THE ENCYCLOPEDIA OF
GEOCHEMISTRY AND ENVIRONMENTAL SCIENCES

ENCYCLOPEDIA OF EARTH
SCIENCES SERIES

Series Editor: RHODES W. FAIRBRIDGE
Columbia University

Published Volumes
THE ENCYCLOPEDIA OF OCEANOGRAPHY (Vol. I)
THE ENCYCLOPEDIA OF ATMOSPHERIC SCIENCES AND ASTROGEOLOGY (Vol. II)
THE ENCYCLOPEDIA OF GEOMORPHOLOGY (Vol. III)
THE ENCYCLOPEDIA OF GEOCHEMISTRY AND ENVIRONMENTAL SCIENCES (Vol. IVA)
THE ENCYCLOPEDIA OF WORLD REGIONAL GEOLOGY, PART 1: Western Hemisphere (Including Antarctica and Australia) (Vol. VIII)

The
ENCYCLOPEDIA
of
GEOCHEMISTRY
and
ENVIRONMENTAL
SCIENCES

ENCYCLOPEDIA OF EARTH SCIENCES SERIES,
VOLUME IVA

EDITED BY

Rhodes W. Fairbridge

Professor of Geology
Columbia University
New York

Dowden, Hutchinson & Ross, Inc.
Stroudsburg   Pennsylvania

Copyright © 1972 by **Dowden, Hutchinson & Ross, Inc.**
Library of Congress Catalog Card Number: 75-152326
ISBN: 0-87933-180-1

All rights reserved. No part of this book may be reproduced or transmitted in any form or by any means—graphic, electronic, or mechanical, including photocopying, recording, taping, or information storage and retrieval systems—without written permission of the publisher.

80 79 78   2 3 4 5
Manufactured in the United States of America.

Distributed world wide by Academic Press,
a subsidiary of Harcourt Brace Jovanovich,
Publishers.

CONTRIBUTORS

Louis H. Ahrens, University of Cape Town, Rondebosch, South Africa. *Geochemical Evolution of the Core, Mantle, and Crust; Geochemistry, Ionization Potentials.*

B. J. Alder, Lawrence Radiation Lab., University of California, Livermore, Ca. 94550. *Core Geochemistry.*

Eugene A. Alexandrov, Dept. of Earth and Environmental Sciences, Queens College, Flushing, N.Y. 11367. *Manganese: Element and Geochemistry.*

J. B. Allen, Bureau of Mineral Resources, Canberra, Australia. *Aluminum Ore Deposits.*

Allen M. Alper, Chemical & Metallurgical Division, Sylvania Electric Products, Inc., Towanda, Pa. 18848. *Titanium.*

G. C. Amstutz, Mineralogisch-Petrographisch Institute der Universität, 69 Heidelberg, Germany. *Copper Ore Deposits; Lead: Economic Geology.*

Michael Anbar, Physical Science Division, Stanford Research Institute, Menlo Park, Ca. 94025. *Sonochemical Processes in the Ocean.*

G. M. Anderson, Dept. of Geology, University of Toronto, Toronto 5, Canada. *Silica Solubility.*

A. C. Andrews, Dept. of Chemistry, Kansas State University, Manhattan, Ks. 66502. *Hydrogen* (with H. Truran).

Ernest E. Angino, State Geological Survey, University of Kansas, Lawrence, Ks. 66044. *Bismuth.*

Jacques Avias, Directeur, C.E.R.G.H. Université de Montpellier II, Montpellier, France. *Nickel: Element and Geochemistry* (with V. Gornitz).

Frédéric Baltzer, Université de Paris, Laboratoire de Sédimentologie, 91, Orsay, France. *Lagoon Geochemistry* (with A. Rivière).

Hubert L. Barnes, Dept. of Geosciences, Pennsylvania State University, University Park, Pa. 16802. *Water: Substance and Solvent* (with D. Langmuir); *Hydrothermal Solutions.*

Isaac Barshad, Dept. of Soil & Plant Nutrition, University of California, Berkeley, Ca. 94720. *Weathering, Chemical.*

Paul B. Barton, Jr., U.S. Geological Survey, Dept. of Interior, Washington, D.C. 20242. *Exsolution; Mineral Genesis—Equilibrium.*

William A. Bassett, Dept. of Geology, University of Rochester, Rochester, N.Y. 14627. *Mass Spectrometry.*

Carrol M. Beeson, Petroleum Engineering Dept., University of Southern California, Los Angeles, Ca. 90007. *Porosity and Permeability* (with G. V. Chilingarian and I. Ershaghi).

Theodore Belsky, Institute of Geophysics, University of California, Los Angeles, Ca. 90024. *Precambrian Hydrocarbons.*

Michael Bender, Lamont Geological Observatory of Columbia University, Palisades, N.Y. 10964. *Carbon Isotope Fractionation; Manganese Nodules.*

J. H. Bennett, Dept. of Nuclear Engineering, University of Missouri, Rolla, Mo. 65401. *Iodine.*

Lloyd V. Berkner (deceased), Graduate Research Center of the Southwest, Dallas, Tx. 75230. *Oxygen—Evolution In the Earth's Atmosphere* (with L. C. Marshall).

Rudolf H. Bieri, Scripps Institution of Oceanographic Office, Washington, D.C. 20390. *Ionium–Thorium Dating.*

Blake W. Blackwelder, Dept. of Geology, George Washington University, Washington, D.C. 20006. *Calcium Carbonate: Economic Geology.*

Enrico Bonatti, Institute of Marine Science, University of Miami, Miami, Fl. 33149. *Authigenesis of Minerals—Marine.*

Kurt Bostrom, Institute of Marine Sciences, University of Miami, Miami, Fl. 33149. *Geochemistry of Sediments —Modern.*

v

CONTRIBUTORS

DONALD R. BOWES, Dept. of Geology, The University, Glasgow W2, Scotland. *Metamorphic Environments—Chemical Mobility.*

GEORGE BOYD, Chemistry Division, Oak Ridge National Laboratory, Oak Ridge, Tn. 37830. *Technetium.*

WILLIAM H. BRADLEY, Pigeon Hill Farm, Milbridge, Maine 04658. *Paleolimnology.*

R. R. BROOKS, Dept. of Chemistry & Biochemistry, Massey University, Palmerston, New Zealand. *Biogeochemistry* (with I. R. Kaplan).

BRUCE BUELL, Union Oil Co. of California, Research Dept., Brea, Ca. 92621. *Flame Spectroscopy.*

DAVED J. BURDON, Food and Agricultural Organization of the United Nations, Via delle Terme di Caracella, Rome, Italy. *Hydrology, Semiarid Regions.*

I. BURKART–BAUMANN, Mineralogisch Institut, 69 Heidelberg, Germany. *Vanadium.*

ROGER G. BURNS, Dept. of Earth and Planetary Sciences, Massachusetts Institute of Technology, Cambridge, Ma. 02139. *Crystal-Field Theory* (with W. S. Fyfe).

S. C. CARAPELLA, JR., Central Research Laboratories, American Smelting & Refining Co., South Plainfield, N.J. 07080. *Arsenic; Thallium.*

DOROTHY CARROLL (deceased), U.S. Geological Survey, Menlo Park, Ca. 94025. *Rainwater.*

RAYMUNDO J. CHICO, 103 Woodlawn Road, Baltimore, Md. 21210. *Epigenesis; Hypogene; Supergene; Syngenesis.*

GEORGE V. CHILINGARIAN, Petroleum Engineering Dept., University of Southern California, Los Angeles, Ca. 90007. *Porosity and Permeability* (with I. Ershaghi and C. M. Beeson).

VEN T. CHOW, Hydrosystems Laboratory, University of Illinois, Urbana, Il. 61801. *Hydrologic Maps.*

GEORGE CLAUS, American Society of Ocean Sciences, Room 1701, 501 Fifth Avenue, New York, N.Y. 10017. *Environmental Pollution—Global Effects* (with G. Halasi-Kun).

PAUL L. CLOKE, Dept. of Geology and Mineralogy, University of Michigan, Ann Arbor, Mi. 48104. *Hydrothermal Solutions—Sulfide Transport; pH–Eh Diagrams; Silver: Element and Geochemistry; Silver: Economic Deposits.*

GERALD A. COLE, Department of Zoology, Arizona State University, Tempe, Az. 85281. *Limnology.*

W. CHARLES COOPER, United Nations Development Programme, Casilla 197-D, Santiago, Chile. *Tellurium.*

S. C. CREASEY, U.S. Geological Survey, Menlo Park, Ca. 94025. *Hydrothermal Alteration of Silicate Rocks.*

EDGAR F. CRUFT, Dept. of Geology, University of New Mexico, Albuquerque, N.M. 87106. *Evaporite Processes* (with S. Feldman); *Oxygen.*

CHARLES D. CURTIS, Dept. of Geology, University of Sheffield, Sheffield 1, England. *Cobalt: Element and Geochemistry; Copper: Element and Geochemistry; Copper: Mineralogy of Compounds; Scandium; Tungsten; Zinc: Element and Geochemistry.*

PAUL E. DAMON, Dept. of Geosciences, University of Arizona, Tucson, Az. 85721. *Trace Elements in Silicate Minerals.*

A. G. DARNLEY, Geological Survey of Canada, Ottawa 4, Canada. *Uranium–Thorium–Lead Age Determination.*

STANLEY N. DAVIS, Dept. of Geology, University of Missouri, Columbia, Mo. 65201. *Hydrogeology.*

DAVID R. DAWDY, U.S. Geological Survey, Engineering Research Center, Fort Collins, Co. 80521. *Water Supply: Economics.*

EGON T. DEGENS, Division of Geological Sciences, California Institute of Technology, Pasadena, Ca. 91109. *Geochemistry of Ancient Sediments.*

WILLIAM F. DENNEN, Dept. of Geology, University of Kentucky, Lexington,

CONTRIBUTORS

Ky. 40506. *Emission Spectrography*.

ROGER J. M. DE WIEST, Dept. of Geological Engineering, Princeton University, Princeton, N.J. 08540. *Hydrology, Subsurface Waters*.

R. C. DOMAN, Technical Staffs Division, Corning Glass Works, Corning, N.Y. 14830. *Phase Equilibria* (with A. M. Alper and R. C. McNally); *Tantalum*.

ISABELLA DREW, Low Library, Columbia University, New York, N.Y. 10027. *Atomic Number and Periodic Table; Spectrophotometry*.

PAUL F. DUBY, Dept. of Engineering, Columbia University, New York, N.Y. 10027. *Clapeyron's Equation; Entropy; Enthalpy or Heat Content; Free Energy and Free Enthalpy; Mass Action and Equilibrium Constant; Thermodynamics*.

ROBERT A. DUCE, Graduate School of Oceanography, University of Rhode Island, Kingston, R.I. 02881. *Bromine*.

R. C. DUGDALE, Department of Oceanography, University of Washington, Seattle, Wa. 98105. *Nitrogen Cycle; Nitrogen Cycle in the Oceans*.

J. DYCK, Geophysics Section, Saskatchewan Research Council, Saskatoon, Sask., Canada. *Geophysical Methods for Hydrologic Search* (with D. J. Gendzwill and T. P. Pepper).

HENRY L. EHRLICH, Dept. of Biology, Rensselaer Polytechnic Institute, Troy, N.Y. 12181. *Manganese Cycle; Mineral Genesis and Transformation—Microbial; Sulfur Oxidation—Bacterial*.

CESARE EMILIANI, School of Marine and Atmospheric Science, University of Miami, Miami, Fl. 33149. *Paleotemperatures—Isotopic Determinations*.

HAROLD E. ENLOWS, Dept. of Geology, Oregon State University, Corvallis, Or. 97331. *Authigenesis of Minerals* (non-Marine).

ROBERT D. ENZMANN, 29 Adams Street, Lexington, Ma. 02173. *Molybdenum*.

IRAJ ERSHAGHI, Petroleum Engineering Dept., University of Southern California, Los Angeles, Ca. 90007. *Porosity and Permeability* (with G. V. Chilingarian and C. M. Beeson).

HANS EUGSTER, Dept. of Geology, Johns Hopkins University, Baltimore, Md. 21218. *Ammonia, In Minerals and Early Atmosphere; Seawater: History*.

ROBERT K. FAHNSTOCK, Dept. of Geology, State University College, Fredonia, N.Y. 14063. *Runoff*.

RHODES W. FAIRBRIDGE, Dept. of Geology, Columbia University, New York, N.Y. 10027. *Cadmium; Cyclic Salts; Geochemical Classification; Halogens; Magnesium Cycle; Natural Resources; Tin: Element and Geochemistry; Rare Gases; Water table*.

FRASER P. FANALE, Planetology Group —Lunar & Planetary Sciences Div., Jet Propulsion Laboratory, La Jolla, Ca. 91103. *Argon; Helium*.

GEORGE T. FAUST, U.S. Geological Survey, Washington, D.C. 20242. *Chemical Mineralogy*.

SANDRA FELDMAN, Dept. of Geology, University of New Mexico, Alburquerque, N.M. 87106. *Evaporate Processes* (with E. F. Cruft).

CYRUS W. FIELD, Dept. of Geology, Oregon State University, Corvallis, Or. 97331. *Isotope Fractionation; Isotope Geology* (Stable); *Sulfur* (Elementary Geochemistry).

R. L. FLEISCHER, Physical Science Branch, General Electric Co., Schenectady, N.Y. 12301. *Fission Track Dating* (with P. B. Price and R. M. Walker).

A. F. FREDERICKSON, King Resources Co., Denver, Co. 80202. *Partition Coefficients*.

C. G. I. FRIEDLAENDER, (Prof. Emer., Dalhousie University, N.S.) 248 Clemow Avenue, Ottawa 1, Canada. *Bubble Motion in Fluid Inclusions*.

MICHAEL J. FROST, Dept. of Mineralogy, British Museum, London, S.W. 7, England. *Crystal Chemistry; Mineral Thermometry; Native Elements* (with L. W. Staples).

CONTRIBUTORS

WALLACE H. FULLER, Dept. of Agriculture, Chemistry and Soils, University of Arizona, Tucson, Az. 35721. *Phosphorus—Element and Geochemistry; Phosphorus Cycle.*

W. S. FYFE, Geology Department, Manchester University, Manchester, England. *Crystal Field Theory* (with R. G. Burns).

RICHARD V. GAINES, Kawecki Berylco Industries, Boyertown, Pa. 19512. *Cesium.*

GEORGE DONALD GARLICK, Dept. of Geology, Humboldt State College, Arcata, Ca. 95521. *Oxygen Isotope Geochemistry.*

D. J. GENDZWILL, Geophysics Section, Saskatchewan Research Council, Saskatoon, Sask., Canada. *Geophysical Methods for Hydrologic Search* (with J. Dyck and T. P. Pepper).

RONALD J. GIBBS, Dept. of Geological Sciences, Northwestern University, Evanston, Il. 60201. *River Geochemistry: Environmental Factors; River Geochemistry: Regional.*

BRUNO J. GILETTI, Dept. of Geological Sciences, Brown University, Providence, R.I. 02912. *Electron Capture; Rubidium–Strontium Dating Method* (with L. Long).

S. S. GOLDICH, Dept. of Geology, Northern Illinois University, DeKalb, Il. 60115. *Geologic Time Scale.*

GORDON GOLES, Dept. of Chemistry, University of California, in San Diego, La Jolla, Ca, 92037. *Precambrian Atmosphere: Geochemical History.*

H. G. GOODELL, Dept. of Environmental Sciences, University of Virginia, Charlottesville, Va. 22903. *Carbon/Nitrogen Ratio.*

VIVIEN GORNITZ, N.A.S.A. Institute for Space Studies, 2280 Broadway, New York, N.Y. 10025. *Acids and Bases; Aluminum; Antimony; Base—Chemical; Chelation; Colorimetry; Columbium; Electrolytes—Flocculation of Colloids; Liesegang Rings; Molecular Sieves; Molecular Weights; Optical Activity; Sodium: Element and Geochemistry; Solubility-product Constant; Stereochemistry and Enantiomorphism; Thermochemistry.*

T. J. GRAY, Atlantic Industrial Research Institute, Halifax, Nova Scotia, Canada. *Catalysis.*

JACK GREEN, California State College, Long Beach, Ca. 90801. *Elements: Planetary, Abundances and Distribution.*

L. PAUL GREENLAND, U.S. Dept. of Interior, Geological Survey, Washington, D.C. 20242. *Gallium.*

WM. H. GROSS, Metal Products Dept., Dow Chemical Co., 2030 Abbott Road Center, Midland, Mi. 48640. *Magnesium.*

O. N. HACKETT, U.S. Geological Survey, Washington D.C. 20242. *Groundwater; Springs.*

DAVID S. HAGLUND, Shell Development Co., Exploration & Product Research Center, P.O. Box 481, Houston, Tx. 77001. *Uranium.*

GEORGE HALASI-KUN, 401 Dodge, Columbia University, New York, N.Y. 10027.../ *Environmental Pollution—Global Effects* (with G. Claus).

FRANCIS R. HALL, Dept. Soil and Water Sciences, James Hall, University of New Hampshire, Durham, N.H. 03824. *Silica Cycle.*

CLIFFORD A. HAMPEL, 8501 Haarding Avenue, Skokie, Il. 60076. *Hafmium.*

STANLEY S. HANNA, Dept. of Physics, Stanford University, Stanford, Ca. 94305. *Mossbauer Effect.*

BRUCE B. HANSHAW, U.S. Geological Survey, Washington, D.C. 20242. *Clay Membrane Phenomena.*

J. F. HARRIS, Technical Research Centre, Cominco Ltd., Trail, British Columbia, Canada. *Indium.*

ROBERT C. HARRISS, Dept. of Oceanography, Florida State University, Tallahassee, Fl. 32306. *Silica—Biogeochemical Cycle.*

STUART A. HARRIS, Geography Department, University of Calgary, Canada.

CONTRIBUTORS

Aluminum Toxicity; Hydrology—Coastal Terrain.

KNUT S. HEIER, *Mineralogisk-Geologisk Museum*, Sars Gate 1, Oslo 5, Norway. *Alkalis, Alkali Metals, and Alkaline-Earth Metals.*

JOHN D. HEM, U.S. Geological Survey, Menlo Park, Ca. 94025. *Aqueous Solutions; Colloids; Connate Water; Juvenile Water; Magmatic Water; Meteoric Water; Oxidations and Reduction; Phreatic Water; Solubility; Suspended Water; Vadose Water; Water—Nonmarine.*

WALTER E. HILL, JR., C. F. and I. Steel Corporation, P.O. Box 316, Pueblo, Water Resources Division, P. O. Box Co. 81002. *Versene (EDTA) Solution Studies.*

W. L. HISS, U.S. Geological Survey, 4369, Albuquerque, N.M. 87106. *Geochemistry of Sedimentary Silica* (with V. A. Kneller).

P. V. HOBBS, Dept. of Atmospheric Sciences, University of Washington, Seattle, Wa. 98105. *Water Molecule: Structure*

GORDON W. HODGSON, University of Calgary, Calgary 44, Canada. *Chlorophyll; Hydrocarbons; Organic Pigments.*

CHARLES M. HOHENBERG, Dept. of Physics, University of California, Berkeley, Ca. 94720. *Xenon.*

H. D. HOLLAND, Princeton University, Dept. of Geological and Geophysical Sciences, Princeton, N.J. 08540. *Hydrothermal Alteration—Non-Silicate.*

DONALD H. HOOD, Institue of Marine Sciences, University of Alaska College, Ak. 99701. *Seawater; Chemistry.*

MARJORIE HOOKER, U.S. Geological Survey, Washington, D.C. 20242. *International Association of Geochemistry and Cosmochemistry.*

R. M. HORDON, Dept. of Geography, Rutgers University, New Brunswick, N.J. 08903. *Hydrologic Cycle; Water Balance.*

GORDON JACOBY, Institute of Geophysics and Planetary Sciences, University of California, Los Angeles, Ca. 90024. *Hydrology, Limestone Terrain.*

LELA JEFFREY, Dept. of Oceanography, Texas A. & M. University, College Station, Tx. 77843. *Organic Geochemistry of Sea Water.*

JOHN T. JENKINS, Dept. of Geotechnical Science, Loyola College, Montreal 262, Quebec, Canada. *Mineral Classes Silicates.*

MEAD LEROY JENSEN, Dept. of Geology, University of Utah, Salt Lake City, Ut. 84112. *Diffusion—Geological Role.*

WILLIAM D. JOHNS, Dept. of Geology, University of Missouri, Columbia, Mo. 65201. *Chlorine.*

IAN R. KAPLAN, Dept. of Geology, University of California, Los Angeles, Ca. 90024. *Biogeochemistry* (with R. R. Brooks); *Sulfur Cycle.*

AARON KAUFMAN, Lamont Geological Observatory of Columbia University, Palisades, N.Y. 10964. *Radioactivity In Rocks.*

ROBERT KAY, Lamont Geological Observatory, Columbia University, Palisades, N.Y. 10964. *Cerium; Dysprosium; Erbium; Europium; Gadolinium; Holmium; Lanthanides; Lanthanum; Lutetium; Neodymium; Praseodymium; Promethium; Samarium; Terbium; Thulium; Ytterbium; Yttrium.*

WALTER D. KELLER, Dept. of Geology, University of Missouri, Columbia, Mo. 65201. *Glacial Milk.*

WILLIAM C. KELLY, Dept. of Geology and Mineralogy, University of Michigan, Ann Arbor, Mi. 48105. *Gold: Economic Deposits.*

JERRY S. KIER, University of Texas, Marine Science Institute, Port Aransas, Tx. 78373. *Barium: Economic Geology.*

WILLIAM A. KNELLER, Dept. of Geology, University of Toledo, Toledo, Oh. 43606. *Geochemistry of Sedimentary Silica* (with W. H. Hiss).

ANNA H. KOFFLER, College of Pharmacy, Ohio Northern University, Ada,

CONTRIBUTORS

Oh. 45810. *Trace Elements in Plants.*

IGHO H. KORNBLUEH, Dept. of Physical Medicine, University of Pennsylvania, Philadelphia, Pa. 19146. *Medicinal Springs.*

KARL KREJCI-GRAF, Geologisch-Paleontologisches Institut der Universität, Senckenberg-Anlage 32, Frankfurt-am-Main, Germany. *Trace Metals in Sediments, Oils and Allied Substances.*

RALPH KRETZ, Dept. of Geology, University of Ottawa, Ottawa 2, Canada. *Beryllium; Lithium: Element and Geochemistry; Potassium; Silicon.*

S. KRISHNA-SWAMI, Tata Institute of Fundamental Research, Colaba, Bombay 5, India. *Thorium.* (with W. Moore).

TEH-LUNG KU, University of Southern California, Dept. of Geological Sciences, Los Angeles, Ca. 90007. *Radium in the Oceans.*

A. W. KUCHLER, Dept. of Geography-Meteorology, University of Kansas, Lawrence, Ks. 66044. *Vegetation Indicators.*

J. LAWRENCE KULP, Teledyne Isotopes, 50 Van Buren Ave., Westwood, N.J. 07675. *Geochronometry.*

D. LAL, Tata Institute of Fundamental Research, Colaba, Bombay, 5, India. *Radionuclides — Cosmic-Ray-Produced.*

ARTHUR M. LANGER, Environmental Science Laboratory, Mt. Sinai School of Medicine, New York, N.Y. 10029. *Mineral Particles and Human Disease* (with A. D. Mackler).

DONALD LANGMUIR, Dept. of Geology, Pennsylvania State University, Pa. 16802. *Water Substance and Solvent* (with H. L. Barnes).

HENRY LEPP, Dept. of Geology, Macalester College, St. Paul, Mi. 55101. *Iron.*

ABRAHAM LERMAN, Isotope Department, Weizman Institute of Science, Rehovot, Israel. *Seawater—Geochemical Balance.*

M. H. LIETZKE, Chemistry Division, Oak Ridge National Laboratory, Oak Ridge, Tn. 37830. *Polonium.*

D. A. LIVINSTONE, Zoology Dept., Duke University, Durham, N.C. 27906. *Ecology; Sodium Cycle.*

WILLIAM LODDING, Dept. of Geology, Rutgers University, New Brunswick, N.J. 08903. *X-Ray Spectroscopy.*

LEON E. LONG, Dept. of Geology, University of Texas, Austin, Tx. 78712. *Lead—Interpretation of Stable Isotope Abundance; Rubidium–Strontium Dating Method; Uranium–Helium Isotopic Age Method.*

JOHN F. LOVERING, School of Geology, University of Melbourne, Parkville, Victoria, 3052, Australia. *Rhenium.*

WILLIAM C. LUTH, Stanford University, Stanford, Ca. 94305. *Silicates—Hyrothermal Systems.*

R. J. P. LYON, Dept. of Geophysics, Stanford University, Stanford, Ca. 94305. *Infrared Analysis.*

ANNE D. MACKLER, Dept. of Community Medicine, Mt. Sinai School of Medicine, New York, N.Y. 10029. *Mineral Particles and Human Disease* (with A. Langer).

FRANK T. MANHEIM, Woods Hole Oceaographic Institution, Woods Hole, Ma. 02543. *Interstitial Waters In Sediments; Natural Brines.*

IAN R. MANNERS, Dept. of Geography, Columbia University, New York, N.Y. 10027. *Environmental Science* (with J. Oliver).

LAURISTAN C. MARSHALL, Southwest Center for Advanced Studies, Dallas, Tx. 75230. *Oxygen Evolution In The Earth's Atmosphere* (with L. V. Berkner).

JACQUES M. MAY, 185 Seaview St., Chatham, Ma. 02633. *Medical Geography.*

W. A. MCBRYDE, Faculty of Science, University of Waterloo, Waterloo, Ont., Canada. *Iridium; Osmium; Palladium; Rhodium; Platinum: Element and Geochemistry; Platinum Metals; Ruthenium.*

CONTRIBUTORS

ALISTAIR W. MCCRONE, University of the Pacific, Stockton, Ca. 95204. *pH–Eh Relations*.

H. L. MCKAGUE, Dept. of Geology, Rutgers University, New Brunswick, N.J. 08903. *Mineralogy*.

ROBERT C. MCNALLY, Corning Glass Works, Corning, N.Y. 14830. *Phase Equilibria; Zirconium: Element and Geochemistry*.

EDWARD MERCY, Dept. of Geology, Lakehead University, Port Arthur, Ont., Canada. *Mantle Geochemistry*.

EMMY BOOY MERRILL, Dept. of Geology and Engineering, Michigan Technical University, Houghton, Mi. 49931. *Bonding; Clay Minerals—Base Exchange*.

J. MURRAY MITCHELL, JR., NOAA, Environmental Data Service, Silver Spring, Md. 20910. *Air Pollution and Global Climate*.

WILLARD S. MOORE, U.S. Naval Oceanography Office, Washington, D.C. 20390. *Cycles—Geochemical: Thorium* (with S. Krishna-Swami); *Radioactive Isotopes; Radium; Radon; Van der Waals Force*.

GEORGE MUELLER, University of Miami, Institute of Molecular Evolution, Coral Gables, Fl. 33134. *Organic Geochemistry; Organic Mineraloids*.

ROBERT F. MUELLER, Goddard Space Flight Center, NASA, Greenbelt, Md. 20771. *Trace Elements—Geochemistry*.

V. RAMA MURTHY, Dept. of Geophysics, University of Minneapolis, Minn. 55455. *Elements*.

RAYMOND L. NACE, Water Resources Division, U.S. Geological Survey, Washington, D.C. 20242, *International Hydrological Decade*.

BARTHOLOMEW NAGY, Dept. of Geosciences, University of Arizona, Tucson, Az. 85721. *Natural Chromatography In Sedimentary Rocks*.

BERT E. NORDLIE, Dept. of Geology, University of Arizona, Tucson, Az. 85721. *Volcanic Gases*.

JOHN E. OLIVER, Dept. of Geography, Columbia University, New York, N.Y. 10027. *Environmental Science* (with I. R. Manners).

ROY OVERSTREET, Professor of Soil Chemistry (deceased), Department of Soils and Plant Nutrition, University of California, Berkeley, Ca. 94720. *Soil Salinity and Alkinity*.

T. P. PEPPER, Physics Division, Saskatchewan Research Council, Saskatoon, Sask., Canada. *Geophysical Methods for Hydrologic Search* (with J. Dyck and D. J. Gendzwill).

F. G. PERCIVAL, O.B.E., Sadlers End, Haslemere, Surrey, England. *Iron: Economic Deposits*.

ORRIN H. PILKEY, Dept. of Geology, Duke University, Durham, N.C. 27708. *Barium; Calcium*.

J. R. POSTGATE, Agricultural Research Council, University of Sussex, Brighton, Sussex BNI9QJ, England. *Sulfate Reduction—Microbial*.

P. BUFORD PRICE, Dept. of Physics, University of California, Berkeley, Ca. 94720. *Fission Track Dating* (with R. L. Fleischer and R. M. Walker).

RICARDO M. PYTKOWICZ, Dept. of Oceanography, Oregon State University, Corvallis, Or. 97331. *Calcium Cycle*.

GEORGE R. RAPP, JR., Dept. of Geology and Geophysics, University of Minnesota, Minneapolis, Mi. 55455. *Selenium*.

A. L. REESMAN, Dept. of Geology, Vanderbilt University, Nashville, Tn. 37203. *Abrasion pH*.

ROBERT C. REYNOLDS, JR., Geology Dept., Dartmouth College, Hanover, N.H. 02755. *Boron; Rubidium*.

A. GLENN RICHARDS, Dept. of Entomology, Fisheries and Wildlife, St. Paul, Mi. 55101. *Chitin and Chitinous Cuticles*.

ARTHUR S. RITCHIE, Dept. of Geology, University of Newcastle, Newcastle, 2308, N.S.W., Australia. *Chromatography*.

CONTRIBUTORS

A. RIVIERE, Université de Paris, Laboratoire de Sédimentologie, 91, Orsay, France. *Lagoon Geochemistry* (with F. Baltzer).

C. J. ROBINOVE, U.S. Geological Survey, Washington, D.C. 20242. *Hydrology.*

HAROLD W. ROBINSON, Assay Dept., Englehard Industries, Inc., Newark, N.J. 07115. *Gold: Element and Geochemistry.*

EDWARD ROEDDER, U.S. Dept. of the Interior, Geological Survey, Washington D.C. 20242. *Fluid Inclusions.*

ROBERT H. ROSE, Chief Geologist, National Park Service, Washington D.C. 20242. *Conservation.*

ALEXANDER B. RONOV, Institute of Geochemistry, Academy of Science, Moscow, B-334, U.S.S.R. *Earth's Crust Geochemistry* (with A. A. Yaroshevsky).

J. N. ROSHOLT, U.S. Geological Survey, Federal Center, Denver 25, Co. 80225. *Protactinium Thorium.*

ROBERT V. RUHE, Dept. of Geology, Indiana University, Bloomington, In. 47401. *Pedology (Soil Science).*

WILLIAM M. SACKETT, Oceanography Dept., Texas A&M University, College Station, Tx. 77843. *Protactinium.*

R. T. SANDERSON, Dept. of Chemistry, Arizona State University, Tempe, Az. 85281. *Electronegativity.*

FRANCIS R. SAUPÉ, Centre de Recherches Petrologie et Geochimiques, 54 Nancy, France. *Mercury.*

JEAN-GUY SCHILLING, Graduate School of Oceanography, University of Rhode Island, Kingston, R.I., 02881. *Rare Earths in Basalts.*

ROBERT SCHMALZ, Dept. of Geology, Pennsylvania State University, University Park, 16802. *Calcium Carbonate: Geochemistry.*

FRED W. SCHONFELD, Los Alamos Scientific Laboratory, University of California, Los Alamos, N.M. 87544. *Plutonium.*

STANLEY A. SCHUMM, Dept. of Geology, Colorado State University, Fort Collins, Co. 80521. *Paleohydrology.*

NICHOLAS M. SHORT, NASA, Goddard Space Flight Center, Greenbelt, Md. 20771. *Actinium.*

MELVIN P. SILVERMAN, Pittsburgh Energy Research Center, U.S. Bureau of Mines, Pittsburgh, Pa. 15213. *Sulfide Mineral Oxidation—Microbial.*

DALE R. SIMPSON, Dept. of Geological Sciences, Lehigh University, Bethlehem, Pa. 18015. *Fluorine; Phosphatization.*

H. JAMES SIMPSON, JR., Mauna Loa Observatory, Hilo, Hi. *Complexes.*

LAWRENCE SLOTE, Environment Techniques, Grumman Aerospace Corp. Bethpage, L.I., N.Y. *Urban Air Pollution.*

LAWRENCE F. SMALL, Department of Oceanography, Oregon State University, Corvallis, Or. 97330. *Photosynthesis.*

CHARLES SMITH, Geological Survey of Canada, Ottawa, Canada. *Chromium.*

R. KARL SMITH, Corning Glass Works, Corning, N.Y. 14830. *Tungsten* (with C. I. Curtis).

RICHARD L. STANTON, Dept. of Geology, University of New England, Armidale, N.S.W., Australia. *Sulfides In Sediments.*

LLOYD W. STAPLES, Geology Dept. University of Oregon, Eugene, Or. 97403. *Antimonates; Antimonites and Arsenites; Carbonates; Chromates; Halides; Iodates; Mineral Classes: Non-Silicates; Molybdates and Tungstates; Native Elements; Organic Compounds; Oxides and Hydroxides; Phosphates, Arsenates and Vanadates; Selenates and Tellurates: Selenites and Tellurites; Sulfates, Sulfides; Sulfo-Salts.*

HAROLD T. STEARNS, P.O. Box 158, Hope, Id. 83836. *Hydrology, Volcanic Terrain.*

F. J. STEVENSON, Dept. of Agronomy, University of Illinois, Urbana, Il. Il. 61801. *Biochemicals; Nitrogen: Element and Geochemistry.*

JOHN S. STEVENSON, Dept. of Geological

CONTRIBUTORS

Sciences, McGill University, Montreal, Canada. *Medical Geology* (with Louise S. Stevenson).

LOUISE S. STEVENSON, Curator of Geology, Redpath Museum, McGill University, Montreal, Canada. *Medical Geology* (with J. S. Stevenson).

R. E. STEVENSON, O.N.R., Scripps Institute of Oceanography, La Jolla, Ca. 92037. *Estuarine Hydrology*.

MINZE STUIVER, Depts. of Geological Sciences and Zoology, University of Washington, Seattle, Wa. 98105. *Carbon-14 Dating*.

EDWARD STURM, Dept. of Geology. Brooklyn College, Brooklyn, N.Y. 11210. *Differential Thermal Analysis; X-Ray Diffraction Analysis*.

F. M. SWAIN, Dept. of Geology, University of Minnesota, Minneapolis, Mi. 55455. *Lake Geochemistry*.

S. R. TAYLOR, Dept. of Geophysics, Australian National University, Canberra, Australia. *Rare Earths (Lanthanide Series)*.

LELAND L. THATCHER, Division of Water Resources, U.S. Geological Survey, Federal Center, Denver, Co. 80225. *Deuterium; Tritium*.

H. G. THODE, McMaster University, Hamilton, Ontario, Canada. *Sulfur Isotope Fractionation*.

C. P. THORNTON, Dept. of Geochemistry and Mineralogy, Pennsylvania State University, Pa. 16802. *Paragenesis*.

JAMES THORP, Dept. of Geology, Earlham College, Richmond, In. 47374. *Soil Erosion*.

SPENCER R. TITLEY, Dept. of Geology, University of Arizona, Tucson, Az. 35721. *Geochemistry (Testing for Elements)*.

JOZSEF TOTH, Groundwater Division, Research Council of Alberta, Edmonton, Alberta, Canada. *Groundwater Motion in Drainage Basins*.

SOLCO W. TROMP, Biometeorological Research Centre, Haarlemmertrekvaart 33, Oegstgeest (Leiden), The Netherlands. *Water Divining*.

ALFRED H. TRUESDELL, U.S.G.S., Experimental Geochemistry and Mineralogy Branch, Menlo Park, Ca. 94025. *Ion Exchange*.

JAMES W. TRURAN, Belfer Graduate School of Science, Yeshiva University, New York, N.Y. 10033. *Hydrogen*.

KARL TUREKIAN, Dept. of Geology, Yale University, New Haven Ct. 06520. *Outgassing Of The Planet Earth*.

WILLIAM G. TURNER, Dept. of Geology, Duke University, Durham, N.C. 27708 *Calcium: Economic Geology*.

ALEXIS VOLBORTH, Dept. of Geology, Dalhousie University, Halifax, N.S., Canada. *Neutron Activation Analysis*.

G. A. WAGNER, Max-Planck-Institut f. Kernphysik, 69 Heidelberg, Germany. *Charged Particle Tracks*.

C. T. WALKER, Dept. of Geology, California State College, Long Beach, Ca. 90804. *Boron Geochemistry in Marine Environments: Paleosalinity*.

ROBERT M. WALKER, Dept. of Physics, Washington University, St. Louis, Mo. 63130. *Fission Track Dating* (with R. L. Fleischer and P. B. Price).

J. MARION WAMPLER, Dept. of Geophysical Sciences, Georgia Institute of Technology, Atlanta, Ga. 30332. *Actinide Series; Lead: Element and Geochemistry*.

J. M. WARDE, 270 Park Avenue, New York, N.Y. 10017. *Economic Deposits; Niobium (Columbium)* (with V. Gornitz).

SAYED A. EL WARDANI, Lockheed Marine Laboratory, San Diego, Ca. 92101. *Carbon: Element and Geochemistry*.

SLADE ST. J. WARNE, Dept. of Geology, University of Newcastle, Newcastle, N.S.W. 2308, Australia. *Derivative Differential Thermal Analysis; Differential Thermogravimetric Analysis; Effluent Gas Analysis; Gas Evolution Analysis; Thermogravimetric Analysis; Oxyluminescence; X-Ray Diffraction —Variable Atmosphere; X-Ray Diffraction—Variable Temperature*.

CONTRIBUTORS

JON N. WEBER, Pennsylvania State University, Mineral Science, University Park, Pa. 16802. *Germanium*.

JOHN WEHMILLER, Lamont Geological Observatory, Columbia University, Palisades, N.Y. 10964. *Carbon Cycle; Strontium Cycle; Strontium: Element and Geochemistry*.

PHILIP R. WHITNEY, Geological Survey, New York State Museum and Science Service, Albany, N.Y. 12224. *Potassium; Potassium—Rubidium Ratio in Geology*.

E. J. W. WHITTAKER, Dept. of Geology and Mineralogy, Oxford University, Oxford, England. *Solid Solution*.

KEMBLE WIDMER, N.J. Dept. Environmental Protection, Trenton, N.J. 08625. *Thermal Pollution*.

LEE WILSON, Dept. of Geology, Columbia University, New York, N.Y. 10027. *Specific Gravity*.

JOHN W. WINCHESTER, Dept. of Oceanography, Florida State University, Tallahassee, Fl. 32306. *Buffe-Systems; Geochemistry*.

KENNETH M. WOLGEMUTH, Lamont Geological Observatory, Columbia University, Palisades, N.Y. 10964. *Oxygen Cycle*.

ROBERT J. WRIGHT, American Metal Climax, Inc., Denver 3, Co. *Uranium: Economic Deposits*.

ALEXEI A. YAROSHEVSKY, Institute of Geochemistry, Academy of Science, Moscow, B-334, U.S.S.R. *Earth's Crust Geochemistry* (with A. B. Ronov).

PREFACE

This volume is the fourth in a series, *Encyclopedia of Earth Sciences*. Earlier ones dealt with Oceanography (I), Atmospheric Sciences and Astrogeology (II), and with Geomorphology (III).

Geochemistry means the chemistry of the planet earth and its compositional evolution. Every element in nature and all the more important chemical cycles and processes are treated in this alphabetically arranged volume. All the major mineral groups and mineral species are noted and there are summaries on the valuable economic ore deposits. Originally it was planned to take in both geochemistry and mineralogy in one volume (IV) but the material proved too voluminous, so it was divided, Vol. IVA to treat primarily geochemistry and environmental science. A separate volume is being prepared specifically dedicated to minerals and mineralogy (Vol. IVB).

Geochemistry is coupled with Environmental Science in this volume because it is the chemical pollution of our planet's air and water that is claiming the attention of many geologists and chemists today. To understand environments of today, it is necessary to look at the environments of the past. Many of the appropriate topics will be found in this volume, although certain details are contained in the other volumes of this series; these are here noted by cross-references. Engineering and sedimentational topics will come in Vol. VI. Especially biological and ecological aspects of the environment are set aside for yet another volume (VII), although our present collection does include such topics as "Medical Geology" and "Mineral Particles and Human Disease."

The Nature and History of Geochemistry

Although geochemistry is generally considered to be one of the younger branches of the earth sciences, the term itself goes back well over a century. It was first used in 1838 by the Swiss chemist C. F. Schonbein, who in 1842 stressed that to comprehend the true nature of our globe it was equally important, beside understanding for example the historical messages of stratigraphy and paleontology, to grasp the fundamental chemical nature and evolution of the constituent materials.

It is only within the last decade or so that we have come to learn how the whole basis of organic evolution rests on a progressive and irreversible biogeochemical evolution of the environment, specifically of the soil and atmosphere. These in turn are intimately bound up in the history of weathering (dominated by bacterial action), and progressive changes in the total biota. In the pre-biologic era of the earth's development, about which we admittedly know very little, it is suspected there existed an atmosphere comparable in some ways to that of the planet Jupiter, which has a reducing environment devoid of oxygen. It was only with the development of photosynthetic life (probably around 3.8 billion years ago) that oxygen started to be generated on an appreciable scale. This was the earth's first great geochemical revolution, initiated by the photosynthesis of living organisms. Nevertheless, the oxygen would not have been freed, but must have been entirely earmarked for the weathering budget. Photochemical dissociation of water vapor also liberates some oxygen, but apparently not on the scale required to oxidize surface rocks and still provide the surplus necessary for the evolution of animal life.

The second environmental revolution was a threshold phenomenon, the result of a gradual build-up of oxygen, which only became available when all the near-surface rocks had completed their first weathering cycle. The first oxygen-breathers, i.e., the first primitive organisms of the animal kingdom, probably did not emerge before 2.9 billion years ago.

Yet a third geochemical revolution is

recognized; it occurred at the beginning of the Paleozoic Era, i.e. the start of the Cambrian, about 600 million years ago. This involved the first appearance of animal fossils with calcium carbonate shells. The process is still rather mysterious. Undoubtedly soft-bodied animals had been widespread in Late Precambrian times and some critical physiologic barrier was overcome at the close of that Era which permitted the soft clams, crabs, jelly-fish, or wormlike organisms to develop rigid exoskeletons. The process may have been related to the rising pH in ocean water, for the partial pressure of carbon dioxide must have been initially much higher than that of oxygen, or it may have been connected with physiologic evolution, an enzyme development, for example, among the higher invertebrates.

Still a fourth geochemical revolution occurred in the late Paleozoic, again a gradational process, around 250 million years ago. This was related to the development of highly organized plants, which flourished in the coal swamp environments. Their burial led to a tremendous removal of elemental carbon from the planetary budget, again modifying the partial pressure of carbon dioxide. Progressive addition of carbon dioxide, of course, continued throughout geological time due to volcanic emanation, but the degree of recycling as time goes by cannot yet be estimated. Its production rate has probably been cyclic, related to the volcanic-orogenic revolutions of plate tectonics, sea-floor spreading, and continental drifting. Relative orogenic quiescence on a global scale alternates with phases of continental collision and violent upheaval.

In this catalog of geochemical revolutions, there is a fifth, again related to physiologic changes, in this case to the evolution of calcium carbonate-secreting nannoplankton, the pelagic coccoliths, and foraminifera. These organisms float freely in the surface waters of the oceans and, on dying, their tests (shells) fall to the bottom and, if not redissolved, become incorporated in the deep-sea sediments. As a result, for cycles of the order of 200 million years, this carbon dioxide sink plays a significant role in the global geochemical budget. A sixth revolution may just be starting, under the influence of man and his burning of hydrocarbons. Time will tell.

The object of this brief review of the earth's five great geochemical revolutions is to stress that geochemistry is far from a study of a static, modern earth. It has to consider not only the processes of present day, but reconstruct the situations of many former environments. Tremendous changes must be recognized between one major cycle and the next. For this reason, rigid rules of geochemical behavior cannot be codified with assurance "for all time." Basic laws of chemistry are, of course, changeless in time and space, applying equally to this and other planets, but they must be interpreted against a secular, changing environmental background. Critical in the understanding of that background is the basic principle: the earth's biota is evolving continuously, and as it evolves it not only changes the physiologic setting for its successors, but also the global environment. This environment in turn is not only significant as an ecologic control for the earth's organisms, but also for many inorganic processes.

During the last century it was customary to look at geochemistry essentially as an analytic science, and often merely as the analytic side of mineralogy, as opposed to the physical or optical procedures. A few simple studies of natural reactions were also included. A classical textbook of this mid-19th century era was that by K. G. Bischof, "Lehrbuch der physikalischen und chemischen Geologie." Later in the century came Justus Roth's "Allgemeine und chemische Geologie" and one by Reinhard Brauns —"Chemische Mineralogie," J. J. Berzelius in Sweden also made a significant contribution. By and large, the field was dominated by the German tradition of analytic chemistry.

Dynamic studies were initially very limited owing to the lack of analytic equipment capable of creating high-temperature and -pressure conditions. Nevertheless, pioneering strides in the field were possible in connection with the evaporite minerals, which undergo complex reactions at relatively low temperatures and pressures. The natural salts early attracted the attention of German chemists and mineralogists because of the fine development and great economic importance of these deposits in the Permian and Triassic of central Germany, e.g., at Heilbrunn, Fulda, and Stassfurt. A journey by Ochsenius to the celebrated Kara Bogaz Gulf off the Caspian Sea (see our Vol. III: *Encyclopedia of Geomorphology*, 1968, p. 579) presented to geologists the evidence of how a complex contemporary evaporite basin operated, under conditions of seasonally contrasting climatic conditions, with varying inflow and evaporation rates. In the United States during the last century the work of Sterry Hunt played an important role. Hunt was particularly interested in weathering, and carried out analyses of river water in the major drainage discharges, such as the Mississippi. He was instrumental in pointing out the major role of chemical erosion—at a time when mechanical erosion was commonly assumed to be far more important. Up until quite late in the 19th century, the mechanical work of rivers and waves was considered to be the key factors in base-leveling or "peneplanation," as W. M. Davis was to call it. Ferdinand von Richthofen, in 1886, wrote that eventually, under stable tectonic conditions the continents would be reduced to a uniform surface of "marine abrasion." As long ago as 1866 this view was opposed in England by G. Maw and by W. Whitaker. According to Mellard Reade, after lengthy experiments (1877), the work of waves was merely limited to a physical removal of debris, already prepared and loosened by chemical weathering. But this was a minority view, and the mechanical school adhered to their dogma until well into the present century.

When it comes to global geochemistry, the name that stands supreme is F. W. Clarke. As a celebrated Bulletin of the U.S. Geological Survey, his "Data of Geochemistry" first appeared in 1908 and was to go through many editions. A new series with the same title has begun to appear in the larger format of "Professional Papers" during the last decade. Clarke's role was primarily to gather in one volume the immensely scattered data on the chemistry of rocks, sediments, and waters then available from all parts of the accessible earth. Particular attention was paid to the "wet analyses" of igneous rocks. The high point of this approach was probably the publication of Arie Poldervaart's "Chemistry of the Earth's Crust" in the Geological Society of America volume (spec. paper 62, 1955) on the "Crust of the Earth" that marked the 200th anniversary of Columbia University.

Meanwhile in Europe, in the early part of the present century, two important schools were growing up. The first was the Russian group, founded by V. I. Vernadsky and A. E. Fersman. Following the tradition of Mendeleyev there was a great emphasis on the chemical elements of the earth's crust, indeed of the globe in toto, and on the reactions observable in nature and reproducible in the laboratory. Vernadsky rightly stressed the interactions of geochemistry with the field of biochemistry, thus establishing the science of biogeochemistry, an area largely ignored by the American-Canadian-British schools, who formerly concentrated on the geochemical role in "hard-rock" petrology.

The second great European school of

PREFACE

the early 20th century was led by the Norwegian, Victor M. Goldschmidt, working partly in Oslo, partly in Göttingen. Essentially a development of the Germanic analytic tradition, this school evolved, as exemplified by the work of Barth, mainly in the field of igneous thermodynamics. Goldschmidt attempted, above all, to determine the relationship of the different elements to the various "spheres" of the planet, recognizing the "siderophile," "chalcophile," "lithophile," "atmophile," and "biophile" elements, according to their common geochemical associations (see article in this volume: "Geochemical Classification"). Goldschmidt's major work "Geochemistry" was originally published in German, but was to play an important role in the English-speaking world after it was issued in translation (ed. by A. Muir, 1954). The major work by K. Rankama and T. G. Sahama, from Finland, evolved later in the same tradition. From a volume in Finnish by Sahama (1947), following a period at the University of Chicago by Rankama, emerged their joint work "Geochemistry" in 1950. The tasks of modern geochemistry, according to Goldschmidt, and summarized by Rankama as a "letter to the editor," in the *American Journal of Science* (1947), are:

(1) "The determination of the abundance of the elements and of the atomic species in the earth.
(2) "The description of the distribution of the individual elements in the different spheres of the earth, and in minerals and rocks.
(3) "The detection of the laws dominating the abundance and distribution of the elements."

The "natural history" of a given element and its budgetary "cycle" must therefore include such things as its mode of occurrence in mineral lattices, its role in the crystallization and differentiation of a magma and in metamorphism (*endogenic history*), its decomposition, weathering, and transport at the surface (*exogenic history*), its geochemical behavior pattern, and its concentration in the crust, mantle, core, meteorites, and in other extraterrestrial bodies (see our *Encyclopedia of Atmospheric Sciences and Astrogeology,* 1967). There is here a transition into "cosmochemistry," a field usually taken as very closely related to geochemistry.

Note on Natural Resources

The Greek philosopher Democritus, who lived from about 460 to 370 BC wrote: "Out of nothing arises nothing; nothing that is can be destroyed. All change is only a combination and separation of atoms." In more recent times this thought is formally codified as the "Law of Conservation of Matter." It explains the Universe. In considering natural resources and pollution, it has great implications. If the planet earth has a certain fixed supply of metal X, that is it. When concentrated deposits are exhausted, we must ask: What happened to it? Could we possibly recycle the wastes or residues?

In the editor's opinion one of the major tasks of geochemists today, and indeed of geologists in general, should be the education of the public in general, and of politicians and statesmen in particular, concerning the nature of the earth's resources. We, the profession, know that natural resources are positively finite. But the general public quite commonly believes that the cornucopia will never run dry. There is no upward limit to "progress," "development," and "expansion." Exploitation of nature can go on to infinity . . . This is the most dangerous fallacy that besets the civilized world. Geologists and geochemists can play a vital role in this plea for sanity.

How to use this Encyclopedia

Throughout this encyclopedia there is extensive cross-referencing from one article to another. The reader will probably look up a key word directly, which is usually a name (such as an element) or the noun in a process identification; in such titles adjectives will normally follow the noun—separated by a comma. If one fails to find the sought-after term, there is an index of perhaps 10,000 entries which may provide the clue.

Some of the desired information will not be found in this volume, but others in our Earth Science series may fill the gap: for oceanographic geochemistry, see Vol. I (*Encyclopedia of Oceanography*, 1966); for meteorites and astrogeology, see Vol. II (*Encyclopedia of Atmospheric Sciences and Astrogeology*, 1967); for geomorphology and weathering processes, see Vol. III (*Encyclopedia of Geomorphology*, 1968). A separate Vol. IVB for *Mineralogy and Economic Geology* is planned. Separate volumes are also in preparation for Petrology and Structure and for Sedimentology, for Stratigraphy, and for World Regional Geology.

Wherever possible, extensive lists of *References* are appended to each article. We must apologize for our failure to ensure that certain articles were brought up to date at the last minute. The whole Encyclopedia of Earth Science project, begun in 1964, has been educational for its editor—proving all sorts of old adages about things that can go wrong, will go wrong, and taking more time than we expected. For the reader, it would be a good plan to familiarize himself, or herself, with the major journals in the field and refer to their indexes for continuing developments. The same specialists frequently continue work in the same fields, so that it is not difficult as a rule to watch for new developments.

Apart from the major science journals like *Nature* and *Science,* the reader will find new material treated, specifically in *Geochimica et Cosmochimica Acta,* and *Chemical Geology* (Vol. 1, 1966–) as well as general geological journals such as the *Geological Society of America Bulletin,* the *Journal of Geology,* and the *American Journal of Science*. The more specialized journals in petrology are sometimes useful: *Journal of Petrology, Lethaea,* and the *Journal of Sedimentary Petrology*. Russian work will be found translated in *Geokhimia,* in the *Doklady of the U.S.S.R. Academy of Sciences,* and as selections in the *International Geology Review*.

Abstracts of papers will be found in the bibliographic volumes of the *Geological Society of America* (Exclusive of N. America), the *U.S. Geological Survey* (N. America only), in French in the *Bulletin Signaletique,* in German in the *Neues Jahrbuch* or *Centralblatt,* and in Russian in the *Referatevni Zhurnal*. More specifically, chemical bibliographies can be found in *Chemical Abstracts*. Most of these abstract volumes are severely limited in their potential usefulness by cumbersome arrangements and lack of intelligent indexes. Since most references are identified by an author and date, it is obvious to any serious researcher that this should be the basic form of bibliographic arrangement, as exemplified by the *Geological Society of America* ("Exclusive of N. America") volumes up to Vol. 30 (1968).

Without the aid of computerized search and retrieval systems, the present anarchy in abstracting services is not merely regrettable, but represents a gross waste of national and international funds.

Standard References. For the general reader, it may be helpful to list here certain basic references in geochemistry and mineralogy, some stressing one point of view, some another.

Ahrens, L. H. (editor), 1968, "Origin and Distribution of the Elements" (Pergamon). (A fundamental, international symposium volume of wide-ranging type.)

Breger, I. A. (editor), 1963, "Organic Geochemistry," (Pergamon). (A symposium volume.)

PREFACE

Broecker, W. S., and Oversby, V. M., 1971, "Chemical Equilibria in the Earth" (McGraw-Hill). (A thermodynamic approach to the chemistry of global systems.)

Degens, E. T., 1964. "Geochemistry of Sediments" (Prentice-Hall). (A brief survey and introduction to weathering products and low temperature aqueous reactions.)

Garrels, R. M., and Christ, C. L., 1965, "Solutions, Minerals and Equilibria" (Harper and Row). (A pioneering introduction to the field as a whole, with stress on low-temperature environments.)

Helgeson, H. C., 1964, "Complexing and Hydrothermal Ore Deposition" (Pergamon and Macmillan). (Another aqueous approach in the medium-range temperatures.)

Krauskopf, K. B., 1967, "Introduction to Geochemistry" (McGraw-Hill). (Leading modern introduction to the whole field.)

Manskaya, S. M., and Drozdova, T. V., 1968, "Geochemistry of Organic Substances" (Pergamon). (A translation from the Russian.)

Mason, B., 1966, "Principles of Geochemistry" (J. Wiley & Sons). (A very well written introduction, that has proved itself by going to three editions.)

Miyake, Y., 1965, "Elements of Geochemistry" (Maruzen Co., Tokyo). (A Japanese volume, translated into English, with an elementary but excellent global approach.)

Nagy, B., and Colombo, U., 1967, "Fundamental Aspects of Petroleum Geochemistry" (Elsevier, Amsterdam). (Collected articles by U.S., Canadian, British, French, and Italian writers. The Elsevier Co. also has a dozen volumes or more in its series: "Methods in Geochemistry and Geophysics.")

Polanski, A., 1965, "Geochemistry of Isotopes" (Warsaw, transl. by N. S. F., Washington). (A specialized treatment, coming from Poland, and presenting different viewpoints and data sources.)

Smith, F. G., 1963, "Physical Geochemistry" (Addison-Wesley). (A Canadian approach, concentrating on the "hard-rock," petrologic aspects.)

Wedepohl, K. H., 1969–, "Handbook of Geochemistry" (Springer-Verlag). (A continuing, modern, very detailed "Germanic" data handbook, bound loose-leaf for updating. Uniquely valuable for the specialist.)

A number of volumes include translations from the Russian, which your editor has helped to bring out, thanks to the continued cooperation of the Plenum Publishing Co., New York. Selected references from these Plenum Press "Monographs in Geoscience," that deal with aspects of geochemistry include:

- L. M. Lebediv, 1967, "Metacolloids in Endogenic Deposits."
- A. I. Perel'man, 1967, "The Geochemistry of Epigenesis."
- S. J. Lefond, 1969, "Handbook of World Salt Resources."
- A. D. Danilov, 1970, "Chemistry of the Ionosphere."
- G. S. Gorshkov, 1970, "Volcanism and the Upper Mantle: Investigations in the Kurile Island Arc."
- B. Persons, 1970, "Laterite—Genesis, Location, Use."
- D. Carroll, 1970, "Rock Weathering."
- R. E. Wainerdi and E. A. Oken, 1971, "Modern Methods of Geochemical Analysis."
- A. S. Povarennykh, 1971, "Crystal Chemical Classification of Minerals."

It may be useful further to indicate here some basic references to mineralogy, essentially more appropriate to our projected Vol. IVB (*Mineralogy and Economic Geology*), but not out of place because of the overlap between Geochemistry and Mineralogy:

- Kaplan, R. (editor) 1965, "A Guide to Information Sources in Mining, Minerals, and Geosciences" (Intersciences Publ.).
- Hey, M. H., 1955, "An index of Mineral Species . . ." (British Museum, 2nd ed., with supplements).
- Strunz, H., 1957, "Mineralogische Tabellen," Leipzig (Akadem. Verlag, 3rd ed.).
- Palache, C., Berman, H., and Frondel, C., 1944–1962, Dana's "System of Mineralogy" (John Wiley & Sons).
- Deer, W. A., Howie, R. A. and Zussman, J., 1962, "Rock-forming Minerals," in 5 vols. (Longmans Green, and John Wiley & Sons).
- Vanders, I., and Kerr, P. F., 1967, "Mineral Recognition," New York (John Wiley & Sons), 316 pp. (49 color plates).

Acknowledgments

A gigantic work of the nature of this series of Encyclopedias of Earth Sciences is not achieved without the generous assistance of very many persons. Although the work has been centered at the editor's home institution, cooperation has been obtained from specialists in all parts of the world. In many cases the internationally recognized specialists have been kind enough to lend help and advice.

During the preparation of this volume, three of our old friends have died: Dr. Roy Overstreet (of the Dept. of Soils, University of California), Dr. Dorothy Carroll (of the U.S. Geological Survey, at Menlo Park, California) and Dr. Lloyd V. Berkner (retired president of the Graduate Research Center of the Southwest in Dallas). We are grateful to colleagues who have checked their proofs.

Certain persons have been particularly helpful in giving advice, and checking over our tables of contents as well as preparing valuable entries: Dr. George Halasi-Kun, organizer of the Columbia University Seminars in Hydrology; Dr. H. D. Holland of Princeton University; Dr. Michael Frost of the British Museum, London. Items signed B.C.C. acknowledge the work of my former assistant Mrs. Charlotte Schreiber.

It is a pleasure to acknowledge the devoted services of "Dotty" Spiro, to whom work on these Encyclopedias has become a labor of love. It is highly appropriate to note the care and attention devoted to our Encyclopedias by Alberta Gordon and Aria Ruks at Van Nostrand Reinhold.

Rhodes W. Fairbridge
New York, N.Y.

A

ABRASION pH

The term *abrasion pH* was defined by Stevens and Carron, its originators (1948), "to designate the pH values obtained by grinding minerals in water." They proposed abrasion pH to be useful in the field identification of minerals, and they tabulated, as determinative references, abrasion pH measurements for about 280 different minerals. Abrasion pH ranges from pH 1 for ferric sulfate minerals, such as coquimbite, konelite, and rhomboclase, to pH 12 for calcium–sodium carbonates, such as gaylussite, pirssonite, and shortite (Table 1).

The recommended technique for determining abrasion pH is to grind, in a nonreactive mortar, a small amount of the mineral in a few drops of water for about 1 min, and then to draw through the mineral-water slurry a pH test paper whose color change indicates the abrasion pH of that mineral.

Abrasion pH effectively differentiates between some of the carbonate minerals: e.g., siderite, pH 5; calcite, pH 8; dolomite, pH 9–10; magnesite, pH 11; and the sodium–calcium carbonates, pH 12. Dioctahedral and trioctahedral phyllosilicates, although similar in appearance, but containing, respectively, dominant aluminum and magnesium, can also be differentiated: pyrophyllite, pH 6; talc, pH 9; muscovite, pH 7–8; and phlogopite pH 10–11. Unfortunately, abrasion pH is less discriminatory for identifying many minerals that yield abrasion pH between 6 and 7; among these are most of the oxides and many silicates, including quartz.

The abrasion pH, as a fundamental property of a mineral, is a function of: (1) the constituent ions of the mineral; (2) the ease with which these ions pass into aqueous solution; and (3) the resultant effects that these ions have in the solution phase of the hydrogen and hydroxyl balance of the solution.

The rate of solution of ions from the mineral is accelerated by abrasion of the mineral in water—this is due to the energy added to the mineral-water system by the mechanical work of pulverizing, which increases the surface area of mineral particles. Many minerals, especially silicates, consist of tightly bound groups of ions (such as oxygen-coordinated silicon and aluminum) joined by relatively soluble cations. These cations are commonly dissolved from the surfaces of the mineral grains more readily than are the more tightly bound groups, and, as a result of the incongruent rate of solution, a thin coating of the more insoluble material may thus be developed on the particles. Further abrasion tends to destroy some of this coating and exposes additional reactive ions at the surface.

TABLE 1. SELECTED MINERALS AND THEIR ABRASION pH[a]

Mineral	pH	pH
Coquimbite	1	
Melanterite	2	
Alum	3	
Glauconite	5	5.5*
Kaolinite	5,6,7	5.5*
Anhydrite	6	
Barite	6	
Gypsum	6	
Quartz	6,7	6.5
Muscovite	7,8	8.0
Calcite	8	8.4
Biotite	8,9	8.5
Microcline	8,9	8.0 9.0*
Labradorite		8.0 9.2*
Albite	9,10	
Nepheline		8.8
Dolomite	9,10	8.5
Enstatite		9.6
Tremolite	10	
Augite	10	8.8
Hornblende	10	8.9
Leucite	10	
Diopside	10,11	9.9
Olivine	10,11	9.6*
Magnesite	10,11	
Phlogopite	10,11	
Actinolite	11	
Wollastonite	11	

[a] The pH values given in the left column are from Stevens and Carron (1948), and were obtained by the method outlined in the article. The pH values in the right column are from Keller et al. (1963), and represent electronic pH measurements obtained from grinding 10 g of crushed mineral in 100 ml of double-distilled water for 1 hr. Note that these pH values are often lower than those determined by Stevens and Carron, and may reflect nonattainment of a chemical equilibrium or in some cases variation in chemical composition. The asterisk indicates pH values from more recent unpublished results. The variation in pH values seems to indicate that establishment of a uniform procedure and a personal reference group of pH values may be necessary for consistent results.

ABRASION pH

The mineral surfaces, and ions dissolved from them, can interact with the aqueous solvent, with each other, or with both in solution. Reaction of the mineral ions with the water, i.e., hydrolysis, overshadows other effects, and is the dominant reaction determining the abrasion pH (Stevens, 1934).

Stevens and Carron (1948) give the following as examples of the hydrolysis reactions of the minerals with water:

$Fe_2(SO_4)_3 \cdot 9H_2O$ (coquimbite) + HOH
$\rightleftarrows 3H_2SO_4 + 4H_2O + 2Fe(OH)_3$ (pH 1)

$Na_2B_4O_7 \cdot 10H_2O$ (borax) + HOH
$\rightleftarrows H_2B_4O_7 + 2NaOH + 9H_2O$ (pH 10)

$PbCO_3$ (cerussite) + 2HOH $\rightleftarrows H_2CO_3 + Pb(OH)_2$ (pH 6)

Grant (1969) showed that the abrasion pH of granitic rocks decreases with increasing chemical weathering and related the rock-derived abrasion pH to the following:

$$\text{Abrasion pH} \approx \frac{f(Na + K + Ca + Mg)}{(\text{clay material})}$$

The abrasion pH of a mineral appears to be closely related to the solubility product constant of the corresponding mineral and, if so, can then be related to the free energy of formation of the mineral from its elements. Insofar as this relationship is valid, the abrasion pH is fundamental by being descriptive of the environment in which a mineral may be potentially formed. Although true equilibrium may not always be reached in abrasion pH, the departure of the abrasion pH from that at true equilibrium tends to be small, since pH is expressed as a logarithmic function. Thus, abrasion pH validly describes the pH available in the process of solution and hydrolysis of a mineral in water.

A. L. REESMAN

References

Grant, W. H., 1969, "Abrasion pH, an index of chemical weathering," *Clays and Clay Minerals,* **17**(3), 151–155.

Keller, W. D., Balgord, W. D., and Reesman, A. L., 1963, "Dissolved products of artificially pulverized silicate minerals and rocks," part 1, *J. Sediment. Petrol.,* **33**(1), 191–204.

Stevens, R. E., 1934, "Studies on the alkalinity of some silicate minerals," *U.S. Geol. Surv. Profess. Paper,* **185A**, 1–13.

Stevens, R. E., and Carron, M. K., 1948, "Simple field test for distinguishing minerals by abrasion pH," *Am. Mineralogist,* **33**, 31–49.

Cross-references: *pH-Eh Diagrams; pH-Eh Relations; Silica Solubility; Solubility; Solubility Product Constant; Weathering.* Vol. IVB: *Phyllosilicates.*

ABSOLUTE AGE DETERMINATION—
See GEOCHRONOMETRY

ACIDS AND BASES

Acids and bases are two closely interrelated classes of chemical compounds, the definitions of which have varied considerably in the development both of chemistry and petrology.

The term *acid* was initially applied to substances that had a sour taste, dissolved many metals, turned the color of certain plant indicators red, and reacted with alkalis to give salts. *Bases* (alkalis) were defined as substances that tasted bitter, felt smooth and slippery on the skin, turned certain plant dyes blue, and neutralized the effects of acids.

Then, in the 18th century, Lavoisier advanced the idea that oxygen was the essential element of acids. His theory survives in the word oxygen, which means acid former. The German word for oxygen is *Sauerstoff,* meaning the substance of acid. In 1840, von Liebig defined acids as hydrogen-containing chemicals that would liberate H_2 gas upon reacting with metals. Studies of electrolytic conductance and ionization of acids and bases led Arrhenius to propose yet a different definition of acid and base. According to him, acids ionized in water to yield *hydrogen ions* (H^+) and bases reacted to give *hydroxyl ions* (OH^-). Neutralizatiton involved the combination of H^+ and OH^- to form H_2O and a salt.

Although the Arrhenius theory had many attractive features, it failed to explain acid-base behavior in nonaqueous solvents and in substances that lacked H^+ and OH^- ions. Therefore, Brönsted and Lowry extended the definition of acids and bases to include reactions in which H^+ and OH^- did not participate. A Brönsted-Lowry acid is a proton donor; a base is a proton acceptor. Acid-base reactions consist of the transfer of a proton from one base to another. In fact, every acid can form a base by losing a proton. Acids and bases related in this way are called *conjugate acid base pairs*. Neutralization involves proton transfer.

In the reaction:

$HCl + NaOH \rightarrow NaCl + H_2O$
acid 1 + base 2 $\rightarrow$ base 1 + acid 2

conjugate acid-base pairs are HCl–NaCl and NaOH–H_2O. In the *hydrolysis* of NH_4Cl:

$NH_4Cl + H_2O \rightarrow NH_3 + H_3O^+ + Cl^-$

NH_4^+ is the acid 1, H_2O is base 2, NH_3 is the base 1 and H_3O^+ is the acid 2. Thus, NH_4^+–NH_3 and H_2O–H_3O^+ are conjugate pairs. A strong Brönsted acid implies a weak conjugate base and vice versa.

The Brönsted-Lowry theory can be applied to nonaqueous solvents as well. The strength of acids and bases, which appears to be the same in water, varies in other solvents. In liquid ammonia, for example, acetic and benzoic acids are as strong as nitric, hydrochloric, and perchloric acids. Only the last three are leveled to the same extent in water, which is less alkaline than ammonia. In pure acetic acid, which is even less basic than water, only the strongest acid will react to any extent, e.g., $HClO_4$.

$$HClO_4 + HC_2H_3O_2 \rightleftharpoons H_2C_2H_3O_2^+ + ClO_4^-$$

Because HF and H_2SO_4 are highly acidic, strong acids are only weakly dissociated in them. Because of their highly acidic nature, they exert a leveling effect on bases and bring out basic properties in what would normally be an acid.

In the reaction:

$$H_2SO_4 + HC_2H_3O_2 \rightarrow H_2C_2H_3O_2^+ + HSO_4^-$$

acetic acid is acting as a base. Proton transfers occur in all these solvents, and the Brönsted-Lowry concept of acids and bases gives these reactions a unified treatment.

A more general concept, the *electronic theory*, was proposed by G. N. Lewis in 1923 and revised in 1938. Lewis defined an acid as a substance that can accept an electron-pair, a base as one that can donate an electron-pair. Although the Lewis definition helps to systematize reactions and is very useful in organic chemistry, the less comprehensive Brönsted-Lowry concept is often more effective because of its simplicity and directness. However, the use of the Lewis definition is increasing. Since there seems to be no simultaneous tendency to reject the Brönsted-Lowry definition, it will become necessary to use a qualifying adjective such as "Lewis acid" or "Brönsted base" to clarify the meaning.

Recently, Pearson (1966) has proposed classifying acids and bases into "hard" and "soft" groups. Hard acids are atoms of small size, high positive oxidation state, and no outer electrons, properties that lead to low polarizability. Soft acids, on the other hand, are large atoms, have low or zero valence, contain easily excited outer (*d*-orbital) electrons, and are therefore highly polarizable. Among bases, a hard base has low polarizability and high electronegativity, whereas a soft base is highly polarizable, has low electronegativity, and is easily oxidized. Hard acids include the alkali and alkali earth ions, H^+, Al^{3+}, Fe^{3+}, and Si^{4+}, among others; hard bases are H_2O, OH^-, F^-, Cl^-, NH_3; soft acids are Cu^+, Ag^+, Au^+, Hg^+, Pt^+; soft bases are I^-, Br^-, S^{2-}. This classification provides a qualitative explanation of the stabilities of various compounds and complexes, because, in general, hard acids associate with hard bases and soft acids unite with soft bases.

For many practical purposes, one can still use the older definition of acids and bases, namely, that acids furnish H^+ ions and bases OH^- ions. This is useful in discussing acid-base equilibria and relative strengths. A weak acid is one that is only slightly dissociated into ions. The ionization is represented by:

$$HA \rightleftharpoons H^+ + A^-$$

where A^- indicates any anion, or, better yet, by

$$HA + H_2O \rightleftharpoons H_3O^+ + A^-.$$

The *equilibrium constant* K_a is called the ionization constant for the acid, and is a fixed quantity at any given temperature.

$$K_a = \frac{[H_3O^+][A^-]}{[HA][H_2O]}$$

but since H_2O is the solvent, its concentration barely changes and $[H_2O]$ can be incorporated into the equilibrium constant.

$$K_a = \frac{[H^+][A^-]}{[HA]}$$

where $[H^+]$ actually represents $[H_3O^+]$. Similarly, for a weak base

$$BOH \rightleftharpoons B^+ + OH^-$$

$$K_b = \frac{[B^+][OH^-]}{[BOH]}$$

A polyprotic acid (one with several replaceable H atoms) ionizes in steps:

$$H_2S \rightleftharpoons H^+ + HS^- \quad K_a(H_2S) =$$
$$K_1 = \frac{[H^+][HS^-]}{[H_2S]} = 1 \times 10^{-7}$$

$$HS^- \rightleftharpoons H^+ \pm S^{2-} \quad K_a[HS^-] =$$
$$K_2 = \frac{[H^+][S^{2-}]}{[HS^-]} = 1.3 \times 10^{-13}$$

For any conjugate pair of acid and base, the two ionization constants K_a and K_b are related by the ion product of water.

$$H_2O = H^+ + OH^- \quad K_w = [H^+][OH^-]$$
$$K_a K_b = K_w$$

Since geochemical reactions frequently occur at elevated temperatures and pressures, it is of interest to know how the equilibrium constants and K_w vary with these parameters. The effect of temperature on the ionization constant of acids and bases is given by the van't Hoff equation:

$$\left(\frac{\partial \ln K_a}{\partial T}\right)_p = \left(\frac{\Delta H^0}{RT^2}\right)$$

If H^0 remains constant over the desired temperature range, then

$$2.303 \log \left(\frac{K_{T2}}{K_{T1}}\right) = \frac{-\Delta H^0}{R}\left(\frac{1}{T_2} - \frac{1}{T_1}\right)$$

A plot of log K_a vs $1/T$ will yield a straight line of slope $-\Delta H^0_{\text{reaction}}/2.303R$.

The effect of pressure changes is expressed by the relation:

$$\left(\frac{\partial \ln K_a}{\partial P}\right)_T = -\frac{\Delta V^0}{RT}$$

When both T and P vary

$$dP/dT = \frac{\Delta H^0_{\text{reaction}}}{T \Delta V_{\text{reaction}}} \text{ (Clapeyron equation)}$$

Empirically, increasing pressure promotes the ionization of water, so that at 2000 atm the K_w is four times its value at 1 atm (Hamann, 1963). Increasing temperature results in a 100-fold increase in K_w ($K_{w(25°C)} = 10^{-14}$, $K_{w(100°C)} = 10^{-12}$; Sillén, 1964).

The pH Concept

At 25°C and a concentration of 1 mole per liter, the ion product of water is 1×10^{-14}. The interrelation of [H+] and [OH−] concentration has led to the introduction of pH as a useful measure of acidity or alkalinity.

The pH is most commonly defined as −log [H+], or, as chemists generally prefer, −log a_{H^+}. A material is acidic in a pH range of 1 to 7, basic in a range of 8 to 14, and neutral at a pH of exactly 7, at normal temperature and pressure. The above definition of pH, although exact, is not very practical, since the activity of a single ion cannot be measured directly. The pH is therefore determined by the use of an EMF cell, consisting of a glass electrode, which dips into the test solution, and a reference electrode (usually a mercury-mercurous chloride electrode in saturated KCl) to complete the circuit. The reference electrode is connected to the unknown solution by a salt bridge of a saturated KCl solution. The glass electrode is constructed so that its potential varies with the logarithm of the H+ ion activity, while the reference electrode has a constant potential. The modern operational definition of pH is given by the equation

$$pH_{\text{unknown}} = pH_{\text{standard}} - \frac{F(E_{\text{unknown}} - E_{\text{standard}})}{2.303\, RT}$$

where F is the Faraday constant, R is the gas constant, E is the electrode potential, and T is the temperature.

Geological Applications

The hydrogen ion concentration is of great significance in chemical reactions accompanying geological processes. The pH of the medium is especially important in regulating the precipitation of hydroxides from solutions, as is seen in the following list:

pH	Precipitation of Hydroxides	Natural Media
11		
	Magnesium, Mg(OH)$_2$	
10		Alkali soil, playa lakes
9		
	Bivalent manganese, Mn(OH)$_2$	
8		Seawater
7	Zinc, Zn(OH)$_2$	River water
6	Copper(II), Cu(OH)$_2$	Rainwater
5	Bivalent iron, Fe(OH)$_2$	
4	Aluminum, Al(OH)$_3$	Peat water
3		Mine waters
	Trivalent iron, Fe(OH)$_3$	
2		Acid thermal springs
1		Oxidizing pyrite

The $CaCO_3$–CO_2–H_2O *buffer system* controls the pH of many terrestrial waters and also governs the formation of calcite and aragonite (Garrels and Christ, 1965).

The transportation and deposition of Fe(OH)$_3$ depends upon the solubility, which, in turn, depends on pH.

$$K_{sp} \text{ Fe(OH)}_3 = [\text{Fe}^{3+}][\text{OH}^-]^3$$

$$[\text{Fe}^{3+}] = \frac{K_{sp}}{[\text{OH}^-]^3} ; \text{ in water } [\text{OH}^-] = \frac{K_w}{[\text{H}^+]}$$

$$\therefore [\text{Fe}^{3+}] = K_{sp}[\text{H}^+]^3 / K_w^3$$

At a fixed temperature, K_{sp} and K_w are constant so that the concentration of iron depends on the cube of the hydrogen ion concentration (Mason, 1966).

Transportation and Deposition of Al$_2$O$_3$ and SiO$_2$

In an acid medium (pH $<$ 4), alumina is quite soluble, silica is not. Thus, in such an environment, e.g., a tropical acid soil, alumina

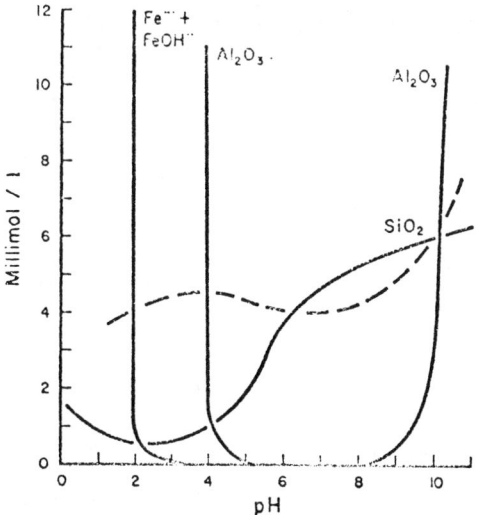

FIG. 1. Solubility of silica gel, aluminum hydroxide, and ferric iron hydroxide in relationship to pH (after Correns, 1939; from Fairbridge, 1967). Broken line indicates a modified SiO_2 solubility, following the work of Krauskopf, 1959.

would be removed and quartz would remain. It is rare to have such a low pH in a sedimentary environment. In the pH range 5–9, the alumina solubility curve is at a minimum, whereas that of silica increases slightly but remains higher than that of alumina. This may account for the development of laterites and bauxite in the weathering of subtropical soils. The pH is especially important in the formation of clays. At pH 4 the solubilities of Al_2O_3 and SiO_2 are such as to favor the formation of kaolinite, whereas in alkaline solutions (pH 8–9), montmorillonite is formed (Mason, 1966).

Hydrothermal alteration of ore deposits is interpreted by Hemley and Jones (1964) as a process of hydrogen metasomatism, by which hydrogen ions are added to the rock, a mole equivalent of cations is released, and the (OH^-/H^+) ratio in the solution is increased. The H^+ or H_3O^+ ions react extensively with O atoms of the original silicate minerals to form hydrous silicates containing OH groups (such as sericite, kaolinite, montmorillonite, biotite, etc.). In general, alteration zones display a net loss of base equivalents (metal cations).

Petrologic Terminology

The terms *acidic* and *basic* have previously been used in petrology to indicate relative concentration of SiO_2, based upon the outdated hypothesis that silicate minerals of the igneous rocks were salts of the silicic acids. By convention those with more than 66% SiO_2 were classed as acidic, those with less than 50% were basic, and the rest were intermediate. Incongruously the more "acid" rocks contain more alkalis (sodium, potassium), although "basic" rocks generally contain more calcium. Such petrologic use of the terms "acidic" and "basic" thus has nothing to do with acid-base relations in geochemistry, and should be avoided.

VIVIEN GORNITZ

References

Barnes, H. L. (editor), 1967, "Geochemistry of Hydrothermal Ore Deposits," Chap. 6, New York, Holt, Rinehart and Winston, Inc., 670pp.

Bell, R. P., 1952, "Acids and Bases: Their Quantitative Behavior," London, Methuen; New York, John Wiley & Sons, 90pp.

Fairbridge, R. W., 1967, "Phases of diagenesis and authigenesis," in (Larsen, G., and Chilingar, G. V., editors), "Diagenesis in Sediments," Amsterdam, Elsevier, pp. 19–89.

Garrels, R. M., and Christ, C. L., 1965, "Solutions, Minerals and Equilibria," New York, Harper and Row, 1450pp.

Hamann, S. D., 1963, "The ionization of water at high pressures," *J. Phys. Chem.*, **67**, 2233–2235.

Harned, H. S., and Owen, B. B., 1958, "The Physical Chemistry of Electrolytic Solutions," Third ed. (*Am. Chem. Soc. Monograph* **137**), New York, Reinhold Book Corp.

Hemley, J. J., and Jones, W. R., 1964, "Chemical aspects of hydrothermal alteration with emphasis on hydrogen metasomatism," *Econ. Geol.*, **59**, 538–369.

King, E. J., 1965, "Acid-Base Equilibria," New York, Macmillan, 341pp.

Krauskopf, K. B., 1959, "The geochemistry of silica in sedimentary environments," in "Silica in Sediments," *Soc. Econ. Paleontologists Mineralogists, Spec. Publ.*, **7**, 4–19.

Krauskopf, K. B., 1967, "Introduction to Geochemistry," Chap. 2, New York, McGraw-Hill Book Co., 721pp.

Mason, B., 1966, "Principles of Geochemistry," Third ed., New York, John Wiley & Sons, 329pp.

Pearson, R. G., 1966, "Acids and bases," *Science*, **151**, 172–177.

Sillén, L. G., 1964, "Stability Constants of Metal-ion complexes," Sect. I, "Inorganic Ligands," *Chem. Soc. London, Spec. Publ.* **17**.

Cross-references: *Abrasion pH; Aqueous Solutions; Base—Chemical; Buffer System; Calcium Carbonate; Electronegativity; Hydrothermal Alteration; Ion Exchange; pH-Eh Diagrams; pH-Eh Relations; Solubility Product Constant; Weathering, Chemical.* Vol. IVB: *Bauxite.* Vol. VI: *Laterites.*

ACTINIDE SERIES

The actinide series comprises the chemical elements with atomic numbers 89–103. The

ACTINIDE SERIES

TABLE 1.

Atomic Number	Name	Origin of Name	Year Discovered	Symbol	Atomic Weight[a]	Half-Life of Longest-Lived Isotope	Electronic Configuration[b]	Valence[b]
89	Actinium	Gr. *aktis, aktinos*	1899	Ac	227.03	22 yr	$6d^17s^2$	3
90	Thorium	Scandinavian god *Thor*	1828	Th	232.038	1.39×10^{10} yr	$6d^27s^2$	4
91	Protactinium	Gr. *protos* and actinium	1913	Pa	231.036	3.48×10^4 yr	$5f^26d^17s^2$	4,**5**
92	Uranium	Planet Uranus	1789	U	238.03	4.51×10^9 yr	$5f^36d^17s^2$	3,4,**5**,6
93	Neptunium	Planet Neptune	1940	Np	237.05	2.20×10^6 yr	$5f^46d^17s^2$	3,4,**5**,6
94	Plutonium	Planet Pluto	1940	Pu	239.05	8×10^7 yr	$5f^67s^2$	**3**,4,5,6
95	Americium	The Americas	1944	Am	[243]	8.0×10^3 yr	$5f^77s^2$	**3**,4,5,6
96	Curium	P. and M. Curie	1944	Cm	[247]	1.6×10^7 yr	$5f^76d^17s^2$	**3**,4
97	Berkelium	Berkeley, Calif.	1949	Bk	[247]	7×10^3 yr	$5f^97s^2$	**3**,4
98	Californium	California (Univ. and State)	1950	Cf	[251]	~800 yr	$5f^{10}7s^2$	3
99	Einsteinium	A. Einstein	1952	Es	[254]	480 days	$5f^{11}7s^2$	3
100	Fermium	E. Fermi	1952	Fm	[257]	94 days	$5f^{12}7s^2$	3
101	Mendelevium	D. Mendeleev	1955	Md	[257]	3 hr	$5f^{13}7s^2$	3
102	Nobelium?	A. Nobel	1957?	No?	[253]	1.58 min	$(5f^{14}7s^2)$	(3)
103	Lawrencium	E. O. Lawrence	1961	Lw	[256]	45 sec	$(5f^{14}6d^17s^2)$	(3)

[a] Atomic weights for Ac, Pa, Np, and Pu, are based on the masses of nuclides that occur naturally on earth. For elements 96–103, the mass number of the longest-lived isotope is given in brackets. Boldface indicates predominant oxidation state under normal conditions.
[b] Parentheses indicate predicted configuration or valence state.

name is derived from the first element of the series, *actinium,* and is used in the same sense as the term *lanthanide series* is used for the rare earth elements. The similarity in chemical properties among the lanthanides is a result of the addition of electrons to the inner $4f$ orbitals, in the fourteen elements following lanthanum, in such a way that the availability of valence electrons is rather uniform throughout the series. Similarly, electrons are added to the $5f$ orbitals in the actinide series (see Table 1), and it is on this basis that the actinides are considered to be a second example of an "inner transition" series.

Occurrence and Origin

All nuclides of the actinide elements are radioactive, and only three of these nuclides are sufficiently long-lived to exist in large quantity on earth. These are two isotopes of *uranium,* U^{238} (half life 4.51×10^9 years) and U^{235} (7.13×10^8 years), and one of *thorium,* Th^{232} (1.39×10^{10} years). Shorter-lived isotopes of uranium and thorium occur naturally among the daughter products of the long-lived nuclides. Notable among these are U^{234} (2.48×10^5 years) and Th^{230} (*ionium,* 7.5×10^4 years), both descendants of U^{238}. Also found among the daughter products of uranium and thorium are isotopes of two other actinide elements, *protactinium* and *actinium.* Natural protactinium consists almost exclusively of Pa^{231} (3.48×10^4 years), a descendent of U^{235}. Pa^{231} decays to Ac^{227} (22 years), the preponderant constituent of natural actinium.

Two other actinide elements have been found to occur naturally on earth. These are *neptunium* and *plutonium,* the first two transuranium elements, and their occurrence in minute traces is a result of natural nuclear reactions on uranium. Pu^{239} (2.44×10^4 years) occurs in uranium ores at a concentration up to about 10^{-11} that of the uranium. Np^{237} (2.20×10^{-6} years) occurs in even lower concentration in uranium ores (see also Table 2).

The original discovery of neptunium and plutonium was a result not of their natural occurrence, but of nuclear reactions in the laboratory. The discovery of these elements near the beginning of World War II, and the subsequent production of kilogram quantities of Pu^{239} within a few years of its discovery, are major parts in the drama of the original development of nuclear weapons. Neptunium was discovered by E. M. McMillan and P. H. Abelson among the products of neutron irradiation of uranium. Plutonium was first detected by G. T. Seaborg and others as a

ACTINIDE SERIES

TABLE 2.

Element	Melting Point (°C)	Density (g/cm³)	Crystal Structure	Ionic Radii (Angstroms) +3	+4	Oxidation Potentials[a] (volts) 0–III	0–IV	III–IV
Ac	1050		FCC	1.11		(~2.6)		
Th	1750	11.72	FCC		0.99		1.90	
Pa	1600	15.37	Tetragonal		0.96		(0.9)	
U	1132	19.04	Orthorhombic	1.03	0.93	1.80		0.631
Np	637	20.45	Orthorhombic	1.01	0.92	1.83		−0.155
Pu	639	19.74	Monoclinic	1.00	0.90	2.03		−0.982
Am	994	13.8	Hexagonal	0.99	0.89	2.32		(−2.7)
Cm	1340	13.5	Hexagonal	0.98				
Bk								(−1.6)

[a] Standard oxidation potentials in acid aqueous solution. Estimated values are given in parentheses. See literature for values involving oxidation states higher than IV.

product of deuteron bombardment of uranium. In the two decades following the discovery of the first transuranium elements, the remaining members of the actinide series were produced by a variety of nuclear reactions, and were identified by Seaborg and his co-workers at the University of California at Berkeley. (Discovery of element 102 was first claimed by workers elsewhere, but their experiments have not been substantiated, so their claim to discovery and the suggested name, *nobelium,* are in dispute.) *Americium* was first produced by neutron irradiation of plutonium. *Curium, berkelium, californium,* and *mendelevium* were produced by *helium ion bombardment* of elements having, respectively, two protons fewer. *Einsteinium* and *fermium* were first isolated as products of *multiple neutron capture* by uranium during the first thermonuclear explosion. Element 102 was produced by bombardment of curium with carbon ions, and *lawrencium* by bombardment of californium with boron ions.

None of the transplutonium elements have been found to occur naturally on earth. Their existence in measurable quantity is not expected since none of the nuclides has a half-life comparable with the age of the earth and since there are no simple nuclear reactions by which they might be formed from uranium. On the other hand, the transplutonium elements are certainly produced by stellar processes, in particular by the supernovae explosions thought to be primarily responsible for the formation of thorium and uranium. It is expected that a tremendous burst of neutrons produced during the supernova explosion is sufficient to build up nuclides through and perhaps beyond the mass range of the heaviest actinides.

Chemical and Physical Properties

The chemistry of the actinide series is similar in some respects to that of the lanthanide series, but there are important differences (see Tables 1 and 2). Outstanding among these are the absence of the +3 oxidation state for thorium and protactinium; the many oxidation states of uranium, neptunium, plutonium, and americium; and the absence of the +2 oxidation state for any member of the series. (In fact, the differences between actinium and the five elements that follow it are so pronounced that some authors reject the concept of an actinide series.) These differences are generally attributable to lower binding energies for the 5f electrons of the actinides in comparison with the 4f electrons of the lanthanides.

Actinium and the transplutonium actinides occur in the tripositive state under normal conditions, and it is these actinides that are most similar to the lanthanides. This similarity is displayed in the composition and structure of compounds and also, to a large extent, in the structures and properties of the metallic states. Although uranium, neptunium, and plutonium do not occur in the tripositive state under normal conditions, their tripositive compounds are similar to those of the other actinides and to those of the lanthanides. Ionic sizes of the actinides become progressively smaller with increasing atomic number (for a given oxidation state), a contraction analogous to the well-known *lanthanide contractiton*. The ionic radii of the actinides are in fact only a few Angstroms larger than those of the corresponding lanthanides, so that the similarities in structures and properties of compounds among the two groups are to be expected.

The properties of thorium are similar to those of its analog cerium, insofar as the latter occurs in the +4 state. Thorium has, however, no 5f electrons, and its properties are more like those of hafnium (periodic Group IV) than those of cerium.

The five elements following thorium have many properties not seen in the lanthanides. These differences are a result of the multiplicity of oxidation states and the stability of some of the higher oxidation states. The complex behavior of these elements is a result of similarity in energy level of electrons in the $7s$, $6d$, and $5f$ orbitals, and is manifested not only in complex oxidation-reduction behavior, but also in complicated structures and phase relationships in the metallic state. The complexity reaches a peak in the element plutonium, which exhibits six different phases in the metallic state, and which, in solution, may have species of all four oxidized states in equilibrium at appreciable concentrations.

In acid aqueous solution, the actinides may exist as M^{+3}, M^{+4}, MO_2^+ (except Pa(V), whose behavior in solution is very complex), or MO_2^{+2} ions, depending on oxidation state. Solutions of the different ions of uranium, neptunium, plutonium, and americium show a variety of colors. These elements may be separated by their oxidation-reduction behavior. Separation of the other actinides, as in the case of the lanthanides, is best achieved by ion-exchange techniques involving the M^{+3} ions.

Nuclear Properties

All nuclides of the actinide elements are unstable, i.e., undergoing *alpha decay*. In many cases *beta decay* processes are predominant over alpha decay, and in a few cases among the heaviest elements the predominant mode of decay is *spontaneous fission*.

For elements of odd atomic number, only one or two isotopes are stable with respect to beta decay, so most isotopes of the odd numbered actinides decay predominantly by β^- emission or *electron capture*. In some proton-rich isotopes, however, alpha decay is predominant over electron capture.

Each of the even-numbered actinide elements has many beta-stable isotopes, so alpha decay is the most common decay mode in this group. Considering only the longest-lived isotope of each (even-numbered) element, there is a consistent increase in alpha decay rate with increasing atomic number (see Table 1). Alpha half-lives range from 10^{10} years for thorium to less than one year for fermium, and perhaps only minutes for element 102. The odd-numbered elements show less regularity in alpha decay rates, but in general their isotopes are shorter-lived than those of their even-numbered neighbors.

Spontaneous fission is an alternative decay mode for a number of nuclides within the actinide series. Those with even numbers of protons and neutrons have higher spontaneous fission rates than neighboring nuclides with unpaired nucleons. Spontaneous fission rates are extremely low among the lighter actinides, but the rates increase rapidly with increasing atomic number, so that for a few of the heaviest nuclides spontaneous fission predominates over alpha decay.

Fission of excited nuclei is an extremely important property among the actinide elements. A number of these nuclides undergo fission upon capture of thermal neutrons, a feature that has been of paramount importance in the exploitation of nuclear energy. U^{235}, Pu^{239}, and U^{233} are of practical importance as nuclear fuels. The excited nuclei formed in production of the heaviest actinides readily undergo fission, often with much higher probability than other modes of de-excitation. This process severely lowers the efficiency of production of the heaviest elements.

Geochemistry

Discussions of the geochemistry of actinium, plutonium, thorium, protactinium, and uranium are given in separate articles. The geochemistry of actinium, neptunium, and plutonium is of little importance because of their exceedingly low terrestrial abundances. Actinium, with no isotope having a half-life of more than 22 years, is normally found only in association with its parent nuclides. Neptunium and plutonium are even more restricted in their occurrence, for not only can they be produced only from uranium, but the nuclear reactions by which they are formed must be initiated by decay of another atom of uranium or other heavy element. Formation of neptunium and plutonium is thus essentially limited to environments where the concentration of uranium (or thorium) is high, and such environments are rare.

There has been some speculation that remnants of original transuranium elements might be found on earth. Unidentified lines in nuclear spectra of some natural materials have been attributed to Cm^{247} (1.6×10^7 years) or to Pu^{244} (8×10^7 years), the longest-lived nuclides among the transuranium elements, but there has been no confirmation of such occurrences. Unless the half-lives are considerably in error, it is overwhelmingly probable that these nuclides are extinct. The role of Pu^{244} in the early history of the solar system and the earth is of some interest. Heavy isotopes of xenon in the earth's atmosphere and in meteorites have been attributed to spontaneous fission of this nuclide, as have an excess of fission tracks found in some meteoritic minerals. Decay of Pu^{244} may have contributed significantly to the early heating of the earth.

Because of the quantity of plutonium that has been artificially produced, the geochemistry of the transuranium elements in the sedimentary environment is of some interest. With the exception of neptunium, which under normal conditions forms the NpO_2^+ ion, the highly charged (+3 or +4) ions of this group are expected to show considerable hydrolysis in natural waters, and thus to be associated with particulate or colloidal material rather than in solution.

Uses

The uses of thorium and uranium are described in separate articles. Of the other actinides, only plutonium (as Pu^{239}) has been used in quantity—as a constituent of nuclear explosives and as a fuel for nuclear reactors. In research, the actinides are used as tracers, and, in a unique use, the lighter actinides have served as the starting materials from which heavier elements are produced. Currently, curium and some heavier elements are being produced in relatively large quantities with the hope of extending the list of elements beyond the actinide series.

A potentially very important use of certain actinides is as a highly concentrated, long-lasting energy source for space vehicles or for remote terrestrial installations. Alpha-emitting nuclides with half-lives of years or tens of years are best suited for this purpose. Pu^{238} (86.4 years), Cm^{242} (162 days), and Cm^{244} (18 years) are of practical importance in this regard.

J. M. WAMPLER

References

*Cunningham, B. B., 1964, "Chemistry of the Actinide elements," *Ann. Rev. of Nucl. Sci.,* **14**, 323–346.
Haissinsky, M., and Adloff, J-P, 1965, "Radiochemical Survey of the Elements," Amsterdam, Elsevier Publ. Co.
*Katz, J. J., and Seaborg, G. T., 1957, "The Chemistry of the Actinide Elements," London, Methuen and Co. Ltd.
Seaborg, G. T., 1963, "Man-Made Transuranium Elements," Englewood Cliffs, N.J., Prentice-Hall, Inc.

* Additional references may be found in this work.

Cross-references: *Actinium; Aqueous Solutions; Cerium; Charged Particle Tracks; Electron Capture; Fission Track Dating; Hafnium; Ion Exchange; Lanthanides; Neutron Activation Analysis; Oxidation and Reduction; Plutonium; Protactinium; Rare Earths (Lanthanide Series); Rare (Noble, Inert) Gases; Thorium; Uranium; Xenon.*

ACTINIUM: ELEMENT AND GEOCHEMISTRY

Properties

Actinium, a radioactive element, atomic number 89 and atomic weight 227, was first identified in 1899 by A. Debierne shortly after the discovery of radium and polonium. This element, placed in Group IIIA of the 7th period of the periodic table, is one of the actinides. The actinides ($Z = 89$–103) show chemical behavior analogous to the lanthanides because electrons are added to the $5f$ shell after the outer $7s$, $6s$, and $6p$ shells are filled and the $6d$ shell contains one electron. Actinium, in its normal trivalent form with an ionic radius of 1.18 Å, has a coordination number of 8 with chlorides (Green, 1959). Actinium can also assume the (IV) oxidation state.

Two natural isotopes of actinium, Ac^{227} and Ac^{228}, are known (Rankama, 1954). Ac^{227} is a principal member of the actinium series, which, together with the uranium and thorium series, are composed of naturally occurring radioelements of high atomic weight. The parent element of the actinium series is U^{235}, sometimes called actinouranium. This series, whose isotopes have mass numbers calculated from $4n + 3$, where n is an integer varying from 58 to 51, follows the decay scheme given in Table 1, leading to Pb^{207r} as the stable end product. Ac^{228} is a short-lived daughter in the thorium series. Artificial actinium isotopes of mass numbers 221, 222, 223, 224, 226, 229, and 230 have been produced.

Minerals

No actinium minerals are known. The element is a natural daughter product of uranium and hence is found in minute quantities in uranium and thorium minerals or in other minerals containing U and Th as trace constituents.

Abundance (Green, 1959)

The parent isotope, U^{235}, comprises only 0.79% of the three natural uranium isotopes. Because of the short half-life (22 years) of actinium, this element will be present in only infinitesimal amounts in uranium-bearing minerals. Thus, a metric ton of pitchblende contains less than 0.20 mg of actinium. Igneous rocks contain an average of 5×10^{-10} ppm of actinium, with granites having up to 7–8×10^{-10} ppm and ultrabasics from 0.2 to 2.0×10^{-10} ppm. Abundances of 0.3×10^{-10} and 0.1×10^{-10} ppm have been assigned to stony and iron meteorites respec-

ACTINIUM: ELEMENT AND GEOCHEMISTRY

TABLE 1. THE ACTINIUM SERIES (FROM GLASSTONE, 1958)

Radioelement	Corresponding Element	Symbol	Radiation	Half-Life
Actinouranium	Uranium	U^{235}	α	7.13×10^8 yr
Uranium Y	Thorium	Th^{231}	β	25.6 hr
Protactinium	Protactinium	Pa^{231}	α	3.43×10^4 yr
Actinium 98.8% | 1.2%	Actinium	Ac^{227}	β and α	21.8 yr
Radioactinium	Thorium	Th^{227}	α	18.4 days
Actinium K	Francium	Fr^{223}	β	21 min
Actinium X	Radium	Ra^{223}	α	11.7 days
Ac Emanation	Radon	Rn^{219}	α	3.92 sec
Actinium A ~100% | ~5×10^{-4}%	Polonium	Po^{215}	α and β	1.83×10^{-3} sec
Actinium B	Lead	Pb^{211}	β	36.1 min
Astatine-215	Astatine	At^{215}	α	~10^{-4} sec
Actinium C 99.7% | 0.3%	Bismuth	Bi^{211}	α and β	2.16 min
Actinium C'	Polonium	Po^{211}	α	0.52 sec
Actinium C''	Thallium	Tl^{207}	β	4.78 min
Actinium D (end product)	Lead	Pb^{207}	Stable	—

tively. A cosmic abundance of 0.5×10^{-14} ppm has been estimated for actinium. Actinium concentrations are determined by sensitive alpha-ray spectrometers or scintillation counters.

Geochemical Behavior

Actinium, dependent on uranium and thorium for its existence, follows these elements in mode of occurrence. Thus, as a lithophile element, it becomes enriched in the late stages of magmatic crystallization and can concentrate in pneumatolytic or hydrothermal phases (Rankama and Sahama, 1950).

AcC (Bi^{211}) and RaC' (Po^{214}) provide the basis of a pleochroic halo method of dating mica-bearing rocks (Rankama, 1954). These two isotopes produce rings of different radii and relative densities of discoloration. The ratio of AcC to RaC' is equal to the ratio of radiogenic Pb^{207} to Pb^{206}. However, discoloration densities may be altered by higher temperatures, limiting the method to rocks that have not been strongly reheated.

Protactinium, the immediate parent of Ac^{227}, has been used because of its half-life (34,300 years) to measure rates of marine sedimentation.

NICHOLAS M. SHORT

References

Glasstone, S., 1958, "Sourcebook on Atomic Energy," second ed., Princeton, N.J., D. Van Nostrand Co., 641pp.

Green, J., 1959, "Geochemical table of the elements for 1959," *Bull. Geol. Soc. Am.*, **70**, 1127–1184.

Rankama, K., 1954, "Isotope Geology," New York, McGraw-Hill Book Co., 535pp.

Rankama, K., and Sahama, T. G., 1950, "Geochemistry," Chicago, Univ. of Chicago Press, 911pp.

Cross-references: Actinide Series; Elements, Planetary *Abundances and Distribution; Lanthanides; Protactinium; Spectrophotometry.*

AEROSOLS—*See* Vol. II

AIR POLLUTION—*See* **POLLUTION**

AIR POLLUTION AND GLOBAL CLIMATE

Global Trends of Climate

Meteorological data reveal a systematic fluctuation of global climate in the past century [see Vol. II, *Climatic Variation (Instrumental Data),*]. This fluctuation has consisted in part of a net worldwide warming of about 0.6°C between the 1880s and the 1940s, followed thereafter by a net cooling which to date (1970) has accumulated to about 0.3°C (Fig. 1). The temperature fluctuation is thought to reflect a change in the planetary heat budget; it has been accompanied also by changes in the large-scale atmospheric circulation and in other climatic elements. It is likely that the bulk of the fluctuation is ascribable to natural causes, for example, variable volcanic dust loading of the upper atmosphere (Table 1 and Fig. 2). A part of the fluctuation, however, may be related to increasing atmospheric pollution in the period.

Air Pollutants Potentially Involved

Of the many substances entering the atmosphere as a by-product of human activities, specifically two—carbon dioxide and particulate materials—are of special concern as possible causes of disturbances in global climate. With regard to CO_2, the average longevity (residence time) of a molecule of this gas in the atmosphere is several years. This is a long enough time for new atmospheric CO_2 supplied through combustion of fossil fuels (now exceeding 10^{10} tons per year worldwide) to become well mixed by air circulations into all parts of the atmosphere. Precision CO_2 measurements in remote sites (Hawaii and Antarctica) verify that such mixing has taken place very efficiently, and that the global atmospheric reservoir of CO_2 has been growing by about 0.2% per year since 1958 when the observations began (Fig. 3). With regard to particulate materials (or *aerosol*), average residence times of those which enter the atmosphere near ground level are measured in days or weeks, rather than years as in the case of CO_2. However, the sources of particulates are so ubiquitous that these materials also tend to be distributed through very large expanses of the atmosphere, albeit in much lower concentrations than those found in the immediate vicinity of urban or industrial centers. Evidence for the geographically extensive spread of particulates, including most continents and the North Atlantic Ocean, has accrued in recent years from a variety of sources. The worldwide total atmospheric particulate loading derived from all human activities is estimated to have increased by a full order of

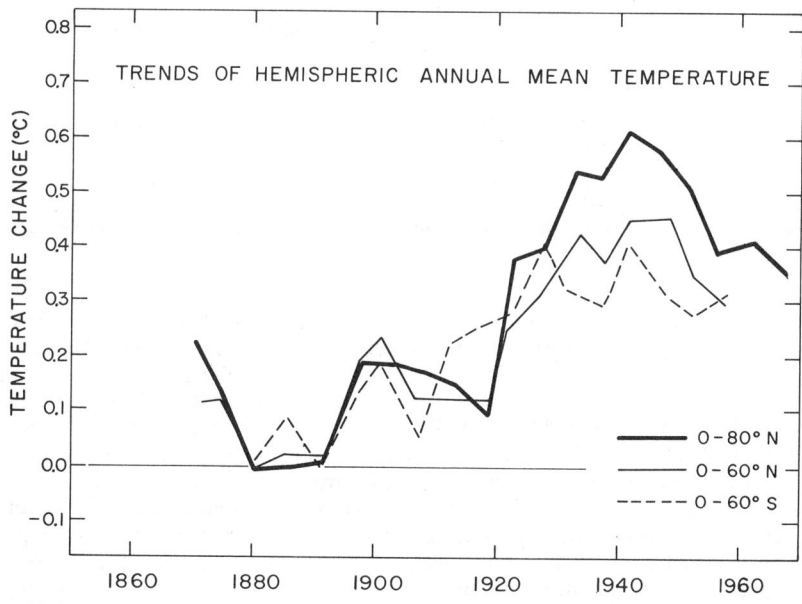

FIG. 1. Changes of mean annual temperature, 1870–1967, integrated over various latitude bands of the earth. Data after 1960, shown for 0–80°N band only, indicates continuation of cooling trend that began in 1940s.

AIR POLLUTION AND GLOBAL CLIMATE

TABLE 1. VOLCANIC ERUPTIONS SINCE 1855

Date	Name and Location			Severity Class[a]
1855	Cotopaxi, Ecuador	1°S	78°W	1-1/2
1856	Awu (Awoe)	3.5°N	125.5°E	2
1861	Makjan, Molucca Is.	.5°N	127.5°E	2
1870	Ceboruco, Mexico	21°N	105°W	3
1872	Vesuvius	41°N	14°E	3
1872	Merapi, Java	7.5°S	110°E	3
1875	Askja (Vatna Jökull) Iceland	65°N	17°W	2
1877	Cotopaxi, Ecuador	1°S	78°W	3
1883	Krakatoa	6°S	105.5°E	1
1883	St. Augustine, Alaska	59.5°N	153.5°W	3
1883	Bogoslov, Aleutians	54°N	168°W	3
1885	Falcon Island	20°S	175°W	3
1886	Tarawera, N.Z.	38.5°S	176.5°E	2
1886	Niafu, Tonga Is.	16°S	175.5°W	3
1888	Bandai San, Japan	38°N	140°E	2
1888	Ritter Is.	5.5°S	148°E	2
1890	Bogoslov, Aleutians	54°N	168°W	3
1892	Awu (Awoe)	3.5°N	125.5°E	2
1902	Mont Pelée, Martinique	15°N	61°W	2
1902	Soufrière, St. Vincent	13.5°N	61°W	2
1902–04	Santa Maria, Guatemala	14.5°N	92°W	1-1/3
1907	Shtyubelya, Kamchatka	52°N	157.7°E	2
1911	Taal, Luzon	14°N	121°E	3
1912	Katmai, Alaska	58°N	155°W	2
1913	Colima, Mexico	19.5°N	104°W	3
1914	Sakurashima, Japan	31.5°N	131°E	3
1921	Andes (Chile-Argentina border)	≃30°S	≃70°W	3
1929	Asama, Japan	36.5°N	138.5°E	4
1931	Kluchev, Kamchatka	56°N	160.5°E	4
1932	Quizapu, Chile	35.5°S	70.5°W	3
1947	Hekla, Iceland	64°N	19.5°W	2
1953	Mt. Spurr, Alaska	61°N	153°W	2
1955	Ranco Puyehue, Chile	40°S	72°W	3
1956	Bezymyannaya, Kamchatka	56°N	160.5°E	2
1960	Puntiagudo et al., So. Chile	39–45°S	72°W	3
1963	Gunung Agung, Bali	8.5°S	115.5°E	1-1/2
1963–65	Surtsey, Iceland	63°N	20.5°W	3
1966	Awu (Awoe)	3.5°N	125.5°E	2
1968	Fernandina I., Galapagos	0.5°S	92°W	2

[a] Severity class reflects the estimated order of magnitude of the total mass of material ejected:
  Class 1: 10–1 $km^3 \sim 10^{10}$ metric tons
  Class 2: 1–0.1 $km^3 \sim 10^9$ metric tons
  Class 3: 0.1–0.01 $km^3 \sim 10^8$ metric tons
  Class 4: 0.01–0.001 $km^3 \sim 10^7$ metric tons
  Fractional classes represent compromises believed to be appropriate. Data based primarily on table by Lamb (1970).

magnitude in the past century, to a present value (smoke-sized particles only) on the order of 10^6 tons (see Fig. 2).

Effect of Pollutants on Atmospheric Heat Budget and Temperature

Both CO_2 and atmospheric particulates (aerosol) are of concern to climate because both are capable of altering the terrestrial heat budget. In the case of CO_2, the gas absorbs long-wave thermal radiation emanating from the earth's surface toward space and re-emits part of that radiation back toward the surface again. Thus it tends to warm the climate near the ground by a mechanism usually referred to as the "greenhouse effect." In the case of atmospheric particulates, the situation is more complicated. The particulates absorb a part of the solar radiation passing through the atmosphere toward the earth's surface (warming effect), and scatter another part back toward space (cooling effect). Depending *(1)* on the relative efficiencies of the aerosol in absorbing and backscattering solar radiation, *(2)* on the altitude of the aerosol, and *(3)* on certain physical properties of the underlying surface, the *net* thermal effect near the ground may be one of either warming or cooling. In

AIR POLLUTION AND GLOBAL CLIMATE

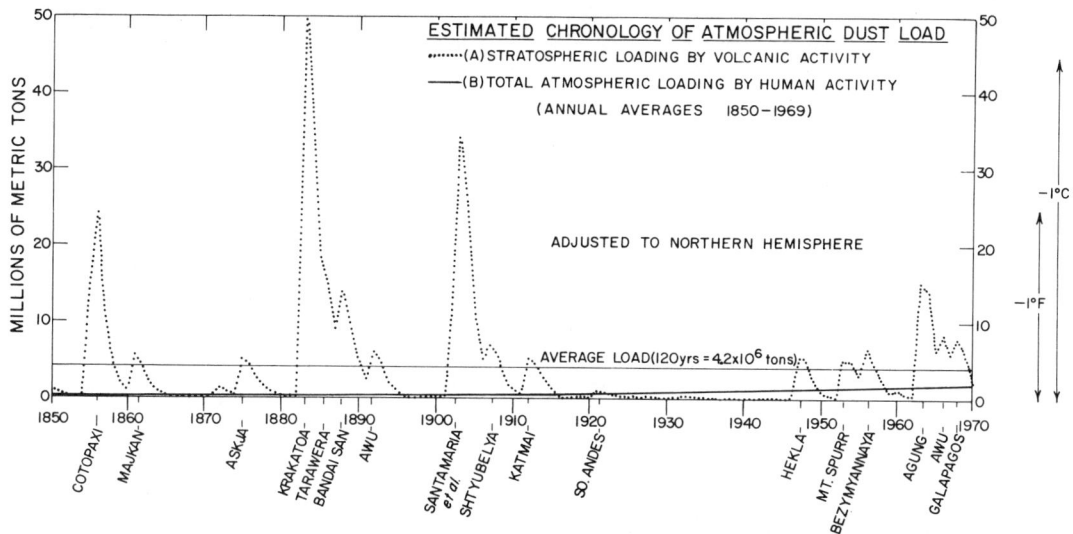

FIG. 2. Estimated chronology of worldwide atmospheric particulate loading, in particle-size range 0.1–1μ, by volcanic activity (stratospheric loading only, dotted curve) and by human activity (heavy solid curve), 1850–1970. Average of volcanic loading added for comparison (thin solid line). Rough calibration of volcanic loading in terms of planetary average temperature effect is shown in the outside margin at right. Volcanic chronology derived from data in Table 1 (see Singer (1970) for details).

most studies of the problem, it has been tacitly assumed that the backscatter effect of aerosol dominates over the absorption effect, inferring a net cooling of surface climate. Surface cooling is clearly indicated in the case of high-altitude aerosol such as volcanic dust veils in the stratosphere. Very recent investigations of the problem of low-altitude aerosol, such as that associated with most human activities, point to a surface warming effect, however. Confirmation of the latter conclusion, with the aid of better observations of the optical properties of aerosol than are presently available, is highly desirable.

The best available determination of the temperature effect of changes of atmospheric CO_2 is that of Manabe and colleagues (see references in Singer, 1970). Assuming constant relative humidity in the atmosphere, and conditions of surface albedo, cloudiness, solar radiation intensity, and other parameters chosen as typical of middle latitudes in an equinoxial season, these investigators calculated that a 10% increase of CO_2 concentration (which approximates the total CO_2 increase since the mid-nineteenth century) would result in a warming of about 0.3°C at all levels in the lower atmosphere, and a substantial cooling at higher stratospheric levels (above about 10 km).

Quantitative estimates of the temperature changes attributable to the long-term increase of atmospheric particulates from human sources have been attempted (see Singer, 1970). These, however, are based on very insecure assumptions and will not be cited here. As noted earlier, the temperature effect of such low-altitude aerosol is now thought likely to be one of warming and not cooling, and therefore supplements the warming effect of fossil CO_2 increases. Thus it is possible to ascribe to air pollution at least a part of the global climatic warming of 0.6°C between the 1880s and 1940s. By the same token, it is difficult to account for the global cooling that set in after the 1940s to any pollution effect.

Indirect Large-Scale Climatic Effects of Pollutants

It should be emphasized that the ultimate climatic response of global atmospheric increases of CO_2 and particulates may extend beyond the basic thermal response discussed above. Concomitant changes of cloudiness, for example those conceivably stimulated by the temperature changes or by possible cloud-nucleating effects of particulates (see *Air Pollution and Urban Climate*), might introduce important revisions to the ultimate thermal adjustment of global climate to the pollution changes. Moreover, the basic thermal effects of pollution might at some point either past or future begin to modify the pattern of the global atmospheric circulation itself. The circulation change, in turn, could then alter the worldwide pattern of climate in such a manner as to cause contrary temperature changes as

13

AIR POLLUTION AND GLOBAL CLIMATE

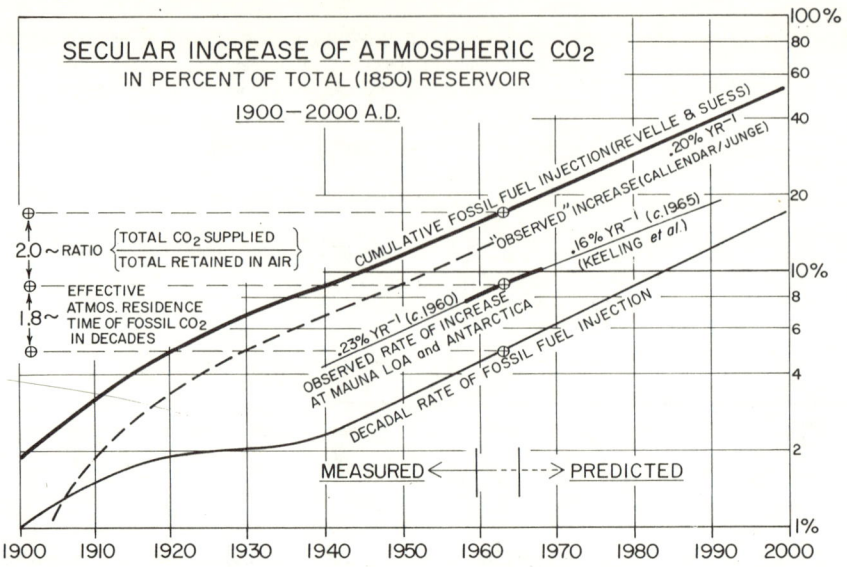

FIG. 3. Secular increase of atmospheric carbon dioxide due to fossil fuel combustion, as percent excess over 19th century base level, 1900–2000 AD. Top curve is limiting increase, assuming that all fossil CO_2 is retained in the atmosphere. Dashed curve is illustrative of early estimates of actual increase in atmosphere. Trend segments in the middle of the figure identify mean level and rates of increase of CO_2 based on precision measurements since 1958. The best estimate of actual increase is about half the increase represented by the top curve.

well as other climatic aberrations in some geographical areas. It will be necessary to await the development of more advanced numerical modeling techniques for the study of atmospheric phenomena, now on the way, before such matters can be investigated intelligently.

In perspective, it should be added that the global climatic fluctuation of the past century is not unlike similar fluctuations known to have occurred in earlier centuries and millennia. From this viewpoint we are amply justified in attributing all past climatic change to natural geophysical forces lying beyond man's control. At the same time, the impact of air pollution on global climate, along with other effects of human activities, clearly threatens to become more competitive with natural climate-driving forces in the future. The problem merits very close monitoring, and improved understanding.

J. Murray Mitchell, Jr.

References

Lamb, H. H., 1969, "Climatic Fluctuations," in "World Survey of Climatology," Vol. 2, Elsevier Publ. Co., Chap. 5.

Lamb, H. H., 1970, "Volcanic dust in the atmosphere; with a chronology and assessment of its meteorological significance," *Phil. Trans. Roy. Soc. London,* **266,** 425–533.

Robinson, G. D., 1970, "Long-term Effects of Air Pollution—A Survey," Hartford, Center for the Environment and Man, Inc.

Singer, S. F. (editor), 1970, "Global Effects of Environmental Pollution," Springer-Verlag New York Inc./D. Reidel Publ. Co.

Cross-references: *Air Pollution and Urban Climate; Environmental Pollution; Thermal Pollution; Urban Air Pollution.* Vol. II: *Aerosols; Atmospheric Circulation—Global; Atmospheric Nuclei and Dusts; Carbon Dioxide Cycle in the Sea and Atmosphere; Climatic Variation (Historical Record); Climatic Variation (Instrumental Data); Energy Budget of the Earth's Surface; Greenhouse Effect.* Vol. VI: *Volcanic Ash Deposits.*

AIR POLLUTION AND URBAN CLIMATE

Nature of Urban-Rural Differences of Climate

It has long been recognized that cities—which represent the most concentrated form of environmental modification by man—differ appreciably from rural areas as to local climatic conditions. Urban climatic anomalies extend to temperature, humidity, cloudiness, visibility, radiation, wind speed, and apparently precipitation as well. The typical direction and magnitude of such anomalies in an average large city is as indicated in Table 1.

AIR POLLUTION AND URBAN CLIMATE

TABLE 1. CLIMATIC CHANGES PRODUCED BY CITIES

Element	Comparison with Rural Environments
Temperature	
Annual mean	1.0 to 1.5 °F higher
Winter minima	2.0 to 3.0 °F higher
Relative humidity	
Annual mean	6% lower
Winter	2% lower
Summer	8% lower
Dust particles	10 times more
Cloudiness	
Clouds	5 to 10% more
Fog, winter	100% more
Fog, summer	30% more
Radiation	
Total on horizontal surface	15 to 20% less
Ultraviolet, winter	30% less
Ultraviolet, summer	5% less
Wind speed	
Annual mean	20 to 30% lower
Extreme gusts	10 to 20% lower
Calms	5 to 20% more
Precipitation	
Amounts	5 to 10% more[a]
Days with < 0.2 in.	10% more[a]

[a] Precipitation effects are relatively uncertain. (see text)

Air Pollution as a Contributory Cause

In accounting for these observed anomalies of urban climate, the role played by air pollution is not altogether clear. Other causative factors include (a) the local addition of heat by combustion and other energy expenditures in the city, or so called *thermal pollution;* (b) the effect of buildings and pavement in altering the disposition of solar heat; and (c) disturbances in the natural wind produced by the mechanical turbulence and obstacle effects of large buildings. Each of these factors gives rise to climatic effects which are similar to the presumed effects of locally concentrated air pollution, and are thus difficult to distinguish.

Urban air typically contains many gaseous and particulate materials, in concentrations substantially higher than those found in "clean" rural air. Those materials thought to have significant effects on climate include water vapor, carbon dioxide, combustion particles, and possibly nitrogen dioxide. Other pollutants, for example, sulfur oxides, nitrogen oxides other than NO_2, ozone, and metallic compounds, may also produce indirect climatic effects of one kind or another; these, however, are thought to be relatively minor and limited to special meteorological conditions.

Urban heat sources (thermal pollution) account for part of the observed excess of temperature in cities (the so-called urban "heat island"), especially in winter when the energy used for space heating and for transportation is comparable per unit city area to the energy received from the sun. The locally added water vapor and carbon dioxide may also contribute to the urban heat island, especially at night, by its ability to absorb and reradiate long-wave thermal radiation escaping from the surface toward the overhead sky. In other words, these gases may contribute to a locally intensified "greenhouse effect."

Effects of Atmospheric Particulates

The basic effects of airborne particulate materials (aerosol) are to scatter and absorb solar radiation, to deplete solar radiation reaching the urban surface, and to reduce visibility through the atmosphere. Further effects on urban temperature and other climatic conditions are uncertain; these depend on many physical properties of the aerosol which must be known in detail, and which tend to vary greatly from one city to another according to the type of local industry and heating fuel, and from one season to another. Such properties include the number, density and size distribution of the particles, the vertical distribution of the particles, and the chemical makeup of the particles, which governs the relative strengths of absorption and scattering by the aerosol.

Urban Pollution and Rainfall

The efficacy of urban pollutants in modifying precipitation, both in and downwind of cities, is not yet clearly established. Observed anomalous patterns of rainfall in cities are difficult to interpret with reliability, partly because of the extreme variability of natural precipitation which erodes the statistical significance of the observed patterns, and partly because of systematic gage catch errors which are largely unavoidable in urban rainfall measurements. Furthermore, such patterns as do appear to be real may be attributable to thermal or mechanical factors in the urban environment, rather than to local excesses of pollution (see Fig. 1). On the other hand, certain air pollutants (or by-products of chemical reactions taking place in polluted atmospheres) are credited with adding significant concentrations of either *ice nuclei* or *cloud condensation nuclei* to the natural background levels of such nuclei. In certain meteorological conditions the addition of such nuclei to urban air might alter the precipitability of clouds either in the immediate vicinity or

AIR POLLUTION AND URBAN CLIMATE

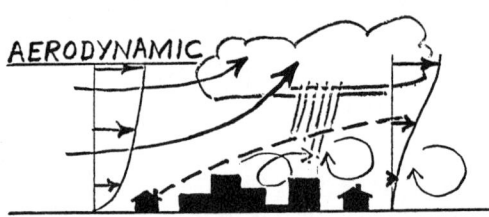

FIG. 1. Schematic diagram showing three kinds of effects of cities on local cloudiness and precipitation. (a) *Thermal* effect refers to enhancement of air convection by thermal pollution and by altered solar heating of the urban surface due to thermal properties of buildings and pavement. (b) *Pollution* effect refers both to enhancement of air convection by absorption of solar radiation in pollution layer, and to possible cloud nucleating effects of pollution particles. (c) *Aerodynamic* effect refers to the rising air motion over the city owing to high turbulent "roughness" and obstacle effect of urban structures. Each effect tends to increase convective cloudiness and precipitation both in and downwind of cities, but the relative importance of the three is not yet known.

downwind of the city. Observational evidence of such an effect is equivocal at best. Some support for it comes, however, from a study by Frederick (1970) which shows a rather systematic tendency for cold-season precipitation at twenty-two urban stations in the eastern United States to average several percent greater on weekdays than on weekends. Further investigations of this problem are needed.

J. MURRAY MITCHELL, JR.

References

Frederick, R. H., 1970, "Preliminary results of a study of precipitation by day-of-the-week over eastern United States," Preprint, Second National Conf. on Weather Modifications, Boston, Mass., *Am. Meteorol. Soc.*, pp. 209–214.

Lowry, W. P., 1969, "Weather and Life," New York, Academic Press, 306pp.

Ludwig, J. H., Morgan, G. B., and McMullen, T. B., 1970, "Trends in Urban Air Quality," *EOS* (American Geophysical Union, Washington, D.C.), **51**, 468–475.

Peterson, J. T., 1969, "The Climate of Cities: A Survey of Recent Literature," National Air Pollution Control Administration, Publication No. AP-59.

U.S. Dept of Health, Education, and Welfare, National Air Pollution Control Administration, 1969, "Air Quality Criteria for Particulate Matter," Publication No. AP-49.

Cross-references: *Air Pollution and Global Climate; Environmental Pollution; Thermal Pollution; Urban Air Pollution.* Vol. II: *Aerosols; Atmospheric Nuclei and Dusts; Greenhouse Effect.*

ALKALIS, ALKALI METALS, AND ALKALINE EARTH METALS

The alkali and alkaline earth metals are listed in Table 1 with the most important parameters controlling their geochemical distribution. Two additional elements in these groups, $Fr^{223}(22m)\beta^-\gamma,\alpha$, and $Ra^{226}(1620y)\alpha,\gamma$, are short-lived radioisotopes in the Ac^{227} and U^{238} series, respectively. Studies on the marine geochemistry of Ra^{226} have given much information on sedimentation rates, growth of corals and other organisms with calcium carbonate skeletons, and circulation of sea-

TABLE 1.

	Group IA				Group IIA			
Alkali Elements	Ionic Radius (1^+)	First Ionization Potential	Electronegativity	Alkaline Earth Elements	Ionic Radius (2^+)	Second Ionization Potential	Electronegativity
Li (3)	0.68	5.4	0.97	Be (4)	0.35	18.2	1.47
Na (11)	0.97	5.1	1.01	Mg (12)	0.65	15.03	1.23
K (19)	1.33	4.34	0.91	Ca (20)	1.01	11.9	1.04
Rb (37)	1.45	4.18	0.89	Sr (38)	1.18	11.03	0.99
Cs (55)	1.67	3.89	0.7	Ba (56)	1.34	10.00	0.97

water. These elements are not considered further here.

Sodium (Na), potassium (K), magnesium (Mg), and calcium (Ca) are major elements (>1%), strontium (Sr) and barium (Ba) minor elements (100–1000 ppm), and lithium (Li), rubidium (Rb), cesium (Cs), and beryllium (Be) trace elements (<100 ppm) in the earth's crust. The abundances of the elements in geological materials are given in the entries relating to individual elements. The geochemistries of the alkali and alkaline earth elements are dominated by their ease of oxidation and the ionic nature of their metal-oxygen bonds. These bonds have more than 76% ionic character on Pauling's scale; exceptions are Be (63%) and Mg (70%). Cs forms the most ionic bond of any element.

With the exception of Mg, and to a much lesser extent Ca and Na, all the alkali and alkaline earth elements are concentrated in the continental crust. This upward concentration is most marked for Cs, which is nearly quantitatively removed from the mantle. The ionic radius of Mg makes it stable in six-coordination with oxygen; Mg is thus a major element in the dense mineral phases of the mantle (olivine, clinopyroxene, spinel). The coupled substitution of Na and Al in pyroxenes becomes important at high pressures, and Ca is a major element in clinopyroxenes and garnets of the upper mantle.

The most important minerals of the alkali and alkaline earth elements are given in Table 2. With the exception of Rb, all the elements may be mineral forming. The trace elements Li, Cs, and Be form separate minerals only at stages of pegmatite formation; the minor elements Sr and Ba do so only in sediments and evaporites. In major silicate rock types the minor and trace elements are incorporated in the mineral structures of the major elements. The most important single parameters controlling these substitutions are ionic size and charge (Table 1). Thus Li substitutes for Mg; Rb, Cs, and Ba for K; and Sr for Ca and K. In cases of charge differences the substitution must be coupled with substitution of some other elements in order to maintain electrostatic neutrality of the structure. Thus when Li^+ substitutes for Mg^{2+} a simultaneous substitution of Fe^{3+} or Al^{3+} for Fe^{2+} and/or Mg^{2+} must take place, and when Ba^{2+} substitutes for K^+, Al^{3+} must replace Si^{4+} in the structure.

The minerals orthopyroxene-clinopyroxene-amphibole-mica form a series in which there is a regular increase in Li concentration and of the Li/Mg ratio. Increase in Li/Mg ratio with differentiation is a common feature in all magmatic rock series but the absolute amount of Li may decrease in the most differentiated rocks, reflecting the decreasing concentration of the Mg host mineral. Greisens and granitic rocks associated with tin and tungsten mineralization are particularly high in Li, at the hydrothermal stage of pegmatite formation the Li concentration may be high enough for independent Li minerals to crystallize.

Be^{2+} is so small that it cannot replace other alkali and alkaline earth elements in mineral structures. Because of its small size Be^{2+} can be expected to occur in magmas only as $(BeO_4)^{6-}$, which may substitute for $(SiO_4)^{4-}$ in silicate structures. Be always occurs in fourfold coordination with oxygen. The substitution of $(BeO_4)^{6-}$ for $(SiO_4)^{4-}$ must involve coupled substitutions involving high-charge cations such as Ti^{4+}, Zr^{4+} etc., or an excess of F^-, Cl^-, or OH^- anions in order to maintain charge balance. This explains why maximum Be concentrations are observed in alkaline rocks, and why amphiboles and micas, where coupled substitutions involving elements of high valency are possible, have higher Be contents than feldspars.

When Rb and Cs substitute for K no coupled substitution is necessary to maintain charge balance. Cs^+ is significantly larger than K^+ and not as easily incorporated in the potassium sites as is Rb. Cs is therefore con-

TABLE 2. MOST IMPORTANT MINERALS

Li	Lepidolite, spodumene, petalite, amblygonite
Na	Plagioclase (albite), amphibole, halite
K	Potassium feldspar, mica, sylvite
Rb	(lepidolite), (amazonite)
Cs	Pollucite, lepidolite
Be	Beryl
Mg	Olivine, pyroxene, amphibole, phlogogpite, garnet, spinel, magnesite, dolomite
Ca	Plagioclase (anorthite), clinopyroxene, amphibole, garnet, apatite, sphene, calcite, dolomite
Sr	Celestite, strontianite
Ba	Barite, witherite (Ba-feldspar)

centrated in the magma during fractional crystallization and its concentration in the residual liquid may be high enough for independent Cs minerals to form toward the late stages of pegmatite formation. This contrasts with the behavior of Rb, which, though it is about thirty times as abundant in the earth's crust as Cs, never forms independent minerals. Because of their more similar ionic sizes, Rb is more perfectly camouflaged by K in minerals than is Cs. The concentration of Rb also increases in the more differentiated rocks, and potassium feldspars of granites and pegmatites are higher in Rb than the feldspars of granodiorites and monzonites. Both Rb^+ and Cs^+ are larger than K^+ and prefer the larger lattice sites (higher coordinatiton with oxygen) of the mica structures to those of the feldspars. Both Rb/K and Cs/K ratios are significantly higher in micas than in the coexisting feldspars.

The K/Rb ratio in minerals, rocks, and meteorites has been studied intensively. Only rarely does this ratio fall outside the limits of 160 and 300 for crustal silicate rocks. Basic rocks tend toward and beyond the upper limits of this ratio, whereas low values are observed in strongly differentiated granites and pegmatites. The universality of this ratio was emphasized by the similarity of the K/Rb ratio (about 300) in ordinary chondritic meteorites and terrestrial rocks. However, the K/Rb ratio of achondrites is about 1500 and recent work has shown high K/Rb ratios in oceanic basalts, particularly the low-potassium tholeiites dredged from the ocean floors. This has led to some speculation about the universality of this ratio and K/Rb ratios of above 1000 have been postulated for the mantle even though most ultrabasic rocks and eclogites have K/Rb ratios around 300. So far, high K/Rb ratios have been found to characterize oceanic tholeiites, oceanic and continental alkali basalts, some syenites and nepheline syenites, anorthosites, and other rocks in the high-pressure granulite facies suite. These rocks also appear to be very low in Cs.

The ionic size of Sr is intermediate between that of Ca and K, and Sr may enter the structural sites, particularly in the feldspars, of either of these elements. The substitution of K by Sr must be accompanied by a coupled substitution between some other elements in order to maintain electrostatic neutrality of the structure. The bulk of the Sr in the crust resides in the feldspars. In many rocks, the distribution ratio of Sr in plagioclase to Sr in potassium feldspar is close to unity. Potassium feldspars of intermediate rocks, monzonites, and syenites may contain above 1000 ppm Sr. In magmatic rocks the Sr contents of the K-feldspars decrease with differentiation and may be very low in feldspars from pegmatites. Other major calcium- and potassium-bearing minerals in the earth crust, i.e., clinopyroxene, amphibole, garnet, contain relatively small amounts of Sr. Consequently, the chemically similar alkaline earth elements, Ca and Sr, do not form a very coherent pair of elements in geological processes. The common accessory minerals apatite and sphene have suitable structures for Sr.

Extremely high Sr concentrations are found in some alkaline and peralkaline rock types. Carbonatites may contain up to 1% Sr, and the high Sr (and Ba) content of carbonatites may serve to distinguish them from metalimestone and marble.

The radioactive decay of Rb^{87} to Sr^{87} is important in rock dating and the Rb/Sr ratio of geological materials has been studied in detail. The abundance of Sr^{87} is conventionally discussed in terms of the Sr^{87}/Sr^{86} ratio. The primitive terrestrial value of this ratio is believed to be 0.698 and increases with time as more Rb^{87} decays to Sr^{87}. Since Rb and Sr behave differently in geological processes, the increase in Sr^{87}/Sr^{86} ratio will be different for the various geospheres. When a magma is generated it inherits the Sr^{87}/Sr^{86} ratio of the source region, and the initial Sr^{87}/Sr^{86} ratios of magmatic rocks could potentially be used to indicate the site of magma generation. Table 3 gives average Rb/Sr ratios in different magmatic rocks. The concept has been used for defining a crustal vs mantle origin of magmas on the basis that the Rb/Sr ratio of the mantle is considerably lower than that of the crust (0.02 vs 0.2). However, the value of 0.2 for the crust is valid only for the upper continental crust, and indications are that this ratio is considerably less in the deep crust. In fact Rb and Sr are not significantly fractionated during basalt magma generation from mantle material, but both elements are removed nearly quantitatively from the source rock. The real fractionation of Rb from Sr takes place during granite formation. Granitic magmas cannot form directly from mantle material except, perhaps, under very special conditions, but may form easily through partial melting of the intermediate rocks constituting the deep crust. The granitic magmas generated in the deep crust will rise, leaving a residuum with extremely low Rb/Sr ratios. Indications are that components of high pressure granulite facies terranes, which may approximate the composition of the deep crust, have lower Rb/Sr ratios than the mantle.

The geochemistry of Ba is completely governed by its replacement of K in K-minerals.

TABLE 3. APPROXIMATE Rb AND Sr CONCENTRATIONS, AND Rb/Sr RATIOS
IN SOME MAJOR IGNEOUS ROCKS

	ppm Rb	ppm Sr	Rb/Sr
Chondrite	2.8	11	0.25
Dunite	0.1	10	0.01
Oceanic tholeiite	1.2	115	0.01
Basalt (continental)	10	230	0.04
Andesite, diorite, monzonite	60	600	0.10
High-calcium granite	110	170	0.65
Low-calcium granite	440	100	4.4

The sizes of the Ba^{2+} and K^+ ions are very similar but the double charge of the barium ion makes a coupled substitution necessary. Virtually all the Ba in the crust is contained in potassium feldspars and to a lesser extent in the micas. The concentration ratio of Ba in K-feldspar to Ba in biotite is about 2 to 1. The maximum Ba concentrations are found in potassium feldspar of intermediate igneous rocks, which may contain more than 1% Ba. Although Ba^{2+} in this way behaves according to the classical capture principle it is not depleted until very late in the differentiation sequence, and is often enriched towards the end stages of magmatic crystallization. The ratios of Ba/Sr, Ba/Rb, and Ba/K all decrease with fractionation of magmatic rocks.

The geochemistry of the major elements Mg, Ca, Na, and K in igneous rocks is very well known and understood. Mg^{2+} enters into six-coordinated structures with silica; it is a major element in the early ferromagnesian minerals and is largely removed from the crystallizing magma with the separation of olivine and pyroxene. Mg is also a major element in biotites of the more acidic rocks, but as the abundance of biotite is low, the overall Mg content of these rocks is low. The concentration of Mg in a crystallizing magma therefore rapidly decreases and the early cumulative rocks are strongly enriched in this element.

Calcium enters clinopyroxene and plagioclase. The calcium-rich plagioclases are the first feldspars to crystallize out of a basic magma and the Ca concentration is at a maximum in the basic rocks (gabbro and basalt).

Sodium in igneous rocks is mainly incorporated in the plagioclases and to a lesser extent in the potassium feldspars. Potassium feldspar does not start to crystallize before the intermediate stages and the plagioclase feldspars become increasingly sodium-rich as fractionation proceeds. Though Na is incorporated into basic plagioclases, its concentration in magmatic rocks increases with fractionation until the late granitic stage when it suddenly decreases. The Na content appears again to build up in the late hydrothermal vapors and the late stages of pegmatite formation are often associated with a sodium metasomatism (cleavelandite).

In peralkaline rocks Na is incorporated in nepheline, and alkaline and peralkaline rocks may be strongly enriched in sodium.

Potassium is concentrated in potassium feldspar and mica, both mineral phases that increase in importance in the more differentiated rocks. Potassium therefore builds up rapidly in the magma until the intermediate stage when K-feldspar starts to crystallize, and the transition from basic to intermediate rocks shows a sudden increase in K content. Monzonites and syenites may be particularly high in potassium. Potassium concentrations in magmatic rocks continue to increase all through to the final granitic stages. The end of fractional crystallization appears in some rocks to be marked only by the addition of quartz; the potassium concentration, along with the concentration of most other elements, will decrease in these rocks.

Potassium is also concentrated in granitic pegmatites, which may consist nearly entirely of potassium feldspar and quartz.

Metamorphic Rocks

The concentrations of alkali and alkaline earth metals in metamorphic rocks are largely inherited from their igneous and sedimentary parent materials. The movement of these elements during metamorphism (metasomatism) is disputed but general agreement prevails in that the alkali elements Na, K, Rb, and Cs are mobile during the process of regional metamorphism. Work on Rb/Sr dating has shown that radiogenic Sr^{87} is easily moved out of some lattice positions in the early stages of metamorphism, and the constant "initial" Sr^{87}/Sr^{86} ratio in rocks of different composition and origins in many high grade metamorphic areas could indicate considerable movement of Sr in rocks.

Albitization and the formation of albite porphyroblasts in low-grade metamorphic rocks show that Na is mobile in the early stages of regional metamorphism. Considerable movement of potassium is evident under high temperature and pressure in the lower amphibolite facies during regional metamorphism. The amphibolite facies is the regional environment of in situ granite formation in the crust (granitization), which involves movement and concentration of potassium.

Because of the general coherence between Rb^+ and K^+ ions the movement of Rb could be expected to be similar to that of K. This is undoubtedly the case to a first approximation, but greater mobility and stronger upward concentration of Rb in the crust is indicated by data from high-pressure granulite facies rocks, many of which have K/Rb ratios close to 1000 compared with the typical ratio of between 200 and 300 in common upper crustal silicate rocks.

As could be expected, Cs is strongly concentrated upward in the continental crust and its concentration is low in high-grade metamorphic rocks. The near-surface concentration of Cs is illustrated by its high concentration in oil brines and many hot spring waters. Cs concentrations of 2 ppm have been found in the water of the Wairakeri hydrothermal system of New Zealand, and the water from Takaanu in Japan contains 4.7 ppm Cs.

Sediments

The processes of weathering, erosion, transport, and sedimentation can cause considerable fractionation of the alkali and alkaline earth elements. In this process, calcium can be separated from the other elements with which it occurs in the magmatic host rocks, and may be deposited as calcium carbonate (calcite) in extensive monomineralic deposits. Most of the deposition is caused by the settling of dead sea animals, corals, etc., which extract calcium carbonate from seawater for their shells. The calcium carbonate in many shells is in the form of aragonite, which may later invert to calcite. However, inorganic deposition of calcite from seawater is also possible as exemplified by the existence of limestone beds from the early Precambrian.

Mg may precipitate with Ca from seawater and form dolomite. The primary formation of dolomite in this way has been established but dolomite commonly forms secondarily, as Mg from seawater replaces Ca in calcite.

Shales have high Sr/Ca ratios, indicating that the absorption sites on clay particles prefer Sr over Ca. This could indicate that Sr is left behind relative to calcium in the weathering products of igneous rocks. However, Sr is concentrated with Ca in carbonates, and deep-sea carbonates are particularly high in Sr (2000 ppm vs 600 ppm in shelf carbonates).

Ba concentrations tend to be low in carbonates but may reach some 100 ppm in deep-sea carbonates. Contrary to Sr, Ba is concentrated in the clay fractions of sediments; normal shales may contain about 580 ppm Ba vs 2300 in deep-sea clays.

K, Na, and the trace elements Li, Rb, Cs, and Ba are all contained in shales. The movement and fractionation of these elements during weathering is caused by an interplay of their high solubilities and absorption on colloidal materials. The alkali metals are absorbed on colloidal clay minerals from their aqueous solution in the order Cs > Rb > K > Na > Li, which could be predicted from the relative values of the ratio of ionization potential/hydrated ionic radius (I/r_h).

Seawater

Na, K, Mg, Ca, and Sr are the major cations in seawater. Their abundances are given in Table 4 relative to a chlorinity of 19‰. The trace alkali and alkaline earth elements are also among the most abundant cations in seawater. Even the least abundant (Cs) is only exceeded in concentration by silicon (Si), aluminum (Al), molybdenum (Mo), zinc (Zn), lead (Pb), iron (Fe), tin (Sn), uranium (U), and vanadium (V).

K. S. Heier

References

Ahrens, L. H., 1964, "The significance of the chemical bond for controlling the geochemical distribution of the elements," *Phys. Chem. Earth*, **5**, 1–54.

Armstrong, R. L., 1968, "A model for the evolution of strontium and lead isotopes in a dynamic earth," *Rev. Geophys.*, **6**, 175–199.

Heier, K. S., and Adams, J. A. S., 1964, "The geochemistry of the alkali metals," *Phys. Chem. Earth*, **5**, 253–381.

Hurley, P. M., 1968, "Absolute abundance and distribution of Rb, K and Sr in the Earth," *Geochim. Cosmochim. Acta*, **32**, 273–283.

Taylor, S. R., 1965, "The application of trace element data to problems in petrology," *Phys. Chem. Earth*, **6**, 133–213.

Table 4. Alkali and Alkaline Earth Elements in Seawater[a]

Major Elements, %		Trace Elements, ppm	
Mg	1.272	Rb	0.12
Ca	0.400	Li	0.1
Sr	0.008	Ba	0.006
K	0.380	Cs	0.002
Na	10.556		

[a] Relative to chlorinity of 19‰.

Turekian, K. K., and Wedepohl, K. H., 1961, "Distribution of the elements in some major units of the earths crust," *Geol. Soc. Am. Bull.*, **72**, 175–192.

Cross-references: *Calcium; Calcium Carbonate; Earth's Crust Geochemistry; Electronegativity; Elements: Planetary Abundance and Distribution; Hydrothermal Solutions; Mantle Geochemistry; Potassium/Rubidium Ratio in Geology; Seawater; Trace Elements; Weathering, Chemical.* Vol. II: *Meteorites.* Vol. V: *Alkalic Rocks; Carbonatites; Magmatic Differentiation; Metamorphism*, Vol. IVB: *Alkali Feldspars; Amphibole Group: Clinopyroxenes; Plagioclase Feldspars; Pyroxeue Group.* Vol. VI: *Evaporites; Sediments.*

ALUMINUM: ELEMENT AND GEOCHEMISTRY

Physical Properties. Aluminum (L. *alumen*), symbol Al; atomic weight 26.98; atomic number 13; electronic configuration $1s^2 2s^2 2p^6 3s^2 3p^1$; melting point 659.7°C; boiling point 2057°C; specific gravity 2.699 (at 20°C); valence 3; crystal structure; face-centered cubic (Cu structure). Only one stable nuclide exists in nature, Al^{27}. Aluminum is a white metal that strongly resists corrosion by forming a protective coating of the white oxide. It is very malleable and fairly ductile. The electrical conductivity is only 60% that of copper, but its lightness makes it useful for transmission lines.

Wöhler was the first to obtain the pure metal in 1827, although Oersted probably isolated an impure form two years earlier. The method of producing the metal by electrolysis of pure alumina, Al_2O_3, dissolved in cryolite, Na_3AlF_6, was patented independently by Hall in the U.S. and Heroult in France, soon after its discovery in 1886.

Compounds. Aluminum occurs in nature as the oxide, *corundum*, Al_2O_3, in cryolite, Na_3AlF_6, in *spinel*, $MgAl_2O_4$, and in *bauxite*, which is a mixture of hydrated aluminum hydroxides, consisting of *gibbsite*, $\gamma\text{-}Al(OH)_3$ *diaspore*, $\alpha\text{-}AlO(OH)$, and *boehmite*, $\gamma\text{-}AlO(OH)$. A rare, but potentially economic mineral is *dawsonite*, $Na_3Al(CO_3)_3 \cdot 2Al(OH)_3$. In addition, aluminum is widespread in rock-forming silicates such as *feldspars, micas,* and *clays.* Important commercial products are the metal itself, alloys, sulfates, $Al_2(SO_4)_3 \cdot 9H_2O$ or $Al_2(SO_4)_3 \cdot 18H_2O$, and the alums, which are double sulfates of Al and either potassium (K), solium (Na), or ammonium (NH_4), e.g., $KAl(SO_4)_2 \cdot 12H_2O$, $NaAl(SO_4)_2 \cdot 12H_2O$, and $NH_4Al(SO_4)_2 \cdot 12H_2O$.

Chemistry. Aluminum loses its 3 valence electrons readily, forming Al^{3+} ions, with an inert gas configuration [that of neon (Ne)]. Its strong affinity for oxygen results in its natural occurrence within silicate rocks. Al^{3+} ions are precipitated in the presence of hydroxides:

$$Al^{3+} + 3NaOH \rightarrow Al(OH)_3\downarrow + 3Na^+$$

Aluminum hydroxide is amphoteric, in that it reacts with both acids and bases:

acid: $Al(OH)_3\downarrow + 3H^+ \rightarrow Al^{3+} + 3H_2O$

base: $Al(OH)_3\downarrow + OH^- \rightarrow Al(OH)_4^-$

The relation of solubility of $Al(OH)_3$ to pH may influence the formation of bauxites and lateritic soils. Because of the amphoterism of $Al(OH)_3$, its solubility is greatest in highly acidic (pH 1–5) and highly basic (pH 9–11) solutions, whereas in the pH range 5–9, the solubility of $Al(OH)_3$ is at a minimum. SiO_2 (an acidic oxide), however, shows little increase in solubility with increasing pH, except above pH 9. In the pH range 5–9, the solubility of alumina is even less than that of silica, and therefore under tropical conditions in nature, SiO_2 is preferentially leached from soils, leaving alumina behind (see the solubility curve in the article *Acids and Bases*).

Al^{3+} ions form complexes with fluoride AlF_6^{3-} and oxyanions $Al(C_2O_4)_3^{3-}$ but not with NH_3 or Cl^-. Hydroxide complexes are present in all but strongly acid solutions.

Aluminum is a very active metal in either acidic or basic solutions:

$$Al^0 \rightarrow Al^{3+} + 3e^- \quad E^0 = 1.66V$$

$$Al^0 + 4OH^- \rightarrow Al(OH)_4^- + 3e^-$$
$$E^0 = 2.35V$$

Geochemistry. Aluminum, after oxygen and silicon, is the most abundant element in the earth's crust. If one considers cosmic abundances, iron (Fe) and magnesium (Mg) are more widespread than Al, as indicated by the average chemical composition of meteorites. Aluminum is concentrated in the sialic crust (the lithosphere) and is thought to be absent from the core and scarce in the mantle. This suggests a pronounced chemical differentiation of the earth. In the upper lithosphere, Al normally occurs combined with oxygen (O) and silicon (Si) (mainly in feldspars).

During the early stages in the magmatic differentiation of igneous rocks, aluminum forms an essential constituent of plagioclase feldspars and spinels. In the main stages of crystallization, Al is somewhat enriched in the first rocks to solidify, but with progressive differentiation, its content decreases. For example, the average Al content of gabbros is higher than that of silicic rocks.

All silicate minerals consist of a framework of $(SiO_4)^{4-}$ tetrahedra that combine differently in various mineral groups to form regular structures. Part of the Si^{4+} can be replaced by Al^{3+} in the (SiO_4) tetrahedra. The substitution of Al^{3+} for Si^{4+} is never complete because of the difference in size of the two ions. Furthermore, Al^{3+} can exist in sixfold coordination with oxygen in micas, amphiboles, and pyroxenes. Thus it can replace Mg^{2+} and Fe^{2+} in spite of its smaller size. The importance of Al in silicate minerals arises from the fact that it can occupy two different structural positions, one of fourfold coordination, the other of sixfold coordination, in the crystal lattices.

The *feldspars,* which are the most abundant constituents of igneous rocks, are aluminosilicates of K, Na, and calcium (Ca). Therefore they carry the bulk of lithospheric aluminum. Anorthite, $CaAl_2Si_2O_8$, contains twice as much aluminum as albite, $NaAlSi_3O_8$ and orthoclase, $KAlSi_3O_8$. Consequently, rocks of basaltic composition, whose plagioclase is richer in the anorthite component, contain more Al than the granitic rocks that are high in K and Na feldspar. For this reason, Al is enriched in the early products of the main stage of magmatic differentiation.

The *micas* are another group of Al silicates. Biotite, $K_2(Mg,Fe^{II})_{6-4}$ $(Fe^{III},Al,Ti)_{0-2}$ $(Si_xAl_yO_{20})(OH,F)_4$, more significant geochemically than muscovite, contains from 10 to 20% Al_2O_3. Muscovite, $K_2Al_4(Si_6Al_2O_{20})$ $\cdot(OH,F)_4$ is higher in aluminia—up to 30%. But muscovite is an essential constituent only of granitic rocks and is absent in mafic ones. Moreover, the bulk of muscovite in granites is of secondary origin, being derived from the alteration of feldspars.

Some *pyroxenes,* e.g., spodumene [$LiAl(Si_2O_6)$] and jadeite [$NaAl(Si_2O_6)$] and *amphiboles,* e.g., hornblende and glaucophane [$Na_2Mg_3Al_2(Si_8O_{22})(OH)_2)$] are also aluminum bearing.

In addition, aluminum is found as a simple silicate, Al_2SiO_5, in *sillimanite, kyanite, andalusite,* and *mullite.* The first three are important constituents of metamorphic rocks, but they seldom occur in igneous ones.

Of lesser geochemical significance are other Al minerals such as *corundum,* Al_2O_3, *chrysoberyl,* Al_2BeO_4 (found in pegmatites, aplites, and mica schists), *topaz,* $[Al(F,OH)]_2SiO_4$, *staurolite* $(Fe^{II},Mg)_2(Al,Fe^{III})_9O_6(SiO_4)_4(O,OH)_2$, *wavellite* $4AlPO_4\cdot2Al(OH)_3\cdot9H_2O$, and *alunite* $K_2Al_6(OH)_{12}\cdot(CO_4)_4$.

Under normal conditions of chemical weathering the feldspars and some other silicates, e.g., leucite, are decomposed into clays and soluble material. During diagenesis the reaction may be reversed, leading to authigenic feldspar.

$2KAlSi_3O_8 + 2H_2O + CO_2 \rightleftarrows$
orthoclase
$\qquad Al_2Si_2O_5(OH)_4 + K_2CO_3 + 4SiO_2$
$\qquad$ kaolinite $\qquad$ soluble, colloidal

The *clay minerals* (kaolinite, montmorillonite, illite, and chlorite) are all hydrous aluminosilicates with layer-lattice structures. The grain size of clays in sediments is very small ($< 5 \times 10^{-3}$mm down to colloidal dimensions). The clays are phyllosilicates (or sheet silicates) that differ in the stacking arrangements of the two basic units consisting of (a) aluminum hydroxide groups in which Al has sixfold coordination and in which Al lies between two layers of O^{2-} or OH^- and (b) linked SiO_4 groups in hexagonal array, which give rise to $(Si_4O_{10})^{4-}$ sheets. Clays also differ in the replaceability of Si and Al by other elements.

The structure of *kaolinite* consists of one $(Si_4O_{10})^{4-}$ sheet linked to one Al–OH sheet. Only two out of the three available sites are occupied by Al ions. No replacement of Al and Si by other elements occurs. Four polymorphs of kaolinite exist, differing chiefly in the stacking sequence. *Montmorillonite* consists of a sheet of (Al–OH) sandwiched between two Si_4O_{10} sheets. These groups are stacked one above the other in the direction of the c axis, with H_2O molecules adsorbed between structural layers. The water content is variable, causing montmorillonite to swell when water-saturated. Considerable atomic substitution is possible in the crystal lattice. Therefore the chemical composition of the montmorillonite groups is exceedingly variable. Furthermore, the three-layer unit is negatively charged so that exchangeable cations (Ca^{2+},Na^+,K^+) may be adsorbed between the unit layers.

Illite or *hydromica* is closely related to muscovite and to montmorillonite but contains more water, less potassium, and a higher Si/Al ratio than muscovite. Illite exhibits less stacking regularity than muscovite, because of fewer interlayer cations. On the other hand, illite differs from montmorillonite in that some Si is always replaced by Al, and the resulting charge deficiency is balanced by K^+ ions, which prevent swelling or expansion. *Chlorite* is derived from montmorillonite by the insertion of a (Mg or Al)–(OH) layer between each montmorillonite triple layer. It has been recently found as an important argillaceous mineral in sediments.

The clays are transported by streams and rivers to the seas where they are deposited. The enrichment of aluminum in hydrolyzates is of geochemical significance. In clays formed under a cold climate, the Al content is comparable to that of igneous rocks. Under tropical weathering, clays become enriched in alumina (laterites, bauxites).

Biogeochemistry. Although aluminum is widely distributed in living organisms, it is not an essential constituent in them. However, some plants may require Al. A certain tree, *Orites excelsa*, has a basic aluminum succinate in its woody tissue. The ashes of *Lycopodium alponium* may contain up to 33% Al_2O_3. Many plants are sensitive to the element, because Al^{3+} ions flocculate negatively charged colloids. This may affect the properties of cultivated soil. Aluminum poisoning of soils is fairly widespread.

Uses of Aluminum. Aluminum is in wide demand and has many industrial uses. Aluminum is required in the manufacture of airplanes, automobiles, industrial machinery, and ships. In the electrical industry, it is used for transmission wires. As Al_2O_3, in artificial rubies and sapphires, it is used as jewels in watches, scales, and other precision instruments. It is also employed in the rust-proofing of iron.

VIVIEN GORNITZ

References

Chinner, G. A., 1966, "The significance of the aluminum silicates in metamorphism," *Earth Sci. Rev.,* 2(2), 111–126.

Goldschmidt, V. M., 1954, "Geochemistry," Oxford, Clarendon Press, 730pp.

Khiterov, N. I., *et al.,* 1963, "Relations between andalusite, kyanite and sillimanite at moderate temperatures and pressures," *Geochemistry Intern.* No. 3, 235–244.

Rankama, K., and Sahama, Th. G., 1950, "Geochemistry," Chicago, University of Chicago Press, 911pp.

Cross-references: *Acids and Bases; Aluminum Ore Deposits; Core Geochemistry; pH-Eh;* Vol. II: *Meteorites.* Vol. IVB: *Amphibole Group; Bauxite; Clays and Clay Minerals; Feldspars; Plagioclase Feldspars; Pyroxene Group.* Vol. VI: *Laterites.*

ALUMINUM ORE DEPOSITS

Natural Occurrence of Aluminum. Although aluminum is the most abundant metal in the earth's crust, the element's strong affinity for oxygen prohibits its natural occurrence in the metallic state. Aluminum has a widespread distribution and is present as an essential component in most rock-forming minerals. The more important of these mineral groups with high alumina content, which occur frequently in natural concentrations, include the aluminosilicates of soda, potash, and lime, such as the feldspars and feldspathoids characteristic of igneous rock masses; the aluminum silicates of the kyanite-andalusite-sillimanite group, which occur in the gneisses and schists of metamorphic areas; hydrated aluminum silicates, such as the clay minerals formed by weathering and sedimentation processes; and, less abundantly, aluminum sulphates, such as alunite, generally considered to be hydrothermal in origin. Some of these potential sources have been utilized as aluminum ores, e.g., nepheline in the U.S.S.R., alunite in Japan, labradorite-feldspar in Norway, andalusite in Sweden, and clay in the U.S. (see details in Bracewell, 1962); however, owing to the comparative ease with which purified alumina can be extracted from gibbsitic and boehmitic bauxites by the Bayer and other processes, the predominant ore of commerce is bauxite. Considerable interest is being shown, however, for the development of new processes, in particular those based on acid leaching, for extracting alumina from low-grade and nonbauxitic materials. In view of the increasing world demand for aluminum, it is reasonable to anticipate that at least part of the world's future supply will be derived eventually from materials such as clays and shales or aluminum phosphates. Dawsonite, $NaAl(OH)_2CO_3$, with 35% acid-extractable alumina, has also been suggested as a potential aluminum ore, although once regarded as a rare hydrothermal mineral of low-temperature origin. Dawsonite is present in relative abundance in the Eocene Green River Formation oil shales of Colorado, where it occurs through 700 ft of rich oil shale, averaging 25 gal/ton, over hundreds of square miles. In places dawsonite may make up to one-fourth by weight of the oil shale, which might thus be utilized as a source of oil and alumina. Dawsonite also occurs in substantial quantities in weathered Pleistocene ash beds in Olduvai Gorge, Tanzania. Smith and Milton (1966) consider that this dawsonite has been formed by reactions between permeating solutions of sodium carbonate and original volcanic glass, nepheline, or clay minerals.

Aluminum is present in high concentrations in many other minerals that are too disseminated or infrequent to be considered as aluminum ores. Apart from the naturally occurring form of alumina, viz., corundum and its varieties emery, ruby, and sapphire, alumina contents around 70% occur in the aluminum spinels; other minerals high in alumina are garnet, epidote, mica, staurolite, chloritoid, scapolite, and dumortierite. High alumina con-

ALUMINUM ORE DEPOSITS

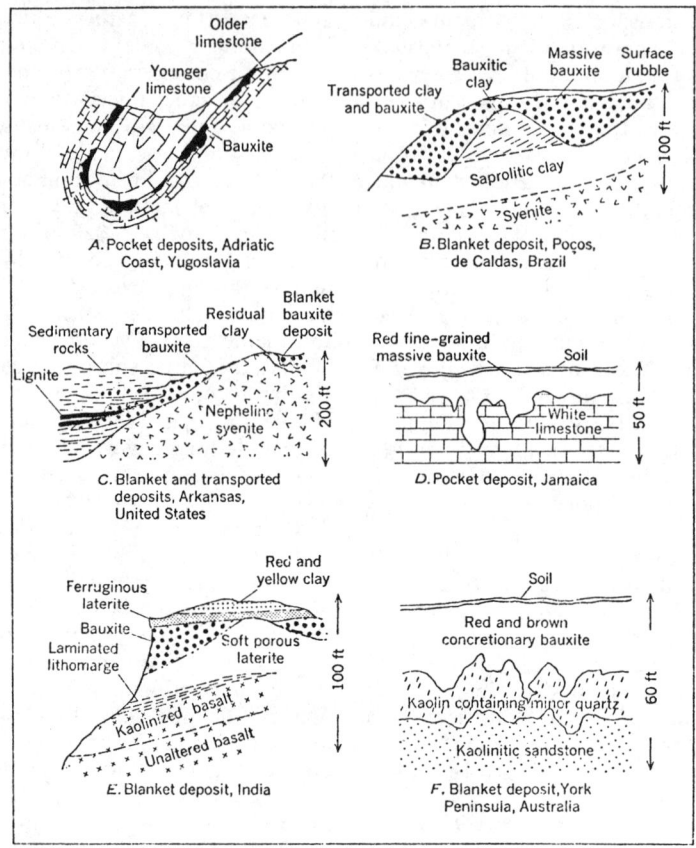

FIG. 1. Typical occurrences of bauxite (from S. H. Patterson, *U.S. Geol. Surv. Bull.* **1228**, 16).

tents also occur in certain pegmatitic minerals, e.g., chrysoberyl ($BeAl_2O_4$) contains 80.2%. Cryolite, (Na_3AlF_6) a pegmatitic mineral occurring in commercial quantity only at Ivigtut in Greenland, contains 24.3% Al_2O_3, but owing to its fluxing properties has been of particular importance to the aluminum industry in the electrolytic process of extracting aluminum.

Bauxite: Definition and Mineralogy. Bauxite was discovered by Berthier at Les Baux near Arles in southern France, and the term was first used by Dufrenoy and Deville, who appear to have considered bauxite to be a mineral of definite composition. Later work demonstrated the mixed mineralogy of bauxite, which can be broadly defined as a rock in which one or more of the hydrated alumina minerals—gibbsite, boehmite, and diaspore, as well as the amorphous and chemically variable cliachite—are present in sufficient abundance for the rock to be usable either as an economic source of alumina or aluminum, or for the manufacture of refractories, abrasives, chemicals, etc.

The term "bauxite" is used to describe a fairly wide range of materials showing considerable variations in mineralogical composition, physical appearance, and mode of occurrence. Figure 1 shows some typical bauxite occurrences. The mineralogy, chemistry, etc., of bauxite have been detailed by Hose (1964), Pearson (1955), and Beneslavsky (see Bushinsky, 1958). Table 1 summarizes the chemical compositions of the hydrated aluminum oxides, one or more of which must be present in any bauxite; Deer et al. (1962) have given a detailed account of their mineralogy.

In addition to gibbsite, two other naturally occurring polymorphs of alumina trihydrate

TABLE 1. ALUMINA HYDRATE MINERALS IN BAUXITE

Mineral	Chemical Composition	Alumina (%)	Water (%)
Gibbsite	$\gamma - Al_2O_3 \cdot 3H_2O$	65.4	34.6
Boehmite	$\gamma - Al_2O_3 \cdot H_2O$	85.0	15.0
Diaspore	$\alpha - Al_2O_3 \cdot H_2O$	85.0	15.0

have been reported, namely nordstrandite from Sarawak and Guam in 1962 (Schlanger, 1964) and bayerite from Israel in 1963; however, these minerals would appear to be infrequent and have not been recorded in bauxite.

Bauxite also contains varying proportions of oxides and hydroxides of iron, manganese, titanium, and other metals, and may also contain some free silica and clay minerals, notably kaolinite, halloysite, or both. With increasing iron content, a transition occurs into the "lateritic bauxites"; these will perhaps become ores of aluminum, and possibly also of iron, titanium, and other metals, as the world's high-grade bauxite reserves are depleted.

Reference is often made in bauxite literature to the amounts of "available" alumina and "reactive" silica in a particular bauxite. The "available" alumina in a bauxite depends largely on how the alumina is distributed between the alumina hydrates and the other mineral constituents. The alumina hydrates themselves show considerable differences in their behavior when treated by the Bayer process. Thus, alumina present as gibbsite is more easily recoverable by the Bayer process than that present as boehmite. Diaspore is so difficult to treat that it is not regarded as an economic source of alumina, and is used principally in the manufacture of refractories. In the Bayer process, alumina is not recoverable when combined with silica since the aluminosilicate minerals react with caustic soda to form insoluble hydrous sodium aluminum silicates. The silica that is in chemical combination with the aluminosilicates is often termed the "reactive" silica to distinguish it from the nonreactive" silica (present as quartz, chalcedony, etc.) that does not react with caustic soda.

Origin of Bauxite. *Residual Weathering.* Following early work by Buchanan, Bauer, Lacroix, Fox, Harrison, and Harder, summarized in a classic study of the Arkansas bauxite region by Gordon et al. (1958), there is general agreement that most bauxite deposits are primarily of "residual" origin and that their formation is a particular case of the laterization process. In brief, bauxitization is attributed to in situ chemical weathering during which such components as soda, potash, lime, magnesia, and silica are removed in solution from the parent rock. Gordon et al. (1958, p. 145) considered that bauxitization takes place if the following conditions are satisfied:

1. A warm humid climate of continuous moisture, in which rainfall considerably exceeds evaporation most of the time.
2. A high temperature, probably exceeding 77°F most of the time, provides an environment in which the microflora can destroy humus faster than the macroflora can manufacture it. (Ground water rich in humic organic material or strong acids deposits silica and kaolin and takes alumina and iron oxides into solution.)
3. Pure rainwater acts upon a porous aluminous rock, preferably coarse-grained, that will permit free movement of water.
4. The rock is located in a topographically elevated, well-drained place, but not so far above sea level that a temperate climate and conditions more favorable for podsolization will prevail.
5. The rock is above the level of the permanent water table.
6. The conditions listed above last for a long period of time.

However, the sequence of events and precise conditions attending the formation of bauxite are still controversial. For example, the kaolinite bodies, which frequently separate the bauxite from the underlying parent rock, are regarded by some investigators as having been derived by resilication of the lower portion of the overlying bauxite, whereas others consider the bauxite to represent the desilicated upper portion of the kaolinite bodies. Thus, several workers, notably Allen, Harder, and Hose (1964) have proposed that bauxite does not form directly from the parent rocks, but that weathering results in the initial formation of a quartzose clay residuum, which is subsequently desilicated to give hydrated residuum of alumina hydrates, etc. Although this process has been convincingly demonstrated on a small scale for naturally occurring materials, there is insufficient evidence to postulate the direct formation of gibbsite on a large scale by the breakdown of kaolinite or halloysite. On the other hand, there is ample evidence of the direct formation of bauxite from the parent rock (see for example, Bleackley, 1964, p. 137).

In discussing the origin of bauxite, two broad genetic types may be distinguished:

1. Bauxites associated with rocks containing aluminosilicate minerals from which the alumina hydrate minerals of the bauxite appear to have been derived, e.g., basic igneous rocks.
2. "Terra rossa" bauxites, which are developed on or in close association with limestones or dolomites.

The "terra rossa" bauxites, which are well represented in the Mediterranean and Caribbean regions, pose a particular problem in that they are considered to represent the bauxitized

25

insoluble clay residuum developed on the limestone surface after prolonged weathering. However, in the case of the Caribbean bauxites, the underlying limestones are exceptionally pure and their low alumina content appears inadequate to provide the material for the great mass of overlying bauxite. For the case of the Jamaican bauxite, Zans (1956) has suggested that the bauxitized material has been derived in part from argillaceous material contributed by adjacent areas of andesitic volcanic material that has been transported by intermittent karst waters and deposited in solution depressions on the limestone surface and then subjected to bauxitization.

Detrital. Subsequent to its formation, the bauxite material may become disaggregated by normal erosion processes and deposited elsewhere as part of a sedimentary series (Gordon, et al., 1958). Detrital material may undergo further enrichment by leaching of impurities subsequent to deposition, as well as by the later addition of hydrated aluminum oxides in veins and impregnations.

Chemical Precipitation. This theory, which was proposed by Archangelsky and revised by Bushinsky (1958), is favored by few geologists outside the U.S.S.R. Following the removal in solution or colloidal suspension of aluminum or iron released by weathering processes, it is postulated that these components are subsequently precipitated in a swampy or coastal lagoon environment.

Volcanic Exhalations. Davidson (1964) has commented on the possible development of bauxite (and hematite) deposits due to metalliferous solutions from submarine hot springs related to volcanic activity. Davidson's account (1964, p. 244) includes an excellent example of this process, which is taking place at present in the volcanic province of the Kurile Islands: "Thus the rivulent Yur'yeva arising from Ekebo volcano on Naramushir island, debouches into the Okhotsk Sea with a volume of 1.8 cubic meters per second and pH 1.72 and brings daily 35 tons of iron and 65 tons of aluminum to the ocean waters; at its mouth iron and aluminum hydroxides form a belt of thick yellow mud stretching seawards for 3 km, with an average width of 300 meters and a thickness of 2 meters."

Present-Day Formation of Bauxite. Bauxite is currently forming in a number of localities, e.g., on metamorphosed acid volcanics and sediments in Malaya, on karst-limestone in Jamaica, and on the wet side of alkaline basalt and andesitic volcanic domes in the Hawaiian Islands. Hose (1964, p. 6) considers gibbsite to be the predominant alumina hydrate forming at the present time, and reports that variations in chemical composition are small although the bauxites are derived from widely different rock types. Hose concludes that bauxites derived from alkaline and intermediate volcanic, igneous, metamorphic, and sedimentary rocks tend to be lower in Fe_2O_3 and TiO_2 than bauxites derived from basic volcanic rock, whereas bauxites derived from limestones are usually higher in Fe_2O_3 and TiO_2 than those derived from basic volcanic rock.

Geographical-Stratigraphical Distribution of World Bauxite Deposits. Bauxite deposits are situated mainly in tropical and subtropical regions although deposits do occur in higher latitudes, for example, as far as 60°N in the U.S.S.R. and at 42°S in Tasmania. A stratigraphical summary of world bauxite deposits is given below; that the stability fields of the hydrated alumina oxides can be broadly correlated with geological history is significant. Thus, gibbsite and boehmite tend to occur in the younger deposits whereas diaspore tends to occur in the older deposits.

Quaternary (Recent and Pleistocene). These include the predominantly gibbsitic tropical lateritic bauxites of Panama, Costa Rica, Hawaii, Fiji, and the Solomon Islands. In Brazil, Venezuela, French Guinea, the Republics of the Ivory Coast and Cameroun, Ghana, Malawi, Malaya, Indonesia, the Caroline Islands, and Western Australia, surface bauxite overlies older rocks and may have been formed before the Pleistocene.

Tertiary. In the Cape York Peninsula of Australia, the vast gibbsite-boehmite deposits under present development are considered to be Tertiary. Other deposits regarded as Tertiary are the following: the gibbsite-boehmite deposits of Jamaica and Hispaniola; the gibbsitic bauxites of the Guianas and Surinam; the Australian deposits in Victoria; the lateritic bauxites of Oregon; the gibbsitic bauxites of the southern United States, the Deccan, Northern Ireland, Hungary, Dalmatia, and Montenegro; and the diaspore deposits of Kashmir.

Mesozoic. Boehmitic deposits, which constitute one-sixth of world bauxite reserves, are confined mainly to southern Europe, where they are associated with a broad belt of Mesozoic and early Tertiary limestones running from northern Spain through southern France, Italy, Yugoslavia, Austria, Hungary, Greece, and Turkey. Varying amounts of gibbsite and diaspore are associated with these beds or lenses of "terra rossa" bauxites, which have been developed on a weathered limestone surface of Lower Cretaceous or Jurassic age. Subsequently, the deposits became submerged and were overlain by younger limestones, whereas later earth movements resulted in their present faulted, folded, and sometimes overturned attitudes.

Palaeozoic. Diasporic deposits are largely confined to areas in the temperate zone, e.g., Russia, China, northern India, northern Turkey, Hungary, Rumania, Scotland, and the United States (Pennsylvania and Missouri). Apart from some deposits in Kashmir, Hungary, Greece, and China, they are all classified as Palaeozoic. Davidson (1964, p. 183) has drawn attention to what appear to be the earliest known bauxite deposits definitely identifiable with a lateritic crust of weathering. The deposits consist of two zones rich in hematite and gibbsite and are situated in the northern Onega district of the U.S.S.R.; they are classed as laterites of Upper Devonian and Lower Carboniferous age, respectively. "The first development of lateritic ores can therefore be equated with the first invasion of the land by a forest flora, a coincidence which lends weight to the commonly held view that organic processes play an essential part in laterization."

Perhaps the oldest known deposit occurs in the Eastern Sayansk region of the Buryat Republic, where a thick bauxite horizon containing diaspore and boehmite is developed at a disconformity in dolomite between (?) uppermost Sinian and Lower Cambrian strata.

Precambrian. Although some Russian geologists (see Bushinsky, 1958) have suggested that the corundum-bearing rocks in the U.S.S.R. associated with early Precambrian metamorphic rock may represent metamorphosed bauxites, the oldest alumina-rich sediment occurs in Swaziland. Here a bed containing diaspore has been worked in the Insuzi Series, which is over 3100 million years old.

World Bauxite Reserves. A bibliography of the world's bauxite deposits has been compiled by Fischer (1955), and a world survey of bauxite deposits has been made by Bracewell (1962). World bauxite reserves, including inferred bauxite, have been estimated by Patterson (1967) at 5800 million tons, with potential bauxite resources of 9600 million tons. In recent years, world production of bauxite has been rising at an average annual rate of 8% and an estimated 45 million tons of bauxite were produced in 1968. Reserves appear to be adequate for the next century.

J. B. ALLEN

References

Bleackley, D., 1964, "Bauxite and laterites of British Guiana," *Geol. Surv. Guiana Bull.*, **34**.

*Bracewell, S., 1962, "Bauxite, Alumina and Aluminum," Mineral Resources Division, Overseas Geol. Surv., London, H. M. Stationery Office.

Bushinsky, G. I., 1958, in (N. M. Strachov, chief editor), "Bauxites, Their Mineralogy and Genesis," 488pp., Moscow. Review by V. A. Zans in *Econ. Geol.*, 1959, **54**(5), 957–965.

Davidson, C. F., 1964, "Uniformitarianism and ore genesis," *Mining Mag.* **110,** 176–185, 244–253.

Deer, W. A., Howie, R. A., and Zussman, J., 1962, "Rock-forming Minerals," Vol. 5, "Non-Silicates," pp. 93–117. London, Longmans Green & Co.

Gordon, M., Tracey, J. I., and Ellis, M. W., 1958, "Geology of the Arkansas bauxite region," *U.S. Geol. Surv. Profess. Paper* **299,** 268pp.

Hose, H. R., 1964, "Bauxite Mineralogy," in "Extractive Metallurgy of Aluminum," Vol. 1, "Alumina," pp. 3–20, New York, Interscience Publishers, a division of John Wiley & Sons.

Moses, J. H., and Mitchell, W. D., 1961, "Bauxite deposits of British Guiana and Surinam in relation to underlying unconsolidated sediments suggesting two-step origin," *Geol. Soc. Am. Spec. Paper*, **68,** 233.

Patterson, S. H., 1967, "Bauxite reserves and potential aluminum resources of the world," *U.S. Geol. Surv. Bull.* **1228,** 176pp.

Schlanger, S. O., 1964, "Petrology of the limestones of Guam," *U.S. Geol. Surv. Profess. Paper* **403-D,** 14–15, 44, 49, 50.

Smith, J. W., and Milton, C., 1966, "Dawsonite in the Green River Formation of Colorado," *Econ. Geol.*, **61,** 1029–1042.

U.S. Bur. Mines, 1953, "Materials Survey on Bauxite," Washington, D.C., U.S. Govt. Printing Office, 309pp.; after Edwards, J. D., et al., 1930. "The Aluminum Industry," Vol. 1, New York, McGraw-Hill Book Co., 358pp.

Williams, L. R., 1965, "Bauxite," *Minerals Yearbook* (U.S. Bur. Mines) Vol. 1.

Zans, V. A., 1956, "The Origin of the Bauxite Deposits of Jamaica," C. R. 20th Int. Geol. Congr. Mexico, 1956, p. 108.

* Additional references may be found in this work.

Cross-references: *Oxides and Hydroxides; Weathering, Chemical.* Vol. IVB: *Bauxite; Clays and Clay Minerals; Green River Mineralogy.* Vol. VI: *Laterization; Leaching; Podzols; Terra Rossa; Weathering, Organic.*

ALUMINUM TOXICITY

Aluminum toxicity is probably as big a problem in highly leached or wet soils as are salinity and alkalinity in the soils of arid regions. Soils are said to exhibit aluminum toxicity when crops grown on them show signs of poor yield and growth due to the content of extractable aluminum in the soil. The evidence for this condition consists of stunted growth and coralloid roots in the cases of maize, cowpeas, and sugarcane

ALUMINUM TOXICITY

(Evans, 1956). The short main roots and secondary roots are dead and further branching gives a mat of stumpy roots. The dead root tips contain much aluminum and also phosphorus. The leaves show marked phosphorus deficiency even when phosphatic fertilizer has been applied to the soil. Leguminous plants do not develop nodules containing nitrogen-fixing bacteria. Similar symptoms appear to be shown in the cases of other crops.

Distribution and Origin

Geographically, aluminum toxicity is widespread in humid regions, ranging from the peats of Canada and Finland to the equatorial coastal gleys and peats of the Guianas. It occurs on the forested slopes of the Appalachians and in the "cat clays" of Holland.

The types of soil most commonly exhibiting toxicity are podzols, red-yellow podzolic soils, peats and mucks, and humic gleys. It may also occur occasionally in gray-brown podzolic soils. Thus it is to be correlated with soils which are permanently wet as well as those which are highly leached. The highly leached soils which dry out during the year rarely show aluminum toxicity, which supports the idea that the aluminum tends to be precipitated irreversibly as gibbsite when the soil is wetted and dried regularly.

The aluminum originates in the primary minerals of the unweathered rock. During the chemical weathering processes, it is freed as the primary minerals decompose, and it then may exist in the soil in one of many different forms. Jackson (1963) has pointed out that this element plays a very important role in the chemistry of the soil processes, particularly in the more highly leached soils. As leaching takes place, the aluminum concentration increases since the other more soluble substances are removed by soluviation. Provided that the soil does not dry out, the aluminum concentration is not altered by appreciable irreversible precipitation in an inactive form.

Not only does the aluminum accumulate in the soil but it may also be absorbed from the soil water into plant tissues and accumulate there as water is lost by transpiration. This has been known since at least 1743. When the plant dies, and organic matter which has been enriched in extractable aluminum is returned to the soil, this can result in extremely large concentrations of extractable aluminum in organic soils (Harris, 1961). Values may be as high as 1.4% extractable aluminum in the peat, even though the underlying soil is only slightly leached. The aluminum comes into the soil in chelated form (say, combined with fulvic acids) in the groundwater. Actual studies of distribution of extractable aluminum in plants and soils show that the two do not correlate closely in a given area, probably due to the variety of forms in which the aluminum may exist (Harris, 1963).

Plant behavior in regard to aluminum concentration various enormously. The natural vegetation consists of species which can tolerate tremendous quantities of extractable aluminum in most cases. The plants grown as crops often grew originally in areas where aluminum toxicity was unimportant or absent. One of the most sensitive plants is jute. It is sensitive to about 300 ppm aluminum (soluble in $N/2$ acetic acid). Rice, cocoa, citrus, black-eye peas, maize, mung, coconuts, castor bean, sesame, and sugar cane will tolerate 300–500 ppm aluminum, i.e., they are fairly tolerant. Finally there are the tolerant crops such as oil palm, cassava, eddoes, plantains, bananas, yams, sweet potatoes, certain varieties of pineapple, and coffee.

Treatment

Various treatments have been suggested. Peats may be drained and placed under improved pasture using a grass such as *Brachiaria purpurascens* which will tolerate soils containing large quantities of aluminum. Within about five years, the peat will have weathered and the quantity of extractable aluminum may drop to levels which will permit other kinds of agriculture. If sufficiently large quantities of lime and fertilizer are added, the soil may be rendered fertile but this is highly uneconomical. Flooding by seawater, first tried by H. Evans on sugar estates in 1960 in British Guiana (now Guyana), is proving a very cheap way of increasing the base status of the soils by replacing some of the ions leached out of the soil. Removal of the excess salt is no problem in a humid climate, and only moderate applications of lime and fertilizers then turn the soils into excellent farm land.

Failure to recognize aluminum toxicity has proved very costly. The Dutch overcame toxicity problems in the Middle Ages by reclaiming the intertidal zone along the shores of the Guianas. Their settlements further inland were generally unsuccessful. In modern times the Dutch have had considerable trouble in utilizing the cat clays in Holland. Failure to survey the soils of the area before commencing construction of irrigation and drainage projects has proved very costly to the Government of Guyana, e.g., their Boerasiree Project on the west coast of the Demerara River. Unfortunately we still have much to learn about aluminum toxicity and its detection, and without numerous expensive soil analyses it is impossible to predict reliably whether wet, leached soils will exhibit this property when drained.

STUART A. HARRIS

References

Evans, H., 1956, in (Crocker, C. D., editor), "Boerasiree West Soil Report," Dept. of Agr., British Guiana (in cooperation with Univ. Maryland) Georgetown. Mimeo., 85pp.

Harris, S. A., 1961, "Soluble aluminium in plants and soils," *Nature*, **189**, 513–514.

Harris, S. A., 1963, "Aluminium in soils and plants on the coastlands of British Guiana," *J. Sci. Food Agr.*, **14**, 259–263.

Jackson, M. L., 1963, "Aluminium bonding in soils: A unifying principle in soil science," *Soil Sci. Soc. Am. Proc.*, **27**, 1–10.

Cross-references: *Aluminum; Chelation; Groundwater; Pedology; Weathering, Chemical.* Vol. VI: *Leaching; Podzols.*

AMMONIA, IN MINERALS AND EARLY ATMOSPHERE

It has been estimated (Mason, 1966; Rubey, 1951; Wedepohl, 1963) that the earth's atmosphere, hydrosphere, and crust contain a total of 44.3×10^{20} g nitrogen. Of this amount, 87% is present in the atmosphere as N_2, and most of the remainder is contained in organic compounds trapped in sedimentary rocks (see Table 1). Nitrogen fixation, most of it through organic processes, accounts for about 4.8×10^{20} g of the nitrogen present in the crust (Hutchinson, in Kuiper, 1954). Thus, a prebiological atmosphere may have contained as much as 98% of the nitrogen present in the outer shells of the earth, indicating a very efficient chemical separation mechanism. Minerals rich in nitrogen are rare, and rock-forming minerals usually contain less than 100 ppm N_2 (see Wlotzka, 1961, reference in Eugster and Munoz, 1966). Nitrogen migrating toward the surface of the earth, through degassing during geologic time (Rubey, 1951), accumulated in the atmosphere as N_2 gas. In contrast, most of the hydrogen arriving at the surface (discounting the probably large amount lost to space) is present in the oceans, whereas the bulk of the oxygen remains locked in the mantle in silicates and oxides. Table 2 gives the comparison between these three gases.

The presence of nitrogen in the atmosphere either as N_2 or as NH_3 was essential to the origin and evolution of life. Without nitrogen, amino acids and hence proteins could not have formed. But where did this nitrogen come from? To answer this question, we must look more closely at the occurrence of nitrogen and ammonia in minerals.

Nitrogen and Ammonium in Minerals

The only commonly occurring nitrogen minerals are the nitrates soda-niter ($NaNO_3$) and niter (KNO_3). They are frequently found in efflorescent crusts of arid regions and in caves. The most massive deposits are those in Chile. Both minerals are water-soluble and owe their existence to very special circumstances. Two naturally occurring nitrides, siderazot (Fe_5N_2) and osbornite (TiN), have been described and both are extremely rare. Siderazot was found on Etna and Vesuvius, and osbornite in the enstatite achondrite Bustee. Small amounts of other nitrides, such as BN and Si_3N_4 have been suspected to be present in some igneous rocks, but their existence has never been verified.

The ammonium feldspar buddingtonite ($NH_4AlSi_3O_8$) was found in hot-spring deposits in California by Erd, White, Fahey, and Lee (1964; reference in Eugster and Munoz, 1966).

None of these minerals can account for the bulk of the nitrogen present in the inorganic parts of the crust. Wlotzka (1961, reference in Eugster and Munoz, 1966) found igneous rocks to contain an average of 20 ppm nitrogen. Most of this nitrogen is located in muscovite (av 60 ppm), biotite (av 33 ppm), or feldspars (av 20 ppm). Average values for sedimentary rocks are considerably higher: limestones, 70 ppm; sandstones, 135 ppm; shales, 580 ppm; clay minerals, particularly illite and glauconite, contain the largest amounts. The distribution of nitrogen among the different minerals indicates that nonbiogenic nitrogen must be present as $(NH_4)^+$ ions, in place of K^+ in the silicate lattices. Some of this nitrogen may originally have been contained in organic compounds and later become fixed in clay minerals.

The presence of ammonium ions in the interlayer sites of micas was demonstrated through infrared studies by Vedder (1964; reference in Eugster and Munoz, 1966), and through synthesis of pure ammonium muscovite, $NH_4Al_2AlSi_3O_{10}(OH)_2$ and ammonium

TABLE 1. NITROGEN CONTENT OF ATMOSPHERE, HYDROSPHERE AND CRUST OF THE EARTH

Unit	Amount $\times 10^{20}$ g	Total Nitrogen (%)
Atmosphere	38.6	87.1
Hydrosphere	0.3	0.7
Crust, sediments	5.2	11.7
Crust, igneous rocks	0.2	0.5
Total	44.3	100.0

TABLE 2. DISTRIBUTION OF NITROGEN, HYDROGEN AND OXYGEN AMONG THE OUTER SHELLS OF THE EARTH

Shell	Thickness (km)	Total Mass (10^{20} g)	Nitrogen		Hydrogen		Oxygen	
			10^{20} g	Wt %	10^{20} g	Wt %	10^{20} g	Wt %
Atmosphere	>100	51.1	38.6	75.6	0.0000023	0.0000045	11.8	23.1
Hydrosphere	3.8	14,100	0.3	0.002	1600	11.3	12,500	88.7
Crust	17	240,000	5.4	0.002	310	0.13	110,000	45.8
Mantle	2883	40,750,000	?	?	?	?	15,700,000	38.6

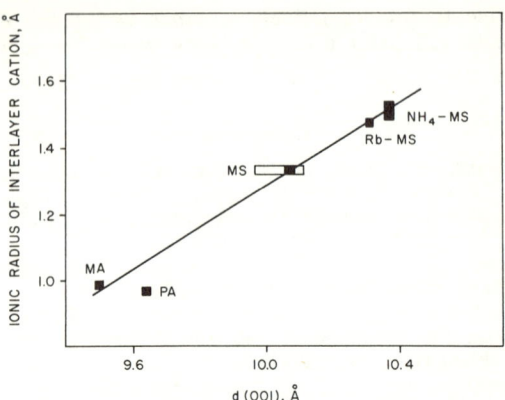

FIG. 1. Spacing of (001) for 1M polymorphs of a number of synthetic dioctahedral micas as a function of the ionic radius of the interlayer cation: $d(001)$ for celadonite varies with f_{O_2}. The radius of NH_4^+ in silicates is uncertain: an average value is plotted. MA, margarite; PA, paragonite; MS, muscovite; Cel, celadonite; Rb–Ms, rubidium muscovite; NH_4–MS, ammonium muscovite (for source of data see Eugster and Munoz, 1966.)

phlogopite, $NH_4Mg_3AlSi_3O_{10}(OH)_2$ (Gruner, 1939; Barrer and Denny, 1961; Eugster and Munoz, 1966). NH_4^+ occupies slightly more space than K^+ and hence the basal spacings of the ammonium micas are larger than those of the potassium analogs. The relationship between ionic radius and the basal spacing, $d(001)$, is shown in Fig. 1 for dioctahedral micas.

If ammonium silicates form readily, why are they so rare in the crust and why is the ammonium content of nonbiogenic rocks so low? In view of the large amount of nitrogen present in the atmosphere, the cause cannot be lack of nitrogen, but must be connected with the stability of ammonium silicates in a near-surface environment.

Stability of Ammonium Silicates

The thermal stability of ammonium silicates compares favorably with that of their potassium and sodium counterparts. Ammonium muscovite, for instance, was found to be stable to at least 670°C at 2kb gas pressure (Eugster and Munoz, 1966). In an alkali-rich environment, however, much of the ammonium may be lost at much lower temperatures through cation-exchange reactions.

$$NH_4AlSi_3O_8 + K^+ \rightleftharpoons$$
buddingtonite
$$\quad\quad\quad\quad KAlSi_3O_8 + NH_4^+ \quad (1)$$
$\quad\quad\quad\quad$ K-feldspar

The ammonia released in this way may be oxidized.

$$4NH_3 + 3O_2 \rightleftharpoons 2N_2 + 6H_2O \quad (2)$$

Oxidation of the silicate may also occur directly.

$$4\ NH_4AlSi_3O_8 + 3\ O_2 \rightleftharpoons 2\ Al_2SiO_5 +$$
buddingtonite   e.g., kyanite
$$5\ SiO_2 + 2N_2 + 8\ H_2O \quad (3)$$
quartz

In either case, water will escape to the hydrosphere and nitrogen gas to the atmosphere, leaving behind potassium and aluminum silicates. The position of the equilibrium Eq. (1) is governed by P, T and the activity ratio a_{NH_4}/a_K. Because of the abundance of Na and K in the crust, the equilibrium is displaced far to the right in most near-surface environments.

If Eq. (2) is deducted from Eq. (3), Eq. 4 results.

$$4\ NH_4AlSi_3O_8 \rightleftharpoons 2\ Al_2SiO_5$$
buddingtonite   e.g., kyanite
$$+ 5\ SiO_2 + 4\ NH_3 + 2\ H_2O \quad (4)$$
quartz

Assuming pure solids, this equilibrium is governed by

$$(K_4)_{P,T} = f_{NH_3}^4 \times f_{H_2O}^2$$

where f_{NH_3} is the fugacity of ammonia, and f_{H_2O} is the fugacity of H_2O. Hence, ammonium silicates are unstable in an anhydrous environment (low f_{H_2O}) and in an environment in which ammonia is unstable. Since the core and mantle of the earth are probably virtually anhydrous, we should not expect ammonium silicates to be present there. Nevertheless, f_{H_2O} is usually high enough in the crust, and the stability of the ammonium silicates then depends on the magnitude of f_{NH_3}, which in turn is governed by

$$3H_2 + N_2 \rightleftharpoons 2NH_3 \quad (K_5)_T = \frac{f_{NH_3}^2}{f_{H_2}^3 \cdot f_{N_2}} \quad (5)$$

In Fig. 2 the ratio $f_{NH_3}^2/f_{N_2}$ has been plotted against T for a range of hydrogen fugacities using thermodynamic data to calculate the value of K of Eq. (5). High hydrogen fugacity and low temperature stabilize ammonia. Probable ranges of f_{H_2} in crustal environments can be evaluated from mineral assemblages present.

Hydrogen Fugacities of Crustal Environments

Mineral assemblages often define and buffer the magnitude of the oxygen fugacity, f_{O_2}, of a particular environment (see, for instance,

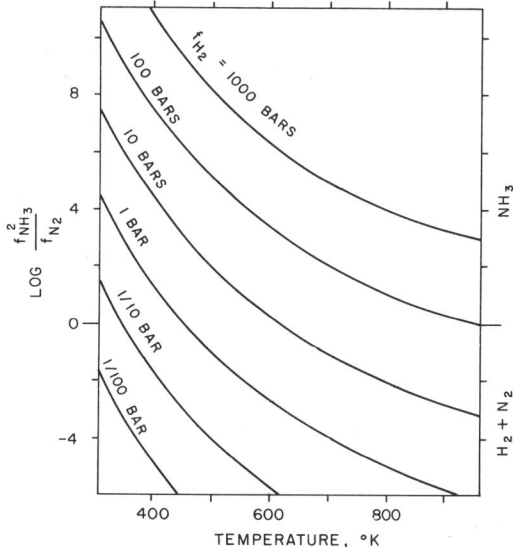

FIG. 2. Stability of ammonia versus nitrogen + hydrogen as calculated from thermodynamic data. Ammonia is more abundant in the upper portion of the diagram, as hydrogen + nitrogen is in the lower part. Hydrogen fugacities have been contoured between 1/100 and 1000 bars.

Eugster and Wones, 1962, reference in Eugster and Munoz, 1966). If f_{O_2} is defined, and the gas phase consists of a mixture of $H_2O + H_2$, f_{H_2} can be calculated for a given P and T from

$$2H_2 + O_2 \rightleftharpoons 2H_2O \quad (K_6)_T = \frac{f_{H_2O}^2}{f_{O_2} \cdot f_{H_2}^2} \quad (6)$$

and from the requirement that

$$P_{gas} = P_{H_2O} + P_{H_2} + P_{O_2}$$

Oxidized mineral assemblages, such as magnetite and hematite, have very low values of f_{H_2}. Values of f_{H_2} of a more reducing assemblage, quartz + magnetite + fayalite (QFM) are shown in Fig. 3 for the two gas pressures $P_{gas} = 1$ bar and $P_{gas} = 250$ bars (dashed curves). Repeated in Fig. 3 are the hydrogen isobars of Fig. 2 for reference. In a gas equilibrated with quartz + magnetite + fayalite, $N_2 + H_2$ predominates over ammonia, except at low temperature and very high pressure. Crustal environments are usually not much more reducing. For comparison, the curves for magnetite + iron (MI, dotted curves), more reducing than either average crust or mantle, are also shown. This assemblage corresponds essentially to pure hydrogen and hence favors ammonia.

If other gases are present, such as CO_2, equilibria are more in favor of the decomposi-

AMMONIA, IN MINERALS AND EARLY ATMOSPHERE

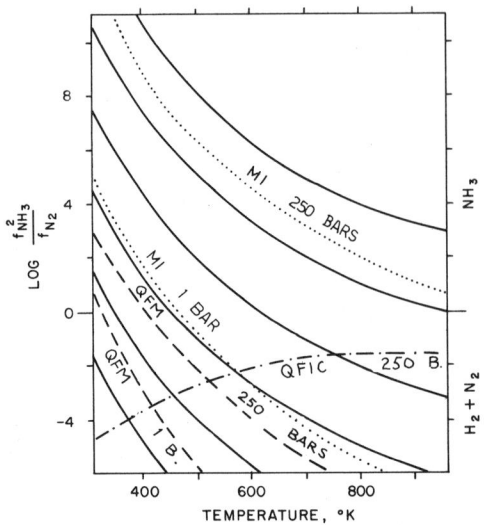

FIG. 3. Data of Fig. 2, to which have been added curves for three mineral assemblages: quartz + magnetite + fayalite (QFM, dashed curves) for gas pressures of 1 and 250 bars; magnetite + iron (MI, dotted curves) for 1 and 250 bars; and quartz + iron + fayalite + graphite (QFIC, dash-dot curve) for 250 bars.

tion of NH_3. For instance, if graphite is present, even an assemblage of metallic iron + quartz + fayalite (QFIC), dot-dash curves) does not stabilize NH_3 (data from Eugster and Skippen, 1967). The data show that NH_3 predominates over $N_2 + H_2$ only in unusually reducing environments, thus accounting very simply for the scarcity of pure ammonium silicates in the crust.

So far only the stability of pure ammonium silicates has been considered. But what about the stability of an alkali silicate containing some ammonia, such as a solid solution between potassium feldspar and buddingtonite? The equilibrium constant K_4 for Eq. (4) in this case is

$$(K_4)_{P,T} = \frac{f_{NH_3}^4 \times f_{H_2O}^2}{\left(a_{NH_4AlSi_3O_8}^{\text{feldspar ss}}\right)} \quad (4)'$$

where $\left(a_{NH_4AlSi_3O_8}^{\text{feldspar ss}}\right)$ is the activity of the buddingtonite composition in the feldspar solid solution, with an average value of about 10^{-4} in igneous rocks. Since K_4 remains unchanged, f_{NH_3}, and hence f_{H_2}, may be much smaller in Eq. (4)' than in Eq. (4). In other words, the ammonium content of igneous rocks is generally low, because the average value of f_{NH_3} in crust and mantle is very low. As more data on synthetic and natural systems become available, we may be able to use the ammonium content of silicates as a quantitative measure of the ammonia fugacity of a particular environment.

Ammonia and Nitrogen in the Early Atmosphere

The present atmosphere of the earth is generally considered to have accumulated by degassing of the earth throughout its history (Rubey, 1951; Holland, 1962). If this is correct, nitrogen would have had to be present in the solid matter that formed the protoplanet (Urey, 1952), probably either as ammonium silicates or as iron nitride. Subsequent to aggregation, the earth differentiated into core, mantle, and crust, in part driven by the heat produced through radioactive decay. Ammonium silicates probably lost most of their ammonia early in this development except in very unusual local environments. The nitrogen released would have reached the atmosphere either as NH_3 or N_2.

It has often been stated (Holland, 1962), that the early atmosphere consisted essentially of ammonia and methane, based on the persuasive argument that a reducing atmosphere was essential for life to evolve. However we have shown earlier, that $N_2 + H_2$ predominate over ammonia in a normal crustal environment. The ammonia fugacity may have been large in a predifferentiated crust, if metallic iron was still present. However, abundant CH_4 would have caused graphite to precipitate, and in an iron + graphite + silicate crust, nitrogen would have predominated over ammonia (Fig. 3). Since life originated long after a granitic crust formed, the atmosphere at that time surely must have already contained much more nitrogen than ammonia. It can be argued that upper levels of the atmosphere were out of equilibrium with the lowest layer. This is probably fallacious, since most of the free oxygen present in the prebiological atmosphere was derived from photodissociation of H_2O in the upper atmosphere.

In summary, the nitrogen present in the atmosphere may have been contained in ammonium silicates and metal nitrides during the accumulation of the protoearth. Some metal nitrides may still be locked in the core or mantle, but ammonium silicates probably released most of their nitrogen during early stages of the differentiation of the earth. At first, this nitrogen may have reached the atmosphere as ammonia, but this was oxidized to nitrogen long before the origin of life. The mass of ammonium silicates necessary to provide the nitrogen present in atmosphere and crust at the present time is small compared with the total mass of the earth. Some ammonium silicates still persist in the crust and perhaps in the mantle in unusual environments, but most of the nitrogen once present in the crust and perhaps also in the mantle, has been swept out into the atmosphere, primarily through oxidation.

The three ions K^+, Na^+, and NH_4^+ occupy similar posititons in alkali silicate structures. It is interesting to contemplate the differences in their fates during differentiation of the earth's surface. K^+, through preferential adsorption on clay minerals, is accumulated in sediments and hence returned to the crust. Na^+, by default, and because of the great solubility of its compounds, is enriched in the hydrosphere, whereas NH_4^+ is oxidized and the nitrogen passes into the atmosphere as N_2. Equally dangerous but tantalizing generalizations can be made by comparing the earth's atmosphere with those of Venus and Jupiter. Apparently, the cytherean atmosphere is richer in N_2 (Dayhoff et al., 1967), perhaps because of more rapid photodissociation of H_2O and loss of H_2. The oxygen would have to be trapped at the surface, and it has been suggested that very active volcanism connected with the high surface temperature (Davidson et al., 1967) may provide the necessary sink. In contrast, the jovian atmosphere still seems to contain NH_3 and CH_4 (see, for instance, Greenspan and Owen, 1967), perhaps representing a stage the earth's atmosphere passed through long ago.

H. P. EUGSTER

References

Davidson, G. T., and Anderson, A. D., 1967, "Venus: Volcanic eruptions may cause atmosphere obscuration," *Science*, **156**, 1729–1730.

Dayhoff, M. O., Eck, R. V., Lippincott, E. R., and Sagan, C., 1967, "Venus: Atmospheric evolution," *Science*, **155**, 556–558.

*Eugster, H. P., and Munoz, J., 1966, "Ammonium micas: Possible sources of atmospheric ammonia and nitrogen," *Science*, **151**, 683–686.

Eugster, H. P., and Skippen, G. B., 1967, "Igneous and metamorphic reactions involving gas equilibria," in (Abelson, P. H., editor), "Researches in Geochemistry," Vol. 2, New York, John Wiley & Sons.

Greenspan, J. A., and Owen, T., 1967, "Jupiter's atmosphere: its structure and composition," *Science*, **156**, 1489–1494.

Holland, H. D., 1962, "Model for the evolution of the earth's atmosphere," *Buddington Vol., Geol. Soc. Am.*, 447–477

Kuiper, G. P. (editor), 1954, "The Earth as a Planet," Chicago, Univ. of Chicago Press, 751pp.

*Mason, B., 1966, "Principles of Geochemistry," Third ed., New York, John Wiley & Sons, 329pp.

Rubey, W. W., 1951, "Geologic history of sea water. An attempt to state the problem," *Geol. Soc. Am. Bull.* **62**, 1111–1148.

Urey, H. C., 1952, "The Planets, Their Origin and Development," New Haven, Yale Univ. Press, 245pp.

Wedepohl, K. H., 1963, "Einige Überlegungen zur Geschichte des Meerwassers," *Fortschr. Geol. Rheinland Westfalen*, **10**, 129–150.

*Additional references may be found in this work.

Cross-references: *Core Geochemistry; Earth's Crust Geochemistry; Hydrogen; Nitrogen; Outgassing of the Planet Earth; Oxidation and Reduction; Oxygen; Water. Vol. II: Atmosphere. Vol. IVB: Nitrates. Vol. VI: Volatiles.*

ANION EXCHANGE—
See **ION EXCHANGE**

ANTIMONATES: ANTIMONITES AND ARSENITES

This is a class of relatively rare minerals found in many different geological environments. Among the minerals in this class, antimony and arsenic sometimes take the place occupied by sulfur in the more common sulfides and sulfates. Some of the minerals can be considered as multiple oxides rather than anisodesmic compounds. Often these minerals are not placed in a separate class but are included with the phosphates and arsenates (see Table on p. 34).

LLOYD STAPLES

Reference

Dana, J. W., and Dana, E. S., 1951, "The System of Mineralogy," Seventh ed., Vol. II, 1017, (revised by Palache, C., Berman, H., and Frondel, C.), New York, John Wiley & Sons.

Mason, B., and Vitaliano, J., 1953, "Antimony oxides and antimonates," *Mineral., Mag.* **30**, 100.

Cross-reference: *Mineral Classes: Nonsilicates.*

ANTIMONY: ELEMENT AND GEOCHEMISTRY

Physical Properties

Antimony (L. *stibium,* mark), symbol Sb, has atomic weight 121.75; atomic number 51; electron configuration [Kr core] $4d^{10}5s^25p^3$; melting point, 630.5°C; boiling point, 1635°C; specific gravity, 6.691 (at 20°C); valence, 3 or 5. Crystal modifications: a hexagonal (rhombohedral) metallic form, reactive yellow allotrope (polymorph), and an amorphous black modification. Isotopic abundances: Sb^{121} 57.25%; Sb^{123} 42.75%. Ionic radii: Sb^{3+} 0.76Å; Sb^{5+} 0.62Å; Sb^{3-} 2.45Å. Antimony is a brittle, blue-white metalloid and a poor conductor of heat and electricity.

ANTIMONATES: ANTIMONITES AND ARSENITES

ANTIMONATES

AX_2O_6 Type

Ordonezite	$ZnSb_2O_6$	⎫
Bystromite	$MgSb_2O_6$	⎬ Isostructural
Tripuhyite	$FeSb_2O_6$	⎭

$A_2X_2O_6(OH,F)$ Type

Bindheimite	$Pb_2Sb_2O_6(O,OH)$	⎫ *Bindheimite Group:* crystal
Romeite	$Ca_2Sb_2O_6(O,OH,F)$	⎭ structure of pyrochlore type

Miscellaneous

Derbylite	$Fe_6Ti_6Sb_2O_{23}$ (?)
Swedenborgite	$NaBe_4SbO_7$
Katoptrite	$Mn_{13}Al_4Sb_2Si_2O_{28}$

ACID AND NORMAL ANTIMONITES AND ARSENITES

Miscellaneous

Armangite	$Mn_3(AsO_3)_2$
Trigonite	$MnPb_3H(AsO_3)_3$
Trippkeite	$CuAs_2O_4$
Schafarzikite	$FeSb_2O_4$

BASIC OR HALOGEN-CONTAINING ANTIMONITES, ARSENITES

Miscellaneous

Ecdemite	$Pb_6As_2O_7Cl_4$ (?)
Heliophyllite	$Pb_6As_2O_7Cl_4$ (?)
Finnemanite	$Pb_5(AsO_3)_3Cl$
Nadorite	$PbSbO_2Cl$
Melanostibian	Mn_2SbFeO_6

Minerals and Compounds

Important antimony minerals are the native element, Sb; *stibnite*, Sb_2S_3; *kermesite*, Sb_2S_2O; *senarmontite*, Sb_2O_3; *jamesonite*, $2PbS \cdot Sb_2S_3$; *boulangerite*, $5PbS \cdot 2Sb_2S_3$; and the sulfantimonides of copper (Cu), silver (Ag), and nickel (Ni). Of these, stibnite is the most common mineral and the chief ore. Magnificent crystal specimens were once obtained from the Island of Shikoku, Japan. Artificial compounds include stibine, SbH_3, an obnoxious, poisonous gas, the chlorides $SbCl_3$ and $SbCl_5$, the sulfides Sb_2S_3 and Sb_2S_5, and the oxides Sb_2O_3 and Sb_2O_5.

Chemistry

The chemical properties of antimony resemble those of bismuth and arsenic. Sb loses its $2(s)$ and $3(p)$ electrons readily, and may exist in the oxidation states -3, $+3$, $+5$, and 0. The existence of simple Sb^{3+} or Sb^{5+} ions is improbable. Sb^{3+} forms complexes: $Sb(OH)_2^+$, $Sb(OH)_4^-$, SbS_3^{3-}, $SbCl_5^{2-}$, and $SbCl_3OH^-$. Sb^{3+} oxide is slightly soluble in dilute acids, more so in strong bases:

$$Sb_2O_3 + H_2O + 2H^+ \rightleftarrows 2Sb(OH)_2^+$$
(in acids)

$$Sb_2O_3 + 3H_2O + 2OH^- \rightleftarrows 2Sb(OH)_4^-$$
(in bases)

Sb^{5+} oxide dissolves in strong acids, forming complexes: $Sb_2O_5\downarrow + 10H^+ + 12Cl^- \rightarrow 2SbCl_6^- + 5H_2O$. It is also soluble in strong bases (KOH), yielding antimonides:

$$Sb_2O_5\downarrow + 2OH^- + 5H_2O \rightleftarrows 2Sb(OH)_6^-$$

The Sb^{3+} sulfide found in nature (stibnite) is dark gray. The Sb_2S_5 precipitate obtained in the laboratory by the action of H_2S in a dilute hydrochloric acid solution of the antimony compound is deep orange. The difference in color has been attributed to differences in particle size, but the orange precipitate is probably metastibnite, an amorphous form of Sb_2S_3.

The chlorides of Sb hydrolyze in water, yielding a whitish precipitate:

$$SbCl_3 + H_2O \rightleftarrows SbOCl\downarrow + 2H^+ + 2Cl^-$$
$$SbCl_5 + H_2O \rightleftarrows SbOCl\downarrow + 2H^+ + 4Cl^-$$

Geochemistry

Antimony is a relatively rare element with cosmic abundances on the order of 0.1–0.2 (Si = 10^6). It has chalcophile properties, combining readily with sulfides. Chondritic meteorites contain about 0.1 ppm Sb, whereas achondrites have 0.02–0.15 ppm of the element. Siderites show up to 0.96 ppm Sb. Antimony shows no marked preference for mafic or silicic rocks. The average concentration of Sb in igneous rocks ranges between 0.1 and 0.9 ppm. Antimony becomes enriched in the early stages of magmatic differentiation in sulfide bodies, as in the pyrrhotite-pentlandite deposits associated with gabbroic rocks. In addition, antimony accumulates in late-stage granite pegmatites, together with niobium and tantalum oxides, in granodiorites, and hydrothermal sulfide deposits. The chief minerals of hydrothermal deposits are stibnite and bournonite, $PbCuSbS_3$. Antimony is also present in galena, in amounts up to 1%, where it replaces either Pb^{2+} (ionic radius 1.20Å) or S^{2-} (1.84Å). Antimony may substitute for arsenic in many minerals, despite a rather large difference in ionic radii (Sb^{3+}, 0.76Å; As^{3+}, 0.58Å). Thus, solid-solution series of tetrahedrite-tennantite and famatinite-enargite are common. Numerous sulfantimonides, such as *pyrargyrite*, Ag_3SbS_3, *tetrahedrite*, Cu_3SbS_3, and *bournonite*, $PbCu(Sb,As)S_3$, are of secondary importance as minerals of sulfide ore deposits. Factors that may induce precipitation of antimony sulfides from alkaline solutions include neutralization by CO_2, oxidation, falling temperatures, and dilution. Antimony is also present in minor amounts in certain mercury deposits (Huitzuco district, Guerrero, Mexico).

Little is known about the behavior of Sb during weathering and erosion. The Sb sulfides are oxidized to the corresponding oxides (e.g., senarmontite, Sb_2O_3). Antimony probably occurs in both hydrolysate and oxidate sediments. Limestones and sandstone are low in antimony (0.2 ppm). Antimony may accumulate with other heavy elements in carbonaceous shales or by adsorption on clays and hydrous oxides. Thus it becomes enriched in MnO_2 sediments and black shales like the Mansfeld Kupferschiefer. Oceanic red clays show considerable enhancement of Sb over igneous rocks. An average of 1 ppm Sb is indicated. The Sb concentration in sea water is extremely low. It is found in trace amounts in certain marine organisms such as actinians, echinoderms, and fishes.

Economic Geology

Most antimony deposits have precipitated from low-temperature hydrothermal solutions at shallow depths. Two different antimony occurrences exist: (*1*) Stibnite is the principle ore mineral, but lesser amounts of cinnabar, pyrite, quartz, and chalcedony are also present (e.g., Hunan, China); (*2*) Complex vein deposits containing other base metals [copper (Cu), lead (Pb), zinc (Zn), silver (Ag), arsenic (As)]; the Sb is recovered as a by-product. Ore minerals are sulfantimonides of the other elements (e.g., Coeur d'Alene, Idaho). The antimony has been deposited as part of the main period of mineralization during which the other metals were emplaced.

China is the world's major antimony producer (31% of the world's total, 1962). Important mining districts are in Hunan Province, and also Kwangsi, Kweichow, Kwantung, and Szechan Provinces. The Chinese deposits consist of fissure fillings of stibnite, cinnabar, and pyrite, and replacements of stibnite-galena in limestones. The Republic of S. Africa ranks second in Sb production. The Murchison Range, Transvaal, furnishes antimony as a by-product of gold ores. The antimony deposits of Bolivia occur in the same general area as the tin-silver deposits. In general, the ore consists of stibnite in quartz gangue, with some galena or gold. Productive regions include Challapata district, department of Oruro, and the Tupiza district, department of Potosi. Yugoslavia, Mexico, U.S., and Czechoslovakia also yield significant quantities of the element. In the U.S., antimony is recovered as a by-product of lead-silver ores at Coeur d'Alene, Idaho. Lesser amounts are extracted from mines in Nevada, Utah, and Alaska.

Antimony has numerous applications. Its property of expansion upon freezing makes it suitable for type-metal alloy. Usually, antimony is alloyed with lead to harden it and increase its resistance to corrosion. Antimony-lead alloys are utilized in storage-battery plates, pipes, and sheathings for electrical cables. Other alloys include Britannia metal (Pb–Sb–Cu), pewter (Pb–Sn–Sb), Queen's metal (Sb–Sn–Cu–Zn), and Sterline (Cu–Sb–Zn–Fe). It also is used in munitions. Antimony of extremely high purity (99.999%) finds applications in semiconductor technology.

VIVIEN GORNITZ

References

Dui-Yin, W., and Saukov, A. A., 1961, "Physicochemical factors in the genesis of antimony deposits," *Geochemistry*, **6**, 510–516.

Ehmann, W. D., and Tanner, J. T., 1966, "The abundance of antimony in meteorites," *Earth and Planet, Sci. Letters*, **1**, 276–279.

Goldschmidt, V. M., 1954, "Geochemistry," Oxford, Clarendon Press, 730pp.
Lamey, C. A., 1966, "Metallic and industrial mineral deposits," Chap. 12, "Antimony and Mercury," New York, McGraw-Hill Book Co., 567pp.
Lawrence, E. F., 1963, "Antimony deposits of Nevada," Reno, Mackay School of Mines, University of Nevada, 248pp.
Quiring, H., 1945, "Die metallischen Rohstuffe, ihre Lagerungsverhältnisse und ihre wirtschaftliche Bedeutung No. 7 Antimon," Stuttgart, Ferdinand Enke Verlag, 158pp.
Rankama, K., and Sahama, Th. G., 1950, "Geochemistry," Chicago, University of Chicago Press, 912pp.
Tunell, G., 1964, "Chemical processes in the formation of mercury ores and ores of mercury and antimony," *Geochim. Cosmochim. Acta* **28**, 1019–1038.

Cross-references: *Acids and Bases; Arsenic; Bismuth; Hydrothermal Solutions; Niobium; Tantalum.* Vol. II: *Meteorites.* Vol. VI: *Red Clay.*

AQUEOUS SOLUTIONS

Water is plentiful in the general vicinity of the interface between the lithosphere and the atmosphere, i.e., at the outer surface of the earth's crust. The portion of this water that is in the liquid state participates in chemical reactions both with rock minerals in the lithosphere and with gases from the atmosphere. The properties of water make it an excellent solvent for most inorganic compounds. Water molecules are dipolar, the liquid acts as an effective insulator for electrical charges, and it persists in the liquid state over a wide temperature range. In addition, it has a high surface tension, and is attracted to mineral surfaces.

The oceans constitute a large reservoir of liquid water (1,370,323,^{000}km^3). Radiant energy from the sun continuously evaporates water from the oceans (about 93cm/yr). The vapor is transported in the atmosphere and is partly condensed and precipitated as rainfall over the land surfaces, which are thus bathed intermittently or continuously in relatively pure water, which returns toward the oceans by gravity flow.

The processes of solution and chemical alteration of the rocks near enough to the land surface to come into contact with circulating water have been going on ever since the surface temperature of the earth became low enough to permit water to condense. It seems logical to assume that these processes have strongly influenced the composition of the outer crust of the earth, and have tended to concentrate the elements in those environments where their chemical behavior naturally would tend to place them. Thus chlorine, which forms few compounds of low solubility, has been concentrated in the ocean, and elements such as silicon and most of the metals, which form compounds resistant to attack by water, are concentrated in rocks and soils.

The geochemist is interested in aqueous solutions that occur in nature because of the part they play in establishing the chemical composition of the earth's crust, and because of the information they can provide on the processes of chemical alteration of rocks from studies of the solutions that are produced by these processes. The composition of the water at the earth's surface is also of interest in itself, as a necessary part of the study and description of the hydrosphere. Published literature on the composition of the ocean, and of rivers, lakes, and groundwater is plentiful. Recent summaries include discussions of marine chemistry by Goldberg (1963), selected analyses of underground waters by White, Hem, and Waring (1963), and data on rivers and lakes by Livingstone (1963).

The processes controlling the composition of river, lake, and underground water are in a general way subject to limiting chemical and geologic conditions. The composition of these waters varies from place to place and from time to time in response to complex interrelationships depending on the amount of water available, the nature of the rock minerals with which the water has had contact, the length of time and degree of intimacy of contact, and such factors as temperature and biologic activity, which tend to promote certain reactions or speed reaction rates. The composition of the ocean is affected by most of the same processes, but their relative importance in marine environments is not the same.

The term "weathering" is applied to those processes by which rocks at or near the earth's surface are altered or broken down chemically and physically. Water is involved in most weathering processes and aqueous solutions result from them. The composition of these solutions is of course an indication of the kinds of processes taking place. Detailed study of weathering processes requires knowledge of solids and solutes participating in chemical reactions and of the conditions under which the reactions take place. Not all of the rather extensive literature on weathering has given consideration to solution chemistry. This omission is usually not present in the more modern discussions of the subject, such as that of Keller (1957) and, in areal studies, that of Miller (1961).

Solution reactions involved in weathering of rock minerals can be classified in general in three types:

1. Direct solution without prior chemical breakdown, as in the solution of an evaporite, e.g., sodium chloride.
2. Complete solution accompanying chemical breakdown by hydrolysis or oxidation or reduction, as in the solution of calcite in water containing H^+.
3. Partial solution accompanying chemical breakdown of mineral structures to yield solutes and sparingly soluble alteration products, as in the decomposition of silicate minerals to yield metal cations, silica, and clay minerals or metal hydroxides.

Reactions of the first type are reversible to yield the original mineral, most commonly by exceeding saturation through removal of water from the system by evaporation, or through changes in temperature. Reactions of the second type are readily reversible as a result of changes in solution characteristics such as pH, redox potential, or activities of ions, and commonly approach a state of chemical equilibrium in natural water. Reactions of the third type are generally not reversible without major changes in temperature and pressure, and often proceed at a very slow rate.

Because rocks and soils are complex mixtures of solids, the waters that have been in contact with them generally have been influenced by reactions of all three types. Often, however, one type of solution reaction may clearly predominate.

Processes of rock alteration deep below the earth's surface may occur in the presence of water at high temperature and pressure. The composition of solutions that participate in these processes is often different from that of solutions produced in normal weathering reactions. Indications of these hydrothermal processes can be obtained from the composition of water from hot springs.

The reactions of weathering are often strongly influenced by plants and animals. Vegetation releases carbon dioxide in respiration and as a decay product, and water passing through the soil obtains from this carbon dioxide a reinforced supply of hydrogen ions with which to attack and bring into solution certain rock minerals or mineral components. Bacteria may influence oxidation or reduction reactions, probably most often by increasing the reaction rate.

The composition of natural aqueous solutions is best reported in terms of concentrations of constituents in the form in which they occur in the solution. Cations and anions are reported in weight-per-weight or weight-per-volume units such as parts per million or milligrams per liter, or in weight-per-volume units calculated to a chemical equivalent basis (Hem, 1970, p. 81). Certain constituents, such as silicon, are present as hydrated oxide; the form of others may not be known, and their concentrations are reported as undissociated element or oxide. A satisfactory understanding of the geochemistry of solutions is best attained by study of these data along with the composition of the solids involved. In some of the literature of geochemistry, the composition of solutions is reported in terms of the percentage composition of the residue remaining when the solutions are evaporated. Data of this kind are misleading on at least two counts; as pointed out here, the complications of weathering reactions may result in residues from weathering solutions that have little resemblance to the original rock, and analyses expressed in terms of per cent of dry residue may foster erroneous conceptions regarding the form in which elements are carried in solution.

The overall results of solution of rock minerals are displayed in the dissolved-solids loads of streams. The principal constituents of these loads are the cations: calcium, magnesium, sodium, and potassium; the anions: bicarbonate, sulfate, chloride, fluoride, and nitrate; and the nonionized silica. The total amount dissolved in water discharging from a square mile of drainage basin during a year is reported by Livingstone (1963) to average 85 metric tons for the whole North American continent and to range from 110 tons for Europe to 6 tons for Australia. Where conditions for rock solution are unusually favorable, as in limestone terranes in humid temperate regions, rates of several hundred tons per square mile per year can be observed.

Figures of this kind have often been used to indicate rates of land erosion, and assuming they represent rates of accretion of dissolved substances to the ocean, they were supposed to be a basis for computing the age of the ocean. These computations meet with difficulties, partly because the dissolved solids yields include material derived from the ocean and the atmosphere, and because a considerable thickness of the upper part of the earth's crust has probably been subjected to repeated erosion and reconstitution of minerals. A more modern view of the circulation of elements into and out of the ocean has been expressed by Barth (1961), who suggested that currently the rate of supply of elements to the ocean is balanced by rate of withdrawal into solid minerals.

In groundwater systems a rather prolonged and intimate contact of water with rock minerals takes place. Chemical reactions that can attain a state of equilibrium are considerably

AQUEOUS SOLUTIONS

TABLE 1. CHEMICAL COMPOSITION OF WATER FROM VARIOUS SOURCES (PPM)

	1	2	3	4	5	6	7	8	9	10	11	12
Silica (SiO_2)	6	9.1	11	3.4	4.1	—	Trace	363	270	8.4	—	13
Aluminum (Al)	0.16	—	—	—	—	—	—	0.2	93	—	—	0.85
Iron (Fe)	0.01	0.04	0.03	0.04	0.36	—	—	0.06	250	0.04	—	0.04
Manganese (Mn)	0.002	—	—	—	—	—	—	0.0	0.4	—	—	0.00
Calcium (Ca)	400	37	12	37	14	346	17,300	0.8	150	46	3210	9.6
Magnesium (Mg)	1272	11	1.6	8.4	3.7	730	41,400	0.0	68	4.2	1690	15
Sodium (Na)	10,560	23	2.7	7.0	3.4	3174	14,300	352	121	1.5	70,900	1740
Potassium (K)	380	—	1.0	0.9	—	85	4400	24	105	0.8	279	20
Strontium (Sr)	13	—	—	—	—	—	—	0	—	—	—	.9
Bicarbonate (HCO_3)	142	115	49	125	50	215	Trace	70	—	146	75	2650
Carbonate (CO_3)	—	—	—	—	—	—	—	0	—	—	—	84
Sulfate (SO_4)	2560	55	0.4	28	4.8	3008	600	23	2061	4.0	—	223
Chloride (Cl)	18,980	23	0.4	7.5	1.5	5338	175,000	405	983	3.5	120,000	830
Fluoride (F)	1.4	0.4	0.0	0.0	—	—	—	25	1.2	0.0	—	5.8
Nitrate (NO_3)	<2.5	3.4	0.2	0.5	0.52	—	—	1.8	—	7.3	—	0.00
Bromide (Br)	65	—	—	—	—	—	7000	1.5	0.3	—	127	3.5
Boron (B)	4.6	—	0.00	—	—	—	—	4.4	32	—	35	1.4
Dissolved solids	34,400	226	65	162	57	12,800	260,000	1310	4160	139	196,000	4300
pH	—	—	7.0	8.0	—	—	—	9.6	1.7	7.0	—	8.6

1. Mean composition of seawater. pH generally ranges from 7.4 to 8.4 (Goldberg, 1963).
2. Mississippi River near St. Francisville, La. Average of daily samples Oct. 1, 1955–Sept. 30, 1956 (*U.S. Geol. Surv.* 1960).
3. Mekong River at Mukdaharn Village, Nakorn Penom Province, Thailand, Oct. 18, 1959. Flow approximately 1,200,000 cu ft/sec (Durum, Heidel, and Tison, 1960).
4. McKenzie River at Arctic Red River, North West Terr., Canada. Flow 450,000 cu ft/sec, July 24, 1958 (Durum, Heidel, and Tison, 1960).
5. Lake Superior. Mean of six samples [Dissolved solids recalculated (Livingstone, 1963)].
6. Caspian Sea. Mean of ten analyses, Aug. 11–21, 1933 [Dissolved solids recalculated (Livingstone, 1963)].
7. Dead Sea, 5 mi east of Ras Fesch Ka, at depth of 300 m (Livingstone, 1963).
8. Spring in Upper Geyser Basin, Yellowstone Park, Wyo.; temp. 94° C, Oct. 16, 1957 (White, Hem, and Waring, 1963).
9. Lower Mendeleev Spring, Kunashir Island, U.S.S.R. (Sulfate includes 911 ppm HSO_4), Oct. 30, 1954 (White, Hem, and Waring, 1963).
10. Big Spring, Huntsville, Ala. From limestone, Mar. 28, 1952 (Hem 1970).
11. Oil-field brine, North Makpat, Kazakh District, U.S.S.R. Well 1750 ft deep, producing from Permian and Triassic rocks (White, Hem, and Waring, 1963).
12. Oil-field water, Ellis Pool, Alberta, Canada. Jan. 8, 1958. Well 3026 ft deep, from Ellis sandstone, Jurassic age (White, Hem, and Waring, 1963).

more likely to do so in a groundwater system than in a river or lake. Because of the many factors that may influence their composition, in addition to the effects of the mineral array present, even in groundwater systems, the solute content may be only rather indistinctly related to the mineralogic composition of the aquifer. Some of these relationships have been pointed out by White, Hem, and Waring (1963).

JOHN D. HEM

References

Barth, T. F. W., 1961, "Abundance of the elements, areal averages and geochemical cycles," *Geochim. Cosmochim. Acta*, **23**, 1–8.

Durum, W. H., Heidel, S. C., and Tison, L. J., 1960, "World-wide runoff of dissolved solids," *Intern. Assoc. of Scientific Hydrology, I.U.G.G. Publ. No. 51*, 618–628.

Goldberg, E. D., 1963, "The Oceans as a Chemical System," in (Hill, M. N., editor) "The Sea," Vol. 2, pp. 3–25, New York, Interscience Publishers, a division of John Wiley & Sons.

Hem, J. D., 1970, "Study and interpretation of the chemical characteristics of natural water," *U.S. Geol. Surv. Water Supply Paper*, **1473**, 363pp.

Keller, W. D., 1957, "The Principles of Chemical Weathering," revised ed., Columbia, Mo., Lucas Bros., 111pp.

Livingstone, D. A., 1963, "The Data of Geochemistry," Sixth ed., Chapter G, "Chemical composition of rivers and lakes," *U.S. Geol. Surv. Profess. Paper*, **440G**, G1–G64.

Miller, J. P., 1961, "Solutes in small streams draining single rock types, Sangre de Cristo Range, N. Mex.," *U.S. Geol. Surv. Water Supply Paper*, **1535F**, F1–F23.

U.S. Geological Survey, 1960, "Quality of surface waters of the United States 1956," Parts 7 and 8, *U.S. Geol. Surv. Water Supply Paper*, **1452**, 250.

White, D. E., Hem, J. D., and Waring, G. A., 1963, "The Data of Geochemistry," Sixth Ed., Chapter F, "Chemical composition of subsurface water," *U.S. Geol. Surv. Profess. Paper*, **440F**, F1–F67.

Cross-references: *Oxidation and Reduction; Water; Weathering, Chemical.* Vol. II: *Evaporation; Solar Energy.* Vol. VI: *Groundwater; Hot Springs; Weathering, Organic.*

AQUIFER

An aquifer is a lithologic unit or combination of such units which has appreciably greater water transmissibility than adjacent units. It stores and transmits water, commonly recoverable in economically usable quantities. Impervious layers or beds of very low permeability which bound an aquifer are termed *confining* beds or *aquicludes* (e.g., the two shale horizons confining a porous sandstone). Any rock of low porosity and permeability is termed an *aquitard*. An *aquifuge* is one so impervious that it neither stores nor transmits water.

The yield or capacity of an aquifer is termed its *storage coefficient*, which is defined as the volume of water released or taken into storage per unit surface areas, per unit change in the component of head normal to the surface of the aquifer (Chow, 1964). In the case of an unconfined aquifer the storage coefficient is equal to the *specific yield*.

B.C.S.

Reference

Chow, Ven Te, 1964, "Handbook of Applied Hydrology," New York, McGraw-Hill Book Co., 1445pp.

Cross-reference: *Groundwater; Groundwater Motion in Drainage Basins; Hydrogeology; Hydrology, Limestone Terrain; Hydrology, Subsurface Waters.*

ARCHIMEDES PRINCIPLE—See SPECIFIC GRAVITY

ARGON: ELEMENT AND GEOCHEMISTRY

Physical Properties

Argon (Gr. *Argos,* inactive); symbol Ar; atomic weight 39.944; atomic number 18; electronic configuration ($1s^2 2s^2 2p^6 3s^2 3p^6$). Argon is one of the rare or "noble" gases: it is a colorless, odorless, and essentially inert gas. Owing to its chemical inactivity, chemical properties such as valence and atomic radius are of relatively little importance in determining its distribution within the earth.

Geochemistry

Earth as a Whole. Argon is by far the most abundant rare gas in the earth's atmosphere, constituting approximately 1% of it by volume (He $\sim$ 5 ppm and Ne $\sim$ 20 ppm). Von Weizsäcker first postulated the formation of Ar^{40} from its unstable isobar K^{40} and suggested that this must be the major source of the argon in the atmosphere. Whereas Ar^{36} is thought to be the most abundant argon isotope cosmically; virtually all the argon in the atmosphere is Ar^{40}, and the preponderance of this one isotope strongly supports this proposed special radiogenic origin. Thus it is normally presumed that the Ar^{40} in the atmosphere has been radiogenetically produced in rocks and subsequently released to the atmosphere by processes of crustal and subcrustal fusion, metamorphism, erosion, etc., throughout the history of the earth. Argon is chemically inactive; it merely remains in the atmosphere rather than being reincorporated into sediments. Also, it is too heavy to escape from the earth's atmosphere into outer space as does helium. Hence the amount of argon in the atmosphere accurately represents the sum of all that is released by the above processes. Wasserburg has pointed out that the amount of argon in the atmosphere is reasonably close to the amount that might be expected from crustal outgassing based on the mass of crustal rocks, their potassium content, age, etc.

In crustal rocks. The argon in the crust is virtually pure Ar^{40}, in contrast to atmospheric argon, which has an $Ar^{40}:Ar^{36}$ ratio of somewhat less than 300:1. The Ar^{40} in the crust has been produced by the decay (by K-capture) of K^{40} in potassium-bearing minerals and rocks. The total half-life of K^{40} (for both K-capture and β-decay) is approximately 1.3 billion years. Even though K^{40} represents only about 0.01% of the total potassium, the amount of Ar^{40} generated by its decay in a potassium-bearing mineral, such as biotite, over a period of a few million years is easily measurable. Most mineral lattices appear to retain their radiogenic argon quantitatively, and the $K^{40}:Ar^{40}$ method of age determination has been routinely applied to rock dating for over a decade. Ages thus obtained are regarded as reliable minimum estimates of the time of crystallization of the rock from which the biotite, muscovite, or hornblende has been extracted. However, if the rock has been subjected to even a mild, pervasive metamorphism and its temperature raised to about 300°C or over for geologic periods of time, extensive diffusive loss of argon may have occurred. Therefore, in metamorphic terrain, such ages are normally presumed to record the time of last major metamorphism and may be virtually independent of the time of initial crystallization of the rock.

Special Occurrences. The argon content of a mineral is not always explicable in terms of its potassium content and age; certain minerals, such as beryl, trap "initial" argon (made by potassium decay in pre-existing rocks) in channels defined by rings of silica tetrahedra. Some "excess" argon (over that predicted on the basis of potassium content and age) has also been found in common rock minerals from deep-seated rocks in cases where the potassium content happened to be low enough to allow extremely low levels of "initial" argon to be discerned. Such argon might be trapped in liquid-gas inclusions or randomly distributed.

In Meteorites

In meteorites also, most argon has been produced by radioactive decay of potassium. Many K:Ar ages have been determined on "stony" meteorites and seem to cluster around 4.5 billion years in agreement with various other estimates of the time of meteorite formation although some are lower, presumably owing to argon loss. In addition to this "radiogenic" argon, considerable amounts of "cosmogenic" argon have been produced in meteorites during their history in space by interactions between cosmic rays and iron and, in "stones," between cosmic rays and calcium nuclei. In this "cosmogenic" argon, the isotopes Ar^{36}, Ar^{38}, and Ar^{40} are directly produced in roughly similar abundances. Assuming a projected rate for these interactions during the meteorite's lifetime in space, it is then possible to determine the duration of its effective exposure to cosmic rays. The unstable nuclei Ar^{37} and Ar^{39} are also produced by such interactions. By studying their relative abundances, much can be learned concerning the constancy of cosmic rays in space and time. Still another "type" of argon has been detected in meteorites; "carbonaceous chondrites," in particular, have been found to contain considerable amounts of argon that has both high $Ar^{36}:Ar^{40}$ and $Ar^{36}:Ar^{38}$ ratios. Argon of this isotopic composition cannot be attributed to radioactive decay, cosmic-ray interactions, or atmospheric contamination. Since, on other grounds, these meteorites are thought to represent the most undifferentiated and unaltered solid solar system material examined to date, and since the observed ratios agree well with independent estimates of what the original "cosmic" ratios should have been, this argon is often referred to as "primordial" argon. It is interesting to note that the $Ar^{36}:Ar^{38}$ ratio in the earth's atmosphere, which should have been unaffected by subsequent radioactive decay, is identical ($Ar^{36}:Ar^{38} = 5.3:1$ in both cases) to that in this particular type of meteorite, confirming that the Ar^{36} and Ar^{38} in the earth's atmosphere are (in contrast to the Ar^{40}), indeed, of "primordial" origin.

FRASER P. FANALE

References

Damon, P. E., and Kulp, J. L., 1958, "Inert gases and the evolution of the atmosphere," *Geochim. Cosmochim. Acta*, **13**, 280.
Pepin, R. O., and Signer, P., 1965, "Primordial rare gases in meteorites," *Science*, **149**(3681), 253–265.
Suess, H., 1949, "The abundance of rare gases in the earth and in the cosmos," *J. Geol.*, **57**, 600.
Turekian, K. K., 1959, "The terrestrial economy

of helium and argon," *Geochim. Cosmochim. Acta*, **17**, 37.

Wasserburg, G. J., and Mazor, E., and Zartman, R. E., 1963, "Isotopic and chemical compositions of some terrestrial natural gases," in (Geiss, J., and Goldberg, E. D., editors). "Earth Science and Meteoritics," Amsterdam, North Holland Publ. Co. (div. of John Wiley), 219–240.

Zahringer, J., 1962, "Isotope effects and abundance of inert gases in stone meteorites and on the earth," *Z. Naturforsch*, **17a**, 460.

Cross-references: *Electron Capture; Helium; Outgassing of the Planet Earth; Potassium; Potassium-Argon Age Determination; Rare Gases.* Vol. II: *Carbonaceous Meteorites; Meteorites.*

ARSENATES—See PHOSPHATES, ARSENATES, AND VANADATES

ARSENIC: ELEMENT AND GEOCHEMISTRY

Arsenic (As) compounds were known as early as 400 B.C. Albert Magnus is credited with preparing elemental arsenic in 1250.

Physical Properties

The atomic number of arsenic is 33, its atomic weight is 74.91, density is 5.72 (20°C), melting point (28 atm) is 817°C, and boiling point is 613°C (sublimes). In dry air elemental arsenic is steel gray in appearance; however in the presence of moisture it is coated with oxide and is black in appearance.

Abundance

Arsenic is considered to be a rare but ubiquitous element of the upper lithosphere. The average abundance of arsenic in igneous rocks is estimated by Goldschmidt to be of the order of 5 g/T.

Meteorites have been found to contain arsenic and, according to the determinations of Nodak and Nodak quoted by Rankama and Sahama, the most significant quantity, 1020 g/T, was found associated with the triolite phase, whereas 360 g/T was associated with the nickel-iron phase.

Some uncertainty exists regarding the detection of arsenic in the solar spectrum and as a result no reliable data exist on the presence of arsenic in the solar atmosphere.

Arsenic is present in seawater; however data regarding its level are conflicting, and values in the range of 0.003 to 0.02 ppm have been reported.

Occurrence

Arsenic is occasionally observed in the native state in nature. However it is more frequently found combined with sulfur, selenium, tellurium, and as sulfo salts and arsenides of various heavy metals such as copper, iron, nickel, and cobalt. It also forms a number of pentavalent arsenate minerals that bear a close geochemical relationship with phosphates and vandates with which it can form some isomorphic compounds. Trivalent arsenic as As_2O_3 is experienced in the two mineral forms, arsenolite and claudetite.

Minerals

Slightly over 160 minerals bearing arsenic have been identified. A characterization of these minerals by Strunz, which appears in Gmelin, is as follows:

Mineral Class:	Elemental	Sulfide
Number of minerals	4	56
	Oxide	Arsenate	Silicate
	8	91	5

A number of these minerals have been tabulated from Dana in Gmelin, and are given in Table 1.

During the course of geochemical separations, the concentration of arsenic is low in the early magmatic sulfides, whereas its concentration becomes more significant in the late magmatic sulfides. Further concentration of arsenic occurs during the pegmatitic stage; however, it is not until the pneumolytic and hydrothermal stages are reached that the most predominant number of arsenic minerals are formed.

Arsenopyrite, which is the most abundant and widespread mineral of arsenic, is found in pegmatites but more frequently in high-temperature gold-quartz veins, high-temperature tin veins, and in contact with metamorphic sulfide deposits.

Arsenic sulfide (realgar) and arsenic trisulfide (orpiment), recognized since antiquity, are found present in hydrothermal veins and volcanic sublimations.

The numerous oxidic minerals of arsenic observed in nature are a result of the oxidation of sulfide and arsenide deposits in contact with the free oxygen of the atmospheres. The arsenate minerals are the most preponderant of the oxide minerals. Some arsenate minerals have been observed in metamorphic rocks deep in the earth. These were formed from sulfide or arsenide minerals reacting with super oxides, which provide the necessary oxidation potential.

Source

Arsenic for commerce is not recovered primarily from an independently mined ore but as

TABLE 1.

Mineral	Formula	Theoretical % As	Crystal
Dimorphite	As_4S_3	75.70	Orthorhombic
Realgar	AsS	70.00	Monoclinic
Orpiment	As_2S_3	60.91	Monoclinic
Skutterudite	$CoAs_3$	79.23	Isometric
Loellingite	$FeAs_2$	72.85	Orthorhombic
Rammelsbergite	$NiAs_2$	71.85 (?)	Orthorhombic
Pararammelsbergite	$NiAs_2$	71.85	Orthorhombic
Safflorite	$CoAs_2$	71.77	Orthorhombic
Niccolite	$NiAs$	56.08	Hexagonal
Maucherite	Ni_3As_2	45.98	Tetragonal
Sperrylite	$PtAs_2$	43.42	Isometric
Domeykite	Cu_3As	28.21	Isometric
Algodonite	Cu_6As	16.42	Hexagonal
Arsenopyrite	$FeAsS$	46.01	Monoclinic
Gersdorffite	$NiAsS$	45.23	Isometric
Cobaltite	$CoAsS$	45.15	Isometric
Lautite	$CuAsS$	43.92	Orthorhombic
Tennantite	$Cu_{12}As_4S_{13}$	20.26	Isometric
Enargite	Cu_3AsS_4	19.04	Orthorhombic
Vrbaite	$Tl_2S \cdot 2As_2S_3 \cdot Sb_2S_3$	23.55	Orthorhombic
Lorandite	$Tl_2S \cdot As_2S_3$	21.87	Monoclinic
Sartorite	$PbS \cdot As_2S_3$	30.87	Monoclinic
Gratonite	$9PbS \cdot 3As_2S_3$	11.33	Hexagonal
Proustite	Ag_3AsS_3	15.14	Hexagonal
Xanthoconite	Ag_3AsS_3	15.14	Monoclinic
Smithite	$AgS \cdot AsS$	30.34	Monoclinic
Trechmannite	$AgS \cdot AsS$	30.34	Hexagonal
Seligmannite	$2PbS \cdot CuS \cdot As_2S_3$	16.99	Orthorhombic
Pearceite	$8(Ag,Cu)_2S \cdot As_2S_3$	6.72	Monoclinic
Arsenolite	As_2O_3	75.74	Isometric
Claudetite	As_2O_3	75.74	Monoclinic

a by-product from the treatment of sulfidic lead and copper ores. The arsenic, when present in trace quantities in these ores, is often a replacement for sulfur, and in higher concentrations it may be present as inclusions of independent minerals such as arsenopyrite, enargite, and tetrahedrite.

S. C. CARAPELLA, JR.

References

Gmelins Handbuch der Anorganischen Chemie, 1952, Teil 18, Weinheim, Germany, Arsen, Verlag Chemie, G. M. B. H.

Goldschmidt, V. M., 1954, "Geochemistry," London, Oxford University Press, 730pp.

Palache, C., Berman, H., and Frondel, C., 1951, "Dana's System of Mineralogy," Seventh ed., Vol. 1, New York, John Wiley & Sons.

Rankama, K., and Sahama, T. G., 1950, "Geochemistry," Chicago, University of Chicago Press, 911pp.

Cross-references: *Oxidation/Reduction*. Vol. IVB: *Arsenic, Native; Meteoritic Minerals.*

ARTESIAN WATER

The name comes from *Artois*, the name of a former province in France, and is used in reference to confined groundwater under a hydrostatic or pressure head. The water is carried by a permeable bed, or *aquifer* (q.v.), sandwiched between impermeable zones above and below it (Fig. 1).

The height of a potential rise (h_f) of artesian water may be higher than the ground surface, in which case it is termed a "fountain of water." If the height of rise is less than the ground elevation, it is spoken of as "semi-artesian water" (i.e., water which moves up in a well, but not to ground level).

The height of rise (h_f) in any given case (see Fig. 1) may be calculated from the following relationship:

$$h_f = p/\gamma_w$$

where p = pressure and γ_w = unit weight of water (62.41 b/ft³ or 1000 kg/m³).

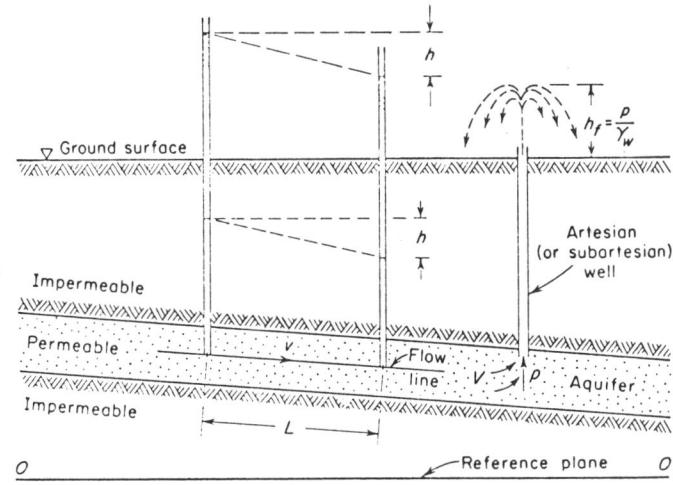

FIG. 1. Artesian (or subartesian) groundwater flow (Jumikis, 1962, p. 256).

A map which shows the surface contours of an imaginary surface to which water would rise in a drilled well is termed a *piezometric contour map* (see *Groundwater*).

B. C. S.

References

Dapples, E. C., 1959, "Basic Geology for Science and Engineering," New York, John Wiley & Sons, pp. 332–334.
De Wiest, R. J. M., 1965, "Geohydrology," New York, John Wiley & Sons, 366pp.
Jumikis, A. R., 1962, "Soil Mechanics," Princeton, N.J., Van Nostrand, 791pp.

Cross-references: *Aquifer; Groundwater.*

ASPHALT, ASPHALTITES, WAXES—See Vol. IV B

ATMOSPHERE, EVOLUTION AND HISTORY—See GEOCHEMICAL EVOLUTION OF THE CORE, MANTLE, AND CRUST; OUTGASSING OF THE PLANET EARTH; OXYGEN—ITS EVOLUTION IN THE EARTH'S ATMOSPHERE

ATMOSPHERIC CHEMISTRY—See Vol. II

ATMOSPHERIC NUCLEI AND DUST—See Vol. II

ATMOSPHERIC POLLUTION—See POLLUTION

ATOMS—See "Encyclopedia of the Chemical Elements" (C. A. Hampel, editor, Van Nostrand Reinhold Company)

ATOMIC NUMBER AND PERIODIC TABLE

The *atomic number*, Z, of an element is equal to the number of protons in the nucleus. Since atoms are electrically neutral, the atomic number is also equal to the number of electrons distributed in energy levels around the nucleus. The mass number or *atomic weight*, A, equals the number of protons plus the number of neutrons in the nucleus.

This distinction between atomic number and atomic weight is significant in the consideration of the development of the periodic table.

During the nineteenth century many chemists were engaged in seeking an arrangement for the then known elements which would emphasize similarities in properties and facilitate their systematic study. The first truly periodic arrangements of the elements should be credited to Beguyer De Chancourtois, professor of geology at the École des Mines, Paris, who emphasized in 1863 the fact that when the elements are arranged in order of increasing atomic weights, there is a recurrence of elements with similar properties. De Chancourtois plotted the values of the atomic weights on a helical curve about a vertical cylinder. This "telluric screw" resulted in an arrangement in which related elements fell on the same vertical line.

Without knowledge of the work of De Chancourtois, the Englishman John A. R. Newlands worked out a similar system of classification between 1863 and 1865. When he arranged the known elements in order of their atomic weights, he noticed in many cases a repetition of chemical properties in each eighth element. Unfortunately, Newlands called his discovery

FIGURE 1. MENDELEYEV'S PERIODIC SYSTEM OF THE ELEMENTS (MOORE, 1963)

PERIODS	I	II	III	IV	V	VI	VII	VIII			0
1	1 H Hydrogen 1.0080										2 He Helium 4.0026
2	3 Li Lithium 6.939	4 Be Beryllium 9.012	5 B Boron 10.811	6 C Carbon 12.0111	7 N Nitrogen 14.007	8 O Oxygen 15.9994	9 F Fluorine 18.998				10 Ne Neon 20.183
3	11 Na Sodium 22.990	12 Mg Magnesium 24.312	13 Al Aluminum 26.982	14 Si Silicon 28.086	15 P Phosphorus 30.974	16 S Sulfur 32.064	17 Cl Chlorine 35.453				18 Ar Argon 39.948
4	19 K Potassium 39.102	20 Ca Calcium 40.08	21 Sc Scandium 44.956	22 Ti Titanium 47.90	23 V Vanadium 50.942	24 Cr Chromium 51.996	25 Mn Manganese 54.938	26 Fe Iron 55.847	27 Co Cobalt 58.933	28 Ni Nickel 58.71	
	29 Cu Copper 63.54	30 Zn Zinc 65.37	31 Ga Gallium 69.72	32 Ge Germanium 72.59	33 As Arsenic 74.9216	34 Se Selenium 78.96	35 Br Bromine 79.909				36 Kr Krypton 83.80
5	37 Rb Rubidium 85.47	38 Sr Strontium 87.62	39 Y Yttrium 88.905	40 Zr Zirconium 91.22	41 Nb Niobium 92.906	42 Mo Molybdenum 95.94	43 Tc Technetium 99	44 Ru Ruthenium 101.07	45 Rh Rhodium 102.905	46 Pd Palladium 106.4	
	47 Ag Silver 107.870	48 Cd Cadmium 112.40	49 In Indium 114.82	50 Sn Tin 118.69	51 Sb Antimony 121.75	52 Te Tellurium 127.60	53 I Iodine 126.9044				54 Xe Xenon 131.30
6	55 Cs Cesium 132.905	56 Ba Barium 137.34	57 La* Lanthanum 138.91	72 Hf Hafnium 178.49	73 Ta Tantalum 180.948	74 W Tungsten 183.85	75 Rh Rhenium 186.2	76 Os Osmium 190.2	77 Ir Iridium 192.2	78 Pt Platinum 195.09	
	79 Au Gold 196.967	80 Hg Mercury 200.59	81 Tl Thallium 204.37	82 Pb Lead 207.19	83 Bi Bismuth 208.980	84 Po Polonium 210	85 At Astatine 210				86 Rn Radon 222
7	87 Fr Francium 223	88 Ra Radium 226.05	80 Ac** Actinium 227								

6	*58-71 Lanthanide Series	58 Ce Cerium 140.12	59 Pr Praseodymium 140.907	60 Nd Neodymium 144.24	61 Pm Promethium 147	62 Sm Samarium 150.35	63 Eu Europium 151.96	64 Gd Gadolinium 157.25	65 Tb Terbium 158.92	66 Dy Dysprosium 162.50
		67 Ho Holmium 164.93	68 Er Erbium 167.26	69 Tm Thulium 168.93	70 Yb Ytterbium 173.04	71 Lu Lutetium 174.97				
7	**90-101 Actinide Series	90 Th Thorium 232.04	91 Pa Protactinium 231	92 U Uranium 238.03	93 Np Neptunium 237	94 Pu Plutonium 239	95 Am Americium 243	96 Cm Curium 245	97 Bk Berkelium 249	98 Cf Californium 249
		99 Es Einsteinium 255	100 Fm Fermium 255	11 Md Mendelevium 256	102 No Nobelium 	103 Lw Lawrencium				

the "Law of Octaves," in analogy with the musical scale. Chemical comedians had a field day, and the importance of Newlands' observation was not given recognition until 1887 when he was awarded the Davy Medal by the Royal Society.

Credit for the development of the periodic law is usually given jointly to Dimitri Mendeleev of Russia, who based his observations on the chemical properties of the elements, and Lothar Meyer of Germany, who based his on their physical properties. Working independently, in 1869 and 1870, both men published quite similar tables for the periodic arrangement of the elements.

Because of his slightly prior claim, the name of Mendeleev is most commonly connected with the Periodic Table of the Elements. Therefore, March of 1969 marked the centennial of one of the most important concepts in chemistry. The contribution of the Periodic Table is that it brings order to the seemingly chaotic mass of materials composing the universe. Hundreds of chemical compounds are seen to fit into relatively few groups based on elements of similar properties.

A modern version of Mendeleev's periodic table (also known as the "short form" of the periodic table) is shown in Fig. 1. Inspection of this table reveals a fault that became evident at an early date: the inversions among certain atomic weights. Argon, $A = 39.948$, precedes potassium, 39.102; similar reversals occur in the positions of cobalt and nickel, tellurium and iodine, thorium and protactinium. In addition, the location of the group VIII elements, the rare earths, and the inert gases (after the turn of this century) offered problems. It became clear that something in the fundamental structure of the elements was responsible for the observed periodicity, but that atomic weights were not the ultimate answer.

The solution to the problem lay in the results of observation of the characteristic radiation of x rays, the frequencies of which are a function of the target material. In 1913, H. G. J. Moseley made the first detailed study of the characteristic x ray spectra emitted by thirty-eight different elements used as targets. He found that the vibration frequencies, v, of the x ray lines were related to the atomic number Z (which was at this time defined only as the numerical position of the element in the periodic chart) by $\sqrt{v} = a(Z-b)$, where a and b are constant for all the elements. Utilizing the treatment developed by Bohr for the hydrogen atom, Moseley came to the conclusion that the atomic number is a measure of the positive charge on the nucleus of an atom. Since the number of positive charges or protons in an atom is a more fundamental quantity than the atomic weight, the modern statement of the *periodic law* is *the properties of the elements are periodic functions of their atomic numbers.*

Because the nucleus of an atom remains unaffected by chemical transformations, it is now recognized that the number and arrangement of the electrons determine the periodic character of the elements. This is illustrated by the extended or "long form" of the Periodic Table found in most modern textbooks (Fig. 2). An abridged discussion of the electronic basis for the periodic classification of elements follows; a more complete account can be found in any good inorganic or physical chemistry text, several of which are listed among the references following this article.

Electrons are arranged in energy levels around the nucleus of the atom. Each energy level is made up of sublevels, or orbitals. From spectroscopic nomenclature these orbitals have been designated *s, p, d,* and *f* in order of increasing energy. Quantum mechanical calculations have shown that *s* orbitals contain a maximum of 2 electrons, *p* orbitals 6, *d* orbitals 10, and *f* orbitals 14. The maximum number of electrons possible in each energy level is $2n^2$, where n is the number of the energy level, but the outermost energy level of any atom can contain no more than 8 electrons. Thus, the first period, corresponding to the first energy level, contains only 2 elements, hydrogen and helium. Hydrogen, with one proton and one electron, does not fit conveniently into the periodic chart. Many authors place it in either Group IA or VIIB (or both), since it has certain properties in common with other elements of these groups. Other authors remove it from the chart completely because it is unique. Helium, and the other five elements of Group O, have completely filled outer energy levels. These *inert* or *rare gases* (q.v.) are chemically unreactive because of the high stability associated with completed energy levels.

The second period, consisting of elements with outer electrons in the second energy level, contain 8 members. The third period is made up of 8 elements also; the additional 10 electrons that the third energy level can contain in its *d* orbitals are not added until the fourth period. The first row of *transition elements,* Sc through Zn, $Z = 21-30$, represent filling of the *d* orbitals (Fig. 3). Because empty, half-filled, or completely filled orbitals represent extra stability, there are some apparent anomalies in the electronic arrangements of certain elements: chromium (Cr), molybdenum (Mo), Copper (Cu), silver (Ag), and gold (Au) contain only one (rather than the expected 2)

ATOMIC NUMBER AND PERIODIC TABLE

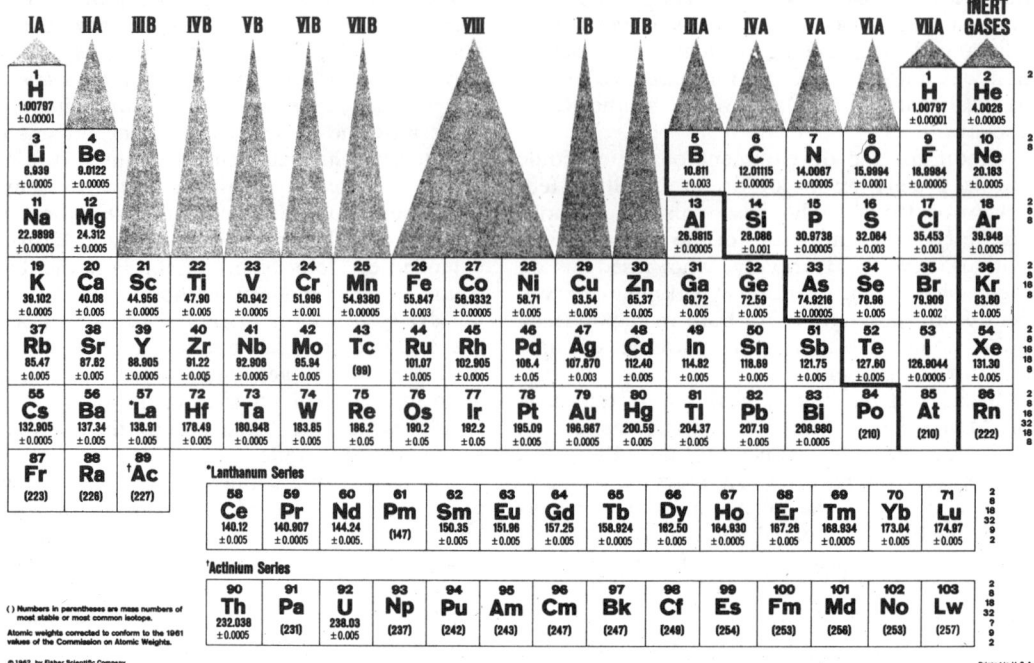

FIG. 2. Periodic chart of the elements (by permission of Fisher Scientific Co.).

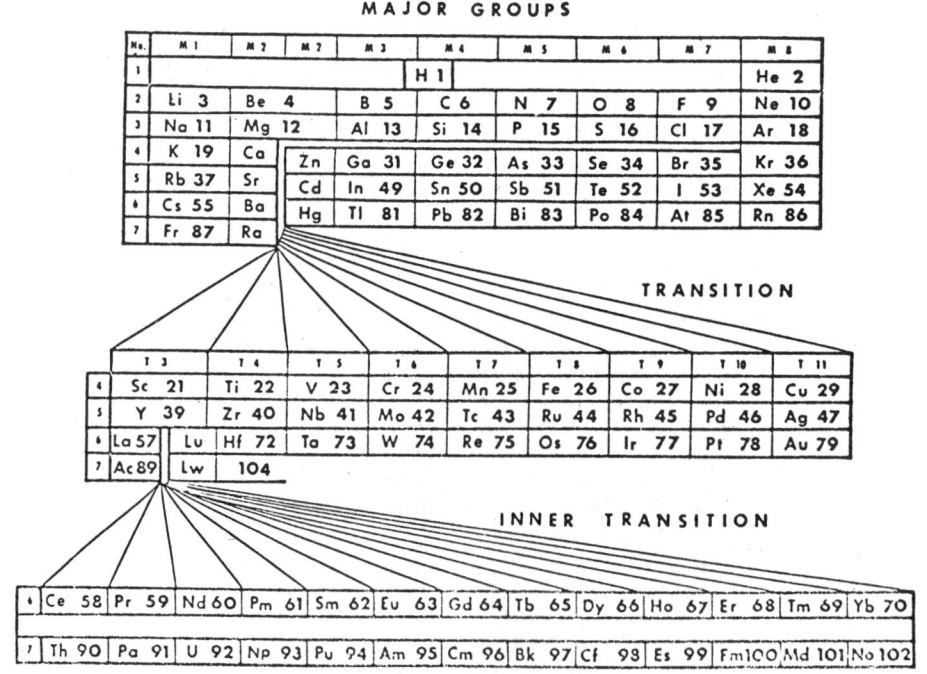

FIG. 3. Transition and inner transition elements.

ATOMIC NUMBER AND PERIODIC TABLE

electrons in their outer energy levels. The unfilled *d* orbitals are responsible for the numerous oxidation states and highly colored compounds characteristic of the transition elements.

The fourth energy level can contain 32 electrons, but because of energy considerations, the *f* orbitals of the fourth energy level do not fill until the sixth period, after 2 electrons are in the sixth energy level. Filling of the *f* orbitals produces the *inner transition elements* (also known as the *actinide series* (q.v.) and the *lanthanides* (q.v.) or *rare earths,* (q.v.). The inner transition elements follow after lanthanum, $Z = 57$, but are usually placed at the bottom of the table (Fig. 3). The remarkable chemical similarity of the lanthanides is explained by the fact that the arrangement of electrons in the two outer energy levels is the same for all fourteen elements.

The geochemical character of an element is largely governed by the electronic configuration of its atoms and hence is clearly related to its position in the periodic table (Fig. 4). *Lithophile elements* (Goldschmidt's term for elements with an affinity for silicates) are those that readily form ions with 8 electrons in the outermost level (Groups IA and IIA). *Chalcophile elements* (with affinity for sulfur) are those in the B subgroups with 18 electrons in the outer levels. *Siderophile elements* (with affinity for metallic iron) are those of group VIII, together with gold and possibly rhenium, with outer electron levels incompletely filled. The inert gases (Group O) are *atmophile elements*.

Goldschmidt also pointed out the marked correlation between geochemical character and atomic volume. Fig. 5 is based on Lothar Meyer's original diagram which appeared in 1870 and indicates the periodic variation of atomic volume with atomic number. Siderophile elements are near the minima on the curve. Lithophile elements are near the maxima and on the portions of the curve with negative slope. The chalcophile elements appear on sections in which atomic volume is increasing with atomic number, followed by the atmophile elements.

The *density* of the elements in the solid state reaches a maximum in the central members of each period.

The *melting point* occurs as a minimum in the inert gases and as a maximum in Group IV (for the lighter elements) and Group VI (for the heavier elements).

The *ionization potential* (q.v.) increases from left to right in each period and from

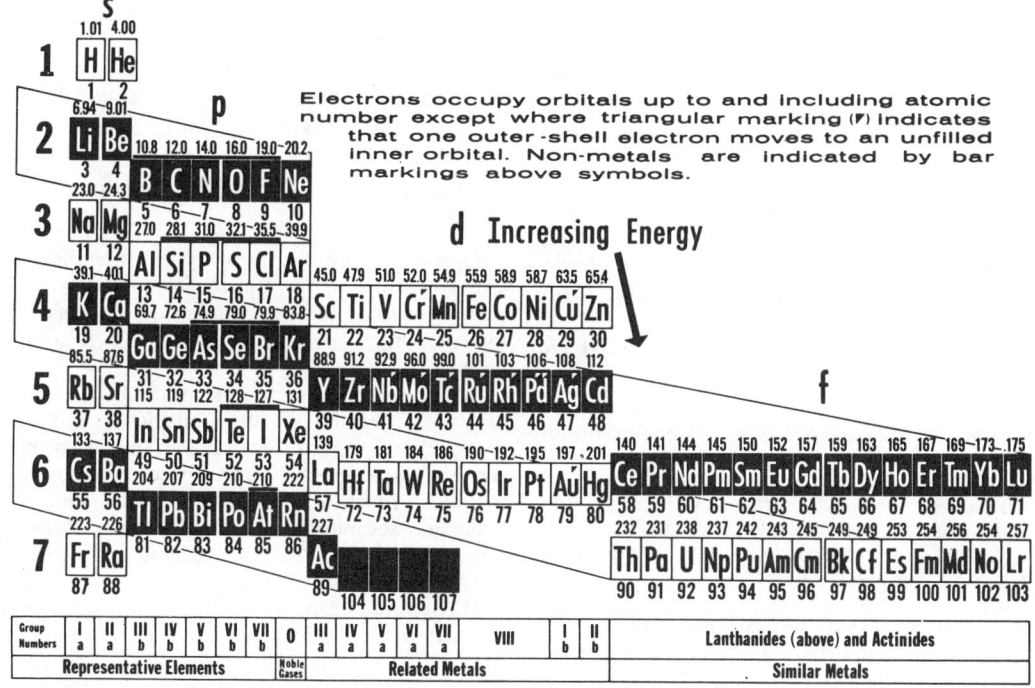

Fig. 4. Electron configurations of the elements (by courtesy of J. W. Eichinger, Jr., and W. T. Lippincott).

ATOMIC NUMBER AND PERIODIC TABLE

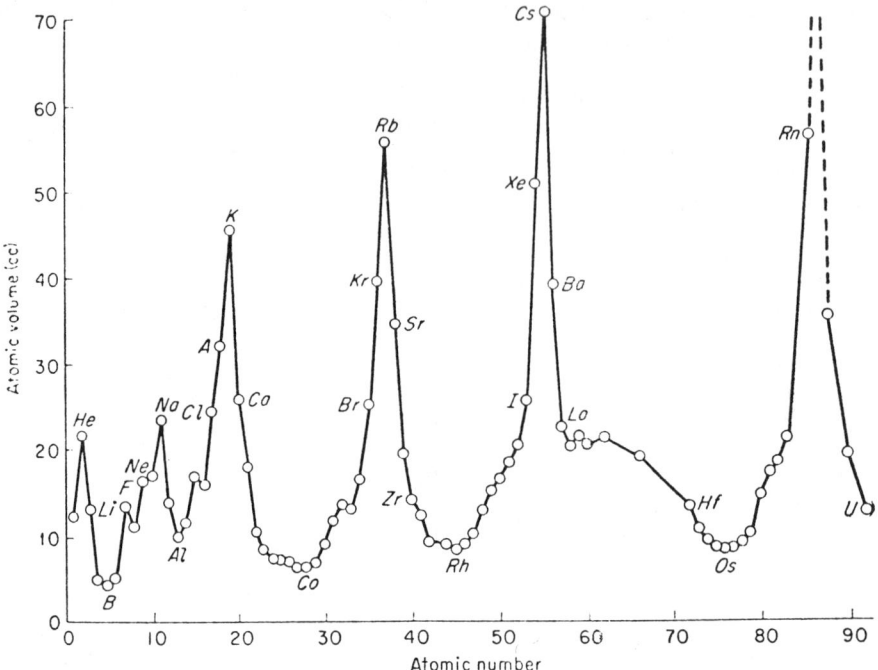

FIG. 5. The atomic volumes of the elements (Gilreath, 1958).

bottom to top in each group. Helium has the highest ionization potential.

The *electronegativity* (q.v.) increases from left to right in each period and from bottom to top in each group (the rare or inert gases are excepted). Fluorine is the most electronegative element; francium the most electropositive.

The *metallic character* decreases from left to right across each period and increases from top to bottom in each group. Francium is the most metallic element. Metallic character and electronegativity are inversely related. The metals, on the left of the periodic table, form basic oxides, and the nonmetals, on the right, form acidic oxides.

I. M. Drew

References

Besançon, R. M. (editor), 1966, "Encyclopedia of Physics," New York, Reinhold Publ. Corp.
Clark, G. L. (editor), 1966, "Encyclopedia of Chemistry," Second ed., New York, Reinhold Publ. Corp.
Day, M. C., and Selbin, S., 1962, "Theoretical Inorganic Chemistry," New York, Reinhold Publ. Corp.
Gilreath, E. S., 1958, "Fundamental Concepts of Inorganic Chemistry," New York, McGraw-Hill Book Co.
Mason, B., 1966, "Principles of Geochemistry," Third ed., New York, John Wiley & Sons.
Mendeleyev, D., 1869, *J. Russ. Chem. Soc.*, **1**, 60.
Meyer, L., 1870, *Ann. Chem., Suppl. Band* **7**, 354.
Moore, W. J., 1963, "Physical Chemistry," Third ed., Englewood Cliffs, Prentice-Hall.
Moseley, H., 1913, *Phil. Mag.*, **26**, 1024.
Van Spronsen, J. W., 1969, "The Periodic System of Chemical Elements," New York, Amer. Elsevier, 368pp.

Cross-references: *Actinide Series; Electronegativity; Lanthanides; Rare Earths; Rare Gases.*

AUTHIGENESIS OF MINERALS—MARINE*

Deposits from the floor of the oceans consist generally of multimineral mixtures. The components of such mixtures may be classified on the basis of their origin, as described below.

Terrigenous. Terrigenous minerals are those originally formed on the continents and subsequently transported into ocean basins through the hydrosphere, the atmosphere, or both. For instance, most of the quartz grains found in deep-sea sediments are terrigenous, deriving originally from the weathering of igneous rocks on the continents.

Biogenous. Biogenous minerals are those formed by biological reactions in the sea. An example is calcite, which can be produced by various planktonic and benthonic organisms,

*Contribution No. 1338 from the Institute of Marine Sciences, University of Miami.

such as foraminifera and coccolithoforidae, as part of their test.

Volcanogenous. These minerals are introduced directly on the ocean floor by submarine volcanic activity and not significantly altered by seawater; e.g., feldspars and pyroxenes are found in some pelagic sediments in the Pacific where they were introduced locally by recent submarine basaltic eruptions.

Cosmogenous. Cosmogenous particles are those derived from outer space (i.e., Fe–Ni spherules). They are quantitatively unimportant.

Authigenous. Authigenous (or authigenic) minerals are those that were formed directly in the ocean by means of inorganic reactions involving seawater or interstitial solutions. Equivalent terms are *hydrogenous*, employed by Goldberg (1954), and *halmeic*, by Arrhenius (1963). Mineral authigenesis can take place in the ocean as a consequence of two types of reactions: (a) When the solubility product of a certain compound is exceeded within a body of seawater, the compound tends to precipitate. However, it is not uncommon that actual precipitation of a certain compound does not take place, even though its solubility product is exceeded, due to kinetic reasons; (b) When terrigenous, volcanogenous, or biogenous solids are unstable under the physicochemical conditions prevailing at the ocean floor, they tend to react with seawater or interstitial water, and form more stable phases. *Halmyrolysis* is a term used to indicate such reactions when they take place in the water column or in the preburial stage of diagenesis.

In addition to those mineral-forming reactions, which take place in the water column or during halmyrolysis, most authors include in the definition of authigenesis the formation of minerals during the early stages of diagenesis (*syndiagenesis* according to Fairbridge, 1967).

A brief review of some common oceanic authigenous minerals and of their mode of formation is given below. This review does not include deposits formed from enclosed bodies of seawater, such as evaporites.

Oxides and Hydroxides

Deposits of iron and manganese oxides and hydroxides cover large areas of the floor of the oceans, as nodules, crusts, and concretions or as small (a few microns) particles dispersed in the sediments. These deposits consist generally of microaggregates of various poorly crystalline oxides and hydroxides. Among the iron minerals cryptocrystalline goethite is prevalent. The manganese minerals include birnessite, a species identical to a synthetic form known as δMnO_2, and todorokite, a hydrous manganese oxide with the approximate formula: $(Mn,Ca) Mn_3O_7 \cdot 2H_2O$. Ramsdellite, γMnO_2; ranciete, $(CaMn)Mn_4O_8 \cdot 3H_2O$; and psilomelane, $(Ba,Mn)Mn_8O_{16}(OH)_4$, have also been reported. In addition, substantial quantities of Fe–Mn hydroxides, which are amorphous at x-ray diffraction, are contained in these aggregates. Hausmannite, which is common in ancient sedimentary manganese deposits on land, has not been reported in modern marine deposits.

The bulk chemical composition of marine Fe–Mn aggregates is quite variable. The majority of the deep-sea concretions tend to have a Fe/Mn ratio close to one; however, deposits where Fe reaches concentrations of 30% by weight are known in the Red Sea and in the southeast Pacific. Yet, deep-sea manganese deposits where iron is almost totally absent are also known, particularly from hemipelagic areas close to the continents. Various other elements are commonly hosted within the structure of the Fe–Mn oxides and hydroxides, particularly Ni, Co, Cu, Pb, etc. A study by electron probe of the distribution of some of these elements within the various phases has shown that Ni and Cu are covariant with Mn, whereas Co is covariant with Fe (Burns and Fuerstenan, 1966).

The Eh and pH conditions of seawater are such that iron and manganese tend to be oxidized and to precipitate. Three major mechanisms of formation of marine iron-manganese minerals can be outlined, depending on the source of the elements in question.

1. Iron and manganese normally contained in seawater (introduced originally by streams and deriving from the weathering of continental rocks) are oxidized and precipitate slowly on the ocean floor, forming small particles dispersed in the sediment or, in areas of low or no sedimentation, crusts and concretions. Much of the iron does not reach far into the open ocean, being less soluble than manganese in seawater. These types of deposits are especially common on topographic highs, which are swept by strong currents, and where detrital sedimentation is scarce or absent, and are often associated with deposits of phosphorite. An example is the manganese pavement found on the Blake Plateau, off the coast of Florida and Georgia (Fig. 1 and Pratt and McFarlin, 1966).

2. Iron and manganese may be introduced in seawater from below through the sea floor in areas of submarine volcanic and hydrothermal activity and from the alteration of igneous minerals. Both elements are subse-

FIG. 1. Cut sample of the manganese-phosphorite pavement on the Black Plateau. The dark regions are manganese oxides and the light regions are phosphorite.

quently oxidized by oxygen contained in seawater, and thus will tend to precipitate. Iron will be deposited first, near its source and give rise to iron deposits (with cryptocrystalline goethite as main component), such as those found in some south Pacific sea mounts (Bonatti and Joensuu, 1966). Manganese has longer residence time in seawater, and will be deposited over wider areas of sea floor. A similar mechanism is probably responsible for the formation of colloidal sediments with unusually high Fe and Mn contents on the East Pacific Rise and other active oceanic ridges (Boström and Peterson, 1968, Fig. 2).

3. In hemipelagic sediments relatively rich in organic matter, reducing conditions are commonly established below the top of the sedimentary column. Under such conditions,

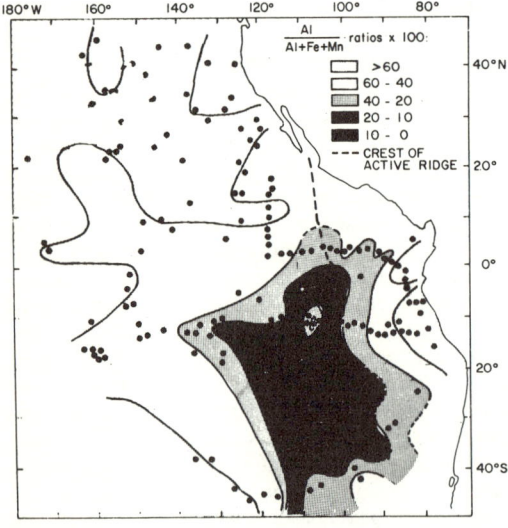

FIG. 2. Ratio Al/Al+Fe+Mn in pelagic sediments from the east Pacific. Sediment poorest in aluminum and richest in manganese and iron found along the rest of the east Pacific Rise (Boström and Peterson, "Marine Geology," 1968).

manganese oxide particles originally contained in the sediments are dissolved; Mn migrates in the pore solution and is reprecipitated near the oxidizing water-sediment interface. The aggregates resulting from such processes are usually very poor in iron, which is mobilized to a lesser extent than manganese. They are found in the Gulf of California, in the Panama and Guatemala basins (east Pacific), in the Baltic, etc.

The rate of accretion of marine Fe–Mn minerals is highly variable; it is generally very low (on the order of 1 mm per 1000 yr) in deposits of the I type, according to Io/Th age determinations (Bender et al., 1966). It is probably much more rapid in precipitates of the II type.

The mineral maghemite has been reported from sediments of the Indian Ocean where it probably originated from the submarine alteration of magnetite crystals (Harrison and Peterson, 1965).

Sulfates

Barite ($BaSO_4$) is a common authigenic component of deep-sea sediments in the Pacific. Its distribution is shown in Fig. 3. Concentrations of $> 8\%$ by weight (on a carbonate-free and manganese oxide-free basis) are found in sediments close to the East Pacific Rise in the south Pacific Ocean. Deep-sea barite occurs often as prismatic crystallites, which may occasionally reach up to 50μ in size but are usually not larger than 10μ. Sr and Pb substitute to some extent for Ba in oceanic barites; according to Arrhenius (1963), a sample of typical Pacific barite contains 5.4 mole % celestite and 0.5 mole % anglesite.

The high concentration of sulfate in seawater probably limits the solubility of barium. This element may become relatively concentrated in deep water due to various causes, such as sinking of remains of barium-rich planktonic organisms or the introduction of barium-bearing hydrothermal solutions through the sea floor; as a consequence, barite will precipitate. Most of the barite present on the East Pacific Rise is probably related to hydrothermal activity on the Rise itself (Arrhenius and Bonatti, 1965). Some marine protozoans, of the group of the xenophyophora, are known to secrete minute (a few microns) crystals of barite as part of their skeleton, and may be responsible for the presence of barite in biogenous sediments from the East Equatorial Pacific. Barite is rare in the north Pacific and the Atlantic. Barite concretions found off the California coast have been ascribed to local hydrothermal activity on the sea floor (Revelle and Emery, 1951).

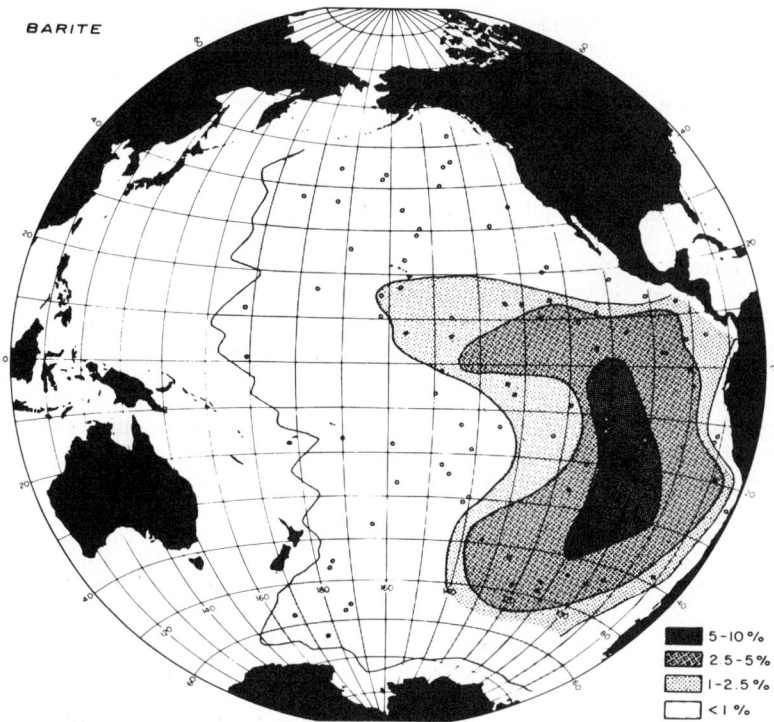

Fig. 3. Distribution of barite (on carbonate- and MnO_2-free basis) in pelagic sediments from the east Pacific (Arrhenius and Bonatti, 1965).

Sulfides

Sulfides are usually formed in the post-burial, early stage of diagenesis in sediments containing initially more organic matter than can be oxidized before burial. These sediments are found in basins close to continents (with high influx of land-derived organic debris), in areas below water masses with high biological productivity, or in basins with poorly oxygenated bottom waters. Often in such environments, reducing conditions (negative Eh) are observed below the sediment surface. Sulfate reduction takes place mainly because of bacterial action; as a result H_2S is produced, which reacts with iron in the pore solution or with iron hydroxides, producing metastable hydrotroilite ($FeS-nH_2O$) and subsequently stable phases such as pyrite and marcasite. In basins where the bottom water contains no oxygen (i.e., the Black Sea) authigenic sulfides may form at the sediment-water interface.

Carbonates

Depending mainly on the temperature and pH of a given body of seawater, $CaCO_3$ is apt to precipitate or dissolve. Precipitation of $CaCO_3$ may occur where seawater is warm and CO_2 is being lost, for instance, through photosynthesis. These conditions are likely to be approached in relatively shallow, tropical waters; classical areas are banks in the Bahamas and off the eastern coast of Florida. Whether precipitation of $CaCO_3$ from seawater always requires biological agents, or whether purely inorganic precipitation is possible, is still a matter of debate. The holders of the latter view point to the fact that the product of the activities for Ca^{2+} and CO_3^{2-} measured in seawater near the surface generally exceeds the activity product for both calcite and aragonite.

According to Cloud (1962), aragonite precipitates in inorganic compounds in shallow banks of the Bahamas where aragonitic ooliths are commonly found. It is not entirely clear why aragonite precipitates rather than calcite, since the former polymorph is metastable under ordinary conditions. From experimental studies it appears that aragonite precipitates preferentially at high temperatures and if Mg^{2+} is present in the solution. During diagenesis aragonite gradually transforms into the stable phase, calcite.

Some authors believe that biological activity is involved to some extent in all cases of $CaCO_3$ precipitation from seawater. According to these authors, inorganic "spontaneous" nucleation of calcite or aragonite does not occur at any appreciable rate even though seawater may be supersaturated with respect to $CaCO_3$. Lowenstam and Epstein (1957) questioned the inorganic formation of aragonite in the Bahamas on the basis of oxygen isotopic measurement.

Submarine inorganic recrystallization of foraminiferal ooze appears to occur when the original biogenous ooze is maintained in contact with circulating bottom waters for long periods of time; that is, when the accumulation rate is exceedingly low. The O^{18}/O^{16} ratio of some recrystallized limestones dredged from guyots in the northern Atlantic indicates that the recrystallization took place in relatively deep water (Milliman, 1966). Magnesium-rich calcite appears to be produced in the first stages of such recrystallization.

Deep ocean water is undersaturated with respect to carbonates due to its low temperature and high CO_2 concentration; in fact, some dissolution of $CaCO_3$ generally takes place only a few hundred meters below the surface. A carbonate compensation depth can be defined, generally between 3000 and 4000 below the surface, below which $CaCO_3$ dissolves at high rates. As a consequence, direct precipitation of carbonates is absent in the deep sea. Exceptions are known: authigenic crystals of calcite and dolomite have been found in red clays from the south Pacific (Fig. 4); they are probably formed as a result of hydrothermal activity in the sea floor (Bonatti, 1966).

Dolomite has not been reported to precipitate directly from open seawater; dolomitization is generally believed to be a diagenetic process (Fairbridge, 1967). Manganocalcite $(Ca,Mn)CO_3$ has been reported as forming in reducing interstitial solutions in hemipelagic sediments from the east Pacific, off Central America (Lynn and Bonatti, 1965). Siderite, $FeCO_3$, can be formed in bottom environments where reducing conditions are prevalent, especially as a result of diagenetic reactions.

Phosphates

Authigenic phosphate minerals in the ocean are essentially limited to two forms of carbonate apatite, dahllite, $Ca_5(PO_4,CO_3OH)_3F$, and francolite, $Ca_5(PO_4,CO_3OH)_3OH$. These two minerals are the main components of deposits of phosphorite, together with cryptocrystalline phases known as collophane. Phosphorites are found as slabs and units of various sizes, or as small nodular-oolithic aggregates. Deposits of phosphorite are generally concentrated in relatively shallow water at the edge of the continental shelf, on banks and other topographic highs where detrital sedimentation is scarce or absent. Examples of these types of deposits are found off California and northern Mexico (Dietz et al., 1942; D'Anglejan, 1967), on the Blake Plateau and other banks off the eastern coast of Florida (Fig. 1), etc. Sea mounts and guyots located far from continents may also provide an environment favorable to the formation of phosphorites.

Three major hypothesis have been proposed to explain the formation of authigenic carbonate-apatites: They may be formed: (1) by chemical precipitation from supersaturated seawater; (2) biogeochemically, for instance, by the action of microorganisms; or (3) by replacement of PO_4^{3-} units in pre-existing carbonates.

The third hypothesis is favored at present, due to the fact that carbonate apatites do not precipitate directly from aqueous solutions, but are formed instead by phosphate replacement of pre-existing calcite (Klement, et al., 1942; Ames, 1959). According to Degens (1965) the environmental conditions required for the deposition of marine phosphorites include: (1) a pH greater than 7, which allows the metasomatic replacement of calcite to proceed at appreciable rates; (2) the presence of calcareous materials; (3) PO_4^{3-} concentrations in seawater in excess of 0.1 ppm; and (4) an environment with scarce or no detrital deposition. The redox potential of the solution apparently does not influence the formation of carbonate apatite.

Fragments of fish bone up to 200μ in size are found occasionally in some deep-sea sediments of the Pacific (Fig. 5). They consist of aggregates of microcrystalline apatite which is slowly dissolving in contact with deep seawater. Gradual replacement of Ca by rare earths appear to take place in these apatites, which have

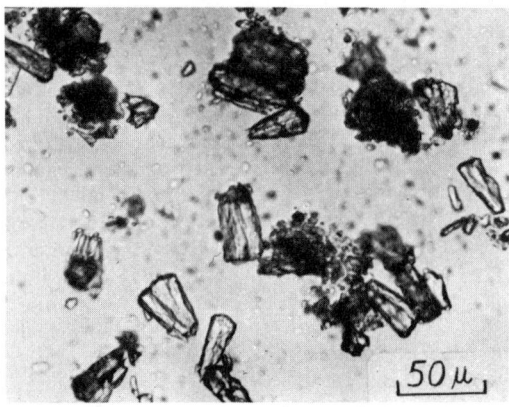

Fig. 4. Crystals of authigenic calcite in a pelagic sediment from the south Pacific (Bonatti, 1966).

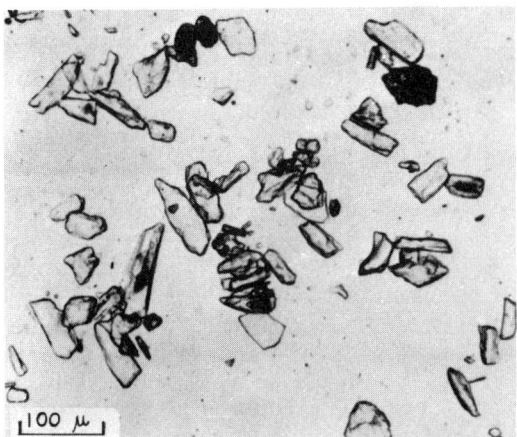

Fig. 5. Fragments of skeleton apatite from Pacific deep-sea sediment.

been found to contain several thousands ppm of lanthanum (La), yttrium (Y), ytterbium (Yb), and other rare earths. (Arrhenius and Bonatti, 1965).

Silicates

Sillén (1961), on the basis of theoretical considerations, has suggested that reactions of detrital silicates with seawater, and precipitation of silica and silicates from seawater, should be important factors in the chemical budget of the oceans. There is suggestive evidence that silica and poorly crystalline silicates may be common in the finest size fraction of many pelagic sediments; whether they are precipitates from seawater or degradation products of detrital silicates is not known. Most of the clearly identified authigenic silicates from the sea floor are products of halmyrolitic reactions rather than of direct precipitation from seawater.

Silica. Inorganic precipitation of silica, mainly in the form of opal, may take place off the mouths of rivers or wherever large quantities of silica are introduced in seawater; for instance, in the vicinity of sites of submarine volcanic-hydrothermal activity. Opal may also precipitate from interstitial solutions within deep sea sediments, especially where the concentration of silica has been maintained at a high level by the presence of biogenous, relatively unstable opal (tests of diatoms and radiolaria). Authigenic growth of quartz in the deep sea has not been reported, even though it has been observed often in shallow water and littoral sediments.

Feldspars. Marine authigenic feldspars have been reported in a few cases; for instance, Mellis (1952) has shown that orthoclase may replace plagioclase in deposits from the Atlantic.

Zeolites. Vast areas of the sea floor are covered by sediments containing the zeolite phillipsite. The distribution of phillipsite in the Pacific Ocean floor is indicated in Fig. 6. Deep-sea phillipsites are generally in the form of elongated, twinned crystals reaching up to 100 μ in length; these crystals may constitute more than 50% (by wt) of the sediment, and are generally associated with smectites, grains of basaltic glass, and Fe–Mn hydroxide particles (Fig. 7). Oceanic phillipsites differ in composition from continental, hydrothermal phillipsites; oceanic phillipsites lack calcium and other bivalent cations and have a relatively high Si/Al ratio (Fig. 8). Phillipsite forms slowly on the sea floor as the end product of low-temperature reactions between grains of basaltic glass and seawater. The Si/Al framework of the zeolite derives probably from the glass, whereas the alkalies are contributed to a large extent by seawater (Arrhenius, 1963).

Oceanic phillipsite grows in areas of low sedimentation rate and is probably a stable phase in deep seawater. Harmotome, a barium-rich zeolite structurally similar to phillipsite, has been reported to be present together with phillipsite in some Pacific sediments (Arrhenius, 1963).

Clinoptilolite has been found sparsely in the Atlantic (Biscaye, 1965; Hathaway and Sachs, 1965; Bonatti and Joensuu, 1968), as crystals usually $<$ 40 μ in size. Marine clinoptilolite may be formed as one of the products of alteration of volcanic glasses more silicic than basalts. In some instances, clinoptilolite is associated with magnesium silicates such as sepiolite (Hathaway and Sachs, 1965) or palygorskite (Bonatti and Joensuu, 1968) and is probably one of the products of the reaction of magnesium-rich solutions with volcanic glass or with minerals of the smectite family.

Layered Silicates

One of the most common marine authigenic layered silicates is *glauconite*, which has a mica-type structure, and is observed to be often disordered and interlayered. Compositionally, glauconites are distinguished from other micas by their high iron content (about 20–25%), a significant proportion of which is in the bivalent form. The term glauconite is not restricted to the specific mineral mentioned above, but is also used as a morphological term to indicate subspherical green pellets common in recent and ancient sediments. Burst (1958) has pointed out that so-called "glauconitic" pellets often do not contain the min-

53

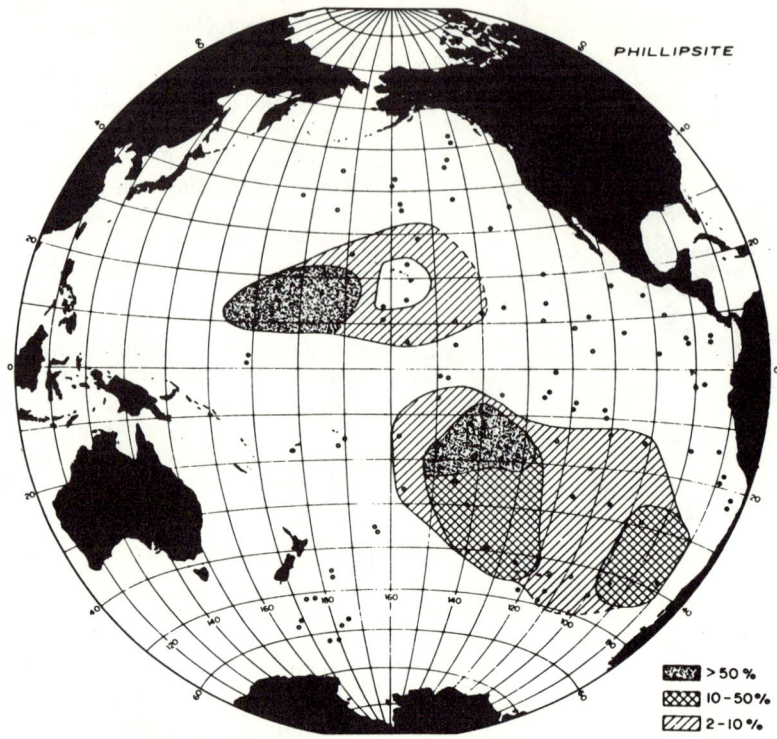

Fig. 6. Distribution of phillipsite (on a carbonate- and MnO_2-free basis) in pelagic sediments from the east Pacific (Bonatti, 1963).

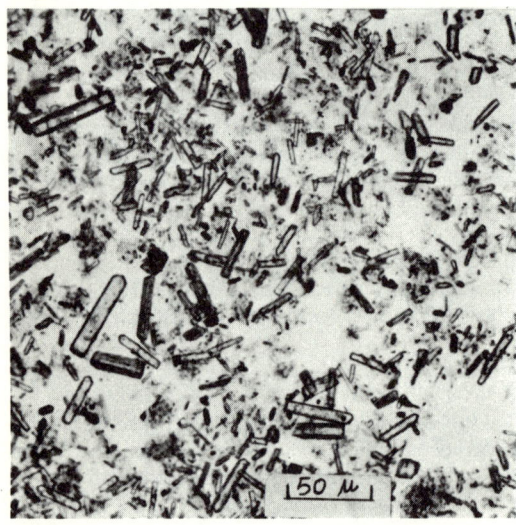

Fig. 7. Phillipsite crystals in a pelagic sediment from the south Pacific.

alteration *(halmyrolysis)* of detrital micaceous clay minerals. In addition to the presence of micaceous clays as a source material, the formation of glauconite requires reducing conditions, which can be provided in local microenvironments by the presence of lumps of organic matter in the sediment or by the activity of mud-feeding organisms. The common pelletal form of glauconitic aggregates

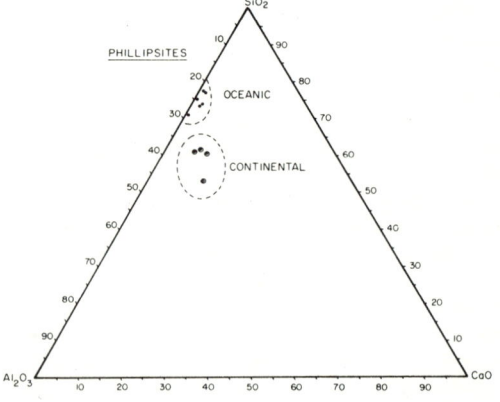

Fig. 8. SiO_2–Al_2O_3–CaO ratios in oceanic and continental phillipsite (Bonatti and Joensuu, unpublished).

eral glauconite, but may consist of aggregates of other clay minerals.

Glauconite is found in a variety of environments within the continental shelf, and it has been found at depths exceeding 2000 m. The formation of glauconite is due to submarine

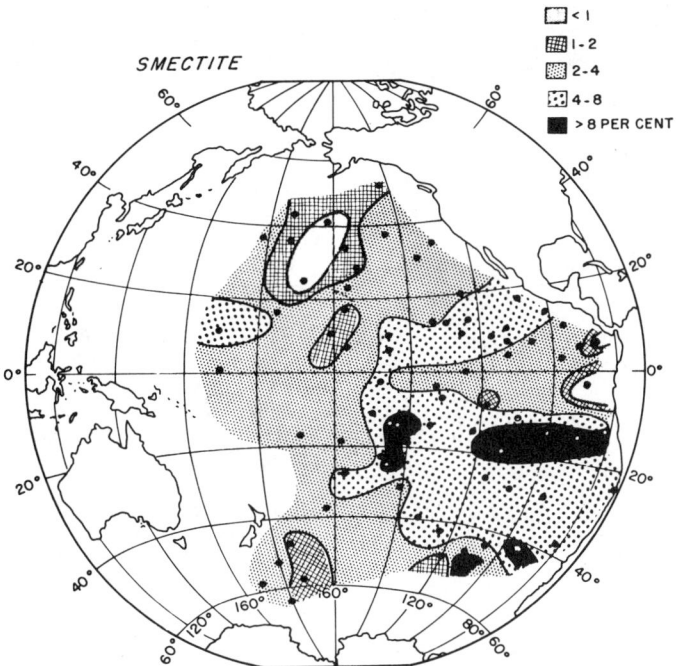

FIG. 9. Distribution of smectites (on a carbonate- and MnO_2-free basis) in pelagic sediments from the east Pacific.

is due to the fact that such aggregates may have been originally fecal pellets of sediment feeders, or internal fillings of foraminiferal shells, or mechanical agglomerates of clay minerals. *Muscovite* is one of the most abundant clay minerals found in the floor of the oceans. Of the various polymorphs of muscovite the high-temperature, $2M$ form appears to be the most abundant. This knowledge and the fact that oceanic micas have been shown to have quite old K/Ar ages (Hurley et al., 1963) indicates that they are terrigenous to a large extent. Some degradation of $2M$ muscovite in seawater appears to take place, resulting in the formation of $1M$, $1Md$, and mixed layered forms, which are found in the fine-size fraction of deep-sea deposits. Authigenesis of $1M$ and $1Md$ micas may result from the slow alteration of kaolinite on the ocean floor.

Minerals of the *Smectite* (montmorillonite) group are widespread in oceanic deposits. Their distribution in the Pacific is shown in Fig. 9. Oceanic smectites are partly terrigenous, partly authigenous. The latter are formed by reaction of volcanic glasses with seawater on the sea floor, and are found in the vicinity of volcanic centers. Laboratory experiments indicate that the alteration of natural glasses and of some igneous silicate minerals in solutions with pH and composition similar to that of seawater results in the formation of montmorillonites. In the southeast Pacific authigenous smectites are iron-rich (nontronites) and rather poorly crystallized; they are closely associated with phillipsite.

Chlorites in deep-sea sediments are generally terrigenous; however, instances where chlorite grows authigenically have been reported in the Pacific (Bonatti and Arrhenius, 1965; Swindale et al., 1967). In both cases chlorite appeared to be formed by diagenetic reactions of iron and aluminum hydroxides with Si and Mg contained in the pore solution of the sediment.

Minerals of the *serpentine* group are formed on the sea floor by alteration of olivine and other femic igneous silicates.

Palygorskite and sepiolite, magnesium silicates, have been found recently in some deep-sea sediments from the Atlantic: sepiolite by Hathaway and Sachs (1965); and palygorskite by Bonatti and Joensuu (1968) (Fig. 10). They are probably formed by reaction of magnesium-rich solutions with volcanic glasses or with minerals of the smectite group.

ENRICO BONATTI

References

Ames, L. L., 1959, "The genesis of carbonate apatite," *Econ. Geol.*, **54**, 829–841.

Arrhenius, G., 1963, "Pelagic Sediments," in (Hill,

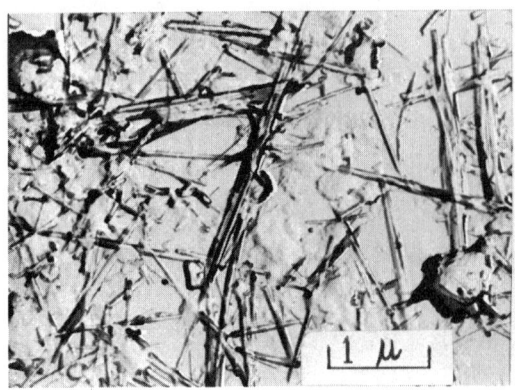

FIG. 10. Electron micrograph of palygorskite from the western Atlantic (Bonatti and Joensuu, 1968).

M. N., editor) "The Sea," Vol. 3 pp. 655–727, New York, Interscience Publ.

Arrhenius, G., and Bonatti, E., 1965, "Neptunism and Volcanism in the Ocean," in (Sears, M., editor), "Progress in Oceanography," Vol. 3 pp. 7–22, New York, Pergamon Press.

Bender, M. L., Ku, T. L., and Broecker, W. S., 1966, "Manganese nodules: their evolution," *Science,* **151,** 325–328.

Biscaye, P. F., 1965, "Mineralogy and sedimentation of the deep sea sediment fine fraction in the Atlantic Ocean," *Geol. Soc. Am. Bull.,* **76,** 803–832.

Bonatti, E., 1963, "Zeolites in Pacific pelagic sediments," *Trans. N.Y. Acad. Sci.,* **25,** 938–948.

Bonatti, E., 1966, "Deep sea authigenic calcite and dolomite," *Science,* **153,** 534–536.

Bonatti, E., and Arrhenius, G., 1965, "Eolian sedimentation in the Pacific off Northern Mexico," *Marine Geol.,* **3,** 337–348.

Bonatti, E., and Joensuu, O., 1968, "Palygorskite from Atlantic deep sea sediments," *Am. Mineralogist,* **153,** 975–983.

Boström, K., and Peterson, M. N. A., 1969, "The origin of Al-poor sediment in areas of high heat flow on the East Pacific Rise," *Marine Geol.* **7,** 427–447.

Burns, R. G., and Fuerstenan, D. W., 1966, "Electron probe determination of inter-element relationships in manganese nodules," *Am. Mineralogist,* **51,** 895–902.

Burst, J. F., 1958, "Glauconite pellets," *Bull. Am. Assoc. Petrol. Geol.,* **42,** 310–327.

Cloud, P. E., 1962, "Environment of calcium carbonate deposition west of Andros Island, Bahamas," *U.S. Geol. Surv. Profess. Paper* **350.**

D'Anglejan, B. F., 1967, "Origin of marine phosphorites off Baja California, Mexico," *Marine Geol.,* **5,** 15–43.

Degens, E. T., 1965, "Geochemistry of Sediments," Englewood Cliffs, N.J., Prentice-Hall, 342pp.

Dietz, R. S., Emery, K. O., and Shepard, F. P., 1942, "Phosphorite deposits on the sea floor off Southern California," *Bull. Geol. Soc. Am.,* **53,** 815–848.

Enlows, H. E., and Oles, K. F., 1966, "Authigenic silicates in marine Spencer Formation at Corvallis, Oregon," *Bull. Am. Assoc. Petrol. Geol.* **50**(9), 1918–1926.

Fairbridge, R. W., 1967, "Phases of Diagenesis and Authigenesis," in (Larsen, G., and Chilingar, G. V., editors), "Diagenesis in Sediments," Amsterdam, Elsevier Publ. Co., pp. 19–89.

Goldberg, E. D., 1954, "Chemical scavengers of the sea," *J. Geol.,* **62,** 249–265.

Goldberg, E. D., 1961, "Chemistry in the Oceans," in (Sears, M., editor), "Oceanography," *Am. Assoc. Advan. Sci. Publ. No. 67,* 583–597.

Harrison, C. G. A., and Peterson, M. N. A., 1965, "A magnetic mineral from the Indian Ocean," *Amer. Mineralogist,* **50,** 704–712.

Hathaway, J. C., and Sachs, P. L., 1965, "Sepiolite and clinoptilolite from the Mid Atlantic Ridges," *Amer. Mineralogist,* **50,** 852–867.

Hurley, P. M., Heezen, B. C., Pinson, W. H., and Fairbairn, H. W., 1963, "K/Ar age values in pelagic sediments of the North Atlantic," *Geochem. Cosmochim. Acta,* **27,** 393–400.

Klement, R., Hüter, F., and Köhrer, K., 1942, "Bildet sich Karbonat-Apatit in wassrigen Systemen?," *Z. Elektrochem. angen. phys. Chem.,* **48,** 334–336.

Lowenstam, H. A., and Epstein, S., 1957, "On the origin of sedimentary aragonite needles of the Great Bahama Bank," *J. Geol.,* **65,** 364–375.

Lynn, D. C., and Bonatti, E., 1965, "Mobility of manganese in diagenesis of deep sea sediments," *Marine Geol.,* **3,** 457–474.

Mellis, O., 1952, "Replacement of plagioclase by orthoclase in deep sea deposits," *Nature,* **169,** 624.

Milliman, J. D., 1966, "Submarine lithification of carbonate sediments," *Science,* **153,** 994–997.

Pratt, R. M., and McFarlin, P. F., 1966, "Manganese pavements on the Blake Plateau," *Science,* **151,** 1080–1081.

Revelle, R., and Emery, K. O., 1951, "Barite concretions from the ocean floor," *Bull. Geol. Soc. Am.,* **62,** 707–724.

Sillén, L. G., 1961, "The Physical Chemistry of Sea Water," in (Sears, M., editor), "Oceanography," *Am. Assoc. Advan. Sci. Publ. No. 67,* 549–581.

Swindale, L. D., and Pow Foong Fan, 1967, "Transformation of gibbsite to chlorite in ocean bottom sediments," *Science,* **157,** 799–800.

Cross-references: *Aqueous Solutions; Authigenesis of Minerals—Nonmarine; Geochemistry of Sediments; Interstitial Waters in Sediments; Mineral Classes: Silicates; Oxidation and Reduction; pH-Eh; Rare Earths; Seawater; X-Ray Diffraction; Weathering, Chemical.* Vol. I: *Black Sea; Mineral Potential of the Oceans; Submarine Plateaus.* Vol. III: *Weathering.* Vol. IVB: *Clays and Clay Minerals; Electron Microscopy; Order-Disorder; Polymorphism.* Vol. VI: *Authigenesis; Deep-sea Sedimentation; Eolian Sediments; Globigerina Ooze; Halmyrolysis; Pelagic Sediments; Siliceous Ooze; Weathering, Organic.*

AUTHIGENESIS OF MINERALS—NONMARINE

Definition

The term *authigene* or *authigenic* (Gr. *authi* "there," or "on the spot," plus *genes* "birth" or "origin") was applied particularly to the minerals of clastic rocks that formed as crystalline units at the place where they were found. The term is now used more broadly to designate all sedimentary rock minerals that form at the site of origin of the rock, either at the time of deposition or after. The contrasting term is *allothigenic,* meaning minerals derived from previously existing rocks.

Authigenic minerals are formed by and are representative of the physical and chemical conditions of the sedimentary environment during and after accumulation of the sediment. The allothigenic minerals are representative of external physical and chemical environments and may well be unstable in the sedimentary environment during and following accumulation.

Authigenesis is the result of an attempt to establish an equilibrium assemblage of minerals by the elimination of the unstable, growth of the stable, and the production of new and stable minerals. Reactions are controlled by the nature of the allothigenic minerals and the interstitial fluids and by the physical and chemical condititons of the environment. The effect of temperature and pressure, the chemical parameters of the acidity or alkalinity (pH), and the oxidation-reduction potential (Eh) seem to be the most important factors controlling the appearance of new minerals and the disappearance of old.

Authigenesis is considered a sedimentary process and should be clearly separated from metamorphism and igneous activity. However, since sedimentary rocks may be deeply buried and subjected to relatively high temperatures and pressures, differentiation of the processes is not easy and the distinction is always an arbitrary one. Attempts to estimate an upper temperature limit for authigenesis are usually found to be in the range 150–200°C, but may be higher. If confining or hydrostatic pressure alone is involved, the upper pressure boundary for authigenesis would be very high, probably thousands of atmospheres, but the application of a shearing stress drastically changes the environment. Although it is known that many mineralogical changes are dependent upon the interaction of temperature, hydrostatic pressure, and shearing stress, the importance of the last is hard to state quantitatively. Changes caused by the pressure of simple overburden should probably be considered authigenic, not metamorphic.

The use of minerals as environmental indicators is a useful but far from perfect guide Feldspar, amphibole, and pyroxene, long thought of as indicative of high-temperature environments, are also produced during authigenesis. It has been suggested that conversion of unstable rocks (usually rocks rich in glass) into mineral assemblages rich in zeolites be considered the first step in metamorphism, and a "zeolitic facies" of regional metamorphism has been proposed. Other students of the subject classify as authigenic all mineral assemblages produced in sedimentary rocks under conditions in which zeolites are formed. The clay minerals recrystallize during authigenesis as well as during low-temperature metamorphism. Such a development is considered most typical of metamorphism, and it has been suggested that the complete recrystallization of the clay minerals in pelitic sediments might serve as an indicator for metamorphism.

Mineral changes in sedimentary and pyroclastic rocks in close proximity to igneous intrusions may be the result of late magmatic fluids rather than authigenesis. Zeolites, commonly found filling vesicles and fractures in igneous rocks and filling pores and replacing the calcite of fossils in sedimentary rocks, are often attributed to late magmatic hydrothermal activity rather than authigenesis if geologic conditions warrant such an interpretation. The evidence of igneous origin is not always clear and the decision to term minerals igneous or authigenic is also an arbitrary one.

Authigenic Processes

Authigenic minerals may be formed by any one of a variety of processes:

1. Deposition may take place about a nucleus formed by a stable allothigenic grain of the same mineral. Growth of this type is termed secondary enlargement, and the new growth is commonly in optical continuity with the nucleus. Deposition may also take place about a nucleus of an allothigenic mineral of slightly different composition, perhaps a different member of an isomorphous group. Under such circumstances the new growth is often slightly out of optical continuity.

2. Authigenic minerals may develop as crystals or crystalline masses in pores as the result of chemical reactions or precipitation from a saturated solution, or they may develop in mineral grains as chemical alteration products. Reactions between fluids and solids are most characteristic, but in some instances reactions between solids and solids take place.

3. Minerals may result from metasomatism,

TABLE 1. COMMON AUTHIGENIC MINERALS (EXCLUDING HALIDES)[a] (FROM FAIRBRIDGE, 1967)

Mineral	Formula	Frequency[b]	Syn-diagenetic	Ana-diagenetic	Epi-diagenetic
Anatase	TiO_2 (tetr.)	R		×	×
Anhydrite	$CaSO_4$	C		×	
Ankerite	$Ca(Mg, Fe)(CO_3)_2$	R		×	×
Aragonite	$CaCO_3$ (orth.)	C	×	×	×
Azurite	$Cu_3(CO_3)_2(OH)_2$	R		×	×
Barite	$BaSO_4$	C	×		×
Bornite	Cu_5FeS_4	R	×		
Brookite	TiO_2 (orth.)	R		×	
Calcite	$CaCO_3$ (rho.)	C	×	×	×
Celestite	$SrSO_4$	C	×		×
Cerussite	$PbCO_3$	R		×	
Chalcedony	SiO_2	C	×	×	×
Chalcopyrite	$CuFeS_2$	R	×		×
Chamosite	$[(Fe, Mg)O]_{15}[Al_2O_3]_5[SiO_2]_{11}[H_2O]_{16}$	R	×	×	
Chlorite group	—	C	×	×	
Collophane	$Ca_3P_2O_8 \cdot H_2O$	C			×
Dahllite	$Ca_{10}(PO_4)_6(CO_3) \cdot H_2O$	C	×		×
Dolomite	$CaMg(CO_3)_2$	C	×	×	×
Galena	PbS	R	×		
Glauconite	$K_2(Mg, Fe)_2Al_6(Si_4O_{10})_3(OH)_{12}$	C	×	×	×
Greenalite	$Fe_6Si_4O_{10}(OH)_8$	R		×	
Gypsum	$CaSO_4 \cdot 2H_2O$	C	×		×
Halite	$NaCl$	C	×		×
Hematite	Fe_2O_3	C			×
Hydromagnesite	$4MgCO_3 \cdot Mg(OH)_2 \cdot 4H_2O$	R			×
Illite	$(K,Na)(Al,Mg,Fe,Li)_2(AlSi_3O_{10})(OH)_2$	C	×		
Kaolinite	$Al_2Si_2O_5(OH)_4$	C			×
Leucoxene	TiO_2 complex	C		×	
Limonite	$FeO(OH) \cdot n(H_2O) + Fe_2O_3 \cdot n(H_2O)$	C			×
Magnesite	$MgCO_3$	R	—	—	×
Malachite	$Cu_2CO_3(OH)_2$	R		×	×
Marcasite	FeS_2 (orth.)	C	×		×
Montmorillonite	$(Al, Mg)_8(Si_4O_{10})_3(OH)_{10} \cdot 12H_2O$	C	×		
Muscovite	$KAl_3Si_3O_{10}(OH)_2$	C			
Natron	$Na_2CO_3 \cdot 10H_2O$	R			×
Nesquehonite	$MgCO_3 \cdot 3H_2O$	R	×		×
Opal	$SiO_2 \cdot nH_2O$	C	×	—	×
Orthoclase	$KAlSi_3O_8$	C		×	×
Phillipsite	$(Ca, Ba, K, Na)_6Al_8(Al, Si)_2Si_{10}O_{40} \cdot 15–20H_2O$	R	×		
Plagioclase	$(Ca, Na)(Al, Si)AlSi_2O_8$	C		×	×
Psilomelane (Wad)	$(Ba, H_2O)Mn_5O_{10}$	C	×		
Pyrite	FeS_2 (iso)	C	×		×
Pyrolusite	MnO_2	C	×		
Quartz	SiO_2	C		×	×
Rhodochrosite	$MnCO_3$		×	×	
Rutile	TiO_2 (tetr.)	R		×	
Siderite	$FeCO_3$	C	×	×	
Sphalerite	ZnS	R	×		×
Strontianite	$SrCO_3$	R		×	
Sulphur	S	R	×		×
Tourmaline	$Na(Mg, Fe)_3Al_6(BO_3)_3Si_6O_{18}(OH)_4$	R		×	
Witherite	$BaCO_3$	R		×	×
Zeolites (phillipsite, heulandite, laumontite, chabazite, natrolite, analcime)		C	×		×
Zircon	$ZrSiO_4$	R		×	

[a] In part, after Twenhofel (1950, p. 288), and Teodorovich (1961). [b] R = rare; C = common.

one mineral being replaced by another of a different composition by the action of migrating solutions.

4. Authigenic minerals may also be due to inversion, when one polymorph is replaced by another in response to the new environment.

5. Recrystallization also effects mineral development, when allothigenic mineral grains change their form, size, or orientation in response to the environment.

6. Volcanic glass in contact with pore fluids commonly alters to a wide variety of clay minerals and zeolites, but simple devitrification is also possible. Devitrification is thought of as the process whereby a glass develops crystalline minerals from elements contained within the glass without addition of outside material. Since the majority of glasses found in sediments are persilicic, the formation of α-cristobalite and feldspar is the usual result of devitrification. However, these minerals usually represent an intermediate stage in authigenesis with quartz and clay minerals or zeolites developing later.

Authigenic Minerals

A considerable array of authigenic minerals has been recognized and described (see Table 1). The most common are those in equilibrium in a physical environment of low temperature and pressure and a chemical environment that is mildly reducing and mildly alkaline. The more unusual authigenic minerals are the result of atypical physical and chemical environments. Fig. 1 is a graph showing the stability fields in terms of oxidation-reduction potentials (Eh) and acidity or alkalinity (pH) for many common sedimentary minerals. Although this graph was originally designed to show the primary sedimentary environment, later authigenic environments can be thought of in terms of a shift in position over the diagram.

Authigenic minerals most commonly encountered are quartz, calcite, and dolomite, and the illite, kaolinite, montmorillonite, and sedimentary chlorite groups. Feldspars, certain iron minerals, zeolites, and even metallic oxides and hydroxides, sulfides, sulfates, and phosphates are not uncommon.

Alpha Quartz. Alpha quartz (SiO_2) occurs commonly as secondary overgrowths on grains, serving as the most common cement in sandstones, or as microcrystalline masses resulting from the recrystallization of opal, or from the solution, transportation, and reprecipitation of SiO_2. Such microcrystalline masses are termed chert.

The solution of such an insoluble material and the reprecipitation of a material that crystallizes so very slowly is probably due to a variety of interacting factors, such as an increase in temperature with burial, increased solubility at pressure points on quartz grains, and the development of local pH gradients within the sediment. Several possible processes have been pictured as developing local pH gradients. Concentrations of organic material may develop carbonic acid or organic acids; or a clay layer between sand grains, if subjected to CO_2-rich solutions, might liberate K_2CO_3, a strong base, which would dissolve SiO_2 at the clay contact. These processes are aided by the vast amount of time available for most geologic processes.

Carbonates. *Calcite* ($CaCO_3$) as found in sedimentary rocks is almost wholly authigenic, the result of the inversion of aragonite or the recrystallization of smaller calcite grains. It may also occur as cement precipitated in pores in the rock or as overgrowth on detrital calcite, chiefly fossil debris. Calcium carbonate as aragonite is deposited in huge amounts biochemically, and probably inorganically as well, but this isomorph is unstable at the temperatures and pressures found in the depositional and postdepositional environments of the sediment and soon inverts to calcite. The calcite itself, especially calcite in extremely small fragments, is very susceptible to recrystallization.

Dolomite ($CaMg(CO_3)_2$) also is almost entirely authigenic, the result of the replacement of calcite. A metasomatic process, under the influence of migrating magnesium-rich fluids, probably causes dolomitization. Such fluids appear to be the result of evaporation concentration of seawater.

Clay Minerals. The principal clay mineral groups of the sedimentary rocks are:

Kaolinite ($Al_2Si_2O_5(OH)_4$)

Montmorillonite
(($Al,Mg)_8(Si_4O_{10})_3(OH)_{10}$ $12H_2O$)

Illite
(($K,Na)(Al,Mg,Fe,Li)_2(AlSi_3O_{12})(OH)_2$)

Sedimentary chlorites
(($Mg_6Si_8O_{20}(OH)_4$ plus $Mg_6(OH)_{12}$)

The clay minerals found in recent sediments or even young sedimentary rocks are probably allothigenic, a function of the source region rather than the depositional environment. The authigenic changes these allothigenic clay minerals experience are not thoroughly understood, but it is clear that such changes do occur. In the older sedimentary rocks, montmorillonite and kaolinite disappear, giving way to illite and chlorite, a possible result of their alteration to a more mica-like structure. Undoubtedly, authigenic kaolinite has been found in desiccation

AUTHIGENESIS OF MINERALS—NONMARINE

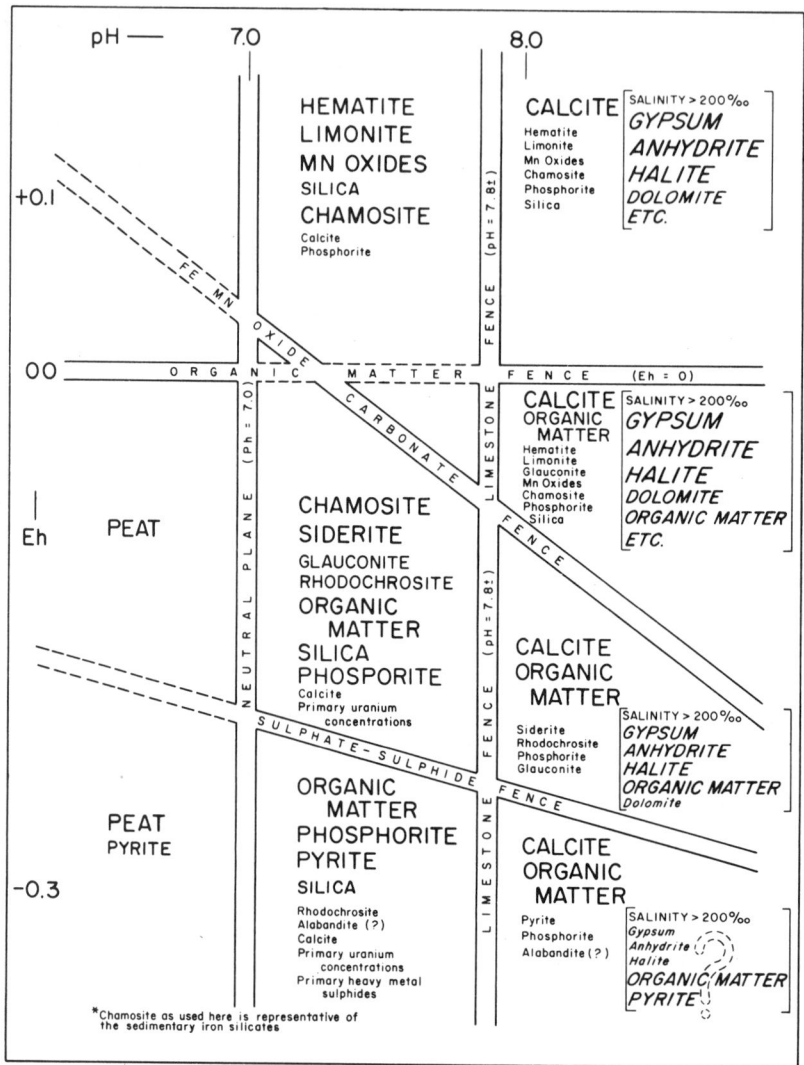

Fig. 1. Sedimentary chemical end-member associations in their relations to environmental limitations imposed by selected Eh and pH values. Associations in brackets refer to hypersaline solutions (Krumbein and Garrels, 1952).

and joint fractures in coal and as vermicular crystals and nodules in shales. Certain pure kaolinite flint clays are thought to be formed by removal of silica and alkalies from mixtures of illite and mixed-layer clay. The leaching is thought to have been accomplished by acid swamp waters after deposition. Relatively pure, well-crystallized 1 M illite has been found in sandstones and is thought to be due to in-place alteration of feldspar in the presence of saline waters.

Feldspars. The principal alkali feldspars are:

    Orthoclase (KAlSi$_3$O$_8$)
    Microcline (KAlSi$_3$O$_8$)
    Albite (NaAlSi$_3$O$_8$)

These three alkali feldspars, occurring in very pure form, are the common authigenic feldspars. They occur in sandstones, shales, and limestones of all ages. They occur as overgrowths on allothigenic minerals, as euhedra, or as grains so small as to be detected only by the x-ray methods. They may be present in very small amounts or they may constitute a high proportion of the rock. Most typical of overgrowths is the case of orthoclase rimming allothigenic microcline, although orthoclase will also rim grains of plagioclase. In these instances the rim has a composition different from the core and does not grow in precise optical continuity. All three alkali feldspars occur as very small (generally less than 1 mm), very simple euhedra exhibiting a prism (110) and basal pinacoid (001).

Details of the formation of authigenic feldspar are poorly understood. It was once thought probable that a high alkali concentration and the presence of CO_2 were required, but more recently the authigenesis of feldspar has been thought of in more simple terms as the result of the leaching and dehydration of illite or montmorillonite.

Iron Minerals. The common iron minerals are:

Glauconite $(K_2(Mg,Fe)_2Al_6(Si_4O_{10})_3(OH)_{12})$
Chamosite
$((FeMgO)_{15}(Al_3O_3)_5(SiO_2)_{11}(H_2O)_{16})$
Pyrite (FeS_2)
Marcasite (FeS_2)
Siderite $(FeCO_3)$

The iron minerals listed above, along with other less common iron minerals, occur in some rocks as important authigenic constituents. The major factor in the formation of these iron minerals is the oxidation-reduction potential. Because of the decomposition of included organic material and the isolation from oxygen-bearing fluids, the burial environment appears to possess a much lower oxidation-reduction potential than the depositional environment. Unless the sediment is flooded with ferric iron, the amount of organic matter in normal sediments is probably enough to reduce all the iron from the ferric state, producing the ferrous iron minerals indicated.

Zeolites. The common zeolites are:

Heulandite-Clinoptilolite
$((CaNa_2)Al_2Si_7O_{18}\ 6H_2O)$
Phillipsite $((\tfrac{1}{2}Ca,Na,K)_3(Al_2Si_5O_{16})\ 6H_2O)$
Thomsonite $((NaCa_2)((Al,Si)_5O_{12})_2\ 6H_2O)$
Analcime $(Na(AlSi_2O_6)\ H_2O)$

The zeolites listed above and others, along with the related mineral analcime, are common authigenic minerals. The presence of volcanic glass in a sediment appears necessary for their formation. Reaction between volcanic glass and alkaline, often saline lake or marine pore water, seems adequate to account for their origin.

HAROLD E. ENLOWS

References

Degens, E. T., 1965, "Geochemistry of Sediments, a Brief Survey," Englewood Cliffs, N.J., Prentice-Hall, Inc., 342pp.

Fairbridge, R. W., 1967, "Phases of diagenesis and authigenesis," in (Larsen, G., and Chilingar, G. V., editors), "Diagenesis in Sediments," pp. 19–89, "Developments in Sedimentology 7," Amsterdam, Elsevier Publishing Co.

Kalkowsky, E., 1880, "Uber die erforschung der archäischen formationen," Neues Jahrb. Geol. Palaeontol. Abhandl., **1**, 1–28.

Krumbein, W. C., and Garrels, R. M., 1952, "Origin and classification of chemical sediments in terms of pH and oxidation-reduction potentials," J. Geol., **60**, 1–33.

Pettijohn, F. J., 1957, "Sedimentary Rocks," Second ed., New York, Harper and Row, 718pp.

Teodorovich, G. I., 1961, "Authigenic Minerals in Sedimentary Rocks," New York, Consultants Bureau, 120pp.

Cross-references: *Authigenesis of Minerals—Marine; Calcium Carbonate; Crystal Chemistry; Interstitial Waters in Sediments; Oxidation and Reduction; pH-Eh.* Vol. IVB: *Clays and Clay Minerals; Feldspars; Glass: Devitrification of Volcanic Glass; Isomorphism; Polymorphism; Zeolites.* Vol. VI: *Authigenesis; Coal; Dolomite.*

B

BACTERIA, BACTERIOLOGY—*See* Vol. I; MARINE MICROBIOLOGY

BACTERIAL SULFATE REDUCTION—*See* SULFATE REDUCTION—MICROBIAL

BACTERIAL SULFUR OXIDATION—*See* SULFUR OXIDATION—BACTERIAL

BARIUM: ELEMENT AND GEOCHEMISTRY

Physical Properties. Barium (Gr. *barus*, "heavy"), is an alkaline earth element: symbol, Ba; atomic weight, 137.36; atomic number, 56; electronic configuration, $1s^2, 2s^2, 2p^6, 3s^2, 3p^6, 3d^{10}, 4s^2, 4p^6, 4d^{10}, 5s^2, 5p^6, 6s^2$; melting point, 714°C; boiling point, 1640°C; and valence, +2. The mass numbers of its stable isotopes and their relative abundances are 138 (71.66%), 137 (11.32%), 136 (7.81%), 135 (6.59%), 134 (2.42%), 130 (0.101%), and 132 (0.09%). Barium isotopes of mass numbers 127, 128, 129, 131, 133, 135, 137, 139, 140, 141, 142, 143, 144, and 145 have been produced artificially.

V. Casciorolus noted the presence of barium in the mineral *barite* in 1602. Metallic barium, which does not occur in nature is silvery white in color, malleable, and slightly harder than lead. It can be produced by the electrolysis of fused barium chloride.

Compounds. The most important naturally occurring compound of barium is *barite* (*baryte*), $BaSO_4$. *Witherite*, $BaCO_3$, is the most common barium-bearing carbonate. *Barytoanglesite*, $(Ba,Pb)SO_4$, and the double salt, *bromlite*, $CaBa(CO_3)_2$, are much less common barium minerals. The most important occurrence of barium in the silicates is in the form of trace impurities. Barium silicates are uncommon and include *celsian*, $BaAl_2Si_2O_8$, the barium feldspar. Several rare barium silicate minerals are found in alkalic rocks including *benitotite*, $BaTiSi_3O_9$, and *leucosphenite*, $Na_4Ba(TiO)_2(Si_2O_5)_5$. Other rare barium minerals are *sanbornite*, $BaSi_2O_5$, *gillespite*, $BaFeSi_4O_{10}$, *banalsite*, $Na_2BaAl_3Si_4O_{16}$, and *nitrobarite*, $Ba(NO_3)_2$.

Chemistry. The barium atom readily looses its two outer electrons, forming the Ba^{2+} ion. The first ionization potential is 5.19V and the second is 9.95V. With the exception of radium, barium is a stronger reducing agent and has more pronounced base-forming tendencies than the other alkaline earths. The atomic radius of barium is 2.22Å, its ionic radius is 1.35Å, and its covalent radius is 1.98Å. Radius ratio predictions indicate that barium should occur in 10 to 12 coordination with oxygen. Based on electronegativity differences, the barium-oxygen bond should be about 82% ionic and 18% covalent in nature. Barium has a relatively low ionic potential (Z/r, in effect a measure of charge per unit area of the ion), hence is taken into aqueous solution in the form of hydrated ions. Of particular importance in the geochemistry of barium is the highly insoluble nature of $BaSO_4$ in natural waters.

Geochemistry. Barium is not a major cation, but is a ubiquitous trace element in the earth's crust. The average igneous rock contains 480 ppm BaO and the average sedimentary rock 430 ppm. In igneous and metamorphic rocks, Ba^{2+} is characterized by its ability to substitute for the K^+ ion. A similar diadochic relationship commonly exists between lead and barium. Barium does not substitute for calcium in significant quantities because the size difference is too great. In terms of the overall geochemical cycle, barium most closely resembles strontium, another important alkaline earth trace element of slightly smaller size.

In igneous and metamorphic rocks the most important occurrence of barium is as a trace element in the tectosilicate mineral, *orthoclase feldspar*, $KAlSi_3O_8$. *Celsian*, $BaAl_2Si_2O_8$, is a relatively rare barium feldspar. Because of the higher charge of Ba^{2+} relative to K^+, barium is commonly emplaced in the first orthoclases to crystalize from the magma. *Orthoclases* from the late stage pegmatitic rocks are typically very low in barium. In general the felsic igneous rocks contain more barium than basic rocks (because more orthoclase is present in the felsic varieties). For example, typical dunites may contain less than 10 ppm barium whereas granites have between 400 and 500 ppm. Alkalic rocks such as syenites contain the most barium (1600 ppm). In the alkalic rocks barium occasionally forms independent minerals but these are of little quantitative importance.

Next in importance in incorporating barium in igneous rocks are the phyllosilicates, *muscovite mica* and *biotite mica*. Muscovites contain as much as 9% barium; biotites have up to 6% barium. In both cases barium substitutes for potassium. As in the case of the feldspars, the barium rich micas are formed at the earliest stages of mica formation, generally at high temperatures.

Barium becomes sufficiently concentrated at the hydrothermal stage of magma evolution or during hydrothermal processes to form independent minerals. The most important of these and in fact the most important of all barium minerals is *barite (baryte)*, $BaSO_4$, a common vein mineral. Other hydrothermal barium minerals include *witherite*, $BaCO_3$, *psilomelane*, a barium manganese oxide, and *barytoanglesite*, $(Ba,Pb)SO_4$.

During weathering, the barium-bearing orthoclases and micas of igneous rocks are readily attacked because of their relatively high temperature of formation, hence relative instability under weathering-zone conditions. Barium is an unimportant and quite variable constituent of river waters, probably averaging about 0.05 ppm. Only a small fraction of the barium contributed to the oceans during geologic time has remained in the water. Typical barium concentrations in oceanic surface water are about 0.01 ppm and are undersaturated with respect to barium. Although the solubility product of barium sulfate decreases with depth, it has been found that the barium content of seawater increases with depth in areas of high organic productivity, even approaching saturation values. At depths between 4000 and 5000 m, concentrations as high as 0.06 ppm have been noted. Deep-sea sediments below such areas tend to be markedly enriched in barium relative to sediments from other areas. The depth increase in barium is related to the production of sulfate ions from oxidizing organic matter. The sulfate ions combine with barium to form the insoluble barium sulfate, which sinks, resulting in a net transfer of barium to deeper waters. Another important means of removal of barium from the sea is adsorption on clay minerals, a process favored by the element's large size and low ionic potential. The third important means of oceanic barium removal is adsorption on negatively charged manganese precipitates and ultimate incorporation into manganese nodules, which typically contain about 0.2% barium. Overall, most barium in recent sediments is tied up with clay minerals.

Biogeochemistry. Barium is not taken up in significant quantities by any animals. It apparently serves no useful organic function and in large amounts it is poisonous to higher animals. *Calcitic* or *aragonitic* shells of marine invertebrates generally contain about 10 ppm barium.

Uses. Metallic Barium is used to aid in the evacuation of gaseous impurities in vacuum tubes. It is also a constituent of various alloys. Barium sulfate is used as a pigment for weighting paper and as a component of drilling muds. A solution of barium hydroxide in water (baryta water) has numerous uses, including the analysis of CO_2 in air. Barium hydroxide is also used in the sugar-refining industry.

ORRIN PILKEY

References

Day, F. H., 1963, "The Chemical Elements in Nature," London, G. G. Harrap and Co., 372pp.

Goldberg, E. D., 1963, "The Oceans as a Chemical System," in (Hill, M. N. editor), "The Sea," Vol. 2, pp. 3–25.

Goldschmidt, V. M., 1954, "Geochemistry," Oxford, Clarendon Press, 730pp.

Ketner, K. B., 1963, "Bedded barite deposits of the Shoshone Range, Nevada," *U.S. Geol. Surv. Profess. Paper* **475B**, 38–41.

Rankama, K., and Sahama, Th. G., 1950, "Geochemistry," Chicago, Univ. of Chicago Press, 712pp.

Cross-references: *Aqueous Solutions; Bonding; Seawater; Trace Elements; Trace Metals.* Vol. IVB: *Clays and Clay Minerals; Phyllosilicates.* Vol. VI: *Pelagic Sediments.*

BARIUM: ECONOMIC GEOLOGY

Minerals. *Barite*, $BaSO_4$, and *witherite*, $BaCO_3$, are the two chief barium-bearing minerals. Barite is abundant and has a worldwide distribution. Witherite is much less common and is not mined in the United States now. At one time it was mined at El Portal, Mariposa County, California. Now, the world's supply comes from deposits in Northumberland County in the north of England.

Barite is also called barytes, tiff, cawk, or heavy spar. It is usually orthorhombic, colorless, white or gray (although it is known in many colors, owing to impurities), transparent to opaque. The streak is white, the luster vitreous to resinous or pearly. The specific gravity of pure $BaSO_4$ is 4.5. This may be reduced by impurities. Well-formed crystals of barite are tabular, have cleavage in three directions and have a hardness of 2.5–3.5 on the Mohs' scale. Barite is relatively insoluble in water and acid. It is chemically inert. Witherite, $BaCO_3$, is orthorhombic. It has an uneven fracture, a luster that is vitreous to resinous, a hardness of 3–3.5 on the Mohs' scale, and a specific gravity of 4.29. Witherite is usually white, gray, yellow, brown, or green, and is transparent to opaque.

Geology. Barite is found in igneous, sedimentary, and metamorphic rocks. The three main types of Barite ore deposits are: (*1*) vein- and cavity-filling deposits, (*2*) bedded deposits, and (*3*) residual deposits.

Vein and Cavity Filling Deposits. Barite and associated minerals occur along faults, joints, and bedding planes, and in breccia zones and solution channels. This type of deposit also occurs in collapse and sink structures in limestone. Sharp contacts of the veins and cavity fillings with the wall rock are common. It often cements faults and collapsed breccia by replacement of the matrix or by filling voids.

Barite in veins and cavity fillings generally is dense, hard, and white to gray. Associated minerals are fluorite, calcite, ankerite, dolomite, quartz, and many sulfide minerals, especially pyrite, chalcopyrite, galena, and sphalerite. Gold, silver, and some rare earth elements occur in some deposits of barite in the United States. Barite is often a gangue mineral in metalic ore deposits.

Host rocks for vein and cavity-filling deposits of barite date from Precambrian to Tertiary time. Deposits dating from the Mesozoic are uncommon in the United States.

Bedded Deposits. Barite is sometimes restricted to certain beds, or a sequence of beds, in sedimentary rocks. Commercially important deposits of this type contain fine-grained, massive, gray barite or abundant crystals and masses of barite with quartz, chert, and carbonates such as calcite, dolomite, siderite, strontianite, and witherite. Pyrite and secondary iron oxides are common.

The beds are generally lenticular in shape and may pinch out abruptly or gradually diminish in barite content vertically and horizontally. The thickness of the bed may be as great as 200 feet, the length greater than 1 mile, and the width as great as a half mile.

Bedded deposits occur almost exclusively in rocks of Late Paleozoic age. There is conflicting evidence as to the origin of these deposits. One theory is that the barite originated either as a primary sedimentary deposit or as a replacement of the host rock by an aqueous solution. Dunbar and Rodgers (1957) hypothesize that the origin of barium in a carbonate environment is due to its expulsion from the metastable aragonite lattice when it converts to calcite. Barium is more abundant in the aragonite lattice than the calcite lattice. During recrystallization of aragonite to calcite in the presence of SO_4^{2-} ions, barite would be formed. If expelled, barium moves and is concentrated by some means, forming a deposit of barite.

Residual Deposits. This type of ore body occurs in unconsolidated material that has formed by the weathering of pre-existing deposits. These deposits are abundant in Missouri, Tennessee, Georgia, Virginia, and Alabama where deposits commonly are in a residuum formed from limestone and dolomite of Cambrian and Ordovician age. Most barite of this type is white and translucent to opaque. It occurs chiefly as mammillary, fibrous, or dense, fine-grained masses 1–6 in. across; less commonly it occurs as subhedral to euhedral crystals. Some pyrite, sphalerite, and galena may occur with the barite. The remainder of the material in these deposits is chert, quartz, and rock fragments in a matrix of red and brown clay.

The barite is either disseminated randomly or concentrated in irregular runs or streaks at various intervals in the residuum. In Missouri the deposits are generally within 15 ft of the surface; in Georgia, some deposits are as deep as 150 ft beneath the surface.

Mining and Milling Procedures. Methods of mining and preparing barite depend on the type of deposit and the ultimate use of the mineral. Residual deposits are mined by *open-pit* methods. The clay is removed in log washers and fine barite contained in the overflow may be recovered by tabling and froth flotation. In some locations, barite occurring in veins is mined by *shrinkage stoping* methods and recovered by *jigging*.

Barite is ground either wet or dry depending on whether it is going to be bleached or not. Material to be bleached is wet ground as well as is material that must be concentrated by flotation. Material that is not to be bleached is usually dry ground. The bleaching operation is customarily considered a trade secret. However, the procedure is well understood in general: the impurities are brought into solution by treating ground barite with sulfuric acid, the pulp is then settled and separated and the impurities are washed out, and finally the residue is dried and packed.

For lithopone or barium chemicals, the barite is converted into an intermediate soluble form. This is accomplished by roasting crushed barite in a kiln with carbon to reduce the $BaSO_4$ to the more soluble barium sulfide, commonly called black ash. The black ash is then leached with hot water. The liquor thus produced is then treated in various ways to make lithopone, blanc fixe, and barium carbonate.

Domestic and Foreign Barite Ore Deposits. The two principal barite producers in the

United States are Baroid Sales Division, National Lead Company, and the Magnet Cove Barium Corporation, a subsidiary of Dresser Industries. A total of forty-one firms produced crude barite in 1963. In addition, there were thirteen producers of barium chemicals and two producers of lithopone. Arkansas and Missouri were the two leading producers of barite; Georgia and Nevada were also major producers.

Arkansas. The largest barite deposit being mined in Arkansas is in the Magnet Cove area near Malvern, Hot Springs County. The deposit is found in a group of replaced beds near the base of the Stanley shale of Pennsylvanian age in a valley formed by a plunging syncline. The ore body is 60 ft thick, ¾ mile long, and nearly a half mile wide. The grade of the ore varies; the richest parts contain 70% $BaSO_4$. Associated minerals are quartz, chert, pyrite, iron oxides, clay minerals, calcite, and alstonite. Other bedded deposits are known in the Stanley shale and the Arkansas novaculite in Montgomery, Pike, and Polk Counties. Deposits of concretionary barite occur in Howard and Sevier Counties in the Trinity sand of Cretaceous age.

Missouri. The major deposits are in the Washington County district. This area has yielded more barite than any other district in the United States. The barite is concentrated in the residuum overlying the Potosi and Eminence dolomites of Cambrian age. Soft white barite is found with small amounts of galena, chert, drusy quartz, and fragments of dolomite in a matrix of red clay. Barite fragments are often ½–6 in. across. The concentrations are generally within 15 ft of the surface although the residuum may be 30 ft thick. Single areas of many acres may be ore bearing, but the richest parts of the deposit form runs or leads 10–20 ft wide and several hundred feet long. Some deposits contain 20–25% barite.

In the Central district, chiefly in Cole, Miller, Moniteau, and Morgan Counties, the deposits occur in clays and solution structures in the Gasconade, Roubidoux, and Jefferson City formations of Ordovician age. These deposits are deeper and more restricted in horizontal extent than those in the Washington County district.

Nevada. Here, barite production has increased greatly since 1950. Most of the mining is in the Battle Mountain area in the north central part of the state. The barite is found in large bedded deposits in the Late Paleozoic and Devonian sediments of several mountain ranges.

The barite is usually fine grained, gray to black, and fetid. Barite appears to have selectively replaced the original constituents in some layers. Some ore contains as much as 97% $BaSO_4$. Deposits of lenticular shape are mined in Devonian cherty limestone and shale in Lander County. A bedded deposit northwest of Dunphy, Eureka County, is also being mined.

Recent studies by the United States Geological Survey indicate that there is no obvious connection between the deposits and igneous activity in the Shoshone Range. The age of the mineralization is not known, but several of these deposits of vein and bedded types are known in Elko, Mineral, Clark, and Pershing Counties.

Georgia. Residual deposits of barite are found in the Cartersville district, Bartow County. The ore is associated with the residuum of the Weisner, Shady, and Rome formations of Cambrian age. White barite occurs in yellow to brown clay that is over 100 ft thick in places. Barite fragments are irregular and average less than 6 in. in diameter. These fragments are often coated with iron oxides and contain small sulfide inclusions. The clay matrix contains rock fragments and jasper. Veins containing barite in fault zones in the bedrock are considered the source of the barite in the residuum. Exhaustion of shallow deposits and higher costs of mining deeper deposits have curtailed production since World War II.

Polk, Floyd, Cherokee, Gordon, Murray, and Whitfield Counties have barite deposits, some of which have been mined in past years.

Tennessee. The Sweetwater district in parts of McMian, Monroe, and Loudon Counties yields most of the barite in this state. It occurs with fluorite and pyrite in veins or shatter zones in the upper part of the Knox group of Ordovician age. The barite is concentrated in overlying residual clays, and only these have been mined. The barite occurs in three separate belts separated by barren areas. Each belt is in a fault block. The large deposits cover several acres and may be as much as 60–80 ft thick. The barite is white and the fragments are usually greater than 6 in. in diameter. Chert and iron oxides are also abundant in the residuum.

In the Del Rio district, Cocke County, barite occurs in veins and replacements of fault breccia associated with thrust faults involving sedimentary rocks of Precambrian and Cambrian age.

California. The largest barite deposits are along the Merced River at El Portal, Mariposa County. Barite, with commercial witherite, replaced beds in the Calaveras formation of Mississippian age. These deposits were worked from 1910 to 1949.

Vein deposits of the Mountain Pass district in San Bernadino County are estimated to have large reserves of ore containing about 20% barite and 10% rare earth minerals.

Other States. Scattered residual deposits, and to a lesser extent vein deposits, have been mined in Virginia and Alabama. North Carolina, South Carolina, Kentucky, Montana, Washington, Idaho, New Mexico, Arizona, and Wisconsin have scattered barite deposits.

Foreign Deposits. Most foreign countries have deposits of barite, but only West Germany, Greece, U.S.S.R., Italy, Canada, Mexico, and Peru mined greater than 100,000 tons in 1958. The world's supply of witherite comes from Northumberland County in the north of England.

Small deposits of barite are mined in Japan, India, Korea, Australia, Algeria, and French Morocco, and lesser amounts in several other African countries.

Industrial Specifications. The many industries that use barite require that it conform to certain specifications. The drilling industry requires the barite to be fine grained, heavy, and chemically inert. Thus the barite must contain a minimum of 92% $BaSO_4$, be free of soluble salts, and have a minimum of 4.2 specific gravity; 90–95% of the material must pass through a 325-mesh screen. Several percent of iron oxide is not objectionable.

The chemical manufacturers require that the barite be a minimum of 94% $BaSO_4$, that there be a maximum of 1% each of Fe_2O_3, $SrSO_4$, and only a trace of fluorine. If the barite is used for lithopone, the $SrSO_4$ content may be higher. The mesh size is also important; if it is too fine the dust may be lost and if it is too coarse the mixing with carbonaceous material is poor. A size specification range of 4- to 20-mesh is requested by most consumers.

The glass manufacturers require that the barite be a minimum of 98% $BaSO_4$, containing a maximum of 1.5% SiO_2, 0.15% Al_2O_3, and 0.15% Fe_2O_3. A mixture of mesh sizes from 30-mesh down to 140-mesh is usually preferred. Iron is the most objectionable impurity.

Uses. Barite has many uses, the most important of which is a weighting agent in well-drilling muds. These muds lubricate and cool the bit, plaster the walls of the drill hole to prevent caving, carry the cuttings up the well to the surface, and restrain high gas and oil pressure to their formation levels to prevent blowouts (the main purpose for which barite is used).

Crushed barite, 16- to 20-mesh, is used by the glass industry. When added to the glass melt it fluxes the heat-insulating froth that forms on the surface of the melt, thus saving fuel. It also acts as an oxidizer and decolorizer, and increases brilliance.

Barite is used as a filler or extender in paint, inks, oilcloth, linoleum, rubber, and other materials. It is also used in the manufacturing of lithopone, a white pigment for paints.

Barite is used to make various barium chemicals such as barium carbonate, precipitated barium sulfate, barium chloride, barium oxide, barium peroxide, barium hydroxide, barium nitrate, and others. Precipitated barium sulfate, blanc fixe, is used as a white paint filler. Barium chloride is used in case hardening leather and cloth, in making magnesium metal, in preventing scum from forming on brick, and in water treatment. Fused barium chloride may be electrolyzed to produce barium metal. Barium carbonate is used as a component in ceramic glazes and enamels and also to prevent the formation of scum on ceramics. It diminishes porosity and prevents discoloration in bricks and it is used in making crown and flint glasses. Barium oxide is used as a component in glass and in the manufacturing of barium peroxide. In electric ferrous metallurgy, it is used to increase the life of acid furnace linings, give a quieter and steadier arc, and reduce sulfur and slag viscosity. Barium hydroxide prevents scumming in ceramics, and is used in lubricating oils and to remove sugar from molasses by the barium saccharate process. Barium nitrate is used in green signal flares, in primers and detonators, and in enamel. Barium metal is a deoxidizer of copper. It is also used as a "getter" to remove the last traces of gases in vacuum tubes, and because of its high electron emission rate when subjected to an electrical potential, it is used in alloys for spark plugs and electron mission elements in electronic tubes. Barium titanate is important in the electronics industry because of its high dielectric constant and its piezoelectric properties. The compound is well suited for use in miniature electronic and communication equipment. Finely ground barite is mixed with synthetic rubber powder and added to asphalt for road, airstrip, and parking-lot construction. The advantages of adding barite to the asphalt are that it results in a more flexible seal and less cracking of the surface, thereby giving longer life to the road surface.

Witherite, natural barium carbonate, is used in refining sugar, and in the manufacturing of barium chemicals, case hardening, and pigments.

Reserves and Production. Domestic commercial reserves were estimated by the United States Geological Survey in 1958 to be 285 million tons of measured and indicated ore, containing about 46 million tons of barite. These reserves are principally in Missouri and the Southern Appalachian states; the remainder is found in Arkansas, Nevada, and California.

The Nova Scotia reserves are estimated to be 3 million tons, and those in the region of Camamu Bay, Brazil, in excess of 800,000 tons.

Domestic production of barite was a record of 1,352,000 short tons, but the market made a comeback in 1959 to 867,000 tons and has remained stable since.

Production of lithopone in the United States has declined due to its replacement in the pigment field by titanium dioxide. In 1947, eight firms produced lithopone, and in 1963 there was only one firm which produced it.

Consumption of barite as a drilling mud has decreased since 1956, but in other industries, its use has increased.

The duty on crude or otherwise unmanufactured barite is $2.55 per long ton. Ground or otherwise manufactured barite has a tariff of $6.50 per long ton levied upon it. Domestic crude barite sells for $12.00–$16.00 per short ton, f.o.b. shipping point, in carload lots.

Future Outlook. Since the major consuming industry of gas and oil-well drilling is falling off from a peak of exploration activity, this should brake the growth of the barite industry. Other industries are picking up some of the slack.

JERRY S. KIER

References

Brobst, D. A., 1960, "Barium Minerals," in "Industrial Minerals and Rocks," Seeley W. Mudd Series, Am. Inst. Mining Engineers, Third ed., pp. 55–64.

Dean, B. G., and Brobst, D. A., 1955, "Annotated bibliography and index map of barite deposits in U.S.," *U.S. Geol. Surv. Bull.* **1019-C**, 41pp.

Dunbar, C. O., and Rodgers, J., 1957, "Principles of Stratigraphy," New York, John Wiley & Sons, 356pp.

Kroll, N. J., 1945, "Processes for making barium and its alloys," *Bur. Mines Inform. Circ.* **7327**, 16pp.

Lewis, R. W., 1965, "Barium," in "Mineral Facts and Problems," *Bur. Mines Bull.* **630**, 91–99.

Cross-references: *Aqueous Solutions; Barium: Element and Geochemistry; Rare Earths.* Vol. IVB: *Crystal Growth.*

BASE—CHEMICAL

Centuries ago, the name "alkali" (from an Arabic word meaning alkaline ashes of plants) was applied to substances which feel soapy, cut grease, and reverse the effects of acids. Later a base was defined as any hydroxyl compound that yielded hydroxyl OH^- ions in aqueous solution. Neutralization was interpreted as a combination of OH^- from the base with H^+ from the acid to form water and also a salt.

The Brönsted-Lowry theory extended the scope of the word base to include any chemical species, ionic or molecular, capable of accepting a proton (hydrogen ion) from another substance. The other substance acts as an acid in giving up a proton. A substance may therefore act as a base only in the presence of an acid. The greater the affinity for protons, the stronger the base. The OH^- acts as a strong base in the reaction

$$H^+ + OH^- \rightarrow H_2O$$

Water in undergoing the reaction

$$H^+ + H_2O \rightarrow H_3O^+$$

behaves as a weaker base.

G. N. Lewis proposed a more general concept of bases in 1923. In order to be considered a base a substance must exhibit the following properties:

1. Neutralization: the base must combine rapidly with an acid.
2. Titration with indicators: the base will change the color of indicators in a characteristic manner.
3. Displacement: a base will replace a weaker base from its compounds.
4. Catalysis: the base will promote chemical reactions which require base catalysts.

A base, in the Lewis concept, is any substance which can donate a pair of electrons to an acid, e.g.,

$$\text{base} + \text{acid} \rightarrow \text{neutralization}$$

$$\overset{H}{\underset{H}{H:\ddot{N}:}} + \overset{Cl}{\underset{Cl}{B:Cl}} \rightarrow H_3N:BCl_3$$

Examples of inorganic bases include $NaOH$, KOH, Na_2CO_3, K_2CO_3, $Ca(OH)_2$ and NH_3.

Examples of organic bases are methylamine, ethylenediamine, aniline, and pyridine.

VIVIEN GORNITZ

References

Bell, R. P., 1952, "Acids and Bases: Their Quantitative Behavior," London, Methuen; New York, John Wiley & Sons, 90pp.

Lewis, G. N., 1923, "Thermodynamics," New York, McGraw-Hill Book Co., 653pp.

Cross-references: *Acids and Bases; Buffer Systems; Ion Exchange; pH-Eh.*

BASE EXCHANGE

BASE EXCHANGE—See **ION EXCHANGE**

BERYLLIUM: ELEMENT AND GEOCHEMISTRY

Properties

Beryllium, of atomic number 4, was isolated from the mineral beryl by L. N. Vauquelin in 1797. In pure form it is a hard gray metal of atomic weight 9.013, density 1.8 (20°C), high melting point, (1283°C), and high boiling point (about 2500°C). The chemical behavior of this element is similar to that of magnesium. In compounds, beryllium forms bivalent ions of relatively small radius (0.35Å). Two natural nuclides of beryllium are known, ^{9}Be and ^{10}Be, the latter being produced by the action of cosmic rays in the atmosphere.

Minerals

Of the following selection of beryllium minerals, bery is by far the most common: beryl, $Be_3Al_2Si_6O_{18}$; bromellite, BeO; chrysoberyl, $BeAl_2O_4$; beryllonite, $NaBePO_4$.

Abundance

Beryllium is found in a variety of rocks, and is appreciably concentrated in certain veins and pegmatite bodies. The following abundance data are by Sill and Willis (1962):

Rock	Be (ppm)
13 chondrites + 1 achondrite	0.038
iron meteorite	<1
granite (G–1)	2.74
diabase (W–1)	0.623
shale	6.14
basalt	1.00

The average abundance in the earth's crust is estimated to be 2.5 ppm.

Geochemical Behavior

Beryllium is concentrated in silicate minerals relative to sulfides. In common crystalline rocks, the element is enriched in the feldspar minerals relative to ferromagnesian minerals, and apparently replaces the silicon ion; 85–98% of the total beryllium may be bound in the feldspar structures (Beus, 1961).

During the process of magmatic differentiation, beryllium is thought to be concentrated in the later stages.

Beryllium is locally concentrated in metasomatic rocks. Thus the skarn iron ores of Central Sweden contain 100 ppm Be.

The greatest known concentrations of beryllium are found in certain pegmatite bodies, where crystals of beryl comprise a few percent of the total pegmatite volume, and may be found in different zones of zoned dykes. The element is also concentrated in certain hydrothermal veins, and some granitic rocks contain sufficient amounts to cause the crystallization of small amounts of beryl.

Beryllium is enriched in the country rocks about certain beryl-containing pegmatite bodies, though the aureoles are normally narrow compared with those for lithium. Thus Kalita (1959) found 30 ppm Be in an aureole, less than 2m thick.

During the weathering of crystalline rocks, and during sedimentation processes, beryllium appears to follow the course of aluminum, and is enriched in some bauxite deposits, clays, and deep-sea deposits.

Beryllium is concentrated by some plants, but its biological role is still unknown.

RALPH KRETZ

References

Beus, A. A., 1961, "Distribution of beryllium in granites," *Geochemistry,* No. 5, 432–437 (translation of Russian).

Beus, A. A., 1966, "Geochemistry of Beryllium and Genetic Types of Beryllium Deposits," San Francisco, W. H. Freeman & Co., 401pp.

Day, F. H., 1963, "The Chemical Elements in Nature," London, George Harrap & Co., 372pp.

Kalita, E. O., 1959, "The problem of the dispersion aureoles of lithium rubidium, and beryllium," *Mat. Geol. Ore dep. Petr., Min., Geochem, Akad Sci., U.S.S.R.* (translation of Russian), 205–211.

Rankama, K., and Sahama, Th. G., 1950, "Geochemistry," Chicago, Univ. of Chicago Press, 911pp.

Sill, C. W., and Willis, C. P., 1962, "The beryllium content of some meteorites;" *Geochim. Cosmochim. Acta,* **26,** 1209–1214.

Cross-references: *Elements; Hydrothermal Alteration.* Vol. IVB: *Bauxite; Clays and Clay Minerals.* Vol. V: *Metasomatic Pegmatite.*

BIOCHEMICALS

Approximately 10^{20} metric tons of organic matter, roughly an amount equivalent to the weight of the earth, has been synthesized by living organisms during the biological history of the earth (Abelson, 1957). Whereas most of the carbon has been recycled innumerable times through various biotic cycles (e.g., photosynthesis and respiration), a small but highly significant fraction has escaped the influence of the biosphere and is preserved in sediments, sedimentary rocks, and fossil fuels.

Approximately 10^{16} tons of organic matter is believed to exist in the earth; practically all of this (>99%) is in fossil forms, namely, as kerogen, coal, and petroleum. Kerogen is by

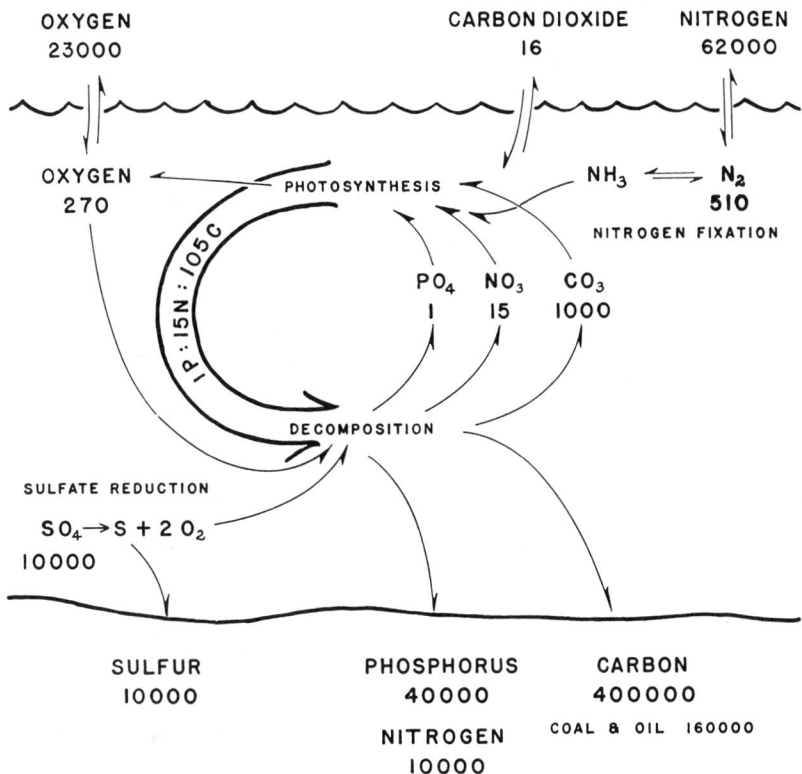

Fig. 1. The biochemical cycle of the sea. The numbers represent the quantities of the respective elements in the atmosphere, the ocean, and the sedimentary rocks, relative to the atoms of phosphorus in the ocean (Redfield, 1958).

far the most abundant form of organic matter, there being about 10^{15} tons in the earth. Coal reserves account for 10^{12} tons. The carbon tied up in fossil organic matter ($\sim 10^{15}$ tons) greatly exceeds that present as CO_2 in the atmosphere ($\sim 10^{10}$ tons).

The total amount of organic matter in living tissues is 10^{12} tons, and an equal quantity exists in terrestrial humus and sea-bottom organic compounds. The annual production of organic matter is $146 \pm 87 \times 10^9$ tons, two-thirds of which is produced in the sea (Mason, 1958, p. 217).

The history of carbon dioxide, oxygen, nitrogen, phosphorus, and sulfur in the earth is bound to organic matter synthesis and decomposition in geologic environments, as illustrated in Fig. 1 for the biochemical cycle of the sea. During the synthesis process, phosphorus, nitrogen, and carbon are selected in proportions of about 1:15:105, respectively. The carbon/oxygen ratio of newly formed organic matter, which, in the sea, is produced largely by photosynthetic microscopic plants called phytoplankton, is generally lower than 1. During the decay process, the various elements are released in about the same proportions in which they were selected, and the transformed organic matter that ultimately becomes incorporated in the bottom muds has elemental ratios similar to the initial photosynthetic product. Subsequent diagenetic changes, resulting from complex physical, chemical, and biological processes, alter considerably the character of the organic matter. The final metamorphic products (kerogen, coal) have considerably higher carbon contents, but considerably lower nitrogen and oxygen contents, than the organic materials from which they were derived.

A detailed account of the geochemistry of organic matter is covered in recent books by Breger (1963) and Colombo and Hobson (1964).

Organic Matter Production in an Aqueous Environment

Except for a relatively few sterile areas, marine and lake waters support a complex floral and faunal population. Most organic production is in the form of plants called diatoms. These and other phytoplankton live in near-surface waters, where they use the energy of sunlight for photosynthesis. They serve as the diet of grazing zooplankton, which in turn are used as food by larger carnivorous animals. As the various organisms die, their remains gradually sink through the water, largely to be consumed by scavengers. Nevertheless, considerable organic detritus reaches the bottom, to be decomposed first by worms, halothurians, crinoids, and other macroorganisms, and finally, by microscopic forms of life, such as bacteria. The first few centimeters of a typical bottom mud is a zone of intense microbial activity; several hundred million bacteria may be present per gram of sediment.

Because of the high biological activity of surface muds, the oxygen content and, hence, the Eh value drop rapidly in passing from the water into the sediment. When the organic content is high and the bottom is stagnant, Eh values of zero or less will be obtained, even in the upper layers of the sediment. A low Eh is important in preventing complete oxidation of organic matter by microorganisms.

As indicated earlier, very little of the organic matter in geologic materials resembles the organic compounds that make up plant and animal tissues; most of it occurs in metamorphized derivatives. Highly fossilized forms are represented by lignite, kerogen, coal, tar, and crude oil. The amorphous brown to black humic and fulvic acids, which are the most characteristic components of soil and marine humus, have properties intermediate between those of lignin and kerogen or coal. Biochemicals such as amino acids, carbohydrates, and lipids occur in deposits of all ages, but they usually comprise only a small fraction of the total organic matter.

The conditions in sediments formed in brackish waters appear to be particularly suitable for the formation of hydrocarbon and hydrocarbon precursors; these are transported by any of several mechanisms to porous reservoir rocks, where they are deposited. Some of the petroleum hydrocarbons may represent the segregation and accumulation of hydrocarbon material synthesized by microorganisms (see *Petroleum,* Vol. VI).

Preservation of Organic Matter

The organic content of sediments, as measured by either nitrogen or organic carbon, decreases with increasing depth in the sedimentary column. The decline is greatest in the region immediately below the water-sediment interface, where oxidative processes are high because of the presence of large numbers of aerobic organisms. After burial to the extent of only a few decimeters in a fine-grained sediment, loss of organic matter is reduced; at greater depths, where the Eh value is low, the loss becomes negligible. The anaerobic state of deep sediments is an excellent environment for the preservation of organic matter. A factor of some significance in reducing biological activity in deep-buried sediments is the low temperature of these environments. Some biochemicals may escape destruction because they occupy pores or cavities inaccessible to microorganisms.

The ability of biochemicals to survive ultimate destruction in soils and sediments is due partly to their association with metals and clay minerals (Stevenson, 1960). Some biochemicals, such as amino acids, can be held within the expanding layers of certain clays, notably montmorillonite. The sorption of organic molecules by clay is a highly selective process leading to the retention of compounds which would otherwise be consumed by microorganisms, or modified chemically. The stability of biochemicals to aging and thermal destruction may be increased through their association with mineral matter.

In terrestrial environments, where oxidative processes predominate, extensive decomposition of organic matter occurs and long-term preservation is unlikely. However, in a dry climate, such as is typical of desert regions, organic compounds may persist for long periods of time. The occurrence of organic remains in prehistoric caves, in catacombs, and in ancient ruins can be attributed to the dry conditions under which they have been preserved. The biochemicals in shells and bones are free from direct attack by microorganisms because of their location within the dense inorganic matrix.

Soil and Marine Humus

The organic matter of marine sediments and terrestrial soils consists largely of a series of high-molecular-weight, brown, nitrogenous polymers collectively referred to as humus. The most characteric fraction is humic acid, an alkali-soluble, acid-insoluble material whose acidic nature is due to carboxyl (COOH) and phenolic OH (C_6H_5—OH) groups. A detailed description of the dark colored pigments in sediments and soils is given in the entry *Humus and Humic Acids* in Vol. VI.

Biochemicals in Soils, Sediments, and Natural Waters

There is reason to believe that any biochemical which occurs as a structural component of plant or animal tissues will be present in the mineral and aquatic habitats of living

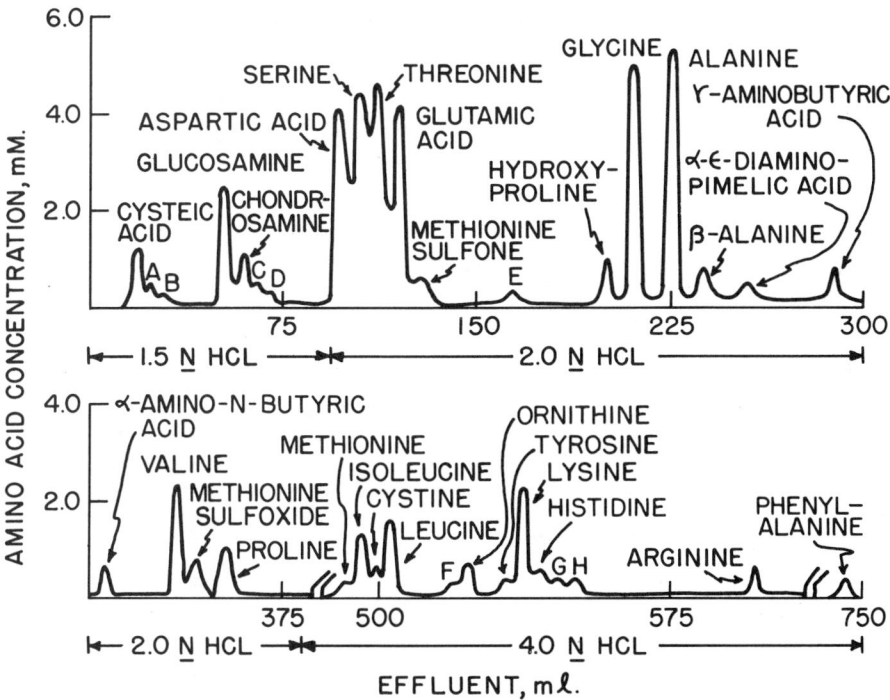

Fig. 2. Separation of amino compounds from soil using ion-exchange chromatography (Stevenson, 1960).

organisms. Thus, it is not surprising that more than 500 organic constituents have been isolated from sediments, soil, and water. They include many alcohols, aldehydes, amines, amino acids, purine and pyrimidine bases, steroids, sugars, and vitamins. Vallentyne (1957) has summarized our knowledge of the biochemicals in geologic materials.

Chromatographic techniques, such as paper partition chromatography, ion-exchange chromatography, and gas chromatography, have greatly facilitated the separation of biochemical compounds from complex geologic materials. Fig. 2 gives the results of a study in which ion-exchange chromatography was used to separate the amino compounds from a hydrolyzate of terrestrial soil. The amino acids and amino sugars that have been found in seston, fresh water, soils, and fresh-water and marine sediments are listed in Table 1.

Porphyrin Pigments. The pioneering work of Treibs in the 1930s established the presence of porphyrin pigments in crude oil and bituminous shales. These compounds were subsequently shown to be derived from chlorophyll, to which they are structurally related. Chlorophyll-degradation products have also been found in recent and ancient marine and fresh-water sediments.

Our knowledge of porphyrin pigments (see Orr et al., 1958) can be summarized as follows: (1) They are widespread in fossil organic matter, but they are more abundant in petroleum and oil shales than in coal. (2) They are preserved in fossil fuels in the form of metal complexes, but the metals are not those originally present in the plant; Mg has been replaced by Va, Ni, and other metals. (3) They are derived largely from chlorophyll. (4) The dihydrophorphyrin structure of chlorophyll has been converted to the porphyrin structure by reduction and dehydrogenation. The general nature of the chemical changes involved in the conversion of chlorophyll A to the prophyrin pigments found in sediments is shown in Fig. 3.

With regard to the metal-porphyrin complexes in petroleum, research has shown that they originated from sedimentary organic matter and that they were formed in the period of diagenesis before the petroleum moved from its source bed. This discovery is among the most significant in showing that petroleum is of biological origin.

Carotenoids. The carotenoids comprise a group of colored lipoidal substances having forty carbon atoms per molecule and a system of conjugated double bonds. About twenty different carotenoids have been found in sediments, of which only two (β-carotene and rho-

BIOCHEMICALS

TABLE 1. AMINO COMPOUNDS IN HYDROLYZATES OF SESTON, FRESH WATER, TERRESTRIAL SOILS, AND FRESH-WATER AND MARINE SEDIMENTS[a]

Amino Compound	Seston	Lake Water	Sea-water	Terrestrial Soils	Fresh-water Sediments	Marine Sediments
Neutral amino acids						
α-Alanine	+	+	+	+	+	+
β-Alanine				+		+
δ-Amino-*n*-butyric acid				+		
γ-Aminobutyric acid				+		
δ, ε-Diaminopimelic acid				+		
Glycine	+	+	+	+	+	+
Hydroxyproline	+			+		
Isoleucine				+	+	
Leucine	+		+	+	+	+
Phenylalanine	+		+	+		+
Proline	+		+	+		+
Serine			+	+	+	+
Threonine			+	+		+
Tyrosine	+	+	+	+	+	+
Tryptophane	+	+				
Valine	+		+	+	+	+
Acidic amino acids						
Aspartic acid	+	+	+	+	+	+
Glutamic acid	+	+	+	+	+	+
Basic amino acids						
Arginine	+		+	+	+	+
Histidine	+	+	+	+	+	+
Lysine			+	+	+	+
Ornithine			+	+		+
Sulfur-containing amino acids						
Cysteic acid[b]				+		
Cystine		+		+		+
Methionine			+	+		
Methionine sulfone[b]				+		
Methionine sulfoxide[b]				+		
Amino sugars						
Galactosamine				+		
Glucosamine				+		

[a] Compiled from data of Degens et al. (1964), Stevenson (1960), and Vallentyne (1957).
[b] Artifacts.

dovioloscin) have been identified with certainty. The structure of β-carotene is shown in Fig. 4.

The original discovery of carotenoids in sediments came as a surprise, because these compounds are rather unstable and can be oxidized easily. The low temperature and the absence of light and oxygen in the sedimentary environment are believed to be responsible for their

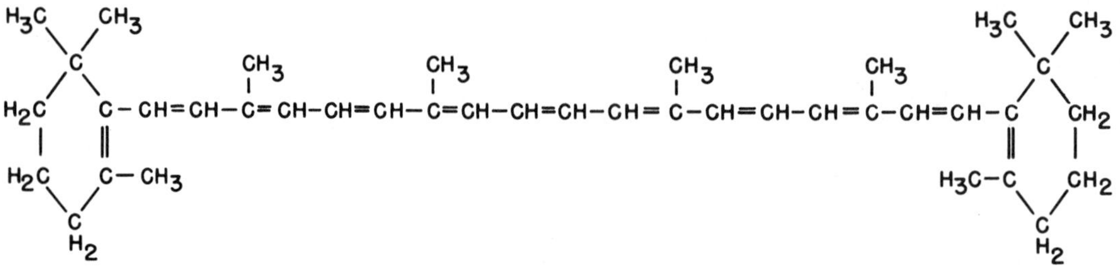

FIG. 3.

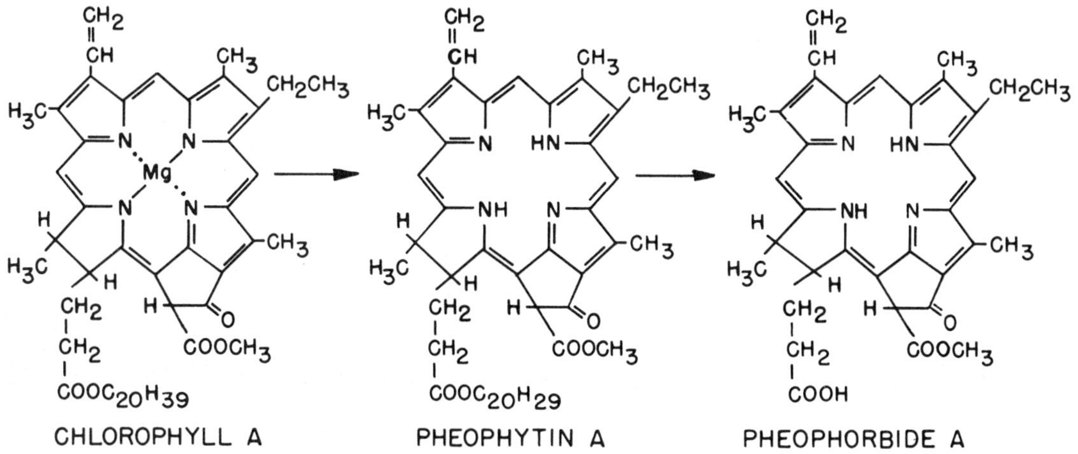

CHLOROPHYLL A → PHEOPHYTIN A → PHEOPHORBIDE A

FIG. 4. The structure of β-carotene.

ability to escape destruction. Carotenoids may be the precursors of certain aromatic hydrocarbons in petroleum.

Paleobiochemistry

Biochemicals typical of those occurring in present-day organisms have been found in fossils and sediments over 300 million years old. For example, protein and polysaccharide derivatives have been detected in sedimentary rocks, deep marine sediments, and fossil shells and bones; cellulose has been found in early Tertiary lignites; and chitin has been found in Eocene remains of insects. The biochemicals in fossils and ancient sediments are expected to provide information about the biochemistry of ancient organisms, the age and thermal history of sediments, and the depositional environments of lakes and oceans during their long histories. Precambrian fossil biochemicals may provide clues to the origin of life. The subject of paleobiochemistry has been reviewed by Abelson (1957) and by Jones and Vallentyne (1960).

It is likely that most, if not all, extinct creatures used amino acid building blocks in performing their biochemical functions. Because most of these compounds have appreciable chemical and thermal stabilities, they are able to survive rather drastic conditions of deposition and burial. Experiments conducted at the Geophysical Laboratory of the Carnegie Institute of Washington, D.C. show that the pattern of amino acids in fossils follows a stability sequence. For example, in a study (Abelson, 1957) in which the amino acids of a recent shell of the clam *Mercenaria mercenaria* were compared with those of a fossil representative of the same species that was 25 million years old, the fossil representative was found to contain only those amino acids which had been shown by thermal studies to be relatively stable; namely, alanine, glutamic acid, glycine, leucine or isoleucine, proline, and valine. In addition to these amino acids, the recent shell contained aspartic acid, lysine, phenylalanine, serine, threonine, and tyrosine. All of the last-named amino acids were found to have relatively low thermal stabilities.

F. J. STEVENSON

References

Abelson, P. H., 1957, "Some aspects of paleobiochemistry," *Ann. N.Y. Acad. Sci.* **69**, 276–285.

Bernfeld, P. (editor), 1963, "Biogenesis of Natural Compounds," New York, Macmillan, 930pp.

Breger, I. A., 1963, "Organic Geochemistry," Oxford, Pergamon Press Ltd.

Colombo, U., and Hobson, G. D., 1964, "Advances in Organic Geochemistry," Oxford, Pergamon Press Ltd.

Degens, E. T., Reuter, J. H., and Shaw, K. N. F., 1964, "Biochemical compounds in offshore California sediments and sea waters," *Geochim. Cosmochim. Acta,* **28**, 45–66.

Jones, J. D., and Vallentyne, J. R., 1960, "Biogeochemistry of organic matter. I. Polypeptides and amino acids in fossils and sediments in relation to geothermometry," *Geochim. Cosmochim. Acta,* **21**, 1–34.

Mason, B., 1958, "Principles of Geochemistry," Second ed., New York, John Wiley & Sons, Inc.

Orr, W. L., Emery, K. O., and Grady, J. R., 1958, "Preservation of chlorophyll derivatives in sediments off Southern California," *Bull. Am. Assoc. Petrol. Geol.,* **42**, 925–962.

Redfield, A. C., 1958. "The biological control of chemical factors in the environment," *Am. Scientist,* **46**, 205–221.

Stevenson, F. J., 1960, "Some aspects of the distribution of biochemicals in geologic environments," *Geochim. Cosmochim. Acta,* **19**, 261–271.

BIOCHEMICALS

Vallentyne, J. R., 1957, "The molecular nature of organic matter in lakes and oceans, with lesser reference to sewage and terrestrial soils," *J. Fisheries Res. Board Can.*, **14**, 33–82.

Cross-references: Calcium Cycle; Carbon Cycle; Cycles—Geochemical; Hydrocarbons; Nitrogen Cycle; Oxygen Cycle; pH-Eh; Phosphorus Cycle; Sulfur Cycle. Vol. I: Marine Microbiology; Marine Sediments; Phytoplankton. Vol. IVB: Clays and Clay Minerals. Vol. VI: Coal; Diagenesis; Fossil Fuels; Humus and Humic Acids; Kerogen; Pelagic Sediments; Petroleum.

BIOGEOCHEMICAL PROSPECTING—
See GEOBOTANICAL AND BIOGEOCHEMICAL METHODS OF PROSPECTING FOR MINERALS *in* Vol. IV B

BIOGEOCHEMISTRY

The term "biogeochemistry" was introduced by the Russian geochemist Vernadsky to describe the study of organisms in relation to mineral formation. The original interpretation has been broadened by Hutchinson (1954), Baas Becking, Vinogradov, and other workers in the field to mean the chemical study of interactions between the biosphere on the one hand and the atmosphere, lithosphere, and hydrosphere on the other. At one extreme, the science merges into ecology and at the other it becomes a specialized form of geochemistry, soil science, or oceanography. In the last decade, biogeochemical investigations have been undertaken on meteorites to determine if living processes may have occurred in the parent planet from which the meteorite fragment was derived.

The mass of the biosphere is smaller than the atmosphere, hydrosphere, or upper lithosphere by factors of 300, 7×10^4, and 10^6 respectively. Since organisms are in dynamic equilibrium with their environment, the actual mass is in no way an indication of their importance for surface processes. Just as a parasite, having only an infinitesimal weight as compared with the host, may cause the death of its host, organisms acting as catalysts may completely change the characteristics of an environment. We must consider not only the assimilatory processes which directly accumulate components into cells, but also dissimilatory processes, generally related to energy metabolism, which may cause a thousandfold turnover of a particular element compared to total cell weight, during the lifetime of the organism.

Accumulation of Elements in Organisms

Marine Biosphere. Oceans cover 71% of the earth's surface and the euphotic zone extends to about 100 m beneath the air-water interface. Total productivity in the oceans has been considered as greater than that in the terrestrial environment. Various estimates give the total annual biomass in the ocean as varying between 300 and 400 (metric) tons/km^2.

Organisms living in the ocean are able to selectively enrich certain elements in their protoplasm. Table 1 gives a comparison of analytical data for seawater and an estimate for the marine biosphere based on the elemental analysis of copepods and lamellibranchs. It can be seen that in this case lead, cadmium, zinc, and manganese are highly enriched. Other groups may preferentially enrich other elements. For example, it is well known that tunicates accumulate unusually high amounts of vanadium.

The mechanism of uptake of trace elements is generally complex. It may be due both to an active metabolic function, for example the formation of hemocyanin in crustacea or hemovanadin in tunicates, or a passive indirect process. It is now known that many enzymes require trace elements as cofactors, and these may accumulate in the cell. On the other hand, many organisms secrete gelatinous membranes rich in organic ligands (proteins of marine animals are high in amines) which may form complexes with transition elements by scavenging them out of solution. For these reasons, transition elements are enriched in invertebrates much more than are alkali or alkaline earth elements.

Another way in which organisms may concentrate trace elements, is by physical adsorption on surfaces. This is important in the case of bacteria and nanoplankton having a total small volume but large surface area. Larger animals, both benthic and nectonic may further concentrate trace elements in waste products deposited on the sediment surface.

After death of an organism, the portions not consumed by predators or decomposed by microorganisms settle to the sediment-water boundary. Some elements are fixed tightly, and subsequently are buried by normal sediment accumulation. Others may be released and will be resolubilized, thus enriching the bottom ocean waters. Such a case has been found for barium, where an increased concentration in bottom water has been measured. Elements that are rapidly removed from seawater will have a short residence time in the ocean, and Table 1 shows this to be true for most transition elements.

Terrestrial Biosphere. This biosphere can be subdivided into a solid component (lithosphere) and aqueous component. Since the volume of fresh water is only about 0.04% that of sea-

TABLE 1. ELEMENTARY COMPOSITION OF THE MARINE BIOSPHERE COMPARED WITH THE HYDROSPHERE

Element	Abundance in Marine Biosphere[a] (ppm)	Abundance in Seawater[b] (ppm)	Residence Time[b] ($\times 10^6$ yr)	Enrichment Factor
O	800,000	857,000	—	0.93
H	102,100	108,000	—	0.95
Cl	10,500	19,000	—	0.55
Na	5,400	10,500	260	0.51
Mg	300	1,350	45	0.22
S	1,400	885	21	1.58
Ca	400	400	8	1.0
K	2,900	380	11	7.63
B	9	65	—	0.14
C	61,000	28	—	2,178
Si	70	3	0.008	23.3
N	15,200	0.5	—	30,400
P	1,300	0.07	—	7.00
I	2	0.06	—	33.3
Fe	70	0.01	0.00014	7,000
Zn	250	0.01	0.18	25,000
Mo	1	0.01	0.5	100
Cu	15	0.003	0.18	5,000
V	3	0.002	0.01	1,500
Ni	5	0.002	0.018	2,500
Mn	50	0.002	0.0014	25,000
Cd	10	0.0001	0.5	100,000
Ag	0.5	0.00004	2.1	125
Pb	5	0.00003	0.002	167,000

[a] Data from Mason (1958) and Brooks and Rumsby (1965).
[b] Data from Goldberg (1965).

water, it is relatively unimportant as a medium for biological activity. The processes described in the previous section, however, may also apply here. Terrestrial productivity has been estimated at 160 tons/km² annually. This may be low, since the productivity of soil microflora is difficult to estimate.

Here, as in the marine biosphere, some elements are preferentially concentrated. Most of the elements found in plants and animals serve a specific purpose: for example, calcium and phosphorus in bone material; carbon, hydrogen, nitrogen, oxygen, and sulfur as invariable constituents of organic compounds in general; transition metals as enzyme cofactors. As in marine organisms, some elements occur in land organisms without serving any apparent purpose. Table 2 compares elemental composition of a mammal, man *(Homo sapiens)* and a plant, alfalfa grass *(Medicago sativa)*.

Compared with animals, plants usually show greater concentrations of manganese, nickel, aluminum, titanium, and boron. On the other hand, animals are relatively enriched in calcium, iron, zinc, and copper.

The absence of a single trace element may convert an otherwise arable soil into an infertile desert. Addition of molybdenum to barren soils of South Australia in the 1950s allowed the establishment of legumes. Likewise, a deficiency in cobalt may prevent ruminants from synthesizing vitamin B12 and may result in a disease known as "bush sickness."

Soil Formation

The conversion of a lithified surface into a soil profile is largely due to biological activity. Plants will physically initiate decomposition of rocks by forming root systems. The humus accumulating on the ground will be decomposed leading to the formation of organic acids, often referred to as "humic acids," rich in heterocyclic phenols. Rock-forming minerals may be solubilized, ultimately leading to a change in texture and the formation of clays.

Elements from subsoils will often be solubilized and concentrated in the A(upper) horizon. In humid tropical environments, lateritic soils may thus form through a concentration of iron oxide. It has also been suggested that alumina-rich soils or clays (bauxite) may be residual after the removal of the silica by plants.

Biogeochemical and Geobotanical Prospecting

Since plants are able to accumulate significant quantities of certain elements, attempts

TABLE 2. ELEMENTARY COMPOSITION OF THE TERRESTRIAL BIOSPHERE COMPARED WITH THE LITHOSPHERE

Element	Abundance in Lithosphere (ppm)	Abundance in *Homo sapiens* (ppm)	Enrichment Factor	Abundance in *Medicago sativa* (ppm)	Enrichment Factor
O	466,000	628,100	1.35	779,000	1.67
Si	277,200	40	0.00014	93	0.00033
Al	81,300	0.5	0.000006	25	0.000307
Fe	50,000	50	0.001	27	0.00054
Ca	36,300	13,800	0.38	5,800	0.16
Na	28,300	2,600	0.09	200	0.007
K	25,900	2,200	0.09	1,700	0.07
Mg	20,900	400	0.02	810	0.04
Ti	4,400	—	—	1	0.0002
H	1,400	93,100	66.5	87,200	62.3
P	1,180	6,300	5.33	7,010	5.98
Mn	1,000	1	0.001	4	0.004
S	520	6,400	12.30	1,037	1.99
C	320	193,700	605.3	113,400	354.37
Cl	200	1,800	9.0	700	3.5
Rb	120	9	0.07	5	0.04
V	110	0.03	0.002	0.2	0.0018
Ni	80	0.03	0.00037	0.5	0.006
Zn	65	25	0.38	4	0.06
N	46	51,400	1117	82,500	1793
Cu	45	4	0.09	25	0.56
Co	23	0.04	0.0017	0.02	0.00087
Pb	15	0.5	0.03	0.5	0.03
Sn	3	2	0.67	1	0.33
Br	3	2	0.67	—	—
B	3	0.2	0.067	7	2.33
Mo	1	0.2	0.20	1	1.00
I	0.3	1	0.03	0.03	0.10

have been made in recent years to exploit this fact for prospecting for minerals.

Much of the pioneer work was carried out in the Soviet Union (Malyuga, 1964), in Canada, and in the United States.

Two approaches to this problem have been made. One technique (biogeochemical prospecting) relies on direct analysis of plant material (usually by spectrochemical analysis) to delineate areas of anomalous mineralization. The other technique (geobotanical prospecting) relies on visual observation of the nature and appearance of the vegetation. A classical example of the latter is the ready identification of serpentine areas by the peculiar sparse and often endemic vegetation growing over it. Regions of lead mineralization can often be traced by the appearance of the ground cover. In the United States, geobotanical prospecting has been successfully applied to the search for uranium in the Colorado plateau by studying the distribution of various species of *Astragalus*.

There is much speculation as to the relative merits of soil or plant analyses in biogeochemical prospecting. Some authorities claim that plant analysis, though more difficult, affords a more representative sample of trace elements in the soil and that deep-rooted species can often give indications of buried ore deposits which normal soil analyses would not reveal.

The technique is at present more advanced in the Soviet Union where the fact that a large part of the country is completely frozen over for much of the year tends to give emphasis to methods which do not involve sampling soils or drilling holes.

Biological Control of the Chemical Environment

Beside the direct removal of constituents as outlined in the previous section, organisms play an important role in controlling the chemistry of the environment. One extreme example known to most readers is the pollution of water reservoirs, either by effluent from sewage or large-scale death of the biomass. The latter may occur either through cyclic blooming and death of plankton, or by liberation of toxins, as in the case of blue-green algae and dinoflagellates, which results in the poisoning of specific metazoans.

Because of their numerical abundance, high rate of reproduction and tolerance of varied environmental conditions, microorganisms are the most important biogeochemical agents. Three broad types of metabolism are carried out by algae, fungi, bacteria, and protozoa which fall into this class. The primary metabolic function is *photosynthesis,* using solar energy to fix carbon dioxide into cellular components and liberating molecular oxygen through the decomposition of water. In *respiration,* molecular oxygen is used to oxidize organic compounds to CO_2 and H_2O, by which process the organism derives energy and synthesizes the necessary structural building blocks. The third type of metabolism is *fermentation,* where the organisms use oxidized forms of organic matter and convert them to less oxidized forms, liberating a variety of organic compounds (e.g., alcohols, ketones, amines) besides carbon dioxide. Photosynthetic organisms are restricted to shallow environments, respiratory organisms to oxygenated environments, and fermentors to oxygen-free environments such as sediments.

Since metabolism always involves either the formation or decomposition of water and the transfer of electrons, biological activity can often be estimated by measurement of the hydrogen ion concentration (pH) and the oxidation-reduction potential (Eh). The two are related for any one reaction of the type

$$M + H_2O \rightleftharpoons MOH + H^+ + e^-$$

by the following general equation:

$$Eh = E^0 - \frac{0.059}{n} pH + \frac{0.059}{n} \log \left[\frac{\text{oxidized M}}{\text{reduced M}} \right]$$

where M is either a metal or compound, E^0 is the standard electrode potential at zero pH (and 25°C), and n is the number of electrons partaking in the reaction.

Such measurements can readily be made by use of a pH meter with a glass electrode for hydrogen ion concentration and a platinum electrode for oxidation reduction potentials. Fig. 1 taken from Bass Becking, Kaplan, and Moore (1960) shows the relationship between measurements made on laboratory cultures of various microorganisms and those made in natural aqueous environments. The close overlap is very obvious.

Extremes in environmental conditions can be formed by all three major types of metabolic functions. The removal of CO_2 by photosynthesis moves the equilibrium away from bicarbonate to carbonate, causing a rise in pH. In unbuffered tidal pools or lagoons, where photosynthesis has occurred, the pH may rise above 10.0 at the end of a day. In such cases the Eh will remain high. On the other hand, certain bacteria are able to oxidize reduced compounds through the use of oxygen. The thiobacilli (or sulfur bacteria) will oxidize sulfides, such as pyrite, to sulfates liberating H^+ and dropping the pH to 1.0 or below. This process is common in soils and mines, where acid mine waters become a hazard to operations. Under oxygen-free conditions, a variety of reactions occur generally at neutral pH. Reducing substances such as methane, phenols, or hydrogen sulfide react with dissolved oxygen, maintaining a low Eh. Since many elements have solubility properties based on their valence state at specific pH values, mineral mobilization or deposition will occur under biological influence.

Deposition of Minerals and Fuels

Minerals. Indisputable evidence for the biological deposition of minerals comes from direct observation of skeletal material on the ocean floor, which is covered to the extent of about 1.7×10^8 km² by calcium carbonate and silica from protozoa and algae. Silica is dominant in Antarctic Ocean sediments and carbonate is prevalent in equatorial sediments, especially at depths above 4500 m, the "compensation point" where high pressure and low temperature work to redissolve calcium carbonate. Estimates indicate that the limestone and dolomite content of the lithosphere is equal to about 13–14 kg/cm² of the earth's surface. Aragonite and calcite are also formed by algal activity, either as a structural support for cells or by increase in pH during photosynthesis and precipitation. This activity has recently been shown to result in the precipitation of dolomite in arid climate lagoons, such as in South Australia and the Persian Gulf. Freshwater carbonates or tufas are also formed by algae, especially near springs. Opal or geyserite often forms in hot-spring areas associated with blue-green algae.

Sulfides are among the most abundant biogenic minerals. It is definitely known that pyrite (iron disulfide) forms in newly deposited sediments as a result of sulfate reduction by bacteria to form hydrogen sulfide which reacts with iron. Pyrite is a common constituent of shale and limestone and often reaches the quality of an ore. Other metal sulfides exist but the question of whether they are a biogenic product is still open. Elemental sulfur, often associated with limestone, gypsum, and anhydrite is considered to be the product of the oxidation of hydrogen sulfide which was derived from bacterial sulfate reduction in the presence of petroleum. These are often asso-

BIOGEOCHEMISTRY

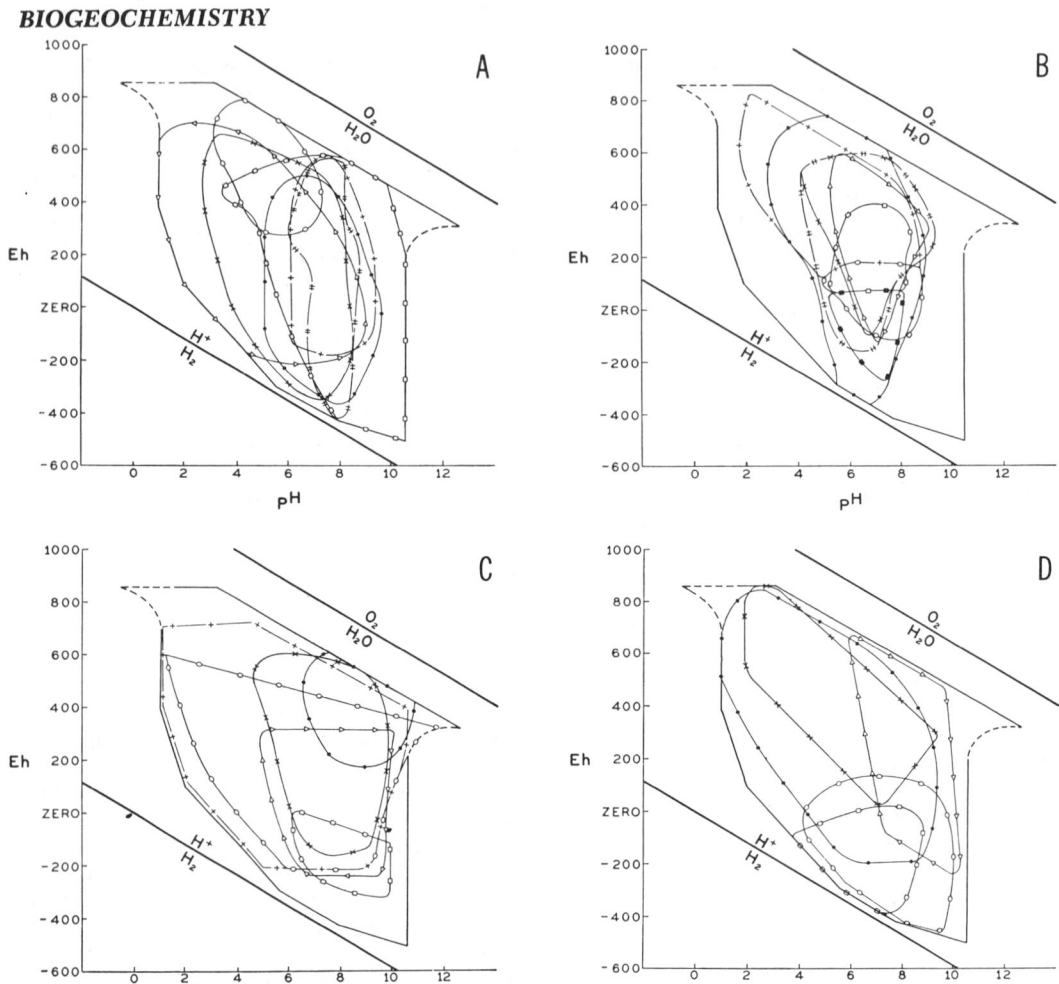

FIG. 1. (A) Approximate "areas" of Eh and pH for some natural environments. The "area" for each environment is bounded by a different symbol: ○ ○ ○ = meteoric water; × × × = peat bogs; ● ● ● = marginal marine sediments; + + + = seawater; ╪ ╪ ╪ = open-sea sediments; □ □ □ = evaporites; △ △ △ = geothermal environment. (B) Approximate "areas" of Eh and Ph for some natural environments. The "area" for each environment is bounded by a different symbol: ● ● ● = soils; △ △ △ = shallow ground water; + + + = oxidized mine water; ○ ○ ○ = primary mine water; × × × = fresh water; ╪ ╪ ╪ = fresh water sediments; □ □ □ = oxidized connate water; ■ ■ ■ = uncontaminated connate water. (C) Approximate "areas" of Eh and pH for some photosynthetic organisms. The "area" for each environment is bounded by a different symbol: ○ ○ ○ = green algae and diatoms; ● ● ● = *Dunaliella*; × × × = *Enteromorpha*; + + + = blue-green algae; △ △ △ = purple bacteria; □ □ □ = green bacteria. (D) Approximate "areas" of Eh and pH for some geologically important bacteria. The "area" for each environment is bounded by a different symbol: ○ ○ ○ = sulfate-reducing bacteria; ● ● ● = thiobacteria; × × × = iron bacteria; △ △ △ = denitrifying bacteria; □ □ □ = heterotrophic anaerobic bacteria.

ciated with "salt domes," such as are common on the Gulf Coast of the southern United States. From 1895–1957 over 130×10^6 tons were mined in this area alone (see Feely and Kulp, 1957).

Several other mineral deposits are thought to have been produced either directly or indirectly by microbiological activity. These include bog iron ore and uranium in the terrestrial environment; phosphate, barite, uranium oxide, and perhaps manganese nodules in marine sediments. No satisfactory evidence is yet available to confirm these claims.

Fuels. Three forms of fossil fuels exist: coal; petroleum; and light hydrocarbon gases. All of these are considered to have been formed through the agencies of biological systems.

Coal is recognized to be the result of rapid burial of plant material. The major coal deposits on earth appear to have been formed in a coastal marsh environment. Rapid burial of the organic matter prevented decay by fungi and other aerobic organisms. Phenols, terpenes, and other components of plant decay further prevented rapid microbial decay leading to preservation of the lignins. Ion-exchange and adsorption processes during the life of the plant and later during decay, enabled several elements to be concentrated in the resulting peat. These include boron, germanium, arsenic, bismuth, beryllium, cobalt, nickel, cadmium, lead, silver, selenium, gallium, molybdenum, and uranium.

Petroleum is generally considered to have been formed in a marine environment, since it occurs in marine sediments associated with marine fossils. Since petroleum-like compounds have been found in old lake deposits and occasionally associated with either volcanic or metamorphic rocks, the marine genesis hypothesis has been questioned as a unique environment of formation. In recent sediments, paraffinic hydrocarbons in the range C_{16}–C_{24} are abundant and are thought to have been derived from the decarboxylation of long-chain fatty acids. Petroleum is thought to have been formed after burial of hydrocarbons and other complex organic components which have escaped decomposition through preservation under reducing conditions.

Light hydrocarbon gases probably have two origins. Those composed of a complex of isomers from methane to C_6 and especially containing benzene are the result of a cracking of heavier petroleum fractions at high temperatures. Methane is often the primary not sole gas in a reservoir. This could have formed through biological degradation of organic matter (cellulose) involving methane bacteria. Heating of coal, however, will also result in the liberation of methane as the major light gas.

Isotope and Organic Geochemistry

Two approaches have contributed greatly to the study of biological deposits (synonyms), making possible a reconstruction of some of the events leading to their formation and accumulation.

Stable-Isotope Fractionation. Experimental evidence has accumulated in the last twenty years to show that organisms are capable of carrying out fractionation of the isotopes of light elements, in particular, hydrogen, carbon, nitrogen, oxygen, and sulfur. Certain marine organisms, in particular brachiopods, will deposit their calcitic shell in equilibrium with the seawater in which they live and the isotopic composition of the calcium carbonate is dependent on the temperature of growth. Hence analysis of the O^{18}/O^{16} ratio of CO_2 in the *unaltered* shell, even hundreds of millions of years after death of the organism, will indicate the temperature of the sea during the period of shell formation. This is the basis of the "O^{18}-carbonate paleothermometer."

Generally, organisms will enrich the light isotope either in their cell tissues or in the metabolic products as a result of a kinetic reaction in which the light isotope reacts most rapidly. The result is that photosynthetic organisms are enriched by about 1 to 2% in C^{12} relative to the starting C^{12}/C^{13} ratio in atmospheric or dissolved carbon dioxide. Degradation of the organic matter will lead to a further enrichment in the light isotope in the regenerated CO_2. Repeated cycles in a closed system may lead to enrichments of 8–9% (80–90‰) in C^{12}. During cell synthesis, the light isotope (either H or C^{12}) is preferentially concentrated in lipids and since petroleum is also isotopically enriched in C^{12} (about 30‰) relative to marine carbonate and bicarbonate, it is thought that petroleum may be largely derived from fats. Table 3 shows isotopic fractionation by bacteria.

Sulfate-reducing bacteria are capable of enriching the product, hydrogen sulfide, in S^{32} by 45‰ during a single-step reduction. This method of enrichment is thought to have left behind the heavier isotope S^{34} in seawater sulfate, enriching it by 20‰ with respect to sulfur in meteorites. Sulfides are enriched on an average by 12‰ in S^{32} relative to meteoritic sulfur. Oxidation of sulfur to sulfate leads to only slight enrichment in S^{32}. In this way it is possible to determine the nature of the starting material in a mixture of sulfur minerals. This technique has been used to show the biological origin of sulfur in salt domes and in Sicilian and several other elemental sulfur deposits (Feely and Kulp, 1957).

Organic Geochemistry. Combined with other studies reviewed in this chapter, organic geochemistry is an important tool for defining the environment of deposition. The nature of organic compounds preserved in the lithified rock may answer the following questions: Did the organic matter originate from biogenic or abiogenic processes? Was the environment fresh water or marine? Did the organic matter form in situ or is it the result of petroleum seepages?

Methods have been developed to extract traces of organic matter on the order of parts per million or less. The presence of amino acids, steroids, photosynthetic pigments, and

BIOGEOCHEMISTRY

TABLE 3. ISOTOPE FRACTIONATION BY ORGANISMS AND IN NATURALLY OCCURRING SUBSTANCES

Isotopic Ratios	Fractionation Factor (°/₀₀)	Standard	Organism or Substance
N^{15}/N^{14}	−8.0 to +8.0	Atmospheric nitrogen	Living matter
	−3.0 to +2.0		Coal
	+1.0 to +16.0		Crude oil
	+1.0 to +17.0		Sedimentary rocks
	−11.0 to +16.0		N_2 in natural gas
O^{18}/O^{16}	−41.0 to +6.0	Mean ocean water	Fresh water
	+6.0 to +15.0		Volcanic waters
	+6.0 to +12.0		Igneous rocks
	+10.0 to +15.0		Sandstones and shales
	+28.0 to +37.0		Marine carbonates
	+28.0 to +32.0		Diatomite
	+40.0 to +42.0		Atmospheric CO_2
	+23.0 to +45.0		Dissolved O_2 in ocean
	+21.0 to +23.0		Atmospheric O_2
C^{13}/C^{12}	+2.5 to −3.5	Marine carbonate	Marine limestones
	+2.0 to −4.0		Recent shells
	−1.5 to −3.0		Ocean HCO_3
	−7.5 to −9.5		Atmospheric CO_2
	−8.0 to −18.0		Marine plants and animals
	−21.0 to −32.0		Trees
	−21.5 to −27.0		Coal
	−22.5 to −27.5		Fossil wood
	−23.0 to −30.0		Petroleum
S^{34}/S^{32}	+10.0 to +35.0	Meteoritic troilite	Evaporite sulfates (marine)
	+28.0 to −40.0		Sedimentary sulfides
	+28.0 to −17.0		Biogenic sulfur
	+23.0 to −17.0		Marine plants and animals
	+19.5 to +20.5		Marine sulfate
	+5.0 to −25.0		Biogenic hydrogen sulfide
	−2.0 to −15.0		Volcanic sulfur
	−10.0 to −45.0		Bacterial sulfate reduction

isoprenoids point to biological origins. The use of optical rotatory dispersion tends to confirm this. Humic acid, derived from lignins, generally suggests the presence of plants and hence terrestrial, rather than marine, conditions of formation. Presence of complex aromatic compounds and light gases would suggest a petroleum seepage. Thus a combination of several chemical techniques can tremendously assist the more established geological methods for reconstructing ancient environments.

Biogeochemical Control of Elemental Cycling

Life on earth as we know it depends on an overall constant composition of the various geospheres. A rapid change in atmospheric composition would probably affect most metazoans. Many marine organisms require strict open-ocean conditions, where variations in environment are at a minimum; plant tolerances have already been described earlier. The reverse is also true; living matter controls the composition of the earth's crust.

It has been estimated that 3.0×10^{10} tons/year of carbon is exchanged. If this figure represented only a fixation of carbon by photosynthesis, the total CO_2 in the atmosphere would disappear in about 30 years and the dissolved ocean bicarbonate in 4000 years. On the other hand, if this value represented a release of carbon dioxide by respiration it would necessitate the removal of oxygen from the atmosphere in 22,000 years. These time spans are of course negligible in the earth's history.

The elements probably most affected by biological cycles are carbon, oxygen, sulfur, nitrogen, phosphorus, and iron. Calcium, silicon, and trace elements are less affected. Nitrogen is very rapidly cycled since it has no large reservoir beside the atmosphere, because most nitrogen minerals stable on the earth are soluble. Phosphorus has only one important valence state under terrestrial conditions (orthophosphate), although it is vitally important as a nutrient for all forms of life, its cycling depends on uptake and release by solution. Both nitrogen and phosphorus may control productivity if they become limiting, and then indirectly may affect other elements. Iron has two valence states that are interchangeable; it may therefore act as a reducing or oxidizing agent. Because of the low solubility of iron oxides, the ionic concentration of iron in natural water is low, but it rarely becomes limiting. Iron may have played an important role in the earth's early history in controlling the atmospheric oxygen content.

Sulfur is important in that it is an abundant element on earth as well as having several valence states from -2 in sulfides to $+6$ in sulfate. The reduced form usually exists in nature as pyrite in shales or other sediments, and the oxidized form is in gypsum or anhydrite. Weathering will leach the former into the ocean converting it to the latter, whereas in the sediments, sulfate is reduced by bacteria and converted ultimately to pyrite.

Oxygen in the atmosphere is generally considered to be largely the result of photosynthesis. Since burial (or conservation) of organic matter must result in a net accumulation of oxygen, the question as to the location of this oxygen is of considerable interest. The amount in the atmosphere, 1.3×10^{15} tons, is too small to be accounted for by the amount of buried carbon in sediment. One possibility is that inorganic reactions which result in sul-

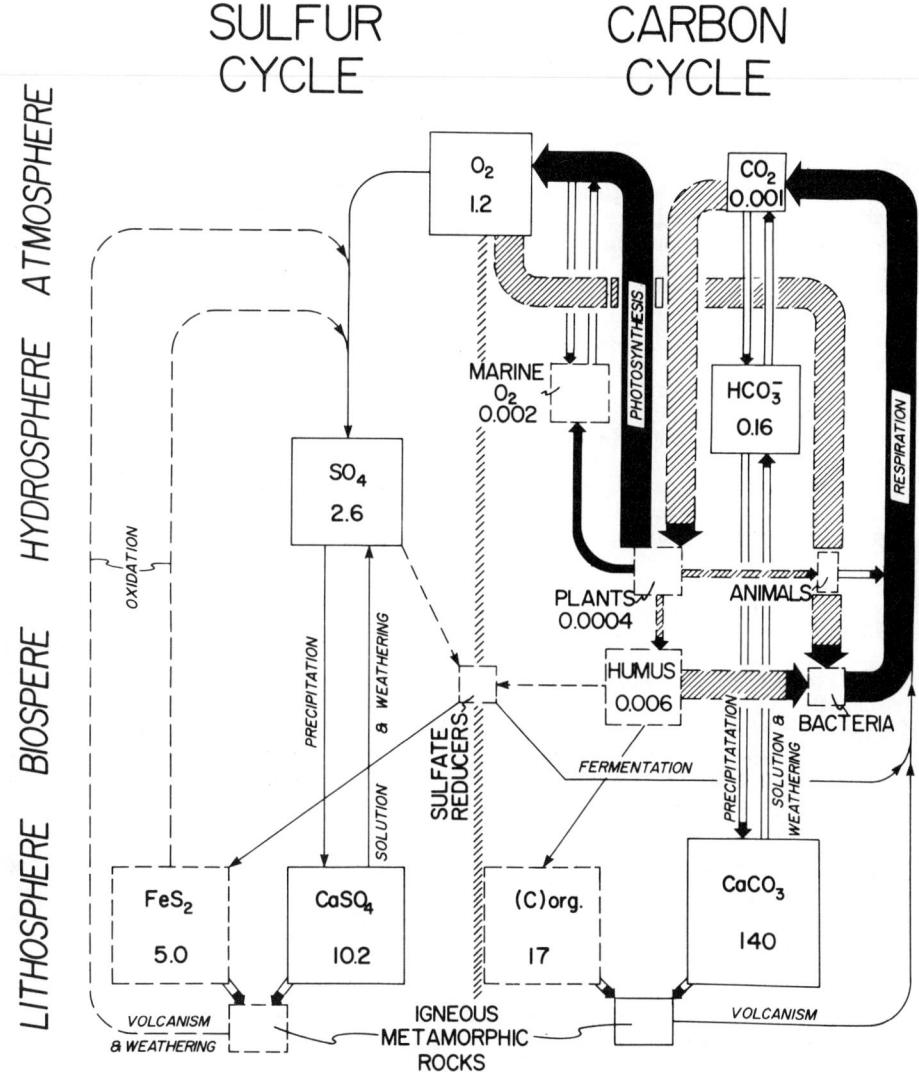

Fig. 2. Interaction between the biogeochemical sulfur and carbon cycles and their mode of controlling the oxygen cycle. Line thickness denotes the approximate magnitude of annual flow rates. Numerical values signify total oxygen contents in 10^{15} metric tons either as molecular oxygen, oxygen sink (FeS_2), or equivalent amount of oxygen released (Corg).

BIOGEOCHEMISTRY

fate and iron oxide formation bind the oxygen. Shifts in any one of the cycles, i.e., the cycles of iron, sulfur, carbon, or oxygen, may affect the others. Thus changes in the earth's history which may have caused excessive leaching of reduced sulfur and iron, added to periods of volcanism, might have placed a heavy demand on atmospheric oxygen; this may have resulted in a decrease. The interaction of the carbon-oxygen-sulfur cycles is shown in Fig. 2.

Meteorites and Exobiology

The search for life on other planets has demanded the application of many of the principles of biogeochemistry. Studies have been carried out on meteorites to identify "organized elements" which may represent fossil organisms and culturing bacteria. Organic components have been studied with the view of determining a biogenic origin. Plans for space exploration demand either the direct observation of life or an attempt to establish if some kind of life ever existed on the extraterrestrial body. Such studies will ultimately help us determine the evolutionary processes that led to the first appearance of life on earth.

R. R. BROOKS and I. R. KAPLAN

References

*Baas Becking, L. G. M., Kaplan, I. R., and Moore, D., 1960, "Limits of the natural environment in terms of pH and oxidation-reduction potentials," *J. Geol.*, **68**, 243–284.
*Brooks, R. R., and Rumsby, M. G., 1965, "The biogeochemistry of trace element uptake by some New Zealand bivalves," *Limnol. Oceanog.*, **10**, 521–527.
Feely, H. W., and Kulp, J. L., 1957, "Gulf coast salt-dome sulfur deposits," *Bull. Am. Assoc. Petrol. Geol.*, **41**, 1802–1853.
*Goldberg, E. D., 1965, "Minor Elements in Sea Water," in (Price, J. P., and Skirrow, G., editors), "Chemical Oceanography," Vol. I, pp. 163–194, New York, Academic Press.
Hutchinson, G. E., 1954, "The biogeochemistry of the Terrestrial Atmosphere," in (Kuiper, G. P., editor). "The Earth as a Planet," pp. 371–433, Chicago, Univ. Chicago Press.
*Malyuga, D. P., 1964, "Biogeochemical Methods of Prospecting," New York, Consultants Bureau.
ᵏMason, B., 1958, "Principles of Geochemistry," New York, John Wiley & Sons.

* Additional references may be found in this work.

Cross-references: Carbon Cycle; Carbon Isotope Fractionation; Cycles—Geochemical; Hydrocarbons; Isotope Fractionation; Lagoon Geochemistry; Organic Geochemistry; Oxidation and Reduction; Oxygen Cycle; pH-Eh; Photosynthesis; Sulfate Reduction—Microbial; Trace Elements. Vol. I: *Marine Microbiology; Marine Sediments; Manganese Nodules.* Vol. II: *Extraterrestrial Life; Meteorites.* Vol. IVB: *Geobotanical and Biogeochemical Methods of Prospecting for Minerals.* Vol. VI: *Coal; Fossil Fuels; Petroleum; Soil Genesis.* pr. Vol. VII: *Algae; Dinoflagellates; Protozoa.*

BIOGEOCHEMISTRY, PELAGIC—See Vol. I

BIOMETEOROLOGY—See Vol. II

BISMUTH: ELEMENT AND GEOCHEMISTRY

Physical Properties

Bismuth (Ger., *Weisse Masse,* white mass, later Wismuth), symbol Bi, atomic weight 208.980; atomic number 83; electronic configuration (Xe core) $4f^{14}5d^{10}6s^26p^3$; melting point 271.3°C; boiling point 1560°C; specific gravity 9.75 (20°C); valence 3 or 5. Crystal modification: hexagonal metallic form (ditrigonal scalenohedral class). Isotopic abundances: Bi^{209}, 100%; several other isotopes occur as daughter products in the decay chains of uranium and thorium. Bismuth is a brittle, diamagnetic, reddish white metal (often with brassy tarnish colors), and is a poor conductor of heat and electricity.

Minerals and Compounds

Bismuth is commonly found in the native state and often contains traces of arsenic (As), sulfur (S), and tellurium (Te). The most important bismuth mineral is bismuthinite (Bi_2S_3) which resembles stibnite (Sb_2S_3) very closely. Bismuth is not abundant and is usually found in veins associated with silver (Ag), cobalt (Co), lead (Pb), tin (Sn), or zinc (Zn) ores. Rarer minerals include bismuthotantalite ($BiTaO_4$), tetradymite (Bi_2Te_2S), guanajuatite [$Bi_2(S,Se)_3$], matildite ($AgBiS_2$), and aramayoite [$Ag(Sb,Bi)S^2$]. Important localities are Frieberg, Saxony; Joachimstal, Bohemia; Bolivia; Cornwall, England; and Great Bear Lake, Canada. The most important compounds are the trioxides and the subnitrates of medicinal use. Its soluble salts are characterized by the formation of insoluble basic salts on addition of water.

Chemistry

The chemical properties of bismuth resemble those of antimony and arsenic. Chemical Bi is stable in air at room temperature, but burns, when heated, to yield Bi_2O_3. At elevated temperatures, it combines with the halogens, sulfur, and a variety of metals. Bi is unreactive toward monoxidizing acids.

$$Bi + H_2O \rightarrow BiO^+ + 2H^+ + 3e^-$$
$$E^0 = -0.32$$

Bi has a strong tendency to assume the $+3$ condition. The $+5$ state is very rare; Bi_2O_5 has not been prepared in the pure condition. Bi^{5+} is a very strong oxidizing agent in acid solution.

$$Bi_2O_5 + 6H^+ + 4e^- \rightarrow 2BiO^+ + 3H_2O$$
$$E^0 = \sim 1.60$$

Treatment of alkali metal bismuthates ($MBiO_3$) with HNO_3 gives a red-brown substance approximating Bi_2O_5 in composition, which compound loses considerable oxygen, even at temperatures below 100°C. Orthobismuthates (M_3BiO_4) are obtained by heating the sesquioxide with appropriate amounts of alkali metal oxide (or peroxide) in air.

A complete series of Bi trihalides is known; all trihalides can be formed by direct combination of the elements. Bi trihalides are partially and reversibly decomposed with the formation of insoluble compounds containing BiO^+ (bismuthyl) cations.

$$BiX_3 + H_2O \rightarrow BiO^+ + X^- + 2H^+ + 2X^-$$

Complexes are known. Bi trihalides yield salts of the types $M[BiX_4], M_2[BiX_5]$, and $M_3[BiX_6]$. Additional compounds, mainly of the 1:1 variety, are formed with organic donor molecules. Bi(III) chloride yields such compounds with tertiary amines. At least two crystalline forms of Bi(III) oxides are known. The compounds exhibit no acidic tendencies and are insoluble in alkalis but dissolve easily in acids to form Bi(III) salts.

The addition of OH^- to solutions of Bi salts gives $Bi(OH)_3$, which is completely basic. The hydrated cationic species $Bi(H_2O)^{3+}$ is present in solution when the sesquioxides are dissolved in concentrated solutions of strong oxyacids. Normal salts such as $Bi_2(SO_4)_3$ or $Bi(NO_3)_3 \cdot 5H_2O$ can be crystallized from such solutions, but they hydrolyze easily.

Geochemistry

There is a dearth of geochemical information concerning Bi and a general outline of the geochemistry of Bi is available only from general principles and mineralogical experience. Bismuth is a distinctly chalcophile element, but it also possesses certain lithophile characteristics. Only small amounts of bismuth are found in rocks and minerals belonging to the early and main stages of magmatic crystallization; it becomes strongly enriched in late magmatic differentiations. Little data is available on the concentration of Bi in the various members of the sequences of magmatic rocks. It is known to be captured in apatites and probably in other early calcium minerals. Bi tends to be concentrated in granitic magmas and is common in granitic pegmatites. In dolerite, concentrations of 0.01–0.80 ppm have been reported. Other rock types and ranges are: syenite (0.02–0.04 ppm), rhyolite (0.02–0.22 ppm), eclogite (0.02–0.10 ppm), and basalt (0.02 ppm).

Bismuth is probably widely distributed in minerals of yttrium and yttrium lanthanides, and in hydrothermal mineral deposits associated with granitic magmas. Bismuth minerals are also known in pneumatolytic deposits associated with volcanic action. Bismuthinite is the most important hydrothermal mineral of Bi. In high-temperature hydrothermal deposits, Bi is often associated with gold (Au), Te, and selenium (Se). It is most commonly present in galena.

Little is known of the behavior of bismuth during weathering. Observations are limited to the oxidation zones of ore deposits. Products of oxidation are bismite, hydrocarbonates, vanadates, and, rarely, silicates of Bi. No oxidation state in Bi minerals higher than 3 is known. Bismuth is enriched to some degree in sedimentary iron ores and in the hydrolyzates and can be detected in some reduzates. Residual sediments are essentially devoid of bismuth.

Various estimates of the abundance of bismuth in the upper lithosphere have been made; however, the few data available cluster around a figure of 0.2 ppm for a crustal abundance.

Economic Geology

In all known deposits of economic import, Bi is found in sulfide minerals or in secondary alteration products of sulfides and related compounds. Commercial ore minerals of bismuth are native Bi, bismuthinite, and bismuth ochre. Few deposits are mined exclusively for Bi. Bi is commonly produced as a by-product of lead refining and is also derived from foreign and domestic ores of Sn, copper (Cu), and Ag.

Bismuth minerals occur in fissure veins and in replacement deposits formed by hydrothermal solutions. Major producing countries are Bolivia, Canada, Japan, Mexico, Peru (about one-quarter of world's supply), South Korea, and the United States.

Extraction of Bi requires a complex process to eliminate other metals. Ores are reduced in furnaces with carbon (C) or iron (Fe) and soda ash. The Bi in Pb is dissolved in HCl as an oxychloride and smelted. Refinery slimes are fused with caustic soda and soda ash. Bismuth metal is consumed in small quantities.

Its grestest uses are in metallurgical, chemical, and medical applications.

ERNEST E. ANGINO

References

Bateman, A. M., 1950, "Economic Mineral Deposits," New York, John Wiley & Sons, 916pp.

Brooks, R. R., and Ahrens, L. H., 1961, "Some observations on the distribution of thallium, cadmium, and bismuth in silicate rocks and the significance of covalency on their degree of association with other elements," *Geochim. Cosmochim. Acta,* **25,** 100–115.

Goldschmidt, V. M., 1954, "Geochemistry," London, Oxford University Press, 730pp.

Kleinburg, J., Argersinger, W. J., and Griswold, E., 1960, "Inorganic Chemistry," Boston, D.C. Heath and Co., 680pp.

Rankama, K., and Sahama, Th. G., 1950, "Geochemistry," Chicago, University of Chicago Press, 911pp.

Cross-references: *Alkalis, Alkali Metals, and Alkaline Earth Metals; Hydrothermal Solutions.* Vol. IVB: *Metallurgy.*

BITUMEN, BITUMINOUS SEDIMENTS—
See Vol. IV B

BONDING

The properties and behavior of minerals, both normally and under conditions of stress, reflect the orientation and strength of bonds between the component atoms. Both of these factors reflect the structure of the individual atoms themselves (see *Atomic Number and the Periodic Table*).

Briefly, an atom consists of a central nucleus composed of protons and neutrons which is surrounded by shells of electrons. In general, bonding in minerals may be examined effectively by concentrating on the role of the valence electrons and merely keeping in mind that the main mass of the atom is concentrated in the small central nucleus. Here are located the positively charged protons and the electrically neutral neutrons, each with an atomic mass of 1. The electrons each have a mass of 1/1835 atomic mass and a negative charge. Only electrons may be detached from their original atom (except in the case of nuclear reactions which form atoms of new materials). Because all electrons are identical in properties they may readily replace one another around any atom.

In any bond formation, the atoms involved strive to complete their outer orbital. This may be accomplished by removing one or more electrons from the outer shell to form a positively charged ion or by filling vacancies in the outer shell to form a negatively charged ion. This removal or addition of electrons may be more or less complete.

There are five types of bonds: (a) ionic (polar), (b) covalent (homopolar), (c) metallic, (d) van der Waals (molecular), and (e) hydrogen. Bonds are rarely completely of one type; however, one form of bonding is usually dominant.

Ionic Bonding

In ionic bonding, an electron from the outer orbital of one atom moves into the outer orbital of another atom. Thereafter, the two atoms are held together by electrostatic attraction.

Characteristically the atoms in groups I and II of the periodic table are donors of electrons and the atoms in groups VI and VII are receptors of electrons. Fig. 1, NaCl, shows how such bonding occurs. The single electron in the third (m) valence shell of Na is readily available for bonding. The Cl atom needs only one electron to complete its outer shell. Bonding occurs when the electron in the third (m) electron shell of Na moves into orbit around the Cl nucleus. The Na$^+$ ion and the Cl$^-$ ion now both possess full outer valence shells. This ionization of the atoms leaves the Na atom with an electrical charge of $+1$ and the Cl atom with an electrical charge of -1.

Because the Na$^+$ has two shells (k and l) and the Cl$^-$ has three shells (k, l, and m) occupied by electrons there is a size disparity between the two ions belonging to the same period. This size disparity resulting primarily from the number of valence shells containing electrons is accentuated by the charged nature of the atoms. Because the eleven protons in the Na nucleus outnumber the ten electrons in the Na$^+$ ion, their attraction pulls the orbiting electrons slightly closer to the nucleus. Conversely, the eighteen electrons orbiting in the Cl$^-$ ion outnumber the seventeen protons in the Cl nucleus and slightly expand the volume occupied by their orbits. The attraction of the

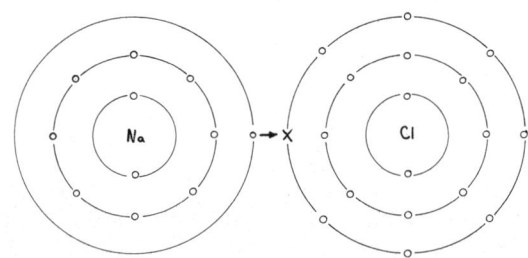

FIG. 1. Ionic bonding of Na to Cl by transfer of one electron from the outer orbital of the Na atom to the vacancy in the outer orbital of the Cl atom.

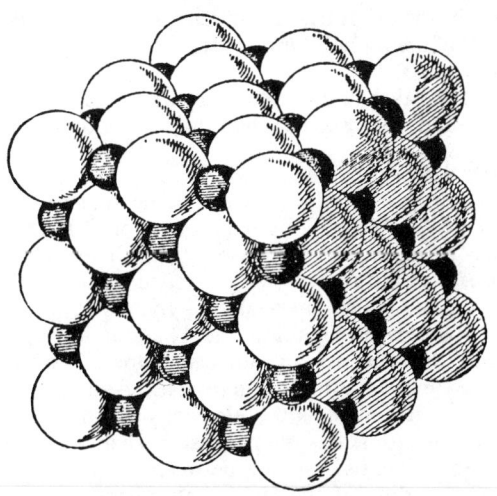

FIG. 2. Halite crystal bonded by ionic bonding. The large spheres are Cl^- ions; the small spheres are Na^+ ions (after Barlow, 1898).

Cl protons cannot completely counterbalance the mutual repulsion of the negatively charged electrons.

Fig. 2 shows the arrangement of Na^+ and Cl^- ions in a crystal of halite. Each ion is a separate entity held in position adjacent to oppositely charged ions only by their electrical attraction. Ionic bonding is found in evaporite minerals and fluorite.

Characteristically, materials formed by ionic bonding may be simple or complex but are usually transparent, brittle, weak, and have low refractive indexes and melting points. The crystals are poor conductors of heat and electricity, but the melts (characterized by low viscosities) are good conductors of heat and electricity (by ionic migration). These properties reflect the nondirectional character of ionic bonds.

Covalent Bonding

Unlike atoms united by ionic bonding, covalently bonded atoms share electrons. These electrons orbit around both the donor and one receptor atom. The location of the shared electron cannot be determined at any given instant, but the relative amount of time spent orbiting each atom can be determined.

This shuttling back and forth of electrons between atoms brings the atoms into much closer proximity than is the case with ionic compounds. In the extreme case of O–H bonding the single proton nucleus of H can be thought of as a bump on the surface of the O atom. This is, of course, not completely accurate; however, it is sufficiently true to account for many properties of OH^- and H_2O.

Atoms of the middle groups of the periodic table, the amphoteric elements, are characterized by covalent bonding. The energy required to completely remove or to add three or four electrons is greater than is commonly available for chemical reactions. By sharing these valence electrons however, the energy needed to satisfy bonding requirements is substantially reduced. Usually these atoms are coordinated with three, four, or six neighboring atoms.

Among the amphoteric elements are Si and C which combine with O to form the basic building units of most of the rocks on Earth. Si and C are both members of Group IV of the atomic table; they may bond either by losing or by gaining four electrons. Si usually bonds with four O atoms in tetrahedral coordination. These SiO_4^{4-} tetrahedra may then satisfy their electrical requirements by sharing varying numbers of O atoms and fulfilling remaining electrical requirements by adding electron donors (e.g., Fe, Na, Ca) to their structures. Similarly, CO_3^{2-} bonds covalently to Ca and Mg in the limestones and dolomites of Earth.

Matter in all three states, gaseous, liquid, and solid, may be organized by covalent bonding. The common molecules in air (e.g., N_2, O_2, CO_2) are united by covalent bonds. Water is composed of covalently bonded H_2O molecules which may be partially ionized by dissolved matter (e.g., Na^+ and Cl^-) in the area.

In general, covalently bonded materials are good thermal and electrical insulators. These transparent materials have high refractive indexes. The crystals are brittle and hard (e.g., diamond) and melt at high temperatures to form very viscous liquids. These crystals are more compressible than those formed by ionic bonding but less compressible than metallically bonded crystals. Covalently bonded crystals frequently are characterized by variation in their properties (e.g., index of refraction) with direction. These variations result from the formation of covalent bonds between atoms located at specific angles to one another.

Metallic Bonding

Unlike ionic and covalent bonding, metallic bonding does not involve temporary or complete loss of valence electrons from orbitals around a particular atom. No single atom holds valence electrons rigidly to itself; all adjacent atoms hold a group of valence electrons communally. Consequently, the electrons are very loosely held in what is sometimes called the "electron cloud." The atoms sharing these electrons approach each other more closely than atoms involved in ionic or covalent bonding. Each metallically bonded atom may be surrounded by a fairly large number of other atoms (i.e., have a high coordination number).

Elements in the lower left region of the periodic table are usually involved in metallic bonding. These elements have low electronegativities (and electropositive character).

When an energy imbalance exists in a metallically bonded system (e.g., one end of a rod receives an input of electrons in an electrical circuit), valence electrons can move readily through the system to attempt to bring about equilibrium. Therefore, the metals are good conductors of heat and electricity. Metallically bonded crystals are soft, malleable, ductile, and opaque to visible and much infrared radiation. They melt at relatively low temperatures to form opaque liquids with low viscosities. These properties reflect the relative weakness of metallic bonding and its lack of specific directional orientation.

Van der Waals Bonding

Van der Waals bonds are weak bonds between atoms and molecules due to electrical attraction. There is neither transfer nor sharing of electrons between atoms.

The force of attraction between two permanent dipoles (materials with a positive and a negative end, e.g., the H_2O molecule is positive at the end where the two H nuclei lie at about $105°$ to one another on the surface of the O atom and negative at the outer end) is inversely proportional to the sixth power of the distance between them. The same relationship obtains for neutral atoms or molecules and for permanent dipoles and dipoles induced by them in other molecules.

The van der Waals attraction between any two atoms, therefore, is very weak. However, between groups of atoms in molecules and particles the attractive force will be the sum of the attractive forces between all the pairs of atoms present. This total force may be of importance.

Van der Waals bonding is found in crystals formed by covalently bonded material (e.g., between layers of talc and pyrophyllite). The direction of these bonds is determined by the shape of the component covalent molecules.

The crystals formed by van der Waals bonds are poor conductors of heat and are electrical insulators. They are soft, weak, plastic, and very compressible. These crystals, as would be expected of weakly bonded materials, have low melting points.

Liquids characterized by van der Waals bonds are electrical insulators and poor conductors of heat. They are highly compressible and have viscosities dependent upon the shape of the component molecules (usually low to moderate).

Clay flakes and sedimentary organic material may both be linked by van der Waals bonding.

Hydrogen Bonding

Hydrogen bonding, as the name suggests, is a bond formed only by hydrogen atoms. Rather than a linking of two atoms by a shared electron, it is linkage of two electronegative atoms (e.g., F,O,N,Cl) by a shared hydrogen atom.

In hydrogen bonding the H atom is covalently bonded to an adjacent atom. At the same time the intense electrostatic field around the H^+ ion attracts the electrons of one other adjacent atom. A system is established wherein the three atoms resonate between covalent and ionic bonding of the H to each adjacent atom. Invariably the H approaches one atom more closely than the other. It therefore is involved in covalent bonding to its nearest neighbor for a greater percentage of time than to the other atom involved.

The bond is weak (the bond energy usually ranges from two to ten kcal/mole) and because of its low bond energy is suited to reactions occurring at temperatures found on the surface of the earth. Hydrogen bonding occurs in ice and diaspore as well as in organic substances.

Summary

Chemical bonding is the linking of atoms into molecules and crystals by the sharing or transfer of electrons or by electrostatic attraction. All types of bonding result from the attraction of oppositely charged particles, both subatomic particles and atoms and molecules.

Although covalent and ionic bonding are the most prevalent in geologic systems, the systems involving hydrogen (e.g., ice), van der Waals (e.g., clays), and metallic (e.g., gold) bonding are also extremely important to the understanding of our mineral surroundings.

EMMY BOOY

References

Barlow, W., 1898, "Geometrische Untersuchung über eine mechanische Ursache der Homogenität der Struktur und der Symmetrie; mit besonderer Anwendung auf Krystallisation und chemische Verbindung," Z. Krist., 29, 433–588.

Moore, W. J., 1962, "Physical Chemistry," Englewood Cliffs, N.J., Prentice-Hall, Inc., 846pp.

Pauling, L., 1939, "The Nature of the Chemical Bond," Ithaca, New York, Cornell Univ. Press, 644pp.

Smith, F. G., 1963, "Physical Geochemistry," Reading, Massachusetts, Addison-Wesley Publ. Co., 624pp.

Cross-references: *Atomic Number and Periodic*

Table; Electronegativity; Van der Waals Force; Water. Vol. IVB: *Evaporite Minerals; Refractive Index.*

BORATES

In recent years the borate minerals have been studied in considerable detail and although complex, much detail is now known about their structures. The most common structure consists of $(BO_3)^{3-}$ anionic groups, with all of the oxygens shared, and the boron situated in triangular coordination, like carbon in the carbonate minerals.

The triangles are often arranged in the form of sheets, which may have water molecules between them. Another arrangement is $(B_2O_4)^{2-}$

Anhydrous Borates

Ludwigite	$(Mg,Fe^{2+})_2Fe^{3+}BO_5$	} *Ludwigite group:* an orthorhombic series with complete substitution of Mg and Fe
Paigeite	$(Fe^{2+},Mg)_2Fe^{3+}BO_5$	
Pinakiolite	$Mg_3Mn^{2+}Mn_2^{3+}B_2O_{10}$	
Hulsite	$(Fe^{2+},Ca,Mg)_4(Fe^{3+},Sn^{4+})_2B_2O_{10}(?)$	
Warwickite	$(Mg,Fe)_3TiB_2O_8$	
Kotoite	$Mg_3(BO_3)_2$	
Jimboite	$Mn_3(BO_3)_2$	
Rhodizite	$NaKLi_4Al_4Be_3B_{10}O_{27}(?)$	
Jeremejevite	$AlBO_3$	
Nordenskioldine	$CaSn(BO_3)_2$	

Hydrous Borates

Pinnoite	$Mg(BO_2)_2 \cdot 3H_2O$
Kernite	$Na_2B_4O_7 \cdot 4H_2O$
Tincalconite	$Na_2B_4O_7 \cdot 5H_2O$
Borax	$Na_2B_4O_7 \cdot 10H_2O$
Priceite	$Ca_4B_{10}O_{19} \cdot 7H_2O\ (?)$
Probertite	$NaCaB_5O_9 \cdot 5H_2O$
Ulexite	$NaCaB_5O_9 \cdot 8H_2O$
Veatchite	$Sr_3B_{16}O_{27} \cdot 5H_2O$
Colemanite	$Ca_2B_6O_{11} \cdot 5H_2O$
Hydroboracite	$CaMgB_6O_{11} \cdot 6H_2O$
Inderborite	$CaMgB_6O_{11} \cdot 11H_2O$
Meyerhofferite	$Ca_2B_6O_{11} \cdot 7H_2O$
Inyoite	$Ca_2B_6O_{11} \cdot 13H_2O$
Kurnakovite	$Mg_2B_6O_{11} \cdot 13H_2O$
Inderite	$Mg_2B_6O_{11} \cdot 15H_2O$
Howlite	$Ca_2SiB_5O_9(OH)_5$
Bakerite	$Ca_4B_4(BO_4)(SiO_4)_3(OH)_3H_2O$
Paternoite	$MgB_8O_{13} \cdot 4H_2O$
Ginorite	$Ca_2B_{14}O_{23} \cdot 8H_2O$
Larderellite	$(NH_4)_2B_{10}O_{16} \cdot 5H_2O\ (?)$
Ammonioborite	$(NH_4)_2B_{10}O_{16} \cdot 5H_2O\ (?)$
Kaliborite	$KMg_2B_{11}O_{19} \cdot 9H_2O$
Sussexite	$Mn_2(B_2O_5) \cdot H_2O$ } *Sussexite group:* an orthorhombic series with probably complete substitution of Mg and Mn
Szaibelyite	$Mg_2(B_2O_5) \cdot H_2O$

Borates With Hydroxyl or Halogen

Fluoborite	$Mg_3(BO_3)(F,OH)_3$
Hambergite	$Be_2(BO_3)(OH)$
Teepleite	$Na_2B_2O_4 \cdot 2NaCl \cdot 4H_2O$
Bandylite	$CuB_2O_4 \cdot CuCl_2 \cdot 4H_2O$
Roweite	$(Mn,Mg,Zn)Ca(BO_2)_2(OH)_2$
Boracite	$Mg_3B_7O_{13}Cl$
Hilgardite	$Ca_8(B_6O_{11})_3Cl_4 \cdot 4H_2O$
Parahilgardite	$Ca_8(B_6O_{11})_3Cl_4 \cdot 4H_2O$

Compound Borates

Lunebergite	$Mg_3B_2(OH)_6(PO_4)_2 \cdot 6H_2O$
Cahnite	$Ca_2B(OH)_4(AsO_4)$
Sulfoborite	$Mg_6H_4(BO_3)_4(SO_4)_2 \cdot 7H_2O$
Seamanite	$Mn_3(PO_4)(BO_3) \cdot 3H_2O$

anionic groups which form chains. A third coordination is in the form of tetrahedra of $(BO_4)^{5-}$.

Because of the different structures present, the properties of the borate minerals vary greatly. Most are white or colorless, light, and soluble in water to some degree and, as a result, they are most common in desert regions. (See Table on p. 87.)

LLOYD W. STAPLES

References

Christ, C. L., 1960, "Crystal chemistry and systematic classification of hydrated borate minerals," *Am. Mineralogist,* **45,** 334.

Dana, J. W., and Dana, E. S., 1951, "The System of Mineralogy," Seventh ed., Vol. II, 320, (revised by Palache, C., Berman, H., and Frondel, C.), New York, John Wiley & Sons.

Schaller, W. T., 1930, "Borate minerals from the Kramer district Mohave Desert, Calif.," *U.S. Geol. Surv. Profess. Paper* **158.**

Tennyson, Ch., 1963, "Eine Systematik der Borate auf kristallchemischen Grundlage," *Fortschr. Mineral.,* **41,** 64.

Cross-references: *Boron; Mineral Classes: Nonsilicates.*

BORON: ELEMENT AND GEOCHEMISTRY

Element Data

Boron was discovered in 1808 independently by Sir Humphry Davy, Gay-Lussac, and Thenard. Symbol B; atomic weight 10.82; atomic number 5; melting point 2300°C; sublimes at 2550°C; specific gravity: crystal 2.34; amorphous 2.37; valence 3; electronic configuration $1s^2, 2s^2, 2p^1$. The element is obtained by heating boron trioxide with magnesium powder. Isotopic abundances: B^{10}, 18.98% B^{11}, 81.02%.

In chemical compounds boron (ionic radius 0.22 Å) is always trivalent, and has properties analogous to those of carbon and silicon. Therefore boron may be classified as biophile and lithophile.

Boron forms nonionic bonds with oxygen, resulting in two types of oxyanions, those in which boron has a coordination number of three (BO_3^{3-}) and those with coordination number four (BO_4^{5-}). The BO_3^{3-} anion occurs principally as soluble alkali and alkaline earth borates, and as the sparingly soluble boric acid (H_3BO_3). Four-coordinated boron is found in small amounts in many silicates (e.g., micas, amphiboles, clay minerals). More extensive boron-silicon diadochy yields the borosilicates, the most important of which is tourmaline.

Boron also forms a series of hydrolyzable, volatile boron trihalogenides, some of which may occur in volcanic gases (e.g., BCl_3).

Boron is important biochemically because it may replace carbon in fats and sugars. It is commonly concentrated in oil field waters, coal, and in the remains of some plants and animals.

Minerals

Silicates. Boron forms few silicate minerals, the more important of which are listed below.
Danburite, $CaO \cdot B_2O_3 \cdot 2SiO_2$; *datolite,* $HCaBSiO_5$, found as a secondary mineral in veins and cavities of basic volcanic rocks often associated with calcite, prehnite, zeolites, and danburite; *axinite,* approx. $(Ca,Mn,Fe)_7(Ca,Mg,H)_4 B_2(SiO_4)_8$, in igneous rocks and contact zones surrounding igneous intrusions; *tourmaline,* $(Na,Ca)(Li,Mg,Fe,Al)_3(Al,Fe^{III})_6 B_3Si_6O_{27}(O,OH,F)_4$, a typical pegmatite mineral and the most important borosilicate; *dumortierite,* $(Al,Fe)_7BSi_3O_{18}$, found in igneous and metamorphic rocks.

Sassolite or Boric Acid $B(OH)_3$. Sassolite occurs as volcanic deposits near fumaroles and in the waters of hot springs.

Evaporites. Evaporites of boron include colemanite, $Ca_2B_6O_{11} \cdot 5H_2O$; borax, $Na_2B_4O_7 \cdot 10H_2O$; kernite, $Na_2B_4O_7 \cdot 4H_2O$; and ulexite, $NaCa_5B_5O_9 \cdot 8H_2O$.

Igneous Processes

During magmatic crystallization boron is incorporated only to a limited extent into the lattices of silicate minerals; most of the common igneous rocks contain rather insignificant amounts (<10 ppm) of boron. (Serpentinized ultrabasics form an exception. It has been proposed that for these rocks boron has been added by autometamorphic processes.) Magmatic boron is enriched in residual fluids, as attested to by the almost ubiquitous occurrence of tourmaline in pegmatites. Magmatic residual fluids may escape and allow boron to find its way to the earth's surface via volcanic exhalations or waters from thermal springs. These processes have contributed most of the boron presently found at the earth's surface, for the combined boron contents of sediments and marine waters exceed by some twentyfold the amount that could have been derived from weathered igneous rocks.

Sedimentary Processes

Goldschmidt and Peters first pointed out that the geochemistry of boron is dominated by marine processes (Goldschmidt, 1954). Seawater contains 4.6 ppm B, largely present as undissociated boric acid. The boric acid and associated sodium borate constitute an important buffer system in marine waters, second in importance only to the carbonate system.

Boron is extracted from marine waters by sedimentation. Small amounts are probably taken out by the formation of syngenetic or authigenic tourmaline, but the most important removal process consists of the substitution of boron for silicon in the clay minerals, particularly in the potassium clays illite and glauconite. Illites deposited from normal marine waters contain from 400 to 600 ppm B; consequently marine shales contain approximately 100 to 200 ppm B. The boron content of carbonate rocks is small (a few ppm) and is controlled by the illite content.

Sandstones contain boron mainly as detrital tourmaline, although authigenic overgrowths about detrital nuclei are probably common. Evaporite sediments contain abundant and varied suites of boron minerals such as hydrated borates of calcium, sodium, and magnesium (colemanite, borax, kernite).

An interesting aspect of boron geochemistry lies in the discovery that the boron content of illite is related to the boron content (and hence salinity) of the water from which the illite was deposited. This principle is the basis for a geochemical method of determining paleosalinity (for a review and bibliography see Reynolds, 1965).

Relative paleosalinities are determined by measuring the boron content of illite from sedimentary rocks. The sample is disaggregated and the clay-size fraction is collected to ensure minimal contamination by detrital tourmaline, mica, and feldspar. The sample is analyzed for boron and potassium. If x-ray diffraction studies indicate the absence of other potassium-bearing minerals, the following relation is used:

$$\text{ppm B in illite} = \text{ppm B in sample} \times \frac{7.7}{\% K_2O \text{ in sample}}$$

In this expression, the value 7.7 is used as the percent K_2O in a pure, hypothetical illite. For some samples, a correction may be required for the presence of clay-sized, authigenic potassium feldspar.

Paleosalinity values (ppm B in illite) measured in this fashion show good correlations with known paleogeographic situations. For example, studies of sediments collected adjacent to ancient reefs commonly show large, consistent differences which allow the clear separation of the fore and back-reef facies. Rocks collected from fossiliferous, presumably normal marine beds, provide a narrow range of values whose mean is approximately 450 ppm B in illite.

Boron contents in the vicinity of 2000 ppm have been observed from illites that are associated with evaporites; values as low as 125 ppm B have been measured in illites from brackish water deposits (Upper Cretaceous coal-clay sequences).

Metamorphic Processes

Diagenesis and metamorphism of boron-containing sediments cause an extensive redistribution of boron. Small amounts of boron are probably lost during diagenesis, and at greenschist facies conditions, boron leaves the illite lattice and appears in the megascopic fraction of the rock as tourmaline porphyroblasts. At metamorphic conditions above the stability limit of white mica (the sillimanite zone) there appears to be a widespread destruction of tourmaline with concomitant boron mobilization and loss from the system.

Economic Geology

Boron is concentrated in evaporite salts of marine origin or in desert playas. The boron that accumulated in continental evaporites was assumed to have been derived from the weathering of volcanic rocks. Recently, it has been suggested that the boron is leached from marine sediments, resulting in localized enrichment of these solutions.

The industrial boron compounds are obtained from bedded deposits beneath old playas, brines of saline lakes and marshes, and also encrustations around playas and lakes. Such encrustations formerly supplied large amounts of boron in Death Valley, California. Today, Searles, Borax, and Kramer Lakes in California are the only borate-producing localities in the U.S. The borates of economic importance include borax, kernite, colemanite, and ulexite. The U.S. produces 93% of the world's boron. In Argentina, second in world production after the U.S., ulexite is mined as layers and nodules in playas. Italian boron deposits are unique. There, boric acid occurs in volcanic steam vents. Evaporation of the steam yields borax.

Uses

Borax and its derivatives have many diverse applications. In the household, it makes a good cleanser, either directly or in soaps. Medically, it is a mild antiseptic. It is an ingredient in baking powder and food preservatives. Many industrial processes require borax. Among them are such unexpected uses as prevention of rancidity in cosmetics, pastes, and glues, and disease in sugar beets and celery. It is also used as a glaze in glazed paper and as fireproofing for wood and textiles. These are just a few examples.

A comprehensive study of the geochemistry of boron has recently been published in a series of papers by H. Harder (see references in Reynolds, 1965).

ROBERT C. REYNOLDS, JR.

References

Christ, C. L., and Garrels, R. M., 1959, "Relations among sodium borate hydrates at the Kramer deposit, Boron, Calif.," *Am. J. Sci.* **257**, 516–528.

Gates, G. R., 1959, "Clay mineral composition of borate deposits and associated strata at Boron, Calif.," *Science*, **130**, 102.

Goldschmidt, V. M., 1954, "Geochemistry," Oxford Univ. Press.

Lerman, A., 1966, "Boron in clays and estimation of paleosalinities," *Earth Sci. Rev.*, **2**(1).

Levinson, A. A., and Ludwick, J. C., 1966, "Speculation on the incorporation of boron into argillaceous sediments," *Geochim. Cosmochim. Acta*, **30**(9), 855–861.

Milton, C., and Eugster, H. P., 1959, "Mineral Assemblages of the Green River Formation (Colorado-Utah-Wyoming)," in (Abelson, P.H., editor), "Researches in Geochemistry," 118–150, New York, John Wiley & Sons.

Muessig, S. J., 1959, "Primary minerals in playa deposits—minerals of high hydration," *Econ. Geol.*, **54**, 495–501.

Reynolds, R. C., 1965, "The concentration of boron in Precambrian seas," *Geochim. Cosmochim. Acta*, **29**, 1.

Schaeffer, R., 1968, Muetterties, E. L. (editor), Review, "The Chemistry of Boron and Its Compounds," *Science*, **160**(3831), 982–983.

Cross-references: *Alkalis, Alkali Metals, and Alkaline Earth Metals; Carbon; Gases, Volcanic; Paleosalinity; Silicon; X-Ray Diffraction Analysis.* Vol. III: *Playa.* Vol. V: *Fumarole.* Vol. VI: *Diagenesis; Evaporites: Hot Springs.*

BORON GEOCHEMISTRY IN MARINE ENVIRONMENTS

Introduction

In 1932 (see references in Shaw and Bugry, 1966), Goldschmidt and Peters demonstrated that boron was concentrated in marine argillaceous sediments. From this observation they argued that clays absorbed boron from seawater during deposition so that boron concentration might be used to discriminate between marine and nonmarine sediments. The pioneer work of Goldschmidt and Peters stimulated other investigations which attempted to prove or disprove that much of the boron variation in sediments was salinity dependent. These investigations, together with general studies of trace elements in sediments, have provided a great wealth of quantitative data. Unfortunately, these data have proved difficult to interpret and many conflicting theories have been advanced to account for boron enrichment and variation in marine sediments. Indeed, the only fact that is generally agreed upon today is that boron is concentrated in marine sediments.

Boron in Tourmaline

The average boron concentration in sandstone is 35 ppm (Turekian and Wederpohl, 1961; see reference in Moore, 1963), but in Recent sands from Buzzards Bay, Massachusetts, average boron concentration is 65 ppm and locally as high as 158 ppm (Moore, 1963). Boron in the bay sands is inversely related to tidal current energy and sorting coefficient. According to Moore, decreasing environmental energy is associated with preferential transport of the light minerals so that tourmaline is enriched in the residual sediment. Thus, the evidence points significantly to mechanical concentration of boron in Buzzards Bay.

The relatively low boron content of many sandstones has led to speculation that significant enrichment of boron in marine sediments is mainly related to incorporation of boron in the argillaceous and organic fractions of sediments (Goldschmidt and Peters, 1932; see reference in Shaw and Bugry, 1966) (Eagar and Spears, 1966; see references in Walker, 1968). The arguments advanced by Moore (1963) are therefore important because they show that boron enrichment in arenaceous sediments may also be related to mechanical processes which lead to concentration of tourmaline.

Boron in Clays

The early work of Goldschmidt and Peters (1932; see reference in Shaw and Bugry, 1966) and later investigations have shown conclusively that clays are mainly responsible for boron fixation in many marine environments. Several of these investigations (Keith and Degens, 1959; Harder, 1959; Frederickson and Reynolds, 1960; Walker and Price, 1963; Potter et al., 1963; see references in Shaw and Bugry, 1966) revealed some correlation between the boron content of shales or clay fractions and depositional salinity. Since the boron content and salinity of sea water are directly related (Frederickson and Reynolds, 1960; see reference in Walker, 1968) speculation exists that clays absorb boron in amounts proportional to boron in solution. However, most authors concede that other environmental parameters also contribute to boron variation in marine sediments.

Experimental Evidence for Boron Sorption. Laboratory investigations show that the concentration of boron absorbed from solution (Harder, 1961; see reference in Walker, 1968) (Lerman, 1966; Porrenga, 1967) is proportional to the concentration of boron in solution, reaction time, and temperature. Boron fixation by several minerals is greatly enhanced by prior grinding to an amorphous state (Porrenga, 1967). The sorption capacities of illite, montmorillonite, and halloysite are similar, but other clay minerals apparently have a lesser sorption capacity.

Location of Boron in the Clay Lattice. Structural investigation of synthetic boron micas and clays (Eugster and Wright, 1960; Stubican and Roy, 1962; see references in Walker, 1968) point to aluminum-boron diadochy in the tetrahedral layers of clays. Harder (1959) and Frederickson and Reynolds (1960; see references in Walker, 1968) stressed that boron associated with illite was firmly held and could not be removed unless the lattice structure was destroyed. Thus, they concluded that in illite at least, boron was held in the tetrahedral sites. On the other hand, uptake of boron by clays in which trivalent substitution in the tetrahedral sites is minimal and the greatly increased uptake which results from prior grinding (Porrenga, 1967) suggests boron adsorption on clay surfaces, as Lerman (1966; see reference in Walker, 1968) has suggested. Both mechanisms probably contribute to boron concentration in marine clays. To what extent adsorbed boron migrates to the tetrahedral sites during lithification is still an open question. Similarly, the stability of adsorbed boron during lithification is also an open question because some of the boron in soils and clays can be removed by boiling water and even greater amounts can be removed by acid treatment (Jackson, 1964; Krasintseva and Shishkina, 1959; see reference in Lerman, 1966).

Boron and Illite Authigenesis. If boron readily displaces aluminum and silicon in the tetrahedral sites of a clay lattice, it is difficult to understand why the reverse reaction is not equally possible. Both Frederickson and Reynolds (1960) and Walker and Price (1963; see references in Walker, 1968) argued that boron entered the tetrahedral sites during illite authigenesis. This theory is attractive because it disposes of those problems associated with an exchange mechanism. However, the evidence for such a mechanism is indirect (Walker, 1968) and the proposed mechanism should be viewed as a working hypothesis which appears to explain the empirical relation between boron in illite and inferred salinity.

Preferential Enrichment of Boron in Illite. It has frequently been observed that illitic clays and shales contain more boron than nonillitic clays and shales. However, in some laboratory experiments, illitic and nonillitic clays absorb comparable amounts of boron from solution (Porrenga, 1967). If part or all of the sorbed boron held on clay surfaces is lost during lithification, much of the boron remaining in the clay fraction would be held in the tetrahedral sites of authigenic illite. It should be noted that acid treatment, commonly used to remove carbonate prior to clay separation, could contribute to loss of sorbed boron and acid soluble clays (chlorites) so that distribution of boron in clay fractions is not necessarily the same as in the untreated sediment. Other factors which may lead to preferential fixation of boron in illite are as follows. Illite tends to be degraded during weathering but other clay minerals are aggraded. Hence illite is commonly more degraded than other clay minerals entering a marine environment, and this promotes boron sorption (Porrenga, 1967). Laboratory data (Lerman, 1966; Porrenga, 1967) show that the amount of boron absorbed by clays is a function of reaction time. Slow deposition rates apparently favor concentration of illite in marine environments. Thus, environments in which illite accumulates are most favorable for boron fixation.

Boron in Recycled Clays. Potassium-argon ages for recent sediments from the Mississippi delta and delta front are consistent with a detrital origin for some or all of the illite (Hurley et al., 1963; see reference in Walker, 1968). As Walker (1968) pointed out, the Gulf Coast is a site of extremely rapid deposition and a detrital origin for illite cannot be assumed if the clastic deposition rate is low. Nevertheless, in some marine sediments much of the illite is probably recycled and a significant part of the boron, therefore, inherited from source. Undisputed proof that some of the boron in marine clays is inherited is lacking, but studies of nonmarine sediments in the North German Plain (Harder, 1959; see reference in Walker, 1968) support this view. In fluvioglacial deposits of the North German Plain, the boron concentration in the clay fraction parallels the boron concentration in the surrounding source rocks. A considerable overlap between boron concentrations reported from marine and nonmarine clays (Keith and Degens, 1959; Potter et al., 1963; see references in Shaw and Bugry, 1966) may be explained in terms of inherited boron (Shaw and Bugry, 1966). For example, boron concentrations are likely to be high in nonmarine clays which contain significant amounts of recycled marine illite. Similarly, low boron concentrations are likely in marine clays containing recycled micaceous clays from metamorphic and nonmarine sedimentary rocks. Presumably, nonillitic clays may also contribute inherited boron, but boron variation due to such clays is likely to be rather small because the boron concentration in nonillitic clays is usually rather low.

Boron in Organic Matter

Boron and Organic Carbon. Eagar and Spears (1966; see reference in Walker, 1968) reported that organic carbon and boron were directly related in British Carboniferous sediments. Boron concentrations in the organic fraction were found to be much less than in living matter. Even so, in organic-rich sediments a considerable part of the total boron was held in the organic fraction. On the other hand, Tourtelot (1964; see reference in Shaw and Bugry, 1966) found that boron and organic carbon were not significantly related in late Cretaceous shales from the western interior of the United States. Walker (1968), working with illitic clay fractions, arrived at a similar conclusion.

Organic carbon in rocks may be present as radicals bonded to clay structures (Eagar and Spears, 1966; see reference in Walker, 1968), as lithified organic fragments, and as hydrocarbons. Thus, conflicting data for boron and organic carbon suggest that boron may be preferentially fixed in one kind of organic material.

Boron and Organic Silica. Siliceous remains of diatoms are reported to contain appreciable amounts of boron (Lewin, 1965; see reference in Porrenga, 1967). However, recent work by the writer of this article suggests that admission of boron into the silica lattice may be an inorganic process at low temperatures. Coarse-size fractions from certain insoluble limestone residues studied by Walker (1963; see reference in Walker, 1968) consist of authigenic quartz crystals. The coarse-size fractions contain extraneous boron which cannot be explained in terms of clay contamination. Heavy mineral separations failed to reveal the presence of tourmaline. Presumably, therefore, boron is held in the authigenic quartz. Possibly, all low-temperature silica in sediments contains boron, whether of organic or inorganic origin, but further work is required before enrichment of boron in silica can be accepted as an established fact.

Conclusions

Many investigations demonstrate conclusively that several mechanisms must contribute to boron enrichment and variation in marine environments. Nevertheless, some writers have argued that boron concentration is a quantitative index of depositional salinity. These writers, however, confined their investigations to illitic clay fractions from stratigraphic sequences dominated by carbonates. The use of boron as a salinity index was based on the assumption that such illites were authigenic and incorporated boron into their structure at the time of deposition. Their conclusions, therefore, do not deny that other mechanisms control the boron content of whole rocks or clay fractions composed of detrital illite and non-illitic clay minerals.

C. T. WALKER

References

Lerman, A., 1966, "Boron in clays and estimation of paleosalinities," *Sedimentology,* **6,** 267–286.

*Moore, J. R., 1963, "Bottom sediment studies, Buzzards Bay, Massachusetts," *J. Sediment. Petrol.,* **33,** 511–558.

Porrenga, D. H., 1967, "Clay mineralogy and geochemistry of Recent marine sediments in tropical areas," Doctoral Thesis, University of Amsterdam.

*Shaw, D. M., and Bugry, R., 1966, "A review of boron sedimentary geochemistry in relation to new analyses of some North American shales," *Can. J. Earth Sci.,* **3,** 49–63.

*Walker, C. T., 1968, "Evaluation of boron as a paleosalinity indicator and its application to offshore prospects," *Am. Assoc. Petrol. Geol. Bull.,* **52.**

* Additional references may be found in this work.

Cross-references: *Hydrocarbons; Paleosalinity; Potassium-Argon Age Determination; Trace Elements; Seawater.* Vol. I: *Marine Sediments.*

BOUNDARY LAYER—*See* Vol. II

BOXWORK—*See* Vol. VI

BROMINE: ELEMENT AND GEOCHEMISTRY

Properties. Bromine, a minor element, is a member of the halogen family. It is a reddish brown liquid under standard conditions, with a melting point of $-7.2°C$ and a boiling point of $58.8°C$. It has a density of 3.12 at $20°C$. It exists as a diatomic molecule with a molecular weight of 159.83. It is moderately soluble in water ($\sim$35g/l at $20°C$) and miscible with nonpolar solvents such as CS_2 and CCl_4. Bromine has two stable isotopes with the following abundances: $^{79}Br = 50.57\%$ and $^{81}Br = 49.43\%$.

Since the bromine atom is only one electron

short of the inert-gas configuration, it forms the uninegative bromide anion or a single covalent bond. Higher covalences are known, but only in oxygen compounds (e.g., BrO_3^-) or interhalogen compounds (e.g., BrF_5).

Abundance. The cosmic abundance of bromine is approximately 0.1 atoms per 10,000 atoms of silicon. The average amount of bromine present in crustal rocks is about 0.0002% or 2 ppm and the amount present in seawater is 0.0065% or 65 ppm. Some representative bromine concentrations in igneous rocks and certain types of chondrites are given in Table 1.

Minerals and Geochemical Behavior. Due to its low abundance and the generally high solubility of its salts in water, minerals containing bromine are not too common. The radius of the bromide ion is 1.96Å, which easily allows it to enter crystals of ionic chlorides, where the majority of bromine is found in minerals. Bromine concentrates, with respect to chlorine, in the later crystallizates of evaporite sediments.

Probably the most common bromine mineral is bromyrite, AgBr. Other, less common, minerals are embolite, Ag(Cl,Br), and iodobromite, Ag(Cl,Br,I). As mentioned above, bromine is often found in significant amounts in chloride minerals such as cerargyrite (AgCl), halite (NaCl), and carnallite ($KMg(H_2O)_6Cl_3$). The latter has been found to contain up to 0.2% Br.

Goldschmidt has calculated that approximately 600 g of rock have been weathered for each kilogram of water in the ocean. Thus the 600 g of igneous rock have been the potential source of the dissolved matter supplied by weathering to one kilogram of seawater. Assuming an average bromine concentration in igneous rocks of 2 ppm, it is seen that the Br concentration in seawater should be 1.2×10^{-3} g Br/kg seawater, or 1.2 ppm. The actual concentration of bromine in seawater is 65 ppm. An explanation for this apparent excess of bromine in the sea suggests that probably only a small fraction of the bromine present in the magma has been fixed in magmatic minerals, most of it remaining in residual solutions and in magmatic gases at high temperatures with eventual solution in the sea. It has also been suggested by other investigators that high-temperature volatilization processes are not necessary for Br, as well as carbon (C), nitrogen (N), chlorine (Cl), and boron (B), to be concentrated in the ocean, and in fact that a high-temperature process is probably not responsible since if it were, certain other elements should also be concentrated in the sea. Urey suggested that Br, as well as C, N, Cl, and B, was removed from the outer parts of the earth through solution in water at low or moderate temperatures (100°C or less).

Economic Geology. The major commercial source of bromine is the sea, although some is also obtained from inland brine deposits and from bitterns. The commercial extraction of bromine from seawater involves acidification of the seawater with H_2SO_4, oxidation of the Br^- by Cl_2, removal of the Br_2 formed by a stream of air, and subsequent purification of the bromine by reduction and reoxidation. The primary use for bromine is in ethylene bromide, a gasoline additive. There are approximately 2 g of bromine per gallon of leaded gasoline. Considerable amounts of bromine are also used in insecticides and for medicinal purposes.

ROBERT A. DUCE

References

Behne, W., 1953, "Untersuchungen zur Geochemie des Chlor und Brom," *Geochim. Cosmochim. Acta*, **3**, 186.

Correns, C. W., 1956, "The Geochemistry of the Halogens" in (Ahrens, L. H. et al., editors), "Physics and Chemistry of the Earth," **1**, 181, New York, McGraw-Hill Book Co.

Goldschmidt, V. M., 1954, "Geochemistry," Oxford, Clarendon Press, 730pp.

Mason, B., 1966, "Principles of Geochemistry," Third ed., New York, John Wiley & Sons, 329pp.

Reed, G. W., and Allen, R. O., Jr., 1966, "Halogens in chondrites," *Geochim. Cosmochim. Acta*, **30**, 779.

Urey, H. C., 1953, "The concentration of certain elements at the earth's surface," *Proc. Roy. Soc. London Ser. A*, **219**, 281.

Cross-references: *Chlorine; Cyclic Salts; Natural Brines; Oxidation and Reduction; Seawater.* Vol. VI: *Salts—Cyclic.*

TABLE 1. BROMINE CONTENT OF SOME IGNEOUS ROCKS AND CHONDRITES

Sample	Br (ppm)
Granites	~1.6
Granodiorites	~2.6
Syenites	~1.1
Gabbros	~2.0
Basalts	~2.7
Hypersthene and bronzite chondrites	~0.15
Carbonaceous and enstatite chondrites	2–11

BUBBLE MOTION IN FLUID INCLUSIONS

Bubbles in *fluid inclusions* (q.v.) may display different types of motion. There is, firstly, the irregular and uninterrupted motion of very small bubbles whose diameters are on the order of the micron, i.e., the *Brownian motion* (see Vol. II, p. 126). The movement will be the more lively the smaller the bubble. If the diameter of the bubble is greater than 0.004 mm, the movement will be hardly perceptible. The movement of bubbles in neighboring inclusions is unrelated.

Then there is a "non-Brownian" mobility, which is a function of outside factors such as change of temperature or inclination of the inclusion containing the bubble. The bubbles with this type of mobility are bigger: their diameter may be on the order of the millimeter. The mobility of bubbles in neighboring inclusions will be parallel also if the bubbles are of different size. This non-Brownian movement is claimed to constitute a very sensitive thermal gradient detector (Roedder 1965, 1967).

The mobility of gas bubbles in the so-called enhydros (Frondel, 1962, p. 230) will be mentioned here (see also Vol. IV. B, *Enhydrate, or Waterstone Enhydro*).

Thirdly, there are gas bubbles in liquid inclusions without any apparent mobility. What causes the inhibition may not always be recognized. The most plausible explanation is that the bubble is in contact with, and in some way adheres to, the wall of the inclusions.

The different types of motion of bubbles in fluid inclusions are frequently lumped together indiscriminately (Deicha 1955). The Brownian motion is probably the most widespread type of bubble motion in fluid inclusions and certainly the most important one with respect to implications in molecular physics and to the discussion of the random process.

The irregular swarming movement of small particles suspended in a liquid has been described by Robert Brown in 1828 (*Ann. Phys. Chem.* 1828, **14,** 294 and *Phil. Mag.* 1828, **4,** 161; cited in synoptic publications such as de Haas-Lorentz, 1913; J. Perrin (English ed., 1923). De Haas-Lorentz gave a concise historical account; he mentioned also observations made prior to Brown, in particular by Brongniart (*Ann. Sci. Nat.* 1827, **12,** 14).

Different interpretations have been formulated to account for the Brownian motion (Wiener, 1863, Jevons, 1869, Gouy, 1889). Gouy explained the Brownian motion as a distant and reduced result of thermal molecular movement. He was the first to give a molecular kinetic interpretation and he attempted also an experimental corroboration, at least a qualitative one. A quantitative formulation was developed in 1905 independently by Smoluchowski (1923) and by Einstein (1956). Einstein spoke of "the movement of small particles suspended in a stationary liquid that might be identical with the so-called Brownian movement"; he reserved judgment as he considered the information on Brownian motion not sufficiently precise. Einstein discussed the phenomenon based on the molecular-kinetic theory. According to this theory a dissolved molecule and a suspended body would differ solely by their dimensions. This means that the assumption has to be made that suspended particles perform an irregular movement even if only a very slow one, because of the molecular movement of the liquid. If the particles can leave a given volume through a partition, they will exert a pressure on the partition, just as molecules do in a solution. The osmotic pressure p will be $p = (RT/V^*)(n/N)$, where n is the number of suspended particles in volume V^*; therefore $n/V^* = n'$ number of particles in unit of volume $p = (RT/N) n'$.

With spherical particles of radius r the coefficient of diffusion is given by:

$$D = \frac{RT}{N} \frac{1}{6\pi kr}$$

The mean value of the displacement Δ of the particles $\bar{\Delta} = \sqrt{2Dt}$; that is, Δ is proportional to the square root of the time.

The formulation developed by Einstein states thus that the square of the average mean displacement of small particles suspended in a stationary liquid is

$$\bar{\Delta} = \frac{RT}{N} \frac{1}{3\pi kr} t$$

where R is the gas constant (8.31×10^7); T is the absolute temperature, in degrees K; N is Avogadro's number (6.02×10^{23}); k is the viscosity of the liquid, in poise; t is the time interval, in sec; and r is the radius of the moving particle, in cm.

The formulation arrived at by Smoluchowski by strictly following gas kinetic procedures differs from the Einstein formula by a numerical factor. Langevin (1908) confirmed the Einstein formulation and gave another derivation for it.

In the Einstein formula, everything can be measured or is known, with the exception of Avogadro's number. Based on this formulation Avogadro's number can be determined. Perrin (1909) arrived at good approximations of N by studying colloidal suspensions.

In the study of fluid inclusions, the unknown is the viscosity of the fluid. Solving the Einstein formula for the viscosity, k:

$$k = \frac{RT}{N\bar{\Delta}^2} \frac{t}{3\pi r}$$

If the time interval t is given, as for instance when using a movie camera with known number of frames per second, and the mean average displacement, as well as the diameter of the moving bubble, are experimentally determined, the viscosity of the liquid in the fluid inclusion can be calculated (Friedlaender, 1970). The value of the viscosity, in turn, may under certain circumstances permit an extrapolation of the composition or concentration of the liquid in the fluid inclusion.

ACKNOWLEDGMENT: Work supported by N.R.C. Grant A-888.

C. G. I. FRIEDLAENDER

References

Deicha, G., 1955, "Les Lacunes des Cristaux et Leurs Inclusions Fluides," Paris, Masson & Cie, 126pp.
Einstein, A., transl. by Cowper, A. D., ed., and Fürth, R., 1956 "Investigations on the Theory of the Brownian Movement" New York, Dover, 120pp.
Friedlaender, C. G. I., 1970, "Brownian movement in liquid inclusions in quartz. Some quantitative observations," *Can. Mineralogist,* **10,** 272–274.
Friedlaender, C. G. I., 1970, "Brownian movement in quartz: an attempt at quantitative evaluation," *Schweiz. Mineral. Petrog. Mitt.,* **50,** 13–20.
Frondel, C., 1962, "Dana System of Mineralogy," Vol. 3, Seventh ed.
Gouy, M., 1889, "Sur le mouvement brownien," *Compt. Rend.,* **109,** 102–105.
de Haas-Lorentz, G. L., 1913, "Die Brownsche Bewegung, Vieweg," Braunschweig, 103pp.
Jevons, W. S., 1869, "On the so-called molecular movements of microscopic particles," *Manchester, Lit. Phil. Soc. Proc.,* **9,** 78–84.
Langevin, P., 1908, "Sur la théorie du mouvement brownien," *Compt. Rend.,* **146,** 530–533.
Perrin, J., 1909, "Mouvement brownien et réalité moléculaire," *Ann. Chim. Phys., 8e sér.,* **18,** 5–114.
Perrin, J., Transl. by D. Le Hammick, 1923, "Atoms," Second ed. in English, London, Constable & Co., of the 11th edition of "Les Atomes," first published in 1913, 231pp.
Roedder, E., 1965, "Non-Brownian bubble movement in fluid inclusions. A thermal gradient detector of extreme sensitivity and rapid response," *Geol. Soc. Am., Special Paper* **87** (Abstr.), 140.
Roedder, E., Device for sensing thermal gradients, U.S. Pat. 3,344,669 (Oct. 3, 1967).
Smoluchowski, M. von, 1923, in (R. Fürth, editor), "Abhandlungen über die Brownsche Bewegung und verwandte Erscheinungen. Ostwald's Klassiker der exakten Wissenschaften," **207,** Leipzig, Akad. Verlagsges.m.b.H., 152pp.
Wiener, C., 1863, "Erklärung des atomistischen Wesens des tropfbarflüssigen Körperzustandes, und die Bestätigung desselben durch die sogenannten Molecularbewegungen," *Pogg. Ann.,* **118,** 79–94.

Cross-references: *Fluid Inclusions.* Vol. II: *Avogadro's Law, Number; Brownian Motion, Movement.* Vol. IVB: *Enhydro, Enhydrite, or Water-Stone.*

BUFFER SYSTEM

A buffer system (so named from "buff," to deaden the shock), is a chemical system where the concentration or thermodynamic activity of one of the components is maintained at an approximately constant level during fluctuations in the system. Such a condition is generally achieved by means of a chemical equilibrium which can absorb or generate the buffered component and compensate for undesired additions or removal of it by external means. In the earth's crust the water vapor content of the atmosphere in contact with a large body of water is controlled or buffered by evaporation or precipitation of liquid water. The acidity of seawater is buffered against short-term changes by solubility equilibria with atmospheric carbon dioxide and sedimentary calcium carbonate, and the content of carbon dioxide in the atmosphere is buffered on a worldwide scale by its solubility in the oceans. In the geochemical laboratory oxygen pressure is buffered during high-temperature synthesis of ferric and ferrous iron minerals by oxides which have definite dissociation pressures. In a sense, the oxygen of the atmosphere, which is continuously being produced by the photosynthesis of green plants, is buffered against unlimited increase by animal and plant respiration and by inorganic oxidation processes.

The operation of a geochemical buffer system can be illustrated by the carbonate system in seawater. Table 1 shows the distribution of carbon in the several reservoirs of the earth's crust, expressed as grams of carbon dioxide

TABLE 1. CARBON IN THE EARTH'S CRUST[a]

Reservoir	$g\,CO_2/cm_e^2$
Carbonate in sediments	13,000
Organic carbon in sediments	4,900
CO_2 in atmosphere	0.46
Living matter on land	0.055
Dead organic matter on land	0.51
Dissolved carbonate in ocean	
$\quad CO_2 + H_2CO_3$	0.16 ⎫
$\quad HCO_3^-$	22.4 ⎬ 25.4
$\quad CO_3^{2-}$	2.8 ⎭
Dead organic matter in ocean	2.0
Living organic matter in ocean	0.006

[a] Revelle and Suess, 1957.

BUFFER SYSTEM

TABLE 2. CARBONATE EQUILIBRIA IN SEAWATER

Equilibrium Reaction	Expression	Equilibrium Constant (20°C) Pure Water	Seawater
(a) CO_2 (gas) $+ H_2O = H_2CO_3$	$\frac{(H_2CO_3)}{P_{CO_2}} = K_p$	$10^{-1.40}$	$10^{-1.47}$
(b) $H_2CO_3 = H^+ + HCO_3^-$	$\frac{a_{H^+}(HCO_3^-)}{(H_2CO_3)} = K_{a_1}$	$10^{-6.38}$	$10^{-6.01}$
(c) $HCO_3^- = H^+ + CO_3^{2-}$	$\frac{a_{H^+}(CO_3^{2-})}{(HCO_3^-)} = K_{a_2}$	$10^{-10.38}$	$10^{-9.17}$
(d) $CaCO_3$ (calcite) $= Ca^{2+} + CO_3^{2-}$	$(Ca^{2+})(CO_3^{2-}) = K_s$	$10^{-8.35}$	$10^{-6.11}$

per square centimeter of earth surface (g CO_2/cm_e^2). Most important for the seawater buffer system are the inorganic carbonate minerals in the sediments, the seawater, and the atmosphere. The sediments contain vast reserves of carbonate to draw upon for solution equilibria, and the seawater contains 55 times more CO_2 (in all its forms) than does the atmosphere. As a result of equilibria between these three reservoirs, *(1)* the acidity of seawater is buffered to pH = 8.0–8.5, independent of arbitrary additions of acid or base, and *(2)* the gaseous carbon dioxide concentration in the atmosphere is buffered to 300 ppm by volume, independent of arbitrary additions of CO_2 by respiration or combustion provided such additions are not too rapid.

Table 2 gives the important equilibrium constant expressions and their values for pure water and seawater, where activities are taken to be the concentrations in moles per liter for H_2CO_3 (including dissolved CO_2), HCO_3^-, CO_3^{2-}, and Ca^{2+} in solution, the activity of hydrogen ion, a_{H+}, as measured by pH = $-\log_{10} a_{H+}$ using a glass electrode, and the partial pressure of gaseous carbon dioxide, P_{CO_2}. The activities of pure solid calcite, $CaCO_3$, and water are each taken as unity. Using these expressions, Figs. 1, 2, and 3 have been constructed showing relationships between the concentrations and pH. It is important to note that HCO_3^- has a much higher concentration than CO_3^{2-} or H_2CO_3 in seawater at pH = 8.0–8.5. Because of this, HCO_3^- acts as a reservoir to replace or absorb CO_3^{2-} and H_2CO_3 if they should be removed from or added to the solution by some process, and this stabilizing effect on the ratios $(HCO_3^-)/(H_2CO_3) = K_{a_1}/a_{H+}$ and $(CO_3^{2-})/(HCO_3^-) = K_{a_2}/a_{H+}$ maintains constancy in pH.

Example 1. Seawater normally contains the concentration $(HCO_3^-) = 2.3 \times 10^{-3}$ moles/liter. Calculate the pH assuming equilibrium with the atmosphere, where $P_{CO_2} = 3.0 \times 10^{-4}$ atm.

$$P_{CO_2} = 10^{-4.00 + 0.48} = 10^{-3.52}$$
$$(HCO_3^-) = 10^{-3.00 + 0.36} = 10^{-2.64}$$
$$a_{H+} = K_{a_1} K_p P_{CO_2}/(HCO_3^-)$$
$$= 10^{-6.01-1.47-3.52+2.64} = 10^{-8.36}$$
$$pH = -\log_{10} a_{H+} = 8.36$$

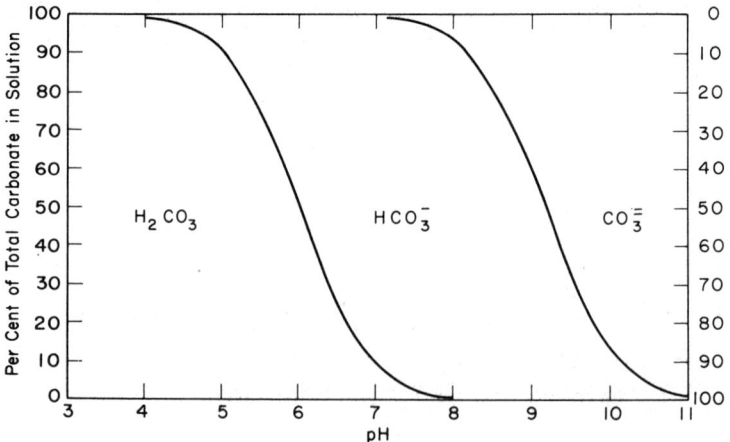

FIG. 1. Relative abundance of carbonate species in seawater as a function of pH.

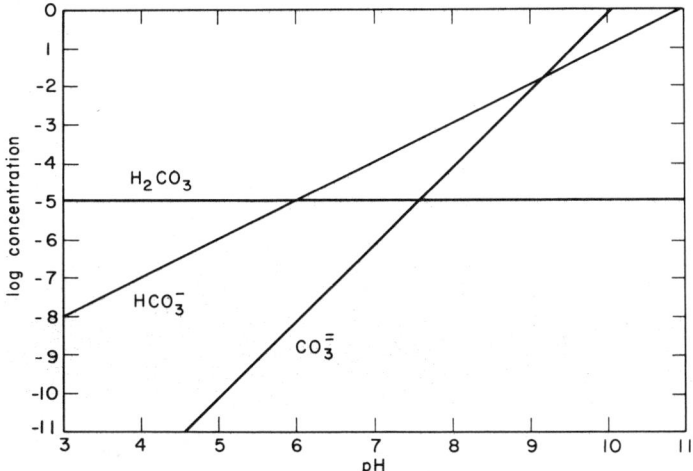

FIG. 2. Concentration of carbonate species in seawater that has a fixed value of P_{CO_2} (3.0×10^{-4} atm) as a function of pH.

Surface seawater is generally observed to have a virtual P_{CO_2} somewhat greater than the atmospheric value, and consequently the pH is slightly lower than the value calculated.

Example 2. If the seawater in example 1 is also in equilibrium with calcite, calculate the concentration of Ca^{2+}.

$$(CO_3^{2-}) = K_{a_2}(HCO_3^-)/a_{H^+}$$
$$= 10^{-9.17-2.64+8.36} = 10^{-3.45}$$
$$(Ca^{2+}) = K_s/(CO_3^{2-}) = 10^{-6.11+3.45}$$
$$= 10^{-2.66} = 2.2 \times 10^{-3}$$

Seawater normally contains $(Ca^{2+}) = 1.0 \times 10^{-2}$ moles/liter, and, although (CO_3^{2-}) is actually lower than the value calculated here owing to a lower pH than calculated in example 1, some supersaturation of seawater with respect to calcite is observed in nature.

Example 3. If industrial pollution adds 1.0×10^{-5} moles of strong acid per liter of the seawater of examples 1 and 2, calculate the change in pH and compare with the pH change if there had been no buffering action.

Solution: Reactions (b) and (c) of Table 2 are shifted to the left by the addition of H^+. In view of the relative concentrations at the initial pH (8.36) (see Fig. 3) and because the same value of a_{H^+} must satisfy both equilibrium expressions, (HCO_3^-) changes virtually not at all, and most of the added H^+ reacts to decrease (CO_3^{2-}) slightly. A smaller portion of H^+ leads to an increase in (H_2CO_3) and P_{CO_2} by the same factor as that of the (CO_3^{2-}) decrease. Accordingly, the final concentrations are, to a good approximation:

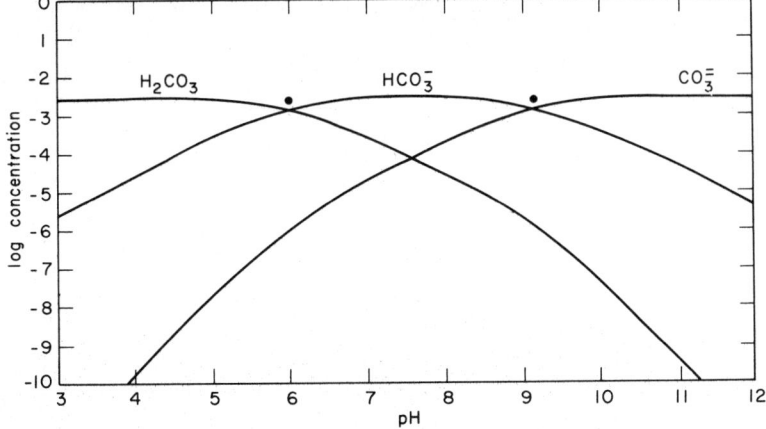

FIG. 3. Concentration of carbonate species in seawater with a fixed value of total carbonate (i.e., $H_2CO_3 + HCO_3^- + CO_3^{2-} = 2.5 \times 10^{-3}$ moles/liter) as a function of pH.

$(HCO_3^-) = 10^{-2.64} + 10^{-5.00} = 10^{-2.64}$
$(CO_3^{2-}) = 10^{-3.45} - 10^{-5.00} = 10^{-3.46}$
$a_{H^+} = K_{a_2}(HCO_3^-)/(CO_3^{2-})$
$= 10^{-9.17-2.64+3.46} = 10^{-8.35}$
$pH = 8.35$
$\Delta pH = 8.36 - 8.35 = 0.01$
$P_{CO_2} = a_{H^+}(HCO_3^-)/K_{a_1}K_p$
$= 10^{-8.35-2.64+6.01+1.47} = 10^{-3.51}$

Without buffering action,

$a_{H^+} = 10^{-8.36} + 10^{-5.00} = 10^{-5.00}$
$pH = 5.00$
$\Delta pH = 8.36 - 5.00 = 3.36$

In the buffered system, equilibrium with the atmosphere and sediments would ultimately be achieved by outgassing a little CO_2 or dissolving a little $CaCO_3$, or both, and the original (HCO_3^-) would be restored exactly.

Example 4. Seawater, initially as in example 1 and in exact equilibrium with calcite, supports photosynthesis leading to removal of 1.0×10^{-5} moles H_2CO_3 per liter. What is the final condition with respect to equilibrium with calcite?

Solution: The concentration relations are best described by the combined reaction

$2HCO_3^- = H_2CO_3 + CO_3^{2-}$
$(H_2CO_3)(CO_3^{2-})/(HCO_3^-)^2$
$= K_{a_2}/K_{a_1} = 10^{-3.16}$

Removal of H_2CO_3 leads to a shift to the right, nearly restoring the original (H_2CO_3) and increasing (CO_3^{2-}) slightly. Owing to its relatively high concentration, (HCO_3^-) undergoes negligible decrease, and the overall acidity decreases slightly because of the removal of H_2CO_3, an acid, and the increase in the ratio $(CO_3^{2-})/(HCO_3^-) = K_{a_2}/a_{H^+}$. Numerically,

$(HCO_3^-) = 10^{-2.64} - 2 \times 10^{-5} = 10^{-2.64}$
$(CO_3^{2-}) = 10^{-3.45} + 10^{-5} = 10^{-3.44}$

Because (CO_3^{2-}) increases, calcite should precipitate because of its slight supersaturation, and the original pH condition would be restored. Alternatively, atmospheric CO_2 could dissolve in the seawater and restore the removed H_2CO_3 and original pH.

Example 5. If the seawater of example 1 became more acid by 0.10 pH unit as an average over the World Ocean, calculate the change in concentration of gaseous CO_2 in the atmosphere.

$pH = 8.36 - 0.10 = 8.26$
$P_{CO_2} = a_{H^+}(HCO_3^-)/K_{a_1}K_p$
$= 10^{-8.26-2.64+6.01+1.47} = 3.8 \times 10^{-4}$ atm

A pH change in seawater, which is small by the standards of laboratory chemistry, could create a change in atmospheric P_{CO_2} which is much greater than has been observed naturally. It appears that the air-sea-sediment carbonate equilibria compose a delicately balanced and stable geochemical buffer system.

JOHN W. WINCHESTER

References

Garrels, R. M., 1960 "Mineral Equilibria at Low Temperature and Pressure," New York, Harper & Row, 254pp.

Garrels, R. M., and Christ, C. L., 1965, "Solutions, Minerals, and Equilibria," New York, Harper & Row, 450pp.

Revelle, R., and Suess, H. E., 1957, "Carbon dioxide exchange between atmosphere and ocean and the question of an increase at atmospheric CO_2 during the past decades," *Tellus* **9**, 18–27

Sillén, L. G., 1961, "Physical Chemistry of Sea Water," in (M. Sears, editor), "Oceanography," Am. Assoc. Advan. Sci. Publ. No. 67.

Cross-references: *Acids and Bases; Boron Geochemistry in Marine Environments; Calcium Carbonate; Calcium Cycle; Carbon Cycle; Oxygen Cycle; pH-Eh; Photosynthesis; Seawater; Thermodynamics; Water. Vol. I: Carbon Dioxide Cycle in the Sea and Atmosphere. Vol. II: Evaporation.*

C

CADMIUM: ELEMENT AND GEOCHEMISTRY

A rare metal element, of silvery white color, cadmium (symbol Cd) has the atomic number 48, an atomic weight of 112.40, and falls into Group IIB of the Periodic Table, between silver (at. no. 47) and indium (49), and with period neighbors zinc (30) and mercury (80). Cadmium was discovered in 1817 by F. Stromeyer, professor of metallurgy in Göttingen, and named after the Greek term "cadmia" which refers to calamine or zinc carbonate. He was studying ores of this nature from Salzgitter, Germany, which contained minor amounts of cadmium.

The metal has a melting point of 321°C and a boiling point of 767°C; its crystal structure is a closely packed hexagon, its density at 20°C being 8.65. Its thermal conductivity is 0.22 cal/cm^2/cm/°C/sec at 20°C. Its valence is +2. In nature it is a mixture of eight stable isotopes, 106, 108, 110, 111, 112, 113, 114, and 116. Radioactive ones are Cd^{106} and Cd^{113}. Artificial isotopes have mass numbers 105, 107, 109, 115 and 117. The metal is non-corrosive in a clean atmosphere, but in urban areas polluted by SO_2 and SO_3 there is rapid corrosion. It will dissolve in most acids but is not soluble in alkalis.

Cadmium forms a sulfide, CdS, used as a yellow pigment and in photoelectric cells; a cyanide, $Cd(CN)_2$; several halides, used in photography; a nitrate, $Cd(NO_3)_2$; an oxide, CdO, used among other things for battery electrodes. The major industrial use of the metal is for electroplating iron and steel. Such plating is resistant to alkalis, which is not so for zinc (galvanized) plate. It is used in many alloys, especially for those needed in high-temperature work. Nickel-cadmium batteries offer multiple charge advantages. Copper with up to 1% Cd has enhanced ductility and tensile strength. It is utilized as rods for controlling nuclear reactors, since it readily absorbs low-energy (thermal) neutrons. It has a potential use in solar cells for conversion of the sun's radiation into electrical power.

Cadmium is highly toxic and particularly dangerous to man when inhaled as dust or in gas form.

Minerals. Minerals that contain cadmium include greenochite, CdS; hawleyite, CdS; xanthocroite, $CdS(H_2O)_x$; cadmoselite, CdSe; monteporrite, CdO; and otavite, $CdCO_3$.

Abundance. Cadmium has a solar abundance of 1.1×10^{-4} ppm (by wt), and in the Earth's crust calculations vary about 0.5 ppm ("Clarke value") to 0.01. In sediments (shale) is abundance averages 0.3 ppm and in igneous rocks, 0.13 ppm. Its association is chalcophile; it is very closely allied to zinc, the average Cn:Cd ratio being estimated at 500:1 up 900:1.

In meteorites cadmium occurs over a range of 1 to $100 \times 10^{-5}\%$ in chondrites; $10^{-3}\%$ in troilite, and $10^{-4}\%$ in kamcite.

Geochemistry. Cadmium behaves very much like zinc, both favoring a combination with sulfur. The bulk of Cd in nature is dispersed as isomorphic impurities in various other minerals, usually sulfides. The principal carrier is sphalerite. Its highest concentration, in zinc minerals, is up to 5%. It is also widespread in lead ores, galena, etc.

In igneous rocks Cd is virtually confined to the principal rock-forming minerals, specifically the ferromagnesian minerals (Sandell and Goldich, 1952), because Cd behaves like iron in the magmatic process. The Cd concentration in basaltic rocks is appreciably higher than in acid types. In the latter it is usually restricted to biotite and to some extent in apatite.

Cadmium does not concentrate during the high-temperature postmagmatic metasomatism and pegmatitization (Vlasov, 1966), except where sphalerite and related minerals are involved. The important source of Cd is relatively low-temperature hydrothermal deposits, silicate-sulfide deposits, iron sulfide deposits, and carbonate-fluoride-barite-sulfide associations.

Supergene concentration of Cd results from oxidation of sulfide deposits, but it tends to be leached out since its sulfate is readily soluble in acid waters. Cadmium in streams has been noted as a pathfinder for geochemical search for zinc (Hawkes and Webb, 1962). Cadmium follows zinc during weathering except in the case of zinc sulfide, in which secondary cadmium sulfide remains after the zinc has been leached. The average Cd concentration in soils is 0.5 ppm. An exceptional concentration, also with zinc, has been noted in oceanic phosphate rock (leached guano), i.e., 100 ppm (Rankama and Sahama, 1950).

Production. World production of cadmium ranges between 25 and 29 million tons annually; U.S. production is around 10 million tons anually. The U.S. government maintains a regular stockpile of 14–16 million tons. The U.S., U.S.S.R., Canada, Japan, and Congo (Leopoldville) accounted for 74% of the world production in 1965. Countries with appreciable production include Mexico, Hondurus, Peru, Austria, Belgium, France, East and West Germany, Italy, Netherlands, Norway, Poland, Spain, Yugoslavia, Zambia, and Australia.

RHODES W. FAIRBRIDGE

References

Hawkes, H. E., and Webb, J. S., 1962, "Geochemistry in Mineral Exploration," New York, Harper & Row, 415pp.

Rankama, K., and Sahama, T. G., 1950, "Geochemistry," Chicago, Univ. of Chicago Press, 911pp.

Sandell, E. B., and Goldich, S. S., 1943, "The rarer metallic constituents in some American igneous rocks," *J. Geol.,* **51,** 167–189.

Vlasov, K. A. (editor), 1966, "Geochemistry and Mineralogy of Rare Elements, and Genetic Types of Their Deposits," Jerusalem, Israel Program for Scientific Translations (from the Russian).

Cross-references: *Hydrothermal Solutions.* Vol. IV B: *Meteoritic Minerals.*

CALCIUM: ELEMENT AND GEOCHEMISTRY

Element and Geochemistry

Physical Properties. Calcium (L. *calx,* chalk), is an alkaline earth element with symbol Ca; atomic weight 40.08; atomic number 20; electronic configuration $1s^2, 2s^2, 2p^6, 3s^2, 3p^6, 4s^2$; melting point 838°C; boiling point 1440°C; density 1.55; and valence +2. The mass numbers of its stable isotopes and their relative abundances are 40 (96.97%), 42 (0.64%), 43 (0.45%), 44 (2.06%), 46 (0.0033%) and 48 (0.185%). Calcium isotopes 39, 41, 45, and 49 have been produced artificially. Some Ca^{40} is derived from the decay of K^{40} thru beta emission.

Calcium was identified by Sir Humphrey Davy in 1808. The metal is produced by the electrolysis of a mixture of $CaCl_2$ and CaF_2. It is silvery white in color and when burned in air forms calcium oxide, CaO, and calcium nitride, Ca_3N_2.

Compounds. Important natural occurrences of calcium are in the form of carbonates, silicates, sulfates, fluorides, and phosphates. Important examples of these types of compounds are *calcite* and *aragonite,* $CaCo_3$; *dolomite,* $CaMg(CO_3)_2$; *anorthite,* $CaAl_2Si_2O_8$; *diopside,* $CaMgSi_2O_6$; *tremolite,* $Ca_2Mg_5Si_8O_{22}(OH)_4$; *gypsum,* $CaSO_4 \cdot 2H_2O$; *fluorite,* CaF_2, and *apatite,* $Ca_5(F,Cl,OH)(PO_4)_3$. Commercially important compounds of calcium include calcium oxide, CaO (quicklime); calcium hydroxide, Ca(OH) (slaked lime); calcium acid sulfide $Ca(HSO_3)_2$; calcium carbide, CaC_2; and calcium cyanamide, Ca NCN.

Vaterite, a rare variety of $CaCO_3$, is apparently deposited during the earliest stages of calcification by some calcareous marine organisms. It rapidly converts to either aragonite or calcite.

Chemistry. Calcium readily looses its two outer electrons forming the Ca^{2+} ion. The first ionization potential is about 6.09V and the second is 11.82V. Its atomic radius is 1.97Å and ionic radius is 0.99Å. On the basis of relative ionic radii of Ca^{2+} and O^{2-}, it is theoretically predicted that calcium, when bonded to oxygen should be surrounded by six oxygen ions, i.e., a sixfold octahedral coordination. However, calcium is very close to the theoretical size for eightfold cubic coordination and, in fact, it often occurs in this state. Based on the electronegativity difference between calcium and oxygen it is predicted that the calcium-oxygen bond should be about 79% ionic and 21% covalent. Calcium exhibits pronounced base-forming tendencies.

Geochemistry. Calcium is the fifth most abundant element in the earth's crust, the seventh most abundant in the whole earth, the eighth most abundant in meteorites, and ranks tenth in the sun. In the universe there are about 490 atoms of calcium per 10,000 atoms of silicon and in the sun's atmosphere there are about 870 atoms of calcium per 10,000 silicon atoms. Calcium is estimated to make up 2.5% of the entire earth and 3.6% of the crust. The average igneous rock contains 5.1% CaO compared to 5.9% CaO for the average sedimentary rock. The latter figure assumes that limestone makes up 12% of all sediments, which is probably low. The average shale contains 3.1%, sandstone 5.5%, and limestone 42.6% CaO.

The basic building block of silicate minerals is the silicate tetrahedron, $(SiO_4)^{4-}$, which can be arranged in various ways by sharing different numbers of oxygens with one another. As a general rule calcium is preferentially located in eightfold coordinated positions in silicate minerals. In the nesosilicates (isolated silicate tetrahedra which share no oxygens) calcium is found in ordered sixfold coordinated positions.

Examples are the metamorphic minerals of the garnet group: *uvarovite*, $Ca_3Cr_2(SiO_4)_3$ and *grossularite*, $Ca_3Al_2(SiO_4)_3$. In the sorosilicates (two isolated tetrahedra which share one oxygen atom) calcium is present mainly in the epidote group; important metamorphic minerals include *epidote*, $Ca_2(AlFe)_3(SiO^4)_3(OH)$. Calcium is an unimportant constituent of the phyllosilicate or sheet silicate group (each tetrahedron shares three oxygens with others), although occasionally this element may occupy interlayer positions. Volumetrically, by far the most important occurrences of calcium in silicates are in the inosilicate and tectosilicate groups. In the inosilicate group (two oxygens in each tetrahedron shared as in the *pyroxenes* and alternately two and three oxygens shared as in the *amphiboles*) both six and eightfold coordinated cation positions are available, but calcium always occupies the eightfold position. Important calcium bearing *pyroxenes* include *diopside*, $CaMgSi_2O_6$, *wollastonite*, $CaSiO_3$, and *augite*, a complex mineral containing Ca, Mg, Fe, Al and other elements in the cation positions. Calcium amphiboles include *tremolite*, $Ca_2Mg_5(Si_8O_{22})(OH)_2$, and *hornblende*, a complex silicate analagous in composition to *augite*. Cations in tectosilicates (all tetrahedral oxygens are shared) are in tenfold coordination and are present to maintain neutrality of charge due to substitution of Al^{3+} for Si^{4+} in the tetrahedra. *Anorthite*, $CaAl_2Si_2O_8$, is the most important calcium tectosilicate.

Fluorite, CaF_2, commonly occurs as a secondary vein mineral in limestones. *Apatite*, $Ca_5(F, Cl, OH)(PO_4)_3$, occurs both as a secondary mineral and an accessory mineral in igneous rocks. *Collaphane* is a microcrystalline variety of the *apatite* group and the principal constituent of many large deposits of *phosphorite* in ancient and Recent sediments. The most common calcium sulfates are *gypsum*, $CaSO_4 \cdot 2H_2O$, and *anhydrite*, $CaSO_4$, both of which can be evaporites or secondary minerals. Other calcium evaporites include *glauberite*, $Na_2Co(SO_4)_2$, *polyhalite*, $K_2Ca_2Mg(SO_4)_4 \cdot 2H_2O$, and the borate *colemanite*, $Ca_2B_6O_{11} \cdot 5H_2O$. The very important carbonate group is represented by *calcite*, $CaCO_3$, *aragonite*, $CaCo_3$, and *dolomite*, $CaMg(Co_3)_2$. All three are sedimentary minerals but they also occur as vein minerals in a variety of rock types. The cations in *calcite* and *dolomite* are in sixfold coordination but in the more dense *aragonite* lattice, calcium is surrounded by nine oxygens. A variety of *calcite* in Recent sediments is *high magnesium calcite* which contains between five and twenty-five weight percent $MgCO_3$. In this structure random substitution of the small magnesium ion for calcium results in a high-energy crystal. By contrast, *dolomite* exhibits a more stable ordered structure. Both aragonite and high-magnesium calcite are unstable on the earth's surface. *Oldhamite*, CaS, is known only from meteorites and is unstable under earth surface conditions. CaO is uncommon in nature because it is reactive and forms $CaCO_3$. The tungstate *scheelite*, $CaWO_4$, is a high-temperature vein ore.

In igneous rocks calcium is an important constituent of both major and accessory minerals. Approximately one-half of the calcium in igneous rocks is present in the *pyroxene* and *amphibole* groups and the other half is present in the *plagioclase feldspars*. During magmatic crystallization, the maximum calcium content is reached at a relatively early stage. This is illustrated by considering the average CaO content of the following calc-alkalic rock types which are arranged in order of crystallization or increasing silica content: *dunites*, 0.7%; *gabbros*, 11.0%; *diorites*, 6.7%; *granodiorites*, 4.4%; and *granites*, 2.0%.

The principal calcium-bearing minerals in sediments are *calcite* and *dolomite*. These minerals not only occur as distinct massive bodies of varying purity but also as cement in other rock types such as sandstone. $CaCO_3$ in the form of *travertine* or *tufa* is deposited at the mouths of springs. Other important occurrences of calcium in sediments are as evaporite and phosphorite deposits.

During metamorphism *limestone* is recrystallized to marble. Other minerals may form, such as the *pyroxene wollastonite* ($CaSiO_3$): $CaCO_3 + SiO_2 = CaSiO_3 + Co_2$. *Dolomite*, on the other hand, may alter to calcite plus MgO or even *olivine*, Mg_2SiO_4. A typical sequence of calcium bearing minerals arranged in order of increasing regional metamorphism of an impure *limestone* is: *amphibole*, *epidote diopside*, and *wollastonite*. Typically, the calcium content of the *plagioclase feldspars* increases with increasing metamorphic grade. The increased calcium in the feldspars is furnished at the expense of other calcium minerals such as *epidote*.

Most calcium minerals, particularly those of igneous and metamorphic origin, are readily attacked by weathering and the calcium re!eased to solution. Because of the equilibrium established between $CaCO_3$ and carbonic acid, $CaCO_3$ is important in regulating the pH of soils. In most lake and river waters calcium is by far the most abundant cation, averaging about 20% of the dissolved solids. However, in seawater calcium makes up only 1.5% of the dissolved solids. The great relative loss of

CALCIUM: ELEMENT AND GEOCHEMISTRY

calcium in seawater is due to the near quantitative precipitation of calcium as $CaCO_3$. It is estimated that only 1.8% of all calcium supplied to the ocean has remained in the water. At the present time one-seventh of the calcium being contributed to the oceans is derived from the weathering of silicates; the remainder comes from calcareous sedimentary rocks.

One of the most important factors controlling the calcium content of both fresh and seawaters is the CO_2 content of the water or ultimately the CO_2 content of the atmosphere. With CO_2 available, calcium remains in solution in equilibrium with the bicarbonate ion rather than forming insoluble $CaCO_3$. In seawater, the solubility of $CaCO_3$ is directly proportional to salinity and the CO_2 partial pressure and inversely related to temperature and pH. Below depths of about 4000 meters, solution of $CaCO_3$ occurs, due in part to lower temperatures and the higher CO_2 content of these waters. Because of the close relationship between CO_2 content and $CaCO_3$ solubility, it is quite possible that the rate of solution or precipitation of $CaCO_3$ has varied significantly in the past depending on volcanic CO_2 productivity. Today, most surface waters of the open ocean are saturated or supersaturated with respect to $CaCO_3$.

At the present time about one-fifth of all sediment being deposited in the oceans is in the form of $CaCO_3$. Most of this calcium precipitation occurs in the deep sea, in middle or lower latitudes. These deposits are almost exclusively organic in origin and the contributing organisms are almost entirely pelagic. Certain algae and foraminifera are the important contributors of $CaCO_3$ and *calcite* is the principal mineral form. Extensive deep-sea deposition of $CaCO_3$ may not have commenced until the Cretaceous period when planktonic foraminifera became abundant. If the assumption can be made that deep-sea material is lost permanently to the continents, the amount of calcium in the geochemical cycle is possibly slowly diminishing.

Of somewhat less quantitative importance in present day oceans are the continental shelf carbonate deposits. Examples of areas of essentially pure $CaCO_3$ deposition in shallow water, where sedimentation rates are much higher than in the deep sea, are the Bahama Banks and the Persian Gulf. In contrast to the dominance of calcite in deep-water sediments, the shallow-water material is dominated by *aragonite* (which typically comprises 40% to 60% of the carbonate material) and *high-magnesium calcite* (10% to 30%). *Low-magnesium calcite* usually ranges in concentration between 10% and 40%. High-magnesium calcite is so unstable under most post-depositional conditions that it is rarely found even in Pleistocene sediments. Unaltered *aragonitic* fossils are not uncommon in Tertiary sediments, but solution or replacement of this mineral occurs readily.

A controversy exists concerning the relative importance of inorganic and organic precipitation of $CaCO_3$ in warm, shallow waters. However in terms of total oceanic $CaCO_3$ deposition, organic precipitation is strongly dominant.

Dolomite, $CaMg(CO_3)_2$, a common mineral in ancient sediments, is relatively uncommon in the Recent. It is known to occur in Recent intertidal and supratidal environments in carbonate areas. Isolated fresh-appearing *dolomite* rhombs have also been observed in deep-sea sediments. More commonly, it is a replacement product of calcareous sediments sometimes occurring while sediment permeability is sufficiently high to allow introduction of magnesium-rich seawater.

Biogeochemistry. Calcium is an indispensable element in the biosphere. The desirable properties which enhance its incorporation into biological systems include its abundance and its

TABLE 1. MINERALOGY OF CALCAREOUS SKELETONS[a]

Organism	Aragonite	Aragonite and Low-Mg Calcite	Low-Mg Calcite	High-Mg Calcite	High-Mg Calcite and Aragonite
Foraminifera	R[b]		C	C	
Sponges				C	
Corals	C			C	
Bryozoa	C			R	C
Brachiopods			C	R	
Echinoderms				C	
Mollusks	C	C	C	R	
Annelids	C			C	C
Arthropods			C	C	
Algae	C		C	C	

[a] Modified after Chave, 1962.
[b] C = common, R = rare.

ability to form a soluble cation in aqueous system. In vertebrate bones, calcium primarily occurs as the phosphate and secondarily as the fluoride and carbonate. In other organisms calcium-bearing internal and external skeletons are formed in such quantities as to have a profound effect on the geochemical cycle of calcium. Calcium is also an important component of plant and animal tissues and fluids serving such functions as an enzyme activator and electrolyte. The calcium cycle has a profound effect on plant growth because of its role in pH control.

The mineralogy of calcareous marine invertebrate skeletal material is summarized in Table 1. Skeletal mineralogy is a group characteristic. In some groups, more than one mineral may be precipitated in discrete layers within a single skeleton.

Uses. Calcium is used as a constituent of certain lead and aluminum alloys and as a reducing agent for the production of certain metals from their oxides. Slaked lime, $Ca(OH)_2$, when mixed with water and sand forms mortar. Portland cement is a mixture of calcium silicates. Plaster of Paris is $CaSO_4 \cdot 1/2\ H_2O$. Various compounds of Ca, particularly calcium phosphates, are widely used as commercial agricultural fertilizers.

ORRIN H. PILKEY

References

Chave, K. E., 1962, "Factors influencing the mineralogy of carbonate sediments," *Limnol. Oceanog.* **7**, 218–223.

Goldberg, E. D., 1963, "The Ocean as a Chemical System, in (Hill, M. N., editor), "The Sea," Vol. 2, pp. 3–25.

Goldschmidt, V. M., 1954, "Geochemistry," Oxford, Clarendon Press, 730pp.

Rankama, K., and Sahama, Th. G., 1950, "Geochemistry," Chicago, Univ. of Chicago Press, 911pp.

Revelle, R., and Fairbridge, R., 1957, "Carbonates and carbon dioxide," *Geol. Soc. Am. Mem.* **67** (1), 239–296.

Cross-references: *Bonding; Calcium Cycle; Earth's Crust Geochemistry; Electronegativity; Elements: Planetary Abundances and Distribution; Evaporite Processes; pH-Eh; Seawater; Weathering, Chemical.* Vol. IV B: *Meteoritic Minerals;* Vol. V: *Diorite; Dunite; Gabbro; Granite.* Vol. VI: *Deep-Sea Sedimentation; Evaporites; Pelagic Sediments; Sedimentation Rates—Deep-Sea; Spring and Crust Deposits; Travertine; Tufa.*

CALCIUM: ECONOMIC GEOLOGY

Natural Occurrence of Calcium. Calcium does not occur in the free state but is widely distributed in compounds, being the fifth most abundant element in the earth's crust. It is more plentiful in igneous (about 3.6%) than in sedimentary rocks and constitutes about 3.2% of the crust of the earth. It is the most abundant metallic element in the human body but ranks third in vegetation, being exceeded by potassium and sodium.

There are over 270 calcium-containing minerals. Those of most importance economically are marble, limestone, iceland spar, calcite, dolomite, fluorspar, anhydrite, gypsum, apatite, asbestos, tachydrite, the plagioclase feldspars, and calcium chloride (in solution in brines). Most of these are treated elsewhere. Only tachydrite, calcium chloride, and metallic calcium will be considered here.

Calcium Chloride. The greater part of commercial calcium chloride is synthetic, derived as a by-product of the Solvay process. However, the percentage recovered from brines and evaporite deposits is growing. The uses of both calcium chloride and calcium magnesium chloride (tachydrite) are increasing.

The chief centers of U.S. production of natural calcium chloride are California, Michigan, and West Virginia.

Calcium chloride is a white, soluble, hygroscopic salt. It is present in tachydrite and as an impurity in carnallite, but its most common natural occurrence is in solution. It forms several hydrates and can be completely dehydrated by heating to 200°C. Anhydrous calcium chloride can absorb more than its weight of water in twenty-four hours. Even in solution it is hygroscopic. In the evaporating pits of Bristol Lake near Amboy, California, when evaporating conditions are poor on winter days the specific gravity of the solution sometimes actually declines.

The conditions under which a calcium chloride-rich brine forms are not known but, unquestionably, unusual geologic conditions are necessary. It has been suggested that a possible source for the high $CaCl_2$ content of the Bristol Lake brine is Bagdad Crater, a recent volcanic cone on the northwest margin of the playa. Another hypothesis is that calcium chloride brines may form by a base-exchange process in which calcium in a contiguous clay bed is exchanged for the sodium in a sodium chloride brine.

The Bristol Lake deposits consist of a number of salt beds interlayered with playa sediment. The area is covered by a five-foot overburden beneath which is a five-foot layer of salt. The calcium chloride is recovered from brine which seeps into pits or ditches sunk below this first salt bed. It is believed, on the basis of evidence from bore holes, that the calcium chloride-rich brine is confined to the top thirty feet of the deposits.

In Michigan and West Virginia, wells sunk into underground saline formations are the source of the brines from which the calcium chloride is recovered.

At Stassfurt, E. Germany, calcium chloride is recovered from carnalite, which contains about three percent calcium chloride as an impurity.

The chief uses of calcium chloride are highway de-icing, dust control, concrete treatment, tire weighting, and brine refrigeration.

Calcium Metal. Pure calcium metal can be machined, sawed, threaded, drawn into wire, pressed, hammered into plates, and cast. It is trimorphous in form, and intermediate to sodium and aluminum in hardness.

Calcium is used to reduce such metals as uranium, thorium, titanium, zirconium, and chromium from their oxides or fluorides; as a deoxidizer and scavenger in the refining of nonferrous metals such as aluminum, nickel, chromium, and copper; to remove bismuth from lead and nitrogen from argon; as a constituent of numerous alloys; as a dehydrating agent for organic liquids; as a decomposing agent for thiophenes and mercaptans; as a desulfurizer for petroleum fractions; to improve the casting uniformity and machinability of iron, some steels, and magnesium and its alloys.

Compounds of calcium are very numerous and widely used in all branches of chemistry.

Until a few years ago, metallic calcium was produced by electrolysis of anhydrous calcium chloride. Today, practically all calcium is produced by the reduction of lime. Powdered lime and commercial-purity aluminum are briquetted; in chrome-nickel iron alloy retorts, under vacuum, and at temperatures of about 1170°C, the aluminum reduces the lime. In the temperature range of 680–740°C calcium vapor crystallizes in water-cooled sections of the retort which project through the furnace wall.

The production and consumption of calcium metal, tachydrite, and calcium chloride have been increasing each year. Many patents for new methods of producing calcium and calcium compounds have been granted in the last few years. However, there are, at present, only two plants in North America producing calcium: one at Canaan, Connecticut and one in Haley, Ontario. Both plants use the *aluminum-thermal vacuum retorts method*.

High-purity lime is required, usually containing over 97% calcium oxide. Silica and iron oxides may be tolerated to 2% or more but the alkali metals must be absent, or nearly so, since they can cause ignition on opening the retort.

WILLIAM G. TURNER

References

Bateman, A. M., 1950, "Economic Mineral Deposits," Second ed., New York, John Wiley & Sons, 916pp.

Bates, R. L., 1960, "Geology of Industrial Rocks and Minerals," New York, Harper & Row, 441pp.

Johnstone, S. J., and Johnstone, M. G., 1961, "Minerals for the Chemical and Allied Industries," New York, John Wiley & Sons.

Lamey, C. A., 1966, "Metallic and Industrial Mineral Deposits," New York, McGraw-Hill Book Co., 567pp.

Wright, L. A. (editor), 1957, "Mineral Commodities," *Calif. Dept. Natl. Resources, Div. Mines Bull.* **176**.

Cross-references: *Calcium Carbonate; Ion Exchange; Natural Brines.* Vol. III: *Playa.* Vol. IVB: *Calcite.* Vol. VI: *Dolomite; Evaporites; Limestone.*

CALCIUM CARBONATE: GEOCHEMISTRY

Mineralogy. Five polymorphs of pure calcium carbonate are known. Three of these, calcite, aragonite, and vaterite, have been assigned mineralogical names although only the first two are known to occur abundantly in nature. The remaining polymorphs, calcite II and calcite III, are structural modifications of the calcite lattice, and are stable only under pressures in excess of 15,000 bars at 300°C (Fig. 1). Both calcite II and calcite III invert rapidly to calcite upon cooling and/or reduction in pressure, and neither is known to occur naturally.

Calcite. The most abundant calcium carbonate mineral is the rhombohedral polymorph, calcite. This mineral, when pure, has a hardness of 3 on Mohs' scale, a specific gravity of 2.710, and a molar volume of 36.93 cm^3. It exhibits nearly perfect rhombohedral cleavage. The crystal structure is essentially a distorted NaCl-type lattice, with the cations in sixfold coordination.

Aragonite. The orthohombric polymorph, aragonite, is as abundant as calcite in Recent deposits, but becomes increasingly rare in high-temperature deposits and in rocks of pre-Pleistocene age. Aragonite is somewhat harder than calcite (3.5–4.0 on Mohs' scale) and has significantly greater specific gravity (2.947) with correspondingly smaller molar volume (33.96 cm^3). It has a distinct cleavage parallel to the c crystallographic axis, and very poorly developed cleavage planes parallel to a and b as well. The structure is a derivative of the nickel arsenide lattice, with the cations in ninefold coordination.

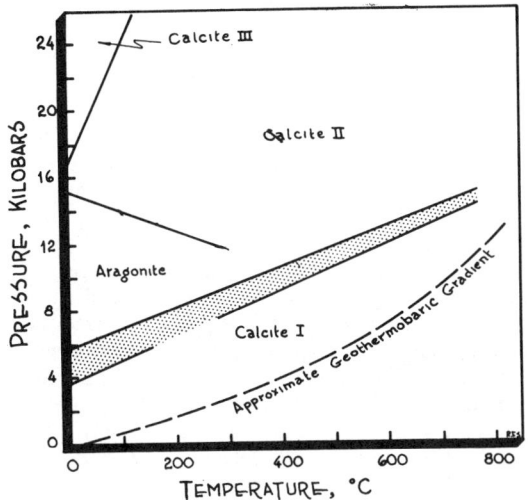

FIG. 1. Pressure-Temperature diagram showing stability relations among the calcium carbonate polymorphs. The shaded area represents the extent of experimental uncertainty in the location of the calcite I-aragonite equilibrium boundary (adapted from Jamieson, 1957; Jamieson, 1958; Simmons and Bell, 1963).

Vaterite. The hexagonal polymorph, vaterite, is very rarely found in nature. Its physical and chemical properties are little known. It is easily prepared in the laboratory, but inverts to either calcite or aragonite upon standing in contact with water or sodium chloride solutions. Its high chemical reactivity explains its scarcity in natural environments, though there is some evidence that vaterite may form as a short-lived initial precipitate in some inorganic and biological systems. The hardness of the mineral is not known; its specific gravity, calculated from x-ray cell size measurements, is 2.645 g/cm³. The corresponding large molar volume, 37.84 cm³, suggests that vaterite, if stable, must be a very low-pressure, and possibly high-temperature, phase.

Calcium Carbonate Hexahydrate. Although a hexahydrate of calcium carbonate ($CaCO_3 \cdot 6H_2O$) has been reported in the literature, little is known of the stability or validity of such a phase. If stable, it must form at low temperature, and presumably under high water pressures.

A calcium carbonate monohydrate is known to exist, and is found in nature intimately admixed with vaterite.

Stability. Calcite I is the stable polymorph of calcium carbonate at 25°C and 1 atm pressure. The Gibbs free energy of formation of calcite under these condittions, $-269.98 \pm .30$ kcal/mole, is only 0.23 kcal/mole less than that of aragonite ($-269.75 \pm .30$ kcal/mole) but because of the large difference between the molar volumes of the two polymorphs, calcite inverts to aragonite at 25°C only under pressures greater than 4000–6000 bars (Fig. 1). Although experimental difficulties lead to some uncertainty about the exact transition pressure at low temperatures, there is no doubt that the high-pressure phase is aragonite (as its small molar volume would suggest) and that the slope of the phase boundary is between 12 bars/degree and 17 bars/degree. It is of interest to note that the near-surface thermobaric gradient for the crust of the earth is close to 8 bars/degree, and even assuming a maximum temperature at the base of a 50 km crust of 800°C, the normal geothermobaric gradient probably nowhere exceeds 12–15 bars/degree Centigrade. An approximate geothermobar is shown in Fig. 1, and its position relative to the calcite-aragonite equilibrium boundary makes it clear that under normal conditions calcite must be the stable polymorph of calcium carbonate throughout the crust.

The free energies of formation of these polymorphs, and their relative stabilities may be changed by admixed chemical impurities, foreign ions held in the lattice in solid solution, elements or complexes adsorbed on the crystal surface ("poisons"), anomalous lattice energies (dislocations, lattice distortions, etc.; see *Order-Disorder*) or by particle-size effects (see *Crystallography*). Such factors, individually or in combination, may be responsible for the widespread occurrence of aragonite in biological systems and for the often reported precipitation of aragonite from seawater under conditions which would otherwise favor calcite as the stable polymorph. It is unlikely, however, that any of these factors will have lasting effects geologically, and aragonite so formed would be expected to have recrystallized to calcite in ancient rocks. The inversion would be favored by moderate temperatures and the presence of interstitial aqueous solutions. This conclusion is in satisfactory agreement with observed natural occurrences of aragonite.

The thermodynamic stability field of vaterite is unknown, though its high reactivity, unusual crystal structure, and very large molar volume suggest that it may be truly stable under no conditions at or near the earth's surface.

The stability fields of the high-pressure polymorphs, calcite II and calcite III, are indicated in Fig. 1.

Crystal Chemistry and Composition. The octahedrally coordinated cation site in the calcite lattice will accommodate divalent cations having radii in the range 0.50–1.00Å without significant distortion. Because the ca-

TABLE 1. CHEMICAL COMPOSITIONS OF NATURALLY OCCURRING CALCITE (2–7) AND ARAGONITE (8–14) (FROM PALACHE, BERMAN, AND FRONDEL, 1951)

	Calcites							Aragonites						
	1	2	3	4	5	6	7	8	9	10	11	12	13	14
CO_2	43.97	43.95	43.55	42.62	43.84	42.08	43.69	44.22	43.95	43.57	42.24	38.98	42.62	43.06
CaO	56.03	55.74	54.41	48.82	34.04	22.15	55.38	53.81	55.96	54.67	51.96	45.77	52.30	52.80
MgO		0.11	0.27		7.28	2.72	0.58	0.12	0.03					
FeO		0.04	0.15		13.05	0.29								
MnO		0.04	0.42	6.21	1.71	16.67								
SrO								0.30		0.67	5.23		3.87	
PbO										0.89		15.08		
ZnO				1.59										3.07
CoO		0.01												
BaO			1.27										tr	
Al_2O_3							0.05	0.32						
Fe_2O_3							0.06	tr						
Na_2O							0.01							
Ce_2O_3		0.007							0.22					
$(La, Sm, Di, Y, Er)_2O_3$		0.025							0.13					
SiO_2		0.032											tr	
H_2O							0tr 0.02							
Insoluble					0.24	0.08		0.30				0.20	1.42	0.53
Total	100.00	99.954	100.07	99.24	100.16	83.99	99.79	99.07	100.29	99.80	99.43	100.03	100.21	99.46

1. Theoretical; 2. Joplin, Mo.; 3. Elba; 4. Langban, Sweden; 5. Toggiano, Modena, Italy; 6. Franklin, N.J.; 7. Marble, Carrara, Italy; 8. Sarajeva, Yugoslavia (hot-spring deposit); 9. Matsushiro, Japan; 10. Tarnowitz, Silesia; 11. Tsumeb, Southwest Africa; 12. Postenje, Serbia; 13. Otago, New Zealand; 14. Friedensville, Penn.

tion is in ninefold coordination in the aragonite lattice, significantly larger cations may be introduced. The compositional variations observed in natural calcites and aragonites reflect this difference (Table 1). With few exceptions, the major impurities reported for calcite (magnesium, manganese, cadmium, ferrous iron, zinc, and cobalt), have radii in the range predicted by crystal chemical theory, suggesting that these foreign ions are present in substitutional solid solution in the calcite lattice. Ions smaller than 0.50Å (e.g., beryllium, radius = 0.35Å) are rarely present in more than trace amounts, even in geochemically favorable environments, and larger ions such as strontium or barium cause distortions of the lattice which preclude their entry into the lattice in significant quantities. Certain elements having suitable ionic radii (trivalent aluminum, r = 0.51Å; chromium, r = 0.63Å; and quadravalent molybdenum, r = 0.69Å) are reported in some calcite analyses, and appear to be present in defect solid solution. In geochemically favorable environments, copper (0.72Å), nickel (0.69Å), vanadium (0.95Å), and certain rare earths enter the calcite lattice, and divalent silver (0.89Å), germanium (0.73Å), and tin (0.93Å) might be found under very unusual circumstances.

As in the case of calcite, the major impurities reported in aragonite analyses appear to be present in substitutional solid solution. Significant amounts of strontium, barium, and lead are reported replacing calcium in aragonite, and in spite of its very small radius, magnesium is often reported as an important impurity. In exceptional environments, trace amounts of radium (1.43Å) and mercury (1.10Å) may occur.

The importance of ionic radius in determining chemical impurities in calcite and aragonite is evident from a comparison of the analyses in Table 1 with the crystal chemical data presented in Table 2. Table 2 also suggests the extent of stable solid solution in each phase at 25°C and 1 atm. It is evident that, in accord with Pauling's rules, the more similar the substituting ion with respect to both size and charge, the more extensive the solid solution observed. At temperatures higher than 25°C the range of solid solution increases, and many systems which show very limited isomorphous substitution at room temperature exhibit unbroken solid-solution series at temperatures of only a few hundreds of degrees. Notable among these are the solid solutions of strontium in aragonite and of magnesium in calcite (Fig. 2).

It is evident from the foregoing that carbonate minerals formed at high temperatures under favorable geochemical conditions may contain solid-solution impurities in amounts which would be unstable under surface conditions. If not too abundant, these excess concentrations may be retained in the calcium carbonate lattice where they may provide a useful indication of the temperature and geochemical ambient of crystallization. Under other cir-

TABLE 2. IONIC RADII, RADIUS RATIO, AND EXTENT OF SOLID SOLUTION EXHIBITED BY MAJOR ELEMENTS WHICH ENTER THE CALCIUM CARBONATE LATTICE IN SOLID SOLUTION

Element	Radius (Ahrens)[a]	Radius Ratio, Cation/CO_3	Extent of Solid Solution at 25°C
Aragonite (C/O = 9)			
Barium	1.34Å	0.87	Very limited
Lead	1.20	0.80	Limited (1:12)
Strontium	1.12	0.75	Partial (1:25)
Calcium	0.99	0.67	
Calcite (C/O = 6)			
Calcium	0.99	0.67	
Cadmium	0.97	0.65	Unknown (presumed complete)
Manganese (2+)	0.80	0.54	Complete
Iron (2+)	0.74	0.50	Nearly complete
Zinc	0.74	0.50	Partial
Cobalt	0.73	0.50	Partial
Magnesium	0.67	0.47	Very limited (see Fig. 2.)

[a] A radius ratio of 0.67 corresponds to the theoretical limit of six-fold coordination in the calcite lattice.

cumstances, the impurities may be exsolved to form secondary carbonate phases, oxides, or hydroxides, or they may be carried away in solution. The resulting equilibrium assemblages may reflect the temperature of exsolution or recrystallization.

By a similar argument, the inversion of aragonite may lead to the formation of calcite abnormally enriched in strontium, lead, or barium, or to polyphase mixtures of calcite and strontianite, witherite, or cerussite.

Isotope Chemistry. Few studies have been directed toward either stable or radioactive calcium isotope distribution in calcium carbonate minerals. Although such studies appear to offer valuable data in igneous and metamorphic rocks, the partition coefficients at low temperatures and the relative scarcity of all isotopes except Ca^{40} and Ca^{44} in nature demand analytical accuracy beyond the limits of instruments presently available.

Coprecipitation studies using radioactively "tagged" isotopes of certain solid-solution impurities (either natural or neutron activated) provide valuable insight into low-temperature solution equilibria, however. Outstanding results have been achieved using neutron-activated Zn to study the calcite-smithsonite system, and Sr^{90} in the aragonite-strontianite system.

Attempts to employ stable isotopes of major impurities to interpret diffusion and metasomatic phenomena have in general yielded ambiguous results. For example, the promise of the Mg^{24}/Mg^{26} method for studying metasomatic dolomitization has not been realized, in part due to the experimental difficulties involved, and in part due to the extremely small amounts of Mg^{26} in natural systems, and the limits imposed by analytical errors.

Isotopic examination of the carbonate anion is, of course, not restricted to the calcium carbonate minerals, but because of their natural abundance, it is here that studies of carbon and oxygen isotopes have achieved their greatest usefulness.

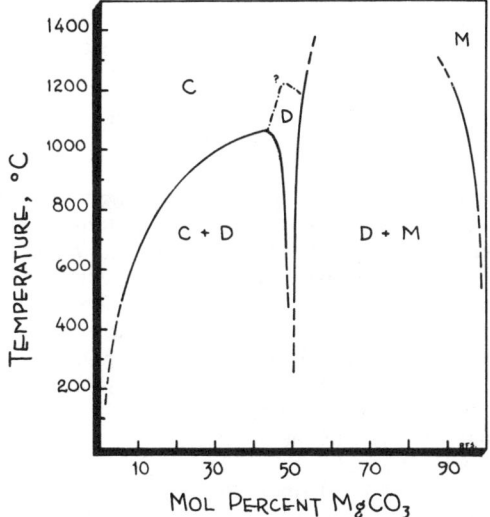

FIG. 2. Temperature-composition diagram for the system $CaCO_3$–$MgCO_3$. Symbols used are as follows: C = calcite phase (including magnesium calcite solid solutions), D = dolomite, and M = magnesite. The section shown is not isobaric, pressures range from 2 kb (at low temperatures) to 15 kb (at 1250°C) (adapted from Goldsmith and Heard, 1961, Fig. 3).

Radiocarbon (C^{14}) dating techniques have met with considerable success when dealing with rocks and fossil material up to 40,000 years old. Such studies have been particularly helpful in interpreting the history of sediment deposition in areas of extensive secondary transport, as for example in the turbidite sequence of the Tongue of the Ocean (Bahamas). Radiocarbon techniques have also been valuable in studies of Recent dolomite formation and in dating calcareous fossil material from late Pleistocene and Recent deposits. Anomalous results occasionally reported from organic carbonates probably reflect the abundance of ancient carbonate material in the environment of the living organism, and serve to emphasize the need for particular care in sampling and interpreting isotopic data from biogenic materials. The errors introduced are small, geologically, but may have great significance when dealing with material less than a few thousand years old. The case of the fresh oyster served on a 500–1000 year old (C^{14}) half-shell at a geological society banquet a few years ago is a familiar and typical example. (For a more detailed discussion, see *Carbon-14 Dating*.)

Because of the nature of the respirational and transpirational processes, terrestrial organisms tend to become abnormally enriched in C^{13} relative to C^{12}. Although the selective concentration of the heavier isotope is less marked in marine organisms, the C^{13}/C^{12} ratio has proved a valuable tool for distinguishing biogenic from inorganic carbonate materials.

A more significant control on the C^{13}/C^{12} ratio is exercised by the loss of volatiles from the sea surface. Seawater is isotopically heavy relative to the atmosphere or to fresh-water bodies on the continents with respect to both carbon and oxygen. Other things being equal, marine carbonate materials are correspondingly heavier isotopically than are fresh-water minerals. Both C^{13}/C^{12} and O^{18}/O^{16} ratios have been used to distinguish marine from fresh-water carbonates. Though in many examples the separation is clear, there exists a large area of isotopic overlap within which interpretation is necessarily ambiguous in the absence of correlative geologic or paleontologic evidence (see *Paleosalinity*).

Isotopic partition between solid materials and adjacent fluid phases is highly sensitive to temperature, and the isotope ratio of carbonate minerals, especially the O^{18}/O^{16} ratio, has been used with outstanding success to estimate the temperature of deposition (see *Paleotemperatures*).

General Geochemistry. *Origin.* Calcium is abundant in mafic igneous rocks, metamorphic rocks, and sedimentary rocks exposed on the continents. It is leached by the action of water (hydrolysis) or by reaction with dilute natural acids (either organic acids or carbonic acid). The calcium is highly mobile, and is carried in solution as ionic calcium or in complexes. Calcium is the most abundant cation in normal surface and ground waters on the continents.

Calcium in solution reacts readily with dissolved carbonate ion (produced by dissolution of carbon dioxide from the atmosphere and by organic decay) and is precipitated as nodules and crusts in soils, as interstitial cement, in vein fillings, and in cave deposits. It is also extracted by organisms in lakes, soils, and streams, and may be deposited as gypsum ($CaSO_4 \cdot 2H_2O$) from solutions rich in sulfate ion. Because of these changes, calcium is much less abundant, relative to sodium, in the sea, though this change reflects in part the massive removal of calcium from seawater by carbonate-secreting marine organisms.

Solubility. The occurrence, distribution, and composition of most deposits of calcium carbonate are directly controlled by the solubility of calcium carbonate minerals in aqueous solutions. The thermodynamic equilibrium constant for the reaction.

$$CaCO^3 (\text{solid}) \rightleftharpoons Ca^{2+}(\text{aq.}) + CO_3^{2-}(\text{aq.})$$

has a value of $10^{-8.41}$ when the solid phase is calcite and $10^{-8.23}$ when the solid phase is aragonite. The values, calculated from the free energies of formation at 25°C and 1 atm, yield the following standard-state solubilities in distilled water and in the absence of carbon dioxide:

Calcite 6.2×10^{-5} molal 6.2 mg/kg(ppm)
Aragonite 7.8×10^{-5} molal 7.8 mg/kg(ppm)

In most natural systems, the activity of carbon dioxide exerts the most important control on the solubility of calcium carbonate minerals. This effect, occasioned by the free hydrogen ion present in carbonic acid solutions, is shown in Fig. 3. In distilled water in equilibrium with the normal atmospheric partial pressure of carbon dioxide ($10^{-3.5}$ atm), the solubility of calcium carbonate increases to nearly ten times that observed in the absence of CO_2; at 1 atm of CO_2, the solubility of both solid polymorphs is close to 1000 ppm. Because surface waters exhibit a range of carbon dioxide activities from 10^{-4} to as great as $10^{-2.6}$ depending upon exposure to the atmosphere, availability of organic decay products, photosynthesis, and organic metabolism, it is evident that rapid local changes in the saturation state of a solution may be observed.

At constant total pressure and partial pres-

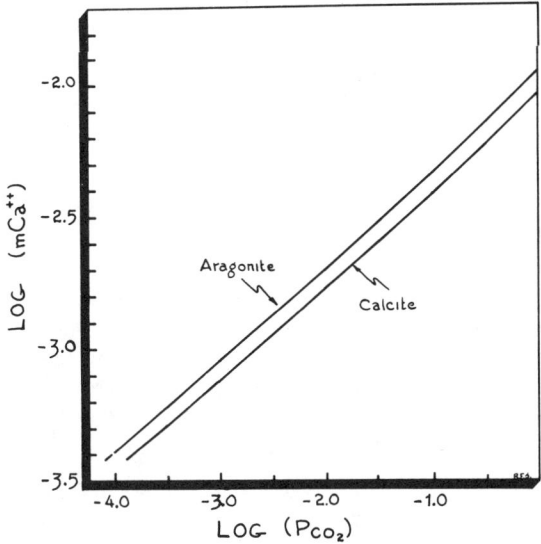

Fig. 3 Solubility of calcium carbonate as a function of carbon dioxide pressure. Solubility of aragonite (upper curve) and calcite (lower curve) expressed as the negative logarithm of the molal concentration of calcium ion in equilibrium with the solid phase plotted against the negative logarithm of carbon dioxide pressure (activity) (unpublished data from Schmalz, 1962).

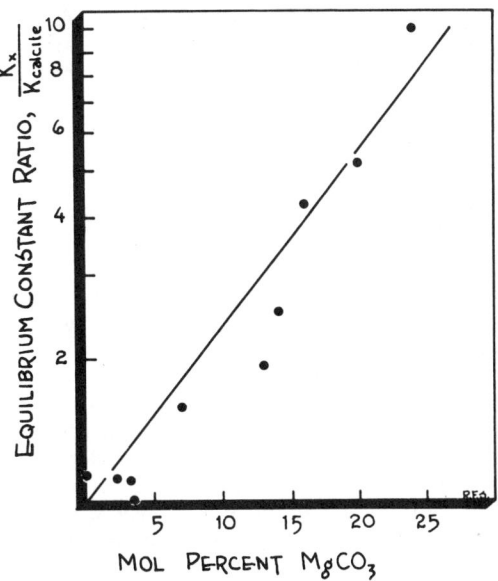

Fig. 4. Solubility of organically secreted magnesian calcites at 25°C expressed as a function of mole percent magnesium and the ratio of equilibrium constants, $K_x/K_{calcite}$ (adapted from Chave et al., 1962, Fig. 1).

sure of carbon dioxide, the solubility of calcium carbonate decreases slightly with increased temperature. This effect is overshadowed, however, by the negative temperature coefficient of CO_2 solubility, but may achieve great significance in high-temperature systems.

Increased total pressure increases the solubility of calcium carbonate minerals to a meaningful extent. Theoretical and experimental estimates of the magnitude of the change, under conditions of constant temperature and carbon dioxide pressure, yield values between 0.15 and 0.3%/atm.

The difference between the solubilities of calcite and aragonite reflect the difference in free energy of formation of the two polymorphs. Any factor which contributes to the free energy of formation of the solid phase other than crystal structure will similarly affect the solubility of the solid. Among the most important contributors to the free energy of the solid phase are impurities present in solid solution, defects and distortions of the crystal lattice, and reduced particle size with corresponding increase in the surface energy contributions. These effects have been studied by many investigators and selected examples are presented in Fig. 4 and Tables 3 and 4.

Most fresh surface waters are undersaturated

TABLE 3. EFFECT OF PARTICLE SIZE UPON THERMODYNAMIC EQUILIBRIUM (SOLUBILITY) CONSTANT FOR CALCITE (AT 25° C)[a]

Average Particle Size		Thermodynamic Equilibrium Constant, $K_{25°}$
(cm)	(microns)	
1	10^4	$10^{-8.41}$
1×10^{-4}	1	$10^{-8.40}$
2×10^{-5}	0.2	$10^{-8.34}$[b]
1×10^{-5}	0.1	$10^{-8.33}$
1×10^{-6}	0.01[c]	$10^{-7.70}$

[a] Data from Chave, and Schmalz, 1966.
[b] It is interesting to observe that the National Bureau of Standards thermodynamic data for calcite yield this value.
[c] Calculated value derived from particle size and specific surface energy.

TABLE 4. EFFECT OF GRINDING-INDUCED STRAIN ON CALCITE SOLUBILITY
FOR VARIOUS PARTICLE SIZE INTERVALS[a]

Size Interval (microns)	$K_{25°}$ Annealed Particles	Grinding Time (min)	$K_{25°}$ Freshly Ground Particles
$> 177 \times > 125$	$10^{-8.41}$	10	$10^{-8.02}$
$> 125 \times > 88$	$10^{-8.41}$	20	$10^{-7.70}$
$> 88 \times > 63$	$10^{-8.40}$	10	$10^{-8.03}$

[a] Data from Chave, and Schmalz, 1966.

with respect to calcium carbonate, but where such waters are warmed, or undergo appreciable reduction in total pressure or partial pressure of carbon dioxide, supersaturation may result, with consequent deposition of calcium carbonate minerals. Hot-spring deposits (travertine) reflect the simultaneous reduction in pressure and evolution of CO_2 where warm subsurface waters emerge; they probably reflect only to a minor extent the loss of water by evaporation. Cave deposits (stalactites, stalagmites, flow stone, etc.) are probably largely due to losses of CO_2 to the cave atmosphere, though evaporation of water may play an important role in some occurrences, and bacterial activity or photosynthesis may play a dominant role in others. The carbonate crusts and nodules of desert soils (caliche) are primarily the result of evaporation. Calcium carbonate minerals are also deposited by direct organic activity, though the mechanisms responsible for such deposits are poorly understood.

In the presence of foreign ions in solution, the solubility of calcium carbonate is increased substantially. In solutions of low ionic strength, this effect is the result of the "foreign ion effect" and may be treated theoretically by the Debye-Hückel equation. In more concentrated solutions (ionic strength greater than 0.1) complexing further reduces the activity of calcium, carbonate, and bicarbonate ions in solution, and thereby increases the amount of calcium carbonate which will dissolve. In seawater (ionic strength = 0.7), the solubility of calcite is nearly ten times as great as in distilled water, temperature, pressure, and carbon dioxide activity being the same. The magnitude of this effect is shown in Fig. 5, in which the apparent solubility product (or ion concentration product) is plotted against salinity. There is some experimental evidence that this trend toward higher solubility with increased salinity reverses after the ionic strength reaches a value close to 1.0 (corresponding to a salinity close to 40‰). This observation has yet to be confirmed, but if valid, it must have important geological consequences.

The factors which affect deposition of calcium carbonate from marine waters are the same as those noted in fresh water, with the added effect of salinity. Organisms which secrete calcium carbonate tests remove large amounts of calcium carbonate directly, and others influence the saturation state of the water mass significantly by changing local activity of CO_2 through respiration or photosynthesis. The surface water of the sea is generally observed to be slightly oversaturated with respect to calcite (about 25–30 ppm) in low latitudes, approximately saturated (in equilibrium) with respect to calcite in mid-latitudes, and slightly undersaturated in high latitudes in both hemispheres. Bottom waters, because of their low temperatures (generally within one or two degrees of zero), high pressures, and high CO_2 activities occasioned by decay of organic matter on the sea floor, are believed to be undersaturated with respect to calcite. The high solubility of calcium carbonate in deep marine waters is perhaps responsible for the reduced abundance of calcium carbonate in deep-water marine sediments (Fig. 6) and its general absence from sediments in waters

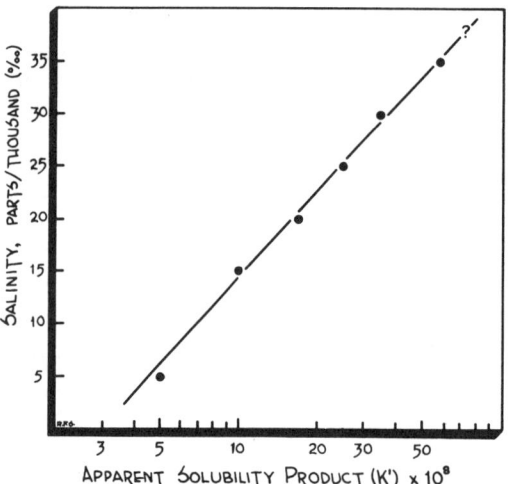

FIG. 5. Apparent solubility product (K') for calcite expressed as a function of salinity (in parts per thousand) (from Wattenberg, 1933).

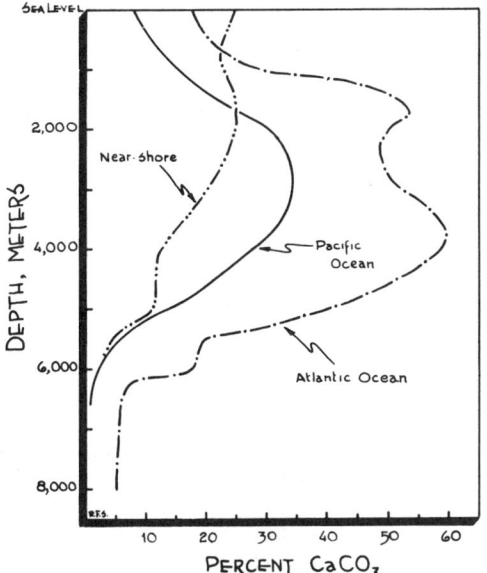

FIG. 6. Average percentage of calcium carbonate in pelagic sediments from the Pacific Ocean, Atlantic Ocean, and near-shore sediments as a function of depth (in meters) (adapted from Sverdrup et al., 1942, Fig. 259).

deeper than 4500 m. Calculations based on known temperature and pressure coefficients of solubility, however, suggest that all marine waters deeper than about 500 m must be undersaturated with respect to calcite, and the absence of calcium carbonate in deep-water sediments may be due to some other effect or to coincidence.

When cool bottom waters are brought to the surface by upwelling, they become warmer and simultaneously the pressure is reduced by as much as several hundreds of bars. The resulting decreased solubility of calcium carbonate may lead to supersaturation, and such a mechanism has been adduced to explain at least part of the carbonate sediment presently accumulating on the Bahama platform. Here, however, additional warming, evaporation, and CO_2 loss as the water spreads westward across the Bahama bank may cause even greater supersaturation and inorganic precipitation. Supersaturation in tropical waters may also be caused by photosynthetic removal of carbon dioxide by green plants. The calcareous algae remove large amounts of dissolved calcium carbonate, which is secreted by the plants and added to the sediment upon their death. Such calcareous algae may in fact be the major contributors to carbonate sediments in tropical waters. The role of bacteria in the deposition of marine carbonate sediments has not been adequately studied, but there is increasing evidence that they play an extremely important part, perhaps being responsible in part for the formation of oolites which have long been considered *prima facie* evidence of inorganic precipitation.

Melting Relations. At one atmosphere pressure, calcium carbonate does not melt, but decomposes to calcium oxide (lime) and carbon dioxide. The more stable polymorph, calcite, breaks down at 895°C, whereas aragonite, being less stable, decomposes at 825°C. Under 1000 bars pressure (CO_2), however, calcium carbonate melts congruently at a temperature slightly above 1300°C in the absence of water. The melting point is lowered by the presence of water, and congruent melting is observed at temperatures as low as 675°C in the presence of excess water at a pressure of 1000 bars. These melting relations, only recently studied in detail, admit the possibility of forming igneous carbonate rocks, and since 1960 a number of occurrences of both extrusive and intrusive carbonate rocks have been described.

Occurrences. Because the composition and other geochemical properties of calcium carbonate differ in accord with the mode of occurrence, it is necessary to treat the various important types of occurrence briefly.

Igneous Rocks. Under normal circumstances in the crust where the necessary volatile pressures (both CO_2 and H_2O) can be developed and maintained, natural carbonate magmas may form and give rise to both extrusive and intrusive igneous rocks. Such rocks, called *"carbonatites"* (q.v. Vol. V), are often found to be made up of more than 90% carbonate minerals, generally a mixture of calcite and sodium carbonate (shortite, nahcolite). Natural carbonatites make up only a small fraction of one percent of the igneous rocks exposed on the continents, but they have now been recognized in Africa, Australia, western North America, and in the U.S.S.R. Carbonatites do not contain aragonite, since the pressures necessary to stabilize aragonite at the melting temperature are rarely attained and if formed, aragonite would rapidly invert to calcite as the rock cooled.

Carbonatites are unusually rich in certain of the rare earth elements (La, Eu, Yb) and Ba, Sr, Zr, Nb, Y, Mn, P, and F have been identified as distinguishing or "index" trace elements.

Carbonatite magmas appear to be immiscible in normal silicate melts, and calcite-filled vugs and amigdules in extrusive and near-surface intrusive bodies, previously interpreted as alteration products, may represent segregations of carbonatite fluids from the cooling silicate mass.

Metamorphic Rocks. The metamorphism of

carbonate rocks, of whatever origin, causes recrystallization of the constituent mineral grains to a single phase or an assemblage of mineral phases stable under the temperature and pressure conditions prevailing at the time of metamorphism. The resulting rock is called a *"marble"* (q.v. Vol. V). Except under unusual conditions of high pressure and low temperature (as in association with the glaucophane schists of California), calcite is the major metamorphic calcium carbonate mineral. The recrystallization is generally accompanied by crystal growth, often with the destruction of primary features (bedding, fossils, etc.) and may, therefore, make genetic interpretation of the rock unit difficult or impossible. Interpretation may be made additionally difficult by ptygmatic folding, flow, etc., resulting from the low elastic limit exhibited by calcium carbonate under pressures characteristic of regional metamorphism or tectonism.

In the recrystallization process, particularly when a polymorphic inversion is involved, trace elements stable in the original mineral may be exsolved. Although slight excesses of certain elements (e.g., strontium) may be retained in metamorphic calcite, suggesting that the original mineral was a strontian aragonite, when present in larger amounts, these exsolved cations may be removed in solution or may crystallize locally as separate phases. Such assemblages as calcite-dolomite, calcite-siderite, and calcite-strontianite are often reported, and wolastonite ($CaSiO_3$) is not unusual in siliceous marbles. The exsolved phases generally appear to be in equilibrium with the primary mineral of the rock (calcite) and as such have been used to estimate the temperatures and pressures at which recrystallization took place (see *Metamorphic Facies, Vol. V*).

Hydrothermal Deposits. Calcite is extremely common in relatively low-temperature hydrothermal deposits. It is the primary gangue mineral in many lead and zinc deposits, and is frequently associated with copper and iron deposits as well. Such hydrothermal calcites are often observed to be abnormally rich in metallic impurities, the abundance of impurities being more or less related to the temperature at which the deposit was formed. Trace elements, especially the rare earths, and those divalent and trivalent elements mentioned earlier are often reported.

Petrographic characteristics suggest that many of these hydrothermal deposits formed by the slow deposition of calcite from circulating calcium- and carbonate-rich solutions, but in some cases small crystallite size suggests rapid deposition, perhaps occasioned by sudden release of carbon dioxide pressure.

Where hot carbonate-rich solutions emerge at the surface, carbonate minerals are deposited as travertine, probably due in part to evaporative concentration of the solution, but due largely to supersaturation occasioned by evolution of CO_2. Travertine deposits are generally associated with late stages in hot-spring development, and consequently with lower-temperature springs. In addition to the impurities mentioned earlier, travertine deposits in volcanic regions commonly carry significant amounts of arsenic, sulfur, and antimony as well as other elements characteristic of fumeroles. Bacterial activity plays an as yet inadequately studied role in the formation of such deposits.

Even at very low temperatures, the increased solubility of calcium carbonate under pressure leads, in the presence of water, to its dissolution in local areas of high stress concentration and redeposition in nearby regions of lesser stress. As a consequence, calcite of exceptional purity is often found filling veinlets, cracks and fractures in even slightly deformed rocks. Such fracture fillings have sometimes been confused with hydrothermal deposits, but their high purity indicates a low-temperature origin, and their distribution often makes their stress-related origin evident.

Evaporite Deposits. Because calcium carbonate is the least soluble mineral present in quantity in solutions of either fresh or seawater, except where sodium and/or sulfate is unusually abundant, it is the first mineral to be deposited upon evaporation of the solvent. Both calcite and aragonite have been identified as primary minerals in early evaporite sequences, and one or both are frequently found interbedded with later evaporite salts such as gypsum, anhydrite, or halite. Because the amount of calcium carbonate in most natural waters is relatively small, in closed (isochemical) evaporite basins, calcium carbonate makes up only a small part of the total salt accumulation, and is generally absent in the later stages. Where reaction between the early-deposited calcium carbonate and the later brines is possible, the carbonate mineral may be altered to dolomite or enriched in strontium, barium, or other elements. In sulfate-rich waters, calcium is first removed as gypsum ($CaSO_4 \cdot 2H_2O$) and the enhanced Mg/Ca ratio in the remaining solution may favor the primary precipitation of a highly magnesium calcite or dolomite. Recent dolomites described from the Dutch Antilles, the Florida and Bahama shelves, and the Red Sea are probably the result of such a process.

Cave deposits are frequently cited as examples of inorganic precipitation of calcium carbonate due to evaporation of water from carbonate-saturated solutions as they flow over a broad surface. The high humidity observed in limestone caverns argues against evaporation as the sole cause of carbonate deposition, however, and the evolution of CO_2 from the solution, coupled perhaps with organic activity, may afford a more satisfactory explanation.

Caliche crusts in soils where evaporative losses from the surface exceed precipitation are clearly evaporite deposits, and carbonate minerals found in such crusts are commonly associated with gypsum.

Upon evaporation (or degassing) of seawater, calcium carbonate is the first mineral to be deposited, though surprisingly aragonite is the polymorph most commonly observed. The deposition of this metastable phase poses a number of interesting problems, but a possible solution is offered elsewhere in this Encyclopedia (see *Evaporite Processes*). If the evaporation takes place in a thermodynamically "open" system, experimental data show that the later sulfate-rich brines redissolve the aragonite, freeing CO_2 and redepositing the calcium as gypsum. This observation may help to explain the relative scarcity of calcite (or aragonite) in many evaporite deposits.

Biogenic Carbonates. Many organisms secrete hard parts which are made up exclusively or in large part of calcium carbonate. The physiologic process or processes by which such organic secretion is accomplished is not yet understood, but organic calcification is the subject of intensive research at this time, and with the aid of modern methods (radioactive tracers, electron microscopy, and microprobe techniques, etc.) substantial progress may be anticipated in this area in the next few years. Organisms are known to secrete both calcite and aragonite, and both mineralogical and chemical composition differ from species to species, and even among individuals of the same species. Many organisms secrete hard parts selectively enriched in particular elements (among which, magnesium is very important) and recent studies have shown that within a single species, the chemical composition, like the isotopic composition, shows variations which reflect ambient temperature and seasonal growth. Within a single individual, different parts of the test (as for example in the Echinoidea) may show significantly different chemical and isotopic compositions. The problem is evidently extremely complex, but it affords promise of important applications in paleoecological studies in particular, and is now being investigated.

One of the most abundant elements in organically secreted carbonate materials is magnesium, and the distribution of magnesium in calcites of organic derivation has now been sufficiently studied to make a few generalizations possible. First, many organisms (see below) secrete calcite containing magnesium in amounts far in excess of the amount which is known to be stable at surface temperatures (Fig. 2). Mollusks with calcite tests containing up to 5% (mole) $MgCO_3$ are well known, echinoderms secrete tests with between 8% and 15% $MgCO_3$, and many algae, particularly the red algae, secrete calcite with more than 20% $MgCO_3$. Such compositions are clearly metastable, and diagenetic reactions would be expected to alter or remove them in sedimentary deposits (see discussion below). Secondly, the amount of magnesium present in the test of any given species varies directly as the temperature of the growth ambient. Tropical echinoderms may carry up to 14% $MgCO_3$, but the same species collected in high latitudes shows much less abundant magnesium. Thirdly, the magnesium content of the organism's test varies seasonally, and the temperature dependence of magnesium content appears to decrease with increased phylogenetic rank.

A brief summary of mineralogic and chemical properties of the hard parts of organisms follows:

Protista: In addition to the well-known siliceous protozoans (diatoms and radiolarians) many single-celled organisms secrete carbonate tests. Most forms secrete unimineralic tests, and calcitic, aragonitic, and magnesian calcitic tests have been recognized.

Porifera: The calcareous sponges are known to secrete only magnesian calcites. Low magnesian calcites (up to a few percent) are most common in high latitudes; somewhat higher magnesium contents are recorded in tropical forms (up to 8%).

Coelenterata: The coelenterates, among which the corals are the most important group, secrete both aragonite and magnesian calcites of moderate magnesium content. Those corals secreting solid skeletal structures, the hexacorals, generally produce aragonite, enriched in some cases in strontium, whereas the "soft" corals (the octacorals including the "sea fans" and *Gorgonia*) build their skeletons of magnesian calcite (up to 15% $MgCO_3$) spicules bound together by organic integuments.

Bryozoans: The Bryozoa secrete both aragonite and moderately magnesian calcite, the magnesium content of the calcites in general increasing in low latitudes. Some varieties secrete bimineralic skeletons made up of mixtures of magnesian calcite and aragonite.

Brachiopoda: Phosphatic brachiopods are familiar, but calcite and low-magnesium calcite are secreted by modern articulated forms.

Mollusca: Most mollusks secrete a two-layered shell: an inner (nacreous) layer of pearly aragonite, which may be strontian, barian, or magnesian, and an outer (conchineal) layer of prismatic calcite which may be somewhat magnesian (up to 5%) and is often impregnated by collogen-like fibers and sheets. The gastropods generally show the two-layer structure, with low-magnesium calcite in the conchineal layer, though some species have tests made up exclusively of aragonite. The pelecypods include the normal two-layered forms, but also some species which secrete either calcite or aragonite exclusively. The cephalopods generally secrete unimineralic tests of either aragonite or moderately magnesian calcite (up to 10%) though in some forms (notably the straight fossil forms) there is evidence that the main shell was aragonitic and the operculum calcite. The amphineura and scaphopods secrete aragonite exclusively.

Echinoderms: Almost without exception, the echinoderms secrete calcite tests, the only significant impurity in which is magnesium. The magnesium content of tropical forms may range between 8 and 14% depending upon the species, but in colder waters the amount of magnesium present decreases to 2–6%. Within a single individual the amount of magnesium in solid solution shows significant variation from plate to plate, depending upon position in the exoskeleton.

Annelida: The annelids secrete aragonite, magnesian calcite (up to 20%), and both phases together. The mineralogy is determined by the particular species; the serpulids secreting the most magnesian calcites known with the exception of certain red algae.

Decapods: The decapods are characterized by low- to moderate-magnesian calcite or phosphatic skeletons.

Ostracoda: the ostracodes secrete calcitic tests with up to 10% $MgCO_3$. The magnesium content is determined in part by species and in part by ambient water temperatures.

Ciraepoda: The ciraepods, like the ostracodes, secrete calcitic tests containing variable amounts of magnesium determined by species and ambient conditions.

The Protochordates and higher animals in general secrete phosphatic skeletons, though small amounts of aragonite may be concentrated in certain organs, and among the birds, calcite is secreted in egg shells.

Algae: The calcareous green algae secrete aragonite almost exclusively, and this material is of very high purity. Few of the brown algae secrete carbonate, but at least one genus, *Padina,* has fronds covered by minute crystals of calcite which by their systematic arrangement with respect to the plant structure appear to be organically elaborated. As in the case of the aragonite secreted by the green algae, this calcite is of high purity. The red marine algae secrete calcite which is characteristically very rich in magnesium. In tropical waters compositions of up to 21% $MgCO_3$ are reported. One genus, *Goniolithon,* contains up to 29% Mg by wet analysis, but a substantial amount of this magnesium appears to be present as the mineral brucite $(Mg(OH)_2)$ which fills cellular openings in the skeletal structure. Seasonal variations in the magnesium content of high-latitude red algae have been reported.

Sedimentary Carbonate Deposits. Sedimentary carbonate deposits may be divided into three general types: biogenic and bioclastic deposits, secondary carbonate deposits, and deposits of mixed and/or uncertain genesis.

Modern and ancient reef masses, preserved in place, generally contain sufficient fossil material to make their origin quite definite. Examples of biogenic rocks include the Recent coral reefs and atolls of tropical marine waters, the Permian algae reefs of the Paleozoic of the north central states. Depending upon the nature and extent of diagenic and postdiagenic alteration, such deposits reflect by both their mineralogy and composition the characteristics of the organisms responsible for their accumulation. Under conditions that cause extensive diagenetic alteration, aragonite and magnesian calcites may be removed or altered to low-magnesium calcite, and in some examples dolomitization may be observed (see below). Bioclastic sediments are those made up of transported and redeposited debris of unquestionable biogenic origin. They include the cochinas of Florida and the Dover chalk, as well as the "fore-reef" facies associated with reef masses both ancient and modern. Because they are made up of identifiable biogenic clasts, the Pleistocene eolianites of Bermuda and the Bahamas are also termed bioclastic. The bioclastic deposits are mineralogically and chemically similar to biogenic rocks preserved in place, and are subject to the same types of alteration.

Secondary carbonate deposits are those accumulations of carbonate debris derived from some preexisting source. Obviously, many of the bioclastic deposits fall into this category. It is evident also that a sedimentary deposit may be classed as secondary with confidence only under unusual circumstances. The modern carbonate sediments of the Florida shelf, containing abundant Pleistocene and Tertiary fossil fragments derived from exposures on the continent, are typical of known secondary carbonate deposits. As in the case of the bioclastic deposits, the mineralogy and composition of secondary carbonates may reflect diagenetic alteration, but it may also reflect extensive changes occasioned by weathering or diagenetic alteration of the primary source.

Many carbonate deposits, including vast thicknesses of large areal extent, are made up in whole or in major part of extremely fine-grained carbonate crystals which provide no evidence of origin. Though such rocks may include scattered fossils, the origin of the fine-grained matrix is the subject of a continuing controversy. The carbonate sequences of the miogeosynclinal facies of practically any geosyncline include many examples of such problematical rocks. Oolitic carbonates, though often accepted as inorganically precipitated, by one or a combination of the processes mentioned earlier, are also the subject of considerable debate at the present time. The most direct evidence in opposition to their inorganic origin is the observation that oolite-like aragonite deposits form in degassed but unsterilized seawater solutions, whereas no such sediment can be produced from water which has been sterilized as well as filtered. The origin of the fine-grained limestones has been attributed to wholly inorganic precipitation, and in defense of this thesis, attention has been drawn to the morphological similarities between artificially precipitated aragonite crystals and those characteristic of both modern and ancient fine-grained carbonate rocks. Because such crystallites differ very slightly from those forming the skeletal secretions of many of the marine algae, however, some workers believe that they must be considered of biological origin. The argument remains unresolved, and a definite conclusion may be impossible in the case of ancient rocks, and even in modern carbonate environments definitive evidence of inorganic precipitation has yet to be discovered. The application of modern research tools has significantly reduced the number of carbonate deposits which admit of an inorganic precipitational origin, however, by revealing that many of the rocks and sediments previously considered to be inorganic precipitates are in fact made up of minute fragments of clearly identifiable coccolith, protozoan, and algal fragments. The electron microprobe and high-resolution electron microscope have proven to be of particular value in such studies.

Diagenesis, Lithification, and Metamorphism. *Diagenesis.* Because most sedimentary carbonate deposits are composed of particles exhibiting a wide range of solubilities, selective removal of the more soluble phases occurs by dissolution early in diagenesis. High-magnesium calcite is removed first, but upon continued exposure to the action of superjacent or interstitial solutions, aragonite and low-magnesian calcites may also be dissolved. As a result of this process, solutions in contact with mixed carbonate phases tend to approach equilibrium with the most soluble phase present. This process may be responsible at least in part for the supersaturation of tropical marine waters with respect to calcite. The effect of selective dissolution upon the sediment is illustrated by the changing mineralogical makeup of sediments from Campeche Bank as a function of particle size (which is roughly correlative with distance from source) shown in Fig. 7. Inasmuch as particle mineralogy and composition is related to the secreting organism, diagenesis may selectively remove from a bioclastic or biogenic assemblage all traces of those organisms whose hard parts are made up of the more soluble carbonate phases. Significant errors may thus be introduced into pale-

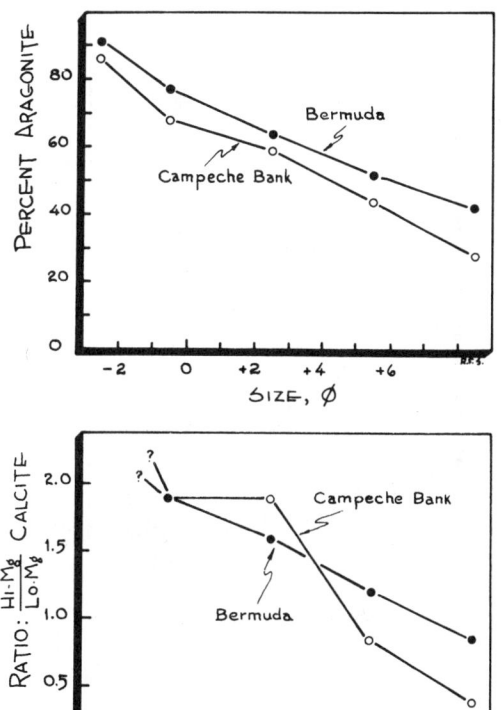

Fig. 7. Aragonite content (weight percent) and high- to low-magnesium calcite ratio in sediments from Bermuda (filled circles) and Campeche Bank (open circles) plotted as average values for the size classes: $> -2\phi$, -2 to $+1\phi$, $+1\phi$ to $+4\phi$, $+4\phi$ to $+7\phi$, and $< +7\phi$ (from Chave, 1962).

ontological and paleoecological interpretations of ancient rocks.

Under certain conditions, metastable phases (high-magnesium calcite, aragonite, etc.) may undergo recrystallization rather than selective removal by solution. Aragonite recrystallizes to calcite; magnesium is leached from the magnesian calcites and its lattice site replaced by calcium. Although these processes of recrystallization and replacement may engender extensive changes in the mineralogical and chemical composition of the rock, they may proceed without destroying the traces of organisms responsible for the original carbonate deposition. Chemical, mineralogic, and paleontologic study of the cores recovered by the Royal Society boring at Funafuti Atoll provide a case in point (Fig. 8). In the upper 220 ft of the core, the bulk material contains highly variable amounts of magnesium, and is found to be made up of several calcite phases of differing magnesium content together with aragonite. The mineralogical and chemical compositions of each fossil fragment correspond with those characteristic of the living forms. Below 220 ft and down to depths of 630 ft, the rock is essentially unimineralic; aragonite and high-magnesium calcites are missing, though many of the species which secrete such materials may still be identified in the core. There can be little doubt that aragonitic forms have inverted to calcite without loss of organic structures, and that magnesium has been leached from the magnesian calcite forms by interaction with the seawater which permeates the reef mass. Below 640 ft, dolomite is universally present; in many samples fossils of organisms which secrete aragonite, calcite, or magnesian calcite are found to be completely dolomitized.

When calcium carbonate minerals are allowed to react with solutions unusually rich in magnesium, dolomitization proceeds in accord with the reaction:

$$2CaCO_3 + Mg^{2+} = CaMg(CO_3)_2 + Ca^{2+}$$

This dolomitization reaction has been studied in the laboratory at low temperatures in tumbling-barrel experiments, and has been postulated to explain Recent dolomitization of carbonate sediments in the Pickelmeer (Netherlands Antilles). Because the dolomitization reaction has a negative volume change it is favored by increased pressure, or in the absence of pressure, dolomitization may lead to increased porosity in a dolomitized carbonate rock. The latter possibility is of considerable significance in studies of oil reservoir rocks.

Lithification. Although both silica and oxide cements are reported, calcium carbonate (generally calcite) is by far the most abundant cement in limestones. Coarse- and fine-grained carbonate sediments may be cemented by simply drying in air, the resulting material often exhibiting surprising strength. Air-dried eolian sands are used as building materials in many tropical areas, and in Bermuda. The nature of the cementation process is not entirely clear, but probably depends in large part upon the deposition of carbonate material in solution in interstitial fluids as desiccation proceeds. In response to increased pressure, calcium carbonate will migrate through intergranular fluids from points of high local stress to points of lower stress, thus filling cavities and pores with carbonate derived from points of intergranular contact. Although many investigators believe that subaerial desiccation is the most important single factor in lithification, most consolidated limestones probably result from a combination of recrystallization and desiccation, be it subaerial or under large lithostatic loading.

Beach rock, which may form in very short periods at the interface between salt- and fresh-water-permeated carbonate sands, repre-

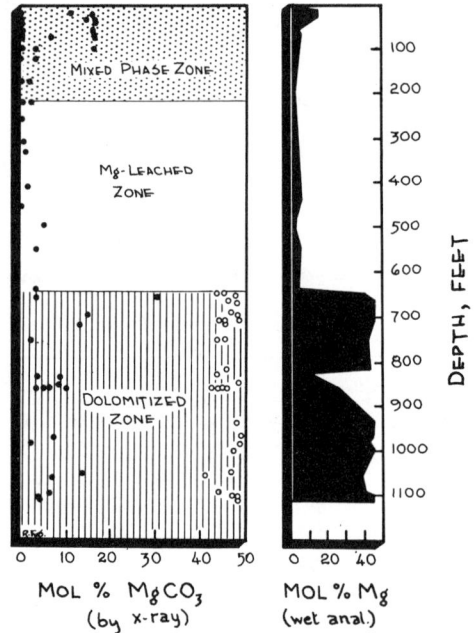

FIG. 8. Mineralogical and chemical compositions of core samples from Funafuti atoll. At left, compositions of mineral phases determined by x-ray measurements of cell size. Calcite phases shown as filled circles; dolomitic phases as open circles. At right, bulk chemical compositions of the same samples determined by wet chemical analysis (adapted from Schmalz, R. F., 1962, unpublished preprint for the American Chemical Society; analytical data from Sollas et al., 1904).

sent a special case of lithification. Here again, though the speed at which lithification proceeds has attracted a great deal of attention, the process is little understood. It is probably related to the mixing of two solutions in which the solubility of calcium carbonate is very different.

Metamorphism. Metamorphism is essentially an extension of the processes of diagenesis carried out under more extreme conditions of temperature and pressure. In addition to the phenomena mentioned earlier in connection with metamorphic carbonate rocks (exsolution, recrystallization, segregation of exsolved phases, etc.) metamorphism may lead to pressure-induced changes in grain shape and size, development of grain overgrowths, exsolution of included organic impurities, cementation, and dolomitization. Most of these have been discussed briefly in earlier parts of this article, and many are treated in detail elsewhere in this Encyclopedia.

ROBERT F. SCHMALZ

References

Abelson, P. H., (Editor), 1959, "Researches in Geochemistry," New York, John Wiley & Sons, 511pp.

Brooks, R., Clark, L. M., and Thurston, E. E., 1950, "Calcium Carbonate and its hydrates," *Phil. Trans. Roy. Soc. London, Ser. A,* **243,** 145–167.

Brown, W. H., Fyfe, W. S., and Turner, F. J., 1962, "Aragonite in California glaucophane schists and the kinetics of the aragonite-calcite transformation," *J. Petrol.* **3,** 566–582.

Bunn, C. W., 1946, "Chemical Crystallography," London, Oxford University Press, Amen House, 422pp.

Chave, K. E., 1962, *Limnol. Oceanog.,* **7,** 218–223.

Chave, K. E., et al., 1962, *Science,* **137,** 33.

Chave, K. E., and Schmalz, R. F., 1966, "Carbonate-seawater interactions," *Geochim. Cosmochim. Acta* 30(10), 60–67.

* Cloud, P. E., 1962, "Environment of calcium carbonate deposition west of Andros Island, Bahamas," *U.S. Geol. Surv., Profess. Paper* **350,** 138pp.

Crawford, W. A., and Fyfe, W. S., 1964, "Calcite-aragonite equilibrium at 100°C," *Science,* **144** (3626), 1569–1570.

* Graf, D. L., "Geochemistry of Carbonate Sediments and Sedimentary Carbonate Rocks," 1960, Parts I, II, III, IV-A and IV-B (bibliography). Circulars # 297, 298, 301, 308 and 309. *Illinois State Geol. Surv.*

* Gold, D. P., 1964, "The average and typical chemical composition of carbonatites," *Proc. Intern. Mineral. Assoc.,* New Delhi, India.

Goldsmith, J. R., and Heard, H. C., 1961, *J. Geol.,* **69,** 52.

Goto, M., 1961, "Some mineralo-chemical problems concerning calcite and aragonite with special reference to the genesis of aragonite," *J. Fac. Sci. Hokkaido Univ., Ser. IV,* **10** (4), 571–640.

Ingerson, E., 1962, "Problems of the geochemistry of sedimentary carbonate rocks," *Geochim. Cosmochim. Acta,* **26,** 815–847.

Jamieson, J. C., 1953, *J. Chem. Physics,* **21,** 1385–1390.

Jamieson, J. C., 1957, *J. Geol.,* **65,** 334–343.

Kamhi S. R., Aug. 1963, "On the structure of vaterite, $CaCO_3$," *Acta Cryst.* 16(8), 770–772.

* Lowenstam, H. A., 1963, "Biologic problems relating to the composition and diagenesis of sediments," in "The Earth Sciences: Problems and Progress in Current Research," Chicago, Univ. of Chicago Press.

Miller, J. P., 1952, "A portion of the system calcium carbonate-carbon dioxide-water, with geological implications," *Am. J. Sci.,* **250,** 161–203.

Miyake, Y., 1957, "A study on the organic productivity and the solubility product of $CaCO_3$ in the ocean by means of the radiocarbon C^{14}," *UNESCO Intern. Conf. on Radioisotopes in Scientific Research,* No. 138, 4pp.

* Palache, C., Berman, H., and Frondel, C., 1951, "The System of Mineralogy," Vol. II, Seventh ed., New York, John Wiley & Sons, 1124pp.

Peterson, M. N. A., 1966, "Calcite: rates of dissolution in a vertical profile in the central Pacific," *Science,* 154(3756), 1542–1544.

Pytkowicz, R. M., 1965, "Calcium Carbonate Saturation in the Ocean," Oregon State Univ. Dept. of Oceanography, Ref. No. 65-16, 6pp.

Rankama, K., 1954, "Isotope Geology," New York, Pergamon Press, 535pp.

Rankama, K., and Sahama, Th. G., 1950, "Geochemistry," Chicago, The Univ. of Chicago Press, 912pp.

* Revelle, R., and Fairbridge, R., 1957, "Carbonates and carbon dioxide," in (Hedgpeth, J. W., editor), "Treatise on Marine Ecology and Paleoecology," Vol. I, *Geol. Soc. Am. Mem.,* **67.**

Simkiss, K., 1964, "The inhibitory effects of some metabolites on the precipitation of calcium carbonate from artificial and natural sea water," *J. Conseil, Conseil Perm. Intern. Exploration Mer,* 29(1), 6–18.

Simmons, G., and Bell, P., 1963, *Science,* **139,** 1197–1198.

Sollas, W. J., et al., 1904, Report of the Coral Reef Committee of the Royal Society, London.

Sverdrup, H. U., et al., 1942, "The Oceans," New York, Prentice Hall, 1087pp.

Wattenberg, H., 1933, *Wiss Ergebn. Deutsch. Atlantisch. Exped. Meteor., 1925–1927,* 8(2), 122–231.

Wyllie, P. J., and Tuttle, O. F., 1960, "The system $CaO-CO_2-H_2O$ and the origin of carbonatites," *J. Petrol.,* **1,** 1–46.

* Zen, E., 1960, "Carbonate equilibria in the open ocean and their bearing on the interpretation of ancient carbonate rocks," *Geochim. Cosmochim. Acta,* **18,** 57–71.

* Additional bibliographic references may be found in this work.

CALCIUM CARBONATE: GEOCHEMISTRY

Cross-references: *Antimony; Aqueous Solutions; Arsenic; Barium; Beryllium; Calcium; Calcium Cycle; Carbon-14 Dating; Crystal Chemistry; Evaporite Processes; Fluorine; Germanium; Hydrothermal Solutions; Interstitial Waters in Sediments; Magnesium; Manganese; Neutron Activation Analysis; Niobium; Order-Disorder; Paleosalinity; Paleotemperatures Isotopic Determinations; Phosphorus; Polymorphism; Rare Earths; Seawater; Solid Solution; Specific Gravity; Thermodynamics; Trace Elements; Vanadium; Zirconium.* Vol. I: *Marine Sediments.* Vol. III: *Algal Reefs; Atolls; Beachrock; Biological Erosion of Limestone Coasts; Coral Reefs; Duricrust; Karst; Limestone Caves; Limestone Coastal Weathering; Stalactite and Stalagmite.* Vol. IV B: *Aragonite; Crystal Growth; Crystallography; Defects in Crystals; Electron Microscopy; Isomorphism; Moh's Scale.* Vol. V: *Carbonatites; Fumaroles; Marble; Metamorphic Facies; Volatiles.* Vol. VI: *Caliche; Diagenesis; Dolomite; Evaporites; Leaching; Oolites; Pelagic Sediments; Spring and Crust Deposits; Travertine.* Vol. VII: *Algae; Annelida; Brachiopoda; Bryozoa; Cephalopoda; Coccoliths; Corals; Echinodermata; Foraminifera; Gastropoda; Mollusca; Paleoecology; Pelecypoda; Porifera; Radiolaria.*

CALCIUM CARBONATE: ECONOMIC GEOLOGY

Description. The two principal calcium carbonate minerals are calcite and aragonite. Calcite crystallizes in the rhombohedral system and aragonite in the orthorhombic system. Certain animals, such as the echinoderms, secrete skeletal material that is a magnesium-rich calcite (5–25% $MgCO_3$) called high-magnesium calcite. High-Mg calcite and aragonite which form in the present-day marine environment are both unstable and are recrystallized to low-Mg calcite quite readily during diagenesis.

Most calcium carbonate in ancient rocks is in the form of low-Mg calcite. Dolomite, a carbonate rock with more than 40–45% $MgCO_3$, and an ordered Ca–Mg crystal structure, is also an abundant carbonate mineral in ancient rocks. Carbonate rocks with 10–40% $MgCO_3$ are termed magnesian limestone, those with less than 10% are termed limestone, and those with more than 95% $CaCO_3$ are termed high-calcium limestone.

Occurrence. Calcium carbonate at the present time is forming primarily in the marine environment. Twenty percent of all sediments being deposited in present-day ocean basins are calcareous. Economic concentrations of calcium carbonate are chiefly massive, fairly pure limestone deposits. Such deposits are formed in shallow-water conditions of minimum dilution by noncarbonate materials. Such conditions are usually found in tectonically stable areas with adjacent land masses of low relief. At the present time, the formation of extensive pure carbonate sediments is restricted to warm-water areas.

Limestones are usually composed of the recrystallized calcareous skeletons of marine organisms and less commonly may be composed all or in part of chemical precipitates. Since chemical deposition (precipitation) probably occurs only under special conditions (a loss of carbon dioxide in the water is usually responsible), the vast bulk of calcium carbonate formed may be attributed to such skeletal contributors as the molluscs, foraminifera, coralline algae, corals, barnacles, bryozoans, tubicolous annelids, ostracods, sponges, and echinoderms. Important skeletal contributors in the past include such forms as crinoids, stromatoporids, archaeocyathids, tetracorals, and brachiopods. The occurrence of organic reefs in the geologic column is summarized by Twenhofel (1950). Figure 1 indicates the processes involved in the formation of a limestone.

Uses. Uses and specific requirements for the various types of calcium carbonate are discussed by Lamar and Willman (1938), Graf and Lamar (1955), and Gillson (1960). The uses of calcium carbonate fall under two broad categories: dimension stone and crushed stone. Dimension stone is stone cut to a specific size for building, construction, or statuary purposes. Crushed stone is irregularly broken up carbonate rock which is ground up for use either in the natural state (road metal, filter beds, concrete aggregate, poultry grit, coal mine dust, filler, and whiting), or converted for multiple uses such as a fluxing agent in metal refining, a soil conditioner, a source of lime, or a chemical raw material for glass making (Bates, 1960). Calcium carbonate is also used as bleaching powder, in soil stabilization (reduces plasticity and increases compressive strength), in asphalt, in the production of rock wool for insulation, in the production of calcium carbide, in the manufacture of textiles (for boiling out, scouring, and bleaching of vegetable fibers), in ceramics, in sugar refining (precipitates the sugar from impure solutions), and in paper manufacture (Gillson, 1960).

One of the most important uses of calcium carbonate is in the production of cement. Since it reaches a high strength in a short time, portland cement is the main type of cement produced in the United States. Ultimate strength is acquired in only one year. Portland cement is made by burning a finely ground artificial mixture of 75% calcium carbonate, 20% silica, alumina, and iron oxide, and 5%

CALCIUM CARBONATE: ECONOMIC GEOLOGY

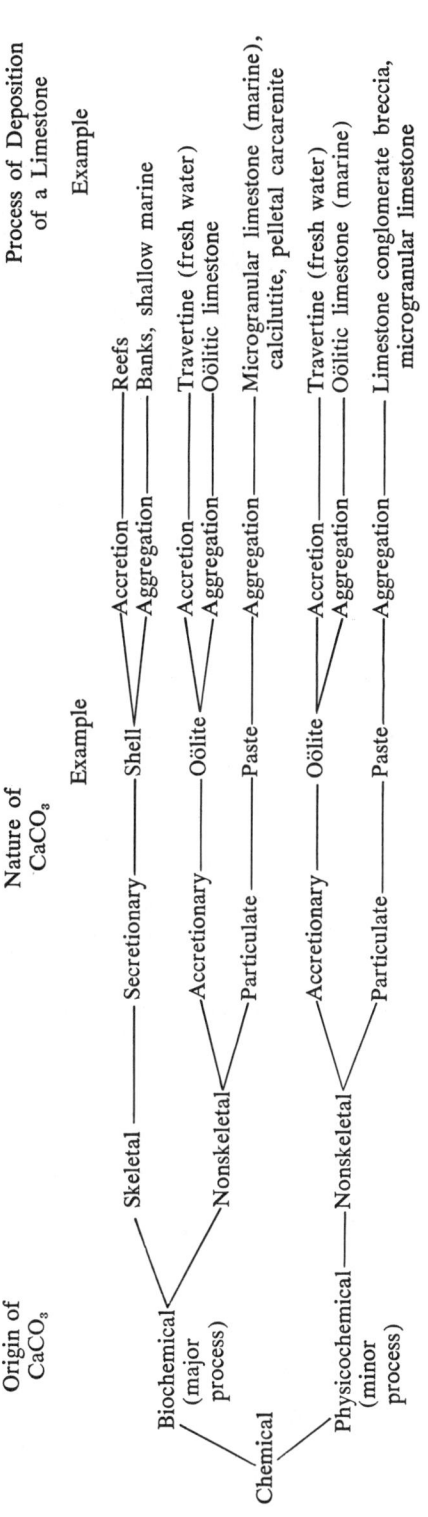

Fig. 1. Origin, Nature, and Process of Accumulation of a Limestone (After Feray, Heuer, and Hewatt, 1962).

miscellaneous materials (magnesium carbonate, sulfur, and alkalies). Limestones, shales, and clays are the commonly used raw materials. Portland cement is a major factor in the national economy, being the structural foundation for all the machinery of the industrial world.

Physical and chemical properties determine the value and possible uses for a calcium carbonate deposit. Sedimentary calcium carbonate may have varying amounts of clay (argillaceous limestone), silica (cherty limestone), iron (ferruginous limestone), or organic matter (carbonaceous limestone). When used as dimension stone, the consistency, shear strength, compactness, color, and presence of irregular or staining matter are important. Chemical composition is crucial when used in glass making or cement preparation. Various comparative measurements to test the suitability of calcareous rocks for the many different uses include measurements on compressive strength, modulus of elasticity, absorption values, unit weight, and abrasion tests.

High-calcium limestones are used to produce sodium carbonate and other chemicals, as metallurgical flux, in glass production, in sugar refining, as mineral feeds for stock, as agricultural limestone, in calcium carbide manufacture, and as whiting substitute. Dolomite is largely used as refractory material and has many uses in common with the calcium limestone. The production of alkalies requires a high-calcium limestone with less than 1% silica. For use as flux, silica and aluminum should be low (less than 5%) and sulfur and phosphorous should be less than 0.1%.

Geologic Record. The distribution of limestones in North America has been briefly outlined by Dunbar and Rodgers (1963). Precambrian calcareous rocks are widespread, but are commonly dolomitic. Paleozoic rocks, in general, contain a larger proportion of calcium carbonate than systems of later eras. The Cambrian and Ordovician carbonate rocks in North America have much lime breccia and finely crystalline limestone. Numerous groups from the Middle Ordovician and successive periods contributed large skeletal elements to the sediments and formed organic-fragmental limestones. Limestones from the Lower Mesozoic are rare but numerous deposits from the Cretaceous occur around the Gulf of Mexico. Cenozoic limestone is common only in extreme southern areas of North America such as Florida. Organic-fragmental limestones seem to dominate these post-Paleozoic deposits.

Example Deposits. Some of the more important United States deposits from an economic viewpoint are discussed by Bates (1960).

119

CALCIUM CARBONATE: ECONOMIC GEOLOGY

In eastern Pennsylvania, the Middle Ordovician Jacksonburg limestone is mined as raw material for cement since it contains about 70% $CaCO_3$, 5% $MgCO_3$, 5% Al_2O_3, 1% Fe_2O_3 and 19% SiO_2. Finished cement should not contain over 4% ferric oxide or over 5% magnesium. The limestone itself should have less than 2% pyrite.

The Columbus limestone (Middle Devonian of Ohio) is used as a source of lime (calcium oxide) and crushed limestone since it is chert-free, is in uniform beds, and is of high-calcium quality which can be used as fluxstone. Other examples of nearly pure calcium limestones include the New Market limestone in Virginia Ordovician, the Valentine limestone (Pennsylvania Ordovician), and the Dundee and Rogers City limestones (Devonian of Michigan).

Much of the limestone dimension stone used in this country comes from the Salem formation (Mississippian of Indiana). This limestone is a calcarenite or detrital rock made up of sand-size grains of both biologic and physiochemical origin.

In Illinois, the Racine dolomite (Middle Silurian) is extensively employed as refractory material, fluxstone, and agricultural rock. This dolomite contains a number of reefs flanked by strata, all of which are dolomitic.

The Holston orthomarble (recrystallized limestone but not metamorphic) of eastern Tennessee is a thick-bedded limestone which takes a highly decorative polish and is used in the construction industry. This orthomarble was formed by the recrystallization of a skeletal hash.

Increasingly during recent years, shell hash, usually oyster shells, from shallow nearshore waters, is being dredged as a source of $CaCO_3$. In the United States, such mining operations are being carried out along the Texas Gulf coast.

Marble. Marble is metamorphosed limestone or dolomite. It lacks rock cleavage, is readily polished, and is highly valued as dimensional stone.

Color and patterns in polished marble are usually due to the composition of the original limestone. Pure marble is white but impurities usually lend a variety of colors. The most important coloring agents in marble include bituminous matter (black marble), iron oxide-hematite (red marble), and iron oxide-limonite (brown marble). Diopside, serpentine, hornblende, or talc inclusions produce a green marble. Marble is the product of regional low-grade metamorphism. Important deposits of marble include the bluish-gray Columbian marble from Vermont. Several other varieties of marble are extensively mined in the Tate District, Georgia. These marbles range in color from rose and green to white. Smaller amounts of marble are mined in Alabama, Colorado, Maryland, and North Carolina.

Mining. Since large quantities of calcium carbonate are usually required by industrial or building concerns, local sources of calcium carbonate are normally exploited in order to minimize transportation costs. Sometimes it is necessary to ship calcium carbonate if special grade or material is required. This is especially true for dimension stone. The current trend in limestone production is toward fewer and larger quarries with savings achieved by using modern, efficient equipment. Stripping and blasting are common removal techniques, but underground mining may be employed where high-quality materials are followed downdip.

BLAKE W. BLACKWELDER

References

Bates, R. L., 1960, "Geology of Industrial Rocks and Minerals," New York, Harper & Row, 441pp.

Dunbar, C. O., and Rodgers, J., 1963, "Principles of Stratigraphy," New York, John Wiley & Sons, 356pp.

Feray, D. E., Heuer, E., Hewatt, W. G., 1962, "Biological, Genetic, and Utilitarian Aspects of Limestone Classification," in (Ham, W. E., editor), "Classification of Carbonate Rocks, a Symposium;" *Am. Assoc. Petrol. Geol. Mem.* **1**, 20–32.

Gillson, J. L., et al., 1960, "The Carbonate Rocks," in "Industrial Minerals and Rocks," *Am. Min. Met. Petrol. Eng. Third ed.*, 123–201.

Graf, D. E., and Lamar, J. E., 1955, "Properties of Calcium and Magnesium Carbonates and their Bearing on Some Uses of Carbonate Rocks," in (Bateman, S. M., editor), *Econ. Geo. Pt. 2*, **1**, 639–713.

Johnstone, S. J., and Johnstone, M. G., 1961, "Minerals for the Chemical and Allied Industries," Second ed., New York, John Wiley & Sons, 293–326.

Lamar, J. E., and Willman, H. B., 1938, "Summary of the Uses of Limestone and Dolomite," *Illinois State Geol. Surv., Rep. Invest.* **49**.

Twenhofel, W. H., 1950, "Coral and Other Organic Reefs in the Geologic Column," *Am. Assoc. Petrol. Geol. Bull.*, **34**, 182–202.

Cross-references: Vol. I: *Marine Sediments.* Vol. V: *Marble.* Vol. VI: *Diagenesis; Dolomite; Limestone; Travertine.*

CALCIUM CYCLE

The calcium carbonate cycle through and within the oceans is involved in the weathering and sedimentation of carbonate minerals and igneous rocks, the formation of calcareous organisms, photosynthesis and oxidation, the exchange of carbon dioxide across the sea surface, and the short-time buffering of the oceanic pH. The carbon dioxide-carbonate system in the oceans was reviewed by Revelle and Fairbridge (1957), Skirrow (1965), and Pytkowicz (1968).

The first problem in studying this system is to determine the various species present, that is, the concentrations of CO_2, H_2CO_3, HCO_3^-, and CO_3^{2-}, as well as that of Ca^{2+}. The determination of the anions hinges upon a knowledge of the dissociation constants of carbonic acid and of boric acid (which contribute to the titration alkalinity and must be accounted for in seawater). These constants were determined at atmospheric pressure by Buch et al. and by Lyman (see Skirrow, 1965) and at high pressures by Disteche and Disteche and by Culberson and Pytkowicz (1968).

Once the species are known, it is possible to determine the degree of calcium carbonate saturation and, hence, the fate of carbonates in the oceans. Results so far suggest supersaturation in the upper layers, with biogenic removal in calcareous tests. Inorganic precipitation, except in special environments, is inhibited by magnesium and possibly by organic coatings. In the intermediate layers, between about 200 and 2000 meters, saturation occurs near the high-latitude sources of intermediate waters. As these waters progress through the oceans, oxidation of organic matter increases the carbon dioxide content, decreases the pH, and causes undersaturation of calcium carbonate by converting carbonate into bicarbonate. At depths greater than 3000–4000 meters, there is undersaturation everywhere in the oceans. Thus, calcium carbonate is removed from the near-surface layers and, as calcareous tests sink, there is partial resolution at depth. The dissolved carbonate, which is over 50% of the amount that settles, is returned to the surface at high latitudes where deep waters rise and by vertical mixing of the waters.

As a result of these processes, the vertical distribution of calcium carbonate often shows a minimum at about 50–200 meters, because of incorporation into calcareous organisms, but below that it increases with depth. The horizontal distribution at depth shows an increase in the direction of motion of the waters. The greatest concentrations are found in the tropical east Pacific where the oldest waters, originating in the north Atlantic and Indian Oceans, are found. At the surface the greatest carbonate concentrations, expressed as specific carbonate alkalinities [the equivalents of inorganic carbon, $(HCO^-_3) + 2(CO_3^{2-})$, per liter divided by the chlorinity to compensate for evaporation and dilution by rains and runoff], are found at high latitudes. This is the result of the compensatory effects of return of carbonate by the upwelling of deep waters and utilization. Utilization alone causes a decrease in the carbonate content of surface waters at lower latitudes.

The carbonate content of sediments does not reflect the high transport of weathered products by rivers into the north Atlantic because of solution at depth. The dissolved carbonate is then carried into the south Atlantic by the Atlantic deep water and finds its way into the Indian and Pacific Oceans via the Antarctic circumpolar waters. Rivers also bring carbonate into other oceans but not to the same extent as in the north Atlantic.

The calcium carbonate cycle must be studied in conjunction with the carbon dioxide cycle because they are related by the reaction $CO_3^{2-} + H_2O = 2HCO_3^-$. An estimate of the fluxes in these cycles was made by Pytkowicz (1967) and is shown in Fig. 1. The flux at the extreme right results from the weathering of igneous rocks. It does not constitute a cycle because this weathering produces primarily sodium carbonate which, being soluble, does not result in precipitation followed by solution. The middle cycle represents calcium carbonate brought by rivers after the weathering of sedimentary rocks, incorporated into shells, and redissolved in part after these shells settle. The cycle at the left is that of carbon dioxide which is utilized in photosynthesis and released during oxidation. It couples with the middle cycle through the reaction which was mentioned earlier.

The fluxes are expressed in mg $CO_2/cm^2/$1000 yr for mass balance purposes and the actual species are shown in parenthesis. The procedure used for the estimates is shown in the original paper (Pytkowicz, 1967). Not all the fluxes could be estimated by the model used; however, there are independent studies of the photosynthetic and CO_2 exchange rates. Carbon dioxide appears to enter the oceans at high latitudes, where low temperatures enhance its solubility, and is released to the atmosphere at low latitudes.

Two interesting results came out of this study. It was found that over half of the calcium carbonate removed from the near-surface layers is redissolved rather than being permanently incorporated into the sediments. Also, not enough silicates were found to go through and be present in the water column relative to existing dissolved carbonate to affect the pH of the oceans in terms of a few thousand years. Carbonates regulate the oceanic pH

CALCIUM CYCLE

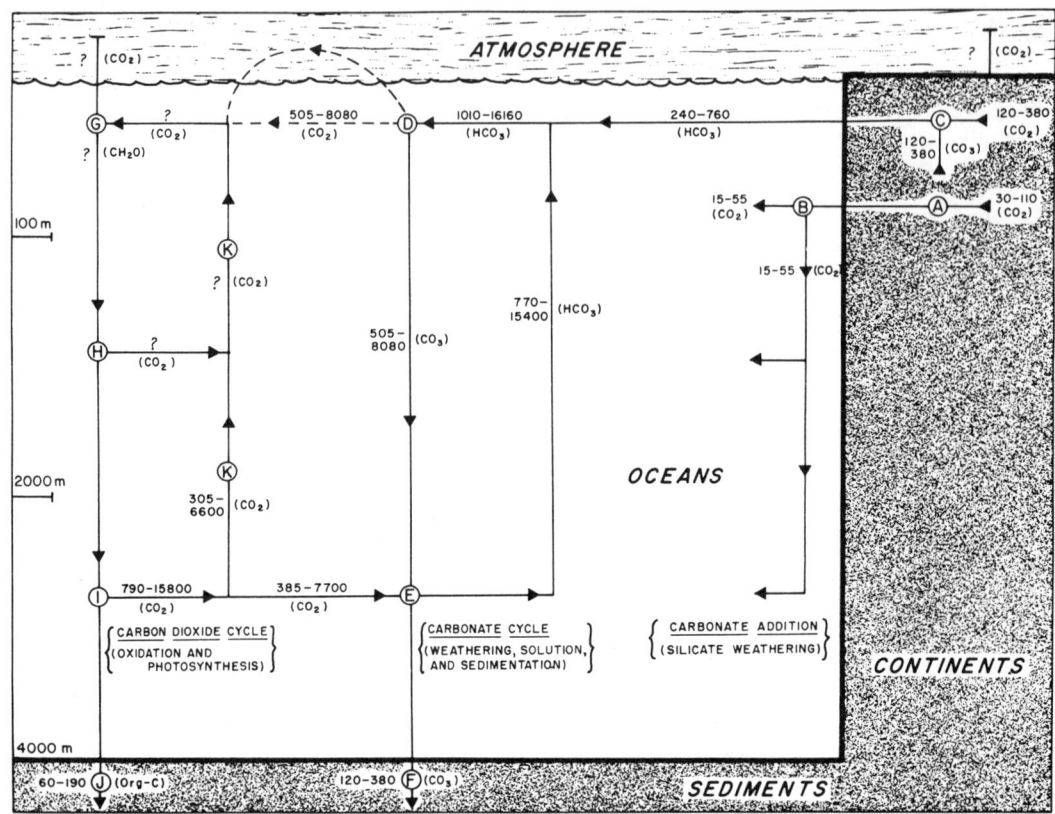

FIG. 1. The fluxes of carbon dioxide and carbonate through and within the oceans. The fluxes are in mg $CO_2/cm^2/1000$ yr, for the species represented in parentheses.

for geologically short-term processes. However, the weathering of igneous rocks, which was a precursor to the formation of sedimentary rocks, controls the long-term pH.

Mackenzie and Garrels (1966) examined the geological history of calcium carbonate in the oceans in its relation to the weathering and sedimentation of carbonate and silicate minerals. The reaction between primordial igneous crust and volatile acid gases produces the sedimentary rock mass and primary oceanic salinity. The chemical composition of the ocean since its early formation has been governed by a dynamic equilibrium between the water and solids carried into it or precipitated from it owing to continuous evaporation and renewal by streams. Hydrogen ion buffering in the oceans and the ion activity ratios of major dissolved species, including calcium, are governed on a long-term basis (million years or so) by reactions involving aluminosilicates.

Consideration of early Precambrian sediments suggests that calcareous algae, of the primitive categories known collectively from their stromatolitic fossil structures as *Collenia*, must have developed as far back as nearly 3×10^9 years ago. Since the primeval atmosphere was continuously enriched in CO_2, due to volcanic exhalation, in spite of early buffering, there is a possibility that the pH would have remained depressed (i.e., an "acid ocean") until a certain threshold level had been achieved. The organic record shows that animal evolution had reached a relatively high level of sophistication prior to the dawn of the Cambrian period. At this point in time (about 600×10^6 years ago) a remarkable and as yet enigmatic event took place: the sudden appearance of $CaCO_3$ shells on diverse, unrelated, and highly organized creatures. Fairbridge (1967) has proposed that this calcification of the mantle or carapace of invertebrate organisms reflected an environmental crescendo or threshold that could be related to the ability of the organism to maintain osmotic excretion in an environment of rising pH and alkalinity.

R. M. PYTKOWICZ

References

Culberson, C., and Pytkowicz, R. M., 1968, "Effect of pressure on carbonic acid, boric acid, and the pH in seawater," *Limnol. Oceanog.* **13**, 403–417.

Fairbridge, R. W., 1967, "Carbonate rocks and paleoclimatology in the biochemical history of the planet," in (Chilingar, G. V., Bissell, H. J., and Fairbridge, R. W., editors), "Developments in Sedimentology," 9A, Amsterdam, Elsevier Publ. Co.

Mackenzie, F. T., and Garrels, R. M., 1966, "Chemical mass balance between rivers and oceans," *Am. J. Sci.*, **264**, 507–525.

Pytkowicz, R. M., 1967, "Carbonate cycle and the buffer mechanism of recent oceans," *Geochim. Cosmochim. Acta* **31**, 63–73.

Pytkowicz, R. M., 1968, "Carbon dioxide-carbonate system at high pressures in the oceans," in (Barnes, H,, editor), *Oceanog. Mar. Biol. Ann. Rev.*, **6**, 83–135, London, G. Allen & Unwin.

Revelle, R., and Fairbridge, R. W., 1957, "Carbonates and carbon dioxide," in (Hedgpeth, J. W., editor), "Treatise on Marine Ecology and Paleoecology," Vol. 1, pp.239–294; *Geol. Soc. Am. Mem.* **67**.

Skirrow, G., 1965, "The dissolved gases-carbon dioxide," in (Riley, J. P., and Skirrow, G., editors), "Chemical Oceanography," New York, Academic Press.

Cross-references: *Calcium Carbonate; Oxidation and Reduction; pH-Eh; Photosynthesis; Seawater Chemistry. Vol. V: Volatiles.*

CAPILLARITY—*See* Vol. II

CAPILLARY WATER—*See* Vol. VI

CARBON: ELEMENT AND GEOCHEMISTRY

Properties of Carbon

The geochemical behavior, cycle, and mineral association of an element depend on, among other things, its chemical properties as dictated by its atomic structure, ionic radius, ionization potential, and chemical affinity. The origin and abundance of an element is determined by its nuclear structure and unclear stability. It is, therefore, fundamental to the understanding of the geochemistry of any element to familiarize oneself with such properties. Table 1 summarizes the numerical properties of carbon.

Carbon exists in two crystalline forms, diamond and graphite, the latter being of lower density. The element carbon is found in more compounds than are any other elements.

Geochemical Behavior

Geochemistry and geochemists tend in general, to deal with the accessible portions of our planet Earth; the lithosphere, the hydrosphere, the atmosphere, and the biosphere. In these environments carbon plays a very significant geochemical role which will be illustrated later. Carbon is also important in extraterrestrial environments. One of the theories advanced to account for the energy of the stars such as our sun is the *carbon cycle* which is simplified in the following series of nuclear reactions:

$$_6C^{12} + {_1H^1} = {_7N^{13}} + \text{Energy}$$
$$_7N^{13} \rightarrow {_6C^{13}} + \beta^+$$
$$_6C^{13} + {_1H^1} = {_7N^{14}} + \text{Energy}$$
$$_7N^{14} + {_1H^1} = {_8O^{15}} + \text{Energy}$$
$$_8O^{15} = {_7N^{15}} + \beta^+$$
$$_7N^{15} + {_1H^1} = {_6C^{12}} + 2He^4$$

The net reaction of this cycle is obtained by addition:

$$(4_1H^1) = {_2He^4} + 2\beta^+ + \text{Energy}$$

The source of energy is, therefore, basically the conversion of hydrogen to helium through carbon, nitrogen, and oxygen intermediates.

Radiocarbon–14

The radioactive species, carbon-14, is formed in the earth's atmosphere as a result of bombardment of nitrogen with cosmic neutrons in the upper atmosphere:

$$_7N^{14} + n \rightarrow {_6C^{14}} + p$$

TABLE 1. PROPERTIES OF CARBON

Property	Numerical Value
Atomic number	6
Atomic weight	12.010
Density of solid at 20°C (g/cm³)	3.51
Melting point (°C)	3570
Boiling point (°C)	3470 (sublimation)
Outer electron configuration	$2s^2\ 2p^2$
Stable isotopes (mass numbers)	12, 13
Most radioactive isotope (mass number)	14
Covalent radius (Å)	0.77
Ionization potential (eV)	11.264
Abundance in earth's crust (%)	0.032

CARBON: ELEMENT AND GEOCHEMISTRY

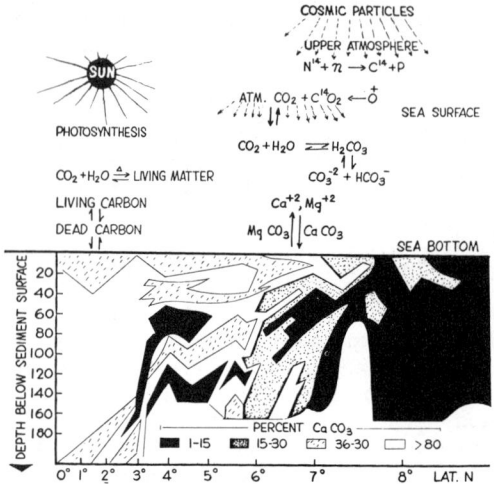

FIG. 1. Path of carbon between the atmosphere and the oceans. Organic and inorganic processes are illustrated.

The radioactive form of carbon, C^{14}, finds its way to the life cycles through photosynthesis, and to the inorganic geochemical cycles through the solution of its compound, carbon dioxide, in the oceans, and subsequent precipitation of carbon minerals and compounds (Fig. 1). Radioactive carbon-14 provides a tool to date the organic and inorganic processes involving carbon over periods on the order of three to five times the half life of C-14, which is about 5600 years.

Stable Carbon Isotopes

Carbon has two important stable isotopes, C^{12} and C^{13}. The ratio C^{12}/C^{13} in nature varies, depending upon the source and origin of carbon and the physical-chemical conditions under which the carbon compounds were formed. Geochemists refer to these variations as fractionation processes. For example, petroleum and natural gas, during their formation, tend to fractionate carbon isotopes and concentrate the lighter isotope, C^{12}, relative to C^{13}, the heavier isotope. Diamonds favor the heavier C^{13} and vegetable carbon favors the lighter C^{12}. Isotope geochemists can, by analyzing for the ratio of C^{12} to C^{13}, derive information indicative of the source and origin of the carbon, whether it is of vegetable and animal origin, igneous, or meteoritic origin.

Carbon Reservoir on Earth

The distribution of carbon in the earth's geospheres is shown in Table 1 in the article *Carbon Cycle* (q.v.); this suggests that the oceans are the reservoir of the element carbon.

SAYED A. EL WARDANI

References

Degens, E. T., 1965, "Geochemistry of Sediments," Englewood Cliffs, N.J., Prentice-Hall, 342pp.
Holland, H. D., 1962, "Model for the evolution of the earth's atmosphere," in "Petrologic Studies," Buddington Volume, Nov., *Geol. Soc. Am.*
Menzel D. W., 1967, "Particulate organic carbon in the deep sea," *Deep-Sea Res.*, **14**(2), 229–238.
Park, R., and Epstein, S., 1960, "Carbon isotope fractionation during photosynthesis," *Geochim. Cosmochim. Acta*, **21**, 110–126.
Ronov, A. B., 1968, "Probable changes in the composition of seawater during geologic time," *Sedimentology*, **10**, 25–43.

Cross-references: *Carbon Cycle; Carbon Isotope Fractionation; Photosynthesis.* Vol. IV B: *Diamond Genesis.*

TABLE 1. CARBON AND CARBON DIOXIDE IN SEDIMENTS, ATMOSPHERE, AND HYDROSPHERE[a]
(all quantities computed as CO_2)

	Total on Earth (10^{17} kg)	Amount per Unit Area (kg/m²)
CO_3^{2-} in sediments	670.0	131000.0
Organic carbon in sediments	40.0 (Degens) 250.0	7900.0 49000.0
CO_2 in atmosphere	0.0235	4.6
$CO_2 + H_2CO_3$ in oceans	0.008	2.2
HCO_3^- in oceans	1.147	318.0
CO_3^{2-} in oceans	0.142	39.5
Total in ocean	1.297	360.0
Dead organic matter on land	0.026	17.0
Living organic matter on land[b]	0.003	2.0
Dead organic matter in ocean[c]	0.100	27.8
Living organic matter in oceans[b]	0.0003	0.08

[a] Data from Revelle and Fairbridge, 1957.
[b] See Tables 4 and 5 for varying estimates of productivity.
[c] This number varies, and may be lower than that given (see Menzel, 1967).

CARBON CYCLE

TABLE 2. FLUXES OF CARBON DIOXIDE IN ATMOSPHERE, HYDROSPHERE, AND BIOSPHERE[a]

	Total on Earth (10^{15} g/yr)	Amount per Unit Area (mg/cm²-yr)
Consumption of fossil fuels, CO_2 produced	9.1	1.6
Weathering of silicates	0.125 ± 0.065	0.035 ± 0.02
Weathering of carbonates	0.90 ± 0.45	0.25 ± 0.13
Total weathering	1.03 ± 0.52	0.29 ± 0.15
Photosynthesis on land[b] (CO_2 consumed)	73 ± 18	55 ± 14
Photosynthesis in ocean[b]	462 ± 303	128 ± 84
Average rate of volcanic production of CO_2 during 3×10^9 yr	0.03	0.006
CO_2 exchange across ocean atmosphere interface	210 ± 130	58 ± 36

[a] Data from Revelle and Fairbridge, 1957.
[b] See Tables 4 and 5 for more estimates of productivity.

CARBON CYCLE

The cycle of carbon in nature is of great interest to biologists, chemists, and geologists. The ability of carbon to exist in gaseous, liquid, and solid forms, as well as its usefulness in both organic and inorganic processes, account for its widespread distribution in various reservoirs. In order to understand the cycle of carbon, the exact nature of these reservoirs and the rates of exchange between them must be understood. Data pertinent to this discussion are included in Tables 1 and 2 and summarized in Fig. 1. It must be emphasized that a great deal of the discussion involves the estimation of the sizes of the various carbon reservoirs; the literature contains many different estimates of these reservoirs, many of which are alarmingly inconsistent, particularly in estimates of biological productivity and biomass compositions.

There is much debate about the ultimate

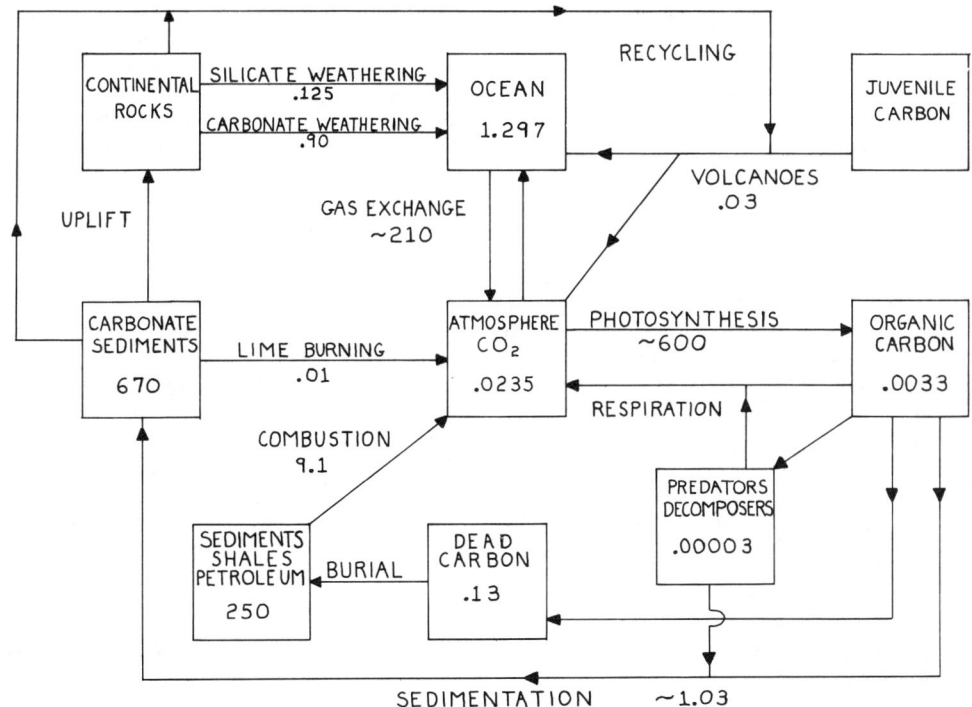

FIG. 1. The carbon cycle. Numbers in boxes refer to total amount of carbon in reservoir, in kg × 10^{17}. Numbers by arrows refer to total fluxes, in units of 10^{15} g/yr. Absent numbers imply that estimates are unavailable.

source of the carbon which is now at the earth's surface. Rubey (1951) argued for continuous outgassing of carbon dioxide from volcanoes and hot springs at a virtually constant rate throughout much of geologic time. His principal argument is that there is little in the geologic record to suggest any large variations in the carbon dioxide content of ocean water. Urey (1952) has argued more in favor of an atmosphere initially of methane; early in earth history this would have been oxidized to carbon dioxide. Holland (1962) also postulates an early atmosphere of methane and ammonia which began being oxidized about 0.5 billion years after the earth's formation. The development of the early atmosphere somehow had to generate the nitrogen that is the major component of today's atmosphere, and at the same time an initial source of oxygen was required. The development of oxygen in the atmosphere has direct bearing on the history of the carbon cycle. Photodissociation of water vapor to give oxygen and ozone has been cited as a possible major source of oxygen. The earliest oxygen in the atmosphere would be used in the oxidation of gases (NH_3 and CH_4) and of various surface minerals. Any excess oxygen or ozone would suffice to protect any primitive "life" from damaging ultraviolet radiation. Once life was initially protected, photosynthesis could have begun in earnest. Oxygen evolution in photosynthesis is a sophistication of the higher plants, and earlier forms of life could have been photosynthetic but not oxygen producing; anaerobic bacteria are a present-day counterpart. It is not accurately known when free oxygen became available in the atmosphere; estimates range from 2.5 billion years ago to 600 million years ago, the beginning of the Cambrian. Berkner and Marshall (1964) argue for continuous growth during the Paleozoic of atmospheric oxygen. On the basis of known calcareous algal deposits (stromatolites) in dated Precambrian rocks, it is reasonable to say that photosynthesis and carbonate secretion were occurring roughly one billion years ago.

The abiotic synthesis of organic matter has been frequently discussed: Miller and Urey (1959) pioneered the experimental work in proving the synthesis of amino acids and other essential organic compounds from a "primitive atmosphere" of methane, ammonia, water, and hydrogen. Electrical discharges were used as the energy source for these reactions. Ponnamperuma (1963, 1966) has extended this work and obtained quite remarkable syntheses of biologically significant molecules such as adenosine triphosphate (ATP), various nucleotides and peptides, and hydrocarbons. Once these compounds are formed, though, the actual step to living organisms is difficult to visualize. Weyl (1968) and others have discussed this topic; most workers call upon some sort of organic aggregate of biologically active molecules as a start of "life."

There is now about 925×10^{17} kg of carbon in the reservoirs at the earth's surface. Because carbon is utilized biologically, it is estimated that the turnover time for photosynthetic carbon may be as little as four or five years, extremely rapid on any geologic time scale. This same estimate would then predict that as much as 95% of all carbon on earth has been used by organisms, and much of this has been recycled. Carbon dioxide in the atmosphere dissolves freely in seawater, but does not remain as CO_2. Instead, it is hydrated to carbonic acid, H_2CO_3, which dissociates to bicarbonate ion, HCO_3^-, and hydrogen ion, H^+. Bicarbonate ion can further dissociate to carbonate ion, CO_3^{2-}, and another hydrogen ion. The relative amounts of these ions are dependent on the pH of the solution; in seawater, pH 7.9–8.2, bicarbonate ion predominates. These equilibria, with dissociation constants (at 25°C), are summarized in Table 3.

In addition, carbonate and bicarbonate ions are complexed by various cations in seawater, mainly magnesium, sodium, and calcium. These complexes exist as neutral or charged species in the water and are not precipitates. Calcium carbonate is precipitated organically by foraminifera, corals, molluscs, etc., and these shells and skeletons are the source of the calcium carbonate on the sea floor. Organisms are able to secrete carbonate shells because the surface ocean in which they live is supersaturated with respect to calcium carbonate, hence the shells do not immediately redissolve. However, calcium carbonate becomes more soluble with depth in the ocean because the solubility product, k_{sp}, increases with increasing hydrostatic pressure and with decreasing temperature. In addition, the oxidation of organic carbon at depth and the finite slow rate of water mass mixing between deep and surface layers of the ocean result in the accumulation of CO_2 in the deep ocean. This accumulation is summarized by the reaction:

$$(CH_2O)_n + nO_2 = nHCO_3^- + nH^+$$

which effectively pushes the system toward undersaturation, whence it begins to dissolve calcium carbonate. The level at which calcium carbonate begins to redissolve is known as the compensation level; the ultimate control on the compensation level is ocean circulation

TABLE 3.

```
atmosphere    CO₂
                ↑↓
ocean           ↕   + H₂O = H₂CO₃ ⇌ H⁺ + HCO₃⁻
                                         ↕ k″
                                         CO₃²⁻ + H⁺
bottom sediments                  Ca²⁺ ↕ k_sp
                                         CaCO₃
```

where
$$k' = \frac{(H^+)(HCO_3^-)}{(H_2CO_3)} = 10^{-6.4}$$

$$k'' = \frac{(H^+)(CO_3^{2-})}{(HCO_3^-)} = 10^{-10.3}$$

$$k_{sp} = (Ca^{2+})(CO_3^{2-}) = 10^{-8.3}$$

and the input of carbonate from continental weathering. The mixing rates between the surface and deep waters of the Pacific are three times slower than those of the Atlantic, and the compensation level as seen in the carbonate content of oceanic sediments is shallower in the Pacific. Turekian (1965) has argued that the ocean is in steady state with respect to carbonate; i.e., the input of carbonate to the oceans is just balanced by the carbonate sedimentation. Using figures for river concentrations of carbonate, this implies that approximately 4 g/(m²)(yr) of calcium carbonate are accumulating on the ocean floor. This is the net rate and is the difference between the carbonate produced at the surface by organisms and that redissolved at depth. Estimates of carbonate productivity are difficult, but a number like 18 g/(m²)(yr) is deduced from carbonate fluxes between surface and deep ocean water masses. Carbonate sediments are the largest reservoir of carbon; through various means besides resolution at depth their carbon is returned to seawater and to the atmosphere and biosphere. Since Cretaceous times pelagic foraminifera have been a major sink for $CaCO_3$ in the deep ocean, but a large percentage of all carbonate sediments have been deposited in shallow-water areas where geologic uplift and subsequent erosion of the limestones serves to recycle the carbonate back into the sea. In addition, weathering of silicate rocks introduces carbonate alkalinity into the ocean by reactions such as the following:

$$Mg_2SiO_4 + 4H_2CO_3 = 2Mg^{2+} + 4HCO_3^- + H_4SiO_4$$

This is actually a charge balance reaction, since the positive magnesium ions must be balanced by an equal negative charge, and SiO_4^{4-} rapidly associates with hydrogen ion to form a neutral species in seawater.

Volcanism and metamorphism are also a significant source of carbon dioxide for the ocean and atmosphere; indeed, outgassing of CO_2 is postulated as the ultimate source of carbon. The absolute rates of recycling of carbonate by burial and metamorphism, and of the introduction of juvenile carbon dioxide, are difficult to estimate. One can calculate an average rate of outgassing by dividing the total amount of carbon on the earth's surface by the age of the earth, but the meaning of this number is rather vague, since it neglects any carbon that may have existed in the primitive atmosphere, and it also fails to consider recycling of carbon by burial.

The combustion of fossil fuels and lime by man during the past 150 years has served to introduce into the carbon cycle a new recycling step which is much more rapid than any geologic process. On the basis of figures for the amounts of various fuels combusted, it would be predicted that the carbon dioxide content of the atmosphere would be about 12% higher now than in 1850. In fact, the pCO_2 now is about 8% higher, and the remaining increase has been taken up by the oceans.

As mentioned above, biologic utilization of carbon is very much more rapid than geologic utilization of carbon. Fig. 1 emphasizes this fact; it should be noted that although predators and decomposers are about 1% of the biomass, they are able to reoxidize 99.9% of photosynthetic production every year. Some of the annual net production is taken out of circulation and is not immediately recycled. This includes the chemically resistant humus fractions of soils (approx. 2.7 kg C/m² land) and undecayed organic matter dissolved or suspended in the sea (0.1–3.0 kg C/m² sea). This also includes the organic matter which becomes entrapped in the sediments of the ocean floor and that entrapped in continental

TABLE 4. NET PRODUCTIVITIES AND NET PRODUCTION OF MAJOR COMMUNITIES

Community	Productivity, (kg)(m^{-2})(yr^{-1})		Production, (kg × 10^{-14})(yr^{-1})	
	Carbon	Dry Matter	Carbon	Dry Matter
Coniferous forest	1.4	2.8	0.14	0.28
Deciduous forest	0.54	1.2	0.025	0.06
Tropical forest	2.3	5.0	0.34	0.73
Other forest	0.54	1.2	0.078	0.17
Arable land	0.91	2.0	0.21	0.46
Grassland	0.91	2.0	0.245	0.54
Desert	0.045	0.1	0.015	0.033
Tundra	0.045	0.1	0.005	0.011
Mean for land	0.71	1.53	1.058	2.284
Ocean	0.072	0.32	0.24	1.070
Continental shelf	0.16	0.73	0.043	0.195
Brown algal zone	1.0	2.9	0.009	0.026
Mean for sea	0.081	0.36	0.292	1.291
Mean for whole earth	0.265	0.70	1.35	3.575

[a] From Bowen, 1966.

sediments. Estimates of the amount of organic carbon in sediments are difficult, but it has been shown that shales average 2.1% organic matter, carbonates 0.29%, and sandstones 0.05% (Degens, 1965). This would mean that about 50 × 10^{17} kg to 250 × 10^{17} kg carbon are tied up in sediments. Again, the literature does not agree on this number, mainly because there are widely differing estimates of the amounts of sediments. The great majority of this organic matter existed as finely disseminated organic carbon in sediments. Probably less than 5% of this has been diagenetically altered to petroleum products. Coal is thought to be a terrestrial deposit, whereas oil most likely is the result of marine diagenesis of organic matter.

Fig. 1 does not distinguish between marine and terrestrial biomasses; Tables 4 and 5 show the productivity of these various biomasses, and it should be obvious how little carbon is not recycled. The tables include several estimates of the various biomasses, and there are some astounding discrepancies which point out the difficulty of such estimates. Of particular interest are the estimates of terrestrial productivity; obviously the ratio of carbon utilization between ocean and land is a poorly known figure which invites further study.

JOHN WEHMILLER

References

Berkner, L. V., and Marshall, L. C., 1964, "The History of Growth of Oxygen in the Earth's Atmosphere," in (Brancazio, P. J., and Cameron, A. G. W., editors), "The Origin and Evolution of Atmospheres and Oceans," pp. 102–126. New York, John Wiley & Sons.

Bowen, H. J. M., 1966, "Trace Elements in Biochemistry," New York, Academic Press.

Degens E., T., 1965, "Geochemistry of Sediments," Englewood Cliffs, N.J., Prentice-Hall, 342pp.

Farmer R. E., 1965, "Genesis of subsurface carbon dioxide," in "Fluids in Subsurface Environments," Am. Assoc. Petrol. Geol. Mem. 4.

Holland, H. D., 1962, "Model for the evolution of the Earth's atmosphere," in "Petrologic Studies," Buddington Volume, Nov. Geol. Soc. Am.

Menzel, D. W., 1967, "Particulate organic carbon

TABLE 5. ESTIMATES OF THE PRODUCTION OF THE BIOSPHERE[a]

Gross			Net		
Land	Sea	Total	Land	Sea	Total
			0.163	0	0.163
			0.151	0.286	0.437
0.20	0.44–2.03	0.64–2.23			
	0.235				
0.24	≤ 1.6	≤ 1.84	0.15		
			0.21	0.32	0.53
				0.53	
0.172	0.25	0.422	0.103	0.25	0.353
			1.06	0.29	1.35

[a] In kg C × 10^{-14}/yr; data from Bowen, 1966.

in the deep sea," *Deep-Sea Res.*, **14**(2), 229–238.
Miller, S. L., and Urey, H. C., 1959, "Organic compound synthesis on the primitive Earth," *Science*, **130**, 245–251.
Miyake, Y., and Matsuo, S., 1963, "A role of sea ice and seawater in the Antarctic on the carbon dioxide cycle in the atmosphere," *Paper, Meteorol. Geophys. (Tokyo)*, **14**(2), 120–125.
Poldervaart, A., 1955, "Chemistry of the Earth's Crust," *Geol. Soc. Amer. Spec. Paper* **62**.
Ponnamperuma, C., 1963, "Synthesis of adenosine triphosphate under possible primitive Earth conditions," *Nature,* **199**, 221–226.
Ponnamperuma, C., 1966, "Possible abiogenic origin of some naturally occurring hydrocarbons," *Nature,* **209**, 979–982.
Revelle, R., and Fairbridge, R. W., 1957, "Carbonates and carbon dioxide," in "Treatise on Marine Ecology and Paleoecology," Vol. 1, *Geol. Soc. Am. Mem.*, **67**.
Ronov, A. B., 1968, "Probable changes in the composition of sea water during geologic time," *Sedimentology*, **10**, 25–43.
Rubey, W. W., 1951, "Geologic history of seawater," *Geol. Soc. Am. Bull.* **62**, 1111–1147.
Turekian, K. K., 1965, "Some aspects of the geochemistry of marine sediments," in (Riley, J. P., and Skirrow, G., editors), "Chemical Oceanography," Vol. 2, New York, Academic Press.
Urey, H. C., 1952, "The Planets: Their Origin and Development," New Haven, Yale Univ. Press, 245pp.
Weyl, P. K., 1968, "Precambrian marine environment and the development of life," *Science*, **161**, 158.

Cross-references: *Ammonia*, in *Minerals and Early Atmosphere; Cycles—Geochemical; Hydrocarbons; Nitrogen; Outgassing of the Planet Earth; Oxygen; Ozone; pH-Eh; Seawater.* Vol. I: *Carbon Cycle in the Oceans; Carbon Dioxide Cycle in the Sea and Atmosphere.* Vol. II: *Earth—Geology of the Planet; Planetary Atmosphere.* Vol. IV B: *Fossil Fuels.*

CARBON DIOXIDE CYCLE IN THE SEA AND ATMOSPHERE—*See* Vol. I or II

CARBON-14 DATING

The carbon-bearing materials that are formed in equilibrium with atmospheric carbon acquire during their formation small amounts of the cosmic-ray-produced C^{14} isotope. When the exchange with the atmospheric carbon comes to an end, the radioactive decay of the C^{14} provides a method for determining the period elapsed since the cessation of the exchange. The method, discovered in 1947 by Libby and his co-workers, has been successfully applied for age determinations up to 70,000 years. For this time interval it is presently the most important dating tool available.

Principles Involved in C^{14} Dating

Upon entry into the atmosphere, the primary cosmic radiation produces neutrons of which a certain number interact with atmospheric nitrogen. The capture of the neutrons by nitrogen results in the production of C^{14} and nonradioactive hydrogen according to the equation $N^{14} + n \rightarrow C^{14} + H$. The C^{14} formed in this way soon combines with atmospheric oxygen and forms carbon dioxide. Since the cosmic-ray-produced neutrons are most abundant at higher altitudes, C^{14} production is mainly confined to altitudes of 15,000 ft and higher. The carbon dioxide derived from the C^{14} atoms arrives within a relatively short time interval at the earth's surface through convection and turbulent mixing in the troposphere.

Of course, the C^{14} produced in this way is not the only source of carbon dioxide; a large quantity is derived from the stable carbon isotopes (C^{12} and C^{13}) present at the earth's surface. As a result the carbon dioxide containing the C^{14} is only a very minor fraction in atmospheric carbon dioxide; the C^{14}/C^{12} ratio $= 10^{-12}$ for atmospheric carbon dioxide.

Through photosynthesis plants acquire small amounts of the C^{14}. Animals acquire it by feeding on plant material and consequently all these living organisms contain a certain amount of the C^{14} isotope. Shells acquire C^{14} because large quantities of the isotope enter the oceans, first as dissolved carbon dioxide, and also as bicarbonate and carbonate due to the exchange of these components with the dissolved carbon dioxide.

When the plant or animal dies, the CO_2 exchange with the atmosphere ceases and C^{14} is no longer added to the organism. From this point on, the C^{14} content (or C^{14}/C^{12} ratio) diminishes with time because of the radioactive decay of the C^{14}. When decaying, the atom ejects one electron (beta particle) and converts again to stable nitrogen. This process is essential in detecting the C^{14}; no instrument can measure the very minute quantities of C^{14} in organic material with sufficient precision but it is possible to detect the charged beta particles ejected by the decaying atoms. The detection is generally done with the aid of a counter tube; the number of decaying atoms per minute is proportional to the amount of C^{14} in the material under investigation.

The C^{14} radioactivity of a given sample decreases exponentially with time with a half-life of 5730 years; after this period only half of the original activity is left. After twice this period (11,460 years) only one quarter is left, after again another 5730 years (17,190 years) only one eighth, and so on. A more general formula is $t = 8270\ ^e\log p$, where p is the ratio of the C^{14} activity of an unknown sample to the C^{14} activity of contemporaneous material and t is the age of the sample. For very old material extremely small amounts of C^{14} are left; a sample that is 46,000 years old will have only a fraction (0.004) left from the original C^{14}.

The basic assumption underlying the whole dating method is that all organisms in a certain reservoir during their life acquire, per gram of carbon, the same amount of C^{14}, whether recently or very long ago. In other words, production of the C^{14} by cosmic radiation should have been constant for at least the last 70,000 years (the time span covered by the method) while the reservoirs (atmosphere, biosphere, oceans) should have kept the same rate of carbon exchange. The presently available evidence indicates that basically these conditions are fulfilled, although deviations occur.

The Measurement of C^{14} Radioactivity

For measurement, the sample has to be converted into a gas or liquid; the first step is the combustion of the sample. By this method carbon dioxide gas is obtained; after an elaborate purification this gas can be used for counting in proportional counters. Other suitable gases for filling proportional or Geiger counters are methane, acetylene, and ethylene; a suitable liquid for scintillation counting is benzene. With the original method used by Libby the carbon in the sample was converted to carbon black and coated on the inside of a metal cylinder that could be inserted in a counter. This method has been abandoned because air-borne fission products (fallout) were often introduced into the counter together with the black carbon sample. Present-day techniques are more suitable for smaller samples because of increased counting efficiencies.

For contemporaneous organic material (not yet influenced by atomic-bomb-produced C^{14}), about 14 beta particles are emitted per minute per gram of carbon. One gram of pure carbon, when converted to carbon dioxide, provides a one-liter counter with a filling of about 2 atms pressure. In order to obtain precise measurements, the counter should have a very low counting rate when filled with "dead" carbon dioxide (containing only the stable C^{12} and C^{13} isotopes).

Several components contributing to the background of a counter are external gamma radiation, cosmic radiation, and the radioactivity of the materials used for the construction of the counter. Most external gamma radiation is absorbed in an iron or lead shield with a thickness of 8 inches or more. Additional shielding with pure mercury or specially selected lead of very low radioactivity is placed around the counter to absorb gamma radiation originating in the iron or lead shield. The top section of the shield is often bulky; thickness of up to 30 inches of iron are used to absorb the low-energy component of the cosmic radiation. Secondary neutrons produced by cosmic radiation in the shield can be absorbed by paraffin plus boric acid. In some instances, the whole counting room is placed underground or underneath a building with many stories. At the University of Groningen, the Netherlands, the counters are situated at the bottom of a 90 ft deep shaft.

The more energetic meson component of the cosmic radiation cannot be eliminated by shielding. This problem is solved by surrounding the sample counter with a ring of counters; this allows electronic cancellation of the pulses produced simultaneously in the central counter and the ring counters. In some instances (Oeschger counter) the guard counters are mounted in the same enclosure as the central counter; here the same filling is used for both the central and ring counters.

For construction of the sample counters the most favored materials are quartz and copper. At present the minimum background obtained for a one-liter counter is about 1 count per minute. Maximum ages of 50,000 years can be measured by counting directly. With the aid of isotopic enrichment another 20,000 years can be added, giving a maximum range of approximately 70,000 years.

The Precision of the Measurement

Owing to the statistical fluctuations in the number of C^{14} disintegrations, the age of a sample can only be determined within certain limits. The statistical error for a measurement depends on the total number of disintegrations counted. The error is inversely proportional with the square root of the total number of disintegrations; the larger the number of disintegrations, the smaller the relative error. The relative error can be made smaller by counting for up to two days and thus accumulating more counts. Younger samples have smaller relative error than older ones, because they have higher activity, which results in a larger number of counted disintegrations.

The counting error, usually given as the standard deviation, is expressed as an age error in the final result. The use of the standard deviation for the calculation of the age error implies that there is still a 32% possibility for the actual age to lie outside the stated limits;

by taking twice the limit this possibility is only 5%. One should therefore bear in mind that for one-third of all published dates the actual C^{14} age lies outside the stated limits.

For one or two days of counting, a typical counter will give a 10,000-year date with an error of about $\pm$ 100 years and a 30,000-year date with an error of about $\pm$ 800 years.

Other sources of laboratory error, e.g., variations in counter background or counter efficiency, are generally small with respect to the statistical deviations.

The half-life of radiocarbon has been determined in various ways; the average of the last three measurements gives 5730 years. However, all existing published dates use the older half-life of 5568 years. To convert an older date to the newer half-life, the age should be multiplied by 1.03. The 5568 years half-life will be kept in use for the publication of dates until agreement has been reached on a definite half-life.

Carbon-14 dates (as well as other isotope dates in geology) are conventionally quoted in years "BP" (Before Present), the year AD 1950 being taken as the fixed datum. Archaeological dates are often adjusted to AD and BC, although one C^{14} year does not need to be equal to one calendar year.

Errors in the Assumption

In order to be able to apply the method, one has to know the initial C^{14} concentration of the specimen under study and the possible post-depositional change in C^{14} concentration by causes other than radioactive decay.

Deviations in the Present Dynamic Reservoir. The C^{14} concentration in contemporary materials is rather uniform, but significant deviations are possible. The surface layers of the oceans have on the average a slightly lower C^{14} concentration than wood, after correction for isotopic fractionation effects. This is caused by the finite mixing rates for carbon dioxide between atmosphere and oceans. Close to the poles, upwelling of old, and consequently less active, water can lower the C^{14} concentration even more, and differences of 10% or more can be found. An approximate correction can be made for these differences and they should not seriously influence the age measurements. It is more difficult to apply a correction for shells formed in more restricted parts of the ocean, i.e., lagoons and estuaries. Fresh waters also give rise to uncertainties due to the introduction of old and C^{14}-free carbonates derived from limestone. Shells and organic material formed at present can have "ages" of up to 2000 years in hard-water lakes.

Wood, one of the most used materials in C^{14} dating, is generally very uniform in C^{14} concentration. Wood of the same age, but from different regions in North America, does not show any measurable differences; if there are differences at all they correspond to age variations of less than 20 years.

Although the distribution of C^{14} is not completely uniform in the dynamic reservoir (biosphere, atmosphere, and hydrosphere) it seldom causes serious problems for C^{14} dating.

Man-Made Processes Influencing the C^{14} Concentration. The study of the contemporary distribution of C^{14} has been complicated by the consumption of fossil fuel. The carbon dioxide obtained from this source dilutes the C^{14} in the atmosphere; around 1950 the C^{14} concentration in the atmosphere was 2% lower than in 1890. More recently, the explosion of thermonuclear bombs added large amounts of C^{14} to the atmosphere, and as a result the 1964 level was nearly twice the natural C^{14} level. The increase in C^{14} activity in the large ocean reservoirs is smaller; in 1966 the carbon-14 in the ocean surface waters was about 20% above its natural levels.

For radiocarbon dating most problems caused by these changes are avoided by comparing unknown samples with material grown before the end of the 19th century.

Variations in C^{14} Concentration During the Past. A more difficult problem is encountered with the assumption that all living organisms in comparable environments have the same initial C^{14} concentration independent of the period in which they were formed.

A large number of precise measurements on tree rings and Egyptian materials of well-known age show that appreciable variations in atmospheric C^{14} content have occurred during the past 6000 years. Basically these variations are of two types: (1) short-term variations of a few percent lasting at most a few centuries, and (2) a long-term increase of about 10% in atmospheric C^{14} content from about 2500 BP to 6000 BP. Measurements of tree-rings older than about 7300 BP are not yet available. The short-term variations result in age errors of maximally 200 years. The long-term variations result in C^{14} ages that are too low for the 2500–6000 BP interval; here 6000 calendar years correspond to about 5200 C^{14} years.

Variations in atmospheric C^{14} content may result from changes in C^{14} production rates, e.g., through variations in cosmic-ray flux, or they may reflect climate-induced changes in the size and exchange rate of reservoirs in which the C^{14} is distributed. Two different mechanisms might be responsible for variations in cosmic-ray flux: (1) changes in the

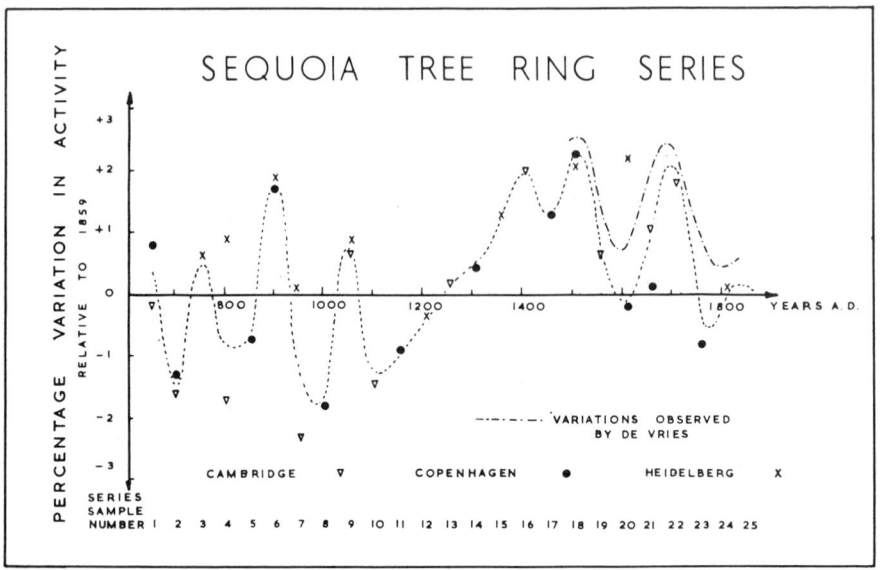

Fig. 1. Curve illustrating variation in C^{14} activity over some 1200 years (from Willis et al., 1960). This curve is approximately in inverse relationship to mean sunspot and eustatic levels. Determinations were carried out from Sequoia tree rings by several laboratories for mutual control.

Earth's magnetic dipole moment and (2) changes in solar wind intensity.

Variations in total magnetic dipole moment of the Earth alter the cosmic-ray flux in the Earth's atmosphere since the deflection of the incident cosmic rays varies with the magnetic-field intensity. The total C^{14} production in the atmosphere is approximately inversely proportional to the square root of the field intensity. Measurements of remanent magnetism in pottery and bricks indicate a variable magnetic-field intensity for the past 8000 years. However, the precision of these paleomagnetic-field measurements is not great, and correlation of magnetic-field changes with the short-term variations in atmospheric C^{14} concentration is not possible.

Variations in solar wind intensity strongly influence the cosmic-ray flux in the Earth's atmosphere. An increase in solar wind intensifies the weak magnetic field carried by the solar wind throughout our planetary system. In the Earth's vicinity a smaller cosmic-ray flux results as cosmic rays directed towards the Earth will be more deflected. Because appreciably greater solar wind intensities are associated with sunspot maxima, a lower C^{14} production is to be expected for extended periods of high sunspot activity. This helio-magnetic modulation mechanism seems to be responsible for the short-term atmospheric C^{14} fluctuations.

No single explanation exists for the long-term increase of C^{14} activity between 2500 and 6000 BP. (See, 1971 Nobel Symposium, XII, Radiocarbon Variations and Absolute Chronology).

Postdepositional Change. Preferential removal of one carbon isotope with respect to another during decomposition can change the C^{14}/C^{12} ratio, but in practice the influence of this effect is unimportant for age measurements. Exchange of carbon with the surroundings seems to have only an important effect for carbonates. More serious is contamination of the sample caused by natural addition of surrounding carbon. Typical instances are organic materials containing rootlets from more recent origin or the precipitation of humic materials leached from overlying soils. This type of contamination is generally important for old samples. Addition of 1% of recent carbon will change the age of a 40,000 years old sample to 33,000 years, but it changes the age of a 1000 years old sample to only 990 years. Rootlets in a sample are normally removed as much as possible, whereas humic acids are dissolved in an alkali solution. In general, an unreasonably large amount of undetected contaminants is required to explain anomalous dates in the 0–5000 years range. Contamination will undoubtedly influence very old dates and an upper limit for the extension of the C^{14}-dating method is mainly governed by contamination problems.

Applications of C[14] Dating

Although the pinpointing of a date within a few years is merely wishful thinking of sample donors, the method has been applied in numerous fields. In geology the most important application has been the establishment of the chronology of climatic changes during the late Pleistocene. Trees and other organic deposits buried by advancing ice masses can be dated and provide a date for the onset of a cold phase; the retreating stage can be dated by dating the earliest organic growth after the ice left the region. Cores from lake sediments, peat bogs, etc., can be analyzed for pollen and since the pollen type distribution reflects climatic changes the determination of the age of critically selected samples establishes an absolute chronology. In desert areas like the Great Basin of western United States, pluvial periods are recorded in lake deposits; work at Searles Lake, California, shows, for example, that pluvial conditions prevailed from 50,000 to 32,000 years and from 24,000 to 10,000 years ago. In general there is a correlation between cold phases in the north and pluvial phases in the southwest. To a large extent there is also synchronism between climatic changes in North America and Europe.

In archaeology the dating method has been important for the establishment of the chronology of different cultures. The method can also help in checking the authenticity of art works for which a certain age is claimed.

Numerous other applications include work on the extinction of animals such as the mastodon, on rates of uplift of shorelines after ice load removal, on postglacial sea level rise, and on volcanic eruptions by dating the preceding organic growth. With the present steep rise in C[14] level due to atomic bomb explosions the C[14] is also used on a large scale to check rates of CO_2 invasion into lakes, rivers, and oceans. It also gives valuable information on the rate of turnover of different biochemical constituents in mammals.

M. STUIVER

References

Aitken, M. J., 1961, "Physics and Archaeology," New York, Interscience Publ.
Bray, J. R., 1967, "Variation in atmospheric carbon-14 activity relative to a sunspot-auroral solar index," *Science,* **156**(3775), 640–642.
Fairbridge, R. W., 1961, "Convergence of evidence on climatic change and ice ages," *Ann. N.Y. Acad. Sci.,* **95**(1), 542–579.
Keith, M. L., and Anderson, G. M., 1963, "Radiocarbon dating: fictitious results with mollusk shells," *Science,* No. 1414, 634–636.
Libby, W. F., 1955, "Radiocarbon Dating," Second ed., Chicago, Univ. of Chicago Press, 175pp.
Suess, H. E., 1965, "Secular variations of the cosmic-ray-produced carbon 14 in the atmosphere and their interpretations," *J. Geophys. Res.,* **70**(23), 5937–5952.
Stuiver, M., 1961, "Variations in radiocarbon concentration and sunspot activity," *J. Geophys. Res.,* **66**(1), 273–276.
Stuiver, M., and Suess, H. E., 1966, "On the relationship between radiocarbon dates and true sample ages," *Radiocarbon* **8**, 534–540.
Vries, H. de, 1959, "Measurement and use of natural radiocarbon," in (Abelson, P. H., editor), "Researches in Geochemistry," pp.169–189, New York, John Wiley & Sons.
Willis, E. H., 1963, "Radiocarbon Dating," in (Brothwell, D., and Higgs, E., editors), "Science and Archaelogy," pp. 35–46.
Willis, E. H., Tauber, H., and Münnich, K. O., 1960, "Variations in the atmospheric radiocarbon concentration over the past 1300 years," *Radiocarbon Suppl.* **2**, 1–4.

Cross-references: *Carbon Cycle; Carbon/Nitrogen Ratio; Geochronometry; Photosynthesis; Seawater Chemistry.* Vol. II: *Climatic Variations; Tree-Ring Analysis.* Vol. VII: *Dendrochronology; Palynology.*

CARBON ISOTOPE FRACTIONATION

Natural carbon is composed of three isotopes, C^{12}, C^{13}, and C^{14}. Carbon-14 makes up a negligible amount of the total carbon reservoir, and is detectable only through its radioactivity. Natural carbon is composed of about 98.9% C^{12} and 1.1% C^{13}.

This C^{12}/C^{13} ratio of about 90:1 is roughly, but not exactly, constant in all natural materials. It varies because both equilibrium and kinetic processes occurring during geochemical events will slightly alter the C^{12}/C^{13} ratio of affected materials. Equilibrium processes, in which the carbon isotopes in two phases are in equilibrium, will cause fractionations (partial separations) of C^{13} and C^{12} because one isotope is relatively more stable in one phase than in another. Kinetic processes, in which C^{13} reacts slightly faster or slower than C^{12}, will also cause fractionations.

The study of the natural variations in the relative abundances of the stable carbon isotopes is interesting in itself; moreover it is an important tool in many geological investigations.

Variations in the ratio of C^{13} to C^{12} are reported as values of δC^{13} (in parts per thousand, ‰, relative to a standard)

$$\delta C^{13} = \left[\frac{(C^{13}/C^{12})_{sample} - (C^{13}/C^{12})_{std}}{(C^{13}/C^{12})_{std}} \right] \times 1000 \quad (1)$$

If the δC^{13} of a substance = 1.5, for example,

CARBON ISOTOPE FRACTIONATION

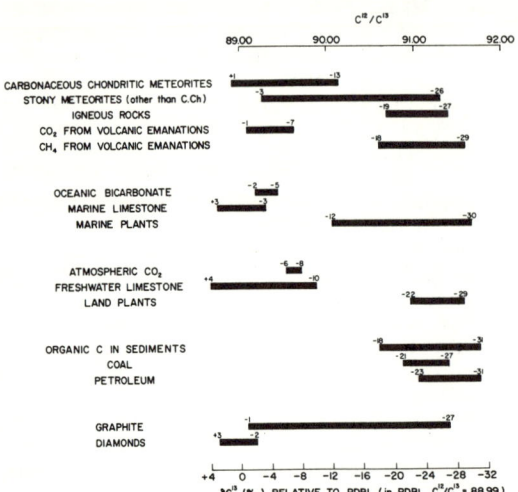

FIG. 1. Variations in the isotopic carbon composition of some natural materials. The variations shown here include typical, but not extreme, values of the different materials. Data are mainly from Craig (1953) and Polanski (1961).

it is enriched in C^{13} by 0.15% relative to the standard. Belemnites from the Peedee formation of South Carolina, having a C^{12}/C^{13} = 88.99, usually serve as the standard for reporting data. All δC^{13} values given here are relative to PDB1. The range of δC^{13} found in some natural materials is shown in Fig. 1.

Fractionating Mechanisms for Carbon Isotopes

Volcanic Gases. In hydrothermal events, the carbon isotopes will be fractionated between CH_4 and CO_2 according to Eqs. (2a) and (2b), where K is a function of temperature.

$$C^{13}O_2 + C^{12}H_4 \rightleftharpoons C^{12}O_2 + C^{13}H_4 \quad (2a)$$

$$k = [C^{12}O_2][C^{13}H_4]/[C^{13}O_2][C^{12}H_4] \quad (2b)$$

Low-Temperature Inorganic Fractionation. Most low-temperature inorganic fractionation of C^{12} and C^{13} is the result of thermodynamic effects. The most important variations can be interpreted in terms of two reactions:

$$HC^{12}O_3^-(aq) + C^{13}O_2(gas) \rightleftharpoons HC^{13}O_3^-(aq) + C^{12}O_2(gas) \quad (3a)$$

$$k \simeq 1.005 \text{ at } 20°C \quad (3b)$$

$$HC^{13}O_3^-(aq) + CaC^{12}O_3 \rightleftharpoons HC^{12}O_3^- + CaC^{13}O_3 \quad (4a)$$

$$k \simeq 1.004 \text{ at } 20°C \quad (4b)$$

Organic Fractionation. Organic fractionation of the stable carbon isotopes is not very well understood, but is apparently due mainly to three kinetic effects. The first fractionating process is the absorption of CO_2 into the immediate reservoir for photosynthesis. The second and largest fractionation occurs during the reaction in which inorganic carbon is fixed into organic carbon. The third major process in which fractionation occurs is the formation of the lipid (fat) fraction from more elementary materials. In the fixation reaction, C^{13} reacts about 1.7% slower than C^{12}; in lipid formation C^{13} reacts about 0.5% slower than C^{12}.

Natural Variations in the Relative Abundance of the Stable Carbon Isotopes

Meteorites and Igneous Rocks. Of all available material carbonaceous chondrites are thought to hold the best record of the isotopic carbon composition of the primitive nebular material from which the solar system condensed. The C^{12}/C^{13} of carbonaceous chondrites ranges from 88.9 to 90.3.

The δC^{13} of stony meteorites of the earth's crust is the result of the still unknown sequence of events in which the planets and meteorites evolved. An understanding of δC^{13} variations in meteorites and planets awaits the development of a satisfactory and accepted theory of meteoritic and planetary evolution.

Volcanic Emanations. Craig (1953) determined δC^{13} of CO_2 and CH_4 from volcanic emanations. Assuming that the CO_2 and CH_4 equilibrated isotopically from the temperature dependence of the "K" in Eq. (2) he calculated that the equilibration had occurred at about 300 C.

Metamorphic minerals. δC^{13} ranges of graphites and diamonds are shown in Fig. 1. Future work on the isotopic carbon composition of these minerals may lead to a better understanding of their origin and genesis. At present, it is not known whether the δC^{13} of these minerals mainly reflects the δC^{13} of the source materials or the fractionation during metamorphism.

Oceanic Bicarbonate. Almost all CO_2 in the oceans exists as the bicarbonate. The δC^{13} of bicarbonate in ocean surface waters is $-2‰$. The thermocline and deep waters are depleted in C^{13} due to the presence of C^{13}-poor bicarbonate originating from the oxidation of organic matter depleted in C^{13}.

Atmospheric CO_2. Carbon dioxide in the atmosphere is in equilibrium with bicarbonate in the oceans. Therefore, from Eq. (3) it is about 5‰ "lighter" than oceanic bicarbonate, or has a $\delta C^{13} \simeq -7‰$.

Seasonal variations have been noted in the

atmospheric δC^{13} over the oceans. Diurnal variations in the δC^{13} over forests have been observed; these are due to the daily local accumulation of isotopically light CO_2 from plant respiration, and its removal due to atmospheric mixing.

Marine Plankton. The isotopic organic carbon composition of marine plants is determined by the total effect of the three causes of organic fractionation listed above. The δC^{13} of oceanic plankton ranges from about $-15‰$ to $-30‰$; the δC^{13} of the fat fraction is about $5‰$ lighter than that of the whole organism. Part of the range of the δC^{13} of oceanic plankton is due to the effect of temperature on the extent of fractionation of the stable carbon isotopes. Other causes of the composition range are not well understood.

Terrestrial Plants and Trees. As in the case of the marine organisms, the δC^{13} of terrestrial plants and trees is determined by the total effect of the three causes of fractionation. The organic carbon in most terrestrial plants is about $20‰$ lighter than atmospheric CO_2. Park and Epstein (1960) found that for tomato plants grown in the laboratory, δC^{13} relative to atmospheric CO_2 decreased by $7‰$ during absorption of CO_2 into the leaf cytoplasm, by $17‰$ during photosynthetic fixation, and by about $3‰$ during partial conversion to fat. Thus, the composition of plants is expected to be roughly $27‰$ lighter than that of atmospheric CO_2. On the other hand, Wickman argued that the isotopic depletion of C^{13} in plants is due mainly to cyclic depletion of that isotope in the air around the plants.

Organic Matter in Marine Sediments. The isotopic organic carbon composition of marine sediments is largely a reflection of the δC^{13} of the organic matter incorporated into the sediments. Thus, there are three main determining factors involved: (1) the δC^{13} of the marine organic matter deposited in the sediment; (2) the δC^{13} of the terrestrial organic matter deposited in the sediment; and (3) the relative amounts of terrestrial organic matter compared to marine-derived organic matter in the sediment.

The value of δC^{13} per mil of petroleum is generally slightly lower than the probable δC^{13} of the source material (Silverman and Epstein, 1960). Thus, the average of δC^{13} for nonmarine oil is about $-31‰$, whereas the δC^{13} for terrestrial plants is about $-26‰$; the average δC^{13} for marine oil is about $-28‰$, whereas that for plankton and recent marine sediments is about $-22‰$. This indicates that in the formation of petroleum either there is an additional depletion of the C^{13} of the source organic matter or only the isotopically light fraction of organic matter in sediments (essentially the fat fraction) is available for conversion to petroleum.

Carbonates. If isotopic equilibrium were attained in the formation of inorganically precipitated carbonates, then from Eq. (4) the carbonates would have an isotopic carbon composition $4‰$ heavier than the dissolved bicarbonate from which they derived. Thus for inorganically precipitated carbonates, the δC^{13} will be $0‰$ to $4‰$ relative to the source carbonate, depending on how closely isotopic equilibrium is reached.

For calcium carbonate deposited as hard parts of organisms, the situation is far more complicated. The primary factor determining the isotopic carbon composition is the δC^{13} of the dissolved carbonate. The processes acting to fractionate this carbon are not well understood. Keith, Anderson, and Eichler (1964) concluded that the carbonate carbon is derived metabolically (at least in mollusks); therefore the δC^{13} in calcareous organisms varies appreciably due to differing biochemical processes occurring in different organisms. On the other hand, Polanski (1965) and others have held that the δC^{13} of organic carbonate is determined by equilibrium with the surrounding waters, but the local inhomogeneities in the surrounding waters spread out the δC^{13} values.

Variations in the δC^{13} of the Oceans and Atmosphere

Studies of the isotopic carbon composition of limestones deposited during the last 600,000,000 years do not show any monatomic variation in δC^{13}. They do, however, show some tendency toward cyclic variations. Whether these variations are real or are the result of an insufficient number of analyses is impossible to say on the basis of the present limited data.

Craig studied variations in the value of δC^{13} for atmospheric CO_2 during the last two millennia by determining the isotopic carbon composition of tree rings. His data showed no consistent variation in δC^{13} except during the first 150 years of tree growth.

Recent studies of δC^{13} of tree rings from trees which grew during the last hundred years have to date not shown any evidence of a change in the isotopic composition of atmospheric CO_2 due to the addition of isotopically light CO_2 from the burning of coal and oil. There probably have been both long-term and short-term variations in oceanic and atmospheric δC^{13}, but such variations are very difficult to detect due to the great variations of the isotopic carbon composition of the various carbon reservoirs.

The Application of Carbon Isotopic Composition Determinations to Studies on Early Fossils

Since the isotopic composition of organic carbon is different from that of sedimentary calcareous carbon, it is theoretically possible to distinguish between an organic and inorganic origin for the carbon in sedimentary rocks. The question of the origin of certain fossil-like structures in very old sediments has been studied. When the isotopic composition of noncalcareous carbon associated with the possible fossil is similar to that of modern organic carbon, it is concluded that the structures are, in fact, fossils.

MICHAEL L. BENDER

References

Craig, H., 1953, "The geochemistry of the stable carbon isotopes," *Geochim. Cosmochim. Acta,* **3**, 53–92.

Keeling, C. D., 1961, "The concentration and isotopic abundances of carbon dioxide in rural and marine air," *Geochim. Cosmochim. Acta,* **24**, 277–298.

Keith, M. L., Anderson, G. M., and Eichler, R., 1964, "Carbon and oxygen isotope composition of mollusk shells from marine and fresh water environments," *Geochim. Cosmochim. Acta,* **28**, 1757–1786.

Park, R., and Epstein, S., 1960, "Carbon isotope fractionation during photosynthesis," *Geochim. Cosmochim. Acta,* **21**, 110–126.

Polanski, A., 1961, "Geochemistry of Isotopes," Wydawnictwa Geologizne, Warsaw, 392pp. (English translation published in 1965 by the U.S. Department of the Interior.)

Rankama, K., 1956, "Isotope Geology," New York, McGraw-Hill Book Co., 535pp.

Sackett, W., M., 1964, "The depositional history and isotopic organic carbon composition of marine sediments," *Marine Geol.,* **2**, 173–185.

Silverman, S. R., and Epstein, S., 1958, "Carbon isotopic composition of petroleums and other sedimentary organic materials," *Bull. Am. Assoc. Petrol. Geologists,* **42**, 998–1012.

Cross-references: Calcium Carbonate; Carbon; Carbon Cycle; Isotope Fractionation; Photosynthesis. Vol. I: Marine Sediments. Vol. II: Carbon Dioxide Cycle in the Sea and Atmosphere; Carbonaceous Meteorites; Planetary Evolution; Tree-Ring Analysis. Vol. VI: Coal; Petroleum.

CARBON/NITROGEN RATIO

The primary biological elements that are essential constituents of the three major types of organic matter are remarkably constant among organisms (Table 1) and consist of the following:

TABLE 1. ELEMENTAL COMPOSITION OF ORGANIC MATTER[a]

Element	Carbohydrates (%)	Lipids (%)	Proteins (%)
O	49.38	17.90	22.40
C	44.44	69.05	51.30
H	6.18	10.05	6.90
P		2.13	0.70
N		0.61	17.80
C/N		113.2	2.9

[a] Rankama and Sahama (1950). For further discussion see Abelson (1959) and Breger (1963).

Plants contain the lowest proportions of protein; invertebrates the highest (Table 2). The effect of the ecologic food chain is therefore to concentrate nitrogen.

The composition of the organic matter in sediments, and thus the C/N, is controlled by the type of organic matter which is deposited in the sediments and upon the state of its degradation. Both of these are determined by the physical, chemical, and biological parameters of the environment. In most marine sediments the phytoplankton and protozoa are the principal contributors to the organic matter so that in the absence of any breakdown the C/N ratio of sediments should be between 5.0 and 6.0. Normally, however, benthonic fauna, including bacteria, and oxidation rapidly deplete the organic matter of energy sources and release CO_2, CO, H_2O, CH_4, NH_3, NO_2, N_2 and other free radicals into the interstitial and overlying water. The result is usually an increase with the depth of burial both in the C/N ratio and of resistant lignin-humus organic complexes and/or of Kerogen (Emery, 1960), although both Bader (1955) and Arrhenius (1950) note rational exceptions. The total organic matter in sediments has been estimated by multiplying either the organic carbon or the organic nitrogen percentage by an appropriate factor. Trask (1939) suggested that the value in recent sediments was 1.8 times the percentage of organic carbon and/or 18 times the nitrogen percen-

TABLE 2. ORGANIC MATTER COMPOSITION IN ORGANISMS[a]

	C/N	Protein (%)	Lipids[b] (%)	Carbohydrates (%)
Marine sessile algae	9.7[c]	7	2	91?
Phytoplankton				
Peridinians	6.0[c]	14	1.5	85?
Diatoms	5.1[c]	29	8	63
Invertebrates				
Copepods	4.4[c]	65	8	22
Higher invertebrates		70	10	20
Vertebrates				
Fish	4.1[c]			
Man	3.8[d]	51.5	47	1.5

[a] Table after Trask (1955) unless otherwise noted. The relative proportions of each of the major categories of organic matter above vary seasonally and with sex in higher organisms. Note the large increase in fat, for energy and insulation, in mamals. All analyses are on a water- and mineral (bone or test)—free basis.
[b] Ether extract consists largely of fats.
[c] Vinogradov (1953).
[d] Rankama and Sahama (1950).

tage, which corresponds to an "average" organic matter of 56% C and 5.5% N. Emery (op. cit.) has used 1.7 and 17, whereas Forsman and Hunt (1958) found 1.22 to be the average correction factor for converting organic carbon percentages in ancient sedimentary rocks to total organic matter.

Arrhenius (1950) showed that the relationship between carbon and nitrogen is not a constant but is of the form $C = aN^b$, where C and N are the carbon and nitrogen concentrations and a and b are empirical constants. Bader (1955) suggested that both a and b are environmentally controlled and are unique to a particular set of environmental conditions.

The C/N ratio and the total organic matter in marine sediments vary widely between environments but one or both are usually an important, sensitive, distinctive environmental characteristic (see Table 3) (e.g., Van Andel, 1964).

H. G. GOODELL

References

Abelson, P. H., 1959, "Geochemistry of organic substances," in (Abelson, P. H., ed.), "Researches in Geochemistry," pp. 79–103, New York, John Wiley & Sons.
Arrhenius, G., 1950, "Carbon and nitrogen in subaquatic sediments," Geochim. Cosmochim. Acta, 1, 15–21.
Bader, R. G., 1955, "Carbon and nitrogen relations in surface and subsurface marine sediments," Geochim. Cosmochim. Acta, 7, 205–211.
*Breger, I. A. 1963, "Origin and classification of naturally occurring carbonaceous substances," in (Breger, I. A., editor), "Organic Chemistry," pp. 50–86, New York, MacMillan Co.
*Emery, K. O., 1960, "The Sea Off Southern California," New York, John Wiley & Sons, 366pp.
Forsman, J. P., and Hunt, J. M., 1958, "Insoluble organic matter (kerogen) in sedimentary rocks of marine origin," in "Habitat of Oil," Tulsa, Oklahoma, Am. Assoc. Petrol. Geologists, pp. 747–778.
Rankama, K., and Sahama, Th. G., 1950, "Geochemistry," Chicago, Univ. of Chicago Press, 912pp.
*Trask, P. D., 1939, "Organic content of recent marine sediments," in (Trask, P. D., editor), "Recent Marine Sediments," pp. 428–453, Tulsa, Am. Assoc. Petrol. Geoogists.
Vinogradov, A. P., 1953, "The Elemental Composition of Marine Organisms," Mem. 2, Sears Foundation for Marine Research.

Cross-references: Carbon: Element and Geochemistry; Nitrogen; Organic Geochemistry. Vol. I: Marine Sediments; Phytoplankton. Vol. IV B: Kerogen. Vol. VI: Globigerina Ooze; Pelagic Sediments; Siliceous Ooze.

References (for Table 3)

Correns, C. W., 1939, "Pelagic Sediments of the North Atlantic Ocean," in (Trask, P. D., editor), "Recent Marine Sediments," pp. 373–395, Tulsa, Am. Assoc. Petrol. Geologists.
Emery, K. O., 1956, "Sediments and water of the Persian Gulf," Bull. Am. Assoc. Petrol. Geologists, 40, 2354–2383.
Emery, K. O., 1960, "The Sea Off Southern California," New York, John Wiley & Sons, 366pp.
Emery, K. O., 1964, "Sediments of Gulf of Aqaba (Eilat)," in (Miller, R. L., editor), "Papers in Marine Geology," pp. 257–273, New York, Macmillan Co.
Emery, K. O., et al., 1957, "Sediments of three bays of Baja, California, Sebastian Viscaino, San Cristobal, and Todos Santos," J. Sed. Petrol., 27, 95–115.
Evans, R. G., 1962, " A Sedimentological Study of Greater Gullivan Bay, Florida," unpublished Master's thesis, Florida State Univ.
Fleece, J. B., 1962, "The Carbonate Geochemistry

CARBON/NITROGEN RATIO

TABLE 3. CARBON, NITROGEN, AND THE CARBON-NITROGEN RATIO IN MARINE SEDIMENTS
(Surface Samples only)

Environment	Carbon (%)	Nitrogen (%)	C/N	Total Organic (%)	Source[a]
Transitional marine					
Delta, Mississippi					
Marsh				31.5(10)[a]	Scruton (1955)
Delta front silts and sands				6.7(13)[a]	Scruton (1955)
Prodelta silty clays				26.5(20)[a]	Scruton (1955)
Offshore clays				8.9(7)[a]	Scruton (1955)
Marginal deposits				6.2(35)[a]	Scruton (1955)
Delta platform, Gulf of Paria	1.06(7)[c]	0.08(7)	13.4(7)	1.80(7)[b]	Van Andel and Postma (1954)
Delta, Gulf of Calif.	0.85–1.44	0.070–0.093			
Foreset-bottomset	0.97(3)			1.7(3)[b]	Van Andel (1964)
Tidal flats	3.8–4.6[b]			6.5–7.8	Stevenson and Emery (1958)
Marshes	10.6[b]			18(91)	Emery (1960)
Marshlands, Newport Bay, California	0.06–16.3[b]			0.1–27.7	Stevenson and Emery (1958)
Mangrove peat, Florida	16.24(22)	0.27(22)	20–32	27.6(22)[b]	Scholl (1963)
Mangrove peat, Florida (1000 Islands)	10.13–23.75	0.04–0.94	60.2(22)	17.2–35.3	Holmes (1962)
1000 Islands, Florida (exclusive of peat)	3.06(88)	0.10(88)	30.6(88)	5.2(88)[b]	Holmes (1962)
	0.11–9.84	0.02–0.57		0.2–16.7	
1000 Islands, Florida	0.40–14.1	0.33–0.38	14–30	0.7–24.0	Scholl (1963)
Whitewater Bay, Florida	6.0–18.4	0.18–0.57	17–33	10.2–31.3	Scholl (1963)
Barataria Bay, La.	0.0–7.6			0–12.9[b]	Krumbein (1939)
San Quintin Bay, Baja, Calif.	0.58(91)	0.031(91)	18.7	0.99(91)[b]	Gorsline and Stewart (1962)
	0.085–2.087	0.005–0.110		0.145–3.548[b]	
Laguna Madre, Texas	0.59–4.12[b]			1–7	Rusnak (1960)
Moriches Bay, N.Y.	1.18–16.8[b]			2–28.6	Nichols (1964)
Apalachicola Bay, Fla.	0.081–2.367	0.007–0.203		0.14–4.02[b]	Kofoed (1961)
Florida Bay, Fla.					
5 Key cores	4.25(47)	1.92(47)	30	7.6	Fleece (1962)
5 Shole cores	2.65(38)	1.69(38)	25	4.7	Fleece (1962)
Bay surface	2.49(139)			4.23(139)	Goodell data
	0.290–6.630			0.49–11.3	
Tampa Bay, Fla.	1.62(530)	0.48(270)	24.5(270)	2.75	Goodell data
	0.030–9.35	0.01–3.28		0.51–15–90	
St. Josephs Bay, Fla.	1.34(102)	0.071	18.9	2.28(102)	Stewart (1962)
	0.5–4.5	0.01–2.4		0.9–7.7[b]	

CARBON/NITROGEN RATIO

Location				Reference
San Cristobal Bay, Baja Calif.	1.3(16) 0.6–2.0		2.21(13)[b] 1.02–3.4	Emery et al. (1957)
Gulf of California		*Normal marine*		
Eastern slope	3.62(29)		6.15(29)[b]	Van Andel (1964)
Basin	2.54(30)		4.32(29)[b]	Van Andel (1964)
Western slope	4.22(15)		7.17(15)[b]	Van Andel (1964)
Gulf of Paria	0.87(33)	1.29(33)	1.48(33)[b]	Van Andel and Postma (1954)
Central gulf	0.18–2.12	0.15–0.189	6.8	
Red Sea	0.19(26)	0.041(26)	4.6	Mohamed (1949)
	0.07–0.39	0.016–0.086		Mohamed (1949)
Gulf of Aqaba	0.127(9)	0.033(9)	3.8	Emery (1964)
Deep water	0.08	0.024	3.33	
Eilat area	0.06–0.12	0.023–0.033		
		0.024(30)		
		0.000–0.042		
Gulf of Suez	0.39(3)	0.051(3)	7.6	Emery (1964)
Persian Gulf	0.8(8)	0.08(28)	10.0	Mohamed (1949)
Arabian Sea	0.47–3.09		10.1–34.2	Emery (1956) Wiseman and Bennett (1940)
Gulf of Aden	2.21–4.76		10.6–13.2	Bennett (1940)
Gulf of Oman	0.57–5.79		4.9–13.5	Bennett (1940)
California Basins	4.1[b]		7.0(80)	Emery (1960)
Gulf of Baja Calif. Basins	2.64(36)		4.49(36)[b]	Van Andel (1964)
Pacific Basins off Baja, Calif.	2.9		5.0	Krause (1964)
Continental Shelf				
Lower California silt-clay	0.96(12)[b]	0.010(70)	1.63(12)	Van Andel (1964)
S.E. Atlantic	0.151(70)	0.079(20)	0.26	Goodell data
Gulf of Paria	0.60(20)	0.018–0.142	1.02(20)	Van Andel and Postma (1954)
Marginal shelf	0.07–1.92		15.1	
California	0.53[b]		7.6	Emery (1960)
W. Gulf Coast	0.53(45)	0.063(149)	0.9	Trask (1953)
West Florida inner shelf	0.33(249)		8.5	Evans (1962)
	0.00–2.55		0.91	
N. Florida inner shelf	4.84(106)	0.22(106)	21(249)	Hyne (1965)
	0.179–1.924	0.008–0.083	0.00–132.50 21.9	
Antarctic			0.304–3.271	
Weddel Sea-Palmer Peninsula	0.241(41)	0.060(41)	4.0(41)	Goodell data
	0.051–1.537	0.001–0.359	0.411	
Bellingshausen-Amundsen Sea	0.414(22)	0.010(22)	4.14(22)	Goodell data
	0.00–0.920	0.005–0.129	0.704	

TABLE 3. (Continued)

Environment	Carbon (%)	Nitrogen (%)	C/N	Total Organic (%)	Source[a]
Ross Sea	0.592(61)	0.082(61)	7.2(61)	1.006	Goodell data
	2.00–0.070	0.013–0.402			
Continental Slope					
S.E. Atlantic	0.543(30)	0.037(30)	14.7	0.92	Goodell data
West Gulf of Mexico	1.02(13)	0.119(79)	8.6	1.73	Trask (1953)
Gulf of Calif.					
(E. and W. slopes)	3.82(44)			6.49(44)[b]	Van Andel (1964)
California	2.41(29)[b]			4.1	Emery (1960)
S. Florida carbonate					
slope	0.84(5)	0.046(5)	18.3	1.43	Milligan (1962)
Antarctic					
Weddell Sea-Palmer					
Peninsula	0.390(10)	0.019(10)	20.5(10)	0.66(10)	Goodell data
	0.174–0.682	0.005–0.077		0.29–1.16	
Bellingshausen-	0.673(35)	0.061(35)	11.0(35)	1.14(35)	Goodell data
Amundsen Sea	0.17–1.18	0.007–0.311		0.29–2.00	
Ross Sea	0.591(84)	0.081(84)	7.3(84)	1.01(84)	Goodell data
	0.07–1.972	0.002–0.439	3.0(27)	0.12–3.35	Emery (1960)
Submarine canyons	1.11[b]				
Deep Sea Floor					
Off California	1.24(11)[b]			2.1	Emery (1960)
Gulf of Mexico	0.63(16)	0.068(16)	9.5	1.07[b]	Trask (1953)
Baja, California	1.71(7)			2.91(7)[b]	Van Andel (1964)
Straits of Florida	0.71(32)	0.05(32)	14.1	1.2	Milligan (1962)
Globigerina ooze					
Arabian Sea	1.11(24)	0.097(24)	11.4	1.89[b]	Wiseman and Bennett (1940)
N. Atlantic	0.222(63)[b]			0.377(63)	Correns (1939)
Western Gulf	0.60(22)	0.065(22)	9.2	1.02[b]	Trask (1953)
Antarctic Ocean	1.150(29)	0.043(20)	26.7	1.96[b]	Goodell data
Diatom ooze					
Antarctic Ocean	0.588(33)	0.062(33)	9.5	1.00	Goodell data
Gulf of Baja, Calif.	2.82(20)			4.79[b]	Van Andel (1964)
Pelagic silts					
Antarctic Ocean	0.809(90)	0.074(90)	10.9	1.38[b]	Goodell data
North Atlantic	0.538(86)[b]			0.916(86)	Correns (1939)
Pelagic clays	0.441(32)	0.057(32)	7.7	0.75[b]	Goodell data
Glacial marine	0.372(86)	0.037(86)	10.0	0.63[b]	Goodell data
(see Antarctic Shelf					
for additional data)					

Oceanic Deeps		
Peru-Chile Trench		
Upper slope	2.43(31)	4.13^b Trask (1961)
Lower slope and basin	0.63(51)	1.07^b Trask (1961)
South Sandwich Trench		
(6000 m and greater)	0.639(9)	0.035(9) 18.3 Goodell data

^{a} Given as wood, grass, etc., and *not* from chemical analyses.
^{b} Calculated from author's data using 1.7 C% = organic matter or organic/1.7 = %C as appropriate.
^{c} All figures in parentheses are the number of observations used to calculate the mean; hyphenated figures indicate the range of data.
^{d} See references for Table 3 at end of article.

and Sedimentology of the Keys of Florida Bay, Florida," unpublished Master's thesis, Florida State Univ.

Gorsline, D. S., and Stewart, R. A., 1962, "Benthic marine exploration of Bahia de San Quintin, Baja, Calif., 1960–61, Marine and Quaternary Geology," *Pacific Naturalist,* **3,** 282–319.

Holmes, C. W., 1962, "Sediments of the 10,000 Islands," unpublished Master's thesis, Florida State Univ.

Hyne, N., 1965, "Sedimentary Environments and Submarine Geomorphology of the Continental Shelf in the Area of Choctawhatchee Bay, Florida," unpublished Master's thesis, Florida State Univ.

Kofoed, J. W., 1961, "Sedimentary Environments in Apalachicola Bay and Vicinity, Florida," unpublished Master's thesis, Florida State Univ.

Krause, D. C., 1964, "Lithology and Sedimentation in the Southern Continental Borderland," in (Miller, R. L., editor), "Papers in Marine Geology," pp. 274–318, New York, Macmillan Co.

Krumbein, W. C., 1939, "Tidal Lagoon Sediments on the Minnesota Delta," in (Trask, P.D., editor), "Recent Marine Sediments," pp. 178–194, Tulsa, Am. Assoc. Petrol. Geologists.

Milligan, D. B., 1962, "Marine Geology of the Florida Straits," unpublished Master's thesis, Florida State Univ.

Mohamed, H. F., 1949, "The distribution of organic matter in sediments from the northern Red Sea," *Am. J. Sci.* **247,** 116–127.

Nichols, M. M., 1964, "Characteristics of Sedimentary Environment in Moriches Bay," in (Miller, R. L., editor), "Papers in Marine Geology," pp. 363–383, New York, Macmillan Co.

Rusnak, G. A., 1960, "Sediments of Laguna Madre, Texas," in (Shepard, F. D., et al., editors), "Northwest Gulf of Mexico," pp. 153–196.

Scholl, D. W., 1963, "Sedimentation in modern coastal swamps, southwestern Florida," *Bull. Am. Assoc. Petrol. Geologists,* **47,** 1581–1603.

Scruton, P. C., 1955, "Sediments of the eastern Mississippi Delta," in "Finding Ancient Shorelines," pp. 21–49, *Soc. Econ. Paleontologists Mineralogists Spec. Publ.* **3.**

Stevenson, R. E., and Emery, K. O., 1958, "Marshlands at Newport Bay, California," Allan Hancock Foundation Publ., Occasional Paper No. 20, 109pp.

Stewart, R. A., 1962, "Recent Sedimentary History of St. Joseph Bay, Florida," unpublished Master's thesis, Florida State Univ.

Trask, P. D., 1953, "The sediments of the western Gulf of Mexico," Pt. 2, "Chemical studies of sediments of the western Gulf of Mexico," "Papers in Physical Oceanography and Meteorology," Mass. Inst. Technol. and WHOI, Vol. 12, 120pp.

Trask, P. D., 1961, "Sedimentation in a modern geosyncline off the arid coast of Peru and northern Chile," XXI Intern. Geol. Congr. Pt. XXIII, pp. 103–118.

Van Andel, T. H., 1964, "Recent marine sediments of Gulf of California," in "Marine Geology of

CARBON/NITROGEN RATIO

Gulf of California," *Am. Assoc. Petrol. Geologists Mem.* **3**, 216–310.
Van Andel, T. H., and Postma, H., 1954, "Recent Sediments of the Gulf of Paria," Reports of the Orinoco Shelf Exp., Vol. 1, Amsterdam, North Holland Publ. Co.
Wiseman, J. D. H., and Bennett, H., 1940, "The distribution of organic carbon and nitrogen in sediments from the Arabian Sea, John Murray Exp., 1933-34," *Sci. Results*, **3**, 193–221.

CARBONACEOUS METEORITES—
See Vol. II

CARBONATE HYDROLOGY—See HYDROLOGY, KARST TERRAIN

CARBONATES

The carbonate class of minerals has the $(CO_3)^{2-}$ anionic group as the fundamental unit of structure. Most of the minerals are anisodesmic oxysalts with carbon situated in the center of an equilateral triangle with oxygen atoms at the corners, producing a planar group. The bonds to the metallic cations are usually ionic.

Two very important groups of carbonates are the calcite group and the aragonite group. The calcite group contains the relatively small metallic cations Mg, Fe, Mn, Co, and Zn, whereas the aragonite group contains the larger cations Ba, Sr, and Pb. Calcium is common to both groups. (See Table below.)

ACID CARBONATES

Nahcolite	$NaHCO_3$
Kalicinite	$KHCO_3$
Teschemacherite	$(NH_4)HCO_3$
Trona	$Na_3H(CO_3)_2 \cdot 2H_2O$
Wegscheiderite	$Na_5(HCO_3)_3CO_3$

ANHYDROUS CARBONATES
$A(XO_3)$ type

Calcite	$CaCO_3$	*Calcite group:* almost complete series between Ca, Mg, Fe, Mn; Zn and Mn, but little substitution by Ba, Sn, and Pb for Ca; hexagonal scalenohedral
Magnesite	$MgCO_3$	
Siderite	$FeCO_3$	
Rhodochrosite	$MnCO_3$	
Cobaltocalcite	$CoCO_3$	
Smithsonite	$ZnCO_3$	
Otavite	$CdCO_3$	
Aragonite	$CaCO_3$	*Aragonite group:* orthorhombic with pseudohexagonal symmetry; partial series at ordinary temperature, but complete substitution at higher temperature
Witherite	$BaCO_3$	
Strontianite	$SrCO_3$	
Cerussite	$PbCO_3$	

$AB(XO_3)_2$ type

Dolomite	$CaMg(CO_3)_2$	*Dolomite group:* rhombohedral; Mg, Fe, Mn substitute for alternate Ca in calcite-like structure
Ankerite	$Ca(Fe,Mg)(CO_3)_2$	
Kutnahorite	$Ca(Mn,Mg)(CO_3)_2$	
Alstonite	$CaBa(CO_3)_2$	
Barytocalcite	$CaBa(CO_3)_2$	
Norsethite	$BaMg(CO_3)_2$	

Miscellaneous

Huntite	$Mg_3Ca(CO_3)_4$
Fairchildite	$K_2Ca(CO_3)_2$
Shortite	$Na_2Ca_2(CO_3)_3$
Benstonite	$Ca_7Ba_6(CO_3)_{13}$

HYDROUS CARBONATES
$A(XO_3) \cdot xH_2O$ type

Thermonatrite	$Na_2CO_3 \cdot H_2O$
Nesquehonite	$MgCO_3 \cdot 3H_2O$
Trihydrocalcite	$CaCO_3 \cdot 3H_2O$
Pentahydrocalcite	$CaCO_3 \cdot 5H_2O$
Lansfordite	$MgCO_3 \cdot 5H_2O$
Natron	$Na_2CO_3 \cdot 10H_2O$

Miscellaneous

Buetschliite	$K_6Ca_2(CO_3)_5 \cdot 6H_2O$
Pirssonite	$Na_2Ca(CO_3)_2 \cdot 2H_2O$
Gaylussite	$Na_2Ca(CO_3)_2 \cdot 5H_2O$

Schroeckingerite	$NaCa_3(UO_2)(CO_3)_3(SO_4)F \cdot 10H_2O$
Voglite	$Ca_2Cu(UO_2)(CO_3)_4 \cdot 6H_2O(?)$
Bayleyite	$Mg_2(UO_2)(CO_3)_3 \cdot 18H_2O$
Swartzite	$CaMg(UO_2)(CO_3)_3 \cdot 12H_2O$
Andersonite	$Na_2Ca(UO_2)(CO_3)_3 \cdot 6H_2O$
Liebigite	$Ca_2U(CO_3)_4 \cdot 10H_2O$
Lanthanite	$(La,Ce)_2(CO_3)_3 \cdot 8H_2O$

CARBONATES WITH HYDROXYL OR HALOGEN

Loseyite	$(Mn,Zn)_7(CO_3)_2(OH)_{10}$
Zaratite	$Ni_3(CO_3)(OH)_4 \cdot 4H_2O$
Hydrozincite	$Zn_5(CO_3)_2(OH)_6$
Aurichalcite	$(Zn,Cu)_5(CO_3)_2(OH)_6$
Rosasite	$(Cu,Zn)_2(CO_3)(OH)_2$
Malachite	$Cu_2(CO_3)(OH)_2$
Phosgenite	$Pb_2(CO_3)Cl_2$
Bismutite	$(BiO)_2(CO_3)$
Artinite	$Mg_2(CO_3)(OH)_2 \cdot 2H_2O$
Azurite	$Cu_3(CO_3)_2(OH)_2$
Hydrocerussite	$Pb_3(CO_3)_2(OH)_2$
Hydromagnesite	$Mg_4(CO_3)_3(OH)_2 \cdot 3H_2O$
Rutherfordine	$(UO_2)(CO_3)(?)$
Sharpite	$(UO_2)_6(CO_3)_5(OH)_2 \cdot 6H_2O(?)$
Dawsonite	$NaAl(CO_3)(OH)_2$
Northupite	$Na_3Mg(CO_3)_2Cl$
Dundasite	$PbAl_2(CO_3)_2(OH)_4 \cdot 4H_2O$
Alumohydrocalcite	$CaAl_2(CO_3)_2(OH) \cdot 2H_2O(?)$
Beyerite	$Ca(BiO)_2(CO_3)_2$
Cordylite	$Ce_2Ba(CO_3)_3F_2$
Ancylite	$Ce_4Sr_3(CO_3)_7(OH)_4 \cdot 3H_2O$
Bastnaesite	$Ce(CO_3)F$
Synchisite	$CeCa(CO_3)_2F$
Parisite	$Ce_2Ca(CO_3)_3F_2$
Roentgenite	$Ce_3Ca_2(CO_3)_5F_3$

COMPOUND CARBONATES

Tychite	$Na_6Mg_2(CO_3)_4(SO_4)$
Bradleyite	$Na_3Mg(CO_3)(PO_4)$
Leadhillite	$Pb_4(CO_3)_2(OH)_2(SO_4)$
Susannite	$Pb_4(CO_3)_2(OH)_2(SO_4)$

LLOYD W. STAPLES

References

Dana, J. W., and Dana, E. S., 1951, "The System of Mineralogy," Seventh ed., Vol. II, 132, (revised by Palache, C., Berman, H., and Frondel, C.), New York, John Wiley & Sons.

Donnay, G., 1953, "The crystallography of bastraesite, parisite, roentgenite, and synchisite," *Am. Mineralogist*, **38**, 932.

Faust, G. T., 1953, "Huntite, $Mg_3Ca(CO_3)_4$, a new mineral," *Am. Mineralogist*, **38**, 4.

Goldsmith, J. R., 1959, "Some aspects of the geochemistry of carbonates," in "Researches in Geochemistry," New York, John Wiley & Sons, 336pp.

Graf, D. L., and Lamar, J. E., 1955, "Properties of calcium and magnesium carbonates," *Econ. Geol.*, 50th Ann. Vol., 639pp.

Kulp, J. L., Kent, P., and Kerr, P. F., 1951, "Thermal study of the Ca–Mg–Fe carbonate minerals," *Am. Mineralogist*, **36**, 643.

Cross-references: *Mineral Classes: Nonsilicates.* Vol. IV B: *Aragonite; Calcite.*

CATALYSIS

The phenomenon of catalysis in both organic reaction systems has been recognized for more than one hundred and fifty years. Chemical reactions involve the interaction between reacting species, redistribution of chemical bonds, and the generation of a product. A substance which facilitates any or all of these various stages can be regarded as a catalyst and there is no restriction, as in the classical definition, that the catalyst itself remain unaltered. In many technologically important catalytic processes the catalyst is consumed, albeit to a very small extent compared to the amount of product. Although the greatest attention has been accorded to gas- and liquid-phase catalytic reactions, solid-state reactions are, in every respect, equally affected by catalysts although their influence is frequently overlooked, particularly since their detailed study is difficult in the extreme.

CATALYSIS

The nature of a catalyst varies very widely and covers both homogeneous and heterogeneous systems. Among the latter, natural clays and other minerals, artificial oxides and sulfides, and metals, either separately or supported on a suitable substrate, are extensively employed. Recently, considerable attention has been accorded the organometallic systems, some of which form the basis of the Ziegler-Natta stereospecific polymerization catalysts. Currently, there is ever-increasing interest in the biological catalytic reactions effected by enzymes. Hardly any type of system has escaped attention and the literature is profuse and becoming more so all the time.

Catalytic activity can most conveniently be characterized by the change in the rate of reaction for a chemical process which occurs when the catalyst is added to the system. In the case of homogeneous systems, comparisons are made on the basis of unit concentration of catalyst whereas in heterogeneous systems it is more appropriate to characterize on the basis of unit surface area. Unfortunately, neither of these specifications are rigidly adhered to, making it extremely difficult to compare data from diverse sources. Although this type of reference is satisfactory for preliminary considerations, in the majority of industrially important systems the specificity of a catalyst, that is, its activity with respect to the preferred reaction as against its activity to undesirable side reactions, is frequently of greater importance. This emphasizes the fact that the overall process is frequently very complex with the catalyst actively participating in the reactions, usually with the formation of intermediate complexes of varying stability.

Among the catalytic reactions which early assumed commercial importance were the fixation of atmospheric nitrogen, the oxidation of ammonia, and the hydrogenation of edible oils to margarine. However, the greatest applications of catalysis have very largely developed from the petroleum and petrochemical industries. The demand for high-octane gasoline during World War II accelerated the development of many catalytic reaction techniques such as cracking or hydrocracking, reforming, alkylation, and many other processes too numerous to consider in detail. Simultaneously, gasoline synthesis from coal or from carbon monoxide and hydrogen was extensively developed and led to a wide variety of products. Particularly since World War II the catalytic production of polymers has attracted enormous attention and is still developing remarkably with the production of textile fibers, rigid polymers, and synthetic elastomers closely resembling natural rubber.

Theories of Catalysis

Perhaps the earliest theory of catalysis was that of Clement and Désormes who, in 1808, postulated the formation of an intermediate compound, a view supported by Ostwald and many others and, in many ways, comparable with the more modern complex or carbonium ion theories. At an early stage, the importance of hydrogen and hydroxyl ions in reaction kinetics was recognized and studied.

During the decade 1920–1930 the concept of acid-base catalysis was extensively developed. Brönsted and Lowry independently correlated catalytic activity with the acidic or basic character of a catalyst, based on the ability of a substance to donate or accept a proton, respectively. The extensive work of Brönsted led to the association of his name particularly with acid sites. Almost simultaneously Lewis defined an acid molecule as one capable of accepting an electron pair into one of its atoms and, conversely, a basic molecule as one which presents an electron pair.

A wide range of phenomena may be correlated in a qualitative manner with Lewis acidity; this has led to interesting speculations regarding reaction mechanisms. However, little reliable quantitative data is available to substantiate these various proposals. The carbonium ion concept for cracking catalyst reactions was introduced by Greensfelder. Tamele has been largely responsible for the correlation of catalytic activity in cracking reactions with the "acidity" of catalyst materials. Friedel-Crafts reactions can also be explained by the intermediate formation of a carbonium ion with the metal halide, and some isomerization reactions have been similarly explained.

Various methods have been used to determine the "acidity" of catalyst surfaces and it is generally fairly simple to define protonic acidity in any one solvent by adsorbing a suitable series of indicators whose strengths relative to water are known. Definition of Lewis acidity is somewhat more difficult and is usually undertaken by the Benesi technique using n-butylamine titrations or by the Phillips technique of ion exchange with ammonium acetate. On alumina silicates it is generally believed that olefins react with Brönsted acid sites to add protons at the double bonds. Alternately, it is believed that isomerization and certain reactions involving paraffins occur at Lewis acid sites. In either case, a carbonium ion is the probable intermediate as established by infrared adsorption studies. In the latter case, the Lewis site abstracts a hydrogen atom.

An alternate theory of analysis was advanced by Balandin in 1929 and has subsequently been developed by his direct associates and designated the "multiplet theory." It emphasizes a geometric arrangement of active sites in a proper ratio to the total surface. The sites are designated singlet, doublet, and so on according to their ability to simultaneously sustain single, double, or multiple bond. Balandin appreciated that if bonding were too strong the surface would be blocked and if too weak the probability of reaction would be small. The concept of directional bonding of the adsorbed species propitious to the formation of specific products was developed.

At about the same time as these direct theories of catalysis were being developed important contributions to the understanding of the catalytic process were being made by the detailed studies of adsorption and desorption kinetics. The heterogeneity of surfaces was established by Taylor and further contribution was made by Polanyi, Roginskii, Rideal, and Garner, whose work could be analyzed theoretically on the basis of the transition-state theory of Eyring.

More recently two important theories have been added. Eley and Dowden have drawn attention to the importance of electron holes in the d-band for a wide number of reactions. At about the same time, a correlation between the defect structure of solids and their catalytic activity was developed simultaneously by Garner and Gray and by Volkenshtein. Catalytic reactions over semiconductors established a direct relationship between the reaction process and the propensity for electron transfer processes to occur between the adsorbed species and the solid. These views are compatible with all previous theories and permit the development of a comprehensive theory of catalysis.

Among the more recent developments in catalysis, the Zieglar-Natta type catalysts, based on a composite derived most frequently from an alkylaluminum and a subhalide such as $\alpha-TiCl_3$, has focused attention once again on the intermediate complex formation. These catalysts are highly active for polymerization and are frequently observed to yield products of high regularity of bond orientation around asymmetric carbon atoms. Such characteristics are generally classified as stereospecific.

No mention can be made at this juncture of the wealth of catalytic reactions occurring in nature and loosely grouped under the heading "biocatalysis." Undoubtedly, the catalytic activity of the enzymes will be accorded even more attention in the future than more conventional catalysis is today.

T. J. GRAY

References

"Advances in Catalysis," Vols. 1–13, New York, Academic Press.
Benesi, H. A., 1957, *J. Phys, Chem.*, **61,** 970.
Brönsted, J. N., 1923, *Rev. Trav. Chim.*, **42,** 718.
Derry, R., Garner, W. E., and Gray, T. J., 1954, *J. Chem. Phys.*, **51,** 670.
Emmett, P. H., editor, "Catalysis," Vols. 1–7, New York, Reinhold Publ. Corp.
Greensfelder, B. S., Voge, H. M., and Good, G. M., 1949, *Ind. Eng. Chem.*, **41,** 2573.
"Kinetics and Catalysis," (English transl.), Vols. 1–7, New York, Consultants Bureau Enterprises.
Lowry, T. M., 1923, *Chem. Ind.*, **42,** 43.
"Soviet Research in Catalysis," (English transl.), Vols. 1–7, New York, Consultants Bureau Enterprises.
Tamele, M. W., 1950, *Discussions Faraday Soc.* **8,** 270.

Cross-references: *Acids and Bases; Bonding; Chelation; Hydrocarbons.*

CATION EXCHANGE—See ION EXCHANGE

CATION SUBSTITUTION—See TRACE ELEMENTS IN SILICATE MINERALS, SUBSTITUTION

CEMENT—See Vol. VI BUILDING MATERIALS

CERIUM: ELEMENT AND GEOCHEMISTRY

Properties. Elemental cerium (symbol Ce) is a very reactive, silvery-white lanthanide, or rare earth metal, with atomic number 58 and outer electronic structure $4f^1 5d^1 6s^2$. It occurs in four stable isotopes: Ce^{136}, 0.193% in nature; Ce^{138}, 0.250% in nature; Ce^{140}, 88.5% in nature; and Ce^{142}, 11.1% in nature. Ce^{142} is weakly radioactive with a half-life of 5×10^{15} years. Trace amounts of cerium have been determined analytically by neutron activation, by x-ray spectrography, and by isotope-dilution mass spectrometry. In geologic environments, cerium forms highly ionic bonds and weak complexes. The most common oxidation state is tripositive, with ionic radius 1.03 Å; tetrapositive cerium has an ionic radius of 0.92 Å. Generally, the chemistry of tripositive cerium is similar to that of the alkaline earth elements and to other light rare earth elements; tetrapositive cerium behaves more like zirconium and thorium.

Berzelius (in 1803) was the first to isolate cerium from other rare earths, in the mineral cerite. Cerium was named after the planetoid Ceres.

Minerals. Cerium is the major rare earth

element in several rare earth minerals, usually with coordination number of ten to twelve. The most common are: monazite (light rare earths, Th) PO_4 and bastnaesite (light rare earths, Th) FCO_3. Numerous other rare earth nitrate, sulfate, carbonate, niobate, and tungstate rare earth minerals exist. Often their structures contain halogens and alkali metals.

Abundance. See *Rare Earths*.

Geochemical Behavior. Schmitt et al. (1964) have established a value of 0.84 ppm cerium in chondrites and 9.7 ppm cerium in eucrite achondrites. The standard deviation within meteorite class was only about 20%. In cosmic processes, high temperatures will not volatilize cerium (boiling point 3469°C).

The cerium concentration in average basalt is 66 ppm, but the values vary by over a factor of ten (Haskin et al., 1966). Alkaline oceanic basalts and most continental basalts have more cerium than oceanic tholeiites and some continental diabases. If mantle garnet peridotite with low cerium content partially melts to yield a basaltic liquid, cerium will concentrate in the liquid.

Igneous differentiation series tend toward higher cerium concentrations. For example, Towell et al. (1965) studied rare earths in the southern California batholith and found 87.6 ± 14 ppm cerium in the Rubidoux Mountain leucogranite, probably a differentiate (by crystal settling) of the San Marcos gabbro, containing 14.5 ± 2.3 ppm. cerium. About half of the cerium in the gabbro is in rock-forming minerals (plagioclase, pyroxene, hornblende), probably substituting for calcium. The remainder is in accessory minerals like apatite. The accessory minerals hold most of the cerium in the leucogranite.

The concentration of cerium is generally higher in granitic rocks (about 85 ppm in granites with 60.70% silica) than in basaltic rocks, and it is even more variable.

Cerium forms complexes in, and is transported by, alkali-, halogen-, and carbon dioxide-rich hydrothermal fluid, but the complexes are not as stable as those formed by heavy rare earths. Metasomatic and pegmatoid deposits often have low ratios of cerium to heavy rare earths compared to granites, indicating low ratios in the hydrothermal solutions.

Cerium has a 48-year residence time in seawater (which is alkaline), the lowest of any rare earth element. Probably it is complexed. Oxidation to tetrapositive cerium and removal by adsorption on clay particles, as well as depletion in surface water in areas of organic productivity, are likely processes. Today, high secondary cerium concentrations in phosphatic fish remains occur only in deep water. Manganese nodules have 0.5% cerium; Pacific Ocean water has only 1.28 ppm cerium (Goldberg et al., 1963).

In soils, the relative amounts of cerium in detrital, clay, and carbonate fractions is not known. Sedimentary rocks have about 75 ppm cerium, with great variation. In a study of soils in Russian platform soils, Balashov (1964) found higher cerium in alkaline soils, the cerium precipitating as hydroxide (probably tetrapositive cerium). Cerium was lower in acid soils, from which it is probably removed as a complex ion.

Biologic concentration of rare earths has been found in hickory leaves; rare earths have been used to trace opium.

Economic Geology. The major source of cerium has been from monazite in beach sands. The accessory monazite originally present in granite rocks resists weathering and is concentrated by sedimentary processes. Australia, Brazil, Ceylon, India, Malaysia, and South Africa mine the largest quantities. Up to 30% (by weight) of monazite may be cerium. Recently, the bastnaesite deposit at Mountain Pass, California has been an important source of cerium, the rare earth element with highest concentration in the ore.

High-purity cerium is used in limited quantities, for example, in optical filters. Most cerium is sold unpurified in rare earth concentrates for use in polishing, petroleum cracking catalysis, lighter flints, arc carbon, and in alloys.

ROBERT KAY

References

Balashov, Y. A., Ronov, A. B., Migdisov, A. A., and Turanskaya, N. V., 1964, "The effects of climate and facies environment on the fractionation of the rare earths during sedimentation," *Geochem. Intern.*, **10**(1), 951–969 (Engl. transl.).

Goldberg, E. D., Koide, M., Schmitt, R. A., and Smith, R. H., 1963, "Rare-earth distributions in the marine environment," *J. Geophys. Res.* **68**, 4209–4217.

Haskin, L. A., et al., 1966, "Meteoric, solar, and terrestrial rare-earth distributions," in (Ahrens, L. H., et al., editors), "Physics and Chemistry of the Earth," Vol. 7, Oxford, Pergamon Press.

Rankama, K., and Sahama, T., 1950, "Geochemistry," Chicago, Univ. of Chicago Press, 912pp.

Schmitt, R. A., Smith, R. H., and Olehy, D. A., 1964, "Rare-earth, yttrium and scandium abundances in meteoric and terrestrial matter," *Geochim. Cosmochim. Acta*, **26**, 67–86.

Towell, D. G., Winchester, J. W., and Spirn, R. V., 1965, "Rare-earth distributions in some rocks and associated minerals of the batholith of southern California," *J. Geophys. Res.*, **70**, 3485–3496.

Cross-references: Alkalis, Alkali Metals and Alkaline Earth Metals; Catalysis; Lanthanide Series; Mass Spectrometry; Neutron Activation Analysis; Rare Earths; Seawater Chemistry; X-Ray Spectroscopy.

CESIUM: ELEMENT AND GEOCHEMISTRY

Cesium (Cs), a rare metal, is the heaviest member of the alkali group of elements. It was discovered by Bunsen and Kirchoff in 1860 through the use of the spectroscope, but the pure metal was not isolated until 1882. Cesium has an atomic weight of 132.9 and atomic number 55. The name is derived from the Latin caesius (sky blue) in allusion to the characteristic line in the blue region of the spectrum.

The metal is silvery white, soft, and melts near room temperature at 28.5°C (83°F). The solid metal has a density of 1.90 and crystallizes with body-centered cubic symmetry. It is the most electropositive of the elements and is extremely active chemically. It burns upon contact with air and reacts explosively upon contact with water. It forms a series of compounds which are similar in properties to those of the other alkali metals, being for the most part very soluble in water. In all its natural compounds, cesium occurs as a monovalent cation. Its ionic radius is 1.69. Only one isotope (cesium-133) is known in nature, but at least eighteen others are known through nuclear reactions. The cesium-137 isotope, with a half-life of 33 years, has been widely publicized, and is one of the most troublesome radioactive wastes to be produced through the fission of uranium.

Minerals

Only three minerals are known in which cesium is an essential component:

Pollucite
$H_4Cs_4Al_4Si_9O_{27}$ Cs_2O, 22–36%

Rhodizite
$CsAl_4Be_4B_{11}(OH)_4O_{25}$ Cs_2O, 5.0–7.5%

Avogadriate
$(K, Cs)BF_4$ Cs, 0–11%

All three of these minerals are rare, but only pollucite is of commercial importance. Although relatively few pollucite localities are known, several of them contain substantial tonnages of pollucite in essentially pure masses, such as the Bernic Lake Mine, Manitoba (over 300,000 tons); Bikita in Rhodesia (150,000 tons); and the Karibib, S. W. Africa deposit (50,000 tons). Probably other deposits will come to light as the presence of pollucite is recognized; however, this mineral is so similar in appearance to white quartz that its distinction from quartz, in the field, is practically impossible except to an experienced mineralogist who is specifically looking for pollucite. Pollucite has recently been found to be isostructural with analcite and a complete natural compositional series exists between the two end members.

The percentage of cesium in its minerals is highly variable because atoms of other alkali metals, particularly rubidium (Rb) and potassium (K), substitute for cesium atoms in the crystal lattice.

Cesium also enters into substitution in the lattice of a number of other minerals, particularly biotite (up to 3.1% Cs_2O), beryl (up to 7.5% Cs_2O), lepidolite (up to 0.3 Cs_2O), and microcline (up to 0.2% Cs_2O). Smaller quantities are occasionally found in about thirty other silicate minerals.

Abundance

The average content of cesium in all igneous rocks is about 4ppm. Most of this is concentrated in the granitic rocks. Among the sediments, the argillaceous rocks have a cesium content of up to about 10 ppm, whereas limestones and sandstones are much poorer in cesium. Most of the cesium in nature is dispersed in minerals such as the micas and feldspars in quantities of a few thousandths or at most a few hundredths of a percent.

Distribution in Rocks

In the magma, cesium is concentrated in the late phase siliceous fraction and the pneumatolytic emanations. Characteristically, the concentration of cesium does not reach a sufficiently high point to form separate minerals except in the pegmatitic phases, and then only, apparently, in conjunction with unusual concentrations of fluorine (F), lithium (Li), rubidium (Rb), tantalum (Ta), beryllium (Be), and similar late-phase elements.

In the granitic rocks, the cesium content decreases markedly and regularly as the series becomes more mafic. Thus, in the ultrabasic rocks, the cesium content is generally about 100 parts per billion or less, whereas the intermediate rocks may have ten times this amount, and the granites from 1 to 50 parts per million.

Sedimentary rocks are very poor in cesium unless they contain large amounts of argillaceous matter. Thus limestones and sandstones may contain 10–100 parts per billion, whereas clays and shales will contain from 1 to 5 parts per million.

In pneumatolytic veins and greisens, cesium is often concentrated, ranging from 10 to 100 parts per million, or more, in quartz-beryl, quartz wolframite, and beryl-molybdenite veins, and in mica greisens and cassiterite greisens.

Geochemical Behavior

In the weathering process, cesium is liberated when the micas, feldspars, or other silicates containing it are broken down. It is taken into solution, but almost immediately resorbed by the clays which also form during weathering. The cesium ions are actively adsorbed on the clay surfaces and penetrate into the intersheet void. Thus it is almost totally removed from the groundwater solution. This explains the low cesium content in seawater (about 5 parts per billion) and in river and lake waters; the only exception to this rule seems to be cesium which is occasionally observed in waters which contain suspended clay particles; however, the cesium is not truly in solution, but is attached to the clay.

Economic Geology

The uses of cesium are at the present time limited to the manufacture of photocells and other small electronic devices; in the degassing of vacuum tubes (now largely superseded by transistors); and in ion engine research. The quantities used are small, and are derived from the processing of pollucite and as a by-product of processing lepidolite for extraction of rubidium.

RICHARD V. GAINES

References

Hampel, C. A., 1961, "Rare Metals Handbook," Second ed., New York, Reinhold Publ. Corp., 715pp.
Rankama, K., and Sahama, Th. G., 1950, "Geochemistry," Chicago, Univ. of Chicago Press, 912pp.
Vlasov, K. A. (editor), 1966, "Geochemistry and Mineralogy of Rare Elements and Genetic Type of Their Deposits," Vol. I: "Geochemistry," 688pp.; Vol. II: "Mineralogy," 945pp. (transl. by Lerman, Z.), Jerusalem, Israel Program for Scientific Translations.

Cross-references: *Alkalis, Akali Metals, and Alkaline Earth Metals; Mineral Classes: Silicates.* Vol. V: *Pneumatolysis.* Vol. VI: *Argillaceous Rocks.*

CHARGED–PARTICLE TRACKS

Charged-particle tracks are linear regions of radiation-damaged material, produced by the passage of charged particles in insulating solids (solid-state track detectors). Charged-particle tracks are visible in the electron microscope; e.g., in micas they appear with a diffraction contrast diameter of ≈ 100 Å. Suit-

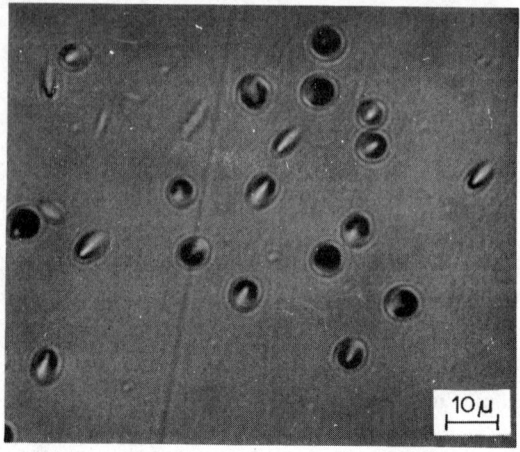

FIG. 1. Fission tracks in tektite glass (australite), etched for 15 sec in hydrofluoric acid.

able chemical etching reagents dissolve the radiation-damaged material faster than the undamaged material. Hence, the submicroscopic tracks can be enlarged to etch channels and *etch pits* (q.v.), now observable with an optical microscope. (See Fig. 1 for fission tracks in tektite glass.)

Solid-state detectors were developed in the last ten years by Young, Silk, Barnes, Price, Walker, and Fleischer (see the references in Fleischer et al., 1965).

The tracks can be formed by different processes: the projectile may either ionize the target atoms, or cause breakup of chemical bonds. Also it may trigger thermal effects, or simply collide with atoms. It is not yet clear to what extent these processes participate in track formation. According to Fleischer et al. (1967a) the registration of charged particles depends on the primary ionization caused along the particle path through a solid. Only then can tracks be formed, when the rate of primary ionization exceeds a critical value. This critical rate differs from solid to solid. Hence, for a given solid the track registration depends upon the mass and energy of the incident particle.

Charged-particle tracks find applications in nuclear science and in the earth sciences. Only the latter application will be discussed in some detail.

The distribution and concentration of *trace elements* (q.v.) may be determined with high sensitivity by means of charged-particle tracks (Fleischer et al., 1968). Elements with strong alpha radiation like radium and plutonium can be identified by registration of alpha-particle tracks in plastic detectors. Small concentrations of fissionable nuclei, such as uranium and thorium, can be determined by registration of the fission fragment tracks in the mineral itself or in an adjacent plastic detector. Elements with large (n, α) cross sections, such as lithium and boron, can be detected in plastics after neutron irradiation by alpha-particle tracks.

The recent discovery of fossil charged-particle tracks in many terrestrial minerals and glasses led to the development of a new dating method: the *fission-track method* (q.v.). Almost entirely, tracks are caused by fragments due to spontaneous fission of uranium-238. The number of fossil tracks in a mineral depends on its age and its uranium content. The uranium content of a sample may be determined by the number of fission tracks of U^{235} induced by a thermal neutron irradiation. Depending on the uranium concentration, this dating method for minerals and glasses may be applied within a large time range, e.g., 100 year old glasses, Precambrian minerals. The fission-track method has been used to date mica, calcite, zirkon, sphene, apatite, allanite, epidote, and glasses (Wagner, 1969). Fission-track ages are apparent ages, mostly being shorter than K/Ar, Rb/Sr, and U/Pb ages. This is due to the sensitivity of the tracks against raised temperatures. The stability of tracks differs from mineral to mineral. Hence, fission-track ages may yield information about the thermal history of rocks.

In mica, albite, and glasses another type of fossil charged-particle tracks were observed in addition to the regular fission tracks. These tracks are produced by the recoil nuclei accompanying alpha decay of the members of the uranium and thorium radioactive series. The etched alpha-recoil tracks can be observed in the phase contrast microscope and/or in the electron microscope. These tracks can be used to date minerals and glasses (alpha-recoil method). However, this method of dating is still in its first stages (Huang et al., 1967).

Other important objects for the application of charged-particle tracks are *meteoritic minerals* (q.v.). In *meteorites* (q.v., Vol. II), fossil charged-particle tracks were recorded in olivine, plagioclase, enstatite, hypersthene, bronzite, diopside, and whitlockite (Fleischer et al., 1967b). Their possible sources are: (a) spontaneous fission of uranium-238 and extinct plutonium-244; (b) induced fission of heavy elements by cosmic rays; (c) slowed heavy nuclei from primary cosmic radiation; and (d) spallation recoils. Tracks of different origin show characteristic properties. They can be identified by their length distributions, angular distributions, depth variations and by correlations of track densities and uranium contents in different mineral species. The fossil charged-particle tracks in meteorites give information on the early history of the *solar system* (q.v., Vol. II), on the elemental abundance, on the energy spectrum of the heavy *cosmic rays* (q.v., Vol. II), and on the space erosion and ablation rates of meteorites.

In studies of the recent elemental abundances in cosmic rays plastic detectors flown in balloons and spacecrafts are used.

In lunar minerals and glasses, recently collected by the Apollo missions, a variety of tracks similar to meteorites were identified: (a) heavy, primary cosmic ray tracks; (b) fission tracks; (c) spallation recoil tracks; and (d) tracks from heavy, solar wind nuclei (Fleischer et al., 1970). From these different kinds of tracks knowledge of cosmic rays, especially their solar component, and of the surface activities, the age and the thermal history of the moon can be derived.

G. A. WAGNER

References

Fleischer, R. L., Price, P. B., and Walker, R. M., 1965, *Ann. Rev. Nucl. Sci.* **15**, 1–28.

Fleischer, R. L., Price, P. B., Walker, R. M., and Hubbard, E. L., 1967a, *Phys. Rev.* **156**, 353–355.

Fleischer, R. L., Price, P. B., Walker, R. M., and Maurette, M., 1967, *J. Geophys. Res.* **72**(1), 331–353.

Fleischer, R. L., Price, P. B., Walker, R. M., Maurette, M., and Morgan, G., 1967b, *J. Geophys. Res.* **72**, 355–366.

Fleischer, R. L., and Lovett, D. B., 1968, *Geochim. Cosmochim. Acta* **32**, 1126–1128.

Fleischer, R. L., Haines, E. L., Hart, H. R., Jr., Woods, R. T., and Comstock, G. M., 1970, *Apollo 11 Lunar Science Conf.* **3**, 2103–2120.

Huang, W. H., Maurette, M., and Walker, R. M., 1967, IAEA Vienna, 415–429.

Wagner, G. A., 1969, *Neues Jahrb. Mineral. Abhandl.* **110**, 252–286.

Cross-references: *Elements; Fission-Track Dating; Geochronometry; Neutron Activation Analysis; Trace Elements; X-Ray Diffraction Analysis.* Vol. II: *Australites; Cosmic Rays; Meteorites; Nuclear Particle Tracks in Meteorites; Solar System; Tektites.* Vol. IV B: *Electron Microscopy; Etch Pits; Meteoritic Minerals.*

CHELATION

Chelates are a class of metal-organic and metal-inorganic coordination compounds in which the metal atom or ion is held between two atoms of a single molecule (the ligand). The word chelate (from the Greek, *chele*, or claw) refers to the pincer-like action of the ligand on the metal atom. The chelating ligand

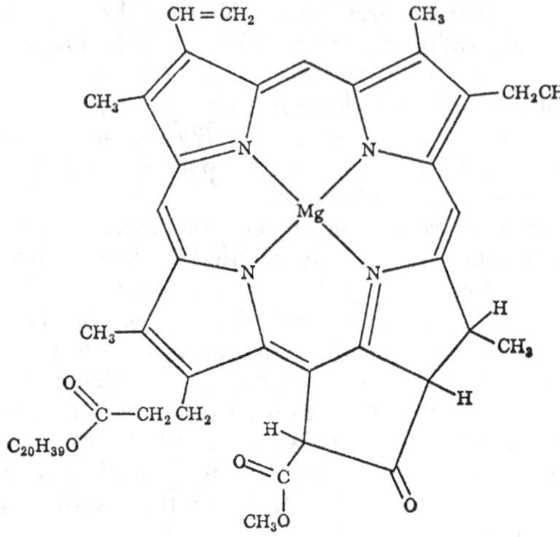

CHELATION

Fig. 1 — Chlorophyll a

Fig. 2 — Hemin

Fig. 3 — Nickel dimethylglyoxime

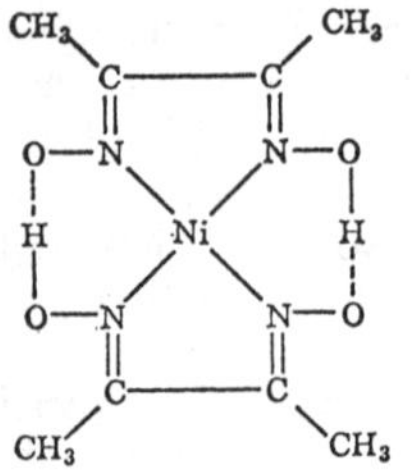

most frequently attaches itself to a metal ion through nitrogen (N), oxygen (O) or sulfur (S) atoms.

Importance and Uses

Chelation plays an important role in biological processes and thus in *organic geo-chemistry* (q.v.). Two chelates fundamental to plant and animal life are chlorophyll and hemin. *Chlorophyll* (q.v.), the green coloring agent of plants, is an example of a porphyrin. The basic unit of the porphyrin system is the pyrrole ring:

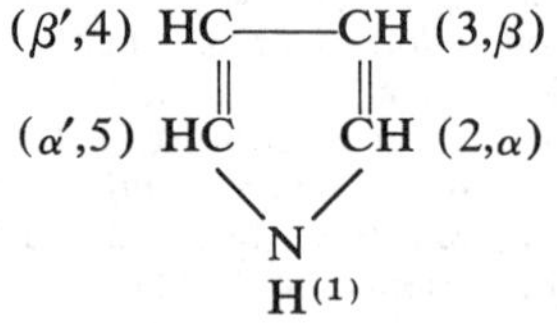

$$(\beta', 4)\ HC\text{===}CH\ (3, \beta)$$
$$(\alpha', 5)\ HC\quad\ CH\ (2, \alpha)$$
$$\diagdown N \diagup$$
$$H^{(1)}$$

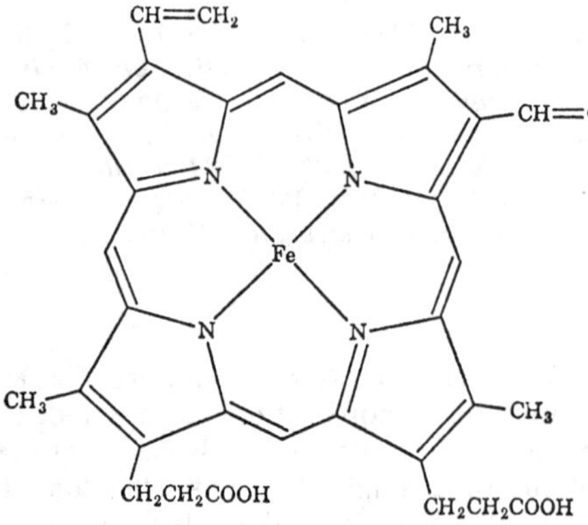

In chlorophyll, four pyrrole rings are bonded to a central Mg atom.

Another porphyrin, hemin, is part of hemoglobin, the red substance in blood. Here four pyrrole rings are joined to iron. Note the similarity in structure to chlorophyll.

Chelates are used in chemical analyses to identify certain elements. The colored chelates of diphenyl thiocarbazone with zinc (Zn) and dimethyl glyoxime with nickel (Ni) are well-known tests for these elements.

Significance for Organic Geochemistry

Very little is known about chelation in the sea, especially as related to the enrichment of trace elements in marine organisms and in some marine sediments. Krauskopf (1956), in a study of the concentration of rare metal elements in seawater, concluded that normal, aerated seawater is undersaturated with respect to these elements. Some may have been removed by precipitation as sulfides but most were probably removed by adsorption or through biologic processes. Black shales (e.g., the German "Kupferschiefer"), asphalts, and other petroleums often contain enrichments of copper (Cu), zinc (Zn), lead (Pb), cadmium (Cd), vanadium (V), uranium (U), molybdenum (Mo) and other transition elements. The possibility that chelation is a factor in the concentration of certain trace elements in sediments is suggested by the observations on chelates and transition elements.

Chelates have the ability to sequester metals (suppress undesirable properties by forming metal complexes). Sequestering agents have been used in treating cases of heavy metal poisoning by suppressing the ionic form which is the actual poison and by easing the excretion of the chelated metal. Metals for which this type of treatment is effective include beryllium (Be), nickel (Ni), mercury (Hg), lead (Pb), copper (Cu), vanadium (V), cadmium (Cd), and manganese (Mn). Sequestration may be used to control the concentration of metals which are either unwanted or tolerable only in small quantities. Thus, sequestration might be a mechanism of "depoisoning" the hydrosphere. Sufficient amounts of certain elements like Cu, Be, arsenic (As), and Pb have been supplied by weathering and erosion during geologic time to have caused serious poisoning of the oceans if some process had not intervened to eliminate those substances.

Trace elements are particularly significant in biological systems, e.g., zinc in carbonic anhydrase, cobalt in vitamin B_{12}, copper replacing iron in the blood pigment of some mollusks, Cu and magnesium (Mg) in chlorophyll, and iron (Fe) in hemoglobin. Plant studies indicate the absorption of small quantities of trace elements in the form of chelates. Chelation of Fe, Mn, Cu, Zn, boron (B), cobalt (Co), and Ni is highly probable, since these elements complex readily with organic molecules, and marine organisms need them for growth. Yet it has not been definitely established whether organometallic complexes are always in the form of chelates.

Chelation of trace metals could affect algae in several ways (Saunders, 1957):

1. Organic chelates may influence the availability of a trace element by lowering its concentration in water to such an extent that the growth of organisms is hindered.
2. If there is an excess of a trace element toxic to an organism, chelation may reduce that ion to a nontoxic level (scavenging, depoisoning).
3. Chelation may remove a metal ion that is antagonistic to a particular metal poison, thus increasing the toxicity.
4. If a trace element tends to precipitate out, a chelate could keep it in soluble form and maintain the concentration above the level that would otherwise prevail.

The effect of chelation on the availability of a metal ion depends upon the nature and quantity of the chelating agent, the nature of the metal, the concentration of the metal ion in water, and the pH.

Chelation affects the *iron cycle* in water. In neutral or alkaline water in which some O_2 is present, iron will be in the Fe^{III} state and will precipitate as $Fe(OH)_3$ or $FePO_4$. This precipitate will settle to the bottom and will deprive the algal organisms of Fe. If there are organic compounds that can chelate iron, it will remain in a soluble form available to the organisms. Amino acids, peptides, and pigments have been found in natural seawater. Although amino acids do not chelate Fe, other organic compounds do. Porphyrins are also effective chelating agents. It appears likely that the availability of trace metals in seawater is to some extent influenced by chelation.

Chelation and Minor Heavy Elements in Sediments

Chelation may have some relation to the accumulation of minor elements in certain sediments and petroleum. Black bituminous shales contain unusually high amounts of trace metals. The Mansfeld Kupferschiefer of Germany is an economically important example. It is a bituminous shale worked as a copper ore, but also enriched in As, Ag, Zn, Co, Pb, V, Mo, antimony (Sb), bismuth (Bi), and gold (Au). Coal also shows enrichment in many uncommon elements such as Be, B, germanium (Ge), arsenic (As), Bi, Mo, and V. In petroleum, U and V commonly occur in trace amounts. An enrichment in trace elements goes together with an increase in the amount of porphyrins. It is known that porphyrins can occur in metal complexes: Fe, Ni, V, gallium (Ga), and titanium (Ti) are tightly bound to such complexes. The metals may have been initially bonded to sulfur. Metal porphyrin complexes are very stable; a V-porphyrin complex is decomposed only in the strongest acids. On the average, asphalt contains 0.1–0.3% V in the pure bitumen, whereas the porphyrin content ranges from 1 to 3%. In an extreme case from the Argentine-Peru border area, pure bitumen had 0.76% V and 8.4% porphyrin. By contrast, the dried matter in green leaves contains only 0.6–1.2% (rarely up to 3%) chlorophyll. Trace elements could migrate with hydrocarbons as organometallic complexes (Goldschmidt, 1954). They could also have been extracted from seawater by porphyrins in marine organisms. Porphyrins are extremely stable, as noted above, and have been identified in shales, asphalts, and petroleum dating back to the Paleozoic. Porphyrins have also been recently detected in carbonaceous meteorites. They apparently withstand the processes of diagenesis, lithification, and deep burial.

Studies by Szalay (1958, 1964, 1967) have demonstrated the high fixation capacity of humic acid for UO_2^{2+}, VO^{2+}, and to a lesser extent the lanthanides, zirconium (Zr), chromium (Cr), Fe, Co, Ni, Cu, and Zn. Humic acid represents a mixture of compounds resulting from the decay of lignin (plant matter) and contains a stable polyaromatic skeleton to which are attached acidic hydroxyl and carboxyl groups. Metallic ions are held to the humic acids by cation exchange (the reversible replacement of ionizable hydrogen by cations).

CHELATION

The polycarboxylic nature of humic acid suggests that it may act effectively as a chelating agent. This property of humic acid is noteworthy because of the frequent association of U and V with organic matter in sediments such as, for example, the sedimentary uranium deposits of the Colorado Plateau.

Although it appears plausible that the extraction of trace elements from seawater and the concentration of the elements in bituminous shales, asphalt, and petroleum may depend on chelation, there is no absolute proof. Although Fe, V, Cu, and Mg do form stable complexes with porphyrin, they could also be attached to other types of organic compounds. In addition, trace metals may be concentrated by other processes such as adsorption on clays, chemical precipitation, or circulation of metal-bearing ground waters after deposition of the sediment.

Role of Chelation in Weathering

Among factors involved in the disintegration and chemical weathering of rocks, minerals, and soils, chelation of metallic elements by organic molecules of plant, animal, and microbial origin plays a significant role. Lichens, for example, contain certain acids (polyhydroxy polycarboxylates) which complex metals (Schatz et al., 1957). The effectiveness of humus and other organic matter in soils in breaking down rock and mineral particles is well established. Experiments have shown that chelating agents may hasten dissolution of a mineral by sequestering metal ions before they pass into solution. Although acids are strong solvents of minerals, their anions themselves may be powerful chelators. Thus dissolution of minerals and rocks may involve a combination of the effects of acid action and chelation (Schalscha, et al., 1967).

VIVIEN GORNITZ

References

Emmons, R. C., and Jones, J. B., 1956, "Chelation, a possible geologic process," *Geol. Soc. Am. Bull.,* **67,** 1692.

Goldschmidt, V. M., 1954, "Geochemistry," Oxford, Clarendon Press, 730pp.

Hodgson, G. W., and Baker, B. L., 1969, "Porphyrins in meteorites: metal complexes in Orgueil, Cold Bokkeveld and Murray carbonaceous chondrites," *Geol. Soc. Am. Spec. Paper* (Abstr. Mexico City Meet., 1968).

Krauskopf, K. B., 1956, "Factors controlling the concentrations of thirteen rare metals in seawater," *Geochim. Cosmochim. Acta* **9,** 1–32.

Krejci-Graf, K., 1959, "Diagnostik der Herkunft des Erdöls," in "Erdölundkohle," pp. 706–712 and 805–815.

Krejci-Graf, K., 1960, "Modern Anschauungen über die Entstehung des Erdöls, Erdöl und Kohle," *Erdgas, Petrochemie,* 836–845.

Saunders, G. W., 1957, "Interrelations of dissolved organic matter and phytoplankton," *Botan. Rev.* **23,** 389–409.

Schalscha, E. B., Appelt, H., and Schatz, A., 1967, "Chelation as a weathering mechanism. I. Effect of complexing agents on the solubilization of iron from minerals and granodiorite," *Geochim. Cosmochim. Acta,* **31,** 587–596.

Schatz, A., Schatz, V., and Martin, J. J., 1957, "Chelation as a biochemical weathering factor," *Bull. Geol. Soc. Am,.* **68,** 1792–1793.

Smith, R. L., 1959, "The sequestration of metals," London, Chapman and Hall, 256pp.

Szalay, A., 1958, "The significance of humus in the geochemical enrichment of uranium," *Proc. Intern. Conf. Peaceful Uses At. Energy, Geneva,* **2,** 182–186.

Szalay, A., 1964, "Cation exchange properties of humic acids and their importance in the geochemical enrichment of UO_2^{++} and other cations," *Geochim. Cosmochim. Acta,* **28,** 1605–1614.

Szalay, A., and Szilagyi, M., 1967, "The association of vanadium with humic acids," *Geochim. Cosmochim. Acta,* **31,** 1–6.

Cross-references: *Bonding; Chlorophyll; Hydrocarbons; Ion Exchange; Organic Geochemistry; pH-Eh; Photosynthesis; Seawater Chemistry; Trace Elements—Geochemistry; Trace Elements in Plants; Trace Metals in Sediments, Oils, and Allied Substances; Weathering, Chemical.* Vol. II: *Carbonaceous Meteorites.* Vol. IV B: *Coal; Petroleum.* Vol. VI: *Diagenesis.*

CHEMICAL BASE—See BASE—CHEMICAL

CHEMICAL MINERALOGY

Chemical mineralogy pertains to the analysis and synthesis of minerals, and to the application of chemical theory to problems in mineralogy and in a more restricted sense to petrology. The alliance of chemistry and mineralogy is an ancient one and derives from a common source in alchemy. The earliest chemists used minerals as raw material for the preparation of their chemicals and as the basic substances from which to extract the then undiscovered chemical elements. Jöns Jakob Berzelius (1779–1848), the Swedish chemist, and Friedrich Wöhler (1800–1882), the German chemist, laid the basis for the analysis of minerals. Although Axel Frederich von Cronstedt, the Swedish Master of Mines, is generally regarded as the founder of chemical mineralogy for his *system of identification of minerals,* it was actually Berzelius who put blowpipe analysis on a sound basis. Berzelius and Wöhler together placed the qualitative and quantitative analysis of minerals on a firm foundation. The science, as the early workers developed it, was mineral chemistry.

Chemical analysis of minerals is a cornerstone of mineralogy for its results give the formulas of minerals. New methods of analysis for minerals are developed as needed in research and they are constantly revised as a result of the development of new techniques and new reagents. Hillebrand's treatise as revised by Lundell, Hoffman, and Bright is a basic reference work. Microchemical tests, spot tests, spectrochemical analysis, staining and contact printing techniques, and thermal analysis are commonly employed in chemical mineralogy.

With the shift of emphasis in chemistry to synthesis, mineral synthesis based on empirical procedures (trial and error) was developed at the hands of many chemists, particularly by the French scientists Ferdinand Fouque (1828–1904) and Michel-Levy (1844–1911). Empirical synthesis reached a sudden maximum and declined rapidly before the turn of the nineteenth century, capitulating to the methods of physical chemistry. Empirical mineral synthesis is still done when making reconnaissance studies and in the search for unstable (and sometimes stable) polymorphic modifications of mineral phases.

The theoretical studies of J. Willard Gibbs (1839–1903), Jacobus H. Van't Hoff (1852–1911), Walther Nernst (1887–1941), and H. W. Bakhius Roozeboom (1945–1907) sounded the death knell of empirical mineral synthesis. Its place was taken by the rigorous physical and chemical approach developed at the Geophysical Laboratory of the Carnegie Institution of Washington under the directorship of the physicist Arthur Louis Day. The synthesis and stability of minerals as a problem in phase rule study and in other physical chemical studies, such as the determination of E_h and pH diagrams, is the major problem in chemical mineralogy.

The application of chemistry to mineralogical problems embraces a wide field of research and includes such topics as the weathering of feldspar to clay minerals; the polymorphism of $CaCO_3$ as aragonite and calcite; the influence of impurities on the morphology of crystals; the chemistry of hornblende (a complex aluminosilicate of calcium, sodium, iron, magnesium, and hydroxyl) crystals from igneous and metamorphic rocks.

ACKNOWLEDGMENT: Copied with permission from the "Supplement to the Encyclopedia of Chemistry" published by the Reinhold Publishing Co., New York, pp. 57–58, 1958, and edited by George L. Clark and G. G. Hawley.

GEORGE T. FAUST

References

Brauns, R., 1896, "Chemische Mineralogie," Leipzig, Tauchnitz, 460pp.
Eitel, W., 1966, "Silicate Science," New York, Academic Press, 5 vols.
Knopf, A., 1941, "Petrology," in "Geology 1888–1938, Fiftieth Anniversary Volume," pp. 333–363, Geol. Soc. Am.
Kraus, E. H., 1941, "Mineralogy," in "Geology 1888–1938, Fiftieth Anniversary Volume," pp. 309–332, Geol. Soc. Am.
Larsen, E. S., 1941, "Geochemistry," in "Geology 1888–1938, Fiftieth Anniversary Volume," pp. 391–413, Geol. Soc. Am.
Von Groth, P., 1926, "Entwicklungsgeschichte der mineralogischen Wissenschaften," Berlin, Springer Verlag, 262pp.
Wöhler, F., 1861, "Die Mineral-Analyse in Beispielen," Göttingen, Verlag Dieterischen Buchhandlung, 234pp.

Cross-references: *Colorimetry; Mineralogy; Thermal Analysis;* Vol. IV B: *Blowpipe Analysis.* Vol. V: *Free Energy and the Gibbs Phase Rule.*

CHEMICAL SEDIMENTS—See PARAGENESIS; also Vol. VI

CHEMICAL WEATHERING—See WEATHERING, CHEMICAL

CHEMILUMINESCENCE—See Vol. VI, LUMINESCENCE

CHEMISTRY, METHODS, etc.—See "Encyclopedia of Chemistry" (Second ed.), Van Nostrand Reinhold Company

CHITIN AND CHITINOUS CUTICLES

Chitin is perhaps best known as a characteristic chemical component in the exoskeletons of arthropods. It is also found in setae, jaws, and gut lining of annelids, in the radula and dorsal shield of mollusks, in the perisarc of medusoid coelenterates, in the stalk wall of bryozoans, in the egg shells of nematodes and acanthocephalans, and less certainly in a few other animal groups. It is also the common constituent of the walls of most but not all groups of fungi.

Chitin is a high-molecular-weight polymer composed of N-acetylglucosamine residues joined together by β-glycosidic linkages between carbon atoms 1 and 4 (Fig. 1). Modern chemical terminology would call it a polymer of 2-acetamido-2-deoxy-α,D-glucopyranose. The molecular chains are long and unbranched. Freed from the protein with which it is normally associated, the chitin chains are parallel to one another and adjacent chains run in

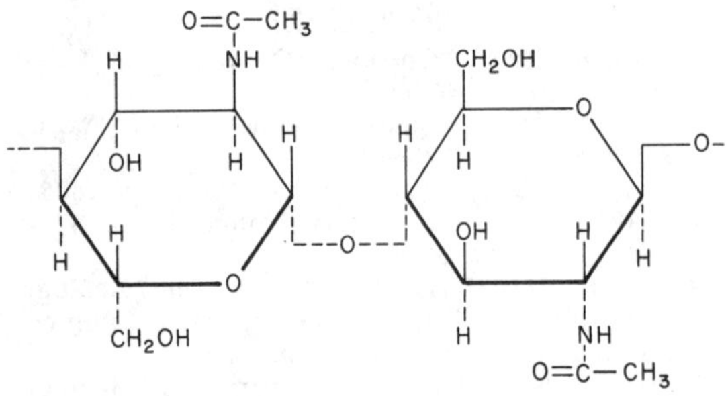

FIG. 1. The repeating unit of a chitin chain.

opposite directions, an arrangement that maximizes the number of interchain hydrogen bonds. In many respects chitin is similar to cellulose, from which it differs by the presence of an acetylamine group on the second carbon atom. Treatment with hot concentrated alkalies removes half or all of the acetyl groups to give a product called *chitosan* which can be made to give a color reaction useful in the identification of chitin in natural objects.

Chitin is not found in natural objects in a pure condition. It is always linked to protein, and hence occurs as a glycoprotein. It is relatively easy to remove the protein moiety with alkali leaving the insoluble chitin as a white residue. This is true both for α-chitin, which is normally linked to arthropodins, and for β-chitin, which is linked to collagen. On the average, chitin accounts for only one-fourth to one-third of the dry weight of noncalcified arthropod cuticles. In calcified cuticles the organic constituents account for 5–60% of the dry weight, and most of the inorganic material is calcite.

The metabolic source of chitin may be glycogen since there are numerous reports of glycogen decrease concurrent with chitin synthesis. However, no direct proof exists. Furthermore, no detailed knowledge is available on the metabolic steps involved in chitin synthesis. Since the naturally occurring substance is a glycoprotein, one would expect that monomers are added in situ to a mixed lattice.

Most of the data on chitinous cuticles pertains to insects although there is a sizable literature on calcified crustacean cuticles. Less is known about other arthropod groups, and very little other than demonstration of the presence of chitin is known for other phyla. At least in the arthropods, chitinous cuticles always have on their outer surface a thin non-chitinous layer called the epicuticle. The epicuticle has been demonstrated to account for the relative impermeability of the cuticle of insects; the much thicker underlying chitinous cuticle provides the arthropod's skeleton. In thickness, cuticles range from about 1μ to several mm.

When first secreted the chitinous cuticle of arthropods is a soft material with a high water content (termed procuticle). This material becomes more compact as much of the water is removed. In membranous areas no further change may occur. In areas that are destined to become sclerites the procuticle becomes modified by the addition of sterols and polyphenols, which react to produce an elastic cuticle (termed mesocuticle). Commonly the modification continues with the addition of more polyphenol derivatives, full dehydration, extrusion of salt molecules, and chemical interactions to produce a tight lattice that gives hard cuticle (termed exocuticle). Usually, but not always, this exocuticle also becomes dark in color or even black (melanized exocuticle). This process of hardening and darkening of sclerite is termed *sclerotization*. Details of the chemical linkages are not known but various types of bonds, including covalent bonds, are involved, and the chitin molecules are not only linked to protein but also crosslinked between molecular chains. It is not yet certain whether the chitin and protein chains are parallel or perpendicular to one another.

Estimates indicate the natural production of billions of tons of chitin annually. Obviously this must be decomposed at a rate approximating the rate of production. Some dozens of kinds of chitin-destroying bacteria and a comparable number of chitin-destroying fungi have been isolated. Chitinase activity has also been recorded for soil amebae, eelworms, and earthworms. The digestive activity of the fluid from the gut of snails has long been known though it is not yet clear whether the chitinase originates in snail tissues or in symbiotic bacteria. The digestive juices of some other animals, both vertebrate and invertebrate, show chitinase activity but this would have little effect on the decomposition of chitin in nature. Natural decomposition must involve primarily the microorganisms of soil and water.

Purified chitinases attack only purified chitin. They have no significant effect on whole cuticle. But this is of only biochemical interest because the relevant digestive juices always contain a mixture of chitinases and proteases as well as other enzymes. The breakdown of cuticle in nature, then, releases not only amino sugars but also amino acids and other organic acids. Commonly the decomposition proceeds further with the release of CO_2 and NH_3, the latter being subsequently partly oxidized to nitrate. Veldkamp reports that up to 60% of the nitrogen of chitin decomposing in soil is recoverable as nitrate.

Decomposition in nature requires considerable time. Under optimum conditions in a test tube the breakdown of chitin by pure chitinase requires days or weeks. In nature, times of one to two months have been recorded for favorable conditions; considerably longer times must be common, especially in dry soils. Anyone can readily demonstrate for himself that the decomposition of hard sclerotized cuticle proceeds much more slowly than that of soft membranous cuticle.

The chemical stability of chitin-containing cuticles results in their ready preservation as fossils. Insects in amber are not uncommon, and some show amazingly fine preservation of microscopic details. Good preservation of microscopic structure including setae, cuticular laminae, and pore canals has been reported from old fossils, the oldest being from a scorpion credited to the Lower Carboniferous period and the trilobites of Cambrian and Carboniferous periods. In only a few cases have valid chemical tests for chitin been performed on fossil specimens. These include a beetle wing from the Eocene and a coelenterate perisarc from the Silurian. Clearly, under favorable conditions chitin can persist for hundreds of millions of years.

Chitin, because of its insolubility, may persist in acid waters long enough for chitinous skeletons to leave an impression or mold whereas structures composed of calcite, aragonite, and even hydroxyapatite may decompose too rapidly. Chitinous materials thus may play an important role in the construction of living organisms in acid lakes and streams, as well as favoring fossilization.

A. GLENN RICHARDS

References

Hackman, R. H., 1965, "Chemistry of the Insect Cuticle," in (Rockstein, M., editor), "The Physiology of Insecta."
Richards, A. G., 1951, "The Integument of Arthropods," Univ. of Minnesota Press.
Travis, D., 1963, "Calcified crustacean cuticles," *Ann. N.Y. Acad. Sci.*, **109**, 177–245.
Veldkamp, H., 1955, "The aerobic decomposition of chitin by microorganisms," *Meded. Landb. Hoogesch., Wageningen*, **55**, 127–174.

CHLORINE: ELEMENT AND GEOCHEMISTRY

Properties

Chlorine (Cl), a gaseous element, of atomic number 17, was isolated by K. W. Scheele in 1774. Not until 1810 did Sir Humphry Davy prove that it was an element; he gave it its name which alludes to its color. It is a yellowish-green gas of atomic weight 35.457, melting point $-102°C$, and boiling point $-35°C$. Its electronegativity value is 3.0. It consists of two stable natural isotopes, Cl^{35} and Cl^{37}. There is no evidence that natural fractionation of these isotopes takes place. Because of its high reactivity chlorine occurs in all naturally occurring compounds as the univalent Cl^- anion, with ionic radius 1.81 Å. The Cl^- anion exhibits limited isomorphous substitution for the hydroxyl ion OH^- (radius 1.40 Å) and is found as a minor constituent of rock-forming silicates.

Minerals

In a small number of minerals chlorine is present as a major constituent. These include:

halite	NaCl
carnallite	$KMg(H_2O)_6Cl_3$
bischoffite	$Mg(H_2O)_6Cl_2$
cerargyrite	AgCl
sal ammoniac	NH_4Cl
atacamite	$Cu_2(OH)_3Cl$
chlorapatite	$Ca_5(PO_4)_3Cl$
sodalite	$Na_4Al_3Si_3O_{12}(Cl)$
Na-scapolite	$Na_4Al_3Si_9O_{24}(Cl)$
zunyite	$Al_{13}Si_5O_{20}(OH)_{18}Cl$

All but the first of these, halite, are quantitatively unimportant. Halite is relatively abundant, occurring in deposits of solid rock salt and in solution as the most abundant dissolved ingredient in seawater and subsurface brines.

If the solid Earth is considered as a whole, most of the chlorine occurs as a minor or trace constituent in hydrous minerals, particularly the micas, amphiboles, and clay minerals, where Cl^- proxies for OH^- in their structures (Correns, 1956).

Abundance

Chlorine has a reported cosmic abundance of about 6 ppm (Green, 1959). This is in contrast to a crustal abundance in the Earth of 180 ppm (Johns and Huang, 1966). In the Earth's crust chlorine ranks as about the twentieth most abundant element.

Table 1 summarizes the average Cl contents of the common rock types (Johns and Huang, 1966).

Chlorine is the most abundant dissolved constituent in seawater, making up 55.04 wt% of the dissolved solids; in contrast, average river water contains only 5.68 wt % Cl (Sverdrup et al., 1942).

Geochemical Behavior

Chlorine is lithophile and highly mobile in character. During the course of magmatic crys-

CHLORINE: ELEMENT AND GEOCHEMISTRY

TABLE 1. CHLORINE CONTENTS OF COMMON ROCK TYPES[a]

Rock Type	ppm Cl
Igneous rocks	
Ultramafic (dunites, peridotites, and pyroxenites)	100
Mafic (basalts, gabbros, diabases)	160
Intermediate (diorites, andesites)	180
Acid	
High-calcium (granodiorites)	220
Low-calcium (granites)	200
Alkalic	
Syenites	430
Nepheline syenites	2170
Sedimentary rocks	
Shales and clays	100
Sandstones	20
Limestones	130
Dolomites	660
Metamorphic rocks	
Schists	354
Gneisses	207
Amphibolites	300

[a] From Johns and Huang, 1966.

tallization, probably most of the chlorine is enriched in the last fraction to crystallize, and especially in the residual liquids and vapors. Alkalic rocks, pegmatites, volcanic emanations, and hydrothermal solutions are characterized by a high concentration of chlorine.

Cl is released during the weathering of silicate rocks, goes into solution as Cl^- ions, and ultimately accumulates in the oceans. It has been shown, however, that the bulk of the Cl which has accumulated in the atmosphere and hydrosphere has been volatilized gradually and continuously throughout geologic time from the Earth's interior (Rubey, 1951). Rubey has estimated the quantity of Cl now at or near the earth's surface, compared with the quantity supplied by the weathering of crystalline rocks as follows:

	Cl (in units of 10^{20} g)
In present atmosphere, hydrosphere and biosphere	276
Buried in ancient sedimentary rocks (previous hydrosphere)	30
Total	306
Supplied by weathering of crystalline rocks	5
Excess Cl, unaccounted for by rock weathering, and derived from Earth's interior	301

Economic Geology

NaCl, either from solid sedimentary deposits of halite, or as obtained from subsurface brines or seawater, is the prime source of chlorine. Salt deposits of Silurian, Permian, and Triassic age are scattered throughout the world.

NaCl is itself an important raw material, and apart from the common household uses, is prerequisite to the production of soda ash, and subsequently glass and many chemical products. Cl_2 is produced by electrolysis of molten NaCl. Cl_2 is important as a germicide, and as a primary reagent in the manufacture of many important organic and inorganic chemicals.

WILLIAM D. JOHNS

References

Correns, C. W., 1956, "The Geochemistry of the Halogens," in (Ahrens, L. H., et al., editors), "Physics and Chemistry of the Earth," Vol. 1, pp. 181–233, New York, Pergamon Press.

Green, J., 1959, "Geochemical table of the elements for 1959," *Bull. Geol. Soc. Am.*, **70**, 1127–1184.

Johns, W. D., and Huang, W. H., 1966, "Distribution of chlorine in terrestrial rocks," *Geochim. Cosmochim. Acta*, **31**, 35–49.

Rubey, W. W., 1951, "Geologic history of sea water," *Bull. Geol. Soc. Am.*, **62**, 1111–1148.

Sverdrup, H. U., Johnson, N. W., and Fleming, R. H., 1942, "The Oceans," New York, Prentice-Hall, Inc., 1087pp.

Cross-references: *Hydrothermal Solutions; Natural Brines; Seawater Chemistry; Weathering, Chemical.* Vol. IV B: *Halite, Isomorphism.* Vol. V: *Volatiles.*

CHLOROPHYLL

Structure

Chlorophyll is a general term referring to a group of photosynthetic pigments including chlorophyll *a*, the structure and absorption spectrum of which are shown in Fig. 1. Chlorophyll comprises a tetrapyrrole pigment with a chelated magnesium atom. The principal structure is that of a dihydroporphyrin (a chlorin) and is marked by an isocyclic fifth ring linking the gamma-methene bridge carbon to the nearest beta carbon of the third pyrrole ring. The nature, properties and distribution of chlorophylls are described by Holt (1965).

Chlorophyll *a* has a vinyl group on ring I. Chlorophyll *b* has the methyl group of ring II of chlorophyll *a* replaced by an aldehyde. Chlorophyll *c* is of unknown structure. Chlorophyll *d* is identical with an oxidation product of chlorophyll *a* in which the vinyl group is replaced by a formyl group. Other chlorophylls have been described (Allen et al., 1960), notably chlorobium chlorophyll *(660)* and chlorobium chlorophyll *(650)*. Many other variations in chlorophyll are indicated by spectral evi-

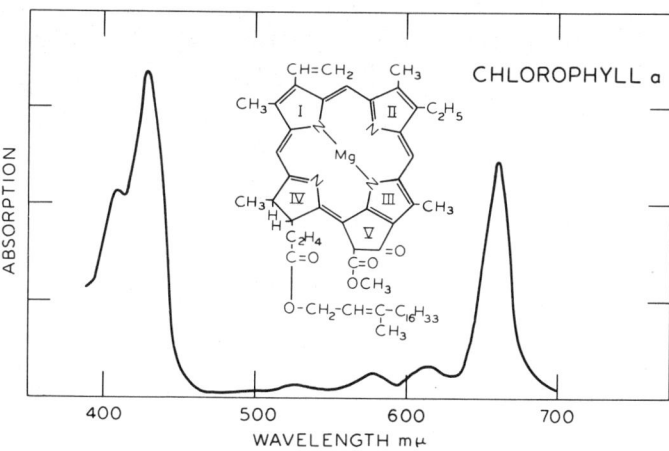

Fig. 1. Structure and absorption spectrum of chlorophyll *a*.

dence. Closer examination of the chlorophyll absorption spectrum in vivo in a large number of organisms has shown that not only does it differ from that in solution and vary from one organism to another, but it may even vary in the same organism, depending on its physiological state. For example, the spectrum of chlorophyll *a* of *Euglena gracilis* has been resolved into three components with absorption maxima at 673, 683, and 695 $m\mu$. The differing relative proportions of these as the cultures age are a function of the light intensity received by the cell as it grows. These forms of chlorophyll *a* are known only as spectroscopic entities. It is not known whether they represent different chlorophyll-protein complexes, different states of aggregation of chlorophyll, or some other phenomenon. There is evidence that the spectroscopically different chlorophyll *a* forms are functional entities. It appears that for efficient photosynthesis it is necessary to activate some accessory pigment such as chlorophyll *b*, a biliprotein, or the form of chlorophyll *a* that absorbs at long wavelengths (695 $m\mu$) (Allen et al., 1960).

The chlorophyll content of phytoplankton falls in the range 4–12% of the ash-free dry weight; of green leaves, about 8%. The classical preparative procedure for chlorophyll is given by Zscheile and Comar (1941). Creitz and Richards (1955) have developed a spectrophotometric method for separate estimations of chlorophylls *a*, *b*, and *c*.

In considering the biosynthesis of chlorophyll pigments, chlorophyll and hemin pigments are structurally very similar, the major differences being that they are chlorin (dihydroporphyrin) and porphyrin structures, respectively, and the complexing metals are magnesium and iron, respectively. In the evolution of living organisms it is probable that chlorophyll arose before heme. The biosynthesis of chlorophyll is comprehensively discussed by Bogarad (1965). Evidently the first building units are glycine and succinyl coenzyme A, yielding δ-aminolevulinic acid, which condenses to form a pyrrolic substance, porphobilinogen:

$$\begin{array}{c} HOOCH_2C-CC-CH_2COOH \\ \parallel \parallel \\ HC C-CH_2NH_2 \\ \diagdown \diagup \\ N \\ | \\ H \end{array}$$

Under mildly acid conditions, this highly reactive material readily forms tetrapyrrole porphyrins, out of which protoporphyrin ultimately appears. Magnesium is chelated, and one of the propionic acid side chains forms the isocyclic fifth ring by oxidative ring closure. Partial reduction of the fourth ring to form the dihydroporphyrin magnesium complex along with esterification of the fourth-ring propionic acid with phytyl alcohol completes the biosynthesis of chlorophyll *a*.

Although the blood pigment heme is clearly associated with a globulin protein of molecular weight 30,000–70,000, the chlorophyll pigment is much less clearly defined in its association with plant proteins and lipids.

Woodward et al. (1960) performed an elegant chemical synthesis of chlorophyll *a* starting with substituted pyrroles, confirming in every respect the structure shown in Fig. 1, established by Fischer and associates (Fischer and Orth, 1947). Fleming (1967) determined the absolute spatial configuration and structure of chlorophyll.

All of the phyla of organisms that carry out green plant photosynthesis, i.e., reduction of carbon dioxide to sugars, contain chlorophyll *a*. This pigment is sometimes accompanied by chlorophyll *b, c,* or *d*. Photosynthetic bacteria contain another chlorophyll, and in green photosynthetic bacteria chlorophyll *a* does not occur, being replaced by other species of chlorophyll. The green algae, like the higher plants, contain chlorophylls *a* and *b*, in the ratio of about three to one. Chlorophylls *a* and *b* are major chloroplast pigments in a different position from the other chlorophylls, which occur in much smaller amounts.

In a review of fossil pigments, Vallentyne (1960) summarized chlorophyll photosynthetic data as follows: (a) the total weight of organic compounds annually synthesized by photosynthetic plants is about 1.6×10^{17} g; (b) photosynthetic organisms have existed on earth for at least 1.6×10^9 years, and possibly for somewhat over 2×10^9 years; and (c) the total mass of organic compounds now present in organic sediments and sedimentary rocks, petroleum, living organisms, etc., is on the order of 6×10^{21} g. This is about $1/10^6$ of the mass of the earth.

The fate of the chlorophylls in geochemical environments is determined to a large extent by the inherent instability of the pigment as a magnesium chelate. The magnesium atom is readily lost from the chlorophylls, forming a series of pheophytins. Hydrolysis of the phytyl esters also takes place readily and results in the corresponding pheophorbides. This appears to be the most common pattern of geochemical degradation. Chlorophylls *a, b,* and *c* of phytoplankton persist into the seston of surface waters, but they are converted to degradation products very soon after deposition. Orr et al. (1958) estimated that in the Santa Barbara basin only 0.2% of the chlorins from phytoplankton reached the underlying sediments. Recent marine and fresh-water sediments contain pheophytins and pheophorbides, the latter becoming more important with increasing depth below the water-sediment interface. Undoubtedly the bulk of the chlorophyll degrada-

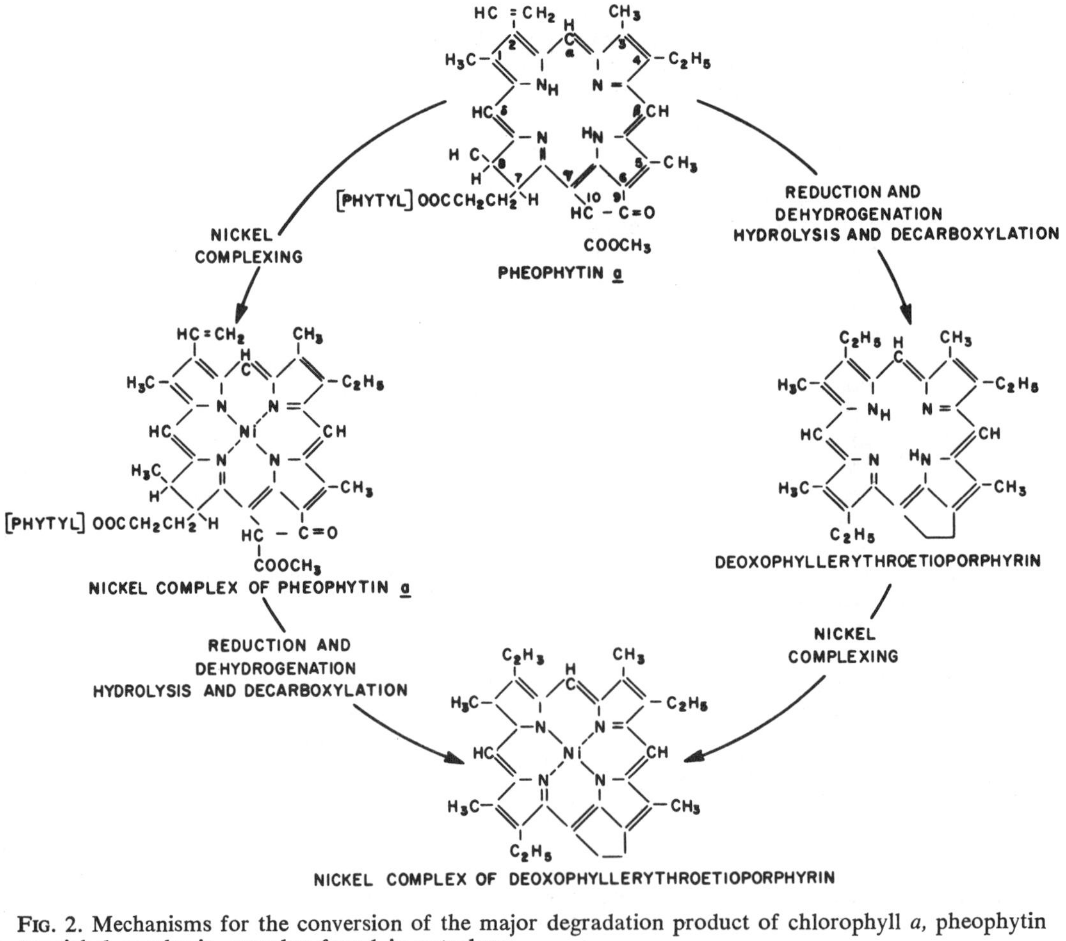

FIG. 2. Mechanisms for the conversion of the major degradation product of chlorophyll *a*, pheophytin *a*, to nickel porphyrin complex found in petroleum.

tion products surviving the water column are destroyed in the sediments. However, some of the products are evidently stabilized through reformation of metal ligand complexes, primarily with nickel. A portion of such metal chlorins are believed to be converted ultimately into metal porphyrins (Fig. 2) commonly found in crude oil and ancient sediments (Hodgson et al., 1968).

The phytol produced from the hydrolysis of the pheophytin esters is fairly resistant to geochemical destruction and may be the source of isoprenoid compounds in petroleum (Bendoraitis et al., 1962); however, more recent studies show the presence of isoprenoid hydrocarbons directly in living plants.

The ingestion of chlorophyll-containing food by the ruminant herbivores on the other hand results in the direct production of the porphyrin phylloerythrin by the bacteria and protozoa of the rumen. Various phorbides and rhodoporphyrin-γ-carboxylic acid have been identified as accompanying phylloerythrin in the feces of herbivores.

G. W. HODGSON

References

Allen, M. B., French, C. S., and Brown, J. S., 1960, "Native and extractable forms of chlorophyll in various algal groups," pp. 33–52 in (Allen, M. B., editor.), "Comparative Biochemistry of Photoreductive Systems," Vol. I, New York, Academic Press, 437pp.

Bogarad, L., 1965, "Chlorophyll Biosynthesis," pp. 29–74, in (Goodwin, T. W., editor), "Chemistry and Biochemistry of Plant Pigments," Academic Press, New York, 583pp.

Bendoraitis, J. G., Brown, B. L., and Hepner, L. S., 1962, "Isoprenoid hydrocarbons in petroleum," Anal. Chem. **34,** 49–53.

Creitz, G. I., and Richards, F. A., 1955, "The estimation and characterization of plankton populations by pigment analysis," J. Marine Res., **14,** 211–216.

Fischer, H., and Orth, H., 1937, "Die Chemie des Pyrroles," Vol. II, Part I, Leipzig, Leipzig Akad. Verlagsges, 720pp.

Fleming, I., 1967, "Absolute configuration and the structure of chlorophyll," Nature, **216,** 151–152.

Hodgson, G. W., Hitchon, B., Taguchi, K., Baker, B. L., and Peake, E., 1968, "Geochemistry of porphyrins, chlorins and polycyclic aromatics in soils, sediments and sedimentary rocks," Geochim. Cosmochim. Acta, **32,** 737–772.

Holt, A. S., 1965, "Nature, Properties and Distribution of Chlorophylla," pp. 3–28 in (Goodwin, T. W., editor), "Chemistry and Biochemistry of Planet Pigments," New York, Academic Press, 583pp.

Orr, W. L., Emery, K. O., and Grady, J. R., 1958, "Preservation of chlorophyll derivatives in sediments off southern California," Bull. Am. Assoc. Petrol. Geologists, **42,** 925–962.

Vallentyne, J. R., 1960, "Fossil Pigments," pp. 83–105, in (Allen, M. B., editor), "Comparative Biochemistry of Photoreductive Systems," Vol. I, New York, Academic Press, 437pp.

Woodward, R. B., et al., 1960, "The total synthesis of chlorophyll," J. Am. Chem. Soc., **82,** 3800–3802.

Zcheile, F. P., and Comar, C. L., 1941, "Influence of preparative procedures on the purity of chlorophyll components as shown by absorption spectra," Botan. Gaz., **102,** 463–481.

Cross-references: *Chelation; Photosynthesis. Vol. I: Phytoplankton.*

CHROMATES

Anisodesmic compounds containing chromate groups are fairly rare in nature. Chromium acts similar to sulfur in the sulfates, with tetrahedral coordination.

The chromates are usually secondary minerals, often compounds of lead. The best known chromate is the beautiful orange-red mineral, crocoite.

	Anhydrous Chromates
	$A_2(XO_4)$ type
Tarapacaite	$K_2(CrO_4)$
Chromatite	$Ca(CrO_4)$
	$A_2(X_2O_7)$ type
Lopezite	$K_2(Cr_2O_7)$
	$A(XO_4)$ type
Crocoite	$Pb(CrO_4)$
(Monazite structure)	
Phoenicochroite	$Pb_3(CrO_4)_2O$
Vauquelinite	$Pb_2Cu(OH)(CrO_4)(PO_4)$
Hemihedrite	$Pb_5ZnF_4(CrO_4)_3O$
	Hydrous Chromates
Iranite	$Pb(CrO_4) \cdot H_2O$

LLOYD W. STAPLES

References

Dana, J. W., and Dana, E. S., 1951, "The System of Mineralogy," Seventh ed., Vol. II, 644, (revised by Palache, C., Berman, H., and Frondel, C.) New York, John Wiley & Sons.

Cross-references: *Chromium; Mineral Classes: Nonsilicates; Sulfur.*

CHROMATOGRAPHY

Chromatography is a set of qualitative and semiquantitative analytical methods, used in organic and inorganic chemistry and related areas, to determine the ions or substances present in a sample, usually a mixture. The constituent substances or ions in a sample are separated by virtue of their ability to move at different rates over or through a suitable medium, such as a porous paper. This mobility is

CHROMATOGRAPHY

achieved by the addition of an appropriate solvent or solvent mixture, the *mobile phase*.

In possibly the first use of the method, the Russian botanist Tswett, in 1903–6, applied an extract of green leaves to the top of a small column of calcium carbonate and allowed a solvent (mobile phase) to run through the column. The original leaf extract divided into two colored zones. Later the column was extruded from its container and the colored zones were cut out for further investigation. Most chromatographic analyses today are concerned with the separation of colorless material but nevertheless the term chromatography has persisted.

Complex Nature of Chromatography. The processes that operate during a chromatographic analysis are often complex and, as yet, are not clearly understood. One or more of the following processes can combine in a chromatographic separation: adsorption (physisorption or chemisorption), diffusion, partition (liquid-liquid, liquid-solid, gas-liquid, etc.), ion exchange, complexation, hydration, and hydrolysis.

Adsorption is the phenomenon whereby substances are held at the surface of a liquid or a solid in such a way that the concentration of the substance(s) adsorbed in the boundary region is higher than in the surrounding region of the contiguous phases. If the adsorptive forces are great (comparable to ionic bonding) *chemisorption* is said to occur. If the forces are weaker (comparable to the forces of cohesion of molecules in the liquid state, i.e., van der Waals forces), then the phenomenon is physisorption. Diffusion is the movement of molecules of liquids and gases to achieve uniform concentration within the confining space. Partition is the phenomenon whereby a solute will become distributed between two immiscible solvents when the three are closely mixed. The concentration of the solute is often much greater in one solvent than in the other. Ion exchange is the reversible exchange of ions between a solid (the ion exchanger) and a liquid (the solution) in such a way that there is no substantial change in the structure of the solid. Modern ion exchangers have open network structures that permit the ion exchange to take place within the exchanger as well as on its surface. Hence absorption functions as well as adsorption. The molecules of many substances, formerly regarded as simple, are now recognized as having complex groupings of ions around a simple molecule (or ion). The process that forms these groupings is complexation. Hydration and hydrolysis are simple examples of complexation.

Natural Chromatography. It is becoming recognized that some natural geologic processes are in reality chromatographic. Natural chromatography operates in some forms of mineral genesis (Ritchie, 1964), in the flow of mixtures of fluids through sedimentary rocks (Nagy, 1960), in soil formation under arid conditions (Yaalon, 1964), and in trace element distribution (McIntire, 1963).

Types of Chromatography. Chromatographic methods are classified in two ways. First, in consideration of the major process thought to be operating, there are adsorption chromatography, partition chromatography, ion-exchange chromatography, and precipitation chromatography. Second, in consideration of the shape or nature of the chromatographic medium, there are column chromatography, paper chromatography, and thin-layer chromatography. Gas chromatography (see below) is a form of either partition chromatography or adsorption chromatography.

Essentials of Chromatography. Fundamentally, the main kinds of chromatography have certain features in common:

1. The materials of the column, the paper, the thin layer, the ion exchanger, etc., which are known as the *support*. The support sometimes incorporates a solvent (usually water). This solvent becomes the *stationary phase* since it is held in the support; the mobile phase moves more or less freely over the support and the stationary phase. For example, silica gel as a support absorbs water into its interstices to form a silica-water complex which is the stationary phase. Likewise, in paper chromatrography a water-cellulose complex forms the stationary phase. In adsorption chromatography the adsorbent is commonly both the support and stationary phase.

2. The mixture to be separated is in liquid solution for most forms of chromatography, but in the gaseous state for gas chromatography.

3. A suitable solvent, the mobile phase, which causes the solutes to separate during what is known as the *development*. The mobile phase in some cases is known as the eluting solution or the eluant.

Column Chromatography. Chromatography columns can be set up and operated to achieve many useful separations. In general, columns can cope with greater masses of solute than can paper or thin-layer chromatography. In the laboratory, columns are usually packed into a buret-type glass cylinder. In the field, polyethylene tubing is more convenient. The column material determines the type of separation achieved (see Table 1 and Ritchie, 1964.)

Column Development. The development can be effected by allowing the solvent to flow ver-

TABLE 1. COMMON COLUMN MATERIALS

Adsorption Columns	Partition Columns	Ion-Exchange Columns	Precipitation Columns
Starch	Cellulose powders	Ion-exchange resins	Ammonium sulfide in agar gel
Talc	Silica gel	Modified cellulose ion-exchangers	Cadmium sulfide cellulose
Calcium carbonate	Kieselguhr	Chelating resins	Zinc sulfide cellulose
Magnesium carbonate			8-Hydroxy-quinoline
Magnesia			
Activated alumina			
Activated carbon			
Activated magnesia			
Zirconium compounds			

tically downwards through the column under the influence of gravity (see Fig. 1). Flow rates can be increased by applying suction at the lower end of the column. However, the solvent can be made to flow upwards through the column by using a solvent reservoir adjusted to a suitable head (see Fig. 2). As an example of column development, an ion-exchange column of a cation exchanger, after packing, can be "regenerated" by allowing a 10% HCl solution to flow downward through the column. Any metal ions held fortuitously on the resin will be removed in the effluent and all replaceable sites on the resin will be occupied by hydrogen ions; the resin is then in the H form. If a 10% solution of sodium chloride is allowed to flow through the column, the sodium ions will replace the hydrogen ions and will themselves be held on the resin. Hydrochloric acid will be contained in the effluent. The capacity of an ion-exchange resin to take metal ions onto its available sites is limited and is expressed in milliequivalents per gram (dry weight) or milliequivalents per milliliter (wet volume). If too great a volume of sodium chloride is allowed to flow, the column will become *overloaded* when all available sites have become occupied by sodium ions. The latter will then begin to appear in the effluent. If the rate of

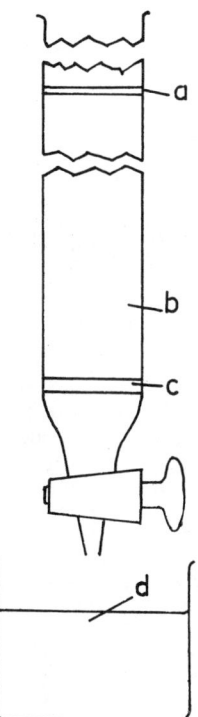

FIG. 1. Downward column development (a) paper pad; (b) column; (c) sintered disc; (d) effluent.

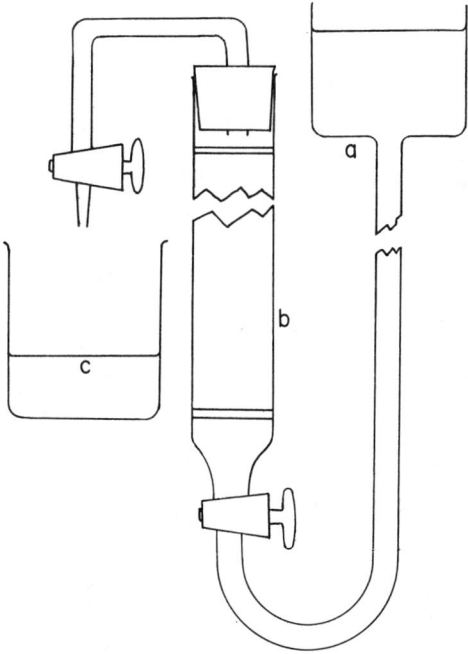

FIG. 2. Upward column development (more often used for "backwashing" column during regeneration); (a) solvent reservoir; (b) column; (c) effluent.

flow is too great, some sodium ions will pass through the column without being held, even though the column is not overloaded. If, then, a 10% solution of calcium chloride is passed down the column holding the sodium ions, the calcium ions will replace the sodium ions on the resin and sodium chloride will appear in the effluent. In another case, if a mixture of 10% sodium chloride and 10% calcium chloride solution is passed down a regenerated column, sodium and calcium ions will be held at first at the top of the column. However, subsequently, on-coming calcium ions will progressively replace the sodium ions. If the volume of the salt solution and the mass and ion-exchange capacity of the resin have been correctly adjusted, all the sodium chloride would have been expelled from the column as effluent (with hydrochloric acid) and all the calcium ions would be held on the resin. Thus the sodium chloride solution would have been separated from the calcium chloride. Subsequently, the calcium chloride could be eluted from the column by passing down a 10% HCl solution.

Although separation of solutes on columns can be achieved in most cases without difficulty, unless the zones into which the solutes are separated are visible (colored or fluorescent) it becomes difficult to decide where the zones are. For inorganic solutes it is sometimes possible to wash the column (after development) with a reagent such as H_2S water (as a colorant) or 8-hydroxyquinoline solution (which colors some zones and renders others fluorescent).

Conductimetric, radiometric, colorimetric, and spectographic examinations of effluent also provide information on the position of the solutes. Once standard procedures for a separation or group of separations have been arrived at, good separations in the effluent can be expected (Ritchie, 1964, p. 99).

Four kinds of column development are used according to the nature of the substance to be analyzed and the degree of separation possible or desired. These are shown briefly in Table 2.

Paper Chromatography. Paper chromatography in its simplest form can be regarded as a kind of partition chromatography. The paper (a variety of grades of "chromatography paper" is available) is the support and the stationary phase is an indefinable "water-cellulose" complex. Whereas in solvent extraction, in which solutes are partitioned between two immiscible solvents, complete separation (even after several extractions) is never achieved, in column and paper chromatography separation is achieved by what amounts to an infinite number of discrete partitions (with infinitely small volumes of solvent) between the stationary and the mobile phases.

TABLE 2. TYPES OF COLUMN DEVELOPMENT

Elution Development	Liquid Chromatogram	Frontal Analysis	Displacement Development
(1) Solutes in minimum of solvent added to column	(1) Solutes in minimum of solvent added to column	(1) Solutes in relatively large volume of solvent added to column	(1) Solutes in a convenient volume of solvent added to the column
(2) Pure solvent added	(2) A series of solvents of increasing eluting power added	(2) No further solvent added	(2) A displacing solution of a substance more strongly adsorbed (or held in some other way) to the column is added
(3) Effluent contains in order: (a) pure solvent; (b) solutes in order of mobility with or without overlap; (c) pure solvent	(3) Effluent contains in order: (a) some of first solvent; (b) a series of effluents each containing a different solute; (c) some of the last solvent (solute zones separated by zones of pure solvent)	(3) Effluent contains in order: (a) pure solvent; (b) solution of the least adsorbed solute increasing in concentration to that of orginal solution; (c) mixed solution of first solute at maximum concentration together with solution of second solute which increases to maximum concentration, and so on for the other solutes	(3) Effluent contains in order: (a) first solvent; (b) least adsorbed solute followed immediately by (c) the next least adsorbed solute and so on; (d) displacing solution as column becomes overloaded

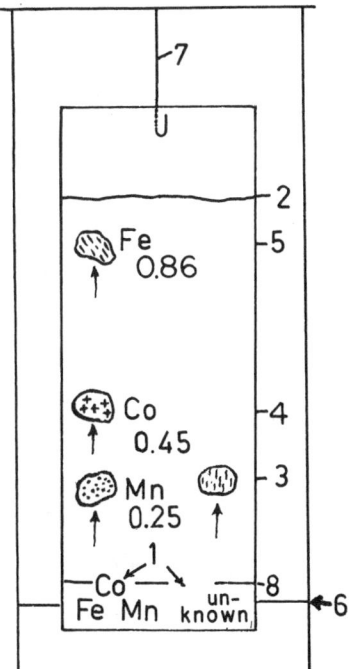

FIG. 3. Ascending development on paper: (1) points of application of solution; (2) solvent front after development; (3) positions of Mn spots; (4) position of Co spot; (5) position of Fe spot; (6) solvent level in closed vessel; (7) hook from lid; (8)–(3) distance moved by Mn ion; (8)–(2) distance moved by solvent.

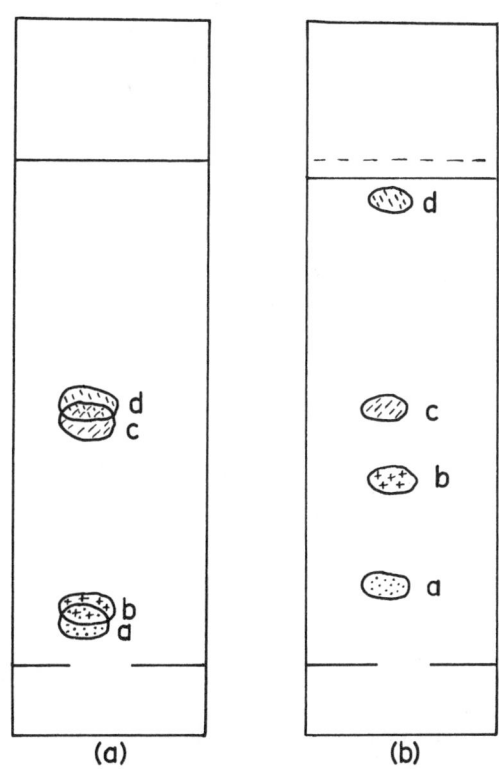

FIG. 4. Multiple development: (a) chromatogram after first development; (b) same chromatogram after second development; (a)–(d) spots of ions.

A solution of the substance to be analysed is applied in a minute quantity (on the order of 5–100 micrograms of solute) at *a point of application* near one end of the paper. Development is achieved by allowing a solvent mixture to flow over and through the paper. If the paper is held vertically the solvent can flow upward by capillarity (Fig. 3) or downward by gravity. If the paper is held horizontally the solvent can be applied slowly at the center and a radial flow direction can be achieved. Thus ascending, descending, and radial development have become common techniques. After development the paper is known as the *chromatogram*. Multiple development can be used in difficult separations by normal ascending or descending development in one direction after which the chromatogram is dried and then subjected to a second different solvent flowing in the same direction (see Fig. 4). Two-dimensional development is achieved if a second solvent flows at right angles to the first (see Ritchie, 1964, pp. 46–54).

Movement of Ions. Ions in the substance to be analysed move in the direction of flow of the mobile phase (or remain stationary) according to their ability to form complexes in the paper-solvent environment. Some ions can move with the solvent front, some lag behind it, and some do not move at all. The movement of an ion relative to the movement of the solvent is called the R_F value of the ion:

$$R_F = \frac{\text{distance moved by ion}}{\text{distance moved by solvent}}$$

The distances are measured from the point of application to the position of the ion (usually a spot) or of the solvent front after development. To minimize the effect of abnormal movement after the solvent reaches the dried solutes (i.e., while equilibrium is being reached), development time should permit a solvent flow of 10–15 cm (4–6 in.). For difficult separations longer travel distances can be permitted. Descending development, which allows the solvent to drip off the lower end of the paper, permits longer travel distances for slowly moving ions. Different solvents move at different rates and hence development times should be adjusted to achieve the desired separation of ions.

CHROMATOGRAPHY

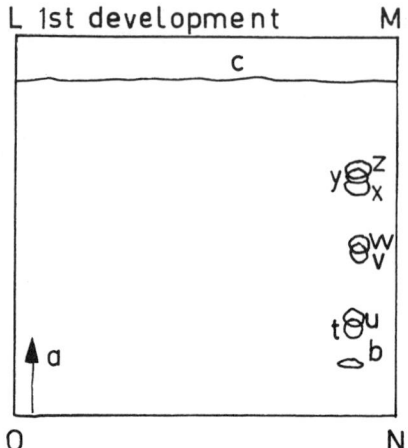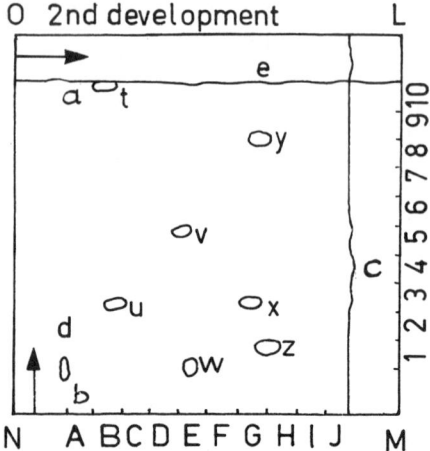

FIG. 5. Two-dimensional development: LMNO, paper square; (a) direction of flow of solvent 1; (b) point of application of sample solution; (c) solvent 1 front; (d) direction of flow of solvent 2; (e) solvent 2 front; (u)–(z) spots produced by ions (overlapping after first development, spaced after second development); (A)–(J) and (1)–(10) grid systm for location of spots.

Sometimes some ions lag behind the main spot to form a tail or comet. Too much solute is often the cause of comets. Small spots formed by the use of a minimum of solute give the most reliable R_F values. In two-dimensional development a grid system for defining the position of the ions may be adopted (Fig. 5).

Special Chromatography Papers. Chromatography papers with special characteristics permit separations on paper in ion-exchange or precipitation chromatography.

Location of Ions on Chromatograms. The positions of the ions on the paper after development are located in a number of ways:

1. Sometimes the ions form colored complexes on the chromatogram. Hence the spots can be marked by soft pencil at the moment the chromatogram is taken out of the solvent.

2. If the ion complex is colorless it can be detected by exposure to the vapor or spray of a suitable detecting reagent which will produce colored or fluorescent spots. Fluorescence, of course, is observed in the dark under ultraviolet light. Specific reagents are available for particular ions.

Identification of the Ions. Sometimes the color of the spots on the chromatogram after development or after treatment with the detecting reagent are distinctive. Specific reagents of course provide positive identification. However, the R_F values provide very reliable identification. For a given ion in a certain solvent mixture, the R_F value will be constant for a given temperature. Many R_F values for metals in a variety of solvents are recorded in the *Journal of Chromatography.* Some, which are suitable for geological analyses, have been recorded (Ritchie, 1964, pp. 77, 79, 86–7, 89, 93–4, 115–117).

Positive identification of metals in a substance can be made by "running" parallel chromatograms (see Ritchie, 1964, pp. 51, 57, etc.). On a rectangle of paper 11 in. (28 cm) × 9 in. (23 cm) six or seven parallel analyses can be conducted. In addition to the "unknown," solutions of probable (or suspected) metal constituents can be applied to the paper (see Fig. 3). Comparison of the spots in the parallel chromatograms confirm or exclude the presence of the suspected metals.

The paper-chromatographic method has some advantages over classical chemical analytical methods and also over the other chromatographic methods. Only minute quantities of rocks or minerals are required. Separations, which are difficult by other means (e.g., of the rare earths or zirconium-hafnium), are achieved with great ease on paper. The separated ion (as a spot) is available for further examination. A selected spot can be cut from the paper and thus becomes available quite free from other ions. Another real advantage is the possibility of its use in the field and under base-camp conditions.

Qualitative, Semiquantitative and Quantitative Analyses. While qualitative analyses can be carried out readily, workers like Agrinier, Nevill and Lever, Spain et al., and Ziegler et al. (all reported in Ritchie, 1964) have shown that the chromatographic method is quite capable of achieving semiquantitative and quantitative analyses that are comparable to the classical methods in accuracy and reliability.

"Batch" Operation Analysis. In addition to

separations by precipitation chromatography or ion-exchange chromatography in columns, several specific separations can be made by the "batch" operation. When one component of the sample is capable of becoming strongly fixed to the ion-exchange resin or the precipitant, while other constituents are not fixed at all, the separation can be made in a beaker in the following way. The regenerated resin or prepared precipitant is placed in a beaker. The solution of the material to be separated is added and stirred. The solution remaining is decanted off and the resin (or precipitant) is washed with a suitable liquid. The metal fixed to the resin (or precipitant) can be recovered by regeneration (in the case of a resin) or by ashing the resin or precipitant. (See the methods of Ziegler et al. reported in Ritchie, 1964, pp. 133–7).

Thin-Layer Chromatography. In thin-layer chromatography (TLC) uniformly thin layers of chromatographic media (as slurry) are spread on glass plates by mechanical spreaders. Thicknesses of 50 to 3000 microns are commonly used. The plates are dried at 110°C for half an hour to activate the medium. In most respects the analysis then proceeds as in paper chromatography. The advantages over paper are that the glass plates are inert whereas the paper rarely is. The medium is also more pure and thickness is more uniform. A greater variety of thin layer media is available. Since the plates can be laid horizontally (and for other reasons) the time of development is much shorter than for paper. Fluorescent media are available and fluorescent spots are much more definite on glass than on paper, on which the patchy fluorescence of the paper itself is confusing.

Gas Chromatography. In gas chromatography a volatile sample (boiling point up to and above 400°C) is carried by a gaseous mobile phase through a column containing a stationary phase. In *gas-liquid chromatography* (GLC), the stationary phase is a liquid solvent distributed on an inert solid support. This is a form of partition chromatography. In *gas-solid chromatography* (GSC) the stationary phase is a surface-active adsorbent of large surface area. This is a form of adsorption chromatography. In each case the gaseous components of the sample are separated by a process of selective and differential retardation acting on the components as partition or adsorption proceeds. Elution development (see Table 2) is the simplest and most widely used method of development. Thus a small gaseous sample is injected into the column at the inlet. A constant flow of *carrier gas* is passed through the column. As in other forms of chromatography, this mobile phase must be generally inert towards the stationary phase and, for this reason, it is usually helium, nitrogen, or hydrogen. The aim is to make the components of the sample separate into bands under the controlled conditions of the experiment (e.g., temperature, pressure, sample volume, carrier-gas flow rate, etc.). When separation is achieved, the bands of the sample components will be separated by pure carrier gas.

If displacement development is used (as it is sometimes in GSC) the small gaseous sample is introduced into the column as before. The carrier gas contains a component that is held more strongly on (or in) the stationary phase than any component of the sample. Thus each component of the sample is displaced forward selectively and, under ideal conditions, passes out of the column in separate bands which are, however, not separated by carrier gas.

In frontal analysis (in GSC) a large volume of uniformly mixed sample and carrier gas is introduced into the column and passed through it. Separation is only achieved within the limits of the method (see Table 2).

Gas chromatographs are usually expensive machines to which are fitted automatic recorders. Percentages of the gaseous components of the sample can be calculated from peaks in the recorder response which can be read off the recording chart. Comparisons with samples of known percentages also permit rapid analysis.

GSC is particularly useful in separating stable gases of low boiling point and light hydrocarbons, which are significant in geology. GLC is used similarly and to some extent is advantageous in that the partition processes give more definite and regular peaks on the recording charts (see Fig. 6).

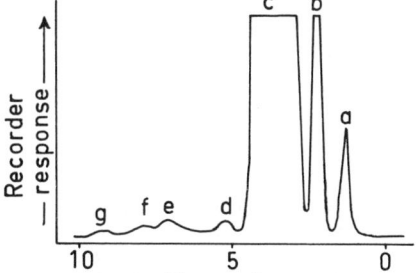

FIG. 6. Gas chromatogram showing separation of a mixture of hydrocarbons: (a) propane + propylene; (b) isobutane; (c) butane + butylenes, etc.; (d) isopentane + 3-methyl-1-butene: (e) 2-methyl-1-butene + *n*-pentane + 1-pentene; (f) 2-pentene; (g) 2-methyl-2-butene (after Coates, V. J. et al., 1958).

CHROMATOGRAPHY

Table 3. Chromatographic Analyses of Geologic Material

Element	Geologic Material	Method[a]	Reference[b]
Pb, Zn, Cd, Mn, Cu, Co	Seawater	PCC	Carritt, p. 66
U	Natural waters	IEC	Coulomb and Goldstein, p. 68
Mo	Natural brine	IEC	Ward et al., p. 70
U	Natural waters	PC	Ward and Marranzino, p. 70
Pb, Cu, Zn, Co, Ni	Surface waters	IEC	Canney and Hawkins, p. 72
U	Natural waters	IEC	Ostle, p. 142
Na, K, Mg, Ca, Ba	Surface waters	IEC	Brown et al., p. 142
U, Be, Mn	Rocks and natural waters	AC, PCC	Sulcek et al., pp. 144–5
U, Th	Rocks and natural waters	IEC	Korkisch et al., pp. 146–7
Be, Mo, As, Bi, Se, Li, B, Ag, Ni, Co, Cu, Nb, Ta, Ti, U, Th, Au	Minerals, rocks, soils, and ores	PC	Agrinier, pp. 74–8, 86–95, 144
Cu, Co, Ni, Nb, Ta, Pb, U, Bi	Soils	PC	Hunt et al., pp. 78–86; Jedwab, p. 141
Se	Silicate materials	PC	Weatherley, p. 95
Ag	Soils and rocks	PC	Almond et al., p. 141
F	Phosphate rock	IEC	Shapiro, p. 142
Co	Soils and rocks	PC	Almond and Bloom, p. 142
Th	Monazite	IEC	Nagle and Murthy, p. 142
Zr, Hf	Minerals	IEC	Street and Seaborg, p. 142
Alkaline earths	Dolomite-limestone	PC	Pollard et al., p. 142
Zn	Meteorites	IEC	Nishimura and Sandell, p. 143
Na, K	Silicate rocks	IEC	Riley, p. 143
Na, Ta	Minerals, ores and concentrates	IEC	Kallmann et al., p. 143
Zn, Fe, Al, Mn, Mg, Ca, etc.	Natural phosphates	IEC	Povondra and Cech, p. 144
Mn, Sc, Cu, Co, Ga, W, Ta, Th, U	Rocks	PC	Coulomb, p. 144
Be	Rocks and natural waters	AC	Sulcek et al., p. 144
Mn, alkaline earths	Rocks	IEC	Povondra and Sulcek pp. 144–5
U	Rocks	IEC and AC	Sulcek et al., p. 145
Ca, Sr, Ba	Minerals	IEC	Sulcek et al., p. 145
Ca, Sr	Ca-rich materials	IEC	Povondra and Sulcek, pp. 145–6
Hydrocarbons	Sediments	AC	Ferguson, p. 147
U	Ores	IEC	Seim et al., p. 147
Th, U	Monazite, etc.	PC	Elbeih and Abou-Elnaga, p. 147
Zn	Iron ores	IEC	Umezaki, p. 148
Amino-acids	Fossils	IEC and PC	Abelson, C.A.[c]
Au (prospecting)	Eluvial, alluvial	PC	Nevill and Lever, pp. 96–8
Fe, Na, K, Ti, Al, Mg, Cu	Coal ash	IEC	Ellington and Stanley, pp. 98–100
55 metal ions	Minerals	PC	Ritchie, pp. 100–128
Sn, Hg, Cu, Pb, Bi, Cd, Sb, As, Zn, Nb, Ta, Co, Ni, Fe	Minerals	PreC	Spain et al., pp. 128–31
Ca, Mg	Mg-limestones	PC	Hjelle, pp. 131–2
Cu, Ag, Au	Ores, minerals	PreC	Ziegler et al., pp. 133–7
In, Tl	G1, W1 and silicate rocks	IEC	Brooks and Ahrens, C.A., 55, 20770d.
Zn	Ores	IEC	Hisoda and Kashikawa, C.A., 55, 26845f.
S	Ores	IEC	Ionescu et al., C.A., 53, 8922c.
Na, K	Rocks and minerals	IEC	Reichen, C.A., 53, 1587a.
Trace elements	Seawater	IEC	Brooks, C.A., 55, 18436g.
Noble metals	Silicate rocks	IEC	Brooks and Ahrens, C.A., 55, 223g.

TABLE 3. (*Continued*)

Element	Geologic Material	Method[a]	Reference[b]
U	Ores and solutions	IEC	Fisher and Kunin, *C.A.*, **51**, 7943h.
Nb	Soils	PC	Grimaldi and Breger, *C.A.*, **55**, 23892i.
Rare earths	Minerals	PC	Lederer and Kertes, *C.A.*, **51**, 9256h.
Rare earths	Monazite	PC	Lederer, *C.A.*, **50**, 3147i.
Sn	Silicate materials	PC	Agrinier, *B.S.*, 25.7.1538.
Fe, Ti	Ilmenite	PC	Elbeih and Gabra, *B.S.*, 25.7.1539.
U	Natural waters	PC	Grassini and Alberti, *B.S.*, 24.7.27036.

[a] PC = paper chromatography; ParC = partition column; PreC = precipitation chromatography; IEC = ion-exchange column; AC = adsorption column.
[b] Page references are from Ritchie (1964); *C.A.* = Chemical Abstracts; *B.S.* = Bulletin Signaletique.
[c] Chap. X in (Breger, I. A., editor), 1963, "Organic Geochemistry," New York, Pergamon Press.

Applications of Chromatography to Geology. In the analysis of geologic materials, the chromatographic methods, either in the laboratory or the field, are at least as attractive as classical methods (Table 3). The efficiency of chromatographic analysis is revealing the presence of hitherto unsuspected matter in geologic materials (e.g., amino acids in fossils and hydrocarbons in soils).

An extension of the chromatographic analysis of minerals is found in their identification on the basis of their major metal components (determined chromatographically). Attempts in this field have been made by Spain et al. (1962) and Ritchie (1962, 1964).

It has become recognized that chromatography is a natural process in geology. B. Nagy (1960) was probably the first to demonstrate that a geologic process was chromatographic in character. Ritchie (1964) and Yaalon (1964) have cited further examples. It is likely that many geologic reactions formerly stated to be due to adsorption, ion exchange, or diffusion will be shown in the future to be more complex in character, that is, chromatographic.

A. S. RITCHIE

References

Coates, V. J., Noebels, H. J., and Fagerson, I. S. (editors), 1958, "Gas Chromatography," New York, Academic Press, 323pp.

McIntire, W. L., 1963, "Trace element partition coefficients—a review of theory and application to geology," *Geochim. Cosmochim. Acta,* **27**, 1209–1264.

Nagy, B., 1960, "Review of the chromatographic 'plate' theory with reference to fluid flow in rocks and sediments," *Geochim. Cosmochim. Acta,* **19**, 289–296.

Ritchie, A. S., 1962, "The identification of metal ions in ore minerals by paper chromatography. I. Opaque ore minerals," *Econ. Geol.,* **57**, 238–247.

* Ritchie, A. S., 1964, "Chromatography in Geology," Amsterdam, Elsevier, 185pp.

Spain, J. D., Ludeman, F. L., and Snelgrove, A. K., 1962, "The use of precipitation chromatography in geochemical prospecting; mineral identification with disposable agar gel columns," *Econ. Geol.,* **57**, 248–259.

Yaalon, D. H., 1964, "Downward Movement and Distribution of Anions in Soil Profiles with Limited Wetting," pp. 157–164 in (Hallsworth, E. G., and Crawford, D. V., editors), "Proceedings 11th Easter School in Agri. Sci., Univ. of Nottingham," London, Butterworths.

* Additional references may be found in this work.

Cross-references: *Bonding; Hafnium: Hydrocarbons; Ion Exchange; Rare Earths; Van der Waals Force; Water; Zirconium.*

CHROMIUM: ELEMENT AND GEOCHEMISTRY

Chromium (Cr) is a hard, steel-white metallic element having the properties listed in Table 1.

Mineralogy. The principal chromium-bearing minerals belong to the chromite or chromian spinel group, with the general formula $(Mg,Fe)O \cdot (Cr,Al,Fe)_2O_3$. Depending on the degree of isomorphic substitution of Al^{3+} and Fe^{3+} for Cr^{3+}, chromian spinels contain from 13 to 65% Cr_2O_3. The dominant zone of composition is shown in Fig. 1.

Chromian spinels form isometric crystals with a black to brownish black color, a brown streak, a density of 3.6–4.8 gm-cm^{-3} and a hardness of 5.5–7.5. Their unit-cell size varies from 8.13 to 8.34Å, the larger sizes representing increased substitution of chromium and iron for aluminum in the crystal lattice. Chromian spinels are primary constituents of ultra-

CHROMIUM: ELEMENT AND GEOCHEMISTRY

TABLE 1. PROPERTIES OF CHROMIUM

Atomic number	24
Atomic weight	52.01
Density (gm-cm^{-3} at 20°C)	7.1
Melting point (°C)	1930 ± 10
Boiling point (°C)	2482
Coefficient of thermal expansion (°C^{-1})	8.2 × 10^{-6} (20°C)
Isotopes (natural) (abundance in wt %)	
Cr^{50}	4.3
Cr^{52}	83.8
Cr^{53}	9.5
Cr^{54}	2.4
Isotopes (artificial)	(half-life)
Cr^{48}	23.5h
Cr^{49}	41.9m
Cr^{51}	27.8d
Cr^{55}	3.5m
Cr^{56}	5.9m
Valence states	
	0 (metal)
	+2 (chromous)
	+3 (chromic)
	+6 (chromate)
Ionic radii (Å)	
Cr	1.28
Cr^{2+}	0.83
Cr^{3+}	0.64
Cr^{6+}	0.35

tarapacaite	4[K$_2$Cr$_2$O$_4$]
crocoite	Pb Cr O$_4$
lopezite	4[K$_2$Cr$_2$O$_7$]
uvarovite (garnet)	Ca$_3$Cr$_2$(SiO$_4$)$_3$
fuchsite (mica)	4[K(Al,,Cr)$_3$Si$_3$O$_{10}$(OH$_2$)]
kämmererite (chlorite)	Mg$_5$(Al,Cr)$_2$Si$_3$O$_{10}$(OH$_8$)

Geochemistry. Chromium occurs in nature principally as the trivalent ion Cr^{3+} with a radius of O.64Å. It readily substitutes for Fe^{3+} (0.67Å) and Al^{3+} (0.56Å) during crystallization. The geochemical history of chromium in the earth's crust starts with its introduction from deeper levels in the mantle as a component of basaltic magmas. Chromium typically is precipitated from magmas at an early stage, either in chromian spinel or in silicate minerals, especially clinopyroxene. These minerals form part of the ultramafic rock suite, and leave a less mafic, residual magma depleted in chromium. An example of the redistribution of chromium as the result of fractional crystallization of basaltic magma is given in Table 2.

Chromite is generally resistant to chemical weathering. Due to its high specific gravity, it may be mechanically concentrated in laterites or heavy mineral placers. The chromium-bearing silicates release chromium which is incorporated into shales and schists. Metamorphism of these weathering products may result in the formation of fuchsite. Little chromium goes into solution, and thus precipitates and evaporates have a low chromium content.

Economic Geology. Since chromite deposits characteristically occur in ultramafic intrusions, an understanding of their mode of occurrence and properties must be based on the characteristics of the host intrusions.

mafic (high magnesia and low silica) rocks and are mined for metallurgical, refractory, and chemical uses.

Other chromium-bearing minerals of lesser abundance, in order of chromium content, include:

eskolite	Cr$_2$O$_3$
daubreélite	FeS·Cr$_2$S$_3$
stichtite	3[Mg$_6$Cr$_2$CO$_3$(OH)$_{16}$·4H$_2$O]

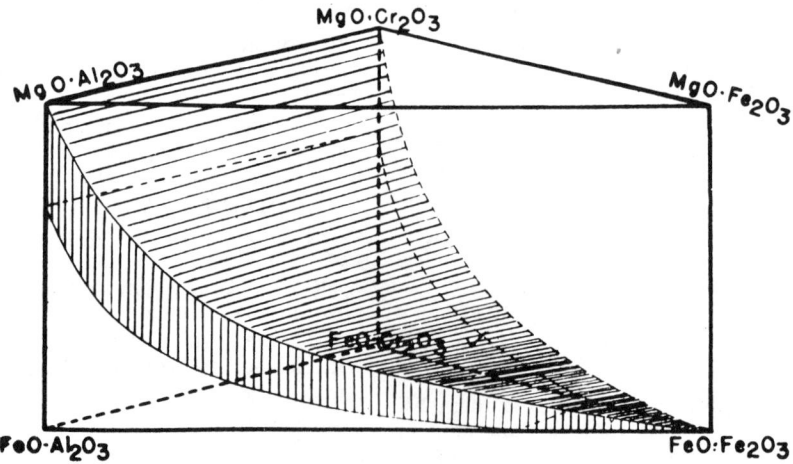

FIG. 1. The spinel triangular prism showing end members and predominant tone of isomorphism of chromian spinel (from Stevens, 1944).

TABLE 2. THE RANGE OF CHROMIUM CONCENTRATION IN VARIOUS ROCK UNITS OF THE MUSKOX INTRUSION, N.W.T., CANADA[a]

Chilled marginal gabbro	110–1000
Lower ultramafic rocks	3000–5000
Upper gabbros	225–1000
Granophyric rocks	25

[a] In ppm; listed in order of solidification.

Ultramafic intrusions are broadly divisible (not without exceptions) into (a) *stratiform complexes* and (b) *"alpine" complexes*. The former are characterized by their layered form, changing from dominantly ultramafic near the base to dominantly gabbroic or granophyric toward the top. Such intrusions are particularly well developed in Precambrian terrains and include the Bushveld Complex of South Africa, the Great Dyke of Southern Rhodesia, the Muskox Intrusion of northern Canada, and the Stillwater Complex of Montana. The "alpine" complexes occur with basic volcanic rocks in old mountain belts (Taconic, Caledonian, Variscan, Cordilleran, Alpine, etc.). They are characterized by a metamorphic fabric related to their orogenic history, which has resulted in a complex arrangement of rock units in contrast to the simple layering of most stratiform intrusions.

The inherent features of the host intrusion are imprinted on the chromite deposits themselves, since the chromite crystallized during the early cooling stage of the intrusion. Thus chromite deposits are divided into (a) *stratiform* types and (b) *podiform* types. Stratiform deposits form extensive layers in stratiform intrusions, some of which can be traced for tens of miles along the strike. Deposits in the Bushveld Complex and Great Dyke form the major chrome resources of the world. The podiform type of chromite deposit occurs in alpine ultramafic intrusions and is characterized by irregular, in places pencil-shaped, form and smaller size. Such deposits are found in Greece, Turkey, the Urals, Cuba, New Caledonia, and many other areas.

The principal geologic events controlling the formation of an economic chromite deposit vary somewhat, depending on whether the deposit is of the stratiform or podiform type. Both chemical and physical processes play an important role. First, the chemical composition of the parent magma, especially its chromium, iron, and aluminum contents and the state of oxidation, will control the partition of chromium between the spinel and silicate phases and the composition of the chromite which will crystallize. Then physical factors, such as the rate of cooling, the rate of settling of chromite crystals relative to silicate crystals in the cooling magma, flowage, and convection currents may decide whether the chromite will concentrate during crystallization into a minable mass, or become disseminated throughout the intrusion in a low grade, noneconomic form. The history of these events is often recorded in the fabric of stratiform deposits and can be readily studied under the microscope because of the undeformed nature of the host intrusion.

The podiform type of chromite deposit may have had an initial history analogous to the stratiform type, but the superposition of a deformational event, related to its formation in an alpine belt, has destroyed most of the original textures on which such an interpretation can be based. These deposits are thus characterized by their contorted or irregular forms and the fractured nature of their component grains. Field evidence suggests that the effect of this metamorphic event has been to deform and tear apart originally larger chromite masses rather than cause a concentration of chromite into larger deposits.

Chromite ores may differ widely in composition, depending on the relative proportions of chromite to silicate minerals in the mass, and the chemical composition of the chromian spinel itself. Ores containing silicate impurities may be enriched by mechanical beneficiation, but spinels low in chromium may only be enriched by chemical methods if warranted. On the basis of their chrome-iron ratio and chromium content, ores are divided according to usage into metallurgical, chemical, and refractory grades. The exact specifications for these grades vary with the user, but premium prices are paid for high-chrome ores (over 48% Cr_2O_3) with a high chrome-iron ratio (over 3:1). Such ores occur principally in the podiform type of deposit, although certain horizons in the Great Dyke are an exception to this rule.

World Distribution. Fig. 2 shows the principal ultramafic belts of the world and the location of the major known chromite reserves. The world's largest reserves occur in the Bushveld Complex of South Africa, and are estimated to be about 2000 million tons of chemical grade chromite. The Great Dyke of Southern Rhodesia contains reserves estimated at 608 million tons, half of which are of metallurgical grade. Large economic deposits occur in Russia, principally in the Ural Mountains. The size of the Russian reserves are not known, but refractory and metallurgical grade chromites are currently being exported to Europe and underselling the chromite mined by western countries. Although chromite deposits are found in North America, none are minable

CHROMIUM: ELEMENT AND GEOCHEMISTRY

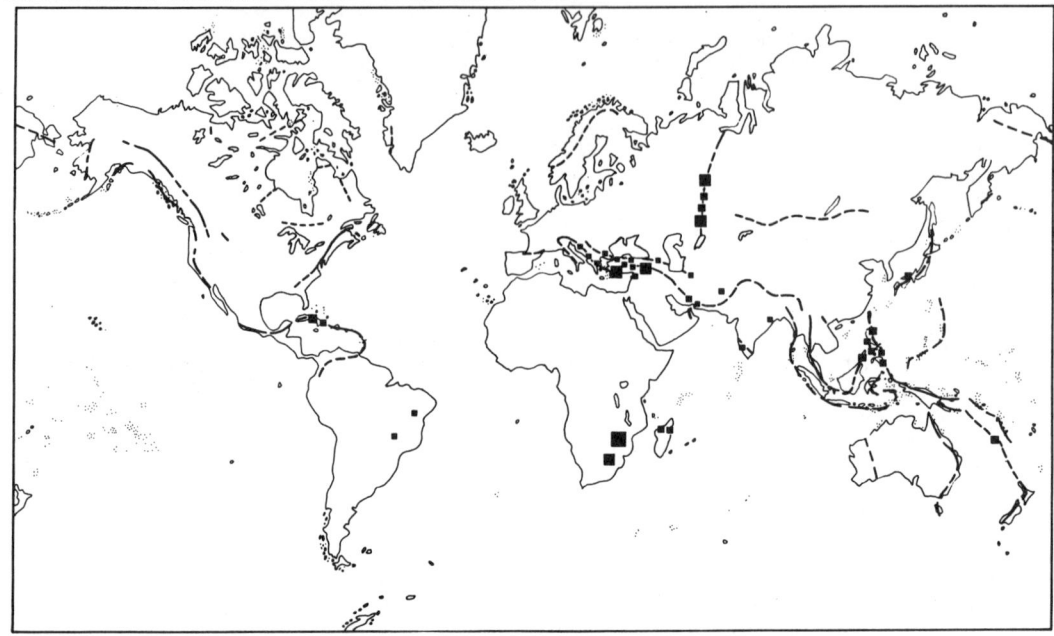

FIG. 2. World map showing principal chromite reserves (■) and trends of ultramafic belts (– – –) containing smaller deposits and possible future discoveries.

under current economic conditions. A summary of potential world reserves is listed in Table 3.

TABLE 3. MAGNITUDE OF POTENTIAL WORLD RESERVES OF CHROMITE[a]

Country	Magnitude (million tons)
Republic of South Africa	500
Southern Rhodesia	100–500
U.S.S.R., Turkey	10–100
Phillipines	
Cuba, New Caledonia, India Yugoslavia, United States Canada, Albania, Greece	1–10
Pakistan, Sierra Leone Japan, Cyprus, Brazil Guatemala, Afghanistan Egypt, etc.	1

[a] After H. A. Heiligman and H. M. Mikami, 1960.

CHARLES H. SMITH

References

Donath, M., 1962, "Die Metallischen Rohstoffe," *Chrom.* **14,** 371.
Fukai, R., and Broquet, D., 1965, "Distribution of chromium in marine organisms," *Bull. Inst. Oceanog.,* **65**(1336), 19pp.
* Heiligman, H. A., and Mikami, H. M., 1960, "Chromite, in Industrial Minerals and Rocks," Third ed., New York, Am. Inst. of Mining and Met. Engrs.
Rankama, K., and Sahama, Th. G., 1950, "Geochemistry," Chicago, Univ. of Chicago Press, 912pp.
Stevens, R. E., 1944, "Compositions of Some Chromites of the Western Hemisphere," *Am. Mineralogist,* **29,** 1–34.
Woodtli, R., (editor), 1964, "Methods of Prospection for Chromite," Paris, Organization for Economic Cooperation and Development, 244pp.

* Additional bibliographic references may be found in this work.

Cross-references: *Earth's Crust Geochemistry; Weathering, Chemical.* Vol. IV B: *Isomorphism; Metallurgy.* Vol. V: *Magma; Ultramafic Rocks.*

CLAPEYRON'S EQUATION

Clapeyron's equation expresses the variation of pressure and temperature for a system in equilibrium at constant composition. It is famous as one of the earliest physiochemical applications of the concept of entropy and the second law of thermodynamics. It is a very important equation for the geologist since it describes the relation between the pressure and temperature of phase transitions such as polymorphic transformations, fusion, vaporization, or sublimation.

At a given *temperature, T,* and *Pressure, P,* the condition for a system to be in chemical equilibrium is that: $\Delta G = 0$, where ΔG represents the change in *Gibbs free energy* (also called free enthalpy or Gibbs function) for the system when a reaction takes place. If the process is the transition of a chemical substance from phase 1 to phase 2, then $\Delta G = G_2 - G_1$, i.e., the difference between the values of Gibbs free energy for that substance in the two phases. For a small change in pressure dP and a small change in temperature dT, the state of equilibrium will be maintained if ΔG remains zero, which means that the differential change, $d(\Delta G)$, must be zero. Taking into account the mathematical relation derived from the second law of thermodynamics, this condition becomes

$$d(\Delta G) = \Delta V dP - \Delta S dT = 0 \quad (1)$$

where ΔV and ΔS are the *volume change* and the *entropy change* for the process. The above equation is usually written:

$$\frac{dP}{dT} = \frac{\Delta S}{\Delta V} \quad (2)$$

The latent heat of transformation, ΔH, (change in enthalpy) is related to the entropy change at the temperature of transformation T by: $\Delta H = T \Delta S$ (since at equilibrium $\Delta G = \Delta H - T \Delta S = 0$) and Eq. (2) can be replaced by the most usual form of *Clapeyron's equation*:

$$\frac{dP}{dT} = \frac{\Delta H}{T \Delta V} \quad (3)$$

Applications

Equilibrium Between a Substance and its Vapor: Clausius-Clapeyron Equation. When considering a transformation into a gaseous phase, the volume of the constituent in the condensed phase can be neglected relative to V_g, the volume of the gaseous phase. If the latter behaves like an ideal gas, then $\Delta V = V_g = RT/P$, where R is the gas constant (1.987 cal/deg-mole). Substituting this into Eq. (2) and (3) (and noting that $dP/P = d\ln P$) leads to the *Clausius-Clapeyron equaiton*:

$$\frac{d\ln P}{dT} = \frac{\Delta \cdot S}{RT} \quad (4)$$

$$\frac{d\ln P}{dT} = \frac{\Delta H}{RT^2} \quad (5)$$

The temperature dependence of the vapor pressure in equilibrium with the solid or the liquid substance can be calculated from thermodynamic data by these equations. Conversely, the latent heat of vaporization or sublimation, and the entropy change, can be obtained by an experimental determination of the equilibrium vapor pressure as a function of temperature. For that purpose, it is convenient to rewrite Eq. (5) in the following way (noting that $d(1/T) = - dT/T^2$)

$$\frac{d\ln P}{d(1/T)} = - \frac{\Delta H}{R} \quad (6)$$

If $\ln P$ (or $\log P$) is plotted against $1/T$, the value of ΔH can be computed from the slope of the line.

Example of a Polymorphic Transformation: Quartz α to Quartz β. (Kern and Weisbrod, 1967, p. 115). At atmospheric pressure α-quartz changes to β-quartz at 846°K. The latent heat of transition is 150 cal/mole and the volume change can be estimated as 0.2 cm³/mole. Straightforward application of Eq. (3) gives $dP/dT = 0.886$ cal/cm³-deg $= 36.6$ aim/deg (1 cal $= 0.0413$ liter-atm). This is the value of the slope of the equilibrium pressure-temperature curve at $P = 1$ atm, $T = 846°K$. Fig. 1 shows a curve derived from direct measurement of equilibrium temperature at various pressures and the calculated slope. The variations of ΔH and ΔV over a small temperature range are usually negligible, and the equilibrium curve can be approximated by a straight line, so that it is possible to predict, by using Clapeyron's equation, the shape of the equilibrium diagram from thermodynamic data at atmospheric pressure.

PAUL DUBY

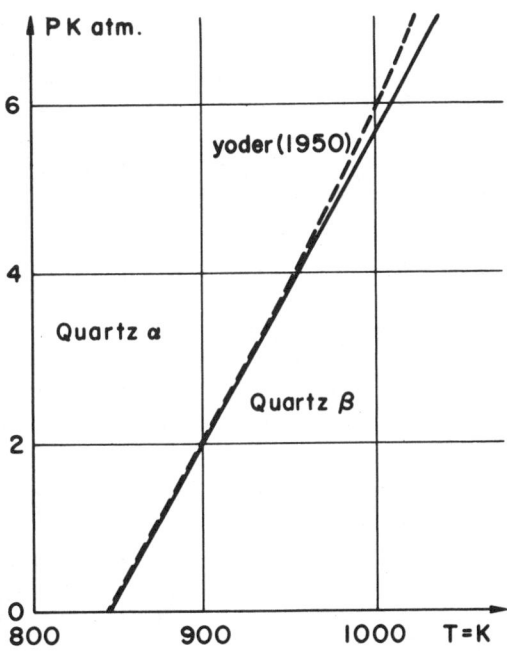

FIG. 1. Equilibrium diagram for the transformation α-quartz to β-quartz. The continuous straight line is the slope predicted by Clapeyron's equation. The interrupted line represents direct equilibrium measurements (from Kern and Weisbrod, 1967, p. 117, by permission of Freeman, Cooper and Co.).

References

Kern, R., and Weisbrod, A., 1967, "Thermodynamics for Geologists," San Francisco, Freeman, Cooper and Co., 304pp. (translated by Duncan McKie from "Thermodynamique de Base pour Minéralogistes, Pétrographes et Géologues," Paris, Masson et Cie, 1964).

Klotz, I. M., 1964, "Basic Chemical Thermodynamics," New York, W. A. Benjamin, 468pp.

Mahan, B. H., 1963, "Elementary Chemical Thermodynamics," New York, W. A. Benjamin, 155pp.

Cross-references: *Enthalpy; Entropy; Thermochemistry; Thermodynamics.*

CLAY MEMBRANE PHENOMENA

Some earth materials behave as natural membranes which retard or prevent the passage of charged ionic species while allowing relatively unrestricted movement of neutral species. Molecules with high dipole moments may likewise be retarded compared with those of lower dipole moment. These membranes are termed semipermeable membranes because of their ion-exclusion properties. Differences in free energy across such membranes can result in the development of osmotic pressure, of differences in electrical potential, and of salt sieving or ultrafiltration.

The exact mechanism by which ion-exclusion occurs is not known but one widely held hypothesis is that an ion-exchange material has

Publication authorized by Director, U.S. Geological Survey. Substantive portions of this entry were extracted from an article by Back and Hanshaw, 1965.

a net electrical charge deficiency; in minerals this deficiency may be caused by any one or a combination of the following ways: (*1*) broken bonds; (*2*) removal of the hydrogen of an exposed hydroxyl group; (*3*) removal of structural cations under certain conditions and; (*4*) substitution of low valence cations for ones of higher valence within the mineral structure. In order to maintain electrical neutrality the material adsorbs a large number of cations and some anions into its pores and produces a diffuse double layer shown diagrammatically in Fig. 1. As the exchange material compacts, the double layers from the sides of a channel overlap so that the pore becomes positively charged. This positively charged pore repels cations and prevents them from entering and, in order to maintain electric neutrality across the membrane, anions also are restricted from moving through the membrane. The streaming potential causes the egress side of a membrane to become positively charged. This charge also tends to retard the passage of cations across the membrane. Recent laboratory investigations have demonstrated that compacted clays may act as semipermeable membranes.

Movement of water across a semipermeable membrane, in the absence of an externally applied pressure differential, may result from the existence of different salt concentrations on opposite sides of the membrane, from the existence of an electrical potential across the membrane, or possibly from the existence of different thermal-energy levels on either side of the membrane. Such phenomena are termed, respectively, chemical osmosis, electro-osmosis,

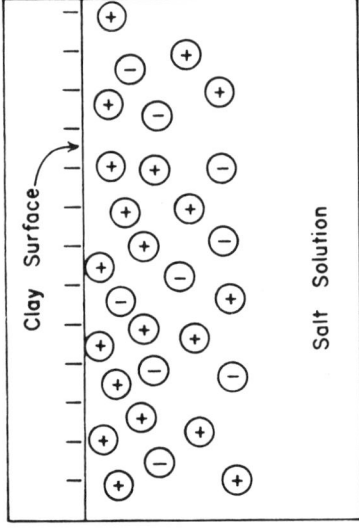 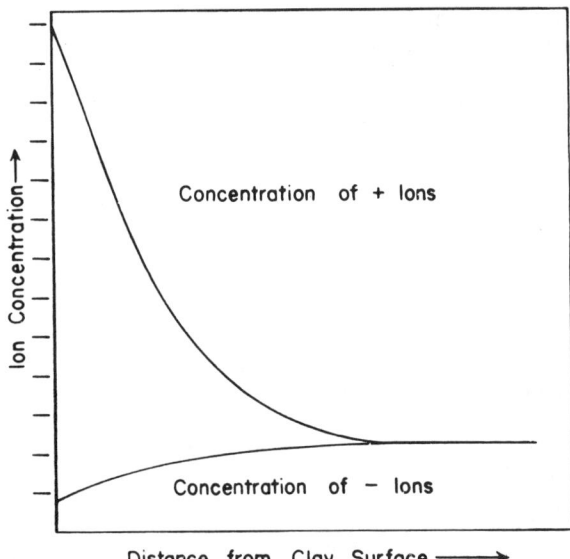

Fig. 1. The diffuse double layer.

and thermo-osmosis. Any one or a combination of these phenomena may cause pressure differentials across a membrane. Little experimental work has been done in the fields of electro- and thermo-osmosis; only chemical osmosis will be discussed below. Ultrafiltration occurs when an electrolyte solution flows through a semipermeable membrane. As a result of ion exclusion the effluent is less concentrated than the original solution.

The phenomenon of chemical osmosis will occur when two aqueous phases having different salt concentrations (that is, different activities of H_2O) are separated by a semipermeable membrane. At constant temperature a greater net transfer of water will occur from the less saline solution (higher activity of H_2O) through the membrane into the more saline solution (lower activity of H_2O), and will cause the pressure to increase in the more saline solution and decrease in the less saline if the two solutions are confined. Osmotic pressure is the equilibrium pressure difference across a membrane which results from a greater net transfer of water from the less saline to the more saline side of the membrane until equilibrium is reached. Alternately, the pressure that must be applied to the more saline side of the membrane to cause the passage of water to be the same in both directions is equal to the osmotic pressure. An equation to express the relation between osmotic pressure and solution activity, derived from thermodynamic arguments, is

$$P_{II} - P_I = \pi = \frac{RT}{V_{H_2O}^0} \ln \frac{\alpha_{H_2O}^I}{\alpha_{H_2O}^{II}} \quad (1)$$

where P_I and P_{II} are pressures corrected for differences in depth below a common datum, π is osmotic pressure, R is the gas constant, T is absolute temperature, $V_{H_2O}^0$ is the molar volume of pure water, and $\alpha_{H_2O}^I$ and $\alpha_{H_2O}^{II}$ are the activities of H_2O in solutions I and II, respectively. Fig. 2 (Hanshaw and Zen, 1965) is a plot of the osmotic pressure as a function of the concentration of NaCl in solution; the reference state is distilled water. In the calculation, $V_{H_2O}^0$ is taken to be 18 ml/mole. For example, at 80°C, corresponding to a depth of about 2–3 km for a normal near-surface geothermal gradient, the activity α_{H_2O} in a halite-saturated binary solution is 0.747 (International Critical Tables, Vol. 3, p. 369) at 1 bar pressure. If α_{H_2O} in saturated halite solution is taken to be about 0.75 for the entire pressure range, then against distilled water ($\alpha_{H_2O} = 1.00$) the value of π is 470 bars. Owing to the presence of other dissolved components, a halite-saturated seawater would cause a greater pressure difference, so that

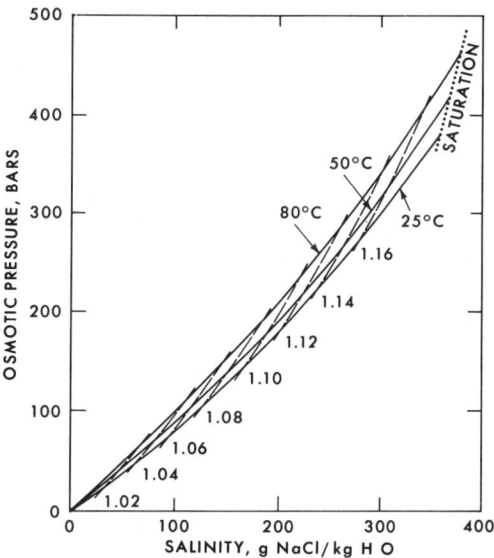

FIG. 2. Calculated osmotic pressure difference across an ideal membrane as a function of NaCl concentration and temperature in the binary system NaCl–H_2O. Dashed lines are constant density lines (from Hanshaw and Zen, 1965).

$\pi = 500$ bars seems a reasonable upper limit.

The flow of water through the membrane to establish osmotic equilibrium is accompanied by ion-exclusion effects which will increase the salinity of the solution on the low-pressure side of the membrane. The solution that passes through the membrane is less saline than that on either side and will tend to decrease the salinity on the high-pressure side of the membrane.

An ion-selective or membrane electrode develops a potential difference when it separates two electrolyte solutions containing the same cation species at different thermodynamic concentrations, or activities. The reproducible static potential that is developed is given by the Nernst equation.

$$\epsilon_{I,II} = C_{I,II} + \frac{RT}{z_i F} \ln \frac{\alpha_{iI}}{\alpha_{iII}} \quad (2)$$

$\epsilon_{I,II}$ is the developed Volta potential, $C_{I,II}$ is an asymmetry potential, z_i is the valence of the ith cationic species, F is the faraday, and α_i is the activity of the ith species. If the activity of the ith species is held constant in one solution and is changed in the other, the equation predicts a 59-mV change per tenfold change in activity of an ion of unit charge at 25°C.

Some pressure and salinity data observed within water-bearing sedimentary rocks cannot be explained by simple gravitational flow of water or by solution of minerals. Certain of these geologic relationships can be explained by osmotically induced cross-formational movement of water, through shales which serve as semipermeable membranes, from an aquifer whose water has a low concentration of dissolved salts to one whose water is more concentrated. Likewise, an externally applied pressure differential on the fluid phase may cause cross-formational water flow through a semipermeable shale membrane, with resultant ion-exclusion effects commonly referred to as salt filtering or ultrafiltration.

Various authors have postulated that salt filtering by clay or shale membranes may account for the formation of brines having concentrations exceeding (in some cases greatly exceeding) the concentration of seawater. These various speculations have been borne out by experimental laboratory investigations on compacted clays by Wyllie (1948, 1949, 1951, 1955), McKelvey and Milne (1962), Kemper (1960), and Hanshaw (1962, 1964). These and other studies are summarized in Hanshaw and Zen (1965). Experiments by Kemper (1961) on shale and compacted clay membranes demonstrated fair agreement between measured and theoretical values of osmotic pressure. The highest value measured by Kemper was more than 100 bars using a 5.4 N NaCl difference across a bentonite membrane. Young and Low (1965) also measured osmotic pressure experimentally in argillaceous rocks. Both of these studies indicate that clay membranes may be less than 100% efficient. Using the Teorell-Meyer-Siever theory, Hanshaw (1962) showed the reason for reduced membrane efficiency at higher solution concentration. At high salt concentrations the clay membranes allow ionic leakage across the membrane; it remains to be determined whether a series of imperfect clay membranes separating solutions of small concentration difference can cumulatively produce a net pressure anomaly corresponding to a single ideal membrane.

White (1965) and Graf et al. (1965, 1966) have suggested that differences in mobility of various species through shale membranes can aid in interpreting the origin of subsurface brines, especially those of anomalous chemical and isotopic composition. These speculations have been only partially substantiated by laboratory studies.

The geologic evidence for the membrane properties of shales has been studied by Berry and Hanshaw (1960). These investigators found closed potentiometric surface contours as low as 400 ft above sea level in the lower Cretaceous Viking sandstone in the central Alberta syncline. The recharge area is about 4000 ft above sea level and the lowest outcrop of the Viking sandstone, where water discharges, is about 1000 ft above sea level and 250 miles away from the closed low contour. The low potentiometric surface in the area of closure cannot be accounted for by gravitational flow of water. Osmotic cross-formational withdrawal of water through the Joli Fou shale into units with high total dissolved solids can account for the observed potentiometric low.

In the Wheeler Ridge anticline, at the southern tip of the San Joaquin Valley in California, the potentiometric surface elevation and formation-water salinity both increase with depth. Water from the lowest sediments tested, sands of Eocene age, is more than twice as saline as water in the Eocene elsewhere in the basin. The piezometric surface of the Eocene sands is 4150 ft above sea level on Wheeler Ridge, whereas the highest Eocene outcrop anywhere in the area is approximately 2200 ft above sea level. The excess head may be caused by high outcrop elevations of basement rocks underlying the Eocene or by tectonic compression. Apparently upward flow of water from the Eocene sands toward the land surface, in areas where the elevation is less than 1000 ft, is salt filtered as it passes through various shales in the stratigraphic sequence. Thus, ultrafiltration would account for the high salinity of waters associated with the various formations on the anticline. Bredehoeft et al. (1963) presented a mathematical model to predict the distribution of ions within a formation undergoing ultrafiltration.

Berry (1959) discussed the occurrence of closed low contours on the potentiometric surfaces of the various formations in the Cretaceous and Jurassic systems of the San Juan basin. Hanshaw and Hill (1969) described potentiometric surface elevations of Pennsylvanian and Mississippian formations in the San Juan basin that are much higher than any of the outcrops of these units. The Gallup sandstone member of the Cretaceous Mesaverde group in the southwestern part of the San Juan basin contains highly saline water in the area where the sandstone pinches out into the Cretaceous Mancos shale. The Gallup is a quartzose sandstone containing lenses of coaly material that establish its nonmarine origin. There are no known evaporite deposits associated with the Gallup sandstone to account for the salinity of the formation water. In discussing the anomalously high salinity in the sandstone, Berry showed that water was moving from outcrop areas toward the pinchout of the Gallup. Thus, water must be moving cross-formationally through the enclosing illitic marine Mancos shale. If the shale is capable of retarding the passage of ionic species then the apparently anomalous salt concentration is explained. Berry postulated also that the closed low piezometric contours within the Cretaceous aquifers in the San Juan basin were caused by downward movement of water cross-formationally from one aquifer to another through the intervening shales. The driving force for this system is presumed to be a core of saline water which must occur in the Jurassic aquifers and be more concentrated than any of the overlying brines. He calculated that a salinity greater than 135,000 ppm in the Jurassic formations would be necessary to make the system work. Each stratigraphically higher aquifer contains water less saline than the one below it at any one geographic point in the basin. Likewise, the size of the potentiometric surface anomaly decreases stratigraphically upward in the Cretaceous aquifers. The areas of high salinity correspond closely with the areas of low potentiometric surface in each aquifer studied. Hanshaw and Hill (1969) postulated that the upper Paleozoic units with known brines and with an otherwise inexplicably high potentiometric surface in the Four Corners area of New Mexico could be the outflow receptors of the San Juan membrane system.

Anomalous water pressures of at least 400 bars above hydrostatic pressure have been measured in oil wells; many of these wells penetrate evaporites and/or shales which separate formation waters of differing salt concentrations. Hanshaw and Zen (1965) suggest that these pressures are explicable by assuming osmotic equilibrium across a membrane which separates saturated halite solution from solutions up to 10% NaCl. Thus, osmotic equilibrium may be an important mechanism for floating thrust sheets. Lubrication of thrust sheets by shales or evaporites and flotation by anomalously high pressure may be simply different manifestations of the same geologic milieu.

In summary, shales acting as semipermeable membranes can cause anomalously high or low potentiometric surfaces and (or) ultrafiltration can cause formation waters to become more saline. Geologically, pressure and salinity anomalies associated with membrane phenomena are most noticeable in rocks of low transmissibility, or in units of moderate transmissibility surrounded by units of low transmissibility. If water is withdrawn from an aquifer by osmosis, the water pressure in that unit decreases if the amount of water removed by osmosis is greater than that replaced by gravitational flow. Pressure increases when water moves cross-formationally into an aquifer; the degree of pressure increase will depend on the relative ease with which gravitational flow dissipates the pressure increase caused by osmosis. Pressure should change very little when water moves into, or out of, a rock unit of high transmissibility because such a unit equalizes pressure rapidly.

BRUCE B. HANSHAW

References

Back, W., and Hanshaw, B. B., 1965, "Chemical Geohydrology," in (Chow, V. T., editor), "Advances in Hydrosciences," Vol. 2, pp. 49–109, New York, Academic Press.

Berry, F. A. F., 1959, "Hydrodynamics and Geochemistry of the Jurassic and Cretaceous Systems in the San Juan Basin, Northwestern New Mexico and Southwestern Colorado," unpublished Ph.D. thesis, School of Mineral Sci., Stanford University.

Berry, F. A. F., and Hanshaw, B. B., 1960, "Geologic evidence suggesting membrane properties of shales," *21st Internat. Geol. Cong. (Copenhagen 1960), Abstr. Vol.*, p. 209.

Bredehoeft, J. D., Blyth, C. R., White, W. A., and Maxey, G. B., 1963, "Possible mechanism for concentration of brines in subsurface formations," *Am. Assoc. Petrol. Geologists Bull.*, **47**, 257–269.

Graf, D. L., Friedman, I., and Meents, W. F., 1965, "The origin of saline formation waters," 2, "Isotopic fractionation by shale micropore systems," *Illinois State Geol. Surv. Circ.* **393**.

Graf, D. L., Meents, W. F., Friedman, I., and Shimp, N. F., 1966, "The origin of saline formation waters," 3, "Calcium chloride waters," *Illinois State Geol. Surv. Circ.* **397**.

Hanshaw, B. B., 1962, "Membrane properties of compacted clays," unpublished Ph.D. thesis, Harvard University.

Hanshaw, B. B., 1964, "Cation-exchange constants for clays from electrochemical measurements," pp. 397–421 in (Bradley, W. F., editor), "Clays and Clay Minerals, 12th Conf.," New York, Pergamon Press, 692pp.

Hanshaw, B. B., and Hill, G., 1969, "Geochemistry and hydrodynamics of the Paradox Basin region, Utah, Colorado, and New Mexico," *Chem. Geol. J.*, **4**, 263–294.

Hanshaw, B. B., and Zen, E-an, 1965, "Osmotic equilibrium and overthrust faulting," *Geol. Soc. Am. Bull.*, **76**, 1379–1386.

Kemper, W. D., 1960, "Water and ion movement in thin films as influenced by the electrostatic charge and diffuse layer of cations associated with clay mineral surfaces," *Soil Sci. Soc. Am. Proc.*, **24**, 10–16.

Kemper, W. D., 1961, "Movement of water as affected by free energy and pressure gradients," II. "Experimental analysis of porous systems in which free energy and pressure gradients act in opposite directions," *Soil Sci. Soc. Am. Proc.*, **25**, 260–265.

McKelvey, J. G., and Milne, I. H., 1962, "The flow of salt solutions through compacted clay," pp. 248–259, in (Ingerson, Earl, editor), "Clays and Clay Minerals," New York, Pergamon Press, 614pp.

Washburn, E. W. (editor), "International Critical Tables, 1926–1933," New York, McGraw-Hill Book Co., 616pp.

White, D. E., 1965, "Saline waters of sedimentary rocks," pp. 343–366, in (Addison Young and J. E. Galley, editors), "Fluids in Subsurface Environments," Tulsa, Okla., Am. Assoc. Petrol. Geologists.

Wyllie, M. R. J., 1948, "Some electrochemical properties of shales," *Science*, **108**, 684–685.

Wyllie, M. R. J., 1949, "A quantitative analysis of the electrochemical component of the S. P. curve," *Am. Inst. Mining Met. Engrs. Trans.*, **186**, 17–26.

Wyllie, M. R. J., 1951, "An investigation of the electrokinetic components of the self potential curve," *Am. Inst. Mining Met. Engrs. Trans.*, **192**, 1–18.

Wyllie, M. R. J., pp. 282–305, 1955, "Role of clay in well-log interpretation," in (Pask and Turner, editors), "Clays and Clay Technology," *Calif. Dept. Nat. Resources, Div. Mines Bull.*, **169**, 326pp.

Young, A., and Low, P. F., 1965, "Osmosis in argillaceous rocks," *Am. Assoc. Petrol. Geologists Bull.*, **49**, 1004–1008.

Cross-references: *Bonding; Electrolytes: Flocculation of Colloids; Natural Brines; Thermodynamics.* Vol. IV B: *Montmorillonite.* Vol. V: *Clay-Shale-Slate-Schist.* Vol. VI. *Evaporites; Oil Wells.*

CLAY MINERALS—BASE EXCHANGE

Base exchange is the substitution of one cation for another on the surface or in the interstices of a suitable host crystal. This phenomenon is not confined to clay minerals, but is common to certain clays, zeolites, and organic resins.

Although it is theoretically possible to have both anion and cation exchange occurring at various sites on clay flakes, the number of available sites for cation exchange is so much greater than those for anion exchange that attention is focused primarily on studies of cation or base exchange. There are two essential requirements for cation exchange (and the less common anion exchange): (a) an imbalance of electrical charge in the crystal; (b) A surficial geometry permitting the adsorption of ions of suitable charge onto the crystal. These requirements are met by some of the clay minerals. Although it is possible to have some ions adsorbed on any clay mineral, base exchange is most commonly encountered in smectites, vermiculites, and, to a much lesser extent, in illites. These minerals all have a residual charge on their unit cell, i.e., they invariably have a deficiency of positively charged atoms in their structures. Atomic substitution may occur in other minerals following Goldschmidt's Rules, but these are minor and merely departures from the ideal case. For example, in kaolinite some substitution of Al^{3+} for Si^{4+} in tetrahedral sites may and does occur, but this is an aberration. The ideal structure of kaolinite is characterized by an exact electrical balance between positively charged aluminum (Al) and silicon (Si) ions and negatively charged oxygen (O) and OH ions.

Smectites have a residual charge on a layer unit cell of 0.7, vermiculite a charge of 1.3, and illite a charge of 1.8. Mica minerals have a residual charge of 2 on a layer unit cell, whereas brittle micas have a residual charge of 4 (Mackenzie and Mitchell, 1966). These unsatisfied negative charges on the crystal are satisfied by ions which locate themselves in appropriate sites on the basal surfaces of the sheet silicates.

The small particle size of clay minerals (generally observed to be less than 2 microns in maximum dimension) is another important contributor to their high base exchange. Any mineral may have a defect structure causing a residual charge on the crystal and may possess surfaces with favorable sites for ion adsorption, but it is the number of such sites in a given quantity of the material which is important. Clays may be loosely described as being "all surface"; smectites may have a surface area as great as $800 m^2/g$ (Van Olphen, 1963). Ions have great difficulty penetrating into the interior of most tectosilicates, for example, but even a large crystal of any clay can be relatively easily penetrated along the (001) surface.

When a clay is formed from preexisting minerals, either by hydrothermal solutions or by low-temperature weathering, it reflects the environment of its formation. "Ex nihil, nihil fit!" The composition of the material altered to form the clay is of great importance. Just as a pure quartz sandstone cannot be metamorphosed into a biotite schist, no clay can be formed which does not contain the elements of the ancestral minerals and altering fluids. It is possible to form a clay which lacks some of the elements originally present in the mineral which was altered. But no clay or other mineral can be formed by elements which were missing from the environment at the time of formation. Therefore, generalizations about the environment of formation of clays must always reflect the necessity of accounting for the source of all the constituent atoms.

Once the crystal is formed, substitution of elements within the layers of the crystal will not take place without destruction of the crystal itself. The small particle size of the clays and relatively high reactivity resulting from large surface areas permits such destruction to take place comparatively easily. Therefore, clays are good indicators of their most recent environment rather than of condititons when the original igneous or metamorphic rock altered.

It is not necessary to destroy completely a clay crystal to alter its character dramatically. The interlayer ions are subject to the *law of mass action* (Toth, 1964). They are most subject to variation with changes in the chemistry of the currently dominant environment. This environment is not necessarily the most recent fluctuation of local conditions, but the conditions which prevailed over a long enough time period for an approximate state of equilibrium to be attained. For example, aqueous solutions passing through limestones will carry calcium ions to adjacent clays (as in the fertile soils overlying limestone bedrock in Pennsylvania, Ohio, etc.). Clays which wash into the ocean will be subject to ion exchange with the dominant sodium (Na) ions of the sea. These are less drastic environmental changes than those which would totally destroy a clay.

No ion exchange can take place unless a medium for transfer is available. Almost everywhere clays are associated with water, which serves as the medium for transfer. Although some water may be held on the surface of clay in a rigid structure, most molecules will be in the liquid state and free to migrate through the soil, sediments, etc. This water will vary in pH and ionic composition with conditions in the environment surrounding the clay particle. There is no single rule for ion exchange which holds for all clays and all conditions. In general, the law of mass action will apply: when an ion predominates in the surrounding fluid, it will tend to become the dominant exchange ion on the clay. Various rules have been propounded at different times concerning the replacing power of various cations and anions on clays, but these vary for different clays, solution concentrations, pH, etc. (Grim, 1968).

Smectites have a high cation-exchange capacity (80–150 meq/100 g, Grim, 1968). It combines a charged unit cell with a very fine (40% less than 0.06 microns, Salmang, 1961) particle size. The basal surfaces of smectites are negatively charged and adsorb cations onto "holes" in the surface hydroxyl sheet. The edges of the smectite flakes are positively charged. These clays form complex gel structures with water. Their behavior can be understood on the basis of the double-layer theory (Verwey and Overbeck, 1948), which suggests that ions are adsorbed on the surface of clay particles to satisfy the electrical imbalance locally present in the crystal. According to Stern's refinement of the work of Helmholtz and Gouy the negatively charged surface of the clay is surrounded by a rigidly held cloud of cations which neutralize the charge on the clay. Steric considerations prevent complete balancing of the charge by this rigidly bound layer, however, and cations are attracted to the neighborhood of the particle to a diminishing extent with increasing distance from the surface of the particle (Fig. 1).

Polar molecules, such as water, are also oriented as a result of the residual charge on the clays. The swelling observed in some clays, e.g., Na smectites, has been attributed by some authors to the formation of solvation hulls of water around the adsorbed exchange ions. Others feel that it is merely dissociation of cations which pries apart clay layers in water, causing them to form double layers around individual clay layers rather than around the several layers composing a crystal.

Geologic Applications

When minerals are weathered to form clays, exchangeable ions are present on the clay surfaces. These ions will change when groundwater brings in other ions to replace the original constituents. Stream erosion transports clays in dilute solutions of ions which vary with local conditions. Deposition in lakes or oceans will introduce further variations in the

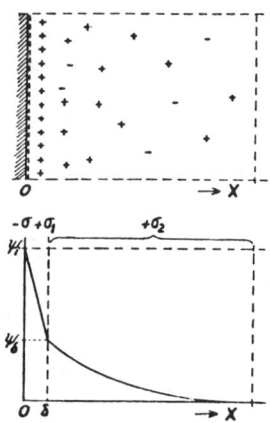

FIG. 1. Schematic representation of ion and potential distribution in the double layer according to the theory of Stern. ψ_t denotes total potential; ψ_δ, zeta potential; χ, distance from particle surface; σ, surface charge density; δ, thickness of Stnern layer.

chemical composition of the environment. Syngenetic reactions, diagenesis, and epidiagenesis will all affect the ions adsorbed on the surfaces of clays. The ions on the edges of clay flakes will be more easily varied than those held between the basal surfaces.

So long as the clays are at relatively low temperatures exchange of cations will take place easily. Temperatures above 100°C will result in stronger bonds forming between the cations and the clays with resultant decrease in cation-exchange capacity. In general, all effects are more pronounced in particles of finer size.

Although pH varies considerably in the environments in which clays are found, the clay mineral type and composition of ions in the water present are much more significant than the effect of the pH (Wiklander, 1964). Formation of new minerals under more severe conditions of change results in kaolinites in leaching acid environments and three-layer clays where alkaline environments lead to free silica in solutions.

The single most striking example of the effect of base exchange in sediments is the fixation of potassium (K) ions on the surfaces of clays. Because K ions fit almost exactly into the "holes" on the hydroxyl surfaces of the clays, they are extremely difficult to dislodge once they are emplaced.

Terrestrial clay minerals are found in primary and secondary deposits. The primary deposits are those of clays in situ, which generally includes both clays formed by low-temperature weathering and those formed by hydrothermal alteration. Secondary clays are those which have been subject to some transportation after formation.

Clays formed by the weathering of igneous or metamorphic rocks are close in chemical composition to the original rock. If there is little movement of water through the soil the clays will contain the alkalis and alkaline earths originally present. But if high rainfall, topography, and temperature permit a relatively rapid flow of soil water, the removal of these ions will permit even a basic rock to form a variety of clays, including kaolinite. Conversely, a granite may form more basic clays if the appropriate ions are introduced by groundwaters. The acid pH caused by plant roots may favor the formation of kaolin, whereas alkaline soils tend to contain smectite or illite depending upon the presence of K, magnesium (Mg), or Ca.

In a cold, wet climate with good drainage and an acid environment (due to plant activity) the first clays formed reflect the composition of the original rock. If these clays contain alkalis or alkaline earths, they are altered to kaolinite by the removal of those ions. Finally, the kaolinite is decomposed; Al and iron (Fe) are removed by groundwater and Si is concentrated near the surface. In a hot wet climate with no organic acids present and a neutral or slightly alkaline environment (lateritic alteration), the early stages of alteration are the same, but it is the Si which is carried away and Fe and Al are residually concentrated.

In arid regions there is little decomposition of silicates and consequently little formation of clays. Because there is little removal of ions by groundwaters, illites and smectites are the clays commonly found in arid soils.

In areas of hydrothermal alteration the types of clays formed reflect the composition of the country rock and the hydrothermal fluid. In general, the zones of more intense hydrothermal alteration are characterized by clays of the kaolinite group whereas smectites and illites are found in zones of less intense alteration.

Secondary deposits of clays contain altered clays. During transportation in rivers alkaline clays may be leached to form kaolin. But unless time in the river is unusually long, the dominant clay in a particular river will tend to reflect the dominant clay in the areas drained by the river.

When deposition takes place in lakes, the clay minerals vary with conditions. In freshwater lakes with little Ca and appreciable movement of water through the sediments, leaching leads to the formation of kaolinite. When Ca is present, kaolinite does not form. In playa lakes the extremely high ion content causes formation of sepiolite-attapulgite rather than illite and smectites.

Glacial deposits of sediments frequently contain unaltered rock flour of colloidal size. Illite and chlorite are the most common clay minerals.

In a marine environment *syndiagenesis* results in the slow alteration of kaolin to illite (by potassium fixation) or to chlorite. The extra Si required is supplied by the Si dissolved in the oceans. According to Harriss (1967), ions in marine sediments are not in equilibrium with the surrounding water. Therefore, this process does not go to completion. The mineral composition of the sea floor reflects the clay mineralogy of the closest land mass: smectites in volcanic areas, illites and chlorites in polar regions, and kaolinites (changing to illites by K-fixation), illites, and smectites near continental land masses. The Ca present in marine waters prevents the formation of kaolinite. The Mg present favors the formation of smectites. Potassium ions are adsorbed onto clays to hasten alteration to illites. Acid conditions should favor formation of kaolinite, but dominant K-fixation negates this.

Anadiagenesis occurs in an alkaline environment which is losing water content. This favors formation of three-layer clays over two-layer clays (smectites and illites over kaolinites). The K-fixation favors formation of illites from smectites. With increasing compaction the tubular minerals (e.g., halloysite, sepiolite-attapulgite) are not found. When conditions approach metamorphism illite and chlorite are formed with increasingly perfect and large crystals.

Under conditions of *epidiagenesis,* clay-rich rocks weather in a manner similar to other rocks. They form new clays when conditions are out of equilibrium with their formational environment. Clay minerals are unchanged when environmental conditions do not fluctuate.

EMMY BOOY

References

Grim, R. E., 1968, "Clay Mineralogy," Second ed. New York, McGraw-Hill Book Co. 596pp.
Harriss, R. C., 1967, "Clay minerals and oceanic evolution," in "Clays and Clay Minerals, Proc. 15th Natl. Conf.," pp. 207–214, New York, Pergamon Press.
Mackenzie, R. C., and Mitchell, B. D., 1966, "Clay mineralogy," in "Earth Science Reviews," pp. 47–91, Amsterdam, Elsevier Publ. Co.
Salmang, H., 1961, "Ceramics, Physical and Chemical Fundamentals," London, Butterworths, 380pp.
Toth, S. J., 1964, "The Physical Chemistry of Soils," in (Bear, F., editor), "Chemistry of the Soil," Second ed., pp. 142–162, New York, Reinhold Publ. Corp.
Van Olphen, H., 1963, "An Introduction to Clay Colloid Chemistry," New York, Interscience Publ., Div. of John Wiley & Sons, 301pp.
Verwey, E. J. W., and Overbeck, J. Th. G., 1948, "Theory of the Stability of Lyophobic Colloids," Amsterdam, Elsevier Publ. Co., 205pp.
Wiklander, L., 1964, "Cation and Anion Exchange Phenomena," in (Bear, F., editor), "Chemistry of the Soil," Second ed., pp. 163–205, New York, Reinhold Publ. Corp.

Cross-references: *Aqueous Solutions; Bonding; Colloids; Hydrothermal Alteration; Hydrothermal Solutions; Ion Exchange; pH-Eh; Weathering, Chemical.* Vol. VI: *Diagenesis; Leaching; Weathering, Organic.*

CLIMATIC VARIATIONS—See Vol. II

COAL—See Vol. IV B

COAL GASIFICATION AND LIQUEFACTION—See Vol. IV B

COASTAL TERRAIN GROUNDWATER—See HYDROLOGY, COASTAL TERRAIN

COBALT: ELEMENT AND GEOCHEMISTRY

Physical and Chemical Properties

Cobalt, symbol Co, atomic number 27, atomic weight, 58.9332, is a hard, bluish-white metal with a high melting point (1490°C) and boiling point (3100°C). The metal is ferromagnetic, as are the other Group VIII "triad" elements, iron and nickel. The classical inclusion of these three elements in Group VIII of the Periodic Table is perhaps misleading; they are all transition elements and exhibit the properties typical of that group. Cobalt has the electronic configuration $1s^2 2s^2 2p^6 3s^2 3p^6 3d^7 4s^2$ and only one stable nuclide (Co^{59}) is known to exist in nature.

The chemistry of cobalt is well-known, it being an important member of the much-investigated first transition series. The majority of its chemical properties can be related directly to the presence of the incomplete d-electron shell in all compounds. Two valence states (Co^{II} and Co^{III}) account for the vast majority of these. In aqueous solution, redox potential data show that complex formation radically affects the relative stabilities of the Co^{II} and Co^{III} states. Two couples are cited as examples:

$$(Co(H_2O)_6)^{2+} = (Co(H_2O)_6)^{3+} + e$$
$$E^0 = -1.84V$$
$$(Co(NH_3)_6)^{2+} = (Co(NH_3)_6)^{3+} + e$$
$$E^0 = -0.1V$$

Co^{III} is thus seen to be greatly stabilized by the presence of nitrogen "donor" molecules. In this valence state it is known to have a considerable affinity for such molecules, and the resultant complexes are particularly inert. Hence, although Co^{3+} cannot exist in simple aqueous solution (water being decomposed), the presence of suitable complexing molecules can cause the trivalent state to become more stable.

Geochemistry

Recent analyses of meteorites show that cobalt follows iron and nickel into the metallic iron and sulfide phases, the nickel:cobalt ratio being about 17:1. In stones, however, the picture is complicated by the fact that much of the metal apparently resides in small troilite inclusions. Sampling is hence difficult and meaningful abundance figures are virtually unattainable.

The crust of the earth as a whole contains about 25 ppm cobalt and this is mostly concentrated in ultrabasic and associated rocks of magmatic origin. Values of over 100 ppm are not uncommon. The bulk of the element is found in the olivines and pyroxenes of these rocks, but the greatest concentrations occur in related sulfides bodies, these providing the majority of workable ore deposits.

In the more acidic series, cobalt concentrations are normally very low, the general trend during differentiation being one of depletion in rocks of increasing silicic composition. This trend is much less obvious, however, than that exhibited by nickel.

In the processes of weathering and sedimentation cobalt shows no very marked differentiation. Since the simple trivalent state is unstable in water, the reactions that lead to the formation of sedimentary oxidate ores of manganese and iron ore are prohibited for cobalt. Conversely, reduzate sediments concentrate cobalt, but this is simply due to the precipitation of insoluble sulfides. Secondary enrichment in sedimentary (and other) sulfides is also plausible, cobalt showing a marked affinity for sulfur.

Resistate sediments are generally poor in cobalt, that which is present presumably representing an initial concentration in magmatic iron minerals.

The mean abundance figure for shales (19 ppm) is near the continental crustal mean of 25 ppm, but deep-sea clay sediments have been reported to contain much more (70–75 ppm). High mobility in solution, with consequent concentration in marine deposits is to be expected as a result of the low ionic potential of the simple Co^{2+} ion (2.8).

Cobalt is the outstanding example of a trace element essential to life. Deficiencies in the available cobalt content of soils lead to the nonactivation of certain enzymes. Vitamin B_{12} is one such enzymic compound and is a co-ordination complex of Co^{III} based on the tetrapyrrole ring structure. Certain marine algae are also known to concentrate cobalt from seawater (enrichment factors of up to 2000 have been reported).

In soils, weathering mechanisms are extremely complicated. Certain bacteria, for example, are known to liberate cobalt already complexed as chelate compounds and this provides a mechanism whereby formally trivalent cobalt can be brought into solution and transported away as stable amino acid complexes. Although very little, as yet, is known of these reactions, they may well play significant roles in the sedimentary cycles of transition elements. If this proves to be the case, the geochemical technique of inferring palaeoenvironments from a thermodynamic study of ancient mineral assemblages must be partially invalidated. Phase stability and solubility relationships are just not obeyed by living communities of microorganisms and again ample evidence is coming forward that ancient sediments were as microbiologically active as are modern ones (Moore, 1964).

Cobalt, although a relatively rare element, is one upon which the world's technical economy is partially based. It is usually obtained as a by-product of the smelting of sulfide copper ores. It is won from the resultant slags. The Katangan Copper belt produces the greater part of the world's current annual production of about fifteen thousand tons.

C. D. Curtis

References

*Mero, J. L., 1965, "The Mineral Resources of the Sea," Amsterdam, Elsevier, 312pp.

Moore, L. R., 1964, "The microbiology, mineralogy and genesis of a tonstein," *Proc. Yorkshire Geol. Soc.* **34,** 235–292.

Warren, H. V., and Delavault, R. E., 1957, "Biogeochemical prospecting for cobalt," *Trans. Roy. Soc. Can., Sect. III,* **51,** 33–37.

Young, R. S., 1957, "The geochemistry of cobalt," *Geochim. Cosmochim. Acta,* **13,** 28–41.

* Additional bibliographic references may be found in this work.

Cross-references: *Aqueous Solution; Chelation; Complexes; Nickel; Organic Geochemistry; Sulfides; Trace Elements.* Vol. VI: *Pelagic Sediments; Weathering, Organic.*

COLLOIDS

The colloidal state represents a condition midway between solution, where the dissolved particles are of ionic or molecular dimensions, a few angstrom units in diameter, and particulate suspensions in which the particles are several microns in diameter and separate from the suspending medium by gravitational settling. No well-defined boundary is possible however, as some ions or molecules are actually in the colloidal particle-size range, and the rate of settling of a suspension can be influenced by various factors. Frequently the colloidal size range is defined as extending from particles about $1\ \mu$ in diameter to particles about $0.001\ \mu$ in diameter. The part of colloid science of interest in geochemistry is mainly that concerned with aqueous suspensions of mineral particles or organic matter, and the interactions between particles and solutes.

Colloidal suspensions in water that are of

importance in geochemistry represent a means by which material is transported in water. Chemical and physical reactions are promoted in such systems, partly owing to the large area of solid-liquid interface characteristic of systems containing many small particles. As a given weight of solid is subdivided into smaller and smaller particles, the surface area increases. The colloidal particles normally carry an electric charge at their surfaces, and a layer of oppositely charged ions from the solution will generally tend to accumulate there.

The electrostatic forces that cause colloidal particles to attract or repel each other have a counterpart in forces of attraction between smaller particles such as ions or molecules. However the interionic forces are much more intense. Attractive forces between molecules, which do not involve effects of chemical bonds, are relatively weak. These intermolecular forces are commonly called van der Waals forces. Many inorganic structures however possess unusually strong surficial charges as a result of the arrangement of their interionic chemical bonds.

The surfaces of minerals may therefore strongly attract ions or molecules. Clay minerals, especially, have areas in their crystal lattices where there is a deficiency of cationic charge. Cations from solutions in contact with clays may be strongly adsorbed on the clay surface at these regions of charge deficiency or along the edges of mineral particles where crystal bonds between ions have been disrupted. Clay minerals characteristically occur in particle sizes near or within the colloidal range. The amounts of adsorbed cations which can be retained by clay surfaces (exchange capacity) depend principally on the nature of the clay, but also to a considerable extent on the amounts and kinds of ions available for adsorption. The exchange capacity of kaolinite, for example, is generally only a few milliequivalents per hundred grams of clay. Montmorillonite, on the other hand, may have an exchange capacity greater than 100 milliequivalents per hundred grams. Ions with a small hydrated radius and high charge are adsorbed preferentially to ions with large radius and low charge. Therefore, for a mineral in contact with solutions that contain cations, adsorbed ions are exchanged for ions in solution until a state of equilibrium is reached. A change in the proportions of calcium and magnesium to sodium in the solution, for example, will be reflected by a change in proportion of adsorbed ions on clays in contact with the solution.

Reactions of this type tend to influence the proportions of monovalent to divalent cations in solution in natural waters in the soil, in groundwater in aquifers, and in streams. The physical properties of soils also are influenced by the proportions of monovalent to divalent adsorbed ions on the mineral surfaces.

A survey of published literature on ion exchange by Robinson (1962) presents many of the methods by which ion-exchange equilibria have been evaluated. An explanation of the preferential exchange behavior of ions and natural adsorbing substances can probably be developed from the model proposed by Eisenman, Rudin, and Casby (1957) to explain the behavior of ions toward glass-membrane electrodes.

In some instances ion-exchange equilibria can be evaluated satisfactorily by equations of the law of mass action. However no fully satisfactory theoretical treatment is available that can be applied over a wide concentration range. Empirical equations of various types can be utilized over shorter concentration ranges. For some kinds of systems, distribution coefficients calculated from the equation

$$K_d = \frac{f_s V}{(1 - f_s) M}$$

where f_s = fraction of ion adsorbed; V = volume of solution, M = mass of adsorbent, and K_d = distribution coefficient, may provide a satisfactory means of summarizing the ion-exchange behavior.

Particles of metal hydroxides in and near the colloidal-size range may carry surface charges that are reversible in sign, depending on the pH of the solution. Ferric hydroxide, for example, forms a suspension of positively charged particles at pH levels below about 5.5, and the charge changes to negative at higher pH. A mechanism is indicated here for coprecipitation of anions such as phosphate and arsenate when ferric hydroxide deposits form at low pH, and for coprecipitation of metal cations such as cooper when the deposits form at a pH near neutrality.

The electrical potential between the outermost adherent layer of water molecules around a charged particle and the adjacent solution is termed the zeta potential. By subjecting colloidal suspensions to an electrostatic field and observing the behavior of the suspended particles through a microscope, the sign of the charge and the relative intensity of the zeta potential can be estimated (see Fig. 1).

Bacteria and organic material found in natural water sometimes behave as colloidal suspensions. The nature of the organic matter which lends a pronounced yellow or brown color to many natural waters is still imper-

COLLOIDS

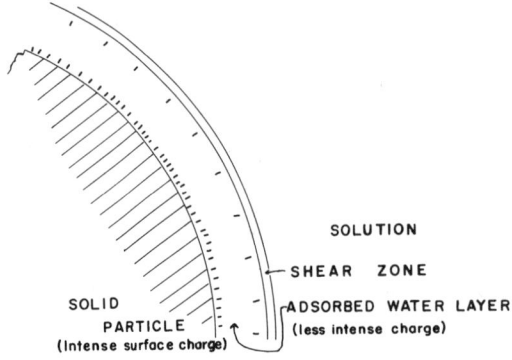

Fig. 1. Section through surface of charged colloidal particle. Adsorbed water layer is rigidly bound to solid; movement occurs at shear zone.

fectly known. However recent studies (Black and Christman, 1963a, 1963b; Lamar and Goerlitz, 1963; Shapiro, 1957) indicate that much of the material consists of high-molecular-weight carboxylic acids. The adsorption or complexing action of this material on inorganic ions is often important. Complexes of oxidizable material, such as ferrous iron, with organic molecules may lend stability to the reduced form of the metal in an oxidizing environment where precipitation would normally be expected.

JOHN D. HEM

References

Black, A. P., and Christman, R. F., 1963a, "Characteristics of colored surface waters," *J. Am. Water Works Assoc.,* **55,** 753–770.
Black, A. P., and Christman, R. F., 1963b, "Chemical characteristics of fulvic acids," *J. Am. Water Works Assoc.,* **55,** 897–912.
Eisenman, G., Rudin, D. O., and Casby, J. U., 1957, "Glass electrode for measuring sodium ion," *Science,* **126,** 831.
Lamar, W. L., and Goerlitz, D. F., 1963, "Characterization of carboxylic acids in unpolluted streams by gas chromatography," *J. Am. Water Works Assoc.,* **55,** 797–802.
Robinson, B. P., 1962, "Ion exchange minerals and disposal of radioactive wastes—a survey of literature," *U.S. Geol. Surv. Water Supply Paper,* **1616,** 132pp.
Shapiro, J., 1957, "Chemical and biological studies on the yellow organic acids of lake water," *Limnol. Oceanog.,* **2,** 161.
Wahlberg, J. S., and Fishman, M. J., 1962, "Adsorption of cesium on clay minerals," *U.S. Geol. Surv. Bull.,* **1140A,** A1–A30.

Cross-references: *Aqueous Solutions; Bonding; Ion Exchange; Oxidation and Reduction; pH-Eh; Van der Waals Force.* Vol. IV B: *Clays and Clay Minerals.* Vol. VI: *Soil Genesis.*

COLOR (OF MINERALS)—See Vol. IV B

COLORIMETRY

Colorimetry is a method of chemical analysis which deals with the measurement of the light absorption by colored solutions. Since light absorption depends upon the concentration of a specific constituent in solution, colorimetry is frequently used by geologists to determine qualitatively the trace quantities of many elements.

The fundamental principle of colorimetry states that the amount of light absorbed by a given substance in solution is proportional to the intensity of incident light and to the concentration of the absorbing species. This is expressed mathematically in the *Lambert-Beer law:*

$$\log I_0/I = abc$$

where I_0 = intensity of incident light
I = intensity of transmitted light
a = absorptivity of the substance
b = light path length
c = concentration of colored substance
I/I_0 = transmittance
$\log I_0/I$ = absorbence.

The term colorimetry is generally restricted to the visual comparison and matching of the color of a standard solution with that of an unknown one, whereas *spectrophotometry* involves the use of a photoelectric cell which measures a narrow band of wavelengths for transmittance.

Visual colorimetry is a simple method and is fairly precise. Essentially it requires the matching of the color of a standard solution with that of an unknown sample so that when they become identical, they must contain the same amount of colored substance in columns of equal cross-section. At this point $C_x b_x = C_s b_s$ and $C_x = C_s b_s / b_x$ where C_x = concentration of unknown solution; b_x = length of light path of unknown solution; C_s = concentration of standard solution; and b_s = length of light path of standard solution. A standard series of solutions is prepared, each with a known concentration, having the same volume as the unknown, and being contained in identical flat-bottomed tubes of equal diameter *(Nessler tubes).* The solutions should be compared in daylight and examined against a white background.

A more refined method uses the Duboscq colorimeter (Fig. 1), which has a dual matched optical system. Uniformly intense light is incident upon both colorimeter tubes and the difference in absorption of the standard and unknown solutions is compensated for by ad-

COLORIMETRY

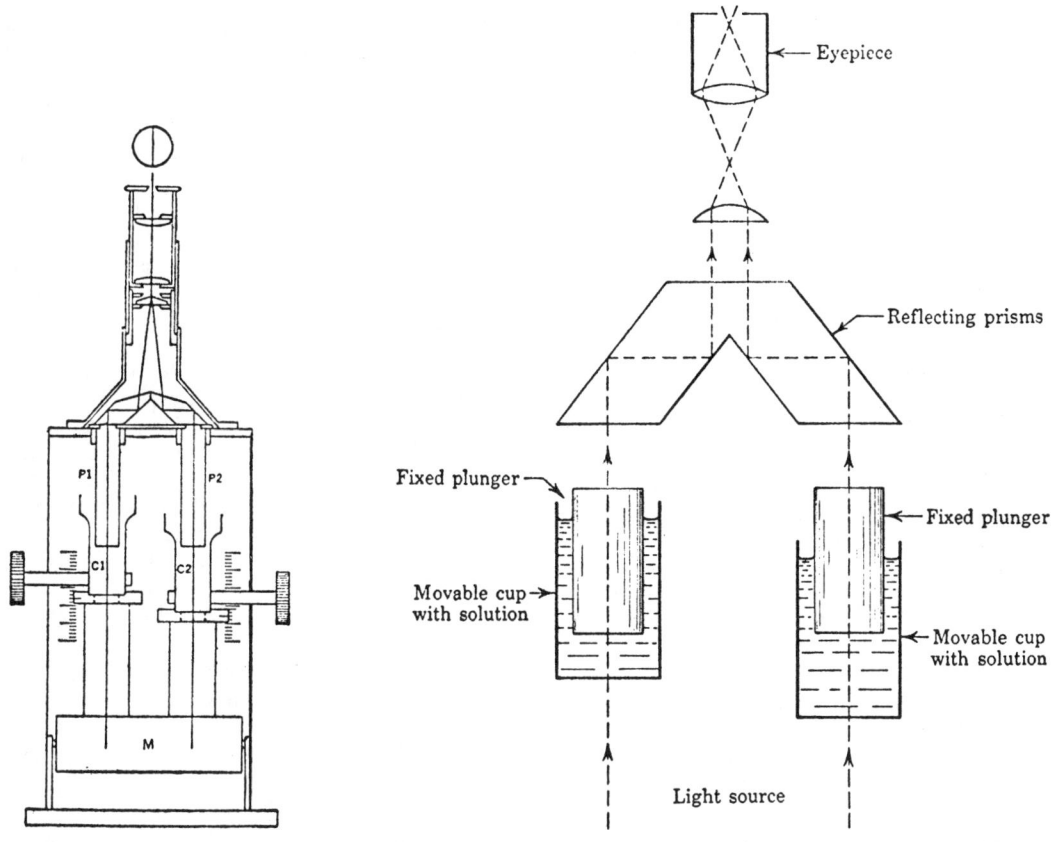

FIG. 1. (Left) Optical path in a colorimeter of the Duboscq type: $P1$, $P2$, plungers; $C1$, $C2$, cups to hold the solutions; M, mirror. The two halves of the field viewed through the ocular appear equally bright when a match has been obtained (from Willard, Merritt, and Dean, 1958). (Right) Simplified diagram of a Duboscq colorimeter (from Hamilton and Simpson, 1958).

justing the thickness of solution through which light passes. When the two colors match $C_x = C_s b_s / b_x$.

Spectrophotometry (q.v.) is the most precise method of measuring light transmittance. It involves the use of a photoelectric cell which measures a narrow band of wavelengths for transmittance, and a prism or grating to give monochromatic light. At each wavelength, three readings are made: (1) sample cell filled with solvent, which gives the value of I_o, (2) the cell filled with standard solution, and (3) cell filled with unknown. The ratio of meter readings of the cell filled with pure solvent and sample is the I_o/I ratio. Spectrophotometry may be used analytically to measure the absorption at a single wavelength, or to determine the whole absorption spectrum of liquids and solutions in the visible and ultraviolet regions.

Turbidimetry and Nephelometry

In a clear solution, the decrease in the intensity of transmitted light is caused mainly by selective absorption of certain wavelengths of light. In colorimetry and spectrophotometry, the loss of incident light by reflection is insignificant. Turbid solutions cause a loss in radiant energy by reflection, refraction, and absorption. The decrease in the intensity of incident light can be related to the concentration of suspended material. This measurement is called *turbidimetry*.

Very dilute suspensions scatter light and the degree of scattering is proportional to the amount of suspended material. *Nephelometry* consists of measuring the light scattered at right angles to the incident beam.

Sources of error in colorimeter methods include fading of color with time, changes in temperature and in light, and deviations from Beer's law. The reason for such deviation lies in the interaction of light-absorbing molecules in solution with each other or with foreign substances.

Colorimetry has widespread application in analytical chemistry and mineralogy. It is used

COLORIMETRY

especially to determine trace amounts of metals in metallic ores, silicate rocks, phosphate rocks, soils, seawater, iron, steel, and other alloys.

VIVIEN GORNITZ

References

Boltz, D. F. (editor), 1958, "Colorimetric Determination of Nonmetals," New York, Interscience Publ., 372pp.

Hamilton, L. F., and Simpson, S. G., 1958, "Quantitative Chemical Analysis," Eleventh ed., New York, Macmillan Co., 566pp.

Pierce, W. C., Sawyer, D. T., and Haenisch, E. L., 1958, "Quantitative Analysis," Chap. 21, New York, John Wiley & Sons.

Sandell, E. B., 1959, "Colorimetric Metal Analysis," Third ed., New York, Interscience Publ., 1032pp.

Snell, F. D., and Snell, C. T. (assisted by C. A. Snell), 1959, "Colorimetric Methods of Analysis—Including Photometric Methods," Vol. 2A, Princeton, N.J., Van Nostrand Co., 803pp.

Willard, H. H., Merritt, L. L., Jr., and Dean, J. A., 1958, "Instrumental Methods of Analysis," Third ed., Princeton, N.J., Van Nostrand Co., 626pp.

Cross-references: *Mineralogy; Spectrophotometry.* Vol. IV B: *Color in Minerals.*

COLUMBIUM—See NIOBIUM

COMPENSATION DEPTH—See Vol. I, Calcium Carbonate Compensation

COMPLEXES

A complex can be viewed simply as the statistical association of two or more ions in solution. This definition is chosen because of the lack of agreement about the bonding mechanism of aqueous inorganic complexes. A working restriction of this definition requires that the individual components of the complex be able to exist independently in solution. Usually a single cation and one or more anions or anionic radicals are involved. The coordination number of the central cation controls the maximum number of anions involved. The anions are mono- or polydentate, depending on the number of atoms in the anion which donate electrons (e.g., CO_3^{2-} is a bidentate ligand). There is no restriction on the resulting charge of a complex species.

Although complexes have long been invoked in analytical chemistry (e.g., transition metal ammine complexes), their significance in geochemical processes has only recently been appreciated. Problems such as the very low apparent activity of CO_3^{2-} in seawater (Garrels and Thompson, 1962), transportation of sufficient relatively insoluble metal ions by hydrothermal ore solutions to form ore deposits (Helgeson, 1964), and solution of calcium carbonate by saline solutions apparently supersaturated with respect to calcite (Revelle and Emery, 1957), have all been analyzed in terms of the role of complexes in reducing the number of free ions considered in the equilibria.

The current analysis of seawater chemistry demands the inclusion of complex species involving the major ions present, such as $MgSO_4^0$, $CaHCO_3^+$, and $NaCO_3^-$ (Garrels and Thompson, 1962). Transportation of Pb^{2+} by hydrothermal solutions has been explained by the formation of soluble high-temperature complexes of Pb^{2+} and Cl^-, especially species such as $PbCl^+$ and $PbCl^{2-}$ (Helgeson, 1964).

The existence of complex species in solution is well accepted, although the bonding mechanisms are not as well agreed upon. The analysis of equilibria usually involves standard thermodynamic parameters, such as *enthalpy* and *entropy*.

JAMES SIMPSON

References

Garrels, R. M., and Thompson, M. E., 1962, "A chemical model for seawater at 25°C and one atmosphere total pressure," *Am. J. Sci.,* **260,** 57–66.

Helgeson, H. C., 1964, "Complexing and Hydrothermal Ore Deposition," New York, Macmillan Co., 128pp.

Revelle, R., and Emery, K. O., 1957, "Chemical erosion of beach rock and exposed reef rock," *U.S. Geol. Surv. Profess. Paper 260-T.*

Cross-references: *Aqueous Solutions; Bonding; Enthalpy; Entropy; Hydrothermal Solutions; Seawater Chemistry; Thermodynamics.*

COMPUTER APPLICATIONS IN EARTH SCIENCES—*See* Vol. VI

CONIOLOGY—*See* KONIOLOGY

CONNATE WATER

Water that is trapped in marine sediments at the time they are laid down in the sea is commonly called *connate water.* As the term implies, connate water is produced at the same time as the rock and constitutes a sort of fossil seawater.

When marine sediments are raised above sea level and subjected to the action of circulating meteoric water, their connate water and solutes are removed by leaching and flushing and carried back to the ocean. The flushing process however is slow in fine-grained and deeply buried sediments, and water which almost certainly owes its high content of dissolved ions to remnants of marine solutions is common in such environments. The water associated with petroleum commonly is saline and occurs in formations whose porosity and structure are generally unfavorable for extensive water circulation and coincident removal of solutes. Usually the salt dissolved in brines that occur in deeply buried rocks is considered to be of connate origin.

Many geological processes may have modified the composition of connate brines to produce their wide range. Obviously these alterations have been extensive. Some connate waters are essentially saturated or nearly saturated solutions of sodium chloride. Others contain large proportions of calcium as well as sodium and chloride. Although the composition of connate water gives few useful clues as to the composition of the ocean in past geologic periods, the ocean has evidently been rich in chloride for a very long time and brines in which anions other than chloride are predominant cannot logically be ascribed entirely to a connate origin.

Among the processes which might be expected to alter the concentrations of ions in connate water are the precipitation of solids such as calcite on mineral surfaces, the solution of rock minerals and evaporites, sorption and desorption of ions on solids, and the differential movement of water molecules and ions through clay and shale strata. The latter effect is equivalent to the ultrafiltration effect of certain types of membranes used in the reverse osmosis method of removing solutes from water. Under the high pressures encountered at great depths, water and ion movements may be very different from the ones expected at low pressure. These effects may possibly explain the high concentrations of solutes in some connate brines and the difference in ion content between such brines and ordinary seawater. Another factor of considerable importance in many places appears to be the biochemical reduction of sulfur from S^{6+} as found in sulfate ions to sulfur in the more reduced forms of free sulfur, polysulfides, or sulfide ions. The brines encountered in oil and gas fields, and the gases themselves, are often high in hydrogen sulfide content as a result of sulfate reduction.

Analyses of a variety of connate brines have been published by White, Hem, and Waring (1963). White (1957) has suggested criteria for distinguishing connate water by means of ratios of concentrations of certain of the dissolved ions to one another. In most connate water the ratios of bromide and iodide to chloride are relatively high and ratios of potassium and lithium to sodium are low.

Analyses tabulated under the topic *Aqueous Solutions* (q.v.) include one for an oil-field brine from the U.S.S.R. which is high in sodium and chloride, as is characteristic of many connate waters. However the composition of connate brines has a wide range. Some connate brines are important commercially as sources of certain elements, especially bromine, lithium, potassium, and magnesium.

JOHN D. HEM

References

White, D. E., 1957, "Magmatic, connate and metamorphic waters," *Geol. Soc. Am. Bull.* **68**, 1659–1682.

White, D. E., Hem, J. D., and Waring, G. A., 1963, "Chemical composition of subsurface waters," *U.S. Geol. Surv. Profess. Paper* **440F**, F30–F39.

Cross-references: *Meteoric Water; pH-Eh Relations; Seawater Chemistry; Sulfate Reduction—Microbial.* Vol. VI: *Marine Sediments.*

CONSERVATION

Conservationists warn of dangers ahead which threaten the continued prosperity and well being not only of the more advanced countries but of all nations of the world (Commager, 1963). These dangers evolve from man's misuse and abuse of his environment and the earth's natural resources on which he depends for survival.

From the beginnings of civilization, every nation's basic wealth and progress have stemmed, in large measure, from its natural resources (Anon., Natural Resources, 1962), many occurring in limited quantity and others having fixed maximum capacities for continued renewal and sustained yield. The conservation movement has evolved through a growing recognition of these problems. Conservationists are therefore concerned with the identification and evaluation of problems involving natural resources and man's environment and with seeking out the necessary knowledge and skills for their correct diagnoses and solutions.

Geology supplies much of the basic scientific information upon which sound conservation principles and practices are based. It is quite appropriate, therefore, to consider the contribution geology makes to conservation theory and to explore ways in which geology and the related earth and life sciences can better serve conservation in the future. Since the solutions of conservation problems depend upon various other sciences, particularly ecology, the relation of geology to them also merits attention.

Conservation is not a science, but, rather, an attitude of mind and a way of life embracing human values, human needs, and human morality in a program of conduct with respect to environment. Conservation is, therefore, concerned with both the condition of environment and human attitudes and behavior. It represents a blending of natural sciences with social conduct and responsibility.

Fosberg (1970) defines conservation as essentially the preservation of man's environment in a condition to fulfill his needs for a healthy and satisfying life. The late President Kennedy, in his 1962 Conservation Message to the Congress, conceived of conservation as the wise use of our natural environment and explained that this means the highest form of national thrift—the prevention of waste and despoilment while preserving, improving, and renewing the quality and usefulness of all our resources. Aldo Leopold (1949) envisioned conservation as a state of harmony between men and the land, a concept in which land includes soils, waters, and the biota as well as the people. He believed that this harmony could best be achieved by the practice of a land ethic having to do with what is right and what is wrong for both the land and the people living upon it.

These definitions constitute the fabric of conservation with material and designs representing the handiwork of the geologist, ecologist, sociologist, economist, philosopher, and others who cherish and strive for higher human goals and aspirations. In the aggregate, the definitions reflect a basic concept of man's admiration, respect, and love for the land. Good conservation practice can enrich human life through utilization of the cumulative knowledge and wisdom of the past and some vision of the future with respect to the wisest use of natural resources.

Man's Place in Geologic History

Primitive man, existing as a part of nature and not apart from nature, interfered but little with the relatively stable relationships within his environment. The subsequent rise in the level of his skills and powers of reason endowed man with a growing capability for altering his surroundings and for manipulating the processes of nature, on a scale comparable to that of a geologic force, for his own economic and social purposes. Knowledge and wisdom, however, have not kept pace with man's technological growth. Thus, by his very nature, man himself is the source of much of the difficulty he encounters in striving for successful social and environmental adjustment.

Geology is the history of the earth and its life which embraces a time span of about 4.6 billion years. This history is recorded in the form of fossils and in the composition and relationships of rocks. These yield clues about geo-dynamic processes and biologic evolution. Within any single human life span, the changes wrought by geologic processes, except for those of a catastrophic nature such as eruptions, earthquakes, and floods, are scarcely discernible. Mainly through the processes of weathering, erosion, sedimentation, and crustal deformation, the natural environments of any given moment of earth history have been transitory. Environmental changes are continuous and will continue into the future. Man's appearance on the scene has been a comparatively recent evolutionary occurrence.

Ecology deals with the interaction of all organisms among themselves and with the inanimate materials and forces of nature. Total environments necessarily embrace both their living and nonliving components (i.e., ecosystems). The recent attention to ecosystems has tended to develop stronger ties between geology and ecology, and closer relationships among all the earth and life sciences. All biologic species have achieved their evolution in relation to geologic history. As Bradley (1963) has observed: "Everything found in or on the earth came to be what it is, and where it is, by geological processes."

While organisms have been shaped to some extent by geology, the converse is also true. Much geology is also modified by organisms as illustrated by their role in the breakdown of rocks and their role in the creation of the earth's atmosphere.

Man's total environment is the most critical, but ignored, of our resources (Anon., "Natural Resources," 1962). Fortunately, an increasing awareness of the importance of understanding the balances of nature is reflected in the rapid development of interest in ecological studies. Man is drastically altering the balance of a relatively stable system, and he must concern himself with both the changes caused by human beings and those produced by natural influences and processes. Some changes may be very harmful. Conservationists warn of the foolishness of tampering with the environment without first striving to determine the long-term effects of our actions (Anon., "Natural Resources," 1962). These problems and their solutions are the concern of conservation and the natural sciences, particularly geology and ecology, can contribute no greater service to society than that of providing conservationists with the basic scientific information they require.

Leggett (1963) reminds us that many aspects of applied geologic science were appreciated and understood long before the beginning of the Christian Era. However, the early interest in earth sciences was not maintained during ensuing centuries and the nineteenth century was well underway before widespread interest was awakened.

The Environment in the United States

The development of the geologic and ecologic sciences in the United States paralleled the national growth. Early progress rested in men learned in fields such as medicine, law, literature, and philosophy, but untrained or meagerly self-trained in science (Merrill, 1924). The country itself, two centuries ago, was sparsely populated and young—politically and economically. About three million people, concentrated along the Atlantic seaboard, constituted but a distant frontier of the Old World. With new frontiers to conquer, the seemingly limitless wilderness to subdue, and an apparently inexhaustible storehouse of natural resources to exploit, it is little wonder that pleas for checking the unnecessary and reckless waste and despoilment of soils, forests, waters, and minerals were largely unheeded. Moreover, the body of accurate and adequate scientific information required to give meaning to these pleas was lacking.

Merrill (1924) summarizes the contributions made to geologic science by individual geologists and also by those attached to the various state and national surveys. State surveys were emerging as early as 1830. A milestone in both geologic science and conservation was achieved in 1879 with the consolidation of all national geological surveys under the newly established United States Geological Survey.

One of America's pioneer naturalists, William Bartram, is noted for his use of the ecological approach in field natural history investigations. His "Travels in North America," published in 1794, covers many facets of natural environments, including the soil, minerals, rocks, and climatology as well as his main work on botany and zoology.

A powerful advocate of the ecological approach to land use emerged in the person of George Perkins Marsh. His views are of interest to geologists because he was among the first to expound the thesis of man's role in changing the face of the earth—a thesis which had been relatively neglected until the past few decades. Marsh challenged the myth of unlimited resources which dominated American thought and action in his day (Udall, 1963). During the remainder of the nineteenth century, two geologists were at the forefront as leading spokesmen for conservation. One was Major John Wesley Powell, leader of one of the great National Surveys and second Director of the U.S. Geological Survey he had helped establish. The second was Nathaniel Southgate Shaler, Professor of Geology at Harvard University.

Powell emphasized that "renewable" resources are renewable only through management which prevents waste, preserves, and renews. His major work in this field, "Report on the Lands of the Arid Regions of the United States," published in 1878 was largely ignored until lately (Powell, 1957; Udall, 1963).

Shaler was the conservationist successor to Marsh in the English-speaking world (Thomas, 1956). In his book, "Man and the Earth," published in 1905, Shaler stressed society's moral and social responsibility to future generations with respect to the modification of natural environments and the exploitation of natural resources. Shaler emphasized that primitive man made very few demands on the earth's storehouse of natural resources, but with the expanding economy which accompanies cultural progress, these demands increase vastly. He linked the destruction of resources with impending shortages. Such intangibles as aesthetics and enlightened human attitudes in relation to human welfare and survival also loomed large in his concept of man's relation to environment. Shaler's thesis that the potential of water, wind, tides, and sun should receive attention and study as possible additional sources of energy encompasses much that is appropriate to modern conservation thought.

The basic scientific information and the skills of geologic science developed and improved phenomenally during the last hundred years. Accelerated progress can be expected in the future. During the same period, great pressures have been created by population growth, technological developments, and industrial and economic expansion.

As the accelerated drain on the world's finite supply of fossil fuels and basic minerals continues, symptoms of what has been termed the "climactic approach to exhaustion" are appearing (Ayres, 1956). Continued progress in gaining basic scientific knowledge and continued advances in applied technology, however, have the effect of creating new resources. Aluminum, for example, once of little consequence, has now become a resource of great value. Fossil fuels and minerals which were previously inaccessible, or could not be extracted and processed economically, can now be used. Nolan (1962) explains that with the use of progressively lower-grade ores, the tonnages of material available in the earth's crust increase geometrically with decreasing grade. With respect to many nonrenewable resources vital to man, however, finite limits in the usable resource base will gain in significance. This presents a challenge to geologists and scientists of allied disciplines to continue the scientific and technologic progress required to provide mankind with the necessities of life. In this progress, the importance of maintaining a condition of environment essential to man's physical, cultural, and spiritual well-being tends to be overlooked.

Conservation for the Future

Conservationists need the assistance of geologists and others to help determine just what are the conditions of environment essential for the fulfillment of man's needs for a healthy and satisfying life. These conditions will vary for every situation. However, the determination in each case can be made only if undisturbed samples of all principal types of natural environments are available as standards. Many types of natural environments are rapidly vanishing. Selections should be made from those that remain. Certain categories may have been destroyed already, but in some instances the re-creation of reasonable facsimiles of natural environments may be justified. Samples of natural environments as free as possible from man-induced changes will always be needed if man is to understand from whence he came and whither he is bound (Leopold, 1963).

National Parks and Reserves

Samples of natural environments still relatively free of man-induced changes occur in many of the scenic and scientific areas administered by the National Park Service. Geologic science, in fact, played an important role in their establishment. Yellowstone was so established in 1872 largely because of its geologic features reported on by the Hayden Expedition of the previous year. In like manner, the establishment of Mount Rainier National Park in 1899 was stimulated by geological studies some three years earlier. Other outstanding examples include Yosemite, Grand Canyon, Glacier, Kings Canyon and Hawaii Volcanoes National Parks, and Katmai and Glacier Bay National Monuments.

Today the undisturbed natural environments preserved within the scenic and scientific areas of the National Park System are being sought out by research scientists of many countries as unique outdoor laboratories and classrooms. These reserves provide scientists with unique opportunities. Many nations of the world have followed the lead of the United States by founding national park systems of their own. Thus, the concept is becoming universally recognized that national parks and other lands, preserved as ecological entities, can provide those natural environments so important in supplying the knowledge civilization will continue to need in an ever-increasing measure.

Other land management agencies of the U.S., government, in addition to the National Park Service, are placing increased emphasis on land inventory and classification to ascertain how public lands can best be used to meet the needs of the future. The concept of preserving some of these lands in their natural condition for their aesthetic, scientific and recreational values is now widely recognized. Besides government activity, recognition must be made of the accomplishments of private organizations and foundations, examples of which include The Nature Conservancy, the National Audobon Society, and the Sierra Club.

Education

Conservation education is generally recognized as one of the most important aspects of the movement. The appreciation and the understanding of natural environments and those processes operating within them receive primary emphasis in the interpretive programs available to the public in the units of the National Park System. Interpretive programs are likewise on the increase in the National Forests and in many of the state and municipal parks throughout the U.S. The extensive publications program of the Geological Survey deserves special mention.

Conservation education leading to a better public appreciation and understanding of natural environments and phenomena also engenders admiration, respect, and love for the land. In addition, there are bonus values which accrue to the people in the realms of physical, mental, and spiritual refreshment.

Although geologists can take justifiable pride in the contributions they have made to the cause of conservation, an even greater challenge lies ahead. The task is a continuing one which can never be regarded as finished. Nor can progress be judged as satisfactory until an adequate number of all representative types of primitive or re-created natural environments have been identified, evaluated, and placed under some system or other of adequate protection and preservation. The National Registry of Natural Landmarks, launched in 1964 by the Department of the Interior, is a step in that direction. The objective of the Registry is to encourage the preservation of sites of exceptional value in illustrating the natural history of America.

ROBERT H. ROSE

References

Anon. (National Academy of Sciences), 1962, "Natural Resources," *Acad. Sci. Natl. Res. Council Publ.,* **1000,** 1–400.

Anon., American Geological Institute, 1962, Report of the 1962 Educational Committee.

Anon., 1965, "Conserving the natural heritage," *Nature,* **208,** 1241–1244.

Anon., 1966, "National Parks Commissions," *Nature,* **209,** 1181–1182.

Anon., 1970, "Frontiers in Conservation," *Proc. Soil Conservation Soc. Am.,* Ankeny, Iowa, 162pp.

Ayres, E., 1956, in (Thomas, W. L., editor), "The Age of Fossil Fuels: Man's Role in Changing the Face of the Earth," Chicago, Univ. Chicago Press, pp. 367–381.

Bradley, W. H., 1963, in (Albritton, C. C., editor), "The Fabric of Geology," Reading, Mass., Addison-Wesley Publ. Co., pp. 12–23.

Commager, H. S., 1963, "The Country Beautiful," *Holiday,* p. 29.

Fosberg, F. R., 1970, in (Gray, P., editor), "Encyclopedia of the Biological Sciences," (Second ed.), New York, Reinhold Publ. Co., pp. 270–273.

Highsmith, R. M., Jr., Jensen, J. G., and Rudd, R. D., 1969, "Conservation in the United States" (Second ed.), Chicago, Ill., Rand McNally & Co., 322pp.

Jennings, J. N., 1965, "Man as a geological agent," *Australian J. Sci.,* **28**(4), 150–156.

Leggett, R. F., 1963, in (Albritton, C. C., editor), "The Fabric of Geology," Reading, Mass., Addison-Wesley Publ. Co., pp. 243–261.

Leopold, Aldo, 1949, "Sand County Almanac," Oxford Univ. Press, pp. 201–226.

Leopold, Luna, 1963, "A national network of hydrologic benchmarks," *U.S. Geol. Surv. Circ.* **460-B.**

Merrill, G. P., 1924, "The First One Hundred Years of American Geology," Preface and p.1, New Haven, Yale Univ. Press, 773pp. (1964, Fasc. of 1924 ed., Hafner Publ. Co.).

Nolan, T. B., 1962, "Role of the geologist in the national economy," *Geol. Soc. Am. Bull.,* **73,** 273–279.

Powell, J. W., 1957, "The Exploration of the Colorado River," Chicago, Univ. Chicago Press (1961, New York, Dover Publ. 400pp.).

Rose, R. H., 1966, "Public Seashores—Their Preservation and Use," *Proc. Internat. Union Conserv. Nature and Natural Resources,* Lucerne, Switzerland, 10th Ann. Meet.

Slayter, R. O., 1969, "Man's use of the environment—the need for ecological guidelines," *Australian J. Sci.,* **32**(4), 146–152.

Thomas, W. L., 1956, "Man's Role in Changing the Face of the Earth," Chicago, Univ. Chicago Press, 1193pp.

Udall, S. L., 1963, "The Quiet Crisis," New York, Holt, Rinehart and Winston, pp. 75–82.

UNESCO, 1968, "Conservation and rational use of the environment," *Nature and Resources,* 4(2), 2–5.

U.S. Govt. Printing Off., 1968, "Man . . . An Endangered Species?" Conservation Yearbook No. 4, 100pp.

Cross-references: *Ecology; Environmental Pollution; Environmental Science; Natural Resources; Soil Erosion.* Vol. I: *Marine Ecology.* Vol. III: *Anthropogenic Influences in Geomorphology; Organisms as Geomorphologic Agents.*

CONSTRUCTION MATERIALS—
See Vol. VI

CONTINENTS AND OCEANS—STATISTICS OF AREA, VOLUME, AND RELIEF—See Vol. III

COPPER: ELEMENT AND GEOCHEMISTRY

Element and Geochemistry

Physical and Chemical Properties. Man's knowledge of copper stems from the dawn of the Bronze Age and the metal has played an important part in all subsequent technological advances. It is a tough, soft, and ductile reddish metal with a thermal and electrical conductivity second only to silver. In nature it occurs both in the free state and combined with other elements in a host of different minerals. Two stable nuclides are known to exist: Cu^{63}, 69.1% and Cu^{65}, 30.9%.

The electronic structure of the free element is $1s^2 2s^2 2p^6 3s^2 3p^6 3d^{10} 4s^1$, the unpaired $4s$ electron being a consequence of the high stability of the completely filled $3d$ level.

The divalent oxidation state is best known in copper chemistry, although many sulfides appear to contain Cu^I rather than Cu^{II}. Much of the chemistry of Cu^{II} is governed by the steric effects of the d^9 electronic configuration (Cu^{2+} being the last transition ion in the first row of the transition elements). Regular octahedral complexes are unknown, Cu^{II} taking up either a square planar or an irregular octahedral configuration (with two much elongated axial bonds, the *"Jahn-Teller"* effect). The forces causing these distortions are equally operative in solid structures, and are the cause of much of the confusion that has arisen in the past concerning the radius of the Cu^{2+} ion.

Geochemistry. In meteorites, copper is restricted almost entirely to the metallic iron and the sulfide phases, virtually none having been reported in the silicates.

In the earth's crust, these relationships are confirmed by the well-marked association of copper and sulfur that is found in most rock types. Probably as a result of this, the occurrence of copper in, for example, ferromagnesian minerals has been overlooked too often.

Although concentrations may well be small, they are geochemically significant. The detailed petrological and mineralogical investigation of the Skaergaard intrusion, however, showed that concentrations of even up to 10^3 ppm occurred in pyroxenes. Wager and Mitchell (1951) found that incorporation of copper into silicate minerals increased with the progress of differentiation until the liquid phase became supersaturated with sulfur. They concluded that a sulfide magma separated as an immiscible liquid phase at that stage, removing the bulk of the copper from the silicate magma. Thereafter, the silicate minerals showed much lower copper contents, but the initial trends were again followed as a second sulfide concentration commenced.

The general geochemistry of sulfide magmas separating from differentiate series was admirably elucidated by Goldschmidt (1954). He recognized two main types, the first of which separated contemporaneously with basic rocks rich in olivine. Pyrrhotite, FeS, pentlandite, (Fe,Ni)S, and chalcopyrite, $CuFeS_2$, are the sulfide minerals characteristic of the resultant rocks. In the second type (high hydrothermal range) the resultant assemblages consist mainly of pyrite FeS_2 with subordinate chalcopyrite. Unlike the first series, nickel is absent, and cobalt and zinc are often present in considerable concentrations.

Copper is most commonly enriched in the late magmatic, hydrothermal stages of magmatic crystallization. The deposits of this widespread type provides about 60–70% of the world's copper production. It ranges from auto-hydrothermal deposits and disseminated porphyries to large veins and manto deposits in the surrounding country rock. Along with complex sulfides, the element also may occur in the native state, for example, in the lavas of Upper Michigan. This is one good example of the noble nature of copper, and of the coinage metals in general.

The behavior of copper during weathering and sedimentation may be compared usefully with that of zinc. Copper sulfides are readily oxidized to the sulfate and copper then released as Cu^{2+}. This ion is mobile in solution and, again like zinc, can be removed efficiently from solution as the sulfide in reduzate sediments or as the carbonate when limestones are encountered during transport. In high pH environments, however, precipitation of basic copper oxy-salts may become important and workable deposits produced. The concentration in seawater is fairly high (0.003 ppm) and, moreover, the reported content of deep-sea clays (250 ppm) is almost five times as great as the mean continental figure. This would indicate that much copper is transported to the sea and hence that its overall mobility in solution is high.

All the significant differences between the sedimentary geochemistries of zinc and copper may well stem from the greater stability of the nitrogen donor complexes of the transition ion. Copper is well known to concentrate in organic-rich sediments and the incorporation mechanism is probably initiated by the formation of chelate complexes when copper is adsorbed on particulate matter (e.g., the Kupferschiefer and Red-Bed copper deposits the world over).

Copper is well-known to be an essential trace element both in plants and animals. Soils derived from highly leached sediments are often copper deficient as a consequence of the metal's high mobility in solution. Cereals grown thereon are particularly susceptible to deficiency diseases, all of which are easily remedied by the administration of suitable copper compounds. Herbivorous animals also react quickly to deficiencies, copper being essential in the synthesis of hemoglobin.

Rich copper deposits are found in many parts of the world, and the present annual production exceeds 4.3 million tons. The most commonly mined deposits are probably those of hydrothermal origin, but many oxidative and secondary deposits are also being economically exploited.

CHARLES D. CURTIS

References

* Goldschmidt, V. M., 1954, "Geochemistry," Oxford, Clarendon Press, 730pp.
Wager, L. R., and Mitchell, R. L., 1951, "The distribution of trace elements during strong fractionation of a basic magma," *Geochim. Cosmochim. Acta*, **1**, 129–208.

* Additional bibliographic references may be found in this work.

Cross-references: *Chelation; Cobalt; Complexes; Nickel; pH-Eh; Silver; Sulfides; Trace Elements; Zinc. Vol. IV B: Meteoritic Minerals.*

COPPER: MINERALOGY OF COMPOUNDS

A very large number of copper minerals have been described in the literature. Precise structural information, however, is still surprisingly scarce. This survey is, therefore, restricted to a discussion of a few representative and comparatively well described examples.

In meteorites, Cu is distributed more or less equally between iron and sulfide phases. Silicates contain only minute traces. The same distributional pattern is observed in the primary rocks of the earth's crust; Cu being invariably restricted to sulfides. Two distinct assemblages of primary sulfides are usually recognized. The first (magmatic range sulfides) are mainly mixed Cu–Fe minerals. Chalcopyrite $CuFeS_2$ is quantitatively the most important mineral of this group. The second assemblage (hydrothermal) is characterized by complex sulfides and sulfo salts containing arsenic (As), antimony (Sb), and bismuth (Bi). Zn minerals, as well as silver (Ag), lead (Pb), and tin (Sn) minerals, are important constituents.

Oxidative breakdown (weathering) of primary sulfides and sulfo salts liberates Cu^{2+} in aqueous solution. Reaction with country rock leads to the formations of the variable and spectacular minerals of the "oxidized zone" of copper-ore deposits.

Reaction of primary Cu–Fe sulfides with Cu^{2+} (percolating down from the oxidized zone) in the region of the water table leads to the formation of another distinct mineral assemblage. The simple sulfides covellite CuS and chalcocite Cu_2S typify this zone of "supergene enrichment."

Elemental Cu is generally found in association with basic extrusives. The exact mechanism of derivation from primary compounds remains obscure.

Some better-known Cu minerals are described briefly in Table 1. Physical data are much simplified (many Cu compounds are polymorphic, for example) but are sufficient for the purposes of the ensuing discussion.

Structural Mineralogy. *Oxides and Oxy Salts.* In cuprite Cu_2O each O is surrounded tetrahedrally by metal atoms. Coordination of Cu is unusual, being two-fold (linear). The overall structure is highly symmetrical, unlike that of all cupric compounds. Tenorite CuO is monoclinic. Cu has an approximately coplanar coordination of 4, with oxygens at the centers of very irregular metal tetrahedra. In the common oxy salts and the silicate dioptase $CuSiO_3 \cdot H_2O$, Cu^{II} coordination varies between square coplanar and octahedral. Chalcanthite $CuSO_4 \cdot 5H_2O$, approaches regular octahedral coordination with 4 equatorial oxygens (H_2O) and 2 axial oxygens (SO_4^{2-}). Atacamite $Cu_2(OH)_3Cl$, malachite $Cu_2CO_3(OH)_2$, and dioptase all show marked distortion of the Cu^{II} coordination octahedron, with 4 short equatorial and 2 long axial separations. For example, atacamite has 2 Cu^{II} positions, both with 4 equatorial OH^- at 1.97–2.07 Å. The Cu^{II} in position 1 has 2 axial Cl^- at 2.76 Å; the Cu^{II} in position 2 has 1 axial Cl^- at 2.75 Å and 1 axial OH^- at 2.36 Å.

In azurite $Cu_3(CO_3)_2(OH)_2$ both Cu^{II} sites are again at the center of a square of 4 oxygens (2 hydroxyl, 2 carbonate) at approximately 2.0 Å. In addition, one Cu has a fifth oxygen at 2.38 Å. One Cu^{II} site, therefore, has square pyramidal coordination, the other has square coplanar.

In summary, the generally low symmetry of Cu^{II} minerals can be attributed to the peculiar but marked tendency for Cu^{II} to adopt environments with coordination intermediate between true square planar and regular octahedral.

Sulfides and Sulfo Salts. The structures of many sulfides and sulfo salts can be related to the zincblende ZnS structure in which each Zn is at the center of a regular sulfur tetrahedron. Apart from this broad generalization, the structural chemistry of the copper sulfides is particularly complicated.

Chalcopyrite $CuFeS_2$ is a superstructure of ZnS with Cu and Fe replacing Zn in a regular arrangement. The question of metal valence state, however, is not readily answered. Formulation as $Cu^{II}Fe^{II}S_2$ seems reasonable, but the chalcopyrite structure is adopted by $CuAlS_2$ and $CuGaS_2$. Recent valence investigations using *neutron diffraction* also indicate the $Cu^IFe^{III}S_2$ formula.

Some sulfides (bornite Cu_5FeS_4, digenite Cu_9S_5 etc.) certainly contain univalent copper. Many sulfides and sulfo salts also have regular tetrahedral Cu coordination at 2.32 ± 0.02 Å. By adopting the following formulation: Chalcopyrite, $Cu^IFe^{III}S_2$, Stannite, $Cu_2^IFe^{II}Sn^{IV}S_4$, and Tetrahedrites, $Cu_3^I(Sb,As)^{III}S_3$, it is possible to assign consistent valence and stereochemistry to many of these compounds.

Hexagonal covellite CuS cannot be $Cu^{II}S$. The structure is mineralogically unique, with one-third of the Cu atoms having 3 nearest sulfur neighbors at 2.19 Å, and the remainder having 4 sulfur neighbors in a tetrahedron at 2.32 Å. Two-thirds of the sulfur atoms are present as S_2 groups (as in pyrite FeS_2). The most reasonable formulation on structural grounds is $Cu_4^I, Cu_2^{II}, (S_2)_2S_2$, once more giving consistent Cu^I valence/stereochemistry.

Ferromagnetic cubanite $CuFe_2S_3$ is another structurally unique compound. Direct Fe–Fe bonding is indicated, which has also been suggested to account for ferromagnetism. Cu coordination is regular tetrahedral. Since the simplest valence formulation is $Cu^{II}Fe_2^{II}S_3$, regular coordination of Cu^{II} is indicated. Single electron delocalization between Fe^{III} pairs, however, explains the ferromagnetism and leaves Cu^I in tetrahedral sites.

Some compounds, however, certainly contain Cu^I in other than regular tetrahedral sites. Digenite (formulated $Cu^{II}Cu_8^IS_5$) has

COPPER: MINERALOGY OF COMPOUNDS

TABLE 1. SOME REPRESENTATIVE COPPER MINERALS

Mineral	Formula	System	Color, Luster, Opacity	Mode of Occurrence
Native metal	Cu	Isometric	Light rose, tarnishing rapidly to brown; metallic; opaque	Commonly associated with basic extrusive igneous rocks
Cuprite	Cu_2O	Isometric	Red to black; adamantine to submetallic; red to yellow by transmitted light	Normally massive, granular or earthy; sometimes fibrous; found in the oxidized zone of copper deposits, together with malachite, azurite etc.
Tenorite	CuO	Monoclinic	Black	Usually found as black powder in the oxidized zones of Cu deposits
Atacamite	$Cu_2(OH)_3Cl$	Orthorhombic	Bright green to apple green; vitreous; transparent	Commonly slender prismatic needles; massive fibrous or granular; secondary oxidized zone mineral, especially under saline desert conditions
Malachite	$Cu_2(CO_3)(OH)_2$	Monoclinic	Bright green; silky or dull; translucent	Commonly massive or encrusting; banded botryoidal form well-known; widespread in upper oxidized zone
Azurite	$Cu_3(CO_3)_2(OH)_2$	Monoclinic	Azure blue to very dark blue; vitreous; transparent	Usually massive or stalactitic; upper oxidized zone
Dioptase	$CuSiO_3 \cdot H_2O$	Hexagonal	Emerald green; vitreous; transparent to translucent	Rare mineral of oxidized zone of weathering
Chalcanthite	$CuSO_4 \cdot 5H_2O$	Triclinic	Sky-blue; vitreous; transparent	Uncommon mineral of upper oxidized zone, particularly in desert conditions
Chalcopyrite	$CuFeS_2$	Tetragonal	Brass-yellow, tarnishing; metallic; opaque	Usually massive being most widespread copper mineral; formed under variety of conditions (occurs in most Cu deposits); basic igneous association quantitatively most important
Bornite	Cu_5FeS_4 (variable)	Isometric	Copper red to pinchbeck, brown on fresh fracture surface; tarnishes; metallic; opaque	Widespread and common copper mineral; usually massive; found with chalcopyrite in primary basic, acidic, and sedimentary rocks
Tetrahedrite–tennantite	$Cu_3(Sb,As)S_3$ (variable)	Isometric	Flint grey to iron-black; metallic; opaque	Crystals commonly tetrahedral; usually massive; found in low to moderate hydrothermal veins with Pb, Zn, Ag minerals; most common and important sulfo salt minerals
Enargite	Cu_3AsS_4	Orthorhombic	Grey black to iron black; metallic; opaque	Usually found massive, granular, or prismatic; often vein or replacement deposits
Bournonite	$PbCuSbS_3$	Orthorhombic	Grey black to iron black; metallic; opaque	Usually found massive and granular; one of the commoner sulfo salts, usually in moderate temperature hydrothermal veins together with tetrahedrite, sphalerite, etc.
Chalcocite	Cu_2S	Orthorhombic	Black; metallic; opaque	Occurs most commonly as black, fine grained, massive material; found principally in supergene enrichment zone, together with other supergene minerals, as alteration products of primary sulfides
Covellite	CuS	Hexagonal	Indigo blue or darker; metallic; opaque except in very thin plates	Crystals rare, commonly massive; crystals exhibit perfect basal cleavage; usually occurs in supergene enrichment zone together with the much more common chalcocite

one irregular octahedral (presumably Cu^{II}) site, 4 regular tetrahedral Cu^I sites, 2 irregular tetrahedral Cu^I sites, and 2 triangular coordination sites.

In general, structural data suggest that Cu^I prefers regular tetrahedral coordination and that this unit is the basis of many sulfides and sulfo salts. Cu^{II} seems not to adopt regular coordination, but information is scarce.

Bonding and Stereochemistry. *Oxides and Oxy Salts.* Cupric oxides and oxy salts are essentially ionic solids and may, therefore, be reasonably considered in terms of *crystal-field theory*. The Cu^{2+} ion has the d^9 electronic configuration and nondegeneracy of the d shell in electrostatic fields (i.e., in proximity to other ions) leads to pronounced destabilization of regular octahedral or tetrahedral environments *(Jahn-Teller distortion)*. In fields of low strength, the stable environment is that of a distorted octahedron with 4 short equatorial and 2 long axial internuclear separations. In high-strength fields, distortion may be sufficient to sever bonding with one or both axial groups.

These crystal-field forces, stemming directly from d-electron distribution asymmetry, are clearly responsible for the unusual crystal structures adopted by ionic cupric compounds.

Electronic transitions between different d states are largely responsible for the strong colors of many transition metal compounds. Two factors influence the color of cupric compounds. Strong crystal fields intensify coloration since d-d transitions move from the infrared (very weak fields) into the visible. Increased distortion of the environment causes marked broadening of absorption bands. The weak-field (almost regular) chalcanthite environment produces pale blues. In malachite, atacamite, and dioptase, greater distortion and field strength produce deeper colors. The less symmetrical azurite structure gives intense, deep coloration whereas the highly asymmetric tenorite is a very good absorber in the visible and is thus black.

Sulfides and Sulfo Salts. Bonding in these compounds is much more covalent in character than that in the minerals considered above. Covalent bonding theory for the d^9 Cu^{II} configuration, however, still predicts Jahn-Teller distortion (due to asymmetry in antibonding electron distribution). Cu^{II} would not, therefore, be expected to occur in regular octahedral or tetrahedral environments.

Cu^I and Zn^{II} have the same electronic configuration ($3s^2 3p^6 3d^{10}$) and are not susceptible to crystal-field effects. The very marked preference of Zn^{II} for regular tetrahedral coordination is a feature of zinc chemistry. Cu^I prefers to adopt regular tetrahedral coordination in compounds other than sulfides: it seems reasonable to predict that Cu^I and Zn^{II} would behave very similarly in sulfides.

This simple treatment is satisfactory provided that copper is dominantly present as Cu^I in sulfides and sulfo salts. This, at least, is consistent with structural interpretation. Many of these compounds, however, possess metallic or semimetallic properties (metallic luster, high conductivity, ferromagnetism etc.). It would be unwise to pursue formal valence and bonding considerations further until these properties have been much more extensively investigated.

CHARLES D. CURTIS

References

Orgel, L. E., 1960, "An Introduction to Transition Metal Chemistry, Ligand-Field Theory," London, Methuen.

Ramdohr, P., 1969, "The Ore Minerals and Their Intergrowth," Berlin, Akademie Verlag, 1400pp.

Wells, A. F., 1962, "Structural Inorganic Chemistry," Third ed., Oxford, Clarendon Press.

Wyckoff, R. W. G., 1964, "Crystal Structures," Second ed., London and New York, John Wiley & Sons, 5 vols.

Cross-references: *Aqueous Solution; Bonding; Crystal-Field Theory; Weathering, Chemical.* Vol. IV B: *Crystallography; Meteoritic Minerals.*

COPPER DEPOSITS

The geochemical behavior of copper (see *Copper: Element and Geochemistry*) determines the distribution of copper during the various rock-forming processes, i.e., the various stages of the geochemical cycle (Fig. 1). Likewise, the commercial accumulations of copper are a function of its geochemical behavior, since ore deposits are but more or less unusual accumulations of metals or metalliferous minerals.

Igneous Associations. In igneous rocks copper is an almost ubiquitous element. In basic rocks it is found along with cobalt and nickel (e.g., Insizwa, Sudbury, etc.). Basaltic, andesitic, and quartz-porphyritic rocks in geosynclinal belts contain almost without exception accumulations of copper minerals, often with more or less molybdenum, cobalt, nickel, silver, etc. These deposits often extend or grade into deposits of similar composition in overlying or intercalated tuffs and sediments. In the Keweenawan lavas of Michigan and in some other places, native copper occurs as an integral part of basaltic lavas, accumulated mostly in the late vesicular spaces.

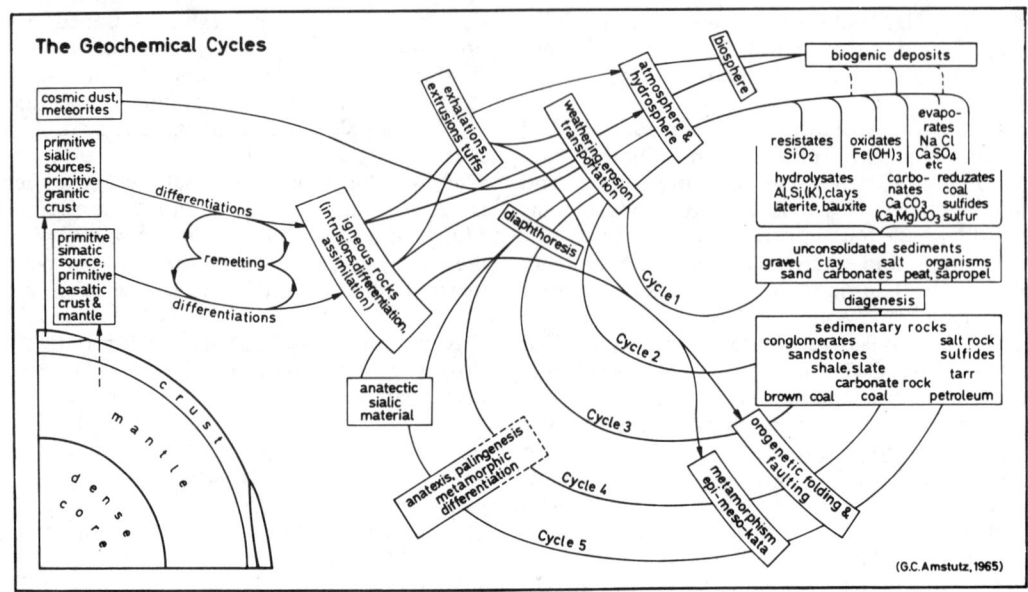

Fig. 1

In intermediate to acidic intrusives (diorites, quartz-monzonites and granites) copper deposits are abundantly found as so-called "porphyry copper" deposits and their derivative or marginal forms (Fig. 2). Examples include Chuquicamata, El Salvador, Chile; Toquepala- Cuajone, Peru; Bingham, Ely, Ray, Miami, Ajo, Morenci, Santa Rita, U.S.A.; Bougainville, New Guinea; Philippines; Yugoslavia; Mamut, Malaysia.

Contact deposits belong to the marginal forms. Their shape and composition is determined more by the intrusive body than by the intruded wall rock (e.g., in some parts of the deposits at Morenci, Arizona, and at Bingham, Utah). Marginal parts of Chuquicamata and almost all of Butte are vein-type deposits. El Teniente (Braden), Chile, is a mineralized volcanic breccia pipe.

Sedimentary Accumulations. In sediments copper precipitates mostly in black shales, in red bed facies and associated sandstones, but can be found occasionally in carbonate rocks, especially as a subfacies within the widespread *stratiform lead-zinc deposits*. Some large representatives of the first types are the "Kupferschiefer" deposits in Europe, the White Pine deposits in Michigan, the Red Bed deposits of Arizona and New Mexico; the Zambian (Rhodesian) Copper Belt; Katanga, etc.

Very numerous are the massive and disseminated deposits in sediments of *orogenic belts* which are normally associated with the volcanic belts mentioned before. Many of these are metamorphic and were, until recently, believed to be of replacement origin, but are now almost universally recognized as primary sediments and/or exhalative products.

In the oxidation and the *groundwater zones* copper is often enriched as oxides, carbonates, silicates, and as secondary sulfides, especially

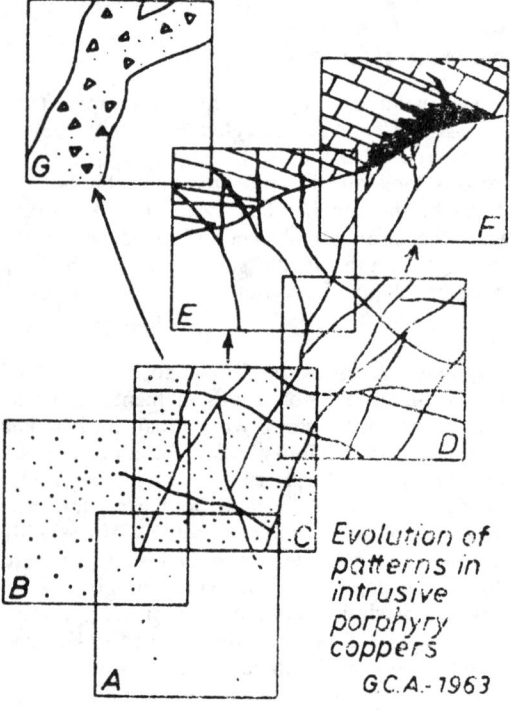

Fig. 2

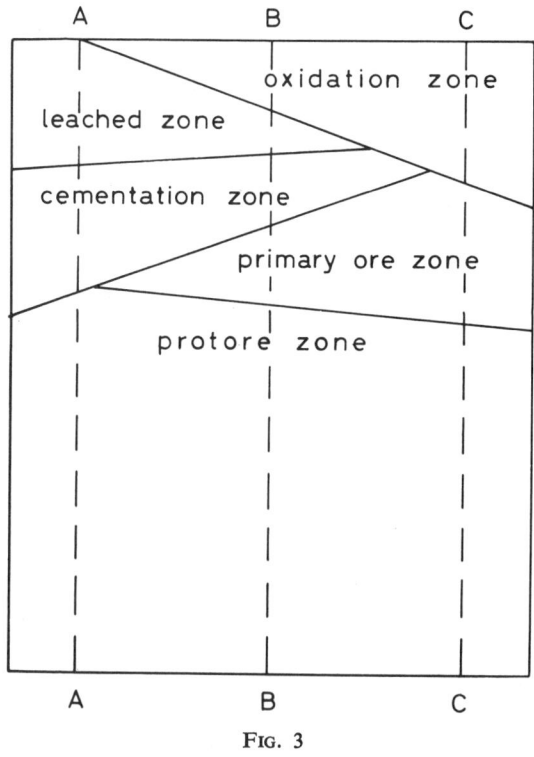

Fig. 3

chalcocite. These *supergene enrichment* zones are often richer than the underlying primary ores. A schematic zoning pattern is given in Fig. 3.

Most abundant among the above-named deposits are the porphyry copper and the black shale deposits. They provide roughly two thirds of the world's copper production. The major copper-producing nations are, in order, the United States, Chile, Zambia (Northern Rhodesia), Congo (Katanga), Peru, and Canada. The copper price has been relatively stable (about 30–35 U.S. cents since 1962), despite significant fluctuations in production and consumption. The annual growth rate of the copper industry is 3–4%. The development of known deposits should satisfy demand until about 1975; after this time, additions to production must come from new discoveries. According to estimates of the U.S. Bureau of Mines the proved copper reserves are 31.5 million tons for the U.S. and 170 million tons for the world.

Mineralogy. From among approximately 150 minerals containing copper, only the following fifteen are important ore minerals: (a) *sulfides:* chalcopyrite, chalcocites, bornite, covellite, enargite, and tetrahedrite-tennantite; (b) *native copper;* (c) *oxidation products:* atacamite, malachite, chrysocolla, cuprite, tenorite, brochantite, and antlerite.

G. C. AMSTUTZ

CORE GEOCHEMISTRY

References

"Gmelins Handbuch der Anorganischen Chemie," 1958, 1961, 1965, 8 Aufl., System-No. 60, Pt. B, Sect. 1–3, Weinheim, Verlag Chemie, 1452pp.

Kraume, E., 1964, "Copper," in "Die metallischen Rohstoffe," **4,** Stuttgart, Enke, 380pp.

Titley, S. R., and Hicks, C. L. (editors), 1966, "Geology of the Porphyry Copper Deposits, Southwestern North America," Tucson, University of Arizona Press, 287pp.

Cross-references: *Supergene.* See also individual minerals in Vol. IV B.

CORE GEOCHEMISTRY

It has been known for some time that seismic velocities experience a sharp discontinuity in the interior of the earth. A detailed analysis has shown that this discontinuity occurs at a radius of 3473 km from the center of the earth whose total radius is 6371 km. The central region is called the core, and the outer region beyond the discontinuity is called the mantle. At the core-mantle interface the density is estimated to jump from 5.5 g/cc to 10.0 g/cc at a pressure of 1.35 mb (1 mb is approximately 1 million atm). From the total mass of the earth and its moment of inertia in conjunction with the seismic velocities, the deduction can be drawn that the core contains about 1.95×10^{27} g, constituting about 1/3 of the mass of the earth while occupying only 1/7 of its volume. Since no shear waves have been observed in the core, it is generally accepted that the core must be in liquid form, although a small seismic velocity discontinuity in the interior of the core indicates that the central 5% in volume of the core is solid.

It is unlikely that the core is made up predominantly of light elements, since at the estimated density (10 g/cc) the observed seismic velocities would not be in agreement with an empirical but systematic relation between velocity and density. This alone would exclude previous suggestions that the core is made of either a metallic form of silicate or an even more dramatic suggestion that the core consists of metallic hydrogen. The first suggestion can be eliminated by consideration of shock-wave experiments (McQueen et al., 1964) in which the mantle silicates have been subjected to pressures as they exist in the core. The metallic hydrogen hypothesis is also made unlikely by theoretical calculations which suggest that the transition from molecular to metallic hydrogen takes place only at considerably higher pressures (18 mb) than those that exist in the core.

The density of the core and general consideration of the chemical abundance of the elements indicate that the core consists mostly of iron. A detailed analysis of shock compression data reveals that the density of the liquid core is about 10% less than that of pure iron and that the density at the bottom of the mantle is about 10% greater than that expected of the mantle constituents. These results suggest that some lighter elements are dissolved in the predominantly liquid iron core and that these light elements are there because the mantle material is soluble in the core. The presence of an appreciable amount of impurity in the iron core could furthermore help to support the dynamo theory of the earth's magnetism. The dynamo theory requires a fairly high electrical resistance of the core material relative to that of pure iron under atmospheric conditions in order that the eddy currents in the liquid core persist sufficiently long. Though little is presently known about the electrical resistance of materials under high-pressure and temperature, the presence of impurities in metals under atmospheric conditions generally raises the resistance.

Returning now to the solubility considerations at the mantle-core interface, they are akin to those that occur in the steel industry when slag dissolves in molten iron and vice versa, except that the pressures and temperatures are considerably higher. The importance of these solubility considerations is that they impose significant constraints on the conditions in the interior of the earth with respect to pressure, temperature, and especially chemical composition, since solubilities are sensitively dependent on these quantities. Because very little is presently known about mutual solubilities at high temperature and pressure, the solubility considerations are based on crude theoretical estimates which permit only general checks on the proposed models of the interior, as, for example, the condition that the proposed constituents of the mantle should not be either completely soluble or insoluble in molten iron.

Solubility considerations (Alder, 1966), were, in fact, undertaken in order to see whether they could explain the above density discrepancies. The suggestion would then be, for example, that the heavier iron replaces the magnesium or silicon or both in the oxides of these elements, which appear to be the major constituents of the lower mantle. The substituted magnesium or silicon could then dissolve in the liquid iron to lower its density. The major support for this hypothesis derives from the fact that iron easily replaces magnesium in the low-pressure form of olivine and magnesium oxide. At higher pressure the change of crystal structure accompanying the transformation of quartz to stishovite makes the low-pressure observation on olivine of doubtful value; however, the crystal structure of magnesium oxide appears to remain unchanged up to the pressures pertaining to the lower mantle. On the other hand, liquid iron or iron oxide could simply be soluble to some extent in the solid oxide system without substituting, and then the lowering of density could be brought about by the solubility of the solid oxides in turn in the liquid iron.

The simple theory of liquids proposed by van der Waals (Hildebrand and Scott, 1962) was used to predict the solubilities. It was first satisfactorily tested in its prediction of the one relevant solubility experimentally available, namely, that of silica in liquid iron under atmospheric conditions. This theory then predicted that magnesium oxide would be soluble to an extent of about 6% in weight in molten iron at the interface conditions while neither stishovite nor the magnesium silicate would be significantly soluble. This magnesium oxide solubility would thus be of the right magnitude to cause the lowering of the density of the core. If these predictions can be trusted, the mantle material must then be a physical mixture of magnesium and silicon oxides rather than a compound like magnesium silicate. Furthermore, the conclusion could be made that a significant amount of oxygen derived from magnesium oxide in the form of a doubly negative charged monatomic ion is dissolved in the liquid core.

In order to increase the reliability of these predictions, an experiment was undertaken to measure the solubility of magnesium oxide in molten iron at 0.2 mb, the most extreme conditions presently accessible in the laboratory but still far removed from the actual conditions of 1.4 mb at the interface. The experiments failed to give any significant data because the equilibrium condition of solubility could not be achieved in the relatively short time (5 min) for which these conditions could be maintained. Nevertheless, further experiments should be carried out, possibly under milder conditions than attempted above, in order to learn more about mutual solubilities of materials of geochemical interest at higher pressures and temperatures.

BERNI J. ALDER

References

Alder, B. J., 1966, *J. Geophys. Res.*, **71**, 4973.
Hildebrand, J. H., and Scott, R. L., 1962, "Regular Solutions," Englewood Cliffs, N.J., Prentice-Hall.
McQueen, R. G., Fritz, J. N., and Marsh, S. P., 1964, *J. Geophys. Res.* **69**, 2947.

Cross-references: *Geochemical Evolution of the Core, Mantle, and Crust; Iron; Van der Waals Force.* Vol. V: *Convection Currents; Core; Dynamo Theory; Geomagnetic Induction and Earth Conductivity.*

CORONA—See Vol. V

COSMIC DUST—See Vol. II

COSMIC RAY-PRODUCED RADIONUCLIDES—See **RADIONUCLIDES, COSMIC RAY PRODUCED**

CRUDE OIL COMPOSITION AND MIGRATION—See Vol. IV B

CRUST—See **EARTH'S CRUST GEOCHEMISTRY; GEOCHEMICAL EVOLUTION OF THE CORE, MANTLE, AND CRUST;** See also Vol. V

CRYSTAL CHEMISTRY

Crystal chemistry is concerned with the relationships between chemical composition, structure, and physical properties in crystalline substances. It became a separate branch of science in 1926 when sufficient structures had been determined, and V. M. Goldschmidt could formulate the general principles underlying the crystal structure of simple inorganic compounds.

Goldschmidt's generalizations were based on the theories and results of a long line of workers including Haüy who, at the start of the nineteenth century, determined the law of rational indexes (see *Crystallography*) by postulating the existence of regularly packed "molecules." This hypothesis was further developed by Barlow and Pope in 1906 who assumed the units to be spherical atoms, and was finally proved in 1912 when Friedrich, Knipping, and Laue discovered that x rays were diffracted by crystals thus enabling W. H. and W. L. Bragg, one year later, to make a crystal structure analysis of sodium chloride.

The basic unit in all crystal structures is either the atom, the ion, or, occasionally, a group of atoms so closely associated as to behave as a single unit in the structure.

An atom consists of a very small, dense, positively charged nucleus about 10^{-12} mm in diameter surrounded by electrons which, when all are present, give electrostatic neutrality to the atom. The electrons sometimes behave as very small particles and sometimes as waves. They occupy a certain region in space around the nucleus rather than travel in a fixed orbit as was once thought. The orbital of the electron comprises the region in which the probability of finding the electron at any given moment is greatest. In the hydrogen atom, the locus of points representing the most probable distance of the electron from the nucleus is a spherical shell, with a radius of 0.53×10^{-7} mm (0.53 Å). This is, of course, the radius of the electron's orbit in the Bohr model of the atom. It is of interest to note that this radius is about 100,000 times that of the nucleus.

Bonding Forces

Crystals are held together by forces of attraction between atoms. These bonding forces are related to the electron distribution in the outer layers of the atoms; they are therefore basically electronic. Four types of bonds which grade into each other continuously are commonly recognized. These are metallic, homopolar, ionic, and Van der Waals bonding.

Metallic Bonding. This type of bonding occurs in crystalline metals. The structure of a metal consists of closely crowded atoms whose outer electrons form a composite orbital or electron cloud which embraces all the positively charged atomic nuclei with their inner electrons. The attraction between the positive ions and the negatively charged electron cloud constitutes the metallic bond.

The high mobility of the outer electrons under an electric field makes metals very good conductors of electricity. Metallic crystals are also good conductors of heat. They usually have simple structures in which the atoms are closely packed thus resulting in high density. Typical examples include the native metals, copper, gold, and platinum, all of which crystallize with cubic close packing. In this structure each layer consists of sheets of tightly packed atoms centered at the points of an equilateral triangular network. Each atom touches six others at the corners of a regular hexagon. The second layer fits over interstices in the first staggered normal to the base of the triangle by two-thirds of the distance from the apex to the base (Fig. 1). Atoms of the third layer occupy positions between the other two. It will be noticed that the closest packed layers are parallel to the face of the octahedron (111). If the third layer is directly over the first, the addition of further layers in the sequence ABABAB . . . results in hexagonal close packing.

Homopolar Bonding. This type of bonding, which is also called *covalent*, occurs in varying degrees in diamond, quartz, and pyrite. The centers of homopolar bonded atoms are further apart than those of metallic bonded atoms,

CRYSTAL CHEMISTRY

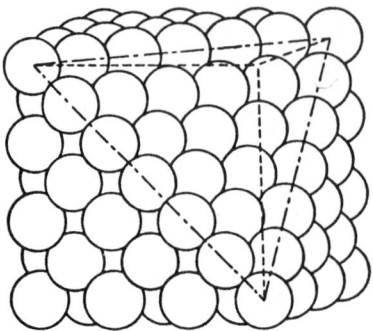

FIG. 1. Structure of native copper (Cu) (after Evans, 1964).

and the valence electrons occupy a molecular orbital that encloses two atoms joined by a single homopolar bond. The most likely position for the electrons at any instant is between the atoms. The concentration of negative charge (electron density) between the two atoms produces the homopolar bond. Homopolar bonds, unlike metallic bonds, are strictly limited to the bonding of one atom to a reasonably small number of others. Also, since the orbitals of many electrons are nonspherical and directed, the homopolar bond is normally directional. For these reasons, homopolar compounds commonly combine atoms to form discrete molecules of definite geometrical shape.

When one electron from each atom is involved in the bond it forms a single homopolar or covalent bond; when two from each are involved, a double bond may result. Single atoms may join with two neighboring atoms to form a linear chain structure. Up to four electrons may be available in certain elements for bonding (e.g., carbon). Atomic positions are complicated by the definite spatial orientation of homopolar bonds and by resonance. Resonance may be explained by saying that when alternative atomic configurations are possible the true configuration is often not one of the alternatives but an intermediate condition to which all the possible configurations contribute. Resonance confers additional stability to the structure. A homopolar bond is also possible when one atom contributes two electrons to the bond while the other contributes none. This variant is sometimes known as a *dative* bond, a *coordinate* bond, or a *donor-acceptor* bond.

Minerals in which all bonds are predominantly homopolar are commonly poor conductors of electricity and heat, transparent with high refractive indexes, brittle and hard (H > 5). In most compounds the homopolar bonds are within groups of molecules which themselves are connected by other types of bonds. Such compounds will be discussed later. One example of a mineral with wholly homopolar bonding is diamond. The diamond structure (Fig. 2) is controlled by the presence in carbon of four bonding electrons whose orbitals are directed toward the corners of a regular tetrahedron.

Pure homopolar bonding is of considerable importance in organic chemistry, but of less importance to mineralogy.

A gradation is possible between metallic and homopolar bonding. Substances such as the semimetals, bismuth and arsenic as well as selenium, pyrrhotite, and most sulfides, all show evidence of having some metallic as well as homopolar bonding. A similar gradation between homopolar and ionic bonding will be discussed under "ionic bonding."

Ionic Bonding. This bond type, which is also called *electrovalent,* heteropolar, or polar bonding, is the bonding of the alkali halides but plays an important part in the bonding of most minerals with the exception of the metals, sulfides, and related groups. Ionically bonded atoms are further apart than homopolar bonded atoms. The ionic bond is produced when an atom acquires one or more electrons from another atom, thus gaining a negative charge while the other becomes positively charged. The two atoms are now attracted to each other by an electrostatic force. The charged atoms are known as *ions,* those having a negative charge being called *anions* and those having a positive charge being called *cations.* In an ionically bonded crystal there are no molecules as such, the whole crystal being a giant molecule in which each ion is equally bonded to those ions of opposite charge at equal distances from it.

Atomic Size. The ionic radius is a concept in which the term radius is used in a special sense. The *ionic radius* is therefore defined

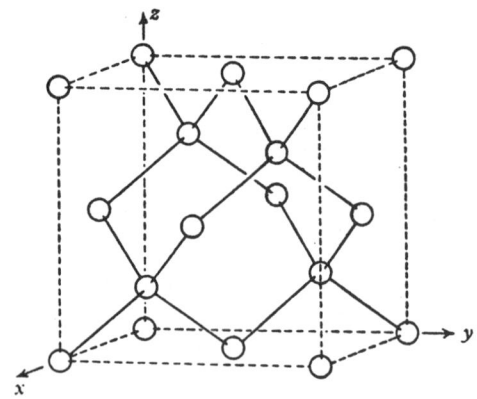

FIG. 2. Structure of diamond (C) (after Evans, 1964).

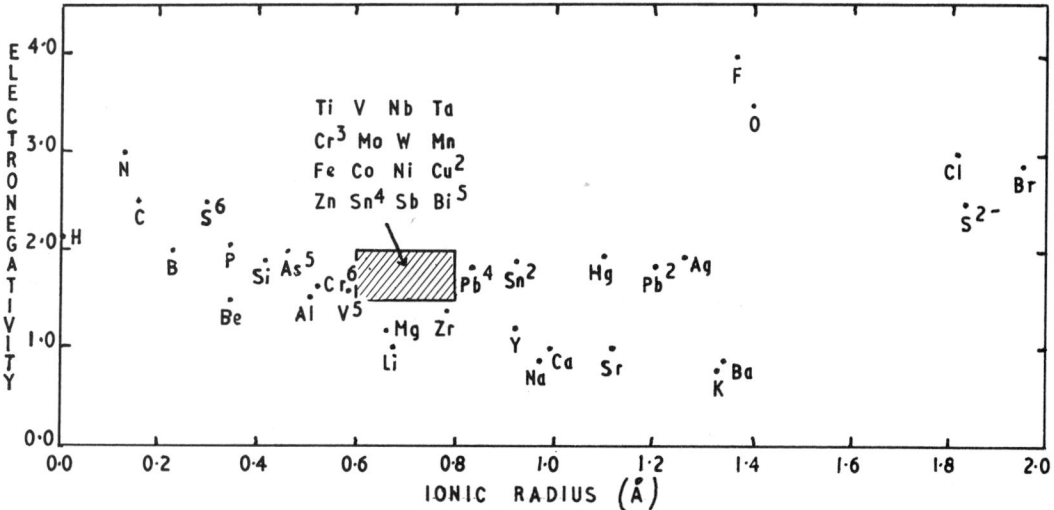

FIG. 3. Ionic radii, adjusted to six coordination, and Pauling electronegativities of the common elements.

as the contribution an ion makes to the length of the bond (bond length) between two ionically bonded ions. This bond length can easily be measured by x-ray methods and since the radii of certain ions may be estimated by various methods, a semiempirical scale of ionic crystal radii may be established. Values of atomic size for other types of bonding may also be determined and, as may be expected, vary considerably even for a single element. Size is of greater importance in ionic crystals and for that reason neither metallic, homopolar, or molecular radii will be discussed further.

The ionic radius of a single element is not constant and varies with the charge, in electron units, with the number of surrounding ions, and even with the ratio between the sizes of the two bonded ions. A diagram which shows the ionic radii of some common elements is given in Fig. 3.

Hydrogen and oxygen are elements of special importance. The hydrogen ion consists of a charged nucleus, the proton, which has no orbital electrons associated with it. It is thus so small that it is effectively a dimensionless positive charge. The OH^- group is effectively a sphere with a radius nearly equal to that of the O^{2-} ion. Oxygen is the commonest element in the earth's crust and most common minerals are oxygen compounds. Since oxygen is a large ion, 1.40 Å (1Å = 10^{-10}m), most of its ionic compounds have their structures controlled by the packing of the oxygen ions, the other ions merely fitting into the interstices between the oxygen ions.

Coordination. The most stable arrangement of spheres of one size around a sphere of smaller size depends on the ratio of the two radii. The number of large cations that surround a smaller anion is known as the *coordination number* of that anion. For the commoner stable arrangements of ions the theoretical coordination numbers for various radius ratios are as follows: 0.15–0.22, 3; 0.22–0.41, 4; 0.41–0.73, 6; 0.73–1.0, 8; and 1.00, 12. Elements with radii giving predicted coordination numbers near the boundary between two different coordination numbers are often able to form compounds with either coordination number. When this occurs it is usually found that the lower coordination number is favored in minerals stable at high temperatures and low pressures.

Pauling's Principles. The coordination numbers of the cations present are one of the factors that control possible structures in ionic compounds. Pauling in 1928 stated a number of principles which could be applied to ionic compounds. He proposed that in any ionic crystal the total positive charge on cations must equal the total negative charge on the anions. The number of anions surrounding each cation is controlled by the ratio of the ionic radii. The contribution of the charge on a cation toward neutralizing the charge on a neighboring anion is equal to the charge on the cation divided by the number of neighboring anions. In stable structures the charge on each ion is neutralized by the contributions from its neighbors.

As an example of the use of Pauling's

principles we may consider the possible stability of a magnesium feldspar with the formula $MgAl_2Si_2O_8$. Each oxygen ion in the group $n(Al_2Si_2O_8)^{2-}$ has an unsatisfied charge of one-fourth. Magnesium ions are six-coordinated with respect to oxygen so that when their double charge is shared among six oxygens they can supply a charge of only one-third to each neighboring oxygen. Thus the charge on a magnesium ion could not be neutralized by the contributions from its near neighbors and it is unlikely that a magnesium feldspar would be stable.

Purely ionic compounds are moderate insulators, poor conductors of heat, transparent with low refractive indexes, brittle, and often only moderately hard, and have fairly high melting points. The structures vary from simple to complex.

Halite, sodium chloride, is a typical ionic compound. The ionic radii of chlorine and sodium are 1.81 Å and 0.95 Å, respectively giving a ratio of 0.52 and indicating that sodium is six-coordinated with reference to chlorine. Since both chlorine and sodium ions bear a single charge (Cl^-, Na^+) the chlorine must also be surrounded by six sodium ions. The simple resultant structure is shown in Fig. 4.

A more complex structure is calcite, calcium carbonate, in which CO_3^{2-} groups having internal homopolar bonding replace simple ions. The arrangement (Fig. 5) is similar to that of sodium chloride but distorted due to the presence of triangular-shaped CO_3^{2-} groups.

Mixed Bonds. A continuous gradation is possible between the fully homopolar bond and the purely ionic bond. Although it is sometimes convenient to treat most common minerals as ionic, the bonding even in a substance

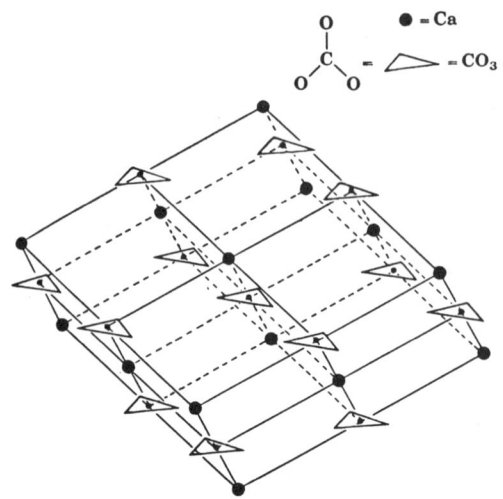

FIG. 5. Structure of calcite ($CaCO_3$) (after Fyfe, 1964).

such as sodium chloride is about 30% homopolar. The percent ionic character of a bond is related to the elements involved in the bond. The *electronegativity* of an element is a numerical expression of its power to attract an electron. The percent ionic character of the bond between two elements is related to the difference between the electronegativities of the two elements involved. The exact relationship is probably complex but one empirical formula giving the approximate percent ionic character is:

$$p = 16\Delta x + 3.5\Delta x^2$$

where p is the percent ionic character of the bond and Δx is the difference between the electronegativities of the two elements involved. Values of the electronegativities are included in Fig. 3. It may be shown that the silicon-oxygen and aluminium-oxygen bonds are predominantly homopolar (37 and 46% ionic, respectively) whereas the bonds between sodium, potassium, calcium, or magnesium with oxygen are predominantly ionic (62, 65, 68, and 85% ionic, respectively).

Hydrogen Bonding. This is a special type of bond that has been considered as a variety of the ionic bond. It occurs in ice and some hydrous minerals, and arises from the linkage of two oxygen atoms through an intermediate hydrogen atom.

Van der Waals Bonding. This type of bonding, which is also called residual bonding, because of its weakness, is found in crystals of the inert gas elements. Van der Waals forces link the homopolar bonded layers in graphite (Fig. 6) and are responsible for the

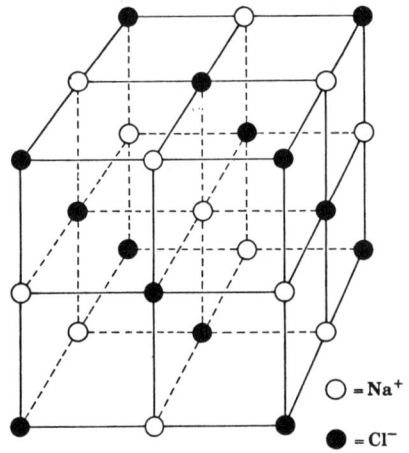

FIG. 4. Structure of halite (NaCl) (after Fyfe, 1964).

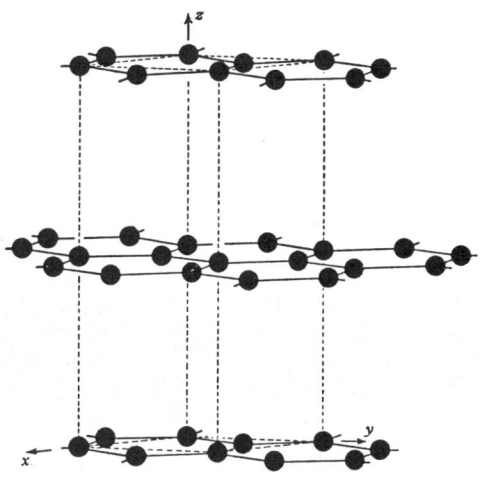

FIG. 6. Structure of graphite (C). The broken lines indicate the unit cell and not bonds (after Evans, 1964).

extreme softness and excellent basal cleavage of this mineral. Van der Waals forces probably contribute a small share to most bonds.

One other possibility exists for an element to be incorporated in a crystal. In what are known as *clathrate compounds,* the atoms are sealed in large cavities in the crystal structure of the compound. The geologically important examples include the trapping of argon (produced by the radioactive decay of K^{40}) in the orthoclase structure and the incorporation of water, helium, and argon into the beryl structure.

Crystal Structures

There are compounds in which all the bonds are identical, as well as the more common compounds in which two or more bond types occur. The terminology used below to express this difference is favored by some workers, but is generally considered rather academic, a direct description of the bonding usually being more informative.

When all the bonds in a crystal are of a similar type and strength the crystal is known as *homodesmic* but when there are significant differences in the bonding forces the crystal is *heterodesmic.* Examples of homodesmic minerals are native copper, fluorite, and diamond, which have essentially metallic, ionic, and homopolar bonding, respectively. Examples of heterodesmic minerals are graphite and calcite. In graphite the carbon atoms are bound into sheets by covalent bonds but the bonding between the sheets is Van der Waals. In calcite, CO_3^{2-} groups (with essentially homopolar bonding within the group) are joined to each calcium by essentially ionic bonds. This formation of radicals within the crystal is characteristic of heterodesmic structures. Common radicals are CO_3^{2-}, SO_4^{2-}, PO_4^{3-}, and SiO_4^{4-}. In heterodesmic minerals the main properties are usually determined by the weakest bond; thus in graphite the weaker Van der Waals bonds between the layers account for its softness, perfect basal cleavage, and extreme anisotropy.

The term *mesodesmic* has been used for those nonhomodesmic crystals in which groupings of atoms occur but which are not as sharply defined as in typical heterodesmic compounds. The silicates are sometimes considered to be examples of mesodesmic compounds. The terms *isodesmic* and *anisodesmic* are sometimes used for compounds in which bond strengths are, respectively, equal in all directions or unequal in different directions. When these terms are used homodesmic and heterodesmic are applied to structures in which one or several bond *types* are present.

The structures found in minerals are extremely numerous. A few examples of the structures of non-silicates have already been given and only the structures of silicates will be discussed further.

Silicate Structures. Silicates consist of silicon and oxygen groups of various arrangements in which the silicon is always surrounded by four oxygens at the corners of a tetrahedron. The high-pressure and high-temperature phase of SiO_2, stishovite, appears to be an exception to this. Stishovite is described as having octahedral rather than tetrahedral coordination. These silicon-oxygen tetrahedra may be isolated or joined by the sharing of one or more oxygens. This joining is analogous to the *polymerization* of carbon in organic chemistry. Silicates are classified according to the manner in which their silicon-oxygen tetrahedra are linked.

Nesosilicates (silicates with isolated $(SiO_4)^{4-}$ groups). These are also known as *orthosilicates.* They include such minerals as olivine [$(Mg,Fe)_2SiO_4$], garnet (e.g., $Ca_2Al_2Si_3O_{12}$), and zircon ($ZrSiO_4$). The structure of the olivine end member, fayalite (Fe_2SiO_4), is shown in Fig. 7. It will be noticed that the silicon-oxygen tetrahedra are linked only by O–Fe–O bonds. The coordination number of iron is six and therefore the iron ions are surrounded by six oxygens lying at the corners of an octahedron.

Sorosilicates and Cyclosilicates (double tetrahedra structures and ring structures, respectively). Silicon-oxygen tetrahedra may combine to give more complicated but still finite groups (Fig. 8). Two tetrahedra may be linked by hav-

CRYSTAL CHEMISTRY

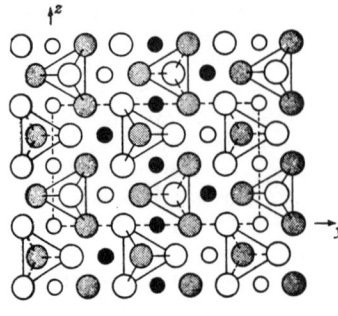

○: Fe in plane of paper ●: Fe at height $\frac{1}{2}a$
◉: O at height $\frac{1}{4}a$ ◯: O at height $\frac{3}{4}a$

FIG. 7. Structure (simplified) of fayalite (Fe_2SiO_4). Silicon, which is in the center of each tetrahedron, is not shown. Small circles represent iron. Lines joining oxygens are not bonds (after Evans, 1964).

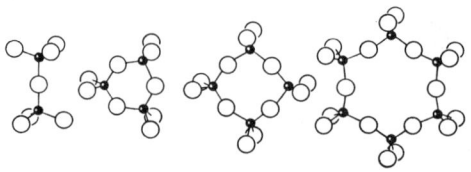

FIG. 8. Examples of finite groups of silicon-oxygen tetrahedra. As in most diagrams of this type the atoms are not to scale with each other or with the bond lengths (after Kostov, 1968).

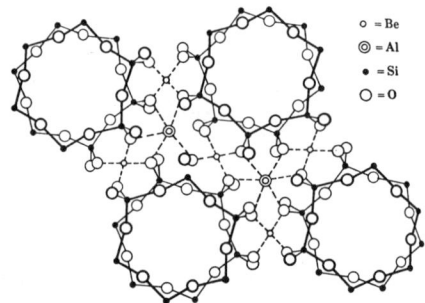

○ = Be
◉ = Al
● = Si
◯ = O

FIG. 9. Structure of beryl ($Be_3Al_2Si_6O_{18}$) looking down on the rings (after Fyfe, 1964).

ing a single oxygen shared between the two tetrahedra, thus giving a group of composition $(Si_2O_7)^{6-}$. Minerals containing such groups (e.g., hemimorphite, $Zn_4(OH)_2Si_2O_7 \cdot H_2O$) have been called diorthosilicates, pyrosilicates, or *sorosilicates*. Three, four, or six tetrahedra may unite by sharing two oxygens to form rings (*cyclosilicate*). Three tetrahedra are linked in benitoite ($BaTiSi_3O_9$), four in axinite ($Ca_2(Fe^{2+},Mn)Al_2(OH)BO_3Si_4O_{12}$), and six in a number of common minerals including beryl ($Be_3Al_2Si_6O_{18}$). By sharing three oxygens a double hexagonal ring containing twelve tetrahedra may be formed as in milarite, $KCa_2AlBe_2Si_{12}O_{30}$. The structure of beryl is given in Fig. 9. Two levels of Si_6O_{18} rings are shown in the diagram; the beryllium and aluminum ions lie in a plane between these two levels.

Inosilicates (chain structures). Infinite linear chains of silicon-oxygen tetrahedra may be produced by the sharing of two or more oxygens. In such minerals the chain structure is commonly reflected in the elongate habit of the crystals and in the presence of prismatic cleavage. Minerals with chain structures are of two types, single chain and double chain, represented by the pyroxenes and amphiboles, respectively. The structure of a pyroxene, diopside ($CaMgSi_2O_6$), is shown, normal to the chains, in Fig. 10. In diopside a single chain is formed by the sharing of two of the oxygens in each tetrahedron, successive tetrahedra facing opposite directions. Such chains have the formula $n(SiO_3^{2-})$. Calcium occupies the sites of eight coordination and magnesium the sites of six coordination. The Ca–O bonds are apparently the weakest and the two cleavages at 88° cut through these. The structure of an amphibole, tremolite ($Ca_2Mg_5Si_8O_{22}(OH)_2$), is given in Fig. 11. In tremolite a double chain is formed when two pyroxene-type chains are joined by sharing suitable oxygens. Such chains have the formula $n(Si_4O_{11})^{6-}$. Again the Ca–O bonds seem to be weakest and the origin of the two characteristic amphibole cleavages at 56° as opposed to the pyroxene cleavages at 88° is obvious from the figures.

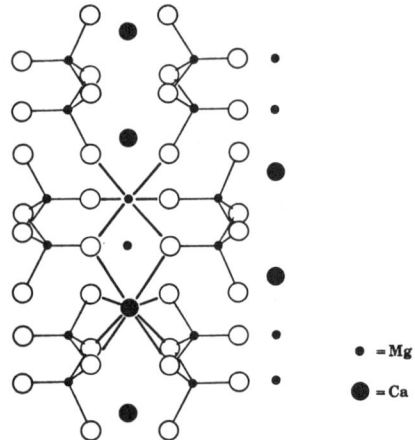

● = Mg
◉ = Ca

FIG. 10. Structure (simplified) of diopside ($CaMgSi_3O_6$) looking along six chains. Small black dots are magnesium ions, large are calcium (after Fyfe, 1964).

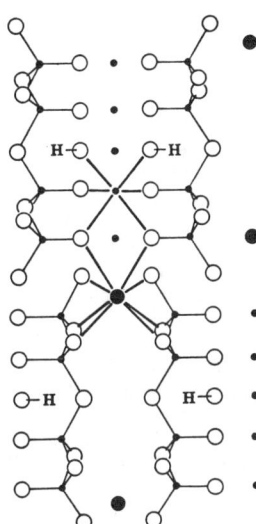

FIG. 11. Structure (simplified) of tremolite ($Ca_2Mg_5Si_8O_{22}(OH)_2$) looking along four double chains (after Fyfe, 1964).

Phyllosilicates (sheet structures). When a chain or ring structure is extended in two dimensions by the sharing of three of the oxygens of the tetrahedra a sheet structure is produced. In minerals with sheet structures the sheets consist of closed six-sided rings, thus accounting for the nearly hexagonal or pseudohexagonal symmetry. The formula of a silicon-oxygen sheet of this type may be written $n^2(Si_2O_5)^{2-}$ where n^2 indicates that the crosslinking takes place in two dimensions. An exceptional sheet structure where the rings are four and eight sided is found in *apophyllite*, which is tetragonal. Minerals with sheet structures predictably exhibit excellent basal cleavage. Among minerals having this structure are the micas, talc, chlorites, and the clay minerals. The structure of a mica, muscovite ($KAl_2(AlSi_3O_{10})(OH)_2$), is shown in Fig. 12. In this structure some of the aluminum (not shown in the figure) replaces the silica in the silicon-oxygen tetrahedra. Cleavage is produced by breaking the relatively weak O–K bonds.

Tektosilicates (framework structures). When all four oxygens of each tetrahedron are shared with adjacent tetrahedra the result is a three dimensional framework. Such frameworks are represented by the structure of quartz, tridymite, and cristobalite and have the formula $n^3(SiO_2)$. The n^3 indicates the crosslinking in three dimensions. The true silicon-oxygen framework structure is structurally neutral and can thus only occur in minerals with the formula SiO_2. If, however, some of the Si^{4+} are replaced by Al^{3+} the structure gains a charge

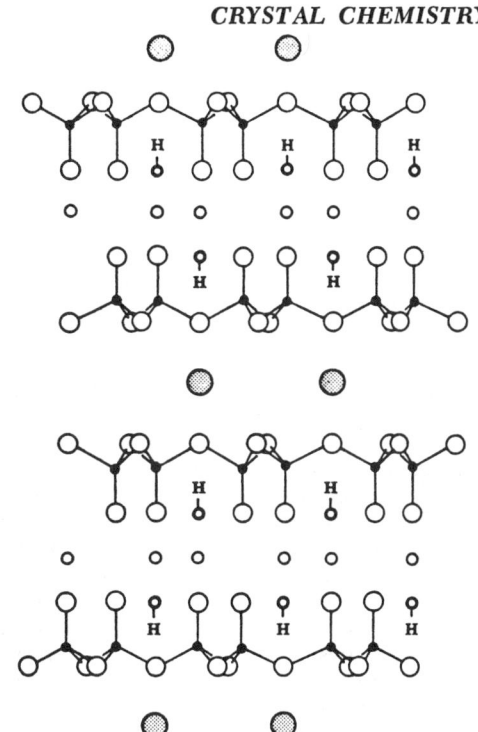

FIG. 12. Structure of muscovite ($KAl_2(AlSi_3O_{10})(OH)_2$) looking along four sheets. Note that Al replacing Si in silicon-oxygen tetrahedra is shown as Si. Shaded circles are potassium ions, small circles aluminum (after Fyfe, 1964).

that must be balanced by the addition of other cations, and a variety of minerals of various composition and structure become possible. Among minerals having this structure are the feldspars. One typical example is the mineral plagioclase, the variety albite having the formula $Na(AlSi_3O_8)$ and containing the structure $n^3(AlSi_3O_8)^-$.

Other groups with framework structures are the feldspathoids and the zeolites. The framework structures of the zeolites are much more open than those of the feldspars. Some zeolites have wide channels in which cations and water molecules are held. This water can be driven out by heat and reversibly replaced by other similarly sized neutral molecules. These zeolites can thus act as molecular sieves, permeable only to molecules smaller than a fixed size.

Properties of Structure Types. Some aspects of the physical properties in each of these groups have already been mentioned. Others are that only nesosilicates and tektosilicates can be cubic and many form equidimensional crystals; phyllosilicates tend to be tabular, and inosilicates and cyclosilicates, prismatic. The sequence: isolated $(SiO_4)^{4-}$ groups; isolated paired groups, rings and chains; sheets; frameworks, is a sequence within which, other things being equal, refractive indexes and density tend to decrease, and the minerals occur later in the crystallization sequence for a cooling magma.

The classification of structures according to the types of groups present is not confined to silicates. Many minerals such as calcite contain isolated groups. Chain groups are present in such minerals as native selenium, layered groups in graphite and molybdenite. In silicate glasses silicon-oxygen tetrahedra are still present but polymerization is variable and irregular. In opal there is evidence of a semirandom framework, derived from a highly disordered form of cristobalite.

Other Structures. A number of aspects of crystal chemistry that can only be mentioned here are dealt with in other articles. It is not uncommon for two substances with analogous formulas, composed of elements having similar ionic radii, to form crystals of similar structure. Such substances, for instance, cerussite and aragonite are said to be *isomorphous* (q.v.). It is also possible for an element or compound to form more than one type of structure. Such substances, for instance, carbon, which occurs as graphite and diamond, are said to be polymorphous, and the phenomena is *polymorphism* (q.v.). It is rare to find minerals in which any one set of identical atomic sites is entirely filled by atoms of a single element. Thus for instance in the olivine structure the site for divalent metals may be filled by almost any ratio of magnesium and iron. This phenomena is known as *solid solution* (q.v.). A more limited version of the same phenomenon is discussed under *Trace Elements in Silicate Minerals*. Crystal structures which differ considerably from the theoretical ideal also occur. Such crystals are known as *defect crystals* (q.v.).

Michael J. Frost

References

Appleman, D. F., et al., 1966, "Chain Silicates," Washington, D. C., Am. Geol. Inst.
Bailey, S. W., et al., 1967, "Layer Silicates," Washington, D.C., Am. Geol. Inst.
* Bragg, L., and Claringbull, G. F., 1965, "Crystal Structures of Minerals," Third ed., London, F. Bell and Sons, 409pp.
Bunn, C., 1964, "Crystals: Their Role in Nature and in Science," New York, Academic Press, 286pp.
* Evans, R. C., 1964, "An Introduction to Crystal Chemistry," Second ed., Cambridge, The University Press, 410pp.
* Fyfe, W. S., 1964, "Geochemistry of Solids," New York, McGraw-Hill Book Co., 199pp.
Goldschmidt, V. M., 1958, in (Muir, A., editor), "Geochemistry," Oxford, Clarendon Press, 730pp.
Grigor'ev, D. P., 1964, "Fundamentals of the Constitution of Minerals," Translated and published in Jerusalem by the Israel Program for Scientific Translations.
Johnson, Q., Smith, G. S., and Kahara, E., 1969, "Automatic determination of crystal structure," *Science,* **164**(3884), 1163–1164.
Khamskii, E. V., 1969, "Crystallization from Solutions," New York, Plenum Press, 106pp.
Klassen-Neklyndova, M. V., 1962, "Plasticity of Crystals," New York, Plenum Press, 196pp.
Kostov, I., 1968, "Mineralogy," Edinburgh, Oliver & Boyd (author's transl. edited by P. G. Embrey and J. Phemister), 589pp.
Pauling, L., 1960, "The Nature of the Chemical Bond," Third ed., Ithaca, Cornell Univ. Press, 450pp.
Wells, A. F., 1962, "Structural Inorganic Chemistry," Oxford, Clarendon Press, 1055pp.

* Additional bibliographic references may be found in this work.

Cross-references: *Bonding; Electronegativity; Ion Exchange; Mineral Classes: Nonsilicates; Mineral Classes: Silicates; X-Ray Diffraction Analysis; Solid Solution; Trace Elements in Silicate Minerals.* Vol. IV B: *Anisotropism; Clays and Clay Minerals; Crystal Growth; Crystallography; Defects in Crystals; Isomorphism; Phyllosilicates; Polymorphism.*

CRYSTAL-FIELD THEORY

Crystal-field theory, developed by Bethe in 1929, has become prominent in recent years in the interpretation of physical and chemical properties of compounds of transition elements. The theory has wide application in geology and accounts for differences in geochemical behavior between transition metal ions and other cations possessing similar charges and sizes.

Crystal-field theory describes the effects of perturbation of d orbitals of a transition metal ion in a lattice. To a first approximation magnetic and exchange forces are disregarded. A transition metal ion in a crystal is subjected to the electric field of surrounding, negatively charged, anions or dipolar groups *(ligands)* which, in this theory, are represented by point negative charges. This "crystalline field" destroys the spherical symmetry possessed by an isolated transition metal ion, and the nature and magnitude of the changes induced within the central cation depend on the type, symmetry, and positions of surrounding ligands.

In an isolated transition metal ion, the five d orbitals are energetically equivalent (fivefold degenerate), and d electrons occupy singly as many orbitals as possible with spins orientated in the same direction so as to minimize interelectronic repulsion *(Hund's Rule)*. However,

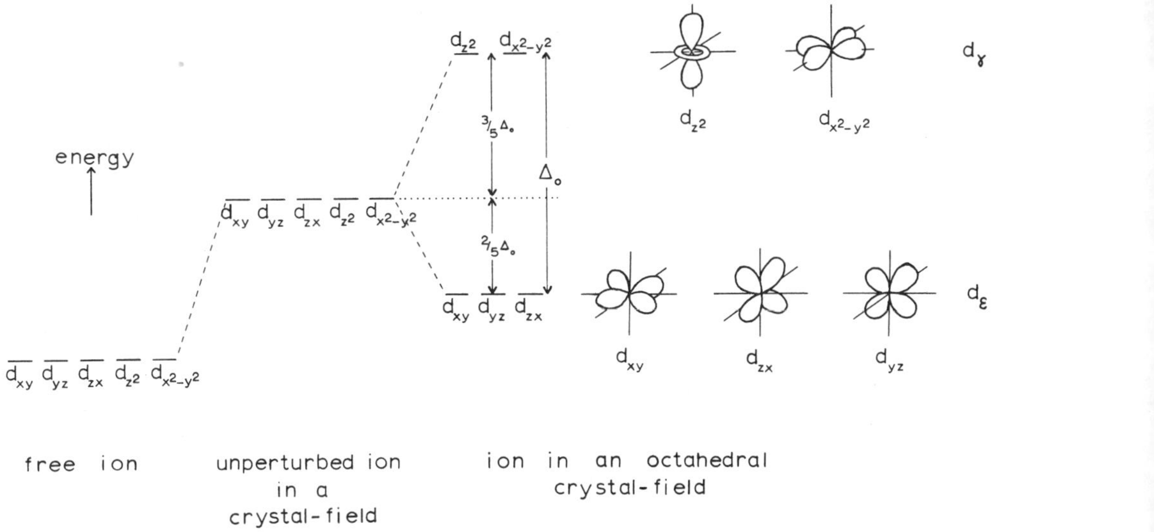

FIG. 1. Energy-level diagram illustrating the splitting of d orbitals by an octahedral crystal field.

the five d orbitals possess different spatial configurations. One group of orbitals, the $d\gamma$ orbitals (alternatively, e_g or T_3 orbitals), consisting of the d_{z^2} and $d_{x^2-y^2}$ orbitals, have lobes which are directed along three cartesian axes (Fig. 1). A second group of orbitals, the d_ϵ orbitals (alternatively, t_{2g} or Γ_5 orbitals), consisting of the d_{xy}, d_{yz}, and d_{zx} orbitals, possess lobes which project between the cartesian axes (Fig. 1). In a crystal field the d orbitals are no longer degenerate and some orbitals are lowered in energy relative to others. By preferentially filling low-energy orbitals, d electrons stabilize certain transition metal ions.

When a transition metal ion is in octahedral coordination with six identical ligands, electrons in all five d orbitals are repelled by the negatively charged ligands, but electrons in the two d_γ orbitals are repelled to a greater extent than are those in the three d_ϵ orbitals (Fig. 1). The energy separation between the d_ϵ and d_γ orbitals is termed "crystal-field splitting" and is denoted by Δ_o (alternatively, $10\ Dq$). Each electron in a d_ϵ orbital stabilizes a transition metal ion by $2/5\Delta_o$ whereas every electron in a d_γ orbital diminishes stability by $3/5\Delta_o$. The resultant net stabilization energy is called *"crystal-field stabilization energy"* (designated by CFSE).

The distribution of d electrons in a given transition metal ion in a crystal field is controlled by two opposing tendencies. Coulomb and exchange interactions between electrons cause them to be distributed over as many orbitals as possible so that there is a maximum number of unpaired electrons with parallel spins (*"Hund's stabilization energy"*), but the crystal-field splitting favors the occupancy of the group of orbitals with the lowest energy *(crystal-field stabilization energy)*. In an octahedral crystal field, ions possessing one, two, or three d electrons (for example, Ti^{3+}, V^{3+}, and Cr^{3+}, respectively) can each have only one electronic configuration and d electrons occupy different d_ϵ orbitals with spins parallel. However, ions possessing four, five, six, and seven d electrons (for example, Cr^{2+} and Mn^{3+}, Mn^{2+} and Fe^{3+}, Fe^{2+} and Co^{3+}, and Co^{2+} and Ni^{3+}, respectively) have a choice of electronic configuration. If the crystal-field splitting is small (the "weak-field" case), d electrons occupy both d_ϵ and d_γ orbitals singly to the maximum extent. The CFSE is reduced by electrons entering the d_γ orbitals, but energy is not expended in pairing of electrons in d_ϵ orbitals already half filled. Alternatively, if the crystal-field splitting is large (the "strong-field" case) it is energetically more favorable for d electrons to fill low-energy d_ϵ orbitals. In this situation, the gain in CFSE outweighs the electron pairing energy. Finally, ions possessing eight, nine, and ten d electrons (for example, Ni^{2+}, Cu^{2+}, and Zn^{2+}, respectively) can each possess only one electronic configuration and d orbitals are filled to completion. In a weak crystal field, an ion generally possesses more unpaired electrons (the *"high-spin"* state) than it does in a strong crystal field (the *"low-spin"* state). The distinction between the low-spin and high-spin configurations is fundamental in understanding magnetic and optical properties of transition metal compounds. Note that ions possessing three, eight, and six (low-spin configuration) d electrons acquire high stabilizations in octahedral crystal fields, whereas ions possessing zero, five, and ten d

electrons have zero CFSE in weak octahedral fields.

When a transition metal ion is in tetrahedral coordination, the d_γ orbitals become the more stable group, but the magnitude of the tetrahedral crystal-field splitting parameter, Δ_t, is smaller than that of the octahedral parameter, Δ_o. If the cation, ligands, and cation-ligand internuclear distances are identical in both octahedral and tetrahedral coordinations, $\Delta_t = 4/9\ \Delta_o$.

The magnitude of the crystal-field splitting parameter, Δ, may be estimated from measurements of absorption spectra. The energy to excite an electron from one d orbital to a vacant position in another of higher energy corresponds to the visible or near infrared region of the spectrum, and absorption of this radiation is the most general origin of color in transition metal compounds. Certain generalizations may be made about the dependence of the numerical value of Δ on the valence and atomic number of the metal ion, the symmetry of the coordinated ligands, and the nature of the ligands:

1. Δ values are higher for trivalent ions than for divalent ions. For the first transition series, the ranges are 7500–12500 cm^{-1} for divalent ions and 14000–21000 cm^{-1} for trivalent ions.

2. The values are about 30% higher for ions in each succeeding transition series.

3. Δ depends on the nature of ligands coordinated about the transition metal. Ligands may be arranged in order of increasing Δ, the *spectrochemical series*," an abridged version of which is I$^-$ < Br$^-$ < Cl$^-$ < F$^-$ < OH$^-$ $\leqslant$ carboxyanions < H$_2$O < NH$_3$ < SO$_3^{2-}$ < NO$_2^-$ << CN$^-$. Ligands at the beginning of the series generate weak crystal fields, whereas those at the end of the series produce strong crystal fields and low-spin electronic configurations in the central transition metal ion. The crossover point from high-spin to low-spin configuration varies from one cation to another and may be ascertained from magnetic measurements and crystal structure analyses of internuclear distances. Note that when an ion changes from a high-spin to low-spin configuration, internuclear distances are drastically reduced.

4. Δ depends on the symmetry of the coordinated ligands, the crystal-field splitting for tetrahedral coordination being 40–50% of the values for octahedral coordination.

Crystal-field stabilization energies of ions produce observable effects on thermodynamic properties (for example, lattice energies and heats of hydration) of transition metal compounds. If the ions were spherically symmetrical and no preferential filling of d orbitals occurred, a given thermodynamic function would display smooth periodic variation across a row of transition elements concomitant with contraction of the ions. However, observed values frequently fall on characteristic two-humped curves with maximum values for ions possessing three and eight d electrons (Fig. 2). Furthermore, the values lie above a smooth curve through points for ions possessing zero, five, and ten d electrons, for which there is zero CFSE. When crystal-field stabilization energies, estimated from absorption spectra of hydrated ions, are reduced from observed heats of hydration, the corrected values conform closely to the smooth curve (Fig. 2). Such an analysis provides convincing evidence in support of the d-orbital energy separation proposed by crystal-field theory.

A further consequence of considerable importance in the thermodynamics of heterogeneous systems, arising from the existence of crystal-field stabilization energies, is that solid solutions (for example, ferromagnesian silicates) containing as one component a compound of a transition element will rarely conform to ideal-solution behavior. A necessary condition for a solution to be ideal is that the heat of mixing of components is zero. This criterion is not fulfilled, however, whenever differences exist between crystal-field stabilization energies of ions in different environments. Measurements have shown that ferromagnesian silicate minerals such as the olivine, pyroxene, and amphibole series, are not ideal solid solutions because the crystal-field stabilization energy of the Fe^{2+} ion varies with composition in each series.

Although configurations in which identical ligands are in regular octahedral and tetrahedral coordination about a transition metal ion exist in aqueous solutions and in melts, such ideal configurations are rarely found in crystal structures. Frequently, cations are located in distorted environments in which metal-ligand internuclear distances are not constant, or surrounding ligands are not identical. Distortion of coordination polyhedra about a transition metal ion is to be expected for theoretical reasons (the *Jahn-Teller effect*). For example, if one of the metal d orbitals is empty while another of equal energy (such as the d_γ orbital group) is half filled, the compound is predicted to distort spontaneously to a different geometry in which a more stable electronic configuration is achieved by making the occupied orbital lower in energy. Sixfold coordination sites in compounds of Cr^{2+} and Mn^{3+} ions, which possess the electronic configuration $d_\epsilon^3 d_\gamma^1$, and Cu^{2+} ($d_\epsilon^6 d_\gamma^3$), are invariably dis-

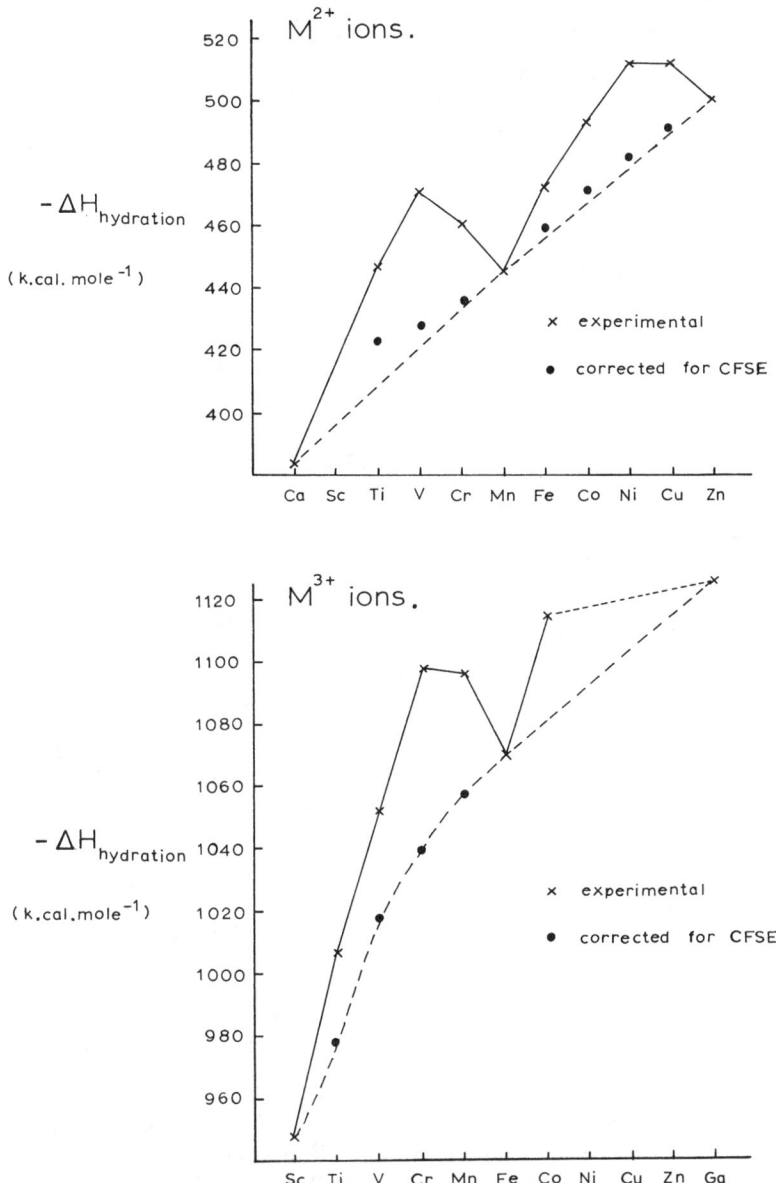

Fig. 2. Heats of hydration of transition metal ions. Experimental values lie on double humped curves. Corrected values, obtained by deducting crystal-field stabilization energies derived from absorption spectra, lie on smooth curves.

torted from regular octahedral symmetry as a result of the Jahn-Teller effect. Minerals in which Jahn-Teller distortion of the crystal structure is observed include the manganese (III) oxides hausmannite, bixbyite, partridgeite, hetaerolite, and manganite, and the copper (II) minerals tenorite and malachite.

Crystal-field theory provides insight into cation distribution in crystal structures and fractionation processes in geological environments. For example, cations in spinels, $X^{2+}Y_2^{3+}O_4$, occur in octahedral and tetrahedral coordination. In "normal" spinels, divalent ions (X^{2+}) occupy tetrahedral sites and trivalent ions (Y^{3+}) fill the octahedral sites. "Inverse" spinels, however, contain half the trivalent ions in the tetrahedral sites, and the remaining trivalent ions plus divalent ions in the octahedral sites. In this and similar cases it is possible to predict the distribution of cations between octahedral and tetrahedral sites, and to account for the type of spinel formed by each transition metal ion, by crystal-field theory. The trends are related to the magnitude of the octahedral "site preference energy" parameter, which is the difference between octahedral and tetrahedral crystal-field stabilization energies of a cation. The tendencies for divalent and trivalent ions to fill octahedral sites and form "inverse" and "normal" spinels, respectively, are related to the magnitudes of the octahedral site preference energies of the ions. Similarly, fractionation patterns of transition metal ions during magmatic crystallization may be correlated with the relative values of site preference energies of cations in the magma (octahedral and tetrahedral sites) and silicate minerals (octahedral sites).

The following geological phenomena may be explained by crystal-field theory:

1. Color and pleochroism of minerals. Differential absorption of polarized light in different directions in a crystal arises from the presence of transition metal ions in distorted environments. Electrostatic repulsion of d electrons varies with direction, and electronic transitions have greater probability (result in higher absorption) when light is polarized in certain directions. Crystal-field theory yields information on the probability of a transition.

2. Fe/Mg ratios in coexisting ferromagnesian silicate minerals. The distributions of Fe^{2+} ions in silicates are governed by the relative orders of CFSE of the Fe^{2+} ion in ferromagnesian silicate crystal structures.

3. Magnetic properties of minerals. Paramagnetism is related to the number of unpaired electrons within transition metal ions in minerals, which is governed by the strength of crystal fields.

4. Internuclear distances in crystal structures. Metal-ligand distances depend on the electronic configuration (high-spin or low-spin) of cations, the low-spin configuration leading to shorter internuclear distances. For example, pyrite (FeS_2) is diamagnetic and possesses abnormally short Fe–S internuclear distances compared to those in paramagnetic pyrrhotite (FeS).

5. Transition metal geochemistry in igneous, sedimentary, metamorphic, and marine environments is rationalized through crystal-field theory applied to both equilibrium and kinetic factors.

ROGER G. BURNS
WILLIAM S. FYFE

References

Burns, R. G., and Fyfe, W. S., 1967, "Crystal field theory and the geochemistry of transition elements" in (Abelson, P. H., editor) "Researches in Geochemistry," John Wiley & Sons, pp. 259–285.

Burns, R. G., 1970, "Mineralogical Applications of Crystal Field Theory," Cambridge University Press, 240pp.

Orgel, L. E., 1966, "An Introduction to Transition-Metal Chemistry: Ligand-Field Theory," London, Methuen and Co., Second ed., 186pp.

Cross-references: *Aqueous Solutions; Entropy; Solid Solution; Thermodynamics.* Vol. IV B: *Color in Minerals; Crystallography; Optical Mineralogy.*

CRYSTALLOGRAPHY—See Vol. IV B

CURIUM—See **ACTINIDE SERIES**

CURRENT BASE—See Vol. VI

CYCLES, GEOCHEMICAL

Chemical Reactions

Chemical reactions tend to reach a state of equilibrium or a point where no change occurs in the concentrations of reactants or products. The earth may be thought of as one large chemical system and geochemistry considers the study of the reactions by which the components of the earth approach equilibrium. The chemical driving forces of these reactions are the free energy differences of the components; the primary physical driving force is the energy of the sun. Also important is the energy derived from radioactive decay within the earth.

Geochemical Cycle

Considered on a grand scale, geochemical reactions recur countless times. Magma is generated within the crust of the earth and erupts on the surface or is emplaced within existing rocks. With time, these rocks are worn away by interaction with the atmosphere and hydrosphere. The reactions proceed through diagenesis, lithification, and orogenesis to metamorphism which, in its highest grade, regenerates a magma (the "pantocycle" of the Termiers, 1963; see Fig. 1). Of course, this never proceeds as a true cycle; at each step components are lost or gained and the secondary magma may not resemble the primary one at all. However, the cycle proceeds with a certain order. Each time the magma crystallizes, definite sequences of minerals are produced. The weathering of igneous and metamorphic rocks produces certain clay minerals whose composition is more a function of the weathering environment than of the type of rock being weathered. Through the cycle the chemical elements obey certain laws which determine the individual geochemical cycles of the elements. A study of the overall geochemical cycle is somewhat cumbersome and unrewarding; studies of the in-

CYCLES, GEOCHEMICAL

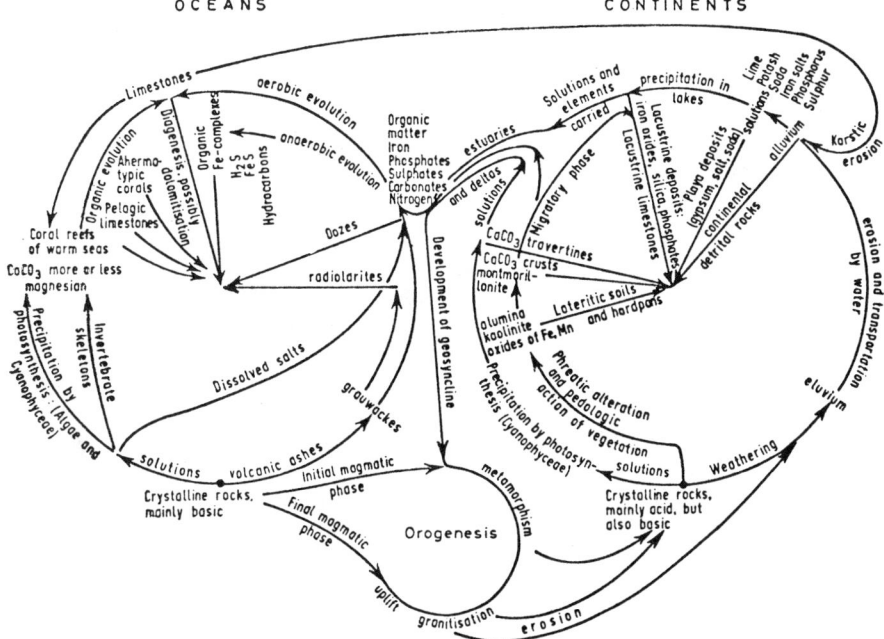

FIG. 1. The geochemical cycle or pantocycle after Termier (1963). The individual cycles are seen as a group with each representing an ordered series of irreversible processes and each proceeding as a result of the operations of the other parts. The overall development of the pantocycle is primarily the result of orogenesis and evolution. Orogenesis, or mountain building, governs the inorganic constituents, and evolution governs the biosphere and its interaction with the surface of the earth.

dividual cycles reveal many intricacies of the system and help give a fuller understanding of the chemical forces at work on the earth. (See *Carbon, Iron, Manganese, Nitrogen, Phosphorus, Sodium, Strontium,* and *Sulfur Cycles.* Also see Figs. 2–8, on pp. 210–216.)

Geochemical Cycles in Seawater

One very important phase of the geochemical cycle for most elements is that which occurs in the ocean. The soluble products of weathering are transported to the ocean and then removed by direct precipitation or by interaction with organic or inorganic substances within the sea. The average length of time an element spends in the ocean gives a measure of its reactivity; geochemists have sought to apply a quantitative measure to this time period. They have defined the residence time of an element as the amount in solution divided by the rate of input or removal:

$$\tau = \frac{A}{dA/dt}$$

There is a primary assumption involved in this concept: that the element is thoroughly mixed throughout the system, a condition necessitating that the residence time be several times the mixing rate of the entire ocean. The rate of supply and removal must be constant for several times the residence time.

The introduction of the residence time concept by T. Barth was a fundamental break with an older school of geochemists headed by V. M. Goldschmidt. The older school held that very soluble elements such as sodium accumulated in the ocean and, thus, that the age of seawater could be calculated from the expression:

$$\text{Age of the sea} = \frac{\text{Amount of sodium in sea}}{\text{Sodium supplied to sea per year}}$$

The age calculated from this equation is only about 200 million years, a figure clearly too low. What the figure represents is not the age of seawater but the mean length of time a sodium ion resides in the ocean. Table 1 lists the residence times of some of the more common elements calculated by E. Goldberg.

The geochemical cycles of elements in the sea may be studied in three major groups. These are: *(1)* major elements, *(2)* trace elements, and *(3)* nutrient elements.

The major elements are characterized by high concentrations and long residence times. These elements are removed from seawater by precipitation in the case of Ca and Sr, by interaction with detrital clay minerals either by

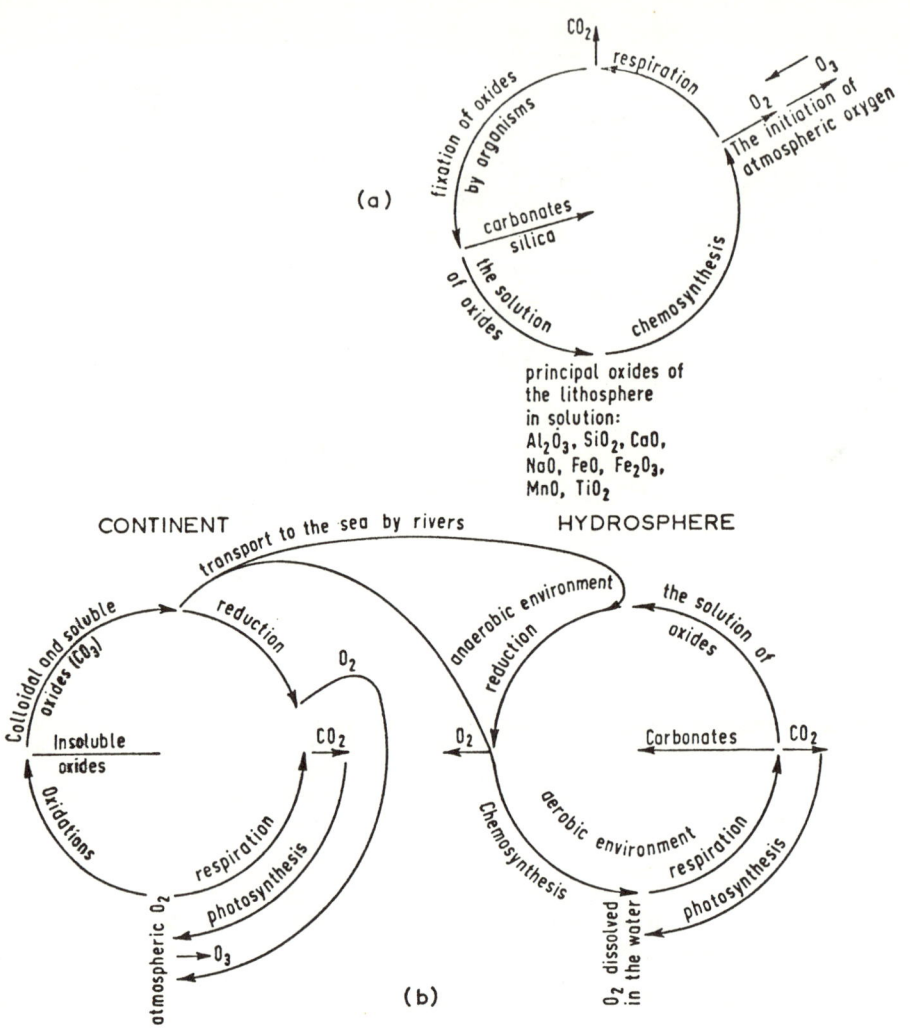

FIG. 2. The evolution of the oxygen cycle after Termier (1963). (a) The hydrosphere during the first two billion years of earth history: The primitive atmosphere was probably composed of H_2O, CO_2, CO, N_2, NH_3, SO_2, H_2S, H_2, CH_4, HCl, HF, and possibly other hydrocarbons and hydrogen halides. It was not an oxidizing atmosphere and could not support life as we know it. (b) The continents and hydrosphere after the first two billion years of earth history: Green plants have produced an abundance of free oxygen and the atmosphere has taken on its present characteristics.

surface absorption or by internal lattice rearrangement, or by the formation of authigenic minerals. Trace elements (e.g., Fe, Al, Ba) are more highly reactive as revealed by their short residence times. They may be removed in the same manner as major ions; surface absorption is probably the major control mechanism. The nutrient elements (primarily P, N, C, O) are extracted from near-surface waters by biological activity. When the organisms die, they sink and there is a strong flux of these life-giving elements from surface to deep water. If these organisms did not decay and release the nutrients to deep waters and if deep waters did not upwell to the surface, the depletion of the nutrient elements from the surface waters would curtail life in the sea. But the cycle is complete and this allows life to flourish in the ocean. (See also *Seawater—Chemistry; Seawater—Geochemical Balance*.)

Fractionations

As elements pass through the geochemical cycle, they tend to concentrate at certain

CYCLES, GEOCHEMICAL

TABLE 1[a]

Element	Type[b]	Concentration in Seawater (mg/liter)	Residence Time (yr)
Li	t	0.17	2×10^7
Be	t	6×10^{-7}	1.5×10^2
Na	m	10,500	2.6×10^8
Mg	m	1,350	4.5×10^7
Al	t	0.01	1×10^2
K	m	380	1.1×10^7
Ca	m	400	8×10^6
Cr	t	5×10^{-5}	3.5×10^2
Fe	t	0.01	1.4×10^2
Si	n	3	8×10^3
Ag	t	3×10^{-4}	2×10^6
Ba	t	0.03	8.4×10^4
Au	t	4×10^{-6}	5.6×10^5
U	t	3×10^{-3}	5×10^5

[a] From Goldberg, 1963.
[b] t = trace element; m = major element; n = nutrient element.

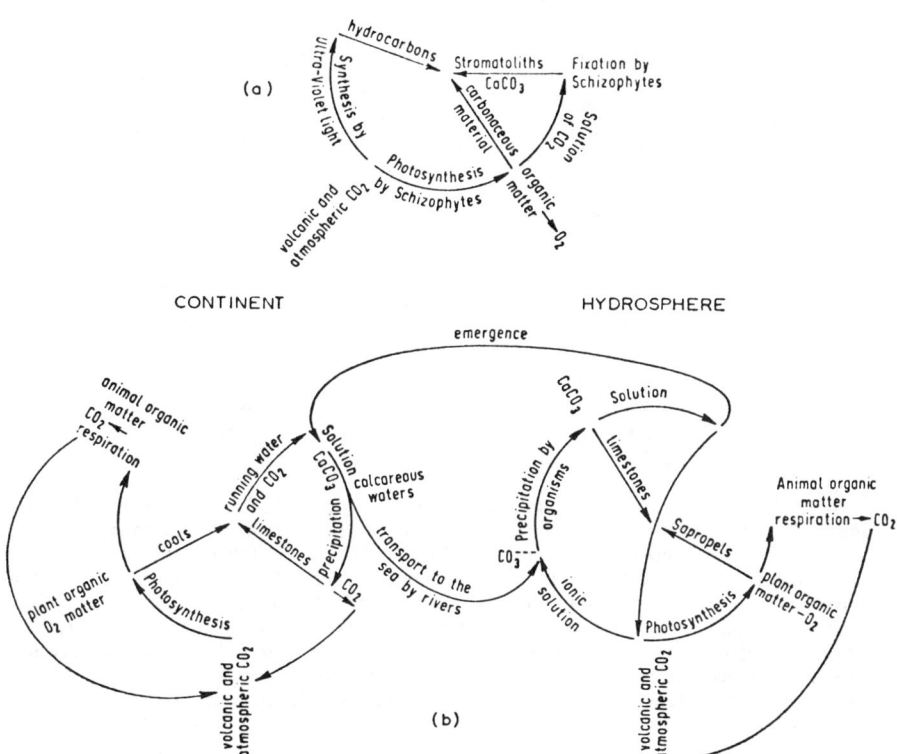

FIG. 3. The evolution of the carbon cycle after Termier (1963). (a) During the primitive atmosphere stage before the appearance of land vegetation: The lower pH of seawater and the lack of abundant $CaCO_3$-secreting organisms left most of the CO_2 in the atmosphere. (b) The present epoch after the appearance of land vegetation: The near saturation of the ocean with respect to $CaCO_3$ and the abundance of $CaCO_3$-secreting organisms decrease the residence time of CO_2 in the various systems. Coupled with extensive photosynthesis, a very active carbon cycle results. Large-scale withdrawal of carbon to form coal and petroleum hydrocarbons has occurred in addition to the deposition of calcareous sediments.

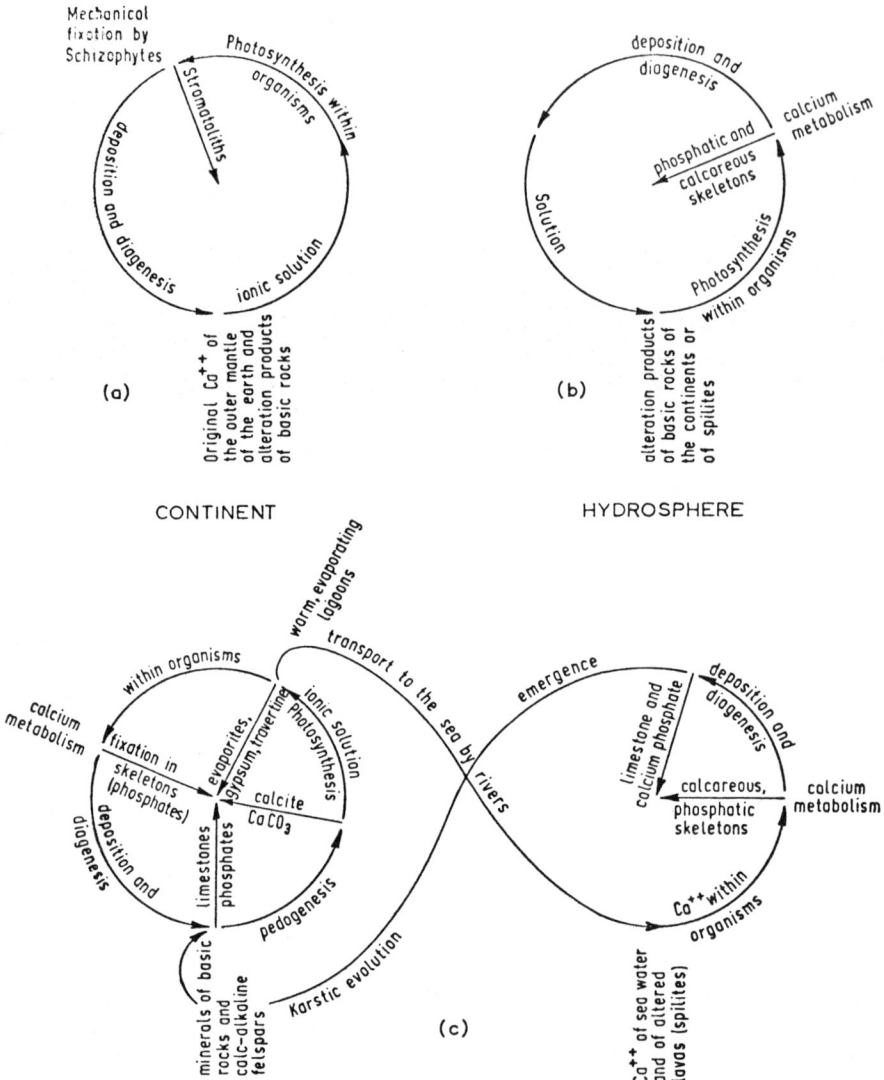

Fig. 4. Evolution of the calcium cycle after Termier (1963). (a) During the Precambrian (>600 million years ago): The low pH of surface waters causes a rapid leaching of Ca^{2+} from rocks. There is limited $CaCO_3$ precipitation. (b) During Cambrian times (600–500 million years before present): Animals with calcareous skeletons appear and cause an increased rate of deposition of $CaCO_3$. (c) During the Present epoch: The appearance of life on the continents has resulted in an increased rate of metabolism of Ca^{2+} as the carbonate and phosphate. The increasing pH of seawater has led to an accumulation of $CaCO_3$ in ocean sediments, aided in particular by the worldwide development of calcareous plankton during and since the late Mesozoic.

stages in association with other elements. Elements that concentrate in the earth's core or with iron deposits are called siderophile; those which are associated with sulfide deposits are chalcophile; those occurring as or with silicates are lithophile; and those which concentrate in the atmosphere are atmophile. Another type of fractionation exhibiting itself in the cycle is the concentration of the different isotopes of the lighter elements. Generally, light isotopes (e.g. C^{12}, O^{16}, S^{32}) are concentrated in organic materials and heavier ones in low-temperature minerals. At high temperatures, isotope fractionation is not as pronounced, providing a geothermometer for certain mineral deposits (see also *Isotope Fractionation*).

Equilibrium

The net effect of the operation of the geochemical cycle is a gradual approach to equi-

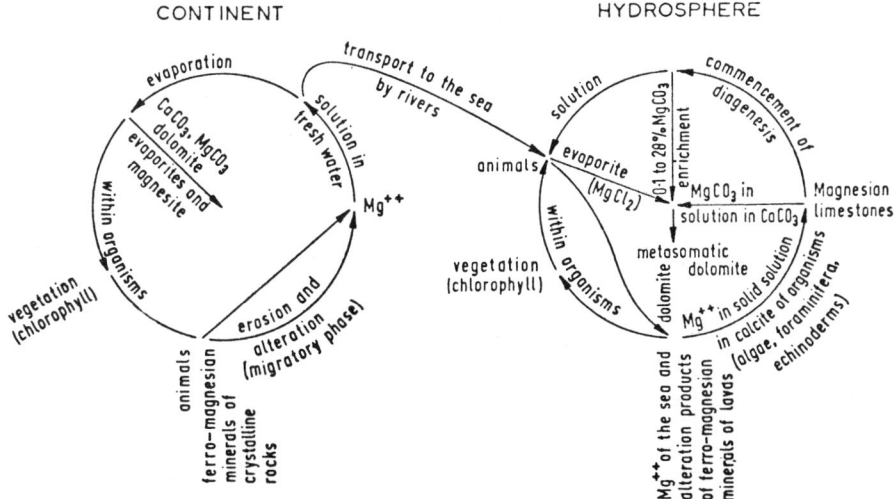

FIG. 5. Present cycle of magnesium after Termier (1963). The magnesium cycle follows rather closely the calcium cycle, because Mg is commonly withdrawn also by certain lime-secreting organisms. Magnesium-rich calcites are metastable and contribute to diagenetic dolomite metasomatism. Magnesium is also a major component in evaporite salts and certain authigenic clays. Although there has thus been a relatively steady withdrawal of magnesium through time, magnesium salts never approach saturation in open seawater.

librium of the constituents of the earth. A major disequilibrium exists between the interior of the earth and the crust. The interior is composed of materials rich in reduced metals; the surface contains oxidized states and is poorer in metals. Gradually, lighter metals are being transferred from the interior and added to the earth's atmosphere, hydrosphere, and lithosphere. The result of this is growth with time of continents and oceans (see *Geochemical Evolution of Core, Mantle, and Crust*).

H. C. Urey (1956) has expressed the basic equilibrium of the earth's crust by the following equation:

$$CaSiO_3 + CO_2 \rightleftarrows CaCO_3 + SiO_2$$

The forward reaction represents weathering of silicates (e.g., $CaSiO_3$) by acidic gases (CO_2); the reverse reaction characterizes metamorphism of a salt ($CaCO_3$) and an acid silicate (SiO_2). As with any chemical reaction, the equilibrium position is determined both by the concentrations of reactants and products and by the temperature and pressure. Weathering proceeds when the concentration of acidic gases increases and when the temperature and pressure are low. Metamorphism requires a high temperature and pressure as well as a concentration of salts and acidic silicates. This reaction more than any other may be considered the controlling equation for the overall geochemical cycle.

WILLARD S. MOORE

References

Barth, T. F. W., 1962, "Theoretical Petrology," Part V, "Geochemical Cycles," pp. 367–379, New York, John Wiley & Sons.

Goldberg, E. D., 1963, "The Oceans as a Chemical System," in (Hill, M. N., editor), "The Sea," New York, Interscience Publ., div. of John Wiley & Sons.

Krauskopf, K. B., 1967, "Introduction to Geochemistry," New York, McGraw-Hill Book Co., 721pp.

Termier, H., and Termier, G., 1961, "L'Evolution de la Lithosphère, III. Glyptogénèse," Paris, Masson et Cie, 471pp.

Termier, H., and Termier, G., 1963, "Erosion and Sedimentation," London, D. Van Nostrand Co. (transl. by D. W. and E. E. Humphries), 433pp.

Urey, H. C., 1956, "Regarding the early history of the Earth's atmosphere," *Geol. Soc. Am. Bull.*, **67**, 1125–1128.

Cross-references: *Calcium; Calcium Cycle; Carbon; Carbon Cycle; Chlorophyll; Geochemical Evolution of Core, Mantle, and Crust; Hydrocarbons; Iron; Magnesium; Nitrogen Cycle; Outgassing of the Planet Earth; Oxygen; Oxygen— Its Evolution in the Earth's Atmosphere; Paleosalinity; pH-Eh; Phosphorus Cycle; Photosynthesis; Seawater, Chemistry; Seawater, Geochemical Balance; Silica—Biogeochemical Cycle; Silica Cycle; Sodium Cycle; Weathering, Chemical.* Vol. I or II: *Carbon Dioxide Cycle in the Sea and Atmosphere.* Vol. I: *Marine Sediments.* Vol. II: *Earth—Geology of the Planet.* Vol. IV B: *Banded Ion Ores.* Vol. V: *Orogenesis; Petrology.* Vol. VI: *Sediments.*

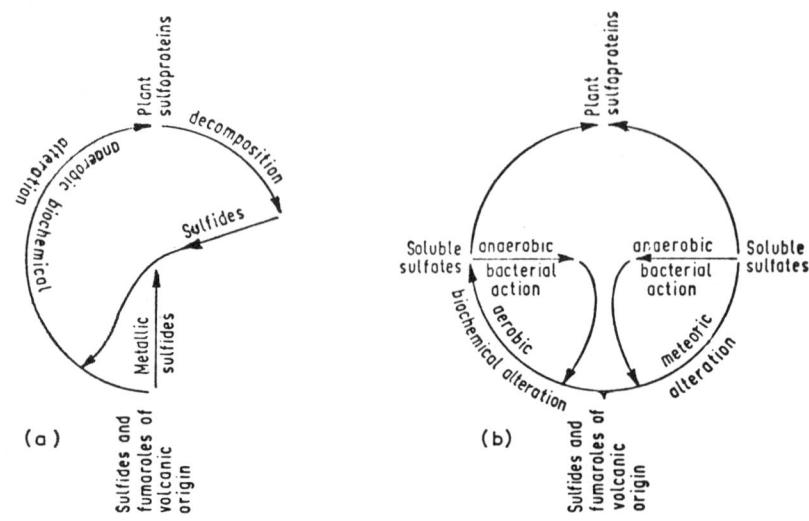
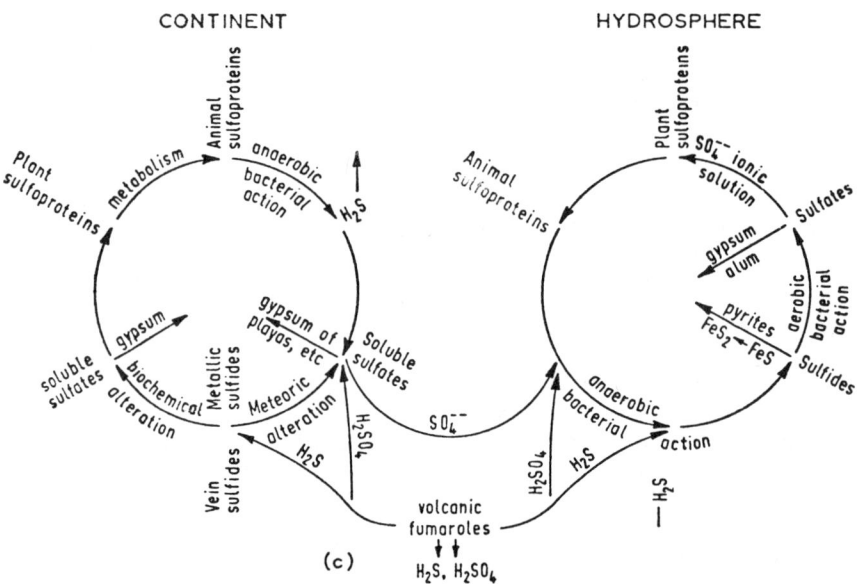

Fig. 6. The evolution of the sulfur cycle after Termier (1963). (a) Primitive atmosphere stage: The release of SO_2 and H_2S from the degassing of the earth results in a high accumulation of acidic sulfur gases in the atmosphere. These react with the lithosphere to produce sulfides. (b) After the development of life, about 3.5 billion years ago, there is anaerobic reduction of sulfates by bacteria. This marks the commencement of isotopic fractionation of S^{32} and S^{34}. (c) Present epoch (since the appearance of life on the continents): The atmosphere is now oxidizing and contains few sulfur gases. Sulfate is the predominant form of sulfur in the ocean and in evaporites. Elemental sulfur is found around volcanoes, and associated with salt domes (the product of bacterial sulfate reduction). Sulfides are found associated with metal ore deposits.

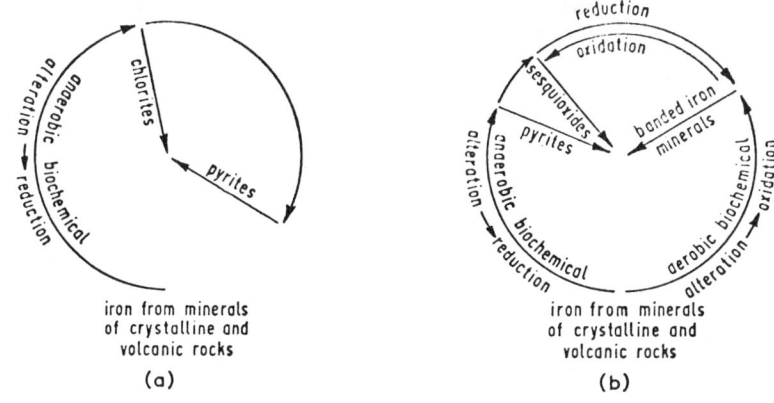

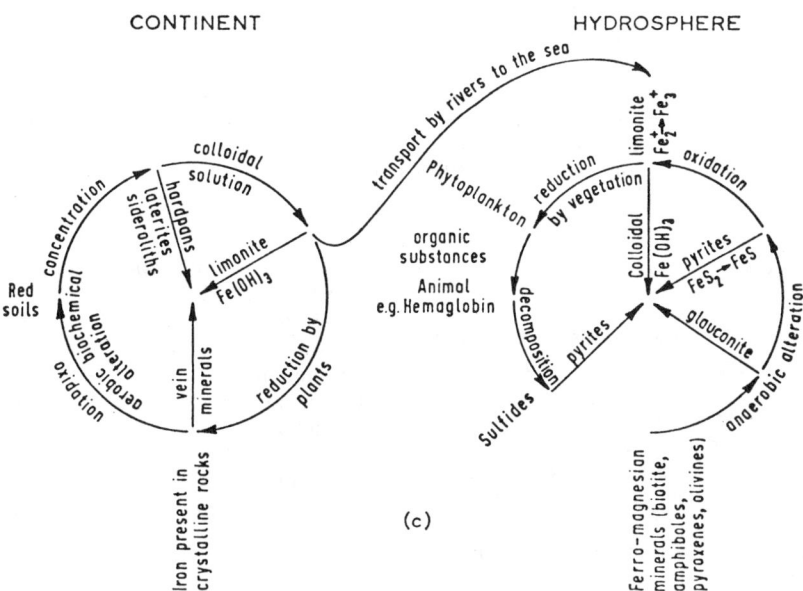

Fig. 7. The evolution of the iron cycle after Termier (1963). (a) During the primitive atmosphere stage when the atmosphere was reducing, iron extracted from rocks existed in the free state or as sulfides. (b) During the Precambrian as the atmosphere became oxidizing due to the activity of aquatic plants, iron oxides were laid down as sedimentary iron ores in vast deposits that have never been matched since then. (c) Present epoch (after the appearance of abundant land plants): Oxygen in the atmosphere has produced iron carbonate and additional iron oxide deposits. The solubility of iron in natural waters has been decreased because Fe^{3+} is much less soluble than Fe^{2+}. In subtropical soils iron is concentrated in laterite and ferruginous duricrust ("ferricrete").

CYCLES, GEOCHEMICAL

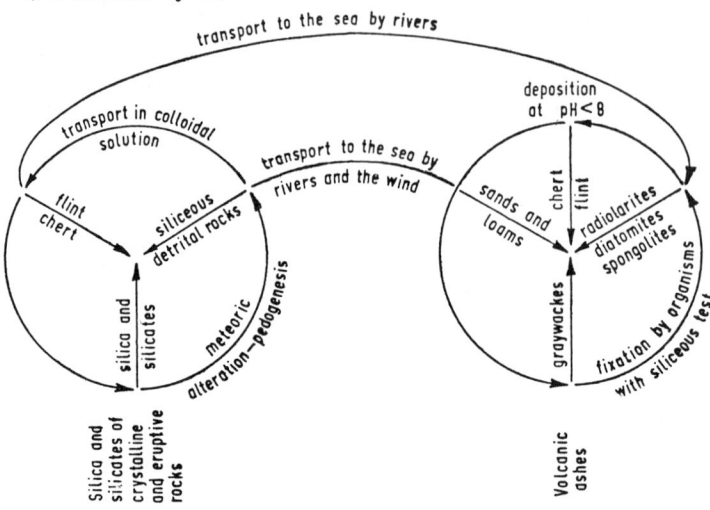

FIG. 8. The silica cycle after Termier (1963). The silica cycle is largely a study of the weathering of the lithosphere. Silica removed from the continents in solution is rapidly withdrawn from seawater where it acts as nutrient for radiolarians, diatoms, and sponges. In semiarid continental soils, silica is often concentrated as a duricrust ("silcrete").

CYCLIC SALTS

Salts that are dissolved in seawater are subject to transfer into the atmosphere in spray and bubbles. They are then liable to be carried by clouds inland and precipitated on the soil along with rain or snow. Eventually they may return, via groundwater and streams, to the ocean. Hence the expression "cyclic salts."

It has become recognized that there are anomalous and variable ratios between the various salts that do not correspond to the normal "conservative" oceanic salt ratio. Japanese workers have found that the bigger and heavier ions and higher valency elements become concentrated in the bubble spray. Thus bromine becomes enriched in contrast to chlorine, and potassium and calcium are enriched in contrast to the sodium. (For details, see *Salts, Cyclic* in Vol. VI of this Encyclopedia.)

The salt particles play a role in nucleation of moisture particles in warm rain. In semiarid lands the cyclic salts tend to concentrate in soils or are transported into playa lakes. They become thus a very serious agricultural problem. Although the total salts in rain tend to diminish in proportion to the distance from the sea, the precipitation-evaporation ratio also drops, so that in many cases the problem is worse farther inland.

A special condition occurs along the coastal desert shores of Chile and Peru where upwelling brings up the cold waters of the Peru (Humboldt) current. These deep waters are enriched in nitrate and the sea breeze carries the nitrate-rich salt spray particles into the Atacama Desert. Rains at higher elevations bring these ions also into the groundwater and a combination of leaching and capillary action concentrates important economic deposits of nitrates.

RHODES W. FAIRBRIDGE

References

Bloch, M. R., Kaplan, D., Kertes, V., and Schnerb, J., 1965, "Ion separation in bursting air bubbles," *Nature*, **209**, 802–803.

Woodcock, A. H., and Blanchard, E., 1955, "Tests of the salt nuclei hypothesis of rain formation," *Tellus*, **7**, 437–448.

Yaalon, D. H., 1964, "The concentration of ammonium and nitrate in rainwater," *Tellus*, **16**, 200–204.

Cross-references: Evaporite Processes. Vol. I: Upwelling. Vol. II: Maritime Climate: Oceanicity; Salt Nuclei. Vol. VI: Salts—Cyclic.

CYCLOSILICATES—See MINERAL CLASSES: SILICATES

D

DEAD SEA—*See* Vol. III

DENUDATION—*See* Vol. III

DERIVATIVE DIFFERENTIAL THERMAL ANALYSIS

Differential thermal analysis (DTA) is the measurement, preferably continuously and automatically, of the amounts of heat taken in (endothermic) or given out (exothermic) by materials as they are heated up or cooled down at a constant rate and undergo some reaction such as dehydration, decomposition, oxidation, melting, or crystallographic changes.

The heat effects which occur during such physical and chemical changes are measured as a differential temperature (ΔT), which when plotted continuously against true temperature (T) or time (t) gives a DTA curve.

The *derivative differential thermal analysis* (DDTA), or "derived differential thermal" curve is the graphical representation of the continuous plot of the first calculus derivative of the differential temperature curve, $d\theta/dt$ (where θ is the differential temperature), against time or temperature.

It therefore represents the first derivative curve of the DTA curve.

History

Although first suggested in 1908 by Burgess, significant applications of this technique have only appeared relatively recently, i.e., Frederickson (1954), de Keyser, 1953, Freeman and Edelman, 1959 (see Wendlandt, 1964), Greenwood, Holdam, and Macrae, 1948, Kelley and Fisher, 1958, Zenchelsky and Segatto, 1957, Borchardt and Daniels, 1957, Kissinger, 1957, and Campbell and Gordon, 1957 (see Campbell, Gordon, and Smith, 1959), and Garn (1965).

Instrumentation and Theory

The DDTA curve of a sample may be obtained by two alternative methods.

1. The first method involves the direct differentiation of the DTA curve, either *manually* (Frederickson, 1954) or *by an R-C (resistance-capacitance) system* (Campbell, Gordon and Smith, 1959). In the former method, successive changes in the value of temperature (T-T') for equal increments of T were observed, recorded, and expressed as $\Delta(T\text{-}T')/\Delta T$, where ΔT was a 10°C temperature increment. In this way the first derivative of the DTA curve is obtained from the manual plot of $\Delta(T\text{-}T')/\Delta T$ vs. T.

In the latter method a recorder containing a retransmitting potentiometer coupled to the slide wire is used; the signal from the potentiometer circuit is recorded on another recorder after differentiation by a simple R-C circuit similar to those of Greenwood, Holdam, and Macrae, 1948, Zenchelsky and Segatto, 1957, and Kelley and Fisher, 1958. Campbell, Gordon, and Smith (1959) also include an evaluation of alternative R-C circuits, calibration procedures, theoretical aspects, and applications. Garn (1965) suggests circuit modifications.

2. The second method involves the simultaneous DTA of two samples in a single sample holder and furnace (Freeman and Edelman, 1959). The temperature differentials of the two samples may be represented by the equations:

$$\Delta T_s = T_s - T_r \quad \text{(sample 1)}$$
$$\Delta T_{s'} = T_{s'} - T_{r'} \quad \text{(sample 2)}$$

where T_s is the temperature of sample (1); $T_{s'}$, the temperature of sample (2); and T_r and $T_{r'}$ the temperatures of the inert reference material used with samples 1 and 2, respectively.

If the temperature applied to each system differs, subtracting the second equation from the first gives:

$$\Delta T_s - \Delta T_{s'} = (T_s - T_{s'}) + (T_{r'} - T_r)$$

However, the heating rates are identical, thus

$$\frac{d\Delta T_{\text{s.s}'}}{dt} = \frac{dT_{\text{s.s}'}}{dt}$$

From the last equation it is clear that derivative curves are obtainable using a single furnace and block, where the inert reference material is replaced by a duplicate sample.

The geometrical arrangement and thermodynamic properties of the substance under investigation cause variations in the temperature differential. If too high, two DTA curves may be obtained with opposite deflections but a common base. Conversely, a true derivative curve, but with reduced sensitivity, is produced where the temperature differential approaches zero (Wendlandt, 1964). However, some temperature differential must be maintained, usually by placing one sample slightly closer to the furnace heating coils. A temperature lag so produced which is satisfactory for one system is often unsatisfactory for another (Freeman and Edelman, 1959).

By a modification of this method (Freeman and Edelman, 1959), using a single three-hole sample block, containing duplicate samples and an inert reference material, simultaneous, but separate and independently recorded DTA and DDTA curves may be obtained. Such pairs of curves determined simultaneously under identical experimental conditions make possible higher quality direct comparison studies of the thermal data being recorded. Thus is obviated the effects of the controllable operational variables on curve configurations (Bayliss and Warne, 1962), which to some extent must occur if the DTA and DDTA curves are obtained separately even though the determinations might be consecutive.

Curve Configuration

Theoretically, the DDTA curve will exhibit two peaks, one below and one above the base line for each individual DTA peak (compare curves A and B in Fig. 1).

In practice this complete complementary paired-peak configuration is not always achieved due to the close proximity of subsequent thermal reactions. These cause the superposition, interference, and resultant modification of the peaks, representing them on both DTA and DDTA curves.

Applications

The configuration of the DDTA curve intensifies even slight changes in slope on DTA curves and individual peaks. It therefore exhibits more detail and aids in the detection, identification, evaluation, and interpretation of features on the DTA curve. Such features, often graphically minor, may be of major importance and might otherwise be overlooked or erroneously attributed to unimportant instrumental peculiarities or instability.

For example, slow transitions occurring over wide temperature ranges show poor DTA definition because of the difficulty of determining accurately the temperatures or positions where individual peaks deviate from and return to the base line. Frederickson (1954) has demonstrated that the degree of precision with which the temperature can be determined,

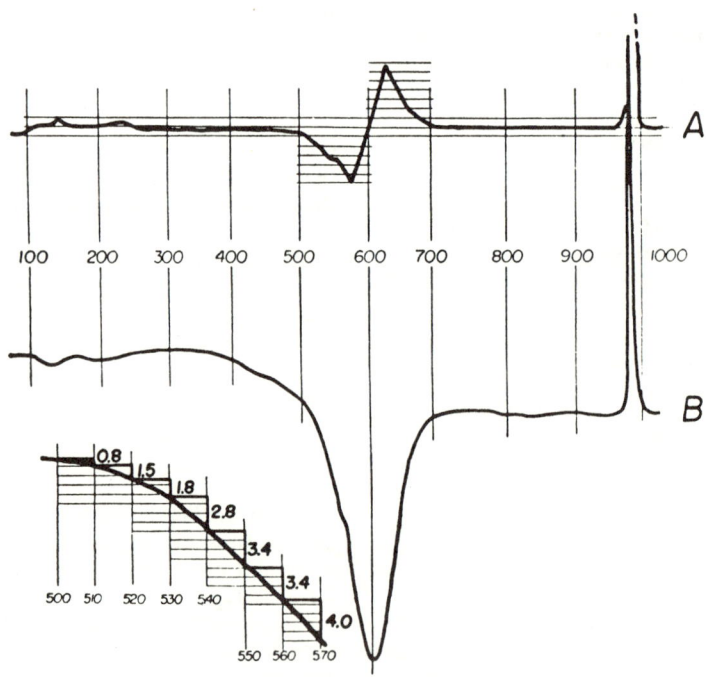

Fig. 1. Differential thermal curve (B) of a sample of kaolinite. Curve A represents the first derivative of the derived differential thermal curve for the same sample (Frederickson, 1954).

where any significant deviation from the base line can be observed, located, or measured, is greatly improved by using DDTA. Furthermore, this leads automatically to a more accurate location of the base line.

Such clarification and refinement of the data represented by the original DTA curve is vital to the accuracy of studies involving decomposition temperatures for peak areas, particularly where the curve is complicated by drifting or displaced base lines.

For zero-order reactions, such as melting or crystalline phase changes, the amount of reactant present is proportional to the magnitude of the DDTA peak representing it. This led to the improved determination of the concentration of individual components in mixtures (Campbell, Gordon, and Smith, 1959).

In respect to this, often the close but not complete overlap of two DTA peaks caused by different components is only evidenced by a slight inflection or kink on one limb of the resultant composite peak. (For example see curves 8, 9, and 10, Fig. 12 in Warne, 1965.) Such minor and easily overlooked features stand out well on the comparable DDTA curve.

Derivative differential thermal analysis may therefore be applied with advantage to studies of base-line location, peak configuration and area, the evaluation of individual components of mixtures, crystallographic inversion and melting temperatures, and kinetics.

S. St. J. Warne

References

Bayliss, P., and Warne, S. St. J., 1962, "The effects of the controllable variables on differential thermal analysis," *Am. Mineralogist,* **47,** 775–778.
* Campbell, C., Gordon, S., and Smith, C. L., 1959, "Derivative thermoanalytical techniques —instrumentation and applications to thermogravimetry and differential thermal analysis," *Anal. Chem.,* **31,** 1188–1191.
Frederickson, A. F., 1954, "Derived differential thermal curves," *Am. Min.,* **39,** 1023–1025.
* Garn, P. D., 1965, "Thermoanalytical Methods of Investigation," New York, Academic Press, 606pp.
*Wendlandt, W. Wm., 1964, "Thermal Methods of Analysis," New York, Interscience Publ., div. of John Wiley & Sons, 424pp.
Warne, S. St. J., 1965, "Identification and evaluation of minerals in coal by differential thermal analysis," *J. Inst. Fuel,* **38,** 207–217.

* Additional bibliographic references may be found in this work.

Cross-references: *Crystal Chemistry; Differential Thermal Analysis.*

DESALINATION PROCESSES—THE U.S. DESALTING PROGRAM

As part of his quest for water, man has long been trying to brew fresh water from the sea. Aristotle, about 350 BC, taught that vapor formed from seawater, when condensed, is no longer salty.

The first practical conversion units did not come until the advent of the steamship with its requirement of fresh water for boilers. For many years, desalination equipment was of major interest only for ships. For this use there were two principal equipment criteria: reliability of operation and space limitations. The fuel cost was a minor consideration. In the present century, land-based conversion plants have been built in very arid regions where little or no natural fresh water is available.

A new dimension was added to the conversion of seawater to fresh water when the U.S. Congress passed the Saline Water Act of 1952. Rising concern over the acute shortage of water in the arid areas and the excessive withdrawal of underground waters motivated government action to provide for the development of practicable low-cost means of producing from seawater, or other saline waters, water of a quality suitable for agriculture, industrial or municipal uses. The Department of the Interior established an Office of Saline Water to conduct a research program for the development of technology for large, low-cost, land-based plants. The 1952 Act provided $2 million for a five-year program.

While it is relatively simple to produce fresh water from ocean water, to do it at low-cost is extraordinarily difficult. The legislation was therefore repeatedly amended, and in 1965 sums were authorized up to $200 million for the period 1962–72. A substantial expansion of the desalting program thus became possible with prototype plant construction.

The increase in water use in the United States has been phenomenal. At the turn of the century, about 40 billion gallons of water were used per day. By 1920, the figure had doubled. It doubled again by 1944 and again by 1965. The current use of water in the United States is now 366 billion gallons per day.

Some areas have ample supplies but in the more arid areas and in some areas where population density continues to rise, the supply of fresh water is being sorely taxed.

When the Office of Saline Water was established, the cost of producing 1000 gallons of fresh water from seawater ranged upward from $4. By 1971 the cost range was about 65–75/ 1000 gal. The short-range goal is to provide

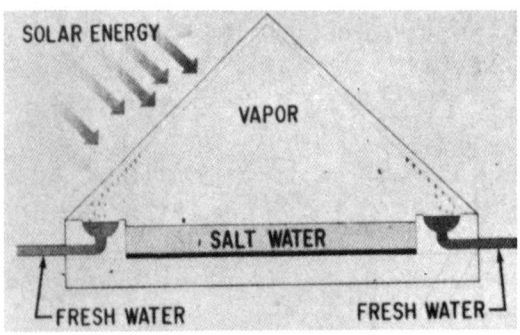

Fig. 1. Schematic diagram of the operation of a saline water conversion solar still. Solar distillation has the advantage of "free" energy. A shallow basin, painted black to absorb the sun's rays, is filled with saline water. Heat from the sun causes the water to evaporate. As the vapor rises, it contacts the cooler glass surface where it condenses and runs down into collecting troughs on each side of the still. Under good solar intensity, approximately one pound of fresh water per square foot of basin surface per day can be obtained.

processes and plant designs with the lowest desalting costs at the earliest possible time, which will permit planning for the intermediate range goal, a plant of 50 million gallons per day. The largest desalting plants built in the United States to date have capacities up to 2.6 million gallons per day. Much of the pilot-plant work is conducted at the O.S.W. Research and Development Test Station at Wrightsville Beach, North Carolina.

The long-range activity is basic research, in part conducted through the award of grants or contracts to agencies, firms and universities.

The simplest and cheapest form of desalination utilizes solar heat, but only small-scale development seems to be feasible (see Figs. 1 and 2). Processes under development can be divided into five categories, with a number of different approaches: *(1)* distillation, using either nuclear or fossil fuel, *(2)* membranes, *(3)* crystallization, *(4)* humidification, and *(5)* chemical.

Long-tube vertical evaporators have been used for many years in the chemical industry to concentrate liquors such as waste sulfite in the pulp industry. Experimental work on the development of the long-tube vertical multiple effect distillation process (LTV) to adapt it for seawater conversion was carried out in a 2000 gallon per day pilot plant at Wrightsville Beach, N.C. After test work this resulted in the design of a 1 million gallon per day LTV plant which was built at Freeport, Texas, in 1960–61 at a cost of $1.25 million, tests indicating a process potential of 1000 gallons of fresh water for less than $1 in the 1 million gallon per day range. The plant at Freeport was initially composed of twelve large evaporators, with 500 tubes, each two inches in diameter and about forty feet long (Fig. 3). Subsequently it was modified to include a multi-stage flash and vertical tube evaporator process (MSF/VTE). Seawater, preheated through heat exchangers is introduced in the

Fig. 2. A solar still of 4000 gallon per day capacity was constructed on the Greek island of Symi by Church World Services. Technical assistance was provided by the Office of Saline Water (O.S.W.).

Fig. 3. The O.S.W. Demonstration Plant at Freeport, Texas, uses a long-tube vertical distillation process to convert seawater into one million gallons of fresh water daily.

top of the first evaporator and falls through the tubes. Steam around the outside of the tubes and within the evaporator shell causes part of the seawater to boil as it falls. Emerging at the bottom of the evaporator is a mixture of vapor and hot brine. The residual hot brine is pumped to the top of the second evaporator where under slightly reduced pressure it again falls through the inside of the tubes. The vapor produced in the first evaporator flows to the outside of the tube bundle in the second unit. Here the vapor is condensed to fresh water by giving up its latent heat of vaporization to the seawater falling through the tubes, which again causes part of the water in the tubes to boil. This same process is repeated through all twelve units or "effects." The condensed vapor, fresh water, is pumped from each effect through heat exchangers to recover as much heat as possible and then into a common line leading to the storage tank.

A second 1 million gallon per day demonstration plant utilizing a multistage flash distillation process was built in 1962 on Point Loma near San Diego, California, at a cost of $1.6 million. Through improved techniques, the operating temperature of the plant was increased from 190 to 250°F, at which it produced 1.4 million gallons per day, the cost being $1 to $1.2 per 1000 gallons. In 1964 the plant was transferred to Guantanamo Naval Base. In a flash distillation plant, the seawater is progressively heated and then introduced into a large chamber where a pressure just below the boiling point of the hot brine is maintained. When the brine enters this chamber, the reduced pressure causes part of the liquid to boil—or flash—into steam. The remaining brine is passed through a series of similar chambers (multi-stages) at successively higher vacuum where the flash process is repeated at progressively lower temperatures. The San Diego plant utilized thirty-six such flashing stages. About 90% of the heat required for boiling is recirculated, and only 10% is supplied by the salt water heater, a steam boiler utilizing fuel oil.

A third 1 million gallon per day plant, using a forced-circulation vapor compression process, was completed in 1963 near Roswell, New Mexico for $1.7 million (Fig. 4). Forced-circulation evaporators have been used successfully in a number of industrial applications. Combining a vapor compressor with two forced-circulation evaporators is a new approach to the saline water problem. The saline feed water is introduced into an exchange unit for removal of hardness and then is heated to about 144°F. At this point, it is treated with acid and then introduced into a vacuum deaerator for the removal of dissolved gasses. It is then further heated to 224°F and pumped into the first effect evaporator. The water is circulated through the tubes under pressure to prevent boiling. As the water emerges into the lower pressure vapor dome, part of the water immediately vaporizes, concentrating the brine 1.6 times. The water that did not vaporize in the first effect is fed to the second evaporator where further heating concentrates it to a factor of 4.0 as it vaporizes in the second vapor dome. The final brine is then sent through a series of heat exchangers to give up its heat to the incoming feed water. The steam generated in the first effect is collected in the vapor dome and pumped to the outside of the tubes in the second evaporator where it gives up its latent heat of vaporization by increasing the temperature of the water in the tubes and thereby is condensed to fresh water. Steam generated in the second evaporator is compressed by the vapor compressor, thus raising its temperature, and is recycled to the first effect to boil more water and to be condensed to fresh water.

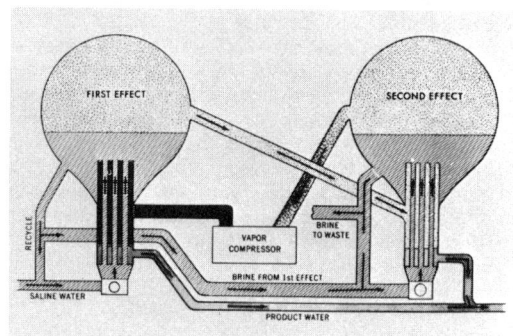

Fig. 4. The O.S.W. Demonstration Plant at Roswell, New Mexico, utilizes a forced-circulation, vapor-compression distillation process to make one million gallons of fresh water a day from brackish water.

A 250,000 gallon per day electrodialysis process demonstration plant was built in 1960 (for $480,000) to demineralize the brackish well water of Webster, South Dakota. An electrodialysis cell (Fig. 5) consists of a sandwich of alternating cation and anion permeable membranes. Upon the application of an electric current, the positively charged ions (such as sodium) pass through the cation permeable membranes and the negatively charged ions (such as chloride) move in the opposite

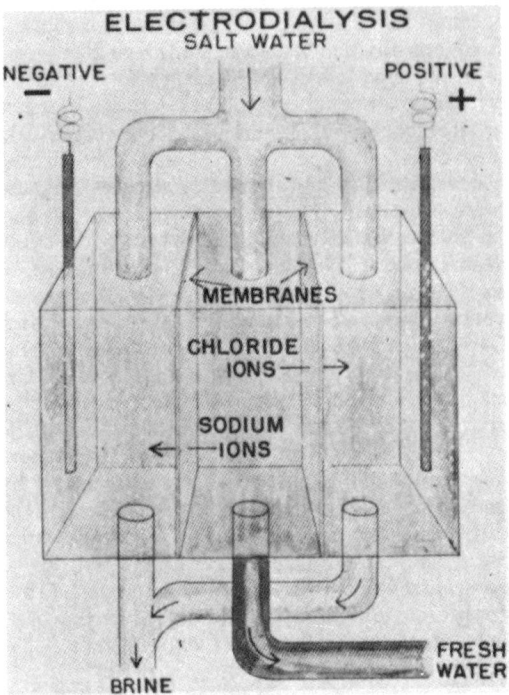

Fig. 5. An electrodialysis process of conversion is utilized to produce 250,000 gallons of fresh water from brackish water daily at the O.S.W. Demonstration Plant in Webster, South Dakota. Saline water enters through the top. As it passes through the middle chamber, chloride ions are attracted by the positive charge and sodium ions are attracted by the negative charge, leaving fresh water in the center chamber. The fresh water is drawn off and used, while the concentrated brines in the side chambers are drawn off and disposed of. This is a simplified diagram; there are 217 membrane cell pairs in each stack in the actual plant.

direction and pass through the anion permeable membranes. The water in the center chamber of each membrane cell is thus depleted of salt while the water passing through the intervening pairs is enriched. The cost of conversion depends on the salinity of the raw feed water. Electrodialysis is now considered to be the most economical process available for mildly brackish water, but it is not competitive with other processes for the conversion of seawater.

A distillation process requires a double phase change: from a liquid to vapor and back to liquid. A crystallization process also requires a double phase change: from a liquid to a solid and back to a liquid. An ice crystal is pure water, thus a freeze-demineralization process involves freezing part of the seawater to form a slush or a slurry of ice crystals, removing the ice from the brine, washing it free of occluded salt, and melting it to fresh water. Work on this promising system is in progress (Fig. 6).

Another promising process under development is known as reverse osmosis (RO), using the salt-rejection properties of cellulose acetate. Specially cast membranes of this material have made possible the development of an inherently simple desalting process. The application of hydrostatic pressure to saline water produces a flow of fresh water through the membranes, the best now available providing a flux of 23 gal of fresh water per day per square foot of membrane. The process has good economic potential because it avoids the costly phase change required by both the distillation and the crystallization processes. RO plants of up to 100,000 gallons per day have been operated.

Studies indicate that substantial cost savings will accrue as modern technology is incorporated in multimillion gallon per day plants, and they indicate that a large nuclear fueled combination power and water plant could produce 150 million gallons of fresh water in the cost range of 22–30¢ per thousand gallons.

As an indication of the growing interest of atomic energy as a source of power, the International Atomic Energy Agency convened panels in 1963 and 1964 to advise it on such power use for desalination. Dual purpose, electricity and water, plants are considered with much favor.

A survey conducted by the United Nations Department of Economic and Social Affairs (Resources and Transport Division) of the water needs of 43 developing countries, shows that the most immediate need for desalting plants is in the size range of 10 million gallons per day or less (see Tables 1 and 2). The U.N. report "Water Desalination in Developing Countries" wisely points out that it still is necessary to conduct extensive feasibility studies. Desalination is not a panacea for all water ills, but there are many locations where today it can represent the cheapest and most reliable source of supply. A second U.N. publication "Water Desalination: Proposals for a Costing Procedure and Related Technical and Economic Considerations" provides an excellent formula to determine realistic costs. Technical data gained through the Office of Saline Water are not for the exclusive use of the United States. At the direction of President Johnson, our desalting technology is being shared with other nations. In 1965 the O.S.W. sponsored the First International Symposium on Water Desalination, held in Washington, D.C.; it was attended by delegates from 58

DESALINATION PROCESSES—THE U.S. DESALTING PROGRAM

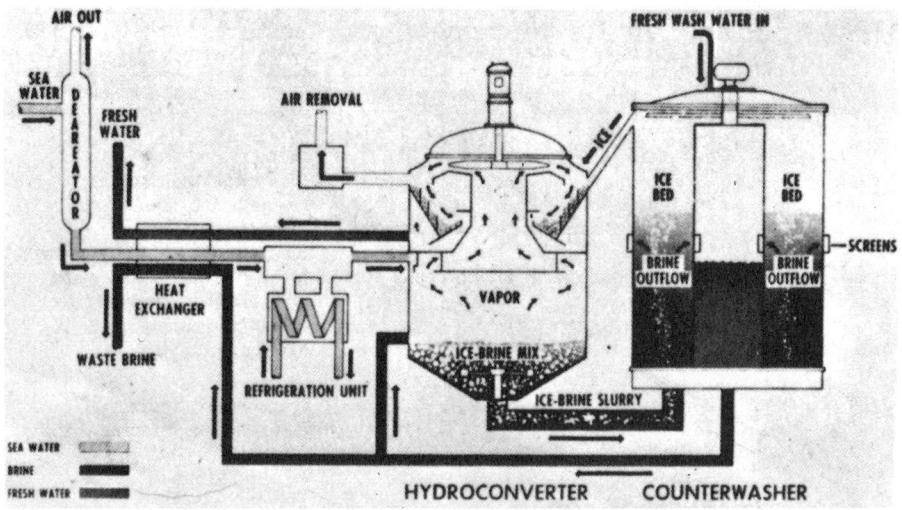

FIG. 6. A vacuum-freezing vapor-compression system. Seawater enters the system at 60–75°F, passes through the deaerator and then through the heat exchangers where it is cooled by product water and waste brine. The feed is introduced into the freezer where it is converted into a 50% ice slurry at 3 mm Hg pressure. The slurry is pumped from freezer to counterwasher where brine is separated and the ice cystals are washed free of salt. From the counterwasher, ice is scraped into a melter where it is melted by the condensation of compressed water vapor from the compressor.

countries, a fair measure of its widespread interest. Its proceedings covered the entire field, both in theory and with respect to worldwide application.

In President Johnson's words: "It would be difficult to exaggerate the power for good, the palliative effect on age-old animosities and problems, that would result from providing an abundance of water in lands which for countless generations have only known shortage."

FRANK C. DiLUZIO

TABLE 1. EXAMPLES OF BRACKISH WATER UTILIZATION

	Total Use (m³/day)[a]
Djibouti, French Somaliland	3,790
Banghazi, Libya	3,788
Zlitan, Libya	1,125
Madagascar	3,800
Mogadiscio, Somalia	2,850
Manama, Bahrain	27,397
Kuwait	23,562
Dhahran, Saudi Arabia	17,000
Ras Tannura, Saudi Arabia	13,622
Famagusta, Cyprus	1,121

[a] 1 m³ = approximately 4000 gallons.

References

Worldwide references may be found cited in the U.N. publications mentioned above and in papers of the First International Symposium (Washington, 1965).

Selected Research and Development reports of O.S.W. listed below may be purchased from the Clearing House for Federal Scientific and Technical Information, 5285 Port Royal Rd., Springfield, Va. 22151.

R & D Nos.

4 1954, "Demineralization of Saline Water with Solar Energy," George Löf, PB 161379.
40 1960, "Saline Water Conversion by Direct Freezing with Butane," Blaw-Knox, PB 161819.
41 1960, "Saline Water Conversion by Freezing: An Integral Processing Unit Using a Secondary Refrigerant," Cornell University, PB 161906.
69 1962, "Investigation and Preparation of Polymer Films to Improve the Separation of Water and Salts in Saline Water Conversion," Monsanto Res. Corp., PB 181467.
* 1963, "Desalination—A Report on Overseas Research of the Water Research Foundation of Australia," PB 181472.
75 "Demineralization of Saline Water through Pressurization Cycles with Ion Exchange Material," Western Independent Research, PB 181485.

DESALINATION PROCESSES—THE U.S. DESALTING PROGRAM

TABLE 2. COUNTRIES AND TERRITORIES WHICH APPEAR TO HAVE POSSIBILITIES FOR THE ECONOMIC APPLICATION OF DESALINATED WATER[a]

Country or Territory	Area	Principal Category of User
Africa		
Ethiopia	Assab	Industry, households
French Somaliland	Djibouti	Households, shipping
Kenya	Scattered	Households
Mali	Lake Faguibine and Goundam area	Industry, mining
Mauritania	Port Etienne	Households, shipping, tourism
South Africa	Mining and stock-raising areas	Mining, households, watering of cattle
Sudan	Port Sudan	Oil refineries, industries
Tunisia	Southern region	Industry, tourism
United Arab Republic	Red Sea region	Mining, industry
Asia		
Bahrain	Manama	Households, industry
India	Madras	Households, shipping, fishing, industry
	Calcutta	
	Kandla	
	Ernakulam Inlet	
Israel	Eilat	Households, industry, shipping
Jordan	Jerusalem	Households, tourism
Pakistan	Makran (West Pakistan)	Households, industry
	Khulna (East Pakistan)	
Qatar	Doha	Households
Saudi Arabia	Eastern Province	Households, industry
	Jedda	
Turkey	Lake Hazar area	Irrigation, households
	Gallipoli Peninsula	Industry, other uses
Europe		
Cyprus	Nicosia, Famagusta, Larnaca	Households, industry
Greece	Syros, Crete, Mykonos, Melos	Households, tourism
Malta	Comino Island	Tourism
Spain	Balearic Islands	Tourism
	Palma de Mallorca	
	Cartagena	Industry, households, tourism
	Alicante	Tourism
	Malaga	
	Benidorm	
	Canary Islands	Tourism, households
	Gran Canarias	
Caribbean and Latin America		
Argentina	Sierra Grande	Mining (Sierra Grande)
	Venado Tuerto	Households (Venada Tuerto)
Bermuda		Tourism
Brazil	Fortaleza, Quixadá	Households
Chile	Taltal	Households, shipping
Ecuador	Santa Elena Peninsula	Tourism, households
Mexico	Cozumel (Quintana Roo)	Tourism
	Carmen (Campeche)	
	Tijuana (Baja California)	Tourism
Netherlands Antilles	Curacao	Households, tourism, industry
	Aruba	
	Bonaire	
Peru	Ite	Fish meal
	Sechura Desert	Mining
	Huasco	Households
Venezuela	Puerto Piritu	Tourism
	Margarita	
Virgin Islands (U.S.)	St. Croix	Households, tourism

[a] Source: United Nations Publication "Water Desalination in Developing Countries."

81 "Absorption—Multistage Flash Distillation Process," Fluor Singmaster and Breyer, PB 181545.
* 1963, "Conversion of Saline Water, a Bibliography," PB 181546.
91 "Mineral By-Products From the Sea," W. R. Grace & Co., PB 181588.
106 1964, "Desalination by Electrosorption and Desorption," Southern Res. Inst., PB 181683.
111 1964, "Reverse Osmosis for Water Desalination," General Atomic, PB 181696.
124 1964, "A Study of Desalting Plants (15 to 150 mgd) and Nuclear Power Plants (200 to 1500 mwt) for Combined Water and Power Production," PB 166396.

* Not numbered.

Bloch, M. R., 1969, "Die Entsalzung des Meerwassers," *Naturwissen.*, **56**, 184–189.

Cross-references: *Hydrology; Hydrology, Semiarid Regions; Natural Brines; Seawater: Chemistry. Vol. I: Salinity in the Ocean.*

DESERTS—See Vol. III

DEUTERIUM

Properties

Deuterium (symbol H^2 or D) is the isotope of hydrogen with two neutrons in the nucleus and atomic mass of 2.014. It was discovered by H. C. Urey, F. C. Brickwedde, and G. M. Murphy in 1931. The doubling of the atomic mass as compared to light hydrogen or protium results in no significant qualitative differences in chemical or physical properties but does result in quantitative differences which are sometimes extreme. For example, both isotopes react with chlorine at 0°C but the reaction rate of protium is about 130 times higher.

Physical properties of the isotopes are compared below:

	Protium	Deuterium
Triple point (°K)	13.96	18.73
Boiling point (°K)	20.39	23.67
Heat of vaporization (cal/mole)	216	293

The principal determining factors in the distribution of deuterium are the properties of the oxide, since deuterium exists principally as heavy water. Physical properties for D_2O are given below:

	H_2O	D_2O
Specific gravity	1.000	1.1078
Melting point (°C)	0	3.81
Boiling point (°C)	100	101.4
Temp. of maximum density (°C)	3.98	11.2
Critical temp. (°C)	374.2	371.5

Abundance

The terrestrial abundance of deuterium is one part in 6700 parts of hydrogen and is controlled essentially by the oceanic abundance. Deuterium may be synthesized in the stars, but is destroyed much more rapidly by thermonuclear reactions. Hence the stellar abundance (D/H) is estimated at 3×10^{-5} to 5×10^{-5} (Severnyi, 1956).

Deuterium abundance is usually expressed as parts per thousand deviation from a standard (SMOW, Standard Mean Ocean Water).

$$\delta D(‰) = 1000 \left[\frac{D/H(Sample) - D/H(SMOW)}{D/H(SMOW)} \right]$$

It is small changes in the δD value caused by fractionating mechanisms, principally evaporation and condensation, that provide the basis for attempts to analyze various atmospheric and hydrological phenomena by deuterium measurements. For example, in the evaporation process, the D/H ratio increases in the liquid residue and decreases in the vapor.

Typical abundance values for various natural waters are given by Friedman et al. (1964) and Dansgaard (1964) (Table 1).

If the delta value for ocean water is taken as zero, it is clear that most other waters will be negative with respect to deuterium abundance. The heavy isotopes precipitate first from oceanic water vapor. Hence, rainfall over the ocean and some coasts has $+\delta D$. The vapor that moves inland must have $-\delta D$. The isotopically lightest water is found in snow and polar ice. This effect is shown by data for the Missouri River which is fed by snow melt. It has the very negative value of -141 in the northern headwaters of South Dakota. As the river flows south the D/H ratio is increased by evaporation and additions of tributaries so that a substantially less negative value is measured at St. Louis. Positive delta values are found in surface waters subject to high evaporation losses, such as the Trinity River in Texas.

TABLE 1.

Location	δD (‰)
Lake Superior	−75
Missouri River (S. Dakota)	−141
Missouri River (St. Louis)	−95
Pecos River (Texas)	−29
Trinity River, East Fork (Dallas)	+16
Stat Nord, Greenland (snow)	−180
Vienna, Austria (precip.)	−65
Bombay, India (rain)	−8

DEUTERIUM

The D/H ratio is affected by climatic conditions and decreases at lower temperatures. This is the basis for the identification of underground water recharged during the Pluvial epoch. Such groundwater with an anomalously low D/H has been found in several northern hemisphere locations, particularly in the Middle East and North Africa.

Geochemical and Hydrological Behavior

Faucher and Thomas and later Roy and Roy (1957) found that deuterium exchanged with hydroxyl hydrogen in clays and developed this as a technique for the investigation of structure. Since the exchange is complete at higher temperatures in a relatively short time, it is assumed that the minerals in most hot spring areas are in equilibrium with the deeper groundwaters.

Boato (1954) determined that D/H in carbonaceous chondrites was in good agreement with the average for terrestrial materials and confirmed that the primeval cosmic ratio was 1/6500, or essentially the same as the present ratio. Edwards (1955) found a somewhat lower ratio in stony meteorites. Friedman (1956) found the ratio in tektites approximately the same as for water dissolved in terrestrial obsidian and fresh water.

Attempts to identify magmatic water by means of the D/H ratio have been inconclusive and most investigations of thermal waters suggest a meteoritic origin. Investigations of 95 brines by Clayton et al. (1966) in the central United States suggest meteoritic origin and appear to show no evidence of deuterium exchange since the time of recharge.

Significant fractionation of deuterium in biological systems, including aquatic life, has long been known. Enrichment of deuterium has been reported in hydrogen from crude oil, methane from natural gas, peat, and curtisite.

Many hydrological applications have been suggested for deuterium but have not as yet been widely accepted. These are reviewed in a guidebook prepared for the International Hydrological Decade (1968).

LELAND L. THATCHER

References

Boato, G., 1954, "The isotopic composition of hydrogen and carbon in the carbonaceous chondrites," *Geochim. Cosmochim. Acta*, **6**, 209.

Brinkmann, R., et al., 1963, "Uber den Deuterium," *Naturwissenschaften*, **19**.

Clayton, R. N., Friedman, I., Graf, D., Mayeda, T,. Meents, W., and Shimp, N., 1966, "Origin of saline formation waters," *J. Geophys. Res.* **71**(16), 3869–3882.

Dansgaard, W., 1964, "Stable isotopes in precipitation," *Tellus*, **16**(4), 436.

Edwards, G., 1955, "Isotopic composition of meteoritic hydrogen," *Nature*, **176**, 109.

Friedman, I., 1956, "Water in tektites," *Proc. 2nd Conf., Nuclear Processes in Geologic Settings*.

Friedman, I., Redfield; A. C., Schoen, B., and Harris, J., 1964, "The variation of the deuterium content of natural waters in the hydrologic cycle," *Rev. Geophys.* **2**(1), 177–224.

International Hydrological Decade, "Guidebook to the Use of Isotopic Techniques in Hydrology," 1968, Vienna, Austria, International Atomic Energy Agency.

Roy, D. M., and Roy, R., 1957, "Hydrogen-deuterium exchange in clays and problems in the assignment of infra-red frequencies in the hydroxyl region," *Geochim. Cosmochim. Acta*, **11**, 72.

Severnyi, A. B., 1956, "Investigation of deuterium on the Sun," *Izvt. Krymsk. Astrofiz. Observ.* **16**, 12.

Cross-references: *Elements: Planetary Abundances and Distribution; Organic Geochemistry; River Geochemistry; Seawater, Chemistry.* Vol. II: *Carbonaceous Meteorites; Tektites.*

DIADOCHY—See TRACE ELEMENTS IN SILICATE MINERALS

DIAGENESIS—See PARAGENESIS; also Vol. VI

DIFFERENTIAL THERMAL ANALYSIS

The method of differential thermal analysis (DTA) makes possible the detection of thermal transitions accompanying many physical and chemical changes taking place in a substance while it is being heated. Energy changes associated with fusion, vaporization, recrystallization, solid-solid phase changes, oxidation, reduction, dehydration, chemical recombination, and decomposition, may be detected and measured. Suitable instrumentation is employed to record exothermic and endothermic reactions as positive and negative deviations from the base line (Fig. 1). The advantage of this method is that it can be carried out with comparatively simple apparatus and that it gives information not easily obtained by other instrumental methods.

Figs. 2 and 3 illustrate common instrumental arrangements. The sample, usually in powdered form, and a thermally inert reference substance, such as calcined α-alumina, are placed in identical specimen holders. The specimen-holder assembly is then placed in a furnace. The furnace temperature is raised at a preset, constant rate. Common heating rates employed in investigations of minerals range

DIFFERENTIAL THERMAL ANALYSIS

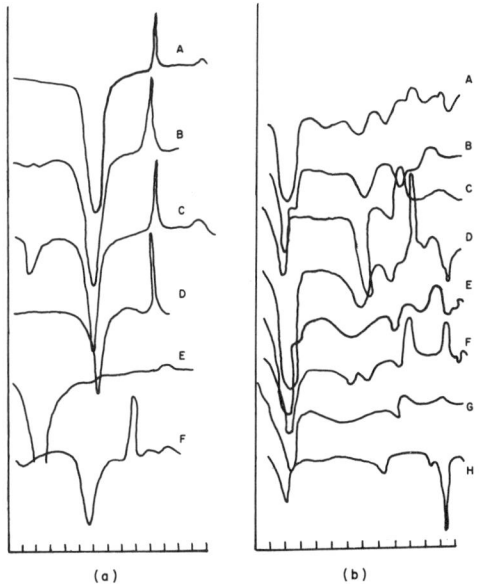

FIG. 1. Thermograms of Clay Minerals (after Grim, 1953).

(a) (A) kaolinite, Georgia, well crystallized
    (B) kaolinite, Illinois, poorly crystallized
    (C) hydrated halloysite, Indiana
    (D) anauxite, Ione, California
    (E) allophane, Bedford, Indiana
    (F) allophane, Iyo, Japan
(b) (A) montmorillonite, Otay, California
    (B) montmorillonite, Tatatilla, Mexico
    (C) montmorillonite, Upton, Wyoming
    (D) montmorillonite, Cheto, Arizona
    (E) montmorillonite, Pontotoc, Mississippi
    (F) montmorillonite, Palmer, Arkansas
    (G) nontronite, Howard County, Arkansas
    (H) hectorite, Hector, California

from 8°/ min to over 20°/ min. Usually, one junction of the differential thermocouple is placed in the center of the sample and the other in the inert reference specimen. The thermocouples are so arranged that the electromotive forces (emf's) produced at the two junctions buck each other. The emf generated by the junction in the reference specimen is mainly a function of the furnace temperature. When no thermal transition takes place in the sample, its temperature will be equal, or almost equal, to that of the inert reference specimen. Therefore, during temperature intervals where no reaction takes place the two emf's are alike and no signal results across the differential thermocouple. When the sample undergoes a transition, accompanied by either the absorption or liberation of thermal energy, the temperature at the center of the sample differs from that prevailing at the center of the reference substance. The imbalance results in an emf proportional to the temperature difference. The signal, usually in tenths of millivolts, may then be amplified and recorded.

Instrumentation. The essential components of a DTA system are: specimen holders, a furnace and furnace temperature programmer, a d.c. amplifier, and a recorder (Fig. 2). Common specimen holders (Fig. 3) are made of nickel, stainless steel, alumina, and other materials. Generally, one sample is studied at a time, but multiple specimen holders, permitting the investigation of several samples simultaneously, may also be used (Fig. 3; Kerr, 1948). Commonly used thermocouples are made of chromel-alumel, or platinum-platinum and 13% rhodium. For low-temperature work, thermistors may be advantageously employed. For ordinary analyses, furnaces should have a maximum temperature of approximately 1200°C. Common electrical resistance heating elements are Nichrome, Kanthal, Globar, and platinum. For specialized investigations, furnaces which permit operation under vacuum, pressure, and inert atmospheres have been described (Wendlandt, 1963; 1964; Garn, 1965).

In order to obtain a constant heating rate, a controller that regulates the furnace temperature is required. The simplest controller is a variable voltage transformer which is advanced by a motor and suitable reduction gears. Usually, a more elaborate temperature controller is used to compensate for the thermal inertia of the furnace. In the early days of DTA, the signal, after having passed through a d.c. amplifier, was recorded manually by noting the voltage at evenly spaced intervals and plotting it vs. time. Today, the signal is recorded more accurately and conveniently by potentiometric recorders. The differential temperature and the furnace temperature may be recorded by two separate strip chart recorders. One records the furnace temperature vs. time, whose curve, ideally, should be a straight line and whose slope is a function of the heating rate. The other records the thermogram, i.e., the differential temperature vs. time. In order to determine the temperature at which a transition occurs, it is necessary to compare the data on the two charts. Alternatively, an X–Y recorder may be employed. The thermocouple sensing the furnace temperature, or the temperature at the center of the inert reference substance (Fig. 2), is recorded on the X axis, while the amplified signal, which is proportional to the differential temperature, is recorded on the Y axis. An X–Y recorder has the advantage of plotting the complete thermogram on a single sheet of graph paper. Further, because the X-axis response is always propor-

DIFFERENTIAL THERMAL ANALYSIS

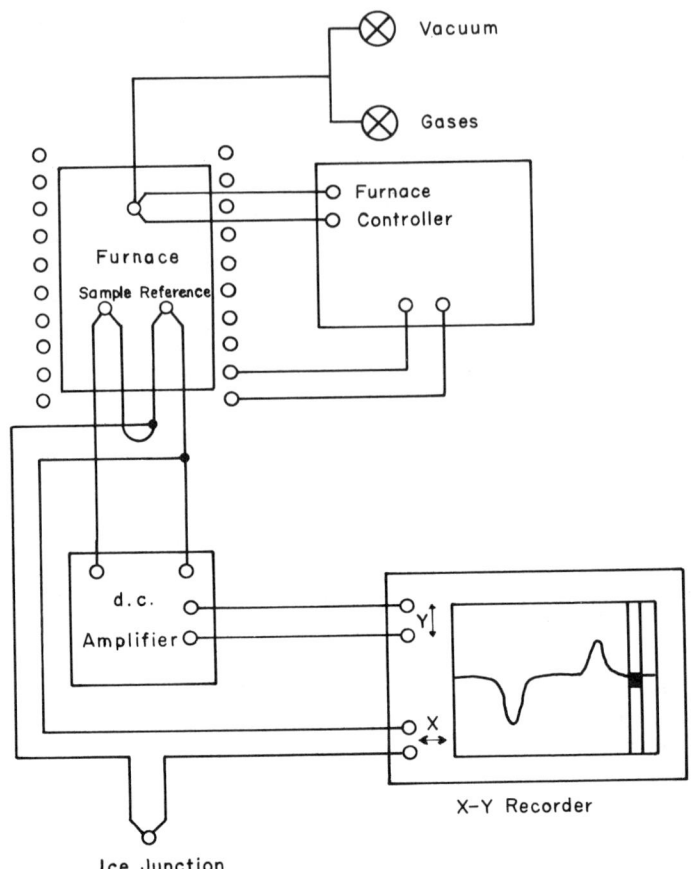

Fig. 2. Schematic diagram of typical differential thermal analysis apparatus.

tional to the furnace temperature, the heating rate need not be perfectly linear with respect to time for ordinary work. An advantage of strip chart recorders is the ease with which the linearity of the heating rate may be checked.

Theory. Conventionally, endothermic and exothermic reactions are recorded as negative and positive deviations from the base line. For example, any of the endothermic reactions recorded on the thermograms of Fig. 1 starts when the curve deviates from the base line in the negative Y direction. The rate of the reaction increases until the peak is reached.

Following the maximum differential temperature, or the peak of the curve, the rate of heat absorption by the sample decreases. Somewhere between the peak and the end of the curve the reaction ceases (Vold, 1949). The area under the curve is approximately proportional to the heat of the transition (Kerr, 1948).

The relationship between the quantity of heat involved and the area under the curve may be expressed as

$$Q = \psi \int_{t_1}^{t_2} \theta \, dt \qquad (1)$$

where Q = quantity of heat, θ = differential temperature, t_1, t_2 = initial and final time of transition, and ψ = proportionality constant to be evaluated experimentally.

The right-hand side of Eq. (1) is the area under the curve times the proportionality constant ψ. Under suitable conditions, an approximate value of ψ may be found from thermograms of substances whose heats of reaction are known. Alternatively, internal standards may be used (Barshad, 1952).

For quantitative work, the ratio of the area under the curve to the heat evolved or absorbed must be known accurately. Several investigators have developed expressions relating the area under the curve to the heat of the reaction. Considering the sample specimen only, and defining the significant area under the curve as the total area, they derived the following expressions (here slightly modified) (Speil, 1945; Kerr, 1948):

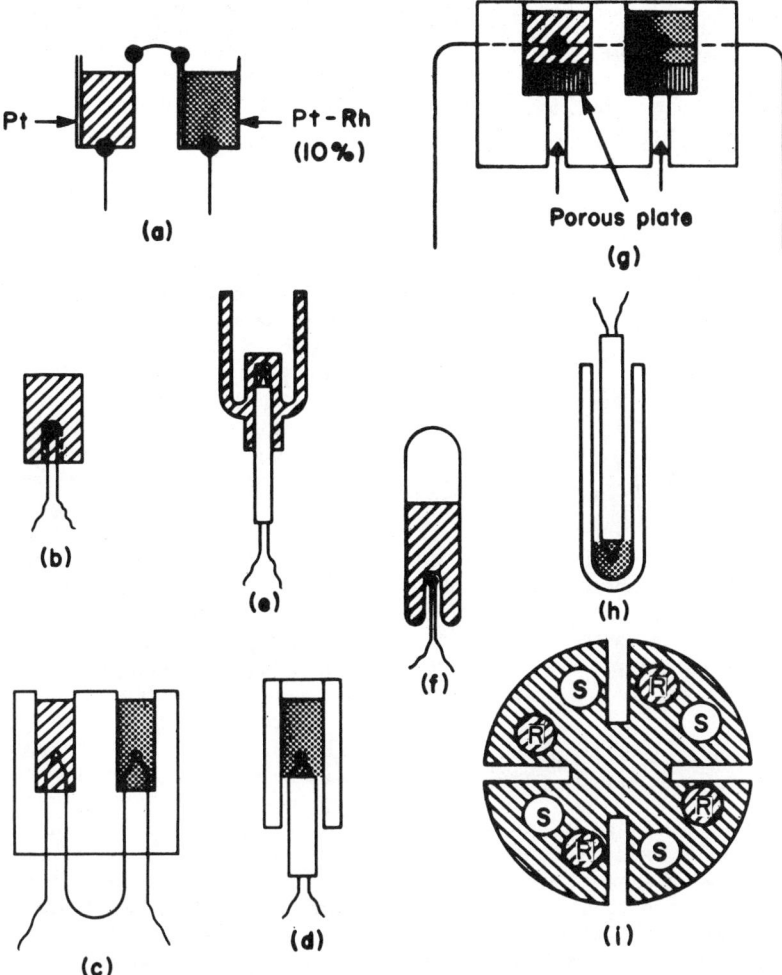

FIG. 3. Various specimen holders. (a) Specimen holders act also as thermocouples. (b) Sample, pressed into shape of cylinder, is placed on thermocouple bead. (c) Sample and reference specimen holder are cut into nickel block. (d) Specimen chamber is a sleeve of ceramic tubing. (e) Small cups, made to fit on top of thin ceramic tubing (two holes), contain specimens. (f) Sealed sample container. (g) Specimen chambers fitted with porous plates to permit circulation of various atmospheres. (h) Specimen chamber is mounted so that thermocouple bead can be lowered into it. (i) Multiple specimen holder. Chambers are cut into a nickel block (modified after Wendlandt, 1963).

$$m \int_a^x \frac{dH}{dt}\, dt + gK_s \int_a^x (T_w - T_s)\, dt = m_s c_s (T_s - T_0) \qquad (2)$$

where m = mass of reactive component in the sample
 m_s = total mass of the sample
 dH = the decrement of heat absorbed from the sample by the endothermic reaction in time interval dt
 g = geometrical shape constant
 K_s = thermal conductivity of sample
 T_w = temperature of nickel block containing specimen at $t = x$
 T_s = temperature at center of sample at $t = x$
 T_0 = temperature at center of sample at $t = a$
 c_s = mean specific heat of sample between $t = a$ and $t = x$

Considering an endothermic reaction, the first factor on the left-hand side of Eq. (2) defines the quantity of heat due to the reaction. The second factor on the left-hand side defines the heat absorbed by the sample. The right-hand side is an expression for the total heat flowing into the sample. Omitting the factor involving the heat of reaction, a similar expression may be written for the inert reference substance. Since only the differential temperature is recorded, the expression for the inert reference substance may be subtracted from that for the sample. Further, by integrating over the entire curve, rather than to the difficult-to-determine point "x," where the reaction presumably ceases (i.e., a to b), the final expression becomes

$$m \int_a^b \frac{dH}{dt} dt = m \, \Delta H$$

$$\Delta H = \frac{gK_s}{m} \int_a^b y \, dt \qquad (3)$$

In attempting to derive an expression for quantitative DTA, other investigators made use of the *divergence theorem*. Ignoring the important effects of the thermocouple wires, the following relationship was developed (Boersma, 1955):

$$\int_{t_1}^{t_2} \theta \, dt = \frac{q}{K_s} \int_0^r \frac{V}{S} dr \qquad (4)$$

where q = heat of reaction per unit volume; r = radius of cylindrical specimen holder cavity; V = volume of sample; and S = surface of sample.

The left-hand side of Eq. (4) is proportional to the area under the curve. For the common cylindrical sample holder $V/S = r/2$, if the surfaces closing the cylinder are ignored. Then Eq. (4) reduces to

$$\int_{t_1}^{t_2} \theta \, dt = \frac{qr^2}{4K_s} \qquad (5)$$

These expressions, as well as others that have been derived, cannot be utilized for quantitative DTA studies because of the following: (a) the unknown thermal conductivity of the sample; (b) the uncontrollable geometrical factors related to sample size, such as shrinkage during the transition; (c) the effects of the thermocouple wires which, under conditions of DTA, perform the double role of temperature transducer and heat sink, and which may account for a reduction of as much as 50% of the area under the curve (de Jong, 1957); (d) the loss of volatile substances and/or water which may change the thermal conductivity of the sample in a manner which would make all theoretical expressions which assume a single-valued thermal conductivity invalid. The use of DTA for calorimetric measurements still awaits a technique which would permit corrections to be made for changes in heat transfer characteristics of the sample taking place during the reaction.

Interpretation of Thermograms. The information obtainable by this method and the interpretation of thermograms may be illustrated with the aid of Fig. 1. The thermograms on the left-hand side were obtained from a series of kaolinite clay minerals. Thermograms for kaolinite are characterized by a large endothermic peak at about 550–600°C. The absorption of thermal energy, which starts at about 400°C, is due to the breakdown of the lattice, accompanied by the liberation of OH. The range of peak temperatures between different samples is believed to be due to differences in particle size and degree of crystallinity. An endothermic reaction at about 110–200°C accompanies the loss of adsorbed water. The intense exothermic reaction taking place at about 950–1000°C is due to the energy liberated when new compounds are synthesized from elements which made up the clay mineral lattice.

The second set of thermograms (Fig. 1) were obtained from a series of montmorillonite clay minerals. The thermograms indicate a loss of adsorbed interlayer water in the temperature range 100–200°C. The size and character of this peak is, among others, a function of the nature of the adsorbed cation. Some of the curves show a distinct break between the temperature ranges at which interlayer water and OH is liberated. Compared to kaolinite thermograms, the endothermic peaks marking OH loss are less distinct and they occur over a wider range of temperatures. Further, unlike the thermograms of kaolinite minerals, there are two distinct endothermic reactions in the high-temperature range. The final exothermic reactions are, as in the case of kaolinite, due to the synthesis of new substances. The interpretation of thermograms is rather difficult. Usually, the investigator must supplement the information from DTA with that obtained by other methods of investigation such as X-ray diffraction analysis, mass spectroscopy, thermogravimetric analysis, gas chromatography, or infrared spectroscopy.

Applications. DTA finds its widest application as an analytical tool of identification (Smothers, 1958; MacKenzie, 1957; Kerr, 1950; Wendlandt, 1964; Garn, 1965). Among the clay mineral groups whose thermal transitions have been investigated by DTA are kaolinite, montmorillonite, illite, vermiculite, chlorite, and palygorskite. Other mineral groups that have been studied are the silica minerals, oxides of iron, aluminum, and manganese, carbonates, sulfides, halides, sulfates, nitrates, borates, phosphates, and some silicates. Interesting applications other than identification have been reported in the literature (Wendlandt, 1964). In some industries, DTA is employed as a quality-control device. For example, it was found that efflorescence of fired ceramic products may be predicted from estimates of gypsum, pyrite, and calcite content as determined from thermograms. Pedologists use thermograms to correlate clay mineral content with soil type (MacKenzie, 1957). Organic chemists are just beginning to use DTA in the study of polymers, starches, carbohydrates, amino acids, proteins, fats, and others. Many thermal decomposition studies have been carried out with DTA. The technique was used successfully in dehydration studies of hydrates. Solid-solid phase transformation studies and Curie point determinations were also carried out with the aid of this method. The study of radiation damage is another interesting application. Equations have been developed which make possible the determination of reaction rates, the order of reaction, and activation energy (Wendlandt, 1963, 1964). Among others, the kinetics of the decomposition of kaolinite minerals were investigated by the DTA method.

Although DTA dates back to 1887, its application to research problems other than identification was made possible only recently by the availability of advanced electronic amplification and recording equipment. The discovery of many new applications of DTA has brought about a renewed interest in this method.

EDWARD STURM

References

Barshad, I., 1952, "Temperature and heat of reaction calibration of the differential thermal analysis apparatus," *Am. Mineralogist,* **37,** 667–694.

Boersma, S. J., 1955, "A theory of differential thermal analysis and new methods of measurement and interpretation," *J. Am. Ceram. Soc.,* **38,** 281–284.

de Jong, G. J., 1957, "Verification of use of peak area for quantitative differential thermal analysis," *J. Am. Ceram. Soc.,* **40,** 42–49.

Garn, Paul D., 1965. "Thermoanalytical Methods of Investigattion." New York, Academic Press, 606pp.

Grim, R. E., 1953, "Clay Mineralogy," New York, McGraw-Hill Book Co., 384pp.

Kerr, P. F., and Kulp, J. L., 1948, "Multiple differential thermal analysis," *Am. Mineralogist,* **33,** 387–419.

Kerr, P. F., Kulp, J. L., Hamilton, P. K., 1950, "Analytical Data on Reference Clay Minerals," Am. Petrol. Inst. Project 49, New York, Columbia Univ. Press.

* Mackenzie, R. C., 1957, "Differential Thermal Analysis of Clays," Aberdeen, Scotland, Central Press.

* Smothers, W. J., and Chiang, Y., 1958, "Differential Thermal Analysis: Theory and Practice," New York, Chemical Publ. Co.

Speil, S. L., Berkelhammer, H., Pash, J. A., and Davis, B., 1945, "Differential thermal analysis. Its application to clays and other aluminous minerals," *U.S. Bur. Mines Tech. Paper,* **664.**

Vold, M. S., 1949, "Differential thermal analysis," *Anal. Chem.,* **21,** 683–688.

* Wendlandt, W. W., 1963, "Differential Thermal Analysis," in (Jonassen, H. B., and Weissberger, A., editors), "Technique of Organic Chemistry," Vol. 1, pp. 209–257, New York, Interscience Publ., div. of John Wiley & Sons.

Wendlandt, Wesley W., 1964. "Thermal Methods of Analysis." New York, Interscience Publ. John Wiley & Sons, 424pp.

* Additional bibliographic references may be found in this work.

Cross-references: *Chromatography; Derivative Differential Thermal Analysis; Organic Geochemistry; Mineral Classes: Silicates; Oxidation and Reduction; Thermogravimetric Analysis; X-Ray Diffraction Analysis; X-Ray Spectroscopy.* Vol. IV B: *Ceramic Technology; Clays and Clay Minerals; Crystal Growth.* Vol. VI: *Pedology.*

DIFFERENTIAL THERMOGRAVIMETRIC ANALYSIS

The *thermogravimetric analysis* (TGA) of a sample is obtained by continuously determining and recording the weight (w) of a sample as a function of temperature (T) or time (t), i.e.:

$$w = f(T \text{ or } t)$$

The resultant curve typically exhibits a stepped configuration, the "flats" and curves of which represent temperature ranges where zero or definite weight changes occur respectively (see *Thermogravimetric Analysis*).

In *differential thermogravimetric analysis* (DTGA) or derivative thermogravimetry (DTG) the differential of the weight change with respect to time (dw/dt), is recorded as a function of temperature (T) or time (t) to produce a graphical representation of the first derivative of the weight-change, i.e.:

$$dw/dt = f(T \text{ or } t)$$

As a result, each curved portion located between "flats" and representing weight variations on the TGA curve now becomes represented by a peak. The DTGA curve therefore

DIFFERENTIAL THERMOGRAVIMETRIC ANALYSIS

consists of one or more peaks departing from a common "horizontal" base line, the areas under which are proportional to the weight changes involved.

The DTGA curve configuration is very similar to the DTA curve of the same substance provided that the individual reactions involve weight changes. Compare curves marked DDTG, DTA, and TGA in Fig. 1. For example, changes such as melting and crystallographic inversions, which show up well on DTA, will be completely absent from the comparable DTGA curve (see *Differential Thermal Analysis*).

History. The initial work described in two papers by De Keyser (1953) was followed by the contributions of Erdey, Paulik, and Paulik (1954), Waters (1956), Erdey (1957), Erdey (1958), Paulik, Paulik, and Erdey (1958),

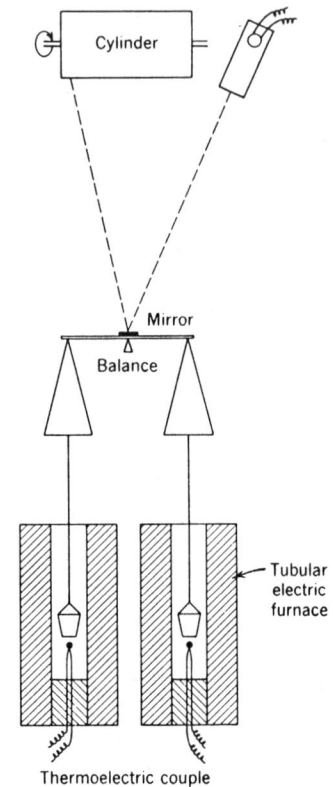

Fig. 2. Schematic diagram of the DeKeyser differential thermobalance (after DeKeyser, 1953; from Wendlandt, 1964). Courtesy John Wiley & Sons, Inc.

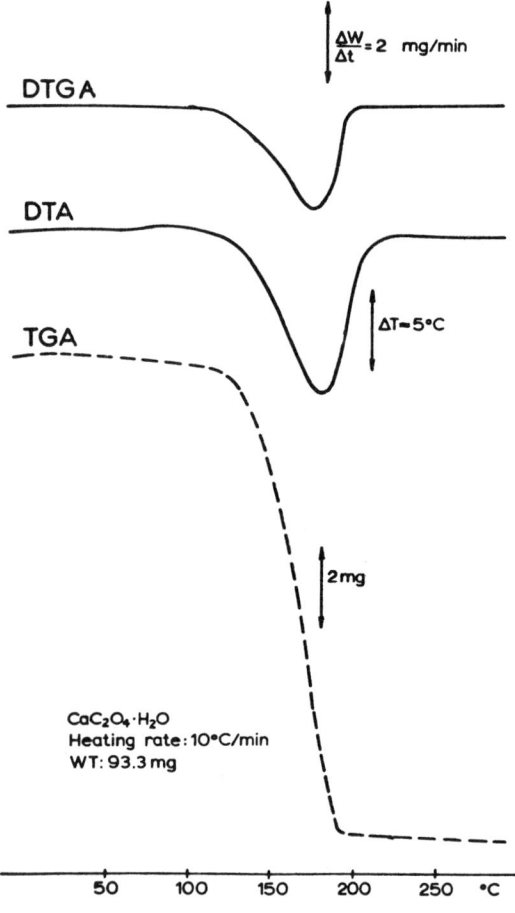

Fig. 1. Simultaneous TGA/DTGA/DTA of the calcium oxalate monohydrate showing the strong relation between the DTGA and DTA peak shapes. Courtesy Mettler Balances, Zurich, Switzerland.

Waddams and Gray (1958), Waters (1958), Campbell, Gordon, and Smith (1959), Erdey, Liptay, Svehla, and Paulik (1962) (see in Wendlandt, 1964) and Lambert (1955), Kofstad (1957), Freeman and Carroll (1958), Soulen (1962), Ingraham and Marier (1964), Wiedemann (1964), Carroll and Manche (1965), Newkirk (1965), Wendlandt (1965) and Lodding (1966) (see in Mettler Instrument Corp. Technical Bulletin No. T-102), and Garn (1965).

Instrumentation and Theory. The DTGA curve may be obtained *(1)* directly, by using a differential recording thermobalance, e.g., De Keyser, 1953 (see Fig. 2) or *(2)* indirectly, from the conventional weight loss or TGA data as obtained from a thermobalance (see *Thermogravimetric Analysis*). The differential (DTGA) curve is obtained from this by manual differentiation and replotting or by automatic differentiation. For the latter an R-C (resistance-capacitance) system is employed (see *Derivative Differential Thermal Analysis*).

DIFFERENTIAL THERMOGRAVIMETRIC ANALYSIS

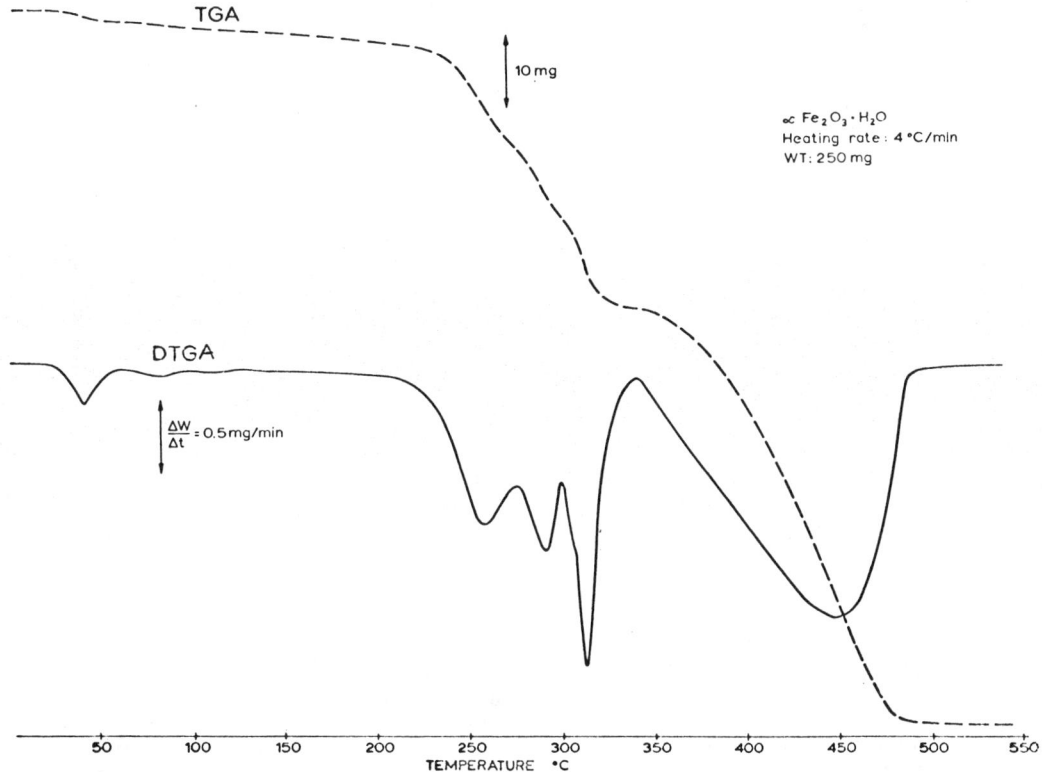

Fig. 3. Reduction in hydrogen of α–$Fe_2O_3 \cdot H_2O$ showing the increase in resolving power of the DTGA compared to the weight loss curve alone. Courtesy Mettler Balances, Zurich, Switzerland.

The advantages of DTGA have been enumerated by Erdey, Paulik, and Paulik (1954), and summarized by Wendlandt (1964).

1. The DTGA curves may be obtained in conjunction with TGA or DTA determinations (for comparisons see Fig. 1).
2. The DTA and DTGA curves are comparable (see Fig. 1), but the results of the former indicate even those changes that are *not* accompanied by weight losses, whereas the latter are more reproducible.
3. Although the curves of DTA extend over a wider temperature range due to subsequent heating of the material after reaction, the DTGA data indicate exactly the temperatures of reaction initiation, termination, and maximum rate.
4. Reactions following each other very closely on TGA curves cannot be easily distinguished since their effects overlap. However the comparable DTGA curves indicate, by sharp peak maxima, the number and location of the thermogravimetric stages present (see Fig. 3).
5. The DTGA curves are exactly proportional to the derivatives of the TGA curves. Thus, because the area under the curves gives the change in weight precisely, the DTGA can give exact quantitative analyses.
6. Derived thermogravimetric analyses may be applied to materials which cannot be subjected to DTA, e.g., some organic compounds, which melt during heating.

S. St. J. Warne

References

Garn, P. D., 1965, "Thermoanalytical Methods of Investigation," New York, Academic Press, 606pp.

* Mettler Instrument Corp. Tech. Bull., No. T-102, Princeton, N.J.

*Wendlandt, W. Wm., 1964, "Thermal Methods of Analysis," New York, Interscience Publ., 424pp.

* Additional bibliographic references may be found in this work.

Cross-references: *Differential Thermal Analysis; Thermogravimetric Analysis.*

DIFFUSION, GEOLOGICAL ROLE

There has been much disagreement among earth scientists pertaining to the degree of importance of the mechanism of diffusion in geological and mineral processes (Bowen, 1921; Gilluly, 1948; and Perrin and Roubault, 1949). Yet a wealth of experimental information has been obtained by metallurgists on specific diffusion rates and diffusion mechanisms for many metal pairs (Barrer, 1941; Jost, 1952; Mehl, 1936; Seitz, 1951), including self-diffusion measurements of similar pairs (Mapother and Maurer, 1948; Seith and Keil, 1933). It is suggested, therefore, that the quantitative laboratory approach developed for the most part by the metallurgists, when tempered with a knowledge of the geological factors, might provide a better understanding of the extent and importance of the geologic role of diffusion.

Diffusion Fundamentals and Factors

Diffusion is defined as the process by which, under the influence of a chemical potential gradient, atoms, molecules, or ions move from one position to another within a solvent phase. When the phase is a gas, a liquid, or a solid, the diffusion process is more closely defined by the modifying adjective gaseous, liquid, or solid, respectively. The "driving force" of diffusion is, of course, energy. This is determined by the chemical potential or activity of the solute atoms, a property that may vary in different solvents.

Mathematical Treatment. Diffusion rates can be expressed quantitatively by use of Fick's first and second laws. Fick (1855) placed diffusion on a rational mathematical basis when he applied, by direct analogy, the equations pertaining to heat conduction derived by Fourier (1822) to the similar process of diffusion.

Fick's first law is as follows:

$$\frac{dm}{dt} = -DA \frac{\partial c}{\partial x} \quad (1)$$

assuming that $D \neq f(c)$. His second law is:

$$\frac{\partial c}{\partial t} = D \frac{\partial^2 c}{\partial x^2} \quad (2)$$

It is from these two basic equations that the general solutions are obtained for the various boundary conditions that exist in the experimental determinations of the diffusion coefficients. Of particular aid in setting up these general solutions are the studies of Carslaw and Jaeger (1959) and Crank (1956). Two of the most common examples that have been used in geological studies for unidirectional diffusion for (a) the semi-infinite case and (b) the finite (thin source of solute) case are given by Garrels et. al. (1949) and Jensen (1952).

In order that the units in Eq. (1) agree, D must be expressed in units of area/time, usually cm^2/sec. The diffusion coefficient (D) therefore is the amount of solute material per unit time (dm/dt) that diffuses through a unit cross-sectional area (A) under a unit concentration gradient $(-\partial c/\partial x)$. D values obtained from the study of diffusion in metals generally vary from about 10^{-5} cm^2/sec at temperatures in excess of 1000°C to even less than 10^{-20} cm^2/sec at temperatures as high as even a few hundred degrees Centigrade. The diffusion rate of Na (sodium) in quartz at 500°C is 5.8×10^{-10} cm^2/sec (Veerhogen, 1952) and that of Na^{22} in microperthite is approximately 10^{-11} to 10^{-12} cm^2/sec at 550°C (Jensen, 1952). The diffusion coefficient of Cu^{64} in steely chalcocite from Kennecott, Alaska, has been determined at temperature above and below the approximate disordering temperature of 105°C. At 130°C, $D \sim 10^{-8}$ cm^2/sec, whereas at 30°C, $D \sim 10^{-10}$ cm^2/sec (Jensen, 1951). A summary of experimental diffusion rates in silicates, as few as they are, is given by Fyfe, Turner, and Veerhogen (1958, p. 63).

In answer to a question that is often asked, viz., how far will a particular substance diffuse through a given rock or mineral, it should now be evident that the diffusion coefficient indicates *only* the amount of material that passes through a given cross-sectional area per unit time. The distance of diffusion is indicated by the concentration curve extending with a negative slope from the interface diffusion plane. Theoretically, therefore, there is a probability that at least one diffusion atom may have traveled an infinite distance from the initial plane!

The Effect of Temperature. The simple concept of the behavior of an atom is that it tends to oscillate about its equilibrium position with a frequency that is determined by its mass and the bond strengths between neighboring atoms. When the atom receives energy, as in the form of heat, its frequency does not change significantly but the magnitude of its oscillations increases. If the magnitude becomes great enough to exceed the bond strengths or "break" the bonds holding the atom in place, the atom may "jump" to a new position. The energy that it must overcome in order to accomplish this is known as the *activation energy*.

It is common to represent this activation energy graphically by a curve of varying amplitude showing energy sumps, or atomic equilibrium locations, and energy peaks of height proportional to the activation energy. Less energy is required for an edge ion to break its bonds than an adjacent surface atom. Movement of interior ions requires even more energy. If the energy barrier is represented by a value of ϵ cal/atom, the probability that the atom will reach the activation energy is proportional to the probability function $e^{-\epsilon/kT}$ where e is the natural logarithm, k is the Boltzmann constant, and T is the absolute temperature.

If $e^{-\epsilon/kT}$ is the probality that ϵ will be reached during one oscillation, then the probability per unit time is $P = fe^{-\epsilon/kT}$ where f is the vibrational frequency. Dushman and Langmuir (1922) have proposed that the diffusion coefficient is approximately equal to the product of the probability per unit time that an atom will jump (P) and the jump distance squared (δ^2):

$$D = P\delta^2$$

therefore,

$$D = \delta^2 f e^{-\epsilon/kT} \quad (3)$$

Since the oscillation frequency and the jump distance are fairly independent of temperature for a given substance, a constant (D_0) is introduced, and:

$$D = D_0 e^{-\epsilon/kT} \quad (4)$$

If both ϵ and kT are multiplied by Avogadro's number (N_0), kN_0 is the gas constant (R), and $N_0\epsilon$ is the activation energy (Q) measured in cal/mole, thus:

$$D = D_0 e^{-Q/RT} \quad (5)$$

This equation has proven to be fairly accurate when checked with experimental diffusion results that are limited to the temperature ranges of a crystal structure that is undergoing very slow disordering rates.

Eq. (5) is useful for the experimental determination of the activation energy of a particular species of solute atoms in a given solvent system over a given temperature range. This is done by first determining experimentally the diffusion coefficients at two different temperatures, at least, and preferably at several intermediate temperatures for improved precision. Threfore, if D_2 and D_1 are the two diffusion coefficients determined at temperatures T_2 and T_1, respectively, it follows that:

$$D_2 = D_0 e - Q/RT_2 \text{ and } D_1 = D_0 e - Q/RT_1 \quad (6)$$

therefore

$$\ln D_2 = \ln D_0 - Q/RT_2 \quad (7)$$
$$\ln D_1 = \ln D_0 - Q/RT_1 \quad (8)$$

Subtracting Eq. (8) from Eq. (7):

$$\ln D_2 - \ln D_1 = \frac{Q}{R}\left(\frac{1}{T_1} - \frac{1}{T_2}\right) \quad (9)$$

therefore

$$\frac{Q}{R} = \frac{\ln D_2 - \ln D_1}{\frac{1}{T_1} - \frac{1}{T_2}} \quad (10)$$

which is the slope of the line shown in Fig. 1. If, therefore, the slope of the line is determined from the diffusion experiments, where the logarithms of diffusion coefficients are plotted versus the reciprocals of the absolute temperatures, the activation energy expressed in cal/mole is:

$$Q = (2.303)(R)(\text{slope}) = (4.583)(\text{slope}) \quad (11)$$

The effect of second-order transformations on diffusion coefficients may now be considered. If the diffusion rates of a solute, such as sodium, were measured in a solvent such as α quartz at various temperatures and the logarithms of the diffusion coefficients were plotted versus the reciprocal of the given temperatures on semilogarithm paper, the result should be approximately a straight line as previously explained. If, however, the line was extended with the same slope to temperature over 573°C, it would be in error. Sodium ions tend to diffuse through β quartz at a higher rate due to the more open structure, and this sudden increase in diffusion would take place at the transformation temperature of about 573°C. In order to determine the activation energy of Na in quartz, two or more diffusion experiments would have to be made at temperatures over 573°C. The resulting curve would probably be similar to that shown in Fig. 2. Presumably, the slope of the two fairly straight lines would be different, since the energy of activation of sodium in α quartz would be very likely less than that in β quartz.

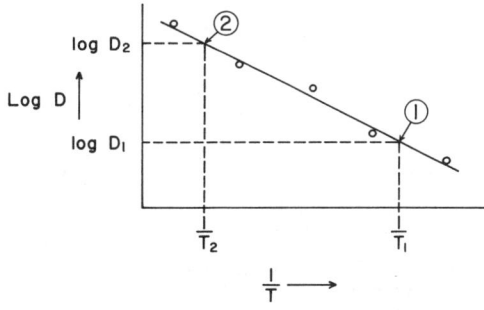

FIG. 1. Plot of experimentally determined diffusion coefficients at different temperatures.

DIFFUSION, GEOLOGICAL ROLE

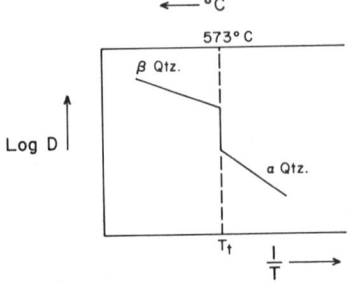

Fig. 2. Variation of diffusion coefficients and activation energies for α and β quartz.

Mechanism of Diffusion. Solid diffusion is usually broken down into three basic mechanisms, especially in the case of metals. These are *interstitial diffusion,* diffusion by *direct interchange,* and the so-called *vacancy* diffusion mechanism. These three mechanisms are shown diagrammatically in Fig. 3.

A series of calculations have been made by several investigators (Huntington, 1942; Huntington and Seitz, 1942, 1949; Zener, 1950) on the energy requirements for the different diffusion mechanisms in the case of self-diffusion of metallic f.c.c. copper. They determined that the activation energy needed to cause a copper atom to jump from its equilibrium site to an interstitial position (known as a *"Frankel defect"*) is approximately 238,000 cal/mole; however, the energy needed for a copper atom to move from one interstitial position to another is only about 11,500 cal/mole.

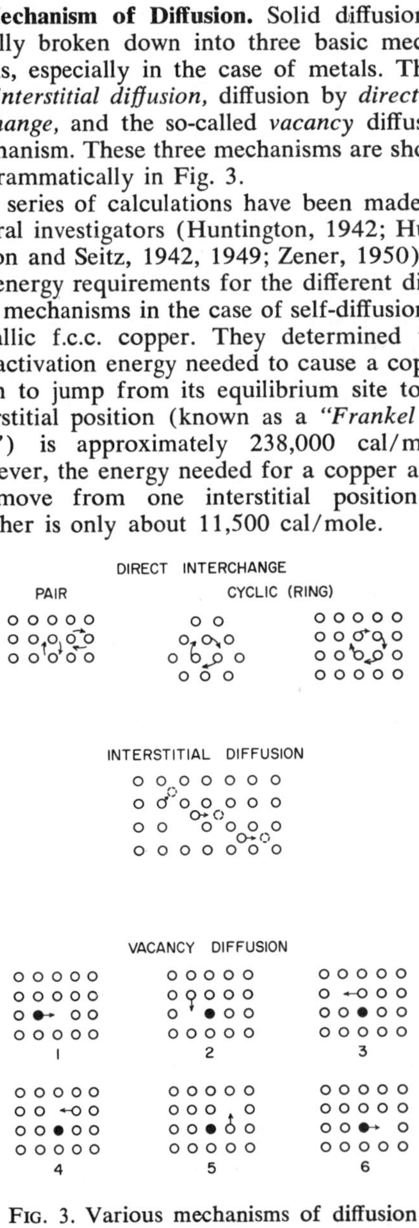

Fig. 3. Various mechanisms of diffusion

Diffusion by direct interchange may occur by the simultaneous exchange of position of two atoms, which is known as pair interchange, or it may occur by a cyclic interchange of three, four, or more atoms. This has been called *ring diffusion* (Zener, 1950), or *"cyclic"* (Buerger, 1934). The energy of activation needed in the copper system for pair interchange is approximately 120,000 cal/mole. For cyclic interchange of four copper atoms, it is lowered to a value of about 91,000 cal/mole. In order to have a direct interchange of two atoms, the crystal structure would need to be strongly distorted during the interchange. Because of this "space problem," the mechanism is not accepted with much favor; however, the cyclic type of interchange need not require as much distortion of the crystal structure. Nevertheless, the probability of four atoms moving in the correct direction at a given instant is rare. Possibly the best reason for belief in the ineffectiveness of diffusion by direct interchange is the so-called *Kirkendall effect* (da Silva and Mehl, 1951; Pfeil, 1929), i.e., the movement of the diffusion interface boundary. It is explained by more rapid diffusion of material across the interface in one direction than diffusion of material in the opposite direction. Diffusion, therefore, by direct interchange cannot account for movement of fiducial markers.

The third mechanism, vacancy diffusion, takes place by an atom moving to a vacant site and thereby creating a vacancy in the site previously occupied. In the case of copper, once again, the activation energy required for a copper atom to move to a vacant site is about 23,100 cal/mole. The energy needed to create a vacancy within the copper structure has already been given as 238,000 cal/mole; however, the energy needed to cause an atom to move on the surface of the crystal and form a vacancy there (known as a *"Schottky defect"*) is much less; it is on the order of 41,600 cal/mole. Through thermodynamic reasoning, furthermore, it is known that any substance has a finite number of vacant atomic sites at any temperature above absolute because of the increase in entropy that results from the disorder that the vacancies introduce. The number of vacant sites, for example, in a sodium chloride crystal, is about one percent of the total sites at a temperature near 800°C (Mehl, 1936), just a few degrees below its melting temperature. In copper, at its melting temperature of 1356°C, it is estimated by approximate calculations (Smigelskas and Kirkendall, 1947) that the proportion of vacancies with respect to the total number of sites available for vacancies may be about 0.1%.

Structural inhomogeneities known as *dislocations* may very well account for many of these vacant sites. In any regard, however, they are thought to be of great importance in further enhancing the mechanism of vacancy diffusion in metals and also in specific structural types of ionically bonded minerals.

The importance of determining the energy requirements of these different mechanisms is significant because the activation energy determined experimentally from self-diffusion studies of metallic copper, including very careful measurements on single crystals, is about 48,000 cal/mole. Apparently vacancy diffusion is the favored mechanism of self-diffusion in f.c.c. metallic copper. It is, of course, possible that the other mechanisms contribute, but to a much lesser extent.

Of primary interest to those of the geologic profession are similar calculations that have been determined for ionically bonded substances, the alkali halides. The computed energies of activation for the various mechanisms of diffusion of Na in NaCl as summarized by Seitz (1951) are given below:

1. The energy needed to produce a pair of positive- and negative-ion vacancies (by moving a positive and a negative ion to the surface of the crystal) is 43,100 cal/mole.
2. The energy needed to move an Na^+ ion vacancy into an adjacent vacant Cl^- site is 11,800 cal/mole.
3. The energy needed to move a Cl^- ion vacancy into an adjacent vacant Na^+ site is 12,900 cal/mole.
4. The energy needed to move vacant Cl^- ion site, that is adjacent to a pair of Na^+ and Cl^- (neutral charge on vacancy pair) ion vacancies is 8,800 cal/mole.
5. The energy needed to force an ion into an interstitial site and create a vacancy is 65,000 cal/mole.

In comparison, experimental values of the activation energy of self-diffusion of Na in NaCl have been determined using Na^{24} as a tracer. In addition, further insight on the prevalent mechanisms of diffusion has been provided by adding given small amounts of the halides of a divalent metal, e.g., $SrCl_2$, which increases the density of vacant Na^+ sites, in order to maintain electric neutrality. The following values have been determined experimentally (Mapother et al., 1948 and 1950):

1. The energy needed to produce a pair of positive- and negative-ion vacancies is 48,500 cal/mole.
2. The energy needed to cause migration of an Na^+ vacancy is 18,500 cal/mole.

The comparison of computed values with the experimentally determined values indicates similarities. Since the NaCl already contains vacancies, a comparatively large additional energy requirement, beyond the energy needed for mobility of Na^+, is not wholly needed to create vacancies.

The probable mechanism of solid diffusion in silicate minerals, which is really of the utmost geologic significance, differs from the above-mentioned methods in very basic manners even though the energy requirements are very likely of the similar order of tens of kcal/mole. Silicates are composed of specific structures of networks based not only upon various arrangements of the silica tetrahedrons but also upon the number of oxygen atoms of one tetrahedron that are shared with adjacent tetrahedrons. The result is a structural framework, formed by the tetrahedrons, and having cations located in the larger openings or "cages" of the framework. When the structure receives energy, in the form of heat, it generally tends to expand and the atoms are then more capable of migrating or diffusing from one open space or cage to another. Basically, this is an interstitial type of diffusion but it might be more descriptive to refer to the mechanism as interstitial diffusion through a network, or interstitial network diffusion.

Interstitial diffusion is also probable in sulfide and oxide minerals. In fact, diffusion seems to be the most reasonable process by which certain types of replacement occur, for example, during replacement of one sulfide by another. In discussing the mechanism of replacement, Buerger (1948) has said, "Lindgren long ago pointed out that replacement occurs on approximately a volume-by-volume basis. While the field evidence for this has been obvious, the mechanism for accomplishing it has been obscure. Diffusion suggests the mechanism. There is a tendency on the part of crystals to have their volumes determined by their largest atoms. Thus, the volumes of the rock minerals are dominated by their oxygen atoms and the volumes of the sulfides are dominated by the packing of the sulfur atoms. Replacement is, therefore, substantially a matter of the diffusion of new metals into the volumes dominated by oxygen or sulfur atoms. Thus, diffusion supplies a mechanism for approximately maintaining volume during replacement." Further corroborating support of this mechanism has been given by studies (Jensen, 1957) which indicate that the S^{32}/S^{34} composition of host sulfides is identical with the S^{32}/S^{34} composition of the replacing sulfide.

Effect of Vapor Pressure of Mineralizer on Diffusion Rates. Considering the abundance of information on diffusion coefficients determined by the metallurgical profession it is surprising to note the lack of information pertaining to the effect of pressure on diffusion rates. Be that is it may, there is considerable indirect field evidence that diffusivity is increased, even appreciably, in minerals and rocks that exist in a medium of high vapor pressure (vapor pressure, for example, of mineralizers such as water or carbon dioxide or both). Some of the indirect evidence is as follows:

1. The rate of crystal growth is controlled by the availability of the appropriate substances to a growing crystal surface. The rate of diffusion of the appropriate substances through a medium such as hydrothermal solutions or a pegmatitic solution lacking effective convection currents may, thus, influence the size of the resulting crystal. The majority of the crystals growing in viscous solutions, such as silica-rich hypabyssal and extrusive magmas, are relatively small. In contrast, the largest crystals generally occur in pegmatites from what is believed to have been an agueous-rich (nonviscous) solution.

2. Determination of the orthoclase-albite binary phase diagram was accomplished (Bowen and Tuttle, 1950) by subjecting the appropriate mixtures to high water vapor pressures in order to increase the rate of crystal growth. At one atmosphere, glass formed upon even slow cooling of the melt; with increased water vapor pressures, only crytoperthites formed that could not be identified microscopically but had to be identified by x-ray techniques. This study is a striking example of the significant increase in the mobility or diffusion and more rapid crystal growth of solute substances under increased vapor pressure of water.

3. Fig. 4 is a photograph of a typical hand specimen from the Bingham porphyry copper stock in Utah. Hydrothermal alteration of feldspar crystals in this specimen has occurred presumably by diffusion of hydrothermal solutions (H_2O and CO_2 for the most part) into and through solid feldspar crystals to form various alteration minerals of clay and sericite. The distance of solid diffusion needed for this alteration is certainly not the diameter (several thousand feet) nor even the radius of the stock but instead, with few exceptions, is only a few centimeters at the most. Hydrothermal solutions moved with relative ease through the myriad of fractures existing in the shattered portion of the Bingham stock. Note that the hand specimen itself (Fig. 4) exhibits without

Fig. 4. View of the fracture surfaces and vugs in a typical hydrothermally altered, quartz monzonite specimen, Bingham porphyry copper deposit, Utah. Note abundance of planar fractures that are minute openings and passageways for movement of hydrothermal solutions (ruler in cm).

exception only vugs and planar surfaces or fracture planes along which the specimen broke when struck with a geologic hammer. Portions of some of these fractures are veinlets that are more than 1 cm wide. Others are made evident only by thin lines on the planar surfaces of the specimen. The higher concentration of quartz and clay minerals along these fractures aids in accentuating the lines. Even so, disseminated sulfide specks and sericitized feldspar crystals do exist within the specimen, not along evident fractures, but near fractures from which diffusion transported the material along grain boundaries and ultimately into the crystals.

Conclusions. The migration or movement of atoms or ions in a solid substance is a fact. The rate of this process of solid diffusion can be measured experimentally and the results treated quantitatively by a rational mathematical approach. Diffusion appears to be of prime aid in the mechanisms by which various geological and mineralogical processes occur, viz., hydrothermal alteration, replacement, crystal growth, and other even more controversial processes.

The distances over which solid diffusion can be effective in these processes is relatively short, rarely more than several centimeters, which is easily attainable by acceptable diffusion rates. Even so, transport of substances by "trunk lines" over considerably greater distances may be explained by relatively rapid movement through rock openings, along fracture zones, around mineral grains, and finally into crystals through crystal flaws and breaks. Solid diffusion transport through the remaining few centimeters is then effected with relative ease, resulting, thereby, in the transport and removal of a relatively large mass exchange of material.

MEAD LeROY JENSEN

References

* Barrer, R. M., 1951, "Diffusion In and Through Solids," Cambridge, University Press, 464pp.
Bowen, N. L., 1921, "Diffusion in silicate melts," *J. Geol.,* **29,** 295–317.
Bowen, N. L., and Tuttle, O. F., 1950, "The system $NA1Si_3O_8$–$KA1Si_3O_8$–H_2O," *J. Geol.,* **58,** 489–511.
Buerger, M. J., 1948, "The role of temperature in mineralogy," *Am. Mineralogist,* **33,** 101–121.
Carslaw, H. S., and Jaeger, J. C., 1959, "Conduction of Heat in Solids," London, Oxford University Press, 386pp.
* Crank, J., 1956, "The Mathematics of Diffusion," London, Oxford University Press.
da Silva, L. C. C., and Mehl, R. F., 1951, "Interface and marker movements in diffusion in solid solutions of metals," *J. Metals,* **191,** 155–173.
Dushman, S., and Langmuir, I., 1922, "The diffusion constant in solids and its temperature coefficient," *Phys. Rev.,* **20,** 113.
Fick, A., 1855, "Ueber diffusion," *(Pogg.) Ann. Phys. Chem.,* **94,** 59–86.
Fourier, J. B., 1822, "Theorie Analytique de la Chaleur," Envres de Fourier.
Fyfe, W. S., Turner, F. J., and Verhoogen, J., 1958, "Metamorphic reactions and metamorphic facies," *Geol. Soc. Am. Mem.,* **73,** 259pp.
Garrels, R. M., Dreyer, R. M., and Howland, A. L., 1949, "Diffusion of ions through intergranular spaces in water-saturated rocks," *Bull. Geol. Soc. Am.,* **60,** 1809–1828.
* Gilluly, J., 1948, "Origin of granite," *Geol. Soc. Am. Mem.,* **28,** 14, 83, 84.
Huntington, H. B., 1942, "Self consistent treatment of the vacancy mechanism for metallic diffusion," *Phys. Rev.,* **61,** 325.
Huntington, H. B., and Seitz, F., 1942, "Mechanism for self-diffusion in metallic copper," *Phys. Rev.,* **61,** 325.
Huntington, H. B., and Eeitz, F., 1949, "Energy for diffusion by direct interchange," *Phys. Rev.,* **76,** 1728.
Jensen, M. L., "Diffusion in minerals," Ph.D. dissertation, 1951, Cambridge, Mass., Mass. Inst. Technol.
Jensen, M. L., 1952, "Solid diffusion of radioactive sodium in perthite," *Am. J. Sci.,* **250,** 808–821.
Jensen, M. L., 1957, "Sulfur isotopes and mineral paragenesis," *Econ. Geol.,* **52,** 269–281.
* Jost, W., 1952, "Diffusion in Solids, Liquids, and Gases," New York, Academic Press, 558pp.
Mapother, D. E., Crooks, N., and Maurer, R. J., 1950, "Self diffusion of sodium in sodium chloride and sodium bromide," *J. Chem. Phys.,* **18,** 1231–1236.
Mapother, D., and Maurer, R. J., 1948 Abstract, "Self-diffusion of sodium in sodium chloride," *Phys. Rev.,* **73,** 1260.
* Mehl, R. F., 1936, "Diffusion in metals," *Trans. AIME,* **122,** 11–56.
Perrin, R., and Roubault, M., 1949, "On the granite problem," *J. Geol.,* **57,** 357–359.
Pfeil, L. B., 1929, "The oxidation of iron and steel at high temperatures," *J. Iron Steel Inst. (London),* **119,** 501–547.
Seith, W., and Keil, A., 1933, "Die selbst Diffusion im festen Blei: *Ziet. Melallkunde,* **25,** 104.
* Seitz, F., 1951, "Fundamental aspects of diffusion in solids," Chap. 4, in (Smoluchowski, R. et al., editors), "Phase Transformations in Solids," New York, John Wiley & Sons.
Smigelskas, A. D., and Kirkendall, E. O., 1947, "Zinc diffusion in alpha brass," *Trans. AIME,* **171,** 130–135.
Van Orstrand, C. E., and Dewey, F. P., 1915, "Preliminary report on the diffusion of solids," *U.S. Geol. Surv. Profess. Paper* **95-G,** 83–96.
Veerhogen, J., 1952, "Ionic diffusion and electrical conductivity in quartz," *Am. Mineralogist,* **37,** 637–655.
Wagner, C., and Hantelmann, P., 1950, "Determination of the concentrations of cation and anion vacancies in solid potassium chloride," *J. Chem. Phys.,* **18,** 72–74.
Zener, C., 1950, "Ring diffusion in metals," *Acta. Cryst.,* **3,** 346–354.

* Additional bibliographic references may be found in this work.

Cross-references: *Aqueous Solutions; Bonding; Entropy; Hydrothermal Alteration; Hydrothermal Solutions; Mineralogy; Thermodynamics.* Vol. II: *Avogadro's Law.* Vol. IV B: *Crystal Growth; Metallurgy.*

DOWSING—See WATER DIVINING

DRAINS AND DRAINAGE—See Vol. VI

DRAWDOWN, CONE OF DEPRESSION

Drawdown is a term applied to the maximum lowering of the groundwater table caused by pumping or artesian flow. It is measured as the difference between the initial level of water in a well before pumping, and the stabilized level of water after a long period of pumping. The stabilized level (steady-state) is achieved

DRAWDOWN, CONE OF DEPRESSION

when the flow into the well from the aquifer is equal to the rate of withdrawal.

The drawdown may be determined from the following formula (Dapples, 1959):

$$d = \sqrt{\frac{D^2 - 2.3q \log_{10} \frac{R}{r}}{\pi P}}$$

where d = thickness of water in well in feet (when pumping is in progress) measured from base of aquifer being pumped
D = thickness of aquifer below static water level in feet (measured to base of well when entirely in aquifer or to base of aquifer if well is deeper)
q = quantity of water pumped in gallons per day
R = radius of base of cone of depression
r = radius of well in feet
P = permeability coefficient (rate of flow in gallons per day through a cross section of one square foot under a hydraulic gradient of 100% at a temperature of 60°F)

Within the *aquifer* (q.v.), immediately around a well, the upper boundary of the water is lowered as water is pumped out (Fig. 1). The water surface boundary, described by an inverted cone, is known as the *cone of depression,* which theoretically extends outward from the well to the limits of the water-bearing bed (Dapples, 1959). The radial distance out from the well over which this water boundary is actually lowered, below that of the static water table, is termed the *radius of influence,* and is approximately one-quarter mile. This must be taken into account for the proper spacing of wells.

B. C. S.

References

Dapples, E. C., 1959, "Basic Geology for Science and Engineering," New York, John Wiley & Sons, 609pp.
Jumikis, A. R., 1962, "Soil Mechanics," Princeton, N.J., D. Van Nostrand Co., 791pp.
Tolman, C. F., 1937, "Ground Water," New York, McGraw-Hill Book Co., 593pp.

Cross-references: *Aquifer; Artesian Water; Groundwater; Water Table.* Vol. VI: *Wells—Water.*

DRILLING FLUIDS, TECHNOLOGY—
See Vol. VI

DROUGHT—See Vol. II

DUST COUNTER—See KONIOLOGY; Vol. II, ATMOSPHERIC NUCLEI AND DUST

DYSPROSIUM: ELEMENT AND GEOCHEMISTRY

De Boisbaudran, in 1886, first isolated dysprosium from other rare earth elements, in the rare earth concentrate holmia. Dysprosium means *difficult of access.*

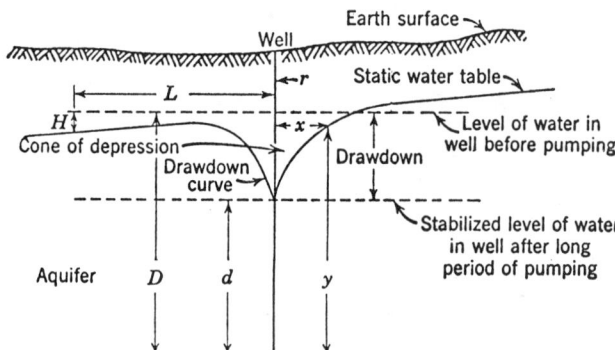

FIG. 1. Diagrammatic representation of a well penetrating an aquifer. Pumping has lowered the water table locally to develop a cone of depression around the well (Dapples, 1959: Modified from Tolman, 1937). Key: r = radius of well in feet; d = thickness of water in well (in feet measured from the base of the aquifer being pumped; x and y = coordinates of any point on the drawdown curve; D = thickness of aquifer below static water level (in feet); H = difference in head between two manometers in path of fluid movement; L = distance along the path of movement between positions of manometers.

Properties

Elemental dysprosium is a very reactive silvery white lanthanide or rare earth metal with atomic number 66 and outer electronic structure $4f^{10}6s^2$. It occurs in six stable isotopes:

Dy^{158}	0.09% in nature
Dy^{160}	2.29%
Dy^{161}	18.8%
Dy^{162}	25.5%
Dy^{163}	25.0%
Dy^{164}	28.2%

One dysprosium isotope found in nature is weakly radioactive: Dy^{156} (0.05%) with a half-life of 2×10^{14} years. Trace amounts of dysprosium have been determined analytically by neutron activation, by x-ray spectography, and by isotope-dilution mass spectometry. In geologic environments dysprosium forms highly ionic bonds and weak complexes. The oxidation state is tripositive, with ionic radius 0.91Å. Generally, chemical behavior is similar to that of the alkaline earths, and almost identical to that of other heavy rare earths.

Minerals

Dysprosium occurs in highest concentration in rare earth minerals that favor heavy rare earths, generally nitrates, sulfates, carbonates, niobates, and tungstates with cation coordination numbers of six to eight, often containing structural halogens and alkali metals. Typical examples are: xenotime (Y, heavy rare earths) PO_4 and fergusonite (Y, heavy rare earths) $(Nb, Ta)O_4$.

Geochemical Behavior

Schmitt et al. (1964) have established a value of 0.31 ppm dysprosium in chondrites and 3.8 ppm in eucrite achondrites. The standard deviation within meteorite class is only about 20%.

The average dysprosium concentration in basalt is probably about 5 ppm, and in granites with 60–70% silica, probably about 6 ppm; these values are interpolated from data in Haskin, et al. (1966). Since dysprosium behaves like other rare earths, the total range in values would also be less than a factor of ten. Partial melting of mantle peridotite yields a basaltic liquid with enriched dysprosium and a mantle source depleted in dysprosium.

Igneous differentiation series tend toward higher dysprosium concentrations, although the total enrichment is less than for light rare earths. Towell et al. (1965) found 3.08 ± 0.12 ppm dysprosium in the Rubidoux Mountain leucogranite phase of the southern California batholith, probably a differentiate (by crystal settling) of the San Marcos gabbro, with 2.79 ± 0.11 ppm.

In basaltic, syenitic, and granitic rocks dysprosium is found mainly in mafic rock-forming minerals, probably substituting for divalent calcium (ionic radius, 0.99Å).

Dysprosium and other heavy rare earths form complexes and are transported by alkali, halogen, and carbon dioxide hydrothermal solutions; metasomatic and pegmatoid deposits formed from such fluids may have high concentrations of dysprosium and higher ratios of dysprosium to light rare earths than granites.

Dysprosium probably has about 450-years residence time in seawater (which is alkaline), about the same as other heavy rare earths. It is probably complexed. Removal by adsorption on clay particles, as well as depletion in surface water in areas of high organic productivity, are likely processes. High secondary dysprosium concentrations in phosphatic fish remains accumulating today occur only in deep water; paleozoic fish remains also have high dysprosium. Manganese nodules have 3,000 ppm dysprosium, whereas Pacific Ocean water has only 0.73 ppm (Goldberg, et al., 1963).

In soils, the relative amounts of dysprosium in detrital, clay, and carbonate fractions is not well known. Sedimentary rocks have about 6 ppm, with a variation of a factor of five. In a study of rare earths in Russian platform soils, Balashov, et al. (1964) found higher dysprosium in alkaline soils, the dysprosium precipitating as hydroxide; lower concentrations are in acid soils, probably due to removal as complexes.

Biologic concentration of rare earths has been found in hickory leaves; rare earths have been used to trace opium.

Economic Geology

The major source of dysprosium has been from monazite in beach sands. The accessory monazite originally present in granitic rocks resists weathering and is concentrated by sedimentary processes. Australia, Brazil, Ceylon, India, Malaysia, and South Africa mine the largest quantities. Less than one percent (by weight) of monazite is dysprosium. Recently, the bastnaesite deposit at Mountain Pass, California has been an important source of dysprosium, again in low concentration.

High-purity dysprosium has a very limited market; it is used for special types of optical filters. Most dysprosium is sold unpurified, in rare earth concentrates used for polishing, petroleum cracking catalysis, lighter flints, arc carbons, and in alloys.

ROBERT KAY

References

Balashov, Y. A., Ronov, A. B., Migdisov, A. A., and Turanskaya, N. V., 1964, "The effects of climate and facies environment on the fractionation of the rare earths during sedimentation," *Geochem. Intern. No. 1,* **10,** 951–969 (English transl.).

Goldberg, E. D., Koide, M., Schmitt, R. A., and Smith, R. H., 1963, "Rare-earth distributions in the marine environment," *J. Geophys. Res.,* **68,** 4209–4217.

Haskin, L. A., Frey, F. A., Schmitt, R. A., and Smith, R. H., 1966, "Meteoric, Solar, and Terrestrial Rare-Earth Distributions," in (Ahrens, L. H., et al., editors) "Physics and Chemistry of the Earth," Vol. 7, pp. 167–321, Oxford, Pergamon Press.

Rankama, K., and Sahama, T. G., 1950, "Geochemistry," Chicago, Univ. of Chicago Press, 912pp.

Schmitt, R. A., Smith, R. H., and Olehy, D. A., 1964, "Rare-earth, yttrium and scandium abundances in meteoric and terrestrial matter—II," *Geochim. Cosmochim Acta* **28,** 67–86.

Towell, D. G., Winchester, J. W., and Spirn, R. V., 1965, "Rare-earth distributions in some rocks and associated minerals of the batholith of Southern California," *J. Geophys. Res.,* **70,** 3485–3496.

Cross-references: *Alkalis, Alkali Metals, and Alkaline Earth Metals; Complexes; Halogens; Hydrothermal Solutions; Mass Spectrometry; Neutron Activation Analysis; Rare Earths.*

E

EARTH, GEOCHEMISTRY—See **EARTH'S CRUST GEOCHEMISTRY; GEOCHEMICAL EVOLUTION OF THE CORE, MANTLE, AND CRUST**

EARTH'S CRUST GEOCHEMISTRY

The earth's crust (lithosphere) is taken here as the outer solid geosphere that is separated from the upper mantle by the Moho boundary. It consists of three shells: a stratified sedimentary shell ("stratisphere"),* composed mainly of sedimentary rocks; a "granitic" shell, distributed only beneath the continents and thinning out at the ocean boundaries; and a "basaltic" shell which has different structures under continents and under oceans. (These layers were identified nearly 100 years ago by Suess as SIAL and SIMA.) The earth's crust was formed during a long period of geological time embracing about 4.5 billion years. It originated as a result of interaction between the differentiates of deep-seated material and the external environment. This interaction, most intensive during the earlier stages of earth history, has ultimately led to the present complicated structure of the crust, which is divided into large blocks varying in their geological history and framework (see Fig. 1).

Within every block or zone there is a tendency toward regular changes of crustal structure with depth, which is manifested in the existence of layers more or less persisting in planetary scale (both in continents and oceans). The mass distribution into blocks and shells within the crust, the chemical composition of the crust, and the conditions under which it was formed are the main problems of quantitative geochemistry.

To resolve these questions it is necessary to know the volume of the crust in the various tectonic zones, the densities, masses, and the quantitative ratios of the various rock types forming the crust, as well as their chemical composition.

---
* Editorial note: "stratisphere" is not used in English; perhaps because it is rather close to "stratosphere" (upper atmosphere).

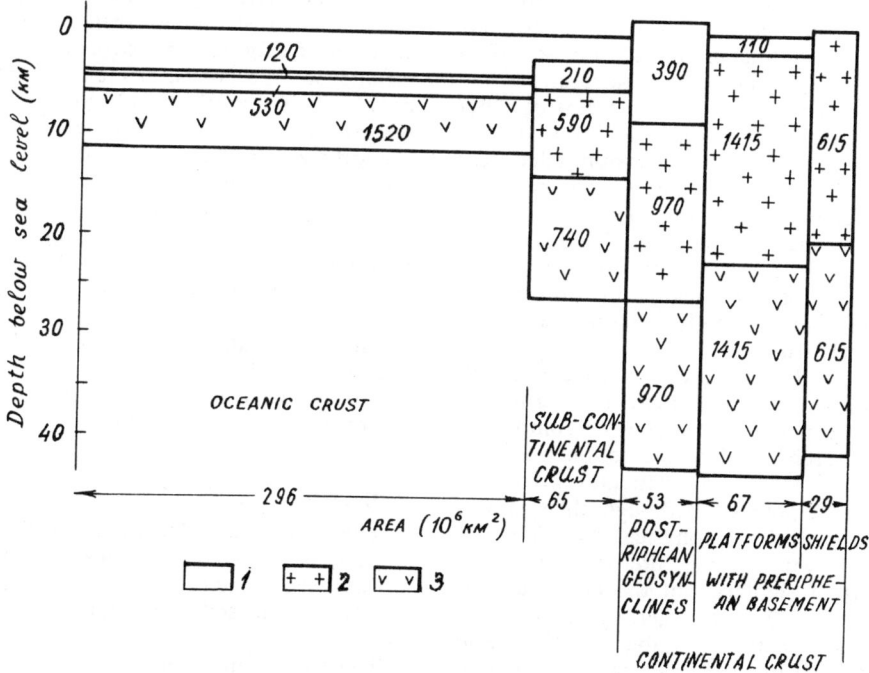

FIG. 1. Schematic subdivision of the earth's crust. Values indicate volume in 10^6 km^3: (1) sedimentary layer; (2) "granitic" layer; (3) "basaltic" layer.

Clarke (1924) and Goldschmidt (1933) calculated the volume of the earth's crust assuming a mean thickness of 16 km (10 miles). Geophysical investigations have determined that the natural boundary between the earth's crust and mantle is the Mohorovičić discontinuity, thus making it possible to obtain measurements of real volume for this solid shell. Such calculations of the crustal volume were performed for the first time by Poldervaart (1955). An analogous attempt, based on the more precise and recent geophysical, geotectonic, and geochemical data, was undertaken by Ronov and Yaroshevsky (1967, 1969).

The determination of the composition of the earth's crust depends largely on the mass ratios of the various types of magmatic, sedimentary, and metamorphic rocks forming its shells. The first attempts to determine the crustal composition were based on the idea that the average compositions of the rock types must be practically the same if one neglects the "excess volatiles" (H_2O, CO_2, etc.), the concentrations of which are higher in the sediments than in other rock types. Such an assumption simplifies the problem, which is now reduced to the determination of the average composition of the magmatic rocks considered to be primary. Several variants of such estimates are available (Clarke, 1924; Goldschmidt, 1933; Vinogradov, 1962; Taylor, 1964). The idea of compositional similarity of genetically different rock types is closely connected with the concept of geochemical balance of weathering, sedimentary, and metamorphic processes. However the first attempts to use the results of estimations of the natural volume ratios of different sedimentary rock types and the data of their average chemical compositions led Kuenen (1950) and Poldervaart (1955) to values quite different from those obtained by geochemical balance calculations. This discrepancy remains the most enigmatic feature of the earth's crustal geochemistry and prevents the derivation of an internally consistent scheme of its chemical structure.

The most difficult problem concerning the chemical composition of the earth's crust concerns the deeper structure and the composition of the "basaltic" shell of the continental crust. The solution of this problem may involve incompatible hypotheses although they may be consistent with the geophysical data. They are limited by two extreme alternatives: *(1)* the hypothesis of similarity of the chemical composition of the "basaltic" and "granitic" shells, and *(2)* the hypothesis of the basaltic (basic) composition of the "basaltic" shell, accepted by Poldervaart (1955) and Pakiser and Robinson (1967) as the basis for geochemical calculations. Ronov and Yaroshevsky (1967, 1969) proceeded from the assumption that the petrographic and chemical composition of the "basaltic" shell gradually changes with depth from a composition similar to that of the "granitic" shell in its upper layers to a composition in the lower horizons corresponding to continental basalt of a geosynclinal type. At present it is impossible to show this tendency in a geochemical section of the earth's crust, and authors have presented it only schematically on the assumption that the basic (basaltic) and acid (granitic) materials take part equally in the formation of the "basaltic" shell. There is no doubt that the uncertainty about the "basaltic" shell composition is the principal cause of possible errors in the calculation of crustal composition.

In the general geochemical balance of the crust the fundamental differences in the structure and composition between continental and oceanic crustal blocks must be considered. For this reason most authors have considered each block separately on the basis of various hypotheses about continental and oceanic tectonic structure. For the continental blocks the accepted hypothesis was usually based on the idea that the crustal material had passed at some time or other through the geosynclinal stage of development.

Proceeding from these assumptions Ronov and Yaroshevsky calculated the volumes, masses, and average chemical composition of the earth's crust and its shells. They used recent data on the tectonic structure of continental and oceanic crusts, Demenitskaya's map of crustal thickness, the results of measurements of sedimentary and volcanic rock volumes in the sedimentary shell, the areal estimates of the magmatic and metamorphic rock ratios on the shields, as well as extensive data concerning the average chemical composition of different rock types. The assumed ratio of the "granitic" and "basaltic" volumes of the continental crust (1:1) was based on average values provided by the latest data from Pakiser and Robinson (1967) and Baliajevsky et al. (1967). The results of the calculations, corrected to 1969, are combined in Tables 1–5.

Sedimentary Shell

The results of calculations of the volumes, masses, and average chemical composition of various sedimentary rock types in continents, recent oceanic sediments, and the stratisphere as a whole are shown in Table 1.

The total volume of the earth's sedimentary shell is equal to 1,050,000,000 km^3 (taking into account the consolidation of recent sediments) and 900,000,000 km³ without volcanic rocks, i.e., about 10% the volume of the crust and 0.1% the volume of the whole earth. The average thickness of the sedimentary shell is 2.0 km; if the area of the shields not covered by sediments is excluded, the average is 2.2 km. On the whole, our values are close to those obtained by Kuenen (1950), whose volume was 71,000,000 km³ and average thickness was 2 km; Poldervaart (1955) gave a volume of 710,000,000 km³ and assumed an average thickness of 1.4 km. The bulk of the sediments have accumulated on the continents (500,000,000 km³) and on the continental margin (210,000,000 km³), whereas only 340,-000,000 km³ is actually in the oceans. This distribution of the sediments, according to Kuenen's (1950) scheme was 200,000,000 km³ for continents and 800,000,000 km³ for oceans including continental shelves; Poldervaart (1955) gave the following values: 180,000,000 km³ for continents, 160,000,000 km³ for the pelagic part of the ocean, and 370,000,000 km³ for the shelf and marginal areas.

An indirect method of calculating the total volume of the sedimentary rocks based on the idea of geochemical balance (Clarke, 1924; Goldschmidt, 1933) resulted in an obviously underestimated figure (330,000,000 km³). Newer estimates by the same method (Horn and Adams, 1966), with the use of an electronic computer, gave a value of 816,000,000 km³, but the estimate of sedimentary rock abundances used in this work differs significantly from measured values.

On the continents, about 75% of the volume of all sedimentary rocks is found in geosynclinal areas and only 25% in the platforms, their average thickness being 10 km and 1.8 km, respectively.

Clay and shale are the most widespread sedimentary rocks on the continents (42%). Arenaceous, volcanic, and carbonate rocks are approximately equally abundant (20, 19 and 18%, respectively). All other rock types, mainly evaporites, comprise about 1%.

A fundamental peculiarity of sedimentary rocks is the pronounced difference between their composition (Table 1 and 6) and the average composition of rocks of the "granitic" shell (Tables 2 and 6), which was the chief source of the sediment material (at least during the last 1.5 billion years). This difference is reflected in a much higher content of water, carbon dioxide, and organic carbon, as well as of sulfur, chlorine, fluorine, boron, and other "excess volatiles" (Rubey, 1951) in the stratisphere and in the hydrosphere. This is an indication of direct release of these excess volatiles from the mantle during the process of its degassing (see *Outgassing of the Planet Earth*).

The other important peculiarity of the composition of sedimentary rocks is their high calcium content, which is a most enigmatic feature of the geochemistry of the outer shell. Quite typical is the displacement of the potassium/sodium ratio in favor of potassium; this is not compensated by the excess sodium in the ocean. This leads to some deficiency of sodium in the sedimentary shell and hydrosphere, taken together, relative to the "granitic" shell. The oxidizing conditions on the earth's surface have determined the higher ferric to ferrous (Fe_2O_3/FeO) ratio in sedimentary rocks, as well as the heightened content of sulfur (sulfate) in them.

These peculiarities are most clearly manifested in platform sediments, since these are products of deep weathering and strongly developed surface differentiation. In contrast, geosynclinal sediments have undergone less extensive alteration (especially sands), and their composition (e.g., graywacke) approaches the composition of the parent rocks.

"Granitic" Shell

The results of calculations of the volumes, masses, and relative abundances of different magmatic and metamorphic rock types in the "granitic" shell are combined in Table 2.

The "granitic" shell is completely restricted to the continents, and its volume and mass are approximately 3,600,000,000 km³ and 9.8×10^{24} g, respectively. The composition and ratio of different rock types in the "granitic" shell are estimated on the basis of the data of average petrographic composition of the various shields (Table 3). As a whole, the data characterizing the various shields are in quite good agreement and are close to the composition of the "granitic" shell which was calculated using the hypothesis that the composition of the metasedimentary rocks of the shields is nearly the same as that of the geosynclinal suites (see Tables 3 and 5). The data of Table 5 were used for the calculation of the average composition of the crust (Table 6).

The "granitic" shell is completely concentrated on the continents. Acidic granitoids and metamorphic rocks are the main rock types of the "granitic" shell; the basic and ultrabasic rocks make up less than 15% of the shell's volume. These volumetric ratios of rocks lead to the acidic chemical composition of the shell, its typical high content of silica, the concentrations of alkalies (K>Na), and of the rare elements (uranium, thorium, rare earths, zirconium, niobium, etc).

TABLE 1. VOLUME, MASS, AND AVERAGE CHEMICAL COMPOSITION OF SEDIMENTARY AND VOLCANIC ROCKS OF CONTINENTS AND SEDIMENTS OF OCEANS

Types of Crust	Large Structural Units of Crust and Layers	Volume, $10^6 km^3$	Average Thickness[a] (km)	Mass $(10^{24}g)$	Types of Rocks and Sediments	Abundance % of Volume on continents and % of Area in Oceans
Continental	Platforms	135	1.8 (1.2)	0.35	Sands	23.5
					Clays	49.3
					Carbonates	21.0
					Evaporites: sulfates, 50% salts, 50%	2.3
					Volcanics	3.9
					Average composition of sedimentary series of platforms	100.0
	Geosynclines	365	10.0	0.94	Sands	18.7
					Clays and shales	39.4
					Carbonates	16.3
					Evaporites: sulfates, 50% salts, 50%	0.3
					Volcanics	25.3
					Average composition of sedimentary series of geosynclines	100.0
	Total	500	4.2 (3.4)	1.29		
Subcontinental	Shelf and continental slope	210	3.2	0.52		
Oceanic	Sediments of seismic layer I	120 (70)[c]	0.4	0.19	Terrigenous	7.3
					Calcareous	41.5
					Siliceous	17.0
					Red deep-sea clays	31.2
					Volcanogenic sediments	3.0
					Average composition of pelagic sediments[d]	100.0
	Sediments of seismic layer II	265	0.9	0.66		
	Sediments of oceanic crust as a whole	385 (335)[c]	1.3	0.85		
Total for sedimentary shell including volcanics		1095 (1045)[c]	2.2 (2.0)[c]	2.65		
Total for sedimentary shell excluding volcanics		900[c]		2.25		

[a] Thicknesses including areas of shields are shown in parentheses.
[b] The sum of analyses exceeds 100.00% by O-Cl_2.
[c] Recalculated for consolidated sediment with a density of 2.5.
[d] In calculating average values, the content of SO_3 and Cl in marine sediments was taken to be the same as in continental ones.

The most important characteristic of the composition of shield metamorphic rocks is their similarity to the composition of the sedimentary suites in geosynclines (Table 5). At the same time the average composition of the "granitic" shell differs appreciably from that of Neogaea (post-Precambrian sedimentary rocks) (Table 6). In all probability this indicates that the processes of metamorphism in the crust take place in a closed system without significant addition of new material. However, the subsequent transformation of the metasediments during the stage of ultrametamorphism and palingenesis apparently developed in an open system and were accompanied by an additional supply of silica and alkali solutions; according to modern concepts, this appears to be the essence of the granitization process.

"Basaltic" Shell

The "basaltic" shell consists of two parts, the continental and the oceanic, differing in structure and apparently in composition. According to the hypothesis assumed here, the "basaltic" shell of the continental crust is formed of strongly metamorphosed rocks of both basic and acid composition together with a significant portion of magmatic rocks (see

EARTH'S CRUST GEOCHEMISTRY

TABLE 1. (Continued)

						Content of Components, % by Weight, and their Masses in Stratified Shell as a whole (10^{24}g)									
SiO_2	TiO_2	Al_2O_3	Fe_2O_3	FeO	MnO	MgO	CaO	Na_2O	K_2O	P_2O_5	Corg	CO_2	SO_3	Cl[b]	H_2O^+
75.75	0.49	6.90	2.55	1.31	0.06	1.43	3.40	0.58	1.83	0.16	0.30	2.75	0.29	0.09	2.11
55.09	0.86	16.30	4.17	1.87	0.05	2.46	4.75	0.75	3.01	0.11	0.99	3.92	0.43	0.12	5.16
9.80	0.18	2.54	0.85	0.54	0.06	6.83	38.93	0.24	0.76	0.07	0.33	35.05	2.36	0.08	1.40
0.30	—	0.07	0.04	—	—	0.31	18.07	24.92	0.31	—	—	0.66	24.95	28.74	8.11
49.22	1.53	15.74	3.33	8.02	0.18	6.11	10.00	2.51	0.73	0.18		0.01	0.03	0.005	1.81
49.12	0.64	10.86	2.97	1.65	0.06	3.36	12.17	1.23	2.11	0.11	0.63	10.00	1.34	0.76	3.26
62.93	0.52	12.12	2.30	3.20	0.11	2.28	5.69	1.92	1.69	0.11	0.25	4.22	0.12	0.03	2.47
55.76	0.71	17.56	3.61	3.35	0.08	2.52	4.08	1.27	2.76	0.15	0.78	2.80	0.11	0.03	4.37
13.30	0.14	2.70	0.43	0.94	0.09	2.92	42.40	0.56	0.49	0.10	0.35	34.44	0.03	0.02	1.05
0.30	—	0.07	0.04	—	—	0.31	18.07	24.92	0.31	—	—	0.66	24.95	28.74	8.11
55.62	1.00	16.12	4.17	4.52	0.24	4.22	6.91	3.33	2.02	0.34		0.01	0.03	0.007	1.46
50.00	0.65	13.69	2.98	3.21	0.13	2.97	11.49	1.86	2.00	0.18	0.42	7.51	0.14	0.03	2.75
49.76	0.65	12.93	2.98	2.79	0.11	3.05	11.67	1.69	2.03	0.16	0.48	8.18	0.46	0.21	2.89
Composition considered to be similar to that of sedimentary rocks of continents															
52.92	0.81	15.73	5.04	1.92	0.33	3.82	4.68	1.59	3.09	0.08	0.30	3.35	—	—	6.34
18.84	0.29	6.09	2.55	0.46	0.19	1.98	35.61	0.70	1.03	0.11	0.28	28.90	—	—	2.97
62.80	0.66	13.21	4.92	1.15	0.14	3.02	1.47	1.52	2.70	0.16	0.26	1.08	—	—	6.91
54.51	0.80	16.40	6.93	0.84	1.05	3.41	2.02	1.27	2.78	0.21	0.22	2.45	—	—	7.11
44.30	2.48	12.41	5.83	6.40	0.41	8.85	9.79	1.88	0.78	0.20	0.32	3.36	—	—	2.99
40.73	0.61	11.41	4.60	0.97	0.47	2.94	16.29	1.13	2.01	0.15	0.26	13.27	—	—	5.17
Composition considered to be similar to that of sediments of seismic layer I															
Composition considered to be similar to that of sediments of seismic layer I															
46.63	0.64	12.44	3.50	2.21	0.22	3.01	13.15	1.51	2.02	0.16	0.41	9.81	0.46	0.21	3.62
1.240	0.017	0.331	0.093	0.059	0.006	0.080	0.350	0.040	0.054	0.004	0.011	0.261	0.012	0.006	0.096
46.62	0.62	10.84	3.24	2.00	0.18	2.71	13.62	1.28	2.09	0.13	0.49	11.60	0.53	0.27	3.78
1.049	0.014	0.244	0.073	0.045	0.004	0.061	0.306	0.029	0.047	0.003	0.011	0.261	0.012	0.006	0.085

Tables 2 and 6). The ratio of basic rocks reaches 50% of the total volume, while their composition is considered to be similar to that of geosynclinal basalts.

The continental crust has considerable thickness and a diversity in composition; the more homogeneous oceanic crust is 86% original oceanic theoleiitic basalts and their metamorphic equivalents. These basalts are characterized by a low content of potassium, rubidium, strontium, barium, phosphorus, uranium, thorium, and zirconium, and high ratios of K/Rb and Na/K, which strongly distinguish them from analogous continental rocks.

The oceanic crust is essentially characterized by the occurrence of ultrabasic rocks, seen in the zones of deep faulting (mid-ocean rift valleys), these rocks being considered outcrops of mantle material. These rocks differ from analogous continental rocks by heightened content of lithophylic elements (Si, Ti, U, Th, Zr, Be, etc.) according to Vinogradov et al. (1969). Probably these peculiarities in the chemistry of basic and ultrabasic rocks of the oceanic crust attest to considerably less differentiation of the oceanic mantle material. The existence of such rocks on the ocean floor implies a possible participation of ultrabasic rocks in the formation

EARTH'S CRUST GEOCHEMISTRY

TABLE 2. DISTRIBUTION OF VOLUMES AND MASSES OF MAGMATIC AND METAMORPHIC ROCKS OF CONTINENTAL, SUBCONTINENTAL, AND OCEANIC TYPES OF THE CRUST AND CHEMICAL COMPOSITION OF ROCKS

Types of Crust	Shells	Volume (10^6 km^3)	Mass (10^{24} g)	Types of Rocks	Abundance, % of Total Volume of Shell	Mass of Rock Types (10^{24} g)
Continental and Subcontinental	Sedimentary	710	1.81	Traps and plateau-basalts of platforms	1.1	0.02
				Basalts of geosynclines	9.2	0.20
				Andesites of geosynclines	7.4	0.16
				Rhyolites of geosynclines	1.8	0.03
	"Granitic"	3590	9.81	Granites	18.1[a]	1.75
				Granodiorites, diorites	19.9[a]	1.93
				Syenites, nepheline syenites	0.3[a]	0.03
				Gabbro	3.7[a]	0.38
				Dunites, peridotites	0.1[a]	0.01
				Gneisses	37.6[c]	3.66
				Crystalline schists	9.0[c]	0.88
				Marbles	1.5[b]	0.15
				Amphibolites	9.8[c]	1.02
				Average composition of magmatic rocks of "granitic" shell	42.1	4.10
				Average composition of metamorphic rocks of "granitic" shell	57.9	5.71
	"Basaltic"	3740	10.85	Acid magmatic and metamorphic ortho- and para-rocks	50.0	5.24
				Basic magmatic and metamorphic ortho- and para-rocks	50.0	5.61
Oceanic	Basalts plus volcanic rocks of seismic layer II	1785	5.18	Oceanic tholeiitic basalts	99.0	5.13
				Alkaline differentiates of oceanic basalts	1.0	0.05

[a] The estimate of relative abundance of magmatic rocks is based on the results of measurement on the maps of Ukrainian and Baltic shields and crystalline basement of Russian platform (acid, 38.3%; base, 3.8%), as well as the data obtained by Fleischer and Chao (1960) (Daly's figures) on the relations of granites, granodiorites, and syenites in the acid group of rocks and basic and ultrabasic rocks in the basic group.

[b] The percentage of marbles was reduced to 1/3 the amount of carbonate rocks in geosynclines, because of their less widespread occurrence in Early Archean rocks of the Precambrian.

[c] A quantitative estimate of the abundances of different types of metamorphic rocks is obtained on the basis of the concept of the geosyncline origin of metamorphic "para-"rock series. The second assumption is that the metasediments and "ortho-"rocks have nearly the same volumetric abundances.

of the lower horizons of the oceanic crust. By analogy, they may also be located at the base of the "basaltic" layer of the continental block, but at present it is practically impossible to estimate quantitatively their contribution to the composition of the crust. Perhaps their abundance is greatly underestimated (0.2%, Table 2).

The Crust as a Whole

About 64% of the whole crustal volume is continental, or 79% when the shelf and subcontinental ("quasicratonic") crust are included; 21% is strictly oceanic crust. The average thicknesses of the various crustal types decrease from 43.6 km for continental, to 23.7 km for subcontinental, and to 7.3 km for oceanic crust. The average thickness of the entire crust amounts to about 20 km.

The chemical composition of the crust (Table 6) as a whole approaches that of intermediate rocks, though it is impossible to find its close analog among them. In rough approximation, the crust's composition can be described as a mixture of the two prevailing types of rocks: granite and basalt (geosynclinal basalt plus oceanic theoleiite) at a ratio of 2:3. The average chemical composition of the crust changes with depth from the sedimentary shell to the "basaltic" one (Table 6), with a continuous increase in the content of iron, magnesium, and alumina and a decrease in the amount of combined water (H_2O^+); the contents of alkalis (potassium) and silica first in-

Table 2. (Continued)

					Content of Components (% by weight)									
SiO_2	TiO_2	Al_2O_3	Fe_2O_3	FeO	MnO	MgO	CaO	Na_2O	K_2O	P_2O_5	Corg	CO_2	S	Cl	H_2O^+
49.22	1.53	15.74	3.33	8.02	0.18	6.71	10.00	2.51	0.73	0.18	0.01	—	0.03	0.005	1.81
49.04	1.36	15.69	5.38	6.37	0.31	6.17	8.94	3.11	1.52	0.45	0.01	—	0.03	0.005	1.62
59.57	0.77	17.30	3.33	3.13	0.18	2.75	5.79	3.58	2.04	0.26	0.01	—	0.03	0.005	1.26
72.74	0.33	13.47	1.45	0.88	0.08	0.38	1.20	3.38	4.45	0.08	0.03	—	0.04	0.02	1.47
72.33	0.31	14.00	0.95	1.50	0.05	0.53	1.36	3.14	5.07	0.15	0.03	—	0.04	0.02	0.61
65.66	0.55	15.72	1.54	2.63	0.10	1.73	3.71	4.10	3.05	0.23	0.03	—	0.04	0.02	0.98
57.73	0.66	18.70	2.51	2.02	0.15	1.77	3.80	4.85	6.97	0.20	0.01	—	—	—	0.61
48.73	1.17	16.80	2.93	7.09	0.21	7.38	11.13	2.35	0.70	0.24	0.01	—	0.03	0.005	1.24
43.54	0.81	3.99	2.51	9.84	0.21	34.02	3.46	0.56	0.25	0.05	—	—	—	—	0.76
69.6	0.5	14.7	1.7	2.2	0.1	1.3	2.1	3.2	3.7	0.2	—	—	—	—	0.7
62.0	0.9	18.5	2.7	4.8	0.1	2.8	1.5	1.9	3.8	0.2	—	—	—	—	0.8
12.38	—	1.18	0.55	—	—	17.89	26.74	0.03	0.05	0.11	—	40.71	—	—	0.36
49.6	1.6	15.5	3.5	7.7	0.2	6.9	9.4	2.9	1.1	0.3	—	—	—	—	1.3
66.91	0.50	15.07	1.41	2.54	0.09	1.78	3.34	3.52	3.72	0.19	0.03	—	0.04	0.02	0.84
63.72	0.74	15.12	2.13	3.48	0.11	2.91	3.88	2.87	3.21	0.22	—	1.05	—	—	0.56
49.74	1.36	16.57	2.33	6.95	0.17	7.54	11.50	2.79	0.19	0.13	—	—	—	—	0.73
47.83	2.84	16.39	4.13	7.50	0.16	5.22	9.07	3.71	1.87	0.48	—	—	—	—	0.80

crease from the sedimentary shell toward the "granitic," and then decrease toward the "basaltic" shell of the continents and oceans.

A comparison between our calculations of the crustal mass (28.50×10^{24} g) with those obtained by Poldervaart in 1955 (23.67×10^{24} g) reveals some differences caused by different initial data and calculation methods. Some differences are also noted in the determination of the average chemical composition of the crust; our estimate is intermediate between those of Poldervaart (1955) and Vinogradov (1962) (Table 4).

Quantitative estimates (using the volumetric method) of the abundances of minor (rare) elements in the continental crust and earth's crust as a whole are still unavailable due to the lack of necessary data. However, rough estimates of their abundance in the continental crust and the earth's crust as a whole could be obtained based on a mixture of the proportions of the main rock types, i.e., granites and basalts. For such estimates we adopted the figures from the latest compilations by Vinogradov (1962) and Taylor (1964) for continental granites and basalts. The results are given in Table 7, where abundances of the main elements are shown for comparison; these were obtained by two methods: quantitative, based on volume measurements, and semiquantitative, computed by the same method as the abundances of the minor elements.

A comparison between the average content

TABLE 3. COMPARISON OF THE AVERAGE CHEMICAL COMPOSITION OF SHIELDS AND OF "GRANITIC" SHELL OF THE CONTINENTS

	Baltic and Ukrainian Shields		Canadian Shield				Average for Shields (2 and 6)	"Granitic" Shell (two variants of calculations)[b]	
Oxides	Seder-holm (1925)[a]	Ronov and Migdisov (1970)	Grout (1938)	Shaw et al. (1967)	Fahring and Eade (1968)	Average for Canadian Shield			
	1	2	3	4	5	6	7	8(1)	9(2)
SiO_2	68.4	66.7	63.9	66.7	66.1	66.4	66.6	65.6	66.0
TiO_2	0.4	0.5	0.8	0.5	0.5	0.5	0.5	0.6	0.5
Al_2O_3	14.8	15.1	17.0	15.0	16.1	15.6	15.3	15.3	15.0
Fe_2O_3	1.3	2.1	2.4	1.4	1.4	1.4	1.7	1.9	1.4
FeO	3.2	3.5	3.0	2.8	3.1	2.9	3.2	3.1	2.8
MnO	—	0.1	—	0.1	0.1	0.1	0.1	0.1	0.1
MgO	1.7	2.3	1.8	2.3	2.2	2.2	2.3	2.3	2.4
CaO	3.4	3.4	4.1	4.2	3.4	3.8	3.6	4.1	3.7
Na_2O	3.1	3.1	3.7	3.6	4.0	3.8	3.4	3.1	3.2
K_2O	3.6	3.1	3.1	3.2	2.9	3.1	3.1	3.4	3.5
P_2O_5	0.1	0.1	—	0.2	0.2	0.2	0.2	0.2	0.2
	100.0	100.0	100.0	100.0	100.0	100.0	100.0	100.0	100.0

[a] Baltic Shield in Finland.
[b] Two variants are based on the different ways of calculation of the metasedimentary rock composition (see Table 5).

TABLE 4. AVERAGE CHEMICAL COMPOSITION OF THE CRUST (wt %)

Oxides	Clarke (1924)	Goldsch-midt (1933)	Vinogradov (1962)	Taylor (1964) Continental crust	Poldervaart (1955) Continental crust	Poldervaart (1955) Entire Lithosphere	Pakiser and Robinson (1967) Continental crust	Ronov and Yaroshevsky (1969) Continental crust	Ronov and Yaroshevsky (1969) Entire Lithosphere
SiO_2	60.3	60.5	63.4	60.4	59.4	55.2	57.8	61.9	59.3
TiO_2	1.0	0.7	0.7	1.0	1.2	1.6	1.2	0.8	0.9
Al_2O_3	15.6	15.7	15.3	15.7	15.5	15.3	15.2	15.6	15.8
Fe_2O_3	3.2	3.3	2.5	7.2[a]	2.3	2.8	2.3	2.6	2.6
FeO	3.8	3.5	3.7		5.0	5.8	5.5	3.9	4.4
MnO	0.1	0.1	0.1	0.1	0.1	0.2	0.2	0.1	0.2
MgO	3.5	3.6	3.1	3.9	4.2	5.2	5.6	3.1	4.0
CaO	5.2	5.2	4.6	5.8	6.7	8.8	7.5	5.7	7.2
Na_2O	3.8	3.9	3.4	3.2	3.1	2.9	3.0	3.1	3.0
K_2O	3.2	3.2	3.0	2.5	2.3	1.9	2.0	2.9	2.4
P_2O_5	0.3	0.3	0.2	0.2	0.2	0.3	0.3	0.3	0.2
	100.0	100.0	100.0	100.0	100.0	100.0	100.0	100.0	100.0

[a] Fe_2O_3 and FeO given as FeO.

TABLE 5. COMPARISON OF AVERAGE CHEMICAL COMPOSITION OF GEOSYNCLINAL
SEDIMENTARY SERIES AND AVERAGE CHEMICAL COMPOSITION
OF THE METAMORPHIC ROCKS

Oxides	Average Chemical Composition of Geosynclinal Sedimentary Rocks (Weight %)	Average Chemical Composition of the Shield Metamorphic Para-Rocks (Weight %)
SiO_2	56.6	58.4
TiO_2	0.7	0.9
Al_2O_3	15.5	15.4
Fe_2O_3	3.4 ⎫ 6.7[a]	2.5 ⎫ 6.6[a]
FeO	3.6 ⎭	4.4 ⎭
MnO	0.2	0.1
MgO	3.4	4.3
CaO	5.5	5.2
Na_2O	2.1	2.5
K_2O	2.3	2.8
P_2O_5	0.2	0.2
Corg	0.5	—
CO_2	2.7[b]	2.4
S	0.1	—
Cl	0.1	—
H_2O+	3.1	0.9
Total	100.0	100.0

	Abundance (% of volume)			
Acidic volcanic rocks	2.8 ⎫		33.1 ⎫	Gneisses
Sandstones	21.0 ⎬ 68.2		68.2 ⎨ 35.1 ⎭	Crystalline shists
Clays and shales	44.4 ⎭			
Carbonates	5.5 ⎫ 5.8		5.8	Marbles
Evaporites	0.3 ⎭			
Basic volcanic rocks	26.0		26.0	Amphibolites
Total	100.0		100.0	

[a] Fe_2O_3 and FeO given as FeO.
[b] See footnote [b] in Table 2.

of elements in the earth's crust and that in chondrites (see Table 7) indicates an enrichment of the earth's crust by the elements forming the most mobile compounds and mixtures in the earth's thermodynamic conditions (Si, Al, Na, K, Rb, U, Th, TR (rare earths), Ba, Sr, Zr, Nb, etc.) and its impoverishment with respect to refractory compounds of such elements as Mg, Fe, Cr, Ni, etc. According to recent theories these chemical features of the crust were created in the process of differentiation of the earth's mantle by fusion and degassing (Rubey, 1951; Vinogradov, 1959).

ALEXANDER B. RONOV
ALEXEI A. YAROSHEVSKY

References

Beliajevky, N. A., Borisov A. A., and Volvovsky, I. S., 1967, "The depth structure of the U.S.S.R. territory," *Sov. Geol.*, **11**, 56–84 (in Russian).
Clarke, F. W., 1924, "The data of geochemistry," *U.S. Geol. Surv. Bull.*, **770**, 841.
Demenitskaya, R. M., 1961, "Principal Features in Structure of the Earth's Crust According to Geophysical Data," Moscow (in Russian).
Engel, A. E. J., and Engel, C. G., 1968, "Rocks of the ocean floor," 2nd Internat. Ocean Congr., Reports of Plenar Meetings; Fundamental Problems of Oceanology, Nauka, Moscow, pp. 183–217.
Fahring, W. F., and Eade, K. E., 1968, "The chemical evolution of the Canadian shield," *Can. J. Earth Sci.*, **5**(5), 1247–1252.
Fleischer, M., Chao, E.C.T., 1960 "Some problems in the estimations of abundances of elements in the earth's crust." Rept. XXI, Intern. Geol. Congress, Copenhagen.
Goldschmidt, V. M., 1933, "Grundlagen der quantitativen Geochemie," *Fortschr. Mineral. Krist. Petrogr.*, **17**, 112–156.
Grout, F. F., 1938, "Petrographic and chemical data on the Canadian shield," *J. Geol.*, **46**, 486–504.
Horn, M. K., and Adams, J. C. S., 1966, "Computer-derived geochemical balances and element abundances," *Geochim. Cosmochim. Acta*, **30** (3), 279–298.
Kuenen, P. H., 1950, "Marine Geology," New York, John Wiley & Sons. 568pp.
Masuda, A., 1966, "Genesis of the earth's crust and the structure within the mantle: a unified chemical theory," *Chem. Geol.*, **1**(2).

EARTH'S CRUST GEOCHEMISTRY

TABLE 6. VOLUMES, MASSES, AND AVERAGE CHEMICAL COMPOSITION OF THE CRUST

Types of Crust	Shell	Volumes (10^6 km^3)	Average Thickness (km)	Mass (10^{24} g)	SiO$_2$	TiO$_2$	Al$_2$O$_3$	Fe$_2$O$_3$
Continental	Sedimentary	500	3.4	1.29	49.90	0.65	12.97	2.99
					0.645	0.008	0.167	0.039
	"Granitic"	3000	20.1	8.20	63.94	0.57	15.18	2.00
					5.243	0.047	1.245	0.164
	"Basaltic"	3000	20.1	8.70	58.23	0.90	15.49	2.86
					5.066	0.078	1.348	0.249
	Total Continental	6500	43.6	18.19	60.22	0.73	15.18	2.48
					10.954	0.133	2.760	0.452
Subcontinental	Sedimentary	210	3.2	0.52	49.90	0.65	12.97	2.99
					0.258	0.003	0.067	0.016
	"Granitic"	590	9.1	1.61	63.94	0.57	15.18	2.00
					1.029	0.009	0.244	0.032
	"Basaltic"	740	11.4	2.14	58.23	0.90	15.49	2.86
					1.247	0.019	0.332	0.061
	Total Subcontinental	1540	23.7	4.27	59.35	0.73	15.07	2.55
					2.534	0.032	0.643	0.109
Oceanic	Sedimentary (layer I)	120	0.4	0.19	40.73	0.61	11.45	4.60
					0.077	0.001	0.022	0.009
	Volcanic-sedimentary (layer II)	530	1.8	1.45	45.62	1.02	14.24	3.37
					0.661	0.015	0.206	0.049
	"Basaltic"	1520	5.1	4.40	49.72	1.37	16.57	2.35
					2.188	0.060	0.729	0.103
	Total Oceanic	2170	7.3	6.04	48.44	1.26	15.85	2.67
					2.926	0.076	0.957	0.161
Total Crust		10,210	20.0	28.50	57.60	0.84	15.30	2.53
					16.414	0.240	4.360	0.722

[a] For each shell the upper line is weight percent; the lower line is mass (10^{24} g).
[b] The mean composition of the sedimentary shell differs from that of Table 1 by recalculation of SO$_3$ to S.

TABLE 7. ABUNDANCES OF CHEMICAL ELEMENTS IN THE EARTH'S CRUST AND CHONDRITES[a] (in ppm) (recalculation of the data of A. Vinogradov, 1962, and S. Taylor, 1964)

Element	Continental Crust	Crust as a Whole	Chondrites
H	(1,530)	(1,520)	—
Li	20	18	3
Be	3	2	3.6
B	10	9	2
C	200 (5,070)	180 (5,220)	400
N	19	19	1
O	461,000 (472,650)	456,000 (468,780)	350,000
F	585	544	28
Na	23,550 (22,060)	22,700 (21,370)	7,000
Mg	23,300 (18,500)	27,640 (23,400)	140,000
Al	82,300 (80,210)	83,600 (80,950)	13,000
Si	281,500 (279,380)	273,000 (277,970)	180,000
P	1,050 (1,040)	1,120 (960)	500
S	350 (530)	340 (400)	20,000
Cl	145 (620)	126 (500)	70
K	20,850 (23,620)	18,400 (19,420)	850
Ca	41,500 (40,000)	46,600 (49,960)	14,000
Sc	22	25	6
Ti	5,650 (4,380)	6,320 (5,040)	500
V	120	136	70
Cr	102	122	2,500
Mn	950 (1,110)	1,060 (1,240)	2,000
Fe	56,300 (46,810)	62,200 (50,880)	250,000
Co	25	29	800

Table 6. (Continued)

			Components, wt % and Mass $(10^{24}g)^a$								
FeO	MnO	MgO	CaO	Na$_2$O	K$_2$O	P$_2$O$_5$	Corg	CO$_2$	S^b	Cl	H$_2$O$^+$
2.80	0.11	3.06	11.70	1.70	2.04	0.16	0.48	8.20	0.18	0.21	2.90
0.036	0.001	0.039	0.151	0.022	0.026	0.002	0.006	0.106	0.002	0.003	0.037
2.86	0.10	2.21	3.98	3.06	3.29	0.20	0.17	0.84	0.04	0.05	1.53
0.234	0.008	0.181	0.326	0.251	0.270	0.016	0.014	0.069	0.003	0.004	0.125
4.78	0.19	3.85	6.05	3.10	2.58	0.30	0.11	0.51	0.03	0.03	1.00
0.416	0.016	0.335	0.526	0.270	0.224	0.026	0.009	0.044	0.003	0.003	0.087
3.77	0.14	3.05	5.51	2.99	2.86	0.24	0.16	1.20	0.05	0.06	1.37
0.686	0.025	0.555	1.003	0.543	0.520	0.044	0.029	0.219	0.008	0.010	0.249
2.80	0.11	3.06	11.70	1.70	2.04	0.16	0.48	8.20	0.18	0.21	2.90
0.015	0.001	0.016	0.061	0.009	0.010	0.001	0.002	0.043	0.001	0.001	0.015
2.86	0.10	2.21	3.98	3.06	3.29	0.20	0.17	0.84	0.04	0.05	1.53
0.046	0.002	0.036	0.064	0.049	0.053	0.003	0.003	0.013	0.001	0.001	0.025
4.78	0.19	3.85	6.05	3.10	2.58	0.30	0.11	0.51	0.03	0.03	1.00
0.102	0.004	0.082	0.130	0.066	0.055	0.006	0.002	0.011	0.001	0.001	0.021
3.82	0.16	3.14	5.97	2.90	2.79	0.23	0.16	1.57	0.07	0.07	1.43
0.163	0.007	0.134	0.254	0.124	0.119	0.010	0.007	0.067	0.003	0.003	0.061
0.97	0.47	2.94	16.29	1.13	2.01	0.15	0.26	13.27	—	—	5.17
0.002	0.001	0.006	0.031	0.002	0.004	0.0003	0.0005	0.025	—	—	0.010
4.23	0.31	5.43	13.67	2.04	1.03	0.14	0.12	6.04	—	—	2.74
0.061	0.0045	0.079	0.198	0.030	0.015	0.002	0.002	0.088	—	—	0.040
6.85	0.17	7.52	11.48	2.80	0.21	0.13	0.01	—	0.03	0.03	0.66
0.306	0.0075	0.331	0.505	0.123	0.009	0.006	0.0004	—	0.001	0.001	0.029
6.11	0.21	6.89	12.15	2.57	0.46	0.13	0.04	1.87	0.02	0.02	1.31
0.369	0.013	0.416	0.734	0.155	0.028	0.008	0.003	0.113	0.001	0.001	0.079
4.27	0.16	3.88	6.99	2.88	2.34	0.22	0.14	1.40	0.04	0.05	1.37
1.218	0.045	1.105	1.992	0.822	0.667	0.062	0.040	0.399	0.012	0.014	0.389

Table 7. (Continued)

Element	Continental Crust	Crust as a Whole	Chondrites
Ni	84	99	13,500
Cu	60	68	100
Zn	70	76	50
Ga	19	19	3
Ge	1.5	1.5	10
As	1.8	1.8	0.3
Se	0.05	0.05	10
Br	2.4	2.5	0.5
Rb	90	78	5
Sr	370	384	10
Y	33	31	1
Zr	165	162	30
Nb	20	20	0.3
Mo	1.2	1.2	0.6
Ru	—	—	—
Rh	—	—	—
Pd	0.015	0.015	1
Ag	0.075	0.08	0.1
Cd	0.15	0.16	0.1
In	0.25	0.24	0.01
Sn	2.3	2.1	1
Sb	0.2	0.2	0.1
Te	—	—	0.5
J	0.45	0.46	0.04
Cs	3	2.6	0.1

TABLE 7. (Continued)

Element	Continental Crust	Crust as a Whole	Chondrites
Ba	425	390	6
La	39	34.6	0.3
Ce	66.5	66.4	0.5
Pr	9.2	9.1	0.1
Nd	41.5	39.6	0.8
Sm	7.05	7.02	0.2
Eu	2.0	2.14	0.08
Gd	6.2	6.14	0.4
Tb	1.2	1.18	0.05
Dy	—	—	0.35
Ho	1.3	1.26	0.07
Er	3.5	3.46	0.2
Tu	0.52	0.5	0.04
Yb	3.2	3.1	0.2
Lu	—	—	0.03
Hf	3.0	2.8	0.5
Ta	2.0	1.7	0.02
W	1.25	1.2	0.15
Re	0.0007	0.0007	0.0008
Os	—	—	0.5
Ir	—	—	0.5
Pt	—	—	2
Au	0.004	0.004	0.17
Hg	0.085	0.086	3
Tl	0.85	0.72	0.001
Pb	14	13	0.2
Bi	0.0085	0.0082	0.003
Th	9.6	8.1	0.04
U	2.7	2.3	0.015

[a] In brackets are shown the abundances of the main elements obtained by a quantitative method based on the volume measurements.

Pakiser, L. C., and Robinson, R., 1967, "Composition of the continental crust as estimated from seismic observations; in The Earth beneath the Continents," *Am. Geophys. Union, Monograph,* **10**, 620–626.

Poldervaart, A., 1955, "Chemistry of the earth's crust," *Geol. Soc. Am. Spec. Paper,* **62**, 119–144.

Ronov, A. B., and Migdisov, A. A., 1970, "Evolution of the chemical composition of rocks of the shields and the sedimentary cover of the Russian and the North American platforms," *Geokhimia,* No. 4, 403–438.

Ronov, A. B., and Yaroshevsky, A. A., 1967, "Chemical composition of the earth's crust," *Geokhimiya,* No. 11, 1285–1309 (in Russian).

Ronov, A. B., and Yaroshevsky, A. A., 1969, "Chemical composition of the Earth's Crust," *Am. Geophys. Union, Monograph,* **13-D**.

Rubey, W. W., 1951, "Geologic history of sea water: an attempt to state the problem," *Bull. Geol. Soc. Am.,* **62**, 1111–1147.

Sederholm, J. J., 1925, "The average composition of the earth's crust in Finland," *Comm. Geol. Finlande Bull.,* **12**, 70.

Shaw, D. M., Reilly, G. A., Muysson, J. R., Pattenden, G. E., and Campbell, F. E., 1967, "An estimate of the chemical composition of the Canadian Precambrian Shield," *Can. J. Earth Sci.,* **4**, 829–853.

Taylor, S. R., 1964, "Abundance of chemical elements in the continental crust: a new table," *Geochim. Cosmochim. Acta,* **28**(8), 1273–1285.

Vinogradov, A. P. 1959, "Chemical Evolution of the Earth," Izd. Akad. Nauk SSSR, Moscow, 43pp. (in Russian).

Vinogradov, A. P., 1962, "Average content of chemical elements in main types of igneous rocks of the earth's crust," *Geokhimiya,* No. 7, 555–571 (in Russian).

Vinogradov, A. P., Udintsev, G. B., Dmitriev, L. V., Kanaev, V. O., Neprochnov, J. P., Petrova, G. N., Rycunov, L. N., Kogan, L. I., 1969, "The structure of the rift zones of the Indian Ocean and their place in the world system of rifts," *Izv. Akad. Nauk SSSR, Ser. Geol.,* **9**, (in Russian).

Cross-references: *Elements: Planetary Abundances and Distribution; Geochemical Evolution of the Core, Mantle, and Crust; Outgassing of the Planet Earth; Rare Earths; Rare Earths in Basalts; Weathering, Chemical.* Vol. V: *Basalt; Geosynclines; Granite; Igneous Rocks; Magmatic Differentiation; Mohorovičić Discontinuity; Tholeiites.*

ECOLOGY

Ecology is the study of organisms in relation to their environment. The term was adopted by Haeckel in 1869 from the Greek word *oikos* (a house or home), although it had been used by H. D. Thoreau fifteen years earlier (Ager, 1963). Ecology emerged as a self-conscious discipline around the beginning of the 20th century.

Life History Studies

The oldest and largest division of ecology deals with the life history of organisms—the places where they live, what they eat or are eaten by, how they grow, reproduce, and die. The findings of this branch provide much of the raw material from which generalizations are constructed. The first step in dealing with a practical problem, such as the outbreak of a serious agricultural pest, is often to gather simple descriptive information about its life cycle.

Plant Communities

The first branch of ecology to develop beyond the stage of life-history study was the description of vegetation. Its basic method is to study the detailed distribution of vascular plants in terms of communities of various types, the pattern and complexity of which depend largely on the climate and soil.

The major theoretical contribution of this school is the idea of *seral succession* toward a stable climax. According to this idea, if a new environment is created for terrestrial plants, or an old one drastically changed, the vegetation that first develops on it does not remain unchanged for eternity, but rather alters the environment so that it becomes more suitable for some new and different kind of vegetation. This in turn changes the environment still more and is replaced by a third kind of vegetation and so on. Ultimately a *climax* vegetation develops which is stable under the prevailing climatic conditions and remains until the climate changes to favor something else, or until some new catastrophe—perhaps the advent of agricultural man—changes the environment drastically once again.

Further study of seral succession showed that vegetation patterns were seldom so simple. At present the climax community is regarded as a useful concept, not something to be identified and examined in the field. Even under constant climatic conditions some sorts of vegetation are not stable, but undergo local cyclical changes. At the other extreme, a few kinds of vegetation replace themselves after disturbance without intervening seral stages. It is clear that the sort of climatic constancy that was implied in the climax idea has not prevailed at least since the beginning of the Pleistocene, and that the climax is best considered an ever-changing end-point toward which vegetation develops rather than a state it actually attains.

Biogeochemistry

In any community the interactions of the organisms with each other and with their environment are so complex as to defy complete understanding. One way of asking answerable questions is to restrict attention to the chemical aspects of ecological processes, and to regard organisms as the causes or effects of geochemical processes. This approach has led to an understanding of the quantitative role of organisms in major chemical cycles, such as the carbon cycle and the nitrogen cycle, and is now being strengthened by the use of isotopic methods. The introduction of isotopic tracers into biogeochemical cycles permits the estimation of reservoir sizes and exchange rates and seems likely to lead to new theoretical developments.

Not all geochemical phenomena are of equal biological importance. Only about a third of the elements of the periodic table are known to have a structural or physiological function in even a single organism. Another third are present in abundances as great as some of those with a demonstrated role, and it is possible that most of the naturally occurring elements have a biological function that cannot yet be demonstrated because of their widespread occurrence as contaminants in culture media. The relative abundance of the elements in organisms is much closer to that in the accessible universe than it is to that in the earth, the earth's crust, or the *biosphere*. This might be construed as evidence for the extraterrestrial evolution of the basic biochemical mechanisms, but it is more likely a coincidence brought about by the combined effects of the chemical fractionations that produced the biosphere and the chemical behavior of the elements enriched in it. Only the elements that produce soluble cations or oxyanions under normal conditions are readily available to organisms, and it is these elements of the biosphere that are important in the composition of living things.

Although generally organisms are composed of the common cosmic materials, there are some essential elements that are often scarce enough in the biosphere to restrict organic growth. For example, phosphorus is chemically very suitable for constructing molecules to transfer energy in cellular metabolism and to carry genetic information. Apparently because of these properties the demand for phosphorus outweighs the supply in many environments.

Physiologists have shown in the laboratory that many other elements, as well as vitamins and inhibitory compounds, are important to

ECOLOGY

controlling or limiting the growth of algal populations, and field enrichment experiments have shown the inadequacy of the old idea of one or a few elements acting as limiting factors for plant growth.

Energy Flow

More recently a field has developed in which organisms are considered in terms of energy transfer instead of chemical reactions (see Fig. 1). Its principal contribution has been to demonstrate that organisms, like most machines, are not efficient converters of energy, so that an organism will incorporate into its body only a small part of the stored chemical energy of its food. This places a severe practical limit on the number of steps that can be maintained in a food chain and has important practical consequences for an expanding human population trying to increase its food supply.

The measurement of energy flow in natural communities presents many difficulties, only some of which are technical ones. Though the subject is suitable for theoretical development, this has been limited by the lack of methods for dealing with the thermodynamics of open systems. Now that a better understanding of the thermodynamics of open systems is being attained by physical chemists, the study of energy flow seems likely to have a productive future.

Populations

By far the greatest body of coherent ecological theory has been created by students of populations. The study of field populations has uncovered many interesting phenomena, such as the periodic oscillation of arctic small vertebrate populations and the seasonal changes in abundance of planktonic organisms. These observations have stimulated both laboratory experiments and the development of deductive mathematical theory.

If a small number of organisms is provided with a new unexploited environment, the ensuing population growth curve exhibits a roughly sigmoid form, with an initial phase of very rapid, almost logarithmic increase and a later phase in which the rate of growth gradually declines to zero. Such a population history can be described by a curve of the form:

$$dN/dt = rN \frac{(K - N)}{K}$$

where dN/dt is the instantaneous rate of growth, r the intrinsic rate of natural increase in the absence of crowding, N the population size at any time, and K the maximum population size. A formula of this kind is simple to use and easy to understand, and although few populations justify the assumptions on which it is based, it has been widely employed not

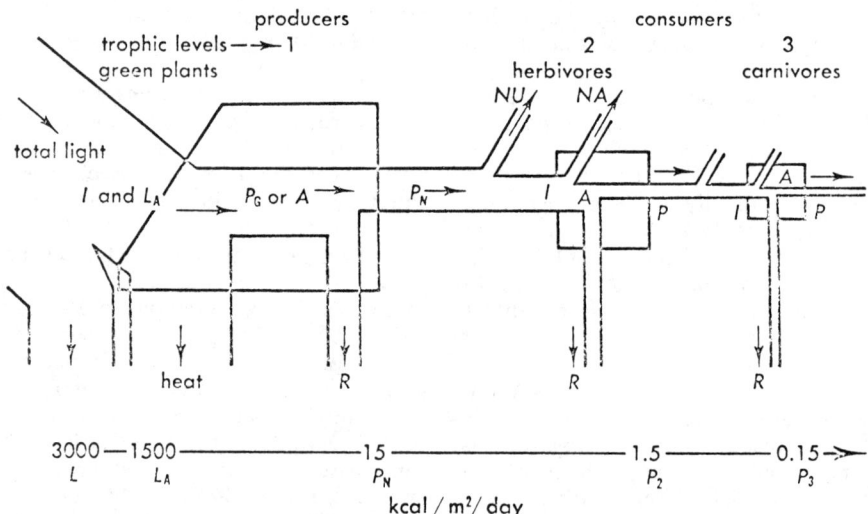

FIG. 1. A simplified energy flow diagram. The boxes represent the standing crop of organisms: (1) producers or autotrophs; (2) primary consumers or herbivores; (3) secondary consumers or carnivores) and the pipes represent the flow of energy through the biotic community. L = total light; L_A = absorbed light; P_G = gross production; P_N = net production; I = energy intake; A = assimilated energy; NA = nonassimilated energy; NU = unused energy (stored or exported); R = respiratory energy loss. The chain of figures along the lower margin of the diagram indicates the order of magnitude expected at each successive transfer starting with 3000 kcal of incident light per m² per day (from Odum, 1963).

only for curve-fitting and the description of population growth, but as a point of departure for population theory. The principal developments of importance from it have been the prey-predator equations of Volterra:

$$\frac{dN}{dt} = r_1 N_1 - \alpha_1 N_1 N_2$$

$$\frac{dN_2}{dt} = \alpha_2 N_1 N_2 - d_2 N_2$$

and the Gause equations of species interaction:

$$\frac{dN_1}{dt} = r_1 N_1 \frac{(K_1 - N_1 - \alpha N_2)}{K_1}$$

$$\frac{dN_2}{dt} = r_2 N_2 \frac{(K_2 - N_2 - \beta N_1)}{K_2}$$

In these equations N_2 is the size of one population and N_1 that of another, α_1 expresses the effect of predation on the prey population, and α_2 expresses its effect on the predator. In the absence of prey the predator should die at the rate of d_2; K_1, K_2 are the saturation values of two species grown alone; r_1 and r_2 are the intrinsic rates of natural increase of species N_1 and N_2; and α and β express the inhibitory effects of these species on each other. If $\alpha > K_1/K_2$ and $\beta > K_2/K_1$ then either N_1 or N_2 may win out in competition, the result depending on the initial concentration of the two species. If $\alpha < K_1/K_2$, and $\beta < K_2/K_1$, the species will coexist. If $\alpha < K_1/K_2$ and $\beta > K_2/K_1$, N_1 will be the sole survivor of competition, and if $\alpha > K_1/K_2$ *and* $\beta < K_2/K_1$, only N_2 will survive.

Work with models of this sort and with experimental laboratory populations that behave more or less in the ways that the models predict has led to current ideas about the *ecological niche*, or way in which an organism fits into the ecological system of which it forms a part. Some people hold that Gause's axiom, which states that two species cannot live indefinitely the same way in the same place under constant conditions, is essentially trivial, while others regard it as one of the grandest generalizations of ecology. Probably its principal value lies in raising the question of how the niches of apparently similar organisms differ, and so compelling ecologists to examine their material very closely. In this, Gause's axiom is similar to the idea of seral succession, which directs attention to the environmental requirements of species as well as to the ways in which the environment is changed by them. Both ideas stimulate the acquisition of useful information although they imply an environmental constancy that is seldom to be found in nature.

Niche theory has led to renewed interest in the taxonomic diversity of natural communities. MacArthur, assuming that there must be some way in which the niches of organisms in an ecological system did not overlap, developed a model of the relative abundance of the individual species in a natural community. According to this model, the expected abundance of the rth rarest species, where there are n species and m individuals and i is the species rank, is given by

$$\frac{m}{n} \sum_{i=1}^{r} \frac{1}{(n - i + 1)}$$

The predictions of the model have been borne out well in a number of taxonomically homogeneous cases, but do not seem to hold generally for natural communities.

A different line of approach has been developed by ecologists interested in practical problems of *population management*, particularly the exploitation of fish populations. The method has been to start with a formula that is essentially a simple equation of continuity:

$$S_2 = S_1 + (A + G) - (M + C)$$

where S_1 and S_2 are the total weight of the population at the beginning and end of the time under consideration, A is recruitment of new individuals to the population, G is growth of the population, M is natural mortality, and C is capture by fishing.

Such an equation is modified as data become available for a specific case and gradually refined until it is a very powerful predictive tool. The equations become cumbersome in the process but the mathematics involved is essentially simple, and electronic computation avoids the tedium and human errors that would have hampered their use twenty years ago.

Unfortunately such models are limited to the populations for which they have been formulated or for others similar to them in essential respects. There seems little immediate likelihood of the development of simple and accurate general models for the dynamics of natural populations. Many ecologists working with organisms other than fish despair of producing useful models without recourse to stochastic theory. The stochastic approach is appealing, for the processes of population change are certainly not deterministic ones, but such a shift involves a great increase in mathematical complexity. The sophistication necessary to handle stochastic models is rare among ecologists, and it has usually been applied to new methods of constructing a deterministic theory. Cole, for example, used finite difference equations to demonstrate that the reproductive potential of a species depends very greatly upon the timing of reproduction in its life cycle and to point out the critical importance of pieces of information, such as the age at first reproduction, which might otherwise be overlooked in life-history studies.

Behavior

The experimental analysis of animal behavior patterns has led to renewed interest among ecologists in the social and psychological aspects of their subject. In particular, studies have focused on territoriality, social stress, hierarchies and other behavioral mechanisms that control population density, on the transmission of traditional information about nesting and feeding sites, and on feeding behavior as it affects an animal's role in the community. Although some behavioral studies have relevance to species diversity, community stability, and population growth rates, much of this work can be carried on profitably outside an ecological context, and behaviorists seem closer to establishing an independent discipline than most other ecologists.

Outlook

It is doubtful that such a heterogeneous collection of approaches to the study of organisms in relation to their environment can be contained indefinitely within a single discipline. At present only the theory of evolution serves as a weak cement to hold the bits together, and few ecologists are able to see clearly its relevance to their activities. Perhaps a major generalization will emerge from a combination of the theories that have been developed in several of the disparate fields. Unless it does, academic ecology seems almost certain to fragment, although the ecological approach may persist for a long time among foresters, game conservationists, and other biological technologists concerned with large areas of the world.

D. A. LIVINGSTONE

References

Ager, D. V., 1963, "Principles of Paleoecology," New York, McGraw-Hill Book Co., 371pp.
Allee, W. C., Emerson, A. E., Park, O., Park, T., and Schmidt, K. P., 1949, "Principles of Animal Ecology," Philadelphia, W. B. Saunders Co. 837pp.
Beverton, J. H., and Holt, S. J., 1958, "On the dynamics of exploited fish populations," London, Fishery Investigations, Ser. II, Vol. XIX, Her Majesty's Stationery Office, 533pp.
Cole, L. C., 1954, "The population consequences of life history phenomena," *Quart. Rev. Biol.,* **29,** 103–137.
Hutchinson, G., 1943, "The biogeochemistry of aluminum and of certain related elements," *Quart. Rev. Biol.,* **18**(1), 1–29; **18**(2), 128–153; **18**(3), 242–262; **18**(4), 331–363.
Odum, E. P., 1954, "Fundamentals of Ecology," Philadelphia, W. B. Saunders Co., 384pp.
Odum, E. P., 1963, "Ecology," New York, Holt, Rinehart and Winston, 152pp.
Odum, H. T., 1957, "Trophic structure and productivity of Silver Springs, Florida," *Ecol. Monographs,* **27**(1), 55–112.
Oosting, H. J., 1948, "The Study of Plant Communities," San Francisco, W. H. Freeman, 389 pp.
Slayter, R. O., 1969, "Man's use of the environment—the need for ecological guidelines," *Australian J. Sci.,* **32**(4), 146–152.
Slobodkin, L. B., 1961, "Growth and Regulation of Animal Populations," New York, Holt, Rinehart and Winston, 184pp.

Cross-references: *Biogeochemistry; Cycles—Geochemical; Elements; Environmental Science; Organic Geochemistry; Vegetation Indicators.* Vol. II: *Vegetation Classification and Description.* Vol. VI: *Radioactive Isotope Tracer Technology.* Vol. VII. *Biosphere; Evolution.*

ECOLOGY, MARINE—See Vol. I; Vol. VII

ECONOMIC GEOLOGY, MINERALOGY— See Vol. IV B

EDDIES IN FLUID MOTION—See Vol. II, FLUID MECHANICS

EDDY CONDUCTIVITY—See Vol. VI

EDTA—See VERSENE (EDTA) SOLUTION STUDIES

EFFLUENT GAS ANALYSIS

The technique of *effluent gas analysis* (EGA)* is virtually always employed in conjunction with *differential thermal analysis* (DTA) (q.v.). It involves the simultaneous "analysis" of the gaseous products evolved from the sample being subjected to DTA.

The DTA is carried out in furnace conditions of dynamic gas flow. The gas flow is used as a carrier for the gas evolved from the test sample. The resultant gas mixture then travels through a connecting tube to a detector cell which measures the *amount* (not type) of additional gas present. This data is then re-

---

* The alternative terms *evolved gas analysis* (EGA) and *evolved gas detection* (EGD) are under consideration by the International Confederation of Thermal Aanalysis.

corded in the form of what is best termed *a gas detection curve,* which ideally is obtained simultaneously with the DTA curve by a two pen *X-Y* type recorder.

However, although used as above, in the literature this term is something of a misnomer because as used it is a gas detection not analysis technique. For this reason the somewhat more acceptable term *gas evolution analysis* (GEA) (q.v.) is preferable unless the evolved gas is actually analyzed, for example by gas chromatographic, mass spectrographic, or infrared methods.

S. St. J. Warne

References

Bayliss, P., and Warne, S. St. J., 1962, "The effects of the controllable variables on differential thermal analysis," *Am. Mineralogist,* **47,** 775–778.

*Garn, P. D., 1965, "Thermoanalytical Methods of Investigation," New York, Academic Press, 606pp.

Garn, P. D., and Kessler, J. E., 1961, "Effluence analysis as an aid to thermal analysis," *Anal. Chem.,* **33,** 952–954.

*Wendlandt, W. W., 1964, "Thermal Methods of Analysis," New York, Interscience Publ., a div. of John Wiley & Sons, 424pp.

*Additional bibliographic references may be found in this work.

Cross-references: *Differential Thermal Analysis; Gas Evolution Analysis.*

Eh–pH, GEOCHEMICAL APPLICATION —See pH–Eh

ELECTROLUMINESCENCE—See Vol. IV B, LUMINESCENCE

ELECTROLYTES: FLOCCULATION OF COLLOIDS

Electrolytes are aqueous solutions of acids, bases, and salts, which conduct electricity because they are dissociated into ions. Electrolytes may be further divided into two classes: nonassociated electrolytes and associated electrolytes. Nonassociated electrolytes exist in solution as simple ions, which may be hydrated. Salts like NaCl or KCl are believed to be nonassociated or "strong electrolytes." Associated electrolytes may form ion pairs, or complexes, or they may be weakly dissociated (weakly ionized) acids or bases. Ion pairs encountered in nature may include $CaCO_3^0$, $Ca(OH)^+$, and $MgHCO_3^-$. Examples of complex ions are $ZnCl_4^{2-}$ or $B_3O_3(OH)_4^-$. In dilute solutions, the solute particles are likely to exist as simple ions. But, with increasing concentration, the crowding of the particles together promotes chemical interaction (as in the formation of covalent or electrostatic bonds). Thus, increased concentration favors development of ion pairs or complex ions.

The colloidal state is one of small particle size ranging from 10^{-3} to 10^{-6} mm (10,000–10 Å). There are many types of colloidal systems such as *solid-gas* (smoke), *liquid-gas* (aerosol, fog), *liquid-liquid* (emulsions), but in sedimentary processes the most significant is a solid-liquid system (sol) in which the liquid is water. Sols resemble liquids in their properties including the ability to flow. They can be either hydrophilic or hydrophobic. The hydrophobic sols are less stable because there is no attraction between the particles and water.

Colloidal particles bear electrical charges either by adsorption of ions or by being ionized themselves. Adsorption occurs to a great extent because of the large surface area of colloids. Silica, humus, colloids, and clays are all negatively charged, while alumina, ferric oxide, chromic hydroxide, lime, and magnesia are positively charged. Because the surfaces of colloidal particles are charged, they will attract ions of opposite charge around them in solution. These are called *counterions*.

The oppositely charged adsorbed layer and counterion layer constitute an *electrical double layer*. All particles of a sol have like charges. When two such particles approach each other so closely that their double layers overlap, then they begin to repel each other. This repulsion stabilizes the sol and prevents it from flocculation.

In addition to the repulsive force, sols are subject to mutual attraction by *Van der Waals forces* (q.v.). The attraction is weak between single molecules but increases as the particles

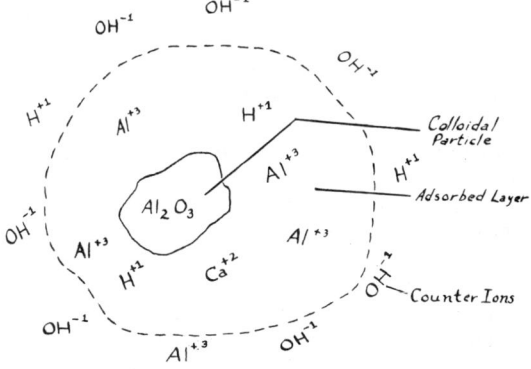

Fig. 1. Schematic diagram of a particle of Al_2C_3 sol, which becomes positively charged due to absorption of plus ions. These in turn attract counterions.

aggregate into larger units. If the repulsion can be overcome, more and more particles will coalesce and the sol will be flocculated.

Repulsion can be decreased by shrinking the double layer, which is accomplished through the addition of electrolytes to the solution. The effectiveness of electrolytes depends upon the concentration and the charge on the ion opposite in sign to the charge on the colloid. The effect of ions of charges 1, 2, and 3 increases by 1^6, 2^6, 3^6 (*Schulze-Hardy rule*).

In geology it is found that silica, ferric hydroxide, aluminum hydroxide, and clays are transported by streams largely in the form of colloids. Particles of colloidal size are so small they do not settle by gravity and thus they remain almost indefinitely in suspension. When the colloidal dispersions reach the sea, flocculation takes place, since seawater is a dilute salt solution and therefore an excellent electrolyte. The result is that seawater greatly accelerates the rate of sedimentation by causing the suspended matter to precipitate. Similar and even more rapid flocculation may occur in supersaline lakes.

Quick Clays

While the presence of electrolytes promotes the flocculation of clays, leaching of salts from freshly precipitated clays, under certain conditions, may lead to the reverse effect; namely the peptization or dispersal of the clay minerals. This phenomenon has been observed in connection with quick clay landslides. A quick clay changes from a normally solid mass into a viscous sol, which is capable of horizontal flowage and has high transporting ability. This comes about as a result of sudden shock, but often without apparent cause. The transformation from clay to fluid occurs within a few minutes over large areas, even on a slope of less than one degree. Quick clays are composed chiefly of layer-lattice silicates. Illite, and to a lesser extent chlorite, are the dominant clay minerals. About half of the dry clay (by weight) consists of minus 2μ particles (colloidal dimensions), although larger particles are also present, which may contribute to the formation of an unstable structure (Osterman, 1964).

Electrolytes cause marine clays to flocculate in an open "card-house" structure. After consolidation, fresh-water leaching removes the electrolyte and weakens the interparticle forces. This results in a metastable condition, which under the right circumstances may lead to collapse of the clay structure, release of adsorbed water, and flowage. Rosenqvist (1952) found the sensitivity of certain Norwegian clays to be inversely proportional to their electrolyte content.

Clays of fresh-water origin may be peptized by humic acids found in the soil.

VIVIEN GORNITZ

References

Drew, I. M., 1966, "Properties of the Bootlegger Cove clay of Anchorage, Alaska," Ph.D. Thesis, Columbia University.

Garrels, R. M., and Christ, C. L., 1965, "Solutions, Minerals, and Equilibria," New York, Harper and Row, 450pp.

Harned, H. S., and Owen, B. B., 1958, "The physical chemistry of electrolytic solutions," Third ed., (ACS Monograph 137), New York, Reinhold Co.

Mason, B., 1966, "Principles of Geochemistry," Third ed., New York, John Wiley & Sons, 329 pp.

Osterman, J., 1964, "Studies on the properties and formation of quick clays," pp. 87–108, Clays and Clay Minerals, Proc. Twelfth Natl. Conf. on Clays and Clay Minerals, New York, MacMillan Co.

Robinson, R. A., and Stokes, R. H., 1959, "Electrolyte Solutions," New York, Academic Press.

Rosenqvist, I. T., 1952, "Investigations in the clay-electrolyte-water system," *Norweg. Geotech. Inst. Publ.,* **9**, 1–107.

Twenhofel, W. H., 1950, "Principles of Sedimentation," Second ed., New York, McGraw-Hill Book Co., 673pp.

Cross-references: *Acids and Bases; Aqueous Solutions; Bonding; Colloids; Complexes; Solubility Product Constant; Seawater, Chemistry; Van der Waals Force.* Vol. IVB: *Clays and Clay Minerals.* Vol. VI: *Quick Clay.*

ELECTRON CAPTURE

Electron capture is a mode of radioactive decay in which an electron in one of the electronic shells of an atom is captured by the atomic nucleus. The result is to convert the atom to one whose atomic number is one less than for the original atom. The most frequent type of electron capture is that where the electron derives from the K shell of orbiting electrons and is consequently known as *K-electron capture.*

The new atom has the same total charge as before the decay, but has a vacancy in one of its electronic shells. This may be occupied by an electron which falls from a higher shell to the lower orbit energy level. As it falls, it emits an x ray of wavelength characteristic of the new atom. This characteristic x radiation is often the best way to establish that electron capture has, in fact, occurred.

The electron capture process has the same net effect as positron emission, that is, the atomic weight remains essentially unchanged as a proton becomes a neutron.

Electron capture is possible because the electron has a finite probability for existing anywhere and, in particular, has a non-zero probability of being in the nucleus. The electrons in the K shell have the highest probability of being in the nucleus, and so are those most often involved in electron capture (K-electron capture).

It is sometimes possible for an atom to achieve a lower energy level either by electron capture or by emission of a negative electron. A case in point is the geochemically important decay of potassium-40. This nuclide may either capture a K-electron to form argon-40 (see *Potassium-Argon Age Determination*) or emit a negative electron from the nucleus to form calcium-40. In any given population of K^{40} atoms, 12% of them will decay to Ar^{40} and 88% to Ca^{40}. It should be noted, however, that the form of branched decay exemplified by K^{40} is not a required characteristic of all electron capture processes.

The stability of a nucleus is determined both by the total number of neutrons and protons and by the ratio of neutrons to protons. Electron capture will be most likely in those cases where there is an excess of protons relative to neutrons. The low-mass isotopes of an element, therefore, are the ones most likely to undergo decay by electron capture.

B. J. GILETTI

References

Friedlander, G., Kennedy, J. W., and Miller, J. M., 1964, "Nuclear and Radiochemistry," Second ed., New York, John Wiley & Sons, 585pp.

Cross-reference: *Potassium-Argon Age Determination.*

ELECTRONEGATIVITY

The term "electronegativity" refers to the relative tendency of an atom to acquire negative charge. This tendency is not precisely defined because no exact theoretical or experimental method of evaluation has yet been devised. Electronegativity exists because the electronic cloud which surrounds each atomic nucleus is inadequate to block off the nuclear charge completely at the periphery. In other words, although every complete atom is electrically neutral as observable from a distance, a small fraction of the total nuclear positive charge can be detected at any point near the surface of the atom. This "effective nuclear charge" at the surface is relatively insignificant in the absence of low energy vacancies capable of accommodating a foreign electron, as in the atoms of M 8 elements (commonly called the *inert* or *noble gases*). However, wherever a low energy electron vacancy occurs in the outer shell of an atom's electronic cloud, the effective nuclear charge as manifest within that vacancy is of major importance. Indeed, it constitutes the cause and means of chemical bond formation, and largely determines both polarity and strength of the bond. Electronegativity is a measure of the force of this effective nuclear charge within an orbital vacancy at the distance of the atomic radius.

Evaluation of relative electronegativity was first accomplished by Pauling. He considered the energy of a heteronuclear single covalent bond to consist of the geometric mean of the homonuclear single bond energies of the two elements, *supplemented* by an electrostatic or ionic energy resulting from uneven electron sharing in the bond, which he attributed to an electronegativity difference between the two elements. He used the difference between the observed bond energy and the average of the homonuclear energies as a measure of the ionic energy. From such differences for various pairs of atoms he established a relative scale in which the most electronegative element, fluorine, has the value 4.0, and within the same period are O 3.5, N 3.0, C 2.5, B 2.0, Be 1.5, and Li 1.0.

Mulliken defined electronegativity as the average of the "valence state" ionization energy and electron affinity. Gordy suggested that electronegativity is a measure of the electrostatic potential at the surface of an atom, expressed as the effective nuclear charge, $Z_{eff}e$ divided by the radius. Allred and Rochow modified this concept by considering the electrostatic force, $Z_{eff}e^2/r^2$, as the measure of electronegativity. Electronegativities have also been estimated from work functions of metals, from force constants determined by infrared spectroscopy, from nuclear magnetic resonance spectroscopy, and by other methods. When adjusted to the same arbitrary scale, conventionally that established by Pauling, these methods give values in surprisingly good agreement, with only a few minor discrepancies that remain controversial. The principal application of *Pauling scale electronegativities,* which are those almost universally given in textbooks, has been qualitative prediction of bond polarity. That is, a given bond between atoms initially differing in electronegativity is polar with a partial negative charge on the initially more electronegative atom. The degree of polarity increases with increasing electronegativity difference.

ELECTRONEGATIVITY

TABLE 1. RELATIVE ELECTRONEGATIVITIES OF SOME ELEMENTS (RELATIVE COMPACTNESS SCALE)[a]

H	3.55	K	0.42	Rb	0.36	Cs	0.28
Li	0.74	Ca	1.22	Sr	1.06	Ba	0.78
Be	2.39	Zn	3.00	Cd	2.59	Hg	2.93
B	2.93	Ga	3.28	In	2.84	Tl(I)	1.89
				Sn(II)	2.31		
C	3.79	Ge	3.59	Sn(IV)	3.09	Tl(III)	3.02
N	4.49	As	3.90	Sb	3.34	Pb(II)	2.38
O	5.21	Se	4.21	Te	3.59	Pb(IV)	3.08
F	5.75	Br	4.53	I	3.84	Bi	3.16
Na	0.70						
Mg	1.56	Sc	1.30	Y	1.05	La	0.88
Al	2.22	Ti	1.40	Zr	1.10	Hf	1.05
Si	2.84	V	1.60	Nb	1.36	Ta	1.21
P	3.43	Cr	1.88	Mo	1.62	W	1.39
S	4.12	Mn	2.07	Tc	1.80	Re	1.53
Cl	4.93	Fe	2.10	Ru	1.95	Os	1.67
		Co	2.10	Rh	2.10	Ir	1.78
		Ni	2.10	Pd	2.29	Pt	1.91
		Cu	2.60	Ag	2.57	Au	2.57

[a] Values for the transitional elements are tentative estimates only.

Far more valuable applications of electronegativity have been made using a different scale based on the relative compactness of the electronic clouds of atoms. Electronegativities thus derived are approximately a linear function of the square root of the Pauling scale values, and in this sense in generally good agreement. They (see Table 1) have been used for quantitative estimation of the partial charges on combined atoms, which in themselves permit correlation of a vast quantity of chemical data and interpretations of many chemical phenomena. The partial charges in turn have been applied to the quantitative calculation of bond energy. Furthermore, more recently, a simple quantitative relationship between homonuclear covalent bond energy and electronegativity has been demonstrated. Experimental homonuclear bond energy can be used to calculate electronegativity, or vice versa.

Space does not permit a detailed description of the concepts and methods mentioned here, but an example of the results obtainable may illustrate the principles involved. Silica, SiO_2, consists of silicon atoms initially of 2.84 electronegativity, tetrahedrally surrounded by oxygen atoms, initially of 5.21 electronegativity, each of which bridges two silicon atoms. The principle of electronegativity equalization states that when two or more atoms initially different in electronegativity combine, their electronegativities become equalized to the geometric mean. For SiO_2 the electronegativity of the compound is $(2.84 \times 5.21^2)^{1/3} = 4.26$. The equalization is brought about through the uneven sharing of the bonding electrons. The oxygen being initially more electronegative, attracts more than a half share of the bonding electrons. By spending more than half time more closely associated with the oxygen, the bonding electrons impart a partial negative charge on the oxygen, expanding the cloud through increased repulsions and decreasing the electronegativity. Simultaneously the silicon atoms shrink because of reduced repulsions, and increase in electronegativity because of reduced shielding between nucleus and bonding electrons. The decrease in oxygen electronegativity from 5.21 to 4.26 is 0.95 If oxygen had acquired an electron completely the electronegativity would have dropped by 4.75. The partial charge on oxygen is defined as the ratio $0.95/4.75 = -0.20$ (minus because the electronegativity decreased). The silicon is left with a partial positive charge of 0.40.

The silicon-oxygen bond, like all heteronuclear bonds, can be treated *as if* its energy were partly covalent and partly ionic. Instead of ionic energy supplementing the covalent energy, as suggested by Pauling, the ionic energy *substitutes* for a part of the covalent energy. The ionic weighting coefficient is half the charge difference: $(0.40 + 0.20)/2 = 0.30$. The covalent weighting coefficient, $1.00 - 0.30$, is 0.70. For the covalent energy one takes the geometric mean, 59.7, of the homonuclear bond energies of silicon (53.4) and oxygen (66.7, as calculated from the O_2 molecule), multiplies by 0.70, and corrects for bond length by the factor (covalent radius sum)/(observed bond length) = 1.90/1.61. This calculation gives 49.1 kcal/

mole of bonds. The ionic energy is simply the conversion factor (to kcal/mole) 332, times the weighting coefficient 0.30, divided by the bond length 1.61, or 61.8 kcal/mole of bonds. The sum, 110.9 kcal, is the Si–O bond energy. Atomization of SiO_2 requires the rupture of four SiO bonds per formula unit; $4 \times 110.9 = 443.6$ kcal/mole for the atomization energy of $SiO_2(c)$. Subtraction of the atomization energies 108.9 for Si and 119.2 for two O gives -215.5 kcal/mole for the calculated standard heat of formation of $SiO_2(c)$. The experimental value is -217.7.

Electronegativity thus permits a quantitative interpretation of the bonding in SiO_2 or any other compound for which appropriate data are known. The same principles allow a superior alternative to the "ionic" model of nonmolecular solids, and offer high hope of eventually elucidating the thermochemistry of mineral substances in general.

R. T. SANDERSON

References

Sanderson, R. T., 1967, "Inorganic Chemistry," New York, Reinhold Book Corp.
Sanderson, R. T., 1971, "Chemical Bonds and Bond Energy," New York, Academic Press.

Cross-references: *Bonding; Elements; Infrared Analysis; Rare Gases.*

ELECTRON MICROSCOPY—See Vol. IV B

ELEMENTS

Of the first 92 elements listed in the Periodic Table, 90 occur terrestrially in abundances which vary by several orders of magnitude. This enormous variation in the abundances of elements in any given sample is the combined result of processes by which elements were produced and the subsequent history in which fractionation of elements may have occurred. Thus, while the abundances of elements in any given sample of the solar system may be fruitfully used to construct its chemical history, the abundance pattern of the total solar system is critical to an understanding of the processes of *nucleosynthesis.*

Cosmic Abundances

Estimates of the abundances of the elements in the Solar System are called the "cosmic abundances." Our present knowledge of these estimates is derived principally from three sources. These are:

(a) *Solar atmosphere:* The detection and abundance estimates of elements in the solar atmosphere are made by spectral analysis of sunlight. It is assumed that the surface layers of the sun preserve the primordial composition of the solar system, although in its interior hydrogen is being burned to helium. Only 66 elements have been recognized in the sun's spectra because some of the elements do not produce detectable spectra under the given conditions, or their spectra are absorbed by the atmosphere of the earth. The relative abundances of the more common elements in the solar atmosphere in units of *atoms per 10^6 atoms of silicon* are given in Table 1. The most noteworthy feature is the overwhelming abundance of hydrogen and helium relative to all other elements by a factor of 100 to 1. Since this is not true for the earth, it is deduced that in the earth these gases were lost, but were retained in their primordial abundances in the sun because of its large gravitational field.

(b) *Meteorites:* These extraterrestrial objects provide an important sample of the non-volatile components of the solar system and since they have not undergone the extreme geochemical differentiation that rocks of the earth have experienced, it is generally believed that meteorites are well suited for elemental abundance determinations for the solar system. Meteorites can be divided into three groups depending on their metal-to-silicate content as follows: *(1)* Irons, principally consisting of a nickel-iron alloy. *(2)* Stony irons consisting of about 50% metal and 50% silicates. *(3)* Stones, containing principally silicate minerals with small amounts of metal. Each of these groups contain variable amounts of a sulfide phase, FeS, called *troilite*. Thus, in order to

TABLE 1. RELATIVE ABUNDANCES OF THE MORE COMMON ELEMENTS IN THE SOLAR ATMOSPHERE[a]

Element	Atomic Number	Relative Abundances (atoms/10^6 atoms of Si)
H	1	3.16×10^{10}
He	2	5.13×10^9
C	6	1.26×10^7
N	7	3.55×10^6
O	8	2.8×10^7
Na	11	6.31×10^4
Mg	12	7.94×10^5
Al	13	5.25×10^4
Si	14	1.00×10^6
P	15	7.94×10^3
S	16	7.08×10^5
K	19	2.09×10^3
Ca	20	4.90×10^4
Ti	22	2.46×10^3
Cr	24	7.59×10^3
Mn	25	4.17×10^3
Fe	26	1.18×10^5
Ni	28	2.82×10^4

[a] From Aller, 1961.

TABLE 2. ESTIMATES OF THE AVERAGE COMPOSITION OF METEORITIC MATTER BY MASON (1966)

Element	Atoms per 10^6 Atoms Si
O	3.4×10^6
Fe	8.5×10^5
Si	1.0×10^6
Mg	9.5×10^5
S	1.07×10^5
Ni	4.74×10^4
Al	6.8×10^4
Ca	5.8×10^4
Na	4.9×10^4
Cr	9.6×10^3
Mn	6×10^3
K	4.3×10^3
Ti	2.8×10^3

estimate the mean composition of meteoritic matter, it is necessary to know the actual relative proportions of these three groups, in addition to estimating the average composition of each group. Since elemental determinations are usually made separately on the metal, silicate, and sulfide phases of meteorites, different investigators in the past have assumed different relative proportions of these three phases to compute the average chemical composition of meteoritic matter. For example, the earlier estimates of Goldschmidt and others were based upon the ratio of the weights of core to the mantle of the earth, but later estimates such as those of Urey and Mason were based on the relative abundances of those three phases actually observed in chondritic meteorites. The average abundances of the more common elements in meteoritic matter estimated by Mason (1966) are given in Table 2.

(c) *Earth:* The most easily available samples from the earth are the crustal rocks which have undergone extensive fractionation with respect to their elemental concentrations. Because of the diversity of rock types found in the crust of the earth, estimate of the mean composition of the crust is difficult and is based upon some assumed distribution of rock types, or by some way of obtaining a sample that is expected to be nearly an average for the entire crust. The crust of the earth, defined as the layer of material above the Mohorovičić discontinuity, is less than 0.5% of the mass of earth. Therefore all estimates of the mean composition of the earth must be based on models regarding the constitution of the core and the mantle and their relative proportions. By analogy with meteorites, and in accord with the seismic data, it is commonly accepted that the core of the earth is composed of iron (or iron-nickel alloy) with subordinate amounts of the sulfide troilite. The composition of the mantle on the other hand is not a well settled matter. Various authors have used materials like peridotite (Washington, 1925; Mason, 1959), eclogite (Niggli, 1928), and silicate material similar to stone meteorites (Smith, 1963), and have calculated the mean composition of the earth. Results of some such calculations are shown in Table 3. In spite of the model-dependent nature of the calculations, the fact is obvious that more than 90% of the earth is made up of four elements, iron (Fe), oxygen (O), silicon (Si), and magnesium (Mg).

Since none of the above three sources gives complete information on cosmic abundances, it has become necessary to combine all the information to construct a table of cosmic abundances. Suess and Urey (1956) made use of the astrophysical measurements of the sun

TABLE 3. AVERAGE COMPOSITION OF THE EARTH

Element	Washington (1925)	Niggli (1928)	Mason (1959)	Smith (1963)
O	27.71	29.3	27.8	29.26
Na	0.39	0.90	0.14	0.56
Mg	8.69	6.73	17.0	11.28
Al	1.79	3.01	0.44	1.24
Si	14.53	14.9	12.64	14.67
P	0.11	0.15	0.03	0.15
S	0.64	0.73	2.74	3.29
K	0.11	0.29	0.07	0.14
Ca	2.52	2.99	0.61	1.40
Ti	0.02	0.54	0.04	0.07
Cr	0.20	0.13	0.01	0.26
Mn	0.07	0.14	0.09	0.22
Fe	39.76	36.9	35.39	34.82
Co	0.23	0.18	0.20	0.17
Ni	3.16	2.94	2.7	2.43

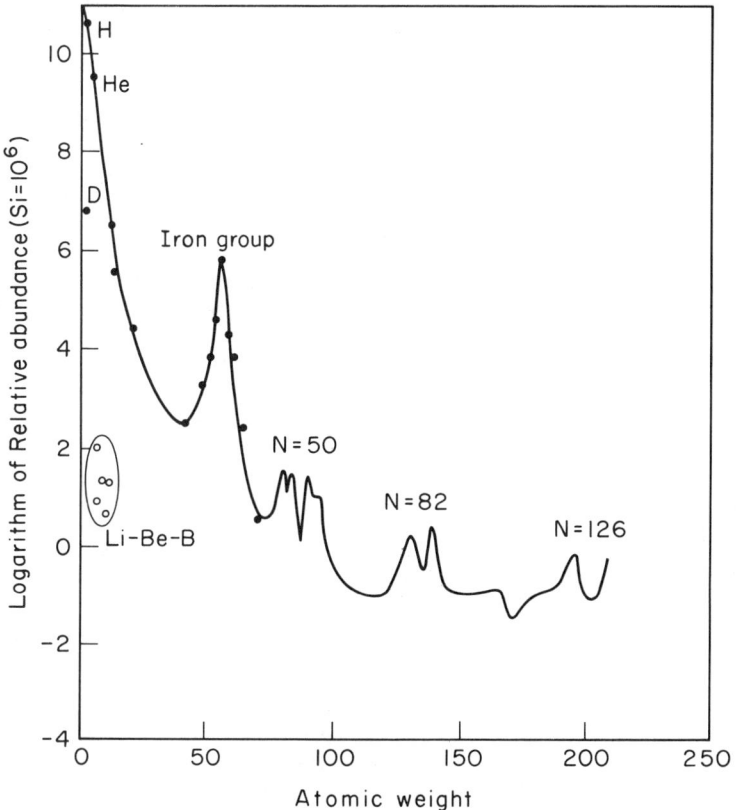

FIG. 1. Schematic diagram of the cosmic abundance curve constructed from the data of Suess and Urey (1956).

for many light elements and data from chondritic meteorites for the heavy elements to construct the first comprehensive table of cosmic abundances. In some instances where analytical data were not available, Suess and Urey made use of well established geochemical ratios of elements to compute the abundances. For example, the bromine abundance was calculated by using the well established Cl/Br ratio in seawater and the Cl abundance. From this table of abundances (Table 4) it was possible to construct a cosmic abundance curve for the elements. The Suess-Urey cosmic abundance curve is shown schematically in Fig. 1.

Since the construction of the Suess-Urey cosmic abundance table, refinements and additions of new data have occurred in the abundance measurements for the sun, the meteorites, and the earth, particularly for many trace elements. Increased knowledge of nuclear physics controlling the synthesis of elements coupled with the new determinations has led to revisions and modifications of the Suess-Urey cosmic abundances. One such calcula-

TABLE 4. ABUNDANCES OF THE ELEMENTS IN THE SOLAR SYSTEM[a]

Element	Suess-Urey (1956)	Aller (1961)
H	4.0×10^{10}	3.16×10^{10}
He	3.08×10^9	5.13×10^9
Li	100	100
Be	20	20
B	24	24
C	3.5×10^6	1.26×10^7
N	6.6×10^6	3.55×10^6
O	2.15×10^7	2.8×10^7
F	1600	3.16×10^4
Ne	8.6×10^6	1.59×10^7
Na	4.38×10^4	6.31×10^4
Mg	9.12×10^5	7.94×10^5
Al	9.48×10^4	5.25×10^4
Si	1.00×10^6	1.00×10^6
P	1.00×10^4	7.94×10^3
S	3.75×10^5	7.08×10^5
Cl	8850	5.62×10^4
Ar	1.5×10^5	2.40×10^5
K	3160	2.09×10^3
Ca	4.90×10^4	4.90×10^4
Sc	28	22
Ti	2440	2.46×10^3
V	220	2.09×10^2

TABLE 4. (Continued)

Element	Suess-Urey (1956)	Aller (1961)
Cr	7.8×10^3	7.59×10^3
Mn	6.85×10^3	4.17×10^3
Fe	6.00×10^5	1.18×10^5
Co	1.8×10^3	1.78×10^3
Ni	2.74×10^4	2.82×10^4
Cu	2.12×10^2	1.00×10^3
Zn	4.86×10^2	6.03×10^2
Ga	11.4	8.9
Ge	50.5	50
As	4.0	4.0
Se	67.6	67.6
Br	13.4	14.0
Kr	51.3	51.3
Rb	6.5	7.1
Sr	18.9	15.9
Y	8.9	8.9
Zr	54.5	10.0
Nb	1.00	1.00
Mo	2.42	2.4
Ru	1.49	0.87
Rh	0.214	0.20
Pd	0.675	0.58
Ag	0.26	0.21
Cd	0.89	0.89
In	0.11	0.18
Sn	1.33	1.18
Sb	0.246	0.28
Te	4.67	3.55
I	0.80	0.71
Xe	4.00	3.63
Cs	0.456	0.46
Ba	3.66	3.80
La	2.00	0.40
Ce	2.26	0.62
Pr	0.40	0.14
Nd	1.44	0.72
Sm	0.664	0.25
Eu	0.187	0.095
Gd	0.684	0.355
Tb	0.0956	0.055
Dy	0.556	0.380
Ho	0.118	0.077
Er	0.316	0.219
Tm	0.0318	0.038
Yb	0.220	0.190
Lu	0.050	0.036
Hf	0.438	0.079
Ta	0.065	0.178
W	0.49	0.126
Re	0.135	0.251
Os	1.00	0.794
Ir	0.821	0.501
Pt	1.625	1.59
Au	0.145	0.145
Hg	0.284	0.178
Tl	0.108	0.112
Pb	0.47	1.00
Bi	0.144	0.100
Th	—	0.032
U	—	—

[a] The numbers in both columns represent abundance in atoms relative to 10^6 atoms of Si.

tion which is based on a composite of astrophysical measurements and meteoritic data is that of Aller (1961) and is shown in Table 4.

General Features of the Cosmic Abundance Curve

1. More than 75% by weight of all matter is hydrogen. Together with helium, it comprises more than 99% by weight of all atoms in the solar system.
2. The abundances drop off exponentially with increasing atomic weight up to $A \sim 100$ and remain reasonably constant for elements with $A > 100$.
3. There is a marked depletion of the elements lithium (Li), beryllium (Be), and boron (B) compared to their neighboring elements. Deuterium is likewise depleted relative to protium.
4. A pronounced peak in abundance exists at mass 56 corresponding to iron.
5. Nuclides of even atomic numbers have higher abundance than those with odd atomic numbers (*Oddo-Harkins rule*).
6. There are peaks in abundances at $A = 80, 90, 130, 138, 196,$ and 208.

Theories of Origin of Elements

Any theory of origin of the elements will necessarily have to account for the general features of the cosmic abundance curve. The salient points of the cosmic abundance curve, mentioned above, have provided an empirical framework for the evaluation of theories of origin of elements.

There have been many theories regarding the origin of elements. Some of the earlier, termed the *equilibrium theories,* proposed that the present distribution of elements represents a "frozen" distribution of nuclides produced at high temperatures and densities. In general, these theories have failed to account for the general features of the cosmic abundance curve, and hence are not in common use at present, having been replaced by *nonequilibrium theories.*

(a) Gamow, Alpher and Herman proposed that at the beginning all matter was composed of neutrons, and as this matter expanded, the neutrons decayed to protons and electrons. The protons then absorbed neutrons and element synthesis was on its way, by successive additions of neutrons and β^- decays. This theory enjoyed widespread success until it was clear that at several places it could not satisfy the features of the cosmic abundance curve. For example, there is no explanation in this theory as to why there is a peak in abundance at iron, nor does it explain how the synthesis continues beyond masses of 5 or 8, because of the fact that He^5 and Be^8 which are produced by neutron capture have extremely short life times, decaying back to He^4. Thus, neutron capture synthesis will only result in formation of elements up to helium and no further.

(b) In an attempt to satisfy the general features of the cosmic abundance curve, Burbidge, Burbidge, Fowler, and Hoyle (1957) proposed that the present elements in the solar system have been manufactured in nuclear reactions occurring in stars. Element synthesis was supposed to start from hydrogen which formed the primitive matter in the galaxy, and which underwent various nuclear reactions in the interior of stars. These authors proposed specifically that several types of nuclear reactions must be invoked to explain the cosmic abundances. These are:

1. Hydrogen burns by successive neutron captures to produce helium in stars with central temperatures of 10^7 degrees absolute and densities of ~ 100 g/cm^3. This reaction can be expressed by the simplified notation $4H^1 \rightarrow He^4$.
2. At higher temperatures on the order of 10^8 degrees in the central core of the star, and at densities on the order of 10^5 g/cm^3, the helium produced in process (1) begins to burn. Three helium atoms coalesce to produce carbon. We can characterize the reaction by the notation $3He^4 \rightarrow C^{12}$. The carbon produced in turn will capture other particles and in succession synthesis of O, Ne, and perhaps Mg takes place.
3. At still higher temperatures of about 10^9 degrees and densities near 10^7 g/cm^3, nuclear reactions produce α particles which interact with material produced in process (2) resulting in Mg^{24}, Si^{28}, S^{32}, A^{36}, and Ca^{40}. This has been termed the α-process.
4. As the star evolves further and reaches the central temperatures of about 3×10^9 degrees, the rate of all nuclear processes is greatly increased and interactions between nuclei produced in (1), (2), and (3) protons and neutrons are supposed to attain a statistical equilibrium so that the iron group elements are synthesized at this stage. This is called the e-process.
5. The synthesis of heavy elements beyond iron occurs principally by capture of neutrons in the iron group nuclei. The neutrons themselves may be produced by α-particle reactions on existing nuclei such as C^{13}, O^{17}, or Ne^{21}. When the rate of neutron capture is slow compared to the beta-decay life times of the nuclei produced this is termed the s-process. The path of synthesis in this process leads up to Bi^{209}, but many nuclei are bypassed. The presence of these bypassed nuclei and the heavy elements beyond bismuth (Bi), uranium (U), and thorium (Th) are attributed to neutron capture processes on a rapid scale, termed the r-process, in which neutron capture occurs on a time scale short compared to intervening β decays of the nuclei produced. These neutrons at high fluxes are produced in supernova explosions.

The hypothesis of element synthesis in stars is well documented both by laboratory studies of nuclear reactions and astrophysical observations. Elements are synthesized in stars and, as nuclear processes take place, the composition of a star changes until at some point an instability arises dispersing the "processed" material into interstellar space where it is mixed with uncondensed hydrogen. Second and later generation stars condense from this interstellar matter and the cycle is repeated many times over. There is evidence, mainly drawn from the abundances of U^{235} and U^{238}, that stellar activity and element synthesis have gone on in our galaxy for periods on the order of 10×10^9 years. The abundance distribution in the solar system is the result, then, of a continuous synthesis of elements in stars during some billions of years before our own solar system condensed.

V. RAMA MURTHY

References

Ahrens, L. H., 1965, "Distribution of the Elements in our Planet," New York, McGraw-Hill Book Co., 110pp.
Aller, L. H., 1961, "The Abundance of the Elements," New York, Interscience Publishers, 283pp.
Burbidge, E. M., Burbidge, G. R., Fowler, W. A., and Hoyle, F., 1957, "Synthesis of elements in the stars," *Rev. Mod. Phys.*, **29**, 547.
Fowler, W. A., 1961, "The Origin of Nuclear Species by Means of Nuclear Reactions in Stars," in "Modern Physics for the Engineer," Second Series, New York, McGraw-Hill, pp. 177–239.
*Mason, G. H., 1966, "Principles of Geochemistry," Third ed., New York, John Wiley & Sons, 329pp.
Suess, H. E., and Urey, H. C., 1956, "Abundances of the elements," *Rev. Mod. Phys.*, **28**, 53–74.

*Additional bibliographic references mentioned in the text may be found in this work.

Cross-references: *Atomic Number and Periodic Table; Elements: Planetary Abundances and*

ELEMENTS

Distribution; Geochemical Evolution of the Core, Mantle, and Crust; Outgassing of the Planet Earth. Vol. II: *Cosmogony; Gravitational Collapse of a Star; Meteorites; Nucleogenesis; Solar System; Stars and Stellar Interiors; Sun.*

ELEMENTS, GEOCHEMICAL CLASSIFICATION—See GEOCHEMICAL CLASSIFICATION OF ELEMENTS

ELEMENTS: PLANETARY ABUNDANCES AND DISTRIBUTION

Definition of Categories

The abundances of elements in a sample do not change, only the data do. Concentrations of an element in a geosphere may vary in space and time because of differentiation, viscosity, solutional, radioactive, shock, gravity, aeolean, pH, Eh, pressure, thermal, mechanical, or other effects. In terms of man's stay on earth only lead, carbon, and certain radioactive isotopes of H, C, Sr, and the transuranium elements have experienced change in the atmosphere, hydrosphere, surface soils, and human artifacts.

The major geospheres in which elements are concentrated are more or less agreed upon but tens of categories within these spheres can be created by the geochemist depending upon the relationship he wishes to quantify. The geochemist should define the limits of his category; in some cases the data he collects may more precisely define these limits. But the category must make geochemical sense. For example, in shales, elemental abundance data for red versus black facies would be desirable from the standpoint of prospecting for sedimentary ore deposits. Likewise of significance in petroleum exploration would be compilations of abundances of "chelatable" metals in the following categories: iron oxides of shallow water oxidizing environments to the iron carbonates of the less oxidizing to the iron sulfides of the deep basinal reducing environments.

Table 1 lists over fifteen conditions upon which to build abundance tables. General categories include entries 1–4; specific categories from 5 on. Usually the geochemist combines categories. In this paper, categories of some major units within subgeospheres (No. 3) are refined by including items 5 and 7; for example, high iron chondrites and pelagic clays. Sixteen categories for the chemical elements are included in Table 7. Columns 1–10 are from Turekian and Wedepohl (1961), column 11 from Goldberg (1965, pp. 163–195), columns 12–13 recalculated from Turkevich et al. (1969), and columns 14–16, recalculated from Aller (1967, pp. 287–288). Sources of information for modifications to this table are given below.

As more compilations are made of categories 5 through 15 of Table 1, the more valuable they may be in quantifying geochemical concepts of a specific nature. Tables of the type exemplified by Table 7 are good for quantifying geochemical concepts of a general nature. Let the user be aware.

General Sources of Information. Information on the behavior of the elements is exponential in its growth. We note philosophically that the study of the properties of the elements has lagged behind their discovery just as the study of the laws governing the distribution of elements has lagged behind their analysis in naturally occurring materials. A comprehensive treatise on the properties of the elements is by Samsonov (1968). The usual atomic structure, crystallochemical, thermodynamic, thermal, optical, electrical, magnetic, and nuclear physical properties are given as they are in Weast (1968). But in addition, Samsonov has compiled two sections of geochemical interest; one detailing oxidation-reduction potentials in water at 298°K and the other on the reactions of elements with H, B, C, Si, N, P, O, S, Se, Te, and refractory materials in the solid phase. Clark (1966) supplements these two handbooks with data on solubility, ionization constants, and phase equilibria of rock and minerals. A crystal chemical classification of minerals is thoroughly covered by Povarennykyh (1969).

More recent data on phase equilibria in silicate systems, especially those including volatiles, are critically reviewed by Eitel (1966). Many systems involving water-silica-and the major oxides are described together with details on the apparatus required for their study. Analytical methods for the study of silicates are also given by Bennett and Hawley (1965) and by May and Cutitta in Abelson (1967, p. 112).

Elemental Abundances Within Geospheres and Subgeospheres

Geochemistry Texts and Reviews. Besides the papers cited in the caption to Table 7, general texts by Mason (1966), Krauskopf (1967), Ahrens (1965), and Wedepohl (1969) have appeared which cover one or more geospheres. To these must be added the geochemical series books edited by Abelson (1967), Ahrens et al. (1966, 1971), and Hurley (1964). The 1963 and 1967 geochemical symposia proceedings edited by Vinogradov (1966) and Ahrens (1968) deserve special mention. Rankama

ELEMENTS: PLANETARY ABUNDANCES AND DISTRIBUTION

TABLE 1. GEOCHEMICAL ABUNDANCE CATEGORIES

Category Based on	Examples
1. Geosphere	Lithosphere, hydrosphere, atmosphere
2. Subgeosphere	Metamorphic, igneous, sedimentary rocks
3. Major unit within subgeosphere	Basalt
4. Minor unit within subgeosphere	Tholeiitic basalt, alkalic basalt
5. Geochemical "break"	
(a) Major element	Low and high calcium "granites"
	Low and high iron concentrations in chondrites
(b) Volatile element	Low to high S, C, and H_2O
(c) Minor or trace element	Low to high rare earth basalts
6. Time interval	Early Paleozoic low Ca/Mg ratios in shales
7. Major province	Basinal, alpine, volcanic, orogenic
	Oceanic: pelagic, tectonically active margins; plateau, abyssal
	Continental: shield, shelf, geosynclinal margins; deltaic
	Lunar mare, lunar highland
	Metallogenic
8. Position	
(a) Level or depth	Aqueous or solid bodies, atmosphere, core samples, soil horizons
(b) Structural	Fold axes and limb zones
(c) Boundary	Dike or batholith margin and interior, sample exterior and interior, zoned crystals, banded rocks
9. Facies	Mineralogical, in subgeospheres
	Regressive and transgressive
	Oxidizing to reducing
	Quiet water to turbulent water
	High Eh to low Eh
	Abiotic and biotic
10. Heat and pressure history including metamorphic grade index	High to low Al in clinopyroxenes
	Lignite to graphite
11. Other environmental conditions	Fractured to massive materials
	Altered to fresh materials
	Acid to basic fluids
	Mineralized to barren rocks
	Low to high gravity
12. Texture	Vesicular to nonvesicular
	Welded to nonwelded
	Coarse to fine grained
	Columnar to noncolumnar
13. Size fraction	Coarse to fine mesh
14. Constituent	Phenocryst and matrix
	Mineral and bulk sample
	Leached and unleached sample
15. Miscellaneous	Radiation responsive to radiation unresponsive
	Biologically processed or produced to nonbiologically processed or produced
	Shocked to nonshocked materials

(1964) updates his earlier book on isotope geology on an element by element basis. A supplementary monograph on isotope geochemistry is by Hart (1969).

General texts dealing with abundances in the subgeospheres are as follows: the mantle by Gaskell (1967) and Runcorn (1967); in igneous rocks by Rodionov (1965), in sediments by Perelman (1967), Degens (1965), and Larsen and Chilingar (1967); and in the oceans by Riley and Skirrow (1965) and Dietrich et al. (1966).

A continuing compilation of papers edited by Fleischer (1963–19–) is intended to provide the sixth edition of Clarke's "Data of Geochemistry." This compilation will provide a comprehensive review of most abundance data in a projected 34-volume treatise.

Besides providing usual abundance data on the distribution of elements in the earth's crust, chapters in print cover volcanic emanations (by D. E. White and G. A. Waring), chemical composition of subsurface waters (by D. E. White, J. D. Hem, and G. A. Waring), chemical composition of rivers and lakes (by D. A. Livingstone), chemical composition of sandstones—excluding carbonate and volcanic sands (by F. J. Pettijohn), nondetrital siliceous sediments (by E. D. Cressman), and marine evaporites (by F. J. Stewart). Another compilation in tabulated form for the major elements is provided by Poldervaart and Green (1959).

Mantle and Crustal Data. Over one hundred individually authored papers on the geochemistry of the solar system, meteorites, mantle, igneous and metamorphic rocks, sediments, ocean water, and the atmosphere are found in the series or symposium proceeding books cited above. Cameron (1968) reviews the chemical relationships among the planets. Abundances in the mantle, not listed in Table 7 because of the uncertainties involved, are defined on the basis of the chondritic model by Mercy in Gaskell (1967, pp. 421–444), on the basis of ultramafic segregation model by Harris in Runcorn (1967, pp. 305–317), and on the basis of an amphibolite model by Bose (1967). Nichols in Runcorn (1967, pp. 285–304) has calculated possible variations in the composition of the upper mantle (Table 2).

General syntheses of these data on the geosphere or subgeosphere level have also been made by over one hundred authors in the last five years. Several examples may be cited. Ringwood (1966a) begins with the cold accretion of the earth from a dust cloud with the composition of a Class I carbonaceous chondrite. He traces the origin of the core from thermodynamic considerations. Subsequent defluidization of the volatiles by the heat of accretion produced a mantle of chondritic composition and a surface crust subsequently altered by continued defluidization in geological time energized by radioactive heat-up.

Ringwood assumes that most of the primordial Na, K, Rb, Cs, S, Zn, Cd, Hg, Ge, Tl, Pb, F, and Cl have been lost by volatilization in the process of the earth's formation. The reviewer does not accept Ringwood's (1966b) hypothesis that the origin of the moon came about as a result of the condensation of a "sediment-ring" of non-volatile silicates following the disruption and escape of a primordial atmosphere from the earth. Ronov (1964, 1968) traces the composition of seawater in geological time based on the composition of sediments and metasediments. The results indicate that higher Ca, Mg, and K and lower Na were present in Precambrian oceans. CO_2 resulting from volcanic and biological activity began to deplete the oceans in Ca and Mg in the early Proterozoic. Major element (Al, Si, Ca, Mg) ratios are used to define the increases or decreases in graywackes or arkoses as a function of time. Ronov and Yaroshevsky (1967) have compiled the average thickness, composition, and mass of these and other rock types.

Relative to abundance data as they relate to the origin of the earth's crust assuming a chondritic mantle, interpretation of major and trace element data is reviewed by Tauson (1965), Shaw (1964), Ringwood (1966b), and Taylor (1964b, 1965); Taylor (1964b) has compiled a new table of continental crustal abundances using a ratio of basalt to granite of 1:1. Recall that the average igneous rock as compiled by A. P. Vinogradov cited in Green (1959) is based on a basalt to granite ratio of 2:1. Rare earth distributions in terrestrial, meteoritic, and solar material are reviewed by Haskin et al. in Ahrens et al. (1968). A key to the evolution of the crust from the mantle may be provided by the behavior of the halogens, especially chlorine (Iwasaki et al., 1968).

Hydrosphere. Table 7 lists values for oceanic abundances taken from Goldberg (1965, pp. 164–165). The numbers, as is the case with all averages, do not disclose the variations in concentration which tell a more interesting geochemical story. For example, Ba, Hg, Pb^{210}, and the rare earths increase with depth in ocean waters. Except for Ba, the enrichment of the other elements with depth may be due

TABLE 2. POSSIBLE VARIATIONS IN THE COMPOSITION OF THE UPPER MANTLE[a]

	(1)[b]	(2)	(3)
SiO_2	45.1	45.2	40.3
TiO_2	0.5	0.0 (2)	0.4
Al_2O_3	4.1	0.8	3.7
Cr_2O_3	0.3	0.4	0.3
Fe_2O_3	2.0	1.5	1.8
FeO	7.9	7.2	7.1
MnO	0.2	0.2	0.1
NiO	0.2	0.3	0.2
MgO	36.7	44.2	32.7
CaO	2.3	—	2.1
Na_2O	0.6	0.1 (5)	0.5
K_2O	0.0 (2)	—	0.0 (2)
P_2O_5	0.1	0.0 (5)	0.1
H_2O	—	—	9.7
CO_2	—	—	0.8
Cl	—	—	0.2
Total	100.0	100.0	100.0

[a] After G. D. Nichols; cited in Runcorn (1967, p. 303).
[b] Column (1), composition of material from which volatiles have been lost, but from which the basaltic fraction has not been removed by partial fusion. Column (2), composition of material from which both volatiles and the basaltic fraction have been removed by partial fusion. Column (3), composition of volatile-rich parts of the upper mantle, such as may occur beneath the midoceanic ridge system. Figures for the chemical constituents are given as weight percentages.

to solid phases resulting from the death of organisms or from the formation of metabolic waste products. Alternatively, the dissolved metals may be absorbed on organic debris at depth. For Ba, bacterial oxidation of detrital organic phases might cause precipitation of barium sulfate. Dissolution of barium sulfate at great depth and consequent return of Ba to the overlying water might help explain the higher concentrations at depth. Hanor et al. (1968) show that in sea-floor cores that are remote (2000 km) from the East Pacific Rise, the deeper sediments are enriched in Ba (9000 ppm) relative to upper portions of the core analyzed (5000 ppm). However, this is presumed to be caused by removal of Ba-rich sediments from the crest of the rise with subsequent burial by Ba-poor sediments. This distribution of Ba in pelagic sediments is consistent with the hypothesis of sea-floor spreading. Would detailed Ba measurements in ocean water at depth and on the surface over rise crests and flanks show a similar pattern to the sedimentary concentrations?

As with many geochemical problems, the interface between geospheres is the last to be analyzed. Consider the ocean and the atmosphere interface where Bruyevich and Kulik (1967) have shown significant chemical interactions to exist. When mechanical evaporation occurs as in sea spray, the chemical composition of the evaporated spray is not the same as the seawater. Experimental data on Black Sea distillates show that phosphorus is preferentially transferred to the atmosphere as an aerosol. However, the principal difference between the composition of seawater and rainwater is in the ratio of sulfates to chlorides; it is 0.14 for the ocean and sometimes greater than 1 for rainwater. In atmospheric moisture absolute concentrations are $HCO_3 > SO_4 > Cl$ which is opposite to the concentrations in ocean water.

Thus the excess of sulfate in river water discharge is not due to continental weathering processes but due to ocean-atmosphere interactions which concentrate sulfates in rainwater.

Also, abundances of sulfates in seawater against which river water discharge concentrations are compared may vary in areas of active volcanism. Gorshkov and Tovarova (1957) have measured extremely high sulfate concentrations in waters leached from fresh pyroclastics at Bezymianny in 1956. A serial decrease in sulfate content was observed from the water extracted from ash expelled in the initial eruptive phase to the main eruptive phase to the later agglomerate flow. Similar concentration patterns were observed for Cl^-, SiO_2, SO_4^{2-}, Fe, Mg^{2+}, Ca^{2+}, Na, and K (Table 3). Submarine eruptions could thus produce local and abnormal concentrations of salts such as fluorine (Wilkniss and Linnenbom, 1968) in ocean water, as they in part may account for the anomalies in Red Sea brines (Degens and Ross, 1969).

Just as sulfate concentrations in river water cannot be explained by the weathering of exposed rocks, neither can sodium concentrations. Without taking space to amplify this fundamental problem, the reader is referred to Perrin (1960) and Livingstone (1963). Perhaps it is enough to say that the puzzle of the sodium balance is still with us and unsolved.

Planetary defluidization discussed by G. D. Nichols in Riley and Skirrow (1965, Vol. 2, pp. 277–294), Vinogradov (1968), and Brancazio and Cameron (1964) remains as a dominant process which is applicable not only to the oceans but to all geospheres, including the atmosphere.

Atmosphere. The presence of banded iron ores, pyritic sandstones, uranium ores, and carbonates in the Precambrian attests to anaerobic conditions. With the first fossils appearing over 2.7 billion years ago, some biogenic oxygen must have been present at that time. Prior to the appearance of early life forms, pre-life organized elements were probably formed and continued into the Middle Precambrian according to Rutten (1968). Thus, early life forms and pre-life forms probably co-existed for one billion years. According to Rutten, the primeval atmosphere was superseded by the oxygenic

TABLE 3. CONCENTRATION OF SALTS IN BEZYMIANNY ASH WATER EXTRACTS[a]

Sample[b]	Salt (mg/100 g of ash)							
	Cl^-	SO_4^{2-}	SiO_2	Fe	Mg^{2+}	Ca^{2+}	Na	K
A	95.88	400.4	2.95	10.57	21.5	157	8.1	3.11
B	55.04	198.4	2.71	8.83	10.2	81.2	5.15	1.68
C	22.7	165	2.4	5.6	3.11	54	6.67	1.7

[a] After Gorshkov and Tovarova, 1957.
[b] A = water extract from ash fall in the neighborhood of the volcano during the initial period of eruption.
B = water extract from ash fall during the main explosion (March 30, 1956).
C = water extract from agglomerate flow.
(Mean values from analysis of five samples.)

atmosphere 1.45 billion years ago. The 1/10 present atmospheric level of oxygen was reached in the Ordovician about 0.4 to 0.5 billion years ago.

Other factors consistent with an early anaerobic reducing atmosphere have been proposed by Berkner and Marshall contributing to Brancazio and Cameron (1964). Berkner and Marshall identify the Precambrian-Paleozoic transition with the rise of oxygen concentration in air to 1/100 of the present level and the appearance of land animals in the late Silurian with a further rise in oxygen level to 1/10 of its present level. These conclusions are not too inconsistent with those of Rutten. However, rather than assume biogenic contributions and the effect of the Pasteur point to increase the oxygen concentrations, Berkner and Marshall assume that interactions of various frequencies of solar radiation with probable constituents of the Earth's early atmosphere influence the increase in oxygen. Brinkmann (1969) extends this mechanism to explain the presence of oxygen in the atmosphere of Venus.

The evolution of a favorable atmosphere on Earth and a toxic one on Venus from the degassing of presumably the same volatiles from the planet's interior has been discussed by Palm (1969). She argues that differences in surface thermal histories are responsible for the divergence in the nature of terrestrial and Venusian atmospheres. The presence of liquid water would determine the course of reactions of the lithosphere with the atmosphere. In this connection, the salts responsible for producing an acid ocean (Table 3) could have been leached from pyroclastics early in the history of a planet and the evolution of an atmosphere would be controlled by the rate at which the ocean waters were buffered.

One of the more detailed studies of the evolution of the atmosphere is by H. D. Holland, contributing to Engel et al. (1962, pp. 447–477). A three-stage phasing is postulated. The first stage, of relatively short duration (one-half billion years), is the reaction of hydrogen-rich volcanic gases with magma and native iron producing a highly reducing atmosphere consistent with all other studies. The second stage evolved as the native iron was removed, resulting in an atmosphere similar to that of volcanic gases with an oxidation state predicted to be close to present day Hawaiian emanations ($P_{H2O}/P_{H2} = 137$ and $P_{CO2}/P_{CO} = 31$). According to Holland, nitrogen would be the most abundant constituent of the second-stage atmosphere. At this time, carbon dioxide took the place of methane and began to participate in processes of weathering and chemical precipitation. The end of the second stage may have been reached some one billion years ago. The third stage began when the rate of production of oxygen by photosynthesis exceeded the rate needed to completely oxidize volcanic gases. Since the late Paleozoic, the rate of accumulation of oxygen has probably gone on at a roughly constant rate.

The present concentration of oxygen and other gases in the atmosphere is given in Table 4, which is taken from F. A. Richards in Riley and Skirrow (1965, p. 199). The variable and constant concentrations of the constituents

TABLE 4. COMPOSITION OF THE ATMOSPHERE[a]

		ppm (or as indicated)
Ammonia	NH_3	0 – Trace (tr)
Argon	Ar	9340 ± 10
Carbon dioxide	CO_2	330 ± 10
Carbon monoxide	CO	0 – tr
Formaldehyde	CH_2O	?
Helium	He	5.24 ± 0004
Hydrogen	H_2	0.5
Iodine	I_2	$< 10^{-4}$ g/m^3
Krypton	Kr	1.14 ± 0.01
Methane	CH_4	2.0
Neon	Ne	18.18 ± 0.04
Nitrogen	N_2	780,840 ± 40
Nitrogen dioxide	NO_2	0 – 0.02
Nitrous oxide	N_2O	0.5 ± 0.1
Oxygen	O_2	209,460 ± 20
Ozone	O_3	0 – 0.07 (summer)
		0 – 0.02 (winter)
Sodium chloride	NaCl	$\sim 10^{-4}$ g/m^3
Sulphur dioxide	SO_2	0 – 1
Xenon	Xe	0.087 ± 0.001

[a] After F. A. Richards in Riley and Skirrow (1965, p. 198).

shown indicate the complexity of atmospheric chemistry. Local chemical variations in the atmosphere certainly exist in active volcanic areas. Naughton et al. (1969) has measured the volume concentrations of gases from Kilauea using remote sensing spectrometry. Although atmospheric contamination may contribute a large absolute error, the first experiments of Naughton and co-workers yield $H_2O:CO_2:SO_2 = 95\%:4\%:1\%$.

One constituent not included in Table 4 is lead. The rate of introduction of lead into the world's oceans is 27 times greater than it was in the Pleistocene. The source of this lead is tetraethyllead in industrial combustion products. As cited in E. D. Goldberg (1965, p. 181), 3.5×10^{11} grams of lead are washed out of the atmosphere per year.

Identification of rates of influx of various elements from the atmosphere into the oceans would be a desirable approach to determining the origin of atmospheric aerosols. Mossop (1965) has identified ammonium sulfate collected as minute particles at an altitude of 20 km. The origin of these aerosols, whether terrestrial, industrial, or extraterrestrial, is being studied. Cadle et al. (1969) report ammonium sulfate, calcium sulfate, and sulfur particles in the fumes from Kilauea. The effect of atmospheric aerosols in producing temperature changes on the surface of the earth by solar blanketing effects is unknown.

Igneous Rocks. Rare earth distributions within smaller abundance categories have provided new light on magmatic differentiation processes, the localization of ore deposits (Vlasov, 1968), and the geological environment (Fleischer and Altschuler 1969). For nonvolatile elements, such as Pd, data by Crockett and Skippen (1966) are also instructive. A wide range of Pd values exists for continental basalts but not for oceanic basalts; the latter average is cited in Table 7. Likewise the abundance of Pd in ultramafics varies widely. The number selected is the average of their dunite and olivine nodule values only. The geochemical break in rare earth distributions cited in Table 1 is documented by Schilling and Winchester (1966). Not only can basaltic rock types in Hawaii be defined by absolute abundances of the rare earths, but the greatest negative slope for the rare earths, La–Eu, may correlate with the highest degree of differentiation. Other correlations in alkalic and tholeiitic basalts have been made for both trace and major elements by M. Prinz and V. Manson in Hess and Poldervaart (1967) and by Philpotts et al. (1968), who endorse partial to complete fusion processes for the distributions observed. Because of widely differing concentrations of trace elements in fresh and altered basalt (Hart, 1969), sample selection especially of submarine basalts should be made with care.

Calc-alkaline volcanics may evolve from partial melting of "wet" basalt or quartz eclogite (derived under dry conditions from basalt in the upper mantle) (Green and Ringwood, 1968). Again, trace element distributions and isotope ratios support the experimental high-pressure data obtained by these authors. The access of water to the upper mantle might result from convective crustal overturn in areas of sea-floor spreading. Green and Ringwood's paper is an example of dozens of papers attempting to correlate geochemically major zonations of rock types in the earth's crust with geochemical abundance anomalies. The oceanic trough-to-continent transition of tholeiites to alkalic basalts, for example, is marked by increasing K/Na ratios (Challis, 1968) in New Zealand. The increase of Al in clinopyroxenes in the presumably more deep-seated alkalic basalts in Japan may reflect the higher temperature-pressure conditions obtaining during genesis of this rock type. A more sensitive indicator of this source environment for alkalic basalts may be the Ga and Ge content of clinopyroxenes (S. R. Taylor, personnel communication). Symposia on phase transformations and the earth's interior (A. E. Ringwood, Canberra, Australia, January 6–10, 1969) have dealt with aspects of these relationships, as did the one on andesites (A. L. McBirney, Bend, Oregon, July 1–5, 1968).

Magmatic processes involving the upper mantle or lower crust may not be the only ones involved in the production of intrusives and extrusives. Major and minor elements (including the rare earths and volatiles in the rhyolitic rocks of New Zealand (Ewart 1967, 1968, Ewart et al., 1968) support the hypothesis that rhyolitic magmas are derived by the partial fusion of Triassic-Jurassic eugeosynclinal graywacke-argillite sequences. New data on the abundances of volatiles such as Hg (cited in Table 7; from Ehmann and Lovering, 1967) may be as important as that for the rare earths in providing additional information on magmatic differentiation processes and fusion.

Abundances data for the minor elements in alkalic rocks are subject to wide variations. Factors of five exist between the abundances of the following elements in the Lovozero and Ilimaussaq agpaitic nepheline syenites of the Kola Peninsula and West Greenland, respectively: Li, Cl, Sr, Mo, Cs, Ba, and Pb (Gerasimovskii, 1968, p. 260). The presence of the mineral lamprophyllite in the Lovozero massif contributes to its abnormally high abundance of Sr and soerensenite in the Ilimaussaq massif contributes to its high abundance of Sn. Comparing Gerasimovskii's data on syenites with those from Turekian and Wedepohl cited in Table 7, factors of five deviations from the Lovozero abundances include: Be, Cr, Co, Zr, Nb, Hf, Ta, and U; factors of five from the Ilimaussaq abundances are: Li, Be, Cl, Cr, Ni, Zn, Zr, Nb, Mo, Sn, Cs, Ba, Hf, Ta, Pb, and U.

Two books which bring together the geochemistry—albeit limited to a chapter apiece—of the end members of the rock clans, the ultramafics and the carbonatites are by Wyllie (1967) and Heinrich (1968), respectively. The role of carbonatites in producing shock-induced mineralogies is unknown at present but may yet tie together widely differing categories into one (see Currie and Shafiqullah, 1967).

We note that the trend for igneous rocks is to show that certain trace elements may prove to be more sensitive indicators than major elements for understanding mantle and crustal processes. The distribution functions of the elements—their lognormality or "abnormality"—are still receiving keen appraisal (Ahrens, 1966; Rodionov, 1965; Butler, 1964). Finally, the actual sampling of raw magma in the crusted lava lakes of Kilauea is giving new experimental insight into changes in composition correlative with their geophysical properties, viscosity, and seismic behavior. The change in composition of Kilauean lavas with time can be explained by the separation of hypersthene either at the magma source or in a storage chamber located higher in the mantle or crust (Wright, 1968).

Metamorphic Rocks. The modern trend in igneous petrology to use geochemistry to understand a specific process also characterizes metamorphic petrology. In folded hornblende granulite facies rocks in Norway, for instance, the fold zones or hinges are relatively enriched in Ti, Fe, Mn, and Mg and impoverished in Al, Na, and K; Si and Ca showing no significant change. The relationships are explained in terms of crystal stability in stress by Parker (1968). Other factors resulting in variations in elemental abundances within metamorphic isograds may involve dehydration reactions as detailed by E. O. Lacy and M. P. Atherton in the proceedings volume of "Controls of Metamorphism" edited by Pitcher and Flinn (1965). W. S. Pitcher summarizes the contributions in this volume by noting that although metamorphism is temperature controlled, deformation produces the fabric, triggers the reactions, and facilitates diffusion. According to Pitcher, the role of diffusion in regional metamorphism is not well known. Vacancy diffusion apparently permits the high bond strengths of silicates to be bypassed in geologically long time periods. The crux of many problems in metamorphic geochemistry is to determine which of two effects is the more important—rate phenomena or equilibria phenomena. While there is abundant evidence for differing rates of nucleation and growth, there is also abundant evidence for coexisting minerals; the latter circumstance strongly suggests near attainment of chemical equilibrium during regional metamorphism.

Other mechanisms may also produce coexistence of otherwise incompatible minerals. Shock is one of these. Whether meteoritic impacts or possibly carbonatitic explosions produce all the isotropism, kink bands, planar structures, shatter cones, or coesite is open to debate. Of significance is the laboratory synthesis of metastable coesite in flint at pressures and temperatures (5–20 kb and 450–900°C) far below the coesite stability field (H. Green, 1968). Evidently the high density of dislocations developed in highly compressed quartz ($> 50\%$) at a strain rate of $10^{-4}/$ sec increases the free energy of the deformed quartz to a higher value than the unstrained coesite. Thus, metastable coesite forms.

Sedimentary Rocks. General texts dealing with the geochemistry of sediments have been cited above. Specific aspects of geochemical variations in sandstones, shales, and carbonates are discussed by various authors in "Developments in Sedimentology—Diagenesis in Sediments" edited by Larsen and Chilingar (1967) followed by Chilingar et al. (1967) which deals exclusively with carbonates. In the latter book, Wolf et al., have compiled data on the elemental composition of carbonate skeletons, minerals, and sediments. Fundamental aspects of petroleum geochemistry are covered by Nagy and Columbo (1967), organic geochemistry by Eglinton and Murphy (1969) and Schenck and Havenaar (1969), salines geochemistry by Mattox (1968) and Rau (1966), and geochemistry of organic substances by Manskaya and Drozdova (1968).

In Table 7, the abundances of the elements in terrigenous sediments only cover sandstones, shales, and carbonates, but as Turekian and Wedepohl (1961) state, conglomerates, arkoses and graywackes are genetically and volumetrically important sediments for which little quantitative data were available (in 1961). These rocks are not simply weathering products because they have in many cases undergone chemical changes from their parent igneous or metasedimentary rocks. Much quantitative data on the major and trace element distributions in arkoses and graywackes have been analyzed by Ronov (1964).

The reader is referred to Turekian and

Wedepohl (1961) from whom the deep-sea sedimentary abundances were obtained for a discussion of their choice of categories. The end-members, clay and carbonate-rich sediments, vary in their sedimentation rates from the Pacific to the Atlantic. The reader is cautioned that wide variations in elemental abundances occur between the two oceanic basins and the averaging technique used by Turekian and Wedepohl for the abundances shown may be less than ideal. Lead in Atlantic clays averages 45 ppm whereas lead in Pacific clays is around three times greater. Also, they have assumed that the abundance of Br and Se in pelagic carbonates is identical to that reported for pelagic clays. The reviewer would prefer an order of magnitude similarity.

Table 7 contains data on Pd, Au, and Ir in pelagic sediments from Crocket et al. (1968) who used neutron activation analysis. Unlike the behavior of the other elements such as barium, Pd or Au are not concentrated along the crest of the East Pacific Rise. Another curious fact is the 100-fold enrichment of Ir in manganese nodules over pelagic sediments although no such enrichment occurs with Pd and Au in these nodules.

Extraterrestrial Abundances

Lunar. In this report no space is available to discuss fully the implications of the lunar alpha scattering data obtained by Turkevich and his co-workers (1969). Suffice to say that it warrants space on Table 7 as the first return of controlled extraterrestrial data. The raw data in terms of atomic percentages as given by Turkevich have been recomputed using the mean values to parts per million by weight, making the following assumptions.

Carbon is assumed to be zero. The reported value is < 3 atomic percent. If present as elemental carbon, no darkening effect was observed on the soil sampler on repeated contact with the surface soil. The general consensus of the results of the alpha-scattering data is toward silicate rocks, which, with the exception of carbonaceous chondrites, are extremely low in carbon. Carbon concentrations, as diamond, produced by impact shock, are also considered to be far below the detection limits of the alpha scatterer.

The maximum value of sodium is used assuming some potassium; the ratio of Na to K assumed is 2:1. This is an unsatisfactory assumption, yet potassium must be present if the other elemental abundances suggest a rock of basaltic composition. An atomic weight of 28 derived from the ratio assumed was used to compute the weight percent of "sodium." Although the sum of sodium and potassium is more reliable than the absolute abundances of sodium and potassium cited in Table 7, these values should be qualified by the assumptions stated.

As Turkevich et al. (1969) state, the values reported for calcium and iron are not specifically related to these elements but rather to groups of elements in the calcium and iron mass ranges. However, the weight percentages are cited assuming that the masses are only calcium and iron. The other elements in silicate rocks, perhaps with the exception of Cr, are relatively rare compared to those of calcium and iron.

For any interpretation of these numbers, corrections based upon the background of the geochemist are employed. The first correction to be made within the limits of the errors stated might be to normalize the data to a possibly more realistic oxygen percentage of around 44%.

Interpretation of these mare and highland data also rests in the mind of the beholder. The author prefers a tholeiitic basalt or basaltic andesite for the mare analysis and a mixed volcanic ash, both mafic and felsic constituents, for the highland analysis. Surveyor VII actually did not land in the true "highland" province but in the dark halo surrounding Tycho. The author assumes that Tycho is a caldera and that volcanic ash of differing composition was deposited on the flanks of Tycho; for an opposite interpretation see Urey and Marti (1968). Contrasting chemistry of pyroclastics is common in many calderas on earth —Askja and Teneriffe, for example.

Meteorite. Summary articles on meteorites are by Vinogradov (1965), Anders (1964), and Wood (1967). Texts by Geiss and Goldberg (1963), Mason (1962), and Moore (1962) include much data on the chemistry of meteorites. The field of tektites is well-covered in a text by O'Keefe (1963) and two symposia proceedings (Vols. 14 and 28) of *Geochimica et Cosmochimica Acta*.

The lunar analyses will probably cut down on the volume of future papers that contend that meteorites and tektites originate on the moon. Arguments for eucritic meteorites being similar to the material of the lunar analyses are less cogent than the argument for basalts or basaltic andesites of a terrestrial type, possibly tholeiites. This is not to argue that the lunar surface will be one of the best collecting grounds for meteorites, especially with regard to the troilite types. The relatively soft dust or ash layer on the lunar surface may cushion low velocity meteoroid impacts and preserve more of the fragile carbonaceous chondrites.

The presence of organized elements and

volatiles in carbonaceous chondrites has focussed attention on their origin. Are they the panspermatic agents for the formation of life on earth? Do they more closely represent the average composition of the solar system than noncarbonaceous chondrites? Table 7 lists abundances in carbonaceous chondrites taken from Aller (1967, pp. 287–288) and recalculated to parts per million based on a silicon content of 14.9% of Class II carbonaceous chondrites analyzed by Wiik (1956) omitting his analyses #9 and 10. Silicon abundances in other stony meteorite groups are given by Vogt and Ehmann (1965). The abundances listed by Aller are taken from data prepared by Drs. R. A. Schmitt and G. A. Goles who have modified the Class I carbonaceous chondrite data with Class II carbonaceous chondrite abundances. These Class II data values differ from the Class I abundances cited by Urey (1964) for F, Na, P, Cl, K, Sc, Ti, Cr, Zn, Ge, Rb, Sr, Y, Cd, Cs, Ba, La, Tl, Pb, Bi, and U (See Haskin et al., 1965). Others, except for Br and C, are from Urey (1964) for Class I carbonaceous chondrites. Br is a new value reported by Liebermann and Ehmann (1967) and C is from Wiik (1956) for the same analyses of Class II carbonaceous chondrites used for calculating the silicon content.

Large differences in Hg content have been measured in carbonaceous chondrites. The differences range from 0.69 to 22.2 ppm from Class III to Class I, omitting the analysis of Orgueil IA, which contains 114 ppm (Ehmann and Lovering, 1967). This high value may represent terrestrial or possibly extraterrestrial contamination, according to Ehmann and Lovering, by adsorption of Hg on the minute S crystals present in Orgueil. A less dramatic increase in Br from Class III (2.45 ppm) to Class I (5.72 ppm) is documented by Liebermann and Ehmann (1967).

In the ordinary chondrite listing of Table 7, new values include Hg averaged from eleven analyses, from Ehmann and Lovering (1967). The high and low iron group chondrite bromine values are from Liebermann and Ehmann (1967).

The direction of differentiation or alteration of the composition of the three classes of carbonaceous chondrites varies depending upon the authority cited, as does the evolutionary position of all carbonaceous chondrites to ordinary chondrites. Ringwood (1966) hypothecates Type I carbonaceous chondrites as a starting material for the origin of planetesimals. This material was subsequently devolatilized or differentiated by either internal heating after condensation and accretion or by collision of Type I carbonaceous chondrite bodies. See also J. F. Lovering, contributing to Moore (1962, pp. 179–198). Anders and Goles (1961) disagree, arguing that carbonaceous chondrites are differentiates of chondrites. The reviewer is impressed with the generally operative process of planetary defluidization that has formed our crust, oceans, and atmosphere, and believes with Anders and Goles that ordinary chondrites can defluidize to yield carbonaceous chondrite compositions and that the classes within the carbonaceous chondrites shown in Table 5 provide an index to the intensity and/or duration of the defluidization process. The Class I carbonaceous chondrites are the most enriched in the defluidized products. Possibly this Class I may represent hydrothermally altered portions of primitive crusts of planetesimals. The preference of the carbonaceous chondrites for water, the halogens, S, C, P, Ge, In, Tl, Cu, V, Cd, Hg, Au, Sn, and Pb and the major elements Na and K make this chondrite group as intriguing as the origin of chondrules themselves.

Whatever origin of chondrules is proposed it should satisfy the dichotomy of iron found in chondrites, the shape of the chondrule, its size, and the internal textures. Supercooling of molten droplets, which was first suggested by J. Wood (1967) and later elaborated upon by Blander and Katz (1967), may provide part of the answer. Diverging bladed crystals on chondrule rims, similar to those observed in borax beads, are clearly explained as supercooling phenomena. Blander and Katz provide a mechanism for the formation of chondrules at temperatures and pressures much lower and probably more plausible than those proposed by Wood, together with an explanation for the high and low iron categories.

Most workers in tektite research no longer believe tektites originate on the moon. Microtektites have been observed in nine out of

TABLE 5. VOLATILES AND CARBON IN CLASSES OF CARBONACEOUS CHONDRITES[a]

Class	Percent C	Percent S	Percent H_2O
I	2.7 – 5.0	5.2 – 6.7	18 – 22
II	1.1 – 2.8	2.3 – 3.7	8 – 17
III	0.2 – 0.6	1.8 – 2.4	0.1 – 1.5

[a] After Wood, 1967a, p. 563.

TABLE 6. COMPOSITION OF PROBABLE EXPLOSION-GENERATED GLASSES[a]

	Microtektites (53)	Philippinite tektites (21)	Tunguska spherule #753 (1)
SiO_2	68.0	70.9	60.0
Al_2O_3	15.0	13.5	17.4
FeO	5.4	5.1	3.4
MgO	4.0	2.7	2.4
CaO	3.2	3.1	5.9
Na_2O	1.1	1.4	1.5
K_2O	1.9	2.3	8.5
MnO	0.1	0.1	0.3
TiO_2	0.7	0.8	0.4
Total	99.4	99.9	99.8

[a] Analysis of Tunguska spherule by microprobe by L. Walters; of microtektites exclusive of bottle green types by microprobe by W. Cassidy, B. Glass, and B. Heezen. Philippinite analyses by C. C. Schnetzler and W. H. Pinson. All iron reported as FeO. Numbers in parentheses represent number of analyses. Data from Glass (1968, 1969).

forty cores in ocean sediments. The similarity of these objects with Australasian and Ivory Coast tektites on the basis of morphology, chemistry (assuming more vapor fractionation for the microtektites), and age (700,000 or 800,000 years) suggests a common origin (Glass, 1968, 1969). A further chemical and morphological similarity of the microtektites with one of four spherules recovered from the Tunguska impact area is given in Table 6.

A cometary explosion on "impact" with the earth's upper atmosphere is suggested by Glass as an explanation for tektites and microtektites.

Solar. Abundance values recalculated to parts per million from atoms per one million silicon atoms are from Aller (1967, pp. 287–288). These values, similar to the cosmic abundances cited in Green (1959), are not the same as meteorite abundances because the volatiles, notably hydrogen and helium, are calculated in with the totals. The values cited are the parts per million by weight of the solar atmosphere of which hydrogen, for example, composes over 70% and helium perhaps a little under 30% by weight. The value for helium is not given by Aller but is assumed by this author to be one order of magnitude less than the hydrogen abundance which is stated by Aller to be 10^{12} relative to 10^6 silicon atoms. This hydrogen value is two orders of magnitude greater than that cited by Cameron (1966, p. 8). The order of magnitude value for helium of 10^{11} atoms per 10^6 silicon atoms is based on the assumptions of Biswas et al. (1963) and arguments by Suess and Urey (1956, p. 57). However, the value for cosmic helium abundances is still quite uncertain and is under investigation by Dr. J. Greenstein at the California Institute of Technology. See also Peebles and Wilkinson (1967).

Abundances for carbon and argon are taken from Suess and Urey (1956). The value for fluorine of 10,000 atoms is an order of magnitude guess calculated from Biswas et al. (1963, p. 3118). The value is an order of magnitude greater than that estimated by Suess and Urey and an order of magnitude less than the stellar abundance cited by Aller. The chlorine value is an order of magnitude adjustment from Suess and Urey; As, Se, and Br and Kr are also taken in slightly modified form from the latter authors, and Ar is cited as given by them. The Ge solar abundance quoted by Aller appears to be low.

Errors inherent in the determination of solar abundances rest in the analytical technique. The analysis requires the measurement of intensity of absorption lines of the elements in question and a knowledge of the structure of the solar atmosphere. For each line the transition probability of the f-value must be determined. As Aller points out, for strong lines one must also know the coefficient of line broadening. All of these factors involve uncertainties. Furthermore, there may be inadequacies in the theory of spectral line formation. Some elements such as gallium, germanium, rubidium, ruthenium, and rhodium, which are represented by weak lines, can be masked or blended with lines of the more abundant elements. For a discussion of stellar abundances, see Fowler and Stephens (1968).

Conclusions

Many phases of experimental geochemistry, both basic and applied, rely on the validity of interlaboratory data. What constitutes a geochemical standard is moot. For most geochemical work, silicate standards of the type discussed by Fleischer (1965) and Flanagan (1967) are appropriate. As Taylor and Kolbe (1964) indicate, the selection of the standard depends on its use. For example, the determination of trace elements in sulfur in the parts per million range would require a hermetically sealed sample for reproducible results because of the possible adsorption of mercury on sulfur at room temperatures. The problem of contamination becomes increasingly important as the concentration levels go down and as a category becomes more specific, i.e., parts per billion of iridium in enstatite chondrules.

The processing of the standardized analytical data obtained from different laboratories has also received much emphasis over the last decade. Machine processing of geochemical data early regarded by Vistelius (1967) as a

ELEMENTS: PLANETARY ABUNDANCES AND DISTRIBUTION

TABLE 7a. ELEMENTAL CONCENTRATIONS (in ppm by wt) IN SIXTEEN GEOSPHERES[a,b]

Element-Symbol-At. No.	Ultramafic (1)	Igneous Basaltic (2)	Igneous Granitic High Ca (3)	Igneous Granitic Low Ca (4)	Syenite (5)	Sedimentary Shale (6)	Sedimentary Sandstone (7)	Sedimentary Carbonate (8)
Actinium-Ac-89	—	—	—	—	—	—	—	—
Aluminum-Al-13	2.0×10^4	7.8×10^4	8.2×10^4	7.2×10^4	7.2×10^4	8.0×10^4	2.5×10^4	4.2×10^3
Americium-Am-95	—	—	—	—	—	—	—	—
Antimony-Sb-51	10^{-1}	2×10^{-1}	2×10^{-1}	2×10^{-1}	10^{-1}	1.5	10^{-2}	2×10^{-1}
Argon-Ar-18			(2.2 × 10⁻⁵cc per gram of rock)					
Arsenic-As-33	1	2	1.9	1.5	1.4	13	1	1
Astatine-At-85	—	—	—	—	—	—	—	—
Barium-Ba-56	4×10^{-1}	3.3×10^2	4.2×10^2	8.4×10^2	1.6×10^3	5.8×10^2	10	10
Berkelium-Bk-97	—	—	—	—	—	—	—	—
Beryllium-Be-4	10^{-1}	1	2	3	1	3	10^{-1}	10^{-1}
Bismuth-Bi-83	—	7×10^{-3}	—	10^{-2}	—	—	—	—
Boron-B-5	3	5	9	10	9	10^2	35	20
Bromine-Br-35	1	3.6	4.5	1.3	2.7	4	1	6.2
Cadmium-Cd-48	10^{-1}	2.2×10^{-1}	1.3×10^{-1}	1.3×10^{-1}	1.3×10^{-1}	3×10^{-1}	10^{-2}	3.5×10^{-2}
Calcium-Ca-20	2.5×10^4	7.6×10^4	2.53×10^4	5.1×10^3	1.8×10^4	2.21×10^4	3.91×10^4	3.02×10^5
Californium-Cf-98	—	—	—	—	—	—	—	—
Carbon-C	—	—	—	—	—	—	—	—
Cerium-Ce-58	10^{-1}	48	81	92	1.6×10^2	59	92	11.5
Cesium-Cs-55	10^{-1}	1.1	2	4	6×10^{-1}	5	10^{-1}	10^{-1}
Chlorine-Cl-17	85	60	1.3×10^2	2×10^2	5.2×10^2	1.8×10^2	10	1.5×10^2
Chromium-Cr-24	1.6×10^3	1.7×10^2	22	4.1	2	90	35	11
Cobalt-Co-27	1.5×10^2	48	7	1.0	1	19	3×10^{-1}	10^{-1}
Copper-Cu-29	10	87	30	10	5	45	1	4
Curium-Cm-96	—	—	—	—	—	—	—	—
Dysprosium-Dy-66	10^{-1}	3.8	6.3	7.2	13	4.6	7.2	9×10^{-1}
Einsteinium-Es-99	—	—	—	—	—	—	—	—
Erbium-Er-68	10^{-1}	2.1	3.5	4.0	7.0	2.5	4.0	5×10^{-1}
Europeum-Eu-63	10^{-1}	8×10^{-1}	1.4	1.6	2.8	1.0	1.6	2×10^{-1}
Fermium-Fm-100	—	—	—	—	—	—	—	—
Fluorine-F-9	10^2	4×10^2	5.2×10^2	8.5×10^2	1.2×10^3	7.4×10^2	2.7×10^2	3.3×10^2
Francium-Fr-87	—	—	—	—	—	—	—	—
Gadolinium-Gd-64	10^{-1}	5.3	8.8	10	18	6.4	10	1.3
Gallium-Ga-31	1.5	17	17	17	30	19	12	4
Germanium-Ge-32	1.5	1.3	1.3	1.3	1	1.6	8×10^{-1}	2×10^{-1}
Gold-Au-79	6×10^{-3}	4×10^{-3}	4×10^{-3}	4×10^{-3}	10^{-3}	10^{-3}	10^{-3}	10^{-3}
Hafnium-Hf-72	6×10^{-1}	2.0	2.3	3.9	11	2.8	3.9	3×10^{-1}

ELEMENTS: PLANETARY ABUNDANCES AND DISTRIBUTION

Element	Col 1	Col 2	Col 3	Col 4	Col 5	Col 6	Col 7	Col 8
Helium-He-2	10^{-1}	—	—	—	—	—	—	3×10^{-1}
Holmium-Ho-67	—	1.1	1.8	2.0	3.5	1.2	2.0	10^{-2}
Hydrogen-H-1	10^{-2}	2.2×10^{-1}	$(6 \times 10^{-5}$ cc per gram of rock$)$	2.6×10^{-1}	10^{-2}	10^{-1}	10^{-2}	1.2
Indium-In-49	—	5×10^{-1}	5×10^{-1}	5×10^{-1}	5×10^{-1}	2.2	1.7	b
Iodine-I-53	5×10^{-1}	5×10^{-1}	b	b	b	b	b	—
Iridium-Ir-77	b	b	—	—	—	—	—	—
Iron-Fe-26	9.43×10^{4}	8.65×10^{4}	2.96×10^{4}	1.42×10^{4}	3.67×10^{4}	4.72×10^{4}	9.8×10^{3}	3.8×10^{3}
Krypton-Kr-36		$(4.2 \times 10^{-9}$ cc per gram of rock$)$						
Lanthanum-La-57	10^{-1}	15	45	55	70	92	30	1
Lead-Pb-92	1	6	15	19	12	20	7	9
Lithium-Li-3	10^{-1}	17	24	40	28	66	15	5
Lutetium-Lu-71	10^{-1}	6×10^{-1}	1.1	1.2	2.1	7×10^{-1}	1.2	2×10^{-1}
Magnesium-Mg	2.04×10^{5}	4.6×10^{4}	9.4×10^{3}	1.6×10^{3}	5.8×10^{3}	1.5×10^{4}	7×10^{3}	4.7×10^{4}
Manganese-Mn-25	1.62×10^{3}	1.5×10^{3}	5.4×10^{2}	3.9×10^{2}	8.5×10^{2}	8.5×10^{2}	10^{2}	1.1×10^{3}
Mendelevium-Md-101	—	—	—	—	—	—	—	—
Mercury-Hg-80	(4×10^{-3})†	(7×10^{-3})†	(2.1×10^{-2})†	(3.9×10^{-2})†	10^{-2}	4×10^{-1}	3×10^{-2}	4×10^{-2}
Molybdenum-Mo-42	3×10^{-1}	1.5	1.0	1.3	6×10^{-1}	2.6	2×10^{-1}	4×10^{-1}
Neodymium-Nd-60	10^{-1}	20	33	37	65	24	37	4.7
Neptunium-Np-93	—	—	—	—	—	—	—	—
Neon-Ne-10			$(7.7 \times 10^{-8}$ cc per gram of rock$)$					
Nickel-Ni-28	2×10^{3}	1.3×10^{2}	15	4.5	4	68	2.0	20
Niobium-Nb-41	16	19	20	21	35	11	10^{-2}	3×10^{-1}
Nitrogen-N-7	6	20	20	20	30	—	—	—
Nobelium-No-102	—	—	—	—	—	—	—	—
Osmium-Os-76	b	b	b	b	b	b	b	b
Oxygen-O-8	4.3×10^{5}	4.4×10^{5}	4.8×10^{5}	4.8×10^{5}	4.8×10^{5}	4.8×10^{5}	5.0×10^{5}	5.0×10^{5}
Palladium-Pd-46	5×10^{-3}	2×10^{-3}	10^{-3}	10^{-3}	—	—	—	b
Phosphorus-P-15	2.2×10^{2}	1.1×10^{3}	9.2×10^{2}	6.0×10^{2}	8.0×10^{2}	7.0×10^{2}	1.7×10^{2}	4.0×10^{2}
Platinum-Pt-78	b	b	b	b	b	b	b	b
Plutonium-Pu-94	—	—	—	—	—	—	—	—
Polonium-Po-84	—	—	—	—	—	—	—	—
Potassium-K-19	40	8.3×10^{3}	2.52×10^{4}	4.2×10^{4}	4.8×10^{4}	2.66×10^{4}	1.07×10^{4}	2.7×10^{3}
Praseodymium-Pr-59	10^{-1}	4.6	7.7	8.8	15	5.6	8.8	1.1
Promethium-Pm-61	—	—	—	—	—	—	—	—
Protactinium-Pa-91	—	—	—	—	—	—	—	—
Radium-Ra-88	—	—	—	—	—	—	—	—
Radon-Rn-86	—	—	—	—	—	—	—	—
Rhenium-Re-75	b	b	b	b	b	b	b	b
Rhodium-Rh-45	b	b	b	b	b	b	b	b
Rubidium-Rb-37	2×10^{-1}	30	1.1×10^{2}	1.7×10^{2}	1.1×10^{2}	1.4×10^{2}	60	3
Ruthenium-Ru-44	b	b	b	b	b	b	b	b
Samarium-Sm-62	10^{-1}	5.3	8.8	10	18	6.4	10	1.3
Scandium-Sc-21	15	30	14	7	3	13	1	1

TABLE 7a. (Continued)

Element-Symbol-At. No.	Igneous					Sedimentary		
	Ultramafic (1)	Basaltic (2)	Granitic		Syenite (5)	Shale (6)	Sandstone (7)	Carbonate (8)
			High Ca (3)	Low Ca (4)				
Selenium-Se-34	5×10^{-2}	5×10^{-2}	5×10^{-2}	5×10^{-2}	5×10^{-2}	6×10^{-1}	5×10^{-2}	8×10^{-2}
Silicon-Si-14	2.05×10^5	2.30×10^5	3.14×10^5	3.47×10^5	2.91×10^5	7.3×10^4	3.68×10^5	2.4×10^4
Silver-Ag-47	6×10^{-2}	1.1×10^{-1}	5.1×10^{-2}	3.7×10^{-2}	10^{-2}	7×10^{-2}	10^{-2}	10^{-2}
Sodium-Na-11	4.2×10^3	1.8×10^4	2.84×10^4	2.58×10^4	4.04×10^4	9.6×10^3	3.3×10^3	4.0×10^2
Strontium-Sr-38	1.0	4.65×10^2	4.4×10^2	1.0×10^2	2.0×10^2	3.0×10^2	20	6.1×10^2
Sulfur-S-16	3.0×10^2	3.0×10^2	3.0×10^2	3.0×10^2	3.0×10^2	2.4×10^3	2.4×10^2	1.2×10^3
Tantalum-Ta-73	1.0	1.1	3.6	4.2	2.1	8.0×10^{-1}	10^{-2}	10^{-2}
Technetium-Tc-43	b	b	b	b	b	b	b	b
Tellurium-Te-52	—	—	—	—	—	—	—	—
Terbium-Tb-65	10^{-1}	8.0×10^{-1}	1.4	1.6	2.8	1.0	1.6	2.0×10^{-1}
Thallium-Tl-81	6×10^{-2}	2.1×10^{-1}	7.2×10^{-1}	2.3	1.4	1.4	8.2×10^{-1}	10^{-2}
Thorium-Th-90	4×10^{-3}	4	8.5	17	13	12	1.7	1.7
Thulium-Tm-69	10^{-1}	2×10^{-1}	3×10^{-1}	3×10^{-1}	6×10^{-1}	2×10^{-1}	3×10^{-1}	4×10^{-2}
Tin-Sn-50	5×10^{-1}	1.5	1.5	3	1	6.0	10^{-1}	10^{-1}
Titanium-Ti-22	3.0×10^2	1.38×10^4	3.4×10^3	1.2×10^3	3.5×10^3	4.6×10^3	1.5×10^3	4×10^2
Tungsten-W-74	7.0×10^{-1}	7.0×10^{-1}	1.3	2.2	1.3	1.8	1.6	6.0×10^{-1}
Uranium-U-92	10^{-3}	1.0	3.0	3.0	3.0	3.7	4.5×10^{-1}	2.2
Vanadium-V-23	40	2.5×10^2	88	44	30	1.3×10^2	20	20
Xenon-Xe-54		(3.4×10^{-10} cc per gram of rock)						
Ytterbium-Yb-70	10^{-1}	2.1	3.5	4.0	7.0	2.6	4.0	5.0×10^{-1}
Yttrium-Y-39	10^{-1}	21	35	40	20	26	40	30
Zinc-Zn-30	50	1.05×10^2	60	39	1.3×10^2	95	16	20
Zirconium-Zr-40	45	1.4×10^2	1.4×10^2	1.75×10^2	5.0×10^2	1.6×10^2	2.2×10^2	19

Note: See footnotes in Table 7b.

TABLE 7b. ELEMENTAL CONCENTRATIONS (in ppm by wt) IN SIXTEEN GEOSPHERES[a,b]

Element-Symbol-At. No.	Pelagic Carbonate (9)	Oceanic Clay (10)	Ocean Water (11)	Lunar Maria (12)	Lunar Highland* (13)	Chondrite (14)	Carbonaceous Chondrite (15)	Solar (16)
Actinium-Ac-89	—	—	—	—	—	—	—	—
Aluminum-Al-13	2.0×10^4	8.4×10^4	10^{-2}	—	1.1×10^5	1.1×10^4	1.2×10^4	9.3×10^{-1}
Americium-Am-95	—	—	—	—	—	—	—	—
Antimony-Sb-51	1.5×10^{-1}	1.0	5×10^{-4}	—	—	8.6×10^{-2}	—	6.9×10^{-6}
Argon-Ar-18	—	—	6×10^{-1}	—	—	—	—	4.3
Arsenic-As-33	1	13	3×10^{-3}	—	—	2.3	—	2.1×10^{-4}
Astatine-At-85	—	—	—	—	—	—	—	—
Barium-Ba-56	1.9×10^2	2.3×10^3	3×10^{-2}	—	—	4.0^b	2.8	9.9×10^{-4}
Berkelium-Bk-97	—	—	—	—	—	—	—	—
Beryllium-Be-4	10^{-1}	2.6	6×10^{-7}	—	—	3.3×10^{-2}	—	4.4×10^{-5}
Bismuth-Bi-83	—	—	2×10^{-5}	—	—	2×10^{-3}	2.0×10^{-1}	1.9×10^{-4}
Boron-B-5	55	2.3×10^2	4.6	—	—	4.3×10^{-1}	—	7.4×10^{-4}
Bromine-Br-35	70	70	65	—	—	$(1.7 \times 10^{-1})_L$† $(1.7 \times 10^{-1})_H$†	2.5†	—
Cadmium-Cd-48	10^{-2}	4.2×10^{-1}	1.1×10^{-4}	—	—	5.9×10^{-2}	8.3×10^{-1}	1.1×10^{-4}
Calcium-Ca-20	3.2×10^5	2.9×10^4	4×10^2	—	1.1×10^4	1.3×10^4	1.5×10^4	1
Californium-Cf-98	—	—	—	—	—	—	—	—
Carbon-C	35	3.45×10^2	28	—	—	1.1	(2.8×10^4)†	1.4×10^2
Cerium-Ce-58	4×10^{-1}	6	5.2×10^{-6}	—	—	1.2×10^{-1}	8.7×10^{-1}	—
Cesium-Cs-55	2.1×10^4	2.1×10^4	5×10^{-4}	—	—	$(1.6 \times 10^2)^b$	1.7×10^{-1}	2.5×10^{-1}
Chlorine-Cl-17	11	90	1.9×10^4	—	—	2.4×10^3	3.8×10^2	9.3×10^{-2}
Chromium-Cr-24	7	74	5×10^{-5}	—	—	7.2×10^2	3.3×10^3	6.6×10^{-2}
Cobalt-Co-27	30	2.5×10^2	10^{-4}	—	—	99	6.9×10^2	4.5×10^{-3}
Copper-Cu-29	—	—	3×10^{-3}	—	—	—	1.3×10^2	—
Curium-Cm-96	—	—	—	—	—	—	—	—
Dysprosium-Dy-66	2.7	27	2.9×10^{-6}	—	—	3.5×10^{-1}	3.1×10^{-1}	—
Einsteinium-Es-99	—	—	—	—	—	—	—	—
Erbium-Er-68	1.5	15	2.4×10^{-6}	—	—	2.5×10^{-1}	2.0×10^{-1}	—
Europeum-Eu-63	6×10^{-1}	6	4.6×10^{-7}	—	—	8.0×10^{-2}	7.3×10^{-2}	—
Fermium-Fm-100	—	—	—	—	—	—	—	—
Fluorine-F-9	5.4×10^2	1.3×10^3	1.3	—	—	88^b	1.0×10^3	1.4×10^{-1}
Francium-Fr-87	—	—	—	—	—	—	—	—
Gadolinium-Gd-64	3.8	38	2.4×10^{-6}	—	—	3.5×10^{-1}	4.6×10^{-1}	—
Gallium-Ga-31	13	20	3×10^{-5}	—	—	5.5	—	9×10^{-4}
Germanium-Ge-32	2	2	6×10^{-5}	—	—	9.6^b	19	5.2×10^4
Gold-Au-79	(8×10^{-3})†	(2.6×10^{-3})†	4×10^{-6}	—	—	1.7×10^{-1}	1.9×10^{-1}	—

ELEMENTS: PLANETARY ABUNDANCES AND DISTRIBUTION

TABLE 7b. (Continued)

Element-Symbol-At. No.	Oceanic Pelagic Carbonate (9)	Oceanic Pelagic Clay (10)	Ocean Water (11)	Extraterrestrial Lunar Maria (12)	Extraterrestrial Lunar Highland* (13)	Extraterrestrial Chondrite (14)	Extraterrestrial Carbonaceous Chondrite (15)	Extraterrestrial Solar (16)
Hafnium-Hf-72	4.1×10⁻¹	4.1	—	—	—	1.9×10⁻¹	2.9×10⁻¹	—
Helium-He-2	—	—	5×10⁻⁶	—	—	—	—	2.85×10⁵
Holmium-Ho-67	8×10⁻¹	7.5	8.8×10⁻⁷	—	—	8.2×10⁻²	7.9×10⁻²	7.14×10⁵
Hydrogen-H-1	—	—	1.08×10⁵	—	—	—	—	7.3×10⁻⁵
Indium-In-49	10⁻²	8×10⁻²	<2×10⁻²	—	—	8.6×10⁻⁴	—	—
Iodine-I-53	5×10⁻²	5×10⁻²	6×10⁻²	—	—	4.1×10⁻²	3.1×10⁻¹	49
Iridium-Ir-77	b	(<10⁻⁴)†	b	—	—	4.2×10⁻¹	—	—
Iron-Fe-26	9×10³	6.5×10⁴	10⁻²	—	5.1×10⁴	(2.1×10⁵)_L (3.0×10⁵)_H	2.4×10⁵	—
Krypton-Kr-36	—	—	3×10⁻⁴	—	—	—	—	3.1×10⁻³
Lanthanum-La-57	10	1.15×10²	1.2×10⁻⁵	—	—	3.6×10⁻¹	3.6×10⁻¹	—
Lead-Pb-92	9	8.0	3×10⁻⁵	—	—	1.9×10⁻¹	1.8×10⁻¹	2.1×10⁻⁴
Lithium-Li-3	5	57	1.7×10⁻¹	—	—	2.4	—	5.4×10⁻⁶
Lutetium-Lu-71	5×10⁻¹	4.5	4.8×10⁻⁷	—	—	3.6×10⁻²	3.2×10⁻²	—
Magnesium-Mg	4×10³	2.1×10⁴	1.35×10³	—	4.4×10⁴	1.5×10⁴	1.3×10⁴	1.4
Manganese-Mn-25	10³	6.7×10³	2×10⁻³	—	—	2.7×10³	1.9×10³	7.9×10⁻²
Mendelevium-Md-101	—	—	—	—	—	—	—	—
Mercury-Hg-80	10⁻²	10⁻¹	3×10⁻⁵	—	—	(1.2×10⁻¹)†	12	—
Molybdenum-Mo-42	3	27	10⁻²	—	—	1.6	—	4.3×10⁻⁶
Neodymium-Nd-60	14	1.4×10²	9.2×10⁻⁶	—	—	6.1×10⁻¹	5.9×10⁻¹	—
Neptunium-Ne-93	—	—	—	—	—	—	—	—
Neon-Ne-10	—	—	10⁻⁴	—	—	—	—	64
Nickel-Ni-28	30	2.25×10²	2×10⁻³	—	—	(1.1×10⁴)_L (1.7×10⁴)_H	1.4×10⁴	1.1
Niobium-Nb-41	4.6	14	10⁻⁵	—	—	—	—	4.2×10⁻⁴
Nitrogen-N-7	—	—	5×10⁻¹	—	—	—	—	30
Nobelium-No-102	b	—	b	—	—	—	—	—
Osmium-Os-76	—	—	—	—	—	8.8×10⁻¹	6.7×10⁻¹	—
Oxygen-O-8	4.7×10⁵	4.7×10⁵	8.57×10⁵	—	4.2×10⁵	3.5×10⁵	—	3.3×10²
Palladium-Pd-46	7×10⁻³	2.8×10⁻³	—	—	—	(7×10¹—)ᵇ	—	4.5×10⁻⁵
Phosphorus-P-15	3.5×10²	1.5×10³	7×10⁻²	—	—	(1.1×10³)ᵇ	1.3×10³	1.5×10⁻¹
Platinum-Pt-78	b	b	b	—	—	1.2	1.7	—
Plutonium-Pu-94	—	—	—	—	—	—	—	—
Polonium-Po-84	—	—	—	—	—	—	—	—
Potassium-K-19	2.9×10³	2.5×10⁴	3.8×10²	d	(1.3×10⁴)ᵇ	(9.2×10²)ᵇ	5.2×10²	4.4×10⁻²

ELEMENTS: PLANETARY ABUNDANCES AND DISTRIBUTION

Element	1	2	3	4	5	6	7	8	9	10	11
Praseodymium-Pr-59	3.3	—	—	—	—	—	2.6×10^{-6}	—	—	1.3×10^{-1}	—
Promethium-Pm-61	—	—	—	—	—	—	—	—	—	—	—
Protactinium-Pa-91	—	—	—	—	—	—	—	—	—	—	—
Radium-Ra-88	—	—	—	—	—	—	1.1×10^{-10}	—	—	—	—
Radon-Rn-86	—	—	—	—	—	—	6×10^{-16}	—	—	—	—
Rhenium-Re-75	b	—	—	—	—	—	b	—	6×10^{-2} $(1.6 \times 10^{-1})_L$ $(2.2 \times 10^{-1})_H$	—	—
Rhodium-Rh-45	b	—	—	—	—	—	b	—	—	—	5.4×10^{-5}
Rubidium-Rb-37	10	—	—	—	—	—	1.2×10^{-1}	—	2.5^b	1.9	5.8×10^{-4}
Ruthenium-Ru-44	b	—	—	—	—	—	b	—	$(7.2 \times 10^{-1})_L$ $(1.1)_H$	8.0×10^{-1}	1.5×10^{-4}
Samarium-Sm-62	3.8	—	—	—	—	—	1.7×10^{-6}	—	2.3×10^{-1}	1.8×10^{-1}	—
Scandium-Sc-21	2	—	—	—	—	—	4×10^{-5}	—	8.9	9.6	6.4×10^{-4}
Selenium-Se-34	1.7×10^{-1}	—	—	—	—	—	4×10^{-4}	—	8.5	—	3.8×10^{-3}
Silicon-Si-14	3.2×10^4	—	—	—	—	2.3×10^5	3	—	1.85×10^5	1.49×10^5	20
Silver-Ag-47	10^{-2}	—	—	—	—	—	4×10^{-5}	—	$(3.6 \times 10^{-2})^b$	—	1.5×10^{-5}
Sodium-Na-11	2.0×10^4	—	—	—	—	$(2.5 \times 10^4)^d$	1.05×10^4	b	7.0×10^3	4.6×10^3	1
Strontium-Sr-38	2.0×10^3	—	—	—	—	—	8.0	—	12	12	9.3×10^{-3}
Sulfur-S-16	1.3×10^3	—	—	—	—	—	8.85×10^2	—	2.3×10^4	8.6×10^4	14
Tantalum-Ta-73	10^{-2}	—	—	—	—	—	—	—	2.4×10^{-2}	1.9×10^{-2}	—
Technetium-Tc-43	—	—	—	—	—	—	—	—	—	—	—
Tellurium-Te-52	b	—	—	—	—	—	b	—	5.6×10^{-1}	2.2	—
Terbium-Tb-65	6.0×10^{-1}	—	—	—	—	—	6.0	—	5.3×10^{-2}	3.1×10^{-2}	—
Thallium-Tl-81	1.6×10^{-1}	—	—	—	—	—	8×10^{-1}	—	$(9.2 \times 10^{-4})^b$	1.1	—
Thorium-Th-90	1	—	—	—	—	—	7	—	4.1×10^{-2}	(6.2×10^{-2})†	—
Thulium-Tm-69	10^{-1}	—	—	—	—	—	1.2	—	3.4×10^{-2}	3.1×10^{-2}	—
Tin-Sn-50	10^{-1}	—	—	—	—	—	1.5	—	4.3×10^{-1}	1.1	4.8×10^{-4}
Titanium-Ti-22	7.7×10^2	—	—	—	—	—	4.6×10^3	—	7.9×10^2	8.0×10^2	4.1×10^{-2}
Tungsten-W-74	10^{-1}	—	—	—	—	—	1	—	$(6.1 \times 10^{-2})^b$	—	—
Uranium-U-92	10^{-1}	—	—	—	—	—	1.3	—	$(1.4 \times 10^{-2})^b$	2.0×10^{-2}	—
Vanadium-V-23	20	—	—	—	—	—	1.2×10^2	—	66	80	1.5×10^{-2}
Xenon-Xe-54	—	—	—	—	—	—	—	—	—	—	—
Ytterbium-Yb-70	1.5	—	—	—	—	—	2.0×10^{-6}	—	2.0×10^{-1}	1.9×10^{-1}	$(1.3 \times 10^{-4})^b$
Yttrium-Y-39	42	—	—	—	—	—	3×10^{-4}	—	2.0	2.2	3.2×10^{-4}
Zinc-Zn-30	35	—	—	—	—	—	10^{-2}	—	55	2.0×10^2	9.3×10^{-3}
Zirconium-Zr-40	20	—	—	—	—	—	1.5×10^2	—	12	11	9.1×10^{-3}

[a] Columns 1 — 10 (Turekian and Wedepohl, 1961, their Table 2); Column 11 (Goldberg, 1965, p. 164-165); column 12, see Table 8 (recalculated from Turkevich, et al., 1969); and columns 14 — 16 (recalculated from Aller, 1967, p. 287-288).
[b] Uncertain values as reported by various authorities.
[c] † Data not in original compilations; referenced in text.
— No data.
L and H refer to low and high iron classes of chondrites.
V, VI, and VII refer to Surveyor missions.
The elements promethium, technetium, and astatine do not occur in nature.
[d] See qualifications cited in text.

ELEMENTS: PLANETARY ABUNDANCES AND DISTRIBUTION

Table 8. Lunar Mare Analyses (Apollo 11)[a]

Element[b]	A-Vesicular	B-Crystalline	C-Breccia	D-Fines
Aluminum	$51000_4{}^{8000}$	$62000_4{}^{8000}$	$58000_2{}^{0}$	$69000_1{-}$
Barium	$90_4{}^{60}$	$95_4{}^{30}$	$98_2{}^{10}$	$68_1{-}$
Calcium	$68500_4{}^{3400}$	$72000_4{}^{2000}$	$79000_2{}^{0}$	$86000_1{-}$
Chromium	$4000_4{}^{190}$	$4150_3{}^{650}$	$2800_2{}^{350}$	$2500_1{-}$
Cobalt	$13_4{}^{8}$	$9_4{}^{2}$	$13_2{}^{1}$	$18_1{-}$
Copper	$5_2{}^{0}$	$5_3{}^{3}$	$8_1{-}$	$+$
Gallium	$5_1{-}$	$6_2{}^{3}$	$+$	$+$
Iron	$146000_4{}^{14000}$	$143000_4{}^{10500}$	$136000_2{}^{17000}$	$124000_1{-}$
Lithium	$15.5_4{}^{4.5}$	$17.5_4{}^{6}$	$12.5_1{-}$	$15.0_1{-}$
Magnesium	$48000_4{}^{2300}$	$48000_4{}^{3000}$	$50000_2{}^{6400}$	$48000_1{-}$
Manganese	$2770_4{}^{760}$	$3250_4{}^{1000}$	$2100_2{}^{500}$	$1750_1{-}$
Nickel	$86_4{}^{160}$	$14_2{}^{8}$	$225_2{}^{14}$	$250_1{-}$
Oxygen	$416000_4{}^{8000}$	$418000_4{}^{7900}$	$426000_2{}^{7000}$	$424000_1{-}$
Potassium	$1360_4{}^{560}$	$1020_4{}^{170}$	$1350_2{}^{210}$	$1000_1{-}$
Rubidium	$4.7_3{}^{3.0}$	$2.6_4{}^{4.0}$	$3.1_1{-}$	$2.2_1{-}$
Scandium	$94_4{}^{32}$	$110_4{}^{50}$	$62_2{}^{8}$	$55_1{-}$
Silicon	$189000_4{}^{19400}$	$190000_4{}^{9800}$	$194000_2{}^{9200}$	$20000_1{-}$
Sodium	$4000_4{}^{660}$	$4100_3{}^{470}$	$2600_2{}^{1600}$	$4000_1{-}$
Strontium	$120_4{}^{80}$	$110_4{}^{80}$	$105_2{}^{65}$	$90_1{-}$
Titanium	$68000_4{}^{6600}$	$56000_4{}^{7500}$	$53000_2{}^{3500}$	$42000_1{-}$
Vanadium	$32_4{}^{9}$	$46_4{}^{23}$	$27_2{}^{7}$	$42_1{-}$
Ytterbium	$4.4_4{}^{2.5}$	$3.0_3{}^{1.9}$	$3.2_2{}^{1.8}$	$2.5_1{-}$
Yttrium	$230_4{}^{50}$	$190_4{}^{100}$	$200_2{}^{130}$	$130_1{-}$
Zirconium	$>1200_3{}^{>530}$	$730_4{}^{410}$	$950_2{}^{780}$	$400_1{-}$

[a] Apollo 11 return data for Mare Tranquillitatis samples (categories A, B, C, D) reported by Anderson et al. (1969) are given above. The subscript indicates the number of samples involved in the average and the superscript is the standard deviation. All data are in parts per million by weight.
For additional geochemical data on lunar samples see NASA SP-235, Apollo 12 Preliminary Science Report, 1970, Washington, D.C., 227pp. and NASA 1971 Lunar Science Conference Abstracts, Houston, 318 pp.
[b] In calculating the standard deviation for copper, the Cu value in original sample 20 was taken at 5 ppm instead of the reported 4.5. The greater than symbol for Zr results from the reported value of >2000 ppm in original sample 57. Nickel was assumed to be zero where reported "not detected" in samples 17, 20, 28, 45, and 72. $+$ indicates no data.

means of correlation with other geological parameters is now being undertaken on a large scale by the United States Geological Survey (Miesch, 1969).

The trends shown by machine analysis of analytical data either in support of lognormal (Ahrens, 1966) or "normal" (Butler, 1964) element distributions in igneous rocks are sometimes controversial not because of the machine technique of analysis or the quality of the analyses but because of the nature of the sample selection. This is refreshing in that the geochemist becomes as important as his data. Machine analysis of abundance values can define a minor unit within a subgeosphere as Manson (cited in Hess and Poldervaart, 1967, Vol. I, pp. 215–269) has done or in revising geosphere and subgeosphere abundances as demonstrated by Horn and Adams (1966). The importance of manganese in the defluidization cycle is emphasized by the computer technique of Horn and Adams.

After better defining abundance values by machine techniques in categories of interest to the academician, the oil and mining geologists have become increasingly aware of the importance of statistical analysis of elemental distributions in prospecting. Only recently have regional geochemical maps been published. In the "1:5,000,000 Hydrogeochemical Map of the U.S.S.R." published in 1966, the abundance of certain trace elements in thermal waters can be correlated with thermal anomalies, loci of structural folds, seeps, and mud volcanoes.

Data on geochemical prospecting are probably equally divided between the open and proprietary literature. Competitive methods for detection of geochemical anomalies usually remain unpublished. In many cases where the institutional objective is fulfilled, the data still do not become available because of the time and effort to compile, organize, and interpret the data for publication.

Two texts on geochemical prospecting of general interest include the one by Ginzburg (1960) and Kvalheim (1967). Both authors cite case histories of the concentration of a given element (P, Au, Fe, Pb, etc.) in the finest size fraction of the host material. On the other hand, agglutination processes may concentrate certain elements in the "coarse" fraction, i.e.

ELEMENTS: PLANETARY ABUNDANCES AND DISTRIBUTION

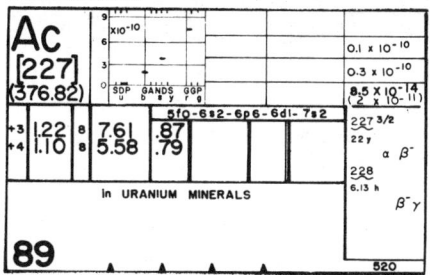
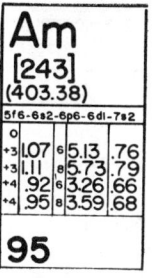
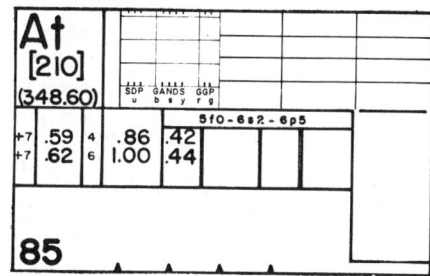

ELEMENTS: PLANETARY ABUNDANCES AND DISTRIBUTION

ELEMENTS: PLANETARY ABUNDANCES AND DISTRIBUTION

CHROMIUM
EINSTEINIUM
COBALT
ERBIUM
COPPER
EUROPIUM
CURIUM
FERMIUM
DYSPROSIUM
FLUORINE

ELEMENTS: PLANETARY ABUNDANCES AND DISTRIBUTION

FRANCIUM
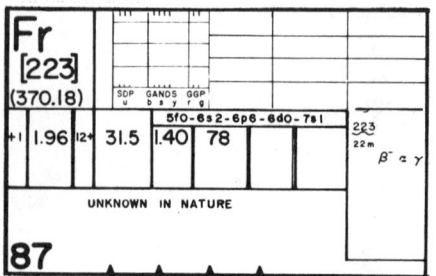

GADOLINIUM
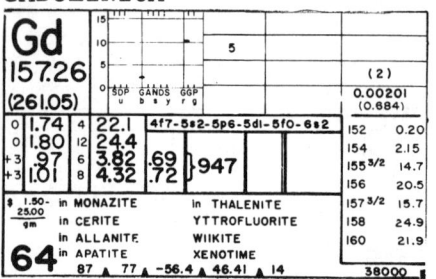

GALLIUM
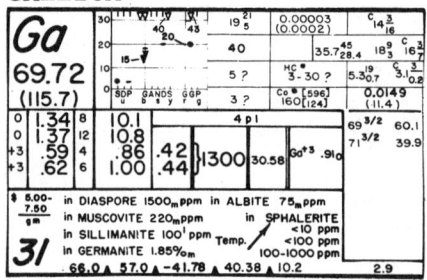

GERMANIUM
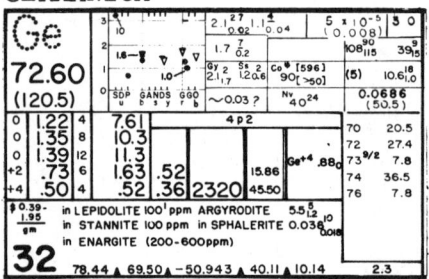

GOLD
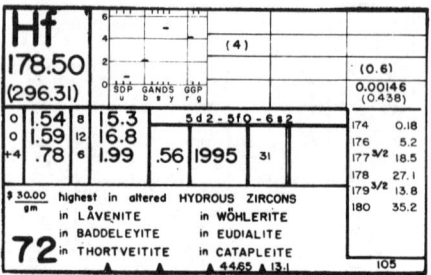

HAFNIUM
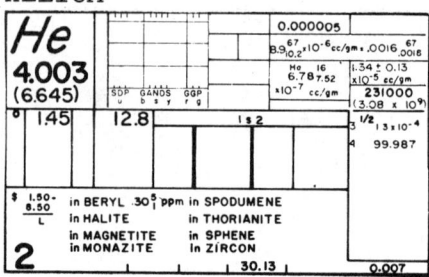

HELIUM
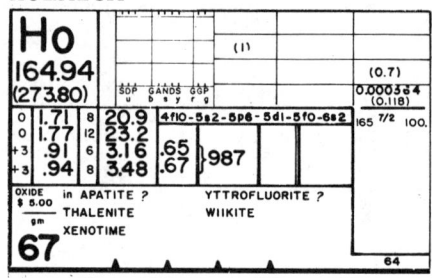

HOLMIUM
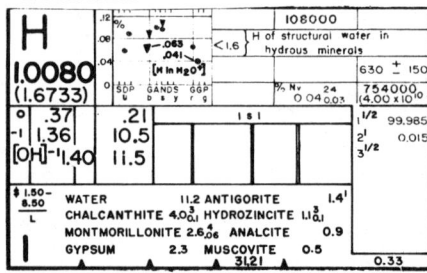

HYDROGEN

INDIUM

ELEMENTS: PLANETARY ABUNDANCES AND DISTRIBUTION

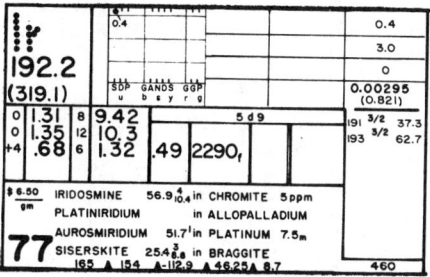

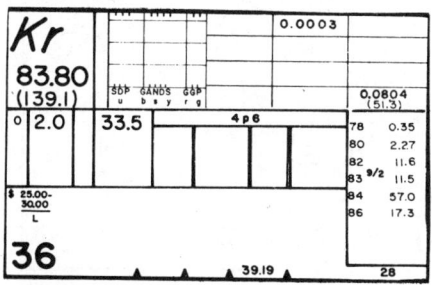

289

ELEMENTS: PLANETARY ABUNDANCES AND DISTRIBUTION

ELEMENTS: PLANETARY ABUNDANCES AND DISTRIBUTION

OSMIUM
PLUTONIUM
OXYGEN
POLONIUM
PALLADIUM
POTASSIUM
PHOSPHORUS
PRASEODYMIUM
PLATINUM
PROMETHIUM

ELEMENTS: PLANETARY ABUNDANCES AND DISTRIBUTION

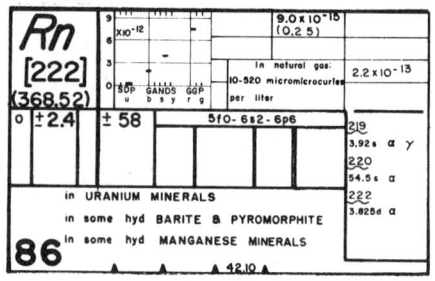

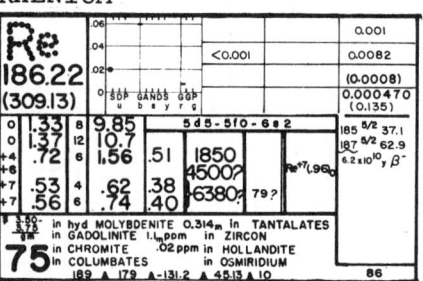

ELEMENTS: PLANETARY ABUNDANCES AND DISTRIBUTION

SILICON
SILVER
SODIUM
STRONTIUM
SULFUR

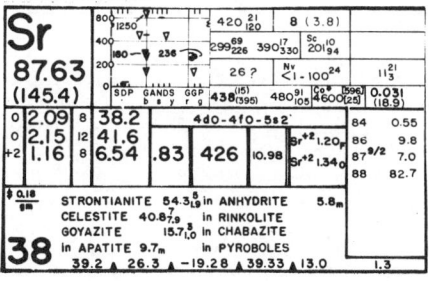

TANTALUM
TECHNETIUM
TELLURIUM
TERBIUM
THALLIUM

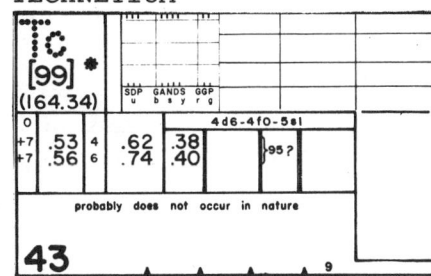
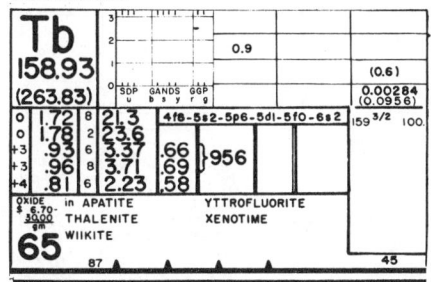

ELEMENTS: PLANETARY ABUNDANCES AND DISTRIBUTION

THORIUM

URANIUM

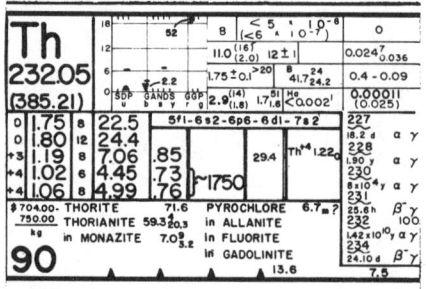

THULIUM

VANADIUM

TIN

XENON

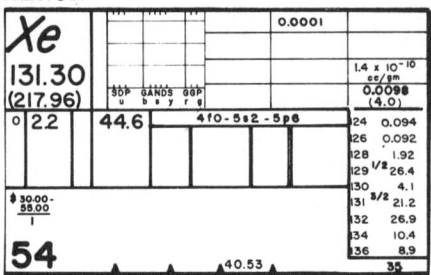

TITANIUM

YTTERBIUM

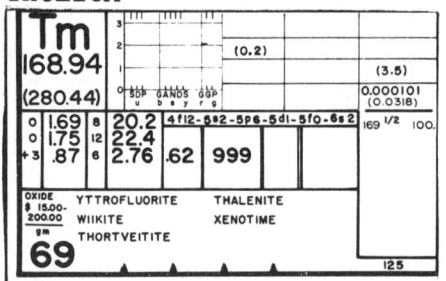

TUNGSTEN

YTTRIUM

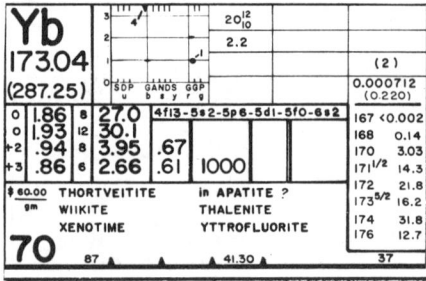

ELEMENTS: PLANETARY ABUNDANCES AND DISTRIBUTION

ZINC

Zn	150 8500	~38	66 42	0.01 (0.021)	1530		
65.38	100 81		80 17 200 100	10 35 12	115		
(108.5)	50		Sn 5 Gy 7 In 2 19 15 25 23 18 16	Co 2	~100?		
	0 SDP GNDS GSP		b y r g	19 11 20 11 73	40 96 410 [>24]	0.594 (486)	
				4s2		64	48.9
0	1.33	8	9.85			Zn⁺² .85	66	27.8
0	1.36	12	1.05			Zn⁺² 1.05	67 5/2	4.1
+2	.71	4	1.50	.51	610	17.89	Zn⁺² 1.19	68	18.6
+2	.74	6	1.70	.53			Zn⁺² 3.42	70	0.63

$ 0.93 – NATIVE (?) Zn 100 HEMIMORPHITE 54.2
1.19   ZINCITE 76.1 2.3 SMITHSONITE 40.0 10
 kg   SPHALERITES 58.7 17 FRANKLINITE 16.7 3.4
         WILLEMITE 58.6 GOSLARITE 15.7 3.5

30 | 31.19 | 22.69 | -16.63 | 38.45 | 9.95 | | 1.10

ZIRCONIUM

Zr	300 700	3000	GI 3 16 2 170 40	132 2 4	0.004 (0.002)		
91.22	200 190		200 137 24 57 63 1080	Sc 4 352 Q 2 11	8		
(151.4)	100 110		Gy 4 Ss 11 Er 2 596 Co 3 14 89 17 9 15 3 210 [10] 51 47	33	37 21 13		
	0 SDP GNDS GSP		b y r g	Do 2 La 14 Ha 4 An 5 17 1 20 11 0.2 0.2 7 16	0.0930 (54.5)		
				4d2-4f0-5s2		90	51.5
0	1.55	8	15.6				91 5/2	11.2
0	1.60	12	17.2			Zr⁺⁴ 1.08a	92	17.11
+4	.79	4	2.07	.56	1980	33.83		94	17.4
+4	.82	6	2.31	.58				96	2.8

$ 35.20 – BADDELEYITE 72.2 4 0.7 in PYROCHLORE 4.2 m
  kg   ZIRCON 49.7 EUDIALITE
          ZIRKELITE 39.2 1 LÅVENITE
          CATAPLEITE ±23 in THORTVEITITE

40 | 125 | 115 | -84.29 | 43.31 | 9.18 | | 0.18

⟺ KEY ⟺

	ABUNDANCES	Deep Sea Sediments	Ocean (Transfer Percentage)	Meteoritic Sulfides C = sulfide phase of chondrites
Element S-SIDEROPHILE C-CHALCOPHILE A-ATMOPHILE L-LITHOPHILE B-BIOPHILE and LITHOPHILE	In ppm unless indicated otherwise (FOR SPECIFIC ROCK TYPES SEE ABBREVIATIONS) Arithmetical Average — Ab — Number of Analyses C — Standard Deviation (variance ratio in brackets) General overall average — Specific { intrusive • GI ● extrusive • WI ▼ IGNEOUS ROCK TYPES	Shales	Metamorphic Rocks	Meteoritic Irons C = iron phase of chondrites
Atomic Weight Mass of isotope of longest half life; [*]ᵃ – Mass of isotope of best known half life		Sandstones	Other	Chondrites C = silicate phase of chondrites
(#)Weight of one atom x10⁻²⁴ grams		Carbonates	Other	Universe including volatiles (Atoms per 10⁶ Si Atoms)

Charge	Radius (Å) Radii listed as function of coordinations. Ionic radii based on 1.40 Å for O⁻²	Coordination	Volume (cubic Å)	Electrons in Each Quantum Group QUANTUM GROUPS 1s thru 7s CORRESPOND TO SHELLS K thru Q; QUANTUM GROUPS FILLED BEFORE 1st GROUP LISTED AND EMPTY AFTER LAST GROUP LISTED (5g zero throughout)				Naturally Occurring Isotopes
				Radius Ratio cation/oxygen	Lattice Energy Coefficient	Ionization Potential (volts)	Miscellaneous Radii Subscript keyed to list of abbreviations ( )-Extrapolated value [ ]-Calculated value	A# % Relative Abundance Nuclear spin in multiples of h/2π indicated by superscript. RADIOACTIVE RADIATION ISOTOPE DECAY ∼∼∼ HALF-LIFE d – Days α – Alpha particle decay h – Hours m – Minutes β – Negative beta decay s – Seconds y – Years γ – Gamma ray (subsequent radiation) e⁻ – Prominent internal conversion electron IT- Isomeric transition K – K electron capture L – L electron capture

Cost Range (reagent purity) | Common, independent, geochemically significant and/or host minerals with % element contained Arithmetic average indicated where superscript indicates number of individual analyses and subscript indicates standard deviation. All other values calculated from idealized mineral formula.

| Atomic Number (slanted for "dispersed" elements) | STANDARD HEAT OF FORMATION ΔHf° (GAS) kg-cal/mole | STANDARD FREE ENERGY OF FORMATION ΔFf° (GAS) kg-cal/mole | LOGARITHM OF EQUILIBRIUM CONSTANT OF FORMATION (25°C) log Kf (GAS) | ENTROPY AT 25°C S° (GAS) cal/degree-mole | ENTROPY AT 25°C S° (SOLID) cal/degree-mole | Thermal neutron absorption cross section in barns |

Pt during laterization of ultramafics (Augustithis, 1967, p. 324). Other "rules" in prospecting are derived by the empirical approach, e.g., that the distribution of vanadium, chromium, nickel, cobalt, and copper increases from sandstones to argillites and decreases with the dilution of sediments by carbonates. The use of remote sensors in disclosing the subtle tonal changes in vegetation and drainage that might reflect these petrographic changes would thus lead to prediction of favorable elemental concentrations.

While not discounting the importance of empiricism in geochemical prospecting, a more sophisticated approach is that presented by Barnes (1967) and Helgeson (1964), who develop the geochemical thermodynamics of ore genesis. The changing stabilities and degrees of formation of the complexes HCl and NaCl and probably KCl and the chloride complexes of magnesium and calcium with temperature and pressure, appear to be of primary importance in the transport and deposition of ore-forming metals (Hegelson, 1964). Of significance in analysis of lead ore deposits is the fact that the precipitation of galena may have four controls:

1. decreasing temperature and total NaCl concentration
2. increasing ratios of total NaCl to total HCl
3. increasing solution pH
4. increasing temperature in the range where the complex HCl becomes sufficiently stable to form at the expense of H_2S in high chloride solutions

The science of geochemistry laid its academic roots in the early 1900s owing to the work of F. W. Clarke, H. Fersman, and W. Vernadsky. It is now a permanent part of industrial research involving oil, ores, and salines. The use of geochemical research in human health and welfare, e.g., study of the effects of nuclear bomb contamination or water pollution, will become more important as man changes elemental abundances of lead, strontium, and carbon. The subject of the December 1968 symposium (AAAS, Dallas, Texas) was on environmental geochemistry in relation to human health and disease. This may be a clue to a trend in applied geochemistry.

JACK GREEN

References

Abelson, P. H. (editor), 1967, "Researches in Geochemistry," Vol. 2, New York, John Wiley & Sons, 663pp.

Ahrens, L. H., 1965, "Distribution of Elements in our Planet," New York, McGraw-Hill Book, 110pp.

Ahrens, L. H., 1966, "Element distributions in specific igneous rocks; VIII," Geochim. Cosmochim. Acta, 30, 109–122.

Ahrens, L. H., (editor), 1968 "Origin and Distribution of the Elements," Elmsford, N.Y., Pergamon Press, 1178pp.

Ahrens, L. H., Press, F., Runcorn S. K., and Urey, H. C. (editors), 1966, "Physics and Chemistry of the Earth," Vol. 7, New York, Pergamon Press, 337pp.

Ahrens, L. H., Press, F., Runcorn, S. K., and Urey, H. C. (editors), 1971, "Physics and Chemistry of the Earth," Vol. 8, New York, Pergamon Press, 510pp.

Aller, L. H., 1967, "Earth, chemical composition of, and its comparison with that of the sun, moon, and planets," pp. 285–291 in S. K. Runcorn, et al., editors), Vol. 1, New York, Pergamon Press, 784pp.

Anders, E., 1964, "Origin, age, and composition of meteorites," Space Sci. Rev., 3, 583–714.

Anders, E., and Goles, G. C., 1961, "Theories on the origin of meteorites," J. Chem. Ed., 38, 58.

Anderson, D. H. et al., 1969, "Preliminary examination of lunar samples from Apollo 11," Science, 165, 1211–1227.

Augustithis, S. S., 1967, "On the phenomenology and geochemistry of differential leaching and elemental agglutination processes," Chem. Geol., 2, 311–329.

Barnes, H. L., 1967, "Geochemistry of Hydrothermal Ore Deposits, New York, Holt, Rinehart and Winston, 670pp.

Bennett, H., and Hawley, W. G., 1965, "Methods of Silicate Analysis," Second ed., New York, Academic Press, 334pp.

Biswas, S., Fichtel, C. E., Guss, D. E., and Waddington, C. J., 1963, "Hydrogen, helium and heavy nucleii from the solar event on November 15, 1960," J. Geophys. Res. 68, 3109–3122.

Blander, M., and Katz, J. L., 1967, "Condensation of primordial dust," Geochim. Cosmochim. Acta, 31, 1025–1034.

Bose, M. K., 1967, "The upper mantle and alkalic magmas," Norsk Geol. Tidsskr., 47, 121–129.

Brancazio, P. J., and Cameron, A. G. W. (editors), 1964, "The Origin and Evolution of Atmospheres and Oceans," New York, John Wiley & Sons.

Brinkmann, R. T., 1969, "The dissociation of water vapor and evolution of oxygen in the terrestrial atmosphere," J. Geophys. Res., 74, 5355–5368.

Bruyevich, S. V., and Kulik, Ye. Z., 1967, "Chemical interaction between the ocean and the atmosphere (salt exchange)," Oceanology, 7(3–4), 279–292.

Butler, J. R., 1964, "Concentration trends and frequency distributions for elements in igneous rock types," Geochim. Cosmochim. Acta, 28, 2013–2024.

Cadle, R. D., Lazrus, A. L., and Shedlovsky, J. P., 1969, "Comparison of particles in the fume from eruptions of Kilauea, Mayon, and Arenal volcanoes," *J. Geophys. Res.*, **74**, 3372–3378.

Cameron, A. G. W., 1966, "Abundances of the elements," in (S. P. Clark, Jr., editor) "Handbook of Physical Constants," *Geol. Soc. Am. Mem.* **97**, 8–10.

Cameron, A. G. W., 1968, "Fundamental problems of the solar system," presented at Am. Geophys. Union Annual Meet., San Francisco, Dec. 2, 1968.

Challis, G. A., 1968, "The K_2O: Na_2O ratios of ancient volcanic arcs in New Zealand," *New Zealand J. Geol. Geophys.*, **11**, 200–211.

Chilingar, G. V., Bissell, H., and Fairbridge, R. (editors), 1967, "Carbonate Rocks," New York, Elsevier Publishing Company, 471pp.

Clark, S. P. (editor), 1966, "Handbook of Physical Constants," *Geol. Soc. Am. Mem.*, **97**, 587pp.

Crockett, J. H., and Skippen, B., 1966, "Radioactive determination of palladium in basaltic and ultramafic rocks," *Geochim. Cosmochim. Acta*, **30**, 129–141.

Crocket, J. H., Hariss, R. C., and MacDougall, J. D., 1968, "Some aspects of the marine geochemistry of palladium, gold, and iridium," presented at Geol. Soc. Am. Annual Meet., Mexico City, Nov. 11–13, 1968.

Currie, K. L., and Shafiqullah, M., 1967, "Carbonatite and igneous rocks in the Brent Crater, Ontario," *Nature*, **215**, 725–726.

Degens, E. T., 1965, "Geochemistry of Sediments, A Brief Survey," Englewood Cliffs, N.J., Prentice-Hall, Inc., 342pp.

Degens, E. T., and Ross, D. A., 1969, "Hot Brines and Recent Heavy Metal Deposits in the Red Sea," New York, Springer Verlag, 700pp.

Dietrich, G., Duing, W., Grasshoff, K., and Koske, P., 1966, "Meteor Research Results," Berlin, Gebruder Borntraeger, 149pp.

Eglinton, G., and Murphy, M. T. J., 1969, "Organic Geochemistry, Methods and Results," New York, Springer Verlag.

Ehmann, W. D., and Lovering, J. F., 1967, "The abundance of mercury in meteorites and rocks by neutron activation analysis," *Geochim. Cosmochim. Acta*, **31**, 357–376.

Eitel, W., 1966, "Silicate Science," Vol. IV, New York, Academic Press, 617pp.

Engel, A. E. J., James, H., and Leonard, B. F. (editors), 1962, "Petrologic Studies; A Volume in Honor of A. F. Buddington," Boulder, Colo. Geol. Soc. Am.

Ewart, A., 1967–1968, "The petrography of the Central North Island rhyolitic lavas; Part 1, Correlations between the phenocryst assemblages; Part 2, Regional petrography including notes on associated ash-flow pumice deposits," *New Zealand J. Geol. Geophys.* **10** and **11**, 182–197 and 478, respectively.

Ewart, A., Taylor, S. R., and Capp, A. C., 1968, "Trace and minor element geochemistry of the rhyolitic volcanic rocks, Central North Island, New Zealand," *Contrib. Mineral. Petrol.*, **18**, 76–104.

Flanagan, F. J., 1967, "U.S. Geological Survey rock standards," *Geochim. Cosmochim. Acta*, **31**, 289–308.

Fleischer, M. (editor), 1963–19—, Data of Geochemistry," Sixth ed., *U.S. Geol. Surv. Profess. Papers*, Washington, D.C.

Fleischer, M., 1965, "Summary of new data on rock samples G-1 and W-1, 1962–1965," *Geochim. Cosmochim. Acta*, **29**, 1263–1283.

Fleischer, M. and Altschuler, Z. S., 1969. "The effects of geological environment on the rare earth composition of minerals," *Geochim. Cosmochim. Acta*, **33**, 65–79.

Fowler, W. A., and Stephens, W. E., 1968, "Resource Letter OE-1 on Origin of the Elements," *Am. J. Phys.*, **36**, 289–302.

Franzgrote, E. J., Patterson, J. H., and Turkevich, A. L., 1968, "Surveyor VII, A preliminary report," National Aeronautics and Space Administration, report NASA SP-173, pp. 207–232.

Gaskell, T. F., 1967, "The Earth's Mantle," New York, Academic Press, 509pp.

Geiss, J., and Goldberg, E. D. (editors), 1963, "Earth Science and Meteorites," Amsterdam, North Holland Publishing Co., 312pp.

Gerasimovskii, V. I., 1968, "Geochemistry of agpaitic nepheline syenites," *23rd Intern. Geol. Congr.* **6**, 258–265.

Ginzburg, I. I., 1960, "Principles of Geochemical Prospecting," New York, Pergamon Press, 311 pp.

Glass, B., 1968, "Microtektites and the Origin of the Australasian Tektite Strewn Field," Tempe, Arizona, Center for Meteorite Studies, Arizona State Univ., 22pp.

Glass, B., 1969, "Electron microprobe analysis of four silicate spherules from the Tunguska impact area," *Science*, **164**, 547–549.

Goldberg, E. D., 1965, "Minor elements in sea water," (J. P. Riley and G. Skirrow editors), in "Chemical Oceanography," Vol. 1 New York, Academic Press pp. 163–196.

Gorshkov, G. S., and Tovarova, I. I., 1957, "Geochemical effect of the Bezymianny volcano eruption," *Proc. Ninth Pacific Science Congr.*, **12**, 244 (published 1961).

Green, H. W., 1968, "Metastable growth of coesite in highly strained quartz aggregate," presented at Am. Geophys. Union Natl. Fall Meet., San Francisco, Dec. 2–5.

Green, J., 1959, "Geochemical Table of the Elements for 1959," *Bull. Geol. Soc. Am.* **70**, 1127–1184.

Green, T. H., and Ringwood, A. E., 1968, "Genesis of the calcalkaline igneous rock suite," *Contrib. Mineral. Petrol.*, **18**, 105–162.

Hampel, C. A. (editor), 1968, "The Encyclopedia of Chemical Elements," New York, Reinhold Publishing Corp., 849pp.

Hanor, J. S., and Garrett, W. B., 1968, "Stratigraphic variation in barium content in sediment cores from the East Pacific Rise," presented at

Geol. Soc. Am. Annual Meet., Mexico City, Nov. 11-13, 1968.

Hart, S. R., 1969, "K, Rb, Cs contents and K/Rb, K/Cs ratios of fresh and altered submarine basalts," *Earth Planetary Sci. Letters,* **6,** 295-303.

Hart, S. R., 1969, in V. V. Belousov et al., (editors), "Isotope geochemistry and geochronology: in the Earth's Crust and Upper Mantle," *Am. Geophys. Union, Geophys. Monograph* **13.**

Haskin, L. A., Frey, F. A., Schmitt, R. A., and Smith, R. H., 1965, "Meteoritic, solar, and terrestrial rare earth distributions," National Aeronautics and Space Administration Accession No. N 66-34044, Report No. NASA—CR-77136, 270pp.

Helgeson, H. C., 1964, "Complexing and Hydrothermal Ore Deposition," New York, Macmillan Company, 128pp.

Heinrich, E. W., 1966, "The Geology of Carbonatites," Chicago, Rand McNally and Company, 555pp.

Hess, H. H., and Poldervaart, A. (editors), 1967, "Basalts," Vol. 1, New York, Interscience Publishers, a div. of John Wiley, 482pp.

Holland, H. D., 1962, "Model for evolution of the earth's atmosphere," in (A. E. J. Engel et al., editors) "Petrologic Studies" Geol. Soc. Am. Boulder, Colo., pp. 447-478.

Horn, M. K., and Adams, J. A. S., 1966, "Computer derived geochemical balances and elemental abundances," *Geochim. Cosmochim. Acta,* **30.** 279-298.

Hurley, P. M. (editor), 1964, "Advances in Earth Science," The M.I.T. Press, Cambridge, Massachusetts, 502pp.

Iwasaki, I., Katsura, T., Ozawa, T., Yoshida M., and Iwasaki, B., 1968, "Chlorine content of volcanic rocks and migration of chlorine from the mantle to the surface of the earth" (pp. 423-427), in (L. Knopoff, C. L. Drake, and P. J. Hart, editors), "The Crust and Upper Mantle of the Pacific Area," *Am. Geophys. Union Monograph* **12,** 522pp.

Krauskopf, K. B., 1967, "Introduction to Geochemistry," New York, McGraw-Hill Book Co., 721pp.

Kvalheim, A., (editor) 1967, "Geochemical prospecting in Fennoscandia," New York, John Wiley, 352pp.

Larsen, G., and Chilinger, G. V. (editors), 1967, "Diagenesis in Sediments," New York, Elsevier Publishing Co., 539pp.

Liebermann, K. W., and Ehman, W. D., 1967, "Determination of bromine in stony meteorites by neutron activation," *J. Geophys. Res.,* **72,** 6279-6287.

Livingstone, D. A., 1963, "The sodium cycle and the age of the oceans," *Geochim. Cosmochim. Acta,* **27.** 1055-1069.

Majmundar, H. H., 1968, "Trace element distribution in pyroxenes," presented at Geol. Soc. Am. Annual Meet., Mexico City, Nov. 11-13, 1968.

Manskaya, S. M., and Drozdova, T. V., 1968, "Geochemistry of Organic Substances," Monogr. in Earth Sciences, Vol. 28, Oxford, Pergamon Press (transl. and ed. by Shapiro, L. and Breger, I.A.).

Mason, B., 1962, "Meteorites," New York, John Wiley & Sons, 274pp.

Mason, B., 1966, "Principles of Geochemistry," Third ed., New York, John Wiley & Sons, 329pp.

Mattox, R. B. (editor), 1968, "Saline deposits," A Symposium based on Papers from the Internat. Conf. on Saline Deposits, Houston, Texas, 1962, Geol. Soc. Am. Spec. Paper 88, 701pp.

Miesch, A. T., 1969, "The Constant Sum Problem in Geochemistry in Computer Applications in the Earth Sciences," New York, Plenum Press, pp. 161-176.

Moore, C. B., 1962, "Researches on Meteorites," New York, John Wiley & Sons.

Mossop, S. C., 1965, "Stratospheric particles at 20 km, altitude," *Geochim. Cosmochim. Acta,* **29,** 210-207.

Nagy, B., and Columbo, U., 1967, "Fundamental Aspects of Petroleum Geochemistry," New York, Elsevier Press, 388pp.

Naughton, J. J., Derby, J. V., Glover, R. B., 1969, "Infrared measurements on volcanic gas and fume; Kilauea eruption, 1968," *J. Geophys. Res.,* **74,** 3273-3277.

O'Keefe, J. A. (editor), 1963, "Tektites," Univ. Chicago Press, 228pp.

Palm, A., 1969, "Evolution of Venus' atmosphere," *Planet. Space Sci.,* **17,** 1021-1028.

Parker, R. B., 1968, "Major element variation in folded granulite facies rocks," presented at Geol. Soc. Am. Annual Meet., Mexico City, Nov. 11-13, 1968.

Peebles, P. J. E., and Wilkinson, D. T., 1967, "The Primeval Fireball," *Sci. Am.* **216,** 28-37.

Perelman, A. I., 1967, "Geochemistry of Epigenesis," New York, Plenum Publishing Corp. 266pp.

Perrin, R., 1960, "The puzzle of the sodium balance," *Compt. Rend.,* **270,** 1766-1769.

Philpotts, J. A., Schnetzler, C. C., Thomas, H. H., Schuhmann, S., and Kouns, C. W., 1968, "Oceanic tholeiites; trace element clues to their origin," presented at Geol. Soc. Am. Annual meet., Mexico City, Nov. 11-13, 1968.

Pitcher, W. S., and Flinn, G. W. (editors), 1965, "Controls of Metamorphism," New York, John Wiley & Sons, 368pp.

Poldervaart, A., and Green, J., 1959, "Abundance of major elements in the earth's crust," *Geotimes,* May-June, Am. Geol. Inst., Washington D.C., pp. 25-27.

Rankama, R., 1964, "Progress in Isotope Geology," New York, John Wiley & Sons, 705pp.

Rau, J. L. (editor), 1966, "Second symposium on salt," Vols. 1 and 2, Cleveland, Ohio, Northern Ohio Geological Soc.

Riley, J. P., and Skirrow, G. (editors), 1965, "Chemical Oceanography," Vols. 1 and 2, New York, Academic Press, 712 and 508pp., respectively.

Ringwood, A. E., 1966a, "The chemical composition and origin of the earth," in (P. M.

ELEMENTS: PLANETARY ABUNDANCES AND DISTRIBUTION

Hurley, editor), "Advances in Earth Science" Cambridge, Mass., The M.I.T. Press.

Ringwood, A. E., 1966b, "Chemical evolution of the terrestrial planets," *Geochim. Cosmochim. Acta*, **30**, 41–104.

Rodionov, D. A., 1965, "Distribution Functions of the Elements and Mineral Contents of Igneus Rocks," Consultants Bureau Special Research Report, Consultants Bureau, New York, 80pp.

Ronov, A. B., 1964, "Common tendencies in the chemical evolution of the earth's crust, ocean and atmosphere," *Geochem. Internat.*, **4**, 713–737.

Ronov, A. B., 1968, "Probable changes in the composition of sea water during the course of geological time," *Sedimentology*, **10**, 25–43.

Ronov, A. B., and Yaroshevsky, A. A., 1967, "Chemical structure of the earth's crust," *Geokhimiya*, **11**, 1285–1309.

Runcorn, S. K. (editor), 1967, "Mantles of the earth and terrestrial planets," New York, Interscience Publishers, a div. of John Wiley, 584pp.

Rutten, M. G., 1968, "The history of atmospheric oxygen," presented at Geol. Soc. Am. Annual Meet., Mexico City, Nov. 11–13, 1968.

Samsonov, G. V. (editor), 1968, "Handbook of the Physiochemical Properties of the Elements," New York, IFI/Plenum Press, 941pp.

Schenck, P. A., and Havenaar, I., 1969, "Advances in Organic Geochemistry, 1968," Elmsford, N.Y., Pergamon, Press, 617pp.

Schilling, J. G., and Winchester, J. W., 1966, "Rare earths in Hawaiian basalts," *Science*, **153**, 867–869.

Shaw, D. M., 1964, "Interpretation geochimique des elements en traces dans les roches cristallines," Monographies Scientifiques, Paris, Masson et Cie, 238pp.

Suess, H. E., and Urey, H. C., 1956, "Abundances of the elements," *Rev. Mod. Phys.* **28**, 53–74.

Tauson, L. V., 1965, "Factors in the distribution of trace elements during the crystallization of magmas," *Phys. Chem. Earth*, **6**, 215–249.

Taylor, S. R., 1964a, "Abundance of chemical elements in the continental crust—a new table," *Geochim. Cosmochim. Acta*, **28**, 1273–1286.

Taylor, S. R., 1964b, "Trace element abundances as applied to a chondritic earth model," *Geochim Cosmochim. Acta*, **28**, 1989–1998.

Taylor, S. R., 1965, "The application of trace element data to the problems in petrology," *Phys. Chem. Earth*, **6**, 133–213.

Taylor, S. R., and Kolbe, P., 1964, "Geochemical standards," *Geochim. Cosmochim. Acta*, **28**, 447–454.

Taylor, S. R., and White, A. J. R., 1966, "Trace element abundances in andesites," *Bull. Volcanologique*, **24**, 177–194.

Turekian, K. K., and Wedepohl, K. H., 1961, "Distribution of the elements in some major units of the earth's crust," *Bull. Geol. Soc. Am.* **72**, 175–192.

Turkevich, A. L., Franzgrote, E. V., and Patterson, J. H., 1969, "Chemical composition of the lunar surface in Mare Tranquillitatis," *Science*, **165**, 277–279.

Urey, H. C., 1964, "A review of atomic abundances in chondrites and the origin of meteorites," *Rev. Geophys.* **2**, 1–34.

Urey, H. C., and Marti, K., 1968, "Surveyor results and the composition of the moon," *Science*, **161**, 1030–1032.

Vinogradov, A. P., 1965, "The composition of meteorites," International Union of Pure and Applied Chemistry, Vol. 10, London, Butterworths, pp. 459–493 (reprinted 1966).

Vinogradov, A. P. (editor), 1966, "Chemistry of the Earth's Crust," Vols. 1 and 2, Soviet Acad. Science in translation, Israel Program for Scientific Translations, Jerusalem.

Vinogradov, A. P., 1968, "Geochemical problems in the evolution of the ocean": *Lithos*, vol. 1, pp. 169–718.

Vistelius, A. B., 1967, "Studies in Mathematical Geology," New York, Plenum Publishing Corp., 294pp.

Vlasov, K. A. (editor), 1968, "Geochemistry and mineralogy of rare elements and genetic types of their deposits," New York, Daniel Davy and Co., 916pp.

Vogt, J. R., and Ehmann, W. D., 1965, "Silicon abundances in stony meteorites by fast neutron activation analysis," *Geochim. Cosmochim. Acta*, **29**, 373–383.

Weast, R. C. (editor), 1968, "Handbook of Chemistry and Physics," Forty Ninth ed., Cleveland, Ohio, The Chemical Rubber Company.

Wedepohl, K. H. (editor), 1969, "Handbook of Geochemistry," New York, Springer Verlag, Vol. I. 442pp; Vol. II, Pt. 1, 586pp; Vol. II, Pt. 2, 667pp.

Wiik, H. B., 1956, "The chemical composition of some stony meteorites," *Geochim. Cosmochim. Acta*, **9**, 279–289.

Wilkniss, P. E., and Linnenbom, V. J., 1968, "Radiochemical experiments in fluorine marine geochemistry," presented at Am. Geophys. Union Natl. Fall Meet., San Francisco, Dec. 2–5.

Wood, J. A., 1967a, "Meteorites," in (R. Fairbridge, editor) "Encyclopedia of Atmospheric Sciences and Astrogeology," New York, Reinhold, pp. 561–564.

Wood, J. A., 1967b, "Chondrites: their metallic minerals, thermal histories, and parent planets," *Icarus*, **6**, 1–49.

Wright, T. L., 1968, "Comparison of Kilauea and Mauna Loa lava composition in space and time," presented at Geol. Soc. Annual Meet., Mexico City, Nov. 11–13, 1968.

Wyllie, P. J. (editor), 1967, "Ultramafic and Related Rocks," New York, John Wiley & Sons, 464pp.

Cross-references: *Calcium Carbonate; Chemical Mineralogy; Chlorine; Crystal Chemistry; Earth's Crust Geochemistry; Gases—Volcanic; Geochemical Evolution of Core, Mantle, Crust; Hydrosphere; Lake Geochemistry; Neutron Activation Analysis; Outgassing of the Planet Earth; Oxidation and Reduction; Oxygen—Evo-*

ELEMENTS: PLANETARY ABUNDANCES AND DISTRIBUTION

lution in the Earth's Atmosphere; Phase Equilibria; Pollution; Rainwater; Rare Earths; River Geochemistry; Seawater, Chemistry; Sulfate Reduction—Microbial; Trace Elements; Water—Nonmarine. Vol. I: *Manganese Nodules; Marine Sediments.* Vol. II: *Atmosphere; Australites; Carbonaceous Meteorites; Meteorites; Moon—Lunar Soil; Solar System; Tektites; Venus.* Vol. IVB: *Banded Iron Ores; Clinopyroxene; Coesite and Stishovite; Geochemical Prospecting.* Vol. V: *Carbonatites; Igneous Rocks; Lithosphere; Metamorphic Rocks; Shock Effects in Rocks and Minerals; Volatiles.* Vol. VI: *Methane; Pelagic Sediments.*

EMISSION SPECTROGRAPHY

Light contains a great deal of information as to the nature, motions, temperature, etc., of an emitting source and the decoding of this information can be accomplished by spectrographic means. Factors which enter into the applications of optical spectrographic procedures to chemical analyses, as practised, are a wedding of science and art requiring more than superficial understanding on the part of the practitioner.

About seventy of the chemical elements can be determined by routine spectrographic means. Table 1 shows those elements excited by a direct current arc together with their approximate lower limit of detection. Since simultaneous excitation of many elements in the same sample occurs, a spectrum will provide information as to the elements present and their relative abundance. Successive spectra of selected samples can then provide an immediate chemical reconnaissance of a particular problem. Procedural refinements make the analytical system quite reproducible, and quantitative determinations of selected elements are commonly made.

The Spectrographic System

The fundamental steps in spectrochemical analysis are the excitation of a sample to emit light, dispersion of this light by an appropriate optical system, the recording of its wavelength and intensity by one of several means followed by appropriate calculations to correlate optical data to concentration. Each step involves a number of phenomena which must be controlled to attain analytical precision and accuracy.

A spectrographic source is essentially a radiant volume containing a controlled amount of sample. This may be a fragment, powder, liquid, or gas, and each, of course, requires a different electrode design to introduce the sample into the light source. The more common devices are conducting pellets or rods which act as self-electrodes, graphite or metal rods provided with a crater into which powders may be packed, aspirators to introduce liquids into a flame, and discharge tubes for gases.

TABLE 1. SPECTROGRAPHICALLY DETECTABLE ELEMENTS

Element	Approx. Lower Limit of Detection (ppm)	Element	Approx. Lower Limit of Detection (ppm)	Element	Approx. Lower Limit of Detection (ppm)
Aluminum	5	Indium	1	Ruthenium	10
Antimony	10	Iridium	100	Samarium	500
Arsenic	100	Iron	1	Scandium	10
Barium	5	Lanthanum	10	Selenium	500
Beryllium	5	Lead	2	Silicon	10
Bismuth	10	Lithium	1	Silver	1
Boron	10	Lutecium	10	Sodium	1
Cadmium	10	Magnesium	1	Strontium	5
Calcium	1	Manganese	1	Tantalum	100
Cerium	500	Mercury	100	Tellurium	100
Cesium	100	Molybdenum	1	Terbium	10
Chromium	1	Neodymium	10	Thallium	1
Cobalt	2	Nickel	2	Thorium	100
Copper	1	Niobium	10	Thulium	10
Dysprosium	10	Osmium	100	Tin	10
Erbium	10	Palladium	10	Titanium	5
Europium	10	Phosphorous	100	Uranium	100
(Fluorine)	100	Platinum	5	Vanadium	10
Gadolinium	500	Potassium	5	Wolfram	10
Gallium	5	Praseodymium	10	Ytterbium	10
Germanium	10	Rhenium	100	Yttrium	10
Gold	10	Rhodium	10	Zinc	100
Hafnium	100	Rubidium	10	Zirconium	10
Holmium	10				

In most sources (sparks, flames, and discharge tubes) the excitation process does not deplete the sample and concentration is directly proportional to exposure (light intensity × time). However, highest sensitivity is usually attained by direct current arc excitation in which the sample is consumed and the number of excited atoms and ions changes in both space and time. Temperature is not constant along the length of the arc column so the efficiency of excitation varies and the fractional volatilization of material from the sample electrode causes the arc gas to change in composition with time. The pole-to-pole variation may be eliminated by illuminating the entrance slit from a point in the arc column or allowing each point in the source to illuminate each point on the slit. This can be done either by appropriate diaphragms or lens combinations or by having no lenses (or mirrors) between source and slit. The direct current arc is also inherently variable in radiant flux, sometimes to the degree that reciprocity failure of emulsions becomes an important source of analytical error. Stabilization by carefully chosen arcing mixtures or external devices such as flowing gas or rotating magnets are usually required.

Spectrographic instruments are essentially simple devices. However, because of the requirements of mechanical stability, optical alignment, and flexibility of use, good instruments are both expensive and delicate. In essence, a spectrograph consists of a narrow (~ 0.01 mm) entrance slit which is illuminated by the light to be decoded and is the light source of the internal optical path. Light diverges as it leaves the slit and is collimated by a lens or concave mirror so as to arrive at the diffraction element in parallel rays. The diffraction of the incident light, i.e., its separation into its component wavelengths, may be accomplished either by a diffraction grating or prism. After diffraction, rays of the same wavelength are brought to focus on the focal curve by passage through a focussing lens. The usual optical spectrographs are capable of analyzing light from the short ultraviolet, ~ 2000 Å, through the visible region, ~ 4000–7000 Å, and up to about 10,000 Å in the infrared. The lower limit is imposed by light absorption in the optical pieces, film gelatin, air, etc., and the upper limit by the insensitivity of the usual recording devices.

The design and construction of spectrographic instruments is beyond the scope of this article. The choice of optical materials and their arrangement is very wide. Each system has its peculiar advantages and each manufacturer offers what he considers to be the most generally useful instrument. Pertinent parameters to consider are speed, dispersion, wavelength coverage, stigmatism, and resolving power.

Line images of the entrance slit produced by different wavelengths of light are brought to focus at different positions on the focal curve of a spectrograph where they may be observed visually, sensed by appropriately placed photoelectric devices, or integrated by a photographic emulsion. Visual (spectroscopic) observation of line position and color may be used for rapid qualitative identification of elements. Photoelectric sensing can be coupled with electronic systems to provide immediate quantitative results for preselected elements by use of prepositioned and calibrated sensors. Photographic recording is probably the most common and certainly the most flexible method. It is also inherently complex. Light striking a photographic emulsion excites a latent conversion of silver halides to silver metal either directly or through the intermediate activity of fluorescent dyes. Photographic development and fixation are then used to bring out the latent image. The amount of silver deposited is dependent upon the amount of light received (exposure), rate of receipt, wavelength of incident light, and photographic processing. All must be held within narrow tolerances to attain high reproducibility.

The amount of silver deposited on a plate or film is only an indirect measure of element concentration in the source. Excitation in a spectrographic source is (usually) a steady-state phenomenon in which atomic concentration (N) and light intensity (I) are related by

$$I \sim NPe^{-E/kT}$$

where P is the transition probability, E the excitation energy of the line, k the Boltzmann constant, and T the absolute temperature. The transition probability and excitation energy are constant for a given line, and hence the intensity is a function of the concentration of atoms and the effective temperature.

To determine intensity and thence concentration, it is first necessary to calibrate the photographic emulsion in order to convert optical density of a line to light intensity. This may be accomplished if the incident relative intensity of several lines or portions of the same line is known or can be fixed. A common procedure is to rotate a stepped sector in front of the slit in such a way that successive portions of the slit receive 1, 1/2, 1/4, 1/8, etc. of the incident radiation. Curves of relative intensity vs. some measure of the optical density of a line for each wavelength region may then be constructed.

EMISSION SPECTROGRAPHY

Optical density (or percent transmission) is measured with a photometer whose usual design allows the movement of a spectral line through the focus of a narrow light beam which is incident upon a photoelectric sensor. As the line blocks light from the sensor, its signal to a galvanometer decreases to a minimum, which is taken as a measure of its opacity.

Analytical Methods

The general goal of spectrochemical analysis is to relate the relative intensity of light of a particular wavelength to the concentration of a particular element in a sample. The determination of relative intensity is a spectrographic problem, but the selection or preparation of analytical standards for control of the concentration is usually a geological one. Reference standards must, in general, contain accurately known amounts of the elements of interest at concentrations near the levels sought and in matrixes equivalent to the unknowns. These requirements are not always met with ease. Some rock materials with analytical certification are available, but not uncommonly the spectrochemist must devise his own standards.

The addition method is a procedure whereby either a single quantitative determination for a given element in a given material or the establishment of a standard series may be made. The usual procedure is to mix the element to be measured into an appropriate base in the proportion, say, 1:9, then dilute this mixture with base 1:9, ad infinitum, thereby providing a series of aliquants of the base material containing known amounts of the added element. A linear plot of amount added vs. relative intensity will then give the amount of element sought by extrapolation.

Analytical curves for spectrochemical purposes are usually plotted as log concentration vs. log relative intensity. Reference standards define the curve, which may then be entered with the intensity measured for an unknown to find concentration. The inherent variations of the optical spectrographic system usually limit the reproducibility of analytical results to $\pm$ 15-25% of the amount determined unless some special attention is paid to the improvement of precision. The most usual course is to refer the intensity of the unknown element to that of a carefully selected, spectrochemically similar element either present in or added to the sample in known amount. Since systemic variations affect unknown and internal standard elements equally, their ratio has rather good precision, usually about $\pm$ 5% of the amount determined.

Analytical curves for methods employing internal (variable internal, mutual) standardization are plotted as log concentration vs. log relative intensity ratio of unknown to standard.

Applications

Spectrochemical analysis lends itself readily to most geological problems in which it is desired to find the distribution of elements in space or time. A few milligrams of material suffice for an analysis and many samples may be studied in a comparatively short time. Some examples of geological applications are: (a) quality control of raw materials and processing; (b) the abundance and distribution of particular elements in general or particular geological environments; (c) correlation and provenance studies; and (d) geothermometric and geobarometric measurements.

WILLIAM H. DENNEN

References

Ahrens, L. H., 1961, "Spectrochemical Analysis," Second ed., Reading, Mass., Addison-Wesley Publ. 454pp.

Brode, W.R., 1945, "Chemical Spectroscopy," Second ed., New York, John Wiley & Sons, 677pp.

Gerlach, W., and Schweitzer, E., 1931, "Foundations and Methods of Chemical Analysis by the Emission Spectrum," London, Adam Hilger.

Harrison, G. R., 1939, "Massachusetts Institute of Technology Wavelength Tables," New York, John Wiley & Sons, 429pp.

Harrison, G. R., Lord, R. C., and Loofbourow, J. R., 1948, "Practical Spectroscopy," Englewood Cliffs, N.J., Prentice-Hall.

Harvey, C. E., 1950, "Spectrochemical Procedures," Glendale, Calif., Applied Res. Labs.

Nachtrieb, N. H., 1950, "Principles and Practice of Spectrochemical Analysis," New York, McGraw-Hill Book Co.

Cross-references: *Neutron Activation Analysis; Spectrophotometry; X-Ray Spectroscopy.* Vol. IVB: *Optical Mineralogy.*

ENANTIOMORPHISM—See STEREOCHEMISTRY AND ENANTIOMORPHISM

ENERGY—See Vol. II, SOLAR ENERGY; Vol. IV B, COAL; GEOTHERMAL ENERGY; OIL SHALES; PETROLEUM; Vol. VI, GEOLOGY—ENERGY; TIDAL POWER; WATER POWER

ENERGY BUDGET OF EARTH'S SURFACE—See Vol. I or II

ENTHALPY OR HEAT CONTENT

The *enthalpy*, H or *heat content*, of a substance is a thermodynamic property defined as the *internal energy*, E, plus the product of the *pressure*, P, times the *volume*, V, of the substance:

$$H = E + PV \quad (1)$$

The enthalpy is an extensive state function; its value depends only on the state and the amount of the substance and not on its previous history. It has the units of energy and it is usually expressed in calories (or kilocalories).

For a process at *constant pressure* ($\Delta P = 0$), in which the only work performed is the mechanical pressure-volume work ($P\Delta V$), the *change in enthalpy*, ΔH, is equal to the heat adsorbed by the system, q (hence the name heat content):

$$\Delta H = \Delta E + P\Delta V = q \quad (2)$$

This relation is a direct consequence of the definition of enthalpy by Eq. (1) and of the mathematical statement of the first law of thermodynamics, namely that the change in internal energy, ΔE, is equal to the heat adsorbed minus the work done ($q - P\Delta V$). It is clear that this thermodynamic relation does not define absolute values of enthalpy or internal energy. Changes in enthalpy, however, are readily measured by calorimetric techniques, and the relative enthalpy values are sufficient for all thermochemical calculations.

Enthalpy-Temperature Relation and Heat Capacity

When heat is adsorbed by a substance, under conditions such that no chemical reaction or state transition occur and only pressure-volume work is done, the temperature, T, rises and the ratio of the heat adsorbed, over the differential temperature increase, is by definition the heat capacity. For a process at constant pressure (following Eq. (2)), this ratio is equal to the partial derivative of the enthalpy, and it is called the *heat capacity at constant pressure*, C_p, (usually in cal/deg-mole):

$$\left(\frac{\partial H}{\partial T}\right)_p = C_p \quad (3)$$

The temperature dependence of H for a substance remaining in the same physical state can be expressed as a function of C_p by integration of equation 3.

If a substance undergoes a transformation from one physical state to another, such as a polymorphic transition, the fusion or sublimation of a solid, or the vaporization of a liquid, the heat adsorbed by the substance during the transformation is defined as the *latent heat of transformation* (transition, fusion, sublimation or vaporization). It is equal to the enthalpy change of the process, which is the difference between the enthalpy of the substance in the two states at the temperature of the transformation. For the purpose of thermochemical calculations, it is usually reported as a molar quantity with the units of calories (or kilocalories per mole (or gram formula weight). The symbol L or ΔH, with a subscript t, f (or m), s, and v is commonly used and the value is usually given at the equilibrium temperature of the transformation under atmospheric pressure, or at 25°C. For a substance undergoing one phase transformation, with a latent heat ΔH_t at a temperature T_t, the enthalpy change between two temperatures, T_1 and T_2, such that $T_1 < T_t < T_2$, is given by

$$H_T - H_T = \int_{T_1}^{T_t} C_p' dT + \Delta H_t + \int_{T_t}^{T_2} C_p'' dT \quad (4)$$

where C_p' and C_p'' are the heat capacities of the substance in the two different physical states. For several successive transformations, additional terms are added. Figure 1 illustrates the temperature dependence of enthalpy and heat capacity.

Very precise measurements of the heat capacity of liquids and solids can be obtained

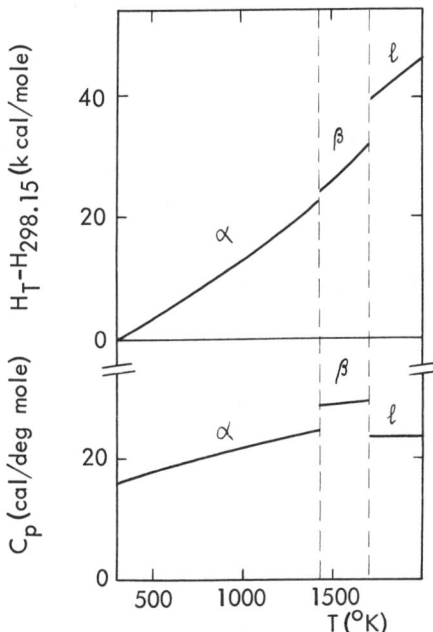

Fig. 1. Example of the temperature dependence of enthalpy relative to 25°C, $H_T - H_{298.15}$, and heat capacity, C_p. The data are for fluorite, CaF_2 (K. K. Kelley, 1960). The discontinuities in the lines correspond to the α to β transition (1424°K) and the fusion (1691°K).

by calorimetric techniques at relatively low temperatures (below 200°C) and they can be extrapolated down to the absolute zero of temperature ($-273.15°C$) by reliable theoretical expressions. In that temperature range, heat capacity data are usually very accurate and enthalpy values are obtained by integration (Eq. (4)). The most reliable method for determining high-temperature enthalpies and heat capacities is the dropping method (or method of mixtures) which consists of dropping the substance under investigation from a furnace at a known temperature into a calorimeter at room temperature. This method determines directly the change in enthalpy (or heat content) of the substance between the temperature of the furnace and that of the calorimeter. Heat capacities are obtained by differentiation (Eq. (3)). The measurement of heat capacity of gases is usually more difficult, and their thermodynamic properties can be more accurately calculated by methods of statistical mechanics based upon energy level of gas molecules obtained from spectroscopic data, or upon the knowledge of the molecular configuration and the vibration frequencies of the molecules.

Molar enthalpy data for elements and inorganic compounds above room temperature are usually tabulated in the form of the heat content above a reference temperature, usually $298.15°K = 25°C$. They are represented by: $H_T - H_{298.15}$ in cal/mole. The data are correlated over a range of temperature by empirical equations such as a series of powers of the absolute temperature or such as the following expression adopted by K. K. Kelley (1960) for his extensive compilation of data on inorganic compounds:

$$H_T - H_{298.15} = aT + bT^2 + c/T + d \quad (5)$$

where T is the absolute temperature (°K) and a, b, c, d are constants determined from experimental data. The corresponding equation for heat capacity is:

$$C_p = a + 2bT - c/T^2 \quad (6)$$

Standard Enthalpy of Formation

For the convenience of tabulation and computation of thermodynamic data, it is essential to present them in a commonly accepted form relative to a single standard state of reference. At all temperatures, the *standard state* for a *pure liquid or solid* is the *condensed phase under a pressure of one atmosphere*. The standard state for *a gas* is the *hypothetical ideal gas at unit fugacity* (equivalent to a "perfect gas" state), in which state the enthalpy is that of the real gas at the same temperature when the pressure approaches zero. Values of thermodynamic quantities for standard-state conditions are identified by a superscript ^{0}, and H^0, for instance, is the enthalpy change of a reaction when reactants and products are in the standard state.

The *standard enthalpy of formation*, ΔH_f^0 (also represented by $\Delta \bar{H}_f^0$ or simply H_f^0), of a substance at a given temperature is by definition, the enthalpy change when one mole of the substance in its standard state is formed, isothermally, at the indicated temperature from the elements, each in its standard state. Usual units are kcal/mole. *For all elements in their stable form at 25°C* (298.15°K), the *enthalpy of formation is zero*. If solid substances have more than one crystalline form, the most stable one is taken as the standard state, and the others have slightly different enthalpies. This convention about zero enthalpy is arbitrary but universally accepted, and it may be compared to the arbitrary choice of zero for terrestrial altitudes. The combination of enthalpies of formation, enthalpies of transition, and heat capacities makes possible the calculation of the enthalpy of a substance, in a given state at a given temperature, relative to a commonly accepted reference.

Enthalpy calculation for mixtures is more complex than for pure substances and its discusion is beyond the scope of the present article. *Aqueous solutions* (q.v.), however, are very important from a geochemical point of view, and reliable data are usually available. The *enthalpy of solution* is the enthalpy change resulting from the dissolution of a substance; it is a function of the solute concentration and its values are reported accordingly in the literature. For a *solute in aqueous solution*, a *standard state* is defined as the *hypothetical ideal solution of unit molality* (1 mole of solute per 1000 grams of water). In this state, the partial molal enthalpy and heat capacity of the solute are the same as in the infinitely dilute real solution. Although it is impossible to prepare a solution of only one ionic species (since the system must remain neutral), it is convenient to apportion the enthalpy (and other thermodynamic properties) between the various ions. This apportionment is not unique and an additional convention has to be made, namely that the *standard enthalpy of hydrogen* ion in aqueous solution (aq) at unit activity, ΔH_f^0 for H^+(aq), *is zero*. The properties of a neutral electrolyte in the standard state are equal to the algebraic sum of the values corresponding to the individual ions.

Heat of Reaction and Gibbs Free Energy Change

For a chemical reaction, at constant pressure with only pressure-volume work performed, the heat adsorbed by the process, q, is equal to the enthalpy change, ΔH, or the sum of enthalpies of the products of the reaction minus the sum of enthalpies of the reactants (taking into account the amount of each).

$$q = \Delta H = \Sigma H_{\text{products}} - \Sigma H_{\text{reactants}} \quad (7)$$

Those heat effects can be easily calculated when the enthalpies of formation and the enthalpy-temperature relations are available for the substances considered. Usually, the *heat of reaction* is defined as the heat evolved by the process, and it is equal to the enthalpy change but opposite in sign, while heats of fusion or vaporization always refer to the heat adsorbed, and for heats of solution the usage varies. In order to avoid any confusion, it is recommended to express heat effects of chemical process by reporting the enthalpy change, ΔH.

Early chemists thought that the heat of reaction, $-\Delta H$, should be a measure of the *"chemical affinity"* of a reaction. With the introduction of the concept of *entropy* (q.v.) and the application of the second law of thermodynamics to chemical equilibria, it is easily shown that the true measure of chemical affinity and the driving force for a reaction occurring at constant temperature and pressure is $-\Delta G$, where ΔG represents the change in thermodynamic state function, G, called *Gibbs free energy* or *free enthalpy*, and defined as the enthalpy, H, minus the entropy, S, times the temperature, T ($G = H - TS$). For a chemical reaction at constant pressure and temperature:

$$\Delta G = \Delta H - T\Delta S \quad (8)$$

and the Gibbs free energy change can be obtained by calculating the enthalpy and entropy change and applying Eq. 8. The criterion for *a spontaneous chemical reaction* is that ΔG be *negative*, and a chemical equilibrium corresponds to the condition $\Delta G = 0$.

Conversely, if the Gibbs free energy change is known as a function of temperature at constant pressure, the enthalpy change can be obtained by a relation which is an alternate form of the *Gibbs-Helmholtz equation*, and which can be derived from equation 8.

$$\left(\frac{\partial(\Delta g/T)}{\partial(1/T)}\right)_P = \Delta H \quad (9)$$

This means that ΔH is the slope of the line representing $\Delta G/T$ versus $1/T$ at constant pressure.

PAUL DUBY

References

Kelley, K. K., 1960, "High-Temperature Heat-Content, Heat-Capacity, and Entropy Data for the Elements and Inorganic Compounds," *U.S. Bur. Mines, Bull.* **584**, Washington, U.S. Govt. Printing Off., 232pp.

Kern, R., and Weisbrod, A., 1967, "Thermo dynamics for Geologists," San Francisco, Freeman, Cooper and Co., 304pp (translated by Duncan McKie from "Thermodynamique de Base pour Minéralogistes, Pétrographes et Géologues," 1964, Paris, Masson et Cie.)

Klotz, I. M., 1964, "Basic Chemical Thermodynamics," New York, W. A. Benjamin, Inc., 468pp.

Mohan, B. H., 1963, "Elementary Chemical Thermodynamics," New York, W. A. Benjamin, Inc., 155pp.

*Rossini, F. D., Wagman, D. D., Evans, W. H., Levine, S., and Jaffe, I., 1952, "Selected Values of Chemical Thermodynamic Properties," *NBS Circular* **500**, Washington, U.S. Govt. Printing Off., 1268pp.

*Presently being revised. Some revised sections have already been issued in advance as NBS Technical Notes 270–1 (1965) 270–2, 270–3 (1967).

Cross-references: *Aqueous Solutions; Entropy; Free Energy and Free Enthalpy; Thermochemistry; Thermodynamics.* Vol. IVB: *Polymorphism.* Vol. V: *Free Energy and the Gibbs Phase Rule.*

ENTROPY

The *entropy*, S, is a thermodynamic function introduced by Clausius to describe the natural evolution of a system and to provide a simple quantitative expression of the *second law of thermodynamics*. It is best defined, for mathematical convenience, by expressing the *differential entropy change*, dS, as the amount of *heat adsorbed* in an *infinitesimal reversible process*, dq_{rev}, divided by the *absolute temperature*, T, at which it is transferred.

$$dS = \frac{dq_{\text{rev}}}{T} \quad (1)$$

A reversible process, such as the one considered here, is a process in which the direction of all exchanges, and especially the flow of heat, can be reversed by a small change of the variables such as temperature and pressure. This is an idealized concept of a limiting case where the temperature of the surroundings equals that of the system, and heat can flow in either direction following infinitesimal temperature changes. This is, in other words, a system in *thermodynamic equilibrium* with the surroundings.

The entropy is a state function of the system, and equation 1 can be used to define the entropy change, ΔS, or the difference between the entropy in state A and state B, $S_B - S_A$, by considering a succession of elementary reversible processes (or a sequence of equilibrium states), and making a summation of the effects, or in mathematical term taking the integral:

$$\Delta S = S_B - S_A = \int_A^B \frac{dq_{\text{rev}}}{T} \quad (2)$$

The *entropy change* is thus the integral of dq_{rev}/T calculated along a *reversible path*, such that at each point the system is in equilibrium with the surroundings at the temperature T. Although this equation does not add very much to the physical understanding of the concept of entropy, it provides a way to define the entropy of a substance and to show that it is a function only of the amount of the substance and its state (as defined by temperature, pressure, composition).

Entropy and the Second Law of Thermodynamics

Reversible, or equilibrium, processes are very important for the purpose of thermodynamic arguments, but they are idealizations, and daily experience indicates that actual or natural processes tend to occur in a natural direction. For instance, the flow of heat from a hot to a cold body or the dissolution of sugar in water are spontaneous or irreversible processes. The purpose of the *second law of thermodynamics* is to predict the direction in which natural processes occur, and the concept of entropy enabled Clausius to state it in a very concise form: "The entropy of the world strives towards a maximum." An alternate statement is that: for any *spontaneous* (or natural) *transforamtion* of an *isolated system, the entropy can only increase*. For an open system, which exchanges heat, work, or matter with the surroundings, it is possible to reverse a natural process but only at the expense of some change in the environment so that the total entropy increases. The same experimental observation can be stated in the following way: the total energy of the universe remains constant but part of it is being transformed into such form which is less available for useful purposes. In any actual process transforming heat into work, some of it must be dissipated at a lower temperature than the original one, and it is no longer available for further transformation. Similarly, when the original energy is of another kind (mechanical, electrical, chemical, etc.), part of it is always dissipated by some irreversible effect, such as friction, which results in heating of the surroundings. The entropy change can be considered a measure of the degradation of available energy, and for any naturally occurring process, it should be positive.

The sign of the entropy change for a given process is the criterion for spontaneity, but it applies only to an isolated system or to the entire universe. This means that, in order to use this criterion to actual processes, all entropy changes must be evaluated, not only those within the system under study, but also those in the surroundings. In general, the changes in the environment are not known, or they are difficult to evaluate, so that the criterion is not practical. For an open system, it is convenient to split the total change in entropy into two parts, the *entropy* which is *exchanged with the surroundings*, $(dS)_e$, and the *entropy* which is *produced by the internal irreversible process*, $(dS)_i$. The former can have either sign, while the latter is always positive or zero:

$$dS = (dS)_e + (dS)_i \quad (3)$$

$$(dS)_i \geqslant 0 \quad (4)$$

One of the purposes of thermodynamic calculations is to express the second law so that the internal entropy production, $(dS)_i$, can be related to other thermodynamic variables. For instance, in predicting the direction in which a chemical process will occur spontaneously, under conditions of constant temperature and pressure, the free energy function introduced by Gibbs, and also called free enthalpy, G, is the most useful criterion. It is equal to the enthalpy, H, minus the product of entropy times absolute temperature ($G = H - TS$); it is minimum for a state of chemical equilibrium and its change is always negative for a spontaneous process.

Statistical Interpretation of Entropy

The definition of entropy, which is given above, is quite adequate to serve as the foundations upon which the whole body of classical thermodynamics is built. The statistical interpretation, however, adds to the understanding of its physical meaning, and also provides a basis for the calculation of absolute values.

(1) *Boltzmann's Relation*. The macroscopic state of a system is fully described by the values of macroscopic variables such as temperature, pressure, volume, chemical composition, or total energy. A microscopic state, on the other hand, is specified only when the loca-

tions, velocities, and energies of all atoms and molecules are known. Usually, many different microscopic states can be described by the same values of thermodynamic variables, and therefore, they are not distinguishable from a macroscopic point of view. The number of so-called microscopic realizations of a macroscopic state of fixed energy is usually represented by Ω. The statistical probability of a given macroscopic state is related to the number of microscopic realizations, and it will be greater if Ω is larger. The second law of thermodynamics can be stated in statistical terms, namely that *every system that is left to itself will, on the average, change towards a state of maximum probability.* Accordingly, S and Ω must behave in a similar way.

It was shown by Boltzmann that, for an isolated system, the entropy, S, is proportional to the natural logarithm of the number of microstates Ω.

$$S = k \ln \Omega \qquad (5)$$

The proportionality factor, k, is Boltzmann's constant ($3.2996 \; 10^{-24}$ cal/deg), and it is equal to the gas constant, R (1.9872 cal/deg mole) divided by Avogadro's number, N (or $R = kN$). The evaluation of Ω is the domain of statistical and quantum mechanics, and, so far, it can be done only in relatively simple cases, such as ideal gases, for which spectroscopic data can be interpreted in terms of energy states and relative probability of these states. Nevertheless, Boltzmann's relation is useful in understanding the meaning of entropy, by representing it as a measure of molecular disorder. An isolated system tends to assume a state of maximum probability and this natural process is accompanied by an increase in entropy which corresponds to an increasing disorder.

(2) *Third Law of Thermodynamics.* While Clausius' definition of entropy provides a method for calculating the difference in entropy between two states, the statistical interpretation of Boltzmann leads to the concept of absolute entropy. A macroscopic state that has only one microscopic realization can be assigned zero entropy: $S = 0$ for $\Omega = 1$. Such a state can be realized by a perfectly ordered crystal at absolute zero (0°K); each atom has a well-defined equilibrium position and it is in its state of lowest energy. This principle is often referred to as the Nernst's theorem or the third law of thermodynamics. It can be stated concisely that *the entropy of all perfect crystalline solids is zero at absolute zero.*

Entropy Calculations

The most useful purpose of entropy data, from a geologist's viewpoint, is in the calculation of geochemical equilibria. Two different relations must be considered for that application, namely the evaluation of entropies from heat capacity data, and the use of entropy changes to calculate free energy changes.

(1) *Relation Between Entropy and Heat Capacity.* The *heat capacity, C,* of a system, is defined as the ratio of the heat absorbed by the system to the temperature increase which results from this absorption of heat. For the purpose of thermochemical calculations, it usually refers to the gram atomic or molecular weight of the substance and it is given as the *molar heat capacity* in cal/deg mole. The term, *specific heat,* generally describes the heat capacity per gram of a substance (although some authors use molar specific heat, or simply specific heat, to designate the molar quantity).

For a reversible process, $dq_{\text{rev}} = CdT$, and taking into account the third law, the absolute entropy, S_T, of a substance at a temperature T, can be calculated from heat capacity data by integration of the equation defining S, from absolute zero to the temperature T.

$$S_T = \int_0^T \frac{dq_{\text{rev}}}{T} = \int_0^T \frac{C}{T} dT \qquad (6)$$

The heat capacity varies with temperature and it depends on the conditions under which heating takes place. The *heat capacity at constant volume,* C_v, is equal to the rate of change of internal energy, E, with temperature, T, for heating at constant volume, V. The *heat capacity at constant pressure,* C_p, is equal to the rate of change of enthalpy, H, with temperature, T, for heating at constant pressure, P.

(2) *Calculations from Heat Capacity Data.* Heat capacity arises from the thermal motion of atoms and molecules, and theoretical calculations, based upon a detailed description of molecular properties, contribute reliable values in favorable cases. The thermodynamic properties of gaseous substances can be evaluated by the interpretation of spectroscopic data or from molecular constant data. For solid substances, the heat capacity at constant volume is related to the frequencies of atomic vibrations in the crystal, and at low temperatures, the theoretical expressions are approximated by C_v equal to a constant times T^3. With increasing temperatures, however, thermal energy is also absorbed by the electrons, and C_v increases faster than predicted by the theory of crystal vibration.

Most thermochemical measurements and computations are performed at constant pressure and values of heat capacity at constant pressure (1 atm) are commonly reported in the literature. Experimental values of C_p are generally not available at very low temperatures (below about 50°K), but in that range, the difference between C_p and C_v is negligible for solids, and expressions such as

$$C_p = \alpha T^3 + \gamma T \qquad (7)$$

where α and γ are constants and the first term corresponds to atomic vibration and the second one (often negligible) is the electronic contribution, can be used to evaluate the entropy from absolute zero to the temperature of the lowest measurement. The entropy, in the range of experimental values of C_p, is obtained by plotting C_p/T versus T and computing the area under the curve. This procedure is illustrated by Fig. 1. (An alternate technique consists of plotting C_p against $\ln T$, since $C dT/T = C_p d\ln T$).

When a substance undergoes a phase transformation such as a polymorphic transition, fusion, vaporization, or sublimation, there is a finite change in entropy. At the temperature of transition, T_t, the process is reversible ($\Delta G = 0$), and the entropy change for the transformation, ΔS_t, is equal to the latent heat of transformation, or enthalpy change, ΔH_t, divided by T_t.

$$\Delta S_t = \frac{\Delta H_t}{T_t} \qquad (8)$$

The entropy of the substance at a temperature T greater than T_t is given by:

$$S_T = \int_0^{T_t} \frac{C_p'}{T} dT + \frac{\Delta H_t}{T_t} + \int_{T_t}^{T} \frac{C_p''}{T} dT \qquad (9)$$

where C_p' and C_p'' are the heat capacities of the substance in the two different states. For several successive transformations, additional terms are added.

(3) *Standard Entropies.* For convenience of calculations, entropy values, computed as described above, are tabulated for substances in their *standard reference state* (solids and liquids under 1 atm pressure and gases in the hypothetical state at unit fugacity) at 298.15°K (25°C). Unlike values of enthalpy, for which the reference is arbitrary, the *standard entropies* are *absolute values* and they should not be confused with entropies of formation. Standard entropies at 298.15°K are designated by $S°_{298.15}$ and reported in cal/(deg)(mole). Above room temperature, heat capacity data are generally given by empirical equations, often derived from enthalpy measurements, and entropy data are calculated and tabulated accordingly (for instance as $S_T - S_{298.15}$).

(4) *Entropy and Gibbs Free Energy Changes.* As mentioned above, the *change in Gibbs free energy*, also called *free enthalpy*, ΔG, for a chemical reaction under constant pressure and at a given temperature is the criterion to predict the direction of a spontaneous process or the conditions of a stable equilibrium. If the enthalpy change, ΔH, and the entropy change, ΔS, are known or can be calculated, the Gibbs free energy change, ΔG, is obtained by the following equation, for a process at constant temperature, T.

$$\Delta G = \Delta H - T \Delta S \qquad (10)$$

The variation of ΔG with temperature is given by the partial derivative at constant pressure, P.

$$\left(\frac{\partial \Delta G}{\partial T}\right)_P = -\Delta S \qquad (11)$$

In other words, $-\Delta S$ is the slope of the line representing ΔG versus T at constant pressure.

Most applications for geochemical purposes involve the calculation of ΔG from ΔH and ΔS, but the same relations can be used to calculate the entropy change, ΔS, from experimental values of ΔH and ΔG at one temperature, or from the temperature dependence of ΔG. The data are usually reported for reactants and products in their standard states at that temperature (superscript ^{0} and subscript T).

$$\Delta S_T^0 = \frac{\Delta H_T^0 - \Delta G_T^0}{T} \text{ or}$$

$$\Delta S_T^0 = -\left(\frac{\partial \Delta G_T^0}{\partial T}\right)_P \qquad (12)$$

In general, the standard Gibbs free energy change for a reaction is determined by experimental measurement of the equilibrium constant for that reaction, or in some cases, by the standard emf of a reversible galvanic cell in-

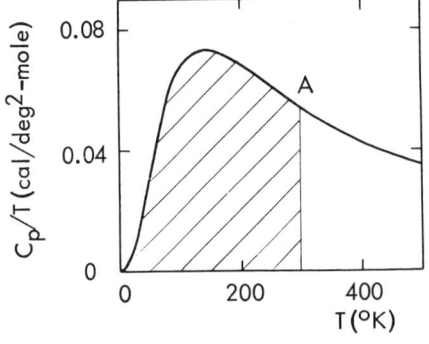

FIG. 1. Calculation of entropy from heat capacity data. Values of C_p/T for fluorite, CaF_2, are plotted versus T. The entropy at 298.18°K (25°C) equals the area under the curve from the origin to point A: $S_{298.15}° = 16.06$ cal/deg-mole (data from Kelley and King, 1961).

volving the reaction. When such data are available, the standard entropy of one of the reactants or products can be calculated from the standard entropy of reaction, if the entropies of all other reactants and products are known.

In particular, this last method is the basis for calculating entropies of ions in aqueous solutions. Absolute values of ionic entropies are not required for equilibrium calculations (since the electric charges must balance) and relative entropies based upon the convention that $S°_{298.15} = 0$ for hydrogen ion, H^+(aq), are commonly used. All ionic entropies are usually tabulated for hypothetical ideal solutions of unit molality (1 mole of solute per 1000 g of water).

PAUL DUBY

References

Kelly, K. K., 1960, "High-Temperature Heat-Content, Heat-Capacity, on Entropy Data for the Elements and Inorganic Compounds," *U.S. Bur. Mines, Bull.* **584**, Washington, U.S. Govt. Printing Off., 232pp.

Kelley, K. K., and King, E. G., 1961, "Entropies of the Elements and Inorganic Compounds," *U.S. Bur. Mines, Bull.* **592**, Washington, U.S. Govt. Printing Off., 149pp.

Kern, R., and Weisbrod, A., 1967, "Thermodynamics for Geologists," San Francisco, Freeman, Cooper and Co., 304pp. (translated by Duncan McKie from "Thermodynamique de Base pour Mineralogistes, Petrographes et Geologues," Masson et Cie, Paris, 1964).

Lewis, G. N., and Randall, M., 1961, "Thermodynamics," Second ed., revised by Pitzer, K. S., and Brewer, L., New York, McGraw-Hill Book Co., 723pp.

Rossini, F. D., Wagman, D. D., Evans, W. H., Levine, S., and Jaffe, I., 1952, "Selected Values of Chemical Thermodynamic Properties," *NBS Circular* **500**, Washington, U.S. Govt. Printing Off., 1268pp. (Presently being revised. Some revised sections have already been issued in advance as NBS Technical Note, **270-3**, 1968).

Waser, J., 1966, "Basic Chemical Thermodynamics," New York, W. A. Benjamin, Inc., 278pp.

Cross-references: *Enthalpy or Heat Content; Free Energy and Free Enthalpy; Phase Equilibria; Thermodynamics.* Vol. IVB: *Polymorphism.* Vol. V: *Free Energy and the Gibbs Phase Rule.*

ENVIRONMENTAL POLLUTION

It will be useful to consider whether the commonly accepted definitions of the words "environment" and "pollution" are adequate or sufficiently precise to serve this discussion. Webster's unabridged dictionary defines environment as "the aggregate of all the external conditions and influences affecting the life and the development of an organism" (2nd intern. ed.) The terrestrial environment is a composite of air, water, and land. Each can be polluted individually or compositely. Since the environment represents not only "conditions" but also "influences" which affect life, such influences on human performance as noise levels also have to be considered.

Pollution is defined by the Committee on Pollution of the United States National Research Council in the following way:

"Pollution is an undesirable change in the physical, chemical, or biological characteristics of our air, land, and water that may or will harmfully affect human life or that of other desirable species, our industrial processes, living conditions, and cultural assets; or that may or will waste or deteriorate our raw material resources." (1965)

This rather broad definition has been narrowed down somewhat by the California State Water Control Board in their characterization of water pollution. "Pollution of water is any impairment of its qualities that adversely and unreasonably affects its subsequent beneficial use." (1963). One notes that the change must be both adverse and unreasonable. The belief that a change from the natural state is not necessarily adverse, but may on the contrary be beneficial, is implicit in this definition. Furthermore, it is also implied that any substance introduced into the water might be considered as a potential pollutant if its concentration is large enough to affect the subsequent uses of the water.

Both definitions of pollution quoted above are based on anthropomorphic criteria. Although in the first definition reference is made to "other desirable species" the wording of the statement implies that the desirability of change is evaluated primarily according to its effect on the human species. The fact is that in a broad sense all of man's activities contribute in some degree to the pollution of the environment. Human existence necessarily leads to the accumulation and elimination of waste products, and all of these products entering the environment are pollutants. Thus, to define precisely what constitutes environmental pollution becomes an almost impossible task. Fresh water entering estuaries is a serious pollutant from the standpoint of the marine environment. Salt water intrusions into loose sedimentary strata and the subsequent replacement of underground fresh-water flows is a type of pollution, which causes great concern to populations inhabiting shore areas. In view of these examples, which show that seemingly unobtrusive natural forces can be considered as pollutants, one is forced, in order to have some point of departure, to accept artificial definitions similar to those quoted above.

Pollution problems can be analyzed from various standpoints, as outlined below, and this outline will be followed in the article.

I. Human uses of the three principal resources: air, water, and land. Consideration of both consumptive and nonconsumptive uses.
II. General categories of pollutants: conservative and nonconservative.
III. Types of pollutants: biological, chemical (organic and inorganic), and physical.
IV. The consistency of the pollutant: gaseous, liquid, solid, or energy.
V. Origin of the pollutant: domestic, agricultural, industrial, or commerce and transportation waste.
VI. Effects of the pollutant: immediate-gross, perceptible, imperceptible.
VII. Possibilities of abatement of pollutants: prevention, treatment (conventional and advanced), disposal, ultimate disposal, recycling, recovery.
VIII. Geophysical and geochemical processes by which the environment regenerates itself: dispersion, oxidative degradation-mineralization, anaerobic degradation.

In view of the interactions between various types of pollution and the several points of view from which any type of pollution can be described, there has to be some repetitions and overlap. Furthermore, some items of information have been deliberately repeated under several headings, with a view to speedy reference.

I. Human Uses of the Environment

The quality and purity of the environment depends upon the ways in which it is used, which may be either consumptive or nonconsumptive.

Consumptive Uses. (*1*) Direct human consumption of any environmental resource such as the drinking of water, the eating of salts, or the breathing of air will remove certain quantities of materials from the natural environment. These materials will not be immediately returned. (*2*) Agricultural uses of the environment, both in plant production and in animal husbandry, will also consume quantities of naturally available air, water, and mineral resources, in disproportion to their rate of replacement. (*3*) In industrial manufacturing processes the three principal environmental commodities are likewise consumed, although the consumption of mineral resources predominates.

As an example of industrial consumptive use of resources, one can consider the case of steel. Approximately 600 kg of new steel per capita are consumed each year in the United States. One-third of this is returned to the furnaces as scrap. About 410 kg of new steel enters the inventory of steel in use, whereas at the same time 350 kg of steel becomes obsolete or can be written off as a loss due to corrosion or other naturally occurring degrading processes. Of these 350 kg approximately 140 kg, or 40%, is recovered and returned to the steel furnaces in the form of junked materials. The balance, that is, 210 kg, is lost, never to be recovered. Most of this becomes buried in dumps; some of it is widely dispersed. This amount of lost steel obviously serves as an eventual source of pollution of the environment with metallic iron. During a one year period, the steel inventory in the United States increases by approximately 80 kg for each inhabitant. The mean lifetime of all steel products is approximately thirty years. In order to support one individual in our society, some 25 tons of materials have to be extracted from the earth and processed each year, including minerals and stone, sand, and gravel. This tremendous demand on the earth's resources leads not only to widespread barren areas and possible future exhaustion of the resources, but also creates a significant problem of waste disposal, both in the manufacturing and mining industries.

The total figures for the annual per capita consumption of metals in the United States show the following distribution: 150 kg each of copper and lead, 100 kg of aluminum and zinc, and 20 kg of tin. To meet the need for these raw materials in products, the nation has to transport 15,000 ton-km of freight per capita per year. Therefore, it seems that man has become an important factor in geologic changes.

The amount of rock and earth he moves, both for mining and construction purposes, has already reached great proportions in the industrialized areas of the world. Since population growth will undoubtedly increase and the developmental processes of the poor countries will also be accelerated, the total potential demand for mineral consumption will increase at remarkable rates. If the whole world population were to be brought up to the per capita steel inventory of the ten richest nations, some 200 billion tons of iron would have to be extracted from the earth. Since iron ores usually contain no more than 3% of usable metal, the disposal of the mining waste of this quantity must result in the creation of mountains of slag.

During consumptive utilization of environmental resources the transformation of materials, either within the body of an organism or through an industrial process, will produce wastes, which when conducted back into the environment will invariably lead to pollution.

Nonconsumptive Uses. The utilization of environmental resources in a nonconsumptive manner is most evident in the varying uses to which natural waters are put. Water serves as routes for navigation: cooling in power production; and habitats for aquatic life. The nonconsumptive use of every resource should, by definition, produce no pollution since in principle no transformation of the resource will occur. However, the fact is that even nonconsumptive uses within the present technological society invariably lead to increased levels of pollution. This fact is a consequence of both laxity in the enforcement of pollution-control laws and thoughtless or nonchalant attitudes toward the environment on the part of the users.

Nonconsumptive Uses of Water. That navigation pollutes waterways stems mostly from the following factors: (*1*) Fuel in outboard gasoline engines is only 65% utilized. The rest is directly discharged with the cooling water. (*2*) Bilgewater containing high levels of hydrocarbons and other compounds is routinely pumped into navigable waters. (*3*) Human waste products, including used waters and garbage, are usually emptied without treatment into the water.

About 65% of all industrial usage of water in the United States serves cooling purposes for power-generating plants. The cooling water effluents are usually 10 to 20°C warmer than the intake waters. Thus this nonconsumptive use results in thermal pollution.

Waters which serve as habitats for aquatic life also become polluted through efforts to harvest the life forms. Since harvesting of shell or fin fish invariably involves boating activity, and boating causes pollution of natural waters, these endeavors also lead to pollution. The common practice of processing the harvested life forms on board will also result in waste products, further adding to pollution levels. Finally, the practice of artificially fertilizing certain restricted areas for the purpose of increasing the catch may produce undesirable levels of both inorganic and organic substances in the waters.

Nonconsumptive Uses of Air. In this category there are two major uses: cooling purposes and flight pathways for aircraft and missiles. When air is deployed for cooling purposes, it usually does not become polluted. On the other hand, when air serves as the flight medium of aircraft, considerable quantities of pollutants are discharged into it. The exhaust gases of airplanes pollute the air directly and grossly. High-flying SSTs, through an increase of vapor condensation and ozone dissipation in the stratosphere, are bringing about secondary changes, the detrimental effects of which cannot as yet be adequately appraised. Take-off and landing noises of jet aircraft and the sonic booms generated by the breaking of the sound barrier cause serious noise pollution; this will be discussed later.

Nonconsumptive Uses of Land. National Parks, Forests, and Seashores used for recreation, can be seen as a source of pollution. The recurrence of man-made forest fires causes air pollution, as well as silting, a form of water pollution, following the erosion of the burned areas. And it is an unhappy fact that human visitors to wilderness preserves, created so that the city-dweller might vacation in a "pure" natural environment, leave behind appalling quantities of waste and garbage.

II. General Categories of Pollutants

In general, pollutants can be classified as either conservative or nonconservative.

Conservative Pollutants. Conservative pollutants are compounds or products which remain inherently unchanged when introduced into the environment. Carbon monoxide in the air, resulting from the burning of fossil fuels, will be retained there for prolonged periods of time without changing its nature. Certain pesticides such as DDT, when introduced into water, will not undergo degradation and will tend to accumulate. Plastic or metal containers, glass bottles, and other scrap materials, when buried on the terrain, may not decompose rapidly and will be retained in land fills for decades.

Nonconservative Pollutants. Nonconservative pollutants will sooner or later disappear from the environment at a rate dependent upon the character and quantity of the starting material. Their volume usually decreases in a logarithmic fashion, as they are degraded by biological or nonbiological forces. The concept of half-life, worked out originally for radioactive decay, can be utilized in estimating the rate of their disappearance. The concentration will thus be reduced to one thousandth (0.1%) of the original in ten half-lives or to one hundredth (1%) in seven half-lives.

III. Types of Pollutants

Depending upon their origins, pollutants can be classified as biological, chemical, or physical.

Biological Pollutants in Water. Biological pollutants in water are usually parasitic on humans; examples are certain water-borne viruses (hepatitis, cocsackie, echo); bacteria, especially Enterobacteriaceae, causing various diseases (dysentery, paratyphus) and even *Vibrio* (cholera); parasitic fungi, most commonly *Candida albicans* (responsible for thrush and athlete's foot); protozoa, which produce such widespread diseases as amoebic dysentery and leishmaniasis; and a host of parasitic worms which cause such common human diseases as schistosomiasis and tapeworm anemias.

One of the most common biological pollutants is the bacterium *Eschericia coli*. This organism is not pathogenic per se, but is a normal constituent of the intestinal flora of vertebrates. It enters into the waters through fecal matter, and is routinely used as an indicator for the presence of fecal contamination.

Biological Pollutants in Air. Many types of biological materials enter into the air. The most commonly found are spores of fungi, pollen grains of flowering plants, cysts or desiccated invertebrates, and also their eggs. Furthermore, certain portions of higher plants or animals may be quite abundant in agricultural and industrial areas, producing air pollution. Plant fibers are present above cotton fields: fur fragments (hairs) abound above both farms and leather factories. Any of these materials may cause allergic reactions in sensitive people. The incidence of hay fever for example in the present population of the United States is about 12%. Agricultural weeds such as ragweed are the most common contributors to the pollen enrichment of the air. Ragweed pollens can travel hundreds of miles and can reach concentrations of over 100 particles per liter. When their number is above 10 per liter, they usually cause serious allergic manifestations in sensitive persons. Some microorganisms, such as certain fungi indigenous to the soil of selected areas, may reach concentrations in the air at times of dust storms sufficient to cause severe mycotic infections. Thus, in the San Joachim Valley of California, *Coccodioides immitens*, an endemic organism, when breathed in large quantities, will produce in the exposed populace the frequently fatal infection of coccidiomycosis. At the present time there are no preventive measures available for the elimination of these air pollutants.

All of the biological pollutants are nonconservative, and if their source is eliminated their population numbers rapidly decline.

Chemical Pollutants. The chemical pollutants fall into two large categories: inorganic and organic materials.

Inorganic Pollutants. Inorganic materials may enter the air or the waters, either as by-products of industrial activity, transportation, or agricultural land use. Most of the components of stack emissions are inorganic gases such as carbon monoxide, sulfur dioxide, and nitrogen oxides. Many of the industrial waste waters contain quantities of inorganic acids, salts, or particulates.

Mining activities produce a particular type of pollutant from acid mine drainage. There are over 68,000 inactive mines in the Appalachian region of the United States alone, in which surface water seepage is producing underground waters with pH as low as 3. These waters, on surfacing, denude the environment, since very few organisms can tolerate such high levels of acidity.

Agricultural activities also produce considerable inorganic wastes. Nitrate and phosphate fertilizers are used in copious quantities in developed countries, frequently in unwarranted amounts. Parts of these inorganic salts which are not utilized by plants will eventually enter the underground water systems and end up in the rivers, lakes, and oceans. Small quantities of fertilizers cannot be considered as detrimental to the environment, since many of the oligotrophic rivers or lakes will be fertilized by them only to the point of higher productivity, but after certain levels have been reached, further addition of these inorganic nutrients will result in overfertility of the receiving waters. Such *eutrophication* of the waters will manifest itself in the occurrence of nuisance organisms, perpetual algal blooms, and the general deterioration of the aqueous environment. A prime example of this process is evidenced by the conditions prevailing in Lake Erie.

Organic Pollutants. The biggest source of organic pollution is domestic sewage. The organic material content of domestic waste waters is usually expressed in terms of oxygen consumption, which is the amount of oxygen required to oxidize the organic matter in the sewage into inorganic materials: carbon dioxide, sulfur oxide, nitrous oxides, and water. The generally accepted expression for oxygen consumption is the *biological oxygen demand* (BOD).

In addition to domestic wastes, many industrial processes involve the discharge of large quantities of organic materials into the environment. Paper and sugar mills produce waste waters loaded with carbohydrates; oil refineries discharge large quantities of hydrocarbons in their effluents; slaughter houses, meat packing, and dairy companies add fatty acids and proteins to the environment. The gasoline engine enriches the atmosphere with gaseous hydrocarbons, many of which contain lead in organic binding. Even waste-disposal facilities produce quantities of organic pollutants which are discharged into the atmosphere or into recipient waters. Incinerators pollute the atmosphere (due to incomplete burning) with hydrocarbons and highly toxic organofluoride compounds, which result from the incineration of plastics. Secondary treatment plants will remove only approximately 90% of the BOD, resulting in a residue which still contains 10% of oxidizable organic matter.

Agricultural land use increases organic pollution in two ways. Animal husbandry gives rise to large amounts of manure, the rational disposal of which is an unsolved problem. Further, the agricultural use of pesticides is a major contributor to organic pollution. The present practice of spreading manure for its fertilizing properties greatly adds to the enrichment of groundwaters through seepage of organic compounds, which usually are not degraded in the top layers of the soil, but enter unaltered the receiving bodies of water. Indiscriminate use of pesticides likewise enriches water through seepage and run-off, and these will either become accumulated in aquatic life forms or undergo slow decomposition.

Physical Pollutants. Physical pollutants result from changes brought about by the physical alteration of the properties of the environment.

Particulates in Water. Sediments entering lakes, rivers, or oceans due to natural erosion are usually not considered to be pollutants. Nevertheless, because of their turbidity-producing effect, they may sometimes modify the qualities of the water to a degree where it no longer can serve consumptive purposes. If the particles are sufficiently fine and are widely dispersed, their sedimentation may take years and they will remain in the water, traversing large distances.

Agricultural use of land areas greatly increases sediment production. In the United States, nearly four billion tons of sediment are reaching receiving bodies of water. These originate from farm lands, streambanks, highways, and urban constructions. These types of pollution from eroded soil are always detrimental to the beneficial uses of water, not only in terms of water supply, recreational usage, flood control, and waterway commerce, but also because they destroy reservoir storage capacities, silt up channels and harbors, and deflect streamflows. The degree of silt (i.e., suspended) load varies from river to river and in the upper range it might be as much as 200,000 mg/liter, whereas in the lower range it is only about 5 mg/liter. The average is between 10 and 30,000 mg/liter. Studies by the United States Department of Agriculture have shown that the clearing of forests and the continued use of land for row crops increased erosion 10,000-fold. Plowing of grassland for continuous row crop cultivation increased erosion twenty to 100 times. Erosion rates in areas of urban construction projects are approximately three times the averages of those found in agricultural areas. This increased erosion becomes a permanent feature of urbanized communities because of the consistently larger quantities of surface runoffs. This is mainly due to the reduction of surface permeability from paving.

The silting of harbors and channels requires the continual dredging of these bottom areas. The produced spoil represents a further problem for waste disposal. Since spoil is greatly enriched with inorganic materials, some of which are highly toxic, dumping grounds for spoil disposal have to be carefully selected. Indiscriminate dumping of spoil, as practiced in the New York bight area, will lead to the complete abolition of marine life.

The building up of sediments and debris in water beds, lagoons, or shallow estuaries may seriously impede the flow of water. The regenerative ability of natural waters is greatly dependent upon the rapidity of exchange. New water has to replace the old either through flow or tidal exchange, and currents are essential to transport away the organic pollutional load. Turbulences and waves will enhance the oxygenation of water, reducing, through degradation, the level of organic pollutants. If flow rates decrease due to accumulation of sediments, or tidal exchanges are diminished because of building up of sediment barriers, the self-regenerating activities of the waters decrease considerably and may eventually be abolished. Thus, the building up of barriers resulting in reduced flows will add to the pollution load in an indirect way by limiting the capacity of the waters for natural removal or degradation of the pollutants.

Particulates in Air. Physical aspects of air pollution include the emission of particulate materials from smoke stacks of industrial establishments, buildings, vehicles, or incinerators. Carbon particles and fly ash may enter the air in large quantities, producing dust covers over industrialized areas. Interaction of these particles with other constituents of the air such as carbon monoxide, and sulfur dioxide may, however, be beneficial. Speculations have been put forth to explain the seeming paradox of the nontoxicity of sulfur dioxide emissions. Supposedly, both the carbon and fly-ash particles serve as absorbents for the gas and through their precipitation they also tend to remove toxic concentrations of sulfur dioxide from the atmosphere. The construction industry, both by building and wrecking, contributes considerably to the enrichment of the air with dust particles. The moving of earth, sand blasting, the utilization of asbestos, all add to dust pollution. Wrecking especially contributes to air pollution, since no really good procedures have yet been developed for the prevention and precipitation of dust particles entering the atmosphere during the destruction of buildings. The present practice of water hosing the surfaces of structures built from brick or stone to bring down the resulting dust are highly inefficient. Thus, at sites of wrecking activities, dust concentrations may reach levels attained in nature only during serious dust storms.

Storms crossing open land areas and even cities, may enrich the air with large quantities of particulate matter through turbulences. Some of these finely dispersed materials do not settle easily and may be kept in the atmosphere for considerable periods of time, adding to the air-pollutional load. Explosives used as part of excavation procedures further increase the dispersion of solids in the air. Agriculture produces physical pollutants of the atmosphere in the form of plant fibers. Pollen grains and spores, resulting mainly from weed growth, may be considered as both biological and physical pollutants. Finally, atmospheric nuclear explosions will project into the air particulate materials that settle slowly and are enriched with radioactive particles, such as strontium-90, which eventually will reach the ground and become assimilated by plants. This element is capable of replacing calcium; therefore, it will appear both in the skeletal systems of mammals and in their high-calcium content products, such as milk. After the testing of nuclear devices it has been shown that the strontium-90 contents of milk are considerably increased. This eventually may lead to incidences of genetic disturbance in the exposed segments of the human population.

Solid Wastes on Land. Solid wastes resulting from households (garbage) and from industrial production represent the largest mass of physical pollutants that require disposal. Their accumulations on land areas not only create an esthetic nuisance but pose an ever-increasing problem as to their ultimate burial. Larger cities are rapidly running out of land-fill areas and new methods have to be developed for getting rid of the accumulating garbage from the increasing population. In 1970 every single individual in the United States population generated an average of seven pounds of garbage per day. The collection, transportation, and disposal of this material cost 4.7 billion dollars, making this the third largest municipal expenditure in the country.

Large parts of the nondegradable solid wastes end up eventually in waters in the form of floating debris. This type of pollution may ruin the esthetic recreational value of shores and beaches, hamper navigation, and decrease fishing areas.

Thermal Pollutants. Another physical type of pollution originates from the heating of the natural environment. Heating of air above large cities may produce convections during cold periods or may result in inversions, in some cases leading to the death of heart and lung patients.

Heating of water resulting from its use for cooling in connection with power generation will produce thermal pollution, eliminating stenothermic species in the recipient waters. Usually, organisms having wider tolerance towards temperature will replace the original fauna and flora and these species may then become over-abundant, creating at the least a nuisance, and other severe pollutional problems.

Heated power-plant effluents are usually enriched with heavy metal ions, leading to inorganic pollution of the discharge areas. Many of these metals are toxic for aquatic life or they are accumulated by certain organisms, such as shellfish, making them unpalatable or dangerous for human consumption.

Radioactive Pollutants. Some infinitesimal amounts of radioactive nuclides also enter receiving waters in the cooling water discharges. Atomic generating plants will supply 60% of the power requirements for the United States by 1980; therefore the amount of radioactive contaminants entering waters will quadruple by that date. These predicted levels are still not considered to be dangerous when the nuclides are widely dispersed, but again certain organisms have a tendency to accumulate up to 100,000 times the quantity of radioactivity in their bodies than that present at any given period of time in one liter of water. Consumption of such organisms, among them oysters, may lead to the accumulation of dangerous levels of radioactivity in the human populations.

Noise Pollution. With the advance of civilization, urban populations are increasingly exposed to higher levels of continuous or intermittent noise. The sources of noise pollution fall into five major categories: surface vehicular traffic, aviation, industry, construction and recreational activity. Tolerable levels of noise cannot be established objectively. Exposure to ninety decibels per eight hours is considered to be the upper limit which a human being can endure without permanent ill effects. This standard is equally as arbitrary as any of the standards set for tolerable limits of irradiation. High noise levels will not only produce physiological changes, such as deafness, but will result in psychological problems, such as increased irritability, lowered reaction time, and a general fatigue of the organisms exposed. Whereas no true noise-control measures have been established and, therefore, noise levels from the four first-mentioned categories do not seem to be reducible at this time, noise from recreational purposes, which represents one of the most damaging manifestations of this type of pollution, could be easily brought under control. Racing cars, jazz bands, and TV and radio stations continuously produce noise levels which are either at the threshhold of tolerability or represent a subliminal irritant, to which a great majority of the urban population is continuously exposed. Some twenty-five decibels per day can produce psychological damage in terms of fatigue. Thus, for the well-being of urban populations, stricter noise-control measures should be initiated.

Sound energy or noise has increased since World War II at a rate of about one decibel per year. As a consequence, major portions of the urban population are already exposed to deafening noise levels. It is estimated that approximately sixteen million industrial workers in the United States suffer some form of occupational hearing loss. Since several additional millions of people live close to airports, or in large cities, noise-induced hearing loss and other ill effects, mainly of a psychological nature, have already developed into a major national problem. Since at the present time there is no comprehensive legislation on noise control, except for the Walsh-Healy Public Contracts Act, which sets maximum permissible noise exposure levels for industries fulfilling contracts for the Federal government in excess of $10,000, not much hope exists that the effects of noise pollution will be brought under control. The Walsh-Healy regulation sets the permissible noise level at ninety decibel for an eight-hour exposure. On the other hand, construction equipment noise, such as that caused by air compressors, jack hammers, and pile drivers, may produce a combined noise output of 3000 or more decibels. Aircraft operations on the ground, such as engine runups for overhaul and repair, actually produce higher and more steady noise levels than the general aviation noise.

Attempts at noise control were instituted in ancient times and in the Middle Ages. For example, in 50 BC Caesar banned all chariot traffic at night in the city of Rome. In 1560 in response to complaints in London about the nocturnal noise due to the practice of husbands beating their wives Queen Elizabeth passed a law prohibiting males from beating their women after 10 o'clock in the evening.

IV. Consistency of Pollutants

Gaseous Wastes. Gaseous wastes appear most frequently in the atmosphere, but small quantities may be found in water, and even trapped in the soil. They originate from emissions of fossil-fuel power plants, industrial complexes, or from the exhaust of automobiles and other vehicles utilizing internal-combustion engines. Water vapors may also be considered, in certain instances, to be polluting modifiers of the environment. The deployment of cooling towers for dissipating excess heat (mainly from power generation) into the atmosphere will produce, especially in the cooler winter months, a saturation of the atmosphere with vapors which give rise to a lasting fog. For a power plant with a one million megawatt capacity, a cooling tower is required whose base is the size of a football field and which reaches about forty to fifty stories into the air. The vapors produced by such a tower could saturate the air in a fifteen mile district with a semi-permanent fog.

Gaseous wastes in the hydrospace occur as a consequence of anaerobic decomposition of organic matter. Bubbles of hydrogen sulfide may be dispersed in oxygen-depleted water, and its toxic effects may completely eliminate biotic forms. Hydrogen sulfide may also be trapped in soil which is not well aerated, especially in areas near garbage dumps and other refuse sites. Another somewhat less toxic gas, methane, is also produced from decomposing organic matter and may pollute waters or can be found above dumping sites.

Finally, some gaseous radionuclides, such as tritium or radon, are commonly present both in waters and in soil, either as naturally occurring elements or as a consequence of improper waste disposal. Atomic power plants produce and discharge some infinitesimally small quantities of tritium into their cooling waters, the half-life of which, fortunately, is only about thirteen days; thus its decay and disappearance from the environment is relatively rapid.

Table 1 shows air pollutant emissions broken down according to the different sources which contribute to it and the estimated levels these pollutants will reach by 1980 if no adequate abatement measures are instituted by that time.

Liquid Wastes. Liquid wastes originate mainly from domestic and industrial consumption of water, and as such they are considered waste waters. A compilation of waste water

ENVIRONMENTAL POLLUTION

TABLE 1. AIR POLLUTANT EMISSIONS[a]

Source	Million Ton	%
Automotive	86	60
Industry	23	17
Power plants	20	14
Space heating	8	6
Refuse disposal	5	3
Particulate matter		
Now	145 million tons	
1980	200 million tons	
Sulfur dioxide		
Now	30 million tons	
1980	45 million tons	
Solid waste		
Now	250 billion lb	
1980	750 billion lb	

[a] "Sources of Air Pollution," PHS Publication 1548, 1966.

according to its sources in the United States (in 1968) is shown in Table 2. To this list, however, one has to add the huge quantities of agricultural waste waters, which have their origin largely in animal husbandry. These waters are removed and spread together with manure on land but quite frequently they enter nearby receiving bodies of water without any purification.

Special industrial liquid wastes originate from the use of inorganic and organic chemicals in certain manufacturing processes. Spent sulfite wastes of the paper and textile industries; dyes of fabric manufacturers; nondegradable organic solvents used in plastic manufacture and by the cleaning industry; toxic electrolytes, such as chromium from electroplating; and large quantities of acids and alkalis employed in varied industrial applications, form liquid wastes which require special disposal. If any of these compounds should enter regular municipal sewage-treatment plants, they would ruin the biological abilities of the plant required for adequate treatment because of their toxic nature, temporarily "poisoning" the plant. Since the treatment of the above-mentioned wastes is most difficult and expensive, certain offshore areas have been designated as acid-waste dumps. In the New York Bight area, the industrial acid waste dumped is composed of 9% hydrogen sulfide, 7% $FeSO_4$, and 25% inert solids. Recent studies have shown that the sediment in this area is completely devoid of life, whereas in the waters of the dump sites large numbers of fish schools can be seen. This seeming discrepancy between the toxic effects of the dumped material on the bottom and its apparent fertilizing impact on the water itself requires further investigation.

Solid Wastes. Solid wastes are posing an ever-increasing problem. The United States produced 207 million tons of trash, garbage, scrap, and other debris in 1969. If compacted, it would have formed a wall fifty feet thick, four stories high, extending from New York City to Los Angeles. A breakdown of the sources of solid wastes is shown in Fig. 1. Although urban waste constitutes only about 7% of the total of solid waste production, it represents the biggest problem of disposal because of the scarcity of available dump sites in highly developed areas. The other 93% of the nation's solid waste originates largely from mining and agricultural activities, and is generally close to potential disposal sites. With the passing of time, however, even the disposal of this waste approaches a critical level. It is not only that the per capita genera-

TABLE 2. SOURCES OF WASTE WATER[a]

Source	Billion gal.
Domestic sewers	5,300
Primary metals	4,300
Chemical, allied	3,700
Paper, allied	1,900
Petroleum, coal	1,300
Food and kindred	690
Other manufacturing	450
Transportation	240
Rubber, plastics	160
Machinery	150
Textiles	140
Electrical machinery	91

[a] "Cost of Clean Water", U.S. Dept. of Interior, 1968.

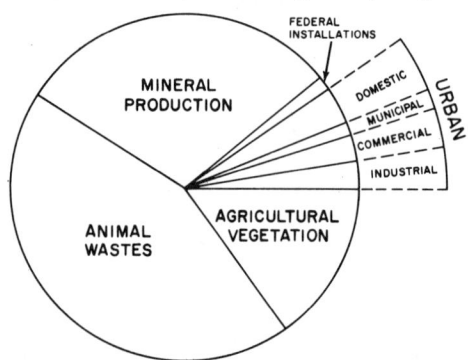

FIG. 1. The solid wastes generated in urban communities, although the crux of most disposal problems, constitute only a small portion of the total produced in the United States. The figures are given in an Ad Hoc Group report to the Office of Science & Technology, Executive Office of the President. In 1967, agriculture generated 58% (animal wastes, 43%; vegetation, 15%); mineral production, 30.8%; federal installations, 1.2%; urban, 7% (domestic, 3.5%; municipal, 1.2%; commercial, 2.3%) and industrial, 3%.

tion of waste has almost doubled during the past twenty-five years, but its composition has changed markedly, in such a way as to cause increased problems of disposal. Concentration of the population in cities and the manufacture of nonreturnable and nondegradable containers has brought about this increase in solid waste. Although glass or even aluminum containers do not pollute the environment, per se, except for causing esthetic damage, plastics may release highly toxic fluorides when they undergo slow degradation or incineration.

Energy Wastes. Thermal ionizing radiation and sound energy comprise this group of pollutants. Thermal energy is the result either of heating or of cooling activities. Both of these will dissipate considerable amounts of waste heat into the environment. Although large-scale temperature changes in extended areas have not yet resulted from this type of pollution, the heating of certain rivers or even lakes to a few degrees above ambiant level has already been noticed in connection with power-plant cooling water discharges. This problem becomes especially acute in subtropical climates, where the summer temperatures are too high to permit the efficient cooling of thermal effluents. Some nuclear power plants in Florida, the effluents of which raised the temperature of the receiving waters to 90°F in the warm summer months, managed to eradicate completely the biotic communities from their lagoons.

The effects of an accumulation of radioactive nuclides producing ionizing radiation, and a general increase on a global scale of the level of radioactivity as a consequence of nuclear testings, atomic generating plants, and the increased industrial-research and medical use of radioisotopes, will be discussed in detail under the section headed Imperceptible Pollution.

V. Origin of Pollutants

In discussing environmental pollution it is convenient to give a detailed analysis of the total pollution picture in reference to the origin of the polluting materials.

Domestic Waste. Domestic waste is either liquid or solid. The liquid waste will invariably end up in sewage-treatment plants, either private (cesspools) or municipal; whereas the solid waste, commonly called garbage, is usually disposed of by municipalities.

Liquid Domestic Waste. Private sewage treatment in the form of utilization of cesspools produces enrichment of the groundwater through seepage of both inorganic and organic materials. If the soil around a cesspool is not sandy but contains quantities of humates, its absorbtive-filtering ability will not be high enough to prevent the dissemination of bacteria in an area fifteen to twenty yards around a cesspool. Frequently (especially in clay-containing soils) the continuous use of the disposing facility will cause the formation of small channels in the soil, in which the discharged materials will flow unhindered, spreading fecal organisms in areas several hundred yards away from their source of origin. Even where the structure of the soil is adequate to retain bacteria, both the inorganic and dissolved organic matter will eventually enter groundwater flows from the cesspools and through acceding into the recipient bodies of water, they will fertilize it. In view of the inadequate treatment provided by cesspools, their installation should either be completely banned or restricted to areas where the nearest recipient waters are at least 500–600 yards away.

Where municipal collection of domestic waste waters exists, the sewers lead into a treatment plant which may provide for one, two, or three stages of treatment. In most instances the technology developed around the turn of the century is still being used and new methods of advanced waste treatment have not been put into practice. Both conventional and advanced treatment techniques will be discussed in Section VII.

The primary pollutants in domestic sewage are the human disease-causing organisms carried in the fecal material; the large quantities of organic fecal matter, creating a heavy oxygen demand during decomposition; and certain toxic substances, such as household poisons, pesticides, and some detergents. Domestic sewage is especially high in nutrients, originating from human fecal matter (organic compounds), urine (inorganic compounds), and the phosphates which have become liberated as a result of degradation of modern detergents. All of these compounds have fertilizing value, and when they reach natural bodies of water will produce favorable conditions for growth and/or overgrowth of aquatic plants. The insoluble residue of the sewage will result in a different problem, similar to that caused by suspended solids and floating debris in streams or lakes. Residue and debris will reduce the availability of light, increase oxygen demand, and impede water flow or flushing action in estuaries.

Solid Domestic Wastes. Solid household wastes, such as trash and garbage, lead to a form of pollution somewhat similar to that produced by untreated domestic sewage. Often these wastes are used as land fill in low wet areas and will form "drainage basins" for small streams and marshes. Great portions of the garbage itself may harbor pathogenic organisms, both for men or for rodents (rats) feeding on it. Rodents are vectors of several serious diseases which can cause epidemics in human populations. The decomposition of the solid organic waste may exert a large oxygen demand on the environment because in many cases the dump area is at least partially submerged; in addition, it will create foul odors noticeable for miles around. Used containers may release some toxic or otherwise harmful materials into the streams which drain the garbage dumps. The corrosion of metallic objects will lead to a heavy accumulation of metal ions in the waters, whereas the decomposition of the organic garbage will enhance the nutrient content of the receiving bodies. Large rainfalls or occasional floodings of the dumping sites will carry floating debris and suspended solids into nearby rivers or estuaries.

Agricultural Wastes. *Manure.* Agricultural waste, while very similar in composition to municipal and domestic wastes, presents an even greater problem than that caused by the human population. It is seldom realized that the farm animals in the United States produce ten times as much waste as people. The magnitude of this problem can be illustrated by the following statements. A cow generates sixteen and one-half times as much manure per day as a human, whereas one hog is equal to two human beings in waste production, and the waste from seven chickens equals that of one man. With the advent of intensive animal husbandry, the practice of maintaining animals in confined areas, such as poultry houses, has made the problem even more acute. On a modern poultry farm it is not uncommon that 100,000 birds are kept in one house. The wastes from such a production unit are equivalent to a human community of 15,000 people. Since agricultural land around cities, on which animal manure could be spread, is becoming especially scarce, this common mode of disposal of animal wastes, practiced for millennia, will have to be changed. Almost irreparable damage has already been caused by the disposal of untreated duck manure into estuarine waters along the southern shore of Long Island, where the over-fertilization of these bays has eliminated a once flourishing oyster industry.

Fertilizers. Another way in which agricultural usage of land contributes to the heavy pollution of receiving waters is through the indiscriminate use of fertilizers. In certain shore areas where intensive farming is practiced, as typified by the potato farms of eastern Long Island, phosphate fertilizers are used in concentrations roughly fifty times higher than the national average. Ninety percent of these fertilizers are not utilized by the plants and reach the estuaries as soluble phosphates, adding to the already high quantities of inorganic and organic nutrients brought in by sewer discharges, dumpings, and seepage.

Biocides. Within the past twenty years a whole array of synthetic organic compounds has been developed for use as biocides to control insects and plant pests. Since 1943, when DDT was first marketed, more than 90,000 pesticide products and formulations have been officially registered in the United States. According to the estimate of the United States Department of Agriculture, over one billion pounds of the major chemical pesticides were produced in this country in 1969. As many of these compounds resist degradation and can accumulate in wildlife, frequently concentrations of 80 to 100,000 times the amounts found in the environment can be detected in predacious fish or birds. By 1965, in the major river basins in the United States, dieldrin reached concentrations between 116 and 214 ppt, and DDT between 4 to 118 ppt. Today these two compounds are common and accepted constituents of all waters, soils, and the products grown in or on them. Predatory fish contain the highest levels. Chlorinated hydrocarbon pesticides are so widespread that even the wildlife of Antarctica, including penguins, sea birds, most of the fish, snails, and other invertebrates show varying amounts of DDT and DDE in their tissues. The imminent toxic effects of these compounds (in spite of extensive surveys) are not well known; and even less understood is their possible chronic toxicity. Only DDT levels have been measured for a significant length of time in humans and it appears to have reached a constant concentration in the body fat of the population of the United States. However, no clear-cut evidence of its deleterious effects has as yet been established. A report by the American Chemical Society (1969, p. 235): states that "There is no evidence at present that long-term low-level exposure to pesticides at concentrations approximating those found in the diet or the environment in the U.S. has any deleterious effect on man. At this time, therefore, the net effect of pesticides on human health in the broad sense is positive."

Industrial Wastes. Industry is one of the greatest producers of pollutants. According to Lacy and Cywin (1969): ". . . Industrial waste contains all known pollutants of concern in water pollution abatement as well as some unidentified factors. Wasteload estimates, based upon an estimate of the "average" quantity of pollutant per product unit, indicate that chemical paper, and food and kindred industrial groups generate about 90% of the BOD in industrial wastewaters before treatment, and that this amount is nearly three times as much as from the sewered pollution of the United States."

In both amount and composition the industrial wastes will vary with the different types of industry and with different factories. Føyn (1965) classifies the main forms of industrial wastes and their chemical types as compiled in Tables 3 and 4.

Treatment of industrial waste is usually carried out by municipal waste treatment plants. Such industrial establishments which produce only a high BOD can safely discharge their waste effluents to sewage systems because the plants probably will be able to handle their waste load. Occasionally, overloading of plant facilities and subsequent poisonings of the secondary treatment beds may occur as the consequence of an unusually high organic load discharge, but such instances are relatively rare. Naturally, when sewage-treatment facilities are utilized, new industrial plants producing high BOD levels will necessitate the expansion of local treatment facilities. In the United States only New York City levies a surcharge on industrial establishments for treating their effluents. The average industrial contribution to pollution is shown in Table 5. Table 6 is the calculation used by the City to establish the surcharge sale, whereas Table 7 shows the excess pollutants treated and the industries billed for them in 1967. This novel approach to "fining" or charging the pollut-

ing industry for the treatment of waste is a concept which probably could be of substantial help in financing the enormous cost of pollution abatement.

Industrial wastes which are strictly toxic, such as nondegradable biochemicals, acids, concentrated heavy metals, etc., obviously cannot be channeled through a municipal treatment facility. Such wastes either have to be treated on the site, or special provisions have to be made for their disposal. Large industrial complexes usually have their own waste-treatment facilities, the effluents of which can enter municipal treatment plants. Thus, acids are neutralized with lime; electrolytes (chromium, mercury, lead, etc.) are precipitated in the form of insolubles; other toxic chemicals are reclaimed and recycled. Where no such facilities exist, special dumping sites are allocated for the disposal of toxic waste materials, and they are either barged to the sea or injected into underground cavities (natural or artificial). Abandoned mines are being filled and sealed with waste products. This latter practice, however promising, produces as yet unsolved technological problems. Underground flows in the mine tunnels eventually exert such pressure on the dumped material that the closure ceilings break open and the outgushing waters wash along the disposed waste. Such outbreaks will then produce acute environmental pollution, leading occasionally to the complete denudation of the area covered by the water flow.

In connection with industrial waste, two specific industries, the mining and the petroleum industry, have to be discussed.

Wastes from the Mining Industry. A special type of industrial waste results from mining activities, and these are particularly damaging pollutants because they introduce large concentrations of both highly soluble inorganic matter and sediments, debris, and other solids that do not settle, such as colloids, into waters. Slurries are frequently used to transport ores through pipe lines where they are mixed with large quantities of water. Usually it is less than half of the slurry water which is recovered for reuse; the rest is released into either settling ponds or streams. These slurry waters will introduce into the local water flows both their dissolved and suspended matter. Many heavy metal ions are highly soluble in water, and arsenics,lead, and even silver salts frequently are carried in the rivers. Brines of common dissolved salts, if released in high enough concentration, will even change the osmotic characteristics of the water, being toxic to the organisms both on a chemical and physical level. Radium-226, a long half-life

TABLE 3. INDUSTRIAL WASTES

Origin	Type of Waste
Slaughter houses Fish and fish-meal factories Breweries Fat and margarine industry Sugar and starch factories Fruit and beet sugar Dairy products Fuel	Organic compounds
Chemical pulp mills, paper mills Fiber and plastics industry Textile industry Leather industry Soap works Photographic Chemistry	Organic compounds and inorganic compounds
Coal iron metal Mining Dye-stuff High explosives	Inorganic compounds

ENVIRONMENTAL POLLUTION

TABLE 4. DRINKING WATER STANDARDS

Determination	USPHS	WHO	Rational Limit (McKee)
Bacterial			
Coliform bacteria, per 100 ml	1.0	0.05[a] 1.0[b]	
Physical			
Turbidity, silica scale units	5	—	
Color, cobalt scale units	15	—	
Odor, maximum threshold number	3	—	
Chemical (mg/liter)			
Alky benzene sulfonate	0.5	—	
Ammonia	—	0.5[a]	
Arsenic	0.05[c]	0.2[a,b]	0.05
Barium	1.0[c]	—	
Cadmium	0.01[c]	0.05[a]	
Calcium	—	200[b]	200
Carbon chloroform extract	0.2	—	
Chloride	250	350[a]	500
Chromium (hexavalent)	0.05[c]	0.05[a,b]	
Copper	1.0	3.0[a]	1.0
Cyanide	0.2	0.1[a,b]	
Fluoride	1.6–3.4[c]	1.5[a]	1.5
Iron	0.3	1.0[b]	1.0
Lead	0.05[c]	0.1[a,b]	0.1
Magnesium	—	125[a]	125
Magnesium + sodium sulfate	—	1000[b]	
Manganese	0.05	0.1[a]	0.1
Nitrate, as NO_3	45	50[a]	
Phenolic compounds	0.001	0.001[a]	
(Potassium)			500[d]
Selenium	0.01[c]	0.05[a,b]	0.05
Silver	0.05[c]	—	
(Sodium)			500[d]
Sulfate	250	250[a]	
(Sulfur)			500[d]
Total Solids	500	1500[b]	
Zinc	5.0	5.0[a]	5.0
Radiological (pc/liter)			
Radium–226	3[c]	—	
Alpha emitters	—	1[a,b]	
Strontium–90	10[c]	—	
Beta Emitters	1000[c]	10[a,b]	

[a] WHO European Standards of 1961.
[b] WHO International Standards of 1958.
[c] Mandatory. Others are recommended by USPHS.
[d] Does not appear as such in original list.

radio nuclide, is a waste product of uranium mining operations; arsenics are released from silver deposits. Cyanide is used in the extraction processes of gold, and the purification of table salts will produce brines high in potassium chlorides. All of these compounds are highly toxic; therefore the mining industry attempts to remove them in forms of "relatively insoluble precipitates." Radium is converted into radium sulfate, and arsenic into its carbonate. Slight solubility, however, still can introduce significant quantities of toxic materials into the aquatic environment. Furthermore, the heaps of tailings which, in most cases, contain these toxic compounds and other contaminants, are exposed openly to weathering, and rain flows percolating through them will invariably dissolve and carry these materials into open rivers. Treatment of tailing overflows seems at this time to be both impractical and uneconomical.

The suspended solids may be slag, sluice, sands, or clays washed from around the desired ores. They will cause turbidity in the receiving streams and may occasionally fill river beds, slowing water movements. In highly turbid

TABLE 5. INDUSTRY AVERAGES—POLLUTANT CONCENTRATIONS[a]

Industry	Firms Sampled	Concentration (mg/l) BOD	SS
Slaughter house	2	1,785	1,920
Meat processor—no pretreatment	5	762	414
Meat processor—with pretreatment	7	485	255
Frozen desserts—open refrigeration	3	537	157
Frozen desserts—closed refrigeration	3	982	440
Milk processor	6	964	266
Juice and fruit drink bottler	5	1,813	116
Brewer's spent grain processor	1	23,762	4,848
Commercial baker (sweet products)	7	3,745	1,530
Candy manufacturer	7	1,635	121
Brewery	2	1,090	422
Soft drink bottler	3	488	38
Potato chip manufacturer	4	1,110	1,657
Commissary	9	586	236
Instant coffee manufacturer	2	594	237
Semiindustrial laundry	6	1,045	601
Pharmaceuticals	2	957	137
Fur dresser or dyer	8	791	962
Laundry (family)	10	535	237
Laundry (linen)	12	695	256
Diaper service	4	270	231
Industrial laundry (wiping cloths)	2	2,506	1,661
Rug cleaning	7	121	74
Prepared salads (potato salad, cole slaw)	6	4,386	5,041

[a] After Imbelli et al. (1968).

waters, owing to a decrease in light penetration, oxygen production is simultaneously decreased. Thus the regenerative ability of the waters is impaired, favoring the development of anaerobic bottom processes, and leading eventually to the release of further noxious or outrightly toxic materials, such as hydrogen sulfide, into the water. Clay minerals, however, may have a beneficial effect in the streams, since they serve as natural absorbants or flocculants for many organic and some inorganic compounds, and thus may improve the quality of the water by removing some nutrients from it.

Wastes from the Petroleum Industry. Finally, in the discussion of industrial waste, special attention has to be paid to the petroleum industry. Drilling for oil or gas results in suspended solids, which may be washed away from the drilling site and released into the environment. Brine deposits are frequently associated with petroleum, and the highly concentrated salt solutions are often forced from the drill pipe by the released well pressure. When this brine enters fresh water it can cause osmotic stress or even death to the organisms present. Accidental spills or uncontrolled leaks of the oil and tar floating from the area may be directly toxic to some aquatic life, especially to birds which frequent the water's surface, or may deprive organisms of adequate illumination. The volatile residue will result in a heavy oxygen demand. When these residues are accumulated by commercially valuable organisms, such as oysters, they render the products unmarketable.

Dramatic instances of oil pollution, such as the *Torry Canyon* disaster of 1968 and the Santa Barbara seepage of 1969, represent only a fraction of the damage caused by the

TABLE 6. COMPILATION OF SURCHARGE[a]

$$D_s = CFV[(SS - 350) + (BOD - 300)] \ldots 1$$

where
- D_s = amount of surcharge ($)
- C = cost per pound ($) of removing pollutants, expressed to the nearest tenth of a cent
- F = $\frac{62.4}{1,000,000}$, i.e., the factor for converting mg/l to lb/mil-ft^3
- V = volume in units of cubic feet of wastewater actually discharged into the sewer system
- SS = mg/l of suspended solids
- 350 = allowable mg/l of SS below which there is no surcharge
- BOD = mg/l of biochemical oxygen demand
- 300 = allowable mg/l of BOD below which there is no surcharge

[a] After Imbelli et al. (1968).

TABLE 7. EXCESS POLLUTANTS TREATED AND SURCHARGES LEVIED[a]

Industry	No. of Accounts	Surcharges Billed (1967)	Percent of Total	Flow Billed (mgd)	Excess Pollutants (lb/day)
Bakery (sweet goods)	44	257,663	27.9	2.65	31,514
Laundries	132	150,896	15.3	7.66	64,125
Breweries	4	91,608	9.2	1.52	21,280
Pharmaceuticals	4	72,678	7.3	4.61	35,645
Milk processing	22	61,199	6.2	2.11	21,582
Meat processor	77	82,130	8.3	3.02	26,358
Commissary	10	50,051	5.1	1.81	4,492
Candy manufacturing	28	31,729	3.2	0.56	8,234
Frozen desserts	24	32,109	3.3	1.62	14,161
Brewers' spent grains	1	26,489	2.7	0.02	19,443
Soft drink bottlers	45	21,824	2.2	2.03	8,762
Miscellaneous chemicals	5	13,205	1.5	0.17	2,575
Steel drum processing	6	6,096	0.6	0.14	1,191
Fur dresser and dyer	11	5,406	0.5	0.06	955
Juice and fruit drinks	7	5,022	0.5	0.04	591
Miscellaneous	45	61,257	6.2	1.12	14,578
Total	465	987,362	100.0	29.14	275,466

[a] After Imbelli et al. (1968).

deliberate cleaning of tanks by ship operators in local waters. There are about 100 oil handling terminals, 100 commercial and naval shipyards, and over sixty oil tank cleaning firms in the United States and in these localities oil is liberally disposed of into the seas during cleaning operations. The effects of chronic oil pollution on marine life are not yet well understood, but from the data available, one can infer that a drastic reduction in both invertebrate fauna and algal flora occurs, permitting the survival of only some protista.

Wastes from Commerce and Transportation. Wastes from commerce in general originate mainly from packaging materials and advertisements. With the change brought about by the introduction of nonreturnable packaging goods, the tendency on the part of individuals to reuse packaging materials has declined, whereas throw-away items have increased. In order to sustain a high level of production, deliberate production of wastes became a national enterprise. If goods are not recycled, either because of nonreusable packing materials, or because the products must be replaced after a very limited lifetime, production is sustained at an artificially inflated high level. Thus, the trend to use relatively cheap materials for packaging, such as aluminum or plastic containers, or to reduce artificially the useful lifetime of items of production, creates massive quantities of wastes in the form of scattered junk and discarded products. To the heaps of packaging and advertising materials must be added junked cars and household equipment with brief use spans. Except for sustaining high levels of production, there is no reason why the life span of an automobile should be only four to five years instead of twenty. The same holds true for practically all household equipment: refrigerators, washing machines, and the like. At this time, there are in the United States over sixty million discarded automobiles piled up along highways, on streets, riverbanks, etc. Their number increases by two and a half million each year. Municipalities are unable to cope with this problem. At the same time, the compacting of these junk piles for reuse in the manufacture of steel is economically feasible only on a very large scale. In order to reuse the vehicles, their interiors have to be burnt out before they are compacted; and since plastics are extensively utilized in automobile manufacture, and the burning of plastic produces noxious gases laden with fluorides for which no efficient air pollution abatement methods have yet been devised, the disposal of junk automobiles and other vehicles of transportation will not be achieved easily without one form or another of pollution. The disposal of discarded military equipment represents a special and large aspect of this problem. At one place in the Arizona desert there are over ten thousand decomissioned World War II military airplanes sitting on the ground. These obviously will not decompose simply by exposure to the elements.

The waste produced by advertisement in the form of masses of paper trash amounts to approximately one hundred pounds per person per year of unnecessary and unwanted printed junk. Incineration of these materials alone, leaving aside consideration of real garbage, would produce serious air pollution problems. Overtaxed city sanitation departments cannot cope with the spreading of leaflets, giveaway items, and other advertising gimmicks in the streets, all of which add to the unhealthy levels of garbage accumulation in the cities. Advertising also contributes appreciably to sound pollution (mentioned in Section III).

Transportation Wastes. As a generalization one can state that since the beginnings of man's existence no means of transportation has yet been devised which would not pollute the environment in one way or another. There are only degrees in the amounts of pollution and differences in types. Rates of comparison, however, can be made. Since the waste produced by one horse in a day is equal to that of sixteen and one-half men, and that of a fifty-horsepower engine is equal to that of twenty-two men, a return to the horse as a means of transportation would hardly seem to be an improvement on our present situation. With increased technological development serving a much larger population, more efficient methods of transportation than the horse and buggy had to be developed, but with them came waste.

All vehicles utilizing internal-combustion engines will pollute primarily the air. This point has been discussed under the section headed Gaseous Pollutants. Land-based vehicles, such as cars, trucks, buses, military vehicles, etc., will pollute not only the air but also secondarily the water. Gasolines containing lead, after burning, will enter into the atmosphere, from whence the lead-containing organic molecules become precipitated and will end finally in waters. Recent studies have shown that highways running parallel to the lake shores will produce a substantial accumulation of lead in the lake water. Although toxic levels for the biotic community have not yet been reached, these accumulations will probably manifest themselves in long-range, imperceptible pollution. Lead, a heavy metal ion, is accumulated selectively by many aquatic organisms; it is passed along the food chain, at the end of which concentrations several thousand times those present in the environment can be found.

Wastes originating from transportation by trains, dependent upon the type of power used, will be somewhat different. Coal and diesel engine trains will pollute the atmosphere and the hydrosphere more or less in the same manner as automobiles. The big advantage, however, lies in the much smaller amount of pollution per unit of transported goods produced by them. Furthermore, their pollutants are more easily dispersed and diluted, since railroad lines often run through open countrysides where they may be the only pollution contributors. Passenger trains produce another type of pollution: that of human waste. Trains are not equipped with waste containing or waste-treatment facilities, and the human waste products are openly discharged onto the tracks. Admittedly, such discharges on infrequently traveled routes do not seem to cause substantial pollution; but on commuter tracks or between closely placed large cities, the accumulation of human waste products not only creates an esthetic nuisance, but can serve as well as a breeding ground for insects and flies, which participate in the dissemination of diseases.

Electric trains do not pollute the atmosphere, and they would seem therefore to be the ideal mode of transportation. One should not forget, however, that they are a prime example of the situation where consumption on a secondary level does not cause pollution, but is shifted back to the primary energy producer. Power plants happen to be one of the major contributors to both air and thermal pollution.

Transportation by boats (mentioned earlier) pollutes waters mainly from the discharge of human waste, bilge waters, and unused fuels. The air polluting effects from powered boats are similar to those produced by internal-combustion engines on land, but their direct discharges into water are much more serious. A report to Congress provides a tentative estimate of the size of the problem and its remedial costs (U.S. Department of Interior, 1968). The study shows that in any given year 110,000 commercial and fishing vessels, about 1,500 Federally owned vessels, and about 8,000,000 recreational watercraft use the navigable waters of the United States. In addition, approximately 40,000 foreign ships enter United States waters each year. Very few of these ships are equipped with proper waste-disposal systems and they usually discharge their untreated or inadequately treated effluents into the estuaries. This causes a further pollutional load which these slow-moving bodies of water can no longer cope with. Laws are being enacted to force existing American watercraft to be outfitted with waste treatment or at least with waste-storage facilities, which would reduce the pollution they cause in the estuaries. However, the cost of complying even with present-day regulations is estimated to be on the order of 600 million dollars, an expenditure the individual boat owners probably will try to avoid.

Airplanes—private, commercial, and military—pollute the air primarily with their exhaust gases. To these one has to add the human waste generated in passenger planes, the disposal of which eventually will either lead to water or land pollution. What has been said about other means of transportation, such as cars or diesel engine trains, holds true for airplanes too, except that their air polluting effects are probably less dangerous because of the greater dispersion and dilution of the pollutants. High-flying supersonic planes cause an additional pollution problem by creating an atmospheric condensation of vapors. An SST burns approximately seventy tons of fuel per hour, the exhaust gases of which are adequate to condense the vapors in the stratosphere, producing an eventual dissipation of ozone. This problem of stratospheric ozone may have serious consequences as far as the total ecological balance of the earth is concerned.

The noise produced by planes at takeoff and landing and the breaking of the sound barrier by supersonic planes increase considerably the level of noise pollution. In addition, sonic booms may be outrightly damaging to terrestrial structures.

Special pollutional problems are posed by the firing of missiles. It is not only the exhaust gases of these projectiles which produce pollution, but, more importantly, our inability to sterilize these delicate instruments. The possibility of contaminating the higher atmosphere with terrestrial microorganisms seems to be remote, but missile landings on extraterrestrial bodies or the invasion of space by astronauts must be seen in the light of the potential hazard of contaminating these hitherto virgin spaces. Outgassing, both from space ships and from the space suits worn by the astronauts, is apparently a technologically unresolvable and unavoidable problem. The microbes leaving the astronauts bodies on the lunar surface travel approximately twelve meters before reaching the ground. Thus in a circle with a twelve-meter radius the lunar soil is contaminated around each astronaut. Each step made by the space traveler increases the radius of the contaminated area. If for no other reason than to ascertain the original conditions existing on the moon, it is important to obtain uncontaminated surface samples. Since, however, collections of rocks or soil are carried out within the twelve-meter circle surrounding the astronauts, occasionally it might become difficult to ascertain whether a certain microorganism is indigenous to the lunar terrain or is the result of terrestrial contamination through outgassing. In this context one should again note that all of man's activities result in some form of pollution, and even our first step on an extraterrestrial body has led to an unavoidable pollution of its surface, attesting to man's presence by leaving behind his waste.

VI. The Effects of Pollutants

Immediate Gross Pollution. *From Domestic Wastes.* Gross pollution of the environment often results from industrial waste discharge, or more frequently from accidents. Since the initiation of waste-treatment practices, massive or imminent pollution through domestic discharges no longer occurs, with one important exception. Apparently working on the assumption that the oceans have an unlimited capacity to assimilate waste, there are still nine United States counties, each with populations of over 50,000, which empty their sewage without any treatment into the open sea. In the rest of the country, domestic wastes undergo at least primary treatment, which partially alleviates the gross pollutional picture that untreated domestic waste may produce.

From Industrial Discharges. Industrial discharges can lead to immediate kills of large numbers of the biotic community. Areas completely denuded of vegetation, the end result of acid mine waste seepage, are frequently found in the Appalachian regions. Elimination of certain desirable species from the environment as a result of a sudden influx of toxic wastes from industrial processes has been noted throughout the country, but most obviously in the marine environment. In the dump areas of the New York bight, the only fauna and flora to survive are in the form of some microorganisms. Sudden releases of thermal pollution produce fish kills by the millions. The dead animals are then carried downstream and their rotting bodies not only cause a nuisance, but also seriously interfere with further consumptive uses of the water.

Perceptible Levels of Pollution. *In Water.* Most gross pollutional effects can be ascribed to accidents, whereas perceptible levels of pollution are the end result of slowly accumulating factors which, during the course of years, bring about a change in the biotic composition of the environment. Even ignoring the directly toxic manifestations of pollutants, environmental degradation will occur inevitably through fertilization. This phenomenon, which is basically a natural process, is only speeded up by human activity. Many lakes or rivers can safely be assumed to have been oligotrophic at the time of their origin, meaning that the amount of inorganic nutrients was minimal and practically all the dissolved organics were absent. At this stage, only very limited biological production can take place; after the death of the photoautotrophic organisms their bodies undergo rapid decomposition and mineralization, and thus there is no fast increase in the nutrient value of the waters. Over a period of thousands of years, however, this situation gradually changes. Both through the accumulation of nutrients within the water, by indigenous production, and through exogenous processes, such as washing in, the lakes and rivers undergo a process called eutrophication, that is, a nutrient enrichment giving rise to higher and higher levels of production. Eventually, the balance becomes shifted towards the accumulation of such great quantities of organic matter that they can no longer be decomposed aerobically, and instead of fast mineralization through oxidative processes, anaerobic sediments will accumulate, leading to the development of swamps or marsh lands. Erosion naturally plays an important role in the filling up of these areas. Man's activity in dumping or otherwise releasing quantities of nutrients into these originally less-productive bodies of water helps to speed up the natural and unalterable process of eutrophication. As a result of such processes, many of the desirable species disappear from the water and become replaced by forms not utilized by man. Lake trout have become practically extinct in Lake Erie, and during the past twenty years have been replaced by a small, herring-like fish which is not harvested for either commercial or sport fishing purposes. It can be questioned whether this so-called deterioration of Lake Erie represents a severe detriment as far as the natural balance of the environment is considered. If the sewage from the large industrial centers around the lake had not entered it, the same eutrophic conditions which are now present in it would have most probably developed in any case in another 200 years. The speeded-up eutrophication has resulted in conditions which are considered to be detrimental only as far as the human population is considered. Although the lake trout have disappeared, the productivity of the lake has increased thirty times, being able to support an animal population ten times larger than that of about twenty years ago. There is no reason why these smelts could not be harvested with appropriate modern techniques and utilized commercially for human consumption, perhaps in the manner of sardines.

Other perceptible types of pollution occur in water, air, or on land as a consequence of nonconsumptive utilization of these resources. Outcries about the pollution of harbors or frequently travelled waterways are raised everywhere. But, again, the issue should be weighed with respect to what is more important: having cheap facilities for the transportation and distribution of both agricultural and industrial products to supply the needs of a growing world population, or restricting commercial boating in waterways in order to restore them to their more or less natural condition. In other terms, is it more important to clean up the East River to the degree that one could fish in it, or to have an easy and cheap access for supplying a population of eight million people with goods?

In the Air. Perceptible pollution of the air is evident only above and around large cities or industrial establishments. Air pollution problems, therefore, become critical only in less than one millionth of the air space immediately surrounding the earth. On the other hand, about 5% of the earth's population is living in these urban areas, and, in their interest, adequate measures of air pollution control should be put into effect.

On the Land. Perceptible pollution of the land is a consequence chiefly of inadequate garbage disposal or the scattering of trash. Problems involved here are mainly esthetic, both in terms of foul odors and of unsightly views.

Imperceptible Pollution. The most serious problems are represented by the imperceptible, long-range, unpredictable manifestations of pollution which may lead to slow but drastic environmental changes. Four factors have to be discussed in this context.

(1) Biocides. The influence of biochemicals (that is, organic compounds such as biocides) including the large-scale application of defoliants, cannot be precisely predicted in advance (see paragraph on pesticides). Since these chemicals by now are present throughout the entire surface of the earth—owing to dispersion, both by water and air—they may influence profoundly the whole global ecosystem. It has been pointed out that many of the pesticides kill off oceanic algae, which are not only primary producers in the marine food chain, but are also the sustainers of an oxidizing environment, Planktonic algae in the oceans produce twice as much oxygen as all the land plants on the earth and if these forms become eliminated, or reduced in number, the ratio of atmospheric oxygen to the other gaseous elements may slowly change, leading theoretically to the eventual extermination of all higher life forms. The predictive figures presented relative to this possibility are controversial, but the global oxygen reserve is large and the depletion rate rather small.

Nevertheless, it must be urged that further studies be undertaken to clarify the problem.

(2) Burning of Fossil Fuels. Another global change which may result in a disturbance of the earth's ecosystem is thought to be brought about by an increase in the utilization of fossil fuels. The burning of coal and petroleum products, and the incineration of waste, results in an atmospheric enrichment of CO_2. The liberation of this gas into the air may increase its amounts relative to other gases to the degree that it will cause a so-called "greenhouse effect" and may eventually lead to an increase in the average temperature of the earth. This could result in the melting of the polar ice caps, a rise in the level of the world's oceans, the submergence of some areas of the continents, and certain changes in climate which could be intolerable for many species both of plants and of animals. Such a global change could have profound effects on the human race.

(3) Depletion of Ozone. Recently some warnings have been put forth regarding the serious consequences on the climate of the earth which might result from the regularly scheduled flights of SST. These planes during each hour of flight will burn, at cruising altitudes in the lower part of the stratosphere, approximately 70 tons of fuel. From this, 83 tons of water, 72 tons of carbon dioxide, and 4 tons each of carbon monoxide and nitric oxide will be produced. Assuming 550 SST flights per day, the mean residence time of water vapor in the stratosphere will be one and a half years. According to calculations based on this model, water vapor in this limited region of the stratosphere would increase from 3 ppm to 5 ppm. This would cause a decrease of $1.5°C$ in temperature of the stratosphere by radiating heat into space and an increase of $0.6°C$ temperature at sea level, as a result of the greenhouse effect. Furthermore, the increased water vapor levels in the stratosphere would decrease its ozone concentration due to photochemical decomposition of water and the subsequent reduction of ozone to molecular oxygen. A reduction of ozone, which is the earth's most efficient shield against lethal ultraviolet and cosmic radiation, could eventually lead to increased numbers of mutations and the possible extermination of life. On the other hand, the most important greenhouse agent in the stratosphere is ozone and a reduction of ozone would lead to a cooling of the lower atmosphere, thus an opposite effect. The question is therefore controversial and as yet unsolved.

(4) Induced Mutations. The fourth possibility of long-range effects from pollution is an induced mutation rate due to the presence of increasing amounts of radionuclides in the atmosphere and the water, and possibly even in the soil. The introduction of radioactivity into surface waters may stem from numerous causes. Both atmospheric contaminants resulting from nuclear explosions and the leaching of uranium deposits will increase the level of radioactivity in water. Aquatic or marine life accumulate radionuclides in their body and through their concentrating effect further enrich the natural environment. Utilization of radioisotopes in medical therapy, research, and industrial processes, the irradiation of people with x-rays and emitters, and the preservation of foods through irradiation will all produce some active nuclides which eventually will turn up in water. Low levels of radioactive wastes are discharged in laundries and hospitals or poured into the sinks of laboratories. All of these eventually will reach sewage systems and end up in natural waters. These isotopes do not decay rapidly, and thus can be disposed of only by dilution. Nuclear power plants also produce a low but constant level of radioactive waste. Both the U.S. Public Health Service and the World Health Organization have set standards for safe levels of emitting materials in drinking water. Alpha emitters cannot be present in higher quantities than 1 pico-curie per liter (pc/l) and beta emitters more than 10 pc/l. This situation, however, is greatly complicated by the occurrence of natural radioactivity. Some hot springs may contain large amounts of radium emanation through which the receiving waters acquire considerable radioactivity. Whereas the River Thames has as slow an activity as 0.01 pc/l, some spring waters in Japan show over 700,000 pc/l. It has been pointed out many times that there is absolutely no safe limit for exposure to ionizing radiation. The probability that a single medical x-ray could produce a genetic change, with consequent lethal mutations in the offspring, is relatively high. Exposure even to permitted levels of ionizing radiation will eventually result in genetic changes, the effects of which may be deleterious in forthcoming generations. Since man's activities, no matter in how small a degree, add to the level of radionuclides in the environment, an increased rate of mutations should be expected.

For the most part, mutations are not useful for the survival of any particular organism, and it can thus be speculated that a general decline of survivability may ensue throughout the whole biotic community. From this point of view, man will not present an exception, and his degeneration in connection with the slowly increasing levels of radioactivity may lead to his eventual extinction.

All four of these pessimistic forecasts, if fulfilled, would see the widespread extermination of life on the planet as a direct consequence of man's activity through the pollution of his environment. Whether or not the predictions, which range from 100 to several thousand years before the final effects would manifest themselves, are realistic will have to be rigorously investigated. Suffice it to say that although concrete proofs are still lacking, the speculations presented above warrant sufficient attention and concern to show the need for large-scale research to ascertain the probability of survival.

VII. Possibilities of Abating Pollutants

Prevention. Although it is recognized that prevention of pollution is the most efficient way to get rid of the problem altogether, in many instances this is not the most feasible answer. It is possible to prevent wastes generated by households, industry, or commerce from entering the environment, but it is almost impossible to prevent pollution from natural phenomena. Unless global environmental control is achieved, which is not foreseeable in the near future, factors such as storms, floods, or even sun spot activities will add to environmental pollution. Prevention, therefore, has to be restricted mainly to those activities of man which speed up the processes of natural pollution. Adequate sewage-treatment facilities may completely clean their effluents, thus eliminating the entry of pollutants into the environment; but it is impossible to prevent storms from washing nutrients into rivers or estuaries. Through the institution of adequate air pollution abatement measures, it may be feasible to remove all atmospheric contaminants from emission, but it still will not be possible to prevent an increase in carbon dioxide in the atmosphere through forest fires. By deep burial on land, or through injection into safe underground cavities, one could remove all radioactive waste produced by humanity from the surface of the terrain, but it will never be possible to control sun spot activities, the ionizing flux of which produces occasionally 100,000 times higher levels of cosmic radiation than what is considered normal and tolerable.

In view of these facts, preventive efforts can be seen to be mainly concerned with the slowing down of phenomena over which man does not have control and to which mankind's activities are merely additive as far as "detrimental" effects are concerned. One has to emphasize, however, that there is no real basis for the evaluation of naturally occurring pollutional factors with respect to their supposed detrimental effects, nor any real basis for supposed detrimental effects, nor any real basis for the evaluation of mankind's activities in terms of changes brought about in the environment. It is important to consider all effects in terms of the whole earth as an entity. Some environmental changes may appear clearly detrimental in a specific area or in relation to a certain species, and yet far-reaching beneficial effects on a global scale may still favor the practice leading to these effects. For example, the use of DDT is known to be destructive to certain bivalves and fish regarded in the United States as desirable species, but the effectiveness of DDT, both in combating malaria and in aiding in the production of higher crop yields in underdeveloped countries faced with serious problems of malnutrition, may be seen as a strong argument in favor of the benefits ensuing from DDT application. One may examine in the same way arguments against the installation of nuclear power plants. Admittedly, limiting the numbers of nuclear power plants will help to limit increases in the amounts of radioactive nuclides which enter the environment; but the advantages of the inexpensive source of energy which such plants can provide may in the future be more than a matter of luxury; energy derived from nuclear power may be necessary for the basic needs and well-being of our burgeoning populations. It seems to be a question of where to draw the line between human progress for the benefit of all humanity and the preservation and restoration of the environment for the enjoyment of a relatively small fraction of humanity.

Treatment. Since, at least at the present time, prevention of pollution on a grand scale seems to be impossible, attention is focused more on the possibilities of waste treatment. However, the yearly increase of industrial waste production in the United States is about 4.5%, three times higher than the population increase. Furthermore, rapid urbanization generates increases in domestic waste at the rate of 6% per year, whereas disposal and treatment facilities are only growing at a 2.1% rate annually. Obviously, even to catch up with these disproportions would require efforts patently unfeasible from the economic point of view, if one is restricted to conventional treatment methods. It is more or less evident that the building of sewage treatment plants or the incineration of solid waste is not a solution. Conventional techniques will not only not be able to overcome the lag but will be a source in themselves of further deterioration of the environment.

Conventional Treatment. Sewage treatment, both that of domestic and of industrial waste, consists of three distinct steps. Primary treatment refers to the simple settling and screening of solids. The effluent contains the solids that do not settle and all the inorganic and organic nutrients dissolved in the water. Its oxygen demand may be as high as 3 g/liter, and it is loaded with both pathogenic and other microorganisms. Its inorganic content is high in terms of fertilizing ability. Secondary treatment is aimed to reduce the amount of solid matter and to remove approximately 90% of the oxygen demanding substances. Most of the microorganisms are killed through chlorination, and an effluent from a secondary treatment plant will have only 10% organic contents, but it will still contain almost the same amount of nitrates and phosphates in solution as the original sewage. The composition of the average secondary treatment effluent is presented in Table 8. Finally, tertiary treatment aims at a reduction of another 7% of the oxidizable compounds, and concomittantly, the removal of some of the inorganic nutrients, especially of phosphates. The establishing of primary and secondary sewage-treatment plants to handle the domestic and industrial waste produced in the United States alone in 1968 would have required an immediate investment of 30 billion dollars. This sum does not include the treatment of other types of waste such as those originating from mining activities or from thermal pollution. It would not cover the disposal of radioactive wastes and does not take into account the restoration of the already deteriorated environment. Costs of disposing of garbage are also not included in this sum. If the total waste problem in the United States in 1968 had been dealt with and some provision had been included for environmental restoration the cost would have been over 100 billion dollars. Obviously, no country can afford such a price. Conventional methods of waste treatment, therefore, will have to be replaced with advanced techniques. Several new and proven procedures exist, none of which, however, has as yet been put into practice.

Advanced Waste Treatment. Table 9 enumerates the more promising advance waste-treatment schemes. These techniques are aimed at the recovery of water and its reuse more or less in its original state. A comparison of waste-removal capacity of conventional and advanced waste treatment plants is given in Tables 10 and 11. Calculations show that unless these techniques are instituted by 1980, the water resources of this country will be totally exhausted (see Fig. 2). On the other hand, if advanced waste treatment techniques are employed, the gap between resources and usage will show a favorable trend (see Fig. 3).

Ultimate Disposal. In areas where a shortage of water is not a limiting factor and it therefore need not be reclaimed, the best possibility for ultimate disposal seems to be to utilize the almost unlimited capacity of the oceans to ab-

TABLE 8. AVERAGE COMPOSITION OF MUNICIPAL SECONDARY EFFLUENT[a]

Component	Average Concentration in Secondary Effluent (mg/l)	Average Increment Added During Water Use[c]	
		mg/l	lb/day/ 1000 pop.
Gross organics	55	52	64
Biodegradable organics (as BOD)	25	25	31
Methylene blue active substance (MBAS)[b]	6	6	7
Na^+	135	70	86
K^+	15	10	12
NH_4^+	20	20	25
Ca^{2+}	60	15	18
Mg^{2+}	25	7	9
Cl^-	130	75	92
NO_3^-	15	10	12
NO_2^-	1	1	1
HCO_3^-	300	100	120
CO_3^{2-}	0	0	0
SO_4^{2-}	100	30	37
SiO_3^{2-}	50	15	18
PO_4^{3-}	25	25	31
Hardness ($CaCO_3$)	270	70	86
Alkalinity ($CaCO_3$)	250	85	100
Total dissolved solids	730	320	390

[a] After Weinberger et al. (1966).
[b] Apparent alkyl benzene sulfonate.
[c] Concentration increase from tap water to secondary effluent.

TABLE 9. SUMMARY OF AWT PROCESS DEVELOPMENT STATUS[a]

Process	Current Scale
Separation Processes—	
Adsorption—granular activated carbon	75,000; 100,000; and 300,000 pilot plants in operation
	5 mgd, 1 mgd, and 0.5 mgd, plants under design
	7.5 and 0.5 mgd plants under construction
Adsorption—powdered activated carbon	15,000 gpd pilot plant in operation
Adsorption—on inorganics	—
Adsorption—on coal	—
Adsorption—on polymers	—
Oxidation—by H_2O_2	—
Oxidation—by O_3	Bench-scale studies ready to start
Oxidation—by corona	—
Oxidation—by autoxidation	Laboratory tests underway
Electrodialysis	75,000 gpd pilot plant in operation
Foam separation—simple	500,000 gpd pilot plant in operation
Foam separation—with additives	Bench-scale studies ready to start
Emulsion separation	—
Electrolysis	—
Electrochemical degradation	—
Biodenitrification	1 mgd pilot plant
Foam harvesting	—
Evaporation	1,000 gpd pilot plant under construction
Extraction—of contaminants	—
Extraction—of water	—
Freezing	Initial bench-scale studies completed
Eutectic freezing	—
Hydration	Desk-top studies
Reverse osmosis	5,000 gpd pilot plant in operation
Ion exchange	5,000 gpd and two 10,000 gpd pilot plants
Coagulation—inorganic	35,000 gpd pilot plant in operation
	75,000 and 100,000 gpd pilot plants under construction
	0.5, 1.0, 5.0, and 7.5 mgd pilot plants under construction
	200,000 gpd pilot plant being designed
Coagulation—organic polyeloctrolytes	100,000 gpd and 200 mgd pilot plants being designed
Filtration—diatomaceous earth	75,000 gpd pilot plant in operation
Filtration—rapid sand and dual media	Bench-scale studies in progress
	0.5, 5.0, and 7.5 mgd pilot plants being designed
Flotation	35,000 gpd pilot plant in operation
Biomass treatment	Initial pilot-scale studies completed
	1 mgd pilot plant being designed
Activated sludge (control)	100,000 and 200,000 gpd pilot plants being designed
Bionitrification	200,000 gpd and 1 mgd pilot plants being designed
Biological PO_4 removal	In-plant studies being planned
	Two, 100,000 gpd pilot plants being designed
Biological PO_4 removal by mineral addition	200 gpd initial pilot scale tests
	3 mgd (with hydraulic short circuiting) pilot tests in field
Coagulant regeneration	Bench-scale and pilot studies being planned
By-product recovery	Desk-top
Foam recycle	Laboratory tests completed

[a] After Weinberger and Stephan (1967).

sorb wastes. There are some twelve billion cubic feet of seawater available for the disposal of wastes for every individual on the earth. According to more or less well established calculations, one part of sewage should be diluted with 200 parts of water in order to avoid damage to the ecology of the recipient body. Thus, given proper dumping and dispersing methods, the oceans could assimilate without harm the combined waste products of all the inhabitants of the earth for several hundred years to come. That this is not the case is

TABLE 10. REMOVAL CAPABILITY OF CONVENTIONAL PRIMARY—SECONDARY TREATMENT[a]

	Percent
BOD[b] removal	90
Total organic removal	80
Suspended solids removal	90
Total phosphate removal	30
Total nitrogen removal	50
Dissolved mineral removal	5

[a] After Weiss and Okun (1967).
[b] Biochemical oxygen demand.

largely due to two factors: first, wastes are not carried far enough out into the open ocean and second, even if this is done, provisions are rarely made for their proper dispersion.

Transportation by Barge to Sea. Sewage-treatment plants produce a residue—sludge—which has to be disposed of and incinerators leave noncombustible solids—ash—which have to be transported somewhere. The disposal of sludge or ash is an expensive problem for which no really good solutions exist in practice. Barging to the sea and dumping as carried out in New York City is not only costly, but because of the proximity of the dump sites to the shore, ends in serious damage to the estuarine environment. Land-fill areas are rapidly running out for both the dumping of sludge and for the deposition of ash. Thus, such practices will have to be modified.

Barging could, however, be a means for rational sludge disposal, if it were carried out properly. The barges would have to empty their loads approximately 100 miles from the shore around the edge of the continental shelf where the ocean currents could disperse the materials into deeper waters, carrying them only very gradually towards the shore. The nutrients rep-

TABLE 11. REMOVAL CAPABILITY OF SEVERAL AWT TERTIARY PROCESSES[a]

	Foam Separation (%)	Coagulation-Sedimentation (%)	Adsorption (%)	Electrodialysis (%)
BOD removal	93	93	99	99
Total organic removal	85	85	99	99
Suspended solids removal	92	99	99	99
"Hard" detergent removal	85	55	95	98
Total phosphate removal	30	95	95	97
Total nitrogen removal	50	50	55	75
Dissolved mineral removal	5	10	15	50

[a] After Weiss and Okun (1967).

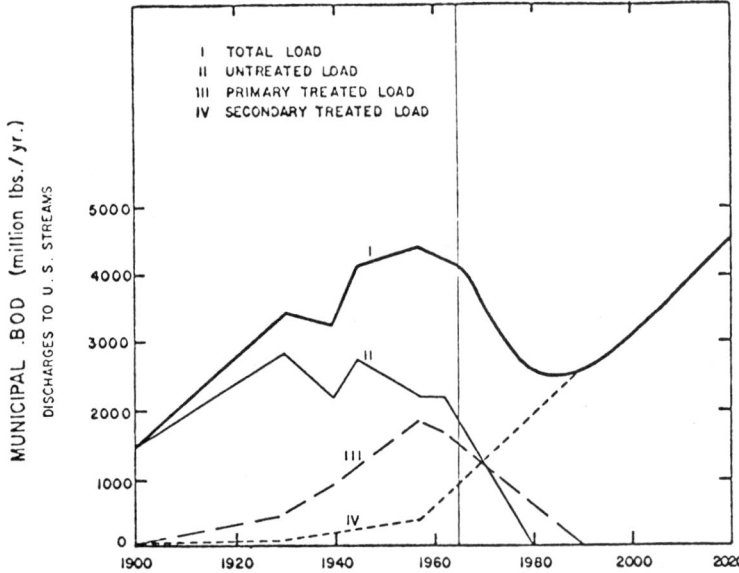

FIG. 2. Estimate of BOD discharges to United States streams from municipal outfalls (AD 1900–2020). After Weinberger et al. (1966).

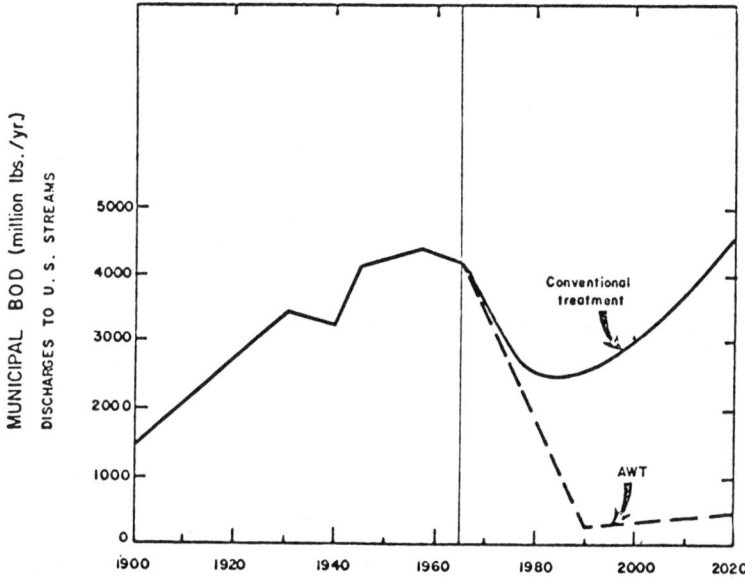

FIG. 3. Potential impact of "Advanced Waste Treatment" (AWT) on municipal discharges of BOD to United States streams. After Weinberger et al. (1966).

resented in the wastes would fertilize the sea, enriching planktonic growth and ultimately resulting in better fish catches. However, the purchase and maintenance of large-capacity barges fast enough to travel such distances seems uneconomical. Thus, it is unlikely that barging will ever serve as a practical solution for this problem.

Incineration at Sea. Another method recently proposed would be to equip large boats with modern incinerating facilities. These ships would run out to the open seas, incinerate their waste loads there, and pollute the air far enough from shore not to have any effect on the terrestrial atmosphere. The ash resulting from incineration would be released to the open sea and become adequately diluted. Old, Liberty-type cargo vessels could be equipped with modern, high-temperature incinerators, and three such ships could actually take care of the garbage-disposal problem of the whole city of New York. The almost nine million inhabitants of this metropolis produce some twenty-three thousand tons of garbage per day. The capacity of a Liberty ship is roughly ten thousand tons, thus it is safe to assume that three such ships making alternate rounds to the open ocean could adequately service the city's daily garbage-disposal problem.

Piping to the Open Ocean. The idea of laying large diameter underwater pipes reaching to the edge of the continental shelf and carrying all the waste material of a large city out beyond the shelf has been pioneered by scientists of the Franklin Institute of Philadelphia. Detailed cost analyses have been performed for the United States Department of Interior, FWPCA Advanced Waste Treatment Research Group. Such a project seems to be economically feasible, in either of two ways: (*1*) The pipeline does not have to be longer than thirty miles or, (*2*) it collects wastes from an area serving more than one community. In the first case, such pipelines could be laid on the West Coast of the United States where the continental shelf is relatively narrow, seldom exceeding twenty miles. On the East Coast, on the other hand, it would be necessary to build up a continuous collecting network from the large metropolitan centers to gather all the wastes of the cities and their suburbs corresponding to populations of over a few million in order to pay for the costs of laying a pipe running over the 100 some miles of continental shelf on the East Coast. According to cost calculations, incineration of 1000 gallons of sewage would cost 2.5 cents, to which the transportation and final disposal of the ash have to be added. Direct disposal to the sea of the same amount of sewage would be less than 1 cent even if it is collected from a 100 mile distance. The cost would be 5 cents per 1000 gallons if the conveyance distance is expanded to 1000 miles. If, however, the whole area covered by a radius of 1000 miles is serviced by a large diameter pipe then the cost would drop considerably below that of other methods of disposal. One also has to take into account that the cost of ash disposal resulting from incinerator residues will increase the original amount required, the total of which would more than balance the higher cost presented by long-distance conveyance.

There is considerable resistance, both in scientific circles and among sanitary engineers, to the idea of piping waste out to the sea. Admittedly, the ecological effects of concentrated wastes, even in the open sea, may be harmful, but this is more a presumption than a proven fact. In the modern design of large-diameter pipelines adequate care is taken for the proper dilution of wastes so that at the time of their emergence into the sea they no longer represent a concentrate, but have already undergone the required 1 to 200 times dilution. Indeed, the apparently more progressive West Coast of the United States has already installed two large-diameter pipe outfalls, one serving an area in Washington State and another in San Diego, California. Two pipelines are being laid in Los Angeles, California. The results of two years of operation have shown the predicted beneficial effect in terms of a greater diversity of the bottom fauna.

Injection into Underground Cavities. Modern schemes for ultimate disposal of treatment residue include land spread and injection into underground cavities, in addition to ocean dumping. Although both the ocean dumping and land spread seem to be relatively cheap solutions to the problem, the first one invariably results in the pollution of estuaries, at least when the dumping sites are too close to the shore, and the second one cannot be efficiently utilized in view of the growing lack of available land areas. Furthermore, the spreading of sludge or ash, although in some instances it may be beneficial by fertilizing sandy or clay-dominated soils which lack nutrients, in the long run, through leaching or seepage, will result in over-fertility of the ultimate receiving bodies of water. Thus, for final disposal, the only feasible solution besides piping to the open ocean seems to be the injection of waste into natural or artificial underground cavities. Approximately one-third of the United States terrain is composed of rocks fractured adequately for disposal of waste by injection. The naturally occurring underground cavities could also be greatly expanded by underground explosions of atomic charges, thereby creating large systems of subsurface cavities, as well as by drilling and mining, adequate to receive waste for long periods of time to come. According to costs analysis, disposal of either concentrated or diluted wastes into underground cavities is cheaper than any treatments presently employed. Nevertheless, particularly in seismic areas or in structurally unstable situations, serious earthquake potentials are involved, as demonstrated by experiences in the Denver region.

Recovery and Recycling. Both ocean disposal through piping of waste to the edge of the continental shelf or injecting it beneath the earth's surface, would eventually result in a loss of fresh water. Since the fresh-water reserve of the earth is less than 1% of the total water supply, such a loss to the oceans or to the deeper layers of the terrain cannot be permitted unless inexpensive methods of producing fresh water out of the oceanic masses are developed. Until desalination becomes economically more feasible, conservation of water resources will have to play a role in the planning of future waste-treatment and disposal techniques.

Closely connected with water recovery is the utilization of other resources which are wasted at the present through discharges or dumpings. Recycling of materials will become essential for the healthy economic growth of this country in the next two decades. In certain industries, the recovery and recycling of utilizable raw materials from wastes is already an established practice. Thus, paper manufacturers are not only recovering cellulose fibers for further utilization of their effluents, but also obtain the sulfites spent in their manufacturing processes. Some power plants burning coals with a high sulfur content, which would enrich the environment with quantities of the noxious gas sulfur dioxide, can recover industrially useful sulfates from their smoke-stack emissions by new scrubbing techniques. The waste heat generated by cooling either the water or the air of power plants, smelters, or other industrial complexes, has been used experimentally to benefit the environment. For example, on Long Island, favorable environmental conditions for year-round growing of shellfish were established by using thermal effluents. In Hungary, the heating of certain sections of cities has been achieved from thermal wastes. Collection and recycling of nondegradable products, such as aluminum cans, has led to the lowering of prices in the manufacture of this metal. The recycling of these raw materials in the economy postpones the exhaustion of the natural resources. Whereas the heaping of used cars throughout this country not only results in ugly sights, but also, through slow corrosion, leads to a general enrichment of the environment with potentially toxic heavy metals; their reuse as raw material for the production of steel or the recent practice of sinking them in the estuaries for the building up of artificial reefs—breeding grounds for fish—makes them useful. The recycling of used cars would delay the gradual exhaustion of natural iron resources. The second practice could yield an increase in fish production, providing nutrition for a growing population. Several other new and ingenious schemes have been worked out for the reutilization of waste products, two examples being prefabricated building materials made out of fly ash, and the use of dried sludge as fertilizer. In a time of decreasing resources, growing population, and a worldwide demand for higher living standards, the recycling of resources seems to have taken on significance.

VIII. Geophysical and Geochemical Processes in Environmental Regeneration

The natural environment has a significant, but not unlimited, capacity to cope with pollution. Ecological cycles can utilize pollutants, returning them to the resources of the earth, but when these cycles become overburdened by the sheer amount of pollutants the natural processes no longer can exercise their beneficial effects. Both physical and chemical processes take place in the three areas of the biosphere, that is, air, water, and land; these processes will help to reduce the accumulation of polluting material.

Dispersion. The dispersion of polluting material in the air is achieved by diffusion or, more efficiently, by wind action. Dispersion is basically a process of dilution, by which local accumulations of high concentrations of polluting materials are thinned out. Frequently dilution alone neutralizes the toxic effects of noxious pollutants. Also, by removing these materials from the site of their production and dispersing them over large areas, more time is provided for their degradation and elimination from the atmosphere. Man's activities may help in the naturally occurring process of dispersion, since explosions, blasting, or even local heating of the air produce convective currents which will bring about the dispersion of polluted materials.

Energy dissipation through cooling is also a form of dispersion by which the local accumulation of polluting material, in this case heat, is dispersed throughout an area large enough to absorb it without detrimental effect to the environment.

The process of dispersion in water is very similar to that occurring in the air. Currents and waves induced either by winds or by the disproportionate heating of water layers will help in carrying away and dispersing locally accumulated pollutants. The processes of diffusion, especially in the case of liquid wastes, will also serve the same purpose, that is, they will give rise to ultimate dilution of polluting materials. It is estimated that a 1- to 200-fold dilution of any toxic material will result in an ultimate concentration so low that it will no longer have a detrimental effect on biotic communities. On the other hand, one should keep in mind that in the case of liquids that are not miscible with water, such as oil, dispersion may be more detrimental than useful in reducing the biologically harmful effects of the pollutant. One gallon of oil spilled on water, when formed into a monomolecular layer, would cover a surface of over 100 acres. This surface, which is the interface boundary between the hydrosphere and the atmosphere, is a most important habitat for many earth organisms. Bacteria, algae, insects, and even birds, live at this interface. When the habitat of these organisms is destroyed by an oil slick, primary- and secondary-production abilities of the area become restricted. On the other hand, when dispersed, the polluting material more easily succumbs to final degradation, therefore, dispersion should be viewed on the whole as a positive factor.

Processes of natural dispersion of waste on the land are not as useful as in either the air or water. Waste can be blown by wind or carried away by streams, or may be washed into the earth by rainfalls, but these are relatively unimportant factors in the dilution of waste concentrates. Animals also may contribute to the spreading of waste materials, by carrying away or scattering garbage. In this respect, man's unintentional or intentional activities play a more important role. The scattering of trash throughout the countryside is one not very helpful way of diluting the local accumulation of waste products, whereas the spreading of manure on land or sprinkling of sludge for fertilizing purposes will serve to disperse waste materials, while at the same time supposedly contributing to the enrichment of the environment through positive use of wastes.

Oxidative Degradation. By now it is generally accepted that the terrestrial atmosphere was originally reducing. This was transformed into the present-day oxidative environment in the middle Precambrian period—between 2.9 and 1.8 billion years ago—as a result of plant activity. Undoubtedly, the most dramatic pollutional effect ever suffered by the earth environment was this drastic change in its atmosphere. The liberation of oxygen—a most toxic substance for anaerobic life forms—eradicated most of the biotic community existing at that time. New types of life had to develop, capable of survival in an oxidative environment. Geochemical processes took on new characteristics, based on the slow oxidative degradation of both organic and inorganic materials.

ENVIRONMENTAL POLLUTION

Chemical oxidation of reduced compounds (many metals for example) occurs constantly on the surface of the earth when the compounds are exposed either to the hydrosphere or to the atmosphere. When biological oxidation is also involved the processes are greatly speeded up through the enzymatic degradation of the organic materials. These processes are most evident in the water and on land. The efficiency of biological oxidation is very high in comparison to the simple chemical process; thus the fact that decaying organic matter does not accumulate in intolerable quantities either in water or on the land is due mainly to microbial oxidative activity. With a few exceptions, all organic materials are attacked by extracellular or intracellular enzymes. The final products of such degradation are carbon dioxide, sulfur dioxide, nitrous oxide, water, and some minerals. The whole process is called mineralization, meaning that through degradation, the mineral resources of the earth are reconstituted.

Oxidative degradation in the air, on the other hand, may result in the accumulation of some toxic substances. Ultraviolet irradiation reaching the earth will produce ozone. This gas, being an extremely active oxidant, will interact with sulfur dioxide—liberated both from natural oxidative degradation, but more pronouncedly from the burning of sulfur-containing fossil fuel—and together they will produce the highly corrosive gas sulfur trioxide. This gas, further combining with atmospheric vapors, will form sulfuric acid, which will exert a corrosive action on practically everything on the earth's surface.

The degree of oxidative degradation in water is directly proportional to the amount of dissolved oxygen. Oxygen in water occurs either as an effect of diffusion from the atmosphere or as a result of the photosynthetic activity of aquatic plants. If the level of organic material in the water is relatively low, the dissolved oxygen content is usually sufficient to oxidatively degrade the organics without causing oxygen depletion. However, if the organic load is too high or the productivity of the water is such that an overabundance of plant cells occurs (which, during the nights, are no longer oxygen producers but consumers), either temporary or complete oxygen depletion of the water may take place. Since aquatic fauna are dependent on the oxygen content of water for their survival, even temporary oxygen depletion results in mass mortality of animal life. The decaying remains will add to the already existing organic load, and then the chances for the natural recovery of the water become almost zero. It is estimated that at the current rate of waste production and discharge into the natural bodies of water of the United States, given the presently existing inadequate waste-treatment facilities, by 1980 all the available oxygen in the waters will be used for the mineralization of the organic load. In other words the level of the biological oxygen demand (BOD) by 1980 will be 100% for the total fresh surface waters of this country. Should this happen, none of the freshwater bodies will be able to fully assimilate the organic load, and because of their oxygen-deprived state, they will be turned into anaerobic sewers, incapable of supporting higher forms of life. It is obvious that drastic measures will have to be taken in the next decade to prevent the occurrence of such a catastrophic alteration of the environment. Fouling of underwater structures by organisms is also part of the natural phenomenon of oxidative degradation. Many microorganisms and higher invertebrates require substrates for their life. Any underwater structure can serve such a purpose. Attached organisms will interfere with the integrity of the structure and will present an unpleasant task of either replacement or cleaning. Fouling organisms usually secrete extracellular products which penetrate and eventually bring about the corrosion and disintegration of the underwater structures. For the existence of these organisms, certain levels of dissolved oxygen must be present in the water. That in heavily polluted waters, such as those represented by New York Harbor or the East River, fouling of underwater structures does not represent a catastrophic problem, is due to the low dissolved oxygen content of these waters. Whereas the heavy organic pollution present is instrumental in the elimination of healthy biotic communities, by the same token the exceedingly high oxygen demand exerted by the organic load also produces depletion of dissolved oxygen to the degree that fouling organisms can no longer survive. Thus, there is little destruction of the underwater structures.

Oxidative degradation on land manifests itself mainly in corrosion. Corrosion is also mostly due to microbial activity. For most microbes, a relative humidity of at least 30% is necessary for survival; thus in hot, arid climates corrosion seldom occurs, or the process is very slow. In climates with high humidity and favorable temperatures, such as those of the coasts of the United States, land corrosion of materials and structures attributable to microbial oxidative processes is a serious problem. With the exceptions of some plastics or glass, all natural or man-made materials are susceptible to microbial attack. Given proper humidity levels, such compounds will be rapidly decomposed and mineralized. This natural process, while on the one hand exceedingly helpful in removing and returning into the resources the waste products of humanity, on the other hand poses serious problems regarding the preservation of structures used by man.

Anaerobic Degradation. Degradation of products through anaerobic activity obviously does not occur in the atmosphere. These processes are restricted to water and to a lesser degree to land environments. Incidence of anaerobic processes in open waters is relatively infrequent. Some bogs, marshes, and swamps show exclusively anaerobic degradation of organic materials, but such phenomena occur normally in the subsurface layers of sediments.

One of the most common products of anaerobic microbial decomposition of organic matter is hydrogen sulfide. This highly reactive gas either will be produced in such copious quantities that it will bubble to the surface of the waters, or it will be consumed in the downward migration of metallic ions and form sulfides. If hydrogen sulfide is produced either in the water or in the sediment, it will poison the biotic communities, leading to a depletion of the oxygen supply in the water. The formation of sulfide minerals in sediments will turn them into a blackish ooze which in geology is generally called "sapropel." Mineral sulfides are usually toxic for higher plants; therefore, even if these bodies of water become filled through natural sedimentation processes, a considerable time must elapse before true vegetation can again develop on them. In marshlands or swamps the vegetation utilizes only a fraction of the minerals available in the bottom sediment, their main inorganic supply being the result of uptake through their stems from the surrounding water. Because the lack of oxygen in these bodies of water prevents the complete mineralization of the organic matter, the dead remains of the biota become only partially degraded. Thus, swamps or marshes will produce quantities of peat or sapropel which eventually may undergo further carbonification and in time will serve as lignite or coaly deposits. Small quantities of the organic carbon in these deposits will be reduced to the degree that either gaseous or liquid hydrocarbons will be produced. Most prominent among these gases is methane, which usually bubbles out from the swamp and when ignited will give rise to straying lights. Many of the other gaseous hydrocarbons become trapped in the sediments and eventually will form natural gas, whereas the liquid hydrocarbons are the basis of petroleum.

Anaerobic degradation processes on the land occur only in the deeper subsurface strata. These processes are relatively unimportant compared with those found in watery environments. Anaerobic microorganisms can degrade organic compounds which occasionally are washed down or buried in these deep layers of soil. The end products of such degradation will be the same as those found in sediments.

Both the geophysical and geochemical processes—both organic and inorganic—will eventually recycle and purify, in the human, utilitarian sense of the word, the main resources of the earth: air, water, and land. This phenomenon of purification, which must be considered as being the normal or natural process in an oxygen-rich environment, is often interrupted by man's activity, either through poisoning or through overloading the system. In this way we can describe man's activities as unnatural interferences with nature's workshop. Such interference will lead eventually to a profound modification of the environment. New cycles are likely to arise, either through "natural processes" or by the activity of man. These cycles may be radically different from those that nature normally provides in a given setting. Man, as a major organism—whether passive or aggressive—in the global ecosystem, must by nature modify his own environment. The change that produced oxygen in the Precambrian period due to the activity of photosynthesizing organisms was undoubtedly more profound, as far as the earth's environment is concerned, than any of the changes which mankind's polluting activities have produced thus far. And there have been several other revolutionary events in the biogeochemical history of the planet (Fairbridge, 1967). Since the Precambrian change was attributable to natural forces, no one would question its moral correctness, in human ethics. The present-day environment modification, which is the result of man's own productive activity, is seen by some observers as an "unnatural" change, and in accordance with certain unquestioned moral principles is thus condemned as wrong, as an a priori evil. The current outcry by emotionally, rather than rationally directed ecologists, is for a return to earlier conditions, pristine conditions assumed to be better and more natural. But man is also a natural organism—albeit somewhat ruthless—and given the magnitude of the problems we have described and the impossibility of turning back the clock of history, may it not be that the solutions to these problems lie in the "unnatural" human technology, in further alterations of existing conditions by imaginative technical manipulations? Does not man have the right, and could he not show the foresight, to induce, on a global scale, changes which could eventually serve his own betterment?

GEORGE CLAUS
GEORGE J. HALASI-KUN

ENVIRONMENTAL POLLUTION

References

Anon., 1967, "How to stop air pollution," *The Sciences* (N.Y. Acad. Sci.), **6**(12), 1–9.

Anon. (Nat. Acad. Sci.—N.R.C.), 1969, "Resources and Man," San Francisco, W. H. Freeman & Co., 259pp.

American Chemical Society, 1969, "Cleaning Our Environment: The Chemical Basis for Action," Washington, D.C., 249pp.

Brady, N. C. (editor), 1967, "Agriculture and the Quality of our Environment," Washington, D.C., Am. Assoc. Advance. Sci. (AAAS), 476pp.

Chow, T. J., and Earl, J. L., 1970, "Lead aerosols in the atmosphere: Increasing concentrations," *Science*, **169**(3945), 577–580.

Coale, A. J., 1970, "Man and his environment," *Science*, **170**(3954), 132–136.

Commoner, B., 1967, "Science and Survival," New York, Viking Press, 150pp.

Cooley, R. A., and Wandesforde-Smith, G. (editors), 1970, "Congress and the Environment," Seattle, Univ. Washington Press, 284pp.

De Bell, G. (editor), 1970, "The Environmental Handbook," New York, Ballantine Books, 360pp.

Dixon, J. P., 1965, "Air Conservation," Report of the AAAS Air Conservation Comm., 348pp.

Edwards, C. E., 1970, "Persistent Pesticides in the Environment," CRC Monoscience Series, 70pp.

Ehrlich, P. R., and Ehrlich, A. H., 1970, "Population, Resources, Environment," San Francisco, W. H. Freeman & Co., 400pp.

Eipper, A. W., 1970, "Pollution problems, resource policy, and the scientist," *Science*, **169**(3940), 11–15.

Fairbridge, R. W., 1967, "Carbonate Rocks and Paleoclimatology in the Biogeochemical History of the Planet," in (Chilingar, G. V., Bissell, H. J., and Fairbridge, R. W., editors), "Carbonate Rocks," Amsterdam, Elsevier Publ. Co., pp. 399–432.

Fortune Magazine Editors, 1970, "The Environment," New York, Harper & Row (Perennial Library).

Føyn, E., 1965, "Disposal of Waste in the Marine Environment and the Pollution of the Sea," in (Barnes, H. and Allen, G., editors), "Oceanography and Marine Biology—An Annual Review," London, Unwin Ltd.

Halasi-Kun, G. J., 1970, "Correlation between Precipitation, Flood, and Windbreak Phenomena of the Mountains . . . ," pp. 77–87, in (Halasi-Kun, G. J., and Widmer, K., editors), "Proceedings of University Seminar on Pollution and Water Resources," Vol. I, 1967/68, New York-Trenton, Columbia University-New Jersey Department of Environmental Protection, 91pp.

Imbelli, C., Pressman, W. B.; and Radiloff, H., 1968, "The industrial wastes control program in New York City," *J. Water Pollut. Control Fed.*, **40**(2), 1981–2012.

Interstate Sanitation Commission of New York, New Jersey and Connecticut, 1969, "Report on the Water Pollution Control Activities and the Interstate Air Pollution Program," 43pp.

Koenig, L., 1967, "Ultimate Disposal of Advanced-Treatment Waste," Public Health Service #AWTR-3 (999-WP-3), 68pp.

Kraybill, H. F. (editor), 1969, "Biological effects of pesticides in mammalian systems," *Ann. N.Y. Acad. Sci.*, **160**, Art. 1, 1–422.

Lacy, W. J., and Cywin, A., 1969, "Federal grants available for industrial pollution control," *Water and Sewage Works*, **4**, 12–15.

Leinwand, G. (editor), 1969, "Air and Water Pollution: Problems of American Society," New York, Washington Square Press, 160pp.

Mawson, C. A., 1965, "Management of Radioactive Wastes," Princeton, N. J., D. Van Nostrand, 196pp.

McKee, J. E., and Wolf, H. W. (editors), 1963, "Water Quality Criteria," Sacramento, Calif., State Water Quality Control Board, 355pp.

Miller, M. W., and Berg, G. G. (editors), 1969, "Chemical Fallout," Springfield, Ill., Thomas, 532pp.

Miller, R. S., et al., 1970, "Man and His Environment: The Ecological Limits of Optimism," New Haven, Conn., Yale Univ. Press, 78pp.

Moncrief, L. W., 1970, "The cultural basis for our environmental crisis," *Science*, **170**(3957), 508–512.

O'Malley, C. K., 1970, "Proceedings of the Conference on International and Interstate Regulation of Water Pollution, March 1970," New York, Columbia University School of Law, 1970, 321pp.

President's Science Advisory Committee, 1965, "Restoring the Quality of the Environment—Report of the Environmental Pollution Panel," Washington, D.C., The White House, 172pp.

Rose, J., 1969, "Technological Injury," London, Gordon and Breach Sci. Publ., 224pp.

Silvester, R., 1967, "Jet mixers in sewage outfalls," *J. Inst. Engineers* (Australia), 33–37.

Singer, S. F., 1970, "Will the world come to a horrible end?" *Science*, **170**(2954), 125.

U.S. Dept. of the Interior, 1968a, "The Cost of Clean Water-Summary Report," Federal Water Pollution Control Admin., **1**, 1–39.

U.S. Dept. of the Interior, 1968b, "Summary Report on Advanced Waste Treatment, July 1964 to July 1967," Federal Water Pollution Control Admin., #WP-20-AWTR-19, 1–96.

U. S. Depts. of the Interior and Transportation, 1968, "Oil Pollution—A Report to the President," #298-767, 1–31.

Van Dyne, G. M. (editor), 1969, "The Ecosystem Concept in Natural Resource Management," New York, Academic Press, 386pp.

Wallauschek, E., and Lützen, J., 1964, "General Survey on Radioactivity in Sea Water and Marine Organisms," Paris, Org. Econ. Coop. and Devel., European Nuclear Energy Agency, Part 2, 96pp.

Weinberger, L. W., Stephan, D. G., and Middleton, F. M., 1966, "Solving our water problems—water renovation and re-use," *Ann. N.Y. Acad. Sci.* **136**(5), 131–154.

Weinberger, L. W., and Stephan, D. G., 1967, "Technology of Advanced Waste-Treatment,"

Internat. Conf. Water for Peace, May 23–31, 1967, Vol. 2, pp. 1, 547–555.

Weiss, C. M., and Okun, D. A., 1967, "Water Quality Technology and Future Prospects," Internat. Conf. Water for Peace, May 23–31, Vol. 4, pp. 195–207.

Woodwell, G. M., 1970, "Effects of pollution on the structure and physiology of ecosystems," *Science*, **168**(3930), 429–433.

Cross-references: *Air Pollution and Global Climate; Air Pollution and Urban Climate; Conservation; Desalination Processes; Environmental Science; Hydrocarbons; Lagoon Geochemistry; Lake Geochemistry; Medical Geography; Medical Geology; Medical Geology—Trace Metals in Mammals; Medicinal Springs; Mineral Genesis and Transformation—Microbial; Mineral Particles and Human Disease; Outgassing of the Planet Earth; Oxygen Cycle; ph-Eh; Phosphorus Cycle; Photosynthesis; River Geochemistry; Sulfate Reduction—Microbial; Sulfide Reduction—Microbial; Sulfide Mineral Oxidation—Microbial; Thermal Pollution; Urban Air Pollution.* Vol. II: *Greenhouse Effect; Inversion; Ozone.* Vol. III: *Denudation; Erosion.* Vol. IVB: *Fossil Fuels; Methane; Natural Gas; Petroleum.* Vol. VII: *Microorganisms.*

ENVIRONMENTAL POLLUTION—GLOBAL EFFECTS—See AIR POLLUTION AND GLOBAL CLIMATE; AIR POLLUTION AND URBAN CLIMATE

ENVIRONMENTAL SCIENCE

The emergence of environmental science as a field of study in its own right during the last decade is a reflection of the inadequacy of existing disciplines to deal with the complexity of the total environmental system. Essentially, environmental science attempts to identify, define, and analyze those physical and biotic processes that actively influence, or are influenced by, man's actions. Of necessity, the subject crosses many disciplines and it becomes the task of the environmental scientist to extract pertinent material from the natural, physical, and biological sciences such that the environment can be viewed holistically. Thus, environmental science seeks to combine the approach of such disciplines as biology, geology, pedology, and climatology, and attempts to avoid the artificial division of the environment into quite distinct and separate compartments.

In some ways the modern approach to environmental science resembles that used by the natural scientists of the 18th and 19th centuries. At that time, however, it was possible for an individual to encompass the realms of many disciplines and resolve them into a meaningful whole. The works of Lamarck, Humboldt, and Darwin are graphic evidence of this. But with the accumulation of data and the development of science, such an interpretation is made exceedingly difficult today. Because of this, a major problem for environmental science is the selection of a suitable framework within which the total environmental system can be adequately studied.

While the problems of devising such a framework are formidable, a number of writers have suggested possible approaches. Miller (1968), for example, uses the energy and mass budgets as the basic premise to study the interface between earth and atmosphere. One of the objectives of his approach is "To demonstrate cooperative action among sciences, by showing that physical and chemical phenomena occur at biological surfaces that usually exist in economic and cultural frameworks." To meet such a demand, Miller suggests a two-part approach. Budgets are firstly evaluated in a systematic way thereby permitting comprehension of individual systems and laying the groundwork for analysis of their interaction. Such analysis is completed at a second stage through the study of synthesizing units which vary from energy and mass budgets of organisms to synthesis on a world scale (Table 1). Miller further provides apt examples of systems that exemplify ". . . the ideas that the landscape is a functioning entity, that the inflows and outflows of energy and matter to and from it are measurable, and that they can be brought into a single framework if we cast a budget in terms of calories of energy and grams of water or other matter." Such treatment is the essence of studies in environmental science. Although other authors have offered useful summary approaches to the treatment of environmental science (e.g. Strahler, 1970), Miller's approach is, to date, the most comprehensive available.

TABLE 1. THE ENERGY AND MASS BUDGET APPROACH[a]

Systematic units
I. Atmospheric motion near the earth's surface
II. Exchange of matter (except water) at the earth's surface
III. Exchanges of energy at the earth's surface
IV. Exchanges of water at the earth's surface
Synthesizing units
V. Synthesizing the energy and mass budgets of organisms
VI. Synthesizing the budgets of elements of the landscape
VII. Synthesizing the budgets of mosaic landscapes
VIII. Synthesizing the budgets of regions into a world pattern

[a] From Miller, 1968.

ENVIRONMENTAL SCIENCE

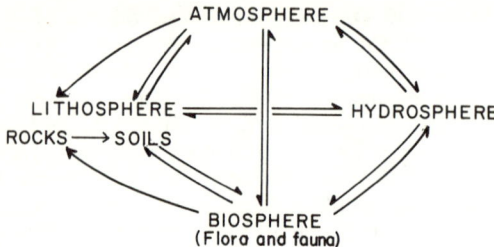

Fig. 1. The environmental system.

The Focus of Environmental Science

Irrespective of the methodology used, the starting point, and indeed the whole focus of environmental science, is the complex of interacting processes which together comprise the total environment (Fig. 1). The prime concern of the environmental scientist, however, is not the components per se but the various ecological processes and cycles that activate the environmental system. These processes and cycles demonstrate the functional interrelationship between the nonliving, or abiotic, environment and the biotic community of plants, animals, and microbes that is fundamental to environmental dynamics. Such interrelationships can be considered at several levels.

At one level the abiotic environment, including such physical parameters as moisture, insolation, and winds, affects the conditions for existence within the biosphere. Thus all living organisms have certain tolerance limits with respect to such things as temperature, moisture, and nutrients. Climate, for example, will exert a considerable influence on the distribution and composition of plant communities, which will in turn affect the distribution of animal species, since each species must remain within access of its food supply. Less obvious is the reciprocal influence of the vegetation and plant cover on local climate and weather (Fig. 2). A similar reciprocal relationship exists between vegetation and fauna, either through selective grazing, which will favor certain plants at the expense of others, or through overgrazing, which may ultimately lead to the destruction and displacement of the natural vegetation cover, exposing soil to accelerated erosion.

At a second level, the functional interdependence of the abiotic and biotic environments is exemplified by the food chains through which energy flows in the biosphere and by the biogeochemical cycles whereby the basic inorganic elements and compounds essential to life, such as water, oxygen, carbon, and nitrogen, are recycled through the environment (Fig. 3).

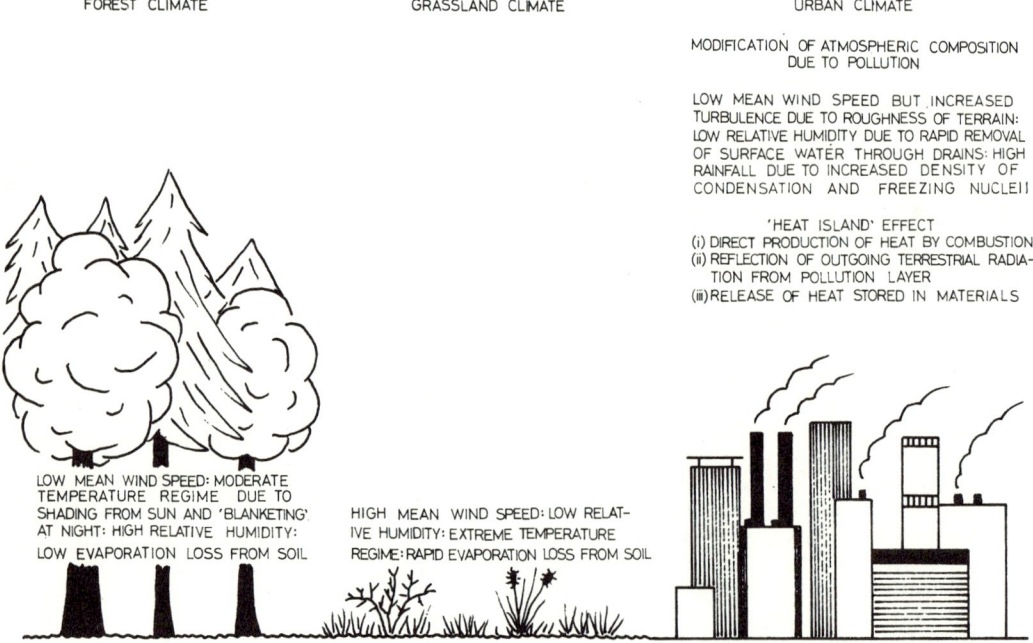

Fig. 2. Influence of surface cover on local climatic regimes.

ENVIRONMENTAL SCIENCE

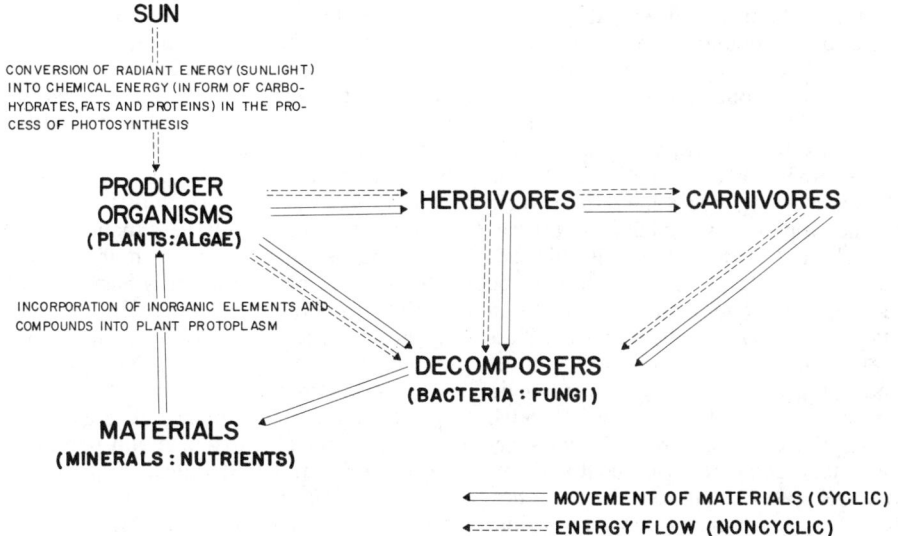

FIG. 3. A simplified model of energy flow and mineral cycling in the environment (after Kormondy, 1969).

Disruption of Environmental Systems through Natural Agencies

It should appear quite evident that the modification or interruption of any single process or cycle will trigger off a chain reaction affecting the total environmental system. For the greater part of earth history such modifications have resulted from the operation of entirely natural processes. As an example of this, one explanation of the cycle of glaciation which has characterized the Quaternary Period invokes changes in the carbon dioxide content of the atmosphere (Plass, 1956, 1959). As illustrated in Fig. 4, the carbon dioxide balance concerns

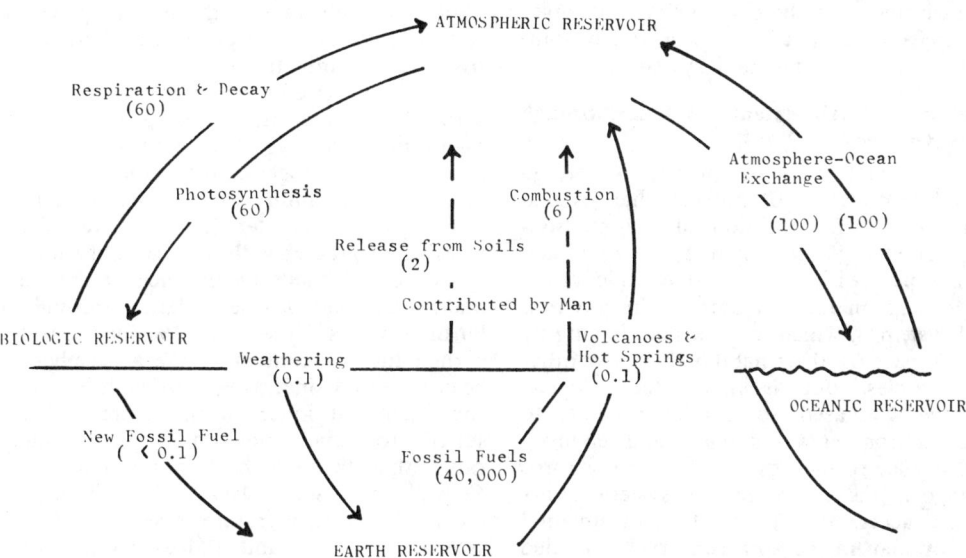

FIG. 4. The Carbon Dioxide Balance. Equilibrium is attained through the exchange that occurs between the great natural reservoirs. Exchange is indicated through the arrows indicating the processes. Figures in parentheses refer to estimated number of billion tons ($\times 10^9$) that are used in each process each year (after Plass, 1959).

339

many parts of the environment. The interchange of carbon dioxide between atmosphere and ocean is of particular significance since the net direction of exchange is dependent upon the relative concentrations of carbon dioxide in the atmosphere and ocean reservoirs. Thus an increase in atmospheric carbon dioxide would result in a movement of carbon dioxide to the oceans, while a decrease would have the reverse effect. But the apparent simplicity of the atmospheric carbon dioxide balance is misleading, since a considerable period of time may be required before the equilibrium is restored through the self-regulating mechanism. Plass suggests that a change in pace of one or all of the input-output processes (e.g., a decrease in volcanic activity, increased withdrawal through the process of photosynthesis by a dense vegetation cover, etc.), may have served to trigger an absolute decrease in atmospheric carbon dioxide, temporarily upsetting the equilibrium and initiating a glacial epoch. At the height of the glacial periods, ice sheets may have absorbed up to 0.3% of the oceans' waters. Inasmuch as only a small volume of carbon dioxide can be retained by ice, carbon dioxide must have gradually accumulated in the ocean reservoir. As a result of the self-regulating mechanism, the excess would be released over a long period of time to the atmosphere, thereby drawing the cycle to a conclusion.

Although carbon dioxide change is not accepted as the sole cause of the glacial epochs, the nature of the reasoning behind the theory provides insight into the complexity of changes resulting from modification of a single component of the environmental system.

Disruption of Environmental Systems through Human Agencies

Environmental scientists must, however, go beyond interpretation of natural changes and look to human modification of the environment. Primitive societies—the hunting and gathering peoples of the pre-Neolithic era—functioned as an integral part of the environmental system in which they existed, having no greater impact on their habitat than any other animal species. But as man's technological skills have increased, so has his capacity to initiate environmental changes, and environmental science recognizes man as a major force of change in the environmental system. However, the accelerated pace of environmental change by man in the 20th century has tended to obscure the fact that man's modification of the environment has been proceeding for many centuries. The discovery and deliberate use of fire, for example, had far-reaching ecological effects, as has the alteration of the biotic community through the replacement of the natural flora and fauna by more useful plant and animal species. In this process, man has essentially simplified the trophic structure of the environment over wide areas, substituting a monoculture of crops for the diversity of the natural environment. At the same time, man has attempted to eliminate his direct competitors among the insect and vertebrate populations. The net result is a man-made system; even the decomposer element in the system is altered since plant growth is harvested and soil fertility maintained through the application of fertilizers. Such systems have hitherto proved capable of supporting the growing human population, but there is, in the artificial and highly specialized systems man has created, the intrinsic danger of imbalance and potential catastrophe (Elton, 1958; Bates, 1955, 1969).

It is precisely because so many human activities have an unpredictable, degradational impact on the total environment system that man as a major force of change is the concern of the environmental scientist. It is unnecessary to look further than man's use of persistent insecticides such as DDT to appreciate the potential synergistic effect of his actions. The decimation of certain carnivore bird populations as a result of the interference of DDT with normal breeding and reproductive processes (Peakall, 1967, 1970) and perhaps reduced rate of photosynthesis by marine phytoplankton when concentrates of only a few parts per billion are present (Wurster, 1968) are only two of the unanticipated consequences of the widespread use of chlorinated hydrocarbon insecticides.

As far as the environmental scientist is concerned, however, the basic problem is that man's understanding of the environment and of environmental processes is less than perfect at a time when his ability to alter the environment has never been greater. It is, for example, impossible to predict with any certainty the consequences of man's disturbance of the atmospheric carbon dioxide balance through the burning of fossil fuels. A 12% increase of the carbon dioxide content of the atmosphere has been observed since 1880, although it has been suggested that much of this increase has resulted from the oxidation of plant materials rather than through the burning of fossil fuels (Revelle and Suess, 1957). Yet although the mean global temperature rose by 0.9°F between the 1880s and 1940s—an increase of the correct magnitude if carbon dioxide is assumed to be the causal factor—there has been a noticeable decline in global temperatures in the two decades since 1940 (Mitchell, 1961). There appears to be some agreement among meteorologists that this decline tendency may have resulted from the increasing turbidity of the atmosphere (i.e., the solid particulate matter such as dust and soot), which has increased the reflectivity or albedo of the earth-atmosphere system. As a result, a greater proportion of the incoming solar radiation is being reflected back into space, offsetting the effect of continued increase in the carbon dioxide content of the atmosphere. These examples serve to illustrate (a) the extent to which man's actions can initiate a chain reaction affecting every aspect of the environmental system—in this case the entire exogenic energy budget of the earth—and (b) the fact that there is still a great deal of guesswork involved in the interpretation of environmental processes; quite evidently the input of vast amounts of carbon dioxide as pollutants into the atmosphere is affecting the climate, but the extent and significance of that effect is quite uncertain.

Numerous similar examples could be cited and, in most instances, similar uncertain conclusions would be reached.

The Status of Environmental Science

The present status of environmental science is evident from the nature of this article. Unlike descriptive appraisals of established disciplines, the content has been concerned with what environmental science is, not what it does or has accomplished. As a discipline it is still in a formative stage and papers are still calling for the establishment of a more realistic, holistic approach to environmental study (Hare, 1970; Dansereau, 1970). In most cases the call has been for deemphasis of the traditional discipline approach and for the development of programs in environmental science.

The noticeable deterioration of the environment in recent years emphasizes the need. It is true that individual environmental problems may be solved on a piecemeal, short-term basis through research in individual disciplines. All too frequently, however, such narrowly focussed approaches have produced adverse effects in other areas. In many instances the technology is already available and solution depends upon socioeconomic actions. As Dansereau notes "Legislation is always behind the times; it is the grammar that must follow general usage." But the concern of the environmental scientist transcends such short-term problems and looks to long-term, more insidious repercussions. In order to fulfill such a role, it is evident that (a) new thinking must occur in traditional educational approaches to the environment and (b) a training program must be established such that what might be termed "specialized generalists," capable of working with and understanding the multidisciplinary nature of the environment, can be educated.

Such developments are slowly occurring in universities and public administration areas but, if environmental science is to develop into an integrating discipline capable of contributing significantly to the present-day environmental crisis, much remains to be completed.

JOHN E. OLIVER
IAN R. MANNERS

References

Bates, M., 1955, "The Prevalence of People," New York, Scribner.

Bates, M., 1969, "The Human Ecosystem" in "Resources and Man," National Academy of Sciences-National Research Council, San Francisco, Freeman, pp. 21–30.

Dansereau, P., 1970, "Challenge for Survival," New York, Columbia Univ. Press.

Elton, C., 1958, "The Ecology of Invasions by Animals and Plants," New York, John Wiley & Sons.

Hare, F. K., 1970, "How should we treat environment?" *Science,* **167,** 352–355.

Kormondy, E. J., 1969, "Concepts of Ecology," Englewood Cliffs, N.J., Prentice-Hall.

Miller, D. H., 1968, "The Energy and Mass Budget at the Surface of the Earth," Assoc. Am. Geographers, Comm. on College Geography, Publ. No. 7, Washington, D.C.

Mitchell, J. M., 1961, "Recent secular changes of global temperature," *Annals N.Y. Acad. Sci.,* **95,** 235–250.

Peakall, C. F., 1967, "Pesticide-induced enzyme breakdown of steroids in birds," *Nature,* **216,** 505–506.

Peakall, D. B., 1970, "Pesticides and the reproduction of birds," *Sc. Amer.* **222,** 73–78.

Plass, G. N., 1956, "The carbon dioxide theory of climatic change," *Tellus,* **8,** 140–153.

Plass, G. N., 1959, "Carbon dioxide and climate," *Sci. Amer.* **201,** 41–47.

Revelle, R. and Suess, H. E., 1957, "Carbon dioxide exchange between atmosphere and ocean and the question of an increase of atmospheric CO_2 during the past decades," *Tellus,* **9,** 18–27.

Strahler, A. N., 1970, "The life layer," *J. Geogr.,* **69**(2), 70–76.

Wurster, C. F., 1968, "DDT reduces photosynthesis by marine phytoplankton," *Science,* **158,** 1474–1475.

Cross-references: *Air Pollution and Global Climate; Air Pollution and Urban Climate; Biogeochemistry; Ecology; Environmental Pollution; Hydrologic Cycle; Photosynthesis; Thermal Pollution.* Vol. I: *Carbon Dioxide Cycle in the Sea and Atmosphere; Marine Ecology; Ocean-Atmosphere Interaction.* Vol. II: *Energy Budget of the Earth's Surface.* Vol. III: *Anthropogenic Influences in Geomorphology; Quaternary Period.*

EPIGENESIS

Noun. (*1*) In rocks and ore genesis processes, minerals or mineral deposits are epigenetic if their formation (deposition, crystallization, etc.) took place after the formation of the enclosing (host) rock. The term epigenetic only refers to time emplacement relations between the host rock and guest minerals (i.e., introduced matter); however, contrary to syngenetic deposits, the host of epigenetic rock and mineral deposits is *always* subject to replacement (metasomatism) and/or cavity filling.

(*2*) Epigenetic deposits are "those introduced into a pre-existing rock" (Lindgren, 1930).

(*3*) Sometimes used to indicate the noncongruency of mineralization with the host rock (e.g., carnotization of sandstones in the Colorado Plateau).

(*4*) In mineral zoning used to indicate mineralization front relations with the host rock.

(*5*) Opposite to *syngenesis* (q.v.) (see also *Ore Genesis*).

(*6*) In sedimentology (lithification), an almost equivalent term is *epidiagenesis*.

Adj. Epigenetic: Of, relating to, or produced by epigenesis (Examples: the ——— nature of the chert, Van Hise, 1904; the ——— nature of carnotization in the Colorado Plateau sandstones).

Compound Adjective: (*1*) *Epigenetic-Supergene* (e.g., some secondary enrichment cases: mineral genesis taking place because of mineral matter redistribution within the epigenetic deposit, due to descending migratory processes.

(*2*) *Epigenetic-Hypogene* (e.g., most vein deposits described in the geological literature, i.e., mineral genesis taking place in the host rock (epigenetically) from ascending mineral fluids).

RAYMUNDO J. CHICO

References

Lindgren, W., 1930, "Pseudo-eutetic textures," *Econ. Geol.*, **25**, 1–13.
Van Hise, C. R., 1904, "A Treatise on Metamorphism," *U.S. Geol. Surv. Monograph* **47**.

Cross-references: *Syngenesis.* Vol. IVB: *Economic Geology; Classification; Ore Genesis.* Vol. V: *Metasomatism; Petrology.*

EPILIMNION

Epilimnion is the term applied to the upper layer of a stratified lake, having relatively uniform warm, well-mixed, and aerated characteristics, which first develops after the spring turnover and continues on through the summer. In late autumn it is cooled by cold winds up to 4°C (maximum density of water). At this point the upper layer becomes more dense than the water below, sinks, and displaces the lower layers. This displacement is termed the *autumn turnover.* The sinking water is not only more dense, it is also well aerated, thereby replenishing the oxygen supply in the bottom waters.

The formation of ice, in the winter, provides protection from further wind action and typically a lake has a temperature profile of 0° just at ice level, down to +4°C in the bottom waters. Spring warming causes melting of the ice cover and as the surface water temperatures rise up to 4°C, convective sinking, aided by the stirring action of the wind, again produces a uniform vertical thermal gradient and a thorough vertical convective circulation. This spring turnover ends when the surface water becomes warmer and less dense, preventing the mixing action of the wind from destroying the vertical density gradient. The stratification proceeds from this point and the *hypolimnion* (q.v.) and *thermocline* (q.v.) are also developed.

B.C.S.

References

Chow, Ven Te, 1964, "Handbook of Applied Hydrology," New York, McGraw-Hill Book Co,. 1445pp.
Hutchinson, G. E., 1957, "A Treatise on Limnology," Vol. I, New York, John Wiley & Sons, 1015pp.
Welch, P. S., 1952, "Limnology," Second ed., New York, McGraw-Hill Book Co.

Cross-references: *Hypolimnion; Limnology; Thermocline.*

EQUILIBRIUM CONSTANT—See MASS ACTION AND EQUILIBRIUM CONSTANT

ERBIUM: ELEMENT AND GEOCHEMISTRY

Discovered in 1843 by Mosander who named it *terbia* after the town of Ytterby, Sweden. The names terbia and erbia were later interchanged. In 1879 Cleve separated erbium chemically from other rare earths.

Properties

Elemental erbium is a very reactive silvery-white *lanthanide,* or rare earth metal, with atomic number 68 and outer electronic structure $4f^{12}6s^2$. It occurs as six stable isotopes:

Er^{162}	0.136% natural abundance
Er^{164}	1.56%
Er^{166}	33.4%
Er^{167}	22.9%
Er^{168}	27.1%
Er^{170}	14.9%

Erbium is determined analytically by neutron activation, x-ray spectrography, absorption spectrometry (it has very sharp spectral lines), and by isotope dilution mass spectrometry. In geologic environments erbium forms highly ionic bonds. It forms weak complexes. The oxidation state is tripositive, with ionic radius 0.88 Å. Generally, chemical behavior is similar to the alkaline earths, and almost identical to other heavy rare earths.

Minerals

Erbium occurs in highest concentration in rare earth minerals that favor heavy rare earths. Generally nitrates, sulfates, carbonates, niobates, and tungstates, with cation coordination number of six to eight concentrate heavy rare earths, and often contain structural halogens and alkali metals. Typical examples are: xenotime (Y, heavy rare earths) PO_4 and fergusonite (Y, heavy rare earths) $(Nb,Ta)O_4$.

Geochemical Behavior

Schmitt et al. (1964) established a value of 0.21 ppm erbium in chondrites and 2.3 ppm in eucrite achondrites. The standard deviation within a meteorite class is only about 20%. The erbium concentration in average basalt is 3.3 ppm, and the values range over less than a factor of ten (Haskin et al., 1966); oceanic and continental basalts have similar ranges. Partial melting of mantle peridotite yields a basaltic liquid with enriched erbium and a mantle source area depleted in erbium.

Igneous differentiation series tend toward higher erbium concentrations, although the total enrichment is less than for light rare earths. The relative abundance of erbium and other heavy rare earths remains unchanged. Haskin et al. (1966) found 2.8 ppm erbium in a Gough Island basalt and 12.1 ppm erbium in a trachyte differentiate. Average granite with 60–70% silica has 3.8 ppm, but the range of values for granites is even greater than for basalts.

In most basaltic, syenitic, and granitic rocks, erbium is mainly in mafic rock-forming minerals, probably substituting for divalent calcium (ionic radius 0.99 Å).

Erbium and other heavy rare earths form complexes and are transported in alkali, halogen, and carbon dioxide-rich hydrothermal fluids may have high concentrations of erbium and higher ratios of erbium to light rare earths than granites.

Erbium has a 430-year residence time in seawater (which is alkaline), about the same as other heavy rare earths. Removal by adsorption on clay particles, as well as depletion in surface water in areas of high organic productivity, are likely processes. High secondary erbium concentrations in phosphatic fish remains today occur only in deep water; paleozoic fish remains also have high erbium content. Manganese nodules have 1380 ppm erbium, whereas Pacific Ocean water has only 0.61×10^{-6} ppm (Goldberg et al., 1963).

In soils, the relative amounts of erbium in detrital, clay, and carbonate fractions is not well known. Sedimentary rocks have about 4.0 ppm erbium, with a variation of a factor of five. In a study of rare earths in Russian platform soils, Balashov et al. (1964) found higher erbium in alkaline soils, the erbium precipitating as hydroxide; lower concentrations are in acid soils, probably due to removal as complexes.

Biologic concentration of rare earths has been found in hickory leaves; rare earths have been used to trace opium.

Economic Geology

The major source of erbium has been monazite in beach sands; the accessory monazite originally present in granitic rocks resists weathering and is concentrated by sedimentary processes. Australia, Brazil, Ceylon, India, Malaysia, and South Africa mine the largest quantities. Less than one percent of the monazite is erbium. Recently, the bastnaesite deposit at Mountain Pass, California, has been an important source of erbium, again in low concentration.

High-purity erbium has a very limited market. Most erbium is sold unpurified in rare earth concentrates used for polishing, petroleum cracking catalysis, lighter flints, arc carbons, and in alloys.

ROBERT KAY

References

Balashov, Y. A., Ronov, A. B., Migdisov, A. A. and Turanskaya, N. V., 1964, "The effects of climate and facies environment on the fractionation of the rare earths during sedimentation," *Geochem. Internatl.* **10**(1), 951–969 (English transl.).

Goldberg, E. D., Koide, M., Schmitt, R. A., and Smith, R. H., 1963, "Rare-earth distributions in the marine environment," *J. Geophys. Res.*, **68**, 4209–4217.

Haskin, L. A., Frey, F. A., Schmitt, R. A., and Smith, R. H., 1966, "Meteoric, Solar, and Terrestrial Rare-Earth Distributions," pp. 167–31, in (Ahrens, L. H., et al., editors), "Physics and Chemistry of the Earth," Vol. 7, Oxford, Pergamon Press.

Schmitt, R. A., Smith, R. H., and Olehy, D. A., 1964, "Rare-earth, yttrium and scandium abundances in meteoric and terrestrial matter—II," *Geochim. Cosmochim. Acta*, **28**, 67–86.

Cross-references: *Complexes; Hydrothermal Solutions; Mass Spectrometry; Neutron Activation Analysis; Rare Earths; X-Ray Spectroscopy.*

EROSION—*See* Vol. III

ESTUARINE HYDROLOGY

An estuary is the wide mouth of a river, or arm of the sea, where the tide meets the river current, or flows and ebbs. It may also be defined as "a body of water in which the river water mixes with and measurably dilutes sea water" (Ketchum, 1951). These definitions do not overlap completely, because a lagoon connected with the sea may also be affected by the tide.

Some scientists prefer to describe the environment in terms of the salinity of the water (saline, brackish, or fresh), but saline water is not restricted to marginal marine areas. Such a description does not consider, therefore, the most characteristic aspect of the estuarine environment—that it is a region of steep and variable gradients in the environmental conditions (Fig. 1).

If the physiography of estuaries is a sole consideration, they can be defined as "bodies of water bordered by and partly cut off from the ocean by land masses that were originally shaped by nonmarine agencies. They are usually perpendicular to the coast line and most of them occupy the drowned mouths of stream valleys and are, therefore, usually considered as evidence of submergence" (Emery and Stevenson, 1957).

Classification

There is no system which is universally used to classify estuaries. A broad classification separates normal (or positive) estuaries, in which fresh-water inflow exceeds evaporation, from inverse (hypersaline) estuaries in which evaporation exceeds fresh-water inflow. Neutral estuaries are those in which neither evaporation nor river discharge dominates.

Estuaries along most coastlines have been formed partly by the subsidence of the land mass and partly by the rise in sea level. These embayments are usually elongate indentures of the coastline with rivers flowing in from the landward ends. Deep estuaries are known as *rias* (q.v. Vol. III). In eastern North America most estuaries are shallow with irregular, or dendritic, shore lines and are normal estuaries.

Along the Gulf Coast of the United States, marine processes have built a series of barrier islands parallel to the coastline. Most of the islands extend across the mouths of estuaries, forming a lagoon and decreasing the width of the estuarine entrance to the open sea.

The exchange of water, in such cases, between the estuary and the open sea is modified by the intervening lagoon in which evaporation may exceed fresh-water inflow. The waters in the estuary, then, have salinities higher than normal as a result of the exchange with the lagoonal water.

Water Characteristics and Circulation

The important feature in an estuary is the intermixing of seawater with the fresh water from land drainage. This interaction usually produces a variation, both horizontal and vertical, in the salinity of estuarine waters. In normal estuaries, salinities range from nearly zero at the river's mouth, to approximately 30‰ at the seaward extremity. In addition, there is generally an increase in salinity with depth.

An inverse estuary also has greater salinities at depth, but the highest salinities are at the head of the embayment rather than at the mouth. There may be a difference of several parts per thousand between the salinity at the head and that of normal seawater (Fig. 2).

Temperature. The water in estuaries, and especially that overlying the tidal flats, is relatively thin, so it follows the variations in temperature of the atmosphere more closely than does the water of the open sea. The water is much colder in winter and warmer in summer than is the sea. The diurnal variation is also greater than in the sea.

There are pronounced variations in water temperature with depth. During the winter, the water is cold and nearly isothermal at all depths. In some instances, in response to cold weather, the surface water may become a degree or so colder than the deeper water. During the summer, solar radiation and minor wind mixing produce a high temperature at the surface with less change at depth. The difference between surface and bottom temperatures may also be influenced by warm or cold river water flowing over the dense seawater.

Where evaporation exceeds river inflow, the summer surface water may become so saline as to sink to the floor of the estuary and flow out of the entrance beneath the incoming sea-

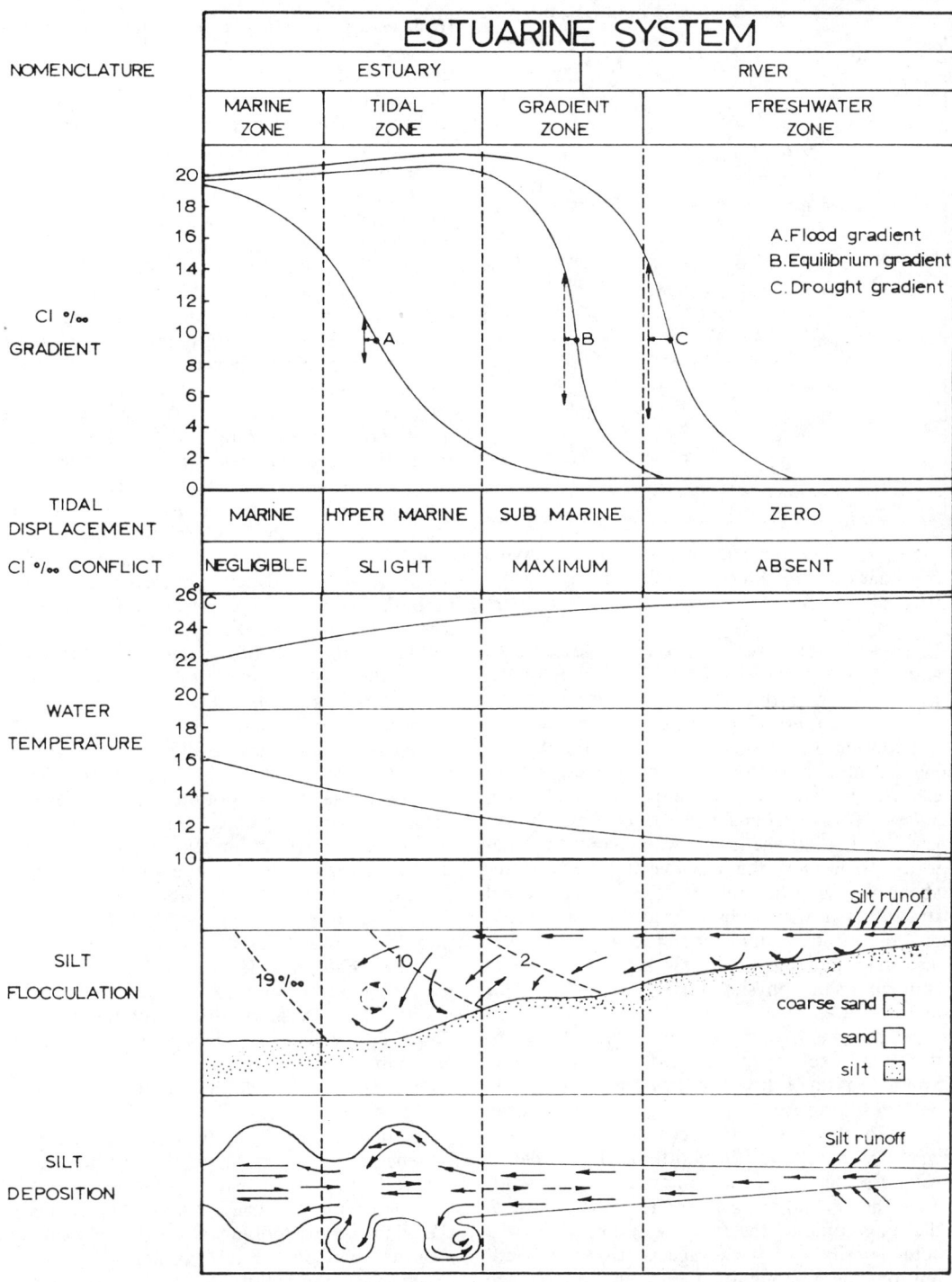

Fig. 1. Some zonal features of a composite Australian estuarine system (modified from Rochford, D.J., 1951, "Studies in Australian estuarine hydrography. I. Introduction and comparative features," *Australian J. Marine Freshwater Res.*, **2**, No. 1).

water. The temperature of the deep water may then be higher than that at the surface.

Circulation. Water movements in estuaries result mainly from the interaction of tides, river flow, and wind. The tides and river flow are usually the dominant factors. In Gulf

ESTUARINE HYDROLOGY

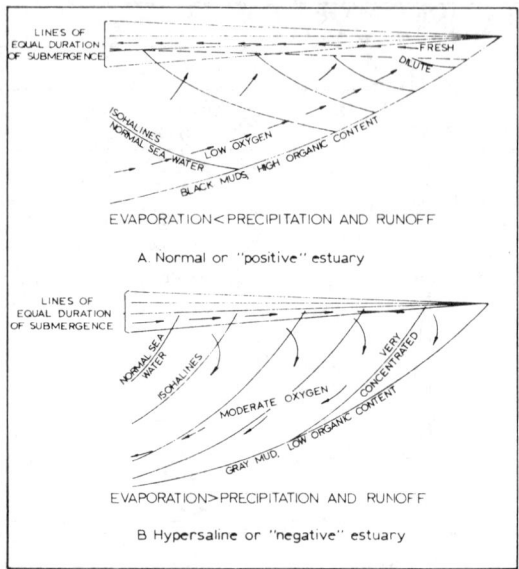

FIG. 2. Schematic sections of the two basic types of estuaries (after Emery and Stevenson, 1957).

Coast estuaries, where the tide range is small and river discharge is at times negligible, wind-induced currents are most important.

Stable Estuaries. In normal estuaries, the distribution of temperature and salinity, and the circulation pattern, are controlled almost exclusively by the tide range and the river inflow. Tidal currents tend to produce turbulent mixing of the river water and the seawater. However, the low-density fresh water above the seawater results in a stable vertical stratification which resists mixing. As a consequence, the relative magnitudes of the river discharge and the tidal flow are significant in controlling the physical structure of the water in the estuary.

Where the river flow is large in relation to the tidal exchange, the seawater enters the estuary as a salt-water "wedge" along the bottom. However, there is frictional drag between the overlying fresh water, the saltwater wedge, and the bottom. The relative velocities of the seaward-flowing fresh water and the intruding salt-water wedge control the magnitude of the friction factor. Thus, the actual position of the wedge is closely dependent on the volume of the river flow. When the volume of river discharge is great, the wedge extends only a short distance into the estuary and, of course, vice versa (Pritchard, 1952).

The salinity of the salt-water wedge remains similar to that of the open sea because there is only minor mixing with the seaward-flowing fresh water. However, at the interface between the two types of water, waves form and sometimes intrude into the surface water. Thus, the salt content in the upper layers increases slightly as the water moves seaward. Even so, throughout the estuary, a sharp salinity gradient exists between the two water layers.

The loss of salt water from the wedge to the upper layer is compensated by a flow of water from the sea (Fig. 2). The exchange from below takes place all along the upper interface of the wedge. As a result, there is a flow directed upstream at all positions within the wedge. The landward-moving water in the salt wedge is minor, however, and of the two, the seaward flow of surface water is far greater.

Partly Mixed Estuaries. Where tidal movements are great as compared to the volume of river discharge, mixing between the seawater and fresh water is sufficient to destroy sharp interfaces. The salt wedge, in such cases, does not exist as an identifiable feature, but a transition layer of definitely increasing salinity does occur. In such an estuary, however, the salinity in both the upper and lower layers decreases toward the head of the estuary.

The chief cause of currents in estuaries in which the waters are partly mixed is the tide. As in the stratified estuary, there is a net water movement superimposed on the tidal currents—a net seaward flow at the surface and a net flow toward the head in the deeper layers. These water motions are not as well defined as in a stratified estuary, and there is no sharp current interface. The flow from the deeper layers toward the surface decreases toward the head of the embayment (Fig. 2). The volume rate of seaward flow increases, therefore, toward the mouth (Pritchard, 1952).

Mixed Estuaries. In wholly mixed estuaries, the movements induced by the tide are far greater than those produced by the river inflow. The waters are completely mixed and are isohaline from the surface to the bottom. At all depths, the salinity decreases from the mouth to the head.

In such estuaries, the outward flowing water is deflected to the right, in the northern hemisphere, because of earth rotation (Fig. 3). Thus, in wide estuaries, the salinity is less on the right side (looking toward the estuarine mouth) than on the left. A net seaward flow exists along the right side and a net landward flow on the left. Water also moves laterally across the estuary from the left to the right side resulting in horizontal mixing.

In narrow, well-mixed estuaries, mixing induced by tidal action may be great enough to eliminate any lateral salinity gradient. There is a net seaward flow in all waters and the

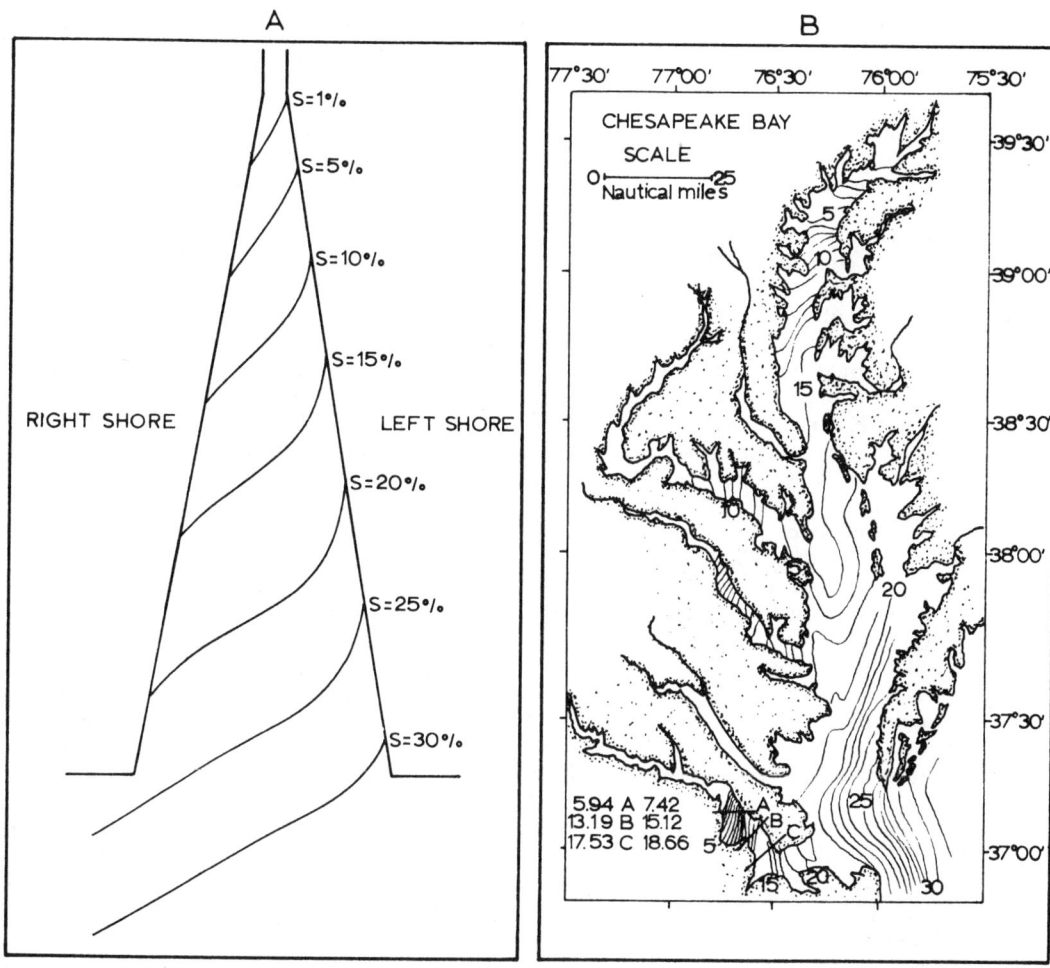

FIG. 3. Hypothetical and actual isohalines in a northern hemisphere estuary (modified from Pritchard, 1952).

only difference in salinity is the normal decrease toward the head.

Estuaries Bordered by Lagoons. Along coastal regions where barrier islands extend across the mouths of estuaries, the water bodies are usually so shallow that mixing by winds is sufficient to produce homogeneous water. Tidal currents are only significant through the inlets between the barrier islands. The total volume of water which flows in and out is relatively small. As a consequence, the rise and fall of the tide and tidal currents are minor within the estuary, and the most significant currents are from wind action.

There is, necessarily, a net flow of water out of these shallow estuaries sufficient to remove the water added by fresh water discharge. The large cross-sectional area of the estuary and the dampening effect of the coastal lagoon reduce the flow so that it is normally not directly measurable. The net motion may be completely modified by wind action to the extent that high water and constant, net inflow may occur during times when prevailing winds blow up the estuary. Strong winds blowing from the land reverse this effect and result in extremely low water levels in the estuary and the extrusion of estuarine waters many miles to sea.

Carbon Dioxide and Oxygen. The oxygen content of estuarine waters reaches a minimum in the pre-dawn hours and a maximum in the later daylight hours in response to the respiration of plants and animals and photosynthesis by the plants. This general trend is complicated by the tidal regime and the river discharge, but annual means indicate such maxima and minima exist (Stevenson and Emery, 1958).

The range of oxygen concentration is

greatest in shallow water where the volume of organisms is greatest. Ranges of from nearly zero at sunrise to 260% of saturation have been measured in areas where marsh grasses are concentrated. In the main body of estuarine water, oxygen values usually lie between 80 and 120% saturation.

A close approach to the variation of carbon dioxide is that of the pH because, basically, the higher the content of CO_2 the lower is the pH. Thus, the photosynthetic process of plants should tend to make the pH low at night and high during the day. The range is usually not great with maximum measured at slightly more than one pH unit (7.5–8.6; Stevenson and Emery, 1958).

There is also a seasonal cycle of carbon dioxide and oxygen. Greater ranges occur in the summer in response to increased plant growth. Increased river flow dilutes the water and reduces the carbon dioxide.

The oxygen concentration usually decreases with depth. Where fresh water flows over seawater, there is only a slow replacement of bottom water from the open sea which may already be low in oxygen. Accordingly, the bottom water may become increasingly more stagnant and less hospitable to oxygen-consuming life. Hypersaline estuaries may have highly oxygenated waters throughout their depth because the bottom water has only recently sunk from the surface.

Seiches in Estuaries. In some of the larger estuaries, the periodic flooding by the tides is supplemented by seiches, long stationary waves. The simplest seiche is one whose node is at the mouth of the estuary and the antinode near the head. The period of the seiche is controlled by the length and depth of the body of water, and where its natural period nearly coincides with that of the tide, as at the Bay of Fundy, a great fluctuation in sea level occurs (about 15 meters). In most estuaries, the seiche is only a few centimeters and is obscured by the much greater tidal amplitude.

Waves. Wind waves are small in estuaries because of the short fetch and the shallow water. They usually cause little erosion although when the tide is high, waves may stir up the muddy sediment on tidal flats. Waves may transport some sand and, because the largest waves come across the widest part of the estuary, they form sand spits pointing upstream in tidal channels.

A tidal bore (a wave of translation) is common in narrow estuaries and tidal channels. As a result of the shape of the entrance and bottom friction, the flooding tide is held back for a time until the water finally rushes up the channel as a steep wall of water. Bores may be from a few centimeters to several meters in height and move at velocities as great as 10 knots. The character of a tidal bore is determined, in part, by the river discharge which must be sufficient to hold back the tide for a period of time.

Estuarine Sediments

The inorganic sediments of estuaries are derived from inflowing rivers bordering sea cliffs, the sea floor outside the estuary, and the reworked deposits of tidal flats and marshes along the shores. Regardless of the source, much reworking of sediment occurs within estuaries. Erosion, too, is evident from the migration of tidal channels and the muddy color of the water when no river inflow is taking place. Some estuaries have entrances narrow enough so that tidal currents scour the bottom locally, leaving rocky or gravelly bottoms. The prevailing condition, however, must be one of deposition, and the average rate of deposition is greater than that of the open sea.

Distribution of Grain Sizes. The coarsest sediment in most estuaries is on the barrier or bay-mouth bar, and consists of sand and cobbles. Generally, this material is too coarse to have been transported across the tidal flats, but is derived from erosion of a sea cliff, then transported and deposited by longshore currents and waves. The excellent sorting and absence of much silt and clay may result from the turbulence of the waves.

The flat portions of the floors of estuaries that are deeper than about 6 m are usually covered by sediment which becomes progressively finer with depth of water. A smooth concentric pattern of sediments may occur, ranging from sand along the shore to fine mud at depth. Such a distribution occurs only where the bottom is relatively flat and current conditions are mild. In estuaries where the deeper areas are extremely irregular, mud occurs only in depression and coarse sediments characterize the shallower bottoms.

The sediment distribution in shallow areas, mostly the tidal flats, is more complex but usually follows a systematic pattern. Most of the flow of water is confined to well-defined channels which slowly migrate over the tidal flat (as shown by remapping at intervals of several years). The velocity of the water is such that the finer grains are swept out, leaving the coarse sediment in the channel. The areas between the channels consists of poorly sorted mud which becomes finer with distance from the tidal channels. Probably most, but not all, of the reworking of sediment in estuaries takes place on the shallow flats where the ebbing and flooding current erode and redeposit the sediment.

Organic Constituents. The sediments contain the remains of all phyla of animals and much plant debris. Even though the remains become scattered by scavenging, decomposition, and diagenesis, the organisms still have enriched the sediments in organic matter, calcium carbonate, silica, nutrients, and other constituents.

Sediments in estuaries located in areas where precipitation exceeds evaporation have organic nitrogen contents from 0.2 to 0.6%, and sediments in hypersaline areas, below 0.2%. The percentage used to differentiate between these areas, or 0.2% organic nitrogen, corresponds to about 1.7% organic carbon, or 2.9% total organic matter. Phosphorus is also abundant in sediments of normal environments, ranging from 0.1 to 0.4%.

Calcium carbonate is variable because of the presence or absence of shells and because of solution induced by acidic conditions. In coastal bays in temperate and arctic regions, calcium carbonate ranges between 0 and 6%, whereas in bays of tropical regions it is 10–47%.

Manner of Deposition. In the tidal section of a fresh-water river, a transition takes place and the distribution of sediments may be quite variable and confused. When the estuary proper is reached, there is some admixture of sea salts, and where the net upstream flow in lower layers occurs, there is a distinct change in sediment distribution. Finer sediments tend to be deposited in the channel (the reverse of conditions commonly found in river channels). In most streams, the bulk of suspended material probably is silt which is deposited directly out of suspension. Clay sizes, however, may be deposited through flocculation. The clays then fall to the deeper floors of the estuaries. Sediment may also travel down rivers at or near the surface in large floating floccules containing organic debris. When these settle to the bottom or are stranded by lowering water level, they are held by capillary action. Near the mouth of the estuary, coarser sediments are again found in the channel as a result of wave action and because much of the silt load has already been deposited in the channel further upstream.

R. E. STEVENSON

References

Arons, A. B., and Stommel, H., 1961, "A mixing-length theory of tidal flushing," *Trans. Am. Geophys. Union,* **32**(3), 419–421.

Emery, K. O., and Stevenson, R. E., 1957, "Estuaries and Lagoons," *Geol. Soc. Am. Mem. 67,* **1**, 673–750.

Ketchum, B. H., 1951, "The flushing of tidal estuaries," *Sewage Ind. Wastes,* **23**(2) 198–209.

Lauff, G. H. (ed.), 1967, "Estuaries," Washington, D.C., A.A.A.S., Publ. 83, 757pp.

Pritchard, D., 1952, "Salinity distribution and circulation in the Chesapeake Bay estuarine system," *J. Marine Res.,* **11**(2), 106–123.

Pritchard, D. W., 1955, "Estuarine circulation patterns," *Proc. Am. Soc. Civil Engrs.,* **81**(717).

Stevenson, R. E., and Emery, K. O., 1958, "Marshlands at Newport Bay, California," *Allan Hancock Foundation Publ. Occas. Paper,* **20**.

Cross-references: *Calcium Carbonate Geochemistry; Hydrology; Lagoon Geochemistry; Oxygen; pH-Eh; Photosynthesis; River Geochemistry; Seawater, Chemistry.* Vol. I: *Bay of Fundy; Coriolis Force; Fetch; Seiche; Tides; Waves.* Vol. II: *Evaporation; Winds.* Vol III: *Baer-Babinet Law; Barriers; Coastal Lagoon Dynamics; Estuary; Fjärd; Fjord; Rias; Rivers.* Vol. VI: *Estuarine Sedimentation; Lagoon Sedimentation.*

EUROPIUM: ELEMENT AND GEOCHEMISTRY

Demarcay, in 1901, was the first to isolate europium in its pure form from other rare earths.

Properties

Elemental europium is a very reactive silvery-white lanthanide, or rare earth, metal with atomic number 63 and outer electronic structure $4f^7 6s^2$. It occurs in two stable isotopes, Eu^{151} (47.8%) and Eu^{153} (52.2%). Europium is determined analytically by neutron activation (its neutron cross section, 4300 barns, is very high), and by isotope dilution mass spectrometry. In geologic environments europium forms highly ionic bonds. It forms weak complexes. The oxidation states are tripositive, with ionic radius 0.95 Å and dipositive, with ionic radius 1.09 Å; chemical behavior is similar to the alkaline earths, and to other rare earths. Its boiling point (1439°C) is lower than most rare earths.

Minerals

Europium occurs in highest concentration in rare earth minerals that favor light rare earths. Generally nitrates, sulfates, carbonates, niobates, and tungstates with high cation coordination numbers concentrate the light rare earths, and often contain structural halogens and alkali metals. Typical examples, with cation coordination of 10 and 11, are: monazite (light rare earths, Th) PO_4 and bastnaesite (light rare earths, Th) FCO_3.

Geochemical Behavior

Schmitt et al. (1964) have established a value of 0.074 ppm europium in chondrites, and 0.72 ppm europium in eucrite achondrites. The standard deviation within a meteorite class is only about 20%.

The europium concentration in average basalt is 2.7 ppm, but the range in values is a factor of ten (Haskin et al., 1966), and no clear trends are noted. However, in some oceanic and continental basalts, europium is enriched relative to its neighbors *gadolinium* and *samarium,* probably due to reduction to a dipositive oxidation state. The relative enrichment could occur during the partial melting of a peridotite source rock in the mantle, or during igneous differentiation.

Igneous differentiation series tend toward lower europium concentrations. Haskin et al. (1966) report 4.3 ppm in a Gough Island basalt, and 0.65 ppm in a trachyte differentiate. Towell et al. (1965) studied rare earths in the Southern California batholith and found 0.629 ± 0.025 ppm europium in the Rubidoux Mountain leucogranite, probably a differentiate (by crystal settling) of the San Marcos gabbro, with 1.05 ± 0.04 ppm. Trivalent light rare earths increase during the same differentiation series, but europium is probably divalent and is removed by concentration in settling crystals. In both gabbro and granite, europium is mainly in rock-forming minerals, substituting for calcium in mafic minerals, but also significantly in plagioclase. Average granite has lower europium (about 1.7 ppm in granites with 60–70% silica) than average basalt, but the concentration is even more variable.

Europium forms complexes and is transported by alkali, halogen, and carbon dioxide-rich hydrothermal fluids, but the complexes are not as stable as those formed by the heavier rare earths. Metasomatic and pegmatoid deposits often have low ratios of europium to heavier rare earths compared to granites, indicating low ratios in the hydrothermal solutions.

Europium has a 160 year residence time in seawater (which is alkaline), about the same as light rare-earths. Probably it is complexed. Removal by adsorption on clay particles, as well as depletion in surface water in areas of high organic productivity, are likely processes. Today, high secondary europium concentrations in phosphatic fish remains occur only in deep water; paleozoic fish remains also have high europium. Manganese nodules have 1040 ppm europium, whereas Pacific Ocean water has only 0.113×10^{-6} ppm (Goldberg et al. 1963).

In soils, the relative amounts of europium in detrital, clay, and carbonate fractions is not well known. Sedimentary rocks have about 2.0 ppm, with a variation of a factor of five.

In a study of rare earths in Russian platform soils, Balashov et al. (1964) found higher europium in alkaline soils, the europium precipitating as hydroxide; lower concentrations are in acid soils, probably due to removal as complexes.

Biologic concentration of rare-earths has been found in hickory leaves; rare-earths have been used to trace opium.

Economic Geology

The major source of europium has been monazite in beach sands. The accessory monazite originally present in granitic rocks resists weathering and is concentrated by sedimentary processes. Australia, Brazil, Ceylon, India, Malaysia, and South Africa mine the largest quantities. Only 0.04% (by weight) of monazite is europium. Recently, the bastnaesite deposit at Mountain Pass, California, has been an important source of europium; the ore has about 0.11% europium.

High-purity europium is expensive and is used in selective absorptiton of light in optical filters, and in control rods of atomic reactors (to absorb neutrons). The development of yttrium-europium oxide red television phosphors has greatly expanded the market. Europium is sold unpurified in rare-earth concentrates used for polishing, petroleum cracking catalysis, lighter flints, arc carbons, and in alloys.

ROBERT KAY

References

Balashov, Y. A., et al., 1964, "The effects of climate and facies environment on the fractionation of the rare earths during sedimentation," *Geochem. Internatl.,* 10(1), 951–969 (English transl.).

Goldberg, E. D., et al., 1963, "Rare-earth distributions in the marine environment," *J. Geophys. Res.,* 68, 4209–4217.

Haskin, L. A., et al., 1966, "Meteoric, Solar, and Terrestrial Rare-Earth Distributions," pp. 167–321 in (Ahrens, L. H., et al., editors), "Physics and Chemistry of the Earth," Vol. 7, Oxford, Pergamon Press.

Rankama, K., and Sahama, T. G., 1950, "Geochemistry," Chicago, University of Chicago Press, 912pp.

Schmitt, R. A., et al., 1964, "Rare-earth, yttrium and scandium abundances in meteoric and terrestrial matter—II," *Geochim. Cosmochim. Acta,* 28, 67–86.

Towell, D. G., et al., 1965, "Rare-earth distributions in some rocks and associated minerals of the batholith of Southern California," *J. Geophys. Res.,* 70, 3485–3496

Cross-references: *Complexes; Gadolinium; Hydrothermal Solutions; Mass Spectrometry; Neutron Activation Analysis; Rare Earths; Samarium.*

EVAPORITE PROCESSES

Fig. 1a. Worldwide evaporite distribution through geologic time (from Kozary et al., 1968).

EVAPORITE PROCESSES

Evaporite minerals, initially formed by chemical precipitation from natural waters, are to be found on every continent and from the Precambrian to the Recent (Fig. 1a). The most extensive evaporite deposits are Permian in age (Fig. 1b), and the present period of geological history represents a period of minimal deposition.

Evaporites form in relatively arid climates where evaporation exceeds precipitation plus runoff, but a high temperature is not essential. High winds over restricted salinas or basins can produce a sufficiently arid environment, and Recent evaporite deposits are known from the coldest regions of the earth, such as Antarctica.

Depositional environments can be formed in a number of ways. Grabau (1920) listed four possibilities: marginal salt pans, marine salinas, lagoons, and relict seas. Wind-carried salt sea spray can also blow large quantities of salt inshore ("cycle salts"; see Vol. II, *Salt Nuclei*), and there is also the possibility of hydrothermal accumulation (Borchert, 1969) and groundwater deposition (Cagle and Cruft, 1969). It is generally considered that the most

EVAPORITE PROCESSES

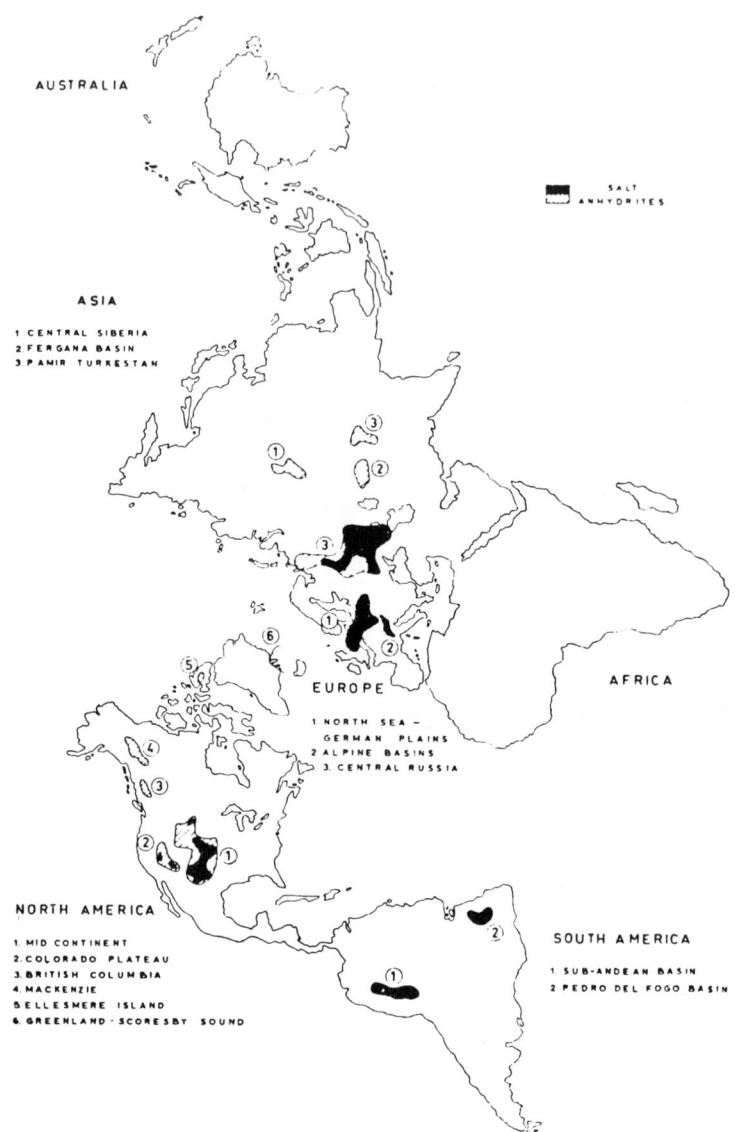

Fig. 1b. Evaporite distribution during the Permian (from Kozary et al., 1968).

extensive and thickest evaporite deposits are of marine origin. In most instances this may well be the case, although extensive deposits formed from groundwater action are recorded in South West Africa (Cagle and Cruft, 1969), and similar deposits occur in Argentina (Cortelezzi and Kilmurray, 1965).

Physical Conditions of Evaporite Deposition

Environments of large-scale evaporite deposition are rare today. The Gulf of Karabogaz in the Caspian Sea is the only large basin of contemporary lagoonal evaporite deposition, but many examples from the geologic past have been recorded. The Permian evaporites of west Texas and New Mexico extend for tens of thousands of square kilometers, with a combined thickness of 1356m for the Castile and Rusler Formation. The Salina Formation of the Silurian extends from Michigan to New York, an evaporite basin covering 400,000 km² (150,000 square miles), and extensive evaporites from the Permian Zechstein Sea once covered much of western Europe.

Since a 427m (1400-ft) column of seawater on evaporation would yield only a 30 cm (1 ft) layer of calcium sulfate and about a 6.7m (22-ft) layer of halite, the evaporation of a huge volume of seawater would be necessary to account for these thick ancient evaporite beds. To explain such extensive deposits. Ochsenius (1888) postulated the barred or restricted basin. The bar, being a reef, clastic barrier, or sill, serves to separate the basin from the normal environment. This allows water to enter through small channels, but with little or no outward leakage. Normal seawater enters at a rate equivalent to the amount being evaporated from the basin surface and becomes more and more concentrated. Some crystallization may take place at the surface at this time but much or all of this precipitate is dissolved in settling. When the density of the evaporating influx water reaches a greater density than the brine above which it is flowing, it begins to sink. The salts crystallize out from this brine in order of their solubility products and settle to the bottom of the basin, where their preservation is dependent on the environment.

Any evaporite theory has to explain an important discrepancy between the volumetric ratio of $CaSO_4$ to $NaCl$ in seawater and in evaporite rocks. In seawater the ratio is 1:30 but in the ancient deposits it is 1:1 or even greater. Also, a theory has to provide an adequate explanation for the extreme rarity of the "bitterns" or late stage precipitates which are found in many evaporite deposits. It is necessary to postulate a return to the ocean of dense, highly saline brines so that precipitation of halite and "bittern" salts frequently does not occur within the basin. Branson (1915) postulated the existence of a series of isolated basins separated by low barriers. The marginal basins develop concentrated brine and with an influx of further water, there is an overflow to supply brine to the basins in the interior. In this way by fractional crystallization in the individual basins, halite and the more soluble salts are deposited in one basin and gypsum in another. Branson calculated that it would be necessary to evaporate 17,375m (57,000 ft) of seawater to get a 12m (40-ft) bed of gypsum, but this is not required if these deposits can form in interior basins or can accumulate in depressions through current or wave action.

Early theories of large-scale evaporite formation discounted much outflow of basin waters, and King (1947) proposed a modification making allowances for it. Normal seawater enters over the barrier, with the surface layer becoming more concentrated by evaporation and eventually sinking. The barrier, however, is incomplete, and either the bottom saline brine flows back out to sea, or the barrier is semipermeable, permitting "reflux" of the solutions back out to the sea. Much of the halite and the more soluble magnesium and potassium salts are thereby removed. This "leaky reef" or "permeable barrier" was used as a model for evaporite deposition in the Michigan Basin by Briggs and Pollack (1967).

It is generally assumed that the restricting element in the evaporite basin is a physical barrier such as a tectonic sill, an organic reef, or a series of bars which have been formed prior to the deposition of the evaporites. From a study of estuaries and pressure surfaces Scruton (1953) recognized the possibility of dynamic barriers to inhibit flow. Movement of water toward the distal end of the basin is induced by hydrostatic head and this potential exists throughout the water column. As the density of the brine increases in the lower water layer, the pressure due to density overbalances the hydrostatic head and a seaward flow of brine commences. If the water stratification in the entrance to the basin is well developed and accompanied by a large inflow, a dynamic barrier can thus form in addition to the physical barrier. The escape of an excessive amount of high-density brine is prevented by friction at the interface between the large amount of surface water and the deep outflowing brine, and between the deep brine and the bottom of the channel.

If the restricting element in the basin is a reef or if carbonates are contemporaneously deposited with the evaporites, a severe problem is introduced because highly saline waters are not generally able to support abundant life, certainly not of reef fauna type. Sloss (1969) therefore postulated an evaporite basin, not necessarily very deep, containing a dense brine layer lying below a layer of water of near-normal marine salinity and surrounded by a large shelf area. Dense brine is concentrated by evaporation on the shelf and flows downslope to the basin bottom where it collects. The living parts of the surrounding reefs can then maintain themselves in upper normal marine waters. As dense brine rather than normal marine water is being supplied to the basin, it is not necessary to evoke very great depths of water to produce thick evaporite sequences in this model.

In Ojo de Liebre, Mexico, a Recent salt pan, organic production is surprisingly high in the saline waters, and large concentrations of algae, calcareous foraminifera, and gastropods are found (Phleger, 1969).

Chemical Conditions of Evaporite Deposition

Laboratory workers attempting to determine the chemical conditions of evaporite deposition have started with both natural seawater

EVAPORITE PROCESSES

TABLE 1. SALT CONCENTRATIONS IN SEAWATER[a]

Density	Volume	Fe_2O_3	$CaCO_3$	$CaSO_4 \cdot 2H_2O$	NaCl	$MgSO_4$	$MgCl_2$	NaBr	KCl
1.0258	1.000								
1.0500	0.533		0.0642						
1.0836	0.316	0.0030	Trace						
1.1037	0.245		Trace						
1.1264	0.190		0.0530	0.5600					
1.1604	0.1445			0.5620					
1.1732	0.131			0.1840					
1.2015	0.112			0.1600					
1.2138	0.095			0.0508	3.2614	0.0040	0.0078		
1.2212	0.064			0.1476	9.6500	0.0130	0.0356		
1.2363	0.039			0.0700	7.8960	0.0262	0.0434	0.0728	
1.2570	0.0302			0.0144	2.6240	0.0174	0.0150	0.0358	
1.2778	0.023				2.2720	0.0254	0.0240	0.0518	
1.3069	0.0162				1.4040	0.5382	0.0274	0.0620	
Total deposit		0.0030	0.1172	1.7488	27.1074	0.6242	0.1532	0.2224	
Salts in last bittern					2.5885	1.8545	3.1640	0.3300	0.5339
Sum		0.0030	0.1172	1.7488	29.6959	2.4787	3.3172	0.5524	0.5339

[a] Grams of salt laid down in Usiglio's experiment recalculated to an initial volume of 1 liter by Clarke (1924).

and synthetic salt systems. Many discrepancies still exist, however, between laboratory data and the thick sequences of evaporite seen in ancient deposits.

Usiglio (1849) evaporated Mediterranean seawater and found that calcium carbonate was precipitated when the water had been evaporated to approximately one-half the original volume and gypsum crystallized when the volume was one fifth of the original. At one tenth the initial volume the solubility products of halite, magnesium sulfate, and magnesium chloride are exceeded, and these salts precipitated. Continued reduction in volume of the brine by evaporation caused the precipitation of sodium bromide and potassium chloride with a salt residue still remaining (Table 1). Although Usiglio precipitated only eight salts, over seventy minerals have been identified from evaporite deposits, of which only twelve, however, are present in significantly large quantities (Table 2).

Since seawater is the medium from which many large-scale evaporites precipitated, the most important ions forming evaporites are Na^+, Mg^{2+}, Ca^{2+}, K^+, Cl^-, SO_4^{2-}, HCO_3^-, and CO_3^{2-} (Table 3). Most of the carbonate crystallizes out as calcium carbonate during the early stages of evaporation and the majority of the calcium and some sulfate is precipitated as gypsum and/or anhydrite. Whether gypsum or anhydrite precipitates depends on several factors including temperature, presence of impurities in the solutions, and degree of supersaturation (Cruft and Chao, 1969) (see *Anhydrite*). When saturation with respect to NaCl is reached, halite will crystallize along with calcium sulfate in the early stages and with magnesium and potassium salts in the later stages of crystallization. The later stages of crystallization can be represented by a ternary diagram with K^+, Mg^{2+}, and SO_4^{2-} in an aqueous solution saturated with NaCl.

Experimental work on this system from 0 to 110°C has been conducted by van't Hoff (1905, 1909, 1912), Jänecke (1923), Borchert (1940) and others (see Borchert and Muir, 1964). At 25°C (Fig. 2) increasing evaporation of the last stages of the solution would lead to the following sequence of salts: spsomite, kieserite-epsomite-hexahydrite, carnallite, kierserite, carnallite, and bischofite. This assumes that later precipitates do not form and crust over the early precipitates and that reaction of early precipitates with the solution is permitted, which is not always the case in natural solutions. Sequences of crystallization in this system at other temperatures are given in Fig. 3.

TABLE 2. EVAPORITE MINERALS

Chlorides

Halite[a]	$NaCl$
Sylvite	KCl
Bischofite	$MgCl_2 \cdot 6H_2O$
Koenenite	$Mg_9Al_4Cl_8(OH)_{22} \cdot 7H_2O$
Chlorocalcite (hydrophilite)	$KCaCl_3$
Carnallite	$KMgCl_3 \cdot 6H_2O$
Tachyhydrite	$CaMg_2Cl_6 \cdot 12H_2O$
Douglasite	$K_2FeCl_4 \cdot 2H_2O?$
Erythrosiderite	$K_2FeCl_5 \cdot H_2O$
Rinneite	NaK_3FeCl_6

Fluorides

Fluorite	CaF_2
Sellaite	MgF_2

Sulfates

Aphthitalite (glaserite)	$(K,Na)_3Na(SO_4)_2$
Thenardite	Na_2SO_4
Barite	$BaSO_4$
Celestite	$SrSO_4$
Anhydrite	$CaSO_4$
Vanthoffite	$Na_6Mg(SO_4)_4$
Glauberite	$Na_2Ca(SO_4)_2$
Langbeinite	$K_2Mg_2(SO_4)_3$
Mirabilite	$Na_2SO_4 \cdot 10H_2O$
Syngenite	$K_2Ca(SO_4)_2 \cdot H_2O$
Loeweite	$Na_4Mg_2(SO_4)_4 \cdot 5H_2O$
Blödite (astrakanite)	$Na_2Mg(SO_4)_2 \cdot 4H_2O$
Leonite	$K_2Mg(SO_4)_2 \cdot 4H_2O$
Picromerite (schoenite)	$K_2Mg(SO_4)_2 \cdot 6H_2O$
Polyhalite	$K_2Ca_2Mg(SO_4)_4 \cdot 2H_2O$
Görgeyite	$K_2Ca_5(SO_4)_6 \cdot H_2O$
Bassanite	$CaSO_4 \cdot \frac{1}{2}H_2O$
Kieserite	$MgSO_4 \cdot H_2O$
Sanderite	$MgSO_4 \cdot 2H_2O$
Gypsum	$CaSO_4 \cdot 2H_2O$
Starkeyite (leonhardtite)	$MgSO_4 \cdot 4H_2O$
Pentahydrite (allenite)	$MgSO_4 \cdot 5H_2O$
Hexahydrite	$MgSO_4 \cdot 6H_2O$
Epsomite (reichardtite)	$MgSO_4 \cdot 7H_2O$
Kainite	$KMg(SO_4)Cl \cdot 3H_2O$
Anhydrokainite	$KMg(SO_4)Cl$
D'Ansite	$MgNa_{21}(Cl_3SO_4)(SO_4)_9$

Carbonates

Calcite	$CaCO_3$
Magnesite	$MgCO_3$
Siderite	$FeCO_3$
Aragonite	$CaCO_3$
Strontianite	$SrCO_3$
Dolomite	$CaMg(CO_3)_2$
Ankerite	$Ca(Fe,Mg)(CO_3)_2$

Borates

Pinnoite	$Mg(BO_2)_2 \cdot 3H_2O$
Kurgantaite	$(Sr,Ca)_2B_4O_8 \cdot H_2O$
Priceite (pandermite)	$Ca_4B_{10}O_{19} \cdot 7H_2O$
Ulexite	$NaCaB_5O_9 \cdot 8H_2O$
p-Veatchite	$SrB_6O_{10} \cdot 2H_2O$
Colemanite	$Ca_2B_6O_{11} \cdot 5H_2O$
Hydroboracite	$CaMgB_6O_{11} \cdot 6H_2O$
Inderborite	$CaMgB_6O_{11} \cdot 11H_2O$
Inyoite	$Ca_2B_6O_{11} \cdot 13H_2O$
Kurnakovite	$Mg_2B_6O_{11} \cdot 15H_2O$
Inderite	$Mg_2B_6O_{11} \cdot 15H_2O$
Howlite	$Ca_2SiB_5O_9(OH)_5$

TABLE 2. (Continued)

Paternoite	$MgB_8O_{13} \cdot 4H_2O$
Ginorite (cryptomorphite)	$Ca_2B_{14}O_{23} \cdot 8H_2O$
Kaliborite	$KMg_2B_{11}O_{19} \cdot 9H_2O$
Szaibelyite (ascharite)	$(Mg)(BO_2)(OH)$
Boracite	$Mg_3B_7O_{13}Cl$
Ericaite	$(Fe,Mg,Mn)_3B_7O_{13}Cl$
Hilgardite	$Ca_8(B_6O_{11})_3Cl_4 \cdot 4H_2O$
Parahilgardite	$Ca_8(B_6O_{11})_3Cl_4 \cdot 4H_2O$
Strontiohilgardite	$(Ca,Sr)_2(B_5O_8(OH)_2Cl)$
Heidornite	$Na_2Ca_3Cl(SO_4)_2 \cdot B_2O_3(OH)_2$
Lueneburgite	$Mg_3B_2(OH)_6(PO_4)_2$
Sulphoborite	$Mg_6H_4(BO_3)_4(SO_4)_2 \cdot 7H_2O$
Danburite	$CaSi_2B_2O_8$

[a] Italicized names indicate most important minerals.

Comparing natural evaporite successions with experimental systems, however, it can be seen that in natural systems there is a deficiency of sulfate at the time when the brine should be saturated with respect to sodium, potassium, magnesium, chloride, and sulfide. This disparity is accompanied by the high ratio of calcium sulfate minerals to halite seen in natural deposits compared to what might be expected from the evaporation of seawater (Table 4 and Fig. 4). Although D'Ans (1947) has attempted to relate the sulfate deficiency in the potash salts to the excess of gypsum and anhydrite, Borchert and Muir (1964) attribute it to the reduction of sulfate to sulfide by bacteria during the early stages of formation of the basin. It is possible that the excess of calcium sulfate in the early stages can be explained by the inflow of seawater, and therefore need not have any relation to sulfate deficiencies in the bitterns. From the great rarity of the bittern salts in most natural successions it appears that only very infrequently does evaporation proceed to these last residual stages.

Dolomite is usually found in association with ancient evaporite deposits, although the volume of dolomite forming today is minor and is formed diagenetically (Kinsman, 1966; Bissel and Chillingar, 1962). Many unsuccessful attempts have been made to precipitate this mineral in the laboratory at low temperatures (See *Dolomite*, this volume; also Vol. VI.)

An association of particular interest is that between euxinic black shales and evaporites. Ferric iron is normally deposited from seawater and may be associated with evaporites indicating neutral or slightly oxidizing conditions at the site of deposition. If, however, a change in climate occurs or there is a sudden influx of terrigenous silt with much organic matter, and active sulfate-reducing bacteria, it is possible that the brine above the sediment–water interface in the basin may become reducing. Salinity changes accompanying evaporite deposition may also produce mass mortality in the basin with decomposition of the organic remains at the bottom by microorganisms (see Vol. III, *Kara-bogaz Gulf*). The relatively impermeable overlying evaporite layers will tend to seal off the organic layers, thus preserving them, and giving rise to the excellent preservation of organics in some evaporite deposits.

TABLE 3. DISTRIBUTION OF DISSOLVED SPECIES IN SEAWATER[a]

Ion	Total Molality	% Free Ion	% Me-SO$_4$ pair	% Me-HCO$_3$ pair (%)	% Me-CO$_3$ pair	
Ca^{2+}	0.0104	91	8	1	0.2	
Mg^{2+}	0.0540	87	11	1	0.3	
Na^+	0.4752	99	1	—	—	
K^+	0.0100	99	1	—	—	
			% Ca-anion pair	% Mg-anion pair	% Na-anion pair	% K-anion pair
Cl^-	0.56	100	—	—	—	—
SO_4^{2-}	0.0284	54	3	22	21	0.5
HCO_3^-	0.00238	69	4	19	8	—
CO_3^{2-}	0.000269	9	7	67	17	—

[a] At 25°C and 1 afm. Chlorinity is 10‰ and pH is 8.1. After Garrels and Christ, 1965.

EVAPORITE PROCESSES

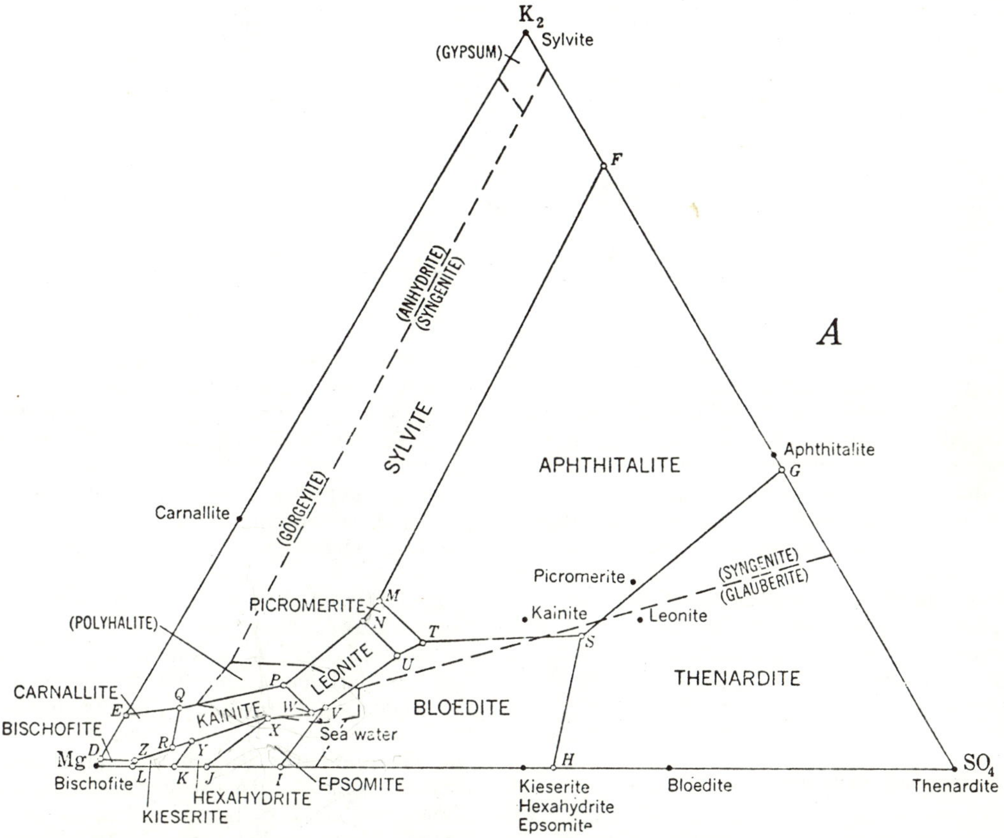

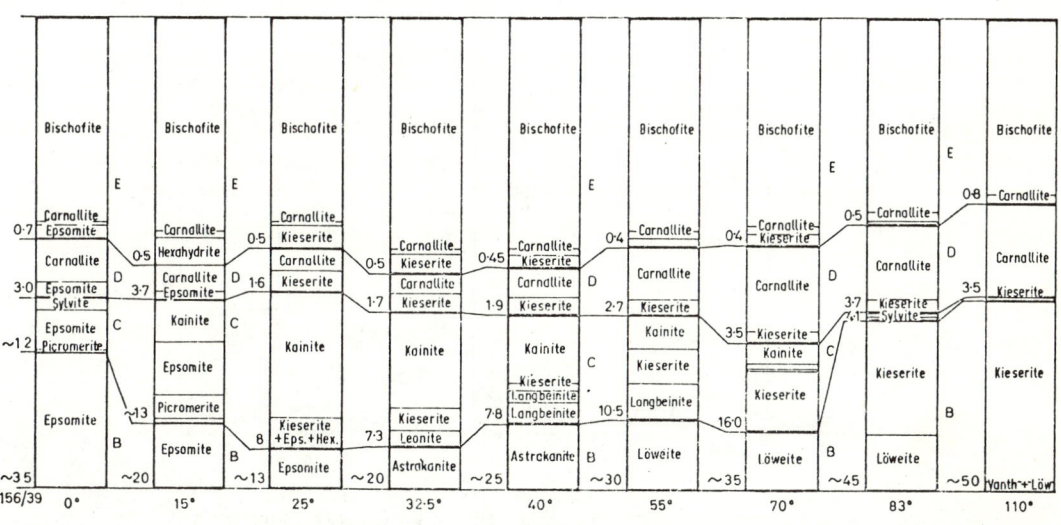

FIG. 2. Ternary diagram of the late stages of evaporite mineral formation when the solution is saturated with NaCl. The fields of Ca minerals are shown by dotted lines (from Stewart, 1963).

FIG. 3. Successions of late-stage evaporite minerals precipitated from seawater at temperatures ranging from 0 to 110°C. (from Borchert and Muir, 1964).

EVAPORITE PROCESSES

TABLE 4. COMPARATIVE PRECIPITATION PROFILES[a]

Components	Thickness in 100 m Evaporite Succession			
	Theoretical from Seawater	Zechstein	Other Marine Evaporites (and Potash)	Gypsum-Halite Deposits
$MgCl_2$	9.4	0.5	0.1	0
KCl	2.6	1.5	1.4	0
$MgSO_4$	5.7	1.0	0.2	0
$NaCl$	78	78	66	23.5
$CaSO_4$	3.6	16	26	58
$CaCO_3$ and $CaMg(CO_3)_2$	0.4	3	6.3	18.5

[a] From Borchert and Muir, 1964.

The thin annual calcite-anhydrite laminae of the Permian Castile formation of West Texas contain pyrite at the boundaries of the layers (Dean, 1967) suggesting perhaps a spring influx of stream water containing microorganisms and algae which normally could not exist in the high salinities necessary for the precipitation of calcium sulfate. Recent examples of euxinic and evaporite facies have been recorded in Peru by Morris and Dickey (1957) who found black mud in close association with deposits of gypsum and halite. Although the surface water pH was 7.8, bottom waters above the muds were at a pH of 6.6, indicating the possible presence of sulfate-reducing bacteria.

Natural Evaporite Deposits

Evaporite deposits can be divided into two major categories, marine and continental, although all variations between these two extremes are possible. Marine deposition of evaporites in deep marginal basins or relict seas are better known in the past than in present analogs. The Zechstein lagoons and the Permian deposits of west Texas and New Mexico were much greater in extent than the present marginal salt pans or supratidal flats in the Trucial Coast of the Persian Gulf and the Gulf of Mexico.

In terms of composition and morphology nonmarine evaporite formations show, as might be expected, much more variation (see Vol. III, *Lakes*). Carbonates and sulfates are generally the major components, but this is dependent to a large extent on source area and climate. In lake and river deposits in humid regions precipitation will give rise first to calcium carbonate and then to sodium and potassium sulfates as concentration proceeds. In arid regions where salts are derived from groundwater, so-

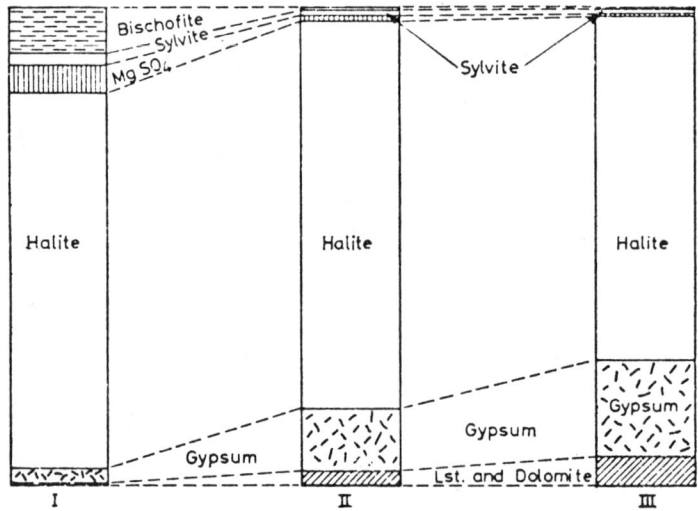

FIG. 4. Comparative thicknesses of evaporite salts precipitated from seawater (I), from the Zechstein evaporites (II), and from the average of other marine evaporites (from Borchert and Muir, 1964).

EVAPORITE PROCESSES

TABLE 5. RELATIONSHIP BETWEEN BRINE COMPOSITION AND CLIMATE[a]

Type of Water	Environment	Relative Ionic Abundance		Hydrochemical Facies
		Cations	Anions	
River and lake	Humid-continental	$Ca > Mg > Na + K$	$HCO_3 > Cl > SO_4$	Low salinity
		$Na + K > Ca > Mg$	$SO_4 > Cl > HCO_3$	High salinity
	Humid-subtropical	$Na + K > Ca > Mg$	$SO_4 > HCO_3 > Cl$	Low salinity
			$HCO_3 > Cl > SO_4$	High salinity
	Steppe	$Ca > Na > K > Mg$	$HCO_3 > SO_4 > Cl$	Low salinity
		$Na + K > Ca > Mg$	$SO_4 > Cl > HCO_3$	High salinity
Ground-water	Desert	$Na + K > Ca > Mg$	$Cl > SO_4 > HCO_3$	Arid
	Active circulation	$Na + K > Ca > Mg$	$HCO_3 > Cl > SO_4$	Low salinity
			$Cl > SO_4 > HCO_3$	High salinity
	Restricted circulation	$Na + K > Ca > Mg$	$Cl > HCO_3 > SO_4$	At all salinities
	Stagnant conditions	$Na + K > Ca + Mg$	$Cl > SO_4 + HCO_3$	At all salinities

[a] After I. Chebotarev, 1956; from Borchert and Muir, 1964.

dium and potassium chlorides and sulfates are predominant (Table 5) (Borchert and Muir, 1964). Continental basins can derive their water indirectly from seawater, from rivers, groundwater, or connate water. "Cyclic salts" are formed seasonally by spray blown onshore by the wind and collecting in basins (see Vol. II, *Salt Nuclei*). Littoral salina deposits (see *Lagoon Geochemistry*) may originate with the percolation of seawater inshore to fill depressions below base level. As a result of evaporation and prior precipitation, Ca^{2+}, CO_3^{2-}, and SO_4^{2-} is low and only the more soluble salts appear. Streams flowing from mountainous regions to arid plains may discharge water as well as sediment to catchment areas in their alluvial fans, or develop underground courses, periodically appearing above ground to form salt pans (see Vol. III, *Playa*). Continental evaporite formation also occurs by deposition from thermal springs, groundwater and connate brines trapped in ancient evaporite sequences. Recycling of earlier evaporite deposits by streams, groundwater, and wind, as in the White Sands gypsum deposits in New Mexico, is not uncommon.

The Permian Basin in west Texas and New Mexico is an example of a large-scale ancient marine evaporite basin, of which no modern analogs are known. Depth of water in the evaporite basin is not known although the presence of thin anhydrite-calcite laminae which can be individually correlated over distances as great as 110 km (Kirkland and Anderson, 1969, personal communication) suggests a period of quiet deposition, probably under deep-water conditions. Applications of the modified "Ochsenius bar theory" to these types of deposits requires a tectonic, climatic, and chemical balance that has to be maintained over a very long period of time. In view of the widespread recurrence of these deposits throughout the geological column, this might be somewhat unreasonable. The mode of origin of these deposits in terms of water chemistry and detailed physiography of the basin is still not adequately known.

The supratidal coastal flat or "sabkha" environment of evaporite deposition on the Trucial Coast of the Persian Gulf (see Vol. III, *Sabkha or Sebkha*) is in an area of low rainfall and high evaporation, with the area of evaporite deposition covering approximately 1000 square miles (Fig. 5) (Kinsman, 1966). The sabkha, composed of Holocene carbonates, is infrequently flooded with brine, but most evaporite deposition is controlled by groundwater. Sabkha accretion seaward has been at the rate of three to six feet for the last 4000 to 5000 years (Kinsman, 1969). A minor amount of aragonite is precipitated interstitially, along with abundant gypsum, but farther away from the coast nodular anhydrite is found. Pseudomorphs of anhydrite after gypsum are found near the coast and with time, and regression of the sea, primary syndiagenetic anhydrite is de-

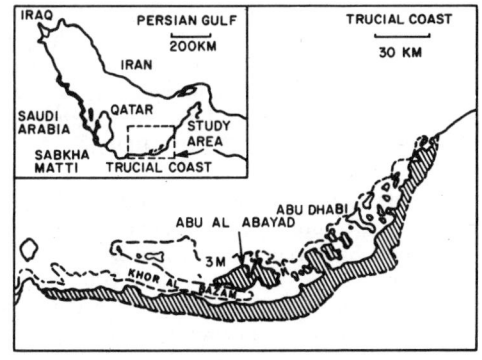

FIG. 5. Map of the Persian Gulf and surrounding countries with black areas indicating coastal sabkha extent, and continental sabkha. More detailed map shows sabkha area by hachures (from Kinsman, 1969).

posited around these pseudomorph nuclei to form the nodular anhydrite (Murray, 1964). After the aragonite, gypsum, and anhydrite are precipitated from the brine, the Mg/Ca ratio must increase and accompanying dolomitization of the carbonate sediments proceeds until the Mg/Ca is sufficiently reduced. Release of calcium from the carbonates promotes the precipitation of more gypsum. Some halite is formed on the sabkha surface, but it is removed either by the wind or is dissolved during flooding (Kinsman, 1969). Pseudomorphs of halite are seen in ancient sabkha deposits, e.g., the Purbeckian of England. Other environments similar to the Trucial Coast are found in Baja California and Australia. The recent evaporite environment north of San Felipe in the Gulf of California is discussed by Kinsman (1969); information on Laguna Ojo de Liebre on the Pacific coast of Baja may be found in Phleger and Ewing (1962), Kinsman (1969), and Phleger (1969). The evaporites of northern Queensland are discussed by Coleman, Sherwood, and Smith (1966).

The movement of evaporite materials by onshore winds to form continental dunes has been recorded in the Trucial Coast and in Baja California, and an example of aeolian deposition of continental evaporites is found in the Tularosa Basin in southern New Mexico. Permian evaporites have been dissolved by streams and groundwater and redeposited in Pleistocene sediments and playa lakes. During dry periods, persistent westerly winds pick up gypsum from surface crusts, and carry it east to White Sands National Monument, an area of approximately 690 square miles of white gypsum dunes.

Extensive evaporites, primarily gypsum and anhydrite, occur on and near the desert surface near high dunes in the Namib Desert of South West Africa. These have been attributed to wind-blown mist bringing in sulfate from the Atlantic Ocean (Martin, 1963). Another explanation is that the sulfur was derived from nearby igneous and metamorphic rocks, and that the evaporites are related to groundwater movement (Cagle and Cruft, 1969).

In the Atlantis II deep of the Red Sea, below 2000 meters, iron oxides, dolomite and anhydrite are forming. Temperatures as high as 56°C, have been recorded, which is 34° higher than normal Red Sea water below 1000 meters. Salinities reach 310 g/liter, which is ten times higher than those in normal seawater (Miller, et al., 1966). This abnormal brine is not widespread and is absent from other nearby deeps. High concentration of heavy minerals such as zinc (Zn), copper (Cu), iron (Fe), and manganese (Mn) and also low concentrations of magnesium (Mg), sulfate (SO_4), and bromine (Br) relative to total salt in seawater support the conclusion that the brine is intimately related to tectonic activity in the Red Sea rift. Older (Tertiary) evaporite deposits bordering the Red Sea are known, and some dolomite and anhydrite may be land derived. Although submarine discharge of thermal waters can be responsible for certain evaporitic minerals, deposits of this sort are apparently very limited in extent at the present time although they might be of much greater importance in ancient deposits.

SANDRA FELDMAN
EDGAR F. CRUFT

References

Bissel, H. J., and Chillingar, G. V., 1962, "Evaporite type dolomite in salt flats of western Utah," *Sedimentology*, **1**, 200–210.

Borchert, H., 1940, "Die Salzlagerstätten des deutschen Zechsteins," *Arch. Lagerstättenforschung*, **67**, 196pp.

Borchert, H., 1969, "Principles of oceanic salt deposition and metamorphism," *Geol. Soc. Am. Bull.*, **80**, 821–864.

Borchert, H., and Muir, R. O., 1964, "Salt deposits," London, Van Nostrand, 338pp.

Branson, E. B., 1915, "Origin of thick gypsum and salt deposits," *Geol. Soc. Am. Bull.*, **26**, 231–242.

Briggs, L. I., and Pollack, H. N., 1967, "Digital model of evaporite sedimentation," *Science*, **155**, 453–456.

Cagle, F. R., and Cruft, E. F., 1969, "Gypsum deposits of the coast of South West Africa," Third Intern. Salt Symp. Proc., Northern Ohio Geol. Soc., Vol. 1, pp. 156–165.

Clarke, F. W., 1924, "The data of geochemistry," *U.S. Geol. Surv. Bull.*, **770**, 841pp.

Coleman, J. M., Sherwood, M. G., and Smith, W. G., 1966, "Chemical and physical weathering on saline high tidal flats, Northern Queensland, Australia," *Geol. Soc. Am. Bull.*, **77**, 205–206

Cortelezzi, C. R., and Kilmurray, J. O., 1965, "Surface properties and epigenetic fractures of gavels from Patagonia, Argentina," *J. Sediment. Petrol.*, **35**, 976–980.

Cruft, E. F., and Chao, P. C., 1969, "Nucleation kinetics of the gypsum-anhydrite transition," Third Intern. Salt Symp. Proc., Northern Ohio Geol. Soc., Vol. 1, pp. 109–118.

D'Ans, J., 1947, "Über die Bildung und Umbildung der Kalisalzlagerstätten," *Naturwissenschaften*, **34**, 259–301.

Dean, W. E., 1967, "Petrologic and geochemical variations in the Permian Castile varved anhydrite, Delaware Basin, Texas and New Mexico," Ph.D. dissertation, Univ. New Mexico.

Garrels, R. M., and Christ, C. L., 1965, "Solutions, minerals, and equilibria," New York, Harper & Row, 450pp.

Grabau, A. W., 1920, "Geology of the nonmetallic

mineral deposits other than silicates," Vol. 1, "Principles of salt deposits," New York, McGraw-Hill Book Co., 435pp.
Jänecke, E., 1923, "Die Enstehung der deutschen Kalisalzlager," Second ed., Brunswick.
King, R. H., 1947, "Sedimentation in the Permian Castile sea," *Am. Assoc. Petrol. Geol. Bull.*, **31**, 470–477.
Kinsman, D. J. J., 1966, "Gypsum and anhydrite of Recent age, Trucial Coast, Persian Gulf. Second Symposium on Salt. " Vol. 1, Northern Ohio Geol. Soc., Cleveland, Ohio, pp. 302–326.
Kinsman, D. J. J., 1969, "Modes of formation, sedimentary associations and diagnostic features of shallow-water and supratidal evaporites," *Am. Assoc. Petrol. Geol. Bull.*, **53**, 830–840.
Kozary, M. T., Dunlap, J. C., and Humphrey, W. E., 1968, "Incidence of saline deposits in geologic time," *Geol. Soc. Am. Spec. Paper*, **88**, 43–57.
Martin, H., 1963, "A suggested theory for the origin and a brief description of some gypsum deposits of South West Africa," *Trans. Geol. Soc. S. Africa*, **66**, 345–359.
Miller, A. R., Densmore, C. D., Degens, E. T., Hathaway, J. C., Manheim, F. T., McFarlin, P. F., Pocklington, R., and Jokela, A., 1966, "Hot brines and recent iron deposits in deeps of the Red Sea," *Geochim. Cosmochim. Acta*, **30**, 341–360.
Morris, R. C., and Dickey, P. A., 1957, "Modern evaporite deposition in Peru," *Am. Assoc. Petrol. Geol. Bull.*, **41**, 2467–2474.
Murray, R. C., 1964, "Origin and diagenesis of gypsum and anhydrite," *J. Sediment. Petrol.*, **34**, 512–523.
Ochsenius, C., 1888, "On the formation of rocksalt beds and mother liquor salts," *Proc. Acad. Nat. Sci. Phila.*, **40**, 181–187.
Phleger, F. B., 1969, "A modern evaporite deposit in Mexico," *Am. Assoc. Petrol. Geol. Bull.*, **53**, 824-829
Phleger, F. B., and Ewing, G. C., 1962, "Sedimentology and oceanography of coastal lagoons in Baja California, Mexico," *Geol. Soc. Am. Bull.*, **73**, 145–182.
Scruton, P. C., 1953, "Deposition of evaporites," *Am. Assoc. Petrol. Bull.*, **37**, 2498–2512.
Sloss, L. L., 1969, "Evaporite deposition from layered solutions," *Am. Assoc. Petrol. Geol. Bull.*, **53**, 776–789.
Stewart, F. H., 1963, "Marine evaporites," *U.S. Geol. Surv. Profess., Paper*, **440-Y**, 52pp.
Usiglio J., 1849, "Analyse de l'eau de la Méditerranée sur les côtes de France," *Ann. Chem.*, **27**, 92–107; 172–191.
Van't Hoff, J. H., 1905, "Die Bildung der ozeanischen Salzablagerungen," Vol. 1, Leipzig, Viewig und Söhn.
Van't Hoff, J. H., 1909, "Die Bildung der ozeanischen Salzablagerungen," Pt. 2, Leipzig, Viewig und Söhn.
Van't, Hoff, J. H., 1912, "Untersuchungen über die Bildungsverhältnisse der ozenischen Salzablagerungen, insbesondere das Stassfurter Salzlagers," Leipzig, Precht and Cohen, 374pp.

Cross-references: *Cyclic Salts; Groundwater; Lagoon Geochemistry; Natural Brines; pH-Eh; Seawater, Chemistry; Solubility Product Constant; Sulfate Reduction—Microbial.* Vol. I: *Red Sea; Salinity in the Ocean.* Vol. II: *Salt Nuclei.* Vol. III: *Dead Sea; Great Salt Lake; Kara-bogaz Gulf; Lakes; Playa; Sabkha; Salton Sea.* Vol. IVB: *Anhydrite; Dolomite; Evaporite Minerals; Gypsum; Halite.* Vol. VI: *Dolomite; Evaporites; Salt Deposits; Salt—Cyclic.*

EVAPORITES—PHYSICOCHEMICAL CONDITION OF ORIGIN—See Vol. VI

EXSOLUTION*

Exsolution (or unmixing or precipitation) is the process of isochemical separation of one or more phases from an originally homogeneous phase without the complete disappearance of the original phase. The term exsolution is most frequently used in connection with solid-state reactions and is frequently considered to occur with falling temperature, but these are not necessary restrictions. The exsolved phases are termed *daughter* phases. The phase which maintains phase continuity is the *parent* phase.

Many minerals are capable of taking appreciable amounts of other components into solid solution and this ability is much more pronounced at high temperatures, as suggested by the schematic phase diagrams shown in Figs. 1 and 2. On cooling there will be a tendency for the high-temperature phases to unmix. Whether or not unmixing will occur, and what form it will take if it does occur, depends upon the rate of cooling and the characteristics of the particular system. As examples: (*1*) Above about 660°C the alkali feldspars, $(K, Na)AlSi_3O_8$, form a complete solid solution series from albite, $NaAlSi_3O_8$, to sanidine, $KAlSi_3O_8$. On comparatively rapid cooling (as is experienced by volcanic rocks which may be erupted at 800–1000°C) there is no exsolution. On the other hand, alkali feldspars from plutonic rocks, which have cooled very slowly from high temperature, frequently show pronounced exsolution to the two-phase assemblage known as perthite. (*2*) At high temperature (900°C) there is complete solid solution between PbS and PbTe, whereas at 200 or 300°C there is very little solid solution. This system reacts so rapidly that it is extremely difficult to quench-mix crystals in the laboratory from above 750°C. These two examples are similar to the system illustrated in Fig. 1. (*3*) Nonstoichio-

* Publication authorized by the Director, U.S. Geological Survey.

EXSOLUTION

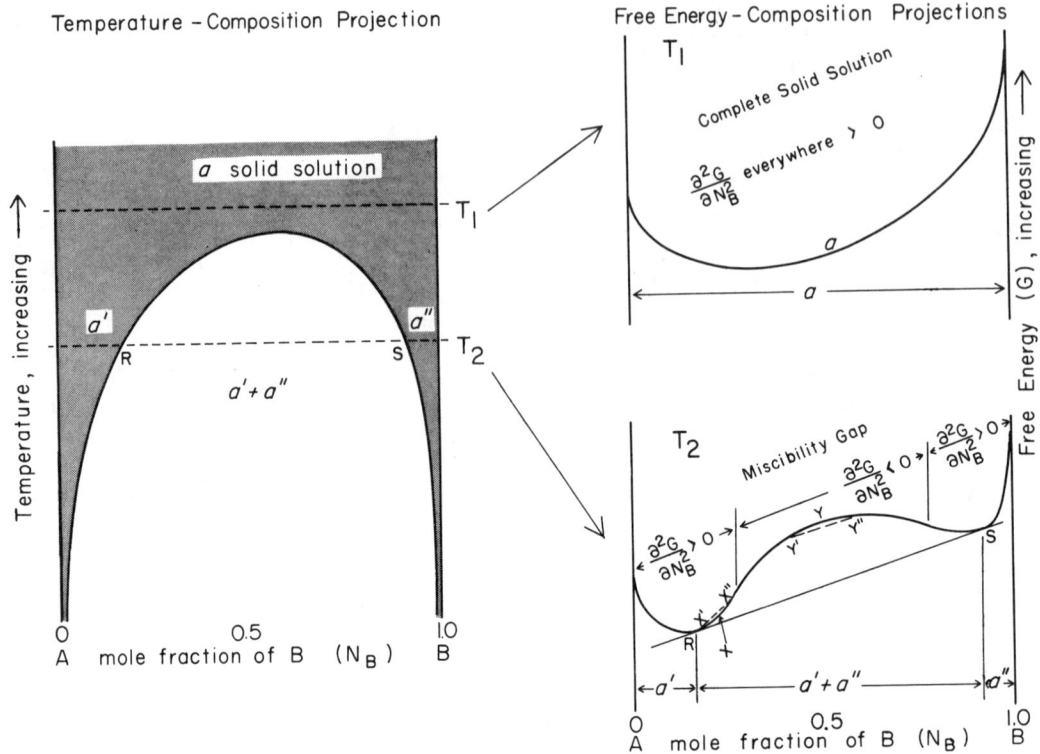

FIG. 1. Left side: temperature—composition diagram showing complete solid solution at temperature T_1 and a miscibility gap at T_2. Right side: free energy—composition diagrams for T_1 and T_2. Note that only one free-energy curve is present at each temperature. Spinodes are indicated (but not labeled) where $\partial^2 G/\partial N_B{}^2 = 0$.

metric compounds such as pyrrhotite, $Fe_{1-x}S$, generally have a more restricted compositional range at lower temperatures (as suggested by the c phase in Fig. 2 which is schematically similar to part of the Fe–S system) so that falling temperature would cause some compositions that had been in the stable one-phase region at higher temperatures to enter the two-phase field at lower temperature and thus to have a tendency to exsolve.

The boundary between the single- and multiple-solid phase regions in a phase diagram is commonly referred to as the *solvus*. The solvus may have any slope though, as mentioned above, it usually reflects the decrease in size of the stability fields of the individual solid phases at lower temperature. The assignment of the terms "parent" and "daughter" to the original and exsolved phases is ambiguous only at the crest of the solvus (Fig. 1); however, situations can occur in which a' exsolves a'' and then a'' further exsolves a' and thus it is possible for a phase to be both a daughter and a parent (and even grandparent).

Pressure as well as temperature may play a role in governing the position of the phase boundaries. However, since the range of temperatures encountered in the crust of the earth (0 to 1200°C) has vastly more influence on most systems than the usual range in pressures (1 to 10,000 bars), pressure is of comparatively minor importance in this connection.

Exsolution is isochemical, that is, there is no change in the bulk composition of the parent + daughter phases from that of the original parent. For example, the commonly observed "exsolution" of ilmenite from a magnetite-ulvospinel solid solution is not simple exsolution since it involves the addition of oxygen to the ulvospinel component as shown by the following simplified equation:

$$3Fe_2TiO_4 + \tfrac{1}{2}O_2 = Fe_3O_4 + 3FeTiO_3$$

Thermodynamic and Kinetic Description of Exsolution

The most stable assemblage of phases under a given set of conditions is that in which the free energy for the particular composition under consideration is the lowest. It is possible to construct diagrams, such as those shown on the right sides of Figs. 1 and 2, in which the

EXSOLUTION

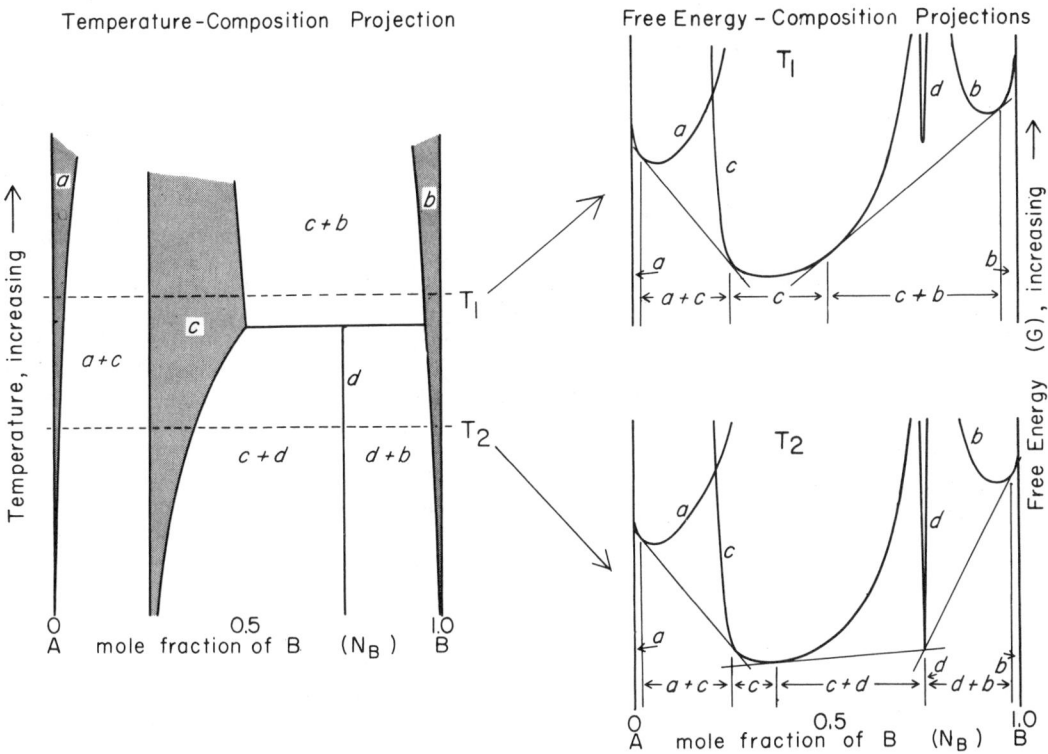

FIG. 2. Left side: temperature—composition diagram showing incomplete solid solution among phases *a, b, c,* and *d*. Right side: Free energy—composition diagram for temperatures T_1 and T_2. Note how the relative lowering of the free-energy curve for phase *d* from T_1 to T_2 changes the width of the stability field for phase *c*. Note also the absence of spinodes and hence the absence of attainable unstable regions.

Gibbs free energy, G, is plotted against composition, N, at constant temperature and pressure. Each phase will have its own particular free energy curve. Those phases with wide compositional ranges will have broad free energy curves, as *a* in Fig. 1. Phases with narrow compositional ranges will have narrow free energy curves, as *d* in Fig. 2. When the bulk composition falls between two curves (Fig. 2), or between two lobes of a single curve (Fig. 1 at T_2), the compositions of the solid solutions represented by the solvus are given by the points of tangency for the common tangent between the two free energy curves (as at R and S, Fig. 1). Thus the composite free energy curve for the stable phases in any system must be concave upward (in one-phase fields) or on a line connecting tangency points (for two-phase fields). The free energy curves for unstable or metastable phases (or for unstable or metastable compositions of phases for which other compositions may be stable) will lie above the minimum free energy curve.

Unmixing reactions involve two related phenomena: nucleation and crystal growth. Crystals possess a surface energy which is proportional to the amount of surface and which is thus an inverse function of the crystal size (assuming uniform crystallite morphology). Because stability is favored by a decrease in energy it is evident that large crystals are more stable than small ones, and when one considers that the initial nuclei must be extremely small it is evident that some appreciable degree of supersaturation is necessary for nucleation. However, although thermodynamics is based on the statistics of very large numbers (i.e., a macroscopic observation of a system composed of many millions of particles), the initial nuclei are thought to be of only a few hundred to a few thousand atoms and thus local aberrations in the overall statistical distribution of particles may permit nucleation without introducing a very high degree of supersaturation to the system as a whole.

In systems such as that illustrated in Fig. 1, the composition of the parent phase is particularly critical. If a homogeneous solid solution, *a*, containing 0.5 mole fraction B (composition Y) is placed at temperature T_2 it will be unstable relative to $a' + a''$ (compositions R and S) as shown by the phase and free energy diagrams. Infinitesimally small statistical aberrations may result in local compositional variations Y' and Y''. Because the free energy curve in this part of the diagram is concave downward ($\partial^2 G/\partial N^2 < 0$) the $Y' - Y''$ chord is now energetically lower than the original Y position. Thus the phases will nucleate easily (the activation energy is very low) and the nuclei can grow spontaneously at the expense of Y. If the diffusion rates are rapid enough Y' and Y'' will eventually reach the stable compositions R and S indicated by the solvus. There is no problem of nucleation because the crystal itself provides the nucleus.

On the other hand, consider the behavior of composition X, which, like Y, is not stable relative to $R + S$. In this portion of the diagram the free energy curve is concave upward ($\partial^2 G/\partial N^2 > 0$) and thus incipient phase separation such as $X' + X''$ is less stable than X and would therefore tend to rehomogenize rather than to dehomogenize. To break down to a more stable assemblage X composition would have to nucleate a phase having 0.6 or more mole fraction B. It is evident that the probability of nucleation is considerably lower for such compositions than for compositions in the region where $\partial^2 G/\partial N^2 < 0$.

The points where $\partial^2 G/\partial N^2 = 0$ are referred to as *spinodes*. The spinodes separate metastable regions (where $\partial^2 G/\partial N^2 > 0$ and where the system will tend to return to its original state if displaced only a small amount, as by $X \rightarrow X' + X''$) from unstable regions (where $\partial^2 G/\partial N^2 < 0$ and where the system will tend to proceed further from the original state if displaced by even an infinitesimal amount, as by $Y \rightarrow Y' + Y''$). Of course, the metastable fields terminate and are replaced by the stable ones beyond R and S (i.e., at the solvus). Experimental studies show a marked decrease in exsolution rates as compositions are shifted from the unstable to the metastable fields, in good agreement with the theory.

The problem of the nucleation of the potential daughter phase is particularly acute in systems such as that shown in Fig. 2 in which there are several phases and thus several free energy curves, and in which there may be no spinodes (and no unstable regions) at all. If there are no structural similarities between the host and daughter phases, nucleation may be difficult. Nucleation most frequently begins at grain boundaries, twin planes, or slip planes, for these are localized areas of higher energy due to surface or strain effects.

Textural Description of Exsolution

The growth of daughter nuclei is of particular importance to the metallurgic and ceramic industries and a very considerable amount of effort has been put forth toward understanding the growth process (see bibliography). Once a stable nucleus forms, the daughter phase tends to grow in such a way as to be as stable as possible. An important way to accomplish this is to decrease the surface energy that must be generated as the mass of the daughter phase increases; and this in turn may be done either by (1) forming crystallographically oriented intergrowths in which the energetic contribution per unit area of interface may be quite small, or (2) by permitting the daughter phase to grow as large and as equidimensional as possible (thereby decreasing the surface area).

If the parent and daughter phases each contain a similar two-dimensional array of atoms that can be shared by both structures, exsolution may take place and yet some degree of crystallographic continuity may be maintained across the interface (see Fig. 3). This behavior is referred to as *coherent* exsolution. The structurally shared interface may be much more stable than a randomly oriented interface between parent and daughter would be, the energy of the coherent interface being not too different from that contributed by a twin plane

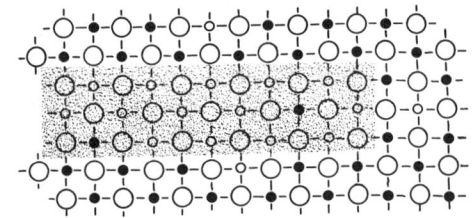

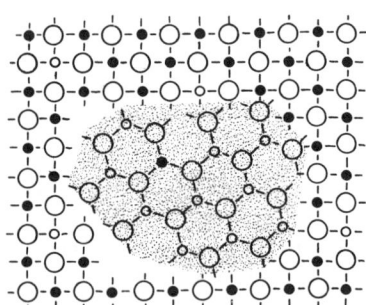

FIG. 3. Schematic diagram of coherent (upper) and noncoherent (lower) exsolution of ($\circ$,•) $\bigcirc$ from (•,$\circ$) $\bigcirc$ on an atomic scale. Area of daughter phases is shaded.

within a single phase. Because the structural matching is usually in only two dimensions (there are more complex relations, see bibliography) the resulting exsolution texture is usually one of plates or lamellae oriented parallel to one or more of the prominent structural patterns of the parent phase. The interface at the ends or edges of such plates may not be coherent.

Even slight disparity in the repeat dimensions between the parent and daughter lattices will eventually lead to sufficient strain to cancel the stability gained by the coherent intergrowth and a complete phase separation (noncoherent exsolution, see lower Fig. 3) takes place. The system will continue to decrease in interfacial energy by permitting the lamellae to take more equant forms, and eventually a granular texture can develop with no textural evidence of the presence of the former solid solution. In some systems coherent exsolution may be maintained to such an extent that the lamellar textures are megascopically observable, whereas in some others the first observable daughter phases are noncoherent.

Once the daughter phase has been nucleated the rate of exsolution is controlled by solid-state diffusion. The rate of solid-state diffusion is a function of temperature (though there may be anomalously high diffusion rates in the vicinity of phase transitions).

Diffusion rates are strongly influenced by the concentration gradients (degree of supersaturation) existing in the exsolving phase, and may also be affected by the presence of components that are not essential to the exsolution process (i.e., impurities).

Diffusion rates vary greatly from some silicate systems which do not equilibrate in weeks at 1000°C to some sulfide systems that react rapidly even at room temperature (see article on *Mineral Genesis, Equilibrium*). In structurally complex minerals there may be drastically differing diffusion rates for atoms in different structural sites; for example, sodium and potassium can readily be exchanged for one another in alkali feldspar whereas the aluminum and silicon exhibit very little mobility.

Problems of Recognition of Exsolution in Minerals

There are two other common processes which produce textures that may be difficult to distinguish from exsolution features. First, partial replacement of a pre-existing mineral by a later one may take place along crystallographically favorable directions leading to the development of lamellae. Transitions to more complete replacement or to veining textures and random orientation of the "daughter" phase are suggestive of replacement rather than exsolution. Replacement lamellae are thought to widen where they intersect other lamellae whereas those formed by exsolution tend to pinch. Second, a growing crystal of some other, structurally similar, pre-existing mineral may provide a foster parent nucleus for the epitaxial growth of oriented crystals that may later become swallowed by further growth of the original mineral. Such epitaxial growth is very common in sulfide ores where, for example, chalcopyrite frequently nucleates on sphalerite crystals. In many cases the preservation of original delicate growth banding in the host sphalerite demonstrates that solid-state diffusion within the sphalerite has been essentially nil and therefore that such lamellae are unlikely to have originated through exsolution.

PAUL B. BARTON, JR.

References

Bradley, R. S., 1951, "Nucleation in phase changes," *Quart. Rev. (London)*, **5**, 315–343.

Garrels, R. M., and Christ, C. L., 1965, "Solutions, Minerals, and Equilibria," New York, Harper & Row, 435pp.

Geisler, A. M., 1951, "Precipitation from Solid Solutions of Metals," in (Smoluchowski, R., Mayer, J. E., and Weyl, W. A., editors), "Phase Transformations in Solids," New York, John Wiley & Sons, pp. 387–544.

Symposium on Nucleation Phenomena, 1952, *Ind. Eng. Chem.*, **44**, 1269–1338.

Cross-references: *Enthalpy; Entropy; Metallurgy; Mineral Genesis—Equilibrium; Solid Solution; Stoichiometry; Thermodynamics.* Vol. IV B: *Alkali Feldspars; Ceramic Technology; Crystal Growth.* Vol. V: *Free Energy and the Gibbs Phase Rule.*

F

FERTILITY OF THE OCEAN—*See* Vol. I

FERRITE—*See* Vol. VI

FISSION TRACK DATING

The fission track method is a dating technique based on the discovery that insulating substances register and store tracks of heavy, energetic nuclear particles. The tracks are revealed by immersing a sample in an etching solution that preferentially attacks the submicroscopic regions of radiation-damaged material produced by the passage of the nuclear particles. In many natural materials, the etched channels thus formed give a faithful record of the trajectory of the particles and, indeed, look much like tracks in a nuclear emulsion (see Fig. 1). In other materials, such as glass (see Fig. 2), the etched tracks give rise to characteristic conical figures.

Most nuclear particles produced in natural radioactive decay processes do not produce enough radiation damage to give tracks that can be revealed by etching. In fact, it has been shown (Price and Walker, 1963) that, in terrestrial samples, only fission fragments from the spontaneous fission of uranium impurities give rise to the fossil (ancient) particle tracks which are a common feature of many minerals. This statement is strictly true when referring to large tracks several microns or greater in length. It has recently been found, however (Huang and Walker, 1967), that the recoil nuclei accompanying alpha particle decay also produce fossil tracks. Such tracks are very short ($\sim 100\text{Å}$) and must be studied by special techniques such as phase contrast microscopy. Since they are much more numerous than fission tracks, they may in the future prove useful as a dating tool.

Other sources of tracks exist in extra-terrestrial materials, and these are discussed in a separate article (see Vol. II, *Nuclear Particle Tracks in Meteorites*).

The number of stored spontaneous fission tracks increases with increasing age of the sample according to the relation:

$$S = (\lambda_F/\lambda_D)N_U k[\exp(\lambda_D T) - 1] \quad (1)$$

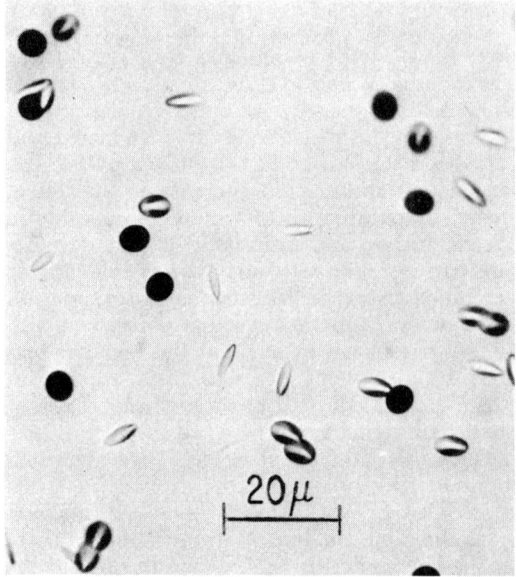

Fig. 2. Etched fission-fragment tracks in a borate glass. Round pits are from fragments crossing normal to the surface; elliptical pits result from fragments passing into glass at acute angles.

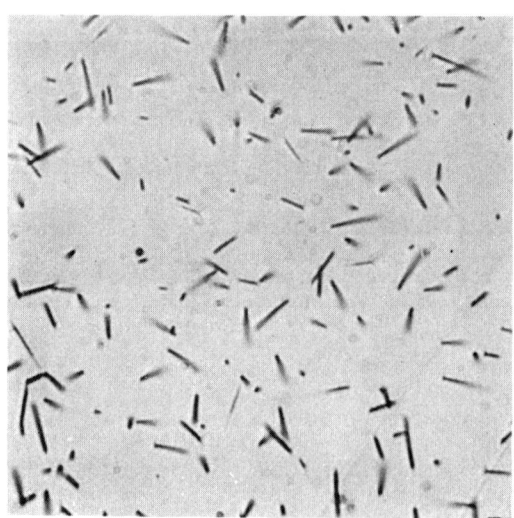

Fig. 1. Naturally occurring fission-fragment tracks in a zircon crystal from Nyasaland, revealed by etching for one minute in phosphoric acid at 480°C (560×).

where S is the density of stored tracks (number per cm^2), λ_D and λ_F are respectively the total and spontaneous fission decay constants of U^{238}, N_U is the number of U^{238} atoms/cm^3, and k is a geometric factor approximately equal to the range of fission fragments in the sample. A uranium concentration of 10^{-6} atom/atom gives an easily measurable track density of ~ 2000/cm^2-million years.

The uranium concentration is inferred from the increase in track density produced by an irradiation with a known dose of neutrons in a nuclear reactor. The density of tracks before and after irradiation is used with Eq. (1) to give the age, T, of the sample. It should also be noted that, even in the absence of fossil tracks, this induced track production provides an extremely sensitive way of measuring the uranium contents of a sample, concentrations lower than 10^{-14} atom/atom being measurable in principle. Proper choice of neutron bombarding energies can also be used to measure Th/U ratios.

To be suitable for fission track dating, a sample must have enough uranium to give a measurable fossil track density during the lifetime of the sample. The tracks must also be stable against thermal effects, and the appropriate etchant must be known. Silicate glasses, mica, apatite, sphene, epidote, zircon, and hornblende are common materials in which tracks are very stable and in which the uranium contents are often high enough to be useful.

Although the fission track method is still in its infancy, it has already yielded several important results, notably the verification of K-Ar ages of tektites, the identification of impactite ages with those of tektites (Fleischer, et al., 1965a), the confirmation of the K-Ar age of Bed I, Olduvai Gorge (Fleischer, et al., 1965b). The method has also been used to show that there is no large discrepancy between the time of formation of tektites and their arrival time on earth (Fleischer and Price, 1964) as had been proposed by some investigators. More recently the simultaneous use of fission track dating (Fleischer et al., 1968) and K-Ar dating (Aumento et al., 1968) has given direct chronological evidence for ocean bottom spreading and confirmation of the hypothesis that the magnetic anomalies on the ocean bottoms are chronologically significant.

The fission track method has the following unique features: (a) it can be used to date materials not hitherto measurable; (b) it can be used to work with very tiny samples including thin sections of rock; (c) it is useful for dating of the Pleistocene; (d) because tracks are removed by heating, fission track dating can be used to date natural heating events; and (e) it is inherently simple. Because of these unique features, and in light of its demonstrated utility for certain problems, this method is becoming a standard tool of geochronology.

For a fuller description of the method the reader is referred to two review articles (Fleischer et al., 1965c and d).

ROBERT L. FLEISCHER
P. BUFORD PRICE
ROBERT M. WALKER

References

Aumento, F., Wanless, R. K., and Stevens, R. D., 1968, "Potassium-Argon Ages and Spreading Rates on the Mid-Atlantic Ridge at 45° North," *Science*, **161**, 1338.

Fleischer, R. L., and Price, P. B., 1964, "Fission Track Evidence for the Simultaneous Origin of Tektites and other Natural Glasses," *Geochim. Cosmochim. Acta*, **28**, 755.

Fleischer, R. L., Price P. B., and Walker, R. M., 1965a, "On the Simultaneous Origin of Tektites and Other Natural Glasses," *Geochim. Cosmochim. Acta*, **29**, 161.

Fleischer, R. L., Price, P. B., and Walker, R. M., 1965b, "Fission Track Dating of Bed I, Olduvai Gorge," *Science*, **148**, 72.

Fleischer, R. L., Price, P. B., and Walker, R. M., 1965c, "Tracks of Charged Particles in Solids," *Science*, **149**, 383.

Fleischer, R. L., Price, P. B., and Walker, R. M., 1965d, "Solid State Track Detectors: Applications to Nuclear Science and Geophysics," *Ann. Rev. Nuclear Science*, **15**, 1.

Fleischer, R. L., Viertl, J. R. M., Price, P. B., and Aumento, F., 1968, "Mid-Atlantic Ridge: Age and Spreading Rates," *Science*, **161**, 1339.

Huang, W., and Walker, R. M., 1967, "Fossil alpha-particle recoil tracks: A new method of age determination," *Science*, **155**, 1103-1106.

Price, P. B., and Walker, R. M., 1963, "Fossil Tracks of Charged Particles in Mica and the Age of Minerals," *J. Geophys. Res.*, **68**, 4847.

Cross-references: *Geochronometry; Potassium-Argon Age Determination; Uranium-Thorium-Lead Dating.* Vol. II: *Nuclear Particle Tracks in Meteorites; Tektites.*

FLAME PHOTOMETRY—See FLAME SPECTROSCOPY

FLAME SPECTROSCOPY

Flame spectroscopy is the utilization of a flame excitation source in the optical emission spectroscopic analysis of elements with excitation potentials and monoxide dissociation energies low enough for flame excitation. This now includes over sixty elements, and has many applications in geochemistry.

FLAME SPECTROSCOPY

The first use of the flame as a spectroscopic source for analyzing elements is generally attributed to Bunsen and Kirchoff in 1860. Lundegårdh (1929, 1934) developed quantitative procedures beginning in 1928 and is considered the father of modern flame spectroscopy. Although his techniques appeared very promising, flame spectroscopy was not used extensively until after 1945 when commercial instrumentation was developed. The wide acceptance of flame spectroscopy as an analytical technique is so indicated by a bibliography (Mavrodineanu, 1967) containing 1987 references to analytical procedures (5113 total references), many chapters in books, and a number of books (see Mavrodineanu, 1967). Three of the most recent books are those by Dean (1960), Herrmann and Alkemade (1963), and Mavrodineanu and Boiteux (1965).

For analysis with the flame spectrometer, a sample in solution form is atomized into a flame, and chemical elements contained in the sample are vaporized and excited to emit characteristic electromagnetic radiation—light rays. This radiation is deflected into a selective filter, a prism, or a diffraction grating so that only a narrow portion of the spectrum is separated and directed onto a suitable detector for measuring light intensity. The detector converts the light into electrical energy which is measured with a galvanometer or recorder.

Flame spectrometers and photometers consist of various combinations of components which include the type of flame used for excitation, the atomizer burner, the light dispersing element or monochromator, the detector, and the measuring device.

Flames used for excitation burn fuel mixtures varying from natural gas-air at a temperature of $2200°K$ to cyanogen-oxygen at $4800°K$. Hydrogen-oxygen provides an excellent flame for general use with a sufficiently high temperature, $2900°K$, to excite many elements and gives low background emission. A nitrous oxide-acetylene flame (Fleming, 1967) has been used more recently to obtain better sensitivity for elements forming monoxides with higher dissociation energies such as aluminum. A special, premixing chimney has also been used with a rich oxygen-acetylene flame to produce similar results (D'Silva et al., 1964). The utilization of flames such as these and chemical reactions (along with thermal excitation) in the lower part of the flame has increased applications to the elements that are more difficult to excite.

For introduction of the sample solution into the flame two types of arrangements are used. One type is an integral, atomizer-burner constructed so the sample solution is injected directly into the flame. Another type has a nebulizer-spray chamber where larger droplets condense and finer droplets are swept into the flame (Herrmann and Alkemade, 1963, pp. 21–34).

The type of dispersing element employed in an instrument determines if it is a flame photometer or flame spectrometer. Flame photometers employ a light filter which limits their use due to interference from spectral overlap. Modern interference filters can now be obtained which transmit narrow light bands but even with their use, applications for flame photometers are limited. Spectrometers employing prisms or diffraction gratings are preferred because of the much wider variety of elements they can determine (ibid., p. 427, also table 15). More recently, the development of better instruments has led to high dispersion, high resolution spectrometers that are utilized to improve line-to-background ratios and provide better detection limits (Buell, 1966).

For detecting the emitted light barrier-layer cells, phototubes and multiplier phototubes can be used. The latter provide better sensitivity, allowing the use of narrower slits and/or less instrument gain. This increases the ratio of line signal to background signal and fluctuations, thus lowering detection limits. The technology of producing excellent multiplier phototubes has improved in recent years making additional improvements in detection limits possible. Cooling devices for such tubes, which decrease dark current fluctuations by orders of magnitude, are also being supplied commercially.

For measuring instrument response, galvanometers or recorders are generally employed. Recorders are becoming more popular because they can be used to average out flame fluctuations, are easier to read, and can be used in conjunction with wavelength scanning to record spectra (Watanabe and Kendall, 1955). More recently digital recording and computer averaging devices have become available and offer possibilities of improving precision.

Quantitative determinations with the flame spectrometer are very rapid and precise. A single measurement, after preliminary instrument adjustments, can be made in ten seconds with a precision as good as $\pm\ 0.2\%$ of the emission obtained (Beckman Instruments, Bull. 259). Because flame methods are empirical, calibration curves must be prepared by plotting the net emission of light versus concentrations of the element to be determined.

Although precision of measurements is excellent, certain errors may be encountered which can lead to inaccuracy (Herrmann and Alkemade, 1963; Gilbert, 1959). These errors are caused by differences between the composition of samples and standards and are often called matrix errors. In certain instances, a large excess of an unwanted element (concomitant) may prohibit the determination of the wanted element (analyte). Depending on the severity and type of interference, various procedures are used to compensate for matrix errors. The most troublesome types of interferences are spectral and chemical interferences. Spectral interferences occur when radiation from concomitants, sample solvents, or flame components overlaps that of the analyte. An entire chapter devoted to spectral interferences and techniques of compensating for them is given in Buell (1969). Chemical interferences are caused by concomitants which have entered the flame and cause reactions that influence the analyte emission. The literature on such interference is extensive and is also included in Chapters 10, 11, and 12 of the works edited by Dean and Rains (cited under Buell, 1969) and other books. When other procedures fail, a good technique is to separate the analyte by solvent extraction (Dean, 1960); this also increases sensitivity.

Most applications of flame spectrometry have been for the determination of sodium, potassium, and calcium in a large variety of samples. In more recent years techniques and instrumentation have improved markedly so that applications for determining many other elements have increased proportionately. Now about fifty elements, shown in Table 1, can be determined with a detection limit of 1 ppm or better and about eighteen more can be determined with less sensitivity. With proper instrumentation, flame sources are now competitive with spark sources for many elements including the transition elements (Buell, 1966; Fassel and Golightly, 1967).

Applications to geology and geochemistry have been mainly for determining the more sensitive alkali and alkaline earth elements in a large variety of samples including geological waters, rocks, minerals, and ores. In more recent years, as the scope of flame techniques has grown, applications to other elements have increased. There are 170 references listed in Mavrodineanu's bibliography (1967) for the last three sample types and over 100 references for water analysis. In the early 1860s, Kirchhoff and Bunsen applied flame spectroscopy to mineral analysis and to the discovery of certain elements. In 1897, Hartley and Ramage applied flame methods to a study of the dissemination of elements in a variety of common ores and minerals. Later (1901) they describe a simplified flame spectrographic method for the analysis of minerals. In this manuscript they also presented an interesting introduction concerning the early history of mineral analysis beginning with extensive blow-pipe applications by Gahn and Bergman in 1818. These earlier applications led to better quantitative procedures such as the determination of alkali metals in minerals by Bossuet (1933). By 1950 a combined spectrographic method including a flame determination of total Fe, Mn, Mg, Ca, Na, and K in silicate rocks was developed by P. H. Lundegårdh. After 1955, applications to elements other than Na, K, and Ca began increasing. The elements that have been determined quantitatively in minerals and rocks now include Ga, In, Tl, Fe, Cu, Co, Mn, La, V, and even PO_4 indirectly. Methods of eliminating interferences have also been improved. A good example is the use of releasing agents developed by Dinnin (1960) to eliminate chemical interferences. Flame spectroscopy has been applied extensively to water analysis, including samples such as oil field brines and seawater. The determination of Sr has been studied extensively and some of the more unique determinations include Cl via CuCl bands and Cs (after concentration) in seawater. Flame spectroscopy has been utilized for various geological studies which includes determining K in *Potassium-Argon age dating* (q.v.). Selected references for some of these more unique applications and for some of the comprehensive studies of routine determinations are compiled in Table 1.

In addition to flame emission spectroscopy, atomic absorption spectroscopy has been applied to the analysis of geological materials and provides some advantages. Atomic absorption will be discussed briefly here because it is closely related to flame emission and because many commercial instruments are now designed to use either technique. A bibliography of atomic absorption references is given in Mavrodineanu (1967).

For atomic absorption measurements, a source of monochromatic radiation (usually a hollow cathode lamp made from the element to be analyzed) is inserted in the optical path of a flame spectrometer so that its light beam passes through the flame and into the instrument. The light from this beam is adjusted to 100% light (O absorbance). When the sample is atomized into the flame, ground-state atoms are formed. These atoms absorb light at their corresponding wavelengths of emission in the excited state. Sample concentrations are calculated from a calibration of absorbance versus concentration. In actual practice, longer flames are used to increase the absorption path, thereby increasing sensitivity, and an AC system is utilized to eliminate most *spectral interfer-*

FLAME SPECTROSCOPY

TABLE 1. SELECTED REFERENCES FOR THE APPLICATION OF
FLAME SPECTROSCOPY TO GEOLOGICAL MATERIALS

Reference No.[a]	Sample Type	Elements Determined	Year
1	Minerals and rocks	Bibliography	1963
2	I Rocks and minerals	Na,K	1963
	II Rocks and minerals	Ca,Sr,Rb,Li	
3	Water, plants, soils and rocks	Alkali and alkaline earth elements	1956
4	W-1 and G-1 Standards	Li,Na,K,Rb,Ca,Cs,Sr	1961
5	Silicate rocks	Li,Rb,Cs	1956
6	Rocks	Li,Rb,Cs	1963
7	Bauxite, tonerde, and zinc-blende	Ga,In,Tl via extraction	1959
8	Mats and concentrates	Fe,Cu,Co via solvent extraction	1959
9	Mineral pulps and rocks	Mn	1955
10	Minerals and glass	La via bandhead	1955
11	Magnetite-ilmenite	V via extraction and bandhead	
12	Phosphate rocks	PO_4 via Ca	1954
13	Phosphate rocks	PO_4 via Ca	1964
14	Standard muscovite for K–Ar age dating and Std G-1	K	1965
15	K–Ar age dating studies in Scotland	K	1965
16	Seawater—review of six methods	Ca,Mg,Na,K,Sr	1954
17	Seawater	Sr	1955
18	Seawater	Cs after concentration	1964
19	Seawater	Cl via CuCl bands	1955
20	Water	Ca,Mg,Na,Fe	1956
21	Oil field water	Modern instrumental methods—45 pp.	1961
22	Oil field water	K,Li,Sr,Ba,Mn,Cs,Rb	1962–1965
23	Formation brines as to source	Flame identification (qualitative)	1961

[a] See references at end of article.

ences derived from sample concomitants, solvents, and flame components. This elimination of most spectral interferences plus better sensitivity for certain elements provides an advantage in many instances for atomic absorption compared to flame emission. On the other hand flame spectroscopy is more sensitive for some elements, provides greater flexibility, can

TABLE 2. DETECTION LIMITS FOR ATOMIC ABSORPTION AND
FLAME EMISSION SPECTROSCOPY[a]

Range for Det. Limit (ppm)	Elements	
	Atomic Absorption	Flame Emission
< 0.01	Be, Ca, Cd, Cu, K, Li, Mg, Mn, Na, Rb, Zn	Ca, Cs, Cu, In, K, Li, Mn, Na, Rb, Sr
0.01–0.1	Ag,[c] Al,[b] Au,[b] Ba,[b] Bi, Co,[c] Cr,[c] Cs, Fe,[c] In, Ga, Mo,[b] Ni,[c] Pb, Rh, Sb, Sc,[b] Sr,[c] V, Yb	Ag,[c] Al, Ba, Co, Cr,[c] Eu, Fe,[c] Ga, Mg, Mo,[b] Ni,[c] Rh, Sc,[b] Tl,[c] V, Y, Yb
0.1–1	As, Dy, Er, Eu, Ge,[b] Hg, Ho, Pd, Pt, Ru, Sb, Si,[c] Sn,[c] Te, Tl, Ti, Tm, Y	Au, B,[e] Be, Cd, Cl_2, Dy, Er, Gd, Ge,[b] Ho, La, Lu, N,[e] Nb, Nd, Pb, Pd, Ru, Sb, Sn,[c] Tb, Ti, Tm, U
> 1	B, Gd, Hf,[d] Ir, La, Lu,[d] Nb,[d] Nd, Pr, Re, Sm, Ta, Tb, U, W, Zr	As, Bi, Ce,[d] Hg,[d] Sr,[d] Os,[d] P, Pr, Pt, Re, Si, Sm, Ta,[d] Te,[d] Th,[d] U,[d] Zn,[d] Zr

[a] Based on optimum conditions for the best published detection limits.
[b] Borderline poor for this range.
[c] Borderline good.
[d] Very poor.
[e] Based on bandhead emission.

utilize emission of bandheads, and does not require a hollow cathode lamp for each element analyzed. *Chemical interferences* are essentially the same for both techniques. Sensitivities for atomic absorption are compared to those for flame emission in Table 2. There should be little difficulty developing methods applicable to almost any sample type for the elements that have detection limits under 0.1 ppm. As detection limits approach 1 ppm and more, limitations will increase.

Atomic absorption has been applied by Billings and Adams (1964) to the determination of many elements, including Ni, Co, and Cd in various rock and mineral types. More recently, a book by Angino and Billings (1967) has been published on this subject. One excellent application for atomic absorption has been for gold assaying described in the *Atomic Absorption Newsletter*, 4(5), 1965. A new technique using direct vaporization of solid samples with rocket propellant mix shows some promise for gold prospecting and other applications. Venghiattis and Whitlock (1967) have applied the technique to analyzing ore concentrates for Cu, Fe, Pb, and Hg. Although the solid-propellant technique may not be as precise as solution techniques it should have some excellent time-saving applications, especially in the field.

By the full utilization of flame spectroscopy and atomic absorption many new applications to geological materials should be possible in the near future.

BRUCE E. BUELL

References

Angino, E. E., and Billings, D. K., 1967, "Atomic Absorption Spectrometry in Geology," New York, Elsevier Publ. Co., 144pp.

Beckman Instruments, Inc., Bull. 259, Instructions for the Beckman Flame Spectrophotometer.

Billings, G. K., and Adams, J. A. S., 1964, "The analysis of geological materials by atomic absorption spectroscopy," *Atomic Abs. Newsletter,* **23**, 1.

Bossuet, R., 1933, "Alkali metals in minerals," *Compt, Rend.,* **196**, 1381; *Chem. Abstr.,* **27**, 4501.

Buell, B. E., 1966, "Comparison of spark excitation to high-and-low-dispersion flame spectrometry," *Anal. Chem.,* **38**, 1376.

Buell, B. E. 1969, "Spectral Interferences," in (Dean, J. A., and Rains, T. C., editors), "Flame Emission and Atomic Absorption Spectroscopy," Vol. 1, Chap. 9, New York, Marcel Dekker.

Dean, J. A., 1960, "Flame Photometry," New York, McGraw-Hill Book Co.

Dinnin, J. I., 1960, "Releasing effects of flame photometry-determination of calcium," *Anal. Chem.,* **32**, 1475.

D'Silva, A. P., Knisely, R. N., and Fassel, V. A., 1964, "The premixed fuel-rich, oxyacetylene flame in flame emission spectrometry," *Anal. Chem.,* **36**, 1287.

Fassel, V. A., and Golightly, D. W., 1967, "Detection limits of elements in the spectra of premixed, oxy-acetylene flames," *Anal. Chem.,* **39**, 466.

Fleming, H. D., 1967, "Use of mixed air nitrous oxide (N_2O)/acetylene flames in atomic absorption spectroscopy," *Spectrochim. Acta,* **23B**, 207.

Gilbert, P. T., Jr., 1959, "Analytical flame photometry: new developments," *Am. Soc. Testing Mater. Spec. Tech. Publ.,* **269**, 73.

Hartley, W. N., and Ramage, H., 1897, "The wide dissemination of some rarer elements, and the mode of their association in common ores and minerals," *J. Chem. Soc.,* **71**, 533.

Hartley, W. N., and Ramage, H., 1901, "A simplified method for the spectrographic analysis minerals," *J. Chem. Soc.,* **79**, 61.

Herrmann, R., and Alkemade, C. T. J., 1963, "Chemical Analysis by Flame Photometry," New York, Interscience Publishers, a div. of John Wiley & Sons (translated by P. T. Gilbert, Jr.).

Libby, W. F., 1956, "Putting on the Heat," *Chem. Eng. News,* **34**, 3442 (contains an illlstrated table of flame temperatures).

Lundegårdh, H., 1929, 1934. "Die Quantitative Spektrolanalyse der Elemente," Jena, Fisher, Part I, 1929; Part II, 1934.

Lundegårdh, P. H., 1950, "Rapid analysis of rocks. Some viewpoints with special emphasis on the possibilities of use of the flame method," *Geol. Foren. Stockholm Forh.,* **72**, 151.

Mavrodineanu, R., 1967, "Bibliography on Flame Spectroscopy, Analytical Applications 1800–1966," *Natl. Bur. Std. (U.S.) Misc. Publ.,* **281**.

Mavrodineanu, R., and Boiteux, H., 1965, "Flame Spectroscopy," New York, John Wiley & Sons.

Venghiattis, A. A., and Whitlock, L., 1967, "Determination of metal ores by solid sampling," *Atomic Abs. Newsletter,* **6**(6), 135.

Watanabe, H., and Kendall, K. K., Jr., 1955, "Flame spectrograms. I. Common metals," *Appl. Spectroscopy,* **9**, 132.

*Complete coverage can be obtained from additional references in above works.

Cross-references: *Alkalis, Alkali Metals, and Alkaline-Earth Metals; Emission Spectrography; Potassium-Argon Age Determination; X-Ray Diffraction Analysis; X-Ray Spectroscopy.* Vol. IVB: *Blowpipe Analysis.*

References to Table 1

1. I, Camacho-Calderon, H. A. Mottola, and C. A. Vallecilla-Risscos, 1963, *Rev. Univ. Ind. Santander,* **5**, 447–449. Bibliography on the determination of distinct elements in minerals and rocks by flame photometry.
2. L. Biagi, R. Pirani, and G. Simboli, 1963, Parts I and II, *Mineral, Petrog. Acta,* **9**, 111-162, 163–178; *Chem, Abstr.,* **63**, 1211 (1965).

3. T. F. Borovik-Romanova, 1956, "Spectral-Analytical Determination of Alkali and Alkaline Earth Elements in Water, Plants, Soils, and Rocks," Moscow, Izdatel. Akad. Nauk, SSSR, 184pp; *Chem. Abstr.*, **52**, 7034 (1958).
4. E. E. Vainshtein and V. I. Lebedev, *Geokhimiya*, **1961**, 362–363; *Chem. Abstr.*, **56**, 10889 (1962).
5. E. L. Horstman, 1956, *Anal. Chem.* **28**, 1417.
6. R. Pouget, 1963, CEA-R-2285, 5pp.; *Chem. Abstr.*, **60**, 1107 (1964). cf CEA-R-2176, 28pp. (1962); *Chem Abstr.* **58**, 5024 (1963), "Analysis of silicate rocks."
7. H. Bode and H. Fabian, 1959, *Z. Anal. Chem.* **170**, 387.
8. N. Mc N. Galloway, 1959, *Analyst*, **84**, 505.
9. R. Ishida, 1955, *Dept. Govt. Chem. Res. Inst., Tokyo*, **50**, 35–39; *Chem. Abstr.*, **50**, 8381 (1956).
10. R. Ishida, 1955 (Part III) *J. Chem. Soc. Japan, Pure Chem. Sect.*, **76**, 60–63; *Chem. Abstr.* **49**, 13012 (1955); *cf ibid.* 56–50 (Part II).
11. C. M. Stander, 1960, *Anal. Chem.*, **32**, 1296.
12. W. A. Dippel, C. E. Bricker, and N. H. Furman, 1954, *Anal. Chem.* **26**, 553.
13. R. Ratner and D. Scheiner, 1964, *Analyst*, **89**, 136.
14. M. A. Lanphere and G. B. Dalrymple, 1965, *J. Geophys. Res.*, **70**, 3497.
15. J. A. Miller and P. E. Brown, 1965, *Geol. Mag.*, **102**, 106; cf J. A. Miller, 1960, *Nature*, **187**, 1019 for ages of rocks in the South Atlantic.
16. R. D. Hitchcock and W. L. Starr, 1954, *Appl. Spectry.*, **8**, 5.
17. T. J. Chow and T. G. Thompson, 1955, *Anal. Chem.*, **27**, 18.
18. T. R. Folsom, C. Feldman and T. C. Rains, 1964, *Science*, **144**, 538.
19. M. Honma, 1956, *Anal. Chem.*, **28**, 1417.
20. G. E. Marsh, 1956, *Appl. Spectry.*, **10**, 8.
21. D. M. Gullikson, W. H. Caraway, and G. T. Gates, 1961, "Applying modern instrumental techniques to oilfield water analysis," *U.S. Bur. Mines R. I.* No. **5737**, 45pp.
22. A. G. Collins (see Mavrodineanu, 1967, entries 337–343), 1962, beginning with: "Methods of analyzing oilfield water: flame-spectrophotometric determination of K, Li, Sr. Ba, and Mn," *U.S. Bur. Mines R. I.*, **6047**, 18pp.
23. R. G. Mihram and K. A. Catto, Jr., 1961, "New method leads to quick identification of formation brines,"*Oil Gas J.*, **59**, 126.

FLOW NETWORK

The flow network, used in groundwater studies, is a construction of streamlines (flowlines) and equipotential lines showing an approximation of the groundwater flow pattern. The first use of a flow net was made by Forchheimer (1914) to study the seepage underneath a concrete dam. It is an imperfect, but useful, tool in the analysis of groundwater flow in a porous medium, utilizing general graphic methods. It may be constructed whenever the flow of groundwater is steady or time independent. It is built on the basis of Laplace's equation:

$$\nabla^2 h = 0$$

where h is the energy per unit weight of fluid (the hydraulic or piezometric head). To make the solution of this equation possible, it is necessary that the value of h, or its normal derivative $\partial h/\partial n$ (n indicating the normal to the boundary of the flow region), or a linear combination of both be known along the entire boundary of the flow region. However, it is sufficient to know h along some parts of the boundary and to have $\partial h/\partial n$ along the remaining parts (Davis and DeWiest, 1966).

The flow network may be conceived of as a two-dimensional graph established by two families of mutually orthogonal lines: (a) flowlines or streamlines which indicate how the water is traveling; (b) equipotential lines (lines which join points of the same potential).

The construction of the flow net, solving Laplace's equation graphically, is relatively simple in the case of regions with fixed boundaries and for confined flows. It becomes more complex for free surface flows or unconfined flows where the location of the water table is not known. The streamlines (a) usually terminate upon the potential lines (b), with exception for places where seepage surfaces are present.

The construction of a flow net is limited to groundwater movement which satisfies the criteria of steadiness and homogeneity for which Darcy's law is valid (see *Hydrology, Subsurface Waters*). It is restricted to examination of two-dimensional cross sections of porous medium which are representative of the main flow and to the analyses of three-dimensional problems with either axial or radial symmetry (Davis and DeWiest, 1966).

B.C.S.

References

Casagrande, A., 1937, "Seepage through dams," *J. New Engl. Water Works Assoc.*, **51**, 131–172.
Chow, Ven Te, 1964, "Handbook of Applied Hydrology," New York, McGraw-Hill Book Co., 1445pp.
Davis, S. N., and DeWiest, R. J. M., 1966, "Hydrogeology," New York, John Wiley & Sons, 463pp.
Forchheimer, P., 1914, "Hydraulik," Leipzig, B. G. Teubner, 566pp.

Cross-references: *Groundwater; Hydrology, Subsurface Waters.*

FLUID INCLUSIONS

Fluid inclusions are the small droplets of fluid, generally water rich, that are found within many natural, as well as synthetic, mineral samples. Most inclusions are two-phase, containing liquid and a bubble of gas or vapor. In size they are rarely as large as one centimeter, and most are less than a millimeter, with those in the range of one to ten micrometer by far the most abundant. They occur in samples from many, but not all, geologic environments. Ordinary water-rich fluid inclusions have not been found in meteorites, nor do they occur in those igneous and metamorphic rocks that have been formed in a strongly water-deficient environment, although inclusions filled with other fluids such as CO_2, or silicate liquid (now a glass), are found in some of these rocks. The occurrence of fluid inclusions in a wide variety of samples indicates that a great many of the geological processes that have yielded the rocks we now see at the surface of the earth have taken place in the presence of these fluids. The inclusions trapped in rocks preserve for us samples of these ancient fluids, and hence enable the geologist to tell a number of things about events of the past that would otherwise be completely hidden.

Fluid inclusions have been described as mineralogical oddities by scientists as far back as Robert Boyle (1672), but actual investigations of their nature and significance started much later, with the studies of Breislak, Davy, and Brewster, around 1820. Since then they have occasionally been studied rather intensively. Interest in them has stemmed mainly from the evidence that they give as to the temperature, pressure, and composition of those geologic environments in which they were trapped millions or billions of years ago. Although they are not a panacea for geological problems, they do provide much valuable information, some of which is otherwise unobtainable.

Observation of Inclusions

Although large fluid inclusions are obvious to the naked eye, ordinary inclusions can be observed only with special techniques. Most crystals that appear white are actually colorless and transparent. The whiteness comes from the presence of large numbers of fluid inclusions (and less frequently, cracks or solid inclusions) that disperse light. Ordinary white quartz is very rich in fluid inclusions, and frequently will have as many as a billion inclusions per cubic centimeter. These are so small that they may make up only 0.1 volume percent of the sample. Such inclusions are normally visible only on examination of a thin section of the rock using the microscope. The standard petrographic thin section is useful, but not optimum, as the larger inclusions are frequently opened in the preparation of the section; somewhat thicker polished plates are generally preferable. Smooth crystal or cleavage faces suffice for many observations, and the effect of a polished surface may be obtained readily, even for irregular specimens, by immersion of the specimen in a fluid of the same index of refraction.

If the gas bubble is small enough to be free of contact with the inclusion walls, it may move when the specimen is tilted, and if only a few micrometers in size, it will be in continuous movement, similar to normal Brownian movement. Smaller numbers of inclusions occur as flaws in many gemstones, and several of the synthetic crystals used in various electronic devices may be ruined by the presence of such defects. The common phenomenon of decrepitation (flying apart) upon rapid heating of minerals is usually due to the pressure developed in fluid inclusions.

Causes of Trapping of Fluid Inclusions

Careful studies of both synthetic and natural minerals have shown that inclusions can be trapped in several different ways. Practically all inclusions represent imperfections in the crystal structure of the host mineral, so any process that will cause an imperfection is a possible cause of trapping. Inclusions are generally categorized as *primary,* when formed simultaneously with the surrounding crystal, or *secondary,* when formed at a later time, as by fracturing and rehealing. Skeletal or dendritic crystal growth, followed by solid perfect growth that covers the imperfect region, is one of the most obvious methods of trapping primary fluid inclusions. If there is interference in the growth of any portion of a crystal by the presence of another mineral grain, a droplet of an immiscible liquid, a gas bubble, or any other such cause, further growth will attempt to cover over this imperfection, and in so doing will trap whatever material or materials are present in the reentrant. Even more commonly, inadequate material supply causes certain portions of the crystal (e.g., centers of faces) to grow at a slower rate. Upon covering over such areas, very large inclusions may be trapped. Occasionally partial solution of a preexisting crystal will cause deep etch pits or etch tubes to be formed, which may then be covered with new growth, yielding inclusions.

Probably the bulk of the small inclusions

in natural minerals results from the rehealing of fractures to form *secondary* inclusions. Crystals embedded in a solid matrix may be fractured by tectonic forces, and even those sticking out into open cavities filled with fluid frequently are fractured owing to stresses set up in the crystal during growth. Any such fracture represents an increase in surface energy, and hence will tend to become eliminated by solution and growth processes on the adjoining sides of the fracture. The fluid trapped in the healed fracture will thus eventually be transformed into a group of rounded or faceted negative crystal cavities filled with liquid, surrounded by an apparently perfect crystal. This fracturing and rehealing may take place almost contemporaneously with the growth of the crystal, or at any later time. In some cases there is evidence that adjoining fluid inclusions have different compositions, suggesting formation during two epochs of fracturing and rehealing, perhaps billions of years apart.

Composition of Inclusions, and Phases Present

In most ordinary fluid inclusions, the major component is water, with lesser amounts of various cations and anions. Their composition has been studied by a wide variety of methods, but there is no single optimum analytical procedure, and for many constituents, satisfactory quantitative procedures have not been devised. A surprisingly large amount of useful data may be obtained by rather simple microscopical techniques. More elaborate microscopical methods may give inexact but useful compositional data even on very small inclusions. Thus the depression of the freezing point of the fluid, i.e., the *freezing temperature* obtained with a microscope freezing stage, yields a semi-quantitative measure of the concentration of salts in the fluid of inclusions as small as ten micromicrograms (10^{-11} g).

The major problems in actual quantitative chemical analysis are not in the analytical procedures themselves, although skillful work is needed since micro- and even millimicrogram quantities are involved in some methods, but rather in the finding and selection of suitable, unambiguous samples, and in the extraction of the inclusion contents without gross contamination or loss. Ball milling or crushing in a vacuum system will yield the water and other volatile components from any inclusions that are opened. If the resulting powder is leached with water or acids, followed by filtering or electrodialysis, the soluble ions that were present in the inclusions may be analyzed by sensitive photometric, colorimetric, and neutron activation techniques. Such procedures yield the most complete data we have on the composition, but some of the results are necessarily ambiguous, owing to a variety of problems of contamination and loss. Crushing in vacuo, followed by mass spectrometry, may provide a powerful tool for analysis of the gaseous constituents, but truly appropriate procedures for the measurement of such important constituents as H^+ and e^- (i.e., pH and Eh) have not been devised.

The analyses that have been made show sodium and chlorine to be the major ions present in most inclusions, with smaller amounts of potassium, calcium, magnesium, sulfate, bicarbonate, and carbonate. Many other constituents, such as borate, fluoride, and H_2S, are found in fluid inclusions. These are usually present as very minor constituents, but may occasionally be present in major amounts. Concentrations of heavy metals are very low, seldom exceeding one percent. The total amount of salts in solution may vary widely, from nearly fresh water, to very concentrated brines.

Commonly the fluids that were trapped at elevated temperatures had more sodium chloride and other salts dissolved in them than they could maintain in solution upon cooling to surface temperatures, and hence crystals of these salts are found in the fluid inclusions. Such crystals, formed from the inclusion fluids *after* trapping, are called "daughter minerals" to distinguish them from incidental solid inclusions of minerals trapped during the growth of the host. As many as eight different daughter minerals have formed in some inclusions.

Since the coefficient of thermal expansion of a water-rich fluid is about thirty times that of most minerals, when cooled from the temperature of trapping to room temperature, the fluid contracts far more than the container surrounding it, and a vapor space is formed. This gas or vapor bubble may be nearly a vacuum, having only the vapor pressure of water at room temperature, or it may contain large quantities of gases that were originally in solution in the fluid. These gases are under high pressure. Carbon dioxide, the most common of these gases, under pressures up to a thousand pounds per square inch (70 bars) at room temperature, is found in fluid inclusions from many types of occurrences. It is found both as liquid and in the gas bubbles. Inclusions may not form a second phase if cooling since trapping has been small, as in many of the salt deposits of the world, or if the trapped fluid is relatively noncondensable, as in some gas inclusions.

In addition to the normal saline water inclusions mentioned above, many other types of fluids may be trapped. Thus it is not rare to find droplets of petroleum-like compounds trapped in minerals formed in some low-temperature environments, such as the fluorite deposits in limestones in southern Illinois. In most cases, these petroleum-like inclusions are not the fluid from which the crystal grew. Widely varying ratios of "petroleum" to water solution are found in adjoining inclusions formed simultaneously, so it is apparent that the "petroleum" inclusions represent droplets of an immiscible liquid, suspended in the water solution from which the crystal grew, which were accidentally trapped by growth of the crystal.

Just as inclusions of water solution are trapped in crystals growing from a water solution, so minerals growing from silicate melts (magmas) may trap droplets of the liquid silicate melt. These liquid silicate inclusions may precipitate one or more crystals (daughter minerals) upon cooling to surface temperature, or they may supercool to form a solid glass inclusion, usually with a shrinkage bubble in it. If droplets of another, immiscible fluid are present in the silicate melt, these may also become trapped in a manner analogous to the trapping of drops of "petroleum" in minerals growing from water solution. Thus in some igneous rocks there are inclusions that prove the presence, at the high temperatures and pressures of crystallization of the rock, of immiscible droplets of an aqueous, highly saline fluid in some granitic melts, and of immiscible droplets of a dense, almost pure carbon dioxide fluid in some basaltic melts.

The occurrence, at room temperature, of two immiscible fluids in inclusions does not always imply the existence of immiscibility under the condititons of trapping. At high temperatures water solutions and carbon dioxide mix in all proportions, and such homogeneous solutions are sometimes trapped as fluid inclusions. At lower temperatures the mutual solubilities are very low, so the originally homogeneous fluid separates to form two immiscible fluids, one rich in carbon dioxide, and the other rich in water and salts. On further decrease in the temperature to below the critical temperature for carbon dioxide, 31°C, the carbon dioxide phase will separate into a liquid and a gas, giving a total of three phases. The actual temperature at which this occurs will depend on the density of the carbon dioxide fluid, and hence on the pressure, temperature, and compositional conditions of trapping.

Geological Thermometry Using Inclusions

One of the most intriguing and valuable applications of fluid inclusions to geological problems is their use as "recording thermometers," registering the temperatures of events long since discontinued. If the bubble in an inclusion represents differential shrinkage, from the temperature of trapping to the temperature of observation, its relative size should give us at least an estimate of the temperature of trapping. A simple inspection of the bubbles in various samples provides qualitative confirmation of the idea. Inclusions in rocks known from independent geological evidence to have formed at relatively low temperatures have small bubbles or none at all; inclusions in rocks formed at high temperatures have large bubbles. In order to quantify this relationship, one merely heats the inclusion with a microscope heating stage until the bubble disappears. This *homogenization temperature* (sometimes called "filling temperature") may or may not be equal to the true *trapping temperature*. Although basically a very simple procedure, there are many problems in the interpretation of such heating experiments. Some of these are illustrated in Fig. 1. One of the fundamental problems, that may or may not be quantitatively significant, is the problem of pressure. If an inclusion traps a homogeneous liquid that is in equilibrium with its vapor (i.e., it is on the "boiling curve," line A-C.P. on the diagram), or traps a homogeneous vapor that is in equilibrium with liquid (line B-C.P. on the diagram), the experimentally determined homogenization temperature will be the true trapping temperature. If, however, as is generally true, the pressure on the fluid at the time of trapping is greater than that given by the line A-C.P.-B, the homogenization temperature will be lower than the trapping temperature. An independent estimate of the pressure at the time of trapping may sometimes be made from the available geological facts, and hence permit a determination of the pressure correction to be added to the homogenization temperature. From the spacing of the isobars on the diagram it is apparent that even a crude estimate of pressure will suffice for inclusions with small bubbles, but will not for those with large bubbles. Actually, it is possible that the four inclusions shown, I through IV, could *all* have been trapped at the same indicated temperature (540°C), although they homogenize at four different temperatures from 170 to 374°C (points 1 through 4). Although the pressure needed to trap inclusion IV at 540°C is geologically not very probable, the other three cases are all very reasonable, geologically, and may occur commonly. Thus the homogenization temperature gives only a minimum temperature. Unfortunately, the position of

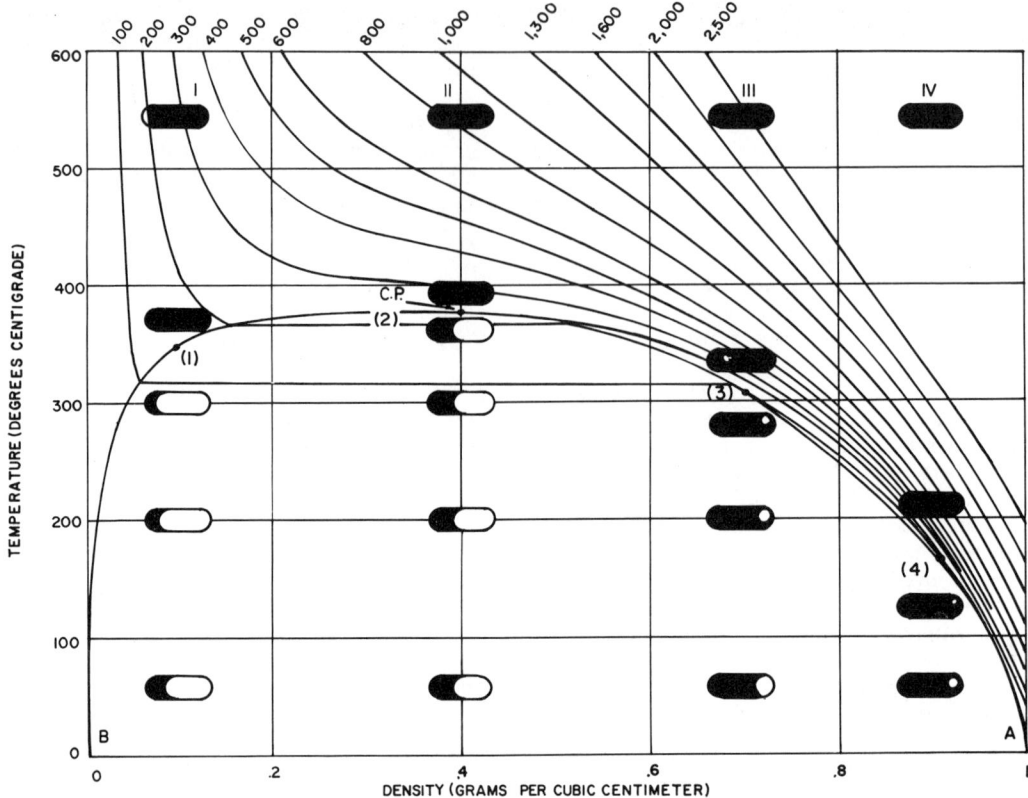

Fig. 1. Thermal behavior of four inclusions (four vertical rows of "capsules") that have different liquid-to-gas ratios is superimposed on the phase diagram for pure water (see text for explanation). C.P. refers to critical point. The thin curving lines are isobars, connecting points of equal pressure (given at the top, in atmospheres).

the curve A-C.P.-B (shown for pure water) is raised very significantly by the salts present in inclusions. These salts also make serious changes in the position and shape of the isobars, so that unless the composition of the inclusion is known, and $P-V-T$ data are available for such compositions, the pressure corrections themselves may be subject to error.

An additional problem stems from changes that may occur in the inclusion after trapping. If leakage of material occurs, either in or out, the homogenization temperatures will change accordingly. If it occurs during the experimental heating, it is readily detectable, but if slow movement of materials occurs along crystal imperfections in response to pressure gradients, over geological time, it may escape detection. There is evidence, however, that such leakage seldom occurs. Another change that may occur is the "necking down" of flat or long inclusions to more equant shapes, thus decreasing the total surface energy. The change in inclusion shape with time is almost ubiquitous, and its progress can be watched with inclusions in soluble salts in the laboratory; there is good evidence for its occurrence in many natural samples. Such changes may affect geological thermometry only if two or more inclusions are formed from what was originally a single individual. As these changes usually occur at less than the original trapping temperature, the inclusion contents are generally no longer homogeneous. Let us assume the simplest case, wherein an elongated two-phase inclusion, with fluid and a shrinkage bubble, splits to form two inclusions. The bubble, representing shrinkage for the entire inclusion contents, is trapped along with some fluid in one inclusion, and only fluid is trapped in the other. On further cooling to surface temperatures the trapped gas bubble in the one will increase in volume, and a new bubble will form in the other. Reheating experiments on these would give too high a temperature for the one, and too low a temperature for the other. If, however, many different inclusions give nearly identical homogenization temperatures, such phenomena need not be considered.

In spite of all the limitations that must be placed on the interpretation of the results of homogenization of gas-liquid inclusions (similar experiments and reasoning can be applied to the redissolving of crystals of daughter minerals), fluid inclusions provide one of the most useful and accurate geologic thermometers, capable of application to a wide variety of samples.

Other Uses of Inclusions

One of the most important geologic processes, from man's point of view, and one of the least well understood, is the formation of ores. There must have been a number of different mechanisms whereby metals were removed from dilute sources (rock or magma), transported, and finally deposited as a concentrated mass. In many cases water solutions were involved, and here inclusions have been able to offer invaluable help in reconstructing events long since discontinued. It is obvious that an effective and intelligent search for ore deposits would benefit from a thorough understanding of the environment of ore deposition. Not only do fluid inclusions provide geothermometers, they also provide, in some samples, useful *geobarometers*, and the only evidence we have on the *density* of the ore fluids.

Perhaps the most difficult problem is to understand how the ore minerals, which in general have low solubility in water, were carried by the fluids in concentrations large enough to produce ore deposits. They must have combined with other materials to form complex ions with higher solubilities, but the nature of many of these complexes is not known. Their formation would partly depend, however, on the composition of the fluids. Therefore the analysis of inclusions may throw some light on the whole chemical cycle involving the formation and transport of the complexes and the conditions under which they eventually broke down, causing the precipitation of ore minerals to form valuable deposits.

EDWIN ROEDDER

References

Gubelin, E. J., 1953, "Inclusions as a Means of Gemstone Identification," Gemological Inst. Am., 220pp.
Ingerson E., 1947, "Liquid inclusions in geologic thermometry," *Am. Mineralogist,* **32** (7–8), 375–388.
Roedder, E., 1962, "Ancient fluids in crystals," *Sci. Am.,* **207**(4), 38–47 (Oct.).

Roedder, E., "Composition of Fluid Inclusions," in (Fleischer, M., editor), "Data of Geochemistry," Chap. JJ, *U.S. Geol. Surv. Profess. Paper* **440.**
Roedder, E., et al., 1962–1963, "Studies of fluid inclusions I, II, and III," *Econ. Geol.,* **57,** 1045–1061; **58**, 167–211 and 353–374.
Smith, F. G., 1953, "Historical Development of Inclusion Thermometry," University of Toronto Press, 149pp.
Sorby, H. C., 1858, "On the microscopical structure of crystals, indicating the origin of minerals and rocks," *Quart. J. Geol. Soc. London,* **14,** part 1, 453–500.
Yermakov, N. P., et al., 1965, "Research on the Nature of Mineral-forming Solutions," Oxford, Pergamon Press (vol. 22 of Internat. Series Mono. Earth Sciences), 743pp.

Cross-references: *Colorimetry; Exsolution; Mass Spectrometry; Natural Brines; Neutron Activation Analysis; pH-Eh; Spectrophotometry.* Vol. II: *Brownian Motion.* Vol. IVB: *Crystal Growth; Defects in Crystals; Electron Microscopy; Etch Pits; Ore-Forming Fluids.*

FLUID MECHANICS—*See* Vol. II

FLUORESCENCE—*See* Vol. IV B, **LUMINESCENCE**

FLUORINE: ELEMENT AND GEOCHEMISTRY

Fluorine, symbol F, atomic number 9, atomic weight 18.9984, is in group VIIA of the Periodic Table, and it has 7 isotopes. Hydrofluoric acid was first prepared in 1771 by Scheele and named from the Latin *fluo,* "I flow." The presence of a new element in it was recognized by Davy in 1813 and Ampère proposed for it the name "fluorine."

At room temperature, fluorine is a yellow gas, somewhat paler than *chlorine* (q.v.). It is a highly reactive gas and difficult to handle. The boiling point is $-188.14°C$, the density of the liquid being 1.108 g/cc. Its critical temperature is $144°K$.

Chemical Features

Fluorine is an exceedingly reactive element of the *halogen group* (q.v.) and is known to combine with most other elements, including some of the inert gases. It is the most powerful oxidizing agent known and highly exothermic in its chemical reactions. Salts of fluorine are called fluorides, of which CaF_2, fluorite, is the most abundant and important. Sodium fluoride, villiaumite, is a rare mineral; however, this salt, artifically prepared, has important uses. Fluoride salts have been used as a flux for more than two centuries.

FLUORINE: ELEMENT AND GEOCHEMISTRY

Minerals

Fluorite (fluorspar)	CaF_2
Cryolite (Greenland spar)	$3NaF \cdot AlF_3$
Fluorapatite	$CaF_2 \cdot 3Ca_3(PO_4)_2$
Villiaumite	NaF
Sellaite	MgF_2
Nocerite	$Ca_3Mg_3O_2F_8$
Fluellite	$AlF_3 \cdot H_2O$

Geochemistry

The ion of fluorine, fluoride, has a single negative charge and a radius of 1.33 Å. In many minerals fluoride substitutes for hydroxyl. Fluoride is widely distributed in the lithosphere and hydrosphere. In average seawater the concentration of fluoride is about 1.3 mg/liter; however, the activity of fluoride in seawater is reduced significantly as a result of complexing. Even with correction for *complexes* (q.v.), the activity of fluoride in seawater exceeds the activity of hydroxyl ions, illustrating the importance of the fluoride ion in regard to biological activity and inorganic precipitates even though the concentration appears small. In most rivers and lakes fluoride concentration is less than 1 mg/liter, or 1 ppm.

Fluoride is present in nearly all igneous and metamorphic rocks, where it substitutes for the hydroxyl in mineral structures, principally in apatite, mica, and amphiboles. Other fluoride-bearing minerals are rare in igneous and metamorphic rocks. Clastic sedimentary rocks commonly contain fluoride-bearing apatite as detrital and subspherical phosphate aggregates, some with an oolitic structure.

Bedded phosphate deposits containing several percent fluoride extend over large areas; for example, North Africa, Florida, and Colorado and adjacent states. With the world resources of phosphate rock being estimated at hundreds of billions of tons, this material is likely to be the major future source of fluorine. Appreciable quantities of CaF_2, fluorite, are present in some limestones and dolomite as fissure veins and replacement beds. Such deposits in southern Illinois and western Kentucky are the principal domestic source of fluorine. Data concerning production, utilization, and sources are presented by Ambrose (1965).

Uptake of fluoride by apatite of teeth and bone has been of interest for more than a century. Early interest centered on the finding that hydroxylapatite of fossil teeth and bone slowly incorporates fluoride as a substitute for hydroxyl of the mineral structure. Thus the fluoride content of such apatite was considered to give a rough age date; however, failure to consider the microarchitecture and the total composition of the teeth and bone resulted in errors in much of the work. The method is only of historical interest because of more precise age date methods. Recently there has been great interest in fluoridation of water supplies and topical application of fluoride to teeth as a means of reducing dental caries. An extensive review on this subject is given in "Fluorine Chemistry," Vol. 4, edited by 299—Encyclo of Earth Sci.—9x10x16 T.R. J. H. Simons (1965). Other volumes of the same series (1–5) consider other aspects of fluorine chemistry.

Economics

Only three minerals are exploited: cryolite, fluorite, and fluorapatite. *Cryolite* ore deposits are uniquely found in Greenland, which exports about 50,000 tons a year. The mineral goes partly into the aluminum industry. *Fluorite* is the most important commercial source of fluorine, with a world production of 2 million tons, mainly from the United States, Mexico, U.S.S.R., China, and a number of European countries. Its main use is in the steel industry, for making hydrofluoric acid, in ceramics, and for synthetic manufacture of cryolite. *Fluorapatite* is utilized for fertilizer but not as a fluorine source.

DALE R. SIMPSON

References

Allen, R. D., 1952, "Variations in chemical and physical properties of fluorite," *Am. Mineralogist*, **37,** 910–930.

Ambrose, P. M., 1965, "Fluorine," in "Mineral Facts and Problems," *U.S. Bur. Mines Bull.*, **630,** 339–349.

Clark, H. C., 1968, "Fluorine," in (Hampel, C. A., editor), "Encyclopedia of the Chemical Elements," New York, Reinhold Publ. Corp., 214–222.

Fleischer, M., and Robinson, W. O., 1963, "Some Problems of the Geochemistry of Fluorine," in (Shaw, D. M., editor), "Studies in Analytical Geochemistry," University of Toronto Press.

Glover, E. D., and Sippel, R. F., 1962, "Experimental pseudomorphs; replacement of calcite by fluorite," *Am. Mineralogist*, **47,** 1156–1165.

Simons, J. H. (editor), 1950–1964, "Fluorine Chemistry," New York, Academic Press, 5 vols.

Cross-references: *Chlorine; Complexes; Electronegativity; Halogens; Rare Gases; Seawater, Chemistry.*

FOG, SMOG, MIST—See Vol. II, **FOG, SMOG, MIST; BIOMETEOROLOGY**

FORMULA CALCULATION—See **MOLECULAR WEIGHTS**

FOSSIL FUELS—See Vol. IV B

FOSSIL MICROBES—See Vol. VII, MICROBES

FREE ENERGY AND FREE ENTHALPY

Definition

The determination of the conditions for a chemical equilibrium or the prediction of the direction of a spontaneous transformation, following the second law of thermodynamics, is made easier by the use of two free energy functions. The *Gibbs free energy*, or *free enthalpy*, G (also called Gibbs function or simply free energy and also designated by F), is defined as the enthalpy, or heat content, H, minus the product of the entropy, S, times the absolute temperature, T.

$$G = H - TS \qquad (1)$$

The *Helmholtz free energy*, or *work content*, A (also called free energy and also designated by F), is defined as the internal energy, E, minus the product, TS.

$$A = E - TS \qquad (2)$$

As a consequence of the definition of enthalpy as the internal energy plus the product of pressure, P, times the volume, V,

$$H = E + PV \qquad (3)$$

a similar equation relates G and A.

$$G = A + PV \qquad (4)$$

The four functions of state, E, H, A, and G, are called *thermodynamic potentials*. Like the internal energy and the enthalpy, the Gibbs and Helmholtz free energies depend only on the state of the substance (which is well defined when state variables such as temperature, pressure, and composition are fixed), and they do not depend on the previous history. They have the units of energy and they are extensive quantities (proportional to the extent of the system and therefore additive). For the purpose of thermochemical calculations, they refer to one gram atomic or molecular weight of the substance, and they are given in calories (or kilocalories) per mole (or per equivalent formula weight).

Significance of Free Energy Changes

The second law of thermodynamics states that the entropy can increase only during a spontaneous or natural transformation of an isolated system. Most of the systems of interest to the chemist or the geologist, however, are not isolated, and they do exchange energy with the surroundings. In that case, it can be shown that the change in entropy of the system, dS, cannot be smaller than the heat received from the environment, dq, divided by the absolute temperature, T. The difference, $dS - dq/T$, is equal to the change in entropy resulting from internal irreversible heat exchanges and it is represented by $(dS)_i$. It must be positive for a spontaneous or natural process and it is zero for an equilibrium or reversible transformation.

$$dS - \frac{dq}{T} = (dS)_i \geqslant 0 \qquad (5)$$

(1) Free Energy Change and Equilibrium. The Gibbs or Helmholtz free energy can be introduced into Eq. (5) by applying the first law of thermodynamics, namely that the change in internal energy equals the heat absorbed minus the work done by the system, and assuming that the latter consists only of pressure-volume work, PdV.

$$dE = dq - PdV \qquad (6)$$

Taking the differential of Eqs. (1) and (3) and combining it with (5) and (6) gives Eq. (7) for the Gibbs free energy

$$dG = VdP - SdT - T(dS)_i \qquad (7)$$

According to this last equation, the change in Gibbs free energy (or free enthalpy), at constant pressure, $dP = 0$, and constant temperature, $dT = 0$, equals the negative of the product of the internal entropy change by the absolute temperature. Following the second law, dG must be negative for a spontaneous process.

$$dG = -T(dS)_i < 0 \qquad (\text{at constant } P, T) \qquad (8)$$

A similar calculation for the Helmholtz free energy (or work content) leads to:

$$dA = PdV - SdT - T(dS)_i \qquad (9)$$

For a process at constant volume, $dV = 0$, and constant temperature, $dT = 0$, the *criterion for spontaneity* is that dA be negative.

$$dA = -T(dS)_i < 0 \qquad (\text{at constant } V, T) \qquad (10)$$

The conditions for an equilibrium or reversible transformation are:

$$dG = 0 \qquad (\text{at constant } P, T) \qquad (11)$$

and

$$dA = 0 \qquad (\text{at constant } V, T) \qquad (12)$$

The physical meaning of these equations is that the *Gibbs free energy and the Helmholtz*

FREE ENERGY AND FREE ENTHALPY

free energy exhibit a minimum under conditions of thermodynamic equilibrium. The significance of the functions G and A for a thermodynamic system is analogous to that of the potential energy for a system consisting of an object in the earth's gravity field.

(2) Driving Forces and Maximum Available Work. For a finite process, at constant temperature, the following relations are derived directly from Eqs. (1) and (2):

$$\Delta G = \Delta H - T\Delta S \quad \text{(isothermal)} \quad (13)$$

and

$$\Delta A = \Delta E - T\Delta S \quad \text{(isothermal)} \quad (14)$$

where the Δs represent finite changes of the thermodynamic functions. In particular, for a chemical reaction, they represent the sum of the values of function for the products each multiplied by a stoichiometric coefficient minus the same sum for the reactants. It is usually given per mole and it is equal to the change of the value of the function for the whole system when the reaction proceeds once as written.

In addition to its use in practical calculations, Eq. (13) provides a more intuitive interpretation of ΔG. The decrease in Gibbs free energy required for a spontaneous process ($\Delta G < 0$) will be greater if there is a decrease in enthalpy ($\Delta H < 0$, or exothermic reaction) and an increase in entropy ($\Delta S > 0$), and the influence of the entropy change becomes more important when the temperature increases. In other words, two driving forces govern the behavior of a chemical system at constant temperature and pressure: the tendency towards minimum enthalpy and the tendency towards maximum entropy. The Gibbs free energy, or free enthalpy, combines those two driving forces. A similar statement applies to the Helmholtz free energy, or work content, which combines the internal energy and entropy. No prediction can be made about the kinetics, however, and even a large negative value of ΔA or ΔG is no assurance that the reaction will proceed at an observable rate.

When the system performs some work other than pressure-volume work, an additional term, $-dw'$, must be added to the energy balance (Eq. (6)), and the mathematical statements of the second law become

$$dG + dw' = -T(dS)_i < 0 \quad (\textit{at constant P, T}) \quad (8')$$

and

$$dA + dw' = -T(dS)_i < 0 \quad (\textit{at constant V, T}) \quad (10')$$

The maximum available work corresponds to a reversible process, $(dS)_i = 0$, and it is equal to the decrease in Gibbs free energy at constant P and T, or the decrease in Helmholtz free energy at constant V and T.

The application of Eq. (8') to an electrochemical system leads directly to the important relation between ΔG and the *electromotive force, ϵ, of a reversible galvanic cell* at a given temperature and pressure,

$$nF\epsilon = -\Delta G \quad (15)$$

where n is the number of electric charges involved in the electrochemical reaction, and F represents the Faraday (96,487 coulombs per equivalent).

Influence of Temperature and Pressure on Equilibrium Conditions

Among the most interesting questions that the geologist can answer with the help of thermodynamics is the effect of temperature and pressure on equilibrium conditions. This involves the determination of the temperature and pressure dependence of the Gibbs free energy change and the use of tabulated thermodynamic data for that purpose.

(1) Temperature and Pressure Dependence of the Gibbs Free Energy Change. For a small reversible change of a system, the internal entropy change, $(dS)_i$, is zero, and following Eq. (7) the variation of Gibbs free energy with temperature and pressure is $dG = VdP - SdT$. The rate of change of G with pressure at constant temperature (or partial derivative) is equal to the volume, V. The rate of change of G with temperature at constant pressure is equal to minus the entropy, $-S$. A similar equation, $dA = PdV - SdT$, applies to changes of Helmholtz free energy with volume and temperature.

Most chemical equilibria are investigated under conditions of constant temperature and pressure (meaning that the final temperature and pressure of the products is the same as the initial values for the reactants), and the question is: knowing the change of Gibbs free energy, ΔG, for a given set of T and P, can the new value be predicted for another set of T and P, or in other words, how does ΔG vary with temperature and pressure? Since G, V, and S are extensive properties (function of the amount of substances), while T and P are intensive variables (not depending on the extent of the system), it can be shown that: $d(\Delta G) = (\Delta V)dP - (\Delta S)dT$, from which it results that the change of ΔG with pressure at constant temperature is ΔV:

$$\left(\frac{\partial \Delta G}{\partial P}\right)_T = \Delta V \quad (16)$$

and the change of ΔG with temperature at constant pressure is $-\Delta S$:

$$\left(\frac{\partial \Delta G}{\Delta T}\right)_P = -\Delta S \quad (17)$$

The Gibbs free energy or free enthalpy change, ΔG, can also be related to the enthalpy change, ΔH.

$$\left(\frac{\partial \Delta G}{\partial T}\right)_P = -\frac{\Delta H}{T^2} \text{ or } \left(\frac{\partial \Delta G/T}{\partial 1/T}\right)_P = \Delta H \quad (18)$$

(2) Standard Free Energy Changes as a Function of Temperature. For the convenience of computations involving thermodynamic data, it is important to tabulate them for substances in a well-defined and widely accepted standard state of reference. At all temperatures, the *standard state* for a *pure liquid or solid* is the *condensed phase under a pressure of one atmosphere*, and the standard state for a gas is the *hypothetical ideal gas at unit fugacity* (equivalent to a "perfect gas" at atmospheric pressure). Values of thermodynamic quantities for standard state conditions are identified by a superscript °, while the temperature T is usually indicated as a subscript. For instance, $\Delta G_T°$ is the Gibbs free energy change, or free enthalpy change, for a reaction when reactants and products are in the standard state, at the temperature T. $\Delta H_T°$ and $\Delta S_T°$ are the standard enthalpy and entropy changes and the following relation applies at any temperature T:

$$\Delta G_T° = \Delta H_T° - T\Delta S_T° \quad (19)$$

The *standard Gibbs free energy* or *free enthalpy of formation*, $\Delta Gf°$ (or $\Delta G_f°$) of a substance at a given temperature is the change in free enthalpy when one mole of the substance in its standard state is formed isothermally at the indicated temperature from the elements, each in its standard state. Usual units are kcal/mole. For all elements in their standard state, the free enthalpy of formation is zero.

$\Delta H_T°$ and $\Delta S_T°$ are functions of temperature and they are generally obtained by their relation with *heat capacity at constant pressure*, $C_p°$.

$$\left(\frac{\partial \Delta H_T°}{\partial T}\right)_P = \Delta C_p° \quad (20a)$$

$$\left(\frac{\partial \Delta S_T°}{\partial T}\right)_P = \frac{\Delta C_p°}{T} \quad (20b)$$

In Eqs. (20a) and (20b) $\Delta C_p°$ is equal to the sum of heat capacities of the products minus that of the reactants when all substances are in the standard state. By integration of Eq. (20) and substitution into Eq. (19) the following relation is obtained, assuming that all reactants and products remain in the same physical state:

$$\Delta G_T° = \Delta H_{298}° - T\Delta S_{298}° + \int_{298}^{T} \Delta C_p° \, dT - T \int_{298}^{T} \frac{\Delta C_p°}{T} dT \quad (21)$$

This is the most useful relation since the standard enthalpy and entropy changes at 25°C (298.15°K) can be obtained from tabulated values, and heat capacities above room temperature are usually available. Taking into account the fact that entropies are equal to zero at absolute zero of temperature, the equation can be rearranged as follows:

$$\Delta(G_T° - H_{298}°)/T = \frac{1}{T}\int_{298}^{T} \Delta C_p \, dT - \int_{0}^{T} \frac{\Delta C_p°}{T} dT \quad (22)$$

The right-hand side of the equation is function of heat capacities only and it provides an alternate way of tabulating data. The expression $-(G_T° - H_{298}°)/T$ for a given substance is called the *free energy* (or *free enthalpy*) *function based on* $H_{298}°$ and it is available in tabular form (for instance at 100° increments of temperature). The Gibbs free energy change is computed easily by combining the change in enthalpy at 298.15, $\Delta H_{298}°$, with the change in free energy function, $\Delta[-(G_T° - H_{298}°)/T]$, for the reaction. The free energy function, $-(G_T° - H_0°)/T$, based on the enthalpy at 0°K, $H_0°$, has also been tabulated, and it is used with $\Delta H_0°$.

Some heat capacity data above room temperature are reported as empirical functions of T, such that $\Delta C_p°$ can also be obtained by combining the coefficients of the products and reactants as Δa, Δb, Δc.

$$\Delta C_p° = \Delta a + 2\Delta bT - \Delta c/T^2 \quad (23)$$

This leads to another useful expression, which is referred to as the five-term equation:

$$\Delta G_T° = (\Delta H_{298}° - I_H) - \Delta aT\ln T - \Delta bT^2 + \Delta c/2T - T(I_S - \Delta a) \quad (24)$$

where I_S and I_H are integration constants.

The free enthalpy change at a pressure P, ΔG_T^P, is usually related to the standard value of atmospheric pressure, $\Delta G_T°$, by integrating Eq. (16) from 1 to P.

$$\Delta G_T^P = \Delta G_T° + \int_1^P \Delta V \, dP \quad (25)$$

(3) *Application to the Polymorphic Transformation Graphite to Diamond* (Kern and Weisbrod, 1967, p. 109). The respective stability fields of graphite and diamond can be determined by the calculation of the change in Gibbs free energy, $\Delta G_T{}^P$, for the reaction graphite → diamond at various temperatures and pressures. If the resulting value is negative, the transformation is spontaneous as written, and diamond is the stable phase. Conversely, if it is positive, graphite is thermodynamically stable. If it is zero, the two phases are in equilibrium, and the locus of values of P and T for which $\Delta G_T{}^P = 0$ represents the boundary between the two fields of stability.

The values of heat capacity for graphite and diamond are well known and $\Delta C_p{}^\circ = C_p{}^\circ$ (diamond) $- C_p{}^\circ$ (graphite) can be obtained to calculate the two integrals of Eq. (22) by a graphical method (or the integrals of Eq. (21) and $\Delta S_{298}{}^\circ$). The heat of reaction, $\Delta H_{298}{}^\circ$ is also known. The Gibbs free energy change, $\Delta G_T{}^\circ$, is readily obtained for any temperature and one atmosphere pressure.

The molar volumes, $V_T{}^P$, of graphite and diamond at any temperature and pressure can be represented by an equation such as

$$V_T{}^P = V_{298}{}^\circ (1 - \chi P)[1 + \alpha(T - 298)] \quad (26)$$

where $V_{298}{}^\circ$ is the volume at 25°C and 1 atm, α is the coefficient of thermal expansion, and χ the isothermal compressibility. Then $\Delta V_T{}^P = V_T{}^P$ (diamond) $- V_P{}^T$ (graphite), and it can be substituted in Eq. (25) to obtain $\Delta G_T{}^P$. The equilibrium pressure at any temperature can be calculated by putting $\Delta G_T{}^P = 0$ and solving for P. The resulting curve is shown in Fig. 1

If the temperature dependences of enthalpy, entropy, and volume are neglected, then the equation reduces to $\Delta G_T{}^P = \Delta H_{298}{}^\circ - T \Delta S_{298}{}^\circ + P \Delta V_{298} = 0$. It is represented by the interrupted straight line in Fig. 1. It has a slope $\Delta S_{298}{}^\circ / \Delta V_{298}{}^\circ$ in agreement with Clapeyron's equation.

At room temperature, diamond can only be formed at pressures in excess of 15,000 atm. A comparison of the equilibrium curve with the mean geothermal gradient shows that it cannot form under ordinary conditions, and its presence in a rock implies extremely high pressures. Once it has been formed, however, it can exist metastably since the rates of transformation are extremely small.

Influence of Composition on the Gibbs Free Energy Change

In addition to equilibria between phases of constant composition such as the one described

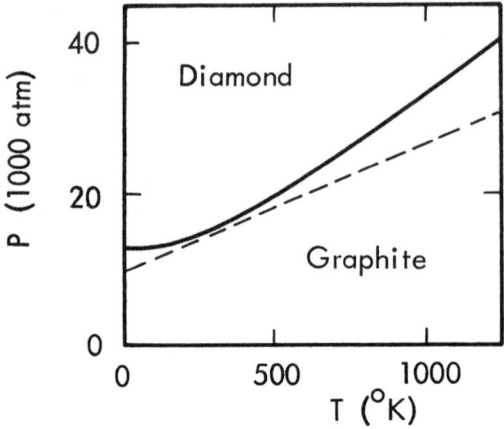

FIG. 1. Equilibrium diagram for the transformation of graphite to diamond. The solid curve represents the equilibrium pressure-temperature relation obtained by solving the equation $\triangle G_T{}^P = 0$. The dashed straight line corresponds to Clapeyron's equation (data from Kern and Weisbrod, 1967, p. 112).

above, the geochemist must often consider reactions between phases of variable composition such as solid or liquid solutions or gas mixtures.

(1) Partial Molal Gibbs Free Energy (or Chemical Potential) and Activity. The change in Gibbs free energy or free enthalpy corresponding to a change in the amount of one constituent when all others, as well as temperature and pressure, remain constant is by definition the *partial molal free energy* or *partial molal free enthalpy* of that constituent and is usually expressed in cal/mole and represented by $\bar{G}_i$. Mathematically, it is a partial derivative, and because of its importance it is also called *chemical potential*, μ_i.

$$\left(\frac{\partial G}{\partial n_i}\right)_{T,P,n_j \neq i} = \bar{G}_i = \mu_i \quad (27)$$

(It can be shown that μ_i is also equal to the partial derivative of the Helmholtz free energy, A, when V and T are constant.)

The *thermodynamic activity*, a_i, of a constituent is defined by the relation:

$$RT \ln a_i = \mu_i - \mu_i{}^\circ = \bar{G}_i - \bar{G}_i{}^\circ \quad (28)$$

where R is the gas constant (1.9872 cal/deg-mole), and $\mu_i{}^\circ$ is the *chemical potential in a reference or standard state*. The choice of a reference state is arbitrary, but it is usually the standard state defined above and, in that case, $\mu_i{}^\circ = \bar{G}_i{}^\circ = \Delta Gf^\circ$. Another reference state may be chosen, however, if it is more convenient for a particular application (e.g., a complex mixture of minerals or a system at a pressure different from one atmosphere). The value of the activity depends on that choice; it must be determined experimentally, but the form of its relation to concentration can often be predicted by theory.

The equilibrium of a typical chemical reaction

$$bB + cC = mM + nN \quad (29)$$

is characterized by the *equilibrium constant, K,* defined as the ratio of activities at equilibrium.

$$K = \left(\frac{a_M^m \, a_N^n}{a_B^b \, a_C^c} \right)_{\text{at equilibrium}} \quad (30)$$

The important relation

$$\Delta G° = -RT \ln K \quad (31)$$

is obtained by using Eq. (28) and (30) in expressing the equilibrium condition, $\Delta G = 0$, for the chemical reaction (29).

(2) Influence of Brine Composition on the Gypsum-Anhydrite Equilibrium (Kern and Weisbrod, 1967, p 275). The relative stability of anhydrite, $CaSO_4$, and gypsum, $CaSO_4 \cdot 2H_2O$, in the presence of pure water at various temperatures is determined by calculating the standard Gibbs free energy change, $\Delta G°$, for the reaction.

$$CaSO_4 + 2H_2O = CaSO_4 \cdot 2H_2O \quad (32)$$

$$\Delta G° = \Delta G f_{\text{gypsum}}° - \Delta G f_{\text{anhydrite}}° - 2\Delta G f_w° \quad (33)$$

For $T = 313°K$ (40°C), $\Delta G° = 0$, and the reaction is reversible (or at equilibrium). At lower temperatures, $\Delta G° < 0$, which means that the reaction is spontaneous from left to right, and gypsum is the stable form of calcium sulfate. Above 40°C, $\Delta G° > 0$, and anhydrite in the stable species.

If the water contains dissolved salts, it is no longer in its standard state, and $\Delta G f_w°$ must be replaced in Eq. (33) by the partial molal free enthalpy, $\bar{G}_w$.

$$\bar{G}_w = \Delta G f_w° + RT \ln a_w \quad (34)$$

The equilibrium condition becomes

$$\Delta G = \Delta G° - 2RT \ln a_w = 0 \quad (35)$$

For instance, the activity of water in a solution containing 5 moles NaCl per liter is known, and it can be used in Eq. (35) to predict an equilibrium temperature of 25°C for the transformation of gypsum to anhydrite in a brine of that composition.

PAUL DUBY

References

Kern, R., and Weisbrod, A., 1967, "Thermodynamics for Geologists," San Francisco, Freeman, Cooper and Co., 304pp (translated by Duncan McKie from "Thermodynamique de Base pour Minéralogistes, Pétrographes et Géologues," Paris, Masson et Cie, 1964).

Klotz, I. M., 1964, "Basic Chemical Thermodynamics," New York, W. A. Benjamin, 468pp.

Lewis, G. N., and Randall, M., 1961, "Thermodynamics," Second ed. (revised by Pitzer, K. S., and Brewer, L.) New York, McGraw-Hill Book Co., 723pp.

*Rossini, F. D., Wagman, D. D., Evans, W. H., Levine, S., and Jaffe, I., 1952, "Selected Values of Chemical Thermodynamic Properties," *NBS Circular 500,* Washington, U.S. Govt. Print. Off., 1268pp.

Waser, J., 1966, "Basic Chemical Thermodynamics," New York, W. A. Benjamin, 278pp.

* Presently being revised. Some revised sections have been issued as NBS Technical Note, 270-3 (1968).

Cross-references: *Clapeyron's Equation; Enthalpy or Heat Content; Entropy; Mass Action and Equilibrium Constant; Phase Equilibria; Thermochemistry; Thermodynamics.*

FUELS—*See* Vol. IV B

FULVIC ACIDS—*See* Vol. VI

G

GADOLINIUM: ELEMENT AND GEOCHEMISTRY

Properties

Elemental gadolinium is a very reactive, silvery-white *lanthanide* or *rare earth* (q.v.) metal with atomic number 64 and outer electronic structure $4f^7\ 5d\ 6s^2$.

Gadolinium occurs as one weakly radioactive isotope: Gd^{152} (0.20% natural abundance, half-life 1.1×10^{14} years), and as six stable isotopes:

Gd^{154}	2.15%
Gd^{155}	14.7%
Gd^{156}	20.5%
Gd^{157}	15.7%
Gd^{158}	24.9%
Gd^{160}	21.9%

Gadolinium is determined analytically by neutron activation (its thermal neutron cross section, 46,000 barns, is one of the largest of any element), x-ray spectrography, absorption spectrometry (it has very sharp spectral lines) and isotope dilution mass spectrometry. In geologic environments, gadolinium forms highly ionic bonds. It forms weak complexes. The oxidation state is tripositive, with ionic radius 0.938Å. Generally, chemical behavior is similar to the alkaline earth elements, and more similar to heavier rare earths than to lighter ones.

Minerals

Gadolinium occurs in highest concentration in rare-earth minerals that favor heavy rare earths. Generally nitrates, sulfates, carbonates, niobates and tungstates with cation coordination number of six to eight concentrate the heavy rare earths, and often contain structural halogens and alkali metals. Typical examples are:

Xenotime (Y, heavy rare earths) PO_4

Fergusonite (Y, heavy rare earths) (Nb, Ta)O_4

Abundance. *See Rare Earths*

Geochemical Behavior

Schmitt et al. (1964) have established a value of 0.32 ppm (parts per million) gadolinium in chondrites and 3.1 ppm in eucrite achondrites. The standard deviation within a meteorite class is only about 20%. In meteoritic matter, volatization may have redistributed alkali metals but not rare earths. Gadolinium has a boiling point of 3000°C.

The gadolinium concentration in average basalt is 5.9 ppm, and the range is less than a factor of ten (Haskin et al. 1966). Oceanic and continental basalts have similar ranges. Partial melting of marine peridotite yields a basaltic liquid with enriched gadolinium and a mantle source area depleted in gadolinium.

Igneous differentiation series may tend toward higher or lower gadolinium concentrations, although the total variation is less than that for light rare earths. The relative abundance of gadolinium and other heavy rare earths remains almost unchanged. Haskin et al. (1966) report 9.0 ppm gadolinium in a Gough Island basalt and 21 ppm gadolinium in a trachyte differentiate. Olivine basalts from Ascension Island have 11.3 ppm, and their soda trachyte differentiates only 6.3 ppm.

Average granite with 60 to 70% silica has 6.7 ppm but the range of values for granites is even greater than for basalts.

In most basaltic, syenitic and granitic rocks, gadolinium is mainly in mafic rock-forming minerals, probably substituting for divalent calcium (ionic radius 0.99 Å).

Gadolinium and other heavy rare earths form complexes and are transported in alkali, halogen, and carbon dioxide-rich hydrothermal fluids; metasomatic and pegmatoid deposits formed from such fluids may have high concentrations of gadolinium and higher ratios of gadolinium to light rare earths than granites.

Gadolinium has a 280 year residence time in seawater (which is alkaline), about the same as other heavy rare earths. Removal by adsorption on clay particles, as well as depletion in surface water in areas of high organic productivity, are likely processes. High secondary gadolinium concentrations in phosphatic fish remains today occur only in deep water; paleozoic fish remains also have high rare earths. Manganese nodules have 4,600 ppm, whereas Pacific Ocean water has only $0.61 \times 10^{6-}$ ppm (Goldberg et al., 1963).

In soils, the relative amounts of gadolinium in detrital, clay, and carbonate fractions is not well known. Sedimentary rocks have about 6.1 ppm, with a variation of a factor of five. In a study of rare-earths in Russian platform soils, Balashov et al. (1964) found higher concentrations in alkaline soils, the gadolinium precipitating as hydroxide; lower concentrations are in acid soils, probably due to removal of complexes.

Biologic concentration of rare earths has been found in hickory leaves; rare earths have been used to trace opium.

Economic Geology

The major source of gadolinium has been from a light rare earth mineral, monazite, in beach sands; the accessory monazite originally present in granitic rocks resists weathering and is concentrated by sedimentary processes. Australia, Brazil, Ceylon, India, Malaysia and South Africa mine the largest quantities. Less than one per cent of monazite is gadolinium. Recently, the bastnaesite deposit at Mountain Pass, California has been an important source of gadolinium, again in low concentration.

High purity gadolinium has a very limited market. Most gadolinium is sold unpurified in rare earth concentrates used for polishing, petroleum cracking catalysis, lighter flints, arc carbons, and in alloys.

ROBERT KAY

References

Balashov, Y. A., et al., 1964, "The effects of climate and facies environment on the fractionation of the rare earths during sedimentation," *Geochem. Internat. No. 1*, **10**, 951–969. (English transl.).

Goldberg, E. D., et al., 1963, "Rare-earth distributions in the marine environment," *J. Geophys. Res.*, **68**, 4209–4217.

Haskin, L. A., et al., 1966, "Meteoric, Solar, and Terrestrial Rare-Earth Distributions," in (Ahrens, L. H. et al., editors), "Physics and Chemistry of the Earth," Vol. 7, Oxford, Pergamon Press, pp. 167–321.

Rankama, K., and Sahama, T. G., 1950, "Geochemistry," Chicago, University of Chicago Press, 912pp.

Schmitt, R. A., et al., 1964, "Rare-earth, yttrium and scandium abundances in meteoric and terrestrial matter—II," *Geochim. Cosmochim. Acta*, **82**, 67–86.

Cross-references: *Complexes; Hydrothermal Solutions; Mass Spectrometry; Neutron Activation Analysis; Rare Earths.* Vol. IVB: *Fergusonite; Monazite; Xenotime.*

GALLIUM: ELEMENT AND GEOCHEMISTRY

Properties

Mendeléeff and de Boisbaudran independently predicted the existence of an aluminum-like element. This element, gallium, was first isolated in 1875 from zinc blende by de Boisbaudran who named it after his native land—Gallia.

The physical properties of gallium are described in Table 1. Unusual features of gallium include: (*1*) the wide range of temperature over which it is a useful liquid (30–1200°C);

TABLE 1. PHYSICAL CONSTANTS OF GALLIUM

Atomic number	31
Atomic weight	69.72
Isotopes and abundances (%)	
Mass No. 69	60.4
Mass No. 71	39.6
Color	Metallic gray
Crystal structure	Orthorhombic, pseudotetragonal
Density (g/cm^3)	
20°C (s)	5.907
29.65°C (s)	5.904
29.8°C (l)	6.095
Melting point (°C)	29.75
Boiling point (°C)	1983
Latent heat of fusion (g-cal/g)	19.16
Vapor pressure (mm Hg)	
1315°C	1
1726°C	100
Specific heat, 29–127°C (cal/g/°C)	0.0977
Electrode potential $Ga^{3+} + 3e = Ga$ ($H_2 = 0.0$ V)	−0.52
Ionic radius for Ga^{3+} (Å)	0.62
Ionization potential for Ga^{3+} (V)	30.7
Ionic potential (Ga^{3+})	4.8

TABLE 2. GALLIUM ABUNDANCES IN VARIOUS MINERALS

Mineral	Formula	Ga (ppm)
Albite	$NaAlSi_3O_8$	40
Almandine	$Fe_3Al_2(SiO_4)_3$	20
Bauxite	$Al_2O_3 \cdot 2H_2O$	90
Beryl	$Be_3Al_2(SiO_3)_6$	40
Biotite	$K(Mg,Fe)_2AlSi_3O_{10}(OH)_2$	40
Blende	ZnS	1–1000
Calcite	$CaCO_3$	0.1
Cancrinite	$3H_2O \cdot 4Na_2O \cdot CaO \cdot 4Al_2O_3 \cdot 9SiO_2 \cdot 2CO_2$	40
Chromite	$FeCr_2O_4$	18
Corundum	Al_2O_3	100
Cryolite	Na_3AlF_6	3
Diopside	$CaMg(SiO_3)_2$	5
Franklinite	$Zn(Fe,Mn)_2O_4$	10
Hematite	Fe_2O_3	1
Hornblende	$(Fe^{3+},Fe^{2+},Al,Mg,Mn)_5(Ca,Na,K)_2Si_8O_{22}(OH)_2$	10
Ilmenite	$FeTiO_3$	5
Lepidolite	$KLiAl_2Si_3O_{10}(OH,F)_2$	100
Limonite	$2Fe_2O_3 \cdot 3H_2O$	3
Magnetite	$FeO \cdot Fe_2O_3$	30
Microcline	$KAlSi_3O_8$	40
Muscovite	$KAl_3Si_3O_{10}(OH)_2$	200
Natrolite	$Na_2Al_2Si_3O_{10} \cdot 2H_2O$	100
Nepheline	$(Na,K)AlSiO_4$	20
Olivine	$(Mg,Fe)_2SiO_4$	2
Oligoclase	$(NaAlSi_3O_8)_{7-9}(CaAl_2Si_2O_8)_{3-1}$	10
Orthoclase	$KAlSi_3O_8$	10
Phlogopite	$KMg_3AlSi_3O_{10}(OH)_2$	50
Quartz	SiO_2	< 1
Siderite	$FeCO_3$	< 1
Sodalite	$3NaAlSiO_4 \cdot NaCl$	100
Spodumene	$LiAl(SiO_3)_2$	60

(2) the exceptionally low vapor pressure of the liquid even at high temperatures, i.e., less than 0.0004 mm of Hg at 900°C, 0.003 mm at 1000°C, and 0.02 mm at 1100°C; (3) the density of the liquid, which is greater than that of the solid (expands on solidification); (4) the property of wetting most surfaces; and (5) its high heat conductivity and thermal stability (useful as a heat-exchange medium).

The ionic radius and ionization potential of gallium are very similar to those of aluminum, resulting in a very high degree of geochemical coherence between gallium and aluminum. To a lesser extent, gallium also follows trivalent iron, which has a similar ionic radius. A notable difference between gallium and aluminum is in the chalcophilic and siderophilic propensities of gallium as demonstrated by its concentration in the sulfide and metal phases of meteorites and in zinc blende.

Minerals

Gallium is a widely dispersed element and is usually present in the 5–200 ppm range in most minerals. Table 2 presents some typical analyses. Gallium is generally concentrated in aluminum minerals (e.g., bauxite, corundum) although cryolite (Na_3AlF_6) is an exception. The camouflage of gallium by trivalent iron is demonstrated in magnetite and franklinite; however, hematite, goethite, and limonite formed by sedimentary or hydrothermal processes contain little gallium.

The mineral germanite $Cu_3(Fe,Ge)S_4$ from Tsumeb, in Southwest Africa, has been found to have the highest known gallium content, showing a maximum of 1.85% Ga. The only other terrestrial sulfide mineral that contains appreciable gallium is blende. The concentration of gallium in this mineral appears to be temperature dependent; low-temperature samples contain 100–1000 ppm Ga and high-temperature samples contain < 10 ppm Ga.

Of the common rock-forming minerals, micas contain the most gallium; muscovite and phlogopite are usually richer than biotite. Feldspar contains more gallium than does pyroxene or olivine. Of the feldspars, albite contains more gallium than does anorthite.

Abundance

The abundance of gallium in typical units of the earth's crust is given in Table 3. It is apparent from these figures that gallium is widely

TABLE 3. ABUNDANCE OF GALLIUM IN UNITS OF THE EARTH'S CRUST

Location	Ga (ppm)
Earth's crust	17
Igneous rocks	
Ultramafic	2
Basalt	18
Granodiorite	20
Granite	17
Sedimentary rocks	
Shale	25
Sandstone	12
Carbonate	4
Seawater	3×10^{-5}
Deep-sea clay	20

dispersed through the earth's crust with little tendency to concentrate in a particular geologic environment.

Geochemical Behavior

The geochemistry of gallium closely parallels that of aluminum. The typical Ga/Al ratio is about 2.4×10^{-4}. Geochemical processes seldom alter this ratio by as much as a factor of 2.

Magmatic differentiation produces some enrichment of gallium in the final crystallate. Such enrichment over the original magma content of gallium seldom exceeds a factor of 2.

Pelagic sediments have the same Ga/Al ratio as granitic rocks implying little fractionation of gallium from aluminum during weathering, transport, and sedimentation. Gallium is concentrated in the hydrolysate fraction of sediments.

The geochemical coherence of gallium and aluminum persists through metamorphism. Metamorphic rocks derived from hydrolysate sediments contain more gallium than do those derived from sandstones and other sediments poor in aluminum.

Gallium is an essential micronutrient of the fungus *Aspergillus niger* and the duckweed *Lemma minor*.

Economic Geology

Most commercially produced gallium is derived as a by-product from zinc ores and from bauxite. It has also been produced from coal flue dust. The demand for gallium is too small to warrant the full potential production rate. Nevertheless, gallium has been found useful: (*1*) in electronic devices; (*2*) in pressure-volume-temperature studies because of its wide liquid range and low vapor pressure; (*3*) as a heat-exchange medium because of its thermal stability, high heat conductivity, low vapor pressure, and wide liquid range; and (*4*) in dental amalgams.

L. PAUL GREENLAND

References

Borisenko, L. F., 1963, "Some characteristics of the distribution of gallium in ultramafic rocks," *Geochemistry*, **8**, 778.

Burton, J. D., Culkin, F., and Riley, J. P., 1959, "The abundances of gallium and germanium in terrestrial materials," *Geochim. Cosmochim. Acta*, **16**, 151.

Cuttitta, F., Clarke, R. S., Jr., Carron, M. K., and Annell, C. S., 1967, "Martha's Vineyard and selected Georgia tektites: new chemical data," *J. Geophys. Res.*, **72**(4), 1343–1349 (Feb.).

Dalton, I. M., and Pringle, W. J. S., 1962, "The gallium content of some Midland coals," *Fuel*, **41**, 41–48.

Greenland, L., 1965, "Gallium in chondritic meteorites," *J. Geophys. Res.*, **70**, 3813–3817.

Nockoids, S. R., and Allen, R., 1956, "The geochemistry of some igneous rock series-III," *Geochim. Cosmochim. Acta*, **9**, 34.

Rankama, K., and Sahama, Th. G., 1950, "Geochemistry," Univ. Chicago Press, 912pp. (1950).

Vorobiev, G. G., 1951, "Gallium in minerals and rocks of Mongolia," *Geochemistry*, **8**, 835.

Wasson, J. T., 1967, "The chemical classification of iron meteorites: I. A study of iron meteorites with low concentrations of gallium and germanium," *Geochim. Cosmochim. Acta*, **31**, 161–181.

Yeremenko, G. K., Walters, A. A., and Klimenchuk, V. I., 1963, "On the distribution of gallium in alkalic rocks, Azov Region," *Geochemistry*, **2**, 145.

Cross-references: *Aluminum; Zinc*. Vol. IVB: *Bauxite; Corundum*.

GASES—VOLCANIC

The major goal, and challenge, of the study of volcanic gases is to determine the composition and behavior of the gaseous (volatile) phase associated with magmas. Magmatic gases are fundamentally related to a wide variety of earth processes and are the controlling factor in many cases. A clear understanding of atmosphere and ocean evolution, rock crystallization, ore deposition, volcanic and hot spring activity, and many other phenomena requires a knowledge of the composition and behavior of magmatic gases.

The problems involved in the acquisition and interpretation of data led many scientists to conclude, as late as the 1940s, that little information could be obtained from the study of volcanic gases beyond the knowledge that certain constituents are present in apparently variable amounts. Modern chemical theory and data and more advanced analytical methods now make it possible to obtain valuable information from volcanic gas samples.

Definition

Volcanic gases are those constituents which are emitted from volcanoes in a gaseous state or chemically associated with the gaseous constituents. Some of these constituents may be deposited around volcanic vents and fumaroles as sublimates while others escape into the atmosphere. Ash, scoriae, cinders, etc., may be propelled from the throat of the volcano by the upward movement of volcanic gases and, in some cases, may be carried long distances through the atmosphere. These solids are not strictly a part of the gaseous phase. There is evidence that some sublimates are formed from material carried as an aerosol in the gaseous phase. If these constituents owe their transport and origin to the truly gaseous constituents, they should be considered as part of the volcanic gas.

Collection Methods

Volcanic gas collections are subject to the caprice of volcanic activity. Collection sites are often inaccessible, especially the most favorable ones, and collecting is usually hazardous. Ideal collecting conditions and sites rarely occur and faultless collections have seldom, if ever, been achieved. Undoubtedly the finest series of gas collections were made by Jaggar and Shepherd from Kilauea, Hawaii during the years 1917–1919. Field conditions and collecting techniques are such that contamination by the atmosphere, water, etc., is always present and postcollection reactions of various types partially or completely destroy the high-temperature compositions.

Collections are usually made in one of two ways: (1) the vacuum tube method and (2) the "pump through" method. The vacuum tube method uses a previously evacuated and sealed glass tube which has one slender, drawn-out end. This end is broken open at the collecting site and the gas is sucked into the tube. The tube is then resealed by melting the glass near the entrance or by capping with an appropriate stopper. The pump through method involves pumping the gas into one or more collection tubes which are located between the collecting site and the pump. In this way the gases may be continually moved through the collection device to condense certain species, obtain a large sample, and seal off the collected portion at the desired time.

Volcanic gas collections have been carried out most extensively in Hawaii, Iceland, Italy, Japan, and Kamchatka; many other areas have been sampled on a less regular basis. Some improvements in collecting devices have been made over the past fifty years, such as packing collection tubes with absorptive materials. Further improvements are needed, however, so that the infrequent opportunities to make good collections can be used to greater advantage.

Analysis

Volcanic gases have been analyzed by a variety of techniques. Most gaseous species can be readily detected qualitatively if they are present in significant amounts; quantitative determinations of trace amounts of gas or of the total gas composition presents a challenge to the analyst. The most successful analyses early in this century were performed by E. S. Shepherd. Shepherd designed an analytical train in which the gas sample could be moved through a series of compartments, each of which contained an appropriate compound to selectively absorb a specific gas species. The volume loss resulting from each absorbtion indicated the amount of each species present. Modern methods of analysis make use of titration, Orsat apparatus, gas chromatograph, emission and absorption spectrography, and standard quantitative chemical analysis procedures. Several of these methods may be used in concert for a complete analysis.

Composition and Behavior

Chemical Composition. The most abundant molecular species in volcanic gases are H_2O, H_2, CO_2, CO, SO_3, SO_2, H_2S, S_2, HCl, HF, O_2, and Ar; these gases are almost always present (see Table 1). Other species may also be present in greater or lesser amounts, but the factors controlling their occurrence are not often defined. Included in this latter group are NH_3, CH_4, CS_2, COS, SO, and others. Theoretically, all ionic and molecular species occurring in the system C–H–S–O–N–F–Cl–X under the given temperature and pressure conditions will be present to some degree. The amounts of most of these species are below the detection limits of the analytical equipment used or may be absent due to disequilibrium reactions. Some species not commonly reported may actually be present but are not sought out by the analytical techniques used.

Analyses of fumarole sublimates and the condensed portions of volcanic gases show that most of the common elements which form rocks and mineral deposits are also carried by the gases (see Table 2). Bicarbonate, bisulfate, nitrate, borate, and bromide have also been detected, sometimes in relatively large quantities. The presence of such elements suggests a genetic relationship between the gases emitted from magmas and ore deposits, but the extent

TABLE 1. SELECTED ANALYSIS OF VOLCANIC GASES

	Compositions (Vol. %) After Deletion of Minimum Probable Atmospheric Contamination				Corrected for Losses and Contaminations:
	Kilauea, Hawaii (Basaltic)	Surtsey, Iceland (Basaltic)	Sheveluch, Kamchatka (Andesitic)	Katmai, Alaska (Rhyolitic-Andesitic)	Kilauea, Hawaii (1200° C)
CO	1.46	0.68	0.000	3.44×10^{-4}	9.78×10^{-3}
CO_2	47.68	9.18	0.0493	0.0184	0.356
H_2	0.48	4.56	0.000	—	2.88×10^{-3}
H_2O	36.18	79.20	10.9	99.94	0.268
S_2	0.04	1.6 mg	—	—	5.25×10^{-4}
SO_2	11.15	5.40	0.0755	—	0.364
SO_3	0.42	—	0.000	—	2.22×10^{-6}
H_2S	—	—	0.408	0.0282	2.71×10^{-4}
HCl	0.04	0.80	0.000	—	2.97×10^{-4}
HF	—	—	0.000	—	—
CH_4	—	—	—	2.16×10^{-3}	8.10×10^{-10}
$N_2 + Ar$	2.55	0.18	83.5	5.02×10^{-3}	5.18×10^{-4}

Surtsey column additional values: 0.36, 6.47, 4.70, 86.16, 18.3 mg, 1.84, —, —, 0.40, —, —, 0.07

Sheveluch column additional: 0.000, 0.680, 0.000, 16.4, —, 1.22, 0.000, 0.408, 0.000, 0.000, —, 86.5

Katmai column additional: 6.35×10^{-4}, 0.0381, —, 99.90, —, —, —, 0.0247, —, —, 7.12×10^{-3}, 5.40×10^{-3}

TABLE 2. CHEMICAL ANALYSIS OF FUMAROLIC GASES, SHOWA-SHINZAN, JAPAN

Component	ppm by wt
SiO_2	253
Al	15
Fe	1.3
Ca	4.6
Mg	32
Na	22
K	15
CO_2	29,200
CO	50
SO_2	1,490
SO_3	21
H_2S	8.0
S	3.7
Cl	728
F	238
Br	1.1
B	39
PO_4	2.8
NO_2	0.01
O_2	51
H_2	685
NH_3	1.3
N_2	567
Ar^{40}	0.6
CH_4	1.5
Ni	0.01
Cu	0.03
Zn	0.5
Ge	0.01
As	0.7
Mo	Trace
Ag	0.003
Sn	0.03
Sb	0.1
Pb	0.03
Bi	0.05

Rn-curies 10^{-7} (1.3)

to which volcanic gases resemble ore fluids is a subject of much debate. It is generally agreed that volcanic gases have been the major source of the ocean waters and atmospheric gases; however, a correlation between volcanic gas compositions and the rate of hydrosphere and atmosphere production is not easily made.

Equilibrium and Contamination. Additions and losses of certain gases due to contamination, cooling, sublimation, etc., lead to considerable doubt as to how accurately collected samples resemble the gas which accompanies the volcanic magma. Furthermore, contamination, reactions, and drastic changes in the physical environment occurring during the ascent of magmas to the surface make inferences about magmatic gases at their source even more uncertain.

Recent work (Ellis, 1957; Heald et al., 1963; Nordlie, 1967) has shown that volcanic gas

GASES—VOLCANIC

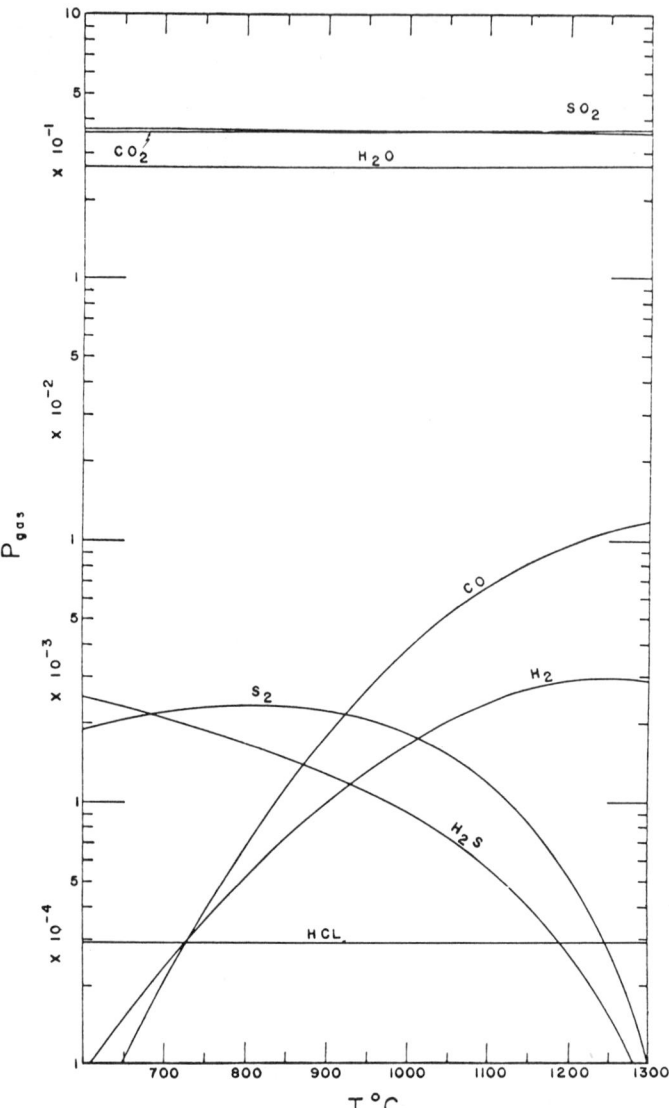

FIG. 1. Equilibrium composition of a basaltic volcanic gas. This graph shows how composition changes as temperature decreases. (Total pressure is 1 atm.)

samples have compositions which approach those predicted by chemical equilibrium calculations (see Table 1 and Fig. 1). The major problems involved in such comparisons are (1) the contamination of the samples before and during collection and (2) the reactions occurring among the gases during and after collection.

It has been established that most of the free O_2, N_2, and Ar in the samples results from atmospheric contamination. Atmospheric oxygen also reacts with the volcanic gases (notably SO_2 and H_2S) to form more oxidized species such as SO_3 and H_2O. Much of the H_2O present in many analyses may also be a contaminant. Water can be added by rain, groundwater, seawater, and from condensation of the volcanic gas itself on rocks around the volcanic vent. Sulfur is abundant around many vents and in some areas has been deposited in large quantities beneath the surface by the rising gases. Therefore, the sulfur content of gas samples often may be lower than the amount originally present.

The extremely rapid rates of reaction of gases at high temperatures and low pressures favors the formation of equilibrium compositions, but this also makes it difficult to "freeze in" high-temperature compositions. Contaminants may react quickly to partially destroy the original compositions. Even without contamination, a collected sample will alter its total composition to one more stable at lower temperatures. Most gaseous species will cease to react at significant rates when they reach room temperature, but the temperatures at which the various gas species become essentially "non-reactive" vary. Thus, the composition of a cooled volcanic gas will reflect this spectrum of quenching temperatures; the final composition will be determined by the conditions under which cooling took place.

Variation in Time and Location. The composition of volcanic gases may vary with time, location on the volcano, and type or composition of the volcano. In some cases the gases emitted during the early portion of an eruptive cycle are rich in sulfur, while the late gases are more rich in H_2O and CO_2. The percentage of halogen gases has been noted to decrease from the beginning to the end of some eruptions and the ratio HCl/HF also may change.

Since hot gases will react with the rocks through which they pass, the distance from the lava source to the collecting site not only cools the gases but also causes a change in total composition. The acid gases such as HCl, HF, and H_2S are notably reactive with rocks. At the same time, the deposition of various sublimates on successively cooler rocks alters the content of these constituents.

Volcanic magmas vary widely in composition from rhyolitic to basaltic and the solubilities of the various gases will vary with magmatic composition. This will be reflected in a variation of volcanic gas compositions among volcanoes and may also result in different time-related compositional changes. The source and history of acidic magmas is distinctly different from basic magmas in many cases; thus the source and history of their associated gases also varies. The evidence available indicates that basaltic volcanoes produce gases more rich in carbon and sulfur while water and acid gases are more typical of rhyolitic volcanoes.

BERT E. NORDLIE

References

Allen, E. T., and Zies, E. G., 1932, "A chemical study of the fumaroles of the Katmai region," *Natl. Geog. Soc. Washington, Contr. Tech. Papers, Katmai*, Ser. 1(2), 75–155.

Basharina, L. A., 1953, "Study of the gaseous products of the volcanoes Kliuchevskii and Sheveluch in 1946–47," *Acad. Nauk SSSR, Lab. Vulkanol., Vulkanol. Stantsii Byull.*, **18**, 31–40.

Ellis, A. J., 1957, "Chemical equilibrium in magmatic gases," *Am. J. Sci.*, **255**(6), 416–431.

Heald, E. F., et al., 1963, "The chemistry of volcanic gases," Pt. 2, "Use of equilibrium calculations in the interpretation of volcanic gas samples," *J. Geophys. Res.*, **68**(2), 545–557.

Nordlie, B. E., 1967, "Composition of the basaltic gas phase," *Geol. Soc. Am. Program 1967 Annual Meet.*

Shepherd, E. S., 1938, "The gases in rocks and related problems," *Am. J. Sci.*, **35A**, 311–351.

Sigvaldason, G. E., and Elisson, G., 1968, "Collection and analysis of volcanic gases at Surtsey, Iceland," *Geochim. Cosmochim. Acta*, **32**, 797–805.

White, D. E., and Waring, G. A., 1963, "Volcanic emanations," in "Data of geochemistry," *U.S. Geol. Surv. Profess. Paper* **440-K**.

Cross-references: *Emission Spectrography; Sulfur.* Vol. IVB: *Exhalative Sedimentary Ore Deposits.* Vol. V: *Fumaroles; Magma; Thermal Activity; Volatiles; Volcanoes.* Vol. VI: *Hot Springs; Spring and Crust Deposits.*

GAS EVOLUTION ANALYSIS

Gas evolution analysis (GEA)* is the alternative but preferred name for *effluent gas analysis* (EGA) under which heading it has already been described and should be referred to. Briefly, dynamic furnace atmosphere differential thermal analysis (DTA) is required (for further details see *Differential Thermal Analysis*) so that any gas evolved from the sample under test will be flushed or transported over a detector cell. As this inert flushing or carrier gas flows at a constant rate, the detector measures only the amount of additional gas which passes over it from time to time. This gas is released as the test sample suffers single or multiple decomposition or dehydration during DTA.

The amount of increase and then decrease in gas flow produced as gas-producing reactions start, progress to their maximum decomposition rate, and then eventually reach completion, is recorded continuously as a function of temperature on an X–Y recorder in the form of a peak (see GEA curve in Fig. 1).

The complete permanently recorded trace is called a GEA curve or gas detection curve.

As the additional gas is produced by reactions within the test sample, a close correlation exists between the GEA peaks and the DTA peaks which represent these gas-yielding endothermic reactions (compare the two curves in Fig. 1).

To minimize operational errors and variables the GEA and DTA curves should be deter-

---

* Alternative terms include "Evolved gas analysis" (EGA) and "Evolved gas detection" (EGD).

GAS EVOLUTION ANALYSIS

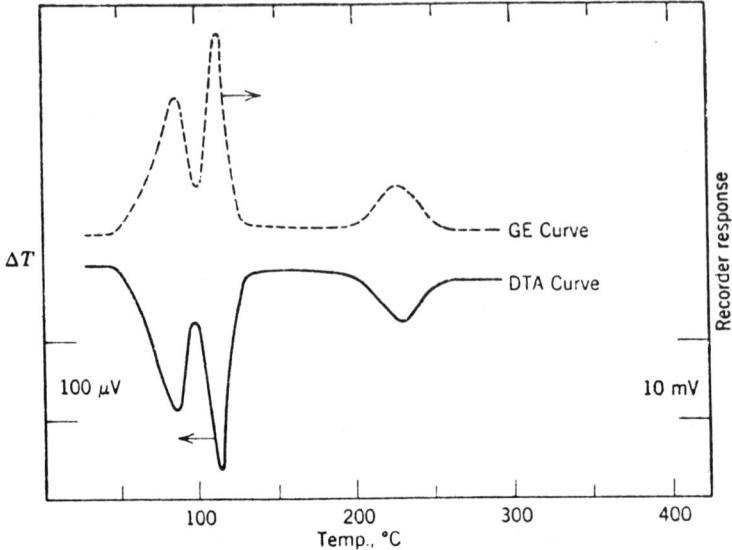

Fig. 1. Simultaneous DTA-GEA curves of $CuSO_4 \cdot 5H_2O$ (35.7 mg of sample) (from Wendlandt, 1964).

mined continuously and simultaneously, preferably on a two-pen recorder although two separate recorders may be used (Bayliss and Warne, 1962).

Two quite separate types of determinations may be made:

1. The GEA curve can be used directly to provide data on the temperatures at which the "gas" peaks begin, reach a maxima, and terminate, peak areas, and the number of peaks present.
2. The gaseous decomposition products, after being utilized to produce the GEA curve, can be further transported to an analysis unit where their actual identity is established by vapor phase gas chromatography, infrared spectroscopy, or mass spectrometry.

History

Significant contributions to the establishment, development, and application of GEA have been made by Gordon and Campbell, 1955, Felton and Buehler, 1958, Rogers, Yasuda, and Zinn, 1960, Murphy, Hill, and Schacher, 1960, Stone, 1960, Ayres and Bens, 1961, Vassollo, 1961, Wendlandt, 1962, Wendlandt, 1963 (see Wendlandt 1964), Garn and Kessler, 1961, and Garn, 1965.

Instrumentation and Theory

Instruments for GEA have been described by Rogers, Yasuda, and Zinn, 1960, and Vassallo, 1961. Simultaneous DTA-GEA was probably initiated by Stone (1960) but was first produced by Ayres and Bens (1961) and Wendlandt (1962).

The schematic layout of Wendlandt's combined DTA-GEA apparatus with detail of the furnace and sample assembly is shown in Fig. 2.

The detection of the evolved gas has been achieved by means of a thermal conductivity cell employing the model airplane glow plugs of Rogers, Yasuda, and Zinn (1960), or the thermistor thermal conductivity cell of Vassallo (1961) and a Felton and Buehler (1958) type bridge circuit (see Wendlandt, 1964).

In practice the purge gas is made to pass through the unknown sample and reference materials. This gas flow is used as the reference gas in the thermal conductivity detector cell. Thus the presence of any gaseous decomposition product will produce a different cooling effect in the detector cell while it is passing through the detector, provided its thermal conductivity differs from that of the carrier gas. The intensity of the differential signal from the detector is proportional to the amount of evolved gas passing through at any instant, provided the concentrations are low. Thus a measure of the quantity of this additional gas is given by the integration of the area under the GEA peak in question (Garn, 1965).

Subsequently, Garn and Kessler (1961) and Garn (1965) have described the replacement of the thermal conductivity detector by a gas density detector, the operation and advantages of which they describe in some detail.

GAS EVOLUTION ANALYSIS

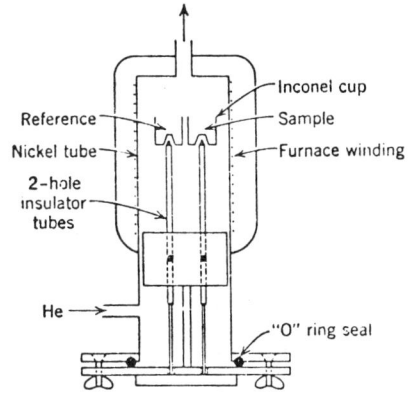

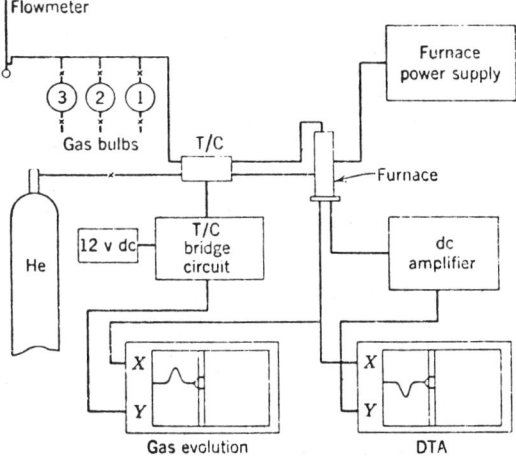

FIG. 2. Combined DTA-GEA apparatus (from Wendlandt, 1964).

The effects on the GEA curves of volatile and nonvolatile compounds, of the operating variables such as gas pressure, flow, and heating rates, the type of carrier gas, sample weight, and detection cell sensitivity were determined by Rogers, Yasuda, and Zinn (1960). The main parameters affected were maximum peak temperatures and heights, while excellent agreement was found between theoretically and experimentally derived GEA curves of pentaerythritol tetranitrate.

Curve Configuration

If the accepted practice of recording endothermic DTA peaks in the direction of the bottom of the chart is followed then the majority of workers prefer to record the GEA peaks in the opposite sense (see Fig. 1).

In this way, *provided the changes causing the DTA peaks are accompanied by gas evolutions*, the resultant DTA and GEA curves will form virtual mirror images of each other. This is exactly the case in Fig. 1, where the following three decomposition reactions not only produce the DTA peaks, but also liberate water vapor, which in turn causes the comparable GEA peaks (Wendlandt, 1962).

85°C peak

$CuSO_4 \cdot 5H_2O(s) \rightarrow$
$\qquad CuSO_4 \cdot 3H_2O(s) + 2H_2O(\ell)$
$2H_2O(\ell) \rightarrow 2H_2O(g)$ (from saturated solution)

115°C peak

$CuSO_4 \cdot 3H_2O(s) \rightarrow CuSO_4 \cdot H_2O(s) + 2H_2O(g)$

230°C peak

$CuSO_4 \cdot H_2O(s) \rightarrow CuSO_4(s) + H_2O(g)$

Alternatively if a given sample reaction releases *no gas* then the resultant DTA peak will have no counterpart on the GEA curve, which will remain featureless over this temperature region.

As in DTA, the actual peaks may be described as small, medium, large, sharp, blunt, well or poorly defined, symmetrical or asymmetrical. These terms are broadly indicative of the type of gas liberating reactions involved, e.g., slow or fast, large or small, between large or small temperature ranges, with rapid or slow decomposition initiation and termination.

The DTA and GEA reactions for particular physical and chemical changes have been summarized by Ayres and Bens, 1961 (see Table 1).

Applications

Firstly, each GEA peak indicates a gas-producing sample reaction and the temperature range in which it occurs. Furthermore each peak may represent the evolution of one or more gases, the compositions of which are unknown and relatively unimportant in most cases (Wendlandt, 1964).

TABLE 1. DTA AND GEA EFFECTS DURING PHYSICAL OR CHEMICAL SAMPLE CHANGES

Physical or Chemical Change	DTA Endothermic Reaction	GEA	
		Yes	No
Decomposition[a]	×	×	×
Fusion	×		×
Crystal transition	×		×
Desorption	×	×	×
Vaporization			
Desolvation	×	×	
Ebullition	×	×[b]	
Sublimation	×	×[b]	

[a] May also undergo an exothermic reaction.
[b] Condensation before reaching gas detector is possible.

GAS EVOLUTION ANALYSIS

Such information may be used directly for sample identification, where the GEA curve configuration is sufficiently diagnostic.

In cases where the actual composition of the evolved gas is required, true gas analysis units (see above), such as a mass spectrograph, must be employed in addition to the GEA apparatus.

The semiquantitative detection of individual components in mixtures may be obtained under reproducible conditions if gaseous products are released but not where the individual components have similar decomposition or boiling points.

Sample changes such as melting, crystallographic modifications and inversions, and melting and chemical reactions which involve no gas production are not recorded on the GEA curve and so remain completely undetected by this method.

These facts form the basis of a most important application of GEA, namely in direct association with DTA, which it complements and the interpretation of which it aids. Thus from the evaluation of GEA and DTA curves of the same sample determined under identical conditions it can be clearly seen which DTA peaks were produced by gas-yielding reactions and which were not and are therefore of the different types enumerated above such as crystallographic inversions. In this way, to some extent, it obviates the use of a thermobalance by indicating DTA peaks produced by at least one type of reaction, always associated by weight losses.

The temperature range over which this "weight loss" occurs and an indication of its magnitude is shown by the relevant GEA peak. Sometimes this may indicate that the temperature range is narrow and/or the peak is unexpectedly small compared with the related DTA peak. In such cases the GEA data clearly indicates that such a DTA peak represents two overlapping reaction peaks, only one of which is associated with gas evolution.

Gas evolution analysis also makes possible studies of the thermal behavior of materials under isothermal conditions and under various gaseous atmospheres and pressures.

S. St. J. Warne

References

Bayliss, P., and Warne, S. St. J., 1962, "The effects of the controllable variables on differential thermal analysis," *Am. Mineralogist,* **47,** 775–778.

Garn, P. D., and Kessler, J. E., 1961, "Effluence analysis as an aid to thermal analysis," *Anal. Chem.,* **33,** 952–954.

*Garn, P. D., 1965, "Thermoanalytical Methods of Investigation," New York, Academic Press, 606pp.

*Wendlandt, W. Wm., 1964, "Thermal Methods of Analysis," New York, Interscience Publishers, a div. of John Wiley & Sons, 424pp.

* Additional bibliographic references mentioned in the text may be found in this work.

Cross-references: *Chromatography; Differential Thermal Analysis; Effluent Gas Analysis; Infrared Analysis; Mass Spectrometry.*

GAS INDUSTRY—See Vol. IV B, **NATURAL GAS**

GEL, GEL MINERALS—See **COLLOIDS; MINERALOIDS**

GEMOLOGY—See Vol. IV B

GEOBOTANICAL AND BIOGEOCHEMICAL METHODS OF PROSPECTING FOR MINERALS—See Vol. IV B

GEOCHEMICAL CLASSIFICATION OF ELEMENTS

A consideration of the planetary (and meteoritic) abundance and relationships of the elements persuaded Victor Goldschmidt (1923) that certain specific geochemical rules of behavior can be established. Basically the elements in the solid planet are distributed between molten iron and silicates according to their various chemical affinities and solubility in the two liquid phases. Following Goldschmidt's words (translation of 1954, p. 11): "Generally, those elements can be expected to be concentrated in the fused iron which have a lower affinity for oxygen than has iron, i.e. those elements which are more easily reduced to the metallic state, and which are readily soluble in molten iron." Table 1 shows the free energy of formation of the principal electropositive oxides. There are uncertainties about some of the thermodynamic criteria so that a few of the figures given by Goldschmidt were only approximate and a better table is that of Rankama and Sahama (1950).

Since the planet Earth has undergone a fundamental differentiation into a silicate-rich mantle and crust, and an iron-nickel (alloy) core, there is a tendency toward a corresponding polarization of the various elements. Thus metals of the palladium and platinum groups are found to be relatively enriched in the iron meteorites; correspondingly these metals, likewise gold and silver, are relatively scarce in the silicate crust. Experimentally these char-

GEOCHEMICAL CLASSIFICATION OF ELEMENTS

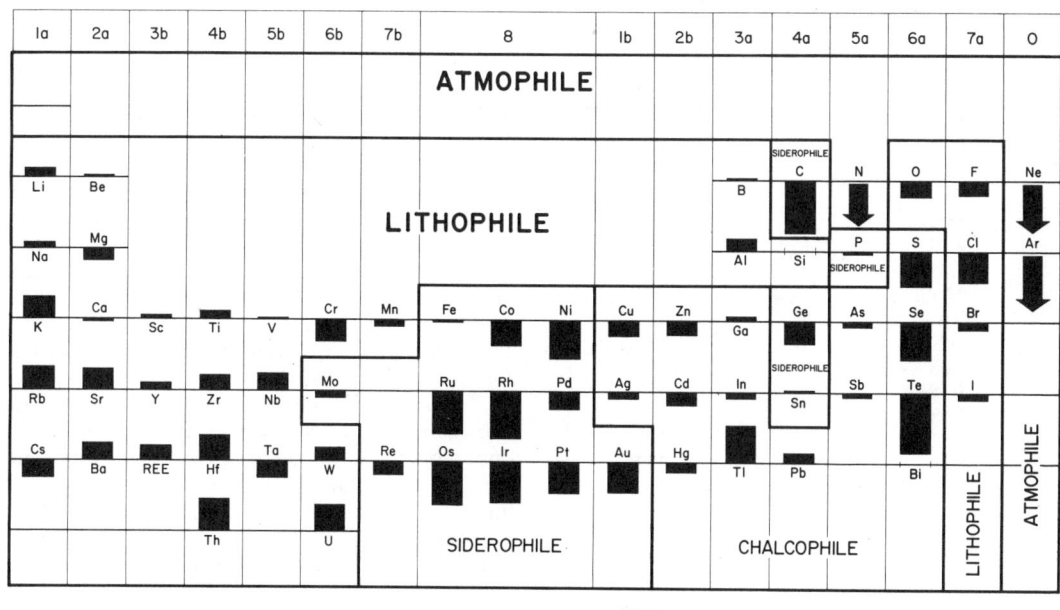

FIG. 1. Excess elements in the earth's crust over cosmic abundances.

TABLE 1. FREE ENERGY OF FORMATION ($-\Delta F°$) OF SOME OXIDES FROM STANDARD STATES OF THE ELEMENTS[a]

Oxide	$-\Delta F°$ (kg-cal/g-atom O_2)
CaO	144.3
MgO	136.2
Al_2O_3	125.6
ZrO_2	122.7
V_2O_3	103.7
TiO_2	102.5
SiO_2	95.2
Na_2O	93.6
Ta_2O_5	89.4
MnO	86.9
Nb_2O_5	85.2
Cr_2O_3	85.1
Ga_2O_3	78.3
ZnO	75.8
In_2O_3	72.7
SnO	60.8
FeO	59.3
WO_3	59.1
CdO	55.2
NiO	53.0
MoO_3	52.7
Sb_2O_3	49.7
PbO	45.1
As_2O_3	44.9
Bi_2O_3	38.9
Cu_2O	35.1
Ag_2O	2.6

[a] At room temperature. From Rankama and Sahama, 1950.

acteristics can be tested to some extent by studies of smelted iron and silicate slags. The latter find analogs in the stony meteorites, which tend to concentrate the alkali and alkaline earth metals. Goldschmidt's faith in metallurgical analogs was not, however, always justified, and a sulfide-oxide sphere is not now considered at all likely (see "Chalcosphere," below).

Under reducing conditions it is sulfur which must be considered in relation to the electropositive elements. As Goldschmidt said (ibid., p. 16) "We may imagine a competition between sulfur and oxygen to combine with an electropositive element and consider the differences between the free energies of sulfide and oxide formation . . ." (Table 2).

Although many other reactions and phenomena are involved, a broad classification on the above basis was offered by Goldschmidt (subsequently modified slightly by Rankama and Sahama, and other workers), as indicated in Table 3. The categories are as follows:

Siderophile Elements. Literally "iron-loving" (from the Greek), these elements tend to concentrate along with iron, nickel, etc., in the earth's core (referred to by Rankama and Sahama as the *Siderosphere*). Conversely these elements have a weak affinity for oxygen and sulfur.

Chalcophile Elements. Literally "copper-loving," these elements have a strong affinity

GEOCHEMICAL CLASSIFICATION OF ELEMENTS

TABLE 2. FREE ENERGY OF FORMATION ($-\Delta F°$) OF SULFIDES[a]

Sulfide	$-\Delta F°$ (kcal/g-atom S)
CaS	109.98
SrS	109.02
BaS	101.28
Na_2S	86.6
MnS	45.52
ZnS	40.37
CdS	33.49
MoS_2	27.10
RuS_2	22.05
WS_2	23.08
PbS	22.73
$FeS\alpha$	23.39
$FeS\beta$	23.04
Tl_2S	22.54
NiS	(21.33)
PtS	18.55
SnS	18.18
FeS_2	18.04
CoS	21.06
Cu_2S	19.22
Bi_2S_3	13.01
PtS_2	12.14
Sb_2S_3	12.31
CuS	11.7
HgS	8.8
As_2S_3	8.68

[a] Goldschmidt, 1954.

for sulfur, typically copper, silver, zinc, lead, etc. All are fairly commonly found in the earth's crust but only localized as in sulfide ore deposits. They are soluble in FeS (troilite, as found in meteorites) and are enriched in the mantle. Goldschmidt believed there was a distinct sulfide-oxide layer (*"Chalcosphere"*) at the base of the mantle, but this is not accepted today.

Lithophile Elements. Literally these are the "rock-loving" elements that are concentrated in the crust, in slag, and in stony meteorites. They are the ones that have a lower free energy of oxidation than iron (per gram-atom of oxygen), and include such elements as magnesium, aluminum, sodium, and silicon. These are the major crustal components (thus, *Lithosphere*).

Atmophile Elements. Literally these are the "air-loving" elements, which exist either in the uncombined state, such as oxygen, nitrogen, etc., or as volatiles such as H_2O, CO_2, and the rare gases, helium, neon, argon, etc. (*Atmosphere*).

Biophile Elements. Literally these are the "life-loving" elements, which are concentrated in living organisms; e.g., H, C, N, O, P, etc. (*Biosphere*).

(It may be mentioned that in the original Goldschmidt work, in German, the terms were spelled "siderophil," "chalcophil," etc., but the suffix "-phile" is more customary in English and is adopted by most authors.)

Periodic System

It has long been known that there was a close relationship in the geochemical behavior

TABLE 3. GEOCHEMICAL CLASSIFICATION OF THE ELEMENTS[a]

Siderophile	Chalcophile	Lithophile	Atmophile	Biophile
Au	Cu Ag	Li Na K Rb Cs	H C N	H C N O P
Ge Sn (Pb)	Zn Cd Hg	Fa	O I Hg	(Na) (Mg) (S) (Cl)
C P (As)	Ga In Tl	Be Mg Ca Sr Ba	He Ne A	(K) (Ca) (Fe)
Mo (W)	(Ge) (Sn) Pb	Ra	Kr Xe Rn	(B) (F) (Si)
Re	As Sb Bi	(Zn) (Cd)		(Mn) (Cu) (I)
Fe Co Ni	(Mo)	B Al Sc Y		
Ru Rh Pd	S Se Te	La Ce Pr Nd Sm		
Os Ir Pt	Fe (Co) (Ni)	Eu Gd Tb Dy Ho		
	(Ru) (Pd) (Pt)	Er Tm Yb Lu		
		Ac Th Pa U Np		
		Pu Am Cm		
		Ga (In) (Tl)		
		C Si Ti Zr Hf		
		(Ge) (Sn) (Pb)		
		V Nb Ta		
		P (As)		
		O Cr W Mn		
		(Fe) (Co) (Ni)		
		H F Cl Br I		

[a] Rankama and Sahama, 1950.

GEOCHEMICAL EVOLUTION OF THE CORE, MANTLE, AND CRUST

TABLE 4. PERIODIC SYSTEM AND THE GENERAL GEOCHEMICAL CHARACTER OF THE ELEMENTS[a]

H									He		
Li		Be	B	C	N	O	F	Ne			
Na		Mg	Al	Si	**P**	S	Cl	A			
K		Ca	Se	Ti	V	Cr	Mn		Fe	Co	Ni
	Cu	Zn	Ga	**Ge**	**As**	Se	Br	Kr			
Rb		Sr	Y	Zr	Nb	**Mo**	Te?		Ru	Rh	Pd
	Ag	Cd	In	**Sn**	**Sb**	Te	I	Xe			
Cs		Ba	La-Lu	Hf	Ta	W	**Re**		Os	Ir	Pt
	Au	*Hg*	Tl	**Pb**	**Bi**	**Po?**	At?	Rn			
Fa		Ra	Ac-Cm								

[a] *Au*: siderophile; **Cu**: chalcophile; Li: lithophile (and atmophile). From Rankama and Sahama, 1950.

between elements and their respective position in the periodic system (Niggli, 1928; Rankama and Sahama, 1950, p. 91; see Table 4).

As pointed out by Rankama and Sahama, these categories are useful for many geochemical purposes, but they should not be taken as strictly correlative with any exclusive earth shell ("sphere").

RHODES W. FAIRBRIDGE

References

Day, F. H., 1964, "The Chemical Elements in Nature," New York, Reinhold Publ. Corp., 372pp.

Goldschmidt, V. M., 1923, "Geochemische Verteilungsgesetze der Elemente," *Videnskapsselskapets Skrifter* (Oslo), I, Math.-nat. Kl. 10.

Goldschmidt, V. M., 1954, "Geochemistry," Oxford, Clarendon Press, 730pp.

Niggli, P., 1928, "Geochemie und Konstitution der Atomkerne," *Fennia*, **50**(6).

Rankama, K., and Sahama, T. G., 1950, "Geochemistry," Chicago, Univ. Chicago Press, 912pp.

Cross-references: *Atomic Number and Periodic Table; Elements; Elements: Planetary Abundance and Distribution; Oxidation and Reduction; Thermodynamics.* Vol. IVB: *Mineral Classification.*

GEOCHEMICAL EVOLUTION OF THE CORE, MANTLE, AND CRUST

The evolution of the earth into its present form is clearly linked with early events in the history of the universe, in particular those involving our galaxy. Much of this earliest history is shrouded in uncertainty. For our purpose, we will assume that at the dawn of time, perhaps some 10^{10}–10^{15} years ago, matter was dispersed in the form of hydrogen, the simplest of all elements. Astrophysicists and nuclear physicists are able to provide a plausible account of how stars may evolve from hydrogen and how at high temperatures—produced by the release of gravitational potential energy as the protostar condenses from hydrogen—atoms, or, rather, nuclei, of heavier elements can be synthesized; some elements are created when a star reaches a supernova stage. Stars do not, however, all have the same age. Some are only in their embryonic stage of evolution and some, including the sun, are regarded as "middle-aged." Younger stars, plus possible planetary systems, are presumed to evolve from cosmic dust and gases which have been produced by older stars, either by escape from the gravitational field of the older stars, or through stellar explosion (nova or supernova). Though H and He may predominate in the primitive cosmic dust, distinct quantities of heavy elements will also be present. Let us now turn to our solar system which is believed to have originated some 5000 million years ago.

At that time a great mass of cosmic dust and gas appear to have accumulated in the region of space we occupy in our galaxy. The total mass of the dust cloud together with its composition (relative abundances of the elements) and physical nature would determine the fate of the evolving embryonic earth. The physical nature of the components of the dust cloud is conjectural. We could either assume that the cloud consisted more or less of a gas of dispersed atoms of different elements with H and He predominant—a situation physically analogous to that obtaining when the oldest stars first evolved from a gaseous condensate of H atoms—or that the dust was composed of a mixture of discrete grains plus gases (notable H and He). Most workers support the second possibility.

Eddies in primitive cosmic material apparently led to the accretion of the various protocomponents (protosun and protoplanets) of our solar system. Much of the mass, > 95%, was concentrated in one embryonic major component, the protosun. Because of its great mass, the relatively cool protosun became intensely hot through the release of gravitational potential energy and perhaps other sources of heat. Temperatures became so high in fact that nuclear reactions developed, some self-sustaining, and the protosun became a hot, fiery star. The masses of the other protocomponents were far less than that of the protosun and, unlike it, the protoplanets evolved into relatively cool objects—planets—as we know them.

Each of the planets has either a distinct structure or may be without structure. Whereas it is difficult to ascertain the structures of the other planets, we have a reasonably clear idea of the structure of the planet earth. According to geophysical evidence, the solid earth comprises three principal components; the core, mantle, and crust (Fig. 1; from Birch, 1965) and we may now go into the problem of their geochemical evolution. As noted, this will depend primarily on the mass, physical nature, and composition of the parent solar dust. As we are concerned particularly with geochemical evolution, let us consider the composition of the primitive starting material. H and He were probably the principal components, but eventually escaped from the evolving protoearth and were not necessarily of great significance in influencing its fate. Particular significance should be attached to (a) those elements which were both abundant and chemically active and (b) those which are radioactive and may have supplied heat.

Estimated Abundances of Elements in the Primitive Solar Cosmic Dust

Meteorites provide us with a good indication of the composition of the primitive starting material. In the first place, the elements which compose both the earth and meteorites had a common origin. Were this not so, the isotopic composition of elements extracted from terrestrial material would have differed substantially from the isotopic composition of elements from meteorites. Some significant differences exist, but these are due either to the effects of cosmic rays or to radioactivity. Difficulties nevertheless arise when attempting to use meteoritic data. The reason is that these objects have had a history different from that of the earth. In addition, their composition is not uniform and we have always to face the problem of which type, or types, of meteorites to use for obtaining abundance data. The chondritic meteorites are by far the most abundant but, on the other hand, the individual chondrites are comparatively small relative to the metal meteorites. For our purpose we shall make use of all meteorite varieties and take into consideration the proportions of the different kinds and their approximate weight relationships. Table 1 gives a list of estimated abundances of the most common elements and Table 2 a list of radioactive nuclides; these are given in two parts: (a) those with comparatively long lives which are

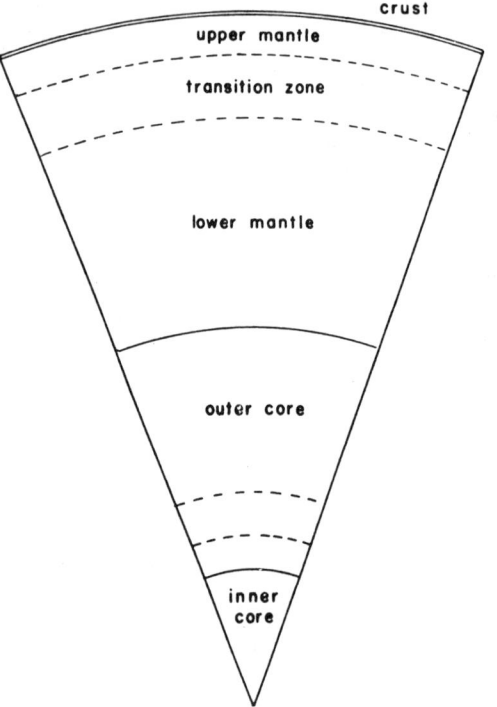

FIG. 1. Structure of the earth, according to geophysical evidence (from Birch, 1965).

TABLE 1. ESTIMATED ABUNDANCES OF SOME ABUNDANT ELEMENTS IN AVERAGE METEORITIC MATTER[a]

Element	Abundance (%)	Atoms relative to Si = 10^6
O	33.0	3,420,000
S	2.1	107,000
Na	0.68	49,000
Mg	13.9	945,000
Al	1.1	68,000
Si	16.95	1,000,000
Ca	1.4	58,000
Fe	28.4	849,000
Ni	1.7	47,500

[a] Expressed as percent and number of atoms relative to Si = 10^6.

TABLE 2. RADIOACTIVE ISOTOPES

Isotope	Half-Life (yr)
Abundant long-lived isotopes	
U^{235}	7.1×10^8
U^{238}	4.5×10^9
K^{40}	1.33×10^9
Some longest-lived extinct isotopes	
Al^{26}	8×10^5
Pd^{107}	7×10^6
I^{129a}	1.6×10^7
Sm^{146}	5×10^7
Pu^{244}	7.5×10^7

Though not in existence, evidence of the daughter (Xe^{129}) has been found in some meteorites.

comparatively abundant and which throughout the lifetime of our planet have produced a considerable amount of heat and (b) a few examples of the longest-lived nuclides which are now extinct but which may have produced intense heat in the very earliest stages of the evolution of the earth.

Evolution of the Core

We will accept Daubrée's early (1863) suggestion that the core (Fig. 1) of the planet is metallic. Density and hydrodynamical vs density relationships (Fig. 3; from Birch, 1965), together with chemical and abundance considerations, lead to the conclusion that the core is composed mainly of iron plus a significant proportion of nickel. If we use the distribution relationships of elements between the three principal phases in meteorites, metal, sulfide, and silicate as a guide, it is likely that several rare elements [gold (Au), silver (Ag), platinoids, molybdenum (Mo), and germanium (Ge), for example] may also be concentrated in the core. Also, there is the possibility that distinct amounts of silicon (Si) in the form of an Fe(Ni)Si alloy, may be present in the core. This arises from the fact that the density of the core appears to be about 15% less than that expected if the core were composed entirely of iron; hence, the need to postulate the presence of the light element silicon. Bearing in mind the above suggestions, we now go into the problem of the geochemical evolution of the core.

Though the primitive cosmic dust was cool, heat was required to produce the core, mantle, and crust. In fact, their evolution depended primarily on the thermal history of the earth (Birch, 1965; Donn, et al., 1965). During the very earliest stages, some heat may have been produced by the release of gravitational potential energy as the dust condensed. A more important source is probably radioactivity, either by comparatively short-lived nuclides whose radioactivity is now exhausted (Table 2) or by the longer-lived nuclides also listed in Table 2. It may be borne in mind that 4-5000 million years ago, U^{238} was approximately twice its present abundance, K^{40} eight or more times its present abundance, and U^{235} as much as sixty times the amount present now.

It will be assumed that the grains in the primitive cosmic dust were composed of SiO_2, Fe_2SiO_4, Mg_2SiO_4, FeS, and metallic Fe; the proportions of these compounds would be such that the relative abundances of the elements are approximately equivalent to that corresponding to a weighted average of the meteorites. We will assume further that the accreting dust was fairly uniform in composition. Once condensed, radioactivity, together with other sources would gradually heat the mass. The ultimate maximum temperature which was reached is a matter of speculation. It was undoubtedly greater than the melting point of iron and many of the metals and may have approached or exceeded the melting points of some silicates. Molten iron would tend to diffuse and coalesce into droplets. Because of their high density, these would tend to descend toward the center of the earth. Whether they could do so, however, depends on the viscosity of the matrix material, assumed mainly to be silicates with some sulfides. A most important fact is that the melting point curve of silicates increases more rapidly with depth (greater pressure) than does temperature. Some relevant information is given in Fig. 2 from MacDonald 1959). In deriving these curves, it has been assumed that the composition of the protoearth was equivalent to that of chondritic meteorites and that the earth reached an initial temperature of 1300°C. Though the silicate in Fig. 2 is diopside and not mantle material, the curve is nevertheless instructive.

For the sake of further discussion, let us assume that softening of the silicate matrix occurs first in the outermost layers (because of the melting point-temperature relationships), beginning at about 200-300 km depth. Provided the whole silicate mass ultimately became comparatively soft (Elsasser, 1963), the coalescing iron droplets would continue to sink and eventually form the earth's core. The possible significance of a single giant convective cell in the protoearth leading to core formation has been suggested.

During the descent to the center, some other metals would tend to follow iron and nickel. Typical examples are the siderophile elements, i.e., metals which are chemically inert and which are at the same time either form substitutional solutions or alloys with iron; some examples have already been noted.

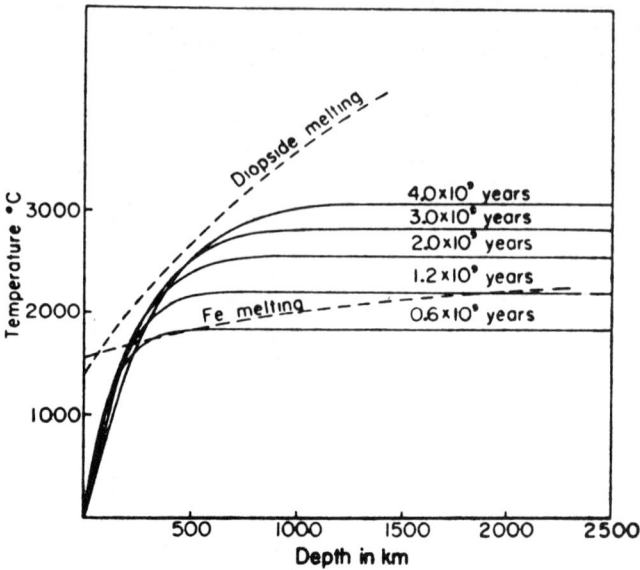

Fig. 2. Temperatures in a homogeneous Earth heated by radioactivity corresponding to that in average chondrite (from MacDonald, 1959).

Core formation may have taken place some 500 million years or so after accretion. One suggested important consequence was an apparent rise in temperature to perhaps 4000–5000°C in the earth's deep interior. Such a temperature rise would strongly influence the geochemical evolution of mantle (including possible sulfides) and crust. There is the interesting possibility, too, of a growing core (Runcorn, 1964) and the consequences this may have had on the evolution of the crust.

Before turning to mantle and crust, brief mention will be made of the sulfides. Sulfur is a fairly abundant element (Table 1) and it is possible that a few percent of FeS may have been present as a constituent in the primitive cosmic dust. If so, it should be borne in mind that the melting point of FeS ($\sim$1195°C) is lower than that of the Fe–Mg silicates. It is possible, therefore, that a high proportion of FeS, together with other metal sulfides, may have been in a molten state. The possible consequences have not been adequately studied and it is still conjectural as to whether FeS and other sulfides are in a dispersed state in the earth's interior or whether the sulfides might to some extent have concentrated and segregated into layers. Let us now turn to the mantle and crust.

Geochemical Evolution of Mantle and Crust

It seems doubtful whether the whole of the silicate matrix became completely molten at the time the core evolved. If this were so and if, on subsequent cooling, the molten mass differentiated thoroughly, a decidedly thicker crust than now exists would have evolved. It is possible, however, that a molten silicate matrix did not differentiate completely. If we are to assume that only "softening" occurred, the original dust components would still exist in their original form as chemical reactions would not have taken place. Developments in the mantle about 4000 million years ago are however indeed speculative. The present crust is about 5 km thick in the oceanic areas and 30–40 km thick in the continental areas. If we assume the mantle to be composed of ultrabasic material fairly similar to that of chondritic meteorites, it is necessary only to melt and differentiate a 700–800 km thick outer mantle in order to produce the present crust. If we assume this to be the case, we may ask the question as to when this happened, i.e., the formation of the crust, and what geophysical and geochemical considerations should be taken into account? According to age data based on the various radioactive age techniques (lead, argon, and strontium), the oldest well-established mineral ages go back $\sim 3500 \times 10^6$ years (Fig. 3). Older ages have been reported, but require confirmation.

Let us assume that 3500–4000 million years ago the outer part of the mantle was completely molten, cooled, and differentiated. Crystallization may have begun at the base because of the melting point-temperature relationship

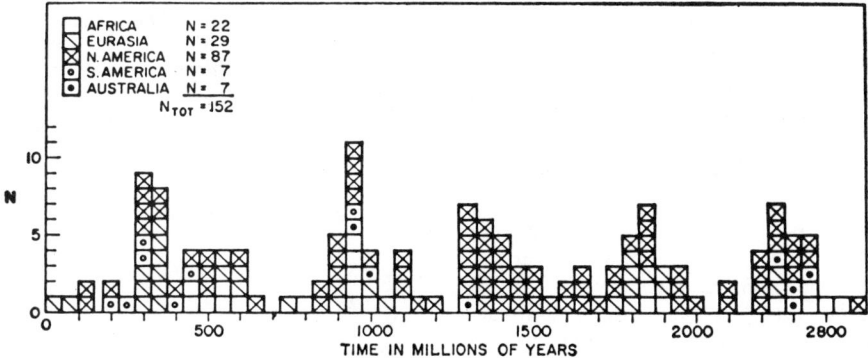

FIG. 3. Distribution of estimated ages of rocks as determined by radioactive techniques (where N = frequency) (adapted from Birch 1965, who refers to original sources). Note: The data refer to pre-1960 values and since then a few well-established ages of 330×10^6 years have appeared in the literature.

(see, for example, Fig. 2) and gradually moved upward. Convection, perhaps in the form of convection cells, may have significantly influenced the path of fractional crystallization (differentiation). In any event, high melting point Fe–Mg silicates would be the first minerals to crystallize and these would tend to segregate and settle toward the base of the cooling mass because of their high density. It is possible that the olivine nodules which are found in some basalts, together with some dunites, may represent such segregates from a cooling upper mantle. A striking feature about these nodules is their uniformity of composition. Thus, for example, the ratio of the two antipathetic components SiO_2 and MgO is remarkably constant; this suggests that these objects were formed under closely similar conditions of temperature, pressure, cooling history, and identical parental (upper mantle) materials.

A lighter fraction of basaltic composition rose to the surface and consolidated there as the primitive crust which might have covered the whole of the earth's surface. Several elements (for example, Li, Na, K, Rb, Cs; Be, Ca, Sr, Ba; B, Al, Ga, In, Tl, Sc, rare earths; Si, Ge, Sn, Pb, Ti, Zr, Hf, Th; As, Sb, Bi, Nb, Ta; U) are enriched in material of basaltic composition relative to mantle material assumed to approximate the composition of chondritic meteorites. It is possible that in some areas fractionation may have gone a stage further, leading to the formation of the first granites or granodiorites some 3300 million years ago. If so, comparatively small volumes were evidently produced, because only a very few of the many hundreds of rock ages now available exceed 3300 million years of age. These rocks are relatively highly enriched in most of the elements noted above, for example, Li, K, Rb, Cs; Be, Sr, Ba; B, rare earths; Si, Zr, Hf, Th; Nb and Ta; U. Table 3 compares the abundances of elements in possible mantle material (average chondrite), together with average basalt and granite.

TABLE 3. ESTIMATED ABUNDANCES OF SOME ELEMENTS IN CHONDRITES, BASALTS, AND GRANITES[a]

Element	Chondrites[b]	Basalt[c]	Granite[c]
Li	2.5	10	30
Be	0.04	0.5	5
B	0.43	5	15
Na	0.68%	1.94%	2.77%
Mg	14.4%	4.5%	0.16%
Al	1.3%	8.76%	7.7%
Si	17.8%	24.0%	32.3%
K	0.09%	0.83%	3.34%
Ca	1.4%	6.7%	1.85%
Sc	8.5	38	5
Ti	850	0.9%	0.23%
Ge	9.5	1.5	1.5
As	2.2	2.0	1.5
Rb	3.0	30.0	150
Sr	11.0	465	285
Y	2.0	25	40
Zr	12.0	150	180
Nb	?	20	20
Sn	0.43	1.0	3.0
Sb	0.10	0.2	0.2
Cs	0.1	1.0	5.0
Ba	4.5	250	600
La	0.34	10	40
Hf	0.19	2	4
Ta	0.023	0.5	3.5
Pb	0.18	5.0	20
Th	0.40	2.2	17
U	0.014	0.6	4.8

[a] As ppm unless otherwise indicated.
[b] From Ahrens (1965).
[c] From Taylor (1964a).

The variation of concentration of each of the different elements in these materials is quite apparent. In these compilations, attempts have not been made to differentiate between different types of basalt. Differences exist, but for our purpose they are minor relative to the major differences in Table 2.

Selective geochemical fractionation, notably the enrichment in the crustal rocks basalt and granite, relative to undifferentiated mantle—assumed to approximate the composition of average chondrites—has been discussed in detail by Taylor (1964b). Such relative enrichment is partly explained by ionic radius and charge considerations.

As a result of the distribution of ages in Fig. 3, the *continental* crust has been in a continuous state of evolution since the primitive crust consolidated; the possible causes are many. Primitive crustal material may erode. The products (sediments) accumulate in the deep oceans and eventually are heated, become molten, and differentiate, leading to a fresh crop of rocks. Zonal fusion (Vinogradov, 1964) is another possible cause for the development of younger crustal rocks. There is the possibility also that developments in the core may, because of thermal effects, also have repercussions on the evolution of the crust. Such postulated events have, however, little bearing on the *geochemical* evolution of the crust beyond the formation of the primitive crust and we won't consider them further. It may be noted, however, that small *changes* in composition between the earliest and youngest continental crust rocks of the same type have been observed; further work is, however, required to establish whether these small differences are indeed so or not.

L. H. Ahrens

References

Ahrens, L. H., 1965, "Distribution of the Elements in our Planet," New York, McGraw Hill Book Co., 110pp.

Birch, F., 1965, "Speculations on the Earth's thermal history," *Bull. Geol. Soc. Am.,* **76,** 133–153.

Donn, W. L., Donn, B. D., and Valentine, W. G., 1965, "On the early history of the Earth," *Bull. Geol. Soc. Am.,* **76,** 287–306.

Elsasser, W. M., 1963, "Early history of the Earth," in "Earth Science and Meteoritics," Amsterdam, North Holland Publ. Co., pp. 1–30.

MacDonald, G., 1959, "Calculations on the thermal history of the Earth," *J. Geophys. Res.,* **64,** 1967–2000.

Runcorn, S. K., 1964, "A growing core and a convecting mantle," Chap. 21 in "Isotopic and Cosmic Chemistry," Amsterdam, North Holland Publ. Co.

Taylor, S. R., 1964a, "Abundance of chemical elements in the continental crust: a new table," *Geochim. Cosmochim. Acta,* **28,** 1273–1285.

Taylor, S. R., 1964b, "Trace element abundance and the chondritic Earth model," *Geochim. Cosmochim. Acta,* **28,** 1989–1998.

Vinogradov, A. P., 1964, "Geochemical aspects of the Earth's crust and upper mantle," I.C.S.U. Review, **6,** 131–136, Amsterdam, Elsevier Publ. Co.

Cross-references: *Core Geochemistry; Elements: Planetary Abundances and Distribution; Geochronometry; Potassium-Argon Age Determination; Radioactive Isotopes; Rubidium-Strontium Dating Method; Uranium-Thorium-Lead Age Determination.* Vol. II: *Cosmic Dust; Cosmogony; Galaxy; Gravitational Collapse of a Star; Meteorites; Planetary Evolution; Planet Earth—Origin and Evolution; Solar System; Universe.*

GEOCHEMICAL PROSPECTING—
See Vol. IV B

GEOCHEMISTRY

Geochemistry is the study of the chemical constitution of the earth and its chemical changes, either taking place now or having taken place in the past. *Cosmochemistry* (see Vol. II), the study of the earth in relation to the solar system and the universe, including the study of the ultimate origin of the earth, overlaps geochemistry, and several fields of *geology, oceanography,* and *meteorology* include chemical considerations which are part of geochemistry. Unlike subfields of the earth sciences defined in terms of the actual materials under investigation, geochemistry may include any inquiry about the earth involving the theoretical or experimental methods of chemistry. In principle, geochemistry is as broad as all of chemistry and all of the earth sciences—geochemistry is what a geochemist does!

Abundances of the Elements

The composition of the materials of the earth's crust is variable, and to determine the true average composition is a major experimental and theoretical challenge. It is, however, of considerable importance to determine the average composition of the earth and of the solar system, for the pattern of relative abundances sets important boundary conditions on theories for the creation of the elements and the evolution of the solar system.

The average composition of the earth as a whole is believed by many to be similar to the composition of the sun and to the composition of chondritic stone meteorites, at

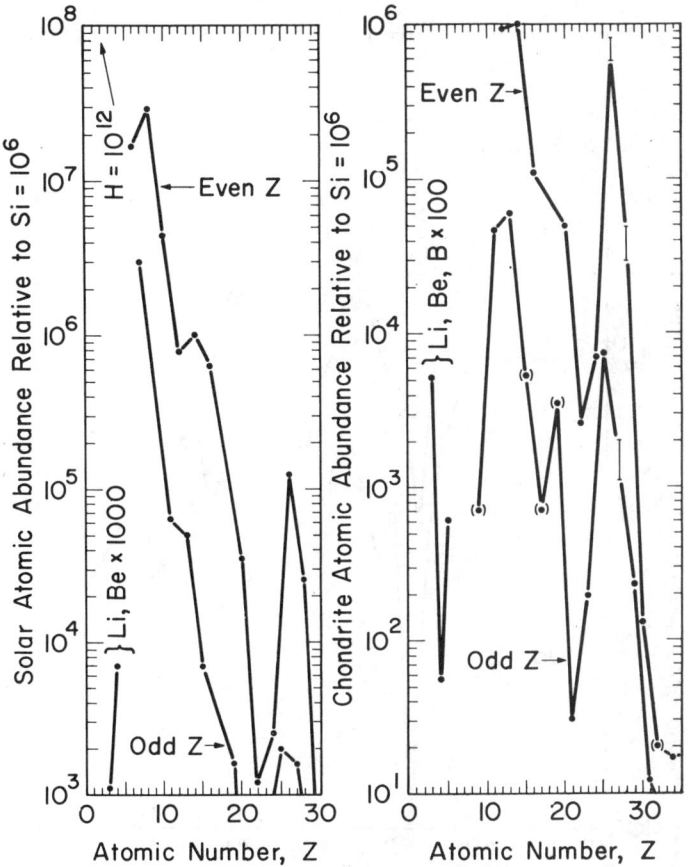

FIG. 1. Solar and chondritic abundances expressed as atoms per 10^6 atoms of Si, after L. H. Aller, 1964, and H. C. Urey, 1964. Volatile elements are not contained appreciably in chondritic meteorites; ( ) is more than two-fold uncertain; vertical bars near $_{26}$Fe indicate values for two different groups of chondrites.

least with respect to the elements which form relatively nonvolatile compounds under earth conditions. Fig. 1 shows the atomic abundances, in the solar atmosphere and in the chondritic meteorites, of the elements as a function of atomic number up to the vicinity of iron, $Z = 26$. Four features are common to both abundance curves: (1) There is a general trend of sharply decreasing abundance with increasing Z. (2) The abundances of elements with odd Z are regularly lower than those of neighboring elements with even Z. (3) A sharp minimum occurs at Li, Be, and B. (4) A sharp maximum occurs at Fe and adjacent elements.

Fig. 2 shows the meteorite abundance curve for middle and heavy weight elements, and we observe: (1) Abundance tends to continue decreasing with increasing Z. (2) The odd-Z abundance curve lies everywhere lower than the even-Z curve. (3) Maxima occur near $Z = 56$ and $Z = 78$. (4) A "shelf" seems to exist in the region $32 \leq Z \leq 40$.

More regularities in abundances appear if we consider the separate isotopes of the elements. Because of the physics of stability of the atomic nucleus, elements of odd Z usually have only one stable isotope and never more than two, and $_{43}$Tc and $_{61}$Pm have no nonradioactive isotopes at all. In contrast, even Z elements generally have several stable isotopes, and $_{50}$Sn has ten. If we presume that the nuclear reactions which took place to create the elements, in a star or wherever it was, created the nuclides (isotopes) in smoothly decreasing abundance with increasing atomic weight, then the sum of isotopic abundances for each element should be greater for even-Z elements, because of the larger number of isotopes, than for odd-Z elements. The second regularity pointed out above for Figs. 1 and 2 supports this presumption.

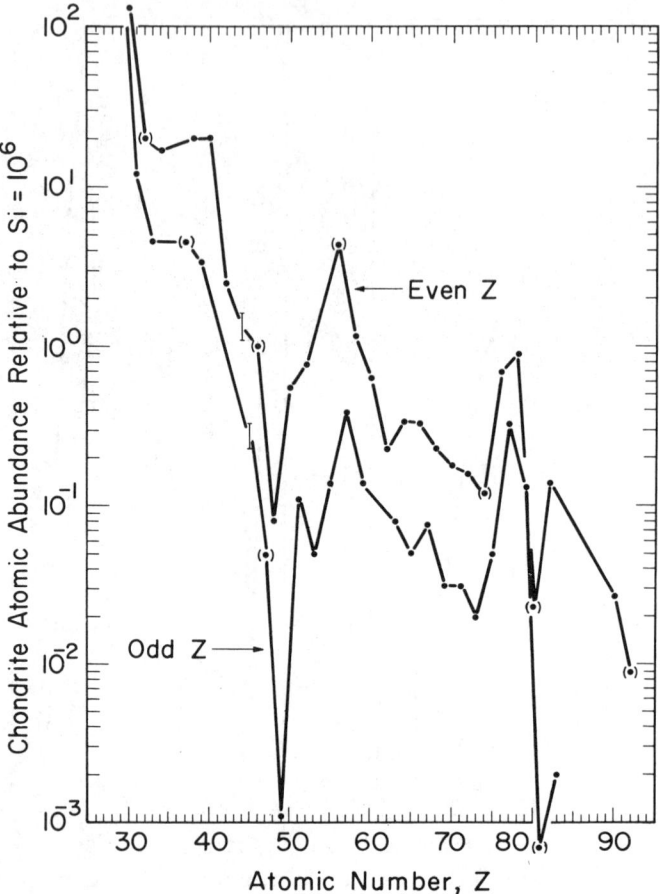

Fig. 2. Chondrite abundances, continuing Fig. 1.

Fig. 3 depicts the patterns of relative isotopic abundance in various parts of the periodic table for several elements of even Z as they occur in the earth's crust; isotopic abundances elsewhere in the solar system are expected to be very nearly the same. Three fairly distinct groupings may be seen: *(1)* The light elements from $_6$C and $_8$O to $_{20}$Ca have by far their greatest abundance in the lightest isotope. (The apparent exception of argon is due to $_{18}$Ar^{40} in the earth's atmosphere produced by $_{19}$K^{40} decay.) *(2)* The principal isotope for $_{22}$Ti, $_{24}$Cr, and $_{26}$Fe is the middle one. *(3)* The general pattern for all heavier elements is a skewed distribution with the maximum abundance among the heavier, but not the heaviest isotopes. The regularities are quite general, although some singularities exist such as $_{18}$Ar and $_{28}$Ni, which require special explanation. Several elements occur in the region of nuclear closed shells, and a closed shell "magic" number of neutrons is associated with enhanced abundance. For example, $_{56}$Ba^{138}, with magic neutron number $N = 82$, has an anomalously high abundance in a pattern otherwise normal for barium. Another cause for unusual isotopic abundance patterns is production of some isotopes from a long-lived radioactive parent, such as $_{18}$Ar^{40}. Three of the four isotopes of lead are produced by decay of uranium or thorium, and the pattern shown for $_{82}$Pb is an average of a variety of abundances for lead taken from different sources in the earth.

Radioactive elements, such as $_{92}$U and $_{19}$K, experience steadily changing relative isotopic abundance patterns throughout geologic time, but at any one instant of time all samples of the element anywhere in the earth will have identical patterns. The patterns for these elements at a time 4.5×10^9 years ago, when the earth was formed, are shown with open circles and dotted lines in Fig. 3. The two uranium isotopes were of similar abundance at that time, in keeping with the view that original nuclide abundances varied smoothly

GEOCHEMISTRY

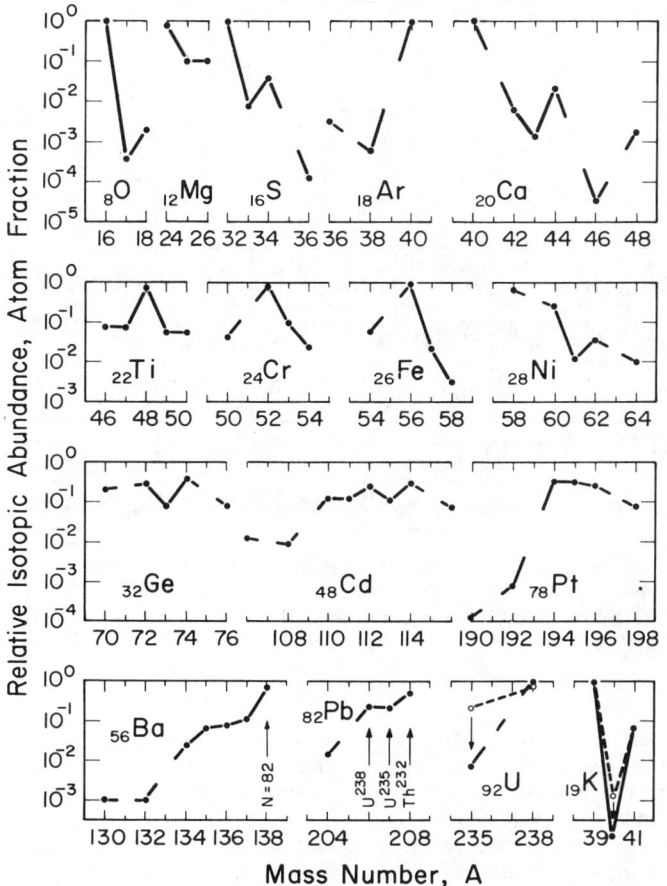

FIG. 3. Relative isotopic abundances in the earth's crust. Lines connecting points are interrupted at unstable isotopes.

with atomic weight, but the abundance of K^{40} was still low, explained by assuming that most of the mass 40 abundance was contained in $_{20}Ca^{40}$.

The theoretical explanation of the origin of the elements depends heavily on these elemental and isotopic abundance relations. Present thinking favors origin by nuclear reactions in a star over a very long period of time ending 4.5×10^9 years ago, whereupon the material of the star was dispersed and shortly thereafter condensed into the earth and planets. Nuclear reactions of different kinds account for the formation of light, medium, and heavy elements: *(1)* Conversion of hydrogen, $_1H^1$, into helium, $_2He^4$. *(2)* Condensation of $_2He^4$ into carbon, $_6C^{12}$, bypassing Li, Be, and B. *(3)* Successive reaction of $_2He^4$ with $_6C^{12}$, and related reactions, to produce the lightest isotopes of the even Z elements up to $_{20}Ca^{40}$. *(4)* A massive fusion of nuclear material to form elements up to $_{26}Fe$, in the region of maximum nuclear stability. *(5)* Neutron capture, gradual and/or cataclysmic, to form the heavier elements, causing enhanced abundance at magic numbers owing to enhanced stability against further neutron capture at the nuclear closed shells. Using the abundances and these broad outlines of nucleosynthesis, experimental and theoretical work is now directed to detailed understanding of the nuclear reactions and of the gigantic physical processes which must have been the ultimate cause.

Lithosphere

The study of the chemical composition of the silicate rocks of the dry land portions of the earth's surface has been central in geochemistry for many decades. *Petrology* and *mineralogy* are concerned with the evidence of minerals as they are found in rocks and its relationship to the possible geologic condititons of formation of rocks and of the entire earth's crust. Geochemists have been especially interested in the distribution of elements of similar chemical properties in rocks and minerals and its bearing on these geologic problems. Fig. 4 is a periodic chart which groups the elements

GEOCHEMISTRY

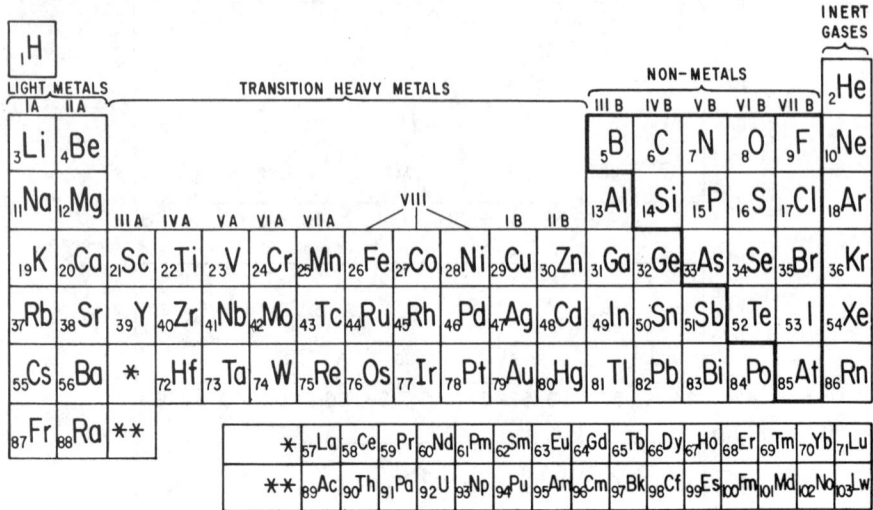

FIG. 4. Periodic arrangement of the elements showing groups having similar chemical properties. Similarity within a group is most evident for the light metals, the lanthanide rare earths, the halogens, and the inert gases, but this chart, based on electron configuration of the atoms, is the basis for interpreting all chemical properties.

according to chemical properties. Most elements are metallic in character, and much geochemical investigation has been directed to these elements. Because the majority of elements are rather rare, their occurrence in minerals is governed largely by crystal chemistry and their ability to replace more abundant elements during crystallization.

In the isomorphic substitution of a *trace element* (q.v.) for a major element of similar properties in a mineral, the ease of substitution depends on (1) the relative abundances of the trace and major elements, (2) the relative inherent thermodynamic stabilities of the corresponding isomorphic trace and major element compounds, and (3) the exactness of "fit" of the trace element in the crystal site vacated by the major element. Factor (3) is clearly dependent on relative ionic sizes, but factor (2) is also affected by the relative sizes of the trace and major element ions. We can see this by examining solubilities of isomorphic compounds, for instance the sulfates of calcium, strontium, and barium which decrease in that order in their solubilities in water at room temperature, corresponding to a monotonic increase in cationic radius. Thermodynamic stability of crystals and solutions, for example as measured by solubility, is intimately related to the strength of chemical bonding and therefore to the closeness of approach of ions and molecules. Simple regularities which depend on ionic size can often be seen, but also very often trends are obscure owing to the several forces acting simultaneously in solids and liquids. Current geochemical research is directed to unraveling this picture and applying it to the trace element distributions in mineralogy and petrology.

Fig. 5 shows some trends in sizes of cations of periodic groups I and II. The alkali and alkaline earth metals are geochemically important in their 1+ and 2+ oxidation states, respectively. The groups K, Rb, Cs and Ca, Sr, Ba, which occur at the beginning of the transition metals, tend to cohere geochemically, and their reciprocal radii vary linearly with atomic number. Divalent lead, because of its similar ionic radius, is geochemically very similar to the alkaline earth elements.

Although the alkali and alkaline earth ions of the same charge increase in size with increasing Z, Fig. 6 shows that the first transition series of ions with the same number of electrons, from K^+ to Mn^{7+}, shrink in size with increasing nuclear charge. Also shown is the lanthanide series of rare earth elements which progressively fills an inner electron shell in such a way that the total atomic size is not increased as electrons are added; the radii of the 3+ ions steadily contract with increasing Z owing to the increasing nuclear charge.

Fig. 7 is a representation of the abundances of the elements in the earth's crust. The abundances, based on the average for many analyses of rocks of different types, have been normalized to the corresponding abundances in chondritic stone meteorites in the belief that the latter approximate the average over the entire volume of the earth for most elements. Thus, Fig. 7 gives a qualitative picture of the degree of enrichment or depletion of the elements in the crust compared to the interior of the earth, and the trends bear scrutiny in detail. Some of the prominent patterns are the following: *(1)* Most elements are more abundant in the earth's crust than in chondrites. *(2)* For middle and heavy weight metals, abundance usually increases with increasing Z and therefore with ionic radius. *(3)* The abundances of the rare earth elements also increase with increasing ionic radius, i.e., decreasing Z. *(4)* Li, Be, and B are strongly enriched in the crust. *(5)* S, Se, and Te are depleted in the crust and the depletion becomes more pronounced with increasing Z. *(6)* The group VIII metals, including Fe, the Pt metals, and Au, are depleted in the crust, and the depletion tends to become more pronounced with increasing Z. Fine detail in the patterns of Fig. 7 may or may not be significant depending on the quality of analytical data in specific cases and the validity of striking an average for the earth's crust or of regarding the meteorites as an average for the whole earth.

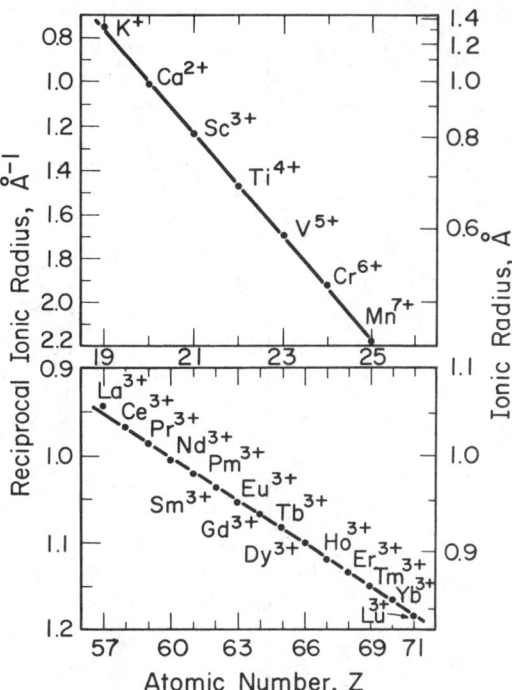

FIG. 6. Ionic radii for ions having the same number of electrons as $_{18}$Ar and for the trivalent lanthanides having the $_{54}$Xe configuration and additional inner 4f electrons.

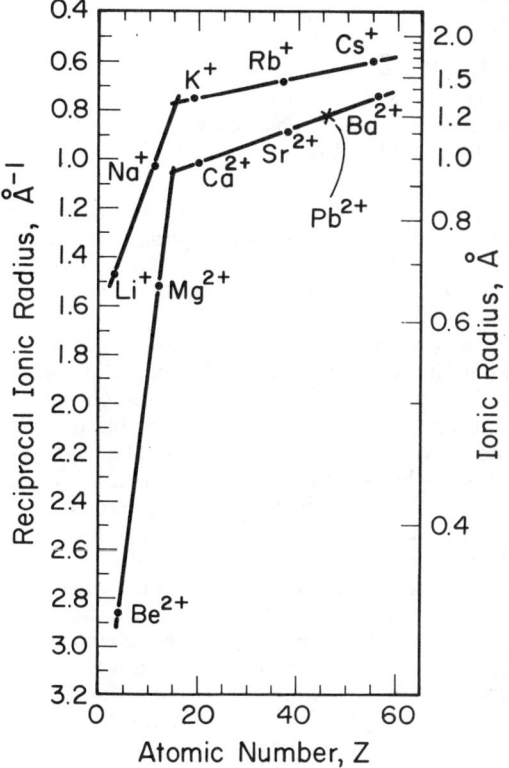

FIG. 5. Ionic radii of alkali and alkaline earth metals.

Exchangeable Reservoirs

The atmosphere, oceans, and upper layers of marine sediments, together with the inorganic and biological materials of the land areas of the earth, comprise a system where transfer of material from one region to another takes place rapidly enough to be observed by geochemists. Current research is directed toward specific chemical questions, and some of this research is described in other articles in this encyclopedia. The most interesting findings occur where detailed observations of nature can be interpreted in terms of known chemical properties, and often discoveries in nature stimulate further research in basic chemistry as well as vice versa.

The Oceans. The chemical analysis of *seawater* (q.v.) for all the elements in the periodic table is not yet complete. Nearly all elements have been detected in at least one sample of seawater, but, apart from ten "major" elements dissolved in seawater, only a dozen others have been studied by several investigators and in waters from many locations. Another twenty or so have been studied enough to establish a likely order of magnitude concentration in average seawater, but for the remaining half of the elements of the periodic table no gen-

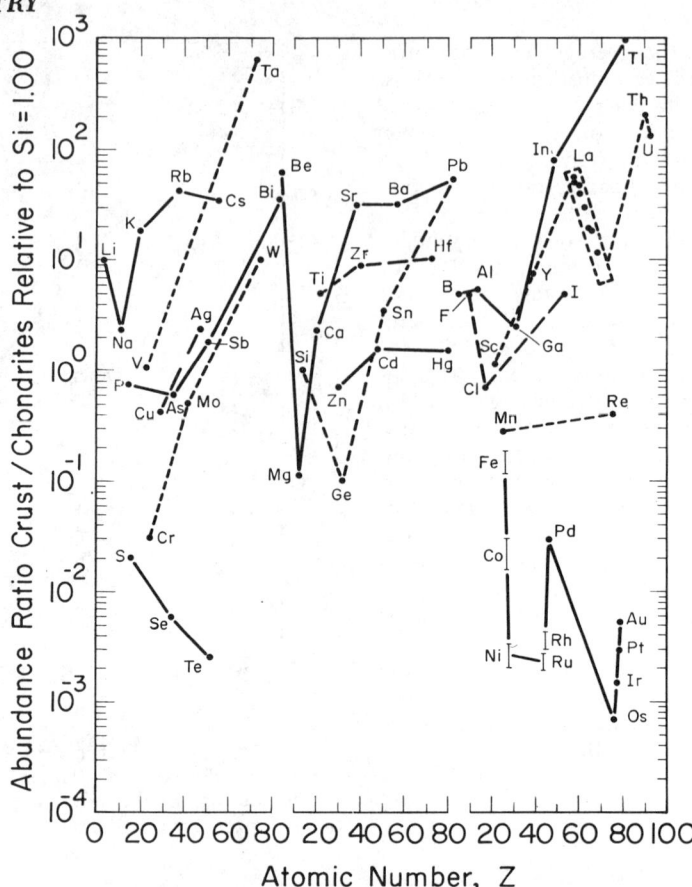

Fig. 7. Abundances of the elements in average earth's crust material, mostly after Mason (1958), normalized to abundances in chondritic stone meteorites and adjusted to Si = 100.

eral geochemical knowledge can be said to exist. In spite of the paucity of exact information on element concentrations in seawater, however, it can be said that the mean residence times of most elements in the oceans, between supply by weathering of rocks and removal by sedimentation, is short compared to geologic time. As a result, concentrations are at very low and roughly steady-state values, and analytical chemistry is accordingly difficult. A few elements have very long residence times and correspondingly high seawater concentrations. Sodium appears not to be removed efficiently by any sedimentation process, and its concentration is probably increasing gradually with the passage of time. Chlorine is so abundant in the sea that it could not have been supplied at the present weathering rate during the entire history of the earth, and the principal addition of chlorine to the sea must have occurred at some earlier time, perhaps by volcanic activity in the primitive earth. Dissolved organic compounds together with living material, which comprise on the order of 1 ppm of the seawater, are limited by the supply of the essential nutrients phosphorus and nitrogen, and research is now beginning to characterize these substances and their genetic relationships.

Some elements in the sea are controlled by solubility equilibria with sediments. Calcium and carbonate ions occur at saturation values, and many animals and plants are able to grow shells from seawater by precipitation of calcium carbonate as calcite or aragonite. Strontium replaces calcium in crystals, and its concentration in the sea is governed by calcium carbonate equilibria. Other elements may also be controlled by simple inorganic equilibria, but many are undoubtedly affected in the main by biological chemistry. Marine chemists aspire to clean-cut chemical descriptions of the processes determining the composition of the sea, but this goal has been realized so far only for a few specific cases.

The Atmosphere. The *atmosphere* (see Vol. II) is believed to be uniform in composition throughout the troposphere, stratosphere, and mesosphere, up to an altitude of 80 km, with 78.09% N_2, 20.95% O_2, 0.93% Ar, and 0.03% CO_2. The concentration of water vapor may be as high as 3% in saturated air at room temperature or as low as 0.01% or less in very cold or very dry areas. Vaporization and condensation of water vapor underlies the most obvious weather changes and is governed largely by adiabatic temperature increase and decrease due to falling and rising air masses. Meteorology is the detailed study of weather, placing special emphasis on thermodynamics. Geochemists have focused attention on the exchange of atmospheric gases with living materials and with the sea, and CO_2 is especially interesting because its rate of movement can be determined by the radioactivity of C^{14}. Solid particles in the atmosphere occur as stable suspensions (aerosols) in still air if their radii are less than about 10 microns, and this dust serves to nucleate raindrops and snowflakes. In recent years research on the chemistry of atmospheric aerosols has been directed particularly to the migration of radioactive debris from nuclear weapons tests, the behavior of industrial pollutants, and the relation between natural aerosols and precipitation.

The Sediments. Sedimentary geochemistry research has been largely inorganic but is now including considerable inquiry into organic chemistry and biochemistry occurring under natural conditions. Questions related to the formation of valuable *mineral deposits* (q.v.), *petroleum* (q.v. Vol. VI), and *coal* (q.v. Vol. VI) has led to many of the recent findings, and the present highly developed state of molecular biology has made timely the geochemical inquiry into the natural conditions leading to the origin and evolution of life (see Vol. VII).

Methods of Geochemical Investigation

In principle, all the methods of chemistry and of geology, oceanography, and meteorology are potentially valuable in geochemical investigation, and geochemists use to advantage every skill they possess.

Many important geochemical findings have been made recently through new methods of chemical analysis. Gas source and solid source *mass spectrometry* (q.v.) has made possible absolute age determination of rocks and minerals by the potassium-argon, rubidium-strontium, and uranium-thorium-lead isotope methods, and this study has led to the detailed investigation of the diffusion of these elements at elevated temperatures, a study which allows us to infer the temperature history of rocks and meteorites. The same techniques of mass spectrometry are used to study natural variations in isotopic abundances caused by chemical reactions in the earth. For instance, the oxygen-18 content of fossil carbonate shells is found to vary with temperature, and these variations have been used to trace the temperature changes in the seas during the ice ages.

Low-level radioactivity measurement of cosmic-ray-produced nuclides (especially carbon-14 and tritium) and of short-lived members of the uranium and thorium decay series has extended age determination to very young ages of biological and sedimentary materials and has given extensive information about the geochemistry of uranium and thorium and their ore deposits. The circulation of gaseous and particulate components in the atmosphere has been investigated taking advantage of cosmic-ray- and bomb-produced radioactivities as tracers, and the same is being done for several elements in the oceans.

Several new analytical methods for determining element concentrations with accuracy and sensitivity have successfully been applied to geochemical research. Rapid silicate analysis for the major elements now includes colorimetric and flame photometric procedures. Trace analysis by colorimetric and neutron-activation techniques is suitable for the low concentrations at which most elements occur. *Chromatography* (q.v.) is a major tool of the organic geochemist, and emission spectrography, x-ray fluorescence, and *electron probe microanalysis* (q.v.) have contributed significantly to petrology. In short, geochemists are alert to all new analytical techniques as they are developed.

Experimental methods to simulate natural processes under controlled laboratory conditions potentially include all of experimental chemistry. Minerals are synthesized in high-pressure apparatus, aqueous equilibria are studied potentiometrically, trace element behavior is studied using radiotracers, and biochemicals are produced in culture media, to mention a few examples. The role of the geochemist is to ask intelligent questions, that is, to look at nature with a hypothesis in mind for the cause of what is being observed and relate the observations quantitatively to known chemical properties. Often the important chemical properties are only poorly understood, and laboratory experimental work is a not inconsiderable part of the geochemical research effort today.

JOHN W. WINCHESTER

References

Ahrens, L. H., Press, F., and Runcorn, S. K., (editors), "Physics and Chemistry of the Earth," serial volumes, New York, Pergamon Press.
Aller, L. H., 1961, "The Abundance of the Elements," New York, Interscience, 283pp.

GEOCHEMISTRY

Fleischer, M., (editor), 1964, "Data of Geochemistry," Sixth edition, *U.S. Geol. Surv. Profess. Paper* **440** (chapters published separately, 1964 et seq.).
Mason, B., 1958, "Principles of Geochemistry," John Wiley & Sons, 310pp.
Miyake, Y., 1965, "Elements of Geochemistry," Tokyo Maruzen, 475pp.
Volborth, A., 1969, "Elemental Analysis in Geochemistry. Pt. A; Major Elements," New York, American Elsevier, 373pp.
Wedepohl, K. H. (editor), 1969, "Handbook of Geochemistry," Berlin, Springer-Verlag, Vol. 1, 442pp.; Vol. 2, pt. 1, 586pp.
See also *Geochimica et Cosmochimica Acta*, journal of the Geochemical Society.

Cross-references: *Biogeochemistry; Chromatography; Colorimetry; Crystal Chemistry; Elements: Planetary Abundances and Distribution; Emission Spectrography; Flame Spectroscopy; Geochemical Evolution of the Core, Mantle, and Crust; Geochemistry of Sedimentary Silica; Mass Spectrometry; Mineralogy; Organic Geochemistry; Potassium-Argon Dating; Radioactive Isotopes; Radionuclides; Rare Earths; Rubidium- Strontium Dating Method; Seawater, Chemistry; Silica Solubilities; Thermodynamics; Trace Elements; Uranium-Thorium-Lead Age Determination; X-Ray Spectroscopy.* Vol. II: *Aerosols; Atmosphere; Cosmochemistry; Cosmogony; Meteorites; Nucleogenesis.* Vol. IVB: *Electron Probe Microanalysis.* Vol. V: *Petrology.* Vol. VI: *Coal; Petroleum; Radioactive Isotope Tracer Technology.* Vol. VII: *Evolution.*

GEOCHEMISTRY OF THE EARTH—See EARTH'S CRUST GEOCHEMISTRY; GEOCHEMICAL EVOLUTION OF THE CORE, MANTLE, AND CRUST

GEOCHEMISTRY, IONIZATION POTENTIALS

The manner in which the elements have distributed themselves in our planet depends mainly on their respective abundances and on their chemical properties. Their chemical properties depend in turn on certain fundamental atomic properties; notably, state of oxidation (valence), size (radius), the magnitude and nature of the forces associated with the atom, and finally, sundry other properties such as polarizability, shape (orbital electron distribution), and whether or not the atom participates in crystal-field (ligand-field) splitting.

One of the most important of all these properties is size, the significance of which was first clearly demonstrated by V. M. Goldschmidt

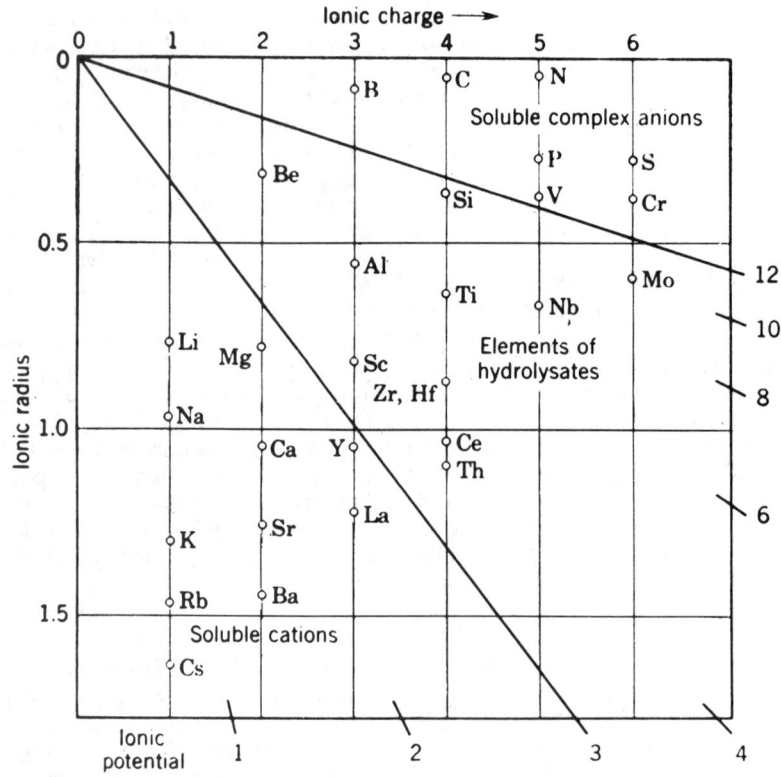

FIG. 1. Properties of chemical elements in relation to ionic potentials

and his co-workers in the late 1920s and the 1930s (Goldschmidt, 1937). Of the other properties, oxidation state is particularly important, but this in turn depends in part on the magnitude and nature of the forces operating within an atom, a general property which itself is of paramount significance for geochemistry. The importance of ionization potentials for geochemistry is in their bearing on the "magnitude and nature" of atomic and ionic forces. In fact, one may say that the role of the ionization potential is as important as that of size. We will consider some seven aspects of geochemistry for which the ionization potential appears to find a useful application. Some of these involve V. M. Goldschmidt's suggested geochemical classification of the elements according to their characteristic tendencies. In this respect we may recall the statement by Neumann (1949); "The primary principle of geochemical classification of the elements must obviously be their tendency to form ionic, covalent and metallic bonds," and again, "A method of measuring such a tendency would have a wide application in geochemistry and one of the first results would probably be the establishment of a sound theoretical basis for the classification of chemical elements as siderophile, chalcophile, lithophile, and atmophile, which classification is at present purely empirical."

The ionization potential is defined as the energy required to remove an electron from a neutral gaseous atom.

$A(gas) \rightarrow A^+ + e^-$ (1st ionization potential)

All ionization potentials are positive because work is done in pulling off an electron. This potential depends upon atomic charge, atomic size, the screening effect of the nucleus by inner electrons, and the closeness of the electronic orbital to the nucleus. The degree of penetration of the orbital is in the order $s > p > d > f$. The ionization potential of an s electron is greater than that of a p electron, because an s electron is closer to, and more influenced by, the nucleus.

Various sets of ionization potentials have appeared from time to time in the literature; those used here are from Kizer (see Ahrens, 1964, for details).

Siderophile Tendency

The typically siderophile elements include several metals which are sometimes described as noble because of their tendency to be chemically inert. Examples are gold (Au), platinum (Pt), palladium (Pd), and to a lesser extent, silver (Ag), copper (Cu), and nickel (Ni). Chemical inertness is controlled mainly by electronic structure and also the firmness of binding of an electron, as indicated by the ionization potential (Ahrens, 1964).

The relative inertness of the noble gases helium (He), neon (Ne), argon (Ar), krypton (Kr), xenon (Xe), and radon (Rn) is due mainly to the fact that the outermost group of electrons in the atoms of each of these elements is complete. In addition, we should bear in mind that their ionization potentials are conspicuously high, thereby indicating that their outermost electrons are tightly bound and not readily available for chemical binding purposes.

Let us return to the metals and compare in the first place the coinage metals (Group IB or 18-electron elements) with the alkali metals. Both the alkali metals and the coinage metals contain a single s electron outside completed electron groups, as indicated in Table 1, and the extremely high reactivity of the alkali metals is evidently due to the ease with which the s electron is removed, as indicated by the low magnitude of the first ionization potential (Table 1). This is due, presumably to efficient shielding (screening) of the nuclear charge by the electron structure of the alkali metals and comparatively feeble positive forces act therefore on the outermost valence s electrons. On the other hand, the s electrons outside the 18-electron group ($\ldots\ldots s^2p^6d^{10}$) in the coinage metals are much more tightly held, as indicated by their respective first ionization potentials (Table 1); compare particularly the pairs, K–

TABLE 1. ELECTRON ASSIGNMENTS (OUTERMOST ELECTRON ONLY) AND FIRST IONIZATION POTENTIALS OF THE GROUP 1 ELEMENTS

Element	Electron Assignment	First Ionization Potential (eV)	Element	Electron Assignment	First Ionization Potential (eV)
Li	$2s^1$	5.39			
Na	$3s^1$	5.14			
K	$4s^1$	4.34	Cu	$4s^1$	7.72
Rb	$5s^1$	4.13	Ag	$5s^1$	7.57
Cs	$6s^1$	3.89	Au	$6s^1$	9.22

Cu, Rb–Ag, and Cs–Au, where in each pair the principal quantum number of the outermost electron is the same. The greater firmness of binding in Cu, Ag, and particularly Au, is evidently due to poor shielding efficiency by the 18-electron group in these elements (Ahrens, 1964).

The striking inertness and strong siderophile tendency of Au is particularly noteworthy and is, in a sense, quite anomalous. Ionization potentials usually fall as we proceed down a vertical group in the Periodic Table (see for example, Li, Na, K, Rb, and Cs in Table 1), but in the group Cu–Ag–Au, there is a large increase of almost 2 eV when passing from Ag to Au, instead of a drop. This increase in ionization potential is apparently due to the effect of the entry of the $4f$ and/or $5d$ electrons (Ahrens, 1964), and were it not for this "anomaly," Au would not have been so siderophile and chemically inert. It probably would have existed in a combined state in geological environments, having a chemistry and geochemistry somewhere between that of Ag and Tl.

When considering the Group II elements, ionization potential data are again helpful with regard to chemical inertness and siderophile tendency, though perhaps the various relationships are not quite so well developed.

If the $3d$ transition metals are considered, chemical inertness and siderophile tendency may follow ionization potential trends quite closely. Thus in the sequence, Mn (d^5s^2), Fe (d^6s^2), Co (d^7s^2), and Ni (d^8s^2), siderophile tendency increases from Mn to Ni; that is, with increase in firmness of binding, as indicated by the respective ionization potentials. The strongly siderophile elements, Pt and Pd, have comparatively high ionization potentials.

Though the magnitude of the ionization potential often provides us wth a useful measured guide as to why an element has siderophile tendencies, some exceptions may be found (Ahrens, 1964).

Cationic Forces

The geochemical significance of cationic forces is discussed under some of the other headings; here we will concern ourselves essentially with the problem of how such forces may be indicated, with particular reference to the use of the ionization potential.

Several terms which are more or less synonymous have been suggested for cationic forces; for example, *polarizing power, effective force of attraction of a cation, binding power of a cation, anion affinity, effective nuclear charge,* and others. Polarizing power is perhaps the most popular of these terms.

It has been recognized for a long time that polarizing power depends primarily on cationic charge (formal ionic charge) and radius. As an indication of the magnitude of the cationic force Goldschmidt suggested the function Z/r^2; Cartledge suggested a similar quantity Z/r, referred to as the ionic potential, for a related purpose; Dietzel has also discussed this property. A somewhat more complex function involving charge and radius is the so-called EK value suggested by Fersman in his studies on the geochemical distribution of the elements. The EK value is defined as,

$$\mathrm{EK} = k\,\frac{Z^2}{2r}$$

where Z = charge and r = radius. Szadezky-Kardoss employed a function (the Verbindungspotential) which embraced radius and charge. This function refers to all constituent ions in a crystal and was suggested as a useful fundamental property; it does not, however, refer to a specific ion as such and will not be considered further here. (For details of references to the works of Goldschmidt, Cartledge, Dietzel, Fersman, and Szadezky-Kardoss, see Ahrens, 1964).

Functions involving the ratio of charge to radius work reasonably well as a measure of cationic force provided only ions with similar electronic structures are considered; thus, for example, Ramberg (see Ahrens, 1964) has pointed out that the ionic potential is a useful measure of cation field strength for the noble gas type cations. (Note: *Ionic potential* should not be confused with *ionization potential*. Ionic potential is simply the ratio of ionic charge to radius.) If, however, different electronic structure types are involved, serious anomalies may arise (Ahrens, 1964; Goldschmidt 1954). Thus for example, the ionic potentials of each component in the pairs Na^+–Cu^+; K^+–$Ag(Au^+)$; Mg^{2+}–Zn^{2+}; Ca^{2+}–Cd^{2+}, and Sr^{2+}–Hg^{2+} are similar, whereas, in fact, the polarizing power of the 18-electron cation (the second in each pair) is far greater than that of the noble gas ion. In order to explain this anomaly, a third property must be considered, namely that of variable screening (shielding): different electron arrangements screen the positive charge on the nucleus with different efficiency and accordingly influence the resultant forces acting on the periphery of the atom or ion and hence the polarizing power of a cation.

Slater, Pauling, and Coulson among others (see references in Ahrens, 1964), have discussed variable screening by different electrons (s, p, d, and f) and have assigned arbitrary and fixed values. In his definition of *screening efficiency* Ahrens (1954) took ionic radius, charge, and ionization potential into consideration.

According to Lakatos, Bohus, and Medgyesi (see Ahrens, 1964) the d^5 half subshell and the complete d^{10} subshell screen the positive charge on the nucleus more effectively than do the incomplete configurations d^4, d^6, and d^9. These workers refer to particularly weak screening by $4d$ $4f$, and $5d$ electrons; in this respect see also Ahrens (1954).

Consider now quantities which have been suggested for indicating the magnitude of cationic forces, but which unlike those noted above, make allowance in one way or another for variabe screening.

Several workers (Pauling, Finkelnburg and Lakatos, Bohus and Medgyesi; see references in Ahrens, 1964) have referred to the quantity $Z - S$ which they define as the effective nuclear charge or effective charge number ($Z =$ total nuclear charge; S is screening) and which is usually designated Z_{eff}. The quantity $Z - S$ is, however, not itself usually used but rather some other function which is supposed to be equal to $Z - S$ or related to it. Finkelnburg suggested nI_1/R where n is the principal quantum number, I_1 the first ionization potential, and R the Rydberg constant. This quantity does not, however, appear to be particularly useful and has been criticized (Ahrens, 1964).

Lakatos, Bohus, and Medgyensi (see reference in Ahrens, 1964) refer to the effective field strength (F^*) defined as

$$F^* = \frac{Z^*}{r}$$

where r is ionic radius and Z^* the effective nuclear charge, $Z - S$. According to Lakatos et al., S and hence Z^* may be obtained from ionization potential data.

The quantity F^* is closely related to $I_{n/r}$, the cationic field function (F) of Ahrens (1964), and I_{n/r^2}, a quantity which Goldschmidt (1954) has suggested in place of his earlier suggested Z/r^2 (see above) as a measure of polarizing power. (I_n is the nth ionization potential; that is, first for singly charged cations, second for doubly charged cations, and so on). As a measure of the magnitude of cationic forces, Ahrens (1964) has suggested the nth ionization potential as such. For the purpose of the present review the nth ionization potential will be used as a measure of the polarizing power of the cation.

An indication of the sensitivity dependence of I_n on radius and charge may be obtained by inspecting the data in Table 2.

In (a) atoms of a similar kind are considered and it may be seen that for an approximately twofold increase in radius (Mg^{2+}, 0.65Å to Ba^{2+}, 1.34Å) the second ionization potential drops by one-third (or increases by one-half); for details about the radius-ionization potential-charge relationships see Ahrens (1956). It is clear from Table 2 (b) that the magnitude of I_n is very sensitively dependent on charge and increases by more than a factor of two if charge is doubled. (Y and Th are transition-type ions and should for strict comparison not be included in Table 2. Because, however, non-transition type ions of high charge and having a radius magnitude of 1Å do not exist and as the shielding efficiency in Y and Th appears to be quite high and similar to that obtaining in 8-electron atoms, Y and Th serve reasonably well for our purpose.)

Chalcophile and Lithophile Tendencies

The metal-sulfur bond in a large number of sulfide minerals is usually classed as essentially covalent whereas the bonds between oxygen and several of the common elements (Na, K, Ca, Mg, Fe, Al, and Si) in silicate and other oxy-minerals are usually considered as ionic. Lack of miscibility between sulfides and silicates and oxides is evidently due largely to the big difference in bond character (Ahrens 1964, for example).

Bond difference apparently stems from differences in the properties of the atoms of the elements in question, sulfur and oxygen, or rather their respective anions, S^{2-} and O^{2-}. S^{2-}, the larger of the two, is more easily polarized and ease of polarization is probably the reason that for a given metal, the metal-sulfur bond is invariably more covalent than the metal-oxygen

TABLE 2. SOME SELECTED IONIZATION POTENTIALS

Cation	r(Å)	I_2(eV)	Cation	r(Å)	Charge	I_n(eV)
	(a)			(b)		
Be^{2+}	0.35	18.2	Na^+	0.97	1	(I_1) 5.14
Mg^{2+}	0.65	15.03	Ca^{2+}	1.01	2	(I_2) 11.87
Ca^{2+}	1.01	11.87	Y^{3+}	0.92	3	(I_3) 20.5
Sr^{2+}	1.18	11.03	Th^{4+}	1.02	4	(I_4) 29.38
Ba^{2+}	1.34	10.00				

(a) Charge constant. (b) Radius approximately constant.

Table 3. Ionization Potentials of Some Cations

	\multicolumn{7}{c}{Singly charged cations}						
	Rb^+	K^+	Na^+	Tl^+	Ag^+	Cu^+	Au^+
Radius (Å)	1.45	1.33	0.97	1.45	1.27	0.97	1.37
Ionization potential (eV)	4.13	4.34	5.14	6.1	7.57	7.72	9.22

	\multicolumn{9}{c}{Doubly charged cations of medium size[a]}								
	Mg^{2+}	Mn^{2+}	Fe^{2+}	Co^{2+}	Zn^{2+}	Ni^{2+}	Pt^{2+}	Pd^{2+}	Cu^{2+}
Radius (Å)	0.65	0.80	0.74	0.72	0.69	0.69	(0.80)	(0.80)	(0.80)
Ionization potential (eV)	15.03	15.64	16.18	17.05	17.96	18.15	18.56	19.43	20.29

[a] Size is 0.65–0.90 Å; see discussion by Ahrens (1964).

bond. We will arrange cations (given charge and radius) in order of increasing ionization potential in order to ascertain whether threshold values exist which if exceeded would give rise to sulfide bonds of such covalency that the metals may be accepted into the sulfide minerals (Ahrens 1964). Singly charged cations and doubly charged ions, each arranged in order of increasing ionization potential, are listed in Table 3.

In the univalent elements, thallium sits neatly on the fence (hence the dashed line) as this element has geochemical affinities both with the heavy alkali metals and with some of the coinage metals. It seems that provided the first ionization potential of a univalent element is greater than about 6 eV, the degree of covalency of the bond formed with sulfur is sufficiently high to enable the metal to be accepted into the structures of the covalent sulfide minerals.

A similar situation seems to hold for the medium-sized divalent cations. Manganese (Mn) is located more or less at the threshold value ($\sim$ 15.6 eV). This element does not usually form sulfide minerals, but traces of manganese are found in many sulfides and the Mn–S bond evidently has sufficient covalency for Mn to be accepted into the structure of various sulfide minerals. For both univalent and divalent metals, those located beyond the respective threshold values may form sulfide minerals.

The above approach to chalcophile-lithophile tendencies based on bond type and ionization potential, has been discussed in some detail by Ahrens (1964). Electronegativities have been used for a like purpose: see for example Ahrens (1964).

Cationic Radii

All published sets of ionic radii are based at least in part on internuclear distance measurements in crystals which are classed as ionic. Some sets of radii are based entirely on internuclear distance measurements. In these sets, the assumption is made that provided the radii of certain anions (O^{2-} and F^- for example) are known, subtraction of their respective radii from the total internuclear distance gives the radius of the cation. It is assumed further that the radii of the anions remain virtually unchanged in various crystals, and herein lies a possible cause of error because in the presence of an intense field, anions may deform and effectively shrink in size. Radii proposed by Pauling (1962) do not suffer from this difficulty as they depend on relationships between screening and radius in isoelectric sequences. That is not to say, however, that such radii are necessarily superior. Ahrens (1952) has compared different sets of radii and finds that smooth relationships exist between Pauling-type radii and ionization potential in isoelectronic sequences, whereas radii based entirely on measurement tend to show irregularities. As one would expect to find a smooth relationship between ionization potential and the radii of isoelectronic ions, it has been concluded (Ahrens, 1952) that Pauling-type radii are usually superior to those based *entirely* on internuclear distance measurements. Ahrens (1952) has prepared a list of radii based in part on such smooth relationships and on internuclear distance measurements.

Structure

Ionic structures are normally those which are said to comply with the so-called ionic radius ratio rules. The structure as such is determined essentially by the manner in which the cations and anions can be arranged into coordination groups according to the different cationic and anionic radii. The situation with regard to covalent crystals is more complex. It is quite a common custom to explain such structures in terms of the orbital hybridization which the electrons of the atoms of some elements are supposed to undergo. There is the problem, however, of whether orbital-hybridization represents a form of "descriptive scientific quackery," far removed from physical reality. Orbital hybridization is connected with a theory of atomic bonding known as the *valence bond theory*. In the process of hybridization there is a rearrangement of the energy levels of the outer electron shell so that new equivalent bonding orbitals are formed. These orbitals are geometrically equivalent and are directed in space as octahedra (d^2sp^3), tetrahedra (sp^3), or square planar (dsp^2). The theory has been criticized because it does not explain the spectra of complex ions, and because of an arbitrary division of bonding types into ionic and covalent extremes.

An alternative approach, based partly on the magnitude of cationic forces (polarizing power) and anionic polarizability has been discussed by various workers, in particular Ahrens (1964) and Ahrens and Morris (see Ahrens, 1964), who utilized the ionization potential as a measure of polarizing power. These workers paid special attention to the structures of the halides, oxides, and sulfides of a large number of metals, and found both continuous and discontinuous changes between structure and ionization potential. In a sense, the approach is akin to that proposed by Bernal and Megaw (see Ahrens and Morris in Ahrens, 1964) in their study of the hydroxyl bond. It may be recalled that in this approach, OH^- is supposed to undergo various changes in shape (charge distribution) as it comes under greater and greater cationic fields and it is supposed that the different structures of the hydroxyl compounds arise as a direct consequence of such changes. Similar ideas to those propounded above are also evident in some of the writings of V. M. Goldschmidt, in particular those on pp. 64–106 of his book (Goldschmidt, 1954).

Covalency and Geochemical Coherence in Silicates

Several pairs or larger groups of elements are geochemically associated; for example, K–Rb–Tl, Zr–Hf, Al–Ga, Si–Ge, and the rare earth elements. In each such pair or larger group, cationic charge is the same and radius similar. However, not all pairs of elements with cations of like charge and similar radius are geochemically associated. Examples are given in Table 4, which lists pairs of ions together with their respective ionic radii and ionization potentials, first for singly charged cations and second for doubly charged cations.

It has been suggested (see Ahrens, 1964, who refers to various workers) that lack of coherence in the common silicate rocks and

TABLE 4. GROUPS OF IONS OF LIKE CHARGE AND SIMILAR RADIUS WHICH ARE NOT CLOSELY ASSOCIATED

Ions	Ionic Radii (Å)	Ionization Potentials (eV)
Na^+	0.97	5.14
Cu^+	0.97	7.72
K^+	1.33	4.34
Ag^+	1.27	7.57
Au^+	1.37	9.22
Mg^{2+a}	0.65	15.03
Zn^{2+a}	0.69	17.96
Ca^{2+}	1.01	11.87
Cd^{2+}	0.97	16.9
Sr^{2+}	1.18	11.03
Hg^{2+}	1.10	18.7

^{a} These two elements may show moderate coherence in some silicates.

minerals is due to the fact that bonding to oxygen by one of the components in a group may be distinctly covalent; consequently ionic substitution to a significant degree does not take place.

Ionization potentials have been successfully used as a guide to whether or not significant covalency is likely to be present (Ahrens, 1964); the higher the ionization potential is, the greater is the degree of covalency. Thus, the Au–O and Hg–O bonds would be particularly covalent. It will be realized that bonding differences may vary very considerably, all the way from such extremes as gold (Au) and potassium (K) and also mercury (Hg) and strontium (Sr) to the very slight differences which exist between the bonds formed by aluminum (Al) and gallium (Ga). For the latter pair only detailed study can show whether in fact the slightly greater degree of covalency in the Ga–O bond, as compared with that of Al–O, is geochemically significant. (The third ionization potential of Ga is a little greater than that of Al.) These problems have been studied in detail by Ahrens (1964).

Electronegativities have been used by Ringwood (see reference in Ahrens, 1964) in place of ionization potentials as a measure of the covalency in silicate minerals.

Metal Binding by Organic Molecules in Geological Environments

The uptake of metals in geological environments by naturally occurring organic molecules may significantly influence the fate of metal distribution in the sedimentary cycle. The general problem of metal biogeochemistry is complex and here only one aspect will be considered, i.e., the problem as to why in the first place, only certain elements form metal-organic compounds, and secondly and more particularly, why the stabilities of such compounds vary so considerably from metal to metal.

The formation and stability of meta-organic compounds and complexes depends on many factors, notably, properties of the metal and of the organic ligand, and the physicochemical conditions (pH and eH, for example) of the system under consideration. Our concern here will be with properties of the metal, in particular the importance of the ionization potential. Before considering the importance of this property we shall briefly indicate the importance of two other properties, charge and size, in order to illustrate the need to maintain these properties constant when we study the significance of the ionization potential.

The effect of variation of charge and of radius is clearly illustrated by data in Table 5. In the first case (variation of charge), the well known reagent EDTA is the ligand and in the second (variation of ionic radius) the oxalic acid ion is the ligand. In both cases, the effect of variation of the property concerned is quite clear. As a generalization, it may be stated that for many, but not all organic ligands, stability tends to increase both with increase in charge and decrease in radius. The ligands considered so far do not have geological significance and have been selected merely for the sake of illustration.

We come now to consider the possible significance of the ionization potential for influencing the stability of metal-organic chelate complexes. For this purpose, some amino acids have been chosen as ligands. These acids readily form metal-chelate complexes and have been found in various sediments; moreover, they represent the degradation products of protein, a principal component of the living cell of both the animal and the vegetable kingdom. Because of the dependence of stability on both charge and radius, only cations of given charge (two) and radius magnitude (0.65–0.80Å) have been chosen.

TABLE 6. LOG STABILITY CONSTANTS OF M^{II}-AMINO ACID COMPLEXES OF MEDIUM SIZED CATIONS WITH GLYCINE AND α-ALANINE

M^{2+} Cation	Second Ionization Potential	Log Stability Constant	
		Glycine	α-Alanine
Mg^{2+}	15.03	3.44	1.96
Mn^{2+}	15.64	5.5	6.05
Fe^{2+}	16.18	7.95	7.0
Co^{2+}	17.05	8.91	8.8
Zn^{2+}	17.96	9.3	9.5
Ni^{2+}	18.15	11.12	10.7
Cu^{2+}	20.29	15.38	14.8

These cations are listed in Table 6, in order of increasing second ionization potential, together with the log stability constants for two of the amino acids.

It is quite clear that in general, stability tends to increase with ionization potential. This generalization, with minor exceptions, holds for many organic ligands. The general inference is that as the cationic force (binding power) increases, the organic ligand becomes more tightly bound by the metal. Exceptional behavior can in part be explained by so-called ligand-field splitting. [The ligand-field theory is another theory of bonding in coordinate compounds of transition elements. The ligands (polar molecules or anions) are assumed to create an electrostatic field (also called *ligand field* or *crystal field*) in which the five normally equivalent d electronic energy levels are split. The amount of splitting depends on the central cation, the strength of the applied field, and the geometrical arrangement of the ligands about the metal atom.] A fuller account of the importance of the ionization potential in biogeochemistry is to be published later.

L. H. AHRENS

TABLE 5. STABILITY CONSTANTS

Cation (and Radius)	Log Stability Constant
EDTA complexes of R^+, R^{2+}, R^{3+}, and R^{4+} cations (radius = 1.0 Å)	
Na^+	1.7
Ca^{2+}	10.7
Re^{3+}	15.5–19.8 (av 17.5)
Th^{4+}	23.2
M^{IIa} oxalate complexes	
Be^{2+} (0.35 Å)	~4.0
Mg^{2+} (0.65 Å)	3.43
Ca^{2+} (1.01 Å)	3.00
Sr^{2+} (1.15 Å)	2.54
Ba^{2+} (1.34 Å)	2.33

References

Ahrens, L. H., 1952, "The use of ionization potentials," Part 1, "Ionic radii of the elements," *Geochim. Cosmochim. Acta*, **2**, 155.

Ahrens, L. H., 1954, "Shielding efficiency of cations," *Nature*, **174**, 644.

Ahrens, L. H., 1956, "Some ionization potential variations and relationships," *J. Inorg. Nucl. Chem.*, **2**, 290.

*Ahrens, L. H., 1964, "The significance of the chemical bond for controlling the geochemical distribution of the elements," Part 1, *Phys. Chem. Earth*, **5**, 1.

Goldschmidt, V. M., 1937, "The principles of distribution of chemical elements in minerals and rocks," *J. Chem. Soc.*, p. 655.

Goldschmidt, V. M., 1954, in (A. Muir, editor) "Geochemistry," London, Oxford University Press.

Neumann, H., 1949, "Notes on the mineralogy and geochemistry of zinc," *Mineral Mag.,* **82,** 575.

Pauling, L., 1962, "The Nature of the Chemical Bond," Third ed., Ithaca, N.Y., Cornell University Press.

* Additional bibliographic references may be found in this work.

Cross-references: *Alkalis, Alkali Metals, and Alkaline Earth Metals; Atomic Number and Periodic Table; Biogeochemistry; Bonding; Crystal-Field Theory; Electronegativity; Elements: Planetary Abundances and Distribution; Geochemical Classification of Elements; Oxygen; pH-Eh; Rare Earths; Rare Gases; Sulfur; Versene (EDTA) Solution Studies.*

GEOCHEMISTRY OF PELAGIC SEDIMENTS, ANTARCTIC—See Vol. I, ANTARCTIC PELAGIC SEDIMENTS

GEOCHEMISTRY OF PRECAMBRIAN ATMOSPHERE AND OCEAN—See PRECAMBRIAN ATMOSPHERE, GEOCHEMISTRY

GEOCHEMISTRY OF SEDIMENTS: ANCIENT

The interpretation of geochemical data of sediments is complicated by the fact that a great variety of both organic and inorganic constituents may coexist which, chemically speaking, are not formed in equilibrium with each other. Actually, the various mineral and organic fractions can enter a sediment at different geologic times and under different environmental circumstances. Consequently, the coexisting compounds in a sediment are mostly cogenetic only in a petrographic but not a physical-chemical sense. In contrast, the igneous minerals closely follow the Bowen's reaction series of high-temperature petrogenesis.

In order to give some information on the kind of constituents that are present in sediments of all geologic periods, the leading mineral and organic end-members are listed in Fig. 1. The data are arranged on the basis of genetic type and chemical and mineralogical nature of the individual compounds. In principle, one can distinguish between four major types of end-members: (*1*) weathering residues of igneous and metamorphic rocks; (*2*) minerals formed at low temperatures in the presence of water; (*3*) mobile phases; and (*4*) organic constituents.

Single species of each of these four groups can occur alone or in association with each other. However, monomineralic sediments such as pure limestones or quartz-sandstones are exceedingly rare. For the most part, sediments represent complex mixtures of a great number of compounds which, from a physical-chemical point of view, are unrelated. Consequently, the composition of a single mineral or organic fraction generally has more geochemical meaning than the sum total chemistry of the whole sediment.

The amount of geochemical information one

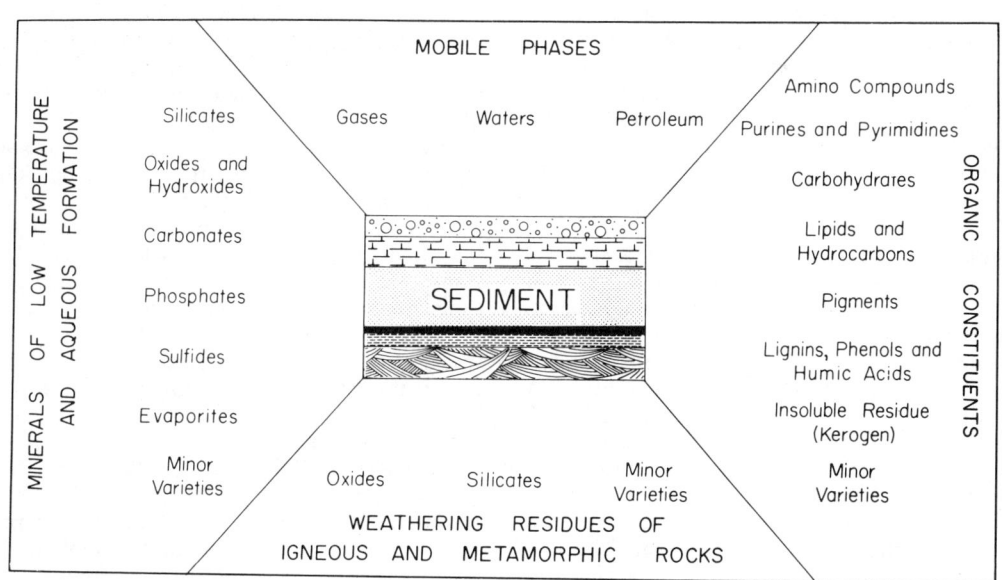

FIG. 1. Major organic and inorganic end members of sediments.

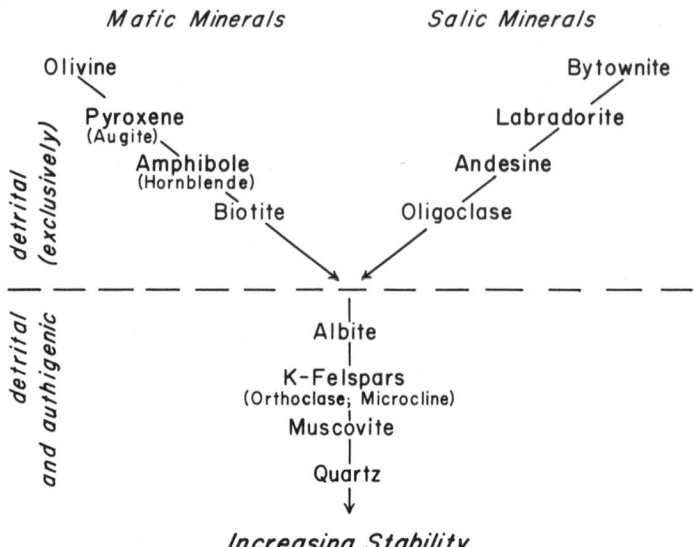

FIG. 2. Weathering stability series of primary rock-forming minerals.

may draw from ancient sediments is so enormous that only a few basic concepts and a small but representative percentage of the published literature could be included within the limits of this article. For more details on various aspects of the geochemistry of sediments, one may consult Braitsch (1962), Breger (1963), Clarke (1924), Degens (1965), Garrels (1960), and Mason (1958).

The information is presented as following the pattern outlined in Fig. 1 and starting with the weathering residues of crystalline rocks.

Weathering Residues of Igneous and Metamorphic Rocks

During weathering and subsequent denudation, the individual igneous and metamorphic minerals respond differently to chemical, biochemical, and physical attack. As a result, some mineral species will be eliminated rather radically and in short periods of time, whereas others are little or not at all affected.

The relative stability of the more common high-temperature minerals under conditions of weathering is illustrated in Fig 2. Interestingly enough, the sequence in Fig. 2 is identical to the reaction series of high-temperature petrogenesis. Thus, the least stable mineral in low-temperature environments at or near the earth surface is olivine and the most stable one is quartz. It is noteworthy that only the four most stable minerals of the stability series may also occur as authigenic sedimentary mineral species.

The presence, the absence, or the ratio of primary rock-forming minerals in a sediment may be used in a number of ways for the interpretation of syngenetic and diagenetic processes. Although morphological, paragenetic, and occasionally chemical relationships among the coexisting mineral constituents can best be recognized and evaluated by conventional petrographic tools such as thin-section studies, there are also geochemical techniques available which can supplement petrographic observations. The weathering residue of igneous rocks, either in the form of isolated minerals or of rock fragments, can be dated, for instance, by the potassium-argon method, as seen in Fig. 3. It becomes apparent that the fluvioglacial sediments which have been transported over hundreds of miles correspond timewise with their original crystalline source in southern Sweden and Finland. Similarly, the age of the Recent Black Sea deposits is identical to the age of their apparent crystalline sources in Transcaucasus (50 million years), Northern Caucasus and Russian platform (500 million years), Ukranian Shield (1020 million years), and the Danube River Basin (300 million years). The data further indicate that there is no preferential loss in argon over potassium during weathering and transportation, even in the case where the

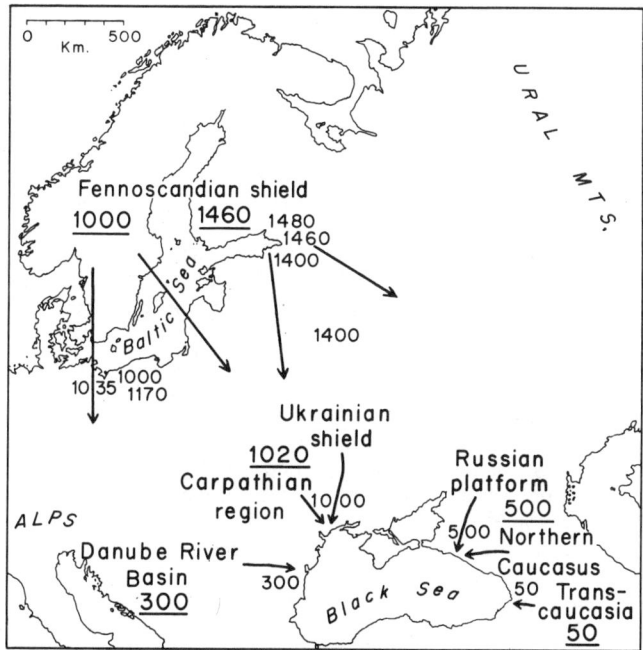

Fig. 3. Potassium-argon ages of detrital rock constituents in some sediments of northern Europe and the Black Sea area in relation to the ages of their source rocks (after Krylov, 1961; for reference, see Degens, 1965).

mineral particles are of subcolloidal size. Thus, the age of the primary detrital constituents in ancient sediments should reflect the age of the crystalline source material. The same applies to clay minerals which actually inherit the potassium-argon age of their crystalline precursors. This then may allow studies on source aspects, paleogeography, paleotectonics, and even on the mechanism of clay mineral formation. Investigations of this type may well be extended to detrital minerals that do not contain potassium, such as zircon, or may involve other methods of absolute age dating.

Other means of investigating the crystalline detritus of ancient sediments include stable isotope studies, e.g., O^{18}/O^{16} ratios in coexisting mineral pairs, and electron-probe microanalysis. The last technique, for example, allows the chemical analysis of individual mineral grains. Thus, source aspects and a great variety of other geologic problems may be tackled by utilizing the isotope distribution and chemistry of the igneous and metamorphic mineral residues in a sedimentary rock.

Minerals Formed at Low Temperatures in the Presence of Water

The bulk of sedimentary rocks is composed of materials that either have been formed during weathering from a crystalline precursor or that have originated in syngenetic or diagenetic environments from compounds dissolved in water. Among these low-temperature minerals, clays are by far the most important ones, quantitatively; they actually represent nearly 50% of the total sedimentary matter (on the dry weight basis). Oxides and carbonates comprise about 10–15% each, while the remaining portion of the low-temperature minerals, i.e., phosphates, sulfides, evaporites, and authigenic silicates other than clays amount to less than 5% of the total. The water content in ancient sediments can be as low as 1% and as high as 50–60%.

1. Silicates. *A. Clay Minerals.* The most obvious way to use clay minerals as a geologic tool is to determine quantitatively the distribution of the various clay minerals present in a sediment. More sophisticated are those investigations that consider in addition the slight variations in crystallographic properties of the various clay minerals involved. Another line of endeavor is the chemical analysis of clay minerals for both their major and minor elements.

Concerning the possible application of clay mineral analysis for environmental evaluations, i.e., in terms of fresh water versus marine, much controversy is established. Whereas some investigators favor the concept that clay minerals are more "at home" in certain environments than in other environments, others conclude that the great majority of clay minerals in sedimentary rocks are not significantly affected both structurally and chemically (bulk composition) by the environment of deposition; namely, clay minerals are formed during continental weathering and consequently are largely detrital in origin.

Considering the distribution of clay minerals with geologic time and their abundance ratio in different environments, there appears to be a greater probability for the accuracy of the second hypothesis, according to which environment has only little to do with the mineralogical transformation of clay minerals. This feature, however, does not necessarily mean that the average clay minerals cannot partially adjust geochemically to the environment of deposition by exchanging some of their elements with those present in the surrounding aqueous environments. It should be emphasized that this adjustment is mainly on a small scale and on an ion-exchange level, unaccompanied by marked structural and chemical changes in the detrital clay composition.

The possible application of clay petrology for geological problems using the relative proportions of coexisting clay minerals in ancient sedimentary rocks is well illustrated in Fig. 4. The clay mineral suite of late Paleozoic sediments from the Oklahoma-Arkansas area changes gradually or abruptly with geologic time or by moving from one sedimentary basin to the other. This relationship in time and space may help considerably in the evaluation of source aspects, paleogeography, and regional tectonics at the time of sedimentation.

In case clay minerals are participating in exchange reactions during their exposure to the environment of deposition, certain geochemical differences on the trace element level are expected to occur if the potential trace elements are more abundant in one than in another environment. Among trace elements suitable as environmental criteria to distinguish between marine and fresh-water sediments, boron has been studied most intensively and successfully so far. This is so, first, because boron is highly abundant in the sea and virtually absent from most fresh waters and, second, boron can be fixed rather tightly and effectively to certain clay minerals. Thus, clays deposited under marine conditions are generally higher in boron by a factor of about 2–10 than the clays formed or deposited in the adjacent continental source area. The usefulness of boron and other trace elements such as lithium, rubidium, vanadium, and others to distinguish between fresh-water and marine sediments has been shown in clay mineral studies in many ancient sedimentary basins all over the world.

An interesting application of clay minerals to geology is their use for the absolute age dating of rocks. Until recently, estimates of the duration of geologic events in the earth's development were made exclusively by determining the age of minerals of intrusive rocks that intersected sediments of known biostratigraphic age. In order to circumvent the uncertanties that necessarily are introduced by

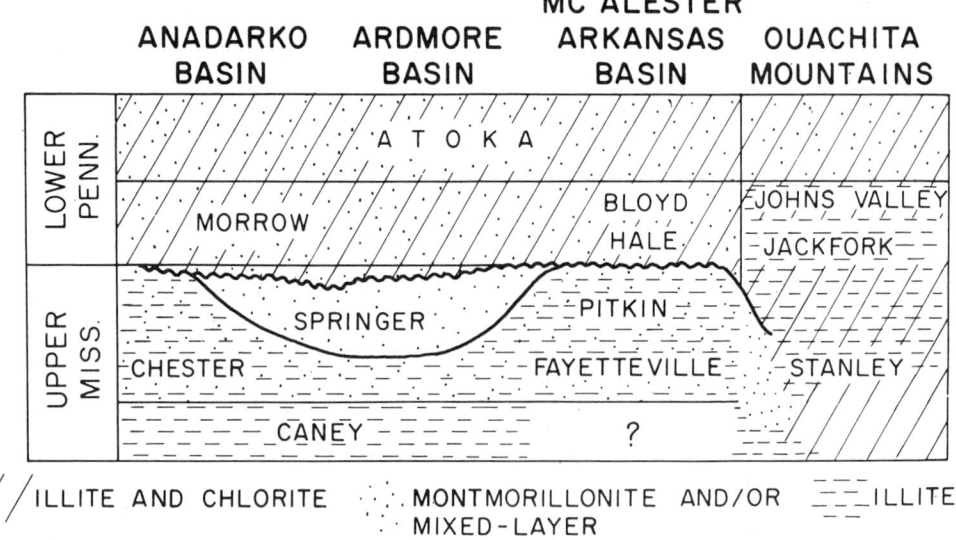

Fig. 4. Clay mineral suite of late Paleozoic sediments from the Oklahoma-Arkansas area (after Weaver, 1958; for reference, see Degens, 1965).

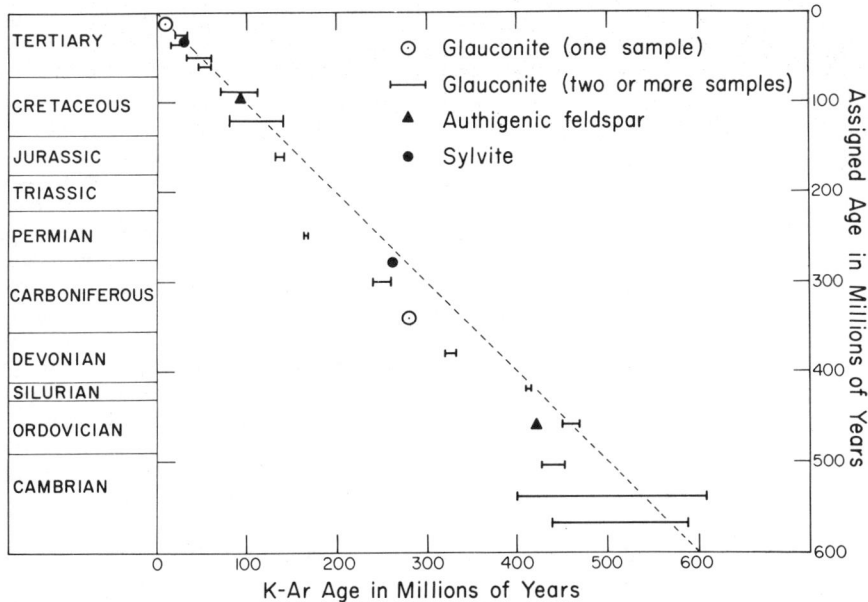

FIG. 5. Relation of potassium-argon age analyses on glauconites, authigenic feldspars, and sylvite to assigned age in millions of years.

such a deductive approach, geochemists looked for authigenic sedimentary minerals which do contain radioactive isotopes and their stable daughter products in measurable quantities. The possibility of such an absolute age determination became feasible with the development of the potassium-argon method through the utilization of glauconites and other potassium-bearing minerals such as sylvite and authigenic feldspars. In particular the worldwide occurrence of glauconites in marine sediments of all ages, and the apparent synchronisms of their formation with that of marine sediments, opened to the potassium-argon method a real possibility for dating sedimentary formations throughout geologic time. A graphical representation of potassium-argon age analyses on glauconites and a few other minerals from the Cambrian and up is given in Fig. 5. Precambrian glauconites have also been dated but, unfortunately, the apparent stratigraphic position is less accurately known. Despite some uncertainties that have been expressed concerning the reliability of glauconite dating, the potassium-argon ages published so far agree surprisingly well with the ages obtained by other methods.

Recently, illites and certain mixed layer clays have also been dated by potassium-argon measurements. The dating of illites, however, faces so many pitfalls that it is to be avoided unless the investigator is well aware of all the problems and uses the method on a basis of analogy only and with limited objectives. Recent illites from the Atlantic, for example, yield ages ranging from 170–470 million years, indicating a detrital origin of the clay minerals under consideration. Age dating of illites, however, may reveal interesting details on the mechanism of clay mineral formation, if ages of the parent materials are available for comparison. From the published work on this subject it appears that clay minerals reflect the age of their crystalline precursor, which means that there is no preferential loss of potassium over argon or vice versa during the alteration of high-temperature minerals in the direction of clays.

Another challenging area of clay mineral research lies in the field of "organo clays." Organo clays are reaction products of specific minerals with specific organic molecules involving the formation of chemical bonds. Clays of the montmorillonite group are the most thoroughly studied ones for their ability to retain organic constituents. The reason for this preference is obvious, considering the fact that montmorillonites offer a large internal surface (nearly 7×10^6 cm^2 $-$g^{-1}). From this area of research valuable information can be obtained regarding the conservation of organic materials with time and the puzzling stability of certain thermodynamically unstable organic compounds in ancient rocks. Light may also be thrown on the origin of petroleum since many data suggest that clay minerals are involved and de facto essential to obtain crude oils.

B. *Feldspars.* Feldspars are the most abundant group of minerals in the lithosphere and occur almost in any crystalline or sedimentary rock. Although sedimentary feldspars range only second in abundance to quartz and clay minerals, the majority of them are of high-temperature origin and thus merely survivors of weathering and diagenesis. Feldspars, however, may also form contemporaneously with deposition, or else at one time during diagenesis and epigenesis. Specimens of this type are generally termed authigenic feldspars so as to separate them clearly from their more common detrital counterparts.

The chemistry and isotopic composition of authigenic feldspars may reveal interesting details regarding paleotemperatures in diagenetic environments or the absolute age of the host strata. For instance, the potassium-argon ages of two authigenic feldspars from the Cretaceous and Ordovician are plotted in Fig. 5. The data agree well with the assigned ages for these periods of 95 and 400 million years, respectively.

C. *Zeolites.* The presence of zeolites in ancient sediments, even in subordinate amounts, can be geochemically rather rewarding because of the fact that some zeolites may serve as thermal indicator minerals. Encouraged by the close agreement that exists between field criteria and experimental data on the mode of zeolite formation, the concept of a zeolite "facies" was developed to bridge the wide gap between low-temperature diagenesis and the hitherto recognized metamorphic "facies." Laumontite, one of the more common zeolite species, is regarded in particular as a potential temperature indicator mineral.

2. Oxides and Hydroxides. Silica and the aluminum, iron, and manganese oxides and hydroxides are geochemically the most interesting members of the oxide family for reasons of their wide abundance in ancient rocks. Because of the fact that a number of crystalline and amorphous modifications exist, most of which respond rather sensitively to changes in pH and Eh, these minerals are extremely useful as sensitive indicators for the Eh/pH relationships that existed at the time of their formation.

A. *Silica.* The only crystalline modification of SiO_2 stable in low-temperature environments is α-quartz. Very commonly it appears in ancient sediments as microcrystalline quartz, which is also known as chalcedony. Chert, one of the most common silicious sediments, is principally composed of microcrystalline quartz. Theoretically, one might expect that quartz crystallizes from any solution that is supersaturated with regard to its solubility product. This, however, is not the case as evidenced by the lack of syngenetic quartz-precipitates, despite the supersaturation stage of SiO_2 in most aqueous environments.

Quartz deposition, indicated by overgrowths on detrital grains or occurrences of interstitial cement, however, can be observed in many ancient rocks. Apparently diagenetic environments are more favorable toward quartz deposition than syngenetic environments. The wide abundance and close association of carbonates and diagenetic quartz in many ancient rocks suggests that the kinetics of quartz deposition and solution is perhaps affected by the presence of certain coexisting carbonate species.

Silica frequently enters sediments in the amorphous form, for instance upon death of silica mineralizers such as radiolaria, sponges, or diatoms. Inasmuch as amorphous silica is metastable, it will become diagenetically mobilized and will migrate in true ionic solution in the direction of a concentration gradient to reprecipitate as quartz and accumulate at points of least solubility. The absence, during Precambrian time, of large populations of silica-secreting organisms that would biochemically extract silica and maintain low concentrations of silica in the oceans suggests that the silica content of the ancient sea was probably much higher than that of the modern ocean. Consequently, inorganic precipitation of amorphous silica might have been possible. Particularly in areas of upwelling waters, the conditions for the deposition of amorphous silica were favorable. Subsequent reorganization of the amorphous precipitate may have resulted in the formation of chert beds, and may account for their worldwide occurrence in shallow marine deposits of the Precambrian.

Cherts and other silicious sediments are poor in trace elements. This is so because the small ionic radius of silicon allows only a small number of elements, e.g., boron, germanium, and beryllium, to proxy for silicon.

From the few data published it appears that variations in the isotopic composition of silicon are considerably smaller ($\sim$ 3 per mil) than those for carbon ($\sim$ 80 per mil) although the mass ratios of C^{13}/C^{12} and Si^{30}/Si^{28} are not significantly different from each other. The type of bonding (ionic versus covalent) has been suggested as a possible explanation for this phenomenon.

More information is available on the oxygen isotope ratios in cherts and diatomites. In general, one can say that factors that will influence the O^{18}/O^{16} ratio in carbonates will similarly affect the O^{18}/O^{16} in crystalline and amorphous silica. Oxygen isotopes have been successfully used for studies on the formation

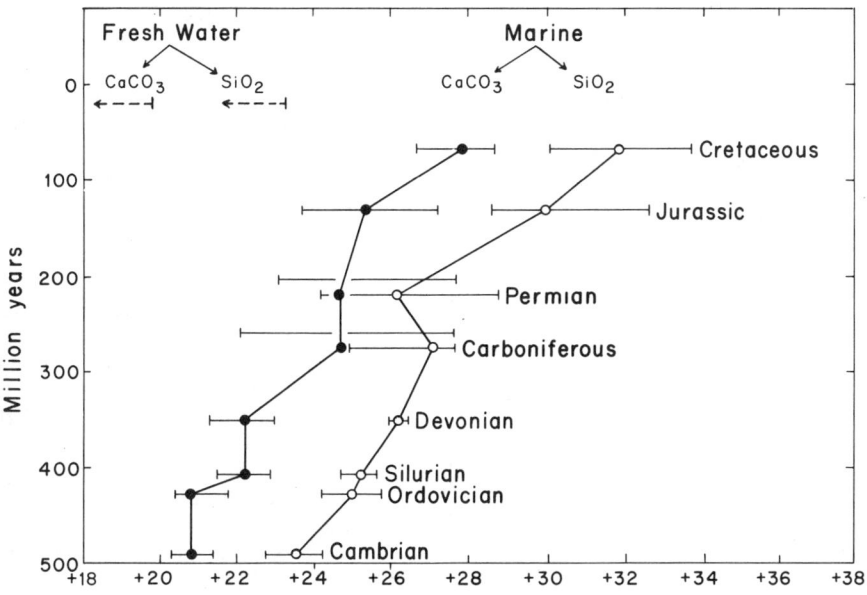

Fig. 6. Variations in O^{18}/O^{16} ratio of carbonates and cherts with geologic age.

mechanism of chert deposition, paleotemperature, environmental aspects, and diagenesis. The systematic change of oxygen isotopes in both carbonates and cherts with geologic age is illustrated in Fig. 6. These data suggest that oxygen isotopes have to be used rather cautiously for the interpretation of paleotemperatures in ancient environments.

B. Aluminum Hydroxides. Rocks containing aluminum hydroxides as their major constituents are generally termed bauxites. Bauxites may develop on any aluminum containing source rock. A warm tropical and alternating wet and dry climate is essential to reduce the prospective parent material to the final aluminum hydroxides. The presence of lignites or high organic activities stimulates the formation of CO_2 and results in the generation of aggressive bicarbonate solutions. Sulfide deposits have the same effect by producing strong acid conditions during weathering. A high redox potential favors the gibbsite formation, whereas boehmite and diaspore may also be found in coexistence with sulfides and carbonaceous materials. Half of the world's aluminum ores have been mined from bauxite deposits associated with limestones and dolomites. The dependence on factors such as abundant rainfall and tropical environments makes the aluminum hydroxides potential indicators for paleoclimates. It should finally be added that aluminum hydroxides are enriched in certain trace elements, in particular gallium, chromium, niobium, molybdenum, zirconium, and titanium.

C. Iron and Manganese Oxides and Hydroxides. The close similarity between iron and manganese is reflected in their common association in igneous rocks. During weathering, transportation, and sedimentation, however, processes are operating that allow an effective fractionation of these two elements. It is common knowledge that iron and manganese ions respond differently to changes in pH and redox potential. The relationships may be quantitatively expressed by means of diagrams showing stability field of possible iron-manganese compounds at room temperature and at different values of pH and Eh. A representative Eh-pH diagram, showing the stability relations of iron oxides, sulfides, carbonates, and silicates at 25°C and one atmosphere total pressure in the presence of water is presented in Fig. 7. The diagram was selected because it bears great significance in studies on sedimentary iron ores, in particular the geographically extensive iron ore formations of the Precambrian.

Calculations and experiments by Krauskopf, Huber, Garrels, Krumbein, and Marchandise (see references in Garrels, 1960) indicate that iron oxides and hydroxides precipitate at lower oxidation potentials than manganese oxides at any given pH. Similarly, under fixed Eh, iron starts precipitating as the oxide at a considerably lower pH than manganese. The formation

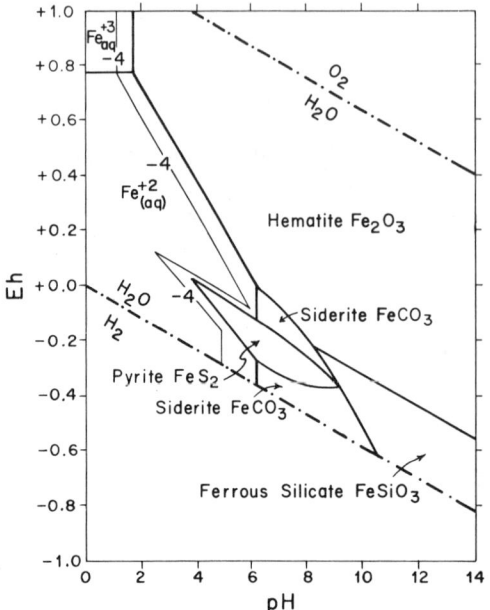

FIG. 7. Eh–pH stability relations among iron oxides, carbonates, sulfides and silicates at 25°C, and 1 atm total pressure in the presence of water. Other conditions: total $CO_2 = 10^0$; total sulfur $= 10^{-6}$; amorphous silica is present. With slight undersaturation in amorphous silica, magnetite enters the ferrous silicate stability field (after Garrels, 1960).

of iron-rich sediments low in manganese as those from the Lake Superior District (Precambrian) and the occurrence of high-purity manganese oxides, e.g., the Eocene deposits of Tschiatura Kutais in South Russia, are a result of the pH/Eh characteristics of iron and manganese compounds. Noteworthy is the high concentration of trace metals in many manganese and iron oxides.

3. Carbonates. Processes causing deposition and solution of carbonate minerals in syngenetic environments are reasonably well understood. This knowledge on the other hand is extremely helpful to reconstruct the general environmental conditions that have led to deposition, solution, or metasomatic replacement of carbonates throughout geologic history. Studies in the area of carbonate analysis for major and trace elements as well as oxygen and carbon isotopes.

Among the carbonates, the two polymorphs of $CaCO_3$, i.e., calcite and aragonite, and the dolomites $CaMg(CO_3)_2$ are of special geochemical interest. It is generally accepted that ancient limestones are largely a product of biological activities, i.e., calcium carbonate has been initially extracted from the sea by organisms, in form of shell materials, etc. In contrast, inorganic precipitation appears to be only a minor contributor. During diagenesis, however, recrystallization and cementation of the calcareous ooze is a common feature. This may result in the formation of secondary calcite and the disappearance of the metastable aragonite which actually is only rarely found in sediments older than the Permian.

Dolomite, on the other hand, is more common in older than in younger sediments. This was interpreted by some authors to mean that dolomites are exclusively a product of diagenesis, i.e., dolomite was formed by metasomatic replacement of calcite. Others, however, suggested that the chemistry of the ancient sea was different and thus favored the precipatation of dolomite, whereas the oceans of today can produce only calcite and aragonite. The occurrence of Recent dolomites in a number of places from all over the world, and where the salinity is somewhat higher than that of the present ocean, seems to support the second statement that seawater that has been slightly altered chemically will yield dolomite rather than calcite or aragonite. In contrast, oxygen isotope data indicate that dolomites cannot precipitate at room temperature and that even the Recent dolomites are products of diagenesis. All dolomites independent of their stratigraphic age have been formed under solid-state conditions from a crystalline calcium carbonate. It is at present not fully understood what kind of environmental conditions in terms of alkalinity, organic activity, salinity, temperature, or other parameters have to be met in order to form dolomite. Most probably, the various $CaCO_3$ polymorphs are metastable in the majority of the diagenetic environments, and they become gradually transformed into the more stable dolomite by straightforward solid diffusion mechanisms. The fact that the Ca/Mg ratio of carbonates generally decrease with increasing geologic age may have the simple explanation that progressively older rocks have, statistically speaking, better chances of becoming dolomitized than comparatively younger sediments. The speed of dolomitization appears to be greatly increased in hypersaline environments, as can be seen by the presence of Recent dolomites in certain evaporite basins. It may be pointed out that these "Recent" dolomites give carbon-14 ages of a few thousand years.

A voluminous literature has accumulated on the trace element composition of limestones and shell carbonates. For instance, a number of studies have been made on the biogeochemistry of magnesium and strontium in the skeletons of calcareous organisms. Several relationships have been suggested between shell composition and chemical and physical environmental factors, such as salinity and temperature. One rather remarkable feature is that certain carbonate layers of a shell are sensitive to changes in temperature or salinity, while other layers do not respond at all. A great number of ancient shell carbonates have preserved their original Sr/Ca or Mg/Ca ratio over millions of years and may thus reveal interesting details on the physics and chemistry of paleoenvironments.

The amount of data on the distribution of oxygen and carbon isotopes in ancient carbonates is rather plentiful. The majority of these studies were concerned with the paleo-temperatures of the ancient sea, or the differentiation of fresh-water and marine-sediments. The pioneering work on oxygen and carbon isotopes by Craig, Emiliani, Epstein, Lowenstam, and Urey (for references see Degens, 1965) has indeed opened new vistas in the area of carbonate geochemistry.

4. Phosphates. Of the more than 150 different phosphates known in nature, only the calcium phosphates are widespread in sedimentary rocks; among these, the carbonate apatites are the most common. Sediments in which carbonate apatites are a major mineral constituent are generally termed phosphorites. The formation of carbonate apatite is a metasomatic phenomenon, i.e., the CO_3^{2-} groups in calcium carbonates are incompletely replaced by PO_4^{3-} groups. It is suggested that, if the diagenetic environment permits, the chemical composition of a completely phosphatized limestone consisting of carbonate apatite will gradually change in the direction of a regular apatite, which means a loss in CO_3^{2-} groups and again in OH^- or F^- with time.

The phosphorite facies is restricted to marine environments, and here to areas not too shallow and not too deep (platform facies), but having access to the open sea at one side. In general, with increasing distance from the shore line, the phosphorus content in the sediments will increase up to a certain point from which it falls off sharply. This feature coincides with the distribution of the associated organic matter, and a cause-effect relationship might be anticipated in the sense that most of the phosphorus was initially collected by photosynthesizing marine organisms. Upon death of the organisms, the phosphorus was carried to the ocean floor via organic detritus. During decay, much of the phosphorus became released to the next environs and available for the phosphate formation.

The total amount of collophane within a single phosphorite deposit can be rather significant. For instance, the Permian phosphorite deposits of the western United States contain approximately 1.7×10^{12} metric tons of P_2O_5 which represents more than five times the amount of P_2O_5 presently distributed in the ocean. Aside from phosphorite deposits of economic interest, phosphatization of all kinds of organic remains such as shell materials, wood, fecal pellets, teeth, etc., can be observed in many ancient sediments.

Phosphate rocks owe some of their rare metals to coexisting organic matter and sulfides, rather than to phosphates themselves. Only a few elements are enriched in significant quantities; these are strontium, arsenic, uranium, the rare earth elements, lead, and zirconium. Of economic interest is the high abundance of uranium and rare earth elements in some ancient phosphate deposits.

5. Sulfides. A number of sulfides occur in sedimentary rocks in a manner compatible with the assumption that they were formed more or less contemporaneously with sedimentation. Quantitatively most important are the two polymorphs of FeS_2, i.e., pyrite and marcasite.

Of the more than thirty ionic and molecular species of sulfur that are chemically important, only five are thermodynamically stable at room temperature in quantities that are geologically significant. These are elemental sulfur, H_2S, HS^-, HSO_4^-, and SO_4^{2-}. In Eh–pH diagrams, the distributions of these five individual species at equilibrium have been determined.

The geochemistry of sulfides is rather well understood. Microbial sulfate reduction (*Desulphovibrio desulfuricans* or *Clostridium desulfuricans*) is the major phenomenon in the early stages of diagenesis. In general, ancient marine sediments contain more sulfides than fresh-water sediments of the same lithology. This feature can be explained by the fact that seawater sulfate represents an unlimited sulfur reservoir for the sulfide formation in marine sediments, whereas no sulfur source equivalent in magnitude is present in fresh-water environments.

The dimorphs of FeS_2 are products obtained by the interaction of $FeS_2 \cdot nH_2O$ (hydrotroilite) and elemental sulfur. Thus, the highly metastable hydrotroilite has to be regarded as the first-formed iron sulfide which in the presence of elemental sulfur will be rapidly altered into more stable sulfide modifications; depending on the acidity of the environments, either pyrite (alkaline) or marcasite (acid) will be the end product. The elemental sulfur essential for the formation of FeS_2 will be obtained by the oxidation of free sulfide. The iron has to come from host strata minerals such as iron hydroxides, which are unstable in Eh–pH environments where sulfides generally form. Also sulfate-reducing bacteria are known to mobilize iron from clay minerals and detrital constituents (amphiboles, pyroxenes, biotites, etc.).

Sulfides are also known to incorporate a number of trace elements, in particular, arsenic, cobalt, nickel, copper, and lead. Like iron, most of these trace elements come from the coexisting mineral and organic matter in the host sediment. This may explain why sulfides present in sandstones have significantly less trace elements than the same sulfides incorporated in a bituminous shale.

The isotope geochemistry of sulfides has been thoroughly studied and the distribution of S^{34}/S^{32} in sedimentary sulfides can be used as a significant geochemical tool for studies on the formation of sulfide minerals, paragenesis, facies, and certain aspects of biogeochemistry.

6. Evaporites. Evaporation of natural waters will ultimately result in the precipitation of its dissolved mineral constituents. The sequence of precipitation is hereby principally determined by the solubility characteristics of its saline compounds. The least soluble material will precipitate first, and the most soluble compound last of all. The processes, however, are not as simple as they appear from this statement. They are complicated by the fact that the solubility of a salt is affected by the nature and amount of coexisting solutes, the water temperature, and the vapor pressure of the system. In a qualitative manner, the Gibbs phase rule expresses the phase relationship in complex salt systems, i.e., the pressure-temperature conditions, and the number and con-

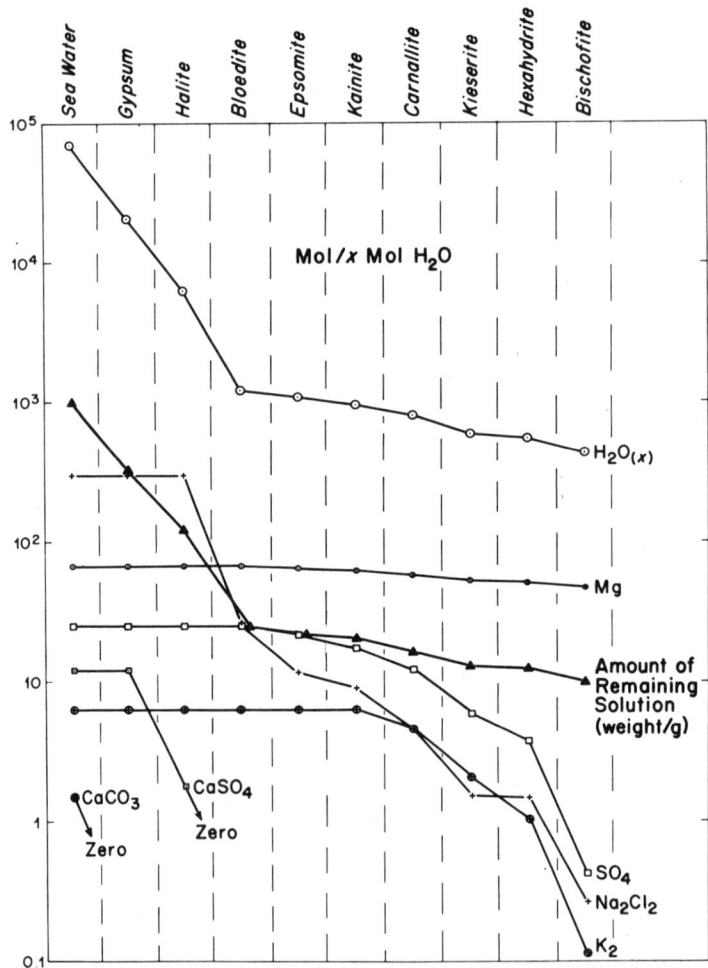

FIG. 8. Chemical composition of ocean water under isothermal evaporation at 25°C, and at the start of precipitation of the individual mineral species (after Braitsch, 1962).

centration of all participating components will determine the number of phases. Although any solution will yield salt deposits upon evaporation, the geologically most significant and best studied from a physical-chemical point of view are the various mineral systems of the sea (for references see Braitsch, 1962).

The reaction products obtained under static isothermal evaporation of ocean water at 25°C, assuming stable equilibrium conditions, are illustrated in Fig. 8. It is the nature of the seawater system, however, that is responsible in many cases for the deposition of metastable mineral phases. This is due to the fact that the stable mineral phases are frequently not precipitated within their stability field for kinetic reasons. For instance, gypsum ($CaSO_4 \cdot 2H_2O$) will even precipitate in the anhydrite ($CaSO_4$) stability field, and syngenetic anhydrite formation has nowhere been observed. In other words, anhydrite can only form during diagenesis from preexisting gypsum.

The bulk of the ancient marine evaporites fall into two major categories: (1) deposits which are slightly and (2) others which are completely impoverished in $MgSO_4$. Factors that determine the type and the sequence of ancient evaporite minerals to be formed during syngenetic and diagenetic events have been summarized in Braitsch (1962).

Mobile Phases

1. Water. One of the more significant topics of modern theoretical and applied geology concerns the water cycle in nature. Inasmuch as water is the main agent during physical and chemical weathering, the carrier and transporter of most sediment materials in the dissolved or particulate state, and the environment of most life processes, the complexity of the problem at hand is well illustrated. Furthermore, water constitutes about 10–15% of the total sedimentary matter.

Studies on the chemical and isotopic composition of water present in ancient sediments are a challenging research topic in the area of sediment geochemistry. This viewpoint rests largely on the consideration that, in a broad sense, water present in the sedimentary rock formation displays a role analogous to magmatic fluids in crystalline rocks. Both types of fluids represent the mobile phase and act as the universal solvent and the carrier and transporter of material from one spot to the other. The solid rock compounds must be considered stationary when compared to the agility of their interstitial waters.

Recent studies by Chave, Chillingar, DeSitter, von Engelhardt, McKelvey, Münnich, Siever, Vogel, White, and Wyllie (for reference see Degens, 1965) have shown the wide application of water studies to age determination, facies aspects, and problems such as the origin of petroleum and high saline formation waters. Not only the chemical parameters of waters but also their stable and radioactive isotopes are useful as geochemical tools.

2. Gases. A variety of gases, both of organic and inorganic origin, have been found in ancient sediments. Abundant and widespread are carbon dioxide, ammonia, hydrogen sulfide, hydrogen, atmospheric air, and the gaseous hydrocarbons among which methane is the most dominant one. The major sources for gases after organic activity and decay has passed its climax are the various radioactive decay products such as argon and helium. Under certain conditions, they may be tightly fixed to the parent material and can be used in some instances for radioactive age determination (glauconites, authigenic feldspars).

Organic Constituents

Sediments have been the principal wastebaskets of posthumous organic debris throughout earth's history. It has been estimated that a total of about 3.8×10^{15} metric tons of organic matter are entrapped in sediments. Most of it is present in a finely disseminated form and associated with shales (3.6×10^{15} metric tons). In contrast, coals comprise only 6×10^{12} metric tons and the ultimate primary petroleum reserves have been estimated at 0.2×10^{12} metric tons.

More than 500 organic constituents have so far been extracted from sediments, soils, and natural waters. This number does not include all the individual organic chemicals that have been isolated from coal tars and crude oils.

The range of investigation which can be performed by studying the organic fractions of ancient sediments is rather wide. For example, there is the opportunity of a direct approach to hitherto inaccessible problems, such as the sequence of origin of the key components of life by studying Mohole-type sediment cores from the deep-sea or well-preserved drill-core sediments available on the continents. Investigations on the interactions of clays and organic matter may throw some light on the generation of biochemically interesting polymers such as peptides or polysaccharides. A study of the mechanism of silification and calcification in biological systems is another challenging topic that will eventually assist in understanding evolutionary trends in earth history. Hydrocarbon generation and accumulation, coal formation, postdepositional microbial activities, diagenetic and thermal stabilities of organic molecules, condensation phenomena, and interactions of organic and inorganic materials are a few additional areas of biogeochemical research which bear great research potentialities in the field of geology, paleontology, and related sciences.

From the published data it can be inferred that most of the biogeochemical alteration of organic matter takes place in the environment of deposition and during early stages of diagenesis, i.e., at a time when microbes and burrowing animals are still alive in soils and sediments. During later stages of diagenesis when organic activity in the strata has more or less ceased, alteration of organic matter predominantly proceeds in the direction of slow nonbiological maturation processes, with thermal degradation being the number one alteration factor. Also, geographical redistribution of organic constituents such as observed during migration and accumulation of petroleum deposits may account for changes in the organic spectrum of ancient sediments by either extracting or supplying certain organic compounds.

Much has to be learned about the distribution pattern and fate of organic constituents in present-day environments, modern shell materials, soils, and Recent sediments, before one can hope to find the key to the understanding of the complex organic spectrum of ancient rocks. To tackle some of the most challenging areas of biogeochemical research, namely, the synthesis of organic matter in the early stages of the earth and other planets, and the biochemical evolution of life and organic matter, one first has to know what kind of organic molecules can survive early and late diagenesis, and which ones are merely products of diagenesis.

Studies of the type performed by Abelson and Hare on amino acids, by Vallentyne and Swain on sugars, by Blumer and Hodgson on pigments, or by Hoering and Hood on fatty acids (references in Breger, 1963; Degens, 1965) may eventually help to outline the path of synthesis and decay of organic matter throughout earth history. A full account on the geochemistry of organic sediments has been presented by Hunt (see Vol. VI).

EGON T. DEGENS

References

Braitsch, O., 1962, "Entstehung und Stoffbestand der Salzlagerstätten. Mineralogie und Petrographie in Einzeldarstellungen," Vol. 3, Berlin, Springer Verlag, 232pp.
Breger, I. A. (editor), 1963, "Organic Geochemistry," New York, The MacMillan Co., 658pp.
Clarke, F. W., 1924, "The Data of Geochemistry," Fourth ed., Gov. Print. Off., Washington, D.C., 821pp.
Degens, E. T., 1965, "Geochemistry of Sediments," Englewood Cliffs, N.J., Prentice-Hall, Inc., 342pp.
Garrels, R. M., 1960, "Mineral Equilibria at Low Temperature and Pressure," New York, Harper and Brothers, 254pp.
Mason, B., 1958, "Principles of Geochemistry," Second ed., New York, John Wiley & Sons, Inc., 310pp.

Cross-references: *Authigenesis of Minerals, Marine; Biogeochemistry; Calcium Carbonate Geochemistry; Carbon: Element and Geochemistry; Carbon-14 Dating; Free Energy and Free Enthalpy; Geochemistry; Hydrocarbons; Interstitial Waters in Sediments; Ion Exchange; Organic Geochemistry; Oxidation and Reduction; Paleosalinity; Paleotemperatures, Isotopic Determinations; pH-Eh; Potassium-Argon Age Determination; Radioactive Isotopes; Seawater, Chemistry; Sulfate Reduction—Microbial; Syngenesis; Trace Elements; Water; Weathering, Chemical. Vol. III: Denudation; Weathering. Vol. IVB: Clays and Clay Minerals; Electron Probe Microanalysis; Evaporite Minerals; Feldspars. Vol. V: Free Energy and the Gibbs Phase Rule; Metasomatism. Vol. VI: Chert; Diagenesis; Evaporites; Limestones; Methane; Organic Sediments and Rocks; Petroleum: Origin and Evolution; Siliceous Ooze; Weathering, Organic.*

GEOCHEMISTRY OF SEDIMENTS: MODERN

Sediments are products of more or less intense mechanical and chemical weathering, erosion, and deposition of primitive and recycled constituents of the earth's crust (see Fig. 1). Sediments eventually form sedimentary, metamorphic, and igneous rocks. Much evidence now indicates that a large fraction of the earth's crust has passed through a sedimentary state at one time or another. Sediments, therefore, play a larger role in the development of the earth's crust than their relatively small mass would indicate (Table 1).

The geochemistry of the different sediments reflects the variations in sediment-forming processes due to the distribution of land and water on the earth, air and water currents, rivers, and differences in climate, biological activity, source rocks, volcanism, degassing of the earth, etc. For many of these geochemical processes the time factor is also important.

Weathering Processes

The original constituents of most rocks at the earth's surface are unstable with respect to the atmosphere and hydrosphere. Due to mechanical weathering, rocks are fragmented

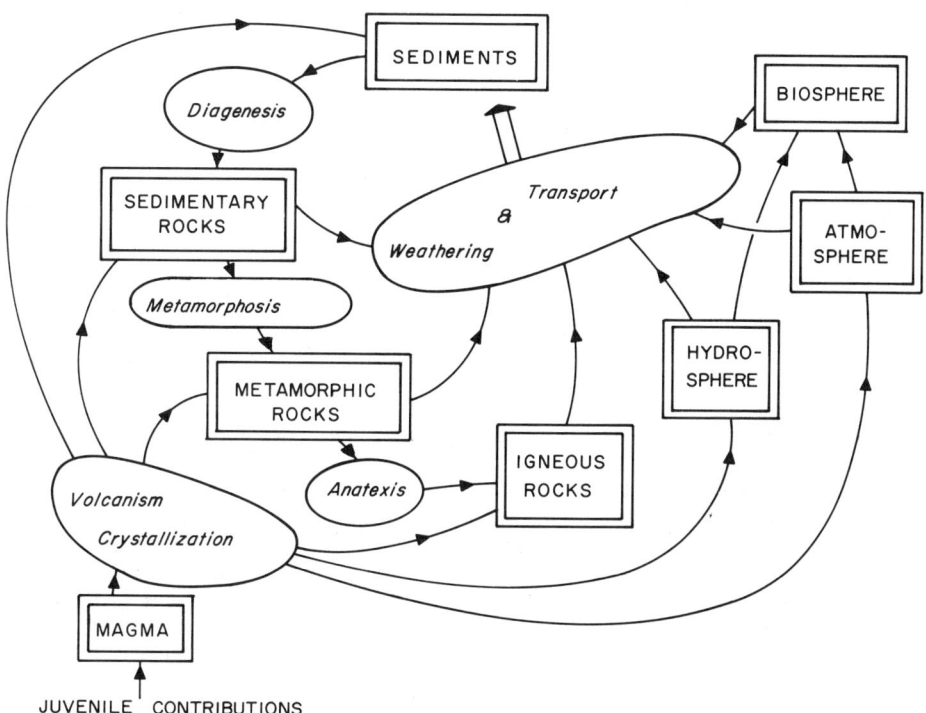

FIG. 1. The role of the sediments in the geochemical cycles in the earth's crust

TABLE 1. MASS OF EARTH SHELLS (%)

Atmosphere	0.00009
Hydrosphere	0.02
Sediments and sedimentary rocks	0.01
Crust (including sediments and sedimentary rocks)	0.4
Mantle and core	99.6

into successively smaller particles. This fragmentation is primarily caused by the differences in volume changes of various phases during temperature variations and by the abrasive action of the solid particles carried by water, ice, and wind. The fragmentation process causes an increase of the surface area of the original rock which enhances the rate of chemical weathering, and vice versa, chemical weathering enhances mechanical breakdown. Main agents in chemical weathering are water, carbon dioxide, and oxygen which tend to react with unstable minerals. Since feldspars, micas, and iron-magnesium silicates such as olivine, pyroxene, and amphibole are the important constituents in most igneous and metamorphic rocks, reactions such as (1)–(4) are occurring extensively during chemical weathering.

These weathering processes tend to control the pH of the aqueous phase (see also *Abrasion pH*). Alkaline solutions (such as in semiarid soils) tend to promote the formation of montmorillonites, whereas acid solutions (such as in the humid tropics) promote the formation of kaolinite and silica. However, the course of chemical weathering, even of the same rock,

(1) $2H_2O + CaAl_2Si_2O_8 + CO_2 \rightarrow CaCO_3 + Al_2Si_2O_5(OH)_4$
  anorthite   calcite   kaolinite

(2) $2KAlSi_3O_8 + 2CO_2 + 3H_2O \rightarrow Al_2Si_2O_5(OH)_4 + 2K^+ + 2HCO_3^- + 4SiO_2$
  orthoclase   quartz

(3) $2KMg_3AlSi_3O_{10}(OH)_2 + 14CO_2 + 7H_2O \rightarrow Al_2Si_2O_5(OH)_2 + 2K^+ + 6Mg^{2+} + 4SiO_2 + 14HCO_3^-$

(4) $(Fe, Mg)_2 SiO_4 + H^+ + CO_2 + \frac{1}{2}H_2O + \frac{1}{4}O_2 \rightarrow FeOOH + SiO_2 + Mg^{2+} + HCO_3^-$
  olivine

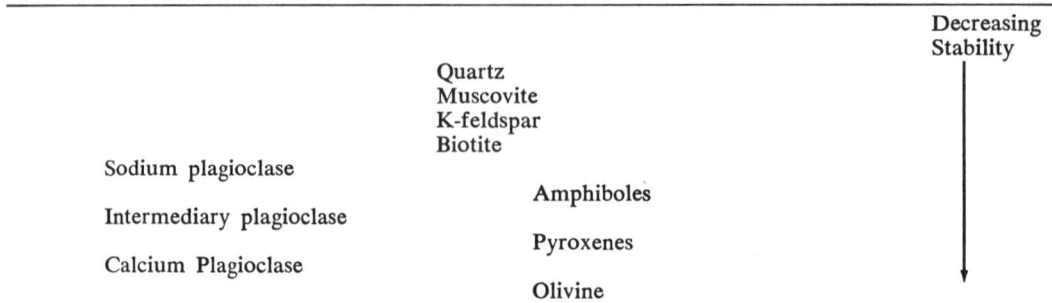

Fig. 2. Weathering stability of minerals in igneous and metamorphic rocks.

varies considerably due to environmental factors such as climate, biological activity, topography, and hydrology. Thus, due to variations in rainfall, basalt can weather to either montmorillonite clays, or to kaolinite clays or again to hydroxides of iron and aluminum such as gibbsite, boehmite, and goethite.

Different minerals are attacked with more or less vigor under various conditions of chemical weathering. The order of stability of igneous and metamorphic minerals is approximately opposite to their order in the reaction series of Bowen (see Fig. 2). This diagram indicates that, for instance, olivine is more unstable than augite and weathers faster into clays, iron oxides, carbonates, etc., than augite does.

In addition to quartz and muscovite, there are some other minerals that are resistant to chemical weathering, namely tourmaline, rutile, zircon, monazite, garnet, magnetite, ilmenite, hornblende, apatite, gold and platinum metals, and diamonds. These minerals tend to be enriched in so-called *resistate sediments*.

Mineralogy of Sediments

The result of weathering, particularly of metamorphic and igneous rocks, is commonly a complete breakdown and reorganization of original phases into new minerals, which are more stable in contact with the atmospheriles. In Table 2 the more important minerals that are present in sediments are given.

The mineralogical composition of a sediment commonly determines the distribution of *trace elements* (q.v.), since these rarely form distinct minerals of their own. Trace elements may be present as diadochic replacement of one or several major elements, as adsorbate on phases with suitable surfaces, or as other undefined constituents, for instance, in organic complexes (see *Chelation;* Vol. IVB: *Organo-Hydrolysates*). Thus, minerals of lead, cobalt, and nickel are rare in sediments; more commonly, these trace elements occur in solid solution or as absorbates on ferromanganoan oxides and hydroxides. Vanadium and uranium are common in carbonaceous shales in which they are associated with organic matter. Boron is still another trace element that rarely forms independent minerals as in some salt deposits and resistates; the major part of boron in sediments is associated with clay minerals, opaline silica, and ferromanganoan precipitates.

Products of Weathering and Sedimentation

The products of weathering may remain in situ as saprolite or as more or less ill-sorted products in which the structure of the original rock is obliterated, forming a soil. The product of weathering may also be redistributed by water, wind, etc., leading to the formation of more or less layered sediments.

Soils. The character of soil depends on climate, drainage, biological and human activities, and parent rock material. Often similar end products are formed from different source rocks as is indicated in Table 3, in which analyses 1, 2, and 3 are soils formed from respectively basalt, limestone, and granite gneiss.

In humid regions the residual materials are enriched in hydroxides of aluminum and ferric iron and depleted in lime, magnesia, and alkalies. Under even higher rainfall and temperatures silica is also removed, leading to the formation of laterite. A mature soil will show a soil profile of zonal arrangement (see Vol. VI, *Soil Genesis*). Typical components in soils are, beside the minerals, also air, water, and organic matter. Many of the agricultural properties of soils are due to the colloidal clays and organic substances present.

Sediments. Sediments vary considerably more in chemical composition than igneous rocks. This is due to many factors. Constituents that at high temperatures form complete solid solutions can often only exist in separate solid phases at low temperatures. This is well known for many carbonates (e.g., dolomite). Differ-

TABLE 2. SEDIMENTARY MINERALS

Mineral or Mineral Groups	Chemical Composition
Clay minerals	
Kaolinite	$Al_4(OH)_8[Si_4O_{10}]$
Halloysite	$Al_4(OH)_8[Si_4O_{10}] \cdot (H_2O)_4$
Montmorillonite	$(Al,Fe^{3+},Mg)_3(OH)_2[Si_4O_{10}]\,Na_x \cdot nH_2O$
Illite	$(K,H_3O)Al_2(H_2O,OH)_2[AlSi_3O_{10}]$
Septechlorite	$(Fe^{2+},Fe^{3+},Al,Mg)_6(OH)_8(Al,Si)_4O_{10}$
Chlorite	$(Al,Mg,Fe)_3(OH)_2(Al,Si)_4O_{10}Mg_3(OH)_6$
Vermiculite	$(Mg,Fe)_3(OH)_2[AlSi_3O_{10}]\,Mg(H_2O)_4$
Micas	
Muscovite	$KAl_2(AlSi_3)O_{10}(OH)_2$
Biotite	$KMg_3(AlSi_3)O_{10}(OH)_2$
Sepiolite	$Mg_2Si_3O_6(OH)_4$
Talc	$Mg_3(OH)_2Si_4O_{10}$
Feldspars	
Orthoclase	$K[(Si,Al)_4O_8]$
Microcline	$KAlSi_3O_8$
Albite	$NaAlSi_3O_8$
Zeolites	
Phillipsite	$(1/2Ca,Na,K)_3Al_3Si_5O_{16} \cdot 6H_2O$
Harmotome	$BaAl_2Si_6O_{16} \cdot 6H_2O$
Clinoptilolite	$(Ca,Na_2)[Al_2Si_7O_{18}] \cdot 6H_2O$
Zircon	$ZrSiO_4$
Tourmaline	$Na(Li,Mg,Fe,Al)_3Al_6B_3Si_6O_{24}(OH)_4$
Amphiboles	$Na_x(Fe,Mg,Ca,Mn)_ySi_8O_{22}(OH)_2$
Pyroxenes	$(Ca,Mg,Fe,)SiO_3$
Oxides	
Quartz	SiO_2
Opal	$SiO_2 \cdot nH_2O$
Rutile	TiO_2
"Manganites"	$MnOOH \cdot nH_2O$
Pyrolysite	$\beta\text{-}MnO_2$
Magnetite	Fe_3O_4
Hematite	$\alpha\text{-}Fe_2O_3$
Hydroxides	
Gibbsite	$\gamma\text{-}Al(OH)_3$
Diaspore	$\alpha\text{-}AlOOH$
Goethite	$\alpha\text{-}FeOOH$
Boehmite	$\gamma\text{-}AlOOH$
Lepidocrocite	$\gamma\text{-}FeOOH$
Limonite	$FeOOH \cdot nH_2O$
Carbonates	
Calcite	$CaCO_3$
Aragonite	$CaCO_3$
Dolomite	$CaMg(CO_3)_2$
Soda	$Na_2CO_3 \cdot 10H_2O$
Trona	$Na_3H(CO_3) \cdot 2H_2O$
Others	
Apatite	$Ca_5(PO_4)_3(F,Cl,OH)$
Pyrite	FeS_2
Marcasite	FeS_2
Hydrotroilite	$FeS_2 \cdot nH_2O$
Barite	$BaSO_4$
Anhydrite	$CaSO_4$
Gypsum	$CaSO_4 \cdot 2H_2O$
Celestite	$SrSO_4$
Epsomite	$MgSO_4 \cdot 7H_2O$
Astrakhanite	$Na_2Mg(SO_4)_2 \cdot 4H_2O$
Bischofite	$MgCl_2 \cdot 6H_2O$
Carnallite	$KMgCl_3 \cdot 6H_2O$
Halite	$NaCl$
Sylvite	KCl

TABLE 3. CHEMICAL COMPOSITION OF SOILS

	1	2	3
SiO_2	40.7	55.42	55.07
TiO_2	7.3	T	1.03
Al_2O_3	30.9	22.17	26.14
Fe_2O_3	8.7	8.30	3.72
FeO	—	T	2.53
MnO	—	—	0.03
MgO	—	1.45	0.33
CaO	1.0	0.15	0.16
Na_2O	0.4	0.17	0.05
K_2O	0.3	2.32	0.14
H_2O+	11.0	7.76	9.75
H_2O-	—	2.10	0.64
P_2O_5	—	T	0.11
CO_2	—	—	0.36
Total	100.3	99.84	100.11

T = trace.

ences in chemical and physical weathering due to climatic variations are other important causes for variability in the ultimate sediment; thus, kaolinite-rich sediments are extensively occurring in the tropical regions, whereas in boreal regions with slow weathering, feldspar detritus (arkose) may occur in the sediments. Furthermore, at the surface of the earth, large variations exist in pH and Eh (redox potential) due to variations in oxygen supply and deposition rates of organic matter, causing some elements to be precipitated whereas others are transported away after the weathering. Organisms preferentially incorporate elements such as calcium (Ca) and silicon (Si) and leave others behind, which further promotes a differentiation of elements. Water and air transport can preferentially sort particles by weight and shape; thus, flaky micas may show a strong tendency to be transported considerable distances by wind and water.

Beach sands are commonly enriched in heavy minerals or minerals resistant to mechanical weathering. A bathymetric zonation of sediments exists; the increased rate of dissolution of calcium carbonate with depth in the oceans explains the absence of significant carbonate deposits in the abysso-pelagic region. Extensive evaporation in basins with limited supply of seawater leads to the formation of *evaporites* (q.v.).

These processes form sediments of a large chemical and mineralogical variety. Thus, some extensively occurring sediments consist largely of either $CaCO_3$ (limestones), SiO_2 (quartz sands), halides (evaporites), iron oxides (iron ores), or aluminum oxide (bauxites).

On the other hand, igneous rocks that are rich in one of these constituents, for instance, $CaCO_3$ (sövites), are exceptionally rare, or may be of a disputable origin. Thus, some high-grade magnetite ore bodies such as Kirunavaara may possibly be ultrametamorphosed iron-rich sediments. For this reason, granite and basalt, which represent the chemical extremes among the common igneous rocks, differ much less from each other than do the major products of sedimentation (see Table 4), namely *resistates* (average sandstone), *carbonates* (average limestone), and *hydrolystates* (average shale) in which the elements tend to be well separated from each other (see also Table 5). The major products of sedimentation are the following:

Residual sediments (or resistates) consist of minerals that are resistant to chemical weathering. Quartz, therefore, tends to become enriched in such sediments, as also to some extent do various heavy minerals such as zircon, tourmaline, amphiboles, apatite, magnetite, ilmenite, and gold.

TABLE 4. CHEMICAL COMPOSITION OF MAJOR CONSTITUENTS OF THE EARTH'S CRUST

	Granite	Tholeiitic Basalt	Average Shale	Average Sandstone	Average Limestone
SiO_2	70.72	50.25	58.10	78.33	5.19
TiO_2	0.41	1.56	0.65	0.25	0.06
Al_2O_3	13.11	16.09	15.40	4.77	0.81
Fe_2O_3	1.85	2.72	4.02	1.07	0.54
FeO	1.97	7.20	2.45	0.30	—
MgO	0.50	7.02	2.44	1.16	7.89
CaO	1.36	11.81	3.11	5.50	42.57
Na_2O	3.35	2.81	1.30	0.45	0.05
K_2O	5.60	0.20	3.24	1.31	0.33
H_2O	—	—	5.00	1.63	0.77
P_2O_5	0.23	0.15	0.17	0.08	0.04
CO_2	—	—	2.63	5.03	41.54
SO_3	—	—	0.64	0.07	0.05
BaO	—	—	0.05	0.05	—
C	—	—	0.80	—	—
MnO	—	0.19	—	—	—

GEOCHEMISTRY OF SEDIMENTS: MODERN

Table 5. Enrichment of Elements in Sediments

Element	Enriched in
Si	Resistates, radiolarian, and diatom oozes
Al	Bauxites, hydrolysates
Ti	Hydrolysates
Fe	Oxidates
Mn	Oxidates
Mg	Carbonates, evaporites
Ca	Carbonates, evaporites
Na	Evaporites
K	Evaporites
P	Phosphate deposits
S	Reduceates, evaporites
Cl	Evaporites
Ga	Hydrolysates
Cu	Oxidates
Co	Oxidates
Ni	Oxidates
V	Reduceates, oxidates
U	Reduceates
B	Hydrolysates
As	Oxidates
Hg	Oxidates

Hydrolysates consist of clays derived from breakdown of silicates. Aluminum, therefore, tends to be enriched in this fraction, as also does potassium by incorporation into the clay minerals. Titanium also tends to be incorporated in these sediments. However, intimately mixed with clays derived from weathering, there are also micas and clays derived intact from preexisting igneous and sedimentary rocks; a sharp distinction usually cannot be made.

Many trace elements are incorporated in hydrolysates; e.g., boron (B) is incorporated by adsorption on the clay minerals and gallium (Ga) by substitution for aluminum (Al).

Oxidates consist largely of oxides of iron and manganese, and in minor or trace quantities also arsenic (As), mercury (Hg), copper (Cu), cobalt (Co), nickel (Ni), lead (Pb), vanadium (V), etc. In extreme cases iron ores or manganese ores result from oxidate deposition.

Carbonates consist largely of carbonates of calcium (Ca) and magnesium (Mg) and only to a small extent of manganese (Mn) or iron (Fe). A typical trace constituent is strontium (Sr).

Evaporites consist largely of chlorides and sulfates of sodium (Na), potassium (K), magnesium (Mg), and calcium (Ca). Depending on the conditions of formation (marine or continental, and the source areas of material) a large variety of evaporites are formed. Thus, for instance, sodium carbonates and boron minerals may be much more enriched in some types of evaporites than others.

Reduceates are strongly enriched in organic matter. The reducing properties of organic matter lead to the formation of sulfides; many trace elements are also enriched in reduceates, the most conspicuous ones being uranium (U) and vanadium (V).

Resistates, hydrolysates, etc., rarely occur in the extreme forms discussed above; most sediments are mixtures of these extremes. In particular, hydrolysate matter is dispersed among all other constituents; even in evaporites some clayey matter settles. Oxidates and reduceates, however, exclude each other but can occur in the same locality in alternating layers.

Other classifications of sediments and sedimentary components have also been proposed; these include *cosmogenous* (for meteorite debris and other extraterrestrial matter), *lithogenous* (from rock material; incorporates quartz, feldspar, clay minerals, micas, pyroxenes, amphiboles, volcanic glass), *hydrogenous* (deposited from aqueous phase: includes constituents such as ferromanganoan minerals, phosphorites, barite, pyrite, phillipsite), *biogenous* (derived from organisms; includes calcites, aragonite, opal, organic matter), and *pore water* (from seawater and its interreaction with deposited matter; includes sea salts). Other terms used are *terrigenous* (of continental origin) or *authigenic* (formed in situ by interaction of various aqueous and solid phases). Often it is impossible to establish to which of these classifications or groupings a constituent should be referred. A mica of igneous origin that enters the sea in a partly decomposed state and incorporates some elements from seawater, for instance Mg, could be called terrigenous or authigenous, it could be the minor constituent of a resistate, reduceate, or evaporite, or be the major constituent of a hydrolysate.

Other classifications are based on environment of deposition or *"facies"* such as shelf sediment, lake sediment, or deep-sea sediments. Deep-sea sediments for instance, can be divided into two major groups; *pelagic sediments* with terrigenous matter of colloidal size or less, and various *muds,* etc., in which the terrigenous matter exceeds colloidal dimension (see Table 6).

Geotectonic Cycles

The oceans have probably existed on the earth's surface for at least three billion years. Yet no sediments from Paleozoic or older are known in the deep sea. Several explanations have been suggested for this in the past, but recent findings indicate that the ocean floors are spreading out from some major oceanic

GEOCHEMISTRY OF SEDIMENTS: MODERN

TABLE 6. CLASSIFICATION OF DEEP-SEA SEDIMENTS

	Pelagic
Brown (red) clay	< 30% Biogenous matter; no terrigenous matter exceeding colloidal size
Diagenetic deposits	(Much phillipsite, Mn nodules)
Biogenous deposits	(> 30% Biogenous matter)
Foraminiferal ooze	(Globigerina ooze); > 30% calcareous matter
Diatom ooze	> 30% Siliceous biogenous matter, largely diatoms
Radiolarian ooze	> 30% Siliceous biogenous matter, largely radiolaria
Coral reef debris	(Slumped coral debris around atolls and reefs, etc.)
Vulcanogen deposits	Pyroclastics and other products of volcanic activity
	Terrigenous
Terrigenous muds	(> 30% Silt and sand of definitive terrigenous origin)
Green muds	
Black muds	
Red muds	
Turbiditites	Formed by turbidity currents
Slide deposits	Formed by slumping
Glacial marine	Formed by activity of ice, often transported by ice

ridges (such as the East Pacific Rise and Mid-Atlantic Ridge) and are thrust under island arcs such as Japan, Kuriles) or continents (as South America). It has been suggested that the under-thrust sediments become involved in the formation of new igneous rocks. Such recycling processes could indeed explain the variability of igneous rocks, since magmatic differentiation, for instance in Hawaii, does not lead to the formation of large volumes of silicic rock. Crude estimates indicate that the amounts of sediments deposited in the past in geosynclines before an orogenic event, and the amount of deep-sea sediment underthrust, are equivalent in mass to the matter now present in batholiths. Fusion and incorporation into a magmatic event, therefore, seems to be the ultimate fate of most sediments. Geochemical balance calculations suggest that this is possible, with the important exception of sodium which cannot be balanced (that is, the amount of sodium in sediments and pore waters is less than is present in batholithic rocks). It has furthermore been suggested that in underthrust or downfolded sediments, many constituents are mobilized and are deposited in shallower sections of the Earth's crust, sometimes as ore bodies.

KURT BOSTRÖM

References

Goldschmidt, V. M., 1954, "Geochemistry," Oxford, Clarendon Press, 730pp.

Rankama, K., and Sahama, Th., 1950, "Geochemistry," University of Chicago Press, 912pp.

Degens, E. T., 1965, "Geochemistry of Sediments," Englewood Cliffs, N.J., Prentice-Hall, 342pp. The section on organic geochemistry is particularly good.

Clarke, F. W., 1924, "Data of Geochemistry," Fifth ed., *U.S. Geol. Surv. Bull.* **770**.

"Data of Geochemistry," Sixth ed., *U.S. Geol. Surv. Profess. Paper* **440, A**.

A large number of articles on the geochemistry of sediments also appear in periodicals such as: *Geochimica et Cosmochimica Acta* (Vol. 1, 1950 through succeeding volumes to the present day).

Geokhymia (available in Engl. transl. as *Geochemistry* from 1956–1963; superseded by *Geochemistry International,* 1964 to present).

Cross-references: *Abrasion pH; Authigenesis of Minerals—Marine; Authigenesis of Minerals—Nonmarine; Calcium Carbonate Geochemistry; Chelation; Earth's Crust Geochemistry; Organic Geochemistry; Oxides and Hydroxides; Solid Solution; Sulfides; Trace Elements; Weathering, Chemical.* See also *Individual Minerals.* Vol. I: *Marine Sediments Mid-Oceanic Ridge.* Vol. III: *Weathering.* Vol. IVB: *Clays and Clay Minerals; Evaporite Minerals; Feldspars; Organo-Hydrolysates; Quartz; Zeolites.* Vol. V: *Sea-floor Spreading; Sövite.* Vol. VI: *Heavy Minerals; Muds; Pelagic Sediments; Saprolite; Soil Genesis; Weathering, Organic.*

GEOCHEMISTRY OF SEDIMENTARY SILICA

Geochemistry

The primary source of silica in solution at or near the earth's surface is the chemical weathering of silicate minerals (Lovering, 1923; Correns, 1940; Siever, 1957). In the decomposition of these minerals by aqueous solution, varying quantities of silica are contributed at different rates depending upon climate, relief, and characteristics of the parent material. The most effective weathering of silicate minerals is accomplished by acidic aqueous solutions, whereas the opal and chalcedony are more readily attacked by alkaline solutions (Lovering, 1923). Quartz, jasper, and taconite are more soluble in solutions of magnesium, calcium, and sodium bicarbonates. Lisitsyn (1967) related the distribution of modern siliceous sediments to climatic zonation. He delineated three major zones: northern, equatorial, and southern. He concluded that the positions of these zones are controlled by the distribution and settling of siliceous phytoplanktonic organisms and sources of terrigenous dilution.

The weathering of silicate minerals is primarily a function of two variables: (1) the mobilities of major cations and (2) the mineral structure (Loughnan, 1962). It appears that in the normal pH range of 4 to 10, the leaching potential is the most important factor influencing the mobility of the cations, through Eh, the redox potential, affects some cations. Parent mineral structures are instrumental in controlling the accessibility of percolating waters to the bonding cations. The higher temperature, metastable silicates of igneous and metamorphic origin decompose rapidly during soil formation and supply the greatest proportion of H_4SiO_4 or monomeric silica. According to Siever (1957) and Goto (1955), if the pH is less than 9.5 the silica in solution will be monomeric. If the pH is greater than 9.5, polymeric silica forms, but if this alkaline solution comes in contact with the near neutral or slightly acid stream water, it will depolymerize to the monomeric form, H_4SiO_4. Okamoto et al. (1957) illustrated that seawater is also an excellent depolymerizer of colloidal silica. Where magnesium salts (found in seawater) and weak acids are present, the lower the silica content in the mineral, the greater the dissolution of the silica (Gruner 1922).

Organic matter plays an important role in the dissolution of amorphous silica (Siever, 1962). If organic adsorbates on the silica surface lower the solubility, one may expect silica will not reach equilibrium with the surrounding waters until the organic matter is removed by decomposition.

Siever (1957, 1962) discussed the importance of alteration of clay to produce free silica during weathering. Alteration of one clay type to another with different silica-alumina ratios will lead to silica liberation to, or absorption from, the surroundings. The reactions which result in the production of silica are those involving the alteration of montmorillonite to muscovite, illite, and kaolinite, or alteration of kaolinite to gibbsite.

The silica in the interstitial waters may originate from the diagenetic alteration of clay or, during compaction, from the dissolution of the small grains of quartz dispersed throughout a shale (Emery and Rittenberg, 1952) or carbonate rock (Hixon, 1964). There is accordingly a relative absence of quartz in sediments of <0.1 micron size. There is additional silica in pore water because of the fact that temperatures associated with burial and compaction are high enough to cause considerable dissolution of silica. However, Kennedy (1950) illustrated that below 140°C the solubility of quartz remains low and he considered it unlikely that a sufficient temperature for silica solution would be reached before the shale was completely compacted. During the compaction of muds the pore fluids are forced upward, due to a vertical gradient of compaction. The compacted fine-grained sediment acts as a semipermeable membrane which selectively concentrates silica immediately below the shale beds by retarding the passage of ions and permitting the passage of water.

The behavior and volume of silica available during diagenesis has been examined by Dapples (1959), Siever (1962), Emery and Rittenberg (1952) and others. Fine-grained silica is unstable in most depositional environments; minor but persistent solution of silica occurs resulting in the pitting and rounding of quartz grains. Early burial is characterized by precipitation of the minor amounts of silica. With increasing depth of burial, carbonates begin to replace quartz or, if carbonate is absent, the quartz will develop deep intergranular penetrations. This dissolution of silica and replacement by carbonate is dependent on temperature and pH values greater than 8.0. The dissolution of siliceous organisms in the early diagenesis of sediments is considered a major source of silica in solution. These organic forms of low-ordered silica have a solubility of about 140 ppm in contrast to crystalline (well-ordered) quartz, the solubility of which in seawater ranges from 6 to 8 ppm. The silica-secreting organisms have an amazing ability to precipitate large quantities of silica out of seawater, which is already undersaturated in amorphous silica. Jorgensen (1953), experimenting with diatoms, found that regardless of the initial concentration of silica, which ranged from 0.65 to 125 ppm, the organisms in a few days reduced the silica in solution to as little as 0.065 ppm. In view of this efficient extraction of silica, it appears that these siliceous organisms are responsible for keeping the oceans undersaturated in silica (seawater contains from 2.0 to 14.0 ppm). Analysis of interstitial water in modern sediments of the continental shelf reveal an increase of concentration of dissolved silica a short distance below the sediment-water interface (Emery and Rittenberg, 1952). Bruevich (1953), Bezruhkov (1955), and Lisitsyn (1955) have made similar observations. They also observed corroded and partially dissolved diatom shells in the sediments of the Bering Sea. The dissolution of amorphous silica is nevertheless very slow (Krauskopf, 1956).

Thermal springs and magmatic waters are an additional source of dissolved silica (Taliaferro, 1933). This source is very important locally, although it probably plays a minor role in the overall silica supply. The silica content of hot springs commonly ranges from 200 to 300 ppm, and occasionally as high as 500 ppm (White, 1947).

This supersaturation accounts for the siliceous sinter and geyserite deposits associated with hot springs. Submarine thermal springs probably would not lead to such concentrations of silica unless the hot springs were below or isolated from wave action. Davis (1918) attributed the origin of the bedded cherts of the Franciscan formation of California to silica supplied by submarine thermal springs (for reference, see Cressman, 1962).

Silica derived from the infalls of volcanic ash and breccia is another possible source for silica in seawater. Volcanic ash and breccia are often intercalated with bedded cherts (Taliaferro, 1933; Goldstein, 1959). Silica may be supplied both by extrusive volcanism and submarine weathering of volcanic ash. Many writers, on the other hand, attribute the silica in bedded cherts to siliceous organisms that would propagate rapidly in such a silica-rich environment. The field relationships and petrographic evidence are broad enough to support both theories.

Eugster (1967) found two new hydrous sodium silicates, $NaSi_7O_{13}(OH)_3 \cdot 3H_2O$ (magadiite) and $NaSi_{11}O_{20.5}(OH)_4 \cdot 3H_2O$ (kenyaite), in the bottom sediments at Lake Magadi, Kenya. Concretions within the magadiite bed consist of chert surrounded by kenyaite. In dilute acids magadiite and kenyaite are converted to $6SiO_2 \cdot H_2O(SH)$, the first known crystalline hydrate of silica. Eugster regards the magadiite as probably a chemical precipitate from alkaline brines which may then be converted by percolating water to kenyaite and eventually to chert. He concluded that this sequence also may provide a mechanism for the genesis of bedded cherts by inorganic precipitation.

Silica Solubility

Silica complexes with water as it dissolves, but the stoichiometry of the hydrated species at any temperature and pressure is not fully known. It is suspected that the complex is $SiO_2 \cdot 2H_2O$ or $Si(OH)_4$ because silicon exhibits a strong tendency to share tetrahedrally four oxygen ions in silicate structures.

Silica may occur in water as molecular or colloidal dispersions. Most of the early investigators believed that natural silica existed largely in the colloidal state, based on work by Kahlenberg and Lincoln (1898), who studied the properties of silica in aqueous solutions without making a single analysis of natural waters. They concluded that dilute solutions in nature (5–30 ppm SiO_2 in most rivers) were also colloidal. This conclusion had been accepted by almost every geologic author concerned with silica solubility until the work of Roy in 1945. Most early investigators were not aware of the ammonium molybdate colorimetric tests to determine silica. The monomeric nature of silica is established by a rapid reaction with ammonium molybdate to form silicomolybdates. Silica in true solution has been referred to by Roy (1945) as molecular dispersed, crystalloid, monosilicic, and monomeric states. Studies by Okura (1951), Iwasaki et al. (1951, 1954), and Krauskopf (1956) suggest that silica in natural waters may occur either in colloidal dispersion or true solution (for Okura and Iwasaki references, see Krauskopf, 1959). However, the micelles of the dispersion are unstable and will be rapidly dispersed if the silica content is less than the amount required for a saturated solution (which is rarely achieved).

The point of saturation is dependent not only upon the temperature and pH condition of the water but on the clay minerals present. These aluminosilicates rapidly release silica to seawater low in silica. Thus, by releasing or combining with silica, aluminosilicates exert a major control on the silica concentration in oceans (MacKenzie and Garrels, 1965).

Krauskopf (1959) discussed the depolymerization and dissolution of silica and the tendency for colloidal suspension and gels to revert to monomeric silica. These reactions are so slow that supersaturated solutions of molecular silica can exist for weeks or months before excess silica coagulates, and dilute suspensions of colloidal silica completely disaggregate into monomeric acid. At ambient temperatures and pressures amorphous silica dissolves forming a true solution (100–140 ppm) in either fresh or marine water, whereas crystalline forms of silica have lower solubilities; e.g., the comparable solubility for quartz is only 6–14 ppm. The solubility increases as the temperature rises but is little affected by changes of pH in the range of 0 to 9. In natural waters silica ranges from approximately 0.1 ppm in snow to 4000 ppm in a California mineral spring. However, it is less variable than any other major dissolved constituents of natural water. Its concentration is less than the amount at equilibrium with the amorphous silica because it is derived largely from materials less soluble than amorphous silica and because it is removed from solution by several factors.

The factors which influence the solubility and rate of reactions between colloidal and monomeric silica are as follows (Krauskopf, 1959): (*1*) temperature; (*2*) pH condition; (*3*) presence of other ions in solution; (*4*) organisms, such as phytoplankton, radiolaria, siliceous sponges, and certain algae; (*5*) degree of crystallinity of silica in solution; (*6*) presence of amorphous silica; (*7*) amount of initial concentration of silica; and (*8*) presence of other solids.

The influences of these various factors are summarized as follows: The equilibrium solubility is increased by a rise in temperature and by the increase of a pH condition greater than 9. It is decreased by the presence of trivalent aluminum and Al_2O_3 (Okamoto et al., 1957); of sodium and large concentrations of carbon dioxide (Lovering and Patten, 1962). The equilibrium solubility is also decreased by the presence of organisms such as sponges (Votinsev, 1948), radiolaria (Arrhenius, 1952), phytoplankton (Jorgensen, 1953), and algae (White et al., 1956). It is also affected by the type of amorphous silica present because the surface area available for reaction varies considerably in the colloidal, gel, opaline, or glassy states.

The equilibrium solubility is decreased by the presence of amorphous silica. White, et al. (1956) reported that the addition of opaline silica caused the immediate precipitation of the colloidal silica from supersaturated hot-spring waters. The rates of dissolution and polymerization are both increased by a rise in temperature, by an increase in pH, and by increased concentrations of electrolytes. Rapid dissolution is also favored by an increase in surface area of the silica, and by an increase in the degree of initial supersaturation. Suspended solids such as ferric hydroxide and certain aluminosilicates may liberate or absorb silica from the surrounding aqueous solution (Siever, 1957, 1962; MacKenzie and Garrels, 1965).

Precipitation of silica can occur by evaporation, cooling neutralization of strongly alkaline solution, reaction with cations, adsorption, and life process of siliceous organisms.

From aqueous solutions of monomeric silica, which is below its equilibrium solubility, crystalline quartz can precipitate but the rate of crystallization at ambient temperature is extremely slow. Colloidal silica with a trace of aluminum ion may assist the precipitation. Siliceous organisms carry out biogenic precipitation. According to Krauskopf (1959) and Bruevich (1953), seawater is an excellent depolymerizer because of the presence of electrolytes. However, one would not expect direct precipitation by this means alone, because the concentrations in normal seawater are far below the equilibrium solubility. Rapid direct precipitation of silica would occur where the equilibrium solubility is exceeded, as in areas of submarine volcanic activity, and where the pH is above 7.

Studies by Bien et al. (1958) around the Mississippi Delta showed that most of the soluble silica was removed from the river water by some process other than mixing with seawater. Although biological uptake by diatoms could account for part of this removal, it is more likely that a major portion is removed by inorganic precipitation. As the chlorosity increased, the silica content decreased. However, the silica content dropped off more rapidly than predicted. Further experiments by Bien et al. showed that inorganic precipitation would occur when suspended solids such as bentonite or aluminum oxide were added to a mixture of seawater and fresh water. They concluded that this was probably an adsorption or coprecipitation process where ions in seawater and suspended matter both are necessary for removal of silica from water.

Correns (1941, 1950) observed an inverse relationship in the solubility of calcium carbonate and silica. In a pH range of 7.5 to 9.0, calcium carbonate becomes more soluble if the pH is slightly lowered; and silica solubility is increased if the pH is raised. Thus changes in pH through a small range could lead to dissolution of calcium carbonate and simultaneous precipitation of silica. This inverse relationship explains the origin of chert nodules in limestone by replacement of the limestone by silica. Sakamoto (1950) suggested that the fluctuation in the pH factor of stream and lake water due to seasonal variation in rainfall would be enough to account for the banded chert of Precambrian iron formation.

Chert

Precipitated silica is disseminated widely through sediments. Most sedimentary silica is classified as *chert* (see Vol. VI), either concretionary ("*flint*") or stratified. In order for chert to form, it requires a very abundant supply of silica, early diagenetic redistribution of freshly precipitated silica, or else late or postdiagenetic replacement of carbonates.

GEOCHEMISTRY OF SEDIMENTARY SILICA

Mineralogy and Crystal Chemistry of Chert

Chert is a common fine-grained siliceous sediment characterized by a wide compositional range of various forms of cryptocrystalline, microcrystalline, and low-ordered quartz. Common varieties and synonyms for chert include jasper, flint, novaculite, silexite, hornstone, lydite, and phthanite. Chert may be composed entirely of opal, chalcedony, and quartz of several species, or may contain varying proportions of each. A general definition by Frondel (1962, p. 221) is that chert is a rock which consists primarily of chalcedony with random interlocking grains of microcrystalline granular and cryptocrystalline quartz. Many cherts are extremely fine-grained and possess many minute pores that make the cherts appear isotropic under the polarizing microscope because of aggregate polarization (light absorption by the many small grains). Unreplaced remains of fossils are common, as are disseminated carbonate rhombs, pyrite, organic carbon, chlorite, and other clay minerals. *Jasper* is chert stained yellow, brown, green, or red. *Flint* is cryptocrystalline quartz which exhibits a conchoidal fracture and is considered by the U.S. Geological Survey to be a variety of chert.

A study of texture and composition was made by Folk and Weaver (1952) who concluded that chert is composed of two petrographic end members: microcrystalline quartz and optically fibrous chalcedonic quartz. No opal-chalcedony admixtures were noted. The composition of chalcedony is still controversial. Correns and Nagelschmidt (1933) and Donnay (1936) believed that chalcedony is composed of a submicroscopic admixture of fibrous quartz and interstitial amorphous silica, probably opal (for references see Siever,* 1962). Later workers have disputed the validity of this admixture. Midgley (1951) applied x-ray diffraction techniques to this problem, however, and concluded that chalcedony seems to be microcrystalline and cryptocrystalline quartz with submicroscopic pores. If *opal* (q.v.) is present its concentration must be considerably less than 10%. These pores are filled with water or air and apparently are the cause of the lower refractive index and density of microcrystalline quartz and chalcedony as compared with normal quartz (see Table 1). Midgley found that the values of the physical properties seem to change in proportion to the abundance of the cavities and the distribution of textural types of fine-grained silica. These fine-grained textural types of silica, when free of voids, have the same properties as quartz. Other fibrous species of quartz, which include quartzine and lutecine, have been recognized in petrographic thin sections by two European workers, Braitsch (1957) and Wilson (1966). In addition, they have recognized lussatite and lussatine, which are varieties of cristobalite. These varieties of silica have not been generally distinguished by American petrographers.

Folk and Weaver (1952), utilizing the electron microscope, distinguished two major types of surface morphologies which they designated *novaculite* and *spongy chert*. The novaculite type is composed of polyhedral blocks with curved surfaces, whereas the spongy type consists of chalcedonic quartz with numerous spherical cavities. The novaculite and spongy morphologies grade into one another. Folk and Weaver suggest that the primary factor which determines the surface morphology is the spacing between sites of crystal nucleation. The work of Darragh et al. (1965) on the origin of precious opal also supports the conclusions of Folk and Weaver. Recent electron optical studies of fine-grained silica by Iwao et al. (1953), Bates (1958), Pittman (1959), Hixon (1964, 1964a), Monroe (1964), and

TABLE 1. PORE DATA ON ELECTRON OPTICAL MORPHOLOGIES OF CHERT[a]

Type of Surface Morphology	Specific Gravity	Average Pore Size (μ)	Pore Densities (pores/μ^2)	Average Total Porosity
Porous (secondary)	2.10–2.40	0.50	300–400	20%
Crystallite	2.40–2.55	0.22	Void space primarily interstitial	10%
Polygonal	2.55–2.65	0.14	Void space primarily interstitial	4%
Intermediate	2.45–2.60	0.46	40–100	6%
Spongy (primary)	2.35–2.45	0.1–0.0015	400–1000	12%
Gel (SMR-55-9112*)	1.9 estd.	0.003	800–2400	Variable, but usually high

[a] From Kneller, 1968, p. 80.

GEOCHEMISTRY OF SEDIMENTARY SILICA

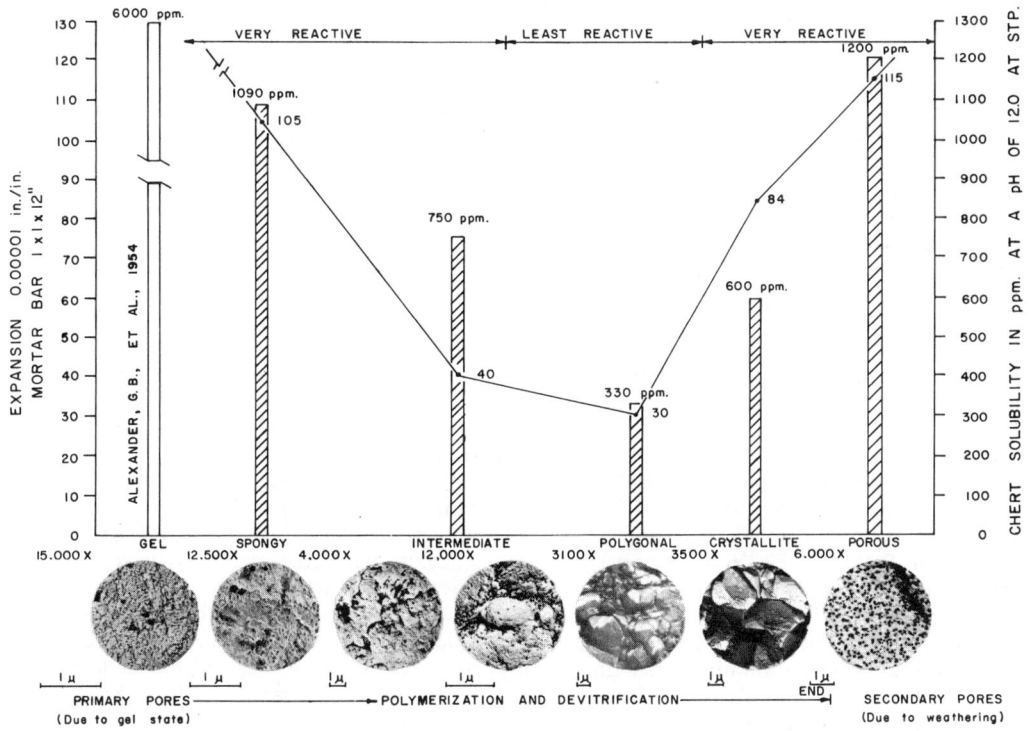

FIG. 1. Electron optical surface morphologies of chert (from Kneller, 1968, p. 79).

Kneller (1965, 1967, 1968) have disclosed similar findings (for references see Kneller, 1968).

Sargent (1923) noted that some chalcedony in thin section may appear brownish and exhibit a lower refractive index. He ascribed this to zones of greater hydration. Folk and Weaver (1952) and Pelto (1956) suggested that they are caused by unspecified dispersion effects between the bubble-filling material and quartz.

Pelto (1956) suggested that the chemical reactivity of chalcedony and chalcedonic cherts is related to the imperfection in the chalcedony crystal structure, similar to grain boundary phenomena in metals. If true, the presence of dislocations and a possible thin vitreous layer interface between the fibrous grains should increase considerably the surface area available for chemical reaction. In addition, the reactivity of chert may be attributed to greater solubility of imperfectly formed, transitional vitreous zones at the ultramicroscopic grain boundaries within fine-grained silica. These transitions are marked by dislocations or islands of bad fit which accommodate impurities such as water. The presence of such regions would increase the solubility of chert.

Kneller (1968) compared the solubility and potential alkali reactivity with the chert morphologies of twenty-two varieties of chert as observed by the electron microscope. He suggested that these morphologies represent possible stages of devitrification and polymerization of fine-grained silica. He also noted that a change in surface area and crystallinity occurs with the grain growth and influences the physical and chemical behavior of the chert. He concluded that degree of devitrification and polymerization is a function of the following factors: (1) time; (2) depth of burial; (3) tectonics; and (4) diagenesis. Fig. 1 and Table 1 summarize this work.

Chert is tough and brittle, and exhibits a splintery to conchoidal fracture. Its hardness approaches that of quartz (7 on Mohs' scale). The abrasive properties of chert are not like those of quartz due to the fibrous nature of chalcedony, and to the variable amount of pore space between the fibers and crystallites. The solubility of chert is higher than that of quartz since solubility is affected by grain size, disordering, and other surface chemistry phenomena.

Crystal chemistry of silica minerals is more complex than often assumed. Summaries of recent developments and references are given by Eitel (1964–66). Especially important are the data of Buerger (1954), who demonstrated that small amounts of trivalent aluminium replace quadrivalent silicon in tridymite. The excess valence charge is made up by alkali or alkaline ions which are stuffed into the large interstitial spaces in the structure. Flörke (1955) observed that most low-temperature opal is not truly amorphous but contains inner stratified layers of cristobalite and tridymite. The tridymite layers result from the disordering of the structure by aluminum, alkali, and alkaline earth ions. The amount of the disordering cations may be considerable. Buerger and Lukesh (1942) report the approximate formula of tridymite from Las Plumas, California, as $NaCaAl_3Si_{15}O_{36}$, and Roy and Roy (1959) have synthesized cristobalite-tridymite solid solutions at high temperature that contain more than 2 mole % $NaAlO_2$. The quartz structure, however, can be stuffed only by univalent lithium, and possibly by divalent beryllium and trivalent boron, because the interatomic space in quartz is much smaller (Buerger, 1954). Keith and Tuttle (1952) have found that aluminum probably substitutes for silicon in the quartz structure with the valence charge compensated by lithium. They suggest from consideration of valence requirements and ionic radii, that titanium can also be substituted for silicon, but their analyses of natural quartz indicate that any substitution by aluminum or titanium structure is very minor. Eitel (1954) states that in addition to these substitutions, ferric iron might substitute for silicon, and hydroxyl for oxygen in chalcedony (see Cressman, 1962, for most of the references cited in this paragraph). Table 2 summarizes the most generally accepted replacement ions in α-quartz and the effect of their admittance into the structure upon the cell parameters (Wilband and Kneller, 1967).

Mielenz et al. (1949) and Swineford and Franks (1959) demonstrated that the silica in siliceous organisms is in the form of opal and that in diatoms, sponges, and radiolara it is amorphous. The index of refraction in opal of siliceous organisms is approximately 1.440 ± 0.002 which indicates, according to Bramlette (1946), a water content of about 9%. The opal of diatoms must be of high purity because when diatomite is cooled from high temperature, tridymite does not form unless cations necessary to disorder the cristobalite have been added (Eitel, 1957). Mielenz, et al. (1949) report the occurrence of cristobalite in an opaline shale, whereas Swineford and Franks (1959) observed from diffraction patterns low cristobalite and tridymite in massive opal, opaline cement in sandstone, and silicified wood from the Pliocene of Kansas.

Most of the siliceous shells and the silica cement in deposits of Tertiary age are opaline, but nearly all of the organic and inorganic silica in Paleozoic cherts is present as microcrystalline quartz and chalcedony. Hoss (1957) advanced evidence for the presence of small amounts of opal in cherts of early Carboniferous age. Opal, quartz, and chalcedony may all be found in Mesozoic cherts. It has long been concluded from these facts that most chert was originally opaline and that opal devitrifies with age to chalcedony and quartz. However, factors other than age are involved, and according to Bramlette (1946) hydrostatic load and deformation (stress) can accelerate the conversion of opal to chalcedony and eventually to quartz. The structure and composition of the opal probably has an effect on the ease of devitrification transformation; e.g., an opal with a tridymite structure might be difficult to convert to quartz because of the large cations stuffed in the interstices, thus stabilizing the crystal structure (Eitel, 1954).

TABLE 2. GENERAL REPLACEMENT IONS IN QUARTZ AND THE EFFECT OF THEIR ADMITTANCE INTO THE STRUCTURE UPON THE CELL PARAMETERS[a]

Type of Impurity	Balancing Ions (position)	Change in Cell Parameters
A. Substitutional impurity replaces Si^{4+}		
(a) Ti^{4+}, Ge^{4+}, B^{4+}		Expansion in α_o and c_o
(b) Al^{3+}, Fe^{3+}, Mn^{3+}, B^{3+}	H^+, Li^+, Na^+ (interstitial)	Unequal expansion in α_o and c_o; greatest in α_o
(c) Al^{3+}	OH^- for O^{2-} (tetrahedra)	Expansion in α_o and c_o
(d) 2 Al^{3+}	1 Mg^{2+}, 1 Ca^{+2} (interstitial)	Unequal expansion in α_o and c_o; greatest in α_o
(e) 3 Al^{3+}	1 Al^{3+}, 1 Fe^{3+} (interstitial)	Expansion in α_o and c_o
B. Interstitial impurity "stuffing"	Most rock-forming elements (K^+, Ca^{+2}, Mn^{+2}, Mg^{+2}, Fe^{+2}) and others[b]	Expansion mainly in α_o

[a] From Wilband and Kneller, 1967, p. 115. Data compiled from Dennen (1966), Frondel (1962), and Cohen and Sumner (1958).
[b] Frondel (1962) reports Rb, Cs, Ba, Pb, Ag, Sn, Cu, Zn, V, Cr, Zr, and U.

Problem of Chert Genesis

There are essentially three theories of chert genesis: primary or syngenetic deposition, secondary or epigenetic replacement of a host (commonly a carbonate rock), and early diagenetic or penecontemporaneous deposition. The syngenetic or contemporaneous theories involved the deposition of silica during sedimentation by detrital, biochemical, magmatic, or direct chemical processes. The epigenetic or secondary theories assume that silica was introduced after sedimentation and was precipitated either in the zone of cementation or in the zone of weathering, Siever (1962) and Dapples (1959) prefer a compromise between these two views. They propose a theory of penecontemporaneous precipitation where the silica is deposited after sedimentation but before consolidation of the host rock. In the past there has been no agreement on how silica was deposited or transformed into chert. In 1965, Peterson and von der Borch (1965) found chert precipitating as a gelatinous opal-cristobalite in lakes associated with the Coorong Lagoons of South Australia. They also observed the deposition of dolomite, magnesite, and magnesian calcite. They concluded that a high pH (9.5 to 10.2) causes the dissolution of detrital silicates, and a lowering of pH (7.0 to 6.5) concomitant with drying of the lakes causes the direct inorganic precipitation of chert. Other investigators who support the thesis that chert can be directly precipitated from natural waters in the form of gelatinous masses on a sediment floor include Tarr (1917), Trefethen (1947), Tarr and Twenhofel (1932), Eugster (1967), and Eugster and Jones (1968). Those authors who suggest that chert is due to secondary replacement of the host rock are Van Tuyl (1918), Fowler et al. (1934), Bastin (1933), White (1947), Emery and Rittenberg (1952), Walker (1962), and Lovering and Patten (1962).

The origin of thick, extensively bedded cherts such as the Monterey (Bramlette, 1946) and Franciscan series of California (Davis, 1918; see reference in Cressman, 1962) and the siliceous shales such as the Mowry formation of the Black Hills region (Rubey, 1929) is more difficult to explain. Several authors postulate that the source for the silica may be by direct precipitation from seawater which had been enriched in silica by submarine volcanic emanations or from river water rich in colloidal silica. Other authors suggest that an abundant source of silica in seawater is the dissolution, migration, and subsequent precipitation of biochemical silica (Bramlette, 1946; Correns, 1950; Krauskopf, 1956, 1959; Cressman, 1962; Siever, 1957, 1962).

Evidently cherts do not have a single mode of origin, but are polygenetic.

WILLIAM A. KNELLER
WILLIAM L. HISS

References

Arrhenius, G., 1952, "Sediment cores from the East Pacific," *Rept. Swedish Deep Sea Exped. 1947-48*, **5**(1), 227pp.

Bastin, E. S., 1933, "Relations of cherts to stylolites at Carthage, Missouri," *J. Geol.*, **41**, 371–381.

Bezrukov, P. L., 1955, "Distribution and rate of sedimentation of silica-rich silts in the sea of Okhotsk," *Dokl. Akad. Nauk S.S.S.R.*, **103**, 473–476.

Bien, G. S.; Contois, D. E.; Thomas, W. H., 1958, "Removal of soluble silica from fresh water entering the sea," *Geochim. Cosmochim. Acta*, **14**, 35–54. (Reprinted in *Silica in Sediments*, Spec. Publ. No. 7, Soc. Econ. Paleontologists and Mineralogists, pp. 20–35, 1959).

Braitsch, O., 1957, "Über die natürlichen Faser- und Aggregationstypen beim SiO$_2$ ihre Verwachsungsformen, Richtungs-statistik, und Doppelbrechung," *Heidelberger Beitr. Mineral. Petrog.*, **5**, 331–372.

Bramlette, M. N., 1946, "The Monterey formation of California and the origin of its siliceous rocks," *U.S. Geol. Surv. Profess. Paper* **212**, 57pp.

Bruevich, S. V., 1953, "K geokhimii kremniga v morye" (Geochemistry of silicon in the sea), *Izv. Akad. Nauk S.S.S.R., Ser. Geol.*, **4**, 67–79.

Buerger, M. J., 1954, "The stuffed derivatives of the silica structures," *Am. Mineralogist*, **39**, 600–614.

Cohen, A. J., Sumner G. G., 1958, "Relationships among impurity contents, color centers and lattice constants in quartz," *Am. Mineralogist*, **43**, 58–68.

Correns, C. W., 1940, "Die chemische Verwitterung des Silikats," *Naturwissenschafte*, **28**, 4.

Correns, C. W., 1941, "Über die Löslichkeit von Kieselsäure in schwach sauren und alkalischen Lösungen," *Chemie der Erde*, **13**, 92–96.

Correns, C. W., 1950, "Zur Geochemie der Diagenese," *Geochim. Cosmochim. Acta*, **1**, 49–54.

*Cressman, E. R., 1962, "Nondetrital siliceous sediments," Chap. T, "Data of Geochem.," Sixth ed., *U. S. Geol. Survey Profess. Paper* **440T**, 23pp.

Dapples, E. C., 1959, "The behaviour of silica in diagenesis," in "Silica in sediments," Soc. Econ. Paleontologists and Mineralogists, Spec. Publ. No. 7, pp. 36–54.

Darragh, P. J., and Gaskin, A. J., et al., 1965, "Origin of precious opal," *Nature*, **209**(5018), 13–16.

Dennen, W. H., 1966, "Stoichiometric substitution in natural quartz," *Geochim. Cosmochim Acta*, **30**, 1235–1242.

Eitel, W., 1954, "Physical Chemistry of the Silicates," Chicago, III., Univ. of Chicago Press, 1592pp.

Eitel, W., 1957, "Structural anomalies in tridymite and cristobalite," *Am. Ceramic Soc. Bull.*, **36**, 142–148.

Eitel, W., 1964–66, "Silicate Science," Vols. 1 (1964), 3 (1965), 4 and 5 (1966), New York and London, Academic Press.

Emery, K. O., and Rittenberg, S. C., 1952, "Early diagenesis of California basin sediments in relation to origin of oil," *Am. Assoc. Petrol. Geol. Bull.*, **36**, 735–806.

Eugster, H. P., 1967, "Hydrous sodium silicates from Lake Magadi, Kanya: precursors of bedded chert," *Science*, **157**, 1177–1180.

Eugster, H. P., and Jones, B. V., 1968, "Gels composed of sodium-aluminum silicate, Lake Magadi, Kenya," *Science,* **161**, 160–163.

Flörke, O. W., 1955, "Zur Frage des "Hoch" Cristobalit in Opalen, Bentoniten, und Gläsern," *Neues Jahrb. Mineral. Monatsh.*, **1955**, 217–223.

Folk, R. L., and Weaver, C. E., 1952, "A study of the texture and composition of chert," *Am. J. Sci.*, **250**, 498–510.

Fowler, G. M., Lyden, J. P., Gregory, F. E., and Agar, W. M., 1934, "Chertification in the tri-state (Oklahoma-Kansas-Missouri) mining district," *Am. Inst. Mining Metl. Engrs. Tech. Publ.* **532**, 1–50.

Frondel, C., 1962, "The System of Mineralogy," Vol. III, "Silica Minerals," New York and London, John Wiley & Sons, 334pp.

Goldstein, A., Jr., 1959, "Cherts and novaculites of Ouachita facies," in "Silica in sediments," Soc. Econ. Paleontologists and Mineralogists, Spec. Publ. No. 7, pp. 135–149.

Goto, K., 1955, "States of silica in aqueous solution," Pt. II, "Solubility of amorphous silica," *J. Chem. Soc. Japan, Pure Chem. Sect.*, **76**, 1364–1366.

Gruner, J. W., 1922, "The origin of sedimentary iron formations: The Biwabik Formation of the Mesabi range," *Econ. Geol.*, **17**, 407.

Hixon, S. B., 1964, "Petrography of the Middle Devonian Bois Blanc Formation of Michigan and Ontario," Unpub. Ph.D Thesis, Univ. of Michigan, 114pp.

Hoss, H., 1957, "Untersuchungen über die Petrographe kulmischer Kielschiefer," *Beiträge Mineral. Petrog.*, **6**, 59–88.

Jorgensen, E. G., 1953, "Silicate assimilation by diatoms," *Physiol. Plantarum*, **6**, 301–315.

Kahlenberg, L., and Lincoln, A. T., 1898, "Solutions of silicates of the alkalies," *J. Phys. Chem.*, **2**, 77–90.

Kennedy, G. C., 1950, "A portion of the system silica-water," *Econ. Geol.*, **45**, 629–653.

*Kneller, W. A., 1968, "A study of chert aggregate reactivity based on observations of chert morphologies using electron optical techniques," in Proc. 17th Ann. Highway Geol. Symposium, Iowa State Highway Comm., Ames, Ia., pp. 73–83.

Krauskopf, K. B., 1956, "Dissolution and precipitation of silica at low temperatures," *Geochim. Cosmochim. Acta*, **10**, 1–26.

*Krauskopf, K. B., 1959, "Geochemistry of silica," in "Silica in sediments," Soc. Econ. Paleontologists and Mineralogists, Spec. Publ. No. 7, pp. 4–19.

Lisitsyn, A. P., 1955, "Distribution of authigenic SiO_2 in the bottom sediments of the western Bering sea," *Dokl. Akad. Nauk S.S.S.R.*, **103**, 479–482 (in Russian).

Lisitsyn, A. P., 1967, "Basic relationships in the distribution of modern siliceous sediments and their connection with climatic zonation," Pts. 1–4, *Internat. Geol. Rev.*, **9**(5–8), 631–652; 842–865; 980–1104; 1114–1130.

Loughnan, F. C., 1962, "Some considerations in the weathering of the silicate minerals," *J. Sediment. Petrol.*, **32**(2), 284–290.

Lovering, T. C., and Patten, L. E., 1962, "The effect of CO_2 at low temperature and pressure on solutions supersaturated with silica in the presence of limestones and dolomite," *Geochim. Cosmochim. Acta*, **26**, 787–796.

Lovering, T. S., 1923, "The leaching of iron protores: Solution and precipitation of silica in cold water," *Econ. Geol.*, **18**, 523–540.

Mackenzie, F. T., and Garrels, R. M., 1965, "Silicates: reactivity with sea water," *Science*, **150**(3692), 57–58.

Midgley, H. C., 1951, "Chalcedony and flint," *Geol. Mag.*, 179–184.

Mielenz, R. C., Witte, L. P., and Glantz, O. J., 1949, "Effect of calcination on natural pozzolans," *ASTM Spec. Tech. Publ.* **99**, 43–91.

Okamoto, G., Okura, T., and Goto, T., 1957, "Properties of silica in water," *Geochim. Cosmochim. Acta,* **12**(1–2), 123–132.

Pelto, C. R., 1956, "A study of chalcedony," *Am. J. Sci.*, **254**, 32–50.

Peterson, M. N. A., and von der Borch, C. C., 1965, "Chert: modern inorganic deposition in a carbonate-precipitating locality," *Science,* **149**(3691), 1501–1503.

Roy, C. J., 1945, "Silica in natural waters," *Am. J. Sci.*, **243**, 393–403.

Rubey, W. W., 1929, "Origin of the siliceous Mowry shale of the Black Hills region," *U.S. Geol. Surv. Profess. Paper* **154-D**, 153–170.

Sakamoto, T., 1950, "The origin of the Precambrian banded iron ores," *Am. J. Sci.*, **248**, 449–474.

Sargent, H. C., 1923, "The massive chert formation of North Flintshire," *Geol. Mag.*, **60**, 168–183.

Siever, R., 1957, "The silica budget in the sedimentary cycle," *Am. Mineralogist,* **42**, 821–841.

Siever, R., 1962, "Silica solubility, 0°–200°C., and the diagenesis of siliceous sediments," *J. Geol.*, **70**, 127–150.

Swineford, A., and Franks, P. C., 1959, "Opal in the Ogallala formation in Kansas," in "Silica in Sediments," Soc. Econ. Paleontologists and Mineralogists, *Spec. Publ.* No. 7, pp. 111–120.

Taliaferro, N. L., 1933, "The relation of volcanism to diatomaceous and associated siliceous sediments," *California Univ., Dept. Geol. Sci. Bull.*, **23**(1), 1–56.

Tarr, W. A., 1917, "Origin of the chert in the Burlington limestone," *Am. J. Sci., 4th Ser.,* **44**, 409–452.

Tarr, W. A., and Twenhofel, W. H., 1932, "Chert and Flint, Treatise on Sedimentation," Second ed., Baltimore, Williams and Wilkins, pp. 519–546.

Trefethen, J. M., 1947, "Some features of the cherts in the vicinity of Columbia, Missouri," *Am. J. Sci.,* **245**, 56–58.

Van Tuyl, F. M., 1918, "The origin of chert," *Am. J. Sci., 4th Ser.,* **45**, 449–456.

Votinsev, K. K., 1948, "Role of sponges in silica cycle in the waters of Lake Baikal," *Dokl. Akad. Nauk S.S.S.R.,* **62**, 661–663 (in Russian).

Walker, T. R., 1962, "Reversible nature of chert-carbonate replacement in sedimentary rocks," *Geol. Soc. Am. Bull.,* **73**(2), 237–242.

White, D. E., 1947, "Diagenetic origin of chert lenses in limestone at Soyatal, State of Queretaro, Mexico," *Am. J. Sci.,* **245**, 49–55.

White, D. E., Brannock, W. W., and Murata, K. J., 1956, "Silica in hot-spring waters," *Geochim. Cosmochim. Acta,* **10**, 27–59.

Wilband, J. T., and Kneller, W. A., 1967, "Relationships between α-quartz cell parameters and some physio-chemical properties of chert aggregates," in A symposium on industrial mineral exploration and development, Spec. Distribution Publ. No. 34, Kansas Geol. Surv., pp. 111–119.

Wilson, R. C. L., 1966, "Silica diagenesis in upper Jurassic limestones of southern England," *J. Sediment. Petrology,* **36**(4), 1036–1049.

Cross-references: *Calcium; Clay Membrane Phenomena; Colloids; Electrolytes: Flocculation of Colloids; Interstitial Waters in Sediments; Lake Geochemistry; Organic Geochemistry; River Geochemistry; Seawater, Chemistry; Silica Solubility, Stoichiometry; Weathering, Chemical; X-Ray Diffraction Analysis.* Vol. I: *Phytoplankton; Submarine Springs.* Vol. IVB: *Chalcedony; Clays and Clay Minerals; Cristobalite; Electron Microscopy; Order-Disorder; Opal; Quartz; Refractive Index.* Vol. VI: *Chert and Flint; Diagenesis; Hot Springs; Jasper; Leaching; Magmatic Waters; Novaculite; Soil Genesis; Volcanic Ash Deposits.* Vol. VII: *Algae; Diatoms; Porifera; Radiolaria.*

GEOCHEMISTRY: TESTING FOR ELEMENTS

Developments in electronic instrumentation, sample handling and preparation, and analytical techniques during the past quarter century make available a wide variety of analytical tools to the geologist and geochemist. The most significant advances in analytical techniques during this period have been those that permit nondestructive analysis of very small amounts of material, and development and refinement of techniques that detect extremely low levels of concentration. The methods now available to the geologist and geochemist range from the simple and easily applied blowpipe techniques to the sophistication of neutron activation analysis.

The classical wet-chemical methods of qualitative analysis still find, and will probably continue to find, extensive utilization in problems of geology. These methods, however, have been augmented by the newer techniques that analyze for an element in terms of certain of its physical properties rather than its chemical properties and do so without loss of the sample or serious modification of its properties. Further advantage of providing quantitative chemical data with the same operations used to identify, many of the new methods have the added composition. Many of the instrumental analytical methods now in common use provide, through some sort of "read-out," a permanent record of the analysis, through either electronic recording or exposure of photographic emulsion.

To the geologist or geochemist, the identification of the mineral, or of the phases present in a substance, is sufficient to establish elemental composition for many purposes. For this important reason, certain of the analytical tools used to identify phases or qualitative mineralogy are included in this discussion.

The information that follows has been derived from many specialized sources to which the reader is referred for more detailed data on equipment, method, application, and cost. In addition, a number of generalized source books and articles are available that provide summary information related to analytical methods. Noteworthy among such works are those of Fletcher (1966), Smales and Wager (1960), and the summary of Melnechuk (1962).

Wet-Chemical Techniques

Most of the easily applied techniques useful in the field and laboratory involve chemical reactions with resultant loss of sample and no permanent record of results. For the most part the various chemical techniques span a broad range of sensitivity. In comparison with many of the techniques of instrumental analysis they require only minimal initial expense in equipment and reagents. They have the disadvantage, in many instances, of requiring a longer time to perform but for many uses, such as certain of the trace element studies, they have greater sensitivity, are less expensive, and are more easily carried out than instrumental analyses. When properly applied by skilled chemists, wet-chemical techniques can be used with a fair degree of confidence in accuracy and precision, dependent mainly on the care exercised in sampling. Wet-chemical methods can be used for either qualitative or quantitative analysis but only rarely can one analytical scheme be useful for both. It is usually necessary that separate methods be used to test specifically for an element and to evaluate specifically its concentration unless analytical schemes are utilized to carry out quantitative chemistry alone.

Although not strictly a wet-chemical method, *blowpipe analysis* (q.v.) is perhaps the most convenient and easily applied means to a rapid qualitative identification. The method involves simple tests by heating or fusing a sample under the blowpipe, the nature of reactions being observed visually. Galbraith (1963) has summarized, with appropriate references, much of the information related to this means of analysis. Largely overshadowed by the more sophisticated analytical tools, the blowpipe remains a fundamental method for use in the field by the lone geological party or in the remote laboratory. Its sensitivity for the elements is variable and depends to a great extent on methods used and elements being tested for.

With the advent of the use of geochemical prospecting techniques in mineral and petroleum exploration, a variety of analytical methods for trace elements has evolved. Used chiefly for quantitative evaluation of trace elements, the methods are nonetheless useful for qualitative analysis through variation of pH, buffers, and extraction methods. While techniques for analysis of this nature continue to evolve, a useful summary of methods used by the U.S. Geological Survey has been published by Ward, et al. (1963). The trace element analyses described are colorimetric determinations using organic dyes such as diphenylthiocarbazone and EDTA. Useful information related to their application can be found in Sandell (1959) and in the book on spot tests by Feigl (1946). The general field of colorimetric analysis has been treated by Snell and Snell (1959).

Wet-chemical qualitative analysis still finds widespread use in the study of minerals and rocks. Among the standard reference works in the literature are those by Scott (1950) and Anderson and Hazelhurst (1946). Of specific applicability to geochemical-geological problems are the analytical methods outlined in detail by Hillebrand et al. (1953) and by Bennett and Hawley (1965). Both works present much detailed information on the analysis of rock and minerals. Although much of the information presented treats the methods of quantitative analysis, the methods are useful for qualitative work and provide information concerning extraction methods.

A number of microchemical methods have been developed which serve as tests for either elements or minerals. Any laboratory that has access to a microscope and a few inexpensive reagents can use these techniques for rapid and accurate qualitative determinations. The standard reference work on microchemical methods is that by Chamot and Mason (1938, 1940). Application of microchemistry to determinative mineralogy has been treated by Short (1940) and by Uytenbogaardt (1951).

Instrumental Analysis

Instrumental techniques are applied in a variety of different ways to measure some physical or chemical property of a solid, gas, liquid, or chemical reaction. Many of the instrumental methods may be considered as transitional between the wet-chemical methods and the nondestructive analysis of physical properties of a substance.

Methods of analysis of a liquid or reaction may involve, as a first step, solution or destruction of the sample, although a few techniques permit re-use of the solution or sample. Among the "transitional" techniques are the application of a variety of methods of measurement of electrical properties. One of the most important of these electrical methods is *polarography,* the measurement of current at different potential drops. All of these "electrical" methods utilize conductivity of the sample to determine qualitative and/or quantitative composition. With polarography, both can be determined simultaneously under proper conditions. Electrical methods are reviewed by Lingane (1958), and polarography is outlined in detail by Kolthoff and Lingane (1952).

Instrumental techniques are applied in the qualitative and quantitative study of colorimetry. Photoelectrometers, which measure absorbency or transmittance at selected wavelengths, or spectrophotometers, which measure the same properties over continuously changing wavelengths of transmitted light, are used to carry out precise measurement of color produced during chemical reactions. Photoelectrometers are best suited to quantitative determinations; spectrophotometers are capable of acquiring both qualitative and quantitative data. Absorption in the visible and ultraviolet are useful for studies of organic compounds. A variety of instruments is available for studies of the spectrum from ultraviolet through the infrared.

Spectroscopic techniques are widely used to make qualitative analysis, and quantitative analyses of variable accuracy. The method measures characteristic wavelengths of emission lines of an element when the element is excited in some manner. Among the useful references on the topic are those related to flame spectrometry, useful in determination of the alkali metal and alkali earth elements (Maurodineanu and Boiteux, 1965), to D-C arc techniques for analysis of minerals and rocks (Bastron et al. (1960) and general references by Ahrens and Taylor (1961) and Brode (1943). Instrumentation available for the method is diverse and ranges from simple, low-cost spectroscopes useful in a small laboratory for rapid determinations to high-quality spectrographs for more precise and accurate work.

Two analytical tools that have proven extremely valuable for a wide variety of elemental determinations are the x-ray fluorescence spectrometer and the electron probe. These instruments analyze a sample nondestructively by exciting it with high-energy radiation and measuring wavelengths and energy of the resultant fluorescence. Largely limited in the early stages of development to heavier elements, the technique is now evolving to use with light elements through use of longer wavelength of exciting radiation. Several excellent references to the general subject are available, noteworthy among which are those by Birks (1959) and Liebhafsky et al. (1960) on x-ray spectrometry, and on the electron probe by McKinley et al. (1966). Another method employing measurement of excitation by high-energy radiation is the method of fluorimetry utilizing ultraviolet light (Radley and Grant, 1954).

X-ray diffraction (q.v.) is probably one of the most widespread methods in use to determine phase or mineralogical composition. With proper calibration, the methods of diffraction can be used to make certain qualitative and quantitative determinations. Among the useful references to the method and its application are the book by Klug and Alexander (1954), and the ASTM diffraction data card files.

Activation analysis involves determination of elemental composition by means of analysis of characteristic beta- and gamma-ray energy and half-lives of artificially produced isotopes. The method is useful for rapid and sensitive qualitative determination. Use of activation, however, requires a considerable amount of equipment and the facilities for handling an irradiating source and the samples. (It is presently in widespread use with larger research organizations but should not be considered as an inexpensive means of analysis.) Useful information may be found in Lenihan and Thomson (1965), and Lyons (1964).

One additional type of instrumental technique deserves considerable emphasis. The petrographic microscope can be used in a great variety of geological problems to carry out qualitative analysis of elements, phase identification, and studies of minerals. Use of the petrographic microscope should be more widespread because of its capabilities but it has been overshadowed by more exotic techniques. It remains a fundamental tool. Application of the technique involves determination of the variation of optical properties of a substance with variation in elemental composition. Reference is made to a variety of good treatments of the subject, foremost of which is probably Winchell (1951). In addition, Larsen and Berman (1934) present useful material and a discussion of the law of Gladstone and Dale treating variation of optical properties with composition. The reader is further referred to the mineralogical periodical literature for past and current applications.

SPENCER R. TITLEY

References

ASTM Diffraction Data Cards, American Soc. for Testing and Materials, Phila. Pa.

Ahrens, L. H., and Taylor, S. R., 1961, "Spectrochemical Analysis," Reading, Mass., Addison-Wesley, 454pp.

Anderson, H. V., and Hazelhurst, 1946, "Qualitative Analysis," New York, Prentice-Hall, 266pp.

Bennett, H., and Hawley, W. G., 1965, "Methods of Silicate Analysis," New York, Academic Press, 334pp.

Birks, L. S., 1959, "X-ray Spectrochemical Analysis," New York, Interscience Publ., 137pp.

Brode, W. R., 1943, "Chemical Spectroscopy," New York, John Wiley & Sons, 677pp.

Bastron, H., Barnett, P. R., and Murata, K. J., 1960, "Method for the quantitative spectrochemical analysis of rocks, minerals and ores and other materials by a powder DC ore technique," *U.S. Geol. Surv. Bull.*, **1084g**, 165–182.

Chamot, E. M., and Mason, C. W., 1938, "Handbook of Chemical Microscopy," New York, John Wiley & Sons, 2 vols.

Feigl, Fritz, 1946, "Qualitative Analysis by Spot Tests," Amsterdam, Elsevier Publ. Co., 574pp.

Fletcher, F. J. (editor), 1966, "Standard Methods of Chemical Analysis," Princeton, N.J., D. Van Nostrand. vols. 3.

Galbraith, F. W., 1963, "Chemical tests for the elements," *Geotimes*, **7**(8), 35–36; **8**(1), 35–36 (AGI Data Sheets 42a and 42b).

Hillebrand, W. F., Lundell, G. E. F., Bright, H. A., and Hoffman, J. I., 1953, "Applied Inorganic Analysis," New York, John Wiley & Sons, 1034pp.

Klug, R. C., 1960, "Activation Analysis Handbook," New York, Academic Press, 219pp.

Klug, H. P., and Alexander, L. E., 1954, "X-Ray Diffraction Procedures," New York, John Wiley & Sons, 716pp.

Kolthoff, I. M. and Lingane, J. J., 1952, "Polorgraphy," New York, Interscience Publ., 2 vols.

Larsen, E. S., and Berman, H., 1934, "The microscopic determination of the non-opaque minerals," *U.S. Geol. Surv. Bull.* **848**, 266pp.

Lenihan, J. M. A., and Thomson, S. J., 1965, "Activation Analysis," New York, Academic Press, 211pp.

Liebhafsky, H. A., Pfeiffer, H. G., Winslow, E. H., and Zemany, P. D., 1960, "X-Ray Absorption and Emission in Analytical Chemistry," New York, John Wiley & Sons, 357pp.

Lingane, J. J., 1958, "Electroanalytical Chemistry," New York, Interscience Publ.

Lyons, W. S. Jr. (editor), 1964, "Guide to Activation Analysis," Princeton, N.J., D. Van Nostrand, 186pp.

Maurodineanu, Radu, and Boiteux, Henri, 1965, "Flame Spectroscopy," New York, John Wiley & Sons, 721pp.

McKinley, T. D., Heinrich, K. F. J., Wittry, D. B., (editors), 1966, "The Electron Microprobe," New York, John Wiley & Sons, 1051pp.

Melnechuk, T., 1962, "Tools for analytical chemistry," *Internat. Sci. and Tech.*, Aug. 1962, Internat. Comm. Inc., 14–25.

Radley, J. A., and Grant, J., 1954, "Fluorescence Analyses in Ultraviolet Light," Fourth ed., Princeton, N. J., D. Van Nostrand.

Sandell, E. B., 1959, "Colorimetric Determination of Traces of Metals," New York, Interscience Publ., Third ed., 1032pp.

Scott, W. W., 1950, "Standard Methods of Chemical Analysis," Vol. 1, Fifth ed., Princeton, N.J., D. Van Nostrand, 1234pp.

Short, M. N., 1940, "Microscopic determination of the ore minerals," *U.S. Geol. Surv. Bull.* **914**, 314pp.

Smales, A. A., and Wager, L. R., 1960, "Methods in Geochemistry," New York, Interscience Publ., 464pp.

Snell, F. D., and Snell, C. T. (assist. by C. A. Snell), 1959, "Colorimetric Methods of Analysis—Including Photometric Methods," Vol. 2A, Princeton, N.J., D. Van Nostrand, 803pp.

Uytenbogaardt, W. V., 1951, "Tables for Microscopic Identification of Ore Minerals," Princeton, N.J., Princeton Univ. Press, 242pp.

Ward, F. N., Lakin, H. W., Canney, F. C., et al., 1963, "Analytical methods used in geochemical exploration by the U.S. Geological Survey," *U.S. Geol. Surv. Bull.* **1152**, 100pp.

Winchell, A. N., 1951, "Elements of Optical Mineralogy," New York, John Wiley & Sons, 3 vols.

Cross-references: *Chromatography; Colorimetry; Flame Spectroscopy; Neutron Activation Analysis; Spectrophotometry; X-Ray Diffraction Analysis; X-Ray Spectroscopy. Vol. IVB: Blowpipe Analysis; Electron Microscopy; Electron Probe Microanalysis; Luminescence; Optical Mineralogy; Thermoluminescence.*

GEOCHEMISTRY OF TRACE ELEMENTS
—See **TRACE ELEMENTS, GEOCHEMISTRY**

GEOCHRONOMETRY

Geochronometry, the measurement of the age of geological and archeological materials, has developed with ever-increasing rapidity in both scope and accuracy during the last decade. It is now possible to date the oldest rock or the youngest paleolithic human site to within a few percent of the absolute age. The methods may be applied to the study of the formation and evolution of rock systems, the existence of metallogenic epochs, or the arrival of man in Hawaii. The purpose of this brief article is to summarize the current status of the primary age methods with regard to their areas of application and their limitations.

Quantitative geochronometry is based on the rate of decay of radioactive nuclides. This rate is independent of external conditions to which geological materials have been subjected. If a specimen contains radioactive isotope A which decays eventually to the stable isotope B, an absolute age may be computed by measuring the present quantities of A and B provided the following conditions apply:

1. The rate of decay of A is known.
2. The initial concentrations of A and/or B are known.
3. The sample has remained a closed chemical system since its inception or since it was fully reconstituted.

At present there are four geochronometric systems which have been demonstrated to be widely applicable. They are based on the decay of radioactive isotopes of uranium, potassium, rubidium, and carbon. For each system, pertinent data, including the age range of application and typical materials are given in Table 1. The first three isotopic clocks are used for the dating of minerals and rock complexes. The natural radiocarbon method is used to date materials that have been in equilibrium with the carbon dioxide-photosynthetic cycle in the past 50,000 years.

There are numerous other methods of geochronometry based on radioactive decay which have less value due to the limited range of application, inherent physical or chemical difficulties, or to incomplete understanding of the processes. Included in these are: systems applicable to sediment or rock minerals, namely U–He, Th–He, U^{238}–U^{234}, U^{238}–Th^{230}, U^{235}–Pa^{231}, Re^{187}–Os^{187}, and spontaneous fission; e.g., cosmic-ray-produced Cl^{36}, H^3, Al^{26}, Be^{10}, etc.; Cl^{36} in surface rocks; He^3 in groundwater, wine, and grain; and radiation damage in metamict minerals.

Only those methods listed in Table 1 will be considered in this discussion. A more compre-

GEOCHRONOMETRY

Table 1. Major Methods in Geochronometry

Nuclides	Half-Life (yr)	λ (yr^{-1})	Effective Range (yr)[a]	Materials
$U^{238}-Pb^{206}$	4.50×10^9	1.54×10^{-10}	10^7-T_0	Zircon, uraninite, pitchblende
$U^{235}-Pb^{207}$	0.71×10^9	9.72×10^{-10}	10^7-T_0	Zircon, uraninite, pitchbende
$Rb^{87}-Sr^{87}$	4.7×10^{10}	1.47×10^{-11}	10^7-T_0	Muscovite, biotite, lepidolite, microcline, glauconite, whole metamorphic rock
$K^{40}-Ar^{40}$	1.30×10^9 (total)	$\lambda\beta\ 4.72 \times 10^{-10}$ $\lambda e\ 5.83 \times 10^{-11}$	[b]10^5-T_0	Muscovite, biotite, hornblende, phlogopite, glauconite, sanidine, whole volcanic rock, sylvite (arkose, sandstone, siltstone)[c]
C^{14}	5710 ± 30	1.21×10^{-4}	0–50,000	Wood, charcoal, peat, grain, tissue, charred bone, cloth, shells, tufa, ground water, ocean water

[a] T_0 = age of the earth, i.e., $\sim 4.6 \times 10^9$ yr.
[b] Under certain favorable conditions the lower limit of this method can be extended to approximately 10^4 yr.
[c] For paleogeographic studies.

hensive review of the methods of dating rock systems has been given recently by Tilton and Hart (1963). The C^{14} method has been discussed extensively by Broecker and coworkers (1959, 1961; Olson and Broecker, 1958).

The U–Pb System

The U–Pb geochronometric system is based on the decay of U^{238} to Pb^{206} and U^{235} to Pb^{207}. It is applicable to uranium-bearing minerals as indicated in Table 1. A less useful but commonly related scheme is the decay of Th^{232} to Pb^{208}. The latter can be used on well-preserved theorium minerals, but in uranium minerals this scheme generally gives a low age due to preferential Pb^{208} loss.

The half-lives of U^{238} and Th^{232} are known to better than 1%; that of U^{235} to about $\pm$ 2%. The lead and uranium concentrations can generally be determined by the isotope dilution method with an error of less than 2%. Thus, the absolute age may be obtained to about 2% for a uranium-bearing mineral if it has remained a closed system and if it has adequate antiquity so that enough radiogenic lead has accumulated. With sufficient analytical effort the relative ages can be obtained within 1%.

The U–Pb system is particularly attractive for geochemical study since chemical alteration of the sample during its geological history will be reflected in a discordance among the isotopic ages calculated from the isotopic ratios (Pb^{206}/U^{238}, Pb^{207}/U^{235}, Pb^{207}/Pb^{206}). The existence of Pb^{204} which is not radiogenic and the isotopic composition of the common lead in the environment at the time of mineral formation makes it possible to subtract the nonradiogenic component from the total abundances of Pb^{206} and Pb^{207} in the mineral. The age equations for the U–Pb system are as follows:

$$T_{238} = 6.50 \times 10^9 \ln (1 + Pb^{206}/U^{238})$$

$$T_{235} = 1.03 \times 10^9 \ln (1 + Pb^{207}/U^{235})$$

where T = age in million years.

Examples of two concordant uraninites and a discordant zircon are given in Table 2. Concordant results generally indicate an absolute time of crystallization for the uranium-bearing mineral. Discordant results require interpretation but may give valuable geochemical information concerning the physiochemical history of the sample. In general, for zircon, the $Pb^{207}-Pb^{206}$ age is a reliable minimum age of crystallization, at least in the Precambrian. In post-Cambrian time the $Pb^{207}-Pb^{206}$ age is so sensitive to error in the Pb^{207} abundance and the U^{235} half-life that it becomes less useful. In this case the $U^{235}-Pb^{207}$ age is generally a reliable minimum age for zircons. A better approximation to the true age of crystallization can be obtained by analyzing a suite of zircons from the same rock sequence. Generally the degree of discordance will increase with uranium concentration. Regardless of the theory of the alteration mechanism (Russell and Ahrens, 1957; Tilton, 1960; Wetherill,

GEOCHRONOMETRY

TABLE 2. EXAMPLES OF CONCORDANT AND DISCORDANT ISOTOPIC AGES[a]

Locality	Mineral	Method	Isotopic Age (million years)
Black Hills, S.D.	Uraninite	$U^{238}-Pb^{206}$	1610 ± 20
		$U^{235}-Pb^{207}$	1615 ± 20
		$Pb^{207}-Pb^{206}$	1620 ± 15
	Microcline	$Rb^{87}-Sr^{87}$	1630 ± 45
	Muscovite	$K^{40}-Ar^{40}$	1590 ± 40
Wilberforce, Ont. (Grenville Province)	Uraninite (Pegmatite)	$U^{238}-Pb^{206}$	1020 ± 10
		$U^{235}-Pb^{207}$	1025 ± 15
		$Pb^{207}-Pb^{206}$	1035 ± 30
	Zircon (Pegmatite)	$U^{238}-Pb^{206}$	900 ± 10
		$U^{235}-Pb^{207}$	930 ± 15
		$Pb^{207}-Pb^{206}$	1000 ± 30
	Biotite (Pegmatite)	$Rb^{87}-Sr^{87}$	970 ± 30
		$K^{40}-Ar^{40}$	940 ± 30
	Biotite (Gneiss)	$K^{40}-Ar^{40}$	850 ± 25
Pikes Peak, Colo.	Zircon (Granite)	$U^{238}-Pb^{206}$	625 ± 25
		$U^{235}-Pb^{207}$	705 ± 20
		$Pb^{207}-Pb^{206}$	980 ± 40

[a] Data from G. R. Tilton et al., 1960 and 1957; Eckelmann and Kulp, 1957.

1963), if the individual zircon samples fall on a nearly straight line on a concordia plot (Fig. 1), the intercept of this line with the concordia curve indicates the time of crystallization.

The isotopic composition of lead in minerals containing negligible uranium, such as galena and pyrite, can be used to estimate the time of ore deposition or it may give information on the crustal history in an area (Russell and Farquhar, 1960). Assuming that the initial lead isotope composition of the earth was the same as that of the troilite phase of meteorites, that the age of the earth is 4.6×10^9 years, and that the average U^{238}/Pb^{204} ratio in the mantle is about 9, a lead isotope growth curve can be calculated. The lead isotope composition of nonradioactive minerals formed from a normal (average mantle-crust) environment will give an approximation to the time of formation ($\pm$ 100 million years). If the environment represents reconstituted granitic

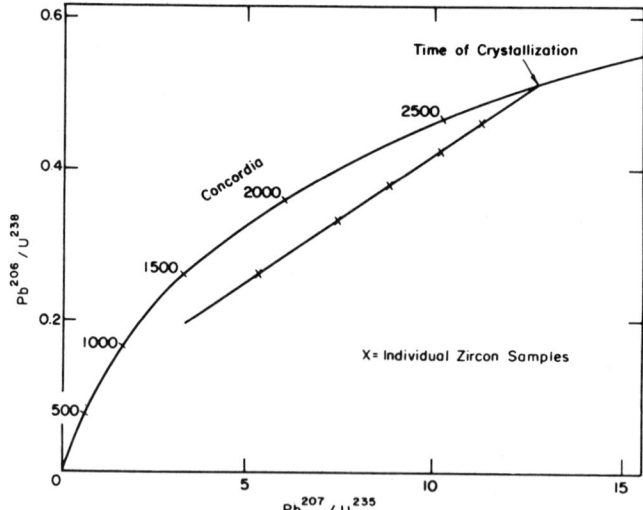

FIG. 1. Concordia plot for the U–Pb system. Points are isotopic ratios from the suite of discordant zircons from one rock complex. The intersection of the straight line with the concordia indicates time of formation or last recrystallization.

crust, the calculated age will be too young due to the addition of excess radiogenic lead. Thus, if the formation is known independently, information on the prior history and composition of the part of the crust may be derived. Using these techniques, the Precambrian origin of the Coeur d'Alene lead ores has been demonstrated (Long et al., 1960). Primary and secondary (later) deposition of lead ores have also been differentiated (Russell and Farquhar, 1960). Similarly, the black schists of the Karelian belt of Finland were shown to have been deposited at least 500 million years prior to the regional metamorphism which produced the present rocks (Wample and Kulp, 1964).

The K–Ar System

The decay of K^{40} to Ar^{40} has proven to be one of the most widely used methods of geologic age determination. It has been applied primarily to micas, but can be used on a variety of sample types (Table 1). The analytical measurements by the isotope dilution methods for potassium or argon can be made to $\pm 1\%$ so that the isotopic age has an analytical error that need not exceed $\pm 2\%$. The decay constant is probably accurate to $\pm 3\%$ so that relative ages on the K–Ar scale can be measured with slightly higher accuracy than absolute ages. The formula for calculating K–Ar isotopic age is as follows:

$$\text{Age (in million years)} = 1885 \ln (1 + 9.10\ Ar^{40}/K^{40}$$

$$K^{40} = 1.19 \times 10^{-4} \text{ at. \% of natural K}$$

One of the attractive characteristics of the K–Ar method is that it virtually always gives a reliable minimum age. The only exceptions to this are found in minerals of extremely low potassium content (such as pyroxene) that were formed under high pressure. These may incorporate sufficient magmatic Ar^{40} during crystallization to overshadow the Ar^{40} from potassium decay. This effect has never been found in high potassium minerals such as biotite, muscovite, hornblende, or glauconite.

The primary limitation in the K–Ar method lies in the diffuse loss of argon at elevated temperatures. For extended periods of time, measurable argon loss can occur in glauconite at temperatures as low as 50°C, in biotite from schists and gneisses at 150°C and in pegmatic muscovite at 200°C, depending on crystal size and perfection. For this reason, the K–Ar age on a biotite from volcanic ash or a shallow intrusion may give the absolute date of the event, but in a metamorphic rock the K–Ar date may represent the last time the particular rock passed upward through the 150°C thermal plane.

Since the Ar^{40}/K^{40} ratio gives a reliable minimum age, many geologic problems can be solved by analyses of whole rock samples. Thus the minimum age for a low-grade metamorphic event can be determined from the analyses of a phyllite, since under the conditions of phyllite formation, all inherited argon would be lost. The minimum age for a welded tuff or a basalt flow could likewise be obtained from whole rock analysis.

Finally, an entirely new application of the K–Ar method is in geomorphology and paleogeography. From whole rock analyses of siltstones, arkoses, sandstones, or glacial drift, it is possible to identify sources of the detrital debris, and in favorable cases, to estimate the percentage contribution from two primary sources.

The K–Ar method has been extended to extremely young samples, particularly by the efforts of the Berkeley group (Reynolds, 1956; Evernden, et al., 1957). The ultimate limitation is not the measurement of the radiogenic Ar^{40} but the accuracy with which the atmospheric correction for Ar^{40} can be made. The fraction of atmospheric argon is determined by measuring the absolute quantity of Ar^{36} in the sample and assuming a certain Ar^{40}/Ar^{36} ratio for atmospheric argon. In older samples the ratio of radiogenic Ar^{40} to total Ar^{40} is 0.9 or greater. In very young samples this value may be less than 0.1. Therefore, uncertainty in the measured isotopic composition of Ar^{40}/Ar^{36} in air, small fractionation effects, and possible incorporation of primary argon in the sample become major sources of error. Nevertheless, by adequate controls Evernden and Curtis (1957) have reported ages as low as 28,000 years by this method.

The Rb–Sr System

The decay of Rb^{87} to Sr^{87} has also played an important role in the modern developments of geochronology. Here again, the elements can be analyzed by the isotope dilution method with an accuracy of 1 to 2%. The primary analytical limitation on the method is the accuracy of the measurement of the Sr^{87}/Sr^{86} ratio. In minerals with low Rb/Sr ratio or of young age, the total analytical error is dominated by the uncertainty in the correction of the common strontium component of the total Sr^{87}. The actual isotopic composition of any primary strontium incorporated into the mineral lattice with high Rb/Sr ratio can be experimentally determined by analyzing the strontium in a Rb-free phase such as apatite. Fig. 2 is a diagram indicating the error expected for various ages and Rb/Sr ratios in the sample.

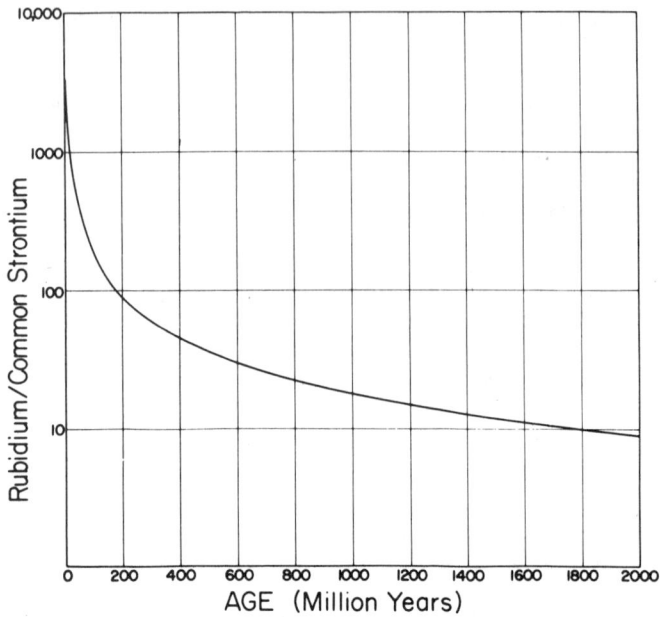

Fig. 2. Diagram showing the rubidium/common strontium ratio needed to obtain an isotopic age to ± 5% provided the raw isotopic ratios can be measured to within ± 1%.

The decay constant of Rb^{87} has now been determined by Flynn and Glendenin to be 1.47×10^{-11} yr^{-1} (1960; Glendenin, 1961). The uncertainty of this determination is probably less than ± 2%. This value is also very close to the best value inferred from comparison of dates on identical minerals by the K–Ar method (Kulp and Engels, 1963). The Rb–Sr isotopic age is calculated from the following equation:

Age (in millions years)
$$= 6.78 \times 10^4 \ln (1 + Sr^{87}/Rb^{87}).$$

The method has been applied successfully to biotite, muscovite, lepidolite, and potassium feldspar from the oldest rocks to those as young as 10^7 years in favorable cases (Table 1). In addition, it can also be applied to a rock as a whole. If Sr^{87}/Sr^{86} (ordinate) is plotted agains Rb^{87}/Sr^{86} (abscissa) for a suite of samples of the same age, the resultant line (known as an isochron) can give information regarding the metamorphic history of the system. The slope of the isochron is a function of the age of the samples. The Sr^{87}/Sr^{86} ratio when Rb^{87}/Sr^{86} equals 0 gives the isotopic composition of the strontium in the environment at the time of the metamorphic event. If the Sr^{87}/Sr^{86} ratio is about 0.71 and the rock suite is granitic in character, this is evidence that the rocks have only experienced a single generation of metamorphism and that the rock may represent a primary portion of the crust. If the Sr^{87}/Sr^{86} intercept lies above 0.71, a second-generation rock is inferred. Furthermore, some indication of the time of the primary metamorphism may be obtained by assuming a Sr^{87}/Sr^{86} ratio of 0.71 and recalculating the age of the sample with the highest Rb/Sr ratio. With such techniques a more comprehensive understanding of the evolution of parts of the continents is possible (Riley and Compston, 1962; Hurley, et al., 1963; Schreiner, 1958).

Two types of alteration cause modification of the isotopic ratios, and hence, the calculated Rb–Sr ages. Elevated temperatures may cause appreciable diffusion of strontium in the system resulting in the dilution of the radiogenic strontium with common strontium from the environment. In general, this is most likely in the case of glauconite and least likely in pegmatitic muscovite or microcline. If a rock has been reheated in its history, but otherwise remained a closed chemical system, the Rb–Sr age will be higher than or equal to the K–Ar age. The other alteration that can particularly effect biotite is base exchange due to groundwater. The net result of this alteration is to lower the Rb–Sr age while leaving the K–Ar age unchanged. This effect has not been found in the case of muscovite. Therefore, to detect subsequent heating or base exchange effects it

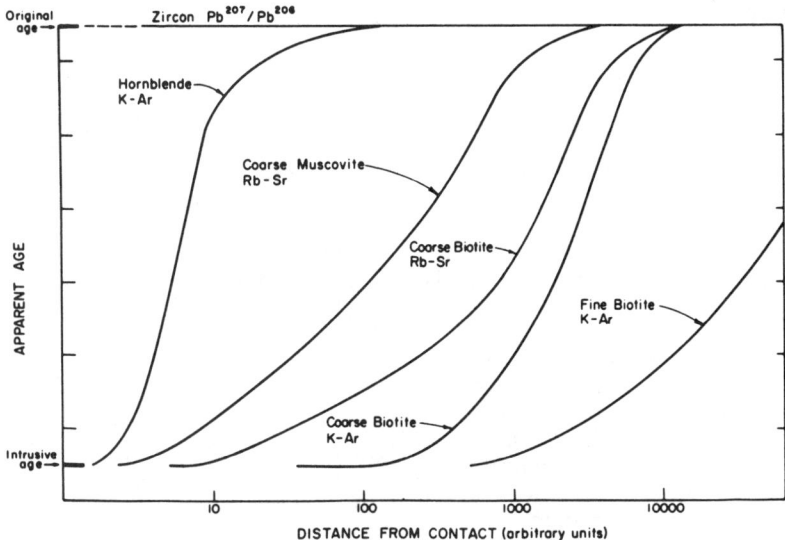

FIG. 3. The effect of heating by an intrusive rock on older minerals. The intrusive rock studied was 10,000 units thick, had a rectangular shape, and provided contact metamorphism in a basement metamorphic complex (from the studies of Hart, Kulp, and Catanzaro; Tilton and Hart, 1963; Hart, 1966; Catanzaro and Kulp, 1964).

is necessary to analyze the mica to obtain both Rb–Sr and K–Ar isotopic ages.

In Table 2, the Black Hills samples show the type of concordance obtainable among all methods. Apparently rebound of the crust occurred in this area in a reasonably short interval after metamorphism. In contrast, the samples from the Grenville area show the greater sensitivity of the mica system to temperature. The uraninite is unaffected, but the apparent age of the coarse biotite from pegmatites is lowered by about 100 million years and the less-well-crystallized biotite from the gneiss lowered by nearly 200 million years. This is interpreted as indicating a crystallization time of about 1030 million years but cooling to about 150°C over 180 million years.

The effect of contact metamorphism on the isotopic ages of a system of minerals is shown in Fig. 3. The zircon system is least affected, followed by hornblende. Fine biotite is more sensitive than coarse showing the effect of crystal size. The most sensitive mineral in this array recovers its primary age within one width of the intrusive. Obviously, samples taken between the contact and the distance of negligible alteration will merely give minimum ages for the original crystallization. The alteration of the K–Ar in biotite is detected ten times as far from the contact as any alteration detectable by the petrographic microscope. The isotopic relationships are therefore the most sensitive parameters in the study of the petrologic and tectonic history of an area.

Radiocarbon Dating

The C^{14} method of age determination depends on the assumptions *(1)* that the rate of formation of C^{14} by cosmic rays on the atmosphere is constant, *(2)* that the rate of mixing of the C^{14} in the atmosphere-biosphere-surface ocean reservoir is rapid relative to the rate of decay, and *(3)* that once material is removed from this reservoir, e.g., death of an animal, completion of annual tree ring, or deposition on ocean floor of foraminifera, no further C^{14} is added and that which is present decays at a known and constant rate.

The method was discovered by Libby and his associates (1961) who demonstrated that these assumptions were at least approximately valid. Subsequently, thousands of measurements have been made in various laboratories and applied to geological and archeological problems. Results have been reported annually since 1959 in the Radiocarbon Supplement published by the *American Journal of Science*.

Carbon from the sample is converted to pure carbon dioxide or methane and counted in an internal gas proportional counter in which background is minimized by massive shielding and anticoincidence circuitry (Fergusson, 1955; Broecker et al., 1959). Other techniques such as liquid scintillation counting have been investigated, but, in general, are less sensitive or less efficient than the internal gas proportional counter method.

The half-life of C^{14} that has been used in reporting most of the data is 5568 years. New work, however, has indicated that the best value is about 5700 years with an uncertainty of less than 1% (Olsson and Karlen, 1963). All laboratories are intercalibrated on the NBS oxalic acid standard. If the second-order correction for carbon isotope fractionation is made, results from these laboratories are strictly comparable (Broecker and Olson, 1961).

If adequate sample is available (5-10 g of carbon) modern or near-modern samples can be counted routinely to $\pm 0.5\%$, which is equivalent to ± 30 years on the date (Oeschger, 1963). The date carries a greater uncertainty than the counting error due to local short-term and worldwide long-term variations in the atmospheric C^{14} concentration (Ralph and Stuckenrath, 1960; Libby, 1963; Broecker et al., 1959). Therefore an isolated radiocarbon date in the range of 0 to 4000 years may have an error from these sources of about ± 160 radiocarbon years. In order to reduce the absolute error in age, it is necessary to know something of the probable concentration of atmospheric C^{14} at the time the sample was in equilibrium with the reservoir.

In the case of old samples, the problems concern the sensitivity of the counting device and the contamination of the sample with younger carbon. Current techniques make it possible to attain a limit of 40,000 years and 50,000 years in special cases. It is conceivable that counting systems could be designed to approach a limit of 100,000 years (Oeschger, 1963). The problem of contamination has been investigated in some detail (Olson and Broeker, 1958). It appears that samples younger than 20,000 years seldom show evidence of detectable contamination whereas samples with ages in excess of four half-lives (23,000 yr.) should be subject to question until the absence of contamination has been demonstrated. Reliable minimum ages will generally be obtained, because in most cases the error will make the sample appear too young. If a sample whose true age is 57,000 years is contaminated with 0.1% modern carbon, the apparent age would be 51,000 years. If contaminated with 1% modern carbon, the age would be about 35,000. This indicates the magnitude of the problem.

The natural radiocarbon method has been successfully applied to a great variety of substances. These include wood, peat, charcoal, leaves, tissues, manuscript, mummy cloth, rope, and unrecrystallized marine shells. Under special conditions it can be applied to bone, fresh-water snails, calcium carbonate terrace and cave deposits, groundwater, and ocean water.

Application of these methods of geochronometry to samples associated with fossil assemblages makes it possible to define geological development in terms of absolute time. A recent review of this geological time scale has been published (Kulp, 1961).

Geochronometry has already caused major developments in the historical sciences but its utilization has barely commenced.

J. Laurence Kulp

References

Broecker, W. S., and Olson, E. A., 1961, "Lamont radiocarbon measurements VIII," *Radiocarbon*, **3**, 176-204.

Broecker, W. S., Tucek, C. S., and Olson, E. A., 1959, "Radiocarbon analysis of oceanic CO_2," *Intern. J. Appl. Radiation Isotopes*, **7**, 1-18.

Catanzaro, E. J., and Kulp, J. L., 1964, "Discordant zircon from the Little Belt (Montana), Beartooth (Montana) and Santa Catalina (Arizona) Mountains," *Geochim. Cosmochim. Acta*, **28**, 87-124.

Eckelman W. R., and Kulp, J. L., 1957, "Uranium-lead method of age determination," *Bull. Geol. Soc. Am.*, **68**, 1117-1140.

Evernden, J. F., Curtis, G. H., and Kistler, R., 1957, "Potassium-argon dating of Pleistocene volcanics," *Quaternaria IV*, 13-17.

Fergusson, G. J., 1955, "Radiocarbon dating system," *Nucleonics*, **13**(1). 18-23.

Fleischer, R. L., et al., 1967, "Origins of fossil charged-particle tracks in meteorites," *J. Geophys Res.*, **72**(1), 331-353.

Flynn, K. F., and Glendenin, L. E., 1960, "Half-life and beta spectrum of Rb_{87}," *Phys. Rev.*, **116**, 744-748.

Glendenin, L. E., 1961, "Present status of the decay constants," *Ann. N.Y. Acad. Sci.*, **91**, 166-180.

Hamilton, E. I., 1965, "Applied Geochronology," New York, Academic Press, 283pp.

Hamilton, E. I. (editor), 1968, "Radiometric Dating for Geologists," New York, John Wiley & Sons, 506pp.

Hart, S. R., 1961, "Mineral ages and metamorphism," *Ann. N.Y. Acad. Sci.*, **91**, 192-197.

Huang, W. H., and Walker, R. M., 1967, "Fossil alpha-particle recoil tracks: a new method of age determination," *Science*, **155**(3765), 1103-1106.

Hurley, P. M., Fairbairn, H. W., Faure, G., and Pinson, W. H., Jr., 1963, "New Approaches to Geochronology by Strontium Isotope Variations in Whole Rocks," *Radioactive Dating*, pp. 201-217, Int. Atomic Energy Agency, Vienna.

Kulp, J. L., 1961, "Geologic time scale," *Science*, **133**, 1105-1114.

Kulp, J. L., and Engels, J., 1963, "Discordances in K-Ar and Rb-Sr Isotopic Ages," *Radioactive*

Dating, pp. 219–238, Int. Atomic Energy Agency, Vienna.
Libby, W. F., 1961, "Radiocarbon dating," *Science,* 133, 621–629.
Libby, W. F., 1963, "Accuracy of radiocarbon dates," *Science* 140, 278–280.
Long, A., Silverman, A. J., and Kulp, J. L., 1960, "Isotopic composition of lead and Precambrian mineralization of the Coeur d'Alene District, Idaho," *Econ. Geol.,* 55, 645–658.
Oeschger, H., 1963, "Low-level counting methods," *Radioactive Dating,* pp. 13–34, Int. Atomic Energy Agency, Vienna.
Olson, E. A., and Broecker, W. S., 1958, "Sample contamination and reliability of radiocarbon dates," *Trans. N.Y. Acad. Sci.,* 20, 593–604.
Olsson, I. U., and Karlen, I., 1963, "The half-life of C^{14} and the problems which are encountered in absolute measurements on beta-decaying gases," Radioactive Dating, pp. 3–13, Int Atomic Energy Agency, Vienna.
Pasteels, P., 1968, "A comparison of methods in geochronology," *Earth-Sci. Rev.,* 4, 5–38.
Ralph, E. K., and Stuckenrath, R., 1960, "Carbon-14 measurements of known age samples," *Nature,* 188, 185–187.
Reynolds, J. H., 1956, "High sensitivity mass spectrometer for noble gas analysis," *Rev. Sci. Instr.,* 27, 928–934.
Riley, G. H., and Compston, W., 1962, "Theoretical and technical aspects of Rb-Sr geochronology," *Geochim. Cosmochim. Acta,* 26, 1255–1281.
Russell, R. D., and Ahrens, L. H., 1957, "Additional regularities among discordant lead-uranium ages," *Geochim. Cosmochim. Acta,* 11, 213.
Russell, R. D., and Farquhar, R. M., 1960, "Lead Isotopes in Geology," New York, Interscience Publishers.
Schreiner, G. D. L., 1958, "Comparison of the Rb^{87}-Sr^{87} ages of the red granite of the Bushneld Complex from measurements on the total rock and separated mineral fraction," *Proc. Roy. Soc. London, Ser. A* 245, 112.
Tilton, G. R., 1960, "Volume diffusion as a mechanism for discordant ages in zircons," *J. Geophys. Res.,* 65, 2933.
Tilton, G. R., Davis, G. L., Wetherill, G. W., and Aldrich, L. T., 1957, "Isotopic ages of zircon from granites and pegmatites," *Trans. Am. Geophys. Union* 38, 360–371.
Tilton, G. R., and Hart, S. R., 1963, "Geochronology," *Science,* 140, 357–366.
Tilton, G. R., Wetherill, G. W., Davis, G. L., and Bass, M. N., 1960, "1000 million-year-old minerals from the eastern United States and Canada," *J. Geophys. Res.,* 65, 4173–4179.
Wampler, J. M., and Kulp, J. L., 1964, "An isotopic study of lead in sedimentaery pyrite," *Geochim. Cosmochim. Acta,* 28, 1419–1458.
Wasserburg, G. J., Burnett, D. S., and Frondel, C., 1965, "Strontium-rubidium age of an iron meteorite,," *Science,* 150(3705), 1814–1818.
Wetherill, G. W., 1963, "Discordant uranium-lead ages," *J. Geophys. Res.,* 68, 2957–2965.

Cross-references: *Carbon-14 Dating; Fission Track Dating; Ionium-Thorium Dating; Potassium-Argon Age Determination; Radionuclides; Rubidium-Strontium Dating Method; Uranium-Helium Isotopic Age Method; Uranium-Thorium-Lead Age Determination. Vol. VI: Radioactive Isotope Tracer Technology.*

GEOLOGIC THERMOMETRY—
See Vol. IV B

GEOLOGIC TIME SCALE

The geologic time scale (Table 1) is a relatively recent development which combines the classical time classification of rocks long used by geologists with numerical time bound-

TABLE 1. PHANEROZOIC TIME SCALE

ERA	PERIOD	EPOCH	BEGINNING OF INTERVAL (MILLION YEARS)
CENOZOIC	QUATERNARY	PLEISTOCENE / PLIOCENE	1.5–2 / 0
CENOZOIC	TERTIARY	MIOCENE	26
CENOZOIC	TERTIARY	OLIGOCENE	37–38
CENOZOIC	TERTIARY	EOCENE	53–54
CENOZOIC	TERTIARY	PALEOCENE	65
MESOZOIC	CRETACEOUS	Upper	100
MESOZOIC	CRETACEOUS	Lower	136
MESOZOIC	JURASSIC	Upper	162
MESOZOIC	JURASSIC	MIDDLE	172
MESOZOIC	JURASSIC	Lower	190–195
MESOZOIC	TRIASSIC	Upper	205
MESOZOIC	TRIASSIC	Middle	215
MESOZOIC	TRIASSIC	Lower	225
PALEOZOIC	PERMIAN	Upper	240
PALEOZOIC	PERMIAN	Lower	280
PALEOZOIC	CARBONIFEROUS / PENNSYLVANIAN		325
PALEOZOIC	CARBONIFEROUS / MISSISSIPPIAN		345
PALEOZOIC	DEVONIAN	Upper	359
PALEOZOIC	DEVONIAN	Middle	370
PALEOZOIC	DEVONIAN	Lower	395
PALEOZOIC	SILURIAN		430–440
PALEOZOIC	ORDOVICIAN	Upper	445
PALEOZOIC	ORDOVICIAN	Lower	500
PALEOZOIC	CAMBRIAN	Upper	515
PALEOZOIC	CAMBRIAN	MIDDLE	540
PALEOZOIC	CAMBRIAN	Lower	570
	PRECAMBRIAN		600

aries based on radiometric age measurements. This time scale is sometimes referred to as the "absolute" time scale, but more appropriate are the terms "geochronologic" or "radiometric" time scale to distinguish it from the older relative time classification. In the latter, the principles of stratigraphic and faunal successions have played predominant roles, and three major time divisions, the Cenozoic, Mesozoic, and Paleozoic eras, were named for recent, middle, and ancient life, respectively. This classification with subdivisions into periods and epochs was well worked out prior to 1850 (see entries under each era in Vol. VII).

Early speculations about the time rates of geologic processes led to attempts to place limits in years on subdivisions of geologic time as well as on the age of the earth. Various methods were tried to calculate intervals of geologic time based on the rates of deposition, erosion, development of life, accumulation of salt in the ocean, and so forth. In all these calculations various assumptions had to be made, commonly on the most meager information. It is not surprising, therefore, that time intervals calculated by different individuals varied greatly.

Many prominent geologists, physicists, and chemists have concerned themselves with the problem of measuring geologic time with the objective of attaining some quantitative values for the intervals represented in the geological classification. One of the foremost is Arthur Holmes whose papers on the subject span half a century (1911–1960). His outstanding work in this field has won him wide recognition. In 1956 Holmes was awarded the Penrose Medal of the Geological Society of America, and in 1964 he shared the prize of the G. Unger Vetlesen Foundation at Columbia University for his contributions to the geologic time scale. A special volume, "The Phanerozoic Time Scale," was dedicated to Holmes by the Geological Society of London, and Table 1 is adapted from a time scale that resulted from the Holmes Symposium.

Phanerozoic Time Scale

In 1913, in a small volume entitled "The Age of the Earth," Holmes outlined how age determinations based on the principles of radioactive decay, in conjunction with geological data on the maximum known thicknesses of rocks assigned to the various geological periods, might be used to construct a quantitative time scale. The ratios of the daughter products, helium and lead, to the parent uranium, were used to calculate these early radioactivity ages. This approach was used by Joseph Barrell in a monumental paper "Rhythms and the Measurements of Geologic Time" in 1917, in which he presented a time scale (Table 2). Holmes' first extended time scale for the Phanerozoic was published in 1933 (Table 2). European scientists who made important contributions up to this time include, among many others, such illustrious names as Charles Lyell, Charles Darwin, Archibald Geikie, Lord Kelvin, Lord Rayleigh, Lord Rutherford, W. J. Sollas, John Joly, and A. de Lapparent. In addition to Barrell, Americans who made significant contributions were chemists such as B. B. Boltwood and F. W. Clarke, and geologists including G. F. Becker, J. D. Dana, G. K. Gilbert, Charles Schuchert, and C. D. Walcott.

The Barrell and Holmes time scales were based on U-Pb age calculations based on chemical determinations. In 1933, F. W. Aston

TABLE 2. EVOLUTION OF THE POST-PRECAMBRIAN (PHANEROZOIC) TIME SCALE
(MILLIONS OF YEARS)

Geologic Division	Barrell (1917)	Holmes (1933)	Holmes (1947)	U.S.S.R. (1960)	Kulp (1961)
Pleistocene	1–1.5	1	1	—	1
Pliocene	7–9	15	12–15	10	13
Miocene	19–23	32	26–32	25	25
Oligocene	35–39	42	38–47	—	36
Eocene	55–65	60	58–68	70[a]	63[a]
Cretaceous	120–150	128	127–140	140	135
Jurassic	155–195	158	152–167	185	181
Triassic	190–240	192	182–196	225	230
Permian	215–280	220	203–220	270	280
Carboniferous	300–370	285	255–275	320	345
Devonian	350–420	350	313–318	400	405
Silurian	390–460	375	350	420	425
Ordovician	480–590	440	430	480	500
Cambrian	550–700	510	510	570	600

[a] Paleocene.

showed by mass spectrographic analyses that lead is composed of a number of isotopes whose abundance ratios he determined in several samples of common lead. This work was followed in 1939–1941 by papers by A. O. Nier. His U-Pb age calculations based on mass spectrometric measurements heralded modern geochronology. The precise isotopic measurements of Aston and Nier provided Holmes with the information he needed for his 1947 geologic time scale (Table 2).

Since World War II, great progress has been made in geochronology with the introduction of the K-Ar and Rb-Sr techniques for age determinations and with the application of U-Th-Pb isotopic age determinations to minerals such as zircon (see *Geochronometry*). New data from a number of geochronology laboratories were used in Kulp's (1961) time scale. Table 2 also includes the geochronologic scale compiled by the Commission on Absolute-Age Determination of Geologic Formations of the U.S.S.R. Academy of Sciences. The committee consisted of 17 persons, and this may be considered indicative of the broad interest in geochronology. Among the contributors to geochronology in the U.S.S.R. may be mentioned E. K. Gerling, N. I. Poleyava, N. P. Semenenko, I. Ye Starik, A. I. Turgarinov, and A. P. Vinogradov.

Refinements in the time scale will follow as more data are accumulated and as critical points are intensively studied. Detailed studies of the Cenozoic chronology have been conducted by J. F. Evernden, G. H. Curtis, and co-workers at the University of California at Berkeley. The greatly improved precision in K-Ar dating achieved by the California group, together with extensive geological field studies, has permitted the dating of intervals differentiated on the basis of fossil land mammals. Consistent radiometric correlations between North American and European mammal localities are obtained, and the mammal ages are also correlated with Pacific Coast foraminiferal ages. Epoch-age correlations, however, are so imperfectly known that Evernden and Curtis have not attempted to place time limits on the Tertiary epochs.

Precambrian Time Scale

Although geologists surmised that a great deal of time is represented in the Precambrian rocks, the immensity of this interval was not fully comprehended or accepted until isotopic age determinations became available. The age of the earth is now commonly taken as 4.55 billion (10^9) years, the age obtained in 1956 by C. C. Patterson by comparing the abundance ratios of lead isotopes for meteorites with terrestrial lead. Many isotopic ages in the range from 2500 to 3600 million years have been determined on mineral and rock samples from the Precambrian shield areas of the Americas, Africa, Australia, and Eurasia (see *Geochronometry*).

The lack of fossils in the Precambrian rocks and the metamorphic changes which they have undergone have made extremely difficult the task of deciphering the stratigraphic succession. Locally, as for example, in the Lake Superior region, the succession and a classification of Precambrian rocks have been worked out, but there is no universally accepted classification. Similarly, the radiometric time scale for the Precambrian succession is still in an elementary form compared to that for the Phanerozoic. The metamorphic processes which have affected in varying degree the minerals of the Precambrian rocks also affected the parent-daughter nuclide ratios. Isotopic ages, therefore, are difficult to interpret in areas of complex metamorphic history and reflect metamorphic events rather than the time of first emplacement or first crystallization. Most of the progress that has been made in Precambrian geochronology has come through the dating of major periods of orogeny.

The progress in worldwide dating of the Precambrian is summarized in papers that were presented at the International Conference, "Geochronology of Precambrian Stratified Rocks," Edmonton, Alberta, June 1967. The papers were published by the Canadian Journal of Earth Sciences (1968) and include work on all the continents. Three proposed Precambrian time scales discussed at the conference are given in Table 3 (on p. 456).

The development of an ordered stratigraphic succession with the use of fossils to correlate rocks in widely separated areas is one of the remarkable achievements of the geological profession. Paleontological and stratigraphic methods remain the most applicable and reliable for correlation in Phanerozoic rocks. Isotopic age measurements now provide a long-needed method for deciphering the succession of Precambrian rocks. In addition, radioactivity age measurements make possible a quantitative approach to the study of geologic history and processes.

SAMUEL G. GOLDICH

References

Barrell, Joseph, 1917, "Rhythms and the measurements of geologic time," *Geol. Soc. Am. Bull.*, **28**, 745–904.

Evernden, J. F., Savage, D. E., Curtis, G. H., and James, G. T., 1964, "Potassium-argon dates and

GEOLOGIC TIME SCALE

TABLE 3. PRECAMBRIAN TIME SCALES

10^6 YRS	CANADIAN SHIELD (STOCKWELL, 1968)		LAKE SUPERIOR REGION (GOLDICH, 1968)	UKRAINIAN SHIELD (SEMENENKO ET AL., 1968)
600		— 570 —	— 600 —	— 550 —
800	PROTEROZOIC	HADRYNIAN		PRECAMBRIAN V
1000		— 880 —		
1200		HELIKIAN	LATE PRECAMBRIAN	— 1200 —
1400		NEOHELIKIAN / — 1280 — / PALEOHELIKIAN		PRECAMBRIAN IV
1600		— 1640 —		— 1700 —
1800		APHEBIAN	— 1800 —	PRECAMBRIAN III
2000				— 2000 —
2200			MIDDLE PRECAMBRIAN	PRECAMBRIAN II
2400		— 2400 —		
2600			— 2600 —	— 2700 —
2800				
3000	ARCHEAN		EARLY PRECAMBRIAN	PRECAMBRIAN I
3200				
3400				
3600				

the Cenozoic mammalian chronology of North America," *Am. J. Sci.,* **262,** 145–198.

Geological Society of London, 1964, "The Phanerozoic time-scale. A symposium dedicated to Professor Arthur Holmes," *Quart. J. Geol. Soc. Lond.,* **120s,** 458pp.

Goldich, S. S., 1968, "Geochronology in the Lake Superior region," *Can. J. Earth Sciences,* **5,** 715–724.

Goldich, S. S., Nier, A. O., Baadsgaard, H., Hoffman, J. H., and Krueger, H. W., 1961, "The Precambrian geology and geochronology of Minnesota," *Minn. Geol. Surv. Bull.,* **41,** 193pp.

Holmes, Arthur, 1913, "The Age of the Earth," London, Harper Bros., 196pp.

Holmes, Arthur, 1933, "The thermal history of the earth," *Wash. Acad. Sci. J.,* **23,** 169–195.

Holmes, Arthur, 1947, "The construction of a geological time-scale," *Trans. Geol. Soc. Glasgow,* **21,** 117–152.

Holmes, Arthur, 1960, "A revised geological time-scale," *Trans. Edinburgh Geol. Soc.,* **17,** 183–216.

Kulp, J. L., 1961, "Geologic time scale," *Science,* **133,** 1105–1114.

Patterson, C. C., 1956, "Age of meteorites and the earth," *Geochim. Cosmochim. Acta,* **10,** 230–237.

Semenenko, N. P., Scherbak, A. P., Vinogradov, A. I., Tougarinov, G. D., Eliseeva, G. D., Cotlovskay, F. I., and Demidenko, S. G., 1968, "Geochronology of the Ukrainian Precambrian," *Can. J. Earth Sciences,* **5,** 661–671.

Stockwell, C. H., 1968, "Geochronology of stratified rocks of the Canadian Shield," *Can. J. Earth Sciences,* **5,** 693–698.

U.S.S.R. Academy of Sciences, Commission on Absolute-Age Determination of Geologic Formations, 1960, "The absolute geochronologic age scale from 1960 data of U.S.S.R. laboratories," *Izv. Akad. Nauk SSSR,* No. 10 (October 1960).

Cross-references: *Carbon-14 Dating; Geochronometry; Potassium-Argon Age Determination; Protactinium-Thorium Dating Method; Radioactive Isotopes; Rubidium-Strontium Dating Method; Uranium-Thorium-Lead Age Determination.* Vol. VII: *Cenozoic Era; Mesozoic Era; Paleozoic Era.*

GEOPHYSICAL METHODS FOR HYDROLOGIC SEARCH

Definition and Scope

Geophysical methods for hydrologic search, broadly speaking, entails the application of the principles of physics as aids in the location and study of underground waters (ground water). Geophysical methods, when applicable, are usually employed when conventional methods, such as surface mapping or test drilling, are inadequate or too costly to find ground water economically.

Geophysical Methods. Geophysical methods for hydrologic investigations may be conveniently, though arbitrarily, grouped into surface methods and subsurface methods. The former rely upon the interpretation of some physical measurement made at the surface to provide subsurface information, whereas the latter make subsurface measurements directly, usually down a borehole. Geophysical measurements may involve, singly or in combination, magnetic, electric or gravity fields, radiation, nuclear, seismic, or thermal properties of the earth (Kelly, 1962; Parasnis, 1962).

Hydrologic Search. Hydrologic search involving the above methods will usually include one or more of the following: (a) The location and delineation of a geological situation which may control the occurrence of ground water. (b) The location and delineation of usable ground water in the aquifer which contains it. (c) The study of quality or quantity of available ground water in a locality.

The use of any geophysical method for hydrologic purposes will depend first on the ability of the geophysical methods to detect the desired situation in the locale and secondly on the relative importance of cost versus reliability. The geophysical method will generally be less reliable but considerably cheaper than other methods of exploration such as drilling.

Applications and Limitations

Applications. Geophysical methods have been successfully applied to hydrological investigations in many parts of the world in igneous, metamorphic, and sedimentary environments. Electrical, seismic, and gravitational techniques have been applied in North America, Europe, Africa, Australia, Asia, and elsewhere. A vast amount of work has been done. A few examples are listed below.

Electrical resistivity methods have been more universally applied to hydrologic search than any other geophysical method because the relative cost is least (Wiebenga, et al., 1962). Electrical resistivity methods have been applied to problems in salt-water encroachment (the Netherlands, the coastal regions of Northwest Germany) and to problems in locating sweet-water sources for the farming industry in Australia (Wiebenga, 1955). Electrical resistivity methods have been used in South Africa and in North America for the location of drill sites for water wells in igneous and in sedimentary environments. The Winkler aquifer, in Manitoba, was discovered by resistivity surveys (Hobson et al., 1962). A Review in *Geophysics,* August 1963, outlines electrical resistivity surveys conducted in West Africa, France, Italy, and in the Middle East to solve hydrological problems (Breusse, 1963).

Seismic methods too have been successfully applied in regions where velocity contrasts are sufficient to give hydrological information. Pleistocene aquifers have been traced by refraction surveys in New Jersey (Gill, et al., 1965) by mapping depth to bedrock. Across the ocean, a water supply for the town of Conakry in Guinea was found in a lateritic complex overlying an eruptive basement of dunite. A study of wave propagation velocities in the laterite overburden indicated a measure of the permeability, whereas seismic refraction methods indicated depressions in the dunite related to permeable zones (Valentin, 1950).

Examples of the application of subsurface, magnetic, gravity, and other methods are given in the appropriate sections below.

Limitations. Although geophysical measurements can be very valuable, their usefulness is subject to limitations:

(a) The object of the search must offer a detectable contrast in the value of the physical parameter (e.g., electrical resistivity) measured by the specific geophysical tool.

(b) There are practical and theoretical limitations to the interpretation techniques employed.

With relation to hydrologic search in particular the following limitations should be pointed out:

(a) Water, as such, is not detected. The contrast useful for the geophysical measurement is offered by the aquifer and the country rock combination and is provided by a range of properties due to, for instance, the concentration of dissolved salts in the water itself.

(b) Water has been and still is a cheap commodity and financial returns are low for hydrological investigations by contrast with those for oil or mineral deposits. A compensating factor, however, is the shallowness of many water supplies and the frequent successes of empirical and qualitative analyses of data for small, local regions.

Surface Methods

Geophysical measurements made at the surface are most often used as a prospecting device to locate a favorable site for production of groundwater. Occasionally more specific information is sought such as areal extent, depth, or quality of an aquifer. Other special

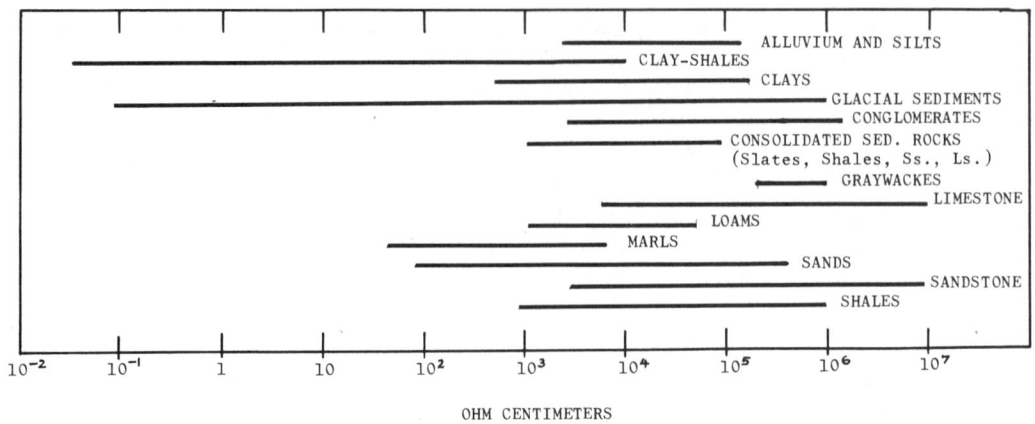

Fig. 1. Sedimentary rock resistivity in ohm-centimeters (after Jakosky, 1950).

problems are sometimes approached. Surface methods are valuable insofar as they reduce the amount of test drilling required to evaluate the groundwater potential of an area.

Resistivity Measurements. To appreciate the value of resistivity measurements it is necessary to realize that the electrical properties of rocks (see Vol. V, *Electrical Properties of Rocks*) have the greatest range of all the physical properties that a geophysicist has to deal with. Of these, the direct current conductivity is the most variable with a range of 10^{20} in naturally occurring minerals. Fig. 1 shows a bar graph indicating the ranges of d.c. (direct current) resistivities of several common sedimentary rocks found in North America, compiled by Jakosky (1950, p. 441), illustrating some of the rocks in which groundwater is often found.

Several factors affect the electrical conductivity of minerals and rocks. Included are such things as porosity, chemical composition, fractures, and the salinity of the interstitial pore fluids. All of these factors are associated with groundwater. Since most rock-forming minerals (except clays) are nonconductors, the conductivity of the rock is determined mainly by the fluid in the pores. The factor most affecting the resistivity, therefore, is the quality and quantity of water contained in the rock.

The electrical resistivity methods, whether they are a.c. or d.c., are intended to detect the direct current resistivity parameters of the earth as a function of depth (see Vol. V, *Resistivity Methods in Geophysics-Galvanic*).

These are popular exploration methods for hydrologic search in many countries. Depending upon the purpose of the survey, and depending upon the geological conditions, any one of a great variety of techniques may be applied when conducting an electrical hydrological exploration program.

The underlying principles in resistivity surveys are the same whether they are for metallic minerals or for groundwater; only the type of anomalies and observable phenomena differ.

Electrical hydrological surveys are used to detect or give information on one or more of the following hydrological problems; the depth to the water table, depth to bedrock, permeability of strata, the detection of buried aquifers, location of potable water supplies, and to specialized problems such as salt-water encroachment as found in northwestern Germany and the Netherlands.

In each of these instances, a careful study of the surface and subsurface geology should be made before conducting a hydrological investigation. This is necessary to assist in the interpretation of the "apparent resistivity" data. In North America, the Wenner configuration or some variation of it is most popular; in Europe, the Schlumberger configuration is most often used. Regardless of the configuration, the apparent resistivity calculated from the current and potential differences due to a specific electrode separation is used as the usual data. Characteristic apparent resistivity results depend on the geology, the location of the survey, and the type of survey conducted. In a glaciated environment for example, as found in many places in North America, increases in resistivity or resistive anomalies are often associated with sweet-water aquifers.

By contrast, in Europe, salt-water encroachment is determined by indications of resistivity lows. Geophysicists in the Far East also make use of the fact that sweet-water aquifers may at times have an apparent resistivity lower than that of the overlying country rock.

Electrical surveys are also conducted, as noted above, for the location of depressions or buried channels in bedrock surfaces. In this

GEOPHYSICAL METHODS FOR HYDROLOGIC SEARCH

FIG. 2. The ABEM Company's Terrameter for Earth Resistivity Measurements, Showing Power Supply, Volt Meter, reels and stakes.

case the physical contrast in resistivity must be between the loose overlying material and the bedrock.

Because of the host of hydrological problems, a great variety of analysis methods have been developed (see Fig. 2). Many of these are empirical and are successful only in very local specific areas. For this reason a careful investigation and correlation with the local geology must be made and all information carefully weighed before deciding upon the best method of interpretation. Analytical and numerical methods for the interpretation of apparent resistivity data are also possible. These are based upon certain simplifying assumptions which make the method tractable. If these necessary electrical and physical conditions are met, thus allowing close agreement with the mathematical assumptions, then very useful hydrological information may be obtained.

Seismic Methods. Although both refraction and reflection seismic techniques have been used in geophysical prospecting for groundwater (see Vol. V, section on *Seismology*), the more common use is seismic refraction in the determination of depth to shallow bedrock where the bedrock is covered by unconsolidated material and soil. Seismic methods may be used to locate depressions or channels in the bedrock surface which may contain deposits of sand and gravel. Occasionally, the seismic method may be used to locate or trace a specific aquifer.

In some areas, reflection seismic methods may be used to locate deep geological structures which could control the occurrence of groundwater, if sweet water may be expected at depth.

As with other geophysical methods a contrast in physical properties is necessary. The velocity of compressional (longitudinal) seismic waves in rock is usually the critical parameter. (See Vol. V, *Elasticity and Rigidity of Rocks*). Table 1 shows some typical velocities of rocks and materials of interest in hydrological work. Seismic velocities usually increase with the specific gravity of the rock, and with its age, depth of burial, and hardness. Rocks which are wet or saturated with water show higher velocities than dry rocks.

This last property makes it possible to estimate the depth to the water table in homogeneous ground.

Some recent work has attempted to use the velocity of seismic shear waves together with compressional waves in order to identify the water table or presence of water in nonhomogeneous ground.

Induced Polarization. The method of induced polarization measures the ability of the ground to store electric charge in a polarization phenomenon when excited by direct or low-frequency current. Although polarization is usually caused by metallic particles in the rock, certain types of clays also cause polarization, though of smaller amplitude than the metallic polarization. This effect has been used to distinguish between sands and clays or mixtures of sand and clay and in estimating the water quality (Vacquier et al., 1957; see also Vol. V, *Electrical Properties of Rocks*).

Magnetic Methods. Measurements of the magnetic field may be helpful in hydrologic search in cases where magnetic rocks have some influence in controlling the location or

TABLE 1. TABLE OF REPRESENTATIVE LONGITUDINAL SEISMIC WAVE VELOCITIES[a]

Material	Velocity (ft/sec)
Weathered surface material	500– 2,000
Gravel, rubble, sand (dry)	1,500– 3,000
Sand (wet)	2,000– 6,000
Clay	3,000– 9,000
Sandstone	6,000–13,000
Shale	7,000–14,000
Limestone	7,000–20,000
Granite	15,000–19,000
Metamorphic rocks	10,000–23,000
Glacial till (Saskatchewan)	5,000– 7,000
Ice	12,050
Fresh water	4,700– 4,900
Seawater	4,800– 5,000

[a] Data compiled from Jakosky (1950) and the Saskatchewan Research Council (private communication).

flow of groundwater (see Vol. V, *Geophysical Methods in Explorations; Rock Magnetism*). This might be exemplified by the case in which an intrusive dyke of magnetic igneous rock has formed an impermeable barrier, trapping groundwater. A magnetic survey could locate such a feature (Enslin, 1955).

Gravity Methods. Careful measurements of the force of gravity can, in some instances, be helpful in hydrologic search. This is because variations in the density of near surface materials can cause local small but measurable changes in the force of gravity (see Vol. V, *Gravity; Gravity Exploration*). This effect has been used in the search for bedrock depressions or buried valleys where, in dense bedrock material, these latter have been filled with deposits of relatively light sand and gravel. The result is a slight decrease in the gravity force over the filled valley which can be detected by careful observation with a sensitive gravity meter (Hall et al., 1962). In common with geophysical methods generally, this method depends on a sufficiently large contrast in physical properties, namely bulk density of the rocks involved.

Airborne Methods. Some recent techniques have been developed which can make geophysical measurements for hydrology, from flying aircraft. The most outstanding is the use of infrared measurements to estimate water temperatures. This has been made use of in the Hawaiian Islands to locate undersea "springs" where cool groundwater has been escaping into the warm ocean.

A second new technique is the measurement of resistivity from the air using powerful electromagnetic pulses. This experimental method promises to supplement and speed up surface resistivity measurements.

Analysis of aerial photos now taken in color can yield valuable information on the hydrology of many areas.

Subsurface Methods

Subsurface measurements are useful in determining depth of water-bearing formations, determining their salinity, and estimating their yield. The geophysical methods include electric, radiation, and thermal logging, as well as special logging techniques (see Vol. VI, *Well Logging*; also Guyod, 1957).

Electric Logging. Most electric logs function by measuring the apparent resistivity of the formations from a borehole. This is done by lowering one or more electrodes on a probe into the fluid-filled hole. A continuous log of resistance vs. depth is produced. The three curves on the right side of Fig. 3 show typical response for different electrode configurations from a test hole in glacial deposits. Note the strong response opposite the sections of sand and gravel. Logs such as this are invaluable to a driller who wishes to know exactly where certain formations are in the hole.

A self-potential log is also shown on the left side of Fig. 3. This represents the variation of electric potential in the hole with respect to a reference electrode at the surface. The self potential (S.P.) log usually distinguishes be-

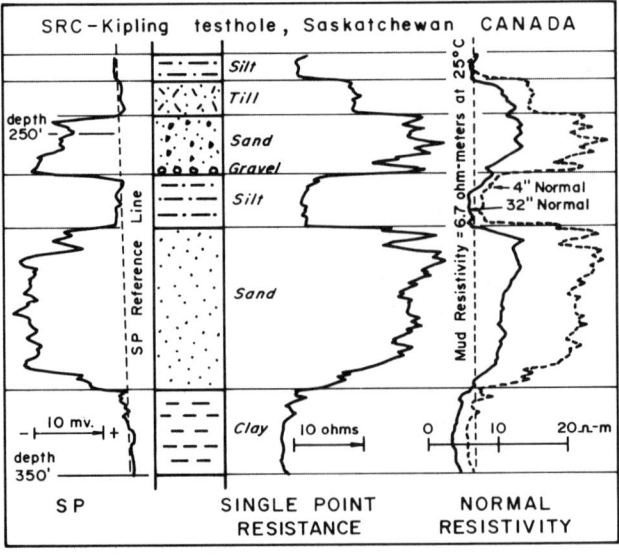

Fig. 3. Geological and electrical logs in glacial drift (from Saskatchewan Research Council).

tween clean sandy formations and formations containing clay or shale. Together with the resistivity log, it may provide an estimate of the freshness of the formation water.

Radiation Logging. A gamma-ray radiation counter may be lowered down a borehole to measure the natural radioactivity of the rocks. Since most normal radiation in rocks is due to potassium, the counter responds to clay minerals which contain much potassium, thereby providing a tool for differentiating between clays and clean sands and gravels which may be good aquifers. The gamma ray log, therefore, appears rather like the S.P. log.

Thermal Logging. Careful temperature measurements in the borehole are sometimes useful in estimating the natural flow of groundwater. The measurement is made possible because moving groundwater carries heat with it and upsets the normal temperature distribution.

Nuclear Magnetic Reasonance Logging. Hydrogen atoms contained in water have a characteristic response when energized by a powerful magnetic field. The measurement and study of this phenomenon may assist the geophysicist in estimating the quantity of recoverable groundwater from a borehole.

Conclusion

In the practical application of geophysics to hydrological search any one, or any combination, of the above methods may be applied to solve a particular problem in a given region. Geophysical methods usually provide more information when used jointly or in combination with other data. As in all geophysical surveys, hydrological surveys require as much supplementary information as is available. Surveys, using more than one method, frequently avoid ambiguities associated with any one method.

D. J. Gendzwill
J. H. Dyck
T. P. Pepper

References

*Breusse, J. J., 1963, "Modern geophysical methods for subsurface water exploration," *Geophysics*, 28(4), 633–657.

Enslin, J. F., 1955, "Some applications of geophysical prospecting in the Union of South Africa," *Geophysics*, 20(4), 886–912.

Gill, H. E., Vecchioli, J., and Bonini, W. E., 1965, "Tracing the continuity of Pleistocene aquifers in northern New Jersey by seismic methods," *U.S. Geol. Surv.* 1965.

Guyod, H., 1957, "Electric detective. Investigation of groundwater supplies with electric well logs," *Water Well Journal* (March and May).

Hall, D., and Hajnal, Z., 1962, "The gravimeter in studies of buried valleys," *Geophysics*, 27(6), Pt. II. 939–951.

Hobson, G. D., Wyder, J. E., and Brandon, L. V., 1962, "Aquifer exploration in Canada by geophysical methods," *Can. Dept. Mines Tech. Surv., Geol. Surv. Can., Reprint No. 54.*

*Jakosky, J. J., 1950, "Exploration Geophysics," Los Angeles, Calif., Trija Publishing Co.

*Kelly, S. F., 1962, "Geophysical exploration for water by electrical resistivity," *J. New Engl. Water Works Assoc.*, 76(2).

*Parasnis, D. S., 1962, "Principles of Applied Geophysics," New York, John Wiley & Sons.

Vacquier, V., et al., 1957, "Prospecting for groundwater by induced electrical polarization," *Geophysics*, 22(3), 660–687.

Valentin, J., 1950, "Seismic Exploration Applied to the Water Supply Problems of Conakry (Guinea)," Paris, Compagnie Générale de Géophysique.

Wiebenga, W. A., 1955, "Geophysical investigations of water deposits, Western Australia." *Australia Bur. Mineral Resources, Geol. Geophys. Bull.*, 30.

Wiebenga, W. A., and Jesson, E. E., 1962, "Geophysical exploration for groundwater," *Australia Bur. Mineral Resources, Geol. Geophys. Bull.*

*Additional bibliographic references may be found in this work.

Cross-references: *Aquifer; Artesian Water; Groundwater; Interstitial Water in Sediments; Water Table. Vol. V: Elasticity and Rigidity of Rocks; Electrical Properties of Rocks; Geophysical Exploration—Electrical Methods; Geophysical Methods in Exploration; Resistivity Methods in Geophysics—Galvanic; Rock Magnetism; Seismology. Vol. VI: Well Logging.*

GEOPHYSICS—See Vol. V

GEOTHERMAL ENERGY—See Vol. IV B

GERMANIUM: ELEMENT AND GEOCHEMISTRY

Germanium, with atomic number 32, is a metallic, silvery-colored, brittle material, with the diamond-type structure (A_4), and the following properties: atomic weight 72.59; atomic volume 13.5 cm^3/g-atom; atomic radius 1.37 Å; melting point 934°C; density at 25°C, 5.323 g/cm^3; hardness on Mohs' scale, 6; opaque to visible and ultraviolet light but transparent to infrared.

Natural germanium consists of five stable isotopes with masses 70 (20.5%), 72 (27.4%), 73 (7.8%), 74 (36.5%), and 76 (7.8%). Although measurable isotope fractionation effects are predicted by theory (Brown and Krouse, 1964), no variations in isotope ratios have been found in terrestrial or meteoric Ge (Shima, 1963). This element exhibits siderophile properties with only minor lithophile and chalcophile tendencies. Thus, Ge is probably abundant in the Fe–Ni core of the earth, but is present in very low concentrations in the upper crust (av. 1.0–1.5 ppm).

Minerals

Minerals with appreciable Ge are exceedingly uncommon:

germanite	$GeS_2 \cdot 7CuS \cdot FeS$
argyrodite	$GeS_2 \cdot 4Ag_2S$
renierite	$Cu_3(Fe, Ge)S_4$
ultrabasite	$28PbS \cdot 11Ag_2S \cdot 3GeS_2 \cdot 2Sb_2S_3$
canfieldite	$4Ag_2S \cdot (Sn, Ge)S_2$

The first three are associated with complex sulfides in low-temperature hydrothermal sulfide deposits, especially in the Cu–Pb–Zn ore body at Tsumeb, S.W. Africa. Stottite, fleischerite, and itoite are rare Ge minerals found in the supergene oxidation zone at Tsumeb (Frondel and Ito, 1957). The Ge^{4+} ion can substitute for Si^{4+} in silicate minerals. Complete replacement can be effected synthetically (e.g., Ge-feldspars, -micas, -garnets, -pyroxenes, etc., Goldsmith, 1950; Tauber et al., 1958; Ringwood and Seabrook, 1963), but in nature, silicates rarely contain more than a few ppm (Harris, 1954; El Wardani, 1957, 1958; Burton et al., 1959). Substitution of Ge for Si occurs preferentially in neso- and ino-silicates (e.g., olivines, pyroxenes) compared to tectosilicates (e.g., quartz).

Because of the analogous properties of silicates and germanates, germanium is of interest to geologists studying phase changes in the earth's mantle. The olivine-spinel transition of Mg_2SiO_4 at 600°C and 150 kilobars, for example, can be compared with the phase transition of Mg_2GeO_4 at 900°C and 0 kilobars (with a 9–10% change in density).

Abundance

Germanium is not significantly concentrated in the rock-forming minerals at any stage of magmatic differentiation; average abundances in the major types of igneous rocks are: granites 1.3, intermediate rocks 1.5, basalts and diabases 1.3, gabbros 1.2, ultrabasic rocks 1.0 ppm.

Geochemistry

Most pegmatite veins contain about as much Ge as granites but higher concentrations are found in pegmatites with abundant Li, Cs, and F minerals; principal Ge carriers are topaz (up to 300 ppm), lepidolite, spodumene, petalite, pollucite, zircon, garnet, feldspars, and tourmaline. During hydrothermal ore-mineral deposition, Ge enters sphalerite (trace to few tenths of 1%; higher Ge in ZnS crystallized at lower temperatures), cinnabar, and enargite. In the weathering process, Ge enters newly formed clay minerals or remains in detrital mineral grains.

Sedimentary rocks as a whole contain about the same amount of Ge as igneous rocks. Exceptions are limestones, evaporites, manganese nodules, and authigenic phillipsite, apatite, silica, etc., in which Ge is usually undetected. Coals contain considerable Ge (up to 250 ppm in coal ash), especially those rich in vitrain and low in ash. There are large regional variations and a single bed may show a fifteen-fold range in Ge content over a lateral distance of several miles. Ge in coal is associated with organic matter, partly by adsorption and partly as an organometallic compound, rather than with mineral matter.

In seawater, germanium is probably present as germanate ion ($GeO_3{}^{2-}$); av. concentration is 5×10^{-5} ppm. Marine scavengers which concentrate Ti, Co, Ni, Cu, Ra, V, Zr, etc., do not concentrate Ge.

Germanium is distributed in extraterrestrial matter in: (1) iron meteorites, from <1.0 ppm in fine octahedrites to 520 ppm in coarse octahedrites; (2) chondrites, fairly uniform, from 7 to 10 ppm, concentrated in the magnetic fraction (El Wardani, 1957; Shima, 1964).

Commercially, Ge is obtained as a by-product of zinc production and base metal refining. This element is of considerable importance to the transistor industry because of its semiconductor properties.

JON N. WEBER

References

Brown, H. M., and Krouse, H. R., 1964, "Fractionation of germanium isotopes in chemical reactions," *Can. J. Chem.*, **42**, 1971–1978.

Burton, J. D., Culkin, F., and Riley, J. P., 1959, "The abundances of gallium and germanium in terrestrial materials," *Geochim. Cosmochim. Acta*, **16**, 151–180.

El Wardani, S. A., 1957, "On the geochemistry of germanium," *Geochim. Cosmochim. Acta*, **13**, 5–19.

El Wardani, S. A., 1958, "Marine geochemistry of germanium and the origin of Pacific pelagic clay minerals," *Geochim. Cosmochim. Acta* **15**, 237–254.

Frondel, C., and Ito, J., 1957, "Geochemistry of germanium in the oxidized zone of the Tsumeb Mine, South-west Africa," *Am. Mineralogist* **42**, 743–753.

Goldsmith, J. R., 1950, "Gallium and germanium substitutions in synthetic feldspars," *J. Geol.*, **58**, 518–536.

Harris, P. G., 1954, "The distribution of germanium among coexisting phases of partly glassy rocks," *Geochim. Cosmochim. Acta*, **5**, 185–195.

Prewitt, C. T., and Sleight, A. W., 1969, "Garnet-like structures of high-pressure cadmium germanate and calcium germanate," *Science*, **163** (3865), 386–387.

Ringwood, A. E., and Seabrook, M., 1963, "High pressure phase transformations in germanate pyroxenes and related compounds," *J. Geophys. Res.*, **68**, 4601–4609.

Shima, M., 1963, "Isotopic composition of germanium in meteorites," *J. Geophys. Res.*, **68**, 4289–4292.

Shima, M., 1964, "The distribution of germanium and tin in meteorites," *Geochim. Cosmochim. Acta*, **28**, 517–532.

Tauber, A., Banks, E., and Kedesdy, H., 1958, "Synthesis of germanate garnets," *Acta Cryst.*, **11**, 893–894.

Cross-references: *Crystal Chemistry; Gallium; Geochemical Evolution of the Core, Mantle, and Crust; Silicon; Zinc.* Vol. II: *Meteorites.* Vol. IVB: *Diamond; Rare Metal Ore Deposits.*

Vol. V: *Pegmatites; Phase Transformation.*
Vol. VI: *Coal.*

GHYBEN-HERZBERG THEORY—See **GROUNDWATER MOTION; HYDROLOGY; COASTAL TERRAIN; HYDROLOGY, SUBSURFACE WATERS**

GLACIAL GEOLOGY: PERIGLACIAL AND GLOBAL EFFECTS—See Vol. III

GLACIAL MILK

Glacial milk is defined as glacial meltwater containing suspended light-colored rock particles the sizes of silt or clay, including the colloidal state. Although not expressed specifically in the definition, the suspended rock particles are pulverized dominantly by abrasive glacial action, and the milk contains ions and compounds in solution, in addition to suspended material.

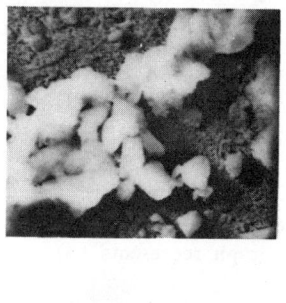
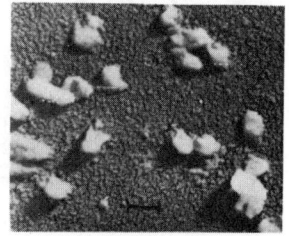
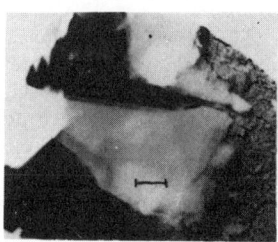
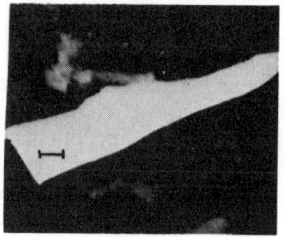
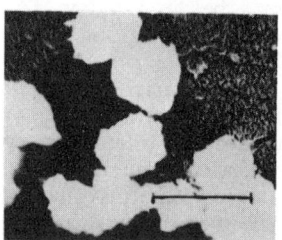
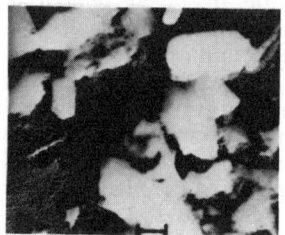

Fig. 1. Electron micrographs of rock flour from glacial milks (bar scale in each micrograph represents 1μ).

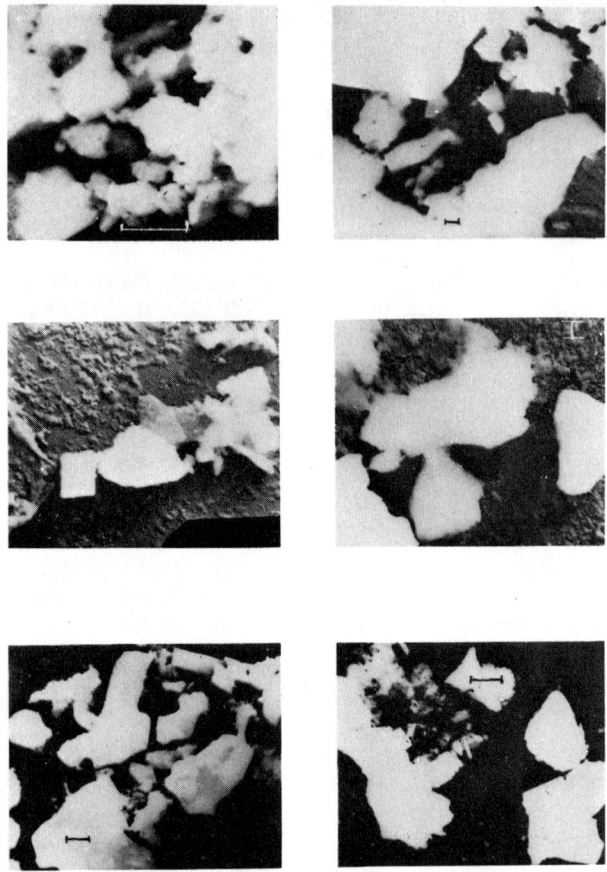

Fig. 2. Electron micrographs of rock flour from glacial milks (bar scale in each micrograph represents 1μ).

In Table 1 are recorded analyses of the mobile load of representative glacial milks, and the dominant type of rock beneath the glacier. In general, the dissolved content of the milk reflects the chemical composition of the rock over which the glacier scours. Calcium (Ca) and magnesium (Mg) from limestone and some silicates are dissolved perhaps to half saturation, judged by dissolving laboratory-pulverized representative rocks from the moraines. Other common rock-forming elements are generally but little dissolved relative to their dissolution upon abrasion in the laboratory. The pH of glacial milk ranges from 6.4 to 6.85, except from Norway glaciers, where it is 4.1–4.5. The cause for this relatively high acidity is not known, but rain along the west coast of Europe commonly yields pH 4.8–5.4.

Electron micrographs of suspended rock particles are shown in Figs. 1 and 2. Shapes of the particles reflect cleavage of minerals from phanerites, but volcanic rocks and argillaceous sediments yielded irregular shapes.

The chemical composition of freshly pulverized glacial rock flour—the suspension particles in glacial milk—is expected to be essentially that of the abraded rock. After long contact with water, silica and the common univalent and divalent metal ions of the rock flour are slowly dissolved, and alumina enriched in it. These changes are suggested from the dissolved ions in the milks, and from a chemical analysis of rock flour which was collected from deposits that settled on tree limbs in a beaver pond in Teton Creek about three miles below Teton Glacier, No. 11 (see Table 2).

It is inferred that glacial milk is a potentially favorable source of inorganic nutrients, both macro and micro, for plants. Its dissolved content, and high specific surface of fine particles, provide readily available ions for plant rootlets. The well-demonstrated fertility of soils in glacial drift, and on alluvial plains deposited by glacial streams, in the United States, Europe, Asia, and Africa have long been a monument

TABLE 1. MOBILE LOAD AND pH OF GLACIAL MILKS[a]

| No. | Locality | Dissolved Major Ions (ppm) | | | | | | | Suspended Solids (ppm) | pH |
		Si	Al	Ca	Mg	K	Na	P	Total		
1-milk	Emmons Glacier Mt. Rainier, Wash.; andesitic volcanics and ash	8.00	0.011	2.1	0.4	0.85	1.51	3.05	12.87	7100	7.5
2-milk	Nisqually Glacier Mt. Rainier, Wash; andesitic volcanics and ash	4.28	0.007	2.07	0.74	0.57	1.54	1.89	9.81	2900	7.4
3-milk	Zermatt, Switz.; greenstone, diabase, and ophiolite	1.18	0.001	10.6	1.11	0.74	0.40	0.49	14.03	173	6.7
4-milk	Les Diablerets, Switz.; limestone	1.35	0.001	26.6	3.83	0.18	1.30	0.60	33.26	5	7.5
5-milk	Grinnell Glacier Glacier Nat'l. Park, Mont.; calcareous argillite	0.55	0.001	8.5	1.42	0.10	0.22	0.27	10.79	6	7.05
6-milk	Stechelberg, Switz.; limestone and crystalline schists	0.61	0.026	19.4	1.51	0.26	0.44	0.36	22.25	87	7.5
7-milk	Grindelwald, Switz.; limestone and crystalline schists	0.66	0.001	17.5	0.525	0.62	0.22	0.33	19.53	34	6.8
8-milk	Trient, Switz.; granite and limestone (?)	1.49	0.001	10.6	7.62	1.05	0.46	0.59	21.22	5	6.9
9-milk	Rhone Glacier Gletsch, Switz.; granite, some mafics	1.14	0.001	4.1	2.56	0.95	0.34	0.49	9.09	109	6.6
10-milk	Bossons Glacier Chamonix, France; granite	0.45	—	7.0	0.14	1.72	0.21	0.16	9.52	5	6.7
11-milk	Teton Glacier Teton Nat'l. Pk. Wyo.; granite and gneiss	0.62	0.003	0.82	0.24	0.53	0.31	0.22	2.52	38	6.4
12-milk	Dinwoody Glacier Wind River Mts., Wyo.; granodiorite and gneiss	0.68	0.003	4.5	2.2	0.59	0.65	n.d.	8.62	5	6.85
13-milk	Nigards Glacier Gaupne, Norway; dark gneiss and schist	0.42	0.001	0.585	0.04	0.13	0.24	0.16	1.58	3.2	4.1
14-milk	Fanarak Glacier Lom, Norway; gneiss and schist	1.06	0.38	1.25	0.08	0.41	0.24	0.49	3.42	5	4.2
15-milk	Bøver Glacier Lom-Skjolden, Norway; gneiss and schist	0.48	—	0.70	0.08	0.38	0.16	0.25	1.80	32	4.5
16-melt-water	Taylor Glacier, West end of Lake; Bonney, Antarctica	0.2	0.09	17.9	0.94	1.14	6.9	—	27.17	—	7.4
17-melt-water	Wright Glacier, Wright tongue of Wilson Piedmont Glacier, Antarctica	0.2	0.03	—	0.09	0.4	2.14	—	2.86	—	7.2

[a] After Keller and Reesman, 1963b.

GLACIAL MILK

TABLE 2. CHEMICAL COMPOSITION OF TETON GLACIAL ROCK FLOUR[a]

Material	Value (wt %)
SiO_2	57.13
Al_2O_3	19.15
TiO_2	0.34
Fe_2O_3	3.55
FeO	1.81
CaO	2.34
MgO	1.69
K_2O	3.21
Na_2O	2.34
P_2O_5	0.22
SO_3	0.10
MnO_2	0.25
Loss igneous	7.82
Total	99.95

[a] From Keller and Reesman, 1963b. Analysis by Bruce Williams Laboratory, Joplin, Missouri.

to the utility of glacial grinding and glacial milk. Glacial milk nourishes the soil as does typical milk the animal world.

W. D. KELLER

References

Coombs, H. A., 1936, "The geology of Mt. Rainier National Park," *Geology*, **3**, 131–212.
Keller, W. D., Balgord, W. D., and Reesman, A. L., 1963, "Dissolved products of artificially pulverized silicate minerals and rocks: Part 1," *J. Sediment. Petrol.*, **33**(1), 191–204.
Keller, W. D., and Reesman, A. L., 1963a, "Dissolved products of artifically pulverized silicate minerals and rocks: Part 2," *J. Sediment. Petrol.*, **33**(2), 426–437.
Keller, W. D., and Reesman, A. L., 1963b, "Glacial milks and their laboratory-simulated counterparts," *Geol. Soc. Am. Bull.*, **74**, 61–76.
Stevens, R. E., and Carron, M. K., 1948, "Simple field test for distinguishing minerals by abrasion pH," *Am. Mineralogist*, **33**, 31–50.

Cross-references: *Abrasion pH; Colloids; ph-Eh.* Vol. III: *Glacial Scour.* Vol IVB: *Electron Microscopy.*

GLASS—*See* Vol. V

GLASS, DEVITRIFICATION—*See* Vol. IV B

GOLD: ELEMENT AND GEOCHEMISTRY

The chemical element gold, Au, atomic number 79, is a heavy, rare metal. It falls into group Ib of the periodic table along with copper and silver. It has only one stable isotope, 197, and twenty radioactive isotopes, 187–203, the lives of which range from seconds to days. Its atomic weight is 196.967. Its melting point is 1063°C. It belongs to the isometric crystal system, but commonly occurs in plates, scales, or masses (hardness = 2.5–3; sp gr = 19.3). It is very malleable and ductile. It is found in various shades of yellow, becoming paler with increasing silver content. Gold is insoluble in ordinary acid but soluble in *aqua regia*.

Because of its malleability, relative chemical inertness, and beauty, gold has long been highly prized by man. Gold ornaments have been found in Stone Age tombs, over 10,000 years old. Since the demand has always outstripped the supply, gold has gradually assumed an artificial value, centered about its use in currency.

Abundance. Gold is quite widely distributed, appearing dispersed in most common rocks, in seawater, and in the lithosphere. The estimated average content in crustal rocks is 0.004 ppm. Assays of meteorites indicate gold to be siderophile and concentrated in the nickel iron core. Goldschmidt reports an average gold content for meteorites of 0.7 ppm. Urey estimates the concentration of gold in the earth as a whole to be 0.25 ppm. Ocean water contains about 0.004 ppm of gold, presumably as $[AuCl_4]$.

Geochemistry. Due to the chemical inertness of gold and because gold compounds are readily reduced to metal, gold is found largely in the free state, i.e., as a "native" metal. Due to its high specific gravity, gold often forms stream-sorted placer deposits; such deposits have been of major economic importance and because of their ease of development were the object of the major gold rushes.

During the main stage of igneous crystallization, gold is enriched in the late magmatic products. The pegmatitic and hydrothermal formations are the most characteristic abodes of gold.

Gold does not combine with either oxygen or sulfur but does react with tellurium. The only compounds of gold known as ore minerals are the tellurides. The most important tellurides are calaverite, $AuTe_2$ and sylvanite, $AuAgTe_4$. Native gold usually contains silver and often small amounts of copper plus traces of platinum-group metals.

Economic Geology. The major typical ore deposits having commercial importance have been condensed by Wise into these four groups:

1. Deposits along or in fractures and fracture systems.
   (a) Quartz veins with only minor amounts of other nonmetallic minerals and sulfides.
   (b) Quartz stringer lodes.

(c) Veins and lodes with less conspicuous quartz and larger proportion of carbonates, silicates, or sulfides in the gangue.
2. Massive deposits.
   (a) Predominantly siliceous replacements.
   (b) Predominantly sulfide replacements.
3. Disseminated deposits.
   (a) Gold in the "porphyry coppers."
4. Residual and mechanical concentrations (placers).
   (a) Gossans and croppings over ore bodies.
   (b) Placers in present streams, alluvials, and beaches.
   (c) Placers in ancient streams, alluvial, and/or beach deposits, unconsolidated or consolidated, incorporated in geologic column.

The largest United States mine is at Homestake, South Dakota. Most of the world's gold comes from Witwatersrand, "the Rand," Republic of South Africa (32 million troy ounces in 1970, 73% of world total). The second largest producer is the U.S.S.R., followed by Canada, the U.S., and Australia.

Gold appears to be concentrated in the seeds of certain plants growing in gold-bearing areas, and is also found in coal ashes and certain marine plants. There is some evidence that certain microorganism have the ability to concentrate gold.

HAROLD W. ROBINSON

References

Anon., 1967, "Geochemical tools used to find and dig gold," *Chem. Eng. News*, **45**(47), 49.

Rankama, K., and Sahama, T., 1950, "Geochemistry," Chicago, Univ. Chicago Press, 912pp.

Robinson, H. W., 1968, "Gold," in (Hampel, C. A., editor), "Encyclopedia of Chemical Elements," New York, Reinhold Book Corp., pp. 244–250.

Ryan, J. P., 1965, "Gold," *Minerals Yearbook*, U.S. Bur. Mines.

Washington, R. A., and Holman, R. H. C., 1966, "A rapid and sensitive method for determining gold in rocks and other geologic materials," *Can. Geol. Surv. Paper*, 65–67.

Wise, E. M., 1964, "Gold: Recovery, Properties and Applications," New York, Van Nostrand.

Cross-references: *Seawater, Chemistry*. Vol. IVB: *Gossan; Placer Deposits; Rare Metal Ore Deposits.*

GOLD: ECONOMIC DEPOSITS

It is uncertain whether copper or gold was the first metal to be discovered by man, but gold was certainly in use in prehistoric times and the goldfields of Nubia were being actively worked by the time of their conquest by the Egyptians in 2700 BC. The first record of gold mining consists of pictorial rock carvings that apparently depict primitive gold-washing operations in Egypt as early as 4000 BC. The total quantity of gold mined by man since these early diggings is obscure, but based on an estimate by Bateman (1950) and more recent data (Ryan, 1968), the total figure is in the neighborhood of 75,000 tons. If all this gold could be gathered and evenly stacked on a regulation basketball court, it would rise to a height of about 26 ft and, at current gold prices, would have a value of about 63 billion dollars.

Types of Deposits. Most of the world's gold has come from placers or from hydrothermal deposits in bedrock in which the gold is either the principal metal of value or a by-product of base metal production.

Beach and alluvial placers derived from erosion of bedrock lodes were the leading source of gold in the United States until 1873 and were exploited in many mining camps in California, Oregon, Alaska, the Yukon Territory, and elsewhere. They served not only as a major source of easily mineable gold, but also as guides to the location of the lode deposits that were soon to surpass them in production. Such placers now account for only 4% of the domestic output, but if the great auriferous conglomerates of the Witwatersrand district (see below) are regarded as placers, this type of deposit is still of overwhelming world importance.

The hydrothermal gold deposits in bedrock occur throughout the world and range in age from Precambrian to Late Tertiary. They are generally confined to orogenic belts in which they are closely associated with stocks, batholiths, and other igneous intrusives of intermediate to acid composition. Important gold deposits are notably absent from regions of flat-lying sediments that have had a mild deformational history. The majority of these deposits are very probably of magmatic hydrothermal origin, but in some cases where the deposits occur in metamorphic terrane, there is usually debate as to whether the gold and associated minerals were of igneous parentage or derived from the country rocks as a result of regional metamorphism.

The gold deposits of hydrothermal origin are so numerous and varied that only a few important types and examples, chiefly domestic ones, can be pointed out here. In many deposits, open filling predominated over replacement and these generally take the form of fissure veins (e.g., Mother Lode, California; Cripple Creek, Colorado; Porcupine, Ontario), of breccia bodies (e.g., Bull Domingo, Colorado), or of saddle reefs (e.g., Bendigo, Australia). Where replacements was the dominant ore-forming process, the resulting deposits may be of the massive type (e.g., Morro Velho, Brazil; Noranda, Quebec), of the lode type (e.g., Homestake, South Dakota; Porcupine and Kirkland Lake, Ontario; Kolar, India), or of the disseminated type (e.g., gold in the porphyry copper type deposits as at Bingham, Utah; "invisible gold" in carbonaceous sediments as at Carlin, Nevada).

Most by-product gold comes from porphyry copper deposits (e.g., Utah Copper, Bingham, Utah), but important amounts are also obtained in the mining of some lead-zinc ores (e.g., Iron King, Arizona; Lark, Utah) and some copper-lead-zinc ores (e.g., Mayflower, Utah; Idarado, Colorado).

Mineralogy. The native metal is by far the most important ore mineral of gold in both placer and bedrock deposits. It is commonly combined with 5–15% silver, and pale yellow alloys containing over 20% silver ("electrum") occur in some deposits. In many ores, the gold is intimately intergrown with sulfides like pyrite and arsenopyrite, but does not itself form a sulfide. Gold does combine in nature with tellurium and in some deposits (e.g., Kalgoorlie, Australia, Cripple Creek, Colorado; Vatukoula, Fiji; Acupan, Philippines), the gold-bearing tellurides are important ore minerals. Chief among these are calaverite ($AuTe_2$), krennerite (($Au,Ag)Te_2$), sylvanite ($AuAgTe_4$), and petzite (Ag_3AuTe_2). A great variety of gangue minerals may accompany gold, but quartz, carbonates, sericite, tourmaline, and fluorite are especially common.

Tenor. Gold mining is an unusual economic enterprise in that the price of the product cannot be adjusted in response to changing costs of production. As a result, the tenor of workable ore tends to increase in times of inflation, and marginal mines operating at maximum efficiency may be forced to close. For the last several years, the average tenor of all placers worked in the United States fluctuated between 23 and 33¢ in gold per cubic yard. Through highly efficient mining methods, the Homestake mine in South Dakota (the leading U.S. producer and only major underground mine in the country mining primarily gold) is profitably mining ore that averages only 0.32 oz of gold per ton. In 1967, ores of the South African goldfields averaged 0.39 oz/ton and those of the open pit mine near Carlin, Nevada, 0.48 oz/ton. In the same year, the major lode mines of Canada produced ores whose average grade ranged from about 0.25 to 0.69 oz/ton. Where mined as a by-product from domestic ores in 1967, gold averaged 0.003 oz/ton in copper ores and 0.025 oz/ton in lead-zinc and copper-lead-zinc ores.

Geochemistry. Under normal conditions of weathering, gold is chemically inert and tends to persist as the native metal in leached ore outcrops and placers. Due to selective leaching of silver, the fineness of placer gold is generally somewhat higher than the lode gold from which it is derived and tends to increase with the distance of transport. Under special conditions of very high oxidation potential combined with high acidity and chloride ion activities, gold may be dissolved as the auric chloride complex in the zone of oxidation, but this behavior is unusual.

Little is known of the hypogene geochemistry of gold. By extrapolation of room-temperature thermodynamic data, Helgeson and Garrels (1968) predict that hydrothermal solutions could dissolve sufficient gold as the aurous chloride complex to account for typical pyritic gold-quartz veins. Precipitation of the gold would be explained simply by cooling of the ore fluids and their reaction with wallrocks. Their model presupposes highly acid ore fluids. In neutral or alkaline ore fluids, gold might be transported as a sulfide or hydrosulfide complex but the stability of such gold complexes at high temperatures and in the presence of chlorides that probably prevail in most hydrothermal fluids has neither been calculated nor verified experimentally. Likewise, the role of organic ligands and carbonaceous matter in the transport and deposition of gold has not been fully explored and may have been underestimated in the past. At least in some deposits (e.g., Carlin, Nevada), it appears that carbonaceous material in the country rocks can serve as an effective precipitant of hydrothermally transported gold.

At the present time, small quantities of ore-grade precipitates are actually forming in several thermal pools of the Taupo Volcanic Zone, New Zealand (Weissberg, 1969). These precipitates are a red-orange mixture of amorphous sulfides and opaline silica containing up to 2.7 oz of gold per ton. The "ore fluids" in this case are slightly acid sodium-potassium-chloride-bicarbonate waters which contain, at most, 0.04 parts per billion gold. The extent to which the present conditions in the Taupo Zone simulate past ore-forming processes elsewhere is speculative.

Domestic Production. Based on recent statistics of Koschmann and Bergendahl (1968), the total production of gold in the United States in the years 1799 through 1965 amounted to

GOLD: ECONOMIC DEPOSITS

TABLE 1. GOLD PRODUCTION IN THE UNITED STATES

Rank	State	Number of Districts	Total Ounces Produced	Percent of U.S. Total
1	California	97	106,130,214	35
2	Colorado	44	40,775,923	13
3	South Dakota	7	31,207,892	10
4	Alaska	43	29,872,981	10
5	Nevada	71	27,475,395	9

307,182,000 troy ounces (present value $10,751 million) and more than three-fourths of this gold came from only five western states (Table 1).

One-half of the total domestic production has come from only twenty-five major districts among which the five leaders were in order of rank (1) Lead, South Dakota, (2) Cripple Creek, Colorado, (3) Grass Valley-Nevada City, California, (4) Bingham, Utah, and (5) the Comstock Lode, Nevada.

The total domestic production of gold from all sources in 1967 was 1,584,000 troy ounces having a value of $55,447,000. Of this total, 69% came from gold and gold-silver mines, 27% as a by-product of base metal ores, and only 4% from placers. According to data of Ryan (1968), almost all of the 1967 production could be credited to five lode gold mines, one gold-silver mine, three placer mines, twelve copper mines, two copper-lead-zinc mines, and two lead-zinc mines. Two mines, the Homestake in South Dakota and Carlin in Nevada, accounted for almost two thirds of the domestic production.

The Witwatersrand Districts. Space available here will not permit description of the world's major gold deposits, but no discussion of gold would be complete without some special reference to the great auriferous conglomerates of the Republic of South Africa. These deposits have dominated world gold production since 1905 and now contribute about two-thirds of the total world output. The gold is disseminated in these ores and tends to occur in pay streaks at selected stratigraphic levels in thin, persistent conglomerates ("reefs") of the Precambrian Witwatersrand system. This system is approximately 24,000 feet thick and has been folded into a broad structural basin roughly 75 × 170 miles in size. Where gold-bearing, the reefs are called "bankets." Forty-nine mines were producing in the district in 1967, some mining at depths over 11,000 feet, and the average tenor of all ores produced was 0.39 oz of gold per ton. Local authorities do not agree on the origin of the ores, some contending that they are epigenetic ores in which the gold was introduced by magmatic hydrothermal solutions after burial of the reefs, and others that the ores are of modified syngenetic origin or essentially placer deposits in which the gold has suffered recrystallization and very local redistribution as a result of mild metamorphism. Whatever their origin, there is no question that these deposits will continue to be the world's leading source of gold for many years to come.

World Production (data from Ryan, 1968). After a steady rise for thirteen years, world gold production decreased slightly in 1967 to a level of 45.6 million ounces valued at $1596 million. The same year saw a 14% decline in U.S. production and a slight cutback in South African gold mining. The ten leading countries in 1967 are listed in Table 2 (Ryan, 1968).

For additional production statistics, the reader is referred to the work of Ryan (1968) and for descriptions of the major gold deposits

TABLE 2. WORLD GOLD PRODUCTION

Rank	Country	Production (troy ounces)	Percent of World Total
1	Republic of South Africa	30,532,880	67
2	U.S.S.R.	5,700,000	13
3	Canada	2,961,999	7
4	United States	1,584,187	3
5	Ghana	762,609	2
6	Australia	627,171	1
7	Philippines	500,417	1
8	Colombia	258,186	0.5
9	Japan	252,769	0.5
10	Mexico	181,491	0.4

in each of these countries to the works of Bateman (1950), Emmons (1937), and Lindgren (1933).

WILLIAM C. KELLY

References

Bateman, A. M., 1950, "Economic Mineral Deposits," New York, John Wiley & Sons, 916pp.

Emmons, W. H., 1937, "Gold Deposits of the World," New York, McGraw-Hill Book Co., 562pp.

Helgeson, H. C., and Garrels, R. M., (1968), "Hydrothermal transport and deposition of gold," *Econ. Geol.,* **63,** 622–635.

Koschmann, A. H., and Bergendahl, M. H., 1968, "Principal gold-producing districts of the United States," *U.S. Geol. Surv. Profess. Paper* **610,** 283pp.

Lindgren, W. T., 1933, "Mineral Deposits," New York, McGraw-Hill Book Co., 930pp.

McLaughlin, D. H., and Wise, E. M., 1964, in (Wise, E. M., editor), "Gold; Recovery, Properties, and Application," Princeton, New Jersey, D. Van Nostrand Co., 367pp.

Ryan, J. P., 1968, Preprint on "Gold" for 1967 *Minerals Yearbook,* United States Bureau of Mines, 22pp.

Weissberg, B. G., 1969, "Gold-silver ore-grade precipitates from New Zealand thermal waters," *Econ. Geol.,* **64,** 95–108.

Cross-references: *Hydrothermal Solutions.* Vol. IVB: *Gangue Minerals; Mining and Mineral Centers of the World; Ore-Forming Fluids; Placer Deposits.*

GREAT SALT LAKE—*See* Vol. III

GROUNDWATER*

The uppermost rocks of the earth's crust constitute a porous medium in which water is stored and through which it moves. Up to a certain level these rocks are saturated with water that is under hydrostatic pressure (atmospheric pressure or greater) and is free to flow laterally under the influence of gravity. Subsurface water in this saturated zone is, by definition, *groundwater* (Meinzer, 1923a, pp. 38–39; 1923b, p. 22), and the uppermost level of the zone is the water table. At some places, such as at the edges of many but not all streams and lakes, the water table intersects the land surface. At most places, however, it is overlain by a zone of aeration in which the water in the pores is under less than atmospheric pressure—it is held in the pores by capillarity. The zone of aeration ranges in thickness from a few inches to several hundred feet. Subsurface water in this zone is termed *vadose water;* it includes water in the soil belt, the capillary fringe, and the intermediate belt between them (Fig. 1). Only the groundwater can be extracted readily for man's use, but the vadose water is important also; it supports plant growth, and the zone of aeration is the medium through which passes most of the water that contributes to groundwater recharge.

From ancient times man has utilized groundwater, drawing it from springs and wells, but only recently has he begun to appreciate its significance with respect to the world's water supply and to amass the scientific and technical knowledge needed to develop and manage it intelligently. According to Nace (1964, p. 18), the volume of groundwater in the upper half mile of the continental crust is perhaps twenty times greater than the combined instantaneous volume of water in all the lakes and rivers of the world. Furthermore, groundwater is almost universally present beneath the land areas of the world, including many areas, such as parts of the arid lands, where no other water is available during much of the time.

Especially significant is the concept of groundwater as a phase of the hydrologic cycle and the corollary concept that groundwater is a renewable resource. Groundwater reservoirs constitute a major mechanism for natural regulation and artificial management of the world's water supply. Also significant is the fact that the huge reserves of groundwater have accumulated over periods of years or centuries. Especially in some arid and semiarid regions, these reserves are not renewable in a practical sense; only recently has attention been given to the implications of planned development of this water as a minable commodity.

Source and Movement

Groundwater can be broadly classified by origin as juvenile, connate and meteoric. Juvenile water is derived from the interior of the earth and includes water of magmatic origin. "It is invariably high in mineral content, however; and, so far as scientists have been able to determine, it is emitted in quantities that are insignificant in comparison to those involved in the hydrologic cycle" (McGuinness, 1963, p. 18). Connate water is water trapped in sediments at the time of deposition. Connate marine water is salty and is avoided as a source of water supply; however, fresh water trapped with nonmarine sediments commonly is usable. Meteoric groundwater is water which is derived from the atmosphere and which moves in the hydrologic cycle; subsequent reference to groundwater concerns only that of meteoric origin.

---

* Webster's Dictionary, Third ed., makes "groundwater" one word. U.S. Geological Survey practice favors it in two words.

GROUNDWATER

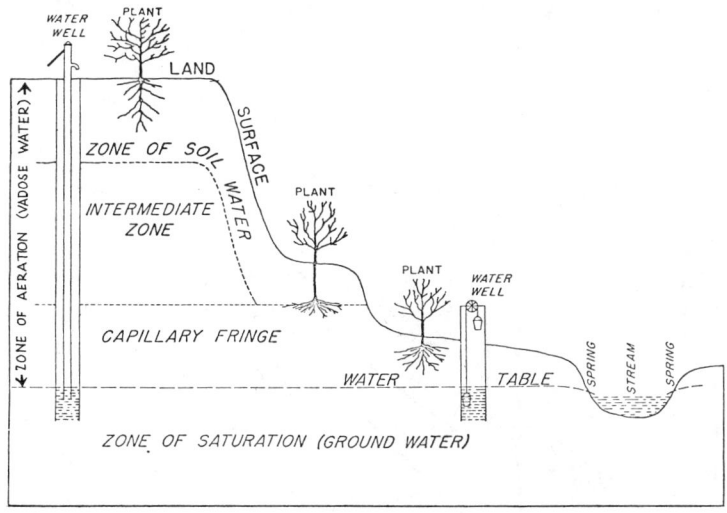

Fig. 1. Zones of subsurface water

Groundwater, like all water in the land phase of the hydrologic cycle, is derived from precipitation. Upon reaching the land surface, the water of precipitation, except for small amounts returned immediately to the atmosphere by evaporation, either flows overland to streams, lakes, marshes or the ocean, or infiltrates the soil. Of the water that enters the soil, a large part is stored temporarily as soil moisture and subsequently is returned to the atmosphere by evaporation or by transpiration of plants; the remainder percolates downward in the porous rocks to the water table, where it enters the zone of saturation.

Storage in the zone of saturation is temporary, the groundwater moving under the force of gravity laterally in the direction of the hydraulic gradient toward discharge areas at the land surface. Ordinarily movement is laminar and in accordance with Darcy's law, which states that the velocity of groundwater movement varies directly with the hydraulic gradient and the permeability of the deposits. In nature, gradients generally are gentle; consequently, the rate of groundwater movement is very slow, commonly ranging from a few feet per year to a few feet per day. In detail, the movement is complex and three-dimensional, owing principally to differences in the permeability of the rocks.

En route to areas of liquid discharge, some of the groundwater may be evaporated or transpired back to the atmosphere (Fig. 2). The remainder seeps into streams or other bodies of surface water in humid regions, sustaining the flow of the streams even during extended periods of fair weather. During times of flood, the direction of groundwater movement adjacent to streams may be reversed; then, usually for only a short time, stream water may recharge—i.e., replenish—the groundwater reservoir. Some streams in arid regions habitually lose water, particularly in certain reaches, and thus replenish underlying groundwater reservoirs.

From the above it may be seen that the occurrence and availability of groundwater are governed by the interaction of numerous environmental factors, especially climate, topography, vegetation, soils and geology. Of these factors, climate, topography, soils and vegetation play mainly an antecedent role; i.e., they affect the distribution of water before it enters the groundwater reservoirs. In contrast, geologic factors—principally the hydrologic properties of the rocks and their stratification and structure—control the movement and storage of groundwater directly.

Hydrologic Properties of Rocks

Porosity, the property of containing openings or pores, explains the ability of rocks to store water and is a measure of the volume that can be stored. There are many kinds of rocks, and their porosities differ greatly according to the number, size and shape of the contained openings. For example, granular unconsolidated rock materials such as sand and gravel, which if well sorted have numerous openings between particles, can store relatively large volumes of water—up to 30 or 40% of the total volume. On the other hand, compact consolidated rocks such as granite and quartzite, whose openings consist of narrow fractures ordinarily few in number, commonly have porosities ranging from only a fraction of a per

GROUNDWATER

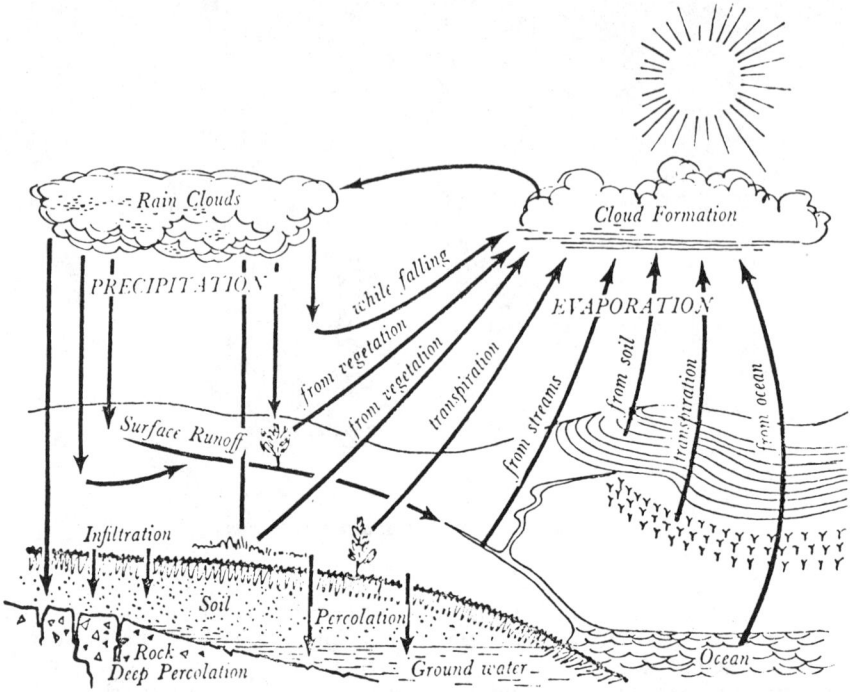

FIG. 2. Hydrologic cycle (from U.S.D.A. Yearbook, 1955).

cent to a few per cent, and therefore can store little water.

Permeability and specific yield are the hydrologic properties of rocks that are of most concern with respect to the flow and storage of groundwater. The first determines the ease with which a rock transmits water under a hydraulic gradient; the second determines the capacity of a saturated rock to yield groundwater by drainage under the force of gravity. Both properties are related to the resultant of two opposing forces—gravity, which tends to move groundwater through the openings in the rock in the direction of the hydraulic gradient, and molecular attraction, which tends to oppose the movement by creating internal friction and by holding water to the walls enclosing the openings. These properties are primarily a function of the size and shape of interconnected openings, inasmuch as size and shape largely determine the effectiveness of molecular attraction. Also, in clays, the mineralogy may exert some control over the retention of water.

Although rocks of high porosity might be expected to have a high permeability and specific yield, some do not. For example, clay with its myriad interlamellar openings commonly has a high porosity, but because the openings are minute, molecular attraction may hold the water to the particles so strongly that the rock is virtually impermeable and will yield little or no water.

Aquifers

Rocks of different types have water-bearing properties that are more or less distinctive. Stratification and regional structure impart orderliness to the distribution of the various rock types and in effect control the extent, subsurface boundaries and hydraulic continuity of aquifers. An aquifer is a formation, group of formations, or part of a formation that contains sufficient saturated permeable material to yield appreciable quantities of water to wells and springs. The term *groundwater reservoir* often is used interchangeably with aquifer, but it is used also in a broader sense to include several related aquifers and confining beds which together may form a hydrologic unit.

Groundwater commonly is classed as occurring under *water-table* (unconfined) or *artesian* (confined) conditions. Water-table conditions exist if the top of the zone of saturation is at atmospheric pressure—i.e., if a zone of aeration freely connected with the atmosphere lies above the water table. Artesian conditions exist if the water in the aquifer is confined beneath less permeable material under sufficient pressure to rise in wells above the base of the confining layer. A water-table aquifer may receive recharge throughout the area it underlies; an artesian aquifer is recharged in an area where the upper confining layer is absent, and therefore where it is not artesian, or via leakage through the confining beds.

Whether water is under artesian or water-table conditions is significant with respect to the spacing of wells and the manner of developing an aquifer. An artesian aquifer is completely saturated from bottom to top and will remain so until the artesian-pressure surface, or *potentiometric* surface, has been so lowered as to dewater the upper part of the aquifer. The withdrawal of water from a well results at the point of withdrawal in a decline of the potentiometric surface. The water that enters the well must be derived by compaction of the aquifer and associated aquicludes and by slight expansion of the water itself. The quantity of water so derived, expressed by the coefficient of storage, is very small—generally only 0.001–0.00001 cubic foot of water per square foot of aquifer for each foot of lowering of the potentiometric surface. Thus, if water is taken from the well at a substantial rate, the potentiometric surface declines quickly for a considerable distance from the well.

In contrast, a body of rock containing a water-table aquifer generally is unsaturated in its upper part. The withdrawal of water from a well results in a lowering of the water table at and near the point of withdrawal and in actual removal of water from storage there. The coefficient of storage of an aquifer under water-table conditions is approximately equal to the specific yield; it may range from 0.01 to as much as 0.30 or even more. Thus the effect of withdrawal is transmitted slowly to other parts of the aquifer.

Aquifers perform a twofold function: they serve as natural storage reservoirs and as distribution conduits. As storage reservoirs, they accumulate water over periods of time ranging from days to centuries, providing the reserve from which nature sustains streamflow during periods of no precipitation and from which man can draw water directly by means of wells. As conduits, they transport water from recharge areas to points of interception by wells or to discharge areas, including bodies of surface water such as streams and lakes, and areas where water discharges by evaporation and transpiration (see Vol VI, *Evapotranspiration*).

Under natural conditions, an aquifer takes in, transports and discharges water at a rate which varies from wet to dry seasons, years, and periods of years but which approaches an average over the long term. This rate may have little to do with the rate at which water can be extracted perennially by man, however. Perennial extraction by man is possible only to the extent that his withdrawal reduces (salvages, or captures) natural discharge without causing an excessive lowering of water levels or depreciation of water quality, plus the extent—if any —to which recharge is increased as a result of the withdrawal.

Especially in extensive aquifers, the points or areas of natural discharge may be so distant that salvage could not be effected even if water levels were drawn down to the bottom of the aquifer at the points of use. Thus, injection of water into the aquifer at or near points of use (*artificial recharge*) at an average rate equivalent to the rate of withdrawal may be necessary if the development is to be perennial. Such practices may be possible in many areas where surplus surface water is available at a time of year other than that during which the need for water is greatest. This planned cyclical storage of water in, and withdrawal of water from, aquifers, to take advantage of their large storage capacity, will be an essential feature of water management in many areas in the future.

Principal Types of Aquifers

On the basis of rock types, aquifers may be divided into six principal types (Thomas, 1952, pp. 10–13): sand and gravel, limestone and other soluble rocks, basalt and other volcanic rocks, sandstone and conglomerate, crystalline and metamorphic rocks, and porous but poorly permeable materials.

Sand and Gravel. Of the several types of aquifers, sand and gravel, which yield water from intergranular openings, are the most widely distributed and yield the bulk of the water taken from wells. They occur principally in glacial outwash, in marine deposits underlying coastal plains, and in alluvial deposits along existing streams and in the fill of intermontane basins.

Sand and gravel aquifers along streams commonly are situated favorably for recharge by induced infiltration. As a consequence, many wells near perennial streams have exceptionally large sustained yields. Notable examples of producing aquifers so situated include alluvium along the Ohio River and the other major rivers of the eastern United States, along the Rhone, Rhine, and Elbe rivers in western Europe, and along the Nile in Africa.

Many of the plains areas of the world are underlain by sand and gravel aquifers. These areas include interior alluvial plains, such as the High Plains of the western United States, the Chad basin in central Africa, and broad plains flanking the east slope of the Andes in

South America. They include *coastal plains*, such as the Atlantic and Gulf Coastal Plains of the United States. They include also lowland *alluvial plains*, such as those of the Tigris and Euphrates Rivers in the Middle East, of the Mekong, Chao Phraya, and Chindwin Rivers in the Far East, and of the vast Indo-Gangetic plain of southern Asia. The Indo-Gangetic alluvial plain is especially notable; sand containing minor amounts of gravel forms here what is perhaps the most productive groundwater reservoir in the world. Wells yielding 1500 gpm (gallons per minute) or more can be developed almost anywhere in an area of more than 400,000 square miles. Beneath the Punjab region alone, there is an estimated 2 billion acre-feet of usable groundwater in storage.

Sand and gravel aquifers are significant producers of groundwater in many intermontane valleys also, such as those of the western United States, along the Andes in South America, and in Iran and Afghanistan—the latter famous for their *ghanats*, tunnels that serve a dual function, as infiltration gallery and conduit.

Limestone and other Soluble Rocks. Limestone and dolomite constitute between 5 and 10% of all sedimentary rocks. They vary widely in ability to store or yield groundwater. The permeability of some aquifers, including many reef limestones, is related closely to the original porosity of the rocks, but the permeability of most of the important aquifers was developed after the rocks were deposited. Under favorable environmental conditions, subterranean waters circulating through openings in limestone and dolomite dissolve out the carbonate minerals. The solution cavities and channels thus created impart to these rocks zones of high permeability, which have set them apart from other common rock types. Some limestones with large solution channels may permit a sizeable river to run through, without storing much or yielding much to wells.

Notable limestone aquifers include those of the southeastern United States, the upper Mississippi Valley, and New Mexico, those of the Antilles, especially Puerto Rico, Jamaica, Haiti and western Cuba, those of the eastern Mediterranean and parts of coastal North Africa, the classic karst region of Yugoslavia, the limestone basins of the Iberian peninsula, the Cretaceous Chalk of England, and the limestones of the Shan plateau of northern Burma.

Individual wells produce large quantities of water from limestone aquifers; for example, yields of 2500 gpm per well from the Biscayne aquifer in southeastern Florida are common. Many of the world's largest springs issue from limestone. Of 65 "first-magnitude" springs [average flow of 100 cfs (cubic feet per second) or more] in the United States, 24 issue from limestone (Meinzer, 1942, p. 425).

Basalt and Other Volcanic Rocks. Basalt and other volcanic rocks also vary widely in their ability to produce groundwater. Although most volcanic rocks are dense and impermeable, they commonly yield water from permeable zones, such as interbeds of sand and gravel, roughened interflow surfaces, flow breccias, lava tubes and fractures. Many massive lavas, and ash-fall deposits generally, are poorly permeable, but they form aquifers in a few areas.

Perhaps the most productive basalt aquifers in the world underlie the Columbia Plateau in the western United States; especially important is the basalt aquifer beneath the Snake River Plain of Idaho. Volcanic rocks are important sources of groundwater also in Central America, along the flanks of the Andes in South America, in Italy, and in some of the island arcs of the central and western Pacific, including the Hawaiian and Philippine Islands and Japan. Older plateau lavas, such as those of southern Brazil, the Abyssinian (Ethiopian) plateau of east Africa, and the Deccan plateau of India, generally yield only small quantities of water.

The more productive basalt aquifers yield large amounts of groundwater; e.g., wells drawing water from the basalt beneath the Snake River Plain commonly yield 2200–4500 gpm. In the United States, 38 "first-magnitude" springs issue from volcanic rocks or associated gravel (Meinzer, 1942, p. 425).

Sandstone and Conglomerate. Sandstone and conglomerate yield water in part from intergranular openings only partly filled with cement and in part from fractures. Their permeability commonly is small to moderate though wells yield as much as a thousand gallons per minute from some sandstone aquifers.

Sandstone aquifers of note include those of the great artesian basin of eastern Australia (Queensland and New South Wales); the sedimentary basins of northeastern and northern Africa, including in northern Africa the well-known sandstones of the Nubian system; the Arabian peninsula; the Khorat plateau of northeastern Thailand; local areas in Spain and France; parts of central and western United States including especially the well-known Dakota Sandstone and the Cambro-Ordovician sandstone.

Crystalline and Metamorphic Rocks. Except where weathered or highly fractured, crystalline and metamorphic rocks have very low porosity and permeability, and yield little water. Where weathered, as near the surface in humid and sub-humid regions, the decayed or disintegrated rock may be slightly more permeable than the fresh rock and may yield small but reliable supplies of groundwater.

Crystalline and metamorphic rocks underlie the great shield areas of the world and are associated with many of the mountain ranges. Where the surface of these rocks has been glaciated, such as in the Canadian shield in North America and the Baltic shield in northern Europe, the mantle of weathered rock is absent and the rocks yield water largely from joints. Saprolite, the decomposed rock of the weathered zone and the immediately underlying fractured rock together form an important source of water in the Piedmont Plateau of eastern United States, the Brazilian shield in South America, the African shield, the Peninsular shield of India, and the Australian shield. In some of these areas, the saprolite may be as much as 100 feet thick and individual wells may yield as much as 100 gpm.

Porous but Poorly Permeable Materials. Clay, silt, shale, glacial till, poorly sorted alluvial and mud-flow deposits, and other porous but poorly permeable rock materials ordinarily are not thought of as aquifers. However, in large areas of the world, where more productive aquifers are lacking or comparatively inaccessible, or where the poorly permeable materials underlie the land surface, millions of wells yield small supplies of water from such rocks, especially for rural use.

Quality of Groundwater

Water circulating in the subsurface phase of the hydrologic cycle contacts a wide variety of soluble materials in soils and rocks. Consequently, natural groundwater contains a great variety of dissolved constituents in various concentrations. The principal constituents are the bicarbonates, sulfates and chlorides of sodium, calcium and magnesium. Other constituents often present, but generally in lesser amounts, are silica, potassium, iron, aluminum, manganese, strontium, fluoride, boron, nitrate, phosphate, and commonly the gases hydrogen sulfide and carbon dioxide.

A general relationship may exist between the mineral composition of groundwater and the mineral composition of the rock types from which it is derived. Thus, calcium bicarbonate water may be associated with limestone or other rocks containing calcite, calcium sulfate water may be associated with rocks containing gypsum, and groundwater having a high chloride content may be associated with rocks containing halite or other salt. Hem (1970, p. 208) notes that:

"A general relationship between mineral composition of a natural water and that of the solid minerals with which the water has been in contact is certainly to be expected. This relationship may be comparatively simple and uncomplicated, as in the case of an aquifer receiving direct recharge by rainfall and from which water is discharged without contacting any other aquifer or other water. Or the situation may be rendered very complex by influence of one or more interconnected aquifers of different composition, mixing of unlike waters, chemical reactions such as cation exchange, adsorption of dissolved ions, and biological influences. Processes involved in soil formation and the soil composition of the area may have a considerable influence on composition of both surface and underground water.

The utility of groundwater for many purposes depends upon its quality. Salty or otherwise highly mineralized water may be unfit to drink. Calcium and magnesium make water hard, impairing the effectiveness of soap as a cleansing agent and forming scale in pipes and boilers. Acid water is corrosive. Excessive iron or manganese causes discoloration. Particular mineral substances may render water unsuitable for processing foods and beverages, many synthetics, and other commodities requiring pure water. Excessive sodium and boron impair the usefulness of water for irrigation. On the other hand, the nearly constant mineral composition and temperature of groundwater make it advantageous for many uses.

It is noteworthy that whereas in the past, recharge to the groundwater reservoirs was derived from water in a virtually natural state, recharge is now derived more and more from water that has undergone some change in dissolved constituents as a result of man's activities. For example, the high nitrate concentrations in some groundwater may be a consequence of recharge in agricultural areas that are fertilized heavily. Likewise, nondegradable, synthetic detergents have been added to ground water in areas where household waste water is discharged into the ground, although the situation in the United States has improved since 1965 when the detergent industry shifted to biodegradable types. Heavy use of salt for deicing roads has resulted in local contamination of groundwater in the northern United States.

Use of Groundwater

Although no authoritative source of data concerning world use of groundwater is available, there is little doubt that the use of groundwater has increased greatly during the past few decades.

GROUNDWATER

In the United States, the total estimated withdrawal of ground water in 1965 (Murray, 1968) for all uses, including generation of fuel-electric power, was about 61 bgd (billion gallons per day). The largest withdrawal was for irrigation—about 42 bgd. Groundwater supplied by industry for its own use, including fuel-electric power generation, was about 8 bgd. Groundwater withdrawn for public supply amounted to about 8 bgd, distributed roughly according to population. Rural use of groundwater was only about 3 bgd, but this constituted most of the water supplied for rural use from all sources. Altogether, in 1965 ground water sources supplied about 1 in every 5 gallons of water withdrawn from all sources for all uses in the United States.

The use of groundwater has not been without problems, of which overdraft accompanied by increasing pumping lifts and lower yields, salt-water encroachment (not only in coastal areas but at some places inland as well), depletion of streamflow and land subsidence are but a few.

A growing problem is the preservation of the quality of the water as it passes through the aquifers. This problem is especially serious because, once the quality of water in an aquifer has deteriorated, remedial action may be effective only after a period of many years.

Groundwater Investigations

Groundwater has been a subject of interest and speculation by man throughout historical time. The meteoric origin of groundwater was advanced by Marcus Vitruvius at about the time of Christ and subsequently was reinforced by the arguments of others, notably Bernard Palissy in the middle of the sixteenth century. In the latter part of the seventeenth century the work of Pierre Perrault (1608–1680), Edmé Mariotte (1620–1684), and Edmund Halley (1656–1742) put *hydrology* for the first time on a quantitative basis, but not until the nineteenth century did groundwater hydrology as a branch of science receive its first great stimulus. Then, largely because of the intense interest in the artesian conditions and the great activity in drilling artesian wells in France, significant results of research relating to groundwater were published, chiefly by French and German scientists and engineers. These included, among others, the results of work by Jules Dupuit (1804–1866) and by Henry Darcy (1803–1858) on the movement of groundwater, by Gabriel Auguste Daubrée (1814–1896) relating geologic structure to the occurrence and movement of groundwater, and by Adolph Thiem (1836–1908), who introduced and applied field methods for tests of groundwater flow.

Since 1900 scientific endeavors in ground water hydrology have increased greatly, owing to the universal need for information to guide water-supply development. A listing of the many contributions by scientists and engineers during this period is beyond the scope of discussion here, but especially outstanding are the contributions of O. E. Meinzer, who, beginning shortly after the turn of the century, pioneered studies of groundwater for the United States Geological Survey. He recognized that aquifers are functional components of the hydrologic cycle and insisted that groundwater should be studied with this fact in mind. As a consequence, he drew together the two traditional approaches to groundwater investigations—that of the geologist, concerned principally with describing the earth medium, and that of the engineer or physicist, concerned principally with fluid mechanics. As a further consequence, he broadened groundwater studies to include all related facets of the hydrologic cycle. This contribution paved the way toward modern ground water hydrology as we know it. A contribution of comparable significance was the mathematical model derived by Theis in 1935, from analogy with an equivalent thermal system, to describe the effect of a well withdrawing groundwater from storage (the "nonequilibrium formula"). This was a first step in quantitative methods of study of transient conditions in groundwater hydraulics.

Scientific groundwater investigations differ in scope according to the problem involved and the complexity of geologic and hydrologic features to be studied. Most early investigations were concerned principally with exploring for, and simply describing, the groundwater resources. Much effort continues to be expended in this direction, but modern studies are directed increasingly toward an understanding of the behavior of ground water as it passes into, through, and out of aquifer systems.

The variety of basic data essential to describing aquifer systems is much more imposing, and the techniques for collecting and interpreting these data are much more complex, than for other phases of the hydrologic cycle. The body of essential basic data commonly includes geologic information, such as surface and subsurface geologic maps, and associated data on the physical and chemical properties of the water-bearing rocks. It includes hydrologic information, such as measurements of hydrologic properties of rocks, maps of groundwater surfaces and measurements of water-level fluctuations, records of discharge from wells and of groundwater accretion to or infiltration from streams, and measurements of precipitation. It includes measurements of water quality and, less commonly, measurements of soil-moisture storage and rates of infiltration into the soil. These data are then subjected to a coordinated analysis whose objectives are to determine the quantity of water in storage, the rate and direction of ground water movement, the quantities of recharge and discharge, and variations in the quality of water in various parts of the aquifer system, and the relations between the aquifer system and the overall hydrologic system.

One of the most promising advances during recent yeears is the development of simulation modeling and its application to analyzing aquifer systems and resolving groundwater problems. Commonly used is the *electric analog model* which, based on the analogy between the flow of electricity and the flow of water, simulates the flow of water through the aquifer and its response to water development. Even more recently, *digital* and *hybrid analog-digital* models have come into use as the capacity and speed of digital computers have increased. Necessarily, the validity of a model is determined by the adequacy and accuracy of the hydrologic and geologic information on which it is based.

The use of simulation models in analyzing performance of aquifers shows great promise and has been highly effective in favorable field situations. However, much additional research in methodology will be required before the performance of complex groundwater systems can be predicted with a precision comparable to that of surface-water flow.

O. M. HACKETT

References

Davis, S. N., and DeWiest, R. J. M., 1966, "Hydrogeology," John Wiley & Sons, 463pp.

DeWeist R. J. M., 1965, "Geohydrology," John Wiley & Sons, 366pp.

Ferris, J. G., and Sayre, A. N., 1955, "The quantitative approach to ground-water investigations," *Econ. Geol.*, 50th Anniversary Volume, pp. 714–747.

Hantush, M. S., 1964, "Hydraulics of wells," in (V. T. Chow, editor), "Advances in Geosciences," Academic Press, pp. 281–429.

Heath, R. C., and Trainer, F. W., 1968, "Introduction to Ground Water Hydrology," John Wiley & Sons, 284pp.

Hem, J. D., 1970, "Study and interpretation of the chemical characteristics of natural water," *U.S. Geol. Surv. Water Supply Paper*, **1473**, Second ed., 363pp.

Hubert, M. King, 1940, "The theory of groundwater motion," *J. Geol.*, **58**(8), 785–944.

Landström, O., and Wenner, C. G., 1965, "Neutron Activation Analysis of Natural Water Applied to Hydrogeology," Univ. Stockholm.

McGuinness, C. L., 1963, "The role of ground water in the national water situation," *U.S. Geol. Surv. Water Supply Paper*, **1800**, 1121pp.

Meinzer, O. E., 1923a, "The occurrence of ground water in the United States: with a discussion of principles," *U.S. Geol. Surv. Water Supply Paper*, **489**, 321pp.

Meinzer, O. E., 1923b, "Outline of ground-water hydrology, with definitions," *U.S. Geol. Surv. Water Supply Paper*, **494**, 71pp.

Meinzer, O. E. (editor), 1942, "Hydrology," Physics of the Earth series, Vol. 9, New York, McGraw-Hill Book Co., 712pp. (see particularly Chaps. 9, 10, 14, and 15).

Murray, C. R., 1968, "Estimated use of water in the United States, 1965," *U.S. Geol. Surv. Circ.*, **556**, 44pp.

Nace, R. L., 1964, "Water of the world," *Natural History*, **73**(1), 10–19.

Pinder, G. F., and Bredehoeft, J. D., 1968, "Application of the digital computer for aquifer evaluation," *Water Resources Res.*, **4**(5) 1069–1093.

Polubarinova-Kochina, P. Ya., 1962, "Theory of Ground-water Movement," Princeton, N.J., Princeton University Press, 613pp. (translated by J. M. Roger DeWiest).

Robinove, C. J., 1962, "Ground-water studies and analog models," *U.S. Geol. Surv. Circ.*, **468**, 12pp.

Schoeller, H., 1962, "Les Eaux Souterraines," Paris, Masson et Cie, 642pp.

Simmons, G., 1962, "On Darcy's Law," *J. Geophys. Res.*, **67**(11), 4516.

Theis, C. V., 1935, "Relation between the lowering of the piezometric surface and the rate and duration of discharge of a well using groundwater storage," *Am. Geophys. Union Trans.*, **16**, 519–524.

Theis, C. V., 1940, "The source of water derived from wells, essential factors controlling the response of an aquifer to development," *Civil Eng.*, **10**(5), 277–280.

Thomas, H. E., 1951, "The Conservation of Ground Water," New York, McGraw-Hill Book Co., 327pp.

Thomas, H. E., 1952, "Ground-water Regions of the United States—Their Storage Facilities," U.S. Congress, House Interior and Insular Afairs Committee, Physical and Economic Foundation of Natural Resources. Vol. 3, 78pp.

Thomas, H. E., and Leopold, Luna B., 1964, "Ground water in North America," *Science*, **143**(3610), 1001–1006.

Thomas, H. E., and Peterson, D. F., Jr., 1967, "Ground water supply and development," in (R. M. Hagan et al., editors). "Irrigation of Agricultural Lands," *Am. Soc. Agronomists Monograph*, **11**, 84–91.

Todd, David K., 1960, "Ground Water Hydrology," New York, John Wiley & Sons, 336pp.

Walton, W. C., 1970, "Groundwater Resource

GROUNDWATER

Evaluation," New York, McGraw-Hill Book Co., 664pp.

Cross-references: *Aquifer; Artesian Water; Connate Water; Drawdown, Cone of Depression; Geophysical Methods for Hydrologic Search; Hydrologic Cycle; Hydrology, Subsurface Waters; Ion Exchange; Juvenile Water; Meteoric Water; Porosity and Permeability; Springs; Vadose Water; Water Table.* Vol. II: *Capillarity; Evapotranspiration; Precipitation.* Vol. III: *Alluvium; Coastal Plains.* Vol. IVB: *Clays and Clay Minerals.* Vol. VI: *Limestone; Sands and Sandstone; Saprolite.*

GROUNDWATER IN COASTAL, KARST, SEMIARID, VOLCANIC TERRAIN—See resp. HYDROLOGY, COASTAL TERRAIN; HYDROLOGY, KARST TERRAIN; HYDROLOGY, SEMIARID REGIONS; HYDROLOGY, VOLCANIC TERRAIN

GROUNDWATER, THERAPEUTIC ROLE —See MEDICINAL SPRINGS

GROUNDWATER MOTION IN DRAINAGE BASINS*

Theoretical and practical studies of groundwater motion generally pertain to either one of two main fields of inquiry: engineering and hydrogeology. Common engineering problems are: artificially created differences involving rapid changes of fluid potentials; high flow rates and small masses of moving water; small areas of influence; and so on. These aspects are found in connection with projects such as: evaluation and development of wells, aquifers, and gas and oil reservoirs; seepage under dams, dikes, and surface water reservoirs; construction of roads, bridges, and large buildings; draining and irrigating agricultural land; tracing and controlling the spread of contaminants; and dewatering of mines. On the other hand, groundwater movement in relation to hydrogeology is usually characterized by naturally existing and slowly changing fluid potentials involving low flow rates but large masses of moving water; large areas of influence, and so on. Accordingly, hydrogeological studies comprise topics such as ground water budgets of drainage basins; type and distribution of chemical facies of groundwater in geologic formations and drainage basins; relation between soil types and quality of groundwater; migration and accumulation of hydrocarbons; the effects of moving groundwater on secondary mineralization, etc.

Whereas the transition between the engineering and hydrogeological problems may be conceived as a matter of scale of time and space, no general formulation of groundwater flow is as yet available out of which either type of problem could be derived as a special case.

Basic Laws of Fluid Motion

The motion of the fluid must be subjected to the principle of the conservation of matter, which is expressed by the *equation of continuity*:

$$\text{div}\,(\rho\vec{q}) = \frac{\partial \rho q_x}{\partial x} + \frac{\partial \rho q_y}{\partial y} + \frac{\partial \rho q_z}{\partial z} = -\frac{\partial \rho}{\partial t} \quad (1)$$

where ρ is density, $\vec{q}$ is the specific volume discharge, defined as the volume of fluid moving through a unit area in unit time, and $\partial \rho/\partial t$ is the variation of fluid density with respect to time at any fixed point in space. Eq. (1) prescribes that a change in density with respect to time must accompany any change in the amounts of masses entering and leaving a volume element at any given point in the flow region. That this change in mass may originate both from the transportation of inhomogeneous fluids at constant speeds and from changing velocities of compressible fluids, becomes evident upon expansion of Eq. (1) into:

$$\text{grad}\,\rho \cdot \vec{q} + \rho\,\text{div}\,\vec{q} = -\frac{\partial \rho}{\partial t} \quad (2)$$

where grad ρ is the variation of ρ in space at any given time, and div $\vec{q}$ is the net increase or decrease of specific volume discharge at any given point of the flow region. If, however, the fluid is homogeneous and incompressible, i.e., grad $\rho = 0$, and $\partial \rho/\partial t = 0$, then

$$\text{div}\,\vec{q} = 0 \quad (3)$$

which is also the condition of incompressibility expressing the fact that the balance of outflow and inflow for a given volume element is zero at any time. For an incompressible inhomogenous fluid, where grad $\rho \neq 0$, and div $\vec{q} = 0$:

$$\frac{\partial \rho}{\partial t} = -\text{grad}\,\rho \cdot \vec{q} \quad (4)$$

describing the temporal fluctuation of ρ at a fixed point of observation.

The *equation of state* gives a quantitative relation between the density (ρ), pressure (p), and absolute temperature (T) of a volume element of the fluid. The general form of this equation is:

---

* Contribution No. 335, Research Council of Alberta, Edmonton, Alberta, Canada.

$$\rho = f(p,T) \quad (5)$$

and for ideal gases:

$$\rho = p\frac{w}{RT} \quad (6)$$

where w is the molar weight, and R is the universal gas constant. For incompressible fluids:

$$\rho = \text{const} \quad (7)$$

The thermal nature of the flow is either adiabatic or isothermal. The general equation set up in reference to a normal state $p_0 \, \rho_0$ is:

$$\frac{p}{p_0} = \left(\frac{\rho}{\rho_0}\right)^n \quad (8)$$

From a combination of (6) and (8) then it follows that for isothermal flow, where $n = 1$,

$$T = \text{const} \quad (9)$$

and for adiabatic flow, where $n \neq 1$:

$$T = \text{const} \cdot \rho^{n-1} \quad (10)$$

The forces acting upon a small volume element dV are:

$$\vec{dF_d} + \vec{dF_a} + \vec{dF_v} = 0 \quad (11)$$

where $\vec{dF_d}$ is the outside driving force; $\vec{dF_a}$ is the force of inertia; and $\vec{dF_v}$ is the viscous resistance of the fluid.

The relation between these forces is given by the Navier-Stokes equations which are fundamental to the theory of fluid flow (Hubbert, 1957, p. 37):

$$\rho\left[\vec{g} - \left(\frac{1}{\rho}\right)\text{grad } p\right]dV = \rho\left(\frac{\vec{Dv}}{Dt}\right)dV$$
$$- \mu[\nabla^2\vec{v} + \tfrac{1}{3}\nabla\nabla\cdot\vec{v}]dV \quad (12)$$

where

$$\frac{\vec{Dv}}{Dt} = \frac{\partial\vec{v}}{\partial t} + \vec{v}\nabla\vec{v} \quad (13)$$

It is seen from the combination of Eqs. (1), (7), (12), and (13) that for steady motion of incompressible fluids, i.e., for $\partial\vec{v}/\partial t = 0$ and $\nabla\vec{v} = 0$, Eq. (12) simplifies to:

$$\rho\left[\vec{g} - \left(\frac{1}{\rho}\right)\text{grad } p\right]dV$$
$$= \rho(\vec{v}\cdot\nabla\vec{v})dV - \mu(\nabla^2\vec{v})dV \quad (14)$$

Hubbert has shown (1957), that in porous media and for low enough velocities for the inertial forces to be negligible the following relation may be derived from Eq. (14):

$$\vec{q} = Nd^2\frac{\rho}{\mu}\left[\vec{g} - \left(\frac{1}{\rho}\right)\text{grad } p\right] \quad (15)$$

where N is a shape factor, d is a characteristic length of the pore structure, ρ and μ are the density and viscosity of the fluid, respectively, and $\sigma = Nd^2(\rho/\mu)$ is the specific fluid conductivity of the system.

An analysis of Eq. (15) shows (Hubbert, 1957) that the quantity $\vec{g} - (1/\rho)\text{grad } p = \vec{E}$ is a force, which can be derived from a potential ϕ provided that $\rho = \text{const}$, or $\rho = f(p)$. Thus

$$\vec{E} = -\text{grad }\phi \quad (16)$$

except for cases of thermal convection, $[\rho = f(p,T)]$, and inhomogeneous fluids, $\rho \neq \text{const}$).

If, however, ϕ is a potential, its value at any point (P) may be calculated by

$$\phi(P) = \phi(P_0) - \int_{P_0}^{P} E_s ds = \phi(P_0)$$
$$+ gz + \int_{p_0}^{p}\frac{dp}{\rho} \quad (17)$$

where $\phi(P_0)$ is the value of ϕ at a point P_0, z and $(p-p_0)$ are the differences in elevation and pressure, respectively, between the points P_0 and P. By relating the potential to an arbitrary standard datum, where $\phi(P_0) = 0$, the following expression is obtained:

$$\phi(P) = \phi = gz + \int_{p_0}^{p}\frac{dp}{\rho} \quad (18)$$

which reduces to

$$\phi = gz + \frac{p}{\rho} \quad (19)$$

anywhere within the flow regime of homogeneous, incompressible fluids, if p_0 is atmospheric pressure. This further gives

$$h = \frac{\phi}{g} = z + \frac{p}{g\rho} \quad (20)$$

the height with respect to a standard datum to which the liquid will rise in a manometer tapped in any point of the system. Thus Eq. (15) may now be written as

$$\vec{q} = \sigma\vec{E} = \sigma\text{ grad }\phi = \sigma g\text{ grad }h \quad (21)$$

or

$$\vec{q} = -k\text{ grad }h \quad (21a)$$

When applied to a straight tube with cross section A, total volume discharge $Q = Aq$, with readings h_2 and h_1 at manometers spaced apart at a distance l, Eq. (21) becomes

$$Q = -KA \frac{h_2 - h_1}{l} \quad (22)$$

which is the same as the famous equation of Darcy (1856). It is equivalent to a special case of Eqs. (21) and (21a). In Eq. (22) K is a factor of proportionality between the fluid discharge and the hydraulic gradient, and it is dependent on the pore geometry of the medium, the fluid properties, and the acceleration due to gravity. Since water may be considered incompressible and having relatively constant viscosity, and since g is constant for all practical purposes at or near the surface of the same planet, the most important variables in K are the rock properties N and d, which, lumped together determine the "hydraulic conductance," k of any porous material. From Eqs. (21), (21a), and (22) it is seen that Darcy's law provides a link between a flow field $\vec{q}$ and a force field $\vec{E}$. If k is isotropic, i.e., if the fluid-conducting properties of the rock are the same in all directions at any given point within the flow regime, then σ is a scalar and the vectors $\vec{q}$ and $\vec{E}$ will coincide; otherwise σ is a tensor resulting generally in different directions of $\vec{q}$ and $\vec{E}$. The magnitude of k is constant at all points in a homogeneous medium, resulting in a constancy in space of σ, and consequently in an equidistant spacing of the equipotential lines at a given flow rate. In nonhomogeneous materials through regions of relatively low k, i.e., low σ, grad ϕ must increase in order for $\vec{q}$ to remain constant. An increase in grad ϕ means a steep slope of the hydraulic gradient, i.e., an increase in the difference of hydraulic head over a unit length of flow path.

Since the basic relation between the field of flow and the field of force is given by Darcy's law, for a study of groundwater flow in a drainage basin it is sufficient to determine the distribution of the fluid potential ϕ. Since ϕ is a scalar potential its steady-state distribution in a region free of sources and sinks is found by the Laplace equation:

$$\nabla^2 \phi = \frac{\partial^2 \phi}{\partial x^2} + \frac{\partial^2 \phi}{\partial y^2} + \frac{\partial^2 \phi}{\partial z^2} = 0 \quad (23)$$

To solve the Laplace equation it is necessary to know ϕ or its first and second derivatives at the boundaries of the flow region, which requires a knowledge of the geometry of the drainage basin. The information necessary to define the geometry can usually be obtained from topographic maps.

The Environment for Groundwater Flow in Drainage Basins

It has been shown that in any area groundwater motion is controlled by the fluid potential ϕ. The distribution of ϕ is determined by the environment, which may be characterized by three groups of condititons. These conditions may be defined in terms of (1) geology, (2) hydrology, and, (3) topography. Accordingly, the quantitative or parametric expressions of the above-mentioned conditions are, for instance, permeability distribution in the rocks, the amount and distribution of water available for infiltration, geometry of the flow region considered, and so on.

Geology. Since the direction and rate of the groundwater flow depends to a large extent on the ease with which water moves through the rocks in any given area, understanding of the type, magnitude, and distribution of the permeability of the geologic formations is important. Permeability has a rough correlation with the rock types, and the degree of cementation, for which reason a stratigraphic or lithologic picture is often used to delineate areas or layers of different permeabilities.

Among the several units of permeability the "darcy" is the most widely used. Expressed in darcys, the permeability of unconsolidated sediments varies between values of 10^{-6} and 10^{-4} for clays; 10^{-4} and 1 for very fine sands, silts, and mixtures of silts and clays; 1 and 10^3 for clean sands and gravels; and 10^3 and 10^5 for clean coarse gravels (Todd, 1959). The most common order of magnitude for permeability of sandstones of oil reservoirs is 10^{-3} and 10^{-1} darcys, but values as low as 10^{-5} and as high as 1 have been reported (Muskat, 1946). Except where they are fractured, crystalline rocks have very low permeabilities, whereas the permeability in limestones and dolomites may vary from virtually nil in tight, dense rock, to very high values in the porous rocks of reefs and in the cavernous areas of karsts.

Depending on the type and geological history of the formations the permeability values in any given area may be quite uniform, as is the case in thick uniform marine deposits, but it also may be very variable as, for example, in floodplain and other continental sediments. Whether an area is described as having a homogeneous or nonhomogeneous distribution of permeability depends to a large extent on the scale of investigation. A detailed study is likely to discover large and abrupt variations of permeability even within a single layer, whereas for general purposes whole formations or even drainage basins may be deemed homogeneous.

The effect of scale on the choice of the appropriate concept can be illustrated by the example of a rock specimen through which water is moving. The relation between the macroscopic flow rate and the pressure drop across the specimen is very accurately given by Darcy's law, despite the tortuosity of the flow paths and the greatly varying velocities of the individual fluid particles. Similarly, a lenticular rock body of permeability contrasting with its surroundings may have a strongly modifying effect on the fluid flow in the near vicinity, but it might be insignificant regarding its influence on the flow in the drainage basin as a whole.

The rock units and geologic features which are most likely to have a strong influence on the flow distribution of entire drainage basins are: buried river valleys that are long and wide compared to the drainage basin within which they are considered, and either with highly permeable valley fill of fractured, permeable banks; layers of strongly contrasting permeabilities having an areal extent which is large compared to the area investigated; relatively extensive faults representing either high or low permeability areas; and so on.

Anisotropy of permeability results in the direction of the flow being different from that of the driving force. Most consolidated sediments are more permeable parallel to the bedding than normal to it. In unconsolidated sediments, or in rocks with solution permeabilities this contrast is usually minor. Due to anisotropy of permeability, groundwater flow lines tend to follow the direction of the bedding more closely than do the force lines.

Hydrology. That surface in the groundwater regime at each point of which the pressure is one atmosphere is referred to as the *water table*. Since this surface is closely related to the upper limits of the regime of saturated flow of groundwater, its configuration is of fundamental importance to the overall development of the flow patterns. The depth of the water table below ground level, however, also represents a position of dynamic equilibrium between recharge from the land surface by precipitation or by a body of surface water, discharge onto the surface by evaporation, vegetal transpiration, or saturated flow of groundwater, and recharge of discharge by groundwater moving from or toward other points in the flow regime. Thus the amounts and distribution of precipitation, evaporation, and bodies of surface water are important factors in defining the environment for the groundwater regime.

In an area, for instance, where evaporation greatly exceeds precipitation, the water table is nearly level even if there is a marked relief in the topography simply because the amount of infiltrating surface water cannot balance the amount of water moving away from the hills as groundwater. This deficiency in precipitation then results in relatively deep water tables under high land, consequently, in low hydraulic gradients (Léczfalvy, 1963).

A stream usually occupies the lowest points of its surroundings and is, therefore, considered as a place where groundwater moves toward the surface. In an arid region, however, where the water table is lower than the stream bed, water may be percolating down to the water table thereby reversing the normally expected direction of groundwater flow.

Topography. The main agent in creating and maintaining initial differences of fluid potential within a drainage basin by intercepting water of precipitation at different altitudes is the topography. A unit mass of water falling on relatively highly elevated areas will retain a larger amount of its potential energy than the one that falls on low grounds. Due to this difference in potential energies of infiltrating water a hydraulic gradient is established which is the actual force causing the groundwater to move. It is thus obvious that the distribution of the topographic features must have a decisive influence on the flow distribution.

Also, an irregular topography will promote an irregular distribution of infiltrating water due to an increase in surface runoff on steep slopes, relative to that on flat and low-lying parts of a drainage basin.

Methods of Flow Studies

It is not possible here to give a detailed account of all methods and techniques used or suitable for studying groundwater movement. A brief review of such methods, however, seems to be appropriate.

Groundwater movement may be studied in the *field* or by *models*.

Field studies may rely on direct observation of the water and its movement, or on observation of phenomena *associated* with the presence of water (Meinzer, 1927; Tóth, 1966b). Particularly when the movement of groundwater is inferred from indirect observation, a good knowledge of the environment is very important.

The environment is studied by means of topographical surveys; geological mapping, subsurface geology, geophysical exploration, permeability determinations by pumping tests;

hydrological and climatological observations such as stage and variation of water levels in lakes and steams, precipitation, evaporation, and so on.

Direct observations of groundwater and its movement are carried out by means such as tracers (dyes, radioactive and nonradioactive isotopes, electrolytes, and so on; Todd, 1959; Schoeller, 1962; International Atomic Energy Agency, 1963); composition and changes with respect to space and time of the natural chemistry of groundwater (Chebotarev, 1955, Back 1960; Schoeller, 1962; position and variation of groundwater levels; and field mapping of seepages, springs, and other occurrences of water obviously of groundwater origin.

Much information regarding the configuration of groundwater flow may be obtained by inference from such indirect observations as vegetation, soil types, soil moisture, salts deposited by groundwater on the land surface, and topographic features related to groundwater.

The main types of groundwater flow models are scale, electrical, membrane, and mathematical models. In *scale models* water or other fluids move through porous material such as sand, lead shots, and so on, or between parallel glass plates. Dyes may be used to facilitate a visual observation of the flow. *Electrical models* may make use either of a network of resistances and capacitors simulating the transmitting and storage characteristics of the water-bearing formations, or they may employ conducting materials such as electrolytes, conducting gelatine, or conducting papers. The principle of *membrane models* is that the shape of a stressed elastic membrane around small deflections is similar to that of the free surface of groundwater around corresponding irregularities. For the development and solution of *mathematical models* the methods of numerical analysis and calculus are most often used. Numerical methods are particularly useful, and sometime the only way open to mathematical treatment in cases where the flow is unsteady, or the boundaries and the initial conditions are irregular.

Whereas certain models may be preferable to others in specific cases, all models have one aspect in common, namely that they cannot perfectly match the highly complicated conditions encountered in nature. Therefore, the results of any model study are necessarily approximations whose correspondence to reality depends on the accuracy of the basic assumptions.

The problem may, however, be reversed by establishing strictly valid basic principles for ideal, i.e., model environments. These principles are then applied to real situations with an accuracy of the results comparable to the closeness of the real case to the model. Whereas a superficial perusal does not reveal a basic difference between the two types of approach, it is evident that much better field information is required for the construction of a working model than for the practical verification of predicted theoretical results.

Distribution and Characteristics of Groundwater Flow Systems

Generally the environment for which the groundwater flow distribution is sought is defined by the following topographic, geologic, and hydrologic conditions:

1. The deepest points of the drainage basin lie along a straight line whose slope is much less than the slopes of the valley flanks.
2. The valley flanks are parallel with and symmetrical with respect to the valley bottom.
3. The water divides on both sides of the basin are symmetrical with each other and at equal distance from the valley bottom.
4. The surface of the valley flanks is a harmonic function of the distance measured horizontally from the valley bottom.
5. The valley is underlain by a horizontal, effectively impermeable layer at an arbitrary depth.
6. The geology of the basin above the impermeable boundary is isotropic and homogeneous.
7. The climatologic conditions are similar at all parts of the basin with a sufficient surplus of precipitation over evapotranspiration, so that the water table is maintained close to the land surface everywhere.

If transitional phenomena are neglected and only the long-term average of the flow is investigated, the problem is that of a steady-state motion. For this reason, and since water may be assumed as incompressible for the present purposes, the fluid-potential distribution in a cross-sectional plane perpendicular to the bottom line of the valley may be obtained by the two-dimensional form of the Laplace equation:

$$\frac{\partial^2 \phi}{\partial x^2} + \frac{\partial^2 \phi}{\partial z^2} = 0 \qquad (24)$$

The boundary conditions are specified by stating that, by virtue of symmetry, no flow occurs across the two vertical planes under the valley bottom and the water divide (Fig. 1)

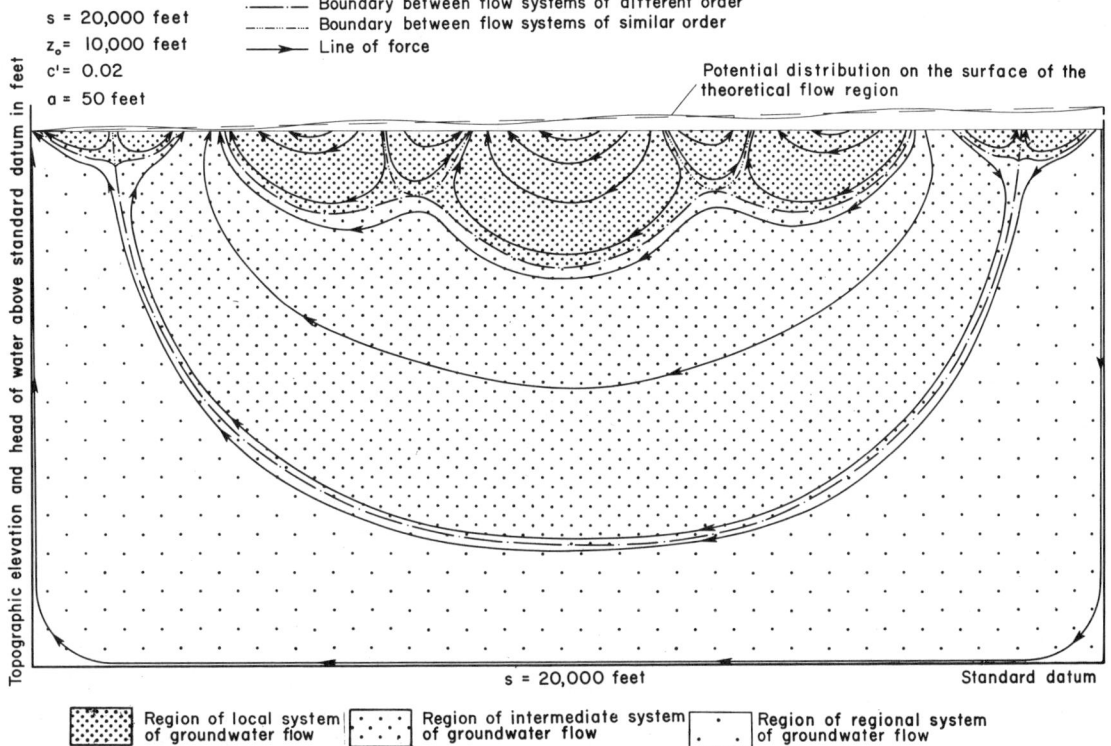

FIG. 1. Theoretical flow pattern and boundaries between different flow systems (after Toth, 1963 *J. Geophys. Res.*, **68** (16), 4795–4812 (August); see Fig. 3).

and also, that no vertical flow occurs across the horizontal impermeable boundary. The fourth condition relates to the water table and specifies that the potential of the water table, similarly to the land surface, is a harmonic function of the horizontal distance from the valley bottom.

The solution of Eq. (24) with the above boundary conditions yields Eq. (25) for the fluid potential:

$$\phi = g \left\{ z_0 + \frac{c's}{2} + \frac{a'}{sb}(1 - \cos b's) + 2 \sum_{m=1}^{\infty} \left[\frac{a'b'(1 - \cos b's \cos m\pi)}{b'^2 - m^2\pi^3/s^2} + \frac{c's^2}{m^2\pi^2} \right] (\cos m\pi - 1) \right] \frac{\cos (m\pi x/s) \cosh (m\pi z/s)}{s \cdot \cosh (m\pi z_0/s)} \right\}$$
(25)

where z_0 is the elevation of the valley bottom with reference to the horizontal impermeable boundary as a standard datum, $a,' b,' c,'$ and s are parameters of the drainage basin, x is distance measured horizontally from and at right angles to the valley bottom, and m's are integers.

Fig. 1 shows the flow distribution calculated with Eq. (25) for one flank of a drainage basin with a regional slope: $c' = 0.02$ and with amplitudes of the local hills and depressions $a = 50$ ft.

Based on similar theoretical and practical investigations a schematic illustration is presented, showing the main features of groundwater flow in a simple drainage basin. Fig. 2 illustrates a vertical cross section of a valley flank between the bottom of the valley and the water divide. The slope of the valley along its thalweg is negligible compared to the slope of the flank, resulting in a flow direction which is nearly perpendicular to the valley bottom. For a circular basin this condition is fulfilled automatically.

The tributaries to the main stream are represented by the subbasins in Fig. 2. These tributaries may again be portrayed by a diagram similar to Fig. 2, the subbasins, however, now consisting of small potholes of hummocky moraine area, and so on. In other words, Fig. 2 is dimensionless, and wells 100 and 1000m in depth for the first case may be taken to illustrate, for instance, 10 and 100m depths, respectively, in the second.

The arrows in Fig. 2 represent the average

GROUNDWATER MOTION IN DRAINAGE BASINS

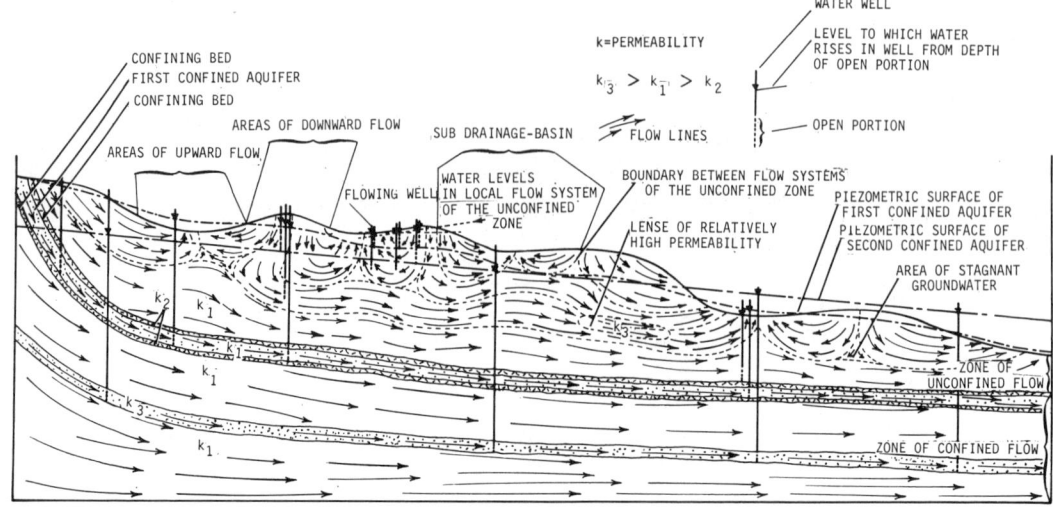

FIG. 2. A schematic illustration of possible groundwater flow systems and associated water levels in a dimensionless drainage basin.

direction of the saturated flow of groundwater. The density of the arrows is approximately proportional to the intensity of the flow at any point of the basin. The drainage basin is divided into two vertical zones, namely the zone of unconfined flow and the zone of confined flow. In the zone of unconfined flow the flow lines may be grouped into systems as indicated by the separating lines. A *flow system* may thus be defined as: "a set of flow lines in which any two flow lines adjacent at one point of the flow region remain adjacent through the whole region; they can be intersected anywhere by an uninterrupted surface across which flow takes place in one direction only." The same definition may be applied to the confined zone, where, however, the boundaries between flow systems are formed by layers of low permeability. A flow system may be called *unconfined* if its flow distribution is controlled by the configuration of the overlying water table; it is *confined* when its flow distribution is influenced only slightly, or not at all, by the potential of the water table. Several independent confined systems may exist above each other, but all superimposed unconfined systems develop in a simultaneous mutual interdependence.

From an analysis of the flow distribution in Fig. 2 a number of general conclusions may be arrived at which are summarized below:

1. Groundwater flow in the unconfined zone of a drainage basin is distributed in *flow systems* of different order, namely: local, intermediate, and regional.
2. A *local system* has its recharge and discharge areas at adjacent topographical highs and lows, respectively. Although the recharge and discharge areas of an *intermediate system* do not occupy the highest and lowest parts in a drainage basin, one or more local systems may be located between them. The recharge area of a *regional system* occupies the water divide while its discharge area lies at the bottom of the basin.
3. The position of the line of separation between unconfined flow systems is continuously shifting in accordance with the fluctuations of the potential field, i.e., of the hydraulic gradient at the water table; these changes are, however, usually insignificant compared to the absolute differences in the potential at different parts of the basin.
4. Along lines where two, and at points where three or more flow systems move in opposite directions, areas of stagnant water form, which may be indicated by increased amounts of total dissolved mineral matter.
5. In case of a linear water table the average direction of groundwater flow is away from the land surface (downward) in the upper half, and it is toward the land surface (upward) in the lower half of the valley. These areas may be referred to as "area of downward flow" and "area of upward flow," respectively.
6. An assymetry of the slope of a basin with respect to the "mid-line" will cause a corresponding assymetry in the flow distribution.

7. In basins with a complex topography (5) and (6) apply to the unit or sub-basins.
8. Due to local hills and depressions alternating areas of downward and upward flow may occur in a basin. Each type of area occupies approximately 50% of the total surface of the basin.
9. By definition a *piezometric surface* can be associated with only one, distinct, and confined aquifer; flow in such an aquifer is independent of the overlying topography. Fig. 2 shows the positions of the piezometric surfaces and the non-pumping water levels as measured in, and contsructed on the basis of, hypothetical water wells of different locations and at different depths of the basin.
10. An imaginary line connecting water levels obtained in shallow wells at different parts of a unit basin has a steeper slope than the same type of imaginary line obtained for deep wells. The two lines cross approximately midway on an even slope. Although the presence of flowing wells is often interpreted as the result of confined flow, when it is associated with the above phenomenon the existence of an actually unconfined situation is likely. Consequently, upward movement of groundwater, i.e., an increasing hydraulic head with increasing well depth, in the lower half of the drainage basin may be a consequence of the natural distribution of fluid potential in an *isotropic* and *homogeneous* medium, and it does not necessarily require the presence of confined conditions.
11. The more pronounced the local topographic irregularities, the deeper are the associated local flow systems.
12. Flow systems occurring in the unconfined zone do not cross major valleys or water divides.
13. The intensity of the flow decreases with depth in a homogeneous, unconfined basin. In a homogeneous confined layer it is uniform across the cross section of the flow.
14. The largest seasonal decline of the water table occurs at the water divide, with potentially the largest seasonal rise at the valley bottom, owing to the natural distribution of the fluid potential. Such a situation will prevail during the winter under cold climates, when, because of freezing of the ground, the effects of evapotranspiration and precipitation are virtually nil.
15. The largest seasonal variations take place in the water levels of small, local systems whereas these variations in deeper and more extended regional flow systems will be gradually decreasing with increasing depths.
16. The amount of mineral matter dissolved and contained by the groundwater increases along and in the direction of the flow paths. The total dissolved solids content is generally high under extended flat areas where the flow velocity is low.
17. Due to the presence of local flow systems waters in adjacent areas may have their origins at distant or geologically unrelated places: a diversified chemical composition of groundwater may be expected within a drainage basin.
18. The effect on the flow distribution of lenses of contrasting permeability presents itself by anomalously low or high water levels in and in the immediate vicinity of the lenses.
19. Water entering the groundwater regime at atmospheric temperature has a tendency to cause a decrease in the geothermal gradient under areas of downward flow, whereas the relatively warm, upward-moving groundwater will increase it in the lower portions of the drainage basin.
20. An accumulation of mineral salts dissolved from the rocks and transported by the water may be expected in the soils of areas of upward flow. The formation of playas, alkali soils, surficial deposits of sodium sulfate and sodium chloride, and so on, may be explained by this mechanism at several localities.
21. Various other field phenomena, such as springs, seepages, flowing wells, phreatophytic vegetation, soap holes, and perennial bodies of shallow surface waters, may be indicative of areas of upward-moving groundwater.

Application and Case Histories of Groundwater-Flow Studies

Studies of the groundwater-flow distribution and its controlling factors in drainage basins may yield revealing information for general geological questions. They may also be useful, or even indispensable, in solving certain practical problems. Obtaining an insight into matters such as the existence or nonexistence of connate waters (Chebotarev, 1955), diagenesis of rock-forming minerals (Bredehoeft et al., 1963; Graf et al., 1965, p. 65); or migration and accumulation of hydrocarbons (Hubbert, 1953) may be expected from this type of investigation. Good examples for the application and, at the same time, improvement of the understanding of regional groundwater motion have been produced in connection with water-supply problems by Nautiyal (1958), Greenman et al. (1961), Alföldi (1963), Pluhowsky and Kantrowitz (1964), Wood and Dale (1964), and Tóth (1966). The relation between the flow distribution and chemistry of groundwater is well demonstrated in the water-quality and contamination studies by Back (1960), Geraghty (1960), Merritt (1961), Schoeller (1962), Hitchon (1963), and Seaber (1965). Studies of the water budget in drainage basins both contribute to and require a good knowledge of the motion pattern of groundwater, as apparent from studies by Plotnikov and Bogomolov (1958), Befani (1961), Rónai (1963), and Szebényi (1965). Regional motion of groundwater may play an important part in the development of soil types and the surface accumulation of mineral salts as indicated in papers by Govett (1958), Sett (1958), and Sadler (1962). Interesting studies of regional flow of groundwater are reported in papers dealing with mining problems by Jones and Subramanyan (1961), and Schmeider (1965). Several of the above-mentioned features are presented in a uniquely compiled form in the Hydrogeological Atlas of Hungary (Schmidt, 1962).

J. Tóth

References

Alföldi, L., 1963, "Lefolyástalan területek vizföldtani kérdései a Mongol Népköztársaság sivatagi és félsivatagi területein" (Hydrogeological problems in closed drainage basins of desert and semi-desert areas in the Mongolian Republic), *Hidrol. Közl.*, **42**(3), 233–240.

Back, W., 1960, "Origin of hydrochemical facies of groundwater in the Atlantic Coastal Plain," report 21st Session, Intern. Geol. Cong., Part 1, pp. 87–95 (Copenhagen).

Befani, A. N., 1961, "Basic principles of the theory of processes of surface and underground runoff," *Bull. Intern. Assoc. Sci. Hydrol.*, **6**(3), 18–31.

Bredehoeft, J. D., Blyth, C. R., White, W. A., Maxey, G. B., 1963, "Possible mechanism for concentration of brines in subsurface formations," *Bull. Am. Assoc. Petrol. Geol.*, **47**(2), 257–269.

Chebotarev, I. I., 1955, "Metamorphism of natural waters in the crust of weathering," *Geochim. Cosmochim. Acta*, **8**, 22–48, 137–170, 198–212.

Darcy, H., 1856, "Les fontaines Publiques de la ville de Dijon," Paris, Victor Dalmont.

Dooge, J. C. I., 1960, "The Routing of Groundwater Recharge through Typical Elements of Linear Storage," I.A.S.H. General Assembly of Helsinki, Comm. of Subterrenean Waters, Publ. No. 52, pp. 286–300.

Freeze, R. A., 1966, "Theoretical Analysis of Regional Groundwater Flow," Ph.D. thesis, Engineering Science, Univ. California, Berkeley, 304pp.

Geraghty, J. J., 1960, "Movement of Contaminants through Geologic Formations," Tech. Div. Activities, Natl. Water Well Assoc., Urbana, Ill., 33–49.

Govett, G. J. S., 1958, "Sodium sulfate deposits in Alberta," *Res. Council Alberta, Prelim. Rept.*, **58-5**, 34pp.

Graf, D. L., Meents, F. F., and Shimp, N. F., 1965, "Chemical composition and origin of saline formation waters in the Illinois and Michigan Basins," G.S.A. Program 1965 Annual Meetings, Kansas City, Mo., 106pp. (Abstract).

Greenman, D. W., Rime, D. R., Lockwood, W. N., Meisler, H., 1961, "Groundwater resources of the coastal plain area of southeastern Pennsylvania," *Penn. Topog. Geol. Surv. Bull.*, **W-13**.

Hitchon, B., 1963, "Composition and Movement of Formation Fluids in Strata above and below the Pre-Cretaceous Unconformity in Relation to the Athabasca Oil Sands," K. A. Clark Volume, Edmonton, Res. Council of Alberta, 63–74.

Hubbert, M. King, 1940, "The theory of groundwater motion," *J. Geol.* **48**(8), 785–944.

Hubbert, M. King, 1953, "Entrapment of petroleum under hydrodynamic conditions," *Bull. Am. Assoc. Petrol. Geol.*, **37**(8), 1954–2026.

Hubbert, M. King, 1957, "Darcy's Law and the field equations of the flow of underground fluids," *Bull. Intern. Assoc. Sci. Hydrol.*, **5**, 24–59.

International Atomic Energy Agency, 1963, "Radioisotopes in Hydrology," Proc. Series, Publ. 71, Vienna, 457pp.

Jones, P. H., and Subramanyan, V., 1961, "Groundwater control in the Neyveli lignite field, South Arcot district, Madras State, India;" *Econ. Geol.*, **56**, 273–298.

Léczfalvy, S., 1963, "A források osztályozása" (Classification of springs; with Russian and German summary), *Hidrol. Közl.*, **43**(1), 46–57.

Meinzer, O. E., 1923, "Outline of Ground-Water Hydrology, With Definitions," U.S. Geol. Survey Water-Supply Paper 494, Washington, D.C., 71pp.

Meinzer, O. E., 1927, "Plants as Indicators of Groundwater," U.S. Geol. Survey, Water-Supply Paper 577, 95 pp.

Merritt, W. F., 1961, "Movement of Radioactive Wastes through Soil II, Measurements of Direction and Effective Velocity of Groundwater Movement," Atomic Energy of Canada, Ltd. CRER-972, Chalk River, Ontario, 10pp.

Muskat, M., 1946, "The Flow of Homogeneous Fluids through Porous Media," Ann Arbor, Mich., J. W. Edwards, Inc., 763pp.

Nautiyal, S. P., 1958, "Artesian Water Supply of the Taraiaand Bhabar, Naina Tal District, Uttar Pradesh, Proc. Symp. Groundwater, 1955, Central Board of Geophysics, Publ. No. 4, Calcutta, pp. 49–66.

Norvatov, A. M., Popov, O. V., 1961, "Laws of the formation of minimum streamflow," *Bull. Intern. Assoc. Sci. Hidr.*, **6**(1), 20–27.

Plotnikov, N. A., and Bogomolov, G. B., 1958, "Classification of underground waters resources and their reflection on maps," *Intern. Assoc. Sci. Hydrol. (General Assembly of Toronto)*, **2**(44), 525pp.

Pluhowsky, E. J., and Kantrowitz, I. H., 1964, "Hydrology of the Babilon-Islip area, Suffolk County, Long Island, New York," U.S. Geol. Survey Water-Supply Paper 1768, Washington, D.C., 119pp.

Rónai, A., 1961, "Az Alföld talajvíztérképe" (Groundwater Map of the Hungarian Plains), *M. Áll. Földtani Int. Budapest*, 102pp.

Rónai, A., 1963, "Az Alföld negyedkori rétegeinek vízföldtani vizsgálata" (Hydrogeological studies of the Quaternary Formations of the Hungarian Plains), *Hidrol. Közl.*, **43**(5), 387–391.

Sadler, L. D. M., 1962, "The shallow groundwater table and its relationship to irrigation development," Proc. Hydrol. Symp. No. 3, Groundwater, Queen's Printer, Ottawa, pp. 335–363.

Schmidt E. R., (editor), 1962, "Magyarország Hidrogeológiai Atlasza" (Hydrogeological Atlas of Hungary), Magyar Állami Földtani Intézet (Hungarian Geological Institute) Budapest, 73pp.

Schmieder, A, 1965, "A rétegvíz utánpótlódásának mennyiségi vizsgálata a Mátra—és Bükkkalján" (Quantitative investigation of the replenishment of formation waters in the foothills of the Mátra and Bükk mountains), *Hidrol. Közl.*, **45**(8), 362–370 and **45**(10), 447–458.

Schoeller, H., 1962, "Les Eaux Souterraines," Paris, Masson Cie, 642pp.

Seaber, P. R., 1965, "Variations in Chemical Character of Water in the Englishtown Formation, New Jersey," *U.S. Geol. Surv. Profess. Paper* **498-B**, B1–B35.

Sett, D. N., 1958, "Geological investigation of the subterranean brines in the Farrukhnagar—Sultanpur area (Gurgaon District), Punjab (1)," Proc. Symp. on Groundwater, New Delhi, 1965, Central Board of Geophysics, Publ. No. 4, Calcutta, pp. 67–78.

Szebényi, L., 1965, "Az artézi víz forgalmának mennyiségi meghatározása" (Quantitative Determination of Artesian Water Circulation), *Hidrol. Közl.*, **45**(3), 125–130.

Todd, D. K., 1959, "Ground Water Hydrology," New York, John Wiley & Sons, Inc., 336pp.

Tóth, J., 1962, "A theory of groundwater motion in small drainage basins in Central Alberta, Canada," *J. Geophys. Res.*, **67**(11), 4375–4387.

Tóth, J., 1966a, "Groundwater geology, movement, chemistry and resources near Olds, Alberta, Canada," *Res. Council Alberta, Bull.* **17**, in press.

Tóth, J., 1966b, "Mapping and Interpretation of Field Phenomena for Groundwater Reconnaissance in a Prairie Environment, Alberta, Canada," *Bull. Intern. Assoc. Hydrol.* **11**(2), in press.

Wood, P. R., and Dale, R. H., 1964, "Geology and ground-water features of the Edison-Maricopa area, Kern County, California," U.S. Geol. Survey Water-Supply Paper 1656, Washington, D.C., 108pp.

Cross-references: *Aquifer; Connate Water; Groundwater; Hydrogeology; Hydrology; Pedology; Porosity and Permeability; Runoff; Water Table.* Vol. II: *Equation of State; Fluid Mechanics.* Vol. VI: *Engineering Geology; Hydraulics.*

H

HAFNIUM: ELEMENT AND GEOCHEMISTRY

Properties

Hafnium is a strong ductile metal located in Group IVB of the Periodic Table below zirconium and above thorium. Its symbol is Hf, its atomic number 72, its atomic weight 178.49, its density 13.29, its melting point 2230°C, and its boiling point 5200°C. The density and melting point of hafnium are among the highest dozen such values for all the metals. Hafnium was discovered at the University of Denmark in 1922 by Coster and de Hevesy, who used x-ray spectra to identify it in zirconium-containing minerals.

Six stable isotopes are present in naturally occurring hafnium: 174, 0.18%; 176 5.15%; 177, 18.39%; 178, 27.08%; 179, 13.78%; and 180, 35.44%. The thermal neutron absorption cross section of natural hafnium is 105 barns/atom, a value that makes hafnium most useful as a control material for nuclear reactors.

The most noted characteristic of hafnium is its chemical similarity to zirconium; in fact, they are almost identical. This factor vitally affects the geochemistry, geology, and exploitation of hafnium.

Minerals

Hafnium always occurs camouflaged in zirconium minerals, and its presence in them was not recognized by chemists until about 1923. The presence of hafnium in zirconium minerals, at an average Hf/Zr ratio of 0.02, is due to the same valence, +4, and nearly identical atomic radii: 1.442 Å for Hf and 1.452Å for Zr. The similarity of the radii, even though their atomic numbers are 72 and 40, respectively, is accounted for by the lanthanide contraction (i.e., the filling of the inner $4f$ level of electrons) among the rare earth elements of atomic number 58 to 71. This factor also accounts chiefly for the remarkably close chemical similarity of hafnium and zirconium, a similarity that makes the separation of these elements very difficult.

There are no minerals containing hafnium independently of zirconium, so the sources of hafnium are the minerals containing zirconium (Table 1). Although some few scarce minerals, usually altered zircons such as alvite and cyrtolite, contain more than 2% Hf, referred to the Zr + Hf content, the principal commercial sources of hafnium are the minerals zircon, $(Zr,Hf)SiO_4$, and baddeleyite, $(Zr,Hf)O_2$, which are processed chiefly for their content of zirconium.

The most important zirconium-hafnium mineral, zircon, is found worldwide in a variety of rocks: crystalline rocks such as nepheline syenites and granites, volcanic rocks such as basalt and quartz porphyry, sedimentary rocks such as sandstone and limestone, and meta-

TABLE 1. SOME MINERALS CONTAINING HAFNIUM[a]

Mineral[b]	$ZrO_2(\%)$	$HfO_2(\%)$	Hf/Zr
Nepheline syenite minerals			
Baddeleyite, ZrO_2	97.7	1.2	0.014
Catapleite, $(Na_2Ca)ZrSi_3O_9 \cdot 2H_2O$	31.5	0.3	0.011
Elpidite, $Na_3(Zr,Ti)Si_6O_{15} \cdot 3H_2O$	20.3	0.2	0.011
Eudialyte, $(Ca,Na)_5Zr_2Si_6(O,OH,Cl)_{20}$	13.5	0.12	0.011
Polymignite, $(Ca,Fe,Y,Zr)(Nb,Ta,Ti)O_4$	29.1	0.6	0.023
Rosenbuschite, $(Ca,Na)_3(Zr,Ti)Si_2O_8F$	19.8	0.3	0.017
Wöhlerite, $NaCa_2(Zr,Nb)Si_2O_8(O,OH,F)$	15.6	0.5	0.034
Zircon, $ZrSiO_4$	64.23	2.0	0.03
Granitic minerals			
Alvite, $ZrSiO_4$ (contains rare earths, Be, U, Th)	48.0	5.5	0.13
Cyrtolite, see alvite	44	17	0.44
Malacon, see alvite	53	4	0.086
Naegite, see alvite	48.3	7	0.17
Zircon, $ZrSiO_4$	62.3	2.7	0.05

[a] Data taken chiefly from Fleischer (1955).
[b] In the formulas, Zr signifies zirconium plus hafnium.

morphic rocks such as slate, quartzite, and marble.

Alkalic rocks such as nepheline syenites have lower Hf/Zr ratios than do granitic rocks, and minerals from granitic pegmatites have the highest Hf/Zr ratios. However, the average Hf/Zr ratio for all minerals is not far from 0.02.

Abundance

The concentration of hafnium in the earth's crust has been estimated to be in the range of 3 to 4.5 ppm. Whatever the absolute value, hafnium is about as abundant as cesium, beryllium, and uranium, less abundant than praseodymium, samarium, and gadolinium, and more abundant than bromine, tin, and tantalum.

Geochemical Behavior

Hafnium follows zirconium in geochemical behavior and the Zr/Hf ratio remains practically constant in any process of magmatic fractional crystallization. In igneous rocks zirconium does not enter into any of the common rock-forming minerals but remains as a separate phase, usually zircon, which also contains a small amount of hafnium.

Zircon is one of the most persistent of minerals and among sedimentation products is classed, along with quartz, as a resistate product of the chemical breakdown of rocks by weathering and hydrothermal processes. For this reason the main ore deposits of zircon are beach sands; hafnium undergoes the same course as the zirconium in sedimentation processes.

Economic Geology

The only commercial source of an altered zircon containing hafnium at a Hf/Zr ratio of more than about 0.02 is a by-product of an alluvial deposit exploited in columbite and tin mining operations in Nigeria. The zircon content of one fraction of the deposit is 70% and the HfO_2 content is 3.5–5%. This fraction, which amounts to about 2500 tons/year in quantity, is used by at least one producer in the United States for the preparation of zirconium and hafnium metals. Unfortunately, other commercial deposits of altered zircons containing high percentages of hafnium are rare or unknown.

The much more prevalent zircon ores of lower hafnium content are found principally in alluvial placer deposits (beach sands and stream deposits) of Australia, India, Brazil, and the United States (Florida, North Carolina, and Idaho). The zircon (of about 0.02 Hf/Zr ratio) obtained is usually a by-product of the mining or dredging of these placer deposits for such minerals as rutile, monazite, and ilmenite, and is the chief source of both zirconium and hafnium. A second source of the metals is the baddeleyite ores of the Pocos de Caldas plateau of Brazil which are reported to contain as much as 93% ZrO_2 and 1.4% HfO_2. The United States is self-sufficient with respect to zircon sources (deposits) for both zirconium and hafnium metals and compounds.

CLIFFORD A. HAMPEL

References

Eilertsen, D. E., 1965, "Hafnium," in "Mineral Facts and Problems," U.S. Bur. Mines Bull., **630**.
Fleischer, M., 1955, "Hafnium Content and Hafnium-Zirconium Ratio in Minerals and Rocks," *U.S. Geol. Surv. Bull.*, **1021-A**.
Gottfried, D., and Waring, C. L., 1964, "Hafnium content and Hf/Zr ratio in zircon from the southern California batholith," U.S. Geol. Survey Profess. Paper **501B**, 88–91.
Hampel, C. A., 1968, "Hafnium," in C. A. Hampel (Editor), "Encyclopedia of the Chemical Elements," New York, Reinhold Publishing Corp., pp. 251–256.
Martin, D. R., and Pizzolato, P. J., 1961, "Hafnium," pp. 198–219 in C. A. Hampel (editor), "Rare Metals Handbook," Second ed., New York, Reinhold Publishing Corp., pp. 198–219.
Miller, H. W., 1960, "The Occurrence of hafnium minerals and ores," in D. E. Thomas and E. T. Hayes (editors), "The Metallurgy of Hafnium," Washington, D.C., U.S. Atomic Energy Commission, pp. 35–40.
Parker, J. G., 1955, "Zirconium and Hafnium," *Minerals Yearbook* (U.S. Bur. Mines).
Vlasov, K. A. (editor), "Geochemistry and Mineralogy of Rare Elements and Genetic Types of Their Deposits," Vol. 1, "Geochemistry of Rare Elements," published in Russian 1964; translation published by Daniel Davey and Co., New York, 1966 (pp. 314–334 on Hafnium).

Cross-references: *Lanthanides; Rare Earths; Zirconium.* Vol. IVB: *Placer Deposits.*

HALIDES

The minerals in the halide class have the electronegative halogens, fluorine (F), chlorine (Cl), bromine (Br), or iodine (I), as the anion. Most of the halide minerals are compounds of chlorine.

The halite (NaCl) structure is of interest historically because it was the first to be determined by using x-ray diffraction. The halite structure is also found in other minerals, such as the galena group, where the AX radius ratio is between 0.41 and 0.73, yielding octahedral coordination. (See Table on pp. 490–491.)

HALIDES

NORMAL HALIDES
AX type

Halite	NaCl	
Sylvite	KCl	*Halite group:* isometric, with halite structure (interpenetrating, face-centered lattices); substitution poor in halite-sylvite series, but complete in cerargyrite-bromyrite series
Villiaumite	NaF	
Cerargyrite	Ag(Cl,Br)	
Bromyrite	Ag(Br,Cl)	
Salammoniac	NH_4Cl	
Nantokite	CuCl	*Nantokite group:* isometric with sphalerite (hextetrahedral) type of structure; secondary minerals
Miersite	(Ag,Cu)I	
Marshite	CuI	
Iodyrite	AgI	
Calomel	HgCl	

AX_2 type

Fluorite	CaF_2	
Sellaite	MgF_2	
Lawrencite	$FeCl_2$	*Lawrencite Group:* hexagonal platy minerals, occurring as sublimates from fumaroles; deliquescent
Scacchite	$MnCl_2$	
Chloromagnesite	$MgCl_2$	
Hydrophilite	$CaCl_2$	
Coccinite	HgI_2	
Cotunnite	$PbCl_2$	
Eriochalcite	$CuCl_2 \cdot 2H_2O$	
Bischofite	$MgCl_2 \cdot 6H_2O$	

AX_3 type

Molysite	$FeCl_3$
Fluocerite	$(Ce,La,Nd)F_3$
Chloraluminite	$AlCl_3 \cdot 6H_2O$

OXYHALIDES AND HYDROXYHALIDES
$A_m(O,OH)_pX_q$ type

Eglestonite	Hg_4OCl_2	
Terlinguaite	Hg_2OCl	
Lorettoite	$Pb_7O_6Cl_2$	
Mendipite	$Pb_3O_2Cl_2$	
Matlockite	PbFCl	*Matlockite group:* tetragonal with perfect basal cleavage due to layer structure of quadratic-packed anions
Bismoclite	BiOCl	
Daubréeite	BiO(OH,Cl)	
Laurionite	Pb(OH)Cl	
Paralaurionite	Pb(OH)Cl	
Penfieldite	$Pb_2(OH)Cl_3$	
Fiedlerite	$Pb_3(OH)_2Cl_4$	
Atacamite	$Cu_2(OH)_3Cl$	*Atacamite group:* orthorhombic, isostructural
Kempite	$Mn_2(OH)_3Cl$	
Paratacamite	$Cu_2(OH)_3Cl$	
Botallackite	Basic Cu chloride	
Cadwaladerite	$Al(OH)_2Cl \cdot 4H_2O$	

$A_mB_n(O,OH)_pX_q$ type

Boleite	$Pb_9Cu_8Ag_3Cl_{21}(OH)_{16} \cdot 2H_2O$?
Cumengite	$Pb_4Cu_4Cl_8(OH)_8 \cdot H_2O$?
Pseudoboleite	$Pb_5Cu_4Cl_{10}(OH)_8 \cdot 2H_2O$?
Percylite	$PbCuCl_2(OH)_2$
Diaboleite	$Pb_2CuCl_2(OH)_4$
Chloroxiphite	$PbCuO_2Cl_2(OH)_2$?
Nocerite	$Ca_3Mg_3F_8O_2$

Koenenite	$Mg_5Al_2(OH)_{12}Cl_4$
Zirklerite	Al,Fe basic chloride
Kleinite	Hg,NH_4,Cl,SO_4
Mosesite	Hg,NH_4,Cl,SO_4

HALIDE COMPLEXES

$A_mBX_3 \cdot xH_2O$ type

Chlorocalcite	$KCaCl_3$
Carnallite	$KMgCl_3 \cdot 6H_2O$
Tachyhydrite	$CaMg_2Cl_6 \cdot 12H_2O$

A_mBX_4 type

Pseudocotunnite	K_2PbCl_4
Avogadrite	$(K,Cs)BF_4$
Ferruccite	$NaBF_4$
Cryolithionite	$Na_3Li_3Al_2F_{12}$

$A_mBX_4 \cdot xH_2O$ type

Douglasite	$K_2FeCl_4 \cdot 2H_2O$?	
Mitscherlichite	$K_2CuCl_4 \cdot 2H_2O$	
Erythrosiderite	$K_2FeCl_5 \cdot H_2O$	*Erythrosiderite group:* orthorhombic, isomorphous series
Kremersite	$(NH_4,K)_2FeCl_5 \cdot H_2O$	

A_mBX_6 type

Hieratite	K_2SiF_6	*Hieratite group:* isometric series; fumarolic sublimates
Cryptohalite	$(NH_4)_2SiF_6$	
Malladrite	Na_2SiF_6	*Malladrite group:* hexagonal; fumarolic sublimates
Bararite	$(NH_4)_2SiF_6$	
Rinneite	NaK_3FeCl_6	
Chlormanganokalite	K_4MnCl_6	

ALUMINO-FLUORIDES

The alumino-fluorides are a special group of halides which have aluminum in six-fold coordination with fluorine as $(AlF_6)^{3-}$ anions.

Isolated octahedra

Cryolite	Na_3AlF_6	
Elpasolite	K_2NaAlF_6	
Pachnolite	$NaCaAlF_6 \cdot H_2O$	Dimorphous, but both monoclinic prismatic.
Thomsenolite	$NaCaAlF_6 \cdot H_2O$	
Jarlite	$NaSr_3Al_3F_{16}$	

Chain structures

Gearksutite	$CaAl(OH)F_4 \cdot H_2O$

Sheet structures

Prosopite	$CaAl_2(F,OH)_8$
Chiolite	$Na_5Al_3F_{14}$

Network structures

Fluellite	$AlF_3 \cdot H_2O$
Ralstonite	$Na(Mg,Al)_6F_{12}(OH)_6 \cdot 3H_2O$
Weberite	Na_2MgAlF_7

COMPOUND HALIDES

Creedite	$Ca_3Al_2F_4(OH,F)_6(SO_4) \cdot 2H_2O$
Trudellite	$Al_{10}Cl_{12}(OH)_{12}(SO_4)_3 \cdot 30H_2O$?

LLOYD W. STAPLES

HALIDES

References

Dana, J. W., and Dana, E. S., 1951, "The System of Mineralogy," Seventh ed., Vol. II, 1, (revised by Palache, C., Berman, H. and Frondel, C.), New York, John Wiley & Sons.

MacMillan, R. T., 1960, "Salt," in "Industrial Minerals and Rocks," Third ed., AIME, Chap. 40.

Pabst, A., 1950, "A structural classification of fluoaluminates," *Am. Mineralogist,* **35,** 149.

Palache, C., Berman, H., and Frondel, C., 1951, "Halides," in "Dana's System of Mineralogy," New York, John Wiley & Sons.

Cross-references: *Electronegativity; Halogens; Mineral Classes: Nonsilicates.*

HALOGENS

From the Greek *halo* (salt) and *-gen* (bearing), the halogens are a family of the related elements fluorine (F), chlorine (Cl), bromine (Br), iodine (I), and astatine (At). All members of the halogens are rather similar, each being nonmetallic and monovalent; each forms anions (negative). They represent a continuous series.

Fluorine (q.v.) is the lightest halogen, a gas, and is most reactive, displacing the others in any reaction, and even oxygen from water; it is the most powerful oxidizing agent known. Iodine is the most stable. Going up the series, the atomic weight increases, as do density, melting and boiling points, and heats of fusion and vaporization. Color deepens from F at pale yellow, Cl yellow green, Br dark red, I deep violet. Each member forms an acid with hydrogen and forms salts with metals (hence the name).

Astatine (also formerly called "alabamine") is a radioactive element, but its longest-living isotope, At^{210}, has a half-life of only eight hours so that its geological role is small; in the outer mile of the Earth's crust at any one time the astatine concentration is calculated to be only 6.86 mg.

The crustal concentration of the stable halogens is estimated as follows: 770 ppm F, 550 ppm Cl, 1.6 ppm Br, 0.3 ppm I. In the ocean the comparable figures are 1.4 ppm F, 18,980 ppm Cl, 85 ppm Br, and 0.05 ppm I. The enormous accumulation of Cl (and to a smaller extent Br) is of course reflected by the NaCl of sea salts, but remains one of the problems of geochemistry.

R. W. F.

References

Anon., 1951, "Iodine, its properties and technical applications," New York, Chilean Iodine Educational Bureau.

Bloch, M. R., 1963, "The social influence of salt," *Sci. Am.,* **209,** 110.

Jolles, Z. E., 1966, "Bromine and its Compounds," London, Ernest Benn Ltd.

Pauling, L., 1959, "General Chemistry," San Francisco, W. H. Freeman Co.

Sconce, J. S. (editor), 1962, "Chlorine," New York, Reinhold Publ. Corp.

Stacy, M., Tatlow, J. C., and Sharpe, A. G. (editors), 1960, "Advances in Fluorine Chemistry," London, Butterworths Ltd.

Cross-references: *Bromine; Chlorine; Elements: Planetary Abundances and Distribution; Floruine; Iodine; Seawater, Geochemical Balance; Seawater, History; Gases, Volcanic.* Vol. IVB: *Evaporite Minerals.* Vol. V: *Volatiles.*

HEAT CONTENT—See ENTHALPY

HEAVY WATER—See DEUTERIUM

HELIUM: ELEMENT AND GEOCHEMISTRY

Physical Properties

Helium (Gr. *helios,* the sun); symbol He; atomic weight 4.003; atomic number 2; electronic configuration $(1s^2)$; isotopic abundances (atmospheric He), $He^4 \sim 100\%$, He^3 $1.3 \times 10^{-4}\%$. Since, as one of the rare or "noble" gases, helium is essentially inert chemically, chemical and crystal chemical considerations such as electronic structure and atomic size are of relatively little importance in determining its distribution in the earth.

Geochemistry

Earth as a Whole. Helium is the second most abundant element cosmically (next to hydrogen) and is continually being formed in the sun and other stars by hydrogen fusion. Despite its high cosmic and solar abundance (the cosmic H/He ratio is thought to be about 10) it is extremely rare on earth. Its abundance in the earth's atmosphere is only five parts per million by volume whereas that of argon (much less abundant cosmically) is two thousand times greater. This absence of helium in the atmosphere is attributed to the fact that since it has such a low atomic weight, it tends to escape from the earth's gravitational field into space, whereas the probability of argon doing so at normal upper atmosphere temperatures is negligible. In fact the mean residence time of helium atoms in the atmosphere is only a few million years. Since the earth is over 4 billion years old it is obvious that virtually no helium can be residual from an initial atmosphere. It is thought that even the small amount of helium which *is* present is not residual but rather the product of alpha decay in the earth's crust. Thus, at least qualitatively, a state of dynamic equilibrium exists between the continual loss of helium into outer space and the supply of new, radiogenic helium which has leaked from the crust and possibly the mantle.

Crustal Rocks. In general, the distribution of helium in the earth's crust is controlled by two factors: (a) the distribution of alpha emitters in minerals and rocks; (b) the ability of those minerals and rocks to retain helium once it has been produced in them by alpha decay.

The source of alpha particles in ordinary rocks are the parts-per-million levels of uranium (U^{238} and U^{235}), Th^{232}, and their shorter-lived alpha-emitting daughters (such as Ra^{226}, Rn^{222}, etc.). The alpha particles from decay of these nuclides have an average range of 20–30 μ in solids and, near the ends of their tracks, pick up two electrons to become helium atoms. Some minerals appear to be incapable of retaining helium quantitatively for geological periods of time, even at surface temperatures. Others, such as magnetite and non-metamict zircons, appear to be retentive. Recently, the helium method has found application to the dating of aragonitic fossil corals and shells. Regardless of mineralogy, if a rock is extensively metamorphosed, or partially melted, a large fraction of its accumulated He will be released. This He, together with radiogenic Ar^{40} (from K-capture of K^{40}) and other radiogenic gases, migrates through the continental crust and the ocean floor into the atmosphere. Although in natural gases, organic compounds are usually dominant, helium concentrations can reach levels of a few percent in certain nitrogen-rich gases and these constitute the largest commercial helium sources.

Special Occurrences. Certain minerals, such as beryl, have the ability to trap magmatic gas rich in helium when they crystallize. This gas is localized (in beryl) in the channels defined by the six-membered rings of silica tetrahedra. In beryls, helium concentrations can reach 10^{-2} scc/g in contrast to the levels of radiogenic helium expected in beryls or major rock-forming minerals based on their alpha activities and ages alone ($\sim 10^{-4} - 10^{-5}$ scc/g). Although highly radioactive minerals (such as uraninite) do show unusually high helium content, they frequently contain only a negligible fraction of the helium produced in them during their histories. This is thought to reflect the extensive damage done to the lattice (hence to its ability to retain He) by the prolonged alpha bombardment.

In Meteorites

Helium ages as high as 4.3 billion years have been obtained on stony meteorites despite the apparent lack of retentivity exhibited by terrestrial samples of their constituent minerals. On the other hand, helium ages on some stony meteorites are much lower than their K–Ar, Rb–Sr, or Pb–Pb ages, indicating that preferential helium loss (relative to argon) may have occurred in these cases. Besides the "radiogenic" helium (pure He^4), significant amounts or "cosmogenic" or cosmic-ray-produced helium from spallation reactions is produced in all meteorites in their orbits prior to impact on the earth. This helium has a very different isotopic composition: a third as much He^3 as He^4 is normally produced in these spallation reactions. The duration of the meteorite's exposure to cosmic rays can be calculated based on the amount of cosmogenic helium and some estimate of the rate at which these reactions took place during its history as a small body in space. These "exposure ages" average about 400 million years for the "irons" and considerably less than 100 million years for stony meteorites, although the two ranges overlap slightly. Carbonaceous chondritic meteorites have been found to contain (along with generally high concentrations of volatiles) considerable quantities of "primordial" helium which was trapped during the earliest stages of the formation of the solar system. Finally, certain "brecciated" stony meteorites contain very large amounts of helium (up to 10^{-2} scc/g) which is thought to have been implanted in their constituent grains by solar wind bombardment.

Economics

Helium is said to constitute 23% by weight of the known universe, but only 0.000001% of the earth's crust. The only significant source is from certain petroleum (natural gas) reserves. Production in the United States is around 4 billion cubic feet annually, being extracted by private companies and at five federally operated plants (Bureau of Mines) in Texas, Oklahoma, Kansas, and New Mexico. A large volume is held in "stock-pile" reserve in underground storage. The only other source in the free world is a plant in Saskatchewan, Canada.

Helium is used for shielded arc welding, gas chromatography, and other industrial purposes.

Helium has been recognized in volcanic gases in eruptions of Vesuvius and other volcanoes, as well as in radioactive rocks and mine gases, but never in commercial quantities. It is believed that the gas found in petroleum sources is a product of alpha radiation from granitic rocks in the Precambrian basement; the gas has migrated upward to be trapped in the same reservoirs as the petroleum gases,

HELIUM: ELEMENT AND GEOCHEMISTRY

reaching percentages of 1–8%. Estimated U.S reserves exceed 200 billion cubic feet.

FRASER P. FANALE

References

Brancazio, P. J., and Cameron, A. G. W., 1963, "The Origin and Evolution of Atmospheres and Oceans," New York, John Wiley & Sons, 314pp.

Damon, P. E., and Kulp, J. L., 1958, "Inert gases and the evolution of the atmosphere," *Geochim. Cosmochim. Acta,* **13,** 280–292.

Fanale, F. P., and Schaeffer, O. A., 1965, "Helium-uranium ratios for Pleistocene and Tertiary fossil aragonites," *Science,* **149** (3681), 312–317.

Konig, H., Warke, H., Bien, G. S., Rakestraw, N. W., and Suess, H. E., 1964, "Helium, Neon and Argon in the Oceans," *Deep-Sea Res.,* **11**(2), 243–247.

Mayne, K. I., 1956, "Terrestrial helium," *Geochim. Cosmochim. Acta,* **9,** 174–182.

Morrison, P., and Pine, J., 1955, "Radiogenic origin of helium isotopes in rocks," *Ann. N.Y. Acad. Sci.,* **62,** 69.

Nicolet, M., 1957, "Aeronomic problem of helium," *Ann. Geophys.,* **13,** 1–21.

Seibel, H. C. W., 1968, "Helium, Child of the Sun," University Press of Kansas.

Signer, P., and Suess, H. E., 1963, "Rare Gases in the Sun, in the Atmosphere and in Meteorites," in (Geiss, J. and Goldberg, E. D., editors), "Earth Science and Meteoritics," Amsterdam, North-Holland Publ. Co. (div. of John Wiley), pp. 241–272.

Verniani, F., 1966, "The total mass of the Earth's atmosphere," *J. Geophys. Res.,* **71**(2), 385–391.

Cross-references: *Argon; Electron Capture; Gases, Volcanic; Hydrogen; Rare Gases; Thorium; Uranium.* Vol. II: *Carbonaceous Meteorites; Meteorites.*

HELIUM: OIL CHEMISTRY—*See* Vol. IVB

HOLMIUM: ELEMENT AND GEOCHEMISTRY

Cleve, in 1879, was the first to isolate holmium from other rare earth elements; he separated it out of the impure heavy rare earth concentrate, holmia. Holmium was named after the city of Stockholm.

Properties

Elemental holmium is a very reactive silvery white *lanthamide* (q.v.), or *rare-earth* (q.v.) metal with atomic number 67 and outer electronic structure $4f^{11}6s^2$. It occurs as one stable isotope Ho^{165}.

Holmium is determined analytically by neutron activation, x-ray spectrography, and by absorption spectrometry (it has very sharp spectral lines). In geologic environments, holmium forms highly ionic bonds. It forms weak complexes. The oxidation state is tripositive, with ionic radius 0.894 Å. Generally, chemical behavior is similar to the alkaline rare earths, and almost identical to other heavy rare earths.

Minerals

Holmium occurs in highest concentration in rare earth minerals that favor heavy rare earths, generally nitrates, sulfates, carbonates, niobates, and tungstates with cation coordination number of six to eight, often containing structural halogens and alkali metals. Typical examples are:

xenotime (Y, heavy rare earths) PO_4
fergusonite (Y, heavy rare earths) $(Nb,Ta)O_4$

Abundance—*See* RARE EARTHS

Geochemical Behavior

Schmitt et al. (1964) have established a value of 0.073 ppm holmium in chondrites and 0.80 ppm in eucrite achondrites. The standard deviation within a meteorite class is only about 20%. In meteoritic matter, volatization may have redistributed alkali metals but not rare earths: holmium has a boiling point of 2600°C.

The holmium concentration in average basalt is 1.11 ppm, and the range is less than a factor of ten (Haskin et al., 1966). Oceanic and continental basalts have similar ranges. Partial melting of mantle peridotite yields a basaltic liquid with enriched holmium and a mantle source area depleted in holmium.

Igneous differentiation series tend toward higher or lower holmium concentrations, although the total variation is less than that for light rare earths. The relative abundance of holmium and other heavy rare earths remains almost unchanged. Haskin et al. (1966) report 2.2 ppm holmium is an Ascension Island basalt and 1.12 ppm holmium in a trachyte differentiate. Towell et al. (1965) found 0.742 ± 0.030 ppm in the Rubidoux Mountain leucogranite phase of the Southern California batholith, probably a differentiate (by crystal settling) of the San Marcos gabbro, with 0.569 ± 0.023 ppm.

Average granite with 60 to 70% silica has 2.2 ppm but the range of values for granites is even greater than for basalts.

In most basaltic, syenitic, and granitic rocks, holmium is mainly in mafic rock-forming minerals, probably substituting for divalent calcium (ionic radius 0.99 Å).

Holmium and other heavy rare earths form complexes and are transported in alkali, halogen, and carbon dioxide-rich hydrothermal fluids; metasomatic and pegmatoid deposits formed from such fluids may have high concentrations of holmium and higher ratios of holmium to light rare earths than granites.

Holmium has a 440-year residence time in seawater (which is alkaline), about the same as other heavy rare earths. Removal by adsorption on clay particles, as well as depletion in surface water in areas of high organic productivity, are likely processes. High secondary holmium concentrations in phosphatic fish remains today occur only in deep water; paleozoic fish remains also have high rare earths. Manganese nodules have 580 ppm, whereas Pacific Ocean water has only 0.22×10^{-6} ppm (Goldberg et al., 1963).

In soils, the relative amounts of holmium in detrital, clay, and carbonate fractions is not well known. Sedimentary rocks have about 1.40 ppm, with a variation of a factor of five. In a study of rare earths in Russian platform soils, Balashov et al. (1964) found higher holmium in alkaline soils, the holmium in groundwater precipitating as hydroxide; lower concentrations are in acid soils, probably due to removal of complexes.

Biologic concentration of rare earths has been found in hickory leaves; rare earths have been used to trace opium.

Economic Geology

The major source of holmium has been from a light rare earth mineral, monazite in beach sands; the accessory monazite originally present in granitic rocks resists weathering and is concentrated by sedimentary processes. Australia, Brazil, Ceylon, India, Malaysia, and South Africa mine the largest quantities. Less than one percent of the monazite is holmium. Recently, the bastnaesite deposit at Mountain Pass, California has been an important source of holmium, again in low concentration.

High-purity holmium has a very limited market. Most holmium is sold unpurified in rare earth concentrates used for polishing, petroleum cracking catalysis, lighter flints, arc carbons, and in alloys.

RORERT KAY

References

Balashov, Y. A., et al., 1964, "The effects of climate and facies environment on the fractionation of the rare earths during sedimentation," *Geochem. Internat. No. 1,* **10,** 951–969 (English transl.).

Goldberg, E. D., et. al., 1963, "Rare-earth distributions in the marine environment," *J. Geophys. Res.,* **68,** 4209–4217.

Haskin, L. A., et al., 1966, "Meteoric, solar, and terrestrial rare-earth distributions," in (Ahrens, L. H., et al., editors), "Physics and Chemistry of the Earth," Vol. 7, Oxford, Pergamon Press, pp. 167–321.

Schmitt, R. A., et al., 1964, "Rare-earth, yttrium and scandium abundances in meteoric and terrestrial matter—II," *Geochim. Cosmochim. Acta,* **28,** 67–86.

Towell, D. G., et al., 1965, "Rare-earth distributions in some rocks and associated minerals of the batholith of Southern California," *J. Geophys. Res.,* **70,** 3485–3496.

Cross-references: *Complexes; Hydrothermal Solutions; Mass Spectrometry; Monazite Neutron Activation Analysis; Rare Earths; Seawater, Chemistry; X-Ray Spectroscopy.*

HOLOCENE—*See* Vol. III

HOT SPRINGS—*See* SPRINGS

HYDROCARBONS*

Hydrocarbons are compounds composed solely of hydrogen and carbon. They occur as gases, liquids, and solids, and are widely distributed in the lithosphere, hydrosphere, and atmosphere, as well as specifically in the biosphere.

Structure and Classification

The existence of hydrocarbons is possible because of shared-electron (covalent) bonding between carbon and hydrogen atoms and between carbon and carbon atoms. The latter bonding gives rise to chains of carbon atoms, and the chains commonly range in length from two to forty carbon atoms. Since a given carbon atom can form covalent bonds with more than two carbon atoms, the carbon chains can be not only straight (normal), but also branched. This makes possible ordinary branched hydrocarbons, commonly with methyl groups (CH_3) positioned along the main carbon chain. A special form of branching is the formation of cyclic structures comprising rings of five or six carbon atoms. If the potential bonding of the carbon atoms is saturated with hydrogen in the foregoing instances, the compounds are termed *aliphatic hydrocarbons,* and they may occur as straight-chain, branched, or cyclic structures, or as combinations of these types.

Hydrocarbons with less hydrogen than that of the foregoing compounds (with the molecular formula C_nH_{2n+2}) are termed unsaturated and may have the formula C_nH_{2n}, C_nH_{2n-2}, C_nH_{2n-4} and so on, with increasing degrees of

* The author acknowledges the support of National Aeronautics and Space Administration Grant NGR-05-020-296.

HYDROCARBONS

TABLE 1. SELECTED HYDROCARBONS IDENTIFIED IN PETROLEUM[a]

Formula	Name	Type	B.P. (°C)	% in Petroleum
CH_4	Methane	Alkane	−161.49	—
C_2H_6	Ethane	Alkane	− 88.63	0.004–0.01
C_3H_8	Propane	Alkane	− 42.07	0.01 –0.7
C_4H_8	Cyclobutane	Cycloalkane	—	0.001
C_4H_{10}	n-Butane	Alkane	− 0.50	0.03 –2.5
C_4H_{10}	Isobutane	Isoalkane	− 11.73	0.001–0.90
C_5H_{10}	Cyclopentane	Cycloalkane	49.26	0.04 –0.32
C_5H_{12}	2,2-Dimethylpropane	Isoalkane	9.50	0.000–0.017
C_5H_{12}	2-Methylbutane	Isoalkane	27.85	0.11 –1.35
C_5H_{12}	n-Pentane	Alkane	36.07	0.12 –2.30
C_6H_6	Benzene	Aromatic	80.10	0.014–1.26
C_6H_{12}	Cyclohexane	Cycloalkane	80.74	0.10 –1.05
C_6H_{14}	2-Methylpentane	Isoalkane	60.27	0.11 –1.20
C_6H_{14}	n-Hexane	Alkane	68.74	0.14 –2.03
C_7H_8	Toluene	Aromatic	110.62	0.09 –1.02
C_7H_{14}	Methylcyclohexane	Cycloalkane	100.93	0.29 –1.85
C_7H_{16}	n-Heptane	Alkane	98.43	0.08 –2.32
C_8H_{18}	n-Octane	Alkane	125.66	1.9
C_9H_{10}	Indan	Aromatic-cycloalkane	177.8	0.003
C_9H_{20}	n-Nonane	Alkane	150.80	1.8
$C_{10}H_{18}$	Naphthalene	Aromatic	217.96	0.06
$C_{10}H_{16}$	Adamantane	Polycycloalkane	—	—
$C_{10}H_{22}$	n-Decane	Alkane	174.12	1.8
$C_{15}H_{32}$	n-Pentadecane	Alkane	270.61	0.8
$C_{20}H_{12}$	Perylene	Aromatic	268–269[b]	—
$C_{20}H_{42}$	n-Eicosane	Alkane	343.8	0.37
$C_{25}H_{52}$	n-Pentacosane	Alkane	401.9	0.20
$C_{30}H_{62}$	n-Triacontane	Alkane	449.7	0.12

[a] Data from Whitehead and Breger (1963).
[b] Melting point.

unsaturation. To overcome the deficiencies in hydrogen, specific carbon-carbon bonds are established by covalently sharing two electron pairs, thus forming double bonds giving rise to *olefins*. If several double bonds occur in a particular molecule, the energy of the molecule is at a minimum if the bonds are in a conjugated pattern, i.e., alternating with single bonds. This pattern, when present in cyclic structures such as benezene (C_6H_6), is particularly stable, and the hydrocarbons involved are aromatic hydrocarbons. A number of aromatic rings may be present in a fused state where certain carbon-carbon bonds are shared by neighboring rings.

The simplest hydrocarbon is methane (CH_4) and at normal temperatures and pressures it exists as a gas. Increasing molecular size reduces the vapor pressure, so that pentane (C_5) to tetradecane (C_{14}) exist as liquids, and higher straight-chain homologs are solids. A number of typical hydrocarbons are listed in Table 1.

Relative to most organic compounds, hydrocarbons are chemically unreactive, except in the cases of olefins and more highly unsaturated (nonaromatic) compounds which are susceptible to chemical attack at the double-bond positions. In general, only severe oxidation or thermal destruction is effective in modifying or destroying the hydrocarbons.

Analytical Methods

The analysis of earth science samples for hydrocarbons is based on extraction, separation, and identification of the hydrocarbon compounds such as indicated in Fig. 1.

Extraction of hydrocarbons from a solid geochemical sample commonly involves crushing followed by solvent extraction for hydrocarbons above about C_{10}. For hydrocarbons of lower molecular weight, the sample is usually evacuated with gentle heating, and the volatilized hydrocarbons caught in a cold gas trap.

Separation methods are based almost entirely upon chromatographic procedures, although distillation methods are still used for gross separations, usually in the case of petroleum samples. Column chromatography is used for separation of classes of compounds, e.g., separation of aliphatics from aromatics through the preferential elution of aliphatics from silica gel with n-hexane or n-heptane. Size differences are used to separate normal aliphatics from branched aliphatics, e.g., molecular sieves

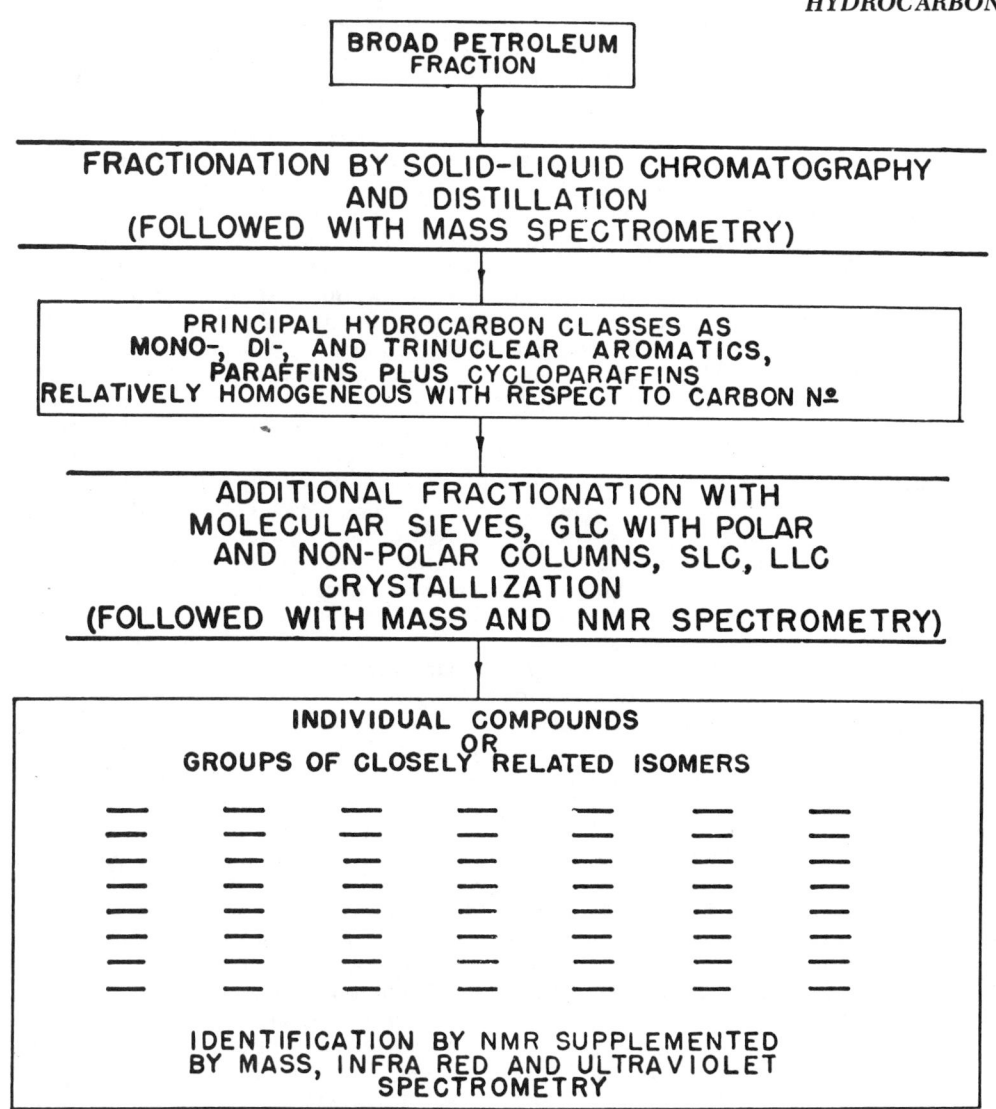

Fig. 1. Operations involved in the isolation and identification of hydrocarbons from a broad petroleum fraction (Mair, 1965).

are capable of accommodating straight-chain hydrocarbons while rejecting the branched isomers. Similar, though less specific, separations can be accomplished in the case of polycyclic aromatics using gel permeation (Edstrom and Petro, 1968).

Gas-liquid chromatography is a much superior method of separating hydrocarbons because its resolving power is so great that in many instances individual compounds are separated completely from neighboring hydrocarbons. The resolution offered by gas chromatography is commonly adequate for identification purposes. Identification is based on comparison of the retention time of the unknown with that of known compounds for chromatography on preferably more than one type of chromatographic column.

Identification is further strengthened through a coupling of gas chromatography with molecular mass measurements. Thus, the compounds resolved by a gas chromatograph are introduced in turn into a mass spectrometer which determines not only the mass of the unknown molecule but also gives a good indication of its structure through its fragmentation pattern.

Other spectral methods are important in hydrocarbon identification. Ultraviolet-visible spectroscopy is particularly useful with aromatic compounds, as is spectrofluorometry. Infrared absorption is most important in the identification of individual hydrocarbon isomers by comparison with reference spectra. It is particularly useful for olefin-type analysis. Raman spectroscopy is important for examining multisubstituted olefins, and complements the corresponding infrared data. Nuclear magnetic resonance (NMR) analysis of hydrocarbons is vital for determination of structure through chemical shift and multiplet spectral patterns. Final definitions of chemical structure for hydrocarbons is sometimes attainable only through the application of x-ray crystalograhpy.

Occurrence

Lithosphere. Hydrocarbons occur commonly up to a few parts per million in soils. The distribution of structural types is wide, including normal, branched, cyclic, and aromatic hydrocarbons. The normal alkanes commonly exhibit odd-carbon number preference. Branched aliphatic hydrocarbons usually include members of the isoprenoid series, i.e., compounds including one or more units with the branched isoprene skeleton of five carbon atoms. Polycyclic aromatic hydrocarbons are probably present, but evidently at levels less than 0.01 ppm.

Hydrocarbons in modern sediments are more fully known. Odd-carbon preference in normal alkanes is common. Aliphatic hydrocarbons occur in the range of 10 to 1000 ppm, and aromatics are frequently equally abundant. Isoprenoid hydrocarbons are common, with most falling in the C_{15} (farnesane) to C_{20} (phytane) range. Polycyclic aromatics are present in small amounts in modern fresh-water and marine sediments.

Hydrocarbons of ancient sedimentary rocks are more fully described. The level of concentration is generally the same or greater than that for modern sediments (Bray and Evans, 1961). Alkanes of all structures are found, including isoprenoids, simple aromatics, and polycyclic aromatics. Of the isoprenoids, pristane (C_{19}) is usually the most abundant, followed by C_{16} and phytane (C_{20}), for overall abundances of up to about 10 ppm of the rock (Hodgson et al., 1968). Isoprenoids are roughly 5–50% as abundant as the n-alkanes. Normal alkanes in sedimentary rocks commonly show odd-carbon preference, usually most strongly in the C_{20}–C_{30} range. Alkanes occur in rocks of all ages, even older than three billion years.

Hydrocarbons in petroleum accumulations are special occurrences of hydrocarbons in the lithosphere and have been studied extensively, as reviewed comprehensively by Smith (1966) and Hunt (1968). Normal alkanes range from one carbon atom (methane) to more than forty, with chains of as many as seventy-four and seventy-eight carbons reported. High-melting waxes containing about twenty-five carbon atoms are largely normal alkanes, but branched and cyclic structures are probably also included. Nearly 100 branched alkanes in the C_4–C_{13} range have been identified in crude oil; in the higher ranges, principal attention has focused on particular series of compounds, mainly the isoprenoids, which are most abundant in the C_{15}–C_{20} range. In general, the branched alkanes with the fewest and simplest branches are most prevalent. Unsaturated hydrocarbons are uncommon in petroleum, but C_6, C_7, and C_8 olefins have been identified.

Cyclic alkanes are ring compounds, usually comprising five- or six-membered rings. All cycloalkanes are saturated, and all except the parent members have alkyl side chains. Thirty-one monocyclic pentanes and forty-two monocyclic hexanes have been identified in petroleum. Two- and three-ring cycloalkanes have been identified, and recently several pentacyclic triterpanes have been isolated (Hills et al., 1968).

Aromatic hydrocarbons are common in petroleum. The simplest are benzene and its alkylated homologs, which are generally less prevalent than either the corresponding alkanes or cycloalkanes. Toluene is almost always more abundant than benzene, generally by a factor of 3 to 6. Polycyclic aromatics with two to five fused rings have been found in petroleum. These occur commonly in alkylated homologous series. As molecular weight increases, more and more compounds are found that are mixtures of alkanes, cycloalkanes, and aromatics. Included are the indanes, tetralins, fluorenes, and cyclopentaphenanthrenes. Asphaltene and resin portions of the heavy gas-oil fraction of petroleum consist primarily of graphite-like clusters of condensed aromatic rings joined together by alkyl groups. Naphthalenic ring systems are also present but are not fused to the aromatic clusters (Erdman, 1965). Hydrocarbon waxes are very common in petroleum and range widely in content from 0 to 33% (Hedberg, 1968).

Biosphere. Hydrocarbons are widely distributed in the biosphere as reviewed by Whitehead and Breger (1963). They are present in primitive and complex organisms, in both internal and external locations. The most common examples are the waxy coatings of plants. Compound types have not been studied extensively but normal alkanes and branched isoprenoid alkanes, as well as iso- and anteiso-alkanes, are well known.

Hydrosphere. Hydrocarbons exist in marine and fresh waters at very low levels. In seawater, C_1 to C_4 hydrocarbons range from about 10^{-7} ml/l to 3×10^{-4} ml/l (Swinnerton and Linnenbom, 1967). Higher hydrocarbons range from 0.01 to 0.2 μg/l, and in fresh surface waters, from 0.2 to 2 μg/l; subsurface oil field waters contain about 10 μg/l (Peake and Hodgson, 1966). The solubility of hydrocarbons in water was thoroughly studied by McAuliffe (1966) for C_{12}–C_{36} hydrocarbons with solubilities ranging from 0.01 to 0.001 μg/g. Hydrocarbons are also accommodated in water in complex dispersed systems showing values in the range of 0.01 to 100 mg/l with preferential accommodation of C_{16}–C_{20} n-alkanes (Peake and Hodgson, 1967).

Atmosphere. Hydrocarbons occur in airborne particulates, probably because of natural or artificial pollution. Terpenes from plants are thought to represent one major source (Went et al., 1967). In urban areas, polycyclic aromatic hydrocarbons are common, with benzo(a)pyrene a typical hydrocarbon at about 1 μg/mg of airborne particulate matter (Stanley et al., 1967).

Origin of Hydrocarbons

Primordial Origin. Hydrocarbons existing in crustal rocks of the earth may have come in part from the primordial matter from which the earth accumulated, through continuing degassing of the planet. No reliable criteria exist for the identification of primordial hydrocarbons, and no clear-cut examples of accumulations of such hydrocarbons are known.

Biogenic Origin. Hydrocarbons are synthesized by living organisms. Normal alkanes in the range from C_{25} to C_{35} appear to be widely distributed in the plant kingdom, occurring in roots, stems, leaves, flowers, fruits, and seeds of a variety of plants (Oro et al., 1965). For example, early classical work on plant waxes showed the presence of n-hentriacontane from flowers of common red, white, and crimson clovers. Odd-carbon preference was observed, and more recent work has defined it more clearly in pyrethrum wax and string bean wax, sugar cane cuticle wax, Gramineae, Crassulaceae, tobacco leaves, rose petal wax, and other plant products. Alkanes are probably common in most plants, having been found in extracts of clover leaves and apple cuticle wax in the range of C_{16} to C_{29}. Alkanes C_{30}–C_{38} are present in cactus leaves.

Normal, iso, and anteiso long-chain hydrocarbons are present in tobacco plants (Kaneda, 1967). The amounts of branched-chain hydrocarbons are almost equal to those of the normal alkanes. The total hydrocarbons make up about 0.1% of the dry weight of the plant. The normals were C_{27}–C_{35}; iso, i-C_{27}, i-C_{29}, i-C_{31}, i-C_{33}, i-C_{35}; anteiso, C_{28}, C_{30}, C_{32}, and C_{34}. Clovers contain normal alkanes ranging from C_{18} to C_{35}, with C_{31} as the major component. In the range C_{24}–C_{34} the odd-carbon abundance is approximately eight times that of the hydrocarbons with even numbers of carbon atoms. Hydrocarbons in the leaves have an odd-carbon preference of five. Bacteria synthesize hydrocarbons, but the alkanes are present in similar amounts than in higher plants and show no marked variations in their distributions and no significant predominance of odd-carbon alkanes. Isoprenoid hydrocarbons in the C_{15}–C_{20} range are present in living organisms, e.g., cactus and Swiss chard. Carotenoids are widespread.

Aromatic hydrocarbons are derived from biogenetic sources, for example, from pinewood tar which contains about 20% retene, 1-methyl-7-isopropylphenanthrene ($C_{18}H_{18}$) (Jönsson, 1968). Crude retene in pine tar was found to contain about seven akyl-substituted phenanthrenes besides retene itself, including 1-methylphenanthrene, 2-ethylphenanthrene, 1,7-dimethylphenanthrene, 1-methyl-7-ethylphenanthrene, and 9,10-dihydroretene.

The biosynthesis of long-chain hydrocarbons is revealed by the metabolic pathways for the biosynthesis of iso and anteiso hydrocarbons. A close relationship is believed to exist between fatty acid and hydrocarbon metabolism. The common fatty acids, myristic, palmitic, and stearic, or biologically equivalent forms, are involved as precursors at a certain stage in the biosynthesis of long-chain hydrocarbons (C_{25}–C_{35}). Similarly, branched-chain fatty acids of C_{14}–C_{18} are involved in the biosynthesis of branched-chain hydrocarbons. Going back one step further, amino acid substrates, L-valine, L-leucine, or L-isoleucine in a tobacco plant are incorporated into the specific long-chain hydrocarbons which are structurally related to the amino acid substrate (Kaneda, 1967). Experiments with both higher plants and microorganisms show that the amino acid is first deaminated to form the α-keto acid which is oxidatively decarboxylated and converted to the corresponding branched-chain acyl coenzyme-A ester. Biosynthesis of the hydrocarbon then proceeds by chain elongation involving multiple C_2 additions or by n-C_{16} or n-C_{18} acid addition, with a final decarboxylation step.

Carotenoids follow the general pattern of terpenoid biosynthesis in that they are derived from the biological isoprene precursor, isopentyl pyrophosphate, which originates from acetyl coenzyme-A (Goodwin, 1968). Chain elongation begins by isomerization of the isoprene precursor to which further molecules are added sequentially to form the C_{10} compound (geranyl pyrophosphate), the C_{15} compound (farnesyl pyrophosphate, the precursor of sterols and other triterpanes), and the C_{20} compound (geranylgeranyl pyrophosphate, the precursor of carotenoids). Two molecules of the C_{20} compound condense to form phytoene, the first C_{40} compound formed in the biosynthetic sequence to the colored carotenoids such as lycopene. The formation of cyclic carotenoids such as α- and β-carotenes is a late step in the chain and takes place either at the neursporene or lycopene level.

Aromatic compounds are commonly synthesized as quinones in microorganisms and larger plants (Thomson, 1965). Benzoquinones are fungal metabolites. Naphthaquinones are common in flowering plants. The largest group of naturally occurring quinones are the anthraquinones. Biosynthesis is probably by acetate-derived aromatic compounds. Naphthacenequinones and phenanthraquinones are biogenic. Perylene quinones are biosynthesized by oxidative dimerization of naphthalene precursors by fungi. Similarly, eight-ring quinones are elaborated by fungi through dimerization of anthraquinones. Such aromatic quinones are readily reduced to the corresponding hydroquinones, and further reduction to the aromatic parent is indicated by the ubiquity of aromatics in geochemical environments.

Hydrocarbons from Immediate Precursors. Hydrocarbons may be formed from specific immediate precursors such as fatty acids and alcohols, and from nonspecific sources such as kerogens and coals. Geochemical conversions are thought to occur both thermally and catalytically.

The most intensively studied reactions pertain to the conversion of normal fatty acids to normal alkanes in sediments and sedimentary rocks. The odd-carbon preference observed for the abuadance of fatty acids in plants persists in fatty acids in sediments and frequently in sedimentary rocks. Thus, in most sediments even-carbon normal fatty acids are much more abundant than odd-carbon fatty acids. Normal fatty acids have been postulated as precursors for normal alkane hydrocarbons because of (1) structural similarities between the two kinds of molecules, (2) ubiquity of fatty acids in biological materials, and (3) fatty acid-normal alkane relationship in sediments (Kvenvolden, 1967). Analytical evidence suggests that normal fatty acids may be precursors for many normal alkanes of intermediate and high molecular weights.

Normal fatty acids are common and essential constituents of living organisms. They are found in oils, fats, and waxes of flora and fauna of marine and terrigenous habitations. Palmitic acid (normal C_{16}) is the most abundant and widespread saturated fatty acid. All even-carbon normal fatty acids from C_2 to C_{38} have been found in nature, most commonly as esters. Natural fats contain predominantly even-carbon acids ranging from C_4 to C_{26}; insect and plant waxes, from C_{26} to C_{38}. Odd-number acids are much less abundant, but chain lengths from C_7 to C_{23} have been reported.

Soils and modern sediments contain fatty acids, and the even-carbon preference is clear. The carbon preference index for the acids appears to be somewhat higher than for the corresponding alkanes in the sediments, e.g., 7.06 vs 4.85, for sediments off the coast of California (Kvenvolden, 1966). Limited data show that the abundance of fatty acids is generally much higher than that of the alkanes in modern sediments, e.g., for the above sediment, normal fatty acids were 132 μmoles/kg, compared with 5.9 for normal alkanes.

Normal fatty acids are present also in sedimentary rocks. The even-carbon preference is generally much less pronounced. In some instances a close correspondence is evident between the fatty acid and alkane patterns of chain length, commonly with the alkane number one unit less than that of the fatty acid of similar position in the abundance pattern, suggesting a simple decarboxylation process under geochemical conditions.

To account for the decarboxylation of fatty acids to hydrocarbons under geochemical conditions, Jurg and Eisma (1964) have shown that fatty acids are decarboxylated under relatively mild conditions (200°C for a number of days) when in the presence of bentonite. Not only is the odd-carbon alkane produced from an even-carbon fatty acid, e.g., C_{21} alkane from C_{22} acid, but discernible yields of alkanes both larger and smaller than C_{21} are also observed. The effect of water suggests that free radicals are involved. The data indicate that such processes might well account for the initial establishment of an odd-carbon preference in alkanes, and secondly, for a dilution of the preference through the random generation of alkanes of all chain lengths from each even-dominant fatty acid.

While the foregoing discussion was limited to normal fatty acids and the corresponding n-alkanes, similar considerations probably apply to branched and cyclic structures. For example, branched-chain fatty acids (iso and anteiso) are isolated from sediments from several environments with concentrations ranging from about 0.1 to about 3 ppm of the sediment. The corresponding normal fatty acids are in some instances less abundant, but generally are an order of magnitude more abundant. The branched C_{14}, C_{15}, and C_{16} are the most abundant iso and anteiso fatty acids. The relatively high ratio of branched-chain to straight-chain fatty acids for the even-numbered carbon molecules suggests a bacterial origin for the branched-chain isomers.

Isoprenoid acids are common in zooplankton, fish, and in marine and terrigenous mammals. They occur also in petroleum, ancient sediments, and modern sediments. The paucity of isoprenoid acids in number and abundance relative to those in ancient sediments has been interpreted as indicating that the acids are formed, in part at least, as a result of geochemical diagenesis, rather than the reverse process of decarboxylation of the organic acids to form hydrocarbons.

The derivation of triterpenoid hydrocarbons of petroleum from naturally occurring steroids and triterpenes is widely accepted. For example, a simple triterpane, gammacerane, is present in the Green River oil shale, and it is believed to have arisen from the corresponding alcohol known to exist in primitive protozoa. More complex starting materials or more complex molecular adjustments are indicated by the discovery of a spirotriterpane in a Nigerian crude oil (Hill et al., 1968). The structure could have arisen from a natural triterpene through a molecular rearrangement induced under acidic reducing conditions. A number of adamantanes and related structures have been identified in petroleum. Lewis acid catalysts readily rearrange cyclanes to alkyladamantanes. The rearrangement has been demonstrated with cholestane, cholesterol, cedrene, caryophyllene, and abietic acid.

Many aromatic compounds found in soils, sediments, and sedimentary rocks appear directly related to naturally occurring pigments in living organisms. Quinoid pigments are the likely precursors of perylene and its alkylated homologs, and for 1,2-benzyperylene and coronene found in modern sediments.

More complex transformations are pictured for other hydrocarbons occurring in the lithosphere. Series of processes are postulated for fragmentation and polymerization of lipid substances, as well as general organic matter, of sedimentary rocks. Such processes are commonly referred to as maturation and are thought to depend largely on thermal effects. Radiation processes probably produce similar results. Carbon isotope studies indicate a hypothetical starting substance for petroleum with an intermediate molecular size (Silverman, 1964). Thermal degradation of this material would be expected to spin off low-molecular-weight fragments enriched in the light carbon isotope C^{12}. The residual material would tend to polymerize and become enriched in the heavy isotope, C^{13}. This pattern is observed in the case of many petroleums, particularly in the case of associated natural gases which are enriched in the light isotope.

The abundance of low-molecular-weight hydrocarbons in the lithosphere is consistent with the maturation concept. Specific direct precursors of hydrocarbons in the C_2–C_9 range are not clearly evident, and yet these hydrocarbons are much more abundant than those clearly derived from biogenic sources. Amino acids, however, may be a source for the aliphatic low-molecular-weight hydrocarbons, through maturation involving both deamination and decarboxylation. In general, thermal degradation of kerogen is acknowledged to yield significant quantities of a variety of low-molecular-weight aliphatic and aromatic hydrocarbons. Such processes would be expected to yield equal quantities of even- and odd-carbon-number alkanes, thus diluting the odd-carbon preference evident in normal alkanes from direct biogenic sources.

Diagenesis of Hydrocarbons

While hydrocarbons are stable compounds relative to most organic substances, they are still subject to destruction, and the abundance of hydrocarbons in the lithosphere is determined by the relative rates of production and destruction. Chemically, hydrocarbons are destroyed by a variety of processes. Oxidation is perhaps one of the most important, and oxidation through direct contact with air as a gas, or in solution in natural waters undoubtedly has a profound influence on the balance of hydrocarbons in the lithosphere. Abstraction of hydrogen from hydrocarbons through chemical attack, e.g., by sulfur, leads to polymerization of hydrocarbons, ultimately to natural high-carbon asphaltites and a variety of unusual hydrocarbon accumulations. Thermal treatment tends to give similar products, with light fragments being volatilized leaving high-carbon residues. Mechanical diagenesis of hydrocarbons is evident in the sense that they are transported through the lithosphere, hydrosphere, and atmosphere of the earth.

Biological destruction of alkanes is widely recognized (Kallio et al., 1961). The mechanisms in the microbial oxidation of alkanes are based on a primary enzymatic attack at the terminal carbon atom which results in the early formation of a fatty acid equal in length to the chain length of the original alkane. Rupture of the hydrocarbon structure and introduction of oxygen into the molecule may occur in at least two ways: (1) dehydrogenation of the alkane to an olefin, followed by addition of water to the double bond, or (2) by hydroperoxidation of the alkane, leading to the formation of an alkyl hydroperoxide. Analysis of microbial alkane oxidation literature indicates fatty acids, normal alcohols, and methyl ketones predominate as identifiable products from n-alkanes of low molecular weight (C_{10} or less). Acids, alcohols, and esters are the major products of microbial oxidation of longer n-alkanes. Peroxides have been detected in culture fluids of bacteria utilizing alkanes, and reduction of hydroperoxides by bacteria has been reported. Likely mechanisms for the microbial utilization of alkanes are as follows: Short-chain alkanes (C_3–C_5) produce methyl ketones, probably via rearrangement of 1-alkyl free radicals; intermediate-chain-length alkanes (C_6–C_{10}) are dehydrogenated to the corresponding 1-alkane; long-chain alkanes (C_{12}–C_{18}) are oxidized by the way of 1-alkyl hydroperoxides, and the formation of n-alcohols which in turn are oxidized to n-fatty acids.

Branched alkanes are probably less readily degraded by microorganisms than are normal alkanes. Bacterial microorganisms grown at the expense of branched alkanes indicated that not only size and position of substituents on the alkane chain but also the number of branches and the degree of branching contributed to the utilizability of the particular hydrocarbon. Branched structures are less available for growth than linear structures: increase in branch size or movement of the branch toward the middle of the chain decreases the availability, as does an increase in the number of branches on a given carbon atom in the chain (McKenna, 1968). Soil microorganisms metabolize pristane and other isoalkanes by two pathways: β-oxidation, and ω-oxidation followed by β-oxidation. Diterminal oxidation of isoalkanes appears to resemble that of n-alkanes in which the oxygenase system attacks first a terminal methyl group to form a monocarboxylic acid and then the antipodal methyl group to form a dicarboxylic acid which is further metabolized by β-oxidation.

GORDON W. HODGSON

References

Bray, E. E., and Evans E. D., 1961, "Distribution of n-paraffins as a clue to recognition of source beds," *Geochim. Cosmochim. Acta*, **22**, 2–15.

Brooks, J. D., and Smith, J. W., 1967, "The diagenesis of plant lipids during the formation of coal, petroleum and natural gas—I. Changes in the n-paraffin hydrocarbons," *Geochim. Cosmochim. Acta*, **31**, 2389–2397.

Edstrom, T., and Petro, B. A., 1968, "Gel permeation chromatographic studies of polynuclear aromatic hydrocarbon materials," *J. Polymer Sci.*, Part C, No. 21, 171–182.

Erdman, J. G., 1965, "The molecular complex comprising heavy petroleum fractions," Hydrocarbon Analysis, *Am. Soc. Testing Mater., Spec. Tech. Publ.*, **389**, 259–300.

Goodwin, T. W., 1968, "Recent developments in the biosynthesis of carotenoids," *J. Sci. Ind. Res.*, **27**(3), 103–105.

Hedberg, H. D., 1968, "Significance of high-wax oils with respect to genesis of petroleum," *Am. Assoc. Petrol. Geol. Bull.*, **52**, 736–750.

Hills, I. R., Smith, C. W., and Whitehead, E. V., 1968, "Hydrocarbons from fossil fuels and their relationship with living organisms," *Am. Chem. Soc. Preprints*, **13**(4), B5–B20.

Hodgson, G. W., Hitchon, B., Taguchi, K., Baker, B. L. and Peake, E., 1968, "Geochemistry of porphyrins, chlorins and polycyclic aromatics in soils, sediments and sedimentary rocks," *Geochim. Cosmochim. Acta*, **32**, 737–772.

Hunt, J. M., 1968, "How gas and oil form and migrate," *World Oil*, **167**, 140–150.

Jönsson, R., 1968, "Separation and identification of some naturally occurring alkylphenanthrenes," *Talanta*, **15**, 425–431.

Jurg, V. W., and Eisma, E., 1964, "Petroleum hydrocarbons: Generation from fatty acid," *Science*, **146**, 1451–1452.

Kallio, R. E., Finnerty, W. R., Wawzonek, S., and Klimstra, P. D., 1961, "Mechanisms in the Microbial Oxidation of Alkanes," in (Oppenheimer, C. H., editor), "Symposium on Marine Microbiology," C. C. Thomas, pp. 453–463.

Kaneda, T., 1967, "Biosynthesis of long-chain hydrocarbons. I. Incorporation of L-valine, L-threonine, L-isoleucine, and L-leucine into specific branched-chain hydrocarbons in tobacco," *Biochemistry*, **6**, 2023–2032.

Kvenvolden, K. A., 1966, "Molecular distributions of normal fatty acids and paraffins in some lower Cretaceous sediments," *Nature*, **209**, 573–577.

Kvenvolden, K. A., 1967, "Normal fatty acids in sediments," *J. Am. Oil Chem. Soc.*, **44**, 628–636.

McAuliffe, C., 1966, "Solubility in water of paraffin, cycloparaffin, olefin, acetylene, cycloolefin, and aromatic hydrocarbons," *J. Phys. Chem.*, **70**, 1267–1275.

McKenna, E. J., 1968, "Microbial oxidation of iso-alkanes," *Am. Chem. Soc. Preprints*, **13**(4), B23.

Mair, B. J., 1965, "Aromatic hydrocarbons in high-boiling fractions," "Hydrocarbon analysis," *Am. Soc. Testing Mater., Spec. Tech. Publ.*, **389**, 214–258.

Naughten, J. J., Heald, E. F., and Barnes, I. L. Jr., 1963, "The chemistry of volcanic gases," *J. Geophys. Res.*, **68**, 539–551.

Olson, R. J., Oro, J., and Zlatkis, A., 1967, "Organic compounds in meteorites, II, Aromatic hydrocarbons," *Geochim. Cosmochim. Acta,* **31,** 1935–1948.

Oro, J., and Nooner, D. W., 1967, "Aliphatic hydrocarbons in meteorites," *Nature,* **215,** 1085–1087.

Oro, J., Nooner, D. W., and Wikstrom, S. A., 1965, "Paraffinic hydrocarbons in pasture plants," *Science,* **147,** 870–873.

Peake, E., and Hodgson, G. W., 1967, "Alkanes in aqueous systems, II, The accommodation of C12-C36 n-alkanes in distilled water," *J. Am. Oil. Chem. Soc.,* **44,** 969–702.

Peake, E., and Hodgson, G. W., 1966, "Alkanes in aqueous systems. I. Exploratory investigations on the accommodation of C20-C33 n-alkanes in distilled water and occurrence in natural water systems," *J. Am. Oil Chem. Soc.,* **43,** 215–222.

Silverman, S., 1964, "Investigations of Petroleum Origin and Evolution Mechanisms by Carbon Isotope Studies," pp. 92–102 in (Craig, H., Miller, S. C., and Wasserburg, G. J., editors), "Isotope and Cosmic Chemistry," Amsterdam, North-Holland Publ. Co., 578pp.

Smith, H. M., 1966, "Crude Oil: Qualitative and Quantitative Aspects," U.S. Bur. Mines Information Circular 8286, 41pp.

Stanley, T. W., Morgan, M. J., and Meeker, J. E., 1967, "Thin layer chromatographic separation and spectrophotometric determination of benzo(a)pyrene in organic extracts of airborne particulates," *Anal. Chem.,* **39,** 1327–1329.

Swinnerton, J. W., and Linnenbom, V. J., 1967, "Gaseous hydrocarbons in sea water: determination," *Science,* **156,** 1119–1120.

Thomson, R. H., 1965, "Quinones: Nature, Distribution and Biosynthesis," pp. 309–332 in (Goodwin, T. W., editor) "Chemistry and Biochemistry of Plant Pigments," London, Academic Press, 583pp.

Went, F. W., Slemmons, D. B., and Mozingo, H. N., 1967, "The organic nature of atmospheric condensation nuclei," *Proc. Natl. Acad. Sci.,* **58,** 69–74.

Whitehead, W. C., and Breger, I. A., 1963, "Geochemistry of Petroleum," in (Breger, I. A., editor, "Organic Geochemistry," MacMillan, Chap. 7, pp. 248–332.

Cross-references: *Carbon; Chromatography; Geochemistry of Sediments—Ancient; Geochemistry of Sediments—Modern; Organic Geochemistry; Oxidation and Reduction.* Vol. IVB: *Coal; Kerogen; Methane; Natural Gas; Petroleum.*

HYDROCLIMATE—See Vol. II

HYDRODYNAMICS—See Vol. II

HYDROELECTRIC ENERGY—See Vol. VI, WATER POWER

HYDROGEN: ELEMENT AND GEOCHEMISTRY

Hydrogen is the lightest of all the known chemical elements, with an atomic weight of 1.008 on the C^{12} scale of masses. It was first discovered in 1766 by Cavendish, who also devised several methods for its preparation and determined the composition of water. The name "hydrogen," meaning "water former," was given to it by Lavoisier in 1783.

Hydrogen in its free elemental state occurs, terrestrially, in relatively small quantities compared to many other elements. It has been detected in volcanic gases and in gases associated with certain salt beds (outgassing from the earth's interior), as well as in a few meteorites. The earth's lower atmosphere contains trace amounts of hydrogen ranging from 1 part in 15000 to 1 part in a million. The relative concentration of hydrogen in the atmosphere increases with altitude until, at higher altitudes, the escape of the lighter hydrogen atoms takes place rapidly.

Hydrogen is present in substantial quantities on the earth in the form of chemical compounds; for this reason it is the ninth most abundant terrestrially occurring element by mass. The most important such compound is water; clearly, one ninth of the mass of all water in the earth's crust, in the oceans, and in the form of water vapor in the earth's atmosphere is hydrogen. Compounds composed primarily of hydrogen, carbon, oxygen, nitrogen, and lesser abundances of heavier elements form the major constituents of animal and vegetable matter. Studies in organic chemistry have revealed enormous numbers of such compounds, ranging from simple hydrocarbons like methane, CH_4, to the large protein-like virus molecules and the important genetic materials, DNA and RNA.

Astrophysical Properties

It is of interest to consider as well the role of hydrogen in the evolution of stars and the history of our galaxy. Hydrogen is by far the most abundant element in our solar system; it constitutes 92% of the total atoms and 74% by mass of the matter in the sun and planets (Cameron, 1968). In contrast, hydrogen atoms constitute only 15% by number of the atoms forming the earth and its atmosphere and a significantly lower fraction of the mass. This discrepancy can be explained in part by the escape of free hydrogen atoms from the earth's atmosphere. It is also possible that the physical conditions which existed at the time of formation of our planetary system had resulted in a nonuniform distribution of the elements in the matter from which the planets were formed. In this regard, the atmosphere of Jupiter is known to contain large amounts of hydrogen both in its molecular form (H_2) and in the form of methane (CH_4) and ammonia (NH_3). It must be noted, however, that due to the significantly larger planetary mass of Jupiter, the lifetime for the escape of free hydrogen from the Jovian atmosphere is appreciably longer than the corresponding lifetime for the earth.

It is the nuclear properties of hydrogen which dictate its fundamental role in providing the energy output of the sun. At temperatures slightly in excess of ten million degrees, hydrogen is converted to helium in the deep interior of the sun by thermonuclear reaction mechanisms. This fusion process is accompanied by an energy release, per helium nucleus formed, of 26.730 MeV (1 MeV = 1.6 × 16^{-6} ergs). This energy release is sufficient to maintain the present luminosity of the sun for a lifetime in excess of 10 billion years while converting only 10% of the mass of the sun from hydrogen into helium.

Observations of the physical characteristics of stars and gas both in our own galaxy and in other galaxies suggest that hydrogen is universally the most abundant constituent. The question as to whether the primordial composition of our galaxy was entirely hydrogen is still unanswered, and a great deal of research relevant to this problem is currently under way. Although there are some stars whose atmospheres appear to contain no appreciable helium, observations of the oldest unevolved stars in our galaxy reveal that the primordial abundance of helium may have been on the order of 20% by mass. This is consistent with current cosmological theories which predict that a substantial conversion of hydrogen to helium will take place in the expansion of the universe immediately following the initial "big-bang" (Peebles, 1966). The synthesis of elements heavier than helium presumably takes place in stellar interiors throughout the history of the galaxy (Burbidge, Burbidge, Fowler, and Hoyle, 1957; Cameron, 1963; Clayton, 1968).

Isotopes of Hydrogen

There are three known isotopes of hydrogen possessing mass numbers of 1, 2, and 3. The isotopes of mass 1 and 2 occur naturally. Hydrogen of mass number 3 is made artificially by bombardment of other species. These isotopes have been named protium, deuterium, and tritium for the mass numbers 1, 2, and 3, respectively. The chemical properties of these three forms are practically identical with the natural isotopic mixture. Tritium has one property unshared by the other two, namely, radioactivity. It is a beta emitter. These various isotopes react with oxygen to give water in similar type chemical reactions. The waters produced have the formulas H_2O, D_2O, and T_2O, and separately are called protium oxide or light water, deuterium oxide or heavy water, and tritium oxide, respectively. These oxides are chemically identical except for properties that might be dependent on atomic masses, such as bond energies. The molecular weights of these oxides are, respectively, 18, 20, and 22. The various possible mixed isotopic hydrogen water species are generally also present in mixtures of these oxides. The relative amounts of deuterium to light hydrogen in naturally occurring waters is approximately one deuterium atom to 6500 protium atoms. This occurrence ratio seems remarkably constant also in other natural sources such as petroleum, hydrates, and organic matter.

To complicate the hydrogen systems is the phenomena of ortho and para nuclear behavior. If the angular momentum of a hydrogen nucleus is ± 1/2 (expressed in $h/2\pi$ units), then the two hydrogen nuclei in combination to produce the hydrogen molecule are able to combine in two ways, either with their spins additive or with their spins opposed. The molecules with their spins additive, and consequently with their total nuclear spin equal to unity, are said to be the symmetrical ortho form and possess the odd quantum states 1, 3, 5, etc. The molecules with their nuclear spins opposed are the antisymmetric para form, and possess the even quantum states 0, 2, 4, etc., and a total nuclear spin equal to zero.

Since for hydrogen, the para forms occupy only the even rotational energy levels and the ortho form occupies only the odd levels, the ratio of the functions for the even- and odd-numbered rotational quantum numbers, J, gives the ratios of the concentrations, 1:3, of ordinary hydrogen due to the para and ortho forms, respectively, which vary with the temperature. At 20°K the percentage of para is 99.82% and of ortho 0.18. At a sufficiently high temperature, where the number of odd terms equals the number of even terms, the proportions of para- to the ortho-hydrogen in ordinary hydrogen systems will be constant at 1 to 3. However, as the temperature is decreased (as $T \to 0°K$) the energy of all molecules must decrease. To do so, the energy must decrease, level by level, to the lowest level where $J = 0$. But $J = 0$ is an even number; so all forms must tend to crowd into this lowest quantum level of $J = 0$. Ortho-hydrogen has its lowest quantum level at $J = 1$; and thus for the ortho form to attain the $J = 0$ level it must change its nuclear spin from odd to even values. Below 20°K this reversible conversion of o-H_2 to p-H_2 does take place, but there is an appreciable reluctance of o-H_2 to do so.

The rate of conversion of the o-H_2 into p-H_2 can be catalyzed by certain forms of charcoal, by some metals, salts, metal oxides, and substances possessing paramagnetic moments. The conversion can be followed by methods involving measurements of vapor pressure, thermal and electrical property changes, conductivity, and heterogeneous adsorption, etc. At high temperature there is always a 1:3 para-to-ortho mixture. Ortho and para forms of hydrogen do show differences in their total energies and their heat capacities.

Production of Hydrogen

In the laboratory, hydrogen is usually prepared by displacement reactions. The hydrogen in compounds like acids, water, and certain bases is replaced with more reactive elements capable of forming positive cations in the replacement compound. Many of the metals are active enough to replace the hydrogen in acids. Some of the most active of the metals will replace the hydrogen in cold water as well as hot water. Certain nonmetals, like white-hot carbon or silicon, can replace the hydrogen in water in the form of steam. This latter preparation is primarily an industrial one.

A common laboratory method uses mossy zinc. When the zinc is placed in nonoxidizing acids, hydrogen gas is evolved. Hydrochloric acid reacts with zinc to yield hydrogen according to the equation

$$Zn + 2HCl \rightarrow ZnCl_2 + H_2$$

This is typical of the type of reactions by which metals replace hydrogen from acids. The ranking of the metals that will liberate hydrogen from nonoxidizing acids is shown in the so-called "electrochemical" (replacement) series. This is an arrangement of the metals in the order of their decreasing ability to displace the hydrogen in acids. The alkali metals, lithium, cesium, rubidium, and potassium, are the most reactive of the metals, and top the list. They are so reactive that they will replace the hydrogen from cold water so that the heat of reaction often will cause the hydrogen to ignite.

Electrolysis of water will produce hydrogen and oxygen and is of importance in the industrial production of oxygen and hydrogen, and especially important where electric power is unusually cheap. An electric current is passed through water containing an electrolyte, usually dilute sulfuric acid, between electrodes housed in separated compartments to prevent the gaseous products from mixing.

Water gas is made by heating steam and white hot coke (carbon) or white hot carbon compounds (such as methane, or even cellulose or wheat straw). This produces a mixture of gaseous CO and H_2 which together is called water gas. Water gas, because of its combustible components, makes a fuel of fair heating efficiency. Many industries, and even municipalities, have used it for industrial and domestic heating.

Coke oven gas serves as a source of additional industrial hydrogen. Bituminous coal is subjected to destructive distillation in retorts at high temperature. The compounds are thermally broken down to simpler ones and eventually gaseous molecules are formed. The gaseous products are separated by distillation leaving the nonvolatile residue behind as coke. Among the volatile products are found ammonia, hydrogen, hydrocarbon gases, tar, carbon monoxide, carbon dioxide, and nitrogen. A ton of bituminous coal produces on the average about 10,000 ft^3 of coke oven gas, of which about 5000 ft^3 is hydrogen. This hydrogen furnishes over a fourth of the world's production.

Physical Properties

Hydrogen is only one-sixteenth as heavy as oxygen and less than one-fourteenth as heavy as air. One liter of hydrogen gas weighs 0.08987 grams at 0°C and 1 atm pressure. It is so light that it can be poured upward through the air, from one container to another. It has no color, taste, or odor. At standard pressure and room temperature about 2 ml of hydrogen will dissolve in 100 ml of water. Physical properties are listed in Table 1. Certain metals, like palladium and platinum, (especially when finely divided) have the capacity of rapidly adsorbing large quantities of hydrogen. The hydrogen is changed from the diatomic to the monatomic condition by the time the molecules become located on the metal surface. In this monatomic state it is in a highly activated or energized state, and consequently it is chemically highly reactive. If given the correct conditions it is immediately oxidized to water with oxygen from the air. Ordinary hydrogen and oxygen mixtures need to be ignited before interaction.

Atomic hydrogen, once called "nascent" hydrogen, comes from the dissociation of ordinary hydrogen diatomic molecules into its constituent atoms. Increasing the energy of each molecule sufficiently to exceed the bond energy dissociates the molecules into separate atoms. The energy of the H–H bond, 103.2 kcal/mole, is large enough to cause the molecules to be very stable. Thus, the separated hydrogen atoms, when formed, are extremely activated. Hydrogen dissociation can be ac-

HYDROGEN: ELEMENT AND GEOCHEMISTRY

TABLE 1. PHYSICAL PROPERTIES OF HYDROGEN

Atomic number	1
Atomic weight	1.00797
Melting point, 1 atm (°C)	−259.1
Triple point, 1 atm (°C)	−259.1
Boiling point, 1 atm (°C)	−252.7
Critical temperature (°C)	−240
Critical pressure (atm)	12.8
Critical density (g/l)	31.2
Density	
gas, 0°C, 1 atm (g/l)	0.0899
liquid, −253°C (g/l)	70.8
solid, −262°C (g/l)	76.0
Specific gravity, air = 1.0	0.0695
Specific heat	
C_p, 0–200°C (cal/g/°C)	3.44
C_v, 0–200°C (cal/g/°C)	2.46
C_p/C_v, 0–200°C	1.40
Heat of combustion, gross (cal/g)	33.940
Latent heat of vaporization, −253°C (cal/g)	107
Latent heat of fusion, −259°C (cal/g)	13.89
Thermal expansion coefficient (per °C)	0.00356
Thermal conductivity, 0°C (cal/cm^2/sec/°C/cm)	0.00038
Viscosity, 15°C, 1 atm (centipoise)	0.0087
Minimum ignition temperature (°C)	574
Thermal neutron absorption cross section (barns)	0.332
Dielectric constant at 2 × 10^6 cycles/sec	1.000264
Velocity of sound, 0°C (m/sec)	1269.5
Diffusion coefficient, into air, 0°C (cm^2/sec)	0.634

complished by other methods in addition to diffusion into metals like palladium. Atomic hydrogen can be prepared by passing ordinary hydrogen through an electric discharge. Hydrogen atoms are also generated in small, but important amounts, in chemical reactions that are initiated and/or propagated through complicated mechanisms involving multiple steps. The initiating energy could be light, heat, high-energy particles, etc. Generally, however, due to the high bond energy, hydrogen molecules do not become involved in the primary initiation steps in such mechanisms, but rather they propagate the reaction chains by replacement of one of the hydrogen atoms in H_2 by other atom types like halogen. It requires light of a wavelength of 2776 Å to furnish sufficient energy to break the bond in the hydrogen molecule.

Second only to helium, hydrogen is the most difficult of all gases to be liquefied. The boiling point of liquid normal hydrogen is 20.40 ± 0.2°K. (At present there is a successful laboratory refrigerating unit that uses liquid-gaseous hydrogen as the refrigerant. It operates continuously at about 24°K.) Upon rapid evaporation of the liquid, the temperature of the residual liquid rapidly decreases and eventually begins to solidify to a snow-white solid which melts at 13.95°K. The critical temperature of liquid hydrogen is 33.2°K with an accompanying critical pressure of 12.8 atm. Hydrogen has its triple point at 13.95°K, and this triple point is also its melting point.

An interesting property of hydrogen, demonstrating its extreme lightness, is the increase in the pitch of reed instruments, including the human voice larynx, when the surrounding air is replaced by this (or other) light gas. The frequency of vibration of such reeds varies inversely as the density of surrounding gas. This gas, because of its extreme lightness, has been used for inflating balloons, but is not favored because of the extreme combustibility of hydrogen in the presence of oxygen. The lifting power of 500,000 ft^3 of hydrogen at 0°C and 760 mm is 38,000 lb. Hydrogen gas has a density of 0.08995 g/l at 0°C, 760 mm in a latitude 45° at sea level, and one gram will occupy 11.117 l at STP. Liquid hydrogen has a density of 0.07 g/cm^3 at −253°C, whereas the density of the solid is 0.076 g/cm^3 at −262°C.

Chemical Properties

Many of the more active metals react directly with hydrogen, forming hydrides. The Group IIA metals, especially beryllium and magnesium, react so violently with hydrogen that they burst into visible flame. The alkali metals are less energetic in their direct union with hydrogen.

The metal hydrides are vigorous reducing agents due, in part at least, to the ionic nature of compounds in which the hydride ion carries a negative one charge.

Hydrogen is capable of forming many compounds with nitrogen. Ammonia, NH_3, is made industrially by direct union of the nitrogen gas obtained by the most economical methods available. The reaction requires a catalyst. Especially in the petroleum industry in the area of petrochemicals, hydrogen is united with many organic compounds. Such synthetic reactions involving hydrogen are called hydroforming.

Bonding of Hydrogen

The bonding in hydrogen and hydrogen-containing molecules represents the three bond types in which the single electron of the hydrogen atom and the two electrons of the hydrogen molecule can participate. The hydrogen atom has a structure consisting of a nucleus surrounded by a $1s$ electron. This is the atomic structure of the protium, deuterium, or tritium atoms. This $1s$ electron is basic in the bond formation of the three bond types, namely, *(1)* bonding through loss of the $1s$ electron, *(2)* bonding through the acceptance of electrons from another atom to form an electron pair, and *(3)* the procurement of an extra electron to give a $1s^2$ structure as in the negative hydride ion, H^-.

When the $1s$ electron of a hydrogen atom is lost to another atom (bond type 1), a bare proton, or the H^+ ion remains. This H^+ ion is about 10^{-6} times the radius of ordinary atoms. The mass of the H^+ ion is essentially the same as that of an H atom, but it has a charge of plus one. The large ratio of charge to mass in such a small volume, makes its polarizing effect on electron clouds surrounding other neighboring atoms so large that opposite ionic-like charges are established. This separation of charges then gives rise to ionic bonds or to dipole attractions. To remove the electron from a hydrogen atom in the gas phase requires about 313 kcal/mole which is so large a requirement that seldom can the H^+ ion form outside of solutions of protonic solvents which can solvate the hydrogen ion. The solvation process then furnishes the extra energy for the solvates to form. Approximately 268 kcal/mole of H^+ ions are required for the process of solvation to be complete. Water, concentrated H_2SO_4, and liquid ammonia are typical protonic solvents. Such solvents undergo self-ionization within their pure liquid states.

The solvated hydrogen ion or *hydronium* ion, is the acidic species in water solutions of acids. The concentration of the H_3O^+ is often expressed in units of pH invented in 1909 by S. P. Sorenson. Sorenson defined pH as pH = $-\log_{10} C_{H^+}$. C_{H^+}, here, is now replaced by the symbol $[H_3{}^+O]$ where the concentration of the hydrated proton is measured in moles per liter. Applying this to pure water which undergoes self-ionization

$$H_2O + H_2O \rightleftarrows H_3O^+ + (OH)^-$$

the ion product is $[H_3{}^+O] \times [OH^-] = 10^{-14}$ at 25°C, and since one H_3O^+ is produced for each $(OH)^-$ then $[H_3{}^+O] = [OH^-] = 10^{-7}$ moles per liter. From this value the pH of pure water is 7. Solutions of lower pH are acid; those of higher pH are basic. Strong acids, because they are assumed to be 100% ionized, lower the pH of their water solutions toward zero. Strong bases raise the pH of their water solutions toward 14.

Hydrogen forms many bonds with many elements, including itself, by the sharing of its electron with an electron from another atom. The $1s$ electron with its + or − spin can pair off with an electron of another atom possessing the opposite spin. The stable electron pair acts as the bonding agent between the two atoms. This is the typical normal covalent bond. The $1s$ orbital of the hydrogen atom is now full at $1s^2$, and the orbital of the shared electron is also full. Such electron pair normal covalent bonds are numerous in the field of organic chemistry. Methane, CH_4, and ethane, CH_3CH_3, are good examples of this type of bonding. Small inorganic molecules often are bonded by the shared pair as in ammonia and HCl. A majority of the bond energy of HCl is probably due to covalent shared pair bonding.

The molecule of H_2 is probably a mixed bonded structure. Each of the hydrogens contribute their $1s$ to the pair and a covalent bond results. But because of resonance between the possible bond structures, the two electrons could, and do, find themselves at times on the same hydrogen atom at the same time. Then ionic bonds are formed between the electron-deficient atom, which is positive, and the other atom, with the extra electron, which is negative. Thus in such a homonuclear molecule as H_2, the bonding is not pure one class or another.

Another bonding method that the proton can undergo is coordinate covalent bonding. Here the H^+ ion with its $1s$ shell completely empty can accept an already formed electron pair from another atom. This establishes a new electron pair bond, very similar to a normal covalent pairing, but different in its method of formation. When the proton becomes solvated in water it shares an electron pair from an oxygen in a water molecule to form H_3O^+

$$H^+ + :\overset{..}{\underset{..}{O}}:H \rightleftharpoons H^+ :\overset{..}{\underset{..}{O}}:H = H_3O^+$$
$$\phantom{H^+ + :\overset{..}{\underset{..}{O}}:}H \phantom{\rightleftharpoons H^+ :\overset{..}{\underset{..}{O}}:}H$$

This is the method of coordination.

Hydrogen reacts with multiple-bonded molecules to produce saturated molecules. Such *hydrogenation reactions* usually require a catalyst along with careful temperature and pressure control to make the process efficient. Mineral, plant, or animal oils are all subject to hydrogenation in various industrial operations.

Petroleum (mineral) oils in the presence of catalysts like copper or nickel, or metallic mixtures, react by opening the double or triple bonds and adding hydrogen across these bonds. This multiple bond saturation changes the physical and chemical properties. The viscosity and boiling points increase. Very viscous or hard oils are produced which can then be processed into lubricants, or can be tailor-made to fit specialized specifications. Finely pulverized coal can be hydrogenated to produce a variety of hydrocarbons. In Europe this reaction is used to produce motor fuel.

The hydrogenation of edible oils and fats is performed by essentially the same type chemistry. Soybean, fish, whale, peanut, cottonseed, hog, and beef oil or fat upon hydrogenation produce the edible fats.

A very unusual bond, involving hydrogen atoms, is frequently found where a single atom of hydrogen apparently is bonded between two electronegative atoms. Oxygen atoms commonly are found participating in a bond where the hydrogen atom is bonded between two oxygens. Fluorine forms these peculiar bonds with hydrogen as a bridge. Such bonds are found in HF and in the $(FHF)^-$ anion of KHF_2. Hydroxyl and carbonyl groups are often involved in such hydrogen bonds, and thus become of special importance in many organic and biological compounds and reactions.

The importance of hydrogen bonds lies in their ability to tie up electronegative atoms in unexpected rings, chains or matrix of bonded structures not normally anticipated from the graphical formula. Although the energy of the bond is about 5 kcal/mole, it is sufficient to cause these unusual structures to be present. Ice shows such hydrogen bonding, especially if the ice is cooled to low temperatures. The water molecules form a nearly complete coordination lattice structure with the hydrogen bonds well formed and maintained. Many liquids are molecular associations of several molecules held together by hydrogen bonds.

A. C. ANDREWS
J. W. TRURAN

References

Andrews, A. C., 1968, "Hydrogen," in (Hampel, C. A., editor), "The Encyclopedia of the Chemical Elements," New York, Reinhold Book Corp.
Burbidge, E. M., Burbidge, G. R., Fowler, W. A., and Hoyle, F., 1957, "Synthesis of the elements in stars," *Rev. Mod. Phys.,* **29**, 547.
Cameron, A. G. W., 1963, "Nuclear Astrophysics," Yale Univ. lecture notes.
Cameron, A. G. W., 1968, "Abundances of the Elements in the Solar System," in (Ahrens, L. H., editor), "Origin and Distribution of the Elements," Oxford, Pergamon Press.
Clayton, D. D., 1968, "Principles of Stellar Evolution and Nucleosynthesis," New York, McGraw-Hill Book Co.
*Farkas, A., 1935, "Orthohydrogen, Parahydrogen and Heavy Hydrogen," London, Cambridge
Laidler, K. J., 1950, "Chemical Kinetics," New York, McGraw-Hill Book Co.
Univ. Press.
Pauling, L., 1943, "Nature of the Chemical Bond," Ithaca, N.Y., Cornell Univ. Press.
Peebles, P. J. E., 1966, "Primeval helium abundances and the primeval fireball," *Phys. Rev. Letters,* **16**, 410.
Pegram, G. B., Huffman, J., and Urey, H. C., 1936, *Phys. Rev.,* **49**, 883.
Sidgwick, N. V., 1950, "The Chemical Elements and Their Compounds," Vol. I, Oxford, Clarendon Press.
Urey, H. C., Brickwedde, F. G., and Murphy, G. M., 1932, *Phys. Rev.,* **39**, 864; *ibid.,* **40**, 1.
Verniani, F., 1966, "The total mass of the Earth's atmosphere," *J. Geophys. Res.,* **71**(2), 385–391.
Washburn, E. W., and Urey, H. C., 1932, *Proc. Natl. Acad.,* **18**, 496.

* Additional bibliographic references may be found in this work.

Cross-references: *Ammonia; Bonding; Catalysis; Deuterium; Electron Capture; Electronegativity; Elements: Planetary Abundances and Distribution; Helium; Hydrocarbons; Outgassing of the Planet Earth; Oxygen; Radioactive Isotopes; Tritium; Gases—Volcanic Water.* Vol. II: *Atmosphere; Galaxy; Jupiter; Nucleogenesis; Solar System; Stars and Stellar Interiors; Sun.* Vol. VI: *Methane.*

HYDROGEN SULFIDE—See Vol. IV B

HYDROGEOLOGY

Definition of Hydrogeology

The meaning of the term hydrogeology is not fixed within modern science. Some researchers use the term in a broad sense to include the occurrence, geochemistry, and geologic activity of all water on or below the surface of the land areas of the world. Internationally, however, a more restricted use is now favored and is here followed, employing hydrogeology to signify the study of *groundwater* with special emphasis on its geologic occurrence, chemistry, and migration (see discussion by Davis and De Wiest, 1965).

Origins of Groundwater

Ancient philosophers generally regarded groundwater, particularly spring water, as the end result of the regional migration of seawater or as the product of the condensation of vapors from deep within the earth. Although some chemical changes do take place in seawater as it migrates through permeable media and condensation of vapor takes place in nature, seawater is not transformed into fresh water and the amount of condensed vapor is far too small to account for even a small fraction of the groundwater found throughout the earth. Researchers now agree that almost all potable groundwater comes from snow, rain, river, or lake waters migrating into permeable zones in the subsurface (see *Hydrologic Cycle*). The depth of active fresh-water circulation is generally less than two kilometers (1.24 miles). Saline and brackish groundwater and some potable water originate as water enclosed between particles of sand, silt, and other sediments as they are deposited in lakes and in oceans. This trapped water is called *connate water* (q.v.). Although it migrates slowly and undergoes some chemical changes, connate water remains isolated from surface water and retains many of the gross chemical characteristics of the original lake or ocean water. A minor amount of saline water is given off by molten rock material, or magma. This water contributes to hot springs, geysers, and fumaroles. Magmatic water has only been detected in regions where there are active volcanoes or where there have been volcanoes within late geologic time. Another source of groundwater is from sedimentary rocks that have been metamorphosed. The differential pressures as well as the high temperatures that constitute part of the process of metamorphism will drive off almost all water from the rocks. The water driven off by metamorphism, or metamorphic water, in a few places finds its way to the surface as saline springs in regions of current or geologically recent mountain building.

Scope of Topic

Hydrogeology can be subdivided into four major activities, namely the study of (*1*) the chemical and physical nature of groundwater, (*2*) the geologic work accomplished by groundwater, (*3*) the relation between rock types and the occurrence of groundwater, and (*4*) the geological, hydrogeological, and geophysical techniques employed in the development of groundwater resources. In addition, the hydrogeologist commonly becomes involved with legal, political, economic, and engineering of groundwater development problems. Only the scientific aspects of hydrogeology will be considered in this article.

Physical and Chemical Nature of Groundwater

Water Temperature. Well water seems cold during hot summer days when water use is at a maximum. This is deceptive inasmuch as the well water almost everywhere is actually warmer than the mean annual air temperature. The reason for the higher temperature is due to the fact that earth temperatures increase with depth, and water tends to be at the same temperature as its surroundings.

Much, if not locally all, of the heat from the earth is transferred by conduction. Radiation is negligible near the earth's surface and mass transport is only important in zones of rapid, nonhorizontal groundwater flow. Fourier's law covering the conduction of heat states that flow is a linear function of the thermal conductivity, cross-sectional area, and thermal gradient. In nearly homogeneous material the thermal conductivity is almost constant so that the increase in temperature with depth, or the geothermal gradient, is constant also, provided the heat flow is normal to the ground surface. The geothermal gradient, and consequently the increase in water temperature per depth, will commonly vary from a maximum of about 1°C for every 20 m of depth in some alluvium to a minimum of about 1°C for every 80 m of depth in some consolidated rocks. Thus, groundwater from a depth of 325 m (roughly 1000 ft) might be expected to have temperatures of from 4 to 16°C higher than the temperature of the ground surface.

Groundwater temperatures may depart widely from values that might be predicted on the basis of a linear geothermal gradient. Some reasons for departures are (*1*) local sources of heat such as large heated buildings, warm waste water, and molten volcanic rocks with associated steam, (*2*) local departures from vertical heat flow caused by mountainous topography, heterogeneous rocks, local cooling at the surface such as exists along many coastlines, and (*3*) mass transfer of heat due to rapid recharge or discharge of groundwater. Areas of rapid, natural groundwater recharge are areas of unusually low temperatures. Areas of rapid, natural groundwater discharge are areas of unusually high water temperatures.

Most hot springs, with water temperatures near boiling, are associated with active or recently active volcanoes. Warm springs, with water temperatures from 5 to 25°C higher than the mean annual temperature, mostly owe their origin to the slow heating of water as it migrates into the earth followed by a rapid rise of the water along very permeable zones in the earth. The upward flow is so fast that the water does not have a chance to cool.

Nuclides in Water. Water is mostly made up of the stable nuclides O^{16} and H^1, however O^{17}, O^{18}, and H^2 are also present in significant amounts. Under favorable circumstances, variations of the heavier nuclides can be used to help interpret the genesis of certain waters. For example, large concentrations of O^{18} and H^2 may suggest that groundwater originated as entrapped lake water which had undergone a long history of partial evaporation thus concentrating the heavier nuclides of the water. The only unstable nuclide that is an essential part of water is tritium (H^3). This nuclide decays fairly fast (half-life 12.4 yr) so that it forms an effective tracer for water movements. Tritium is only introduced into groundwater from the surface since it is produced partly by natural cosmic ray reactions in the air and partly by nuclear explosions. From 1954 to the present, nuclear explosions have been by far the most important source of tritium. If groundwater is essentially free of tritium, it means that the water has not had contact with the atmosphere for at least three or four decades. Rainfall from the past decade can be detected easily in the subsurface inasmuch as tritium concentrations are from 10 to 1000 times natural tritium concentrations that existed in rainwater prior to 1954.

Major Dissolved Constituents. Most groundwater contains only seven major dissolved constituents. These are Ca, Mg, Na, SO_4 HCO_3, and Cl which are ionized as well as H_4SiO_4 which is essentially nonionized under normal conditions. The rather restricted number of constituents may seem a bit surprising in light of the more abundant elements in the earth's crust. The elements Al, Fe, Ti, and Mn which are abundant in rocks are virtually absent in normal groundwater (see Table 1).

Natural controls that limit dissolved constituents in groundwater are given in Table 2. Unusual waters are found, however, that depart rather widely from the average conditions described in Tables 1 and 2. Low pH values will allow the presence of more than 15 ppm of Fe, local sources of contamination will produce shallow groundwater with as much as 600 ppm NO_3, and various gases can be dissolved in concentrations of more than 20 ppm if they occur in abundance in the subsurface.

Minor and Trace Constituents. Most minerals are soluble enough so that modern techniques of chemical analyses can detect small amounts of the mineral components in water. Concentrations are commonly in the range of a few parts per billion. For example, titanium is probably present in groundwaters in concentrations of from 0.1 to 2.0 parts per billion, lithium in concentrations from 1.0 to 500 parts

TABLE 1. COMPOSITION OF NATURAL WATERS (ppm)

Element[a]	Seawater	Potable Groundwater (Average Values for the U.S.)[b]	Groundwater of Excellent Chemical Quality (Salco, Alabama)[c]	Oil-field Brine (Plaquemines Parish, La.)[c]
Si (H_4SiO_4)	2.0	18	9.4	16
Al	0.01	0.04	—	3.1
Fe	0.01	0.07	0.14	110
Ca	400	45	0.6	9210
Na	10,500	35	2.5	63,900
K	380	2.5	0.5	869
Mg	1350	11	0.2	1070
Ti	0.001	0.001	—	—
P (PO_4^{3-})	0.07	0.03	—	—
Mn	0.002	0.02	—	30
F	1.3	0.2	0.0	1.4
S (SO_4^{2-})	2650	36	2.5	153
C (HCO_3^-)	140	215	6	115
Cl	19,000	16	2.8	124,000

[a] Listed in order of abundance in the earth's crust. Actual ion or compound present and analyzed is given in brackets.
[b] Davis and DeWiest, 1966.
[c] U.S. Geol. Survey Analysis.

HYDROGEOLOGY

Table 2. Factors Controlling Natural Composition of Groundwaters

Factors	Constituents Most Commonly Affected	Comments
Low solubility of one or more natural compounds		
Temperature variations most important	H_4SiO_4, various gases	H_4SiO_4 is reported as SiO_2 in most analyses; solubility increases with temperature; solubility of gases decrease with temperature
Partial pressure of CO_2 a critical factor	Ca^{2+}, Mg^{2+}, HCO_3^-	High partial pressure of CO_2 favors dissolution of Ca and Mg compounds; Fe^{2+} is the most abundant if Eh and pH are in their normal ranges; Mn, U, and other less-abundant constituents are also affected by changes in oxidation states
Oxidation state of the ions, or Eh–pH relationships most important	Fe^{2+}, Fe^{3+}	
Surface phenomena of clays, colloids, and organic components		
Exchange of cations	Na^+, K^+, Ca^{2+}, Mg^{2+}	All minor and trace constituents that occur as cations are also affected; abundant ions in solution will displace other ions on clay surfaces; reactions are reversible if ion concentrations are changed
Exchange and sorption of anions	I^-, PO_4^{3-}	
Membrane effects	All constitutents	High concentrations developed behind membranes
Biological activity		
Bacteria	SO_4^{2-}, H_2S, NO_3^-, NH_4^+	Bacteria in the subsurface will reduce SO_4^{2-} to H_2S and NO_3^- to NH_4
Plants	HCO_3^-	Plant roots will take up or give off CO_2, depending on the local physical conditions; much CO_2 is also produced from the decay of organic material
Animals	NO_3^-	Nitrates are one of the products of the decay of animal wastes

per billion, and germanium from 0.1 to 10 parts per billion. Much work is currently being done on analyses of minor and trace constituents but few general summaries of results have been published.

Interest in minor and trace constituents has generally been centered on either the possibility of using water analyses as a guide to mineral deposits or as a guide to the presence of constituents that affect the health of animals, humans, or plants. Just a few parts per million of lithium and boron have been found harmful to many plants. Iodide in concentrations of less than 1.0 ppm will inhibit the formation of goiters. Concentrations of cadmium of less than 0.1 ppm may have adverse effects on humans and some animals.

Radionuclides. In addition to H^3 which forms a part of the water molecule, many other radioactive nuclides, or radionuclides, have been found in water. Three rather distinct sources of these radionuclides exist; these are (1) primordial radionuclides and their daughter products, (2) products of natural activation, mostly in the atmosphere, and (3) artificially produced radionuclides. Some of the radionuclides of importance to hydrogeology are listed in Table 3.

A number of uses have been made of radionuclides in hydrogeologic studies. Radon, a gas, is moderately abundant in groundwater but is lost rapidly to the atmosphere once the water reaches the surface. The points at which groundwater enters streams can, therefore, be correlated with locally high radon contents of streams. Both tritium, H^3, and C^{14} can be used to give the approximate length of time that water has been out of contact with the atmosphere. Artificially produced radionuclides such as Co^{60} and Br^{82} can be introduced into

TABLE 3. RADIONUCLIDES OF INTEREST IN HYDROGEOLOGY

Nuclide	Origin	Half-Life	Relative Biological Hazard
C^{14}	Cosmic-ray activation of atmospheric gas	5600 yr	Moderate
H^3	Nuclear fusion, also some from natural cosmic-ray activation	12.4 yr	Low
I^{131}	Nuclear fission, entirely artificial	8.04 days	High
Ra^{226}	Decay product of uranium	1620 yr	Very high
Rn^{222}	Gaseous decay product of radium	3.83 days	Moderate
Sr^{90}	Nuclear fission, entirely artificial	29 yr	Very high
U^{238}	A primordial radionuclide, entirely of natural origin	4.5×10^9 yr	Low

groundwater and will serve as effective tracers of direction and rate of groundwater movement.

Geologic Work of Groundwater

Dissolution of Rock-Forming Minerals. Perhaps one of the most spectacular aspects of hydrogeology is the study of caverns (See Vol. III, *Speleology*). The work of groundwater is everywhere evident. The initial solution of calcite and dolomite within the rock commonly produces long interconnected caverns that ultimately are modified by roof collapse and deposition of dripstone. If collapse of caverns reaches the surface or if dissolution by water at the soil-rock interface advances irregularly, a rough hummocky topography with numerous depressions is formed.

Less spectacular but no less important is the general formation of small interconnected openings in limestone and dolomite or the enlargement of closely spaced fracture zones in many types of rocks. The selective dissolution of part of the initially solid rock can produce zones which will transmit water, oil, or gas through long distances. More commonly, however, such zones are restricted to a few cubic meters of rock in any single locality.

The dissolution of calcite, dolomite, gypsum, and halite in sedimentary rocks produces the most obvious effects. Some of the silica-rich minerals can also dissolve in ordinary groundwater, so the long-term effects of water circulation must be to increase permeability even though the rocks may be almost chemically inert by ordinary standards.

Deposition of Minerals. Groundwater may become supersaturated with respect to many common minerals. Under proper environmental conditions these minerals will precipitate. For example, warm water forced out of compacting clays may migrate upward into cooler horizons. Silica which is highly soluble only at moderate to high temperatures will be precipitated in the pore spaces of the cooler rocks. Common cementing minerals transported by groundwater are quartz, calcite, limonite (a complex of several minerals), opal, and dolomite.

Springs coming to the surface or entering bodies of water may be supersaturated owing to changes in temperature of the water or partial pressures of various gases originally present in the water. Large masses of travertine are deposited by spring water which may be near saturation with respect to $CaCO_3$. The water becomes saturated as small amounts of CO_2 are lost to the atmosphere.

Economic geologists also study the migration of underground water with the hope that more can be learned about the transportation and deposition of ore deposits. Several elements, notably lead, zinc, mercury, antimony, copper, uranium, and manganese, can be transported in significant amounts by groundwater that is relatively near the earth's surface, although both the salinities and temperatures of these waters can be well above those of ordinary potable water (see Vol. IV B, *Mineral and Ore Deposits*).

Mechanical Erosion. Groundwater does not normally remove particulate material by mechanical means. This is owing to the slow velocity of groundwater movement. Exceptions are found in some caverns which are modified by swiftly flowing streams and at or near springs where the water is concentrated in large openings.

Relation between Rock Types and Groundwater Occurrence

Metamorphic and Plutonic Igneous Rocks. Metamorphic rocks such as slate and gneiss and plutonic igneous rocks such as granite and gabbro will not be able to transmit significant amounts of groundwater unless the rocks have been fractured or modified by weathering. Fracturing by faulting, landslides, and other causes will produce cracks with much wider openings near the ground surface than at depth where confining pressures of overlying rocks are effective. Also, weathering is a surface phenomenon that commonly penetrates only 10–100 m into the rocks. As a consequence, almost all water wells in metamorphic and plutonic igneous rocks obtain water from depths of less than 100 meters. Deeper wells are only justified where open fault zones are known to exist at greater depths. Experience with mines and tunnels has shown that large amounts of water may be encountered in these fault zones but that the chances of penetrating such zones without proper geological guidance is almost negligible.

Groundwater from shallow wells in metamorphic and plutonic igneous rocks is almost everywhere of excellent chemical quality for human use. Total dissolved solids are less than 1000 ppm and commonly less than 200 ppm. Hardness is low to moderate and other potentially troublesome constituents are all but lacking. Nevertheless, where rocks are thoroughly fractured, openings may be wide enough to allow rather easy contamination of water supplies from artificial sources.

Volcanic Rocks. Basalt, rhyolite, and other volcanic rocks are generally much more permeable than plutonic igneous rocks. Indeed, in areas of recent volcanic rocks, the hydrogeologist concentrates his attention on the impervious zones as these must be present to block the rapid subsurface drainage of the groundwater. In areas without natural barriers, wells have penetrated more than 1000 m of permeable rock without encountering saturated zones.

Features that increase the permeability of volcanic rocks are cooling cracks, bubble holes, molds of trees and other debris, buried stream gravel, lava tubes, and most important, open spaces between successive lava flows. These open spaces owe their origin to the fact that fresh lava tends to cool at its base so rapidly that it solidifies before it has a chance to flow into all the minor cracks and depressions of the underlying, older lava flow.

Impermeable zones in volcanic rocks are fine-grained volcanic ash, buried soils between lava flows, thick lava flows, and vertical dikes. Groundwater is either perched on top of the more-or-less horizontal zones or is blocked behind intersecting vertical dikes that are tabular in form.

Fragmental material cast out from a volcano may mix with mud and soil to form a porous but rather impermeable deposit. Such materials only become permeable where the fine sediments have been washed out as the material is transported in local streams.

Quality of water from volcanic rocks is good except near areas of active volcanoes or hot springs. Generally the water is somewhat harder and total dissolved solids are greater than in metamorphic or plutonic igneous rocks. Water withdrawn from basalt in many coastal areas has been contaminated by seawater that moves readily into the rocks.

Sedimentary Rocks. Groundwater occurrence in sedimentary rocks can be predicted with greater accuracy than in igneous or metamorphic rocks. The reason for the greater predictability lies in the fact that most sedimentary rocks have uniform characteristics over many miles or even hundreds of miles and the water-bearing nature of the rocks is related to the rock itself rather than fractures or weathering zones that may be almost randomly distributed.

Of the sedimentary rock types, shale will transmit the least water although it may hold large amounts in minute openings. Some shale is brittle and may become fractured; in which case water will pass rather easily through the fractures.

Sandstone is generally able to transmit significant amounts of water unless it is firmly cemented so that interconnected pore space is absent. Famous flowing wells, or artesian wells, of eastcentral Australia and northcentral United States obtain their water from sandstone. Initial pressures encountered by the artesian wells in the United States were so large that small flour mills and early electrical generators were driven by the underground water.

Geologically young limestones are among the best water-producing rocks known. Many individual wells in the young limestones of Florida yield more than four million gallons each day. Older limestones tend to be much more compact and yield water only from solution openings. Median yields of wells in very old limestones are more than two orders of magnitude less than wells in young limestones.

Some of the groundwater obtained from sedimentary rocks is actually a brine that is almost saturated with respect to NaCl and less commonly $CaCl_2$. Subsurface layers of salt are partially dissolved to yield some of the brine. This origin, however, seems to apply to only a small part of the brines. Many researchers postulate that slowly moving water molecules will eventually pass through shales but that hydrated ions will be left behind to accumulate in the deeply buried sandstones and limestones.

Brines have mostly been formed in sedimentary rocks that are more than 100 million years old. Younger sedimentary rocks commonly contain water with about the same salinity as ocean water. This is reasonable because the majority of sedimentary rocks were first formed in the ocean. Migration and interaction with the local environment, however, has modified the original ocean water. Specifically, more silica, calcium, and bicarbonate goes into solution but magnesium, sulfate, and some sodium is removed.

Potable groundwater is found in sedimentary rocks that were originally deposited in fresh-water lakes and streams or in rocks that have had their saline water flushed out by circulating fresh water. In some areas such as northeastern Kansas, the saline water has been flushed out to depths of less than 20 meters; in other areas, such as southern Texas, the removal of saline water has been effective to more than 1000 meters.

Nonindurated Sediments. Sand, gravel, silt, boulders, and other nonindurated sediments are found at the surface in most of the world. In many regions these sediments form a thin veneer which only contains water temporarily after heavy precipitation. In other regions the sediments have thicknesses that exceed 1500 meters. Some of the world's largest reserves of potable groundwater are found in these regions of thick sediments. In the United States, the Gulf Coastal Plain is potentially the most important, although the highest intensity of development has been in the San Joaquin Valley of California. Much smaller developments that are exceedingly important economically have occurred in the alluvial valleys of the major rivers in central United States. Industrial complexes in St. Louis, Peoria, Kansas City, Louisville, and other cities are heavily dependent on groundwater from river alluvium.

Water from nonindurated sediments is commonly potable except in desert regions where evaporation has concentrated salts in the central parts of closed basins or in coastal regions where connate water of marine origin may be present. Buried organic debris may produce reducing conditions favorable for the mobilization of large amounts of iron or manganese in the water. Although not toxic, such water will have a disagreeable taste and will stain clothing and bathroom fixtures.

Development of Groundwater

Exploration for Groundwater. Five questions should be answered in the exploration for groundwater, these are: *(1)* Is water present? *(2)* What is its biological and chemical quality? *(3)* Are the rocks able to transmit the water easily? *(4)* Is a significant amount of water in storage? *(5)* Are local sources of groundwater recharge available so that supplies will not be eventually exhausted? Various methods of exploration will give varying degrees of refinement to the answers to these questions. Ideally, all methods of exploration should be interwoven. In practice, economic considerations as well as availability of personnel and equipment limit the exploration effort.

Geologic and hydrologic exploration are the fastest as well as the cheapest methods available. Mapping rock types, measuring spring discharges, tracing the positions of faults, and other field activities may be able to cover several square kilometers each day. The ability of the rock to transmit fluids, or its permeability, and its total pore space, or porosity, are estimated (see Table 4). The positions of water-bearing horizons, or aquifers, are then determined. Water quality and the location of hidden sources of recharge, however, are difficult to estimate.

Geophysical methods of groundwater exploration depend on accurate measurement of the resistivity of earth materials to electrical

TABLE 4. WATER-BEARING PROPERTIES OF COMMON EARTH MATERIALS

Relative Porosities (in decreasing order)	Ease of Water Movements (Permeability) (in decreasing order)
Fresh-water clay	Gravel
Volcanic tuff	Basalt, geologically young surface flows
Shale, poorly indurated	Sand
Sand	Limestone, with solution opening
Gravel	Sandstone, poorly indurated
Sandstone, poorly indurated	Volcanic tuff
Basalt, geologically young surface flows	Limestone, dense
Limestone, compact with solution openings	Fresh-water clay
Sandstone, thoroughly cemented	Sandstone, thoroughly cemented
Limestone, dense	Shale, poorly indurated
Granite, gneiss, and other metamorphic and plutonic igneous rocks	Granite, gneiss, etc.

currents, small variations of the gravity field, response of rocks to vibrations from explosions (seismic methods), variations in the earth's magnetic field, and heat radiation given off by the earth's surface. All methods are most successful where natural contrasts in physical properties are the greatest and where geologic structure is relatively simple. In general, electrical resistivity methods and seismic methods have proved most useful. Recently, infrared surveys of coastlines and rivers have been able to pinpoint the presence of subaqueous springs by sharp contrasts in water temperature indicated on scan images. Also, temperatures in shallow test holes have been reported to correlate with heat redistribution by rapidly flowing waters in aquifers.

Test drilling is certainly the most diagnostic of all exploration techniques. It is, unfortunately, also the most expensive and the slowest. Much of the time and cost can be eliminated, however, if drilling sites have been located by geological and geophysical means. Once drilling is undertaken, numerous modern methods of bore-hole geophysics are available to help interpret groundwater quality as well as permeability and porosity of the enclosing rocks.

Production of Groundwater. The hydrogeologist commonly assists the engineer in the design and testing of wells. Aquifers to be tapped by the well are chosen, hardness and mechanical stability of the geologic units are estimated, water quality is estimated, and recommendations for testing procedures are given (See Vol. VI, *Geohydrology*).

STANLEY N. DAVIS

References

Back, W., and Hanshaw, B. B., 1965, "Chemical Geohydrology," in "Advances in Hydroscience," Vol. 2, New York, Academic Press, pp. 49–109.
Davis S. N., and DeWiest, R. J. M., 1966, "Hydrogeology," New York, John Wiley & Sons, 463pp.
Landström, O., and Wenner, C. G., 1965, "Neutron Activation Analysis of Natural Water Applied to Hydrogeology," Univ. Stockholm.
McGuinness, C. L., 1963, "The role of ground water in the national water situation," *U.S. Geol. Surv. Water-Supply Paper,* **1800,** 1121pp.

Cross-references: *Aquifer; Artesian Water; Connate Water; Elements; Geophysical Methods for Hydrologic Search; Groundwater; Hydrologic Cycle; Hydrology, Dissolved Substances; Hydrology, Limestone Terrain: Hydrology, Subsurface Waters; Hydrosphere; Interstitial Water in Sediments; Medicinal Springs; Natural Brines; pH-Eh; Radionuclides: Cosmic Ray Produced; Seawater, Chemistry; Silica Solubility; Springs; Trace Elements.* Vol. I: *Radionuclides in Oceans and Sediments.* Vol. III: *Limestone Caves; Speleology; Stalactite and Stalagmite.* Vol. IVB:

Mineral and Ore Deposits; Vol. V: *Fumarole; Igneous Rocks; Metamorphic Rocks; Volcanic Rocks.* Vol. VI: *Geohydrology.*

HYDROLOGIC ASSOCIATIONS—See Vol. VI

HYDROLOGIC CYCLE

General Concepts

The hydrologic cycle (or water cycle) is the never-ending *circulation* of water and water vapor over the entire earth (Chow, 1964). This circulation penetrates the three parts of the total earth system: the *atmosphere* (the gaseous envelope above the hydrosphere), the *hydrosphere* (the water covering the surface of the earth), and the *lithosphere* (the solid rock beneath the hydrosphere). Solar energy and gravity provide the energy for the circulation.

The hydrologic cycle has no beginning or end. Water is evaporated from the oceans and the land, with the former providing the largest amounts. The evaporated water is carried into the atmosphere, usually drifting tens to hundreds of miles before being returned to the earth as rain, snow, hail or sleet (McDonald, 1962). This precipitated water may be intercepted and transpired by plants, may run over the ground surface and into streams, or may infiltrate into the ground. A considerable part of the intercepted and transpired water and the surface runoff returns to the air by evaporation. The infiltrated water may seep down to deeper zones of the earth, forming groundwater storage which may later flow out to streams as runoff and finally evaporate into the atmosphere to complete the hydrologic cycle (Chow, 1964).

Thus, the hydrologic cycle is involved with the processes of evaporation, precipitation, interception, transpiration, infiltration, seepage, storage, and runoff.

The quantity of water going through the hydrologic cycle during a given period for an area can be evaluated by the hydrologic equation (or continuity equation):

$$I - O = \Delta S$$

Where I is the total inflow of surface runoff, groundwater and total precipitation; O is the total outflow, which includes evapotranspiration, and subsurface and surface runoff from the area; ΔS is the change in storage in the various forms of retention and interception.

The hydrologic cycle may be illustrated qualitatively (Fig. 1), descriptively (Figs. 2 and 3), and quantitatively (Fig. 4).

HYDROLOGIC CYCLE

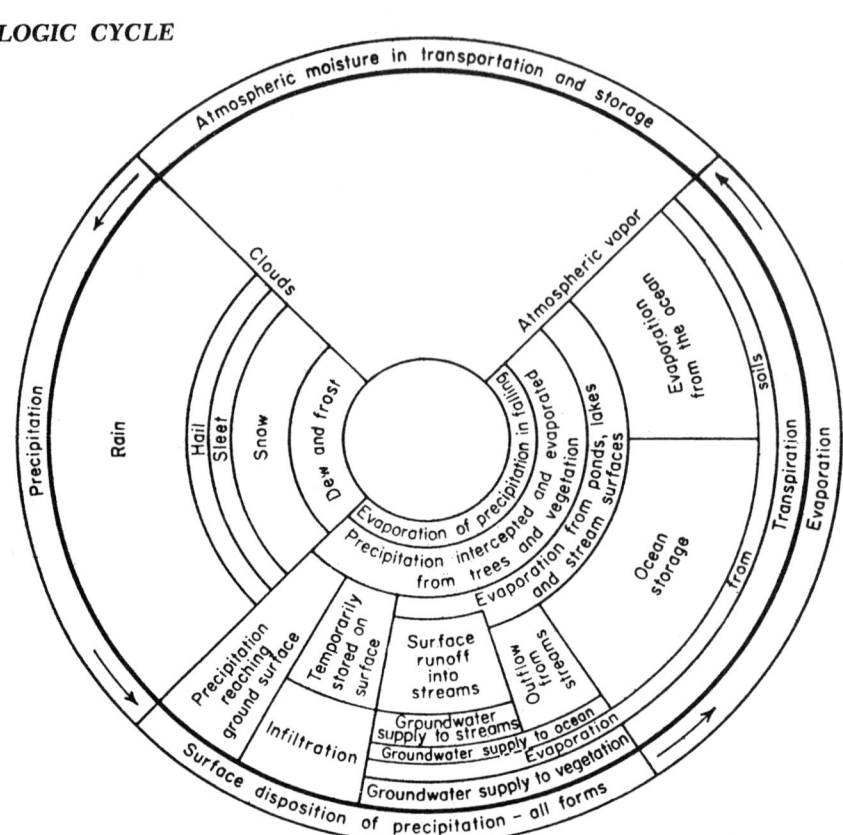

Fig. 1. The hydrologic cycle—a qualitative representation (Horton, 1931).

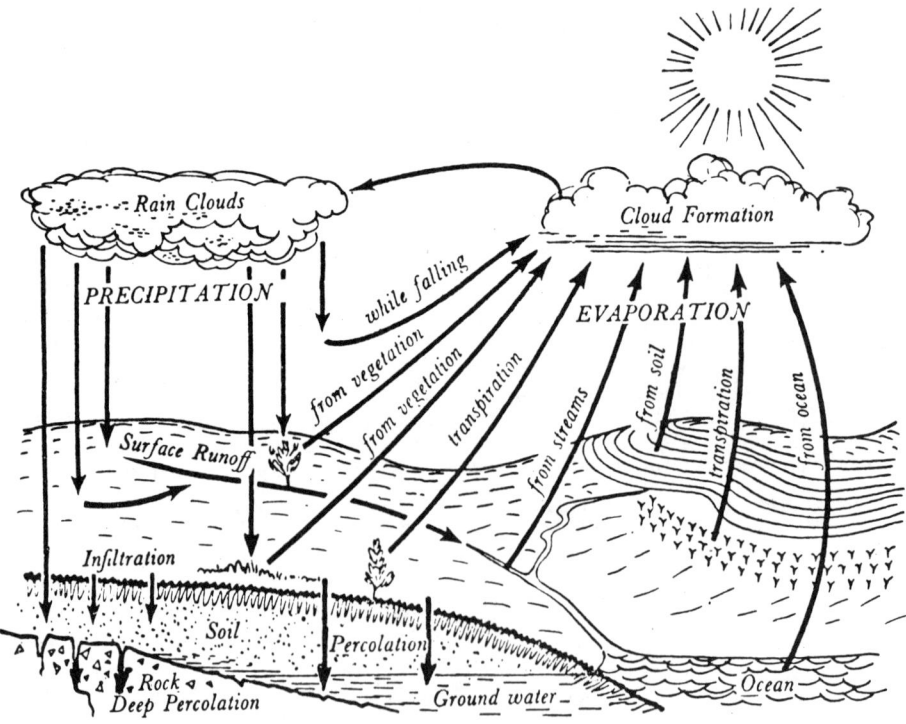

Fig. 2. The hydrologic cycle— a descriptive representation (Ackermann, Colman and Ogrosky, 1955).

HYDROLOGIC CYCLE

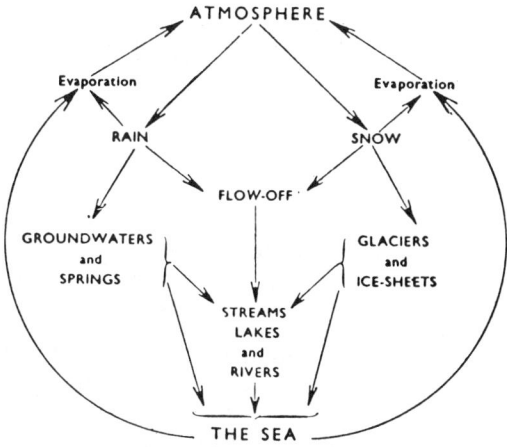

FIG. 3 The circulation of meteoric water. In addition to the main evaporations here indicated it should be noted that evaporation takes place from all exposed surfaces of water and ice (e.g., lakes, rivers, glaciers, and ice sheets) and also from the soil and from plants and animals. Part of the water that ascends from the depths by way of volcanoes reaches the surface for the first time; such water is called *juvenile water* to distinguish it from the *meteoric water* already present in the hydrosphere and other outer zones of the earth (Holmes, 1965). (Note: the word "runoff" is generally used rather than "flow-off.")

The Magnitude of the Hydrologic Cycle

Each year approximately 96,000 cubic miles (4×10^5 km^3 or 4×10^{20} g^3) of water is evaporated from the earth's surface. Of this amount, the oceans account for 81,000 cubic miles (84.4%), and inland water bodies and wet soils provide the remaining 15,000 cubic miles (15.6%; Landsberg, 1945).

Most of the inland evaporation occurs into relatively dry air masses. Much of the water evaporated from the oceans is transported by maritime air masses (which can hold considerably more water vapor than continental air masses) to the continents, where total precipitation amounts to 24,000 cubic miles/yr (100,000 km^3/yr). This is enough water to cover the entire state of Texas (267,339 square miles, 692,408 km^2) to a depth of 475 ft (144.8 m). Of the 24,000 cubic miles of water precipitated, 9000 cubic miles (38,000 km^3) (37.5%) returns to the sea as runoff to balance the excess of precipitation over evaporation inland (Landsberg, 1945).

Reichel (1952) calculates that the mean annual precipitation for the entire world is 34 in. (86.4 cm), which is balanced by a comparable amount of evaporation. It is estimated that 97% of all the water in the world, or over one quadrillion (10^{15}) acre-ft (1234×10^{15} m^3, is contained within the oceans. If the earth were a uniform sphere, this volume

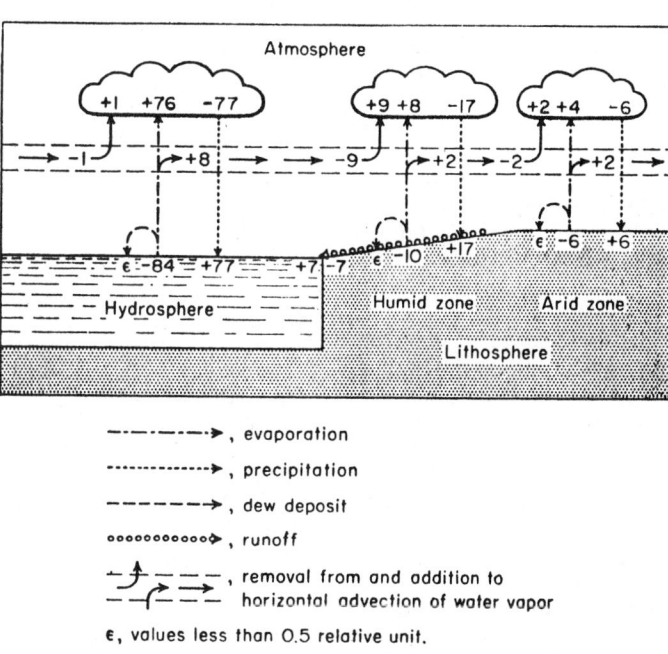

FIG. 4. The hydrologic cycle—a quantitative representation. 100 relative units = 85.7 g/cm^2/yr or 85.7 cm (33.8 in.), mean annual precipitation for the world (Lettaü, 1954).

of water would cover the earth to a depth of 800 ft (243.8 m; Wolman, 1962).

The total volume of fresh water on the earth is estimated at 33 trillion acre-ft (4.1 × 10^{15} m³ or 4.1 × 10^{21} g³), approximately distributed as follows (Chow, 1964):

	Percent
Polar ice and glaciers	75
Groundwater between 2500 and 12,500 ft (762 and 3810 m)	14
Groundwater less than 2500 ft (762 m)	11
Lakes	0.3
Soil moisture	0.06
Atmosphere	0.035
Streams	0.03

Note that these figures are stationary estimates of distribution. Thus, huge amounts of water pass through the atmosphere while the water content is relatively small at any given moment.

The average annual precipitation over the continental United States would amount to 30 inches (76.2 cm) if it were spread evenly. However, topographic configurations and patterns of atmospheric circulation result in uneven distribution of precipitation, ranging from a few inches in the arid southwest to over 100 in. (254 cm) in parts of the Pacific northwest. The 17 western states receive only 25% of our total precipitation but contain 60% of the land area.

The 30 in. (76.2 cm) of water for the United States represents 4,800,000,000 acre-ft/yr, or 4300 billion gal/day. Of this amount, 21½ inches (54.6 cm), or 71.7% is returned to the atmosphere by the processes of evapotranspiration. The remaining 8½ in. (21.6 cm), or 28.3%, becomes surface and groundwater runoff into the oceans (Robinove, 1963).

The Role of the Hydrologic Cycle

If the atmosphere and the earth are considered as separate entities, radiation and conduction fail to provide balanced heat budgets, as the earth's surface has a net gain and the free atmosphere a net loss. The link between the gain and loss is the hydrologic cycle.

Some of the heat absorbed by the earth's surface is expended in evaporation, and therefore transferred to latent heat (the quantity of heat absorbed or emitted without change in temperature during a change of state of unit mass of a material, "hidden heat"), which is later realized as sensible heat (the heat added to a body when its temperature is changed) and released to the atmosphere when the vapor condenses to clouds. Evaporation is high where relatively cool air sweeps over warmer oceans. The highest evaporation values found in the northern hemisphere occur in the Atlantic and Pacific trade wind belts south of 30°N. High values also occur over the northwestern Pacific and North Atlantic oceans during winter when cold, dry continental air masses (cP and cA) move over warmer waters (Petterssen, 1964).

The average life of water vapor molecules in air varies from an hour to several days. Latent heat is usually liberated far from the regions where evaporation occurred. This is particularly true of evaporation in the trade wind belts, which supply much of the vapor that eventually precipitates in middle and high

TABLE 1. ESTIMATED RUNOFF OF MAJOR WORLD RIVERS[a]

	1000 cfs	100 m³/sec
Rivers discharging into the Atlantic		
Eastern North America		
Mississippi	620	175.5
St. Lawrence	500	141.5
South Atlantic slope	325	92.0
North Atlantic slope	210	59.4
	1655	468.4
Europe		
Danube	225	63.7
Rhine	76	21.5
Rhone	59	16.7
Dnieper	59	16.7
Elbe	24	6.8
Garonne	24	6.8
Don	24	6.8
	491	139.0
South America		
Amazon	3600	1018.8
Orinoco	600	169.8
Parana	526	148.9
Uruguay	136	38.5
	4862	1376.0
Africa		
Congo	1600	452.8
Niger	326	92.3
Orange and Zambezi	352	99.6
Nile	100	28.3
	2378	673.0
Total for Atlantic Ocean	9386	2656.4
Rivers discharging into the Pacific		
Columbia	345	97.6
Colorado	23	6.5
Yukon	180	50.9
Australia	354	100.2
Japan and Korea	225	63.7
Middle latitude Asian rivers	2250	636.8
	3377	955.7

[a] From Livingstone (1963).

TABLE 2. OCEANIC AND LAND-DRAINAGE AREAS, IN 1,000,000 SQUARE MILES AND 1,000,000 KM[a]

Ocean	Area square miles	Area km²	Land Area Drained square miles	Land Area Drained km²	% of Total Land Area	% of Area Drained to Ocean Area
Atlantic	37.8	98	25.9	67	45.3	68.5
Indian	25.3	65.5	6.6	17	11.5	26.1
Antarctic	12.4	32	5.4	14	9.4	43.5
Pacific	63.7	165	6.9	18	12.1	10.8
Interior drainage			12.4	32	21.7	
Total	139.2	360.5	57.2	148	100.0	

[a] Murray (1888).

latitudes. Thus, the circulation of water is a key part of heat transfer from low to high latitudes and from oceans to continents (Petterssen, 1964).

The Return of Water to the Oceans

In spite of the relatively uniform pattern of evaporation in the various latitudinal belts of the ocean, there is marked regional imbalance in the return flow of water to the oceans. The explanation lies in the concentration of major rivers (Amazon, Mississippi, Congo, Niger, St. Lawrence, Danube, Po, Nile and Rhine) which drain into the Atlantic Ocean and its marginal seas (Gulf of Mexico, Black Sea). In contrast, the Pacific has only a limited number of major discharge outlets (Yangtze, Hwang-Ho, Yukon, Columbia, and Colorado).

Table 1 provides a listing of the mean annual discharges of the world's major rivers. Note the paucity of data for specific Asian rivers. Table 2 provides further evidence that the Atlantic not only drains the largest portion of the earth's land surface, but has the highest proportion of land area drained to ocean area.

ROBERT M. HORDON

References

Ackermann, W. C., Colman, E. A., and Ogrosky, H. O., 1955, "From ocean to sky to land to ocean," in "Water." *Yearbook Agr. (U.S. Dept. Agr.)*, pp. 41–51.

Carroll, D., 1962, "Rainwater as a chemical agent of geologic processes," *U.S. Geol. Surv. Water Supply Paper* **1535-G**, 18pp.

Chow, V. T., 1964, "Hydrology and Its Development," in "Handbook of Applied Hydrology," New York, McGraw-Hill Book Co., pp. **1**-1 to **1**-22.

Erikson, E., 1952, "Composition of atmospheric precipitation," *Tellus*, **4**, 215–232; 280–303.

Goldberg, E. D., 1960, "Hydrosphere, Geochemistry of," in "Encyclopedia of Science and Technology," New York, McGraw-Hill Book Co., pp. 565–568.

ogy," Second edition, New York, Ronald Press,
Holmes, A., 1965, "Principles of Physical Geology," 1288pp.

Horton, R. E., 1931, "The field, scope and status of the science of hydrology," *Trans. Am. Geophys. Union*, **12**, 189–202.

Hutchinson, G. E., 1957, "A Treatise on Limnology," New York, John Wiley & Sons, 1015pp.

Kuenen, P. H., 1955, "Realms of Water," New York, John Wiley & Sons, 327pp.

Landsberg, H., 1945, "Climatology," in "Handbook of Meteorology," New York, McGraw-Hill Book Co., pp. 927–997.

Lettau, H., 1954, "A study of the mass, momentum and energy budget of the atmosphere," *Arch. Meteorol. Geophys. Bioklimatol., Ser. A*, **7**, 131–153.

Livingstone, D. A., 1963, "Chemical composition of rivers and lakes," *U.S. Geol. Surv. Profess. Paper* **440-G**.

McDonald, J. E., 1962, "The evaporation-precipitation fallacy," *Weather, London*, **17**(5), 168–170; 172–177 (May 1962).

Murray, J., 1888, "On the height of the land and the depth of the ocean," *Scottish Geogr. Mag.*, **4**, 1.

Petterssen, S., 1964, "Meteorology," in "Handbook of Applied Hydrology," New York, McGraw-Hill Book Co., pp. **3**-1 to **3**-39.

Reichel, E., 1952, "Der Stand des Verdunstungsproblems" (The status of the evaporation problem), *Ber. Deut. Wetterdienst, Bad Kissingen*, **35**, 155.

Robinove, C. J., 1963, "What's happening to water?" *Smithsonian Inst. Ann. Rept.* 1962, 375–389.

White, D. E., Hem, J. D., and Waring, G. A., 1963, "Chemical composition of subsurface waters," *U.S. Geol. Surv. Profess. Paper* **440-F**, 67pp.

Wolman, A., 1962, "Water resources, a report to the Committee on Natural Resources of National Academy of Sciences—National Research Council. *Natl. Acad. Sci. Natl. Res. Council, Publ.* **1000-B**.

Cross-references: *Groundwater; Hydrology; Juvenile Water; Meteoric Water; Rainwater; Runoff; Water Balance.* Vol. I: *Ocean-Atmosphere Interaction.* Vol. II: *Atmosphere; Energy Budget of the Earth's Surface; Evaporation; Hydrosphere; Precipitation.* Vol. V: *Lithosphere.*

HYDROLOGIC MAPS

Hydrological maps may be any maps that display hydrologic information. Because of a great variety of types, scales, and legends of such maps, a complete standardization of the maps is practically impossible. Nevertheless, there is a need to aid worldwide understanding and coordination of hydrological facts as presented by the many hydrologists and hydrological agencies over the world by affording more easy comparison of maps from different countries, and to help map preparation in general, for example, by eliminating detailed explanations for some symbols. In response to this need, several international organizations have developed programs on hydrological maps.

The UNESCO Coordinating Council of the International Hydrological Decade (UNESCO/IHD Coordinating Council) recognized the problems, and the expected results on hydrological maps are as follows:

1. The mass of data concerning hydrological phenomena is so great that summarized pictorial representation is essential for ready comprehension.

2. Maps are an effective form of presenting hydrologic data in readily understandable form. Numerous proposals have appeared, many examples are available, and several hydrological maps of various types are currently in preparation.

3. The Coordinating Council of the Decade was urged by the intergovernmental meeting of experts in April 1964 to prepare, in cooperation with governmental and nongovernmental organizations, climatological, hydrological, and hydrogeological maps.

4. Hydrologic agencies in many countries have prepared water balance maps of selected river basins. These have variously been classified as hydrologic atlases, hydro-climatic maps, and hydrometeorological maps.

5. One of the most useful types initially would be a hydrological atlas of the world. These maps would be prepared on the basis of those published for different regions or continents.

The first task on hydrological maps would be to agree on basic standards. These include: base-map scales, sheet sizes, standard symbols, etc. All those symbols referring to hydrology in the broad meaning of the term should be extracted from the international legends already laid down for hydrometeorological and hydrogeological maps while new symbols should be added in respect to other particular types of maps which will be prepared. This would mean making arrangements with competent organizations to obtain the necessary data.

The UNESCO/IHD Coordinating Council therefore decided at the first Council meeting in Paris on 24 May 1965 to establish a working group that will make proposals regarding the objectives of hydrological maps, the types of maps to be prepared on a priority basis, and the standards, scales and symbols to be adopted. It recommended that all countries prepare bibliographies and catalogs of maps and send samples to the UNESCO/IHD Coordinating Council Secretariat, and urged the Secretariat to maintain cooperation with the World Meteorological Organization, the International Association of Scientific Hydrology (IASH), and the International Association of Hydrologists (IAH).

At the same Council meeting, the UNESCO/IHD Coordinating Council also recommended that a hydrogeological map of the arid zones be drawn up, providing a simplified representation of the basic features of groundwater tables from the standpoint of scientific knowledge and exploitation possibilities. Also recommended was that a small-scale hydrogeological map of Europe be prepared, to provide a simplified representation of the location and extent of the main groundwater tables.

TABLE 1. SCALES FOR CLASSIFICATION OF HYDROGEOLOGICAL MAPS PROPOSED BY MARGAT[a]

Purpose	Large-Scale Maps (1/20,000 to 1/200,000)	Small-Scale Maps (1/200,000 to 1/1,000,000)
Maps for scientific purposes	Detailed hydrogeological map Specific hydrogeological map Hydrogeochemical map and specific hydrogeochemical maps Map of productivity	Synthetical hydrogeological map Specific hydrogeological map Hydrogeochemical map Map of availabilities
Maps for practical purposes	Map of groundwaters Map of productivity Applied hydrogeochemical map	Map of groundwaters (popularization); hydrogeological map of reconnaissance or orientation Map of availabilities Hydrogeochemical map of reconnaissance

[a] Margat, 1966.

The UNESCO/IHD Working Group on Hydrological Maps first met in Paris in 1967. It agreed that the scales proposed by Margat (Table 1) would serve as a suitable basis for classification of hydrological maps. It studied a proposed hydrogeological map of Europe and set up a Panel to examine the draft legend for hydrogeological maps which had been made by IASH and IAH (IASH, 1962). The general work plan was given as follows:

1. Collect representative hydrological maps, select or determine types of maps needed, and assign priorities and give guidance for preparation on the basis of map-use practices and need for scientific, governmental, management, and other uses.

2. Update the existing exhibit of hydrological maps and arrange for the distribution of samples to the working group.

3. Prepare a guidebook on the preparation of hydrological maps.

4. Maintain liaison with other working groups of the Council and other international organizations with known or potential hydrological map-making needs and be prepared to guide, assist, and coordinate their efforts in this regard.

5. To provide, as required, advice to IHD national committees for the preparation of international (regional and continental) maps by bi- and multilateral governmental groups.

Organized efforts on the preparation of hydrological maps began before the UNESCO/IHD programs. The IASH started studying the international legend for hydrogeological maps at the General Assembly of the International Union of Geodesy and Geophysics (IUGG) in Rome in 1954 and has had it on the agenda in all subsequent Assemblies, which met every three years. The IAH has studied the preparation of a legend and the drawing of hydrogeological maps since 1952. At the request of a UNESCO Advisory Committee, the Commission of Subterranean Waters of the

TABLE 2. INTERNATIONAL LEGEND FOR HYDROGEOLOGICAL MAPS (IASH, 1962)

Subject	Symbol recommended	Suitability of use. G = large scale maps, P = small scale maps, S = special maps.
A. *Topography*		
Topography	Symbols conform as far as possible with international usage (grey)	G, P, S
B. *Geology*		
1. Geological formation	Only if age is essential to hydrogeological understanding, colours should be used and conform as far as possible with international geological usage	G, P, S
2. Stratigraphy	Letters, symbols and patterns should conform with international usage (black) contour line, broken where uncertain	G, P, S
3. Height or depth of formation (top or base) relative to the national reference level	——— 78 ——— − − − (black)	G, P, S
4. Contact between permeable and impermeable or semipermeable formations	line of contact (black)	G, P, S

TABLE 2. (*Continued*)

Subject	Symbol recommended	Suitability of use. G = large scale maps, P = small scale maps, S = special maps.
5. Strike and dip		G, P, S
6. Axis of anticline, with direction of axial plunge	or (black)	G, P, S
7. Axis of syncline, with direction of axial plunge	or (black)	G, P, S
8. Flexure, with direction of downthrow side	(black)	G, P, S
9. Idem, not affecting covering layers	(black)	G, P, S
10. Fault, with direction of downthrow side	(black)	G, P, S
11. Idem, not affecting covering layers	(black)	G, P, S
12. Overthrust fault	teeth un upper plate (black)	G, P, S
13. Abnormal contact	(black)	G, P, S

IASH held a symposium on hydrogeological maps at the IUGG General Assembly at Helsinki, Finland in 1960. The result was an outstanding display of carefully selected hydrogeological maps of a great variety. Within the Commission of Subterranean Waters of the IASH, a standing committee was then created and met in Helsinki. Its functions were divided into two parts: firstly to deal with small-scale maps (for example, from 1:200,000 to 1:500,000), and secondly to study large-scale, or detailed, maps. The committee was

TABLE 2. *(Continued)*

Subject	Symbol recommended		Suitability of use. G = large scale maps, P = small scale maps, S = special maps.
C. *Lithology* For lithology the standard international letters, symbols and patterns, in brown colour or in the colour of the geological formation (see B. 1), are recommended. In areas with a complicated lithology a mixture of the single symbols may be used. Semi-permeable and impermeable formation are to be omitted.			
1. Gravels, gravelly deposits		(brown or in the colour of the geological formation B.1)	G, P, S
2. Sand		,,	G, P, S
3. Sandstone		,,	G, P, S
4. Conglomerate		,,	G, P, S
5. Limestones		,,	G, P, S
6. Dolomites		,,	G, P, S
7. Calcareous sinter		,,	G, P, S
8. Porous volcanic ejecta		,,	G, P, S
9. Sodium salt		,,	G, P, S

to prepare a report for each, with recommendations on the making of maps. The legend and recommendations were to be flexible enough for worldwide adoption. After its subsequent annual meetings and coordination with the annual meetings since 1959 of the representatives of the IAH, a group within the International Association of Geological Sciences, the committee made considerable progress. As a result, more maps were added to the original display in 1960 until over 200 maps were available for study and compari-

HYDROLOGIC MAPS

TABLE 2. (Continued)

Subject	Symbol recommended	Suitability of use. G = large scale maps, P = small scale maps, S = special maps.
10. Gypsum	(brown or in the colour of the geological formation B. 1)	G, P, S
11. Chemical properties of the formation	The lithological symbols may also indicate chemical properties	G, P, S
D. *Hydrography*		
All natural waters in blue		G, P, S
1. Perennial stream, with direction of flow	(blue)	G, P, S
2. Idem, highly polluted	See F. 7	
3. Idem, with high chloride content	See F. 8	
4. Seasonal stream, with direction of flow	(blue)	G, P, S
5. Intermittent stream, with direction of flow		G, P, S
6. Disappearance point of stream	(blue)	G, P, S
7. Gauging station, with yearly average flow and area of catchment	83 / 1251 (blue)	G, P, S
8. Marsh, seasonal marsh	(blue)	G, P, S
9. Flood-stage area, area inundated during floods	(blue)	G, P, S

son. By its fourth meeting in 1962, an outline legend had been prepared and most major problems resolved. The color controls being agreed upon are as follows: gray for topography, black for geological symbols, brown for lithological symbols, blue for natural waters, violet for chemistry, and red for artificial works. Both the Food and Agricultural Organization, which was represented at the meeting, and IAH agreed on the legend proposed and that the legend be published in black and white in the IASH Bulletin and as a UNESCO

HYDROLOGIC MAPS

TABLE 2. (*Continued*)

Subject	Symbol recommended	Suitability of use. G = large scale maps, P = small scale maps, S = special maps.
10. Surface water divide	• • • • • • • • • • (blue)	G, P, S
11. Spring	⚲ (blue or darker blue) The inside of the symbol should be reserved for hydrochemical data (in colours according to F. 3 and 5), the outside for hydrodynamical data. The example given shows one of the possibilities. 4 ⚲ 1 3 2 The symbol can be used as the basis of symbols for a further classification of springs	G, P, S *e.g.*: 1 = filing number 2 = temperature 3 = altitude 4 = discharge
12. Group of springs	The same symbol as for a spring but larger) (blue or darker blue)	G, P, S
13. Thermal or thermomineral spring	The same symbol as given in 11 but whit thicker outline (blue or darker blue)	G, P, S
14. Natural pond or waterhole with no outlet	⌣ (blue or) darker blue)	G, P, S
15. Salt lake	See F. 9	

publication. After a period of use, a publication in color would be issued.

The International Legend for Hydrogeological Maps (Table 2) was first published in both English and French in the Bulletin of IASH in 1962 (IASH, 1962) and then in English, French, and Spanish as a special UNESCO document in 1963 (UNESCO, 1963). The latter document was presented at the IUGG General Assembly in Berkeley, California in 1963, with the recommendation that the member countries of the IUGG adopt

HYDROLOGIC MAPS

TABLE 2. (*Continued*)

Subject	Symbol recommended	Suitability of use. G = large scale maps, P = small scale maps, S = special maps.
E. *Groundwater hydrology*		
1. Height or depth of water level at a given time and relative to the national reference level	isohypses, isopiezometric lines or ground water contours; broken line where uncertain —450— — — - - (blue)	G, P, S
2. Direction and actual velocity of the ground-water flow (eg. in m/day)	↗10 (blue darker blue)	G, S
3. Ground-water divide	○ ○ ○ ○ ○ ○ ○ ○ ○ ○ (blue)	G, P, S
4. Boundary of area with confined ground water	+ + + + + + + + × × (blue)	G, P, S
5. Boundary of area of artesian flow	////////|||||||||||||||||||| (blue)	G, P, S
6. Boundary of waterbearing formation	See B. 4	G, P, S
7. Ground-water barrier	⊥⊥⊥⊥⊥⊥⊥⊥⊥⊥ (blue)	G, S
8. Average depth of top of saturated part of waterbearing formation, confined or inconfined, below groundsurface	contour lines e.g. in the colour of the formation —-150—	G, S
9. Height or depth of top and/or base of waterbearing formation relative to the national reference level	See B. 3	

the legend and encourage its use. A by-product of the committee's work was the large number of maps collected and reviewed as a basis for symbols for the international legend. More than 150 of these maps and the UNESCO document have been displayed at several international meetings and exhibited at universities and water resources and hydrology agencies in many countries. It is the intent of the Committee and the UNESCO that the International Legend for Hydrogeological Maps be distributed as widely as possible for

TABLE 2. (Continued)

Subject	Symbol recommended	Suitability of use. G = large scale maps, P = small scale maps, S = special maps.
10. Isopachytes, lines of equal thickness of the water-saturated bed at a given time, with thickness (in m)	⌒ 18 ⌒ (blue line, figure in red)	S
11. Different ground-water horizons (aquifers)	To be shown by cross sections or planimetrically (by colour left to the discretion of the author)	G, S
12. Infiltration conditions of covering layers. Qualitative description, e. g. — good — moderate — poor	Patterns, at the discretion of the author	S
13. Transmissibility	Contours or colours, at the discretion of the author	
14. Average yield of wells. Order of magnitude represented by areas of equal well yield, or for selected wells of approximate specific capacity (discharge divided by drawdown) or by total discharge of the wells for a specific drawdown	A range of shades of one colour, greater intensity of colour indicating greater yield	G, S
15. Exploitable yield per unit of the development area of the aquifer	A range of shades of blue	G, S
16. Annual rainfall (depth in mm)	isohyets ⎯⎯300⎯⎯ (blue)	S

review and constructive criticism and that the worldwide use of the symbols be encouraged.

V. T. Chow

References

Margat, J., 1966, "La Cartographie d'Hydrogéologie," *Chronique d'Hydrogéologie*, No. 9, Bur. Rech. Géol. Minières, Paris, France.

IASH, 1962, "A Legend for Hydrogeological Maps." VIIe Année, *Bull.* No. 3, September.

UNESCO, 1963, "International Legend for Hydrogeological Maps," Document NS/NR/20, January.

HYDROLOGIC MAPS

TABLE 2. *(Continued)*

Subject	Symbol recommended	Suitability of use. G = large scale maps, P = small scale maps, S = special maps.
F. *Hydrochemistry*		
1. Total concentration or total chloride or total hardness, etc. of ground water	isocone or isochloride etc. contours; broken line where uncertain ——— 2 ——— — — ✓ (violet) or a range of shades in cross sections or on special maps	G, P, S
2. Depth of interface between fresh and salt ground water below the national reference level	contours, broken where uncertain ══════18══════ ══ ══ (violet)	G, P, S
3. Chemical composition of the ground water	colour representing predominant property; bi-coloured streaks representing mixed features. Concentration is indicated by different shades of the colour or by isocones	S
Bicarbonate water		
calcium	clear blue	
magnesium	violet-blue	
sodium	dark (Prussian) blue	
Sulphate water		
calcium	yellow	
magnesium	orange	
sodium	yellow-brown	
Chloride water		
calcium	green-brown	
magnesium	blue-green	
sodium	green	
4. Temperature in degrees Centigrade	(violet)	G, S, P
5. Mineral or thermal water	Symbol of spring (D.11) or well (G.2 etc.) or pond (D.14) with thicker outline, (blue) the inside of the symbol being reserved for hydrochemical data in colours according to F.3 or symbols as shown below	G, P, S

Cross-references: *Hydrologic Associations; Hydrogeology; Hydrology; International Hydrological Decade; Water Balance.* Vol. III: *Geomorphic Maps.*

HYDROLOGIC MAPS

TABLE 2. (*Continued*)

Subject	Symbol recommended	Suitability of use. G = large scale maps, P = small scale maps, S = special maps.
not fit for use	⊘	
< 2 gr/l	○	
2-4 "	⊙	
4-8 "	◐	
> 8 "	●	
not determined	⊖	
6. Chemical properties of the water-bearing formation	See C.11	
7. Highly polluted stream (organic pollution)	(blue line with grey shading on each side)	G, P, S
8. Stream with high chloride content	(blue line with violet shading on each side)	G, P, S
9. Salt lake	(blue line with violet shading along contour of lake)	G, P, S
G. *Boreholes, Wells and other Works* All artificial works are indicated in red		
1. Borehole	(red)	G, P, S
2. Dug well *)	○ (red)	G, P, S
3. Dug well, dry	● (red)	G, P, S
4. Drilled well *)	⊕ (red)	G, P, S
5. Drilled well, dry	⊕ (red)	G, P, S

(*) The inside of the symbol should be reserved for hydrochemical data, the outside for hydrodynamical data. The example given shows one of the possibilities

```
  6 1 2
   ⊕
  5 3
    4
```

e.g.:
1 = number
2 = static level
3 = depth
4 = temperature
5 = drawdown
6 = yield

Subject	Symbol recommended		Suitability of use. G = large scale maps, P = small scale maps, S = special maps.
6. Artesian well, flowing *)		(red)	G, P, S
7. Artesian well, non-flowing *)		(red)	G, P, S
8. Recharge well *)		(red)	G, P, S
9. Group of wells	The same symbol as for a well but larger	(red)	G, P, S
10. Cistern		(red)	G, S
11. Storage reservoir for surface water		(red)	G, P, S
12. Catchment of a spring		(red square, symbol for spring in blue)	G, S, P
13. Drainage gallery		(red)	G, P, S
14. Pipe line		(red)	G, P, S
15. Dam (with capacity of reservoir e.g. in million m³)		(red)	G, P, S
16. Underground dam		(red)	G, P, S
17. Canal, irrigation canal (perennial waters)		(red)	G, P, S
18. Idem, (flood waters)		(red)	G, P, S
19. Drainage canal or artificial drain		(red)	G, P, S
20. Gauging station at a stream	See D. 7		
21. Hydro-electric station		(red)	G, P, S
22. Mine, used		(red)	G, P, S
23. Id., not used		(red)	G, P, S
24. Quarry		(red)	G, P, S

HYDROLOGY

Hydrology has been defined in several ways. One of the best definitions is that of Meinzer (1942) who said, "Hydrology is, etymologically, the science that relates to water. It is, however, an earth science. It is concerned with the occurrence of water in the earth, its physical and chemical reactions with the rest of the earth, and its relation to the life on the earth. It includes the description of the earth with respect to its waters. It is not concerned primarily with the physical and chemical properties of the substance known as water. Like geology and the other earth sciences, it uses the basic sciences as its tools, but in doing so, it has developed a technique and subject matter that are distinct from those of the basic sciences."

Meinzer's definition of hydrology places the subject in the field of earth science. Although it is concerned, for example, with the effects of plants on the chemical constituents in soil water, it is not usually concerned with the chemical use of water within plant cells, the special province of the botanist.

The Hydrologic Cycle

Let us examine briefly the broadest aspect of hydrology, the hydrologic cycle. The major reservoirs of water on the earth are the oceans, and they form one link in what is known as the hydrologic cycle. The hydrologic cycle is the circulation of water from the ocean into the atmosphere; the movement of atmospheric moisture across oceans and over the continents; the precipitation of moisture as snow and rain; the flow of water in streams and lakes; the evaporation and transpiration of water from the surface back into the atmosphere; the movement of water beneath the surface of the ground; and the discharge of water back into the ocean to continue its endless journey. The hydrologic cycle is continuous, and its various phases are made up of endless and complex details. Water in any one phase of the cycle cannot be treated as a single subject. It must be considered in relation to its total natural environment and its use by man, and as a function of time so that it may be properly utilized as a renewable and conservable resource.

Let us begin by considering the moisture in the atmosphere. Water evaporates from the surface of the ocean. It rises in the air and is borne by winds over the land masses, where a part of it condenses and falls as rain or snow. In the western United States, most precipitable moisture is brought from the Pacific Ocean and carried eastward. When winds from the Pacific Ocean reach the west coast, the warm, moisture-laden air is lifted high over the mountains and becomes cooler. Cool air cannot hold as much water vapor as warm air and so the moisture falls as rain or snow on the western slopes of the coastal ranges and the Sierra Nevada. East of the coastal ranges and the Sierra Nevada, the air is drier; there is less moisture to precipitate and fall, and as a result, the region between the coastal areas and the Rocky Mountains is arid to semiarid. The lifting and drying effect of the continental highlands is again shown as the air reaches the Rocky Mountains. Precipitation is greatest on the western slopes of the Rocky Mountains; the drier air moves across the mountains to the eastern slope, where precipitation is less.

Precipitation in the central part of the United States increases as additional air masses moving from the Gulf coast, the Atlantic, and the Arctic region bring moisture-laden air over the continent.

Water is abundant in the United States. If all the precipitation that fell within the limits of the 48 conterminous states during an average year was spread evenly over the country, it would stand 30 inches deep. However, as the pattern of air movement indicates, this precipitation is uneven. Some areas receive only a few inches of rainfall during the year, while others receive more than 100 inches.

The water represented by the average 30-inch depth is about 4800 million acre-ft/yr, or about 4400 billion gal./day. This is an enormous amount of water, but unfortunately not all of it is available for use. Of the 30 inches of water, about 21 inches is evaporated from open water areas and the soil or is transpired from the leaves of plants and thus returned to the atmosphere before it can be diverted for our use. Only part of this 21 inches supports cultivated crops, native grass, and forests; the rest is evaporated or used by nonbeneficial plants.

The remaining 9 inches of rainfall moves over the surface to streams as "direct runoff," or seeps to the water table to become "groundwater," later to discharge into streams as groundwater runoff." Of this 9 inches, man withdraws the equivalent of about 2 inches from streams, lakes, reservoirs, springs, and wells (but in part this represents the same water used over again) and uses it for municipal and rural water supplies, industry, and irrigation. About half an inch evaporates or is transpired, in part naturally and in part as a result of the activities of man (principally irrigation). The remainder joins the "unused" water to make a total of about 8 inches flowing into the oceans. Actually, the "unused" water is used too, though not "withdrawn"—for hydropower (to an extent equivalent to nearly twice the average stream flow), for dilution of sanitary and industrial wastes, for navigation, for recreation, and by fish and wildlife.

Basic Principles of Hydrology

The basic principle of hydrology is that water is a dynamic, moving resource. The concept of the hydrologic cycle was formed by the fourth century, but quantitative proof that precipitation was the major source of water on and below the land surface was not forthcoming until the seventeenth century.

The development of practical and usable principles of hydrology was greatly hampered in the past (and to a slight extent this is true even today) by those who attempted to explain some of the more spectacular evidences of water action as isolated phenomena and not as part of the unified hydrologic cycle. Groundwater, in particular, became the subject of much mystical speculation.

Since the eighteenth century, the development of hydrologic principles has been aimed at refinement of the understanding of each distinct phase of the hydrologic cycle and of the relationships between the phases. A few of the more important principles are listed below, not necessarily in order of importance or discovery.

(1) The recognition that groundwater moves from points of high pressure to points of low pressure (down gradient) and that gradients are often, but not exclusively, related to rock type and structure.

(2) The fact that the velocity of flowing water, on the surface or underground, is governed by the differences in pressure head, or slope, and the resistance of the confining channel or of the aquifer.

(3) The knowledge that water is capable of dissolving and carrying large amounts of mineral matter that changes composition as the water comes in contact with various types of potential solutes.

(4) The geologic recognition that water transports and deposits vast quantities of solid rock waste and is a major agent in the modification of land forms and in chemical alterations underground.

(5) The fact that natural (underground) or artificial (surface) storage of water modifies the regimen of water in an area by changing the time of flow.

Hydrologic Studies

With these basic principles in mind, the scientists, geographers, and engineers, who are known today as hydrologists, set out to accomplish two basic tasks, neither of which is near completion at the present time. The first is an inventory or description of the water resources of the world—the amount of water in storage, rates and volumes of precipitation, recharge and discharge, the quantitative availability and suitability of water for man's use, and the effects of water, such as floods and droughts.

According to a calculation by Nace (1964), the world's estimated water supply is distributed as shown in Table 1.

The second basic task is research in hydrology—the refinement of understanding water in all its phases. A brief review of our knowledge of the hydrologic cycle and some of the remaining problems is in order here.

Much is known about the fundamentals of hydrology and the laws that govern the occurrence and movement of water on and below the surface of the earth. Research directed toward discovering these facts has continued and been accelerated throughout the United States and elsewhere, and has been carried on by government agencies, universities, and foundations. A great deal has been learned, but a great deal more remains to be understood.

The mechanics of precipitation of water from atmospheric vapor as rain, hail, and snow are fairly well known. However, we still do not understand what governs distribution of precipitation in time and space well enough to predict how much rain or snow will fall where and when—to say nothing of being able to influence them. Stations for the recording of precipitation and temperature are scattered throughout the United States, but most of them are in heavily populated areas. However, much of the precipitation falls where the population is small and scattered, such as the mountainous areas of the West, and records from these areas are spotty and inadequate. More must be learned about the distribution of precipitation in these areas before we can understand and predict the distribution of the primary source of water (Langbein and Hoyt, 1959, p. 41).

When water reaches the land surface, a portion soaks into the ground and is stored as soil water which is available for the growth and nourishment of plants. This *zone of soil moisture* may at times be completely saturated with water—i.e., all the pore spaces between the grains of soil and rock may be filled with water, or they may be only partially filled. We need to know more about the mechanism of the filling and draining of the soil-moisture zone. We also need to know a great deal more about how much water is extracted by crop and native plants from this soil-moisture zone and how much water is evaporated from the land and lost to further use.

TABLE 1. DISTRIBUTION OF WORLD'S ESTIMATED WATER SUPPLY

Location	Surface Area (square miles)	Water Volume (cubic miles)	Percentage of Total Water
Surface water			
Fresh-water lakes	330,000	30,000	0.009
Saline lakes and inland seas	270,000	25,000	0.008
Average in stream channels	—	300	0.0001
Subsurface water			
Soil moisture and vadose water	50,000,000	16,000	0.005
Groundwater within depth of half a mile	50,000,000	1,000,000	0.31
Groundwater —deep lying	50,000,000	1,000,000	0.31
Total liquid water in land areas	50,600,000	2,070,000	0.635
Icecaps and glaciers	6,900,000	7,000,000	2.15
Atmosphere (at sea level)	197,000,000	3,100	0.001
World ocean	139,500,000	317,000,000	97.2
Totals (rounded)		326,000,000	100

Between the zone of soil moisture and the water table is the *zone of aeration*. Water in excess of the amount that the soil can hold (field capacity) moves downward under the force of gravity. Water must move through this zone, which is not saturated with water, in order to enter the saturated zone where the groundwater moves laterally through completely filled pore spaces. The mechanics of the movement of water in this zone are complex and not understood as well as the movement of water in some other phases of the hydrologic cycle.

Fluid movement in the zone of aeration involves movement in three phases; water, water vapor and air. Water is retained in the zone of aeration as films of water surrounding particles of the rock or completely filling some of the void spaces between the grains. Water completely filling the voids can move downward until it reaches the water table as groundwater recharge. Downward movement of water in the zone of aeration is primarily in the water phase; the transfer of water vapor does not contribute significantly to groundwater recharge.

Particular *groundwater reservoirs,* such as the deep sandstone beds that underlie the northern Great Plains, may underlie tens of thousands of square miles, while another groundwater reservoir, such as the sand beds underlying Long Island, may be confined to relatively small areas. The movement of groundwater in small aquifers (water-bearing beds or strata) may be only part of a large pattern of groundwater movement throughout a larger area. The effects of water use and development must be studied in both large and small areas in order to understand fully the regimen of groundwater.

Water moves through groundwater reservoirs until it is discharged as springs, by seepage into streams and lakes, and through withdrawal by man. Such discharge allows a continual movement of water through aquifers and provides room for recharge. The amount of water moving through the ground and the total amount of water removed from the groundwater reservoir in any specific period of time are known only approximately. The water that seeps into streams and lakes provides the base flow of the streams—i.e., the low flow that is sustained through the driest part of the year. If water is diverted from a groundwater system and withdrawn for use, such as irrigation, or the water is evaporated back into the atmosphere, the base flow may be reduced substantially or even eliminated. The complex interrelationship of water on the surface and under the ground is one part of the hydrologic cycle which we need to study more intensively in order to make the best use of both sources of water.

Runoff occurs on the land surface after soil

moisture deficiencies have been satisfied and infiltration of precipitation to the ground is proceeding at its maximum rate. When a storm occurs, the rate of surface runoff starts at zero and gradually increases until it becomes a relatively constant percentage of the rainfall rate. The amount of rainfall, its intensity and duration, and the antecedent moisture deficiency determine the total quantity of runoff from a given storm. Rain falling on snow-covered ground may hasten melting and thus cause a higher runoff than would be expected for the same storm falling on bare ground. The relations between the variable factors contributing to runoff are not completely known, but it seems likely that more precise measurements, rather than theoretical analyses, will lead to an improved understanding of the phenomena.

The volume of water moving in a well-defined channel is governed by the gradient of the channel and water surface, the roughness of the channel, and the cross-sectional area of the water body.

Computation of this volume (*discharge*) may be made directly by measuring the cross-sectional area and the velocity, or indirectly by measuring each of the above factors. Continuous monitoring of discharge at key locations on streams is a vital factor in water management.

Storage of surface water in *reservoirs* regulates the flow of streams so that water supply can meet the demand throughout the year, flood peaks can be suppressed, power can be generated, or recreational use can be made of the water. Natural lakes may be considered to be stream flow in storage. This is well illustrated by the contrast between the relatively steady discharge of the St. Lawrence River which drains the Great Lakes, and the natural discharge of the Missouri River, which drains much of the Great Plains.

The problems of regulating the flow of streams to provide beneficial use and control has always been a challenging task to the hydrologist and will continue to be one of his major tasks.

Evaporation (q.v.) and *transportation* are two phases of the hydrologic cycle that are difficult to study quantitatively. Only in the last few years have instruments and mathematical techniques been developed to measure and calculate the rate and amount of evaporation from open water and land surfaces and the transpiration of plants. Evaporation may be estimated by measuring the loss of water from open pans on the land surface, but application of the evaporation rates to lakes and swamps can be misleading. Recent studies of energy budgets and heat transfer provide more reliable means of calculating total *evapotranspiration* (q.v.) than we have had in the past. In the future we may look to the use of instruments carried in aircraft to give us gross figures on the total evaporation from a particular area.

The water on and below the surface is not pure; it contains varying amounts of different chemical substances in solution (see *Rainwater*). The amount of material that is carried in solution by the water depends upon the solubility of the rocks with which the water comes in contact and the length of time of contact. Research into the physical and chemical properties of water and rocks and the interrelations of the water and the dissolved mineral matter are of extreme importance because of the uses to which we put water. Industries, municipalities and irrigation all require water that is within certain but different limits of chemical, physical and bacteriological quality. Groundwater at a particular place generally has a fairly constant chemical quality, but water in streams in the same area may vary greatly in chemical quality during the year.

Along with these two basic tasks is the necessity for using water economically and solving serious water problems. The practical hydrologist has provided the means for controlling and using water, and because of his need for better techniques of control, he has aided and given impetus to the descriptive and theoretical studies of water.

Summary

Thus, it can be seen that in reality there are two hydrologic cycles—one describing the endless global movement of water and another cycle of hydrology that encompasses the description of water resources, the development of better understanding of water through description and research, the application of scientific and engineering techniques to the control and use of water, and the feedback of the results of experience to provide more comprehensive knowledge of water.

CHARLES J. ROBINOVE

References

Chow, V. T. (editor), 1964, "Advances in Hydroscience," New York, Academic Press.

Chow, V. T., 1967, "New trends in hydrology," *Nature and Resources,* 3(2), 4–8.

Langbein, W. D., and Hoyt, W. G., 1959, "Water Facts for the Nation's Future," New York, The Ronald Press.

Linsley, R. K., Jr., Kohler, M. A., and Paulhus, J. L. H., 1949, "Applied Hydrology," New York, McGraw-Hill Book Co.

McGuinness, C. L., 1963, "The Role of Ground Water in the National Water Situation," *U.S. Geol. Surv. Water-Supply Paper* **1800**.

Meinzer O. E. (editor), 1942, "Hydrology," Physics of the Earth Series, New York, Mc-Graw-Hill Book Co.

Nace, R. L., 1964, "Water in the World," *Natural History*, **73**, No. 1, 10–15.

Nace, R. L., 1970, "World Hydrology: Status and Prospects," *Internat. Assoc. Sci. Hydrol., Publ. No.* **92**.

U.S. Dept. of Agriculture, 1955, "Water, The Yearbook of Agriculture," U.S. Govt. Printing Office, Washington, D.C.

Cross-references: *Desalination; Groundwater; Hydrogeology; Hydrologic Cycle; Hydrology, Semi-arid Regions; Porosity and Permeability; Rainwater; Runoff; Vadose Water; Water—Substance and Solvent.* Vol. II: *Energy Budget of the Earth's Surface; Evaporation; Evapotranspiration.*

HYDROLOGY, COASTAL TERRAIN

In coastal districts, the fresh water in the water table migrates slowly downhill to the sea. Because of their different densities, the fresh water and salt water do not generally mix, except in the ocean where the tides, waves and currents do the mixing. In the aquifers in coastal districts, the less dense fresh water tends to float on the more dense saline water just like an iceberg. Figure 1 shows the shape of the fresh water lens on a sandy island assuming that the fresh water is being replenished by rainfall. The relationship between the thickness of the fresh water body (a) and the depth of the lowest part of the freshwater body below sea level (b) is:

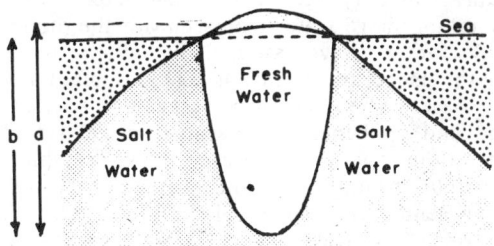

FIG. 1. The characteristic shape of the freshwater lens in islands made of uniformly permeable materials in humid areas.

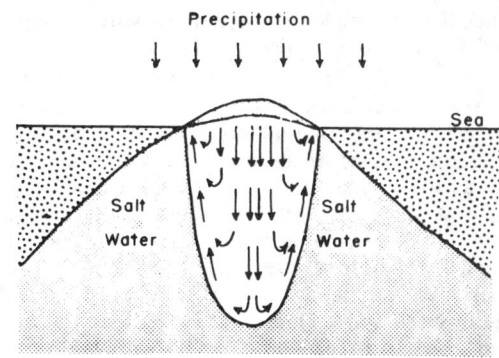

FIG. 2. Flow lines within the fresh-water lens on the island shown in Fig. 1. The lens is recharged by infiltration of rain and snow into the ground but loses water by diffusion and by flowage into the sea.

$$\frac{b}{a} = \frac{\text{Specific gravity of fresh water}}{\text{Specific gravity of seawater}} = \frac{40}{41}$$

Thus for every foot the fresh water stands above sea level, the surface of the salt water lies some forty times as many feet below sea level. These are, of course, only approximate figures, depending mainly on the salinity of the seawater and the purity of the fresh water. Figure 2 shows the flow lines, i.e., the paths of water movement, for the fresh water contained within the lens. Both the lens and the underlying salt water will rise and fall with the tide unless there is a barrier between the underground water and the sea. The time of the peaks and troughs of the fluctuations becomes later as traced inland, just as the time of high and low tide becomes progressively later as it is traced up the tidal part of a river. The time between the peaks and troughs will remain the same, while the time lag will be constant for a given well.

Some mixing of the fresh and salt water does take place at the interface. Usually this is negligible, but it can reach appreciable proportions under certain favorable conditions. This produces a brackish water zone which may be quite thick. This zone occurs where there are considerable fluctuations in the level of the interface due to tidal action or irregular heavy rains. Thus a strong development of a brackish zone is found in the basalt aquifers along the coast of Oahu in Hawaii (Visher and Mink, 1964). It is also increased by pumping the wells in these regions as we shall see below. Like the fresh water, this brackish water lens moves slowly downslope.

Variations in Water Table Pattern with Climate and Form of the Aquifer

At coastal sites on permeable materials where several large rivers cross the area and there is continuous replenishment by precipitation, the floating fresh water and its flow lines are as in Fig. 3a. With a seasonal reduction in the replenishment, seawater tends to enter the rivers in low-lying districts. Even in areas of tropical rainforest, seawater may penetrate nearly to the headwaters of coastal rivers in the drier seasons. Thus in May and early June, 1961, seawater reached some 116 miles up the Barima River in British Guiana, i.e., almost to its headwaters. In this case, the freshwater

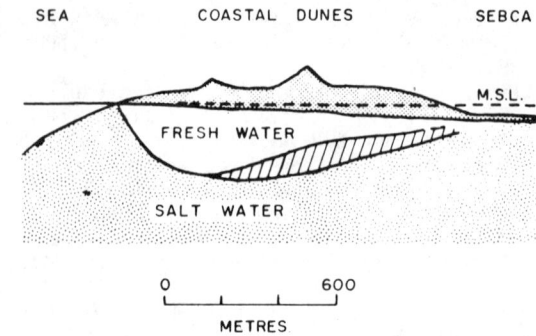

FIG. 4. Cross section of the groundwater zones below a sand dune belt at Zaura, Libya (adapted from Underhill and Atherton, 1964). Note the water table sloping inland, the brackish water zone (cross-hatched), and the localized freshwater lens. Vertical exaggeration, 1:10,000.

lens and its attendant flow lines take the form shown in Fig. 3b. Tropical lowland rivers where this may occur are marked by a line of mangroves, usually *Rhizophora* species, lining the bank. Behind this line occurs the normal swamp or rainforest.

In desert areas on permeable materials where there is negligible recharge from local precipitation, the fresh water takes the form of a thin, almost horizontal sheet (Fig. 3c).

This undergoes local modification where coastal dunes separate the sea from a zone below sea level as at Zaura in Libya. The relationship between the Mediterranean Sea, the dunes, the salty depression or sebca (sabka), and the groundwater are shown in Fig. 4 (after Underhill and Atherton, 1964). A fresh-water lens occurs beneath the dunes and is replenished by rainfall. The water table slopes downhill inland to the sebca. There the intense evaporation disposes of fresh water flowing toward the coast from further inland, fresh water and brackish water flowing downslope from the dunes, and seawater which has traveled inland beneath the fresh water. Thus Zaura can only expand its fresh-water supply by taking more water from beneath the dunes.

So far, we have assumed that all the rocks on the shore are permeable. This is often not the case. Where artesian basins dip seaward and end up beneath the sea, a special set of conditions applies. Figure 5 shows a typical example from the Baltimore area, Maryland, after Bennett and Meyer (1953). Where a well is sunk in the fresh water, the well is artesian, but where it penetrates the saline water, it is not. This is due to the different densities and hence piezometric surfaces of the two kinds of water.

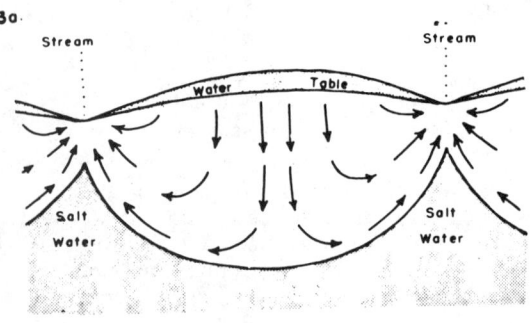

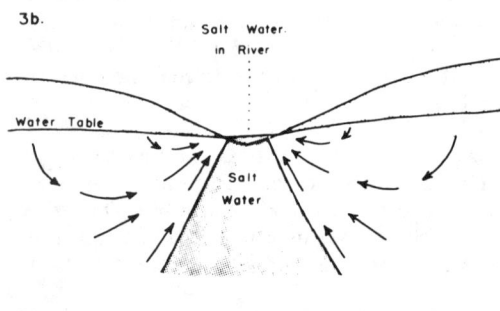

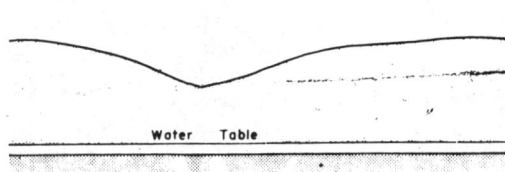

FIG. 3. Idealized common situations on coasts (partly after Parker, 1955). Fig. 3(a) represents a coastal area with uniformly permeable materials under a humid climate. Fig. 3(b) shows what happens in the dry season when salt water backs up the rivers. Fig. 3(c) shows the nature of the freshwater table along an arid coast.

HYDROLOGY, COASTAL TERRAIN

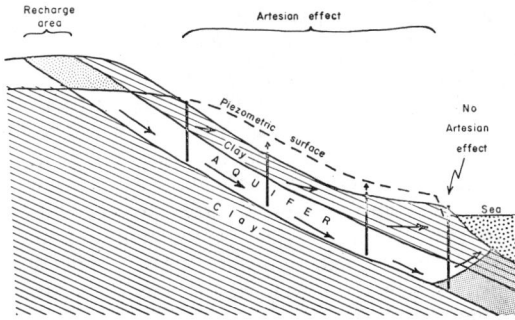

FIG. 5. Idealized cross section of a coast where an artesian aquifer outcrops some distance offshore as well as on the land surface (after Parker, 1955). This applies to most of the Atlantic coast of the United States from Long Island to Florida, parts of the Gulf coast and Pacific coast of the United States, parts of the Hawaiian islands, as well as to the Australian artesian basins.

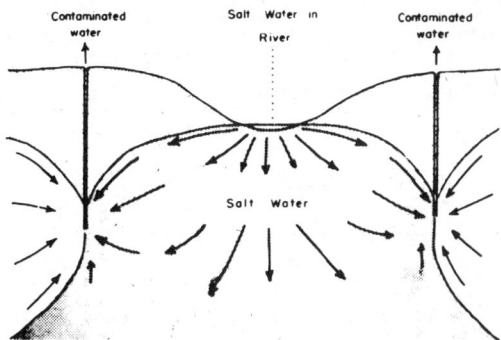

FIG. 6. Effect of pumping wells along the margins of streams after seawater has backed up the rivers as in Fig. 3(b). The contamination will last until fresh water again fills the river and the salt water has sunk to the position in Fig. 3(a).

Water Tables and Eustatic Sea Level Changes

The worldwide rise of sea level as the glaciers melt has an important effect on the water tables in coastal areas. As the sea level rises, so does the water table and the saline water. This increases the tendency of tides to sweep saline water into rivers and it tends to push the fresh water shoreward. In the case of the Atlantic seaboard of the United States, the sea is rising at a rate equivalent to approximately 2 ft/century. It has been estimated that this small rise will cause the fresh water in the artesian aquifer of New Jersey to recede inland at the rate of 1–4 miles/century, depending on the dip of the aquifer (Long, in Parker, 1955). Thus a small rise in sea level can have quite a large effect.

Exploitation of the Fresh Water

The thin layer of fresh water underlain by salt water means that great care must be taken in exploiting the fresh water. The usual method of exploitation today is by a well from which the water flows or is pumped. When water is taken from a well, the surface of the water table is lowered close to the well by an amount depending on the output of the well and the porosity of the aquifer, among other factors. As we have already stated, the thickness of the fresh water below a point is some forty-one times its height above sea level. Thus for every foot of lowering of the surface of the water table, the fresh water-saltwater boundary moves 40 ft nearer the surface. Thus it does not take a great output of water to cause the bottom of the fresh water layer to rise to the bottom of the well. Thereafter the water produced by the well will be saline. By limiting the production, contamination can be stopped.

Particular care must be taken in the case of artesian basins in coastal regions. The quantity of fresh water that is stored is finite, and the amount of recharge is limited. Overpumping will cause influx of saline water, as is happening in the Savannah area of Georgia and South Carolina in the United States (McCollum and Counts, 1964). Only restricted pumping or artificial recharge will prevent the eventual salinization of this currently productive freshwater source.

In the case of salt water backing up a river in the dry season, pumping water near the river produces the result in Fig. 6. Once this has happened, the state cannot be altered until fresh water again fills the river. Then the salt water subsides to its former level and the wells can start producing fresh water once again.

This emphasizes the advantage of the Libyan collecting galleries used by the Romans. These are in a region where there is little precipitation and the fresh water layer is only a few inches deep. Any well would be contaminated by the saline waters underneath. However, the collecting galleries are about 400 meters long and skim off the fresh water from a large area, thus producing a large quantity of fresh water with negligible downdraw.

Compared with the collecting galleries, pumping out water always has the disadvantage of causing greater mixing of fresh and salt water (Fig. 7). This is greater in the case of the standard single-pumped well than from the double-pumped well (Underhill and Antherton, 1964). In the latter case, water is pumped from the brackish and saline water

537

HYDROLOGY, COASTAL TERRAIN

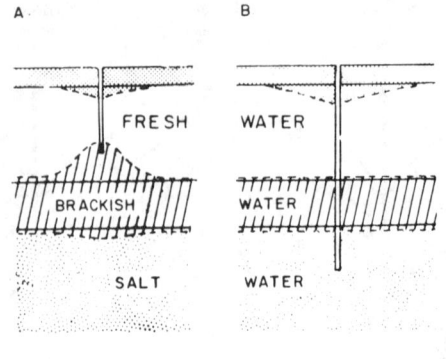

FIG. 7. Results of intermittent pumping of a single pumped well (case A) contrasted with the effect of constant pumping of a double pumped well (case B). In the latter case, water is abstracted from the brackish and salt water zones at a rate calculated to maintain a horizontal interface between the fresh-water and brackish water (after J. H. Edelman in Underhill and Atherton, 1964).

zones in just the right quantities to keep the brackish water–fresh water interface horizontal. However, some means of disposing of the brackish and saline water must then be found, and this is not always easy.

Constantly pumped normal wells on shores such as at Oahu, Hawaii, show thickening of the brackish water lens on the seaward side only. Similar wells which are used intermittently produce a thickening of the brackish water lens at the expense of the freshwater lens on all sides of the well. This is due to the water movements involved in the alternate thickening and thinning of the freshwater lens.

Draining of Swamps and Marshes

Many tropical shores have large areas of freshwater swamps and marshes behind them. The shores of the Guianas in South America are but one example. It is very tempting for agriculturists and land development corporations to try to drain these swamps and put them to some more profitable use. Unfortunately such swamps are the source of the fresh water in the coastal districts. Thus when there was an attempt made to drain part of the Everglades behind Miami, the drains proved unsuccessful, but they caused a marked thinning of the freshwater layer. When salt water entered the drainage canals in the dry season, it promptly contaminated the freshwater lens upon which Miami relied for its water supply. Thus if the city of Miami is to use ground-water for drinking purposes, the Everglades must remain a swampy wilderness.

In the case of the Guianas, the early Dutch settlers realized the problem, and instead of reclaiming the swamps, they reclaimed the intertidal zone. This required the building of hundreds of miles of embankments or polders, together with numerous tide-operated sluices so that the land lying below high-tide mark could be drained, thus ensuring an ample supply of irrigation water from the swamps behind the polders.

STUART A. HARRIS

References

Bennett, R. R., and Meyer, R. R., 1953, "Geology and ground-water resources of the Baltimore area, Md.," *Maryland Bd. Nat. Resources Dept. Geol. Mines Water Resources Bull.*, **4**.

Cooper, H. H., Jr., 1964, "Sea water in coastal aquifers," *U.S. Geol. Surv., Water Supply Paper* **1613-C**, 84pp.

McCollum, M. J., and Counts, H. B., 1964, "Relation of salt-water encroachment to the major aquifer zones, Savannah Area, Georgia and South Carolina," *U.S. Geol. Surv., Water Supply Paper* **1613-D**, 26pp.

Parker, G. G., 1955, "The encroachment of salt water into fresh," in "Water," *Yearbook Agr. U.S. Dept. Agr.*, 615–635.

Underhill, H. W., and Atherton, M. J., 1964, "A coastal ground water study in Libya and a discussion of a double pumping technique," *J. Hydrol.*, **2**, 52–64.

Visher, F. N., and Mink, J. F., 1964, "Ground-water resources in Southern Oahu, Hawaii," *U. S. Geol. Surv. Water Supply Paper* **1778**, 133pp.

Cross-references: *Aquifer; Ground Water; Hydrology; Hydrology, Subsurface Waters; Water Table.* Vol. I: *Mean Sea Level Changes.* Vol. III: *Mangrove Swamps; Sabkha or Sebkha.* Vol. VI: *Polder.*

HYDROLOGY, LIMESTONE TERRAIN

The primary factor in the hydrology of limestone terrains is the solubility of carbonate rocks in *aqueous solutions* (q.v.), which leads to the (largely) underground network of pipes and channels known as *karst* (see Vol. III). For the purpose of this discussion, limestone terrains are defined as regions where carbonate rock formations extend from near the surface to below the *water table* (q.v.). The soil, and the material directly beneath, strongly influence the hydrology of these terrains. An impermeable cover overlying a thick limestone sequence can cause most of the precipitation to pass out of the area as surface runoff and the effect of the limestone is minimized. A pervious soil horizon, or erosion of the impermeable cover, will facilitate infiltration to the limestone subtrate.

Most precipitation has a pH of 5.5 ± 1.0. As the water moves through a soil, the pH can be decreased due to the fact that the partial pressure of CO_2 is usually higher in the soil than in the atmosphere, and/or by the addition of humic acids from organic sources. The acid aqueous solution then enters the underlying carbonate rocks.

The hydrology of a carbonate region can range from a karst area with many interconnecting passages and caves, readily absorbing surface waters and transmitting these waters fairly rapidly from source area to discharge or storage areas, to a situation where a limestone formation of low permeability acts as an aquiclude and the terrain has high surface runoff and little available groundwater. The variability of the hydrology is as significant as the unique hydrology of a fully developed karst terrain and makes the concept of an average hydrology rather hypothetical. This variability can be demonstrated by comparing an area in Texas where the Edwards limestone absorbs an estimated 156,000 acre-feet per day from surface stream flow (Swinnerton, 1942, p. 669) with a chalky portion of the Cooper Marl in South Carolina which is used as an unlined public water supply conduit for Charleston (Stringfield and LeGrand, 1969, p. 368).

The permeability of carbonate rocks varies greatly as a function of its purity, Ca:Mg ratio, texture, structure, and history. After deposition, a limestone has primary porosity and permeability controlled by the size, shape, and distribution of the sediment. Under compaction, consolidation, and recrystallization, the primary permeability is decreased. Secondary porosity and permeability, caused by joints, fractures, and enlargement of openings by solution, increases with the age of the formation.

Given a surface or cover that will absorb precipitation and permit infiltration to the limestone, the water then enters a system of interconnected openings. In the formative stages of a subsurface drainage network the permeability, whether primary or secondary, will be higher in some zones than in others. This produces maximum flow in such zones and the greatest solution of the limestone occurs here, thus further increasing the permeability in the same zone. The fissures will often be smaller and more evenly distributed in the upper portion of the formation and larger and fewer with depth. This distribution results in the areal source of water, precipitation, entering the system and flowing toward a linear or point discharge zone. The zones of discharge are due to the presence of one or more barriers to continued subsurface flow. These barriers may be an impermeable rock formation or structure, usually a fault, or another body of water, usually the ocean.

The concentration of groundwater flow in conduits of this sort which are enlarged by solution, as opposed to the more uniform flow distribution in other porous formations, is peculiar to carbonate rock terrains. Enlarged joints and fractures usually act as the vertical conduits, and solution along bedding planes provides for lateral movement. Sinkholes can be major recharge points or, if lined by impermeable clay, often residual, may form lakes perched above the regional water table. A natural well is formed where a vertical solution conduit reaches from the surface to the water table, e.g., the sacred Mayan well at Chichen Itza in the state of Yucatan, Mexico.

The water table in limestone terrains which have undergone extensive karstification fluctuates greatly, rising in rapid response to precipitation input and falling due to rapid flow of water through solution passages to the discharge zones. Because there may be perched flow or storage on impermeable layers and along some bedding planes, some hydrologists question the validity of the water table concept in limestone terrains. However, there is a recognizable piezometric surface in most of the wide variety of carbonate terrains.

If the base level is low near a limestone area, full karstification can take place. The zone of most solution widening is believed to be at or just below the water table. As the table gradually lowers, the zone of solution also lowers and the network can develop through all levels of a formation down to the regional base level. The interconnecting hydrologic system can be extended through the whole area but the ultimate regional water table would be close to base level. With a higher base level the system is unlikely to extend to any great depth but groundwater would be more available for use. The paleohydrology of a limestone terrain is important because a system of solution-enlarged fissures often depends on past flow conditions. After the system is developed, any raising of base level, tectonically or by a change in sea level, will induce storage and circulation through the entire system.

With an impermeable cover or a dense, impermeable limestone most of the flow will occur on the surface. With a permeable cover or limestone with well-developed permeability, surface flow will vary greatly as water enters and leaves the underground circulation system.

HYDROLOGY, LIMESTONE TERRAIN

TABLE 1. TYPES OF CARBONATE AQUIFER SYSTEMS IN REGIONS OF LOW TO MODERATE RELIEF[a]

Flow Type	Hydrological Control	Associated Cave Type
I. Diffuse flow	Gross lithology	Caves rare, small, have irregular patterns
	Shaley limestones; crystalline dolomites; high primary porosity	
II. Free flow	Thick, massive soluble rocks	Integrated conduit cave systems
A. Perched	Karst system underlain by impervious rocks near or above base level	Cave streams perched—often have free air surface
1. Open	Soluble rocks extend upward to level surface	Sinkhole inputs; heavy sediment load; short channel morphology caves
2. Capped	Aquifer overlain by impervious rock	Vertical shaft inputs; lateral flow under capping beds; long integrated caves
B. Deep	Karst system extends to considerable depth below base level	Flow is through submerged conduits
1. Open	Soluble rocks extend to land surface	Short tubular abandoned caves likely to be sediment-choked
2. Capped	Aquifer overlain by impervious rocks	Long, integrated conduits under caprock; active level of system inundated
III. Confined flow	Structural and stratigraphic controls	
A. Artesian	Impervious beds which force flows below regional base level	Inclined three-dimensional network caves
B. Sandwich	Thin beds of soluble rock between impervious beds	Horizontal two-dimensional network caves

[a] After White, 1969.

Data from stream gages, for example in the Nashville Basin, underlain by Ordovician limestones, show low runoff rainfall ratios and substantial variation depending on whether the gaging point is above or below a discharge zone (Jacoby, 1970). There can be a total absence of surface runoff as in the Yucatan Peninsula of Mexico. A large river can disappear, as the Lesse in southern Belgium, or a large spring can issue forth, e.g., the Fountain of Vaucluse in southern France. The latter is the largest recorded spring flow from a single opening and has a mean flow of 800–1000 ft^3/sec.

Carbonate aquifers can be the most productive of any rock type and, at the other extreme, there are dense, often impure, limestones which are aquicludes. The main control is secondary porosity and permeability. The yield to wells can be prodigious as in the case of the Tertiary limestone Floridian Aquifer. Because of the frequent anisotrophism of secondary permeability in carbonate rocks, in most cases a high-yielding well must be sited at the intersection of fractures and/or encounter solution passages in the limestone. If the water table is unstable, as in some karstic areas, well yield will be unreliable. It may be prolific after rain but go dry in a short time.

In 1969 W. B. White categorized aquifers into the types in Table 1.

Many of the most productive aquifers in the United States are type IIB or IIIA where a karst region has been buried or lowered below base level and the circulation is deep.

There are two problems associated with wells in carbonate areas. First, the high transmissibility of some areas reduces the usual effectiveness of an aquifer in purifying the water. Chemical pollutants and pathogenic bacteria that are removed in most aquifers may persist through the large solution passages in many limestone aquifers. This is critically important in considering deep-well disposal of contaminates in a limestone area. Second, the water is often saturated or even supersaturated with calcium carbonate. At a well head the pressure is reduced and the $CaCO_3$ can precipitate to a degree that will seal some of the fissures and encrust the well, causing a decrease in yield.

GORDON C. JACOBY, JR.

References

Back, W., and Henshaw, B., 1965, "Chemical Geohydrology," in (Chow, V. T., editor), "Advances in Hydroscience," Vol. 2, New York, Academic Press, pp. 49–109.

Castany, G., 1967, "Traité Pratique des Eaux Souterraines," Second ed., Paris, Dunod, 661pp.
Davis, S. N., and DeWiest, R. J. M., 1966, "Hydrogeology," New York, John Wiley & Sons, 463pp.
Jacoby, G. C., Jr., 1970, "Fluvial Flow as a Function of Drainage Basin Morphology in the Southeastern United States." Ph.D. Thesis, Columbia University (unpublished).
Stringfield, V. T., and LeGrand, H. E., 1969, "Hydrology of carbonnte rock terraines—a review," *J. Hydrol.,* **8**(3 and 4), 349–376, 377–417.
Swinnerton, A. C., 1942, "Hydrology of Limestone Terraines," in (Meinzer, O. E., editor), "Hydrology," New York, Dover Publ.
White, W. B., 1969, "Conceptual models for carbonate aquifers," *Ground Water,* **7**(3), 15–21.

Cross-references: *Aqueous Solutions; Aquifer; Calcium Carbonate; Ground Water; Hydrology; Subsurface Waters; Paleohydrology; pH-Eh; Porosity and Permeability; Runoff; Water Table.* Vol. III: *Karst; Limestone Caves.* Vol. VI: *Limestones.*

HYDROLOGY, SEMIARID REGIONS

General Concepts

The semiarid regions of the earth's surface occur as transition zones between the arid deserts and the subhumid belts. Water movement will shape the landscape, according to its geology and past topography, and will work in conjunction with wind erosion, solar insolation, temperature changes, and soils (stable or in movement), as well as with the vegetation and the animals which live thereon, to produce an ecological balance of all factors, either in a temporary or a permanent sense.

Such a balance will incorporate past balances, achieved under past conditions, and may preserve some of them as "fossil" types; the balance itself will have been changed by man's activities, living on the water, the vegetation and the animals. Under the rigorous conditions of life in the semiarid regions, the successful intervention of man has generally tilted the balance in the direction of diminishing returns from those products on which he can live.

The Semiarid Regions of the Earth. Strict definitions of climatic regions vary according to the purpose for which they are made; plant ecology, faunal distribution, living conditions for man and domesticated flocks impose different boundaries for the semiarid regions of the world. However, the controlling factors in the arid zones are, clearly, low precipitation and the amount of heat, and these have been the basis of many scientific classifications, since that of Martonne and Aufrère in 1925. The usual classification today is based on Thornthwaite (1948), and has been modified by Meigs (1952) who produced maps of the distribution of the arid homoclimates (Fig. 1).

In defining arid and semiarid areas, Meigs (1952) used only three factors: *humidity, season of precipitation* and *temperature.* Extremely arid (E) is defined as an area with at least one entirely rainless twelve months; arid (A) and semiarid (S) are regions where precipitation is less than potential evaporation after the 1948 Thornthwaite formulas. Precipi-

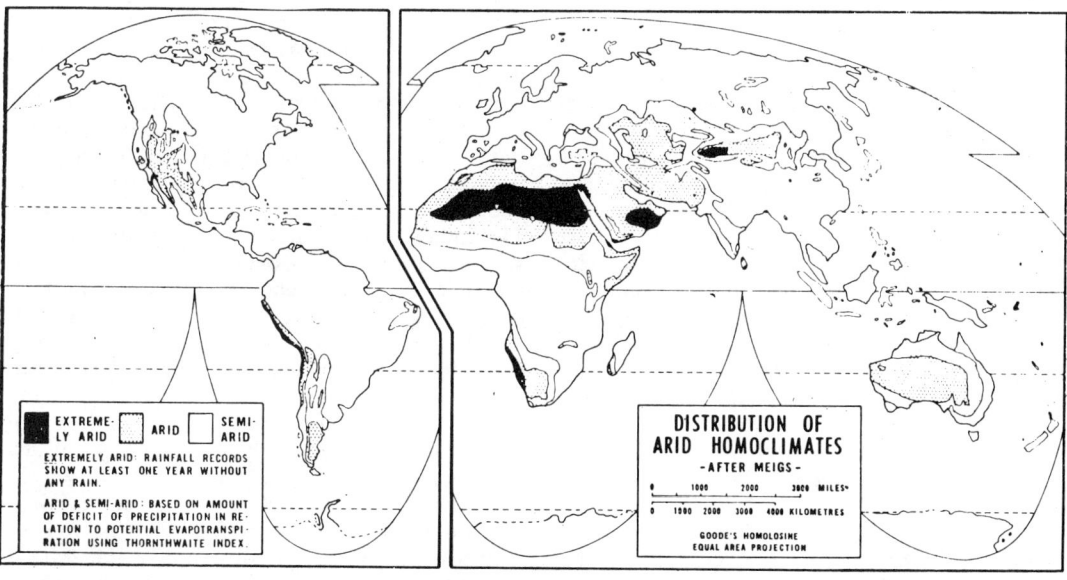

Fig. 1. World distribution of arid and semiarid climates (after Meigs, 1952; from White, 1961).

tation may be irregular and non-seasonal (a); it may occur in summer (b), as on the savanna and pampas under a monsoon or tropical semiarid climate; or it may occur in winter (c), as on the steppe or even tundra (not further considered here) under the Mediterranean semiarid climate. Finally, Meigs gives indices for the temperatures of the coldest and hottest month, with (0) for $-0°C$; (1) for $0-10°C$; (2) for $10-20°C$; (3) for $20-30°C$; (4) for $+30°C$.

Thus "S.c.l.3." would indicate a semiarid area, precipitation occurring in a cold winter, with the coldest month in the $0-10°C$ average range and the hottest month in the $20-30°C$ average range; this would be a typical Mediterranean semiarid climate, occurring in Morocco, Algeria, Lebanon, northern Iran and also on the western coast of the United States around latitude 35°N.

Arid and semiarid lands account for over one-third of the land surface of the earth, while cultivated lands account for but one-tenth of the whole. The greatest belt of arid and semiarid regions extends across North Africa as the Sahara, through the Arabian Peninsula with the "Empty Quarter" of extreme aridity, into the Salt Desert of Iran, and the Takla Makan of Central Asia. In North America, the Great American Desert has thus been described by Shreve (1942) in plant ecological terms:

"Beginning in the north, it will be seen that the desert extends southwards from central and eastern Oregon, embracing nearly all of Nevada and Utah except the higher mountains, into southwestern Wyoming and western Colorado, reaching westwards into California to the eastern bases of the Sierra Nevada, San Bernardino and Cuyamaca mountains. From southern Utah the desert extends into northeastern Arizona at the same time that it occupies the western and southwestern parts of that State. On the highlands of southeastern Arizona and southern New Mexico the continuity of the desert is broken by 'Desert Grassland Transition.' At a slightly lower elevation, it reappears in the valleys of the Rio Grande and Pocos rivers, extending as far east in Texas as the lower courses of Devil's River. In the north, an isolated area of desert occupies part of the Columbia River basin in eastern Washington."

Hydrology. In connection with the International Hydrological Decade of 1966–76, hydrology has been defined as follows: "Hydrology is the science which deals with the waters of the earth, their occurrence, circulation and distribution on the planet, including their response to human activity" (White, 1963).

A stricter definition is that given in the *Journal of the Institution of Water Engineers* (White, 1963, p. 381): "Hydrology is the science of the occurrence and movement of water over and under the earth's surface from the moment of precipitation to the moment of entry into the ocean or of evaporation into the atmosphere."

The Hydrologic Cycle. The limited amounts of water which enter and which circulate within the semiarid and arid regions are the main subject of this presentation. But to clarify the position, it is desirable to glance at the general circulation of water which takes place above, on or at limited depths below the surface of the earth.

The hydrologic cycle describes the circulation of water in nature. It traces the water from its evaporation from the surface of the earth, through its movement as water vapor in the atmosphere to its condensation as clouds and mist. If the minute droplets of this initial condensate can grow by coalescing, then precipitation will fall, as rain, hail or snow, according to initial and subsequent temperature conditions. Of the water which thus returns to the surface of the earth, much will be evaporated or transpired through the vegetation almost at once, depending on the heat available for such vaporization. Precipitation which falls on the sea short-circuits much of the hydrologic cycle.

The non-evaporated portion will first sink into the soil and fill its pores. Thereafter, the remaining non-evaporated water may run off on surface or sink underground to reach the porous and permeable rocks, known as "aquifers," which can transmit and store water. Surface runoff will form streams, rivers and lakes; the main river of the region may flow to the sea, or evaporation may be so great, and the quantity of water so small, that no outlet to the sea can be achieved. Then the surface drainage takes place in a closed basin, very common in the semiarid regions of the world. The underground water will move through the aquifers according to hydraulic conditions, remaining free of evaporation losses, until it reaches the springs or seepages through which it is discharged from the aquifer. It then once more forms part of surface flow and is again subject to evaporation.

The Water Balance Sheet. A watershed is an area of land which "sheds" or discharges all its surface runoff through a common point, which is often the estuary of the main river draining the watershed to the sea. If there is no movement of groundwater into or out of the watershed beneath its encircling water divides, then the watershed forms a hydrological unit for which the following balance holds for any unit period of time, which is usually the hydrologic year of twelve months:

Total Precipitation on the Watershed equals
- Evapotranspiration to the Atmosphere
+ Surface Runoff, including Spring Discharge
+ Infiltration to the Groundwater
± Changes in Water Storage (surface and underground)

If all the items in this balance sheet can be measured directly, then the amount of errors incurred can be determined insofar as the equation does not balance. If some major constituent is not, or cannot be, measured, then its size is found by difference, and all errors of measurement tend to become indeterminate. Very often, evapotranspiration is determined by difference, since it is difficult to measure directly; since evapotranspiration is of major importance in the semiarid and arid zones, its determination by difference greatly reduces the accuracy of the wather balance sheet (Fig. 2).

The Heat Balance Sheet. The sun supplies heat to the surface of the earth, and this heat is again dissipated, so that over periods of say a year, the surface of the earth neither gains nor loses heat; as described by Drummond (1958), radiation maintains this thermal balance. The heat reaching the earth from the sun amounts to 2 g-cal/cm²/min, and in the arid zones more than 80% of this heat reaches the surface of the earth—89% at Yuma, Arizona, and 84% at Helwan in Egypt (Thornthwaite, 1958).

In order to avoid heating up, the earth's surface must return this incident heat to the atmosphere. The rate of return is determined by the reflectivity and emissivity of the different surfaces—bare soil, vegetation, water and others. The ability of a surface to reflect incident heat is known as its albedo; it is high in desert surfaces and accounts for the brilliant light and high air temperatures.

Water is a good conductor of heat, and the wetting of a dry soil will double its heat capacity and increase tenfold its heat conductivity. So, the intake of heat into a wet soil is high, yet the heat is used to vaporize part of the water; due to the high latent heat of vaporization of water, such water takes back much heat to the atmosphere. Where the heat supplied is generally greater than the heat required to vaporize all water on the surface of

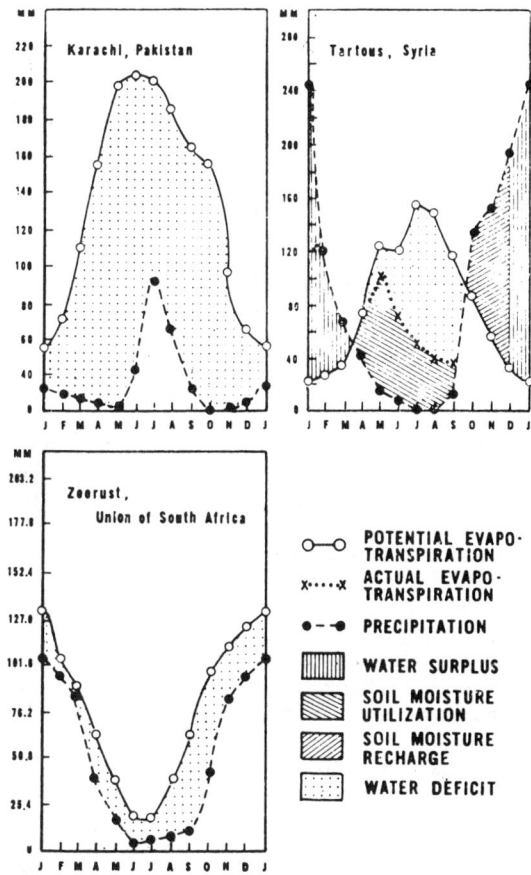

FIG. 2. Water balance diagrams for Karachi, Tartous and Zeerust. (These averages were obtained using the Thornthwaite method for computation; White, 1961).

the area (including water rising by capillary action), there will be found the arid and semiarid zones of the earth.

Collection of Hydrologic Data in the Semiarid Regions

Conditions within the arid and semirarid zones are generally unfavorable to the collection of hydrologic data. Lack of permanent settlements in the past has resulted in almost no long-term records, while the scattered settlements of today provide insufficient meteorological posts for wide regions of sharply differing precipitation.

Within the semiarid regions of the Eastern Mediterranean, for example, the longest record of precipitation is that maintained by the American Colony in Jerusalem; this record commences in 1844 and gives some idea (Burdon, 1959) on precipitation variations over the past 120 years. Such a glance into the past is valuable, but other indicators, such as the geomorphology, plant ecology and historical accounts of floods and other natural phenomena, must be used when information on earlier climates is required.

Instruments. The difficulties regarding the use of instruments to measure precipitation, heat, humidity, wind strength and direction, as well as dew and dew point and evaporation from measuring pans, are clear. There are too few permanent habitations at which to establish stations, still fewer trained and conscientious people to man such stations. For simple precipitation stations, there is also a lack of interest; rain may fall only 10–20 days a year. The application of radar to determine areal rainfall (Kessler, 1966) and the use of photographs from satellites to show cloud cover, areas wetted by precipitation and vegetative growth offer new possibilities of overcoming some of the data collection difficulties in arid regions and are now being tested.

Gauging stations for surface runoff present different problems. The presumed river bed is often only a wide expanse of boulders with numerous channels down any one of which the water may flow. Such flow may be lost over quite short distances by infiltration into the gravels. To choose a gauging site is difficult, but not impossible. Seldom can the flow of the river at the gauging station be calibrated, so that a height-discharge relationship is known; each flow must be measured for depth and speed, possibly over the first twenty years. So, a man must be at the gauge station when the river flows, and such flow may, like the rain, occur but a few times each winter. In the rare cases where reservoirs have been constructed, they can be used to measure runoff whose existence, but not amount, was known when the dam was planned.

Measurements of the amount of groundwater in storage can be achieved by recorders or depth measurements at wells or boreholes; such points are often centers of settlement. Soil moisture measurements are seldom possible on a permanent base, but are made for specific investigations. This also applies to the composition of the waters, though regular chemical sampling of groundwater is feasible.

Ecological Conditions, Past and Present. In the northern temperate zones, the present ecology has been greatly influenced by the glacial and interglacial periods of the Quaternary era. Further south, the glacials were represented by arid periods, more or less alternating with the "Pluvials," during which precipitation was greater over what are now the semiarid and arid zones; the present deserts may have greatly narrowed or disappeared, though some investigators hold that they shifted toward the equator and remained as deserts. The last Pluvial began some 15,000 years ago, with a small increase in pluviosity from 5000–3000 BC. There is much evidence for such increased pluviosity in the past, including the deeply incised wadis of the Syrian, Arabian and North African deserts, the types of soil which have developed, many under light forests, as well as the great quantities of groundwater stored in their aquifers, not a little of which may be fossil water, infiltrated in periods of much higher precipitation.

Attempts have been made to date this presumed "fossil" groundwater by carbon-14 methods; the age as so determined shows a curious and indeed suspicious convergence on an age of $\pm 25,000$ years. Such convergence suggests an equilibrium of C^{14} and C^{12} in the water-aquifer system and so a false dating by the carbon-14 method.

Again, geomorphological studies can reveal much of the past climates of the dry regions of the earth. Christian and Jennings (1957) speak of the detection of lost rivers and river channels, determination of places where floods have occurred and may occur again, as well as of the origin of piedmont deposits where vanished rivers once lost their carrying power as they swept sediment down from the hills and deposited it at their feet.

Since "the fundamental concept in plant ecology is that the vegetation of an area taken as a whole is the result of a long period of adaptation to the whole environment—rainfall, evaporation, dew deposition, wind, insolation, temperatures, soil, etc." (Pichi Sormolli, 1955, p. 9), it follows that present day plant ecology not only will give data on present-day climate, but will also assist in understanding the past climates which have brought about the present ecological pattern. For the plant ecology of the arid and semiarid zones of North America, a summary has been made by McGinnies (1955) showing the relationships between plant associations and climates, and indeed defining some climatical areas in terms of plant ecology.

Presentation of Data. Hydrologic data is mainly presented in tables and on maps; most atlases cover the main features of the arid and semiarid zones. The special maps prepared by Meigs (1952) have already been mentioned, as defining and classifying these zones.

Also of special interest in defining the water balance for the semiarid and arid zones of the eastern Mediterranean regions are the maps prepared by Carter (1957). These show, for the regions bordering the Red Sea and the Persian Gulf, the mean annual potential evapotranspiration, water deficiency and water surplus.

Many more types of map exist, each showing different aspects of the semiarid and arid zones. As more data are accumulated and analyzed, so will more detailed maps reveal present position and future development potentials for these zones where life has been difficult in the past but where modern development methods should produce great improvements in the future.

Precipitation in Semiarid Regions

Precipitation in the semiarid regions is restricted and kept low by the inability of moisture-bearing winds to penetrate into, and cool down within, such regions. Zones of high pressure may prevent the entry of winds, and the great desert areas are mainly associated with this meteorological phenomenon. Such winds as do enter arid and semiarid regions may have had no opportunity of acquiring moisture by passage over oceans or sea, or they may have been forced to lose their moisture in passing over high mountains, as in the "rain-shadow" deserts of Imperial Valley and the Jordan-Syrian steppe. Again, lack of orogenetic effects within the regions, combined with high heat reflected from the ground, may prevent cooling of the incoming winds so that no moisture condenses to form clouds or precipitation, as in the coastal deserts of Chile, southern California, Morocco and western Australia.

Average figures are misleading, but the semiarid regions of the world tend to receive between 400 and 100 mm of precipitation; below 100 mm, the regions become truly arid. Within the Mediterranean semiarid climatic zone, a well-distributed cold weather precipitation of 250 mm is sufficient to mature a crop of winter-sown barley; an additional 100 mm is required for a good wheat crop. However, precipitation within the semiarid zones is distinguished by great variations from year to year, while the intensities of such rains as occur tend to be high; light showers may be evaporated before they reach the ground. As one goes from semiarid to arid regions, the regular seasonal precipitation tends to be entirely replaced by irregular showers. They may be of high intensity and very restricted extent, and often occur when cold air overlies warm air and some factor (orographic or meteorological) permits a sudden uprush of warm air which cooling condenses its moisture to produce a cloudburst.

Condensation over the seas and coasts tends to take place mainly on nuclei of sea salts, as well as volcanic and cosmic dust (Erickson, 1958). Over deserts and semi-deserts, aeolian erosion processes introduce much dust into the atmosphere—the famous red sunsets and dust storms of the deserts and wrongly cultivated marginal lands—which form condensation nuclei. In these zones, precipitation tends to contain material of continental as well as marine origin.

Almost all precipitation in the semiarid and arid regions occurs as rain, though snow is not unknown, especially on ground over 1000 meters in elevation. Dew and even hoarfrost are also of importance, and are due to the great differences between day and night temperatures. Where infiltration conditions are good, as over coastal sand dunes, or suitable vegetation exists, such dew may make a permanent addition to the useful water resources of the area. Some plants, such as tomatoes, appear to be able to take in dew directly through the stomata on the leaves (White, 1961, p. 50); such intake by desert vegetation is under study.

Wind-wells, mounds of high-albedo stones through which the wind can blow, are reported by many authors from areas such as the Crimea and the south of France. They supply water by night condensation on the rapidly cooling stones. The author has not seen any such wells in operation. But in the Negev of Sinai, it was found that the Byzantines (?) had irrigated vines by planting them at the base of an octahedron of open stones, the upper pyramid above ground surface to condense moisture and the lower inverted pyramid leading the condensate down to the vine root.

Evaporation and Transpiration

High evaporation—and high transpiration in vegetated areas—is the dominant hydrological characteristic of the semiarid and arid zones. Both transpiration and evaporation are high because abundant heat energy is supplied to change the limited amounts of liquid water into water vapor, either directly or through the biological processes. In this way, the heat balance of the area is maintained.

Evaporation may be considered (King, 1961, p. 55) as involving three dynamic processes which occur simultaneously:

(1) A flow of water vapor by turbulent and molecular diffusion from the evaporating surface to the atmosphere:

(2) A flow of heat by radiation, convection and conduction to the evaporating surface and its removal therefrom as latent heat of evaporation;

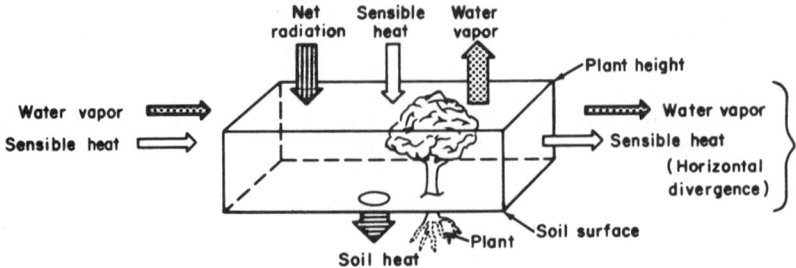

Fig. 3. Complete energy balance for a land surface (diagram modified after K. M. King, 1961).

(3) A flow of water through the soil and plants to the evaporating surface.

King has presented his conception of these processes in terms of complex equations and has simplified them in diagrams, e.g., Fig. 3. It will be noted that the evaporating surface extends from the top of the vegetation down to the surface of the soil; evaporation may also take place within the pores of the soil, but a cover of dust will develop and is a good retarder of evaporation.

With diffuse rainfall, as much as 100% evaporation loss may occur, so that there is no surface runoff, no replenishment of the groundwater and indeed no increase even in the soil moisture; the amount of heat stored in the ground and supplied during and immediately after precipitation has been sufficient to vaporize all the water supplied by the precipitation. Since the heat supply is greater during summer (the hot season), precipitation of an amount and intensity which would be completely evaporated in summer may produce increased soil moisture, infiltration and even surface runoff in the cold winter season. Night rainfall may suffer less evaporation than day rainfall, though such diurnal changes are less important than seasonal effects.

On the other hand, the intensity of the precipitation may enable much more water to be supplied to a surface than can be evaporated by the heat then available. If the water were to remain *in situ* just as it fell, then subsequent heat supplies would, almost by definition within the semiarid zone, evaporate all the precipitation. In nature, however, the nonevaporated precipitation is first used to increase soil moisture or flow off at surface as overland flow. Where the soil cover is nil or negligible, infiltration takes place directly into the pores and cracks of the rock, or overland flow occurs over impermeable outcrops.

The water stored as soil moisture remains within the influence of the heat supply; if the soil is planted, this heat can be used beneficially for plant growth, but if the soil is barren, most of the water is evaporated. However, if the soil is deep, a dry dust cover can develop, with an insulating effect which reduces or stops evaporation from the deeper soil layers; the effect of capillary rise of water is of particular importance in this stage of the hydrological cycle.

The problems of evaporation are not difficult to describe in general terms, but in detail they are extremely complex and indeed have not been solved either insofar as the design of instruments to collect the necessary data is concerned or with regard to the different theoretical thermodynamical approaches which are possible. Much detailed information will be found in Deacon et al. (1958), but the position can only be summed up in the words of Hare (1961): "We have woken up to the fact that natural evaporation is one of the most bafflingly difficult processes to study, and that we get nowhere unless we consider the evidence gathered by five or six different groups of workers—hydrologists, meteorologists, soil scientists, plant physiologists, agronomists and others." And so, evapotranspiration still remains a major problem in the hydrology of the semiarid regions of the earth.

Surface Runoff

The water which runs off on the surface also tends to remain within the influence of surface heat and thus evaporates. In the immediate sense, this leads to rivers with dwindling volumes of waters or to temporary lakes found only after rains and dry through most of the year. Evaporation from the surface of such bodies of standing water will be very high, and it is for this reason that artificial surface storage in ponds and reservoirs is subject to many disadvantages in semiarid regions. Attempts are being made to reduce evaporation by covering the water surface with a monomonecular layer of chemicals (fatty alcohols), but such layers can be broken by wind or contaminated by dust and so rendered impermanent.

The inability of surface waters to maintain themselves against evaporation has a long-term effect in that it permits the formation of basins of inland drainage. Any basin of closed drainage will cease to exist if the average annual storage of surface water exceeds evaporation from its central lake system, for then the lake will rise and spread each year till it overtops the lowest point of the encircling water divide over which it will discharge to the ocean level and also cut its way down so as to reduce the size of the lake. Such a general conception may be nullified by heavy infiltration (as in the closed karst basins of humid and other regions) or by rising mountain barriers and other geogenetic phenomena which may produce closed basins of surface drainage; but where such extraneous factors are not present, basins of closed drainage are characteristic of the semiarid lands of the world. The ability of the Nile, the Euphrates-Tigris, the Indus, the Colorado and similar rivers, to keep open their basins is due to the fact that the amount of incoming surface waters (originating in non-arid regions) exceeds the evaporation losses.

Infiltration and Groundwater

Finally, turning to the water which infiltrates to the groundwater, this is the only water which places itself effectively beyond the reach of further heat supply and, thus, further evaporation. Within the surface zone, it would appear that heat changes are sufficient to produce some evaporation and condensation, as in certain sand dunes in eastern Saudi Arabia. However, in general, evaporation losses are nil from water which is stored underground in the rock aquifers, whose temperatures are unaffected by surface insolation, though the geothermal gradient in the earth's crust (and other causes) will heat up groundwaters to much higher temperatures than that of the original infiltration.

Since precipitation stored in the rock aquifers by natural infiltration is immune from evaporational losses, and is often cool, it is of particular importance in the semiarid zone; in the past, it has been tapped by wells and galleries (kherazes, khanats, fogarras, chains-of-wells, etc.) of numerous types and is now exploited from boreholes. Extraction by modern methods has often exceeded the small natural replenishment, and this has led first to the idea of inducing additional infiltration or recharge to depleted aquifers and recently to the more positive approach of storing water underground, just as in many places natural gas is stored underground.

The different types of groundwater which can occur in the aquifers needs to be distinguished carefully in the semiarid zones, since only truly meteoric groundwater is a renewable resource, and extraction of other types is equivalent to the mining of a non-renewable mineral. Groundwater originating by current infiltration of precipitation is known as "meteoric" water, whereas groundwater of similar origin, but which has formed in the past and is no longer forming today, is known as "fossil" groundwater The latter may have formed during the Pluvial periods of the Quaternary, but attempts to date it by the time of decay of radioactive constituents have not yet proved completely reliable. Water which was entrapped in the sediments at the time of formation is known as "connate" groundwater; in well-flushed aquifers, it is not of importance, but low water put-through under semiarid conditions may permit saline connate water to affect the composition of the normal meteoric groundwater. Finally, there is "magnetic" or "juvenile" groundwater, said to consist of water given off from cooling magma; current thinking minimizes the amount of such water and emphasizes the heating of meteoric water which comes into the zone of high geothermal gradients associated with recent granitic and volcanic activities.

The temperature of groundwater is influenced by the temperature of the infiltration, by the geothermal gradient in the aquifer, by the depth and time the groundwater takes to pass through the aquifer and by certain chemical reactions between the water and the aquifer, of which the most important are exothermic oxidation processes, such as the conversion of sulfides (marcasite, pyrite, etc.) to sulfates. And in certain areas, such as the rift valley of Jordan, warm and hot springs at Hamme in the Yarmouk tributary must owe some of their heat to the effect of recently intruded and extruded basaltic magma.

Hydrochemistry and Soil Chemistry

Precipitation is never pure H_2O, but contains salts and gases in solution. The salts are dissociated into cations, mainly Ca, Mg, Na and K, while the anions are HCO_3, Cl and SO_4; carbon dioxide is the main dissolved gas. These elements in solution in precipitation may be of marine or terrestrial origin. Ericksson (1958, pp. 170–174) estimates the annual precipitation of sea salts as perhaps three kilograms per hectare (kg/ha) for the drier steppe regions south of the Sahara, as 2 kg/ha in the Kalahari, and 1 kg/ha for the righ plateaux of Iraq and Iran. Full evaporation of the water which carries these salts will result in their deposition more or less where they fell; surface runoff will concentrate them in the central evaporating pans in basins of closed drainage, while infiltration to the aquifers may be with water which already is far from pure. Thus, in Syria, "the chemical composition of the precipitation may be changed from an initial figure of some 20 ppm to concentrations of from 100–200 ppm by evapotranspiration and leaching of precipitates in the zones of precipitation; thus recharge waters to aquifers in Syria may contain from 50–200 ppm of total soluble salts" (Burdon and Mazloum, 1958, p. 87).

Precipitation of salts from immediate, full evaporation of precipitation can, over long periods, produce an appreciable *in situ* salt content in the soil. Soil formation itself in the semiarid zone is often more mechanical than chemical or organic; hence, primary soils reflect the composition of the underlying rock without much change by leaching. Where rainfall is higher however, solution effects become more important. Thus in limestone regions, soil formation is almost nil with very low rainfall; solution and redeposition (caliche, havara, nari) occur in regions of intermediate rainfall, while under strong rainfall (upper limit of arid zone) solution is dominant and karst topography, with residual terra rossa soils, marks the true Mediterranean climatic regions.

However, it is in areas of closed drainage where waters have been concentrated by surface flow, and subsequently their salts have been precipitated by evaporation, that the main effects of chemical precipitation can be seen. In this way are formed the salt pans and salt lakes of Australia and South Africa, the salinas of North America and the sebkhas of North Africa and the Middle East. "The floors are true solonchak soils, with salts flocculating the silt and clay fractions into grains readily moved by saltation and so erosion" (Christian and Jennings, 1957, p. 62).

While the composition of the infiltrating waters has a strong influence on the composition of underground water, nevertheless the composition of the aquifer, as well as the volume of water and its rate of movement, will dominate the chemical composition of the underground waters in the semiarid zone.

The solution of low-soluble carbonates by conversion to bicarbonates will continue as long as there is some carbon dioxide in solution in the groundwater; such action will produce bicarbonate waters with up to 600 ppm of total dissolved salts. Thereafter, mineralization of the groundwater will continue by hydration of sulfates, halite (NaCl) and other soluble salts in the aquifers. In some aquifers, such as the Fars Formation of Iraq–Syria (of lagoonal facies), such soluble salts will be very abundant, while in other aquifers, such as the continental, arkositic sandstones of the Sahara and Arabia, soluble minerals are almost completely absent. When the amount of groundwater flowing through the aquifer is large, such soluble salts tend to be removed and the aquifer flushed and cleaned out; likewise, fast-moving groundwater will flush an aquifer quicker than slow-moving water. Since the amount (and sometimes even the rate of movement) of groundwater in the semiarid zone tends to be small, mineralization by dissolution of the aquifer tends to be high.

At the point of natural discharge from aquifers, springs or marshy ground occurs. If the spring is large, a perennial river carries off the discharge, and an oasis is formed, or else a great city such as Damascus (fed by the Barada River flowing mainly from Ain Figeh) comes into existence. If the discharge is small or diffuse, a saline marsh tends to form, of which one of the greatest is the Qatarra Depression in Egypt, the probable discharge zone for the sandstone aquifer of the Western Desert of Egypt.

Water Development in the Semiarid Zones

Studies of the hydrology of the arid and semiarid zones are of much interest to pure science, but they are undertaken mainly to facilitate the proper development of the most critical resource of such regions—the water. Accordingly, a few notes are given on some of the lines along which development of water is taking place—for domestic supplies, including urban development and tourism, for watering the pasturing animals, for irrigation and for industry.

Surface Management. Surface management in the semiarid zones is, from the water viewpoint, directed to making use of the water before it is lost by evaporation. It is directed to increasing transpiration through useful vegetation.

Thus, it covers the maintenance of the optimum plant cover over the region, and in this sense optimum plant cover is that which gives maximum grazing to the pasturing herds and flocks. In addition to transpiring much of the direct precipitation, such vegetative cover reduces soil erosion while its shade conserves moisture in the soil. Its root system will help infiltration and generally build up the soil structure.

Control and Storage of Surface Runoff. When surface runoff occurs, it is liable to be intense but short-lived; the amount will be in excess of the requirements of the vegetation at that time. One method of control is to spread the water over large areas, by diverting it from the wadi or stream bed and controlling it behind earth banks in such a way that its flow velocity is never sufficient to erode the retaining structures. In order to control the flood above such water-spreading areas, conditions may warrant the construction of a spate-breaker, a dam with a permanent but restricted underflow, whose reservoir will fill but temporarily after floods. In such schemes, the surface runoff is mainly stored in the soil, from which it can be extracted by the vegetation; recharge to aquifers may occur.

Water may be stored for longer periods of time in surface reservoirs, but in such cases the water surface is subject to heavy evaporation losses. The stored water may be for watering animals, for domestic supplies, for irrigation, for power generation; the longer it is stored, the more is lost. To reduce such losses, it is possible to cover the water surface with a protective layer or film one molecule thick, whose action McArthur (1959) has described as follows: "When a long-chain fatty alcohol such as hexadecanol is placed on the surface of the water, because the work of adhesion between the hydroxyl group of the molecule and the water is greater than the work of cohesion between alcohol molecules, the latter separate from the mass and move across the water surface to form an oriented surface film one molecule thick." This film substantially reduces water evaporation, but it liable to be destroyed by dust, broken by wind and in need of constant replenishment.

Control of Aquifers. Aquifers may be controlled in many ways, of which the most obvious are extraction from galleries, wells and bore holes. But loss of water from the shallow portion of aquifers can be reduced by the control of phreatophytes whose long roots can reach down over 20 meters to tap the groundwater. Some such phreatophytes as alfalfa are beneficial and should be encouraged; others, such as mesquite, cottonwood, etc., are of little value yet use up considerable quantities of valuable groundwater.

Seepages and springs may be developed so as to minimize evaporation losses and make the full discharge available for beneficial use. When the maximum discharge of a spring occurs when its waters are least required, it may be possible to overpump the aquifer when water is required so that all recharge is stored below the spring discharge level and there is no flow to waste during the period when water is not required.

Underground Storage of Groundwater. While engineering works have been carried out to use surface waters to recharge aquifers, the approach has been negative in that it is used to correct the ill-effects of overpumping from the aquifers. There is now a tendency to take a more positive approach and to attempt to store surplus water underground, in the way in which surplus natural gas, and even oil, is stored. Water so stored is free of evaporation losses, though the location and the use of such underground reservoirs call for detailed geotechnical investigations.

Such storage has not yet been attempted in many places, but in Morocco it has been tried successfully at Tafilalet, in the Dra Valley and in the Souss valley.

Weather Modification. Attempts to increase precipitation by various forms of cloud seeding (silver iodide, dry ice, even water drops) have been made over many of the semiarid regions of the world, often with what appears to be success. However, all such methods of increasing precipitation depend for true success on the ability of the atmospheric circulation to bring more moisture into the region; in this, cloud seeding does not appear to be very effective.

Consideration has also been given to modifying climate by allowing large surfaces of water to form upwind of the semiarid area under consideration. Would the introduction of the Mediterranean Sea to the Qatarra depression increase precipitation along the Alexandrian coast in addition to generating power from a fall of some 70 meters?

Desalting of Brackish Waters. The desalting of brackish waters is of much importance to the semiarid regions, for not a little of their existing water resources are rendered unusable due to salinities lying above the limits of human, irrigation or animal use, yet well below the salinity of seawaters. Solar energy is generally available for the desalting of such waters, but efficiency is low, and costs are comparatively high. On some oil fields, such as those of Libya and Kuwait, waste gases offer a source of energy which would otherwise go to waste.

In considering modern engineering methods, it must not be forgotten that natural adaptation processes have evolved vegetation and animals which are able to convert brackish waters into liquids and foods which can be readily consumed. Dates flourish in high saline water; some rices will grow almost in the sea. Again, the halitophyte vegetation of the coasts and salt marshes can be eaten by grazing animals, while waters with salinities of up to 10,000 ppm are efficiently converted by sheep into nourishing milk, one of the mainstays of life for many types of nomad.

DAVID J. BURDON

References

Burdon, D. J., 1959, "Handbook of the Geology of Jordan," Amman, Jordan, Government Printer.

Burdon, D. J., and Mazloum, S., 1959, "Some Chemical Types of Groundwater from Syria," in "Salinity Problems in the Arid Zones," UNESCO—Arid Zone Research XIV, Paris, pp. 73–90.

Carter, D. B., 1957, "World Climatic Atlas—Special Sheets for Red Sea and Persian Gulf," Laboratory of Climatology, Centerton, N. J.

Christian, C. S., and Jennings, J. N., 1957, "Geomorphology," in "Guide Book to Research Data for Arid Zone Development," Chap. 5, UNESCO—Arid Zone Research IX, Paris, 191pp.

Deacon, E. L., Priestley, C. H. B., and Swinbank, W. C., 1958, "Evaporation and the Water Balance," in "Climatology: Review of Research," UNESCO—Arid Research X, Paris, pp. 9–34.

Drummond, A. J., 1958, "Radiation and the Thermal Balance," in "Climatology: Review of Research," UNESCO—Arid Zone Research X, Paris, pp. 56–74.

Eriksson, E., 1958, "The Chemical Climate and Saline Soils in the Arid Zone," in "Climatology: Review of Research," UNESCO—Arid Zone Research X, Paris, pp. 147–180.

Hare, F. K., 1961, Summing up of Symposium on Evaporation—Cat. No. R32-361/2, the Queen's Printer, Ottawa.

Kessler, E., 1966, "Radar Measurements for the assessment of areal rainfall: Review and outlook," *Water Resources Res.*, **2**, 413–425.

King, K. M., 1961, "Evaporation from Land Surfaces," Proc. of Hyd. Sym. No. 2—Evaporation, Cat. No. R32-361/2, The Queen's Printer, Ottawa, pp. 55–82.

Martonne, E. de and Aufrère, L. 1925, "L'Indice d'Aridité," *C.R. Acad. Sci. Paris*, **180**, 939

McArthur, I. K. H., 1959, "Control of Evaporation Losses from Water Surfaces," Report, International Committee on Irrigation and Drainage, New Delhi.

McGinnies, W. G., 1955, "Plant Ecology—The United States and Canada," in "Review of Research," UNESCO—Arid Zone Research VI, Paris, pp. 250–301.

Meigs, Peveril, 1952, "World Distribution of Arid and Semi-Arid Homoclimates," in "Review of Research on Arid Zone Hydrology," UNESCO—Arid Zone Research I, Paris, pp. 208–214.

Pichi-Sermolli, R. E., 1955, "Plant Ecology—Tropical East Africa (Ethiopia, Somaliland, Kenya, Tanganyika)," in "Review of Research," UNESCO—Arid Zone Research VI, Paris, pp. 302–360.

Shreve, F., 1942, "The Desert Vegetation of North America," *Botan. Rev.*, **8**, 195–246.

Thornthwaite, C. W., 1948, "An approach towards a rational classification of climate," *Geograph. Rev.*, **38**, 55–94.

Thornthwaite, C. W., 1958, "Introduction to Arid Zone Climatology," UNESCO—Arid Zone Research XI, Paris, pp. 15–22.

White, G. F., 1961, "Science and the Future of Arid Lands," Paris, UNESCO, Paper NS (MC). 60/D.30a/A.

White, G. F., 1963, "Preparatory Meeting on the Long-Term Programme of Research in Scientific Hydrology," Paris, UNESCO/NS/181.

White, G. F., 1963, "The Education and Training of Hydrologists," Report of a Committee, *J. Inst. Water Engrs.*, **17**, 381–391.

Cross-references: *Aquifer; Connate Water; Groundwater; Hydrologic Cycle; Hydrology; Hydrology, Limestone Terrain; International Hydrological Decade; Juvenile Water; Meteoric Water; Pedology; Rainwater; Runoff; Water Balance.* Vol. II: *Albedo; Atmospheric Nuclei and Dusts; Climatic Classification; Dew, Dew Point; Equatorial and Tropical Climate; Evapotranspiration; Humidity; Insolation; Orographic Precipitation; Rain Shadow; Rainfall Distribution.* Vol. III: *Arid Cycle; Deserts; Erosion; Quaternary Period; Sabkha or Sebkha.* Vol. VI: *Caliche; Leaching.*

HYDROLOGY, SUBSURFACE WATERS

Scope of Topic

The study of subsurface water generally includes a consideration of its chemical and physical properties, geological environment, natural movement, recovery and utilization. More specifically, the hydrology of subsurface waters concentrates on the study of the laws of the occurrence and movement of subterranean waters (Meinzer, 1942).

History

Accounts of well water and well construction abound in ancient literature and are specially well known from the biblical record of *Genesis*.

Well construction in the Near East was by man and animal power aided by hoists and primitive hand tools, despite great difficulties. Egyptians had perfected core drilling in rock as early as 3000 BC. The ancient Chinese people were able to achieve wells with depths up to 1500 meters through sustaining a slow drilling rate over a period of years.

The greatest achievement in groundwater utilization by ancient peoples was in the construction of long infiltration galleries, or kanats, which collected water from alluvial fan deposits and soft sedimentary rock. These structures, commonly several kilometers long, collected water for both agricultural and municipal purposes. Kanats were probably first used more than 2500 years ago in Iran; however, the technique of construction spread eastward to Afghanistan and westward to Egypt. One extensive kanat system built about the year 500 BC in Egypt is said to have irrigated 3500 km² of fertile land west of the Nile (Tolman, 1937). Many kanats are still in use today in Iran and Afghanistan, the best known of which are in Iran on the alluvial fans of the Elburz Mountains.

Modern percussion methods of well drilling were developed more or less independently in Western Europe. The impetus for this development came largely from the discovery of flowing wells, first in Flanders about AD 1100, then a few decades later in eastern England and in northern Italy. One of the first wells was dug in AD 1126 by Augustinian monks from a convent near the village of Lillers (De Wiest, 1965). In Gonnehem, Flanders, near Bethune, four wells were drilled and were cased to nearly twelve feet above ground level so that they were able to deliver water at sufficient height to drive a water mill. The wells were several hundred feet deep and tapped water under pressure from a formation consisting of fractured chalk that had its outcrop area in the higher plateaus of the Province of Artois. These and other similar wells in the region of Artois became so famous that flowing wells were eventually called *artesian wells* after the name of the region.

The methods of drilling for water have improved rapidly since the end of the nineteenth century, partly owing to knowledge borrowed from oil and gas drilling. The most significant single advance in drilling techniques has been the development of hydraulic rotary methods. Early rotary drilling was done with the aid of an outer casing; however, in about 1890 thick mud was found to be sufficient for holding up the walls of the hole, and the outer casing was no longer used. With this new efficiency and with the successful drilling of the Spindle Top oil field in Texas in 1901 by rotary methods, rotary drilling has steadily gained in popularity. The perfection of the deep-well turbine pump in the years between 1910 and 1930 added a further stimulus to the well-drilling industry (Davis and De Wiest, 1966).

Founders of Hydrology of Subsurface Waters

Although his scientific work was somewhat related to that by Hagen and Poiseuille, Henry Darcy (1803–58) was the first person to state clearly the mathematical law which governs the flow of groundwater. Darcy developed his formula as a result of experimentation with filter sands, and presented it in 1856 in an appendix of a report on the municipal water supply of Dijon, France. Jules Dupuit, of France, was the first scientist to develop a formula for the flow of water into a well. Modern methods of higher mathematics were first applied extensively to ground water flow by Philip Forchheimer of Austria and C. S. Slichter of the United States.

Modern and most significant advances in groundwater hydraulics in the United States were made in the 1930–40 period by Theis, Jacob, Muskat and Hubbert (Muskat, 1937; Hubbert, 1940; Todd, 1959).

Darcy's Law

The hydrodynamic microscopic picture of the flow of groundwater is very complicated and not amenable to mathematical treatment. Indeed, for isothermal flow, the three velocity components, the pressure, and the density at any point of the fluid are the five unknown quantities in problems of groundwater flow. Water is treated as incompressible, except in the calculation of the storage coefficient of water-bearing strata. If the density is assumed to be constant, it would theoretically be possible to solve for the unknown pressure and velocity, if equations of motion (of the Navier-Stokes type) and of conservation of mass were available. Hydrodynamically, such a problem would be tractable if the granular skeleton were a simple geometrical assembly of prismatic, unconnected tubes. The seepage path, far from being a prismatic channel however, is tortuous, branching into a multitude of tributaries and recombining several of them as the flow proceeds.

Darcy's law in its original form avoids the insurmountable difficulties of the hydrodynamic microscopic picture by introducing a doubly averaging microscopic concept. First, it considers a fictitious flow velocity, the Darcy velocity or specific discharge through a given cross section A (Fig. 1) of porous medium rather than the true velocity between the grains. Second, it treats average hydraulic values rather than local hydrodynamic values of this velocity. The basic reason for the introduction of this simplifying concept lies in the nature of Darcy's experiment: it utilized a sand-filled cylindrical pipe which permitted a measurement of only the average hydraulic values. The flow in the pipe of Fig. 1 proceeds from higher to lower head h. The velocity head part in h is negligible so that

$$h = z + \frac{p}{\gamma} + \text{Constant} \qquad (1)$$

HYDROLOGY, SUBSURFACE WATERS

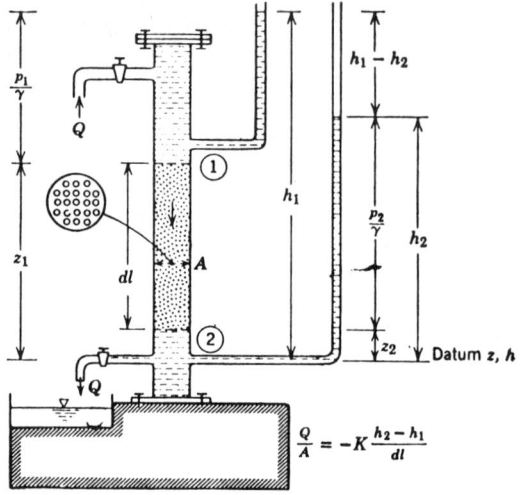

FIG. 1. Apparatus to demonstrate Darcy's law. *Courtesy M. K. Hubbert.*

in which z is the elevation head and p/γ is the pressure head, being the ratio of the water pressure in the pores to the unit weight of the fluid. The flowrate Q is proportional to the head loss, inversely proportional to the length of the flow path, and proportional to a coefficient K which depends on the nature of the sand and of the fluid. Darcy's law may be expressed as

$$Q = KA(h_1 - h_2)/dl = -KA\frac{dh}{dl} \quad (2)$$

The coefficient K is the fluid conductivity and may be expressed as

$$K = k\frac{\gamma}{\mu} \quad (3)$$

in which γ is the unit weight of the fluid, μ is its dynamic viscosity and

$$k = cd^2 \quad (4)$$

is the permeability of the medium; d is a characteristic length, say the average pore size of the sand, and c is a dimensionless constant or shape factor which takes into account effects of stratification, packing, arrangement of grains, size distribution and porosity.

Flow in Confined Aquifers

Aquifers are geologic formations or strata containing water in their voids or pores that may be removed economically and used as a source of water. They are separated from each other by aquicludes or aquitards. *Aquicludes* are geologic formations so impervious that for all practical purposes they completely obstruct the flow of groundwater (although they may be saturated with water themselves) and completely confine other strata with which they alternate in deposition. A shale is an example of an aquiclude. *Aquitards* are geologic formations of a rather impervious and semi-confining nature which transmit water at a very slow rate compared to the aquifer. Over a large area of contact, however, they may permit the passage of large amounts of water between adjacent aquifers which they separate from each other. Clay lenses interbedded with sands, if thin enough, may form aquitards.

Consolidated sandstones are common examples of confined aquifers. The flow in such aquifers (Fig. 2) is governed by the equation

$$\nabla^2 h = \frac{S}{T}\frac{\partial h}{\partial t} \quad (5)$$

in which T is the transmissivity of the aquifer, the product of K and b, the thickness of the aquifer. S is the storage coefficient of the aquifer

$$S = \gamma b(\alpha + n\beta) \quad (6)$$

in which α is the vertical compressibility of the granular skeleton of the medium, treated as a continuum, β is the compressibility of the fluid, and n is the porosity of the medium. The storage coefficient is dimensionless, and it may be conceived of physically (see Fig. 2) as the amount of water in storage that is released from a column of aquifer with unit cross-sectional area and per unit decline of head (De Wiest, 1969). Its two parts may be interpreted as:

$\gamma b\alpha$ = water in storage released due to the compression of the intergranular skeleton per column with unit cross-section and per unit decline of head.

$\gamma bn\beta$ = water in storage released due to the expansion of the water per column with unit cross-section and per unit decline of head.

The coefficients T and S are called the formation constants of the aquifer and the knowledge of these coefficients is indispensable in the planning of the production of an aquifer. S and T are determined in the field by means of pumping tests. Theis proposed a solution of Eq. (5) in the case of an aquifer of infinite extent, initially and uniformly under a constant head H, and then pumped for a constant flow rate Q. If s represents the drawdown $H - h$, the solution proposed by Theis is

$$s = \frac{Q}{4\pi T}W(u) \quad (7)$$

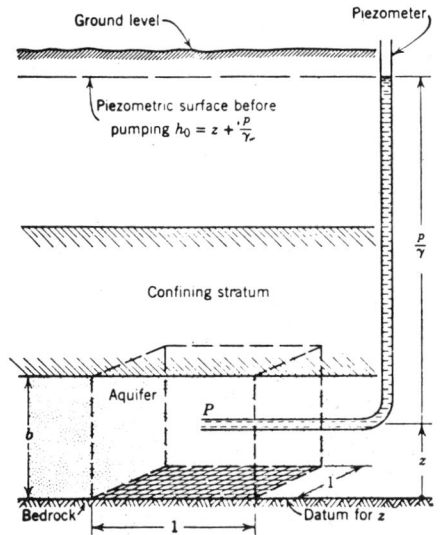 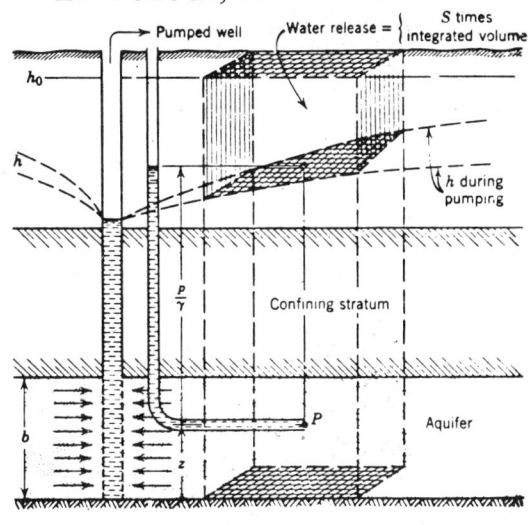

FIG. 2. Physical interpretation of storage coefficient. *Courtesy R. J. M. De Wiest.*

in which

$$u = \frac{r^2 S}{4Tt} \quad (8)$$

where r is the distance from the pumped well to the well where the drawdown is observed and t is the time since pumping started. The function $W(u)$ is tabulated as the exponential integral. S and T are determined by superimposing (Fig. 3) a graph of log $W(u)$ versus log u on a plot of log s versus log r^2/t, so that the pumping data fit the tabulated data. A match point chosen on the overlapping portion of the sheets determines mutual values of s, $W(u)$, r^2/t, and u which may be inserted in Eqs. (7) and (8). The solution of Eqs. (7) and (8) for S and T is then straightforward.

Leakage into and from Aquifers

Not always is a water-producing stratum completely confined between perfectly impervious aquicludes. Often the adjacent strata are semipervious and exchange water with the main aquifer. Consider Fig. 4, where for reasons of simplicity only leakage from above takes place. The semi-confining stratum is overlain by sand, and the capacity for lateral inflow is sufficient to maintain essentially a constant head in spite of the downward leakage into the main aquifer. Impervious bedrock is datum plane for head. Before water was withdrawn from the main aquifer, the head h in the main aquifer was uniform and equal to the height H of the water table above the bedrock. After water is withdrawn from the main aquifer, the head h is represented by a curved line as indicated in Fig. 4, and because the ponded water remains under head H, a head difference is established between the top and the bottom of the semi-confining stratum which will induce leakage through this stratum. The hydraulic conductivity K of the main aquifer is so large com-

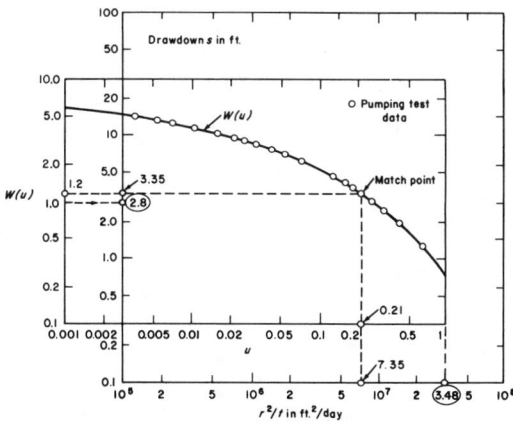

FIG. 3. Theis' graphical method to determine S and T. *Courtesy R. J. M. De Wiest.*

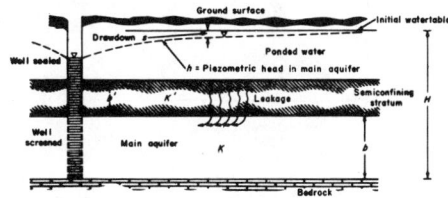

FIG. 4. Leakage into aquifer. *Courtesy R. J. M. De Wiest.*

pared to that of the semi-confining stratum K' that it is safe to assume that water seeps vertically through the semi-confining layer and is refracted over 90° to proceed horizontally in the main aquifer.

In unsteady flow, the drawdown s satisfies the equation

$$\nabla^2 s - \frac{s}{B^2} = \frac{S}{T} \frac{\partial s}{\partial t} \qquad (9)$$

in which $B = \sqrt{Kbb'/K'}$ is the leakage factor, b' is the thickness of the semi-confining stratum, and the other symbols are as defined before.

The solution of Eq. (9) for boundary conditions and initial condition as prevailing to obtain Eq. (7) becomes

$$s = \frac{Q}{4\pi T} W(u, \frac{r}{B}) \qquad (10)$$

in which $W(u, r/B)$ is the well function for leaky artesian aquifers and u is defined by Eq. (8). It is evident that the case of perfect confinement of Section (5) can be obtained from the more general case of Section (6) by making $B \to \infty$ in Eqs. (9) and (10).

The formation constants S, T and B are determined graphically from pumping test data by a method similar to Theis' method of Section (5).

Salt-Water Encroachment

Saltwater encroachment or intrusion is the shoreward movement of water from a sea or ocean into confined or unconfined coastal aquifers and the subsequent displacement of fresh water from these aquifers.

Ghyben–Herzberg Theory. The hydrostatic equilibrium between immiscible fresh-water and salt-water bodies in contact with each other along a certain interface was studied first by Ghyben and Herzberg (De Wiest, 1965). The equation for the depth of the interface is (see Fig. 5)

$$z_s = \frac{\rho_f}{\rho_s - \rho_f} z_w \qquad (11)$$

where ρ_f is the density of fresh water, ρ_s is the density of salt water, z_w is the height of the freshwater table above mean sea level, and z_s is the depth of the interface below sea level. This equation is in good agreement with measurements made in the field indicating that for every foot of fresh water above mean sea level, the thickness of the freshwater lens resting on the salt water was about 40 feet. The limitations of the hydrostatic theory are obvious: if both fluids were truly in static condition, the water table would have zero slope and the interface would become horizontal, with fresh water overlying salt water by mere density difference. Furthermore, fresh water is in a continuous state of motion due to changes in the water table, for example, because of replenishment, evaporation and discharge. It has been recognized for a long time that fresh water seeps into the ocean above sea level. Such water was tapped in earlier times as potable water for use on sea-going vessels. The escape of fresh water below sea level was not considered either in the Ghyben–Herzberg theory.

Hubbert, pointing at the dynamic rather than the hydrostatic equilibrium of the freshwater/saltwater interface, showed the discrepancy between the actual depth to salt water and the depth as calculated by the Ghyben–Herzberg formula for flow conditions near the shore line (Fig. 6).

Hubbert's concepts were confirmed and extended by Lusczynski (De Wiest, 1965) who took into account the existence of a zone of dispersion between fresh water and salt water. Contingent upon the reading of water levels in some observation wells, Lusczynski's work

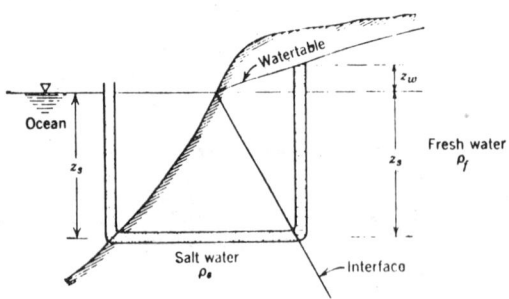

FIG. 5. Saltwater intrusion according to the Ghyben-Herzberg theory. *Courtesy R. J. M. De Wiest.*

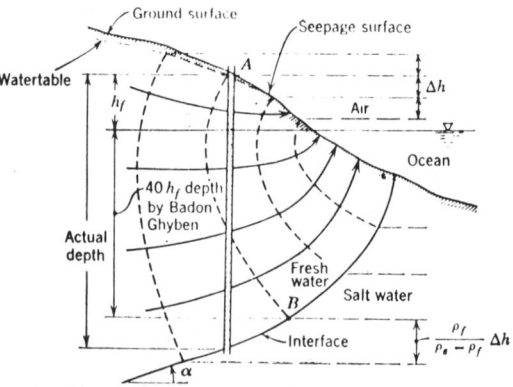

FIG. 6. Discrepancy between actual depth to salt water and depth calculated by Ghyben-Herzberg theory. *Courtesy M. K. Hubbert.*

allows for the computation of a three-dimensional velocity picture in a medium where the salt content of the water gradually varies. This picture may be constructed regardless of the often complicated boundary conditions created by the geological nature of the aquifers which preclude a complete analytical solution of the problem. The value of Lusczynski's work resides in its practical application.

Roger J. M. De Wiest

References

Davis, S. N., and De Wiest, R. J. M., 1966, "Hydrogeology," New York, John Wiley & Sons, 463pp.

De Wiest, R. J. M., 1965, "Geohydrology," New York, John Wiley & Sons, 366pp.

De Wiest, R. J. M., 1969, "Flow through Porous Media," New York, Academic Press, 530 pp.

Hubbert, M. K., 1940, "The theory of groundwater motion," *J. Geol.*, 48(8), 785–944.

Meinzer, O., 1942, "Hydrology," New York, Dover, 712pp.

Muskat, M., 1937, "The Flow of Homogeneous Fluids Through Porous Media," New York, McGraw-Hill Book Co., 763pp.

Todd, D. K., 1959, "Ground Water Hydrology," New York, John Wiley & Sons, 336pp.

Tolman, C. F., 1937, "Ground Water," New York, McGraw-Hill Book Co.

Cross-references: *Groundwater; Hydrology; Hydrology, Coastal Terrain; Hydrology, Limestone Terrain; Porosity and Permeability; Water Table.* Vol. II: *Hydrodynamics.* Vol. VI: *Wells, Water.*

HYDROLOGY, VOLCANIC TERRAIN

Types of Rocks

The development of groundwater in volcanic terrains has increased enormously in the last twenty-five years. Volcanic terrains are made of rocks erupted from volcanoes and intrusive rocks which congealed below the surface. The eruptives fall in three main categories according to composition: *(1)* basalt, a dark colored rock low in silica and high in ferromagnesian minerals; *(2)* rhyolite, a light colored rock high in silica and low in ferromagnesian minerals, and *(3)* a whole series of rocks of intermediate composition between basalt and rhyolite, such as andesite, dacite, latite, and trachyte. The silica-rich magmas are commonly more explosive and tend to form steep cones close to their vents. Such vents are Mt. Lassen, California; Mt. Hood, Oregon; and Mt. Ranier, Washington. The basalt, being more fluid, forms plains, such as the Snake River Plain of Idaho and the Columbia River plateau of Washington, Oregon, and California. Basalts form the high islands of the Central Pacific of which Mauna Loa and Kilauea volcanoes on Hawaii are the best known. If poured out molten, they are flows; if blown out, they are pyroclastics; if solidified in cracks or other voids in the crust, they are intrusives; and if deposited as fragments in a vent, they are throat breccias.

Permeability

Aa basalt flows and block silicious lavas are among the most permeable rocks on the face of the earth. Before weathering practically no surface run-off occurs. The same is true of pahoehoe, the billowy form of basalt commonly filled with caverns. In contrast, the massive parts of both silica-rich and silica-poor lavas where not broken by faulting, have low permeability.

Dikes, sills, plugs, and bosses are the chief shallow intrusive rocks. All have low permeability, but sills will perch water at high level if intruded into permeable rocks, and dikes confine great quantities of water at high level in the Hawaiian Islands. Dikes also direct the movement of water and in some mines cause serious drainage problems.

Ash and welded tuffs have low permeability but commonly serve as perching formations but, where interbedded with permeable beds and warped, give rise to artesian water. Pumice deposits are highly permeable and where extensive, as around Crater Lake, Oregon, give rise to numerous large springs and spring-fed perennial streams with exceptional steady flow (Stearns, 1931). Pumice beds interbedded with muds give rise to artesian springs and wells in the Valle Grande Caldera, New Mexico (unpublished report by the writer to the Atomic Energy Commission, 1948).

Springs from Buried Drainages

Basaltic lava flows which buried drainage systems are common in the western United States. Generally groundwater flows along the buried channels and, at the terminous of the lava, issues as large springs. Usually pillow lava formed by the hot lava entering water lies in the stream channels at the base of the basalt. It is exceedingly permeable in the Pliocene and Quaternary lavas but in the Miocene and older lavas interstitial clays reduce permeability. Most of the Thousand Springs near Twin Falls, Idaho, issue from pillow lava. They discharge about 5000 cubic feet per second, and large quantities of electricity are generated from the water as it falls over numerous dams constructed below the springs. Of the sixty-seven first-magnitude springs in the United States discharging more than 100 ft³/sec (Meinzer, 1927) thirty-six issue from basalt. One spring in the Deschutes Basin has been added to Meinzer's list by the writer.

Big Springs, Idaho, near Yellowstone Park, discharges about 180 ft³/sec from spherolitic obsidian at the terminus of a blocky silica-rich lava flow in an ancient caldera (Stearns, Bryan, and Crandall, 1939). Several other large springs issue in the same area from silicious lavas presumably filling ancient valleys.

Occurrence of Water

The water table is only 2–3 ft/mile in basalts in the Hawaiian Islands, indicating their high permeability (Stearns and Vaksvik, 1935). (See Fig. 1 for the relation between geologic structure and occurrence of water for a typical volcanic island.) It is 4.5–25 ft/mile in the basalts of the Snake River Plain, Idaho (Stearns, Crandall, and Steward, 1938). The Miocene Columbia River basalts were spread out as thick sheets and many beds resemble intrusive rocks and have low permeability. At the Brownlee Dam on the Snake River near Baker, Oregon, the writer found a syncline which yielded abundant artesian flows to drilled wells. The permeability was largely the result of fracturing during the folding. Upon excavation it was found that the upper confining member of the artesian system was the dense upper part of the lava flow. Thus, it is possible to find water moving through the lower part of a basalt and confined under pressure by the upper part. From 1949 to 1958, under the writer's direction, the Dole Corporation developed enough water by deep wells and shafts from water confined at high levels by tight gauge along faults to irrigate 10,000 acres of pineapple on the island of Lanai, Hawaii. Similar fault barriers to the movement of groundwater have been described in the Columbia River basalt (Newcomb, 1961). The Triassic traps (basalts and dolerites) of New England and the Palisades of the Hudson River have low permeability.

Huge quantities of water have been developed in the Hawaiian Islands by tunnels penetrating dike complexes (Stearns and Vaksvik, 1935). Many miles of tunnels have been driven to develop water for sugar cane by driving along ash beds interstratified with basalts on the island of Hawaii (Stearns and Clark, 1930) and at the bottom of lava-plastered valleys on Maui (Stearns and MacDonald, 1942).

The greatest development of groundwater from basalt has been done by wells and shafts in Oahu. The total consumption of groundwater from all wells, tunnels, and shafts on Oahu amounted to 487 million gallons per day in 1962 according to U.S.G.S. records. Honolulu is entirely supplied by water from basalt. Most of the water is from artesian basins where the upper confining member of the basaltic aquifer is lateritic soil and alluvium and no lower confining bed exists. In basaltic is-

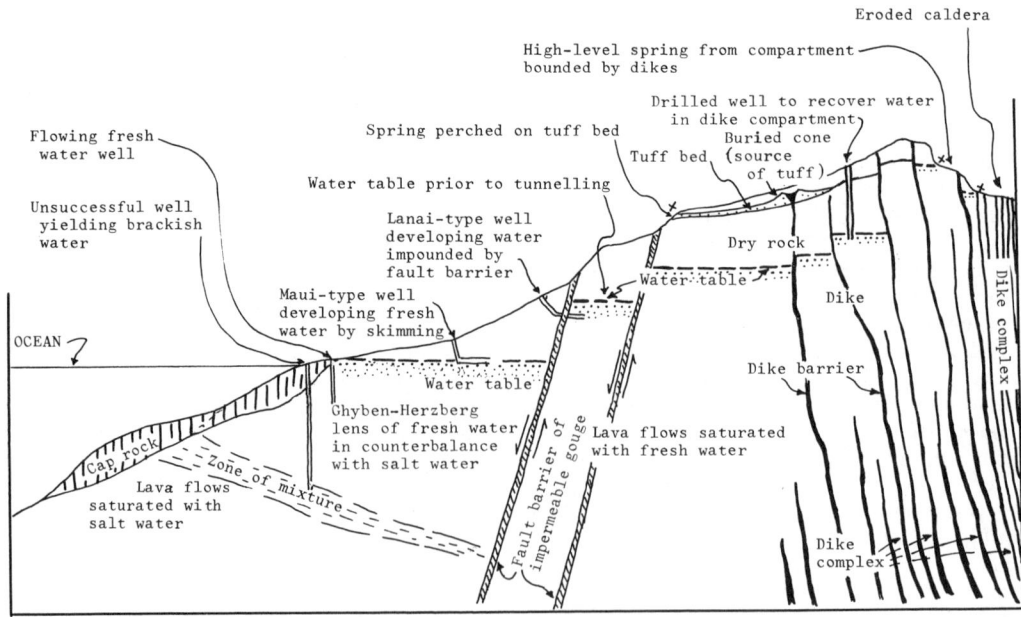

FIG. 1. Cross section of typical eroded volcanic island showing relation of geologic structures to occurrence of water and methods for development.

lands the fresh water floats on seawater according to its density or the Ghyben-Herzberg principle. For each foot fresh water stands above sea level it extends 40 feet below sea level, like cream floating on a pan of milk. Excessive pumpage causes salt contamination.

Natural basaltic reservoirs have proven to be excellent disposal areas for atomic wastes. The well-known Hanford plant, Washington, and Arco Reactor, Idaho, are located in basaltic lava fields. Several of the underground atomic explosions were set off in volcanic tuffs in Nevada.

Dams and reservoirs located in lava fields always should be built after very thorough study of water table conditions and permeability. Some dams have failed completely to hold water even with an inflow of 3000 ft^3/sec, and others have caused extensive damage because of leakage into adjacent drainage areas (Stearns, 1929). A few lakes in the western United States have been dammed by lava flows and all their inflow escapes underground through their natural dams.

For more information the reader is referred to the writer's chapter "Water in Volcanic Terranes" in "Hydrology" (Meinzer, 1942) and its bibliography.

HAROLD T. STEARNS

References

Meinzer, O. E., 1927, "Large springs of the U.S.," *U.S. Geol. Surv. Water Supply Paper* **557**, 94pp.

Meinzer, O. C. (editor), 1942, "Hydrology," New York, Dover Publ. Inc., 712pp.

Newcomb, R. C., 1961, "Storage of ground water behind subsurface dams in the Columbia River basalt, Washington, Oregon, and Idaho," *U.S. Geol. Surv. Profess. Paper* **383A**, 15pp.

Stearns, H. T., 1929, "Success and failure of reservoirs in basalt," *Am. Inst. Min. Met. Eng. Trans., Tech. Publ.* **215** (Class 1, Min. Geol. No. 26), 111–112 (abstract).

Stearns, H. T., 1931, "Geology and water resources of the middle Deschutes River Basin, Oreg.," *U.S. Geol. Surv. Water Supply Paper* **637**, 125–212.

Stearns, H. T., Bryan, L. L., and Crandall, L., 1939, "Geology and water resources of the Mud Lake region, Idaho," *U.S. Geol. Surv. Water Supply Paper* **818**, 125pp.

Stearns, H. T., and Clark, W. O., 1930, "Geology and water resources of the Kau District, Hawaii, including parts of Kilauea and Mauna Loa volcanoes," *U.S. Geol. Surv. Water Supply Paper* **616**, 194pp.

Stearns, H. T., Crandall, L., and Steward, W. G., 1938, "Geology and ground-water resources of the Snake River Plain in southeastern Idaho," *U.S. Geol. Surv. Water Supply Paper* **774**, 268pp.

Stearns, H. T., and MacDonald, G. A., 1942, "Geology and groundwater resources of the island of Maui, Hawaii," *Hawaii Div. Hydrogr. Bull.*, **7**, 334pp.

Stearns, H. T., and Vaksvik, K. N., 1935, "Geology and groundwater resources of the island of Oahu, Hawaii," *Hawaii Div. Hydrogr. Bull.* **1**, 479pp.

Cross-references: *Artesian Water; Groundwater; Hydrology; Porosity and Permeability; Springs; Water Table.* Vol. V: *Flood Basalts; Volcanic Rocks.*

HYDRONIUM—See HYDROGEN

HYDROSPHERE—See Vol. II

HYDROTHERMAL ALTERATION—NONSILICATE

Hot, aqueous solutions frequently alter rocks through which they pass. Minerals in these rocks may recrystallize, dissolve, or be replaced by other minerals, and the physical and chemical properties of rocks may be changed dramatically during such hydrothermal alteration. This article is concerned primarily with replacement reactions. These may involve any mineral, and may produce either a mineral or minerals in the same or in different mineral groups. The replacement of calcite ($CaCO_3$) by dolomite [$CaMg(CO_3)_2$] is an example of a common replacement reaction within the same group; the replacement of calcite ($CaCO_3$) by sphalerite (ZnS) is an example of replacement of a mineral in one group by a mineral in another group. Table 1 lists the seven major mineral groups and their 49 possible types of replacement reactions. There are, of course, many other reactions involving members of minor mineral groups, but Table 1 can serve as a convenient framework for the discussion of most wall-rock alteration reactions. Silicates are by far the most abundant and complex mineral group. A great deal of the literature on wall-rock alteration therefore deals with reactions which fall along row 7, and in particular with the processes and products belonging to box 7/7. The article on *Hydrothermal Alteration of Silicate Rocks* (by S. C. Creasy) is entirely devoted to these matters, and this article touches on silicates only as products of the hydrothermal alteration of minerals in rows 1 through 6.

The mineralogy and geometry of hydrothermal alteration have been studied intensively for more than a century. Ramdohr's classic book (1960) summarizes a great deal of the data pertaining to the ore minerals themselves. Lindgren's "Mineral Deposits" (1933) is still an unsurpassed source of information for megascopic aspects of hydrothermal altera-

TABLE 1. SOME MAJOR TYPES OF REPLACEMENT REACTIONS[a]

Original Mineral \ Replacement Mineral	Native elements (1)	Sulfides and sulfosalts (2)	Oxides and hydroxides (3)	Halides (4)	Carbonates (5)	Sulfates (6)	Silicates (7)
(1) Native elements	1/1	1/2	1/3	1/4	1/5	1/6	1/7
(2) Sulfides and sulfosalts	2/1	2/2	2/3	2/4	2/5	2/6	2/7
(3) Oxides and hydroxides	3/1	3/2	3/3	3/4	3/5	3/6	3/7
(4) Halides	4/1	4/2	4/3	4/4	4/5	4/6	4/7
(5) Carbonates	5/1	5/2	5/3	5/4	5/5	5/6	5/7
(6) Sulfates	6/1	6/2	6/3	6/4	6/5	6/6	6/7
(7) Silicates	7/1	7/2	7/3	7/4	7/5	7/6	7/7

[a] Category n/m contains all reactions in which a mineral of type n is replaced by a mineral of type m.

tion. But the detailed interpretation of the chemistry of hydrothermal alteration has become possible only recently.

The large amount of thermochemical data on metals, sulfides, and oxides obtained by metallurgists has been summarized by Kelley in a series of Bulletins of the U.S. Bureau of Mines, and by Kubaschewski and Evans (1958), Holland (1959, 1965), Barton and Skinner (1967), Robie (1966) and Robie and Waldbaum (1968). With these data, the stability field of most of the native metals, and of the common sulfide and oxide minerals, can be plotted in log fs_2-log fo_2 diagrams (Holland, 1959, 1965; Barton and Skinner, 1967). As was to be expected, minerals commonly associated with each other turned out to have overlapping stability fields in such diagrams, and the alteration of minerals and mineral assemblages could be interpreted in terms of changes of temperature and/or of the fugacity of sulfur and oxygen in hydrothermal solutions. The fugacity of sulfur and oxygen in these solutions is, however, so small that the concentrations of S_2 and O_2 of themselves cannot exert strong controls on the mineralogy of the rocks through which hydrothermal solutions are flowing. In this sense they function as indicator variables, much like the hydrogen ion activity in most near-surface solutions.

The relationship between the fugacity of sulfur and oxygen, and the quantitatively important sulfur species in solution has been described by Barnes and Czamanske (1967) and by Raymahashay and Holland (1968, 1969). In Fig. 1, the stability field of the iron sulfides and oxides are shown on a log fs_2-log fo_2 diagram together with the contours of total dissolved sulfur concentration at 350°C, at a total pressure of 270 atm, a solution pH of 5.0, and a total ionic strength of 1.0. To the left of the maximum of the total dissolved sulfur contours, sulfur is present largely in the reduced state as H_2S, HS^-, and S^{2-} complexes; to the right of the maximum, sulfur is present largely in the oxidized state as HSO_4^-, SO_4^{2-}, and SO_4^{2-}-complexes. The replacement of one sulfide of iron by another (pyrite by pyrrhotite or vice versa), of one oxide by another (magnetite by hematite or vice versa), of a sulfide by an oxide (pyrite or pyrrhotite by magnetite or hematite), or of an oxide by a sulfide (magnetite or hematite by pyrite or pyrrhotite) can be related to changes in total dissolved sulfur or to changes in the oxidation state by sulfur via such a diagram. But as the position of the ΣS contours depends on other variables as well, it is usually impossible to define uniquely with the data of diagrams such as Fig. 1 the changes in temperature and in the chemistry of hydrothermal solutions that have led to a particular replacement. Supplementary information, particularly from the composition of *fluid inclusions* (q.v.), is often extremely helpful (Roedder, 1967).

The reactions discussed above are typical of those reactions of categories 1/1 through 3/3, of Table 1, in which the cation remains unchanged. Frequently, however, this is not the case. If a sulfide AS is replaced by a sulfide BS

$$AS + B^{n+} \rightarrow BS + A^{n+} \qquad (1)$$

the replacement process is controlled by the relationship between the activity ratio $a_{A^{n+}}/$

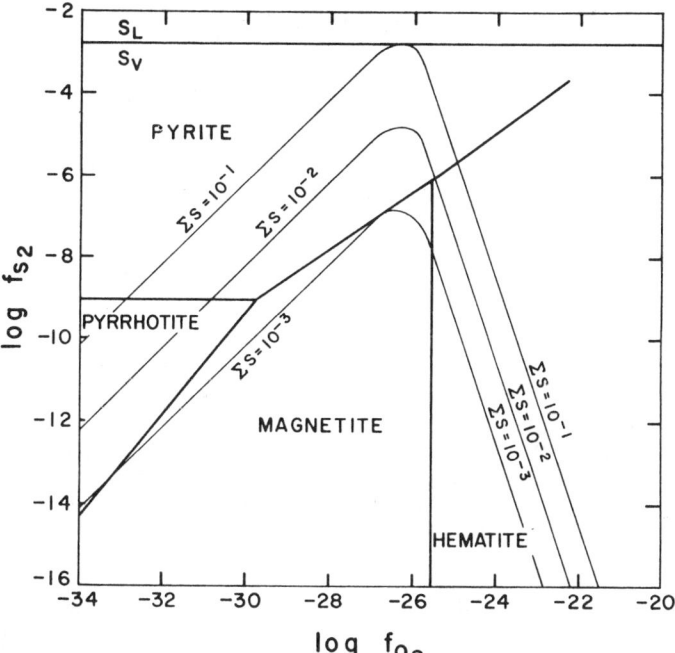

Fig. 1. Contours of total dissolved sulfur (ΣS) in moles/kg solution at 350° C. The total pressure is 270 atm, the pH is 5.0, and the ionic strength is 1.0 (Raymahashay and Holland, 1969).

$a_{B^{n+}}$ to the equilibrium constant for Eq. (1). It is related to the fugacity of S_2 and O_2 only in so far as changes in these fugacities influence the activity ratio $a_{A^{n+}}/a_{B^{n+}}$ as, for instance, via the formation or destruction of complex ions.

The importance of cation activity ratios extends beyond categories 1/1 through 3/3, to all categories in Table 1 along the upper-lower right diagonal n/n. Its importance for silicate replacement reactions in box 7/7 is discussed in the entry *Hydrothermal Alteration of Silicate Rocks*. Its importance for the dolomitization of limestones by the reaction

$$2\ CaCO_3 + Mg^{2+} \rightarrow CaMg(CO_3)_2 + Ca^{2+} \quad (2)$$

and for the replacement of dolomite by magnesite

$$CaMg(CO_3)_2 + Mg^2 \rightarrow 2Mg(CO_3) + Ca^{2+} \quad (3)$$

has been studied by Rosenberg and Holland (1964), Rosenberg, Burt, and Holland (1967), and Johannes (1966). Fig. 2 shows the relationship between the mole fraction of calcium in solutions in equilibrium with calcite, dolomite, and magnesite between 275 and 420°C. Solutions brought in contact with carbonates with which they are not in equilibrium will react with such carbonates until either the carbonate has been completely replaced, or until the composition of the solution has been changed by the replacement process so that it is no longer out of equilibrium with the solid carbonate phases.

Toward higher temperatures progressively greater CO_2 pressures are required to maintain carbonates that are stable with respect to oxides, hydroxides, and silicates. The replacement of limestones and dolomites by a variety of silicate and oxide minerals in skarns is eloquent testimony of the insufficiency of CO_2 pressures at temperatures above about 450°C to forestall replacement reactions in categories 5/3 and 5/7. Greenwood (1967) and Eugster and Skippen (1967) have discussed in some detail the theory of phase equilibria in the presence of gas mixtures containing CO_2.

The systematics of nonsilicate hydrothermal alterations are now quite well understood. The importance of the fugacity of S_2, O_2, CO_2, and H_2O, of the ionic strength of hydrothermal solutions, and of the ratio of cation and anion activities has been demonstrated for a large number of reactions, and stability limits have been determined for a number of geologically important phases. However, much more work remains to be done on the chemistry of aque-

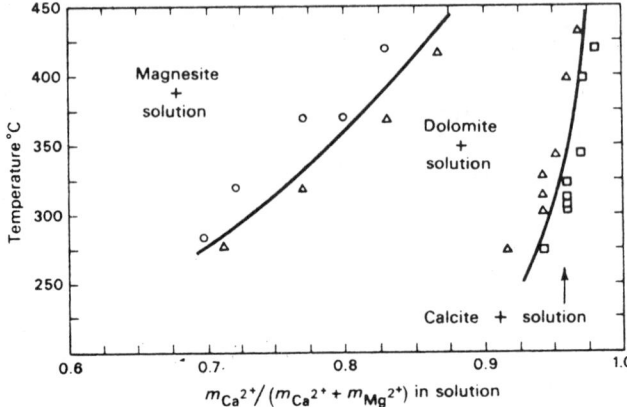

FIG. 2. The mole fraction $m_{Ca^{2+}}/(m_{Ca^{2+}} + m_{Mg^{2+}})$ in solutions in equilibrium with calcite + dolomite and dolomite + magnesite at temperatures between 275 and 420°C (Rosenberg and Holland, 1964).

ous solutions at high temperature before any degree of completeness can be claimed for our knowledge of the hydrothermal alteration of nonsilicate minerals.

HEINRICH D. HOLLAND

References

Barnes, H. L., and Czamanske, G. K., 1967, "Solubilities and transport of ore minerals," Ch. 8 in (H. L. Barnes, editor) "Geochemistry of Hydrothermal Ore Deposits," New York, Holt, Rinehart, and Winston.

Barton, P. B., Jr., and Skinner, B. J., 1967, Sulfide mineral stabilities, Ch. 7 in (H. L. Barnes, editor), "Geochemistry of Hydrothermal Ore Deposits," New York, Holt, Rinehart, and Winston.

Eugster, H. P., and Skippen, G. B., 1967, "Igneous and metamorphic reactions involving gas equilibria," in (Ph. H. Abelson, editor) "Researches in Geochemistry," Vol. 2, New York, John Wiley & Sons, pp. 492–520.

Greenwood, H. J., 1967, "Mineral equilibria in the system $MgO-SiO_2-H_2O-CO_2$," in (Ph. H. Abelson, editor) "Researches in Geochemistry," Vol. 2, New York, John Wiley & Sons, pp. 542–567.

Holland, H. D., 1959, "Some applications of thermochemical data to problems of ore deposits. I. Stability relations among the oxides, sulfides, sulfates and carbonates of ore and gangue Minerals," *Econ. Geol.*, **54**, 184–233.

Holland, H. D., 1965, "Some applications of thermochemical data to problems of ore deposits. II. Mineral assemblages and the composition of ore-forming fluids," *Econ. Geol.*, **60**, 1101–1166.

Johannes, W., 1966, "Experimentelle Magnesitbildung aus Dolomit + $MgCl_2$," *Contrib. Mineral. Petrol.*, **31**, 51–58.

Kelley, K. K., 1960, "Contributions to the data on theoretical metallurgy. XIII. High temperature heat-content, heat-capacity and entropy data for the elements and inorganic compounds," *U.S. Bur. Mines Bull.* **584**.

Kubaschewski, O. and Evans, E. Ll., 1958, "Metallurgical Thermochemistry," Third ed., Pergamon Press.

Lindgren, W., 1933, "Mineral Deposits," Fourth ed., New York, McGraw-Hill Book Co., 930pp.

Ramdohr, P., 1960, "Die Erzmineralien und ihre Verwachsungen," Berlin, Akademie Verlag, 1089pp.

Raymahashay, B. C., and Holland, H. D., 1968, "Composition of aqueous solutions in equilibrium with sulfides and oxides of iron at 350°C," *Science*, **162**, 895–896.

Raymahashay, B. C., and Holland, H. D. 1969, "Redox reactions accompanying hydrothermal wallrock alteration," *Econ. Geol.*, **64**, 291–305.

Robie, R. A., 1966, "Thermodynamic properties of minerals," in (S. P. Clark, Jr., editor) "Handbook of Physical Constants," Rev. ed., Geol. Soc. Am. Memoir 97, Sect. 20, pp. 437–458.

Robie, R. A., and Waldbaum, D. R., 1968, "Thermodynamic properties of minerals and related substances at 298.15°K (25.0°C) and one atmosphere (1.013 Bars) pressure and at higher temperatures," *U.S. Geol. Surv. Bull.* **1259**.

Roedder, E., 1967, "Fluid inclusions as samples of ore fluids," in (H. L. Barnes, editor), "Geochemistry of Hydrothermal Ore Deposits," New York, Holt, Rinehart, and Winston, Chap. 12.

Rosenberg, P. E., and Holland, H. D., 1964, "Calcite-dolomite-magnesite stability relations in solutions at elevated temperatures," *Science*, **145**, 700–701.

Rosenberg, P. E., Burt, D. M., and Holland, H. D., 1967, "Calcite-dolomite-magnesite stability relations, the effect of ionic strength," *Geochim. Cosmochim. Acta*, **31**, 291–296.

Cross-references: *Complexes; Fluid Inclusions; Hydrothermal Alteration of Silicate Rocks; Phase Equilibria.*

HYDROTHERMAL ALTERATION OF SILICATE ROCKS—GENERAL PRINCIPLES*

Hydrothermal alteration results from the chemical reaction between the mineral phases of rocks and the components of an aqueous solution. For a thorough understanding of any hydrothermal alteration the temperature, pressure, and composition of the altering solution should be known, as well as the compositions of the mineral phases in the rock before and after alteration. Unfortunately, only fragments of the desired data are generally available. Traces of solutions that may have been involved in alteration can be recovered from fluid inclusions; alteration temperatures, which seem to range from $<100°C$ to about $650°C$, may be deduced from the fluid-vapor relations in fluid inclusions and from phase equilibrium relations (melting points, inversions, dissociations and decompositions, exsolutions, eutectics, and others); compositions of the unaltered rocks can be approximated through geologic correlations between fresh and altered equivalents; lastly, and perhaps the only unequivocal data, the chemical composition of the alteration mineral phases and the bulk composition of the altered rock can be obtained readily by chemical analyses.

Investigations on thermal springs (White, 1957), and on fluid inclusions (Barton, 1959; Smith, 1954; and Hall and Friedman, 1963), suggest that the aqueous solutions responsible for hydrothermal alteration are relatively concentrated in sodium, calcium, and potassium chloride and contain some calcium, bicarbonate, and sulfate ions. Smith (1954), in addition, found liquid carbon dioxide which he considered to be a component of hydrothermal solutions. A symposium, which gathered to review research on the chemistry of ore-forming fluids, included advocates for transport of metals in brines, in bisulfide solutions, and in polysulfide solutions (Roedder, 1965). Most of the fluid inclusion data, however, support brines as the ore solution.

Bulk changes in chemical composition between fresh and altered rocks, assuming constant volume, have been determined for some types of deposits (Lovering, 1949; Anderson, 1950; Tooker, 1963; and Creasey, 1965). In the proper chemical environment, almost any element may pass into solution and be moved from its original site. Greatest changes seem to involve sodium, calcium, hydrogen, sulfur, and potassium, the first two being leached and the last three added. Variations between different alterations, however, are so great that generalizations have little meaning unless identified with a particular type of alteration.

Table 1 lists some of the common primary and secondary minerals found near ore deposits and the most common products formed from their hydrothermal alteration. In some deposits, the products of early reactions become the reactants of later reactions. Some minerals therefore occur on both sides of the list.

Chemical Considerations

Hemley and Jones (1964, p. 539) emphasized that in the chemistry of rock alteration the *phase rule* and the *equilibrium constant principle* are fundamental concepts. Goldschmidt (1911) first pointed out that the phase rule could be applied to metamorphic rocks in which mineral phases were in equilibrium. Hydrothermal alteration is a type of metamorphism; the phase rule is as applicable

* Publication authorized by the Director, U.S. Geological Survey.

TABLE 1. COMMON MINERALS AND THEIR HYDROTHERMAL ALTERATION PRODUCTS

Existing Minerals (Primary and Secondary) = Reactants	Possible Alteration Minerals = Products
1. Plagioclase	1. Kaolinite, montmorillonite, muscovite, epidote, carbonate, K-feldspar
2. K-feldspar	2. Kaolinite, muscovite
3. Biotite	3. Chlorite
4. Amphibole and/or pyroxene	4. Chlorite, carbonate, epidote
5. Chlorite	5. Biotite, muscovite
6. Fe-Mg biotite	6. Mg-biotite
7. Kaolinite and/or muscovite	7. Pyrophyllite
8. Magnetite	8. Sulfide
9. Kaolinite and/or montmorillonite	9. Muscovite
10. Montmorillonite	10. Kaolinite
11. Muscovite	11. Kaolinite
12. Quartz	12. None

to it as to contact or regional metamorphism. The phase rule can be written as follows:

$$P + V = C + 2 \qquad (1)$$

where P is the number of phases, V is the variance or number of degrees of freedom, such as pressure, temperature, and concentration of each phase, and C is the number of components. Goldschmidt reasoned that any common metamorphic rock must be stable over a range in both temperature and pressure, otherwise it would not occur commonly. In such a metamorphic assemblage, the variance is at least 2, and consequently P equals C; that is, the maximum number of mineral phases in equilibrium equals the number of components. This is Goldschmidt's mineralogical phase rule.

The equilibrium state of the reaction of one substance in contact with an aqueous solution and the formation of another is defined by the equilibrium constant. Because any hydrothermal alteration is the sum of many reactions, the individual equilibrium constants and the reactions they define are the building blocks for the entire alteration process.

The equilibrium constant is expressed in terms of the activity of the individual constituents. Activity, which is the concentration as modified by the effects of other dissolved substances and the solvent, is expressed by $a = km$, where a is the activity, m the concentration, and k the activity coefficient. The equilibrium constant is expressed as the ratio of the products to the reactants in a chemical reaction. Where b and c moles of reactants B and C have come to equilibrium with d and e moles of products D and E, the equilibrium constant, K, can be expressed as follows:

$$K = \frac{a_D^d \cdot a_E^e}{a_B^b \, a_C^c} \qquad (2)$$

Many of the reactions that characterize hydrothermal alteration are hydrolytic. During hydrolysis of silicate minerals, H^+ is selectively consumed and cations, chemically equivalent in amount, are released to the hydrothermal solution. The following equation showing the transformation of K-feldspar to muscovite exemplifies a hydrolytic reaction.

$$3KAlSi_3O_8 + 2H^+ = KAl_3Si_3O_{10}(OH)_2$$
$$+ 2K + 6SiO_2 \qquad (3)$$

Such reactions are so common in hydrothermal alteration that Hemley and Jones (1964) describe the hydrothermal process that produces clays and K-mica as essentially hydrogen metasomatism.

Hydration is the chemical combination of water and another substance; it is common in some types of alteration, but should not be confused with hydrolysis. During hydration, there is no selective consumption of H^+, and cations are not released to the altering solutions. The reaction of aluminum oxide plus water to form gibbsite is hydration.

$$Al_2O_3 + 3H_2O = 2Al(OH)_3 \qquad (4)$$

Experimental Studies

Laboratory studies of two types have produced most data pertinent to the understanding of hydrothermal alteration: (*1*) dehydration studies that deal with temperatures and H_2O pressure in an otherwise isochemical system; (*2*) reactions involving changes in bulk composition and therefore dealing with the composition of the aqueous altering solution, temperature (pressure), and the chemical composition of solid phases. Both types of investigations have been fruitful, but rock alteration is a type of metamorphism and therefore studies in metasomatic chemistry, such as those of Hemley, Meyer, and others (Hemley, 1959; Meyer and Hemley, 1959; and Hemley, Meyer, and Richter, 1961) on the composition of the aqueous phase in equilibrium with the mineral phases, constitute a significant advance in alteration studies.

Fig. 1 shows some of the experimental univariant dehydration curves and divariant phase diagrams most applicable to hydrothermal alteration. These data, however, do not take into account the composition of the aqueous solutions in equilibrium with solid phases. These univariant dehydration curves show the maximum pressures and temperatures under which kaolinite, montmorillonite, pyrophyllite, quartz-muscovite, muscovite, and phlogopite can form and be stable in the isochemical systems investigated. For example, at a water pressure of 12,000 psi, which is equivalent to a depth in the earth's crust of about 10,000 feet, and on the basis of present information, kaolinite is not stable above 400°C, montmorillonite perhaps 400–480°C (depending on composition), pyrophyllite above 575°C, the muscovite-quartz pair above 640°C, and the muscovite above 655°C; these data are presented in Fig. 1. More recently pyrophyllite was produced from quartz + kaolinite at 410–415°C and 15,000 psi P_{H_2O} (Carr and Fyfe, 1960), and at 300°C and 14,500 psi in a run of a years' duration (Reed and Hemley, 1966). Carr and Fyfe (1960) found that pyrophyllite + quartz reacted at about 420°C and 15,000 psi P_{H_2O} to produce X-andalusite and quartz.

Above a pressure of about 1,000 psi, the upper stability limit of phlogopite (Mg-biotite) exceeds the minimum melting curve of a "wet"

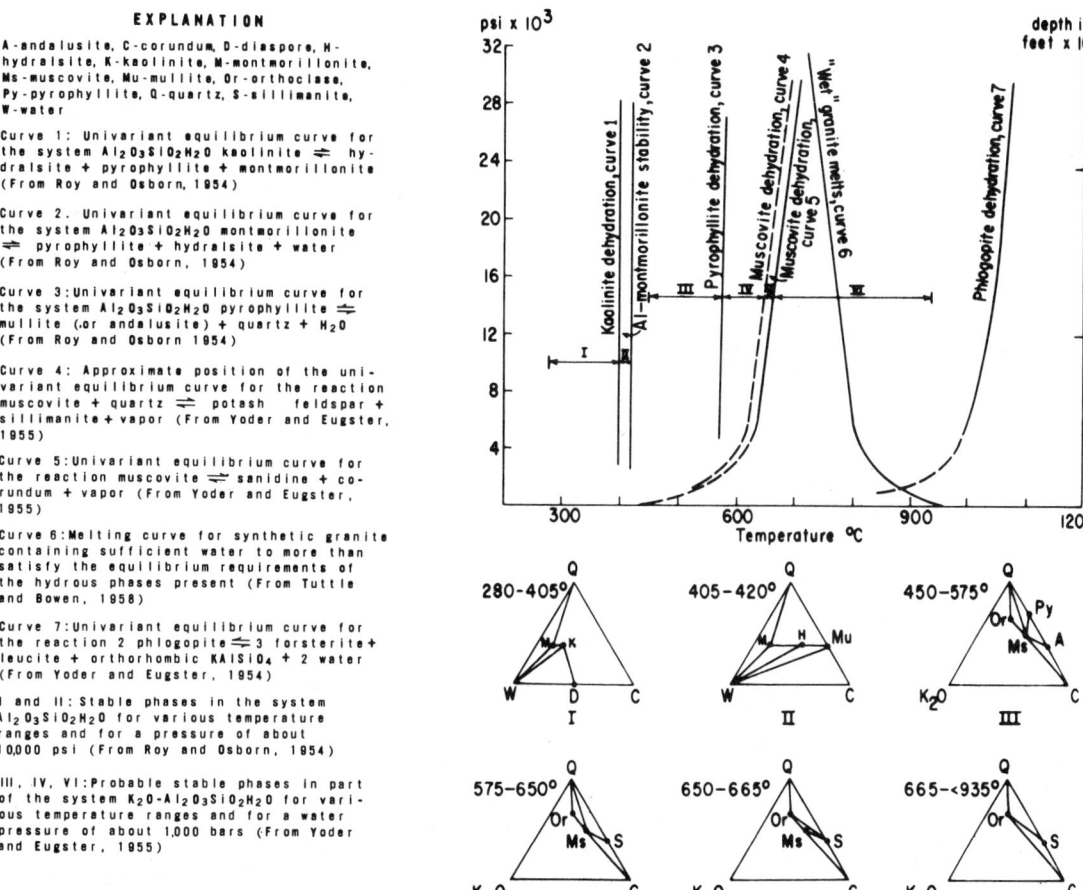

Fig. 1. Experimental dehydration curves for the stability limits of kaolinite, Al-montmorillonite, muscovite, and phlogopite; the reaction curve for the quartz-muscovite pair; and the melting curve for a wet granite. Divariant equilibria in the areas between the dehydration curves for the systems Al_2O_3–SiO_2–H_2O and K_2O–Al_2O_3–SiO_2–H_2O (From Creasey, 1966).

granite (water content about 9%). Yoder and Eugster (1954, p. 179) have indicated that iron-bearing biotites can be grown at temperatures lower than the Mg varieties, but the general relation shown in Fig. 1 holds for all biotites. This relation shows that the presence or absence of biotite gives no information on the maximum temperature of formation of a hydrothermally altered rock, since it can safely be assumed that hydrothermal alteration occurs at temperatures below the minimum melting of a wet granite.

The phase assemblages in triangles I to VI in Fig. 1 show the stable and probably stable minerals for the temperature ranges given for each triangle. The suggested stability fields for these assemblages are bounded by the univariant dehydration curves. It should be pointed out, however, that the mineral assemblages are determined by the components used. Because triangles I and II do not have K_2O as a component in the system, muscovite cannot be shown as a phase, although it also is stable under the P–T conditions. Possible phase assemblages of the system K_2O–Al_2O_3–SiO_2–H_2O in a point projection on the K_2O–Al_2O_3–SiO_2 plane over the temperature range of 200 to 575°C at about 1,000 bars are shown in Fig. 2. Note that Fig. 2 is essentially the same as triangle III of Fig. 1, except that montmorillonite and kaolinite have been added because the temperature is now within the suggested stability range of these minerals.

Hemley (1959) and Hemley, Meyer, and Richter (1961) investigated the systems K_2O–Al_2O_3–SiO_2–H_2O and Na_2O–Al_2O_3–SiO_2–H_2O in which concentration in the aqueous phase and temperatures were varied while the pres-

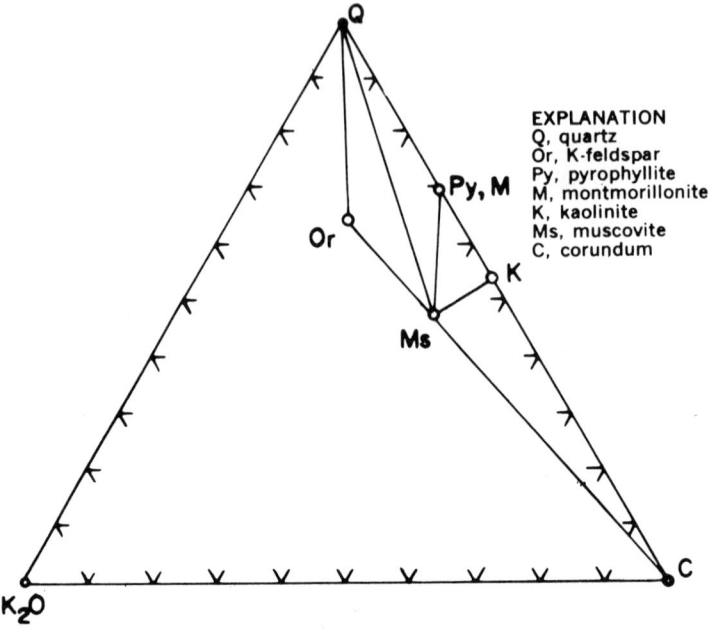

Fig. 2. Composition diagram from the projection on the K_2O–Al_2O_3–SiO_2 plane part of the system K_2O–Al_2O_3–SiO_2–H_2O for the temperature range 200–575°C and a pressure of about 1,000 bars (modified from Yoder and Eugster, 1955).

sure was held constant. The concentration parameters are the quotients $mKCl/mHCl$ for Fig. 3 and $mNaCl/mHCl$ for Fig. 4. Equilibrium values were produced by varying the initial molalities of KCl, NaCl, and HCl, with solid phases greatly in excess, and allowing the system to reach equilibrium. The solutions were in equilibrium with quartz. Hemley's experimental method gives the ratios of total molar KCl or NaCl rather than ion ratios, but for purposes of simplicity these are commonly referred to as K^+/H^+ or Na^+/H^+ ratios. Fig. 3 shows the equilibrium curves for the system with K_2O and Fig. 4 for the system with Na_2O. The pertinent reactions for the two systems are shown in Eqs. (5)–(12).

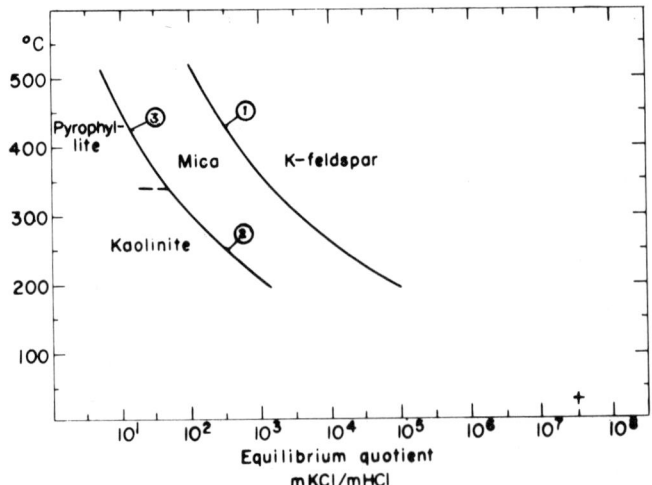

Fig. 3. Some stability relations in the system K_2O–Al_2O_3–SiO_2–H_2O as a function of temperature and the $mKCl/mHCl$ ratio. Quartz presents; 15,000 psi total pressure (from Hemley, 1959).

$$3/2 \text{ KAlSi}_3\text{O}_8 + \text{H}^+ \rightleftharpoons 1/2 \text{ KAl}_2\text{AlSi}_3\text{O}_{10}(\text{OH})_2 + 3\text{SiO}_2 + \text{K}^+ \quad (5)$$
K-feldspar muscovite quartz

$$\text{KAl}_2\text{AlSi}_3\text{O}_{10}(\text{OH})_2 + \text{H}^+ + 3/2 \text{ H}_2\text{O} \rightleftharpoons 3/2 \text{ Al}_2\text{Si}_2\text{O}_5(\text{OH})_4 + \text{K}^+ \quad (6)$$
muscovite kaolinite

$$\text{KAl}_2\text{AlSi}_3\text{O}_{10}(\text{OH})_2 + 3\text{SiO}_2 + \text{H}^+ \rightleftharpoons 3/2 \text{ Al}_2\text{Si}_4\text{O}_{10}(\text{OH})_2 + \text{K}^+ \quad (7)$$
muscovite pyrophyllite

$$\text{Al}_2\text{Si}_2\text{O}_5(\text{OH})_4 + 2\text{SiO}_2 \rightleftharpoons \text{Al}_2\text{Si}_4\text{O}_{10}(\text{OH})_2 + \text{H}_2\text{O} \quad (8)$$
kaolinite pyrophyllite

$$3/2 \text{ NaAlSi}_3\text{O}_8 + \text{H}^+ \rightleftharpoons 1/2 \text{ NaAl}_3\text{Si}_3\text{O}_{10}(\text{OH})_2 + 3\text{SiO}_2 + \text{Na}^+ \quad (9)$$
albite paragonite

$$\text{NaAl}_3\text{Si}_3\text{O}_{10}(\text{OH})_2 + \text{H}^+ + 3\text{SiO}_2 \rightleftharpoons 3/2 \text{ Al}_2\text{Si}_4\text{O}_{10}(\text{OH})_2 + \text{Na}^+ \quad (10)$$
paragonite pyrophyllite

$$1.17 \text{ NaAlSi}_3\text{O}_8 + \text{H}^+ \rightleftharpoons 0.5 \text{ Na}_{0.33}\text{Al}_{2.33}\text{Si}_{3.67}\text{O}_{10}(\text{OH})_2 + 1.67 \text{ SiO}_2 + \text{Na}^+ \quad (11)$$
albite montmorillonite

$$3\text{Na}_{0.33}\text{Al}_{2.33}\text{Si}_{3.67}\text{O}_{10}(\text{OH})_2 + \text{H}^+ + 3.5 \text{ H}_2\text{O} \rightleftharpoons 3.5 \text{ Al}_2\text{Si}_2\text{O}_5(\text{OH})_4 + 4\text{SiO}_2 + \text{Na}^+ \quad (12)$$
montmorillonite kaolinite

Garrels and Christ (1965), using data from Hemley (1959), Hemley, Meyer, and Richter (1961), and Orville (1963) constructed a three-dimensional diagram showing the approximate phase relations in the system Na_2O–K_2O–Al_2O_3–SiO_2–H_2O at 300°C and 15,000 psi (Fig. 5). The mineral phases shown in this diagram are the same as those shown in Fig. 3 and 4, so that the data for all three diagrams are essentially the same, the presentations being somewhat different.

Studies of Altered Rocks

Studies of the hydrothermally altered rocks associated with ore deposits are virtually as old as historical records on ore deposits. However, until the mid-1940s the petrographic microscope was the principal tool used in the study of altered rocks. Unfortunately, the resolving power of the petrographic microscope is inadequate for the fine-grained intergrown mineral phases of many altered rocks; x-ray diffraction equipment nicely covers some of the deficiencies of the microscope, and in conjunction with a microprobe, permits identification of minerals, distribution of elements within individual minerals, and interpretation of textural relations between mineral phases. Hence the last two decades have produced several excellent accounts of hydrothermally altered rocks in individual deposits or districts, and several attempts have been made to synthesize types of alteration from many districts. Space does not permit citation of all the studies that have contributed to recognition of principles of alteration. The following are selected contributions summarizing nature's complex experiments on alteration of silicate rocks.

Studies of successive alteration envelopes outward from quartz and quartz-metal veins that cut igneous rocks of intermediate composition have revealed an inner envelope of sericite passing outward into clay minerals. Adjacent to tungsten veins in Boulder County, Colorado, Lovering (1941) recognized an inner (veinward) envelope of sericite that passed outward into a clay zone composed of dickite adjacent to the sericite and allophane and montmorillonoid against the fresh quartz monzonite host rock.

Sales and Meyer (1948) in their classic study of alteration at Butte, Montana, describe quartz-sulfide veins flanked by successive envelopes of sericite, kaolinite, and montmoril-

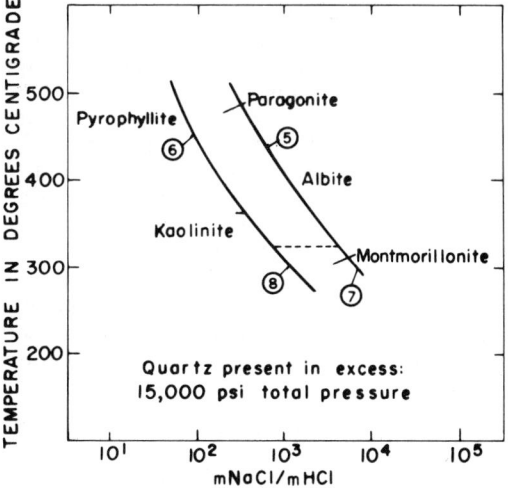

Fig. 4. Some stability relations in the system Na_2O_3–Al_2O_3–SiO_2–H_2O as a function of temperature and the $m\text{NalC}/m\text{HCl}$ ratio (from Hemley and others, 1961).

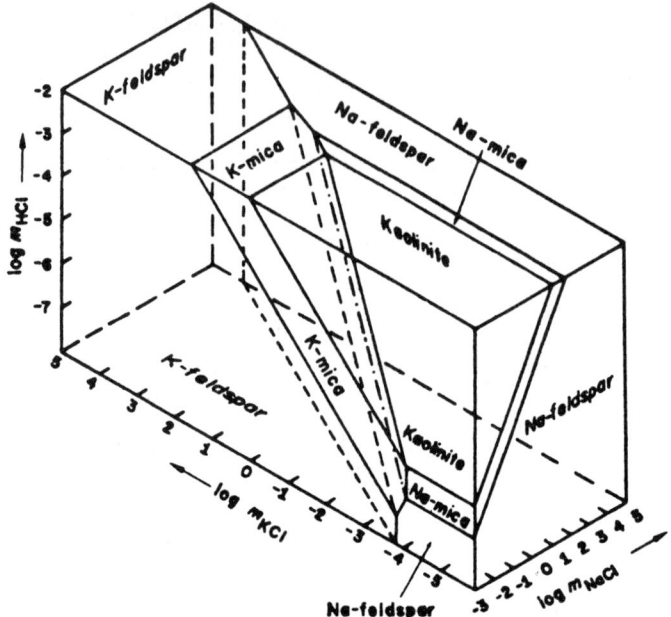

FIG. 5. Approximate phase relations in the system Na_2O–K_2O–Al_2O_3–SiO_2–H_2O at 300°C and 15,000 psi; quartz is assumed to be present in excess (from Garrels and Christ, 1965).

lonite; the host rock is quartz monzonite. Tooker (1963) found similar alterations adjacent to veins in the Central City district, Colorado. Sericite was abundant next to the vein and was associated with quartz and pyrite. Clay minerals predominated beyond the sericite: where alteration is strong, kaolinite somewhat intergrown with montmorillinite and illite-sericite predominates, whereas where alteration is weak, montmorillonite occurs along fractures or as an alteration product of plagioclase.

The observed alteration envelopes adjacent to the veins described above appear to agree with the experimental data in Figs. 3 and 4. If temperature decreased outward from a vein (as seems probable in a shallow environment), kaolinite could be stable in the cooler environment while sericite was stable veinward at higher temperatures (Fig. 3). If the cation/H^+ ratio increased outward from the vein owing to hydrolytic reactions, kaolinite would tend to be stable in the environment of lower cation/H^+ ratio and montmorillonite in the environment of higher ratio. The observed agrees remarkably well with the predicted.

Reaction (11) and reactions (13)–(15) (Hemley and Jones, 1964) are applicable to the alterations described from the vein outward where M = base cation equivalent to the small charge deficiency x.

In deposits such as the disseminated copper deposits of the southwestern United States where large masses of silicate rocks have been hydrothermally altered, the sharp changes and consistencies noted adjacent to the quartz-metal veins are not always apparent. Perhaps this is because the alteration histories of these deposits are highly complex, as Fournier (1967) indicated for the porphyry copper deposit near Ely, Nevada. Here porphyry intrusions of slightly different ages had different alteration histories.

Nevertheless, considerable information on mass alteration of large volumes of rock is available. Alteration similarities are common, but so are alteration differences; no two deposits are exactly alike, but no deposit is en-

$$1/2\ Al_2Si_2O_5(OH)_4 + 3H^+ = SiO_2 + 5/2\ H_2O + Al^{2+} \quad (13)$$
$$\text{kaolinite} \qquad\qquad\qquad \text{quartz}$$

$$KAl_3Si_3O_{10}(OH)_4 + H^+ + 3/2\ H_2O = 3/2\ Al_2Si_2O_5(OH)_4 + K^+ \quad (14)$$
$$\text{sericite} \qquad\qquad\qquad\qquad \text{kaolinite}$$

$$Na_2CaAl_4Si_8O_{24} + 4H^+ \rightarrow 2M_xAl_{2+x}Si_{4-x}O_{10}(OH)_2 + 2Na^+ + Ca^{2+} \quad (15)$$
$$\text{andesine} \qquad\qquad\qquad \text{montmorillonite}$$

tirely unique. Creasey (1966) recognized three principal types of alteration in the igneous rocks associated with the disseminated copper deposits (The term "facies" was used in an earlier paper (Creasey, 1959) to describe these types of alteration, but because of inconsistency with proper metamorphic usage the term should not be used in this context.): propylitic, argillic, and potassic, all of which are based on the assumption of a general approach to equilibrium at the time of alteration.

The propylitic type of alteration is distinguished from the argillic and potassic by the prominence of lime-bearing minerals of the carbonate and epidote groups. The agrillic type of alteration is distinguished by members of the kaolinite and montmorillonite groups of clay

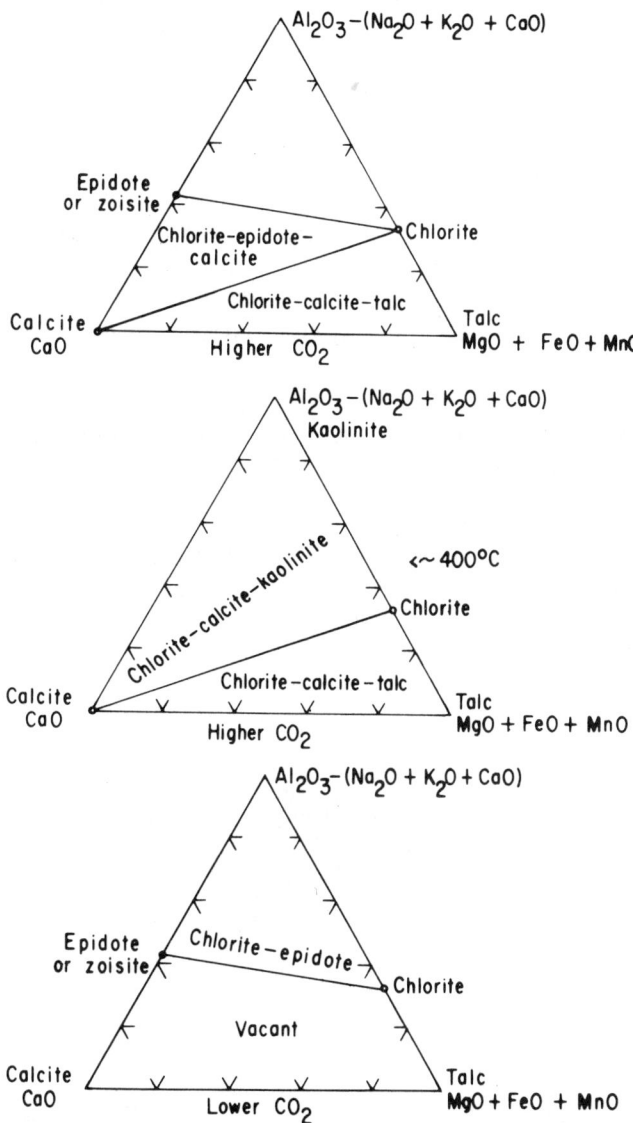

FIG. 6. Compatibility diagrams for the propylitic type of hydrothermal alteration. SiO_2 and H_2O occur in excess, and quartz is always a phase in the final assemblage. Muscovite and albite are common additional phases when K_2O and Na_2O are components of the system. Sphene or leucoxene and apatite are also stable mineral phases. The true plots of some minerals are within the large triangles instead of along the margins as shown (from Creasey, 1966).

minerals in conjunction with strong leaching of CaO. Lime may be important only to stabilize montmorillonite. At least there seems to be no other stable mineral in argillized rocks with lime as a component. The transformation of the clay minerals to mica or some other mineral marks the appearance of potassic alteration. In this type, both secondary muscovite and biotite are prominent, and secondary K-feldspar is abundant. The appearance of secondary K-feldspar is often concomitant with secondary biotite; they are a stable pair.

Propylitic alteration occurs either alone or peripheral to the argillic and potassic types. Fig. 6 contains three idealized compatibility diagrams for propylitic alteration based on observational data reported in the literature and by the author. Two of the diagrams include three possible assemblages that form in the presence of considerable CO_2, and the third diagram contains a fourth assemblage that forms in a lower CO_2-pressure environment. The four assemblages are as follows: (1) chlorite-calcite-kaolinite; (2) chlorite-calcite-talc; (3) chlorite-epidote-calcite; and (4) chlorite-epidote. As indicated in Fig. 6, the assemblage chlorite-epidote-calcite may form at temperatures above the stability temperatures of the clay minerals. Equations (16)–(19) are pertinent to the above mineral assemblages.

$$6Ca_2(Mg, Fe)_5Si_8O_{22}(OH)_2 + 12CO_2 + 14H_2O = 5(Mg, Fe)_6Si_4O_{10}(OH)_8$$
$$\text{actinolite} \qquad\qquad\qquad\qquad\qquad\qquad\qquad \text{chlorite}$$
$$+ 12CaCO_3 + 28SiO_2 (1)(2)(3) \qquad (16)$$
$$\text{calcite} \qquad \text{quartz}$$

[Eq. (16) is taken from Fyfe, Turner, and Verhoogen (1958, p. 220).]

$$Na_2CaAl_4Si_8O_{24} + 4H^+ + 2H_2O = 2Al_2Si_2O_5(OH)_4 + 4SiO_2 + 2Na^+ + Ca^{++} (1)$$
$$\text{andesine} \qquad\qquad\qquad\qquad \text{kaolinite} \qquad\quad \text{quartz}$$
$$(17)$$

$$4CaAl_2Si_2O_8 + 5MgSiO_3 + 5H_2O = 2Ca_2Al_3Si_3O_{12}(OH) + Mg_5Al_2Si_3O_{10}(OH)_8 + 4SiO_2$$
$$\text{anorthite} \qquad \text{pyroxene} \qquad\qquad\qquad \text{epidote} \qquad\qquad \text{chlorite}$$
$$(18)$$

[Eq. (18) is taken from Deer, Howie, and Zussman (1962, p. 199)]

$$Ca_2Mg_5Si_8O_{22}(OH)_2 + 4CO_2 = 2CaMg(CO_3)_2 + Mg_3Si_4O_{10}(OH)_2 + 4SiO_2 \qquad (19)$$
$$\text{tremolite} \qquad\qquad\qquad\qquad \text{dolomite} \qquad\quad \text{talc} \qquad\quad \text{quartz}$$

[Eq. (19) is taken from Turner (1948, p. 133)]

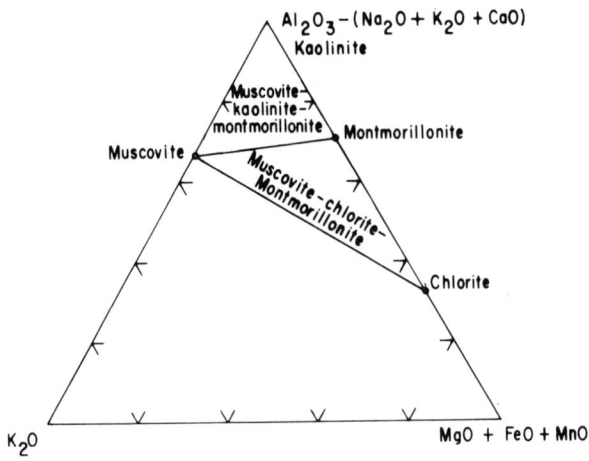

FIG. 7. Compatibility diagram for argillic alteration. SiO_2 and H_2O occur in excess, and quartz occurs in every assemblage. The true plots of some minerals are within the large triangle instead of along the margins as shown, and CaO and Na_2O may be accessory constituents of montmorillonite (modified from Creasy, 1966).

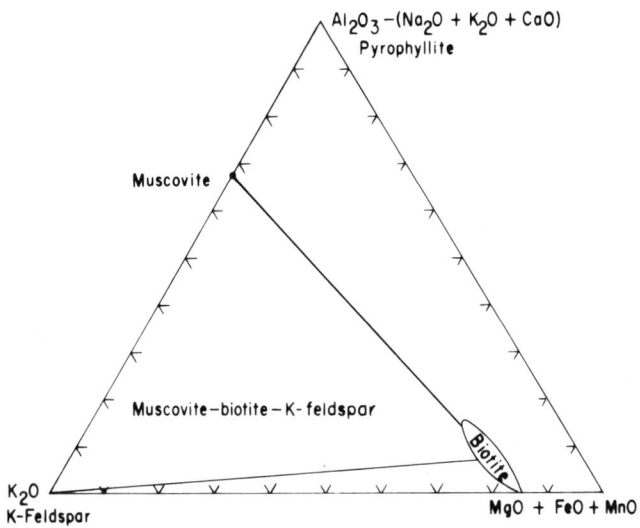

FIG. 8. Compatibility diagram for potassic and quartz-muscovite alterations. SiO_2 and H_2O occur in excess, and quartz occurs in the biotite-muscovite-K-feldspar assemblage (from Creasy, 1966).

Argillic alteration is distinguished by the presence of clay minerals (members of the kaolinite or montmorillonite groups) and by indications of strong leaching of lime. In some argillized rocks other elements such as sodium, iron, and magnesium are also leached. Mineralogically, the loss of lime is reflected chiefly by the absence of members of the epidote and carbonate groups. The amphiboles are unstable here as well as under propylitic alteration. Fig. 7 is a compatibility diagram of argillic alteration. It includes two stable mineral assemblages: muscovite-kaolinite-montmorillonite and muscovite-chlorite-montmorillonite. Reactions (5), (15), (17), and the reaction (20) are applicable to the mineral assemblages.

In addition, quartz is a phase in every assemblage, and other phases may exist if other components occur. A sulfur-bearing component, possibly FeS_2, is reflected by the common presence of pyrite. TiO_2 is reflected in rutile, and a copper-bearing component in chalcopyrite. Possibly the four-phase assemblage muscovite-kaolinite-chlorite-montmorillonite is also stable, and if so, the presence of montmorillonite probably indicates that another component, such as CaO or Na_2O, is definitive for the system. However, the maximum permissible number of phases is not necessarily present.

The potassic alteration is distinguished by the assemblage muscovite-biotite-K-feldspar or any two of the three phases. The key minerals are new biotite and K-feldspar. Since muscovite also occurs in the propylitic and argillic alterations, it has no special significance, but so far as this author knows, it always occurs with the biotite and K-feldspar. Fig. 8 is a compatibility diagram for the potassic alteration. Reaction (20) and reaction (21) are pertinent to the potassic alteration assemblage.

In addition, chlorite is known to react with K^+ to form biotite, but the reaction is difficult to write because the compositions of the chlorites and biotites involved are inadequately known.

The clay minerals are not stable during potassic alteration. Their presence or absence is the most striking difference between the two types of alteration. Since the clay minerals are unstable above about 300–450°C, some products of potassic alterations form above that temperature range, although favorable concen-

$$3/4\ Na_2CaAl_4Si_8O_{24} + 2H^+ + K^+ = KAl_3Si_3O_{10}(OH)_2 + 3/2\ Na^+ + 3/4\ Ca^{2+} + 3SiO_2 \quad (20)$$
andesine — muscovite — quartz

$$Na_2CaAl_4Si_8O_{24} + 4K^+ + 4SiO_2 = 4KAlSi_3O_8 + 2Na^+ + Ca^{++} \quad (21)$$
andesine — K-feldspar

trations of K^+ and H^+ can result in potassic alteration at lower temperature.

Quartz-sericite-pyrite, without either a clay mineral or K-feldspar associated, is a common assemblage that does not fit into any of the three previously described alteration types. If clay were present, the assemblage would belong to the argillic alteration, and if K-feldspar were present, it would belong to the potassic. It seems best to consider it a distinct type of alteration. It can be represented, however, on the compatibility diagram for the potassic alteration, Fig. 8. Reactions (5), (9), and (20) are pertinent for the formation of sericite (muscovite) from feldspars. Where sericitic alteration impinges on earlier clay alteration, the reserve of reaction (14) would be applicable.

Silicification is a type of alteration that is commonly associated with the more complex alterations previously described. It originates in several ways: (*1*) by complete leaching of all cations other than silica, leaving a porous mass of quartz (Eq. (13)); (2) by deposition in open spaces, such as in some quartz veins; and (*3*) by replacement, which can be considered a combination of *1* and *2*, that is, simultaneous leaching and deposition. The intense hydrolytic leaching implies an altering solution of low cation/H^+ ratio.

1967   S. C. Creasey

References

Anderson, C. A., 1950, "Alteration and metallization in the Bagdad porphyry copper deposit, Arizona," *Econ. Geol.* **45**(7), 609–628.

Barton, P. B., Jr., 1959, "The chemical environment of Ore Deposition and the problem of low-temperature ore transportation," in (Abelson, P. H., editor) "Researches in Geochemistry," New York, John Wiley & Sons, pp. 279–300.

Carr, R. M., and Fyfe, W. S., 1960, "Synthesis fields of some aluminum silicates," *Geochim. Cosmochim. Acta*, **21**(1–2), 99–109.

Creasey, S. C., 1959, "Some phase relations in hydrothermally altered rocks of porphyry copper deposits," *Econ. Geol.*, **54**(3), 351–373.

Creasey, S. C., 1965, "Geology of the San Manuel area, Pinal County, Arizona," with a section on "Ore deposits," by J. D. Pellitier and S. C. Creasy, *U. S. Geol. Surv. Profess. Paper* **471**, 64pp.

Creasey, S. C., 1966, "Hydrothermal Alteration," in (Titley, S. R., and Hicks, C. L., editors), "Geology of the Porphyry Copper Deposits, Southwestern North America," Tucson, Univ. Arizona Press, pp. 51–74.

Deer, W. A., Howie, R. A., and Zussman, J., 1962, "Ortho- and Ring-Silicates," Vol. 1 of "Rock-Forming Minerals," New York, John Wiley & Sons, 333pp.

Eitel, W., 1966, "Hydrothermal Silicate Systems," Vol. 4 of "Silicate Science," New York, Academic Press, 617pp.

Fournier, R. O., 1967, "The porphyry copper deposit exposed in the Liberty open-pit mine near Ely, Nevada; Pt. 1, Syngenetic formation; Pt. 2, The formation of hydrothermal alteration zones," *Econ. Geol.*, **62**(1, 2), 57–81, 207–227.

Fyfe, W. S., Turner, F. J., and Verhoogen, J., 1958, "Metamorphic reactions and metamorphic facies," *Geol. Soc. Am. Mem.*, **73**, 259pp.

Garrels, R. M., and Christ, C. L., 1965, "Solutions, Minerals and Equilibria," New York, Harper and Row, 450pp.

Goldschmidt, V. M., 1911, "Die Kontaktmetamorphose in Kristianiagebiet," *Norsk Vidensk. Skrifter, I. Math.-Naturv. Klasse*, **1**, 483pp.

Hall, W. E., and Friedman, I., 1963, "Composition of fluid inclusions, Cave-in-Rock fluorite district, Illinois, and Upper Mississippi Valley zinc-lead district," *Econ. Geol.*, **58**(6), 886–911.

Hemley, J. J., 1959, "Some mineralogical equilibria in the system $K_2O-Al_2O_3-SiO_2-H_2O$," *Am. J. Sci.*, **257**(4), 241–270.

Hemley, J. J., and Jones, W. R., 1964, "Chemical aspects of hydrothermal alteration with emphasis on hydrogen metasomatism," *Econ. Geol.*, **59**(4), 538–569.

Hemley, J. J., Meyer, C., and Richter, D. H., 1961, "Some alteration reactions in the system $Na_2O-Al_2O_3-SiO_2-H_2O$," in "Geological Survey Research 1961," *U. S. Geol. Surv. Profess. Paper* **424-D**, D338–D340.

Lovering, T. S., 1941, "The origin of the tungsten ores of Boulder County, Colorado," *Econ. Geol.*, **36**(3), 229–279.

Lovering, T. S., 1949, "Rock alteration as a guide to ore—East Tintic district, Utah," *Econ. Geol. Monograph*, **1**, 64pp.

Meyer, C., and Hemley, J. J., 1959, "Hydrothermal alteration in some granodiorites," in (Swineford, A., editor), "Clays and clay minerals," 6th Nat. Conf. on Clays and Clay Minerals, Berkeley, 1957, Proc., 89–100.

Orville, P. M., 1963, "Alkali ion-exchange between vapor and feldspar phases," *Am. J. Sci.*, **261**(3), 201–237.

Reed, B. L., and Hemley, J. J., 1966, "Occurrence of pyrophyllite in the Kekiktuk Conglomerate, Brooks Range, northeastern Alaska," *U.S. Geol. Surv. Profess. Paper*, **550-C**, C162–C166.

Roedder, E., 1965, "Report on S.E.G. symposium on the chemistry of the ore-forming fluids," *Econ. Geol.*, **60**(7), 1380–1403.

Roy, Rustum, and Osborn, E. F., 1954, "The system $Al_2O_3-SiO_2-H_2O$," *Am. Mineralogist*, **39**(11–12), 853–885.

Sales, R. H., and Meyer, C., 1948, "Wall rock alteration at Butte, Montana," *Am. Inst. Mining Metall. Engrs. Tech. Publ.*, **2400**, 25pp.

Smith, F. G., 1954, "Composition of vein-forming fluids from inclusion data," *Econ. Geol.*, **49**(2), 205–210.

Tooker, E. W., 1963, "Altered wallrocks in the central part of the Front Range mineral belt,

Gilpin and Clear Creek Counties, Colorado," *U.S. Geol. Surv. Profess. Paper,* **436**, 102pp.

Turner, F. J., 1948, "Mineralogical and structural evolution of the metamorphic rocks," *Geol. Soc. Am. Mem.,* **30**, 342pp.

Tuttle, O. F., and Bowen, N. L., 1958, "Origin of granite in the light of experimental studies in the system $NaAlSi_3O_8$–$KAlSi_3O_8$–SiO_2–H_2O," *Geol. Soc. Am. Mem.,* **74**, 153pp.

White, D. E., 1957, "Thermal waters of volcanic origin," *Geol. Soc. Am. Bull.,* **68**(12, pt. 1), 1637–1657.

Yoder, H. S., Jr., and Eugster, H. P., 1954, "Phlogopite synthesis and stability range," *Geochim. Cosmochim. Acta,* **6**(4), 157–185.

Yoder, H. S., Jr., and Eugster, H. P., 1955, "Synthetic and natural muscovites," *Geochim. Cosmochim. Acta,* **8**(5–6), 225–280.

Cross-references: *Aqueous Solutions; Exsolution; Fluid Inclusions; Hydrothermal Solutions; Natural Brines; Phase Equilibria; X-Ray Diffraction Analysis.* Vol. IVB: *Clays and Clay Minerals; Mineral and Ore Deposits.* Vol. V: *Metamorphic Rocks; Metasomatism.* Vol. VI: *Leaching.*

HYDROTHERMAL SOLUTIONS

Any hot, aqueous fluid, whether found in near-surface or deep geologic environments or formed experimentally, is properly described as *hydrothermal;* because hot water is a strong solvent, such fluids inevitably are *solutions.*

Compositions of hydrothermal fluids have been determined directly by analysis of samples from fluid inclusions (summarized by Roedder, 1968), from hot springs (Waring, 1965), and from wells drilled for geothermal power (Ellis, 1967). Experimental and thermodynamic studies of mineral-solution equilibria by Burnham, Hemley, Holland, and others (see Barnes, 1967) provide indirect evidence of compositions. Typically, chloride content is in the range 10^3–10^4 ppm, and pH varies from moderately acidic to moderately alkaline (except where exposure to air supports oxidation of H_2S to produce sulfuric acid). Total solute concentration in fluids from inclusions is commonly about 4–20% by weight. Ellis and Mahon (1967) have shown that solution composition is fixed by geologically rapid equilibration with host rocks and that exotic constituents such as Li, B, F, and As provide no indication of the origin of such fluids. Consequently, to use "hydrothermal" descriptively to imply a magmatic origin is inappropriate, regardless of composition, because once outside the magma chamber, magmatic fluids promptly lose their identity. The older literature frequently contains obsolete, nearly synonymous usage of "hydrothermal" and "magmatic solutions."

The origin of hydrothermal solutions in geothermal wells and hot springs is now commonly attributed to deep circulation of meteoric water with only a presently indetectable fraction originating from magmatic emanations. This conclusion results principally from isotopic investigations of Yellowstone Park, of geothermal areas of New Zealand, of wells near the Salton Sea, and of brines in the Atlantis Deep of the Red Sea; see, for example, Craig, 1966, or Uzumasa, 1965.

The physical state of hydrothermal solutions in the earth's crust may be as a simple liquid or a supercritical fluid. The critical point of aqueous solutions decreases slightly as CO_2 content rises but, more importantly, it rises steeply with salt concentration from 374°C for pure water to 600°C for a 20% NaCl solution, an effect which results in an expanded range for the liquid state at typical high solute contents. Where supercritical fluids are present, for example during the crystallization of a magma, fluid densities may approach those of a liquid, i.e., 0.5–1.0. Evidence for the existence of a gas phase (at subcritical temperatures) in subsurface environments is rare; fluid inclusions, for example, consistently show a uniform filling temperature improbable if gas bubbles were also present and trapped in inclusions.

H. L. BARNES

References

Barnes, H. L. (editor), 1967, "Geochemistry of Hydrothermal Ore Deposits," New York, Holt, Rinehart, and Winston, 670pp.

Craig, H., 1966, "Isotopic composition and origin of the Red Sea and Salton Sea Geothermal Brines," *Science,* **154**, 1544–1548.

Ellis, A. J., 1967, "The chemistry of some explored geothermal systems," in (Barnes, H. L., editor), "Geochemistry of Hydrothermal Ore Deposits," New York, Holt, Rinehart, and Winston, pp. 465–514.

Ellis, A. J., and Mahon, W. A. J., 1967, "Natural hydrothermal systems and experimental hot water/rock interactions (Part II)," *Geochim. Cosmochim. Acta,* **31**, 519–538.

Roedder, E. (in press), "Composition of fluid inclusions," in "Data of Geochemistry." *U.S. Geol. Surv. Profess. Paper,* **440 JJ**.

Waring, G. A., 1965, "Thermal springs of the United States and other countries of the world —a summary," *U.S. Geol. Surv. Profess. Paper,* **492**, 383pp.

Uzumasa, Y., 1965, "Chemical Investigations of Hot Springs in Japan," Tokyo, Tsukiji Shokan, 189pp.

Cross-references: *Aqueous Solutions; Fluid Inclusions; Springs; Water.* Vol. V: *Water: Solubility in Rocks.* Vol. VI: *Magmatic Water.*

HYDROTHERMAL SOLUTIONS—SULFIDE TRANSPORT

The nature of hydrothermal solutions has been discussed for many years, and is still a subject of much controversy and uncertainty. Virtually all the relevant geological and experimental data have been interpreted in several ways; the interpretations characteristically involve opinions on other controversial questions, such as the origin of granite, the physical state of the fluid, and the sources of the water and other constituents of the solutions.

Origin

Postmagmatic Solutions. Most authors evidently believe that hydrothermal solutions originate as emanations from magmas. By analogy to gases evolved during volcanic activity, it is often thought that at least the earliest of the evolved solutions contain much HCl, SO_2, SO_3, CO_2, HF, and other acid gases together with the predominating water. The physical state of the emanations is in doubt (cf Krauskopf, 1967), and surely depends on the depth and temperature of the intrusion. Under suitable conditions the solution may well be gaseous (cf Tunell, as referred to in Weissberg, et al., 1966), but of moderate density. Water, above its critical point and of appreciable density, e.g., 0.2 g/cm^3 is a good solvent for many substances, some of which are believed to be present in hydrothermal solutions and to act as complexing agents facilitating the dissolution of sulfides. Whether such a solution is above or below its own critical point, if it exists, is a moot question which would seem to be unanswerable until its chemical composition is known. Some solutes, e.g., NaCl, increase the temperature of the critical point of aqueous solutions, and others, as HCl, lower it. Moreover, dissolved silicates may cause intersection of the critical and solubility curves so that at the particular composition of the emanation no critical point may exist (cf Smith, 1963, as referred to in Krauskopf, 1967). Unless the hydrothermal solution undergoes boiling or condensation after it has formed, there is no advantage to classifying it as gaseous, liquid, subcritical, supercritical, or pneumatolytic (cf Ingerson in Štemprok, 1965). Some authors have thought that condensation does occur as the solution migrates toward the surface (Krauskopf, 1967, and Lindgren, 1933, referred to therein), but some recent opinion favors no such discontinuity of state during the history of the solution (Ingerson, ibid.).

Metamorphic Solutions. Suggestions have also been made that at least some hydrothermal solutions arise through the expulsion of aqueous solutions from sedimentary rocks during progressive metamorphism. In its simplest form, this hypothesis presumes that all substances in the solution arise from within the sedimentary column and its contained pore water; many metamorphic reactions involve the release of water from silicate minerals and this water together with that present originally in the pores will tend to dissolve a portion of the trace metals from the minerals until a state of near equilibrium is reached. As with postmagmatic solutions, complexing is to be expected. On migration to other regions, the equilibrium conditions will change, and ore deposits may be formed. In more complicated situations this process may be combined with contact metamorphic effects, postmagmatic emanation, and deep circulation of meteoric waters. The difference between this process and postmagmatic solutions arising from granitic bodies resulting from partial melting of sediments is largely one of timing and degree.

Meteoric Solutions in Thermal Areas. A number of studies have indicated that a high percentage to nearly all of the water issuing from hot springs and geysers in volcanic areas is meteoric in origin, i.e., stemming ultimately from rainwater (cf Krauskopf, 1967, and chapter by Elder in Lee, 1965). This suggests that intrusions at depth may act primarily as an energy source in establishing convection in the region, and not as the source of water. Moreover, it may be possible for the circulating water to leach trace metals and sulfur from rocks through which it permeates in the same manner as proposed for the metamorphic origin of hydrothermal solutions.

Isotopic Studies. Determinations of the isotopic ratios of hydrogen (i.e., hydrogen to deuterium to tritium), oxygen, and sulfur have been made in attempts to answer some of the preceding questions. Unfortunately, the data may be interpreted in a number of ways. For instance, as a magma approaches the surface it may be greatly undersaturated in water and dissolve large amounts of meteoric water contained in nearby rocks. As the magma crystallizes, this water would be evolved and be classified "postmagmatic," but might still retain its meteoric isotopic characteristics. In the case of granites, the water may have been meteoric initially and dissolved during partial melting of sediments.

The isotopic ratios for sulfur species in the hot springs of volcanic areas are close to meteoritic values and thus to those expected in mantle-derived magmas. It may be, however, that this sulfur enters the solution by decomposition of traces of sulfides in the volcanics under the action of circulating meteoric water. Thus, although the sulfur comes indirectly from the magma, the solution would not be "postmagmatic."

Other isotopic studies on sulfur and lead suggest that some ore deposits arise from postmagmatic or mantle-derived solutions, some from sedimentary sources, and some from regeneration processes within preexisting igneous and metamorphic rocks (cf chapter by Ault in Abelson, 1959).

Composition

Bulk Chemical Composition. Data on the composition of hydrothermal solutions comes largely from fluid inclusions, hot springs, and bore-hole waters. A large range of compositions exists among these data, and some geologists, on the basis of theoretical considerations, believe the initial range to have been even greater. Some data are given in the accompanying table. If the initial emanations from magmas contained much HCl, SO_3, and other acid gases, there must have been extensive reaction with rocks to produce the mixture of alkali and alkaline earth chlorides, sulfates, and bicarbonates now observed.

Views differ as to how representative are the fluids in inclusions. As a crystal grows from its mother liquor, some additional phase (sometimes gaseous) may grow on the surface of the crystal, and together with some of the mother liquor become part of an inclusion; in this way an inclusion that is not representative of the hydrothermal solution may be formed. Some students of fluid inclusions believe that entrapped materials represent the exhausted solution, and thus do not indicate compositions of pristine fluids (Yermakov, 1965). Because ore deposition probably always involves some reaction with wall rocks or admixed waters, and the dissolved ore minerals may remain in colloidal suspension or supersaturated solution for some time, the included fluids may differ from the original due to oxidation of sulfide to sulfate (and consequent decrease of pH), loss of sulfide to wall rocks, changed ratios of alkali ions, etc. It may also be possible that the composition of the inclusion changes subsequent to formation, e.g., by diffusion of hydrogen. For the most part, however, fluid inclusions are accepted as being reasonably good samples of the ore-forming fluid.

A number of sulfides, e.g., orpiment, cinnabar, metacinnabar, mercury, and stibnite, are apparently being deposited from active hot springs, examples being noted in California and New Zealand. The solutions observed at the surface undoubtedly differ from those at depth due to admixture with surface waters and partial oxidation and acidification. Comparison of near surface waters with those from deep bore holes in some thermal areas, notably near Wairakei, New Zealand, provides convincing evidence of this. Although the geologic evidence is sufficiently obscure to preclude proof that ore bodies have been formed by such solutions, the evidence argues strongly in favor of such an origin for at least some epithermal deposits.

A possible example of hydrothermal solutions related to metamorphism is represented by the remarkable waters issuing from deep bore holes near the Salton Sea, California (cf White, Anderson, and Grubbs, 1963, as referred to in Krauskopf, 1967). A partial analysis is included in the accompanying table, and, in addition, high concentrations of Cu, Zn, and other metals were found. Deposits in the exit pipe are very high in Cu, Zn, Au, Ag, and other metals and contain fine-grained copper and iron sulfides.

pH of Solutions. Little is known regarding the pH of hydrothermal solutions at high temperature, although much discussion has been devoted to this topic. It is more important to know the chemical composition; with this knowledge the extent of reaction with rocks could be estimated. It is primarily for this purpose and to aid in calculations of mineral solubilities that geologists have been interested in the pH. Hemley and Jones (1964, cf reference in Krauskopf, 1967) have shown, however, that the ratios of acid concentrations to the concentrations of the corresponding salts, e.g., the HCl/KCl ratio, are far more significant than the pH by itself.

Until recently (cf Franck as referred to in Krauskopf, 1967) it was not realized that acids become quite weak at high temperatures and that a moderate concentration of HCl in aqueous solution may not produce a very low pH. Accordingly, nearly neutral hydrothermal solutions may contain appreciable HCl and not alter feldspars or other silicates common in igneous rocks. As the solution cools, however, the acid becomes stronger and produces alteration (cf Helgeson, 1964; Hemley and Jones, 1964, as referred to in Krauskopf, 1967). An essential feature of this alteration is the removal of hydrogen ion from the solution and alkalis and alkaline earths from the rock.

If enough information is available on complex formation, dissociation constants, concentrations, and activity coefficients, the pH can be calculated. Most of the requisite data is unknown, although some estimates can be made. Even in a comparatively simple case, such as the PbS–$NaCl$–HCl–H_2O system studied by Helgeson (1964), the simultaneous consideration of many equilibria is required.

HYDROTHERMAL SOLUTIONS—SULFIDE TRANSPORT

Complex Formation. Knowledge of the bulk chemical composition alone provides only limited insight into the nature of an aqueous solution at high temperature. Most electrolytes, as noted previously for acids, become less dissociated with increasing temperature, and experimental and thermochemical data indicate more extensive formation of complexes, such as $PbCl_4^{-2}$ and $ZnS \cdot nH_2S$. A good understanding of these fluids requires better data on the complexes, their degrees of formation, and their mutual relations than exist at present. Estimates by Helgeson (1964) of the relations in the $PbS-NaCl-HCl-H_2O$ system are consistent with many geologic observations.

Transport of Ore Minerals

The consensus of opinion is that sulfide minerals are carried as complexes in migrating solutions, rather than as colloids or by diffusion. A number of types of complexes may be involved, and one hydrothermal solution may differ from another in this respect. Attention has, to date, been focused on complexes containing only one ligand, such as chloride or bisulfide, and one metallic ion. In the actual

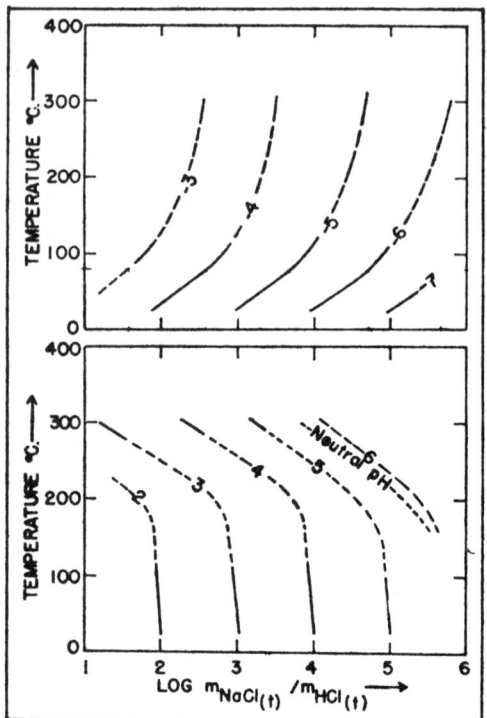

Fig. 2. Calculated "B" isosolubility curves for galena (top) and iso-pH curves for the solutions (bottom) in the system $PbS-NaCl-HCl-H_2O$ at fixed $m_{NaCl(t)} = 1.0$. The numbers indicate log $m_{Pb(t)}$ (top) or pH (bottom).

case it may be that several different ligands are coordinated to a single metallic ion or that several metallic ions occur in the same complex.

The order of stability or degree of formation of complexes of different metals presumably accounts for the commonly observed zoning of mineral deposits. As conditions change during deposition the least stable complexes under the new conditions break down and the corresponding sulfides are deposited. More stable complexes persist to a later stage thereby producing at least a partial separation of different metals.

Chloride Complexes. Figs. 1 and 2 show the general nature of solubility relations for PbS calculated by Helgeson (1964). Due to uncertainties in the data, two different approximations were used for the first and second ionization constants of H_2S, giving rise to the A and B models. A limited amount of experimental data tends to confirm the general magnitude of solubility, but is insufficient to verify the computed curves. It is also known that copper, silver, and gold can be dissolved and redeposited as native metals by chloride solutions.

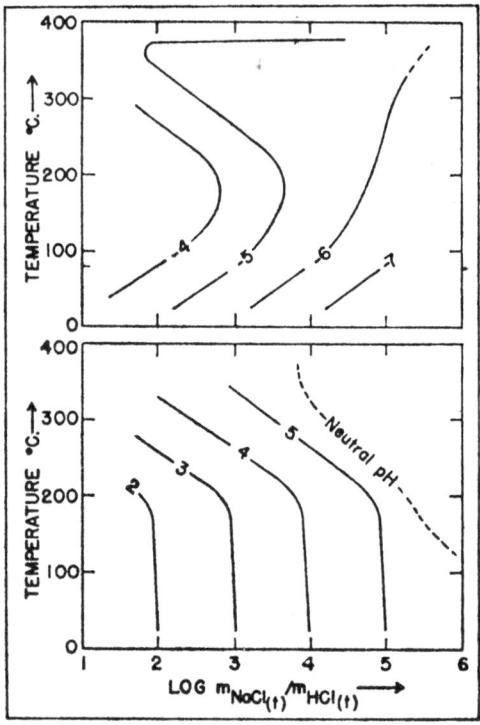

Fig. 1. Calculated "A" isosolubility curves for galena (top) and iso-pH curves for the solutions (bottom) in the system $Pbs-NaCl-HCl-H_2O$ at fixed $m_{NaCl(t)} = 1.0$. The numbers indicate log $m_{Pb(t)}$ (top) of pH (bottom).

The nature thus deduced for hydrothermal solutions is consistent with fluid inclusion and hot-spring data, the requirements of rock alteration, and geologic concepts of controls on deposition of ores. This concept is far easier to apply to deposits of native copper and silver than to deposits involving complexes containing sulfide. As with rock alteration, the concentration ratio of NaCl to HCl is more significant than pH.

Complexes with Bisulfide Ion and Hydrogen Sulfide. A different point of view is the opinion that the principal complexes are with HS^- and H_2S [cf Anderson, Barnes, Hinners, and Holland as referred to in Helgeson, (1964), and Barnes in Štemprok (1965)]. In this case concentrations of sulfide species up to one molar are required to achieve solubilities considered geologically significant. These relations are supported primarily by experimental studies. Except for comparison with fluid-inclusion data, which only rarely show detectable concentrations of sulfide, this concept seems as satisfactory as that with chloride complexes for many deposits. Observations of H_2S at hot springs provide support for the strengths required, if appreciable dilution or oxidation of the original solution by meteoric waters can be shown. Likewise the large amounts of sulfide sometimes added to altered wall rocks argue for this hypothesis.

Sulfide Complexes. In some instances it seems likely that solutions more alkaline than those just discussed may be important. Alkaline hot springs containing sulfide species, and depositing cinnabar, metacinnabar, and mercury are known in California. Experimental studies of relevant systems support this concept (cf Weissberg et al., 1966). These solutions appear to have been more alkaline at depth and to have contained more sulfide. As they approach the surface, oxidation to sulfate occurs by uptake of oxygen, producing a decrease in pH and colloidal suspensions or supersaturation of dissolved heavy metal sulfides. The waters at the surface contain principally Na_2SO_4 and NaCl, and would thus conform to the view that fluid inclusions are likely to represent the exhausted solution. Otherwise the arguments for or against this hypothesis are much the same as for bisulfide complexes.

Other Complexes. Complexes with thiosulfate, polysulfide, or polythionates have been suggested, but these seem likely only under near-surface conditions and would result from partial oxidation of sulfide. These species are all known to disproportionate to sulfide and sulfate at temperatures near 200°C.

Complexes with sulfate may be important

TABLE 1. COMPOSITION OF SOME NATURAL WATERS (ppm)

Description	Li	Na	K	Rb	Cs	Ca	Mg	F	Cl	Br	I	SO$_4$	HBO$_2$	HCO$_3^-$ or CO$_3^{2-}$
Drillhole 44 Wairakei, New Zealand	14.2	1320	225	2.8	2.5	175	0.035	8.3	2260	6.0	0.3	36	117	—
Drillhole Salton Sea, California	321	54,000	23,800	150	20	40,000	800	—	184,000	700	—	0	520	—
Liquid inclusions in quartz, Volynia[a]	—	21,700	1,100	—	—	290	—	—	34,800	—	—	—	—	235
Liquid inclusions in quartz, Aar Massif[b]	2,200	22,100	7,700	—	—	3,300	—	—	17,700	—	—	5,500	—	19,900

[a] Gas in inclusions contains 10–95% H_2S, 3–74% CO_2, and other undetermined gases.
[b] Percent of CO_2 in inclusions was 9.5; analysis recalculated to 100% excluding CO_2 to obtain approximate composition of aqueous phase.

under unusual conditions or for metals not normally found in vein deposits as sulfides, but generally are not stable enough to be significant in sulfide-bearing solutions. Other types of complexes, as with bicarbonate, are thought to be important only for other kinds of deposits.

Deposition from Solution

Physicochemical Factors. For all the types of complexes discussed previously, reaction with wall rocks should change the chemistry of the ore-bearing solution. These changes would result in solubility decreases and deposition. Examples of such reactions include reduction of the acid content of the fluids caused by rock alteration, and of the sulfide content by sulfidization of iron minerals in the walls. In most cases decreases in temperature and pressure or changes in Eh will also cause deposition. Reaction with admixing meteoric waters would likewise tend toward precipitation of ore minerals.

Localization. The chemical relations applicable to ore solutions are generally consistent with geologic observation on the localization of ores. Thus, permeable rocks and open structures are favorable because they allow the solutions to flow through and react with the walls or other groundwaters. Nearly impermeable barriers along a channel will tend to promote longer periods of reaction and more complete exhaustion of the solutions nearby. Highly reactive rocks, such as limestone, will cause marked changes in the fluids and be good localizers of ores. Changes in pressure and temperature as the solutions migrate will also be influential, particularly near the surface.

Variations as deposits clog channelways, and slow or divert flow are to be expected. This may lead to different mineralogy in diverse parts of the vein, with good ore in some places and barren quartz in others. Fluctuations in chemical composition, pressure, and temperature are also likely, especially near the surface. These are likely to produce periods of supersaturation or colloid formation with concomitant deposition and times of dissolution of previous deposits.

PAUL L. CLOKE

References

*Abelson, P. H. (editor), 1959, 1967, "Researches in Geochemistry," New York, John Wiley & Sons, 2 vols.
*Barnes, H. L. (editor), 1967, "Geochemistry of Hydrothermal Ore Deposits," New York, Holt, Rinehart and Winston, 670pp.
Helgeson, H. C., 1964, "Complexing and Hydrothermal Ore Deposition," International Ser. Monographs in Earth Sciences, Vol. 17, Oxford, Pergamon Press, 128pp.
*Krauskopf, K., 1967, "Introduction to Geochemistry," New York, McGraw-Hill Book Co., 721pp.
Lee, Wm. H. K. (editor), 1965, "Terrestrial Heat Flow," Geophysical Monograph Series, No. 8. Washington, Am. Geophys. Union.
*Štemprok, M. (editor), 1965, "Symposium: Problems of Postmagmatic Ore Deposition," Vol. II., Prague, Geol. Survey of Czechoslovakia.
*Weissberg, B. G., Dickson, F. W., and Tunell, G., 1966, "Solubility of orpiment (As_2S_3) in Na_2S-H_2O at 50–200°C and 100–1500 bars, with geological applications," *Geochim. Cosmochim. Acta*, 30, 815–827.
Yermakov, N. P., et al., 1965, "Research on the Nature of Mineral-Forming Solutions," International Series of Monographs in Earth Sciences, Vol. 22, London, Pergamon Press.

* Additional bibliographic references may be found in this work.

Cross-references: *Colloids; Complexes; Fluid Inclusions; Geochemistry; Groundwater; Hydrology; Hydrothermal Solutions; Isotope Geology—Stable; Isotopic Variations in Mineral Deposits; Magmatic Water; Mineral Genesis; Mineral Thermometry; Oxidation and Reduction; pH-Eh; Solubility; Sulfides in Sediments; Sulfur Cycle; Trace Elements; Gases—Volcanic; Water.* Vol. II: *Hydrosphere.* Vol. IVB: *Mineral and Ore Deposits.* Vol. V: *Basalt; Bathygenesis; Geosynclines; Hydrothermal Processes and Differentiation; Igneous Rocks; Magma; Metasomatism; Orogenesis; Volatiles; Volcanoes and Volcanology.*

HYDROTHERMAL SYSTEMS—SILICATES—See SILICATES: HYDROTHERMAL SYSTEMS

HYDROXIDES—See OXIDES AND HYDROXIDES

HYPOGENE

Adj. 1. Epigenetic or syngenetic mineral formation caused by ascending intramagmatic or intratelluric fluids (also broadly known as hydrothermal mineralization).
2. Opposite to supergene.
3. In geomorphology, used to indicate endogenous factors (diastrophism and volcanism) affecting landscape evolution.
4. Originally used for any deep-seated process, e.g., leading to magmatism and metamorphism (Lyell, 1833).

RAYMUNDO J. CHICO

Reference

Lyell, C., 1833, "Principles of Geology," London, 3 Vols.

Cross-references: Vol. III: *Geomorphology.* Vol. IVB: *Economic Geology; Mineral Deposit Classification; Ores, Ore Genesis.* Vol. V: *Magmatism; Metamorphism.*

HYPOLIMNION

Hypolimnion is the term applied to the lower, or bottom, layer of a thermally stratified lake or artificial water body. It is made up of cool, dense, stable water with a restricted oxygen supply which remains all summer much as it was when replenished by the spring turnover, if the lake is deep and clean. If the lake is shallow and dirty, it may suffer a depletion of its oxygen supply and an accumulation of H_2S and other undesirable gases. The temperature of this water is fairly uniform from its upper surface (at the *thermocline*) to the lake's bottom, and is largely cut off from the atmosphere and from wind action by the *epilimnion* (q.v.) and the *thermocline* (q.v.).

B. C. S.

References

Chow, Ven Te, 1964, "Handbook of Applied Hydrology," New York, McGraw-Hill Book Co., 1445pp.

Hutchinson, G. E., 1957, "A Treatise on Limnology," New York, John Wiley & Sons, Vol. 1, 1015pp.

Welch, P. S., 1952, "Limnology," Second ed., New York, McGraw-Hill Book Co.

Cross-references: *Epilimnion; Limnology; Thermocline.*

I

ICE-AGE THEORY, METEOROLOGY—
See Vol. II

IGNEOUS PARAGENESIS—*See*
PARAGENESIS; *also* Vol. V

INDIUM: ELEMENT AND GEOCHEMISTRY

Chemical Character and Affinities

Indium, a silvery-white metal of atomic weight 114.82, specific gravity 7.3 (20°C), melting point 155°C, and boiling point 1450°C, was discovered by Reich and Richter in 1863. Its atomic number is 49 and it occupies the third group of the Periodic Table along with gallium (Ga) and thallium (Tl).

Indium has two stable isotopes, mass numbers 113 and 115, the relative abundances of which are 4.2 and 95.8%, respectively. A number of short-lived radioactive isotopes are also known.

Physicochemical considerations suggest (Vlasov, 1966) that In is most closely related to Sn^{2+} and Cd, followed by Fe, Ga, Tl, and to a lesser degree Zn, Cu, and Pb; and that In^{3+} will be the most stable oxidation state under natural conditions.

Indium is dominantly a chalcophile element, though its dependence upon Fe (when in the dispersed form) indicates partial siderophilic affinities. The few data on the indium content of meteorites indicate preferential concentration in the metal and sulfide phases.

Abundance and Mode of Occurrence

The crustal abundance of indium has been estimated by most workers at the strikingly low figure of 0.1 ppm.

The existence of independent indium minerals (the sulfides $InFe_2S_4$ and $InCuS_2$, the native metal, and the hydroxide) has only recently been established (Picot et al., 1963; Genkin, 1963; Zalashkova et al., 1963; referenced in Vlasov, 1966). These are very rare and the bulk of the indium in nature occurs as a trace constituent in minerals of other elements, the highest contents being found in certain sulfides (notably sphalerite, chalcopyrite, and stannite), sulfosalts, and cassiterite, particularly of the wood-tin variety. Some iron sulfides and iron-rich silicates also contain indium but in much lower concentrations.

Indium is obtained commercially as a by-product from the smelting of certain zinc ores. Current world production (exclusive of the U.S.S.R.) has been estimated at between 1.0 and 1.5 million troy ounces per year.

Concentration of Indium in Hydrothermal Processes

Significant concentration of indium appears to take place only in particular phases of the hydrothermal stage of igneous activity, notably in tin-rich associations such as the cassiterite-polymetallic ores. Average indium contents of sphalerite and chalcopyrite from this environment are 1440 and 445 ppm (Vlasov, 1966); associated stannite and cylindrite are also rich in indium. A slightly lesser degree of concentration is characteristic of the dark iron-rich sphalerite of cassiterite-sulphide skarns and cassiterite-lead-zinc ores.

The large pyritic base metal deposits contain considerably lower concentrations (average 18 ppm in chalcopyrite: Ivanov et al., 1962 in Vlasov, 1966) but in view of their size constitute important potential reserves of indium.

Although the concentration of indium in sulfide ores from any province depends ultimately on its abundance in the primary ore-forming fluid and on the conditions of deposition, it consistently shows a remarkable dependence on the tin content of the host ores.

Indium in Other Hypogene Environments

No significant concentration of indium occurs in either low-temperature environments such as telethermal Pb–Zn or Hg–Sb–As deposits or in very high-temperature (magmatic) sulfides of the ultrabasic Cu–Ni association. Similarly in the pegmatitic and postmagmatic metasomatic processes, indium, with rare exceptions (certain tin greisens), is dispersed without concentration.

In igneous rocks acid varieties show a slightly higher average indium content ($n.\ 10^{-5}\%$) than more basic ones ($n.\ 10^{-6}\%$).

Indium in Supergene Processes

The behavior of indium under supergene conditions is inadequately known. Oxidized zinc minerals like smithsonite and goslarite have indium contents 10 to 100 times lower than primary sphalerite; particularly low contents appear characteristic of sulfates.

Concentrations of up to tenths of a percent have occasionally been recorded from iron hydroxides, which is a result, presumably, of their high sorptive capacity (Vishnevskii et al., 1958 in Vlasov, 1966).

The indium content of sediments ranges generally from $n.10^{-6}$ to $n.10^{-5}\%$, with highest contents in argillaceous and some sandy sediments and lowest in chemical carbonates (Vlasov, 1966). The concentration of indium in some black shales and anthraxolite is apparently connected with organic processes (Shaw, 1952).

J. F. HARRIS

References

Ludwick, M. T., 1959, "Indium," New York, Indium Co. of America, 770pp.

Shaw, D. M., 1952, "The geochemistry of Indium," *Geochim. Cosmochim. Acta,* **2,** 185–206.

Shaw, D. M., 1957, "The Geochemistry of Gallium, Indium and Thallium. A review," in "Physics and Chemistry of Earth," Vol. 2, London, Pergamon Press, pp. 164–211.

Vlasov, K. A., 1966 (editor), "Indium," in "Geochemistry of the Rare Elements," (translation from the Russian) Israel Program for Scientific Translations, Jerusalem, pp. 459–487.

Cross-references: *Gallium; Hypogene; Supergene; Thallium; Trace Elements; Zinc.* Vol. IVB: *Ore-Forming Fluids.*

INERT GASES—See RARE GASES

INFRARED ANALYSIS

The infrared portion of the electromagnetic spectrum was discovered by W. Herschel in 1800, but the first absorption experiments were not made until 1881 when Abney and Festings measured the absorption spectrum of organic molecules in the 0.7–1.2 μ region. Julius in 1892 followed this with the first assignment of an absorption band (near 3.33 μ) to a characteristic chemical structure in the methyl group (see references in Brugel, 1962, p. 7). The first catalogs of infrared spectra of minerals, in the region up to 15 μ, were published by W. W. Coblentz between 1905 and 1910. However, the big expansion in infrared spectrophotometry awaited the development of automatic-recording instrumentation during World War II.

For many years infrared has been used predominantly in structural analysis of organic materials. It has been used sparingly in inorganic structural and mineralogical studies, and its value as a quantitative tool in the compositional analysis of minerals and rocks has been but little utilized. The analysis of reflected infrared radiation from polished surfaces of minerals, gems, and rocks is an almost unexplored field. Apart from the works of Coblentz (1905–1910) and Pfund (1945) (on gems), the most significant contributions have been the classical studies of Simon and McMahon (1953), and of Gardon (1956) on the radiative cooling properties of glass slabs. Spectral analysis of emitted infrared radiation has been given predominance, influenced by the studies of nose-cone reentry and the attendant problems of heat dissipation from refractory coatings. The wavelength range of this work is in the near infrared (1–5 μ), and only little investigation has been made in the region of diagnostic analysis for rock and mineral composition, that is, from 10 to 25 μ and beyond.

The requirements for satellite construction are principally total emissivity and total absorptivity, generally in the region of the solar energy spectrum, that is, 0.5–2.5 μ. The designer of refractory coatings for nose cones is similarly interested in the high-temperature, short-wavelength regions. It is only when scientists become interested in the sources of infrared emission at low ambient temperatures (300–400°K) that studies probe the 10–30 μ region. This is the region of the fundamental vibration of aluminum-oxygen and silicon-oxygen bondings in minerals. It is also the region of the fundamental vibration for most of the inorganic salts (carbonates, nitrates, sulfates, phosphates, etc.) which occur in terrestrial rocks and sediments. Thus, it is the region of paramount importance to the mineralogist.

Infrared analytical techniques are capable of distinguishing minerals within four major groups (Lyon, 1967).

1. Minerals of relatively constant chemical composition (e.g., quartz).
2. Minerals that exhibit marked differences in composition, as between two end members (e.g., plagioclase feldspars and olivines), or more numerous components (e.g., pyroxenes and amphiboles). The infrared spectra clearly show small differences in the amounts of major elements but not of trace elements or low-level impurities.
3. Minerals of constant chemical composition but of different structural modifications (e.g., SiO_2 as quartz, coesite, stishovite, cristobalite, or tridymite; $CaCO_3$ as calcite or aragonite; microcline and orthoclase feldspars).

INFRARED ANALYSIS

Table 1. Wavelengths and Frequencies of the Principal Anion Absorptions[a]

Group	Absorption Peaks	
	Wavelength (μ)	Wave Number (cm^{-1})
CO_3^{2-}	6.90– 7.09	1450–1410
	11.36–12.50	880–800
NO_2^-	7.14– 7.70	1400–1300
	11.90–12.50	840–800
NO_3^-	7.09– 7.46	1410–1340
	11.63–12.50	860–800
PO_4^{3-}	9.09–10.50	1100–950
SO_4^{2-}	8.85– 9.26	1130–1080
	14.71–16.40	680–610
All silicates	9.09–11.10	1100–900

[a] See Fig. 1.

4. Minerals that vary both in structural modification and chemical composition (soda-plagioclase with high- and low-temperature forms which may range from An_0 to An_{20} in composition).

The inorganic anion groups have strong, usually simple, absorption spectra. Some of these are listed in Table 1 and shown in Fig. 1. A strong absorption within one of these bands implies that a given functional anion group (e.g., CO_3^{2-}) is present; either the wavelength of this strong absorption peak or that of smaller peaks elsewhere in the spectrum will indicate to which metal cation the group is bonded (such as to calcium as $CaCO_3$ or to magnesium as $MgCO_3$). Intermediate values for the principal absorptions may be considered diagnostic of solid solutions (crystalline solutions). Details for the carbonate groups appear in Table 2.

Tektosilicates

A typical example of the use of infrared techniques is shown in the analysis of some of the tektosilicates (framework silicates), e.g., quartz and its polymorphs.

Silica is the simplest of the tektosilicates and consists of a framework of SiO_4 tetrahedra with each oxygen linked to another in a neighboring tetrahedron. Silica can exist in various structural groups, and spectra of several members are shown in Fig. 2. Quartz and coesite retain the tetrahedral coordination, whereas stishovite shows the rutile (sixfold coordination) structure; these features are readily observable in their infrared spectra.

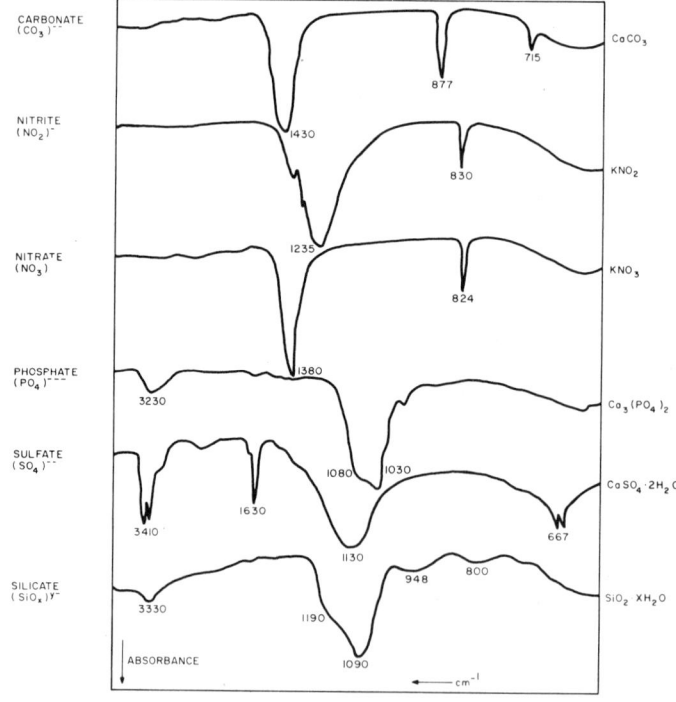

Fig. 1. Absorption spectra of the principal inorganic anions. (This figure is redrawn from Miller and Wilkins, 1952.) Frequencies are given in wave numbers (reciprocals × 10⁴ of wavelengths in microns).

INFRARED ANALYSIS

TABLE 2. CARBONATE GROUP ABSORPTION FREQUENCIES
SHOWING VARIATION WITH BONDED METAL

Mineral	Wavelength (μ)			Wave Number (cm^{-1})		
	Calcite subgroup					
Calcite $CaCO_3$	6.97	11.45	14.04	1435	873	712
Rhodochrosite $MnCO_3$	6.98	11.53	13.76	1433	867	727
Siderite $FeCO_3$	7.03	11.55	13.57	1422	866	737
Magnesite $MgCO_3$	6.90	11.27	13.37	1450	887	748
	Dolomite subgroup					
Ankerite $Ca,Fe(CO_3)_2$	6.90	11.27	13.77	1450	877	726
Dolomite $Ca,Mg(CO_3)_2$	6.97	11.35	13.70	1435	881	730
	Aragonite subgroup					
Cerussite $PbCO_3$	6.94	11.89	14.77	1440	841	677
Witherite $BaCO_3$	6.82	11.61	14.43	1445	860	693
Strontianite $SrCO_3$	6.80	11.61	14.14	1470	860	707
Aragonite $CaCO_3$	6.90	11.61	14.04	1450	860	712

Stishovite is considered to be the first mineral discovered in which silicon occurs with oxygen in a sixfold coordination (octahedral) instead of tetrahedral coordination.

Tarte and Ringwood (1962) have briefly drawn attention to the absence of SiO_6 octahedra in silicate minerals; however, on the basis of other work in XO_4 tetrahedral, and octahedral, groups, they predicted the infrared absorption frequencies for this octahedral configuration to be between 14.3 and 16.7 μ (700 and 600 cm^{-1}). The stishovite infrared absorption curve (Fig. 2) shows a major absorption at 13.0–15.9 μ (769–628 cm^{-1}), effectively supporting their prediction. The data show a marked change in the position of the major absorption peaks when passing from the SiO_4 tetrahedral coordinations of quartz, coesite, and fused silica at 9.1–9.29 μ (1098–1077 cm^{-1}), to the SiO_6 octahedral coordination in stishovite at 13.0–15.9 μ; 10.54–11.30 μ (769–628 cm^{-1}; 949–885 cm^{-1}). A parallel to this has been recorded by Dachille and Roy (1959) for GeO_2-quartz and GeO_2-rutile structures, which show absorption peak shifts from 11.5 μ (870 cm^{-1}) for the "quartz" structure to 14.0 μ (715 cm^{-1}) for the "rutile" polymorph.

General assignments for spectral peaks in quartz are:

(Si–O) stretching
    9.10–12.5 μ (1100–800 cm^{-1})

(Si–Si) stretching
    12.5–16.67 μ (800–600 cm^{-1})

(Si–O–Si) bending and distortion
    21.74–23.3 μ (460–430 cm^{-1})

Three typical illustrations from a recent report (Lyon, 1967) are shown here. Fig. 3 shows absorption spectra for the O–H stretching region in minerals observed between 2.7 and 3.2 μ (3700 and 3125 cm^{-1}). These spectra were prepared from samples at 0.15%

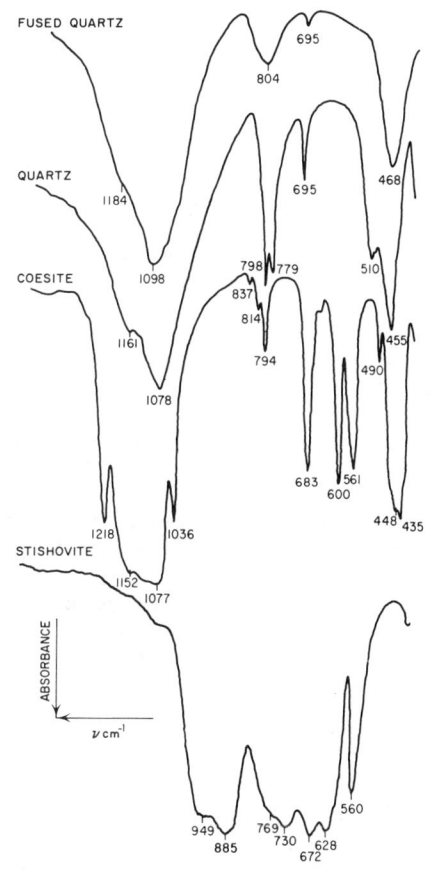

FIG. 2. Absorption spectra for polymorphs of SiO_2—quartz, coesite, stishovite, and fused silica. The curves are displaced vertically.

INFRARED ANALYSIS

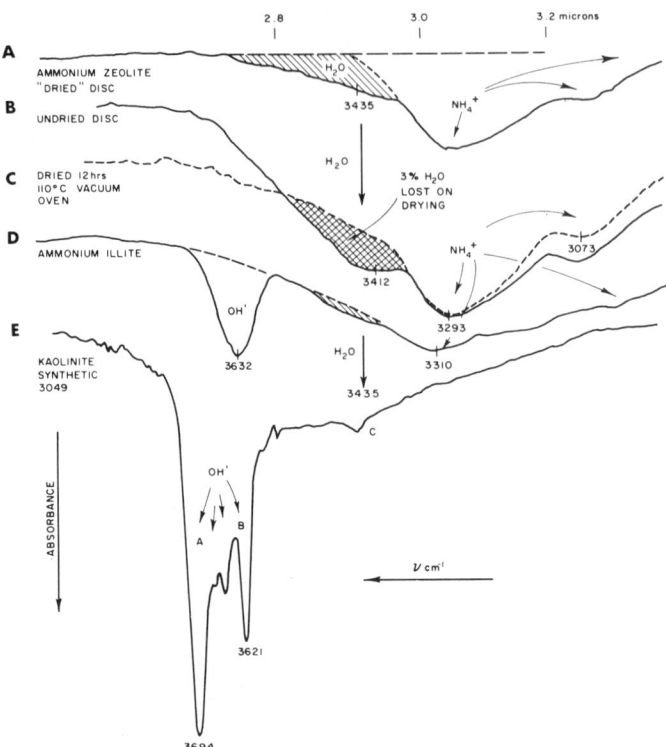

Fig. 3. Absorption spectra for H_2O^-, NH_4^+, and several types of OH^-.

embedded in KBr pellets and examined on a Perkin-Elmer 221 grating infrared spectrophotometer. Excellent resolution in this particular wavelength region is indicated. Spectra A, B, and C indicate the effects of drying overnight at 110°C in a vacuum oven, and it is clear that all of the water physically absorbed on the potassium bromide pellet can be removed by this simple preliminary technique. Spectrum D indicates the presence of the NH_4^+ ion and shows the O–H stretching frequencies of an ammonium illite. Spectrum E indicates that with the degree of resolution available, four O–H stretching frequencies are observable in a kaolinite specimen. At the wavelength of maximum absorption for H_2O^- (approximately 2.91 μ; 3435 cm^{-1}) the spectrum indicates that little water is retained on the kaolinite specimen after this treatment.

Fig. 4 shows the absorption spectra for a group of clinopyroxene insolicates, including acmite, omphacite, jadeite, diopside, and hedenbergite. It is of mineralogical interest to note the similarity of the infrared spectra of omphacite, acmite, and jadeite, indicating the presence of a solid solution between the acmite and jadeite end members. It is also quite clear that although diopside and hedenbergite show comparable spectra, they differ extensively from the other three pyroxenes.

Fig. 5 indicates the reflection spectra for a series of rock samples on which polished surfaces were prepared. There is a peak shift of 160 cm^{-1} in these spectra for the suite of rock compositions ranging between tektite and chondritic meteorites. The rocks of acid composition, i.e., tektite, granite, obsidian, and rhyolite, appear in the lower half of the figure. Crystalline structure is indicated by the complex fine structure in the spectrum of the granite specimen, but the *peak position* is not changed even if the material is glassy (as in obsidian). A comparable pattern is shown by the basic materials, i.e., basalt, gabbro, and dunite. The dunite and gabbro again show spectral peaks from the individual crystals in the rock.

Available Spectral Compilations

An excellent detailed compilation of spectra is that of Horst Moenke (1962), scientific collaborator at the Jena Optical Works. This extensive compilation contains over 350 photographic reproductions of spectra and was pre-

INFRARED ANALYSIS

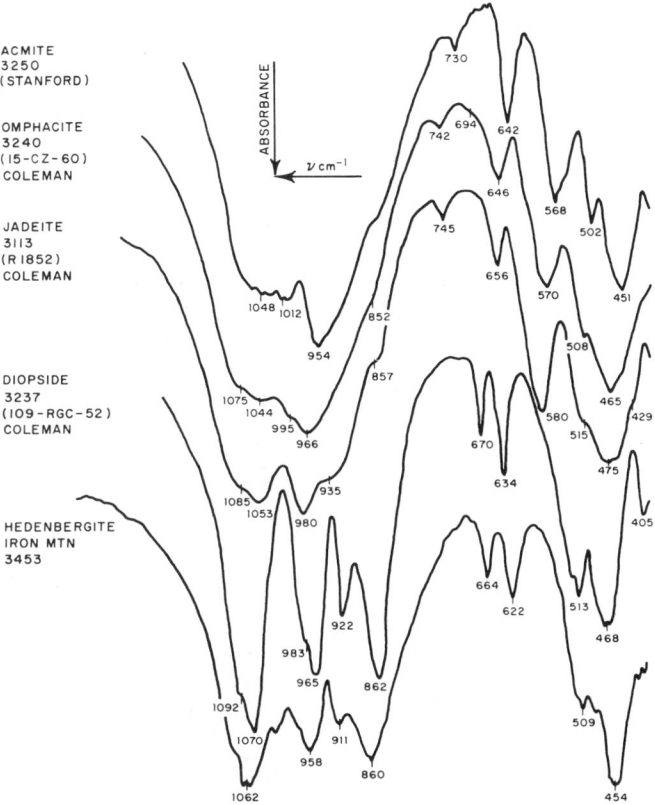

Fig. 4. Absorption spectra for the clinopyroxene inosilicates (single chains of tetrahedra each sharing two oxygens, SiO_3). Examples shown are acmite, omphacite, jadeite, diopside, and hedenbergite. The curves are displaced vertically.

pared for the Commission for Spectroscopy of the German Academy of Sciences in Berlin. Unfortunately, copies have been extremely difficult to obtain in the U.S.A., but it is hoped that they will be available in the near future.

This compilation contains infrared spectra of the most abundant and economically important halides, oxides, hydroxides, carbonates, nitrates, borates, sulfates, chromates, tungstates, molybdates, phosphates, arsenates, vandates, and silicate minerals in the spectral region of 4000 to 400 cm^{-1} (2.5 to 25 μ). It also contains forty pages of text (in German) describing the mineral groups investigated. The manner of *photographic* reproduction of the spectra is instructive and indicates the high quality of the original data and the low noise levels of the instrument during preparation of the spectrum.

Another source (Lyon, 1962) is a bibliography with a detailed index. This seventy-six-page booklet embraces the fields of absorption, reflection, emission, and transmission in minerals and related materials. More than 440 references are contained in this text. Annotations have not been made, but the "Chemical Abstracts" listings (volume, number, and index pagination) are given for more than 95% of the references. The subject index contains approximately 1200 listings. Mineral spectra listings include borates (29), carbonates (30), halides (3), phosphates and vanadates (30), sulfates, tungstates, and arsenates (30), sulfites (8), and oxides and hydroxides (53). The silicates alone comprise 269 spectral listings.

It is hoped that this note will alert mineralogists to some of the atlases and bibliographical collections of data which are becoming available. Mineralogists are interested in the chemical and structural composition of minerals, whether of natural or synthetic origin. Infrared spectrophotometry represents a somewhat neglected but highly promising avenue for investigation. The study of crystal structure has been greatly advanced by the use of one particular instrument, i.e., the x-ray diffractometer. Not enough notice has been paid by

INFRARED ANALYSIS

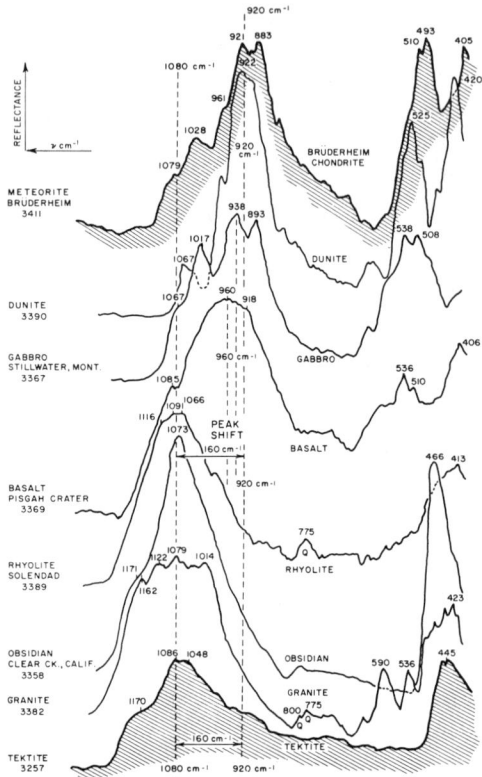

FIG. 5. Peak shift of 160 cm^{-1} for the reflection spectra for rock compositions between tektite and chondritic meteorites. Samples are chondrite from Bruderheim, Alberta; dunite from Twin Sister, California; gabbro from Stillwater, Montant; basalt from Pisgah crater, California; rhyolite from Soledad, California; and others as noted. The reference beam was attenuated about 40%. The curves are displaced vertically.

mineralogists to other definitive tools. X-ray analysis is sensitive to long-range order, or to the periodic arrangement of atoms in a crystalline structure. Infrared analysis, on the other hand, is much more sensitive to short-range ordering, or ordering on a nearest-atom basis. The two are essentially complementary. Infrared analysis has the ability to bridge the gaps between crystalline solid, liquid, and gas. It is extremely valuable to the mineralogist to study the effects of major element substitutions so common in the mineralogical realm.

1965
R. J. P. Lyon

References

Brugel, W., 1962, "An Introduction to Infrared Spectroscopy" New York, John Wiley & Sons, 420pp. (translated by A. R. and A. J. D. Katritzky).

Coblentz, W. W., 1905, 1906, 1908, "Investigations of Infra-red Spectra," Publications 35, 65, 97 of The Carnegie Institution of Washington (now republished in a single volume with same title by joint sponsorship of the Coblentz Society and the Perkin-Elmer Corporation, 1962, 660pp).

Dachille, F., and Roy, R., 1959, "The Use of Infra-red absorption and molar refractivities to check coordination," Z. Krist., 3(6), 462–470.

Gardon, R., 1956, "The emissivity of transparent materials," J. Am. Ceram. Soc., 39, 278–285.

*Lyon, R. J. P., 1962, "Minerals in the Infrared," Menlo Park, Calif., Stanford Res. Inst., 76pp.

*Lyon, R. J. P., 1963, "Infrared absorption spectroscopy," chap. 8 in Zussman, J. (editor), "Physical Methods in Determinative Mineralogy," London, Academic Press, pp. 371–403.

*Miller, F. A., and Wilkins, C. H., 1952, "Infrared spectra and characteristic frequencies of inorganic ions," Anal. Chem., 24, 1253–94.

Moenke, H., 1962, "Mineral Specktren," Deutsche Akad. Wissen. Berlin, Akademie-Verlag, 42pp.

Pfund, A. H., 1945, "The identification of gems," J. Opt. Soc. Am., 35, 611–614.

Simon, J., and McMahon, H. O., 1953, "Study of the structure of quartz, cristobalite, and vitreous silica by reflection in infrared," J. Chem. Phys., 21, 23–30.

Tarte, P., and Ringwood, A. E., 1962, "Infrared spectrum of the spinels Ni_2SiO_4, Ni_2GeO_4, and their solid solutions," Nature, 193, 971–972.

* Additional bibliographic references may be found in this work.

Cross-references: Crystal Chemistry; Mineral Classes: Silicates; Mineralogy; Spectrophotometry; X-Ray Diffraction Analysis. Vol. II: Meteorites; Tektites. Vol. IVB: Coesite and Stishovite; Quartz. Vol. V: Rhyolite.

INOSILICATES—See CRYSTAL CHEMISTRY; MINERAL CLASSES: SILICATES

INTERNATIONAL ASSOCIATION OF GEOCHEMISTRY AND COSMOCHEMISTRY

The objectve of the International Association of Geochemistry and Cosmochemistry (IAGC), organized at UNESCO headquarters in Paris in November 1965, is to advance these sciences through international cooperation, particularly through the organization of working groups and commissions, holding of conferences, and issuance of publications. The business of the Association is directed by member organizations who alone have voting power determined by the size of the membership. Other organizations and individuals may adhere but have no voting power. The Association has no permanent headquarters and operates through the officers. As of January 1969, the officers are Earl Ingerson, President (University of Texas, Austin, Texas); Louis H. Ahrens, Vice-President (University of Cape Town, South Africa); Ken Sugawara, Secretary (Nagoya University, Japan); John F. Lovering, Treasurer (Australian National University, Canberra). Working groups have been organized to coordinate activity concerned with the geochemistry of sediments, extraterrestrial chemistry, geochemical nomenclature, geochemical documentation, isotope geochemistry, and applied geochemistry.

M. H.

INTERNATIONAL HYDROLOGICAL DECADE

The International Hydrological Decade (IHD) which began in January 1965, is an international effort by hydrologists in the principal countries of the world to upgrade the science of hydrology and to find ways for the science better to serve man's welfare. This article summarizes the general nature of the IHD. A more detailed statement is that of Nace (1964), who also published an account (1965) for the general reader.

The variety and complexity of unsolved water problems that exist today, the many problems for which only partial or unsatisfactory solutions have been devised, and the vast array of new problems that will arise in the readily foreseeable future, all lend emphasis to the fact that hydrology will have a major role in the future welfare of mankind.

Hydrology—the scientific study of water in its endlessly self-repeating cycle of global circulation from ocean to atmosphere to land and back to sea again—is a laggard science in some respects. Hydrologists have not, for example, equaled some other earth scientists, in the adaptation of twentieth-century instruments, methods and basic scientific knowledge to their specific purposes. Also, large-scale aspects of the hydrological cycle are not well understood, for work has centered chiefly on local and river-basin phenomena.

However, water problems in many areas already have outgrown water-resources plans based on local and river-basin unit areas. The time is at hand for planning and development on regional and continental scales. In order to provide the necessary scientific base for rational planning and development, hydrologists recognize the need to develop a broader and deeper understanding of water as a substance and of its behavior in the water cycle at all scales—local, continental, hemispheric and global. Thus, while hydrologists as scientists are motivated by the desire to understand, as members of society they are also spurred by a desire to make water science and technology more widely available and effective in all parts of the world.

Water data are grossly inadequate for two-thirds of the land area of the world, containing two-thirds of its human population, chiefly in the so-called underdeveloped countries, many of which have practically no scientists. Some have no hydrologists. Much of the work by these countries in the IHD will be a bootstrap operation. They must learn by doing, while more of their qualified people are obtaining formal education and training in hydrology. Moreover, even elementary data collection and hydrologic studies will constitute constructive and essential progress in these areas and will be an indispensable contribution to global hydrology. The global situation cannot be described or understood adequately while basic data are lacking for two-thirds of the world.

In the program of the IHD, each country is responsible for its own work at its own expense, though some may obtain financial support from sources such as the technical assistance program of the United Nations or the foreign aid programs of advanced countries. Nearly 100 nations have already announced their intentions to participate in the program, and about 85 have formed national committees to coordinate their activities. The United Nations Educational, Scientific and Cultural Organization (UNESCO) has accepted the responsibility for primary international sponsorship of the program on the intergovernmental plane and has appointed a Coordinating Council for the IHD composed of representatives from 21 of its member states. Other intergovernmental agencies also have expressed strong interest and have important roles in the program. These include the World Meteorological Organization, the United Nations Food and Agriculture Organization, the International Atomic Energy Agency, the World Health Organization, and various regional economic commissions of the United Nations. More than a dozen international scientific agencies have also expressed interest. The varied nature of these associations bespeaks the wide interdisciplinary nature of hydrology. They will assist in the planning of activities. At UNESCO's request, the International Council of Scientific Unions has appointed a Scientific Committee on Water Research to serve as international advisor on the IHD program.

Activities by participating nations will include establishment of basic data networks and collection of essential data in countries which have yet to develop advanced hydrological competence; analytical studies of water budgets and balances, local to global in scope, by countries whose scientists have the necessary competence; and advanced and specialized applied and basic research by workers from scientifically advanced countries. Instruments used will range from simple staff gauges on rivers through electronic analogs and computers in laboratories to special instruments in rockets and orbiting satellites for remote sensing of hydrologic phenomena on a global scale. In addition, international agencies and organizations will sponsor programs of education and training, preparation of training manuals and handbooks, standardization of instruments and methods, scientific symposia and seminars, and related activities.

UNESCO staged two important intergovernmental meetings of experts in 1963 and 1964 to lay the groundwork for the IHD. UNESCO'S General Conference (governing body) late in 1964 declared the Decade open effective January 1, 1965. In June 1965, UNESCO convened the first meeting of the Coordinating Council for the IHD, which drafted a consolidated preliminary world program based on program proposals for the various Member States.

As was expected, the early stages of the program contained no spectaculars. No rocket whizzed at the push of a button when the new year chimed in. That may come later, but, in general, hydrology is not a spectacular science, nor is it widely recognized. For example, about 0.0015% of the United States population is classifiable as "hydrologist"—not enough to constitute even a recognizable minority group. Increase by a factor of 10 would not alter the proportion much, but it would alter greatly the social, economic and scientific impact of hydrology. This is a major objective of the IHD—to focus attention on the social and economic importance of hydrology and hydrologists and thereby increase the number of scientists who are working on problems of water—the central problems of Century 21.

Raymond L. Nace

References

Anon., 1970, "Hydrology: mid-decade conference," *Nature and Resources*, **6**(2), 11–12.
*Nace, R. L., 1964, "The International Hydrological Decade," *Trans. Am. Geophys. Union*, **45**, No. 3, 413–421.
Nace, R. L., 1965, "New age for hydrology." *Nat. Hist. Magazine*, **74**, No. 1, 63–68.
Nace, R. L., 1970, "World Hydrology: Status and Prospects," *Internat. Assoc. Sci. Hydrol., Publ. No.* **92**.
UNESCO, *Nature and Resources* (Bull. Intern. Hydrologic. Decade), publ. quarterly, Vol. 1(1), June 1965.
UNESCO, 1969, "Hydrological Forecasting," *Proc. WMO/UNESCO Symp., Australia, Nov.–Dec., 1967*, Geneva, World Meteorol. Org. (U.S. distributor, UNIPUB, New York), 322pp.

* Additional bibliographic references may be found in this work.

Cross-references: *Conservation; Hydrologic Cycle; Hydrology; Natural Resources; Water Balance; Water—Substance and Solvent; Water Supply: Economics.* Vol. VI: *Water Engineering Associations and Societies; Water Power.*

INTERSTITIAL WATERS IN SEDIMENTS*

Interstitial waters are aqueous solutions that occupy the pore spaces between grains of soils, sediments, and rocks. They are sensitive indicators of particle-fluid reactions and equilibria in sediments, reflect migration pathways and origin of fluids, and supply most of the nutrient salts to terrestrial plants (via soils); they have often been suggested as important sources of reactive constituents for the formation of economically valuable deposits such as heavy metal sulfides, phosphorites, and iron and manganese ores. Their role in the formation of oil and gas deposits is certainly significant, but still poorly known. In the larger sense, they are involved in most diagenetic (post-depositional) reactions in sedimentary and metamorphic rocks (see *Fluid inclusions; Natural Brines, Clay Membrane Phenomena;* also in vol. VI: *Diagenesis; Groundwater.* Discussion below is limited largely to Holocene and subrecent deposits.

Early Studies

Perhaps the first major study of interstitial waters was published by the French agronomist, T. Schloesing, in 1866. Schloesing displaced pore fluids from a quantity of soil, and carefully analyzed the effluent. Since that time, studies of sediment-water-plant inter-

*Publication approved by the director, U.S. Geological Survey.

actions have continued to be a major area of interest in soil science (Mehlich and Drake, 1958).

The first geological-oceanographic investigation of interstitial waters was published by Sir John Murray, the pioneer British oceanographer, and R. Irvine in 1895. Murray and Irvine squeezed fluids from a "blue mud" from the Scottish coast and found that whereas the major components of seawater remained in fairly constant proportions in the mud, oxygen had been depleted, sulfate was lost, and bicarbonate alkalinity increased. The relative proportions of the constituents indicated that the (largely bacterial) oxidation of organic carbon to CO_2 had balanced most of the reduction of sulfate to form sulfide. Moreover, on realizing that there was an appreciable amount of dissolved manganese in the pore water, Murray revised his earlier concept of a largely volcanic origin for deep-sea manganese concretions. He suggested that, along with river-borne and volcanic supplies of Mn, a major source might be manganese diffusing up into seawater from terrigenous continental sediments undergoing syngenetic reduction.

Interstitial chlorinity of coastal, and especially intertidal sediments, was studied by British biologists from the late 1920s (see references in Smith, 1955). Some coastal sediments have lower interstitial chlorinities than average local bottom water because of the influence of groundwater discharge; however, higher salinities also arise through gravitational convection and downflow of heavier solutions through permeable sediments (Callame, Scholl; see Scholl, 1965). Valyashko (1963) has emphasized the importance of gravitational downmovement of brines in fossil sedimentary basins as well (see also Vol. I: *Submarine Springs*).

Interest in the geochemistry of pore waters revived in the middle-late 1930s partly through the writings of the Russian geochemist, V. I. Vernadsky. Soviet earth scientists, led by A. P. Vinogradov, S. W. Brujewicz, and L. A. Shchukarev studied pore waters from the Caspian and Black Seas, as well as fresh and saline lakes. Following World War II, Soviet oceanographers, including Brujewicz, Zaitseva, Shishkina, Tageeva, Tikhomirova, Starikova, and Gorshkova took advantage of effective sediment squeezers developed by P. A. Kriukov and expanded interstitial water studies to encompass major and minor ions, nutrients, organic, and absorbed constituents in sediments from the Russian northern seas, the Black, Caspian, and Baltic Seas, and the Atlantic, Pacific, and Indian Oceans.

Outside the Soviet Union, Emery and Rittenberg made detailed studies of biogenic cycles involving interstitial waters from Southern California offshore basins. Kullenberg, the Swedish oceanographer who had developed the piston corer, utilized cores up to ten meters long to determine whether marine sediments might retain a record of the chlorinity of the water in which they were deposited. Marked decreases in chlorinity with depth in Baltic Sea sediments (approaching glacial-equivalent sediments), and parallel observations by Brujewicz in the Black Sea seemed to contradict the smoothing of salt gradients expected as a result of molecular diffusion.

Enrichments of trace elements in interstitial waters were noted by Tageeva and Tikhomirova in the Caspian, Black, and Russian northern seas, by Hartmann in the Baltic Sea, by Gorshkova in several northern seas, and by Presley, Brooks, and Kaplan (1967) in Pacific red clays. Abnormal concentrations of copper (Cu), zinc (Zn), lead (Pb), and other heavy metals are known from pore fluids in sediments of the Red Sea hot brine deeps (Brooks et al., and Hendricks et al., in Degens and Ross, 1969).

Extensive references to studies of interstitial waters in sediments, including those not cited with date, above and subsequently, are given in Brujewicz (1966), and in a Russian monograph devoted to interstitial waters of oceanic sediments (Shishkina, 1970).

Ammonia, carbon dioxide, and methane dominate gases dissolved in interstitial waters in Southern California offshore basins (Emery and Hoggan, 1958). Similar studies were made on lake and paddy muds in Japan by Koyama and his co-workers (Koyama, 1955).

Carbonate equilibria in sediments are revealed sensitively by analysis of interstitial magnesium (Mg), calcium (Ca), alkalinity, pH, and saturometry (equilibration of pore fluid with solid $CaCO_3$ followed by pH measurement), as shown by Berner (1966) in studies of sediments from Bermuda and south Florida. However, more work is needed to understand the interrelation of puzzling phases such as high magnesian calcite, aragonite, and dolomite.

Sillen's suggestion that the ultimate composition of seawater may be controlled by equilibrium between breakdown of silicate rocks and formation of new silicates lends new importance to pore water studies. Powers (1957) demonstrated loss of Mg in interstitial waters of estuarine sediments, which he attributed to uptake by clay minerals such as montmorillonite and chlorite. On the other hand, potassium (K), as well as silica, has been shown to

INTERSTITIAL WATERS IN SEDIMENTS

TABLE 1. ANALYSIS OF PORE WATER FROM DEEP-SEA HEMIPELAGIC SEDIMENTS[a]

Depth (cm)	Br	Cl	SO_4	Alk	Anions	Na	K	Ca	Mg	Cations	Total Solids (g/kg)
Bottom water	0.83	543	55.6	2.4	602	469	9.7	20.5	104	603	—
25-51	0.87	547	57.6	3.2	609	478	13.0	19.9	99	610	35.2
75-100	0.90	547	56.5	4.4	608	478	12.7	19.8	98	609	35.2
160-184	0.89	546	54.8	4.4	606	477	13.1	20.1	96	606	35.2
223-248	0.83	551	56.1	4.5	612	482	13.3	20.6	97	613	35.6

[a] Piston core, Sta. 2164, northeast Pacific Ocean, approximately 51°N, 161°E, 400 km east of southern Kamchatka. Depth 5570 m. The sediment is gray-brown, fine-grained, clayey, having less than 0.1% CO_2 (carbonate), organic C decreasing from 0.50% at 25-51 cm depth to 0.22% at 223-248 cm depth in core. Water content decreased from 58% at 25-51 cm to 48.5% at 223-248 cm. All units except total solids are in mg-equivalents per kg. Data from Shishkina (1959) in Brujewicz (1966).

TABLE 2. ANALYSIS OF PORE WATERS FROM TERRIGENOUS-BIOGENIC SEDIMENTS[a]

Depth (cm)	Br	Cl	SO_4	Alk	Anions	Na	K	NH_4	Ca	Mg	Cations	Total Solids (g/kg)
Bottom water	0.79	526	54.3	2.3	583	451	9.5	—	21.0	102.5	584	—
10-60	0.84	523	49.8	10.3	584	454	11.3	0.7	20.5	97	584	33.8
60-117	0.86	526	37.4	17.3	582	—	11.0	1.5	19.0	95	—	34.0
117-172	0.88	524	28.5	24.8	578	456	10.7	2.4	17.8	93	580	33.3
172-226	0.90	532	13.4	32.8	579	—	—	3.1	15.6	95	—	33.2
409-420	0.93	532	5.5	40.5	579	456	10.7	2.8	14.2	95	579	33.1
470-480	0.95	535	2.9	42.8	582	461	10.4	2.9	12.9	94	581	33.5

[a] Piston core, Sta. 1780, Okhotsk Sea, approximately 56°N, 154°E, depth 964 m. The sediment is a gray-green, diatomaceous clay with black sulfide staining in part. Organic carbon content is approximately 1.5%, amorphous silica about 40%, and carbonate CO_2 less than 1%, pH 8.1-813; water content about 70%. Concentrations (except total solids) in mg-equivalents/kg. Data from Shishkina (1959) in Brujewicz (1966).

be enriched in interstitial waters of some deep-sea sediments by Siever et al. (1965), as well as by the Russian authors previously mentioned (see Tables 1 and 2). For other diagenetic reactions involving interstitial waters see von Engelhardt (1967) and references therein.

Ocean Drilling

The recent development of deep-ocean drilling techniques offers greatly expanded scope for interstitial studies. Rittenberg, Emery, and collaborators (1963) and Siever, Beck, and Berner (1965) found only minor chemical changes associated with diagenesis in interstitial waters from deep-ocean sediments in the Experimental Mohole, taken from nearly 3600 m depth and cored to 170 m below the sediment-water interface. In contrast, a variety of phenomena have been observed in interstitial waters from sediments near continents.

Off eastern Florida, fresh waters in Eocene carbonates several hundred meters below sea bottom, and more than 120 km seaward of the coast, were revealed as a result of JOIDES (Joint Oceanographic Institutions Deep Earth Sampling Program) drillings in 1965. These were attributed by Manheim (1967) to land-connected submarine aquifers (see Vol. I: *Submarine Springs*).

Increases in interstitial salt concentrations which approached saturation with NaCl were encountered on the continental slope of the northern Gulf of Mexico in fine-grained, clayey sediments which were drilled above diapiric intrusive features (Fig. 1). Penetration of evaporites in some of the holes confirmed that the salinity gradients were the result of upward diffusion of salt leached from salt plugs (Manheim and Bischoff, 1968).

Interstitial waters in cores from a 1100 m

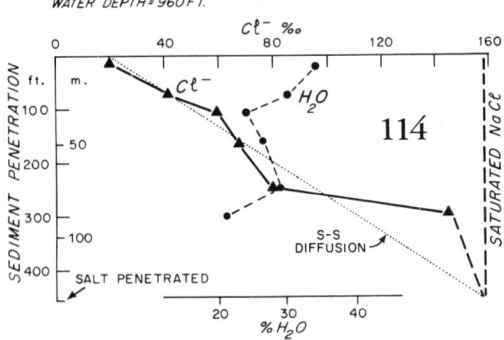

FIG. 1. Chloride concentration and water content with depth in cores from 27°30′N, 95°50′W in Gulf of Mexico (Chloride in g/kg). Dotted line refers to theoretical steady-state distribution of salt (diffusion) between saturated salt and seawater concentrations (from Manheim and Bischoff, 1968).

deep hole, drilled in shallow waters of the Caspian Sea (Pushkina, 1967), revealed a transitional increase in salt concentration to a maximum of 140 g/l (total solids) at 600–800 m. The potassium content of the pore waters remained nearly constant, whereas sodium and calcium, as well as chloride, increased sharply. Extraordinary concentrations of heavy metals, especially Cu (to 50 mg/l), were found in the deeper brines.

Many questions involving the fate of pore waters in earlier deep oceanic sediments may be clarified as results of the (JOIDES) deep-drilling program in the Atlantic and Pacific Oceans. Theoretical questions concerning the properties of interstitial fluids include diagenesis and the stability constants of minerals, migration mechanisms of fluids, diffusion, membrane, and osmotic phenomena and ion exchange behavior. Controversial studies on these problems are scattered through the soil science, biological, geological, civil engineering, hydrological, and desalination literature.

F. T. MANHEIM

References

Berner, R. A., 1966, "Chemical diagenesis of some modern carbonate sediments," *Am. J. Sci.*, **264**, 1–36.

Brujewicz, S. W., 1966, "Khimiya gruntovykh rastvorov (Chemistry of interstitial waters)," in Khimiya Tikhogo Okeana, Izdatel'stvo "Nauka," Moscow, Pt. 2, 263–358.

Degens, E. T., and Ross, D. A., 1969 (editors), "Hot Brines and Recent Heavy Metal Deposits in the Red Sea," Springer Verlag, New York, 600pp.

Emery, K. O., and Hoggan, D., 1958, "Gases in marine sediments," *Bull. Am. Assoc. Petrol. Geol.*, **42**, 2174–2188.

Koyama, T., 1955, "Gaseous metabolism in lake muds and paddy soils," *J. Earth Sci., Nagoya Univ.*, **3**, 65–76.

Manheim, F. T., 1967, "Evidence for submarine discharge of water on the Atlantic continental slope of the southern United States, and suggestions for further research," *Trans. N.Y. Acad. Sci., Ser. II*, **29**, 839–853.

Manheim, F. T., and Bischoff, J. L., 1968, "Geochemistry of pore waters from Shell Oil Co. drill holes on the continental slope of the northern Gulf of Mexico," in Symposium on subsurface brines, Spec. Issue (Angino, E. E., and Billings, G. K., editors), *Chemical Geology*.

Mehlich, A., and Drake, M., 1958, "Soil chemistry and plant nutrition," in (Bear, F.E., editor), "Chemistry of the Soil," *Am. Chem. Soc. Monograph*. 126, pp. 286–327.

Murray, J., and Irvine, R., 1893, "On the chemical changes which take place in the composition of the sea waters associated with blue muds on the floor of the ocean," *Trans. Roy. Soc. Edinburgh*, **37**, 481–507.

Powers, M. C., 1957, "Adjustment of clays to the marine environment," *J. Sediment. Petrol.*, **27**, 355–372.

Presley, B. J., Brooks, R. R., and Kaplan, I. R., 1967, "Manganese and related elements in the interstitial water of marine sediments," *Science*, **154**, 906–909.

Pushkina, Z. V., 1967, "Geochemistry of pore solutions in the Quaternary and Pliocene sediments of the southern Caspian Sea," *Dokl. Akad. Nauk, Earth Sci. Sect.* **148**, 921–924.

Rittenberg, S. C., Emery, K. O., Hulsemann, Jobst, Degens, E. T., Fay, R. C., Reuter, J. H., Grady, J. R., Richardson, S. H., and Bray, E. E., 1963, "Biogeochemistry of sediments in Experimental Mohole," *J. Sediment. Petrol.*, **33**, 140–172.

Schloesing, M. T., 1866, "Sur l'analyse des principes solubles de la terre végétale," *Compt. Rend.*, **63**, 1007–1012.

Scholl, D. W., 1965, "High interstitial water chlorinity in estuarine mangrove swamps, Florida," *Nature*, **207**, 284–285.

Shishkina, O. V., 1970, "Geokhimiya okeanskikh i morskikh porovykh vod (Geochemistry of oceanic and marine interstitial waters)," Izdatel'stvo "Nauka," Moscow, in press.

Siever, R., Beck, K. C., and Berner, R. A., "Composition of interstitial waters of modern sediments," *J. Geol.*, **73**, 39–73.

Smith, R. I., 1955, "Salinity variations in interstitial water of sand at Karnes Bay, Millport, with reference to the distribution of *Nereis diversicolor*," *J. Marine Biol. Assoc.*, **34**, 33–46.

Valyashko, M. G., 1963, "Genesis rassolov osadochnoi obolochki ("Genesis of brines in sedimentary rocks)," in (Vinogradov, A. P., editor), "Khimiya Zimnoi Kory," Vol. 1, Izdatel-stvo "Nauka," Moscow, pp. 253–277.

Von Engelhardt, 1967, "Interstitial Solution and Diagenesis in Sediments," in (Larsen G., and Chilingar, G. V., editors), "Diagenesis in Sediments," Amsterdam, Elsevier, pp. 503–521.

Cross-references: Clay Membrane Phenomena; Fluid Inclusions; Groundwater; Natural Brines; Pedology; Seawater. Vol. I: Submarine Springs. Vol. VI: Diagenesis; Pelagic Sediments.

INVERSION—See Vol. II

IODATES

Iodates are rare minerals occurring only in very arid regions, such as the west coast of South America. The structures are not completely understood.

These minerals are similar chemically to the arsenites, selenites, and tellurites and sometimes are listed with them because they have similar pyramidal structural groups (AsO_3, SeO_3, TeO_3, IO_3).

IODATES

Anhydrous and Hydrous Iodates
$A(XO_3)_2 \cdot xH_2O$ type

Lautarite — $Ca(IO_3)_2$
Bellingerite — $Cu_3(IO_3)_6 \cdot 2H_2O$

Iodates with Hydroxyl or Halogen

Salesite — $Cu(IO_3)(OH)$
Schwartzembergite — $Pb_5(IO_3)Cl_3O_3$

Compound Iodates

Dietzeite — $Ca_2(IO_3)_2(CrO_4)$

LLOYD W. STAPLES

References

Berman, H., and Wolfe, C. W., 1940, "Bellingerite, a new mineral from Chuquicamata, Chile," *Am. Mineralogist*, **25**, 505.

Dana, J. W., and Dana, E. S., 1951, "The System of Mineralogy," Seventh ed., Vol. II, 312, (revised by Palache, C., Berman, H., and Frondel, C.), New York, John Wiley & Sons.

Cross-reference: *Mineral Classes, Nonsilicates.*

IODINE: ELEMENT AND GEOCHEMISTRY

Iodine is of considerable importance because of its presence and necessity in living organisms and for determination of early geological history. The most comprehensive report on iodine is by the Chilean Iodine Education Bureau (1956). This report relies heavily for data on rocks from von Fellenberg's laboratory (see Chilean Report references). Goles and Anders (1962) found von Fellenberg's data too high by factors of 5–30. Thus a redetermination of iodine in the major rock types is being undertaken using the more reliable neutron activation analysis.

Table 1 shows previous iodine abundances as summarized by Turekian and Wedepohl (1961) and recent data from Manuel's laboratory.

Iodine is enriched with organic material in sediments as observed by Itkina's laboratory (see Gulyayeva and Itkina, 1962). Goldschmidt (1962), however, using von Fellenberg's data, suggested an "atomic dispersion" theory for iodine. This is not upheld by recent data.

Manuel's data for igneous rocks indicate low iodine content of around 40 ppb; for a metamorphic rock <10 ppb I; for ultrabasic rocks around 100 ppb I with enrichment of iodine in carbonatites (0.5–1 ppm); for sedimentary rocks widely varying iodine content of 20–8000 ppb, probably associated with organic content; and for deep-sea sediments about 50 ppm I (see Becker et al., 1968; Bennett, 1968a, b).

In summary, iodine has an extremely variable content in crustal material, being enriched with organic content. From crustal iodine information an I^{129}–Xe^{129} formation interval for the earth can be calculated (Becker et al., 1968).

J. H. BENNETT

References

Becker, V., Bennett, J. H., and Manuel, O. K., 1968, "On iodine and uranium in ultrabasic rocks and carbonatites," *Earth Planet. Sci. Letters*, **4**, 357–362.

Bennett, J. H., 1968a, "Geochemical Abundances of Iodine," Master's Thesis, University of Missouri at Rolla.

Bennett, J. H., 1968b, unpublished data.

Chilean Iodine Education Bureau, 1956, "The Geochemistry of Iodine," London, Stone House.

Goldschmidt, V. M., 1962, "Geochemistry," Oxford, Clarendon Press.

Goles, G. G., and Anders, E., 1962, "Abundances

TABLE 1. IODINE CONTENTS OF TERRESTRIAL MATERIALS

Type	Turekian and Wedepohl (1961) (ppb)	Manuel's Laboratory (ppb)
Igneous rocks		
Ultrabasic	500	≈ 100
		Range 7–500
Basaltic	500	≈ 20 (one sample)
Granitic	500	≈ 40
		Range < 50
Metamorphic rocks	No value	< 10 (one sample)
Sedimentary rocks		
Shales	2200	≈ 10,000
		Range 7,000–38,000
Sandstones	1700	≈ 50
		Range 20–140
Limestones	1200	≈ 15,000
		Range 3,000–29,000
Deep-sea sediments	50	≈ 50,000
		Range 10,900–49,000

of iodine, tellurium and uranium in meteorites," *Geochim. Cosmochim. Acta*, **26**, 723–737.

Gulyayeva, L. A., and Itkina, E. S., 1962, "Halogens in marine and fresh-water sediments," *Geochemistry*, **6**, 610–615.

Turekian, K. K., and Wedepohl, K. H., 1961, "Distribution of the Elements in Some Major Units of the Earth's Crust," *Geol. Soc. Am. Bull.*, **72**, 175–191.

Miyake, Y., and Tsunogai, S., 1963, "Evaporation of iodine from the ocean," *J. Geophys. Res.*, **68**(13), 3989–3993.

Cross-references: *Halogens; Neutron Activation Analysis; Seawater, Chemistry.*

ION EXCHANGE*

In many natural materials, one component may be replaced by another without major change in the structure of the material. When the component replaced is an ion, the process is known as *ion exchange*. Nearly all natural ion exchangers consist of an aluminosilicate or silicate network held together by predominately covalent bonds but with excess negative charges at well-defined sites that are satisfied by cations held by electrostatic attraction. These cations may be replaced by an equivalent number of other cations from an external reservoir without destroying the structure of the material. Since cations are exchanged, the process is called *cation exchange*. If negatively charged ions are exchanged on positive sites the process is *anion exchange*. The ion-exchange reaction may be written as follows:

$$A_{aq.\ soln.}^+ + BX = AX + B_{aq.\ soln.}^+ \quad (1)$$

where A^+ and B^+ are exchangeable cations and X^- represents the negatively charged framework.

The number of X^- sites in a given weight of exchanger is a fixed characteristic of such materials as zeolites, feldspars, natural glasses, and some clay minerals such as montmorillonites and illites. In these materials, the exchangeable cations are an essential component of the material and occupy positions in the structure where excess negative charges have resulted from the substitution of, for example, Al^{3+} for Si^{4+} or Mg^{2+} for Al^{3+}. In Fig. 1, schematic structures of glass and illite are shown, and the relation of exchangeable cations to lattice substitutions may be seen. In other ion exchangers, such as quartz or kaolinite the negative charges result from broken bonds on the edges of the particles, and the

* Publication authorized by the Director, U. S. Geological Survey.

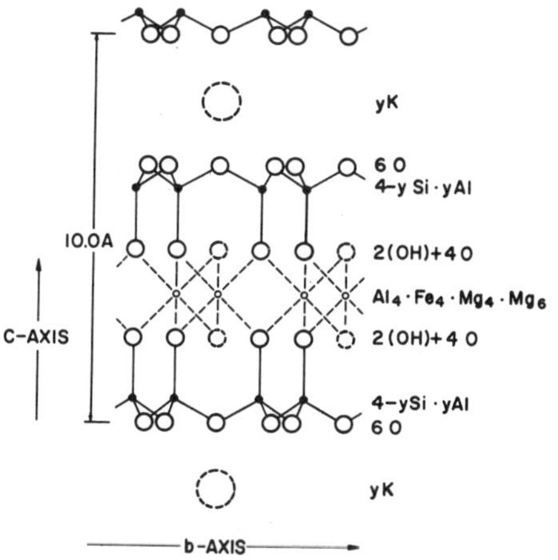

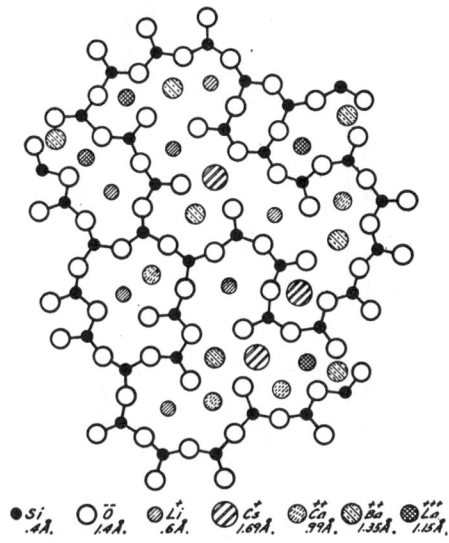

Fig. 1a. Schematic presentation of the crystal structure of illite $(OH)_4K_y(Al_4 \cdot Fe_4 \cdot Mg_4 \cdot Mg_6)(Si_{8-y} \cdot Al_y)O_{20}$ (after Grim, 1942).

Fig. 1b. Schematic representation in two dimensions of the structure of silica glass. Shaded circles represent exchangeable cations (after Perley, 1949).

number of X^- sites depends on the state of subdivision. The usual measure of *ion-exchange capacity* is milliequivalents per 100 grams, wich is the number of moles (gram molecular weight) of singly charged negative sites in one hundred grams of dried exchanger. It is usually determined from the number of moles of NH_4^+ ion taken up by the Na^+ form of the exchanger, since NH_4^+ readily displaces Na^+ on most natural exchangers. The exchange capacity, formula, and type of exchange site for some natural exchangers is given in Table 1.

591

ION EXCHANGE

Table 1. Some Natural Exchangers

Exchanger	Approximate Formula	Origin of Negative Sites	Exchange Capacity (meq/100 g)
Montmorillonite group	$(½Ca,Na)_{.33}(Mg,Al,Fe^{3+})_{2-3}(Si,Al)_4O_{10}(OH)_2 \cdot nH_2O$	Al–Si and Mg–Al substitution	70–100
Illite (hydromica)	$K(Al,Mg)_{2-3}(Si,Al)_4O_{10}(OH)_2$	Al–Si and Mg–Al substitution and broken bonds on particle edges	10–40
Quartz	SiO_2	Broken bonds on particle edges	Low
Kaolinite	$Al_4Si_4O_{10}(OH)_8$	Broken bonds on particle edges	3–15
Zeolite (chabazite)	$CaAl_2Si_4O_{12} \cdot 6H_2O$	Substitution of Al^{3+} for Si^{4+}	400
Natural glass	Indefinite, mainly K, Na aluminosilicate	Substitution of Al^{3+} for Si^{4+}	200

The tendency for the exchange of ion B^+ for ion A^+ as given in Eq. (1) may be exactly expressed by the thermodynamic exchange constant

$$K_B{}^A = \frac{[AX][B^+]}{[BX][A^+]} \quad (2)$$

where brackets denote activities (or thermodynamic concentrations) and K is related to the standard change of free energy in a process by $\Delta G^0 = -RT \ln K$. A negative value of ΔG^0 and therefore a value of $K > 1$ means that the reaction shown in Eq. (1) will go toward the right.

The practical use of Eq. (2) requires knowledge of activities of species in the aqueous solution and in the solid exchanger. Although the activities of ions in the solution can be evaluated by established theory (e.g., by the use of the Debye-Hückle equation), the activities of the ions in the exchanger can only be evaluated by making special assumptions. For natural exchangers, the most useful assumption is that the ions in the exchanger form binary solid solutions in which the rational activity coefficient, λ, of each component is a function only of the mole fraction of the component with the relationships:

$$\lambda_{AX} = \frac{[AX]}{N_{AX}}, \quad \lambda_{BX} = \frac{[BX]}{N_{BX}} \quad (3)$$

$$\ln \lambda_{AX} = C(1 - N_{AX})^2,$$
$$\ln \lambda_{BX} = C(1 - N_{BX})^2 \quad (4)$$

where N is the mole fraction and C is a constant for each pair of ions. Such binary solutions are called symmetrical regular solutions. Using Eqs. (3) and (4), Eq. (2) can be written as:

$$\ln K_B{}^A = \ln \frac{[B^+]N_{AX}}{[A^+]N_{BX}} + C(1 - 2N_{AX}) \quad (5)$$

This relation has been successfully applied by Barrer and Falconer (1956) to zeolite ion exchanges, by Christ and Truesdell (1964) to clay ion exchanges, and by Garrels et al. (1962) to natural glass ion exchanges.

The relative tendency of an exchanger to hold various ions on its echange sites is known as its *selectivity* and is expressed by the relative values of $K_B{}^A$, $K_C{}^A$, etc. Considerable progress has been made by Eisenman (1962) in explaining the origin of selectivity on an atomistic level. Selectivity arises, in his view, from the competition for each cation between the water dipoles in aqueous solution and the negative sites in the exchanger.

The ion exchange reaction may be broken

up into partial reactions to make a thermodynamic cycle.

$$AX = A_g^+ + X_g^- \quad \Delta U_{diss.\ AX}$$
$$B_{aq.}^+ = B_g^+ \quad -\Delta U_{hyd.\ B^+}$$
$$B_g^+ + X_g^- = BX \quad -\Delta U_{diss.\ BX}$$
$$A_g^+ = A_{aq.}^+ \quad \Delta U_{hyd.\ A^+}$$
$$\overline{AX + B_{aq.}^+ = BX + A_{aq.}^+} \quad \Delta U_{exchange}$$

If the standard free energy change, ΔG^0, for each of the partial reactions could be evaluated, then ΔG^0 and therefore K_B^A of exchange could be calculated. However, although the hydration energies of the cations are known, the exchange site, X^-, cannot be isolated, and the dissociation energies cannot be determined experimentally. If the site is represented by a rigid spherical anion of -1 charge, the electrostatic energy required to separate the site and a cation from the distance of closest approach, $(r^+ + r^-)$, to infinity may be calculated as $322/(r^+ + r^-)$ kcal/mole. This is equivalent to the internal energy of dissociation, $\Delta U^0_{diss.}$

$$\Delta U_{diss.} = \int_{r^+ + r^-}^{\infty} F dx = \int_{r^+ + r^-}^{\infty} \frac{e^2}{x^2} dx = \frac{322}{r^+ + r^-} \text{ kcal mole}$$

Free energy is equal to internal energy, plus energy of expansion, less unavailable thermal energy. Since both expansion and thermal energies are very small in ion exchanges, internal energies are fairly good approximations to free energies, and calculated behavior based on internal energy corresponds well with experimental results. A series of calculations of $\Delta U_{exch.}$ for a pair of cations, for example Cs^+ and Na^+, using differing values of r^- generates a curve of $\Delta U_{exch.}$ as a function of r^-. When this is done for a series of cations paired with Cs (as in Fig. 2), selectivity orders result. For the five alkali cations, Li^+, Na^+, K^+, Rb^+, and Cs^+, eleven selectivity orders have been calculated by Eisenman, most of which have been demonstrated to occur on a natural ion exchanger (Fig. 2).

Ion exchange is important in geologic processes in several ways. When clay minerals and zeolites are produced by weathering processes, their exchange sites are occupied by the alkali or alkaline earth cations available at the time. These minerals may be transported into a different environment in which other cations are available, in which case ion exchange modifies the composition of both the clay mineral and the new environment. Thus, these ion exchangers are a major reservoir of exchangeable cations. Natural glasses as found in vitric tuffs have less well-known ion exchange properties; however, it is very likely that ion exchange on natural glass is an important process in the formation of saline groundwaters by exchange of sodium ions from the glass for hydrogen ions (and to some extent potassium ions) from the groundwater. Ion exchange is also involved in the formation of zoned feldspars in a crystallizing magma. Here the external reservoir is the still liquid magma rather than an aqueous solution but the process is similar. Ion exchange has been used in the study of solid solution in alkali feldspars.

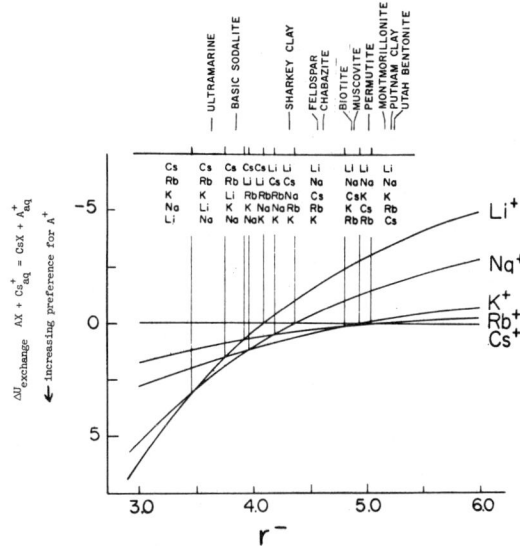

FIG. 2. Calculated selectivity orders for alkali cations as a function of effective anionic site radius and the orders observed on natural exchangers. Each curve was separately calculated from a thermodynamic cycle (see text). Different exchange orders result from crossovers of the curves. At each crossover of two curves there is a reversal of selectivity and therefore a new order (after Eisenman, 1962).

The fertility and mechanical properties of soils depend in a large part on what exchangeable cations are held by clay minerals in the soil. Clay minerals are a major reservoir of ions used in plant nutrition. For example, a soil with principally K^+ in the exchange positions of the clay constituents will be more fertile than the same soil with Ca^{2+} on the clays. Exchangeable cations on clays can be artificially changed to improve soils for agriculture or construction.

Ion-exchange membranes may be formed where a relatively impermeable shale layer separates two sandstone aquifers. Shale is composed largely of clay minerals which are ion exchangers. Cations can exchange freely from one clay particle to another and thus have a

ION EXCHANGE

high mobility relative to anions which are electrostatically excluded from the vicinity of the fixed negative sites of the clay. Thus, the shale may be a *semipermeable membrane*. In addition, the clay minerals show selectivities among the exchangeable ions, thus the membrane is *permselective* since each ion will have a different mobility in the shale. The effect of such a membrane is to modify the composition of subsurface waters which may contribute to the observed *hydrochemical zoning* of deep aquifer basins.

A. H. TRUESDELL

References

Banin, A., 1967, "Tactoid formation in montmorillonite: Effect on ion exchange kinetics," *Science*, **155**(3758), 71–72.

Barrer, R. M., and Falconer, J. D., 1956, "Ion exchange in feldspathoids as a solid slate reaction," *Proc. Roy. Soc. London Ser. A*, **236**, 227–249.

Carroll, D., 1959, "Ion exchange in clays and other minerals," *Geol. Soc. Am. Bull.*, **70**, 749–779.

Carroll, D., 1964, "Ion-exchange capacity of sediments from the Experimental Mohole, Guadalupe site," *J. Sediment. Petrol.*, **34**, 537–542.

Christ, C. L., and Truesdell, A. H., 1963, "Cation exchange in clays interpreted by regular solution theory," *Abstr., Geol. Soc. Am. Spec. Paper* **76**.

Eisenman, George, 1962, "Cation selective glass electrodes and their mode of operation," *Biophys. J.*, **2**, 259–323.

Garrels, R. M., and Christ, C. L., 1965, "Solutions, Minerals, and Equilibria," New York, Harper and Row, 450pp.

Garrels, R. M., Sato, M., Thompson, M. E., and Truesdell, A. H., 1962, "Divalent ion sensitive glass electrodes," *Science*, **135**, 1045–1048.

Grim, R. E., 1942, "Modern concepts of clay minerals," *Illinois State Geol. Surv. Rep. Invest.*, **80**, 50pp.

Helfferich, Friedrich, 1962, "Ion Exchange," New York, McGraw-Hill Book Co., 624pp.

Perley, G. A., 1949, "Glasses for the measurement of pH," *Anal. Chem.*, **21**, 394–401.

IONIUM–THORIUM DATING

Introduction

The ionium/thorium (thorium 230/thorium 232) method of dating deep-sea sediments was proposed by Picciotto and Wilgain (1954) and was studied in detail by Goldberg and Koide (1962). In order to be consistent with previous terminology, thorium 230 will be referred to as ionium in this article. As shown in Fig. 1, the primary source of ionium in the oceans is from the radioactive decay of dissolved uranium 238. Some ionium is also introduced into the oceans as a result of rock weathering on the continents. Thorium 232 originates entirely from rock weathering. Ionium has a half-life of 7.52×10^4 years; thorium 232 is longer lived with a half-life of 1.39×10^{10} years. Since both isotopes have the same chemistry in seawater, they should be

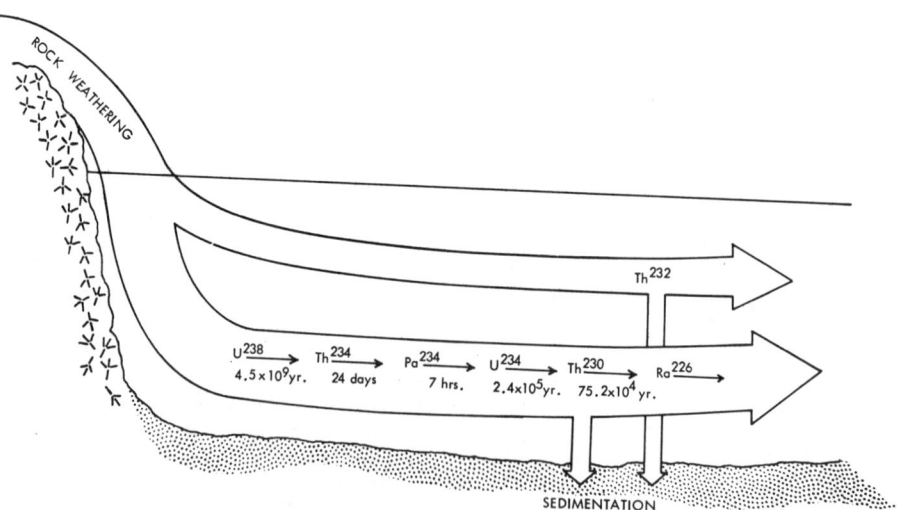

FIG. 1. Schematic representation of thorium-230 (ionium) and thorium-232 isotopes in the oceans. The long horizontal arrows represent the entrance of these isotopes into the oceans via rivers and runoffs after their release from rocks by weathering processes. Most of the thorium-230 (ionium) originates from the decay of the uranium-238 series; however, some of it is also released into the oceans through rock weathering. The short vertical arrows represent precipitation of these isotopes into sediments. Half-lives of the isotopes are given under arrows.

simultaneously removed from the water column into the sediments. Once these isotopes enter into the sediments, the amount of ionium will decrease to an unmeasurable quantity in a half million years or so, due to radioactive decay, while the amount of thorium 232 will remain unchanged. By measuring the decrease in the ionium/thorium 232 ratio with depth in the sediments, a radioactive "age" of the sediments can be obtained.

The decrease in the ratio of ionium to thorium is used rather than the decrease in the ionium concentration alone for age determinations, inasmuch as variations in the rate or type of sedimentation can cause variations in the concentration of ionium. However, such variations should affect equally the concentration of thorium 232 so that the ratio ionium/thorium 232 should not be affected. Therefore, any changes in this ratio should be due solely to changes resulting from radioactive decay.

Since uranium also precipitates in the same sedimentary phases that accommodate ionium and thorium 232, a correction should be made for the ionium that grows in due to the decay of uranium. Usually this correction is not made because the amount of uranium-supported ionium only becomes significant after several half-lives of ionium. This is 300,000 to 400,000 years, which is the upper age limit or maximum age of the sediments that can be dated by this method.

Distribution of Thorium

Calculations of the amount of ionium and thorium 232 in seawater and their rate of entry into the oceans indicate that these isotopes have relatively short residence times as a result of their chemical reactivity. Their residence times are of the order of 100 years which is short relative to the mixing time of the oceans. This leads to the expectation that the ionium/thorium 232 ratio will vary in different areas of the oceans depending primarily on the input of thorium 232. Ionium should have a fairly uniform production throughout the oceans as its parent uranium, with a residence time of approximately a half million years, has a uniform distribution in the oceans.

Thorium 232 enters the oceans via rivers and runoff and because of its high reactivity does not become uniformly distributed. Thus, coastal waters which receive a large volume of river and runoff waters should have a low ionium/thorium 232 ratio relative to open ocean water. This agrees with observational data (Somayajulu and Goldberg, 1966).

There also appears to be variations in the vertical distribution of the ionium/thorium 232 ratio in seawater. Goldberg and Koide (1962) suggested that there is a transfer of thorium 232 from surface to deeper waters through biochemical and inorganic processes taking place in the water column resulting in surface waters having a high ionium/thorium 232 ratio. This idea is supported by the observed relationship between the calcium carbonate content and ionium/thorium 232 ratios of surface sediments. Sediments with a high calcium carbonate content have high ionium/thorium 232 ratios. This high calcium carbonate content of sediments is due to their high content of planktonic foraminifera which live in the upper few hundred meters of the ocean. Upon death, these foraminifera sink to the bottom bringing with them the high ionium/thorium 232 ratio of the surface waters.

Thus, the ionium/thorium 232 ratio of the surface sediments is controlled by the source and rate of input of thorium 232, biochemical and inorganic processes, and surface and deep currents which distribute these isotopes. Nevertheless, there appears to be a distributional pattern in the ionium/thorium 232 ratios of surface sediments which is consistent with the ideas given above (Goldberg and Koide, 1962).

Methods

In order for the ionium/thorium 232 method to be used successfully, the following four assumptions must be valid:

1. The ionium/thorium 232 ratio of the seawater overlying the sediments that have been dated has remained constant over the dating interval. A necessary but not a sufficient condition to establish the validity of this assumption is the observation that ionium/thorium 232 ratios of seawater generally match the ionium/thorium 232 ratio of the surface sediments underlying the seawater.

2. Ionium and thorium 232 isotopes have the same chemical behavior so there is no fractionation of the isotopes between the seawater and the sediments accumulating them.

3. There is no migration of these isotopes within the sediments.

4. The analyzed materials do not contain thorium isotopes of detrital origin.

If the above assumptions can be met, then the age of the sediments can be obtained from the following relationships:

$$R = \frac{\text{Io} - \text{Io}_u}{\text{Th}}, \frac{R}{R_0} = e^{-\lambda t}$$

where Io = ionium that precipitated from seawater; Io_u = ionium that grew in from uranium in the sedimentary phases; Th =

thorium 232 precipitated from seawater; R_0 = ionium/thorium 232 activity ratio in the surface sediments; λ = decay constant for ionium; and t = time.

A half-life for ionium of 80,000 years was used prior to 1962 so that correction to the currently accepted half-life of 75,200 years is necessary for earlier age determinations. The relationship between the half-life, $t_{1/2}$, and the decay constant, λ, is given by $t_{1/2} = 0.693/\lambda$.

The chemical technique for separation of the authigenic thorium isotopes is to leach the sediments with concentrated hydrochloric acid. This treatment removes the absorbed isotopes leaving the detrital ionium and thorium isotopes behind.

The ionium and thorium 232 isotopes are extracted from the acid solution by electrodeposition on a platinum disc and assayed using alpha spectrometry. This ratio is often given in units of disintegrations of ionium per disintegration of thorium 232.

Results

Most deep-sea cores show a constant exponential decrease in the ionium/thorium 232 ratio with depth (Fig. 2).

Some cores show breaks in the rate of decrease of the ionium/thorium 232 ratio with depth (Fig. 3). The interpretation given by Goldberg (1965, 1968) is that these changes in the rate of decrease of the ionium/thorium 232 ratio with depth represent changes in sedimentation rates. Ku, Broecker, and Opdyke (1968) believe that the accuracy of the method is not sufficient to justify the detailed interpretation of Goldberg. Instead they draw a best-fit line through all data points assuming a constant rate of sedimentation for the entire core.

In a few cores the ionium/thorium 232 ratio does not decrease with depth but either increases or remains constant. Possible explanations are that these cores represent sediments that have been disturbed either by burrowing organisms and currents or that there is excessive uranium-supported ionium.

Discussion

Of the four assumptions that must be made for this dating technique to be valid, the assumption that only authigenic ionium and thorium 232 are analyzed is the most difficult to accept. The reason is that many cores which

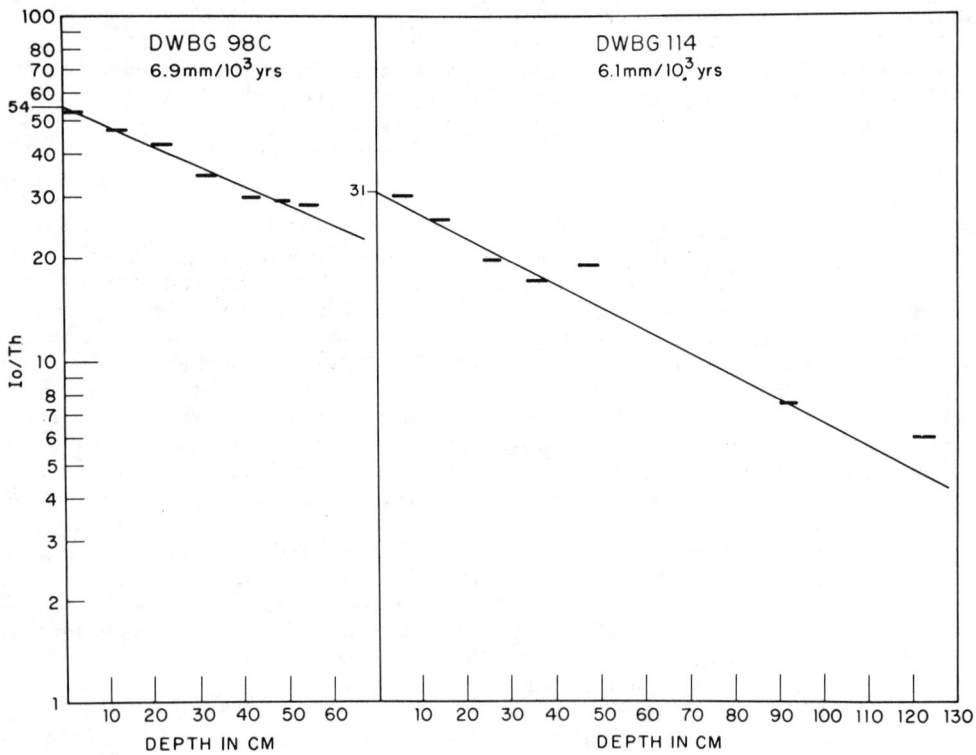

FIG. 2. Ionium/thorium-232 ratios as a function of depth in two cores, DWBG 98C and DWBG 14. Sedimentation rates have remained constant as indicated by the straight-line relationship of ionium/thorium-232 ratio with depth. Sedimentation rates are given beneath core numbers (Blackman and Somayajulu, 1966).

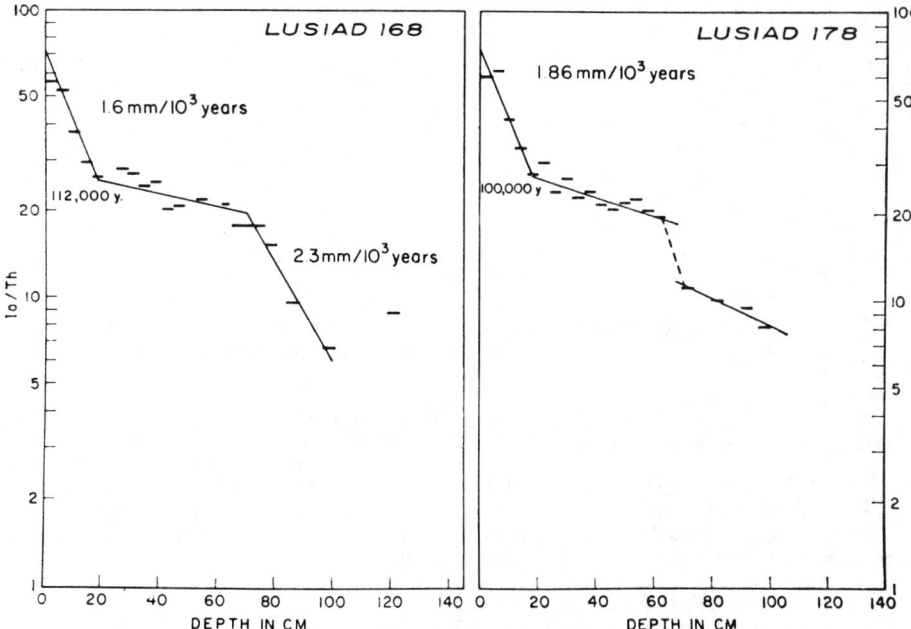

FIG. 3. Ionium/thorium-232 ratios as a function of depth in two cores Lusiad 168 and Lusiad 178. Sedimentation rates have changed as indicated by breaks in the slope of the line. Sedimentation rates are given (Goldberg and Griffin, 1964).

have few authigenic minerals nevertheless have high thorium 232 concentrations. This thorium 232 must therefore be detrital in origin or sorbed on surfaces by ion exchange. However, as has been pointed out by Goldberg and Griffin (1964) if, instead of using the decrease in the ionium/thorium 232 ratio, the decrease in activity of ionium alone is used, much the same results are obtained.

The accuracy and validity of this method can best be checked by independently dating cores on which ionium/thorium 232 determinations have been made.

One core dated by the ionium/thorium 232 method was also dated by the protactinium/ionium method and the two methods gave the same results. (For a discussion of the protactinium/ionium method, see *Protactinium/Thorium Dating* in this volume).

Two cores from the Pacific Ocean dated by the ionium/thorium 232 method were compared to three Atlantic cores dated by the protactinium and protactinium/ionium methods. The cores from the Pacific Ocean have a series of well-defined Pleistocene events based on foraminiferal changes. These Pleistocene events were correlated with similar events that took place in the three Atlantic cores. The ages of these Pleistocene events dated by the two different methods can be compared. Thus, through paleontologic correlations, comparisons between the ionium/thorium 232 method and the protactinium, protactinium/thorium methods can be made. These results are presented in Fig. 4. In general, there is good agreement in the various boundaries of the Pleistocene events. Thus, there is good agreement between the two different dating methods.

Good agreement is also obtained between sedimentation rates based on the ionium/thorium 232 method and magnetic reversals (Goldberg, 1968; Ku et al., 1968) which provides another independent check on the accuracy and validity of the method.

However, there are many cores in which the ionium/thorium 232 method gives ages older by a factor of 10 than ages based on *carbon-14 dating* (q.v.). These cores are mainly from the mid-Atlantic ridge valleys and abyssal hills (Goldberg and Griffin, 1964). No explanation has been given for this discrepancy.

Summary

The ionium/thorium 232 method affords a means of dating sediments up to 300,000 to 400,000 years old. Cores dated by this method usually show an exponential decrease in their ionium/thorium 232 ratio with depth, less frequently they show changes in the rate of decrease of this ratio with depth. The first result is interpreted as representing sediments with constant rates of accumulation, the latter re-

IONIUM–THORIUM DATING

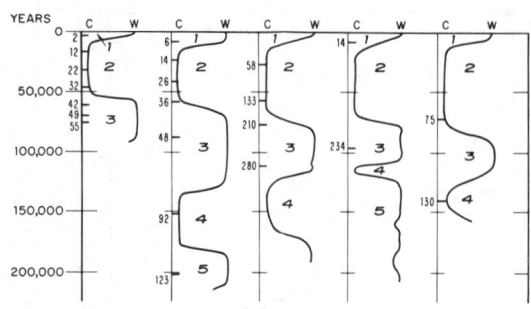

FIG. 4. Plot of five cores showing division into warm (W, interglacial stages) and cold (C, glacial stages) on the basis of foraminiferal analyses. Vertical scale is time in years. The small dashes at the left of the plots represent sections of the cores dated and the numbers beside them are the depths of these sections. The numbers in the center of the plots identify the glacial stages. The two cores on the left are from the Pacific Ocean and have been dated by the ionium/thorium-232 method. The others are from the Atlantic and have been dated by the protactinium or protactinium/ionium method. As can be seen from the plots the ages of the boundaries between the glacial and interglacial stages in the different cores are fairly close together. This indicates that the ionium/thorium-232 and protactinium dating methods are in agreement (Blackman and Somayajulu, 1966).

sult is interpreted as representing sediments with changing rates of sedimentation. Comparison of this method with the protactinium/thorium method gives good agreement. However, in some cases results of this method do not agree with results obtained by the carbon-14 method.

ABNER BLACKMAN

References

Blackman, A., and Somayajulu, B. L. K., 1966 "Pacific Pleistocene Cores: faunal analyses and geochronology," *Science,* **154**(3751), 886–889.

Goldberg, E. D., 1965, "An observation on marine sedimentation rates during the Pleistocene," *Limnol. Oceanog.,* **10**, R125–R128.

Goldberg, E. D., 1968, "Ionium/thorium geochronologies," *Earth and Planetary Sci. Letters,* **4**, 17–21.

Goldberg, E. D., and Griffin, J. J., 1964, "Sedimentation rates and minerology in the South Atlantic," *J. Geophys. Res.,* **69**(20), 4293–4309.

Goldberg, E. D., and Koide, M., 1962, "Geochronologic studies of deep sea sediments by the ionium/thorium methods," *Geochim. Cosmochim. Acta,* **26**, 417–450.

Kigoshi, K., 1967, "Ionium dating of igneous rocks," *Science,* **156**(3777), 932–934.

Ku, T. L., Broecker, W. S., and Opdyke, N., 1968, "Comparison of sedimentation rates measured by paleomagnetic and the ionium methods of age determination," *Earth and Planetary Sci. Letters,* **4**, 1–16.

Picciotto, E., and Wilgain, S., 1954, "Thorium determination in deep-sea sediments," *Nature,* **173**, 632–633.

Somayajulu, B. K. L., and Goldberg, E. D., 1966, "Thorium and Uranium isotopes in seawater and sediments," *Earth and Planetary Sci. Letters,* **1**, 102–106.

Cross-references: Authigenesis of Minerals, Marine; Biogeochemistry; Carbon-14 Dating; Ion Exchange; Isotope Fractionation; Protactinium-Thorium Dating Method; Radioactive Isotopes; Seawater; Spectrophotometry; Uranium; Weathering, Chemical. Vol. VI: Sedimentation Rates, Deep-Sea.

IONIZATION POTENTIAL—See GEOCHEMISTRY: IONIZATION POTENTIALS

IONS—See TRACE ELEMENTS IN SILICATE MINERALS

IRIDIUM: ELEMENT AND GEOCHEMISTRY

Properties

Iridium, a member of the platinum group of metals, was first identified by Smithson Tennant in 1804. The element has atomic number 77 and atomic weight 192.2. Two stable isotopes are known with mass numbers 191 (38.5%) and 193 (61.5%). It is the chemical element of greatest density, 22.65 g/cc at 20°C. In crystalline form the metal is based on a face-centered cubic lattice with $a = 3.8394$ Å. The metallic radius for 12- coordination is 1.354 Å. The melting point of the pure element is 2443°C. Of all metals, iridium displays the greatest resistance to corrosion; another important physical property is the great mechanical strength of the metal at high temperatures.

Occurrence

Like all the platinum metals, it is characterized by indifference of the element itself to chemical attack. Accordingly, it is generally found in native form and alloyed with other platinum metals, usually together with iron, some copper and nickel, and sometimes other metals. In most deposits of contemporary economic significance this element is so attenuated that confident statements about its chemical form in these deposits are not justified. However, the absence of known mineral compounds of iridium supports the view that its natural occurrence is uncombined.

Another interesting form in which iridium occurs is in alloys with osmium, of which two forms are described: *iridosmine* has a hexagonal crystal structure (like osmium) and *osmiridium* has a cubic structure (like iridium). Each of these embraces a wide range of composition, including other platinum metals; the transition from cubic to hexagonal form takes place at about 32% osmium.

Abundance and Geochemical Features

The platinum metals, including iridium, are present in the earth's crust at very low concentration; for instance, iridium has been estimated at 0.001 ppm. Such low content is consistent with the notion that these metals are siderophilic in character, and have concentrated at the time of the earth's condensation in the iron-nickel core in preference to the rocky crust. Support for this hypothesis is also drawn from the abundance of iridium and the other metals in meteorites, among which the average measured concentrations are significantly greater in siderites than in chondrites; also, in chondrites, iridium is preferentially concentrated in the nickel-iron phase. It seems well established also that the platinum metals are generally found more abundantly in ultrabasic rocks than in silicic rocks, and are often associated with sulfide minerals such as pyrrhotite and nickeliferous pentlandite (see also *Platinum Metals*).

Economic Geology

From this point of view the platinum metals are conveniently treated together as a group. In all commercially worked deposits of these metals platinum and palladium together account for something like nine-tenths of the total platinum metals, so that iridium is clearly a minor component. Its commercial value derives from its importance as a hardening and strengthening agent for commercial platinum.

W. A. E. McBryde

References

Hampel, C. A. (editor), 1961, "Rare Metals Handbook," Second ed., New York, Reinhold Publ. Corp., pp. 304–335.

Hampel, C. A. (editor), 1968, "Encyclopedia of the Chemical Elements," New York, Reinhold Publ. Corp., pp. 299–305.

Rankama, K., and Sahama, Th. G., 1950, "Geochemistry," Chicago, University of Chicago Press, pp. 688–694.

Standen, A. (editor), Kirk Othmer "Encyclopedia of Chemical Technology," Vol. 15, Second ed., New York, John Wiley & Sons, pp. 832–878.

Wright, T. L., and Fleischer, M., 1965, "Geochemistry of the Platinum Metals," *U.S. Geol. Surv. Bull.* **1214-A.**

Cross-references: *Earth's Crust Geochemistry; Geochemical Evolution of the Core, Mantle, and Crust; Osmium: Element and Geochemistry; Platinum: Element and Geochemistry; Platinum Metals.* Vol. II: *Meteorites.*

IRON: ELEMENT AND GEOCHEMISTRY

Iron is the most important metal in the universe. The symbol is Fe; atomic number 26; Group VIII of the Periodic Table; atomic weight 55.847; electron configuration (A core) $3d^6, 4s^2$; melting point 1535°C; boiling point 2800°C; sp. gr. 7.874 (20°C); valence 2, 3, 4, or 6.

Pure iron is bright silvery white, malleable, and magnetic. The metal has four allotropic forms known as alpha ("ferrite"), beta, gamma, and delta with transition points at 770, 928, and 1530°C. Both alpha and beta forms have a body-centered cubic crystal structure, the difference involving a loss of magnetic permeability, whereas the gamma allotrope has a cubic close-packed structure. Iron has four stable isotopes in nature: Fe^{54}, Fe^{56}, Fe^{57}, and Fe^{58}.

Mineralogy. Iron is a common element in the earth's crust and it is found as a major or minor constituent in representatives of all of the mineral classes. In igneous rocks it is chiefly found associated with magnesium in the silicates olivine, pyroxene, amphibole, and mica. Minor amounts of iron-bearing pyrite, pyrrhotite, magnetite, and ilmenite also occur in these rocks. In sedimentary and other near-surface deposits, iron occurs chiefly in the minerals hematite, goethite, siderite, magnetite, pyrite, pyrrhotite, and in various secondary iron silicates such as chamosite, glauconite, greenalite, stilpnomelane, and minnesotaite.

Chemistry. Iron is very reactive chemically and the pure metal tarnishes rapidly in air or water. The elements titanium, vanadium, chromium, manganese, cobalt, and nickel are similar to iron chemically because they have similar electron configurations. These elements, including iron, have been called the ferrides by Landergren.

Iron is unique among the more abundant crustal elements in that it occurs in several valence states. In most minerals formed near the surface of the earth it is combined with oxygen as the ferric (Fe^{3+}) ion. In deep-seated rocks it occurs chiefly in the ferrous

IRON: ELEMENT AND GEOCHEMISTRY

TABLE 1. AVERAGE IRON CONTENTS OF VARIOUS ROCKS[a]

Material	Fe (%)
Crust	5.0
Igneous rocks	
Gabbro	8.8
Diorite	5.6
Granodiorite	3.3
Granite	2.5
Sedimentary rocks	
Limestone	0.38
Sandstone	0.99
Shale	4.7
Iron Formation	28.0
(Lepp and Goldich, 1964)	

[a] From Rankama and Sahama, 1950.

(Fe^{2+}) state in silicate minerals. The ferrous/ferric ratio is thus roughly a function of depth of origin and time of atmospheric interaction.

The ability of iron to be oxidized or reduced in natural environments markedly affects its geochemical cycle, and the cycles of other elements as well. Goldschmidt (1958) stated two rules which govern the fixation and mobilization of iron in aqueous solutions, namely:

Rule 1: Oxidizing conditions promote the precipitation of iron, reducing conditions promote the solution.

Rule 2: Acid conditions generally promote the solution of iron, alkaline conditions promote the precipitation of iron.

The susceptibility of iron to surface oxidation is one of the main reasons that it tends to become separated from such elements as Ca and Mg in natural waters.

Ferric oxide may act to oxidize organic matter and thus become reduced to the ferrous state. The ferrous solutions thus formed may be reoxidized by the oxygen of the atmosphere. Iron thus acts as a catalyst in the cycle of carbon and the cycle of iron is closely bound to the oxygen cycle. Iron plays an important role in the biosphere. In animals it acts in transporting oxygen from air or water to animal tissue. In green plants it is necessary for the formation of chlorophyll.

Geochemistry. *Abundance.* Iron is believed to be the most abundant element in the earth as a whole. In the early history of the earth much of the iron became concentrated in the core and mantle. In the crust or lithosphere it is the fourth most abundant element after oxygen, silicon, and aluminum. Table 1 shows the abundance of iron in various crustal units. The greatest variation in iron content is seen to be in the sedimentary rocks because these form near the surface where oxidation is prevalent.

The group of sedimentary rocks called iron formations or ironstones consisting chiefly of iron minerals and chert, or of iron minerals and clay or calcium and magnesium carbonate represent the most extensive concentrations of iron in the crust. Fig. 1 shows the areas of iron formation in northeastern North America. Similar rocks are found on most continents.

Exogenic Cycle. The exogenic or surface cycle involves the action of water and air on earth materials. When deep-seated rocks with predominately ferrous minerals are exposed to weathering, the iron is oxidized and ferric oxide is formed. Because of its relative insolubility, ferric oxide tends to be separated from the more mobile weathering constituents such as solutions containing Ca^{2+}, Mg^{2+}, Na^+, K^+, and other ions.

Large quantities of ferric oxide (hematite) or hydrated ferric oxide (goethite) are trapped in tropical to subtropical soils called laterites. Lateritic weathering differs from weathering under temperate climatic conditions in that even silica tends to be removed in solution. The weathering residuum, therefore, is made up mainly of aluminum and iron oxides with minor titanium and phosphorous. The proportion of aluminum to iron depends upon the composition of the rocks being weathered and also upon the length of time that the laterite has existed. The ferric oxides in a laterite may be locally reduced by organic matter, resulting in the remobilization of iron. In time, some laterites thus tend to become more aluminum rich. Some laterite deposits contain sufficient iron to be classed as ores.

Where weathering, and particularly lateritic weathering, occurs on an iron-rich sediment such as an iron formation, extensive concentrations of relatively pure iron oxide may be formed. These residual deposits do not contain appreciable aluminum or titanium because the iron formations from which they formed did not contain these constituents.

Although iron is concentrated above its crustal average of five percent in many ways, by far the most extensive concentrations are in sedimentary rocks called iron formations or ironstones. These iron-rich sediments have been studied by many investigators, including Alexandrov, Alling, Dorr, Dutton, Goodwin, Gross, Gruner, Huber, James, Krishnan, Moore, Percival, Sakomoto, Tyler, Van Hise, White, Woolnough, and many others (see Lepp and Goldich, 1964). Opinions as to their origin have changed in time from the once prevailing view that they are related to

IRON: ELEMENT AND GEOCHEMISTRY

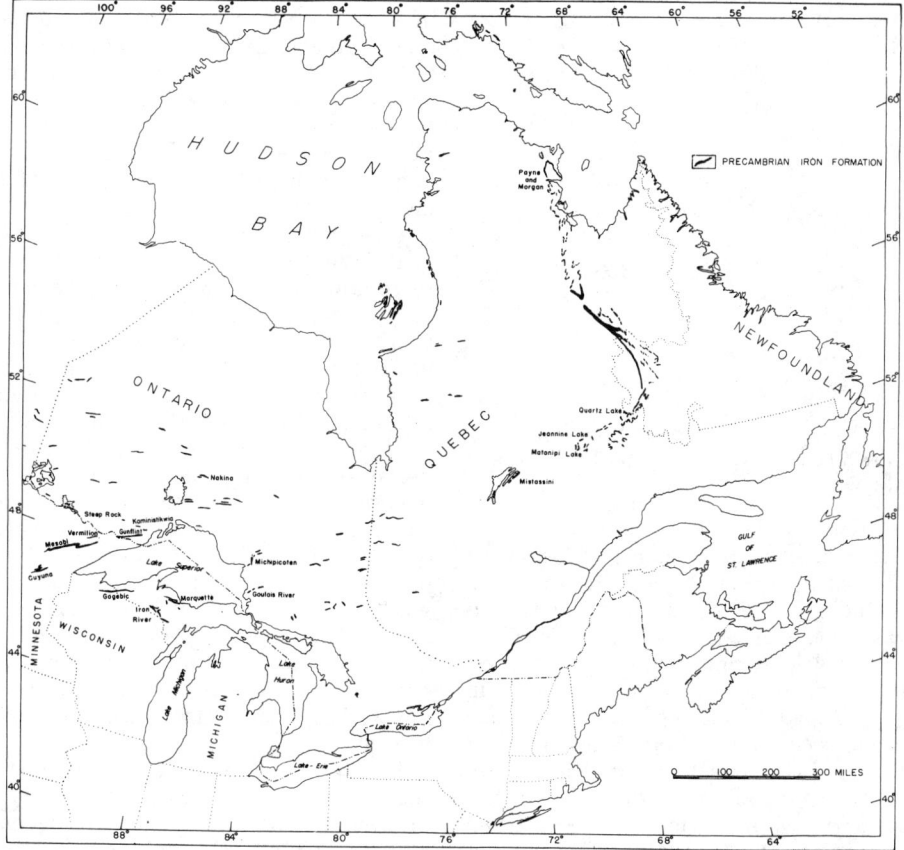

Fig. 1. Map showing Precambrian iron formations (black) in northeastern North America.

volcanism, to the currently most accepted view that they are true sedimentary rocks that have acquired their constituent ions through weathering of other rocks.

It was noted earlier that iron is frequently concentrated as insoluble ferric oxide at the site of weathering. Under nonoxidizing or acid conditions such as exist in certain humose soils, iron may go into solution as ferrous bicarbonate. Some iron is also moved in streams as colloidal ferric oxide. Ferrous carbonate is precipitated from solutions containing $Fe(HCO_3)_2$ when carbon dioxide is removed. Iron may also precipitate as ferric oxide from solutions of $Fe(HCO_3)_2$ as a result of a change in the oxidation potential.

Small deposits of iron carbonate or iron oxide in swamps and lakes, known as bog ores, are formed from weathering solutions. Most of the iron transported by rivers reaches the oceans, however, where it is precipitated. Very little iron stays in seawater which contains only 0.002–0.02 g/ton. If the iron precipitates over a long period of time in an area where other sediment is not accumulating, an iron-rich sediment will result. Most of the time the iron is mixed at the site of deposition with other erosional products and no enrichment of iron occurs. Note (Table 1) that the most abundant type of sedimentary rock, shale, has about the same iron concentration as the average crust.

There appears to have been a change in the nature of iron sedimentation with time. Precambrian iron formations the world over are composed chiefly of chert and one or more iron minerals such as magnetite, siderite, hematite, or iron silicates. Paleozoic and younger ironstones, on the other hand, usually contain calcium and magnesium carbonates or sand and clay with such iron minerals as hematite, goethite, siderite, or chamosite. Fig. 2 shows the chemical differences between samples of the two types.

Lepp and Goldich (1964) interpret this difference in the nature of iron sedimentation with time to indicate an oxygen-deficient atmosphere in the Precambrian. Other investi-

IRON: ELEMENT AND GEOCHEMISTRY

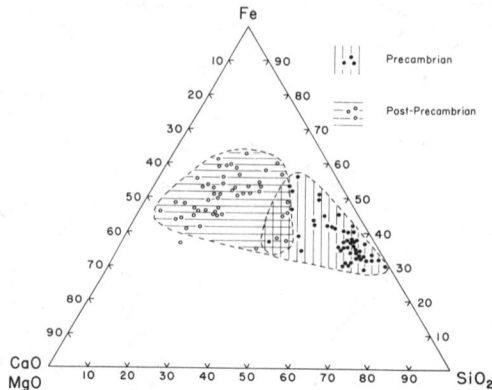

FIG. 2. Composition of Precambrian and later iron-rich sediments in terms of total Fe, SiO$_2$, and CaO + MgO.

gators have reached a similar conclusion on different evidence (see Cloud, Macgregor in James, 1966). Fig. 3 shows how iron would travel with the elements Ca, Mg, and Si in an oxygen-deficient atmosphere. In the present atmosphere iron has a different cycle in that it is trapped in laterites as ferric oxide and thus it is separated from SiO$_2$. Most of the younger ironstones contain from 5 to 10% Al$_2$O$_3$ as opposed to the 1% or less in Precambrian iron formations. As shown in Fig. 3 oxidizing conditions at the site of lateritization produce Fe^{3+} which behaves like aluminum. Reducing or neutral conditions, on the other hand, allow iron to move in solution as Fe^{2+} and thus to behave much like calcium and magnesium.

Endogenic Cycle. The endogene processes are those that take place beneath the surface of the earth, including melting, metamorphism, action of rising fluids in forming new minerals in open spaces or by replacing existing minerals (metasomatism). Landgren (1948), following V. M. Goldschmidt, emphasized that the endogene processes generally lead toward homogeneity of earth materials, whereas the exogene processes tend toward the separation and, consequently, the enrichment of the elements.

Whereas the largest concentrations of iron in the crust are the result of surface processes, there are certain iron-rich deposits such as the Kiruna ores of Sweden that have characteristics suggesting a deep-seated origin. The difficulty of interpreting the endogene concentrations of iron becomes apparent when it is realized that these could be either surface concentrations that became remobilized after burial, or they may be true magmatic or other deposits that owe their existence solely to endogene processes. Scientists are not agreed on the origin of many deep-seated iron deposits. Landergren (1948), for example, proposes that the primary concentration of iron in the Kiruna deposits must have taken place in the surface cycle. Many other investigators, however, believe that these and similar deposits are purely the result of magmatic processes.

Fig. 4 is modified after Landergren and it shows the surface cycle to be the major site of iron concentration. Exogenic deposits may become deeply buried and the iron remobilized to form metasomatic deposits or contact deposits near an igneous intrusive. The exogene deposits may in some instances be melted and the iron redistributed as magmatic, metasomatic, or contact metamorphic deposits.

Iron in laterites and residual deposits is present as Fe$_2$O$_3$ and hence these deposits involve the fixation of oxygen along with iron. Barth (in Rankama (1950)) calculated that

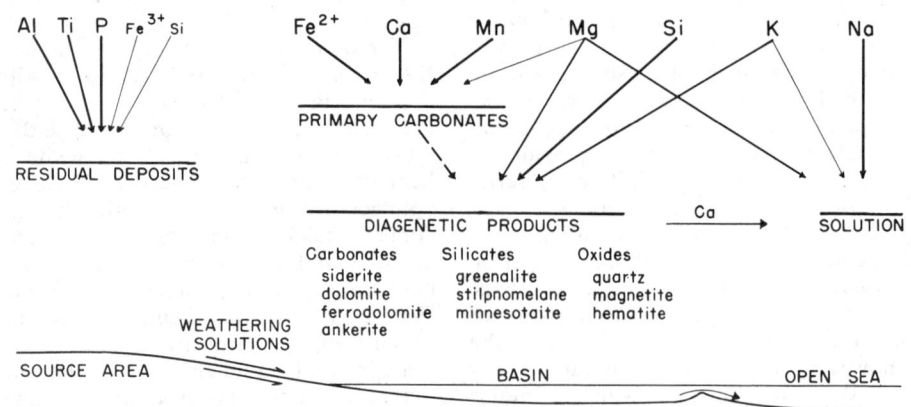

FIG. 3. Lateritic weathering model showing chemical differentation necessary to produce Precambrian iron formations (Lepp and Goldrich, 1964).

IRON: ECONOMIC DEPOSITS

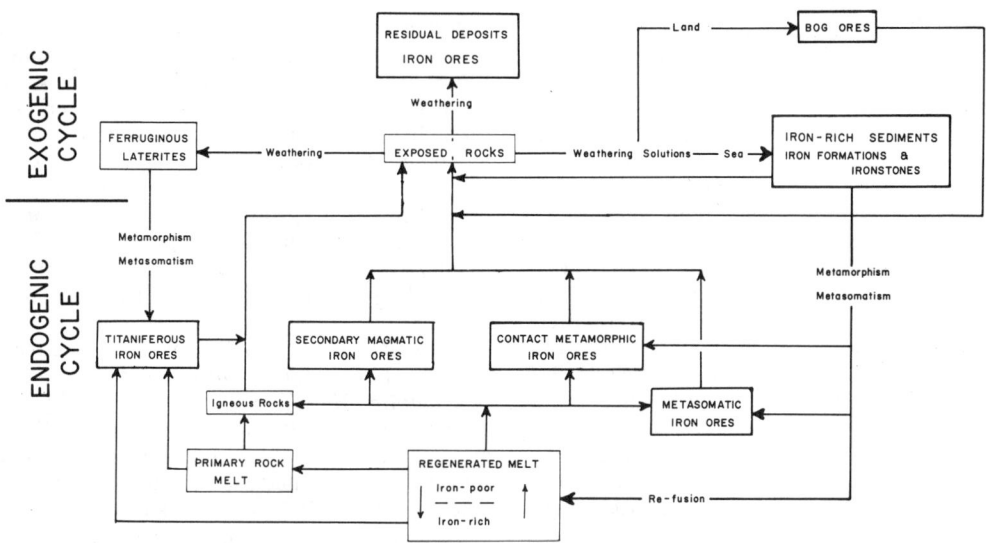

FIG. 4. The iron cycle (modified after Landergren, 1948).

all of the atmospheric oxygen would suffice to oxidize the iron in only 1.43% of the upper crust. The fact that the oxygen of the atmosphere has not been depleted suggests that there must be some process that returns such trapped oxygen to the atmosphere. Barth and Landergren suggest that the exogene ferric deposits release oxygen when they are brought to great depths by orogenic movements. Most plutonic iron deposits contain magnetite ($FeO \cdot Fe_2O_3$) rather than hematite (Fe_2O_3) as the principal iron mineral. Iron may thus act as a respiratory agent in the lithosphere trapping oxygen near the surface and releasing it at depth.

Extensive deposits of iron as magnetite with ilmenite and sometimes chromite are found in some gabbroic bodies. These deposits appear to be the result of magmatic differentiation but the details of their origin are not always clear. The oxides of the ferrides are among the first minerals to crystallize from a melt. Yet many of the titaniferous iron deposits exhibit textures indicating that the iron minerals formed later than the silicate minerals of the rock. In some deposits of this type magnetite is obviously late because it is squeezed into rocks surrounding the gabbroic intrusive. Much more work is needed to clearly distinguish true magmatic iron enrichments from reworked surface deposits.

H. LEPP

References

Goldschmidt, V. M., 1958, "Geochemistry," Oxford University Press, pp. 647–676.

James, H. L., 1954, "Sedimentary facies of iron formation," *Econ. Geol.*, **49**, 235–293.

*James H. L., 1966, "Chemistry of the iron-rich sedimentary rocks," *U.S. Geol. Surv. Profess. Paper* **440-W**, 60pp.

*Landergren, Sture, 1948, "On the geochemistry of Swedish iron ores and associated rocks," *Sveriges Geol. Nndersokn. Årsbok*, **42**(3), 182pp.

*Lepp, H., and Goldich, S. S., 1964, "Origin of Precambrian iron formations," *Econ. Geol.*, **59**, 1025–1060.

Rankama, K., and Sahama, T. G., 1950, "Geochemistry," Chicago, Univ. of Chicago Press, pp. 657–676.

* Contains extensive bibliographies.

Cross-references: *Carbon Cycle; Earth's Crust Geochemistry; Geochemical Evolution of Core, Mantle, and Crust; Oxidation and Reduction; Oxygen Cycle; Weathering, Chemical.* Vol. IVB: *Goethite; Hematite.* Vol. V: *Magmatism; Metasomatism.* Vol. VI: *Bog Iron Ore; Ironstone; Laterites.*

IRON: ECONOMIC DEPOSITS

The simple definition given in the "Survey of World Iron Ore Resources" (United Nations, 1955), is that "if the material is, or may be considered as being, exploitable for extraction of metallic iron, it is iron ore." As this definition has an economic basis it concedes that particular deposits may or may not be iron ores at different times. Thus the former black-band iron ores of Britain, worked and smelted for many years, are no longer iron ores, and the Taconites of the Lake Superior region and of the Quebec-Labrador iron belt are now important iron ores, though they were not iron ores prior

IRON: ECONOMIC DEPOSITS

to the 1950–1960 decade. Equally the location of deposits is significant in this definition; Taconite-type rocks in many countries are still not considered as iron ores.

Factors Affecting Viability. (a) Grade. An iron content over 60% in the saleable product is desirable, and ores may range up to 69% Fe, or even slightly over, but many other factors are to be considered.

(b) Reducibility is important. Hematite is preferred to magnetite in this regard, and the hydrated ore, goethite (limonite) is favored, even with a lower iron content.

(c) Location is vital. Distances to smelting plants (particularly by rail with high transport costs), local availability of coal and flux, and a reservoir of skilled labor are important factors.

(d) Presence of lime in a deposit otherwise of high-iron grade may assist fluxing, but lime will not necessarily commercialize a low-iron deposit which has a long transport distance to the smelter, since adequate supplies of flux are commonly available.

(e) For export ores, short rail transport is desirable, but port facilities for large ore boats may mean rail transport up to 400 or perhaps 500 miles.

(f) Deposits of greater distances than 500 miles from a port may be worked if a local industry is, or can be, established.

(g) Cheap fuel or power may enable a low-iron deposit to be improved in grade, or to be reduced locally.

(h) Physical condition is important. High-iron material of very fine grain size may be unsuitable for transport without costly pretreatment.

(i) Political factors may be influential, e.g., the nearest port to the large deposits of Tindouf in Algeria is in Morocco. Again, countries may be willing, for national prestige, to establish reduction plants for ores that are not normally economic to smelt.

Chief Iron Ore Minerals. *Magnetite* Fe_3O_4 (considered as spinel—$FeO \cdot Fe_2O_3$); Fe = 72.4%, O = 27.6%; isometric system, commonly in octahedra; strongly magnetic (natural "lodestone"); specific gravity 5.18; color black with black streak.

Hematite, Fe_2O_3; Fe = 70%, O = 30%; hexagonal-rhombohedral-scalenohedral; thin to tabular crystals; may be in micaceous or brilliant flaky particles (specularite or oligist); may be botryoidal, as kidney ore or pencil ore; may be in octahedra, replacing magnetite, and then called martite; specific gravity 5.26; color blue-black when crystalline, red in powder and streak. Earthy red variety is "red ochre."

Maghemite (gamma hematite); Fe_2O_3. Magnetic; isometric with red streak; contains some ferrous iron; specific gravity 4.876. Occurs naturally with hematite at temperatures below about 530°C, but tends to alter to hematite from 280°C upward. Above 530°C the stable phases are magnetite and hematite. It is probably more common in nature than is generally recognized, and is a readily reducible ore. There is a (rarer) corresponding gamma form of the hydrated ferric oxide.

(*Wustite,* FeO, can be made synthetically, in slags, but is not a natural mineral.)

Goethite, $Fe_2O_3 \cdot H_2O$; Fe = 62.9%; O = 27.0%; H_2O = 10.1%; orthorhombic, dipyramidal, but commonly in massive, or botryoidal, reniform or stalactitic fibrous aggregates; color light to dark brown; streak yellowish-brown; specific gravity 4.37; adsorbed moisture gives a variable water content.

Lepidocrocite has the same composition as goethite and is dimorphous with it. Streak is red.

Limonite is a rock rather than a mineral, mainly composed of goethite; sometimes called brown hematite. With fine clay admixture it forms yellow ochre. Specific gravity is variable according to water content and impurities. Limonite is a convenient ore term when composition is doubtful or variable, but it should not be confused with laterite, though both are mainly composed of goethite. Limonite may contain *nontronite,* an earthy, unctuous hydrated iron-alumina silicate allied to montmorillonite, and *jarosite,* $KFe_3(OH)_6(SO_4)_2$, as a secondary alteration mineral.

(*Stilpnosiderite* is a term used for the gel of iron oxide that forms a colloidal element in the oölites of the iron ores of Lorraine.)

Siderite, $FeCO_3$ (chalybite, spathose iron); Fe = 48.2%; hexagonal-rhombohedral; specific gravity 3.7–3.9; color light to dark brown; streak white, other carbonates may be combined; *ankerite* is the dolomitic iron carbonate, occasionally with some manganese.

Ilmenite, $FeO \cdot TiO_2$; Fe = 36.8%; Ti = 31.6%; hexagonal-rhombohedral; specific gravity 4.7; mainly mined for its titanium content, but iron can be recovered as a by-product.

Pyrites, FeS_2; isometric, diploidal or pyrite class; Fe = 46.6%. Alters readily to hydrated iron oxide. Normally mined for sulfur, not for iron, but the burnt residue is marketed as iron ore and used as sinter feed.

Silicates of Iron. The complex iron silicates characteristic of the Lake Superior taconites contribute by their breakdown to the enriched iron ores to some extent, but as silicates they are not in themselves desirable as ore minerals. The chief of these silicates are:

(a) Minnesotaite, in light greenish-grey needles and fibers, with a talc structure (Gruner, 1946).

(b) Stilpnomelane, brown to green, in radiating needles or minute grains.

(c) Greenalite, commonly in dark olive-green granules whose analysis is difficult because of the small size and impurity of the grains. Gruner found it similar to serpentine, but containing some Fe^{III} in addition to the Fe^{II} of serpentine.

(d) Grunerite, $(OH)_2Fe_7(Si_4O_{11})_2$—an iron-rich amphibole.

Two silicates are locally iron ore constituents, of diminishing acceptability:

(a) *Chamosite,* approximately $2SiO_2 \cdot Al_2O_3 \cdot 3FeO$ aq.; $Fe = 30\%$; $SiO_2 = 25\%$; $Al_2O_3 = 19\%$ (Taylor, 1949). This is a greenish, kaolin-type alumino-silicate of iron that is a notable constituent of the British Jurassic oolitic iron ores. It cannot be economically beneficiated, and its iron is extracted in the blast furnace.

(b) *Thuringite* has a similar chemical composition, but with a chlorite structure. It is a constituent of some German ores.

Glauconite is a hydrated silico-aluminate of iron and potash and is of variable composition. It is common in Cretaceous greensands, often as casts of foraminifera, and decomposes to limonite. It is not at present of importance in iron ores.

Associated Impurities. The impurities in iron ores are partly from the gangue (including overburden) and partly intimately mixed with the ore itself. The gangue materials may often be removable to a large extent by simple types of pretreatment before smelting, preferably at the mine to reduce the cost of freight per unit of iron. Washing and wet screening will remove sticky clay material. A massive siliceous gangue may call for a gravity separation (such as heavy media separation), and an objectionable laterite cap might need this also. Carbon dioxide and excessive water could be reduced by heat treatment if the local cost is not excessive. Free moisture is generally reduced as much as possible by natural drying before shipment. Goethitic ores may be acceptable for their good reducibility.

Intermixed fine-grained deleterious matter may in some cases be reduced when a lower grade ore is finely ground for concentration of the iron, but in direct-shipping ores of good grade such materials remain to be removed in the furnace processes.

Silica and alumina combined totals are in any case low when the iron content is really high—over 65%—as in the rich magnetite ores and the higher-grade hematites of the Lake Superior type. To avoid the risk of getting pasty high-alumina slags, a silica/alumina ratio higher than unity is generally preferred.

Manganese rarely presents any difficulty. Often a small amount of Mn is desired in the pig, and even a high-Mn ore can usually be blended with other ores to give a suitable all-in proportion.

Lime and magnesia help to flux the ore; small amounts of soda and potash are undesirable, but may be accepted. If a high-lime, low-iron ore with low silica and alumina is available at no great distance from a reduction plant it can be regarded as in effect a ferruginous flux.

All the phosphorus of the ore goes into the pig iron, and has to be removed in the steel furnace. Formerly, ore used to produce hot metal for the acid Bessemer process was called "Bessemer ore," and had an upper limit of 0.0009% P for each 1% of Fe, but this process is of declining importance, and a ratio up to 0.002% P per unit of iron in the ore is generally suitable for modern steel-making practice.

Sulfur should be low in the iron ore. The coke for the blast furnace introduces sulfur, and the ore should add as little as possible. However, if the ore is used for sinter-production the sulfur can be largely eliminated, and acts as a fuel, in the sinter process.

Barytes is not a common impurity, but may be even slightly advantageous in special cases with sintering. The sulfur of the barytes is driven off on the sinter strand, and the residual BaO assists in de-sulfurizing the iron in the blast furnace.

Vanadium, if present, goes mainly into the pig iron in the blast furnace, but passes into the slag in the steel process, and the vanadium can then be recovered as a valuable by-product.

Titanium can be tolerated but should not exceed 1% TiO_2 in the ore. If the TiO_2 content is small most of it goes into the blast furnace slag. A little goes into the pig, but is eliminated in the steel furnace.

Cr_2O_3 should be low, and if consumers are asked to state their maxima for other impurities, such as nickel, copper, zinc, lead, tin, and arsenic they will specify "nil," but in practice very small amounts, say on the order of 0.01% are ignored.

Origins of Iron Ore Deposits. One should bear in mind that few things are more disputed than the genesis of ore bodies.

Magmatic Origin. The primary source of iron is the igneous magma, and its concentration may commence in the magma. Large magnetite masses may have been formed by segregation within a magma that is high in iron by early crystallization, possibly with gravity separation, to give a mass of solidified magnetite within the original magmatic rock. The indications are, however, that more frequently the segregation is injected either into the country rock, or into the cooling intrusive rock itself. The magnetite of Kiruna in northern Sweden is generally considered to be an injection into syenite and quartz porphyries, and the Adirondack, and many other titaniferous magnetites, are presumed to have similar origins. "Filter pressing," the squeezing-out of residual liquid from the crystal mush of a partly crystallized magma, may be a stage in the process. The large magnetite deposits of El Laco, Northern Chile, resemble lava flows in places (Park, 1961) and squeezed-out segregations may have been extruded as flows more commonly than is recognized.

Contact-Metasomatic Origin. The later residual liquid and volatile constituents of a crystallizing magma may convey iron to more distant host rocks, and magnetite or hematite or mixed deposits may result, as in some of the mines in the Urals (Mount Magnitnaya for example): in Malayan ores; Iron Springs, Utah, and many others.

Hydrothermal Origin. The hot residual solutions, mixed in some cases with meteoric waters, may give rise to disseminated or to vein deposits. The latter are only exceptionally of importance in iron ores. These solutions may also effect the replacement of limestones and other rocks on a large scale, occasionally as magnetite (oxidizing to hematite in part), as in Uixan, Morocco and Marmora, Canada, or as goethite, in Ouenza, Algeria and Djerissa, Tunisia.

A special case of this type of origin is that of submarine exhalations, as in the German Lahn-Dill ores. These are called exhalative-sedimentary ores, and of course they do have a sedimentary aspect, but one may also regard the sea as the host rock.

The foregoing are deposits with direct association with the magmatic source of the iron. Secondary processes act on these existing concentrations, and also on a larger scale on the iron present in smaller percentages in rocks in general. The average iron content of the earth's crust is 4.7% (James, 1966). Ultrabasic rocks such as peridotites contain 8 to almost 12% of iron, and their contained silicates break down readily under tropical conditions. The thin films that color so many sandstones red also offer an easily accessible source of iron. Although ferric iron is not readily dissolved it can be reduced by natural agencies to a more soluble form for transportation. Weathering exposes minerals to mechanical erosion, leading to dispersal or concentration. The removal of other rock constituents such as alkalis and silica in solution may leave residual masses that are very high in iron.

Sedimentary Origin. Syngenetic sedimentary beds of iron ore derive their iron from the chemical and physical weathering of pre-existing rocks of all types. Although there are detrital iron sands, sandstones and grits, and high iron conglomerates (e.g., Yampi Sound, Western Australia), yet transportation of the iron is mainly in solution, by the action of carbonated waters or by humic or other organic acids. More rarely iron may be transported as sulfate.

Deposition from bicarbonate solutions is by loss of carbon dioxide, in which increasing temperature may be a factor. Castaño and Garrels (1950) attributed the formation of the Clinton ores of New York to the transportation of ferrous iron in solution in aerated river waters of pH 7 or lower, and subsequent precipitation as ferric oxide in aerated ocean waters; solid calcium carbonate is in equilibrium with the seawater. The precipitation takes place both in the water and by replacement of the calcium carbonate. Moore and Maynard (1929) made laboratory experiments indicating that iron transported as ferric oxide hydrosols stabilized by organic matter would be precipitated by the electrolytes present in seawater.

When iron solutions are in contact with an aluminous mud on the sea floor, silicates such as greenalite or chamosite may be formed. The chamosite forms oölites in suitable shallow waters. In reducing conditions with organic debris siderite may be deposited. The British Jurassic oölitic iron ores contain mainly chamosite oölitic grains in a matrix that is largely sideritic. Much of the siderite is not directly precipitated, but results from carbonation of chamosite (Taylor, 1949). When the environment alternates between oxidizing and reducing conditions, alternating shells of different minerals may be produced in the oölitic grains. In the Jurassic ores of Lorraine the oölitic grains were largely oxidized before their consolidation by calcareous cement.

Major oölitic deposits are summarized in Table 1. The Tertiary pisolitic limonites of the Hamersley province of Western Australia (Robe River, Mount Enid etc.), estimated at 6,000 million tons with 50 to 60% Fe content, are not oölitic. They are almost unique,

IRON: ECONOMIC DEPOSITS

TABLE 1. MAJOR OÖLITIC SEDIMENTARY IRON ORES

Location	Reserves[a] (million tons)	Average Fe (%)	Type	Age
S. Africa—Pretoria	3000	40–50	Magnetite; hematite; limonite; chamosite	Precambrian
Queensland, Constance Range	257	51.5	Hematite and siderite	Precambrian
Alabama, Tennessee, and Georgia	1200	35	Hematite with lime	Silurian
Algeria—Tindouf	400	57	Hematite, with some magnetite	Devonian
Argentina	148	56	Do.	Devonian; Carboniferous
Lorraine and Luxembourg	6000	28–35	Limonite; hematite	Jurassic
Great Britain	4000	24–35	Limonite; siderite; chamosite	Jurassic
Egypt	180	50	Hematite	Cretaceous
U.S.S.R. Kazakhstan, Lisakov	1611	36	Limonite	Oligocene
U.S.S.R.—Kerch	2000	33–40	Limonite	Oligocene
W. Australia—Robe River	6000	50–57	Limonite (pisolitic)	Late Tertiary

[a] The estimates of reserves are of unequal validity, and are given chiefly as indications of relative size. The use of the lower-grade ores is declining, except in cases where they can be pretreated before feeding to the blast furnace. Old established plants, now fully amortized, may be able to continue using the Jurassic ores for some time.

but have some similarity with the Oligocene Lisakov deposits.

In low-iron sedimentary rocks such as glauconitic sandstones the glauconite may be altered to limonite, as at Diest in Belgium, where the iron content is from 19 to 28%. These are not now worked, but it should be possible to beneficiate them in future if required.

The former blackband ores (clay ironstone) of Great Britain were formed by siderite precipitated in reducing conditions along with carbonaceous matter in coal measures, but these are no longer worked. Bog iron ores are usually of minor importance and extent. Where decaying vegetation in bogs and swamps is available, iron is deposited as siderite, but oxidation to limonite readily follows. Iron bacteria may be effective in precipitating ferric hydroxide.

The "banded iron formation" rocks (jaspilites, taconites, itabirites etc.) with an iron content commonly around 30 percent are also sediments, chemically precipitated, but are not "direct feed" ores. Their high silica content may be naturally leached out to give very high grade ores, or they may in favorable circumstances be crushed and concentrated before use in the blast furnace. They are described later.

Residual Origin. Laterites have a high content of ferric and alumina hydroxides. They form as a carapace over vast areas in tropical countries that have alternating wet and dry seasons. They are produced by the leaching away of silica, alkalis, and alkaline earths from a wide range of source rocks. Ultrabasic peridotites, and preexisting rich iron deposits, yield laterites with over 50% Fe content, the iron being mainly as goethite. Many of these high-iron types are spoiled as iron ores on account of their high content of alumina, and in some occurrences by deleterious amounts of chrome, nickel, and cobalt. Titanium may also be present. Rarely the nickel content may be high enough for the rock to be worked as a nickel ore. Australia has imported nickel-bearing laterite as iron-ore from New Caledonia. The lateritic iron ore of Conakry, in the Guinea Republic, was exported in moderate quantities for over a decade, with a chrome content of 1.15%. If an economic method of recovering the chrome, nickel, and alumina as by-products can be devised, these huge deposits in the world's monsoon areas may become valuable mixed-metal ores. At present, where they form cappings over iron deposits of other types, they contaminate these ores with alumina, and are a nuisance.

Sedimentary-residual Origin. The rich ores formed mainly by the leaching of silica from the primary Precambrian banded iron formation contain far more iron than the whole of the other types combined. They form very large iron ore fields in all continents, and their reserves are reckoned in billions of tons.

The primary rocks are characterized by alternate iron-rich and silica-rich bands, with rare oölitic and algal developments. The iron is largely in hematite, but may be in magnetite, siderite, iron-silicates, or pyrite. The hematite is frequently martite, and banded hematite rocks outcropping at the surface may cover banded magnetite rocks at depth, and the magnetite may be of primary origin. Magnetite and siderite in alternating bands of about a quarter-inch thickness occur in Swaziland. Magnetite-carbonate ores of the Kursk magnetic anomaly include a magnesite-siderite carbonate.

There is a large controversial literature regarding the origin of both the primary banded rocks with iron content ranging from about 20 to 35%, and the leaching process producing massive friable or powdery ores with iron contents from 60 to 68% or even more.

It is now generally agreed that the banded rocks originated as marine chemical precipitates, with the iron and silica derived from land masses. However, volcanic sources were suggested by Van Hise and Leith (1911). Moore believed that igneous activity on a large scale was necessary to form the very siliceous chemical deposits (Moore and Maynard, 1929). More recently volcanic exhalations have been suggested by Oftedahl (1958). The very common association of purple phyllites with the banded rocks should not be ignored; they could perhaps be altered tuffs. There are pyrolastics associated with the ores of Hamersley Province, Western Australia (Laberge, 1966).

Moore and Maynard (1929) showed that ferric hydroxide is precipitated in a few days when colloidal iron and silica solutions stabilized by organic matter come in contact with electrolytes of the sea, whereas the silica is not completely coagulated after several months. Thus intermittent supplies of solutions, possibly seasonal, would cause repetition of banding.

Sakamoto (1950) attributed the banding to a cyclical deposition of colloids due to periodic changes of pH in shallow lake waters, in areas with a monsoon climate. Hough (1958) also postulated large fresh-water lakes, but of sufficient depth to permit seasonal density stratification, causing variation in Eh and pH. During slightly reducing and acid conditions iron remained in solution but silica was deposited, but in alkaline and oxidizing phases iron was precipitated. Both these authors note that the conditions they postulate have not occurred since Precambrian times. Given suitably large solutions of iron and silica (and these could be derived from lateritized areas as suggested by Sakamoto) these explanations are plausible. The restriction of the deposits mainly to the Precambrian (and Lower Palaeozoic) suggests that here we have an exception to Lyell's principle of uniformitarianism, and this could be linked up with a high CO_2 content in the Precambrian atmosphere, as has been suggested by MacGregor (1927). The fine banding may be seasonal, but Cullen (1963) considers that the thicker composite divisions of more siliceous and more ferruginous phases are related to recurrent isostatic adjustments in geosynclines.

Possibly the least altered type of primary rock is that of the banded hematite jaspers of Bihar and Orissa, India, and of the Hamersley province in Western Australia. The occasional presence of rhombic pseudomorphs suggests that some, at least, of the original iron may have been as siderite. In most occurrences the siliceous bands have been recrystallized to quartzite, and the Brazilian term "itabirite" might be usefully restricted to this banded hematite quartzite type (Dorr and Barbosa, 1963). Amphiboles, and occasionally garnets, have been produced in metamorphosed banded iron silica rocks, e.g., in the Sula Range of Sierra Leone. James (1954) considers that of the Lake Superior silicates only greenalite appears to be definitely primary.

James states that the evidence indicates that "deposition took place in restricted basins, which were separated from the open sea by thresholds that inhibited free circulation and permitted development of abnormalities in oxidation potential and water composition." He divides the primary iron-rich rocks of the Lake Superior region into four major facies—sulfide, carbonate, silicate, and oxide. The sulfide type is of major importance in the Iron River-Crystal Falls district of Michigan. In many areas of these rocks in other countries it is absent or of minor importance. Carbonate facies exist in some other countries, e.g., at depth in the Middleback Range of South Australia The silicate type is not prominent in many of the largest fields, and, where present, may be due to metamorphic reactions in some cases, as James states, though primary silicates no doubt existed (as also in Jurassic ores).

Though there are exceptional occurrences where silica is absent in the banding, yet in the general world picture it is the iron oxide-silica banded types that predominate. The sulfide, carbonate, and silicate facies are of minor significance in many of the very large

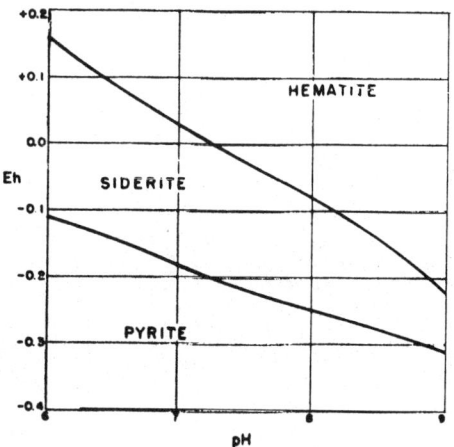

FIG. 1. Fields of stability of hematite, siderite, and pyrite (from Krumbein and Garrels, 1952).

deposits of Brazil, South Africa, India, Australia, and elsewhere.

Krumbein and Garrels (1952) discussed chemical sediments of marine origin within (a) a normal open-circulation environment, and (b) environment-restricted controls such as submerged sills involving at least partial stagnation in enclosed basins, with resultant variations in pH and Eh. Their diagram (Fig. 1) gives the fields of stability of hematite, siderite, and pyrite in terms of Eh and pH.

In earlier years the comparatively low-iron banded rocks were not considered as ores in themselves, and this still prevails in many countries with large deposits of the high-grade enriched ores, but in the United States and Canada, chiefly, the banded rocks (taconites) are now being crushed; the iron is separated out (magnetically in the magnetite types, or, in the case of hematite, by gravity methods such as spirals), and the concentrates containing Fe from 64 to 67 or 68% are agglomerated, mainly to pellets, which may be partially prereduced.

Naturally enriched direct-shipping iron ores are produced by the leaching of silica from the banded rocks, in many cases by meteoric waters, but possibly by hydrothermal solutions in some deposits. Where the leached-out silica has been replaced, either metasomatically, or at some time after the leaching, by secondary iron oxide, hard massive ores have been formed, with an iron content little below that of pure hematite. Where leaching alone has occurred without the introduction of secondary iron, friable broken or biscuity ore remains, collapsed or slumped by the loss of the intermediate bands. At the ground surface these slabby fragments, mixed with soil, are called "chapinha" in Brazil. When screened free of siliceous fines these friable ores commonly have from 60 to 64% Fe. Where the original siliceous bands contained a cloudy mass of very small hematite crystals, the almost complete removal of silica by natural processes leaves the fine iron oxide as a powdery ore ("blue dust" of India), in some places with vertical thicknesses exceeding a hundred feet. The powder, in spite of some collapse, commonly still preserves the original banded appearance, and in some occurrences may contain thin bands of kaolinous matter (possibly leached tuffs). The blue dust may be in equidimensional granules (frequently octahedral martite crystals in India), or when metamorphosed may be as flaky specular crystals ("micaceous hematite"). Fairly numerous tabular fragments may be included in the powder, but the grain size is generally very fine. 40% of the powder may be of a size below 200-mesh. Dorr and Barbosa (1963) report that much powdery ore in Brazil has been formed by disaggregation of hard hematite, but this is not general.

These various physical types of ore pass laterally into each other, or into the unaltered primary banded iron-oxide-silica rocks, along the strike (Fig. 2). The massive type is related to the surface topography in some Indian deposits. The powdery ore rarely reaches the surface (where it would be rapidly eroded), but may be at a shallow depth, protected by a consolidated cap, as at Mount Nimba in Liberia, and the Simandou Range in Guinea.

Estimates of most of the world's reserves of Lake Superior type high-grade ores have been based largely on surface deposits. The pub-

FIG. 2. Lateral passage from hard banded hematite jasper (J) to residual laminated hematite (lower right), and to powdery friable hematite (upper right), with slumping caused by removal of silica. (Noamundi Iron Mine, Bihar, India.)

IRON: ECONOMIC DEPOSITS

lished figures for the high-grade deposits in Australia, Brazil, Canada, China, Gabon, Guinea, India, Liberia, Mauritania, Rhodesia, Sierra Leone, South Africa, United States, U.S.S.R., and Venezuela total over 130 billion tons. In addition, the low-grade "taconites" that are postulated for development, or are already being mined, mainly in North America and the U.S.S.R., may be put at an even higher total, and the uncalculated banded iron oxide-silica rocks that are not at present considered to be iron ore may add many hundreds (or even thousands) of billions of tons more, with iron contents ranging from 20 to 35% The present day ores are worked mainly as surface deposits by open-cast methods, but the ores of the Kursk magnetic anomaly are located at a depth of 1000 meters. Further deposits hidden at depth in other countries may exist.

The estimates of world resources of presently exploitable iron ores published in 1954 in the United Nations survey (1954) was roughly 24½ billion tons. Within less than a decade of this date important new discoveries in Canada, Chile, West Africa, the U.S.S.R., China, Australia, and elsewhere had increased by many billions of tons the estimates of both high-iron direct-shipping ores and the readily treatable itabirites. These estimates are still largely based on surface examination. It is, however, abundantly evident that there is no lack of iron ore for the future.

F. G. Percival

References

Bateman A. M., 1950, "Economic Mineral Deposits," New York, John Wiley & Sons, 916pp.
Blondel, F., and Marvier, L. (editors), 1952, "Symposium sur les gisements de fer du monde," XIX Congr. Geol. Intern. Alger.
Castano, J. R., and Garrels, R. M., 1950, "Experiments on the deposition of iron with special reference to the Clinton iron ore deposits," *Econ. Geol.*, **45**, 755–770.
Cullen, D. J., 1963, "Tectonic implications of banded ironstone formations," *J. Sediment. Petrol.* **33**, 387–392.
Dorr, J. V. N., and Barbosa, A. L. M., 1963, "Geology and ore deposits of the Itabira district, Minas Gerais, Brazil" *U.S. Geol. Surv. Profess. Paper* **341-C**, 1–110.
Gruner, J. W., 1946, "The mineralogy and geology of the taconites and iron ores of the Mesabi Range, Minnesota," St. Paul, Minn., Off. Comm. Iron Range Res., 127pp.
Hough, J. L., 1958, "Fresh-water environment of deposition of Precambrian banded iron formation," *J. Sediment., Petrol.* **28**, 414–430.
James, H. L., 1954, "Sedimentary facies of iron formation," *Econ. Geol.* **49**, 235–293.
James, H. L., 1966, "Chemistry of the iron-rich sedimentary rocks," *U.S. Geol. Surv. Profess. Paper* **440-W**, 1–61.
Krumbein, W. C. & Garrels, R. M., 1952, "Origin and classification of chemical sediments in terms of pH and oxidation-reduction potentials," *J. Geol.*, **60**, 1–33.
Laberge, G. L., 1966, "Altered pyroclastic rocks in iron-formation in the Hamersley Range, Western Australia," *Econ. Geol.*, **61**, 147–161.
MacGregor, A. M., 1927, "The problem of the Precambrian atmosphere," *South African J. Sci.*, **24**, 155–172.
Moore, E. S., and Maynard, J. E., 1929, "Solution, transportation and precipitation of iron and silica," *Econ. Geol.* **24**, 272–303; 365–402; 506–527.
Oftedahl, C., 1958, "A theory of exhalative-sedimentary ores," *Geol. Foren. Stockholm Forh.*, **80**(1), 1–19.
Park, C. F., Jr., 1961, "A magnetite 'flow' in northern Chile," *Econ. Geol.*, **56**, 431–436.
Sakamoto, T., 1950, "The origin of the Precambrian banded iron ores," *Am. J. Sci.*, **248**, 449–474.
Taylor, J. H., 1949, "Petrology of the Northampton sand ironstone formation," *Mem. Geol. Surv. Gt. Britain*, 111.
United Nations, 1955, "Survey of world iron ore resources," Publication 1954 II.D.5.
Van Hise, C. R., and Leith, C. K., 1911," "The geology of the Lake Superior region," *U.S. Geol. Surv. Monograph* **52**.

Cross-references: *Electrolytes; Hydrothermal Solutions; Iron; Meteoric Water; Oxidation and Reduction; pH-Eh; Titanium; Vanadium.* Vol. IVB: *Banded Iron Ores; Exhalative Sedimentary Ore Deposits.* Vol. V: *Magma; Metasomatism.* Vol. VI: *Laterites; Leaching.*

IRON, OXIDATION AND REDUCTION—MICROBIAL

Microbial Iron Oxidation

Environmental Conditions. Microbial iron oxidation occurs in terrestrial and aquatic environments. The Eh limits within which microbial iron oxidations have been observed in nature range from +60 to +850 mV, and the pH limits range from 2.0 to 8.9 (Baas Becking, Kaplan, and Moore, 1960). Iron oxidizing bacteria differ in their Eh–pH requirements.

Types of Microbial Interactions in Iron Oxidation. Microbes may cause iron oxidation directly by enzyme catalysis, or they may cause it indirectly by forming metabolic end products which promote nonenzymatic iron oxidation.

Enzymatic Iron Oxidation. (a) *Autotrophy.* Enzymatic oxidation of iron (Silverman and Ehrlich, 1964) has been clearly demonstrated with members of the *Ferrobacillus-Thiobacillus* group of bacteria. These organisms possess an enzyme system which transfers electrons from ferrous iron to O_2:

$$2Fe^{2+} + 2H^+ + 1/2\ O_2 = 2Fe^{3+} + H_2O$$

A portion of the free energy released in this electron transfer to O_2 is available to the bacteria for assimilation of CO_2 in protoplasmic synthesis (Chemo-autotrophy), thereby freeing the organisms from a dependence on organic matter as a source of energy and assimilable carbon. Not all of the reducing power from ferrous iron oxidation is transferred to O_2, however. A significant portion is used for reducing CO_2 in protoplasmic synthesis by a process called reverse electron transport (Aleem, Lees, and Nicholas, 1963).

Iron oxidation by the *Ferrobacillus-Thiobacillus* group of bacteria proceeds only at acid pH (below pH 5). Under these conditions, autoxidation of ferrous iron is much slower than enzymatic oxidation. At neutral pH and above, autoxidation of iron is so rapid that it is not easily distinguishable frob biological oxidation. However, lowering the Eh of the milieu may make possible the observation of biological iron oxidation by organisms whose enzymes work at a pH as high as 8.5.

(b) *Mixotrophy*. The literature contains reports by Lieske, Praeve, and Sartory and Meyer, in which it is indicated that iron oxidation by some bacteria is detectable at neutral pH in the presence of low but not high concentrations of organic matter. (Silverman and Ehrlich, 1964). On the basis of such observations, Lieske was the first to suggest that at low concentrations of organic matter, the appropriate organisms oxidize ferrous to ferric iron and use some of the free energy released thereby for assimilation of organic matter. At high concentrations of organic matter, he proposed that these organisms oxidize a portion of the organic matter, instead of ferrous iron, to obtain energy required for assimilation of some of the remainder of the organic matter.

(c)) *Biocatalytic function of iron*. Most aerobic and some anaerobic microbes, and plants and animals have iron-containing enzymes which function in electron-transfer reactions. The catalytic action of these enzymes involves reversible oxidations and reductions of the iron. Since the function of the iron in the enzymes is catalytic and not as a substrate, only minute traces of iron are involved. These reactions of iron are, therefore, not of geological significance with respect to iron accumulation.

Nonenzymatic Iron Oxidation by Microbes. Many kinds of microbes in nature can cause oxidation and precipitation of dissolved ferrous iron through the end-products they produce (Starkey and Halvorson, 1927). Oxygen evolution and CO_2 consumption by photosynthesizing algae, for example, can promote rapid autoxidation of ferrous iron. Similarly, a rise in pH and concurrent rise in Eh due to the production of alkaline substances, such as ammonia, can cause autoxidation of ferrous iron, as can the metabolic production of organic oxidizing agents, such as citric acid.

Microorganisms Promoting Iron Oxidation. *Neutral Environment.* The stalked bacterium, *Gallionella ferruginea*, and the sheathed bacterium, *Leptothrix ochracea*, have been described as facultative autotrophic oxidizers of ferrous iron at neutral pH (Perfil'ev et al., 1965; Silverman and Ehrlich, 1964). Other organisms considered to be active iron oxidizers under these conditions include *Metallogenium, Kusnezovia, Caulococcus, Siderococcus,* and *Ochrobium*. Algae belonging to the Cyanophyceae, Chrysophyceae, Volvocales, Chlorococcales, Euglineae, Conjugales, and Ulothricales have been implicated in autoxidation of ferrous iron and precipitation of ferric iron, because the O_2 they evolve promotes iron oxidation, and the CO_2 they consume promotes iron precipitation. Moreover, many common heterotrophic bacteria, which through their metabolism cause a rise in pH at a relatively high Eh, are thought to promote autoxidation of ferrous iron and precipitation of ferric iron. Not all bacteria associated with iron precipitation seem to be responsible for iron oxidation, however. Some members of the family of iron bacteria, the Siderocapsaceae, are not believed to oxidize iron. Rather, they precipitate ferric iron about their cells, by adsorbing it on their cell surface, or by oxidizing the chelating moiety of an organic ferric complex. This is also believed to be true of some members of the two families of sheathed bacteria, Chlamydobacteriaceae and Crenothricaceae, and some protozoan flagellates, *Anthophysa, Siderodendron, Bikosoeca,* and *Siphomonas*.

Acid Environment. Ferrous iron, iron sulfides and pyrite exposed to acid solution are oxidized by autotrophic bacteria of the *Ferrobacillus-Thiobacillus* group. Heterotrophic bacteria are not known to act on iron under these conditions.

Microbial Iron Reduction

Environmental Conditions. Microbial iron reduction occurs in terrestrial and aquatic environments. The process is favored by acid pH and low Eh, but not necessarily anaerobiosis.

IRON, OXIDATION AND REDUCTION—MICROBIAL

Enzymatic Iron Reduction. Although experimental evidence for enzymatic reduction of ferric to ferrous iron is meager, it is carried on by some bacteria. The role of iron in these biochemical reactions is to replace O_2 or organic compounds as terminal electron acceptors in respiratory metabolism. *Bacillus polymyxa, B. circulans* and other Bacilli have been demonstrated to reduce ferric iron (Perfil'ev et al., 1965; Silverman and Ehrlich, 1964).

Nonenzymatic Iron Reduction. Nonenzymatic ferric iron reduction is thought to result from the chemical action of metabolic end products, such as H_2S or organic reducing agents, or through the metabolic consumption of O_2 and the formation of organic or carbonic acids. It is to be noted that solubilization of insoluble ferric iron is not necessarily a criterion of chemical reduction of ferric iron, since the formation of stable organic chelates of ferric iron with certain metabolic end products can also account for an observed increase in soluble iron. Many different kinds of bacteria can reduce ferric iron with their metabolic end products, or by the consumption of O_2 from the environment (Starkey and Halvorson, 1927). Of special practical interest is *Desulfovibrio desulfuricans,* which produces strong reducing conditions by its formation of H_2S from sulfate.

Significance of Microbial Action on Iron

It is probable that some microbes have played important roles in the accumulation of certain iron deposits by their action on iron. The finding of fossil algae and bacteria in some iron oxide ores points to this. However, it is not always certain that the role of the bacteria was to oxidize ferrous iron. In the opinion of some, bog iron ores were deposited mainly as a result of microbial oxidation of organic ferric iron chelates.

Microorganisms are important in maintaining an iron cycle in nature by oxidizing and precipitating ferrous iron on the one hand, and by reducing and solubilizing ferric iron on the other. This is vital for those forms of life which derive their nutritional trace-iron requirements from soil or water, and which can absorb such iron only in soluble form.

Iron-oxidizing bacteria, especially those belonging to the *Ferrobacillus-Thiobacillus* group, may cause serious problems in stream pollution and metal corrosion because of the acid ferric-iron-containing solutions they produce. Other iron bacteria can cause serious problems in public water supplies by their growth and obstruction of iron pipes and conduits. Sulfate-reducing bacteria can present a serious problem in iron pipe corrosion under anaerobic conditions, in the petroleum and natural gas industry.

H. L. EHRLICH

References

Aleem, M. I. H., Lees, H., and Nicholas, D. J. D., 1963, "Adenosine triphosphate-dependent reduction of nicotinamide adenine dinucleotide by ferro-cytochrome c in chemoautotrophic bacteria," *Nature,* **200**, 759–761.

*Baas-Becking, L. C. M., Kaplan, I. R., and Moore, D., 1960, "Limits of the natural environment in terms of pH and oxidation-reduction potentials," *J. Geol.,* **68**, 243–284.

Perfil'ev, B. V. D., Gabe, D. R., Gal'perina, A. M., Rabinovich, V. A., Sapotnitskii, A. A., Sherman, E. E., and Troshanov, E. P., 1965, "Applied Capillary Microscopy," New York, Consultants Bureau, 122pp. (English transl.).

*Silverman, M. P., and Ehrlich, H. L., 1964, "Microbial formation and degradation of minerals," *Advan. Appl. Microbiol.,* **6**, 153–206.

Starkey, R. L., and Halvorson, H. O., 1927, "Studies on the transformations of iron in nature. II. Concerning the importance of microorganisms in the solution and precipitation of iron," *Soil Sci.,* **26**, 381–402.

* Additional bibliographic references may be found in this work.

Cross-references: Catalysis; Chelation; Iron; Oxidation and Reduction; pH-Eh. Vol. VII: Algae; Microorganisms, Microbes.

"ISO" TERMS—*See* Vol. II

ISOTOPE FRACTIONATION

An element is a chemically discrete substance of which all atoms have the same atomic number (proton number or Z). Atoms are the fundamental building components of matter and those of each of the 103 elements are uniquely defined by the number of protons in their nuclei. For example, the nuclei of hydrogen, oxygen, sulfur, and lead atoms invariably contain 1, 8, 16, and 82 protons, respectively. However, the number of neutrons in the nuclei of atoms of nearly all elements may vary within narrow limits. As a consequence, the atoms of an element may exist as several distinct nuclear species, called isotopes, that necessarily have the same proton number (Z), but have different neutron numbers (N), and, accordingly, have different mass numbers (A, or $Z + N$). Prior to the discovery of isotopes by J. J. Thomson in 1913, their existence had been inferred from the fractional atomic weights exhibited by most elements relative to the masses of their component neutrons and protons.

Isotopes of the elements cited above, for

example, would include *hydrogen* as $_1H^1$ (protium, $_1D^2$ (deuterium), and $_1T^3$ (tritium); *oxygen* as $_8O^{16}$, $_8O^{17}$, and $_8O^{18}$; *sulfur* as $_{16}S^{32}$, $_{16}S^{33}$, $_{16}S^{34}$, and $_{16}S^{36}$; and *lead* as $_{82}Pb^{204}$, $_{82}Pb^{206}$, $_{82}Pb^{207}$, and $_{82}Pb^{208}$. Since the original investigations of J. J. Thomson and his colleague F. W. Aston, over 300 naturally occurring isotopes have been found for the elements ranging from 1 to 92 in atomic number (hydrogen to uranium). They include both stable and unstable (radioactive) isotopes that formed either with the original synthesis of the elements in the solar system or later via the decay of radioactive precursors.

Isotope fractionation refers to the unequal *partitioning* of the isotopes of an element between two or more phases during chemical, physical, and biological processes. The results of early investigations with primarily heavy elements suggested that the chemical properties were uniform for the various isotopes of an element. This preliminary conclusion was partly supported by the fact that the chemical properties of an element are determined by the number and arrangement of its orbital electrons. As the isotopes of an element have the same number of electrons that are arranged in a similar configuration, they should exhibit similar chemical behaviors although their nuclear masses and structures are different. This was only an approximation, however, for investigators later obtained experimental verification that differences in the chemical properties of hydrogen and deuterium compounds did exist and that these agreed with previous calculations based on statistical mechanical methods (Urey and Greiff, 1935). Isotopes of the light elements display significant differences in their properties, as demonstrated by fractionation effects obtained for those of hydrogen, carbon, nitrogen, oxygen, and sulfur, but the magnitude diminishes rapidly for elements of increasingly higher atomic weight. The classic reference, *Isotope Geology*, by Rankama (1954) provides excellent background on the history, theory, instrumentation, and applied and experimental results of isotopic investigations in the earth sciences.

Equilibrium Effects

The isotopes of a light element may be fractionated during equilibrium reactions that are governed by the law of mass action because of slight differences in their chemical properties. These differences, and those of affinity and bond energy, are related to differences in thermodynamic properties such as *enthalpy, entropy,* and *free energy* (q.v.). As the thermodynamic properties of atoms and molecules are related to their vibrational frequencies, which in turn are mass dependent, these interrelationships provide a theoretical foundation for the fractionation of isotopic compounds by chemical and physical processes.

The magnitude of fractionation for an equilibrium reaction may be obtained from the equilibrium constant that is derived from partition function ratios of the isotopic compounds involved. Methods of calculating partition functions and their ratios from spectral data are described by Urey and Greiff (1935), Bigeleisen and Mayer (1947), Urey (1947), and Tudge and Thode (1950). A typical exchange reaction may be illustrated by the equation

$$aA_1 + bB_2 \rightleftharpoons aA_2 + bB_1 \quad (1)$$

where A and B represent molecules with a common element and subscripts 1 and 2 designate the light and heavy isotopes respectively. The *equilibrium constant*, K, is given by

$$\ln K = -\frac{\Delta F^0}{RT} \quad (2)$$

or

$$\ln K = -\frac{aF_{A_2}^0 + bF_{B_1}^0 - aF_{A_1}^0 - bF_{B_2}^0}{RT} \quad (3)$$

where F^0 is the standard free energy. The free energy, F, may in turn be related to the partition function, Q, by the equation

$$F = E_0 + RT(\ln N - \ln Q) \quad (4)$$

where E_0 is the "zero point energy" of the molecule and N is Avogadro's number. Substituting Eq. (4) in Eq. (3), and simplifying, one obtains the relation

$$K = \left[\frac{Q_{A_2}}{Q_{A_1}}\right]^a \bigg/ \left[\frac{Q_{B_2}}{Q_{B_1}}\right]^b \exp \left[\frac{-aE_{0_{A2}} + bE_{0_{B1}} - aE_{0_{A1}} - bE_{0_{B2}}}{RT}\right] \quad (5)$$

If E_0 is taken as the bottom of the potential energy curve, rather than as the "zero point energy" for the molecule, the exponential term becomes unity as the potential energy curves for all molecules are similar, and Eq. (5) reduces to

$$K = \left[\frac{Q_{A_2}}{Q_{A_1}}\right]^a \bigg/ \left[\frac{Q_{B_2}}{Q_{B_1}}\right]^b \quad (6)$$

The partition function ratio of a compound is defined rigorously by the equation

$$\frac{Q_2}{Q_1} = \frac{\sigma_1}{\sigma_2}\left(\frac{M_2}{M_1}\right)^{3/2} \frac{\Sigma e^{-E_2/kT}}{\Sigma e^{-E_1/kT}} \quad (7)$$

where σ_1 and σ_2 are the symmetry numbers of the two molecules, M_1 and M_2 are their molecular weights, E_1 and E_2 are the energy states of the molecules, and the summations are taken over all possible energy states. Because of uncertainties for data needed for computations based on this and related equations, Urey (1947) has defined new partition functions, Q_1' and Q_2', by means of approximations, identities, and substitutions. The new partition functions are given by

$$\frac{Q_1'}{Q_2'} = \frac{Q_2}{Q_1}\left(\frac{m_1}{m_2}\right)^{3/2n} \quad (8)$$

Where m_1 and m_2 are the atomic weights and n is the number of the isotopic atoms being exchanged. The partition functions are equilibrium constants for exchange reactions between the compound considered and the separate atoms. As before, they relate to the equilibrium constant of Eq. (1) simply by

$$K = \left[\frac{Q_{A_2}'}{Q_{A_1}'}\right]^a \Big/ \left[\frac{Q_{B_2}'}{Q_{B_1}'}\right]^b \quad (9)$$

Although the calculation of a single partition function is difficult, by means of the foregoing transformation, the calculation of partition function ratios for isotopic molecules may be readily accomplished from a knowledge of vibrational frequencies alone. However, the calculated equilibrium constants for nearly all polyatomic molecules and ions differ slightly from experimentally determined values. These differences arise through errors and uncertainties that are inherent to the calculation and from the application of statistical mechanical formulas that assume ideal gas behavior to liquid and solid phases (Urey, 1947).

Although the equilibrium constant provides a means of estimating the distribution of isotopes of an element among the components of a reaction, the *fractionation factor* (*enrichment factor* or *separation factor*) is commonly of greater interest to the investigator. According to Urey (1947), the fractionation factor, α, may be defined as the overall ratio of the isotopes of an element in one chemical compound as compared to this ratio in a second chemical compound. The fractionation factor, α, is nearly identical to the equilibrium constant, K, as given by the equation

$$\alpha = K^{1/n} \quad (10)$$

where n is the maximum number of exchangeable isotopic atoms for any one of the compounds under consideration. If all components of an equilibrium reaction have just one exchangeable isotopic atom, the equivalency of terms is obvious ($\alpha = K$). However, Eq. (10) is not valid for isotopic compounds of hydrogen or for those in which the two or more isotopic atoms do not occupy equivalent positions in the compound.

Equilibrium constants for selected isotopic compounds of carbon, oxygen, and sulfur that have been derived from partition function ratios calculated by Urey (1947), Craig (1953), and Ault and Kulp (1960) are listed in Table 1. In general, the fractionation effects are small as is suggested by the predominance of values near unity (most hydrogen compounds, not shown, are exceptions). For equilibrium constants that are larger than unity, the heavy isotope is preferentially concentrated in compounds listed in the left-hand column of Table 1.

TABLE 1. CARBON ISOTOPE EXCHANGE EQUILIBRIA[a]

	$\dfrac{C^{13}O_3^{2-}}{C^{12}O_3^{2-}}$	$\dfrac{C^{13}O_2}{C^{12}O_2}$	$\dfrac{C_d^{13}}{C_d^{12}}$	$\dfrac{C^{13}H_4}{C^{12}H_4}$	T (°K)
$\dfrac{C^{13}O_3^{2-}}{C^{12}O_3^{2-}}$	1.000	1.016	1.023	1.087	273.1
		1.012	1.023	1.075	298.1
		1.004	1.018	1.042	400
		0.999₄	1.014	1.026	500
		0.997₅	1.011	1.016	600
$\dfrac{C^{13}O_2}{C^{12}O_2}$		1.000	1.007	1.070	273.1
			1.010	1.061	298.1
			1.014	1.038	400
			1.014	1.026	500
			1.013	1.019	600
$\dfrac{C_d^{13}}{C_d^{12}}$			1.000	1.063	273.1
				1.050	298.1
				1.024	400
				1.012	500
				1.006	600

[a] Urey, 1947; Craig, 1953.

TABLE 2. OXYGEN ISOTOPE EXCHANGE EQUILIBRIA[a]

	$\left(\dfrac{CO_2^{18}}{CO_2^{16}}\right)^{\frac{1}{2}}$	$\left(\dfrac{CO_3^{18(2-)}}{CO_3^{16(2-)}}\right)^{\frac{1}{3}}$	$\left(\dfrac{O_2^{18}}{O_2^{16}}\right)^{\frac{1}{2}}$	$\dfrac{H_2O^{18}}{H_2O^{16}}$	T (°K)
$\left(\dfrac{CO_2^{18}}{CO_2^{16}}\right)^{\frac{1}{2}}$	1.000	1.022	1.037	1.055	273.1
		1.021	1.033	1.047	298.1
		1.016	1.021	1.027	400
		1.013	1.016	1.017	500
		1.011	1.011	1.011	600
$\left(\dfrac{CO_3^{18(2-)}}{CO_3^{16(2-)}}\right)^{\frac{1}{3}}$		1.000	1.015	1.033	273.1
			1.012	1.026	298.1
			1.004	1.011	400
			1.001	1.004	500
			1.000	1.000$_4$	600
$\left(\dfrac{O_2^{18}}{O_2^{16}}\right)^{\frac{1}{2}}$			1.000	1.017	273.1
				1.014	298.1
				1.006	400
				1.002	500
				0.999$_9$	600

[a] Urey, 1947.

Several generalizations are relevant to the fractionation effects produced by equilibrium isotope-exchange reactions. For example, the extent of fractionation is proportional to the relative mass differences of the isotopic compounds. From a consideration of Eqs. (11) and (12), given by Tudge and Thode (1950),

$$S^{32}O_4^{2-} + H_2S^{34} \rightleftharpoons S^{34}O_4^{2-} + H_2S^{32}$$
$$K = 1.074 \text{ at } 25°C \quad (11)$$

$$S^{32}O_4^{2-} + H_2S^{36} \rightleftharpoons S^{36}O_4^{2-} + H_2S^{32}$$
$$K = 1.145 \text{ at } 25°C \quad (12)$$

it is obvious that the fractionation effect (not the K's) doubles as the percentage mass difference of the isotopes doubles. Accordingly, the equilibrium constants and fractionation effects for isotopic compounds of hydrogen are particularly large because of significantly larger mass differences (for example, 100% between H and D). More importantly, the partition function ratios, as given by Eq. (7), and equilibrium constants derived therefrom are temperature dependent. This temperature effect may be inferred from the values listed in Table 1. In general, the equilibrium constants for most isotopic-exchange reactions rapidly approach unity with increasingly higher temperatures of equilibration. This effect, by which the magnitude of fractionation varies inversely with temperature, is graphically portrayed in Fig. 1 for equilibria involv-

TABLE 3. SULFUR ISOTOPE EXCHANGE EQUILIBRIA[a]

	$\dfrac{S^{34}O_4^{2-}}{S^{32}O_4^{2-}}$	$\dfrac{S^{34}O_2}{S^{32}O_2}$	$\dfrac{H_2S^{34}}{H_2S^{32}}$	$\dfrac{S^{34(2-)}}{S^{32(2-)}}$	T (°K)
$\dfrac{S^{34}O_4^{2-}}{S^{32}O_4^{2-}}$	1.000	1.052	1.091	1.107	273
		1.039	1.073	1.088	298
		1.027	1.050	1.061	373
		1.015	1.027	1.034	523
		1.007	1.013	1.016	723
$\dfrac{S^{34}O_2}{S^{32}O_2}$		1.000	1.038	1.053	273
			1.033	1.046$_5$	298
			1.023	1.032$_9$	373
			1.012	1.018$_7$	523
			1.006	1.009$_3$	723
$\dfrac{H_2S^{34}}{H_2S^{32}}$			1.000	1.015	273
				1.013	298
				1.010	373
				1.006	523
				1.004	723

[a] Ault and Kulp, 1960.

ISOTOPE FRACTIONATION

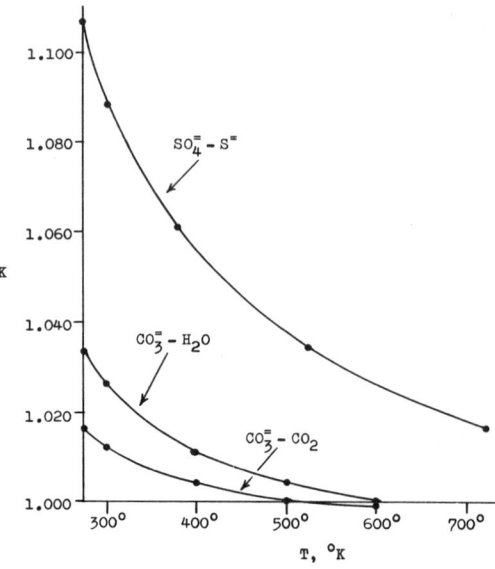

FIG. 1. Variation of equilibrium constants with temperature for the isotopic compounds carbonate ion-carbon dioxide (C^{13}/C^{12}), carbonate ion-water (O^{18}/O^{16}), and sulfate ion-sulfide ion (S^{34}/S^{32}).

ing common isotopic compounds of carbon, oxygen, and sulfur. The temperature dependency of equilibrium constants and attendant fractionation effects for these and other reactions provides the theoretical basis on which methods of isotope geothermometry are founded. Additionally, the fractionation effects for some compounds, notably those of carbon, chlorine, and sulfur, are partly dependent on the valence state of the isotopic element. The equilibrium constant generally increases with increasing valence dissimilarity between the isotopes of the element in the equilibrating compounds. From equilibrium constants given for carbonate ion-methane and sulfate ion-sulfide ion in Table 1, it may be seen that the heavier isotopes of carbon and sulfur are preferentially concentrated in the more oxidized components of these reactions to which they are more strongly bonded. This valence or redox effect partly accounts for the appreciable fractionation that has been observed for isotopes of the relatively heavier elements in nature (for example, sulfur).

In summary, the magnitude of fractionation attributable to isotope-exchange reactions is inversely proportional to temperature and is directly proportional to the relative mass differences of the isotopic compounds, the valence dissimilarity of the isotopic element between the compounds, and the extent to which equilibrium prevails.

Kinetic Effects

Although differences in affinity and equilibrium between isotopes are static, there are also dynamic differences that are related to the kinetics of chemical reactions. Under nonequilibrium conditions, a reaction that proceeds in one direction to completion is unidirectional. Isotope fractionation may arise from such processes provided that the reactants are not entirely consumed. As previously noted, the thermodynamic properties of substances depend on the vibrational frequencies of the molecules, and this property is related to the mass of the component atoms in the molecules. Additionally, the binding energy of atoms in molecules partly depends on the mass of the atoms. At moderate temperatures this mass effect is related to differences in the zero-point energies that are the residual vibrational energies of diatomic and polyatomic molecules at absolute zero. As heavy atoms have lower zero-point energies than light atoms, and consequently smaller frequencies of vibration, it follows that similar although lesser differences exist between the isotopes of an element. Therefore, isotope fractionation may take place during unidirectional chemical reactions because of differences in the reaction rates of the reactant isotopic molecules.

The theory of isotopic fractionation in unidirectional reactions has been treated by Bigeleisen (1949). The equation

$$\frac{k_1}{k_2} = \frac{C_1}{C_2} \frac{Q_1^*}{Q_2^*} \cdot \frac{Q_{A2}}{Q_{A1}} \left(\frac{m_2^*}{m_1^*}\right)^{\frac{1}{2}} \quad (13)$$

where the k's are the specific rate constants, C is the transmission coefficient, Q^* and Q_A are the partition functions of the activated complex and reactant, respectively, m^* is the effective mass of the activated complex, and subscripts 1 and 2 denote the light and heavy isotopic molecules, relates the ratio of rate constants for competitive reactions to chemical differences of the isotopic molecules as determined by their partition function ratios.

Because the identities of the activated complexes are rarely known, and calculations of their effective masses and partition function ratios accordingly are uncertain or impossible, Eq. (13) merely serves to qualitatively approximate the fractionation effect that accompanies unidirectional reactions. However, the $(m_2^*/m_1^*)^{\frac{1}{2}}$ term is always equal to or larger than unity and the activation energy of the reactant molecule is generally larger than that of the activated complex. As a consequence, and except for compounds of hydrogen, the larger reaction rate will normally favor the isotopically lighter reactant molecules (Bigeleisen, 1949). For example, kinetic fractionation factors of up to 1.062, and with S^{32} concentrated in the products, have been obtained experimentally from organic and inorganic reactions of sulfur compounds. In general, the magnitude of kinetic isotope fractionation varies inversely with temperature, the mass of the isotopic element, and the completeness of the reaction.

Physical Effects

In accordance with energy considerations, the mass differences that are inherent to isotopic compounds may similarly give rise to fractionation effects during physical processes. These may occur in response to either equilibrium or unidirectional conditions. The principles are employed in centrifugation, diffusion, distillation, and electromagnetic methods of separating isotopic compounds for industrial and research purposes.

Little is known at present concerning the relative importance of the various physical processes in causing isotope fractionation in most geologic environments. However, the available evidence suggests that isotopic effects related to chemical reactions are more important. Fractionation effects that have been determined for isotopic compounds of hydrogen, nitrogen, and oxygen in processes of condensation, evaporation, and diffusion are in agreement with those predicted from theory. For example, a compound that equilibrates as two distinct phases within a system will preferentially concentrate the isotopically lighter species in the less dense phase (liquid or gas). Accordingly, the nearly consistent depletion of water vapor in O^{18} and D relative to its ocean source implies that equilibrium prevails throughout most evaporation-condensation processes of the meteorological cycle.

Diffusion (q.v.) is the movement of a solute, as ions, atoms, or molecules, in a solvent phase under the driving force of a concentration or chemical potential gradient. Considered in terms of geologic processes, the importance of which remains unresolved, diffusion may be subdivided into two types. The first is solid-state diffusion, which involves the movement of a solute along grain boundaries and surfaces or within a crystal lattice, and the second is solute diffusion, which refers to the movement of a solute in a liquid medium.

The role of diffusion as a mechanism effecting isotope fractionation has been investigated by Sentfle and Bracken (1955). For the diffusion of atoms or ions from an interface under semiinfinite boundary conditions, the solute concentration, C_x, at any distance x from the interface is given by

$$C_x = C_0 \left(1 - erf \frac{x}{2\sqrt{Dt}} \right) \quad (14)$$

where C_0 is the solute concentration at the interface, D is the average diffusion coefficient of the solute, t is the time during which diffusion occurs, and *erf* is the error function. According to Graham's law, the comparative rates of diffusion (effusion), k, for isotopic species of a solute are inversely proportional to the square root of their masses, m_1 and m_2, as follows

$$k = \frac{D_1}{D_2} = \sqrt{\frac{m_2}{m_1}} \quad (15)$$

For hydrogen sulfide k equals 1.029 and, therefore, H_2S^{32} diffuses 2.9 % more rapidly than H_2S^{34}. The isotopic enrichment, δ in percent, at a distance x from the interface is

$$\delta(\%) = \left(\frac{C_{1_x}}{C_{1_0}} \Big/ \frac{C_{2_x}}{C_{2_0}} - 1 \right) \cdot 10^2 = \left(\left[\frac{1 - erf \frac{x}{2\sqrt{kD_2 t}}}{1 - erf \frac{x}{2\sqrt{D_2 t}}} \right] - 1 \right) \cdot 10^2 \quad (16)$$

Although the above considerations apply to both solid-state and solute diffusion, the diffusion coefficients of the latter may be several orders of magnitude larger. Unlike isotopic fractionation in chemical reactions, the extent of enrichment during processes of diffusion is related to the distance of mass transport and generally increases with increasing temperature. Because both concentration and isotopic enrichment at any point are functions of diffusion distance, little fractionation will occur until the concentration of solute at the diffusion front is less than 10% of that at its source. Accordingly, Senftle and Bracken (1955) have concluded that isotopic fractionation by diffusion is likely to be significant only for minor elements of a crystal or mineral, and because diffusion is normally a local phenomenon, fractionation effects for a major element are unlikely to be detected.

Presentation of Analytical Data

Although equilibrium constants given in Table 1 and related fractionation factors are near unity, present analytical methods are capable of resolving isotopic variations as small as ± 0.0001. After conversion of the isotopic element in the sample to a compound amenable for analysis, the ratio of heavy to light isotopes of the element, R, is determined by mass spectrometry. As the absolute abundances of the isotopes are rarely known with great accuracy, the method attains a much higher order of precision by means of comparative ratio analyses between the sample and a standard. The ratio difference between sample and standard, which is a measure of differing isotopic composition, is normally expressed as a deviation from the standard by delta values, δ, in permil, ‰. The ratios, R, are related to the delta permil values, δ, by the equation

$$\delta‰ = \left(\frac{R_{sample}}{R_{standard}} - 1\right) \cdot 10^3 \quad (17)$$

Enrichment or depletion of the heavier isotope in the sample, relative to the standard, is indicated by positive or negative permil values respectively. The fractionation factor, α, and equilibrium constant, K, may be calculated from the measured delta permil values for two mineral phases, A and B, as follows

$$\alpha = K^{1/n} = \frac{1 + \frac{\delta A‰}{1000}}{1 + \frac{\delta B‰}{1000}} \quad (18)$$

Other approximations and identities that are commonly used to portray isotopic analytical data include

$$1000 \ln \alpha \approx \delta A‰ - \delta B‰ = \Delta_{AB} \quad (19)$$

and

$$\alpha \approx 1 + \frac{\delta A‰ - \delta B‰}{1000} \quad (20)$$

These interrelationships (Eqs. (10), (18), (19), and (20)), between the fractionation effects determined from naturally occurring samples (Eq. (17)) and those predicted from theory (Eq. (9) and Tables 1–3), provide a useful foundation on which preliminary interpretations of isotopic abundance data can be made.

Cyrus W. Field

References

Ault, W. U., and Kulp, J. L., 1960, "Sulfur isotopes and ore deposits," *Econ. Geol.*, **55**, 73–100.
Bigeleisen, J., 1949, "The relative reaction velocities of isotopic molecules," *J. Chem. Phys.*, **17**, 675–678.
Bigeleisen, J., and Mayer, M. G., 1947, "Calculation of equilibrium constants for isotopic exchange reactions," *J. Chem. Phys.*, **15**, 261–267.
Craig, H., 1953, "The geochemistry of stable carbon isotopes," *Geochim. Cosmochim. Acta*, **3**, 53–92.
Craig, H., Miller, S. L., and Wasserburg, G. J., 1964, in "Isotope and Cosmic Chemistry," (dedicated to H. Urey), Amsterdam, North Holland Publ., 533pp.
Fontes, J. C., 1966, "Fractionnement isotopique dans l'eau de cristallisation du sulfate de calcium," *Geol. Rund.*, **55**(1), 172–178.
Rankama, K., 1954, "Isotope Geology," New York, Pergamon Press, 535pp.
Senftle, F. E., and Bracken, J. T., 1955, "Theoretical effect of diffusion on isotopic abundance ratios in rocks and associated fluids," *Geochim. Cosmochim. Acta*, **7**, 61–76.
Tudge, A. P., and Thode, H. G., 1950, "Thermodynamic properties of isotopic compounds of sulphur," *Can. J. Res.*, **28B**, 567–578.
Urey, H. C., 1947, "The thermodynamic properties of isotopic substances," *J. Chem. Soc. (London)*, **1947**, 562–581.
Urey, H. C., and Greiff, L. J., 1935, "Isotopic exchange equilibria," *J. Am. Chem. Soc.*, **57**, 321–327.

Cross-references: Atomic Number and Periodic Table; Carbon Isotope Fractionation; Deuterium; Diffusion, Geological Role; Enthalpy or Heat Content; Entropy; Free Energy and Free Enthalpy; Hydrogen; Lead; Mass Action and Equilibrium Constant; Mass Spectrometry; Oxygen Isotope Geochemistry; Partition Coefficients; Radioactive Isotopes; Sulfur Isotope Fractionation; Thermodynamics; Tritium.

ISOTOPE FRACTIONATION IN THE OCEAN—See Vol. I

ISOTOPE GEOLOGY—STABLE

Rankama (1954) has defined isotope geology as the investigation of geologic phenomena by means of stable and unstable isotopes of the elements and of changes in their abundance. The stable isotopes are those whose absolute abundances do not change with time as a consequence of radioactive decay. The stable isotopes of the lighter elements, however, may exhibit measurable variations in abundance from one geologic environment to another. Because isotopic fractionation is both predictable from theory and demonstrable by experiment the observed variations provide qualitative and quantitative insight with respect to the conditions and importance of chemical, physical, and biological processes that have been operative in the various geologic environments. Since K. Rankama's monumental treatise in 1954, isotope geology has contributed materially to a more detailed knowledge of many geologic phenomena through a combination of theoretical, experimental, and applied studies relevant to stable isotope distributions. Reviews by Ault (1959), Ault and Kulp (1959), Craig (1953), Epstein (1959), Epstein and Taylor (1967), Holser and Kaplan (1966), Jensen (1967), Taylor (1967a; 1967b), and Thode and others (1961) offer excellent background concerning these investigations.

Conventions

Isotopic abundances are determined by mass spectrometry from comparative ratio analyses of the samples relative to a standard. The ratio normally measured is that of the heavier, less abundant isotope to the lighter more abundant isotope; for example D/H, C^{13}/C^{12}, O^{18}/O^{16}, and S^{34}/S^{32}. The choice of the heavier isotope for elements such as oxygen and sulfur that consist of more than two isotopes is usually a compromise between largest mass difference and relative abundance. Differences in the measured ratios between sample and standard are a function of differing isotopic abundances. These differences are expressed as delta (δ) deviations of the heavier isotope (D, C^{13}, O^{18}, or S^{34}) in the sample relative to the standard as permil ($^0/oo$ or parts per thousand) values. Positive or negative delta permil values denote enrichment or depletion of the heavier isotope in the sample relative to the standard. The standards on which the comparative analyses are based facilitate interlaboratory comparisons of similar isotope data and they include Standard Mean Ocean Water (SMOW) for hydrogen and oxygen (see Taylor, 1967a), the Chicago PDB standard (Cretaceous belemnite from the Peedee formation, South Carolina) for carbon, and troilite (FeS) from the Canyon Diablo meteorite for sulfur. Delta values for these standards are zero permil by definition. Additional details concerning the isotopic standards and interrelationships of ratios (R), delta permil values ($\delta\permil$), equilibrium constants (K), fractionation factors (α), and common identities may be found in the references previously cited.

General Isotopic Distributions

Isotopic abundances of carbon, oxygen, and sulfur are summarized in Figs. 1, 2, and 3. Although these data are representative of only part of isotopic research in the past two decades, they serve to define the pronounced variability of isotopic abundances and the commonly systematic character of these variations with respect to particular compounds and geologic environments. Thus, the isotopic compositions of minerals and rocks are clearly dependent on their origin and subsequent history.

The isotopic data for a number of carbon-bearing compounds are illustrated in Fig. 1 (Craig, 1953). Carbonates, and especially

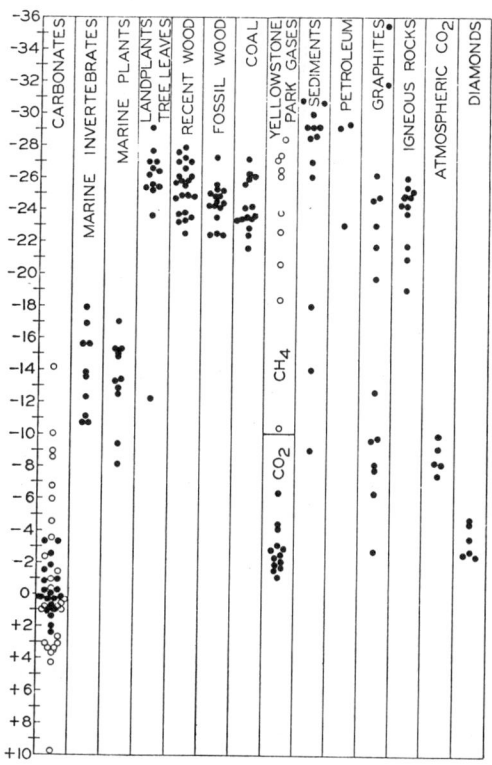

FIG. 1. Isotopic abundances of carbon $\delta C^{13}\permil$ (Craig, 1953, Fig. 3).

those of marine origin (closed circles), are enriched in C^{13} relative to organic carbon that occurs in both recent flora and fauna and ancient fossilized wood and coal. Moreover, isotopically light sedimentary carbon is formed by contributions of C^{13}-depleted organic carbon and the relatively large isotopic variability of carbonates is partly attributable to organic effects. Atmospheric carbon dioxide is slightly depleted in C^{13} relative to carbonate, as predicted from theory and verified by experiment, although contamination by organic fuels may have enhanced the effect as illustrated in Fig. 1. In general, dolomites are isotopically similar to limestones and the metamorphism of carbonates does not produce notable fractionation. Additionally, the isotopic abundance of carbon in minerals, rocks, and organic matter does not exhibit an age effect with respect to geologic time.

Oxygen isotope data are voluminous as is consistent with the abundance and wide distribution of this element. As the O^{18} content of minerals normally increases with increasing strength of the oxygen-cation bond it is preferentially concentrated in quartz and dolomite relative to other silicates and car-

619

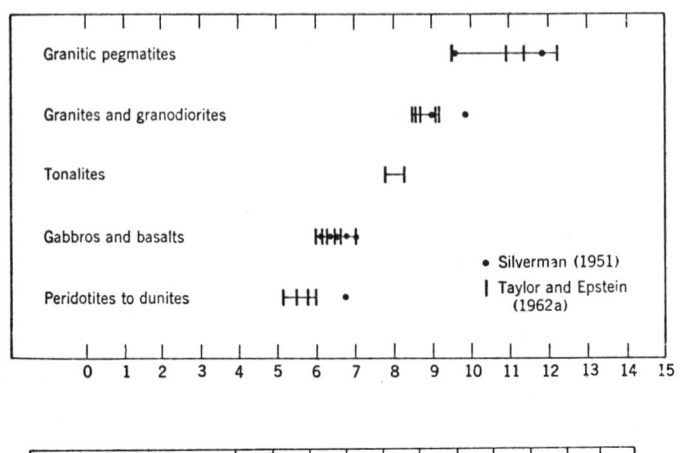

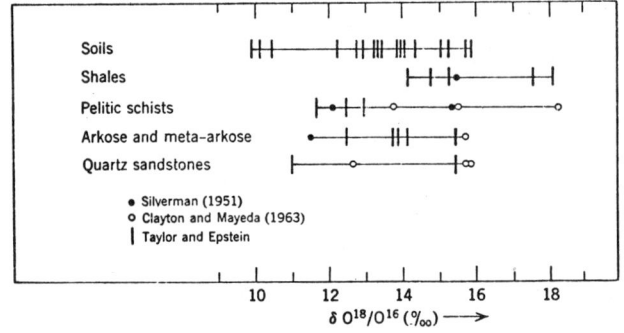

Fig. 2. Isotopic abundances of oxygen as δO^{18}‰ (Epstein and Taylor, 1967, Figs. 3 and 14).

bonates. According to Epstein and Taylor (1967) and Taylor (1967a), the order of progressive O^{18} enrichment for equilibrium exchange reactions involving the common rock-forming minerals is magnetite, chlorite, biotite, olivine, hornblende, pyroxene, muscovite, plagioclase feldspar, calcite, alkali feldspar, dolomite, and quartz for a total range of nearly 11‰. This mineralogical effect is evident from the igneous whole-rock analyses shown in Fig. 2, after Epstein and Taylor (1967), as the more silicic lithologies are progressively enriched in O^{18}. The direction of oxygen isotopic trends with magmatic differentiation, however, may vary locally in response to such factors as composition, disequilibrium, and water content. Sedimentary rocks are isotopically more variable because of the diverse genesis and mineralogy of their components. The temperature dependency of isotopic fractionation is manifested by O^{18}-enriched authigenic minerals as is illustrated by the whole-rock data for shales in Fig. 2. Marine carbonates and cherts are markedly enriched in O^{18} by as much as approximately +27 and +32‰, respectively, although this effect is obscured in samples older than Cretaceous because of contamination by exchange with O^{18}-depleted meteoric waters. With metamorphism, the O^{18} content of igneous rocks may increase whereas that of sedimentary rocks may decrease. However, the isotopic record of a metamorphic event is normally complex as it is governed by both the kinds and proportions of the original country rocks and by the extent to which equilibrium prevails.

Variations in the S^{34} content of naturally occurring sulfur compounds exceed 110‰ (11%). This broad range, as shown in Fig. 3, is attributable to kinetic fractionation effects and reservoir constraints that are associated with the bacterial reduction of sulfate to sulfide in near-surface environments. Isotopic uniformity that characterizes individual occurrences of magmatic and hydrothermal sulfides is consistent with their high temperatures of formation. Gross compositional variations, relative to meteoritic sulfur, for such deposits arise from conditions of disequilibrium or assimilation. Evidence suggests that S^{34} is slightly enriched in sulfides that are formed during successively later stages of magmatic differentiation. Present day seawater is iso-

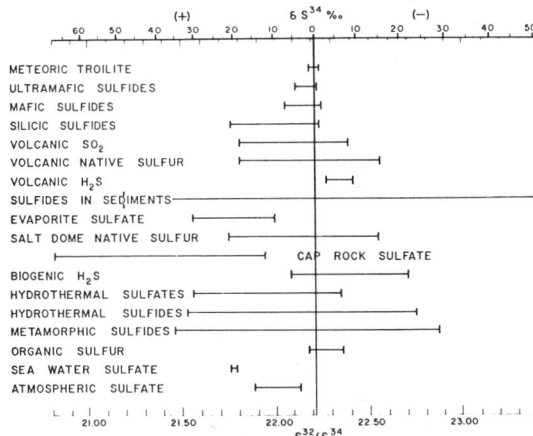

FIG. 3. Isotopic abundances of sulfur as δS^{34}‰ and S^{32}/S^{34} (Field, compiled from literature through 1965).

topically uniform at +20‰ and recent evaporites are compositionally similar. Nonetheless, since Precambrian time the isotopic composition of seawater sulfate, as deduced from evaporites, has ranged from +10 to +30‰ as a result of variable influx and sedimentation rates for S^{34}-enriched sulfate and S^{34}-depleted sulfide.

Applications

Coexisting mineral pairs are potentially useful geothermometers because the fractionation effects between isotopic compounds are temperature dependent. Oxygen-isotope geothermometry has achieved considerable success in studies of paleotemperatures and igneous, metamorphic, and hydrothermal environments because of the ubiquitous distribution of oxygen-bearing minerals, the facility with which they equilibrate, and the experimentally determined calibration curves for many common minerals. Fractionation trends provided by the calibration curves and combined with the measured fractionation factor for two co-existing minerals ideally permit calculation of both the isotopic temperature and composition of the depositional media. Recent studies by Sakai (1968) and other investigators suggest that sulfur-isotope geothermometry may be applicable to ore deposits but this technique must await development of the appropriate calibration curves for the various sulfide minerals.

As the isotopic abundance of an element may grossly characterize a particular geologic environment, such data may serve to index the sources from which and the processes by which minerals and rocks have formed. For example, the large depletion of petroleum in C^{13} points unequivocally to a derivation from organic precursors. Combined studies of carbon and sulfur isotope abundances by Thode and others (1954) and later investigators dramatically proved the biogenic origin of salt dome native sulfur deposits and their associated calcite cap rock hosts; native sulfur depleted in S^{34} relative to its evaporite source and calcite of the cap rock isotopically similar to its C^{13}-depleted hydrocarbon source (Figs. 1 and 3). Laboratory investigations of the bacterial reduction of sulfate to hydrogen sulfide and attendant fractionation effects have verified the effectiveness of this mechanism to produce sulfides that are variably depleted in S^{34} and which are common to recent muds and older sedimentary rocks (Jensen, 1967). According to the review by Taylor (1967b), similarities in the measured O^{18} contents of dolomites and limestones, which are contrary to fractionation effects obtained experimentally from equilibrium assemblages, suggest that much dolomite has formed secondarily by replacement of calcite.

Although isotopically variable, the D and O^{18} contents of surface water and atmospheric precipitation exhibit a linear and covariant relationship. Craig (1966) and other investigators have used this striking correlation to identify the source of geothermal brine and saline formation waters as being of meteoric origin; not of juvenile or connate origin as had been previously assumed. Furthermore, the O^{18}-depleted vein minerals at Butte, Montana, suggest at least partial mixing of the hydrothermal fluids with meteoric waters. Forthcoming hydrogen, oxygen, and sulfur isotope investigations will undoubtedly define the importance of meteoric waters in hydrothermal systems. Moreover, they should provide a means by which the hypogene and supergene constituents of mineralogically complex alteration assemblages may be distinguished. Hydrothermal deposits exhibit weak isotopic gradients for oxygen and sulfur (Field, 1966; Taylor, 1967a and 1967b). Although the expression of these trends is not distinct, additional refinement of existing sampling, mineral separation, and analytical procedures may encourage more widespread application of isotopic surveys to mineral exploration.

Cyrus W. Field

References

Ault, W. U., 1959, "Isotopic fractionation of sulfur in geochemical processes," in (Abelson, P. H., editor), "Researches in Geochemistry," New York, John Wiley & Sons, pp. 241–259.

Ault, W. U., and Kulp, J. L., 1959, "Isotopic

geochemistry of sulphur," *Geochim. Cosmochim. Acta,* **16,** 201–235.

Craig, H., 1953, "The geochemistry of the stable carbon isotopes," *Geochim. Cosmochim. Acta,* **3,** 53–92.

Craig, H., 1966, "Isotopic composition and origin of the Red Sea and Salton Sea geothermal brines," *Science,* **154,** 1544–1548.

Epstein, S., 1959, "The variations of the O^{18}/O^{16} ratio in nature and some geologic implications," in (Abelson, P. H., editor), "Researches in Geochemistry," New York, John Wiley & Sons, pp. 217–240.

Epstein, S., and Taylor, H. P., Jr., 1967, "Variation of O^{18}/O^{16} in minerals and rocks," in (Abelson, P. H., editor), "Researches in Geochemistry," New York, John Wiley & Sons, pp. 29–62.

Field, C. W., 1966, "Sulfur isotope abundance data, Bingham district, Utah," *Econ. Geol.,* **61,** 850–871

Holser, W. T., and Kaplan, I. R., 1966, "Isotope geochemistry of sedimentary sulfates," *Chem. Geol.,* **1,** 93–135.

Jensen, M. L., 1967, "Sulfur isotopes and mineral genesis," in (Barnes, H. L., editor), "Geochemistry of Hydrothermal Ore Deposits," New York, Holt, Rinehart, and Winston, pp. 143–165.

Rankama, K., 1954, "Isotope Geology," New York, Pergamon Press, 535pp.

Sakai, H., 1968, "Isotopic properties of sulfur compounds in hydrothermal processes," *Geochem. J. (Japan),* **2,** 29–49.

Taylor, H. P., Jr., 1967a, "Oxygen isotope studies of hydrothermal mineral deposits," in (Barnes, H. L., editor), "Geochemistry of Hydrothermal Ore Deposits," New York, Holt, Rinehart, and Winston, pp. 109–142.

Taylor, H. P., Jr., 1967b, "Stable isotopes," *Trans. Am. Geophys. U.,* **48,** 686–693.

Thorde, H. G., Monster, J., and Dunford, H. B., 1961, "Sulphur isotope geochemistry," *Geochim. Cosmochim. Acta,* **25,** 159–174.

Thode, H. G., Wanless, R. K., and Wallouch, R., 1954, "The origin of native sulfur deposits from isotope fractionation studies," *Geochim. Cosmochim. Acta,* **5,** 288–298.

Cross-references: *Carbon-14 Dating; Carbon Isotope Fractionation; Geochronometry; Isotope Fractionation; Mass Spectrometry; Meteoric Water; Oxygen Isotope Geochemistry; Paleotemperatures—Isotopic Determination; Potassium-Argon Age Determination; Protactinium-Thorium Dating Method; Rubidium-Strontium Dating Method; Sulfur Isotope Fractionation; Uranium-Thorium-Lead Age Determination.*

ISOTOPE GLACIOLOGY—See Vol. VI, GLACIOLOGICAL ISOTOPE STUDIES

ISOTOPE TRACER TECHNOLOGY—See Vol. VI, RADIOACTIVE ISOTOPE TRACER TECHNOLOGY

ISOTOPE PALEOTEMPERATURES—See PALEOTEMPERATURES—ISOTOPIC DETERMINATION

ISOTOPE VARIATIONS IN MINERAL DEPOSITS—See Vol. IV B

J

JUVENILE WATER

The term *juvenile water* refers to water which has been derived from the crustal rocks or from the interior of the earth, and at the time of its appearance in the circulating water of the hydrosphere represents an accretion to the available water supply. Juvenile water is therefore water which has not previously been a part of the hydrosphere.

Although it is simple enough to define what is meant by the term, the practical application of the definition of juvenile water is another matter. Some geochemists ascribe much of the volume of water now present in the hydrosphere to slow accretion of juvenile water throughout much of geologic time. This implies that a fraction of the water in hot springs and of that present in volcanic gases may even now still be of juvenile origin. A rival hypothesis, which probably has a larger number of supporters, is that the hydrosphere for the most part came into existence in very early geologic time and has not changed greatly in volume since then.

Although many investigators have tried to devise means for identifying juvenile water, none of the procedures have been successful in demonstrating beyond doubt that any significant fraction of the water discharged by hot springs, geysers, or fumaroles is of juvenile origin. It is, of course, very difficult to be certain that a particular water has never been at the surface of the earth before, and it is this feature of the definition that causes the principal difficulty.

Magmatic water is water derived from a magma, or included in a magma (rock melt). There may be a fraction of juvenile water in a magmatic water but the definition is not concerned with this. Thermal water in volcanic regions probably contains a fraction of magmatic water. White (1957) however reported that the contribution from this source was probably less than 5% in the typical thermal springs of the United States.

Craig, Boato, and White (1956) reported that the heavy isotope of oxygen, O^{18}, was enriched in geothermal water, and probably one of the better methods for identifying magmatic water is the determination of the proportion of O^{18} to O^{16}. However, the enrichment of O^{18} is not very great, and the possible exchange of the heavier for the lighter isotope as the water comes in contact with oxygen held in the crystal lattices of rock minerals injects a good deal of uncertainty into the calculation of percentages of magmatic water that might be present in a water sample.

White (1957) suggested that magmatic waters tend to contain high proportions of sodium and chloride. The analysis of a sample from a hot spring in Yellowstone National Park, which is included in a table under the topic *Aqueous Solutions* (q.v.), exhibits some of the chemical properties believed by White to be characteristic of solutions that contain some magmatic water. It seems likely that similar properties would be found in juvenile water. This subject is also discussed in this volume under the topic *Magmatic Water*.

JOHN D. HEM

References

Craig, H., Boato, G., and White, D. E., 1956, "Isotopic geochemistry of thermal waters," *Proc. 2nd Conf. Nuclear Processes in Geologic Settings, Publ.* **400,** Natl. Acad. Sci., Natl. Res. Coun., pp. 29-38.

White, D. E., 1957, "Thermal waters of volcanic origin," *Bull. Geol. Soc. Am.,* **68,** 1637-1658.

Cross-references: *Aqueous Solutions; Gases, Volcanic; Hydrosphere; Magmatic Water; Springs.* Vol. V: *Fumarole.*

K

K CAPTURE—*See* **ELECTRON CAPTURE**

KARA-BOGAZ GULF—*See* Vol. III

KARST TERRAIN, GROUNDWATER, HYDROLOGY—*See* **HYDROLOGY, LIMESTONE TERRAIN**

KEROGEN—*See* Vol. IV B, **KEROGEN; OIL SHALES**

KONIOLOGY (CONIOLOGY)—*See* Vol. II

KRAKATOA WINDS—*See* Vol. II

KRYPTON: ELEMENT AND GEOCHEMISTRY

Like most of the other noble gases, krypton (from the Greek: κρυπτοσ—secret, hidden), chemical symbol: Kr, is a chemically inactive, colorless, odorless, monatomic, and normally gaseous element with atomic number 36. It was discovered in 1898, together with neon and xenon, by W. Ramsay and M. W. Travers as a trace constituent in the troposphere. There are six stable isotopes of Kr with the mass numbers 78, 80, 82, 83, 84, and 86. In addition, two radioactive isotopes of Kr are of importance in geochemistry: Kr^{81} with a half-life of 2.1×10^5 years, produced in the atmosphere by cosmic rays, and Kr^{85} with a half-life of 10.3 years, a fission product introduced into the atmosphere mainly by nuclear bombs. The abundance of Kr in dry tropospheric air is 1.14×10^{-4} percent by volume. Kr, like other noble gases, is utilized in electric discharge tubes for advertising and airfield beacons. In high-power incandescent lamps, a mixture of Kr and Xe is used to prolong filament life and increase light output. Radioactive Kr^{85}, a β emitter, may find future application as a direct energy source for fluorescent light.

Properties

The internationally accepted atomic weight for Kr of tropospheric composition (Kr^n) is 83.80, corresponding to a density of gaseous Kr^n under standard conditions (0°C and 760 mm Hg) of 3.749 g/l.

At the triple point (t.p.), the properties of Kr are: T (t.p.) = 115.78°K; P (t.p.) = 548.7 mm Hg; density of solid, ρ_s (t.p.) = 2.826 g/cm³; density of liquid, ρ_L (t.p.) = 2.451 g/cm³. Critical values: $T_c = 209.4°K$; $P_c = 54.3$ atm. The boiling point is 119.81°K. For a comprehensive review and discussion of noble gas data up to the triple point, see Pollack (1964).

Chemical Behavior

Kr may react with the most electronegative elements, primarily fluoride F and oxygen O. However, only KrF_2 has been positively identified as a (solid) compound, stable up to dry ice temperature. KrF_2 has been prepared from Kr and F_2 by irradiation with photons, electrons, or protons.

Occurrence

Atmosphere. The isotopic composition of Kr^n was determined by A. O. C. Nier (1950) who gives the following percentage abundances:

$Kr^{78} = 0.354 \pm 0.002$ $Kr^{80} = 2.27 \pm 0.01$
$Kr^{82} = 11.56 \pm 0.02$
$Kr^{83} = 11.55 \pm 0.02$ $Kr^{84} = 56.90 \pm 0.1$
$Kr^{86} = 17.37 \pm 0.02$

Kr was probably accreted with the rest of the terrestrial matter.

Hydrosphere. Oceanic waters contain dissolved Kr^n (not distinguishable from tropospheric Kr) in amounts varying from 4.2×10^{-5} ml(STP)/liter to 9.5×10^{-5} ml(STP)/liter (Bieri et al., 1966), determined mainly by the temperature dependence of solubility. However, relatively large deviations may be present where mixing of different water masses occurs; such effects can only be more pronounced for Xe.

Lithosphere. Evidence for Kr produced by spontaneous fission of U-238 and Th-232 has been found in the mineral *euxenite,* and for additional neutron-induced fission in pitchblende from the Congo (Wetherill and Inghram, 1953). Kr of normal composition, but in comparatively large amounts (1.1×10^{-6} cm³/g (at STP)), have been found in a sample of *thucholite* from the Besner mine in Ontario. Fission-produced Kr is not likely to have contributed significantly to the Kr in the atmosphere.

Meteorites. Meteorites contain Kr in widely varying amounts and of clearly different origins. Trapped or "primordial" Kr with a composition close to Kr^n is most commonly found, but spallogenic Kr or Kr produced by (n, α) reactions on Br or I is also fairly well established. The recently discovered cosmic ray-produced Kr^{81} is important for the determination of a radiation age of meteorites. For more details and references, see for instance, Marti et al., (1966).

RUDOLF H. BIERI

References

Bieri, R. H., Goldberg, E. D., and Koide, M., 1966, "The Noble gas contents of Pacific seawaters," *J. Geophys. Res.,* **71,** 5243–5265.

Marti, K., Eberhardt, P., and Geiss, J., 1966, "Spallation, fission and neutron capture anomalies in meteoritic krypton and xenon," *Z. Naturforsch.* **21a,** 398–413.

Nier, A. O., 1950, "A redetermination of the relative abundances of the isotopes of neon, krypton, rubidium, xenon, and mercury," *Phys. Rev.,* **78,** 450–454.

Pollack, G. L., 1964, "The solid state of rare gases," *Rev. Mod. Phys.* **36,** 748–791.

Wetherill, G. W., and Inghram, M. G., 1953, "Spontaneous fission in uranium and thorium," *Nuclear Proc. in Geol. Settings,* 30–32.

Cross-references: *Neon: Element and Geochemistry; Radioactive Isotopes; Rare Gases; Xenon: Element and Geochemistry.* Vol. II: *Meteorites; Troposphere.*

LAGOON GEOCHEMISTRY

A lagoon is a shallow body of salt water, partly isolated from the sea by a sand barrier or a coral reef. It communicates with the sea through inlets or "passes" which may be permanently, seasonally, or exceptionally open. Lagoon water is usually the result of the mixing of continental and sea waters, i.e., brackish water. When seawater or brackish water concentrates under wind or solar evaporation, it becomes a brine. These mechanisms control the rich organic life of brackish environment as well ultimately as the aridity of hypersaline conditions.

Origin of Lagoon Waters

Seawater comes into the lagoon through one or several inlets, or it percolates through the sand barrier. The amount of seawater in a lagoon is controlled by the width, the depth, and the number of inlets, by the permeability of the barrier and by anything liable to change the sea level. Tides, winds, and hurricanes can effect short-period changes in sea level, isostasy and eustasy cause long-period changes. In pure seawater, the ions are in a constant ratio so that salinity of a sample can be stated immediately when its chlorinity is known. This is not true in lagoon water where more reliable values are obtained from conductivity or density measurements.

TABLE 1. PERCENTAGE COMPOSITION OF DISSOLVED SOLIDS IN RIVER AND SEAWATER[a]

Ion	River Water (weighted average)	Seawater
CO_3^{2-}	35.15	0.41 (HCO_3^-)
SO_4^{2-}	12.14	7.68
Cl^-	5.68	55.04
NO_3^-	0.90	—
Ca^{2+}	20.39	1.15
Mg^{2+}	3.41	3.69
Na^+	5.79	30.62
K^+	2.12	1.10
$(Fe,Al)_2O_3$	2.75	—
SiO_2	11.67	—
Sr^{2+}, H_3BO_3, Br^-	—	0.31
	100.00	100.00

[a] From Sverdrup, Johnson, and Fleming, 1959.

Continental waters carry mineral salts from the drainage basins. They reach the lagoon by rivers, or by water table discharge or by underwater karst springs. Rainfall brings in salts which are directly derived from the sea ("cyclic salts"). Fresh water intake as a whole is a function of rainfall. Bicarbonate and Ca^{2+} ions account for more than half the salts of an average fresh water (Table 1). Brackish water does not strictly have the composition of dilute seawater because the ratios between ions are slightly changed by Ca^{2+}, bicarbonate (HCO_3^-), and other ions of fresh water.

When mixing is favored by water movements in the lagoon, a gradation of water composition can be observed between the river mouth and the sea inlet; fresh-water discharge, tides, winds, and longshore currents contribute to the mixing. On the other hand, a stratification of waters of both origins may be observed. Movement of the deepest water may be reduced or stopped altogether by the depth of a lagoon, or a sill across the inlet. Density differences will then result in keeping fresh water lying over salt water, as well as warm water over cold. The higher the temperature, the greater the density difference. In tropical lagoons, stratification is very likely to appear because warm fresh water floats easily over cold seawater (see Vol. I, *Thermocline*). In colder countries, during winter time, ice usually inhibits any movement of the underlying water.

Under arid climatic conditions, evaporation exceeds both rainfall and continental water discharge combined and the lagoon water becomes hypersaline; eventually, it results in a brine. An "antiestuarine" water circulation may appear whereby seawater flows in through the upper part of the inlet and hypersaline water moves out near the bottom, as in the Mediterranean at the Strait of Gibraltar (see Vol. I). In such a case, the salt concentration is limited. Antiestuarine flow is probable in the Laguna Madre of Texas (Phleger and Ewing, 1962).

Where deep outflow is absent, brines become very salty and eventually evaporites are deposited. In an arid environment, the underground water table around the lagoon shows a chlorinity increasing from the lagoon toward the land (Butler, 1967; Baltzer, 1970). In such conditions, evaporation is enhanced by mudcracks and the area becomes a *playa* or a *sebkha* (see Vol. III).

Geochemical Conditions in Lagoon Waters

Chlorinity Changes. Change in chlorinity is very important because it shows the behavior of the original seawater in the lagoon and in the water table. Chloride ion concentration is expressed in grams per liter in the brackish environment. In hypersaline water, the X notation is convenient: for instance, X1 means that chlorinity is seawater chlorinity (19.35g/l) and X3 means 58 g/l (Butler, 1967). The study of changes of salinity in terms of space is made with maps on which curves ("isohalines" or "isochlors") joining points of equal chlorinity are drawn. They show the increase in chlorinity from the river mouth toward the inlet in estuarine lagoons, and how ocean water becomes concentrated from the sea toward the inner lagoon in hypersaline ones. Cross sections show how fresh water floats above seawater.

Changes of salinity with time may occur periodically or incidentally. Tides involve daily changes of chlorinity which reach a maximum at spring high tide and a minimum at spring low tide. Seasonal changes of salinity are related to the amount of sunshine which concentrates brackish waters and to the rainfall which dilutes them. In a dry season, evaporation is sufficient for a salinity change to appear at the end of the day; e.g., in Unare lagoon, Venezuela (Okuda et al., 1965). There, salinity is higher toward the border of the lagoon. Rainfall and salinity show a definite inverse relation in the lagoon of Malamocco, Italy (Franco, 1962).

Incidental changes of salinity are very important. They occur when a barrier is broken by a storm, when man digs a channel through it, or when river discharge increases catastrophically. The chlorinity distribution may then change completely in a short time and it has serious consequences for plant and animal life (Shepard and Moore, 1960).

The distribution of the solutes is similar to salinity distribution (for most of them) and curves of equal content ("isopleths") are similar to chlorinity distribution curves. Salts which are more abundant in fresh water than in seawater will give a similar pattern of curves, but the trends will be opposite. A very good example is given by Kato (1966) from the lagoon of Cananeia, in Brazil. It shows the regular dilution of silica from the river mouth toward the sea.

Lagoon Water Temperature. Lagoon water temperature is related chiefly to air temperature. The relation is closer at low tide than at high tide. It usually reaches a minimum in winter and a maximum in summer (Franco, 1962; Carl, 1940). In the humid tropics, fresh water is warmer than seawater. In arid countries, seawater concentration results in a brine with a very high temperature (37°C in Laguna Tamaulipas of Mexico (Copeland, 1967) and 40°C in an underground brine around a lagoon of the Trucial coast, Persian Gulf (Butler, 1967). The specific heat of pure water is 1. Salt waters have a lower one, e.g., 0.932 for seawater with a salinity of 35 g/l at temperature of 17.5°C. Less heat is therefore necessary to increase the temperature of salt water than of fresh water under the same conditions. In a lagoon environment and in the related water table, the pressure remains close to atmospheric pressure. According to the phase rule, temperature and concentration will be the only agents controlling the nature of simple, double, or triple salts deposited. The very high temperatures possible account for the occasional deposition of unexpectedly high-temperature salts.

pH in Lagoon Waters. The pH of seawater is rather stable, the figure being usually between 7.5 and 8.4 (Sverdrup, Johnson, and Fleming, 1959). In lagoon waters, pH changes are controlled by many agents, so that, according to local environment, trends may be in opposite directions.

Alkaline conditions are usually related to seawater (Carl, 1940). In a brine, a pH as high as 9 may occur when $CaCO_3$ is about to precipitate. Alkalinity (in the oceanographers' sense) is then at a maximum. But in fresh water, a very high alkalinity is also possible because of contact with calcium carbonate (Rivière and Vernhet, 1958). The pH will be as high as 9 if CO_2 pressure is zero (Garrels and Christ, 1965). Quite often a close relationship is found between high pH figures and the amount of sunshine. The CO_2 consumption for photosynthesis obviously controls this relation (Carl, 1940; Okuda et al., 1965). In a euxinic environment, as at the bottom of a humid tropic lagoon (e.g., Abidjan, Ivory Coast) Debyser (1955) found a relatively high pH, due to ammonia evolved from bacterial decay of proteins.

Continental waters discharging into tropical lagoons show low pH figures due to humic acids of the soils and tannic acid of mangrove swamps (Debyser, 1955; Kato, 1966). In mangrove mud, bacterial reduction of sulfates releases H_2S with a slightly acid reaction. In an arid environment, drying up of that H_2S-rich mangrove mud results in a reoxidation of sulfur and strongly acid conditions with new sulfate.

Dissolved Oxygen. Oxygen solubility in salt water decreases as temperature and, to a lesser degree, salinity increase. In this way, a *saturation percentage* can be defined just as in oceanographic practice. Usually, at a given station at the surface of a lagoon, oxygen varies directly with the sunshine, due to photosynthesis. In temperate countries, the most noticeable change is seasonal. In tropical environment, the changes are daily: maximum at 12.00–13.00 hours, minimum at night (Debyser, 1955; Okuda, 1965).

Near the bottom of the lagoon, photosynthesis is limited by light absorption and the respiration of organisms that consume oxygen. If water movements are impeded either by topography or water stratification, oxygen consumption may exceed renewal, and euxinic conditions occur (Debyser, 1955) in a similar way to ectogenic meromictic lakes (see Vol. VI). Decaying organic matter is usually very abundant in brackish water sediments, especially in the humid tropics and sometimes also in the temperate climates. Kato (1966) has shown that the oxygen content in a Brazilian lagoon is an inverse linear function of organic matter in suspension. It explains Okuda's remark that windy weather lowers the oxygen content in the Unare lagoon, Venezuela (Okuda et al., 1965).

Changes of oxygen content with tides, at every station, show the main differences between oxygen distribution under different climates. In temperate countries, high oxygen contents are measured at low tide, mainly because the water influenced by aquatic plants lies only near the edge of the lagoon. At high tide, seawater which is slightly undersaturated with respect to oxygen enters the lagoon and the oxygen content there is at a minimum. Under tropical conditions, water coming from the mangrove swamp predominates at low tide, and at high tide photosynthesis by phytoplankton in the central lagoon plays the major part.

Accidental changes in oxygen content result from the death of aquatic plants due to salinity change or to storms piling them up in large amounts. This suppresses photosynthesis, and more than that, the decay of this organic matter consumes the remaining oxygen (Carl, 1940; Rivière and Vernhet, 1964).

In hypersaline lagoons, the inflow of seawater compensating for evaporation brings in dissolved oxygen. Because the salinity limits biological activity, the environment remains with a fairly high oxygen content.

Oxidation-Reduction. Measurements of Eh show a direct relationship to oxygen. But in addition, they allow investigations into environments completely deprived of oxygen, and are thus very useful in lagoon waters and their sediments. Two measurements may be used: Eh and rH_2 which allow equilibrium calculations in the laboratory because of the easy computation of the free energy of the reaction. For field geology, rH_2 measurements are very convenient. By analogy with pH, rH_2 is the antilogarithm of molecular hydrogen concentration. It is related to Eh and pH:

$$rH_2 = \frac{2\,Eh \cdot F}{2.302\,RT} + 2\,pH$$

where Eh = oxidation-reduction potential between hydrogen reference electrode and platinum electrode; R = perfect gas constant; and T = absolute temperature in °K.

There is a specially devised pH meter (Ponselle) that gives a direct reading of rH_2, by the potential difference between a glass electrode and a platinum electrode (Rivière). The advantage for field geology is that *in situ* measurements of rH_2 give an immediate idea of the oxidation-reduction characteristics of the environment and immediate comparisons may be made without computation (cf. our modification of Garrels' table: Fig. 1).

In lagoon water, rH_2 usually ranges between 12 at the bottom and 30–32 at the surface. When euxinic conditions develop, it becomes lower than 12, reaching 5 and probably less (Rivière and Vernhet, 1964). A zero rH_2 value would indicate a totally reducing environment. A further lowering would take hydrogen out of water molecules.

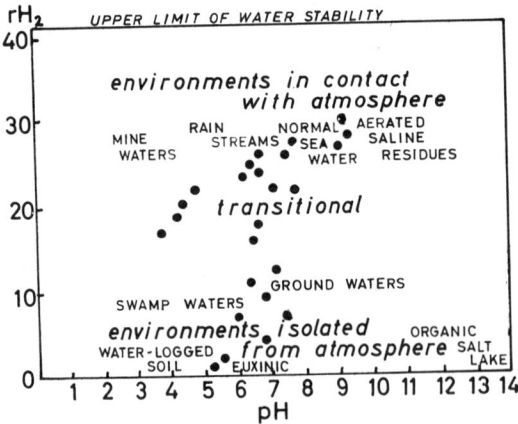

FIG. 1. Examples of environments written in terms of rH_2 and pH, compared with water and sediments of lagoonal origin (modified from Garrels, 1960). Note: $rH_2 = 0$ is also the lower limit of water stbaility.

On the borders of a semiarid tropical lagoon (e.g., west coast of New Caledonia) the tidal and supratidal sediments also show a range of rH_2 from very low figures in the mangrove sediments (lower than 9) to relatively high figures in sun-cracked mud, and under an algal mat, where it is higher than 20. These changes also explain a ferric ion concentration towards the border of the lagoon. The seawater table moving with evaporation from the lagoon toward the dry land has first to pass through the mud of a mangrove swamp. There, it is subject to reducing conditions which provide the iron with a maximum solubility in the ferrous state. The seawater table becomes more and more hypersaline as it passes into more oxidizing areas thanks to mud cracks. Due partly to the excessive concentration, but mainly to oxidation of iron to the ferric state, precipitation of iron occurs near the borders of this lagoon (Baltzer, 1970).

Biological Aspects of Lagoon Geochemistry

A nonhypersaline lagoon environment is usually very rich in nutrients. Fresh-water discharge carries much organic matter (plant and animal debris, plankton, humic acids, and mineral salts), which is consumed on the spot. At the same time seawater inflow carries in some plankton. Both are used for metabolic development in the lagoon (Lillelund, 1960). Excess organic matter sinks as sediment with the remains of the organisms which lived in the lagoon. Bacterial decay gives back phosphates and nitrates (Kato, 1966) in the form of available nutrients in the lagoon water. Estuarine dynamics trap nutrients in a lagoon because of the salt-water wedge which moves slowly toward and under river fresh-water discharge (Pritchard). That wealth in nutrients allows a high productivity of organic matter at every level of organization.

On the other hand, the antiestuarine system, when it has an outflowing undercurrent of hypersaline water, is expected to be very poor in nutrients, although we do not know of a complete study on the subject.

Consequences of Changes in Geochemical Conditions. Some organisms are adapted to periodical changes in tidal and estuarine environments with respect to salinity, water height, and so on. As a result there is a zonation which is most typical in mangrove swamps, but is also evident in salt marshes in temperate countries. In a similar way burrowing organisms penetrate into sediments looking for a suitable salinity. They find it because of the stabilizing effect of the sediment upon salinity. That zonation of organisms results in a local accumulation of various types of organic matter or of calcium carbonate. In turn, that organic matter has purely geochemical consequences, as remarked earlier, with iron concentration due to oxidation-reduction.

Other organisms find in a lagoon an environment which suits them perfectly for a while and they multiply quickly. Later on, a *mass mortality* (q.v. Vol. I) occurs when environment changes, even if a few individuals are able to escape thanks to the salinity gradients allowing them to find a suitable spot for survival. Carl (1940) quotes an annual succession in the plankton population which was observed during almost two years. Abrupt increases or decreases in salinity that lead to the death of aquatic plants have severe consequences on the oxygen content. The greater the hypersaline conditions, the less the organisms can tolerate these changes in oxygen concentration. No survivors are left when halite or other salts begin to crystallize (see *Kara-Bogaz Gulf,* Vol. III).

Microorganisms. In his study of Abidjan lagoon, Ivory Coast, Debyser (1955) points out that all bacteria found in this environment were halophilic, but it is known that very highly saline conditions stop their development. Every one of the main bacteriological functions were found. The *sulfur cycle* (q. v.) is complete, including sulfate reducing and sulfur oxidizing groups. As soon as physical and chemical conditions allow, one or several of the bacterial activities begin to operate.

Chemical Sedimentation in a Lagoon Environment

Calcium Carbonate Deposition. Calcium carbonate precipitation has been frequently observed in lagoon environments, in several forms: "milking" or "whitings" (i. e., clouds of microcrystalline particles in the water), deposits associated with organic matter (algal mats, wood, fishes), or cementation of sands resulting in a calcarenite or beachrock, and finally öolite deposition.

It is convenient to consider separately limestone precipitation in a brackish environment and precipitation in hypersaline water. For brackish water Rivière and Vernhet (1958) give a general answer to the problem. The weak acids that are linked to strong alkalis in seawater represent the so-called alkalinity. They were compared to the Ca^{2+} and Mg^{2+} ion content as a function of salinity (Fig. 2). It shows that the alkalinity of natural brackish waters in equilibrium with limestone of the bottom sediments increases slightly when salinity and Ca^{2+} ions decrease (salinity $\leq$ seawater). In this way, within fresh-water and

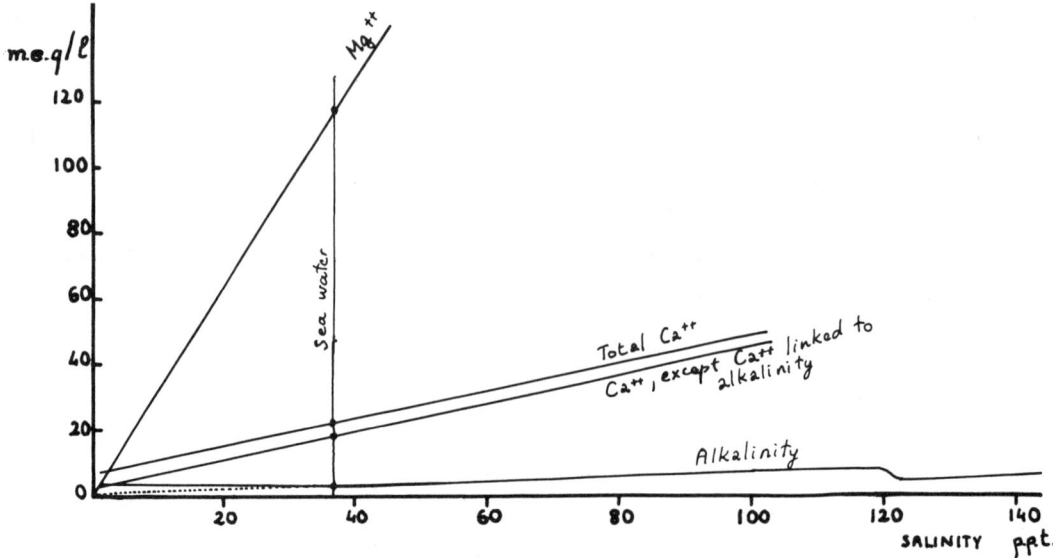

Fig. 2. Mg^{2+}, Ca^{2+} and alkalinity in a Mediterranean lagoon in Languedoc (France) as a function of salinity (Rivière, 1969). Salinity is between 37.5 ppt and 140 ppt. The alkalinity values are from Copeland, 1967.

seawater salinity limits, the ratio between Ca^{2+} ions and acid carbonate and carbonate ions increases as salinity decreases. It is at a maximum in fresh water, it approaches ½ in brackish water (with 5 g/l salinity), and it is only $1/10$ in seawater. The law of mass action may be safely applied for the lower salinities and remains qualitatively true in seawater, in spite of the ionic pairs which appear when salinity increases (Garrels and Christ, 1965).

The equilibrium controlling the reaction between the very soluble bicarbonate ion and slightly soluble calcium carbonate, may be written:

$$Ca(HCO_3)_2 + Ca^{2+} + 2\ OH^- = 2\ CaCO_3 + H_2O \quad (1)$$

By application of the law of mass action:

$$[Ca(HCO_3)_2] \cdot [Ca^{2+}] \cdot [OH^-]^2 = k[CaCO_3]^2 \cdot [H_2O] \quad (2)$$

$[H_2O]$ being constant and equal to 1, and as crystals of $CaCO_3$ are present:

$$[CaCO_3] = 1$$

so that:

$$[Ca(HCO_3)_2] = K \frac{1}{[Ca^{2+}] \cdot [OH^-]^2} \quad (3)$$

Eq. (3) confirms that, at a constant pH, a decrease of Ca^{2+} concentration will lead the equilibrium towards an increase in bicarbonate concentration and a consecutive solution of the carbonate in the sediments, so that the alkalinity rises. Conversely, increasing Ca^{2+} concentration will result in increasing carbonate concentration. Calcium carbonate, being less soluble, will precipitate.

Some examples of limestone precipitation in a lagoon environment illustrate this mechanism. When sunshine concentrates lagoon water near the shore, the water being in equilibrium with the sedimentary calcium carbonate, the salinity increases, and with it the Ca^{2+} concentration. The equilibrium is displaced in favor of calcium carbonate, which precipitates between the sand grains on the shore, where evaporation is at a maximum. This results in a cement which is first pasty but eventually becomes the consolidating agent of a lagoon *beachrock* (Rivière and Vernhet, 1958).

In the same manner, oölites associated with high-energy beaches of lagoons of the French Mediterranean coast (Rivière and Vernhet, 1958) or of the Texas Laguna Madre (Rusnak, 1960) result from strong dry winds. Evaporation increases salinity and Ca^{2+} concentration, and carbonate precipitation occurs. Structure of the oölites shows that their development is not continuous but, on the contrary, interrupted by periods of inactivity.

Bacterial reduction of sulfates in lagoon water produces hydrogen sulfide which is a weak acid, and this augments alkalinity (calcium sulfide is more soluble than bicarbonate). If the H_2S diffuses into the atmosphere, alkalinity decreases and calcium carbonate precipitates. This behavior was observed in vitro by Rivière and Vernhet (1964) in lagoon water naturally enriched with H_2S and Ca^{2+}.

In very small lagoons, or rather brackish water ponds, behind sand barriers on some of the Tuamotu atolls in the South Pacific Trichet (1967) studied the products of the decay of a fast-growing floating algal mat and the fecal pellets due to the digestion of it by invertebrates. Both the decay products and the pellets gradually changed to aragonite. This organic matter contains a large amount of oses and uronic acids, which behave as weak acids with alkalis, and alkalinity is raised. There is next a decrease of alkalinity due to the rising salinity (evaporation, mixing with pure seawater), or to the rising Ca^{2+} content only (bacterial digestion of oses and uronic acids linked with Ca^{2+}); both will result in calcium carbonate precipitation.

It is noteworthy that consumption of CO_2 increases the alkalinity, and $CaCO_3$ must precipitate. But the law of mass action as written in Eq. 3 shows that an increase of OH^- concentration has the same effect as Ca^{2+} increase, so that high pH, by itself, is effective in producing $CaCO_3$ precipitation.

Dalrymple (1966) describes a calcareous micrite deposited on an algal mat due to the bacterial decay of the algae with a development of alkaline conditions with ammonia. In the case of mass mortality of fish in a lagoon, the mechanism observed in vitro by Berner (1968) is likely to involve $CaCO_3$ deposition. The anaerobic decay of fish gives ammonia and amines without any reduction of sulfate. Ca^{2+} ions in the environment do not give $CaCO_3$, but combine with fatty acids giving, for instance, the palmitate of calcium or magnesium. $CaCO_3$ is likely to deposit later on, when the fatty acids are digested.

In humid tropics, sediments have a low calcium carbonate content, 1.04% on the average in the West African Abidjan lagoon sediments (Debyser, 1955). This is a result of acid continental water flowing into the lagoon, with low alkalinity, and of acid conditions within the lagoon, due to organic matter decay. The lagoon water dissolves the small amount of calcium carbonate concentrated in shells of molluscs in this environment.

In hypersaline lagoons, the seawater become concentrated, the alkalinity increases until it reaches a high supersaturation and then $CaCO_3$ precipitates (Copeland, 1967). This reaction is used by salt harvesters to separate $CaCO_3$ from the brine. It occurs also in the concentrating water table around lagoons in arid and semiarid countries.

Magnesium Carbonate Precipitation and Dolomite. Precipitation of both magnesium and calcium carbonate has been widely reported in lagoons of South Australia (e.g., Alderman and Van Der Borch, 1960). After their deposition they change diagenetically to dolomite and calcite. Salinity is usually high, up to 200 g/l but seasonally may be as low as half seawater concentration. Aquatic plants in such an environment at times grow very actively with strong photosynthesis, and high pH readings are reported by the authors to account for the precipitations. Magnesium precipitation is also observed in relation with the concentrating salt water tables around lagoons in arid climatic regions. It is found in the upper layers of some sebkhas. According to Butler (1967) dolomite forming in such conditions around the Trucial coast lagoons of the Persian Gulf is due to a penecontemporaneous replacement of aragonite.

Evaporite Precipitation. When seawater is concentrated in lagoons without any bottom countercurrent flowing out through the inlets, brines are concentrated until salts precipitate. The evaporite processes are described by Feldman and Cruft (this volume), but it is noteworthy that crystals may form either in the lagoon body of water or underground, in the salt water table around the lagoon (sebkha). Depending on local conditions (seawater inflow, depth of the lagoon, temperature, rainfall) evaporite minerals will be deposited after calcium and magnesium carbonates; the sequence will be: hydrated sodium sulfate (glauber salt or mirabilite), gypsum, anhydrite, sylvite, halite, and double and triple salts.

Classification of Lagoons in Terms of Climatic Conditions*

As a conclusion to their study of the lagoons of the central coast of Texas, Shepard and Moore (1960) point out that most of the lagoons of the world show similar features. It is true that the mechanisms at work may be observed more or less in the same way in any lagoon, depending on changes in climate or river pattern, or depending on the place we consider in the lagoon. Shepard and Moore distinguish "bay near river, deep central bay, and bay near inlet"; Debyser distinguishes

---

* See "Climatic classification" in Fairbridge, R. W., editor), 1967, "Encyclopedia of Atmospheric Science and Astrogeology," New York, Reinhold (especially Péguy's classification).

marine, estuarine, brackish, and lacustrine environments. Although we know that a classification based upon climate is much too straightforward, it gives a convenient clue to the problem.

Temperate Climate (some months either arid or cold). Under a temperate climate, salinity ranges from fresh water to seawater with sometimes a hypersaline tendency in summer on the borders of the lagoon farthest from fresh- and seawater influxes. Abundant vegetation traps detrital sedimentation. Limestone precipitation, in forms such as micrite, oölites, or calcarenites, is not rare. In some isolated basins on the borders of these lagoons, gypsum and even a little halite may precipitate in summer, but salinity of the body of water as a whole remains moderate.

Temperate Climate (some months arid and some months cold). This type of climate is typical of that around the Caspian and the Aral seas, and, to a lesser extent, around the Black Sea. On the eastern coast of the Caspian Sea, the aridity is extreme, and wintry cold impedes the development of plants and in some instances it results in a complete desert, even close to the sea. Kara-Boghaz Gulf (*q.v.* Vol. III) belongs to that type. Concentration of seawater is very efficient in lagoons where physiography favors it. The lagoon water is then a "bittern" (a bitter solution that remains after salt has crystallized out of seawater), which is diluted by inflowing seawater. Deposition of the evaporites and calcium carbonate is extremely effective. Such lagoons play a great part in the salt equilibrium of the seas around which they lie. Kara-Boghaz Gulf is the major agent which lowers the salinity of the Caspian Sea. In lagoons of the Black Sea where there is appreciable river inflow salinity is much lower, and stratification of water results in stagnation of the lower layers of water (See Vol. I, *Black Sea*).

Arid Climate (some months possibly tropical or temperate). The body of lagoon water is hypersaline, as in the previous case. High temperature permits the growth of some mangrove around the lagoons, where salinity, although very high, is not excessive. Algal mats and bare salt flats form a large part of the intertidal and supratidal environments: they are known as sebkhas (sabkhas) or coastal playas. The best known are those around the Red Sea and the Persian Gulf; they are under extremely arid conditions. Others are in the Gulf of California and in South Australia. The Laguna Madre of Texas (Rusnak, 1960) shows some similarities with such environment, although it is less extreme.

The temperature of bitterns is sometimes very high in summer. Limestone deposition is abundant. Gypsum and anhydrite are typical, and halite too, mainly in sediments around the lagoon in the lagoon water table. Dolomite is often reported to appear, either directly on the floor of the lagoon (South Australia) or in the water table, by replacement of aragonite.

Lagoons of this arid climate type appear in countries where climates, although hot, are by no means properly considered as arid. This is due to the topographic situation of the environment, and any hills or mountain chain to windward will result in arid conditions in the lagoon to leeward (e.g., in the rainshadow of the southeast trade wind in New Caledonia).

Humid Tropical Climate (some months possibly arid). Hot and humid climates prevail on many lagoons in the world (Africa, Central and South America, etc.). Rainfall is heavy and dilution largely exceeds evaporation. This environment favors mangrove swamps as well as other plant and animal life.

When lagoons are deep enough, it is the most suitable environment for the examination of a stratification of warm fresh water above cold salt water, where euxinic conditions are expected. Calcium carbonate deposition is rare. Even calcium carbonate deposited by molluscs (shells) tends to dissolve.

Coral Lagoons (see Vol. III, **Lagoon—Coral-Reef Type**)

Lagoons in the middle of atolls are a particular case of warm climate lagoons. Environments are similar to coastal lagoons behind sand barriers, except that terrigenous sedimentation is rare. Fresh-water income is almost entirely limited to rainfall. Salinity is usually somewhat higher than in seawater, and sometimes definitely hypersaline. Coralline algae, corals, molluscs, and foraminifera play a major part in the huge calcium carbonate deposition. On the lagoon floor a very thin calcareous mud has an origin which is not certain: it appears to be partly detrital and partly due to fresh (primary) precipitation.

Mangrove swamps commonly develop in the protected bays. Abundant plant debris is deposited and its decay gives rise to much humic and tannic acid which dissolves large holes in the reef limestone.

Extensive lagoons limited by barrier reefs are situated around the Coral Sea, and in the South West Pacific, generally. These include the Great Barrier Reef, Queensland, and the Barrier Reef of Tagula, southeast of New Guinea, and the two parallel Barrier Reefs around New Caledonia. These lagoons are so

LAGOON GEOCHEMISTRY

TABLE 2. CLIMATIC CLASSIFICATION OF LAGOONS

Climate	Temperate Climate (some months either arid or cold)	Temperate Climate (some months arid and some months cold)	Arid Climate (some months possibly tropical or temperate)	Humid Tropical Climate (some months possibly arid)
Examples	Eastern U.S. coast; western and southern France; northeastern Italy	Lagoons of the Caspian sea, e.g., Kara Boghaz; limans of the Black Sea	Lagoons of the Red Sea, Persian Gulf, Gulf of California, southern Australia	Lagoons of the Ivory Coast (West Africa), Venezuela, Brazil
Usual surrounding landscape	Salt marsh and reed bank	Salt marsh; bare mud algal flat when hypersaline; desert in extreme cases	Salt marsh; algal mat; bare mud of sebkhas; a few mangrove swamps	Mangrove swamps; fresh-water swamps and intermediates
Salinity	From 0 to 35 g/l rare hypersalinity	Hypersaline except for lagoons receiving large rivers	Strongly hypersaline	Brackish, from 0 to 35 g/l
Water stratification	Fresh water or brackish water above seawater; ice above water in winter	Salt water above cold hypersaline water in hypersaline lagoons; when large river discharges, stratification is very marked	Possible stratification, but without definite trend	Warm fresh water above cool seawater; stable
Oxidation-reduction conditions	Photosynthesis induces summer supersaturation in oxygen; strong reduction by mass mortality	Seawater replacing evaporated water brings oxygen in hypersaline lagoons; in nonhypersaline ones, oxygenated water in upper layers; stagnation in bottom layers	Seawater replacing evaporated water brings oxygen which is not completely consumed by organisms; strong photosynthesis by specialized plants	Possible supersaturation of surface water by photosynthesis; stagnation of bottom water very common
Typical minerals	Calcite, aragonite (calcarenite, öolite); a little gypsum and halite in summer	Calcite, aragonite (öolites); mirabilite, gypsum, halite in hypersaline lagoon; pyrite in nonhypersaline lagoon	Aragonite, calcite gypsum, anhydrite, halite, dolomite, etc.	Pyrite

633

large that they have the properties of the open sea. Inside these large lagoons, smaller ones may be found, with detrital barriers which bring us back to the general lagoon type.

F. BALTZER
A. RIVIÈRE

References

Alderman, A. R. and Van Der Borch, C.C., 1960, "Occurrence of hydromagnesite in sediments in South Australia," *Nature*, **188**(4754), 931.

Baltzer, F., 1970, "Etude sédimentologique du Marais de Mara (Côte Ouest de la Nouvelle Calédonie) et de formations quaternaires voisines," Expédition sur les Récifs Coralliens de la Nouvelle Calédonie, Vol. IV, Paris, Editions de la Fondation Singer-Polignac, 1–146.

Berner, R. A., 1968, "Calcium carbonate concretions formed by the decomposition of organic matter," *Science*, **159**, 195–197.

Butler, G. P., 1967, "Brines and their evaporites, in the Trucial Coast," Seventh International Sedimentological Congress, London, Mimeographed Communication, 4pp.

Carl, G. C., 1940, "Some ecological conditions in a brackish lagoon," *Ecology*, **21**(1), 65–74.

Copeland, B. J., 1967, "Environmental characteristics of hypersaline lagoons," *Contrib. Marine Sci.*, **12**, 207–18

Dalrymple, D. W., 1966, "Calcium carbonate deposition associated with blue green algal mats, Baffin Bay, Texas," *Inst. Marine Sci., Publ.*, **10**, 187–200.

Debyser, J., 1955, "Etude sédimentologique du système lagunaire d'Abidjan (Côte d'Ivoire)," *Rev. Inst. Franc. Pétrole*, **10**(5), 319–334.

Emery, K. O., 1969, "A coastal pond, studied by oceanographic methods," New York, Elsevier, 100pp.

Franco, P., 1962, "Condizioni fisiche e chimiche delle acque lagunari nel porto canale di Malamocco—I Guigno 1960-Guigno 1961," *Arch. Oceanogr. Limnol., Ital.*, **12**(3), 225–255.

Garrels, R. M., 1960, "Mineral Equilibria at Low Temperature and Pressure," New York, Harper & Row.

Garrels, R. M., and Christ, C. L., 1965, "Solutions, Minerals and Equilibria," New York, Harper & Row.

Kato, K., 1966, "Chemical investigations on the hydrographical system of Cananeia lagoon," Sao Paulo Universidad, *Inst. Ocean. Bol.* **15**(1), 1–45.

Lillelund, K., 1960, "Der Schlendorfer Binnensee—Limnologische Untersuchung eines Strandgewässers an der deutschen Ostseeküste. V—Grundlagen der Sedimentationsvorgänge und ihre Bedeutung für die Bodenproduktion," *Z. Fischerei*, **9**(5–6), 417–424.

Okuda, T., Gomez, J. R., Alvarez, J. B., and Garcia, A. J., 1965, "Condiciones hidrograficas de la laguna y rio Unare," *Bol. Inst. Ocean., Univ. Oriente, Cumana, Venezuela*, **4**(1), 60–107.

Phleger, F. B., and Ewing, G. C., 1962, "Sedimentology and oceanography of coastal lagoons in Baja California, Mexico." *Bull. Geol. Soc. Am.*, **73**, 145–82.

Rivière, A., and Vernhet, S., 1958, "Contribution à l'étude sédimentologique de l'étang de Leucate-Salses (Languedoc-Roussillon)," *Eclogae Geol. Helv.*, **51**(3), 1959.

Rivière, A., and Vernhet, S., 1964, "Contribution à l'Etude de la sédimentologie des sédiments carbonatés," in (Van Straaten, editor), "Deltaic and shallow marine deposits," "Developments in Sedimentology," Vol. 1, Amsterdam, Elsevier pp. 356–361.

Rusnak, G. A., 1960, "Sediments of Laguna Madre, Texas," in "Recent Sediments, Northwest Gulf of Mexico," *Am. Assoc. Petrol. Geol. Tulsa, Oklahoma*, pp. 153–196.

Shepard, F. P., and Moore, D. G., 1960, "Bays of Central Texas coast," in "Recent Sediments, Northwest Gulf of Mexico," *Am. Assoc. Petrol. Geol.*, Tulsa, Oklahoma, pp. 117–152.

Sverdrup, H. U., Johnson, M. W., and Fleming, R. H., 1959, "The Oceans," Englewood Cliffs, N.J., Prentice Hall, 1087pp.

Trichet, J., 1967, "Essai d'explication du dépôt d'aragonite sur des substrates organiques," *Compt. Rend.*, **265**, 1464–1467.

Cross-references: *Cyclic Salts; Evaporite Processes; Free Energy; Mass Action and Equilibrium Constant; Limnology; Natural Brines; Organic Geochemistry; Oxidation and Reduction; pH-Eh; Photosynthesis; Seawater, Chemistry; Sulfate Reduction, Microbial; Sulfur Cycle.* Vol. I: *Black Sea; Mass Mortality in the Sea; Mediterranean; Thermocline.* Vol. II: *Climatic Classification.* Vol. III: *Beachrock; Caspian Sea; Coastal Lagoon Dynamics; Kara-bogaz Gulf; Lagoons; Lagoon—Coral Reef Type; Mangrove Swamp; Playa; Sabkha or Sebkha.* Vol. IVB: *Anhydrite; Evaporite Minerals; Gypsum.* Vol. VI: *Euxinic Basin; Evaporites; Oölites; Swamp, Marsh, Bog; Whitings.*

LAKE GEOCHEMISTRY

The following aspects of this subject will be discussed: (a) inorganic and organic geochemistry of Holocene and Pleistocene lakes of nonglaciated regions; (b) inorganic and organic geochemistry of Holocene and Pleistocene regions; (c) organic geochemistry of a Tertiary lake deposit.

(a) 1. **Inorganic Geochemistry of Lakes of Nonglaciated Regions.** The sediments of lakes of nonglaciated regions range widely in composition depending on climatic conditions, source materials, relief of the land, and the stage of lake development. The sediments of nonglacial lakes of humid regions are strongly influenced in their characteristics by prevailingly high organic productivity. Their clastic fractions are typically detrital quartz, feldspar, clay minerals, volcanic and carbonate frag-

LAKE GEOCHEMISTRY

TABLE 1. ANALYSES OF CORES P-1 AND P-3[a]

Depth, m		Ignition loss	SiO_2	$(Al,Fe)_2O_3$	CaO
P-1	4.70	11.63	40.75	39.09	2.38
	5.60	16.84	40.71	12.51	18.37
	6.80	25.40	48.34	14.33	5.47
P-3	3.0	18.90	40.60	—	0.59
	3.1	16.90	43.00	—	0.92
	3.2	12.45	43.51	39.85	2.49
	3.4	20.00	42.07	—	0.33
	3.9	14.91	40.58	35.75	1.51
	4.0	28.70	36.66	—	1.50
	4.5	30.07	39.30	17.33	5.72
	4.9	23.90	54.96	—	0.93
	5.0	16.33	50.55	24.19	1.22
	5.4	27.40	52.03	—	2.64
	5.5	19.00	34.43	12.94	21.63
	5.9	28.40	44.59	—	8.70
	6.0	19.38	55.51	13.30	4.55
	6.4	22.90	57.64	—	2.76
	6.5	18.60	59.05	—	3.90
	6.9	21.50	52.42	—	3.04
	7.0	11.93	53.58	22.42	5.30
	7.9	26.20	50.15	—	2.15
	8.0	20.80	50.88	19.07	3.10
	9.0	25.60	54.67	—	2.97
	9.1	17.85	54.32	18.64	2.72

[a] Percentage composition of samples dried at 105–110°C (Hutchinson et al., 1956).

ments, and so forth, depending on the source materials. Relatively little authigenic mineral formation takes place.

The lake sediments of arid regions, on the other hand, may, in addition to the detrital fractions, contain important authigenic constituents. An example of the effect of climatic changes on calcium carbonate contents of sediments is that of Lake Patzcuaro, Michoacan, Mexico (Hutchinson et al., 1956). This lake basin of southwestern Mexico lies in a volcanic plateau where late Tertiary and Pleistocene lavas cover Cretaceous and early Tertiary rocks. The present lake water is high in sodium and potassium bicarbonate, low in calcium and magnesium, and moderate in chlorine and sulfite. The lake fluctuates about 60 cm in level yearly and attains about 15 m in depth. A core of the lake sediments is shown in Table 1. The age of the lake sediments based on studies of the pollen record, rates of sedimentation, and carbon-14 analyses at 7.9 is about 3300 years and at 5.5m is about 1700 years. The interpretation of the data is that a moist time interval began about 3300 years ago and was maintained until about 2300 years ago, as is reflected in the low CaO content of the sediments. The interval from 2300 to 1700 years ago is marked by higher CaO contents and suggests a dry low-water episode. Supporting evidence for an alkaline low-water stage of the lake is in certain diatoms (*Navicula oblonga, Anomoeonis sphaerophora*, etc.) that commonly occur in alkaline waters, as well as of dry-climate arboreal pollen. The increase in CaO at 4.5 m may represent another shorter low-water period. The increase in alumina and iron oxide in the upper part of the sedimentary section was caused by changes in erosional details of possible cultural origin in the drainage basin.

Great Salt Lake, Utah and its predecessor, Lake Bonneville, formed in a large complexly block-faulted basin the area of which has decreased over 600,000 years from about 13,300 km² to the present 10,878 km². The water alternated from fresh-water, high-water, overflowing stages to saline, low-water, closed-basin stages about seven times during the history of the lake (Eardley and Gvosdetsky, 1960; Bradley, 1963). In a 600-foot core of the Bonneville-Great Salt Lake sediments, a volcanic ash layer, the Pearlette Ash, occurs at 548 feet in the core (Eardley and Gvosdetsky, 1960). This ash, of the Kansan Glacial Stage has been found from Kansas to the northwestern United States.

The present content of dissolved salts in Great Salt Lake, associated salt crusts, and salt water in the lake clays is about 6×10^9 tons (Bradley, 1963), which amounts to an annual increment of about 1.1×10^6 tons. At this rate the last fresh-water lake stage would have been about 6000 years ago, but C^{14} dating suggests that the last overflow stage was about 12,000 years ago (Eardley et al., 1957). The discrepancy was attributed by Eardley et al. to removal of salt from the Bonneville basin by deflation. Bradley (1963) feels that this explanation may not be plausible and cites the work of Langbein (1962) which indicates that marginal mud flats may store large amounts of salt in closed basins having shrinking bodies of saline waters. Thus the estimated total storage of salt in the water, crusts, and clays of Great Salt Lake may be too low.

The water of Great Salt Lake is similar in proportional composition to that of seawater. Although the actual salinity of the water has varied between about 14% in 1877 to about 27% in 1935 the relative proportions of the salt remain about the same except for $MgCl_2$ and KCl which tend to increase slightly relative to NaCl and Na_2SO_4 as the latter precipitates out (Eardley, 1938). The following composition of the lake water occurred in 1913 (Eardley, 1938): Cl, 55.48%; Br, 0%; SO_4, 6.68%; CO_3, 0.09%; Li, 0%; Na, 33.17%; K, 1.66%; Ca, 0.16%; Mg, 2.76%; Fe_2O_3–Al_2O_3–SiO_2, 0%; salinity, 20.349%.

Eardley (1938) believes that the saline matter was originally derived from marine salt beds in the Jurassic rocks that lie on the eastern side of the Great Salt Lake drainage area in Sanpete Valley, as well as from more disseminated and sparser sources in other rocks of the drainage area.

The chemically precipitated minerals of Great Salt Lake sedimentary deposits are aragonite, $CaCO_3$; dolomite, $2MgCO_3 \cdot CaCO_3$; and parasepiolite (?) $2MgO \cdot 3SiO_2 \cdot 4H_2O$. The floor and margins of the lake at various times become the sites of crystallization of halite (NaCl) and mirabilite or Glauber salt ($Na_2SO_4 \cdot 10H_2O$). The latter forms especially during cold weather. The common sulfates, anhydrite, or gypsum have not formed because of the deficiency of calcium ions in the lake waters. The widespread precipitation of carbonates in the lake waters is of complex nature but appears to be caused by: (1) reduction of CO_2 content of stream waters entering the more saline lake waters in which CO_2 is less soluble than in fresh water; (2) evaporation of lake water following spring floods and resulting precipitation of carbonates; (3) algal photosynthesis which extracts CO_2 and promotes precipitation of $CaCO_3$ from the water; (4) bacterial metabolism which may, in the case of periphyte types, locally increase the pH by producing ammonia and reducing the solubility of the water for CO_2 and bicarbonate. The abundance of Mg as compared to Ca in Great Salt Lake brines is noteworthy as is the primary or penecontemporaneous formation of dolomite.

(a) 2. **Organic Geochemistry of Lakes of Nonglaciated Regions.** An example of a lake in a tropical region, the bottom sediments of which are richly organic, is Lake Nicaragua, western Nicaragua (Swain, 1961b, 1966; Swain and Gilby, 1965). The large lake basin 170 × 75 kilometers, is of tectonic origin and contains uniform profundal volcanic, copropelic (organic-ooze) silts that vary from a few to many meters in thickness. There is active volcanism in and near the lake and this, in combination with active cultivation around the lake, results in a rapid sedimentation rate. The organic geochemistry of the lake sediments is representative of tropical large lakes. Saturated hydrocarbons and aromatic hydrocarbons occur in similar but small amounts. The values are comparable to those of poorly productive (oligotrophic) lakes of cool temperate regions. In an arid-region lake, by contrast (Pyramid Lake, Nevada), the saturated hydrocarbons greatly exceed the aromatics. The rather high content of organic pigments is suggested as accounting for the source of higher aromatics in Lake Nicaragua. The saturated hydrocarbons may be derived in part from the rich diatom population in the lake as well as from the abundant crops of the alga *Botryococcus* ($=$ *Elaeophyton?*).

The protein amino acids of Lake Nicaragua sediments also are related to the algal plankton crops. The amino acids are similar in variety and amounts to those of the sediments of productive (eutrophic) lakes of cool temperature regions. In comparison to the latter, however, the basic amino acids lysine, histidine, and arginine are higher than is typical of low-carbonate lake sediments. Carbohydrate content of Lake Nicaragua sediment is also similar to that of eutrophic cool-temperate lakes and apparently also has an important source in the plantonic organisms of the lake. The organic pigments of the sediments of Lake Nicaragua comprise pheophytin from chlorophyll, carotenes, and yellow-fluorescing pigments of flavinoid character. These compounds are in amounts comparable to those in eutrophic cool-temperate lakes and their source is principally in the planktonic material, or in the degradation products resulting from the action of benthonic, mud-eating (ilyotrophic) organisms.

The protein amino acid residues of peats of Dismal Swamp, Virginia, and North Carolina show a relationship to the environment of accumulation (Swain, Blumentals, and Millers, 1959). The Dismal Swamp peat consists of an upper unit 1.5–2 m thick of reddish brown woody peat derived from a black gum and cypress forest; this overlies 2 m or more of dark copropelic peat related to marsh and lake facies of the Swamp. The amino acids attain their greatest concentration in the lower part of the woody peat. The uppermost part of the peat does not yield amino acids to acid hydrolysis and paper chromatography. The amino acids released by microbial degradation of proteins are thought to become attached to humic acid micelles which, being water soluble, are concentrated downward during fluctuations of water levels in the Swamp. A decrease in total amino acids occurs in the dark copropelic peat as compared to the overlying woody peat; downward migration of amino acids into the copropelic peat is inhibited by its impervious character. Basic amino acids, for example histidine, are more plentiful in acidic Dismal Swamp peats than in more alkaline peats of the north-central United States, owing to their formation as stable salts. Acidic amino acids may be degraded by bacterial decarboxylation more easily in the acidic environment. Flavinoid pigments in Dismal Swamp peat also decrease downward from the reddish brown peat to the dark brown peat, both as a result of natural degradation and a decrease in biological activity in the copropel.

(b) 1. **Inorganic Geochemistry of Lakes of Glaciated Regions.** The sediment-geochemistry of lakes of low-carbonate glaciated areas differs from that of high- or moderate-carbonate bedrock and glacial drift. In Linsley Pond, Connecticut (Hutchinson and Wollack, 1940) and similar low-carbonate lakes, the inorganic constituents are almost completely detrital in origin: the silica is principally detrital quartz, the alumina is in feldspar and clays, and the titania in sphene, rutile, and ilmenite, etc. A little of the silica is represented by authigenic diatom frustules. High silica values at the base and top of the sediment (47–73%) represent the early oligotrophic lake stage and a contribution from agriculture, respectively. The rather high iron contents of the basal sediments (12% Fe_2O_3) may be partly detrital but also includes (bio-) chemically precipitated ferrous sulfide hydrate (melnikovite) and perhaps siderite. The manganese (0.08–1.03% MnO) in the sediment is also probably of (bio-) chemical origin. The low calcium (1.0–2.7% CaO) and rather high Mg/Ca ratios suggest that both are detrital to a certain extent. The phosphrous content (0.055–0.254% P_2O_5) is highest in the lower third of Linsley Pond sediments as a result of sorption with mineral matter (Livingston and Boykin, 1962). The reaction was due to the high ion-exchange capacity of the lake mud in the early stages of lake development. Lake productivity is believed also to be inversely related to exchange capacity of the sediments.

A lake that lies in an area of calcareous glacial drift, Nisswa Lake, Crow Wing County, Minnesota (Roepke, 1959) is characterized by rich marl deposits (15–47% CaO) the amount of which increases upward in the sediments. The lower part of the sediment is high in insolubles and represents an oligotrophic lake stage. There is correlation between M_2O_3 (0.42–2.74%) and insoluble (2.75–41.45%) content of the sediments resulting from the abundant clastic feldspars, clays, and other alumino-silicates in the marl sediments. Quartz grains and diatom frustules account for observed deviation between M_2O_3 and insolubles. There is a correlation between Fe_2O_3 (0.22–8.05%), SO_3 (0.58–2.87%), and ignition loss (5.96–23.60%) that is suggested as reflecting iron-sulfide formation in the presence of abundant organic matter. In a thermally stratified lake, ferric hydroxide undergoes reduction beneath the thermocline during summer stagnation and reacts with hydrogen sulfide evolved from decaying organic matter to form pyrite; below 15 feet in the Nisswa Lake sediments, judged from their sulfur content (0.8–2.8% SO_3), the lake was thermally stratified. Manganese appears to have formed authigenically in the profundal area of lake Nisswa sedimentation but not to any extent in the littoral sediments; again the manganese seems to have been related to subthermocline processes, probably microbial in nature. Carbonate deposition was most extensive in the shallower parts of the Lake Nisswa basin; the deeper parts apparently act as a sump for detrital and organic debris as well as being less favorable for carbonate precipitation. In other lakes of the region, however, carbonate deposition may prevail in the profundal areas both as a result of lesser deposition of other material and of issuance of carbonate-bearing springs in the deeper parts of the lake (Swain, 1961a).

(b) 2. **Organic Geochemistry of Lakes and Bogs of Glaciated Regions.** The organic geochemistry of the deposits of Rossburg Bog, Minnesota is exemplary of this type of deposit (Swain, 1967). This deposit began about 8000 years ago as a postglacial lake that quickly became eutrophic; 5–6 m of copropelic eutrophic lake sediments accumulated in the lake. Following the eutrophic lake stage the bog became filled with organic debris, became dystrophic, productivity declined, and moss and forest peat developed to a thickness of 4–5 m up to the present surface. Saturated hydrocarbons occur in relatively high values in the middle part of the moss peat but in general are smaller in amount there than in the underlying copropelic peat; presumably these are original differences in source material. The aromatic hydrocarbons are somewhat more abundant than the saturated hydrocarbons but have similar distribution. Napthols form a significant part of the aromatic fractions and are believed to have been formed by degradation of naphthylamine, an auxin or growth accelerating substance. Low-temperature pyrolysis of the moss peat yielded phenols as a principal distillation product; the copropelic peat yielded toluene. The phenols are possibly derived from sphagnol or a related peat acid, and the toluene from carotenes which are known to occur in the copropel.

The protein amino acids of Rossburg Bog peat, as has been shown for other peats, do not have their greatest concentration in the uppermost layers of peat but a meter or so below. As in Dismal Swamp, the basic amino acid content indicates that acid conditions have prevailed through much of the history of Rossburg Bog. The ratios of the amounts of certain amino acids in the bog are related to stratigraphy. The valine-to-alanine ratio, for example, is rather uniform and near unity in the moss peat but is higher in the copropelic peat and is lower in the underlying silt. Such variations as this probably are due to differences in source material rather than to postdepositional changes.

The carbohydrate components of Rossburg Bog do not reflect the peat stratigraphy as well as do the amino acids, but certain relationships do exist. There is an increase in carbohydrates about 0.3 m below the surface that may be related to changing water levels. Mannose is rare in the upper 0.6 m of moss peat but shows a small increase below; galactose shows a similar distribution above and below 1.3 m in the moss peat. Ribose, possibly of primary microbial origin, is more abundant in the copropelic peat than in the moss peat. Glucose, xylose, and arabinose are most common in the upper part of the moss peat and decrease to a roughly uniform level in the lower moss peat and underlying copropelic peat. Both mannose and galactose are important constituents in the *Sphagnum* moss that contributes to much of the moss peat, but these two sugars are not abundant in the peat as a whole by comparison with glucose and xylose and are apparently less stable than the latter two in this deposit. The relative stability of sugars in acid lake and peat deposits seems to be less than in slightly alkaline or neutral deposits; the reduction in total carbohydrates from the original source material is about 10^{-1} in neutral and slightly alkaline deposits but is about 10^{-2} in acidic deposits.

Chlorophyll-derived pigments such as pheophytin are present in more or less uniform amounts in both the moss peats and copropels of Rossburg Bog. Carotenes, however, are much more abundant in the copropel than in the overlying moss peat, apparently owing to a primary source of carotenes from planktonic algae in the lake stage that largely disappeared in the bog stage.

(c) **Organic Geochemistry of a Tertiary Lake Deposit.** The Eocene Green River Oil Shale of Colorado, Utah, and Wyoming has been studied for its organic contents because of its importance as a source of petroleum. The Green River Formation was deposited in several former lake basins in southwestern Wyoming, northeastern Utah, and northwestern Colorado. The formation comprises more than 5000 feet of shale, sandstone, fossiliferous freshwater limestone in the lower part and carbonate, organic-rich oil shales, and evaporites in the middle and upper parts. Crude oil of highly paraffinic composition is produced from sandstones and oölitic limestones of the lower Green River, and the middle and upper parts of the formation contain large reserves of oil shale from which oil can be obtained by distillation or hydrogenation. In addition, the Green River Formation is noted for its contents of solid hydrocarbons, several of which can be identified from their infrared absorption spectra (Hunt, Stewart, and Dickey, 1954): ozocerite (absorption maximum at 13.9μ); gilsonite and tabbyite (small λ_{max} at 8.9, and 9.65μ and larger λ_{max} at 13.9μ; albertite and ingramite λ_{max} at 5.9, 10.8, and 13.4μ). There is stratigraphic control on these hydrocarbons which in turn appears to be related to source materials and the sedimentary environment. The ozocerite, high in paraffinic hydrocarbons, originated in the basal fresh-water, mollusk-, and ostracode-rich Green River shales; grahamite and albertite occur and seem to have originated in the lower few hundred feet of the formation. The gilsonite and wurtzilite on the other hand formed in the more alkaline facies of the middle and upper Green River, and probably are related to the pigment-producing planktonic algal organic matter of those stages of the lakes. In support of the latter conclusion is the finding of isoprenoid hydrocarbons in the middle Green River oil shales (Eglinton, et al., 1966). These hydrocarbons, pristane and phytane, probably were derived from the phytyl alcohol side chains of chlorophyll.

The amino acids of the Mahogany Ledge, a commercial-grade Green River Oil Shale include several common protein constituents plus γ-aminobutyric acid of nonprotein origin, amounting to only 0.014% of the total nitrogen (Jones and Vallentyne, 1960). The original extensive suites of amino acids, probably from phytoplankton, have been largely degraded and complexed with the kerogen of the oil shale.

The carbohydrates of the Mahogany Ledge consist of glucose and possibly xylose and glycerol (Palacas et al., 1960) whereas those of the lower Green River are glucose and possibly galactose.

Aromatic carboxylic acids from the Green River Oil Shale include benzoic, propanoic, butanoic, indanoic, tetrahydronaphthoic, and naphthoic acids (Haug et al., 1967). These are suggested as having been derived from cyclic terpenoid precursors. A number of dicarboxylic acids, keto acids, kerogen acids and isoprenoid acids and fatty acids have also been found in the Green River Oil Shale (Abelson and Parker, 1962; Lawlor and Robinson, 1965; Eglinton et al., 1966; Douglas et al., 1968). The isoprenoid acids are thought to have been derived from the phytol side chain of chlorophyll.

In summary, there is a more or less close relationship between the preserved inorganic and organic constituents of lake and bog sediments and the source materials and sedimentary environments under which the materials accumulated. These generally overshadow but do not completely obscure diagenetic changes in the sediments.

F. M. SWAIN

References

Abelson, P. H., and Parker, P. L., 1962, "Fatty acids in sedimentary rocks," *Carnegie Inst. Wash. Yearbook*, **61**, 181–184.

Bradley, W. H., 1963, "Paleolimnology," in (Frey, D. G., editor), "Limnology in North America," Madison, Univ. Wisconsin Press, pp. 621–652.

Douglas, A. G., Douraghi-Zadeh, K., Eglinton, G., Maxwell, J. R., and Ramsey, J. N., 1968, "Fatty acids in sediments . . ." in (Hobson, G. D., and Speers, D., editors), "Advances in Organic Geochemistry," 1966, Oxford, Pergamon Press, 315–334.

Eardley, A. J., 1938, "Sediments of Great Salt Lake, Utah," *Bull. Am. Assoc. Petrol. Geol.*, **22**, 1305–1411.

Eardley, A. J., and Gvosdetsky, V., 1960, "Analysis of Pleistocene core from Great Salt Lake, Utah," *Bull. Geol. Soc. Am.*, **77**, 1323–1344.

Eardley, A. J., Gvosdetsky, V., and Marshall, R. E., 1957, "Hydrology of Lake Bonneville and sediments and soils of its basin," *Bull. Geol. Soc. Am.*, **68**, 1141–1201.

Eglinton, G., Maxwell, J. R., Murphy, Sister M. T. J., Henderson, W., and Douraghi-Zadeh, K., 1966, "Hydrocarbons and fatty acids in algal shales and related materials," *Program Geol. Soc. Am.* 79th Ann. Meet., San Francisco, p. 59.

Haug, P., Schnoes, H. K., and Burlingame, A. L., 1967, "Isoprenoid and dicarboxylic acids isolated from Colorado Green River Shale (Eocene)," *Science*, **158**, 772–773.

Hunt, J. H., Stewart, F., and Dickey, P. A., 1954, "Occurrence of hydrocarbons of Uinta Basin," *Bull. Am. Assoc. Petrol.*, **38**, 1671–1678.

Hutchinson, G. E., Patrick, R., and Deevey, E. S., Jr., 1956, "Sediments of Lake Patzcuaro, Michoacan, Mexico," *Bull. Geol. Soc. Am.*, **67**, 1491–1504.

Hutchinson, G. E., and Wollack, A., 1940, "Studies on Connecticut lake sediments. II. Chemical analyses of a core from Linsley Pond, North Branford," *Am. J. Sci.*, **238**, 493–517.

Jones, J. D., and Vallentyne, J. R., 1960, "Biogeochemistry of organic matter. I. Polypeptides and amino acids in fossils and sediments in relation to geothermometry," *Geochim. Cosmochim. Acta*, **21**, 1–34.

Langbein, W. B., 1962, "The salinity and hydrology of closed lakes," *U.S. Geol. Surv. Profess. Paper* **412**, 1–20.

Lawlor, D. L., and Robinson, W. E., 1965, "Organic acids from the Green River Formation," *Div. Pet. Chem., Am. Chem. Soc.*, Detroit Meet., May 9.

Livingston, D. A., and Boykin, J. C., 1962, "Vertical distribution of phosphorus in Linsley Pond mud," *Limnol. Oceanog.*, **7**, 57–62.

Palacas, J. G., Swain, F. M. and Smith, F., 1960, "Presence of carbohydrates and other organic compounds in ancient sedimentary rocks," *Nature*, **185**(4708), 234.

Roepke, H. H., 1959, in (Schwartz, G. M., et al., editors), "Investigations of the Commercial Possibilities of Marl in Minnesota," St. Paul, Minn., Iron Range Res. and Rehab. Comm., pp. 50–105.

Swain, F. M., 1961a, "Limnology and amino acid content of some lake deposits in Minnesota, Montana, Nevada and Louisiana," *Bull. Geol. Soc. Am.*, **72**, 519–546.

Swain, F. M., 1961b, "Reporte preliminar de los sedimentos del fonds de los lagos Nicaragua y Managua," *Bol. Serv. Geol. Nac. Nicaragua*, **5**, 11–29.

Swain, F. M., 1966, "Bottom sediments of Lake Nicaragua and Lake Managua, western Nicaragua," *J. Sediment. Petrol.*, **36**, 522–540.

Swain, F. M., 1967, "Stratigraphy and biochemical paleontology of Rossburg Bog (Recent) Aitkin County, Minnesota," in (Teichert, C. and Yochelson, E., editors), "Essays in Paleontology and Stratigraphy," R. C. Moore Commem. Vol., Lawrence, Kansas, Univ. Kans. Press, pp. 445–475.

LAKE GEOCHEMISTRY

Swain, F. M., Blumentals, A., and Millers, R., 1959, "Stratigraphic distribution of amino acids in peats from Cedar Creek Bog, Minnesota and Dismal Swamp, Virginia," *Limnol. Oceanog.*, **4**, 119–127.

Swain, F. M., and Gilby, J. M., 1965, "Ecology and taxonomy of Ostracoda and an alga from Lake Nicaragua," *Staz. Zool. Napoli,* **33**(suppl.), 361–386.

Cross-references: *Authigenesis of Minerals, Nonmarine; Biogeochemistry; Carbon-14 Dating; Green River Hydrocarbons; Organic Geochemistry; Organic Pigments; pH-Eh; Seawater, Chemistry; Weathering, Chemical. Vol. I: Phytoplankton. Vol. III: Great Salt Lake; Holocene; Lakes. Vol. IVB: Green River Mineralogy; Mirabilite. Vol. VI: Kerogen; Lacustrine Sedimentation; Marl; Oil Shales; Weathering, Organic. Vol. VII: Algae; Diatoms; Palynology; Pleistocene; Tertiary.*

LAKES—See Vol. III

LANTHANIDES

The lanthanides, sometimes called lanthanons, are the fourteen elements following lanthanum in the Periodic Table. Table 1 lists them, together with lanthanum, yttrium, and scandium.

The chemical properties of the lanthanides are similar because their outer electronic configurations are identical; electrons fill inner, chemically inactive, f orbitals. Adding an electron to an f orbital and a proton to the nucleus of one lanthanide gives the lanthanide of the next higher atomic number. The increased electrostatic attraction of the nucleus on all the f electrons is only partly offset by the mutual repulsion or shielding of the electrons in the highly directional f orbitals, resulting in a decrease in atomic and ionic radius with increased atomic number (the "lanthanide contraction"). However, heavy rare earths form more stable hydration complexes than light rare earths, so that the heavy lanthanides have larger aquated radii than the light lanthanides.

Since many lanthanide compounds are ionic, the decrease in ionic radius from heavy to light lanthanides explains the regular variation in many chemical and physical properties. The relative stability of different lanthanide ions in ionic crystals depends on ionic radius; lanthanides, especially heavy ones, substitute for calcium (ionic radius 0.99 Å). Notable discontinuities occur for europium and cerium in geologic environments they occur in divalent and tetravalent oxidation states respectively, as well as in the trivalent state common to other lanthanides.

Generally the spectral and magnetic properties of rare-earth ions originate from electronic interactions in the shielded f shell, and are not affected by chemical bonding (as are

TABLE 1.

Element	Atomic Number	Electronic Structure of Ion[a] (Charge of Ion)		Ionic Radius (Charge of Ion)
Lanthanum (La)	57	[Xe]	(+3)	1.06Å (+3)
Cerium (Ce)	58	[Xe]	(+4)	0.92Å (+4)
		[Xe]$4f^1$	(+3)	1.034Å (+3)
Praseodymium (Pr)	59	[Xe]$4f^2$	(+3)	1.013Å (+3)
Neodymium (Nd)	60	[Xe]$4f^3$	(+3)	0.995Å (+3)
Promethium (Pm)	61	[Xe]$4f^4$	(+3)	0.979Å (+3)
Samarium (Sm)	62	[Xe]$4f^5$	(+3)	0.964Å (+3)
Europium (Eu)	63	[Xe]$4f^6$	(+3)	0.950Å (+3)
		[Xe]$4f^7$	(+2)	1.09Å (+2)
Gadolinium (Gd)	64	[Xe]$4f^7$	(+3)	0.938Å (+3)
Terbium (Tb)	65	[Xe]$4f^8$	(+3)	0.923Å (+3)
Dysprosium (Dy)	66	[Xe]$4f^9$	(+3)	0.904Å (+3)
Holmium (Ho)	67	[Xe]$4f^{10}$	(+3)	0.894Å (+3)
Erbium (Er)	68	[Xe]$4f^{11}$	(+3)	0.881Å (+3)
Thulium (Tm)	69	[Xe]$4f^{12}$	(+3)	0.869Å (+3)
Ytterbium (Yb)	70	[Xe]$4f^{13}$	(+3)	0.858Å (+3)
Lutetium (Lu)	71	[Xe]$4f^{14}$	(+3)	0.848Å (+3)
Yttrium (Y)	39	[Ar]	(+3)	0.88Å (+3)
Scandium (Sc)	21	[Kr]	(+3)	0.68Å (+3)

[a] [Xe] designates a closed shell of xenon configuration: $1s^2 2s^2 2p^6 3s^2 3p^6 3d^{10} 4s^2 4p^6 4d^{10} 5s^2 5p^6$.
[Ar] argon configuration $1s^2 2s^2 2p^6 3s^2 3p^6$.
[Kr] krypton configuration $1s^2 2s^2 2p^6 3s^2 3p^6 3d^{10} 4s^2 4p^6$.

the properties of transition metals). The visible absorption lines of many lanthanides are sharp.

Lanthanum, the lanthanides, yttrium and sometimes scandium, are often grouped together under the name *rare earths* (q.v.), and further into the heavy rare earths or yttrium group, including yttrium and the lanthanides europium to lutecium, and the light rare earths or cerium group, including lanthanum and the lanthanides cerium to samarium.

The history of the tedious and confusing progress leading to isolation of the rare-earths began with the discovery of a rare black mineral (gadolinite) at Ytterby, Sweden by C. A. Arrhenius in 1787. J. Gadolin isolated a new "earth" or oxide from the mineral; it was thought to contain a single metal, which was named yttria. In 1803 M. Klaproth and, independently, J. Berzelius, isolated a different earth, ceria, from the mineral cerite, found at Bastnaes, Sweden. In the 1840s, the Swedish chemist Mosander used differences in solubility to separate yttria into three earths, yttrium, old erbia, and old terbia; he also separated ceria into cerium, lanthanum, and didymia. Attempts at separation continued, but not until after 1900 was didymia separated into the elements praseodymium, neodymium, samarium, and europium; old erbia and old terbia were separated into their constituent heavy rare earths.

Promethium 147, with a 2.6-yr half-life, was the first definitely characterized promethium. Marinsky et al., (1947) separated it from uranium fission products by ion exchange, an analytical technique (together with solvent extraction) that has been used widely to separate both trace and commercial quantities of lanthanides.

For more on lanthanides see entries on the individual elements and on rare earths.

ROBERT KAY

References

Cotton, F. A., and Wilkinson, G., 1962, "Advanced Inorganic Chemistry: A Comprehensive Text," New York, Interscience Publishers a div. of John Wiley, 959pp.

Haskin, M. A., and Haskin, L. A., 1966, "Rare earths in European shales: a redetermination," *Science,* **154**(3748), 507–509.

Marinsky, J. A., Glendenin, L. E., and Coryell, C. D., 1947, "The chemical identification of radioisotopes of neodymium and of element 61," *J. Am. Chem. soc.,* **69**, 278.

Moeller, T., 1963, "The Chemistry of the Lanthanides," New York, Reinhold Publ. Co., 117pp.

Cross-references: *Atomic Number and Periodic Table; Bonding; Ion Exchange; Rare Earths.*

LANTHANUM: ELEMENT AND GEOCHEMISTRY

Properties

Elemental lanthanum is a very reactive, silvery-white *rare earth* (q.v.) metal, with atomic number 57 and outer electronic structure $5d^1 6s^2$. The elements from atomic number 58 to 71 are chemically similar to lanthanum, and therefore are called *lanthanides* (q.v.).

Natural lanthanum is 99.911% La^{139} and 0.089% La^{138}, a radioactive isotope with a half-life of 2×10^{11} years.

Lanthanum is determined analytically by neutron activation, x-ray spectrography, and isotope dilution mass spectrometry. In geologic environments, lanthanum forms highly ionic bonds; it forms weak complexes. The oxidation state is tripositive, with ionic radius 1.60 Å. Generally, chemical behavior is similar to the alkaline earths, and to other light rare earths.

Mosander (in 1839) first isolated lanthanum from other rare earths; he separated it out of ceria, a mixed light rare earth oxide. The name lanthanum means "to lie hidden."

Minerals

Lanthanum occurs in highest concentration in rare earth minerals that favor light rare earths, generally nitrates, sulfates, carbonates, niobates and tungstates with high cation coordination number, often containing structural halogens and alkali metals. Typical examples, with cation coordination number 10 and 11, are:

monazite (light rare earths, Th) PO_4

bastnaesite (light rare earths, Th) FCO_3

Abundance

Schmitt et al. (1964) established a value 0.30 ppm (parts per million) lanthanum in chondrites and 3.7 ppm in eucrite achondrites. The standard deviation within a meteorite class is only about 20%. In meteoritic matter volatilization may have redistributed alkali metals but not rare-earths: lanthanum has a boiling point of 3468°C.

The lanthanum concentration in average basalts is 17 ppm, but the range is over a factor of ten (Haskin et al., 1966). Alkaline oceanic basalts and most continental basalts have more lanthanum than oceanic tholeiites and some continental diabases. Partial melting of mantle peridotite yields a basaltic liquid with enriched lanthanum and a mantle source area depleted in lanthanum.

The lanthanum concentration in granites with 60–70% silica averages 84 ppm, a value that is higher, but also more variable, than for basalts.

Igneous differentiation series tend toward higher lanthanum concentrations. The relative abundance of lanthanum and other rare earths changes. Haskin et al. (1966) report 47 ppm lanthanum in a Gough Island basalt and 200 ppm in a trachyte differentiate. Towell et al. (1965) found 24.5 ± 1.0 ppm in the Rubidoux Mountain Leucogranite phase of the southern California batholith, probably a differentiate (by crystal settling) of the San Marcos Gabbro, with 4.0 ± 0.16 ppm. About half of the lanthanum in the gabbro is in rock-forming minerals (plagioclase, pyroxene, hornblende), probably substituting for calcium. The remainder is in accessory minerals like apatite. The accessory minerals (probably apatite, zircon, sphene, and monazite) hold most of the lanthanum in the leucogranite.

Lanthanum forms complexes in alkali, carbonate, and halogen-rich hydrothermal fluids, but they are not as stable as heavy rare earth complexes. Pegmatoid and metasomatic deposits, formed from hydrothermal fluids, often have low lanthanum to heavy rare earth ratios compared to granites.

Lanthanum has a 210 year residence time in sea water, about the same as other light rare earths. Probably it is complexed. Removal by adsorption on clay particles, as well as depletion in surface water in areas of organic productivity, are likely processes. Today, high secondary lanthanum concentrations in phosphatic fish remains occur only in deep water; Mesozoic fish remains also have high secondary lanthanum. Manganese nodules have $\sim 11,700$ ppm lanthanum whereas Pacific Ocean water has only 2.9×10^{-6} ppm (Goldberg et al., 1963).

In soils, the relative amounts of lanthanum in detrital, clay, and carbonate fractions is not well known. Sedimentary rocks have about 39 ppm lanthanum, with a variation of a factor of five. In a study of rare earths in Russian platform soils, Balashov et al. (1964) found higher concentrations in alkaline soils, the lanthanum precipitating as hydroxide; lower concentrations are in acid soils, probably due to removal as complexes.

Biologic concentration of rare earths has been found in hickory leaves; rare earths have been used to trace opium.

Economic Geology

The major source of lanthanum has been from monazite in beach sands. The accessory monazite originally present in granitic rocks resists weathering and is concentrated by sedimentary processes. Australia, Brazil, Ceylon, India, Malaysia, and South Africa mine the largest quantities. Usually, about 5% (by weight) of monazite is lanthanum. Recently, the bastnaesite deposit at Mountain Pass, California has been an important source of lanthanum.

High-purity lanthanum has limited application in electronics and in optical filters. Most lanthanum is sold unpurified as misch metal (a mixed rare earth alloy), or as oxide polishing powders. Unpure rare earths are also used as petroleum cracking catalysts and in arc carbons.

ROBERT KAY

References

Balashov, Y. A., et al., 1964, "The effects of climate and facies environment on the fractionation of the rare earths during sedimentation," *Geochem. Internat. No 1,* **10,** 951–969 (English transl.).

Goldberg, E. D., et al., 1963, "Rare-earth distributions in the marine environment," *J. Geophys. Res.,* **68,** 4209–4217.

Haskin, L. A., et al., 1966, "Meteoric, Solar, and Terrestrial Rare-Earth Distributions," in (Ahrens, L. H., et al., editors), "Physics and Chemistry of the Earth," Vol. 7, Oxford, Pergamon Press, pp. 167–321.

Rankama, K., and Sahama, T. G., 1950, "Geochemistry," Chicago, University of Chicago Press, 912pp.

Schmitt, R. A., et al., 1964, "Rare-earth, yttrium and scandium abundances in meteoric and terrestrial matter—II," *Geochim. Cosmochim. Acta,* **28,** 67–86.

Towell, D. G., et al., 1965, "Rare-earth distributions in some rocks and associated minerals of the batholith of Southern California," *J. Geophys. Res.,* **70,** 3485–3496.

Cross-references: *Halogens; Mass Spectrometry; Neutron Activation Analysis; Rare Earths; X-Ray Spectroscopy.*

LEACHING—*See* Vol. VI

LEAD: ELEMENT AND GEOCHEMISTRY

Physical Properties

Lead (from Anglo-Saxon *lead*), symbol Pb (L. *plumbum*); atomic weight 207.19. Natural variations in isotopic composition of lead produce small variations in atomic weight, but almost all samples of common lead have atomic weights in the range 207.17–207.25. Samples of radiogenic lead vary in atomic weight from 206.1 to near 208.) Atomic number 82; electronic configuration (beyond xenon) $4f^{14}5d^{10}6s^26p^2$; melting point 327.4°C; boiling point 1620°C; specific gravity 11.35

(20°C); valence 2 or 4; crystal structure: face-centered cubic (Cu structure). Four stable nuclides occur in nature in varying relative amounts; for almost all samples, the isotopic abundances are within the following ranges: Pb^{204} 1.2–1.6%; Pb^{206} 20–28%; Pb^{207} 20–23%; Pb^{208} 50–54%. Four short-lived radioactive isotopes, Pb^{210}, Pb^{211}, Pb^{212}, and Pb^{214}, occur naturally as decay products of uranium and thorium. Lead is a bluish-white metal, very soft, and malleable. The bright luster of a fresh surface dulls readily by oxidation, but the metal is resistant to corrosion. Metallic lead has poor tenacity, is ductile, and is a poor conductor of electricity.

Lead was one of the first metals to have been used by man (dating back to the 4th millenium BC) and may have been the first obtained by smelting of an ore. In the Ancient World, lead was used for making glazes, pigments, and glass, and writings were inscribed on lead sheets. By Roman times, it was used in large quantities for plumbing and other architectural purposes. Lead has always been obtained almost exclusively by smelting of galena, PbS, which is more abundant than all other lead minerals.

Compounds

The major lead mineral is galena, PbS, but a large number (about twenty) lead sulfates also contribute to the production of this metal. In the oxidized zone of sulfide ore deposits a great variety of lead minerals are found, the most important of which are *cerussite*, $PbCO_3$, and *anglesite* $PbSO_4$. In a rare but interesting occurrence, *cotunnite*, $PbCl_2$, is formed as a sublimate from volcanic fumaroles. Important commercial products are the metal itself; alloys with other low-melting temperature metals; lead tetraethyl, $Pb(C_2H_5)_4$; lead dioxide, PbO_2; and several compounds which are useful as pigments: Pb_3O_4 (red lead), $2PbCO_3 \cdot Pb(OH)_2$ (white lead), $PbSO_4$ (sublimed white lead), and $PbCrO_4$ (Chrome yellow).

Chemistry

Although lead is a member of Periodic Group IV, the bivalent compounds of lead in which the two $6s$ electrons are inert are more usual than the quadrivalent compounds in which all four valence electrons are used in bond formation. A number of colorless salts of Pb^{2+} have fairly strong ionic bonds, but a variety of colored bivalent compounds, including the oxide, the sulfide, and the iodide, have complex bonding relationships. Bivalent lead forms relatively unstable covalent compounds with organic radicals. Lead may be partially oxidized to the tetravalent state by heating in air to form Pb_3O_4. Fully oxidized lead is formed only under very strong oxidizing conditions, and few simple compounds other than PbO_2 are stable.

$$Pb^{2+} + 2 H_2O \rightarrow PbO_2 + 4H^+ + 2e^-$$
$$E° = -1.456V$$

The tetravalent state can be stabilized by the formation of complex anions. Tetrahedrally hybridized lead forms a variety of covalent compounds with organic radicals, of which the lead *tetraalkyl* compounds are of particular importance.

Many of the salts of Pb^{2+} are relatively insoluble in water, in some cases owing to the ionic properties of the large cation, and in other cases owing to the degree of covalent bonding in the crystal. The very low solubility of PbS (except at very low pH) is most important in the geochemistry of lead in solution. Pb^{2+} forms an amphoteric hydroxide which is only slightly soluble in the intermediate pH range but dissolves as $Pb(OH)_3^-$ in the presence of excess alkali (but not with ammonia). The quadrivalent PbO_2 is also distinctly amphoteric, forming $Pb(OH)_6^{2-}$ in alkaline solution. A variety of other complex anions may be formed in aqueous solution, some of which have been used to advantage in ion exchange separation of lead.

Lead lies just above hydrogen in the electrochemical series.

$$Pb° \rightarrow Pb^{2+} + 2e^- \qquad E° = 0.126V$$

It reacts very slowly with nonoxidizing acids, but is readily dissolved by nitric acid or highly concentrated sulfuric acid. Its relative inertness is of particular commercial importance in the manufacture of sulfuric acid. In the atmosphere, a thin film of oxide or carbonate is formed on the surface of lead metal, giving high resistance to atmospheric corrosion.

Geochemistry

Lead is the most abundant of the heavy elements (atomic number greater than 60) in the earth's crust. Its average abundance in the crust is about 15 ppm by weight. Several factors contribute to its relatively high abundance: the cosmic abundance of lead is relatively high because its three major isotopes, Pb^{206}, Pb^{207} and Pb^{208}, are end products of the "slow" neutron capture processes which are responsible for the formation of a portion of the nuclides of the heavy elements. These same isotopes of lead are also the ultimate end products of decay of a considerable portion of the radioactive nuclides produced by "rapid" neutron capture processes. Of overwhelming importance in this regard are the decay of uranium and thorium; these elements are highly enriched in the earth's crust, and their decay throughout geologic time has produced one-third of the lead in the crust today. Finally, lead itself tends to be enriched in the crust; the abundance of lead in ultrabasic rocks and in stone meteorites is roughly 100 times less than its crustal abundance, so it is expected that there is a similar difference in lead content between the crust and the mantle.

The lead content of the metal phase of iron meteorites is also quite low, generally 0.1–0.2 ppm. The sulfide phase (troilite), however, may contain from several ppm to several tens of ppm lead. This, together with the common occurrence of lead as the sulfide (galena) in the earth's crust, has led many authors to classify lead as a *chalcophile* element. However, the major portion of lead in the earth's crust is found, not in sulfide minerals but dispersed in silicates. Clearly lead behaves also as a *lithophile* element in the crust.

The occurrence of lead in magmatic rocks is governed primarily by its relationship to potassium. The divalent lead ion has an ionic radius of 1.20 Å, slightly smaller than that of K^+ (1.33Å), and it is as a substitute for potassium that lead is most abundant. Micas and potassium feldspars have, on the average, 25–30 ppm lead. Lead also occurs in appreciable concentration in minerals with low potassium content—plagioclase, amphiboles, and pyroxenes. Lead may in part substitute for calcium (Ca^{2+}, 0.99 Å) in these cases. Of the common types of igneous rock, lead is most abundant in granite, having an average abundance of 20 ppm.

In spite of its higher charge and smaller size than the K^+ ion, lead tends to be preferentially excluded from early-formed potassium minerals and thus enriched in minerals formed during the final stages of crystallization. This effect is strikingly seen in the case of some pegmatitic K-feldspars which have several hundred ppm lead. (Note however, that the Pb/K ratio may be relatively high in rocks such as basalt because of substitution of lead for calcium.) The failure of lead to follow the accepted rules for ionic substitution is attributed to the moderately high electronegativity of lead (1.60) compared to the alkali and alkaline earth metals. (Strontium, in contrast, with the same charge and about the same size as lead, behaves normally in that the Sr/K ratio is highest in early-formed K minerals.)

There is considerable evidence that lead is not distributed in igneous minerals in a manner as uniform as that of the more electropositive elements, suggesting that the tendency of lead to form bonds with covalent character affects distribution on a submicro scale in ways which are not yet clear.

The behavior of lead during weathering and sedimentation is strongly dependent on environmental conditions, a fact which is reflected by the highly variable lead content of soils (2–200 ppm). Under reducing conditions lead forms the highly insoluble sulfide. Under normal oxidizing conditions lead may be carried in solution in somewhat acidic waters (but not in high concentration because of the low solubility of the sulfate). In neutral or somewhat alkaline water, the lead ion becomes hydrolyzed and is readily co-precipitated with hydroxides of more abundant elements or adsorbed by clay minerals. Owing to this behavior and also to precipitation via biological activity, the concentration of lead in seawater is extremely low—0.03 parts per billion.

Because of the efficiency of the sulfide precipitation of lead, it is sometimes found in fairly high concentration in sediments formed under reducing conditions. The great bulk of sedimentary lead is found, however, in ordinary shale at an average concentration of 20 ppm. It is interesting that deep-sea clays contain considerably more lead than average shale; a number of factors may contribute to this effect.

The abundance of lead in metamorphic rocks is determined primarily by the nature of the original rock. It is found in highest concentration in potassium-rich metamorphic rocks such as mica schists and gneisses.

The biogeochemical role of lead is a minor one. Its most important biological aspect is its toxicity, which is sometimes important in environments where the concentration of lead is abnormally high. The important role of lead in sulfide ore deposits is described in another article. Another very important aspect of the geochemistry of lead—its occurrence as the radiogenic daughter of uranium and thorium, and the importance of this occurrence in the interpretation of geological events and processes—is more fully described in other articles.

J. M. WAMPLER

References

*Ahrens, L. H., 1964, "The Significance of the Chemical Bond for Controlling the Geochemical Distribution of the Elements—Pt. 1," in (Ahrens, L. H., et al., editors), "Physics and Chemistry of Earth," Vol. 5, Oxford, Pergamon Press, pp. 1–54.

*Taylor, S. R., 1964, "The Application of Trace Element Data to Problems in Petrology," in (Ahrens, L. H., et al., editors), "Physics and Chemistry of the Earth," Vol. 6, Oxford, Pergamon Press, pp. 133–213.

*Wedepohl, K. H., 1956, "Untersuchungen zur Geochemie des Bleis," *Geochim. Cosmochim. Acta,* **10,** 69–148.

*Additional bibliographic references may be found in this work.

Cross-references: *Alkalis, Alkali Metals and Alkaline Earth Metals; Bonding; Earth's Crust Geochemistry; Electronegativity; Lead: Interpretation of Stable Isotope Abundances; Oxidation and Reduction; Sulfides in Sediments; Sulfosalts; Thorium; Uranium; Uranium–Thorium–Lead Age Determinations.* Vol. II: *Meteorites.* Vol. IVB: *Galena.* Vol. V: *Fumaroles.*

LEAD AND ZINC: ECONOMIC DEPOSITS

In nature, lead and zinc virtually always display a similar geochemical behavior, as outlined in separate entries (q.v.). These geochemical properties control the distribution and migration of lead and zinc in the course of the various rock-forming processes, i.e., the geochemical cycles (q.v.; see also Fig. 1 of the entry *Copper: Economic Deposits*).

In *igneous* rocks lead and zinc are distinctly more abundant in intermediate and still more in acidic rocks than in basic and ultrabasic ones. In intermediate to acidic extrusive belts such as the greenstone belts of the Caledonides that extend from Finland through Norway to Ireland, and even more pronounced in the Kuroko deposits of Japan, commercial lead zinc contents are found in massive stratiform pyrite deposits closely associated or mixed with copper sulfides. These worldwide stratabound volcanic or volcanic-exhalative-sedimentary deposits occupy positions A to H or even I in Fig. 1. In intrusive igneous rocks the lead and zinc sulfides are normally more clearly separated from the pyrite or copper sulfide portions of veins or mantos. Frequently, the lead-zinc portions of mines form the outer zones of vein systems in which the lower levels display a copper mineralization, a sequence which is called "telescoping" on account of its progressive mineralization. Barite and fluorite often accompany lead and zinc ores or follow as a last generation.

In *sediments* lead and zinc sulfides are found abundantly in all known carbonate provinces, but do occur also in the shale facies and as cements in sandstones and conglomerates, together with a calcareous cement. Barite, fluorite, and less abundantly copper, nickel, and cobalt contents are also found within or in adjacent zones. In these worldwide and very extensive deposits the two metal sulfides,

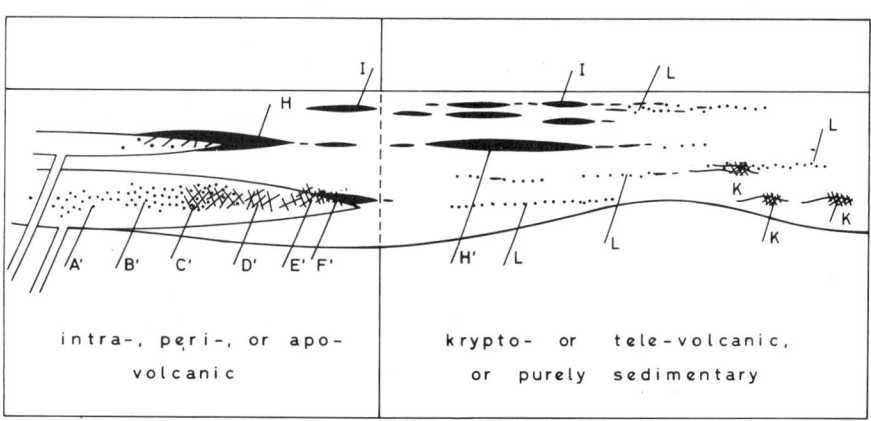

Fig. 1. Stratiform or strata bound types of ore deposits. *Left:* some basic ore patterns in and near extrusive metalliferous porphyries. (Normally with spilitic or porphylitic greenschist or green tuff "facies".) *Right:* some basic patterns in sediments. Some typical examples: Rio Tinto belt, Spain-Portugal; volcanic belt from Ireland to Scotland to Norway; Appalachian and New Brunswick belt; Coast Range belt, Western United States; Noranda belt, Quebec, and other Precambrian belts in shield areas; Ural Mts. belt; Japanese belts of three geologic ages. A', B', C', D', E', F', correspond roughly to the intrusive patterns in porphyry coppers; H and H' are massive deposits adjacent and somewhat removed from the extrusive rock; I are lenticules, layers, or layered patches; K are late diagenetic compaction fractures in and near organic riffs; L represent disseminations. H', I, K, and L may also occur as euxenic sediments in no connection with extrusive rocks or exhalations.

galena, PbS, and sphalerite, ZnS, are invariably late diagenetic, galena postdating sphalerite in most instances. This leads locally to late diagenetic semi-crosscutting accumulations in solid sulfide beds or pockets or in diagenetic compaction fissures, for examples along organic reef structures.

Among these worldwide stratabound lead-zinc deposits, those of the Mississippi Valley, Silesia, and Morocco-Algeria-Tunis are to date the most extensive lead-zinc producers; but identical deposits are present in similar lithologic, paleographic positions in all carbonate provinces. These paleographic loci are apparently zones of reduced lithologic thicknesses, so-called pinch-outs along coasts or over subaqueous highs.

Oxidation zones and oxidation ores of lead and zinc may be of commercial value and are mined in many countries, although only in relatively small quantities. Silver (normally in the native state) often accumulates in such oxidation zones.

Metamorphism has affected all the above-mentioned deposits and may cause recrystallization and very local remobilization. There is no safe proof for the large-scale remobilization of sulfide ores proposed some years ago. The zinc oxide deposits in the metamorphic carbonate beds of Franklin, New Jersey, and a similar occurrence in Russia, are probably metamorphic oxidation products of original stratiform sphalerite deposits. Also the stratabound ores, metamorphic or not, must be considered today as contemporaneous on the basis of the perfect congruency on all scales, the abundant sedimentary textures, and the results of isotopic analyses.

The *world production* of Pb and Zn in 1968 was (for the Western World), in 1000 metric tons, according to estimates of a U.N. study group:

	Pb	Zn
Mine production	2133	3886
Smelter production	2825	3590
Consumption	2630	3694

Most abundant among the deposits described are the stratabound deposits in carbonate provinces (right side on Fig. 1). The production from the so-called massive sulfide belts (left side in Fig. 1) is also considerable (about two-thirds and one-third, respectively, of the total production). The vein-type deposits in or around igneous rocks occupy a relatively unimportant position in the production of these metals. The major Pb- and Zn-producing nations are Canada, the U.S.A., Australia, Mexico, Peru, and Yugoslavia.

The *Mineralogy* of lead and zinc ores is extremely simple except for the oxidation ores, in that sphalerite, ZnS, and galena, PbS, provide about 80 to 90% of the whole production of these two important base metals. Important diadochic by-products of zinc ores are cadmium, indium, and germanium. In many lead ores, silver is an important by-product. The most common economic minerals are (leaving out rarer compounds which nevertheless occasionally contribute metal contents):

Lead
(a) *sulfides: galena,* boulangerite, bournonite, jamesonite
(b) *oxides etc.:* cerussite, anglesite, etc.

Zinc
(a) *sulfide(s): sphalerite,* with wurtzite (β-ZnS), and schalenblende (colloform ZnS)
(b) *oxides etc.:* zinkite, franklinite, smithsonite, etc.

References

Berg, G., Friedensburg, F., and Sommerlatte, H., 1950 "Die Metallischen Rohstoffe. Ihre Lagerungsverhältnisse und ihre wirtschaftliche Bedeutung," 9. Heft, Blei und Zink, Enke, Stuttgart, 468p.

Brown, J. S., Genesis of stratiform lead-zinc-barite-flourite deposits (Mississippi Valley type deposits) Monograph 3, *Econ. Geol.,* **443** (Papers from a symposium organized by Charles H. Behre, Jr., Feb. 1966; sponsored by UNESCO, SEG, IUGS).

Routhier, P., 1963, "Les Gisements Métallifères. Géologie et Principes de Recherche," 2 vols., Masson, Paris, 1282pp.

Schneiderhöhn, H., 1962, "Erzlagerstätten. Kurzvorlesungen zur Einführung und Wiederholung," Fischer, Stuttgart, Fourth ed., 371pp.

LEAD: INTERPRETATION OF STABLE ISOTOPE ABUNDANCES

There are four isotopes of lead: Pb^{204}, Pb^{206}, Pb^{207}, and Pb^{208}. The last three are the stable end products of decay of U^{238}, U^{235}, and Th^{232}, respectively (Table 1). The fourth lead isotope, Pb^{204}, has an extremely long half-life (10^{17} yr) and no known radioactive precursor. The abundance of Pb^{204} serves as an index to which other isotopic abundances may be compared. The analyses are performed by a mass spectrometer.

The present-day ratio of uranium:thorium:lead in the earth's crust is approximately 1:4:6: Roughly one-half of the Pb^{206}, one-third of the Pb^{207}, and one-fifth of the Pb^{208} in average crustal lead has been generated by decay of uranium and thorium since the earth was chemically differentiated, about 4.5 billion years ago. However, on a local scale,

LEAD: INTERPRETATION OF STABLE ISOTOPE ABUNDANCES

TABLE 1. URANIUM AND THORIUM ISOTOPE ABUNDANCES AND HALF-LIVES

Parent	Daughter	Present-Day Atomic Abundance in Parent	Half-Life (yr)
U^{238}	Pb^{206}	99.3%	$4.49 \pm 0.01 \times 10^9$
U^{235}	Pb^{207}	0.7	$7.13 \pm 0.16 \times 10^8$
Th^{232}	Pb^{208}	100	$1.39 \pm 0.02 \times 10^{10}$

large variations in isotopic composition exist depending upon the history of the lead before incorporation into a particular rock or mineral.

Ordinary Lead

The interpretation of Pb isotope distributions is based upon mathematical models developed by Holmes, Houtermanns, Russell, Farquhar, and others. These models assume that at the time of the earth's formation, all lead had an identical isotopic composition that is the same as in certain iron meteorites (the least radiogenic lead known from any source). During a time interval starting with the earth's beginning, coexisting uranium and thorium in the earth contributed radiogenic lead to the primordial lead. At the end of this interval, the lead was mobilized and transported to a new environment, such as a mineral deposit, in which it was effectively separated from the radioactive elements. From that time on, the isotopic composition was "frozen." The present-day composition thus depends upon: (a) the primordial composition (Table 2), (b) the (U + Th)/Pb ratio in the earth during the stage of radiogenic addition, (c) the decay constants and relative abundances of U and Th isotopes, and (d) the time when the lead was isolated. If lead was extracted at different times from a region of uniform U–Pb distribution, a plot of Pb^{207}/Pb^{204} vs Pb^{206}/Pb^{204} ratios for leads of different age will lie on a smooth curve (Fig. 1). The ages calculated from Pb isotopic composition are called "model lead ages." Lead in certain ore deposits, including some of the world's greatest, apparently has experienced the simple one-stage history outlined above; it is termed "ordinary lead."

TABLE 2. LEAD ISOTOPE ABUNDANCES IN IRON METEORITES

Pb^{206}/Pb^{204}	Pb^{207}/Pb^{204}	Pb^{208}/Pb^{204}
9.35	10.2	29.0

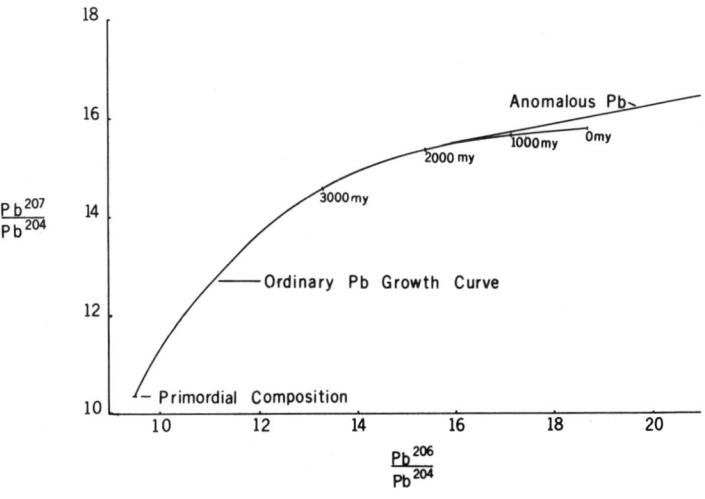

FIG. 1. Diagram of Pb^{207}/Pb^{204} vs. Pb^{206}/Pb^{204} indicating the primordial isotopic composition and a growth curve for ordinary lead corresponding to a source region in which U^{238}/Pb^{204} (today) = 9.08. The isotopic composition of lead varies as shown according to the time of isolation from the source region. The straight line is made up of an array of points derived from anomalous lead from a single locality (in this example, Sudbury, Ontario).

Anomalous Lead

If the Pb isotope ratios were produced by the contribution of two or more distinct U–Th–Pb systems, the lead is termed "anomalous." Anomalous leads may be formed by mixing of lead from regions of differing (U + Th)/Pb composition, or through remobilizing preexisting ore deposits. An age calculated upon the basis of a simple model would be seriously in error if the lead were actually anomalous. Anomalous leads may be recognized if they indicate an age older than that of the enclosing rocks (B type) or an unreasonably young, e.g., negative age (J type). A wide dispersion of isotope ratios from samples from the same ore deposit suggests that the lead is anomalous. Sometimes the ratios form a linear array (Fig. 1) that helps to specify the character of the source of the anomalous component and (within limits) the interval during which it was generated.

Studies of lead in common rocks, of which only a few have been completed thus far, are technically difficult because of the low concentration of Pb (0.01–100 ppm) and ease of laboratory contamination. They promise to contribute important information about the rate of evolution of the continents and ocean basins, and may provide a way to establish genetic connections between intrusive igneous rocks and ore deposits.

LEON E. LONG

References

Cannon, R. S., Pierce, A. P., Antweiler, J. C., and Buck, K. L., 1961, "The data of lead isotope geology related to problems of ore genesis," *Econ. Geol.*, **56**, 1–38.

Doe, B. R., 1970, "Lead isotopes," Berlin, Springer-Verlag Co., 137pp.

Kanasewich, E. R., 1968, "The interpretation of lead isotopes and their geological significance," in "Radiometric Dating for Geologists," New York, Interscience Publ., pp. 147–223.

Patterson, C., 1956, "Age of meteorites and the earth," *Geochim. Cosmochim. Acta*, **10**, 230–237.

Russell, R. D., and Farquhar, R. M., 1960, "Lead Isotopes in Geology," New York, Interscience Publ., 243pp.

Russell, R. D., 1963, "Some recent researches on lead isotope abundances," in "Earth Science and Meteorites," Amsterdam, North-Holland Publ. Co., pp. 44–73.

Tilton, G. R., and Steiger, R. H., 1965, "Lead isotopes and the age of the Earth," *Science*, **150**(3705), 1805–1808.

Cross-references: *Earth's Crust Geochemistry; Geochemical Evolution of the Core, Mantle, and Crust; Lead: Element and Geochemistry; Mass Spectrometry; Uranium–Thorium–Lead Age Determinations. Vol. II: Meteorites. Vol. IVB: Mineral and Ore Deposits.*

LIESEGANG RINGS

When a crystal of silver nitrate is placed on a glass slide covered with a dilute solution of potassium dichromate in gelatin, the resulting precipitate of silver dichromate forms in concentric bands or rings around the crystal. The phenomenon, termed *Liesegang rings* after the discoverer, has been observed with many salts and inorganic compounds. Factors which influence the formation of concentric rings or bands include the nature and concentration of the gel, the electrolyte concentration, and the temperature. Rings have also been produced in aqueous solutions without a gel. The concentration of the outer electrolyte must be higher by several orders of magnitude than that of the inner electrolyte. If the temperature increases, the solubility and diffusion rates increase, fewer bands form and the spacing between them is greater.

Several theories have been proposed to explain Liesegang formation. The supersaturation theory of W. Ostwald is adequate for a description of the gross features. In the case where the $AgNO_3$ diffuses into the gelatin with $K_2Cr_2O_7$, the precipitate ($Ag_2Cr_2O_7$) does not form immediately, but remains in supersaturated solution. Nucleii form a short distance behind the diffusion front and $Ag_2Cr_2O_7$ migrates toward them, leaving a clear band through which $AgNO_3$ must diffuse before the cycle is repeated. Evidence to support the theory includes the rapidity of ring formation and the speed of crystal growth from the time that the rings first appear. Other theories involve effects of adsorption of the inner electrolyte by the precipitate, coagulation of the precipitate as a colloidal dispersion rather than a supersaturated solution, and diffusion waves which interfere to produce the bands.

The best explanation of the phenomenon depends upon the presence of a highly extended, very reactive interface between solid and liquid in the system. The dispersed phase of the system may be formed in situ or be already available, as with colloidal clay or silica gels in nature. The colloidal dispersed phase adsorbs cations of the electrolyte initially dissolved in the medium. Anions are then available to react with the cations from an incoming electrolyte. If the reaction product is less soluble than either of the original electrolytes, a precipitate will form. When the precipitate forms at the boundary of the reacting substances, the gel in the immediate vicinity has been depleted of ions, and there-

FIG. 1. Liesegang rings of hematite in the Noonkanbah Formation, Kimberley Div., Western Australia. *Courtesy R. W. Fairbridge.*

fore the diffusion electrolyte must travel further before encountering a zone rich in the other electrolyte. The rate of band formation is proportional to the base-exchange capacity of the clay.

In geology, the banding of certain agates, malachite, metallic ores, and orbicular granite has been attributed to Liesegang rings. Orbicular granite or gneiss consists of cores of radially oriented plagioclase, biotite, or hornblende, surrounded by shells of plagioclase, biotite, and hornblende. The orbicules of Lonesome Mt. in the Beartooth Mts. of Montana-Wyoming show a progressive enrichment of Na-plagioclase toward the core, and an outward diffusion of Ca, Mg, and Fe. The shell structure has been compared to Liesegang rings because both are a result of an internally controlled rhythmic diffusion process that produces discontinuous, periodic precipitation. If the banding varies exponentially, this is evidence of the Liesegang phenomenon. Exponential progression has been found in metamorphic orbicules, spherulites, and some agates. Arithmetic spacing has been observed in igneous orbicules, concretions, oolites, varved argillites, and zoned plagioclase. Presumably the banding in the latter examples was caused by other processes, such as growth by superposition of successive increments. (For examples of Liesegang rings see Figs. 1 and 2.)

FIG. 2. Ferruginous liesegang rings of limonite alternating with leached zones in a deeply weathered microgranite near Dire-Dawa, Ethiopia. Width of outcrop, about 2 m. *Courtesy S.S. Augustithis.*

Banded precipitation has been carried out experimentally by using a hydrogel made from a very fine particle-sized fraction of Wyoming bentonite to which $K_2Cr_2O_7$ and $CuSO_4$ have been added. Liesegang rings have also developed in a test-tube containing a sample of gyttia from Lake Haruna, Japan. Reddish-brown stripes of iron oxide had formed. Sugawara claims that stratification caused by Liesegang rings exists in lakes. Changes in season and temperature affect the diffusion of Fe^{2+}. While reducing conditions prevail at the lake bottom, the top is oxidizing. An oxidation-reduction reaction produces the rings. Carl and Amstutz (1958) produced Liesegang rings with silver-nitrate, potassium dichromate, in a gelatin-cemented sand. The rings, however, ended at grain boundaries or deviated around them. They suggest that the weathering rings of limonite in natural rocks may be formed in the same way. Such periodic weathering rings frequently accompany exfoliation (see also Augustithis and Ottemann, 1966).

VIVIEN GORNITZ

References

Augustithis, S. S., and Ottemann, J., 1966, "On Diffusion Rings and Spheroidal Weathering," *Chem. Geol.*, **1**, 201–209.

Carl, J. D., and Amstutz, O. C., 1958, "Three dimensional Liesegang rings by diffusion in a colloidal matrix and their significance for interpretation of geological phenomena," *Geol. Soc. Am. Bull.*, **69**, 1467–1468.

Hauser, E. A., 1955, "Silicic Science," Princeton, N.J., D. Van Nostrand Co., Inc., 188pp.

Leveson, D. L., 1959, "Orbicular rocks of the Lonesome Mountain area, Beartooth Mountains, Montana and Wyoming," *Geol. Soc. Am. Bull.*, **70**, 1637.

Ollier, C. D., 1967, "Spheroidal weathering, exfoliation and constant volume alteration," *Z. Geomorphol.*, [N.F.] **11**, 103–108.

Sassen, R., 1967, "Crystals and Liesegang structures in silica gel," *J. Geol. Educ.*, **15**(5), 198–199.

Stern, K. H., 1954, "The Liesegang phenomenon," *Chem. Rev.*, **54**, 79.

Sugawara, K., 1934, "Liesegang's stratification developed in diatomaceous gyttja from Lake Haruna, Japan," *Bull. Chem. Soc. Japan.*, **9**, 402.

Vallentyne, J. R., 1955, "A laboratory study of the formation of sediment bands," *Am. J. Sci.*, **253**, 540–552.

Cross-references: *Electrolytes; Ion Exchange; Oxidation and Reduction.* Vol. III: *Exfoliation; Spheroidal Weathering.* Vol. VI: *Gyttja.*

LIME—*See* Vol. VI

LIMNOLOGY

Limnology is the science of lakes, a synthesis of many disciplines, drawing its devotees from various scientific fields. It includes all aspects of the study of inland waters, although concerned largely with the physicochemical nature of lakes, their flora and fauna. Stream study has lagged behind lake investigation, although the ecological approach to rivers falls within the realm of limnology as now understood.

F. A. Forel is considered by many to have been the first limnologist, and his first volume (1892) on Le Léman (Lake of Geneva) stands as a milestone. Because it dealt with environmental factors rather than biota of the lake, subsequent studies on physicochemical aspects of lacustrine habitat have been termed Forelian limnology. In America, the pioneer work of E. A. Birge, C. Juday and their students in Wisconsin marked the onset of modern limnology and made conditions in Wisconsin lakes a yardstick for later studies in other regions (see Mortimer, 1956). Birge, one of the first Americans to work with the microcrustaceans known as Cladocera, was led from what was essentially a biological study of their spatial and seasonal distribution in Lake Mendota, to a study of physical and chemical factors involved in puzzling fluctuations of cladoceran population—in other words, to Forelian limnology. Since then, limnological research has touched on countless facets of the lacustrine microcosm, but a unifying goal has linked most investigations. This involves the assay of productivity and the exploration of factors which interact (Fig. 1) to make a given lake more or less productive of living material than its neighbor or some distant counterpart.

Volume I of "A Treatise on Limnology" by Hutchinson (1957) is an outstanding modern source for information on geological, physical, and chemical limnology. The original researches pertaining to most of the material presented below are cited in that volume and are not duplicated here.

The Origin and Fate of Lakes

Lake basins owe their origins to diverse causes, many of which are geological "accidents" or "catastrophic." Geologically, they are temporary phenomena in geomorphic evolution (see Vol. III, *Lakes*). Tectonic events have created some of the oldest and deepest lakes of the world: the African rift lakes and Lake Baikal of Siberia with an ancient, largely endemic fauna. Vulcanism, glacial activities, solution of calcareous substrates, aeolian forces, and even meteoritic impact have created lake basins. Hutchinson (1957) has summarized 76 major categories and 8 subdivisions of these events which have resulted in lake genesis.

Because of the nature of their origins, lakes

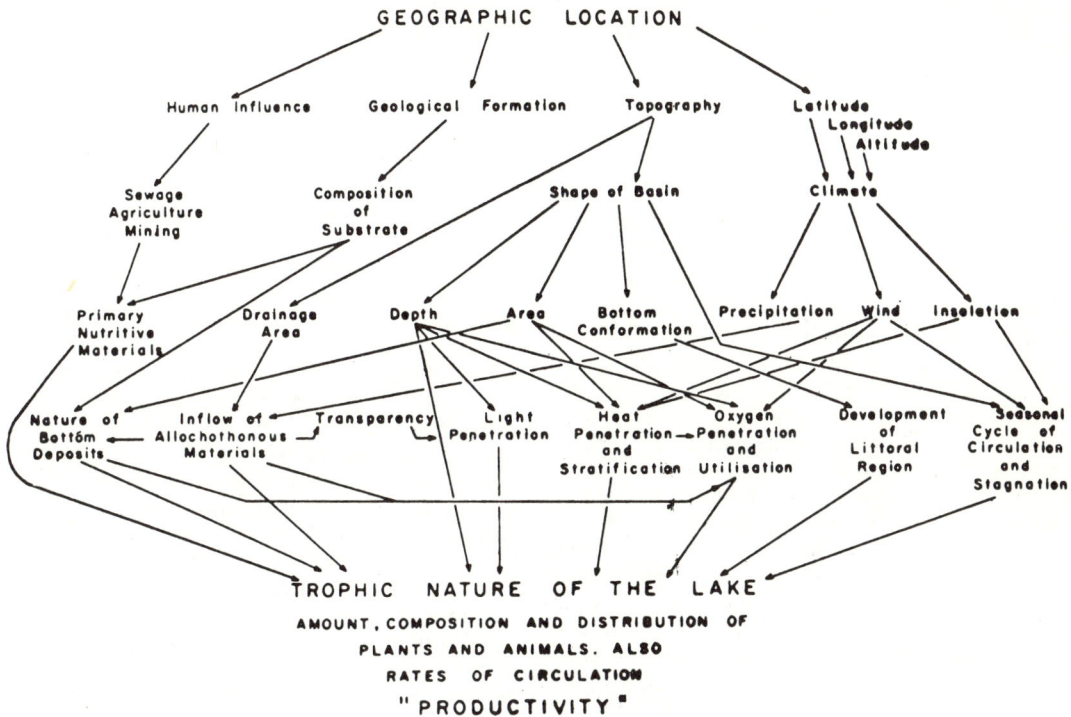

Fig. 1. Interaction of factors governing the trophic nature of a lake (from Rawson, 1939, *AAAS Publ. 10*).

often occur in definite geographic districts of related bodies of water, similar in age, geological history and surroundings. Comparison of lakes within districts, and between lake districts, has contributed much to understanding the factors which determine productivity and the fundamental trophic nature of a lake. Thus the importance of climate and of soils in influencing productivity has been revealed. Moreover, basin shape, often related to its origin, plays a role in determining the ultimate nature of a lake.

No matter what its origin, the lake is doomed to eventual extinction because of its concave nature and accumulation of autochthonous and allochthonous materials which gradually obliterate the depression. Thus, a lake passes from a youthful stage to maturity, senility, and extinction. The rate of succession depends on various factors; for example, introduction of domestic sewage enriches the lake and accelerates the aging process. Although complete definitions of the following terms will be postponed, the youthful lake may be described as *oligotrophic,* the mature lake as *eutrophic.* Many intermediate stages between extreme oligotrophy and extreme eutrophy occur, and the term *mesotrophy* can be applied to them. Senility is characterized by much shallow water and the conspicuous encroachment of large aquatic plants upon the open water. Extinction often involves a marshy meadow which is later colonized by plants typical of terrestrial situations. If drainage is poor and the lake is protected from wind, a floating bog-mat may close over and eventually obliterate the open water. Bog lakes are acid, or at least circumneutral, and are typified by characteristic marginal vegetation contributing to the floating mat. When calcium content is low in bog-lake waters, decay of organic matter is reduced greatly. Plant fragments from the bog mat accumulate in flocculent layers, and the water may become tea colored from humic matter. Flocculated humic colloids contribute to bog sediments to form a characteristic deposit termed *dy* by Scandinavian researchers. Under such conditions, nutrients are not recycled by decay, and the lake approaches extinction as a *dystrophic* lake.

Study of sediment cores has given much information about lake succession and past conditions. The sequence of allochthonous pollen relics in lake deposits tells much of vegetational changes, and hence, ecological conditions of the past—paleoecology. The chemical and physical nature of sediments, and microfossils and plant pigments derived from the lake itself, bespeak lacustrine changes which occurred, and this is the theme of paleolimnology (see Deevey, 1955; Bradley, 1962).

Morphometry

Limnological work on a body of water is enhanced if a bathymetric map is available, showing shore-line and subsurface contours. Important morphometric parameters derived from maps include area, volume, maximum and mean depth (volume/area). The form of the lake may be expressed as a *shore-line index,* comparing length of shore line to that of a theoretical circle with the same area as the lake, and as *volume development* which compares basin shape to a cone with a height equal to the maximum depth, and basal area the same as that of the lake's surface. Related to area is the maximum length, which has particular significance when considered as a wind-effective dimension. Morphometric parameters often reflect the lake's origin, e.g., irregular where glacial scouring was involved, circular and with high volume development in lakes occupying caldera depressions, and high shore-line indexes in lakes formed by damming of stream systems.

Except for small lakes, there is generally an increase in depth as area increases and there is a relationship between depth and productivity. Lakes with mean depths greater than 20 meters produce less per unit area than those with a shallower average depth, other conditions being equal (Rawson, 1955). A surprising number of lakes have maximum depths well below sea level, their basins being called *crypto-depressions.*

Physical Limnology

Light. Radiant solar energy, directly or indirectly arriving at the lake surface, is considered the primary energy source for all productivity within, although in special cases import of organic material may be significant. Common limnological usage expresses solar energy as total energy, g cal/cm^2 of lake surface per day or some longer period of time.

In an ideal system of pure water, various light waves are absorbed at different rates and the penetration is selective (Fig. 2). Blue light penetrates farthest because absorption of wavelengths of about 4700 Å is least. At the red end of the spectrum penetration is reduced greatly because most wavelengths above 7500 Å are absorbed and converted to heat within the first meter of water. This includes more than 50% of the total radiation falling on the surface. Optical properties of lakes vary because of differences in suspended or dissolved materials, with the latter particularly exhibiting selective absorption effects.

Subsurface photoelectric cells are used to determine the depths at which various percentages of surface radiation still persist, with 1% marking the lower limit of the *euphotic zone.* This is the lake region, somewhat arbitrarily chosen, in which photosynthesis can occur.

A standard method of comparing lake transparencies employs the *Secchi disc,* a white disc, usually of 20-cm diameter. The depth at which the disc first disappears from the observer's vision is the recorded transparency value. Secchi disc values of about 50 m represent extremes. Such great transparency goes hand in hand with intense blue color; waters poor in nutrients and organic productivity are usually blue. Some lakes are known in which Secchi-disc readings rarely surpass one meter; such bodies of water are yellow-green to yellow because of abundant suspended microscopic organisms.

Thermics and Lake Classification. Distribution of heat in a lake, evidenced by vertical temperature profiles is quite different from a curve constructed on data from the penetration of light. The difference is brought about by wind action, which distributes heat to greater depths than solar radiation and conduction could do alone. Work of the wind in producing observed temperature distribution can be calculated; it varies from lake to lake, but a reasonable mean approximation is 0.1 g-cm of work necessary to mix a calorie of heat.

Seasonal warming and cooling of lake waters have profound effects because of concurrent density changes. In winter, an ice-covered lake may show an inverse temperature gradient with the warmest water at its greatest depths. This is because of the unique property of water; its greatest density is at 4°C. Just beneath the ice it is colder and therefore lighter than water at 4°C. Although the lake is protected from wind, and the period of ice cover is called the *winter stagnation* period, there are gentle currents beneath the ice. Early inferences about such currents were based on temperature changes, and these have been confirmed by introducing and tracing radio-sodium (Na^{24}) beneath the ice. Horizontal currents are especially striking. These are caused, in part, by a greenhouse effect of solar radiation through the ice especially at the lake margins. When water which is near 0°C is warmed, density increases and it flows down the basin slope.

Vernal ice melting results in a lake exposed to wind action, practically uniform in temperature and, therefore, in density. Wind-generated currents mix the lake thoroughly;

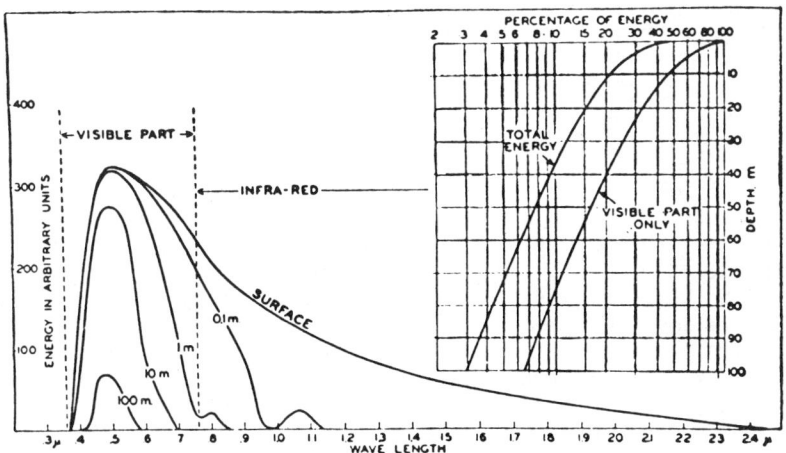

FIG. 2. Spectral penetration of solar radiation through various depths of distilled water.

this is the *spring overturn*. As days become warmer, heat accumulated in the upper strata is distributed downward by the wind, but there comes a time when the warm, upper waters differ so in density from the deeper waters that wind no longer mixes the lake completely. This results in thermal stratification and the period of *summer stagnation*. A nearly uniform layer, the *epilimnion,* mixed by wind, lies at the top; below, lies a region where temperature falls rapidly with depth. This is the *thermocline* or *metalimnion,* where density changes parallel, to some extent, the temperature changes. Recently the thermocline has been defined as a plane where the maximum change in temperature occurs, although once defined as the stratum where temperature drops at least 1°C with each increase in depth of one meter. Below this region lies a colder, dark region perhaps not much warmer than the temperature that prevailed during winter, and protected from wind action by the density barrier above. Thus, the lake is divided into two parts separated by the metalimnion—the well-mixed epilimnion and the cold, stagnant *hypolimnion.*

An important point, however, is that density differences between adjacent temperatures are far greater at high temperatures than at low. Thus, the density change between 24 and 25°C is 31 times that between 4 and 5°C. For this reason, the thermocline, defined as a plane where the maximum temperature change occurs, may be less important than another plane where maximum density change occurs. The summer *stability* of a body of water, defined as work needed to mix the lake uniformly, destroying the stratification with no addition or loss of heat, might, therefore, be as great in a warm tropical lake with a gentle temperature profile, as in a colder lake with pronounced temperature stratification (Fig. 3).

In autumn as air temperatures fall, heat is lost from upper waters, the thermocline lowers, and eventually the lake is homothermal and of the same density. The stability is zero, and the wind mixes the entire lake so chemical and physical conditions throughout approach uniformity. This is the *fall overturn*. Cooling of the entire body of water continues until one calm, cold night a skim of ice forms and *winter stagnation* commences.

Seasonal temperature phenomena described above are typical of a *dimictic* lake, there being two periods of circulation separated by

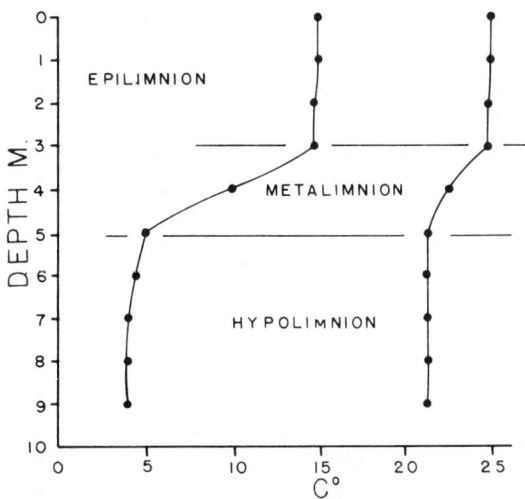

FIG. 3. Vertical temperature profiles from a cold and warm lake. The vertical density-change relations are the same in each curve.

653

two periods when complete mixing is prohibited. Such lakes are common in temperate regions. However, basin shape and area, surrounding topography, climate, and their effects in diminishing or enhancing wind action may modify this annual cycle. Some lakes are *monomictic,* mixing thoroughly but once a year. Cold monomictic lakes with a winter stagnation period circulate freely during the rest of the year, if shallow enough and wind-exposed. Warm monomictic lakes are found at lower latitudes where ice cover does not occur and only summer stratification interrupts circulation. *Polymictic* lakes include small high-altitude lakes which stratify during warm days and become uniform in temperature at night because of surface cooling and ensuing convection currents. This term might apply also to lakes situated near the boundary of warm monomictic and dimictic lakes, where winter ice covers form and disappear several times with circulation occurring between each, and tropical lakes which frequently turn over and stratify. Some shallow lakes in mild climates never stratify and are best described on the basis of older terminology as *third-class lakes.* Conversely, in Antarctica permanently ice-covered lakes occur which must be defined as *amictic.*

All lakes described above except the amictic type, are *holomictic,* their entire water mass circulating at overturn periods. By contrast, *meromictic* lakes occur in which bottom waters are stabilized by the density imparted by dissolved matter and are permanently stagnant. Circulation is limited to the upper waters. Unusual factors combine to create meromixis. Certain plunge-basin lakes formed beneath old Pleistocene waterfalls are so deep relative to surface area that they cannot circulate completely after a critical amount of dissolved material accumulates in the deeps. In such cases, meromixis is caused, to a great extent, by unusual morphometric parameters.

Other meromictic lakes owe their character to events which introduce dense salty water into a fresh-water lake, or conversely, introduce fresh water into a markedly saline lake. In either case, a dense layer is overlain by superficial lighter water. In arid regions where evaporation rates are high, saline meromictic lakes are known which owe their stability to more complex processes. If ice cover occurs, freezing-out effects tend to concentrate various salts in deeper layers, and mirabilite precipitates if the solubility product of Na_2SO_4 is attained. In closed basins especially, these events combined with snow and runoff may effect meromictic conditions.

The upper layer of relatively light water which may stratify thermally at times, and circulate at others, is the *mixolimnion.* The deep layer of permanently stagnant water is the *monimolimnion.* Between the two is a region where salinity, and therefore density, increases markedly with depth. This is the *chemocline,* analogous to the metalimnion although the density gradient is caused by materials in solution rather than by temperature change.

Meromictic lakes usually reveal anomalous temperature profiles. Many show a marked low point within the curve with warm water lying beneath colder water, a condition in no way related to the normal temperature-density relations that occur in the neighborhood of 4°C. For example, 25-degree water may lie below 18-degree water, which in turn lies below 20-degree surface water. This condition is known as *dichothermy.* Another phenomenon, termed *mesothermy,* involves a maximum in the vertical temperature curve. The rare condition, in which one or more maxima and minima occur within the curve, is called *poikilothermy.* The apparent discrepancies in density and stability on the basis of temperature are compensated for by materials in solution which impart greater density to underlying warm waters.

The source of heat in the monimolimnia of meromictic lakes is not always obvious, although bacterial metabolism may account for slow heat accumulation which can be lost only through conduction and diffusion. In shallow meromictic lakes, solar radiation warms the saline bottom waters, and heat is stored as in certain solar furnaces.

The various types of annual temperature cycles, described above, used to categorize lake types are based on intensity factors. A further aspect of lacustrine thermal properties is the capacity factor—the heat gained or lost during a period of time. The *annual heat budget* is commonly calculated, although segments of it such as *summer heat income* can be considered alone. The former is the total amount of heat entering a lake between the times of its lowest and its highest heat content. Each extreme is determined by summing the products of temperature and volume of every stratum. Because total heat contents determined this way are a function of lake volume, comparisons between lakes of different sizes are validated by dividing total heat by surface area. The heat budget is, therefore, expressed in terms of calories per square centimeter.

In general, heat budgets are strongly correlated with mean depths, areas, and volumes. Lake Baikal, the world's deepest lake (1741 meters), has an annual heat budget on the order of 60,000 cal/cm². This is two or three times that of the average dimictic lake in a temperate region. Tropical lakes, which show relatively little annual temperature variation, usually have low heat budgets. A notable exception is the large Atitlán in Guatemala; exceptional wind action distributes an enormous quantity of heat throughout the lake during the summer warming period and the resulting heat income is some 22,000 cal/cm^2, a value much like those derived from many northern and alpine lakes in Europe.

Water Movements. A further aspect of physical limnology deals with the movement of water, including various types of waves, currents and turbulence. The air-water interface is regarded as important in considering the motions within the water, for it is here that wind stress acts and radiant energy enters to produce different densities, which in turn may be modified by the action of wind. In addition, other phenomena are involved and interact with the current patterns: the Coriolis force and even tides can be measured in very large lakes; influent river water carrying dissolved and suspended materials, or of low temperature, may move through the lake as a so-called *density current*.

A lake phenomenon of interest is the standing wave termed the *seiche*. This is periodic oscillation occurring when strong winds cease after having blown water to one side of the lake. The water flows back to the former windward side and then again to the downwind shore, rocking with a period determined by basin shape, length and depth. Surface seiches are especially obvious in large lakes with gentle marginal slopes, for large areas of the shallows may be alternately exposed and flooded with a seiche of relatively small amplitude. These, however, are far smaller than the internal seiche produced in stratified lakes; the accumulation of surface water depresses the thermocline and sets it rocking with an amplitude which surpasses that of the surface seiche above. Internal seiches are detected by the rise and fall of isotherms. These rhythmic currents, alternating in direction, are probably the most important deep-water movements in lakes, creating turbulence which distributes heat and dissolved and suspended materials both vertically and horizontally.

Chemical Limnology

Inorganic Ions and Compounds. The commonest inorganic cations of inland fresh waters are generally these in sequence of abundance: $Ca^{2+} > Mg^{2+} > Na^+ > K^+$. Similarly, the anions are: CO_3^{2-} (usually as HCO_3^-) $> SO_4^{2-} > Cl^-$. There are many exceptions to such arrangements, but most dilute fresh waters are essentially calcium-bicarbonate waters. In some closed basins of arid regions evaporation and concentration leads to the early precipitation of $CaCO_3$ and the relative enrichment of other ions. Further concentration will lead to the precipitation of gypsum ($CaSO_4 \cdot 2H_2O$), and the principal anion becomes chloride. Such a sequence depends on the original composition of the water; if the calcium content is high, almost all the carbonate and sulfate will be removed as the solubility products of $CaCO_3$ and $CaSO_4$ are attained. Even after calcium is gone, sulfate may be removed by the precipitation of mirabilite ($Na_2SO_4 \cdot 10H_2O$). In some lakes, calcium precipitation has not removed the carbonate and the waters are now essentially soda waters, with $NaHCO_3$ and Na_2CO_3 as major constituents. These are less common than those saline lakes where sulfate and especially chloride predominate.

In oxygen-free hypolimnia of some eutrophic lakes and the monimolimnia of meromictic lakes, sulfate is reduced to hydrogen sulfide. This substance is poisonous to most organisms, but the sulfur bacteria require it as a hydrogen donor in chemosynthesis and a modified type of photosynthesis.

Dissolved Oxygen and Lake Typology. Most dissolved oxygen, the distribution of which is of extreme importance to lacustrine fauna, arises as a by-product of algal photosynthesis or that of larger aquatic plants. The basic formula for this synthesis of carbohydrate by green plants in the presence of light is:

$$6H_2O + 6CO_2 \rightarrow C_6H_{12}O_6 + 6O_2$$

Some dissolved oxygen may be derived through atmospheric exchange when the water surface is wind-disturbed, although diffusion of molecular oxygen through undisturbed water appears negligible.

During circulation periods, oxygen is distributed throughout, often approaching or achieving the saturation point. This is because its solubility increases inversely to water temperature, and overturn periods are usually cold; this applies especially to spring circulation following ice melt in a dimictic lake. At 25°C, 5 mg O_2/liter would represent about 90% of saturation, but the same quantity at 5°C is less than 60% of the possible saturation. These data apply to sea level; correction factors are applied to account for altitudes' effects on solubility. During calm periods when intense photosynthesis prevails, upper waters may hold dissolved oxygen in excess of saturation; conversely, if respiration and decay are excessive, the waters may become undersaturated.

During summer, the metalimnion serves as a barrier, prohibiting rapid penetration of wind-driven substances into the hypolimnion. Thus oxygen produced in the well-lighted turbulent epilimnion cannot be distributed to the deeps. In fact, oxygen decreases and CO_2 increases in the stagnant hypolimnion as a result of decomposition, certain bacteria of decay utilizing oxygen and producing CO_2. The rate of hypolimnetic oxygen decrease depends on several factors, one of which is the amount of organic material sinking from the epilimnion, the so-called *trophogenic* zone where most living matter is produced. Lakes can be classified on the basis of their summer oxygen profiles. Fertile lakes with high rates of production show marked oxygen depletion below the thermocline; poorer producers in some instances show almost no decrease. Rich lakes, which show summertime *clinograde* oxygen stratification, are eutrophic. Thermally stratified lakes with abundant summer oxygen throughout reveal an *orthograde* oxygen curve, and are termed oligotrophic (Fig. 4).

A method of comparing productivity of different lakes is based on observing decline in hypolimnetic oxygen between two or more dates and referring it to unit area of hypolimnion surface. Products of oxygen decrease at different depths and the volumes of the strata in which the decreases occur are summed to yield total oxygen lost. This is divided by hypolimnion surface area and the number of days involved in the decrease. One somewhat arbitrary scheme places the rate of hypolimnetic oxygen decrease in mesotrophic lakes between the upper limits for oligotrophy and the lower limits for eutrophy, between 0.025 mg O_2/cm^2/day and 0.055 mg O_2/cm^2/day. The method applies best to lakes at least 20 m deep where light penetration and vertical turbulence play minimum roles in altering hypolimnion oxygen values.

In tropical lakes, a factor other than total organic productivity assumes importance in oxygen deficit and CO_2 production. Higher hypolimnion temperatures increase metabolic rates, and the oxygen curve may reflect this rather than the magnitude of organic productivity in the epilimnion.

The vertical oxygen profile may be modified by factors other than geological surroundings and climate which obviously govern fertility, productivity and rate of decomposition. These include morphometric parameters of the basin. A relatively large hypolimnion volume starts with a greater supply of oxygen to be utilized in microbial metabolism than does a relatively smaller hypolimnion. Therefore, two lakes of identical surface area and productivity might show quite different rates of hypolimnetic oxygen depletion. The English Lake Windermere lies in a region of fertile soil and is productive; its hypolimnion volume is so great, however, that there is not a marked decrease of oxygen there, and its summertime oxygen profile resembles the oligotrophic type.

Unusual oxygen profiles prevail at times. More than 50 lakes are known which show *plus-heterograde* oxygen curves, characterized by a persistent metalimnetic maximum where oxygen is in excess of the theoretical 100% saturation existing at the spring overturn. In many such lakes the blue-green alga *Oscillatoria agardhii* produces dense blooms and high oxygen concentrations at the intermediate temperatures and low light intensities typical of the metalimnion.

Metalimnetic oxygen minima producing *negative-heterograde* curves are not easily explained. In some lakes, a concentration of non-migrating animals in the metalimnion may consume oxygen to such an extent that a minimum occurs within the curve.

An alternative to describing the trophic nature of a lake on the basis of intensity of hypolimnetic oxygen consumption is to measure the rate of CO_2 production. However, more CO_2 appears than can be accounted for by O_2 depletion alone. The CO_2 produced by aerobic oxidation of organic matter can be estimated within reason by multiplying the observed decline in oxygen by a factor of 0.85, a mean value for the respiratory quotient of aquatic organisms. But, in addition to aerobic bacterial action, anaerobic metabolism may occur in tropholytic waters. Anaerobic bacteria

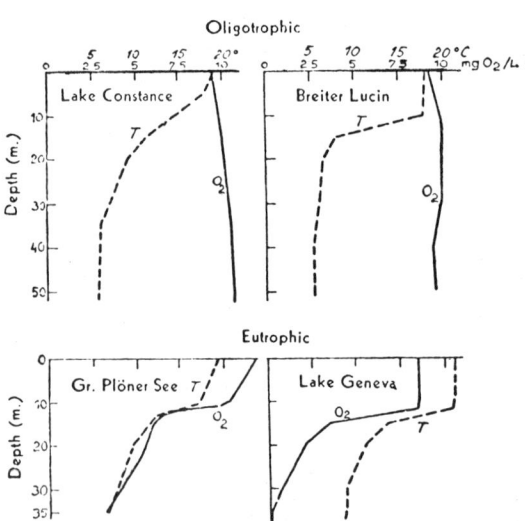

FIG. 4. Examples of eutrophic clinograde and oligotrophic orthgrade oxygen curves in stratified lakes.

break down organic material without utilizing oxygen in the process. About 50% of the organic carbon thus decomposed becomes CO_2; the other half becomes methane, CH_4.

Hydrogen Ion, CO_2, and Alkalinity. For most lakes, the hydrogen ion concentration expressed conventionally as pH, ranges between 6.0 and 9.0, but remarkable exceptions occur. Some volcanic lakes are extremely acid with pH values below 2.0, and the pH of bog lakes is often in the neighborhood of 4.0. In these instances, H_2SO_4 is responsible for the acidity, and oxidation of pyrite is one common source of the acid. A further mechanism for acid production exists in bog lakes. It is well known that sphagnum peat softens the water percolated through it. This involves cation exchange; thus soluble sulfate compounds moving through peaty matter may lose metallic cations and gain hydrogen ions by base exchange. Rainwater alone could account for much of the sulfate ion because it contains a mean of about 2 mg/liter; furthermore, sulfur dioxide is atmospherically derived.

Persistent pH values of 10 or above are usually associated with high concentrations of sodium and magnesium in alkaline lakes, but in the average freshwater lake which shows no extreme acid or alkaline qualities, pH is governed by a CO_2-bicarbonate-carbonate buffering system.

Alkalinity is a rather poorly named component of water chemistry which is commonly determined in limnological procedure. Total alkalinity is acid-combining capacity, arrived at by titration with strong acid to the methyl orange or equivalent end-point at a pH of about 4.5. Under normal conditions, alkalinity is a measure of bicarbonate and carbonate, although at a high pH some hydroxide may be present. Because calcium is one of the most abundant lake cations, and commonly exists as $Ca(HCO_3)_2$, alkalinity titrations in normal waters also reflect calcium content. Perhaps for this reason, total alkalinity serves as a rough index of productivity, those lakes with very low titers being less productive than those with moderate or high alkalinities. In the saline waters of "soda lakes" the carbonates and bicarbonates are compounds of sodium rather than calcium.

Close relationships exist between alkalinity, hydrogen ion content expressed as pH, and the gas CO_2 (Fig. 5). It is customary to consider CO_2 as existing in three states; "bound" as in CO_3; "half-bound" as in HCO_3; and free as gas. In simple terms, CO_2 makes for lower pH and greater acidity as it combines with water to form carbonic acid, H_2CO_3. Limestone, or $CaCO_3$ is insoluble but in carbonic

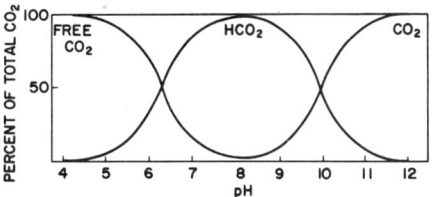

Fig. 5. Relation of pH to the percentage of CO_2 in each of its states.

acid it dissolves as $Ca(HCO_3)_2$. Calcium bicarbonate stays in solution only if there is CO_2 in excess of the equilibrium value. The excess CO_2 prevents the following reaction:

$$Ca(HCO_3)_2 \rightarrow CaCO_3 + H_2O + CO_2$$

This explains events which occur as the issue from a subterranean spring moves downstream. In calcareous regions, spring water is high in $Ca(HCO_3)_2$, free CO_2, and hydrogen ion concentration. Warming and turbulence results in the loss of free CO_2 to the atmosphere, a rise, therefore, in pH and conversion of bicarbonate to insoluble carbonate which precipitates as *marl*. Thus total alkalinity is less in the stream than at the spring source.

Biogenic marl precipitation results as photosynthetic plants remove CO_2 from water or, as most aquatic plants can do, remove CO_2 from the bicarbonate radical converting it to the insoluble carbonate. Marl formation is restricted largely to the shallows, because sinking carbonate comes in contact with carbonic acid in the tropholytic zone and goes into solution. As a result, bicarbonate and, therefore, alkalinity increases in the stagnant hypolimnion.

The vertical distribution of pH in a stratified lake is explained by some of the above remarks. In the trophogenic epilimnion, photosynthesis removes free CO_2 as rapidly as it is formed. The result may be a pH value well in excess of 8.0. Below the thermocline in the tropholytic region, CO_2 is abundant and the pH is lower. Thus, in a eutrophic lake, the pH curve parallels, to some extent, the clinograde oxygen curve, and the CO_2 curve is the reverse of both. Fluctuations in free CO_2 have marked effects on the pH of soft-water lakes, but not so in well-buffered water with high alkalinity.

Iron Cycle. The lacustrine iron cycle shows a close relation to oxygen, pH and CO_2 conditions. Ferric iron, the oxidized form, is insoluble and scarce in well-oxygenated waters. In the absence of oxygen and at a low pH, iron occurs in the ferrous reduced state which is soluble. At overturn periods with abundant oxygen at all depths, iron is scarce because it is precipitated as ferric phosphate and ferric hydroxide. An oxidized microzone is formed at the mud-water interface prohibiting diffusion of substances from the underlying sediments. In lakes with orthograde oxygen curves, this microzone may be permanent. In eutrophic lakes, however, reducing conditions prevail in the stagnant hypolimnion, and ferric compounds are converted to ferrous iron which diffuses into the hypolimnion. If it reaches oxygenated levels, it is converted to the ferric form and reprecipitated. For this reason, the hypolimnion acts as an iron trap and progressive enrichment occurs there. The disappearance of the oxidized barrier at the mud surface allows other substances to diffuse out, and there is not only a marked increase in hypolimnion iron, but in phosphates, bicarbonates, silicates and ammonium nitrogen as well. These may be circulated at overturn periods and utilized by epilimnetic algae, accounting in part for the blooms which occur at the fall overturn.

Sometimes evidence of the relation of pH, CO_2, oxygen and iron is visible near springs. Flocculent masses of the reddish ferric hydroxide may be seen below, but not at the spring mouth. Ferrous iron carried in solution from underground sources is converted to ferric iron and precipitated when the water becomes oxygenated.

Under intense reducing conditions hydrogen sulfide appears in the hypolimnion and, with iron, forms ferrous sulfide. This tends to precipitate and accounts for the black color of the slimy sediment called *sapropel*, which is characteristic of meromictic lakes and others with long periods of anaerobiosis.

Nutrients and Minor Elements. Other chemical constituents of the aquatic environment are classified as nutrient, or in certain instances minor elements. Nitrogen and phosphorus are especially important. Compounds of the former are derived from atmosphere via rain, or from terrestrial decomposition. Furthermore, nitrogen fixation brought about by bacteria and blue-green algae is of some magnitude in lakes. Aquatic blue-greens of the family Nostocaceae, and especially species of *Anabaena*, are important in this fixation. Lake waters contain nitrogen in the combined forms—nitrate, nitrite, and ammonia. Ammonium nitrogen, derived from protein breakdown, may be especially abundant in the hypolimnia of rich lakes during stagnation periods.

Phosphorus is less abundant than nitrogen and is usually a limiting factor in fertility. It is derived largely from soil in the drainage area through weathering of phosphatic rocks, but phosphate-compounds leach less readily than those of nitrogen. Experiments with radiophosphorus, P^{32}, show that inorganic phosphate added to a lake disappears quickly from the water. It is taken up and stored by planktonic algae and sedimentation follows rapidly. Soluble, inorganic phosphate is often difficult to demonstrate in waters with high productivity because planktonic algae are remarkably efficient in taking it up. Soluble organic compounds of phosphorus are probably not useful, but bacteria rapidly reduce them to inorganic phosphates which may be utilized by autotrophic organisms.

Silica is taken up by diatomaceous algae for frustule construction. Its abundance dissolved in epilimnion water usually is negatively correlated with diatom-population fluctuations, and it exhibits stratification.

Cobalt, copper, zinc, aluminum and other metallic elements occur in lake waters in trace quantities and presently are considered of minor importance. Further research will modify this outlook. Cobaltous ion, for example, is a component of B_{12}, and evidence is accumulating that this vitamin may control the species composition of planktonic algae populations. Also, molybdenum seems to affect general productivity, possibly through a role in governing nitrogen metabolism.

Primary Productivity

The rate of photosynthesis carried on by aquatic plants may be referred quantitatively to oxygen produced, CO_2 utilized, carbon fixed or glucose produced per unit surface area over some span of time. In most lakes, the limnetic algae are the important *primary producers*. Various methods have been devised for determining their productivity (Ryther, 1956); most involve the use of replicate dark and light bottles. A water sample with its included planktonic populations is collected from some depth, placed in a clear bottle and a light-tight bottle and immediately returned to the original level where the bottles are suspended for a period of time. Oxygen determinations made on water from that level at this time serve as a base for the experiments. After six or eight hours, oxygen is determined in each bottle. Decrease in the dark bottle represents respiration of plankton and consumption of oxygen in bacterial decomposition of organic substances, and it is assumed the same amount of oxygen was utilized in each bottle. Oxygen increase in the light bottle represents *net production* for the period of exposure. *Gross production* is the sum of oxygen consumed in the dark bottle and the determined increase in the clear bottle.

Because the oxygen method is limited by the sensitivity of the iodometric Winkler test, some workers have determined CO_2 uptake as a measure of productivity. This may be done in two ways. First, initial pH and total alkalinity values are determined; from these data, free CO_2 may be calculated. At the end of the experiment, changes in pH reflect either net decrease in CO^2 because of assimilation or increase due to respiration in the dark bottle. Second, a known amount of radiocarbon, C^{14}, is added to the light bottle whose initial free CO_2 and bicarbonate content is known. Because C^{14} is assimilated at a rate only slightly different from that of C^{12}, later Geiger-counter determination of the C^{14} in algal cells filtered from the bottle water makes it possible to calculate total amount of carbon assimilated from the original $C^{12}:C^{14}$ ratio in the water. The radioactivity of filtered cells probably reflects carbon incorporated in the photosynthate minus that which was respired, and is, therefore, net productivity.

Obvious sources of error in the light-dark bottle method arise with differential increase of algae in the clear bottle making greater surface area available to bacteria, the extra bacterial activity, and any other discrepancy between the two bottles. To escape these inherent errors some workers have studied diel fluctuations in dissolved oxygen at different levels in the lake itself, multiplying by proper volumes to get total amounts. The hours of darkness represent the light-tight bottle; during daylight, the lake is comparable to the clear bottle. Field measurements of oxygen changes used to calculate productivity are probably valid, if day and night respiration are the same and if atmospheric exchange is negligible. Similar methods have been employed in calculating primary productivity in stream environments, where oxygen changes occur as water passes over beds of aquatic vegetation or substrates encrusted with photosynthetic algae.

A further method is based on the relations between light intensities, the amount of chlorophyll *a*, and photosynthesis. Chlorophyll *a* is determined colorimetrically following acetone extraction of residual cells filtered from a known water volume. Concurrent oxygen increases in the lake are observed and an *assimilation number,* the ratio of oxygen produced per milligram of chlorophyll, is established. Assimilation values are extremely variable, but useful means are determined empirically. Knowing the depth and volume of the euphotic zone and its chlorophyll content, the assimilation number is used to convert these data to rate of oxygen production or carbon fixation. Values from many lakes reveal mean assimilation rates of about 4.0 mg C fixed/mg chlorophyll *a*/hour. This method assumes a fixed relation between light and photosynthesis and essentially ignores adaptations of algae to varying light intensities. Winter phytoplankton, for example, may include extreme "shade species"; autotrophic assimilation under ice and snow cover has been demonstrated at 0.06% of incident radiatiton.

Habitats and Communities

Thousands of species of plants and animals may occur in a lake, but relatively few have attracted the attention of limnologists. In general, they have been studied on a spatial community basis involving many taxonomic groups in each instance.

Plankton. The unique plankton community consists of many species, usually minute, which are found floating or feebly swimming in open waters, the limnetic zone. Plankton is collected by straining lake waters through nets of fine silk. Various devices, including the noteworthy Clarke-Bumpus sampler which records the volume of water strained, have been developed for quantitative collecting. Organisms captured by net are the *net plankton*. The *nannoplankton* includes diminutive algae and flagellates which pass through the interstices of the net. Centrifuging and membrane filtration are used for sampling nannoplankters, which surpass the net plankton in total mass and numbers. Freshwater plankton includes five main groups: the various plants known collectively as algae, a term which covers microscopic plants best assigned to several different phyla, and classed as phytoplankton; the Protozoa; the rotifers or Rotatoria; the cladoceran crustaceans; and the copepod crustaceans. The animals are the zooplankton.

To these five groups should be added the interesting dipterous larva, *Chaoborus,* termed the phantom larva. The adult resembles a mosquito and is assigned to a subfamily of the mosquitoes, the Chaoborinae. The larvae are usually found in eutrophic waters, spending the daylight hours on or near the sediment surface in the deeper portions of the lake. At night they rise to prey upon zooplankters, and are conspicuous in plankton collections taken at that time.

Plankton productivity has been examined in the search for indices of lake productivity, and there are many data on numbers, volumes and weights of plankton. A stumbling block has been the lack of precise knowledge concerning life histories and hence the span of time involved in a generation. For this reason, many published data refer only to standing crops; actual turnover rates, which would more closely approximate productivity, are not known.

Cladocera are extremely important members of most plankton communities. In recent years, attention has been focused on cladocerans of the family Chydoridae. Few live in the limnetic region, although at least one, *Chydorus sphaericus,* is a common, cosmopolitan plankter. Chydorid head shields remain well preserved in lake sediments and seem to be distributed more or less uniformly across the lake basin. The study of these relics in lake stratigraphy has become an important part of paleolimnology. Most planktonic Cladocera leave no well-defined microfossils, but species of *Bosmina* constitute an exception and their carapaces and head shields, as in the chydorids, remain preserved in lacustrine deposits (Frey, 1964).

Benthos. Thermal stratification makes possible definition of bottom regions, benthic zones. The *profundal zone* is overlain by hypolimnion water and is composed of fine-grained sediments. In eutrophic lakes this is a planktogenic and coprogenic ooze, only partially oxidized or slightly reduced. This is a grey-brown material called *gyttja*. In oligotrophic lakes profundal sediments are often mineralized.

The *littoral zone* at the lake margins contains coarser sediments. It is often defined as the region from shore to the depth marking the deepest growth of the rooted aquatic plants. Between the littoral and profundal is the *sublittoral zone*.

The community of bottom-dwelling organisms is the *benthos*. In most instances, limnologists have studied the profundal benthos, neglecting the taxonomically richer communities at littoral and sublittoral depths. In eutrophic lakes with summertime oxygen deficits, the profundal benthos is impoverished, made up only of those species which can tolerate anaerobic conditions. Typically, certain larvae of the dipterous family, Chironomidae, occur here. These midge larvae, however, are largely so-called, "red blood worms" of the *Chironomus plumosus* type which includes a handful of species. Evidence is accumulating that they feed largely on settled plankton in the bottom deposits rather than on the organic sediments themselves.

Tubificid oligochaete worms, tiny fingernail clams, especially species of *Pisidium,* and larval *Chaoborus* are also members of the profundal benthos in some eutrophic lakes. The scanty representation in this species list is largely a function of anoxic conditions.

In oligotrophic lakes the profundal fauna is enriched in species, but diminished in total biomass. Chironomid larvae of the *Tanytarsus* type occur in the profundal area where oxygen is adequate. Stratigraphic studies show that they are replaced by *Chironomus* as eutrophication proceeds.

Quantitative benthic sampling utilizes dredges which cover known areas and take sediment samples to depths of 15–20 cm. The Ekman dredge, covering about 125 cm^2, is employed in most studies. The greatest standing crops of profundal benthos occurring in eutrophic lakes are near 500 kg fresh weight/ha, which is some 100 times greater than the mass reported from extremely oligotrophic lakes.

A largely neglected segment of the bottom fauna is the *microbenthos,* the assemblage of microscopic species which pass through the sieves normally employed in screening dredge samples. Included are protozoans, nematodes, rotifers and various microcrustaceans such as ostracods.

Minor Communities. The *Aufwuchs* or *periphyton* is the community encrusting submerged objects. It is comprised of algae and many taxonomic groups of animals.

In the interstitial water among beach sand grains, occurs another community, the *psammon*. Many microscopic forms are found here which are absent from the littoral sediments just a few centimeters away.

A minor community is the *neuston,* associated with the water-air surface film. Several plants and animals live either on the upper or under surface of this film.

The *nekton* is composed of powerful swimmers, an assemblage including fishes, some large crustaceans and insects. Salmonid fish are typical of oligotrophy or mesotrophy and disappear with eutrophy. In North America the bass and sunfishes, centrarchids, replace them and typify eutrophic waters.

Community Metabolism

Limnologists have been concerned with the quantitative flow of energy through lacustrine ecosystems since the important paper of Lindeman (1942) focused attention on this dynamic aspect of communities. The role of organisms as members of trophic levels, utilizing and passing on energy derived from lower levels is the essence of this approach.

Calories derived from solar radiation represent the original source of energy. The first trophic level includes autotrophic organisms of which the phytoplankters are most noteworthy; they are the *producers*. In spite of high annual production, they are relatively inefficient, converting less than 1% of the radiation to energy-containing substance.

The *primary consumers* are herbivorous animals which feed directly on plants, and fall prey to the carnivores, the *secondary* and *tertiary consumers,* which represent the third and fourth levels.

Each level contains less energy than the preceding, the loss by respiration increasing progressively from lower to higher trophic levels. Efficiency of conversion of food energy to protoplasm, however, increase simultaneously. The simplicity of the scheme is misleading, because most carnivores are not restricted to feeding on organisms of the next lower level. Furthermore, detritus and associated bacterial flora are an important energy source.

The *decomposers,* mostly heterotrophic bacteria, break down organic substances in dead plants and animals to an inorganic state, thereby returning nutrients to the green producers for recycling.

GERALD A. COLE

References

Bradley, W. H., 1962, "Paleolimnology," in (Frey, D. G., editor) "Limnology in North America," Madison, Wisc., University of Wisconsin Press, pp. 621–652.

Deevey, E. S., 1955, "The obliteration of the hypolimnion," *Mem. Ist. Ital. Idrobiol., Suppl.* **8,** 9–38.

Forel, F. A., 1892, "Le Léman: Monographie Limnologique," Tome 1, "Géographie, Hydrographie, Géologie, Climatologie, Hydrologie," Lausanne, F. Rouge, xiii + 543pp.

Frey, D. G., 1964, "Remains of animals in Quaternary lake and bog sediments and their interpretation," *Arch. Hydrobiol. Beih., Ergebn. Limnol.,* **2,** 1–114, 2 pl., 1 table.

Hutchinson, G. E., 1957, "A Treatise on Limnology," Vol. I, "Geography, Physics, and Chemistry," New York, John Wiley & Sons, xiv + 1015pp.

Lindeman, R. L., 1942, "The trophic-dynamic aspect of ecology," *Ecology,* **23,** 399–418.

Mortimer, C. H., 1956, "An Explorer of Lakes," in "G. C. Sellery, E. A. Birge—A Memoir," Madison, Wisc., University of Wisconsin Press, pp. 163–211.

Rawson, D. S., 1955, "Morphometry as a dominant factor in the productivity of large lakes," *Verh. intern. Ver. Limnol.,* **12,** 164–175.

Ryther, J. H., 1956, "The measurement of primary production," *Limnol. Oceanogr.,* **1,** 72–84.

Cross-references: *Buffer Systems; Calcium; Calcium Carbonate; Chlorophyll; Environmental Pollution; Epilimnion; Hydrogen; Hypolimnion; Ion Exchange; Lake Geochemistry; Nitrogen; Oxygen; Paleolimnology; pH-Eh; Phosphorus; Photosynthesis; Rainwater; River Geochemistry; Sodium; Sulfur Oxidation—Bacterial; Thermocline.* Vol. I: *Phytoplankton; Seiche; Transparency.* Vol. II: *Coriolis Force.* Vol. III: *Crater Lakes; Cryptodepressions, East African Lakes; Glacial Lakes; Lake Atitlan; Lake Baikal; Lake Geneva.* Vol. VI: *Dy; Gyttja; Lacustrine Sediments; Marl; Sapropel.* Vol. VII: *Algae; Diatoms; Paleoecology.*

LITHIUM: ELEMENT AND GEOCHEMISTRY

Properties. Lithium (Gr. *lithos,* stone), symbol Li, atomic number 3, is one of the alkali elements, and is the lightest of all metals (half the density of water). It comes in Group IA of the Periodic Table. Its oxide was separated from the mineral petalite by A. Arfvedson in 1817. The element was later isolated in minute quantity by H. Davy, and in 1855 by Bunsen and Matthiesson, who determined its properties. In its pure form, lithium is a silver-white metal of hardness 0.6. The atomic weight is 6.940, density 0.53 (20°C), melting point 180°C, boiling point 1326°C, and it possesses a remarkably high specific heat. In compounds, the lithium ion is univalent, with an ionic radius of 0.68 Å.

Two natural isotopes of lithium are known. These are Li^6 and Li^7, their relative abundance being 7.4 and 92.6% respectively.

Minerals. The following is a list of some of the more common lithium minerals, all of which are found in pegmatite bodies:

petalite (Li, Na) $AlSi_4O_{10}$ (2–4% Li_2O)

spodumene $LiAlSi_2O_6$ (about 8% Li_2O)

eucryptite $LiAlSiO_4$ (about 8% Li_2O)

lepidolite (lithium mica) K (Li Al)$_3$ Al, Si)$_4$ O_{10} (OH, F)$_2$ (3–5% Li_2O)

lithiophilite Li (Mn,Fe)PO_4 (up to 22% LiO_2)

amblygonite $LiAlPO_4$(F,OH) (8–10% LiO_2)

Abundance. The average lithium contents of three rock groups are given by Horstman (1957) as follows: granitic rocks, 40 ppm; basaltic rocks, 17 ppm; sedimentary rocks, 53 ppm.

The average abundance of lithium in the earth's crust has been variously estimated to be 25–40 ppm. The element is concentrated in certain pegmatite bodies which contain spodumene and other lithium minerals. In Maine, for example, certain pegmatite bodies contain 0.4 to 1.6% Li_2O (Sundelius, 1963). However, many simple pegmatite bodies contain very low concentrations of this element. Thus certain simple pegmatite bodies in northern Karelia (U.S.S.R.) contain only 4 ppm lithium, compared with 20 ppm in the enclos-

LITHIUM: ELEMENT AND GEOCHEMISTRY

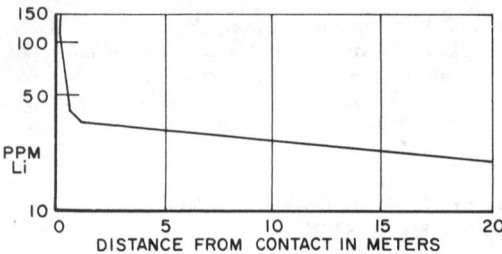

FIG. 1. Lithium content of basic rock as a function of distance from a pegmatite vein (Ryabchikov and Solov'yeva, 1961).

ing gneisses (Ryabchikov and Solov'yeva, 1961).

Geochemical Behavior. Lithium is very reactive chemically and tends to enter silicate minerals rather than sulfide minerals. In silicates, it replaces iron and magnesium rather than other alkalis, sodium, and potassium. Thus in certain gneisses, biotite was found to contain 4.6 times as much lithium as coexisting muscovite (Ryabchikov and Solov'yeva, 1961). Also, it is enriched in ferromagnesian minerals relative to the feldspar minerals. Within the group of ferromagnesian minerals, it is enriched in biotite and amphibole relative to pyroxene and olivine.

The process of magmatic differentiation causes an enrichment of lithium in the later stages. Thus in the Skaergaard mafic complex lithium remains fairly constant at about 2 ppm in the early and middle differentiates, but in the acid differentiate it rises to 6 ppm, and finally to 20 ppm (Wager and Mitchell, 1961).

The process of rock metamorphism does not cause a notable depletion in lithium. However, in certain contact aureoles about granite and pegmatite bodies, lithium was evidently added to the country rocks. Thus the lithium content of a basic metamorphic rock, adjacent to a pegmatite body, has been enriched from about 20 ppm to 100 ppm, as determined by Ryabchikov and Solov'yeva (1961) and is shown in Fig. 1.

During weathering and sedimentation, lithium evidently is readily adsorbed by clay minerals. Hence shales normally contain a relatively high Li content (about 60 ppm, compared with about 40 ppm in granitic rocks), and very little of this element reaches the sea, which contains about 0.1 ppm Li.

RALPH KRETZ

References

Heier, K. E., and Adams, J. A. S., 1964, "The Geochemistry of the Alkali Metals," in (Ahrens, L. H., et al., editors), "Physics and Chemistry of the Earth," Vol. 5, New York, McGraw-Hill Book Co., pp. 255–381.

Horstman, E. L., 1957, "The distribution of lithium, rubidium, and caesium in igneous and sedimentary rocks," *Geochim. Cosmochim. Acta,* **12,** 1–28.

Landolt, P. E., and Sittig, M., 1961, "Lithium," in (Hampel, C. A., editor), "Rare Metals Handbook," Second edition, New York, Reinhold Publ. Corp., pp. 239–270.

Rankama, K., and Sahama, T. G., 1950, "Geochemistry," Chicago, Univ. Chicago Press, 912pp.

Riley, J. P., and Tongudai, M., 1964, "The lithium content of sea water," *Deep-Sea Res.,* **11,** 563–569.

Ryabchikov, I. D., and Solov'yeva, B. A., 1961, "Geochemistry of rubidium and lithium in micaceous pegmatites of northern Karelia," *Geochemistry,* No. 4, 356–365, (transl. from Russian).

Sundelius, H. W., 1963, "The peg claims spodumene pegmatites, Maine," *Econ. Geol.,* **58,** 84–106.

Vlasov, K. A. (editor), 1966, "Geochemistry and Mineralogy of Rare Elements and Genetic Types of Their Deposits," Jerusalem, Israel Program for Scientific Transl., 2 Vols. (transl. by Z. Lerman).

Wager, L. R., and Mitchell, R. L., 1951, "The distribution of trace elements during strong fractionation of basic magma—a further study of the Skaergaard intrusion, East Greenland," *Geochim. Cosmochim. Acta,* **1,** 129–208.

Cross-references: *Alkalis, Alkali Metals, and Alkaline Earth Metals; Lithium Ore Deposits.* Vol. V: *Magmatic Differentiation; Pegmatites; Skaergaard Intrusion.*

LITHIUM: ECONOMIC DEPOSITS

Lithium and its compounds have increased manyfold in interest and importance in the last two decades, largely as a result of research during and since World War II. Among the major industrial uses are lubricants, ceramics, glass, metallurgy, batteries, air-conditioning systems, and welding. Lithium and its compounds also have important applications in nuclear energy and may find use in the manufacture of high energy fuels.

Lithium in the earth's crust is concentrated in the last stages of magmatic crystallization chiefly in granitic pegmatites and such deposits are the principal commercial source of lithium ores. Although granitic pegmatites are not uncommon, lithium-bearing pegmatites of sufficient size, grade, and accessibility to be of economic interest are sparsely distributed. Lithium is also found in certain sediments, mineral springs, and natural brines; however, only one nonpegmatite source, the brine deposit at

TABLE 1. COMPOSITION OF THE MAJOR LITHIUM ORES

Mineral	Formula	Percent Li_2O In Commercial Minerals
Spodumene	$LiAlSi_2O_6$	4–7
Lepidolite	$K_2Li_3Al_4Si_7O_{21}(OH,F)_3$	3–4
Amblygonite	$LiAlFPO_4$	8–9
Petalite	$LiAlSi_4O_{10}$	3.5–4
Dilithium-sodium-phosphate	Li_2NaPO_4	19–21

Searles Lake in California, has been of appreciable commercial interest.

Mineralogy. Minerals containing lithium as a major constituent are most commonly silicates and phosphates. Out of twenty-seven minerals reported to contain in excess of 2% Li_2O only four—spodumene, lepidolite, amblygonite, and petalite—are at the present time of major importance as lithium sources. In addition to these minerals, the compound dilithium-sodium-phosphate recovered from Searles Lake brine is an important lithium source in the United States. The composition of these principal lithium raw materials is given in Table 1.

Spodumene, a lithium pyroxene, is by far the most common lithium ore mineral. It generally occurs as gray to green lathlike crystals ranging in size from fractions of an inch up to many feet in length. *Lepidolite,* a lithium mica, second in abundance, is normally found as compact aggregates of small flakes, varying in color from red to violet and gray to yellow. Some varieties contain significant quantities of cesium and rubidium. *Amblygonite,* a lithium aluminum fluophosphate, has the highest Li_2O content of the ore minerals but its use is limited by lack of known reserves. White in color, it occurs in cleavable masses. *Petalite,* a lithium aluminum silicate, is of special interest for its ceramic properties. Milky white to clear and sometimes pink in color, it superficially resembles quartz with which it usually occurs. Other less common lithium-bearing minerals which have been used commercially to some extent are *zinnwaldite,* a lithium mica; *eucryptite,* a lithium aluminum silicate; *triphylite,* a phosphate of iron and lithium, and *lithiophilite,* a phosphate of manganese and lithium.

Economic Geology. Prior to World War II, the commercial recovery of lithium minerals was confined to pegmatite deposits containing mineral crystals and aggregates of sufficient size and grade to permit selective mining and hand sorting. However, as demand increased in the early forties, suitable beneficiation methods were developed in the United States, which made possible the mass mining and economic recovery of small crystals of spodumene disseminated through large pegmatite bodies. Such deposits because of their outstanding size, number and relatively uniform grade presently constitute the major sources of lithium in the United States and Canada. The largest United States resources of lithium are contained in disseminated spodumene occurring in pegmatites in the Kings Mountain district, North Carolina, which presently furnish most of the domestic output; indicated ore reserves are estimated to contain 40,000 tons of Li_2O in ore averaging 1.7% Li_2O, and inferred ore is estimated to contain 1.28 million tons of Li_2O in ore averaging 1.3% Li_2O. Spodumene is also the principal ore mineral in the pegmatites of the Black Hills, South Dakota, which formerly supplied the bulk of United States lithium ore production; amblygonite, lepidolite, and small quantities of other lithium minerals have also been produced; reserves, both indicated and inferred, are estimated at 16,000 tons of Li_2O in material averaging 1.2% Li_2O. In other pegmatite localities in the United States, indicated reserves in pegmatites are estimated at 2,000 tons of Li_2O in ore averaging 1.2% Li_2O. The foregoing reserve estimates in pegmatites, made by the United States Geological Survey (Norton and Schlegel, 1955), were based on a minimum grade of 1% Li_2O. Elsewhere important deposits of spodumene are found in Canada in Ontario, Manitoba, and the Northwest Territories and in the Republic of Congo.

The major economic source of lepidolite and petalite is the large pegmatite at Bikita, Rhodesia, which also contains subordinate spodumene and amblygonite. Petalite and lepidolite with minor amblygonite are presently being mined commercially from pegmatites in Southwest Africa and to a lesser extent in the Republic of South Africa and Australia.

Other pegmatite sources in the western nations which have produced lithium minerals in recent years are in Argentina, Brazil, Surinam, Spain, Mozambique, Rwanda, Burundi, and Uganda. In the eastern Nations, economic deposits of lithium minerals are reported in East Germany, Czechoslovakia, and the U.S.S.R.

Dilithium-sodium-phosphate obtained from the brines of Searles Lake, California, as a by-product in the production of potash and borax, has been a commercial lithium source since 1938. The "Lake" consists of a solid body of alkali salts 60–90 ft deep having an exposed surface area of 12 square miles. The brines, averaging 0.015% Li_2O, occur in voids in the salt body. Reserves were estimated by the United States Geological Survey (1955) at 90,000 tons of contained Li_2O.

Four sources—the United States, Canada, Rhodesia, and Southwest Africa—currently supply the bulk of the production of lithium ores and concentrates in the western world. In North America, spodumene is the chief ore produced, and substantial quantities of lithium are also recovered from brine at Searles Lake, California. Rhodesian production consists mainly of lepidolite and petalite and Southwest Africa has been a consistent source of the three ore minerals petalite, lepidolite, and amblygonite. In the past, Rwanda-Burundi has furnished substantial quantities of amblygonite.

The chapter "Lithium raw materials" by T. E. Kesler in "Industrial Minerals and Rocks" published by the American Institute of Mining, Metallurgical, and Petroleum Engineers, New York (1960) contains an extensive bibliography. "Lithium—A Materials Survey," Information Circular 8053 published by the Bureau of Mines, U.S. Department of Interior, Washington (1961), also provides comprehensive fundamental information on lithium and its compounds. Statistical data and a review of developments are published annually by the Bureau of Mines in the chapter on lithium in the Minerals Yearbook.

J. M. WARDE

References

Broadhurst, S. D., 1956, "Lithium Resources of North Carolina," No. Carolina Dept. of Cons. and Devel., Div. Min. Res., Inf. Circ., 37pp.
Cameron, E. N., et al., 1949, "Internal Structure of Granitic Pegmatites," *Econ. Geol. Mon.* **2**, 115pp.
Eilertsen, D. E., 1965, "Lithium," *U.S. Bur. Mines Mineral Yearbook,* 1–4.
Norton, J. J., and Schlegel, D. M., 1955, "Lithium resources of North America," *U.S. Geol. Surv. Bull.* **1027-G**, 26pp.

Cross-references: *Natural Brines.* Vol. IVB: *Amblygonite; Ceramic Technology; Lepidolite; Lithium: Element and Geochemistry; Metallurgy; Petalite; Spodumene.* Vol. V: *Pegmatites.*

LITHOSPHERE—See Vol. V

LUMINESCENCE—See Vol. IV B

LUTETIUM: ELEMENT AND GEOCHEMISTRY

Properties

Elemental lutetium is a very reactive silvery-white *lanthanide* (q.v.), or *rare earth* metal with atomic number 71 and outer electronic structure $4f^{14}5d6s^2$. It occurs as one stable isotope, Lu^{175} (97.40% natural abundance); Lu^{176} (2.60% natural abundance) is weakly radioactive, with a half-life of 4.6×10^{10} years.

Lutetium is determined analytically by neutron activation and x-ray spectrography. In geologic environments, lutetium forms highly ionic bonds and weak complexes. The bonds are the most covalent and the complexes the strongest of any trivalent rare earth, however, since the ionic radius (0.848 Å) is the smallest of any trivalent rare earth. Generally, its chemical behavior is similar to the alkaline earths, and almost identical to other heavy rare earths.

Urbain, von Welsbach, and James (in 1907) were the first to isolate lutetium from the other rare earths; they separated it out of ytterbia, a heavy rare earth concentrate. Lutetium takes its name from Lutetia, the ancient name of Paris.

Minerals

Lutetium occurs in highest concentration in rare earth minerals that favor heavy rare earths, generally nitrates, sulfates, carbonates, niobates, and tungstates with cation coordination number of six to eight, often containing structural halogens and alkali metals. Typical examples are:

xenotime (Y, heavy rare earths) PO_4

fergusonite (Y, heavy rare earths)(Nb, Ta)O_4

Geochemical Behavior and Abundance

Schmitt et al. (1964) have established a value of 0.031 ppm lutetium in chondrites and 0.35 ppm in eucrite achondrites. The standard deviation within a meteorite class is only about 20%. In meteoritic matter, volatization may have redistributed alkali metals but not rare earths: lutetium has a boiling point of 3327°C.

The lutetium concentration in average basalt is probably about 0.40 ppm, and the range is less than a factor of ten (these values intepolated from data of Haskin et al., 1966).

Oceanic and continental basalts have similar ranges. Partial melting of mantle peridotite yields a basaltic liquid with enriched lutetium and a mantle source area depleted in lutetium. Igneous differentiation series may tend toward higher or lower lutetium concentrations, although the total variation is less than for light rare earths. The relative abundance of lutetium and other heavy rare earths remains unchanged. Towell et al. (1965) found 0.198 ± 0.012 ppm in the Rubidoux Mountain leucogranite phase of the southern California batholith, probably a differentiate (by crystal settling) of the San Marcos gabbro with 0.256 ± 0.015 ppm.

Average granite with less than 60% silica has 0.45 ppm, but the range of values for granites is even greater than for basalts.

In most basaltic, syenitic, and granitic rocks, lutetium is mainly in mafic rock-forming minerals, probably substituting for divalent calcium (ionic radius 0.99 Å).

Lutetium and other heavy rare earths form complexes and are transported in alkali, halogen, and carbon dioxide-rich hydrothermal fluids; metasomatic and pegmatoid deposits formed from such fluids may have high concentrations of lutetium and higher ratios of lutetium to light rare earths than granites.

Lutetium has a 570-year residence time in seawater (which is alkaline), about the same as other heavy rare earths. Removal by adsorption on clay particles, as well as depletion in surface water in areas of high organic productivity, are likely processes. High secondary lutetium concentrations in phosphatic fish remains today occur only in deep water; paleozoic fish remains also have high rare earths. Manganese nodules have 115 ppm, whereas Pacific Ocean water has only 0.116×10^{-6} ppm (Goldberg et al., 1963).

In soils, the relative amounts of lutetium in detrital, clay, and carbonate fractions is not well known. Sedimentary rocks have about 0.60 ppm with a variation of a factor of five. In a study of rare earths in Russian platform soils, Balashov et al. (1964), found higher rare earths in alkaline soils, the lutetium in groundwater precipitating as hydroxide; lower concentrations are in acid soils, probably due to removal of complexes.

Biologic concentration of rare earths has been found in hickory leaves; rare earths have been used to trace opium.

Economic Geology

The major source of lutetium has been from a light rare earth mineral, monazite in beach sands; the accessory monazite originally present in granitic rocks resist weathering and is concentrated by sedimentary processes. Australia, Brazil, Ceylon, India, Malaysia, and South Africa mine the largest quantities. Less than one percent of the monazite is lutetium. Recently, the bastnaesite deposit at Mountain Pass, California, has been an important source of lutetium, again in low concentration.

High-purity lutetium has a very limited market. Most lutetium is sold unpurified in rare earth concentrates used for polishing, petroleum cracking catalysis, lighter flints, arc carbons, and in alloys.

ROBERT KAY

References

Balashov, Y. A., et al., 1964, "The effects of climate and facies environment on the fractionation of the rare earths during sedimentation," *Geochem. Internat. No. 1,* **10,** 951–969 (English transl.).

Goldberg, E. D., et al., 1963, "Rare-earth distributions in the marine environment," *J. Geophys. Res.,* **68,** 4209–4217.

Haskin, L. A., et al., 1966, "Meteoric, Solar, and Terrestrial Rare Earth Distributions," in (Ahrens, L. H., et al., editors), "Physics and Chemistry of the Earth," Vol. 7, Oxford, Pergamon Press, pp. 167–321.

Rankama, K., and Sahama, T. G., 1950, "Geochemistry," Chicago, University of Chicago Press, 912pp.

Schmitt, R. A., et al., 1964, "Rare-earth, yttrium and scandium abundances in meteoric and terrestrial matter—II," *Geochim. Cosmochim. Acta,* **28,** 67–86.

Towell, D. G., et al., 1965, "Rare-earth distributions in some rocks and associated minerals of the batholith of Southern California," *J. Geophys. Res.,* **70,** 3485–3496.

Cross-references: *Complexes; Lanthanides; Neutron Activation Analysis; Rare Earths; Seawater, Chemistry; X-Ray Spectroscopy.* Vol. IVB: *Meteoritic Minerals; Xenotime.*

M

MAGMA—*See* **PARAGENESIS**; *also* Vol. V

MAGMATIC WATER

The water derived from the melting of rocks, or which is present in rock melt, is termed *magmatic water*. A major part of the gaseous exudations in volcanic regions is water vapor. White and Waring (1963) reported that generally steam constitutes more than 90% of the volatile material coming from volcanoes. However there usually is considerable opportunity for meteoric water to become incorporated in volcanic gases, so not all this can be attributed to magmatic sources. As noted in the discussion of the topic *juvenile water* (q.v.), there is a strong probability that a portion of the water issuing from hot springs in volcanic regions is of magmatic origin. However White (1957) suggested that the proportion of magmatic water in such solutions was probably not generally more than 5%.

Magmatic water may be partly juvenile; that is, water which has never been in the hydrosphere before. However, many magmas probably have had opportunity to pick up water from the hydrosphere at some time in their past history.

Magmatic water, according to White (1957) can contain considerable quantities of some of the more volatile elements and may often be high in chloride, fluoride, or boron, and have a high ratio of lithium to sodium.

JOHN D. HEM

References

White, D. E., 1957, "Thermal water of volcanic origin," *Bull. Geol. Soc. Am.*, **68**, 1637-1658.

White, D. E., and Waring, G. A., 1963, "Volcanic emanations," *U.S. Geol. Survey Profess. Paper* **440K**.

Cross-references: Gases, Volcanic; Juvenile Water. Vol. V: *Volatiles.*

MAGNESIUM: ELEMENT AND GEOCHEMISTRY

Magnesium, symbol Mg, has atomic number 12 and atomic weight 24.312. It falls in Group IIA of the Periodic Table, below beryllium and above calcium. It is a silvery white metal and its crystal system is hexagonal. It has three natural isotopes, Mg^{24} (78.8%), Mg^{25} (10.1%), and Mg^{26} (11.1%). Sir Humphrey Davy discovered the oxide of the new element in 1808, the metal being first isolated by the Frenchman A. Bussy, in 1828.

Geochemistry

The metal magnesium is among the most prevalent elements in the earth's crust and estimates of its content have ranged from 2.0 to 2.5%. It is the eighth most abundant of the chemical elements and the sixth most abundant metallic element. It is the third most abundant structural metal, being exceeded only by aluminum and iron.

According to H. Comstock of the U.S. Department of the Interior, Bureau of Mines, magnesium is a constituent of more than 150 minerals although only about sixty can be classed as magnesium-bearing ores. The naturally occurring sources of magnesium being used for the commercial production of magnesium compounds and/or the primary metal are the minerals brucite, dolomite, magnesite, and olivine, as well as seawater, bitterns, and underground brines. The chemical compositions and magnesium content of these minerals and solutions plus some additional high-magnesium-content minerals are shown in Table 1. Bitterns, which are concentrated brines, are not included because they can vary widely in composition and magnesium content, depending on the source of the original brine.

Magnesium is one of the predominate elements in the mantle surrounding the earth's core as well as in the earth's crust. Surrounding the iron-nickel core of the earth there is a mantle of magnesium-iron silicates referred to by Suess, since late in the last century, as the SIMA (from Si and Mg). Since the core and mantle are estimated to comprise about 99% of the earth's bulk, the earth as a whole can be considered as composed of 33% iron, 25% oxygen, and 25% silicon and magnesium. The balance of 17 percent is comprised of the remaining chemical elements.

Sources of Magnesium

Well over half of the primary magnesium metal produced in the world today is extracted

TABLE 1. CHEMICAL COMPOSITIONS AND MAGNESIUM CONTENT OF VARIOUS NATURAL MINERALS AND SOLUTIONS

Material	Chemical Composition	Magnesium Content (%)
Brucite	$Mg(OH)_2$	42
Magnesite	$MgCO_3$	29
Serpentine	$H_4Mg_3Si_2O_9$	26
Enstatite	$MgSiO_3$	24
Olivine	$(Mg, Fe)_2SiO_4$	19
Kieserite	$MgSO_4 \cdot H_2O$	18
Dolomite	$CaCO_3 \cdot MgCO_3$	13
Langbeinite	$2\,MgSO_4 \cdot K_2SO_4$	12
Epsomite	$MgSO_4 \cdot 7\,H_2O$	10
Kainite	$MgSO_4 \cdot KCl \cdot 3\,H_2O$	10
Carnallite	$MgCl_2 \cdot KCl \cdot 6\,H_2O$	9
Polyhalite	$2\,CaSO_4 \cdot MgSO_4 \cdot K_2SO_4 \cdot 2\,H_2O$	4
Underground brines	Mg, Ca, K, Na sulfates and chlorides	1
Seawater	Na, Mg, Ca, K, etc., chlorides and sulfates	0.13

from seawater which has an average magnesium content of 0.13%. This is equivalent to six million tons per cubic mile. With the volume of seawater on earth estimated at 331 million cubic miles, the total quantity of magnesium existing in the seas adds up to the astronomical figure of two quadrillion tons. Consequently, the supply of available magnesium on earth from seawater alone can be considered inexhaustible. The world's annual production of magnesium currently is only about 200,000 tons, whereas an estimated 137 million tons of magnesium are being washed into the ocean each year by rivers.

Production of Magnesium Metal

Ever since the first commercial production of magnesium began, there have been only two basic reduction methods for obtaining the metal in metallic form—electrolytic and thermal. Throughout the world today approximately 80% of the primary magnesium produced is by the electrolytic process in which anhydrous or partially dried magnesium chloride is broken down by electrolysis into its components of metallic magnesium and chorine gas (see Fig. 1). The raw materials from which the magnesium chloride is obtained are seawater and dolomite or a combination of the two. In the United States, the Dow Chemical Company uses seawater (Fig. 2). Norsk Hydro-Elektrisk in Norway uses seawater and dolomite in combination, whereas the U.S.S.R. is reputed to be using dolomite in combination with carnallite.

Thermal reduction, using ferrosilicon to reduce dolomite, is the method used to produce the remaining 20% of the magnesium produced in the world today (Fig. 3). This

FIG. 1. Primary magnesium transported by conveyor from the ingot casting machine, are stacked on magnesium pallet ingots prior to shipment or storage. Most of the world's production of magnesium today comes from electrolytic cells which convert magnesium chloride obtained from seawater into metallic magnesium and chlorine gas.

method is used in Canada, France, Italy, Japan, and, to a relatively small extent, was formerly used in the United States.

Uses of Magnesium

Magnesium has many important uses in industry. A versatile metal, it has both structural and nonstructural uses. In the latter category are such uses as: alloying ingredient in aluminum, lead, and zinc; deoxidizer and desulfurizer in the manufacture of copper and nickel alloys; additive in the manufacture of ductile iron; reducing agent in the production of titanium and zirconium; catalyst in the production of certain organic chemical com-

MAGNESIUM: ELEMENT AND GEOCHEMISTRY

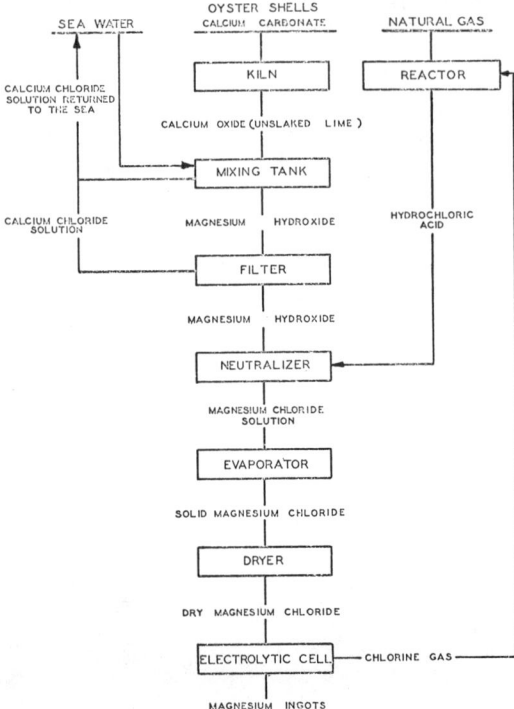

FIG. 2. Dow process for the production of magnesium from seawater.

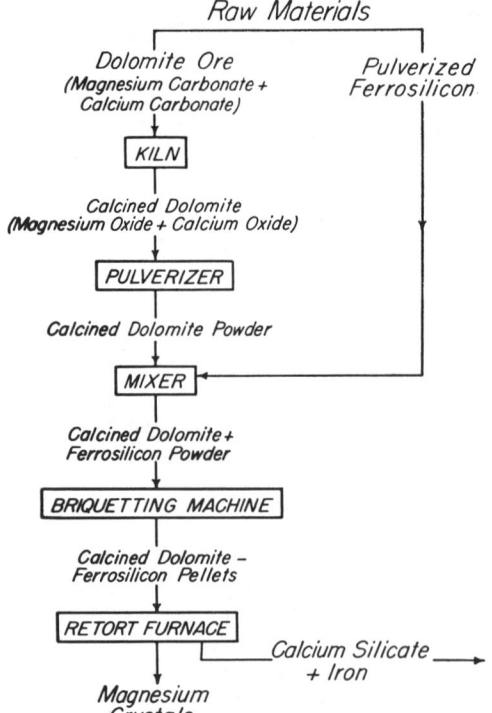

FIG. 3. Flow diagram of the Pidgeon ferrosilicon process for producing magnesium from dolomite.

pounds; photoengraving metal and galvanic anodes for the protection of water heaters and underground structures from corrosion.

Major structural uses usually are based on the extremely low weight of magnesium. The alloys of the metal have an average specific gravity of 1.8 and have been known for many years as industry's lightest structural metal. In addition to its low density, magnesium has excellent machinability characteristics which also contribute to its desirability for many structural applications. The major structural uses of magnesium alloys include: the construction of aircraft, missiles, and space vehicles; materials handling equipment, household appliances and business machines; portable tools of many kinds including lawn mowers and chain saws; automobile die castings and many types of jigs, fixtures, gages, and other items for the tooling industry.

W. H. Gross

References

Beck, A., 1941, "The Technology of Magnesium and Its Alloys," transl. of "Magnesium und seine Legierungen," Second ed., London, F. A. Hughes & Co. Limited, 512pp.

Church, F. L., 1966, "Magnesium: starting to make out in mass markets," *Modern Metals*, **22**(6, July), 57–81.

Comstock, H., 1963, "Magnesium and Magnesium Compounds," a Materials Survey, U.S. Department of the Interior, Bureau of Mines, Washington, D.C., 128pp.

Emley, E. F., 1966, "Principles of Magnesium Technology," London, Pergamon Press Ltd., 1013pp.

Gross, W. H., 1967, "Magnesium and magnesium alloys," in (Standen, A., editor), Kirk-Othmer: "Encyclopedia of Chemical Technology," Vol. 12, Second ed., New York, John Wiley & Sons, Inc., pp. 661–708.

Gross, W. H., 1949, "The Story of Magnesium," Metals Park, Ohio, American Society for Metals, 258pp.

Lerman, A., 1965, "Strontium and magnesium in water and in Crassostrea calcite," *Science*, **150** (3697), 745–751.

Letolle, R., 1962, "Recherches recentes sur la geochimie isotopique du magnesium," *Rev. Geogr. Phys. Geol. Dyn.*, 2nd series, Vol. 5.

Pytkowicz, R. M., Duedall, I. W., and Connors, D. N., 1966, "Magnesium ions: activity in seawater," *Science*, **152** (3722), 640–642.

Roberts, C. S., 1960, "Magnesium and Its Alloys," New York, John Wiley & Sons, Inc., 230pp.

Shigley, C. M., 1951, "Minerals from the sea," *J. Metals (AIME)*, **3**(Jan.) 25–29.

Tangen, T., 1963, "Magnesium Production in Norway—Present State and Future Growth," Paper #63-6, Proceedings of The Magnesium Association.

Thompson, M. E., 1966, "Magnesium in sea water:

an electrode measurement," *Science* **153** (3738), 866–867.

Williams, L. R., 1965, "Magnesium," *U.S. Bur. Mines Mineral. Yearbook.*

Williams, L. R., 1965, "Magnesium Compounds," *U.S. Bur. Mines Mineral. Yearbook.*

Cross-references: *Core Geochemistry; Earth's Crust Geochemistry; Natural Brines; Seawater, Chemistry.*

MAGNESIUM CYCLE

The element magnesium is one of the major constituents of the earth and is also a major element in seawater (concentration: 1.35 g/liter, or 3.7% of the total dissolved ions). Because of its constancy in solution relative to seawater, it is referred to as one of the "conservative ions."

Although the average Mg content of river water is 3.4% of the dissolved salts (Clarke), this does not give a true proportion of the material dissolved from the earth's crust and transported to the sea because of the role of *cyclic salts* (q.v.; see also Vol. II, *Atmospheric Nuclei and Dust*), i.e., those that are carried in small droplets from the ocean spray, incorporated in clouds and so returned as rain and then recycled. Sverdrup, Johnson, and Fleming (1942) gave a corrected estimate for the Mg content of river water (less cyclic salts) as 3.03%, out of a total of 27.35×10^{14} g of all ions carried down annually from the land.

According to the calculations of Goldschmidt (subject to some correction by Rankama and Sahama, 1950), the Mg supplied by the earth's surface erosion to seawater represents 12.54 kg/ton for the whole of geologic time, but of this only 10% remains in solution. Like the other alkali elements, with their low ionic potentials, Mg is largely extracted from seawater and incorporated in *evaporite deposits* (see Vol. VI). Although $MgCO_3$, magnesite, is highly soluble, under diagenesis $MgCO_3$ slowly reacts with $CaCO_3$ to form the double carbonate $CaMg(CO_3)_2$, dolomite, which is hard and very stable in nature in the earth's crust (see Vol. VI, *Diagenesis*). In the older carbonates of the geological record (say, older than 3×10^8 yr) most have become thus "dolomitized." In evaporites, $MgSO_4.H_2O$, $MgCl_2.6H_2O_3$, and other highly soluble salts are commonly precipitated, but a large evaporite body, isolated at shallow depth in the earth's crust, also becomes stabilized because it is impervious to water (owing to its low strength, it flows plastically and seals joints and cracks).

In this way large volumes of Mg compounds are withdrawn from solution in the ocean through geological time. From time to time, dolomites and evaporites become re-elevated by orogenic processes and recycled. Most of the source materials, however, are the *ferromagnesian minerals* of primary igneous rock such as basalt, a typical mineral of which is $(Mg,Fe)_2SiO_4$, olivine. Under ocean-floor spreading the recycling rate of suboceanic basaltic crust is something like $2–3 \times 10^8$ years. How much of the Mg in "primary" igneous rock is slowly recycled and how much comes directly from the mantle is still somewhat problematic.

A diagrammatic summary of the global Magnesium Cycle is given by Termier (and reproduced here in the article, *Cycles—Geochemical*). Termier divides it into two systems, one on the land and one in the ocean, as follows:

(a) *Continents*
1. Solution of Mg^{2+} ions from solid rock by soil processes.
2. Transport in fresh water (ground, river).
3. Inorganic evaporation (playa, lagoons, salt lakes—stabilization in evaporites and dolomite).
4. Organic fixation (incorporated in chlorophyllic plants and thence ingested by animals; much of the organic fraction is directly recycled through soils, but some of it passes via rivers to the ocean).

(b) *Oceans*
1. Solution of solid rocks (dolomite, basalt, etc.) (diurnally variable pH).
2. Transport in solution by currents.
3. Inorganic evaporation (coastal lagoons—stabilization in evaporites and dolomite).
4. Organic fixation (by plants, e.g., calcareous algae, as metastable "high-Mg calcite"; smaller role played by animals such as some mollusca, echinoderms, foraminifera; soft chlorophyllic plants usually decay and thus recycle their Mg content).
5. Burial and diagenesis (metastable calcite combines with Mg-rich connate water by metasomatic processes, both during syndiagenesis and anadiagenesis, to become stabilized as dolomite).
6. Cyclic salt removal (by sea spray, etc., Mg salts are removed into the atmosphere, and transported in clouds, thence into the continental system).

Changes Through Time

There appear to have been few changes in principle to the Magnesium Cycle through time, at least since the appearance on earth of chlorophyllic plants (about $2.9 \pm 0.2 \times 10^9$ years ago). Calcareous algae (*Collenia*, etc.) are known widely on all continents in Precambrian rocks dating back more than 2.7×10^9 years, the resultant *stromatolitic limestones* being found to be almost always dolomitized.

RHODES W. FAIRBRIDGE

References

Fairbridge, R. W., 1967, "Carbonate Rocks and Paleoclimatology," in (Chilingar, G. V., Bissell, H. J., and Fairbridge, R. W., editors), "Carbonate Rocks," Amsterdam, Elsevier Publ. Co., pp. 399–432.

Rankama, K., and Sahama, T., 1950, "Geochemistry," Chicago, University of Chicago Press, 912pp.

Sverdrup, H. U., Johnson, M. W., and Fleming, R. H., 1942, "The Oceans," New York, Prentice Hall, 1087pp.

Cross-references: *Cycles—Geochemical; Cyclic Salts; Evaporite Processes.* Vol. II: *Atmospheric Nuclei and Dust.* Vol. IVB: *Evaporite Minerals.* Vol. V: *Metasomatism; Sea Floor Spreading.* Vol. VI: *Diagenesis; Evaporites; Salts—Cyclic.*

MANGANESE: ELEMENT AND GEOCHEMISTRY

Properties

The atomic weight of manganese is 54.94, its number in the periodic table is 25, and it is represented in nature by one isotope, Mn^{55}. The combined or ionic manganese exists at or near the surface of the earth at oxidation levels +2, +3, +4, and +6 (Hem, 1964). The ionic radius in sixfold coordination of Mn^{2+} is 0.80, Mn^{3+} is 0.70, and that of Mn^{4+} is 0.60. The electronegativity of Mn^{2+} is 1.4. The ionic potential of Mn^{2+} is 2.5, and that of Mn^{4+} is 6.7 (Mason, 1958).

Manganese in iron meteorites is a very chalcophile element, and is lithophile in silicate meteorites (Goldschmidt, 1954). In the upper lithosphere it is an oxyphile element, is not siderophile, and is slightly chalcophile (Rankama and Sahama, 1950). The biosphere, in particular, is favorable for the formation of manganese-bearing soils and sediments, and manganese ores. Manganese is a member of the iron family and both elements are closely associated in geochemical history.

Abundance

The cosmic abundance of manganese is 69 manganese atoms per 10,000 atoms of silica (Suess and Urey in Mason, 1958). The average composition of meteoritic matter ranges from 0.26% in silicate phase and 0.2% in troilitic phase, to 0.01 in nickel-iron phase (Goldschmidt, 1954). Its abundance in ultrabasic rocks is 0.112%, and 0.096% in granite (Rankama and Sahama, 1950). The content of manganese in sedimentary rocks varies from a trace in some sandstones to 0.038% in limestones, 0.089% in phyllites, 0.117% in red clay, 0.260% in marine siderite ores, 7.45% in bog iron ores (Rankama and Sahama, 1950), 0.002 to 10% in soils (Goldschmidt, 1954), and up to 50% in sedimentary manganese ores. The latter figure indicates concentrations up to 500 times the average abundance of 0.1% manganese in the earth's crust.

During magmatic differentiation, the ratio Mn:Fe increases from 1:50 to 1:10 (Goldschmidt, 1954). The relative concentration of manganese continues to increase in the contact zones of magmatic intrusions and in some hydrothermal deposits. Concentrations of great magnitude develop under favorable conditions in the zone of weathering.

Geochemistry

Under tropical and subtropical conditions manganese may be concentrated in residual deposits, while under colder climatic conditions manganese is leached by acid solutions from the soils and transported to the basins of deposition. The manganous ion is here the subject of preferential leaching as compared to leaching of ferrous iron due to smaller ionic radius of the latter (Goldschmidt, 1954). Manganese is removed from the zone of weathering as bicarbonate or as a complex with organic acids derived from decaying plants.

The content of manganese in river water reaches 0.13 ppm, but usually does not exceed 0.1 ppm (Konovalov in Hem, 1964). The amount of manganese present in solution in sea water is only 0.0007 to 0.001 ppm (Harvey in Mason, 1958). Over 90% of manganese introduced by the rivers to the sea is in the form of suspension. The precipitation of dissolved manganese in the form of oxides and carbonates is controlled by the pH and Eh of solutions and follows the precipitation of iron. The inorganic catalysis accelerates precipitation. Bacteria also contribute to precipitation if manganese is present in the form of organic complexes. During sedimentation, the negatively charged sols and gels of $Mn(OH)_4$ and MnO_2 absorb cations of Li, K, Ca, Ba, B, Ti, Co, Ni, Cu, Zn, Tl, Pb, and W. Manganese also may be supplied to the sediments by hot springs. During the process of diagenesis the pH and Eh, in freshly formed sediments, depend on the content of organic matter which reduces manganese, forming carbonates and even hauerite. Any subsequent metamorphism is also accompanied by reduction of manganese to trivalent and bivalent states. In palingenetic and magmatic rocks, manganese is represented exclusively by manganous ion replacing, isomorphically, Mg^{2+}, Ca^{2+}, and Fe^{2+} in silicates.

EUGENE A. ALEXANDROV

References

Goldschmidt, V. M., 1954, "Geochemistry," London, Oxford University Press, pp. 621–642.
Hem, J. D., 1964, "Deposition and solution of manganese oxides," *U.S. Geol. Surv. Water-Supply Paper*, **1667-B**, 42pp.
Mason, B., 1958, "Principles of Geochemistry," New York, John Wiley & Sons, 310pp. (2nd ed. 1966).
Mohr, P. A., and Allen, R., 1965, "Further considerations on the deposition of the Middle Cambrian manganese carbonate beds of Wales and Newfoundland," *Geol. Mag.*, **102**(4), 328–337.
Pratt, R. M., and McFarlin, P. F., 1966, "Manganese pavements on the Blake Plateau," *Science*, **151**(3714), 1080–1082.
Rankama, K., and Sahama, Th.G., 1950, "Geochemistry," Chicago, University of Chicago Press, pp. 640–653.
Roy, S., 1966, "Syngenetic Manganese Formations of India," Calcutta, Jadavpur University, 219pp.
Sapozhnikov, D. G., (editor-in-Chief) et al., 1967, "Manganese Deposits of the Soviet Union," Nauka, Moscow, 460 pp. (in Russian only).
Strakhov, N. M., et al., 1968, "Geochemistry of Sedimentary Deposition of Manganese Ore," Nauka, Moscow, 495 pp. (in Russian only).

Cross-references: *Catalysis; Electronegativity; Iron; Manganese Cycle; Manganese Nodules; Organic Geochemistry; pH-Eh; Seawater, Chemistry; Weathering, Chemical.* Vol. II: *Meteorites.* Vol. IVB: *Meteoritic Minerals.* Vol. VI: *Diagenesis; Hot Springs; Weathering, Organic.*

MANGANESE CYCLE

Chemical Aspects

The abundance of manganese in the earth's crust has been estimated to be 0.09% (Bear, 1955). On the earth's surface, manganese is distributed among the solid phases of igneous, metamorphic, and sedimentary rock, and the aqueous phases of soil-water, ponds, lakes, rivers, and the seas. A cyclic movement of manganese occurs between these various solid and aqueous phases (Fig. 1). Two major cycles can be recognized: a terrestrial cycle, in which the contact between solid and aqueous phases is very large, and a lacustrine and marine cycle, in which the contact between solid and aqueous phases is more limited. In each of the two cycles, manganese in rock minerals is brought into solution by weathering processes, such as acid leaching, oxidation-reduction reactions, and extraction with chelating agents. Once in solution, manganese, usually in the form of manganous manganese, may be subject to redeposition through formation of carbonates, oxides, silicates, or sulfides under appropriate conditions, or through adsorption to preformed oxides of manganese or other adsorbents. Manganiferous sediments may be changed into sedimentary rock by lithification. Since both manganiferous sediments and sedimentary rock are subject to weathering, the cycles are continually repeated. The primary origin of manganese going through these cycles is igneous rock, which was derived from magma that issued from the mineral reservoir of the interior of the earth. Manganese can move from the terrestrial cycle to the lacustrine and marine cycle through drainage from land by rivers and streams, ultimately feeding into lakes and seas. Continental run-off is thought to be a major source of manganese found in ocean waters, the other major source being the weathering of igneous rocks formed as a result of submarine volcanism. A reverse movement of manganese from the lacustrine and marine cycle to the terrestrial cycle occurs through formation of manganiferous deposits in the littoral zones, and through land-building processes such as have taken place over past geological ages.

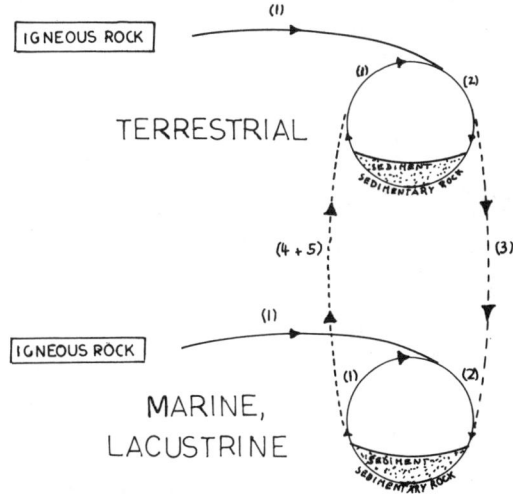

1 WEATHERING
2 REDEPOSITION
3 LAND DRAINAGE
4 LITTORAL DEPOSITION
5 LAND FORMATION

FIG. 1. Manganese cycles.

The movement of manganese from solid to aqueous phases and back to solid phases is influenced by its solubility when in mineral form, and its precipitability when in soluble form (Trost, 1958). Manganese that is strongly held in the crystal lattice of a mineral is less readily dissolved than manganese that is loosely held and easily replaced. Dissolved manganous manganese is precipitated by excess of sulfide, carbonate at alkaline pH, or other basic substances. It may also be precipitated by reaction with suitable oxidizing agents which transform it to an oxide. However, none of the chemical or physical influences on the movement of manganese can be considered without the influence of Eh and pH (Hem, 1963). At low Eh and pH, manganese is more likely to be extracted from a solid phase and either kept in solution or precipitated as a manganous compound than at high Eh and pH. At high Eh and pH, manganese will be more easily precipitated as a higher oxide than at low Eh and pH. Chelating agents, such as the naturally occurring humic acids, bicarbonate and sulfate, may complex manganous manganese in solution so extensively that, in spite of prevailing conditions favoring precipitation, it will remain in solution (Hem, 1963). Chelating agents may also act as extractors of manganese from the solid phase if the binding power of the agent is stronger than that of the crystal lattice of the mineral. Living organisms, particularly microorganisms, play very important roles in the progress of the manganese cycle (Alexander, 1961; Silverman and Ehrlich, 1964).

Microbial Aspects

Microbial Dissolution of Manganese Compounds in Solid Phases. Carbon dioxide and organic acids produced by microbes, particularly in soil, can dissolve manganese in a solid phase such as $MnCO_3$, and if present in excess, may even prevent reprecipitation of the dissolved manganese. Enzymatic reduction of oxidized manganese (Mn^{III} and Mn^{IV}) by microbes (bacteria and fungi) at acid pH and appropriate Eh may bring manganese into solution. Such microbial reduction can occur even in the presence of oxygen (Trimble and Ehrlich, 1968). In instances where iron accompanies manganese in a solid phase, microbes may separate manganese from iron by preferential solubilization. Such separation may also occur nonbiologically, as in thermal action on iron and manganese carbonates (Trost, 1958).

Microbial Precipitation of Dissolved Manganese. Some microorganisms can precipitate manganese by oxidizing Mn^{II} to Mn^{III} and Mn^{IV}, or by forming precipitating reagents like carbonate, sulfide, or certain organic compounds (Ehrlich, 1968; Silverman and Ehrlich, 1964). Other microorganisms may precipitate manganese by adsorbing it to their cell surface in an oxide form.

Special Manganese Accumulations

Sedimented manganese oxides sometimes occur in concretions called nodules. They are reported from terrestrial, lacustrine, and marine environments, but they differ from each other in their chemical and physical properties, depending on their source. The nodules of soil and fresh water have been less well studied than marine nodules. The marine nodules are found on most ocean bottoms at depths from 65 fathoms on down. On the average they may consist of 19% manganese and 14% iron with traces of other metals such as nickel, copper, cobalt, titanium, zinc, lead, phosphorus, aluminum, and zirconium. Their mode of formation has been variously postulated to involve physicochemical processes and biological processes. Recent evidence lends strong support to the idea of microbial participation in nodule genesis (Ehrlich, 1968; Perfil'ev, et al., 1965).

Microbes Active in Manganese Oxidation and Reduction

Manganese Oxidation. Microorganisms reported to be active in manganese oxidation in soil, and in fresh and salt water were summarized by Silverman and Ehrlich (1964). In addition, Perfil'ev and Gabe have indicated that *Kusnezovia* and *Caulococcus* are active in accumulating manganese, presumably by oxidation (Perfil'ev, et al., 1965). Tyler and Marshall have implicated *Hyphomicrobium* sp. in Tasmania.

Manganese Oxide Reduction. Published reports of microorganisms active in manganese oxide reduction include members of the genus *Bacillus, Thiobacillus thiooxidans,* a coccus, and yeast (Perfil'ev, et al., 1965; Silverman and Ehrlich, 1964). However, the ability to reduce oxides of manganese is more widespread among bacteria and fungi than published reports indicate.

Importance of the Manganese Cycle

An active manganese cycle can be beneficial, as in soil fertility (Alexander, 1961). It provides plants, and indirectly animals, with the manganese needed in their nutrition. Manganese, to be available to plants, must be soluble. A concentration of 100 ppm to 1% available manganese is satisfactory. A concentration below this range may cause stunted growth, chlorosis, or necrosis, as in peas and oats. A concentration of more than 1000ppm available manganese is toxic to plants. Management of pH and Eh in the soil can control the level of manganese available to plants.

An active manganese cycle can also be a nuisance, as in the management of public water supplies (Silverman and Ehrlich, 1964). Manganese along with iron is likely to oxidize in aerated, alkaline waters, leading to discoloration of the water. The oxidized manganese and iron may deposit on the walls of conduits and pipes, especially with the assistance of some sheathed and stalked bacteria, and some others. Extensive growth of these bacteria, stimulated by the manganese and iron, can lead to obstruction of the pipes. Remedial and preventive chemical and physical treatments of waters and pipes are possible.

H. L. EHRLICH

References

*Alexander, M., 1961, "Introduction to Soil Microbiology," New York, John Wiley & Sons, Inc., 472pp.
*Bear, F. E. (editor), 1955, "Chemistry of the Soil," New York, Reinhold Publishing Corp., 373pp.
Ehrlich, H. L., 1968, "Bacteriology of manganese nodules. II. Manganese oxidation by cell-free extract from a manganese nodule bacterium. *Appl. Microbiol.*, **16**, 197–202.
Hem, J. D., 1963, "Chemical Equilibria and Rate of Manganese Oxidation," *Geol. Survey Water Supply Paper* **1667-A**, 64pp.
Perfil'ev, B. V., Gabe, D. R., Gal'perina, A. M., Rabinovich, V. A., Sapotnitskii, A. A., Sherman, E. E., and Troshanov, E. P., 1965, "Applied Capillary Microscopy," New York, Consultants Bureau (English transl.), 122pp.
*Silverman, M. P., and Ehrlich, H. L., 1964, "Microbial formation and degradation of minerals," *Advan. Appl. Microbiol.*, **6**, 153–206.
Trimble, R. B., and Ehrlich, H. L., 1968, "Bacteriology of manganese nodules. III. Reduction of MnO_2 by two strains of nodule bacteria," *Appl. Microbiol.*, **16**, 695–702..
Trost, W. R., 1958, "The Chemistry of Manganese Deposits," Mines Branch Research Report R 8, Dept. of Mines and Technical Surveys, Ottawa, Canada, 125pp.

* Additional bibliographic references may be found in this work.

Cross-references: Chelation; Cycles—Geochemical; Manganese: Element and Geochemistry; Manganese Nodules: Organic Geochemistry; Oxidation and Reduction; pH-Eh; Solubility; Weathering—Chemical. Vol. VII: Microorganisms.

MANGANESE NODULES

Manganese nodules are rocks composed largely of ferromanganese oxides formed by precipitation at the bottom of lakes and the oceans (Figs. 1, 2 and 3). They range in size from microns to meters; their morphology is highly variable. They contain up to 55% manganese (Mn), 35% iron (Fe), and 2% nickel (Ni), cobalt (Co), and copper (Cu). Their mineralogy is varied; the Mn and Fe phases are generally poorly crystalline and consist of goethite and varied polymorphs of the manganese oxides; in addition, nodules contain

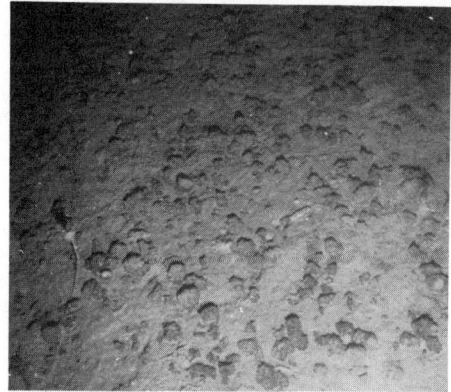

FIG. 1. Manganese nodules on the ocean floor. This picture was taken in the Antarctic, 57°S 83°W, by the USNS *Eltanin*, Water depth, 4850m.

FIG. 2. A manganese nodule from the North Pacific. *Courtesy* Guy Mathieu.

MANGANESE NODULES

FIG. 3. Cross section of a manganese nodule. The light rectangular stone in the center is an olivine fragment around which the nodule grew. *Courtesy Guy Mathieu.*

the detrital and authigenic minerals found in adjacent sediments.

Manganese nodules were first discovered in the open ocean by Thompson, Murray, and Renard during the *Challenger* expedition (1873–1876). In 1878, Buchanan reported the occurrence of nodules in the Firth of Clyde, a shallow-water area, and by the end of the century at least five additional occurrences of manganese nodules in shallow marine environments had been discovered (Manheim, 1965).

Early workers chemically analyzed about a score of manganese nodules, and hypothesized about their mechanism of growth. Two main hypotheses emerged: (a) nodules grow by the slow precipitation of manganese from seawater; and (b) nodules are formed by the rapid precipitation of manganese released in submarine volcanism.

Until the 1950s little additional work was done except for some early measurements of manganese nodule growth rates. The last decade, however, has seen a strong revival of interest in manganese nodules, stimulated both by the expansion of oceanographic facilities and the realization of the economic importance of manganese nodules as ores. Recent work has greatly enlarged our knowledge and understanding of the chemical composition, growth rates, mineralogy, occurrence, and general morphology of manganese nodules.

Occurrence and Morphology

In large areas of the ocean floor, manganese nodules may be absent; in others, they may cover nearly 100% of the area. In the entire Pacific, nodules have been estimated to cover about 10% of the ocean floor; in the Indian and Atlantic the coverage is less. In some places, the weight concentration of nodules ranges up to 5 gm-cm^{-2}. The local variability in manganese nodule concentration is great; two ocean bottom photographs only a few meters apart may show very different nodule concentrations.

Nodules have a wide range of external shapes. They may be spherical, conical, botryoidal, sheetlike, etc. The internal structures of nodules are no less varied than the external forms. Commonly nodules have as nuclei shark and other fish teeth, volcanic shards, pumice, foraminifera, and other hard objects around which the ferromanganese oxides accrete. In some spherical nodules, there are concentric ferromanganese bands around the nucleus, and in flat block-shaped nodules there are horizontal layers. Thin intercalated bands of clay or phosphorite are sometimes observed. This complex internal morphology records the varied conditions which existed during the growth of the nodule.

Mineralogy of Manganese Nodules

Manganese nodules are composed of cryptocrystalline minerals. They are known to consist of three major manganese phases: δMnO_2 (birnessite), 10-Å manganite, and 7-Å manganite. δMnO_2 is the most highly oxidized form, and has a chemical composition of about $MnO_{1.9}$. It has a sheet structure, the randomly oriented sheets being 50–100 Å thick. The two manganites consist of alternating layers of ordered MnO_2 sheets and disordered sheets whose main cations are Mn^{2+} and Fe^{3+}. These ions are coordinated with OH^-, H_2O, and other anions. The 7-Å manganite has the higher O/Mn ratio. Other manganese and iron phases which have been discovered in nodules include todorokite (Mn^{2+}, Mg^{2+}, Ba^{2+}, Ca^{2+}, K^+, $Na^+)_2$ Mn_5^{4+} $O_{12} \cdot 3H_2O$, ramsdelite, psilomelane, and goethite (FeOOH).

Authigenic minerals with cations other than manganese and iron are found in small amounts in manganese nodules. These include barite-celestite, rutile, and anatase. Detrital minerals constitute about 20% by weight of manganese nodules. This fraction, which is intercalated between the ferromanganese oxide bands, includes clay minerals, phillipsite, feldspars, micas, quartz, pyroxene, and hornblende.

Barnes (1967) has examined the depth

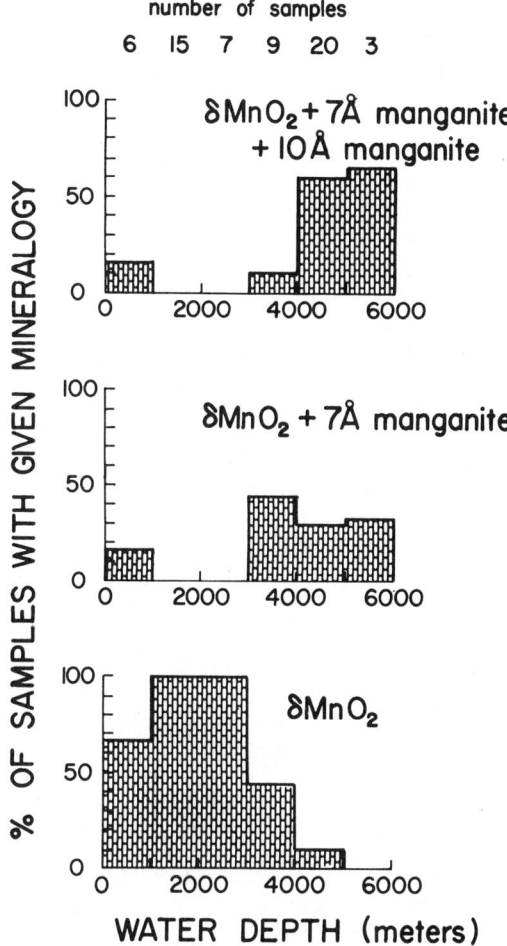

FIG. 4. Depth dependence of the mineralogy of manganese nodules (after Barnes, 1967).

dependence of the mineralogy in nodules taken from the Pacific. His data indicate that above 3500 m in depth the only important manganese phase is δMnO_2; below 3500 m, 10-Å manganite and 7-Å manganite coexist with δMnO_2 (Fig. 4). The depth-related factor thus controlling the mineralogy has not yet been elucidated; perhaps the observed phase change is pressure induced.

The Chemical Composition of Manganese Nodules

A total of about 400 chemical analyses of manganese nodules have been reported. Table 1 shows the average major element compositions of manganese nodules of the different oceans, and the composition ranges in each ocean. The concentration of minor elements in manganese nodules is shown in Table 2.

Mineralogy is one factor controlling the trace element composition of manganese nodules. Lead (Pb) and cobalt (Co), which both have highly oxidized forms, substitute for Mn^{4+} in the highly oxidized δMnO_2 polymorph, whereas they are discriminated against by the less highly oxidized manganites. On the other hand, Ni and Cu, which do not have highly oxidized states, are discriminated against by δMnO_2, but are enriched in the manganites where they substitute for Mn^{2+}. Since most nodules above 3500 m consist mainly of δMnO_2, they are somewhat enriched in Co and Pb, and depleted in Ni and Cu, relative to deeper water nodules containing 7-Å and 10-Å manganite. Superimposed on the mineralogical control of trace element

TABLE 1. MAJOR ELEMENTS COMPOSITION (ppm) OF MANGANESE NODULES FROM THE DIFFERENT OCEANS[a]

Element	Atlantic Ocean		
	Max	Min	Av
Al_2O_3	98,000	26,000	54,000(17)
SiO_2	338,000	23,000	140,000(20)
P	280	280	1,690(5)
Ti	9,100	530	44,000(19)
Mn	189,000	101,000	138,000(16)
Fe	259,000	15,400	164,000(20)
Co	9,100	900	3,300(19)
Ni	5,900	1,600	4,200(19)
Cu	3,000	300	1,300(19)
Zn	710	350	560(15)
Mg	24,000	14,000	17,000(4)
Ca	111,000	4,000	33,000(16)
Sr	1,900	580	1,100(15)
Ba	8,000	1,400	4,900(15)
Pb	2,300	930	1,800(16)
Na	35,000	14,000	23,000(4)
K	9,500	4,900	40,000(8)

MANGANESE NODULES

TABLE 1. (*Continued*)

Element	Indian Ocean Max	Indian Ocean Min	Av
Al_2O_3	106,000	4,000	68,000
SiO_2	371,000	184,000	58,000(6)
P	—	—	2,600(1)
Ti	6,500	2,500	4,200(6)
Mn	210,000	95,000	140,000(5)
Fe	209,000	97,000	140,000(6)
Co	2,900	900	1,900(6)
Ni	11,500	1,300	6,300(6)
Cu	18,100	1,100	5,400(6)
Zn	650	470	570(4)
Mg	—	—	
Ca	18,000	10,400	14,000(4)
Sr	960	640	860(4)
Ba	5,000	2,600	3,700(4)
Pb	1,500	1,000	1,300(4)
Na			
K			

Element	Pacific Ocean Max	Pacific Ocean Min	Av
Al_2O_3	138,000	10,000	58,000(54)
SiO_2	402,000	13,000	94,000(54)
P	4,300	200	1,400(44)
Ti	17,000	1,100	6,700(54)
Mn	530,000	17,000	193,000(216)
Fe	212,000	900	117,000(216)
Co	15,000	10	3,200(210)
Ni	24,600	360	6,600(210)
Cu	15,000	300	3,890(194)
Zn	800	400	470(54)
Mg	24,000	10,000	17,000(54)
Ca	44,000	8,000	19,000(54)
Sr	1,600	240	810(54)
Ba	6,400	800	1,800(54)
Pb	3,600	200	900(54)
Na	47,000	15,000	26,000(54)
K	31,000	3,000	8,000(54)

a Numbers in parentheses indicate number of analyses.

composition is the influence of variable trace element content of the seawater from which nodules derive. The concentrations of all the ferrides vary widely in the oceans, and these variations are reflected in nodule compositions.

Growth Rates of Manganese Nodules

Two theories for the rates of growth of manganese nodules have been prevalent. One proposes that manganese nodules are formed by rapid precipitation from submarine volcanic exhalations. The most important evidences cited for this are the common occurrence of small volcanic rocks as nodule nuclei, and the occurrence of manganese crusts around terrestrial volcanic exhalates. Contrary to this idea, others have held that manganese nodules grow by the slow accumulation of metals from seawater. Any solid object could serve as a nucleus around which the nodule could grow, and movement by sea-floor-dwelling creatures could prevent burial even if nodules accumulate more slowly than the coexisting sediment. During the last few years, growth rates of manganese nodules have been determined by various methods (Barnes and Dymond, 1967). The results all indicate that the nodules measured grow at a rate of a few millimeters per million years. This does not exclude the possibility that nodules in certain areas evolve more rapidly, but it certainly appears that most deep-sea nodules grow slowly.

TABLE 2. MINOR ELEMENTS COMPOSITION OF MANGANESE NODULES[a]

Element	Concentration (ppm)
Be	3(7)
B	290(58)
Sc	10(58)
V	550(58)
Cr	10(58)
Ga	10(54)
Ge	6(2)
Y	310(58)
Zr	620(58)
Nb	85(8)
Mo	500(58)
Ag	3(58)
Cd	2(9)
Sn	267(3)
La	160(54)
Ce	130(8)
Yb	30(58)
W	66(2)
Hg	2(1)
Tl	150(10)
Bi	30(12)
U	4(5)

[a] Numbers in parentheses indicate the number of analyses.

Occurrence of Nodules Within Pelagic Sediments

In a given area beneath the ocean floor, there is about one nodule in each two meters of sediment thickness for each nodule at the sediment top. This is the ratio to be expected considering the (known) relative rates of sediment and nodule accumulation (Bender et al., 1966).

Mechanisms of Nodule Growth

The mechanism by which nodules accumulate manganese and other elements has been another area of debate. The observations that bacteria found in manganese nodules are capable of oxidizing divalent manganese, and that manganese nodules contain appreciable amounts of organic matter, have led some to conclude that nodules are primarily the result of bacterial fixation of manganese. Others have held that nodules are formed by inorganic precipitation of metals supersaturated in seawater. Barnes showed that the hypothesis of inorganic growth is quantitatively tenable by calculating that the rate at which manganous ions in seawater diffuse to the surface of manganese nodules is roughly equal to experimentally determined accumulation rates of manganese in the nodules. Thus, there is some experimentally and theoretically tenable evidence for both mechanisms of manganese accumulation, and it cannot yet be said which one predominates.

MICHAEL L. BENDER

References

Arrhenius, G., 1963, "Pelagic Sediments" in (Hill, M. N., editor) "The Sea," Vol. 3, New York, Interscience, Publ., pp. 655–718.

Barnes, S. S., 1967, "Minor element composition of ferromanganese nodules," *Science,* **157,** 63–65.

Barnes, S. S., and Dymond, J. R., 1967, "Rates of accumulation of ferro-manganese nodules," *Nature,* **213,** 1218–1219.

Bender, L., Ku, T. L., and Broecker, W. S., 1966, "Manganese nodules: their evolution," *Science,* **151,** 325–328.

Bonatti, E., and Nayudu, Y. R., 1965, "The origin of manganese nodules on the ocean floor," *Am. J. Sci.,* **263,** 17–39.

Buser, W., and Grutter, A., 1956, "Uber die Natur der Manganknollen," Schweiz, *Mineral. Petrog. Mitt.,* **36,** 49–62.

Goldberg, E. D., and Arrhenius, G., 1958, "Chemistry of Pacific pelagic sediments," *Geochim. Cosmochim. Acta,* **13,** 153–212.

Manheim, F. T., 1965, "Manganese-iron accumulations in the shallow marine environment," in (Schink and Coreless, editors), "Symposium on Marine Geochemistry," University of Rhode Island, Occ. Publ. No. 3-1965, pp. 217–276.

Sackett, W. M., 1966, "Manganese nodules: Thorium-230: Protactinium-231 ratios," *Science,* **154**(3749), 646–647.

Cross-references: Lake Geochemistry; Manganese: Element and Geochemistry; Seawater, Chemistry; Trace Elements. Vol. I: *Manganese Nodules (Deep-Sea); Mineral Potential of the Ocean.* Vol. VI: *Pelagic Sediments.*

MANTLE GEOCHEMISTRY

The records of observational seismology—the study of the elastic waves produced by earthquakes—have led to the hypothesis that the earth is made up of a series of concentric layers separated from one another by discontinuities of various orders. Together with data concerning certain physical and mechanical properties of the earth, such as the volume, mass, moment of inertia, density, rigidity, incompressibility, pressure, and the gravitational intensity, these observations have resulted in the formulation of mathematical models for the structure of the earth (Bullen, K. E., in Gaskell, 1967). Some consequences of the interpretation of available data are summarized in Table 1.

Definition of the Mantle

The mantle of the earth is that region between the two first-order discontinuities recog-

MANTLE GEOCHEMISTRY

TABLE 1. AN APPROXIMATION TO THE EARTH'S INTERNAL LAYERING AND THE VALUES OF SOME PHYSICAL PARAMETERS[a]

Region	Range of Depth (km)	Name of Layer	Characteristics of P, S Velocities[b]	Features	Fraction of Volume[c]	Density (g/cm³)	Pressure (10¹² dyn/cm²)	Gravity (cm/sec²)	Rigidity (10¹² dyn/cm²)
A	0–33	Crust	Complicated	Heterogeneous	0.0155				
			Moho discontinuity						
B	33–410	Upper Mantle	Normal gradient	Probably homogeneous	0.1667	3.3	0.01	985	0.6
C	410–1000		Greater than normal gradients	Transition layer	0.2131	4.7	0.4	995	1.9
D'	1000–2700	Lower Mantle	Normal gradients	Probably homogeneous	0.4428				
D''	2700–2900	Lower Mantle	Gradient near zero	Transition layer					
			W-G discontinuity			5.7 / 9.7	1.3	1030	3.0 / 0.0
E	2900–4980	Outer core	Normal P gradient	Homogeneous fluid	0.1516	9.7	1.3	1030	0.0
F	4980–5120	Transition region	Negative P gradient	Transition layer	0.0028	(12.5)	3.2	(500)	(0.2)
G	5120–6370[d]	Inner core	Smaller than normal P gradient	Solid	0.0076	(13.0)	3.7	0	(1.3)

[a] After Bullen, in Gaskell, 1967.
[b] The S velocities are effectively zero in the outer core.
[c] Volume ratio, crust:mantle:core = 1:51:10.
[d] The value of 6370 km depth refers to the center of the earth.

nized by analysis of seismological data. The uppermost boundary (top of region B) is known as the *Mohorovičić discontinuity* and corresponds to an abrupt change in the compressional wave velocity, V_p, from about 6.4 to about 8.0 km/sec. The discontinuity is not a sharp boundary but the value of 33 km (Table 1) is accurate to within a few kilometers for the average continental shield area. Under mountain ranges the discontinuity is observed at depths up to 70 km but under the ocean basins at depths as shallow as 5 km below the solid surface. The lowermost boundary of the mantle is referred to as the *Wiechert-Gutenberg discontinuity* (bottom of region D) at a depth of 2900 km.

Gross Chemical Features of the Earth

Many inferences about the nature and composition of the different zones of the earth are speculative to varying degrees because of the inaccessibility of the interior and because of the differentiated nature of the earth. The overall composition may be deduced from *spectroscopic evaluation* of solar and stellar radiation, from nuclear chemical and astronomical theories of the origin of the elements and the evolution of the solar system, and from the analytical study of the meteorites (Mason, 1962 and 1966).

Meteorites (see Vol. II) are part of the solar system and may be fragments of disrupted asteroids. Although the amount of meteoritic matter falling on earth is estimated at between 1,000 and 10,000 tons daily, only the larger and easily accessible ones become the subject of laboratory study. Meteorites are of diverse composition and structure but constitute three broad categories: *(1)* Irons (siderites); *(2)* Stony irons (siderolites); and *(3)* Stones (aerolites). The irons consist essentially of nickel-iron alloy containing between 4 and 20% Ni (rarely greater amounts) together with a large number of accessory minerals of which troilite FeS, schreibersite (Fe, Ni, Co)$_3$P, and graphite are important. The stones consist mainly of silicate minerals (olivines and pyroxenes) with minor amounts of a metal phase similar to that of the iron meteorites. The average composition of *chondrites* (the major class of stony meteorites, distinguished by the presence of rounded bodies of about 1 mm diameter consisting of olivine and/or pyroxene) is about 40% olivine, 30% pyroxene, 10–20% nickel-iron, 10% plagioclase, and 6% troilite.

One of the classical hypotheses of astronomy is that the meteorites are remnants of a defunct planet or planets. This idea seems to have been current more than a century ago and is associated with the names of Daubrée and Boisse. Almost as soon as it was observed that there are only three main classes of meteorite, the concept of a zoned earth became firmly established and the equivalence of iron meteorites with the earth's metal core and the stony meteorites with its silicate mantle was generally assumed (Table 2).

The results of studies during the last few decades (Ringwood, in Hurley, 1964, pp. 287–294) show a convergence of information from the three different fields already mentioned. Such sources lead to the supposition that the Earth consists essentially of an Fe–O–Si–Mg system (plus a minor sulfide phase concentrated in the core) with small amounts (greater than 1%) of Ni, S, Ca, and Al and minor amounts (0.1–1%) of Co, Na, Mn, K, Ti, and P. All the other known elements probably do not amount to more than 0.1% of the

TABLE 2. COMPOSITION OF METEORITE MATTER (wt %)[a]

	Metal (from Irons)	Metal (from Chondrites)	Silicate (from Chondrites)	Average Chondrite
O			43.7	33.24
Fe	90.78	90.72	9.88	27.24
Si			22.5	17.10
Mg			18.8	14.29
S				1.93
Ni	8.59	8.80		1.64
Ca			1.67	1.27
Al			1.60	1.22
Na			0.84	0.64
Cr			0.38	0.29
Mn			0.33	0.25
P			0.14	0.11
Co	0.63	0.48		0.09
K			0.11	0.08
Ti			0.08	0.06

[a] After Mason, 1966.

whole. But the earth is a differentiated body, physically and chemically, and the evidence of geophysical studies combined with the fundamental philosophy of geochemistry requires a *partition of the elements* between the core, mantle, and crust according to their affinities for the major phases, metal and silicate. Thus, the core of the earth is believed to consist of a *nickel-iron alloy* containing a notable quantity of an immiscible FeS phase (Ringwood favors an Fe–Ni–Si alloy) whereas the uppermost parts of the earth and especially the crust are rich in *silicates* and *oxides* containing elements such as Ca, Al, Na, K, and Ti. This major differentiation of the earth, which is far from being a quantitative separation of the elements, has resulted in a mantle dominantly composed of Mg and Fe (ferromagnesian) silicates.

Composition of the Upper Mantle

Basaltic rocks (mixtures of olivine, pyroxene, and plagioclase) are distributed in space and time on a worldwide basis. The depth of generation of *basaltic magma* is known, on the basis of studies of earthquake foci (Kuno, in Gaskell, 1967), to be greater than 60 km below surface, extending downward to about 400 km. Because of the experimentally determined phase relationships among the minerals of basaltic rocks (see the publications of the Geophysical Laboratory, Washington) and knowledge of the *heat flow* through the surface of the earth it is inferred that the generation of basaltic magma involves only the partial melting of the materials of the upper mantle. These materials must have properties in accordance with the observed physical properties (V_p averages about 8.2 km/sec, density is between 3.3 and 3.4 g/cm^3) and be of the correct composition to yield a basaltic magma fraction. Accordingly, some kind of *peridotite* (consisting of olivine, pyroxene, and garnet) must be the principal rock type of the upper mantle (see discussions by Mercy, in Gaskell, 1967, pp. 423–428; and by Ringwood, in Hurley, 1964, pp. 298–306). In order to obtain even an approximate notion of the chemical composition of the upper mantle, two different methods have been used, the synthetic, associated with the names of Ringwood et al., and the analytic, associated with the names of O'Hara and his various co-workers.

The Pyrolite Model. The term '*pyrolite*' (Ringwood, in Hurley, 1964) serves to indicate a mixture of alpine-type peridotite (originally Ringwood used the composition of *dunite*, nearly pure olivine rock) and basalt in a ratio of 3 to 1, the composition then being the mean chemical composition of the upper mantle-lower crust system (Table 3). *Fractional melting* of this system could provide basaltic magma, leaving a residue of peridotite or even dunite. The model pyrolite composition is a close match to the overall composition of the mantle which can be derived from the chondrite model of the earth. The composition of column 4, Table 3 is derived by Ringwood from a consideration of the derivation of a correct mantle/core relationship using the nonvolatile components of the *carbonaceous chondrites* (a small class of the chondrite meteorites believed to have retained primordial abundances of many elements) as a model for the primitive oxidized

TABLE 3. DERIVATION OF PYROLITE COMPOSITION[a]

	Alpine Peridotite	Hawaiian Basalt[b]	Pyrolite[c]	Mantle Composition Derived from Chondrite
SiO$_2$	43.98	48.72	45.16	43.25
TiO$_2$	0.03	2.77	0.71	
Al$_2$O$_3$	0.25	13.40	3.54	3.90
Cr$_2$O$_3$	0.57	—	0.43	
Fe$_2$O$_3$	0.04	1.70	0.46	
FeO	7.42	9.88	8.04	9.25
MnO	0.14	0.18	0.14	
NiO	0.26	—	0.20	
CoO	0.01	—	0.01	
MgO	46.96	8.98	37.47	38.10
CaO	0.33	11.30	3.08	3.72
Na$_2$O	0.01	2.23	0.57	1.78
K$_2$O	—	0.58	0.13	
P$_2$O$_5$	—	0.24	0.06	

[a] Ringwood, in Hurley, 1964.
[b] Average Hawaiian olivine tholeiite.
[c] Mixture of three parts of peridotite and one part basalt.

TABLE 4. MODEL MANTLE COMPOSITION[a]

	1	2	3	4
SiO_2	33.32	35.85	29.84	43.25
MgO	23.50	25.19	26.29	38.10
FeO	35.47	6.14	6.38	9.25
Al_2O_3	2.41	2.59	2.69	3.90
CaO	2.30	2.47	2.57	3.72
Na_2O	1.10	1.18	1.23	1.78
NiO	1.90	—	—	—
Total	*100.00*	*73.52*	*69.00*	*100.00*
Fe		24.88	25.87	
Ni		1.60	1.66	
Si		—	3.47	
Total		*26.48*	*31.00*	

Column 1   Average composition of principal components of Type I carbonaceous chondrites (Orgueil and Ivuna) on a C-, S-, and H_2O-free basis

Column 2   Analysis from column 1 with [FeO/(FeO + MgO)] reduced to be consistent with probable value for earth's mantle (0.12)

Column 3   Analysis from column 2 with sufficient SiO_2 reduced to elemental silicon to yield a total silicate to metal mass ratio of 69/31 as in the earth

Column 4   Model mantle composition: silicate phase from column 3 recalculated to 100%

[a] Ringwood, in Hurley, 1964.

overall composition of the earth (Table 4). Assuming an FeO/(FeO + MgO) molecular ration of 0.12 for the mantle (the available evidence is that the ratio lies between 0.1 and 0.2) then the analysis of column 2 is obtained by reducing the appropriate amount of iron and all of the nickel. This calculation gives a mantle/core mass ratio of 73.5/26.5 and to obtain the correct ratio of 69/31 it is necessary to transfer some SiO_2 to the core as elemental silicon. The analysis of column 3 implies that the core contains about 11 weight percent or 20 atomic percent of silicon. The model mantle composition (column 4) is obtained by recalculating to 100% the silicate analysis of column 3.

The Natural Garnet-Peridotite Model. General petrological research, based on mineralogical, chemical, and structural analyses of magmatic rocks within the crust, and high temperature-high pressure experimental work on the phase chemistry of silicate rocks and minerals (see especially the publications of the Carnegie Institution, Washington and the important review by O'Hara, 1968) have led to a somewhat detailed knowledge of the probable chemical and mineralogical composition of the uppermost parts of the mantle. There is good evidence that the two most important mineral assemblages among natural peridotitic rocks, the spinel-peridotite (olivine + orthopyroxene + clinopyroxene + spinel) and the garnet-peridotite (olivine + orthopyroxene + clinopyroxene + garnet), may be representative of the nature of the lower crust and upper mantle, respectively.

The work of O'Hara and his colleagues (see references in O'Hara, 1968) has led to the supposition that of all the many kinds of ultramafic rocks known to exist within the crust (for general information about such rocks see Wyllie, 1967) the garnetiferous-peridotite nodules in kimberlite *diatremes* (see Chap. 8 of Wyllie, 1967) are the most suitable source material from which the parent liquid giving rise to basaltic magma, as seen in the low-pressure regime of the surface of the earth, may be derived. These particular nodules, then, may be representative of the type of unmodified upper mantle composition. Certainly there is plentiful evidence that kimberlites have erupted into the crust from depths of 125 to 250 km and carry with them a diverse population of nodule fragments, some derived from within the mantle, others derived from the crustal wall rocks through which the pipes have been drilled. The incorporated nodules are almost entirely accidental in character, being either source mantle (garnet-olivine-orthopyroxene-clinopyroxene assemblages = garnet lherzolite) or the crystalline residua of partial melting processes at high pressure (garnet-olivine-orthopyroxene assemblages = garnet harzburgite, and olivine-orthopyroxene assamblages = harzburgite) and cognate accumulates (crystal aggregates separated from magmatic liquids by gravitational and convectional processes) of possibly earlier magmatic cycles.

Chemical analyses of a garnet-lherzolite nodule and its constituent minerals, from a kimberlite pipe, are quoted (Table 5) to show

MANTLE GEOCHEMISTRY

TABLE 5. GARNET-LHERZOLITE, WESSELTON MINE, SOUTH AFRICA, AND ITS CONSTITUENT MINERALS[a]

	Whole Rock	Olivine	Orthopyroxene	Clinopyroxene	Garnet
SiO_2	48.92	41.73	56.76	54.13	41.45
TiO_2	0.07	0.03	0.07	0.20	0.17
Al_2O_3	1.97	0.04	1.52	2.41	22.09
Cr_2O_3	0.34	0.01	0.17	1.56	2.30
Fe_2O_3	0.27	0.04	0.06	1.34	1.90
FeO	6.03	7.91	4.97	1.28	7.02
MnO	0.12	0.10	0.11	0.08	0.45
NiO	0.25	0.42	0.10	0.05	0.02
CoO	0.003	0.003	0.003	<0.001	0.004
MgO	39.39	49.56	35.99	16.46	19.97
CaO	2.40	0.08	0.24	20.65	4.56
Na_2O	0.19	—	—	1.81	0.04
K_2O	<0.01	—	—	0.02	<0.01
P_2O_5	0.04	0.08	0.01	0.01	0.02

[a] Analyses recalculated to 100%.
Volume percent minerals in rock:
  Olivine   45
  Orthopyroxene   40
  Clinopyroxene   10
  Garnet   5
University of Edinburgh, Grant Institute of Geology collection, A3/10596.

the kind of upper mantle composition which can be derived from the hypotheses formulated by O'Hara et al. Caution is desirable in the interpretation of the relatively few such analyses presently available since not only do the compositions of the four principal phases vary but the proportions of the phases are variable between different peridotite nodules. In addition, the concentrations of some minor and trace elements are much more abundant in the conveying and possibly contaminating kimberlite than in the nodules themselves. A comparison of the garnet-lherzolite analysis with Ringwood's pyrolite composition (Table 6) identifies some differences which may be important for petrogenetic theory.

TABLE 6. COMPARISON OF GARNET-LHERZOLITE (1) WITH PYROLITE (2)

	(1)[a]	(2)
SiO_2	48.92	45.16
TiO_2	0.07	0.71
Al_2O_3	1.97	3.54
Cr_2O_3	0.34	0.43
Fe_2O_3	0.27	0.46
FeO	6.03	8.04
MnO	0.12	0.14
NiO	0.25	0.20
CoO	0.003	0.01
MgO	39.39	37.47
CaO	2.40	3.08
Na_2O	0.19	0.57
K_2O	<0.01	0.13
P_2O_5	0.04	0.06
Total	100.00	100.00

[a] Recalculated to 100%.

Radioelements in the Mantle

Investigations of the heat flow through the earth's surface (see relevant chapters in Gaskell, 1967) provide limiting conditions for the distribution of the radioactive elements U, Th, and K within the mantle. The work of Birch, Clark, and MacDonald (references quoted by Mercy, in Gaskell, 1967) has led to an earth model in which the ratios $K/U = 1 \times 10^4$ and $Th/U = 3.7$ (values determined by analysis of crustal rocks) are fixed, and the temperature gradient is assumed to increase linearly from 0°C at the surface to 1000°C at a depth of 600 km. Then the radial distribution of the heat-producing elements is: *Oceanic structure:* all radioactivity above 1500 km; U(1500–465 km):U(564–0 km) = 18.2. *Continental structure:* all radioactivity above 1500 km; U(1500–465 km):U(465–45 km):U(45–0 km) = 1:3.29:40.0. These distributions give the observed heat flow (taken as 65 ergs-Cm^{-2}-sec^{-1}) provided the average U concentration in the mantle lies between 4 and 5×10^{-8} g/g. An average mantle concentration of U about 5.5×10^{-8} g/g would produce melting in the upper mantle and an excess of heat flow.

Recent calculations by Shaw (Table 7), based on analyses for U, Th, and K in rocks exposed in the Canadian Precambrian Shield, show the implications of accepting a meteorite composition for the whole earth. Comments by Shaw are that both the chondrite model and the Orgueil model would produce a mantle with a K content and a ratio of K/U inconsistent with the observed data from possible

TABLE 7. U, Th, AND K MASSES IN THREE EARTH MODELS, AND THEIR CONCENTRATION IN THE WHOLE UPPER MANTLE PLUS THE OCEANIC CRUST, AFTER FORMATION OF THE CONTINENTS[a]

	Continental Crust	Chondrite Earth	Orgueil Earth	Achondrite Earth
U (g $\times 10^{20}$)	0.462	0.657	1.434	2.988
Th (g $\times 10^{20}$)	1.944	2.378	3.884	11.952
K (g $\times 10^{20}$)	4,870	50,500	25,100	20,920
	Concentration of residual U, Th, and K in the whole upper mantle plus oceanic crust			
U (g/g $\times 10^{-8}$)		1.6	8.0	21
Th (g/g $\times 10^{-8}$)		3.5	16	82
K (g/g $\times 10^{-4}$)		37	17	13
10^{-4} K/U		23	2.1	0.6

[a] Data by Shaw, D. M., in Ahrens, L. H., 1968.

mantle materials. Although Shaw argues in favor of the *achondrite model* (achondrites are a special class of stony meteorites—Mason, 1962), the concentration of U so produced in the mantle is inconsistent with MacDonald's hypothesis.

Composition of the Mantle Layers C and D

The investigation of phase transformations in the high-pressure high-temperature environment of the mantle has been made possible by the recent development of static high-pressure apparatus and the application of shock-wave techniques which generate enormous transient pressures (Ringwood, in Hurley, 1964). At a depth of about 400 km (i.e., at at pressure of about 120 kb) magnesian pyroxene ($MgSiO_3$) recrystallizes to forsterite (Mg_2SiO_4) plus stishovite (the rutile polymorph of SiO_2). In layer C, reconstruction of forsterite takes place, involving a change from a hexagonal close-packed structure to a more symmetrical spinel structure having cubic close packing. This process appears to be complete in layer D with the formation of homogeneous matter having the NaCl structure, in which Si, as well as Mg and Fe, occupies octahedral sites. These changes, correlated with the abnormally high rate of increase of velocity with depth in the transition layer, imply only small changes in the concentration of Fe in the deeper layers. Indeed, McQueen et al. have been able to calculate that the probable composition of layer D approximates to 0.56 MgO + 0.26 SiO_2 + 0.18 FeO (see references quoted by Mercy, in Gaskell, 1967; and Ringwood, in Hurley, 1964).

EDWARD MERCY

References

Ahrens, L. H. (editor), 1968, "Origin and Distribution of the Elements," New York, Pergamon Press, 1178pp.

Carnegie Institution, Washington. Annual Reports of the Director, Geophysical Laboratory, in Carnegie Institution Year Books, especially the publications of the last decade.

*Gaskell, T. F. (editor), 1967, "The Earth's Mantle," London, Academic Press, 509pp.

*Hurley, P. M. (editor), 1964, "Advances in Earth Science," Cambridge, Massachusetts Institute of Technology, 502pp.

*O'Hara, M. J., 1968, "The bearing of phase equilibria studies in synthetic and natural systems on the origin and evolution of basic and ultrabasic rocks," *Earth Sci. Rev.,* **4**(2), 69–133.

Mason, B., 1962, "Meteorites," New York, John Wiley & Sons, 274pp.

Mason, B., 1966, "Principles of Geochemistry," Third edition, New York, John Wiley & Sons, 329pp.

Wyllie, P. J., 1967, "Ultramafic and Related Rocks," New York, John Wiley & Sons, 464pp.

* Additional bibliographic references may be found in this work.

Cross-references: *Core Geochemistry; Elements: Planetary Abundances and Distribution; Geochemical Evolution of the Core, Mantle, and Crust; Phase Equilibria; Radioactivity in Rocks; Spectrophotometry.* Vol. II: *Asteroids; Earth—Geology of the Planet; Meteorites; Planet Earth—Origin and Evolution; Solar System—Origin.* Vol. IVB: *Clinopyroxenes; Mantle Mineralogy; Orthopyroxenes.* Vol. V: *Basalt; Core; Gutenberg-Wiechert Discontinuity; Heat Flow; Lherzolite; Mohorovičic Discontinuity; Peridotite; Ultramafic Rocks.*

MARINE ECOLOGY—See Vols. I and VII

MARINE GEOCHEMISTRY—See **SEAWATER CHEMISTRY**; also Vol. I

MARINE MICROBIOLOGY—See Vol. I

MASS ACTION AND EQUILIBRIUM CONSTANT

The *law of mass action* or *law of chemical equilibrium* is a relation between the concentrations of reactants and products of a chemical reaction under equilibrium conditions. It was proposed first by C. M. Guldberg and P. Waage in 1863, as a consequence of experimental results showing that the rate of a chemical reaction is directly related to the concentrations of the reactants. Although this approach is not as general as the derivation based on thermodynamic principles, it provides a more intuitive understanding of the physical meaning of a chemical equilibrium as a dynamic state, which is characterized not by the cessation of all reactions, but by the fact that the rates of the forward and reverse reactions are equal.

A general equation for a typical chemical reaction is

$$b\,B + c\,C = m\,M + n\,N \quad (1)$$

where B, C, M, and N represent chemical formulas and b, c, m, and n are stoichiometric coefficients. The rate of this reaction proceeding forward (from left to right) is equal to the product of a rate constant, k_f, by the concentrations of the reacting substances, x_B and x_C, each raised to a power equal to the stoichiometric coefficient:

$$r_f = k_f\, x_B^b\, x_C^c$$

Similarly, the rate of the backward reaction is

$$r_b = k_b\, x_M^m\, x_N^n$$

At equilibrium, $r_f = r_b$, which leads directly to the mathematical statement of the law of chemical equilibrium relating the concentrations of reactants and products to a constant, K_x, which is called the *equilibrium constant* of the reaction.

$$\left(\frac{x_M^m\, x_N^n}{x_B^b\, x_C^c}\right)_{\text{equilibrium}} = \frac{k_f}{k_b} = K_x \quad (2)$$

The subscript x is used as a reminder of the fact that the equilibrium constant, K_x, and the rate constants, k_f and k_b, depend on the scale and units chosen to express the concentrations.

Thermodynamic Derivation of the Law of Chemical Equilibrium

A straightforward consequence of the *first and second laws of thermodynamics* is that, for a system at constant temperature and pressure, *the condition for equilibrium* is: the Gibbs free energy, or free enthalpy, represented by G, exhibit a minimum, or $dG = 0$ at constant T, P. The application of this result to a chemical system, provides not only a rigorous proof of the law of mass action, but it also leads to the introduction of new thermodynamic functions and relationships which are very useful for practical calculations of the conditions for chemical equilibrium.

Chemical Potential and Equilibrium. The change in Gibbs free energy, or free enthalpy, of a chemical system, which is due to a change in the amount of one constituent when everything else remains constant, is called the *chemical potential* of that constituent in the system and it is represented by μ_i. Its mathematical definition is the partial derivative of G with respect to n_j, the number of moles of the constituent, when the temperature, T, pressure, P, and number of moles of other constituents, $n_{j \neq i}$, remain constant. It is also called partial molal Gibbs free energy, or free enthalpy, and represented by $\overline{G}_i$.

$$\mu_i = \left(\frac{\partial G}{\partial n_i}\right)_{T,\,P,\,n_j \neq i} = \overline{G}_i \quad (3)$$

A consequence of the definition is that the increment in free enthalpy, dG at constant T,P, is

$$dG = \sum_i \mu_i\, dn_i \quad (4)$$

In order to apply the last relation to a chemical reaction, it is convenient to represent the degree of advancement of the reaction by a reaction variable, ξ. A small increment of that variable, $d\xi$, means that $b\,d\xi$ moles of B and $c\,d\xi$ moles of C have reacted to form $m\,d\xi$ moles of M and $n\,d\xi$ moles of N. The corresponding increment in free enthalpy is

$$dG = (m\mu_M + n\mu_N - b\mu_B - c\mu_C)\,d\xi \quad (5)$$

The equilibrium condition at constant T,P is $dG = 0$, and since $d\xi$ is small but not zero, it results that

$$\left(\frac{\partial G}{\partial \xi}\right)_{T,P} = \Delta G = m\mu_M + n\mu_N - b\mu_B - c\mu_C = 0 \quad (6)$$

A curve representing G versus ξ exhibits a minimum for the value of the reaction variable which corresponds to equilibrium. For smaller values of ξ, the derivative is negative and the forward reaction is spontaneous. For larger values of ξ, it is positive and the reverse reaction is spontaneous. The function $-(\partial G/\partial \xi)_{T,P}$ can be considered as a thermodynamic driving force for the reaction and it is called the *affinity* of the reaction. It is usually referred to as: "$-\Delta G$," since it is equal to the change in free enthalpy for the system when the reaction proceeds once as written; it is the sum of the partial molal free enthalpies of the product each multiplied by the stoichiometric coefficient minus the same sum for the reactants.

Activity and Activity Coefficient. For practical use of the general equation describing a chemical equilibrium, it must be expressed in terms of measurable quantities directly related to the composition of the system. This is achieved by expressing the chemical potential, or partial molal free enthalpy, of a constituent, μ_i, in terms of its value in an *arbitrary standard state*, μ_{i0}, and a logarithmic function of *the activity*, a_i which represents the deviation from that standard state, and which is defined by the relation

$$\mu_i = \mu_i^0 + RT \ln a_i \qquad (7)$$

where R is the gas constant and T the absolute temperature.

When pure substances in condensed phases are considered, the pure solid or pure liquid at each temperature and 1 atm pressure is the usual standard state. For gases, the standard state is taken as the ideal gas at unit fugacity (approximately 1 atm pressure). In the case of mixtures, several reference states are available and the choice is usually dictated by its convenience for a particular application. The standard state of the solvent is usually taken as the pure solvent; that of the solute is generally a hypothetical state of unit activity in which the partial molal enthalpy and heat capacity of the solute are equal to the values at infinite dilution. When the standard states are so chosen, it results from Eq. (7) that the activity of a pure solid or liquid is unity, the activity of an ideal gas equals its partial pressure, and the activities of constituents of ideal solutions equal their concentrations.

In the case of many dilute solutions, the activities are nearly equal to concentrations and it is convenient to represent the deviation relative to an ideal behavior by defining an *activity coefficient*, γ_i, as the ratio of the activity, a_i, to the concentration, x_i, or

$$a_i = \gamma_i x_i \qquad (8)$$

The activity coefficient is equal to unity at infinite dilution of the solute, and it deviates from unity as the concentration increases.

Equilibrium Constants. After introducing the activities, as defined by Eq. (7), the condition for chemical equilibrium at constant T,P given by Eq. (6), becomes

$$m\mu_M^0 + n\mu_N^0 - b\mu_B^0 - c\mu_C^0 + RT(m \ln a_M + n \ln a_N - b \ln a_B - c \ln a_C) = 0 \qquad (9)$$

The *thermodynamic equilibrium constant*, K_a, is defined by

$$K_a = \left(\frac{a_M^m \, a_N^n}{a_B^b \, a_C^c}\right)_{\text{equilibrium}} \qquad (10)$$

and ΔG^0 is given by

$$\Delta G^0 = m\mu_M^0 + n\mu_N^0 - b\mu_B^0 - c\mu_C^0 \qquad (11)$$

Then Eq. (9) becomes

$$\Delta G^0 = -RT \ln K_a \qquad (12)$$

It is an extremely useful relation which summarizes the thermodynamic description of a chemical equilibrium.

The relationship between the equilibrium constant in terms of activities, K_a, and the equilibrium constant in terms of concentration, K_x, results directly from the definition of the activity coefficients.

$$K_a = \frac{\gamma_M^m \, \gamma_N^n}{\gamma_B^b \, \gamma_C^c} K_x \qquad (13)$$

K_a is a thermodynamic equilibrium constant, which is function of temperature, pressure, and the choice of standard states, while K_x depends not only on the same factors, but also on the composition of the system. For dilute solutions which approach ideal behavior as the concentration of the solute decreases, the ratio of activity coefficients approaches unity and the value of K_x approaches that of K_a.

Temperature Dependence of the Equilibrium Constant. The relation between the equilibrium constant and the standard free enthalpy change of reaction is very useful for practical calculation of the conditions of chemical equilibrium from thermochemical data. In particular, the temperature dependence of K_a is obtained from that of ΔG^0, and expressed by the Van't Hoff equation

$$\left(\frac{\partial \ln K_a}{\partial T}\right)_P = \frac{\Delta H^0}{RT^2} \qquad (14)$$

namely the partial derivative of $\ln K_a$ with respect to T, at constant P, is related to the standard enthalpy change for the reaction, ΔH^0. The same relation written as

$$\left(\frac{\partial \ln K_a}{\partial 1/T}\right)_P = -\frac{\Delta H^0}{R} \qquad (15)$$

shows that the slope of the plot of $\ln K_a$ versus $1/T$ is equal to $-\Delta H^0/R$.

MASS ACTION AND EQUILIBRIUM CONSTANT

These equations are quantitative statements of the principle of Le Chatelier, that any change in a variable that determines the state of a system in equilibrium causes a shift in the position of equilibrium in a direction that tends to counteract the change in the variable considered. In this instance, upon a temperature increase, the equilibrium of an endothermic reaction ($\Delta H^0 > 0$) is shifted to the right (K_a increases), while it is shifted to the left for an exothermic one.

Examples of the Use of the Law of Mass Action

Most problems of geological interest involve the calculation of equilibrium conditions for complex chemical systems in which several reactions must be considered simultaneously. The mathematical solutions are often time consuming, and the labor can be greatly reduced by programming the calculations for a computer. The following examples are necessarily limited to a single aspect of a practical problem, but their generalization to more complex cases is quite straightforward, and does not require any new basic concept.

Equilibrium in Volcanic Gases. The constituents of volcanic gases are all simple substances about which basic thermochemical data are available. Water vapor is usually the dominant species and the other constituents are compounds of hydrogen and oxygen with carbon, sulfur, chlorine, fluorine and nitrogen (e.g., CO_2, CO, CH_4, CO_2, SO_2, H_2S, HCl, HF, NH_3, . . .). The amount of free hydrogen is variable, while free oxygen is always substracted from the analysis on the assumption that its presence is due to admixed air.

Standard chemical potentials of free enthalpies of formation are easily obtained for the constituents of volcanic gases, and the equilibrium constants or various reactions among them can be calculated over a wide temperature range. Since each substance takes part in several reactions, occurring simultaneously, the calculation of the proportions of gases present at equilibrium requires solving a set of a large number of equations, which express not only the law of mass action for each independent equilibrium, but also the mass balance of each element. The variations of composition of volcanic gases with temperature and pressure can be predicted and results are in fair agreement with some actual gas analyses.

The nature of the lava from which the gases come has an influence on the composition of fumarole gases. The thermodynamic approach to this question is illustrated by the relation between the hydrogen content of a volcanic gas and the ratio of ferrous to ferric iron in the lava (Kern and Weisbrod, 1967, p. 285).

The water vapor in the gases at high temperature is in equilibrium with the products of its dissociation:

$$H_2O = H_2 + 1/2\ O_2 \qquad (I)$$

Assuming that the gases behave as perfect gases, the law of mass action is expressed in terms of partial pressures.

$$\frac{p_{H_2}\, p_O{}^{1/2}}{p_{H_2O}} = K_I \qquad (16)$$

The reaction of oxydation of ferrous to ferric oxide

$$1/2\ O_2 + 2\ FeO = Fe_2O_3 \qquad (II)$$

provides a relation between the state of oxidation of the lava and the partial pressure of oxygen.

$$\frac{a_{Fe_2O_3}}{a_{FeO}{}^2\, p_O{}^{1/2}} = K_{II} \qquad (17)$$

The two reactions have been investigated and values of K_I and K_{II} can be obtained either from direct experimental measurements or from tabulated thermodynamic data. They are combined to describe the oxidation of ferrous oxide by water vapor

$$2\ FeO + H_2O = Fe_2O_3 + H_2 \qquad (III)$$

for which the law of mass action is a combination of equations 16 and 17.

$$\frac{p_{H_2}\, a_{Fe_2O_3}}{p_{H_2O}\, a_{FeO}{}^2} = K_{III} = K_I K_{II} \qquad (18)$$

If ideal behavior of the solution of iron oxides in the mixture of silicates is assumed, the activities, a_i, are equal to the mole fractions, x_i, and the relative amount of hydrogen to water vapor, given by the ratio of partial pressures, is related to the concentrations of ferrous and ferric iron in the lava.

$$\frac{p_{H_2}}{p_{H_2O}} = \frac{x_{FeO}{}^2}{x_{Fe_2O_3}} K_{III} \qquad (19)$$

Since the equilibrium constant, K_{III}, varies with temperature, Eq. (19) can also be used to determine the temperature of a lava with a known oxidation ratio, which is in equilibrium with volcanic gases of a given composition.

Solubility of Minerals in Aqueous Solutions. The dissolution of rock constituents in an aqueous environment is a typical problem that can be treated quantitatively by chemical thermodynamics. The example of the solubility product of a slightly soluble salt illustrates the application of the law of mass action.

Calcium sulfate dissolves in water by producing the free ions Ca^{2+} and SO_4^{2-}:

$$CaSO_4 = Ca^{2+} + SO_4^{2-} \quad (20)$$

The equilibrium constant in terms of activities is K_a

$$\frac{a_{Ca^{2+}}\, a_{SO_4^{2-}}}{a_{CaSO_4}} = K_a \quad (21)$$

For the purpose of thermodynamic calculations, ionic concentrations are usually expressed in moles of solute per 1000 grams of water (molality), and represented by m. The commonly accepted standard state for ions is the hypothetical ideal solution of unit molality. The activity is equal to the product of the molality by an activity coefficient, γ, which approaches unity when the solution is infinitely dilute. The standard state for $CaSO_4$ is the pure solid, and the activity in Eq. (21) is unity. The *solubility product in terms of activities*, K_a, becomes

$$K_a = a_{Ca^{2+}}\, a_{SO_4^{2-}} = \gamma_{Ca^{2+}}\, \gamma_{SO_4^{2-}}\, K_m \quad (22)$$

where the *solubility product in terms of molalities*, K_m, is defined by

$$K_m = m_{Ca^{2+}}\, m_{SO_4^{2-}} \quad (23)$$

If the only source of calcium and sulfate ions is the solid salt, then the concentrations of the ions are equal, and the solubility, S, is given by

$$S = m_{Ca^{2+}} = m_{SO_4^{2-}} = \sqrt{K_m} \quad (24)$$

The solubility product, K_m, varies with the total amount of electrolyte in solution. This is illustrated by the fact that different values are obtained experimentally for the solubility of gypsum ($CaSO_4 \cdot 2H_2O$): 0.015 mole/1000g H_2O in water, and 0.045 mole/1000g H_2O in a 1-molal NaCl solution.

The solubility product, K_a, is a thermodynamic quantity which depends only on temperature and pressure. It is related to the standard chemical potentials or free energies of formation through Eqs. (11) and (12). It can be determined experimentally by extrapolating to infinite dilution the values of solubility at various electrolyte concentrations, and it is often represented by K_{so}.

The ionic activity coefficients in aqueous solutions depend on the total amount of dissolved electrolyte and they differ significantly from unity even for relatively dilute solutions (γ is about 0.4 for the 0.015 molal solution of $CaSO_4$ in water). In order to predict solubilities from thermodynamic data, it is necessary to estimate the values of activity coefficients by comparison with experimental values published in the literature or by the various relations derived from the Debye-Hückel theory of electrolytes.

In general, additional factors must be taken into account for the calculation of the solubility of a single salt. If the solution contains a common ion, the solubility of a slightly soluble salt is usually decreased. Eq. (24) no longer applies and Eqs. (22) and (23) must be used, together with a mass balance for the ions. Consideration must also be given to the form of the solid which is present. In the example of calcium sulfate, the anhydrous form, anhydrite, is slightly more soluble at room temperature than the hydrate, gypsum. Last but not least, the dissolution equations are generally not as simple as Eq. (20), and complex or associated ions must usually be considered.

The calculation of the solubility of minerals in a natural aqueous environment is more complex than the consideration of a single solubility product; it usually involves a large number of species and it requires the simultaneous solution of several equations similar to (21). A very important and quite typical example is the discussion of carbonate equilibria (Garrels and Christ, 1965, p. 74). The solubility of calcium carbonate depends not only on one equilibrium constant defined by temperature and pressure, but it is also a function of the pressure of CO_2, the acidity of the solution and the composition of the solid phase. The relations between the variables result from the application of the law of mass action to all the independent equilibria among the various species in solution.

PAUL DUBY

References

Garrels, R. M., and Christ, C. L. 1965, "Solutions, Minerals and Equilibria," New York, Harper and Row, 450pp.

Kern, R., and Weisbrod, A., 1967, "Thermodynamics for Geologists," San Francisco, Freeman, Cooper and Co., 304 pp. (translated by Duncan McKie from "Thermodynamique de Base pour Minéralogistes, Pétrographes et Géologues," Masson et Cie, Paris, 1964).

Krauskopf, K. B., 1967, "Introduction to Geochemistry," New York, McGraw-Hill Book Co., 721pp.

Waser, J., 1966, "Basic Chemical Thermodynamics," New York, W. A. Benjamin, Inc., 278pp.

Cross-references: *Enthalpy or heat content; Entropy; Free energy and free Enthalpy; Solubility Product; Thermochemistry; Thermodynamics.*

MASS SPECTROMETRY

There are three terms commonly applied to instruments which measure atomic masses. *Mass spectroscope* is used to describe instruments in which information is displayed for immediate visual observation by means of a fluorescent screen or some similar device. *Mass spectrograph* refers to instruments in which the mass differences are recorded on a photographic plate or film. *Mass spectrometer* is the term applied to instruments which collect information by means of an electric signal which is amplified and displayed by galvanometer or strip chart recorder. *Mass spectrography* and *mass spectrometry* apply to mass spectrographs and spectrometers. *Mass spectroscopy* may be limited to mass spectroscopes or it may, especially in British usage, refer to the field of mass measurement.

History

In 1886 E. Goldstein observed that a hole in the plate of a gas discharge tube permitted a luminous beam to pass through to the back side of the plate. He called this phenomenon *Kanalstrahlen*. In 1898 W. Wien found that a magnet caused the Kanalstrahlen to be deflected.

The significance and usefulness of Kanalstrahlen were not realized until J. J. Thomson, the great English physicist, recognized them as beams of positive ions. He conducted experiments between 1907 and 1913 in which he passed such a beam between electric plates and poles of a magnet arranged in such a way that the magnetic and electric fields were parallel. After passing through the fields the positive ions were allowed to impinge on a photographic film where they left images having a parabolic shape. By measuring the sizes of the parabolas, Thomson was able to show that neon is a mixture of atoms having different masses. Thomson's method is analogous to a pin hole camera in that the magnetic and electric fields had no focusing value. They only caused separation of the heavy and light atoms.

In 1918, A. J. Dempster, at the University of Chicago, designed an instrument taking advantage of the ability of a magnetic field to focus a beam of positive ions as well as separate the heavy from the light ions. The optical properties were essentially those of a cylindrical lens focusing a divergent beam onto a line. This is called direction focusing. An instrument capable of velocity focusing, that is focusing a beam containing ions of different velocities, was designed in 1919 by F. W. Aston at Cavendish Laboratory in Cambridge, England. Aston introduced the term *mass spectrograph*.

In 1934 three groups independently developed instruments capable of focusing a beam that is not only divergent but also contains ions of different velocities. The three groups were J. M. Mattauch and R. Herzog of Austria, K. T. Bainbridge and E. B. Jordan of England, and A. J. Dempster in the United States.

During World War II, A. O. Nier, at the University of Minnesota, took advantage of the recent advances in high-vacuum technology to develop a sensitive and reliable mass spectrometer with direction focusing. Nier's design is basic to nearly all of the mass spectrometers used for measuring isotopic abundances in geological investigations.

Principles of Mass Measurement

Although there are various methods and instruments for determining masses of atoms, only those based on the deflection of an ion beam are discussed here. If we wanted to separate ping pong balls from golf balls that are mixed together in a basket, one way of achieving this might be to take the basket to the top of a tall tower in a wind and dump the contents. The golf balls would land much closer to the base of the tower than the ping pong balls. Mass spectrographs and mass spectrometers separate ions of different mass in a similar fashion. A magnetic field is used in place of a wind and an electric field is used instead of gravity to accelerate the ions.

As an ion beam passes through a magnetic field, the force causing an individual ion to be deflected can be expressed by:

$$f = Hev \qquad (1)$$

where H is the strength of the magnetic field, e is the charge on the ion, and v is the velocity of the ion. This force continues to act in a direction perpendicular to the direction of motion, and so the ion travels along a circular path. The centrifugal force which is opposite and equal to the centripetal force is given by:

$$f = \frac{mv^2}{r} \qquad (2)$$

where m is the mass of the ion and r is the radius of curvature.

By combining Eqs. (1) and (2), we can write

$$mv = rHe \qquad (3)$$

The kinetic energy given to an ion by the accelerating potential may be expressed by

$$\tfrac{1}{2}mv^2 = eV \qquad (4)$$

where V is the accelerating potential.

By combining Eqs. (3) and (4), we can relate the mass/charge ratio of the ion to the measurable parameters of the instrument;

$$m/e = \frac{r^2 H^2}{2V} \qquad (5)$$

If atomic mass units, electronic charge units, centimeters, gauss, and volts are used, Eq. (5) may be expressed as

$$m/e = 4.82 \times 10^{-5} \frac{H^2 r^2}{V} \qquad (6)$$

Mass spectrographs and mass spectrometers are generally used for different applications. The mass spectrograph is designed to record many masses with high resolution simultaneously so that the geometry and field strengths are fixed at the time of exposure. The precision of the method lies in our ability to measure distances on a photographic plate accurately. The mass spectrometer shifts from mass number to mass number by changing the accelerating potential or the strength of the magnetic field. The separated beams are detected by an electronic pickup and amplifier. This instrument is not suitable for measuring mass with precision but is capable of much sensitivity and precision in the measurement of beam intensities, hence relative abundances of ions of different mass in an ion beam.

Of the instruments described above, the mass spectrometer is most generally applied to geologic problems. The reason for this is that knowledge of relative abundances of isotopes is most useful in studying geological problems such as age, temperature, and history of a specimen. Therefore, only the mass spectrometer will be discussed under instrumentation.

Instrumentation

There are two different types of mass spectrometers, the gas analysis instrument and the solid source instrument. In the gas analysis instrument the sample, in gas form, is admitted to the spectrometer tube, ionized, and accelerated through a magnetic field. In the solid source instrument the sample in solution is placed on a filament and dried. The filament is placed in the spectrometer and heated electrically. This vaporizes the sample and, in some cases, ionizes it. From that point on both types of mass spectrometer function similarly.

The sample in gas form is ionized by an electron beam traveling from filament F (Fig.

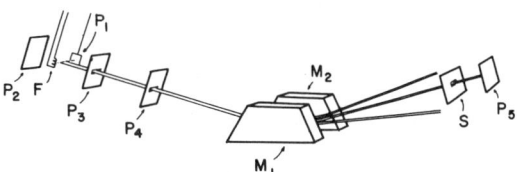

FIG. 1. Basic components of a mass spectrometer. Ions of a sample are produced by an electron beam passing from filament F to plate P_1. These ions are repelled by P_2, drawn through the slit in P_3, and accelerated by a potential on P_4. As they pass between the pole faces of a magnet, their path is bent. Only ions having a particular mass/charge ratio pass through slit S to P_5 for a given accelerating potential and magnetic field strength.

1) to Plate P_1 between two parallel plates P_2 and P_3. The purpose of plates P_2 and P_3, called the repeller and draw out plates, respectively, is to move the ions of sample through a slit in P_3 into the space between plates P_3 and P_4. The potential difference between plates P_3 and P_4 accelerates the ions through a slit in P_4 and down the spectrometer tube between the magnetic poles M_1 and M_2.

There the ions are separated according to mass/charge ratio into beams of different direction. At certain combinations of magnetic field and accelerating potential the various beams of ions enter slit S and impinge on plate P_5. As a result P_5 becomes positively charged. The charge leaks to ground through a very high resistance so that the charge on the plate is a function of the beam intensity while the beam is impinging on it. The charge is then detected by sensitive electronics, and the signal is amplified and read out on a galvanometer or strip chart recorder.

Summary

Although abundances of isotopes can be measured in other ways such as neutron activation and direct counting, the mass spectrometer remains one of the most widely applicable methods for geological purposes. Discussions of the applications of isotope abundance measurements will be found elsewhere in this encyclopedia.

WILLIAM A. BASSETT

References

Clark, G. L. (editor), 1960, "Encyclopedia of Spectroscopy," New York, Reinhold Publ. Corp., pp. 582–647.

Friedman, I., 1954, "Mass Spectrometry," in (Faul, H., editor), "Nuclear Geology," New York, John Wiley & Sons, pp. 64–74.

Rankama, K., 1954, "Isotope Geology," New York, McGraw-Hill Book Co., pp. 46–61.

Cross-references: *Isotope Geology; Neutron Activation Analysis; Radioactive Isotope Tracer Technology.*

MEDICAL GEOGRAPHY

Definition

Medical geography is the science of the distribution of human disease throughout the inhabited world. While this distribution can be shown on maps, medical geography is not limited to map making. The map is a tool which gives a transitory picture of what the phenomenon of disease is at a given time, but medical geography, being a science, purports to explore the causes of the phenomenon under observation.

(A) What is disease? Disease is that alteration of human cells, tissues, and organs which jeopardizes survival in a given environment. This definition brings together the two main terms of the equation of health, namely the host (whose ability to survive implies health) and the environment, whence the challenges to survive arise. The genetic nature of the host governs its responses to environmental challenges, while the geographical nature of the environment determines the challenges themselves.

(B) On the basis of this definition, disease appears as a maladjustment which can involve the host and the physical environment, the host and the biotic environment, the host and the socio-cultural environment or a combination of all.

(1) The Host and the Physical Environment. Little is known of the multiple effects of physical forces such as temperature, pressure, humidity, radiation, and others on the human organism. Their effect has not been measured objectively. A wide range of responses occurs when man moves out of his habitual environment, with death at both ends of the spectrum. In between these extremes, only subjective impressions like comfort or discomfort can be used to gauge their impact on man. Lee (1957) has studied combined effects of temperature and humidity on human activities and has designed a chart which allows one to predict the significance of given conditions.

Motion sickness, mountain sickness and pressure sickness (the bends) are all minor examples of the effects that physical factors in the environment can produce. In all these examples, health (survival) is assured when the tissues of the host "adjust"; if they don't, survival becomes impossible. Some individuals seem to withstand motion and tolerate changes in altitude or pressure better than others. This "threshold of tolerance" which makes James sick in a car while John can fly his stunt plane upside down is governed by James' and John's genetic makeup, i.e., their "genotypes." The genotype will govern not only the basic response to physical stimuli arising in the geographical environment but also the ability to be conditioned and the level which that conditioning may reach.

(2) The Host and the Biotic Environment. The complex formed by host and habitat is referred to as a "niche" by ecologists. The number of "niches" is, of course, infinite (see *Ecology*). They are defined by a complex of geographical features that in turn allow the living things inside the area to survive. The environment provides the wherewithal of life, e.g., food and oxygen; to use it efficiently the living things have to organize themselves into what is known as "society."

To an ecologist, a society of living things is a temporary equilibrium between these things which occurs when the dynamism of reciprocal exclusion has been exhausted. This means that societies of living things are always the result of a compromise and are always temporary. If the environment changes, the balance between the members of the living society will change; some who were at the bottom may rise to the top and vice versa.

This concept of a society of living things is important if the geographical distribution of disease is to be understood. The presence or absence of a pathogenic agent in a "niche" may be the result of social pressures. For example, when a physician injects penicillin into the blood of his patient, he introduces a new living thing which completely transforms the social structure of microorganisms to the benefit of the host. The reverse, of course, may occur and we know now that ill-advised use of antibiotics or toxicants may bring about a "population explosion" of resistant strains hitherto subdued by the susceptible strains. Because pathogenic agents are often parasites of other living organisms, the disease picture of a region is closely related to the presence or absence of the vectors of these parasites (e.g., anopheles and *P. falciparum*, one of the malaria agents), or to the presence or absence of reservoir hosts of the parasites (e.g., rats and *P. pestis*). All the inhabitants of a particular geographic niche, either on the microscopic or on the macroscopic scale, are therefore intimately related to the geographic environment.

(3) The Host and the Cultural Environment. Culture is, to the medical geographer, the sum total of the concepts and techniques used by

individuals or population groups to control their environment. Obviously, culture is linked to geography. The houses of man, his foods, his ways of life in general depend to a large degree upon the location of his residence. Certain techniques used for procuring food and shelter, if they have insured survival and prosperity, become traditional and often remain so long after their original justification has ceased to exist. Disease, or maladjustment, may occur on the occasion of challenges imposed by culture. For example, the habit of smoking is a cultural trait which has known an enormous development under the pressure of a certain kind of life. It is now accepted that the alteration of tissues which we call cancer of the lung is very frequently a consequence of this habit. Excessive intake of certain foods produced in overabundance in certain areas plays a role in obesity, arteriosclerosis, etc. So, in some cases does the artificial coloring that our conditioning demands. It can also be said that certain psychiatric conditions, while not caused by cultural pressures, are conceivably made worse or become apparent in a large number of people because of these pressures. The problem created by a case of schizophrenia in a family varies not only with the severity of the disease but also, and to a considerable degree, with the environment. Undoubtedly, two schizophrenic persons with the same level of disease would make two very different patients if one lived in a two-room apartment in New York and the other in the large household of a wealthy Arab landlord.

(C) In the previous paragraphs we have several times referred to the "host." What is the host? In the present context it is, of course, a human or a population of humans. In the context of medical geography how does the host figure in the distribution of disease? We have seen that the challenges from the environment vary considerably with geography. The hosts, too, vary; the environment make them different, but their inescapable genetic heritage makes them different, too. The "genotypes" of men are many and are inexorably conditioned by the genes which only "mutations" can change.

The genotype plays a role in establishing susceptibility to pathogenic agents or natural immunity. This is obvious when we compare man and animals; it is less obvious, but just as true, when we compare two men. Cows do not catch leprosy, man does not as a rule catch foot and mouth disease. (There are, of course, a number of diseases common to man and animals.) James will be susceptible to motion sickness or to tuberculosis whereas John will be a good sailor and will never catch tuberculosis, perhaps through lack of exposure or because he is naturally immune. One person is allergic to eggs, another is not; one has asthma, another does not. All these individual properties of the individual are basically conditioned by the genotype, and the genotype varies throughout the inhabited world.

Blood groups are considered to be genetic markers. Maps of blood groups made by Mourant (1971) give an idea of the refinement that may be attained. The study of these genetically transmitted blood groups and blood factors may eventually help to further clarify the causes of the geographic distribution of diseases. A classic example is that of the sickle cell trait. This is a form of hemoglobin found in certain African tribes, so called because it gives the red cell a "sickle" shape. It has been found that the bearers of this trait are immune to infection by *Plasmodium falciparum,* the agent of malaria, thus greatly influencing the map of disease and the medical geography of the area. The science of mapping genes is still in its infancy but it is the holder of great promises.

Finally, a study of the mutagenic factors present in the environment may eventually be rewarding, but here we are looking into the distant future. It was mentioned above that in the light of present knowledge, only mutations could change the genes. Thus we are led to consider the factors that bring about these mutations. We know of some of these such as heat, chemicals, radiation, but we do not know what mutations each of them causes. Yet, we know that these mutagenic factors have a geographical distribution and that the fields of irradiation or the shower of cosmic rays are not the same in all corners of the world. Since this action may be felt by microscopic agents as well as by man, since a mutation may change a virulent bacteria into a nonvirulent one, the structure of the pathogenic stimuli present in an environment may change vastly with changed genotypes and mutations. This adds another field for our exploration, another problem for medical geography to explore if disease distribution is to be understood.

Methodology

What method should medical geographers follow to make progress in the fields we have discussed above? Methodology is still in the formative stage. A few textbooks, however, exist which could give guidance to would-be medical geographers (May, 1959). A large series of world maps of medical geography has been published by the American Geographical Society, New York.

MEDICAL GEOGRAPHY

TABLE 1. PROVISIONAL CORRELATIONS BETWEEN GEOGENS AND PATHOGENS[a]

Two-Factor Complexes	Altitude and Latitude	Light and Cloud	Drainage and Humidity	Temperature	Income	Population Density	Sanitation	Diet	Communications	Animal Life	Parasitism	Remarks	Pathogens
Epidemic meningitis	−	−	?	?	−	+	+	?	+	−	?	Rainfall, humidity, and water drainage may influence infection by droplets	Man / Meningococcus
Epidemic cholera	−	?	+	?	+	−	+	?	−	−	?		Man / V. cholerae
Endemic cholera	?	?	+	+	+	+	+	+	+	?	?	The nidification of the disease is governed by unknown factors	Man / V. cholerae
Bacillary dysentery	+	?	+	+	+	+	+	+	−	?	?	Flies may play a part in the spread of the disease but their role is mechanical rather than biological	Man / Shigella and group
Typhoid fever group	−	+	+	+	+	+	+	+	+	?	?	Flies may play a part in the spread of the disease but their role is mechanical rather than biological	Man / Eberthella and group
Tuberculosis	+	+	+	+	+	+	+	+	−	+	−	Infected droplets play a part in the spread of the disease	Man / M. tuberculosis
Diphtheria	?	?	?	?	+	+	+	−	+	+	−	Infected droplets play a part in the spread of the disease	Man / C. diphtheriae
Scarlet fever	+	?	?	+	+	+	+	−	+	+	−		Man / S. scarlatinae
Brucellosis	−	+	+	+	+	−	−	+	−	+	−		Man / B. melitensis and group
Gonorrhea	−	−	−	+	−	−	−	−	−	−	−		Man / N. gonorrhea
Anthrax	?	?	−	+	−	−	+	−	−	+	−	The spores of B. anthracis remain viable in soil indefinitely; geographical factors influencing the soil may influence the spores	Man / B. anthracis
Leprosy	−	?	?	?	−	+	+	?	−	−	−		Man / M. leprae
Tetanus	?	?	−	+	−	+	+	−	−	+	−	C. tetani is widely distributed in the soil and therefore probably sensitive to geographical factors influencing the soil	C. tetani
Poliomyelitis	−	?	?	++	−	+	−	?	++	?	?	Epidemiology still under discussion: infected droplets may play a part in the spread of the disease; presence of virus in sewerage is important	Man / Viruses
Influenza	−	?	++	++	−	++	++	?	++	?	?	Epidemiology still under discussion; the respective roles of the different viruses, the role of animal reservoir, and the passage from endemicity to epidemicity are not clear; droplet infection is certain	Man / Viruses

MEDICAL GEOGRAPHY

Disease													Remarks	Agent/Vector
Measles	−	−	−	−	+++	+++	−	−	+++	−	−	?	Air-borne infection by droplets is probable means of spread	Man / Virus
Smallpox	−	−	−	?	+++	+++	−	?	+++	?	?	?	Air-borne infection by droplets probable; correlation has been found between incidence of smallpox and aqueous vapor tension in India and England	Man / Viruses
Lymphogranuloma venereum	−	−	−	−	−	−	−	+++	−	−	−	−	The virus is destroyed by ultraviolet rays	Man / Viruses
Amboebic dysentery	+	?	+	?	++	−	−	−	−	?	?	?		Man / *E. dysenteriae*
Ascariasis	+	+	+	+	+	+	−	+	+	+	+	+	Seasonal variation of incidence has been observed; humidity seems more important than temperature	Man / *A. lumbricoides*
Ancylostomiasis	++++	+	−	+	++++	++++	+	++	+++++	++	+++	++	Warmth, moisture and oxygen influence the development of the embryo outside the body; ova are resistant to cold and dryness	Man / *A. duodenale* and group
Syphilis	+??	+	?	?	++?	−	−	−	++	−	++	−	Porous, sandy soil and shade favor the growth of the larva	Man / *T. pallidum*
Yaws	+?	?−	?	?	+?	−	−	−	+?	−	?	−	Man does not seem to respond in the same way to *T. pallidum* in different latitudes	Man / *T. pertenue*
Pinta	++	?	−	?	?+	−	−	−	+	−	−	−	Disease limited to low humid tropics; at high altitudes, lesions dry up; the role of flies in transmission is mechanical	Man / *T. carateum*
Dengue	+?	+	+	?	+++++	?	−	?	++++++	++++	++	+	Prevalent in warm humid tropics, especially along stream banks; *Simulium haematopotum* may help spread mechanically	Man / Mosquito / Virus
Malaria	+?	+?	?	+?	++	+−	−	+−	++?	−	+	+?	*Aedes aegypti* has been up to 5000 ft but is limited to 40° N–40° S latitude	Man / Anopheles / *Plasmodium*
Trypanosomiasis	++?	+	+?	+?	+++	++	+?	+?	+++	?	+++	+++	In some regions wild game and the native pig act as reservoirs. High temperatures favor the development of the parasite in flies	Man / Fly or bug / *Trypanosoma*
Filariasis Type: *W. bancrofti*	−	+	?	?	++	+−	−	?+	++	?+	?+	+−	Antelope and buffalo may act as reservoirs for *Onchocerca* in Africa	Man / Fly, mosquito, or tabanid / *Onchocerca*, *Wuchereria Loa*
Filariasis Type: *D. medinensis*	?	?	−	−	−	+	−	?	+++	?	?	−		Man / Copepod / *Dracunculus*
Schistosomiasis	?+	?+	?	?	+++	−	+	?	+++	?	+	−		Man / Snail / *Schistosoma*
Cestode diseases Type: *T. echinorococcus*	−	?	?	?	++	+	−	−	+++	−	+++	+++	In this complex, man is an accidental factor and a dead end for the parasite	Dog / Sheep-Man / *E. granulosus*

MEDICAL GEOGRAPHY

TABLE 1. (Continued)

Two-Factor Complexes	Altitude and Latitude	Light and Cloud	Drainage and Humidity	Temperature	Income	Population Density	Sanitation	Diet	Communications	Animal Life	Parasitism	Remarks	Pathogens
Relapsing fevers	+	+	?	+	+	+	+	+	+	+	+	Malnutrition seems to be a predisposing factor in man. Cold weather favors lousiness. Vector may be transported over long distances. Infection is transmitted to next generation. In California a fourth factor may enter in, since some rodents are a reservoir for spirochetes. Housing and clothing geogens	Man
	+	?	+	+	-	+	+	-	+	+	+		Pediculus or Ornithodoros
	-	?	?	+	-	+	-	-	-	-	+		Spirochete
Leptospiral diseases	-	?	+	?	-	-	+	-	-	+	+	Seasonal correlations have been recorded in various parts of the world. Correlation has been shown between survival of spirochetes and salinity of water in which they live	Rodent
	-	?	+	?	-	-	+	+	-	+	+		Man
	-	?	-	+	-	-	-	-	-	-	+		Leptospira
Epidemic typhus	-	+	-	+	+	+	+	+	+	-	+	Cold weather favors lousiness	Man
	-	?	-	?	-	-	+	?	-	+	+		Pediculus
	-	-	-	+	-	+	+	+	?	-	+		R. prowazeki
Scrub typhus	-	-	-	+	-	+	+	-	?	+	+	The adult mite lives on a certain type of grass, such as the coarse grasses growing on abandoned estates	Man
	+	?	+	?	-	-	+	?	+	+	+		Rodent
	-	?	-	?	-	-	+	?	-	?	+		Larval mite
	-	-	-	+	-	-	+	+	-	+	+		Rickettsia
Tularemia	-	?	-	?	-	-	-	?	-	+	+	Dermacentor is vector and also reservoir, since it remains infected all its life; transmits infection to its eggs. The disease may be transmitted directly	Rodent
	-	-	?	?	-	-	-	?	-	+	+		Dermacentor, Ixodes
	-	-	-	?	-	-	-	?	-	+	+		Man
	-	-	-	+	-	-	-	+	-	+	+		P. tularensis
Trematode diseases Type: Clonorchiasis	+	?	-	?	-	-	-	-	-	+	+	Some trematodes do not require a second intermediate host. The cercariae erupt and attach themselves to certain types of aquatic vegetation. Vegetation and soil geogens	Man
	+	?	-	?	-	-	-	-	-	+	+		Snail
	-	?	+	?	-	-	+	-	-	+	+		Fish
	-	?	+	+	-	-	+	-	-	+	+		Fluke

An inventory of known facts is, of course, the first step—as it is, or should be, in every field of science (see Table 1). The most severe criteria should be applied before calling a fact a fact, as erroneous deductions would soon occur if weak criteria of reliability were applied. This inventory should be made in all of the fields we have mentioned in the above discussion, and no causal relationship should be assumed because of coincidence. On the basis of these inventories, maps of facts can be made that could present a picture of suspected cases (always subject to doubt as diagnoses are not reliable), confirmed cases determined by incontrovertible tests and susceptibilities to certain infective agents. Sample studies of blood antibodies showing previously unrecognized exposure to communicable diseases indicate a latent level of infection in the population. Absence of such antibodies indicates a susceptible population ready to succumb to a newly introduced agent. Yellow fever, for example, has never occurred in Asia so no antibodies against the virus can be found in the blood of Asians; threfore, severe quarantine precautions must be and are taken to avoid introducing the virus lest a terrible epidemic occur. Agents, vectors, reservoir hosts, other animals present, vegetable cover (which governs the animal society), soils (which govern crops and part of the food resource), rain, temperature, radioactive elements, etc., can be plotted on maps revealing certain possible correlations.

Once the inventory has been made, speculation may be in order as to what role each factor plays in the pathologic complex present in a given spot. This speculation may lead to a verifiable hypothesis which further observation or experimentation might confirm or disprove.

Medical geography is a science based on a knowledge of the earth, the climate above the earth, the plant cover of the earth, the fauna related to these plants, the human creatures related to earth, climate, plants and animals. It evaluates the interrelation of these factors and through this evaluating establishes the responsibility of a factor or group of factors in preventing the adjustment of man to his environment. Once the responsible factors of maladjustment, or disease, have been discovered, control of disease is facilitated.

JACQUES M. MAY

References

Lee, Douglas H. K., 1957, "Climate and Economic Development in the Tropics," Published for the Council on Foreign Relations, New York, Harper & Brothers.

MEDICAL GEOGRAPHY

May, Jacques M., 1950, "Medical geography: its methods and objectives," *Geograph. Rev.*, **40**, 9–41

May, Jacques M., 1959, "The Ecology of Human Disease," Vols. 1–2 of "Studies in Medical Geography," New York, M.D. Publications.

May, Jacques M., 1955–1971, "The ecology of malnutrition," Vols. 3–9, of "Studies in Medical Geography," New York, Hafner Publ. Co.

Mourant, A., 1954, "The Distribution of Human Blood Groups," Vol. 1, Oxford, Blackwell Scientific Publications, 438pp.

Cross-references: *Air Pollution and Global Climate; Air Pollution and Urban Climate; Ecology; Environmental Pollution; Environmental Science; Medical Geology; Mineral Particles and Human Disease.*

MEDICAL GEOLOGY

There are two main aspects of medical geology: the study of the total geological environment and its relation to the health of man; and the study of petrographic and mineralogical methods of parts of the human body or foreign substances therein.

The Geological Environment and Health

That environmental factors are important and, at times, crucial in human health and disease has long been recognized by students of human geography (see *Medical Geography*). But whereas geographic studies comprise the entire range of environmental factors, including such variables as climate and sanitation, the geologist confines his attention to the soils and underlying bedrock. Because the relationship between man and his geological environment is frequently less direct than between man and, for example, his climate, it is not surprising that he has been slow in recognizing this relationship. Hence, most significant work in this field has been quite recent.

Warren and Delavault (1949, 1960), and others working in biogeochemical prospecting, drew attention to the large and selective concentrations of metals found in certain plants. This metal concentration appeared to be related to the underlying bedrock even in some cases where the soil cover was different in mineral content from that of the underlying rock. Attention was then focused on the fact that important trace elements found their way to the human body from both soils and the underlying bedrock, and the geological environment was seen to be significant in human health and disease.

Man depends ultimately upon the rocks of the earth's crust for his trace elements, but unfortunately the route that these trace elements may take, in getting from the earth's crust into human organs, is extremely complex. The relationships between trace elements in rocks and in soils, and between soils, plants and animals, are only now beginning to be understood (see *Trace Elements in Plants*).

Many important variations in rock composition point to the need for extremely careful and detailed study if significant applications of geology to health problems are to be made. Limestones and dolomites may be either high or low in lead content, two monzonites closely resembling one another mineralogically may vary greatly in trace elements, and even slates of the same geological age in one small area have been found to differ in content of some trace elements by a factor of as much as 20:1 (Warren, 1964b).

Although soils largely, but by no means entirely, derive their trace elements from the geological formations to which they ultimately owe their origin, there is always the problem of whether the soil is residual or transported. In any event, the total amounts of trace elements, as well as their distribution in soil, is often significantly altered by the soil-forming process. Again, plants may be variable and selective in their trace element content, and plant-animal relationships introduce further variables even before man enters the picture. It has been found that cattle can eat fodder that contains sufficient copper, yet suffer from copper deficiency because there is too much molybdenum in this fodder. Also, some animals can eat without harm a food containing materials that would be poisonous to most other animals.

The need for exact and detailed geological and mineralogical work is underlined by the small amounts of materials often involved in human health. Small deficiencies may lead to serious ailments and, on the other hand, quite modest amounts of materials may be highly poisonous.

Epidemiological studies are shedding some light on these problems, but correlations with trace element studies are still uncommon. Special difficulties arise because medical statistics are ordinarily given for political rather than geologically significant areas, and because most medical statistics concern death rather than illness.

The work of Helen Cannon, who has combined experience in study of uranium mineralization with geobotany, is pointing the way to significant advances in field correlation of geology and health. She has been interested in relating geochemical environments to incidence of cancers and cardiovascular diseases, and this involves careful and detailed geological field studies as a basis for a review of health in the area. Similarly, Warren (1964a) in cooperation with a number of physicians, is making important studies of copper, lead, zinc and molybdenum content of some rocks, soils and vegetal matter from areas reporting a high prevalence of multiple sclerosis.

That dental health is affected by trace elements is also generally acknowledged, with most of the attention centered on the reduction of caries by the addition of fluorine in suitable form and amount to drinking water. This program, now rather extensive, developed from observations that inhabitants of areas with high fluorine content in waters and soils had stronger and healthier teeth than the general population. Widespread research is continuing in the matter of optimum intake of fluorine for dental as well as for general health. It is possible that the keen interest and the positive results from this research have tended to obscure the role played by other trace elements in dental health, and there is need for further information.

Petrographic and Mineralogical Methods in Medicine

Rock Dust Diseases of the Lungs. In cases of rock dust exposure, it is important to know the type of dust and its origin, and this requires a detailed study of the rock being worked and of the minerals that go to make it up. Since it has been found that dust samples taken from different working faces, even in the same mine, show significant mineralogical differences from place to place, very careful petrographic and mineralogical studies are required (Stevenson, 1959).

Most of the dusts usually encountered by man, such as the sands of the desert and the dusts of the street, are too large to reach the lungs and are trapped in the nose and upper respiratory passages. However, particles 5μ in diameter or less can reach the alveoli of the lungs, and many dusts resulting from mining and quarrying are in this size range. In the ashed lung specimens of men who have died from silicosis, Hunter (1959) reports that the most representative particle measured 1μ in diameter.

Considerable confusion has arisen because, in medical literature, silicosis is usually cited as being caused by the inhalation of "free or uncombined silica," and the great danger of many silicate minerals was long overlooked. However, it is true that many rock dusts, such as those from limestones, dolomites, and iron oxide ores, which do not contain siliceous accessory minerals, seem to be inert when taken into the lung and can be inhaled without serious effect.

Careful mineralogical study of ashed lung tissue from patients suffering from silicosis has shown that the most dangerous rocks contain siliceous minerals which have a tendency to shatter into fibrous forms. Thus pure quartz, unless it is very finely ground, is far less dangerous than sericite or chrysotile. The ashed lung tissue of silicotic patients generally reveals a characteristic felting produced by minute, usually fibrous, silicates. Although originally such studies could be made only in autopsies, these contributed greatly to knowledge of the disease and to proof of industrial injury valuable to the heirs in legal compensation cases. Now, however, most ashed lung studies are based on biopsies, and are helpful to the physician in interpreting the patient's symptoms and advising further treatment. They have also been valuable in alerting management to dangerous areas of rocks rich in fibrogenic siliceous minerals, so that special precautions could be taken, particularly in "hard rock" metal mines and asbestos mines.

Coal-miners' pneumoconiosis, rather than classical silicosis, is the most widespread disease in coal mines, and because it does not develop as rapidly as silicosis, its danger was not recognized for many centuries. However, since dust at a coal face is nearly always a mixture of coal and siliceous material, continued inhalation of coal dust generally produces a slowly progressive fatal disease.

The usual dust suppression methods in mines and quarries include ventilation, wet cutting, and extensive spraying of dusty areas. The prevention of silicosis by metallic aluminum has been accomplished in recent years, based largely on research coordinated by the McIntyre Research Foundation in Ontario. It was found that inhalation of small amounts of aluminum dust with the rock dust rendered the siliceous material insoluble and hence benign, and that miners who inhaled aluminum dust before going on shift were protected from the dust hazard. Aluminum prophylaxis is now used extensively in many parts of the world (Irwin, 1964).

Calculi in the Human Body. Medical problems of the cause and recurrence of calculi have been greatly aided by petrographic and mineralogical studies of these stones, using the methods ordinarily employed in the study of true rocks and minerals. This is especially true of such concretions as kidney stones, which are generally composed of materials identical to minerals found in the earth's crust.

Prien and Frondel (1947) have applied the oil immersion method of optical crystallography with great success in the identification of the component minerals in calculi, and they

MEDICAL GEOLOGY

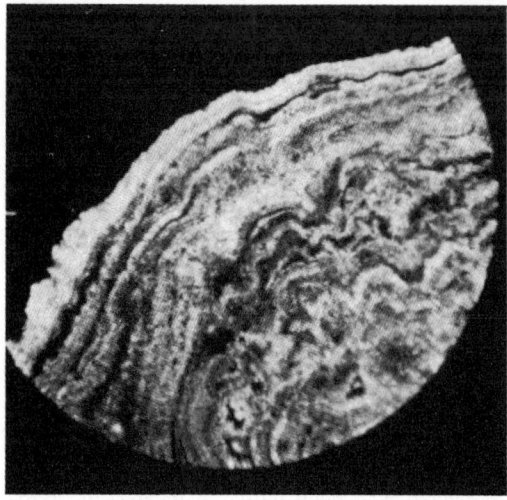

FIG. 1. A portion of a urinary calculus showing colloform structure. Light gray rings are clear, crystalline whewellite; darker grey rings are yellow whewellite; white outer ring is weddellite. The dark lines between some of the rings are organic matter, probably blood (crossed nicols, 20×).

have made significant statistical studies of the various types of stones. Whewellite, weddellite, struvite, apatite (both carbonate-apatite and hydroxyapatite), and brushite were found to be common constituents.

Lucas, Stevenson and Stevenson (1950) have emphasized the value of augmenting oil immersion identification with petrologic study in thin sections (Figs. 1 and 2). The structure of the calculus may then be seen, and the formation of the stone, with the paragenesis of the constituent minerals, traced from the centers of crystallization to the outer encrustation. With sufficient knowledge of the case history of the patient, it may be possible to correlate pathogenesis and petrogenesis.

Mineralogical Studies of Human Bones and Teeth. The "mineral" composition of bones and teeth is similar to that found in rocks of the earth and, hence, may be studied in the same way; indeed it is the mineralogist who has been able to explain to the physiologist the true nature of the "inorganic" component of bones and teeth (McConnell, 1962).

It has been found that dental enamel is composed of a single "mineral" phase which is dahllite (carbonate hydroxyapatite). Dentin and bone have a composition very similar to dental enamel, and their crystalline structure must be essentially isostructural with that of dental enamel. Fossil studies have shown that the usual composition of fossil teeth and bone is also dahllite, although fossilized material may become francolite (carbonate fluorapatite) by fluorine enrichment during metamorphism.

Studies of fluoride in bones and teeth by mineralogical methods suggest that fluoride, when ingested, may change the crystal texture of bone and tooth apatites and stabilize them by increasing crystal size and decreasing crystal strain.

JOHN S. STEVENSON
LOUISE S. STEVENSON

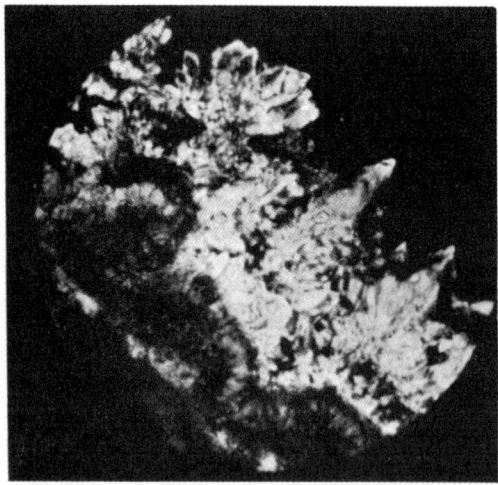

FIG. 2. The rim of the same calculus shown in Fig. 1. Weddellite crystals are encrusted on whewellite "tree-rings"; darker material between the crystals is struvite (crossed nicols, 40×).

References

Hunter, Donald, 1959, "Health in Industry," Pelican Medical Series, Pelican Books, pp. 176–215.
Irwin, D. A., 1964, "The Use of Aluminum in the Prevention of Silicosis." Publ. McIntyre Research Foundation, Toronto, Ont., pp. 1–9.
Lucas, O. C., Stevenson, John S., and Stevenson, Louise S., 1950, "Petrologic study of a urinary calculus," *Trans. Roy. Soc. Can., Ser. III, Section IV,* **44,** 35–40.
McConnell, Duncan, 1962, "The crystal structure of bone," *Clinical Orthopaedics.* No. 23, 253–268, J. B. Lippincott Co.
Prien, Edwin L., and Frondel, Clifford, 1947, "Studies in Urolithiasis: I. The composition of urinary calculi," *J. Urol.* **57,** No. 6, 949–991.
Stevenson, John S., 1959, "Mineralogy and the field geologist," *Can. Mineralogist,* **6,** Part 3, 303–306.
Warren, Harry V., 1964a, "Geology, trace elements and epidemiology," *Geograph. J.* **130,** Part 4, 525–528.
Warren, Harry V., 1964b, Introductory remarks, Medical Geology and Geography Symposium, A.A.A.S., Montreal, Dec. 28, 1964 (Report in *Science,* 1965).
Warren, Harry V., and Delavault, R. E., 1949, "Further studies in biogeochemistry," *Bull. Geol. Soc. Am.,* **60,** 531–559.

Warren, Harry V., and Delavault, R. E., 1960, "Observations on the biogeochemistry of lead in Canada," *Trans. Roy. Soc. Can., Ser. III, Section IV,* **54,** 11–20.

Cross-references: *Biochemicals; Copper; Fluorine; Lead; Medical Geography; Medical Geology—Trace Metals in Mammals; Medicinal Springs; Mineral Particles and Human Disease; Organic Geochemistry; Pedology; Trace Elements in Plants; Zinc.* Vol. IVB: *Asbestos; Illite; Optical Mineralogy; Serpentine.* Vol. VI: *Mines and Mining; Quarries, Quarrying; Soil Genesis.*

MEDICAL GEOLOGY—TRACE METALS IN MAMMALS

Much harm or much good can be done by minute changes in the tiny quantity of trace metals we carry about with us. Everyone is aware of the dangers of lead and the virtues of iron, but we still have a great deal to learn about most of the others. Quest for such knowledge may be greatly facilitated by the nuclear bombardment technique known as neutron activation analysis, which can detect as little as one billionth gram of a metal.

Nutritional supplements contain some trace metals in small amounts on the rather uncertain theory that they may do some good. But when we give these substances to achieve obscure benefits, we may be doing harm without realizing it. Table 1 lists thirty-four trace metals present in the body. Some of them are essential, others are undesirable if not actively toxic.

Copper and zinc are unquestionably as essential as iron. So is cobalt, as a constituent of vitamin B_{12} (cyanocobalamin), although human anemia has never been precipitated by withholding it. Manganese and molybdenum are vital to our physiology, but deficiency states have not been demonstrated. Vanadium might be added to the list, but authorities differ as to whether it should be classed as essential or nonessential.

Tipton (1963) classes all other trace metals as abnormal, whether or not normal physiologic functions can be ascribed to them. Strain (1961), however, points out that there is nothing abnormal about our inheritance of trace element patterns, and that our intercellular fluid may well contain them in about the same proportions as the sea. Normal food supplies significant amounts of most metals, and drinking water may contain more than 50 μg/liter of admittedly toxic substances such as lead, which we consume without apparent harm.

The principal harm done by the less desirable metals, as emphasized by Schroeder (1960), among others, is that they tend to displace physiologically desirable elements belonging to the same periodic group, thereby upsetting what appears to be natural balance. Barium and beryllium inhibit absorption of calcium, for example, and strontium displaces calcium. Furthermore, some of the "unphysiologic" elements accumulate with age, notably aluminum, barium, silver, tin and titanium, but do not normally reach dangerous levels.

Metals with Physiologic Roles

Copper. As a component of red blood cells and many other tissues, copper participates in the action of a number of essential enzymes. Without it, laboratory animals, and presumably man as well, cannot utilize iron; they cannot reproduce, they lose coordination and pigmentation. Many physicians prescribe it in anemias and malnutrition. But human needs are amply met by an intake of 2 mg/day, and we could hardly consume less than this except on a starvation diet. Hypocupremia develops in patients who have a combination of hypoproteinemia, iron deficiency, kwashiorkor and sprue.

TABLE 1. AVERAGE LEVELS OF TRACE METALS[a] (g/70 kg man)[b]

Iron	4	Antimony	<0.09	Cobalt	<0.003
Zinc	2.3	Lanthanum	<0.05	Beryllium	<0.002
Rubidium	1.2	Niobium	<0.05	Gold	<0.001
Strontium	0.14	Titanium	<0.015	Silver	<0.001
Copper	0.1	Nickel	<0.01	Lithium	<0.0009
Aluminum	0.1	Boron	<0.01	Bismuth	<0.0003
Lead	0.08	Chromium	<0.006	Vanadium	<0.0001
Tin	0.03	Ruthenium	<0.006	Uranium	0.00002
Cadmium	0.02	Thallium	<0.006	Cesium	<0.00001
Manganese	0.02	Zirconium	<0.006	Gallium	<0.000002
Barium	0.016	Molybdenum	<0.005	Radium	0.0000000001
Arsenic	<0.01				

[a] The table includes nonmetallic elements such as boron, whose chemical behavior is similar to that of true metals.
[b] After W. H. Strain (AAAS symposium on geochemical evolution, Denver, 1961).

Copper metabolism is disturbed in some diseases, as indicated by its redistribution in the body. Serum levels are high in many infections because the copper content of tissue cells is depleted by tissue damage.

Cobalt. As it is an integral part of vitamin B_{12}, cobalt may speed blood regeneration in radiation anemia, and perhaps in anemia of infection or uremia. Ordinary needs are amply met by any ordinary human diet. Deficiency has not been proved in man, but sheep grazing on pasturage low in cobalt become debilitated. Undesirable overloading with cobalt would increase blood lipids.

Zinc. Occurring in many tissues and enzymes, zinc is indispensable to growth and to the functions of blood cells, liver, kidneys and other organs (probably including the male gonads, where its concentration is relatively high). The hepatic level is low in liver disease, particularly alcoholic cirrhosis, the renal level is low in chronic kidney disease, and the zinc content of white cells is reduced in chronic myeloid and lymphatic leukemia, with return to normal in remission. Deficiency should obviously be guarded against when nutrition is below standard, as it is in many parts of the world, but most people get enough zinc from drinking water, and the supply is well rounded out if they eat animal protein. The body curbs excretion when intake is low.

Halstead and Prasad (1963) studied clay-eating Iranian dwarfs characterized by severe anemia and hypogonadism. Prasad and associates continued the study with a group of Egyptian dwarfs who were anemic because of schistosomiasis and hookworm infection. Zinc deficiency (frequently associated with iron deficiency) was demonstrated by radioactive zinc turnover study, and zinc therapy appeared to promote body growth and development of secondary sex characteristics. This suggests that zinc might also be beneficial in less extreme cases of retarded somatic and sexual development.

As with anything else, an overabundance of zinc is dangerous and sometimes toxic, as when food is contaminated by galvanized containers. A diet chronically high in zinc may aggravate dental caries.

Manganese and Molybdenum. These metals are prerequisite to normal metabolism, bone formation and body growth. Although a well-rounded diet assures adequate amounts, the need for them is so great that supplementation might be considered whenever nutrition is questionable. Manganese especially may be useful in vitamin deficiency, as it has a sparing action on thiamine and relieves a form of paralysis seen in vitamin E-deficient animals. It also counteracts hydralazine toxicity in animals.

A diet rich in molybdenum protects children and experimental animals from dental caries, as has been demonstrated in various areas of New Zealand, Hungary and England (Somersetshire). Since ash from foods with high molybdenum content was more effective than molybdenum compounds given alone, other metals must also be involved. Molybdenum is recommended as an adjunct to iron for treatment of anemia, especially in pregnancy, because it appears to improve tolerance and speed hemoglobin regeneration.

Manganese and molybdenum have low toxicity, but an excess of either disturbs metabolic balance. Manganese can displace iron, and molybdenum sometimes interferes with copper.

Vanadium. Although vanadium has relatively high toxicity, it also has valuable functions, chief among which is inhibition of cholesterol and lipid synthesis. Curran (1960) lowered serum cholesterol as much as 20% by adding vanadium to the diet of young men with normal cholesterol levels; and workmen exposed to vanadium have relatively low cholesterol.

It has been often stated (and occasionally contradicted) that death from cardiovascular disease is in inverse proportion to the hardness of drinking water. The calcium and magnesium content is not responsible for any such effect, but Strain has pointed out that vanadium is almost certainly a significant factor. Soil of the perimeter of the United States is generally poor in vanadium, in contrast to the region west of the Mississippi and the Southwest. Of death rates for coronary disease in 163 metropolitan areas, coastal cities had the highest and cities of the Mississippi Valley and the Southwest the lowest. Fatalities from atherosclerosis were nearly three times as frequent in Savannah, Georgia, as in Lincoln, Nebraska.

Like fluorine and strontium, and in contrast with zinc, nickel, selenium, tellurium and barium, vanadium promotes mineralization of dental enamel. The incidence of caries in man and laboratory animals is inversely proportionate to vanadium intake; fluorine enhances the protective effect of vanadium, and molybdenum could be expected to have a similar additive effect.

Metals with Dubious Roles

Nickel. The physiology of nickel is anomalous: it may be involved in pigmentation and in the action of enzymes and ribonucleic acid, but these roles are putative at best. Nickel is ubiquitous in nature, yet its distribution in the body is bizarre. The kidneys and liver of 69% of New Yorkers and Chicagoans contain it, whereas Miamians seem to store none at all in these organs. The cause and significance of this difference are unknown. Nickel is present in the lungs of all people, and the concentration builds up with advancing age. High intake induces dental caries in rats and appears to be toxic to farm animals.

Chromium. In animal studies, chromium contributes to growth, longevity and resistance to disease, perhaps by figuring in the action of ribonucleic acid; but man is probably better off without much of it. In the rat, it promotes synthesis of cholesterol and fatty acids.

Selenium. Despite its toxicity, selenium is potentially useful to man and demonstrably useful to animals. By inhibiting oxidation of polyunsaturated fatty acids, selenium curtails production of migrant "free radicals." These have the unwanted action of polymerizing body proteins, thereby diminishing the elasticity of tissues. Selenium would accordingly be a desirable element to include in the fountain of youth, except that its antioxidant effect can be more safely supplied by vitamin E. The antioxidant effect of selenium also prevents the muscular dystrophy of sheep known as white muscle disease, but there is no reason to think it could benefit the human disease, which is caused by different mechanisms. Circumstantial evidence incriminates selenium as a cause of dental caries, which is widespread in areas where the element is plentiful, and less so where it is not.

Lithium. In the heyday of the health spas (medicinal springs), multiple physiologic benefits were ascribed to lithium, but it fell into disrepute when a few fatalities resulted from its indiscriminate use in low-sodium diets. Its toxicity can be largely counteracted by sodium.

Many Metals are "Unphysiologic"

We are constantly exposed to elements that accumulate innocuously in the body. The tin content of canned foods, particularly tomatoes that have been stored for some time, may be alarmingly high, but most of it is insoluble, poorly absorbed and rapidly eliminated. Foods that are properly canned are not dangerous. Aluminum ingested from water, food, cooking ware or antacids may decrease absorption of other substances but is not inherently toxic. Progressive accumulation of titanium has no known baneful or beneficial effects.

A number of elements are capable of doing harm by usurping the function of essential ones—beryllium substituting for magnesium, silver and gold for copper, rubidium and cesium for potassium, lithium for sodium, arsenic for phosphorus, tellurium and selenium for sulfur, barium and strontium for calcium, tungsten for molybdenum, cadmium and mercury for zinc. Cadmium is one of the worst offenders, disturbing kidney function and increasing reabsorption of sodium. The cadmium level is high in hypertension and is lowered by antihypertensive treatment. Cadmium is also toxic to the testes. Gunn and colleagues (1963) induced testicular injury in rats by giving cadmium, and prevented such damage by giving zinc simultaneously. Others have found selenium more effective than zinc, because selenium has the capacity for combining with cadmium, as with other metals.

Industrial poisoning occurs with antimony, arsenic, beryllium, cadmium, lead, mercury, strontium, tellurium and thorium. Strontium has had much attention because of its radioactive form in fallout. Fortunately it is not well absorbed and its concentration in bone rises and falls in response to serum levels. Strontium laid down in teeth, however, is a permanent and undesirable acquisition. It has been reported that sodium alginate binds and eliminates strontium without inhibiting absorption of calcium.

Lead poisoning used to be frequent in industry but is now largely limited to pica in children. We all carry lead in our bodies and accumulate it until late middle age, after which we excrete more than we absorb.

Medicinal metal poisoning is still frequent, some through accidental ingestion and some of it occurring as a result of the calculated risk of justifiable treatment with toxic substances. Antimony, arsenic, barium, bismuth, boron, gold and mercury have all been involved.

Most of us get enough of the essential trace metals to meet physiologic needs, and few of us are exposed to them in toxic amounts. In deficiency states, supplementation with certain minerals (and vitamins) is advisable if not mandatory, but in restricted amounts in order to guard against upsetting the delicate balance in their relationship.

Deficiency of copper is rare, and hypercupremia is as undesirable as hypocupremia; but many authorities recommend adding this highly essential element when the diet requires fortifying. The same considerations apply to cobalt, manganese and molybdenum. Recent findings indicate that zinc should be added in areas where natural sources are inadequate. Molybdenum and other metals can be relied upon to counter any tendency zinc may have to aggravate caries. Up to the present, vanadium has not figured in vitamin-mineral preparations. Since it is deficient in many localities, it should perhaps be included for its benefits on the circulation and on the earth.

It has been wisely stated that there are no harmless substances—only harmless ways of using them.

(Reprinted with minor alterations from Pfizer *Spectrum,* by courtesy of the editor.)

References

Alexander, G. V., and Nusbaum, R. E., 1962, "Zinc in bone," *Nature,* **195,** 903.
Asling, C. W., and Hurley, L. S., 1963, "The influence of trace elements on the skeleton," *Clin. Orthop.,* **27,** 213–264.
Butt, E. M., and Nusbaum, R. E., 1962, "Trace metals and disease," *Ann. Rev. Med.,* **13,** 471–480.
Claton, B. E., 1963, "Daily requirements of trace metals and vitamins," *Nutr. Dieta,* **6,** 334–338.
Curran, G. L., 1960, "Vanadium and the Biosynthesis of Cholesterol and Other Lipis," Hahnemann Symposium on Metal Binding in Medicine, Philadelphia, pp. 216–218.
Gunn, S. A., Gould, T. C., and Anderson, W. A., 1963, "The selective injurious response of testicular and epididymal blood vessels to cadmium and its prevention by zinc," *Am. J. Pathol.,* **42,** 685–702.
Halstad, J. A., and Prasad, A. S., 1963, "Zinc deficiency in man," *Israel Med. J.,* **22,** 307–315.
Maynard, L. A., 1962, "Trace elements," *N.Y. J. Med.,* **62,** 544–547.
Morrison, A. B., and Campbell, J. A., 1963, "Trace elements in human nutrition," *Can. Med. Assoc. J.,* **88,** 523–527.
Prasad, A. S., *et al.,* 1961, "Syndrome of iron deficiency anemia, hepatosplenomegaly, hypogonadism, dwarfism and geophagia," *Amer. J. Med.,* **31,** 532–546.
Schroeder, H. A., 1960, "Possible Relationships between Trace Metals and Chronic Diseases," Hahnemann Symposium on Metal Binding in Medicine, Philadelphia, pp. 59–67.
Schroeder, H. A., 1961, "Hardness of local water-supplies and mortality from cardiovascular disease," *Lancet,* **1,** 1171.
Schroeder, H. A., and Balassa, J. J., 1961, "Abnormal trace elements in man; Cadmium," *J. Chronic Diseases,* **14,** 236–258 (August); *Ibid.,* "Lead," 408–425 (October).
Schroeder, H. A., Balassa, J. J., and Tipton, I. H., 1962, "Abnormal Trace Metals in Man; Nickel," *J. Chronic Diseases,* **15,** 51–65 (June); *Ibid.,* 1963, "Titanium," **16,** 55–69 (January); *Ibid.,* "Vanadium," **16,** 1047–1071 (October).
Schutte, K. H., 1964, "The Biology of the Trace Elements; Their Role in Nutrition," London, Lippincott.
Strain, W. H., 1961, "Effects of Some Minor Elements on Animals and People," AAAS Symposium on Geochemical Evolution, Denver.
Tipton, I. H., and Cook, M. J., 1963, "Trace elements in human tissue. II. Adult subjects from the U.S." *Health Phys.,* **9,** 103–145.

Cross-references: *Medical Geography; Medical Geology; Medicinal Springs; Mineral Particles and Human Disease; Neutron Activation: Analysis; Sea Water: Chemistry; Trace Elements: Geochemistry; Trace Elements in Plants.* See also *Individual Elements.*

MEDICINAL SPRINGS

The exploitation of medicinal or mineral springs with curative properties has long attracted attention, particularly in Central Europe during the last century.

The term "spa" for a mineral spring or curative resort (taking its name from the Belgian village of Spa, near Liege) has been in the literature since the sixteenth century.

Balneo-geohydrology is a small and relatively new segment of geohydrology specializing in finding water sources characterized by unusual chemical composition and physical properties which can be exploited for medicinal purposes. Determination of the geological formations through which such water flows or is stored and evaluation of peloids found in some regions, such as peats, fango or muds, are other tasks of this branch of geology. (In medical terminology, peloids are natural substances formed by a geological and/or biological process; they are classified according to origin, chemical composition and radioactivity. Fango is a type of peloid that is characterized by high-inorganic content and low organic content, extensively used as a therapeutic mud in Italy.)

Mineral waters are distinguished from potable (drinking) waters by a number of factors: *(1)* higher contents of mineral matter, *(2)* presence of gas, *(3)* presence of trace elements, *(4)* higher temperature, and *(5)* radioactivity. Various amounts and combinations of these factors can be present. As with other kinds of water, most mineral waters originate in the terrestrial water cycle. On its long upward journey from great depths, water leaches and dissolves many minerals present in the geological formations with which it comes in contact. Such mineral waters are called *vadose.* Primary volatiles from the earth's mantle may sometimes become mixed with it. This apparently newly formed water is known as *juvenile.* Both vadose and juvenile waters may become heated by magmatic or geochemical reactions and emerge as hot springs.

Thermal springs producing water of 20°C (68°F) to 37°C (98.6°F) are classified as warm and those discharging water above body temperature of 37°C are classified as hot springs. Some thermal springs contain relatively large amounts of dissolved mineral matter while others are poorly mineralized. Both may be radioactive to various degrees. It remains still unproved whether there are any mineral springs producing solely juvenile water. The temperature of water changes with the seasonal fluctuation of soil and air temperatures. The chemical composition of mineral springs is by no means constant as the concentration of single elements may vary for a number of reasons.

Mineral water appears on the surface of the earth by way of natural springs, or it may be obtained artificially by sinking of deep wells. Drilling of wells is undertaken to supplement the occasionally insufficient quantity of water from existing mineral springs or to procure it in regions where no natural springs are present. In both instances, location of mineral water depends on a thorough geological survey prior to the contemplated drilling. On the other hand, temperature, radioactivity and the kind of minerals present in water can give important clues to the geological formations through which it passes.

The concentration of minerals in water is expressed in milligrams per liter or parts per million. In older analyses, the term grains per gallon was used. The composition of water is reported in form of ions as cations (+) and anions (−). The undissociated combinations, e.g., silicic acid (H_2SiO_3) are listed separately. The following cations are frequently present: calcium (Ca^{2+}), magnesium (Mg^{2+}), sodium (Na^+), iron (Fe^{2+}), lithium (Li^+), aluminum (Al^{3+}) and potassium (K^+); fluoride (F^-), iodide (I^-), bromide (Br^-), chloride (Cl^-), bicarbonates (HCO_3^-) and sulfates (SO_4^{2-}) appear as anions. Oxygen, carbon dioxide, hydrogen sulfide and radon are the usual gaseous constituents. In some localities, thermal springs carry radioactive elements in relatively high concentrations.

In addition to the dissolved mineral substances, gases and radioactive elements, temperature, taste, color, odor, turbidity and pH determine the medicinal usefulness of mineral water. Spectrographic examinations of mineral waters revealed the presence of many elements in extremely low amounts. The following oligoelements—copper, zinc, cobalt, molybdenum, boron, selenium, chromium and vanadium—seem to be of special therapeutic significance. However, the essentiality, the function and the physiological role of these elements are still not fully understood (see *Trace Elements*). The fact that some microelements ingested accidentally with food or water are deposited in animal and human tissues without having any noticeable effect or function makes this biological issue even more puzzling.

The "safe levels" of radioactive constituents such as uranium, radium, thorium, radon, thoron, tritium, carbon 14 and potassium ($K^{39,40,41}$) in water suitable for internal or external use were tentatively established by health authorities and balneological organizations in Europe. The widely fluctuating and divergent views in regard to the safety and effectiveness of low doses of radiation prevents a uniform agreement. Thus, considerably different levels are quoted in pertinent literature.

Only water containing more than 1 g/kg of dissolved minerals is, in some European countries, classified as mineral water. Although a great many varieties of these waters are extensively used, some of lesser mineralization (the acratopegs) also enjoy a great popularity, probably due to the abundance of trace elements. According to the type, temperature and potability, mineral waters and peloids are employed therapeutically in many parts of the world. Chronic disorders of the integument, the musculoskeletal, gastrointestinal, biliary, respiratory and cardiovascular systems are the most common indications for spa therapy.

Since time immemorial, mineral springs were held in high esteem by all races and strata of the population. Mineral water is probably the only medicinal medium which has outlasted millennia of progress and is today, in spite of vast advances in the healing arts, in even greater demand than before. Keen observations extending over many centuries established empirically the therapeutic specificity of certain types of mineral waters. Depending on temperature, mineral contents, gases, taste and purity (bacteriological safety), mineral water is being used in Europe, Asia, Central and South America very successfully for bathing, drinking, inhalation, gargling, internal irrigation and for filling of therapeutic pools. Peloids are in great demand for treatment of arthritis, rheumatism and allied conditions and are dispensed as hot mud baths or in the form of local applications. Selected kinds of pure or naturally carbonated mineral waters are bottled and widely distributed as table water.

Certain kinds of peloids and mineral waters are important natural assets. We feel that the pinpointing of their exact location and determination of their composition could be of great value in the future. It is certainly not too far fetched to say that earth sciences may help medicine to rediscover an excellent, but neglected, form of therapy.

IGHO HART KORNBLUEH

MEDICINAL SPRINGS

References

Amelung, A., and Evers, A., (editors), 1962, "Handbuch der Baeder und Klimaheilkunde," Stuttgart, W. K. Schattauer Verlag.

Archives of Medical Hydrology, Pisa, Italy, Nistri-Lischi Publishers (official journal of the International Society of Medical Hydrology and Climatology).

"Cyclopedia of Medicine, Surgery, Specialties," 1959, Philadelphia, Pa., F. A. Davis.

Fundamenta Balneo-Bioclimatologica, Stuttgart, F. K. Schattauer Verlag (periodical in German).

Kuenen, P. H., 1955, "Realms of Water," New York and London, John Wiley & Sons.

Kurashova, C. V., et al., 1962, "Kurorti U.S.S.R.," Moscow, State Publishing Office for Medical Literature (in Russian).

Kurortologia (Physical Therapy and Corrective Physical Culture, Transactions), Moscow, Ministry of Health of U.S.S.R. (periodical in Russian).

Licht, Sidney (editor), 1963," Medical Hydrology," New Haven, Conn., Elizabeth Licht Publ.

Therme (Review of European Thermalism), Italy Castrocaro Terme (Forli) (periodical).

U.S. Department of Agriculture, 1955, "Water," *The Yearbook of Agriculture*, Washington, D.C.

Zeitschrift für angewandte Bäder- und Klimaheilkunde, Stuttgart, F. K. Schattauer Verlag.

Cross-references: Hydrology; Hydrology, Limestone Terrain; Juvenile Water; Medical Geography; Medical Geology; pH-Eh; Springs; Trace Elements—Geochemistry; Vadose Water. See also *Individual Elements*.

MEMBRANES—See CLAY MEMBRANE PHENOMENA

MERCURY: ELEMENT AND GEOCHEMISTRY

Mercury (Hg) ranks among the first metals discovered by Man (perhaps as early as 1500 BC) and many references can be found in ancient authors.

The principal physical constants of Hg are:

Atomic number	80
Atomic weight	200.59
Temperature of melting	$-38.87°C$
Heat of fusion (15°C)	2.82 cal/g
Temperature of vaporization	356.58°C
Heat of vaporization (20°C)	73.27 cal/g
Vapor tension (20°C)	12×10^{-4} mm Hg
Density (20°C)	13.5462
Electronic configuration [Xe core]	$4f^{14}5d^{10}6s^2$

Mineralogy

The mineralogy of Hg is comparatively simple. Cinnabar is the only common mineral. It can be primary or an alteration product of more complicated basic metal sulfides containing Hg. The other minerals are mineralogical curiosities, among which the oxysalts reveal arid climatic conditions. Native mercury is frequently an oxidation product of cinnabar.

Some of the general features for mercury minerals are: (*a*) bright color (yellow and red); (*b*) hardness less than 3.5 in the Mohs' scale; (*c*) high refractive index (> 2) trans-

TABLE 1. MERCURY MINERALS[a]

Name	Formula	Symmetry	Typical Occurrence
Native Hg	Hg	—	Almaden, New Almaden, Idria
Cinnabar	Hg S	Rhombh.	Frequent
Metacinnabar	Hg S	Cubic	Idria, New Almaden
Tiemannite	Hg Se	Cubic	Marysville (Utah)
Coloradoite	Hg Te	Cubic	Keystone (Colo.)
Livingstonite	$HgSb_4S_7$	Monocl.	Huitzuco and Guadalcazar, (Mex.)
Moschellandsbergite	Hg_3Ag_x	Cubic	Moschellandsberg (Germany) Challanches (France)
Arquerite	$HgAg_x$	Cubic	Arqueros (Chile)
Kongsbergite	HgAg	Cubic	Kongsberg (Norway)
Potarite	Hg_2Pd_3	Cubic	Potaro River (Brit. Guiana)
Amalgam s. s.	Hg_3Au_2	Cubic	Chanarcillo (Chile)
Calomel	Hg_2Cl_2	Tetr.	Almadenejos (Spain)
Coccinite	$Hg I_2$	Tetr. ?	Casas Viejas (Mexico)
Montroydite	HgO	Orthor.	Terlingua (Texas)
Magnolite	Hg_2TeO_4	?	Keystone (Colorado)
Eglestonite	Hg_2OCl_2	Cubic	Terlingua (Texas)
Kleinite	$Hg_2N(Cl, SO_4) \cdot xH_2O$	Hexag.	Terlingua (Texas)
Terlinguaite	Hg_2OCl	Monocl.	Terlingua (Texas)
Mosesite	$Hg_6(NH_3)_2Cl_2(SO_4)(OH)_4$	Cubic	Terlingua (Texas)
Schuetteite	$Hg SO_4 \cdot 2H_2O$	—	Terlingua (Texas)

[a] See Dana's *System of Mineralogy* or any other Handbook of Mineralogy.

TABLE 2. RADII OF HG AND RELATED METALS[a]

Element	Zn	Ag	Cd	Hg	Tl	Pb	Bi
Metallic	1.37	1.44	1.52	1.55	1.71	1.74	1.82
Tetrahedral	1.31	1.53	1.48	1.48	1.47	1.46	1.46
Ionic charge	2+	1+	2+	2+	1+	2+	3+
Ionic radius	0.83	1.13	1.03	1.12	1.49	1.32	1.1

[a] From Goldschmidt, 1954, p. 275.

parency, and high birefringence (if not isometric).

Traces of mercury may occur in the following minerals: tetrahedrite and tennantite (they are then called schwazite and hermesite), penroseite, argyrodite, clausthalite, altaite, and sphalerite.

Geochemistry

Hg is *chalcophile*. Table 2 shows the great similarity between Hg and Cd, due to the lanthanide contraction.

"It is likely therefore that the geochemistry of mercury will, in many respects, be analogous to that of cadmium, excepting such differences as must result from differences in ionization energy of the elements and the free energy of formation of their compounds" (Goldschmidt, 1954, p. 275). Saukov (1946) points to some other peculiarity: the double bond of monovalent Hg. Calomel is to be written: Cl–Hg–Hg–Cl, or Hg_2Cl_2.

Hg has a high ionization potential (I.P.) and therefore a strong tendency to remain in the native state: 1st I.P., 10.39 eV (Au, 9.20; Ag, 7.53); 2nd I.P., 18.65 eV.

The clarke of Hg is still uncertain:

Earth's crust:
Clarke and Washington (1924) O.n ppm
Fersman (1939) 0.05
Vinogradov (1949) 0.077

Universe:
Mason (1966) 0.3 atom for 10^6 atoms Si

The figures of Ehmann and Lovering tend to lower by about a half the clarke of the crust given by Vinogradov, but to raise the overall figure for the earth, since the mantle would have a higher content.

Isotopes. Dibeler (1955) determined the isotopic composition of Hg as follows:

196	0.156%	201	13.27%
198	10.12	202	29.64
199	16.99	204	6.79
200	23.07		100.36

Deviations from these figures have been explained as follows: *(1)* regional differences in isotopic composition, *(2)* radioactive origin of some isotopes, and *(3)* fractionation.

Geochemical Cycle. Berg stated in 1927: ". . . mercury appears in the geochemical cycle at the end of the hydrothermal phase like a comet, and disappears very soon, becoming undiscoverable." Meanwhile only the adsorption on clay has been discovered. According to Saukov (1946) the transportation of Hg as a sulfate or as a chloride seems possible (see Fig. 1).

Cosmochemistry. In the solar atmosphere Hg is unknown. I. and W. Noddack (1931) report 0.2 ppm Hg in troilite from meteoric iron (Cañon Diablo) and less than 0.01 ppm in chondrites (1934). Carbonaceous chondrites disclosed from 20 to 114 ppm Hg (Ehmann and Lovering). Loss of Hg, through heating, may have been suffered by meteorites on entering the terrestrial atmosphere.

Igneous Rocks. The concentrations of mercury found in various igneous rocks are shown below.

Magmatic rocks (Berg, 1927)	0.027 ppm
(Clarke and Washington, 1927)	0.1/1
(Noddack, 1934)	0.02
(Goldschmidt, 1938)	0.1
Granite (Stock and Cucuel, 1934)	0.058
Granite (Preuss, 1941)	0.01
Acidic intrusives (Saukov, 1946)	0.064
Gabbro (Stock and Cucuel, 1934)	0.079
Gabbro, Preuss, 1941)	0.1
Basalt, Kaiserstuhl (Stock and Cucuel, 1934)	0.1
Basic effusives (Saukov, 1946)	0.09

More recent figures (Vinogradov, Turekian, and Wedepohl, etc.) are of the same order of magnitude, but results obtained by neutron activation analysis disclose much lower values, which decrease from acid to ultrabasic rocks (Ehmann and Lovering): U.S.G.S. standard G-2, granite—0.039 ppm; DTS 1, dunite—0.004 ppm.

Especially noteworthy are the concentrations found by these authors in deep-seated pipes (kimberlite, 0.20 ppm) and especially their various inclusions (up to 1.23 ppm).

In regions of Hg mineralization (Crimea, Bulkin) the clarke may reach 1 to 20 ppm.

Hydrothermal Deposits. Hg appears markedly in "epithermal" deposits, although it also

MERCURY: ELEMENT AND GEOCHEMISTRY

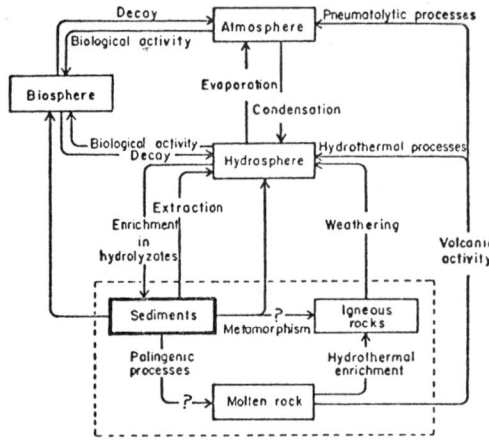

Fig. 1. The cycle of mercury (from Rankania and Sahama, 1950).

appears as trace material in mesothermal sulfides (see *Mineralogy*). Cinnabar is known to deposit in hot springs (Steamboat Springs and Sulfur Banks, e.g.), where it is supposedly carried as a sulfide complex. The genesis of cinnabar deposits is usually explained by a similar process, but these deposits may be syngenetic or epigenetic.

Sedimentary Rocks. It has been shown by Heide and Böhm, G. (1957) that Hg is concentrated in hydrolysates by adsorption. General values are:

Composite of 36 Paleozoic marine shales (Stock and Cucuel, 1934)	0.51 ppm
Later analyses of shales (the same)	0.5
Shale, Trias (Heide and Böhm, 1957)	0.187
Chinle shale (Lausen, 1936)	40
Sandstone (Stock and Cucuel, 1934)	0.033
Sandstone, average (Preuss, 1941)	0.1
Limestone (Stock and Cucuel, 1934)	0.033
Limestone, Muschelkalk (Heide and Böhm, 1957)	0.048

Seawater of the Japanese Sea contains 0.002 ppm Hg, of which 90% is fixed by organisms. The content increases with depth (Hosohara).

Soils. Cinnabar is not easily altered and may be found as a detrital mineral. Its brittleness, however, makes it disappear after short transportation (a few km). Mercury in soil has been determined by Stock and Cucuel (1934) as follows:

Humic horizon of forest soils	$3-8 \times 10^{-2}$ ppm
Forest soils	$10-29 \times 10^{-2}$
Cultivated soils	$3-7 \times 10^{-2}$
Clay soils	$3-3.4 \times 10^{-2}$
Sand	$0.1-2.9 \times 10^{-2}$

In accordance with these figures, the concentration of Hg in the "A" horizon has been found in U.S.S.R. (Dvornikov, et al., 1963).

Atmosphere. The earth's atmosphere contains relatively large amounts of Hg, provided by metallurgical industrial processes, natural evaporation, volcanic eruptions, and coal smoke. Elimination takes place through rain and adsorption on soil. Stock and Cucuel found the following concentrations: air, 2×10^{-8} g/m^3; rainwater, 2×10^{-4} ppm; spring water, 2×10^{-5} ppm.

Biosphere. Enrichment in Hg can be observed in marine animals and algae. The biogeochemistry of Hg is still unknown.

Geochemical Prospecting. *Plants.* Concentration of Hg takes place in *Ledum decumbens* and *Holostium umbellatum*.

Dispersion Patterns (aureoles or coronas). Dispersion aureoles are useful in prospecting. They may either be primary or secondary. The latter may be the result of a mechanical dispersion agent or a dissemination of Hg in a gaseous or dissolved state. This prospecting method can also be used for other metal deposits giving off small amounts of Hg (Hawkes and Williston).

Economy

Uses. According to the U.S. Bureau of Mines Minerals Yearbook the total consumption for the U.S. has been of 65,301 flasks for 1962 (Table 3).

Price and Market. Mercury is regularly quoted at New York and London as follows:

Average quotation N.Y. (1962)	$191.21
Average quotation N.Y. (1965)	$570.75
Lowest since 1910 (1911)	$ 40.07
Highest since 1910 (1965)	$775.−

TABLE 3. MAJOR USES OF MERCURY

Uses	Amount Used (%)
Electrical apparatus	17.1
Redistilled	13.8
Electrolytic preparation of Cl_2 and NaOH	11.2
Industrial and control instruments	7.9
Agriculture (fungicide, bactericide)	6.5
Pharmaceuticals	5.2
Other	31.1
Total	*100.0*

TABLE 4. WORLD PRODUCTION OF MERCURY

Country	Flasks of 76 lb	Percent
North America		
Canada	20	—
Mexico	18,000	6.5
United States	40,000	14.5
South America		
Bolivia	30	—
Chile	370	0.1
Colombia	3	—
Peru	3,280	1.2
Europe		
Czechoslovakia	725	0.3
Italy	57,291	20.8
Rumania	200	—
Spain	82,760	30.1
U.S.S.R.	40,000	14.5
Yugoslavia	16,419	6.0
Africa		
Tunisia	174	—
Asia		
China	26,000	9.5
Japan	4,820	1.8
Philippines	2,500	0.9
Turkey	2,620	1.0
World total (estimate)	275,000	99.8

Hg is sold in iron flasks containing 76 lb of metal and trading is usually done in lots of 25 flasks (Strauss).

Metallurgy. Metallurgy of Hg is very simple, because cinnabar is the only important mineral and it usually appears alone. The metallurgical process is an oxidation followed by distillation. It is necessary to keep the highly toxic Hg vapors from escaping into the air.

World Production. Preliminary figures for 1965 (Minerals Yearbook) are given in Table 4.

Mercury Ore Deposits

Europe. *Italy.* The important mines are around the "Monte Amiata" (Tuscany), an extinct tertiary volcano. Hg appears along the contact between Mesozoic and Cenozoic limestone and shales, especially where the rocks have been fractured and altered. Antimony (Sb) occurs in the same region, but often separately.

Spain. The Almadén (Ciudad Real) mines exploit two, almost vertical, quartzite seams, that are impregnated with cinnabar. According to Saupé, mineralization is not related to tectonics and is anterior to the diagenesis of the quartzite. A connection with Silurian volcanism is probable. These seams are of Silurian age and alternate with bituminous slates. Beck (1909) and Ransome (1921) showed a partial substitution of quartzite by cinnabar, which in fact is due to remobilization. Other minerals are pyrite, baryte, and dolomite. The reserves seem to be important (average grade 3% ?).

At Mieres (Asturias) cinnabar occurs, together with arsenic (As), in a Carboniferous limestone.

Yugoslavia. The Idria deposit occurs in a thrusted series (Carboniferous and Triassic schists, sandstones, and dolomites), above the thrust plain. All rocks are intensively fractured and impregnated with cinnabar, which locally forms veinlets. Bitumen, pyrite, and chalcopyrite are abundant. Berce (1958) showed a relationship to early geosynclinal volcanism, which is not immediately apparent, the volcanic rocks having been removed by thrusting.

Russia. At Nikitowka (Donetz Basin), cinnabar occurs in a Carboniferous sandstone bed, dipping at 50° below shales.

Other deposits are recorded in the Caucasus and in the Fergana in silicified limestone. Features of the Altai deposits are a strong tectonic control and consanguinity with Pb, Zn, Cu mineralization. Hg occurs also in the Maritime Provinces. About 200 prospects are known, and U.S.S.R. is presently exporting Hg.

North America. *United States.* (a) New Almadén (California). The ore occurs as stockworks in silicified and fractured serpentines and metamorphic rocks, along two deep fractures, which were the channel ways for ascending solutions. Together with cinnabar occur pyrite, opaline, and calcite. "On the whole, the ore then lies underneath an impervious capping and practically on the contact with the intrusive serpentine" (Lindgren, 1933).

(b) New Idria (California). "The ore lies along a clayey streak separating sandstones . . . from the serpentinized rocks of the hanging wall. The ores appear in three forms—as normal veins, as irregular stockworks, and as impregnations in sandstone. There are no effusive rocks in the vicinity." (Lindgren, 1933). Metacinnabar is frequent.

(c) Terlingua (Texas). Yates and Thompson (1959) distinguished several types of ore bodies according to decreasing economic interest: (*1*) at limestone-clay contacts, (*2*) connected with karstic phenomena, (*3*) in breccia pipes, and (*4*) in igneous rocks, ranging from rhyolite to syenogabbro. Minerals are cinnabar, chalcedony, gypsum, aragonite, pyrite, and the unique oxysalts (see last group of Table 1).

(d) Red Devil (Alaska). "The Red Devil mine is located on the western limb of an

Upper Cretaceous anticline, that has been subjected to several stages of faulting" (Lyman, 1961). Two perpendicular andesite dikes are the foci of mineralization. About 25 ore shoots are located in ancient openings, due to small faults. The ore is complex: the stibnite-realgar-orpiment content exceeds the cinnabar content.

Mexico. The Huitzuco deposit lies in a strongly fractured limestone crossed by intrusive granite. The limestone is partially metasomatized by gypsum. Minerals are livingstonite, cinnabar, metacinnabar, antimonite, and pyrite.

At Guadalcazar the minerals are the same with some calcite and fluorine, the latter being scarce in Hg deposits.

South America. *Peru.* The Huancavelica deposit is located along a north-south trending anticline cut by Tertiary andesites and basalts. Porous sandstones and fissured limestones are impregnated with cinnabar, pyrite, mispickel, realgar, calcite, baryte, quartz, and bitumen.

Chile and Colombia. Production comes from "Los Mantos" (near Punataqui, Chile) and "La Esperanza" (Colombia).

Asia. *China.* The province of Kveichow, with the surrounding provinces of Szechuan, Hunan, and Kwangsi is an old-time production center for Hg. Connection between fracturation, eruptive rocks, and mineralization is strong. Preferred loci are fractured anticlines. In the same region appear Hg and Sb. Host rocks are Paleozoic limestones and shales (Veichow Juan, 1946).

Japan, Philippines, and Turkey. The important mines of these countries are: Japan: Itomuka and Yamato; Philippines: Palawan Island; Turkey: Izmir and Kutahaya provinces.

Genesis of Hg Deposits. It is no longer true that "the uniform character of the quicksilver deposits points to a common genesis for all of them" (Lindgren, 1933). According to Barnes, Romberger, and Stemprok (1967), "disulfide and sulfide complexes of Hg are potential ore carriers." Deposition could take place through any mechanism that lowers the stability of these complexes, e.g., decrease in temperature or pH, oxidation, or dilution.

World Distribution and Geotectonics. The described Hg deposits belong to either one of the two following belts: (1) Circumpacific or (2) Mediterranean-Himalayan. The confirmation of this distribution is given by the presence of numerous smaller, unmentioned prospects and mines in these belts, and an almost definite absence of Hg outside of them.

The geotectonic approach has been studied by recent Russian geologists. Khain, (1962) showed localization of epithermal Hg deposits, along with As and Sb, in miogeosynclines. Kreiter et al. (1963) found favorable geologic environment for Hg in two types of geosynclinal regions: (1) "regions characterized by sedimentary rocks with local minor intrusions in geosynclines of carbonate-terrigenous type"; (2) "regions located in foredeeps and intermontane basins of volcanic type" containing gold and silver tellurides and selenides, together with Hg, As and B.

Maucher (1965) supposes the existence of a "lower paleozoïc Sb–Hg–W formation related with lineaments parallel to the borders of old continental blocks, which is not connected to an orogenic cycle."

FRANCIS R. SAUPÉ

References

Bailey, E. H., and Smith, R. M., 1964, "Mercury —its occurrence and economic trends," *U.S. Geol. Surv. Circ.* **496.**

Barnes, H. L., Romberger, S. B. and Stemprock, M., 1967, "Ore solution chemistry II. Solubility of HgS in sulfide solutions," *Econ. Geol.,* **62,** 957–982.

Behrend, F., and Berg, G., 1927, "Chemische Geologie," Stuttgart, Enke Verlag, 595pp.

Clarke, F. W., and Washington, H. S., 1924, "The composition of the Earth's crust," *U.S. Geol. Surv. Profess. Paper* **127.**

Engel, G. T., 1965, "Mercury," *U.S. Bur. Mines Minerals Yearbook.*

Fersman, A. E., 1939, "Geochemical methods of prospecting," *U.S. Geol. Surv. Circ.* **127** (transl. from the Russian).

*Goldschmidt, V. M., 1954, "Geochemistry," Oxford, Clarendon Press, 730pp.

Lindgren, W., 1933, "Mineral Deposits," New York, McGraw-Hill Book Co., 930pp.

Mason, B., 1966, "Principles of Geochemistry," Third ed., New York, John Wiley & Sons, 329pp.

*Maucher, A., 1965, "Die Antimon-Wolfram-Quecksilber-Formation und ihre Beziehungen zu Magmatismus und Geotektonik," *Freiberger Forschungsh.* [C]**186,** 173–188.

Preuss, E., 1941, "Beiträge zur spektralanalytischen Methodik, Pt. 2," *Z. Angew. Mineral.,* **3,** 8.

Rankama, K., and Sahama, T., 1950, "Geochemistry," Chicago, Univ. Chicago Press, 912pp.

Reed, G. W., Jr., and Jovanovic, S., 1967, "Mercury in Chondrites," *J. Geophys. Res.,* **72**(8), 2219–2228.

Saukov, A. A., 1946, "Geochemistry of Mercury," *Tr. Inst. Geol. Nauk SSSR,* **78**(17), 129pp.

Schuette, C. N., 1931, "Occurrence of quicksilver ore bodies," *Trans. A.I.M.E. Gen. Vol.,* 403–488.

Stock, A., and Cucuel, F., 1934, "Die Verbreitung des Quecksilber," *Naturwissenschaften,* **22,** 390–393.

Warren, H. V., Delavault, R. E., and Barakso, J., 1966, "Some observations on the geochemistry of mercury as applied to prospecting," *Econ. Geol.,* **61**(6), 1010–1028.

* Additional bibliographic references mentioned in the text may be found in this work.

Cross-references: *Cadmium; Elements: Planetary Abundances and Distribution; Geochemistry: Ionization Potentials; Hydrothermal Solutions—Sulfide Transport; Isotope Fractionation; Mineralogy; Neutron Activation Analysis.* Vol. II: *Meteorites.* Vol. IVB: *Cinnabar; Geochemical Prospecting.*

METALIMNION—See THERMOCLINE

METALS, METALLURGY, METALLOGENESIS—See Vol. IV B

METAMORPHIC ENVIRONMENTS—CHEMICAL MOBILITY

Interpretations of the extent of chemical mobility in metamorphic environments have varied greatly. Zones of *regional metamorphism* (See Vol. V) have been interpreted both (*1*) as the result of isochemical metamorphism with diffusion of material limited to a few mm (Harker, 1939) and (*2*) as the result of regional metasomatism with large-scale bulk transfer of material (Read, 1957). Much geochemical data now suggests that compositions approximating to original compositions can be established for many or most rocks of low metamorphic grade (or facies) and for some of high metamorphic grade. But for some high-grade rocks, chemical mobility has been a significant factor, particularly in the formation of gneisses and migmatites (See Vol. V, *migmatization*) and the progressive development of granitic magma (Read, 1957; Mehnert, 1968). There is considerable debate concerning the mechanism(s) of migration of material and whether, on a large scale, this migration has been related to significant addition and removal of some elements, or to redistribution of elements.

Apart from water and other volatiles, metamorphism adjacent to igneous intrusions is generally isochemical with mineralogical readjustment associated with increase in temperature (See Vol. V, *Aureoles*). In some cases constituents from the magma have invaded rocks adjacent to the intrusion (pneumatolytic metamorphism) with the formation of minerals containing boron (B), fluorine (F), chlorine (Cl), sulfur (S), and sodium (Na) (Harker, 1939). Reaction between magma and limestone produces skarns composed of lime-silicate minerals together with minerals of iron, zinc, lead, copper (sulfides), and tungsten which, in places, are in concentrations of economic value. Suggestions that large layered ultrabasic/basic masses, such as the Bushveld complex of the Transvaal (See Vol. VIII, *Geology of South Africa*), were formed by metasomatic transfusion of a sedimentary sequence are treated with caution.

Many mechanisms have been proposed to explain the migration of material in metamorphic environments. *Solid diffusion* of ions has been demonstrated, on a small scale, in laboratory experiments and used by Lapadu-Hargues (reference in Read, 1957) to explain variations in chemical composition of metamorphic rocks with depth, the mobility of ions being a function of their size. Most mechanisms involve a fluid phase capable of moving through rocks (Ramberg, 1952), particularly along shear planes in foliated rocks. In metamorphic environments this pore solution usually bears little resemblance to the initial pore solution in sediments, the differences reflecting interaction of sediments with their interstitial solutions during diagenesis and metamorphism (Helgeson, 1967). The process of *concretion* is operative in the development of individual minerals. This involves *enrichment in stable constituents,* and its converse, the *extraction and redeposition of the most soluble substances (secretion).* These two, with variations in solubility due to heterogeneous pressure playing an important role, accentuate differences in mechanical properties of rocks and lead to the formation of banded rocks by *metamorphic differentiation* (Ramberg, 1952; Barth, 1962). The experimental evidence of Winkler and von Platen (references in Turner, 1968) indicates that *partial melting (anatexis)* is a mechanism in the formation of migmatites and granitic magma. Metamorphic differentiation caused by gravitation is significant in the formation of planetary structures but has no obvious role in metamorphic petrogenesis (Barth, 1962).

As metamorphic activity over large areas is associated with heat flow in the crust, chemical mobility will be expressed as upward, as well as lateral, movement of material, something that is universally accepted in the case of volatile constituents. Accordingly rocks of high *T-P* facies are considered before those of lower *T-P* facies, corresponding, generally, to an upwards and outwards progression in the crust (Fig. 1). This is the reverse of that generally adopted in mineralogical and fabric studies of progressive metamorphism (see Vol. V, *Metamorphism*) and means that those rocks for which the original composition is the most difficult to determine, are the first to be considered.

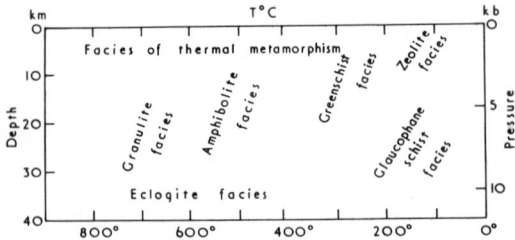

FIG. 1. Metamorphic facies and crustal conditions.

Eclogite Facies

The real significance of rocks of the eclogite facies is still enigmatic, but the consistent compositional similarity of rocks of this facies with those of basic igneous rocks (Turner, 1968) suggests that metamorphic reconstitution was isochemical, or largely so, volatile constituents excluded.

Granulite Facies

Some coarse-grained rocks with granulite facies mineral assemblages have compositions generally similar to those of pelitic, calcareous, and siliceous sediments. Other granulite facies rocks show variations in rock and mineral geochemistry consistent with that shown by an igneous suite (Fig. 2). A volcanic assemblage of basalts, andesites and rhyodacites is the likely original nature of other banded granulites which show structural characteristics of a supracrustal assemblage and have small intercalations of distinctive metasediments (Khoury, see reference in Bowes, 1969). Compared with little-altered Archean lava flows, which are poorer in K_2O and richer in H_2O than post-Archean equivalents, these Archean granulites of the Kylesku group of Scotland are low in K_2O, as well as H_2O— features of granulite facies rocks in general (Ramberg, 1952)—but otherwise generally similar (Fig. 3). Thus the available evidence suggests that, to a large extent, granulite facies metamorphism is isochemical but with a migration of potassium and water into rocks undergoing amphibolite facies metamorphism. Upward migration of cesium (Cs), thallium (Tl), rubidium (Rb), lead (Pb), as well as the radioactive elements uranium (U) and thorium (Th) [and potassium (K)] is also likely (Heier, 1965), while redistribution associated with the development of more mafic and more felsic bands (and to and from hinge zones and limbs of folds) involves limited complementary migration of titanium (Ti), iron (Fe), manganese (Mn), magnesium (Mg), and aluminum (Al), sodium (Na), potassium (K). The presence of small segregations and veins of pegmatitic material represents the very limited quantity of partial melt material which is one of the products of dehydration of semipelitic-pelitic sedimentary and acid-intermediate igneous assemblages under granulite facies conditions

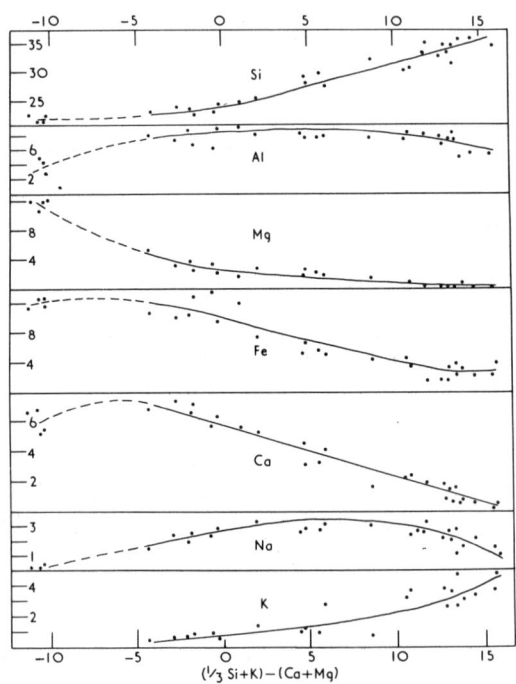

FIG. 2. Variation diagram for granulites of Madras (after Howie; reference in Turner, 1968).

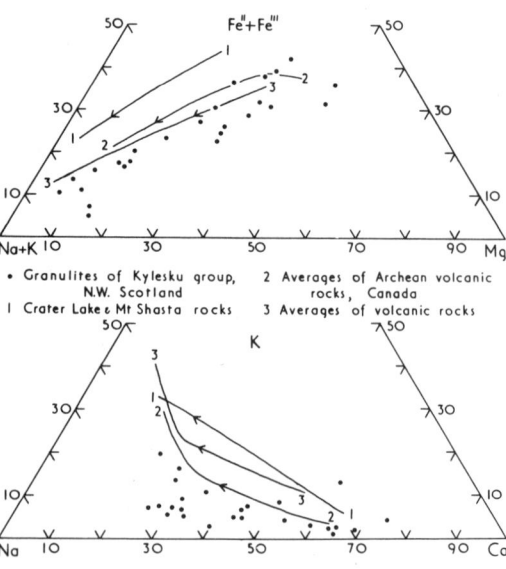

FIG. 3a. Cation percent plots of granulites from Kylesku group of northwest Scotland: comparative trends after (1) Nockolds and Allen (references in Evans and Leake, 1960); (2) Goodwin, 1968; and (3) Daly (reference in Barth, 1962).

METAMORPHIC ENVIRONMENTS—CHEMICAL MOBILITY

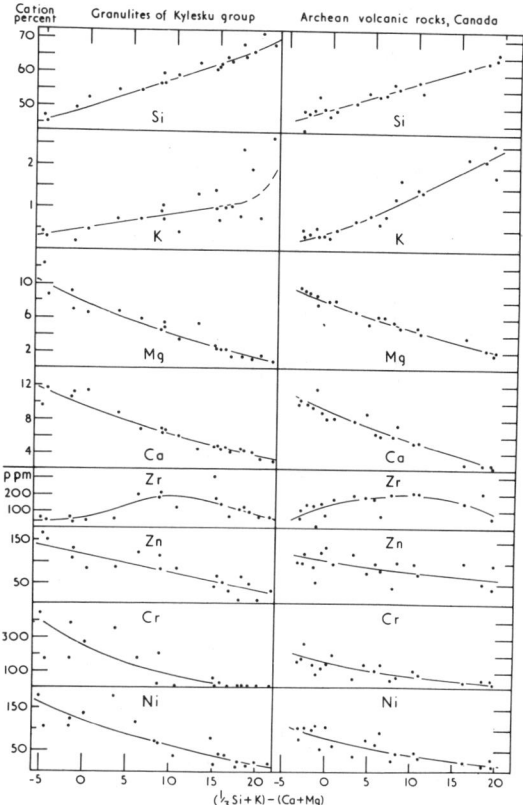

FIG. 3b. Modified Larsen plots of granulites of the Kylesku group and Archean volcanic rocks of the Canadian Shield (from Baragar and Goodwin, 1969).

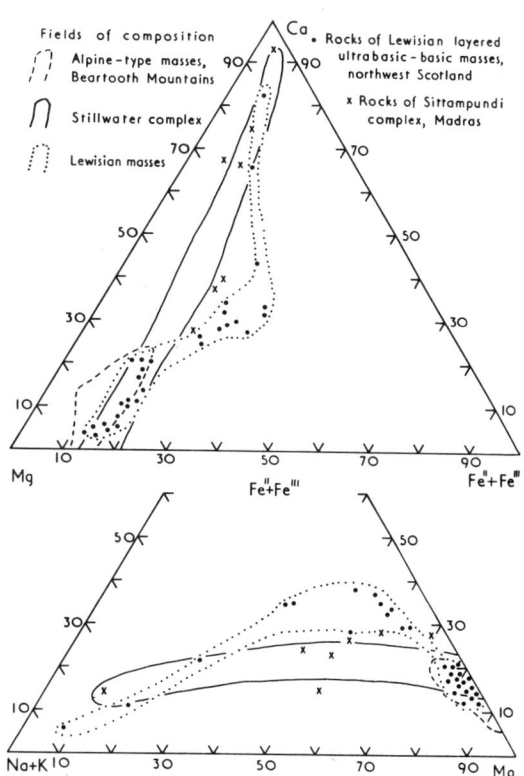

FIG. 4. Cation percent plots of metamorphosed igneous rocks from layered ultrabasic/basic masses in northwest Scotland (cf. Bowes and Wright; references in Bowes, 1969) and Madras (Subramanian, 1956) and field of composition of alpine-type ultrabasic masses, Beartooth Mountains, Montana and Wyoming. The field of composition of the Stillwater complex, Montana, is given for comparison.

(on the order of 600–700°C and 3–6 kb—equivalent to 10–20 km of depth). Some of this quartzo-feldspathic melt may move upwards out of the granulite facies environment.

Some layered ultrabasic/basic igneous masses (see Vol. V, *Intrusions, Layered*) have retained their essential bulk composition and many of their structural characteristics under granulite facies conditions and despite tectonic fragmentation. Both the integrated field of composition and overall compositional trend of Archean metaperidotites, metapyroxenites, meta-anorthosites and garnet-pyroxene-plagioclase rocks (chemically similar to olivine norite/gabbro) in Scotland are indicative of isochemical metamorphism of a tectonically fragmented igneous suite (Fig. 4).

Amphibolite Facies

Rocks undergoing amphibolite facies metamorphism have water and, at least in certain cases, potassium and some trace elements migrating up into them. This migrating material may be concentrated along tectonically controlled channelways. Ultrabasic/basic igneous complexes are reconstituted with the development of hydrous minerals, including potassium-bearing micas, but otherwise have compositions which correspond with those of unmetamorphosed igneous complexes. The alpine-type ultramafic rocks of the Beartooth Mountains of Montana and Wyoming are chemically similar to the basal parts of the nearby Stillwater igneous complex (Fig. 4) while the bulk chemistry of the Sittampundi complex of Madras corresponds with that of parts of large layered igneous complexes (Fig. 4), despite polymetamorphism and the transformation of all primary minerals.

The original basaltic/doleritic nature of many hornblende schists and amphibolites, including strongly striped varieties (Fig. 8) has been established by showing the general correspondence of major and trace element proportions and variation trends with those shown

METAMORPHIC ENVIRONMENTS—CHEMICAL MOBILITY

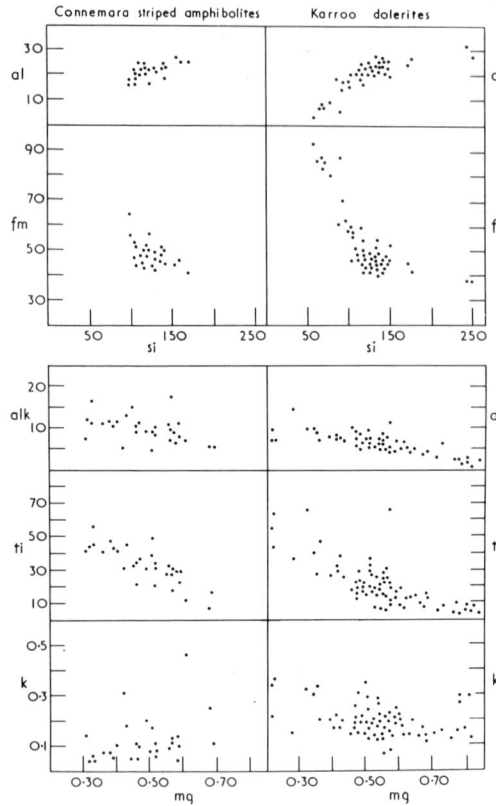

FIG. 5. Plot of Niggli numbers: *al* and *fm* against *si* and *alk, ti* and *k* against *mg* (after Evans and Leake, 1960, pp. 351–352).

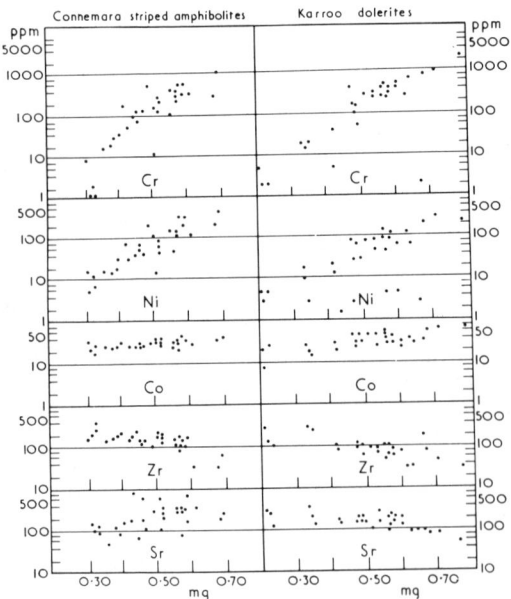

FIG. 6. Plot of Cr, Ni, Co, Zr, Sr against *mg* (Niggli number) (after Evans and Leake, 1960, pp. 358–359).

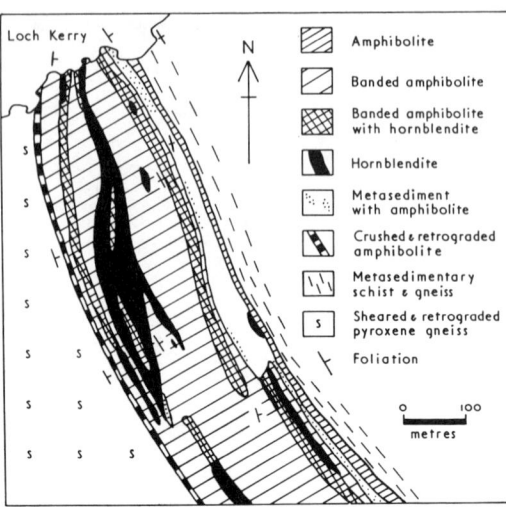

FIG. 7. Map of Loch Kerry basite sheet, Ross-shire, Scotland (after Bowes and Park; reference in Bowes, 1969).

by the products of differentiated basaltic magma (Figs. 5 and 6). Even where extensive metamorphic differentiation has resulted in the development of large masses of contrasting composition (Fig. 7), as well as small-scale bands and lenses (Fig. 8), with compositions differing greatly from that of the original igneous parent (Fig. 9), the process has been essentially one of redistribution of elements: Ti, Fe, Mg, and calcium (Ca) segregated from silicon (Si), Al, and Na with little change of

FIG. 8. Banded amphibolite with plagioclasite bands (white) and hornblendite lenses and bands (black (from Loch Kerry basite sheet, Ross-shire, Scotland).

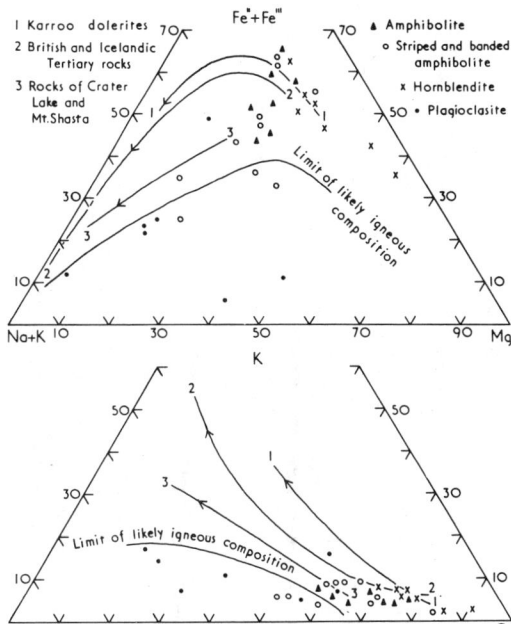

FIG. 9. Cation percent plots of rocks from Loch Kerry basite sheet, Ross-shire, Scotland with trends of some igneous suites for comparison (after Bowes and Park; reference in Bowes, 1969).

bulk composition. Differential solubility under conditions of heterogeneous pressure in foliated rocks plays an important role. The operation of selective partial melting appears to be precluded by the temperature of formation indicated by the mineral assemblages and by the rock compositions. Both the differing behavior of the felsic and mafic masses in a tectonic environment and the greater chemical mobility of the Si, Al, and Na results in the movement of feldspathic material along foliation planes (Fig. 8) and, in places, across these planes, to give discordant relations.

The amphibolite facies corresponds with the medium- to high-grade zones of progressive regional metamorphism such as those described for pelitic rocks in the southeastern Highlands of Scotland (Harker, 1939). The mineral associations are governed by the bulk chemical composition and the temperature-pressure conditions: almandine, staurolite, kyanite and sillimanite are indicator minerals for progressively higher grades. Corresponding sequences of mineral paragenesis in an isochemical environment have been described from many parts of the world (Turner, 1968), variations in mineralogical expression being related to variations in crustal conditions, particularly heat flow (den Tex, 1965). Some lithological bands have acted as closed systems even for oxygen with element migration re-stricted to small-scale diffusion in connection with mineralogical reconstitution (Chinner, see reference in Barth, 1962). Even where metasedimentary schists have an "injected" appearance, this has been explained as the result of small-scale differential mobilization during metamorphism which was isochemical on a regional scale (Butler, see reference in Mehnert, 1968).

The association of quartzofeldspathic rocks with the highest grades of metamorphism in the amphibolite facies is common and related to the development of migmatites and granitic magma. The progression from almandine-, staurolite-, and kyanite-bearing pelitic schists through sillimanite-bearing gneiss to biotite-plagioclase gneiss and biotite-perthite gneiss in New York State is related to large-scale transport of material: Si, Na, Ca, K, Ti added and Mg, Fe, Al, phosphorus (P) subtracted in the transformation of slate, through schist, to augen gneiss (Barth, 1962). Gradations from semipelitic and pelitic schists to quartzofeldspathic gneisses, which are much veined with granitic material, occur, particularly in greatly deformed Precambrian rocks. The compositions of the standard cells (units containing 160 oxygens and approximately 100 cations—Barth, 1962) of averages of metasedimentary schists and quartzofeldspathic gneisses which show these gradational relations in the Precambrian of Scotland are, respectively:

$K_{2.2}Na_{4.9}Ca_{2.8}Mg_{6.4}Fe_{5.2}Al_{14.9}Ti_{0.4}Si_{56.4}P_{0.1}$
$(O_{147.5}[OH]_{12.5})_{160}$

$K_{1.8}Na_{7.6}Ca_{2.4}Mg_{1.2}Fe_{1.4}Al_{15.1}Ti_{0.1}Si_{62.4}P_{0.1}$
$(O_{152.5}[OH]_{7.5})_{160}$

Metasediments are changed to quartzofeldspathic gneisses

	By adding
	2.7 ions of Na
	0.2 ions of Al
	6.0 ions of Si
Total	8.9 cations
representing	27.3 valencies
	By subtracting
	0.4 ions of K
	0.4 ions of Ca
	5.2 ions of Mg
	3.8 ions of Fe
	0.3 ions of Ti
	7.5 ions of H
Total	17.6 cations
representing	27.3 valencies

Averages of trace element proportions (in

parts per million) of metasedimentary schists are:

Ba 411, Ce 37, Co 41, Cr 305, Cu 78, Ga 14.5, La 51, Nb 7, Ni 97, Rb 95, Sr 149, Zn 151, Zr 166

Comparable averages for quartzofeldspathic gneisses are:

Ba 903, Ce 55, Co 36, Cr <50, Cu 44, Ga 14, La 54, Nb <4, Ni 18, Rb 61, Sr 497, Zn 33, Zr 127

This progressive feldspathization (Bowes and Bhattacharjee, see reference in Bowes, 1969) is associated with a complementary development of mafic segregations by residual concentration (see Fig. 8), with the bulk composition of the region remaining essentially constant, apart from water and possibly potassium and some trace elements migrating upwards from granulite facies environments. Under temperature-pressure conditions in the range 400–450°C/4 kb (14 km) to 525°/2 kb (7 km), for temperature gradients of 30 and 70°C, respectively, and during penetrative deformation, the most soluble (felsic) substances are segregated into bands rich in oligoclase, some approaching trondhjemite in composition (see Fig. 11).

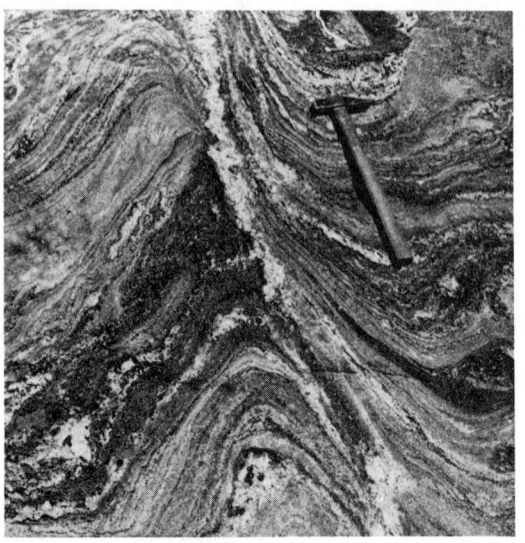

FIG. 11. Folded quartzo-feldspathic gneiss with granitic bands and lenses and granitic vein material in the fold hinge zone from Mingulay, Outer Hebrides; cf. Bowes and Hopgood, reference in Bowes, 1969).

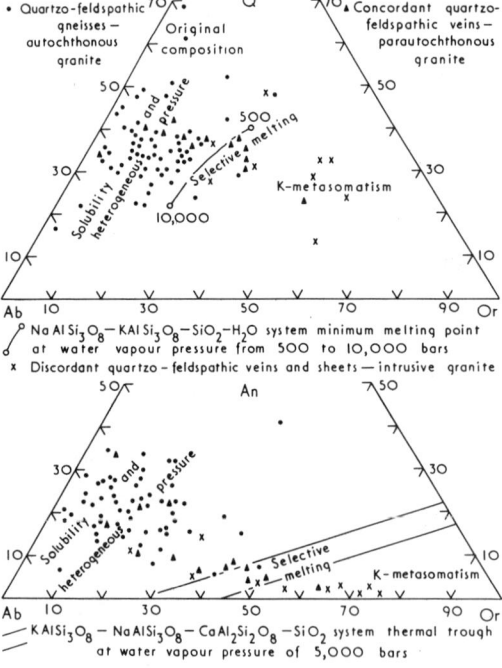

FIG. 10. Plots of normative proportions of Lewisian granitic rocks, northwest Scotland—(after Bowes, 1967; Bowes and Bhattacharjee, reference in Bowes, 1969).

The sodium-dominant nature of many banded quartzofeldspathic gneisses (autochthonous granitic rocks) is well known (Read, 1957; Fig. 10) and there is a general correspondence in the normative Q–Ab–Or and An–Ab–Or proportions for quartzofeldspathic gneisses, felsic bands in quartzofeldspathic gneisses and segregation bands in hornblende schists/amphibolites (Bowes, 1967). Extraction and redeposition of the most soluble substances, differentiation due to heterogeneous pressure and residual concentration of least soluble substances appear to have accentuated existing inhomogeneities (e.g., bedding, tectonic banding), with partial melting not playing an important role (see Mehnert, 1968).

Some quartzofeldspathic bands in gneisses pass laterally and gradually into concordant vein-like masses of granitic and pegmatitic aspect (Fig. 11) resulting in the development of migmatitic (injection) gneisses. There was only local redistribution of material in the formation of some such veins (White, see reference in Mehnert, 1968). However, some concordant quartzofeldspathic veins, which show gradational relationships with felsic bands in gneisses, pass, in the hinge zones of folds, into discordant granitic and pegmatitic veins (Fig. 11; see Bowes and Hopgood, reference in Bowes, 1969). The normative Q–Ab-Or and An–Ab–Or proportions of many such parautochthonous granitic rocks correspond with those of minimum melting point liquids, indicating the operation of selective partial melting (Fig. 10). The required elevation of temperature may be related to the channelled uprise of volatiles, and potassium, from deeper levels, shown by the large grain size of many discordant veins and the replacement of oligoclase by potassium feldspar. Concentration of radioactive elements moving up from granulite facies environments may also play a significant role in the local rise of temperature.

The composition of mobile vein material varies depending upon the relative proportions of more soluble material squeezed into hinge zones, partial melt products, and uprising volatile material. In places the vein complex feeds upwards, using vertically-disposed structures, to give a considerable concentration of granitic magma. In other places there is intimate veining in a migmatite complex (Brown, see reference in Mehnert, 1968). Where the intimate veining (which acts as a heat source) is controlled by horizontal or shallowly dipping structures, local reversals of metamorphic zonation occur (Read, 1957). In such an environment metamorphic differentiation is operative as well as injection.

Many granitic veins in migmatite complexes pass into microcline-rich pegmatitic veins and sheets which show evidence of high water content and replacement of oligoclase by microcline. A corresponding replacement is seen in some gneisses, with the formation of muscovite-microcline gneiss. The effects of this late stage potassium metasomatism has been widely recognized, as has the dominance of potassium-rich rocks in the later stages of the granite series (intrusive granites, Fig. 10), with the normative An proportion having a significant bearing on the composition of partial melt material. Field, textural, and compositional evidence indicate the effects on the products of both metamorphic differentiation, and the crystallization of magma derived by partial melting, of volatiles and potassium. These migrate upward, particularly in structurally controlled belts and appear to represent material expelled from deeper metamorphic levels.

Greenschist Facies

In greenschist facies rocks chemical mobility is only on a local scale. It results from the extraction and redistribution of the most soluble substances with the formation of quartz and quartz-feldspar segregations. This metamorphic differentiation begins at 200–300°C (McNamara, see reference in Turner, 1968). Its effects are seen first in the rocks showing the greatest deformation due to tectonics where abnormally high surface energies in stressed portions of detrital grains lead to a local uprise in the amount of material dissolved and so to supersaturation of the pore-fluid.

Glaucophane-Schist Facies

Mineral assemblages in this facies result from reconstitution at considerable depths but low temperatures. The conditions are the result of such rapid sedimentation and crustal downwarping that normal heat flow from below and heat generated by radioactivity within the sequence was unable to reach normal steady-state conditions. There is considerable very local movement of material, particularly in carbonate-bearing rocks, but no clear regional geochemical gradients.

Zeolite Facies

Characteristic mineral assemblages represent the products of deep *diagenesis* (q. v. Vol. VI) which merge into the lowest grade of schists. There is some very local metasomatic activity and very limited movement of material in mineral reconstitution.

D. R. Bowes

References

Baragar, W. R. A., and Goodwin, A. M., 1969, "Andesites and Archean vulcanism of the Canadian Shield," in (McBirney, A. R., editor), "Proceedings of the Andesite Conference," *Dept. Geol. and Min. Industries, State of Oregon, Bull.*, **65**, 13–44.

*Barth, T. F. W., 1962, "Theoretical Petrology," Second ed., New York, John Wiley & Sons.

Bowes, D. R., 1967, "The petrochemistry of some Lewisian granitic rocks," *Mineral. Mag.*, **36**, 342–363.

*Bowes, D. R., 1969, "Lewisian of northwest Highlands of Scotland," *Am. Assoc. Petrol. Geol. Mem.*, **12**, 575–594.

Evans, B. W., and Leake, B. E., 1960, "The composition and origin of the striped amphibolites of Connemara, Ireland," *J. Petrol.*, **1**, 337–363.

Goodwin, A. M., 1968, "Archaean protocontinental growth and early crustal history of the Canadian Shield," *XXIII Intern. Geol. Congr., Prague*, **1**, 69–89.

Harker, A., 1939, "Metamorphism," Second ed., London, Methuen.

Heier, K. S., 1965, "Metamorphism and the chemical differentiation of the crust," *Geol. Fören. Stockholm Förh.*, **87**, 249–256.

Helgeson, H. C., 1967, "Solution chemistry and metamorphism," in (Abelson, P. H., editor), "Researches in geochemistry," Vol. 2, New York, John Wiley & Sons.

*Mehnert, K. R., 1968, "Migmatites and the origin of granitic rocks," Amsterdam, Elsevier.

Ramberg, H., 1952, "The origin of metamorphic

and metasomatic rocks," Chicago, Univ. of Chicago Press.
Read, H. H., 1957, "The granite controversy," London, Murby.
Subramanian, A. P., 1956, "Mineralogy and petrology of the Sittampundi complex, Salem district, Madras State, India," *Bull. Geol. Soc. Am.,* **67,** 317–390.
den Tex, E., "Metamorphic lineages of orogenic plutonism," *Geol. Mijnbouw,* **44,** 105–132.
*Turner, F. J., 1968, "Metamorphic Petrology," New York, McGraw-Hill Book Co.

* Additional bibliographic references may be found in this work.

Cross-references: *Hydrothermal Solution; Interstitial Waters in Sediments; Mineral Thermometry; Paragenesis; Phase Equilibria; Radioactivity in Rocks; Silica Solubility; Solubility; Trace Elements; Water. Vol. V: Anatexis; Aureoles; Heat Flow; Intrusions, Layered; Magmatic Differentiation; Metamorphism; Metasomatism; Migmatization; Pegmatites; Skarn; Volatiles; Zeolite Facies. Vol. VI: Diagenesis. Vol. VIII: South Africa.*

METAMORPHISM—See PARAGENESIS; also Vol. V

METEORIC WATER

Meteoric water is defined as all the water which is circulating in the *hydrologic cycle* (q. v.). This would include gaseous and condensed water in the atmosphere as well as the liquid water in the rivers and at or below the land surface. In the usual sense in which this term is used, however, the stored water in the hydrosphere, including that of the ocean and the polar ice caps, and connate subsurface water, or water held in combination in rocks (magmatic water) are not included.

The composition of meteoric water is indicated by the chemical analyses tabulated under the topic *Water—Nonmarine* (q.v.), and by the analyses of the more dilute waters in the table under the topic *Aqueous Solutions* (q.v.). Rainwater generally contains only small amounts of solutes, derived from solution of gases, dust, and aerosols. Near the ocean the composition of rain is influenced strongly by particles of salt swept from the surface of the sea, and concentrations of a few ppm of chloride and sodium are commonly observed. The proportion of sulfate to total ions in solution increases inland, and rain in most places a few hundred miles or more from the ocean contains more sulfate than chloride (see *Rainwater; Water—Nonmarine*).

Gorham (1961) has summarized some of the geochemically significant aspects of the chemistry of rainwater. Whitehead and Feth (1964) computed the following average concentrations for major constituents in rain from inland and coastal regions of the United States, using data which had been published by others:

Constituent	Concentration (ppm) Coastal Locations	Inland Locations
Sulfate (SO_4)	2.45	2.14
Chloride (Cl)	4.83	0.22
Sodium (Na)	3.68	0.42
Calcium (Ca)	0.58	1.41

Runoff in rivers and other natural terrestrial waters replenished by rainfall generally has a chemical composition greatly different from that of rainwater, owing to weathering reactions in rock and soil, biochemical effects, and many other influences.

JOHN D. HEM

References

Gorham, E., 1961, "Factors influencing supply of major ions to inland waters, with special reference to the atmosphere," *Geol. Soc. Am. Bull.,* **72,** 795–840.
Whitehead, H. C., and Feth, J. H., 1964, "Chemical composition of rain, dry fallout, and bulk precipitation at Menlo Park, Calif., 1957–59," *J. Geophys. Res.,* **69,** 3319–3333.

Cross-references: *Aqueous Solutions; Connate Water; Hydrologic Cycle; Magmatic Water; Rainwater; Runoff; Water—Nonmarine. Vol. II: Aerosols; Salt Nuclei.*

METEORITES—ORGANIC CONSTITUENTS—See Vol. II, CARBONACEOUS METEORITES

METEOROLOGICAL CYCLES—See Vol. II

METEOROLOGY—See Vol. II

METHANE—See Vol. IV B

MICA GROUP—See MINERAL GROUPS: SILICATES

MICROBIAL IRON OXIDATION AND REDUCTION—See IRON OXIDATION AND REDUCTION—MICROBIAL

MICROBIAL ISOTOPE FRACTIONATION —See CARBON ISOTOPE FRACTIONATION; SULFUR ISOTOPE FRACTIONATION

MICROBIAL MINERAL GENESIS AND TRANSFORMATION—See MINERAL

GENESIS AND TRANSFORMATION—MICROBIAL

MICROBIAL MINERAL SULFIDE OXIDATION—See **SULFIDE MINERAL OXIDATION—MICROBIAL**

MICROBIAL OXIDATION AND REDUCTION OF IRON—See **IRON OXIDATION AND REDUCTION—MICROBIAL**

MICROBIAL SULFATE REDUCTION—See **SULFATE REDUCTION—MICROBIAL**

MICROBIOLOGY—See Vol. I, **MARINE MICROBIOLOGY;** also Vols. VI, VII

MICROORGANISMS, MICROBES—See Vol. VII

MICROSCOPY—See "ENCYCLOPEDIA OF MICROSCOPY" (G. L. Clark, editor, Reinhold Publ.)

MIGRATION—See Vol. IV B, **CRUDE OIL COMPOSITION AND MIGRATION; PETROLEUM GEOLOGY**

MINERAL CLASSES: NONSILICATES

Minerals are here arranged in classes following the Berzelius-Dana plan, according to their anionic groups. The order followed is that of *Dana's System of Mineralogy,* Seventh ed. The classes are listed below in alphabetical order, and appear as separate entries in this volume.

ANTIMONATES, ANTIMONITES, AND ARSENITES
Antimonates
Acid and normal antimonites, arsenites
Basic or halogen-containing antimonites, arsenites

Antimonides (see sulfides, etc.)
Arsenates (see phosphates, etc.)
Arsenides (see sulfides, etc.)

BORATES
Anhydrous borates
Hydrous borates
Borates with hydroxyl or halogen
Compound borates

CARBONATES
Acid carbonates
Anhydrous carbonates
Hydrous carbonates

MINERAL CLASSES: NONSILICATES

Carbonates with hydroxyl or halogen
Compound carbonates

CHROMATES
Anhydrous chromates
Hydrous chromates

HALIDES
Normal halides
Oxyhalides and hydroxyhalides
Halide complexes
Fluoaluminates
Compound halides

HYDROXIDES AND OXIDES CONTAINING HYDROXYL
See Oxides

IODATES
Anhydrous and hydrous iodates
Iodates with hydroxyl or halogen
Compound iodates

MOLYBDATES AND TUNGSTATES
Anhydrous molybdates and tungstates
Basic and hydrated molybdates and tungstates

NATIVE ELEMENTS
Metals
Semimetals and nonmetals

NITRATES
Anhydrous or hydrous nitrates
Nitrates with hydroxyl or halogen
Compound nitrates

ORGANIC COMPOUNDS
Oxalates
Hydrocarbons

OXIDES
Simple oxides
Oxides containing uranium, thorium, and zirconium
Hydroxides and oxides containing hydroxyl
Multiple oxides
Multiple oxides containing niobium (columbium), tantalum, and titanium

PHOSPHATES, ARSENATES, AND VANADATES
Anhydrous acid minerals
Anhydrous normal minerals
Hydrous acid phosphates
Hydrated normal minerals
Anhydrous phosphates with hydroxyl or halogen
Hydrated phosphates with hydroxyl or halogen
Compound phosphates, etc.

VANADIUM OXYSALTS
SELENATES AND TELLURATES; SELENITES AND TELLURITES
Selenides (*see* sulfides, etc.)

717

MINERAL CLASSES: NONSILICATES

SULFATES
Anhydrous acid sulfates
Hydrated acid and normal salts
Anhydrous sulfates with hydroxyl or halogen
Hydrous sulfates with hydroxyl or halogen
Compound sulfates

SULFIDES (WITH TELLURIDES, ARSENIDES, ANTIMONIDES, SELENIDES)

SULFOSALTS
Tellurates and tellurites (*see* selenates, etc.)
Tellurides (*see* sulfides, etc.)
Vanadates (*see* phosphates, etc.)

LLOYD W. STAPLES

References

Dana, J. W., 1971, "Manual of Mineralogy," Eighteenth ed., New York, John Wiley & Sons (revised by Hurlbut, C. S., Jr.).

Dana, J. W., and Dana, E. S., 1944, 1951, "System of Mineralogy," Seventh ed., Vols. I and II, New York, John Wiley & Sons (revised by Palache, C., Berman, H., and Frondel, C.)

Dennen, W. H., 1960, "Principles of Mineralogy," New York, Ronald Press, 453pp.

Hey, M. H., 1955, "Chemical Index of Minerals," Second ed., London, British Museum; also Appendix to the Second ed., 1963.

Mason, B., and Berry, L. G., 1959, "Elements of Mineralogy," San Francisco, W. H. Freeman & Co.

Cross-references: See under each class name; also principal mineral species, individually in Vol. IVB.

MINERAL CLASSES: SILICATES

Silicates, those minerals containing the elements silicon and oxygen, make up about 95% of the earth's crust. Not only are these minerals notable because of their abundance, but also because of their great variety, for of the established 2000 or so mineral species, about 700 are silicates. The diversity of these minerals is explained by the number of different ways in which the same elements can combine, for although one-third of all elements have been identified in silicate structures, yet only twelve of these occur frequently in the common minerals of the class.

Early Classification of the Silicates

From the time that chemical analysis was employed as a tool in mineralogy, the multiplicity of silicates was recognized. In the first decade of the nineteenth century, silicates came to be regarded as the salts of various silicic acids, few of which could be synthesized in the laboratory. More than one hundred years later, in 1914, Clarke (p. 8) expressed the faith of the times when in referring to the silicates he wrote, "They must represent known or probable silicic acids; and any scheme which fails to take the latter consideration into account is inadmissible." He listed the following eight hypothetical silicic acids as being the only ones necessary to account for the natural silicates.

orthosilicic acid H_4SiO_4
metadisilicic acid $H_2Si_2O_5$
metasilicic acid H_2SiO_3
orthotrisilicic acid $H_8Si_3O_{10}$
orthodisilicic acid $H_6Si_2O_7$
trimetasilicic acid $H_6Si_3O_9$
dimetasilicic acid $H_4Si_2O_6$
trisilicic acid $H_4Si_3O_8$

By regarding them as acid, basic, and mixed salts of these acids, with varying degrees of hydration, all silicates could be accounted for. Thus, quartz was considered an anhydride of orthosilicic acid, and plagioclase an isomorphous mixture of trisilicic and orthosilicic acids.

The silicic acid concept persisted until the 1930s, by which time the use of x rays in the examination of crystalline structures *was well* advanced. Primarily due to the work of the Braggs (references in Hurlbut, p. 128), the crystal structures of the silicates began to be revealed, and the belief that silicic acids had any significance in the formation of the silicates was discarded. For a detailed examination of the silicic acid concept, see references to Clarke (1914), Doelter (1912), and Tschermak, contained in Berman (1937).

Crystal Chemistry of the Silicates

In all silicate minerals studied so far, except for stishovite, a high-pressure phase of quartz, x-ray investigations have shown that a silicon atom is invariably surrounded by four oxygen atoms which take up positions at the corners of a tetrahedron centered by the silicon atom (Figs. 1 and 2). The radii of the silicon and oxygen ions make it possible for the silicon ion to fit at the center of this ionic cluster, and the bonding in this arrangement (50% ionic, 50% covalent) is so strong that a breakdown of the group in the presence of other

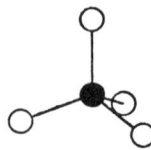

FIG. 1. Nesosilicate Schematic representation of SiO_4 tetrahedron.

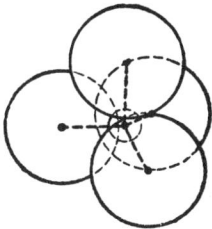

FIG. 2. Nesosilicate SiO$_4$ tetrahedron showing correct relative sizes of atoms.

ions does not occur. Thus, the fundamental unit of structure in all silicates is the SiO$_4$ tetrahedron. Whatever other ions may be present in the structure, silicon and oxygen invariably maintain their tetrahedral linkage. The silicon ion is said to exhibit fourfold, or tetrahedral, coordination with the oxygen ion.

The silicates are essentially ionic compounds. In such structures each cation tends to be surrounded by anions. The number of anions that can fit around a given cation can be calculated from the relative sizes of the ions, and this number is termed the coordination number of the cation. (For a more detailed examination of the concepts referred to here, see the article on *Crystal Chemistry*.)

Many of the cations found in silicates occur only with a single coordination, thus silicon invariably has fourfold coordination, and tetravalent titanium, trivalent chromium and divalent magnesium, iron, and manganese, all have sixfold coordination. Other silicate cations are of a size which permits various degrees of coordination, for example, calcium and sodium ions may occur with sixfold, sevenfold or eightfold, coordination (usually the latter type in silicates), and potassium and barium may occur with sixfold, seven-, eight-, nine-, tenfold, or twelvefold, coordination. Aluminum, the most abundant element in the crust of the earth after oxygen and silicon, has a special role in silicate structure, because its size is such that it may occur with either fourfold, or sixfold, coordination. When it occurs with fourfold coordination, it forms an AlO$_4$ tetrahedron of a size and coherence similar to that of a SiO$_4$ tetrahedron, and may substitute for the SiO$_4$ tetrahedron. When it occurs with sixfold coordination, it occupies a site interstitial to the SiO$_4$ tetrahedra, and serves to link these tetrahedra with simple ionic bonds which are much weaker than those existing amongst the ions of the tetrahedral grouping. In some silicates, aluminum occurs with both types of coordination.

Ions of similar size, which therefore have similar coordination, tend to occupy similar sites in silicate structures, that is, they may substitute for one another within these structures.

Silicate minerals in general are characterized by their capacity to permit a high degree of ionic substitution. The variation in chemical composition within essentially the same crystal structure that results from this substitution confounded early attempts to classify the silicates. A degree of understanding, still imperfect, of the factors controlling the substitution has now been achieved from x-ray studies. The chief factor is the size of the ions; empirically it is found that elements whose ionic radii differ by more than 15% do not commonly proxie each other (the silicon-aluminum pair being a notable exception). Temperature of formation is a second factor; minerals formed at high temperatures exhibit a greater tolerance for substitution than do low-temperature minerals. A third factor is the nature of the structure; e.g., quartz rarely contains aluminum substituting for silicon, but such a substitution is common in the feldspars. A fourth factor is ionic availability; in environments where ions are not available, they obviously cannot enter into the composition of a mineral.

Ions of different valence frequently substitute for each other, but this is possible only if elsewhere in the structure another change is made such that the overall electrical balance between cations and anions is maintained. Paired, or coupled, substitution is very common in silicate minerals. Divalent calcium substitutes for univalent sodium in the plagioclase feldspar series, but for each ion of calcium taking the place of sodium, an accompanying substitution of trivalent aluminum for tetravalent silicon must occur. In other minerals, the coupled substitution may involve a cation and an anion, two anions, or even the elimination of an ion to produce a lattice vacancy. Some structures are sufficiently flexible to accept an increased number of ions during substitution. In certain zeolites, for example, two sodium ions may substitute for a calcium ion without causing a profound change in the structure. Ions whose difference in charge is more than one unit do not commonly proxie each other, even though their radii may be of closely similar magnitude.

Structural Classification of the Silicates

The great variety of the silicates results from the different ways in which the SiO$_4$ tetrahedra may be joined within a crystal structure. The tetrahedra may exist as independent units, or may be linked to each other in various ways by sharing oxygen atoms. Each pattern of linkage yields a structural

MINERAL CLASSES: SILICATES

TABLE 1. SUMMARY OF SOME CHARACTERISTICS OF THE STRUCTURAL SUBCLASSES OF THE SILICATES

Subclass	Linkage Pattern of Tetrahedra	Si:O	Formula Unit and Charge	Example
Nesosilicates	Independent tetrahedra	1:4	$(SiO_4)^{4-}$	Olivine $(Mg,Fe)_2SiO_4$
			$(Si_3O_{12})^{12-}$	Pyrope garnet $Mg_3Al_2(SiO_4)_3$
Sorosilicates	Independent double tetrahedra	2:7	$(Si_2O_7)^{6-}$	Lawsonite $CaAl_2Si_2O_7(OH)_2H_2O$
Cyclosilicates	Tetrahedra in closed rings			
	(a) Trigonal rings		$(Si_3O_9)^{6-}$	Benitoite $BaTiSi_3O_9$
		1:3		
	(b) Tetragonal rings		$(Si_4O_{12})^{8-}$	Axinite $(Ca,Mn,Fe)_3Al_2(BO_3)\text{-}Si_4O_{12}(OH)$
	(c) Hexagonal rings		$(Si_6O_{18})^{12-}$	Beryl $Be_3Al_2Si_6O_{18}$
Inosilicates	Chains of indefinite length			
	(a) Single chains	1:3	$(Si_2O_6)^{4-}$	Diopside $CaMgSi_2O_6$
	(b) Double chains	4:11	$(Si_8O_{22})^{12-}$	Tremolite $Ca_2Mg_5Si_8O_{22}(OH)_2$
Phyllosilicates	Continuous sheets			
	(a) Two-layer structure		$(Si_4O_{10})^{4-}$	Serpentine
		2:5		$Mg_6Si_4O_{10}(OH)_8$
	(b) Three-layer structure		$(Si_4O_{10})^{4-}$	Muscovite $KAl_2(AlSi_3O_{10})(OH)_2$
Tectosilicates	Continuous three-dimensional framework	1:2	$(SiO_2)^0$	Quartz SiO_2
			$(AlSi_3O_8)^{1-}$	Orthoclase $KAlSi_3O_8$

unit possessing a characteristic silicon to oxygen ratio, and having an overall electrical charge which must be balanced by an appropriate number of cations. The present-day classification of silicates is based upon the patterns of linkage amongst the tetrahedra, and six fundamental subclasses, or types, are generally recognized. The names of the subclasses employed here are those proposed by Strunz in 1941 (reference in Strunz, 1970). In their excellent descriptive work, Deer et al., combine two of the subclasses (nesosilicates and sorosilicates in their category 'orthosilicates.'

Table 1 summarizes the chief characteristics of each silicate subclass.

The diagrams of silicate structures used in the following descriptions are after those found in Winchell and Winchell, pp. 239–242.

Subclass Nesosilicates. In this structure the SiO_4 tetrahedra exist as independent anionic groups linked to each other only through the cations of the structure (Figs. 1 and 2). The silicon to oxygen ratio for minerals having this structure is 1:4. Minerals of the olivine and garnet groups are typical nesosilicates.

Subclass Sorosilicates. One oxygen atom is shared by two SiO_4 tetrahedra in this structure, producing an independent double tetrahedral group having a silicon to oxygen ratio of 2:7 (Fig. 3 and 4). The double tetrahedra are joined by means of the cations in the lattice. A few minerals, such as lawsonite and hemimorphite, have structures in which only double tetrahedra are found, but some minerals usually classified as sorosilicates have lattices

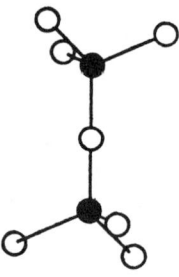

FIG. 3. Sorosilicate. Schematic representation of double tetrahedral linkage.

MINERAL CLASSES: SILICATES

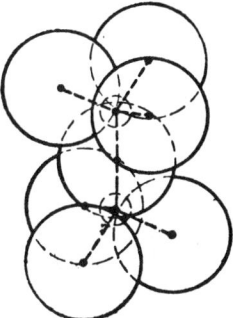

Fig. 4. Sorosilicate. Double tetrahedron showing correct relative sizes of atoms.

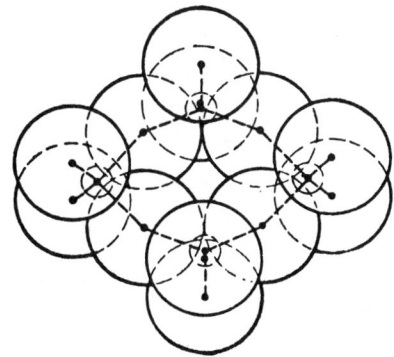

Fig. 6. Cyclosilicate. Tetragonal ring structure (Si_4O_{12}).

containing isolated independent tetrahedra as well as double tetrahedra (e.g., species of the epidote group).

Subclass Cyclosilicates. In this structure, two oxygen atoms of each SiO_4 tetrahedron are shared with adjacent tetrahedra, and the tetrahedra linked by these oxygens form closed, ring-like groups. Three possible configurations of cyclic tetrahedra are known, each having a silicon to oxygen ratio of 1:3; three tetrahedra join to produce a Si_3O_9 ring (Fig. 5), four tetrahedra join to produce a Si_4O_{12} ring (Fig. 6), and six tetrahedra join to produce a Si_6O_{18} ring (Fig. 7). Benitoite, axinite, and beryl, are mineral examples of the trigonal, tetragonal, and hexagonal ring structures respectively. Cations link the individual ring groups in a variety of ways. Rings may be positioned with respect to one another in such a way that vacuities exist in the structure, and these may accommodate ions, uncharged atoms, and molecules which are not essential constituents of the structure.

Subclass Inosilicates. In minerals of this subclass, the SiO_4 tetrahedra are linked together by sharing oxygen atoms to form elongated structures of indefinite length. Two arrangements of these chains are possible. In the single chain structure (Fig. 8), two of the

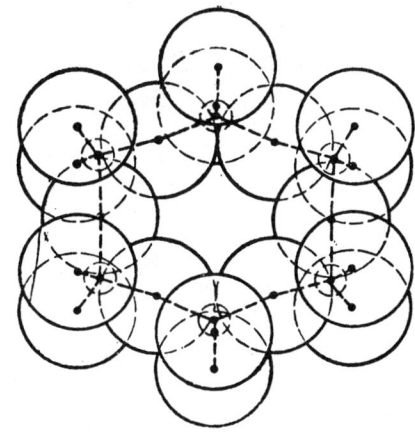

Fig. 7. Cyclosilicate. Hexagonal ring structure (Si_6O_{18}).

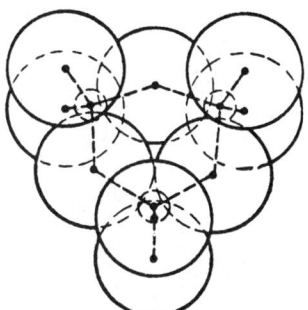

Fig. 5. Cyclosilicate. Trigonal ring structure (Si_3O_9).

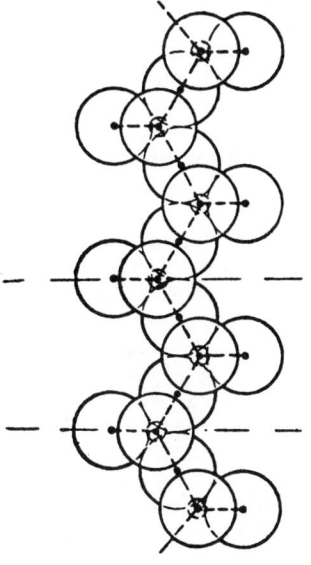

Fig. 8. Inosilicate. Single chain structure.

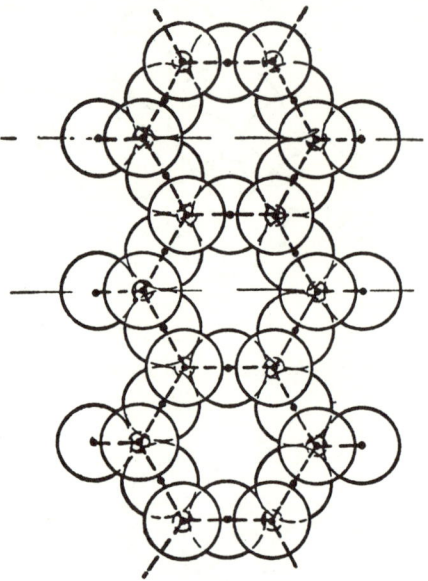

FIG. 9. Inosilicate. Double chain structure.

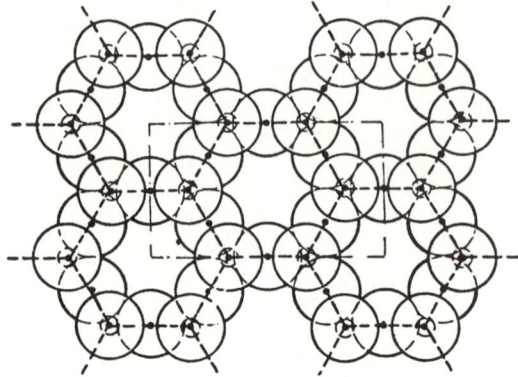

FIG. 10. Phyllosilicate. Continuous sheet structure.

oxygen atoms of each tetrahedron are shared by adjacent tetrahedra, and the individual chains are joined to each other only by the cations of the structure. The minerals of the pyroxene and pyroxenoid groups have this arrangement, in which the silicon to oxygen ratio is 1:3. In the double chain structure (Fig. 9), two single chains occur side by side, and are linked by the sharing of an oxygen atom between alternate tetrahedra. The remaining positions in the structure are filled by cations which serve to complete the linkage between tetrahedra. In this arrangement, the silicon to oxygen ratio is 4:11. Minerals of the amphibole group have this structure. In both single and double chain structures, the elongation is most often parallel to the c-axis direction.

Subclass Phyllosilicates. The SiO_4 tetrahedra of phyllosilicates are linked together to form continuous sheets extending in the plane perpendicular to the c axis. Three of the four oxygens of each tetrahedron are shared by neighboring tetrahedra (Fig. 10). The sheets are termed silica layers, silica sheets, or tetrahedral layers. A second kind of sheet-like layer, consisting of cations in six-fold coordination with oxygen and hydroxyl anions, constitutes an essential part of the phyllosilicate structure. Each layer of this type (termed an octahedral layer) is linked either to a single silica layer, to form a two-layer phyllosilicate, or to two silica layers, to form a three-layer phyllosilicate. The composite layers (consisting of an octahedral layer and either one or two silica layers) are stacked in a direction parallel to the c axis. Only weak bonds hold the composite layers together, hence minerals having this structure, such as the micas, exhibit perfect cleavage in the plane perpendicular to the c axis. The silicon to oxygen ratio in minerals of this subclass is 2:5. A comprehensive treatment of the phyllosilicates is given in a separate article in this volume.

Subclass Tectosilicates. Tectosilicate minerals have a three-dimensional network of SiO_4 tetrahedra, wherein each tetrahedron shares all of its oxygen atoms with adjacent tetrahedra (Fig. 11). In this arrangement the silicon to oxygen ratio is 1:2. The net charge on the formula unit SiO_2 (or Si_2O_4, or Si_3O_6, or Si_4O_8) that represents the makeup of such a structure is zero, hence it might appear that the only tectosilicates possible must have a composition of SiO_2. In fact, many of the

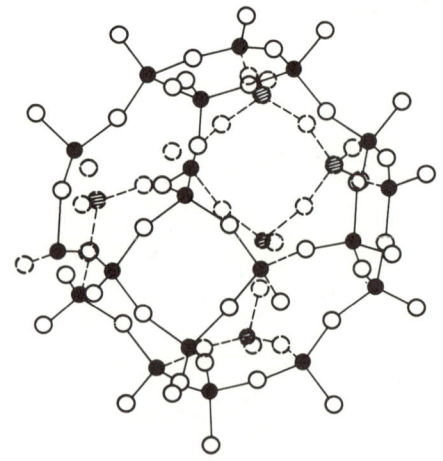

FIG. 11. Tectosilicates. Three-dimensional framework structure.

important rock-forming minerals such as the feldspars, feldspathoids, and zeolites (all aluminosilicates), are tectosilicates. This variety in composition is possible because of the substitution of AlO_4 tetrahedra for SiO_4 tetrahedra in the framework, for up to half of the SiO_4 tetrahedra may be substituted for by AlO_4 groups. For each AlO_4 tetrahedron entering the structure, a negative charge of one is acquired by the structure, and thus cations must be introduced to restore electrical balance.

List of Silicate Minerals

Only the better known silicate minerals appear in the following list. Many of the groups, series, and individual species of the silicates are described in detail elsewhere in another volume (IV B), and the reader is referred to the appropriate articles for more complete treatment of the makeup of a group or series, or for the description of varieties of species.

For ease of reference, the listing within each subclass is completely alphabetical, even though this arrangement sometimes separates species or even groups or series, which for structural and chemical reasons are usually listed together.

In classifying minerals according to structural subclass, the work of Deer et al. (Vols. 1–4) has been closely followed. A few additions to the minerals described by Deer et al. have been made, Mason and Berry, and Hurlbut serving as the source for these.

(a) NESOSILICATES
Aluminum silicate group
   Andalusite
   Kyanite
   Sillimanite
   Staurolite
   Topaz
Chloritoid
Datolite
Dumortierite
Garnet group
   Almandite
   Andradite
   Grossularite
   Hydrogrossularite
   Pyrope
   Spessartite
   Uvarovite
Humite group
   Chondrodite
   Clinohumite
   Humite
   Norbergite
Olivine group
   Fayalite
   Forsterite
   Monticellite
   Olivine
   Tephroite
Sphene
Thorite
Willemite
Zircon

(b) SOROSILICATES
Epidote group
   Allanite
   Clinozoisite
   Epidote
   Piemontite
   Zoisite
Hemimorphite
Lawsonite
Melilite group
   Akermanite
   Gehlenite
   Melilite
Pumpellyite
Vesuvianite (idocrase)

(c) CYCLOSILICATES
Axinite
Benitoite
Beryl
Cordierite
Tourmaline

(d) INOSILICATES
(1) Single-Chain Structures
Pyroxene group
   Aegirine–aegirine–augite series
     Aegirine (acmite)
     Aegirine–augite
   Augite
   Diopside–hedenbergite series
     Diopside
     Ferrosalite
     Hedenbergite
     Salite
   Jadeite
   Johannsenite
   Omphacite
   Orthopyroxene Series
     Bronzite
     Enstatite
     Hypersthene
   Pigeonite
   Spodumene
Pyroxenoid group
   Pectolite
   Rhodonite
   Wollastonite
(2) Double-Chain Structures
Amphibole group
   Alkali–amphibole subgroup
     Arfvedsonite
     Eckermannite

MINERAL CLASSES: SILICATES

    Glaucophane
    Riebeckite
  Anthophyllite–cummingtonite subgroup
    Anthophyllite series
      Anthophyllite
      Gedrite
    Cummingtonite series
      Cummingtonite
      Grunerite
  Calcium–amphibole subgroup
    Basaltic hornblende
    Hornblende series
      Common hornblende
      Edenite
      Ferroedenite
      Ferrohastingsite
      Ferrotschermakite
      Pargasite
      Tschermakite
    Tremolite–actinolite series
      Actinolite
      Ferroactinolite
      Tremolite

        (e) PHYLLOSILICATES

Apophyllite
Chlorite group
  Chamosite
  Clinochlore
  Daphnite
  Penninite
  Ripidolite
  Thuringite
Clay Minerals
  Illite
  Kaolinite group
  Montmorillonite group
  Vermiculite
Mica group
  Biotite
  Glauconite
  Lepidolite
  Margarite
  Muscovite
  Paragonite
  Phlogopite
Prehnite
Pyrophyllite
Septechlorite group
  Amesite
  Greenalite
  Septechamosite
Serpentine group
  Antigorite
  Chrysotile
  Lizardite
Stilpnomelane
Talc

        (f) TECTOSILICATES

Feldspar group
  Alkali-feldspars
    Anorthoclase
    Microcline
    Orthoclase
    Sanidine
  Plagioclase series
    Albite
    Andesite
    Anorthite
    Bytownite
    Labradorite
    Oligoclase
Feldspathoid group
  Cancrinite
  Leucite
  Nepheline
  Sodalite
Scapolite series
  Dipyre
  Marialite
  Meionite
  Mizzonite
Silica minerals
  Cristobalite
  Opal
  Quartz
  Tridymite
Zeolite group
  Analcite
  Chabazite
  Heulandite
  Natrolite
  Stilbite

JOHN T. JENKINS

References

Berman, H., 1937, "Constitution and classification of the natural silicates," *Am. Mineralogist* **22**, 342–408.

Clarke, F. W., 1914, "The constitution of the natural silicates," *U.S. Geol. Surv. Bull.* **588**, 128pp.

Deer, W. A., Howie, R. A., and Zussman, J., 1962, "Rock-Forming Minerals," Vols. 1–4, London, Longmans.

Hurlbut, C. S., Jr., 1959, "Dana's Manual of Mineralogy," Seventeenth ed., New York, John Wiley & Sons, 600pp.

Mason, B., and Berry, L. G., 1968, "Elements of Mineralogy," San Francisco, Freeman and Co., 550pp.

Strunz, H., 1970, "Mineralogische Taballen," Fifth ed., Leipzig, Akad. Verlagsg., 560pp.

Winchell, A. N., and Winchell, H., 1951, "Elements of Optical Mineralogy, Part 2, Description of Minerals," Fourth ed., New York, John Wiley & Sons, 551pp.

Cross-references: *Crystal Chemistry; Earth's Crust Geochemistry; Silicon: Element and Geochemistry; Trace Elements in Silicate Minerals; X-Ray Spectroscopy.* Vol. IVB: *Amphibole Group;*

Crystal Growth; Feldspars; Garnets; Mica Group; Phyllosilicates; Pyroxene Group; Quartz; Rock-Forming Minerals; Zeolites. See also Individual Minerals in Vol. IVB.

MINERAL FUELS—See Vol. IV B

MINERAL GENESIS—EQUILIBRIUM*

The attainment or lack of attainment of equilibrium during the formation of mineral deposits is significant in two respects. First, the equilibrium features may permit one to evaluate some aspects of the physicochemical environment of mineral genesis. As an example, the composition of pyrrhotite ($Fe_{1-x}S$) that formed in equilibrium with pyrite (FeS_2) under certain conditions may be used to interpret the temperature and the activities of sulfur and iron during mineral formation. Second, the nonequilibrium features may enable the geologist to recognize a sequence of changing conditions and thus to decipher geologic history.

Concepts and Terminology

Several modifications of the term "equilibrium" are in common use. *Equilibrium* refers to a state from which a system has no spontaneous tendency to change. Various types of "equilibrium" are illustrated by a gravitational example in Fig. 1. The unmodified term "equilibrium" generally refers to stable equilibrium, in which case the system is stable relative to all possible alternative configurations. A system in *metastable equilibrium* is stable relative to all adjacent states and will revert to the metastable equilibrium position if displaced by only a small amount; if displaced sufficiently, it will proceed to a more stable state. An *unstable equilibrium* is one which is not stable relative to at least one adjacent state.

An important distinction should be made between true equilibrium in which there is no driving force tending to alter the configuration of the system and *frozen* or *false equilibrium* in which a driving force exists but is too small to produce a detectable rate of reaction. When considered on a sufficiently large scale (as suggested in Fig. 1 there may be little distinction between metastable and frozen equilibrium because a series of minor metastable regions (Fig. 1) may impede the approach to equilibrium. The energy required to raise the system from a valley to the top of a divide is the activation energy. In the mechanical system shown in Fig. 1 the activation energy may be provided by vibrations that may jostle the sand sufficiently to permit the block to more closely approach the reference line. In a chemical system the activation energy may be required to break chemical bonds

* Publication authorized by the Director, U. S. Geological Survey.

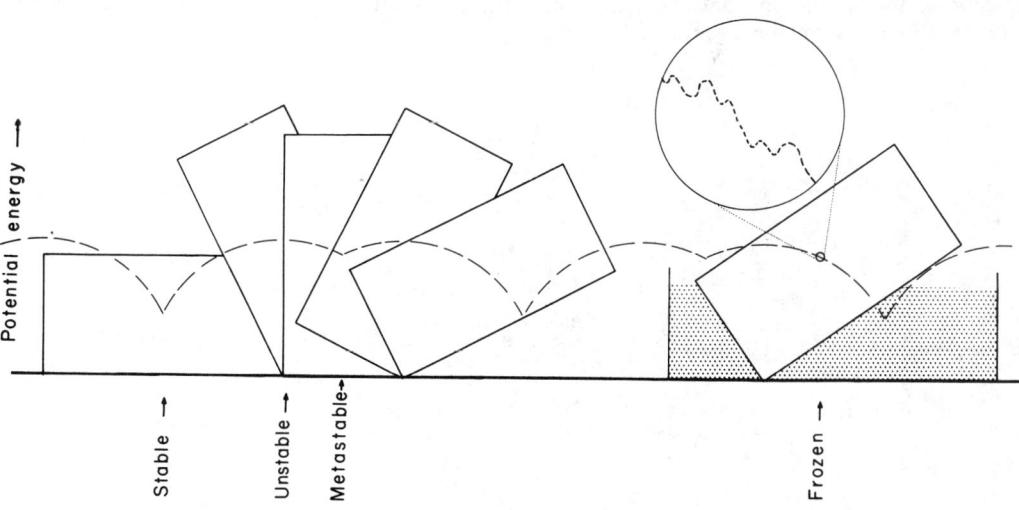

Fig. 1. Gravitational example of several modifications of the term equilibrium. Different positions of the two-dimensional block have various potential energy levels relative to the substrate. The loci of such energy levels for a rolling block are given by the dashed curve. Equilibrium (whether stable, metastable, or unstable) is attained only where the potential energy curve is at a maximum or a minimum. Frozen equilibrium (in the macroscopic sense) is shown at the right as a block whose fall is impeded such as by partial immersion in a sand-filled box (buoyancy neglected). The enlargement shows that the sluggish nature of the reaction can be attributed to numerous microscopic metastable states which do not permit the block to roll smoothly.

and may be supplied by thermal motions of the atoms or molecules. The higher the temperature the greater the thermal energy and the more likely it is that a particle will have sufficient energy to break a chemical bond and move to a new, more stable position. Thus the rate of chemical reaction increases with temperature.

Just as equilibrium may be discussed in terms of relative stability, we may also classify it in terms of the processes driving the system toward equilibrium, the most important of which are *thermal, mechanical* and *chemical*. (Processes such as those arising from magnetic or electrical fields are rarely important in geologic considerations.) At thermal equilibrium, the temperature is uniform throughout the system. At mechanical equilibrium, in the simplest case, pressure is uniform throughout the system. At chemical equilibrium, the chemical potential of each component is uniform throughout the system. Because so many geologic processes take place at essentially constant temperature and pressure we may focus our attention on chemical equilibrium and the factors influencing it.

Textural equilibrium may be considered as a variant of chemical and/or mechanical equilibrium in which the surface energies of the phases are minimized. Surface energies are dependent upon the size, shape, and relative orientation of the mineral grains, and perhaps on the compositions of the very thin regions adjacent to the intergrain contacts. The energetic contributions of the grain surfaces will be minimized by increasing grain size and by segregation of each phase into individual masses in which the grains are parallelly oriented.

Finally, equilibrium may be attained with respect to one reaction and not to others. For example, a solution may be in equilibrium with barite with regard to the amounts of Ba^{2+} and SO_4^{2-} in solution, but the sulfate may be out of equilibrium with other sulfur-bearing ionic species with respect to oxidation-reduction reactions. Such a situation may be termed *partial equilibrium*.

Descriptive Relations Between Minerals

To describe the relations between minerals, as observed in the field, it is useful to have terms that are unencumbered by connotations of equilibrium. Minerals occurring in direct contact with one another, without the presence of an intermediate phase, may be referred to as an *assemblage*. If the minerals are believed to represent either a current or a frozen equilibrium the term *equilibrium assemblage* is used. Minerals occurring together with unspecified spatial and equilibrium relationships are termed an *association*.

Mineral associations may have many complex relationships; some of the more common "end member" types between hypothetical minerals A and B are illustrated in Fig. 2. Let phases A and B both be capable of forming solid solutions with other components so that comparable compositional banding is possible.

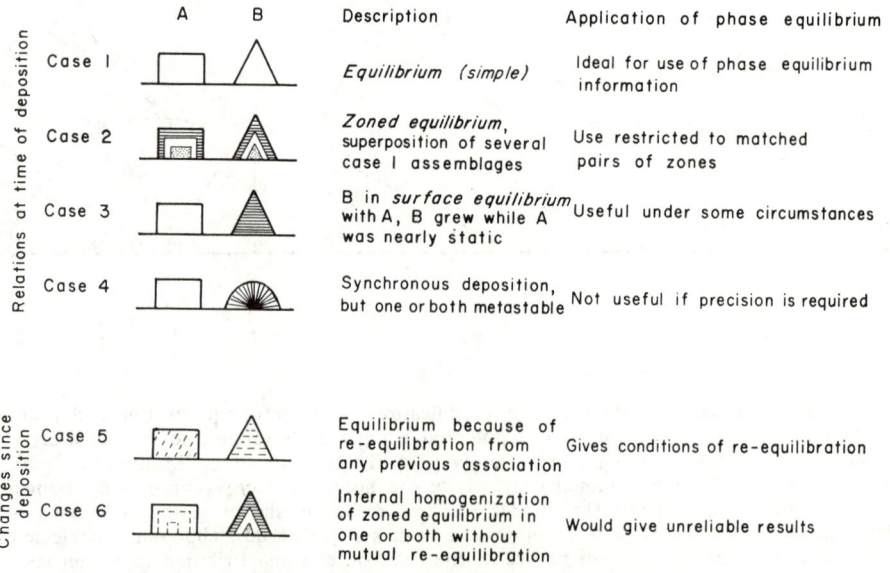

FIG. 2. Some possible relations between coexisting minerals (from Barton et al., 1963).

Conditions at Initial Deposition. Case 1. The crystals formed in complete equilibrium with each other and with their environment, which was constant throughout crystal growth. The grains are uniform throughout. This case is ideal for the application of laboratory equilibrium studies to determine the physicochemical environment in which the minerals formed. It should be clearly realized, however, that not all associations of homogeneous crystals fall into case 1 (note cases 3, 5, and 6). This case may be termed *simple equilibrium* when necessary to avoid ambiguity.

Case 2. The crystals of A and B were deposited simultaneously from solutions whose character changed gradually or spasmodically during crystal growth. A given growth zone in A is matched by a correlative growth zone in B. The array is equivalent to a series of superposed case-1-type equilibria. This relation is known as *zoned equilibrium*. It is an especially useful variation of the partial equilibrium described previously; equilibrium was maintained between the crystal surface and the depositing solution during crystal growth, while the solid state reactions within the crystal were in a state of frozen equilibrium.

Case 3. This is a special variety of case 2 in which mineral A neither grew nor dissolved while mineral B was being deposited. As will be discussed more fully later, equilibration between crystals and solution is frequently rapid. Therefore the fact that mineral A neither grew nor dissolved suggests that its surface was approximately in equilibrium with the solution and so also with mineral B throughout the growth of B. Thus we have the unilateral situation wherein mineral B is in equilibrium with the surface of mineral A, but the bulk of A is not necessarily in equilibrium with B. Mineral B is in *surficial equilibrium* with mineral A. This is an extremely common case and is potentially very useful for the application of phase equilibria.

Case 4. This is synchronous deposition but with one (or both) phases metastable. No detailed inference on the environment of ore deposition may be drawn from such relations except that the rate of deposition was too rapid for equilibrium to be attained.

Postdepositional Changes. In the four previous cases the minerals were assumed to have been preserved in their original condition, but this assumption is frequently not valid. Two significant examples of post-depositional changes are described as cases 5 and 6.

Case 5. The minerals have re-equilibrated under conditions different from those obtaining at the time of deposition. This may be the result of failure to quench as the deposit cools or of a later metamorphic event. Although the end product may be difficult to distinguish from case 1, geologic evidence ranging from regional field relations to microscopic mineral textures may help to make this distinction.

Some minerals react much more readily than others and a special situation that is potentially very misleading is illustrated separately as case 6.

Case 6. The minerals were originally associated in zoned equilibrium but, during postdepositional history one or both of the crystals has homogenized. For example, suppose that A and B are respectively galena and sphalerite and that they originally formed a zoned equilibrium association as case 2. Laboratory studies have shown (Fig. 3) that diffusion rates are much higher in galena than in sphalerite, and thus all vestiges of zoned equilibrium may be annealed out of the galena without the galena equilibrating with the sphalerite. The sluggishly reacting sphalerite may even retain its original compositional banding.

Exsolution (q.v.) although part of case 5, merits special attention. True exsolution should present no problem provided that the original state can be reconstructed. A major problem,

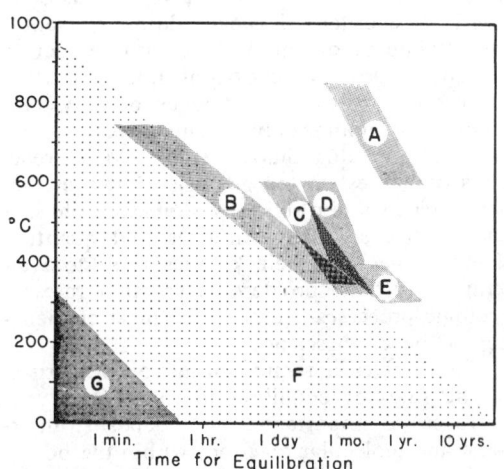

FIG. 3. Plot of time for equilibrium against temperature. A is the general curve for most solid-state reactions involving sphalerite. B is for the distribution of sulfur and tellurium between galena and altaite. C is for the equilibration of pyrite with pyrrhotite. D is for the distribution of selenium between galena and chalcopyrite. E is for equilibration between the ruby silvers. F is the general region within which lie most solution-crystal equilibria. G is the general region for most homogeneous reactions within the aqueous phase (from Barton et al., 1963).

of course, is distinguishing exsolution from replacement or from simultaneous deposition.

Reaction Rates and the Attainment of Equilibrium

The purpose of this section is to note some of the experimental observations on reaction rates and to discuss in general terms the implications of the different rates to the interpretation of processes of mineral genesis.

With very few exceptions, none of which are important geologically, reaction rates increase with temperature. The rates of most reactions in the vicinity of room temperature increase by a factor of from 1.5 to 3 or 4 (and rarely more) for each 10°C rise in temperature. Fig. 3 illustrates times for equilibration of some types of reactions at various temperatures.

Depending on the specific reaction, pressure may either increase or decrease reaction rates; unless a vapor phase is involved, however, the effect of pressure is small and is negligible compared to temperature for most reactions.

The most generally applied criteria of equilibrium in natural systems are based on adherence to the mineralogical phase rule (number of phases does not exceed the number of components) or on the presence or absence of particular minerals in particular associations. These criteria, however, do not guarantee the attainment of equilibrium with respect to the compositions of coexisting minerals.

Reaction rates in homogeneous systems (those containing only a single phase) are amenable to quantitative study. Most reactions of interest to the geologist, however, are heterogeneous (involving two or more phases) and involve so many variables that quantitative treatment of their kinetics is seldom attempted; yet the problem of reaction rates in geologic processes is among the most challenging in the earth sciences.

Fig. 3 gives a semiquantitative comparison of the rates of equilibration for a few types of reactions important to ore deposits. Interionic and molecular reactions within the homogeneous aqueous fluid are generally very rapid, though there are exceptions such as the inertness of sulfate ion toward redox reactions. Reactions between the solution and crystal bathed in the solution are somewhat slower than reactions in the homogeneous fluid, but there is a considerable range in behavior. In general, the more soluble the solid phase, the more rapid the reaction. The range of reaction rates in solid state reactions is particularly variable, and recognition of this fact is of paramount importance in the interpretation of the textures and compositions of minerals. Moreover, some of the solid state reactions are several orders of magnitude more sluggish than the solution-crystal reactions. This means that once a crystal, for example sphalerite, has precipitated from solution it may retain its initial composition even though the nature of the solution bathing it may change drastically. The changing nature of the depositing solution may be reflected in growth bands of different composition within the sphalerite but there will be little tendency for the zoned crystal to homogenize itself internally through solid-state diffusion. Other minerals, such as chalcocite or bornite, have such rapid rates of solid-state diffusion that the probability of the retention of a zoned crystal is remote. Hence, the probability of a bornite retaining information about its conditions of formation is very much less than for a sphalerite.

Aside from the problem of nucleating new phases (see *Exsolution*), solid-state reaction rates are governed principally by the rates of solid-state diffusion, which are in turn related to factors such as bond strengths. High bond strength tends to be associated with hardness, brittleness, high melting point (if congruent or nearly so), and low vapor pressure. Thus sphalerite, pyrite, arsenopyrite, magnetite, hematite, alabandite, and most silicates are far more sluggish in their solid-state reactions than are galena, chalcopyrite, bornite, chalcocite, argentite, the Ag–Au tellurides, electrum, pyrrhotite, and most sulfosalts. In general, the ease of studying a system in the laboratory (by differential thermal analysis, x-ray heating camera, or short-term phase equilibrium investigations) is inversely proportional to the reliability of applying the results to the study of the original conditions of mineral formation.

PAUL B. BARTON, JR.

References

*Barton, P. B., Jr., Bethke, P. M., and Toulmin, P., 3rd., 1963, "Equilibrium in ore deposits," *Mineral. Soc. Am., Spec. Paper*, **1**, 171–185.

Garrels, R. M., 1959, "Rates of geochemical reactions at low temperatures and pressures," in (Abelson, P. H., editor), "Researches in Geochemistry," New York, John Wiley & Sons, pp. 25–37.

Garrels, R. M., and Christ, C. L., 1965, "Solutions, Minerals and Equilibria," New York, Harper & Row, 435pp.

* Additional bibliographic references may be found in this work.

Cross-references: *Bonding; Differential Thermal Analysis; Exsolution; Metamorphic Environments—Chemical Mobility; Mineral Thermometry; Phase Equilibria; Solid Solution.* Vol. IVB: *Crystal Growth.*

MINERAL GENESIS AND TRANSFORMATION—MICROBIAL

Various biological agents have been directly or indirectly responsible for the formation of some major types of mineral accumulations in nature, including coral reefs, certain sulfur and metal sulfide deposits, some iron and manganese oxide deposits, some calcite formations, calcareous and siliceous oozes on some sea bottoms, and fossil fuels like coal and petroleum. Microorganisms have been and are presently extensively involved in the formation of some sedimentary deposits (Kuznetsov et al., 1963; Silverman and Ehrlich, 1964). Mineral accumulations formed by them may be produced actively or passively (Grigor'ev, 1965).

Mechanisms of Microbial Mineral Accumulation

Active—Enzymatic. Microbes, particularly certain bacteria, may change dissolved inorganic substances into insoluble substances by enzymatic, i.e., biologically catalyzed, oxidations or reductions. One example of such activity is the oxidation of soluble ferrous iron to insoluble ferric hydroxide or oxide by the *Ferrobacillus-Thiobacillus* group of bacteria, *Gallionella* spp., or *Leptothrix ochracea* (Kuznetsov et al., 1963; Silverman and Ehrlich, 1964). Another example is the reduction of sulfate to hydrogen sulfide by *Desulfovibrio* spp. or *Desulfotomaculum nigrificans* followed by oxidation of the hydrogen sulfide to elemental sulfur by other bacteria, such as *Thiobacillus thioparus*, green and purple sulfur bacteria, and *Beggiatoa* spp. Some bacteria benefit from oxidations of inorganic matter by deriving reducing power and necessary energy for CO_2 assimilation from these reactions, or by merely deriving reducing power for CO_2 assimilation, the necessary energy then being obtained from sunlight (photosynthesis). Some other bacteria and fungi benefit from reduction of inorganic matter by deriving oxidizing power for energy release from some of the organic matter used by them as food. (Silverman and Ehrlich, 1964).

Enzymatic intervention by microbes in mineral genesis does not always involve an attack of purely inorganic matter. Dissolved, complexed inorganic ions may be enzymatically destroyed by various kinds of bacteria and fungi through use of the organic complexing agent as food. This results in the freeing of the inorganic ion, which, if unstable in solution, is likely to precipitate due to hydrolysis or salt formation. Bog iron may have been precipitated in this manner (Harder, 1919), although enzymatic oxidation of ferrous to ferric iron may also have been involved (Kuznetsov et al., 1963).

Active—Interaction with Metabolic End Products. Not all microbial mineral accumulations are directly due to enzymatic intervention. Metabolic end products formed by microorganisms during their growth may interact nonenzymatically with mineral matter and precipitate it. Large amounts of CO_2, produced during microbial oxidation of organic food, for instance, can precipitate Ca as $CaCO_3$ at neutral or alkaline pH, and Fe and Mn as $FeCO_3$ and $MnCO_3$ at somewhat reduced Eh and slightly acid pH. Oxygen, evolved as a product of photosynthesis by certain algae, can oxidize ferrous iron and manganous manganese to their oxides at appropriate pH. Bacterially produced hydrogen sulfide can precipitate heavy metal ions such as iron, copper, lead, bismuth, and others (Silverman and Ehrlich, 1963).

Passive—Adsorption to Cell Surface. When microbial mineral accumulation is passive rather than active, the following reaction may take place. Iron and manganese oxides may accumulate on the cell sheath or stalks of some bacteria, or encrust certain algae and protozoa (Silverman and Ehrlich, 1964). This accumulation does not seem to involve direct or indirect enzymatic intervention but merely adsorption of preformed iron and manganese oxides to the surface of an organism.

Magnitude of Microbial Reactions with Minerals

The magnitude of microbial action on minerals can be appreciated when it is considered that total energy demand and reducing or oxidizing requirements of microbial metabolism are great, and that the yields of free energy and reducing or oxidizing power per unit of mineral reaction are relatively small. Hence, to meet its total requirements, a microbe has to transform a large amount of mineral. To illustrate, the assimilation of 1 mole of CO_2 by *Ferrobacillus ferrooxidans* has been found to require the free energy released by the oxidation of 51.5 moles of ferrous iron to ferric hydroxide (Silverman and Lundgren, 1959).

Transformation or Degradation of Mineral Accumulations by Microbes

Microbes may alter preformed mineral accumulations of magmatic, hydrothermal, or sedimentary origin. They may accomplish this by enzymatic oxidation or reduction of a susceptible mineral, thereby causing solubilization of the mineral. They may also accomplish this through chemical erosion of susceptible minerals by end-products of their metabolism. Dissolved matter formed by these processes may be carried away from the site of origin and remain in solution for a prolonged period or may soon reprecipitate under chemical or biological influence to form secondary mineral accumulations (Silverman and Ehrlich, 1964).

First Occurrence of Microbial Mineral Action

Fossil records of microbes have been found in sedimentary rocks over 3 billion years old (Engel et al., 1968; Schopf et al., 1965). At present it is not certain when microbes played their first important role in mineral transformation. Current evidence suggests that they were active as geologic agents in Precambrian times (Schopf et al., 1965).

H. L. Ehrlich

References

*Engel, A. E. J., Nagy, B., Nagy, L. E., Engel, C. G., Kremp, G. O. W., Drew, C. M., 1968, "Alga-like forms in Onverwacht, South Africa. Oldest recognized lifelike forms on earth," *Science*, **161**, 1005–1008.

Grigor'ev, D. P., 1965, "Ontogeny of Minerals," Jerusalem, Israel Program for Scientific Transl. (Transl. from Russian), 250pp.

Harder, E. G., 1919, "Iron-depositing bacteria and their geologic relations," *U.S. Geol. Surv. Profess. Paper* **113**, 89pp.

*Kuznetsov, S. I., Ivanov, M. V., and Lyalikova, N. N., 1963, "Introduction to Geological Microbiology," New York, McGraw-Hill Book Co., 252pp. (English transl.).

*Schopf, J. W., Baarghoorn, E. S., Maser, M. D., and Gordon, R. O., 1965, "Electron microscopy of fossil bacteria two billion years old," *Science*, **149**, 1365–1367.

*Silverman, M. P., and Ehrlich, H. L., "Microbial formation and degradation of minerals," *Adv. Appl. Microbiol.*, **6**, 153–206.

Silverman, M. P., and Lundgren, D. G., 1959, "Studies on the chemoautotrophic iron bacterium *Ferrobacillus ferrooxidans*. II. Manometric studies," *J. Bacteriol.*, **78**, 326–331.

* Additional bibliographic references may be found in this work.

Cross-references: *Free Energy or Free Enthalpy; Iron Oxidation and Reduction—Microbial; Organic Geochemistry; Oxidation and Reduction; pH-Eh; Photosynthesis; Sulfide Mineral Oxidation—Microbial; Sulfur Isotope Fractionation —Biological Processes; Sulfur Oxidation—Bacterial.* Vol. III: *Coral Reefs.* Vol. VI: *Fossil Fuels; Pelagic Sediments; Siliceous Ooze.* Vol. VII: *Coral Reef Builders; Microorganisms, Microbes.*

MINERAL GROUPS—ORGANIC—See **ORGANIC MINERALOIDS**

MINERAL PARTICLES AND HUMAN DISEASE

Origins of Mining and Mineral Exploitation

Sometime during man's evolution, between the development of *Australopithecus* as the ground-living ape man (ca. two million BC) and *Neanderthal Man* (ca. 100,000 BC), man learned to use some forms of natural rock and mineral fragments as tools. In the course of implement sophistication, some 250,000 years ago Sinanthropus (Peking Man) "domesticated" fire. With the development of these tools and the use of fire, higher social orders formed and the roots of civilization began (Table 1).

It was not, however, until 6000 BC that Neolithic man acquired tool-making materials by means of direct recovery from the earth. At this same point in human history man first smelted an ore to recover a metal. We could imagine how a fire accidentally set on a malachite bed and a weathered outcrop of cherty limestone along a stream provided the beginnings of mining and metal recovery. Man's first heavy occupational exposure to mineral dust had begun.

Only in relatively recent human history, however, has man become aware that those materials recovered from the earth have given rise to problems which have affected his health and well-being.

Dust Diseases of the Lung

Today, it is recognized that inhalation of many species of mineral particles may produce disease in man. The diseases mainly involve the lungs. The major pathogenic response is collagen formation; that is, the formation of long-chain polypeptide macromolecules of "fibrotic" tissue in areas of the lung where the gas exchange processes take place. The formation of this material decreases O_2–CO_2 exchange capacity and elasticity of the organ. The lung loses its operational capability and decreases in function as an organ. The degree of incapacity may range from slight to severe. This response of the lung to a mineral dust in which fibrotic tissue is formed is called a *pneumoconiosis;* if the pathogenic agent is identified specifically, it may be given a specific name, such as silicosis (from silica dust), talcosis (talc dust), asbestosis (asbestos dust) silicatosis (olivine dust), nephthelosis (nephthelene dust). Although all of these minerals induce collagen formation, the pattern of development differs for each; that is, the distribution and morphology of the fibrotic tissue changes with mineral type. An x ray of the chest will show different patterns of opacities,

Table 1. Human History and Mineral Exploitation

Date	Comments
1870	Industrial Revolution
	Use of steel
	Carbonized steel (ca. 1700)
1200	Wide use of many modern metals
$0_{BC}{}^{AD}$	
1000	First common use of iron implements ("Iron Age"); iron, copper, tin in use
1500	
2500	Bronze in use ("Bronze Age"); copper and tin in use
3000	Beginnings of written history; evidence of silicosis in mummies
4500	Iron in use in Egypt ("Chalcolithic" or "Copper-Stone Age"); grinding of stone
5500	Copper in use in Egypt ("Neolithic Age"); mining and smelting widespread
6000	Flint mined in Great Britain ("Mesolithic Age")
20,000	Upper Paleolithic Age (Magdalenian); widespread use of fashioned stone implements
70,000	Middle Paleolithic Age (Mousterian); use of flint, chert, and glass implements
150,000	Lower Paleolithic; use of fire and some fashioning of natural materials
550,000	First evidence of implements in association with human remains
600,000	Archeolithic Age; some evidence suggests the origin of the tool-implement usage pre-dates Lower Paleolithic
1,000,000	Pliocene-Pleistocene boundary
2,000,000	Australopithicus appears as the ground-living "man-like ape"

characteristic for each mineral type, suggesting that each mineral induces change along different metabolic pathways. It has been reported that the dimensions of collagen chains, as viewed by means of electron microscopy, differ in the different dust diseases. The general term, *pneumoconioses,* applies to "dust diseases of the lung."

Although the lungs are directly affected, other organs may be involved as well. For example, silicotic nodules (local concentrations of silica and altered tissue) have been observed in specific places in the lymphatic system (hilar nodes) and the spleen. Some indirect metabolic effects are also evident, as an increase in "rheumatoid factor" in human serum. The inhalation of mineral dusts invariably includes ingestion as well (swallowing of sputum and mucus).

Direct contact with fumes and dusts in mines may affect the skin, eyes, and nasopharynx directly; e.g., the inhalation of the mineral trona may induce nosebleeds. But these health effects are relatively minor when considering possible systemic disorders. There are a number of mineral dusts which can induce extremely serious diseases, such as malignant neoplasms. These minerals are termed carcinogenic agents. The radioactive minerals, containing uranium and thorium, have been shown to account for the development of lung cancer. Also, the inhalation of the mineral species called asbestos (chrysotile, crocidolite, anthophyllite, and amosite) has been correlated with significant increase of neoplasms of many types. The problem of correlation of mineral dust exposure and disease is related to many factors, which will be discussed below.

Diseases in Early Times

There is archeological evidence which suggests that Neolithic man used grinding as well as chipping techniques to fashion petrified wood, chalcedony, quartize, silicified slate, volcanic glass (obsidian), and fine-grained diorite into instruments. An expert so inclined to grind rather than chip may have suffered some form of pneumoconiosis. This assumption is based on observations made today among "Stone Age Cultures" in which men engaged in grinding quartz-containing rocks develop silicosis. The first "prima facie" evidence of pneumoconiosis in ancient man has been found in Egyptian mummies (ca. 3000 BC). Examination of their residual lung tissue indicated the presence of silica (quartz) particles admixed with residual leathery fibrotic tissue. This exposure was likely "environmental" rather than "occupational." The first written account of an awareness of mineral-dust inhalation and disease is recorded in the 5th century BC. King Solomon presented some twenty mining towns in Gallilee to Hiram, the king of Tyre. As the commentary indicates, Hiram was both an excellent and experienced miner, his knowledge obtained from the mining of gold in Ophir. When Hiram inspected the twenty mines presented to him in the land of Cabul, he reacted by returning them to Solomon! He did so because the gold and silver were admixed with cobalt. He had observed that cobalt ores were injurious to the miners and that they were usually so imbedded with other ores that it tended to make the recovery almost impossible.

In the 4th century BC, the practice of medicine flourished in Greece. Although the Hippocratic doctors largely ignored the diseases of slaves and low freeborn workers, they recognized the fact that miners tended to develop the disease which they called "asthma," which manifested itself by the symptom dyspnea (shortness of breath).

The Roman Empire by the last century BC and the 1st century AD had within it enormous expanses of natural resources. These were exploited to the fullest and, in doing so, great engineering skills in mining and metal recovery were developed. Concomitant with exploitation was the occurrence of disease. Pliny the Elder, the great chronicler of this time, recorded the dangers associated with "vermilion" workers who mined cinnabar (mercuric sulfide): "those employed in the works preparing vermilion covered their faces with a bladder skin of a pig that they may not inhale pernicious powder, yet that they can see through the skin." However, there were often times when crude precautions failed to prevent disease or death. For example, Pliny recorded the death of diggers in pits where "rising vapors pervaded the air."

So it was that more than 2000 years ago the beginnings of a highly ordered civilization demanded the exploitation and the use of natural resources. Exposure to dusts of rocks and minerals was observed to be injurious to the health of those who mined and processed the ores.

Beginnings of Occupational Health Study

The study of diseases associated with mining began during the Renaissance. For example, Paracelsus (late 15th to early 16th century physician) recognized the poisonous effects of mercury as well as toxic effects of many other metals. He also described the high prevalence of lung diseases in miners. Paracelsus' accounts were followed by the first, complete, treatise concerning mining and metal recovery, De Re Metallica, written by Georgius Agricola, published in 1556. Agricola was trained as a physician and practiced in the Austrian mining town of Joachimsthal. However, he was educated in the sciences and letters as well, and therefore his attentions rested upon diverse interrelated subjects. He was directly concerned with the health of the miners, but also gathered and recorded all of the facts known at that time on the subject of mining engineering, mineral exploitation and metal processing. Because of his training, interests, and education, he began the first interdisciplinary study of the diseases associated with mining and mineral processing. Many of the diseases were so well described that physicians today recognize them as specific types of pneumoconioses; e.g., silicosis, as well as differentiating them from tuberculosis and carcinoma of the lung. Agricola recommended methods to improve the working conditions in the mines; he suggested that ventilation would reduce dusts and would correspondingly reduce disease. The condition of extreme dryness produced clouds of dust which were "corrosive, and produced consumption of the lungs." The extent to which mineral dusts might shorten life was indicated by his anecdotal remarks concerning miners of the Carpathian Mountains. ("Women were found to have married seven times because their mining husbands died of 'the terrible consumption' observed in the mines in the area.") Agricola listed specific materials from which severe toxic reactions were observed. For examples, ores containing cadmium and cobalt were observed to eat away the wet feet and hands of workmen and also produced serious injuries to the lungs and eyes. He recommended the wearing of rawhide boots, elbow-length gloves and forms of face masks to protect themselves against such dusts.

Agricola also described diseases occurring among those men who worked in the metal recovery aspects of mining. In one of these smelting processes, which involved sphalerite, wounds and ulcers were caused in workmen who removed the zinc oxide residue, called black pompholyx, from the furnaces. The oxide was so corrosive that the keys used to open the smelting plant doors were made of wood rather than iron.

It was a short time after this period that a document related occupational diseases and air pollution. Authored by John Evelyn, a 17th century London physician, it recognized that soft coal burning in London produced ill effects on those who breathed the fume-laden London air. He had recognized that those who mined the coal suffered similar discomforts. He wrote, "Newcastle coal, as an expert physician affirms, causes consumptions, pthythis and indisposition of the lungs, not only by the suffocating abundance of smoke, but also by its virulency, for all subterranean fuels have a kind of virulence of arsenical vapor rising from it, which, as it speedily destroys those who dig it in the mines, so does it, by little and little, those who use it above them." This remarkable document suggested, on the basis of wrong evidence, that materials which cause disease among those who are occupationally exposed to them may also cause disease among people who use the materials.

It was not, however, until the Industrial Revolution that the application of 18th and 19th century science and technology provided western Europe and North America with widespread mining, smelting, and the associated host of new diseases and occupational disasters.

The Industrial Revolution and Modern Pneumoconiosis Concepts

With the advent of the Industrial Revolution, increased demands for raw materials were met with increased mining and milling operations. Mines employed hundreds of thousands of workers and corresponding numbers were employed in the mineral processing occupations. At this time, medical and social awareness increased and the problems of occupational exposure to mineral dusts took definition in the form of scientific inquiry.

In England in 1831 C. T. Thackrah presented his treatise on industrial diseases and singled out materials and industries where the prevalence of disease was marked. He recognized that miners and metal grinders suffered severe lung disease when sandstone was the principal exposure. He noted that limestone caused no such inconvenience or shortening of life. One factor he considered was that sandstone mines were primarily dry whereas limestone mines had a great deal of water percolating through them. He reasoned that, although the basic materials might have been different, the water in the limestone did decrease the amount of dust generated while working in them. He also recognized the general caustic nature of the sandstone dust; miners who were heavily exposed to this material rarely lived past forty years of age. According to Thackrah the grinders in Sheffield who used dry grindstones seldom lived past the age of about thirty years, whereas those who ground table knives and used wetted stones lived to almost forty and fifty years of age! This indicated to him that although the materials were pernicious, precautions might be used to suppress dusts and prolong lives.

Many of the first such classical observations were made in the mines of Great Britain. The workers in the Cornish copper and tin mines, which were principally in granitic rocks rich in quartz, suffered with very high silicosis rates. In 1863, 10,000 deaths of males between the ages of fifteen and seventy-five were recorded for Great Britain. Of these 10,000 deaths, 5596 were recorded in Cornwall, whereas only 2876 were reported for all the rest of England. Of these deaths in England, 1523 were attributed to "consumption" whereas of the deaths in Cornwall, 4439 were attributed to the same cause. This extraordinarily high consumption rate was attributed to the high dust content of the mines, and suggested to the physicians in the area that the tissue damage of the lungs was produced by the mineral dust. It was also recognized at the time that lungs so attacked by mineral dusts were highly susceptible to attack by microorganisms, especially tuberculosis.

By the end of the 19th century, it was widely recognized that a large number of minerals were pathogens and that inhalation or ingestion of these mineral dusts introduced health hazards. The hazards included shortening of life, or life with marked disability.

20th Century Modern Concepts in Minerals and Human Disease

In the first half of the 20th century, an impressive list of minerals and rocks which were known to cause disease in man had been compiled (Table 2). In order to understand the nature of the interaction mechanisms involved in producing disease, the nature of man's physiology and the dusts are required.

Man's Breathing Physiology. Human lungs are considered to be an extension of the skin. They are in direct contact with the environment. Man is a mammal who breathes through his nose. The nose serves as a filtering device equipped specifically to remove large particles from the air before it enters the lungs. However, man is conspicuous in his almost unique habit of breathing air through his mouth thereby bypassing the nasal passage mucociliary defense mechanism. Particles therefore normally filtered in the upper respiratory tract pass on through into the main bronchial tree. Elimination of particles from human lungs starts with particles impinging upon the mucus in the main bronchi. They are physically removed by a reverse peristaltic motion and ciliary action.

Man's breathing volume is approximately 300 cc of air per breath, at rest, and increases to almost 600–1000 cc per breath at work. He may breathe 12–40 (or more) times a minute, again a function of how much oxygen his tissues need and are getting. At strenuous work, up to 20 liters of air per minute may be exchanged between organism and environment; up to 1 m^3 of air per hour; almost 8 m^3 of air per workday. Taking into consideration the values of aerosol concentration in various mines and mills, this may constitute a considerable number of particles that may enter his lungs. Values for aerosols range from two particles per cc of air to several hundred particles per cc. Therefore, a man may inhale and ingest anywhere from 16,000,000 to al-

TABLE 2. NATURAL MATERIALS ASSOCIATED WITH HUMAN DISEASE

(1) ROCKS

Rock Type	Mineral Components	Comments
Coal	Anthracite Bituminous ± quartz	Coal miner's *pneumoconiosis*; miner's consumption"; focal emphysema; anthacosis; silico-anthacosis; may be related to rank (anth > bitum); "melanosis" or "black lung"; greater incidence of tuberculosis as well; may progress to "massive fibrosis"
Fullers earth	Bentonites and related montmorillonites ± quartz + free silica	Pneumoconiosis without massive fibrosis and nodules (diffuse); mottled x-ray appearance; some related silicosis
Diatomaceous earth	Opaline diatom fragments	Some lung scarring developed; much greater in processing when calcined (conversion of opal to cristobalite); silicosis with progressive massive fibrosis
Granite Quartzite Sandstone Slate	All with large amounts of quartz	Silicosis; silico-tuberculosis; nodular silicosis; fibrosis; enlarged and hardened lymph glands; silicotic nodules in spleen
Pumice	Volcanic glass, some devitrification to quartz	Resembles silicosis; some linear scarring
Limestone	Calcite ± quartz	Some bronchitis; some emphysema; some scarring reported; when calcined for industrial purposes, toxicity increases; caustic burns; dermatitis; ulceration of skin; injury to conjunctiva and cornea
Marble	Calcite ± quartz	
Dolomite	Dolomite ± quartz	
Gypsum	Gypsum ± evaporites	Some bronchitis; when calcined for industrial purposes, toxic effect to skin increases; some irritation to eyes, nose, and pharynx
Anhydrite	Anhydrite ± sulfur	
Bauxite	Hydrated aluminum oxides	Some reports of lung scarring

(2) SILICATE MINERALS

Mineral Name	Major Uses	Comments
Amosite	Used as asbestos	One of the major asbestos minerals; asbestosis; lung cancer; cancer of the gastrointestinal tract; pleural and peritoneal mesothelioma; possible increase in malignancies of the lymphatic system
Anthophyllite	Used as asbestos	See amosite
Biotite	Insulation and filler	Pulmonary fibrosis; silicosis; strong association with free quartz, likely a major factor in producing fibrosis
Chalcedony	Pottery and grinding material (abrasive)	Silicosis; "potters asthma"; "potters consumption"; silicotic nodules in spleen; silico-tuberculosis; progressive pulmonary fibrosis
Chert	See chalcedony	See chalcedony
Chrysotile	See anthophyllite	See anthophyllite; this mineral is the asbestos type which accounts for over 90% of asbestos consumption in the United States
Cristobalite	By-product produced	See chalcedony; cristobalite is more fibrogenic than quartz
Crocidolite	See amosite	See amosite
Diatomaceous earth (opal)	See chalcedony	See chalcedony; the opaline composition is altered when the earth is processed; it is often converted to cristobalite, in whole or in part

Feldspar (K,Na spar)	Ceramics	Silicosis; often attributed to the included quartz content of the pegmatite-derived mineral
Flint	See chalcedony	See chalcedony
Fluorite	Industrial flux	Fluorosis; silicosis; the latter is often attributed to the associated quartz
Graphite	Crucibles; hi T facings; electrodes; paints	Graphite pneumoconiosis; tuberculosis; resembles silicosis
Kaolin	Ceramics; filler	Some lung scarring observed, but only in areas associated with free quartz
Kyanite	Ceramics	Some lung scarring observed
Muscovite	See biotite	See biotite
Nepheline	Industrial uses	Nephelosis; some lung scarring
Olivine	Industrial uses	Silicatosis; some lung scarring
Phlogopite	See biotite	See biotite
Pumice (obsidian and scoria)	Abrasive; insulation	Silicosis
Quartz	Abrasive; industrial uses	See chalcedony; this material is likely responsible for a number of diseases attributed to other minerals
Sericite	Impurity associated with other minerals	Some fibrosis observed in men exposed to sericite dusts; however, attributed by some to the admixed free quartz
Sillimanite (mullite)	See kyanite	See kyanite
Talc	Industrial uses; filler	Talcosis; an asymmetrical fibrosis (one lung involved); talc pneumoconiosis
Tridymite	See cristobalite	See chalcedony; also, more fibrogenic than quartz

(3) Ore Minerals

Metal	Minerals	Comments
Aluminum	Bauxite ore corundum	Aluminosis; some lung scarring; corundum grinders may suffer a severe pneumoconiosis; emphysema; free silica (quartz) associated with corundum deposits, some scarring attributed to this
Arsenic	Cobaltite Enargite Realgar Orpiment Arsenopyrite Smaltite	Local skin and mucous irritant; carcinogen; anemia (hemolytic agent); hemoglobinuria; most severe reactions are observed during the smelting of arsenic-containing ores; miners of the ore are reported to have "excess" lung cancers although the published accounts are few; also, many of the ores are admixed with other materials which confound the data, and represent "mixed dust pneumoconiosis"
Beryllium	Beryl Chrysoberyl Bertrandite	Berylliosis; chronic lung disease; pulmonary lesions; acute poisoning; granuloma; pneumonitis; most of the severe reactions are observed in the processing from the ore; the ore minerals are associated with quartz-rich rocks which add a "mixed dust pneumoconiosis" effect
Cadmium	Greenockite	Most effects are related to mineral processing and recovery; generally a by-product of Pb and Zn smelting; no pneumoconiosis reported for the mineral dust itself; renal and pleural involvement

(3) Ore Minerals (Continued)

Metal	Minerals	Comments
Chromium	Chromite	Some reports of lung cancer among chromite miners, but most of the effects are among the "chromite workers" who process the ore; the ore also includes a number of other active mineral substances such as serpentine phases (chrysotile)
Cobalt	Smaltite Linneite Cobaltite Erythrite	Some reports of excess deaths due to lung cancer and cancer of the main bronchus; "hardmetal disease"; excess cancer among the cobalt ore processors
Iron	Hematite	Siderosis; scarring of the lung increases as the quartz content of the ore increases; some reports of increased lung cancer among the taconite ore miners of the Superior Region and Newfoundland; some experimental work supports this observation
Lead	Galena Cerussite Anglesite	Some reports of pneumoconiosis among the miners of galena, likely produced by associated mineral dusts; most of the severe reactions are observed among the processors and users of the by-products; diseases of the central nervous system; nephritis; "plumbism"; anemia; Pb content of ambient air in smelting towns is generally high
Manganese	Pyrolusite Braunite Manganite Hausmannite Rhodochrosite	Some reports of pneumoconiosis among the miners of manganese ores, but most of the effects are observed among the ore processors; effects the central nervous system (Parkinsonism syndrome); unusually high rate of pneumonia among ore processors
Mercury	Cinnabar	A range of systemic diseases are recognized among the miners and processors; nephrosis (renal lesions); "salivation"; "vertigo"; "paralysis"; "Hatters shakes"; erethism; stomatitis; "mercury poisoning"; perforation of nasal septum
Nickel	Pentlandite Niccolite Millerite Garnierite	Some pneumoconiosis observed among the miners but generally attributed to the admixed gangue minerals; some observations which indicate a higher than normal lung cancer rate among miners of Ni in "hard rock areas"; higher incidence of lung and nasopharynx cancer amongst Ni smelters
Phosphorus	Apatite	Some x-ray changes of lungs with minor scarring; some neurological disorders induced in men who process mineral for P recovery; phosphine and pesticide forms toxic-to-lethal
Platinum	Platinum Sperrylite	No disease associated with water-worked placer deposits; direct mining in mafic rocks generally associated with serpentine minerals; some lung scarring observed which resembles asbestosis (chrysotile?)
Selenium	Tiemannite Guanajuatite Clausthalite Naumannite Eucairite Chalcomenite	Pneumoconiosis observed are attributed to admixed gangue minerals; some observed in hard-rock mining of Pb,Cu,Hg,Ag, ores; most adverse effects observed in ore processors; severe irritation of nose and eyes; gastro-intestinal disorders; dental caries
Tellurium	Montanite Emmonsite Durdenite Tetradymite	As in the case of selenium, the pneumoconiosis observed are attributed to the admixed gangue minerals; some scarring in areas of hard rock mining of Bi,Pb,Ag,Hg; most adverse effects are observed in ore processors; gastro-intestinal disorders; renal disorders (impairment or death)

Element	Mineral(s)	Description
Silver	Silver, Argentite	Some x-ray changes of lungs but attributed to the admixed gangue minerals; some disease observed in men processing ore; argyria—discoloration of the skin to a pale gray or blue-gray
Tin	Cassiterite	Severe x-ray changes in miners; tin pneumoconiosis; some associated silicosis when quartz admixed in as gangue
Titanium	Rutile, Sphene, Ilmenite	Severe lung scarring, titaniosis; scarring regardless of the nature of the gangue minerals
Tungsten	Wolframite, Tungstite	Lung scarring observed among the miners of tungsten; tungsten pneumoconiosis; ore often associated with cobalt minerals which have been suspect as the causative agent
Uranium	Uraninite, Carnotite, Pitchblende, Thorium ores, Vanadium ores	Some silicosis reported for the sandstone deposits of uranium; excess lung cancers reported for the Colorado Plateau
Vanadium	Vanadinite, Carnotite	Some permanent lung scarring; immediate reaction is a respiratory irritant; irritant to eyes; susceptibility to pneumonia
Zinc	Sphalerite	Shortness of breath; some minor lung changes; some deaths reported from pneumoconiosis

most 2,000,000,000 particles from aerosols per workday! Particle retention and elimination rates have been measured and it has been observed that as time progresses, an individual exposed to heavy concentrations of dusts may lose his elimination potency. Therefore a larger number of particles are progressively retained. It is not surprising that in a fresh lung weight of 800 g, for a man who was occupationally exposed to heavy dust concentrations, 20 to 50 g of inorganic dust may be recovered after the individual's death. These particles are for the most part biologically resistant and they stay, permanently lodged in tissue. Some studies have shown that some mineral particles ingested by alveolar macrophages (large cells which phagocytize foreign bodies and remove them from the organ) kill the macrophage and become rereleased into the organ. This may be true for both quartz and asbestos mineral dusts.

Properties of Inorganic Dusts. There are a number of biologically important properties of mineral dusts. The *particle-size* average and range of the particles within the aerosol are most important. As the size of the particle decreases, its ability to enter the terminal bronchioles of the lung increases proportionately. It is generally considered that the particles with less than $5\text{-}\mu$ diameters may enter the terminal bronchioles and enter into the alveolar spaces. It is here that fibrotic tissue develops. However, with fibrous dusts, the fiber diameter may be less than $5\ \mu$ and the overall particle length may go up to several hundred microns. This has been observed in cases of asbestos exposure where fibers on the order of several hundred microns in length have been found beyond the terminal bronchioles. For a given mass, as the particle size of the dust decreases, the *surface area* increases. Therefore chemical potential and surface activity is proportionately increased. The relative proportions of each size class within the aerosol is considered important in disease etiology. It has been observed experimentally that biological response to dusts increases as the fine-size particles increase in number. Also, the nature of the particle *charge* determines how the particles may interact in the biological system. Test systems involving red blood cells show that alteration of surface charge influences degree of hemolysis. The *chemistry* of the particle is important in that compounds made up of toxic elements induce pathogenic responses when dissolved in the lung. This would include both *bulk chemical composition* and the *trace elements* associated with the particle. There are some minerals that have associated with them *trace minerals* on a submicroscopic scale.

One such mineral, chrysotile asbestos, has associated with it trace amounts of magnetite and chromite which are intergrown into the fibrils on a micron scale. These associated minerals may be released in the lung and may in themselves induce disease. The *crystal form* of a material may also induce different disease patterns within a biological host. For example, the different forms of silica such as quartz, tridymite, cristobalite, although all forms of SiO_2, produce different degrees of fibrogenic responses in laboratory animals. It appears that the crystallographic properties induce different responses in different periods of time; the more ordered the crystalline state of a given mineral is, the more severe the biological response.

Interaction Mechanism. There have been a number of theories proposed to describe the interaction mechanism of various types of mineral dusts in the cause of disease. One of these mechanisms involves *mechanical abrasion*. Particles physically rubbed against cell membranes induce rupture. This would either cause cell death or change the metabolic pathways of the cell resulting in a pathogenic response. This was suggested for the interaction of both silica and asbestos fibers. In the latter, it was presumed the needle-like particles punctured cell membranes and caused cell death in that manner.

Another possible interaction mechanism is *chemical*. According to this theory, the particle dissolves in part or totally in the lung and releases chemical agents which either passively or actively enter into biological reactions. Some elements inhibit defensive metabolic responses and in that manner cause disease indirectly. Other theories have proposed that substances eluted from minerals may act as *catalysts* for reactions and in some indirect manner induce diseases. Still other theories regard the *surfaces* of minerals as important factors in causing disease. The particles may act as *nucleating agents* and adsorb either toxic substances or various biological components necessary for cellular response and disease prevention. Others have indicated that surfaces of minerals are active sites for adsorbing proteins and amino acids and thereby altering significantly metabolic pathways.

Methods of Study. Of the many materials man may come in contact with, some may be active and alter normal body metabolism inducing pathologic responses. How can correlation of mineral exposure and disease be achieved? The science which does this is called *epidemiology*, the statistical analysis of disease patterns. One of the bases of the dust exposure-disease relationship is a comparison of the specific working group with the experience of the "general population." Knowing the overall disease incidence and prevalence rates, one can compare the "general experience" with the experience of specific workers. Using this data, one can compare, for example, the asbestos workers' morbidity (disease) and mortality (death) data with the general population. It has been shown that the death rate among asbestos workers is such that on the average they die much earlier than the "general population." Also, the diseases from which they die are quite different. Half of the workmen who are exposed to asbestos dust in their occupations die of malignant tumors. Comparing this with the overall "general population" of the United States, one finds that there is a tremendous excess of malignant disease deaths.

Using the epidemiological principles as outlined above, the occupational experience of any group of working men may be judged to be "hazardous" or "unhealthy."

There have been occasions where physicians working with a specific group of men did not need statistical analysis of morbidity or mortality data to indicate whether or not a particular group of men experienced greater death rates or died from specific diseases. For example, in the Joachimsthal area of Austria, the incidence of lung cancer among these miners was so great that it was readily observable to the local physicians. An important factor to consider in the epidemiological study of disease caused by mineral particles is the *long lapsed time period* between first exposure to the mineral and the first sign of disease. For example, asbestos workers do not develop their diseases until some twenty to thirty years after their initial exposure to the mineral dust. The concentrations are so high that extremely short time periods of exposure may be sufficient to induce pathogenic or carcinogenic responses.

Trace Elements and Disease

There have been numerous studies to indicate that trace elements in soils and minerals may be the active agents responsible for human disease. In the western part of the United States, cattle do not graze in areas where selenium-bearing strata outcrop. There have been studies in Sweden that have shown some correlation between the distribution of lead in rocks and soils and the occurrence of neurological disorders. There is furthermore some correlation between the occurrence of radioactive minerals in clays and soils and the occurrence of birth defects.

Although at the present time some of these correlations may appear to be hypothetical, certainly we know that the hardness of groundwater and in some areas the dry climate influence the endemic occurrence of kidney stones.

Measures Taken to Prevent Disease

Within the Mines. When minerals and rock materials are being mined, numerous measures should be taken to reduce the dust concentration in the ambient air. Ventilation of areas in which the materials are recovered, removed and transported, and the wetting down of dry areas, are the two most common dust suppressant measures used. Application and use of special recovery equipment, which produces little dust during operation, has also reduced the generation of dusts; e.g., special machinery used in the coal industry has limited and controlled coal-dust aerosols in many mines. Respirable dusts have also been reduced by the change in mining techniques, including the reconversion of high-level subsurface mines to open pits.

Measures Used by the Miner. The miner himself may take precautions and reduce the amount of respirable particles to which he may be exposed by wearing an air-filtering device. The air-fed respirator and face mask helps limit his exposure to respirable dust. However, facial comfort and local working practices often reduce the amount of time when these respirators are worn. In wet mines, where the combination of dampness and mineral dust may produce a corrosive effect upon the skin, special protective clothing has been worn by miners.

Regulatory Measures. In many instances decrease of disease associated with mining and mineral processing has been brought about through legislative action. For example, in 1914, the British Royal Commission recognized that the minerals quartz, quartzite, flint, and sandstone were substances which caused injury to the miners. Specific measures were legislated which indicated the maximum permissible levels of mineral dusts to which workmen could be exposed. It is of interest to note that some materials which were considered innocuous to humans at that time have since been found to be pathogens on continued observation and surveillance. Among these are coal dusts, slate, iron ore and talc dusts.

Controlling the Environment and Problem Recognition

Since the Industrial Revolution, man's workplace and environment have changed radically. His biological system has been exposed to ever-increasing amounts of toxic substances. These have been introduced to the earth's surface through mining and milling operations. Not only is the man exposed occupationally involved with these substances, but the exposure in some cases is so widespread that it has become a general environmental problem of global proportions. For example, the asbestos minerals have been so frequently used in a number of products and in a number of industries that their dissemination is almost universal. The use of asbestos spray insulation in New York City has reached such proportions that chrysotile asbestos fibers and fibrils are found in lung tissue of almost every urban dweller who dies within the city. This is also true of the less toxic mineral talc. Particles of this mineral have been found in the upper levels of the Greenland icecap indicating a worldwide dissemination after its use as a pesticide excipient.

What effects these low-level exposures have on human health is unknown at the present time. Certainly these minerals are pathogens and carcinogens if inhaled in large quantities. What makes these materials particularly insidious is that they remain permanently entrapped in tissue. The normal dosage-response relationships and toxicity threshold limit values do not hold for such long-lived materials. What is most important, however, is that the biologically active nature of these mineral substances should be recognized, and that many of them do indeed cause diseases.

ARTHUR M. LANGER
ANNE D. MACKLER

References

Agricola, Georgius, 1556, "De Re Metallica," Basel," translated by Herbert C. Hoover, 1950, New York, Dover Publications, Inc., p. 638.
Davies, C. N., ed., 1961, "Inhaled Particles and Vapours," Proc. 1st Intern. Symp. on Inhalation of Particles and Human Disease, Oxford, Pergamon Press, p. 495.
Davies, C. N., ed., 1967, "Inhaled Particles and Vapours," Proc. 2nd Intern. Symp. on Inhalation of Particles and Human Disease, Oxford, Pergamon Press, p. 605.
Holt, P. F., 1957, "Pneumoconiosis, Industrial Diseases of the Lung Caused by Dust," London, Edward Arnold Ltd., p. 268.
Hunter, Donald, 1969, "The Diseases of Occupations," London, The English Universities Press, Ltd., p. 1259.
International Conference on Pneumoconiosis, Johannesburg, 1969, Republic of South Africa, Department of Mines, p. 338, in press.
Orenstein, A. J., ed., 1960, Proceedings of the Pneumoconiosis Conference, 1959, Johannesburg, Boston, Little, Brown and Company, p. 632.

Cross-references: *Chronium; Cobalt; Environmental Science; Medical Geography; Medical Geology; Medical Geology—Trace Metals in Mammals; Pollution; Silicon; Thorium; Trace Ele-*

ments in Plants; Uranium. Vol. IVB: *Individual Minerals.* Vol. VI: *Mines and Mining.*

MINERALOGY

A mineral is a naturally occurring, inorganically formed chemical element or compound. Except for opal and native mercury, minerals have an orderly arrangement of their constituent atoms, ions, or molecules. This internal order is expressed externally as crystal faces and by various physical properties of minerals such as cleavage and diffraction of x rays.

Mineralogy is the study of minerals, their origin, occurrence, and physical and chemical properties. In 1774 the famous German mineralogist Abraham Gottlob Werner stated the objectives of mineralogy to be the development of: (1) an ideal mineral classification system and (2) more accurate means of mineral identification (Adams, 1938). With the discovery of x-ray diffraction and the development of more rapid and accurate analytical chemical techniques these objectives have been achieved. The current system of mineral classification is well established and unlikely to undergo any major change. With the development of modern instrumentation most mineral identifications became routine. The principal objective of modern mineralogy is to develop a basic understanding of minerals through their occurrence, chemistry, and physics, so that they may be used to better understand the genesis and subsequent history of the minerals themselves, the earth and the rock units of which it is composed. This fundamental understanding of minerals will come from both field and laboratory investigations and will incorporate the modern techniques and theories of chemistry and physics. Such investigations will include fields of study such as: crystallography (q.v.)—the study of the arrangement of the constituent ions, atoms, and molecules in minerals; crystal chemistry (q.v.)—the study of bonding between the chemical components; experimental mineralogy (see *Synthetic minerals*), the synthesis of minerals in laboratories in order to set limits and define as closely as possible the physical and chemical conditions under which given minerals are stable; chemical mineralogy (q.v.), and geochemistry, (q.v.)—study of the chemistry of minerals; and finally crystal physics and solid state technology—study of the physical phenomon associated with minerals. Important contributions will continue to be made by field studies and investigations.

History

The exact time when man began to study minerals is lost in antiquity. It was probably the first time he realized a particular mineral had a special beauty or utilitarian use and he began to search for more. The Greeks were the first to attempt to classify minerals, and did it ineffecually on the basis of color. Most of the Greek, Roman, and Middle Age writings are concerned with the gem minerals. Interest was on their mythical healing properties and the attributes they gave to the wearer. There were no investigations of minerals themselves and each succeeding writer drew from earlier writers. With very few exceptions little was added to the general knowledge of minerals.

It was not until the early sixteenth century that mineralogy began to emerge as a science. At this time the most important mining region in Europe was Saxony. In 1527 Georgius Agricola was appointed to the position of City Physician in the mining town of Joachimsthal in Bohemia. In Joachimsthal and later in Chemnitz, Agricola was closely associated with all aspects of the mining industry. Through his association with mining men and his frequent visits to mines and smelters he accumulated a wealth of practical information. According to Hoover (1950) ". . . he was the first to found any of the natural sciences upon research and observation, as opposed to previous fruitless speculation." It was because of Agricola's practical knowledge that he was able to propose a classification based on properties of minerals other than color alone. His classification was proposed in *De Natura Fossium* published in 1546. Although his classification, by today's standards, is naive it was an enormous improvement over anything previously proposed. Agricola was the founder of the science of mineralogy.

For the next 200 years interest in minerals expanded, principally for economic reasons. The most significant scientific advancement during this period was the discovery and demonstration, by Nicolaus Steno, that the interfacial angles on quartz were constant regardless of the size and shape of the quartz crystals.

By 1800 analytical chemistry was sufficiently developed so that relatively accurate chemical analyses of minerals were possible. It then became possible to classify minerals on something more substantial than appearance and physical properties. The leading mineralogist of the time was Abraham Gottlob Werner. Werner's recognition of the importance of chemical composition is shown by his classification which was based on both chemistry and physical properties. The first classification based entirely on chemical composition was proposed by the Swedish chemist Jons Jakob Berzelius.

Shortly before Werner's death another science began to emerge out of mineralogy, crystallography the science of crystalline material. The earliest leader in its development was Rene-Just Haüy. It was Haüy who first perceived the orderly internal arrangement of minerals while examining cleavage rhombs from a broken calcite crystal. Others who contributed to the growth and development of crystallography were Rome de L'Isle, William H. Wollaston, W. H. Miller, Auguste Bravais and many others. The 1800's were a period of observation and fact gathering. There were important advances such as development of lattice theory by M. S. Frankenheim in 1842 and Bravais in 1850, the development of the petrographic microscope by H. C. Sorby in the eighteen fifties. These advances supported the theory of an orderly internal arrangement within minerals, but revealed nothing concerning the specific arrangements. This was left to Max von Laue in 1912. At von Laue's suggestions, x rays were passed through a crystal of sphalerite. Diffraction of the x rays by the crystal lattice was recorded on a photographic plate. Two years later, William H. Bragg and William L. Bragg worked out the crystal structure of NaCl, halite. Since then the structures of many minerals have been determined.

In this short historical review many mineralogists have been neglected from lack of space, not lack of contribution.

Current Mineralogic Research

It was pointed out earlier that the object of modern mineralogy is to develop a basic understanding of the genesis and subsequent history of minerals. To do this requires both laboratory and field investigations.

Current laboratory studies can be divided into two broad categories. The first, which may be called analytical, is principally the measurement of chemical and physical properties of minerals. The second, which may be called experimental, is the synthesis of compounds having the same chemical and physical properties as minerals. These two types of studies are not unrelated, and very often overlap.

Analytical Studies. Analytical studies can be divided into three areas based on the properties being studied. The first involves studies related to the structure of minerals. The second studies the chemistry of minerals. And the third involves studies which are neither purely structural nor purely chemical.

Structural Methods. The object of structural studies is to determine the arrangement of the constituent particles. Knowledge of the arrangement has been useful in : (*1*) explaining the physical properties of minerals, (*2*) as a guide to the genesis and subsequent history of a given mineral, and (*3*) as a model for synthesizing synthetic material which have desirable physical properties. In a sense structural studies had begun when Haüy recognized the significance of the calcite cleavage rhombs. Although it was soon realized that cleavage fragments were not the basic building blocks of minerals as Haüy supposed, it was not until the development of x rays that serious structural studies became possible.

(a) X-ray diffraction. Although other methods of structural analysis have been developed, x-ray diffraction studies remain the most important. This method is based on the diffraction of x rays by the crystal lattice. This occurs because the wavelength of the x rays and the distance between the lattice planes are of the same magnitude, approximately 1–10Å. The x rays may be either white radiation as used in the Laue method or monochromatic as used in other methods of x-ray diffraction. X-ray diffraction techniques can also be divided into two groups on the basis of sample preparation. A powder sample is used for most routine identifications. For structural analysis and determinations of lattice constants a single crystal is used to diffract the x rays.

(b) Electron diffraction. Electron diffraction techniques are complementary to conventional x-ray diffraction techniques. The electron beam within an electron microscope is focused on the sample. The diffracted beam is recorded on a photographic plate. One essential difference between x-ray diffraction and electron diffraction is the wavelength of the radiation used. The wavelength of the electron beam is shorter by a factor of 1/10 to 1/100. Electron diffraction is especially effective in the study of the fine-grained platy clay minerals. When not being used for diffraction the electron microscope can be used to study the morphology of fine grained materials and to select grains for diffraction. Electron diffraction contrast photographs have been used to study twinning, exsolution lamellae, and defect structures.

(c) Infrared absorption. Absorption of infrared radiation results when a bond in the sample has the same vibration frequency as the incident infrared radiation. Lyon (1967) lists four situations in which infrared techniques are capable of distinguishing minerals. They are (*1*) minerals of constant composition, (*2*) minerals in a solid solution series, (*3*) polymorphs, and (*4*) minerals that are in a solid solution series and are also polymorphs.

Other techniques which have been used, but not extensively, in studying the crystal structure of minerals include: neutron diffraction and nuclear magnetic resonance.

Chemical Methods. The importance of the chemistry of minerals has been recognized since the days of Berzelius. Before x-ray diffraction became a common identification technique many mineral identifications were based on qualitative chemical methods, especially *blowpipe analysis* (q.v.). With the advent of rapid instrumental techniques following World War II, it became possible to make rapid and accurate analyses of minerals. Systematic chemical variations in minerals, chemistry of minerals from different geological environments, and chemistry of new minerals became easier to study. These studies have led to a better understanding of the origin and histories of many minerals.

The chemical elements, which compose a mineral can be divided into three general groups. Major elements are those which are characteristic and essential for the mineral to exist such as Si (silicon) in the mineral olivine (Mg_2SiO_4). The minor elements are those elements which substitute in large amounts ($>1\%$) for the major elements, but which are not necessary in order for the mineral to exist. Trace elements substitute in small amounts, ($<1\%$), for the elements of the previous two groups.

Below is a brief description of some of the techniques of chemical analysis of minerals.

(a) Gravimetric and volumetric analysis. Despite the numerous analytical instruments now available the gravimetric and volumetric wet chemical analysis techniques are more widely used than ever. The development of rapid analysis techniques and the need for carefully analyzed standards has prompted this increase. Nevertheless, these methods work best for experienced analytical chemists. Wet chemical methods are most suitable for the determination of essential and minor elements.

(b) X-ray fluorescence. X-ray fluorescence is perhaps the most widely used instrumental technique for chemical analysis. This technique is based upon the production of secondary x rays by impinging primary x rays on the sample. The wavelengths of the secondary x rays are characteristic of the elements in the sample; the intensity of the secondary x rays, at the characteristic wavelength, are proportional to the amount of the element in the sample. X-ray fluorescence is capable of analysis over a concentration range from 100% to several parts per million on at least sixty elements. The method is rapid and nondestructive.

(c) Electron probe microanalysis. A similar technique is *electron probe microanalysis* (q.v.). However, the source of excitation is a beam of electrons. Because the electron beam can be focused on a spot as small as 0.2μ this method is applicable to analysis of either very small samples or selected small areas within a sample. It is used chiefly for the determination of essential and minor elements.

(d) Emission spectroscopy. *Emission spectroscopy* (q.v.) is used chiefly for trace element analysis. A characteristic spectrum for each element in the sample is produced by exciting the atoms or ions in an electrical arc. This method is most satisfactorily used when the elements sought are present in amounts between 1 ppm and 10,000 ppm (1%).

(e) Atomic absorption spectroscopy. A beam of monochromatic light is passed through a flame into which a liquid sample is aspirated. An amount of light proportional to the amount of the sought after element in the sample is absorbed by atoms of the element in the flame. This method is used chiefly for trace elements, although the ease by which a liquid sample can be diluted makes it suitable for major and minor element analysis.

(f) Mass spectroscopy. Isotopic and trace element analysis can be performed with various mass spectrographs. Isotopic compositional data is used for absolute age determinations and paleotemperature studies. The mass spectrograph allows trace element analysis in the parts per billion range.

There are a number of other techniques which yield compositional data, such as activation analysis, chromatography, and polarography, but these are used to a lesser extent in mineralogy.

Other Methods. Into this category fall the methods and techniques which are neither structural nor chemical. Again the purpose of these techniques is to obtain data to aid in the identification and understanding of minerals.

(a) Optical microscopes. Optical techniques have been used for over 100 years and are one of the most rapid and accurate methods of mineral identification. The minerals may be studied in either transmitted or reflected light. In both methods plane polarized light may be used.

Transmitted light techniques are commonly used in the study of the rock-forming minerals, although it may be used in the study of any nonopaque mineral. When transmitted light is used the rock or mineral under study must be sliced to 0.3 mm or crushed to fragments in the size range of -100 mesh $+120$ mesh. If thin slices or thin sections, as they are called, are used then the textural reactions between the minerals may be observed. Compositional zoning within a mineral can be observed in a number of minerals if it is present. Identification and estimates of composition are easily made on minerals in thin sections.

Grain mounts are commonly used in more difficult identifications and for more accurate estimates of composition.

If the minerals are opaque, reflected light is used to study them. A highly polished plane surface is a prerequisite for use of reflected light. Although not as versatile as transmitted light, reflected light is commonly used in the study of ore minerals.

(b) Differential thermal analysis (q.v.) is used chiefly in the study of fine grained hydrous minerals. It is a means of identification as well as a means of studying solid state reactions induced by heating. With this method the sample and an inert standard are heated at a uniform rate. Exothermic and endothermic reactions take place when hydration and phase changes occur. The temperatures at which these occur are characteristic of many clay minerals, and can be used for their identification.

Experimental Mineralogy. Most minerals form in environments which cannot be directly observed or measured. The object of this area of mineralogy is to simulate these conditions in the laboratory, insofar as possible, in order to define the limits and conditions of stability.

Over the years a number of different experimental techniques have been developed. To treat this area of mineralogy we can divide it into three areas of research which overlap each other. They are (1) aqueous mineral synthesis, (2) moderate- to high-temperature mineral synthesis, and (3) high-pressure mineral synthesis.

Aqueous Mineral Synthesis. Aqueous mineral synthesis is concerned with the stability of minerals in the presence of a solution. Many of the initial studies were carried out at atmospheric conditions; however, it is quite possible to carry on these studies at moderate temperatures (600°C) and pressures. Most commonly studied are the effects pH and Eh on the solubility. Data from such studies is applicable to the formation of minerals on or near the surface of the earth and to the formation of minerals from aqueous solutions within the earth such as hydrothermal ore deposits.

Moderate- to High-Temperature Mineral Synthesis. There are many different types of studies carried on in moderate- to high-temperature mineral synthesis. In most studies the sample is enclosed in an inert metal capsule heated to a desired temperature, with or without an increase in pressure. When equilibrium within the sample has been achieved the capsule is quenched and the results studied. It is possible from such studies to determine the effects of pressure and compositional variations upon the melting points of a mineral or suite of minerals (rock). Polymorphic changes which occur with increasing temperature and/or pressure can be studied also.

Such research is applied to the formation of igneous and metamorphic minerals within the earth's crust.

High-Pressure Mineral Synthesis. High-pressure studies with or without increased temperature are used to simulate conditions below the earth's crust in the mantle. Such research involves pressures up to 200 kilobars (twenty million times the atmospheric pressure) and temperatures up to 2000°C (3992°F) (Bell and England, 1967). Under these conditions many reactions which are difficult to study at lower temperatures and pressures can be speeded up.

Summary. Minerals form in response to the interaction of chemical and physical conditions. Some minerals can form over wide ranges in these conditions, others form in very restrictive ranges. By observing where minerals occur in nature and synthesizing them in the laboratory we begin to set limits on their stability. Detailed chemical and physical investigations allow mineralogists to place greater limits on the environment in which a given mineral formed. Thus a history for that portion of the earth can be developed.

Books on Mineralogy

Because of both the scientific and popular interest in mineralogy a number of books have been written on the subject. They range from the most elementary books to highly specialized treatises on specific mineral groups, occurrences, or localities, and a number of books on gem minerals.

A number of guide books to mineral collecting localities are available. Many states have published guides to collecting areas within the state. The U.S. Geological Survey has published a book on *Gem Stones of the United States* (Schlegel, 1957) which lists collecting localities. *Mining and Mineral Operations in the United States—A Visitors Guide* lists mines, quarries and collecting areas which may be visited in each of the fifty states. In addition this book has a reference section which lists publications pertaining to the mineralogy and geology of a number of states.

H. L. McKague

MINERALOGY

References

Adams, F. D., 1938, "The Birth and Development of the Geological Sciences," New York, Dover Publications.

Bell, P. M., and England, J. L., 1967, "High-Pressure Experimental Techniques" in (Abelson, Ph. H., editor) "Researches in Geochemistry," Vol. 2, New York, John Wiley & Sons.

Correns, C. W., 1949, "Einführung in die Mineralogie," Berlin, Springer, 414pp.

Dennen, W. H., 1960, "Principles of Mineralogy," New York, Ronald Press Co.

Dietrich, R. V., 1969, "Mineral Tables," New York, McGraw-Hill Book Co.

Ford, W. E., 1949, "A Textbook of Mineralogy," New York, John Wiley & Sons.

Hoover, H. C., and Hoover, L. C., 1950, "Georgius Agricola's De Re Metallica" (introduction to translation), New York, Dover Publications.

Hurlbut, C. S., Jr., 1959, "Dana's Manual of Mineralogy," New York, John Wiley & Sons.

Kraus, E. H., and Slauson, C., 1947, "Gems and Gem Materials," New York, McGraw-Hill Co.

Liddicoat, R. T., Jr., 1962, "Handbook, of Gem Identification," Los Angeles, Gemological Institute of America.

Lyon, R. J. P., 1967, in (J. Zussman, editor), "Physical Methods in Determinative Mineralogy," New York, Academic Press.

Mason, B., and Berry, L. G., 1968, "Elements of Mineralogy," San Francisco, W. H. Freeman and Co.

Maxwell, J. A., 1969, "Rock and Mineral Analysis," New York, John Wiley & Sons, 584pp.

Oelsner, H. O., 1961, "Atlas der wichtigsten Mineralparagenesen im mikroskopishen Bild," Bergakad. Freiberg-Fernstudium, 309pp.

Philipsborn, H. von, 1964, "Erzkunde," Stuttgart, F. Enke, 247pp.

Pough, F. H., 1953, "A Field Guide to Rocks and Minerals," Boston, Houghton Mifflin Company.

Sinkankas, S., 1964, "Mineralogy for Amateurs," Princeton, Van Nostrand Co.

Sinkankus, S., 1959, "Gemstones of North America," Princeton, Van Nostrand Co.

Schlegel, D. M., 1957, "Gem Stones of the United States," *U.S. Geol. Survey Bull.* **1042-G**.

Staff, Bureau of Mines, 1967, "Mining and Mineral and Operations in the United States," Washington, U.S. Government Printing Office.

Zin, H. S., and Shaffer, P. R., 1957, "Guide to Rocks and Minerals," New York, Golden Press.

Cross-references: *Chromatography; Crystal Chemistry; Differential Thermal Analysis; Differential Thermogravimetric Analysis; Electron Microscopy; Emission Spectrography; Geochemical Classification of Elements; Geochemistry; Minerals; Infrared Analysis; Mass Spectrometry; Mercury; Neutron Activation Analysis; Radioactive Isotopes; Radioactaive Isotope Tracer Technology; Solid Solution; Spectrophotometry; Trace Elements; X-Ray Diffraction Analysis; X-Ray Spectroscopy.* Vol. IVB: *Blowpipe Analysis; Crystal Growth; Crystallography—Morphological; Defects in Crystals; Electron Probe Microanalysis; Fluorescence Techniques in Microanalysis; Fluorescence Techniques in Mineralogy; Gemology; High-Pressure Minerals; Mineral Classification: Principles; Mineral Deformation; Mineral Properties; Optical Mineralogy; Order-Disorder; Polarization and the Polarizing Microscope; Polymorphism; Synthetic Minerals.*

MINERALOIDS

According to the A.G.I. *Glossary* a mineraloid is "a term used to designate materials that are commonly considered to be minerals, but are *amorphous,* hence excluded by some definitions." An example is *allophane* (q.v.). Glass may also be a mineraloid. Webster's Dictionary does not define mineraloid, but the Oxford English Dictionary gives mineraloid as a "metamict substance derived from a mineral" and indicates it as essentially synonymous with *"Gel mineral,"* i.e., a noncrystalline substance that originated as a gel. However, "metamict" is assumed to apply to minerals where the structure has been destroyed by radioactivity, which would not necessarily be appropriate to all mineraloids. Structurally, the products of colloidal genesis are often colloform, "collomorphic" or reniform aggregates. But not all colloform structures are noncrystalline because in the course of time, under diagenesis, etc., the substance may slowly crystallize. As Lebedev (1967, p. 4) remarked: "Gels cannot exist for an indefinitely long time; generally they become crystallized and recrystallized. The gel textures are thereby transformed into crystallization textures (granular, fibrous, radial, etc.), and the aggregate becomes a metacolloid." In this way a given chemical association in one instance may be a mineraloid, while in another locality a similar combination may have a definite crystallinity. Minerals like limonite, collophane, or opal, for example, seem to show gradations. Most, if not all, mineraloids are low-temperature, low-pressure products related to weathering and hydration. Adsorbtion may play an important role, as is common in the weathering zone, and a wide variety of composition is thus indicated.

Many so-called minerals are thus amorphous mixtures and are only classified with the "mineral kingdom" as distinct from animal or vegetable, but do not conform to a mineral definition that requires crystallinity. Some of these amorphous species are solid solutions, as, for example, is *chrysocolla,* the nature of which was worked out in pioneering studies by Foote and Bradley (1913). Chrysocolla was found to be a solid solution of copper oxide, silica, and water.

Ross and Kerr (1934) pointed out that: "Minerals of this type introduce a problem in mineral nomenclature. Crystalline minerals have definite compositions, or vary only within definite limits, and so differ fundamentally from the amorphous materials. For this reason such amorphous materials have been called *"mineraloids"* to distinguish them from true minerals, although they are commonly listed as minerals in nearly all mineralogies and no doubt will continue to be so classified."

In considering the nature of *allophane* (q.v.), a clay mineraloid, Ross and Kerr concluded: "Optical studies, x-ray diffraction patterns, and dehydration curves all show that allophane and related materials have no definite atomic structure but are amorphous, that is, they are structureless mutual solid solutions of various definite chemical composition, and wide variations in composition are to be expected."

On the other hand, modern x-ray methods and electron microscopy have shown that many of the previously so-called amorphous minerals may be in fact submicroscopically crystalline and in some instances can be established as minerals and identified by their crystal systems. Many early references to amorphous materials may have to be revised.

R.W.F.

References

Foote, H. W. and Bradley, W. M., 1913, "On the solid solution in minerals. IV, the composition of amorphous minerals as illustrated by crysocholla," *Am. J. Sci.,* 33(4), 180–184.

Lebedev, L. M., 1967, "Metacolloids in Endogenic Deposits," New York, Plenum Press (transl. from Russian by J. B. Suthard), 298pp.

Ross, C. S., and Kerr, P. F., 1934, "Halloysite and allophane," U.S. Geol. Surv. Profess. Paper **185-G,** 135–148.

Cross-references: *Colloids; Organic Mineraloids.* Vol. IVB: *Allophane.*

MINERAL SPRINGS—*See* MEDICINAL SPRINGS

MINERAL SULFIDE OXIDATION— MICROBIAL—*See* SULFIDE MINERAL OXIDATION—MICROBIAL

MINERAL THERMOMETRY

Mineral thermometry, the study of the temperatures at which minerals form or were formed, is a branch of the wider study of the temperatures and pressures of the earth's crust. Although attempts have been made from the earliest days of geology to estimate the temperatures at which rocks and minerals were formed, absolute methods of reasonable accuracy have been developed only recently.

Among the more important of these developments has been the development since 1948 of the oxygen-isotope methods, the study during the last half century of silicate-phase-equilibria in the laboratory, the development during the last decade of the sphalerite thermometer, and the renewal during the same period of the study of inclusions as a means of thermometry.

For general reviews of the methods and problems of mineral thermometry see Ingerson (1955) and Smith (1963, pp. 403–524). Recent data on isotope methods is given by Rankama (1963).

Sedimentary Minerals

The carbonate oxygen-isotope thermometer is based on the variation, with temperature, of the distribution of the two oxygen isotopes O^{16} and O^{18} between calcite or aragonite and the water with which it is in equilibrium. If the O^{16}/O^{18} ratio of the water is known, and it can be assumed that the carbonate was deposited under equilibrium conditions and has not been altered since deposition, the method is accurate to $\pm 1°C$. By careful choice of the carbonate, usually shell material deposited by marine organisms such as belemnites, pelecypods, gastropods, and foraminifers, these conditions can be met. An example of the results obtained by this method is given as Fig 1.

It should be noted that the temperature given is that of the *deposition* of the carbonate and, for organic carbonate, some knowledge of the growth habits of the organism is essential before this can be interpreted in terms of regional temperature.

Phosphate oxygen-isotope thermometry has been suggested as a technique which, combined with carbonate thermometry, might enable the isotope ratios of ancient waters to be determined.

Igneous Minerals

Theoretical and laboratory studies of the stability of individual minerals and of phase relations in two, three, and four component systems, including the granite and the basalt systems, have made possible the understanding of the crystallization history of a great many rock types and hence the formation temperature of their constituent minerals.

Other methods that have been used include the study of liquid and of solid inclusions (see

MINERAL THERMOMETRY

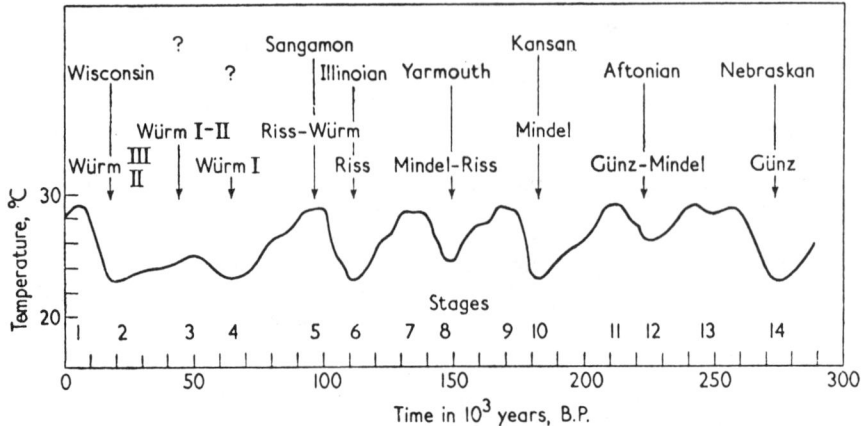

Fig. 1. Generalized Pleistocene temperature variations for low latitudes according to data based on oxygen-isotope thermometry (from Rankama, 1963; after Emiliani, 1955). (Editorial note: the correlations offered above suggest glacials and interglacials, but subsequent chronometric models differ notably. Ed.)

"Hydrothermal Vein Minerals" below), the study of the composition of co-existing titaniferous magnetite and ilmenite and the study of the feldspars.

Metamorphic Minerals

Metamorphic Facies. Metamorphic rocks may, on the basis of their mineralogy, be placed in various *metamorphic facies* (q.v. in Vol. V), each of which has a field of stability which is dependent on pressure, temperature and the chemical potential of the mobile elements. Estimates of the temperature-pressure boundaries of many of the facies have been made (see Vol. V, *Metamorphism: Temperatures and Pressures*).

Solid Solution. An approach which has the advantage of giving a continuous, rather than a discontinuous scale, has been the study of the composition of minerals which form *solid solutions* (q.v.) Among the single minerals which have been used are K-feldspar, plagioclase, garnet, and biotite, and among mineral pairs, plagioclase-zoisite, garnet-hornblende, garnet-biotite, garnet-chlorite, and ilmenite-titaniferous magnetite. Scales based on single minerals may be used only with rocks of uniform composition and within a single area but with this limitation may be very valuable. Thus it is found that the ratio of the refractive index to the length of the unit cell edge in garnet, which is a function of its composition, may be used as an arbitrary estimate of metamorphic temperature.

Study of mineral pairs enables palaeoisotherms to be drawn which are independent of the chemical environment, but the absolute temperatures have usually only been estimated from limited field data. An exception is the ilmenite-titaniferous magnetite pair which has been studied in the laboratory and is claimed to give a temperature accurate to $\pm 50°C$ (Buddington and Lindsley, 1964). This, with the *two-feldspar method* of Barth, seems particularly useful

Other Methods. Among other methods, the study of the fit of crystalline inclusions seems to be particularly promising. The temperature of *no-strain fit*, which, corrected for pressure, is assumed to be the temperature of formation, may best be obtained by the study of decrepitation curves.

In contact metamorphic rocks the existence of a temperature gradient makes a number of methods which give limiting temperatures particularly valuable. Among the more interesting of these is the demagnetization of magnetic minerals at their Curie points and the changing of the thermoluminescent patterns of limestones at fixed temperatures which may be determined in the unaltered rocks.

Hydrothermal Vein Minerals

The Sphalerite-Pyrrhotite Thermometer. The *sphalerite-pyrrhotite* thermometer is based on the observation that when sphalerite crystallizes under equilibrium conditions in the presence of pyrrhotite, the mole percent FeS in the sphalerite is a function of the temperature and pressure. This function has been experimentally determined. An example of the results obtainable is given in Fig 2. The accuracy of the temperature obtained is dependent on a number of factors which have been discussed by Smith (1963).

MINERAL THERMOMETRY

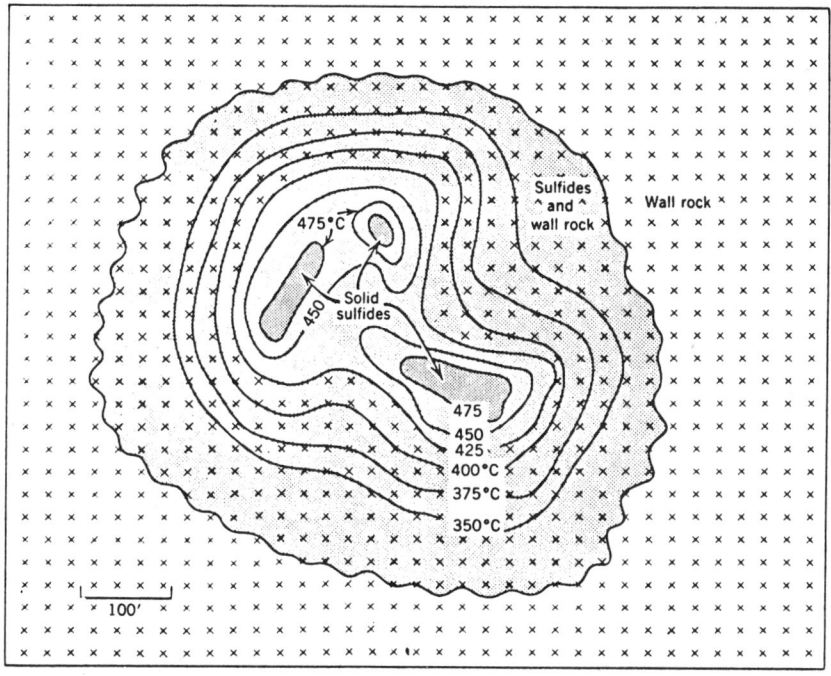

Fig. 2. Schematic plan view of the 1500-foot level of the Balmat No. 2 Mine showing isotherms based on the iron contents of sphalerites determined by Doe (1956) (from Kullerud, 1959).

Phase Changes in Inclusions. Inclusions which now contain more than one phase may, if the system has remained closed and no reaction between the mineral and the included material has taken place, be considered to have contained only one phase at the temperature and pressure of deposition. If the mineral is raised to a temperature at which there is again only one phase, this temperature is a minimum temperature of formation for the mineral and a point on the straight line which passes through the temperature and pressure of formation. The commonest type of inclusion for study is that containing, at room temperature, a saline liquid, and a gas bubble. The temperature at which only one phase is present may be determined by observation under a microscope fitted with a heating stage or by use of equipment to measure and record the decrepitation rate during heating. Such equipment produces graphs of decrepitation rate or decrepidographs which are often complex but with the aid of heating stage observations may be interpreted with some confidence.

Other Methods. Study of hydrothermal vein deposits has indicated that certain mineral assemblages are characteristic of high, medium or low temperature deposits. The temperature pressure boundaries for many of these are not reliably known although studies of synthetic sulfide systems are rectifying this (Kullerud, 1959). Other solid solution thermometers besides the sphalerite-pyrrhotite thermometers are possible, including the pyrrhotite-pyrite thermometer.

The Smith pyrite thermometer is a "thermoelectric potentiometer . . . calibrated to read temperature of crystallization of hydrothermal vein pyrite" (Smith, 1963). It is based on the use of thermoelectric potential as a measure of the size of blocks in the mosaic structure of pyrite which is assumed to be mainly dependent on the temperature of deposition. It is a method of high precision and low accuracy and can only be used safely on quartz-carbonate-sulfide hydrothermal vein pyrite.

Methods based on fit of solid inclusions and the thermoluminescence of transparent carbonates as well as the carbonate oxygen-isotope thermometer, and most of the other previously described methods, have occasionally been used in the study of hydrothermal vein minerals.

MICHAEL J. FROST

References

Buddington, A. F., and Lindsley, D. M., 1964, "Iron-titanium oxide minerals and synthetic equivalents," *J. Petrol.* **5,** 310–357

MINERAL THERMOMETRY

Emiliani, C., 1955, "Pleistocene temperatures," *J. Geol.,* **63,** 538–578.

*Ingerson, E., 1955, "Methods and problems of geologic thermometry," *Econ. Geol.,* 50th Ann. Vol., 341–410.

Kullerud, G., 1959, "Sulfide Systems as Geological Thermometers," in (Abelson, P. H., editor), "Researches in Geochemistry," pp. 301–335, New York, John Wiley & Sons.

Rankama, K., 1963, "Progress in Isotope Geology," New York, Interscience, 705pp.

Smith, F. G., 1963, "Physical Geochemistry," Reading, Mass., Addison-Wesley, 624pp.

*Additional bibliographic references may be found in this work.

Cross-references: *Earth's Crust Geochemistry; Fluid Inclusions; Hydrothermal Solutions; Mineralogy; Solid Solutions;* Vol. IVB: *Refractive Index; Thermoluminescence.* Vol. V: *Metamorphic Facies; Metamorphism: Temperatures and Pressures.*

MINERAL WAXES—See Vol. IV B, ASPHALT, ASPHALTITES, ASPHALTOIDS AND MINERAL WAXES

MINES AND MINING—See Vol. IV B

MODELS—See Vol. II

MOLECULAR SIEVES

A molecular sieve is one of a series of unique adsorbents which are found, for example, in crystalline metal aluminosilicates belonging to the class of minerals known as zeolites. The term molecular sieve generally refers to zeolites although clays, graphite, and feldspathoids are capable of accommodating foreign ions and molecules (see table 1; also *Fluid Inclusions;* (*Complexes*). Zeolites (q.v.) found in nature are secondary minerals occurring most commonly in cavities and veins of basic igneous rocks, particularly near volcanic vents, and in their deep-sea derivatives such as phillipsite. The Ca and Na content has been partly derived from feldspar and feldspathoids. The different varieties of the group are associated with each other and with pectolite, apophyllite, datolite, prehnite, and calcite. Synthetic forms of the minerals, as well as species unknown in nature, can be prepared by a hydrothermal process.

Properties

Zeolites have many unusual properties which are a consequence of their crystal structure. Members of this class of silicates undergo dehydration with little change in crystal structure. Dehydrated crystals are honeycombed with regularly spaced cavities interconnected by passages of molecular dimensions into which foreign ions and molecules can fit. Because of the high surface area of the cavities, the forces of adsorption are strong. The dimensions of the apertures in the channels determine the size of the adsorbed molecule. For example, chabazite may adsorb water vapor, and methyl and ethyl alcohol, but acetone and benzene are largely excluded.

Uses

The adsorption properties of molecular sieves are employed in drying gases and liquids to very low residual concentrations of water. Because of the uniform size of the channels in a molecular sieve those crystal molecules with a larger cross section than the effective diameter of the zeolite pore are excluded from entering the crystal. Molecules with smaller cross sections are adsorbed. This property enables one to separate molecules in fluid mixtures or the basis of shape and size as, for example, in the separation of straight-chain hydrocarbons from branched-chain paraffins and aromatic derivatives in a complex mixture. Hydrocarbon separation finds increasing application in the production of biodegradable detergents, and the improvement of automobile gasoline.

A major commercial use of zeolites is in water softening. Certain ions, especially Ca^{2+} and Mg^{2+}, cause hardness in water. They are undesirable because they form insoluble precipitates with soap (scum or curd). When hard water is permitted to pass through a bed of crushed zeolite, Ca^{2+} and Mg^{2+} ions are attracted by the crystal, and K^+ and Na^+ ions are leached out. For every Ca^{2+} or Mg^{2+}

TABLE 1. STRUCTURAL CLASSIFICATION OF INCLUSION COMPLEXES[a]

Arrangement of Guest Molecules	Examples of Corresponding Host Lattices
1. Molecules in isolated cavities	Nosean, sodalite, ultramarine feldspathoids, scapolite
2. Molecules in parallel channels	Cancrinite, sepiolite, attapulgite
3. Molecules between layers	Montmorillonite, vermiculite, graphite
4. Molecules in channel networks	Zeolites

[a] From Mandelcorn, 1964. p. 311.

atom, two K^+ or Na^+ ions are exchanged. This process is called *ion exchange*. When the zeolite becomes saturated with Ca^{2+} and Mg^{2+}, it is drained and then flooded with a concentrated NaCl solution. The zeolite may then be re-used. Today, synthetic zeolites and artificial resins have largely replaced the natural zeolites as commercial water softeners.

Structure

All naturally occurring zeolites are aluminosilicates with the exchangeable cations Na^+, K^+, Ca^{2+}, and Ba^{2+}. The general formula is (A_2^I, A^{II}) O, $Al_2O_3 \cdot nSiO_2 \cdot mH_2O$ with n and m variable. Artificial zeolites may also contain Li^+, Na^+, K^+, Rb^+, Cs^+, Tl^+, NH_4^+, and Sr^{2+}. The variability of n in the formula arises from isomorphous replacement of the type NaAl $\rightleftharpoons$ Si and CaAl $\rightleftharpoons$ NaSi. The basic structural unit in zeolites is the tetrahedron SiO_4^{4-} and AlO_4^{5-}.

A given tetrahedron is linked to four others through an apical oxygen atom but does not share faces or edges. The resulting network bears a negative charge because of isomorphous replacement of Si^{IV} by Al^{III}. The maximum substitution of Al^{III} for Si^{IV} is the ratio 1:1, and leads to a complete ordering of Al and Si. The charge is neutralized by an electrochemical equivalent of interstitial exchangeable cations. Positions of cations and water molecules are known with less certainty because of their mobile nature, but probably occupy positions in the cavities.

While the fundamental structural unit in zeolites is the tetrahedron, the complex frameworks they form are easier to visualize in terms of polyhedra constructed from groups of tetrahedra. The Al or Si atom is situated at the vertex of the polyhedron with the O atoms centered nearby. Eight tetrahedra may be connected to form a cubic unit and twelve to make a hexagonal prism. The cube as a structural unit is found in phillipsite and laumontite, while the hexagonal prism is found in chabazite, gmelinite, levyite, and erionite. A still more complex polyhedral unit is the cubeoctahedron, which forms the basis of the faujasite structure.

Some zeolites have an even more complex arrangement. The structure of natrolite is based on a cluster of five Al,Si tetrahedra joined in chains running parallel to the c axis, but forming crosslinks which are Si–O–(Si, Al) bonds (see Fig. 1).

The bond density is greatest in the direction of the c axis, which accounts for the fibrous nature of natrolite. Mordenite is interesting in several respects. It displays the minimum substitution of Al for Si (Al/Si = 1:5). Each $(Si,Al)O_4$ tetrahedron is part of one or more five-membered rings, which contribute to the extraordinary stability of this mineral. These five-membered rings are crosslinked to form chains parallel to the c axis. Two systems of intracrystalline channels exist: the large main channels parallel to [001], 6.6Å in diameter, and narrower passages, parallel to [010], 2.8Å across. Mordenite can readily sorb molecules of cross section less than 3Å, but does so more slowly with the large CH_4 and C_2H_6 molecules. Mordenite does not sorb molecules whose critical dimension exceeds 4Å, implying the presence of stacking faults which block the passage of larger molecules (Meier, 1961).

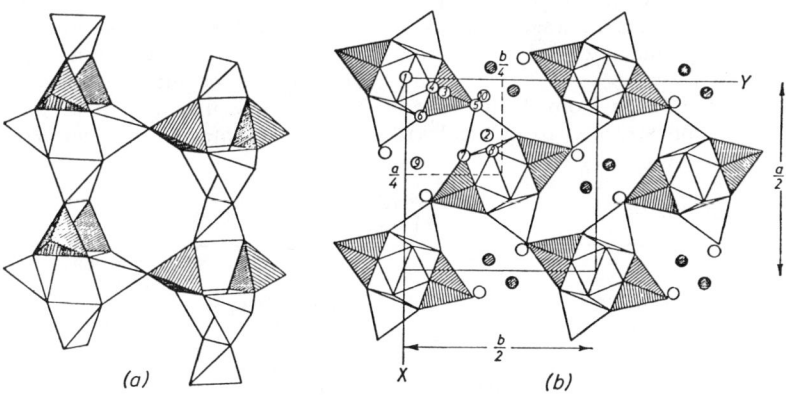

FIG. 1. (*a*) The chains in natrolite. The lower SiO_4 tetrahedra are linked to the upper AlO tetrahedra (shaded) of neighboring chains, (*b*) (001) projection of natrolite structure. Repeat distances are $a/2$ and $b/2$ in this projection. The notation used for atoms of the asymmetric unit is: (1) Si_I; (2) Si_I; (2) Si_{II}; (3) Al; (4) O_I; (5) O_{II}; (6) O_{III}; (7) O_{IV}; (8) O_V; (9) Na; (10) H_2O (*178*) (after Mandelcorn, 1964, Fig. 28).

MOLECULAR SIEVES

The zeolites belong to merely one family in a large number of inorganic inclusion complexes in which "guest" molecules occupy interstices in the crystal lattice of the host. These are summarized in Table 1.

VIVIEN GORNITZ

References

Breck, D. W., 1964, "Crystalline molecular sieves," *J. Chem. Educ.*, **41**, 678–689.
Deer, W. A., Howie, R. A., and Zussman, J., 1963, "Rock-Forming Minerals, Vol. 4, Framework Silicates," New York, John Wiley & Sons.
Mandelcorn, L. (editor), 1964, "Non-Stoichiometric Compounds," New York, Academic Press, Chap. 6.
Meier, W. M., 1961, "The crystal structure of mordenite (ptilolite)," *Z. Krist.*, **115**, 439–450.

Cross-references: Chelation; Complexes; Crystal Chemistry; Fluid Inclusions; Hydrocarbons; Ion Exchange; Mineral Classes: Silicates; Stoichiometry; Vol. IVB: Crystal Growth; Zeolites.

MOLECULAR WEIGHTS

The molecular weight of a compound is the sum of the atomic weights of the atoms in a molecule. The atomic weights (and therefore the molecular weights) are relative, originally being arbitrarily referred to an assigned weight of oxygen $= 16.0000$. However, since 1961 the scale of atomic weights has been based upon $C = 12.000$ for the atomic mass of the principal isotope of carbon (International Union of Pure and Applied Chemistry). A gram-molecular weight, or gram-mole, is the amount of a substance the mass of which, expressed in grams, equals the molecular weight. The number of molecules in a gram-mole is 6.023×10^{23}. This value is called *Avogadro's number*.

Molecular weights range from 2 (hydrogen gas) to several million (large virus molecules).

Calculation of Molecular Weights

1. Find the molecular weight of gypsum, $CaSO_4 \cdot 2H_2O$:

$$
\begin{aligned}
Ca &= 40.08 \\
S &= 32.06 \\
4 \times O &= 64.00 \\
2 \times H_2O &= 36.03 \\
\hline
M.W. &= 172.17
\end{aligned}
$$

2. Find the molecular weight of olivine with the composition $(Mg_{0.6}Fe_{1.4})SiO_4$:

$$
\begin{aligned}
0.6 \times Mg &= 0.6 \times 24.31 = 14.59 \\
1.4 \times Fe &= 1.4 \times 55.85 = 78.19 \\
Si &= 28.09 \\
4 \times O &= 64.00 \\
\hline
M.W. &= 184.87
\end{aligned}
$$

3. Find the % of Cu and S in chalcocite, Cu_2S:

$$
\begin{aligned}
2Cu &= 127.08 \\
S &= 32.06 \\
\hline
M.W. &= 159.14
\end{aligned}
$$

$$\% Cu = \frac{127.08}{159.14} \times 100 = 79.8\%$$

$$\% S = \frac{32.06}{159.14} \times 100 = 20.2\%$$

4. Calculation of the formula of a garnet from chemical analysis. The general formula of garnet is $X_6^{II}Y_4^{III}Si_6O_{24}$, where X represents divalent ions such as Ca, Mg, Fe^{II}, and Mn and Y represents trivalent ions such as Al, Cr, Fe^{III}, and Ti. Some tetrahedrally coordinated Al may substitute for Si. Table 1 shows a typical calculation.

In Table 1, column (1) gives the chemical analysis of the garnet, expressed in weight percent of the oxide. Column (2) is obtained by dividing the weight percentages of column (1) by the molecular weight of the corresponding oxide. Column (3) is derived from column

TABLE 1.

	Wt % Oxide (1)	Molecular Proportion of the Oxide (2)	Atomic Proportion of Oxygen (3)	No. cations Based on 24 Oxygen (4)	No. Ions in Formula (5)
SiO_2	38.03	0.6332	1.2664	11.904	5.952
Al_2O_3	22.05	0.2163	0.6489	6.099	4.066
Fe_2O_3	0.88	0.0055	0.0165	0.155	0.103
FeO	29.17	0.4060	0.4060	3.816	3.816
MnO	1.57	0.0221	0.0221	0.208	0.208
MgO	6.49	0.1610	0.1610	1.513	1.513
CaO	1.80	0.0321	0.0321	0.302	0.302
Total	99.99		2.5530		

(2) by multiplying each entry by the number of oxygen atoms in each oxide. If the garnet formula is based on a total of 24 oxygen atoms, the oxygen atomic proportions must be recalculated to a total of 24, which is done by multiplying all of them by 24/total of column (3) (24/2.553 = 9.400). Results are tabulated in column (4). Column (5) lists the number of cations associated with oxygen in column (4). The column (4) entry is divided by the number of oxygens per metallic element (e.g., for SiO_2, divide column (4) by 2; for Al_2O_3, divide column (4) by 3/2). The formula then becomes

$$(Ca_{0.3}Mg_{1.5}Fe_{3.8}{}^{II}Mn_{0.2})\ (Al_{4.0}Fe_{0.1}{}^{III})$$
$$(Si_{5.95}Al_{0.05}O_{24})$$

5. Find the molecular weight of the garnet in problem 4.

$$M = \frac{D \times V \times N}{Z}$$

$$D = 4.08 \text{ g/cc}$$
$$V = Cl_a{}^3 = (11.53\text{Å})^3$$
$$= 1532.8 \times 10^{-24} \text{ cm}$$
$$Z = 8$$

$$M = \frac{4.08 \times 1532.8 \times 10^{-24} \times 6.023 \times 10^{23}}{8}$$

$$M = 470.8$$

(Data and method adapted from Deer, Howie, and Zussman, 1966.)

Experimental Methods

Molecular weights are determined by a variety of methods, the choice depending upon the properties of the substance, the purpose, and the available equipment. Some of the more familiar methods are the gas-density method, colligative properties, mass spectroscopy, ultracentrifugation, and x-ray diffraction.

The *gas-density method* is based upon Avogadro's law which states that equal volumes of different gases at the same (low) pressure and temperature have equal numbers of molecules, and therefore the weights of gases are proportional to their molecular weights. The method consists of measuring the density as a function of pressure at constant temperature. The ratio d/p is linear at low pressure. The ideal (or perfect) gas law relates molecular weight to the density, pressure, and temperature of gas:

$$M = \frac{w}{V}\frac{(RT)}{p} = \frac{d}{p}RT$$

where w = weight, V = volume, p = pressure, R = universal gas constant, and T = temperature in degrees absolute. Therefore

$$(d/p)_{p=0} = M/RT$$

The ideal gas law is strictly valid only at $p = 0$, but is approximate at atmospheric pressure. The pressure coefficients for different gases at one atmosphere may be quite appreciable, and they differ for each gas.

Colligative properties such as vapor-pressure lowering, boiling-point elevation, freezing-point depression, and osmotic pressure all depend upon the number of solute particles dissolved in solution. Boiling point (bp) elevation and freezing point (fp) depression are commonly employed to compute molecular weights.

The addition of a nonvolatile solute to a solvent causes an increase in the boiling point which is proportional to the molal concentration of solute.

$$K \text{ bp} = RT\text{ bp}^2/\Delta H_{vap}N$$

where $K\ bp$ = ebullioscopic constant; $T\ bp$ = B.P. of solvent; ΔH_{vap} = heat of vaporization of the solvent; R = gas constant; and N = no. of moles solvent/1000 g solvent

To find the M.W. of dilute solutions:

$$M = K \text{ bp}\frac{(1000)}{(G)}\ g/\Delta T \text{ bp}$$

where g = wt in grams of solute; G = wt in grams of solvent; and ΔT = difference in boiling points between the solution and pure solvent.

In a similar manner, it can be shown that addition of solute to solvent causes a lowering of the freezing point. The molecular weight is calculated by corresponding formulas:

$$K \text{ fp} = \frac{RT \text{ fp}^2}{\Delta H_f N}$$

where $K\ fp$ = cryoscopic constant; $T\ fp$ = freezing point of solvent; and ΔH_f = heat of fusion

$$M = K \text{ fp}\frac{(1000)}{G}\ g/\Delta T_f$$

where g and G have the same meaning as before and ΔT_f is the difference in F.P. temperatures between the pure solvent and the solution.

In *mass spectroscopy*, the isotopes of a chemical element are separated according to mass. The molecular weight is found by calculating the physical atomic weights of different isotopes of the same element and their relative abundances.

MOLECULAR WEIGHTS

Ultracentrifugation is applied to the determination of molecular weights of macromolecules that have been subjected to intense gravitational fields.

X-ray methods can be used to find the molecular weight of crystalline compounds. This method is probably the best suited for geologic problems.

X-ray diffraction data yield cell dimensions, from which one can calculate molar volume. Knowing the density, and the number of molecules per unit cell, one can compute the molecular weight; for example:

$$D = \frac{M \times Z}{V \times N}$$

where D = density; M = molecular weight; Z = molecules or atoms/unit cell; V = volume of the unit cell; and, N = Avogadro's number.

VIVIEN GORNITZ

References

Benson, S. W., 1963, "Chemical Calculations," Second ed., New York, John Wiley & Sons, 254pp.

Cameron, A. E., and Wichers, E., 1962, "Report of the International Commission on Atomic Weights (1961), *J. Am. Chem. Soc.*, **84**, 4175–4197.

Clark, G. L. (editor), 1966, "Encyclopedia of Chemistry," Second ed., New York, Reinhold Publ. Corp.

Deer, W. A., Howie, R. A., and Zussman, J., 1966, "An Introduction to the Rock-Forming Minerals," New York, John Wiley & Sons, 528pp.

Cross-references: *Mass Spectrometry; X-Ray Diffraction Analysis*. Vol. II: *Avogadro's Law, Number*.

MOLYBDATES AND TUNGSTATES

This is a mineral class containing $(MoO_4)^{2-}$ or $(WO_4)^{2-}$ anionic groups with tetrahedral coordination, producing anisodesmic compounds. There are two important isostructural groups in the anhydrous division of this class. The wolframite group has complete substitution of iron (Fe) and manganese (Mn). The scheelite group has partial substitution of tungsten (W) for molybdenum (Mo) and partial substitution of lead (Pb) for calcium (Ca). There is a complete solid solution series between scheelite and powellite. (See Table below.)

ANHYDROUS MOLYBDATES AND TUNGSTATES
$A(XO_4)$ Type

Huebnerite	$Mn(WO_4)$	*Wolframite group:* complete solid solution series exists between huebnerite and ferberite
Wolframite	$(Fe,Mn)(WO_4)$	
Ferberite	$Fe(WO_4)$	
Sanmartinite	$(Zn,Fe)(WO_4)$	
Scheelite	$Ca(WO_4)$	*Scheelite group:* partial substitution of Pb for Ca between scheelite and stolzite, and powellite and wulfenite; also partial substitution of W for Mo between scheelite and powellite, and wulfenite and stolzite
Powellite	$Ca(MoO_4)$	
Wulfenite	$Pb(MoO_4)$	
Stolzite	$Pb(WO_4)$	
Raspite	$Pb(WO_4)$	

BASIC AND HYDRATED MOLYBDATES AND TUNGSTATES
Miscellaneous

Cuprotungstite	$Cu_2(WO_4)(OH)_2$
Koechlinite	$(BiO)_2(MoO_4)$
Ferritungstite	$Fe_2(WO_4)(OH)_4 \cdot 4H_2O$ (?)
Lindgrenite	$Cu_3(MoO_4)_2(OH)_2$
Ferrimolybdite	$Fe_2(MoO_4)_3 \cdot 8H_2O$ (?)
Anthoinite	$Al(WO_4(OH) \cdot H_2O$

LLOYD W. STAPLES

References

Dana, J. W., and Dana, E. S., 1951, "The System of Mineralogy," Seventh ed., Vol. II, 1063, (revised by Palache, C., Berman, H., and Frondel, C.), New York, John Wiley & Sons.

Kerr, P. F., 1946, "Tungsten mineralization in the United States," *Geol. Soc. Am. Mem.*, **15**.

Vermaas, F. H, S.. 1952, "South African scheelites and an x-ray method for determining members of the scheelite-powellite series," *Am. Mineralogist*, **37**, 719.

MOLYBDENUM: ELEMENT AND GEOCHEMISTRY

Cross-references: *Mineral Classes: Nonsilicates; Solid Solution.*

MOLYBDENUM: ELEMENT AND GEOCHEMISTRY

An oxide and acid of molybdenum were prepared from the mineral molybdenite, MoS_2, and recognized as belonging to a new element by the Swede K. S. Scheele in 1778. The metal was prepared from the oxide by P. J. Hjelm in 1782. Molybdenum is a very heavy metal with a specific gravity of 10.2 at 20°C; it is malleable, ductile, and exceedingly tough. Its oxide vaporizes (sublimes) at about 1400°C, the pure metal melts at 2622°C±10°C and boils at 5560±20°C when it is shielded from atmospheric oxygen and other substances it could combine with. The heat of fusion (melting) is 6.6 kcal/g-mol; the heat of vaporization is 128 kcal/g-mol; the thermal conductance at 20°C is 0.35 cal/cm²/cm/°C/sec; the specific heat at 20°C is 0.061 cal/g/°C; the electrical conductance from 0 to 20°C is 0.19 microhms −1. The atomic volume of molybdenum is 9.4, and the atomic radius =0.68Å when it is in the metallic state. The atomic weight of molybdenum is 95.94. The atomic number is 42. The metal crystallizes with body-centered cubic symmetry, space group: O_h^9-Im3m, lattice constants: 3,150±0.005, structure type: A2(b.c.c.).

The atomic radius is 0.68Å(+4). The first ionization potential of the metal is 166 kcal/g-mol. Pauling's value of electronegativity is 1.8. In naturally occurring crystals it is most commonly found in fourfold and sixfold coördination. The ionic radius of a 4+ ion is 0.70Å, and the ionic radius of a 6+ ion is 0.62Å.

There are seven stable isotopes of molybdenum. The most common is 98(23.78%); the others in order of decreasing abundance are: 96(16.53%), 92(15.84%), 95(15.72%), 100 (9.63%), 97(9.46%), and 94(9.04%). The most common radioactive isotope generated in pile reactors and as debris in nuclear explosions is 99 with a half-life of 67 which decays by emission of electrons and gamma rays; rarer unstable isotopes are 90 and 91 which emit positrons, 93, 101, and 102, which emit electrons during decay.

Chemical Properties

The most common valence is +6, less common valences (oxidation states) range from 5+, through 4+, 3+, to 2+, the least common. The electron structure is: $(Kr)4d^55s^1$.

At high temperatures molybdenum has a high affinity for iron, and to an extent for Ni, Ti, Mn, and Mg. This suggests that Mo would tend to concentrate in the iron-nickel rich core of the earth and other terrestrial-type planetary bodies. Molybdenum ions tend to be fixed by FeO and MnO_2 radicals. This suggests that Mo could be "ballooned" upward from the interior of the earth. It further suggests that the mantle of the earth may be depleted as compared with the core and the near surface volumes of the planet.

Molybdenum has a high affinity for sulfur. During hydrothermal crystallization (crystallization from very hot water usually under high pressure) sulfides of other metals will only form after all the molybdenum has formed molybdenite, MoS_2. At markedly lower temperatures, particularly in the absence of sulfur, molybdenum ions show a significant affinity for calcium and lead ions.

Where manganese dioxide (MnO_2), ferric oxide (FeO) ochers or their hydrated oxides are present, as in zones of oxidation or sediments, molybdenum tends to enter configurations similar to $FeO \cdot 3MoO_3 \cdot 8H_2O$. The manganese-iron-rich nodules abundant on the ocean floor can typically have molybdenum contents from 200 to 2000 g/ton. In mesothermal (medium-temperature) and epithermal (relatively low-temperature) mineralized veins there is tendency for bluish, *ilsemannite* crystals to form when ferrous iron is low and the concentration of molybdenum ions is high. Beautiful, golden, *wulfenite* crystals tend to form in the proximity of siderite ($FeCO_3$) and as tiny inclusions in the siderite. These tendencies are expressed in the equation

$$Fe^{2+} + Mo^{6+} = Fe^{3+} + Mo^{5+}$$

Minerals

The most common molybdenum mineral is *molybdenite*, MoS_2, found as a primary mineral in metasomatic deposits such as Climax, Colorado, and eastern Greenland; in most acidic (quartz-rich) and often in the much rarer basic (calcium-rich) pegmatites; and in hydrothermal veins, particularly hypothermal (high-temperature), mesothermal (medium-temperature), and not as frequently in epithermal or telethermal (low-temperature) veins. The free energy of formation of molybdenite is −54.19(kcal/g-mol).

Other molybdenum minerals which are occasionally recovered are: *molybdite*, MoO_3, with a free energy of formation of −157.6 (kcal/g-mol), typically formed from molybdenite during the cooling stages of replacement, pegmatite, and hydrothermal-vein deposits. *Wulfenite*, $PbMoO_4$, and *powellite*,

$CaMoO_4$, are typically formed during the cooling stages and weathering of vein and disseminated deposits of copper and lead mineral concentrations. Powellite is often associated with scheelite and thus recovered with tungsten.

Other molybdenum minerals not currently recovered commercially include: *Ilsemannite*, $Mo_3O_8 \cdot nH_2O \pm H_2SO_4$ (molybdenum blue) found in the Valley of Ten Thousand Smokes where the molybdenum ions originate from rhyolite magmas and are precipitated as ilsemannite by the action of H_2S on these ions. Ilsemannite is also found in pelites (muds) of reducing enviornments like the Black Sea where H_2S acts to precipitate molybdenum ions and colloids from the water. *Chillagite*, $Pb((Mo,W)O_4)$ is an isomorphic mix of wulfenite and stolzite formed as veins and pegmatites cool and/or are weathered. The *Scheelite Series* is an isomorphous group with the general formula: $Ca,Sr,Ba, Pb(Cd,MoO_4)$, (WO_4) formed largely as a replacement series during the cooling stages of mineralized areas, and to a much lesser extent by replacements during weathering processes.

Abundance

In nebulae Mo is probably concentrated at about 3 g/ton, in the solar atmosphere, 0.02 g/ton, in Ni-Fe meteorites, up to 16.6 g/ton (maximum), and in troilites, up to 11 g/ton, according to Noddack.

In basaltic rocks on earth the abundance is 2–4 g/ton (Sahama) and in granitic rocks, 2.3 g/ton (Goldschmidt, 1958). In ocean waters it is 0.001–0.005 g/ton and seems to be present as molybdate ions and complex anions, to a very minor degree as colloidal suspensions. Here the Mo originates from continental erosion and river transport; and also from leaching of freshly exposed basalts, gabbros, peridotites, etc., at continually spreading mid-oceanic rifts.

In Sediments and Soils. Precipitates as manganese dioxide nodules are found abundantly spread over certain areas of the ocean floor containing up to 1000 g/ton and even 3000 g/ton Mo.

Saprolites (muds rich in organic matter) in the Dead Sea asphalts contain 100–1000 g/ton Mo; black schistose pyrite-rich shungites of Finland have up to 50 g/ton; black shales in general have 10–50 g/ton Mo.

Silts and muds (pelites) typified by the slightly metamorphosed Kupferscheifer (copper shale) or Mansfeld, Germany, contain 50–200 g/ton Mo, apparently having precipitated from the ocean waters by the hydrogen sulfide-rich reducing environment. Calcareous clays and shales in general have 24 g/ton Mo.

Sandstone (psammites) of mixed quartz, feldspar, calcite grains, and muds carry g/ton; sandstones of only quartz and feldspar have about 2 g/ton Mo. The sandstones of Colorado which contain carnotite deposits contain distributed Mo in the form of ilsemannite; it seems to have been transported there as an oxidation product, colloid, complex anion, or molybdate ion and precipitated as alkali molybdates by waters rich in calcium ions. Conglomerates (psephites) carry about 2 g/ton Mo.

Molybdenum concentration in forest litter is up to 50 g/ton, in resins up to 800 g/ton, and in amber up to 700 g/ton. Fungi and nitrogen-fixing azotobacter tend to concentrate Mo up to 100 g/ton. A cattle condition or "disease" develops due to grazing over alkaline soils in which Mo is concentrated to 24 g/ton or so.

In Metasediments. Massive gabbro and norite bodies (formed from volcanic piles and intercalated calcareous and silicious sediments) carry 3–25 g/ton Mo. Small norite, gabbro, and charnockite bodies (formed from calcareous and silicious sediments) carry 6–50 g/ton. Calcareous granulites (formed from calcareous muddy sandstanes) carry 15 g/ton. Schists (formed from silts, muds, and organic-rich muds) locally have 30–500 g/ton. Cherty quartzites (formed from chert beds and volcanic ash beds) have 1–5 g/ton. Granulites (formed from sandstones and grits) carry about 1 g/ton. Gneisses (formed from sandstones, grits, and conglomerates have about 2 g/ton.

Granitic (rhyolite) magmas which probably incorporate masses of sediments, melting them into the liquid as they move toward volcanic vents, carry 7 g/ton Mo. On the other hand, basaltic magmas (often called oceanites) have only about 2 g/ton. Plateau basalts have about 4 g/ton Mo.

Cosmochemistry of stable and unstable molybdenum nuclides is currently a matter of educated guessing based on very scanty information. According to the theorems of imperfect statistics general principles can be ventured in this fashion; however net information available at this time seems to be far short of firm (1σ, 2σ, $n\sigma$ . . .) statistical thresholds. Molybdenum nuclei are relatively stable and could only be fused into heavier nuclei, or split (fissioned) into lighter nuclei in limited volumes of the universe. These volumes include supernovae, possibly hot, O- and B-type stars, possibly quasers, and to a minor? extent by cosmic rays. Observation of spectral lines radiated from the debris of supernovae such as the Crab Nebula indicate a greater abund-

MOLYBDENUM: ELEMENT AND GEOCHEMISTRY

TABLE 1.

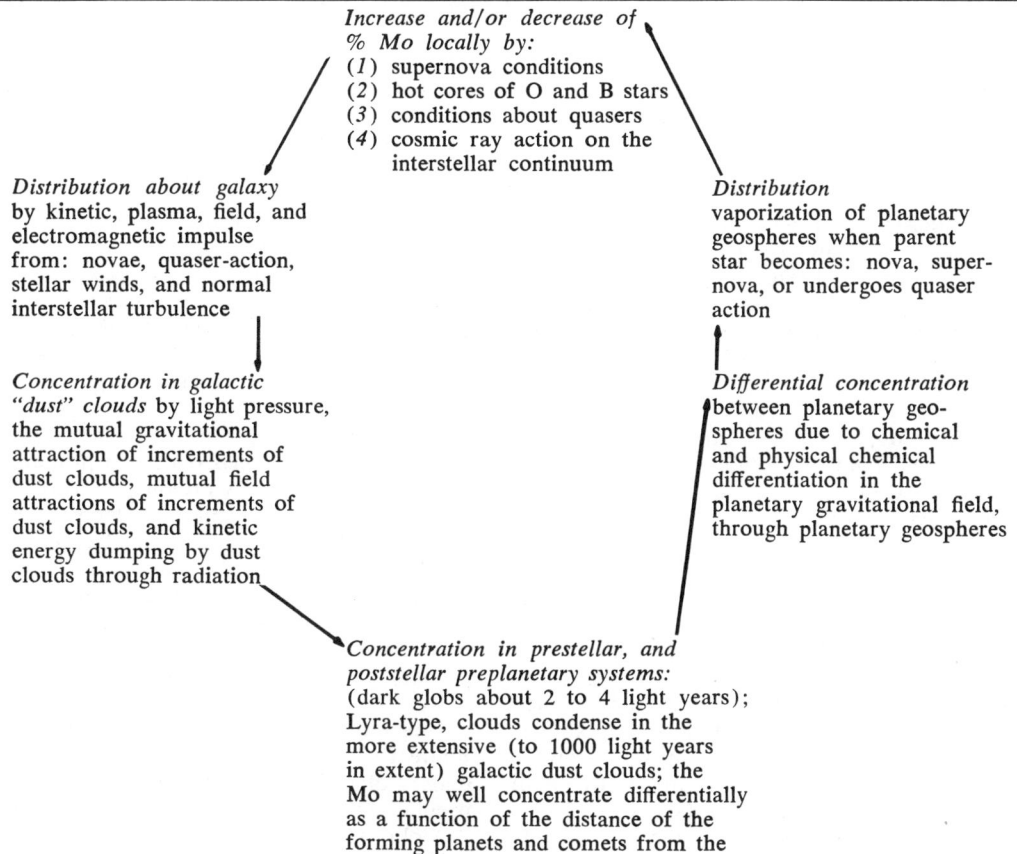

ance of Mo than is observed in stellar atmospheres. Possible trends are diagrammed in Table 1.

Differentiation and cycling of molybdenum between terrestrial geospheres may be surmised from more abundant observations than those available to cosmochemists. Planetologists have available to them meteorites, geological outcrops, blast furnace differentiations, and thermodynamic experiments in the laboratories of the world which suggest the following: There could be a maximum of molybdenum in the metallic? liquid-like (suggested by seismic, meteorite, geological outcrop, and neutrinoe telescope data) central body of the earth. Here, in the central body, pressures and temperatures seem to be high enough to inhibit and to an extent prevent the formation of MoS_2, a mineral which forms with the exothermic radiation of 27.10 kcal/g-atoms of S and FeS with 25.00 kcal/g-atoms of S. Generally Mo would tend to differentiate downward because of its high specific gravity whereas S would tend to differentiate upward because of its relatively lower specific gravity. There would be a tendency superimposed on pure gravitational differentiation of Mo to attach itself to S and therewith be "ballooned" upward out of the central body, if it were not for the high temperatures and pressures in this volume of the earth. In the central body it seems that Mo should tend to enter the metallic phase, as MoO_3 forms with the exothermic radiation of 52.5 kcal/g-atom O and FeO with 59.6 kcal/g-atom O. Differentiation between the central body, lower mantle, and upper mantle of the earth would seem to be dominated by gravitational sulfur-ballooning, and isolated by differential free energies of combination. This is diagrammed in the section suggesting a mode of differentiation between the basaltic (SIMA) rocks of the ocean basins and the granitic (SIAL) rocks of the continental platforms.

There is evidence that there is a three-dimensional pattern of convection currents in the upper mantle of the earth (see 1st and 2nd Conferences on Planetology and Space Mission Planning, Annals N.Y. Acad. Sci., 1966 and 1969). The writer thinks that the continental platforms of the earth and similar second-order geomorphic features of the Moon, Mercury, Venus, and Mars are positioned over descending regions in this pattern. He believes that the universally interconnected ocean basins represent areas beneath which convection cells are rising. Current geophysical measurements indicate that the ocean floors spread away from the universal "mid-oceanic" ridges or rifts to move toward the continents, which with other evidences suggest subcrustal friction.

If there is indeed deep convection including the entire upper mantle of the earth, molybdenum would tend to ascend where the currents ascend; and descend toward the interior of the earth beneath the continents. The convection currents could in this manner provide a mechanism for vertical differentiation. Horizontal differentiation between the basaltic oceanic rocks and the granitic continental rocks would in this theory take place in the geosynclinal structures at the margin, and less frequently within the continental platforms. It is in these volumes of rock that melting and mixing of basaltic rocks that are ever moving toward, then beneath, the continents like a treadmill, would most likely take place (see Table 2).

Cycling of molybdenum in the rocks of the continental platforms seems to be a function of both the *hydrological cycle,* in which rivers, etc., flow to the seas to be evaporated and recycled by rainfall on the lands that feed the rivers; and the *lithological cycle,* in which sediments representing the debris of granitic masses are carried by erosion to active geosynclinal basins, there to be recomposed into massive granitic rocks.

Another theory of granite-mass (batholithic) formation, considers them as differentiates of primary basic magmas. This could come about through the differentiation of rocks brought beneath the continental platforms by the previously mentioned deep convection currents. (The writer is inclined toward a granitization theory, in which most granitic masses are sediments reconstituted into granitic rocks by the extreme physical conditions found in the central volumes of geosynclines.) Table 3 suggests the movement of molybdenum from granitic massif (or other sources) to sedimentary basins. The reconcentration of molybdenum into economic deposits will be considered in the next section.

Economic Geology of Molybdenum

Molybdenite is mined both as a primary ore mineral and as a by-product, mostly from copper mining operations. Deposits from which molybdenum is, or could be, recovered as a primary ore can be classed as (*1*) magmatic, (*2*) metasedimentary and metasomatic, (*3*) pegmatite, and (*4*) hydrothermal. These four classes of deposits are related to either primary magmatic activity or geosynclinal activity.

Magmatic, Metasedimentary, and Metasomatic Ores. The molybdenite deposit at Climax, Colorado, is often considered to be a product of differentiation of the Boulder Batholith. Others regard the "batholith" as a volume of metamorphosed sediments, and contend that the molybdenum and related minerals originated from particular beds within this sedimentary pile.

In Climax MoS_2, mineralization was followed by cooling and oxidation of molybdenite to yellowish molybdite ochre, and small-scale local formation of pegmatitic quartz-rich and hydrothermal veins.

TABLE 2.

Basaltic rocks driven beneath the continental platforms, but locally they react strongly at site of geosynclines	Horizontal movement of basaltic rocks toward continental platforms just below Mohorovicic layer?	Vertical movement of non-solid matter toward surface; the surface cracks; fills with magma, which solidifies
Vertical movement of non-solid matter downward	Horizontal movement of dense, basic, nonsolid matter just above the Weichert-Jeffries discontinuity	

TABLE 3.

Hydrological cycle
(1) land uplifted and erosion, leaching, etc., takes place; lateritic soils and termite nests of laterite are very poor in Mo, which leaches out

(2) Gilpin County Colorado, mine waters are deep blue due to 8 ppm Mo ions; Note: the oxidation/reduction potentials of Mo ions determines the degree to which they will interact with H_2O dipoles, and therefore their solubility in H_2O

Hydrological cycle
(1) Continental fresh waters with Mo ions will often carry them to calcilutites where the Mo tends to be fixed by these calcium-rich muds; in the rocks of the Colorado Plateau MoCa salts are associated with uranium and vanadium minerals, all of which precipitated together

(2) Continental brackish waters are associated with peat bogs and coal swamps in which Mo ions tend to precipitate

(3) Shelf, estuary, and oceanic waters are typical of the geosynclinal cycle; Mo ion content tends to be similar through water volumes of great extent, but with distinct geometric patterns

Abundance in idealized geosynclinal sediments
(1) Molybdenum concentration in psephites (conglomerates), grits (pebbly beds), psammites (sands), and cherts tends to be very low; these would not be source beds of molybdenum for later concentration in favorable structures transecting favorable physical and chemical gradients; molybdenum tends to concentrate in pelites (muds, silts), limestones, and other precipitates

(2) In reducing (oxygen low) waters Mo tends to be fixed in H_2S-rich pelites; Mo tends to be fixed in silica-rich pelites that metamorphose to massive silexite (nearly pure quartz) beds; Mo tends to be concentrated in banded marine iron ores; where such ores have been subjected to pneumatolytic reconcentration, as in Sweden, concentrations of Mo reach 5000g/ton

Other deposits include: (1) The Henderson deposit in Colorado. (2) The enormous concentration in the Schuchert Valley of eastern Greenland at approximately 71° 51′N and 24° 50′W, where the rocks are of granulite facies. (3) Two structures in southern Africa and one in Australia are mineralogically and structurally similar to the above. (4) A deposit in New South Wales seems to be a hydrothermal deposit.

Pegmatites and Hydrothermal Ores. Traces of molybdenite, MoS_2, are almost universally found in quartz-pegmatites but rarely in economic quantities. Economic pegmatite-like deposits from which molybdenite has been produced as the major mineral or as a by-product include: Telemark, Norway; Questa, New Mexico; numbers of smaller pegmatites in Maine; and the pegmatites of the Ontario District, British Columbia; in these areas the MoS_2 tends to be most highly concentrated in the quartz phase.

The O.K. Mine in Beaver Lake, Utah, in monzonite country rock, is a hydrothermal

MOLYBDENUM: ELEMENT AND GEOCHEMISTRY

TABLE 4.

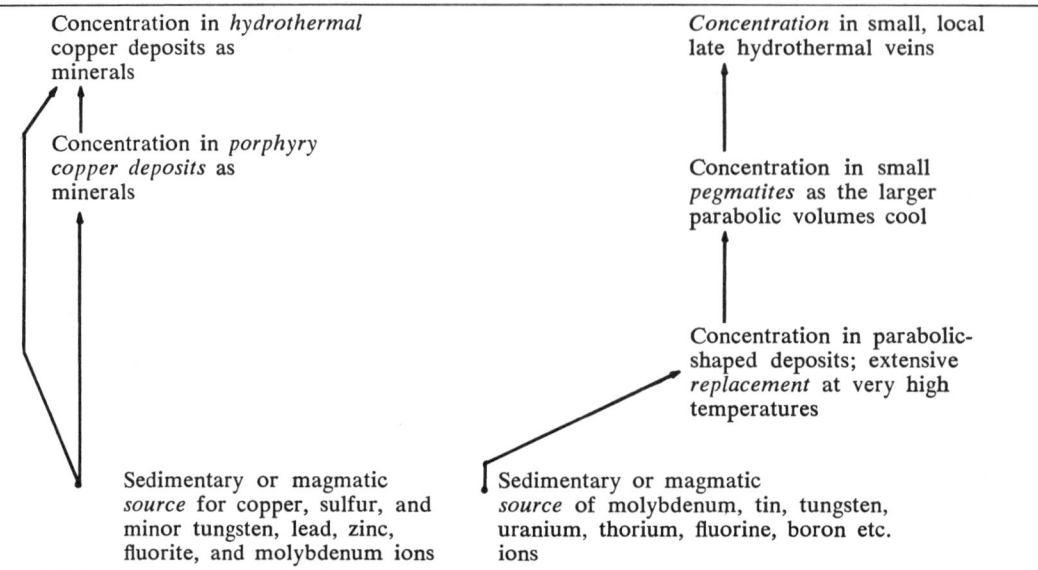

deposit from which copper and molybdenum have been produced. Molybdenite, MoS_2, is most abundant at depths where copper values drop; molybdite, MoO_3, is most abundant where the copper values rise and wall rocks are more highly sericitized. In Lainejaur, Sweden, ReS_2, rhenium sulfide, is found isomorphic with MoS_2 in quantities of up to 0.25%. Hydrothermal lead (galena)-zinc (sphalerite) deposits associated with calcareous tungsten mineralization (scheelite) often contain molybdenum crystallized as wulfenite.

Porphyry Coppers. Molybdenite is found as a minor mineral in all of the porphyry coppers and is often recovered. Porphyry coppers are of such great size that their by-product molybdenum supplies on the order of 30% of the nation's needs. Molybdenite is found in copper schists and shales; where these are worked, molybdenum is often recovered. The percentage of molybdenum is low in the massive copper-nickel mineralizations of great norite-anorthosite suites as at Petsamo, Sudbury, or the Bushveld. It is also low in the multitudes of smaller copper-rich norites, charnockites, and anorthosites of India, Australia, and Namaqualand, South Africa. Molybdenite is more frequently found as a minor mineral with higher-temperature copper ores than with mesothermal and epithermal concentrations. Pyrrhotite-pentlandite-chalcopyrite is a common hypothermal copper ore, tending to be associated with minor amounts of tin (cassiterite), tungsten (wolframite-ferberite), zircon, and later-formed fluorite; and usually minor quantities of molybdenite.

Free world production of Mo is about 100 million pounds annually, about one-third being from the United States. Its main use is in hard steel alloys and high-temperature alloys. Minor uses are in x-ray equipment, pigments, catalysts, etc.

Table 4 suggests paths of molybdenum differentiation applicable to primary magmas, or as the writer is inclined, to deposits derived from granitization of sedimentary piles. It should be noted that according to currently available information, field geologists seeking molybdenum ores should either look for the more abundant copper lodes, or for parabolic structures as described above.

ROBERT D. ENZMANN

References

*Goldschmidt, V. M., 1958, "Geochemistry," Oxford, Clarendon Press, 730pp.
Kirkemo, H., Anderson, C. A., and Creasey, S. C., 1965, "Investigations of molybdenum deposits in the conterminous United States," *U.S. Geol. Surv. Bull.*, **1182-E**, 90pp.
Lindgren, W., 1933, "Mineral Deposits," Fourth ed., New York, McGraw-Hill Book Co., 930pp.
Mason, B., 1966, "Principles of Geochemistry," Third ed., New York, John Wiley & Sons, 329pp.
Morning, J. L., 1965, "Molybdenum," *U.S. Bur. Mines Minerals Yearbook*.
*Rankama, K., and Sahama, T. G., 1950, "Geochemistry," Chicago, University of Chicago Press, 912pp.

*Additional bibliographic references mentioned in the text may be found in this work.

Cross-references: *Geochemical Evolution of the*

Core, Mantle, and Crust; Hydrothermal Solutions; Seawater, chemistry; Weathering, chemical. Vol. II: Cosmochemistry. Vol. IVB: Geobotanical and Biogeochemical Methods of Prospecting.

MÖSSBAUER EFFECT

The resonant absorption and fluorescence of gamma radiation by nuclei has been investigated by a number of different methods over the past thirty years. In 1958 R. L. Mössbauer reported a new phenomenon in resonant nuclear absorption, one which has made possible important advances not only in nuclear physics, solid state physics, and certain fundamental problems of physics, but also in the allied fields of chemistry, biology, and geophysics. In 1961 the Nobel prize in physics was awarded to Mössbauer for his discovery.

In the emission of radiation from a free nucleus of mass M, the nucleus recoils with an energy $E^2/2Mc^2$, where E is the energy of the nuclear transition and c is the velocity of light. Hence the energy of the emitted quantum is $E - E^2/2Mc^2$. Similarly, for absorption the quantum must have an energy $E + E^2/2Mc^2$. Therefore, in a resonance emission and re-absorption process involving the same nuclear species, there is a line shift of magnitude E^2/Mc^2. Since this shift is usually greater than the nuclear line width, the resonance absorption is greatly diminished. Many ingenious techniques have been devised by various investigators to restore the resonance condition in this process. All of these methods, in one way or another, utilize the Doppler effect to produce the compensating shift in energy. A number of investigations of nuclear absorption and fluorescence have also been carried out with sources having a continuous distribution of energy. In this case the energy lost in recoil is not important, but the experiments are difficult because of the large background of radiation.

It was demonstrated by Mössbauer that, under certain conditions, a nucleus embedded in a solid will not recoil on emission or absorption of radiation. Hence resonance absorption can take place with a stationary source and absorber, since it is not necessary to shift the energy of the radiation by means of the Doppler effect. Of more importance, however, is the fact that in this process the emission and absorption lines are not broadened by thermal Doppler motion and one observes the natural energy spread of the radiation. Thus one can obtain extremely narrow resonance.

The effect had already been encountered in the case of x rays. In fact, the fraction of nuclei in the solid which emit (or absorb) without recoil is given by the well-known Debye-Waller factor

$$f = e^{-2W}$$

where

$$W = 3\frac{R}{k\theta}\left[\frac{1}{4} + F\left(\frac{T}{\theta}\right)\right]$$

R is the recoil energy, θ the Debye temperature, and F a function of T/θ which is tabulated in references on x rays. A similar effect occurs in the scattering of neutrons.

It can be seen from the Debye-Waller factor that a large resonant absorption can be obtained if the recoil energy is small compared to the Debye energy $k\theta$. Furthermore the function $F(T/\theta)$ decreases with decreasing temperature, so that the absorption can be large if the temperature is low compared to the Debye temperature.

Resonant Absorption in Fe^{57}

Although the Mössbauer effect has been studied in over thirty nuclei, the most important and wide-spread application is to the nucleus Fe^{57}. The nuclear energy-level diagram is shown in Fig. 1. The great interest and importance of Fe^{57} are due to the following:

1. Because of the low energy of the radiation and fairly high Debye temperature of iron, the resonant absorption is very large at room temperature.

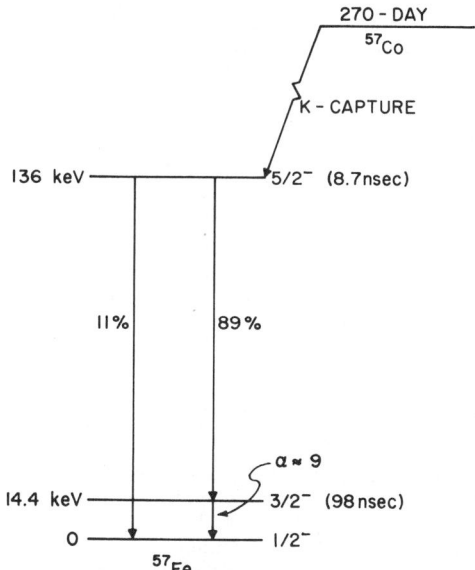

FIG. 1. Energy level diagram of Fe^{57} produced in decay of CO^{57}.

MÖSSBAUER EFFECT

2. The very narrow width of the resonance level ($\Gamma = 4.5 \times 10^{-9}$ eV) provides a very sensitive tool for observing very small shifts and splittings in energy. The fractional width $\Gamma/E = 3.1 \times 10^{-13}$.
3. Because iron and many of its compounds and alloys are magnetic, the resonance absorption provides an important means of studying ferromagnetism.
4. Since iron is an important constituent of many chemical and biological systems and is found in many minerals and meteorites, the Mössbauer effect has found important applications in these fields.

Characteristic Spectra

With relatively simple techniques, it is possible to observe the characteristic Mössbauer spectra in nuclei. In a typical experiment, the energy of the emitted radiation is Doppler shifted by relative motion between a source and observer. An absorption spectrum is then obtained by measuring transmission of gamma rays through the absorber as a function of the source velocity (i.e., energy).

The characteristic features of the spectra are produced by the following shifts or splittings:

1. A magnetic (Zeeman) splitting resulting from the interaction of the nuclear magnetic dipole moment with the magnetic field produced at the nuclear site by the

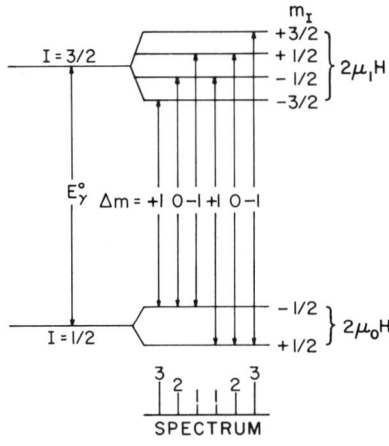

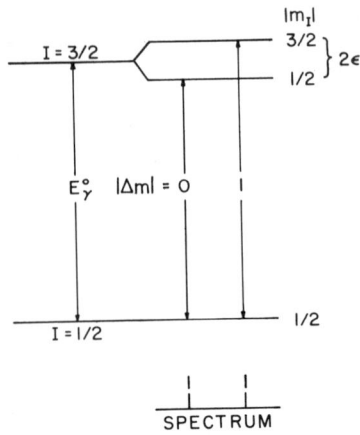

FIG. 3. Energy level diagram of Fe^{57}, illustrating the electric quadrupole splitting resulting from the interaction of the nuclear electric quadrupole moment with the electric field gradient produced at the nuclear site by the atomic electrons.

atomic electrons (Hanna et al., 1960). In ferromagnetic environments, iron nuclei interact with large internal fields (approximately 3×10^5 Oe) to produce the six-line Zeeman pattern illustrated in Fig. 2.

2. An electric quadrupole splitting resulting from the interaction of the nuclear electric quadrupole moment with the electric field gradient produced at the nuclear site by the atomic electrons (Kistner and Sunyar, 1960). In noncubic environments, electric field gradients can exist at the nucleus and the characteristic two-line spectrum (without magnetic splitting) shown in Fig. 3 is obtained.

FIG. 2. Energy level diagram of Fe^{57}, showing the magnetic (Zeeman) splitting of the ground state (nuclear spin $I = 1/2$) and the first excited state ($I = 3/2$). The magnetic-dipole selection rule $\Delta m = 0, \pm 1$ permits only those transitions shown in the diagram. The spectral intensities are those appropriate to nuclei in randomly oriented magnetic fields.

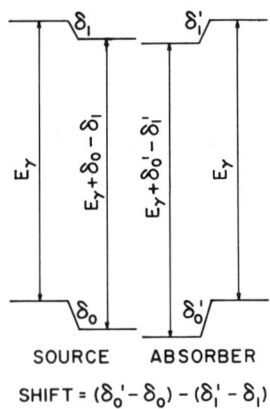

FIG. 4. Energy level diagram of Fe^{57}, illustrating the electrostatic shift (isomer or chemical shift) produced by the Coulomb interaction of the nuclear charge with the overlapping electron cloud.

3. An electrostatic shift (isomer or chemical shift) produced by the Coulomb interaction of the (positive) nuclear charge with the overlapping (negative) electron cloud (Kistner and Sunyar, 1960). If the nucleus rearranges its charge during emission or absorption of the radiation and if the electron cloud at the source nucleus differs from that at the absorber nucleus, the resonant line is shifted, as illustrated in Fig. 4. In practice it is necessary to distinguish a relativistic (second-order Doppler) shift (Pound and Rebka, 1960) from the isomer shift. The relativistic shift depends not only on the temperature difference between source and absorber but also indirectly on the chemical forms of the source and absorber. Ideally, an isomer shift should be determined at absolute zero.

One of the successes of Mössbauer research has been the demonstration (Walker et al., 1961) that the characteristic features of the spectrum can be associated with the oxidation or valence state of the iron ion. Thus, values of the isomer shift corresponding to the Fe^{3+} state range from $+0.1$ to $+0.5$ mm/sec, those for the Fe^{2+} state range from $+1.1$ to $+1.4$ mm/sec, and those for iron atoms in metals and alloys range from -0.2 to 0.0 mm/sec. The quadrupole splittings observed for the Fe^{3+} state range from 0 to 0.5 mm/sec; for the Fe^{2+} state the range is from 0.3 to 1.8 mm/sec—except that for any ion in local cubic symmetry the splitting is zero, as is the case for iron metal and several iron alloys. The hyperfine field at an iron nucleus ranges from about 5×10^5 Oe for some trivalent iron ions to about 3×10^5 Oe for iron atoms in metals and alloys and to zero in most of the divalent compounds.

In order to obtain a quantitative analysis from a Mössbauer spectrum, it is necessary to consider the relationship between the intensity (or area A) of an absorption line and the number n of absorbing nuclei per unit area of the sample. In the simplest case the relationship for a single line is

$$A = \tfrac{1}{2}\pi n f \sigma_0 \Gamma$$

where Γ is the width of the line, σ_0 is its maximum (resonant) absorption cross section, and f is the fraction of absorbing nuclei that absorb without recoil. This relation is valid only for small absorption, $nf\sigma_0 \ll 1$, and if the area is corrected for non-Mössbauer emission from the source.

A comprehensive study of the Mössbauer effect in Fe^{57} has been made by Preston et al. (1962).

Applications of the Mössbauer Effect

The applications of the Mössbauer effect (see e.g., Wertheim, 1964 and Goldanskii and Herber, 1968) may be classified as follows:

1. Use of the very great sensitivity of the effect to study relativistic problems and transient time effects associated with the radiative process.
2. The study of specific nuclear properties.
3. The study of solid state phenomena which produce the characteristic Mössbauer spectra, particularly ferromagnetism and transient spin relaxation phenomena.
4. The investigation of specific chemical properties of atoms and molecules in solids. These studies of atomic and molecular structure have found applications in chemistry, biology, and geophysics.
5. The use of the effect as a means of qualitative and quantitative chemical analysis based on measurements of intensities of Mössbauer spectra. As an

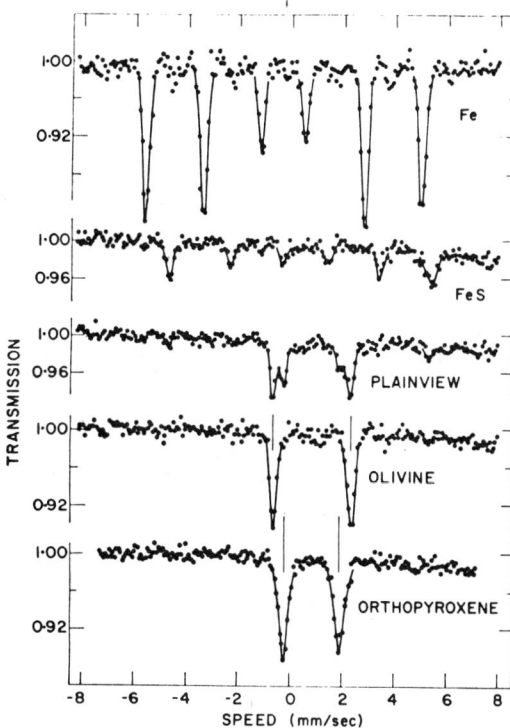

FIG. 5. Mössbauer absorption spectra of the Plainview meteorite and reference minerals present in most stone meteorites (from Sprenkel-Segel and Hanna, 1964).

illustration, the application of this method to the analysis of iron in a meteorite (Sprenkel-Segel and Hanna, 1964) is shown in Fig. 5. Further applications have been made to the analysis of iron-bearing minerals and the Mössbauer analysis of moon dust has been carried out.

STANLEY S. HANNA

References

Faraday Society, 1968, "The Mössbauer Effect," Proc. Symp. London, 1967, New York, Plenum Press, 140pp.

Goldanskii, V. I., and Herber, R. H. (editors), 1968, "Chemical Applications of Mössbauer Spectroscopy," New York, Academic Press.

Hanna, S. S., Heberle, J., Littlejohn, C., Perlow, G. J., Preston, R. S., and Vincent, D. H., 1960, *Phys. Rev. Letters* **4,** 177.

Kistner, O. C., and Sunyar, A. W., 1960, *Phys. Rev. Letters* **4,** 412.

Lafleur, L. D., Goodman, C. D., and King, E. A., 1968, "Mössbauer investigation of shocked and unshocked iron meteorites and fayalite," *Science* **162**(3859), 1268–1270.

Mössbauer, R. L., 1958, *Zeits. Phys.* **151,** 124.

Pound, R. V., and Rebka, G. A., 1960, *Phys. Rev. Letters* **4,** 337.

Preston, R. S., Hanna, S. S., and Heberle, J., 1962, *Phys. Rev.* **128,** 2207.

Sprenkel-Segel, E. L., and Hanna, S. S., 1964, *Geochim. Cosmochim. Acta,* **28,** 1913.

Walker, R. L., Wertheim, G. K., and Jaccarino, V., 1961, *Phys. Rev. Letters* **6,** 98.

Wertheim, G. K., 1964, "Mössbauer Effect: Principles and Applications," New York, Academic Press.

Cross-references: *Neutron Activation Analysis; X-Ray Spectroscopy.* Vol. II: *Magnetic Fields (Stellar); Meteorites; Relativity; Solar Magnetism.*

N

NATIVE ELEMENTS

Pure elements and their alloys as found in nature are referred to as "native"; they are classified into metals, semimetals, and nonmetals. Some of the native elements, found as minerals, occur in commercial quantities, but others (e.g., lead) are rare. Pure nickel has not been found as a mineral and this is also true of other common elements such as silicon and titanium.

In addition to the pure elements and their alloys, members of the solid-solution series, intermetallic compounds, and, by many writers, all carbides, nitrides, silicides, and phosphides are placed in this class. Some writers include, instead of the last group, the arsenides, antimonides, and bismuthides of copper, silver, and gold.

Although only eighteen elements occur as minerals, there are about forty-eight named mineral species in this class. Of these only copper, silver, gold, diamond, graphite, and iron (including kamacite) are at all common.

Native copper and gold may have been collected and used as early as 12000 BC. Meteoritic iron was apparently discovered later, possibly about 4000 BC and, shortly after, native silver may have come into use. Sulfur and graphite also have a long history and seem to have been known well before 1000 B.C.

Table 1 gives, in the order of the periodic

TABLE 1. MINERAL SPECIES INCLUDED IN THE NATIVE ELEMENTS

Metals

Gold	Au	Gold group: These minerals are isometric, and have face-centered cubic lattices (copper structure); common forms are cubes and octahedrons. They all have rather similar physical properties. In addition to the elements themselves there are a number of alloys, and the intermetallic compound *maldonite* (Au_2Bi).
Silver	Ag	
Copper	Cu	
Lead	Pb	
Mercury	Hg	Mercury is unusual in being a liquid at ordinary temperatures, but it crystallizes in rhombohedral crystals below $-39°C$.
Platinum	Pt	Platinum group: Consists of the face-centered isometric species platinum, palladium, and platiniridium. Iridosmine and siserskite are hexagonal and form a series; Ruthenium and rhodium also substitute. They are thus alloys with each other, as well as with gold, iron, and some other metals.
Palladium	Pd	
Platiniridium	Ir, Pt	
Iridosmine	Ir, Os	
Siserskite	Os, Ir	
Iron	Fe	Iron group: Rare terrestrial minerals (e.g. *josephinite*), but important constituents of meteorites, as *kamacite* and *taenite*. There are also iron-nickel alloys and a group of iron carbides, nitrides, silicides, and phosphides, also essentially confined to meteorites. Minerals in this group can only form under extreme reducing conditions.
Nickel-iron	Fe, Ni	
Tantalum	Ta	
Tin	Sn	
Zinc	Zn	

Semimetals and Nonmetals

Arsenic	As	Arsenic group: Hexagonal (rhombohedral lattice) and have one direction of perfect cleavage (0001) and tin white color, often tarnished. An allotropic form of arsenic is arsenolamprite which is orthorhombic. Allemontite is an intermetallic compound.
Allemontite	AsSb	
Antimony	Sb	
Bismuth	Bi	
Selenium	Se	Tellurium group: Rare, hexagonal minerals (trigonal, trapezohedral). Tin white to gray. There is a solid-solution series between them.
Tellurium	Te	
Sulfur	S	Sulfur group: Consists of at least three allotriomorphic forms (alpha, beta, and gamma) in nature, the amorphous form (sulfurite), and several other artificial forms.
Diamond	C	Carbon group: Diamond is isometric whereas graphite is hexagonal, the three minerals showing the greatest diversity of properties of any polymorphous minerals.
Graphite	C	
Lonsdaleite	C	

NATIVE ELEMENTS

table (alloys, etc., being classified according to the element latest in the table), the mineral species included in the native elements.

LLOYD W. STAPLES
MICHAEL J. FROST

References

Dana, J. W., and Dana, E. S., 1944, "The System of Mineralogy," Vol. 1, Seventh ed. (Palache, C., Berman, H., and Frondel, C., editors), New York, John Wiley & Sons.

Fleischer, M., 1966, "Index of new mineral names, discredited minerals, and changes of mineralogical nomenclature," *Am. Mineralogist,* **51,** 1248–1357.

Hawley, J. E., and Rimsaite, Y., 1953, "Platinum metals in some Canadian uranium and sulphide ores," *Am. Mineralogist,* **38,** 463–475.

Hey, M. H., 1955, "Index of Mineral Species and Varieties," British Museum (Natural History), 728pp. plus Supplement.

Hey, M. H., 1968, *Mineral. Mag.,* 1146–1163.

Kraus, E. H. (editor), "First, Second, and Third Symposia on diamonds," *Am. Mineralogist,* **27,** 1942; **28,** 1943; **31,** 1946.

Cross-references: See Vol. IVB: *Individual Elements and Minerals.*

NATURAL BRINES*

Surface Brines

Natural brines are here defined as natural aqueous solutions which contain salts in concentrations significantly greater than seawater (35 g/kg or 36 g/liter total salt since g/liter = g/(kg)(density of brine)).

The largest body of natural brine on the earth's surface is the Gulf of Kara Bogaz, an isolated arm of the Caspian Sea. It had an area (variable) of about 8000 km² and a volume of about 24 km³ in 1956. Its water contained about 30% salt (Urazov and Sedel'nikov, 1959). The Dead Sea, between Israel and Jordan, has a smaller area (940 km²) but is much deeper (average depth 145 m), and contains nearly six times greater volume: 136 km³ (Neev and Emery, 1967). Great Salt Lake (Utah) follows in size, with an area of about 1600 km² and a volume of about 12 km³. The compositions of these and other selected surficial brines are listed in Table 1.

Many smaller coastal marine bays and lagoons in areas of low runoff contain evaporated seawater; examples include the Persian Gulf (sebkhas), Australia (coorongs), the Gulf of Mexico (Laguna Madre, Texas), Baja California (Ojo de Liebre), the Peru-Chile coast (Boca de Virrila), and enclosed atoll lagoons in the Pacific Ocean and Caribbean Sea. Notwithstanding the general conception, a warm climate is not required to produce evaporative brines; for example, brine lakes are known from Antarctica (see Lake Bonney, Table 1).

Whereas the above brines were formed principally through evaporation of seawater and leaching of preexisting marine evaporites, ephemeral lakes, playas, and other partly evaporated water bodies in many arid areas may contain brines derived largely from continental sources. Leaching of igneous and volcanic rocks tends to produce alkaline waters and evaporation of hot spring effluents from such rocks may lead to carbonate-borate-silica rich waters such as nos. 6, 7, and 8 in Table 1. Brines under surface crusts of salt in Searles Lake, California, also contain concentrations of unusual elements such as arsenic, lithium, tungsten, and antimony (Table 1, no. 7).

A different type of brine occurs in crater lakes or springs associated with some active volcanoes. These may be highly acid, owing to emission of fumes of HCl or H_2S, which in the presence of atmospheric oxygen may be oxidized to H_2SO_4 by bacteria. The Ebeco Volcanic lake (Table 1, no. 9), for example, is nearly 2 molar in hydrochloric acid. Such acid solutions may leach out and concentrate substantial amounts of iron, aluminum, or other elements which normally occur in low concentration in surficial brines.

Surface brines tend to be ephemeral, and were more common at times in the past, as attested by vast buried evaporite deposits. A dramatic illustration of the evolution of a major brine-forming area is the Red Sea basin, especially as documented in a current monograph on deep, hot brine pools discovered in 1964–65 (Degens and Ross, 1969). The present Red Sea contains a borderline brine having in excess of 40 g/kg total salt. During Pleistocene time, the lowering of sea level caused shoaling of the inflow channel from the Arabian Sea, which in turn caused the salt concentration to rise to as much as 80 g/kg (W. G. Deuser and E. T. Degens, in Degens and Ross, 1969). Still earlier, during middle Tertiary time, the Red Sea was the site of deposition of thick layers of rock salt and other evaporitic minerals. The cycle is completed by subrecent rifting, which has permitted brines from deeper evaporitic layers to escape upward to form metal-depositing brine pools in the present Red Sea deeps (Table 2, no. 1).

---

* Publication approved by the Director, U.S. Geological Survey.

NATURAL BRINES

TABLE 1. CHEMICAL COMPOSITION OF SELECTED SURFACE BRINES[a]

Name	1. Gulf of Kara Bogaz (Turkmen SSR)	2. Great Salt Lake (Utah)	3. Dead Sea (av.) (Israel-Jordan)	4. Dead Sea E. drainage	5. Lake Bonney (Antarctica)	6. Hot Spring Lake Magadi (Kenya)	7. Searles Lake (Calif.)	8. Alkali Lake Playa (Ore.)	9. Ebeco Volcano Paramushiro (Kuriles)	10. Seawater
Major elements (g/kg)										
H	—	—	—	—	—	—	—	—	1.7	—
Na	50.8	85.7	31.8	25.3	51.4	12.8	88	105.0	0.11	10.8
K	5.0	4.6	5.9	0.98	2.8	0.20	30.2	5.4	0.09	0.39
Ca	0.12	0.33	13.7	33.1	1.65	0.0	0.012	0.0	0.089	0.41
Mg	38.1	8.1	33.0	66.1	24.2	0.0	"trace"	0.0	0.031	1.29
Cl	135.4	146.7	172.5	218.7	162.0	6.0	122	18.6	61.2	19.4
Br	0.60	0.12	4.1	8.9	—	0.11	0.81	—	—	0.065
SO$_4$	0.60	—	0.38	0.04	3.3	0.15	45.7	18.6	0.093	2.7
CO$_3$	78.4	—	—	—	—	3.2	27.2	113.0	0.97	—
HCO$_3$	0.38	0.33	0.18	0.47	0.10	15.8	14.7	15.4	—	0.14
B$_4$O$_7$	—	0.020	0.10	—	0.12	0.032	0.97	0.97	"much"	0.016
Sum	309.0	246.0	262.0	353.6	245.6	38.6	336.8	276.1	64.6	35.2
Density (20°C)	—	1.208	1.231	1.321	1.196	—	1.3	—	1.032	1.025
pH	—	7.5–7.9	6.3	—	6.8	9.0	9.4	9.9	−0.23	8.0
Minor components (ppm)										
SiO$_2$	—	11.9	19	—	3	92	—	1130	45	<1
Al	—	—	—	—	—	—	14	—	60	<1
Fe	—	0.03	2	0.07	—	0.8	14	—	232	0.01
Mn	—	—	3	—	—	—	—	—	3	0.01
Sr	—	—	200	—	—	—	83	—	1	8
Li	—	48	20	—	—	1.0	18	—	2.3	0.2
NH$_4$	—	82	4	—	—	—	19	—	5	<1
F	—	7	—	—	—	156	770	—	68	1.3
PO$_4$	—	—	—	—	0.1	9.2	—	—	—	<1
Isotopic ratios										
δD (‰)	—	−9.6	—	—	—	—	−3.1, −1.7	—	—	0
δO^{18} (‰)	—	−7	+4.4	—	—	—	—	—	—	0
δS^{34} (‰)	—	+16 (SO$_4$)	+12 to +15 (SO$_4$) −21.5 (H$_2$S)	—	—	—	+15.2 (SO$_4$)	—	—	+20

[a] Isotopic notation refers to deviation in indicated units from Standard Mean Ocean Water (SMOW) (oxygen and hydrogen) and meteoritic troilite (sulfur). The numbers at the top of the table refer to the following:

1. Sample taken during 1955. Source: Urazov and Sedel'nikov, 1959; halite, mirabilite (Na$_2$SO$_4$·10H$_2$O), gypsum, and other sulfate minerals precipitating from solution.
2. Average from Hahl, 1964; pH and median value for Br supplied by Robert Madison, U.S. Geol. Survey, Salt Lake City; deuterium supplied by Irving Friedman, U.S. Geol. Survey, Denver; oxygen isotope value from Degens et al. (1964); sulfur isotope value supplied by R. L. Mauger, Univ. Utah, Salt Lake City.
3. Source: chiefly Neev and Emery, 1967; sulfur isotope data from Nissenbaum and Kaplan, 1967; Arie Nissenbaum (personal communication) reports also (in ppm): Sr 270, Li 15, Mn 7, Fe 0.006, Ni 0.007, Co 0.006, Cu 0.14, Zn 0.33, Cd 0.003, Pb 0.032, U 0.0011, in lower water mass.
4. Source: Neev and Emery, 1967.
5. Taken from 32-m depth, at −2.5°C. Source: Angino et al., 1964. Also 4.2 ppm I0$_4^-$.
6. Source: Eugster and Jones, 1968.
7. Compiled from summary by Boiko, 1963, and White et al., 1963. Also (in g/liter): S^{2-} 0.33, WO$_3$ 70, Sb$_2$O$_3$ 6, Mo 0.6, Rb 1, I 30, As$_2$O$_3$ 192 ppm. Sulfur isotopes from Holser and Kaplan, 1966; deuterium data from Irving Friedman, U.S. Geological Survey, Denver (personal communication).
8. Source: Jones, 1966. Temperature at time of recovery 96°F.
9. Central crater of upper crater. Source: Ivanov, 1957 (see in White et al., 1963). Also (in ppm): Co 0.3, Ni 0.1, Cu 0.03, Ti 0.2, I 0.4, As$_2$O$_3$ 1. Temperature 100°C.
10. Average ocean water, mainly from Culkin, 1965.

TABLE 2. CHEMICAL COMPOSITION OF SELECTED SUBSURFACE BRINES[a]

Name	1. Red Sea (Atlantis II Deep)	2. W. Caspian Sea	3. Harris, County, Texas	4. Sweetwater County, Wyoming	5. Eddy County, N.M.	6a. Angara-Lena Basin (East Siberia)	6b. Angara-Lena Basin (East Siberia)	7. Salton Sea (Calif.)	8. Keweenawan Peninsula (Mich.)	9. Arima, Honshu (Japan)
Depth (m)	2000	550–559	2220	460 m below land surface	42	2200–2220	2203–2210	1500	915	—
Age of rocks	Recent	Pliocene	Oligocene	Eocene	Permian	Lower Cambrian	Lower Cambrian	Pliocene	Pre-cambrian	Miocene
Major elements (g/kg)										
Na	92.6	32.2	41.2	84.7	43.9	6.3	31.8	51.0	11.9	19.6
K	1.9	0.23	0.14	0.16	2.1	3.1	12.0	25.0	–0.038	4.4
Ca	5.2	10.5	4.0	0.004	0.097	77.0	104.2	40.0	62.9	3.9
Mg	0.76	3.0	0.90	0.005	38.3	14.0	21.6	0.73	0.18	0.038
Sr	0.048	—	0.13	0.0006	—	4.3	(6)	0.75	0.32	0.002
Cl	156.0	61.5	74.8	21.3	16.8	187.2	297.5	185.0	128.0	41.7
Br	0.128	—	0.062	0.022	—	5.3	7.2	0.15	1.0	0.052
SO$_4$	0.84	3.1	0.21	0.17	222.0	0.18	3.6	0.056	0.088	0.00
HCO$_3$	—	0.024	0.24	13.8	1.4	0.69	0.66	—	0.024	0.57
CO$_3$	—	—	—	88.5	—	—	—	—	—	—
Sum	257.7	110.6	122.1	209.0	324.6	298.1	484.6	310.3	204.6	71.3
Density (20°C)	1.20	1.081	1.087	1.204	1.345	1.286	(1.52)	1.264	1.174	—
pH	(5)	6.6	6.5	10	—	4	—	5–6	6.5	5.8
Minor components (ppm)										
NH$_4$	—	—	27	1.3	—	156	—	482	10	44
I	0.03*	—	15	8.6	—	10	—	22	3.2	0.9
F	0.051*	—	1.1	18	—	17	—	18	—	0.7
B	7.8*	—	17	133	—	0.02–500	—	520	—	535
SiO$_2$	64	—	48	47	—	0.3–3000	—	110	2.5	146
PO$_4$	0.1	—	0.0	110	—	—	—	—	—	1.4
Fe	81	10–100	23	—	0.00	1000	—	3200	1	157
Mn	82	100–1000	1	—	0.00	0.2–2000	—	2000	2.5	42
Cu	0.26	32	0.3	—	—	0.6–29	—	10	0.7	0.1
Zn	5.4	—	0.0	—	—	0.2–28	—	970	—	0.2
Pb	0.63	13	0.0	—	—	0.6–84	—	104	—	0.4
Ag	—	2.5	—	—	—	0.1–6.5	—	1	—	—
Li	0.26*	—	6	0.5	—	0.02–500	—	300	7	53
Rb	—	—	—	—	—	0.7–15	—	169	—	—
Ba	1	10–100	31	—	—	0.5–160	—	200	4	59

[a] Samples 1–6 represent brines from sedimentary rocks. Sample 7 is probably of sedimentary origin, undergoing low-grade metamorphism. Samples 8 and 9 are from igneous rocks, but are of unknown origin.

The numbers at the top of the table refer to the following:

1. Source: Spencer and Brewer, in Degens and Ross, 1969. Values with asterisk refer to Discovery Deep at 2160 m; from Brewer, Culkin and Riley, cited in above paper.
2. Borehole No. 4, Alyat structure, southwestern Caspian Sea. Water depth about 10 m (?). Sum excludes minor components. Source: Pushkina, 1965.
3. Source: White, 1965. Also, Al 6.0 ppm.
4. Seep from shale immediately above Trona deposit in Green River Formation. Source: White et al., 1963. Also, Ti 0.4 ppm.
5. Test well Yates No. 1, reaching Salado Brine, in Castile Formation evaporites. Source: White et al., 1963.
6. Tyretskaya Borehole No. 11. Source: Pinneker, 1966. 6a. refers to water recovered at well bore; 6b. refers to interstitial water extracted from argillaceous sandstone core (in evaporitic-dolomite-clastic sequence). Density is fictive, based on extrapolation of density-total solids plots (in grams per liter), since the salts remain in solution only at the greater temperatures and pressures obtaining at depth. Sum excludes minor components.
7. Water from arkose under temperatures exceeding 270°C. Source: White, 1965. Also (in ppm): Sb 0.5, Al 450, As 15, Cs 20, Sn 0.65, Hg 0.0008.
8. Drip from copper mine in Precambrian lava flows. Source: White, 1965. Also, Al 5.4 ppm. Type sample of "connate" (original) pore fluid. Temperature 94°C.

766

NATURAL BRINES

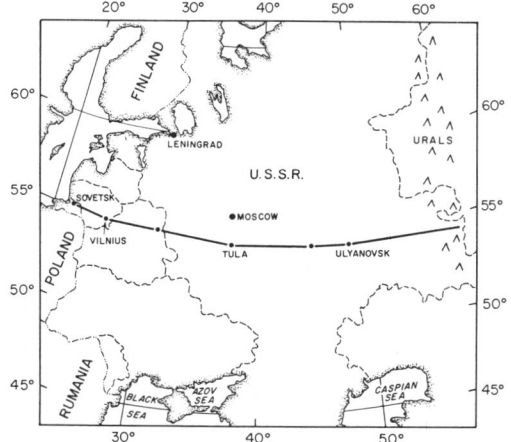

FIG. 1. Location map for traverse across European Russia and Lithuania.

Subsurface Brines

Brines occur in sedimentary strata in many, if not most, of the world's sedimentary basins, in rocks from Pleistocene to Precambrian age. Although some brines are known from springs or other points of discharge from subsurface formations, mixing and flushing by meteoric waters reduces salt concentrations in near-surface strata, and most concentrated brines have been found during deeper boring for oil, or to a lesser extent, coal, metalliferous deposits, or other purposes. The general increase in brine concentration with depth is illustrated with unusual consistency in Figs. 1 and 2, depicting a section across European Russia and Lithuania. Locally, however, all manner of brine distributions may be observed, including fresh and brackish waters beneath more concentrated brines, at depths up to 4000 m, as in Gulf Coast strata of the United States (Myers and Van Siclen, 1964).

The composition of subsurface brines has been the object of much classification and discussion (see Schoeller, 1956; Slavyanov, 1958; Krejci-Graf, 1963; White et al., 1963; and White, 1965 for extensive reviews and references). However, certain generalizations apply broadly. Below the zone of influence of actively recharged groundwater the most common type of brine contains mainly sodium and chloride, followed distantly by calcium. The proportions of magnesium and potassium are generally less than in seawater, and sulfate tends to be reduced by bacteria to sulfide where sufficient organic matter is present. However, the full range of observed natural brines covers a great variety of compositions. Recent investigations have revealed that calcium chloride-dominated brines, frequently associated with dolomite-evaporite sequences are more common than once thought, and contain appreciable amounts of potassium, magnesium, and bromine. The most concentrated brines known belong to this group, reaching a current record of more than 500 g/kg (750 g/liter) in lower Cambrian rocks from the Angara-Lena basin in eastern Siberia

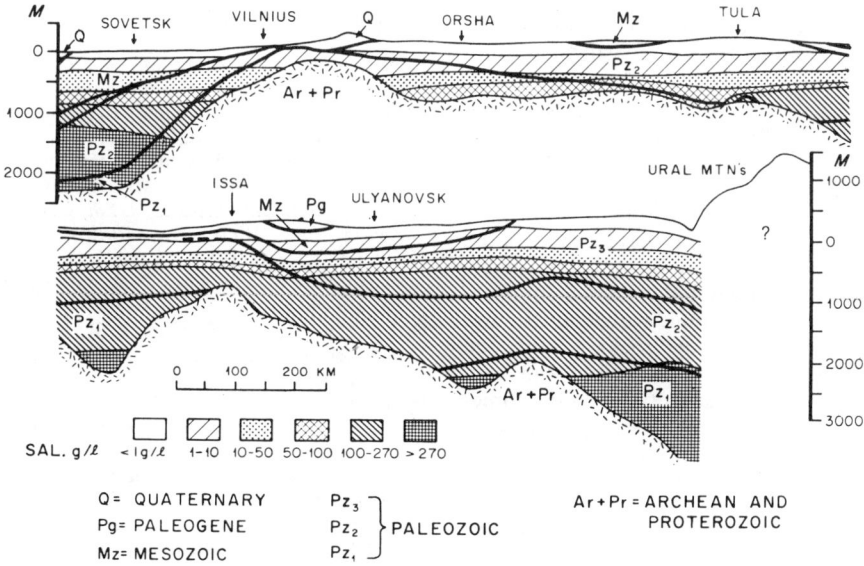

FIG. 2. Hydrochemical section across European Russia and Lithuania. (Modified from B. N. Arkhangelskii, in Kamenskii et al, 1959, p. 58). Heavy lines delineate stratigraphic intervals as indicated.

(Pinneker, 1966; see also Table 2, no. 6a and 6b).

Many subsurface brines show clear associations with evaporite deposits and accompanying redbeds. However, studies on many brines (Degens et al., 1964; Clayton et al., 1966) have so far failed to show the increases in the proportion of heavy isotopes of hydrogen and oxygen which would be expected if the water in the brines itself were a relic of evaporated seawater. Thus, for evaporite-associated brines, the high salt concentrations seem to be better explained by leaching and exchange reactions between meteoric or seawater and evaporite minerals. Leachates of evaporitic salts may move from the vicinity of their origin by gravitational convection (Valyashko, 1963), diffusion, artesian movements and osmotic mixing, subject to the distribution of permeability in the host strata.

The existence of brines in strata carrying normal marine fossils or showing no obvious relationship to evaporite deposits has long stimulated the search for mechanisms capable of forming brines without evaporitic processes. Proposed mechanisms include subsurface evaporation of water by its entrainment in and escape with natural gas, general outgassing of the earth's crust, gravitational settling of salt ions, loss of water through hydration of minerals, kinetic separation of salt from water molecules in a pressure gradient, supply of brines from deeper magma chambers, and relatively recently, selective retention of salts by clay membranes (see Degens and Chilingar, 1967). Several of these mechanisms have not withstood criticism, but conclusive evaluation of the origin and migration of subsurface brines is hampered, in general, by the fact that analytical and other data have been available almost exclusively from permeable rocks such as sandstones and porous carbonates, neglecting the intervening, poorly permeable strata.

Minute fluid inclusions within a variety of minerals (Roedder, 1967 and references cited therein) give evidence that brines of probable sedimentary origin may be involved in processes forming hydrothermal ore deposits, and metamorphic and igneous rocks. Enhanced temperature and pressure increases the mobility and complexing activity of high ionic-strength brines (Helgesen, 1964), promoting interaction with and leaching of the rocks, and tend to enrich heavy metals in the brines. These brines may later, under suitable conditions, release metals to form ore minerals. "Hydrothermal" waters from the Salton Sea area (California) (Table 2, no. 7), and the Cheleken Peninsula (Caspian Sea) (Lebedev, 1967), may provide examples of ore-forming brines from strata in between sediments and metamorphic rocks-in-the-making.

FRANK T. MANHEIM

References

Angino, E. E., Armitage, D. B., and Tash, J. C., 1964, "Physicochemical limnology of Lake Bonney, Antarctica," *Limnol. and Oceanog.*, **9,** 207–217.

Boiko, T. F., 1963, "Ozero Serls i ego litienosnye i vol'framonosnye rassoly (Searles Lake and its lithium and tungsten-bearing brines)," *Tr. Inst. Mineral., Geokhim., Kristallokhim. Redkikh Elementov Akad. Nauk SSSR*, **17,** 47–66.

Clayton, R. N., Friedman, I., Graf, D. L., Mayeda, T. K., Meents, W. F., and Shimp, N. F., 1966, "The origin of saline formation waters. 1. Isotopic composition," *J. Geophys. Res.*, **71,** 3860–3882.

Culkin, F., 1965, "The major constituents of sea water," in (Riley, J. P., and Skirrow, G., editors), "Chemical Oceanography," Vol. 1, London, Academic Press, pp. 121–161.

Craig, H., 1966, Isotopic composition and origin of the Red Sea and Salton Sea geothermal brines," *Science*, **154**(3756), 1544–1548.

Degens, E. T., and Chilingar, G. V., 1967, "Diagenesis of subsurface waters" in (Larsen, G., and Chilingar, G. V., editors), "Diagenesis in Sediments," Amsterdam-New York, Elsevier, pp. 477–502.

*Degens, E. T., and Ross, D. A., 1969, "Hot brines and recent heavy metal deposits in the Red Sea," Berlin-New York, Springer Verlag, pp. 1–600.

Degens, E. T., Hunt, J. M., Reuter, J. H., and Reed, W. E., 1964, "Data on the distribution of amino acids and oxygen isotopes in petroleum brine waters of various ages," *Sedimentology* **3,** 193–225.

Eugster, H. P., and Jones, B. F., 1968, "Gels composed of sodium-aluminum silicate, Lake Magadi, Kenya," *Science,* **161,** 160–163.

Hahl, D. C., and Langford, R. H., 1964, "Dissolved-mineral inflow to Great Salt Lake and chemical characteristics of the Salt Lake brine. pt. 2. Technical Report," *Utah Geol. and Mineralog. Surv. Water Res. Bull.* **3,** 40pp.

Helgesen, H. C., 1964, "Complexing and hydrothermal ore deposition," London-New York, Pergamon Press.

Holser, W. T., and Kaplan, I. R., 1966, "Isotope geochemistry of sedimentary sulfates," *Chem. Geol.,* **1,** 93–135.

Hunt, J. M., Hays, E. E., Degens, E. T., Ross, D. A., 1967, "Red Sea: Detailed survey of hot-brine areas," *Science,* **156**(3774), 514–516.

Jones, B. F., 1966, "Geochemical Evolution of Closed Basin Water in the Western Great Basin" in (Rau, J. L., editor), "Second Symposium on Salt," Cleveland, Northern Ohio Geol. Soc., **1,** 181–200.

Jones, B. F., Rettig, S. L., and Eugster, H. P., 1967, "Silica in Alkaline Brines," *Science,* **158** (3806), 1310–1314.

Kamenskii, G. N., Tolstikhina, M. M., and Tolstikhin, N. I., 1959, "Gidrogeologiya SSSR," Moscow, Gesgeoltekhizdat.
Krejci-Graf, K., 1963, "Diagnostik der Salinitätsfazies der Olwässer," *Forschr. Geol. Rheinland Westfalen,* **10,** 367–448.
Lebedev, L. M., 1967, "Contemporary deposition of native lead from hot Cheleken thermal brine," *Doklady Akad. Nauk SSSR,* **174,** (earth science sections in translation), 173–176.
Myers, R. L., and Van Siclen, Dewitt, 1964, "Dynamic phenomena of sediment compaction in Matagorda County, Texas," *Trans. Gulf Coast Assoc. Geol. Soc.,* **14,** 241–252.
Neev, David, and Emery, K. O., 1967, "The Dead Sea," *Israel Geol. Surv. Bull.* **41.**
Nissenbaum, A., and Kaplan, I. R., 1967, "Stable isotope geochemistry of the Dead Sea, Israel," *Program Geol. Soc. Am. Meet.,* p. 165 (abstr).
Pinneker, Ye. V., 1966, "Rassoly Angaro-Lenskogo artezianskogo basseyna," Moscow, Izdatel'stvo "Nauka."
Pushkina, Z. V., 1965, "Porovye vody glinistykh porod i ikh izmenenie po razrezu" (Pore water in the clayey rocks and their changes with depth), in (Strakhov, N. M., editor), "Postsedimentatsionnye izmeneniya chetvertichykh i Pliotsenovykh glinistykh otlozhenii Bakinskogo Arkhipelaga," Moscow, Izdatel'stvo "Nauka," pp. 160–203.
*Roedder, Edwin, 1967, "Fluid inclusions as samples of ore fluids," in (Barnes, H. L., editor), "Geochemistry of Hydrothermal Ore Deposits," N.Y. Holt, Rinehart, and Winston, pp. 515–574.
Schoeller, H., 1956, "Geochimie des Eaux Souterraines; Application aux Eaux des Gisements de Petrole," Paris, Soc. des Editions.
Slavyanov, N. N. (editor), 1958, "Materialy soveshchaniya po voprosam formirovaniya podzemnykh vod (Symposium on the question of formation of underground water)," *Tr. Laboratorii gidrogeologicheskikh problem,* **16,** 1–359.
Urazov, G. G., and Sedel'nikov, G. S., 1959, "Gidrokhimicheskii rezhim zaliva Kara Bogaz-Gola i perspektivy ego promyshlennogo ispol'zovaniya (Hydrochemical regime of the Gulf of Kara Bogaz and its economic utilization)," *Trudy Okeanogr. Komiteta,* **5,** 328–332.
Valyashko, M. G., 1963, "Genezis rassolov osadochnoi obolochki (Genesis of brines in sedimentary rocks)," in (Vinogradov, A. P., editor), Khimiya zemnoi kory, 1, Moscow, Izdatel'stvo "Nauka," pp. 253–277.
White, D. E., 1965, "Saline waters of sedimentary rocks," *Mem. Am. Assoc. Petrol. Geol.* **4,** 342–366.
White, D. E., Hem, J. D., and Waring, G. A., 1963, "Chemical composition of subsurface waters," *U.S. Geol. Surv. Profess. Paper,* **440-F,** 1–67.

*Additional bibliographic references may be found in this work.

Cross-references: *Aqueous Solutions; Evaporite Processes; Fluid Inclusions; Ground water, Hydrothermal Solutions; Lagoon Geochemistry; Meteoric Water; Seawater, Chemistry; Springs; Sulfate Reduction—Microbial.* Vol. III: *Dead Sea; Great Salt Lake; Kara-Bogaz Gulf; Playa; Sabkha or Sebkha.* Vol. IVB: *Evaporite Minerals.* Vol. VI: *Leaching.*

NATURAL CHROMATOGRAPHY—
See Vol. IV B

NATURAL CHROMATOGRAPHY IN SEDIMENTARY ROCKS

Chromatography is a physicochemical method of separation. The components of a fluid mixture are separated during flow through a suitable porous body which contains fine-grained particles. Recent sediments and clastic sedimentary rocks are porous and permeable bodies which usually contain clay minerals or other colloidal-sized components. Connate water and other formation fluids keep continuously percolating through sedimentary rocks, driven by pressure gradients inherent to the structure and geometry, etc., of the rock strata. The natural rate of water movement through rocks varies; it has been estimated to range between 5 ft/day to 5 ft/year (Meinzer and Wenzel, 1942). Formation fluids, such as connate water and petroleum, contain numerous dissolved components, inorganic salts, and/or organic substances. The flow rate of some of the components may be selectively affected by the colloidal mineral matter or stationary organic constituents of the rocks. This process eventually leads to the separation and concentration of some of the fluid components. This process is called *natural chromatography*. It should be noted that the term chromatography arose from early experiments which involved the separation of colored substances. The term chromatography, as it is used today, has nothing to do with color, and many compounds which are separated by chromatography are indeed colorless.

The principle of natural chromatography has been known for many years, although the name as such has not been in use until recently. The early experiments of Day (1897) and of Gilpin and Cram (1908) with petroleum have shown that asphaltic matter and some high-molecular-weight, polar compounds are separated from the fluid phase during flow through columns packed with certain clay minerals. Subsequently, Kobayashi and his co-workers (Roper, Doscher, and Kobayashi, 1958) found that the components of natural gases moved with different velocities through sandstone that was saturated with crude oil.

These investigators calculated the retention times of the gaseous components in accordance with the chromatographic mechanism. Mileshina et al. (1959) conducted flow experiments in which crude oil was passed through quartz sand and mixtures of quartz sand, clay minerals, feldspars, and calcite. Some of the nonhydrocarbon constituents were selectively adsorbed on the mineral matter. Experiments with water solutions have also been conducted. The behavior of some organic ionic substances, dissolved in seawater and flowing through layers of sand and clay in a geological model system was investigated by Nagy and Wourms (1959). It was noted that the movement of the organic ions in sand and clay resembled the movement of similar substances in ordinary laboratory chromatographic columns. The components of the mixture were separated in the sand and clay layers and increased in concentration in layers rich in clay.

There are geological field observations which seem to support the results of these experiments and which indicate that chromatography may have taken place in rocks. Bonham (1956) suggested that regional variations in the trace element content of crude oils in central Oklahoma may be attributed to selective adsorption processes which occurred within the rocks during petroleum migration. A similar conclusion was reached by Hodgson and Baker (1959) regarding the trace metal distributions within some Canadian oil fields. Brooks (1959) suggested that the asphalt-free crude oils in Pennsylvania are the result of their chromatographic flow through rocks, and Smith et al. (1959) pointed out, in a study of crude oils, that the preferential occurrence of compounds in petroleum may be caused by "natural chromatography."

The experiments and the field observations raised the possibility that chromatography may be a geological process. Consequently, it seemed necessary to examine this possibility in terms of the well-known physicochemical laws of chromatography. It was found (Nagy, 1960) that natural chromatography can be explained and predicted to occur in rocks by the classical chromatographic "plate" theory of Martin and Synge (1941). This finding points out that the physical laws which control fluid flow in rocks and in the chromatographic systems in the laboratory are essentially the same.

Chromatography, as it is used in the laboratory, is a separation process which can be effected by wholly different physical and chemical phenomena. Some components of fluid mixtures can be separated by ion-exchange with the stationary phase (ion-exchange chromatography). Other components may be separated because they have different solubilities in thin films of liquids coating the fine-grained, stationary solid phase (partition chromatography). Still other components may be separated through differential adsorption to the solid (adsorption chromatography) or through differential solubility in immiscible liquid droplets (emulsion chromatography). The commonly used paper chromatography is usually a partition chromatographic mechanism, whereas the chromatographic fractionation of bituminous matter on silica gel, a technique which is often used by organic geochemists, is adsorption chromatography. The classical theory of Martin and Synge defines the separation processes without taking into account the specific physicochemical causes of the separation.

According to this theory, a chromatographic column may be visualized as consisting of several hypothetical layers, or as these investigators called it, "theoretical plates." (This terminology is in analogy to countercurrent extraction.) Within each of these theoretical plates equilibrium in concentration is established between the flowing and stationary phases. The thickness of each of the plates (which is called h.e.t.p. or height equivalent to one theoretical plate) is such that the solution issuing from the plate is in equilibrium with the mean concentration of the solute in the stationary phase throughout the layer. Thus, the thickness of the theoretical plates depends on the minimum quantity of the adsorbent that is necessary to reach a concentration equilibrium with the flowing phase under a given experimental condition.

The concentration of a component in any part of the column may be calculated with the following formula:

$$Q_r = \frac{1}{\sqrt{(2\pi r)}} \, e^{r-V_r} \left(\frac{V_r}{r}\right)^r$$

where

Q_r = total quantity of solute in plate r
r = serial number of the plates; numbered in consecutive order from the top to the bottom of the column

$$V_r = \frac{v}{V_L + \alpha V_S}$$

v = volume of solution that has flowed through the column
V_L = volume of the moving phase in one theoretical plate
V_S = volume of the stationary phase in one theoretical plate

α = partition (distribution) coefficient (gram solute per milliliter of stationary phase per gram solute per milliliter of moving phase, at equilibrium)

This formula can be used to calculate the concentration of a component at a given location, the location where separation between components first begins, the extent of band spreading, etc. It is apparent that a major factor in chromatographic flow is α, the partition coefficient. This factor, as well as the other factors in the formula, can be experimentally determined. A detailed description of this relationship is given by Martin and Synge (1941), and its geochemical application by Nagy (1960).

It follows from the Martin and Synge theory, as expressed by this equation, that chromatographic effects will take place during fluid flow through rocks whenever the partition coefficient has a numerical value greater than zero. In other words, natural chromatography will take place whenever there is a physicochemical interaction between the fluids and the rock. Such interaction may be the adsorption of compounds from the fluid on the mineral surfaces, ion exchange with clay minerals, solubility in bituminous matter, kerogen, etc. Such phenomena may occur repeatedly during intrastratal fluid flow.

Natural chromatography probably plays a role during various stages of diagenesis. When recent sediments are first compacted, the water displaced from the reduced pore volume is forced to move out and up through the layers. Organic compounds derived from decaying biological matter may then be subjected to natural chromatography. It may be interesting to ask a speculative question at this point. Recent sediments contain a multitude of organic compounds including traces of nonpolar hydrocarbons. Petroleum, a formation fluid that has undergone extensive migration and therefore was subjected to prolonged selective filtration (natural chromatography), contains mainly the nonpolar hydrocarbons. One could then ask if the polar components of the original organic debris of muds might not have been selectively removed by mineral matter during intrastratal fluid flow? Natural chromatography might have played a role in petroleum formation as well as in some diagenetic processes leading to the change in the composition of the rocks, as well as of the fluids.

BARTHOLOMEW NAGY

References

Bonham, L. C., 1956, *Bull. Am. Assoc. Petrol. Geol.,* **40,** 897.
Brooks, B. T., 1959, *J. Inst. Petrol.,* **45,** 42.
Day, D. T., 1897, *Proc. Am. Phil. Soc.,* **36,** 112.
Gilpin, J. E., and Cram, M. P., 1908, *Am. Chem. J.* **40,** 495.
Hodgson, G. W., and Baker, B. L., 1959, *Bull. Am. Assoc. Petrol. Geol.* **43,** 311.
Martin, A. J. P., and Synge, R. L. M., 1941, *Biochem. J.,* **35,** 1358.
Meinzer, O. E. (editor), 1942, "Hydrology," New York, McGraw-Hill Book Co., 712pp. ("Physics of the Earth," Vol. 9).
Mileshina, A. G., Safonova, G. I., and Kanaeva, N. A., 1959, *Geol. Nefti i Gaza,* **3,** 55.
Nagy, B., 1960, *Geochim. Cosmochim. Acta,* **19,** 289.
Nagy, B. and Wourms, J. P., 1959, *Bull. Geol. Soc. Am.,* **70,** 655.
Roper, W. A., Doscher, T. and Kobayashi, R., 1958, *J. Petrol. Technol.* **10,** 61.
Smith, H. M., Dunning, H. N., Rall, H. T., and Ball, J. S., 1959, Preprint, 24th Midyear Meeting of API, Division of Refining, New York.

Cross-references: *Chromatography; Hydrocarbons; Phase Equilibria.* Vol. VI: *Crude Oil; Kerogen; Natural Gas; Petroleum.*

NATURAL GAS—*See* Vol. VI

NATURAL RESOURCES

A natural resource is defined as anything in the natural environment, any natural *product, process,* or *feature,* that can be used by man. The object is defined primarily in terms of the geosciences, but its utility is defined in terms of human economy. Examples of such products could be soil or coal or an iron ore; a process example could be a strong prevailing wind (usable as an energy source for windmills); a feature example could be an element of the landscape such as a waterfall (used either as a water-power energy source or as a tourist attraction). All these things are clearly useful to man, although in most cases some *development* or *industrial activity* may be necessary for adapting them to his purposes. Man himself is a resource, but conventionally is classified as "human resources" and, although interacting, is excluded from a strict study of "natural resources."

In classifying natural resources, the first breakdown is into two main categories: (a) *products,* such as minerals, combustible fuels, timber, livestock, fish; and (b) *amenities,* or geographic situation resources, such as a waterfall that affords a hydroelectric potential, or a coral reef or a sandy beach that provides a summer tourist resort. Within each of these there is a further subdivision into *renewable* and *nonrenewable* potentials. Combustible fuels are clearly nonrenewing. Once the world's easily obtained petroleum supplies are used up (in half a century, if not sooner), this nonrenewing source will need to be replaced by something else. Petroleum does re-form slowly beneath the sea floor, but the process takes millions of years. Water power, on the other hand, is completely renewable. The tale, though probably apocryphal, is told of the young lady who, on viewing her first hydroelectric power station, inquired: "And what do they do with the water, after they have squeezed the electricity out of it?"

Various forms of *pollution* (q.v). are serious hazards to the amenity type of resource. As a gross generalization one may say that the leftover residues of category (a), the resource products, constitute the bulk of the pollutants for category (b), the amenities. *Conservation* (q.v.) in the latter is essential. Nature is an open system, a dynamic interrelated whole, so that any interference to any one of its multiple, complex, overlapping cycles, is transmitted eventually to the others. One speaks of the *"ecologic balance"* or the *"balance of nature"* that is easily upset by man's exploitation.

The rational way to try and avoid pollution, and indeed any kind of upset to the ecologic balance, is by means of the integrated survey and multidisciplinary policing of all natural resource development. Too often in the past the policing has been left to the department that has a vested interest in the production of the very thing that is the cause of the trouble. When new surveys are planned to develop natural resources it is highly desirable that specialists in conservation and pollution be incorporated in the study teams. Foresight in such matters is not always easy and long-term consequences are not necessarily visible when the plans are prepared. For this reason, permanent commissions and supervisors are needed to maintain a vigilant watch over all aspects of the natural environment that are liable to be so impinged.

Conservationists tend to polarize into two camps: the *emotionalists* versus the *rationalists*. The extreme form of the emotionalist regards natural resource development as anathema. He would put flowers in his hair and live in a cave. Or, in a less extreme case, he would declare vast areas of the earth as "Natural Parks," surround them with barbed wire fences and exclude man (as an unnatural element, we assume). The writer would prefer not to have to live in a cave. Some compromise is needed. The moment Paleolithic man lit his first fire to warm himself from the Ice Age he started pollution on its way. The modern city is merely a logical development from that stage. It is also the center of the most highly developed modern intellectual activity. The writer urges that the rational approach is to encourage the passionate and emotional dedication to nature, but to recognize that man, as an incredibly powerful and rapacious animal, will destroy it all unless guided and controlled. Calm and rational guidance is therefore the only alternative to anarchy.

Philosophy

A basic question of philosophy, or more precisely, social ideology, has to be considered in contemplating major programs of natural resource development. One may pose the questions: "Does the population of a given area want or need to have its resources developed?" Can we assume that the dictatorial presidents or ruling cliques of certain states have the maximal well-being for their feudal proletariat uppermost on their list of national priorities? Where there is any shadow of doubt about the answers, careful demographic studies of the probable impact of development upon the indigenous population would seem to be indicated. There seems little doubt that certain regions of the earth have already grossly exceeded their optimal population level, so that vigorous steps should be taken to prevent further resource development in such areas and strive for decentralized transfer to the undeveloped regions.

Economics

The concept of human utility in the natural resource definition involves a consideration of *demand*. The extensive coal seams of Antarctica represent potential but at present unexploitable resources, because there is no permanent population there and the demand is lacking. By the same token, some change in man's habits or the discovery of a substitute may reduce or eliminate the demand. Thus the demand for chinchona bark, formerly harvested in great quantity in the East Indies and elsewhere (for pharmaceuticals), is now reduced owing to the manufacture of synthetic quinine. Some minerals, formerly classified as gemstones, have dropped in value because of decreasing rarity and changes in fashion. Thus, resources must be viewed as dynamic and having changing values.

The economic criterion introduces therefore severe restrictions to the theoretical natural resource. The latter may be designated therefore as either:

(a) A *potential resource*. The product may exist in a given region but it may not be exploitable for any one of several reasons: the world market price may be too low in relation to costs of development; the situation may be too remote or transport may be too costly; the local population may be disinterested in development; or it may remain unexploited through lack of capital. A *survey* may show that a resource is probably developable, and can be exploited, provided that certain conditions are met.

(b) A *developed resource*. The product is recognized; man is using it, but is he using it wisely? A mine may be working through a copper lode; a large heap of "tailings" accumulates near the mine, an eyesore on the one hand and a source of potential water pollution on the other. Study of the tailings (mineral and geochemical) may disclose that valuable rare metals are being dumped out with the waste. An oilfield, located far from industrial centers, produces crude oil that is transported to populated areas by pipeline or tanker, but it also produces natural gas that is possibly not economic to transport and in consequence is burned off in giant flares; the energy is wasted; the smoke pollutes the atmosphere. Either a high energy-consuming industry (e.g., a metal smelter) should be built near the gas source, or the gas should be reinjected into the ground to help maintain the oil pressure.

The concept of *wastage of natural resource* is therefore high in the "problem book" of rational thinkers. The desire to conserve has to be matched against the economic framework. If costs of conservation run too high, the resource development may assume the role of the "white elephant." Company or national bankruptcy can result. Wastage can be evaluated against several standards: (a) in terms of conservation of a valuable commodity—on a company, national, or international standard; and (b) in terms of protection against a pollutant—likewise on a company, national, or international standard.

Range and Limits of Natural Resources

The spectrum of knowledge about theoretical or potential resources covers practically the whole span of the natural sciences, but within the restraints of the definition. Some examples within this spectrum are as follows:

Climate (see Vol. II). Although climate is rarely a resource in itself, it is frequently a limiting parameter, and may play an important role in cost analysis. It applies particularly to agriculture, tourism, some energy potentials (e.g. solar), and most developmental activities (need for heating, air conditioning; seasonal stoppage of work due to monsoon rains, etc.). It should be stressed that although satellite surveillance and regional forecasts of world weather are now commonplace, there are many important regions that have either no weather service or no efficient means of communicating to the remote villages such things as storm warnings or drought prediction. On an international basis the W.M.O. (World Meteorological Organization) organizes global work, but many countries totally lack any sort of weather service.

Water Resources (see *Hydrology,* etc., in this volume). Although the geography of surface water (lakes, rivers, swamps, and so on) is well-known, underground water resources are much less known, especially outside of Europe and North America. Since precipitation and evaporation are two of the greatest year-to-year variables in climatology, the matter of seasonal reappraisal is particularly important. Regular air photographic monitoring by satellite or high-flying aircraft can contribute enormously to such knowledge. The International Hydrologic Decade (1965–1974) represents a positive worldwide attack on the problem.

Mineral Resources (see Vol. IVB). Ore deposits are often (erroneously) thought of as the best explored and most thoroughly assessed of the world's natural resources. It is recognized that lonely prospectors have fossicked over an extraordinary large fraction of the earth's land surface, and many mining centers have been established on this basis. However, these hardy souls were often only in search of one thing, be it gold, silver, or copper, but they were not trained to identify minerals of the rare earths or other elements which have perhaps only become important during the last decade or so. And, most important, they could not see beneath carpets of soil or forest, crusts of laterite, or of ice or sand dunes. Modern surveys with airborne magnetometers, seismic parties on land, geochemical surveys, all are needed on a very wide scale in the underdeveloped countries. Even the fundamental geologic mapping is often only of a reconnaissance type. It may be noted that about one-third of the United States still lacks specific state geological survey maps. Although many parts of the U.S. and almost all of northwestern Europe have beautiful geological maps, there are many parts of the world, notably in Asia, South America, and some parts of Africa, that do not even possess an efficient geological survey. The International Union of Geological Sciences has a number of well-developed world mapping and research programs and, jointly with UNESCO in 1969, launched the "International Geological Correlation Programme" which is beginning to play an important role in solving problems and clarifying many areas of confusion. The U.N. Resources and Transport Division has a long-range program for the study of mineral resources, water resources, and energy in the developing countries.

Energy Resources (see Vol. IVB). Conventionally energy resources are taken to include coal, peat, petroleum, natural gas, oil shale, and hydroelectric power. With the exception of the last one, all of these involve *fossil fuel* combustion (and therefore varying degrees of pollution). Classified as "unconventional" energy resources are the more recently developed nuclear (atomic) energy potentials, geothermal energy (volcanic steam, etc.), direct solar energy, and wind power.

Soil Resources (see *Pedology*). Soil surveys are essentially basic to any rational development of agriculture, yet up till recently much of the world's agriculture has been practiced on a hit-or-miss, empirical basis. This is another area, along with water, that can benefit enormously from systematic air photography or geophysical sensing from high-flying aircraft or satellites. Soil reconnaissance mapping of the world on scales of 1:5 million or better are available through the FAO Bureau of World Soil Resources and other groups, but new development often calls for pilot area operations with mapping on 1:10,000 scale, and such coverage is rather exceptional.

Agriculture and Forestry Resources. In this area there are basically two categories of resource: first, the already-existing resource, such as virgin forest, open grassland, native animals, or plants; secondly, the resource region that can be adapted, cleared, or taken over for planting or grazing. These areas of study lie outside the general field of competence of the geoscientists, but must be mentioned here in order to keep them in perspective. Ultimately, many of the surveys of climate, water, minerals, and soils are directed towards "servicing" the needs of the agriculturalists. Thus a basic appreciation of the biological factors is highly desirable in the educational and philosophic makeup of the earth scientists. A somewhat semantic question sometimes arises as to the limit of a natural resource in an area so modified by man (one can think of a Dutch polder, for example) that it is very far indeed from its pristine state. It is suggested here that everything that is basically related to the natural earth (no matter how modified) fits the definition. Industrialization introduces a totally new, man-made element and is therefore excluded.

Oceanic Resources. These are divisible again into two categories, but on a different basis. The mineral resources of the sea (such as by evaporation, or from exploration of the sea bed) are strictly in the geological or geographical area. In contrast, the fisheries resources are clearly in the biological field. Generally, there is no interaction between the two, although funds and ships allocated vaguely to "oceanography" often become an area of competition between these two diverse fields. Integrated surveys are relatively rare although the two disciplines have much to learn from each other; benthic fish are ecologically tied to their substrate, an area understood by the geologist, and today the modern marine geologist is often the best informed person on the distribution and ecology of all the planktonic life forms that have preservable tests (foraminifera, coccolithophoridae, diatoms). The development of fish farming, particularly in semiartificial coastal fishponds and around artificial "reefs" calls for close collaboration. The purely mineral development of the sea floor, such as for manganese nodules, is as yet largely only a potential, but off-shore oil drilling techniques have now reached such a high state of development that the off-shore regions constitute major provinces of the world's petroleum resources (see Vol. IVB).

Inventory and Appraisal of Resources

The key to successful development of resources lies in the word rational: care, forethought, consultation. The operations leading to this end go through the following steps:

(a) **Survey and Inventory.** Scattered through the voluminous files of government laboratories, and the scientific literature of the world journals, there is an immense accumulation of data. Bibliographic catalogs of sources and analyses of data in depth (going back for about a century) are needed on a world basis and reduced to push-button simplicity.

Additional surveys and maps are needed. For almost every state and country in the world (not merely the "underdeveloped" ones) there is a great need for integrated surveys, where five or six disciplines are represented on every survey team. An appropriate method has been developed for tackling the problem. It was first evolved in Australia to deal with the broad, thinly populated spaces there, and is known as the "land system survey" (see detailed discussion in Vol. III of this series).

Inventory implies some sort of central data bank and indexing. With modern electronic *data processing* and retrieval this type of problem presents little difficulty in developed countries, but most regions still lack these basic facilities. Regional map collections and air photographic archives should be associated with the libraries and information banks. In the United States, a relatively advanced country, most of the integral states still lack any such central repositories.

(b) **Services and Training.** The education and technical training of personnel is a key problem, especially in the underdeveloped lands. In the more developed parts of the world, there are often the knowledgeable people available but they are not always in the appropriate positions to guide and direct rational resource development. It is assumed that every state must have been furnished with *scientific and technical service departments* which provided a source of information, advice, research, and training. Such services include: a *Meteorology Service; Geological Survey* (including geophysical and geochemical divisions); *Topographic Survey* (which hopefully maintains a loose-leaf national or state Atlas, embracing all map data from Soils to Population Density); *Agriculture and Soil Service* (which in some regions may be split up into Agronomy, Livestock, Soils, and Forestry divisions, according to demand); *Fisheries and Wildlife Service;* and last but not least an overall *Natural Resource Service* (which may have separate conservation, development, city planning, and pollution divisions.

Often, the most economic way to train technical personnel is through an apprentice program with each service, with possibilities of university scholarships for the most gifted. For professional training, university instruction is essential, although few universities offer general resource-oriented programs. Specialized training in meteorology, soils, geology, and so on is universally available but very few institutions treat the overall picture of total ecology and resources.

There are three critical stumbling blocks in the way of obtaining optimum results from these professional services and training programs. They are:

(a) *The Technological Gap.* The primary and secondary school training in many countries is inadequate, poor, or follows antiquated syllabi: mathematics that ignore the discovery of the slide rule; history that concentrates on the battlelines of long-forgotten wars; language courses that are preoccupied with the archaic. The curriculum frequently does not recognize that manual training would be more appropriate to many personalities than is preparation for a college career that may be totally inappropriate, except on grounds of intellectual or social snobbery. In some underdeveloped countries the instruction is given in an alien language, from textbooks prepared by residents of some unrelated culture group or by a religious sect that devotes more time to dogma memorization than to instruction in carpentry.

What is needed is top-level and well-paid teachers; in New York City, for example, the young teacher's salary is 10% less than that of a rooky garbage collector. What is needed in high schools is a curriculum that includes *general science* for not less than four years, so that everyone has a year each of physics, chemistry, geology, and biology: not in separate, concentrated lumps, but spread out over the whole period, so that the level of instruction can evolve with the student, and so that integration can be attempted at every level. Always the global approach should be stressed; likewise the oneness of humanity and its environment. In the underdeveloped countries the establishment of regional, international colleges can go a long way to promote the spirit of grander unity in place of merely expanded tribalism, often referred to as "patriotism."

(b) *The Information Gap.* This is a different sort of gap: the gap between available information and the people who need it. It is often seen on the political level, where the legislator has not informed himself about the facts available to the specialists. Basically it is a failure to read or consult. "Awareness" is a fashionable word to imply a bridging of the gap, but, alas, it is often employed by those who have made no effort in that direction. In many developing countries there is a desire for "instant development," and sometimes there is an assumption of intuitive know-how. A modest approach is an unpalatable counsel, but, patiently and sympathetically applied, can work wonders.

(c) *The Brain Drain* (or loss of trained specialists). One of the frustrating aspects of training personnel to aid in the rational government of natural resources is the process of siphoning off the cream of the intellectual crop. This process may occur within a highly developed country, where the modestly paid civil service scientist is siphoned off into wealthly industries for higher rewards; or it may be from poor countries to richer countries. It has happened since the days of ancient Rome. The attraction, financial, social, and intellectual, of the imperial city has often attracted the most gifted people: artists, musicians, scientists. As a consequence, the remote areas most needing development may be intellectually impoverished. The solution is not simple, but can be met, in this modern age of rapid transportation, by a combination of financial incentives, frequent trips to and fro, public recognition, scientific opportunities

Evaluation and Utilization of Natural Resources

Finally we come to the critical phase: the resources have been mapped, cataloged, and studied by trained specialists and now something has to be done about them. Action has to follow the following procedure in strict order:

(a) *Evaluation.* Let us take a mineral find, for example. A field geologist discovers specks of gold in a quartz vein. It is impossible to extract the gold without a costly mine and concentrator. Before embarking on this expense, it is essential to *prove reserves* of sufficient volume to justify development. Accordingly, a diamond-drilling program, a costly affair, is a necessary prerequisite. Note also that many such drilling programs prove the gold deposits would be uneconomic and therefore should be dropped. This saves wastage in the long run, but the cost of the drilling has to be written off by the profits of the successful operations. Therefore it is customary for countries to allow oil or mineral development groups a larger profit margin than is usual for industrial development because of the high risks involved.

In an agricultural operation, a *pilot project* is often the best procedure. A small plot or a model farm can demonstrate if the soil and other factors are appropriate. More attention to this aspect might have avoided some of the monumental failures (like the East African groundnuts scheme) of the past.

Economically it is essential to prove feasibility in one way or another before major financing can be obtained. Nevertheless, very large sums can often be raised from banks merely to survey and test for feasibility.

(b) *Utilization.* This is the logical end pont, assuming of course the positive nature of the feasibility study. At this level responsibility generally passes out of the hands of scientists and into the realm of national economics and politics. At this level, however, the scientist may still make some contributions. For instance, how much is this commodity or amenity needed within the regional economy? The priority question. Or again, would it be best to develop a certain item close to the large center of population (to give employment, perhaps?) or would it be best to keep it to some remote spot where its pollution potential will be minimal. In the case of small countries would an international, sectoral, or regional approach be practicable?

These types of questions again underline a theme in this article: that mutual understanding and integration of all scientific knowledge is the only *rational* way of developing natural resources. There are other ways, for instance, by the individualistic, ruggedly independent approach—say, of the ancient, bearded gold prospector and his patient donkey—but in a crowded modern world the procedure is more often on a cooperative and interdependent basis.

Acknowledgment: In preparation of the above article the author wishes to acknowledge the inspiration and help obtained from the United Nations while editing a committee report on *"Natural Resources in Developing Countries: investigation, development and rational utilization"* for the UN Economic and Social Council, approved 5 February 1969; the subject matter has also been offered in lectures at Columbia University.

RHODES W. FAIRBRIDGE

References

Ahmad, J., 1960, "Natural Resources in Low Income Countries: An Analytical Survey of Socio-economic Research," Univ. of Pittsburgh Press, 118pp.

Allen, S. W., 1959, "Conserving Natural Resources," Second ed., New York, McGraw-Hill Book Co.

Alverson, D. L., Longhurst, A. R., and Gulland, J. A., 1970, "How much food from the sea?" *Science,* **168**(3930), 503–505.

Anon. (Scientific Am.), 1963, "Technology and Economic Development,' New York, Alfred A. Knopf, 205pp.

Anon., 1967, "Integrated surveys and applied geomorphology," *Nature and Res.,* UNESCO, **3**(3), 7–9.

Anon. (Nat. Acad. Sci.—N.R.C.), 1969, "Resources and Man," San Francisco, W. H. Freeman & Co., 259pp.

Baer, J. G., 1965, "Teaching about natural resources and conservation problems (with particular reference to Africa)," *Nature and Res.,* UNESCO, **1**(4), 9–14.

Barnett, H. J., and Morse, C., 1963, "Scarcity and Growth: The Economics of Natural Resource Availability," Baltimore, Johns Hopkins Press.

Brown, H., et al., 1963, "Next Hundred Years," New York, Viking Press.

Clawson, H. (editor), 1964, "Natural Resources and International Development," Baltimore, Johns Hopkins Press.

Dasmann, R. F., 1959, "Environmental Conservation," New York, John Wiley & Sons.

Fischer, J. L., and Potter, N., 1964, "World Prospects for Natural Resources," Baltimore, Johns Hopkins Press.

Flawn, P. T., 1966, "Mineral Resources: Geology, Engineering, Economics, Politics, Law," Chicago, Rand McNally, 406pp.

Foncin, M., Froehlich, W., and Sommer, P., 1966,

"Bibliographie Cartographique Internationale, 1964," Paris, A. Colin, 781pp.
Herbert, F. W., 1965, "Careers in Natural Resource Conservation," New York, Walck, Inc.
Huberty, M. R., and Flock, W. L., 1959, "Natural Resources," New York, McGraw-Hill Book Co.
Jackson, N., and Penn, P., 1966 "Dictionary of Resources and Their Utilization," Oxford, Pergamon Press.
Kindsvater, C. E., 1964, "Organization and Methodology for River Basin Planning," Atlanta, Ga., Georgia Inst. Tech., 561pp.
Myrdal, G., 1963, "Challenge to Affluence," New York, Pantheon Books.
Smith, G. H. (editor), 1965, "Conservation of Natural Resources," New York, John Wiley & Sons.
Van Dyne, G. M. (editor), 1969, "The Ecosystem Concept in Natural Resource Management," New York, Academic Press, 386pp.
Ward, R. J., 1967, "The Challenge of Development," Chicago, Aldine Publ. Co., 500pp.
Zimmerman, E. W., 1964, "Introduction to World Resources," New York, Harper.

Cross-references: *Air Pollution and Global Climate; Air Pollution and Urban Climate; Conservation; Ecology; Environmental Science; Hydrology; Manganese Nodules; Pedology; Pollution; Water Pollution.* Vol. I: *Nutrients in the Sea.* Vol. II: *Climatology.* Vol. IVB: *Fossil Fuels; Geothermal Energy Sources; Minerals; Oil Industry; Petroleum.* Vol. VI: *Agricultural Geology; Agrogeology; Agronomy.*

NEODYMIUM: ELEMENT AND GEOCHEMISTRY

In 1885 Von Welsbach first isolated neodymium from other rare earths; he separated the element out of didymia, an intermediate rare earth concentrate of cerite, the first known light rare earth mineral, discovered by Cronstedt in 1751.

Properties

Elemental neodymium is a very reactive, silvery white *lanthanide* (q.v.) or *rare earth* (q.v.) metal with atomic number 60 and outer electronic structure $4f^4\ 6s^2$. It occurs as six stable isotopes:

Nd^{142}	27.1%	natural abundance
Nd^{143}	12.2%	
Nd^{145}	8.3%	
Nd^{146}	17.2%	
Nd^{148}	5.7%	
Nd^{150}	5.6%	

One isotope, Nd^{144} (23.9%), is an alpha emitter, with radioactive half-life of 5×10^{15} years.

Neodymium is determined analytically by neutron activation, x-ray spectrography, absorption spectrometry (it has very sharp spectral lines), and by isotope dilution mass spectrometry. In geologic environments, neodymium forms highly ionic bonds. It forms weak complexes. The oxidation state is tripositive, with ionic radius 0.995Å; dipositive neodymium is oxidized by water, and is unstable in most crystal lattices. Generally, chemical behavior is similar to that of the alkaline earths, and very similar to that of other light rare earths.

Minerals

Neodymium occurs in highest concentration in rare earth minerals that concentrate light rare earth. Generally nitrates, sulfates, carbonates, niobates, and tungstates with high cation-coordination number favor the light rare earths and often contain structural halogens and alkali metals. Typical examples, with cation-coordination number 10 and 11 are: monazite (light rare earths, Th) PO_4; bastnaesite (light rare earths, Th) FCO_3.

Geochemical Behavior

Schmitt et al. (1964) have established a value 0.58 ppm (parts per million) neodymium in chondrites and 7.0 ppm in eucrite achondrites. The standard deviation within a meteorite class is only about 20%. In meteoritic matter volatilization may have redistributed alkali metals but not rare-earths: neodymium has a boiling point of 3027°C.

The neodymium concentration in average basalt is 32 ppm, but the range is over a factor of ten (Haskin et al., 1966). Alkaline oceanic basalts and most continental basalts have more neodymium than oceanic tholeites and some continental diabases. Partial melting of mantle peridotite yields a basaltic liquid with enriched neodymium and a mantle source area depleted in neodymium.

The neodymium concentration in granites with 60 to 70 per cent silica averages 55 ppm, a value that is higher, but also more variable, than for basalts.

Igneous differentiation series tend toward higher neodymium concentrations.

Towell et al. (1965) found 23.5 ± 2.4 ppm in the Rubidoux Mountain Leucogranite phase of the Southern California batholith, probably a differentiate (by crystal settling) of the San Marcos Gabbro, with 7.68 ± 0.77 ppm. About half of the neodymium in the gabbro is in rock-forming minerals (plagioclase, pyroxene, hornblende), probably substituting for calcium. The remainder is in accessory minerals like apatite. The accessory minerals (probably apatite, zircon, sphene, and monazite) hold most of the neodymium in the leucogranite.

NEODYMIUM: ELEMENT AND GEOCHEMISTRY

Neodymium forms complexes in alkali, carbonate, and halogen-rich hydrothermal fluids, but they are not as stable as heavy rare earth complexes. Pegmatoid and metasomatic deposits formed from hydrothermal fluids often have low neodymium to heavy rare earth ratios as compared to granites.

Neodymium has a 180 year residence time in sea water, about the same as other light rare earths. Probably it is complexed. Removal by adsorption on clay particles, as well as depletion in surface water in areas of high organic productivity are likely processes. Today, high secondary neodymium concentrations in phosphatic fish remains occur only in deep water; mesozoic fish remains also have high secondary neodymium. Manganese nodules have 13,800 ppm neodymium whereas Pacific Ocean water has only 2.3×10^{-6} ppm (Goldberg, et al, 1963).

In soils, the relative amounts of neodymium in detrital, clay, and carbonate fractions is not well known. Sedimentary rocks have about 37 ppm neodymium, with a variation of a factor of five. In a study of rare-earths in Russian platform soils, Balashov et al. (1964) found higher concentrations in alkaline soils, the neodymium precipitating as hydroxide; lower concentrations are in acid soils, probably due to removal as complexes.

Biologic concentration of rare earths has been found in hickory leaves; rare earths have been used to trace opium.

Economic Geology

The major source of neodymium has been from monazite in beach sands. The accessory monazite originally present in granitic rocks resists weathering and is concentrated by sedimentary processes. Australia, Brazil, Ceylon, India, Malaysia, and South Africa mine the largest quantities. Usually, about ten percent (by weight) of monazite is neodymium. Recently, the bastnaesite deposit at Mountain Pass, California has been an important source of neodymium.

High purity neodymium has limited application in electronics and in optical filters. Most neodymium is sold unpurified as misch metal (a mixed rare earth alloy), or as oxide polishing powders. Impure rare earths are also used as petroleum cracking catalysts and in arc carbons.

ROBERT KAY

References

Balashov, Y. A., et al., 1964, "The effects of climate and facies environment on the fractionation of the rare earths during sedimentation," *Geochem. Internat.* **10**(1), 951–969 (English transl.).

Goldberg, E. D., et al., 1963, "Rare-earth distributions in the marine environment," *J. Geophys. Res.*, **68**, 4209–4217.

Haskin, L. A., et al., 1966, "Meteoric, Solar, and Terrestrial Rare-Earth Distributions," in (Ahrens, L. H., et al., editors), "Physics and Chemistry of the Earth," Vol. 7, Oxford, Pergamon Press, pp. 167–321.

Rankama, K., and Sahama, T. G., 1950, "Geochemistry," Chicago, University of Chicago Press, 912pp.

Schmitt, R. A., et al., 1964, "Rare-earth, yttrium and scandium abundances in meteoric and terrestrial matter—II," *Geochim. Cosmochim. Acta*, **28**, 67–86.

Towell, D. C., et al., 1965, "Rare-earth distributions in some rocks and associated minerals of the batholith of Southern California," *J. Geophys. Res.*, **70**, 3485–3496.

Cross-references: *Complexes; Emission Spectrography; Lanthanides; Mass Spectrometry; Neutron Activation Analysis; Rare Earths.*

NEON: ELEMENT AND GEOCHEMISTRY

General Information

Neon (from the Greek: $\nu\epsilon o\sigma$ = new, recent) is one of the rare or "noble" gases, with the chemical symbol Ne. It is a chemically inactive, colorless, odorless, monatomic, and normally gaseous element with atomic number 10, discovered in 1898 by W. Ramsay and M. W. Travers as a rare constituent in the troposphere. It has three stable isotopes with mass numbers 20, 21, and 22. In dry air, the abundance of Ne is 0.001818% by volume. Ne belongs to the group of the noble gases which include helium, argon, krypton, xenon, and radon (synonym expressions: rare gases, inert gases). Its characteristic red-orange light emission on recombination with electrons in low-pressure discharge tubes has found wide and popular use in outdoor advertising displays. Modern applications of Ne include its use in the He–Ne gas laser and as a filling gas for spark chambers (an important research tool in high-energy physics, used in the discovery of the μ-neutrino at the Brookhaven National Laboratory in 1962). Some of the Ne lines in the emission spectrum serve as a secondary standard for the definition of the meter.

Properties

The internationally accepted atomic weight for neon of tropospheric composition (Ne^n) is 20.183. This corresponds to a density of gaseous Ne^n under standard conditions (0°C and 760 mm Hg) of 0.9002 g/l.

At the triple point (t.p.), the properties of Ne^n are: $T(\text{t.p.}) = 24.56°K$, $P(\text{t.p.}) = 323.5$ mm Hg, density of solid ρ_s (t.p.) = 1.444 g/cm^3, and density of liquid ρ_L (t.p.) = 1.248 g/cm^3.

Critical values: $T_c = 44.5°K$ and $P_c = 25.9$ atm. The boiling point is 27.07°K. For a comprehensive review and theoretical discussion of noble gas data up to the triple point, as well as for references, see Pollack (1964).

Chemical Behavior

For all purposes, Ne must be considered as chemically inert.

Occurrence

Atmosphere. The isotopic composition of Ne^n has recently been redetermined by Eberhardt et al. (1965). Relative abundances of the three isotopes in percent are: Ne^{20} = 90.5±0.07; Ne^{21} = 0.268±0.002; Ne^{22} = 9.23±0.07. The origin of Ne in our atmosphere is uncertain, but influx of solar wind during periods of low geomagnetic intensity is an attractive possibility.

Hydrosphere. In the oceans, Ne is typically present in concentrations varying from 1.92×10^{-4} ml(STP)/liter to 1.53 ml(STP)/liter (Bieri et al., 1966). Most of the variation is due to the temperature dependence of the solubility of this gas in seawater, but fluctuations in the partial pressure of Ne in the atmosphere, mixing of water masses and dissolution of air bubbles close to the surface may also be involved. The isotopic composition of Ne extracted from seawater is not distinguishable from that of air.

Lithosphere. Isotopic analysis of the Ne extracted from U + Th bearing minerals, notably pitchblende from the Congo, indicates excess Ne^{21} and Ne^{22} which can be attributed to a production from the nuclear reactions (Wetherill and Inghram, 1953):

O^{18} (α, n) Ne^{21} and F^{19} (α, n) Na^{22} $\to^\beta$ Ne^{22}

It was also found in large amounts (7.5×10^{-6} cm^3/g) in a *thucholite* from Ontario. Alteration of the atmospheric composition of Ne due to nuclear reactions, especially for Ne^{21}, over geologic times is considered possible.

Meteorites. Ne concentrations in meteorites have been utilized in developing the cosmic histories of these objects. The highest concentrations occur in the dark phase of brecciated chondrites or achondrites (6×10^{-5} cm^3(STP)/g in Fayetteville). From variations in the isotopic composition with meteorite class and in its concentration relative to other noble gases, it is possible to assign at least two different origins for the element: (a) "Cosmogenic" or "spallogenic" Ne, produced by the interaction of cosmic rays with elements of atomic number > 10, and (b), "primordial" or trapped Ne of unknown but much speculated origin. The first component has a typical isotopic composition (depending on chemical composition, energy spectrum of bombarding particles and diffusion losses): Ne$^{20} \sim 30\%$; Ne$^{21} \sim 33.3\%$ and Ne$^{22} \sim 36.7\%$. The second is similar to air neon.

RUDOLF H. BIERI

References

Bieri, R. H., Goldberg, E. D., and Koide, M., 1966, "The noble gas contents of Pacific seawaters," *J. Geophys. Res.*, 71, 5243–5265.

Eberhardt, P., Eugster, O., and Marti, K., 1965, "A redetermination of the isotopic composition of atmospheric neon," *Z Naturforsch.*, 20a, 623–624.

Pollack, G. L., 1964, "The solid state of rare gases," *Rev. Mod. Phys.*, 36, 748–791.

Wetherill, G. W., and Inghram, M. G., 1953, "Neutron production in rocks: Variations in isotopic abundances in nature due to (α, n) and (α, p) reactions," *Nuclear Proc. Geol. Settings*, pp. 33–35.

Cross-reference: *Rare Gases.*

NEUTRON ACTIVATION ANALYSIS

Neutron activation provides one of the major tools in geochemistry. This form of analysis ("NAA") is an extremely sensitive method for detection and for quantitative determination of most of the elements in the periodic system, often being 100 to 1000 times more sensitive than earlier methods. The sample is placed in a nuclear reactor and irradiated by neutrons. As a result, radioactive nuclides are formed that reflect the stable nuclei of the sample. The radiation spectrum is then studied as a function of energy and time after irradiation, and the original quantities of each element present can be deduced.

This article describes nuclear reactions, excitation, radioactivity, and transmutations induced by neutrons, as applied in neutron activation analysis within the realm of analytical geochemistry.

Activation analysis is considered to originate from the early classical work by von Hevesy and Levi in 1936 and Seaborg and Livingood in 1938. History, theory, and early literature sources on nuclear physics and activation are compiled by Glasstone (1958). The growth rate of literature on activation analysis between the years 1950 and 1968 has been exponential. Presently an activation analyst has a literature approaching 5000 items with many hundreds of articles added each year. A recent literature compilation on this subject covering the period 1957–1968 is by Lutz, Boreni, Maddock, and Meinke (1968).

Nuclear activation is caused by the interaction of elements with neutrons, n, or protons, p. Other nuclear particles, usually accelerated nuclei, electrons, and in some cases high energy gamma rays, can also excite the atom.

Nuclei of elements are built by the fusion of protons and neutrons, collectively called nucleons. When negative electrons, corresponding in count to the number of protons in the nucleus, balance out the sum of the positive charges of these protons and surround the nucleus, an atom is formed. A nuclide is an atomic species with a fixed neutron and proton number.

Neutrons and protons are the basic building blocks of the nucleus—the core of the atom. The positive nucleus of common hydrogen, $_{1}^{1}H^{1+}$, the proton, when combined with one neutron forms the nucleus of the "heavy" hydrogen deuterium, $_{1}^{2}H$, denoted also as D. The addition of a second neutron to the deuterium nucleus results in the formation of a third hydrogen isotope—tritium, $_{1}^{3}H$, or T, which does not occur in nature. The natural element "hydrogen", H, is thus a mixture of two isotopes, or nuclides having the same number of protons, $_{1}^{1}H$ and $_{1}^{2}H$. Naturally occurring elements are mixtures of isotopes which are different in mass due to varying numbers of neutrons in their nuclei. Some elements are composed of a single isotope, but artificially, man has produced by nuclear activation numerous isotopes of all elements. Isotopes have nearly identical chemical properties but they exhibit different nuclear characteristics and may thus be identified. Certain nuclides are radioactive, they disintegrate spontaneously, e.g., $_{92}^{235}U$, $_{92}^{238}U$, $_{92}^{234}U$, $_{19}^{40}K$, $_{90}^{232}Th$. All nuclides can be excited, or transmuted into other nuclides by some form of activation with nuclear particles.

Neutron activation is the most commonly used form of nuclear activation. When a nuclide is activated by an interaction with a neutron, gamma photons (electromagnetic radiation), alpha particles ($\alpha = _{2}^{4}He^{2+}$) or beta particles (β = electron) are emitted. This "radioactivity" can be detected by such devices as the Geiger Müller Counter, and identified by proportional counters, the scintillation crystal-phototube combination, the solid-state germanium or silicon lithium-drifted detectors, etc. The γ rays exhibit a definite energy spectrum and are therefore used to identify the decaying activated nuclide or its daughter products. This process is called *scintillation* or *gamma ray spectrometry*. The energies of the β particles show continuous energy distribution with characteristic maxima but their low penetrating power through counter windows makes counting of beta particles less attractive. Alpha particles are monoenergetic for each emitter but they are stopped by the thinnest sheets of matter and are thus of limited help in practical rapid identification of nuclides for analytical purposes.

Distinction is often made verbally between the "reactor" or "slow neutrons" and the "accelerator" or "fast neutrons" depending on the source. However, this is confusing, since fast neutrons are also generated in reactors and the accelerator neutrons can be slowed (thermalized, moderated) by making them collide with light nuclei, especially those of hydrogen or carbon (graphite). As projectiles, neutrons can have different energies and consequently are propelled at different speeds. Being electrostatically neutral they are not repulsed by the positive nuclei but rather attracted or "captured." A *capture of a neutron* results in a compound nucleus which is usually in an excited or high-energy state, because an addition of a nucleon (p or n) increases the energy of the product by about 8 MeV (million electron volts; one eV equals the energy acquired by a charged particle falling through a potential of 1 volt), plus the kinetic energy of the projectile. Thus even neutrons slowed down to thermal energies of ~ 0.003 eV, when captured, cause usually an ejection of a proton, an alpha particle, or gamma radiation emanating from the excited target nucleus. Activated as well as naturally radioactive nuclides break up (decay) at specific rates, which are typical for each radioactive species.

Decay of a Radioelement

Decay of a radioelement can be described in time units during which one half of the originally present atoms disintegrate. These time intervals are denoted as the half-life of the nuclide. The decay constant, λ, and the half-life constant, $T_{1/2}$, are properties by which a radioactive nuclide can be identified.

When the relative activity of a radioelement is plotted against time units, during which this activity decreases by one half of the original intensity, an exponential decay curve results (Fig. 1). Mathematically, this type of curve means that the number of atoms disintegrating in a unit of time is proportional to the number of atoms of the decaying nuclide present. Since the decay goes on, leaving less and less atoms of the species, the rate, or number of dis-

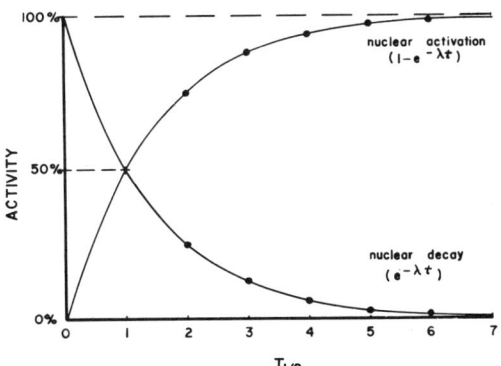

FIG. 1. Exponential decay and activation curves (from Volborth, 1969, p. 284). *Courtesy Elsevier Publ. Co.*

integrations per time unit, also changes continuously.

If we denote as N the total number of atoms of a radionuclide having a decay constant λ, the number of disintegrated atoms dN, and the short time interval that it takes for these atoms to disintegrate as dt, we see that

$$-\frac{dN}{dt} = \lambda N \quad (1)$$

(atoms decay in this reaction, decreasing in number, whence the negative sign for the rate).

We can also write

$$\frac{N}{dN} = -\lambda dt \quad (2)$$

and integrating we get

$$\ln(N_t/N_0) = -\lambda t \quad (3)$$

or

$$N_t = N_0 e^{-\lambda t} \quad (4)$$

in exponential form, where N_t represents the number of undecayed atoms after the time interval t, N_0 is the original number of atoms at any zero time, and e (=1.718 ...) is the base of natural logarithms, ln. To transform to Briggsian or ordinary logarithms, log, we apply the conversion factor 0.4343, thus

$$\log(N_t/N_0) = -0.4343 \lambda t \quad (5)$$

The half-life of a nuclide can be calculated using this equation, if the decay constant, λ, is known, because after one half-life, $T_{1/2}$, the ratio of the number of undecomposed atoms N_t to the atoms at time zero, N_0, will be 1:2, or, substituting in the equation above

NEUTRON ACTIVATION ANALYSIS

$$\log 1/2 = -0.4343 \,\lambda T_{1/2}$$
$$\log 1 = \log 2 - 0.4343 \,\lambda T_{1/2}$$
$$0.3010 = 0.4343 \,\lambda T_{1/2}$$
$$T_{1/2} = \frac{0.693}{\lambda} \quad (6)$$

This notation expresses the relation of half-life to the decay constant. (5) can be rewritten

$$\log N_t = \log N_0 - 0.4343 \,\lambda t \quad (7)$$

which shows that when the logarithm of N_t (of the number of atoms of the decaying species present at any time) is plotted against time, with reference to an arbitrary zero (e.g., starting time of experiment), a straight "curve" (line) will result and its slope will be -0.4343λ. The decay constant, λ, could obviously be calculated from the slope, if such a decay curve were plotted. The decay constants of nuclides, the disintegration of which can be conveniently followed by detector-counter-combinations, can be calculated using Eq. (7). One assumes that the rate (intensity, I) of counted particles or photons in a short time interval is proportional to the number of remaining atoms. Measured after time t this rate of disintegration, I_t, can be substituted for N_t in Eq. (7). If we start our experiment measuring the rate of disintegration at an arbitrary zero time, I_0, we can write

$$\log I_t = \log I_0 - 0.4343 \,\lambda t \quad (8)$$

Determining the counting rate, I_t, at certain time intervals and plotting logarithms of I_t against time, one can graphically determine the slope of the decay curve. Since I_0 is the disintegration rate at an arbitrary time zero it does not need to be known. For example if we have plotted such a curve in seconds and determined the slope to be -0.0412, we can calculate the half-life of the radioactive nuclide. The slope being equal to -0.4343λ, it follows that

$$-0.0412 = -0.4343\lambda$$
$$\lambda = 0.0949$$

and substituting in Eq. (6), we get

$$T_{1/2} = \frac{0.693}{0.0949} = 7.3 \text{ sec}$$

which is the half-life of the radioactive nitrogen–16 isotope, $^{16}N^*$, produced by bombarding oxygen-16, ^{16}O, with 14 MeV neutrons.

If the disintegration of more than one radionuclide is measured simultaneously and the logarithms of total counts, $\log I_t$, plotted at

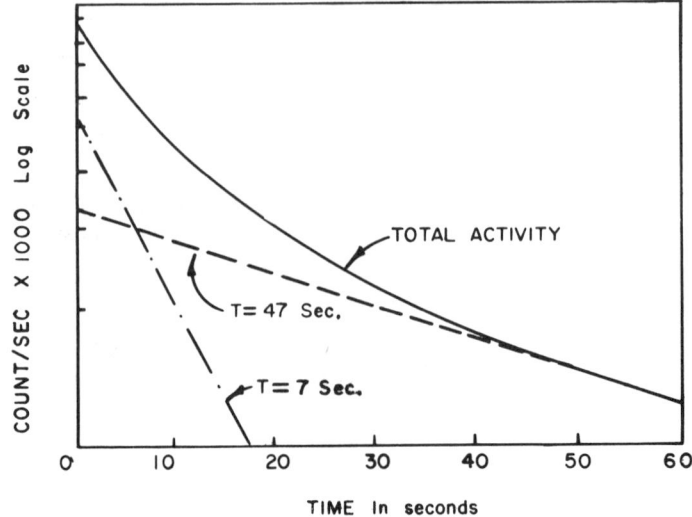

Fig. 2. The effect of decay of short half-life radionuclide on a decay curve of a longer-lived species when measured simultaneously. In the case shown, the composite curve approaches linearity at about 50 sec after the start of the counting. The short-lived isotope has a half-life of 7 sec, corresponding closely to the half-life of N^{16}. If the "oxygen" activity is to decay at least 99%, the waiting period should be 50 sec, preferably 60, when the curve becomes linear. Counts taken after that waiting period would represent the nuclide with the half-life of 47 sec (from Volborth, 1969, p. 321). *Courtesy Elsevier Publ. Co.*

time intervals as above, the emerging plot will be a curve instead of a line, as shown in Fig 2. It is possible to analyze such composite decay curves in order to separate the individual slopes, and determine the appropriate decay constants of the radionuclides present. In *gamma-ray spectrometry* one often approaches this problem from a practical standpoint and waits till the short lived components decay and the curve straightens, after which the slope of the longer-lived nuclide can be graphically determined. Literature and practical methods in gamma-ray spectrometry are given by Crouthamel (1960), Overman and Clark (1960), and Price (1964).

The determination of half life by the method of plotting logarithms of intensity at time intervals becomes impractical when the decay rates decrease because time required to perform a series of measurements and the accumulation of statistically meaningful numbers of counts during relatively short time periods becomes impossible. In order to calculate the decay constants of radionuclides with *long half-lives* one has to know the total weight of the sample, and the fraction of the nuclide investigated in this sample. This permits the calculation of the actual number, N, of atoms of the nuclide, knowing its atomic weight and the Avogadro number. Measuring a total number of disintegrations, ΔN, in such a weighed sample in a specific time period, Δt (e.g., one hour or day), and substituting in Eq. (1), ΔN for dN, and Δt for dt we get

$$-\frac{\Delta N}{\Delta t} = \lambda N \qquad (9)$$

or

$$\lambda = -\frac{\Delta N/\Delta t}{N} \qquad (10)$$

The decay constant, λ, can now be calculated since the other quantities are known.

With *very-long-lived radioelements* where there may be only a few disintegrations per year per sample one has to rely on analytical separations of the daughter elements. Assuming that these are in equilibrium with the parent nuclide, their ratios may be used to determine the radioactive constants λ, and $T_{1/2}$. Reversely one may calculate the time that must have elapsed since the deposition of a certain relative amount of a radioelement or its daughter (see *Geochronometry*).

Nuclear activation is a process reverse to nuclear decay. The product nuclide in a neutron activation reaction will naturally (if radioactive) decay at a rate characteristic for the new species. In time, enough of the product nuclide will form so that the number of decaying atoms will equal the number of atoms newly formed. Thus, one can think of a radioactive equilibrium when the ratio of the daughter (in this case compound nuclide) to its parent (target nuclide in neutron activation) will remain constant. This ratio will naturally depend on the respective decay constants. When one plots the activity of the activated atomic species against its half life (Fig. 1) one gets the *activation curve* which is transposed to the nuclear decay curve. The point in time when the saturation would occur is called the saturation point. From the exponential nature of the activation curve we can see that this point can only be approached asymptotically, but never reached. The exponential function $(1-e^{-\lambda t})$, which defines this curve, is called the saturation factor in neutron activation.

In *radioactive decay* the fraction of the radionuclide remaining after time $nT_{1/2}$ is obviously $(1/2)^n$, expressed in percent it is 50, 25, 12.5, 6.25% . . . *In neutron activation*, saturation acquires supplementary values of 50, 75, 87.5, 93.75%, etc., respectively. The relative production of activated atoms can thus be seen to decrease to insignificant yields after an irradiation (activation) period of four half-lives, making longer exposures to neutron fluxes impractical.

As sources of neutrons the geochemist most frequently uses (a) reactors ("piles"), (b) isotopic sources, and (c) Cockcroft-Walton positive particle accelerators or Van de Graaff electron accelerators.

The Nuclear Reactor as Source of Neutrons

The nuclear reactor is the main source of neutrons and in terms of neutron fluxes, n/cm²/sec, available, the most powerful. Neutrons are produced in reactors as a result of *fission* of uranium, plutonium, and thorium. In this process the target atoms break up into two fragments of approximately equal size after being hit by neutrons and simultaneously, more neutrons are ejected from the fission fragments. Some heavy nuclides, e.g., $^{235}_{92}$U, undergo fission by thermal neutrons. Most heavy nuclides require higher energy neutrons to be split. When the energy of the incident neutron is about 10 MeV the $(n,2n)$ reaction becomes possible, because the system has enough energy (10 + 8 MeV) for two neutrons to be ejected (as mentioned earlier). Whenever one neutron causes fission and more neutrons are formed, self-sustained fission becomes possible, assuming that enough neutrons formed interact with the fissionable isotope. The *nuclear chain reaction* is best understood when one starts with one neutron and assumes that the number will be doubled in geometrical progression: 1, 2, 4, 8, 16, . . .

When fissionable material is piled up in sufficient quantities and the formation of neutrons within is controlled by moderating with heavy water, D_2O, or carbon (graphite rods), and shielded by water, paraffin or concrete, we call this structure a *nuclear* (atomic) *reactor*. From the above, we can also see that the energies of neutrons in nuclear reactors will be spread over a wide energy spectrum. Naturally there are many different types of reactors.

For neutron activation, *samples* are packed in containers ("rabbits") and transferred pneumatically to a desired location within the core of the reactor. Depending on the half-life, the nuclear cross section and the concentration of the nuclide, the sample is irradiated for minutes, hours, days, weeks, or months. After removal, the activated sample is usually treated chemically (Haissinsky and Adloff, 1965; Harvey, 1962; Lavrukhina, Malysheva, and Pavlotskaya, 1967) to separate the radioactive element for further gamma-ray spectrometric identification and counting. In some cases a direct nondestructive gamma-spectrometric analysis and determination are possible.

The nuclear cross section, σ, denotes the probability of interaction between the target nucleus and the projectile. For a given nuclide it varies with the energy of the incident particles and the type of reaction. Thus, because reactors have a wide spectrum of neutron energies (e.g., average energy of 2 MeV, peaked at about 0.6 MeV in ^{235}U fission reactors) one refers loosely to *"thermal" neutron cross sections* (actually for neutron energies of ~0.025 eV, and velocities of about 2200 m/sec). More appropriately one speaks of the "total" cross section, σ_T, which is the sum of the "partial" cross sections, e.g., σ_c, σ_a, σ_s, where σ_c represents the (n,γ) or the neutron capture reaction, the σ_a is the sum of the effects of neutron absorption by such reactions as (n,α), (n,p), etc., and σ_s denotes the cross section for all the scattering processes:

$$\sigma_T = \sigma_c + \sigma_a + \sigma_s \qquad (11)$$

The nuclear chemist uses the *activation cross section*, σ_{act}, which is usually equivalent to the capture cross section, σ_c, unless the activation product is a stable nuclide, in which case the activation cross section is naturally zero, whereas the σ_c can be large, as in the case of ^{114}Cd, for example. Cross-sections which refer to a particular nuclide are called "isotopic" cross sections. It is obvious that nuclear cross sections will depend on the method of actual measurements. When neutron cross sections are plotted as a function of the neutron energy, *"excitation curves"* are obtained. These curves are used as an aide by the nuclear chemist in order to choose the best possible neutron energies (or sources) for given reactions, see "Neutron Cross-Sections" by Hughes and Harvey, 1958. Useful nuclear data are also tabulated by Koch (1960) and data on light nuclei by Lauritsen and Ajzenberg-Selove (1962). The Radiological Health Handbook (1960) compiles information on health hazards and radiation monitoring.

Nuclear cross sections are expressed in cm^2 per nucleus. Since the average diameter of a nucleus is 10^{-12} cm, actual cross-sectional areas will be on the order of 10^{-24} cm^2 per nucleus, which when used as a unit is called a *barn*, 1×10^{-24} cm^2.

The most common interaction of nuclides with reactor neutrons is called the (n,γ) reaction. This results when slow neutrons are captured by the target nucleus and the compond nucleus emits the excess energy as gamma radiation. The product will be an isotope of the target nucleus. This reaction is therefore extensively used in neutron activation analysis and to produce radioisotopes.

Nuclear transmutations can be of different nature, e.g., (n,γ), (α,p), (p,n), (d,n), (d,p), (d,γ), (γ,n), (n,p), $(n,2n)$, (d,t), (p,d), etc. In all these cases the first symbol indicates the projectile, the second the emitted particle (or photon) from the compound nucleus.

These reactions are:

$${}^{63}_{29}\text{Cu} + {}^{1}_{0}n \rightarrow {}^{64}_{29}\text{Cu} + \gamma \quad (n,\gamma)$$

$${}^{1}_{1}\text{H} + {}^{1}_{0}n \rightarrow {}^{2}_{1}\text{H} + \gamma \quad (n,\gamma)$$

$${}^{238}_{92}\text{U} + {}^{1}_{0}n \rightarrow {}^{239}_{92}\text{U} + \gamma$$
$$(n,\gamma) \text{ slow neutrons}$$

$${}^{238}_{92}\text{U} + {}^{1}_{0}n \rightarrow {}^{237}_{92}\text{U} + 2\,{}^{1}_{0}n$$
$$(n,2n) \text{ fast neutrons}$$

$${}^{14}_{7}\text{N} + {}^{1}_{0}n \rightarrow {}^{14}_{6}\text{C} + {}^{1}_{1}\text{H} \quad (n,p)$$

$${}^{9}_{4}\text{Be} + {}^{4}_{2}\text{He} \rightarrow {}^{12}_{6}\text{C} + {}^{1}_{0}n \quad (\alpha,n)$$

Activation with reactor neutrons achieves the highest sensitivity when compared to other analytical methods. Concentrations of elements as low as in parts per billion, and in some cases less, can be determined. Due to the nature of neutron interaction with nuclides, the detection limits are directly proportional to the *neutron fluxes*, which are measured in $n/\text{cm}^2/\text{sec}$.

Most elements can be activated with thermal neutrons. More common cases of neutron capture, (n,γ), are with ^{12}C, ^{23}Na, ^{26}Mg, ^{27}Al, ^{30}Si, ^{35}Cl, ^{41}K, ^{51}V, ^{55}Mn, ^{63}Cu, ^{69}Ga, ^{75}As, ^{31}Br, ^{113}Cd, ^{113}In, ^{123}Sb, ^{127}I, ^{138}Ba, ^{180}Hf, ^{197}Au, ^{232}Th, and ^{238}U. All these isotopes have been determined in trace amounts by reactor neutron activation for geochemical purposes. Thermal neutron activation has been used by geochemists almost exclusively to determine trace amounts of elements in rocks, minerals, water, air, plants, industrial products, and ore, but it can be used also as a tool for major element determination. A cross section of a modern *pulsed nuclear reactor*, the Triga Mark I, built by Gulf General Atomic Incorporated, is shown in Fig. 3. Literature on reactor radioactivation and radiochemical methods in geochemistry is given by Mapper and Moorbath (in Smales and Wager, 1960).

The advantages of thermal neutron activation in reactors are in high thermal neutron fluxes (up to $10^{15} n/\text{cm}^2/\text{sec}$ and up to $10^{16} n/\text{cm}^2/\text{sec}$ in reactors of special design and with pulsing capabilities as compared with maximum fast-neutron levels of $\sim 10 n^{10}/\text{cm}^2/\text{sec}$ in the immediate vicinity of targets in modern laboratory type accelerators) which results in detection limits better by at least four or five orders of magnitude than can be achieved by accelerator activation. The uniformity of exposure of samples, the inherent stability of operation, the ability to irradiate simultaneously a large number of samples of different size and for long time periods, are other advantages of thermal neutron activation reactors. However, nuclear research reactors are too expensive for small laboratories and require constant care and permanent highly trained personnel. Samples to be irradiated have to be transported therefore to nuclear research centers, and as a result of this, the geochemist has to "farm out" a part of his research or spend time working in an unfamiliar environment, especially if short-lived isotopes are being studied. Nuclear reactors represent a constant potential health hazard even when noncritical. Licensing requirements are stringent and strict safety procedures are mandatory. Much of the sample handling and preliminary chemical separation are performed by remote control from behind radiation shields, thus requiring special "hot" laboratories, radioactivity hoods, radioactive waste disposal, decontamination facilities, etc.

Isotopic Neutron Sources

Isotopic neutron sources are used in prospecting for ores by the geochemist and mining geologist, mostly when portability becomes an important factor. Because these devices have fluxes which are orders of magnitude less than

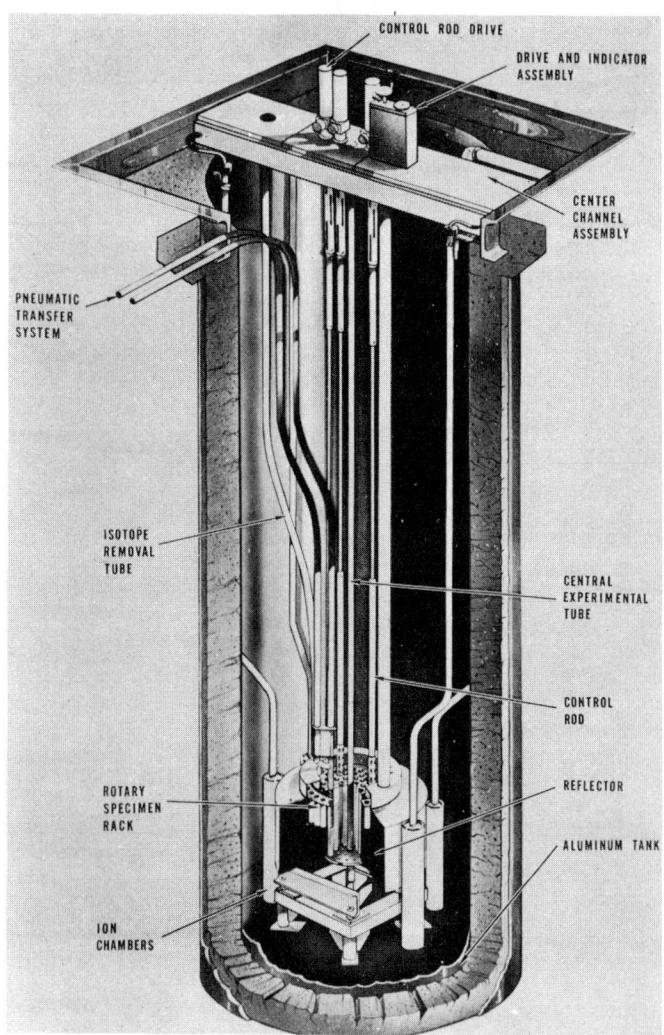

Fig. 3. Triga Mark I Reactor, cross section. *Courtesy Gulf General Atomic, Inc., San Diego, Calif.*

in reactors, the trace element determination with these is often restricted to concentrations above 0.001% in solid samples. These sources are utilizing usually the (α,n) reaction to produce fast neutrons using radioactive isotopes as sources of the α radiation. Activation by these sources is usually of the (n,γ) type. When neutrons produced in these units are slowed down by moderators, such isotopic neutron sources are called *neutron howitzers*. These are mostly used for educational purposes.

Photon-induced reactions of the (γ,n) (γ,p), (γ,np), or (γ,α) type take place with high energy gamma-photons (>10 MeV). In the case of ^{9}Be the reaction $^9\text{Be}(\gamma,n)$ 2^4He occurs with lower gamma energies of 2.04 and 2.6 MeV from artificial radioactive nuclides ^{124}Sb or ^{208}Tl(ThC″), respectively. This reaction has been used to determine beryllium in parts per million in low grade ores. Portable units, "*beryllometers*," have been developed recently for rapid determination of traces of beryllium in the field. Neutrons produced in the reaction above can be counted with gas-proportional neutron detectors of the boron trifluoride type and the concentration of beryllium thus determined (Taylor, 1964).

When neutron sources in combination with radiation detectors are lowered into drill holes the process is called *neutron well-logging* (see Vol. VI, *Well Logging*). It is used extensively in the oil exploration, hydrological investigations, and mining industry. Geophysical applications, nuclear activation in geology, and Russian literature on the subject are given by

Larionov (1963). Reactors, Van de Graaff accelerators, some neutron generators, and isotopic radiation sources are described by Charlesby (1964). Recently high-intensity neutron sources that use the artificial nuclides californium-252, americium-241, and curium-242 have been described (Reinig, 1968; Wahlgren, Wing, and Stewart, 1968).

The most commonly used *gamma source* utilizes the cobalt-60 isotope, ^{60}Co*, which emits 1.17- and 1.33-MeV gamma rays and has a half-life of 5.3 yr. Cobalt irradiation facilities and "cells" are mostly used in medical and biological research. Other radioactive isotopes, e.g., ^{182}Ta* and ^{137}Cs*, are used for similar purposes, and also as tracers and for calibration of gamma counting electronic devices.

Accelerator Neutron Sources

Numerous types of nuclear particle and electron accelerators exist. Neutrons produced by these generators are mostly of high energy measured in millions of electron volts and are therefore commonly referred to as "fast." Naturally these neutrons can be slowed by moderators, e.g., water or paraffin.

Van de Graaff electrostatic high voltage generators have been used as most versatile particle accelerators since 1929. Originally constructed for the acceleration of electrons, these machines were soon adapted to propel positive particles by reversing the polarity within the charging belt. Thus, in addition to high energy electrons (which can be regulated within a wide energy range of about 1–10 MeV and higher), positive particles and ions can be accelerated to desired energies along the tubes (drift tubes) of these devices. This permits the nuclear scientist to experiment with high energy protons, deuterons, tritons, alpha particles, and heavier nuclei and study their interaction with other nuclei placed as targets in the path of the beam. Through the extranuclear interaction of these particle beams with atoms high energy x rays are produced. Neutrons can be generated by bombarding beryllium metal blocks with these x rays. This photo-induced reaction ^{9}Be(γn) 2^{4}He produces polyergic high energy neutrons which can be moderated by surrounding the beryllium block by paraffin. Thermal neutron activation sources of this type are frequently used when reversing polarity of the Van de Graaff accelerator is inconvenient (Burrill in Charlesby, 1964).

Cockroft-Walton type positive particle accelerators are preferred for instant generation of monoenergetic "fast" neutrons. These are essentially Van de Graaff machines with reversed polarity, equipped with hydrogen, deuterium, or tritium gas sources, leak systems and ionizing devices, which are usually either of the Radio frequency or the Penning type. These last mentioned "ion bottles" deliver atomic and ionic particles of the desired gas into the accelerator part of the generator to be propelled toward the "target" which can be chosen so as to produce the desired reaction. A modern Cockroft-Walton type "fast" neutron generator is pictured in Fig. 4. Fast-neu-

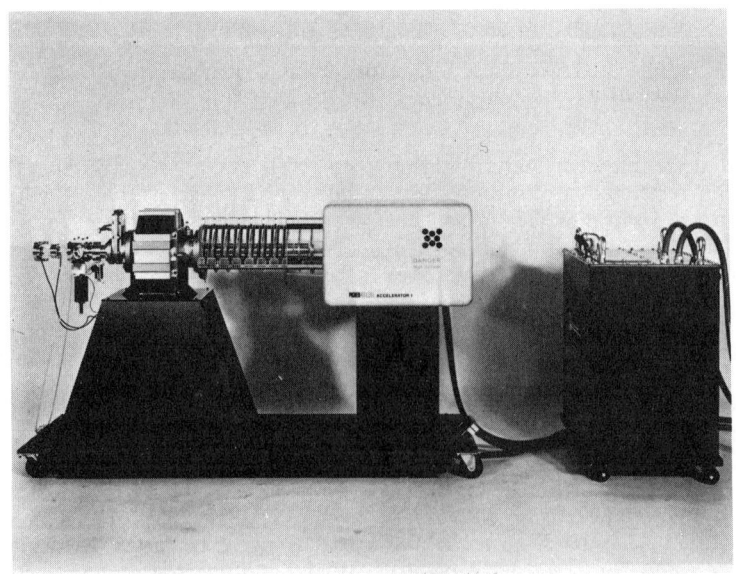

Fig. 4. Fast-neutron generator capable of 10^{11} n/sec output (Accelerator 1). *Courtesy Picker Nuclear, White Plains, New York.*

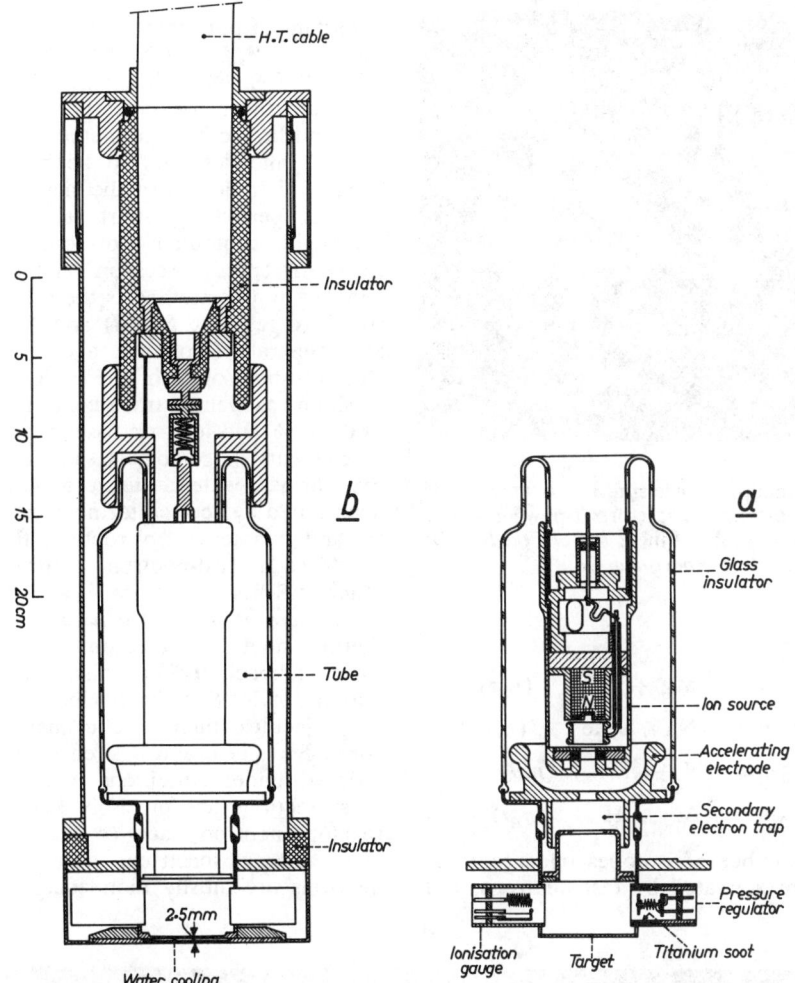

FIG. 5. Philips sealed neutron tube without (*a*) and with (*b*) metal casing. Penning Ion source is at a positive potential of 150 kV, watercooled target is at earth potential. Special hydrogen pressure regulator regulates and stabilizes the gas pressure inside the tube. Neutron output is 3×10^{10} neutrons per second at a target current of 1.5 mA. Self-loading target ensures a life of more than 1000 hours at a constant neutron output. *Courtesy O. Reifenschweiler and Philips Research Laboratories, Eindhoven, Holland.*

tron activation, acceleration, instrumentation, and methods are described by Taylor, 1964, and Volborth, 1968.

Quite recently sealed neutron sources (tubes) capable of high neutron outputs (e.g., 3×10^{10} n/sec) for prolonged time (up to 1000 hours) have been built (Reifenschweiler, 1968; Volborth, 1968, pp. 302–4). Figure 5 shows a cross section of the Philips high-output neutron tube, developed by Reifenschweiler (op. cit.). The long life of this tube is achieved through so-called self-replenishing of the target which is accomplished by accelerating a mixture of ionized deuterium and tritium.

Reactions most frequently used in the production of "fast" neutrons are T (d,n) ^{4}He, D (d,n) ^{3}He, ^{9}Be (d,n) ^{10}B, ^{7}Li (p,n) ^{7}Be, and ^{9}Be (p,n) ^{9}B. With the exception of the last-listed reaction, fast neutrons produced by these reactions are essentially monoenergetic. For example in the deuterium-tritium reaction, which can also be written out

$$^2_1\text{H} + ^3_1\text{H} = ^4_2\text{He} + ^1_0n + 17.6 \text{ MeV}$$

neutrons of (approximately) 14.7 MeV energy are formed.

Most important nuclear reactions with fast neutrons are of the (n,p) and the (n,α) type.

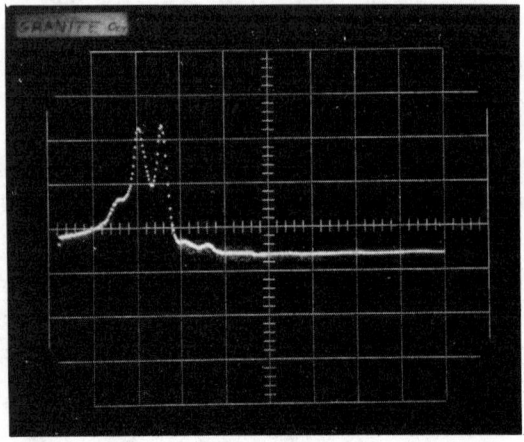

Fig. 6. Gamma-ray spectrum of "oxygen," N^{16}, as obtained from an activated rock-powder sample (polaroid photograph). Linear scale. *Courtesy Dr. H. A. Vincent, The Anaconda Co.*

For example:

$$^{27}_{13}Al + ^{1}_{0}n = ^{27}_{12}Mg + ^{1}_{1}H \quad (n,p)$$

$$^{27}_{13}Al + ^{1}_{0}n = ^{24}_{11}Na + ^{4}_{2}He \quad (n,\alpha)$$

$$^{16}_{8}O + ^{1}_{0}n = ^{16}_{7}N + ^{1}_{1}H \quad (n,p)$$

$$^{28}_{14}Si + ^{1}_{0}n = ^{28}_{13}Al + ^{1}_{1}H \quad (n,p)$$

A large number of isotopes undergo mutations and are activated by fast neutrons. It is important to note that relatively light nuclides, e.g., ^{6}Li, ^{10}B, ^{12}C, ^{14}N, ^{16}O, ^{19}F, ^{20}Ne, ^{24}Mg, ^{28}Si, ^{31}P, ^{32}S, ^{35}Cl, ^{42}Ca, ^{45}Sc, ^{52}Cr, ^{55}Mn, and ^{56}Fe, are most conveniently activated by the fast-neutron activation method, and that the product nuclides tend to be relatively short-lived, which facilitates the neutron activation analysis permitting short activation periods and rapid accumulation of results. A good example of speed, precision, and accuracy obtainable by this method is the activation analysis of oxygen (see Fig. 6) and silicon in rocks and minerals (Volborth and Vincent, 1967; Vincent and Volborth, 1967; Volborth, 1968).

Major, as well as trace concentrations down to 0.001% and in some cases parts per million of elements in geological samples, meteorites, ores, brines, waters, etc., can be conveniently determined by accelerator neutrons. The counting and remote control room of the University of Nevada fast-neutron activation facility, Mackay School of Mines, is shown in Fig. 7.

Accelerator neutron activation complements thermal neutron activation. It represents "instant radioactivity" produced in the "push button" style. Specific nuclei can be selected as projectiles, their kinetic energy varied according to need, and desired nuclear reactions induced under conditions where the neutron cross sections are precisely known. While in reactors neutron energies are spread over a wide energy spectrum, accelerator-produced neutrons are mostly monoenergetic, thus per-

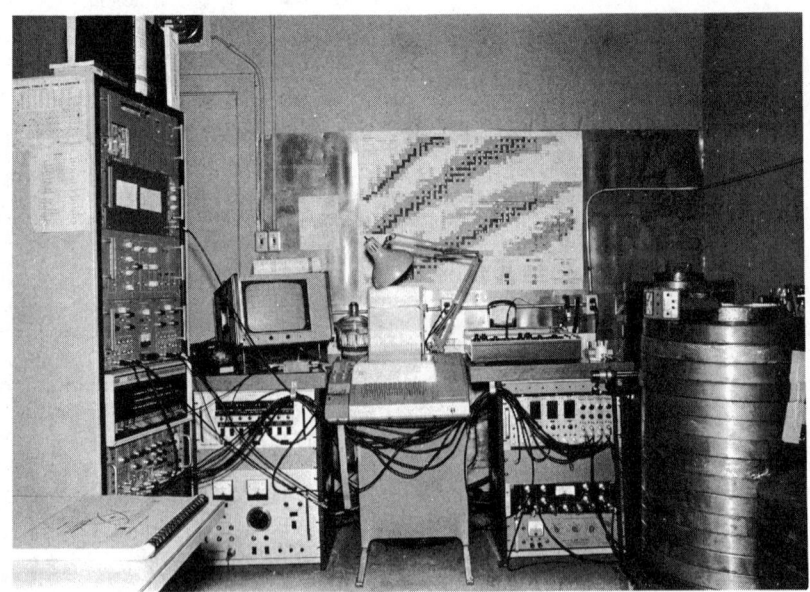

Fig. 7. Gamma-ray counting and remote control room of the University of Nevada fast-neutron activation facility. Mackay School of Mines, Reno, Nev.

mitting a higher selectivity in neutron activation analysis. Modern accelerators can generate neutrons within minutes after the start of operation, their neutron yields (n/sec) can be precisely regulated, they are more convenient to operate than reactors, and their licensing requirements, shielding, and safety precautions are less demanding. Moderating the fast neutrons by water or paraffin, one can achieve thermalized neutron fluxes of about 10^7 n/cm^2/sec. This permits the laboratory to experiment with heavy element neutron activation simulating a reactor for preliminary research or educational purposes. Shielded basement corners, vaults or small concrete wells are sufficient permitting the installation of an accelerator neutron facility in almost any well-planned laboratory.

In conclusion one is tempted to state that for the geochemist, the combination of neutron activation with gamma-ray spectrometry as well as with x-ray fluorescence represents an ability to perform a total rapid nondestructive analysis of rocks. This consists of the determination of all the eight major elements (including oxygen) and most trace elements without having to destroy the sample (Volborth, Fabbi, and Vincent, 1968). However, it must be stressed here that thermal neutron activation still employs essentially chemical separation methods due to interfering reactions and the complex nature of the gamma spectra. Significant advances are presently being made in the perfection of higher resolution lithium drifted germanium and other detectors, making truly nondestructive radioactive assays feasible for additional isotopes. Simultaneous development of multichannel gamma-ray analyzers, integrators, and computers of greater resolution (with 4000 channels and more) makes the nondestructive techniques of neutron activation more versatile every year.

ALEXIS VOLBORTH

References

Charlesby, A., 1964, "Radiation Sources," New York, MacMillan, 268pp.

Crouthamel, C. E., 1960, "Applied Gamma-Ray Spectrometry," New York, MacMillan, 443pp.

Gillespie, A. S., and Hill, W. W., 1961, "Sensitivities for activation analysis with 14-Mev neutrons," *Nucleonics,* **19,** 170-173.

Glasstone, Samuel, 1958, "Sourcebook on Atomic Energy," New York, Van Nostrand, 641pp.

Haissinsky, M., and Adloff, J. P., 1965, "Radiochemical Survey of the Elements," Amsterdam, Elsevier Publ., 177pp.

Harvey, B. G., 1962, "Introduction to Nuclear Physics and Chemistry," Englewood Cliffs, New Jersey, Prentice Hall, 370pp.

Hughes, D. J., and Harvey, J. A., et al., 1955, revised 1958, suppl. 1960, "Neutron Cross-Sections," BNL-325, U.S. Gov't. Printing Office, Washington, D.C.

Koch, R. C., 1960, "Activation Analysis Handbook," New York, Academic Press, 219pp.

Larionov, V. V., 1963, "Yadernaya Geologiya i Geofisika (Nuclear Geology and Geophysics)," Moskow, Gostoptekhisdat.

Lauritsen, T., and Ajzenberg-Selove, F., 1962, "Energy levels of Light Nuclei, Nuclear Data Sheets," U.S. Natl. Acad. Sci., Washington, D.C., 339pp.

Lavrukhina, A. K., Malysheva, T. V., and Pavlotskaya, F. I., 1967, "Chemical Analysis of Radioactive Materials," Radiokhimicheskii Analiz, U.S.S.R. Acad. Sci., CRC Press, Cleveland, 386pp. (transl. from the Russian).

Lutz, G. J., Boreni, R. J., Maddock, R. S., and Meinke, W. W., 1968, "Activation Analysis: A Bibliography," Part I, 511pp., 2 Appendixes, NBS Technical Note 467.

Overman, R. T., and Clarke, H. M., 1960, "Radioisotope Techniques," New York, McGraw-Hill Book Co., 476pp.

Price, W. J., 1964, "Nuclear Radiation Detection," New York, McGraw-Hill Book Co., 430pp.

Radiological Health Handbook, Rev. Edition 1960, U.S. Dept. of Health, Ed., and Welfare, Washington, D.C., PB-121784R, 468pp.

Rankama, Kalervo, 1956, "Isotope Geology," New York, Pergamon Press, 535pp.

Reifenschweiler, O., 1968, "A High Output Sealed-off Neutron Tube with High Reliability and Long Life," *Proc. Intern. Conf. Mod. Trends in Activation Anal.,* Washington, Oct. 7-11.

Reinig, W. C., 1968, "Advantages and applications of Cf^{252} as a neutron source," *Nuclear Applications,* **5,** 1, 24-25.

Smales, A. A., and Wager, L. R., 1960, "Methods in Geochemistry," New York, Interscience Publ., 464pp.

Taylor, D., 1964, "Neutron Irradiation and Activation Analysis," London, George Newnes Ltd., 185pp.

Vincent, H. A., and Volborth, A., 1967, "High precision determination of silicon in rocks by fast-neutron activation analysis," *Nuclear Applications,* **3,** 753-757.

Volborth, A., 1968, "Elemental Analysis in Geochemistry," Amsterdam, Elsevier Publ., 328pp.

Volborth, A., Fabbi, B. P., and Vincent H. A., 1968, "Total non-destructive analysis of CAAS syenite," *Advances X-Ray Analysis,* **11,** 158-163.

Volborth, A., and Vincent, H. A., 1967, "Determination of oxygen in USGS rock standards by fast-neutron activation," *Nuclear Applications,* **3,** 701-707.

Wahlgren, M., Wing, J., and Stewart, D. C., 1968, "A High Intensity ^{241}Am-Be-^{242}Cm Neutron Source," Proc. Internat. Conf. Modern Trends in Activation Analysis, Washington, Oct. 7-11.

Cross-references: *Emission Spectrography; Geochronometry; Isotope Geology; Mass Spectrometry; Radioactive Isotopes; Radioactivity in Rocks; X-Ray Spectroscopy.* Vol. IVB: *Electron Probe Microanalysis.* Vol. V: *Radioactive Iso-*

tope Tracer Technology. Vol. VI: *Radioactivity in Sediments; Well Logging*.

NICKEL: ELEMENT AND GEOCHEMISTRY

Nickel (from the German "Kupfernickel," old Nick copper); symbol Ni; atomic weight 58.71; atomic number 28; electronic configuration: $1s^22s^22p^63s^23p^63d^84s^2$; melting point 1453°C; boiling point 2730°C; specific gravity 8.90 (20°C); valence 2 or 3; crystal structure: the α form is hexagonal close-packed, the β form is face-centered cubic. Nickel was discovered by Cronstedt in 1751.

Nickel is a malleable, ductile, tenacious, slightly magnetic, silvery white metal, which conducts heat and electricity fairly well. It can take a high polish. It is ferromagnetic at ordinary temperatures but becomes paramagnetic at elevated temperatures (Curie point = 353°C). It is closely related in chemical properties to iron and cobalt, all three of which belong to group VIIIb in the periodic table. Five stable isotopes are known in nature:

$Ni^{58} = 67.76\%$ $Ni^{62} = 3.66\%$
$Ni^{60} = 26.16\%$ $Ni^{64} = 1.16\%$
$Ni^{61} = 1.25\%$

Seven radioactive isotopes of nickel have been identified; the mass numbers are 54, 56, 57, 59, 63, and 65.

Minerals

Primary nickel minerals

Breithauptite	NiSb
Gersdorffite	(Ni, Fe, Co)AsS
Millerite	NiS
Niccolite	NiAs
Pentlandite	$(Fe, Ni)_9S_8$
Rammelsbergite	$(Ni, Co)As_2$
Skutterudite	$(Co, Ni)As_3$
Ullmannite	NiSbS
Orcelite	Ni_2As

Secondary nickel minerals

Annabergite	$(Ni, Co)_3(AsO_4)_2, 8H_2O$ (oxidation product of Ni arsenides)
Garnierite	$(Ni, Mg)_3Si_2O_5(OH)_4$ (nickeliferous antigorite)
Nepouite	(nickeliferous chlorite)
Pimelite	$(Ni, Mg)_3Si_4O_{10}(OH)_2$ (nickeliferous saponite)

Nickel forms compounds with sulfates, chlorides, nitrates, carbonates, oxides, hydroxides, and with organic complexes (Ni dimethylglyoxime, used as a qualitative test for nickel) (see Table 1).

Chemistry

Nickel is usually divalent in its compounds and is predominantly ionic in character. The metal is not affected by water or damp air at

TABLE 1. OCCURRENCE OF NICKEL ORE DEPOSITS

	Sulfide and arsenide ores
Canada	Sudbury (Ontario)
	Lynn Lake (Manitoba)
	Pacific Nickel (British Columbia)
U.S.S.R.	Petsamo (Pechanga)
	Monchegorsk (Kola Peninsula)
	Norilssk (Siberia)
Union of South Africa	Rustenburg (Nickel is a by-product of platinum deposits)
U.S.A.	Yakobi Island (Alaska)
	Mount Nickel (Montana)
	Fredericktown, Madison County (Missouri)
Norway	Evje; Hosanger; Hoiaasen; Ringerike; Narvik
Sweden	Boliden, Lainjaur Mine (byproduct of nickel)
Australia	Kambalda
	Silicate ores and nickeliferous lateritic iron ores
Indonesia	Borneo, Celebes (Pomalaa-Kolaka Lakes region), West Irian
Philippine Republic	Surigao, Nonoc Islands
New Caledonia	Thio; Poro; Poum
Cuba	Nicaro, Moa Bay
Venezuela; Guatemala; Colombia	
U.S.A.	Oregon (Nickel Mountain Mine, Riddle, Oregon)
Brazil	Tocantins, Goias, Minas Gerais
Greece	Atalante-Larymna
U.S.S.R.	Urals

ordinary temperatures, nor is it dissolved by the alkali hydroxides. It dissolves readily in nitric acid and aqua regia, and is slowly attacked by sulfuric acid. In the ionic form, nickel has the oxidation state of two ($Ni = Ni^{2+} + 2_e^-$; $E° = 0.250$). Finely divided nickel adsorbs up to seventeen times its volume of hydrogen, thus it has a great catalytic power. In compounds, nickel is a bivalent cation with an ionic radius of about 1.24Å. Bonds with oxygen are partly covalent. Trace elements or minor constituents often have appreciable effects on the properties of the metal. In addition to simple salts, nickel forms a variety of coordination compounds in which the coordination number is 6 or 4 (e.g., $Ni(NH_3)_6^{2+}$, blue). Even in solution, Ni^{2+} ion is really a complex ($Ni(H_2O)_6^{2+}$). Its preference for a coordination number of 6 is seen in the hydrated salts: $NiSO_4 \cdot 6H_2O$, $NiCl_2 \cdot 6H_2O$, $Ni(NO_3)_2 \cdot 6H_2O$, which are all hexahydrates. The anhydrous salts are generally yellow ($NiSO_4$, NiF_2) but the color deepens as the anion becomes more polarizable: $NiCl_2$—yellow-brown; $NiBr_2$—dark brown; NiI_2 and NiS—black. The hydrated salts and aqueous solutions are green. The divalent nickel compounds closely resemble divalent Co^{2+} in chemical behavior.

Geochemistry

Geochemically, nickel is siderophile and will join metallic iron wherever such a phase is present. The Ni in meteorites is strongly concentrated in the metal phase, alloyed with iron (Fe). Ni is only slightly miscible in Fe, and two phases separate out at lower temperatures. Therefore Ni–Fe occurs in meteorites as two distinct minerals, *kamacite* (5.5% Ni) and *taenite* (27–65% Ni), possibly representing evolutionary stages in the evolving core of a now exploded asteroid (see Vol. IV B, *Widmanstätten Figures*). Nickel also occurs in the earth's core, together with the other siderophile elements. The nickel-iron alloy of the core (also called *barysphere*, or NIFE by Suess) is often assumed to have a Ni content comparable to that of iron meteorites, i.e., with a Fe/Ni ratio of 11:1. In chondritic meteorites (representative for meteoritic matter in general) the Fe/Ni ratio is usually about 17:1. In the mantle the Ni concentration is only about 0.02%. Nickel is important in the incandescent gases of all stars with a universal abundance of 1.1.

The high affinity of nickel for sulfur accounts for its frequent occurrence in magnetic or metamorphic segregates of sulfide bodies. In "early magmatic" sulfide segregates, Ni is found associated with a pentlandite-pyrrhotite intergrowth (as at Sudbury, Ontario). *Pentlandite*, $(Ni, Fe)_9S_8$, is the principal metallic ore. Nickel also occurs as sulfide and arsenide minerals in "late magmatic" sulfide segregates and metalliferous veins. Important are: *niccolite, breithauptite, millerite, rammelsbergite, gersdorffite*, and *ullmannite*.

Igneous rocks (average content, 80 ppm), as a rule, contain small amounts of pentlandite, pyrrhotite and pyrite, which are Ni bearing. Yet most of the nickel is incorporated into the crystal structures of the silicates. Nickel tends to concentrate in mafic and ultramafic rocks, mainly in the olivine, pyroxene and hypersthene of igneous origin. In peridotitic rocks the average content in Ni is 0.2%, a part often being in very small masses of natural iron-nickel alloys of awaruite or josephinite type (nickel content up to 77%). Nickel silicates are closely related to corresponding Mg minerals, both structurally and chemically, because the ionic radii of Ni^{2+} and Mg^{2+} are identical to (0.78Å). Nevertheless the ionic radius of Ni (like Co, Fe, Mg) does not permit it to enter even into solid solution in feldspathic structures, but only into silicates with tetrahedra, either isolated or in chains. Nickel silicates are rather rare and only garnierite has any value as an ore.

The weathering of ultrabasic rocks (especially peridotites and serpentinites) gives rise to iron-, nickel- and silica-rich solutions. Unlike Mn^{2+} and Fe^{2+}, Ni^{2+} is very stable in aqueous solutions and is capable of migration over long distances. As the solution sinks, iron oxidizes and precipitates as ferric hydroxide, ultimately losing water to become goethite and hematite in which a small amount of nickel ions are "trapped" (nickel content of this red or yellow material can reach 1.5%). Nickel, magnesium and silicon continue downward and as soon as the water is neutralized, they precipitate as complex hydrous silicates with layer lattices (Ni chlorites: neponite and clay-like minerals, e.g., garnierite, nickeliferous chrysotile).

The nickel content of seawater is quite low (0.0015–0.006 g/ton), or 8 tons of Ni per cubic mile (10^{-7}%), a total of 2500 million tons of Ni in the whole ocean. Most of the Ni released by weathering and carried to the sea remains in the solid products of weathering.

In deep-sea sediments, the Ni content is exceptionally high (20–1000 ppm; average 140 ppm). In fresh water, the average content of Ni is about 10^{-6}%. Nickel is present in all soils, in air dust, and in animal or vegetal material (about 10^{-5}%. Some animal organs are exceptionally rich in nickel, for example the pancreas.

Bog iron ore and oolitic iron ore contain appreciable amounts of nickel. Under reducing conditions, Ni precipitates as the sulfide in marine muds, rich in organisms. The Ni content of petroleum and its derivatives is often quite high. Usually Ni is associated with paraffin hydrocarbons while vanadium (V) is confined to asphalt. Ni, like V and Mo (molybdenum), forms organometallic compounds with hydrocarbons. The three elements probably acted as catalysts in reactions involving H_2S and organic compounds. Some of this nickel may be derived from meteoric dust. Occasional observation of iron globules or chondrules in deep-sea red clay lends support to this hypothesis. Manganese nodules, covering large parts of the Pacific floor, contain about 1% nickel.

Economic Geology

Nickel ore deposits, i.e., rock masses rich enough in nickel to be economically exploited, fall into two classes: (a) sulfides and sulfarsenides and (b) laterites.

Sulfide and Sulfarsenide Nickel Ores. The sulfide ore bodies are typically an association of pyrrhotite, pentlandite with lesser amounts of chalcopyrite. These deposits occur in or near peridotite or norite intrusions and are usually assumed to be genetically related to them (e.g., Sudbury, Ontario). Nickel may also form sulfides and sulfarsenides together with cobalt and iron in hydrothermal veins or veins linked to lateral secretion, but these deposits are generally small and unimportant sources of nickel (for example: Austrian deposits—Schladming).

Lateritic Nickeliferous Ores. These result from weathered ultrabasic rocks, peridotites or serpentinites in tropical to subtropical climates. The laterite formed from the weathering of serpentinite contains 45–55% iron, and about 1–2% nickel (nickeliferous iron laterite). Most of the Ni is probably included in goethite, limonite and serpentine. Examples of such deposits include those in Cuba, the Philippines and Indonesia. Weathering of fresh peridotite yields a nickel silicate laterite, in which garnierite (nickel-antigorite) is the chief nickel mineral (New Caledonia and Indonesia). It occurs at the base of the lateritic cover at the contact with the unaltered bedrock or in a transition zone, where the mineral-rich solutions are neutralized. Nickel-rich ore body outcrops occur mainly at the intersection of old and new lateritic surfaces (Avias, 1952). Factors controlling the location of ore bodies are mainly paleomorphological, tectonic, and paleohydrogeological (Avias, 1969).

Exploitation of Nickel

The first exploitation of nickel ores was reported twenty-two centuries ago in China; in Yunnan a "white copper" contained about 20% nickel and 70% copper. A Bacterian coin of very similar composition found in Turkestan was minted in 235 BC, and is the alloy still widely used today for coinage. Saxon copper miners of the 17th century, in the Schneeberg region of Germany, found that some copper-like ores on being smelted gave a very hard metal they could not use and they called it "Kupfernickel," it is said, after "Old Nick."

It was in 1751 in Sweden that Axel Cronstedt, the chemist, identified a new metal which he named nickel. The ore was further studied by Bergmann in 1775 and Richter in 1804.

The European nickel ores were first sent to China to be treated and set back as "pekfong." In 1825 the German firm Hemminher of Berlin marketed a similar product called "neusilber" or "argentan," mainly used for coins. About 1878 the discovery by Fleitmann that the addition of magnesium to nickel made it malleable opened up the modern use of this metal.

The discovery of nickel in New Caledonia by Garnier in 1863, and the exploitation of garnierite after 1874, made this island for a long time the world's principal source.

The discovery, however, of important copper and nickel ores in Falconbridge township, near Sudbury (Canada) in 1916 made Canadian production the most important in the world (mainly by the "International Nickel Company" founded in 1902). The U.S.S.R., Finland, Cuba, the U.S.A., Brazil, Greece, and Japan have since become nickel-producing nations. Big areas of lateritic ores from the western Pacific (Philippines, Indonesia, New Guinea) are as yet unexploited. The world production of nickel metal exceeds 500,000 tons. The pentlandite-pyrrhotite deposit at Sudbury, Ontario, accounts for over half of the world's production.

Prospecting, exploration, and mining of nickel are today mainly based on geomorphic studies, combined with detailed geological mapping with the help of aerial photographs (for laterite-type deposits) and magnetometer surveys (for sulfide deposits).

Nickel-laterite type deposits are mined by open pit methods; sulfide deposits are mined either by open pit methods or by underground methods. After concentration (by crushing and flotation) sulfide ores are smelted in blast furnaces with coke (as the reducing agent) and limestone. The resulting "matte" separated from the slag goes to converters, and thence to the refinery. Lateritic ores, after drying or pelletization, are sent to electric furnaces, then to converters, and give ferronickel ingots (75% Ni). These are sent to the blast furnaces for smelting with limestone, gypsum, and coke, and separation from the slag; they are then moved to converters, giving a rich nickel matte, and then to the refinery for pure nickel production.

Hydrochemical processes (by leaching nickel from the ore) exist, and were used during World War II for Cuban lateritic ores.

Similar, and other, processes are under current research for new and more economic methods of recovering nickel from the lateritic ore types.

Uses of Nickel

Most of the nickel consumed goes into the production of alloys of various sorts for the aircraft and the plating industries. It also goes into various stainless appliances, for electronic industries (vacuum tube components, electronic tubes, transducers, etc.), or for chemical industries (corrosion-resistant applications). Nickel is also used as a catalyst for hydrogenation of fats and oils or for hydrogen production through the reaction between hydrocarbons and steam, or for desulfurization of petroleum products, or for oxidation of various organic compounds. In organic compounds or nickel carbonyl it is used as an antiknock additive for fuel. Its utilization is also very promising for nickel-iron alkaline batteries or nickel-cadmium storage batteries.

JACQUES AVIAS

References

Amiel, J., and Besson, J., 1963, "Nouveau Traité de Chimie Minerale Sous la Direction de Paul Pascal," Vol. 17, Pt. 2, "Cobalt, Nickel," Paris, Masson et Cie, 896pp.
Avias, J., 1952, "Note sur la genèse des gites nickelifères en Nouvelle Calèdonie," *C.R. 19th Congr. Geol. Int., Alger, sect.* **12**, 271–272.
Avias, J., 1969, "Note sur les facteurs contrôlant la genèse et la destruction des gites de nickel de la Nouvelle Calèdonie. Importance des facteurs hydrologiques et hydrogéologiques," *Compt. Rend.,* **268**, 244–246.
Boldt, J. R., and Queneau, P., 1967, "The Winning of Nickel," Princeton, N.J., D. Van Nostrand, 482pp.
Brocas, J., and Picciotto, E., 1967, "Nickel Content of Antarctic Snow: Implications of the Influx Rate of Extraterrestrial Dust," *J. Geophys. Res.,* **72**(8), 2229–2236.
Cornwall, H. R., 1966, "Nickel Deposits of North America," *U.S. Geol. Surv. Bull.,* **1223**.
Dhavernas, J., 1955, "Histoire du Nickel," Paris, Centre d'Information du Nickel.
Howard-White, R. B., 1963, "Nickel, an Historical Review," Toronto, Longmans Canada Ltd.
Kornilov, I. I., 1963, "Nickel and its Alloys," London, Oldbourne Press, 348pp. (Ageev, N.V., editor; transl. from Russian).
Lombard, J., 1956, Sur la géochimie et les gisements de nickel," *Chronique Mines d'Outremer,* **24**(244).

Cross-references: Catalysis; Core Geochemistry; Geochemical Evolution of the Core, Mantle, and Crust; Hydrocarbons; Solid Solution; Weathering, Chemical; Vol. IVB: Mineral and Ore Deposits; Mining and Mineral Centers of the World; Widmanstätten Figures. Vol. VI: Red Clay.

NICKEL–IRON—See IRON

NIOBIUM (COLUMBIUM): ELEMENT AND GEOCHEMISTRY

Niobium (from Niobe, daughter of Tantalus), also called Columbium, symbol Nb or Cb, atomic weight 92.91, atomic number 41, melting point 2500°C, boiling point 3700°C, specific gravity 8.4 (20°C), electronic configuration: Kr core $4d^45s^1$, valence 3 or 5, crystal structure: body-centered cubic. The element was discovered in 1801 by Hatchett, and was prepared by Blomstand who reduced the chloride by heating in hydrogen in 1864. Niobium is a rare gray metallic element, which forms an acidic oxide from which salts may be derived. There is only one stable nuclide in nature—Nb^{93}. Nb is used as a ferroniobium alloy in steel, in the aerospace industry because of its high melting point, and it may, in the future, serve as sheathing in nuclear reactors because of its superior resistance to corrosion.

Compounds of Nb

Niobium and tantalum generally occur together in nature because of their similarities in chemical properties and ionic radii (Nb^{5+} = 0.69 Å, Ta^{5+} = 0.68 Å). Minerals of niobium are all complex oxides of several metallic elements. The major ore is the isomorphous series—*columbite-tantalite* (Fe,Mn)(Nb,Ta)$_2$O$_6$. Of increasing importance as a source of the metal is the *pyrochlore* (Ca,Na,Ce)(Nb,Ti,Ta)$_2$(O,OH,F)$_7$–*microlite* (Ca,Na)$_2$(Ta,Nb)$_2$(O,OH,F)$_7$ group. In addition to several minerals containing Nb, Ta, U, Th, and other rare earths, such as euxenite, samarskite, fergusonite, and eschynite, the element is

NIOBIUM (COLUMBIUM): ELEMENT AND GEOCHEMISTRY

Table 1. Composition of Some Niobium and Tantalum Ore Minerals

Mineral	General Formula	Nb_2O_5	Ta_2O_5
Columbite	$(FeMn)(NbTa)_2O_6$	47–78	Tr–34
Tantalite	$(FeMn)(TaNb)_2O_6$	2–27	53–84
Pyrochlore	$(Na,Nb)Nb_2O_6F$	26–73	0.2–22
Microlite	$(Na,Ca)_2Ta_2O_6(O,OH,F)$	3–30	33–77

found occupying positions in the crystal lattices of Ti-bearing minerals (Ti^{4+} radius 0.64 Å) such as *ilmenite*, $FeTiO_3$; *rutile*, TiO_2; *brookite*, TiO_2; *sphene*, $CaTiSiO_5$; *perovskite*, $CaTiO_3$; and also *wolframite* $(FeMn)WO_4$. The absence of Nb silicates and sulfides is noteworthy. Nb is distinctly a lithophile element. (See Table 1.)

Geochemistry

Niobium is enriched in the earth's crust relative to its cosmic abundance by a factor of 25:1. The element is widely scattered in the earth's crust, at a concentration of 24 ppm. It occurs naturally only in the pentavalent oxidation state and replaces Ti^{4+} in its compounds. Ionic substitution of Nb and Ta for Zr in zircon is also of geochemical interest. Since zircon is widely distributed as an accessory mineral in igneous rocks, it is an important carrier of Nb.

In accordance with a common rule of geochemistry, a rare ion in a high oxidation state should enter the crystal structure of a common low valence ion in the earliest stages of fractional crystallization. This is not the case of Nb^{5+} with regard to Ti^{4+}. The earliest magmatic minerals (ilmenite in norite) contain less Nb than late stage sphene from syenites or rutile from granites. This reversal of the normal sequence of ionic fractional crystallization may indicate that Ti, Nb, and Ta are not present as free positive ions but rather as the central atoms of complex anions forming chains by polymerization through oxygen bridges.

Niobium becomes concentrated in the later stages of magmatic crystallization in granite pegmatites and especially nepheline syenites where it attains a concentration of 100–300 ppm. In the weathering of rocks, Nb tends to accumulate in hydrolyzate sediments and is therefore found in shale and clay rather than limestone and sandstone.

Economic Geology

Historically, most of the niobium produced has come from granitic sources, but by far the largest resources of niobium are contained in "carbonatites" and associated alkaline igneous rock complexes. (Carbonatites are intrusive carbonate rocks occurring in such complexes). Other significant resources are contained in bauxites resulting from the weathering of nepheline syenite.

The presence of valuable co-products associated with the niobium-tantalum minerals is an important factor in determining the economic potential of a given deposit. Columbite and tantalite are generally recovered as co-products of such minerals as cassiterite (tin), mica, beryl, feldspar, and spodumene. Niobium-tantalum minerals may also contain economically recoverable quantities of rare earths, uranium, thorium, and other geochemically related elements.

Based on genetic relationships, deposits of niobium-tantalum minerals may be classified into two broad categories as follows:

1. Deposits in or derived from *granites and granite pegmatites:* Columbite and tantalite may occur as primary accessory minerals in granite or granitic pegmatites (very coarse grained granite). Primary columbite has been produced from the decomposed granite of the Jos-Bukuru complex in Nigeria; however, pegmatites have been the historical source of most columbite, tantalite, and microlite. Such occurrences are usually relatively small individual deposits; however, notable exceptions are the two large pegmatites at Manono, Republic of Congo, which have furnished most of the world production of tantalite and substantial quantities of columbite as co-products in tin mining. Other important pegmatite sources of niobium-tantalum minerals are in Brazil, Nigeria, Mozambique, and Australia. Residual or placer deposits of columbite-bearing cassiterite gravels derived from the weathering of granitic rocks in the Nigerian tin fields have supplied the major portion of the world output of niobium since World War II, mainly as a by-product of tin mining. The Nigerian concentrates as exported contain 55–68% Nb_2O_5 and 5–13% Ta_2O_5. Columbite-tantalite derived from the weathering of pegmatites and granite is also a by-product of tin mining in Malaysia, particularly at Johore and Kedah. In the United States extensive placer deposits of euxenite with some columbite, monazite and zircon resulting from the weathering of granitic rocks occur in Bear Valley, Idaho,

where the total resources of Nb_2O_5 are estimated at 8000 tons and Ta_2O_5 at 2000 tons.

2. Deposits associated with *carbonatites and alkaline igneous rocks:* Carbonatites and associated alkaline rocks in many areas of the world have been found to contain significant amounts of niobium, uranium, thorium, and rare earths. The principal niobium mineral in such occurrences is pyrochlore; others include secondary columbite, niocalite (calcium niobium silicate), and columbium-perovskite. Common associates include zircon, apatite, rutile, magnetite, and rare earth minerals. Among the niobium-bearing carbonate deposits of economic interest are the deposits associated with the East African rift valleys and those in Norway, Germany, Brazil, the United States, and Canada. A number of these are under investigation and some have been mined commercially. An outstanding example of the latter is the Oka deposit in Quebec with reported reserves of 225 million tons of ore averaging 0.25–0.6% Nb_2O_5 in pyrochlore associated with magnetite, apatite, and pyrite, which commenced production in 1961 and now furnishes a substantial portion of the world niobium supply. Oka concentrates contain 50% to 58% Nb_2O_5 and average 0.4% Ta_2O_5. In Arkansas significant niobium resources occur in niobium-bearing titanium minerals associated with carbonatite and associated alkaline rocks in the Magnet Cove area. Niobium minerals may also occur in alkaline rocks where carbonatites are not associated as in the case of the important deposits of loporite occurring in nepheline syenite massifs in the Kola Peninsula of the U.S.S.R. The loporite, a niobium-perovskite contains 0.3–10.8% Nb_2O_5. Deposits of bauxite formed from the weathering of nepheline syenite in Arkansas contain extensive niobium resources in niobium-bearing ilmenite. Black sands and red mud, intermediate products in the production of alumina from Arkansas bauxite, are also potential sources of niobium. In Nigeria, the riebeckite granites of Kaffo Valley contain substantial niobium resources in disseminated pyrochlore.

Most of the world output of niobium-tantalum minerals have been obtained as a by-product of tin mining. Production of concentrates and slags containing these minerals has fluctuated with the tin output. Since World War II pyrochlore has been growing interest as a source of columbium and by 1964 production of niobium in pyrochlore approximated that from other sources.

The major niobium producers, as of 1965, are *(1)* Nigeria, *(2)* Canada, and *(3)* Brazil.

Reserves of niobium and tantalum in ore in Western Nations (1964) were estimated by the U.S. Bureau of Mines at 9,100,000 tons Nb_2O_5 and 12,000 tons Ta_2O_5. Niobium sources are mainly in pyrochlore in Africa and the Americas.

An extensive bibliography on niobium and tantalum minerals, ores, and ore preparation is contained in Sisco and Epremian (1963). Statistical information on production, consumption, and price is given in the chapters on niobium and tantalum in the U.S. Bureau of Mines, Minerals Yearbook.

VIVIEN GORNITZ
J. M. WARDE

References

Barton, W. H., 1962, "Columbium and tantalum, a materials survey," *U.S. Bur. Mines Inf. Circ.* **8120**, 110pp.

Pecora, W. T., 1956, "Carbonatites: A review," *Geol. Soc. Am. Bull.,* **67**(11), 1537–1555.

Sisco, F. T., and Epremian, E. (editors), 1963, "Columbium and Tantalum," New York, John Wiley & Sons.

Stevens, R. F., Jr., 1965, "Columbium and Tantalum," U.S. Bur. Mines, Minerals Yearbook, 16pp.

Van der Veen, A. H., 1963, "A study of pyrochore," *K. Nederl. Geol. Mijnb., Geol. Ser.,* **22**, 188pp. (Engl.).

Watts, J. T., Tooms, J. S., and Webb, J. S., 1962-63, "Geochemical dispersion of niobium from pyrochlore-bearing carbonatites in northern Rhodesia," *Trans. Inst. Mining Met.,* **72**(11). 729–747.

Cross-references: *Rare Earths; Tantalum. Vol. IVB: Placer Deposits;* see also *Individual Minerals. Vol. V: Carbonatites; Pegmatites.*

NITROGEN: ELEMENT AND GEOCHEMISTRY

The seventh element in the periodic table, nitrogen (N) was first recognized as an element by Lavoisier, but its existence was discovered earlier (in 1772) independently by C. Scheele in Sweden and by Daniel Rutherford in Scotland. It has an atomic weight of 14.0067 and is the most common gas in the atmosphere, where its concentration is 78.03%, in contrast to 20.99% oxygen. Its abundance in nature is 0.3 parts per thousand.

The gas is odorless, colorless, nonflammable, nontoxic, and nonexplosive. In nature it is nonreactive, except under special conditions, e.g., under the electrical discharge of thunderstorms, when nitric acid, HNO_3, may form. N_2 dissolves in water at 0°C to 0.0231 volume N_2/volume of water at 1 atm. In the atmosphere it forms a homogeneous mixture with

NITROGEN: ELEMENT AND GEOCHEMISTRY

oxygen, etc. Nitrogen gas reverts to liquid form at 77.36°K, a clear waterlike liquid, and freezes at 63.14°K. The solid state has two crystalline forms; the alpha form is cubic (low temperature) and the beta form is hexagonal. The liquid is important commercially for its cryogenic properties.

Geochemistry

Nitrogen, an unusual and versatile element, has been of interest to geochemists for more than a century because of its occurrence in the four recognized spheres of the earth, namely, the lithosphere, atmosphere, hydrosphere, and biosphere. The geological significance of nitrogen arises from the fact that the air over each square foot of the earth's surface contains over three tons of nitrogen, and that, after C, H and O, no other element is so intimately associated with the reactions carried out by living organisms. Nitrogen may be an important key to unraveling the mysteries of how the earth was formed, and how life evolved.

Terrestrial Distribution of Nitrogen. From data published in the geologic literature (Emery, et al., 1955; Hutchinson, 1944, 1945; Mason, 1958), the following inventory for nitrogen in the earth can be given:

Sphere	Nitrogen content $\times 10^{13}$ metric tons
Lithosphere	16,360
Atmosphere	386
Hydrosphere	2.3
Biosphere	0.028

Note: The value for nitrogen in the biosphere was calculated from the total amount of carbon in living organisms (2.8×10^{17} g; Mason, 1958, p. 217) and assuming that the average C/N ratio of living tissue is 10.

The survey indicates that the bulk of the nitrogen (about 98%) exists in the lithosphere. Most of the remainder is found in the atmosphere. The amounts that occur in the hydrosphere and biosphere are relatively small.

As far as the validity of the inventory given above is concerned, it should be mentioned that only the nitrogen content of the atmosphere (386×10^{13} metric tons) is known with any degree of accuracy. The point will be made later that the estimate for the lithosphere ($16,360 \times 10^{13}$ metric tons) may be up to 80% too high.

The inventory fails to place into proper perspective the significance of the nitrogen in each sphere. In contrast to the nitrogen in the lithosphere and in the atmosphere, that in the biosphere is relatively reactive. The molecular nitrogen of the atmosphere, although stable chemically, is active to the extent that interchange with the biosphere occurs through biological nitrogen fixation and denitrification (see *Nitrogen Cycle*).

The immense importance of nitrogen in the biosphere is emphasized by the fact that over one-half of the dry weight of nearly all multicellular animals and of many microorganisms is in the form of organic nitrogen compounds, and that practically all of the biochemical processes carried out by living organisms are catalyzed by nitrogen-containing compounds called enzymes. The cycling of carbon, oxygen, phosphorus, and sulfur in the earth is intimately associated with biochemical nitrogen transformations.

Nitrogen Isotopes. Two isotopes of nitrogen are present in the earth. One is N^{14}, which makes up approximately 99.63% of the total; the other is N^{15}, which is relatively scarce. A third isotope, N^{13}, is known, but it is unstable (half life of 10.1 min). In the atmosphere, the distribution of the two stable isotopes is in accord with the following exchange reactions:

$$N^{14} N^{14} + N^{15} N^{15} \underset{\longleftarrow}{\overset{k\,=\,4.0}{\longrightarrow}} 2 N^{14} N^{15}$$

The N^{15} content of the nitrogen in rocks tends to increase with increasing geological age, a result that may be due to preferential diffusion of the lighter isotope.

Nitrogen in the Lithosphere. Nitrogen is distributed extensively throughout the silicate phase of the earth. It is found in soils, sediments, minerals, fossils, and rocks of all types. Many natural products, such as coal and petroleum, contain nitrogen. Except for small quantities of elemental nitrogen that may be occluded in the tubular channels of such minerals as beryl and attapulgite, the nitrogen contained in the lithosphere is in a combined state.

On the basis of data given for the amount of nitrogen in primary rocks (Hutchinson, 1954), in sediments (*ibid.*), and in the earth's core (Scalen, 1958, p. 51), and from a consideration of the relative masses of the crust and mantle (0.024×10^{27} and 4.075×10^{27} g, respectively (Mason, 1958, chap. 3), the distribution of nitrogen in the lithosphere is as follows:

Geosphere	N content $\times 10^{13}$ metric tons
Primary (igneous) rocks	
of the crust	100
of the mantle	16,200
Core of the earth	13
Sediments (fossil N)	35 to 55
Terrestrial humus and sea-bottom organic compounds	0.025

The main point brought out by the inventory is that the bulk of the nitrogen in the lithosphere is held by primary (igneous) rocks of the mantle, even if often exaggerated. This nitrogen can be thought of in terms of "frozen assets," since it is far removed from the geologically-active regions of the earth.

In comparison to the total quantity of nitrogen in the lithosphere, that held in terrestrial humus and sea-bottom organic compounds is negligible. The value given (0.025×10^{13} metric tons) was based on Goldschmidt's (1954, p. 355) estimate that the amount of organic carbon in the pedosphere (soil) and its marine equivalent is 0.29 and 0.19 g/cm^2 of the total surface of the earth, respectively, and that the C/N ratio of soil and marine humus is 10.

The inventory given above must be accepted with caution. Some of the values are based upon inadequate data; others are derived from results published before the advent of modern methods of quantitative chemistry. Reliable values are difficult to obtain for geologic materials because of sampling errors, contamination, and the difficulty of measuring accurately the small amounts normally present.

The value often quoted for the nitrogen content of igneous rocks, and the one used in preparing the inventories listed above, is that given by Rayleigh (1939), who found that the nitrogen content of a series of rocks ranging from the ultrabasic to the acidic was remarkably constant, the average being about 50 ppm or 0.005% by weight. Some recent studies (Scalen, 1958; Stevenson, 1962) have not confirmed these claims. The nitrogen content of rocks now appears to be more variable than Rayleigh's data indicate, and the average may be considerably less than 50 ppm, perhaps as little as one-fifth of that amount.

Very little is known of the chemical forms in which nitrogen exists in rocks. Most of the nitrogen is liberated as ammonia by heating with alkali and can be described as "ammoniacal nitrogen." According to recent reports (Stevenson, 1960, 1962), the nitrogen may exist as ammonium ions held within the structures of such minerals as the micas and feldspars. Contrary to a view expressed in the early literature, little, if any, of the nitrogen in igneous rocks is in organic combination. Oxidized nitrogen is absent in rocks.

In sediments, the nitrogen occurs largely in the form of organic matter (see *Biochemicals;* also Vol. VI, *Humus and Humic Acids*).

The C/N Ratio of Soils, Sediments, and Sedimentary Rocks. A distinguishing feature of soils, sediments, and sedimentary rocks is the constancy of the C/N ratio. For marine and lake sediments, and for the surface layer of terrestrial soils, the ratio generally falls within well-defined limits, usually from about 10 to 20. Deep marine and lake sediments, and subsurface soils, often have substantial lower ratios. The C/N ratio of sedimentary rocks varies widely, both high (> 40) and low (< 5) values having been reported.

Traditionally it has been assumed that the C/N ratio is a characteristic of indigenous organic matter. The practice has been to assume that the nitrogen determined by the Kjeldahl method is in an organic form. Research conducted in the past decade (Stevenson, 1960, 1962) indicates that this procedure may be incorrect, because a significant amount of the nitrogen in some specimens may be in an inorganic form, namely, as ammonium that is contained within the lattice structures of clay minerals. Clays such as montmorillonite, illite, and vermiculite can hold considerable amounts of ammonium between the hexagonal layers formed by linking Si–O tetrahedra; the ammonium neutralizes negative charges arising from the substitution of aluminum for silica at the silica-oxygen tetrahedra. The unusually low C/N ratios reported for clay and marl deposits, for many subsurface soils, and for deep marine and lake sediments now appear to be due to ammonium held by clays.

The relationship between inorganic nitrogen (clay-bound ammonium) and the C/N ratio of several soils is given in Table 1. In the subsoils (B and C horizons), where the C/N ratio is low, rather high amounts of the nitrogen occur in the ammonium form. An interesting point brought out by the data is that the C/organic N ratio of the soil decreases with increasing depth, a phenomenon that soil scientists still have not explained.

Insufficient data are available on the distribution of ammonium in marine and lake sediments. Amounts equivalent to those found in terrestrial soils are to be expected.

Some interesting results were reported recently (Stevenson, 1962) on the ammonium contents and C/N ratios of some paleozoic shales. These data are reproduced in Table 2. In the three shales which had low organic matter (carbon) content, from one-half to two-thirds of the nitrogen occurred in the ammonium form. These shales had rather low C/N ratios (< 13). A considerably lower proportion of the nitrogen in the three shales high in organic matter (carbon) content occurred as ammonium, and they had somewhat higher C/N ratios (> 30). When the nitrogen values were corrected for the amounts of ammonium present, the resulting C/organic N ratios of

TABLE 1. CARBON-NITROGEN RELATIONSHIPS IN SOME TERRESTRIAL SOILS

Soil	Organic C (ppm)	Total N (ppm)	NH_4^+–N[a] ppm	NH_4^+–N[a] % of N	$\dfrac{C}{N}$	$\dfrac{C}{\text{Organic N}}$
Brunizem						
A_1	30,100	2,900	136	4.7	10.4	10.9
B_2	6,600	950	166	17.5	6.9	8.4
B_3	5,000	810	177	21.9	6.2	7.9
C_2	1,400	360	148	41.1	3.9	6.6
Gray-brown podzolic						
A_1	22,800	2,050	123	6.0	11.1	11.8
B_1	4,700	710	148	20.9	6.6	8.4
C_1	2,900	530	186	35.1	5.5	8.4
Red-yellow podzolic						
A_1	11,600	1,220	113	9.3	9.5	10.5
B_1	4,700	670	124	18.5	7.0	8.6
B_3	3,500	620	108	17.4	5.6	6.8
Planosol						
A_1	14,200	1,630	66	4.1	8.7	9.1
B_2	2,100	310	58	18.7	6.8	8.3
B_3	3,800	640	91	14.2	5.9	6.9
C	1,900	330	93	28.2	5.8	8.0

[a] Includes trace amounts of nitrogen as NH_4^+ salts and "exchangeable" NH_4^+.

the six shales were remarkably similar (34.9 to 43.5).

Attempts have been made from time to time to relate the C/N ratio of sediments and sedimentary rocks to such factors as source of organic matter, biological activity, depositional environment, and diagenesis processes. Since some of the nitrogen usually regarded as organic may exist as ammonium, these relationships will require a re-examination. For example, the low C/N ratios of certain sedimentary rocks have been attributed to selective preservation of nitrogen-rich organic compounds, whereas they could be a reflection of the dilution of organic matter with mineral material containing high amounts of ammonium.

With regard to diagenetic changes in sedimentary organic matter, sequential loss of nitrogen compounds undoubtedly occurs, the rate of loss of each component being dependent upon its chemical stability. The proportion of the nitrogen present as amino acids and other labile compounds appears to decrease with increasing age, whereas the biologically-resistant humus forms increase. The insoluble organic matter in sedimentary rocks (kerogen) has a substantially lower nitrogen content than the humus from which it was derived.

Nitrogen Compounds in Petroleum. Most crude oils contain trace amounts of nitrogen compounds. They are remnants of the original source material. The general types of compounds present are similar to those illustrated below:

PYRIDINES

CARBAZOLES

QUINOLINES

VANADIUM-PORPHYRIN COMPLEX

TABLE 2. CARBON-NITROGEN RELATIONSHIPS IN SOME PALEOZOIC SHALES[a]

Sample No.	Orangic C (ppm)	Total N (ppm)	NH_4^+–N ppm[b]	NH_4^+–N % of N	$\dfrac{C}{N}$	$\dfrac{C}{\text{Organic N}}$
1	6,300	670	499	73.1	9.4	36.8
2	6,400	510	344	67.5	12.6	38.6
3	10,500	810	521	64.3	13.0	36.3
4	53,100	1,730	236	13.6	30.7	35.5
5	138,400	3,720	539	14.5	37.2	43.5
6	120,800	4,030	564	14.0	30.0	34.9

[a] From Stevenson 1962.
[b] Includes trace amounts of nitrogen present as NH_4^+ salts and "exchangeable" NH_4^+.

Nitrogen makes up as much as 0.8% of the weight of crude oil. For reasons that are not clear, California oils from Tertiary formations have higher nitrogen contents than other crude oils. Most of the nitrogen in crude oil occurs in the asphaltic fractions.

Trace amounts of nitrogen in petroleum will poison the catalysts that are used in cracking processes. Metals associated with porphyrins (see structure depicted above) are also poisonous to cracking catalysts, and they are deleterious to certain types of fuel combusters.

Atmospheric Nitrogen. Nitrogen comprises 78% by volume (75% by weight) of the gases in the atmosphere. Except for minute amounts of nitrous oxide, ammonia, nitrite, nitrate, and organically bound nitrogen (associated with cosmic dust), this nitrogen exists as diatomic N_2.

The N_2 molecule obtains a triple bond (N≡N), and is very stable chemically. A temperature of over 4000°C is required to decompose it into its constituent atoms. Until the development in 1913 of the Haber-Bosch process for synthesizing ammonia by reacting elemental nitrogen with hydrogen, the only way in which atmospheric nitrogen became available to living organisms was by biochemical nitrogen fixation and the formation of nitrogen oxides through electrical discharge and photochemical reactions in the atmosphere.

The geochemical cycling of nitrogen in the earth is concerned mainly with the passage of molecular nitrogen into and out of the atmosphere. The processes involved are largely biological. For example, Hutchinson (1944) concluded that the amount of nitrogen utilized by nitrogen-fixing microorganisms greatly exceeded that which was fixed by inorganic processes. The value he reported for biochemical fixation was from 0.01 to 0.08 mg of N_2 per cm^2/yr; for the nitrogen fixed by chemical processes the value was 0.0035 mg N_2 per cm^2/yr. A more elaborate treatment of the cyclic migration of nitrogen can be found in *Nitrogen Cycle*.

Geological History of Atmospheric Nitrogen. Our concept of the origin of atmospheric nitrogen is closely allied to the mode of origin that we ascribed to the earth. With regard to the latter, two main schools of thought have existed. The first is that the earth, and the solar system as a whole, was formed by the accretion of small solid particles called planetesimals; the second, that the earth was torn from the body of the sun. Astronomical and geochemical evidence is highly in favor of the planetesimal hypothesis. According to this hypothesis, the nitrogen in the earth's atmosphere was derived largely from compounds initially occluded or chemically combined with the planetesimals.

It is not known whether all of the nitrogen in the present-day atmosphere was formed gradually over geological time or if some of it is a remnant of an original protoatmosphere. According to the school of thought developed by Urey (1952), the planetesimals were vaporized as they arrived at the surface of the protoplanet; therefore, the temperature was too high to allow for the retention of gaseous products. Urey's view is that practically all of the nitrogen in the atmosphere originated by release of nonvolatile compounds (metallic nitrides; ammonium chloride) from the interior of the newly formed earth. This could have been brought about by outgassing of igneous rocks as the temperature of the earth was increased due to heat generated by compression and decay of radioactive elements (U^{234}, U^{232}, and K^{40}).

Other geochemists maintain that only a fraction of the nitrogen in the atmosphere arose from the release of nitrogen compounds from igneous rocks after the earth was formed. Rubey (1951) has concluded that there is far too much nitrogen in the atmosphere, and in sediments and sedimentary rocks, to be accounted for by the simple weathering of igneous materials.

Evidence that the nitrogen content of the atmosphere has increased steadily throughout geological time has come from the finding that nitrogen is a normal constituent of magmatic gases. Small additions of nitrogen have been made to the atmosphere by volatilization of nitrogenous compounds from meteorites during their entry into the earth's atmosphere.

The chemical forms in which nitrogen has existed in the atmosphere over geological time is unknown. Many geochemists are of the opinion that the early atmosphere was reducing and that the nitrogen was present as ammonia; others believe that it occurred as diatomic N_2, the form in which it occurs today. According to the ammonia hypothesis, the early atmosphere contained considerable quantities of hydrogen; consequently, the atmosphere was reducing and the nitrogen was in a reduced state (as ammonia). With the loss of hydrogen (and other light elements) by diffusion into space, and with the appearance of free oxygen in the atmosphere (photochemical dissociation of water vapor; photosynthesis by green plants), the ammonia was oxidized to elemental nitrogen.

Nitrogen in the Hydrosphere. In the ocean, nitrogen occurs as molecular nitrogen, ammonium, nitrite, nitrate, and dissolved and particulate organic matter (Emery et al., 1955). The total quantity is about 2.3×10^{13} metric tons. Molecular nitrogen is the dominant form, amounting to 2.2×10^{13} tons, or over 95% of the total. The combined amount of ammonia, nitrite, and nitrate is about 58×10^{10} tons; most of this (57×10^{10} tons) is in the nitrate form. Approximately 34×10^{10} tons of nitrogen occur in organic matter.

Only the nitrogen present in combined form (as ammonia, nitrite, nitrate, and organic matter) belongs to what might be called the "nitrogen reserve" of the ocean (Emery et al., 1955). Molecular nitrogen occurs as a dissolved gas; therefore, it is virtually unaffected by chemical or biological activity in the water.

For all practical purposes, the nitrogen reserve of the ocean can be considered to be in a state of quasi-equilibrium. Variations in abundance of the different forms of nitrogen may occur with depth, season, biological activity, and other factors, but, in the long run, the amount of each remains constant. In other words, the total quantity of nitrogen in the nitrogen reserve at any one time represents a balance between nitrogen gains and losses.

New sources of nitrogen to the ocean are the land and atmosphere, from which combined nitrogen is carried by rivers and rain. The total amount of nitrogen added each year is believed to be about 78×10^6 metric tons, of which 19×10^6 tons, or about one-fourth of the total, are transported by rivers.

The loss of nitrogen by deposition of organically bound nitrogen is estimated to be 8.6×10^6 metric tons per year. The unaccounted-for nitrogen, amounting to about 70×10^6 tons, is believed to be lost by bacterial dentrification (Emery et al., 1955).

Nitrogen in the Biosphere. Many difficulties are encountered in determining the distribution of nitrogen in living matter (the biosphere). Unlike other spheres, the biosphere is in a constant state of change. Also, the matter of the biosphere is not uniformly distributed, and the nitrogen contents of different organisms vary widely. The value given earlier for total nitrogen in the biosphere (0.028×10^{13} metric tons) is at best an approximation.

An indication of how the nitrogen in the biosphere is distributed can be obtained by consideration of the annual productivity of organic carbon. Mason (1958, p. 217) has examined these relationships. The total annual production of organic carbon is believed to be $20 \pm 5 \times 10^9$ metric tons for terrestrial environments and $126 \pm 82 \times 10^9$ metric tons for marine environments, giving a grand total of $146 \pm 87 \times 10^9$ metric tons for the earth. The superiority of the ocean as a habitat for living organisms is readily apparent. The ocean is twice as fertile as the land, and it predominates in the amount of living matter it supports.

Prebiological History of Organic Nitrogen Compounds. Nitrogen compounds (amino acids, purine and pyrimidine bases, porphyrins) are of paramount importance to the efficient functioning of life processes. From the point of view of science, these compounds, among others, had to be on hand before life arose on this planet.

The principal reactions leading to the development of biochemically reactive substances are summarized below (Fox, 1963):

CO, CO_2, and/or CH_4 Electricity, heat,
NH_3 or N_2 solar radiation
H_2O, H_2, H_2S, metals, ———————————→
and H_3PO_4 β rays, γ rays, x rays

Biochemical staples (organic
acids, amino acids, vitamins,   Heat
carbohydrates, pyrimidines,   ———→
porphyrins, etc.)

Macromolecules (proteins,
nucleic acids, polysaccharides) → Organisms

The initial step was the formation of organic molecules from volatile compounds, presumably in an extralithospheric atmosphere. The compounds became concentrated in the shallow seas which formed by condensation of water vapor during cooling of the earth. Following the polymerization of active biochemicals into proteins, nucleic acids and polysaccharides, molecular systems (enzymes) were developed which had the ability to synthesize protoplasmic constituents from the surrounding medium. Ultimately the living organism developed.

Geochemists have successfully produced biochemical substances in the laboratory under conditions resembling those of the primitive earth. For example, amino acids have been synthesized by subjecting such mixtures of gases as CO_2, H_2, H_2O, and NH_3, and CO_2, H_2, H_2O, and N_2 to an electrical discharge.

F. J. STEVENSON

References

Emery, K. O., Orr, W. L., and Rittenberg, S. C., 1955, "Nutrient budgets in the ocean," in "Essays of the Natural Sciences in Honor of Captain Allan Hancock," Los Angeles, University of Southern California Press, pp. 299–309.

Fox, S. W., 1963, "Prebiological formation of biochemical substances," in (Breger, I. A., editors, "Organic Geochemistry," Oxford, Pergamon Press.

Goering, J. J., and Dugdale, R C., 1966, "Denitrification rates in an island bay in the equatorial Pacific Ocean," *Science,* 154(3748), 505–506.

Goldschmidt, V. M., 1954, "Geochemistry," Oxford, Clarendon Press.

Hutchinson, G. E., 1944, "Nitrogen in the biogeochemistry of the atmosphere," *Am. Scientist* 32, 178–195.

Hersh, C. K., 1968, "Nitrogen," in (Hampel, C. A., editor), "Encyclopedia of Chemical Elements," New York, Reinhold Book Corp., pp. 454–459.

Hutchinson, G. E., 1954, "The Biochemistry of the Terrestrial Atmosphere," in (Kuiper, G. P., editor), "The Solar System," Vol. II, *"The Earth as a Planet,"* Chicago, University of Chicago Press, pp. 371–433.

Lewis, R. W., 1965, "Nitrogen," U.S. Bur. Mines Mineral. Yearbook.

Mason, B., 1958, "Principles of Geochemistry," Second edition, New York, John Wiley & Sons

Moore, C. B., and Gibson, E. K., 1969, "Nitrogen Abundances in Chondritic Meteorites," *Science,* 163, 174–176.

Rayleigh, L., 1939, "Nitrogen, argon, and neon in the earth's crust with application to cosmology," *Proc. Roy. Soc. (London) Ser A,* 170, 451–464.

Rubey, W. W., 1951, "Geologic history of sea water," *Bull. Geol. Soc. Am.* 62, 1111–1147.

Scalen, R. S., 1958, "The isotopic composition, concentration, and chemical state of the nitrogen in igneous rocks," Ph.D. thesis, University of Arkansas. "University Microfilms," Ann Arbor, Michigan, Microfilm No. 59-1379.

Stevenson, F. J., 1960, "Some aspects of the distribution of biochemicals in geologic environments," *Geochim. Cosmochim. Acta,* 91, 261–271.

Stevenson, F. J., 1962, "Chemical state of the nitrogen in rocks," *Geochim. Cosmochim. Acta,* 26, 797–809.

Urey, H. C., 1952, "The Planets: Their Origin and Development," New Haven, Yale University Press

Cross-references: Biochemicals; Carbon Cycle; Carbon-Nitrogen Ratio; Nitrogen Cycle; Nitrogen Cycle—In The Oceans; Organic Geochemistry; Outgassing of the Planet Earth; Oxygen Cycle; Phosphorus Cycle; Seawater, Chemistry; Seawater: Geochemical Balance; Volcanic Gases; Weathering, Chemical. Vol. II: Atmosphere; Cosmogony. Vol. IVB: Clays and Clay Minerals. Vol. V: Lithosphere. Vol. VI: Crude Oil; Diagenesis; Humus and Humic Acids; Hydrosphere; Kerogen; Petroleum

NITROGEN CYCLE

The cycling of nitrogen in the earth is largely the result of activities carried out in the biosphere. An entire sequence of events is involved, some of which take place in the cells of microorganisms and some in the tissues of higher plants. The intermediates include gaseous, mineral, and organic forms of nitrogen.

The biological significance of nitrogen is evident from the realization that all reactions carried out by living organisms require the intervention of organic compounds containing nitrogen. Without nitrogen, life on the earth as we know it would not exist.

The cycle of nitrogen in nature is given in Fig. 1. A feature of particular interest is biochemical *nitrogen fixation,* a process carried out by only a few bacteria and blue-green algae. *Mineralization* is the conversion of organic nitrogen to inorganic forms. The initial reduction to ammonia is referred to as *ammonification;* the oxidation of this compound to nitrate is termed *nitrification.* The utilization of ammonia and nitrate by plants and microorganisms constitutes *assimilation.* Com-

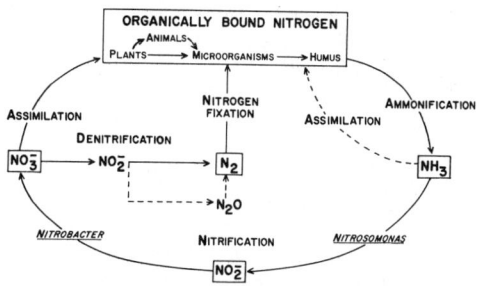

FIG. 1. The nitrogen cycle in nature (adapted from a drawing by Stanier, et al., 1963, p. 536).

bined nitrogen is ultimately returned to the atmosphere through biological denitrification, thereby completing the cycle.

The basic feature of nitrogen transformations centers on oxidation and reduction reactions. The oxidation state of nitrogen ranges from $+5$ for nitrate (NO_3^-) to -3 for ammonia (NH_3). Intermediate oxidation states include $+3$ for nitrite (NO_2^-), $+1$ for nitrous oxide (N_2O) and hyponitrous acid ($H_2N_2O_2$), 0 for elemental nitrogen (N_2), and -1 for hydroxylamine (NH_2OH).

The processes of biological nitrogen fixation, mineralization, assimilation, and denitrification are discussed briefly in the following sections. Detailed information on the various processes can be obtained in the references.

Biological Nitrogen Fixation

Although a vast supply of nitrogen occurs in the earth's atmosphere (386×10^{13} metric tons), it is present as a free, inert gas and cannot be used directly by higher forms of plant and animal life. The covalent triple bond of the N_2 molecule ($N\equiv N$) is highly stable and can be broken chemically only at high temperatures and pressures. Biochemical nitrogen-fixing organisms, on the other hand, perform this seemingly impossible task at ordinary temperatures and pressures.

The ability of a few bacteria and blue-green algae to fix elemental nitrogen can be regarded as being second in importance only to photosynthesis for the maintenance of life on this planet.

The biochemical process of fixing nitrogen is responsible for much of the fertility of agricultural soils. Even with the tremendous expansion in facilities for producing fertilizer nitrogen since World War II, legumes are still the main source of fixed nitrogen for the majority of the world's soils. Nitrogen-fixing processes carried out in prehistoric times created the nitrogen compounds that are present in many commercially important natural deposits, such as coal, petroleum, and the *caliche* of the Chilean desert.

A scheme of the various pathways believed to be involved in the fixation of molecular nitrogen is given in Fig. 2.

The organisms which fix nitrogen are conveniently grouped into two types: (1) the nonsymbiotic fixers, or those which fix nitrogen apart from a specific host; and (2) the symbiotic fixers, or those which fix nitrogen in association with higher plants.

Nonsymbiotic Nitrogen Fixation. The free-living microorganisms capable of utilizing molecular nitrogen include a number of blue-green algae of the family *Nostocaceae*, various photosynthetic bacteria (e.g., *Rhodospirillum*), several aerobic bacteria of the genus *Azotobacter*, and certain anaerobic bacteria of the genus *Clostridium*. A variety of other organisms, including some actinomycetes and fungi, have been reported from time to time to fix nitrogen, but these claims have not all been verified and, in any event, the amounts of nitrogen fixed are too small to be of practical or ecological significance.

The requirement of the photosynthetic bacteria for both irradiation and anaerobiosis restricts their activities to shallow, muddy ponds or estuarine muds. They generally are found as a layer overlying the mud and covered by a layer of algae; fixation of nitrogen is possible because the pigments of the photosynthetic bacteria absorb light rays in the region of the spectra not absorbed by the pigments of the overlying algae.

In contrast to the photosynthetic bacteria, the blue-green algae are widely distributed in nature. They occur in almost every environmental situation where sufficient sunlight is available for photosynthesis, including barren rock surfaces and frozen wastelands. Together with the lichen fungi, they form surface crusts of varying densities in the desert areas of the United States and the semiarid regions of eastern Australia. The initial vegetation on the pumice and ash of Krakatoa after the volcanic explosion of 1883, which completely denuded the island of all visible forms of plant life, was a dark-green gelatinous layer containing blue-green algae. Many scientists believe that the blue-green algae are responsible for the fixation of nitrogen in rice fields.

Azotobacter and *Clostridium* are universally present in terrestrial soils of the world, although the former is not always found in acid soils. The normal condition of *Clostridium* is the spore form, vegetative growth occurring only during anaerobic periods following rains.

In lateritic soils of the tropics, the species of *Azotobacter* indigenous to the temperate zones (*A. Chroococcum, A. vinelandii, A. macrocytogenes, A. agilis*) are supplemented, or replaced, by *A. Beijerinckia*, a nitrogen fixing microorganism discovered only recently (Alexander, 1961, p. 312). This organism is confined largely to soils of southeast Asia, South America, northern Australia, and tropical areas of Africa. A satisfactory explanation for the unique geographical distribution of *Beijerinckia* has not been found.

Under natural conditions, the nitrogen-fixing capabilities of free-living bacteria are greatly restricted. These organisms utilize organic compounds as a source of energy, a factor which limits their activities to environments contain-

NITROGEN CYCLE

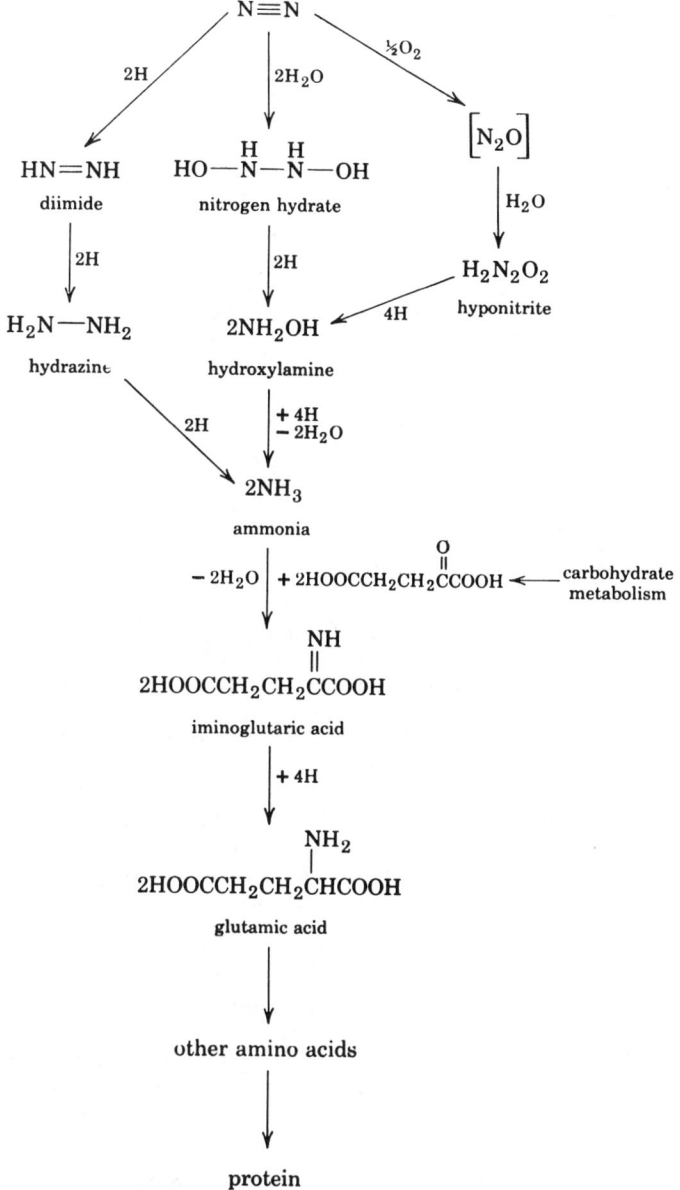

Fig. 2. Pathways of biochemical nitrogen fixation (from Alexander, 1961).

ing rather high amounts of available organic matter. No more than six pounds of nitrogen per acre are believed to be added to the soil each year by the combined activities of non-symbiotic nitrogen-fixing microorganisms.

Symbiotic Nitrogen Fixation. The recognized symbiotic nitrogen fixers belong to the bacterial genus *Rhizobium;* in this case, nitrogen is fixed only when the bacteria are living within nodules on the roots of leguminous plants. The leguminosae family contains from ten to twelve thousand species, most of which are indigenous to the tropics. Thus far, only about 1200 species have been examined for nodulation, of which about 90% have been found to bear nodules. Approximately 200 species are cultivated by man; about 50 species are grown commercially in the United States. Wild legumes are important for the fixation of nitrogen in natural ecosystems.

Six species of *Rhizobium* are recognized: *R. meliloti, R. leguminosarum, R. phaseoli, R.*

NITROGEN CYCLE

TABLE 1. RHIZOBIUM-LEGUME ASSOCIATIONS[a]

Rhizobium Species	Cross-Inoculation Group	Host Genera	Legumes Included
R. meliloti	Alfalfa	Medicago	Alfalfa
		Melilotus	Sweet clover
		Trigonella	Fenugreek
R. trifolii	Clover	Trifolium	Clovers
R. leguminosarum	Pea	Pisum	Pea
		Vicia	Vetch
		Lathyrus	Sweetpea
		Lens	Lentil
R. phaseoli	Bean	Phaseolus	Beans
R. lupini	Lupine	Lupinus	Lupine
		Ornithopus	Serradella
R. japonicum	Soybean	Glycine	Soybean
	Cowpea[b]	Vigna	Cowpea
		Lespedeza	Lespedeza
		Crotalaria	Crotalaria
		Pueraria	Kudzu
		Arachis	Peanut
		Phaseolus	Lima bean

[a] From Stevenson, F. J., 1964.
[b] This group has not attained species status.

japonicum, R. lupini, and R. trifolii. These bacterial-plant groups, often called cross-inoculation groups, include only a small percentage of the leguminous species. The important legumes with which the above-mentioned species of Rhizobium are associated is given in Table 1. The average amount of nitrogen which is fixed varies from 40 to 200 pounds per acre per annum.

Many nonlegumes also form symbiotic relationships with nitrogen fixing microorganisms. They include plants belonging to the families Betulaceae, Elaeagnaceae, Myricaceae, Coriariaceae, Rhamnaceae, and Casuarinaceae. Their geographical distribution is outlined in Table 2.

Contrary to common belief, nodulated nonlegumes are not freak plants of limited distribution, but are important sources of fixed nitrogen for plants in general. The dominant plants on nitrogen-poor soils are often nonlegumes which have the ability to form symbiotic relationships with nitrogen fixing microorganisms. In addition, many nonlegumes are known to contain colonies of nitrogen fixing microorganisms in nodules or special glandular pockets on their leaves.

Nitrogen in Atmospheric Precipitation. The accession of combined nitrogen in rain and snow, consisting of ammonia, nitrite, nitrate, and organic nitrogen associated with cosmic dust, supplements that which is fixed by biochemical agents. The total addition of nitrogen to the soil-plant ecosystem, estimated to be from two to six pounds per acre per year in most localities, is too small to be of agricultural significance except possibly in the humid tropics. Some of the nitrogen in precipitation (ammonia and the organic forms) originates

TABLE 2. DISTRIBUTION OF NODULED NONLEGUMES[a]

Family	Genus	Number of Species Nodulated	Geographical Distribution
Betulaceae	Alnus	15	Cool regions of the northern hemisphere
Elaeagnaceae	Elaeagnus	9	Asia, Europe, North America
	Hippophae	1	Asia and Europe, from the Himalayas to the Arctic Circle
	Shepherdia	2	Confined to North America
Myricaceae	Myrica	7	Temperate regions of both hemispheres
Coriariaceae	Coriaria	1	Discontinuous distribution
Rhamnaceae	Ceanothus	7	Confined to North America
Casuarinaceae	Casuarina	12	Tropics and subtropics, extending from East Africa to the Indian Archipelago, Pacific Islands, and Australia

[a] From Stevenson, F. J., 1964.

Mineralization

Very little of the nitrogen stored in organic compounds through the growth of higher plants is converted to ammonia by animal metabolism. Most of it remains in combined form until acted upon by microbes. During the decay of plant and animal tissues by microorganisms, some of the combined nitrogen is assimilated and becomes incorporated into microbial protoplasm, primarily in the form of proteins and nucleic acids. This nitrogen is eventually converted to ammonia following death of the microorganisms.

The conversion of organic nitrogen compounds to mineral forms, a process referred to as mineralization, encompasses two distinct microbiological processes: (1) ammonification, in which organic nitrogen is converted into ammonia; and (2) nitrification, in which ammonia is oxidized to nitrate. The over-all reaction is shown below.

$$\text{organic nitrogen} \xrightarrow{\text{Ammonification}} NH_3 \xrightarrow{\text{Nitrification}} NO_2^- \longrightarrow NO_3^-$$

Both aerobic and anaerobic organisms participate in the ammonification process, whereas only aerobic organisms oxidize ammonia to nitrate.

In aerobic soils and sediments, the ammonia formed in the first stage of mineralization is rapidly converted to nitrate; this is the main form of nitrogen available to plants. Ammonia accumulates only in environments where oxygen is limiting, such as in swamps.

The nitrification process is carried out by two groups of autotrophic bacteria, one deriving energy through the oxidation of ammonia to nitrite, the other through the oxidation of nitrite to nitrate.

Five genera of ammonia oxidizers (*Nitrosomonas, Nitrosococcus, Nitrosospira, Nitrosocystis,* and *Nitrosogloea*) and two genera of nitrite oxidizers (*Nitrobacter, Nitrocystis*) are recognized. However, only one genus of each group is of ecological importance, *Nitrosomonas* and *Nitrobacter,* respectively.

The reaction catalyzed by *Nitrosomonas* results in the removal of three pairs of electrons from ammonia, with the formation of hydroxylamine, hyponitrite, and nitrite.

$$NH_3 \xrightarrow{\frac{1}{2}O_2} \underset{\text{Hydroxylamine}}{HO-NH_2} \xrightarrow{-2H}$$

$$\tfrac{1}{2}(HO-N=N-OH) \xrightarrow{\frac{1}{2}O_2} \underset{\text{Nitrite}}{HO-N=O}$$
$$\text{Hyponitrite}$$

Members of the genus *Nitrobacter* remove two electrons from nitrite, probably by the following mechanism.

$$\underset{\text{Nitrite}}{HO-N=O} \xrightarrow{H_2O}$$

$$HO-N-(OH)_2 \xrightarrow{-2H} \underset{\text{Nitrate}}{HO-N\underset{O}{\overset{O}{\diagup\!\!\!\diagdown}}}$$

Nitrite is rarely found in soils or sediments, even under conditions of rapid nitrification, because the oxidation of this compound by *Nitrobacter* proceeds at a rate equivalent to, or greater than, the oxidation of ammonia by *Nitrosomonas.* However, nitrites are often found in alkaline desert soils, because of ammonia toxicity to *Nitrobacter.*

During the Napoleonic wars, nitrates were badly needed in France for the manufacture of gunpowder, and with the natural source in Chile cut off by a blockade, the nitrates were produced biologically by composting soil with manure and lime.

Heterotrophic bacteria, actinomycetes, and fungi have all been shown to produce nitrite or nitrate from ammonia and organic nitrogen compounds in pure culture. However, the ecological importance of heterotrophic nitrification has yet to be determined.

Because nitrate carries a negative charge, it is not held by clay minerals in soil and can be lost to the sea. The nitrate not utilized by plants or microorganisms, either in soils, sediments, or water, is ultimately returned to the atmosphere through the activities of denitrifying bacteria (to be discussed). Natural nitrate deposits are rare, the most significant ones being located in Chile, Egypt, Mexico, South Africa, and the United States. Only the Chilean deposit is of commercial significance.

Assimilation

Although green plants are able to use organic forms of nitrogen such as amino acids, practically all of the nitrogen taken up exists in two inorganic compounds, ammonia and nitrate. If nitrate is the form in which nitrogen is assimilated, it must first be reduced in the cell to ammonia, which is subsequently incorporated into two amino acids, aspartic acid and glutamic acid. These amino acids, in turn, serve as the precursors for all other nitrogen compounds in living matter.

The reactions by which nitrate is assimilated

are essentially the reverse of those shown earlier for ammonification and nitrification.

Microorganisms in soils and sediments compete with higher plants for ammonia and nitrate. During the decomposition of low-nitrogen plant remains, mineral nitrogen is incorporated into microbial tissue and temporarily immobilized. This nitrogen may become a part of the stable humus, whereby it is gradually mineralized and recycled as a result of the continuing activities of microorganisms.

Denitrification

In denitrification, nitrate and various intermediate reduction products substitute for oxygen as the terminal acceptor of electrons in metabolic reactions. The end products (N_2 and N_2O) are gases; these eventually become part of the atmosphere.

The geochemical significance of denitrification arises from the fact that the process acts as a balance on biochemical nitrogen fixation. It is analogous to the relation between photosynthesis and respiration in the carbon cycle. Just as organically bound carbon is returned to the atmosphere (as CO_2) through respiration, so is combined nitrogen returned through denitrification.

Some scientists believe that the reason nitrogen is the principal constituent of the earth's atmosphere is because of the continued activity of denitrifying microorganisms throughout geological history. In any event, it is likely that most of the atmospheric nitrogen has passed at least once through the denitrification cycle. The annual exchange of nitrogen between the atmosphere and the biosphere has been reported to range from 0.017 to 0.034 mg/cm^2/yr (Hutchinson, 1944). This corresponds to a cycle length of between 44 and 220 million years, or from one-tenth to one-half of the time span from the Cambrian to the present.

The bacteria responsible for denitrification are all facultative anaerobes, and they are universally distributed in soils, sediments, and water. Both autotrophic and heterotrophic organisms are involved. The autotrophic denitrifiers include *Micrococcus denitrificans*, a facultative autotroph, and *Thiobacillus denitrificans*, an organism which oxidizes sulfur while reducing nitrate under anaerobic conditions. The heterotrophic denitrifiers comprise numerous genera; most of them belong to the genera *Pseudomonas, Micrococcus, Achromobacter,* and *Bacillus*. These organisms are also active in the decay of plant and animal residues in soils and sediments.

The following biochemical pathway represents the probable mechanism of bacterial denitrification:

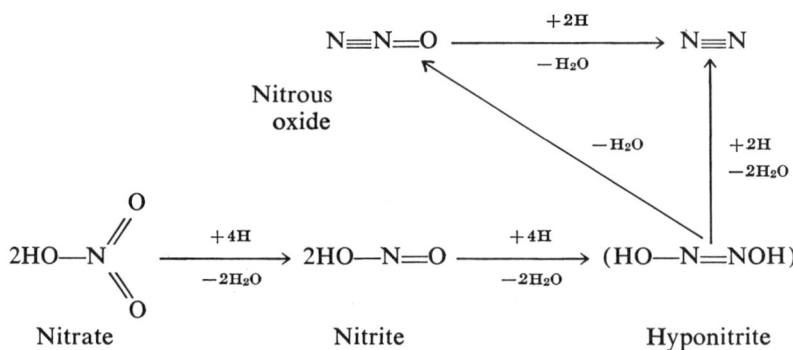

F. J. STEVENSON

References

Alexander, M., 1961, "Introduction to Soil Microbiology," New York, John Wiley & Sons.
Hutchinson, G. E., 1944, "Nitrogen in the biogeochemistry of the atmosphere," *Am. Scientist*, **32**, 178–195.
Stanier, R. Y., Doudorff, M., and Adelberg, E. A., 1963, "The Microbial World," Second ed., Engewood Cliffs, N.J., Prentice-Hall.
Stevenson, F. J., 1964, "Soil nitrogen," in (Sauchelli, V., editor), "Fertilizer Nitrogen, Its Chemistry and Technology, New York, Reinhold Publ. Corp.
Stevenson, I. L., 1964, "Biochemistry of Soil," in (Bear, F. E., editor), "Chemistry of the Soil," New York, Reinhold Publ. Corp.

Cross-references: *Ammonia, in Minerals and Early Atmosphere; Biochemicals; Cycles—Geochemical; Microorganisms; Nitrogen: Element and Geochemistry; Nitrogen Cycle in the Ocean; Organic Geochemistry; Oxidation and Reduction; Photosynthesis; Precambrian Atmosphere—Geochemistry.* Vol. VI: *Caliche; Coal; Humus and Humic Acids; Petroleum; Soil Genesis.* Vol. VII: *Biosphere.*

NITROGEN CYCLE IN THE OCEAN

The variety of forms in which nitrogen exists in the sea as well as the variety of chemical and biological reactions known or suspected to occur makes the study of that element in the marine environment a fascinating one. The subject may be conveniently divided into two portions, one concerned with the internal workings of the nitrogen cycle as it is known today, and the other with the history and fate of nitrogen in the sea from the geological viewpoint.

Structure and Dynamics

A discussion of the nitrogen cycle is conveniently carried out with the aid of a diagram such as that shown in Fig. 1 where the main reservoirs and active pathways of transformations are identified. Nitrogen occurs in dissolved form primarily as N_2, nitrate (NO_3^-), dissolved organic nitrogen, and ammonia (NH_4^+) (see Table 1). All of these are available to a variety of organisms, including the phytoplankton or algae, the latter being responsible for photosynthesis or primary productivity of new organic matter in the sea (see Fig. 1, pathways 1, 2, and 3). A remarkable set of biologically mediated reactions occur within the ammonia-nitrite-nitrate system. The oxidative steps yielding nitrite and eventually nitrate from ammonia are energy yielding; organisms capable of using this energy to synthesize organic compounds are known and belong to the autotrophic bacteria. The number of species is small and specialization is apparently complete, i.e., a given species is able to carry out only the ammonia to nitrite step (path 8a) and another distinct species the nitrite to nitrate step (path 9a). In the better studied land species, both groups are known, while in the sea only recently has the organism carrying

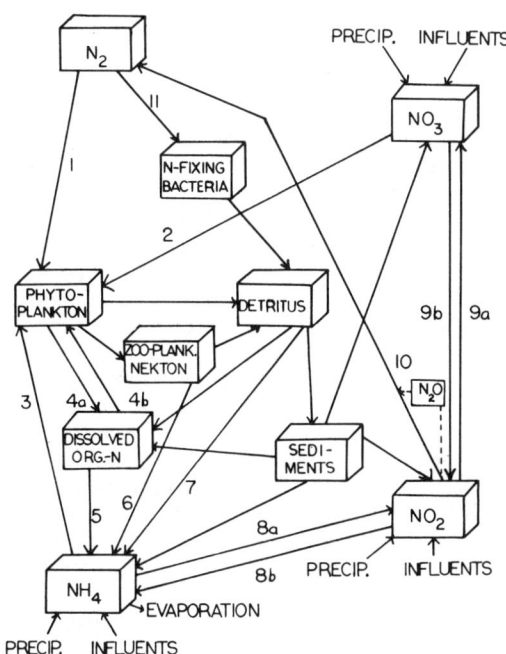

Fig. 1. The nitrogen cycle in the ocean.

out the oxidation to nitrite been isolated. The processes described above are known as nitrification and the organisms as nitrifying bacteria. Recently nitrification has been observed in a heterotrophic organism as well.

The reverse set of reactions beginning with nitrate and leading to the production of ammonia and nitrite may be carried out by a variety of microorganisms under low oxygen conditions. In this case, nitrate or nitrite is used as a terminal hydrogen acceptor in place of oxygen. The process is termed *nitrate reduction* (paths 8b, 9b) except in the special case of *denitrification* (path 10) where gaseous products, N_2, NO, or N_2O, are produced.

TABLE 1. NITROGEN BUDGET IN THE SEA

Reserves	
Inorganic combined nitrogen (NH_4^+, NO_2^-, NO_3^-)	5.8×10^{11} metric tons
Dissolved organic nitrogen	3.4×10^{11} tons
Total	9.2×10^{11} tons
Gains	
Land drainage (inorganic combined, dissolved organic)	19×10^6 tons/yr
Precipitation on sea surface	59×10^6 tons/yr
Biological nitrogen fixation (computed by the author on the assumption that 1% of the nitrogen required by photosynthetic organisms in the tropical and subtropical ocean is supplied from this source)	8×10^6 tons/yr
Total	86×10^6 tons/yr
Losses	
Permanent burial in sediments	8.6×10^6 tons/yr
Unknown (by difference)	77.4×10^6 tons/yr
Total	86×10^6 tons/yr

Relative Importance of Inorganic Nitrogen Sources for the Phytoplankton

In the past nitrate has been assumed to be the fundamental source of nitrogen for algae (phytoplankton) living in the sea. Probably this idea has become fixed as a result of the relatively large amounts of nitrate which accumulate in the surface waters of temperate regions during the winter, and also because reliable analytical methods for that compound have long been available. In contrast ammonia shows a low seasonal variability and is difficult to measure quantitatively. However, it is now recognized that the pool size of a compound existing in the sea, or in other systems, often reflects the activity of the compound in an inverse manner, i.e., the more active the pathways leading to and especially from the pool, the smaller the pool size is likely to be. Modern tracer techniques utilizing compounds labeled with heavy nitrogen, N^{15}, have recently been applied to research on nitrogen in the sea. As a result it is now clear that the low concentration of ammonia is maintained in the sea because it is used at a rapid rate, primarily by the phytoplankton. In fact, only during spring flowering of phytoplankton in nitrate-rich temperate waters does nitrate equal ammonia in the amount of nitrogen supplied. At other times of the year and in the tropical ocean, nitrate levels fall and that compound contributes little to the phytoplankton which then draws its nitrogen primarily from the ammonia pool. Ammonia is apparently supplied at an equal rate to the pool by regeneration from particulate organic matter and probably by breakdown of dissolved forms of organic nitrogen. The turnover time for ammonia (the time required to replace or exhaust the amount present at a given time) varies from less than one day to several days.

Recently investigations have shown high rates of nitrogen fixation in certain areas of the subtropical ocean associated with the presence of a blue-green alga, *Trichodesmium* spp. This organism forms extensive blooms over large areas and those blooms are apparently the oceanic counterpart of the well-known blue-green algal blooms so commonly occurring in fresh waters, some of which are now known to draw heavily on atmospheric nitrogen as a source of that element. It appears that at least on a local basis, it may be of great importance under certain conditions and at certain times of the year.

Although the ability of a variety of marine algae to utilize simple nitrogen-containing organic compounds is well established, little is known about the distribution and utilization of these compounds in nature. Preliminary experiments (conducted at sea by the author and his colleagues) using N^{15} labeled compounds suggest that at least glycine and urea are indeed utilized by photosynthetic organisms. The same experiments show, however, that ammonia is released from these compounds very rapidly under some circumstances. Further investigations along these lines are likely to prove highly interesting.

Origin and Fate of Oceanic Nitrogen Compounds

The sea receives nitrogen from precipitation on the surface, from land drainage, and from nitrogen fixation. Losses are incurred from burial in sediments, from transfer of ammonia (and perhaps other nitrogen compounds) into the atmosphere at the sea surface, and from denitrification.

A number of investigators have drawn up budgets for these gains and losses in order to obtain an overall view of the cycle. Table 1, from a recent balance sheet by Emery, et al., (1955), shows values for the entire ocean.

It is around the "unknown" item that one of the most interesting unsolved problems of the aquatic nitrogen cycle revolves. On land, denitrification is the process by which combined nitrogen is returned to the atmosphere, thereby maintaining a balance against nitrogen fixation. Denitrification is, however, an anaerobic process and although such environments are common in the soil, they are rare in the ocean, usually occurring in fjords or trenches with shallow sills which prevent mixing and circulation of trapped deep water. Organisms capable of denitrification have been isolated from the sea, but until a method is devised for estimating their activity in nature, a figure for the loss of nitrogen from this source cannot be assigned.

The escape of ammonia across the sea-air interface provides another possible mechanism for the loss of nitrogen from the sea. Recent data suggest that the pH of sea water (7.5 — 8.4) is sufficiently basic to allow slow diffusion of this compound into the air. If ammonia is indeed lost from the sea surface, then a portion of the combined nitrogen obtained by the sea from precipitation may be directly recycled.

The construction of such a balance sheet is predicated upon the assumption that the mean composition of sea water in terms of nitrogen compounds is not changing. That this assumption is probably correct can be seen by computing the turnover time for the nitrogen reserves:

total reserves/annual gains = 9.2×10^{11}/
86×10^6 = about 10^4 years

a very short time, indeed, in terms of geological time. One interpretation of this figure would be to the effect that the mean nitrogen reserves of the sea would double in 10^4 years if there were no losses.

Nitrogen in Sediments

A large proportion of the nitrogen incorporated into organic matter is regenerated in the water column, about 90 percent in the upper 200 meters with further regeneration occurring in the deep water. Thus, only a small proportion of the organic matter produced in the photosynthetic zone reaches the sediments.

The subsequent fate of sedimented organic nitrogen depends upon oxygen conditions within the sediments of a given region; however, in all cases ammonia is produced as the first inorganic compound. In the presence of oxygen, bacterial nitrification takes place yielding detectable and often large quantities of nitrite and nitrate. Lacking oxygen nitrification does not occur. Organic nitrogen remains in substantial quantities, usually remaining as the largest single constituent. The activity of denitrifying bacteria in anaerobic sediments may result in loss of fixed nitrogen. A decrease in total nitrogen occurs with depth suggesting that nitrogen is lost to the overlying water through diffusion and perhaps through the activities of burrowing animals.

Nitrogen-fixing microorganisms are also known to live in the upper layers of sediments. *Azotobacter* and *Clostridium,* aerobic and anaerobic bacteria, respectively, have been found in Black Sea sediments and in other areas. Estimates of the nitrogen-fixing activities of this portion of the sediment flora are not available.

R. C. DUGDALE

References

Cooper, L. H. N., 1937, "The nitrogen cycle in in sea," *J. Marine Biol. Assoc U.K.* **22,** 183–204.

Emery, K. O., Orr, W. L., and Rittenberg, S. C., 1955, "Nutrient budgets in the ocean," "Essays in Honor of Captain Allan Hancock," Los Angeles, Univ. of Southern California Press, pp. 299–309.

Harris, E., 1959, "The nitrogen cycle in Long Island Sound," *Bull. Bingham Oceanographic Coll.,* **17,** 31–65.

Rittenberg, S. S., Emery, K O., and Orr, W. L., 1955, "Regeneration of nutrients in sediments of marine basins," *Deep-Sea Res.,* **3,** 23–45.

Sverdrup, H. U., Johnson, M. W., and Fleming, R. H., 1942, "The Oceans," Englewood Cliffs, N.J., Prentice-Hall, 1087pp.

Cross-references: *Nitrogen: Element and Geochemistry; Nitrogen Cycle; Organic Geochemistry; Photosynthesis; Seawater, Chemistry.* Vol. I: *Phytoplankton.*

NOBLE GASES—See **RARE GASES**

NONMETALLIC MINERALS—See **MINERAL CLASSES: SILICATES**

NUCLEAR FUELS—See Vol. IV B

NUTRIENTS IN THE SEA—See Vol. I

OCEAN-ATMOSPHERE INTERACTION— *See* Vol. I

OCEANOGRAPHY—*See* Vol. I

OIL COMPOSITION AND MIGRATION— *See* Vol. IV B, **CRUDE OIL COMPOSITION AND MIGRATION**

OIL SHALES—*See* Vol. IV B

OPTICAL ACTIVITY

Materials said to be "optically active" are those which have the capacity to rotate the plane of polarization as plane-polarized light passes through them. Since plane-polarized light possesses both right and left circularly polarized beams, anything which retards one of these beams more than the other will produce rotation of the original plane of polarization. This phenomenon is known as *optical activity, optical rotation,* or *rotary polarization.* It is caused by a spiral arrangement of atoms in the crystal (as in quartz) or of the molecule (as in natural petroleum).

In Crystals

Many inorganic crystals (notably quartz and cinnabar) and some organic ones are optically active. Most inorganic crystals lose this property on fusion. This indicates a lack of certain symmetry elements in the crystal, namely an *n*-fold rotatory axis ($n=1$, 2, or 4). Thus, the valence electrons bonding the atoms in such crystals are not symmetrically distributed, and a beam of plane-polarized light encounters an unsymmetrical electrical field.

In the microscopic study of rocks and minerals, optical rotation is not important because most rock-forming silicates are optically inactive. Quartz and cinnabar (HgS) are two minerals that rotate plane-polarized light. In fact cinnabar has a greater optical rotation than any other known substance. Its rotation in red light at 325°C is about twenty times that of quartz. (See Fig. 1 in the entry *Stereochemistry and Enantiomorphism* for a diagram of right- and left-handed modifications of quartz.)

In Organic Compounds

Optical activity in organic compounds is displayed by asymmetric molecules, rigid structures with hindered rotation or systems with consecutive double bonds. The chemist defines an asymmetric molecule as being one that is nonsuperimposable on its mirror image. Two molecules, identical in chemical composition and structure, but nonsuperimposable mirror images of each other are called *optical isomers*. The members of such a pair are *enantiomers* (or enantiomorphs). Optical activity may be predicted if four different substituents are attached to a carbon atom. In most cases, such compounds show optical activity. Another test for asymmetry is lack of a plane or center of symmetry. Compounds with two asymmetric carbon atoms (four different substituents on each C atom) but having a plane of symmetry will be optically inactive. These are termed *meso* compounds.

A closs of organic compounds, allenes, having an even number of consecutive double bonds and two different substituents on the end carbon atoms will be optically active (Fig. 1). Similarly, spirenes, or consecutive ring compounds, will be optically active if there is an even number of rings and two different groups attached to the end carbons (Fig. 2).

Many naturally occurring organic compounds show optical rotation. Petroleum is optically active, which is a valuable test for distinguishing it from synthetic oil. Optical activity in organic matter has been considered as proof of its biological origin. Its presence in carbonaceous chondrites (if established) would constitute strong evidence for the existence of extraterrestrial life.

Optical rotation is measured with a polarimeter. It consists of a light source, two nicol prisms, and a tube in between the nicols to contain the substance under examination

Fig. 1. Allene isomers.

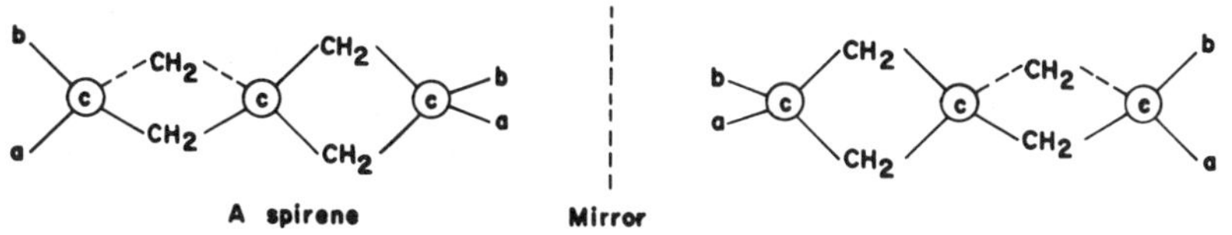

ORGANIC COMPOUNDS

FIG. 2. Spirene isomers.

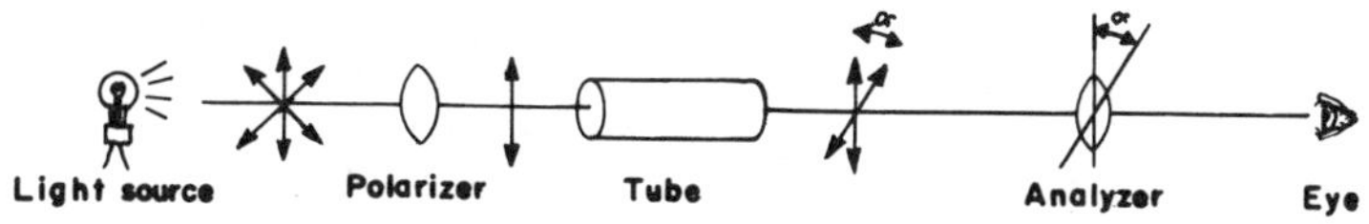

FIG. 3. Schematic diagram of a polarimeter.

(Fig. 3). Light passing through the polarizer is plane-polarized. If the substance is inactive and the analyzer at 90° to the polarizer, no light passes through. If, on the other hand, the substance rotates the plane of polarization, the second nicol prism must be turned through some angle (α) until light is at a minimum. The amount of rotation depends upon the concentration (for a solution) or density (solid), the length of the tube, temperature, and wavelength of light used.

At a given temperature and wavelength of light, the specific rotation is given by

$$\left[\alpha\right]_{-}^{+} = \frac{\alpha}{ld}$$

where α = observed rotation, l = length of the tube, and d = density.

VIVIEN GORNITZ

References

Cameron, E. N., 1961, "Ore Microscopy," New York, John Wiley & Sons, 293pp
Conn, G. K. T., and Bradshaw, F. J., (editors), "Polarized Light in Metallography," London, Butterworths Scientific Publ., 130pp.
Heinrich, E. W., 1965, "Microscopic Identification of Minerals," New York, McGraw-Hill Book Co., 414pp.
Kerr, P. F., 1959, "Optical Mineralogy," Third ed., New York, McGraw-Hill Book Co., 442pp.
Meinschein, W G., Frondel, C., Laur, P., and Mislow, K., 1966, "Meteorites: optical activity in organic matter," Science, 154(3747), 377–380.
Winchell, H., 1965, "Optical Properties of Minerals, a Determinitive Table," New York, Academic Press, 91pp.

Cross-references: *Bonding; Crystal Chemistry.* Vol. II: *Carbonaceous Meteorites.* Vol. IVB: *Optical Mineralogy; Polarization and the Polarizing Microscope.*

ORES, ORE GENESIS—See Vol. IV B

ORGANIC COMPOUNDS

Some mineralogists prefer to consider natural organic compounds as mineraloids rather than true minerals. They are included here for completeness. The two principal divisions are the salts of organic acids, and hydrocarbons. In some classifications the resins (amber) are also treated as a separate division. Many of the organic compounds that have been given mineral names are mixtures, especially in the case of hydrocarbon compounds.

OXALATES

Whewellite	$CaC_2O_4 \cdot H_2O$
Weddellite	$CaC_2O_4 \cdot 2H_2O$
Humboldtine	$FeC_2O_4 \cdot 2H_2O$
Oxammite	$(NH_4)_2C_2O_4 \cdot H_2O$
Minguzzite	$K_3FeC_2O_4 \cdot H_2O$
Stepanovite	$NaMgFe(C_2O_4)_3 \cdot 8\text{--}9H_2O$

HYDROCARBONS

Evenkite	$C_{24}H_{50}$
Kratochwilite	$C_{13}H_{10}$
Simonellite	$C_{15}H_{20}$
Phylloretite	$C_{18}H_{18}$
Fichtelite	$C_{19}H_{34}$
Hartite	$C_{20}H_{34}$
Pendletonite	$C_{24}H_{12}$
Curtisite	$C_{24}H_{18}$
Flagstaffite	$C_{10}H_{18}(OH)_2 \cdot H_2O$

LLOYD W. STAPLES

References

Murdoch, J., and Geissman, T. A., 1967, "Pendletonite, a new hydrocarbon mineral from California," *Am. Mineralogist*, **52**, 611.
Strunz, H., 1970, "Mineralogische Tabellen," Fifth ed., Leipzig.

Cross-references: *Hydrocarbons; Mineraloids; Organic Mineraloids.*

ORGANIC COSMOCHEMISTRY

ORGANIC COSMOCHEMISTRY—See Vol. II

ORGANIC GEOCHEMISTRY

For the past hundred years or so, studies on coals, oils, asphalts, amber, and other organic substances have been carried out within the framework of geology, economic geology in particular, and mineralogy. More recently, a branch of the geosciences has emerged, organic geochemistry, which aims to become a systematic study of the entire range of carbonaceous complexes within the earth's crust, termed by myself as the terrestrial "carbosphere" (see Vol. II, Cosmochemistry—Organic).

Genesis of the Terrestrial Carbosphere

The formation of organic molecules, inclusive of amino acids, was experimentally proved in a wide range of mixtures of H_2, CH_4, NH_3, CO, CO_2, and CN, acted upon by high-energy agencies such as ionizing radiation, electrical discharges, etc. The spontaneous condensation of unsaturated molecules, such as CN, C_2H_2, etc., is also well known.

It is now generally believed that the earth condensed at relatively low initial temperatures from cosmic dust, and that possibly the surface of the earth was, at no stage of its history, at an elevated temperature. Carbonaceous substances may have condensed at the following stages: (a) in the cosmic dust, which may either be a primary condensate, or of solar flare origin; (b) during the condensation of our planet; (c) within the primitive terrestrial atmosphere, hydrosphere, or crust. The studies of cosmic dust samples retrieved by rockets, the determination of uranium ages of hydrocarbons in archaic terrestrial rocks, and other investigations in the future may give us clues which may define the cosmological stage, or stages, in the course of which the original carbosphere condensed.

In the recent literature (Fox, 1963) it has been generally presumed that the condensation of the carbosphere took place subsequent to the condensation of the earth, mainly in the primitive atmosphere and hydrosphere, but also possibly within the crust. The range of organic molecules, including amino acids, may have formed, according to experimental evidence, from nonoxidating atmospheres, over a considerable range of composition. The condensation of amino acids to polypeptides could have proceeded, according to our laboratory investigations, only at relatively elevated temperatures, perhaps in the vicinity of hot springs. On the other hand, it is possible that reactions would occur at lower temperatures too, through a longer "geological" time scale.

The living organisms must have generated at a certain stage of the condensation of the amino acids. Their present role in the changing of what we may term the "prebiological" equilibrium conditions of the carbosphere is difficult to assess. The writer's opinion is that the organisms tend to increase the total volume of the carbosphere through the conversion of atmospheric CO_2 by photosynthesis and, simultaneously, tend to deoxygenate it by splitting off the low-molecular-weight products of their metabolism.

Organic Geochemistry of Isotopes

The isotope ratios of C, H, O, N, and S within the various carbonaceous complexes are of considerable theoretical interest as geological indicators of physicochemical conditions. More important, perhaps, these ratios may furnish evidence as to the biogenic or abiogenic origin of a given hydrocarbon. It was found that the ranges of C^{12}/C^{13} ratios (Rankama, 1954) are the lowest in sedimentary carbonates (88–91), intermediate in substances of magmatic origin, including graphite (88.5–91.5), and highest in the presumably mainly biogenic sediments (89.5–95.5). This is due mainly to the effect of the living organism, which tends to expel C^{13}–enriched carbon in its products of metabolism (including CO_2, which produces the sedimentary carbonates) and to retain a higher proportion of C^{12}. The considerable overlap between the ranges of the biogenically differentiated and undifferentiated juvenile carbons may be caused by the effects of previous geological history of the carbon atoms which constitute the particular carbonaceous substance. The study of individual organic molecules, or phases from the same carbonaceous complex may help to clarify this. Thus, Colombo, and co-workers (1965) found that ethane, propane, and butane contain isotropically heavier carbon than methane from the same gas field in Italy, indicating a process of fractionation in the course of the genesis of the gases.

Very little is known at present regarding the distribution of isotopes of H, O, N, and S within the carbosphere. The S^{32}/S^{36} reach the maximum of 23.0 in biogenic S and the minimum of 21.6 in that of magmatic origin. This rather broad range makes the element a good potential indicator for biogenic or abiogenic origin of a given carbonaceous substance. In the case of H, O, and N, there appears to be little, if any, biogenic separation, but these elements may prove to be useful in the future

TABLE 1. DISTRIBUTION OF THE MAJOR ELEMENTS

Element	Subdivision (in approximate order of decreasing proportions)
H	Hydrosphere, sedimentary rocks (clays) *carbosphere*, atmosphere, igneous rocks (in micas, etc.), biosphere
C	Sedimentary rocks (carbonates), *carbosphere* juvenile carbon (carbonatites, graphite, diamond) biosphere, atmosphere
N	Atmosphere, *carbosphere*, biosphere, sediments (nitrates in evaporates), igneous rocks
O	Igneous rocks, sedimentary rocks, hydrosphere, atmosphere, *carbosphere*, biosphere
S	Igneous rocks (sulfides), hydrothermal deposits (sulfides, sulfates), sediments (sulfur, sulfates, sulfides), *carbosphere*, biosphere, hydrosphere

as indicators for fractionation and other physicochemical changes in the geological history of a given organic phase.

Distribution and Organic Geochemical Cycles of C,H,O,N, and S

None of the five major elements of the "carbosphere" concentrate exclusively in it; their subdivision is shown in Table 1, which is taken, with some minor alterations, from Mueller (1963).

The results of experiments referred to above indicate that organic molecules may condense abiologically at present in the terrestrial atmosphere, hydrosphere, and crust; however, owing to the low concentration of carbon-containing gases on the one hand, and the great biochemical skill of the primitive organism on the other hand, biogenic condensation appears to be the more important factor, and the sole process regarding which direct recent observational evidence exists. The bulk of inorganic C, in the form of CO_2, condenses as carbohydrates within the green plants through photosynthesis, which process also assimilates the O and H, present originally as water. Some inorganic N and S enters the carbosphere mainly through microbiological activities. A high proportion of all the five elements are being recycled. The ratio of *biosynthesized fraction/recycled fraction* may prove to be the highest for C and the lowest for S, for the entire biosphere, but there are no exact estimations of this at present.

With the death of the organisms (and close-to-surface exposure of abiogenic substances) a reversed-cycle stage is reached, leading to partial or total decomposition of the carbonaceous complexes. It appears that O and N tend to liberate at the fastest rate mainly as H_2O and N_2; the geological time scale stability of C and H within the organic molecules is greater; the decomposition products are mainly CO, CO_2, CH_4, and H_2O. Sulfur actually tends to be added to the aging carbonaceous complexes, at least under the more commonly prevailing conditions, perhaps substituting gradually for part of O in the hydroxyl, carboxyl, and other oxygen-containing groups.

Distribution of Trace Elements Within the Carbosphere

The trace elements within prevalently carbonaceous phases of the earth's crust may be present in the following chemical forms:

(a) *Colloidal inorganic impurities.* The high concentrations of Na, Ca, Al, etc., which were spectroscopically detected in the ashes of bitumens from the Uinta Basin, Utah (Bell and Hunt, 1963), may be mainly of this type.

(b) *Inorganic substances absorbed on the carbonaceous phase.* It is possible that the concentrations within coal seams of Ge, Ga, and other metals belong to this group.

(c) *Organic complexes and atoms actually attached to the organic molecules* are of the greatest interest from our point of view. The organic geochemical behavior of elements, which enter the organic molecule, may be tentatively summarized as follows:

(1) *Biophyll elements* are forced into the organic molecule through the living organism, but they have no marked affinity toward the carbonaceous phase. Such are, for example, Mg of the chlorophyll of plants; Fe, of the haemoglobin of animals; and Cu, from the blood of certain lower sea animals. Some elements are concentrated by a single species only, as, for instance, in the enrichment to economically exploitable levels of Se in certain plants.

(2) *Biophyll and carbophyll elements,* These include the four major elements of the carbonaceous phase, that is, H,N,O, and S, although the latter tends to show more pronouncedly "carbophyll" affinities, as indicated by its gradual enrichment

with the aging of biogenic sediments (see above). P seems to be more pronouncedly biophyll. It appears that the organic compounds, which are stable under laboratory conditions, may decompose during geological time. At any rate, no P has been so far recorded in carbonaceous substances, with the exception of the usually young guanos.

(3) *Carbophyll elements.* Porphyrin complexes of V, Ni, and, to a smaller extent, Cu and Fe have been recorded by Dunning (1963) from oil shale extracts and other bitumens, and in petroleum by Whitehead and Breger (1963). These metals substitute for the Mg and Fe in the original biogenic compounds. The chemistry of Co, Sn, Mo, and numerous other heavy metals found in ashes from crude oil has not, so far, been determined. The presumably juvenile tucholites (see below) contain a high percentage of U, Th, rare-earth, and other metals. These may be partially organic or the well-known CO complexes but, unfortunately, the substances in question have not yet been studied in detail. It is suggested by Kronfeld (1963) that the U in marine black shales from North America is partially present as an organouranium complex, and partially as physically adsorbed oxides on the carbonaceous matter.

Finally there are certain "carbophyll" elements which do form stable synthetic compounds, but they have not been detected so far or isolated from any natural hydrocarbon. Thus Siever and Scott (1963) postulate the formation of silicon compounds in the course of silicification of wood, and possibly also in the course of certain biological processes, but the presence of these in any fossil hydrocarbon or recent biological product has not so far been proved. The absence or great scarcity of natural organic compounds of F, Cl, Br, and I, can be explained by the fact that the atmosphere and hydrosphere of our planet has been free, at least since the early stages of its geological history, from halogens in active form. The presence of organic Cl in certain carbonaceous meteorites is indicative of HCl or Cl_2 in the atmospheres of the parent bodies.

The above brief summary clearly indicates the deficiency of our present knowledge regarding both the distribution of trace elements, and their types of association with the carbonaceous phase. Detailed studies of these problems in the future may yield theoretically interesting data on the following aspects:

(a) Distinction between biogenic and abiogenic hydrocarbons.
(b) The presence of V and Ni porphyrins, and the absence of Mg porphyrins from geologically old crude oils, etc., indicates that the original biophyll Mg tends to be substituted for by more pronouncedly carbophyll elements over a rather long geological time period. The analysis of more recent sediments may enable us to estimate the rate of exchange, and even to establish a new method of age determination.
(c) Comparison of trace element contents of carbonaceous sediments may serve as an indicator of diffused hydrothermal activity within certain areas. For example, the author's view has been to interpret the increase of U content of sediments approaching a zone of granitization within the St. Juana area of Chile (Mueller, 1960), as indicative of escape of U-containing hydrothermal solutions from the underlying diorite intrusion.
(d) Comparison of affinities toward trace elements, of differing types of hydrocarbons. Here it was found (Mueller, 1960) that in veins with "differentiated hydrocarbon assemblages" from Derbyshire (see *Organic Mineraloids*) the uranium tends to concentrate in the lowest H/C and highest O/C phase.

Biogeochemistry

The above term is not as yet in general use, although it is a useful collective term for describing the study of the geological history of the individual types of biogenic molecules. These may remain partially preserved, partially degraded and may partially decompose to simple molecules in the course of aging of the sediments. The factors which determine these three procsses are (a) the type of molecule; (b) the age of the sediment; (c) certain specific factors which played a role in the geological history of the sediment in question, such as thermal metamorphism, etc.

The "biogeochemistry" of the most important types of organic molecules can be briefly outlined, as follows below.

Amino Acids. The probable long time scale stability at room temperature was experimentally proven for one of the twenty-one amino acids, namely Alanine (Abelson, 1963). Some of the original amino acid is preserved in the soils, although a high proportion becomes decomposed by the microorganism (Stevenson, 1956; see in Abelson, 1963). It was found (Shacklock and Drakey, 1927; see in Abelson, 1963), that both the diamino and monoamino acids are preserved in peat, lignite, and sub-bituminous coals, but bituminous coals and anthracite contain only the geologically apparently more stable monoamino acids. It was observed (Erdman, Marlet, and Hanson; see in Abelson, 1963), that an Oligocene marine shale from the Anahuav Formation of Fort Bent County, Texas, is depleted of amino acids by a factor of about six, in comparison with recent sediments from the Gulf of Mexico, although the CO_3^{2-}, organic C, and organic N percentages show little change, as seen in Table 2.

A considerable volume of literature deals with amino acids in individual recent species or their fossils. Divergencies in amino acid percentages and compositions have been observed in fossils of different shells and other marine organisms, which originate from the same bed. It was found (Ezra and Cook; see in Abelson, 1963), that in human bones, more than 5000 years old, many of the originally existing amino acids disappear, aspartic acid being the most persistent, but even this disappears in very ancient bones. The pattern of preservation of free and condensed amino acids may depend considerably on the conditions in which the fossil in question is preserved.

Carbohydrates. As in the case of the amino acids, the celluloses, lignins, and sugars, tend to diminish with the aging of closely comparable sediments, and there appear to be divergencies between the geological stabilities of the different individual molecules. For example, it was found, in the course of the preparation of a test bore-hole for the Mohole project (Degens; see in Vallentyne, 1963), that the percentage of sugars in the total organic C fell from 3.8 to 0.2% at depth from the bottom surface of the Pacific to 138 meters beneath it. There were considerable differences observed in the rate of diminution of the diverse sugars with the age of the sediment; thus, galactose showed a lower degree of geological stability than glucose, but a far greater volume of observations would be necessary to establish the precise stability order of sugars arising from certain specific localized conditions, and experimental procedures.

In the course of coalification of cellulose and lignin, the general trend appears to be a partial decomposition and a partial depolymerization. Thus, it was found by Reiff (see in Vallentyne, 1963) that, whereas a molecule of recent cellulose contains an average of 2–3000 glucose units, various Miocene lignites contained only from 40 to 800 glucose units. It was found by Staudinger and Jurisch (see in Vallentyne, 1963), that depolymerization of each type of cellulose proceeds to an apparent maximum at given temperatures, which would enable the diverse fossil celluloses to be used as geological thermometers. However, it is possible that in the course of geologic time other factors than temperature would also influence the degree of depolymerization. Carbohydrates were recorded from Precambrian rocks by Prokopom (1958; see in Vallentyne, 1963).

Lipids. According to the definition of Bloor, (quoted by Bergman, 1963), the term "lipid" is understood to include alyphatic fatty acids and their esters, alcohols and hydrocarbons, and terpinoids and steroids. All these water-insoluble substances of mainly fatty, and occasionally resinous, consistency have the common property of low O/H ratio which, in turn, renders them geologically more stable than the amino acids or carbohydrates, due to the fact that the molecule may remain virtually intact even after the elimination of all the water. According to some authorities, this explains why, of all the biogenic molecules, only the lipids and their degradation products produce major, essentially pure deposits, in the form of petroleum, their ages reaching back to the Paleozoic.

It was found by Abelson (1963) that the chromatograph of extracts of alum shales, Green River Shale, and other old and recent sediments still reveal the maxima of C-14, C-16, and C-18 chains, which are also of the greatest biological significance. A considerable range of lipid-type molecules was detected in seawater by L. M. Jeffery, et al. (1963), uniform distribution being found throughout the oceans; variations in depth have not been

TABLE 2. COMPARISON OF AMINO ACIDS IN RECENT AND OLIGOCENE SEDIMENTS

Content	Recent	Oligocene
Carbonate carbon (%)	0.71	1.49
Organic carbon (%)	0.53	0.27
Organic nitrogen (%)	0.044	0.032
Amino acids (M/g)	3.0	0.51

Principal amino acids in recent sediments: (order of decreasing abundance) valine, leucines, Alanine, glutamic acid, aspartic acid, glycine, proline, tyrosine, phenylalanine
Principal amino acids in Oligocene sediments: Alanine, glutamic acid, glycine, proline, leucines, aspartic acid

studied so far. The possible decarboxylation of biogenic lipids to paraffins and other hydrocarbons was extensively discussed in the problem of genesis of petroleum (Whitehead and Breger, 1963).

Pigments. It appears that the best preserved pigments in old sediments are the porphyrins (Dunning, 1963), which are present mainly as V, Ni, and Cu complexes, the original biogenic Mg and Fe complexes being gradually replaced by the former metals as shown above.

The presence of flavinoid pigments in peats and Quaternary lake sediments was spectroscopically demonstrated by Swain and Venteris (1963).

The occurrence of numerous types of the less stable, or less abundant, pigments has not been proved as yet in any old sediments. Such are, for example, the carotenoids, the anthocyanins, and a considerable number of biological pigments of sporadic distribution.

The pinpointing of certain pigments, characteristic for a given species, has considerable theoretical interest, as this would enable us to trace their degradation in geologic time, under diverse geological conditions. The claim of Blumer (1951; quoted by Dunning, 1963) of separation of alkyl dibenzoperylene and related fossil hydroxyquinoline dyes from Mesozoic sea lilies is of interest in this respect.

The above is, of necessity, a most sketchy outline of our present volume of data dealing with biogenic molecules in recent and old sediments and their degradation products.

Table 3 (from Shabarova 1955; quoted by Vallentyne, 1963) gives some information concerning the abundance of the main types of biogenic molecules in different sediments.

Changes in Constituents of Carbonaceous Complexes

The foregoing is a brief summary of our present-day knowledge regarding the distribution of and changes (mainly degradation) in molecules, which are generally considered as typically biogenic. Very little work has been done so far regarding the analysis of alteration of constituents of biogenic or abiogenic carbonaceous phases in general. From the point of view of what we may term "analytical convenience" we may group these "constituents" as follows; (a) individual organic molecules; (b) types of organic molecules (paraffins, alkenes, aromatics, etc.); (c) groups (hydroxyl, carboxyl, ester, amine, etc.).

One of the aims of the organic geochemist is to reconstruct the "history of alteration" of constituents of a given carbonaceous complex, from the stage of its final settling (sedimentation, condensation, etc.). From the point of view of "geochemical behavior" the constituents of the carbonaceous complexes can be grouped, as a first approximation, as follows:

(1) Stable constituents. These tend to remain chemically unaltered under the main range of reducing and relatively low-temperature conditions within the earth's crust. These include, in order of presumed decreasing stability, the paraffins, the alkenes, olefines, aromatics, and organic acids. It appears that the C—H, C—C,

TABLE 3. PROPORTION OF INDIVIDUAL COMPONENTS IN THE ORGANIC MATTER OF RECENT SEDIMENTS[a]

Sampling Site	Sediment Type	Total Organic Compounds Determined	Lipids	Protein	Carbohydrates	Unhydrolysable Residue	Total Hydrolysable Compound
Continental shelf, eastern U.S.S.R.							
988-m depth	AD	4.57	4.38	6.96	0.74	87.97	12.03
1497-m depth	AD	8.48	5.91	7.75	traces	87.34	12.66
Black Sea							
Costal mud	G	7.37	7.75	56.31	10.83	24.69	75.31
50-m depth	AG	3.89	2.55	9.41	0.00	88.04	11.96
220-m depth	AB	4.89	2.33	10.73	0.00	86.84	13.16
294-m depth	AB	8.05	2.09	3.17	0.00	94.74	6.26
Caspian Sea							
18-m depth	AS	4.98	8.70	7.48	1.03	82.79	17.21
18-m depth	AG	4.56	7.68	9.43	1.54	81.35	18.65
White Sea							
90-m depth	G	2.41	5.60	11.15	5.25	78.00	22.00
Kura river delta							
advance delta	AG	4.69	2.15	7.88	0.00	89.97	10.03
Lake Biserovo							
unstated depth	B	35.77	6.90	34.85	33.25	25.10	74.90

[a] Expressed as a percentage of total organic compound found. A= argillaceous; B= black mud; D= diatomaceous; G= grey mud, S= silty.

C=C bonds, and possibly also the COOH, C—S, and C=S bonds, are of relative high geological stability.

(2) Metastable constituents. These show a trend of gradual microbiological and chemical alteration in geologic time toward a stable constituent. These include all the pronouncedly biogenic molecules, and alcohols, esters, amines, etc., containing C—OH, —CO, —COH, —CH$_2$OCO, C—NH$_2$, etc., groups.

(3) Labile constituents. These decompose under average conditions within a short time, although they might be preserved as inclusions in minerals in the case of juvenile condensates, or in a hermetically closed biogenic phase, like amber. In the abiogenic range the substances in question may include various organometal complexes, whereas compounds of biogenic origin may cover a considerable range of pigments, enzymes, vitamins, etc.

Exposure to localized conditions, such as diverse types of metamorphism, oxidizing conditions along the land surface, excessive microbiological activity, etc., would, of course alter (and diminish) the stabilities of the diverse molecule types and groups. The detailed effects of each of these "localized agencies" of alteration have not been systematically studied so far for a really representative range of conditions. It was found by the author (Mueller, 1963, quoting his work in 1951, 1954, and 1960) that within the Derbyshire orefield of England, in areas with a semianthracitic carbonaceous substance in the lower Carboniferous shale, the mainly calcitic veins contain a highly carbonized, mainly aromatic hydrocarbon, "carbonite." With the presumably higher temperature, fluoritic veins, however, apart from the mainly aromatic mineral, a mainly olefinic one, "Bernalite," and paraffinic oils and waxes (ozocerite), are associated; the side veins surrounding these contain elaterite and mutabilite, which are low-boiling, highly oxygenated organic mixtures. It appears that the originally aromatic coal tar-type distillates became hydrogenated under conditions of higher temperature mineralization, and the resulting high-molecular-weight products subdivided into three unmixable liquids of mainly aromatic, olefinic, and paraffinic constitution, the high volatile, oxygenated molecules escaping into the country rock surrounding the vein. In districts of the Derbyshire orefield, where the shales contain a low-grade kerogen, undifferentiated asphalt and asphaltite-type substances are accumulated in the veins.

Description of Organic Minerals

The term "organic mineral" can be defined as an assemblage of organic molecules with only subordinate (less than, say, 20%) of inorganic contaminants. Every given organic mineral is a homogeneous, or at least quasi-homogeneous, unit of the "carbosphere." Whereas the inorganic minerals are, by definition, composed of a single type of molecule, or an isomorphous mixture, with subordinate impurities, the majority of the "organic minerals" are mixtures of many molecules. In order to stress this difference, the alternative terms "organic mineraloid," or "bitumen" or "hydrocarbon" are used in the literature to denote all or certain types of organic substances of geological age.

"Organic mineralogy" is the branch of organic geochemistry which deals with the problems of classification, description, and genesis of the "organic minerals." In the present encyclopedia, the most important organic minerals are described under appropriate headings, which also contain discussions of the problem of their genesis (see Vol. VI, *Asphalt; Coal; Kerogen; Natural Gas; Petroleum;* etc.). A systematic survey of the theories of genesis of the entire range of organic minerals, including the rarer varieties, is given in the article entitled *Organic Mineraloids.*

GEORGE MUELLER

References

*Abelson, P. H., 1963, in (I. A. Breger, editor), "Organic Geochemistry," Pergamon Press, pp. 431–455.
Abelson, P. H., Hoering, T. C, and Parker P. L., 1963, in (Colombo, U. and Hobson, G. D., editors), "Advances in Organic Geochemistry," Pergamon Press, pp. 169–174.
Bell, K. G., and Hunt, J. M., 1963, "Organic Geochemistry," pp. 333–366.
Bergmann, W., 1963, "Organic Geochemistry," pp. 503–542.
Brock, T. D, 1967, "Bacterial growth rate in the sea: direct analysis by thymidine autoradiography," *Science,* **155**(3758), 81–83.
Clayton, R. K., 1965, "The biophysical problems of photosynthesis," *Science,* **149**(3690), 1346–1354.
Colombo, U., Gazzarrini, F., Sironi, G., Gontiantini, R., and Torgiorni, E., 1965, *Nature,* **205,** 1303.
*Dunning, H. N., 1963, "Organic Geochemistry," pp. 376–430.
Fox, W. S., 1963, "Organic Geochemistry," pp. 36–49.
Jeffery, L. M., Pasby, B. F., Stevenson, B., and Hood, D. W., 1963, "Advances in Organic Geochemistry," pp. 175–198.
Kronfeld, J. A., 1963, "Advances in Organic Geochemistry," pp. 261–262.
Manskaya, S. M., and Drozdova, T. V., 1968, "Geochemistry of Organic Substances," Oxford, Pergamon Press, 345pp (transl. from the Russian).

Mueller, G., 1963, Proc. VI. Int. Petroleum Congress, Frankfort, Sect. 1, pp. 383–396.
Mueller, G., 1960, Int. Geol. Cong. XXI. Ses., Norden, Part 15, pp. 123–132.
Mueller, G., 1963, *Nature*, **198**, 731.
Parker, P. L., Van Baalen, C., and Maurer, L., 1967, "Fatty acids in eleven species of blue-green algae; geochemical significance," *Science*, **155**(3763), 707–708.
Rankama, K., 1954, "Isotope Geology," New York, McGraw-Hill Book Co., pp. 151–160; 181–270, 276–289.
Siever, R., and Scott, R. A., 1963, "Organic Geochemistry," pp. 579–595.
Swain, F. M., and Venteris, G., 1963, "Advances in Organic Geochemistry," pp. 199–214.
*Vallentyne, J. P., 1963, "Organic Geochemistry," pp. 456–502.
Whitehead, W. L., and Breger, I. A., 1963, "Organic Geochemistry," pp. 248–332.

*Additional bibliographic references may be found in this work.

Cross-references: *Carbon; Earth's Crust Geochemistry; Hydrocarbons; Hydrogen; Isotope Fractionation; Isotope Geology; Nitrogen; Organic Mineraloids; Photosynthesis; Sulfur; Trace Elements. Vol. II: Carbonaceous Meteorites; Cosmic Dust; Cosmochemistry—Organic. Vol. IVB: Green River Mineralogy. Vol. VI: Asphalt; Coal; Kerogen; Natural Gas; Petroleum.*

ORGANIC GEOCHEMISTRY OF SEAWATER

As on land, animal and plant life is not uniformly distributed; some areas of the ocean are marine deserts (for example, the Sargasso Sea) and other areas show a profusion of plant and animal life. Of course, physical factors and inorganic nutrients, such as phosphorus, nitrogen, and silicate, profoundly influence the distribution and abundance of marine life, but the observation that in areas of similar physical properties and inorganic composition, a dense plankton bloom exists in one region while a similar adjacent one may be almost devoid of life seems to point to other critical factors, such as organic growth factors, toxins, trace elements, but probably to a combination of these. In support of this view, many plant nutritionists have found that marine algae grown in the laboratory require vitamins and other organic substances as well as certain trace elements for growth.

It has also been suggested that succession of plankton species in a given water mass may be due to organisms using and excreting various organic growth-promoting or inhibiting substances. Lucas (1955) not only emphasized the significance of these so-called "organic metabolites" on the distribution and growth of marine algae, bacteria, fungi, protozoa, and larvae but focused attention on the effects of these materials on the physical responses of marine worms, echinoderms, molluscs, crustaceans, and fish. According to reviews by Johnston (1955), Saunders (1957), Vallentyne (1957), Provasoli (1963), and Hood (1963), dissolved organic components are important biologically from at least five viewpoints: (*1*) organic compounds, even in very dilute solution, may serve as sources of energy for algae, bacteria, fungi, and larvae; (*2*) vitamins and auxins serve as growth stimulators for algae, bacteria, fungi, and larvae; (*3*) organic toxins can kill entire populations; (*4*) the presence of certain organic compounds in seawater can trigger feeding of oysters and other organisms, even though the compounds themselves may not be used for nutrition; and (*5*) nutrient or toxic trace elements may be chelated or "solubilized" with organic compounds, thus indirectly influencing the growth of organisms.

Although most of the investigations made on organic solutes in seawater have been stimulated by biological interests, these organic solutes may have geochemical significance as well. For instance, it has been suggested that proteins, peptides, and organic acids may increase the solubility of calcium carbonate and phosphate in seawater. Consideration of the known physical chemistry of calcium and carbonate components alone is inadequate to explain the dynamic relationships of this buffer system in the marine environment.

In experimental studies on the origin and migration of petroleum, it was noted that the solubility of crude oil in water is increased by naturally occurring high-molecular-weight organic acids. Organic acids adsorbed or occluded to particles suspended in seawater which subsequently settle to the bottom may eventually participate in the transport of petroleum. At present, this concept is pure speculation. However, organic matter of seawater does adsorb in part to clay and sand particles and changes the physical properties of the particles to a certain extent. It has been shown that organic matter absorbed to pure sand lowers its density, so that the grains are more easily transported than before.

There are also experimental and observational evidences that organic solutes in seawater may cause sea slicks in areas of alternate turbidity and calm. Air bubbled through previously filtered seawater caused aggregation of very fine, surface-active particles on the surface to form a slick of reduced surface tension after cessation of the bubbling. Field observations seemed to confirm the laboratory experiments. These studies, indicating the possibility of the formation of organic particles from solutes, are of considerable biological significance since some of the solutes are concentrated on the surface in the form of particles that are then available to filter-feeders. The chemical composition of these colloidal micelles has not yet been determined.

Total Organic Carbon, Nitrogen, and Phosphorus in the Oceans

Organic matter in seawater is divided into particulate and dissolved on the basis of filtering through 0.45-micron millipore filters with the soluble passing through the paper and the particulate remaining on the filter. The soluble organic matter is the subject of this discussion.

The quantity of dissolved organic carbon exceeds that of organic particulate by a factor of 7–8 in waters rich in phytoplankton and, in deep water, by a factor of 1000. Table 1 summarizes the distribution of total soluble organic carbon, nitrogen, and phosphorus in the world oceans. The amount of total carbon in the open ocean varies generally from 0.2 to 2.7 mg/l of carbon. A factor, at present inaccurately known, must be used to multiply the organic carbon values to obtain concentrations of organic material. Higher values are found in landlocked areas: 3.3 mg/l carbon in the Black Sea, 4.6 in the Baltic, 6.0 in the Sea of Azov, and 8.0 in the landlocked coastal area of the Dutch Wadden Sea. Organic nitrogen concentrations are in the range 0.03–0.60 mg/l of nitrogen, and phosphorus concentrations 0.0–29.0 mg/m^3. The distribution of the organic solutes is not homogeneous either horizontally or vertically. However, there is a minimum of organic matter in the oxygen minimum layer. The highest organic carbon concentrations are in water masses of high phytoplankton productivity or recent productivity.

The soluble organic carbon/nitrogen ratio in the sea can often be used to characterize local water masses. The average carbon/nitrogen ratio for seawater is 2.7, whereas for fresh river and lake water, the ratio is 12.6. This ratio in living plankton has a value of 6.0, so it is thought that in seawater carbon is preferentially consumed and in fresh water nitrogen is preferentially consumed. Since the bacterial flora of seawater is quite different from fresh water, decomposition proceeds differently to yield a higher proportion of nitrogen-containing compounds in seawater.

The possible sources of the organic solutes in seawater are: (1) decomposition of dead marine organisms; (2) excretion of these materials by living organisms; (3) run-off from land; and (4) diffusion from sediments on the sea floor during periods of turbidity. Duursma (1960) from his extensive studies of total organic matter in many oceans presented evidence that decomposition of dead organisms is the primary source and that excretion of soluble organics by living organisms is not directly

TABLE 1. SUMMARY OF ORGANIC CARBON, NITROGEN, AND PHOSPHORUS DISTRIBUTION IN THE OCEANS[a]

Locality	Organic Carbon (mg/l)	Organic Nitrogen (mg/l)	Organic Phosphorus (mg/m^3)
North Atlantic	0.20–1.30	0.04–0.40	0.0– 9.0
North Atlantic	1.04–1.97	—	—
Subtropical Atlantic	—	—	2.0– 9.0
Atlantic (near Bermuda)	2.28–2.42	0.16–0.32	—
Atlantic	2.40–2.48	0.24–0.26	1.0–21.0
Pacific	0.6 –2.7	—	—
Pacific	0.98–2.68	0.07–0.11	—
North Pacific	—	0.09–0.32	—
North Sea	0.50–1.80	0.08–0.54	0.0–19.0
North Sea	2.40–4.16	—	—
Baltic	3.52–6.63	—	—
Baltic	2.0 –4.6	—	—
Wadden Sea	1.0 –8.0	0.10–0.60	6.0–27.3
Norwegian Sea	0.45–1.38	0.10–0.21	—
English Channel	—	—	2.0– 8.0
Black Sea	2.83–3.36	—	—
Caspian Sea	12.0	—	—
Sea of Azov	4.63–6.02	—	—
White Sea	2.3 –4.3	—	—
Greenland Sea	2.0 –2.1	0.03–0.38	0.9–29.0

[a] Values given for the same ocean twice were determined by independent investigators. (After Duursma, 1960; and Hood 1963).

obvious in the sea, since the maximum dissolved carbon value occurs after a plankton bloom. In some areas of high run-off from rivers, the presence of fresh-water organic solutes is clearly evident. Flocculation of river-borne organics in seawater is a prolonged process. It has been estimated that 5–36% of fresh-water organic carbon precipitates only after 200 days. The contribution of sediments to the dissolved organic fraction is unknown, but except in rare cases in shallow water areas, it seems to be of minor importance.

Chemical Nature of Marine Organic Compounds

Data on specific organic compounds in seawater are few and limited, yet it is evident that seawater is a very dilute solution of hundreds of organic compounds. However, the relative proportions of organic classes, such as proteins, amino acids, carbohydrates, lipids, humic acids, vitamins, and other trace growth factors have not yet been determined for a given water sample. A summary of the organic solutes which have been identified is given in Table 2 with concentrations whenever available.

Proteins and Amino Acids. Because of the very low concentration and the difficulties involved in separating amino acids, peptides, and proteins from seawater containing an average of 36 g of salt per liter, very few determinations have been made, and even these are not strictly quantitative and represent only minimal values.

By means of a microbiological assay method, isoleucine in the free state was detected in 50% of the Pacific samples tested, but threonine, tryptophan, glycine, methionine, histidine, and arginine appeared only sporadically. Cystine, proline, and leucine did not occur in any of the Pacific samples tested. In deep water of the Gulf of Mexico, estimation of free and combined amino acids were made using standard chemical methods. Amino acids in the combined form predominated by a factor of three over the free amino acids in the samples tested. Eighteen amino acids in the combined form were determined, and eleven in the free form. The concentrations of total free and combined amino acids at four depths in deep water in the Gulf of Mexico ranged from 12 to 38 mg/m^3 with some variation with depth. A minimum value occurred at the oxy-

TABLE 2. DISSOLVED ORGANIC COMPOUNDS REPORTED AS IDENTIFIED IN SEAWATER[a]

Organic Substances	Locality	Concentrations
Free and combined amino acids	Gulf of Mexico	12–38 mg/m^3
Amino acids (proteins)	Gulf of Mexico and Caribbean	Traces to 13 mg/m^3
Free isoleucine, glycine, threonine, tryptophan	Pacific Ocean	Present
Carbohydrate (as arabinose equivalents)	Gulf of Mexico (nearshore water)	0.0–100 mg/l
Carbohydrate (as sucrose equivalents)	Pacific Coast, U.S.A.	0.14–0.45 mg/l
		0.0–2.6 mg/l
Carbohydrate (as arabinose equivalents)	South Atlantic (30°N to 25°N)	Max. value of 12 mg/l
Carbohydrate (as arabinose equivalents)	Gulf of Mexico Continental Shelf	0.0–3.0 mg/l
		50 mg/l in red tide
Dehydroascorbic acid	Gulf of Mexico	Present
Citric acid	Atlantic (off French Coast)	0.025–0.145 mg/l
Malic acid	Atlantic (off French Coast)	0.028–0.277 mg/l
Acetic, formic, lactic, and glycolic acids	Northeast Pacific (surface and inshore)	<0.1 mg/l
Fatty acids	Gulf of Mexico	0.4–0.8 mg/l
Fatty acids	Pacific	0.014–0.12 mg/l
Carotenoids and brown fatty matter	English channel	2.5 mg/l
Complex lipids	Gulf of Mexico (nearshore and offshore)	1.0–8.0 mg/l
Vitamin B$_{12}$	See Table 3	0.0–200 millimicrograms/liter
Biotin	Pacific	Present
Niacin	Pacific	Present
Uracil	Pacific	Present
Purines	Pacific	Present
Plant hormones	North Sea	Present
Volatile sulfur-containing compounds	English Channel	20–50 µg/l

[a] After Provasoli (1963) and Hood (1963).

gen minimum layer and a maximum value just below the minimum layer. The amino acid content with depth was also quite variable. There was also some evidence of free peptides in these samples.

Carbohydrates. Total carbohydrates have been estimated probably more often on more samples than any other organic constituent of seawater except vitamin B_{12}, yet it is not yet known which specific carbohydrates are present. Colorimetric methods have been used with an arbitrary monosaccharide as an artificial standard. These colorimetric methods are not strictly specific for true carbohydrates. In nearshore waters of the Gulf of Mexico, up to 100 mg/l of carbohydrate have been reported, but in areas in an estuary of the Gulf, concentrations ranged from 0.0 to 20 mg/l. Maximum concentrations in Pacific nearshore water approached 7.9 mg/l, but concentrations offshore were in the range 0.1–1.0 mg/l.

This organic fraction reacting with carbohydrate reagents may be algal in origin, since the filtrates of several cultures of marine flagellates and diatoms contain organic solutes reacting with the carbohydrate reagents. These carbohydrates are biologically active, for they affect the pumping rate of oysters. Aside from this, little is known of the significance of these materials.

Lipids and Organic Acids. Since organic acids and lipids are more resistant to biological attack than are the simple amino acids, proteins, and carbohydrates, it is not too surprising to find them in appreciable amounts in seawater, despite the low solubility of lipids in water. Several independent investigations have shown the presence of a number of organic acids. Citric and malic acids were found in the littoral water off Southern France at concentrations of 0.1 mg/l. In Pacific surface water near Washington State, acetic, formic, lactic, and glycolic acids were reported at concentrations of less than 1 mg/l. Since the extraction of the acids with diethyl ether lasted several weeks, the investigators believe that some of this material may have been the result of decomposition of more complex compounds. A mixture of unidentified hydroxylated carboxylic acids with an average molecular weight of 395 was reported by another investigator for some pelagic Pacific water. High-molecular-weight organic acids were also found in the Gulf of Mexico. Their buffer capacity was determined, but actual concentration in either case was not reported.

Lipids, as defined here, are organic compounds preferentially soluble in fat solvents, such as hexane, chloroform, or ethyl acetate. Lipids can be removed from seawater by acidification and extraction with these solvents. Free and combined fatty acids were found by independent investigators in both the Pacific and Gulf of Mexico. In both sets of samples, saturated and unsaturated fatty acids of chain lengths varying from C_{10} to C_{22} were reported in total concentration levels of 0.01–0.12 mg/l in the Pacific and 0.1–0.8 mg/l in the Gulf of Mexico. In both cases fatty acid content and total fatty acid concentration were variable horizontally and vertically.

Subsequently, lipid residues from the Gulf of Mexico have been fractionated on silicic acid columns into several classes of lipids, including hydrocarbons, sterol esters of fatty acids, sterols, free fatty acids, complex glycerides of fatty acids, phospholipids, and several unidentified compounds. In the lipid fraction alone, there are hundreds of compounds. There are at least thirty distinct hydrocarbon compounds, all present in very low dilutions. The constituents and concentrations of the various lipid classes varied somewhat vertically and horizontally in the samples analyzed. Concentrations of the total lipid residue, exclusive of the small solvent residue, ranged from 1 mg/l of total organic material in pelagic waters to as much as 8 mg/l in near-shore productive water in the Gulf of Mexico.

Vitamins and Growth Factors. Vitamin B_{12}, thiamine, biotin, and niacin are indispensable for the growth of many marine algae and bacteria. Thus, these growth stimulators should be present in seawater, and a large number of microbiological assay measurements show that they are indeed present in varying concentrations, usually but not always, in sufficient concentrations to support growth. The assay for vitamin B_{12} is a rather complex one, since vitamin B_{12} activity may represent a number of similar compounds and not simply cyanocobalamin. Fortunately, microorganisms vary in the types of B_{12} compounds to which they respond, so that by selection of the proper combination of organisms, more or less complete characterization of the vitamin B_{12} activity of seawater may be obtained.

For a summary of vitamin B_{12} distribution in the oceans, see Table 3. Assays for vitamin B_{12} in the North Atlantic and northern North Sea show a seasonal variation in concentration in surface samples, i.e., values of 0.1 millimicrograms/l in summer in 1.0 millimicrograms/l in autumn. There was marked variability with depth in the Bay of Biscay, with the lowest values in the euphotic zone (zone of plankton production). A concentration range in a 3600-m cast of 0.3–4.0 millimicrograms/l was reported for the Bay of Biscay. In Long Island Sound a similar

TABLE 3. QUANTITY OF VITAMIN B_{12}-LIKE COBALAMINS IN SEAWATER AS DETERMINED BY MICROBIOLOGICAL ASSAY[i]

Locality and Type of Water	Assay Organism	Concentration Range (millimicrograms/liter)
Vineyard Sound, Woods Hole (coastal)	*Euglena gracilis*[a]	30–200
Northwest Arm, Halifax N.S. (coastal, polluted)	*Stichococcus sp.*[b]	10
Pier at Millport, Scotland (coastal)	*Monochrysis lutheri*[c]	5–10
Aberdeen Bay, Scotland (coastal)	*Ochromonas malhamensis*[e]	4
	Lactobacillus leichmannii[d]	6
Northern North Sea	*L. leichmannii*[f]	0.1 (Aug.)–1.2 (Oct.)
	O. malhemensis	
Butt of Lewis	*L. leichmannii*	0.4 (Apr.)–2.0 (Feb.)
	O. malhemensis	0.5 (Aug.)–2.0 (Apr.)
Norwegian Deeps	*L. leichmannii*	2.26 (mean of 34)
	O. malhemensis	
Bay of Biscay	*Euglena gracilis*	0.57 (av. of 7)
0–30°N along 130°E (N. Pacific) (12 stations)	*Euglena gracilis*	0.3–1.15 (0–100 m)
		0.5–2.5 (below 200 m)
Kagoshima Bay	*Euglena gracilis*	3.2 (av. of 34)
Bahia Fosforescente (Puerto Rico)	*Escherichia coli*[g]	3.0–3.5 (inside bay)
		1.3 (outside bay)
Long Island Sound (coastal)	*Thraustochytrium globosum*[h]	4.5–11.4 (after spring diatom bloom)
		12–14.6 (winter)
Sargasso Sea (upper 50 m)	*Cyclotella nana*	0–0.03 (May to Oct.)
		0.03–0.1 (Nov.–Apr.)

[a] Responds to true B_{12}, pseudo-B_{12}, and factors A,C,C_2.
[b] The specificity of *Stichococcus* toward various cobalamins is unknown.
[c] Responds to true B_{12}, pseudo-B_{12}, and factors A, I and H.
[d] Responds to true B^{12}, pseudo-B^{12}, factors A, I, and also desoxyribosides.
[e] Responds only to true B_{12} and factor I.
[f] The values obtained with the two assay organisms are similar, i.e., true B_{12} may be the major cobalamin (80–90%) in these waters.
[g] Responds to true B_{12}, pseudo-B_{12}, all factors including factor B, and to methionine.
[h] Responds only to true B_{12}.
[i] After Provasoli (1963).

seasonal variation of 4.5–12 millimicrograms/1 was found. In the Pacific near Japan, variability in B_{12} concentrations to the extent of 0.0–26.3 millimicrograms/1 were found. The occurrence of B_{12} in the Sargasso Sea was measured for one year using the diatom *Cyclotella nana*. The quantity of B_{12} in waters above 50 m varied from undetectable to 0.03 millimicrograms/1 from May to October. This lack of the vitamin seemed to control the species of plankton present. The dominant organism was *Coccolithus huxleyi* which does not require B_{12}, but thiamine. Diatom blooms occurred in April after the level of B_{12} had increased to 0.06–0.1 millimicrograms/1.

Other vitamins that have been determined by microbiological assay include thiamine, niacin, and biotin. These methods for these vitamins are not yet strictly quantitative, but allow good rough estimations. In 55% of Pacific water tested, biotin was present in variable concentrations. Niacin was found to be present in 50% of some near-shore samples in the Pacific. Microbiological assays sensitive to 25 millimicrograms/1 of thiamine in Long Island Sound water showed that detectable quantities were not often found in the Sound, but near-shore samples enriched by land drainage showed relatively high concentrations (63–65 millimicrograms/1 at a river outfall and 30–40 millimicrograms/1 near Charles Island).

Phytohormones, biologically active on *Avena* coleoptiles, were found in marine phytoplankton and zooplankton samples as well as in a chloroform extract of seawater from the North Sea.

Toxins. Several genera of flagellates are known to produce toxins which kill other marine organisms, sometimes on a mass mortality basis. Others produce a neurotoxin which does not affect most marine organisms but is toxic to man. The "red tide" forms are *Gymnodinium* sp., *Gonyaulax* sp., *Cochlodinium catena-*

tum, Exuviaella baltica, Pyrodinium phoneus, Thalassiosira decipiens, Horniella marina, and a brackish water organism *Prymesium parvum.* The toxins are generally thermolabile, nondialyzable, readily adsorbed, and sensitive to pH and oxidation. It has been shown that the toxin of *Prymesium parvum* is probably a lipid, since it is soluble in methanol and behaves chromatographically like one. Its nondialyzability indicates that it may exist as a micellar structure in solution. Acrylic acid, present in the algae *Phaeocystis,* inhibits growth of microflora in the intestines of penguins and is also present in seawater. However, its effects on other organisms are not known at this time.

The poison of *Gonyaulax catenella* is not toxic to marine organisms, but it is a very potent poison to mice and men. A dosage of 0.18 μg is lethal to a 20-g mouse in 7 to 15 min; the lethal dose for man is 3–4 mg. This neurotoxin is a strongly basic nitrogenous compound of low molecular weight (372), yielding under various oxidations and hydrolysis guanidopropionic acid, guanidine, urea, ammonia, and carbon dioxide. Shellfish concentrate the poison from a bloom of *Gonyaulax catenella* in the siphon or digestive glands and thus becomes poisonous to man if he eats the contaminated shellfish.

Miscellaneous Compounds. Some sulfur-containing volatile constituents have been reported in English Channel seawater at the level of 20–50 μg/l of carbon and 1.5–4.0 μg/l of sulfur. The compounds were noted in the distillates of seawater and showed strong ultraviolet absorption. Similar behavior was shown in distillates of culture media of actively growing bacteria-free algae. Should these volatile compounds include dimethyl sulfide, which is known to form complexes with heavy metal salts, such elements as copper and mercury may be complexed with the volatile organics in seawater.

Evidence has also been presented that a considerable portion of copper, maganese, and zinc is also bound in some manner with the organic solutes of seawater. Organic compounds containing iron and boron are also assumed to be present on the basis that analyses for these elements are higher when seawater has been strongly oxidized than for untreated samples.

Other compounds shown to be present in seawater are uracil and purines. In microbiological assays of near-shore water samples in the Pacific, uracil was present in 30% of the samples and purines in 2%. In offshore samples uracil appeared in 25% of the samples, whereas purines appeared in only one sample out of forty five. The purine mutant responded to any one of the four purine bases (adenine, guanine, xanthine, or hypoxanthine).

A carbon-protein complex (possibly a pectin or uron) was found to be the end-product of complete decomposition of marine organisms, and is thus thought to be present in seawater. The carbon/nitrogen ratio of this decomposition end-product is about 10, whereas the average value for the ratio of the organic matter in seawater is 2.7, so it appears that this end-product is not a major constituent of seawater.

LELA M. JEFFREY

References

Duursma, E. K., 1960, "Dissolved organic carbon, nitrogen, and phosphorus in the sea," *Neth. J. Marine Res.,* **1**, 1–148.

Hood, D. W., 1963, "Chemical oceanography," *Oceanogr. Mar. Biol. Ann. Rev.,* **1**, 129–155.

Johnston, R., 1955, "Biologically active compounds in the sea," *J. Marine Biol. Assoc. U.K.,* **34**, 185–195.

Lucas, E. E., 1955, "External metabolites in the sea," *Suppl. Deep-Sea Res.,* **3**, 139–148.

Provasoli, L., 1963, "Organic regulation of phytoplankton fertility," in (Hill, M. N., editor), "The Sea," Vol. 2, "The Composition of Sea Water. Comparative and Descriptive Oceanography," New York, Interscience Publ., pp. 165–219.

Saunders, G. W., 1957, "Interrelations of dissolved organic matter and phytoplankton," *Botan. Rev.,* **23**, 389–409.

Vallentyne, J. R., 1957, "The molecular nature of organic matter in lakes and oceans, with lesser reference to sewage and terrestrial soils," *J. Fisheries Res. Board Can.,* **14**, 33–82.

Cross-references: *Buffer Systems; Carbon-Nitrogen Ratio; Cycles—Geochemical; Hydrocarbons; Organic Geochemistry; Seawater: Chemistry.* Vol. I: *Mass Mortality; Nutrients in the Sea; Phytoplankton.* Vol. VI: *Crude Oil; Petroleum.*

ORGANIC MINERALOIDS

An "organic mineraloid" can be defined as an assemblage of organic molecules with only subordinate quantities of inorganic contaminants, termed "ashes." The terms "organic mineral," "bitumen," and "hydrocarbon" are used as alternatives in the literature. Each "organic mineraloid" is a homogeneous or rather quasi-homogeneous unit of the "carbosphere" which was described from the point of elemental and molecular composition and genesis in the entry *Organic Geochemistry.*

Classification of Organic Mineraloids

The systematical classification of organic mineraloids is greatly hampered through our

ORGANIC MINERALOIDS

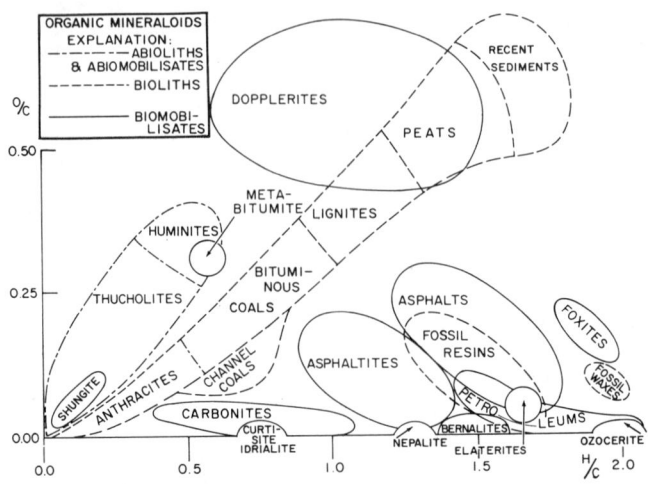

FIG. 1. Chart of organic mineraloids, relating oxygen/carbon ratio and hydrogen/carbon ratio. (Correction: for "channel coals," read "cannel coals.")

lack of knowledge of any one of the suitable properties for the entire range of the terrestrial carbosphere. The hitherto proposed classifications are based on the properties shown below.

Elemental Composition. The system of Owen, quoted by Breger (1963) groups the organic mineraloids according to H% and (O+S+N+H+C)%. It includes an attempt to combine the data of elemental composition with genesis through the derivation of the latter from processes of decarboxylation and dehydration of celluloses. This is to a certain extent true for the coal series, but it is certainly not applicable to other groups of organic mineraloids. Fig. 1, which summarizes the elemental composition of organic mineraloids, is based on the H/C and O/C ratios of each one of the main groups.

Molecular Composition. Our knowledge of the data in question is as yet too incomplete to form a basis of a systematic classification, particularly in the case of the rarer varieties of organic mineraloids. In view of the heteromolecular nature of most of the substances in question, a classification based on individual molecules would be impracticable; however, some advances can be made toward a system based on "dominant molecule types," that is, distinguishing mainly paraffinic, naphthenic, olefinic, aromatic, heterocyclic, lignin, lipid, resin, etc., classes, and also substances which contain some or all types of organic molecules as major constituents or in approximately even proportions. A distinction into substances with predominant or characteristic functional groups such as olefinic, aromatic bonds, OH−, COOH− would lead to results somewhat similar to the above (see Mueller, 1954).

Various Physical and Chemical Properties. The classifications based on what we may term as secondary properties have the disadvantage of occasional inclusion into the same class of substances of fundamentally differing chemical structures; on the other hand, some of them are based on data which are at present known for a greater number and variety of organic mineraloids than chemical data. Abraham (1945, quoted by Bell and Hunt, 1963), distinguished CS_2-soluble *bitumens* and CS_2-insoluble *"pyrobitumens."* The former subdivide into liquid petroleum, fusible mineral waxes and asphalts, and not readily fusible asphaltites; the pyrobitumen are divided into a relatively oxygen-poor class, "asphaltic pyrobitumen," and an oxygen-rich class of "non-asphaltic pyrobitumen," that is, coals. A handy classification through tests feasible in the field was made by Deul and Breger (quoted by Breger, 1963), which consists of distinction between material soluble and insoluble in a mixture of 1 part benzene, 1 part alcohol, and 1 part carbon disulfide: asphalts, asphaltites, oil, coal, and metamorphosed asphaltic material, respectively.

Genetical Classification. The main difficulties in the use of this method are that the mode of formation of a considerable number of organic substances is still being discussed, or even unknown; further, that two or more distinct processes may produce similar assemblages of organic molecules through coincidence. Thus, for example, the possibility of existence of abiogenic, hydrothermally distilled, and primary petroleums was discussed by the writer (in a paper read before the XXIInd International Geology Congress, Delhi, 1964, which is published in the Abstracts of the Congress).

The present-day genetical classifications are based chiefly on the nature of the organic debris from which the carbonaceous matter in question formed. Thus, Down and Himus (1940; quoted by Breger, 1963), distinguish (a) *kerogen rocks,* consisting of the organic sediments of essentially algal sapropelic origin, which include the oil shales, and according to certain theories, also oil, asphalts, and gas, (b) *kerogen coals,* which yield oil on distillation, and are mostly mixtures of algal matter with rafted pollen and plant debris, thus representing sedimentation in coastal areas or in fresh water; (c) *coals,* which contain plant matter exclusively. For reaosns which are discussed below, this classification cannot be extended to the entire range of organic mineraloids.

The scheme of the writer (1954, 1963b, 1970 partly unpublished) distinguishes (a) abioliths and (b) bioliths. Both of these main genetical groups subdivide into primary substances and "mobilisates." The former remain essentially intact after the process of original sedimentation, condensation, etc., losing (or gaining) mainly small groups or fragments of the polymerized molecules present. Subsequent to the stage of emplacement the "mobilisates" undergo a process of mobilization, such as distillation, fractionation, emulsification, etc., which separates the "mobilisates" from the primary "bioliths" or "abioliths."

The bulk of the "bioliths" formed and matured at low temperature, and under rather closely comparable conditions, and therefore, the composition of the original biogenic debris, and its age, determine to a considerable extent its present-day properties. Therefore, the classification based on the properties of the biogenic sediments, as outlined above, seems adequate. The more or less localized and high-temperature agencies which cause "mobilization" at a postsedimentation stage of the organic material produce, in most cases, such drastic chemical transformations that the properties of the resultant organic mineraloid seem to depend more on the nature of these secondary processes than on that of the primary sediment. For this reason, in our brief genetical description of organic mineraloids that follows, stress will be placed on this secondary alteration factor.

In most cases, the H/C ratio of a given solid organic mineraloid decreases with increasing age and/or increasing intensity of the "secondary metamorphosing or mobilizing" agency. Therefore, our description of the members of each genetical type in order of decreasing H/C, coincides, to a certain extent, at least, with a genetical subdivision according to increasing age.

Description of Organic Mineraloids

Abioliths. A considerable degree of uncertainty exists as to the extent and variability of abiogenic hydrocarbons within the earth's crust (see Sylvester-Bradley and King, 1963). The substances, which are more likely to be of abiogenic origin according to the views of the writer (Mueller, 1963b), are briefly described below.

Primary condensates may be anticipated in archaic sediments prior to the advent of the living organism which occurred, according to presently available evidences, 2 to 4 $\times$ 10^9 years ago. Organic molecules were found in numerous archaic sediments from Canada, South Africa, Rhodesia, etc., around this age, but there are as yet no reliable clues which would enable us to distinguish between biogenic and abiogenic origin. The writer believes that the finely divided carbonaceous substances in the meteorites are such "primary," abiogenic condensates (see Vol. II. *Cosmochemistry—Organic*).

Thucholites and other abiogenic mobilisates are associated, as a rule, with pegmatite veins, which often traverse only granite, excluding the possibility of distillation from biogenic sediments. *Huminite* has been described from pegmatites of the Ermlan and Grythytte districts of Sweden (see Ekman and Helland, quoted by Doetler and Leitmer, 1931). It is a brown substance that has the appearance of coal. *Thucholite* is a black bituminous substance found both in pegmatites and in hydrothermal veins which are characterized by a high percentage of uranium, thorium and other carbophyl elements. On distillation there is no fusion, and only gases are liberated; the high percentage of CO indicates the presence of the carbonyl complexes of U and Th (see extensive literature in Doetler and Leitmer, 1931). The localities in which thucholite-type substances were recorded are widespread and appear to include all the shield areas, zones rich in pegmatites: Fennoscandia; Ontario, Canada; South Africa; Colorado, U.S.; Isle of Man, Great Britain, etc.

Miscellaneous products include organic substances of mainly waxy consistency, which were found in fumarole vents in Kamchatka and in vesicles in lavas of Etna. The origin of these and others is doubtful; they may be condensation products of the CH_4 of fumarolic emanations, they may originate through the distillation of biogenic sediments, or they may be microbiological (mainly algal) products. The same genetical uncertainty applies to diverse low O/H, low U% (that is, "non-thucholite type") substances found in innumerable hydrothermal veins, within the British Isles (see Sylvester-Bradley and King, 1963). The majority of these may be biogenic mobilisates as shown below. The possibility of bitumens of hybrid, partially juvenile, and partially biogenic origin was discussed by Robinson (1963) among others.

Bioliths. *Fossil Resins.* Resins are the most stable biogenic products, and they are preserved, without any substantial alteration of elemental composition and structure, right from the Mesozoicum, although a single fossil resin, *"scleretinite,"* was recorded by Mallet (quoted by Hintze, 1933, 1935), from the Coal Measures of Wigan, England.

The *"amber group"* of fossil resins are distinguished from the rest through their compactness, which renders them suitable as ornamental stones, and the chemical distinction of yielding usually over 1% succinic acid on distallation. The color ranges between pale yellow, orange, red, and brownish, green and blue being rare colorations. N is 1,515–1,547; it is optically isotropic except in the case of stress anomalies. Schmidt (quoted by Doetler and Leitmer, 1931), concludes that amber is essentially a mixture of succinic acid; HOOC-CH_2CH_2-COOH; succinoresinolester; and succinoabietinic acid, with some borneolester. The main locality of amber is East Prussia (now U.S.S.R. territory), where it occurs in Eocene or younger, mainly marine, strata as a secondary concentrate from previous land sediments. The perfectly preserved residues of insects, mosses, fungi, and even a lizard in amber present interesting although hitherto unexplored experimental opportunities in the determination of degradation of biogenic substances. Amber-like fossil resins have been found in innumerable localities in North America, South America, Europe, and Asia, mostly associated with deposits of lignite.

The fossil resins dissimilar to amber because of more friable consistency and lack of liberation of succinic acid on distillation, form a rather heterogeneous group of substances, the majority of which have not been investigated with modern methods so far (Doetler and Leitmer, 1931). Such are, for example, *beckerite, stantienite,* and *glaserite,* which are associated with amber in the East Prussian localities. *Cedarite* is a brittle fossil resin from the vicinity of Cedar Lake, Canada. *Copalite* is a Recent or Quaternary resin collected in commercial quantities from the forest soils of the Belgian Congo, Brazil, and other tropical countries; *Highgate resin* from the London Clay appears to be the most similar fossil variety of Copalite. *Kiscellite,* from Budapest, is a variety with the highest sulfur content (3.99%); the sulfur may be mainly of secondary origin through oxidation of pyrite crystals of the Oligocene clays. *Siegburgite* is a chemically interesting aromatic fossil resin, from the Tertiary Sandstone from Siegburg, Germany. Klinger and Pisschki (quoted by Doetler and Leitmer, 1931), separated from it cinnamic acid, styrol, benzene, and toluene through distillation. It is claimed that the resin originated from fossils of Liquidambar Auropaeum, which are found in abundance in the surrounding limestone. Existing members of the same species are known to produce resins with aromatic molecules.

Miscellaneous Biogenic Substances. In addition to the above fossil resins, the preservation of several other types of more or less homogeneous biogenic substances were reported (see, for summary of most of the records, Doetler and Leitmer, 1931). Thus *waxes* are chemically resistant, but their soft consistency tends to cause them to disperse. They are recorded from brown coals, such as *geomicirite* from Gesterwitz, Germany; *pyropissite* from Halle a.d. Sale, Germany, and *denhardite* from British East Africa. *Sea wax* is possibly an algal product, reported from several localities. Waxy substances were also recorded from vertebrate fossils from the Cretaceous period onward.

Under certain conditions of sedimentation, which cannot be delineated as yet, leaf cuticles and cellulose may become preserved in lignites and brown coals. Thus, for instance, an eight-inch thick *paper coal* overlying a twenty-inch thick normal brown coal seam was described from Park Co., Indiana, by Neavel and Guenel (1958). It is also claimed that the chitin of crustaceans tends to remain preserved from the Cambrian age onward, but this has not been substantiated so far.

Coals. The members of the coal series represent relatively less stable terrestrial and aquatic plant substances in a process of gradual carbonization. Their properties and genetical histories are described in the entry *Coals* in Vol. VI.

Kerogens. The finely divided organic sediments which produce appreciable quantities of oil on distillation appear to be derived from mainly sapropelic sediments, and they are described under *kerogen* (q.v. Vol. VI). Very little is at present known regarding those finely divided organic sediments, which do not produce oil on distillation, but which represent the bulk of total carbonaceous matter within the earth's crust.

Products of Localized Metamorphic Processes. The effects of specific conditions in the course of sedimentation have not been studied so far, although abundant carbonaceous matter exists in certain evaporites, phosphate deposits, cherts, etc. Regarding the effects of localized conditions subsequent to sedimentation, we have, at present, little data at our disposal. Tomkeieff (1954) mentions the old miner's terms of "coke coal," "Dundy," "blind coal," "deaf coal," "cinder coal," "columnar coal," "Niggerhead coal," and "coal apples," which types are all produced through the proximity of intrusions within the coal fields of Great Britain. Surface oxidation may also change the composition of coals, and according to Tomkeieff (1954) the miners in England use the terms "crop coal," "crow coal," or "craw coal" for coal of diminished calorific value close to the ground surface.

Biomobilisates. According to the definition of the writer (Mueller, 1954, 1963b), the "biomobilisates" are substances which were separated from the biogenic sediment in a liquid, suspended, or vapor phase. The above definition includes *petroleum* and *natural gas* (qq.v. Vol. VI). However, the details of the genetic history of these products are still very much discussed.

The genetic history of the solid bitumens can be often more readily reconstructed, as these less mobile substances are likely to remain closer to the location of their mobilization from the original sediments. On the other hand, the detailed classification is rendered difficult by the fact that a number of the bitumens are inadequately described; the literature, which originates in great part from the nineteenth century, often includes quite different substances under the same term, or distinguishes similar substances with two or more terms. The genetic types of bitumens which can be distinguished with data available to date are described briefly below.

Mobilisates of Mild Regional Metamorphism. It was concluded by Bell and Hunt (1963) that in the case of semisolid or solid bitumens, which occupy fissures in carbonaceous strata, a process of oozing or distillation (mobilization in our terminology) is more clearly indicated than in the case of petroleum. Examples are occurrences of bitumens in veins in Albert County, New Brunswick, Canada (albetite), Ontario, Canada (anthroxilite), and Uinta Basin of Utah, U.S.A. The properties of the most important members of these "regional mobilisates" can be summarized as follows (Doetler and Leitmer, 1931).

Ozocerite. Varieties and synonyms include: pietrickite, hatchettite, borislavite. Ozocerite is a mixture of mainly straight-chain paraffins. It has a waxy consistency and brownish to greenish color. Density: 0.916–0.971; melting point: 56–85°C. In recent literature, indications of the presence of branched-chain paraffines were reported. The largest economic deposits are within the Borislavsk area of the U.S.S.R., and the Uinta Basin of Utah.

Asphalts. Under this term are generally understood black or dark brown, fusible and CS_2-soluble substances. They are composed mainly of cyclic and heterocyclic molecules of high molecular weight. The "Trinidad Lake Asphalt" originates from a unique lake of some twenty-six million tons capacity situated at the southwest extremity of the island. The presence of hot salt springs in the middle of the asphalt mass indicates that the material may have been melted out or distilled by hydrothermal agencies. Signs of hydrothermal origin appear in other occurrences of asphalt-type substances from Bermuda, Venezuela, St. Juan de Berengera, Peru, Los Angeles, California, the Massif central, France, Siracuse, Sicily, etc.

Asphaltites. These are the rather insoluble substances which usually fill fissures within sedimentary strata, clearly indicating an ultimate mobilized origin. The transition between the less polymerized asphalts and the more polymerized asphaltites is a gradual one. Bell and Hunt (1963) distinguish from the Uinta Basin, U.S., New Brunswick, Canada, and other localities, *gilsonite* (C: 85–86%), *wurzilite* (C: 79.5–80%), *ingramite* (C: 84%), *albertite* (C: 83–87%), and *anthroxilite* (C: 92–94%). A number of other terms were used in the older literature which seem to be unjustified, as according to our presently available data the properties and inferred histories of genesis of the substances in question lie within the broad ranges of the above asphaltites.

Mobilisates Through Localized Agencies. A close inspection and reinterpretation of records in the older literature (for summary, see the previously quoted books of Doetler and Leitmer, 1913–1931, and Hintze, 1933) often reveal the presence of a localized agency which seems to be genetically related to the bitumens in question. It is possible that a proportion of the bitumens included at present under the previous heading may eventually prove to belong to one of the genetic groups which are described below.

Igneous Contact Distillates, Fusates, or Sublimates. *Pyroretine* from Kusig, Czechoslovakia is a brown, brittle resin from which waxes can be extracted (see Dana, quoted by Doetler and Leitmer, 1931). It fills crevices within lignite which was intruded by large basaltic dikes in Ausig, Czechoslovakia. *Piauzite* is a relatively little altered resin from the proximity of porphyry intrusions in lignite of

ORGANIC MINERALOIDS

Trifail, Austria (see Hardinger, quoted by Doetler and Leitmer, 1923). *Schererite* is a camphor-like substance from the proximity of basaltic dikes which cut the lignite of Wilhelmszeche Mine, near Bach Village, Germany (see Noggerath, quoted by Doetler and Leitmer, 1931). The elemental compositions of all the above bitumens lie within the asphaltite range, whereas the species below are characterized by unusually low H/C and high O/C, as shown in Fig. 1. *Metabitumite* is a hard, insoluble substance which appears as globules up to 5 mm in diameter within the Lower Carboniferous Limestone surrounding a basalt vent near Speedwell Mine, Castleton, Derbyshire, England (Mueller, 1954). *Shungite* is, according to Polkanov (1938), a highly carbonized substance which forms major deposits within areas where Paleozoic shales are cut by intrusions at Shunga, Karelia, U.S.S.R. Asphaltum and albertite-type bitumens were also recorded from a granitized Triassic shale within the Sta. Juana district of Central Chile (Mueller, 1960).

Hydrothermal Contact Distillates, Fusates, and Sublimates. A considerable proportion of hydrothermal veins, which cross carbonaceous strata contains apparently mobilized droplets of bituminous substances. To quote only a few examples: asphalt-type bitumens have been recorded by the author (Mueller, 1954) from Ecton Mine, Staffordshire, England. The country rock of the area is a Lower Carboniferous Shale, which proved to yield oil on distillation. Droplets of an ashpaltite-type bitumen appear within calcite chalcopyrite veins traversing a black, Cretaceous dolomite, some 10 km to the west of the Laguna Maule, Central Chile, according to hitherto unpublished observations of the writer.

In certain veins of predominantly mercury mineralization, substances of mainly polycyclic structure appear which may be sublimation products of the organic sediments which surround the deposits in question. The close-to-surface, low-pressure conditions, which are believed to prevail during the formation of mercury deposits, would favor the liberation of the readily sublimable substances with condensed-ring systems. Thus, for example, *curtisite* is a hydrocarbon with properties close to anthracen from Skaggs Springs, Sonora Co., California (see Wright and Allen, quoted by Doetler and Leitmer, 1931). *Idrialite* was interpreted by Goldschmidt, (quoted by Doetler and Leitmer, 1931) as a mixture of substituted phenathrenes from Idria, Yugoslavia. *Nepalite* is a light green resin which yields mainly aromatic oils on distillation; the locality is Phoenix Mercury Mine, Napa Co., California (see Becker, quoted by Doetler and Leitmer, 1931). Finally, *pendletonite* was described by J. Muedoch, in the course of the meeting of the Organic Geochemical Society, during 26–28 September 1966 in London. It was found in a small mercury deposit near New Idria Mine, San Bonito Co., California, and the pale yellow monoclinic crystals proved to have properties identical to the hydrocarbon coronene.

Hydrothermal Differentiates. It was found by the author (Mueller, 1954) that within those areas of the North Derbyshire Orefield, England, in which the Lower Carboniferous Shales contain coal-type primary organic sediments, an insoluble brittle substance, carbonite, which yields mainly aromatic fractions on distillation appears in the presumably low-temperature, mainly calcitic hydrothermal veins. The high fluorspar and ore percentage veins of presumed higher temperature of formation contain a slightly more soluble and resinous variety of carbonite, but in addition to this there are four other types of bitumens, which show sharp phase boundaries against each other, indicating that they originally were present as immiscible liquids. These are: (a) *bernalite* (former term, "olefinite"), a dark red, green to yellow fluorescent resin of mainly olefinic composition. (b) *Ozocerite* and *native vaseline*. (c) *Elaterite*, a dark brown material of elastic consistency, which consists mainly of rather low-molecular-weight and slightly oxygenated olefins, tends to concentrate within the "fractionation series" of side veins, which surround the vein. (d) *Foxite* (former term used by the writer was "mutabilite"), a differentiate of highly oxygenated and mainly saturated molecules which tends to concentrate within the extreme limits of the "fractionation series." This is a tough, brown to buff colored resin, and certain specimens develop a chalk-white surface zone following fracture, presumably due to the escape of moisture from submicroscopic pores.

The I.B.P.s and M.P.s of the bitumens indicate that the temperature within the veins ranged between 300 and 400°C, but it dropped below 100° at a distance of a few feet from the veins (Mueller, in press). The hitherto unparalleled 5 in. (13 cm) long and ¼ in. (7mm) diameter rods of bernalite indicate that this differentiate entered the vein from the country rock through channels seen within the hydrothermal mineral crust, in a plastic state. In contrast, a "liquid injection" is indicated for carbonite, ozocerite, elaterite, and foxite, as these bitumens invariably occupy all the available space between crystals.

The well-defined interrelationships between

the properties of the bitumen assemblage and the hydrothermal mineral composition of the veins were explained by the writer as thermal effects. In case of the low-temperature, calcite-depositing mineralization, an essentially aromatic coal tar-type substance entered the veins, which eventually polymerized to the carbonite. With increasing temperature of mineralization in the higher ore + fluorspar-percentage veins, the original aromatic substances hydrogenated, producing in addition to the aromatic phase, immiscible liquids, which contained mainly olefinic and mainly paraffinic molecules, respectively, and the highly oxygenated molecules were expelled along the thermal gradient surrounding the veins into the "fractionation series" in a low-molecular-weight state.

It appears, through a scrutiny of the older literature, that there are numerous veins in which two or more distinct bitumens have been recorded. Thus, for example, Shrotter (quoted by Doetler and Leitmer, 1931), records three distinct bitumens from Oberhart, Glognitz, Austria, which are within pyritic veins traversing brown coal: the colorless to yellow crystalline *hartite*, an amorphous brown earthy substance called *psartite*, and *ixolite*, found as red, resinous pieces. Bonatti (quoted by Doetler and Leitmer, 1931), distinguishes "two resinous substances from sulfide ore veins, which traverse lignite from Gavilli, Tuscany, Italy: *yellow pyropissite* and *brown pyropissite*. Both of the above two sets of differentiates are within the composition range of asphalts, and the individual members differ mainly from the point of O/C ratio, as is the case between ozocerite and foxite in Derbyshire. Dunitz (1938) described a "thioelaterite" with 2.94% S, and with "minute resinous inclusions," which may be olefinic differentiates.

Devitrificates. The only concrete example of devitrification products was reported by Walker (1934). According to him, a fossil resin, *chemavinite*, from Cedar Lake, Manitoba, Canada, contains monoclinic crystals of an organic substance, *enelectrite*.

Capillary Concentrates. These organic mineraloids are mostly fossil terpenes or their degradation products which apparently concentrated through capillary forces into rents of fossil woods; the most important example is *fichtelite*, from Kolbermoor, Fichtelgebirge, Germany (see Bamberger and Strasser, quoted by Doetler and Leitmer, 1913-31). This is a colorless, soft substance which appears in right- and left-handed monoclinic crystals with melting point 460°C. The compositions of these substances are within that of the range of fossil resins.

Hydrolysates. This group of organic mineraloids consist of substances which underwent hydrolysis, emulsification, and subsequent coagulation, under waterlogged, but still rather oxidative close-to-surface conditions. The most characteristic examples may be mentioned. *Dopplerite* is found in practically all peat deposits as a soft, gelatinous substance as gully fillings, nests, or as veins up to 20 cm thick (Doetler and Leitmer, 1931). This is claimed to be a complex mixture of humic acids with smaller quantities of Fe^{2+}, Mg^{2+}, Al^{3+}, K^+, and Na^+, and some SO_4^{2-} and Cl^- ions. *Coroongite,* from the Coroong District, S. Australia, is found in the form of circular films on marshy soil (see Cumming, 1903; quoted by Doetler and Leitmer, 1931). It is a rubbery, partly soluble and saponificable substance of possibly algal origin. *Scharizite* is a black, colloidal substance from the Drachencave, Mixnitz, Steiermark, Austria (see Schandler, quoted by Doetler and Leitmer, 1931). Its high N content of 2–33% and traces of organic phosphorus indicate origin from the fossil bears and other animals preserved in the cave.

Ionic Precipitates. The hitherto known products of chemical precipitation of biogenic matter are salts of organic acids, although in certain cases the possibility of juvenile magmatic origin of the respective organic mineraloids cannot be discarded. The individual minerals are described in the textbooks of mineralogy (see, for example, Dana, 1951). *Oxanite,* $(NH_4)_2C_2O_4 \cdot H_2O$, is an ammonium oxalate in the form of pulvurent masses from the guano of Guanape Island, Peru. *Humboldtine,* $Fe(C_2O_4) \cdot 2H_2O$, is an iron oxalate usually occurring in capillary, botryoidal masses found in lignites; for example, from Kolosub, Czechoslovakia, as a product of close-to-surface oxidation. *Wedellite*, $Ca(C_2O_4) \cdot 2H_2O$, is a hydrated calcium oxalate, which is found in tiny isolated crystals in the bottom mud of the central Weddell Sea, Antarctica. *Whewellite,* $Ca(C_2O_4) \cdot H_2O$, was reported from bituminous coal in Brux, Switzerland, and elsewhere; on the other hand, its occurrence in high-temperature veins which traverse only igneous rocks from Freiberg, Germany, indicates a possible juvenile magmatic origin for some of the occurrences. *Julienite,* $Na_2CO(SCN)_4 \cdot 8H_2O$, the complex sulfocyanate is reported from schists near Kambove, Katanga, Congo. *Mellite,* $Al_2C_{12}O_{12} \cdot 18H_2O$, aluminum mellate, is a white to yellow aggregate of usually small tetragonal prisms. It is found in lignitic clay from several localities of the Paris Basin, and elsewhere.

GEORGE MUELLER

ORGANIC MINERALOIDS

References

*Bell, G., and Hunt, J. M., 1963, in (Breger, I. A., editor), "Organic Geochemistry," Oxford, Pergamon Press, pp. 333–366.

*Breger, I. A., (editor), 1963, "Organic Geochemistry," Oxford, Pergamon Press, pp. 50–86.

Dana, J. D., and Dana, E. S., 1962, "The System of Mineralogy," Seventh edition (Palache, C., Berman, H. and Frondel, C., editors), New York, John Wiley & Sons, 3 vols.

*Doetler, C., and Leitmer, H., 1913-31, "Handbuch der Mineral-chemie," Vol. 4, Part 3, p. 793 et seq.

Dunitz, B. L., 1938, "Chemie der Erde," Vol. 2, p. 576.

*Hintze, G., 1933, etc., "Mineralogie," Vol. 1, Part 4B, p. 1341, et seq. See also headings on "Organic Minerals" in Erganzuengsband, 1936, and 1954–1955.

Mueller, G., 1954, *Compt. Rend.* **19e.** *Ses. Cong. Geol. Int., Alger,* Ser. 13, Fas. 12, p. 279.

Mueller, G., 1960 Proc. Int. Geol. Congress, 21st Session, Norden, Part. 15, pp. 123–132.

Mueller, G., 1963a, Proc. of the 6th World Petroleum Congress, Frankfurt-am-Main, Sect. 1, paper 29, -PD1, pp. 1–14.

Mueller, G. (1963b). *Nature,* Vol. 198, p. 731.

Mueller, G., 1970, "Indications for High Temperature Processes in Organic Geochemistry," in "Advances in Organic Geochemistry," New York, Pergamon Press, 443pp.

Neavel, R. C., and Guennel, G. K., 1958, Program, Annual Meeting, U.S. Geol. Society, Abstract.

Polkanov, A. A., 1938, "The Karelian Autonomous Soviet Socialist Republic," Proc. Int. Geol. Congress, Moscow, p. 50.

Robinson, R., 1963, "Advances in Organic Geochemistry" (Editors U. Colombo and G. D. Hobson), Oxford, Pergamon Press, pp. 7–10.

Sylvester-Bradley, P. C., and King, R. J., 1963, *Nature,* **198,** 728–31.

Tomkeieff, S. E., 1954, "Coals and Bitumens," Oxford, Pergamon Press.

Walker, T. L., (1934), Univ. Toronto Studies, Geol. Ser., No. 36, p. 11.

*Additional bibliographic references mentioned in text may be found in this work.

Cross-references: *Hydrocarbons; Hydrothermal Solutions; Mineraloids; Organic Geochemistry; Precambrian Hydrocarbons. Vol. II: Carbonaceous Meteorites; Cosmochemistry—Organic. Vol. VI: Asphalts; Coal; Kerogen; Natural Gas; Petroleum; Resin and Amber.*

ORGANIC PIGMENTS

Carotenoids

Carotenoids are terpenoid compounds built up from the isoprenoid structure

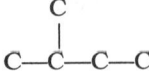

Almost all naturally occurring carotenoids are made up of eight isoprenoid units consisting of forty carbon atoms, in a tetraterpenoid form. Carotenes are carotenoids which contain only carbon and hydrogen; xanthophylls are oxygenated carotenoids. Carotenoids containing elements other than carbon, hydrogen, and oxygen are not known. Weedon (1965) has reviewed the structure, distribution, and function of carotenoids.

Most carotenoids are colored, their color being due to the long series of alternating single and double bonds in the molecule. For example, Fig. 1 shows the light absorption spectrum and chemical structure of β-carotene, a typical carotenoid.

Carotenoids represent a very widespread group of naturally occurring pigments. They are universally present in photosynthetic organisms, sporadically encountered in the non-photosynthetic tissues of higher plants and in fungi and bacteria, and present in most forms of animal life. A major function of carotenoids is to protect the cells of photosynthetic organisms against photosensitization. An ancilliary function relates to the fact that carotenoids can absorb light in the blue region of the spectrum and pass the energy on to the chlorophyll, increasing the efficiency of photosynthesis.

The fate of carotenoid pigments in sedimentary systems has been comprehensively examined by Vallentyne (1960). The content of the pigments in various algal groups generally fall in the range of 200 to 2000 ppm (dry basis). The ratio of xanthophylls to carotenes in algae range from 4:1 to 9:1. In extracts of marine seston the ratio increases to a range of from 7:1 to 14:1. At least twenty different carotenoids occur in wet sediments, and the oldest carotenoid-containing sediments yet discovered are about 100,000 years old. Probably the bulk of the pigments in the sediments come from the deposition of seston from the overlying water, but it is known that some synthesis of bacterial and fungal carotenoids occurs in near-surface sediments. Evidently there is a selective destruction of the xanthophylls in the sediments, since the xanthophyll/carotene ratio in the sediments is lower than in the contributing seston. Carotenoids have not been reported in ancient sediments or in petroleum.

Mulik and Erdman (1963) examined the contribution of carotenoids to the formation of petroleum compounds. While considerable quantities of benzenoid hydrocarbons including toluene, xylenes, and naphthalene are formed by thermal treatment (188°C for 72 hr) of β-carotene in an aqueous sediment,

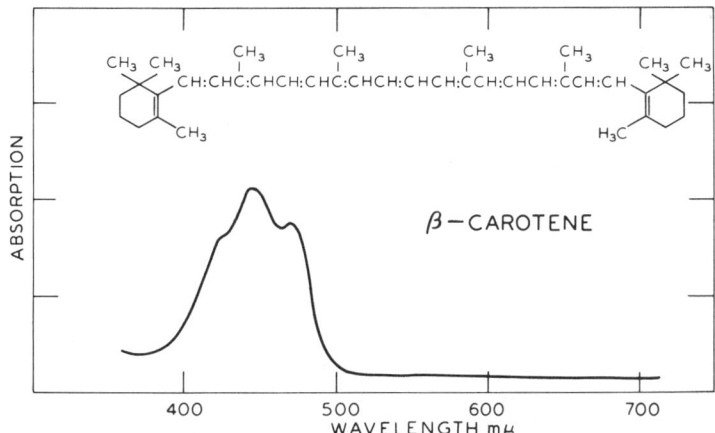

FIG. 1. Structure and absorption spectrum of β-carotene, a typical carotenoid pigment.

the source of the benzene and the bulk of the toluene and xylene of crude oil is still unknown. It is of interest to note that one of the known constituents of petroleum, 2,2,6-trimethylcyclohexane-1-carboxylic acid, may be a residue of β-carotenoid pigment.

Porphyrins

Porphyrins are pigments consisting of four pyrrole units linked by four methene bridges in a planar structure. The spectra of porphyrins generally consist of a major band near 400 mμ and four minor bands, usually in the 500–650 mμ range. Many individual porphyrins result from the substitution of the beta hydrogen atoms with alkyl or alkyl acid groups. Porphyrin compounds form ligand complexes with divalent cations. Such metal complexes are somewhat more stable and are characterized by a major band near 400 mμ and two minor bands in the 500–600 mμ range. The gross structure and chemical behavior of the porphyrin pigments has been well established by Hans Fischer and others (e.g., Fischer and Orth, 1937).

Porphyrins are widely distributed throughout the animal kingdom, principally as metal complexes. Porphyrin synthesis appears to be one of the most fundamental attributes of living cells, as noted by Rimington and Kennedy (1962) in a general review of the distribution, structure and metabolism of porphyrins. The iron complexes, or hemes, play a vital role in oxygen transport in hemoglobin proteins, and in enzyme systems such as the cytochromes.

In sedimentary rocks porphyrins are fairly common as metal complexes, particularly in rocks which are rich in organic matter such as oil shales (Treibs, 1935; Hodgson et al., 1968). Porphyrins are present in petroleums. The porphyrins of oil shale and petroleum are generally metal complexes with nickel and vanadium (as vanadyl). Iron complexes have been reported, but there is little or no evidence to indicate the presence of other cations in such porphyrin complexes. It is generally believed that the major porphyrin component is deoxophylloerythroetioporphyrin, the structure and spectrum of which (in a nickel complex) are shown in Fig. 2. Recent data show the occurrence of geochemical porphyrins in homologous series, both with and without the isocyclic ring of the chlorophylls. Selected ancient shales have vanadyl porphyrins reported in the range of 0.002 to 0.3 ppm.; oil shales, 800 to 4000 ppm.; petroleum 0.1 to 1000 ppm. Coal porphyrins have been reported in the range 0.02 to 17 ppm.

Consideration of the general structure of ancient sediment porphyrins suggests a source from the porphyrin pigments of living organisms, notably the blood hemin pigments and the enzymatic pigments. However, the structure of the ancient sediment porphyrins in most cases is directly related to the chlorophyll pigments, through the fifth ring in the basic structure. Only a small proportion of the porphyrins found in sediments show a direct relationship to the hemin pigments as indicated by the absence of the fifth ring, and even this may have been due to destruction of the chlorophyll fifth ring rather than to a direct hemin source. This seems to be particularly likely in the case of coals, where the chlorophyll source is probable, the hemin-like pigment is evident, and the opportunity of diagenic partial destruction of pigments is high. Vanadyl pigments are related to high-sulfur reducing conditions and appear to be

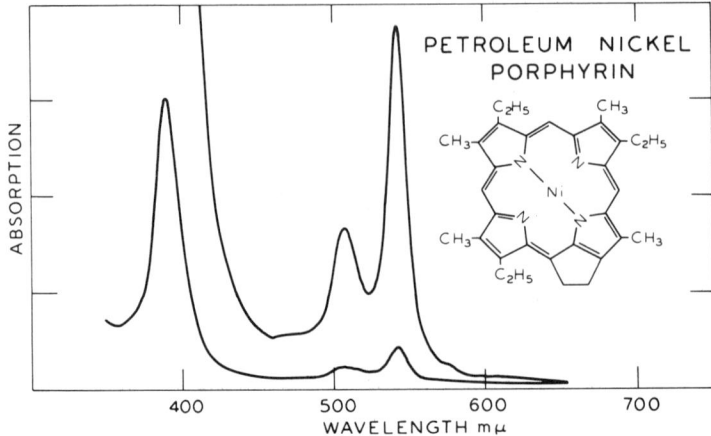

Fig. 2. Structure and absorption spectra (two concentrations) of a petroleum porphyrin pigment.

derived from vanadyl pigments of living organisms. Nickel pigments are evidently formed from chlorophyll-degradation products.

Porphyrin pigments occur in recent sediments, but in very low concentrations relative to chlorophyll-like pigments—0.1 to 1 ppm.

Chlorins

The term chlorin is currently applied to both chlorins and phorbins. These compounds differ from porphyrins in only one structural feature: in chlorins one of the resonating double bonds in the basic porphyrin structure is saturated. The molecule thus has two "extra" hydrogen atoms, and the chlorins (which are green) are therefore sometimes known as dihydroporphyrins. Chlorins, like porphyrins, also form metal complexes. The best known and most widely distributed chlorin pigments are the chlorophylls (see *Chlorophyll*). The chlorins are representative of the plant kingdom. The bulk of the chlorins contain magnesium as the complexing metal in the living organisms, but other metals including vanadium are evidently involved in at least some of the enzyme systems.

The spectrum of a typical chlorin pigment shows a major peak very similar to that of the porphyrins, with the minor peaks shifted into the 500–700 mμ range.

Magnesium complexes of chlorins are very unstable and under sedimentary conditions break down to free chlorins (Orr, Emery, and Grady, 1958). Free chlorins are very common in recent sediments and soils with concentrations commonly in the range 1–10 ppm. The initial degradation products are esters which give rise to free acids. The chlorin pigments generally do not persist into ancient sediments, probably because of an inherent instability of the uncomplexed pigments. A notable exception was the observation of uncomplexed chlorins in Triassic strata such as the Serpiano oil shale of Switzerland, but it was established that these chlorins had been formed through the hydrogenation of porphyrins under extreme reducing conditions, probably in a late diagenetic stage in the sediment, and not directly from chlorophyll. In the diagenesis of chlorins under ordinary sedimentary conditions, the bulk of the chlorin material is lost, with only a trace preserved as porphyrin metal complexes in ancient sedimentary rocks and petroleum.

A further group of pigments related to porphyrins are the tetrahydroporphyrins which are marked by four "extra" hydrogens. The most common examples are the bacteriochlorophylls. The spectra of these compounds are similar to those of the chlorophylls except that the minor peaks occur at still higher wavelengths.

Phycobilins

Tetrapyrrole porphyrin structures undergo degradation in animals to the bile pigments in which the porphyrin ring is ruptured. A similar class of pigments occurs naturally as prosthetic groups of photosynthetically active conjugated proteins. These are phycobilins, and the complexes with proteins are biliproteins. Biliproteins are commonly found in algae (up to 2%, dry weight), but not in higher plants. In algae they are either blue phycocyanins or red phycoerythrins, and they fluoresce red and orange respectively. The binding of the phycobilins to proteins is strong in the case of algal biliproteins, but very weak

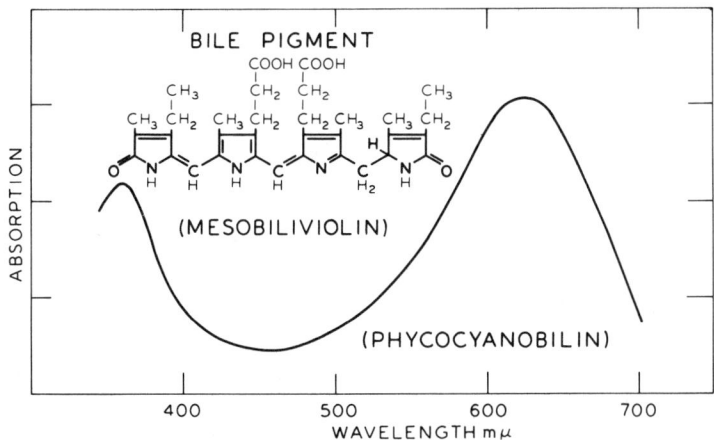

Fig. 3. Structure of a bile pigment with the absorption spectrum of a bile-pigment protein complex.

in biliproteins of other organisms, e.g., butterflies. Linkages are evidently through propionic acid side chains of the pigment and amino groups of the protein.

The absorption spectra of phycobilins differ considerably from those of porphyrins. In aqueous solution phycoerythrins show major absorption between 540 and 570 mμ, and phycocyanins between 610 and 660 mμ. There is no major absorption band corresponding to the intense Soret band of porphyrins at about 400 mμ. Fluorescence is at about 575 and 650 mμ for the red and blue pigments, respectively. Of the phycobilins which are believed to occur in biliproteins, phycocyanobilin and phycoerythrobilin have been most extensively studied (O'hEocha, 1965). Both are soluble in organic solvents and in aqueous acids. Both are unstable in alkali and in concentrated hydrochloric acid. The latter converts phycocyanobilin to a pigment having the spectral properties of mesobiliviolin (Fig. 3). When treated with zinc acetate, phycobilins form zinc complex salts which fluoresce freely. The zinc complexes are readily demetallated under acidic conditions.

The molecular weights of biliproteins range from about 130,000 to 300,000. Seventeen amino acids have been identified, with aspartic and glutamic acids generally most abundant followed by leucine and arginine.

Phycobilins and chlorophylls have different biosynthetic pathways although a heme compound is probably a precursor of phycocyanin. Other work summarized by O'hEocha (1965) suggests that bile pigment precursors of phycoerythrin and phycocyanin are formed photochemically. The protein moieties of biliproteins are formed *de novo,* and do not derive from proteins already existing in the cells.

Phycobilins have not been recognized in geochemical situations, nor have biliproteins. The pigments are moderately stable, however, and might be expected in modern sediments, but not in ancient sediments.

Flavonoids

Flavonoid compounds are primarily responsible for the red and blue pigments of flowers and fruit. Other flavonoid pigments provide the yellow and white colors of flowers, and finally some of the brown and black pigments found in plants are due either to the products of oxidation of flavonoid and related phenolic compounds, or to their chelates with metals. Harborne (1965) recently reviewed their distribution and properties.

The basic structure of flavonoids is that of flavone comprising two benzene rings joined together by a three carbon link which is formed into a gamma-pyrone ring. The various classes of flavonoid compounds differ from one another by the state of oxidation of this three carbon link. In most cases the flavonoid compounds exist in plants as glycosides, involving simple hexoses and pentoses, and di- and trisaccharides.

Four major classes of flavonoid glycosides are the anthocyanins, the flavonols, the chalkones and the aurones. As the conjugation in the structure increases, the flavonoids absorb light at longer and longer wavelengths, with the anthocyanins absorbing at the longest wavelengths of any class of flavonoid compounds, i.e., at about 530 mμ.

The flavonoid pigments, primarily the anthocyanins, are significantly more stable than the chlorophylls and carotenoids in senescent tissue, but no attempts have been made to

Phytochrome

Phytochrome is a naturally occurring blue protein with striking and manifold expressions of physiological control in the growth of plants. The most important examples of such action is the control of flowering, elongation of structures, germination of seeds and spores, formation of anthocyanins and carotenoids, display of crassulaecean metabolism and induction of dormancies. The pigment is a sensor of two of the most important factors of a plant's environment, namely, light and temperature. The characteristics of phytochrome were recently summarized by Butler et al. (1965) and Hendricks and Borthwick (1965).

The structure of the phytochrome chromophore is unknown. The absorption spectrum is similar to that of biliproteins from algae, but there are only one or two chromophore units per protein molecule. The major absorption bands occur in the red region of the spectrum at 660 and 730 mμ, respectively, for two interconvertible forms. The lack of a Soret band precludes a cyclic tetrapyrrole structure for the chromophore.

Phytochrome is isolated from dark-grown seedlings, for example, seedlings of oats, and is measurable in the seedlings of a large number of common garden vegetables. The protein moieties have molecular weights of about 90,000 and 150,000.

The presence of phytochrome or phytochrome products in geochemical situations has not been recognized, and would perhaps be indistinguishable from the biliproteins.

Quinones

Approximately 200 quinones have been isolated from vegetable sources, as reviewed by Thomson (1965a). The majority occur in flowering plants and fungi, and to a lesser extent in bacteria and algae. They range in color from pale yellow to almost black. Structures range from simple benzoquinones to polycyclic compounds containing eight rings, and some quinones bear long alkenyl side chains and are known as isoprenoid quinones. The biosynthesis pathway is commonly that involving acetate-malonate, with the shikimic acid route to aromatics seldom utilized.

Benzoquinones have limited distribution. The majority are fungal metabolites. Naphthaquinones are common in flowering plants, and as many as 20 such compounds have been identified. The largest group of naturally occurring quinones is that of the anthraquinones. Biosynthesis is probably by acetate-derived aromatic compounds. They occur frequently as glycosides and as mixtures of closely related compounds in fungi and higher plants. The most widely distributed is emodin, a trihydroxymethylanthraquinone.

Naphthacenequinones comprise a unique group of red, antibiotic, pigments confined to the Actinomycetales. They occur either as glycosides in combination with unusual amino sugars or as free aglycones. Phenanthraquinones are known, also.

Closely related to the naphthaquinones is a small group of perylene quinones which are evidently biosynthesized via oxidative dimerization of naphthalene precursors by fungi. In a similar manner, eight-ring quinones are elaborated by fungi from the dimerization of anthraquinones.

The carbon skeleton of such compounds are expected to be relatively stable geochemically and polycyclic hydrocarbons are commonly found in soils, sediments and sedimentary rocks. Perylenes in alkylated series with principal absorption at about 430 mμ and corresponding fluorescence at 475 mμ are readily recognized spectrally in extracts of such samples.

Pteridines

The pteridine ring system, shown in Fig. 4 together with the spectrum of a typical pteridine compound, probably occurs in all living organisms. The function of the simple pteridines in living organisms is not well understood, but that of the folic acid compounds comprising a large number of pteridines relates to a number of enzyme reactions. A review of pteridines has been made by Forrest (1962). Again, pteridines would be expected in organic debris in sedimentary systems, but they have not been reported in recent or ancient sediments.

Miscellaneous Organic Pigments

A number of other pigments make up additional, minor, groups of plant pigments, as reviewed by Thomson (1965b). In at least some of these, the appearance of color is often incidental, arising from a minor modification of a chromophore which normally absorbs only in the ultraviolet region.

Nonnitrogenous miscellaneous pigments include cyclic dienones, gamma-pyrones, sclerotirins and vulpinic acid pigments. Many of the nitrogenous pigments are based on pyrollic components reminiscent of bile pigments or of the direct linking of pyrrole rings in the corrin nucleus of the vitamin B_{12} chromo-

ORGANIC PIGMENTS

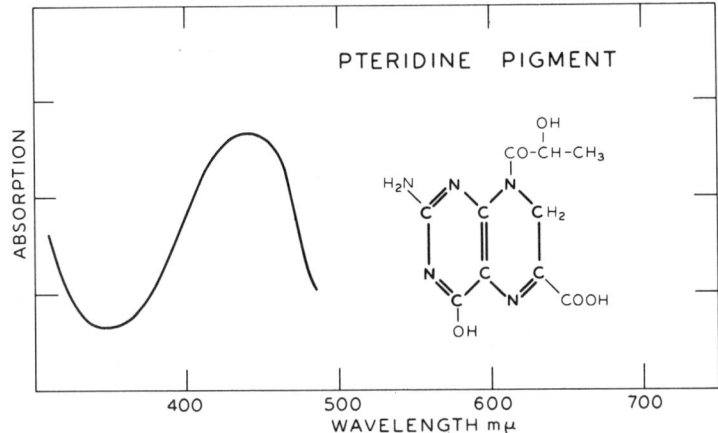

FIG. 4. Structure and absorption spectrum of a typical pteridine pigment.

phore. Phenazine pigments are common, as are phenoxazones which occur as the red chromophore in antibiotic chromopeptides produced by Streptomyces. Melanins are ill-defined polymeric pigments which commonly include derivatives of indole.

Review of Organic Pigments

Dunning (1963) gives a comprehensive review of the geochemistry of organic pigments.

G. W. HODGSON

References

Butler, W. L., Hendricks, S. B., and Siegelman, H. W., 1965, "Purification and Properties of Phytochrome," pp. 197–210, in (Goodwin, T. W., editor), "Chemistry and Biochemistry of Plant Pigments," London, Academic Press, 583pp.

Dunning, H. N., 1963, "Geochemistry of Organic Pigments," pp. 367–430, in (Breger, I. A., editor) "Organic Geochemistry," New York, Pergamon Press, 658pp.

Fischer, H., and Orth, H., 1937, "Die Chemie des Pyrrols," Vol. 11, Pt. 1, Leipzig Akad. Verlagsges., 720pp.

Forrest, H. S., 1962, "Pteridines: Structure and Metabolism," pp. 615–637, in (Florkin, H. and Mason, H. S., editors) "Comparative Biochemistry," Vol. IV, Pt. B, Academic Press, 841pp.

Harborne, J. B., 1965, "Flavonoids; Distribution and Contribution to Plant Color," pp. 247–278 in (Goodwin, T. W., editor) "Chemistry and Biochemistry of Plant Pigments," London, Academic Press, 583pp.

Hendricks, S. B., and Borthwick, H. A., 1965, "The Physiological Functions of Phytochrome," ibid., pp. 405–436.

Hodgson, G. W., Hitchon, B., Taguchi, K., Baker, B. L., and Peake, E., 1968, "Geochemistry of Porphyrins, Chlorins and Polycyclic Aromatics in Soils, Sediments and Sedimentary Rocks," Geochim. Cosmochim. Acta, **32**, 737–772.

Mulik, J. D., and Erdman, J. G., 1963, "Genesis of Hydrocarbons of Low Molecular Weight in Organic-Rich Aquatic Systems," Science, **141**, 806–807.

O'hEocha, C., 1965, "Phycobilins," pp. 175–196, in (Goodwin, T. W., editor) "Chemistry and Biochemistry of Plant Pigments," London, Academic Press, 583pp.

Orr, W. L., Emery, K. O., and Grady, S. R., 1958, "Preservation of Chlorophyll Derivatives in Sediments Off Southern California," Am. Assoc. Petrol. Geol. Bull., **42**, 925–962.

Rimington, C., and Kennedy, G. Y., 1962, "Porphyrins: Structure, Distribution and Metabolism," pp. 557–614, in (Florkin, H. and Mason, H. S., editors) "Comparative Biochemistry," Vol. IV, Pt. B, Acadimic Press, 841pp.

Thomson, R. H., 1965a, "Quinones: Nature, Distribution and Biosynthesis," pp. 309–332, in (Goodwin, T. W., editor) "Chemistry and Biochemistry of Plant Pigments," London, Academic Press, 583pp.

Thomson, R. H., 1965b, "Miscellaneous Pigments," ibid., pp. 333–354.

Treibs, A., 1935, "Chlorophyll—und Hämin Derivate in bituminösen Gesteinen, Erdölen, Erdwachsen und Asphalten," Justus Liebig's Ann. Chem., **510**, 42–62.

Vallentyne, J. R., 1960, "Fossil Pigments," pp. 83–105, in (Allen, M. B., editor) "Comparative Biochemistry of Photoreactive Systems," Vol. 1, Academic Press, 437pp.

Weedon, B., C., L., 1965, "Chemistry of the Carotenoids," pp. 75–125; "Distribution of Carotenoids," pp. 127–142; "The Biosynthesis of Carotenoids," pp. 143–173, in (Goodwin, T. W., editor), "Chemistry and Biochemistry of Plant Pigments," London, Academic Press, 583pp.

Cross-references: *Chelation; Chlorophyll; Complexes; Hydrocarbons; Molecular Weights; Photosynthesis.* Vol. VI: *Coal; Oil Shales; Petroleum.* Vol. VII: *Algae.*

ORGANIC SEDIMENTS—See Vol. VI

OSMIUM: ELEMENT AND GEOCHEMISTRY

Properties

Osmium is a member of the group of six platinum metals which also appear in a block in the periodic table of the elements. It was first identified by Smithson Tennant in 1804. The element is notable for its great density, 22.61 g/cm^3 at 20°C (second only to iridium), its high melting point, and its considerable hardness. Its atomic number is 76 and atomic weight 190.2. There are seven stable isotopes with mass numbers and relative abundance as follows: 184 (0.018%), 186 (1.59%), 187 (1.64%), 188 (13.3%), 189 (16.1%), 190 (26.4%), and 192 (41.0%). The crystalline metal is structured on a hexagonal close-packed lattice whose characteristic constants are $a = 2.7341$ Å and $c/a = 1.5800$. These lead to an average atomic radius for 12-coordination of 1.35 Å. The melting point of the pure metal is given as 3050°C.

Osmium is atypical among the platinum metals for the comparative ease with which it reacts with oxygen. Thus the compact metal heated in air or oxygen to 200°C, and finely divided metal even at ordinary temperature, will form a gaseous oxide OsO_4 which is toxic and has a pungent odor. The metal is not attacked by acids but is susceptible to reaction with alkalis.

Occurrence

Typically among the platinum metals the principal occurrences are as native metal alloys, a number of which have been given mineral names; however, the composition of these varies over appreciable ranges. Osmium appears in alloys with iridium which also contain smaller amounts of the other platinum metals, especially platinum and ruthenium. Of these *iridosmine* embraces alloys containing greater than 32% osmium which crystallize in the hexagonal form, and *osmiridium* with cubic crystal structure contains up to 32% osmium. The terms *nevyanskite* and *siserskite* are sometimes applied to iridosmines which respectively have iridium or osmium as the predominating component. Only small amounts of osmium occur in native platinum or palladium.

There are no mineral compounds of osmium as such, but in the mineral laurite (RuS_2) osmium has been found up to about 3%, and there have recently been one or two other osmium-bearing mineral species characterized by the electron probe microanalyser (Wright and Fleischer, 1965).

Abundance and Geochemical Features

The abundance of osmium in the earth's crust has been estimated as 0.001 ppm. However, in view of the relative amounts of the platinum metals that have actually been produced commercially during the past fifty years, this figure may be too high (Hampel, 1961). On the other hand, in meteorites osmium is relatively more abundant, ranking about third among the platinum metals or roughly equivalent in abundance with palladium. The platinum metals including osmium are usually more abundant in siderites than in chondrites; but in contrast to platinum, osmium does not concentrate preferentially in the nickel-iron phase in chondrites (see also *Platinum Metals*).

Economic Geology

For consideration of economic geology the platinum metals are most appropriately treated as a group since they occur together and are refined together. Commercially much less osmium is produced annually than any of the other platinum metals, and this is applied mainly to the production of hard-wearing alloys (see *Platinum Metals*).

W. A. E. McBryde

References

Hampel, C. A., (editor), 1961, "Rare Metals Handbook," Second ed., New York, Reinhold Publ. Corp. pp. 304–335.

Hampel, C. A., (editor), 1968, "Encyclopedia of the Chemical Elements," New York, Reinhold Publ. Corp. pp. 494–499.

Rankama, K., and Sahama, Th. G., 1950, "Geochemistry," Chicago, University of Chicago Press.

Standen, A., editor, 1963, Kirk-Othmer "Encyclopedia of Chemical Technology," Second ed., New York, John Wiley & Sons, pp. 832–878.

Wright, T. L., and Fleischer, M., 1965, "Geochemistry of the Platinum Metals," U. S. Geol. Surv. Bull., **1214-A**.

Cross-references: Platinum Metals. Vol. IVB: *Electron Probe Microanalysis; Meteoritic Minerals.*

OUTGASSING OF THE PLANET EARTH

The Earth's Atmosphere and Hydrosphere

The three main components of our atmosphere are nitrogen, oxygen, and argon in that order of abundance. The hydrosphere and sedimentary rocks, however, contain a large amount of material which can exist in the vapor phase at temperatures associated with

Table 1. "Excess" Volatile Materials in Present Atmosphere and Hydrosphere and in Buried Sedimentary Rocks (Rubey, 1951) (in units of 10^{20} grams)

H_2O	16,600
Total C as CO_2	910
Cl	300
N	42
S	22
H	10
B, Br, A, F, etc.	4

volcanic action. Rubey (1951) has estimated the total amount of these elements on the earth's surface and in the crust that cannot be derived by the reconstruction of primary silicate rocks through weathering. His estimates of the quantity of the "excess volatiles," as he calls them, are given in Table 1.

The source of such a complexion of potentially volatile components and the mechanism of modification of their proportions are two important areas of research interest.

Volatiles on the surface of the earth may have several origins: *(1)* primordial volatile envelope associated with the accretion of the earth; *(2)* gases derived within a very short time after the origin of the earth as a result of reduction of iron in the earth; *(3)* major episodic degassing of the earth as a result of mainly thermal effects due to radioactive heating within the earth; *(4)* continuous degassing of the earth through volcanoes and similar means; *(5)* accretion from the solar wind.

Aside from modification of the input volatile composition due to oxidation, reduction, and evaporation resulting in solid phases in the crust, loss of hydrogen and helium from the atmosphere through energetic escape from the earth's gravity field must be considered. With the construction of adequate models for each of these mechanisms of supply and modification, the composition of volatiles on the surface of the earth can be reproduced.

Outgassing of the Planet

Carbon Dioxide and Water. Ringwood (1959) has shown that if we choose to start with carbonaceous chondritic material for the construction of the earth and produce the amount of reduced metallic iron found in the earth's core, enormous quantities of CO_2, CO, and H_2O would be released to the atmosphere. Most of this would have to be lost early in the earth's history together with any primordial atmosphere or else the earth would have developed an atmosphere possibly like that of Venus. The loss mechanism is unknown.

Whereas volcanoes and associated phenomena have for a long time been associated with volatile supply to the earth's surface, their quantitative importance cannot be measured since we cannot determine the rate of supply of truly juvenile gases (i.e., gases that have never existed on the surface of the earth or in the crust before) by identification of any one or group of chemical species as truly diagnostic.

Argon and Helium. The only clear indication of possible supply of materials from depth is found in the presence of the large quantity of argon-40 in the atmosphere. This isotope is produced primarily by the radioactivity decay of potassium-40 (0.0119% of potassium isotopes in abundance). Helium-4 is similarly supplied to the atmosphere from the decay series of uranium and thorium. The easy loss of helium from our atmosphere to space precludes the evaluation of the total quantity supplied to the atmosphere through geologic time.

Argon. The difficulties in constructing a degassing model for argon are formidable. We must know: *(1)* the abundance of potassium in the earth, *(2)* the distribution of potassium in the earth as a function of time, and *(3)* the mechanisms for degassing in the crust and the mantle. All we can do at present is use reasonable estimates of these parameters and check for internal consistency.

Within the last ten years, there have been several attempts at quantifying the argon degassing problem. One of the common features of several models is a crust or protocrust which behaves as a closed system except for the loss of argon. The concept of a primordially differentiated earth is generally the basis for such models, although their utility lies in their computational simplicity. Two options remain if we reject a closed-crust model: *(1)* using as evidence contemporary and ancient volcanism, there appears to be addition of material to the crust from the mantle; *(2)* there may be continuous exchange between the crust and the mantle, the gross chemical composition of each remaining the same on the long-time scale.

Since the mechanisms of crustal growth and the degassing processes from both crust and mantle are inadequately known, the following perfectly general model which considers only the solid earth and the combined atmosphere and hydrosphere as the significant features may be useful.

If there were little or no primordial argon-40 when the earth was formed and the loss of argon generated by the decay of potassium-40 from the solid earth obeys a first-order law, we can set up a differential equation describing how the argon content of the solid earth varies with time:

$$\frac{dA}{dt} = \gamma K_0 e^{-\gamma t} - \alpha A \quad (2)$$

where A is the amount of argon in the solid earth at any time t; t is time measured from the beginning of the earth ($t_0 = 0$); γ is the decay constant for K^{40}; K_0 is $= \frac{R}{1+R} K_0'$ where K_0' is the primordial abundance of K^{40} in the earth and R is the branching ratio of K^{40} ($\gamma_{E.C.}/\gamma_{\beta-}$); and α is the "degassing" constant.

With the above model, and if a chondritic earth is assumed as a first approximation, $\alpha = 2.81 \times 10^{-11} \text{yr}^{-1}$, and the growth of argon-40 in the atmosphere is that shown in Fig. 1. The rate of supply of argon to the atmosphere at the present time by this model is 3.02×10^{32} atoms/yr or 1.88×10^6 atoms/(cm^2)(sec).

Helium. The use of the argon degassing constant for other gases is of course dependent on their similarity of behavior. If a chondritic composition is assumed for the earth, crustal abundance estimates require that 67% of the uranium and thorium are in the crust. On the above premise, the mantle degassing constant applies then to 33% of the helium produced in a chondritic earth. Using an equation analogous to that for argon with $\alpha = 2.8 \times 10^{-11} \text{yr}^{-1}$, the present-day supply rate of helium to the atmosphere from the mantle is 0.263×10^{32} atoms/yr.

The crust is obviously more easily degassed of helium than the mantle if we consider the observed high leakage rate from crustal materials. We do not need to consider if the uranium and thorium have always been concentrated in the outer parts of the earth since the residence time (10^7 yr) of helium is short compared to the length of geologic eras (10^8 yr). If we assume for the sake of simplicity that the rate of production of helium-4 in the crust is equal to the rate of loss from the crust to the atmosphere, we arrive at a present-day crustal degassing rate of 1.83×10^{32} atoms/yr. The total supply rate at the present time from both the crust and mantle is then 2.1×10^{32} atoms/yr or 1.3×10^6 atoms/(cm^2)(sec).

Other Volatiles. If the model assuming that most of the argon in the atmosphere is derived from the mantle is correct, then it is conceivable that the same model may be applicable to the supply of other volatiles to the earth's surface. It is not possible, however, to measure the magnitude of this contribution.

KARL K. TUREKIAN

References

Brown, H., 1952, "Rare Gases and the Formation of Earth's Atmosphere," in (Kuiper, G. P., editor) "The Atmospheres of the Earth and Planets," Second ed., Chicago, Univ. of Chicago Press, p. 258.

Cameron, A. G. W., 1963, "The origin of the atmospheres of Venus and the Earth," *Icarus*, **2**, 249–257.

Damon, P. E., and Kulp, J. L., 1958, "Inert gases and the evolution of the atmosphere," *Geochim. Cosmochim. Acta*, **13**, 280–292.

Holland, H. D., 162, "Model for the Evolution of the Earth's Atmosphere," in "Petrologic Studies," New York, Geological Society of America, pp. 447–477.

Ringwood, A. E., 1959, "On the chemical evolution and densities of the planets," *Geochim Cosmochim. Acta*, **15**, 257–283.

Rubey, W. W., 1951, "Geologic history of sea water," *Geol. Soc. Am. Bull.*, **62**, 1111–1148.

Signer, P., and Suess, H. E., 1963, "Rare Gases

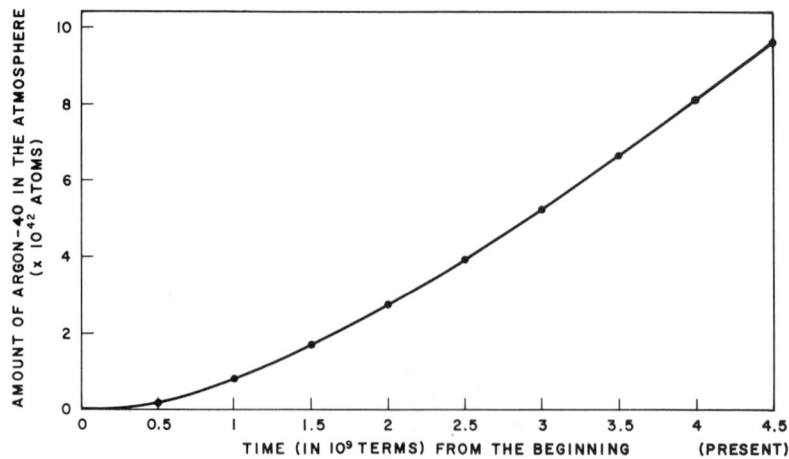

FIG. 1. Growth of argon-40 in the earth's atmosphere with time.

in the Sun, in the Atmosphere, and in Meteorites," in (Geis, J., and Goldberg, E. D., editors) "Earth Science and Meteoritics," Amsterdam, North Holland, pp. 241–272.

Turekian, K. K., 1959, "The terrestrial economy of helium and argon," *Geochim. Cosmochim. Acta,* **17**, 37–43.

Cross-references: *Argon; Gases—Volcanic; Geochemical Evolution of Core, Mantle, and Crust; Helium; Nitrogen; Oxygen—Evolution in the Earth's Atmosphere; Potassium; Thorium; Uranium.* Vol. V: *Volatiles; Volcanic Gases.*

OXIDATION AND REDUCTION

The chemical processes of oxidation and reduction are changes in availability of outer orbital electrons of the chemical elements. These electrons are the ones involved in the formation of chemical bonds between atoms or ions. The loss of electrons from an ion constitutes oxidation; the gain of electrons constitutes reduction. The two processes are concomitant—that is, when one ion is reduced the electrons gained must come from some outside source, usually involving the oxidation of some other ion. An electrical current however may also be involved to supply or remove electrons.

Although oxidation or reduction of any chemical element is possible, the relative ease with which the orbital electron arrangement may be changed differs considerably from one element to another. Those elements, specifically the noble gases, which have a completely filled outer shell, are essentially immune to oxidation or reduction. Recently, however, it has been learned that these elements may enter into chemical combinations under some circumstances.

The noble metals, especially gold and members of the platinum group, are normally found in the uncombined, or native, state and are relatively stable toward oxidation. The more reactive metals such as sodium and potassium normally exist in an oxidized state and are resistant to reduction. These differences in behavior were observed early in the stages of systematizing chemistry and formed the basis of the so-called electrochemical series. In this arrangement, elements listed higher in the series would be capable of displacing from their solutions the ions of any element listed below them. In the process the element going into solution would be oxidized and the one displaced would be reduced. Hence this series could be considered an indication of ease of oxidation; the most readily oxidizable elements being at the top of the list and the least oxidizable at the bottom.

More quantitative and exact interpretations of this important index of the chemical behavior of elements have been developed along with more sophisticated concepts of atomic and molecular structure and chemical bonding. The electrical potential of a single electrode of an element immersed in a solution of its ions represents a convenient index of oxidation behavior. Such potentials can be measured under standard conditions (temperature 25°C, pressure 1 atm, reactants present in unit activity) using the standard hydrogen electrode to complete the electrical circuit. The potential of the standard hydrogen electrode is arbitrarily stated to be 0.00 volts. A list of standard oxidation potentials for oxidation couples is given in Table 1. More complete listings are given in standard references in this field, one of the most widely used being that of Latimer (1952). When written in the form given here, with increasing positive voltages indicating an increase in oxidizing conditions, they conform to the so-called European sign convention, which is generally utilized in geochemical literature. It should be noted that these potentials may be determined for oxidation couples where the more reduced form of the element

TABLE 1. STANDARD OXIDATION-REDUCTION POTENTIALS

Couple	Standard Potential, $E°$
$Li^+ + e = Li$	−3.045
$K^+ + e = K$	−2.925
$Ba^{+2} + 2e = Ba$	−2.90
$Sr^{+2} + 2e = Sr$	−2.89
$Ca^{+2} + 2e = Ca$	−2.87
$Na^+ + e = Na$	−2.714
$Mg^{+2} + 2e = Mg$	−2.37
$Al^{+3} + 3e = Al$	−1.66
$Zr^{+4} + 4e = Zr$	−1.53
$Mn^{2+} + 2e = Mn$	−1.18
$SiO_2 + 4H + 4e = Si + 2H_2O$	−0.86
$Zn^{+2} + 2e = Zn$	−0.763
$Cr^{+3} + 3e = Cr$	−0.74
$Fe^{+2} + 2e = Fe$	−0.440
$Co^{+2} + 2e = Co$	−0.277
$Sn^{+2} + 2e = Sn$	−0.136
$Pb^{+2} + 2e = Pb$	−0.126
$2H^+ + 2e = H_2$	−0.000
$Sn^{+4} + 2e = Sn^{+2}$	+0.15
$Cu^{+2} + 2e = Cu$	+0.337
$Cu^+ + e = Cu$	+0.521
$Fe^{+3} + e = Fe^{+2}$	+0.771
$Ag^+ + e = Ag$	+0.799
$Pd^{+2} + 2e = Pd$	+0.987
$Br_2 + 2e = 2Br^-$	+1.065
$O_2 + 4H + 4e = 2H_2O$	+1.229
$Cl_2 + 2e = 2Cl^-$	+1.3595
$Au^{+3} + 3e = Au$	+1.50
$MnO_4^- + 4H + 3e = MnO_2 + 2H_2O$	+1.695
$F_2 + 2e = 2F^-$	+2.87

OXIDATION AND REDUCTION

has already undergone some degree of oxidation. The standard oxidation potentials in Table 1 presuppose in general that the reactions involved take place in aqueous solution.

For a more general evaluation of the relative stability of ions in various oxidation states, the concept of ionization potential has been developed. The ionization potential represents the work required to dislodge an electron from the outer shell and move it beyond the area of influence of the ion. This is a somewhat more generally applicable value which would indicate the ease of oxidation in any environment. Some applications of the ionization potentials of elements in theoretical geochemistry have been made, for example by Ahrens (1953), but the more specific studies of chemical behavior of elements in rocks, natural water, and related environments have generally utilized electrochemical concepts relating to aqueous solutions.

The relative degree or intensity of oxidizing conditions in a solution can be represented by an electrical potential, controlled by oxidation or reduction equilibria like the couples listed in the table. This potential, commonly referred to in geochemical literature as "redox potential," is represented by the symbol Eh and is expressed in volts. Potentials indicating oxidation power greater than the standard hydrogen electrode are considered to be positive. The redox potential is related to the standard potential for a specific redox couple by the Nernst equation which can be written

$$\mathrm{Eh} = E^0 + \frac{RT}{nf} \ln \frac{[A_{ox}]^a [B_{ox}]^b}{[C_{red}]^c [D_{red}]^d}$$

where R = universal gas constant (1.987 cal/degree-mole); T = temperature in degrees Kelvin; $\ln$ = base e logarithm; n = number of electrons appearing in redox equation; f = Faraday constant (23,060 cal/V); and E^0 = standard potential in volts.

The terms in square brackets represent activities of products and reactants expressed in mass-law form.

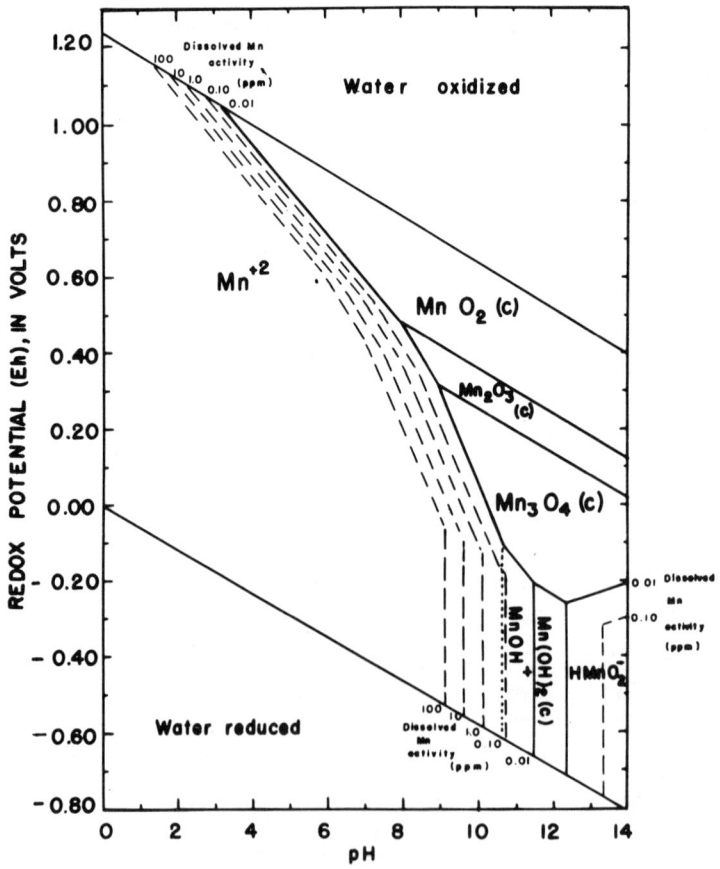

FIG. 1. Fields of stability of manganese species in aqueous solution. Total dissolved manganese activity 0.01 to 100 ppm, bicarbonate and sulfate absent.

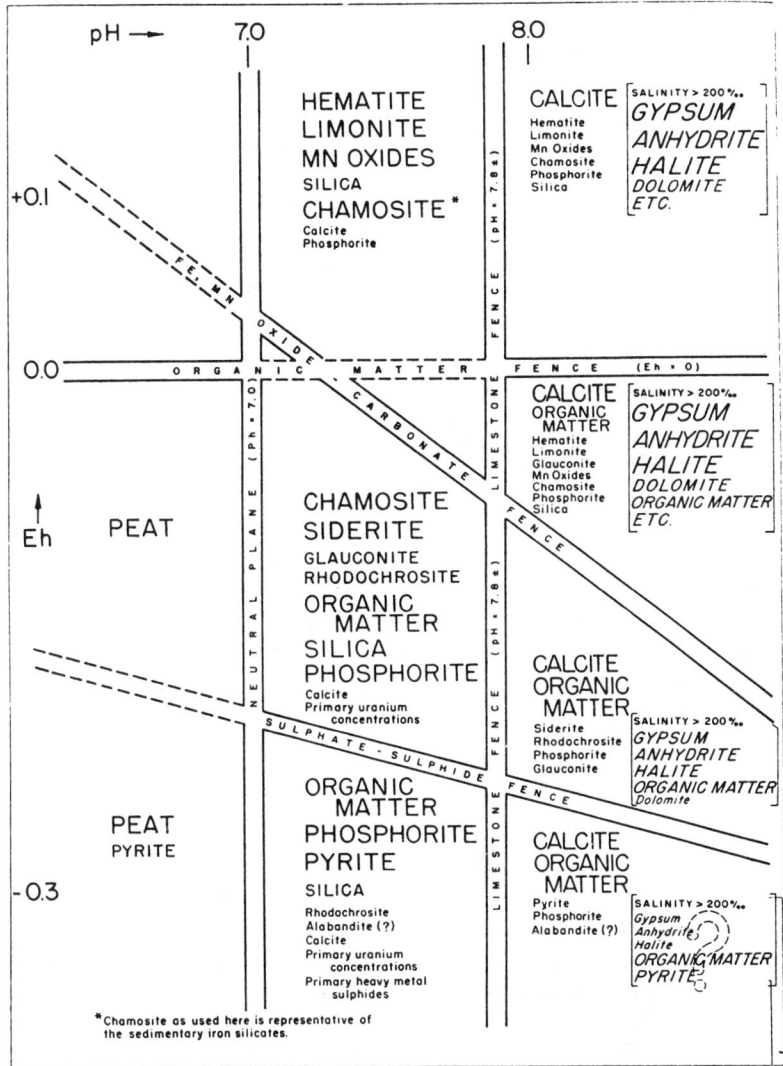

Fig. 2. Sedimentation environments and products as related to pH and redox potential (Krumbein and Garrels 1952).

The net change in standard free energy which occurs in a redox reaction is related to the standard potential by the equation

$$-\Delta G^0 = nfE^0$$

When the standard free energy values of reactants and products are known, it is possible to calculate the standard potential.

Many redox reactions are of interest in geochemistry, especially in such processes as rock weathering, deposition of ores, and in the processes controlling the composition of natural water. The Nernst equation provides a means of quantitatively evaluating some of these reactions and processes, when a state of chemical equilibrium has been reached.

Many redox reactions also involve hydrogen ion activity and if the activities of the other reacting substances are specified, the Nernst equation can be reduced for the reaction in question to an expression relating the two variables Eh and pH. In recent years various writers have used two-dimensional graphs in which pH is plotted as abscissa and Eh as ordinate to represent the conditions required for stability of given forms of solid species or dissolved ions of the elements. These Eh-pH, or stability-field, diagrams seem to have been utilized first in Europe by M. J. N. Pourbaix (1949) in relation to corrosion of metals. The geochemical application of Eh-pH diagrams in the United States has now become extensive.

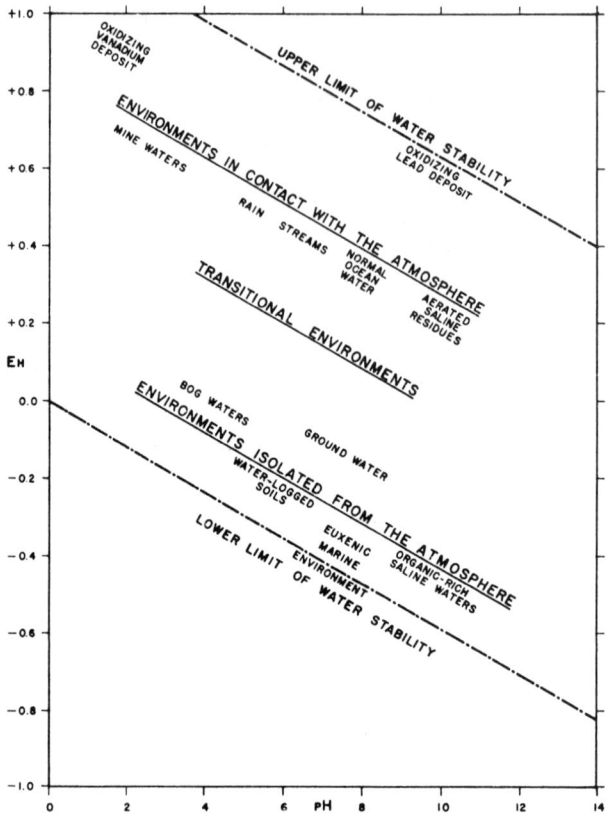

Fig. 3. Approximate position of some natural environments as characterized by Eh and pH (Garrels, 1960).

Garrels' (1960) text contains a thorough discussion of their preparation and use.

Fig. 1 is an Eh-pH diagram for an aqueous solution containing manganous ions (Hem, 1963). The conditions required for stability of the solid oxides and hydroxides are those within the solid lines. These lines bound the fields of stability of the species considered in preparing the diagram, when total activity of dissolved manganous ions is 0.01 ppm. Dashed lines show where these boundary lines would be located at increasing dissolved manganese activities, when equilibrium is established. A solid of the composition indicated should be deposited if the pH and Eh are within the stability field of that solid.

The reserve supply of solids and dissolved ions which can take part in redox reactions serves to stabilize the Eh in a manner that is analogous to the buffering effect associated with pH. The redox stabilizing effect is termed "poising." In natural systems in contact with the atmosphere, the Eh is poised by reactions involving oxygen.

Application of theoretical chemistry in studies of redox systems has improved understanding of a number of problems in geochemistry. Examples include work on chemistry of uranium-vanadium ore deposition by Evans and Garrels (1958), deep-sea sediments by Zobell (1946), estuarine environments and other sedimentary systems by Bass-Becking and others (1960), chemistry of iron by Lapteva (1958) and Hem and Cropper (1959), chemistry of manganese by Krauskopf (1957), chemistry of ground water by Germanov et al. (1959) and by Back and Barnes (1963), and by many others in the United States and abroad.

JOHN D. HEM

References

Ahrens, L. H., 1953, "The use of oxidation potentials, Part 2, Anion affinity and geochemistry," *Geochim. Cosmochim. Acta,* **3,** 1–29.

Baas-Becking, L. G. M., Kaplan, I. R., and Moore, D., 1960, "Limits of the natural environment in terms of pH and oxidation-reduction potentials," *J. Geol.,* **68,** 243–284.

Back, W., and Barnes, I. K., 1962, "Equipment for

field measurement of electrochemical potentials," *U.S. Geol. Surv. Profess. Papers,* **424C,** 366–368.

Evans, H. T., Jr., and Garrels, R. M., 1958, "Thermodynamic equilibria of vanadium in aqueous systems as applied to the interpretation of the Colorado Plateau ore deposits," *Geochim. Cosmochim. Acta,* **15,** 131–149.

Garrels, R. M., 1960, "Mineral Equilibria at Low Temperature and Pressure," New York, Harper and Bros., 254pp.

Germanov, A. I., Volkov, G. A., Lisitsin, A. K., and Serebrennikov, V. S., 1959, "Experiment for the study of the oxidation-reduction potential of ground waters," *Geokhimiya,* **1959,** 259–265.

Hem, J. D., 1963, "Chemical equilibria and rates of manganese oxidation," *U.S. Geol. Surv., Water Supply Paper,* **1667A,** 1–64.

Hem, J. D., and Cropper, W. H., 1959, "Survey of ferrous-ferric chemical equilibria and redox potentials," *U.S. Geol. Surv. Water Supply Papers,* **1459A,** 1–31.

Krauskopf, K. B., 1957, "Separation of manganese from iron in sedimentary processes," *Geochim. Cosmochim. Acta,* **12,** 61–84.

Krumbein, W. C., and Garrels, R. M.,1952, "Origin and classification of the chemical sediments in terms of pH and oxidation-reduction potentials," *J. Geol.,* **60,** 1–33.

Lapteva, O. N., 1958, "Oxidation-reduction potentials of solutions containing ferric and ferrous ions as a function of the pH," *Zhur. Priklad. Klim.,* **31,** 1210–1215.

Latimer, W. M., 1952, "Oxidation Potentials," Second edition, Englewood Cliffs, N.J., Prentice-Hall, 392pp.

Pourbaix, M. J. N., 1949, "Thermodynamics of Dilute Aqueous Solutions," London, Edward Arnold and Co., 136pp.

Zobell, C. E., 1946, "Redox potential of marine sediments," *Am. Assoc. Petrol. Geol. Bull.,* **30,** 477–513.

Cross-references: *Aqueous Solutions; Bonding; Free Energy; Geochemical Classification of the Elements; Geochemistry: Ionization Potentials; pH-Eh; Rare Gases; Water.*

OXIDATION AND REDUCTION OF IRON—MICROBIAL—See IRON OXIDATION AND REDUCTION—MICROBIAL

OXIDES AND HYDROXIDES

Oxygen is the most abundant element in the earth's crust and it is not surprising that the oxides comprise one of the most important classes of minerals. In this class are included isodesmic structural types, whereas the anisodesmic types in which oxygen is in anion radicals, such as carbonates and phosphates, are placed in other classes. Multiple oxides are also included in this class (e.g., the spinels), and for the sake of convenience, the hydroxides and oxides containing hydroxyl are included, although because of their importance they are sometimes placed in a separate class. (See Table below.)

SIMPLE OXIDES

A_2X Type

Cuprite	Cu_2O	
Water (or ice)	H_2O	

AX Type

Periclase	MgO	*Periclase group:* isometric, hexoctahedral; complete series exist between FeO and NiO, CoO and MgO, CaO and CdO; also between CoO, NiO, and MnO
Bunsenite	NiO	
Manganosite	MnO	
Wustite	FeO	
Monteponite	CdO	
Calcium oxide	CaO	
Zincite	ZnO	*Zincite group:* hexagonal, hemisymmetric crystals
Bromellite	BeO	
Tenorite	CuO	
Paramelaconite	$(Cu_{1-2x}Cu_2)O_{1-x}$	Also called "paratenorite"
Montroydite	HgO	
Litharge	PbO	
Massicot	PbO	

A_3X_4 Type

Minium	Pb_3O_4

OXIDES AND HYDROXIDES

A_2X_3 Type

Hematite	Fe_2O_3	⎫ *Hematite group:* rhombohedral; mostly tabular // {0001} and show parting; ilmenite, geikielite, and pyrophanite form an isomorphous series
Corundum	Al_2O_3	
Ilmenite	$FeTiO_3$	
Geikielite	$MgTiO_3$	
Pyrophanite	$MnTiO_3$	⎭
Arsenolite	As_2O_3	⎱ *Arsenolite group:* secondary minerals formed from oxidation of arsenic and antimony ores; isometric or pseudomorphic
Senarmontite	Sb_2O_3	⎰
Claudetite	As_2O_3	
Valentinite	Sb_2O_3	
Bixbyite	$(Mn,Fe)_2O_3$	
Braunite	$(Mn,Si)_2O_3$	

AX_2 Type

Rutile	TiO_2	⎫ *Rutile group:* tetragonal, usually prismatic and vertically striated
Pyrolusite	MnO_2	
Cassiterite	SnO_2	
Plattnerite	PbO_2	⎭
Anatase	TiO_2	
Brookite	TiO_2	
Tellurite	TeO_2	
Cervantite	Sb_2O_4	
Stibiconite	$(Sb)Sb_2(O,OH,H_2O)_{6-7}$	
Bismite	Bi_2O_3	
Sillenite	Bi_2O_3	

A_mX_n Type

Ilsemannite	$Mo_3O_8 \cdot nH_2O$?
Russellite	$(Bi_2W)O_3$
Tungstite	$WO_3 \cdot H_2O$?

Oxides Containing U, Th, Zr

Baddeleyite	ZrO_2	
Uraninite	UO_2	⎱ *Uraninite group:* isometric (fluorite structure)
Thorianite	ThO_2	⎰
Gummite	$UO_3 \cdot nH_2O$	
Clarkeite	$Na_2U_2O_7$	
Becquerelite	$2UO_3 \cdot 3H_2O$?	
Schoepite	$4UO_3 \cdot 9H_2O$?	
Fourmarierite	$PbO \cdot 4UO_3 \cdot 5H_2O$?	
Curite	$2PbO \cdot 5UO_3 \cdot 4H_2O$?	
Vandenbrandite	$CuO \cdot UO_3 \cdot 2H_2O$	
Ianthinite	$2UO_2 \cdot 7H_2O$	

Hydroxides and Oxides Containing Hydroxyl

AX_2 Type

Brucite	$Mg(OH)_2$	⎫ *Brucite group:* hexagonal; flexible, soft, with perfect basal cleavage indicating layered lattice structure
Pyrochroite	$Mn(OH)_2$	
Portlandite	$Ca(OH)_2$	⎭
Lepidocrocite	$FeO(OH)$	⎱ *Lepidocrocite group:* platy crystals with perfect cleavage
Boehmite	$AlO(OH)$	⎰

OXIDES AND HYDROXIDES

Diaspore	AlO(OH)	
Goethite	FeO(OH)	
Groutite	MnO(OH)	*Diaspore group:* orthorhombic pyramidal minerals with exception of manganite, which is monoclinic
Montroseite	(V,Fe)O(OH)	
Paramontroseite	VO_2	
Manganite	MnO(OH)	

Heterogenite	CoO(OH)	

Hydrotalcite	$Mg_6Al_2(OH)_{16} \cdot CO_3 \cdot 4H_2O$	*Hydrotalcite group:* rhombohedral; platy with perfect basal cleavage; soft, low-temperature minerals
Stichtite	$Mg_6Cr_2(OH)_{16} \cdot CO_3 \cdot 4H_2O$	
Pyroaurite	$Mg_6Fe_2(OH)_{16} \cdot CO_3 \cdot 4H_2O$	

Manasseite	$Mg_6Al_2(OH)_{16} \cdot CO_3 \cdot 4H_2O$	*Sjogrenite group:* hexagonal, but in other respects like hydrotalcites.
Sjogrenite	$Mg_6Fe_2(OH)_{16} \cdot CO_3 \cdot 4H_2O$	
Barbertonite	$Mg_6Cr_2(OH)_{16} \cdot CO_3 \cdot 4H_2O$	

Brugnatellite	$Mg_6Fe(OH)_{13} \cdot CO_3 \cdot 4H_2O$
Psilomelane	$(Ba, Mn^{(II)})_2 Mn^{(IV)}_8 O_{16}(OH)_4$

AX_3 Type

Sassolite	$B(OH)_3$
Gibbsite (Bauxite)	$Al(OH)_3$
Hydrocalumite	$Ca_4Al_2(OH)_{14} \cdot 6H_2O$

MULTIPLE OXIDES
ABX_2 Type

Delafossite	$CuFeO_2$

AB_2X_4 Type

Spinel	$MgAl_2O_4$	
Hercynite	$FeAl_2O_4$	
Gahnite	$ZnAl_2O_4$	
Galaxite	$MnAl_2O_4$	
Magnesioferrite	$MgFe_2O_4$	*Spinel group:* isometric, with the A atoms divalent and B atoms trivalent; there are three series included: spinel, magnetite, and chromite with Al, Fe, Cr as the trivalent ion, respectively
Magnetite	$FeFe_2O_4$	
Franklinite	$ZnFe_2O_4$	
Jacobsite	$MnFe_2O_4$	
Trevorite	$NiFe_2O_4$	
Maghemite	$\gamma\text{-}Fe_2O_3$	
Magnesiochromite	$MgCr_2O_4$	
Chromite	$FeCr_2O_4$	

Hausmannite	$MnMn_2O_4$	*Hausmannite group:* tetragonal, but similar to spinels
Hetaerolite	$ZnMn_2O_4$	

Crednerite	$CuMn_2O_4$

AB_4X_7 Type

Hoegbomite	$Mg(Al,Fe,Ti)_4O_7$
Sapphirine	$(Mg,Fe)_{15}(Al,Fe)_{34}Si_7O_{80}$
Plumboferrite	$PbFe_4O_7$
Magnetoplumbite	$Pb(Fe,Mn)_6O_{10}$
Hematophanite	$Pb(Cl,OH)_2 \cdot 4PbO \cdot 2Fe_2O_3$

ABX_3 Type

Quenselite	$PbMnO_2(OH)$
Perovskite	$CaTiO_3$

A_2BX_4 Type

Chrysoberyl	Al_2BeO_4	Olivine-type structure

OXIDES AND HYDROXIDES

A_2BX_5 Type

Pseudobrookite	Fe_2TiO_5

AB_2X_5 Type

Chalcophanite	$ZnMn_2O_5 \cdot 2H_2O$
Zirkelite	$(Ca,Fe,Th,U)_2(Ti,Zr)_2O_5$

AB_4X_8 Type

Coronadite	$Pb_2Mn_8O_{16}$
Hollandite	$Ba_2Mn_8O_{16}$
Cryptomelane	$K_2Mn_8O_{16}$
Todorokite	$(H_2O)_2Mn_8(O,OH)_{16}$
Woodruffite	$(Zn,Mn)_2Mn_5O_{12} \cdot 4H_2O$

Multiple Oxides Containing Cb, Ta and Ti

ABX_4 Type

Pyrochlore	$NaCaCb_2O_6F$ (variable)	*Pyrochlore group:* usually metamict, probably isometric; isomorphous series
Microlite	$(Na,Ca)_2Ta_2O_6(F)$	
Fergusonite	$Y_2Cb_2O_8$ (variable)	*Fergusonite group:* tetragonal series; metamict
Formanite	$Y_2Ta_2O_8$ (variable)	
Yttrotantalite	Complex multiple oxides	
Polymignite	Complex multiple oxides	
Stibiotantalite	$SbTaO_4$	*Stibiotantalite group:* orthorhombic series
Stibiocolumbite	$SbCbO_4$	
Bismutotantalite	$Bi(Ta,Cb)O_4$	
Simpsonite	$Al_2Ta_2O_8$	

$A_mB_nX_p$ Type

Arizonite	$FeTi_3O_9$
Brannerite	$(U,Ca,Th,Y)(Ti,Fe)_2O_6$

AB_2X_6 Type

Tapiolite	$FeTa_2O_6$	*Tapiolite group:* tetragonal series
Mossite	$Fe(Cb,Ta)_2O_6$	
Columbite	$(Fe,Mn)(Cb,Ta)_2O_6$	*Columbite group:* orthorhombic series
Tantalite	$(Fe,Mn)(Ta,Cb)_2O_6$	
Euxenite	$(Y,Ce)(Cb,Ta)_2O_6$	*Euxenite group:* orthorhombic series
Polycrase	$(Y,Ce)(Ti,Cb,Ta)_2O_6$	
Eschynite	$(Ce,Ca,Fe,Th)(TiCb)_2O_6$	*Priorite group:* orthorhombic series
Priorite	$(Y,Er,Ca,Fe,Th)(TiCb)_2O_6$	
Samarskite	$(Y,Er,Ce,U,Ca,Fe,Pb,Th)(Cb,Ta,Ti,Sn)_2O_6$	
Thoreaulite	$SnTa_2O_7$	

$A_mB_nX_p$ Type

Betafite	$(U,Ca)(Cb,Ta,Ti)_3O_9 \cdot nH_2O$	*Betafite group:* isometric (metamict) series?
Djalmaite	$(U,Ca)(Ta,Cb,Ti,Zr)_3O_9 \cdot nH_2O$	

Lloyd W. Staples

References

Dana, J. W., and Dana, E. S., 1944, "The System of Mineralogy," Seventh ed., Vol. I, 489, (revised by Palache, C., Berman, H., and Frondel, C.), New York, John Wiley & Sons.

Evans, H. T., Jr., and Mrose, M. E., 1955, "A crystal chemical study of montroseite and paramontroseite," *Am. Mineralogist* **40,** 861.

Frondel, J. W., and Fleischer, M., 1954, "Glossary of uranium and thorium bearing minerals," *U.S. Geol. Surv. Bull.* **1009-F.**

Heinrich, E. W., 1958, "Mineralogy and Geology

of Radioactive Raw Materials," New York, McGraw-Hill Book Co., pp. 26–64.

Newhouse, W. H., and Glass, J. P., 1936, "Some physical properties of certain iron oxides," *Econ. Geol.*, **31**, 699.

Palache, C., 1935, "The minerals of Franklin and Sterling Hill, Sussex Co. New Jersey," *U.S. Geol. Prof. Paper* **180**.

Cross-references: *Mineral Classes—Non-Silicates; Oxygen.*

OXYGEN: ELEMENT AND GEOCHEMISTRY

Physical Properties

The atomic symbol of oxygen is O. The free element exists as O, most commonly as O_2, and also as O_3 (ozone) and O_4. Physical properties are: atomic weight 15.9994; atomic number 8; electron configuration $1s^2 2s^2 2p^4$; melting point (O_2) $-218.8°C$; boiling point (O_2) $-183.0°C$; density (STP) 1.4285 g/liter; first and second ionization potentials (O^+) 13.614, (O^{2+}) 35.146 eV; electron affinity $O + 2e \rightarrow O^{2-}$ is -7.28 eV; electronegativity (Pauling scale) 3.5. At normal temperatures, O_2 is a colorless, odorless, paramagnetic gas condensing to a bluish liquid at low temperatures and freezing to cubic γ oxygen (α and β modifications are also known). Its solubility in water is 34.3 cm^3/liter at 15°C and 1 atm.

Natural oxygen is composed of three stable isotopes: O^{16} (99.76%), O^{17} (0.04%), and O^{18} (0.2%), although variations in the abundances of these isotopes occur due to fractionation and form the bases of the important method of paleotemperature determination treated elsewhere in this series.

Minerals

Except mainly for the halides and sulfides, oxygen is an important constituent of almost all minerals. In combination with Si and Al as polymerized networks of Si–O and Al–O, oxygen forms the basic structural building blocks of most minerals.

Metallic oxides (i.e., corundum, rutile, and the iron oxides, etc.) are also important minerals. As OH^- or H_2O, oxygen is an important constituent of many minerals (brucite, gypsum); in combination, mainly with C, S, and P, it occurs as complex oxyanions in many other minerals (i.e., calcite, anhydrite, apatite).

Chemistry

Oxygen has a small atomic radius (0.74 Å), although two electrons can fill the $2p$ orbitals to form O^{2-} with a large ionic radius (1.40 Å). The formation of ionic oxides of the type $M^{n+}O^{2-}$ is common, even though the electron affinity for the formation of O^{2-} is high. Ionic oxide stabilities are due to the lattice energies of the compounds evidenced by the standard heats of formation of the oxides, which are of the order of several hundred kcal/mole. The electronegativity of oxygen is second only to fluorine, and strongly electropositive elements (i.e., the alkali metals) form truly ionic oxides.

Covalent oxygen compounds include CO_2 and H_2O, although oxygen bonds with less electropositive elements, i.e., Si–O, can have a partial covalent character.

The role of oxygen in combining with metals to form oxides and silicates has been discussed from a thermodynamic point of view by Goldschmidt (1958) and Latimer (1950).

Geochemistry

The geochemistry of oxygen is really dependent on the geochemistry of other elements forming oxygen-metal compounds. The O^{2-} ion can probably occur in isomorphous substitution with OH^- and F^- in minerals. Because of the analytical difficulties involved in the direct determination of oxygen, some of these substitutions are questionable, i.e., the existence of an oxyapatite molecule substituting for fluor- or hydroxyapatite. Analytical difficulties have also impeded the evaluation of stoichiometry of oxygen-containing minerals, an important but relatively neglected field of analytical geochemistry.

Molecular O_2 is the stable form of the free element, and the equilibrium constant for the reaction $O_2 \rightleftharpoons 2O$ is very small at temperatures up to 3000°C. At 27°C the degree of dissociation at 1 atm is 2.3×10^{-41}. However, strong ultraviolet radiation at 2000–2600 Å will readily dissociate O_2 with subsequent recombination to ozone by a three-body process. Ozone is also formed by oxygen absorption of radiation of wavelength 1760–1925 Å (the *Runge-Schumann bands*). Mass spectrometric determinations of the upper atmosphere from rockets have shown that the $O:O_2$ ratio increases rapidly with height above 100 km, reaching a value of 10 at 200 km. The greatest concentration of ozone in the atmosphere is at 20–30 km, and the existence of the ozonosphere is responsible for absorbing shortwave radiation which would otherwise be lethal to terrestrial life. Oxygen forms 20.99% of dry air by volume at lower altitudes and this percent is relatively constant over the earth's surface. Up to 100 km (within the zone of convection and turbulent mixing), the ratio of oxygen to other gases is relatively constant, but above this altitude there is an increase in oxygen due to diffusive separation of gases.

OXYGEN: ELEMENT AND GEOCHEMISTRY

Earth is the only planet containing appreciable free oxygen in its atmosphere. Venus and Mars contain combined oxygen as CO_2 and water, and the outer giant gas planets have apparently negligible amounts of atmospheric oxygen. The unique position of earth in this respect is probably due to the abundance of life processes on the planet.

Calculations of the mass of oxygen in the various terrestrial environments show that the amount of free atmospheric oxygen on the planet is negligible compared to that bound with other elements.

	Mass of oxygen
Atmosphere	0.116×10^{22} g
Hydrosphere	121×10^{22} g
Lithosphere	2000×10^{22} g
Mantle	$154{,}000 \times 10^{22}$ g

Oxygen, of course, is a primary component of the hydrosphere as H_2O and as complex oxyanions, mainly SO_4^{2-}. The free element occurs dissolved in seawater. In surface waters, where active photosynthesis is taking place, the water is generally saturated or over saturated with free oxygen. The percent saturation drops below the photic zone (see *Seawater, Chemistry*).

The lithosphere contains an average of 46.6 wt % O, mainly bound in silicates and oxides, which makes it the most abundant element. Oxygen shows less variation in atomic percent in different rock types of the lithosphere than any other element (average 60.6 atom percent). The process of surface weathering is essentially an oxidation cycle, under present conditions, so that free atmospheric oxygen is still combining with crustal elements. Crustal weathering processes could readily absorb all the atmospheric oxygen within geological time. Goldschmidt has estimated that between 0.256 and 0.544 kg of free oxygen per cm² of the earth's surface has been fixed in sediments during geological time. This is equivalent to or greater than the present amount of free oxygen. It is difficult, however, to estimate how much oxygen has been recycled back into the atmosphere during metamorphism and volcanic activity. Oxygen is evolved by volcanic gases in combination with C, H, and S mainly as CO_2, H_2O, and SO_2. Fig. 1 represents the inorganic cycle of oxygen after Day (1964).

Because the solid earth is unsaturated in oxygen, and the elements are not in their highest stable valence states, it is improbable that the free oxygen of the atmosphere existed in early geological time (see *Oxygen—Its Evolution in the Earth's Atmosphere*).

Oxygen is essential to terrestrial life processes, either as free O_2 in respiration processes or as CO_2 and H_2O in photosynthesis. Less than 0.0001% of the total oxygen on this planet exists in the free state necessary for maintenance of most life forms. Whether this unique state has led to the abundance of terrestrial life, or whether it is because of it, is a matter of fascinating speculation. Oxygen-containing compounds are important in supplying the energy for bacterial processes, particularly those involving the anaerobic bacteria. In our present civilization oxygen is important not only for the maintenance of life, but also for the energy output involved in the combustion of carbonaceous fuels. The amount of atmospheric oxygen used up in these processes is negligible at the present time, but severe problems exist in the pollution of the atmosphere by the combustion products and in the possibility of increasing the atmospheric CO_2 content to a point where thermal changes become significant (see Fig. 2).

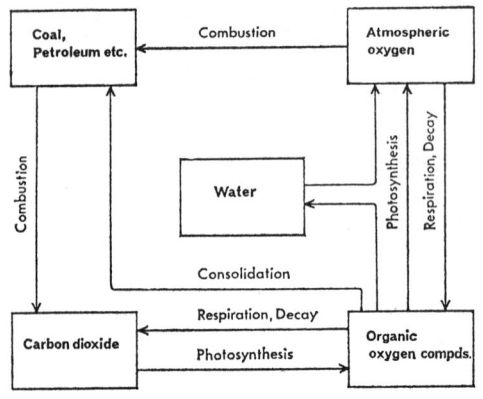

FIG. 2. Organic cycle of oxygen (after Day, 1964).

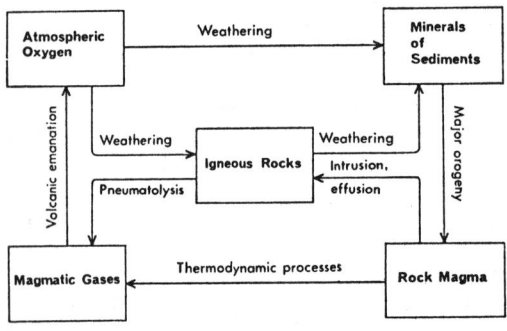

FIG. 1. Inorganic cycle of oxygen (after Day, 1964).

EDGAR F. CRUFT

OXYGEN: EVOLUTION IN THE EARTH'S ATMOSPHERE

References

Ardon, M., 1965, "Oxygen," New York, Benjamin, 106pp.

Day, F. H., 1964, "The Chemical Elements in Nature," New York, Reinhold Publ. Co., 372pp.

Goldschmidt, V. M., 1958, "Geochemistry," Oxford University Press, 730pp.

Latimer, W. M., 1950, "Astrochemical problems in the formation of the earth," *Science*, **112**, 101–104.

Montgomery, H. A. C., et al., 1964, "Determination of dissolved oxygen by the Winkler method and the solubility of oxygen in pure water and sea water," *J. Appl. Chem.*, **14**(7), 280–296.

Sato, M., and Wright, T. L., 1966, "Oxygen fugacities directly measured in magmatic gases," *Science*, **153**(3740), 1103–1105.

Cross-references: *Isotope Fractionation; Oxides; Oxygen Cycle; Oxygen—Its Evolution in the Earth's Atmosphere; Oxygen Isotopes; Seawater: Chemistry; Stoichiometry; Weathering, Chemical. Vol. II: Atmospheric Chemistry; Ozone.*

OXYGEN: EVOLUTION IN THE EARTH'S ATMOSPHERE

From the earliest scientific thought about the earth's surface structure, speculation involving climatic change has been advanced to account in part for lack of uniformity in geologic structure. Early writers such as Aristotle and Pliny, followed by the more precise thought of Hutton and Lyell, have viewed climatic modification as one source of geologic difference. Such views are elaborated in all modern texts on Historical Geology (c.f. Moore, 1933; Dunbar, 1949; Kummel, 1961) following extensive study by many workers.

The idea that the relative composition of the earth's atmosphere has been undergoing significant change throughout its history could be advanced quantitatively only in the present century with the application of modern physics and physical chemistry, biology, and genetics to the processes of the earth, its hydrosphere, and its atmosphere in relation to geologic observation.

Absence of Primordial Atmosphere

Out of the work of Goldschmidt (1937), Harrison Brown (1952), Spencer-Jones (1950), Kuiper (1951), Urey (1952), Alfven (1954), Fesenkov (1959), Vinogradov (1959), and many others, it is generally accepted that upon its agglomeration the earth was without a significant primordial atmosphere. The relative abundance of the rare gases, illustrated in Table 1, shows that their abundance on earth ranges from one ten-billionth to one millionth of their cosmic abundance in order of their ascending molecular weights.

TABLE 1. FRACTIONATION FACTORS OF RARE GASES[a]

Element	Atomic Weight	Fractionation Factor
Neon	20	$\sim 10^{10}$
Argon (36)	36	$\sim 10^8$
Krypton	83	$\sim 2 \times 10^6$
Xenon	130	$\sim 10^6$

[a] After Brown, 1952.

Likewise, the relative abundance of the lighter elements, hydrogen (H) and helium (He), on the sun and the more massive planets, and the paucity of these lighter elements on the inner planets, shows that during their agglomeration, these inner planets lost the greater proportion of mass attributable to the usual abundance of the gaseous elements. Moreover, the accumulation of A^{40}, from decay of K^{40}, corresponds to the estimated age of the solid earth of about five billion years.

Thus, all lines of evidence point to the absence of any primordial atmosphere, and to subsequent growth of an atmosphere primarily from secondary sources self-contained within the earth. This view is consistent with the agglomeration of the earth from planetisimals whose gravitational fields were too small to retain a primordial atmosphere.

The Primitive Reducing Atmosphere

According to Urey (1952), Vinogradov (1959), and others, the physical chemistry of compounds comprising the major portion of the earth's crust indicates that since its agglomeration the earth has never been substantially molten. The earth consequently retains large quantities of gases chemically bound in various ways. These are sufficient to provide for the source of the secondary atmosphere.

The volcanic origin of this secondary atmosphere is developed by Rubey (1955), Urey (1952), Holland (1962), Vinogradov (1959), and others. The continents have been built at an average rate of 1–3 km³/yr from volcanic effluents, based on estimates of Sapper (1927), Verhoogen (1946), Bullard (1962), and Wilson (1954). In evaluating these estimates, on one hand following Vinogradov, volcanic activity should have been considerably greater in earlier eras, precedent to long decay of radioactive elements. On the other hand, following Verhoogen, the existing andesitic volcanoes, particularly in the Pacific andesite ring, represent to some extent merely reworked continental materials. Wilson's estimate of a rate of average continent building of 1.3 km³/yr throughout geologic time, and down to present times, is compatible within an order of magnitude with present estimates of total volume of continental (crustal) materials.

Accompanying these solid effluents are corresponding volumes of gases—primarily primitive water vapor, presumably released from water of crystallization, together with CO_2, N_2, SO_2, H_2, Cl_2, in substantial quantities, accompanied by traces of many other gases. No oxygen is released directly from volcanic effluents.

The absence of a significant content of oxygen in the primitive secondary atmosphere is confined by several lines of evidence. First, there is no suitable source. Second, the incomplete oxidation of early sedimentary materials (~three billion years of age) as demonstrated by Rankama (1955), Ramdohr (1958), Lepp and Goldich (1959), and others, and summarized by Rutten (1962), suggests very early lithospheric sedimentation in a reducing atmosphere. This evidence is in concordance with conclusions of extensive studies of the physical chemistry of early volcanic effluents by Holland (1962), who shows that such activity occurred in a strongly reducing atmosphere.

Finally, the rapidly growing evidence on the origin of life on Planet Earth appears to forbid a significant oxygen concentration until evolution of relatively advanced microorganisms exhibiting organized properties such as photosynthesis. The work of Haldane (1933), Oparin (1924) and Bernal (1949) directed attention to the steps leading to the organization of the simple biological cell, which is now recognized as a very advanced evolutionary entity.

Oparin visualized a logical series of evolutionary syntheses starting from inorganic and very simple organic materials of nonbiological origin (i.e., the simplest compounds of H, C, O, and N) together with some traces of other elements (such as S, P, and Fe), finally ending with the organized cell replete with living function which Bernal defines pragmatically as "the embodiment within a certain volume of self-maintaining chemical process." Oparin made the interesting suggestion that each step in the achievement of the whole process was the consequence of natural experimentation on a large scale guided by natural selection.

The extensive literature on evolution leading to the simple cell is developed in work by Wald (1955), Rabinowitch (1951), Calvin (1959, 1962), Sagan (1957, 1961), Anfinsen (1961), and many others. Synthesis of amino acids and other complex elements of cell structure is demonstrable in a reducing atmosphere, in the presence of ultraviolet light, which provides the energy for chemosynthesis through photoexcitation (c.f. Miller, 1959). Viable chemical precursors to the living cell were selected step-by-step from a thin soup of ever more complex organic compounds (c.f. Florkin, 1959).

During this phase of primitive evolution of preliving compounds, oxygen is a powerful poison, acting to break them down as they are formed. Abelson (1957) has shown, for example, that organic materials such as amino acids which are normally stable for long intervals in anoxygenic atmospheres, are quickly degraded in presence of atmospheres that are significantly oxygenic, particularly in the presence of a catalytic energy source such as visible light.

Photosynthesis can proceed in anoxygenic atmospheres as shown by Hill (1939) and Hill and Scarisbrick (1940), and in the process the whole warehouse of oxygenic components required for growth and reproduction can be synthesized as shown by Calvin and his co-workers. Thus, the evidence from evolutionary experiment and theory, relating to primitive molecular components of life, confirms the anoxygenic (reducing) character of the primitive secondary atmosphere.

In absence of widespread photosynthesis, carbon dioxide will rise as a major constituent of the primitive atmosphere. The physicochemical studies of Rubey (1951) show that CO_2 must be limited to perhaps thirty times its present concentration in the atmosphere (or 1% of present atmospheric partial pressure), due to regulation through chemical precipitation as carbonate from seawater, a chemical reaction, of course, related to temperature.

Oxygenic Concentration in Primitive Reducing Atmosphere

The widespread presence of lithospheric oxides throughout the Archeozoic and Proterozoic eras is indicative of the presence of active quantities of oxygen. Origin of primitive oxygen from photodissociation of water vapor by ultraviolet light of suitable energy has been suggested by many investigators. Urey pointed out in 1959, however, that the concentration of oxygen produced by photodissociation of H_2O in the primitive atmosphere must be self-regulating. This occurs because H_2O and O_2 absorb the photodissociating ultraviolet in almost identical wavebands. H_2O is precipitated from the troposphere, thus severely limiting its concentration at higher levels. O_2 produced by photodissociation of H_2O is distributed exponentially above the surface. Thus, as O_2 increases in concentration, it shadows the H_2O available for photodissociation shutting off the production of O_2. With respect to the small concentration of water vapor above the stratosphere, Nicolet and Bates (1963) have shown

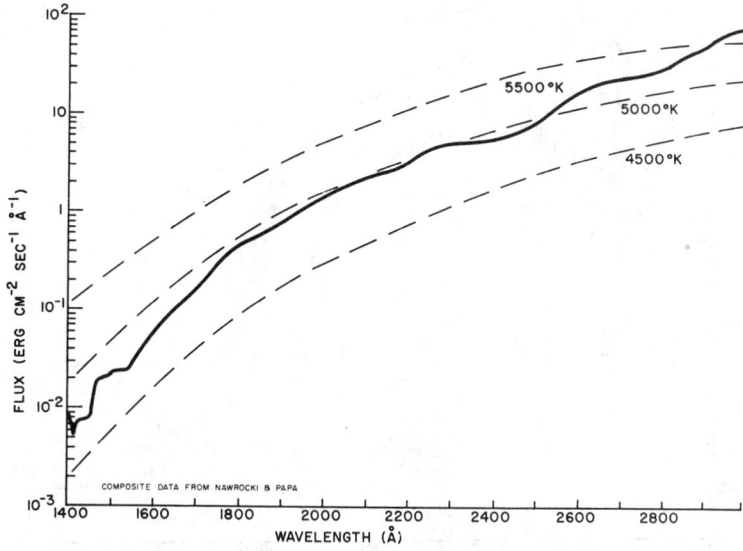

Fig. 1. Solar intensity (1400–3000 Å).

that as a consequence of photodissociation of small quantities of CH_4 by very energetic ultraviolet radiation, water vapor is actually produced at the expense of oxygen, so that no net production of oxygen occurs above the tropopause. Consequently, oxygen in the primitive atmosphere is limited by the Urey effect.

Berkner and Marshall (1964) have calculated the self-regulated level of oxygen as a consequence of the "Urey effect" in the primitive atmosphere. Excellent data on solar radiation in the UV spectrum are now available from the space probes of many workers, including Hall, Damon, and Hinterregger (1963), Detweiler, Garrett, Purcell, and Tousey (1960), to name but a few. Available data on solar UV radiation have been summarized by Nawrocki and Papa (1961), (Fig. 1) after Johnson's (1954) new evaluation of the solar constant.

Radiation down to 1400 Å arises in the upper 100 km of the solar photosphere. As shown from the studies of Wilson (1963) on the evolutionary history of stars of the main sequence, similar to our sun, this photospheric radiation above 1400 Å is extremely stable—probably self-regulating and unvarying over very long periods comparable to the age of the earth. Radiation below 1400 Å arises primarily in the chromosphere and corona, and may have been as much as three times the present average level (thus involving total average fluctuations of as much as 20 to 1) in this spectral region over shorter intervals. From the point of view of total ultraviolet energy, however, by far the major bulk (99.9%) arises from the stable photospheric radiation above 1400 Å. Therefore, in computing oxygen concentrations in the primitive troposphere, only radiation above 1400 Å need be considered.

Absorption of this ultraviolet spectrum by various possible component gases above 1000 Å is now well known and is summarized in Figs. 2a and 2b after the rather complete work of Watanabe and his colleagues (1953) and the contributions of many others.

In the spectral region above 1400 Å, absorption by H_2 and N_2 and by the rare gases He, Ne, Ar, Kr, and Xe is negligible. Of the important probable atmospheric constituents, only H_2O, CO_2, O_2, and O_3 absorb radiation strongly in this region, and only O_3 is important as an absorbing gas above 2200 Å.

In order to interpret the relative contributions of different concentrations of the several possible constituent gases as absorbers over the UV spectrum, Berkner and Marshall combine these data on radiation and on absorption at each wavelength to calculate the NTP equivalent path length that reduces the transmitted radiation to an arbitrary negligible level of 1 erg-cm^{-2}-sec^{-1}-(50 Å)$^{-1}$. This reference energy level is actually 10^{-4} of the level in a 50 Å band at the peak of solar radiation at approximately 4500 Å (i.e., the transmitted energy level, after absorption, which is 0.01% of the average incident radiation in a 50 Å band at the peak wavelength of solar radiation; a transmitted level equivalent to about fifty times the brightness of the full moon). Thus, the path length, x, in cm, of any constituent gas (NTP) required to reduce the radiation

OXYGEN: EVOLUTION IN THE EARTH'S ATMOSPHERE

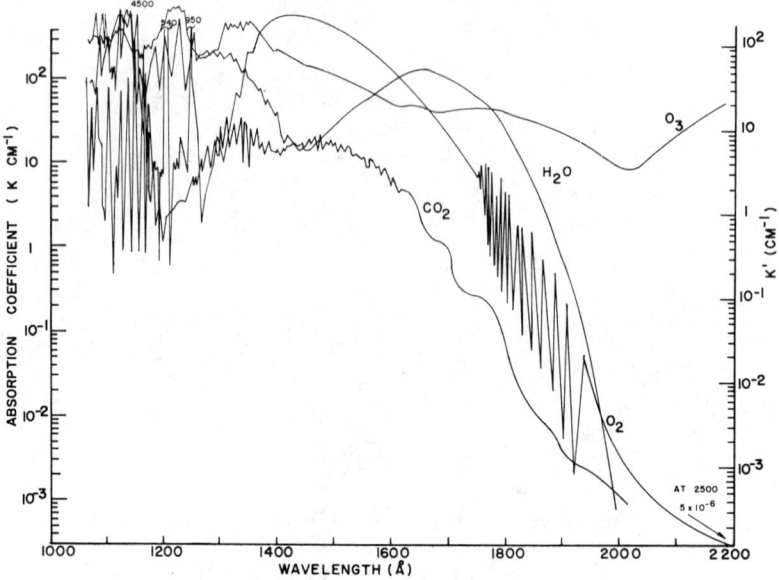

FIG. 2a. Composite of ultraviolet absorption in atmospheric gases.

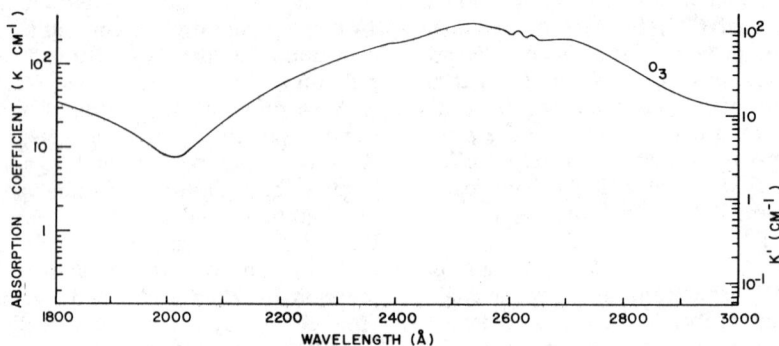

FIG. 2b. Ultraviolet absorption in O_3 (data from Watanabe and Vigroux).

to 1 erg-cm^{-2}-sec^{-1}-(50 Å)$^{-1}$ is expressed as the path length required for its "extinction."

The results of combining radiation and absorption data are shown in Fig. 3 for H_2O vapor.

Here it is apparent that the wavelength, at which significant penetration through H_2O vapor can occur, is not very sensitive to concentration. It drops only about 100 Å for a decrease of three orders of magnitude of H_2O vapor in the most significant range of pressures. (Note also that a change of reference level from 1 to 0.1 erg-cm^{-2}-sec^{-1}-(50 Å)$^{-1}$, for example, will simply shift the curve upward by one decade in Fig. 3.) In the absence of oxygen, and within the range of CO_2 pressures already discussed, the important photochemical reactions that produce oxygen can be summarized by

$$H_2O + 2h\nu \rightarrow 2H + O$$

which provides the source of atmospheric oxygen.

Since oxygen is distributed exponentially above the base of the stratosphere, while water vapor is precipitated out to very low concentrations at the same level, the presence of oxygen will shadow the H_2O vapor in its range of photodissociation (the "Urey effect") as seen from Fig. 4. In this figure it is evident that when the path length of O_2 above the troposphere (10 km) approaches 35 cm, water vapor is completely shadowed from the dissociative radiation band for H_2O up to 1950 Å. This path length of 35 cm of O_2 above 10 km corresponds to a total path length of O_2 above the surface of about 100 cm NTP.

This path length (100 cm NTP) represents

852

OXYGEN: EVOLUTION IN THE EARTH'S ATMOSPHERE

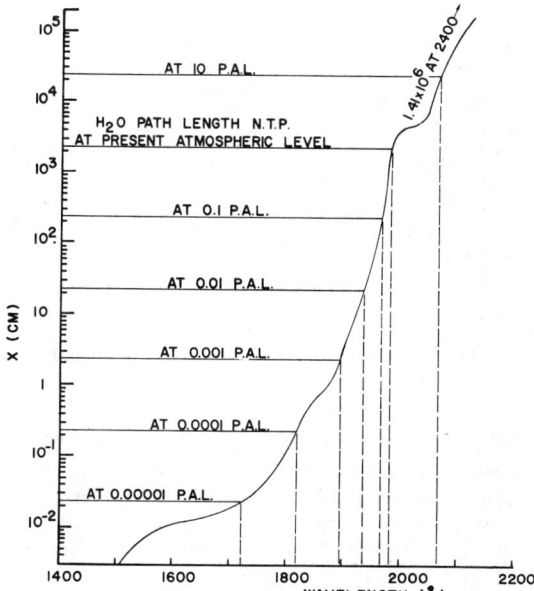

FIG. 3. Thickness of H_2O required to absorb available ultraviolet energy to "extinction" (1 erg-cm^{-2}-sec^{-1}-$(50Å)^{-1}$).

an oxygen level (see Fig. 4) of somewhat less than 0.001 present atmospheric concentration. On the premises of the above discussion, *the upper limit of oxygenic pressure in the primitive atmosphere is less than 0.1% of present atmospheric level (PAL).*

Photodissociation of CO_2 leads to a series of reactions that do not appear to yield a net production of oxygen. However, since CO_2 absorbs UV significantly in the same bands as H_2O and O_2, its presence in excess quantities will lower, somewhat, the estimates of oxygen concentration in the primitive atmosphere. This effect has not yet been quantitatively evaluated over possible ranges of CO_2 to obtain better estimates of oxygenic concentrations in the primitive atmosphere.

Precambrian Lithospheric Oxides

It is often assumed that because of the extensive oxides found in the Precambrian (Proterozoic) lithosphere, that atmosphere must have been significantly oxygenic. This assumption seems unnecessary when the pertinent reactions are reviewed:

$$O + O + M \rightarrow O_2 + M$$
$$O_2 + O + M \rightarrow O_3 + M$$
$$O_3 + M_s \rightarrow \text{surface oxides}$$

coupled with the additional reactions

$$O + O_3 \rightarrow 2O_2$$
$$O_2 + h\nu \rightarrow O + O$$

In the primitive atmosphere the supply of O_3 is maintained close to the surface where it is removed through surface oxidation at very high reaction rates.

Berkner and Marshall have computed the available oxygen supply over geologic time,

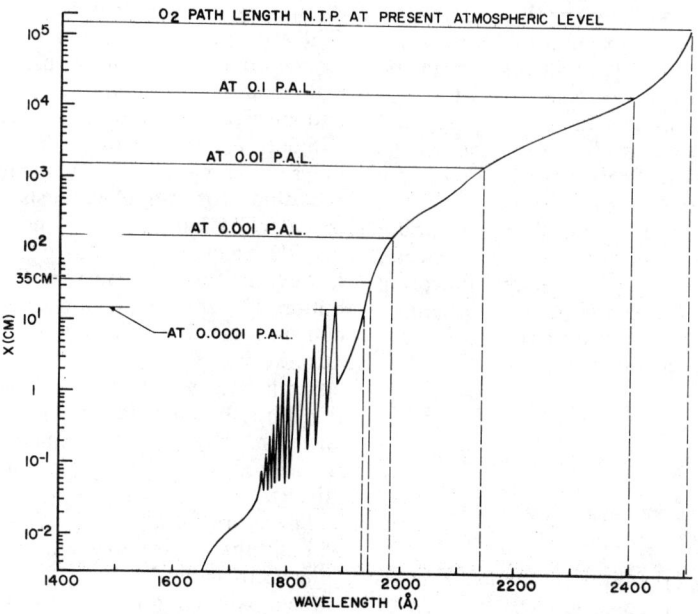

FIG. 4. Thickness of O_2 required to absorb available ultraviolet light to "extinction" (1 erg-cm^{-2}-sec^{-1}-$(50Å)^{-1}$).

from the total energy available for photodissociation. It is found that the available energy is on the order of 100 times that necessary to account for existing Precambrian lithospheric oxides. (The calculations show that for each kilometer of completely oxidized metals in the lithosphere, oxygen equivalent to the oxidation of about one-third of the water of the oceans would be required.) Likewise, sufficient UV energy is available to dissociate the primitive concentration of O_2 to the extent that O_3 will be maintained as a significant constituent near the surface. These calculations lead to the realization that oxidation rate of lithospheric materials in the Archeozoic and in the Proterozoic is dependent not so much on the absolute concentration of oxygen as on its chemical form and the reaction rate in that form. Consequently, the classic assumption is unnecessary that abundance of lithospheric oxides in the Precambrian dictates high oxygenic levels in the primitive atmosphere.

Thus in the primitive atmosphere, the oxygen balance is dominated by a rate of loss which consumes oxygen promptly upon its production. The rapid removal of oxygen from the primitive atmosphere, and its inherent self-regulation by the "Urey" process, together with the other geochemical and biological evidence, leads therefore to the conclusion that the oxygen level in the primitive atmosphere of the earth was $<10^{-3}$ (i.e., $<0.1\%$) of present atmospheric levels.

Rise in Oxygenic Concentration

The subsequent rise of oxygenic level can be attributed only to photosynthesis. The summary of Rabinowitch (1951) shows that in the present atmosphere, all oxygen passes through the photosynthetic process in ~2000 years, all CO_2 in ~350 years, and all H_2O in the oceans in ~2×10^6 years. These intervals are very short compared to geologic periods.

Thus photosynthesis, in oxidizing liquid water and at the same time reducing carbon dioxide to carbohydrate, is the overpowering source of oxygen in the present atmosphere.

At any time after the primitive, the oxygen balance will be determined by:

Plus: Photochemical dissociation of H_2O
         Photosynthesis
Minus: O_2 and O_3 oxidation of surface materials
         Decay and Respiration
         O_2 in H_2O solution

As Holland (1962) points out, the rise of oxygen occurs as a consequence of a small differential between two much larger opposing effects, which relatively speaking increase together. As the initial oxygen from photosynthesis is released, moreover, it simply substitutes for oxygen from photodissociation of H_2O, because of the inherent "Urey" regulation to the level $O_2 < 10^{-3}$ PAL. This balance is illustrated qualitatively in Fig. 5.

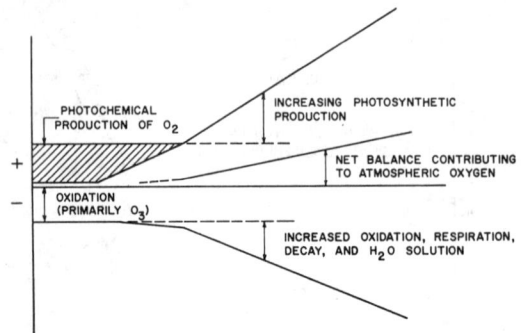

FIG. 5. Factors in early oxygen balance.

Not until the net rate of production of oxygen, primarily by photosynthesis, exceeds the net rate of O_2 dissociation, and its consequent loss as an active oxidant, can equilibrium values of oxygen exceed the levels in the primitive atmosphere and building of a stable oxygenic atmosphere begin. This is the critical criterion for the growth of an oxygenic atmosphere.

This leads to inquiry concerning the ecology for the rise of photosynthesis. Caspersson (1950) and Davidson (1960) show that cell absorption of UV arises from absorption by nucleic acids primarily between 2600 and 2700 Å, and by proteins between 2700 and 2900 Å. Cell absorption in these bands is highly lethal to cell function in all forms, degrading chemical function and halting growth, reproduction, and survival. The only available terrestrial constituents for shadowing lethal radiation and at the same time transmitting energy-rich visible radiation for photosynthesis is atmospheric ozone and liquid water. The thickness of O_3 (NTP) required to absorb the available UV to "extinction" is shown in Fig. 6. The distribution of ozone in idealized form for various concentrations of oxygen is shown in Fig. 7. Ozone is distributed roughly uniformly in a column between its level of maximum production and the surface, to which it is convected and lost. The height of maximum production is lowered as the oxygenic concentration diminishes.

Therefore, the total path length of ozone will diminish with oxygenic concentrations as shown in Table 2.

The path lengths for absorption of varying wavelengths of UV in liquid water have been measured by Dawson and Hulburt (1934).

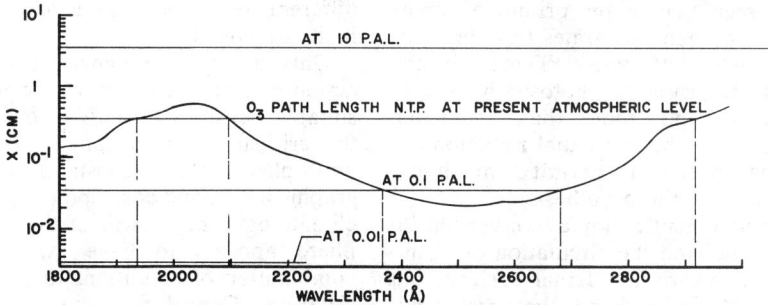

FIG. 6. Thickness of O_3 required to absorb available ultraviolet light to "extinction" (1800–3000Å) (1 erg-cm^{-2}-sec^{-1}-(50Å)$^{-1}$) (after Berkner and Marshall, 1964).

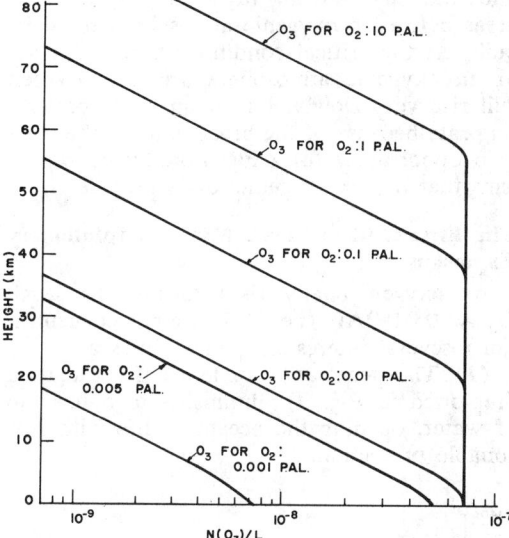

FIG. 7. Estimated distribution of ozone for various levels of oxygen (after Berkner and Marshall, 1964).

Combining the data on path lengths of incident radiation through varying partial pressures of oxygen, corresponding path lengths in ozone, and resultant path lengths in liquid water to reach extinction, yields Fig. 8, which shows the penetration of UV in liquid water, in the presence of various oxygenic atmospheres.

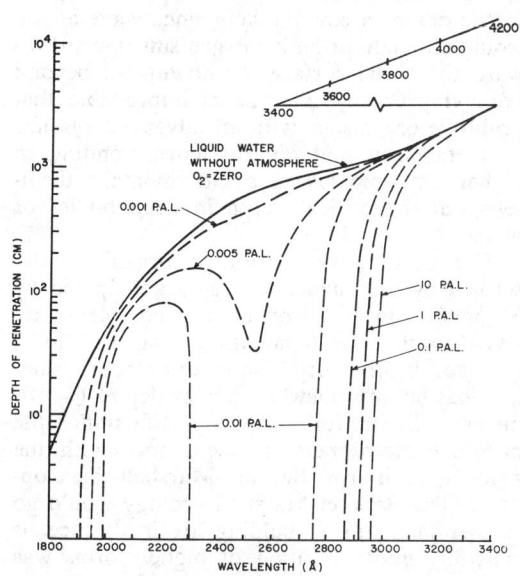

FIG. 8. Path length in liquid water in the presence of O_2 and O_3 required for various concentrations of O_2 to absorb available ultraviolet light to "extinction" (1 erg-cm^{-2}-sec^{-1}-(50Å)$^{-1}$).

TABLE 2. ESTIMATED TOTAL PATH LENGTHS OF O_3 FOR VARYING PRESSURES OF O_2[a]

O_2 (NTP) Proportion of PAL	~ Height of O_3 Column (cm)	Av. % O_3 (NTP) (atm column)	~ Path Length of O_3 (cm, NTP)
10.0	65×10^5	7×10^{-8}	0.5
1.0	47×10^5	7×10^{-8}	0.33
0.1	28×10^5	7×10^{-8}	0.2
0.01	12×10^5	4×10^{-8}	0.05
0.005	12×10^5	1.6×10^{-8}	0.02
0.001	12×10^5	4×10^{-9}	0.005

[a] The crude integration of total path length established in this table is all that present data coupled with basic assumptions will justify. Note particularly that the concentration of O_3 is in higher proportion to O_2 at the surface at the lower oxygenic levels, when surface oxidation more profoundly affects the oxygen balance.

Here it is seen that in the primitive atmosphere, lethal radiation penetrates to a depth of five to ten meters of water. Therefore, the requirements for primitive photosynthesis are:

(a) A water depth more than about ten meters, sufficient to shadow lethal radiation but no deeper than needed to permit a maximum of visible light for photosynthesis.

(b) A limited gentle liquid convection to avoid on the one hand the circulation of primitive organisms toward the lethal surface, but to provide, on the other hand, the organic nutrients synthesized photochemically at the surface in the presence of UV, according to the processes described by Miller.

This rigidly restrictive ecology, anticipated by Berkner (1952) and Sagan (1957, 1961), describes bottom-dwelling organisms (green algae or their evolutionary precursors) in protected shallow lakes or seas. In particular, life in the oceans seems unlikely since wave action would circulate primitive organisms upward toward the lethal surface, or downward beyond life-giving energy. It appears improbable that primitive organisms, without advanced systems of metabolism and control, corresponding to higher oxygenic levels, could maintain themselves at the critical depth in deep bodies of water.

The ecology for primitive organisms described by this model is largely anticipated by Sagan (1961) based on his less complete data on absorption and radiation. Sagan's model calls for benthic organisms restricted to shallow seas but at somewhat greater depths ($\sim 10^2$ meters). Sagan forbids pelagic life under the primitive atmosphere in accordance with the more quantitative Berkner-Marshall development. The Berkner-Marshall ecology would go further in rigidly forbidding life in the oceans generally until evolution of higher forms was made possible by a more oxygenic and protective atmosphere.

Warm pools associated with volcanic hot springs, rich in nutrient minerals and elemental compounds appear prime candidates for the origin of life and photosynthesis. The ancient bioherms, 2.7 billion years old, are known from the work of Hoering and Abelson (1961) to have supported photosynthetic organisms, and could well have been at the base of such pools. Such bioherms must be very close to the seat of life. The rigid ecologic insulation between such pools admits the possibility of multiple origins of living organisms, with natural selection among groups of different origin only at a later era when the permissive ecologic environment became more general. This "admission" may be restricted, however, by the recent recognition that the genetic code among quite different terrestrial organisms may be identical in its organization.

Only as the continents grow with volcanic action can sufficient areas for photosynthesis at suitable densities of activity be found to meet the critical criterion for a growing oxygenic atmosphere. With constantly changing geographic areas, and corresponding fluctuations of climatology, the growth of the oxygenic atmosphere appears to have awaited the proper combination of conditions to satisfy the critical criterion. Considering the lowered levels of light energy below lethal depths of water, Berkner and Marshall tentatively estimate that photosynthetic activity at about present distribution per unit area must have covered between one and ten percent of present continental areas before an oxygenic atmosphere could be built. As the critical conditions for production of an oxygenic atmosphere are met, oxygen will rise very slowly, but at an ever accelerating rate because of the broadening of the ecologic opportunity for oxygen-producing organisms that higher oxygenic levels provide.

The First Critical Level: Marine Evolutionary Explosion

As oxygen finally rises toward the level, $O_2 \sim 0.01$ PAL (i.e., 1% present concentration) several interesting potentialities arise.

(1) The penetration of lethal UV radiation (replotted in Fig. 9) diminishes to a few cm of water, opening the oceans to life with reasonable protection.

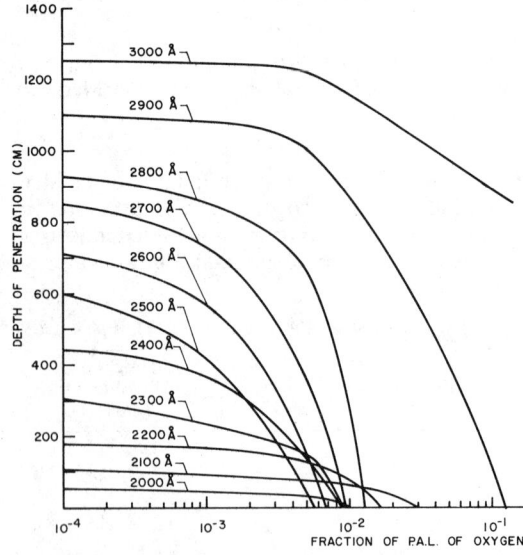

FIG. 9. Penetration of ultraviolet light in liquid water with various combinations of oxygen and ozone atmospheres (intensity at extinction = 1 erg-cm^{-2}-sec^{-1}-(50Å)$^{-1}$).

(2) The oxygenic level reaches the "Pasteur point" where organisms can change their metabolic process from fermentation to respiration. Thus the energy available for chemosynthesis jumps from ~ 20 cal/g-mole to ~ 675 cal/g-mole (c.f. Genevois, 1927, and Rabinowitch, 1951).

(3) At this same oxygenic concentration, many primitive organisms change from anaerobic "photoreduction" to "photosynthesis" through rapid oxidation of an hydrogenase enzyme, thus broadening the base for evolutionary activity.

This leads to a search of paleontologic and geologic history for a radical and explosive change in marine evolutionary forms, corresponding to the opening of entirely new and far more widespread evolutionary opportunities as $O_2 \rightarrow 0.01$ PAL. There is, of course, just one such evolutionary explosion—the Cambrian—beginning 600 million years ago. (This dating follows the most recent geologic and geochronologic conclusions (c.f. Kummel, 1961)). By immediate inference, the oxygenic level, $O_2 \rightarrow 0.01$ PAL is identified by Berkner and Marshall as immediately preceding the opening of the Cambrian, following the earlier suggestion of Berkner.

Prior to the Cambrian there is utterly no evidence in the fossil record of any form of life advanced beyond the elementary algae, fungi, and bacteria; i.e., the simplest forms of thallophyla (cf Rutten, 1962). Since Proterozoic sediments favorable to fossil preservation have been diligently studied for more than a century by a host of workers, the complete absence of fossils representing more advanced forms prior to the Cambrian evolutionary explosion has been considered heretofore as a scientific "puzzle" (cf Kummel, 1961). The usual assumption is that evolutionary Precambrian precursors could have had only "soft" parts that were unfavorable for fossilization (although it should be noted that in subsequent ages, fossils of this general kind are not infrequent).

Under the interpretation dictated by the Berkner-Marshall model, no advanced precursors to the Cambrian evolutionary explosion should be expected until sufficient oxygenic concentrations opened the evolutionary niche that presaged the opening of the Cambrian. In accordance with this model, the geologic record should be read exactly as presented in nature.

Considerable evidence can be adduced to support the Berkner-Marshall model. For example:

(1) In the Cambrian evolutionary explosion, the early emergence of the complex respiratory system as a major organic feature, with corresponding development of the circulatory system to circulate oxygen and remove CO_2, and the nervous system to control the process, is interpreted as evolution awaiting sufficient oxygenic levels; (i.e., $>1.0\%$ PAL).

(2) The diurnal sinking of the pelagic organisms during daylight (c.f. Hardy, 1960) can be interpreted as the early Cambrian evolutionary response of the more advanced organisms, as their explosive evolution began, to the lethal UV radiation near the surface, a response carried by the organism down to the present.

(3) At the opening of the Cambrian, the precedent low levels of photosynthesis had left high CO_2 atmospheric levels, encouraging widespread development of shelled carbonaceous organisms related to the Cambrian limestones, with corresponding reduction of CO_2 levels through subsequent periods.

The Cambrian explosion of marine evolution corresponding to the rise of oxygen to 1% PAL has been designated "the first critical level" of oxygen in the earth's atmosphere.

Strong interactions between the details of the geologic column and the varying atmospheric composition can be anticipated, particularly with respect to the temperature-sensitive surface concentrations of ozone. Rubey (1955) has opened this subject with his exploration of CO_2. Wildt (1942) has suggested the present reddish surface of Mars as due to rapid surface oxidation in the presence of favorable concentrations of ozone. Further analysis of such interactions, particularly in the Proterozoic periods and perhaps in the Archeozoic should prove fruitful.

The Second Critical Level: Evolution Ashore

Following the opening of the Cambrian, the complexity of life is known to have multiplied rapidly. In the next few million years more than 1200 species of different creatures appeared, many of very considerable size and variety of characters. During this time the foundations for all modern phyla were laid. In particular, complex and efficient forms of respiratory apparatus were evolved independently among various phyla as increasing oxygen levels presented favorable opportunities for selection of such evolutionary advances. These advanced respiratory systems provided the mechanistic basis for the concurrent development of circulatory systems, digestive tracts, central nervous systems, bisexual modes of reproduction, and similar characters associated with advanced biological organisms having sufficient metabolic energy through effective supply of oxygen and removal of oxidized carbon.

As oxygen rises with widespread marine photosynthesis to the level of 0.1 PAL (10% present concentration), Fig. 9 shows that lethal radiation will for the first time be largely shielded from dry land. This will open a new ecological niche for evolution above the water and ashore.

Recent evidence indicates the possibility of microscopic palynologic organisms on land as early as mid-Silurian, indicating that plants had reached the surface to release spores or pollen. But the geologic record shows no direct evidence of any "advanced" form of life ashore until the late Silurian period, 420 million years ago. Then a number of different phyla of plants and animals exploded on dry land. By the early Devonian, thirty million years later, great forests had appeared, and the first amphibian vertebrates were found ashore. Thus the mid- or late Silurian is interpreted by immediate inference in the Berkner-Marshall model as the earliest stage at which plants could emerge above the surface without danger of lethal "sunburn." Therefore the period ~420 million years ago is identified with the second critical oxygenic level, $O_2 \sim 0.1$ PAL.

The Rise of Oxygen to Present Quasi-permanent Level

The explosion of evolution ashore increases photosynthesis in a step function by some 20 to 25%, again tilting the oxygen balance radically toward the plus side.

In examining the oxygen balance in light of all factors studied to date, the relation between production rate and final equilibrium may be represented crudely by Fig. 10. Much more refinement of this balance is, of course, to be expected in future studies.

Following the late Silurian, high rates of photosynthesis are induced without corresponding quantities of organic materials immediately available ashore for decay and replenishment of CO_2. This suggests that oxygen may have "overswung" the present level to a somewhat higher value as the lush life of the Carboniferous developed. Then, with reduction of CO_2, the earth would cool, due to loss of the "greenhouse" effect of CO_2, leading to the ice ages of the Permian Period. As the earth cooled, photosynthesis would sharply fall, due to de-adaptation of oxygen-producing organisms, leading to a radical drop of atmospheric oxygen as the balance was upset. Thus, the phase difference of maximum of production of O_2 and of CO_2 suggests that the levels of these two atmospheric components in the post-Silurian atmosphere may have been unstable, the instability being damped by the ever-improving adaptation of organisms to wider environmental ecologies. Pending more analytical

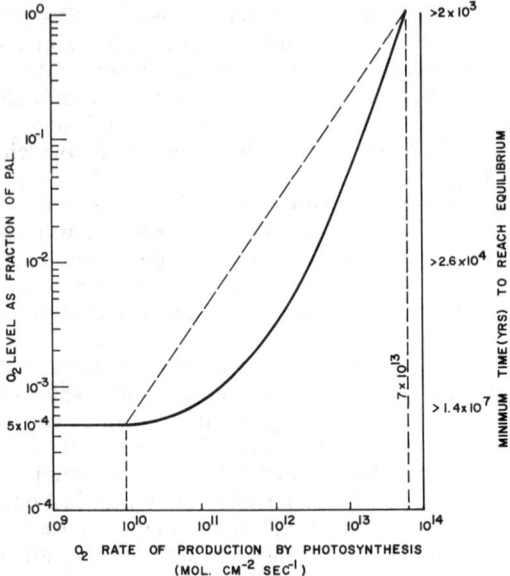

Fig. 10. Estimated atmospheric levels of oxygen as a function of production rates.

study, a tentative estimate has been made on the order of 10^8 years for a complete oscillation above and below the present quasi-permanent level. We must not expect complete symmetry of this function, however, since any drop of oxygen is likely to arise from a series of interrelated events centered around de-adaptation of oxygen-producing organisms, events whose mutual interaction would force an unstable and precipitous drop of oxygenic levels followed by a very slow return.

The tremendous range of geologic events recorded in the rocks in the late Paleozoic and subsequent eras await detailed interpretation in light of the factors affecting the oxygen balance as described by the Berkner-Marshall model. Without such detailed study and interpretation it is unwise to attribute major geologic or paleontologic events during the more recent geologic periods, such as ice ages, or the collapse of the reptilian phyla at the end of the Cretaceous, to fluctuating oxygenic levels however suggestive these may be. Nevertheless, the de-adaptation of any important group of oxygen-producing organisms, over any considerable area, whether it arises from the temperature changes accompanying an ice age, or through extensive defoliation and radiation poisoning as a consequence of thermonuclear war, must have a profound effect on the oxygen balance. The quantitative aspects of such problems await further, and obviously more difficult analysis.

The Berkner-Marshall model is as yet too

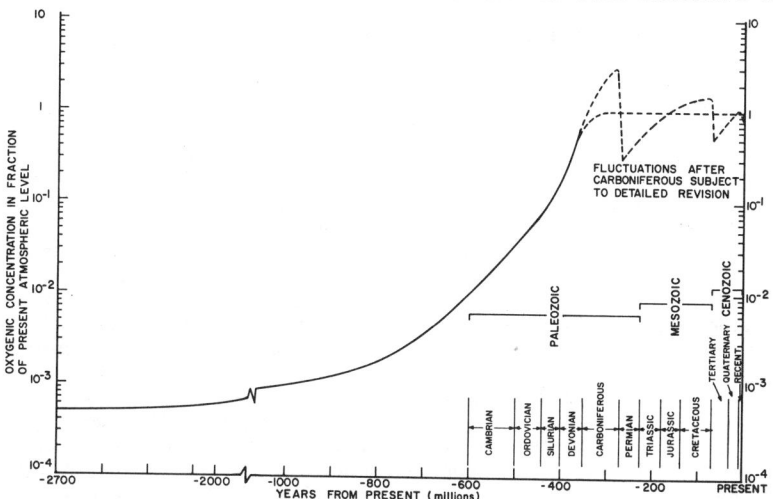

FIG. 11. Model of oxygenic evolution.

recent to have received widespread acceptance. Earlier studies have put oxygen in the Proterozoic at higher concentrations, though now on less firm grounds (c.f. Rutten, 1962). (see Fig. 11.)

In the Berkner-Marshall model it is necessary to consider evolution in two separate roles.

(1) The character and geographic extent of living organisms which at any time must be the major contributors to the balance of the oxygenic atmosphere at any period beyond the primitive.

(2) The identification of a period of explosive evolutionary change as an indicator of the timing of critical oxygenic concentrations in the atmosphere, e.g., as an indicator of major and appropriate physical opening of the environment.

In connecting these two roles, the evolutionary process during any geologic period has been interpreted as an intricate interaction between the oxygenic level made possible by living organisms, and the natural evolutionary exploitation of that level for the new ecologic niches which that oxygen concentration opens to evolution. As such evolutionary opportunities are captured, the oxygen level is then raised as a consequence, thereby permitting a new round of evolutionary development at appropriate oxygenic levels. From time to time critical levels are reached that permit vast physical ecologic opportunities, due to critical responses of organisms to favorable oxygenic levels. Evolution is viewed as a process of rapid scanning of the ecologic scene by the whole range of phyla to exploit quickly any advantages of natural selection that the changing atmospheric composition can afford.

This model opens wide new avenues for geologic and paleontologic study and interpretation in light of the underlying hypothesis.

L. V. BERKNER (deceased)
L. C. MARSHALL

ACKNOWLEDGMENT: This research was supported in part by the National Science Foundation under grant GP-768 and by the United States Weather Bureau under grant Cwb 10531. This paper is based on a report submitted to the National Science Foundation dated November 26, 1963.

References

Abelson, P. H., 1957, "Some aspects of paleobiochemistry," *Ann. N.Y. Acad. Sci.*, **69**, 276.

Alfven, H., 1954, "On the Origin of the Solar System," Oxford Clarendon Press.

Anfinsen, C. B., 1961, "The Molecular Basis of Evolution," New York, John Wiley & Sons.

Berkner, L. V., 1952, "Signposts to future ionospheric research," *Geophys. Res. Papers* No. 12 (from Proc. Conf. Ionosph. Physics, 1950).

Berkner, L. V., and Marshall, L. C., 1964, "The history of oxygenic concentration in the earth's atmosphere," Presented at Faraday Soc. Meet., Edinburgh, Scotland, April 2–3, 1964, Published in *Faraday Soc. Discussions*, **1964**, No. 37, 122–141.

Berkner, L. V., and Marshall, L. C., 1965a, "History of major atmospheric components," *Proc. Natl. Acad. Sci.*, **53**(6), 1215–1226.

Berkner, L. V., and Marshall, L. C., 1965b, "On the origin and rise of oxygen concentration in the Earth's atmosphere," *J. Atmos. Sci.*, **22**(3), 225–261.

Bernal, J. D., 1949, "The physical basis of life," *Proc. Phys. Soc.*, (London), **62A**, 537.

Brown, H., 1952, "Rare gases and the formation of the earth's atmosphere," in (Kuiper, G. P., editor), "The Atmospheres of the Earth and Planets," Chicago, The University of Chicago Press, p. 258.

Bullard, F. M., 1962, "Volcanoes, in History, in Theory, in Eruption," Texas, University of Texas Press.

Calvin, M., 1959, "Evolution of enzymes and the photosynthetic apparatus," report in "The Origin of Life on the Earth," Symp. Intern. Union Biochem., Moscow, 1957, New York, The MacMillan Company, Vol. 1, p. 207.

Calvin, M., and Bassham, J. A., 1962, "The Photosynthesis of Carbon Compounds," New York, W. A. Benjamin, Inc.

Calvin, M., 1962, "Communication: from molecules to Mars," Bull. Am. Inst. Biol. Sci., XII (5), 29.

Caspersson, T. O., 1950, "Cell Growth and Cell Function: A Cytochemical Study," New York, W. W. Norton and Company, Inc.

Davidson, J. N., 1960, "The Biochemistry of the Nucleic Acids," Fourth ed., New York, John Wiley & Sons.

Dawson, L. H., and Hulburt, E. O., 1934, "The absorption of ultraviolet and visible light by water," J. Opt. Soc. Am., 24, 175.

Detweiler, C. R., Garrett, D. L., Purcell, J. D., and Tousey, R., 1960, "The intensity distribution in the ultraviolet solar spectrum," Ann. Geophys., 17, 263.

Dunbar, C. O., 1949, "Historical Geology," New York, John Wiley & Sons.

Fesenkov, V. G., 1959, "Some consideration about the primaeval state of the earth," report in "The Origin of Life on the Earth," Symp. Intern. Union Biochem., M. Florkin, President, Moscow, 1957, New York, The Macmillan Company, Vol. 1, p. 9.

Florkin, M., 1959, selected References from "The Origin of Life on the Earth," Symp. Intern. Union Biochem., Moscow, 1957, New York, The Macmillan Company, Vol. 1, pp. 503 and 578.

Genevois, L. (Bordeaux), 1927, "On the respiration and fermentation in green plants," Beochem. Z., 186, 461.

Goldschmidt, V. M., 1937, "Geochemische Verteilungsgesetze der Elemente. IX. Die Mengenverhaltnisse der Elemente und der Atom-Arten," Norske Videnskaps-akad. Oslo, Skr., Mat.-Nat., Kl., 4, 148.

Haldane, J. B. S., 1933, "The Origin of Life," New York, Harper.

Hall, L. A., Damon, K. R., and Hinteregger, H. E., 1963, "Solar extreme ultraviolet photon flux measurements in the upper atmosphere of August, 1961," Space Research III, Amsterdam, North Holland, pp. 745–759.

Hardy, A. C., 1960, "The Open Sea: Its Natural History," Boston, Houghton Mifflin, 2 vols.

Hill, R., 1939, "Oxygen produced by isolated chloroplasts," Proc. Roy. Soc. London, Ser. B, 127, 192.

Hill, R., and Scarisbrick, R., 1940, "The reduction of ferric oxalate by isolated chloroplasts," Proc. Roy. Soc. London, Ser. B., 129, 238.

Hoering, T. C., and Abelson, P. H., 1961, "Carbon isotope fractionation in formation of amino acids by photosynthetic organisms," Proc. Natl. Sci. U.S., 47, 623.

Holland, H. D., 1962, "Model for the evolution of the earth's atmosphere," in "Petrologic Studies: A Volume to Honor A. F. Buddington," Princeton, N.J., Princeton University, p. 447.

"International Critical Tables of Numerical Data, Physics, Chemistry and Technology," 1926–1930, prepared under the auspices of the International Research Council of the National Acadamy of Sciences, by National Research Council of the United States of America, New York, McGraw-Hill Book Company.

Johnson, F. S., 1954, "The Solar Constant," J. Meteorol., 11, 431.

Kuiper, G. P., 1951, "On the origin of the solar system," Proc. Natl. Acad. Sci., 37, 1.

Kummel, B., 1961, "History of the Earth," San Francisco, W. H. Freeman and Co.

Lepp, H. and Goldich, S. S., 1959, "Chemistry and origin of iron formations," Geol. Soc. Am., Bull. 70, 1637.

Miller, S. L., 1959, "Formation of organic compounds on the primitive earth," report in "The Origin of Life on the Earth," Symp. Intern. Union Biochem., Moscow, 1957, New York, The Macmillan Company, Vol. 1, p. 123.

Moore, R. C., 1933, "Historical Geology," New York, McGraw-Hill Book Co., 673pp.

Nawrocki, P. J., and R. Papa, 1961, "Atmospheric Processes," Geophysics Corporation of America, Bedford, Mass., A.F.C.R.L. Report Contract #AF 19(604)7405.

Nicolet, M., and Bates, D. R., 1963, International Scientific Radio Union, Proc. General Assembly, Tokyo.

Oparin, A. I., 1924, "Origin of Life," London, Uropa Publications.

Oparin, A. I., 1953, "Origin of Life," New York, Dover Publications, p. S213.

Rabinowitch, E. I., 1951, "Photosynthesis and Related Processes," New York, Interscience Publishers.

Ramdohr, P., 1958, "Die uran-und Goldlagerstatten Witwatersrand, Blind River District," Abhandl. Deut. Akad. Wiss. Berlin Kl, Chem., Geol. Biol., 3, 35pp., 19 pl.

Rankama, K., 1955, "Geologic evidence of chemical composition of the pre-Cambrian atmosphere," Geol. Soc. Am., Spec. Paper 62, 651.

Rubey, W. W., 1951, "Geologic history of sea water: an attempt to state the problem," Bull. Geol. Soc. Am., 62(2), 111.

Rubey, W. W., 1955, "Development of the hydrosphere and atmosphere, with special reference to probable composition of the early atmosphere," Geol. Soc. Am., Spec. Paper 62, 631.

Rutten, M. G., 1962, "The Geological Aspects of Origin of Life on Earth," Amsterdam, Elsevier Publ. Co., 146pp.

Rutten, M. G., 1969, "Sedimentary ores of the Early and Middle Precambrian and the history of atmospheric oxygen," Sed. Ores., Feb.

Sagan, C., 1957, "Radiation and the origin of the gene," Evolution, 11, 40–55.

Sagan, C., 1961, "On the origin and planetary

distribution of life," *Radiation Res.,* **15**(2), 174–192.
Sapper, K., 1927, "Vulkankunde," Stuttgart, Engelhorns.
Spencer-Jones, Sir H., 1950, "The evolution of the earth's atmosphere," *Sci. Progr.,* **38**, 417.
Urey, H. C., 1952, "The Planets: Their Origin and Development," New Haven, Yale University Press.
Verhoogen, J., 1946, "Volcanic heat," *Am. J. Sci.,* **244**(11), 745.
Vinogradov, A. P., 1959, "The origin of the biosphere," report in "Origin of Life on the Earth," Symp. Intern. Union Biochem., M. Florkin, President, Moscow, 1957, New York, The Macmillan Company, Vol. 1, p. 23.
Wald, G., 1955, "The Origin of Life," chap. 1, Part I, in "The Physics and Chemistry of Life," New York, Simon and Schuster, p. 3.
Watanabe, K., Zelikoff, M., and Inn, E. C. Y., 1953, "Absorption coefficients of several atmospheric gases," Geophysical Research Papers, No. 21, A.F.C.R.C. Technical Report #52-23, Geophysics Research Directorate, Cambridge, Mass.
Wildt, R., 1942, "The geochemistry of the atmosphere and the constitution of the terrestrial planets," *Rev. Mod. Phys.,* **13-14** (Jan., 1941, Dec., 1942), 151–159.
Wilson, J. T., 1954, "The development and structure of the crust, rates of erosion, deposition and volcanism," in "The Solar System," Vol. II of (Kuiper, G. P., editor), "The Earth as a Planet," Chicago, University of Chicago Press, p. 150.
Wilson, O. C., 1963, "A probable correlation between chromospheric activity and age in main-sequence stars," *Astrophys. J.,* **138**, 832.

Cross-references: *Carbon Cycle; Geologic Time Scale; Organic Compounds; Outgassing of the Planet Earth; Oxygen Cycle; Oxygen: Element and Geochemistry; Photosynthesis; Potassium-Argon Dating; Volcanic Gases.* Vol. II: *Atmospheric Chemistry; Climate and Evolution; Greenhouse Effect; Ice-Age Theory; Ozone; Solar Radiation.* Vol. VI: *Methane.* Vol. VII: *Algae; Cambrian; Evolution; Paleontology.*

OXYGEN CYCLE

Oxygen is the most abundant element of the earth's crust and is more widely distributed in the lithosphere, hydrosphere, and atmosphere than any other major component. This is due to its wide variety of silicates, metallic oxides, its presence in water, its importance in organic compounds and life processes, and its existence as free oxygen in the atmosphere. It is clearly the most significant oxidizing agent.

Although the lithosphere and hydrosphere are the largest reservoirs for oxygen, the most important ones are in the carbon compounds and the atmosphere. Oxygen is very closely associated with carbon in life processes and carbonate geochemistry (see *Carbon Cycle*). For those interested in a nontechnical description of the organic components necessary for life, Adler's *How Life Began* may be useful.

Oxygen on the Earth

Oxygen is ubiquitous due to its varied gaseous, liquid, and solid forms. Its chemical forms include water, free oxygen gas, organic and inorganic compounds of carbon, nitrogen, sulfur, and phosphorus, and numerous ionic oxides and silicates. Mason (1958) suggests that oxygen makes up 47% of the earth's crust but is second in abundance to iron for the earth as a whole. In solar abundances, oxygen is surpassed only by hydrogen and helium.

The distribution of oxygen in its various forms and reservoirs is presented in Table 1. The significant feature to note is that the lithosphere of silicates and oxides is by far the largest reservoir. The hydrosphere includes only about 10% of the oxygen and the atmosphere only about 0.01%. The amount stored in the biosphere is quite negligible in quantity but nevertheless most significant in the oxygen

TABLE 1. AMOUNTS OF DIFFERENT CHEMICAL FORMS OF OXYGEN IN KG PER M² EARTH'S SURFACE[a]

Chemical Form	Lithophere	Hydrosphere	Atmosphere	Plant Biosphere	Animals
H_2O	490,000	2,510,000	30	5.9	0.057
O_2	—	20	2300	—	—
Organic O	530	5	—	0.89	0.0064
CO_2 or CO_3^{2-}	490,000	310	3.4	—	—
PO_4^{3-}	93,000	0.4	—	0.019	0.00046
N_2O	—	0.00035	0.0017	—	—
NO_3^-	0.0017	3.8	—	—	—
SO_2	—	—	0.000004	—	—
SO_4^{2-}	15,000	4,950	—	0.0023	0.000056
Inorganic O	20,400,000	76	—	0.0004	0.00001
Total O	21,500,000	2,515,000	2330	6.8	0.064

[a] After Bowen, 1966.

OXYGEN CYCLE

TABLE 2. EXCHANGE RATES OF OXYGEN BETWEEN BIOLOGICALLY IMPORTANT RESERVOIRS[a]

Exchange	kg O_2/(m²)(yr)
Water transpired by land plants	65
Plant production of free oxygen	0.26
Oxygen used in respiration by predators and decomposers	0.26
Fuel combustion	0.0093
Oxygen from reduction of nitrates	0.019
Oxygen from plant reduction of sulfates	0.0019
Oxygen from bacteria reduction of sulfates	0.00063
Production of organically combined oxygen	0.86
Oxygen lost by biologic carbonate sediments	0.000089

[a] From Bowen, 1966.

cycle. Mason (1958) arrives at similar ratios of the reservoirs by considering the oxygen percentage of total material in each. He gives no value for the biomass because of the large uncertainty.

The oxygen of the lithosphere is involved in many reactions between minerals which are studied by the methods of *petrology* (see Vol. V). Water is taken up or released in many mineral equilibria. A discussion of such reactions is beyond the scope of this article (see, for example, *Serpentine* in Vol. V). The role that water plays in life processes is well known. The cycling of organic compounds and carbonates are discussed under carbon chemistry. Oxides of phosphorus, nitrogen, and sulfur are also necessary in various life processes.

The crucial role of oxygen is its ability to exist in the free gaseous state as a universal oxidizing agent. Although very small in quantity, the biomass plays the major role in the oxygen cycle between the hydrosphere and the atmosphere. Table 2 is a summary of exchange rates of oxygen between reservoirs that are of biological importance. It may be observed that exchange of free oxygen by the biomass is so low that 5000 to 10,000 years would be necessary for the turnover of atmospheric oxygen. Fig. 1 shows the oxygen cycle with the major reservoirs and the dominant exchange rates of gaseous oxygen.

The mechanisms of exchange of gaseous oxygen are production by photosynthesis and loss by biologic respiration and the oxidation of organic material and reduced metals. Photosynthesis that occurs in green plants can be represented by the reaction:

$$CO_2 + H_2O \xrightarrow[\text{chlorophyll}]{\text{solar energy}} \text{organic molecules} + O_2$$

Analogous release of oxygen occurs during the reduction of sulfur and nitrogen compounds by bacteria. Respiration by animal life is the reverse of the above reaction using oxygen to oxidize organic molecules with the associated release of energy.

The primitive atmosphere of the earth was probably devoid or nearly devoid of oxygen gas (Rubey, 1955; Urey, 1952; Poldervaart, 1955; Rankama, 1955). The question arises as to the original form of the oxygen now in the atmosphere. Most of it was probably in the form of water and carbon dioxide that have been outgassed from the earth. From them oxygen gas is released by various mechanisms.

Oxygen in the Atmosphere

This article has stressed the importance of free atmospheric oxygen and the biomass as the crucial features of the oxygen cycle. As suggested above, the earth's primitive atmosphere was devoid of oxygen and thus reducing conditions would have prevailed. Sometime since the formation of the earth 4.5×10^9 years ago the present atmosphere developed. Conditions changed from reducing to oxidizing with the absence and subsequent presence of free atmospheric oxygen. For these reasons it seems appropriate to review briefly the origin of the atmosphere. It will be noted that distinct differences of opinion exist as to when the oxygen in the atmosphere reached its present level.

There are two hypotheses concerning the origin of the atmosphere (Rubey, 1955). The first is that the present atmosphere is a residual of a primitive one occurring when the earth was formed. The second is that both the hydrosphere and atmosphere developed by

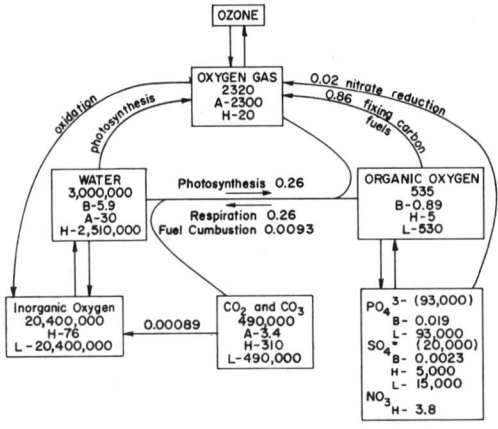

FIG. 1. The oxygen cycle. Numbers inside boxes are kg O per m² and those attached to arrows represent turnover in kg O per m² per year. (B, biosphere; A, atmosphere; H, hydrosphere; L, lithosphere.)

either continual or spasmodic outgassing of the earth. From the volume of the hydrosphere and the composition of the atmosphere, the latter hypothesis seems more tenable.

The primitive atmosphere was certainly quite different in composition. Urey (1952) suggests that its components included water, hydrogen, nitrogen, ammonia, and methane. Rubey (1955) argues that the persistence of methane and ammonia would depend on free hydrogen. However, this element preferentially escapes from the earth's atmosphere since its molecular velocity approaches the escape velocity. He concludes that nitrogen and carbon dioxide were the dominant components. Whatever the major constituents, it seems certain that the primitive atmosphere was reducing and largely devoid of oxygen.

There are two mechanisms which result in free oxygen gas. These are photodissociation and photosynthesis. *Photodissociation* is the interaction of high-energy radiation with water vapor (in the atmosphere) to give free hydrogen and oxygen:

$$2H_2O \xrightarrow{h\nu} 2H_2 + O_2$$

With the earth's surface now largely shielded from this radiation by ozone, such a reaction can only occur high in the atmosphere where radiation is sufficiently intense. *Photosynthesis* releases oxygen by biological processes as discussed previously. These two mechanisms then are responsible for atmospheric oxygen. A juvenile source is quite unlikely since volcanoes, furmaroles, and hot springs have no free oxygen.

Clearly the origin of atmospheric oxygen is closely associated with early life. Poldervaart (1955) suggests that in the early history of the earth the photodissociation mechanism was sufficient to initiate atmospheric oxygen. Other authors feel that free oxygen is entirely due to biological processes and began accumulating in the atmosphere only after life had begun (Berkner and Marshall, 1965; Bowen, 1966; Cloud, 1968; Commoner, 1965). The first living organisms may have been anaerobes living in a reducing environment below about ten meters of water (Commoner, 1965). This environment was necessary for protection from the ultraviolet radiation reaching the earth's surface. (With no atmospheric oxygen, the earth's surface is not shielded by an ozone layer.)

It would be very useful to know the oxygen content of the atmosphere during its accumulation through time. As Holland (1965) points out, lower oxygen pressure unfortunately does not result in direct effects on ocean chemistry, the one environment where reducing conditions could be observed. From studies of the probable origin of life and of sedimentary rocks such as the Precambrian banded iron formations, quite distinct differences of opinion have evolved.

Oxygen began to accumulate about 2×10^9 years ago due to the newly emergent biological systems (Commoner, 1965), but Fairbridge (1967, p. 416) sets the date back to at least 2.7×10^9 years, on the basis of the emergence of the stromatolitic algae. From studying radiation absorption by oxygenated atmospheres and from paleontological evidence, Berkner and Marshall (1965) conclude that oxygen has accumulated largely since the Precambrian. The beginning of the Cambrian may be marked approximately as the time when pO_2 reached 1% of the present level, the so-called *"Pasteur Level."*

Holland (1965) takes the position that atmospheric oxygen has been close to its present level since the Precambrian. This is based on the uniform carbon-13 content of carbonate sediment through the Precambrian and the apparent uniformity of sedimentary rocks. The argument from carbon-13, however, is not applicable since the small differences expected in the carbon isotope ratio would be impossible to detect. The presence of banded iron formations in Precambrian rocks possibly reflects the lower atmospheric oxygen and the delicate balance of evolving life systems (Cloud, 1968).

The question of the accumulation of oxygen comes down to exactly when its partial pressure reached the present atmospheric level or came close to it. No clear evidence is available for either the Pre- or Postcambrian theories.

At the present time oxygen is generally assumed to be more or less in equilibrium. The present control of the oxygen in the atmosphere is by photosynthesis and respiration in biological systems. In ocean sediments the oxygen content is controlled by a feedback mechanism from the fixation of organic material. Photosynthetic processes are controlled by the presence of radiation and the supply of nutrients. If the oxygen level rises, more organic material is oxidized bringing the free oxygen content down again. When the oxygen level drops, more organic material is permanently stabilized until the level rises. As an alternative to this equilibrium hypothesis, there may be a slow accumulation of atmospheric oxygen still in progress but this delicate increment is impossible to measure.

Ozone is the atmospheric component which shields the biosphere from lethal cosmic radiation. In contrast to other important atmospheric components, it is both chemically formed and destroyed within the atmosphere. The important reactions of production are:

$$O_2 + h_\nu \longrightarrow O + O$$
$$O_2 + O \longrightarrow O_3$$

The reactions of destruction are:

$$O_3 + h_\nu \longrightarrow O_2 + O$$
$$O_3 + O \longrightarrow 2O_2$$

Junge (1963) discusses other possible mechanisms but the above ones are the most significant. Most of the atmospheric ozone resides in the stratosphere. The main maximum in concentration ranges between 20 and 25 km above the earth's surface dependent on season and latitude (Junge, 1963).

Conclusions

The important oxygen reservoirs of the earth and the exchange rates between them have been observed. The critical role of oxygen is its occurrence in the free gaseous state and in the biological cycles. Oxygen gas has accumulated in the atmosphere throughout the history of the earth by the processes of photodissociation and photosynthesis, the latter being the dominant mechanism. At present, the biological cycles appear to control the oxygen content of the atmosphere. Ozone, the triatomic species of oxygen, shields the living biosphere from lethal cosmic radiation.

KENNETH M. WOLGEMUTH

References

Adler, I., 1957, "How Life Began," New York, The John Day Co.
Berkner, L. V., and Marshall, L. C., 1965, "History of major atmospheric components," *Proc. Natl. Acad. Sci.*, **53**, 1215.
Bowen, H. J. M., 1966, "Trace Elements in Biochemistry," New York, Academic Press.
Cloud, P. E., 1968, "Atmospheric and hydrospheric evolution on the primitive earth," *Science*, **160**, 729.
Commoner, B., 1965, "Biochemical, biological, and atmospheric evolution," *Proc. Natl. Acad. Sci.*, **53**, 1183.
Fairbridge, R. W., 1967, "Carbonate Rocks and Paleoclimatology," in (Chilingar, G. V., Bissell, H. J., and Fairbridge, R. W., editors), "Carbonate Rocks"(A), Amsterdam, Elsevier Publ., pp. 399–432.
Holland, H. D., 1965, "The history of ocean water and its effect on the chemistry of the atmosphere," *Proc. Natl. Acad. Sci.*, **53**, 1173.
Junge, C. E., 1963, "Air Chemistry and Radioactivity," New York, Academic Press.
Mason, B., 1958, "Principles of Geochemistry," New York, John Wiley & Sons.
Poldervaart, A., 1955, "Chemistry of the earth's crust," *Geol. Soc. Am. Spec. Paper* **62**, 119.
Rankama, K., 1955, "Geologic evidence of the chemical composition of the precambrian atmosphere," *Geol. Soc. Am. Spec. Paper* **62**, 651.
Rubey, W. W., 1955, "Development of the hydrosphere and atmosphere," *Geol. Soc. Am. Spec. Paper* **62**, 631.
Urey, H. C., 1952, "The Planets. Their Origin and Development," New Haven, Yale Univ. Press.
Weyl, P. K., 1968, "Precambrian marine environment and the development of life," *Science*, **161**, 158.

Cross-references: *Carbon Cycle; Cycles—Geochemical; Outgassing of the Planet Earth; Oxidation and Reduction; Oxygen: Element and Geochemistry; Oxygen—Its Evolution in the Earth's Atmosphere; Ozone; Photosynthesis; Sulfate Reduction; Water. Vol. II: Atmospheric Chemistry. Vol. IVB: Banded Iron Ores. Vol. V: Petrology; Serpentine.*

OXYGEN ISOTOPE GEOCHEMISTRY

The abundances of the three stable isotopes of oxygen in ocean water are: O^{16}, 99.763%; O^{17}, 0.0372%; and O^{18}, 0.1995%. (These abundances were derived from Nier's 1950 data for air.) The O^{18}/O^{16} ratio in natural materials varies over a range of 10% (100 permil). This variation is due to equilibrium and kinetic fractionation effects which provide much information on geochemical processes.

Isotopic analyses are usually performed on CO_2 gas in a double-collecting, dual gas-feed mass spectrometer of the type illustrated in Fig. 1.

The enrichment or depletion of O^{18} in substance A, relative to a standard, is reported in terms of the δ (permil) notation:

$$\delta_A = \left[\frac{R_A}{R_{std}} - 1\right] 1000$$

where $R = O^{18}/O^{16}$. The most common standard is Standard Mean Ocean Water, abbreviated SMOW (Craig, 1961). PDB is a carbonate standard commonly used in paleotemperature studies. The δ value of PDB is +30.4 permil on the SMOW scale.

The fractionation factor, $\alpha{}_B^A$, for two substances A and B is defined by

$$\alpha{}_B^A = \frac{R_A}{R_B}$$

It can be shown that $1000 \ln \alpha \approx \delta_A - \delta_B$.

Fig. 2 shows the δ values, relative to SMOW, and the approximate O^{18}/O^{16} ratios of a wide variety of natural substances. Atmospheric CO_2 is roughly 40 permil enriched in O^{18} relative to ocean water. Dissolved oxygen in deep ocean water is also highly enriched in O^{18}. The op-

OXYGEN ISOTOPE GEOCHEMISTRY

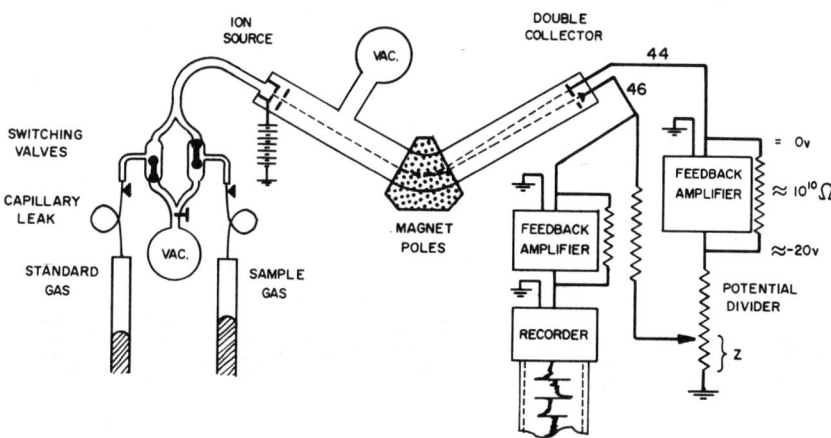

FIG. 1. Schematic diagram of a ratio mass spectrometer. A fraction, Z, of the $CO^{16}O^{16}$ (mass 44) signal is used to balance out the $CO^{16}O^{18}$ (mass 46) ion current. Small departures from null potential on the mass-46 collector are amplified and continuously recorded. The measured δ value (δ_m) is given by

$$\delta_m = \left[\frac{Z_{sample}}{Z_{standard}} - 1\right] 1000 + \text{correction derived from recorder trace}$$

Additional corrections are necessitated by the presence of C^{13}-bearing ions and by various instrumental imperfections (see Craig, 1957).

posite extreme is represented by snow near the South Pole, which has a δ value of -60 permil (falling outside Fig. 2). This snow condensed from vapor depleted in heavy isotopes by previous condensation.

Most igneous rocks fall in the range of $+5$ to $+10$ permil. Sediments are usually enriched in O^{18} relative to igneous rocks due to interaction with surface waters at low temperatures.

It is instructive to compare the natural isotopic variations summarized in Fig. 2 with the experimentally determined equilibrium fractionations plotted in Fig. 3. Equilibration with ocean water obviously plays a major role in determining the isotopic compositions of marine carbonates. The relative isotopic compositions of coexisting minerals in metamorphic rocks are also determined chiefly by mutual isotopic equilibration.

Analytical Methods

The usual methods of preparing carbon dioxide for mass spectrometric analysis are as follows: Oxygen is circulated over red-hot carbon while the CO_2 produced is frozen out in a trap cooled with liquid nitrogen. The recommended technique for most silicates and oxides is fluorination with F_2 or BrF_5. The liberated O_2 is converted to CO_2 as above.

Milligram quantities of water can be analyzed by reaction with BrF_5. However, water is routinely analyzed by an indirect method in which a small quantity of CO_2 is equilibrated with several grams of water at 25°C. The equilibrium fractionation factor at this temperature, as determined by several workers, is 1.04075 ± 0.00016 (Bottinga and Craig, 1968).

Carbonates are reacted at 25°C with 100% phosphoric acid, which yields two-thirds of the oxygen as CO_2. Fractionations involved in this reaction, for a number of carbonates, have been determined by Sharma and Clayton (1965). As an example, α (CO_2/calcite) = 1.01025. Sulfates are converted to $BaSO_4$ and then decomposed by carbon-reduction at 1100°C.

Meteorites

A large number of meteorites have been analyzed by Reuter, et al, (1965) and Taylor et al. (1965). They found that carbonaceous chondrites vary widely from -1 permil (Felix) to $+12$ permil (Orgueil); high- and low-iron chondrites vary from $+4.8$ to $+5.9$ permil; and basaltic achondrites vary from $+4.2$ to $+4.5$ permil (see Figs. 2 and 5). Terrestrial ultramafic rocks are similar in oxygen isotopic composition to the ordinary chondrites. Pallasites, mesosiderites, and basaltic achondrites are isotopically lighter by one to two permil.

Tektites

Taylor and Epstein (1966) report that Ivory Coast tektites and Bosumtwi Crater impactite glasses range from about $+12$ to $+15$ permil, whereas almost all other tektites are restricted

865

OXYGEN ISOTOPE GEOCHEMISTRY

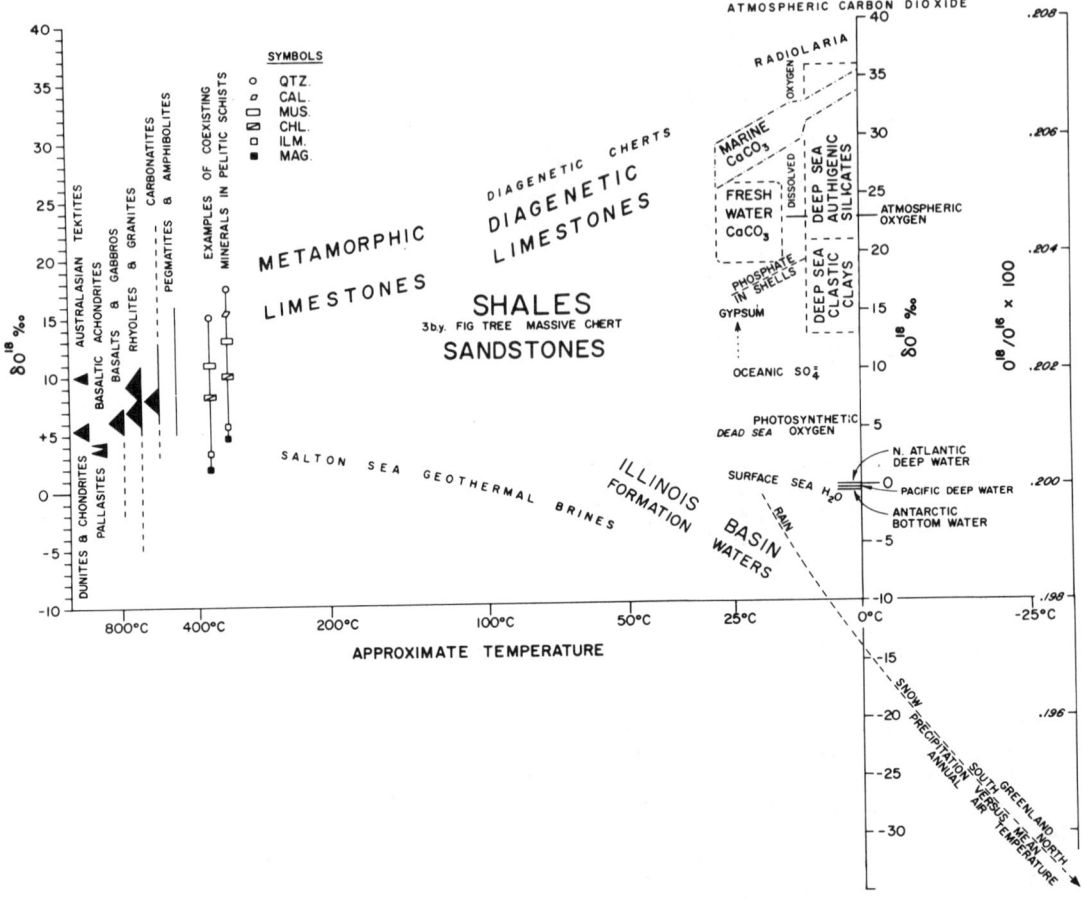

Fig. 2. Oxygn isotope compositions of various substances plotted against their approximate temperatures of formation or deposition. Antarctic snow, not plotted, extends to −60 permil. The temperature scale is chosen to match that of Fig. 3.

to the range +9 to +11 permil. If these latter tektites are terrestrial the data suggest that they were derived from igneous rocks or high-rank metamorphic gneisses of granitic composition, because the δ values of soils and argillaceous sediments are usually higher and more variable.

Igneous Rocks

Isotopic data on igneous rocks, largely from Taylor and Epstein (1962), Taylor (1968) and Garlick (1966), are plotted in Figs. 2, 4, and 5. Fig. 2 shows a gradual increase in mean δ values from ultramafic rocks through gabbros and basalts to rhyolites and finally granites. The derivation of rhyolites from basalts by igneous differentiation with little change in isotopic composition requires that crystal-melt isotopic fractionations are very small. Phenocryst-groundmass fractionations are indeed small when corrected for differences in chemistry (Garlick, 1966).

It is an empirical observation (see Fig. 4) that the O^{18}/O^{16} ratios of coexisting silicate minerals in igneous and metamorphic rocks are approximately related to their chemical compositions according to the following equation:

$$\delta O^{18} = KI + C$$

where

$$I = \frac{Si + 0.58 \text{ Al equivalents}}{\text{total equivalents}}$$

K is a constant determined by the temperature of equilibration, and C reflects the bulk isotopic composition of the rock. This index provides a method for the comparison of isotopic compositions among silicate minerals and rocks of differing chemistry, as is done in Figs. 4 and 5. (The positions on the abscissa of the

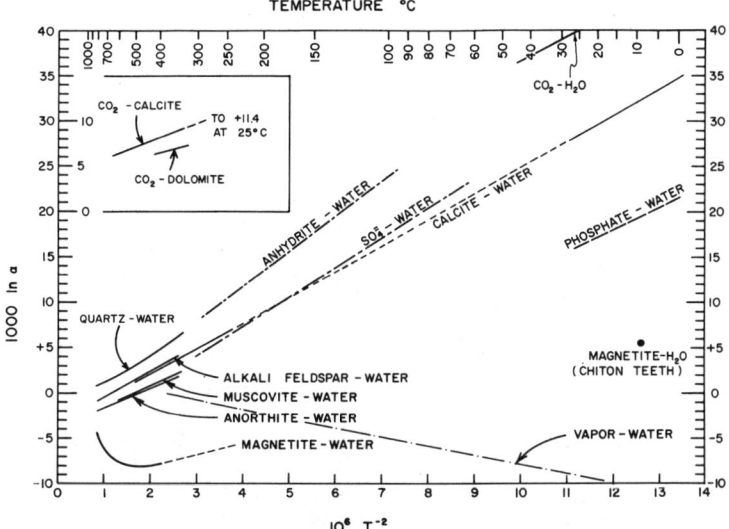

FIG. 3. Experimental fractionation curves. CaCO$_3$–H$_2$O: Epstein, et al. (1953); SiO$_2$–H$_2$O, CaCO$_3$–H$_2$O, Fe$_3$O$_4$$_2H_2$O: O'Neil and Clayton (1964); muscovite–H$_2$O: O'Neil and Taylor (1966); feldspars–H$_2$O: O'Neil and Taylor (1967); sulfates–H$_2$O: Lloyd (1967); phosphate–H$_2$O (indirect): Longinelli (1966); vapor–water: Bottinga and Craig (1968); α (ice–water) = 1.003 (O'Neil, 1968); carbonates–CO$_2$: O'Neil and Epstein (1966); dolomite–H$_2$O $\approx$ quartz–H$_2$O in temperature range 300–500°C: Northrop and Clayton (1966); slope of CO$_2$–H$_2$O: Botinga (1968). The low-temperature carbonate and phosphate curves are based on organic precipitates.

N.B. $1000 \ln \alpha_B^A \approx \delta_A - \delta_B \approx 1000 (\alpha_B^A - 1)$

nonsilicates, magnetite and ilmenite, were arbitrarily chosen in order that these minerals be approximately collinear with the silicate minerals). Temperatures of equilibration, as deduced from the slopes in Fig. 4, evidently decrease in the following sequence: olivine

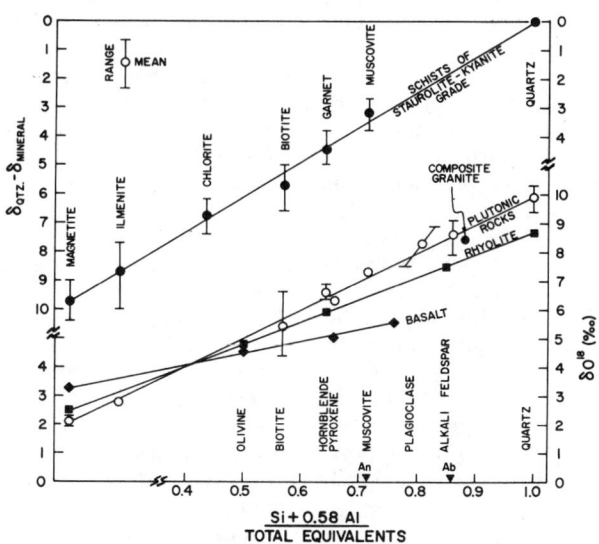

FIG. 4. See text for explanation. Note that δ values are plotted in the lower half of the figure, whereas δ quartz–δ mineral fractionations are plotted in the upper half.

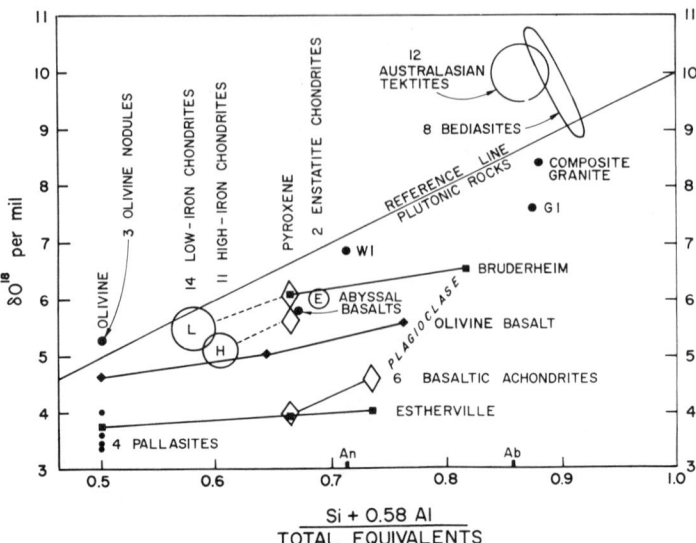

Fig. 5. Comparison between terrestrial and extraterrestrial materials. The reference line for plutonic rocks is from Fig. 4.

basalt; fayalite-bearing rhyolite; seven quartz-bearing plutonic rocks; ten kyanite-grade metamorphic rocks. Evidence of subsolidus isotopic exchange among minerals in plutonic rocks is provided by the Labrieville anorthosite (Anderson, 1966) in which coexisting plagioclase and ilmenite differ by 5 permil, whereas large pockets of pure plagioclase (7.2) and pure ilmenite (4.9) preserve their initial isotopic compositions and differ by only 2.3 permil.

Exceptionally low O^{18}/O^{16} ratios observed in some Scottish Tertiary (and Skaeegaard) igneous rocks are attributed to exchange with light meteoric waters during crystallization and cooling (Taylor, 1968). Eclogitic xenoliths from the Roberts Victor kimberlite range in δ value from 8 to 2 permil. An inverse correlation between their O^{18}/O^{16} ratios and CaO contents has been observed (Garlick and MacGregor, in preparation). It is probable that crystal-melt fractionations are responsible for the observed trends, but there is as yet no adequate explanation of the isotopic disequilibrium between the early crystallized O^{18}-rich eclogites ($\delta = 8$ permil) and the ultramafic xenoliths ($\delta = 5$ to 6 permil) also found in kimberlites.

The majority of carbonatites are similar in isotopic composition to other igneous rocks. Anomalous δ-values are possibly due to post-crystallization alteration (Taylor et al., 1967; Garlick, 1966).

Hydrothermal Deposits

Garlick and Epstein (1966) found the δ values of vein quartz at Butte (Montana) to vary from 0 to +13 permil, and attributed the variation to mixing of O^{18}–poor meteoric water with magmatically derived hydrothermal fluid. On the other hand, Rye (1966) showed that the Providencia (Mexico) deposits were precipitated from solutions having a relatively constant δ-value near +7 permil. It is obviously difficult to make any general statement regarding the isotopic nature of hydrothermal deposits.

Natural Waters

Water condensed from atmospheric vapor is enriched in O^{18} and D relative to the vapor (Fig. 3). The residual vapor is consequently depleted in heavy isotopes, causing subsequent precipitation to be lighter than the initial precipitation (Fig. 2). The worldwide correlation between the O^{18} and D contents of meteoric waters (Fig. 6) is given by:

$$\delta D = 8\ \delta O^{18} + 10\ \text{permil}$$

Evaporation into unsaturated air involves a kinetic effect which causes a slope of 5 rather than 8. This accounts for the north-east African trend in Fig. 6, and the offset of the meteoric water line from the point representing surface ocean water.

Craig (1963) classifies geothermal waters into two types:

1. Near-neutral chloride type waters (and geothermal steam) which have experienced various amounts of oxygen isotope exchange with rocks through which they have circulated. Their deuterium contents are similar to those of local rain or snow. Examples of this type

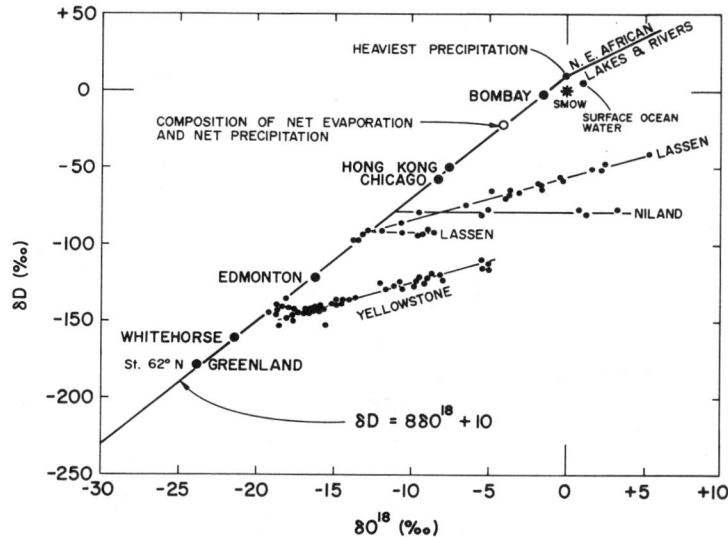

FIG. 6. The isotopic compositions of typical meteoric and geothermal waters (after Craig, 1963; and Dansgaard, 1964).

occur at Larderello, Hekla, Steamboat Springs, and Niland (Salton Sea geothermal brines).

2. Acid type hot springs which differ from local rain or snow in both their oxygen and hydrogen compositions. They are enriched in both O^{18} and D by nonequilibrium evaporation at relatively shallow depths. Most of the water at Yellowstone Park is of this type. Both types are well represented at the Geysers and in Lassen Park.

Clayton et al. (1966) have shown that the oxygen isotope compositions of oil-field brines from several North American basins are correlated with salinity and temperature. The brines are meteoric waters which have interacted with sediments. The range of compositions of Illinois basin brines is shown in Fig. 2. These brines may be in approximate equilibrium with calcites ranging from +22 to +26 permil. The hot brines of the Red Sea, on the other hand, appear to be isotopically identical to ocean water at the southern end of the Red Sea (Craig, 1966).

Ocean water, itself, is not uniform in isotopic composition. The effects of evaporation and precipitation in the Atlantic are shown in Fig. 7. Freezing increases the salinity of ocean water, but has a relatively minor effect on its isotopic composition because $\alpha \frac{\text{ice}}{\text{water}} = 1.003$ (O'Neil, 1968).

Secular changes in the mean isotopic composition of the oceans have occurred as a result of variations in the extent of continental glaciation. The maximum effect during the Pleistocene has been variously estimated at 0.5 to 1.5 permil. Changes on an even longer time scale may result from interactions with the solid earth. Perry (1967) suggests that the low O^{18} contents of ancient cherts (Fig. 8) are due to low O^{18} contents in the ancient oceans.

Free Oxygen

Atmospheric oxygen has a δ value of +23 permil, although calculations indicate that oxygen in equilibrium with ocean water would be isotopically similar to ocean water (Urey, 1947). The high O^{18} content of oxygen is due to the preferential consumption of O^{16} during respiration of plants and animals which, at steady state, must remove oxygen equal in composition to photosynthetic oxygen added to the atmosphere. Photosynthetic oxygen has an average δ value of about +5 permil. The composition of respired oxygen is discussed by Lane and Dole (1956). The O^{18} increase with depth of dissolved oxygen in ocean water is due to the preferential consumption of O^{16}.

Carbon Dioxide

Atmospheric CO_2, having a δ value of roughly +41 permil (Keeling, 1961) is in approximate isotopic equilibrium with ocean water at 25°C. However, isotopic exchange with atmospheric water also influences the composition of CO_2 (Bottinga and Craig, 1968).

Sulfates

The δ values of sulfate dissolved in ocean water, observed by Longinelli and Craig (1967)

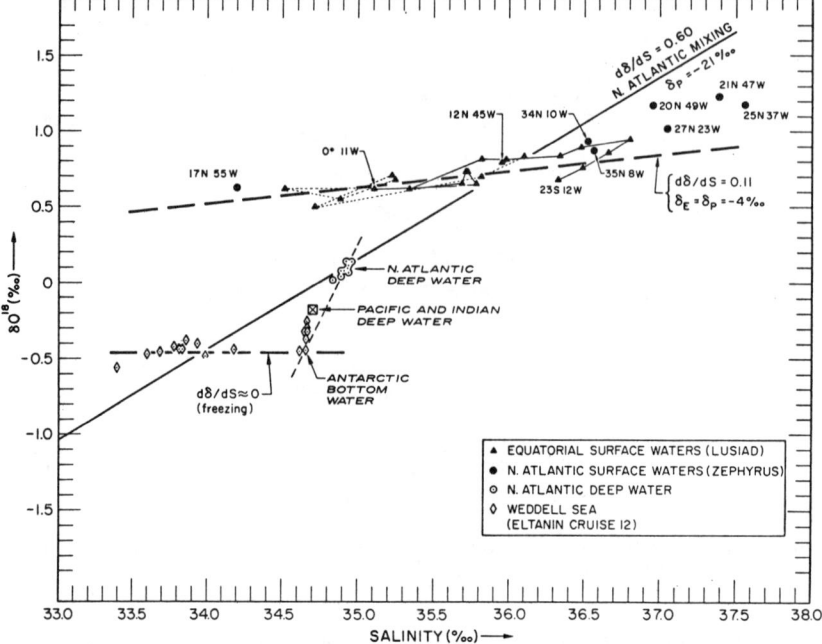

Fig. 7. δO^{18}–salinity relationships in the Atlantic. δ^E and δ^P refer to the isotopic composition of evaporating vapor and precipitation, respectively (Craig and Gordon, 1965).

and Lloyd (1967), are close to +9.7 permil. However, equilibration experiments indicate that oceanic sulfate should be much more enriched in O^{18} (see Fig. 3). Lloyd suggests that the sulfate is out of equilibrium because the turnover in the biogenic sulfur cycle is more rapid than the isotopic exchange rate with water.

Sediments

The isotopic compositions of various sediments are shown in Fig. 2. Detrital sediments

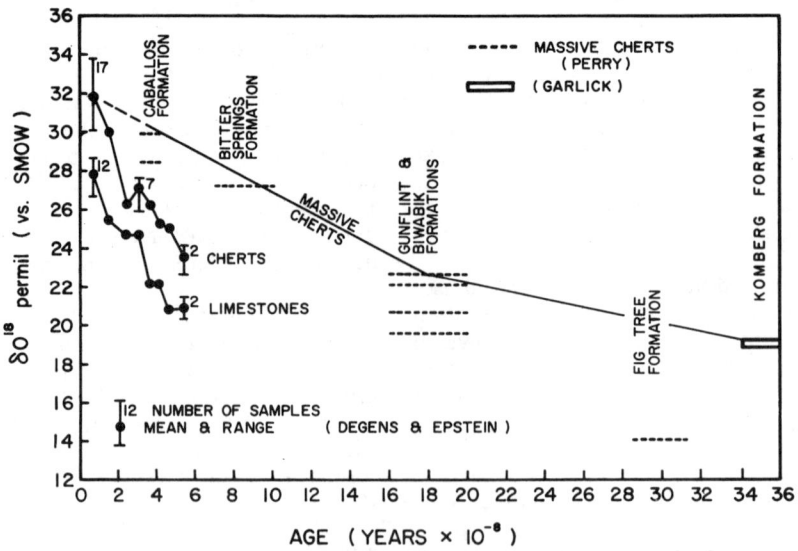

Fig. 8. Isotopic compositions of limestones, cherts, and massive cherts as a function of age (Degens and Epstein, 1962; Perry, 1967; and Garlick, unpublished).

OXYGEN ISOTOPE GEOCHEMISTRY

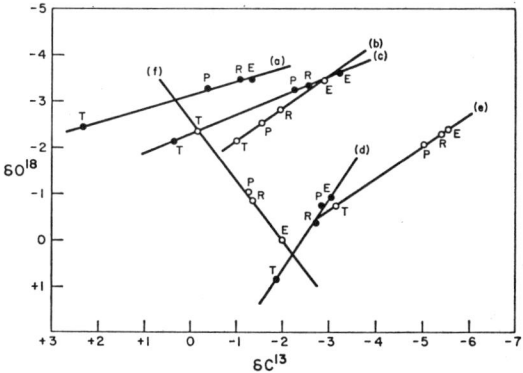

FIG. 9. Isotopic correlations in lantern parts from individuals of six echinoderm species. T, P, R, E represent tooth, pyramid, rotula, and epiphysis. The standard is PDB (Weber and Raup, 1966).

tend to retain the isotopic compositions of their source rocks or soils, whereas most authigenic minerals are 20 to 30 permil enriched in O^{18} relative to the waters from which they precipitate (Savin, 1967). $CaCO_3$ precipitates have received the most attention. Some organisms (notably foraminifera) apparently deposit calcite or aragonite in isotopic equilibrium with their environmental waters and can thus provide information on the temperatures at which the organisms lived, if the water compositions are known (cf. Emiliani's article on *Paleotemperatures* in this volume). In contrast, other organisms (e.g., corals and echinoderms) exhibit vital effects and do not precipitate their carbonates in equilibrium with their environments (see Fig. 9).

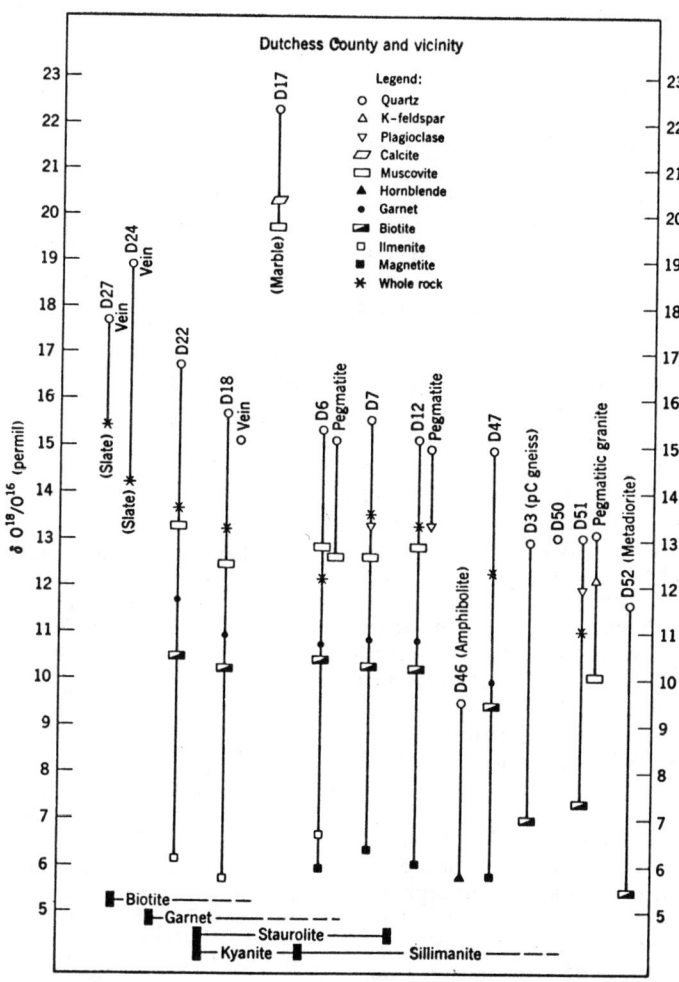

FIG. 10. Isotopic compositions of mineral assemblages in rocks from Dutchess County and vicinity (Garlick and Epstein, 1967). The rocks are pelitic schists unless otherwise noted. Metamorphic grade increases from left to right.

871

OXYGEN ISOTOPE GEOCHEMISTRY

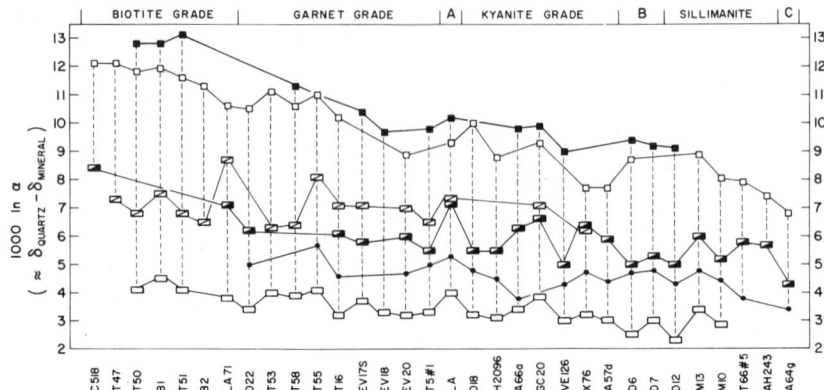

FIG. 11. Isotopic fractionations between quartz and magnetite (solid squares), ilmenite (open squares), chlorite (slashed boxes), biotite (half-filled boxes), garnets (dots), and muscovite (open boxes), in pelitic schists of Dalradian-type metamorphism. A = chloritoid-kyanite grade; B = staurolite-sillimanite grade; C = orthoclase-sillimanite grade (Garlick, 1969).

Diagenesis of Sediments

Recent dolomites usually have isotopic compositions similar to those of coexisting limestones, yet high-temperature dolomites are enriched in O^{18} relative to calcites (see Degens and Epstein, 1964). This is evidence that sedimentary dolomites usually form by diagenesis of calcium carbonate without isotopic change. On the other hand, Clayton et al. (1968) have measured a dolomite-water fractionation factor of 1.035 in Deep Springs Lake which indicates direct crystallization of dolomite from solution.

Variations with age in the compositions of cherts and limestones are shown in Fig. 9. Degens and Epstein (1962) attribute their observed variations to diagenetic processes such as exchange with groundwaters, but Perry (1967) suggests that the lower O^{18} contents of ancient massive cherts reflect lower O^{18} contents in the ancient oceans.

An example of diagenesis within the ocean environment is provided by siliceous pyroclastic layers in deep sea sediments (Garlick and Dymond, 1967). The δ values of ash layers change from about +9 permil in the Pleistocene to about +20 permil in the Eocene.

Metamorphic Rocks

The dimensions of isotopic exchange during metamorphism vary tremendously. Taylor (in press) has shown that the metamorphosed anorthosites of the Adirondack Massif are isotopically homogeneous ($\delta \approx +10$ permil) over thousands of square kilometers, but are 3 permil enriched in O^{18} relative to unmetamorphosed anorthosites from other parts of the world ($\delta \approx +7$ permil). In contrast, Perry and Bonnichsen (1966) and Anderson (1967) have described isotopic heterogeneity over distances of a few centimeters in metasedimentary rocks.

A classic example of regional metamorphism, Dutchess County, is shown in Fig. 10. Isotopic fractionations between coexisting quartz and other minerals in Dutchess County and other Dalradian-type pelitic metasediments are plotted in Fig. 11. When these data are compared with experimentally determined fractionations (Fig. 3) it is possible to obtain estimates of the temperatures at which these rocks crystallized.

G. DONALD GARLICK

References

Aldaz, L., and Deutsch, S., 1967, "On a relationship between air temperature and oxygen isotope ratio of snow and firn in the South Pole region," *Earth Planet. Sci. Letters*, **3**, 267.

Anderson, A. T., 1966, "Mineralogy of the Labrieville anorthosite," Quebec. *Am. Mineral.*, **51**, 1671.

Anderson, A. T., 1967, "The dimensions of oxygen isotopic equilibrium attainment during prograde metamorphism," *J. Geol.*, **75**, 323.

Bottinga, Y., 1968, "Calculation of fractionation factors for carbon and oxygen isotopic exchange in the system calcite-carbon dioxide-water," *J. Phys. Chem.*, **72**, 800.

Bottinga Y., and Craig, H., 1968, "High temperature liquid-vapor fractionation factors for H_2O-HDO-H_2O^{18}" (abstract), *Trans. Am. Geophys. Union*, **49**, 356.

Bottinga Y., and Craig, H., 1968, "Oxygen isotope fractionation between CO_2 and water, and the isotopic composition of marine atmospheric CO_2," *Earth Planet. Sci. Letters*, **5**, 285.

Clayton, R. N., Friedman, I., Graf, D. L., Mayeda, T. K., Meents, W. F., and Shimp, N. F., 1966,

"The origin of saline formation waters, 1, Isotopic composition," *J. Geophys. Res.,* **71,** 3869.

Clayton, R. N., Jones, B. F., and Berner, R. A., 1968, "Isotope studies of dolomite formation under sedimentary conditions," *Geochim. Cosmochim. Acta,* **32,** 415.

Craig, H., 1957, "Isotopic standards for carbon and oxygen and correction factors for mass-spectrometric analysis of carbon dioxide," *Geochim. Cosmochim. Acta,* **12,** 133.

Craig, H., 1961, "Standard for reporting concentrations of deuterium and oxygen-18 in natural waters," *Science,* **133,** 1833.

Craig, H., 1963, "The isotopic geochemistry of water and carbon in geothermal areas," Proc. Spoleto Conference on Nuclear Geology, Ed. Tongiorgi, pp. 17–53.

Craig, H., 1966, "Isotopic composition and origin of the Red Sea and Salton Sea geothermal brines," *Science,* **154,** 1544.

Craig, H., and Gordon, L. I., 1965, "Isotopic oceanography," *Symp. Marine Geochim., Narragansett Marine Lab., Univ. R. I. Publ.,* **3,** 277.

Dansgaard, W., 1964, "Stable isotopes in precipitation," *Tellus,* **16,** 436.

Degens, E. T., and Epstein, S., 1962, "Relation between O-18/O-16 ratios in coexisting carbonates, cherts and diatomites," *Bull. Am. Assoc. Petrol. Geol.,* **46,** 534.

Degens, E. T., and Epstein, S., 1964, "Oxygen and carbon isotope ratios in coexisting calcites and dolomites from recent and ancient sediments," *Geochim. Cosmochim. Acta,* **28,** 23.

Epstein, S., Buchsbaum, R., Lowenstam, H. A., and Urey, H. C., 1953, "Revised carbonate-water isotopic temperature scale," *Bull. Geol. Soc. Am.,* **64,** 1315.

Garlick, G. D., 1966, "Oxygen isotope fractionation in igneous rocks," *Earth Planet. Sci. Letters,* **1,** 361.

Garlick, G. D., 1969, "The stable isotopes of oxygen," in (Wedepohl, editor) "Handbook of Geochemistry," Springer Verlag.

Garlick, G. D., and Epstein, S., 1966, "The isotopic composition of oxygen and carbon in hydrothermal minerals at Butte, Montana," *Econ. Geol.,* **61,** 1325.

Garlick, G. D., and Epstein, S., 1967, "Oxygen isotope ratios in coexisting minerals of regionally metamorphosed rocks," *Geochim. Cosmochim. Acta,* **31,** 181.

Garlick, G. D., and Dymond, J. R., 1967, "Oxygen isotope exchange between volcanic glass and ocean water" (abstract), *Trans. Am. Geophys. Union,* **48,** 236.

Gross, M. G., and Tracey, J. I., Jr., 1966, "Oxygen and carbon isotopic compositions of limestones and dolomites, Bikini and Eniwetok Atolls," *Science,* **151**(3714), 1082–1084.

Keeling, C. D., 1961, "Concentration and isotopic abundances of carbon dioxide in rural and marine air," *Geochim. Cosmochim. Acta,* **24,** 277.

Lane, G. A., and Dole, M., 1956, "Fractionation of oxygen isotopes during respiration," *Science,* **123,** 574.

Lloyd, R. M., 1967, "Oxygen-18 composition of oceanic sulfate," *Science,* **156,** 1228.

Longinelli, A., 1966, "Ratios of O-18/O-16 in phosphate and carbonate from living and fossil marine organisms," *Nature,* **211,** 923.

Longinelli, A., and Craig, H., 1967, "Oxygen-18 variations in sulfate ions in sea water and saline lakes," *Science,* **156,** 56.

Nier, A. O., 1950, "A redetermination of the relative abundances of the isotopes of carbon, nitrogen, oxygen, argon and potassium," *Phys. Rev.,* **77,** 789.

Northrup, D. A., and Clayton, R. N., 1966, "Oxygen-isotope fractionations in systems containing dolomite," *J. Geol.,* **74,** 174.

O'Neil, J. R., 1968, "Hydrogen and oxygen isotope fractionation between ice and water," *J. Phys. Chem.,* **72,** 3683.

O'Neil, J. R., and Clayton, R. N., 1964, "Oxygen isotope geothermometry," in (Craig et al., editors), "Cosmic and Isotopic Chemistry," North Holland, p. 157.

O'Neil, J. R., and Epstein, S., 1966, "Oxygen isotope fractionation in the system dolomite-calcite-carbon dioxide," *Science,* **152**(3719), 198–200.

O'Neil, J. R., and Taylor, H. P., 1966, "Oxygen isotope equilibrium between muscovite and water," *Trans. Am. Geophys. Union,* **47,** 212.

O'Neil, J. R., and Taylor, H. P., 1967, "The oxygen isotope and cation exchange chemistry of feldspars," *Am. Mineralogist,* **52,** 1414.

Perry, E. C., 1967, "The oxygen isotope chemistry of ancient cherts," *Earth Planet. Sci. Letters,* **3,** 62.

Perry, E. C., and Bonnichsen, B., 1966, "Oxygen isotope exchange between quartz and magnetite from the metamorphosed Biwabik iron-formation, Minnesota" (abstract), *Trans. Am. Union,* **47,** 212.

Reuter, J. H., Epstein, S., and Taylor, H. P., 1965, "O^{18}/O^{16} ratios of some chondritic meteorites and terrestrial ultramafic rocks," *Geochim. Cosmochim. Acta,* **29,** 481.

Rosholt, J. N., et al., 1962, "p^{231}/Th^{230} dating and O^{18}/O^{16} temperature analysis of core A254–Br-C.," *J. Geophys. Res.,* **67,** 2907–2911.

Rye, R. O., 1966, "The carbon, hydrogen and oxygen isotopic composition of the hydrothermal fluids responsible for the lead-zinc deposits at Providencia, Zacatecas, Mexico," *Econ. Geol.,* **61,** 1399.

Savin, S. M., 1967, "Oxygen and hydrogen isotope ratios in sedimentary rocks and minerals," Ph.D. Thesis, Cal. Inst. Tech.

Sharma, T., and Clayton, R., 1965, "Measurement of O-18/O-16 ratios of total oxygen of carbonates," *Geochim. Cosmochim. Acta,* **29,** 1347.

Taylor, H. P., 1968, "Oxygen isotope studies of anorthosites," in "Origin of Anorthosites," N.Y. State Museum Spec. Publ.

Taylor, H. P., 1968, "The oxygen isotope geochemistry of igneous rocks." *Contri. Mineral-Petrol.,* **19,** 1.

Taylor, H. P., Duke, M. B., Silver, L. T., and Epstein, S., 1965, "Oxygen isotope studies of minerals in stony meteorites," *Geochim. Cosmochim. Acta,* **29,** 489.

Taylor, H. P., and Epstein, S., 1962, "Relation between O-18/O16 ratios in coexisting minerals of igneous and metamorphic rocks," *Bull. Geol. Soc. Am.,* **73,** 461 and 675.

Taylor, H. P., and Epstein, S., 1966, "Oxygen isotope studies of Ivory Coast tektites and impactite glass from the Bosumtwi crater, Ghana," *Science,* **153,** 173.

Taylor, H. P., Frechen, J., and Degens, E. T., 1967, "Oxygen and carbon isotope studies of carbonatites from the Laacher See district, West Germany, and the Alno district, Sweden," *Geochim. Cosmochim. Acta,* **31,** 407.

Urey, H. C., 1947, "The thermodynamic properties of isotopic substances," *J. Chem. Soc.,* 562–581.

Weber, J. N., and Raup, D. M., 1966, "Fractionation of the stable isotopes of carbon and oxygen in marine calcareous organisms—the Echinoidea," *Geochim. Cosmochim. Acta,* **30,** 681 and 707.

Weber, J. N., 1964, "Oxygen isotope fractionation between coexisting calcite and dolomite," *Science,* **145,** 1303–1305.

Cross-references: *Authigenesis of Minerals; Isotope Fractionation; Isotope Geology—Stable; Mass Spectrometry; Natural Brines; Outgassing of the Planet Earth; Oxygen Cycle; Paleotemperatures; Seawater, Chemistry; Sulfates; Sulfur Cycle. Vol. I: Salinity in the Oceans. Vol. II: Carbonaceous Meteorites; Tektites. Vol. VI: Diagenesis; Meteoric Water.*

OXYLUMINESCENCE

Certain minerals emit light as a function of temperature when slowly heated, at a constant rate, to temperatures below that of their incandescence, to produce the phenomenon of *"thermoluminescence"* (q.v.; see also *Thermal Analysis*). Furthermore, some 75% of the 3000 plus samples examined by Daniels, Boyd, and Saunders (1953), showed visible manifestations of this property.

The term *oxyluminescence* was first proposed by Ashby in 1961 to describe the technique that is closely related to thermoluminescence, where the sample must be heated either in oxygen or air (in contact with oxygen) for light emission to take place. This had been reported previously, but not specifically named, by a number of authors in connection with vapors and solutions (see Ashby, 1961).

Except for the added atmosphere control facilities, the apparatus is the same as for thermoluminescence.

Several factors are vital for the production and intensity of oxyluminescent effects, e.g., oxygen must be present for the production of light emission, the intensity of which is proportional to the concentration of oxygen in contact with the sample, whereas the presence of antioxidants has the effect of decreasing the intensity of the light produced (see Fig. 1).

The present conclusion is that oxylumines-

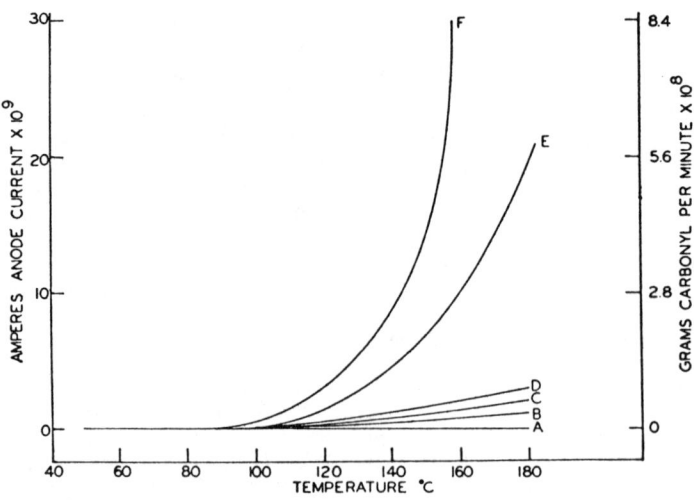

FIG. 1. Variation of oxygen concentration with intensity of oxyluminescence. Each curve was obtained with a different gas mixture flowing over the polymer. (A) Argon without oxygen; (B) 2.6% by volume oxygen in nitrogen; (C) 6.4% oxygen in nitrogen; (D) 11.0% oxygen in nitrogen; (E) air; and (F) oxygen. The polymer was Hercules-type 6501 polypropylene (from Ashby, 1961).

OZONE

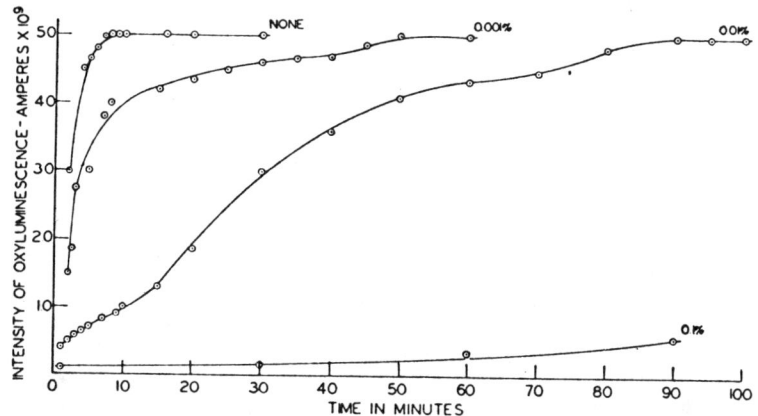

FIG. 2. Effect of antioxidant concentration on oxyluminescence. Each curve was obtained with different concentrations of antioxidant present. The antioxidant was one part 4, 4-thiobis(6-*tert*-butyl-*o*-cresol) to one part dilaurylthiodipropionate. The polymer was Hercules-type 6501 polypropylene (from Ashby, 1961).

cence is related to the incipient oxidation of minerals. This is often modified by the presence of antioxidants, which may be considered as substances which decrease the characteristic light emission of the sample under test (see Fig. 2).

The technique is of some diagnostic value and useful in evaluating substances for oxidant and antioxidant properties.

S. St. J. Warne

References

Ashby, G. E., 1961, "Oxyluminescence from polypropylene," *J. Polymer Sci.,* **50,** 99–106.

Daniels, F., Boyd, C. A., and Saunders, D. F., 1953, "Thermoluminescence as a research tool," *Science,* **117,** 343–349.

Cross-references: *Oxygen; Thermal Analysis; Thermoluminescence.*

OZONE—*See* Vol. II

PALEOHYDROLOGY

Paleohydrology is the study of the waters of the earth, their composition, distribution, and movement on ancient landscapes from the first occurrence of precipitation to the beginning of hydrologic record keeping. It provides a link between the hydrology of the present and the sciences concerned with earth history and past environments.

Paleohydrologic speculations and conclusions are dependent on existing knowledge of the effects of climatic, vegetational and geologic controls on runoff, sediment yield, sediment concentration, water chemistry, groundwater and lake levels, and the nature of flood events. These relations, however, are not directly applicable to the remote geologic past when vegetation was absent or when it was evolving toward its present condition and distribution. This important paleohydrologic consideration has been discussed by Cayeux, Russell, Schwarzbach, and Tricart and Cailleux (see references in Schumm, 1968).

Although runoff and sediment movement from a drainage basin are usually measured at a specific location along a river channel, these data, when averaged for a period of years, provide information on the hydrologic climate or regimen of the drainage system. Relations developed among modern data for average precipitation, temperature, runoff, and sediment yield permit estimates to be made of regional paleohydrology when paleoclimatic information is available. On the other hand, relations that have been developed between channel morphology, sediment load, and runoff provide a basis for the estimation of paleochannel discharge.

Hydrologic Basis

Relations that exist among mean annual temperature, precipitation, runoff, and sediment yield, as based on data obtained for the United States, are shown on Figs. 1 and 2. The curves show in a general way what average hydrologic differences can be expected among the climatic regions. Using a somewhat different line of reasoning, the curves can also be used to demonstrate the hydrologic effects of a climate change (Schumm, 1965).

Only the curve for an average temperature

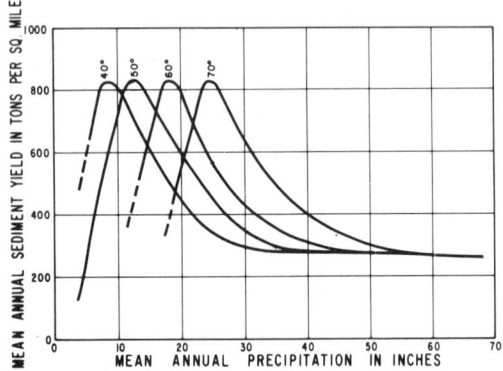

Fig. 1. Curves illustrating the effect of average temperature on the relations between mean annual sediment yield and mean annual precipitation (from Schumm, 1965).

of 50°F will be discussed, but the principles apply to all curves of Figs. 1 and 2. Sediment yield from a drainage basin will increase from a minimum in a region of no precipitation to a peak between 10 and 15 inches of precipitation. The decrease in sediment yield beyond this peak is a result of the increased effectiveness of vegetation in protecting the soil and in retarding erosion, that is, a transition from desert shrubs to grassland. Sediment yield rates decrease further to lower values in forested regions of high rainfall.

The curves of Fig. 2 show how runoff will increase with increased precipitation for uniform monthly precipitation at different mean annual temperatures. For the 50°F curve, below about 20 in. of precipitation, runoff increases slowly with increased precipitation because of relatively high water losses to evaporation and infiltration. The quantity of water lost through transpiration and interception by vegetation increases, as precipitation increases, to about 40 in. of precipitation when the water requirements of infiltration, evaporation and transpiration have been met; above 40 in. of precipitation, runoff increases directly with precipitation (Langbein et al., 1949).

The relations of Figs. 1 and 2 demonstrate that depending on the climate before a climate change, increased precipitation will increase runoff but sediment yield rates may increase or decrease. In very arid regions an increase in

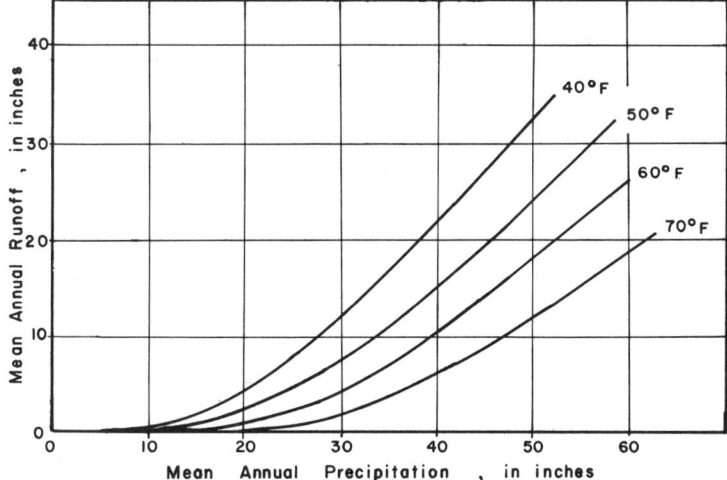

FIG. 2. Curves illustrating the effect of average temperature on the relation between mean annual runoff and mean annual precipitation (after Langbein et al., 1949).

precipitation will increase sediment yield because the vegetational cover will not improve sufficiently to retard erosion effectively, whereas under an initially semiarid climate an increase of precipitation will improve the vegetational cover sufficiently to cause a decrease in sediment yield rates. Therefore, changes in the quantity of sediment moving out of a drainage system cannot be considered as a function of precipitation and runoff alone. The vegetational influence must also be evaluated, but this becomes increasingly difficult as the more remote geologic periods are considered.

Vegetational Effects

Modern hydrologic relationships can be applied at best to only the last few million years of the billions of years of earth history, for it was only during the middle Cenozoic Era (about 25 million years ago) that vegetation evolved to its present status. Many peculiarities of the fluvial and deltaic sedimentary deposits of the early Paleozoic era can probably be attributed to the vastly different hydrology of a phytological desert; that is, a vegetationally barren surface which, nevertheless, received adequate precipitation to sustain present forms of plant life. In addition, the slow colonization of the earth's surface by evolving vegetation undoubtedly caused profound changes in the hydrologic cycle during at least the last 400 million years of geologic time. Four major divisions of geologic time are paleohydrologically significant as follows: (1) prevegetation time; (2) time during colonization of alluvial areas by primitive vegetation; (3) time during colonization of interfluves by flowering plants; and (4) time following appearance of grasses.

Using what is currently known about vegetational influences on sediment yield and runoff, an attempt has been made in Figs. 3 and 4 to show how sediment yield and runoff would have varied with precipitation through geologic time.

It is difficult to establish the limits of the four time periods of paleohydrologic significance on the basis of paleobotanical evidence. However, it is generally agreed that, although vascular land plants have been identified in the Cambrian Period, significant colonization of the land did not occur before the Devonian Period about 400 million years ago and that between the Devonian and Cretaceous Periods (400 million to 135 million years ago) vegetation was progressively expanding its range as new forms evolved.

With the appearance of vegetation capable of surviving abrupt and possibly severe climatic fluctuations, less favorable habitats were colonized, and it was probably between the Cretaceous Period and the Miocene Epoch between 135 and 25 million years ago, that the interfluve or upland areas were colonized. With the appearance of grasses in Miocene time the effects of vegetation on the hydrologic cycle must have been as they are today.

Prevegetation Paleohydrology

During Precambrian and early Paleozoic time, the lack of an effective vegetational cover would have permitted sediment yield rates to increase with increased precipitation to a maximum rate, dependent only on the erodibility or weathering rates of the rocks (Fig. 3, curve 1).

PALEOHYDROLOGY

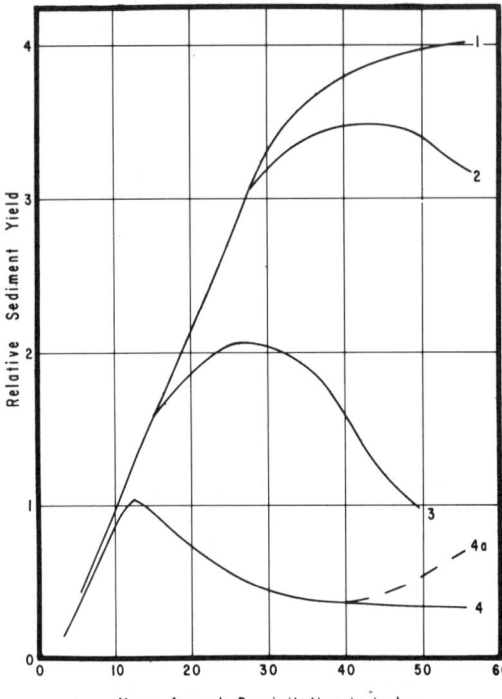

FIG. 3. Hypothetical series of curves illustrating the relations between precipitation and relative sediment yield (maximum average sediment yield of today is 1) during geologic time as follows: before the appearance of land vegetation (curve 1); following the appearance of primitive vegetation on the earth's surface (curve 2); following the appearance of grasses (curve 4). Curve 4a shows the increase in sediment yield rates for tropical monsoonal climates (from Schumm, 1968).

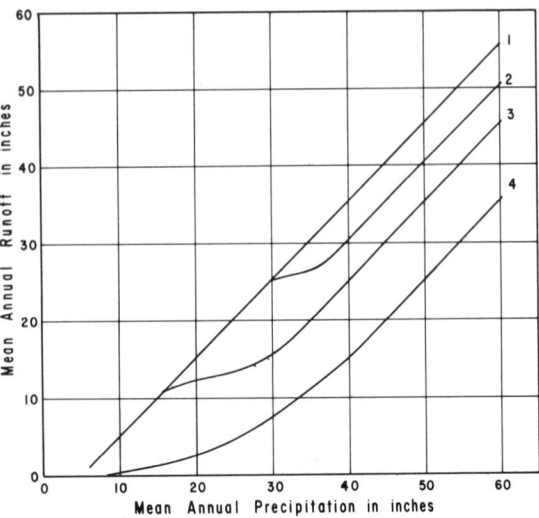

FIG. 4. Hypothetical series of curves illustrating the relations between precipitation and runoff during geologic time as follows: before the appearance of land vegetation (curve 1); following the appearance of primitive vegetation on the earth's surface (curve 2); following the appearance of flowering plants and conifers (curve 3); following the appearance of grasses (curve 4) (from Schumm, 1968).

In Fig. 3 this maximum rate is arbitrarily shown as four times greater than the maximum sediment yield of the present. It is assumed that the severe microclimatic environment, characterized by extremes of temperature and rapid changes of temperature at the ground surface, would have caused rapid weathering. The uninhibited erosion and transport of the weathered material would have produced rapid rates of denudation. Of course, the prevegetation denudation rate for weakly-cemented sedimentary rocks would have been much greater than that shown, but it would have been much less for massive igneous rocks.

Fig. 4, curve 1, indicates that when transpiration and interception losses do not occur, runoff increases directly with precipitation after some loss due to infiltration and evaporation. This loss is assumed to have been about 5 in. because runoff at present is minimal from regions which receive less than 5 in. of precipitation annually. The significant factor during prevegetation time was that runoff would have increased directly with precipitation, and flood peaks would have been much greater in the absence of vegetation. Runoff was about 50% of precipitation at 10 in. of annual precipitation (curve 1, Fig. 4), but it increased to 75% at 20 in. of precipitation. Above 40 in. of precipitation, curve 1 parallels curve 4, but curve 4 is lower, reflecting a 20-in. water loss due to evapotranspiration.

To summarize, during prevegetation time, runoff occurred more frequently as major floods; and both sediment yield and runoff increased with increased precipitation. The sediment yield rates rose to a plateau, depending on the erodibility of the rocks forming the drainage basin, and runoff would have been greater than at present.

Paleohydrology During Evolution of Vegetation

As colonization of the land surface by primitive plants progressed during late Paleozoic and early Mesozoic time, the presence of vegetation undoubtedly decreased erosion in certain parts of the landscape, and the peak of the sediment yield curve would have decreased somewhat (Fig. 3, curve 2). Also, a discontinuity would have appeared in the runoff-precipitation relation for those higher rainfall areas, where vegetation was present and where water was utilized by the primitive vegetation (Fig. 4, curve 2). Primitive vegetation probably occurred only on coastal plains and in humid valleys and, therefore, its effects initially may have been local. Sediment was shed continually from the barren hillslopes and temporarily stored in the valleys as in prevegetation times, but at least in the downstream areas, the vegetation would have tended to stabilize the sedimentary deposits and the stream channels.

As more modern and hardier plants evolved during the Mesozoic and early Cenozoic, less favorable habitats were colonized, and vegetation moved up into interfluve areas, thereby further decreasing sediment yields and compressing the peak of the precipitation-sediment yield curve (Fig. 3, curve 3). The discontinuity in the precipitation-runoff relation would have shifted into drier regions (Fig. 4, curve 3), and, with the appearance of grasses, during late Cenozoic time, erosion and runoff conditions similar to those of the present but without man's influence, would have prevailed (Figs. 2 and 3, curves 4).

Gilluly has concluded from an estimate of the volume of Triassic and younger sediments of the Atlantic coast of the United States that an average rate of Mesozoic and Cenozoic erosion was about equal to present rates of erosion. However, present erosion rates may be on the average twice that of natural rates because of man's influence. For this reason Mesozoic and early Cenozoic erosion rates are shown as being about double that of the present (curve 4, Fig. 3).

Seasonal Precipitation

It should also be recognized that highly seasonal precipitation should increase sediment yields, as suggested by Fournier and Douglas (see references in Schumm 1968). In fact, some information on sediment yields from regions of tropical monsoon climates suggests that the sediment yield curve will rise again as rainfall increases above 50 inches (Fig. 3, curve 4a); however, this increase may, in fact, be due to the rugged topography and intensive land use of the areas from which the data used to establish curve 4a were obtained.

Sediment Character

The curves of Fig. 3, although providing information on the relative volumes of sediment delivered from a drainage system during a given climate, provide no information on the type of sediment to be expected. Nevertheless, as based on geologic investigations, it can be assumed that not only will sediment yields decrease at higher amounts of precipitation but the material will be finer for sediment yield curves 3 and 4 of Fig. 3, whereas fluvial sediments during prevegetation time (curve 1) and during colonization of the land by primitive vegetation (curve 2) should have been coarse grained because complete weathering was precluded by rapid erosion. Therefore, it may be expected that the very different hydrologic characteristics of the geologic past should be reflected in the characteristics of terrestrial sedimentary deposits and paleochannels.

Paleochannel Discharge

Leopold and his associates and Dury, among others (see references in Schumm, 1968), have developed empirical equations that relate the dimensions, shape, meander wavelength and sinuosity of river channels to water discharge, and to type of sediment load moved through a channel (ratio of bedload to total sediment load). Using these relations the discharge of some Quaternary paleochannels of southeastern Australia have been estimated, and deductions concerning the nature of the climatic changes of this region were based on the relations between channel morphology, discharge and sediment type (Schumm, 1968). Similar techniques could be utilized to estimate the discharge rates in pre-Quaternary paleochannels if the dimensions of the channel are known and if an estimate of the nature of sediment load can be made from exposures of bed and bank materials.

Groundwater

Other aspects of paleohydrology are locally important. For example, the change of groundwater levels due to climatic and other hydrologic changes may be of significance geomorphically and of major significance to the interpretation of archaeological sites (Raikes, 1967; see also *Paleolimnology*).

The evidence concerning the paleohydrology of the remote past is as nebulous as that concerning its paleoclimate; nevertheless, modern hydrologic and climatic data, when combined with information concerning the colonization of the land surface by vegetation, provide a scientific basis for paleohydrologic conclusions.

STANLEY A. SCHUMM

References

Langbein, W. B., et al., 1949, "Annual runoff in the United States," *U.S. Geol. Surv. Circ.*, **52**, 14pp.

Raikes, R., 1967, "Water, Weather and Prehistory," New York, Humanities Press, 208pp.

Schumm, S. A., 1965, "Quaternary Paleohy-

PALEOHYDROLOGY

drology," in (Wright, H. E., Jr. and Frey, D. G. editors), "The Quaternary of the United States," Princeton, Princeton Univ. Press, pp. 783-794.

*Schumm, S. A., 1968, "Speculations concerning paleohydrologic controls of terrestrial sedimentation," *Geol. Soc. Am. Bull.,* **79**, 1573-1588.

*Additional bibliographic references may be found in this work.

Cross-references: *Groundwater; Hydrologic Cycle; Hydrology; Paleolimnology; Runoff; Vegetation Indicators.* Vo. II: *Climate and Geomorphology; Climatic Variations (Historical Record); Earth—Geology of the Planet; Paleoclimatology; Planet Earth—Origin and Evolution; Precipitation.* Vol. III: *Denudation; Drainage Basin; Erosion; Quaternary Period.* Vol. VI: *Sedimentation.* Vol. VII: *Cenozoic Era; Mesozoic Era; Paleozoic Era.*

PALEOLIMNOLOGY*

A prime objective of paleolimnology is to illuminate the evolutionary stages through which a lake has passed. The sequence must be established by inference from the stratigraphic record and from the chemical and mineralogical composition of the sediments and the remains of organisms. As long as a lake persists, it undergoes change, as the morphometry changes, as its sediments accumulate, and as the earth's crust deforms. But always in the evolution of a lake or of any geomorphic feature, there is a tendency to come to equilibrium with the controlling factors of climate and tectonic forces. These two dynamic factors are themselves influenced by the antecedent geologic history. Our objective is also to try to reconstruct the paleoclimate and the hydrography and physical geography of the surrounding terrain. But perhaps our most important objective should be to quantify dimensions and the rates at which events took place, the amounts of substances grown, moved, or fixed, the concentrations of dissolved substances, and, if at all possible, the magnitude of the forces involved.

The earliest paleolimnological work in the United States was done by G. K. Gilbert (1891). This was an elaborate study and reconstruction of Pleistocene Lake Bonneville, the ancestor of the present Great Salt Lake of Utah. The fact that this classical investigation was prompted by a desire to understand the geologic history and the fact that it turned out to be also a classic in paleolimnology illustrates how geology grades into limnology along the path of paleolimnology. The reaction is, of course, completely reversible. Lake

* Publication authorized by the Director, U.S. Geological Survey.

Bonneville, at its maximum stage, was a large fresh-water lake with an area of 19,750 square miles and a maximum depth of 1,050 feet. Gilbert considered its morphometry, hydrography, the organisms that lived in it, and the sediments it laid down.

Closely parallel in time and scope was I. C. Russell's study (1885) of Lake Lahontan, another large Pleistocene lake in western Nevada. Russell's work actually followed that of Gilbert but was published six years earlier. After these two comprehensive studies comparatively little was done for several decades.

In trying to reconstruct the history of an extinct lake, one must adjust to the fact that virtually all the evidence is fragmentary. All interpretations draw heavily on inference and analogy. The facts that we have come from observing and measuring stratigraphic sections exposed by accidents of erosion, though, if we are fortunate, there may also be cores taken from holes drilled usually for another purpose.

We have four means by which we can interpret the deposits: *(1)* analogy with known geologic processes, *(2)* inference that the fossil organisms lived in the same kinds of environment as their living counterparts do now, *(3)* analogy with the limnology and sediments of existing lakes, and *(4)* inferences drawn from the chemical composition and mineralogy of the sediments, particularly the authigenic minerals.

Paleolimnological interpretations of a number of extinct lakes are summarized below.

Permian—Wellington Formation

Tasch and Zimmerman (1961) found a considerable number of localities in the outcrop belt of the Wellington Formation in Kansas and Oklahoma that contain identifiable remains of pemphicycliid, estheriid, and leaiid conchostracans. These beds are of Permian age. They concluded that the Permain ponds were temporary, free from currents, that the bottoms of the ponds consisted of limy clay, and that the water was probably very shallow. They inferred, further, that the water in these ancient ponds was probably turbid a good part of the time because the mineral particles enclosing the fossils are so very fine grained.

Jurassic—Todilto Limestone

A very large lake of Late Jurassic age once existed in northwestern New Mexico, southwestern Colorado, and northeastern Arizona (Anderson and Kirkland, 1960). The sediments of this lake, known as the Todilto Limestone, consist of a thin, very extensive limestone member 7-8 ft thick and an overlying massive gypsum member that is generally 70–100 ft thick. On the basis of the varves the authors estimate that the limestone was deposited in about 14,000 years and the gypsum member in about 6,000 years, making the total life of the lake only 20,000 years. The hydrographic basin of this lake is inferred to have had an area of about 150,000 square miles. At its maximum extent the lake occupied approximately 23% of its hydrographic basin. From reasonable evidence they deduce that the average summer temperature was about 85°F and the average winter temperature was about 57°F—a climate somewhat like that of the interior of southern California today. The fossil fish belong to two species, *Pholidophorus americanus* Eastman and *Leptolepis schowei* Dunkle.

Lower Cretaceous—Newark Canyon Formation

Lake beds make up part of the Newark Canyon Formation in Eureka County, Nevada. The formation contains unionids, fossil fish, and fossil plants, all of which indicate an Early Cretaceous age. Many of the beds in this ancient lake deposit have varves. These indicate an average rate of accumulation for this kind of lake sediment of about 1,500 years per foot of compacted rock. The fossil fish, which are small, have been assigned to the genus *Leptolepis* Agassiz.

Paleocene and Eocene—Flagstaff Limestone

La Rocque (1960) has given a reconstruction of the limnology of a large lake that existed for two or three million years in central Utah and in which the Flagstaff Limestone (Paleocene to Eocene) was deposited. The Flagstaff lake formed in a fault depression. Tectonic movement occurred in a succession of pulses, which led to an alternation of limestone and shale with changes in molluscan fauna. In the land snails, which occur sparingly in the lake basin sediments, La Rocque finds some evidence for a warm, semiarid climate. At times it may possibly have had an area as great as 7,000 square miles. The abundance of calcium carbonate in the sediments indicates that this was a hard-water lake. But during the second phase a considerable amount of gypsum was precipitated. During the first phase an extensive growth of submerged and emergent rooted plants flourished around the margin of the lake, on shoals, and around the islands. The pelecypods belong to two groups, the Naiades and Sphaeriidae. Gastropods are represented by gill breathers and lung breathers. One of the most striking aspects of this assemblage is the great abundance of *Hydrobia utahensis* (White), which makes up 32–91% of the individuals. In the second or middle phase, the Flagstaff lake, though about double the size of the first phase, was a barren environment, apparently because of its high sulfate concentration. No molluscan shells, other than a few land snails, were found.

During the third or final phase, the lake retained its large size but returned to very much the same sort of habitat it was during the first phase.

Eocene—Green River Formation

The Green River Formation of southwestern Wyoming, northwestern Colorado, and northeastern Utah was deposited in two large lakes and one somewhat smaller lake that were essentially contemporaneous. They are of early and middle Eocene age. The largest, Lake Uinta, in Utah and Colorado, was in existence between 5 and 8 million years. Gosiute Lake, in Wyoming, lasted some 4 million years, but the smallest lake in the extreme southwest of Wyoming probably had a much shorter life. (See Fig. 1 for an imaginary view of Eocene Gosiute Lake.)

There is good geologic evidence for believing that the lake basins formed by downwarp. The hydrographic basin of Gosiute Lake includes roughly 48,500 square miles. By using all the data available it was deduced that the climate had the following average values: temperature, 65°F; rainfall on the lake, 34 in.; rainfall on watershed, 38 in.; evaporation from lake surface, 52 in.; evapotranspiration from the watershed, 29 in.; runoff from the land, 8.5 in.

Gosiute Lake contained one large island of Upper Cretaceous rocks and several small islands of Jurassic rocks along its western margin. The varves in the Green River Formation consist of one lamina of microgranular carbonates (calcite and dolomite) and one lamina of virtually structureless organic matter in which are particles of carbonates. The varves range in thickness from 0.01 to 10.0 mm, the weighted average being 0.18 mm. In the carbonaceous shale beds, which were deposited in weedy bays and along gently sloping shores, species of the following genera have been identified: *Potamogeton, Nymphaea, Typha, Pontederia, Sparganium, Myriophyllum, Lemna, Salvinia, Equisetum, Pediastrum, Chara,* and a chrysophyte.

Most of the fossil microorganisms are preserved in the structureless organic matter of the beds of rich oil shale. These include a large number and variety of fungus spores, at least one lichen spore, many pollen grains, spores of mosses and ferns, a considerable algal flora, and such things as plant hairs, insect

PALEOLIMNOLOGY

FIG. 1. Imaginary view west along the south shore of the Eocene Gosiute Lake. The animals are small precursers of the modern horse (after S. H. Knight).

parts such as wing scales, hairs, and fragments of wings and carapace. One of the most remarkable of these fossils is a single germling cell of *Spirogyra* in which the spiral chloroplast has been preserved.

No diatoms have been found in the earliest swamp and lake beds, and no fresh-water diatoms as old as the Eocene are known. The commonest fossils in the fluviatile sediments adjacent to the margin of the lake are the remains of turtles and crocodiles, which must have been very numerous. Fish and bird bones are also found in these lake margin beds, along with one snake. Farther back from the lake, remains of marsupials, insectivores, tillodonts, primates, rodents, carnivores, condylarths, perissodactyls, and artiodactyls have been found.

In the beds that formed in shoal water, the commonest fossils are the snails *Goniobasis*, *Viviparus*, the clam *Unio* (?) and the ostracods *Cypridea* and *Candona*. In the carbonaceous shale beds estheriid shells are locally extremely numerous, and caddis fly cases were found (Bradley, 1924). Locally in the less carbonaceous beds mycetophyllid and tabanid fly larvae are found in great abundance. Adult insects are locally common in the Green River Formation of Colorado and Utah, and more than 300 species have been described (Scudder, 1878; Cockerell, 1916). Fragmentary and well-preserved fossil fish, all small, are found locally in Gosiute Lake beds. The famous, beautifully preserved fossil fish represented in many museums come from the small Green River lake that lay west of Gosiute Lake. Shells of *Sphaerium*, *Valvata*, and a small freshwater mussel occur in beds that formed in somewhat deeper water, along with ostracods. Partly broken, clathrate loricae with rather coarse, rounded openings have been found, which are tintinnids, perhaps *Dictyocysta* sp.

Oligocene—Florissant Lake Beds

Despite a voluminous literature on the remarkable fossil plants and insects found in these Colorado beds, very little has been done with the paleolimnology. MacGinitie thinks lakes and ponds formed because volcanic ash falls and mud flows disrupted the streams to cause ponding. The lake beds contain diatoms, sponge spicules, and pollen grains. "In addition to the common fossil plants and insects, there are thin layers covered with enormous numbers of ostracod shells and also ephippia or egg cases of water fleas (Cladocera). In places, flattened shells of fresh water gastropods are abundant, as are fish coprolites with masses of insect fragments. In a few places fish skeletons are common and well preserved." The climate was warm temperate with an annual rainfall not exceeding 25 in. and probably not more than 20 in.

Miocene—Barstow Formation

In a remarkable paleontologic paper Palmer (1957) provided a brief account of an extinct, shallow closed lake in the Mojave Desert of southern California. The beds deposited in this lake are of middle or late Miocene age.

The aquatic fauna "is dominated by three species: dytiscid beetle larvae belonging to

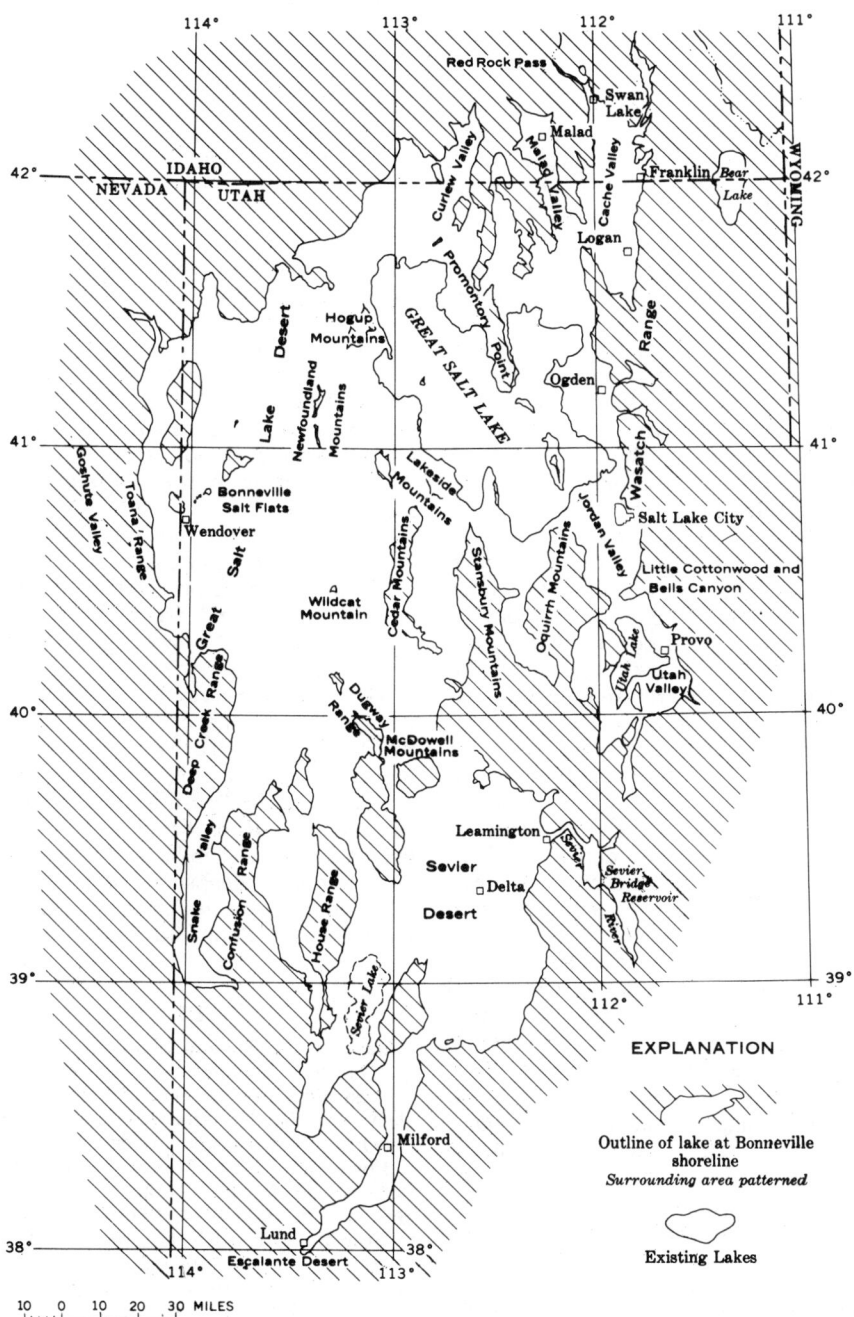

FIG. 2. Map of Lake Bonneville, Great Salt Lake, and Utah Lake.

Schistomerus californense, larvae and pupae of *Dasyhelea australis antiqua*, and an indeterminate fairy shrimp. These constitute 98% or more of the assemblage. Next most abundant are larval water mites." Other aquatic forms include pupae only of *Culicoides* and *Dasyhelea*. Represented only by larve are dragonflies and a tendipedid.

The nonaquatic fauna contains more species but fewer individuals, the most numerous being thrips. Included also are beetles, a leaf hopper, several bugs, and one spider. Many of these arthropods are astonishingly well preserved, some showing a considerable amount of the internal anatomy.

Pleistocene Lakes

Lake Bonneville, Utah. Most of the recent work on Lake Bonneville and its descendant, Great Salt Lake, Utah (see Fig. 2), has been done by, or under the direction of, Eardley (1938; Eardley et al., 1957; Eardley and Gvosdetsky, 1960). At its maximum stage Lake Bonneville occupied a large part of the western half of Utah and extended a little way into Nevada and Idaho. It probably was in existence for much of the Pleistocene epoch—something on the order of a million years. Sharing the wide-swinging changes of the Pleistocene climate, its history has been correspondingly variable. At about the transition from the Pleistocene to the Holocene (15,000–9000 BP) it was large and fresh, but during the interglacial stages it seems to have been smaller and saline, as is the present Great Salt Lake in the semiarid climate of today.

Prior to the late Pleistocene, there were seven comparable fresh-water stages when the lake was large and overflowed. From the characteristics of the sediments in the core, the authors estimate that the lake was fresh and shallow except for repeated brief saline stages.

Eardley made a number of collections of algae growing on calcareous deposits, which had formed in response to the photosynthesis of the algae. The calcareous deposits, bioherms, are restricted to shallow water and make extensive, irregular sheet-like deposits that have an aggregate extent of about 100 square miles. They form in water that ranges from a few inches to 10 or 12 ft in depth. More than 50% of the surface of these mounds is covered with the "brown to pink, blue-green, unicellular, colonial alga *Aphanothece packardii*." The calcareous deposits themselves are nodose, lamellar, or arborescent, and most of them are very porous.

Lake San Augustin, New Mexico. This extinct lake occupied an elongate basin in west-central New Mexico. During late Pleistocene time it had a maximum area of 225 square miles, a maximum depth of 165 ft, and occupied about 13% of its hydrographic basin. During this late Pleistocene stage it cut many shore terraces and built bars and beach cusps (Powers, 1939).

Searles Lake, California. The stratigraphy and the later part of the history of Searles Lake in southeastern California has been presented, along with 33 radiocarbon dates, by Flint and Gale (1958).

The ancestral Searles Lake occupied three adjacent, typical Great Basin block-faulted valleys. At its maximum extent it had an area of about 385 square miles and a maximum depth of at least 750 ft. Because the present-day Searles Lake contains an enormous volume of valuable salts and brine, many exploratory boreholes have been drilled, at least one to 875 ft deep.

Cores show that the "upper salt" is 60 to 90 ft thick and has relatively few, thin mud partings. Below this is a greenish to black mud, the "parting mud," 10 to 13 ft thick, much of it with alternating light and dark laminae. Below is another salt sequence, the "lower salt," 25 to 35 ft thick. This also consists of a complex mixture of salts. Below it is the "bottom mud," which is about 100 ft thick, and grades downward into a unit known as the "mixed layer," which consists of an alternation of mud layers and beds of nahcolite, trona, and gaylussite, but no halite.

Carbon-14 dates on samples taken from the "parting mud" and from the upper part of the "bottom mud" show ages respectively of 13,000 and 14,000 years.

The ancient Searles Lake received most of its water from the overflow of Owens Lake, which, in turn, received its water from the high Sierras. At its highest stages Searles Lake overflowed. The "parting mud" contains fossil fish (a cyprinid, probably *Siphateles*) and pollen grains, which "indicate a cooler and wetter climate during deposition of the Parting Mud than during precipitation of the overlying Upper Salt" (Flint and Gale, 1958, p. 698).

W. H. Bradley

References

Alderman, A. R., and Skinner, H. C. W., 1957, "Dolomite sedimentation in the south-east of Australia," *Am. J. Sci.*, **255**, 561–567.

Anderson, R. Y., and Kirkland, D. W., 1960, "Origin, varves, and cycles of Jurassic Todilto Formation, New Mexico," *Bull. Am. Assoc. Petrol. Geol.*, **44**, 37–52.

Bradley, W. H., 1924, "Fossil caddice fly cases from the Green River Formation of Wyoming," *Am. J. Sci.*, **7**, 310–312.

Bradley, W. H., 1929a, "Algae reefs and oolites of the Green River Formation," *U. S. Geol. Surv. Profess. Paper* **154**, 203–223.

Bradley, W. H., 1929b, "The varves and climate of the Green River epoch," *U.S. Geol. Surv. Eurv. Profess. Paper* **158**, 86–110.

Bradley, W. H., 1931, "The origin of the oil shale and its microfossils of the Green River Formation in Colorado and Utah," *U.S. Geol. Surv. Profess. Paper* **168**, 37–56.

Bradley, W. H., 1963," Paleolimnology, in (Frey, D. G., editor), "Limnology in North America," Madison, Wisconsin, Univ. Press, pp. 621–652.

Clisby, K. H., and Sears, P. B., 1956, "San Augustin Plains—Pleistocene climatic changes," *Science,* **124,** 537–539.
Cockerell, T. D. A., 1916, "Some American fossil insects," *Proc. U. S. Natl. Museum,* **51,** 91–92.
Deevey, E. S., 1955, "The obliteration of the hypolimnion," *Mem. Ist. Ital. Idrobiol. Suppl.* **8,** 11–36.
Eardley, A. J., 1938, "Sediments of Great Salt Lake, Utah," *Bull. Am. Assoc. Petrol. Geol.,* **22,** 1305–1411.
Eardley, A. J., and Gvosdetsky, V., 1960, "Analysis of Pleistocene core from Great Salt Lake, Utah," *Bull. Geol. Soc. Am.,* **77,** 1323–1344.
Eardley, A. J., Gvosdetsky, V., and Marsell, R. E., 1957, "Hydrology of Lake Bonneville and sediments and soils of its basin," *Bull. Geol. Soc. Am.,* **68,** 1141–1201.
Flint, R. F., and Gale, W. A., 1958, "Stratigraphy and radiocarbon dates at Searles Lake, California," *Am. J. Sci.,* **256,** 689–714.
Frey, D. G., 1961, "Developmental history of Schleinsee," *Verhandl. Inter. Ver. Limnol.,* **14,** 271–278.
Gilbert, G. K, 1891, "Lake Bonneville," *U.S. Geol. Surv. Monograph* I, 438pp.
Hutchinson, G. E., 1957, "A Treatise on Limnology," New York, John Wiley & Sons.
La Rocque, A., 1960, "Molluscan faunas of the Flagstaff Formation of Central Utah," *Mem. Geol. Soc. Am.,* **78,** 1–100.
MacGinitie, H. D., 1953, "Fossil plants of the Florissant beds, Colorado," Carnegie Inst. Washington, Publ. **599,** 1–188.
Meinzer, O. E., 1922, "Map of the Pleistocene Lakes of the Basin and Range Province and its significance," *Bull. Geol. Soc. Am.,* **33,** 541–552.
Miller, D. N., Jr., 1961, "Early diagenetic dolomite associated with salt extraction process, Inagua, Bahamas," *J. Sediment. Petrol.,* **31,** 473–476.
Milton, C. M., and Eugster, H., 1959, "Mineral Assemblages of the Green River Formation," in (Abelson, P. H., editor), "Researches in Geochemistry," New York, John Wiley & Sons, pp. 113–150.
Minder, L., 1923, "Uber biogene Entkalkung im Zürichsee," *Verhandl. Intern. Ver. Limnol.,* **1,** 20–32.
Nipkow, F., 1920, "Vorläufige Mitteilungen über Untersuchungen des Schlammabsatzes im Zürichsee," *Rev. Hydrol.,* **1,** 100–122.
Palmer, A. R., 1957, "Miocene arthropods from the Mojave Desert, California," *U.S. Geol. Surv. Profess. Paper* **294-G,** 237–277.
Powers, W. E., 1939, "Basin and shore features of the extinct Lake San Augustin, New Mexico," *J. Gemorphol.,* **2,** 345–356.
Russell, I. C., 1885, "Lake Lahontan, *U.S. Geol. Surv. Monograph,* **11,** 288pp.
Scudder, S. H., 1878, "The fossil insects of the Green River shales," *U.S. Geol. Geogr. Surv. Terr. Bull.* **4,** 747–776.
Tasch, Paul, and Zimmerman, J. R., 1961, "Fossil and living conchostracan distribution in Kansas-Oklahoma across a 200-million year time gap," *Science,* **133,** 584–586.
Vallentyne, J. R., 1957a, "The molecular nature of organic matter in lakes and oceans, with lesser reference to sewage and terrestrial soils," *J. Fisheries Res. Bd. Can.,* **14,** 33–82.
Vallentyne, J. R., 1967b, "Carotenoids in a 20,000-year-old sediment from Searles Lake, California," *Arch. Biochem. Biophys.,* **70,** 29–34.
Welten, M., 1944, "Pollenanalytische, stratigraphische und geochronologische Untersuchungen aus dem Faulenseemoos bei Spiez," *Veröffentl. Geobotan. Inst. Rübel,* **21,** 1–201.

Cross-references: *Geologic Time Scale; Lake Geochemistry; Limnology; Paleohydrology; Paleotemperatures: Isotopic Determination.* Vol. II: *Paleoclimatology (Historical Data).* Vol. III: *Glacial Lakes; Great Salt Lake; Holocene Lakes; Pluvial Lakes; Quaternary; Terraces—Lacustrine.* Vol. IVB: *Green River Mineralogy; Oil Shales.* Vol. VI: *Lacustrine Sedimentation; Varves.* Vol. VII: *Paleontology; Palynology; Palynology of Quaternary Lake and Swamp Deposits.*

PALEOPATHOLOGY—See Vol. VII

PALEOSALINITY

The salinity of water in which a sedimentary rock accumulated has been named paleosalinity. The term was formulated by Frederickson and Reynolds in 1960 (see reference in Walker, 1968) but the concept is much older because fossiliferous rocks have long been recognized as the deposits of ancient seas analogous to those of today.

For a long time, paleosalinity had to be estimated from faunal, mineralogical, and lithological information. However, recent advances in geochemistry have shown that trace element abundance in rocks and minerals is closely linked to depositional environment. Methods for estimating paleosalinity based on this principle are being developed but are not universally applicable. Therefore, geological methods continue to be extremely useful, especially for checking geochemical interpretations.

Geological Methods for Estimating Paleosalinity

Paleoecology. Fossils are often a useful guide to paleosalinity, especially if they are closely related to living forms as in Cenozoic rocks. The environmental significance of many fossils from older rocks, which cannot be similarly compared with living forms, may be less certain. Advanced discussions of this aspect of paleoecology and many additional references are published by Ager (1963) and Ladd (1957; see reference in Ager, 1963).

Mineralogy. The mineralogy of carbonates and evaporites is closely linked to paleosalinity. For example, calcite, anhydrite, and halite normally indicate progressively higher paleosalinities but recrystallization and metasomatism may obscure the primary relationship between mineralogy and salinity. For further information, the recently published chapter on marine evaporites by Stewart (1963) should be consulted.

Lithology and Bedding. Paleosalinity may be indirectly suggested by facies, texture, and bedding structures in cyclothems. In typical Carboniferous cyclothems, marine limestones or shales at the base grade up through silts into cross-bedded sandstones. The upward increase of grain size and directional bedding structures near the top of each cycle are attributed to a transition from marine to deltaic conditions associated with the repeated infilling of a shallow sea. Thus, a cycle of decreasing salinity is inferred. Duff and Walton (1962) in a statistical study of British Coal Measure cyclothems make reference to many earlier investigations.

Geochemical Methods for Estimating Paleosalinity

Different trace elements are enriched in marine and freshwater rocks and this has suggested that trace elements could be environmental indicators. Two methods have been developed: the first depends on measuring the concentration of a large number of elements in a rock; the second depends on measuring the concentration of a single element in a mineral component of a rock.

Estimating Paleosalinity from Abundance of Trace Elements in Rocks. Degens et al. (1957), followed by Keith and Degens (1959; see references in Potter et al., 1963), showed that the abundance of many trace elements in sedimentary rocks was a function of depositional environment. Discrimination between marine and nonmarine was improved if element ratios, rather than absolute abundance was considered. Further work by Potter et al. (1963) demonstrated that most marine and freshwater rocks could be statistically discriminated if several trace elements were considered collectively.

Estimating Paleosalinity from Trace Elements in Minerals. *Early Work.* The concentration of some trace elements in authigenic minerals during sedimentation appears to be a function of paleosalinity. For example, Imbrie (1955; see reference in Ager, 1963) found that the strontium-calcium ratio in the carbonate fraction of the Permian Florena Shale of Kansas reflected depositional salinity. Harder (1959) and Frederickson and Reynolds (1960; see references in Walker, 1968) independently showed that the concentration of boron in illite was partly determined by paleosalinity. Again, Frederickson (1962; see reference in Walker, 1968) suggested that the concentration of bromine in evaporite minerals was controlled by salinity.

The enrichment of trace elements in authigenic minerals during sedimentation has been explained in terms of partition coefficients by Frederickson (1962; see reference in Walker, 1968). According to Frederickson, "Partition coefficients relate the amount of the trace element that enters a growing crystal to the amount of that element in the coexisting fluid. These coefficients are constants and are characteristic of each mineral species. When the coefficient is known, the composition of the fluids in which the crystal grew can be calculated by measuring the level of concentration of that element in the mineral." Thus, Frederickson has outlined a geochemical basis for determining paleosalinity. Several trace elements may prove to be paleosalinity indicators but, as yet, only boron has been extensively investigated.

Boron and Paleosalinity. Frederickson and Reynolds (1960; see reference in Walker, 1968) found that the boron content and salinity of sea water were directly related. They argued that the boron content of illite might be a function of boron in sea water at the time of deposition and so, indirectly, of paleosalinity. An ideal formula for illite was assumed and the boron content of pure illite (adjusted boron) in clay fractions, isolated by size fractionation, was calculated by the following formula:

$$\text{Boron in pure illite} = \frac{\text{Boron in clay fraction} \times \% \text{ K}_2\text{O in pure illite}}{\% \text{ K}_2\text{O in clay fraction}}$$

The boron content of pure illite was thought to be an index of paleosalinity because it was directly related to paleosalinity inferred from other geological evidence. Some boron in sediments is contributed by tourmaline but detrital tourmaline was believed to be absent in the fine size fractions studied by Frederickson and Reynolds. Later work (Walker, 1963; see reference in Walker, 1968) supported this conclusion.

Evidence from Laboratory Studies. Boron is rapidly absorbed from solution by illite; the final concentration being a function of time, temperature and solution concentration

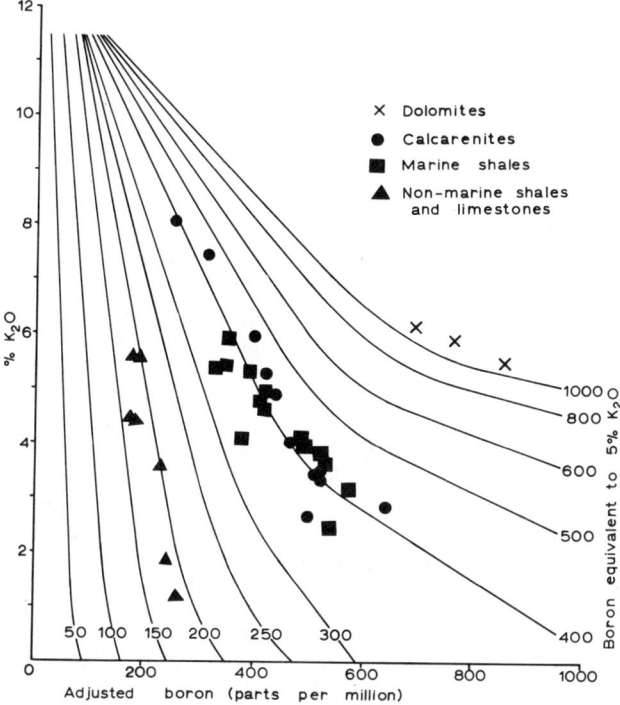

Fig. 1. Departure curves for computing equivalent boron (from Walker and Price, 1963; reference in Walker, 1968). The experimental data used to establish the curves has been superimposed. To compute equivalent boron, plot coordinates for adjusted boron and K20, and the curve on which the intersection falls is equivalent boron.

(Harder, 1961; Lerman, 1966; see references in Walker, 1968). Once absorbed, boron is firmly held and probably enters the tetrahedral sites of the mica lattice (Harder, 1961; Frederickson and Reynolds, 1960; see references in Walker, 1968). Boron-silicon diodochy was confirmed by Stubican and Roy (1962; see reference in Walker, 1968) who studied the X-ray diffraction patterns and infrared absorption spectra of synthetic boron clays.

Recent Developments. Walker and Price (1963; see reference in Walker, 1968) found that the boron content of illite, although reflecting paleosalinity, was also a function of grain size and composition. In illites of constant grain size ($< 0.5\mu$) and similar inferred paleosalinity, adjusted boron or boron in pure illite (a function of $[B]/[K_2O]$) and potassium were inversely related. The concentration of boron [B] and potassium [K_2O] varied sympathetically but the rate of increase of boron $d[B]/d[K_2O]$ was inversely related to potassium [K_2O] so that adjusted boron (B/K_2O) was inversely related to potassium. Departure curves (Fig. 1) based on the experimental data of Walker and Price were used to compensate adjusted boron values for variation related to the concentration of K_2O. These revised boron values (equivalent boron), being unrelated to [K_2O], were regarded as realistic indices of pareosalinity.

Adams et al. (1965) and Walker (1964; see references in Walker, 1968) reported that equivalent boron values from brackish water Holocene illites of the Dovey Estuary, Wales, and from brackish water Carboniferous illites, were similar. Also, the correlation between equivalent boron and mean salinity in the Dovey Estuary was statistically significant (Fig. 2). Such results suggest that a boron-illite equilibrium reflecting salinity is established during deposition and is not substantially changed by diagenesis and lithification. A general relation between the concentration of boron in clay fractions and salinity has also been reported from the Niger Delta, West Africa (Porrenga, 1967). Since illite is not the dominant mineral in these clay fractions, paleosalinity estimates based on boron in non-illitic clays may eventually prove to be possible.

Some authors have questioned the validity

PALEOSALINITY

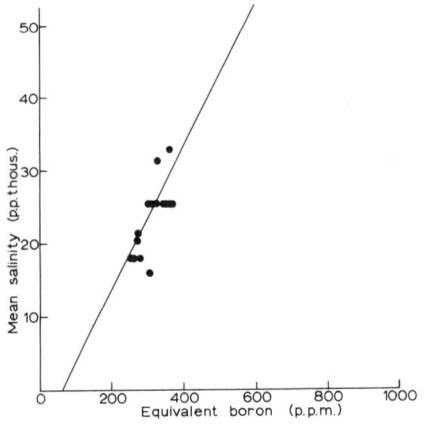

FIG. 2. Correlation between mean salinity and equivalent boron in the Dovey Estuary, Wales (based on data from Adams et al., 1965).

of boron in illite as a salinity index. For example, Eager and Spears (1966; see reference in Walker, 1968) reported that boron was related to organic carbon content rather than salinity in certain British Carboniferous sediments. Levinson and Ludwick (1966; see reference in Walker, 1968) argued that the amount of boron absorbed by illite was an inverse function of grain size. A fortuitous relation between boron and salinity might develop, therefore, if the grain size of illite in marine sediments was less than in nonmarine sediments. Shaw and Bugry (1966; see reference in Walker, 1968) stressed that boron could be held in recycled illite, non-illitic clays and tourmaline, so that in some sediments boron was related to source and mineralogy rather than paleosalinity.

Walker (1968) reviewed much of the previous work on boron in clays and sediments and concluded that in clay size fractions ($< 0.5\mu$) from sedimentary sequences dominated by carbonates most of the illite was authigenic. Hence the boron concentration in such clay fractions was salinity dependent because equilibrium between boron in seawater and illite was established at the time of deposition. In whole rocks and clay fractions from rapidly deposited clastic sequences, however, boron concentration was not a reliable salinity index because it was related in part to parameters other than salinity.

Application to Paleogeography. Frederickson and Reynolds (1960; see reference in Walker, 1968) using the boron content of pure illite (adjusted boron) as an index of paleosalinity showed that the zone of highest paleosalinity in Devonian off-reef shales of southern Alberta, Canada (Fig. 3) coincided with the stage of maximum vertical reef growth. Thus, paleosalinity changes measured geochemically agreed with other geological information.

Walker (1964; see reference in Walker, 1968) applied the concept of equivalent boron to cyclothems in the late Visean Yoredale Formation of northern England. Equivalent boron correlated closely with salinity changes in each cyclothem inferred from geological information (Fig. 4), but was anomalously high in coarse, porous sandstones, previously regarded as deltaic. However, reexamination of

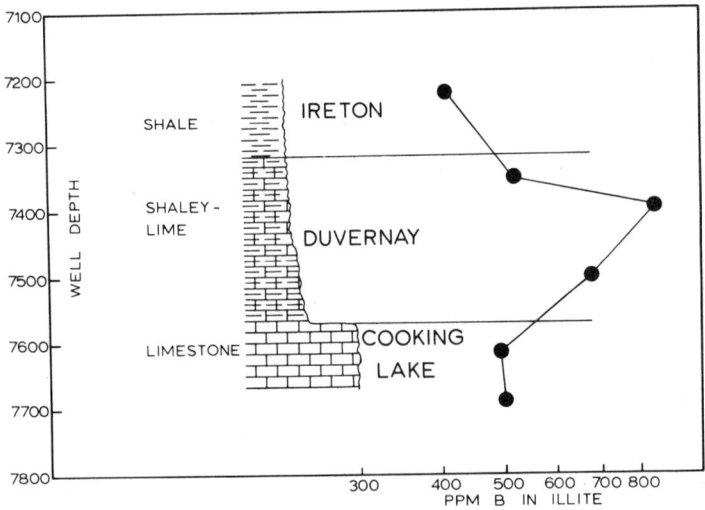

FIG. 3. Paleosalinity variation reflected by the boron content of pure illite in Devonian off-reef shales, Alberta. Maximum reef growth and the highest paleosalinity occurred in Duvernay times.

PALEOSALINITY

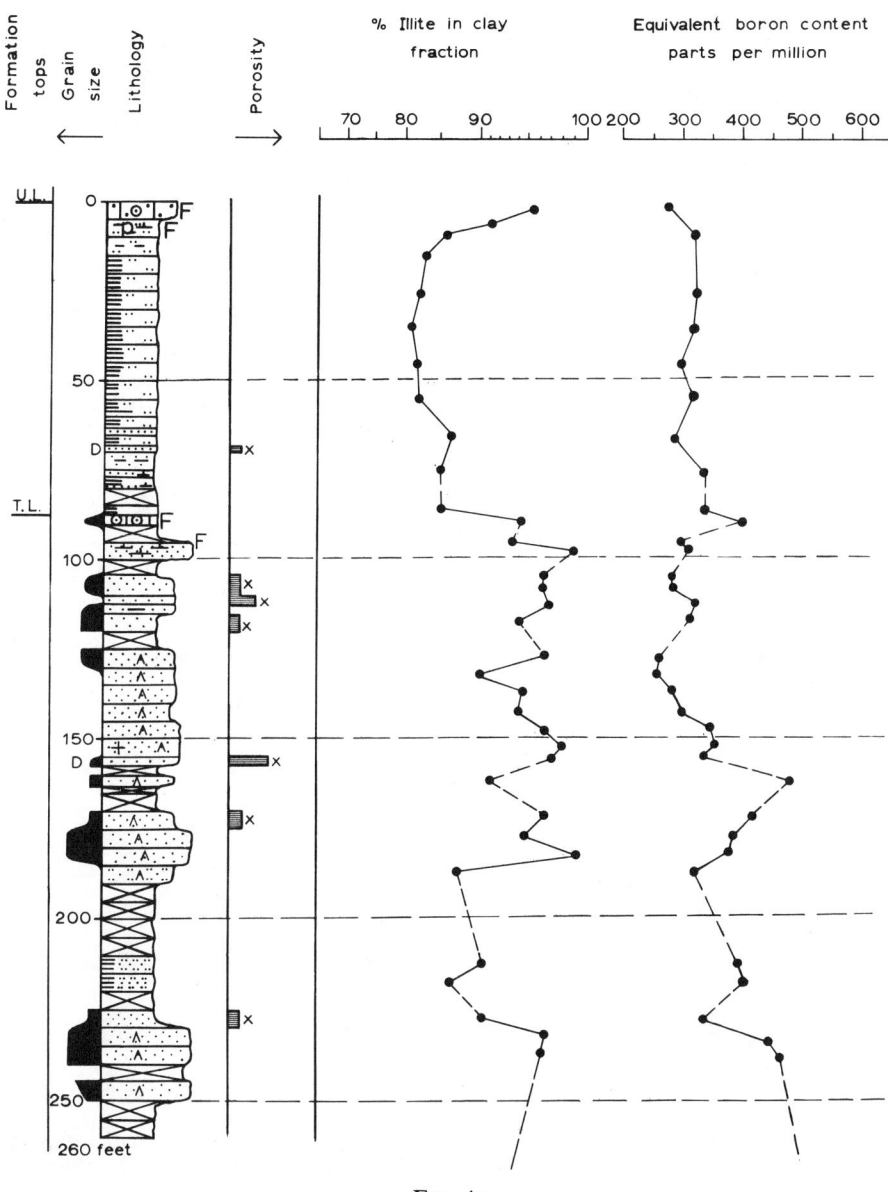

FIG. 4a.

the geological evidence showed that some of these sands were definitely marine and they were therefore reinterpreted as beach and shallow bay sands deposited at the seaward margin of a delta. Toward the south, the cyclic Yoredale Formation passes into a massive carbonate facies. Precise correlation between the two contrasted facies by faunal and lithological methods has not proved possible. Further work by Walker (1967; see reference in Walker, 1968) led to identification of salinity cycles in the massive carbonate facies which could be correlated with the cyclothems in the north.

Reynolds (1965; see reference in Porrenga, 1967) finds that the boron content of pure illite from relatively unmetamorphosed Precambrian carbonates and from post Cambrian marine rocks is similar. Thus the boron content of seawater must have remained constant for 1.9 billion years or more. Even if the ratio of boron and salinity have changed since Precambrian times, boron could still be used as an index of relative salinity changes in Precambrian rocks of similar age.

Conclusions

This review shows that no method for estimating paleosalinity is universally applicable.

PALEOSALINITY

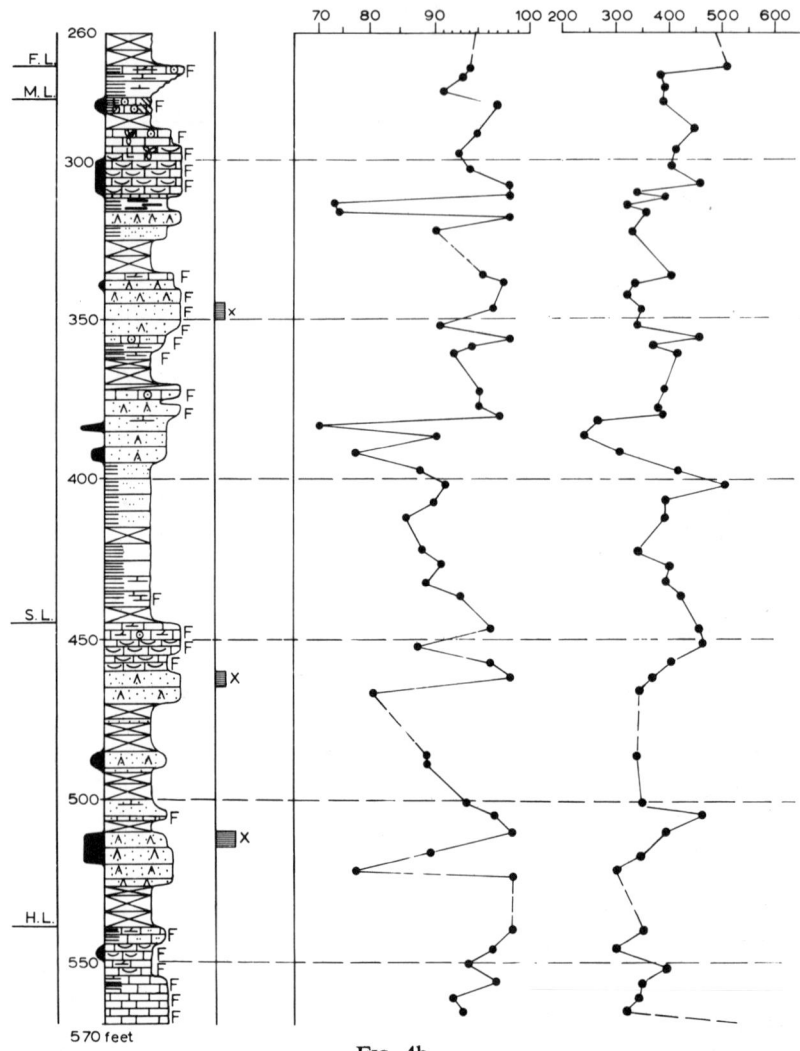

FIG. 4b.

Paleoecology is the best guide in late Cenozoic and Recent rocks and in many other fossiliferous rocks which do not contain authigenic illite as the dominant clay mineral. Mineralogy is a useful guide to the paleosalinity of evaporites. However, in recrystallized evaporites, geochemistry may yield more reliable information. Boron is a reliable guide to salinity in illitic rocks if other geological information indicates that slow erosion, transport and deposition favored growth of authigenic illite. In other rocks which contain few or no fossils, the relative abundance of trace elements may be the only indication of environment.

C. T. WALKER

References

Adams, T. D., Haynes, J. R., and Walker, C. T., 1965, "Boron in Holocene illites of the Dovey Estuary, Wales, and its relationship to paleosalinity in cyclothems," *Sedimentology*, 4(3), 189–195.
*Ager, D. V., 1963, "Principles of Paleoecology: an introduction to the study of how and where animals and plants lived in the past," New York, McGraw-Hill Book Co., 371pp.
*Duff, P. M. D., and Walton, E. K., 1962, "Statistical basis for cyclothems: a quantitative study of the sedimentary succession in the East Pennine Coal Field," *Sedimentology*, 1, 235–255.
Lerman, A., 1966, "Boron in Clays and Estimation of Paleosalinities," *Earth-Sci. Rev.*, 2(1).
Levinson, A. A., and Ludwick, J. C., 1966, "Speculation on the Incorporation of Boron into Argillaceous Sediments," *Geochim. Cosmochim. Acta*, 30(a), 855–861.
*Porrenga, D. H., 1967, "Clay mineralogy and geochemistry of Recent marine sediments in tropical areas," Doctoral Thesis, University of Amsterdam.

PALEOTEMPERATURES—ISOTOPIC DETERMINATIONS

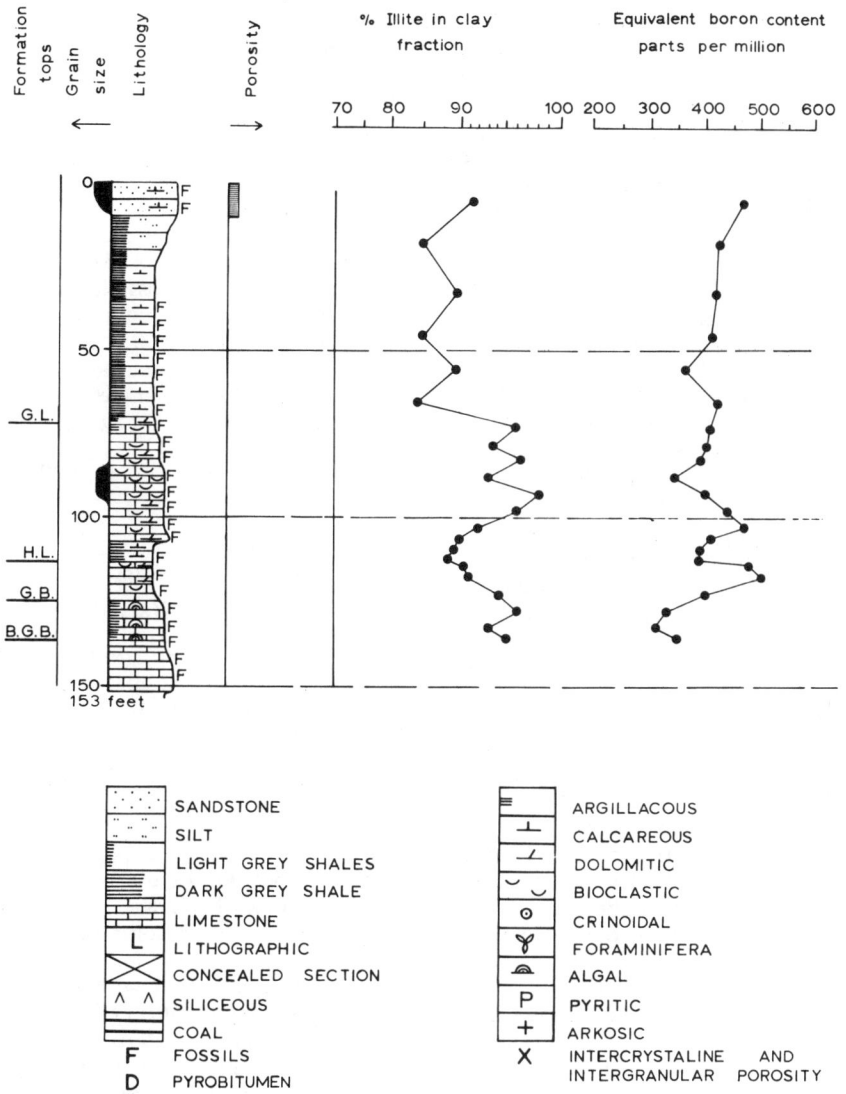

Fig. 4c.

*Potter, P. E., Shimp, N. F., and Witters, J., 1963, "Trace elements in marine and freshwater argillaceous sediments," *Geochim. Cosmochim. Acta*, **27**, 669–694.

Reynolds, R. C., 1965, "The concentration of boron in Precambrian seas." *Geochim. Cosmochim. Acta*, **29**, 1–16.

*Stewart, F. H., 1963, "Marine evaporites," in "Data of Geochemistry," sixth edition, *U.S. Geol. Surv. Profess. Paper* **440-Y**.

*Walker, C. T., 1968, "Evaluation of boron as a paleosalinity indicator and its application to offshore prospects," *Am. Assoc. Petrol. Geol. Bull.*, **52**, 751–766.

*Additional bibliographic references may be found in this work.

Cross-references: *Authigenesis of Minerals; Boron; Geochemistry; Mineralogy; Seawater, Chemistry; Seawater, History; Trace Elements; X-Ray Diffraction Analysis.* Vol. IVB: *Evaporite Minerals; Illite.* Vol. VI: *Diagenesis; Evaporites; Grain Size Studies; Lithology.* Vol. VII: *Cyclothems; Paleoecology; Paleogeography.*

PALEOTEMPERATURES—ISOTOPIC DETERMINATIONS

Molecular energy may be divided into (*a*) translational energy, (*b*) vibrational energy, (*c*) rotational energy, and (*d*) electronic energy. In chemical reactions, the various isotopes of the different elements present in a system tend to concentrate to different extents in different compounds, so as to produce the maximum decrease in the free energy of the system. At ordinary temperatures and above, and for all elements except hydrogen, only the vibrational energy of the molecules affects isotopic distributions.

The hydration reaction

$$CO_2 + H_2O \leftrightarrows H_2CO_3 \quad (1)$$

may be used as an example to illustrate the mechanism of exchange of isotopes. One oxygen atom in 500 is an O^{18}, one in 2700 is an O^{17}. O^{17} may be disregarded because of its scarcity in nature. One couple of CO_2–H_2O molecules in 166 couples of the system in Eq. (1) will contain an O^{18} atom. This may be either in the CO_2 or in the H_2O molecule. Both these molecules combine to form an H_2CO_3 molecule, and when the reverse reaction takes place, the O^{18} will have the choice of entering the CO_2 molecule or the H_2O molecule. The probability of entering either one of these two molecules is not exactly in the ratio 2:1 but is somewhat greater than 2 for the CO_2 molecule. This is because the total vibrational energy of the couple $CO^{18}O^{16}$–H_2O^{16} is somewhat smaller than the total vibrational energy of the couple CO_2^{16}–H_2O^{18}.

The O^{18}/O^{16} fractionation factor (α) between CO_2 and H_2O has been calculated and found equal to 1.045 at 0°C. This means that, at 0°C, the ratio O^{18}/O^{16} in the CO_2 molecules is 45 permil greater than the same ratio in the water molecules.

Reaction (1) provides a mechanism for the attainment of isotopic equilibrium in the CO_2–H_2O system. Isotopic equilibrium may also be reached in the carbonate-water system, as shown by the following general equations:

$$CO_2 + H_2O \leftrightarrows H_2CO_3 \quad (2)$$

$$H_2CO_3 \leftrightarrows H^+ + HCO_3^- \quad (3)$$

$$HCO_3^- \leftrightarrows H^+ + CO_3^{2-} \quad (4)$$

$$CO_3^{2-} + Ca^{2+} \leftrightarrows CaCO_3 \quad (5)$$

The decrease of the free energy of a system, when exchange reactions take place and a given isotope is preferentially concentrated in a given compound, becomes less important with respect to the total energy of the system with increasing temperature, because of the increase of the total energy. Thus, the preferential concentration of a given isotope in a given compound will matter less and less with increasing temperature, and the fractionation factor α will tend toward unity.

The temperature dependence of the oxygen-isotope fractionation factor α was proposed by Urey (1947) as a basis for determining the temperature of precipitation of the calcium carbonate by measuring its oxygen-isotopic composition. With a precision mass spectrometer of the type developed by Nier (1940, 1947) and modified by McKinney and co-workers (McKinney et al., 1950), and with the extraction techniques developed by McCrea (1950), Epstein et al. (1953), and Naydin et al. (1956), it is possible to measure the O^{18}/O^{16} ratio in any calcium carbonate sample with a precision of ± 0.1 permil. The relationship between temperature and isotopic composition of the carbonate was found empirically by Epstein et al. (1953) to be

$$T = 16.5 - 4.3(\delta - A) + 0.14(\delta - A)^2 \quad (6)$$

where

$$\delta = \frac{O^{18}/O^{16}(\text{sample}) - O^{18}/O^{16}(\text{standard})}{O^{18}/O^{16}(\text{standard})} \quad (7)$$

is the difference in isotopic composition between the sample and the Chicago PDB-1 standard, and

$$A = \frac{O^{18}/O^{16}(W') - O^{18}/O^{16}(W)}{O^{18}/O^{16}(W)} \quad (8)$$

is the correction to be applied if the oxygen isotopic composition of the water (W') is different from that of average seawater (W).

The oxygen-isotopic composition of surface water in the open seas is indeed variable. Carbon dioxide equilibrated with ocean water at 25°C ranges from about 0 permil (with respect to carbon dioxide from PDB-1) at the equator to about +1 permil in the tropical evaporation belts, and to about −1 permil in the higher latitudes. The total range (2 permil) is equivalent to a temperature change of 9°C; calcium carbonate deposited from water having an isotopic composition of −1 permil will give an apparent temperature 9°C higher than a carbonate deposited at the same temperature from water having an isotopic composition of +1 permil. The uncertainty resulting from this effect can be eliminated only by measuring directly the isotopic composition of the water (this can be done only for the modern ocean) or by measuring the oxygen-isotopic composition of a compound coprecipitated in equilibrium with the water and the carbonate (a technique which is still in the experimental stage). The uncertainty, however, can be minimized if the oxygen-isotopic composition of the seawater can be estimated within reasonably narrow limits (Craig, 1964). It is also necessary, if meaningful temperature data are to be obtained, that the original oxygen isotopic composition be still preserved in the sample. Although unaltered samples of calcium carbonate are commonly found in Cenozoic and even Mesozoic deposits, unaltered Paleozoic samples are found only rarely. The mineralogical and chemical criteria by which unaltered samples can be recognized as such have been discussed by Lowenstam (1961).

Paleotemperature Analysis of Fossil Shells

The oxygen-isotope method of paleotemperature analysis has been applied extensively to Mesozoic belemnites and Cenozoic mollusks and foraminifera. Belemnites are extinct cephalopods with a thick calcium carbonate rostrum as part of their endoskeleton. Isotopic analysis of these rostra by Urey, Epstein, Lowenstam, Bowen, Naydin, and others revealed two successive major temperature cycles having an amplitude of several degrees Celsius and extending across about 60 million years of Mesozoic time. Although the temperature trend across the Mesozoic-Cenozoic boundary remains unknown, for lack of fossils suitable for analysis, isotopic analysis of Tertiary mollusks by Dorman and Gill (1959) showed additional major temperature fluctuations. Fig. 1 shows the temperature trend during the Mesozoic and Tertiary for the middle northern and southern latitudes. As may be seen, the amplitude of the temperature fluctuations is several degrees Celsius, and the wavelength is some 20 million years. In addition, there seems to be an overall average temperature decrease amounting, again, to several degrees Celsius (dashed line in Fig. 1). While this decrease

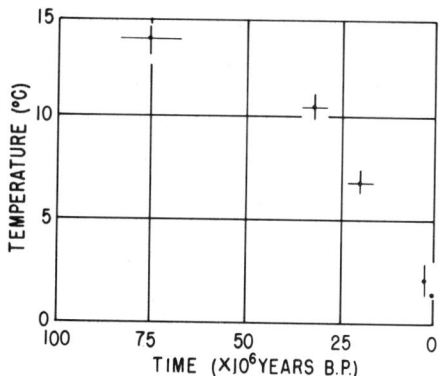

Fig. 2. Temperature decrease, indicated by isotopic analysis, for the high latitudes and the abyssal waters of the ocean during the past 75 million years.

may result from latitudinal changes of the land masses from which the fossils were collected, it is comparable to a temperature decrease of about 12°C for the high latitudes (and the abyssal waters of the oceans) during the past 75 million years, deduced from paleotemperature analysis of Cretaceous belemnites from Alaska and Siberia (Epstein, 1959) and of Tertiary benthonic faraminifera from the equatorial Pacific (Emiliani, 1954, 1961; Lowenstam and Epstein, 1959) (Fig. 2). Thus, the data so far available for the Mesozoic and Tertiary, although still very limited, suggest that the average temperature of the earth's surface decreased 5 to 10°C during the past 150 million years.

A possible explanation for the inferred decrease in overall temperature is a decrease in solar emission. This possibility cannot be excluded, because, in spite of the assumed great secular stability of the sun as a nuclear engine, 150 million years does not represent an entirely negligible fraction of its probable life.

Variation in solar emission could also explain the temperature fluctuations shown in Fig. 1. However, there appears to be a fair correlation between the times of occurrence of the temperature maxima and the times of occurrence of known, major marine transgressions. It would seem, therefore, that the observed temperature fluctuations could be accounted for by changes in the earth's albedo and, therefore, by terrestrial dynamics alone.

The most recent portion of the curve of Fig. 1 represents the Pleistocene. This epoch, characterized by the repeated occurrence of major glaciations, has been defined, by unanimous decision of the 7th INQUA Congress (Denver, Colorado, 1965), as the time that has elapsed since the first appearance of the ben-

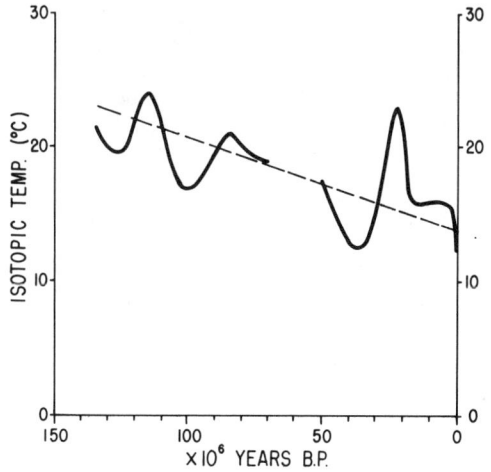

Fig. 1. Temperature variations, indicated by isotopic analysis, for the middle northern and southern latitudes during the past 150 million years.

thonic foraminiferal species *Hyalinea (Anomalina) baltica* (Schroeter) in the late Cenozoic section at le Castella, Calabria, southern Italy (Richmond and Emiliani, 1967). The time of occurrence of this event is unknown at present, and, therefore, the duration of the Pleistocene is also unknown. Various estimates range from 0.7 to 2 million years (see *Quaternary*, in Vol. III).

Shells of planktonic foraminiferal species have been used extensively in isotopic studies of Plio-Pleistocene paleotemperatures. In the low latitudes, two of the more common species, *Globigerinoides rubra* and *G. sacculifera*, deposit their shell material within fifty meters of the ocean surface (Jones, 1967). At these latitudes, therefore, isotopic analysis yields yearly average surface temperature. In the middle and high latitudes, where significant seasonal variations occur in the surface temperature of the ocean, the temperatures obtained by isotopic analysis of *G. rubra* and *G. sacculifera* are essentially summer averages.

Paleotemperature analysis of fossil shells of *Globigerinoides rubra* and *G. sacculifera* collected at close stratigraphic intervals from the section at le Castella revealed marked temperature oscillations and an overall decrease in temperature from the Pliocene into the Pleistocene (Emiliani, 1971).

Analysis of Deep-Sea Cores

The temperature oscillations which occurred during the past million years have been studied by means of isotopic, chemical, and micropaleontological analysis of deep-sea cores of globigerina-ooze facies. These analyses, together with carbon-14 and protactinium-231/thorium-230 age measurements, revealed quasi-sinusoidal temperature oscillations having an apparent range of 9 or 10°C (uncorrected for glacial-interglacial changes in the isotopic composition of seawater) in the Caribbean and the equatorial Atlantic, and an average wavelength of 50,000 years (Fig. 3). Results for all Atlantic and Caribbean cores so far analyzed isotopically were cross-correlated by different criteria, in an effort to detect and account for stratigraphic discontinuities, and an "average core" was reconstructed by averaging, over all cores, the stratigraphic thicknesses of each successive temperature stage (Emiliani, 1966). A generalized curve for temperature relative to time was then reconstructed, based on the O^{18}/O^{16} data (corrected for the glacial-interglacial changes in the isotopic composition of seawater), together with the C^{14} and Pa^{231}/Th^{230} age measurements and reasonable extrapolations therefrom (Fig. 4). This curve, which extends from the present to about 425,000 years ago, shows eight temperature cycles which, with the single exception of the cycle extending from 65,000 to 15,000 years ago, have approximately the same amplitude.

In order to translate the temperature values obtained through isotopic analysis into the approximately true temperatures of Fig. 4, a correction has been applied for the amount and the isotopic composition of (a) the seawater removed during glacial ages and (b) the seawater added, in excess of the present amount, during interglacial ages. The average isotopic composition of continental ice was estimated at -15 permil (Emiliani, 1955, 1966).

The temperature oscillations which occurred during the past 425,000 years, illustrated by the generalized temperature curve of Fig. 4, are established rather firmly with respect to both amplitude and age. Accurate comparisons and correlations of this curve with the glacial and interglacial events of the continents are assured whenever absolute ages of sufficient precision are available for continental events. In part because of sampling uncertainties, the ages so far obtained by K^{40}/Ar^{40} dating of materials from glacial and periglacial areas are generally not sufficiently accurate to allow unequivocal correlation with the oceanic stages. However, some analyses by Frechen and Lippolt (1965), which assign an age of 350,000

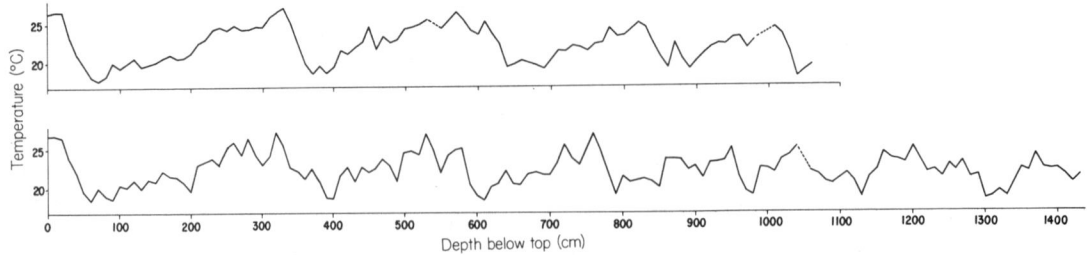

Fig. 3. Temperatures indicated by isotopic analysis of two Caribbean deep-sea cores: (top) P6304-8; (bottom) P6304-9.

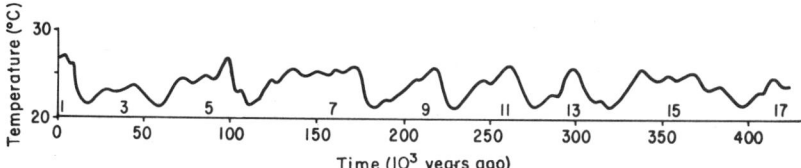

Fig. 4. Generalized temperature curve for the surface water of the Caribbean Sea.

to 400,000 years to the Günz, indicate that the temperature curve shown in Fig. 4 represents the entire glacial Pleistocene as classically understood.

In view of the fact that the continental deposits which have been dated accurately appear to correlate with temperature stages, there is little doubt that close correlations will also be obtained for the continental deposits which have not yet been dated, or which have not been dated with sufficient accuracy. In this context, it appears almost certain that the simple scheme of four or five major glaciations, still widely accepted, will have to be abandoned. In its place is apparently emerging a picture of many alternating high- and low-temperature stages. At least some of these are probably the "interstadials" within the classic glacial and interglacial stages of the Pleistocene.

Causes of Glaciation

The study of the oceanic record has provided significant evidence toward solution of the problem of the cause of the ice ages. When the first cores were analyzed by the oxygen-isotope method, an apparent periodicity of 40,000 to 50,000 years was noted, a periodicity similar to that of the variations of summer insolation in the high latitudes (Emiliani, 1955). Now that additional cores have been analyzed, statistical treatment has become feasible. With this approach, it was found that the correlation coefficient between the astronomically calculated ages of the summer insolation minima at 65°N and the stratigraphic position of the temperature minima in the "average core" previously mentioned is 0.997. Although the correlation coefficient refers to two incremental sequences it is nevertheless highly significant and a causal relationship is strongly suggested (Emiliani, 1967).

Variations in summer insolation in the high northern latitudes have been used by many workers, during the past 100 years, as a basis for theories of glaciation. The modern treatment by Emiliani and Geiss (1959) provides a plausible explanation for the inception of glaciation in the late Cenozoic and the occurrence and timing of the glacial and interglacial events of the Pleistocene. In this treatment, in fact, it was suggested that the inception of glaciation was made possible by Tertiary cooling caused by an albedo increase associated with increased continentality and the Alpine orogenesis. It was further suggested that the successive Pleistocene glaciations were started by quasi-periodic decreases of summer insolation in high northern latitudes but proceeded largely as a self-sustaining process; and that complete deglaciation outside Antarctica and Greenland resulted from surface freezing of the northern North Atlantic, with plastic flow of glaciers, heat absorption by ice melting, and crustal warping providing the necessary impedance. More recently it was noticed that, whereas variations in summer insolation at high northern latitudes apparently modulate the frequency of the Pleistocene temperature oscillations, insolation variations in the northern evaporation belt (about 25°N) may determine the amplitude (Emiliani, 1966). Thus, it is to be expected that high insolation at 25°N coupled with low insolation at 65°N will result in a strong glaciation; and that low insolation at 25°N coupled with high insolation at 65°N will result in a warm interglacial. Intermediate conditions will produce milder glacials and interglacials.

If this treatment is correct, it is to be expected that glacial and interglacial ages will continue to alternate in the future for some millions of years, or until erosion of the modern high relief and marine ingression have sufficiently reduced the earth's albedo to bring glaciation to an end. Meanwhile, it is to be expected that a new glaciation will begin within a few thousand years and reach its peak about 15,000 years from now.

CESARE EMILIANI

References

Blackman, A., and Somayajulu, B. L. K., 1966, "Pacific Pleistocene cores: faunal analyses and geochronology," *Science*, **154**(3751), 886–889.

Bowen, R., 1961, "Paleotemperature analyses of Mesozoic Belemnoidea from Germany and Poland," *J. Geol.*, **69**, 75–83.

PALEOTEMPERATURES—ISOTOPIC DETERMINATIONS

Bowen, R., 1961, "Paleotemperature analyses of Belemnoidea and Jurassic paleoclimatology," *J. Geol.*, **69**, 309–320.

Bowen, R., 1966, "Paleotemperature Analysis," Amsterdam, Elsevier Publ. Co., 265pp.

Craig, H., 1964, "The isotopic geochemistry of water and carbon in geothermal areas," Conf. on "Nuclear Geology on Geothermal Areas," Spoleto, Sept. 9–13, 1963, Consiglio Nazionale delle Ricerche, pp. 17–53.

Dorman, F. H., 1966, "Australian Tertiary paleotemperatures," *J. Geol.*, **74**, 49-61.

Dorman, F. H., and Gill, E. D., 1959, "Oxygen isotope paleotemperature measurements on Australian fossils," *Roy. Soc. Victoria, Proc.*, **71**, 73–98.

Emiliani, C., 1954, "Temperatures of Pacific bottom waters and polar superficial waters during the Tertiary," *Science*, **119**, 853–855.

Emiliani, C., 1955, "Pleistocene temperatures," *J. Geol.*, **63**, 538–578.

Emiliani, C., 1955a, "Pleistocene temperature variations in the Mediterranean," *Quaternaria*, **2**, 87–98.

Emiliani, C., 1958, "Paleotemperature analysis of core 280 and Pleistocene correlations," *J. Geol.*, **66**, 264–275.

Emiliani, C., 1961, "The temperature decrease of surface sea-water in high latitudes and of abyssal-hadal water in open oceanic basins during the past 75 million years," *Deep-Sea Res.*, **8**, 144–147.

Emiliani, C., 1964, "Paleotemperature analysis of the Caribbean cores A254-BR-C and CP28," *Geol. Soc. Am., Bull.*, **75**, 129–144.

Emiliani, C., 1966a, "Paleotemperature analysis of the Caribbean cores P6304-8 and P6304-9 and a generalized temperature curve for the past 425,000 years," *J. Geol.*, **74**, 109–126.

Emiliani, C., 1966b, "Isotopic paleotemperatures," *Science*, **154**(3751), 851–857.

Emiliani, C., 1967, "The generalized temperature curve for the past 425,000 years: a reply," *J. Geol.*, **75**, 504–510.

Emiliani, C., 1971, "Paleotemperature variations across the Plio-Pleistocene boundary." *Science*, **171**, 60–62.

Emiliani, C., and Geiss, J., 1959, "On glaciations and their causes," *Geol. Rundschau*, **46**(1957), 576–601.

Emiliani, C., Mayeda, T., and Selli, R., 1961, "Paleotemperature analysis of the Plio-Pleistocene section at Le Castella, Calabria, southern Italy," *Geol. Soc. Am., Bull.*, **72**, 679–688.

Epstein, S., 1959, "The Variation of O^{18}/O^{16} in Nature and Some Geologic Implications," in (Abelson, Ph.H., editor), "Researches in Geochemistry," New York, John Wiley & Sons, pp. 217–240.

Epstein, S., Buchsbaum, R., Lowenstam, H., and Urey, H. C., 1953, "Revised carbonate-water isotopic temperature scale," *Geol. Soc. Am. Bull.*, **64**, 1315–1325.

Frechen, J., and Lippolt, H. J., 1965, "Kalium-Argon-Daten zum Alter des Laacher Vulkanismus, der Rheinterrassen und der Eiszeiten," *Eiszeitalter Gegenwart*, **16**, 5–30.

Jones, J. I., 1967, "The significance of the distribution of planktonic foraminifera in the equatorial Atlantic undercurrent," *Micropaleontology*, **13**, 109–124.

Lowenstam H. A., 1961, "Mineralogy, O^{18}/O^{16} ratios, and strontium and magnesium contents of recent and fossil brachiopods and their bearing on the history of the oceans," *J. Geol.*, **69**, 241–260.

Lowenstam, H. A., 1964, "Paleotemperatures of the Permian and Cretaceous periods," in (Nairn, A. E. M., editor), "Problems in Palaeoclimatology," New York, Interscience Publ., pp. 227–252.

Lowenstam, H. A., and Epstein, S., 1954, "Paleotemperatures of the post-Aptian Cretaceous as determined by the oxygen isotope method," *J. Geol.*, **62**, 207–248.

Lowenstam, H. A., and Epstein, S., 1959, "Cretaceous paleotemperatures as determined by the oxygen isotope method, their relations to and the nature of Rudistid reefs," Intern. Geol. Cong., XX Session, Mexico, 1956, Symposium del Cretácico, pp. 65–76.

McCrea, J. M., 1950, "On the isotopic chemistry of carbonates and a paleotemperature scale," *J. Chem. Phys.*, **18**, 849–857.

McKinney, C. R., McCrea, J. M., Epstein, S., Allen, H. A., and Urey, H. C., 1950, "Improvements in mass spectrometers for the measurement of small differences in isotope abundance ratios," *Rev. Sci. Instruments*, **21**, 724–730.

Naydin, D. P., Teys, R. V., and Chupakhin, M. S., 1956, "Determination of the climatic conditions of some regions of the USSR during the Upper Cretaceous period by the method of isotopic paleothermometry," *Geochemistry (U.S.S.R.) (Engl. transl.)*, 1956(1960); 752–764.

Naydin, D. P., Teys, R. V., and Zadoroshny, I. K., 1964, "Some new data on the temperatures of Maastrichtian basins of the Russian platform and adjacent areas according to the isotopic oxygen composition in belemnite rostra," *Geochimia*, **1964**, 971–979.

Nier, A. O., 1940, "A mass spectrometer for routine isotope abundance measurements," *Rev. Sci. Instruments*, **11**, 212–216.

Nier, A. O., 1947, "A mass spectrometer for isotope and gas analysis," *Rev. Sci. Instruments*, **18**, 398–411.

Richmond, G. M., and Emiliani, C., 1967, "The Plio-Pleistocene boundary," *Science*, **156**, 410.

Rosholt, J. N., Emiliani, C., Geiss, J., Koczy, F. F., and Wangersky, P. J., 1961, "Absolute dating of deep-sea cores by the Pa^{231}/Th^{230} method," *J. Geol.*, **96**, 162–185.

Rosholt, J. N., Emiliani, C., Geiss, J., Koczy, F. F., and Wangersky, P. J., 1962, "Pa^{231}/Th^{230} dating and O^{18}/O^{16} temperature analysis of core A254-BR-C," *J. Geophys. Res.*, **67**, 2907–2911.

Urey, H. C., 1947, "The thermodynamic properties of isotopic substances," *J. Chem. Soc.*, **1947**, 562–581.

Urey, H. C., Lowenstam, H. A., Epstein, S., and McKinney, C. R., 1951, "Measurement of paleotemperature and temperatures of the Upper Cretaceous of England, Denmark, and the southeastern United States," *Geol. Soc. Am. Bull.*, **62**, 399–416.

Cross-references: *Calcium Carbonate Geochemistry; Free Energy; Geochronometry; Isotope Fractionation; Oxygen Isotope Geochemistry; Seawater, Chemistry; Seawater, Geochemical Balance.* Vol. II: *Climatic Variations.* Vol. III: *Glaciation; Ice-Age Theory; Quaternary.* Vol. VI: *Pelagic Sediments.*

PALLADIUM: ELEMENT AND GEOCHEMISTRY

Properties

Palladium is a member of the platinum group of metals with which it is generally associated in nature. It was discovered as the second new element in native platinum by Wollaston in 1803.

The atomic number of the element is 46 and its atomic weight 106.4. There are six stable isotopes whose mass numbers and relative abundances are as follows: 102 (0.8%), 104 (9.3%), 105 (22.6%), 106 (27.2%), 108 (26.8%), and 110 (13.5%). The metal crystallizes with a face-centered cubic lattice for which the lattice constant $a = 3.8907$ Å. The radius of the atom, for 12-coordination in the metal, is 1.375 Å; slightly smaller covalent radii are observed in complexes with square planar or octahedral coordination. The density of the pure metal is 12.02 g/cm^3 at 20°C, and its melting point is 1552°C.

Occurrence

Palladium is appreciably more reactive chemically than the other platinum metals, and possibly for this reason it is known in more compound mineral forms than any other metal in the group. As with the other members of the platinum group a common occurrence is in native metal alloys. It is the principal component in *palladium* (cubic and *allopalladium* (hexagonal), and a major component in *palladiplatinum;* it is a minor component in many other precious metal alloys such as native *gold, platinum, iridosmine,* etc.

A number of mineral compounds of palladium have been characterized in a recent review (Wright and Fleischer, 1965); several of these have been identified only recently. Among the better known examples are *potarite* (PdHg), found in Guyana; *stibiopalladinite* (Pd$_3$Sb), found in the Bushveld complex of the Transvaal; *froodite* (monoclinic) and *michenerite* (cubic), both PdBi$_2$, found in the Sudbury basin and elsewhere. Platinum metals occur at the level of a few parts per million in sulfide minerals, especially those associated with nickel-bearing ore bodies; in most other minerals the platinum metal content is generally low, though individual instances of a few parts per million have been recorded. As a rule platinum predominates over palladium in such mineral occurrences, but in deposits that have come to be of great economic importance in the Noril'sk region, U.S.S.R., palladium appears to be on the order of five times as abundant as platinum.

Abundance and Geochemical Features

Among the platinum group elements palladium is probably the most abundant member on the earth's crust; current estimates place its abundance there at 0.01 ppm. The abundance data for palladium in meteorites, however, place this element fourth within the platinum group (ahead of rhodium and iridium) in concentration; and deductions from this evidence place it in the same ranking for cosmic abundance.

The general distribution of platinum metals in rocks is discussed in another entry (see *Platinum Metals*) and it will only briefly be summarized here that these elements show a marked preference for ultrabasic rocks in which they probably occur as local disseminations of native metal. Most placer deposits of platinum metals are considered to have resulted from the extensive weathering of such primary deposits.

The platinum metals are distinctly siderophilic in character and this is manifest in their greater abundance in siderite meteorites than in chondrites. However, whereas platinum in chondrites is strongly concentrated in the iron-nickel phase, palladium is more evenly distributed between the iron-nickel and the troilite phases. Such behavior reveals a more chalcophilic character for palladium than for platinum.

Economic Geology

This is discussed in the entry *Platinum Metals* for the group of metals together. As a result of the much increased production of platinum metals from the U.S.S.R. in recent years, palladium is commercially available on a greater scale than platinum. It finds many industrial applications of which the most important is the production of corrosion-free alloys for electrical gear.

W. A. E. McBryde

References

Hampel, C. A. (editor), 1968, "Encyclopedia of the Chemical Elements," New York, Reinhold Publ. Corp., pp. 513–519.

Hampel, C. A. (editor), 1961, "Rare Metals

Handbook," Second edition, New York, Reinhold Publ. Corp., pp. 304–335.
Rankama, K., and Sahama, Th. G., 1950, "Geochemistry," Chicago, University of Chicago Press, 912pp.
Standen, A. (editor), Kirk-Othmer: "Encyclopedia of Chemical Technology," Second ed., New York, John Wiley & Sons, Vol. 15, pp. 832–878.
Wright T. L., and Fleischer, M., 1965, "Geochemistry of the Platinum Metals," *U.S. Geol. Surv. Bull.*. **1214-A**.

Cross-references: *Platinum Metals*. Vol: IVB: *Meteoritic Minerals; Rare Metal Ore Deposits.*

PARAGENESIS

The paragenesis of minerals, the sequence in which they are produced in a mineral deposit, determines which minerals will be associated with one another in or on the earth's crust. These mineral parageneses can be divided into

1. Igneous parageneses which result from crystallization of minerals from relatively high-temperature solutions.
2. Sedimentary parageneses which are composed of the products of the chemical and/or physical breakdown of some preexisting deposit under essentially surface conditions.
3. Metamorphic parageneses which are produced from some preexisting deposit by an increase in temperature and/or pressure.

None of these three major families is sharply distinguished from the others, for transitions between all of them do occur.

Igneous Parageneses

The solutions involved have temperatures in the range 50–1200°C, and the various igneous deposits are produced by crystallization caused by the fall in temperature from the higher to the lower of these values.

At higher temperatures the solution is a complex melt containing various metallic ions [principally aluminum (Al), iron (Fe), magnesium (Mg), calcium (Ca), sodium (Na), and potassium (K)], abundant silicon, and oxygen, with the last two probably linked together as Si-O tetrahedra. Dissolved in this melt are small amounts ($2\pm\%$) of water and other volatile materials, and suspended in it are usually crystals of various refractory minerals such as olivine, magnetite, and chromite. Such a mixture is called a *magma*. During very early stages in its cooling the crystals formed may settle to the bottom of the magma chamber where they become concentrated into layers, lenses, or irregular clots called *crystal segregations*. These are generally formed of various iron-bearing oxides, the early-formed silicates, and a few quantitatively minor native elements. In other cases, these early crystals occur as disseminated grains of minor importance in the deposits formed when the remainder of the magma solidifies.

Except at the very early stages of crystallization, the principal minerals formed are silicates; the mineral deposits composed of these are the normal *igneous rocks*. Among the silicate minerals we can distinguish two principal mineral parageneses, called in this case *reaction series*. The importance of these series in the crystallization of igneous rocks was first pointed out by N. L. Bowen in the 1920s. One of these involves only the iron- and magnesium-bearing (mafic) minerals; the other, those minerals (felsic) that contain no iron or magnesium. Both parageneses result largely from reaction between already crystallized minerals and the remaining liquid portion of the magma, resulting either in a gradual change in composition of a mineral species ("continuous reaction") or in the transformation of one mineral species to another ("discontinuous reaction"). These two series are illustrated in Fig. 1, where the temperature decreases from left to right. The minerals at the far left are the ones most likely to occur in crystal segregations.

As these minerals crystallize from the magma, the relative water content of the remaining liquid increases, for the minerals are anhydrous or nearly so, and this serves to increase the fluidity of the liquid. Eventually the magma becomes saturated in dissolved gases. At this stage the still-liquid remnant of the original magma will be of small amount and will differ chemically from the earlier melts as a result of concentration in it of the less common metallic and nonmetallic ions. The deposits formed from such late melts are called *liquid segregations;* they usually occur as lenses or sheets that intrude into bodies of normal igneous rock or the country rock around them. The minerals present depend on the composition of the parent magma from which the segregation developed, so that one developed from a basic parent magma will usually be composed of different minerals than one derived from an acid parent magma. These variations are also illustrated in Fig. 1.

If the magma continues to crystallize beyond the point where it has become saturated in water, some water must separate to form an additional phase composed of supercritical water "vapor." Under high confining pressure this will have a rather high density and may

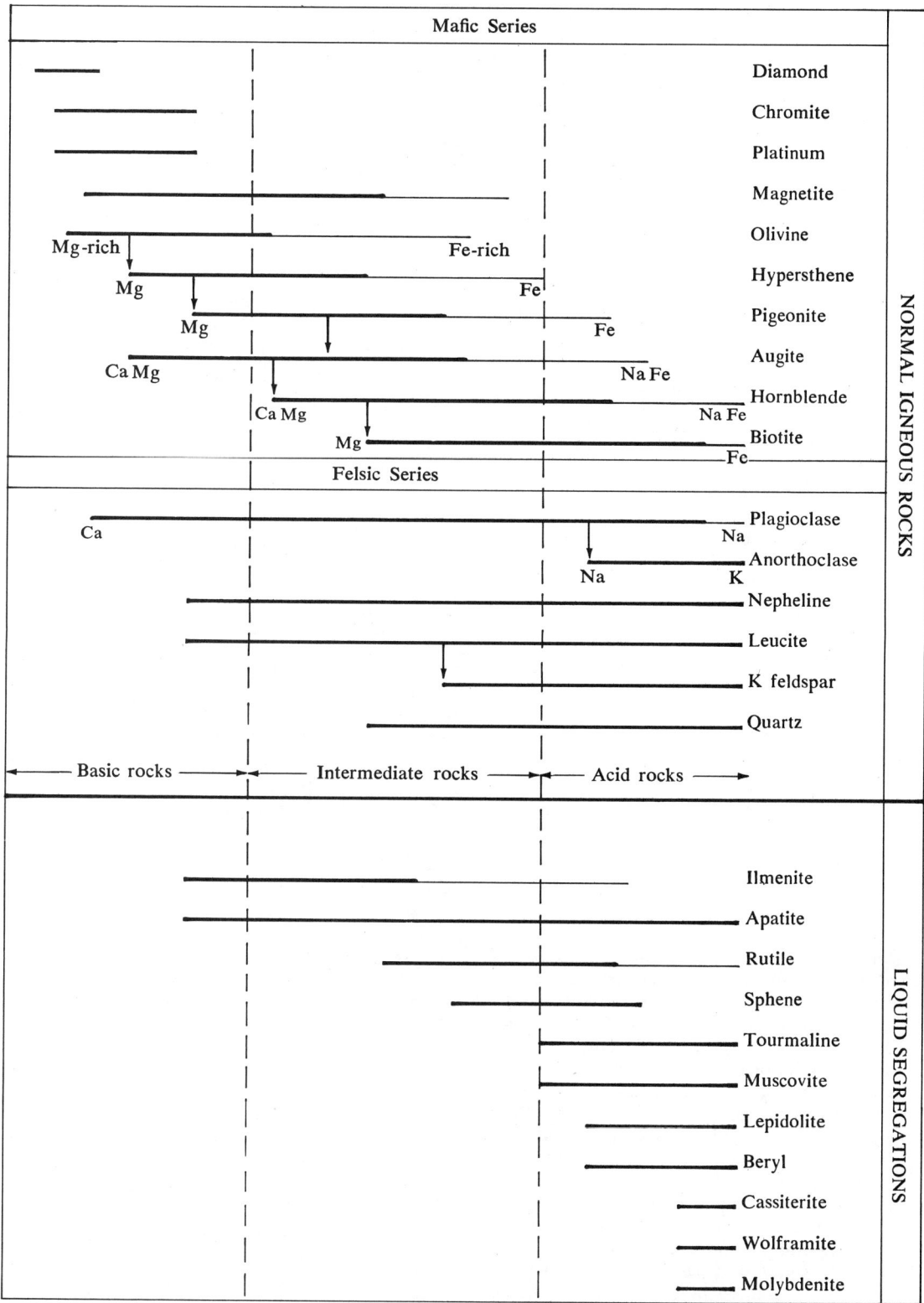

Fig. 1. Mineral paragenesis in the normal igneous rocks and their associated liquid segregations. Vertical arrows indicate discontinuous reactions. Changes in composition produced by continuous reactions are indicated by chemical symbols below lines representing these.

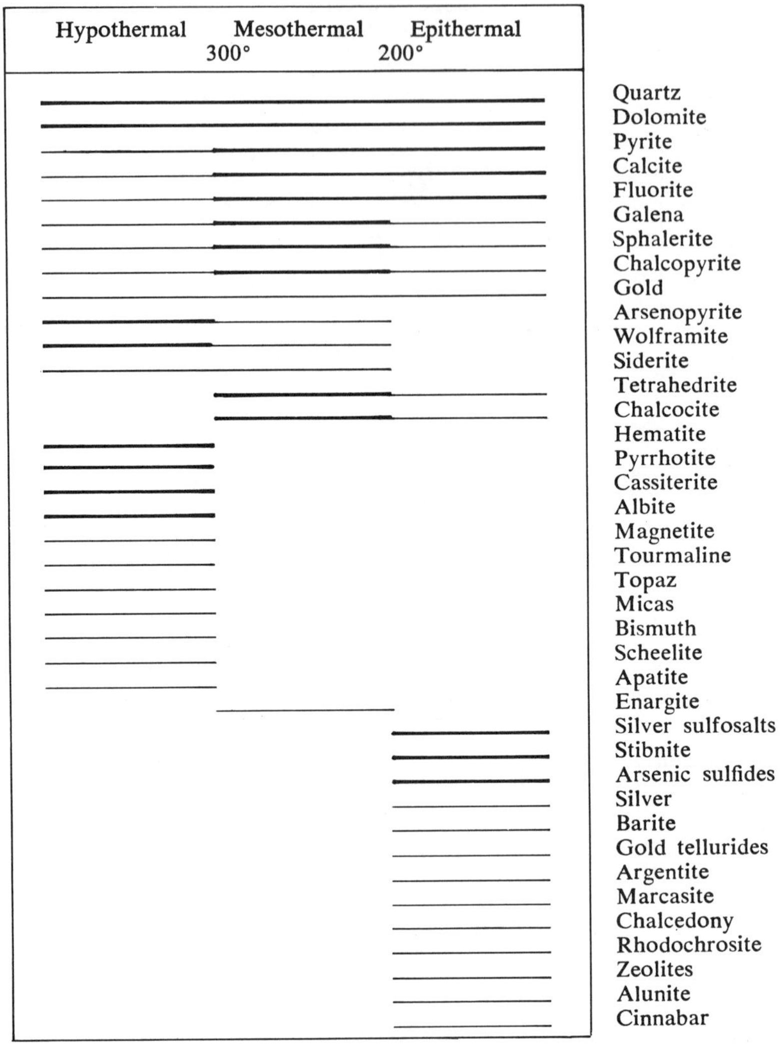

PARAGENESIS

FIG. 2. Mineral paragenesis in the hydrothermal deposits. Temperature decreases from left to right.

Hypothermal 300°	Mesothermal 200°	Epithermal	Mineral
────────	────────	────────	Quartz
────────	────────	────────	Dolomite
────	────────	────────	Pyrite
────	────────	────────	Calcite
────	────────	────────	Fluorite
────	────────	────	Galena
────	────────	────	Sphalerite
────	────────	────	Chalcopyrite
────	────────	────────	Gold
────	────────		Arsenopyrite
────	────────		Wolframite
────	────────		Siderite
	────────	────────	Tetrahedrite
	────────	────────	Chalcocite
────			Hematite
────			Pyrrhotite
────			Cassiterite
────			Albite
────			Magnetite
────			Tourmaline
────			Topaz
────			Micas
────			Bismuth
────			Scheelite
────			Apatite
	────────	────	Enargite
		────────	Silver sulfosalts
		────────	Stibnite
		────────	Arsenic sulfides
		────────	Silver
		────────	Barite
		────────	Gold tellurides
		────────	Argentite
		────────	Marcasite
		────────	Chalcedony
		────────	Rhodochrosite
		────────	Zeolites
		────────	Alunite
		────────	Cinnabar

contain a considerable amount of dissolved mineral matter; with further cooling the supercritical solution may change to a liquid solution. Such supercritical and liquid aqueous solutions can be grouped together as hydrothermal solutions; their deposits are called *hydrothermal deposits*. These are formed by the alteration of preexisting minerals, by the replacement of preexisting minerals, and by the precipitation of new minerals in open cavities in the country rock. The typical associations are those produced by replacement and cavity filling. The minerals present will depend on the temperature and composition of the solutions. A generalized paragenesis for these minerals is represented in Fig. 2, where the temperature decreases from left to right. The paragenesis can be divided into three broad classes of hydrothermal deposits; the names and temperature ranges of each of these is indicated on the diagram.

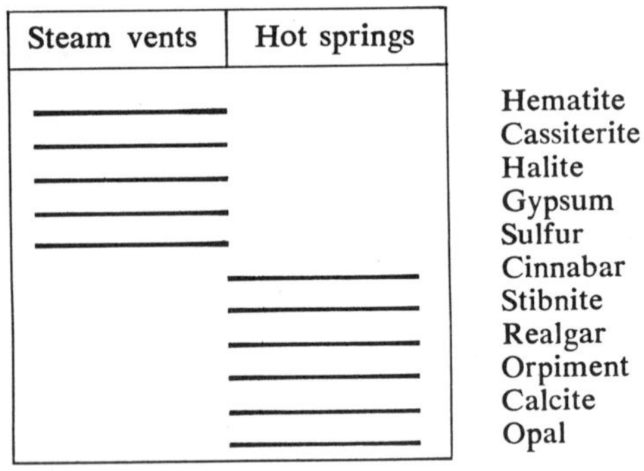

Steam vents	Hot springs	Mineral
────────		Hematite
────────		Cassiterite
────────		Halite
────────		Gypsum
────────		Sulfur
	────────	Cinnabar
	────────	Stibnite
	────────	Realgar
	────────	Orpiment
	────────	Calcite
	────────	Opal

FIG. 3. Mineral paragenesis in deposits around steam vents and hot springs.

If the solutions eventually reach the surface,

either as a vapor or as a liquid that has usually been considerably diluted by water of atmospheric origin, a mineral deposit is commonly formed where they emerge; here also the minerals present depend on the temperature of the solutions. This is shown in Fig. 3, where the minerals formed around steam vents are compared with those formed around hot springs.

Sedimentary Parageneses

Sedimentary minerals are of two types: those that have survived unchanged from the parent material, and those that have been produced during one or the other of the sedimentary processes. These processes are: (a) weathering; (b) erosion, transportation, and deposition; and (c) diagenesis.

The *residual deposits,* produced by weathering alone, may be products of either the mechanical break-up or the chemical decomposition of the parent material. In most cases both processes have contributed, though new minerals are produced only by chemical weathering. The minerals produced and the minerals that survive from the parent material depend on the conditions—mainly climatic—under which weathering occurs, though other factors such as how long the parent material has been subjected to weathering may also be important. The principal minerals produced are the clay minerals and the oxides of iron, aluminum, and manganese; the minerals most resistant to weathering are the micas, the alkali feldspars, and quartz. These relationships are illustrated in Fig. 4, in which the "intensity of chemical weathering" increases from left to right; this can be thought of as combining a climatic factor which increases with rising temperature and rainfall, with a time factor. A deposit at the right-hand end of the diagram would contain no survivors from the parent, whereas one at the left-hand end would contain no weathering products.

If the products and survivors of weathering are picked up by a current of air, water, or ice, transported for some distance, and then redeposited, the deposit formed is called a *transported sediment.*

If transportation has been by mechanical means the deposit will generally be composed of the same *insoluble* minerals as its source residual deposit, although not all these minerals can stand mechanical transportation with the same ease. Fig. 5 attempts to illustrate this; distance of transport increases from left to right. Deposits of minerals formed in this way are called *detrital sediments.* The more resistant minerals in such a sediment will usually occur as relatively large grains; the finer-grained material between them will be composed of the easily disintegrated minerals such as clay and chlorite. Such material often serves to give coherence to the deposit.

On the other hand, some weathering products are transported in solution or as colloidal suspensions and are redeposited by changes in the chemical environment; these are the *chemical sediments.* This redeposition may be

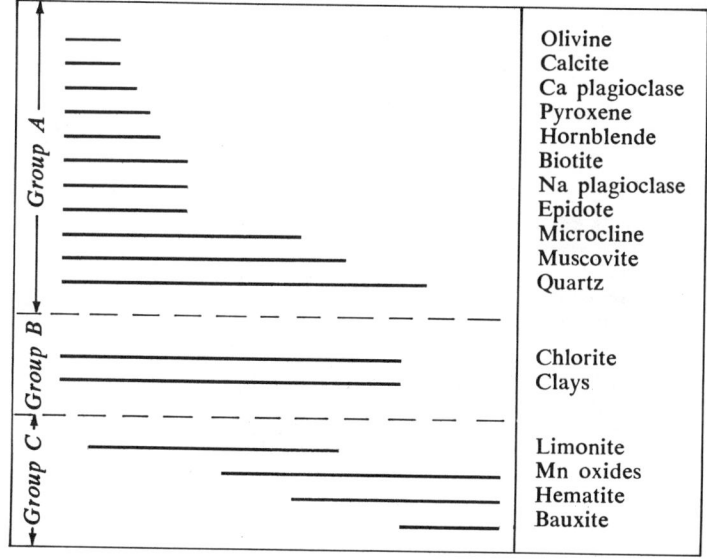

Fig. 4. Mineral paragenesis in residual deposits. Minerals in Group A are all survivors of chemical weathering; those in Group C are products of chemical weathering; those in Group B may be either.

PARAGENESIS

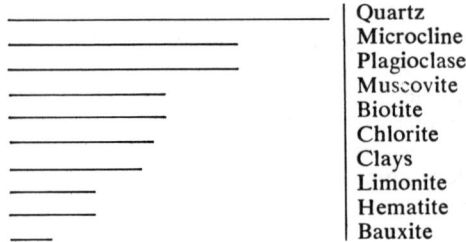

FIG. 5. Ability of the more resistant minerals of residual deposits to withstand mechanical transportation. The distance transported increases from left to right.

brought about by organisms (e.g., oyster shells) or by evaporation (salt beds) or by the addition of some new chemical to the solution (flocculation of silica by seawater). Those deposited from colloidal suspensions are usually oxides (of silicon or iron, for example). Those deposited from true solutions are mostly chlorides, carbonates, and sulfates (mainly of sodium, potassium, and calcium). In Fig. 6 the solubility increases from left to right; if all these minerals were equally abundant in natural solutions, those near the left side would occur most commonly in chemical sediments and those near the right side least commonly. The minerals present in a given chemical sediment will also depend on two other factors: the acidity of the environment (measured by its pH) and the presence of oxidizing or reducing conditions (measured by its Eh). A more complete discussion of these factors will be found elsewhere in this volume.

Both detrital and chemical sediments may be subjected to changes after they have been deposited. If the temperatures involved are low (less than about 100°C) these fall under the heading of *diagenesis;* higher temperature changes fall under the heading of metamorphism (see below). The important changes involved are compaction, cementation, and replacement. A very common and important result is the conversion of a loose sediment into a coherent sedimentary rock. The minerals produced during cementation are essentially the same as those comprising the chemical sediments, whereas replacement may result in the formation of a number of new minerals, particularly from carbonate-rich sediments. The following replacements are most common:

$$\text{Calcite} \rightarrow \text{Chalcedony}$$
$$\text{Calcite} \rightarrow \text{Dolomite}$$
$$\text{Calcite} \rightarrow \text{Hematite}$$
$$\text{Anhydrite} \rightarrow \text{Gypsum}$$
$$\text{Gypsum} \rightarrow \text{Anhydrite}$$
$$\text{Clay} \rightarrow \text{Micas}$$

Metamorphic Parageneses

These may also contain two generations of minerals: some that have survived the changes to which they have been subjected, others that have been produced by these changes. The changes themselves grade imperceptibly into those produced by diagenesis and by hydrothermal replacement (see below). The principal ones are (*1*) increase in pressure, (*2*) increase in temperature, and (*3*) change in chemical environment. A mineralogical change that is promoted by a rise in temperature will commonly be opposed by a rise in pressure; this is particularly true if a fluid (gas or liquid) phase is produced, for high pressure will tend to prohibit the increase in volume caused by this. Thus the reaction

$$CaCO_3 + SiO_2 \leftrightarrows CaSiO_3 + CO_2$$

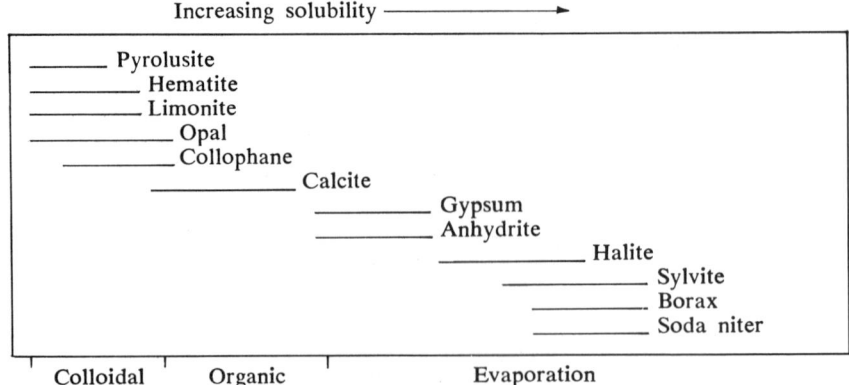

FIG. 6. Paragenesis of minerals in chemical sedimentary rocks. The usual mode of precipitation is indicated along the bottom of the diagram.

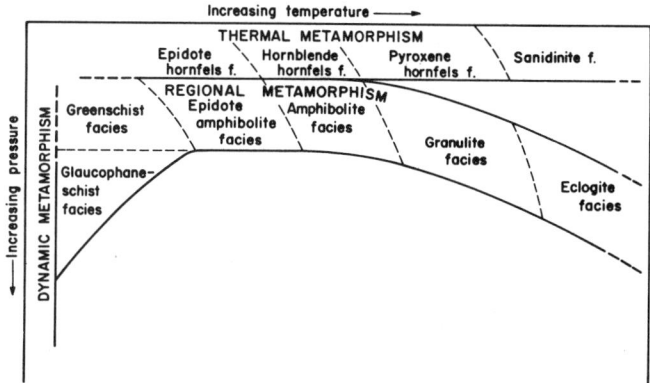

Fig. 7. Physical conditions of metamorphism.

is driven to the right by rising temperature and to the left by rising pressure, for the volume on the right side is greater than that on the left.

Ignoring for the present the effect of chemical environment, we can represent the conditions under which metamorphism occurs by Fig. 7. Three main fields are marked off. *Thermal metamorphism* is characteristically associated with igneous intrusions; *dynamic metamorphism* is characteristically associated with near-surface shearing; and *regional metamorphism* is produced during the deformation of geosynclinal zones to produce alpine mountain chains. Mineralogic changes are produced only by the last two types. In these we can recognize a number of stages of intensity, called *metamorphic facies;* these are named in Fig. 7. Fig. 8 (see p. 904) presents the parageneses in these rocks. Anhydrous higher-density minerals become more abundant with increasing metamorphic grade (from left to right in Fig. 8). In the highest-grade rocks the minerals are essentially the same as those in the igneous rocks, for temperatures in the two cases fall in the same range. At this point partial melting of some assemblages will occur, if water is present, to form a granitic magma, so that the metamorphic rocks also grade imperceptibly into the igneous rocks.

A change in chemical environment usually results from a change in the amount or composition of the interstitial fluids. Such a change is common in metamorphism associated with igneous intrusions, where hydrothermal solutions are introduced from the intrusion itself. Particularly in calcitic and dolomitic rocks, this results in replacement by minerals characteristic of the higher-temperature hydrothermal deposits (*contact metasomatism*). Similar changes on a regional scale, particularly in the lower-grade facies, are perhaps the result of migration of fluids from the deeper higher-grade zones. These appear to be particularly rich in soda, boron, and carbon dioxide; their most common results are the carbonation of calcium-magnesium silicates and the development of tourmaline and albite.

C. P. Thornton

References

Jahns, R. H., 1955, "The study of pegmatite deposits," *Econ. Geol.*, 50th aniv. Vol., pp. 1025–1130.

Niggli, Paul, 1954, "Minerals and Mineral Deposits," San Francisco, W. H. Freeman & Co., 559pp.

Pettijohn, F. J., 1957, "Sedimentary Rocks," New York, Harper and Brothers.

Schneiderhöhn, Hans, 1962, "Erzlagerstätten," Stuttgart, Gustav Fischer Verlag, 371pp.

Turner, F. J., and Verhoogen, J., 1960, "Igneous and Metamorphic Petrology," New York, McGraw-Hill Book Co., 694pp.

White, D. E., 1955, "Thermal springs and epithermal ore deposits," *Econ. Geol.*, 50th anniv. Vol., pp. 99–154.

Oelsner, O., 1966, "Atlas of the Most Important Ore Mineral Parageneses Under the Microscope," Oxford, Pergamon Press, 311pp. (transl. by Hazzard, B. J.).

Cross-references: *Geochemistry of Sediments; Hydrothermal Solutions; Interstitial Water; Metamorphic Environments—Chemical Mobility; Mineral Classes—Silicates; Native Elements; Oxidation and Reduction; Oxides; PH-Eh; Weathering, Chemical.* Vol. IVB: *Alkali Feldspars; Clays and Clay Minerals; Crystal Growth; Evaporite Minerals; High-Pressure Minerals; Mica Group; Mineral and Ore Deposits; Quartz; Refractory Miinerals.* See also Individual Minerals. Vol. V: *Igneous Rocks; Magma; Metamorphism; Metamorphic Facies; Volatiles.* Vol. VI: *Diagenesis; Hot Springs; Spring and Crust Deposits.*

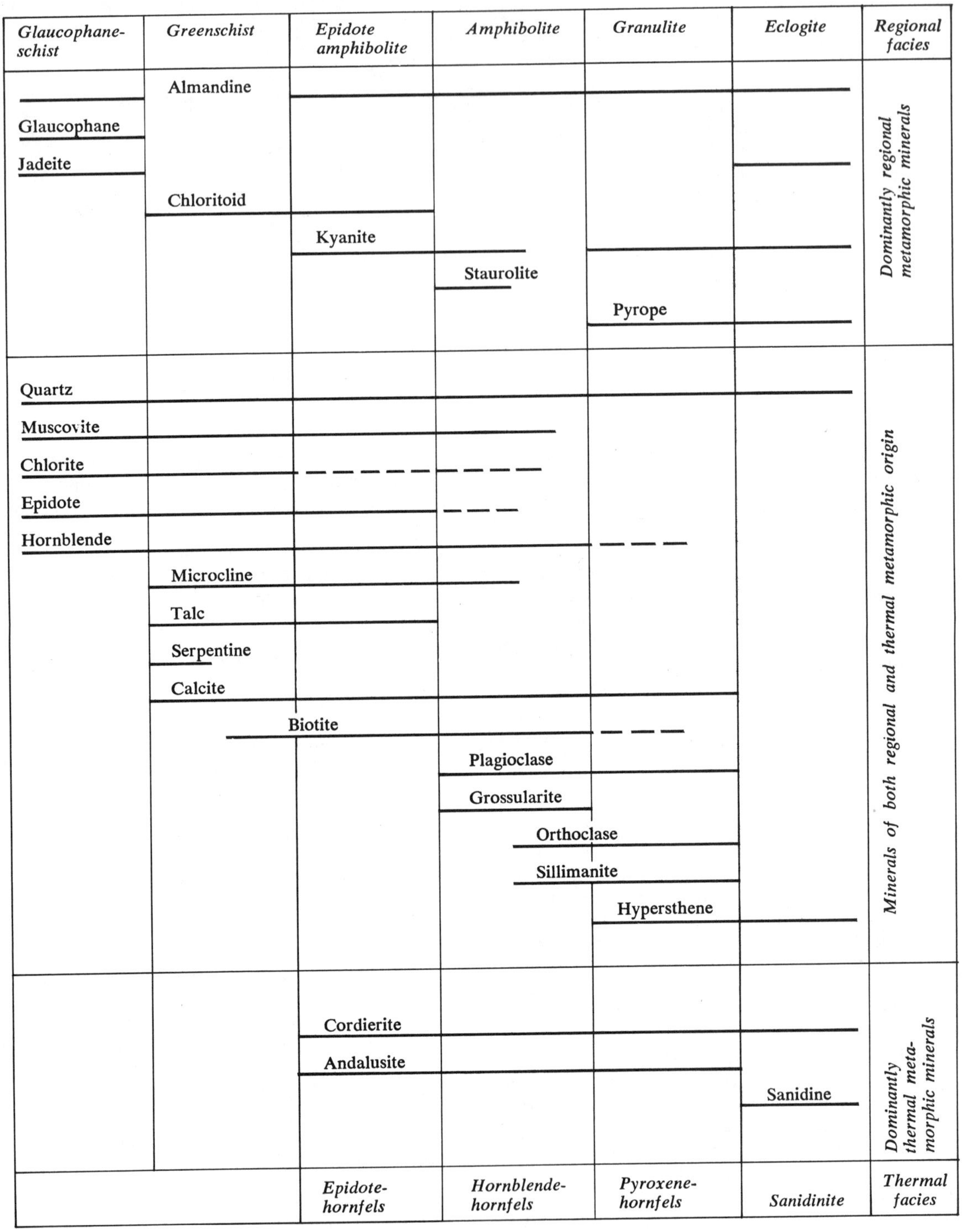

PARAGENESIS

Glaucophane-schist	Greenschist	Epidote amphibolite	Amphibolite	Granulite	Eclogite	Regional facies
	Almandine					*Dominantly regional metamorphic minerals*
Glaucophane						
Jadeite						
	Chloritoid					
		Kyanite				
			Staurolite			
				Pyrope		
Quartz						*Minerals of both regional and thermal metamorphic origin*
Muscovite						
Chlorite		– – – – –	– – –			
Epidote			– – –			
Hornblende				– – –		
	Microcline					
	Talc					
	Serpentine					
	Calcite					
		Biotite		– – –		
			Plagioclase			
			Grossularite			
				Orthoclase		
				Sillimanite		
				Hypersthene		
		Cordierite				*Dominantly thermal metamorphic minerals*
		Andalusite				
					Sanidine	
	Epidote-hornfels		Hornblende-hornfels	Pyroxene-hornfels	Sanidinite	Thermal facies

Fig. 8. Mineral paragenesis in metamorphic rocks. The possible effects of metasomatism have been omitted. Temperature increases from left to right.

PARTICLE TRACKS—See **CHARGED PARTICLE TRACKS**

PARTITION COEFFICIENTS

Partition coefficients may serve as a quantitative tool for solving earth science problems.

Partition coefficients are a measure of the degree to which a given element in a liquid enters a solid growing in the presence of that liquid. This is an old concept dating from 1872 when Berthelot first published his law that was theoretically explained in 1891 by Nernst to give us what is known today as the *Berthelot-Nernst distribution law*:

$$\frac{C_S}{C_L} = K \qquad (1)$$

where (C_S) is the concentration of the element of interest in the solid and (C_L) is its concentration in the liquid; K is the distribution constant or partition coefficient.

Geochemical Importance of Partition Coefficients

Reconstructing, in a quantitative fashion, the chemical environment in living systems of past geological times, and deducing the chemical state of existing systems, all from fragmentary information, are exciting and important major tasks of the future for the fields of geology, paleontology, and many sciences dealing with present and past situations.

Partition coefficients are of great interest because they are a measure of the behavior of elements existing in relatively small quantities that become trapped in growing crystals long before the concentration of these "trace" elements reaches a level where they will separate or precipitate from the system (before their solubility product has been attained). The chemical conditions under which these elements can remain in solution are known. Their presence as trace elements in a crystal, therefore, allows us to reconstruct the chemical and physical parameters that prevailed during the growth of the host crystal in which they are now found.

The pioneering work of Goldschmidt and his students in the 1930s led to a beginning of our understanding of the amounts of trace elements in rocks, minerals, and organisms, as well as in the earth and the universe. Goldschmidt and his students also laid the groundwork for an understanding of the principles governing the structural role of minor constituents in crystals. The chemists and physicists of today, in designing transistors, lasers, and many other new products are exploiting these concepts to determine the role these elements play in modifying the properties of matter.

Partition coefficients enable the earth scientist to investigate quantitatively some of the great problems that have been handled only on a qualitative basis in the past. Some examples of the problems taken from several fields which, in the near future, will yield to this approach are briefly mentioned here.

Because the geologist uses present-day models to reconstruct the environments of the past, some key problems of "present" importance are mentioned first.

Only the chemistry of the body fluids of vertebrates is independent of the chemistry of the fluids in which the organism lives. The ability of an organism to keep the chemistry of its body fluids within viable limits determines its ecological range. The range of some species are different even when the major elemental composition of their body fluids is apparently the same. Although certain trace elements are known to be of critical importance to the healthy functioning of marine organisms, almost nothing is known about the role they play in the organism and its interaction with its environment. The trace elements entering the organism from the external environment via the organism's body fluids become incorporated in its skeleton and, therefore, are a record of both the characteristics of the organism itself and the environment in which it lived. Changes in its skeletal composition, much like tree-ring spacings and character, can be used to detect annual environmental changes during the life of the organism.

Organisms can remain alive and healthy only when their body-fluid composition remains within certain ranges. These ranges are known for some of the major elements but not for the minor elements. Studies of this kind may explain why an organism died: was he poisoned by an influx of new material into his environment or was he forced to move to another environment in which he could not thrive? This factor undoubtedly played a major role in the extinction of many types of organisms in the geological past. Whereas mountain building typifies the early events of a geological period, chemical weathering becomes very important as a process for reducing irregularities in the land surface during the closing stages of a period. During these later stages, great changes in the composition of the waters in seas must have occurred. These changes could easily have led to the poisoning of especially the more highly specialized forms of life that had restricted ecological niches. Distribution coefficient studies on unrecrystallized or unaltered shells will provide an excellent check on this very interesting possibility. Studies of this kind may be valuable in developing chemical means of diagnosing pathological conditions in living organisms including man. The size, shape and condition of oyster shells, for example, is determined in part by the environment in which a given species lives. Paleontologists cannot tell whether the differences in size, shape and complexity of skeletons (shells) of some organisms record evolutionary changes with time or merely record differences caused by environmental situations prevailing in different places at the same time. All of these, and many like them, are problems of fundamental importance to many areas of geology, biology, and ecology involving both living and ancient organisms.

Because so many of our present concepts in geology and associated sciences are related to problems involving the ocean, a better understanding of the chemistry of the ocean throughout geological time would provide new clues to the solution of many old problems. Many such concepts are tied, for example, to changes in the salinity of the sea throughout geological times. Much progress has been made recently to show how variations in the boron content of certain clay minerals can be used to evaluate the paleosalinity of the ocean (Frederickson and Reynolds, 1960). This rough quantitative procedure can be used to provide clues as to the location of old barriers such as coral reefs, a task of great importance in locating petroleum reservoirs. This technique could be an ideal independent check on the interpretation of cyclothems, the salinity level prevailing when certain paleontological assemblages lived, and many related problems.

The word *solution* can refer to both aqueous and molten liquids and solid solutions. The equations that follow are of a general kind, and hence apply equally as well to molten magmas and metamorphic rocks as to aqueous liquids. The partition coefficient concept therefore can be applied to the complete range of geological environments extending from the present to the most ancient past.

Real Crystals

If partition coefficients are to be applied successfully to any problem, a fundamental understanding is required of the principles, processes, and products involved. The product and processes involved are crystals, the manner in which the trace elements become incorporated into the crystals, and the degree to which they remain there to serve as "fingerprints" of the external environment.

Most people conceive of ideal crystals, which do not exist in nature. A brief review of some aspects of crystals important to partition coefficient studies is given below.

Well-formed crystals or ordinary salt (NaCl) can be grown at a very rapid rate, even faster than one-hundredth of an inch per day. The size of the ions making up the crystal are about one hundred-millionth of an inch in diameter; hence, roughly 100 layers of ions can be added to the growing crystal face each second. With all this hustle and bustle at the growing face, it is not at all surprising that many materials present in the solution, in addition to sodium (Na) and chlorine (Cl) ions, get trapped in the growing surfaces and eventually become incorporated within the crystal.

More than any other property, orderliness is the feature that distinguishes the crystalline from other states of matter. Crystals are made up of atoms or ions arranged about fixed average positions in an orderly pattern that is repeated throughout the substance. The degree of orderliness within a crystal, however, is far from perfect. Experience has shown that the larger crystals are made up of many smaller crystals, often called *domains*. From this point of view, even large so-called single crystals are actually a large array of domains stacked together in only a more or less regular fashion. The boundaries or gaps between these domains may vary from a few ion diameters to as many as 8000 diameters or more (at which time they become visible to the eye) and are termed *lineages*. These boundaries are internal surfaces within the crystal.

Anyone attempting to grow a crystal soon discovers that it is much easier to dissolve a crystal than it is to grow one: it is easier to create a disordered array of ions (the liquid state) than an ordered array (the crystalline state). When a crystal is grown from a supersaturated solution, we are referring to a situation where more solid is contained in a liquid than belongs there under equilibrium conditions. This is another way of saying that even in liquids where the ions have a relatively high degree of mobility, enough of them cannot get together easily to form a crystallite having an orderly internal arrangement. As a result, the first crystallites that appear from such a solution are often quite impure and have a very disordered arrangement—a very high number of domains stacked together in a more or less jumbled array. Such crystallites are relatively unstable and frequently dissolve again. If they do not dissolve many times, the interior of the crystallite may have a much different trace element composition than the layers formed at a later time. The importance of this behavior of crystals to distribution coefficient studies will be described later.

Even the chemist's "pure solutions" are not absolutely pure, but may contain many different ions in very small quantities. In nature, the solutions from which crystals grow have a very complicated composition: seawater, for example, contains most of the elements in the periodic table. When crystals are grown from solutions it is not surprising that a pure perfect crystal—a crystal without foreign ions and perfectly ordered—exists only in someone's imagination.

Crystals grow by the addition of material to their surfaces. The impurities in the liquid can be incorporated in the growing solid in several ways. The "impurities": (a) may exist as mixed crystals where the foreign ion can isomorphously substitute, because of its size and charge, for a host ion and become uniformly distributed throughout the crystal; (b) may occur in solid solution because many materials have at least some degree of mutual solubility, again resulting in a random distribution of the trace constituent in the host media; (c) may form chemical compounds within the host crystal; (d) may become attached by adsorption, which is largely a surface phenomenon where the trapped ions may remain at an interface or diffuse into the crystal; or (e) by occlusion or mechanical trapping of ions, other crystallites or the solution itself may become trapped on surfaces, along lineages or other discontinuities in the crystal.

With this number of alternative ways to incorporate a trace constituent in a growing crystal, it is not surprising that trace constituents can be incorporated in growing crystals long before the trace constituent approaches saturation in the growing media.

The law governing the distribution of a trace constituent between the growing crystal and its surrounding liquid, when the trace element becomes uniformly distributed throughout the crystal, is the Berthelot-Nerst distribution or *homogeneous distribution law* [see Eq. (1)], which states that the ratio between the concentration of the trace element in the solid (C_S) and the liquid (C_L) is a constant. The constant (K) is called the partition coefficient or *distribution constant*. The constant is a function of both temperature and pressure. This constant is independent of the amount of trace element present only when the amount is small; it is also independent of the presence of other trace constituents providing also that they exist at low levels of concentration.

Another way of expressing this equation is by relating, in the solid, the ratio of the trace ion and the ion it replaces (the carrier ion) to the ratio of the two in the liquid as:

$$\frac{\left(\frac{Cr}{Tr}\right)_{solid}}{\left(\frac{Tr}{Cr}\right)_{liquid}} = D \qquad (2)$$

The partition coefficient D is related to K in the preceding equation by the relationship

$$D = K \ \frac{\text{grams of carrier per cm}^3 \text{ of saturated solution}}{\text{grams of carrier in the solid}} \qquad (3)$$

When the trace constituents are not homogeneously distributed during precipation or even after considerable aging, the *Doerner-Hoskins* heterogeneous relationship is applied:

$$\left(\frac{Tr}{Cr}\right)_{crystal\ surface} = K \left(\frac{Tr}{Cr}\right)_{liquid} \qquad (4)$$

In this relationship, the ratio of the amounts of trace constituent in the liquid and solid is constant for any composition of liquid, but the ratio in the different layers varies as the composition of the liquid changes. For this to occur, each layer must be quickly covered over by another before it has a chance to react with the liquid and change composition along with it; and from the center to the outside of the crystal the trace element concentration will systematically vary as the composition of the liquid changes. (See appendix I for an expansion of this formula into other useful forms.)

In practice, it is usually found that the behavior of the system is intermediate between the conditions described by homogeneous and inhomogeneous laws.

The major variables that might theoretically affect the distribution coefficient are temperature, pressure, and concentration of the trace element.

The distribution coefficient D will vary with temperature as

$$\ln D = C_1 T^{-1} + C_2 \qquad (5)$$

where C_1 and C_2 are constants and T is the absolute temperature. (See Appendix II for deriviation of this relationship.)

The constants are different for each system; hence, temperature may have an appreciable influence on some systems and very little on others. Unfortunately, little experimental work has been done on systems of interest to the geologist, so the use of distribution coefficients as temperature indicators has not yet been successfully applied.

The partial molar volumes of the trace element and the carrier for which it is substituting are approximately the same; hence, the effect of pressure on the distribution coefficient should be negligibly small.

When the "trace" ion substitutes isomorphously for the carrier ion, the distribution coefficient remains constant throughout the complete range of composition. Even in some nonhomogeneous systems following Eq. (4) the distribution coefficient remains constant over a range of 10^7-fold increase in concentration. In most geological problems, other factors contribute greater errors than an assumption that the coefficient remains constant for all trace constituents.

Influence of Rates on Partition Coefficients

Experimental work has shown that the rate at which a given crystal grows has a marked influence on the amount of trace element incorporated into the growing crystal. Rapidly precipitated material has a much larger surface area and usually is more poorly crystallized (ordered) than slowly grown crystals; hence, they incorporate more contaminants. This fact is of great practical importance in industrial extraction and separation procedures. Fig. 1 shows, for both homogeneous and nonhomogeneous (logarithmic) distributions, the relationship between the percentage of tracer incorporated in a carrier at different precipitation rates. The higher values of D and λ represent the fastest rates of precipitation (Bonner and Kahn, 1958).

Many studies have been made to determine both the amounts of trace elements taken up by carriers precipitated at different rates and aged for various periods of time. Aging is related to the amount of recrystallization the precipitate undergoes in the presence of the fluid from which it separated. Because this recrystallization is relatively rapid, even layers deep within the precipitate appear to reach equilibrium with the surrounding fluid. For this reason, and because most geological systems form over a long period of time, the homogeneous distribution law will be the one most likely to be followed. Departures from it will represent a different process (a new event) affecting the system at a later time than the original crystallization series of events.

Guideposts for Future Work

In contrast to the large amount of excellent work done on partition coefficients in other areas of science, many of the systems of geological interest investigated by geologists have been carelessly handled and have produced few valuable contributions. Much more careful petrographic and x-ray structural work is needed to ensure that the samples being studied are single minerals or phases and not mixtures of minerals. Although this can be done with relative ease with large crystals, it is a very difficult problem with fine-grained minerals like many carbonates and clays, which are minerals of great importance to geological studies. Much more effort is needed for the development of better separation procedures and for the careful application of those already existing before results of the quality needed can be obtained to help solve many of the existing geological problems now facing us. With the advent of the electron microprobe and new analytical techniques, both the amounts and compositions of intergrown mixtures are within the realm of precise measurement. Hence, many of the difficult technical problems of the past may soon be solved.

Influence of Ion-Exchange Processes on Partition Coefficients

In the preceding sections we were concerned with a small amount of solids growing from a relatively large amount of liquid. Here we consider the reverse situation: a volumetrically small amount of fluid moving through a large amount of solid having a correspondingly large amount of surface. This system is very important geologically because large quantities of fluids move through sediments and rocks during the times when they are transformed from simple sediments into metamorphic or magmatic or granitized rocks where the system becomes mobilized in the presence of aqueous or molten fluids.

The exchange of cations between a liquid and solid is stoichiometric, and the equilibria of exchange reactions are governed by the classical absorption isotherm or Freundlich equation.

Ion-exchange processes have become very important in both the laboratory and industry for the separation of ions having very similar chemical properties in solution when present either in large or small concentrations. In the

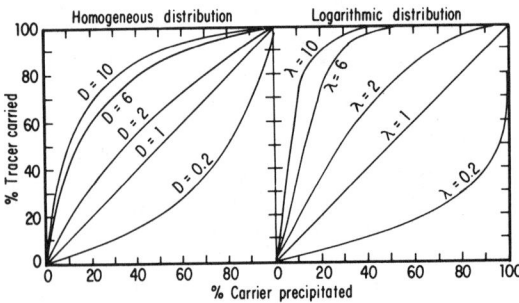

Fig. 1. Efficiency with which a trace element is carried for different values of distribution coefficients, D and λ.

laboratory, a solution containing the cations (or anions) to be separated is poured into the top of a tube containing the ion-exchange material. Under appropriate conditions, all of the cations become adsorbed in a narrow bank at the top of the exchange bed. A buffered solution containing a complexing agent is then slowly passed through the column. The ions are desorbed, carried a short distance down the column and readsorbed. The cation most tightly adsorbed to the exchange material and least soluble in the complexing agent stays behind or moves a lesser distance down the column, hence, becomes separated from its companion ions. Complete separations can be obtained by this procedure.

The pH, composition, and concentration of the eluting solution play key roles in the efficiency of the eluting process. Flow rates, the surface area, and chemical nature or activity of the exchange material, temperature, and column dimensions are also parameters of the process. Total surface activity, flow rates, and temperature affect the degree to which equilibrium is reached between the exchange solid and the ions in the fluid at any time.

That equilibrium is closely approached in some geological systems is highly probable because flow rates are often extremely slow and changes in the bulk composition and pH of the eluting formation fluid or ore fluid may be very gradual as is the temperature gradient. For these reasons, even though the rock formations may be relatively inefficient exchange materials, remarkable separations such as columbium from tantalum occur in certain ore deposits. This process also may account for the concentration of metals from formation fluids to produce ore deposits (the old lateral secretion concept), the concentration taking place where an important change in the composition of the fluid or exchange material (rock type) occurs.

As discussed above, according to the Berthelot-Nerst distribution law, the ratio of the equilibrium concentrations (activities) of a substance in two phases is constant at a given temperature. Once equilibrium conditions prevail, the same principles apply, so that partition coefficients, largely following the homogeneous distribution law, should also be useful in studying the genesis and paragenesis of ore deposits as well as other geochemical separations or concentrations.

APPENDIX I

Nonhomogeneous or The Logarithmic Distribution Law

When the surface layers are covered over before they reach equilibrium with the solution, the concentration of the trace element within the crystal may be higher in the center than near the outer portions of the crystal, giving rise to a nonhomogeneous distribution. If x and y are the amounts of trace and carrier element deposited in a given layer and a and b are the total amounts of trace and carrier element, and since dx and dy, the amounts of trace and carrier element deposited in a given layer, are proportional to their respective concentrations, then

$$\frac{dx}{dy} = \lambda \frac{(a-x)/v}{(b-y)/v} \quad (6)$$

where v is the volume of the liquid from which the crystal is growing and λ is a constant characteristic of the system called the *logarithmic distribution coefficient*. To put the equation in a more useful form, it is integrated to give

$$\log \frac{a}{a-x} = \lambda \log \frac{b}{b-y}$$

or

$$\log \frac{\text{total trace constituent}}{\text{tracer constituent in solution}} = \lambda \log \frac{\text{total carrier}}{\text{carrier in solution}} \quad (7)$$

When λ is greater than unity, the trace constituent element concentrates in the growing crystal and decreases in concentration from the center outward; when λ is less than unity, the trace element concentrates in the liquid and the highest concentrations in the crystal are at its surface.

APPENDIX II

For any mineral of interest (McIntire, 1963), a general formula can be used to express the trace constituents (Tr), the bulk carrier ion being precipitated (Cr), and the remainder of the formula (X) unaffected by the substitution of (Tr) for some (Cr):

$$(\text{Tr},\text{Cr})\text{X}$$

The distribution coefficient can be shown in several forms as

$$D = \underset{(a)}{\frac{(\text{Tr}/\text{Cr})_S}{(\text{Tr}/\text{Cr})_L}} \underset{(b)}{\frac{(M^S\text{TrX}/M^S\text{CrX})}{(M^L\text{Tr}^{m+}/M^L\text{Cr}^{m+})}} =$$

$$\underset{(c_1)}{\frac{K_{\text{Cr}}\text{X}}{K_{\text{Tr}}\text{X}}} \underset{(c_2)}{\left(\frac{\gamma_{\text{Tr}^{m+}}}{\gamma_{\text{Cr}^{m+}}}\right)} \underset{(c_3)}{\exp\left(\frac{-\Delta\mu}{RT}\right)} \quad (8)$$

In expression (a), Tr, Cr, S, and L indicate,

respectively, trace and carrier ions, solid, and liquid. In (b), M is the mol fraction of components in the solid, m is the mol fraction in liquid and m+ are the mol concentrations in the liquid. In expression (c_1) K is the thermodynamic solubility product for pure TrX at a given temperature and pressure, and γ is the molal activity coefficient in the liquid (c_2). In the remainder of expression (c_3), $\Delta\mu$ is the excess free energy of the partial molar free energy involved in transferring one mole to Tr^{m+} from a large quantity of ideal solid solution to the real solid solution at the same mole fraction; R is the gas constant and T is the absolute temperature.

The distribution coefficient is therefore made up of related factors, the first of which (c_1) is the ratio of the solubility products of the two solid end-members; the second (c_2) involving the molal activity coefficients (γ) in the liquid, evaluates the influence of the ions in the liquid; and (c_3) involves the nature of the solid solution. The general case (discussed by Hermann and Suttle, 1961) where the ratio of the carrier with a charge $m+$ (Cr^{m+}) to the remainder of the structural formula (X^{m-}) having a negative sign is expressed as v_+/v_- where the v's are small whole numbers. The above formula then gives

$$D = \frac{(\text{Tr/Cr})_S}{(\text{Tr/Cr})_L} = \underbrace{\left(\frac{K_{\text{Cr}_{v+}X_{v-}}}{K_{\text{Tr}_{v+}X_{v-}}}\right)^{1/v+}}_{(c_1)} \underbrace{\left(\frac{\gamma_{\text{Tr}^{m+}}}{\gamma_{\text{Cr}^{m+}}}\right)}_{(c_2)}$$

$$\underbrace{\exp\left(\frac{-\Delta\mu_{\text{Tr}_{v+}X_{v-}}}{v_- RT}\right)}_{(c_3)} \quad (9)$$

where (c_1) expresses the character of a solid solution. When the carrier and tracer ions are similar in electronegativity and size, the solution will approach ideality and the distribution coefficient will be constant over the entire range of substitution. If the solid solution is ideal, coefficients in (c_2) will be similar and the term will approach unity as will the resultant free energy (in term c_3).

Even when the solid solution is not ideal, and the substitution occurs only in small amounts, the result will be very similar to that of an ideal solid solution and the distribution coefficient, at least within experimental error, will be almost constant.

To evaluate the influence of temperature on the partition coefficient, Eq. (1) is differentiated with respect to temperature, holding pressure (P), and other factors (N) constant; we obtain

$$\left(\frac{\partial \ln D}{\partial T}\right)_{P,N} = -\frac{1}{v_+ RT^2}$$
$$[(h^0_{\text{Cr}_{v+}X_{v-}} - h^L_{\text{Cr}_{v+}X_{v-}})$$
$$- (h^0_{\text{Tr}_{v+}X_{v-}} - h^L_{\text{Tr}_{v+}X_{v-}})$$
$$- (h^S_{\text{Tr}_{v+}X_{v-}} - h^{ID}_{\text{Tr}_{v+}X_{v-}})] \quad (10)$$

Here we see that the temperature coefficient is made up of three terms involving the difference between partial molal enthalpies (the h terms) where h_0 refers to the pure crystalline solid, h^L to the complex liquid solution containing both trace and carrier ions, and h^{ID} refers to the ideal solid solution.

Again, because $\text{Cr}_{v+}X_{v-}$ and $\text{Tr}_{v+}X_{v-}$ are either ideal solid solutions or, when the tracer exists in small quantities, approach ideal solid solutions, their heats of solution will vary in a parallel manner over a range of temperature, and so, for practical purposes, can be considered to be constant. If this is correct, on integration of Eq. (3), the following relationship is obtained:

$$\ln D = C_1 T^{-1} + C_2 \quad (5)$$

where C_1 and C_2 are constants. This is the equation of a straight line when $\ln D$ is plotted against T^{-1}.

Experimental work with a number of salts indicates that the equation is correct. We can, therefore, expect significant variations in a partition coefficient with temperature. This means to the geologist that partition coefficients may provide a tool for studying paleotemperature changes. Such an independent check of O^{16}/O^{18} paleotemperature techniques provides promise of a way to evaluate paleotemperatures in a wide range of geological materials originally formed from aqueous solutions, molten liquids (magmas), and solid solutions (metamorphic rocks), and is thus a great challenge to future geologists.

A. F. Frederickson

References

Bonner, N. A., and Kahn, M., 1958, "Behavior of Carrier-Free Tracers," in (Wahl, A. C., and Bonner, N. A., editors), "Radioactivity Applied to Chemistry," New York, John Wiley & Sons, pp. 102–140.

Frederickson, A. F., 1962, "Partition coefficients—new tool for studying geologic problems," Bull. Am. Assoc., Petrol. Geol., **46**, 518–34.

Frederickson, A. F., and Reynolds, R. C., Jr., 1960, "Geochemical Method for Determining Paleosalinity," in (Swineford, A., editor), "Clays and Clay minerals," Vol. 8, New York, Pergamon Press, pp. 203–213.

Hermann, J. A., and Suttle, J. F., 1961, "Precipitation and Crystallization," in (Kolthoff,

PEDOLOGY (SOIL SCIENCE)

I. M., and Elving, P. J., editors), "Treatise on Analytical Chemistry," Pts. 1 and 3, New York, Interscience Publ., pp. 1367–1409.

McIntire, W. L., 1963, "Trace element partition coefficients—a review of theory and applications to geology," *Geochim. Cosmochim. Acta,* **27,** 1209–1264.

Cross-references: *Crystal Chemistry; Ion Exchange; Paleosalinity; Paleotemperatures; Paragenesis; Phase Equilibria; Seawater, Chemistry; Solid Solution; Stoichiometry; Trace Elements.* Vol. IVB: *Crystal Growth; Electron Probe Microanalysis; Ore Genesis.* Vol. VII: *Ecology; Paleontology.*

PEDOLOGY (SOIL SCIENCE)

The term pedology is from the Greek, *pedon,* meaning ground, and *logos,* meaning a discourse thereof or science. Pedology is the study of the soil in its natural position (Jenny, 1941) in regard to its morphology, characterization, classification, and genesis (U.S., 1951). The study involves the geology of the regolith where most weathering and erosion take place, the physics and chemistry of a complex colloidal system, and the biology of flora and fauna relative to their environment (Robinson, 1949).

Soil

A soil is a naturally occurring three-dimensional body with morphology and properties resulting from effects of climate, flora, and fauna, parent rock materials, topography, and time (U.S., 1951). A soil occupies a portion of the land surface, is mappable, and commonly is composed of horizons that parallel the land surface. A vertical section downward through all the horizons of the soil is called a *soil profile* (Fig. 1).

Soil Mapping, Morphology, and Characterization

First in pedologic work is studying, defining, and mapping soils in the field. A soil map (Fig. 2B) shows distribution of soils relative to other physical or cultural features of the earth's surface (U.S., 1951). Each soil is differentiated from others by number, color, texture, structure, arrangement, chemical composition and thickness of soil horizons, thickness of the true soil, and character and geology of the soil material (Marbut, 1922). Each mapping unit, alike in these characteristics of a soil or combinations of soils, is delineated on a map by a boundary.

Characterization of a soil requires selection of a representative soil profile that is described as quantitatively as possible, utilizing comparative charts for color, structure, and other properties, and accurately measuring soil horizons. The horizons (Fig. 1) are labeled if possible. Soil profiles are described within a standardized nomenclature (U.S., 1951) so that comparison may be made with other soils. Samples

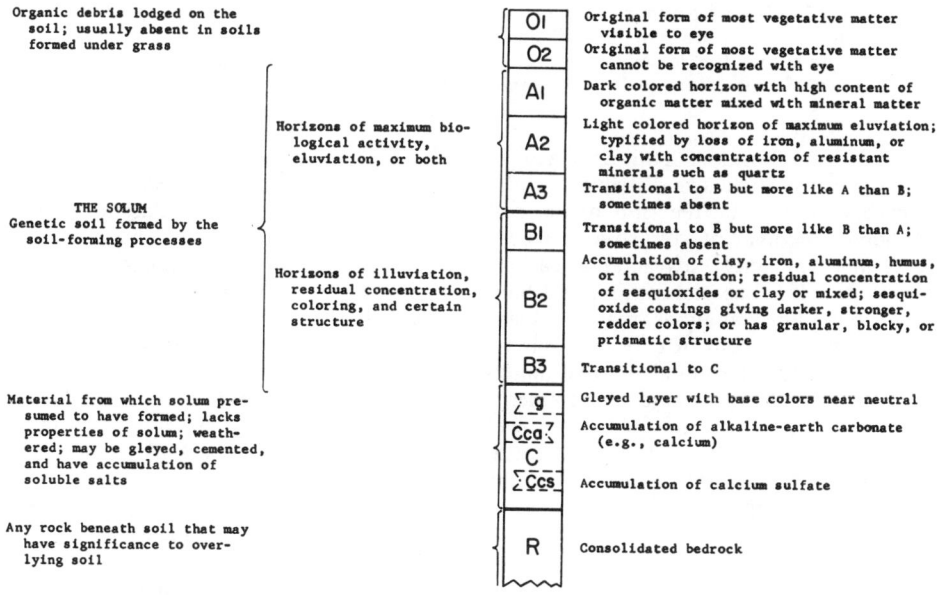

Fig. 1. Hypothetical soil profile that has all the principal soil horizons Not all of these horizons are present in any profile, but every profile has some of them. (From U.S. Dept. Agr., 1951, 1960.)

PEDOLOGY (SOIL SCIENCE)

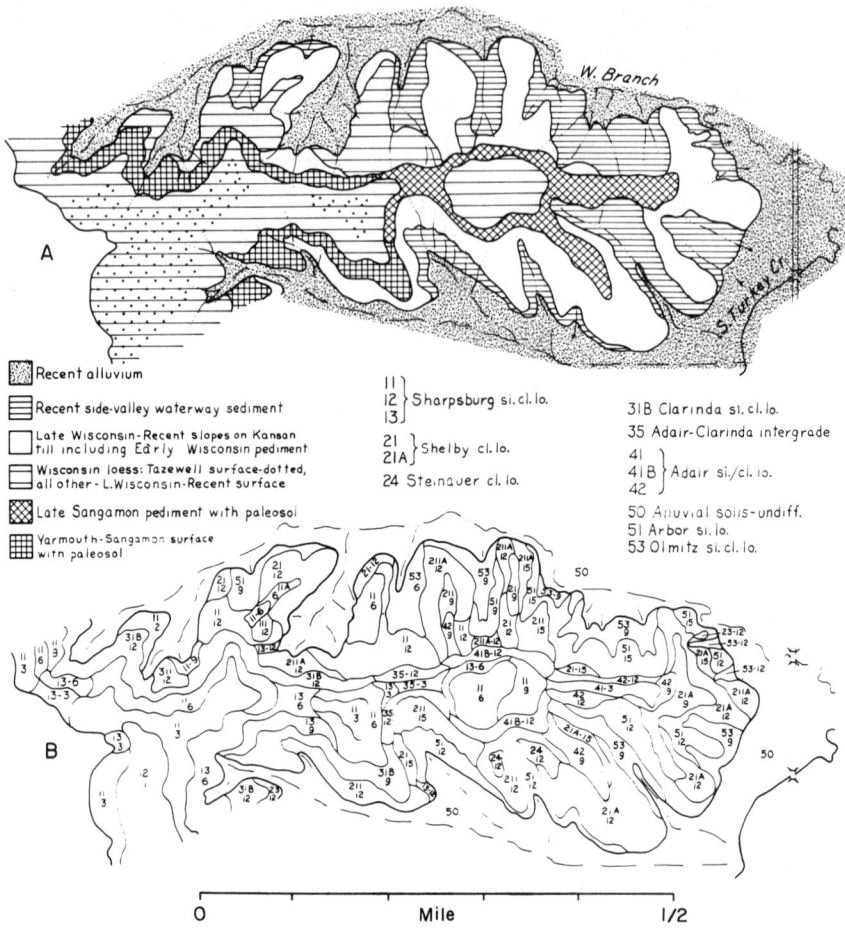

FIG. 2. Map of geomorphic and surficial deposits (A) and soil (B) of area in Adair County, Iowa. On the soil map, the upper or first number is the soil-mapping unit, and the lower or second number is the slope gradient in percent of surface of unit (from Ruhe and Daniels, 1958).

are collected from horizons and analyzed in the laboratory for particle-size distribution, pH, organic carbon, nitrogen, free iron oxide, calcium carbonate equivalent, moisture tension, cation-exchange capacity, extractable cations (Ca, Mg, H, Na, K), base saturation, and bulk density. Among other analyses may be identification and measurement of clay minerals by x-ray diffraction and differential thermal analysis, identification and measurement of elements and compounds by x-ray fluorescence or wet chemistry methods, and grain-mineral identification and study of soil fabric by use of the petrographic microscope.

Soil Classification

Soil classification has been based on soil properties in modern times but always has been tempered with concepts of soil genesis, with external associations, or with use of the soil.

The first systematic classification was by Dokuchaiev in Russia in 1882 (see Simonson, 1962 for history of soil classification). Based on field and laboratory characteristics soils were grouped in three categories, the highest having three and the lowest thirteen classes. The 3 highest classes were "normal" soils of the dry land vegetative zones and moors, "transitional" soils of washed or dry land sediments, and "abnormal" soils. The 13 subclasses, "soil types," were named and related to climatic or vegetative regions. The Dokuchaiev system involved properties of the soil with external associations of climate and vegetation. Later an associate, Sibirtsev, renamed the highest classes, *zonal, intrazonal,* and *azonal.*

In the United States the first soil surveys were made to aid in guiding the selection of crops and the application of soil-management practices. Soils were considered as a medium for plant growth. Although soil properties were utilized in grouping soils, use of soil also was involved. Systematic classification of soils began with Coffey's work (1912) and was continued in the first comprehensive system by Marbut (1936). The final scheme consisted of six categories with two classes in the highest, several thousand in the lowest, and various numbers in intermediate categories. Criteria for classes were mainly characteristics of the soils but in some cases were inferences of genesis of the soil and in other cases were features outside of the soils themselves. Attempts were made to rectify these defects in revised systems (Baldwin, et al, 1938; Soil Science, 1949). Descending categories in the revisions were order, suborder, great soil group, family, series, and type. There were only three classes in the order, but nine suborders, thirty-nine great soil groups, and many thousands of classes in the lower categories.

The soil series is of primary importance in the lower categories and is defined as a group of soils having genetic horizons similar in differentiating characteristics and arrangement in the soil profile, except for the texture of the surface soil, and developed from a particular type of parent material. The soil type is distinguished within the soil series by texture of the surface layer (Riecken and Smith, 1949). This definition differs from the Russian soil type that is analogous to the American great soil group, which is a group of soils having common internal characteristics developed through the influence of environmental forces of broad geographic significance, especially vegetation and climate (U.S., 1938).

The classification system now used in the United States has inconsistencies; e.g., genetic definition of two orders and morphologic definition of the third; vague definition of the classes, particularly the great soil groups; and use of the virgin soil as a basis of definition. Cultivated soils, for example, are classified on presumption of the properties the soil had when virgin. Consequently a new system, the "Seventh Approximation" (U.S., 1960) is currently being tested. It has ten orders, forty suborders, 120 great groups, 400 subgroups, some 1500 families, and more than 7000 series. Soil type has been dropped. The ten orders are differentiated on gross composition, degree of horizonation, presence or absence of certain horizons, and a combined index of weathering and weatherability of minerals (Simonson, 1962). These strictly morphologic criteria were selected because they presumably reflect major differences in the genesis of soils. The same principles are carried downward through the suborder and great group although additional soil characteristics are used for distinguishing classes at each level. At the subgroup level a typifying class is defined that has the definitive properties of a specific great group. Distinguished also are intergrade classes that have some properties of another great group and extragrade classes that have some properties not definitive of any other great group. The soil series remains essentially unchanged. The new classification system is based primarily upon properties of the soil. Genetic bias is permitted only in the selection of properties to define a class. Such bias is based on the present status of knowledge of the genetic pathways in soil formation.

Soil Genesis

Soil genesis is the mode of origin of soil and is evaluated from the basic kinds of changes that take place. These changes are additions, removals, transfers, and transformations of organic matter, soluble salts, alkaline earth carbonates, sesquioxides, silicate-clay minerals, and other substances (Simonson, 1959). These changes are observed directly from laboratory experiments or are inferred by relating measured morphological, physical, and chemical properties of a part to other parts of a soil.

In experiments (Thorp et al., 1957) organic acids and distilled water leachates from tree leaves are passed through columns of different soil materials. Bleached surface layers and subjacent layers of stronger color form in the columns. Effluent solutions from the base of the columns contain detectable amounts of calcium, magnesium, iron, manganese, phosphorus, potassium, and sodium. Very fine silicate clay, e.g., illite, montmorillonite, vermiculite, and chlorite, also were suspended in the effluent. Removal, transfer, and transformation are demonstrable experimentally. Examination of the columns showed that clay was partially removed from the bleached layers and was deposited in voids in the lower layers. In this case removal (eluviation) and addition (illuviation) took place. The net effect of the study was the experimental formation of something similar to a podzolic soil.

Organic matter best illustrates additions to a soil and is formed in the biological decomposition of plant and animal residues by soil microorganisms. Plants supply most of the organic matter as dry matter added to the soil surface and as roots in the subsurface. According to Jenny (1941), short grass prairie in semiarid regions may annually add 0.7 ton/acre of dry matter; tall grass prairie in subhumid regions, 0.8–1.7 tons/acre; pine forest in more humid areas, 2.1 tons/acre; and tropical rain forest, 45–90 tons/acre. Under bluegrass, roots may comprise 2.4 tons/acre in the top 4 in. of a soil (Jenny, 1941).

During decomposition plant materials are converted to carbon dioxide, water, mineral elements, and other chemically altered substances. Less resistant materials are consumed first by soil microbes so that more resistant plant materials remain with the new organic compounds that are synthesized by the organisms (Bartholomew, 1957). At any time the organic matter at a place in the soil reflects an equilibrium state of the addition of new material to the system, removal of more readily decomposable material, and transformation to other forms by microorganisms and other agents. Organic matter also may be transferred within the soil by physical and physicochemical processes. Burrowing animals, worms, and insects turn over the soil and physically mix adjacent portions. Freezing and thawing and wetting and drying do the same thing. Colloidal organic matter may be flushed downward or laterally and coagulate as coatings on structural aggregates in the soil.

Radiocarbon dates of organic matter from surface horizons of soils not only reflect the equilibrium status but point out the relatively rapid turnover in the system. In the Edina soil in southern Iowa organic matter is 410 ± 110 years old in the 6-in. surface layer (Fig. 3). In the next subjacent layer the age is 840 ± 220 years (Broecker, et al., 1956). At depths of 23 to 25 in. the organic carbon is 1545 ± 110 years.* Yet the whole soil is probably 14,000 years old (Ruhe and Scholtes, 1956).

Transfer within the soil system also is inferred from constituents other than organic matter. In the Edina soil total clay ($< 2\mu$) is 23% in the A2 and 52% in the B2 horizon. Fine clay ($<0.06\mu$) is 10% and 33% respectively, whereas coarse ($2-0.2\mu$) and intermediate ($0.2-0.06\mu$) clay are uniform or slightly increase with depth in the profile (Fig. 3). These distributions suggest that fine clay was formed in the lower horizon but also has moved from the A to the B horizon. The latter conclusion is substantiated by fabric studies. Under the petrographic microscope films of oriented clay particles coat surfaces of soil-structure aggregates and fill voids in the lower horizon but progressively decrease to and are absent in the upper horizon. Such coatings are due to movement in suspension and subsequent deposition (Brewer, 1960). Some of the clay coatings are complexly mixed with organic matter. The inferred illuviation is

* Unpublished data, Soil Survey, Soil Conservation Service, U.S. Dept. Agriculture.

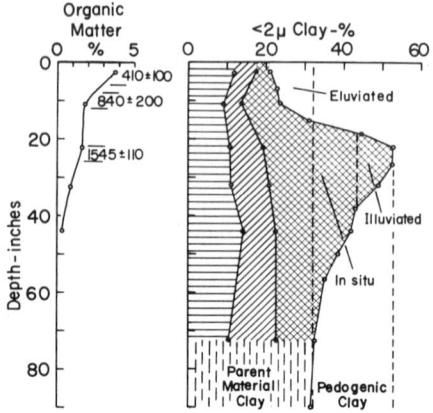

FIG. 3. Organic matter and clay content of Edina soil in Wayne County, Iowa. The radiocarbon age of the organic matter in the upper two horizons is in years before present. Clay fractions are $2-0.2~\mu$ (horizontally lined), $0.2-0.06~\mu$ (diagonally lined), and $<0.06\mu$ (cross hatched). (Data from Ulrich, 1950.)

based on measured morphological, physical, and chemical properties of associated relative parts of the soil. As previously shown very fine clay was transferred through soil columns experimentally (Thorp, et al., 1957). Other compounds in the soil such as soluble salts, sesquioxides, and alkaline earth carbonates can be measured and analyzed and similar conclusions reached.

All changes in soil do not act in the same direction but some act in opposition (Simonson, 1959). For example, additions of plant matter tend to be partially opposed by microbiological activity. As soil organisms convert plant matter to soil-organic matter a part is destroyed. Transfer of material downward is opposed by internal churning of the soil by animals, worms, or insects, freezing-thawing, wetting-drying, growth of large roots, or tree throw. The balance within the system determines the nature of the soil that will evolve (Simonson, 1959).

Combination of changes results in the formation of soil horizons. Dominance of additions and transformation of organic matter at and near the soil surface causes formation of a surface horizon that has high content of organic matter. Transformation of silicate minerals to silicate-clay minerals that are subsequently transferred downward forms an upper horizon depleted of some clay and a subjacent horizon enriched in clay. Soluble salts and alkaline earth carbonates may be dissolved, transferred in solution, and removed entirely from the soil system or may be precipitated and added to a horizon at depth.

The four kinds of changes that develop soil horizons are dependent upon many basic processes such as hydration, oxidation, reduction, solution, precipitation, freezing, thawing, wetting, drying, and the like. These processes, in turn, depend upon the factors of soil formation: parent material, topography, climate, vegetation and other living organisms, and time (Jenny, 1941).

Parent material, the material from which soils form, controls processes and changes from a mineralogic-composition and physical-property sense. Quartz sand subjected to solution and precipitation yields silica. Transformations in the soil are relatively simple, i.e., high-temperature quartz to low-temperature quartz (cristobalite). A parent material of quartz, feldspar, amphiboles, mica, calcite (e.g., loess) subjected to similar processes yields many compounds of calcium, magnesium, iron, potassium, and silicon. Clay minerals, sesquioxides, alkaline earth carbonates, and soluble salts form and can be removed, added, transferred, and further transformed. The composition of original minerals plus weathering by-products is extremely complex.

The physical nature of the parent material also affects processes and changes. A porous, permeable medium gives ready access to gases and liquids. Hydration, oxidation, and solution proceed rapidly. A more impermeable medium impedes access of agents and is affected slower by processes and changes. Parent material not only affects processes but also controls their rate of change.

Topography is the geometric configuration of the soil surface. It can be described quantitatively by slope gradient, nature of contours, elevation, and direction. This factor operates indirectly by controlling other factors, particularly climate. Commonly in mountain ranges rainfall increases and temperature decreases with elevation. Consequently the amount of moisture available for soil-forming processes and the temperatures at which the processes operate are related to climate, which, in turn, is controlled by topography. On a level hilltop descending across convex side slopes to concave foot slopes, moisture in the soils will differ even under the same rainfall and with the same parent material and vegetative cover. On the level hilltop, most rainfall will seep into the soil. On the convex side slope some water will seep into the soil and some will run off downslope. Less moisture is available in the soil than on the hilltop. On the concave foot slope not only will rainfall supply moisture but also runoff from the higher side slope will provide excess moisture. Soil climate varies in these three topographic areas. In some areas, such as a glacial drift plain, topography may control depth to groundwater so that one soil in a closed depression may be saturated for long periods, wheras another soil on an adjacent rise may be above the water table and less wet. Soil climate also varies in these conditions.

Moisture and temperature define climate, which controls processes and changes in the soil system. Although local climatic variations are detectable in their influence on soils, the relationships are best shown on a large geographic basis. The distributions of great soil groups in the United States and the soil types of Russia are related to climatic zones. This is partly due to the direct influence of climate on soil formation and partly to the effect on soils of vegetation, which in turn, is related to climate (Blumenstock and Thornthwaite, 1941).

Vegetative effects on soils can be divided broadly into the processes and changes under prairie, forest, mountain meadow, tundra, tropical rain forest, and others. Under forest, for example, leaf fall is the principal source of organic matter. Litter under conifers is low in mineral constituents and highly acid. Under deciduous trees litter is high in mineral matter and not as acid. In a subhumid zone, prairie litter from grass generally is highest in mineral matter and not acid. Grass also provides a more abundant root mat in the upper part of the soil. As a result such grassland soils have a thicker, darker surface horizon that is higher in organic matter (Fig. 4). Solutions percolating the more acid forest litter alter the subjacent parent material more intensively even though the soils of similar nature, on the same landscape, and formed during the same period of time as the grassland soil. As a result the soils under forest are more acid, more horizonated, and more strongly developed (Figs. 4, 5). Eluvial and illuvial horizons are more pronounced.

Time is the duration of processes acting in a soil. Time usually is determined geologically and involves dating the material in which the soil forms and dating the surface on which the soil forms. For example, soil 12 (Fig. 2B) formed on a level constructional loess surface (Fig. 2A). Post-deposition erosion is negligible. Radiocarbon dates at the base of the loess are 21,000 to 24,500 years. The loess can be traced under a younger glacial till whose base is dated at 14,000 years (Ruhe and Scholtes, 1956). As soil 12 is formed in the upper six feet of the loess that is seventeen feet thick, the soil is younger than 21,000 years. Where the loess is buried by till and undisturbed during burial, only a very weak soil profile is developed in the top of the loess. Consequently most soil properties in soil 12 (not buried by

PEDOLOGY (SOIL SCIENCE)

FIG. 4. Profiles of the Tama (left) and Fayette (right) soils in eastern Iowa. Both soils formed from the same loess parent material, on the same geomorphic surface (of same age and topography), and under similar climate. The soils differ, however, in that the Tama formed under prairie and Fayette under forest conditions. Scale in feet and inches. (Photo R. W. Simonson.)

Soil 221 (Fig. 2B) is formed in glacial till that underlies the loess (Fig. 2A) on slopes that bevel the level loess summit. Slopes are younger than the summit and descend beneath valley fills (Fig. 2A) whose base is dated at 6800 years. This is the maximum age of the slope and soil 21 formed on it. Dating is done by geomorphologic techniques and radioisotope age. When absolute dating in years is not possible, only relative geologic dating can be used.

In soil-genesis studies a soil-forming factor is evaluated by holding four of them constant and analyzing the effects of variability of the fifth. To determine effects of vegetation, soils under forest and grass are studied where parent material, climate, topography, and time are as identical as possible. Evaluation is extremely difficult; five factors control many processes and produce four kinds of changes, some acting in the same and some in opposite directions.

ROBERT V. RUHE

References

Bartholomew, W. V., 1957, "Maintaining organic matter," in "Soil," U.S. Dept. Agriculture Yearbook, Washington, D. C., pp. 245–252.

Baldwin, M., Kellogg, C. E., and Thorp, J., 1938, "Soil classification," in "Soils and Men," U.S. Dept. Agriculture Yearbook, Washington, D.C., pp. 979–1001.

Blumenstock, D. I., and Thornthwaite, C. W., 1941, "Climate and the world pattern," in "Climate and Man," U.S. Dept. Agriculture Yearbook, Washington, D.C., pp. 98–127.

Brewer, R., 1960, "Cutans—their definition, recognition, and interpretation," *J. Soil Sci.*, **11**, 280–292.

till) formed less than 14,000 years ago. The age of the soil as determined by stratigraphic techniques enhanced by radioisotope age.

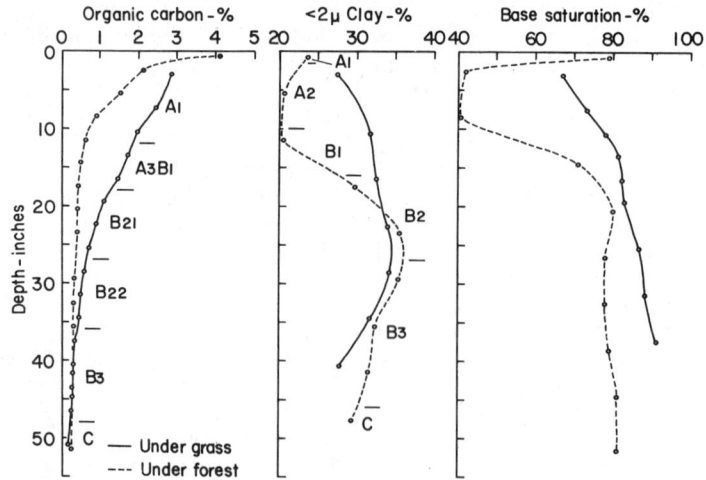

FIG. 5. Profile properties of Tama (solid line) and Fayette soils (broken line), illustrating differences of soils formed under prairie and forest conditions (see Fig. 4). (Data from Smith et al., 1950.)

Broecker, W. S., Kulp, J. L., and Tucek, C. S., 1956, "Lamont natural radiocarbon measurements III," *Science,* **124,** 154–165.

Coffey, G. N., 1912, "A study of the soils of the United States," *U.S. Dept. Agr. Bur. Soils Bull.,* **85,** 114pp.

Glinka, K. D., 1927, "The Great Soil Groups of the World and Their Development," Ann Arbor, Mich., Edwards Bros., 150pp. (translation).

Jenny, H., 1941, "Factors of Soil Formation," New York, McGraw-Hill Book Co., 281pp.

Marbut, C. F., 1922, "Soil classification," *Am. Assoc. Soil Surv. Workers Bull.,* **3,** 24–32.

Marbut, C. F., 1936, "Soils of the United States," in "Atlas of American Agriculture," U.S. Dept. Agriculture, Washington, D.C., Pt. 3, 98pp.

Riecken, F. F., and Smith, G. D., 1949, "Lower categories of soil classification," *Soil Sci.,* **67,** 107–115.

Robinson, G. W., 1949, "Soils, Their Origin, Constitution and Classification," Third edition, London, Thomas Murby and Co., 573pp.

Ruhe, R. V., and Scholtes, W. H., 1956, "Age and development of soil landscapes in relation to climatic and vegetational changes in Iowa," *Soil Sci. Soc. Am. Proc.,* **20,** 264–273.

Ruhe, R. V., and Daniels, R. B., 1958, "Soils, paleosols, and soil-horizon nomenclature," *Soil Sci. Soc. Am. Proc.,* **22,** 66–69.

Simonson, R. W., 1959, "Outline of a generalized theory of soil genesis," *Soil Sci. Am. Proc.,* **23,** 152–156.

Simonson, R. W., 1962, "Soil classification in the United States," *Science,* **137,** 1027–1034.

Smith, G. D., Allaway, W. H., and Riecken, F. F., 1950, "Prairie soils of the upper Mississippi Valley," *Advan. Agron.,* **2,** 157–205.

Soil Science, 1949, Special issue on Soil Classification, **67**(2), 77–191.

Thorp, J., Strong, L. E., and Gamble, E., 1957, "Experiments in soil genesis—the role of leaching," *Soil Sci. Soc. Am. Proc.,* **21,** 99–102.

Ulrich, R., 1950, "Some physical changes accompanying Prairie, Wiesenboden, and Planosol soil profile development from Peorian loess in southwestern Iowa," *Soil Sci. Soc. Am. Proc.,* **14,** 287–295.

U.S. Dept. Agriculture, 1938, "Soils and Men," yearbook of Agriculture, Washington, D.C., 1232p.

U.S. Dept. Agriculture, Soil Survey Staff, 1951, Soil Survey Manual, Handbook 18, Washington, D.C., 503pp.

U.S. Dept. Agriculture, Soil Survey Staff, 1960, "Soil Classification, a comprehensive system, 7th approximation," Soil Conservation Service, Washington, D. C., 265pp.

Cross-references: *Biochemicals; Calcium; Differential Thermal Aanalysis; Geochronometry; Iron; Magnesium; Organic Geochemistry; Oxidation and Reduction; pH-Eh; Potassium; Silicon; Soil Erosion; Soil Salinity and Alkalinity; Vegetation Indicators; Weathering, Chemical; X-Ray Diffraction Analysis.* Vol. II: *Climatic Classification; Vegetation Classification.* Vol. III: *Loess; Organisms as Geomorphic Agents; Regolith; Slopes; Weathering.* Vol. IVB: *Amphiboles; Calcite; Clays and Clay Minerals; Feldspars; Quartz.* Vol. VI: *Soil Genesis; Soil Morphology; Soil Profile; Soil Texture; Soils of the World; World Soil Map.*

PELAGIC SEDIMENTS, GEOCHEMISTRY
—*See* Vol. I, ANTARCTIC PELAGIC SEDIMENTS; PELAGIC BIOGEOCHEMISTRY; Vol. VI. PELAGIC SEDIMENTS

PERIODIC TABLE—*See* ATOMIC NUMBER AND PERIODIC TABLE

PERMEABILITY—*See* POROSITY AND PERMEABILITY

PETROLEUM—*See* Vol. IV B

PHASE EQUILIBRIA

Phase Rule

The phase rule of Willard Gibbs is the basis of all the phase diagrams. These diagrams (graphical expressions of the phase rule) are very useful in understanding what phase changes can take place in a system as either the temperature, pressure, or composition is changed. Phase diagrams are important in interpreting past geological phenomena and in developing new materials.

The phase rule is expressed mathematically by the equation $P+F = C+2$, where $C =$ number of components of the system, $P =$ number of phases present at equilibrium, and $F =$ degrees of freedom (variance) of a system. In a condensed system (systems under a constant, arbitrarily fixed pressure) it is expressed as $P+F = C+1$.

Definitions

System. A system is any part of the Universe which can be isolated and defined. A condensed system is one in which the vapor pressures of the liquid and solid phases present are insignificant in comparison to atmospheric pressure. In the condensed system the vapor phase is not considered important and only the solid and liquid phases are analyzed. Pressure is considered constant under these circumstances. Recently, much more phase equilibrium work has been done in which the gaseous phases (i.e., H_2O, CO_2, SO_2, O_2, Cl, F, etc.) have been considered.

Components. Components are the smallest number of independently variable chemical constituents necessary and sufficient to express the composition of each one present in any state of equilibrium. The components can be elements or compounds, but the latter must not dissociate during the experimental conditions being studied.

Phase. A phase is any part of a system which is physically homogeneous and mechanically separate from any other portion of the system.

A homogeneous system is composed of one phase. A heterogeneous system is composed of more than one phase. The phase rule only applies if there is complete equilibrium.

Degrees of Freedom or Variance. Degrees of freedom are the number of independent variables, usually temperature, pressure, and composition, which must be arbitrarily fixed in order that the system can be completely defined. Degrees of freedom of a system can also be defined as the number of intensive variables which can be altered independently without causing a phase change (either the disappearance or formation of a phase). The degrees of freedom are expressed as univariant (one degree of freedom), bivariant (two degrees of freedom), and invariant (no degree of freedom).

Boundary Line (Curve). A boundary line is the place where adjacent liquidus surfaces intersect in a ternary phase diagram. The region enclosed by boundary lines is the primary phase region. The primary phase is the first phase to crystallize upon solidification of the liquid.

Alkemade Line. An Alkemade line is a straight line connecting two primary phase composition points whose regions are separated by a boundary curve.

Alkemade (Van Rijn van Alkemade) Theorem. The Alkemade theorem is useful for predicting the path of crystallization and compatibility of phases. The point where the Alkemade line intersects the boundary curve is the point of maximum temperature on the boundary curve. From that point, the temperature decreases in either direction along the boundary curve. When the Alkemade line does not intersect the boundary curve, then the maximum temperature is the projected end of the boundary curve that intersects the Alkemade line. This theorem is extremely important in understanding how the liquid composition changes with crystallization in ternary systems. It is also useful in defining composition triangles which are separated by Alkemade lines.

Composition (or Compatibility) Triangle. The composition triangles are the triangular areas in a condensed ternary system which are bound by the three Alkemade lines connecting the composition points of the pure, three primary phases whose liquidus surfaces meet at a point. When all of the Alkemade lines are drawn, a ternary system is separated into compatibility triangles. If the three solid substances at the vertices of the triangles are not soluble, they represent the final equilibrium products of crystallization for any compostion within the triangle. When crystal solution occur between any of the three substances, the final solidified phases may be reduced in number to two or one. When the compatibility triangles in a ternary system (or compatibility tetrahedra in a four-component system) are known and the character of the crystal solution is known, then the final phase assemblage on cooling (under equilibrium conditions) is known. Also, the eutectics and/or peritectics involved can be identified on the phase diagram.

Congruent Melting Point. The congruent melting point is the temperature, at a defined pressure, in which a solid substance changes to a liquid of the same composition.

Incongruent Melting Point. The incongruent melting point is the temperature, at a specified pressure, in which a solid phase transforms into another solid phase plus a liquid, both of different composition than the original solid phase.

Eutectic. The eutectic is the lowest melting point composition, (an invariant point) located at the intersection of the two liquidus curves in a binary system, and three of the surfaces in a ternary system.

Isobar. The locus of all points of constant pressure in a phase diagram is the isobar.

Isopleth. An isopleth is a line or plane of constant composition in a phase diagram.

Isotherm. An isotherm is the locus of all points on the liquidus of constant temperature in a ternary system. Isotherms can be read like contour lines on a map and are used to show the configurations of the liquidus surface.

Liquidus. Any point where equilibrium between liquid and the primary phase exists is the liquidus. At temperatures above the liquidus (a line in binary systems, a surface in ternary systems), the system is completely liquid.

Solidus. The solidus is the locus of all temperature points, below which the system is completely solid. The solidus is either a line or surface, respectively, in binary and ternary systems. In between the liquidus and solidus is a region where solid and liquid are in equilibrium.

Polymorphism. Polymorphism is the term used to classify substances that can have different atomic structures and properties (polymorphs) even when their chemical composition remains constant. Phase transformations occur when either pressure or temperature is changed.

Solid Solution. Solid Solution can be described as a single phase, either crystalline or amorphous, which can vary in composition over a limited range without the formation of an additional phase.

Crystal Solution. Crystal solution, which is a species of solid solution, is a single crystalline phase which can vary in composition over a limited range without the formation of an additional phase. The amount and type of crystal solution, usually temperature and/or pressure dependent, will modify the properties of the phase significantly.

Interpretation of Phase Diagrams

The interpretation or use of a few simple phase diagrams will follow. For a more extensive discussion (see Levin, et al., 1964.) The interpretation of ternary diagrams is much more complicated than binary diagrams, however, many areas can be thought of as binary profiles through the ternary system.

Phase Diagram Showing Complete Crystal Solution Between the End Members, Phases A and B. If a melt of a composition X (50% A—50% B) starts to crystallize at temperature T_4 (X_1), the crystal-solution phase which crystallizes first will have the composition Y_1 (~90% A—10% B). As the melt is cooled to X_2 (T_3), the composition of the crystal-solution phase moves toward Y_2 (~75% A—25% B). At the point X_3 (T_2), the melt X is completely crystallized and the crystal solution has a composition of X (50% A—50% B) (Fig. 1).

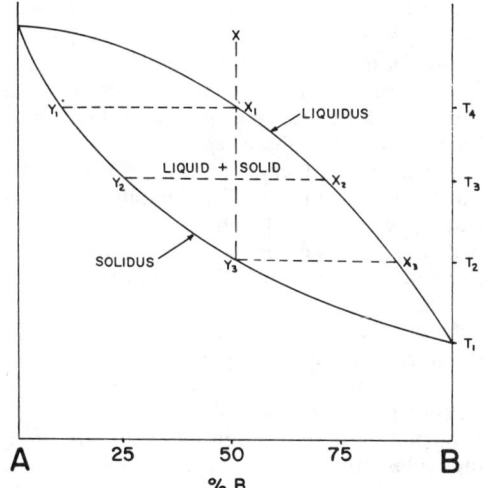

FIG. 1. Phase diagram showing complete crystal solution between the end members, phases A and B.

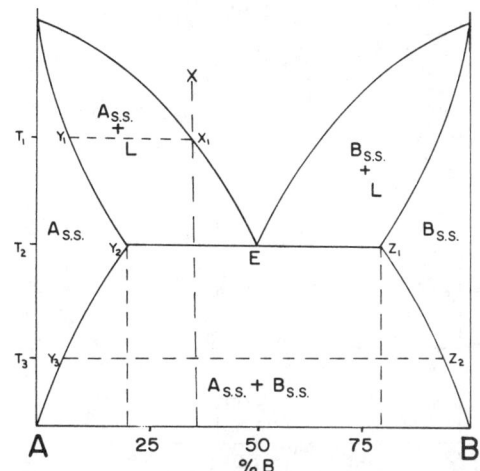

FIG. 2. Phase diagram showing limited crystal solutions of end members plus a eutectic.

Phase Diagram Showing Limited Crystal Solutions of End Members Plus a Eutectic. When a melt of a composition X starts to crystallize at temperature T_1, the composition of the phase will be a crystal solution of Y_1. From X_1 (temperature T_1) to E (temperature T_2) only the crystal solution phase of A will crystallize and the composition of the A solid-solution phase will be between Y_1 and Y_2. At T_2, the eutectic temperature, both phases A crystal solution Y_2 (~80% A − 20% B) and B crystal solution Z_1 (23% A − 77% B) will completely crystallize. On further cooling to temperature T_3, the composition of the crystal-solution phases are changing to $A_{ss} = Y_3$ and $B_{ss} = Z_2$ (Fig. 2).

Phase Diagram Showing Liquid Immiscibility. A melt with the composition X is a homogeneous liquid above temperature T_1. At T_1, two liquids form with the respective compositions L_1 and L_2. As the melt is cooled to temperature T_2, the liquids are changing in composition, respectively, to a and c. At this temperature L_1 is depleted and solid A starts to crystallize until the eutectic temperature T_3 is reached where A and B crystallize. Below this temperature no liquid exists in this system (Fig. 3).

Phase Diagram Containing Peritectic (an Invariant Point). A melt of composition X will start crystallizing at temperature T_1 (X_1). Only B will crystallize until temperature T_2, (the peritectic point) is reached. At this temperature the solid phase B will react with the liquid to form the compound $A_{25}B_{75}$. On further cooling only compound $A_{25}B_{75}$ will crystallize until the eutectic temperature T_3 where the remaining liquid will crystallize as the phase A and compound $A_{25}B_{75}$ (Fig. 4).

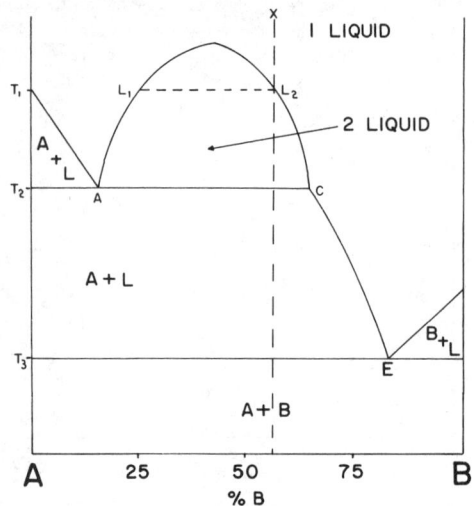

FIG. 3. Phase diagram showing liquid immiscibility.

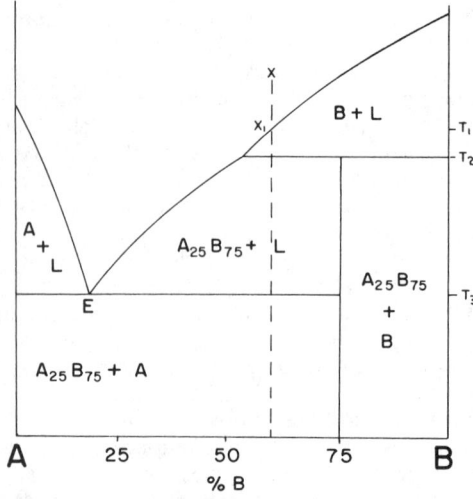

FIG. 4. Phase diagram showing peritectic.

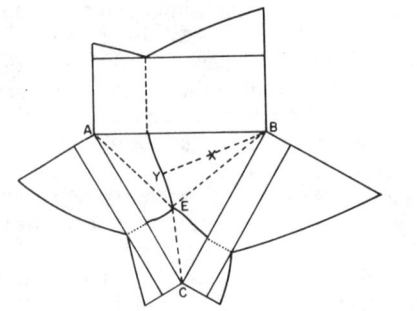

FIG. 5. Ternary diagram (from Masing, 1944).

The above diagrams are quite simple, and only the crystallization paths have been discussed. However, upon heating the specimens, the same paths are used in a reverse manner.

Ternary Diagram. Fig. 5 shows a ternary system in which the corresponding binary systems are simple mechanical mixtures of the components. They possess eutectics, but form neither solid solutions nor compounds. In addition, no solid solutions or compounds occur in this ternary system.

The triangle of concentration of a mechanical mixture with an eutectic is shown. The binary systems AB, BC, and CA are placed along the sides of the concentration triangle. Representation of this kind is often used instead of a perspective arrangement.

The path of crystallization can be demonstrated by considering an alloy of the composition X. The first material to crystallize will be phase B. As crystallization proceeds along the path XY, liquid becomes richer in A and C. At point Y both A and B will simultaneously crystallize and the liquid becomes richer in C. The path of crystallization will now be from Y toward C. When the eutectic E is reached, A, B, and C will crystallize simultaneously at constant temperature until the liquid is gone. (For more detailed discussion of complicated ternary diagrams consult Masing, 1944.)

Construction of Phase Diagrams

Sample Preparation. In phase equilibria studies it is necessary to use as high-purity raw materials as are available, to prevent the effects of contamination (e.g., liquidus suppression, inversion modification, crystal solution, etc.). Wherever possible, high-purity single crystals can be used. The sample should be formed into an intimate mixture which is generally accomplished by exact weighing, milling, and calcination. Other techniques such as hot pressing, isostatic pressing, sol formation, presintering, and premelting are used. The size of the sample used is critical in certain aspects of phase equilibria, particularly in the quench of crystal solution with a minimum of exsolution. The batch ingredients should be thoroughly characterized. Chemical analyses, microscopic, and X-ray examination of samples should be done before and after experiments.

Heat Treatment. In determining phase equilibria at high temperature, furnaces can be used such as gas combustion, electrical resistance, electric arc, electric induction, electron bombardment, high-intensity arc image, and solar furnaces. A typical type of induction-heating furnace used in high-temperature phase equilibrium work is described in an article by McNally, Peters, and Ribbe (1961). In selection of a furnace, it is imperative to consider the atmosphere to be maintained, which can be a neutral, oxidizing, reducing atmosphere, or a vacuum. In systems where the gas phase is important, experiments are done under specified pressures. When the work is done at pressures greater than one atmosphere, the material can be contained in a "bomb" or strong container which permits the study of the system under pressure, temperature, and composition changes. In systems where the amount of oxygen is subject to change with temperature and pressure, the work is done under controlled partial pressure of oxygen. Muan and Osborne show how the PO_2 can be controlled by using CO/CO_2 gas mixtures. Studies of systems where a gas phase is important are being done at the Geophysical Laboratories, Washington, D.C.; Pennsylvania State University; United States Geological Survey, Washington, D.C.; University of Chicago; California Institute of Technology; and Sheffield University.

Even in condensed systems, pressure is important when simulating environments in the earth's lower crust, mantle, or core. These systems are also important in the production of artificial crystals (e.g., diamonds). In this work, a resistance-heated or induction-heated furnace with either a hydraulic press or an anvil is used.

The rapid quenching of samples to prevent unmixing is important, and this can be done by air, water, liquid nitrogen, or molten-metal quenching.

Temperature Measurement. Temperature measurements below 1000°C are generally made with thermometers because of their low cost and simplicity of operation. Expansion and pressure type thermometers are used to about 500°C. Resistance thermometers which are either metallic (e.g., Pt, Ni, or Cu) or non-metallic are generally used to about 1000°C. From 1000 to 2200°C refractory thermocouples (e.g., Pt-Rh, Ir, W-Re (are used. Above 2000°C the optical pyrometers are usually employed. For special studies calorimetric, photographic, electron movement, or spectrographic techniques are used. The advantages and disadvantages of these techniques are discussed by Kingery (1959) and Campbell (1956). Whichever technique is used for temperature measurements, calibration of the device is important. This is accomplished by reference to secondary standards such as melting points of known materials. If black body conditions are not maintained, corrections should be made or approximated for viewing windows and for emissivity of the material and surroundings.

Melting temperatures of points along the liquidus can be determined by such methods as: melting in a suitable crucible or on a setter; observing the melting of a suspended rod or sample in the furnace; by *Differential Thermal Analysis* (q.v.); or by microscopic analysis of samples heated above and below the liquidus.

Petrographic and X-Ray Diffraction Techniques. Petrographic and x-ray diffraction data are extremely important for constructing phase diagrams. Not only are these techniques available for phase identification, but they can also be employed for determining solid solutions, liquid immiscibilities, eutectics, and peritectics. Both x-ray diffraction and petrographic data provide answers for locating solidus and subsolidus boundaries.

Two methods are used in the preparation of specimens for determining phase equilibria. *(1)* In the *direct method* the specimen is observed at the desired temperature. At the present time high-temperature furnaces are available for x-ray diffraction work and, to a limited temperature, for petrographic work. Some furnaces available for x-ray diffraction equipment are able to reach temperatures over 2000°C, but many are limited to approximately 1500°C. *(2)* In the *indirect method,* phase equilibrium data are collected from specimens which are heated to a particular temperature and then rapidly quenched in water, air or some other medium to preserve the assemblages of phases in equilibrium at that temperature (see Fig. 6).

In simple binary systems, maximum limits of solid solution can be delineated by changes in lattice parameters or refractive indices. To do this, specimens of different compositions are heated to a particular temperature, then quenched; accurate lattice parameters and refractive indices are collected and both are plotted against composition. Refractive indexes and lattice parameters normally increase or decrease with solid solution until a maximum amount of solid solution is obtained; then these values remain constant. Several experiments must be done at different temperatures in order to complete the solid-solution boundaries of the solidus and subsolidus.

In many systems, glasses form and are stable when quenched from temperatures above the solidus and liquidus. If glasses exist, their refractive indices can be related to composition; then the liquidus and solidus boundaries may be determined with refractive-index data.

One of the most valuable tools that can be used for working out phase diagrams is the petrographic microscope. Samples can be studied in the form of polished sections or thin sections. They can be prepared

PHASE EQUILIBRIA

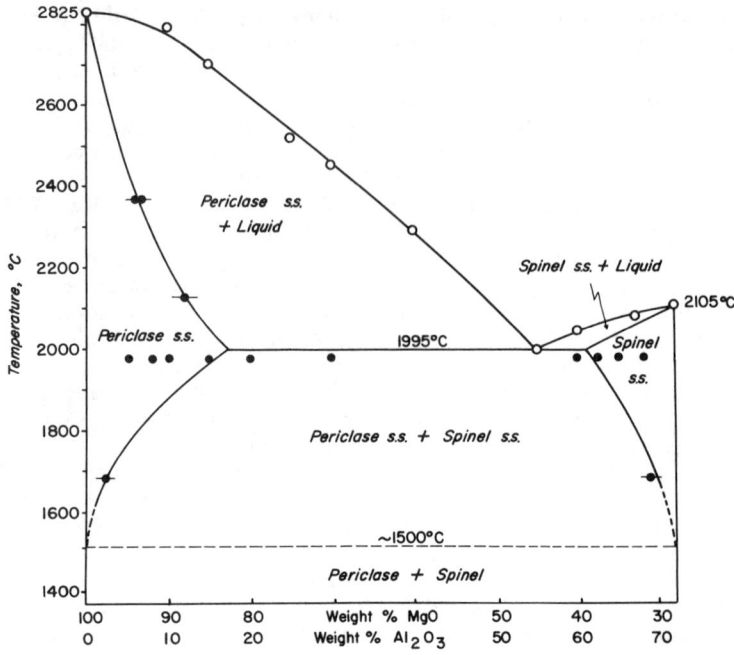

FIG. 6. Phase diagram for MgO–Al$_2$O$_3$ system.

from slowly cooled or quenched specimens; the different cooling rates produce different microstructures that give valuable clues for constructing phase diagrams. When the microstructures of a slowly cooled specimen can be described as a crystalline phase surrounding small blebs of a second phase, it infers that either the second phase was in solid solution in the first phase or that liquid immiscibility of the two phases existed before solidification. It is simple to determine which of these two took place by examining quenched specimens of the same compositions. If liquid immiscibility occurred, the small blebs would still be present; however, if solid solution occurred at elevated temperatures, the quenched specimen would not have any small blebs present. It is also possible to determine subsolidus boundaries by observing the change of microstructures of quenched specimens. A second intergranular phase will appear when the maximum solid solution boundary is passed.

Petrographic data can be useful in locating eutectics. The eutectic composition can be located by finding the composition which lies between compositions in which the first phase to crystallize are different. At the eutectic, the phases often will crystallize in a dendritic manner.

For more detail and better understanding of the construction of phase diagrams, see Alper (1970), Alper, McNally, Ribbe, and Doman (1962), and Brown and Hummel (1964).

The electron microscope and the electron microprobe are also valuable tools in studying phase equilibrium, as are differential thermal analysis and *thermogravimetric analysis* (q.v.).

Predicting an Approximate Phase Diagram for a System

Vorres (1963) has continued some of the work of Dietzel in using differences in ionic field strength, (field strength = $Z/(r_{cation} + r_{anion})^2$, the ratio of the ionic charge of the cation over the square of the sum of the ionic radii of the cation and anion), to predict the number of compounds that might form in simple binary systems. Systems which contain cations with large differences in field strength tend to have more compounds than those with smaller differences in field strength. Differences in field strength are useful in predicting whether liquid immiscibility will take place. Usually, immiscibility occurs when one ion has a high field strength (> 1.0) and when the difference in field strength of two ions is less than a certain minimum (in SiO$_2$ systems < 1.3). Ramberg (1954) has worked out several rules which are useful in predicting what compounds will form in systems that have more than one type of cation and anion.

The use of free-energy composition diagrams in the construction of phase diagrams is discussed by Darken and Gurry (1953). Methods for evaluating thermodynamic data from the position of conjugation lines in ternary systems can be obtained from a paper by Muan (1964).

In general, many of the older, oxide phase diagrams do not show enough crystal solubility at high temperature. Crystal solubility increases as the ions approach the same size and valence state. However, at high temperatures a large amount ($>$ 5 mol %) of ionic substitution can occur between ions which have very different radii ($>$ 40% difference). The solution of Mg^{2+} in CaO and Ca^{2+} in MgO are examples. A large amount of crystal solution can exist at high temperatures when ions with very different radii and sizes are substituted. The solution of Si^{4+} and Al^{3+} in periclase at high temperature are examples.

The position of the eutectic is very important. In simple systems with a binary eutectic, the eutectic composition will be closer to the lower-melting phases if there is a large difference in the melting points of the phases. If the eutectic is not close to the lower-melting-point phase, then there is either an intermediate compound or an area of liquid immiscibility.

The slope of the liquidus often suggests what is occurring in the system. Systems having a simple eutectic and no peritectic tend to have steeper slopes than systems having an intermediate, incongruently melting phase. The liquidus of systems with high crystal solubility appear to have a flatter slope than systems with only a minor amount of crystal solubility. If the lower-melting components form groups, they do not supress the liquidus as rapidly as components with similar melting points which do not form groups. A strong indication of a liquid immiscibility region is a flat liquidus.

If slowly solidified samples ($>$ 10 min) from a system have more phases (contain nonequilibrium phases) than predicted, an incongruently melting phase or a metastable phase may exist.

Since the microstructure is related to the phase diagram, it is important to analyze the microstructure as well as to obtain x-ray patterns, refractive indices, and chemical analyses. If the samples are cooled to slowly, the x-ray pattern might not show the existence of high-temperature crystal solution, whereas, the microstructure will be indicative of crystal solution and other high-temperature phenomena.

Application of Phase Diagrams

Phase diagrams are very useful in interpreting a material's geologic past and in the synthesis of new materials. They give insight into the following natural geologic problems:

1. The relationship between chemical composition, pressure, and temperature in affecting the stabilities of minerals. This gives the geologist insight into the physicochemical environment which may have existed when the minerals were formed naturally. This information is useful in determining the origin of rocks (particularly igneous and metamorphic) and economic mineral deposits.
2. The phase assemblages occurring under equilibrium conditions and how the phase assemblages might change with changing environment (temperature, chemical composition, pressure). Also, they are useful in determining what phase assemblages might occur under nonequilibrium conditions.
3. The investigation of the crystallization sequence of igneous minerals.
4. The predication of what phases may exist in the lower part of the crust, mantle, and core.
5. The origin of meteorites.

Phase diagrams are useful in industrial laboratories in the following ways:

1. Synthesizing materials.
2. Solution and precipitation hardening in metals and ceramics.
3. Sintering of metals and ceramics.
4. Crystallization of metals and ceramics.
5. Predicting how changing chemical and physical environments will affect metals and ceramics.
6. Extraction of metals from their ore; e.g., slags and temperature needed.

A. M. ALPER
R. C. DOMAN
R. N. MCNALLY

References

Alper, A.M., 1970, "Theory, Principles, and Techniques of Phase Diagrams," in "Phase Diagrams: Materials Science and Technology," Vol. 1, New York, Academic Press.

Alper, A. M., 1970, "The Use of Phase Diagrams in Metal, Refractory, Ceramic, and Cement Technology," in "Phase Diagrams: Materials Science and Technology," Vol. 2, New York, Academic Press.

Alper, A. M., 1970, "The Use of Phase Diagrams in Electronic Materials and Glass Technology," in "Phase Diagrams: Materials Science and Technology," Vol. 3, New York, Academic Press.

Alper, A. M., McNally, R. N., Ribbe, P., and Doman, R. C., 1962, "The systems MgO-

PHASE EQUILIBRIA

MgAl$_2$O$_4$," *J. Am. Ceram. Soc.*, **45**(6), 263–268.

Bowen, N. L., 1928, "Evolution of Igneous Rocks," Princeton University Press, 332pp.

Brown, J. J., and Hummel F. A., 1964, "Phase equilibria and manganese—Activated luminescence in the system CdO-P$_2$O$_5$, and Zr$_2$P$_2$O$_7$-Cd$_2$P$_2$O$_7$, summary for the system ZnO-CdO-$_2$O$_5$," *J. Electrochem. Soc.*, **111** (9), 1052–1057.

*Campbell, I. E., 1956, "High Temperature Technology," New York, John Wiley & Sons, 526pp.

Doman, R. C., Barr, J. B., McNally, R. N., and Alper, A. M., 1963, "Phase equilibria in the system CaO-MgO," *J. Am. Ceram. Soc.*, **47**(7), 313–316.

*Darken, L. S., and Gurry, R. W., 1953, "Physical Chemistry of Metals," New York, McGraw-Hill Book Co., pp. 326–342.

Dietzel, A., 1942, *Zeits. Elektrochem*, **48**, 9–23.

Findley, A., Campbell, A. N., and Smith, N. O., 1951, "The Phase Rule and its Applications," Ninth edition, Dover Publ., 494pp.

Gibbs, J. W., 1928, "The Collected Works of J. Willard Gibbs," London, Longmans, Green and Co., Vol. 1, pp. 54–371.

*Kingery, W. D., 1959, "Property Measurements at High Temperatures" New York, John Wiley & Sons, 415pp.

Kracek, F. C., and Clark, S. P., Jr., 1966, "Melting and transformation points in oxide and silicate systems at low pressure," in (Clark, S. P., editor), "Handbook of Physical Constants," *Geol. Soc. Am. Mem.*, **97**, Sec. 13, 301–322.

*Levin, E. M., Robbins, C. R., and McMurdie, H. F., 1964, "Phase Diagrams for Ceramists," Am. Ceram. Soc. 601pp.

Masing, G., 1944, "Ternary Systems," New York, Dover Publ. 173pp.

McNally, R. N., Peters, F. I., and Ribbe, P. H., 1961, "Laboratory furnace for studies in controlled atmospheres; Melting points of MgO in a N$_2$ atmosphere and of Cr$_2$O$_3$ and air atmosphere," *J. Am. Ceram. Soc.*, **44**(10), 491–493.

Muan, A., 1964, "Proc. 5th Int. Symposium on the Reactivity of Solids," Munich, Germany.

*Muan, A., and Osborne, E. F., 1965, "Phase Equilibria Among Oxides in Steelmaking," sponsored by Am. Iron and Steel Instit., Reading, Mass., Addison-Wesley Publishing Co.

Ramberg, H., 1954, *Am. Min.*, **39**, 256.

Roy, R., 1963, "Crystal Chemistry in Research on Ionic Solids," Physics and Chemistry of Ceramics, New York, Gordon & Breach Science Publishing.

Vorres, K. S., 1963, "Estimation of phase diagrams for systems of two binary compounds," *J. Am. Ceram. Soc.*, **46**(187), p 410.

Wahlstrom, E. F., 1950, "Introduction to Theoretical Igneous Petrology," New York, John Wiley & Sons, pp 9–74.

Wetmore, F. E. W., and LeRoy, D. J., 1951, "Principles of Phase Equilibria," New York, McGraw-Hill Book Co., 200pp.

*Additional bibliographic references may be found in this work.

Cross-references: *Crystal Chemistry; Differential Thermal Analysis; Electron Microscopy; Free Energy and Free Enthalpy; Mineral Genesis: Equilibrium; Solid Solution; Solubility; Thermogravimetric Analysis; X-Ray Diffraction Analysis.* Vol. IVB: *Ceramic Technology; Crystal Growth; Electron Probe Microanalysis; Industrial Laboratory Geology; Mineral Properties; Optical Mineralogy; Polymorphism; Refractive Index; Refractory Minerals.*

PHASE RULE—See FREE ENERGY; also Vol. V, FREE ENERGY AND THE GIBBS PHASE RULE

pH–Eh DIAGRAMS

For many years geologists have discussed the effects of acids, bases, oxidants, and reductants on the formation and alteration of rocks and minerals. Most minerals hydrolyze to some extent and are thus affected by acids or bases. Moreover, many minerals contain elements that may acquire more than one oxidation state under natural conditions and are thus affected by oxidation or reduction processes accompanying geologic changes. Among these elements are iron, carbon, copper, sulfur, vanadium, chromium, manganese, arsenic, iodine, mercury, and uranium. A convenient way to summarize many of the relations deduced from geological and chemical studies is by means of pH-Eh diagrams.

Definitions and Mathematical Relations

Activity. The use of *activities* of substances in solution, instead of their *concentrations*, facilitates the exact computation of chemical relationships. Activity may be defined as an "ideal" concentration. For a number of reasons the activity of a solute is not in general the same as the concentration of a solute in a given solution. The ideal case would require the molecules or ions of the solute to have no interactions with the solvent that differ from those with each other. In other words, a solute particle would not attract or repel solvent particles either more or less than it does other solute particles. A solution of methyl alcohol in ethyl alcohol nearly achieves this ideal, but solutions of ionic salts in water deviate greatly from it.

Because the mathematics of ideal solutions are very convenient, it is customary to convert the actual concentrations of non-ideal solutions to activities. For dilute aqueous solutions containing only electrolytes of known strengths the correction factor or activity coefficient can be calculated accurately from theory. For more concentrated solutions it must be determined experimentally. In geochemical applications the solutions are often dilute enough that concentrations do not differ greatly from activities and consequently serious errors do not result from the use of concentrations in place of activities.

The concentrations or activities of pure substances do not vary so long as no impurities are added. For convenience, all such substances are assigned an activity of 1. Thus, pure water or pure magnetite has an activity of 1.

pH. The pH of a solution is defined as the negative logarithm (base 10) of the activity of hydrogen ion. Thus, if the activity of H^+ is $1 = 10^0$, the pH $= -\log_{10} 10^0 = 0$; if $(H^+)^* = 10$, the pH $= -\log 10 = -1$, and if $(H^+) = 0.001 = 10^{-3}$, the pH $= 3$.

The dissociation product of water is 10^{-14} at 25°C and 1 atmosphere pressure. This means that the activity of H^+ multiplied by the activity of OH^- and divided by the activity of H_2O is 10^{-14}; i.e.,

$$\frac{(H^+)(OH^-)}{(H_2O)} = 10^{-14}$$

(The concentration, and thus the activity, of water decreases as the concentration of solute increases.) For the water in most lakes and streams the concentration of dissolved substances is low, and the activity of water can be taken as unity without measurable error. For these waters $(H^+)(OH^-) = 10^{-14}$. For seawater and saline groundwater this approximation sometimes leads to appreciable error.

The neutral point is defined as the circumstance in which the activity of H^+ equals that of OH^-. For dilute aqueous solutions at 25°C and 1 atm, this means a pH of 7. The pH of the neutral point changes slightly with temperature, pressure and salinity.

Eh. The Eh of a solution is a measure of the oxidizing or reducing tendency of the solution. Because oxidation or reduction is fundamentally an electrical process, this tendency is measured by an electrical potential. All measurements are referred to the standard hydrogen electrode, whose potential is taken as zero.

The standard hydrogen potential of zero corresponds to hydrogen gas at 25°C and one atmosphere pressure in equilibrium with an aqueous solution containing hydrogen ion at an activity of 1.0 (the solution, of course, is also at 25°C and 1 atm.). The potential varies with the pressure of the H_2 gas and with the activity of H^+ in accordance with the equation

* Parentheses indicate activity.

$$\text{Eh (in volts)} = \frac{0.059}{2} \log \frac{(H^+)^2}{P_{H_2}} \quad (1)$$

where (H^+) denotes the activity of hydrogen ion and P_{H_2} the pressure of hydrogen gas. The reaction $H_2 = 2H^+ + 2e$ corresponds to this equation.

In Fig. 1 an electrolytic cell is shown with a standard hydrogen electrode on the left and an inert Pt electrode on the right. Hydrogen gas moves into the left electrode and bubbles over the Pt, which is in contact with the solution. If conditions are varied from those shown, the tendency of the Pt in the H_2 electrode to gain electrons is increased as P_{H_2} is increased, and the tendency for the Pt to lose electrons is increased as the pH of the solution is decreased. Thus for fixed values of P_{H_2} and pH this tendency is determined; for $P_{H_2} = 1$ and pH $= 0$, this tendency is arbitrarily called zero. If the inert Pt electrode is inserted in the same solution near the H_2 electrode, the tendency for it to gain or lose electrons will be the same (if equilibrium prevails) as in the H_2 electrode, due to the presence of dissolved H_2 and H^+ ions. Insertion of the Pt electrode in another solution, which is connected electrically and physically to the first by a solution contained in a small bent tube as shown, makes possible the measurement in volts of the tendency of the Pt to lose or gain electrons from whatever substances are present in the second solution as compared to that in the first. (It is recognized that junction potentials may exist between solutions, but here they are assumed to be zero.)

Eq. (1) may be rewritten as

$$\text{Eh} = -0.059 \text{ pH} - \frac{0.059}{2} \log P_{H_2} \quad (2)$$

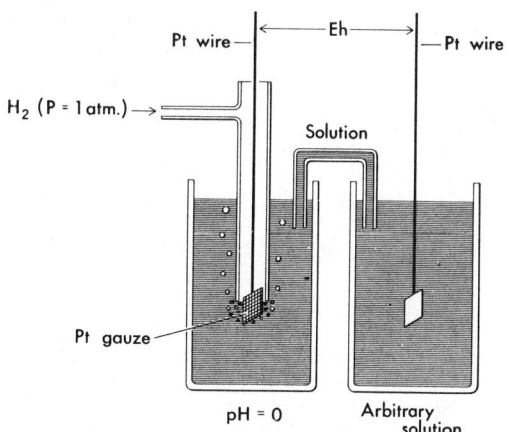

Fig. 1. An electrolytic cell showing a standard hydrogen electrode on the left and a Pt electrode in an arbitrarily chosen solution on the right.

pH-Eh DIAGRAMS

From this equation it is apparent that, if pH is held constant, and reducing substances are added to lower the Eh of a solution, then the equilibrium partial pressure of H_2 must increase. (It is assumed throughout that the total gas pressure is 1 atm., gases not named being inert and decreasing in partial pressure as P_{H_2} increases.) For example, suppose we have a solution of pH = 7 and Eh = 0. Then the logarithm of the equilibrium pressure of H_2 is

$$\log P_{H_2} = -\frac{2}{0.059} \text{Eh} - 2\text{pH}$$
$$= 0 - 2 \times 7 = -14$$

That is, $P_{H_2} = 10^{-14}$. If the Eh is now reduced to -0.118 V, $\log P_{H_2} = -2/0.059 \times (-0.118) - 14 = -10$, or $P_{H_2} = 10^{-10}$. Lowering of Eh to -0.413 V gives $P_{H_2} = 1$. Attempts to lower the Eh further result merely in the production of hydrogen at 1 atm. pressure from the reduction of water because the total gas pressure at the earth's surface may not exceed about 1 atm. Reductants then cause active bubbling of H_2 into the overlying atmosphere at 1 atm pressure. (This picture is complicated, however, by the non-equilibrium phenomenon of overvoltage, which will not be considered here because we are concerned only with equilibrium conditions.)

Extension of these considerations to all other pH's leads to the conclusion that for each pH there is an Eh corresponding to evolution of hydrogen from water. This can be represented mathematically by setting $P_{H_2} = 1$ in Eq. (2):

$$\text{Eh} = -0.059 \text{ pH} - 0.0295 \log P_{H_2}$$
$$= -0.059 \text{ pH} - 0.0295 \log (1)$$
$$= -0.059 \text{ pH} \qquad (3)$$

If we now plot Eh against pH, this equation appears as a straight line as shown in Fig. 2. This line represents the lower stability limit of water.

The upper stability limit of water is related to the reaction

$$2 H_2O = O_2 + 4 H^+ + 4e \qquad (A)$$

The corresponding Eh equation is

$$\text{Eh} = E° + \frac{0.059}{4} \log \frac{(P_{O_2})(H^+)^4}{(H_2O)^2} \qquad (4)$$

This is of the same form as equation (1), but with $E°$ (the standard potential) added. This can be seen more clearly by writing a generalized reaction,

$$nA + mB + \ldots = rG + sH + \ldots + xe \qquad (B)$$

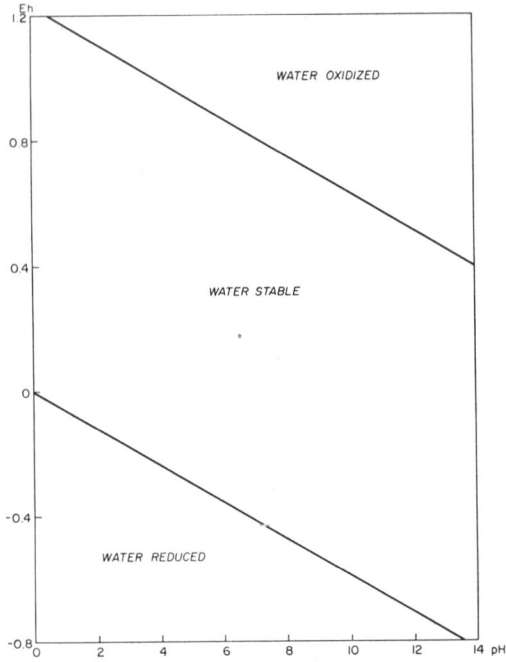

FIG. 2. pH-Eh diagram showing the stability field of water.

where $n, m, \ldots$, and $r, s, \ldots$, represent the number of moles of reactants or products; A, B..... are the reactants; G, H,..... are the products; x the number of electrons; and e the electrons. Corresponding to this is the equation,

$$\text{Eh} = E° + \frac{0.059}{x} \log \frac{(G)^r (H)^s \ldots}{(A)^n (B)^m \ldots} \qquad (5)$$

In the case of hydrogen gas being oxidized to H^+, $E° = 0$. For the oxidation of water to O_2 and H^+, $E°$ is found by suitable experiments and calculations to be 1.23 V.

Equation (4) may be rewritten as follows:

$$\text{Eh} = 1.23 + \frac{0.059}{4} \log \frac{P_{O_2}}{(H_2O)^2}$$
$$+ \frac{4(0.059)}{4} \log (H^+)$$
$$= 1.23 + 0.015 \log \frac{P_{O_2}}{(H_2O)^2} - 0.059 \text{ pH}$$

If the solution is dilute, $(H_2O) = 1$, and if we wish to consider the breakdown of water at room pressure, we set $P_{O_2} = 1$. Then Eh = 1.23 -0.059 pH. This gives a line parallel to that of equation (3) and lying 1.23 V above it as shown in Fig. 2. This is the upper stability limit of water. Since water is almost universally present in geological problems, nearly all surficial geological environments must lie between the two lines shown in this figure.

It is important to note which quantities determine the line slopes, i.e., the coefficients of the pH, in the two cases discussed so far. The slope equals the coefficient 0.059 times the negative of the coefficient of H^+ in reaction (B), (H^+ must for this purpose be considered always on the right hand side of the reaction, with a negative coefficient in some cases), divided by the coefficient, x, of e in reaction (B). The slope may have various values, either positive or negative, depending on the reaction.

Evaluation of Constants

Calculation of $E°$. In order to construct pH-Eh diagrams one must know how to obtain appropriate values of $E°$ for a large number of reactions. Some of these can be found in textbooks, but the most general and satisfactory way is to calculate them from the standard free energy of the reaction. A full understanding of the meaning of free energy and its usefulness can be obtained only by extended study. Space does not permit an explanation of this meaning or of the means of determining the free energy. We are, however, able to look up standard free energies in tables and can use them in simple mathematical formulas. The first of these formulas states that

$$\Delta F° = +23.060\, E°\, x \qquad (6)$$

where $\Delta F°$ is the standard free energy change in kcal/mole of an oxidation reaction (i.e., with the electrons written on the right of the equal sign), x is the number of electrons involved in the oxidation, and $E°$ is the standard potential.

The problem of calculating $E°$ has now become one of finding $\Delta F°$ for the reaction. This is done simply by adding up the standard free energies of formation of the products (taking appropriate account of the relative number of moles), and subtracting from this total the sum of the standard free energies of formation of the reactants. Referring again to reaction (B) this statement may be put in mathematical form as follows:

$$\Delta F°_{reaction} = r\Delta F°_{fm,G} + s\Delta F°_{fm,H} \cdots$$
$$- n\Delta F°_{fm,A} - m\Delta F°_{fm,B} - \cdots$$

where $\Delta F°_{fm,G}$ refers to the standard free energy of formation of one mole of G, etc. The standard free energy of formation of electrons is zero.

As an example we may calculate the $E°$ for reaction (A). By using one of the references listed at the end of this discussion, we find that the standard free energies of formation of everything in the reaction are zero except for water; its $\Delta F°_{fm}$ is -56.69 kcal/mole. The standard free energy of reaction is, therefore,

$$\Delta F°_{reaction} = 0 + 0 - 2(-56.69)$$
$$= 113.38 \text{ kcal}$$

From equation (6) we now get $E° = 113.38/23.060 \times 4 = 1.23$ V, as stated previously.

A more complicated case of geologic interest is presented by the reaction,

$$2\, Fe_3O_4 + H_2O = 3\, Fe_2O_3 + 2\, H^+ + 2e$$
magnetite $\qquad\qquad$ hematite

The tables list -242.4 kcal/mole for magnetite and -177.1 kcal/mole for hematite. (Accurate and consistent data should, of course, be used. Except for more recent better data, the values listed by Garrels and Christ (Table 1) (1965), Latimer (1952), and Rossini et al. (1952) are quite accurate and consistent internally as well as with each other. Those tables in the "Handbook of Chemistry and Physics" (Hodgman, 1959 and later) that are reproduced from Latimer (1952) or from Rossini et al. (1952) are therefore good.) The standard free energy for this reaction is then $\Delta F° = 3(-177.1) + 2\,(0) - 2(-242.4) - (-56.69) = 10.2$ kcal and $E° = 10.2/23.06 \times 2 = 0.221$ V. This value of $E°$ tells us that magnetite is indicative of very reducing conditions during its formation, and that hematite indicates weakly to strongly oxidizing conditions.

Calculation of Equilibrium Constants. One further expression of great use in constructing

TABLE 1. SELECTED DATA FROM GARRELS AND CHRIST[a]

Substance	$\Delta F°$, (kcal/mole)
H_2	0
H^+	0
O_2	0
H_2O	-56.69
e	0
Cu	0
Cu^+	12.0
Cu^{++}	15.53
Cu_2O	-34.98
CuO	-30.4
$Cu(OH)_2$	-85.3
CH_4	-12.14
C (graphite)	0
CO_2	-94.26

[a] Garrels and Christ, 1965, pp. 411–418.

pH-Eh DIAGRAMS

pH-Eh diagrams is that for the equilibrium constant. These constants are needed when reactions are met that involve hydrogen ions, but do not involve electrons, that is, there is no oxidation or reduction.

The equilibrium constant for a given reaction may be calculated from the standard free energy of the reaction. This is done by means of the equation

$$\Delta F° = -1.364 \log_{10} K \qquad (7)$$

where K is the equilibrium constant at the standard conditions of T and P.

For instance, for the reaction

$$H_2O = H^+ + OH^-$$

we read from tables that $\Delta F°_{fm,OH^-} = -37.60$ kcal/mole. This gives $\Delta F°$ reaction $= -37.60 - (-56.69) = 19.09$ kcal. From equation (7) we then get $\log K = 19.09/-1.364 = -14.00$, i.e., $K = (H^+)(OH^-)/(H_2O) = 10^{-14}$ as stated above.

Applications

The Fe–H_2O System. Now that the mathematical background has been explained, it is possible to consider an example of geologic interest. In nature it is found that in the presence of water, iron is either in the form of an oxide or hydroxide or is dissolved in the water. This behavior of iron can be well understood by reference to a pH-Eh diagram. Probably most of the iron in solution is present as Fe^{2+} or Fe^{3+} ions, and these are the only species considered in this discussion. However, the total solubility of iron is equal to the sum of the concentrations of all species present, such as $Fe(OH)^{2+}$, $Fe(OH)_2^+$, $HFeO_2^-$, and others in addition to Fe^{2+} and Fe^{3+}.

To construct the diagram, all known oxidations and pH-dependent precipitations are considered. The first step consists of listing all the iron compounds in the system in order of increasing state of oxidation. Because the only other elements present are hydrogen and oxygen, the only possible compounds are oxides and hydroxides. The list then reads as follows:

Valence of Fe	Solid substances
0	Fe
2	FeO
	$Fe(OH)_2$
2 and 3	Fe_3O_4
3	Fe_2O_3
	$Fe(OH)_3$
	FeOOH

In case of doubt concerning what compounds exist, a table giving free energy values may be scanned to locate all those for which data are available. The tables referred to above include most minerals except silicates; complex basic halides, sulfates, and other salts; and complex sulfides, sulfarsenides, sulfantimonides.

The second step is to write balanced reactions for the oxidation of the most reduced substance (in this case Fe) to each of the other compounds. In the present instance this means six reactions. In doing this, care must be exercised not to use oxygen or hydrogen gas or hydroxide ion, but only water, hydrogen ion, and electrons, in order to balance the reaction. The six reactions are as follows:

(a) $Fe + H_2O = FeO + 2H^+ + 2e$
$E° = -0.037$ V

(b) $Fe + 2H_2O = Fe(OH)_2 + 2H^+ + 2e$
$E° = -0.047$ V

(c) $3Fe + 4H_2O = Fe_3O_4 + 8H^+ + 8e$
$E° = -0.084$ V

(d) $2Fe + 3H_2O = Fe_2O_3 + 6H^+ + 6e$
$E° = -0.051$ V

(e) $Fe + 3H_2O = Fe(OH)_3 + 3H^+ + 3e$
$E° = +0.059$ V

(f) $Fe + 2H_2O = FeOOH + 3H^+ + 3e$
$E° = -0.052$ V

The third step consists of calculating $E°$'s for each of these reactions. The values thus obtained are shown beside each reaction. We are now in a position to write pH-Eh equations analogous to Eq. (5) for each reaction. These are listed below:

(a') $Eh = -0.037 - 0.059 pH$

(b') $Eh = -0.047 - 0.059 pH$

(c') $Eh = -0.084 - 0.059 pH$

(d') $Eh = -0.051 - 0.059 pH$

(e') $Eh = +0.059 - 0.059 pH$

(f') $Eh = -0.052 - 0.059 pH$

Note that equations (a) to (f) are derived from Eq. (5) by substituting activities of 1 for all solids and for water. If these equations are plotted on an pH-Eh diagram they form a series of parallel lines. It is suggested that the reader actually plot these equations to see their relative positions. One of these, namely (c'), lies lower than all the rest. This means that native iron will oxidize to magnetite before it will oxidize to anything else. In other words, only one of these lines represents a stable association—that showing oxidation of iron to magnetite—and the rest metastable or unstable because iron metal will already have disappeared before the Eh can be raised high enough to reach the other lines. Another way of saying this is that reactions a, b, d, e, and f will not take place because iron metal will not be present under conditions in which the reactions can occur. (Exceptions can occur only if the Eh is raised faster than the reactions can take place. This is possible in the laboratory, but generally not in nature.) We then keep the line for equation (c) on the diagram and delete the rest.

In some cases the lines will not all have the same slope and may intersect within the pH range of interest (between pH 0 and 14). In this case the most reduced substance would oxidize first to a different compound at high pH than at low pH. Part of each line, that lying below the other line, is then retained on the equilibrium diagram, and the higher portions deleted. A third line corresponding to the equilibrium between the two different somewhat oxidized substances must then be drawn, starting from the point of intersection of the two lowest lines. The equation for this third line may be obtained by writing the balanced reaction between the two oxidized substances, being careful not to use O_2, H_2, or OH^-, and calculating the pH-Eh equation or equilibrium constant. (This equation may be derived directly from the two previous equations for the lowest lines and consequently the corresponding line must pass through their point of intersection, if the arithmetic is done correctly.)

It is important to note that this line for (c′) lies below the stability range for water. This means that water, one of the reactants in (c) is not stable under conditions where Fe is stable and, consequently, the line (c′) likewise does not represent a stable association. If water and iron metal are placed together, the iron will be oxidized according to reaction (c)—i.e. the most easily produced iron oxide will be formed—and water will be reduced according to the reaction $2e + 2H^+ = H_2$. The net reaction is then

$$3Fe + 4H_2O = Fe_3O_4 + 4H_2$$

This will proceed until either all of the iron is oxidized or all of the water reduced. At the earth's surface the former will prevail and iron will soon be oxidized.

Because lines lying outside the stability field of water are not usually of geologic interest, they are omitted from the diagram. Most of the corresponding reactions involve water, and, consequently, the assemblages are not stable.

Completion of the diagram requires that we next consider possible oxidations to higher valences to ascertain lines representing other stable assemblages. In the event that many oxidation states are possible the procedure must be repeated many times until no more assemblages lying within the stability boundaries of water can be found. Having determined that iron will oxidize to magnetite, we follow this procedure in the fourth step to find out to what the magnetite will oxidize. To do this we proceed as in step two. Magnetite and water are taken as reactants and balanced reactions are written for the oxidation to each more oxidized iron compound. In this case three reactions are found:

(g) $2Fe_3O_4 + H_2O = 3Fe_2O_3 + 2H^+ + 2e$
$E° = 0.221$ V

(h) $Fe_3O_4 + 5H_2O = 3Fe(OH)_3 + H^+ + e$
$E° = 1.21$ V

(i) $Fe_3O_4 + 2H_2O = 3FeOOH + H^+ + e$
$E° = 0.208$ V

The next step is to calculate $E°$'s for each reaction. Considerations analogous to those already discussed show that magnetite will oxidize to FeOOH (goethite) before it will to Fe_2O_3 (hematite) or $Fe(OH)_3$. This is clear from a consideration of Eq. (g′), (h′), and (i′). In this case water will not itself oxidize the magnetite; some other oxidant, such as O_2, must be added to accomplish this.

(g′) Eh = 0.221 −0.059pH
(h′) Eh = 1.21 −0.059pH
(i′) Eh = 0.208 −0.059pH

The only possible oxidation for goethite involves the oxidation of the water which forms part of the structure; the iron is already in its highest oxidation state. This reaction is

$$2FeOOH = Fe_2O_3 + 2H^+ + \tfrac{1}{2}O_2 + 2e$$
$E° = 1.30$ V

The corresponding pH-Eh line lies somewhat above the water boundary, and so is not shown.

From the preceding discussion it is clear that the stability ranges of water and magnetite partly overlap. The overlapped portion may be shown on a pH-Eh diagram as a stability field for magnetite in the presence of water. This is done in Fig. 3. It is also clear that the stability ranges for goethite and water partly overlap. This area is shown as a stability field for goethite in Fig. 3.

Sometimes hematite, Fe_2O_3, forms instead of geothite, however, and because the rate of the reaction, $Fe_2O_3 + H_2O = 2FeOOH$, is very slow, the hematite persists. This production of hematite instead of goethite is possible partly because of the small difference of the $E°$'s of reactions (g) and (i). Hematite is favored by higher temperatures and dry climates.

From this completed pH-Eh diagram a number of interesting conclusions can be drawn. Because oceans, streams, lakes, and near-surface waters filling pore spaces in rocks are more or less in contact with air, we would ex-

pH-Eh DIAGRAMS

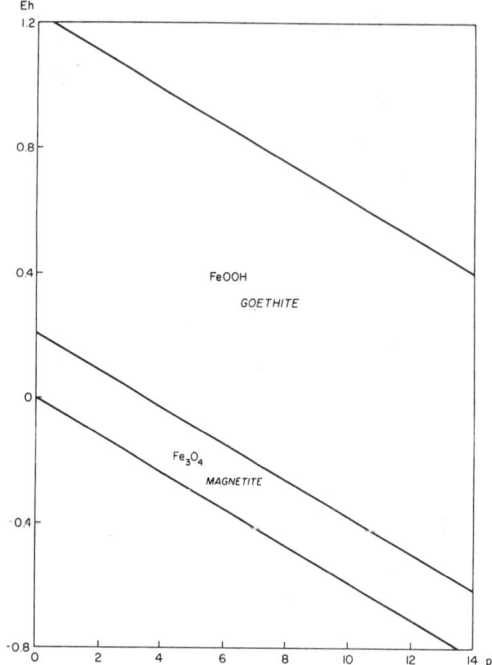

FIG. 3. pH-Eh diagram showing the stability fields for goethite and magnetite in the presence of water.

pect the oxygen of the air to maintain these waters in a rather strongly oxidizing condition. We therefore expect to find goethite or hematite in these environments. This is generally found to be true, and chemically accounts for the brown or red color of many soils, sands and other surface materials.

At greater depth out of contact with the air the environment is almost always more reducing. This is partly because of buried organic material in sediments, and partly because of the pressure and temperature conditions of formation of metamorphic and igneous rocks. In keeping with the reducing environment it is found that these rocks are usually black, gray, or white and contain little or no goethite or hematite.

As a third example, it is clear why iron so readily forms rust (chiefly goethite) on exposure to water and air.

An interesting modification to this diagram can be made by considering the equilibria with ferrous and ferric ions in the solution. Perhaps the easiest of these to understand is that between Fe^{3+} and goethite. This may be represented by the reaction,

$$Fe^{3+} + 2H_2O = FeOOH + 3H^+$$

Because this does not involve electrons, it must be considered in terms of its equilibrium con-

stant. This may be obtained by the method described previously and is

$$6.30 = (H^+)^3/(Fe^{3+}) \qquad (8)$$

This requires that at each pH there will be a corresponding activity of ferric ions. A series of vertical lines, each one of which corresponds to a different activity of ferric ion may, therefore, be drawn in the goethite field.

If we take a large enough amount of goethite and add to it some dilute acid, the goethite will dissolve until the activity of ferric ion rises high enough and that of H^+ falls low enough so that this equilibrium constant is satisfied. If the final solution has a pH of 1.73, one can quickly determine that $(Fe^{3+}) = 10^{-6}$ mole/l. This concentration of Fe^{3+} is so slight that the amount of H^+ consumed during the reaction would have no measurable effect on the initial pH. This procedure gives us a quantitative measure of the solubility of goethite and its dependence on pH.

Another set of lines corresponding to the activity of ferrous ion in the goethite field may also be constructed. This is done by use of the reaction, $Fe^{2+} + 2H_2O = FeOOH + 3H^+ + e$, and the equation,

$$Eh = 0.723 - 0.178\, pH - 0.059 \log (Fe^{2+}) \qquad (9)$$

For each activity of ferrous ion there is a corresponding line of slope -0.178 V per pH unit.

The usefulness of this relation appears if we take sufficient goethite, a little dilute acid, and a reducing agent capable of reducing ferric ion to ferrous but not goethite to magnetite. In this case (Fe^{2+}), pH and Eh will all increase until Eq. (9) is satisfied. This relationship tell us that the solubility depends not only on pH, but also on Eh.

Clearly it is important in this connection to know under what conditions ferric ion dominates over ferrous in solution, and vice versa. This is simply done by writing the reaction,

$$Fe^{2+} = Fe^{3+} + e$$

and the corresponding equation,

$$Eh = 0.771 + 0.059 \log \frac{(Fe^{3+})}{(Fe^{2+})}$$

If $(Fe^{3+}) = (Fe^{2+})$, $Eh = 0.771$ V. If (Fe^{3+}) dominates over (Fe^{2+}) in solution $Eh > 0.771$ V and if (Fe^{2+}) dominates over (Fe^{3+}) in solution $Eh < 0.771$ V. It is convenient to draw a horizontal line on the diagram separating these two regions of dominance. This does not imply that Fe^{3+} does not exist below $Eh = 0.771$ V or that Fe^{2+} does not exist above $Eh = 0.771$ V. Both ions are present throughout the entire stability range of water.

Because magnetite exists stably only at low Eh, we need consider only the equilibrium with Fe^{2+}. This is given by the reaction $3Fe^{2+} + 4H_2O = Fe_3O_4 + 8H^+ + 2e$, and the equation,

$$Eh = 0.962 - 0.236 \text{ pH} - 0.0885 \log (Fe^{2+}) \quad (10)$$

If we decide that a solubility of 10^{-6} moles/l is geologically significant (based on measurements of iron in groundwater, for instance), we may now draw a pH-Eh diagram showing those conditions under which goethite and magnetite will dissolve. This is done by drawing the lines on the diagram for which (Fe^{3+}) and (Fe^{2+}) equal 10^{-6} mole/l. This is done in Fig. 4. All solutions more acid than this will dissolve goethite and magnetite to at least the extent of 10^{-6} mole/l. Therefore, these areas are labeled only as Fe_3^+ or Fe^{2+} in solution. Areas at higher pH are those in which these minerals are insoluble and are shown the same as in Fig. 3.

Substitution of these values into Eqs. (8), (9), and (10) yields

$$(H^+) = 10^{-1.73} \quad (8')$$

$$Eh = 0.369 - 0.178 \text{ pH} \quad (9')$$

and

$$Eh = 0.431 - 0.236 \text{ pH} \quad (10')$$

respectively. The corresponding lines are plotted in Fig. 4.

The lines corresponding to equations (8') and (9') meet the line at $Eh = 0.771$ V at a point, because only two of the three equations are independent. Likewise the lines for Eqs. (i'), (9') and (10') meet at a point. It should be noted that a different choice of activity of (Fe^{3+}) and (Fe^{2+}) will lead to lines parallel to those for Eqs. (8'), (9'), and (10'), and a longer or shorter line at $Eh = 0.771$ V.

This diagram is of use in understanding the behavior of iron in various natural environments. In most surface waters (i.e., lakes, streams and oceans) the Eh is relatively high, but not as high as 0.771 V; pH values range from about 5 to about 8.5. Thus goethite is the stable phase and is quite insoluble; it is not surprising to find that iron released from silicate rocks during weathering precipitates as goethite and remains in this form. The reducing environments at depth, which were mentioned previously, usually have a high enough pH to prevent the dissolution of magnetite or goethite.

A few environments have low enough pH to cause active leaching of iron. These include some stagnant waters in acid bogs and deep lake basins. In these cases the water is both reducing and acid due to the decay of organic material.

Another environment of iron leaching is commonly found in the upper, or A, horizon of many soil profiles. Here decaying organic material keeps the pH low and the Eh weakly reducing. This leads to removal of iron and the production of a grayish-white leached layer just below the surface accumulation of organic matter. This type of soil (a podzol) is common in forests in the northern part of the U.S. In tropical environments where the organic material decays almost at once no humus accumulates and the soil water does not become acid or reducing. Consequently, an accumulation of goethite frequently develops (a laterite). In the southeastern U.S. an intermediate condition prevails so that some leaching occurs but not enough to remove all the iron; this produces a red or brown soil. Chelation of iron is also probably important in soils (see Walton, 1953).

A third environment of interest is that which overlies sulfide ore deposits. Here oxidizing FeS_2, pyrite, produces goethite and sulfuric acid,

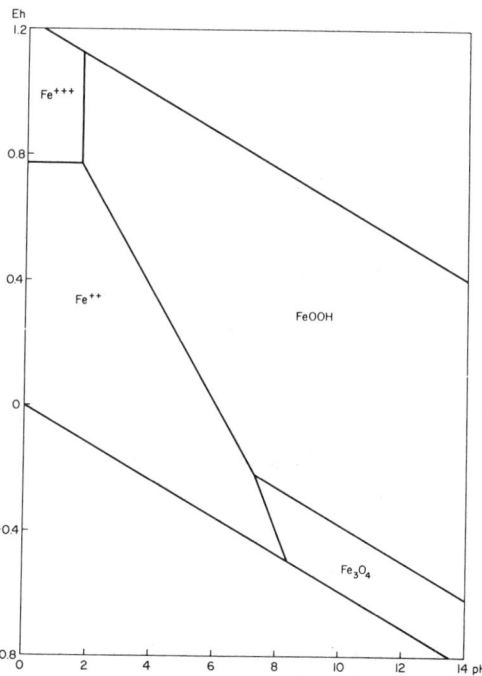

Fig. 4. pH-Eh diagram showing the stability fields for goethite and magnetite in the presence of water and showing fields of dominance for Fe^{2+} and Fe^{3+}. The boundary between the fields for ions and solids is drawn at an activity of 10^{-6}.

pH-Eh DIAGRAMS

$$2FeS_2 + 7\tfrac{1}{2}O_2 + 5H_2O = 2FeOOH + 4H_2SO_4$$

Because of the strong acid and free access to oxygen a very low pH and very high Eh result. This causes the leaching of some of the goethite and the presence of a high concentration of ferric ion in solution. (In all the previous cases it is dominantly ferrous ion in solution.) As this solution sinks into the ground it may cause the oxidation and leaching of other sulfides, e.g.

$$Cu_2S + 10Fe^{3+} + 4H_2O = 2Cu^{2+} + SO_4{}^{2-} + 8H^+ + 10Fe^{2+}$$

In this manner the outcrop overlying an ore deposit may be leached of all metals except the iron in the red to brown rock containing an excess of goethite. The copper or other metals leached in this way may become concentrated at a little deeper level because of reactions such as

$$14Cu^{2+} + 5FeS_2 + 12H_2O = 7Cu_2S + 5Fe^{2+} + 3SO_4{}^{2-} + 24H^+$$

Summary

In this discussion procedures for constructing pH-Eh diagrams have been presented. In the application of these diagrams it has been pointed out how different environments correspond to different parts of the diagram. This in turn indicates that iron, in particular, and other elements, in general, should have different chemical behavior in each environment. The correspondence of the behavior expected from these chemical considerations and the actual behavior in nature show that these diagrams are useful in understanding natural processes.

PAUL L. CLOKE

References

Berger, W. H., 1967, "Foraminiferal ooze," *Science*, **156**(3773), 383–385.

Cloke, P. L., 1966, The geochemical application of Eh-pH diagrams," *J. Geol. Educ.*, **14**(4), 140–148.

Garrels, R. M., and Christ, C. L., 1965, "Solutions, Minerals, and Equilibria," New York, Harper and Row, 450pp.

Hodgman, C. D. (editor), 1959, "Heat of Formation of Inorganic Oxides and Values of Chemical Thermodynamic Properties," in "Handbook of Chemistry and Physics," Forty-first ed., Cleveland, Chemical Rubber Publ., pp. 1820–1837, 1870–1906.

Latimer, W. M., 1952, "The Oxidation States of the Elements and Their Potentials in Aqueous Solution," Second ed., New York, Prentice-Hall, 392pp.

Park, K., 1966, "Deep-sea pH," *Science*, **154**(3756), 1540–1542.

Peterson, M. N. A., 1966, "Calcite: Rates of Dissolution in a Vertical Profile in the Central Pacific," *Science*, **154**(3756), 1542–1544.

Rossini, F. D., Wagman, D. D., Evans, W. H., Levine, S., and Jaffe, I., 1952, "Selected values of chemical thermodynamic properties," *Natl. Bur. Stds. Circ.*, **500**, 1268pp.

Walton, H. F., 1953, "Chelation," *Sci. Am.*, **188**, 68–77.

Cross-references: *Abrasion pH; Acids and Bases; Aqueous Solutions; Chelation; Geochemistry of Sediments; Iron, Economic Deposits; Manganese; Mineral Genesis; Oxidation and Reduction; Phase Equilibria; Solubility; Solubility Product Constant; Sulfides in Sediments; Thermodynamics. Vol. IVB: Gossan. Vol. VI: pH-Eh Relations.*

pH–Eh RELATIONS

Oxidation may involve any of three types of chemical reactions: (*1*) loss of hydrogen, (*2*) gain of oxygen, or (*3*) loss of an electron. Thus, by the loss of two hydrogen atoms, ethyl alcohol is oxidized to acetaldehyde. By addition of oxygen, acetaldehyde is oxidized to acetic acid. By loss of electrons ferrous iron is oxidized to ferric iron, sodium atoms are oxidized to sodium ions, and chloride ions are oxidized to chlorine atoms

$$Fe^{2+} \xrightarrow{-e} Fe^{3+}$$
$$\text{ferrous} \qquad \text{ferric}$$

$$Na \xrightarrow{-e} Na^+$$

$$Cl^- \xrightarrow{-e} Cl$$

Reduction is the opposite to oxidation. That is, it involves (*1*) gains of hydrogen, (*2*) losses of oxygen, or (*3*) gains of electrons.

Oxidation-reduction potential (= Eh = "redox" potential) of natural solutions is related to pH and is a measure of the tendency of ions, atoms, or molecules in such solutions to lose (oxidation) or gain (reduction) electrons under given conditions.

Because of the relationships of pH and Eh to electron exchanges, the same instrument can be used to acquire readings of both parameters. Commonly slide-wire potentiometers (or the more modern vacuum-tube voltmeters), which measure the potential difference between a saturated calomel reference electrode and a glass electrode immersed in an unknown solution, are used for pH measurements. Eh measurements are obtained by replacing the glass electrode with a platinum electrode. For more information on pH-Eh measurement see: Bates, 1954; Gold, 1956; Barnes, 1959; or Garrels, 1960.

pH-Eh RELATIONS

TABLE 1. OXIDATION-REDUCTION POTENTIALS OF SELECTED REACTIONS
OCCURRING IN THE NATURAL ENVIRONMENT[a]

Oxidation $\longrightarrow$	Oxidation-Reduction Potential (mV)
$2H_2O \longrightarrow O_2 + 4H^+ + 4e$	1230
$Fe^{2+} \longrightarrow Fe^{3+} + e$	740
$4OH^- \longrightarrow O_2 + 2H_2O + 4e$	400
$H_2S \longrightarrow S + 2H^+ + 2e$	170
$H_2 \longrightarrow 2H^+ + 2e$	0
$NH_3 + 9OH^- \longrightarrow NO_3^- + 6H_2O + 8e$	−120
$Fe(OH)_2 + OH^- \longrightarrow Fe(OH)_3 + e$	−650

[a] The oxidized form of any reactant can oxidize the reduced form of any reactant lower in the sequence, and suffer reduction in the process (e.g. Fe^{3+} oxidizes H_2).

Relative oxidation or reduction strengths of various reactions are noted as positive or negative, respectively, in reference to the reaction:

$$H_2 \longrightarrow 2H^+ + 2e$$

which is arbitrarily assigned a redox value of 0 (see Table 1).

Table 1 is a list of relative oxidation-reduction potentials of reactions significant in natural sedimentary environments. Materials with high redox potential can oxidize those with a lower value and suffer reduction themselves in the process. Reading upward in the table, the reactions are arranged in order of increasing oxidative power. The oxidized form of any reaction pair can oxidize the reduced form of any pair lower in the list. Oxidation-reduction potentials vary with varying concentrations of the reacting substances, and with pH (see Fig. 1). For pure water, at 25°C, exposed to atmospheric oxygen at one atmosphere of pressure, Eh is theoretically related to pH according to the equilibrium equation:

$$Eh = 1.22 - 0.059 pH$$

However, Garrels (1960, p. 70) notes that, partly because of direct action of dissolved oxygen on the platinum electrode, water systems exposed to and containing dissolved air actually show measurable potentials "according to the approximate relation $Eh = 0.70 - 0.06 pH$."

The stability limits of water as a function of pH and Eh at 25°C and 1 atmosphere total pressure, with appropriate hydrogen and oxygen partial pressures, are shown in Fig. 1. Pure water is theoretically stable only between the sloping boundaries of the figure. That is, it will not be electrolyzed (broken up) into its components hydrogen and oxygen if a voltage less than the difference between the lines (i.e., 1.23 mV) is applied between the instrumental electrodes immersed in the water. Conversely, at greater applied voltages, or at pH-Eh values outside of the boundaries, pure water should dissociate into hydrogen and oxygen. Natural waters behave similarly. The shaded area of Fig. 1 shows the actual pH-Eh field of fluids found in natural environments, as recorded by Baas Becking and others (1960).

ALISTAIR W. McCRONE

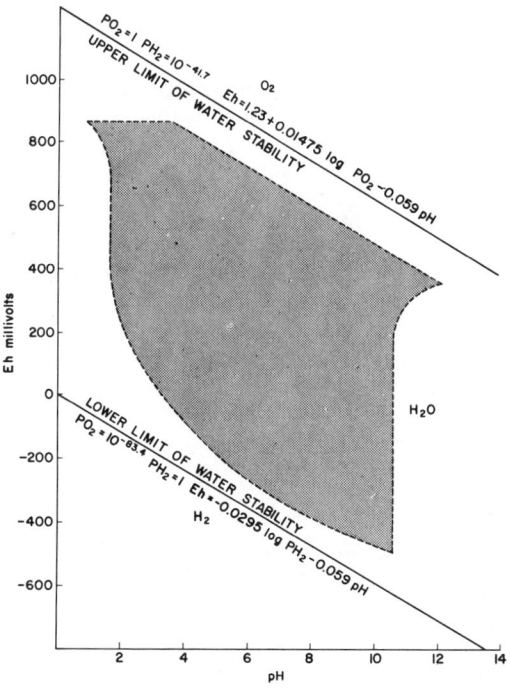

FIG. 1. Theoretical stability limits of pure water in terms of pH and Eh, at 1 atm pressure and 25°C. Shaded area is pH–Eh field of observed natural waters (after Baas Becking et al., 1960). P_{O_2} and P_{H_2} are partial pressures of oxygen and hydrogen, respectively. For explanation of equations see Garrels (1960, pp. 104–110).

References

Baas Becking, L. G. M., Kaplan, I. R., and Moore, D., 1960, "Limits of the natural en-

pH-Eh RELATIONS

vironment in terms of pH and oxidation-reduction potentials," *J. Geol.,* **68,** 243–284.

Barnes, H., 1959, "Apparatus and Methods of Oceanography, Part I, Chemical," London, George Allen and Unwin, 341pp.

Bates, R. G., 1954, "Electrometric pH Determination," New York, John Wiley & Sons, 331pp.

Garrels, R. H., 1960, "Mineral Equilibria at Low Temperature and Pressure," New York, Harpers, 254pp.

Gold, V., 1956, "pH Measurements," New York, John Wiley & Sons, 125pp.

Cross-references: *Ion Exchange; Oxidation and Reduction; Seawater, Chemistry.* Vol. I: *Chemical Oceanography.*

PHOSPHATE CYCLE—See PHOSPHORUS CYCLE

PHOSPHATE AND GUANO—See Vol. IV B

PHOSPHATES, ARSENATES AND VANADATES

The minerals in this class are anisodesmic compounds with $(XO_4)^{3-}$ groups, in which X is P, As, or V. Mutual substitution of P for As and As for V is common. Substitution with other $(XO_4)^{2-}$ groups such as sulfates, molybdates, and chromates is rare. (See table below.)

The most common minerals of this class are those in the apatite group, and pyromorphite, amblygonite, monazite, turquoise, vivianite and wavellite. Although considerable work has been done on the structures of this class of minerals, much remains to be completed to bring our knowledge up to that of the structures of the silicate minerals, to which to some extent they are analogous.

Anhydrous Acid Minerals

Monetite	$CaH(PO_4)$
Schultenite	$PbH(AsO_4)$

Anhydrous Normal Minerals

$AB(XO_4)$ Type

Triphylite	$LiFe(PO_4)$	*Triphylite group:* consists of two isostructural orthorhombic series with Fe–Mn substitution; this is the monticellite type structure.
Lithiophilite	$LiMn(PO_4)$	
Hühnerkobelite	$(Na,Ca)Fe(PO_4)$	
Varulite	$(Na,Ca)Mn(PO_4)$	
Natrophilite	$MaMn(PO_4)$	
Sicklerite	$(Li,Mn^{II},Fe^{III})(PO_4)$	*Sicklerite group:* an orthorhombic series derived from oxidation of triphylite-lithiophilite
Ferri-sicklerite	$(Li,Fe^{III},Mn^{II})(PO_4)$	
Alluaudite	$(Na,Fe^{III},Mn^{II})(PO_4)$	*Alluaudite group:* a series derived from oxidation of hühnerkobelite-varulite
Mangan-alluaudite	$(Na,Mn^{II},Fe^{III})(PO_4)$	
Heterosite	$(Fe^{III},Mn^{III})(PO_4)$	*Heterosite group:* an orthorhombic series with trivalent Fe–Mn substitution
Purpurite	$(Mn^{III},Fe^{III})(PO_4)$	
Beryllonite	$NaBe(PO_4)$	
Hurlbutite	$CaBe_2(PO_4)_2$	
Arrojadite	$Na_2(Fe^{II},Mn^{II})_5(PO_4)_4$	

$A_3B_2(XO_4)_3$ Type

Berzeliite	$(Mg,Mn)_2(Ca,Na)_3(AsO_4)_3$	*Berzeliite group:* an isometric series with garnet structure and Mn–Mg substitution
Manganberzeliite	$(Mn,Mg)_2(Ca,Na)_3(AsO_4)_3$	
Caryinite	$(Ca,Pb,Na)_5(Mn,Mg)_4(AsO_4)_5$ (?)	

$A_3(XO_4)_2$ Type

Whitlockite	$Ca_3(PO_4)_2$
Graftonite	$(Fe,Mn,Ca)_3(PO_4)_2$

$A(XO_4)$ Type

Xenotime	$Y(PO_4)$	Tetragonal, with zircon type structure
Monazite	$(Ce,La,Di)(PO_4)$	

PHOSPHATES, ARSENATES AND VANADATES

Berlinite	$Al(PO_4)$	Monoclinic
Lithiophosphatite	$Li_3(PO_4)$	
Rooseveltite	$Bi(AsO_4)$	

Hydrous Acid Phosphates

$(AB)_m H_n (XO_4)_p \cdot xH_2O$, with $m + n:p > 2:1$ Type

Sterocorite	$Na(NH_4)H(PO_4) \cdot 4H_2O$
Hannayite	$Mg_3(NH_4)_2H_4(PO_4)_4 \cdot 8H_2O$
Hureaulite	$Mn_5H_2(PO_4)_4 \cdot 4H_2O$

$AH(XO_4) \cdot xH_2O$ Type

Brushite	$CaH(PO_4) \cdot 2H_2O$	*Brushite group*: monoclinic and isostructural with gypsum
Pharmacolite	$CaH(AsO_4) \cdot 2H_2O$	
Haidingerite	$CaH(AsO_4) \cdot H_2O$	
Newberyite	$MgH(PO_4) \cdot 3H_2O$	
Forbesite	$(Ni,Co)H(AsO_4) \cdot 3\frac{1}{2}H_2O$ (?)	
Roesslerite	$MgH(AsO_4) \cdot 7H_2O$	*Roesslerite group*: monoclinic; isostructural, and probably isomorphous
Phosphorroesslerite	$MgH(PO_4) \cdot 7H_2O$	

Hydrated Normal Minerals

$AB(XO_4) \cdot xH_2O$ Type

Struvite	$(NH_4)Mg(PO_4) \cdot 6H_2O$

$AB_2(XO_4)_2 \cdot xH_2O$ Type

Dickinsonite	$Na_6(Mn,Fe,Ca)_{14}H_2(PO_4)_{12} \cdot H_2O$	
Fillowite	$Na_6(Mn,Fe,Ca)_{14}H_2(PO_4)_{12} \cdot H_2O$ (?)	
Fairfieldite	$Ca_2(Mn,Fe)(PO_4)_2 \cdot 2H_2O$	*Fairfieldite group*: triclinic; no substitution of Mg and Mn
Collinsite	$Ca_2(Mg,Fe)(PO_4)_2 \cdot 2H_2O$	
Roselite	$Ca_2(Co,Mg)(AsO_4)_2 \cdot 2H_2O$	*Roselite group*: monoclinic, isostructural
Brandtite	$Ca_2Mn(AsO_4)_2 \cdot 2H_2O$	
Reddingite	$(Mn,Fe)_3(PO_4)_2 \cdot 3H_2O$	*Reddingite group*: orthorhombic, and isomorphous with Mn substituting for Fe
Phosphoferrite	$(Fe,Mn)_3(PO_4)_2 \cdot 3H_2O$	
Landesite	$Fe_6Mn_{20}(PO_4)_{16} \cdot 27H_2O$ (?)	
Stewartite	Mn phosphate	
Salmonsite	Mn,Fe phosphate	
Anapaite	$Ca_2Fe(PO_4)_2 \cdot 4H_2O$	
Parahopeite	$Zn_3(PO_4)_2 \cdot 4H_2O$	
Hopeite	$Zn_3(PO_4)_2 \cdot 4H_2O$	
Phosphophyllite	$Zn_2(Fe,Mn)(PO_4)_2 \cdot 4H_2O$	
Trichalcite	$Cu_3(AsO_4)_2 5H_2O$ (?)	
Picropharmacolite	$(Ca,Mg)_3(AsO_4)_2 \cdot 6H_2O$ (?)	
Vivianite	$Fe_3(PO_4)_2 \cdot 8H_2O$	*Vivianite group*: monoclinic; series between erythrite, annabergite and possibly koettigite; bobierrite, hoernesite and symplesite may also belong to this group
Erythrite	$Co_3(AsO_4)_2 \cdot 8H_2O$	
Annabergite	$Ni_3(AsO_4)_2 \cdot 8H_2O$	
Koettigite	$Zn_3(AsO_4)_2 \cdot 8H_2O$	

$A(XO_4) \cdot xH_2O$ Type

Variscite	$Al(PO_4) \cdot 2H_2O$	*Variscite group*: two series, variscite strengite and mansfieldite-scorodite; substitution of (AsO_4) and (PO_4) is small
Strengite	$Fe(PO_4) \cdot 2H_2O$	
Scorodite	$Fe(AsO_4) \cdot 2H_2O$	
Mansfieldite	$Al(AsO_4) \cdot 2H_2O$	

PHOSPHATES, ARSENATES AND VANADATES

Metavariscite	$Al(PO_4) \cdot 2H_2O$	*Metavariscite group:* monoclinic polymorphs of variscite and strengite; also called clinovariscite and clinostrengite
Metastrengite	$Fe(PO_4) \cdot 2H_2O$	
Weinschenkite	$(Y,Er)(PO_4) \cdot 2H_2O$	Gypsum structure
Churchite	$(Ce,Ca)(PO_4) \cdot 2H_2O$	
Rhabdophane	$(Ce,Y)(PO_4) \cdot H_2O$	

ANHYDROUS PHOSPHATES WITH HYDROXYL OR HALOGEN
$A_m(XO_4)_p Z_q$, with m:p > 4:1 Type

Sahlinite	$Pb_{14}(AsO_4)_2O_9Cl$	
Holdenite	$(Mn^{II},Ca)_4(Zn,Mg,Fe^{II})_2(AsO_4)(OH)_5O_2$	
Hematolite	$(Mn^{II},Mg)_4Al(AsO_4)(OH)_8$	
Chlorophoenicite	$(Zn,Mn)_5(AsO_4)(OH)_7$	*Chlorophoenicite group:* monoclinic, isostructural
Magnesium chlorophoenicite	$Mg_5(AsO_4)(OH)_7$	
Synadelphite	$(Mn,Mg,Ca,Pb)_4(AsO_4)(OH)_5$	

$A_7(XO_4)_2 Z_q$ Type

Lacroixite	Na,Ca,Al fluo-phosphate
Morinite	Na,Ca,Al fluo-phosphate
Jezekite	$Na_4CaAl_2(PO_4)_2(OH)_2F_2O$ (?)
Allactite	$Mn_7(AsO_4)_2(OH)_8$

$A_3(XO_4)Z_q$ Type

Clinoclase	$Cu_3(AsO_4)(OH)_3$
Cornetite	$Cu_3(PO_4)(OH)_3$
Georgiadesite	$Pb_3(AsO_4)Cl_3$
Atelestite	$Bi_3(AsO_4)O_2(OH)_2$ (?)
Flinkite	$Mn_2^{II}Mn^{III}(AsO_4)(OH)_4$
Retzian	$Mn_2^{II}Y(AsO_4)(OH)_4$

$(AB)_5(XO_4)_2 Z_q$ Type

Walpurgite	$Bi_4(UO_2)(AsO_4)O_4 \cdot 3H_2O$
Erinite	$Cu_5(AsO_4)_2(OH)_4$
Pseudomalachite	$Cu_5(PO_4)_2(OH)_4 \cdot H_2O$ (?)
Arsenoclasite	$Mn_5(AsO_4)_2(OH)_4$
Andrewsite	$(Cu,Fe^{II})_3Fe_6^{III}(PO_4)_4(OH)_{12}$
Laubmannite	$Fe_3^{II}Fe_6^{III}(PO_4)_4(OH)_{12}$

$AB(XO_4)Z_q$ Type

Adelite	$CaMg(AsO_4)(OH,F)$	*Adelite group:* orthorhombic disphenoidal; isostructural
Conichalcite	$CaCu(AsO_4)(OH)$	
Austinite	$CaZn(AsO_4)(OH)$	
Duftite	$PbCu(AsO_4)(OH)$	
Descloizite	$ZnPb(VO_4)(OH)$	*Descloizite group:* orthorhombic dipyramidal, but similar to adelite group although in different class; arsenic may substitute for V in descloizite group
Mottramite	$CuPb(VO_4)(OH)$	
Pyrobelonite	$MnPb(VO_4)(OH)$	
Calciovolborthite	$CuCa(VO_4)(OH)$	
Turanite	$Cu_2(VO_4)(OH)$ (?)	
Volborthite	Cu vanadate	
Herderite	$CaBe(PO_4)(F,OH)$	*Herderite group:* monoclinic with complete substitution of F for (OH)
Hydroxyl-herderite	$CaBe(PO_4)(OH,F)$	
Amblygonite	$(Li,Na)Al(PO_4)(F,OH)$	*Amblygonite group:* triclinic series with easy substitution of Na and Li, and F for (OH)
Montebrasite	$(Li,Na)Al(PO_4)(OH,F)$	
Natromontebrasite	$(Na,Li)Al(PO_4)(F,OH)$	

Tilasite	$CaMg(AsO_4)F$	
Durangite	$NaAl(AsO_4)F$	
Plumbogummite	$PbAl_3(PO_4)_2(OH)_5H_2O$	
Gorceixite	$BaAl_3(PO_4)_2(OH)_5H_2O$	*Plumbogummite group:* hexagonal and isostructural with alunite group and beudantite group; limited substitution between members
Goyazite	$SrAl_3(PO_4)_2(OH)_5H_2O$	
Crandallite	$CaAl_3(PO_4)_2(OH)_5H_2O$	
Deltaite	$Ca(Al_2Ca)(PO_4)_2(OH)_4H_2O$	
Florencite	$CeAl_3(PO_4)_2(OH)_6$	
Dussertite	$BaFe_3(AsO_4)_2(OH)_5H_2O$	
Chenevixite	$Cu_2Fe_2(AsO_4)_2(OH)_4 \cdot H_2O$ (?)	
Brazilianite	$NaAl_3(PO_4)_2(OH)_4$	
Griphite	$(Na,Ca,Fe,Al)_3Mn_2(PO_4)_{2.5}(OH,F)_2$	
Arseniopleite	Mn basic arsenate	

$A_2(XO_4)Z_q$ Type

Wagnerite	$Mg_2(PO_4)F$	
Triplite	$(Mn^{II},Fe^{II})(PO_4)F$	
Triploidite	$(Mn^{II},Fe^{II})_2(PO_4)(OH)$	*Triploidite group:* monoclinic; complete substitution of Mn and Fe, except in sarkinite which is isostructural
Wolfeite	$(Fe^{II},Mn^{II})_2(PO_4)(OH)$	
Sarkinite	$Mn_2(AsO_4)(OH)$	
Sarcopside	$(Fe,Mn,Ca)_7(PO_4)_4F_2$ (?)	
Olivenite	$Cu_2(AsO_4)(OH)$	*Olivenite group:* orthorhombic; isostructural with partial substitution of As, P, and Cu, Zn
Libethenite	$Cu_2(PO_4)(OH)$	
Adamite	$Zn_2(AsO_4)(OH)$	
Frondelite	$Mn^{II}Fe_4^{III}(PO_4)_3(OH)_5$	*Frondelite group:* orthorhombic; probably complete isomorphous series
Rockbridgeite	$Fe^{II}Fe_4^{III}(PO_4)_3(OH)_5$	
Tarbuttite	$Zn_2(PO_4)(OH)$	
Augelite	$Al_2(PO_4)(OH)_3$	
Dufrenite	$Fe^{II}Fe_4^{III}(PO_4)_3(OH)_5 \cdot 2H_2O$ (?)	
Dewindtite	$Pb_3(UO_2)_5(PO_4)_4(OH)_4 \cdot 10H_2O$	
Phosphuranylite	Ca uranyl phosphate	

$A_5(XO_4)_3Z_q$ Type

Fluorapatite	$Ca_5(PO_4)_3F$	
Chlorapatite	$Ca_5(PO_4)_3Cl$	
Hydroxylapatite	$Ca_5(PO_4)_3(OH)$	
Carbonate-apatite	$Ca_{10}(PO_4)_6(CO_3) \cdot H_2O$	*Apatite group:* hexagonal dipyramidal minerals which fall into the apatite series, pyromorphite series, and svabite series; the last five minerals in the group are isostructural with apatite and there is considerable substitution
Pyromorphite	$Pb_5(PO_4)_3Cl$	
Mimetite	$Pb_5(AsO_4)_3Cl$	
Vanadinite	$Pb_5(VO_4)_3Cl$	
Svabite	$Ca_5(AsO_4)_3(F,OH)$	
Hedyphane	$(Ca,Pb)_5(AsO_4)_3Cl$	
Dehrnite	$(Ca,Na,K)_5(PO_4)_3(OH)$	
Lewistonite	$(Ca,K,Na)_5(PO_4)_3(OH)$	
Fermorite	$(Ca,Sr)_5(P,AsO_4)_3(F,OH)$	
Wilkeite	$Ca_5(P,S,Si,CO_4)_3(OH)$	
Ellestadite	$Ca_5(Si,S,P)_3(Cl,F,OH)$	
Tavistockite	$Ca_3Al_2(PO_4)_3(OH)_3$	
Arsenobismite	Bi basic arsenate	

$(AB)_3(XO_4)_2Z_q$ Type

Lazulite	$(Mg,Fe^{II})Al_2(PO_4)_2(OH)_2$	*Lazulite group:* monoclinic series with complete substitution of Fe and Mg between end members
Scorzalite	$(Fe^{II},Mg)Al_2(PO_4)_2(OH)_2$	

PHOSPHATES, ARSENATES AND VANADATES

Souzalite	$(Mg,Fe^{II})_3(Al,Fe^{III})_4(PO_4)_4(OH)_6 \cdot 2H_2O$
Carminite	$PbFe_2(AsO_4)_2(OH)_2$
Parsonsite	$Pb_2(UO_2)(PO_4)_2 2H_2O$

Hydrated Phosphates with Hydroxyl or Halogen

$(AB)_m(XO_4)_p Z_q \cdot xH_2O$, with $m:p > 3:1$ Type

Borickite	$CaFe_5(PO_4)_2(OH)_{11} \cdot 3H_2O$

$(AB)_3(XO_4)Z_q \cdot xH_2O$ Type

Veszelyite	$(Cu,Zn)_3(As,PO_4)(OH)_3 \cdot 3H_2O$
Tsumebite	$Pb_2Cu(PO_4)(OH)_3 \cdot 3H_2O$
Hemafibrite	$Mn_3(AsO_4)(OH)_3 \cdot H_2O$
Freirinite	$Na_3Cu_3(AsO_4)_2(OH)_3 \cdot H_2O$
Liroconite	$Cu_2Al(AsO_4)(OH)_4 \cdot 4H_2O$
Evansite	$Al_3(PO_4)(OH)_6 \cdot 6H_2O$
Liskeardite	$(Al,Fe)_3(AsO_4)(OH)_6 \cdot 5H_2O$

$(AB)_5(XO_4)_2 Z_q \cdot xH_2O$ Type

Cornwallite	$Cu_5(AsO_4)_2(OH)_4 \cdot H_2O$
Tyrolite	$Cu_5Ca(AsO_4)_2(CO_3)(OH)_4 \cdot 6H_2O$ (?)
Akrochordite	$MgMn_4(AsO_4)_2(OH)_4 \cdot 4H_2O$ (?)
Ceruleite	$CuAl_4(AsO_4)_2(OH)_8 \cdot 4H_2O$
Renardite	$Pb(UO_2)_4(PO_4)_2(OH)_4 \cdot 7H_2O$
Dumontite	$Pb_2(UO_2)_3(PO_4)_2(OH)_4 \cdot 3H_2O$

$A_2(XO_4)Z_q \cdot xH_2O$ Type

Bayldonite	$(Cu,Pb)_2(AsO_4)(OH)$ (?)
Leucochalcite	$Cu_2(AsO_4)(OH) \cdot H_2O$
Tagilite	$Cu_2(PO_4)(OH) \cdot H_2O$
Spencerite	$Zn_4(PO_4)_2(OH)_2 \cdot 3H_2O$
Isoclasite	$Ca_2(PO_4)(OH) \cdot 2H_2O$
Euchroite	$Cu_2(AsO_4)(OH) \cdot 3H_2O$
Delvauxite	$Fe_2(PO_4)(OH)_3 \cdot xH_2O$ (?)

$(AB)_m(XO_4)_p Z_q \cdot xH_2O$, with $m:p = 2:1$ Type

Leucophosphite	$K_2(Fe,Al)_7(PO_4)_4(OH)_{11} \cdot 6H_2O$	
Childrenite	$(Fe^{II},Mn^{II})Al(PO_4)(OH)_2 \cdot H_2O$	*Childrenite group:* orthorhombic with complete series due to substitution of Mn and Fe
Eosphorite	$(Mn^{II},Fe^{II})Al(PO_4)(OH)_2 \cdot H_2O$	
Davisonite	$Ca_3Al(PO_4)_2(OH)_3 \cdot H_2O$ (?)	
Wardite	$Na_4CaAl_{12}(PO_4)_8(OH)_9 \cdot 3H_2O$	
Millisite	$(Na,K)CaAl_6(PO_4)_4(OH)_9 \cdot 3H_2O$	
Lehiite	$(Na,K)_2Ca_5Al_8(PO_4)_8(OH)_{12} \cdot 6H_2O$ (?)	
Mixite	$Cu_{11}Bi(AsO_4)_5(OH)_{10} \cdot 6H_2O$ (?)	

$(AB)_m(XO_4)_p Z_q \cdot xH_2O$, with $m:p = 7:4$ Type

Sampleite	$NaCaCu_5(PO_4)_4Cl \cdot 5H_2O$	
Turquois	$CuAl_6(PO_4)_4(OH)_8 \cdot 4H_2O$	*Turquois group:* triclinic; possibly complete series by substitution of Al and Fe^{III}
Chalcosiderite	$CuFe_6(PO_4)_4(OH)_8 \cdot 4H_2O$	
Ludlamite	$(Fe^{II},Mg,Mn)_3(PO_4)_2 \cdot 4H_2O$	
Arseniosiderite	$Ca_3Fe_4(AsO_4)_4(OH)_4 \cdot 4H_2O$ (?)	
Egueite	$CaFe_{14}^{III}(PO_4)_{10}(OH)_{14} \cdot 21H_2O$ (?)	
Mitridatite	Ca, Fe phosphate	
Richellite	Ca, Fe phosphate	
Englishite	$K_2Ca_4Al_8(PO_4)_8(OH)_{10} \cdot 9H_2O$	

PHOSPHATES, ARSENATES AND VANADATES

$A_3(XO_4)_2Z_q \cdot xH_2O$ Type

Legrandite	$Zn_{14}(AsO_4)_9(OH) \cdot 12H_2O$
Beraunite	$Fe^{II}Fe_4^{III}(PO_4)_3(OH)_5 \cdot 3H_2O$ (?)
Ceruleolactite	$Al_3(PO_4)_2(OH)_3$
Wavellite	$Al_3(PO_4)_2(OH)_3 \cdot 5H_2O$
Sterrettite	$Al_6(PO_4)_4(OH)_6 \cdot 5H_2O$
Troegerite	$(UO_2)_3(AsO_4)_2 \cdot 12H_2O$

$(AB)_m(XO_4)_pZ_q \cdot xH_2O$, with $m:p = 3:2$ Type

Bermanite	$MnMn_2(PO_4)_2(OH)_2 \cdot 4H_2O$
Roscherite	$(Ca,Mn,Fe)_2Al(PO_4)_2(OH) \cdot 2H_2O$
Minyulite	$KAl_2(PO_4)_2(OH) \cdot 3\tfrac{1}{2}H_2O$ (?)
Tinticite	$Fe_3^{III}(PO_4)_2(OH)_3 \cdot 3\tfrac{1}{2}H_2O$
Metavauxite	$FeAl_2(PO_4)_2(OH)_2 \cdot 8H_2O$
Paravauxite	$FeAl_2(PO_4)_2(OH)_2 \cdot 8H_2O$
Vauxite	$FeAl_2(PO_4)_2(OH)_2 \cdot 7H_2O$
Gordonite	$MgAl_2(PO_4)_2(OH)_2 \cdot 8H_2O$
Calcioferrite	$Ca_3Fe_3(PO_4)_4(OH)_3 \cdot 8H_2O$ (?)
Xanthoxenite	$Ca_2Fe(PO_4)_2(OH) \cdot 1\tfrac{1}{2}H_2O$
Montgomeryite	$Ca_4Al_5(PO_4)_6(OH)_5 \cdot 11H_2O$
Overite	$Ca_3Al_8(PO_4)_8(OH)_6 \cdot 15H_2O$
Torbernite	$Cu(UO_2)_2(PO_4)_2 \cdot 8{-}12H_2O$
Autunite	$Ca(UO_2)_2(PO_4)_2 \cdot 10{-}12H_2O$
Uranocircite	$Ba(UO_2)_2(PO_4)_2 \cdot 8H_2O$
Saléeite	$Mg(UO_2)_2(PO_4)_2 \cdot 10H_2O$
Zeunerite	$Cu(UO_2)_2(AsO_4)_2 \cdot 10{-}16H_2O$
Uranospinite	$Ca(UO_2)_2(AsO_4)_2 \cdot 8H_2O$
Metatorbernite	$Cu(UO_2)_2(PO_4)_2 \cdot 8H_2O$
Meta-autunite	$Ca(UO_2)_2(PO_4)_2 \cdot 2{-}6H_2O$
Metazeunerite	$Cu(UO_2)_2(AsO_4)_2 \cdot 8H_2O$
Bassetite	Fe^{II} uranyl phosphate

Torbernite group: layer-type structures parallel to {001} with PO_4 or AsO_4 tetrahedra; stable for values of nH_2O over 8 and ranging to 12

Metatorbernite group: structure similar to torbernite group but stable for values of nH_2O from 6–8 to 5 or $2\tfrac{1}{2}$

$(AB)_m(XO_4)_pZ_q \cdot xH_2O$, with $m:p < 3:2$ Type

Pharmacosiderite	$Fe_3(AsO_4)_2(OH)_3 \cdot 5H_2O$
Cacoxenite	$Fe_4(PO_4)_3(OH)_3 \cdot 12H_2O$
Vashegyite	$Al_4(PO_4)_3(OH)_3 \cdot nH_2O$ (?)
Taranakite	$K_2Al_6(PO_4)_6(OH)_2 \cdot 18H_2O$ (?)

Compound Phosphates, etc.

$AB(XO_4)Z_q$ Type

Beudantite	$PbFe_3(AsO_4)(SO_4)(OH)_6$
Corkite	$PbFe_3(PO_4)(SO_4)(OH)_6$
Hinsdalite	$(Pb,Sr)Al_3(PO_4)(SO_4)(OH)_6$
Svanbergite	$SrAl_3(PO_4)(SO_4)(OH)_6$
Woodhouseite	$CaAl_3(PO_4)(SO_4)(OH)_6$
Lindackerite	$Cu_6Ni_3(AsO_4)_4(SO_4)(OH)_4 \cdot 5H_2O$ (?)

Beudantite group: isostructural with alunite and plumbogummite groups

Miscellaneous

Chalcophyllite	$Cu_{18}Al_2(AsO_4)_3(SO_4)_3(OH)_{27} \cdot 33H_2O$
Ardealite	$Ca_2H(PO_4)(SO_4) \cdot 4H_2O$
Kribergite	$Al_4(PO_4)_2(SO_4)_2(OH)_2 \cdot 8H_2O$ (?)
Diadochite	$Fe_2(PO_4)(SO_4)(OH) \cdot 5H_2O$
Sarmientite	$Fe_2(AsO_4)(SO_4)(OH) \cdot 5H_2O$
Pitticite	Fe sulfate-arsenate
Kolbeckite	Ca,Be,Al silicate-phosphate

Ardealite — Gypsum—type structure

Lloyd W. Staples

PHOSPHATES, ARSENATES AND VANADATES

References

Barr, J. A., 1960, "Phosphate rock," in "Industrial Minerals and Rocks," Third ed., AIME, Chap. 35.

Dana, J. W., and Dana, E. S., 1951, "The System of Mineralogy," Seventh ed., Vol. II, 654, (revised by Palache, C., Berman, H., and Frondel, C.), New York, John Wiley & Sons.

Fisher, D. J., 1958, "Pegmatite phosphates and their problems," *Am. Mineralogist,* **43,** 181.

Richmond, W. E., 1940, "Crystal chemistry of the phosphates, arsenates, and vanadates of the type A_2XO_4 (Z)," *Am. Mineralogist,* **25,** 441.

Cross-references: *Mineral Classes—Nonsilicates.* Vol. IVB: *Individual Minerals.*

PHOSPHATIZATION

Phosphatization is the process by which phosphorus is emplaced in a host material. The process may involve removal of preexisting material by solution or suspension and a later deposition of phosphate material, or it may be a pseudomorphic replacement. Host materials for phosphatization may be carbonate minerals, rocks, or fossils; wood; or silicate minerals or rocks. Generally the product of phosphatization is some variety of apatite; however, other calcium phosphate compounds, aluminum phosphates, and a variety of rare phosphate minerals are known to form by this process.

The following discussion will be concerned first with a consideration of phosphatization in which the products are apatite or other calcium phosphate compounds, and second with the formation of other phosphate products.

Apatite, approximately $Ca_5(PO_4)_3(OH,F,Cl)$, is one of the most ubiquitous minerals in nature. It is found in almost all types of igneous, metamorphic, and sedimentary rock. In the igneous and metamorphic rocks, apatite is generally present in low concentrations as discrete, well-formed crystals. In sedimentary rocks, apatite is present as detrital grains, phosphatic fossils, cement, or as microcrystalline aggregates in the form of pebbles or pellets. Of these, the last three may be produced by phosphatization.

The action of various solutions on phosphate-bearing rocks has been known for many years (Reese, 1892, and Graham, 1925). Phosphate can readily be leached from rock by dilute acid solution, and it can even be preferentially leached from limestone and dolomite, provided the appropriate carbonate ion activity is present in the solution. Under other conditions, the carbonate minerals may be preferentially dissolved thereby concentrating apatite as a residual deposit. Solutions bearing leached phosphate may react with some susceptible material producing phosphate minerals by replacement or precipitation. Where the susceptible material is some form of calcium carbonate, the reaction generally produces apatite, although whitlockite, brushite, and monetite are also known to form. Simplified expressions for reactions involving brushite and apatite are given in Eqs. (1) and (2).

The interaction of alkali phosphate solution and calcium carbonate minerals has been studied by Ames (1959) and Simpson (1964). Ames concludes that the process of formation of the apatite is by replacement, and that such replacement will cease when the concentration of phosphate ions is less than 0.09 ppm with a solution pH of 11.8 at 28°C. Simpson shows that apatite formed by interaction of sodium phosphate solution and calcite has a number of anomalous properties. It contains more than five per cent carbon dioxide and an even greater amount of water. It lacks observable birefringence, has a mean index of refraction of about 1.588, and gives a diffuse X-ray diffraction pattern. These properties show considerable variation as a function of pH and temperature of the solution in which the apatite formed. At a pH below about 6.4, brushite ($CaHPO_4 \cdot 2H_2O$), a mineral isostructural with gypsum, forms by interaction of the alkali phosphate solution and calcite.

Potassium phosphate and ammonium phosphate solutions also interact with calcite to form apatite and brushite.

Simpson (1964) reports that pellets of partially phosphatized calcite were formed in experiments when finely divided fragments of calcite, under an alkali phosphate solution, were only intermittently agitated. In thin section, the pellets show neither a radial nor concentric structure but rather are simply a fine-grained mosaic of apatite and calcite. Such pellets may form by an early cementation of

1) calcite + phosphoric acid + water $\longleftrightarrow$ brushite + carbonic acid

$$CaCO_3 + H_3PO_4 + 2H_2O \underset{\text{pH} < 6.4}{\rightleftharpoons} CaHPO_4 \cdot 2H_2O + H_2CO_3$$

2) calcite + phosphoric acid + water $\longleftrightarrow$ apatite + carbonic acid

$$10CaCO_3 + 6H_3PO_4 + 2H_2O \underset{\text{pH} > 6.4}{\rightleftharpoons} Ca_{10}(PO_4)_6(OH)_2 + 10H_2CO_3$$

the calcite grains with apatite and a later rounding of the aggregate by abrasion. A still later, slow reaction may phosphatize the cemented calcite grains. As mentioned earlier, such phosphatized calcite has anomalous properties for apatite; however, the properties are similar to those of collophane. A similar process may account for some of the natural phosphate pebbles; however, it could not account for pebbles that show a concentric or radial structure. Pebbles or pellets with such internal structures may have formed by a precipitation and growth, by replacement of a preexisting structured pebble, or by a combination of the two processes.

Although dolomite under an alkali phosphate solution is slightly altered, apatite does not form. However, in the same solution, calcite can be phosphatized in the presence of dolomite. The unreactive nature of dolomite to phosphatization is well shown by the abundance of dolomite in sedimentary phosphate rock and the nearly complete absence of magnesium phosphate minerals or magnesite. Although dolomite is relatively unaltered by phosphate solutions, aragonite may be even more susceptible to reaction than calcite because of a greater solubility.

Summaries of suggested processes by which the sedimentary phosphate deposits may have formed are presented by many investigators (Kazakov, 1937; Mansfield, 1927; Blackwelder, 1916; McKelvey et al., 1953; Emery, 1960; Smith and Whitlatch, 1940; Strakhov, 1960). Other important contributions are found in Section XI of the International Geological Congress, 1952, entitled "Origine des Gisements de Phosphates de Chaux." Many investigators agree on the importance of upwelling of cold oceanic waters which are relatively rich in inorganic phosphorus. With upwelling of water and changes in temperature and carbon dioxide pressure, there may be precipitation of apatite. Likewise, the solutions may be rich enough in phosphorus to interact with calcium carbonate and form apatite. The partial pressure of carbon dioxide in the system would be of supreme importance because of its effect on the dissolution of calcium carbonate minerals, and its effect on the composition and therefore solubility of the formed apatite. Many phosphorite deposits contain abundant organic material and sulfide compounds, a feature that suggests that biological activity is important in the formation of some phosphate deposits. Some investigators consider phosphorite deposits to be simply an accumulation of tests and skeletal parts of organisms. Other investigators contend that decomposition of the organisms releases phosphate, ammonia, and other compounds to solution, and apatite is formed from the solution by precipitation or replacement.

Guano, chiefly the excrement of birds, accumulates mostly on arid islands near areas of upwelling waters. Deposits of guano are a rich source of phosphorus and nitrate compounds and, as such, are readily susceptible to leaching by sporadic rains or groundwater. These solutions may then react with underlying rocks producing calcium phosphate, iron or aluminum phosphate, or ammonium-magnesium or sodium phosphate minerals. Rodgers (1948) in considering the origin of some island phosphate deposits concludes that bird droppings were the original source of phosphorus and that rain water dissolves out phosphoric acid which in turn attacks the underlying limestone. Later weathering and solution of the limestone resulted in a concentration of calcium phosphate minerals in residual clays.

Examples of phosphatization producing aluminum phosphate materials are given by Altschuler et al. (1956). They conclude that clay and apatite were dissolved and the resulting phosphoric acid reacted with clay to form crandallite $[CaAl_3(PO_4)_2(OH)_5 \cdot H_2O]$ and wavellite $[Al_3(PO_4)_2(OH)_3 \cdot 5H_2O]$. Other modes of origin are also given for these minerals.

Apatitized wood has been reported in several occurrences. Gulbrandsen et al. (1963) report an occurrence in the Moreno formation, California. This formation is an organic-rich mudstone with intercalated sandstone lenses, and marine fossils are reported to occur throughout the section. It is suggested that the wood itself may have created a local environment favorable for the deposition of apatite.

There are many examples of phosphatization in nature and experiments have been conducted that demonstrate that the process is feasible in nature. However, the importance of such a process to account for large sedimentary phosphate deposits, such as those of the western United States, is still a subject of debate.

DALE R. SIMPSON

References

Altschuler, Z. S., Jaffe, E. B., and Cuttitta, F., 1956, "The aluminum phosphate zone of the Bone Valley Formation, Florida, and its uranium deposits," *U.S. Geol Surv. Profess. Paper* **300**, 495–504.

Amos, L. L., 1959, "The genesis of carbonate apatite," *Econ. Geol.*, **54**, 829–842.

Blackwelder, E., 1916, "The geologic role of phosphorus," *Am. J. Sci.*, 4th series, **42**, 285–298.

Emery, K. O., 1960, "The Sea Off Southern California," New York, John Wiley & Sons, 366pp.

PHOSPHATIZATION

Emigh, G. D., 1958, "Petrography, mineralogy, and origin of phosphate pellets in the Phosphoria Formation," *Idaho Bureau Mines and Geol., Pamphlet*, **114**, 60pp.

Graham, W. A. P., 1925, "Experiments on the origin of phosphate deposits," *Econ. Geol.*, **20**, 319–334.

Gulbrandsen, R. A., Jones, D. L., Tagg, K. M., and Reeser, D. W., 1963, "Apatitized wood and leucophosphite in nodules in the Moreno Formation, California," *U.S. Geol. Surv. Profess. Paper* **475-C**, Art, 85, pp. 100–104.

International Geological Congress, 1953, "Origine des Gisements de Phosphates de Chaux," Section XI (Alger, 1952), 196pp.

Kazakov, A. V., 1937, "The phosphorite facies and the genesis of phosphorite, Geological investigations of agricultural ores USSR," *Trans. Sci. Inst. of Fert. and Insects-Fungicides*, No. 142, pp. 95–113.

Mansfield, G. R., 1927, "Geography, geology, and mineral resources of part of southeastern Idaho," *U.S. Geol Surv. Profess. Paper* **152**, 411pp.

McKelvey, V. E., Cathcart, J. B., Altchuler, Z. S., Swanson, R. W., and Buck, K. L., 1953, "Domestic phosphate deposits," Chapter XI in "Soil and Fertilizer Phosphorus," vol. 4, New York, Academic Press, 492pp.

Reese, C. L., 1892, "Influence of swamp water on the formation of phosphate nodules," *J. Sci.*, **43**, 402–406.

Rodgers, J., 1948, "Phosphate deposits of the former Japanese Islands in the Pacific: A reconnaissance report," *Econ. Geol.*, **43**, 400–407.

Simpson, D. R., 1964, "The nature of alkali carbonate apatite," *Am. Mineralogist*, **49**, 363–376.

Smith, R. W., and Whitlatch, G. I., 1940, "The phosphate resources of Tennessee," *Tenn. Div. Geol. Bull.*, **48**, 444pp.

Strakhov, N. M., 1960, "Climate phosphatae accumulation Economic Geology," USSR (translated) no. 1–2, London, Pergamon Press, p. 1-14.

Cross-references: *pH-Eh; Phosphorus; Weathering, Chemical.* Vol: I: *Upwelling.* Vol. IVB: *Apatite; Phosphates and Guano;* see also *Individual Minerals.* Vol. VI: *Leaching.*

PHOSPHORESCENCE—See Vol. IV B, LUMINESCENCE

PHOSPHORUS: ELEMENT AND GEOCHEMISTRY

Phosphorus, symbol P, is an element with an atomic number of 15 with six reported isotopes. The one stable isotope, $_{15}P^{31}$, has a mass of 31, although the chemical atomic weight is nearer 30.975. The binding energy of the 15 protons and 16 neutrons in the nucleus in the phosphorus atom has been found to be 8.4 MeV or, in another equivalent term, 2×10^8 kcal/gram-atom.

Physical Properties

Elemental Phosphorus. Elemental phosphorus is a solid at room temperature. In the nonmetallic solid form, phosphorus is a *white* or pale yellow, wax-like material that is very reactive. In the metal-like solid form it is dark *red*, more dense, and less reactive than in the white form. The *brown* and *black* solid forms are higher temperature modifications of red phosphorus.

Phosphorus exists in the *vapor* phase as P_4 molecules which have a tetrahedron structure.

Elemental phosphorus is too reactive and too readily oxidized to exist uncombined in nature. The chemistry of phosphorus in nature is largely the chemistry of the phosphate ion.

Chemical Properties

Phosphorus forms compounds to which all the oxidation numbers from -3 to $+5$ can be assigned. It resembles nitrogen in this respect. In the upper lithosphere, phosphorus is an oxyphile; i.e., it appears in an oxidized state. Unlike other elements of the lithosphere, it does not show any well-defined behavior as far as distribution is concerned among the silicates, sulfide, and metal phases in physicochemical systems.

Compounds of phosphorus generally are considerably more reactive than similar compounds of carbon. The explanation for the greater reactivity of phosphorus compared with carbon chemistry is based on a difference in electron structure. The compounds exhibiting triply connected phosphorus are unusually reactive because the unshared pairs of electrons are readily available for the formation of an "activated complex." The compounds having quadruply-connected phosphorus generally are less reactive than those triply connected. The phosphorus and adjacent atoms of all compounds based on triply and quadruply connected phosphorus (tetrahedrally coordinated), appear to obey the octet rule. According to this rule eight electrons are assigned to the valence shell of each atom. (See Van Wazer, 1958, for a complete review of phosphorus chemistry.)

Classification

Though phosphorus exhibits coordination numbers of 1, 3, 4, 5, and 6 in its compounds, none of those having coordination number of 1 are stable, and those having 5 and 6 are very reactive. Almost all of the vast number of phosphorus compounds have coordination numbers of 3 and 4. Classification of phos-

Table 1. Classification of Phosphorus Compounds[a]

Number of Attached Atoms	1	3	4	5	6
σ-Bond hybridization for observed symmetry	p	p^3 with sp^3 character	sp^3	sp^3d	sp^3d^2
Directional characteristics	Linear	Trigonal pyramidal	Tetrahedral	Trigonal bipyramidal	Tetragonal bipyramidal
Geometry	P—X	(trigonal pyramidal diagram)	(tetrahedral diagram)	(trigonal bipyramidal diagram)	(tetragonal bipyramidal diagram)
Number of known structures	ca. 6	Thousands	Millions	ca. 6	2
Stability under normal conditions	Unstable	Stable; but electron donor	Most stable as a class	Very reactive	Very reactive

[a] From Van Wazer (1958) Vol. I, page 71.

phorus compounds may be made, in a gross manner, based on coordination number or number of atoms attached to phosphorus (as suggested by Van Wazer, 1958). Table 1 shows his suggested scheme of classification.

Occurrence and Abundance

Universe and Earth. Phosphorus is abundantly present in the upper lithosphere, in meteorites, and even in the solar atmosphere. Thus phosphorus is believed to be one of the original elements present during the formation of the earth. The gross abundance in the universe is estimated to be very similar to the weight percentage of the phosphorus of the earth's sphere. By knowing both the composition of the stony and iron meteorites and the ratio of these kinds of materials in the earth's sphere, it is possible to arrive at some estimate of the abundance. Phosphorus is estimated to be the eleventh element in igneous rocks on the earth's surface, the thirteenth element in meteorites, and twenty-four in the sun's atmosphere. The "abundance of phosphorus" data as reported by Van Wazer (1961) appear in Table 2. The importance of the composition of meteorites, believed to have their origin in a planet very similar to the earth, in estimating the abundance of phosphorus in the universe, is emphasized by Rankama and Sahama (1950), who summarize work of F. W. Clark, H. S. Washington, V. M. Goldschmidt, and

Table 2. Abundance of Phosphorus[a]

Source	Weight (% P)
In the metal phase of meteorites	0.22
In the sulfide phase of meteorites	0.31
In the silicate phase of meteorites	0.158 ± 0.013
Gross abundance in meteoritic type material	0.18–0.19
In igneous rocks	0.10 ± 0.03
In sedimentary rocks	
Sandstones	0.04
Red clay	0.14
Shales	0.08
Limestones	0.02
Gross abundance on the surface of the earth	0.10–0.12

[a] From Van Wazer (1961) Vol. II, p. 956.

others. Van Wazer (1961) further summarizes and updates our knowledge of abundance of phosphorus.

Natural Form. Phosphorus occurs naturally on the earth's surface primarily in orthophosphates $[PO_4]^{3-}$. Small amounts of gaseous phosphine, PH_3, have been found in polluted springs and in the hypolimnion of lakes or marshes under highly reducing conditions. There is no phosphorus in the atmosphere except as it appears in dust particles and microbial debris. Water-soluble orthophosphorate appears only in very small amounts. Calcium, iron, and aluminum, as well as certain clay minerals of the soil, readily unite with soluble phosphorus to render it insoluble.

Soil. In the soil, phosphorus ranges from 0.022 to 0.83% P. In a soil having 0.064% P, an acre-foot would have approximately $4,000,000 \text{ lb} \times 0.064\% = 2560$ lb of P. Only a small fraction of this is available (soluble) for plant and microbial use.

Water. Ocean and seawater contain meager amounts of phosphorus. The concentration in seawater varies almost as much as in soils, ranging from 0.001 to 0.1 ppm. Lake and river waters contain traces of phosphorus. Recent pollution by man has increased the phosphate concentration in some instances above the natural ranges previously reported.

Rock. Igneous rocks contain phosphorus principally as *apatite*, $Ca_5[(F, Cl, OH)(PO_4)_3]$. Apatite contains between 18.0 and 18.7% P. As indicated by the empirical formula a second anion is an essential part of apatite. This anion may be Cl^-, F^-, OH^-, or CO_3^{2-}. The corresponding apatites are called, *chlorapatites, fluorapatites, hydroxyapatites,* and *carbonate apatites,* depending on the predominant anion (see articles: *Phosphates and Guano; Phosphatization*). Other cations and anions also occur in apatite structure. Pure species of carbonate apatite and hydroxyapatite are not found in nature.

Secondary Minerals. Some secondary minerals formed from phosphate in solution are: *vivianite,* $Fe_3^{2+}[PO_4]_2 \cdot 8H_2O$; *pyromorphite,* $Pb_5[Ce(PO_4)_3]$; *weinschenkite,* $(Y, Er)[PO_4] \cdot 2H_2O$; and *turquois,* $CuAl_6[(OH)_2 PO_4]_4 \cdot 4H_2O$. *Wavellite* $[Al_3(OH_3(PO)_4)_2 \cdot 5H_2O]$ has been found in soil but is less common than calcium, iron, and aluminum phosphates and clay mineral phosphates. The soil forms of phosphate have not been clearly defined.

Rare Deposits. Only a small portion of the phosphorus in the igneous rock of the earth's crust does not occur as apatite. Two other primary phosphates that occur in limited amounts are the rare earth phosphates *monazite,* $Ce[PO_4]$, and *zenotime,* $Y[PO_4]$, which may be found in granites. Nepheline, syenite, pegmatite, and granite also may contain some lithium, beryllium, aluminum, and magnesium phosphates such as $NaBe[PO_4]$, $Li(Fe^{2+}, Mn^{2+})[PO_4]$, $(Fe^{2+}, Mn^{2+})_2[F, PO_4]$ and $LiAl[(F, OH) PO_4]$. These phosphates play only a very small role in the geology of phosphorus.

Sea and Ocean Floor. Phosphorite deposits in the ocean show that collophane, composed of isotropic carbonate fluorapatite, is the principal mineral form in which phosphorus is present (see *Phosphates and Guano*). Francolite, also a carbonate fluorapatite, was found associated with collophane.

Phosphorus—Organic

Origin in Biophilic Systems. The biophilic nature of phosphorus is emphasized by the fact that life could not exist without phosphorus. The multitude of facts concerning biological life processes that have appeared recently in the literature has greatly overshadowed the inquiry into the progressive evolutionary events, if any, that preceded the starting point of the biological portion of evolutionary processes. The postulate that the earth's atmosphere started as a mixture of methane, ammonia, and water vapor requires a wholly new concept concerning phosphorus and "life process." Such an atmosphere would be anaerobic or nearly so (see *Precambrian Atmosphere*). In an atmosphere devoid of oxygen, dissolved phosphorus presumably would be substantially more abundant.

Gulick (1955) visualized the presence of soluble inorganic compounds in the earth's waters in an atmosphere which was anoxic or nearly so, to substantiate his argument that "life originated under conditions in which phosphorus was only partially oxidized." Data in Table 3 show that the fully oxidized phosphoric acids produce quite insoluble salts, whereas the less oxidized phosphites and hypophosphites are more soluble. These two phosphorus species could well satisfy the demand by life processes for soluble phosphorus in an atmosphere devoid of oxygen.

Gulick further postulates that at the redox potentials necessary for phosphorus to appear as soluble salts or ions in the presence of calcium, an emf of 0.3–0.51 V would be necessary. This eliminates a host of substances and reactions that various authors have suggested. In fact, both carbon dioxide and carbon monoxide do not look like possible combiners in the *initial* reactions with phosphorus. Carbon compounds with hydrogen are more logical leading contenders for "first bonding compounds with phosphorus." Assuming an at-

TABLE 3. SOLUBILITIES OF INORGANIC PHOSPHORUS IN WATER

Substance	Formula	Solubility (gram-atoms/ metric ton)	Remarks
Monocalcium phosphate	$(Ca(H_2PO_4)_2)$	—	Ppt. as insol. salt by Fe or Al
Dicalcium phosphate	$CaHPO_4$	1.16	Ppt. as insol. salt by Fe or Al
Tricalcium phosphate	$(Ca_3(PO_4)_2)$	0.16	But 0.004 in seawater because of Ca
Ammonium calcium phosphate	NH_4CaPO_4	Insol.	—
Calcium fluorophosphate	$(Ca_{10}F_2(PO_4)_6)$	Insol.	0.004 in seawater
Calcium pyrophosphate	$(Ca_2P_2O_7)$	Insol.	—
Calcium metaphosphate	$(Ca(PO_3)_2)$	Insol.	—
Calcium hypophosphate	$(Ca_2P_2O_6)$	Insol.	—
Calcium phosphite	$CaHPO_3$	Slightly sol.	—
Calcium hypophosphite	$(Ca(H_2PO_2)_2)$	2000	—

mosphere of a gas mixture of carbon monoxide, methane, cyanogen, and ammonia, together with moisture and hydrogen where photochemical energy is supplied, one could postulate bondings complete to phosphorus, "energy-rich" bonds.

Carbon-to-carbon linkages would occur to form C_2H_6, C_2H_4, C_2H_2, etc., photochemically. Amine formation by NH_2 bonding such as ethylene amine $H_2C=CHNH_2$ may take place. This is substantiated in part by the finding that glycine, alpha alanine, etc., could be detected when a mixture of methane, ammonia, water, and hydrogen were subjected to an electrical discharge. In the presence of an abundance of ammonia as is claimed by many authors, the following reaction could take place (see Gulick, 1955):

Jackson Reaction

$$CO + 2NH_3 \rightarrow NH_4OC\!\!=\!\!N + H_2 \rightarrow H_2NC\!\!=\!\!N + H_2O + H_2$$
$$\text{cyanamid}$$

Further

$$H_2NC\!\!=\!\!N + H_4NOPHOH \rightarrow H_2N\!\!-\!\!C\!\!-\!\!NH_3\!\!-\!\!O\!\!-\!\!PHOH$$

cyanamid   ammonium phosphite   guanidine   phosphite

(with $\|O$, $\|NH$, $\|O$ respectively)

$$H_2N\!\!-\!\!C\!\!-\!\!NH\!\!-\!\!PO(OH)_2 + H_2 \leftarrow$$
$$\|\;NH$$

Phosphoguanidine

Thus the "high energy" phospho-compounds are formed by energy supplied by *dehydrogenation*.

The proposal as represented by the final formation of such a compound as *phosphoguanidine* as an initial phosphorus compound, further must postulate several "geochemical crises" to bring us back to modern times. Some such "crises" are:

(a) The organisms must have had to adapt to a condition where the soluble phosphites ceased to exist, by developing a mechanism for concentrating and utilizing the very *meagerly soluble phosphates* found in waters, rock matrix and soil solution.

(b) Adaptation to the very *meager concentration* of *carbon monoxide* (CO) which presumably was among the chief nutrients of the earliest living forms.

(c) Of great significance was the development of metabolic process to fit the presence of free oxygen to make direct use of oxidative reactions.

Life further had to make certain critical adjustments. For example, it had to adapt to neoabsence of ultraviolet light by developing porphine, an iron-containing enzyme, that can capture and utilize the less energy-yielding light waves in the visible range.

Although Gulick's concept of the role of phosphorus in primitive living systems is intriguing indeed and is well defended, there remain certain factors that have yet to be put into the overall jigsaw puzzle. For example:

(a) PO_4 is the form of phosphorus in igneous rock
(b) PO_4 is the only stable phosphorus state in aqueous systems even at low PO_2
(c) the life that evolved is extremely intolerant of even low concentrations of lower oxides of phosphorus
(d) phosphorus exists almost wholly as phosphate in living systems

Phosphorus is extremely reactive with oxygen as are the lower oxide forms and they all go avidly to PO_4. Moreover, phosphorus originally occurred in igneous rock as apatite and would have to be reduced to other forms to accommodate the Gulick concept.

Phosphorus in the Biosphere. *Availability.* Phosphorus appears in the biosphere as a fully oxidized element. The accessible or "available" phosphorus for biological use in our modern lithosphere is almost exclusively in the form of soluble orthophosphate ions. Although phosphorus shares the important process of life with such elements as nitrogen, oxygen, hydrogen, carbon, and sulfur, it differs from these elements in its availability. Phosphorus is the least available of these elements to most living systems.

Compounds. Numerous organic phosphates are found in plants, animals, and microorganisms. (Inorganic phosphates also occur in living cells.) Such organic compounds containing phosphorus as complex nucleoproteins, phosphoproteins, ortho- and polyphosphate esters, phospholipids and phosphagens are well known to appear in living tissues. The chemistry of organic orthophosphates in biological systems alone is so involved that it would take an entire volume to elucidate. These phosphates include esters of orthophosphoric acid: alkyl orthophosphates, aryl orthophosphates, and esters in which organic radicals possess functional groups. The latter class includes the sugar phosphates, glycerophosphates and the mixed anhydrides of organic acids and orthophosphoric acid.

Perhaps the most unexpected trend to evolve from the vast knowledge recently accumulated on cell biochemistry, is the similarity in processes necessary for life as well as similarity in organic compounds entering into life processes despite great differences among the various species of living organisms. Phosphorus is an indispensable element in the commonly shared chemical patterns of life.

WALLACE H. FULLER

References

Gulick, A., 1955, "Phosphorus as a factor in the origin of life," *Am. Sci.,* **43,** 479–489.
Katchman, B. J., 1961, in (Van Wazer, J. R., editor), "Phosphorus and Its Compounds," Vol. 2, pp. 1281–1343, New York, Interscience Publ.
Lewis, R. W., 1965, "Phosphate Rock," U.S. Bur. Mines Min. Yearbook.
McKelvey, V. E., 1967, "Phosphate deposits," *U.S. Geol. Surv. Bull.,* **1252-D,** D1-D21.
Rankama, K., and Sahama, T. G., 1950, "Geochemistry," Univ. Chicago Press, 912pp.
Van Wazer, R., 1958, and 1961, "Phosphorus and Its Compounds," Vol. I (1958) Vol. 2 (1961), New York, Interscience Publ.
Whaley, T. P., and Currier, J. W., 1968, "Phosphorus," in (Hampel, C. A., editor), "Encyclopedia of the Chemical Elements," New York, Reinhold Book Corp., 524–533.

Cross-references: *Biogeochemistry; Complexes; Geochemical Classification of Elements; Minneraloids; Organic Geochemistry; Phosphatization; Phosphorus Cycle; Precambrian Atmosphere; Seawater, Chemistry; Seawater, History.* See also *Individual Minerals.* Vol. II: *Meteorites.* Vol. IVB: *Phosphates and Guano.*

PHOSPHORUS CYCLE

Phosphorus is a constituent of meteorites, igneous rocks, soil, lakes, rivers, the sea, and all living organisms. It cycles readily between inorganic and organic systems. It is present in the sun's atmosphere where it has been identified quantitatively.

Phosphorus is an essential constituent of all living organisms. In living cells it functions in the conservation and transfer of energy in metabolic reactions. As a limiting factor for plant growth, phosphorus is second in importance only to nitrogen. Phosphorus is apt to be the most important, ecologically, of all the elements contained in living organisms. The reason for this is twofold, because (a) the living organism tends to concentrate phosphorus, raising its level considerably above that of the source, and (b) a deficiency of available phosphorus is more likely to limit *reproduction* and hence productivity of any region on the earth's surface than any other material except water. The biological cycle of growth, reproduction, death, and decay provide a basis for the phosphorus cycle, a cycle of concentration, utilization, and return to soil, lake, and sea.

Since phosphorus performs such a diversity of roles and is essential to the persistence of life on this earth, a few basic facts concerning its chemical and physical characteristics and its occurrence and abundance are necessary as a background to an understanding of this important cycle (see *Phosphorus: Element and Geochemistry*).

Gross Cycle

The most outstanding feature of the overall phosphorus cycle is that it is composed of many interwoven cycles, each carrying out some specific function as a part of earth's dynamic system. The overall cycle is like a wheel with the original phosphorus source, igneous rock, at the hub (see Fig. 1). Oriented around this hub are the many cycles, the most prominent of which are: lithosphere (upper consolidated rock including sediments and ores), pedosphere (soil), hydrosphere (water—fresh and marine), zoosphere (animal biosphere—higher and lower life). The dynamic force that makes the phosphorus turnover within and between cycles is the life processes of the biosphere which, basically, derive energy from the sun. This is depicted in Fig. 1 as the power at the hub of the phosphorus cycle that touches all component cycles.

Another overall phosphorus cycle, representing the present-day conditions in the continental United States and its adjoining coastal waters according to Van Wazer (1961) is presented in Fig. 2. He presents estimated rates of phosphorus transfer taking place between different systems. These figures should not be confused with total amounts. Again, this cycle shows the great interaction between specific segments as well as the dynamics of the biological processes.

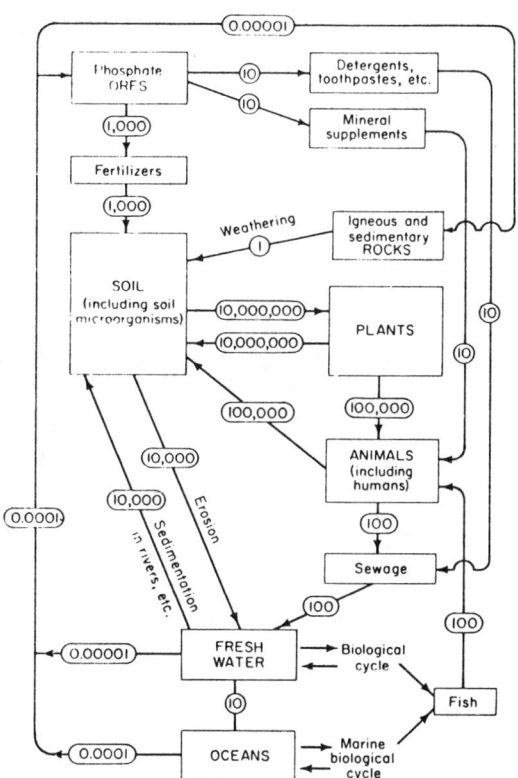

Fig. 2. The overall phosphorus cycle in the present-day continental United States and its connections with the coastal waters. The approximate rate of phosphorus transfer is given in million pounds per year of phosphorus equivalent, as estimated to the nearest order of magnitude (from Van Wazer, 1961, Vol. II, p. 1295).

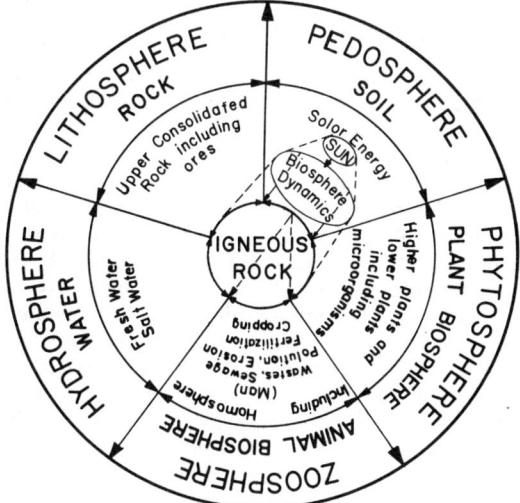

Fig. 1. Gross phosphorus cycle.

Early Phase. All of the phosphorus on the earth's surface originally was present in igneous rock, primarily as apatite. Weathering of this rock liberated phosphorus in a dissolved state. Dissolved phosphorus was then either (a) absorbed into animal and plant systems and cycled, or (b) redeposited in ancient waters as insoluble or slowly soluble apatite rocks. Some, no doubt, united with calcium, iron, and aluminum and other ions forming clay minerals to become part of the early "pedosphere" or soil.

Modern Phase. The early deposits of phosphate formed in waters now represent the majority of phosphate-rich ores which are mined. These phosphate deposits are being distributed to the four corners of the earth because of man's relatively recent, rapid development of uses for phosphorus in industrial and domestic chemicals. As the human population continues to explode over the earth's surface, man's cyclic effect on concentration, utilization, and deposition will increase in geometric proportion to population increases. The *homosphere* as a segment of the total biosphere even now is playing a critical part in the redistribution of phosphorus on the earth. Some examples are:

(a) Accelerating soil erosion with consequent loss of the organic soil layer, the layer of richest accumulation of phosphorus, to river bottoms and sea floors.
(b) Distribution of agricultural chemicals —fertilizers, pesticides, minerals—from phosphate ores to vast areas of the earth's surface, poor in concentration, where much of the phosphorus becomes fixed in the pedosphere in a form unavailable to life processes.
(c) Losses of crop phosphorus through sewage disposal that eventually "dead ends" in river and ocean bottoms.
(d) Industrial and domestic losses through development of phosphorus chemicals, detergents and solvents, and consequent disposal.
(e) Pollution losses as gases or colloidal particles to the atmosphere, and final dilution by thin distribution over the earth's surface.

Biological Dynamics

Rate-Controlling Factor. The biological processes form the common rate-controlling factor for all the cycles. The biological system may be compared to the mainspring on a delicately balanced watch that keeps the wheels of the numerous integrated cycles turning. The rate of turnover of phosphorus in this cycle is a function of the rate of biological activity. The cyclical events of solubilization, growth, decay and deposition are all entered into and rate-controlled by life processes. The biological cycles function similarly in the hydrosphere, pedosphere, and the lithosphere and are a very essential part of the general geochemical cycle of phosphorus.

Assimilation. Assimilation is used here to mean the building up of complex organic compounds from simple compounds. Perhaps the least well-represented parts of the phosphorus cycles shown here are the organic transformations. Biologically-available phosphorus, which includes both water and dilute-acid soluble phosphorus as well as certain organic phosphorus compounds, is taken up from the environment by living organisms. Part is transformed into a diversity of organic compounds and a part is retained as mobile, inorganic phosphorus in the cell fluids. The inorganic form maintains a high degree of mobility within the living organism. Phosphorus is thus concentrated from a very *dilute level* of the organism's habitat to a fairly *high level* in the living organism. Thus all living organisms—animals, plants, and microorganisms—temporarily immobilize phosphorus by incorporating it into their cells.

Dissimilation. Dissimilation is used here to mean the breaking down of complex organic compounds into simpler ones. Upon death of the cell, decay begins and phosphorus is mineralized; i.e., released again in the form of orthophosphates. This is brought about by enzyme action or organisms. The biological phases of the cycle are complicated by the "feeding" of one organism on another. Most organisms, whether plant or animal, can use some of the simple organic phosphorus compounds contained in living tissues as well as the inorganic forms. However, during decay phosphorus is always in excess because CO_2 is lost to the atmosphere, whereas phosphorus continues to recycle until the excess is liberated gradually as inorganic orthophosphate.

Deposition. In mammals most of the phosphorus is centered in the skeletal structure. This is more resistant to decomposition than the less dense tissues. Thus skeletal materials are known to accumulate in waters to form sources of phosphate deposits, generally apatite in nature. Higher organisms generally are richer in phosphorus than are lower forms.

Lithosphere—Rock Cycle

The phosphorus in igneous rocks is slowly solubilized by weathering process. The soluble phosphates, colloidal particles, and fine rock fragments are transported by wind, water, and ice to all segments of the overall phosphate cycle. Carbon dioxide, a product of biological metabolism, when united with water, forms a weak acid which enhances the solubility of phosphorus in rocks. Indeed, the microorganisms in the soil have been found to solubilize inorganic phosphorus of rock and rock fragments. Parts of the soluble phosphorus is absorbed by the biological phases of the pedosphere, phytosphere, zoosphere, and hydrosphere, and a part is reprecipitated in the pedosphere, hydrosphere, and lithosphere as secondary phosphate minerals and calcium phosphates. Notable among secondary deposits is phosphorite, which is found precipitated on ocean floors. An excellent example of such deposits is on the ocean floor off the coast of Southern California (Dietz et al., 1942). Deposition in soils also occurs usually in finely divided particles as calcium, iron and aluminum phosphate. Consolidated sediments such as sandstones, shales, limestones and clay minerals

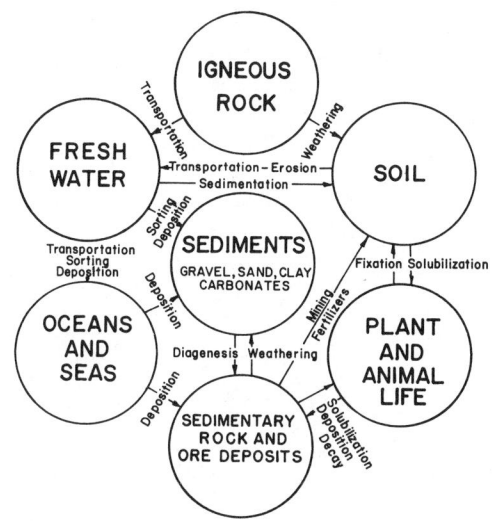

FIG. 3. The phosphorus cycle in the minor lithosphere.

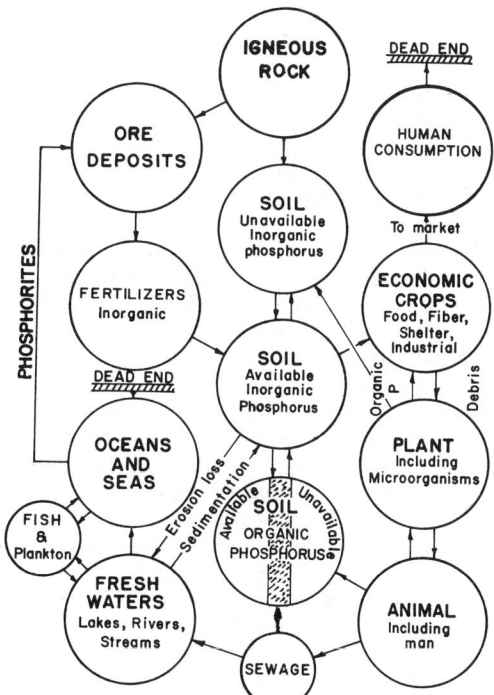

FIG. 4. Phosphorus cycle in soils.

also contain some phosphorus representing redistribution, accumulation, and deposition in the rock lithosphere. Such phosphorus-containing materials, however, do not always represent a dead-end since weathering processes again can return the phosphorus to a more rate-active cycle (see Fig. 3).

Pedosphere—Soil Cycle

The phosphorus cycle in soils touches all other cycles just as does igneous rock. Generally the cycle has been considered to begin with *rock* and proceed as follows: Rock → soil → plant → animal → rock. The soil cycle is much more complicated, however. *Indirectly*, for example, it contributes to fresh water, oceans, phosphate ores, and finally to igneous rock as part of the overall phosphorus cycle. *Directly*, soil phosphorus contributes in a reversible manner, as:

soil ←--→ plant ←--→ animal ←--→ sewage
    ↖ rock ←--------→ fresh water ↗
              (irrigation of soils)

The general cycle of phosphorus in soils has been diagrammed a number of times, some of which may be found by reviewing Alexander (1961), Katchman (1959; in Van Wazer, 1961), and Truog (1953). Alexander emphasized the biological cycle of phosphorus in soils, Katchman the soil-plant interrelationship, and Truog the mineral, particularly calcium, interplay in the phosphorus cycle of the soil. Pierre (1948) did not present a diagram of the cycle but reviewed thoroughly the phosphorus cycle and soil fertility. Fig. 4 is an attempt to incorporate all these views into one phosphorus cycle in soil.

Hydrosphere—Fresh-Water Cycle

Lakes, Streams, and Rivers. The phosphorus cycle in fresh waters is very complicated. Waters contain both undissolved (sestonic) and dissolved phosphorus. Total amounts range from a trace to as much as one part per million. However, stream and river waters usually contain less than 0.1 ppm total phosphorus. The amount of soluble phosphorus is only a fraction of this total. Phosphorus is generally higher in streams and rivers coursing over sedimentary rocks than in the water of granitic and metamorphic rocks. Hutchinson (1957) categorizes lake phosphorus into: (a) soluble phosphorus, (b) acid-soluble sestonic phosphorus, (c) organic soluble (and colloidal) phosphorus, and (d) organic-sestonic phosphorus, as a basis for a discussion on the phosphorus in lakes. In general, surface waters of lakes are uniformly low in phosphorus. Human contamination, however, can greatly change the value. Lake Erie, for example, is notably contaminated with disposal by human beings. Here human intervention is shown to have indispltably upset the natural phosphorus cycle. Lakes high in natural plant debris, peaty materials, contain

PHOSPHORUS CYCLE

higher levels of phosphorus than clear lakes. Agricultural drainage as a result of accelerated erosion of soil, alters the nature and quantity of phosphorus in lakes, rivers and streams. Lakes may vary as much as 1:1 to 1:90 in ratio between soluble phosphorus to other forms of phosphorus.

Variation in undissolved phosphorus is due primarily to biological growth, phytoplankton and algae. Soluble phosphorus may be higher at greater depths due to decomposition and liberation of soluble orthophosphates from decaying organic debris. Iron as well as calcium salts in waters tend to precipitate phosphorus in lake bottoms as minerals of slight solubility. Striking increases in total phosphorus occur at the bottom during periods of stagnation and are attributed to the reduction of iron (III) phosphate to the more soluble iron (II) phosphate. Decreases are associated with the bloom period when the increased plankton and algae population absorb most of the available phosphate.

Shallow muds generally lose phosphorus to deep muds as a result of gravitational drainage, differences in density of phosphate-laden water, and to thermal changes.

In fresh water, phosphorus is depleted from the upper reaches of the streams to lower levels by stream flow of soluble forms and colloidal and debris suspensions. The phosphorus accumulates finally on lake bottoms, sea and ocean floors as rivers and streams ultimately reach these more static levels. The phosphorus cycle in fresh water may be visualized as in Fig. 5.

Hydrosphere—Salt-Water Cycle

Seas and Oceans. The amount of phosphorus in sea- and ocean waters is less than 0.1 ppm, with a range of 0.1 to 0.001 ppm. Early studies by Dietz et al. (1942), and Sverdrup et al. (1942), as reported by Van Wazer (1961), show that the amount rapidly increases with depth from a trace at the surface to a maximum at levels of 900 to 1300 meters, depending on the ocean tested. Only relatively small fluctuation occurs below this depth. Maximum concentrations in the Atlantic, Pacific and Indian Oceans are about 0.060, 0.090, and 0.10 ppm, respectively. Van Wazer (1961) suggests a weighted-average phosphorus factor of about 0.07 ppm to arrive at a figure of about 100 billion tons of dissolved phosphate in the present seas of the world.

Deposits of phosphate ore are principally marine in origin. These sedimentary deposits exceed in size and number the igneous deposits which are high in phosphorus. An excellent classification of the principal kinds of phosphate deposits is given by Van Wazer (1961), along with other data. Soluble phosphorus brought in from rivers and streams along with phosphorus of decaying marine life form the insoluble deposits on the ocean floors. Nodular phosphorite was found by Dietz et al. (1942) as the most abundant single type of rock on banks, escarpments and on walls of submarine canyons off the Southern California coast. This rock was shown to form by direct precipitation from soluble sources in the water. Thus the seas represent the final repositories for phosphorus from all other solubilizing sources in the overall phosphorus cycle.

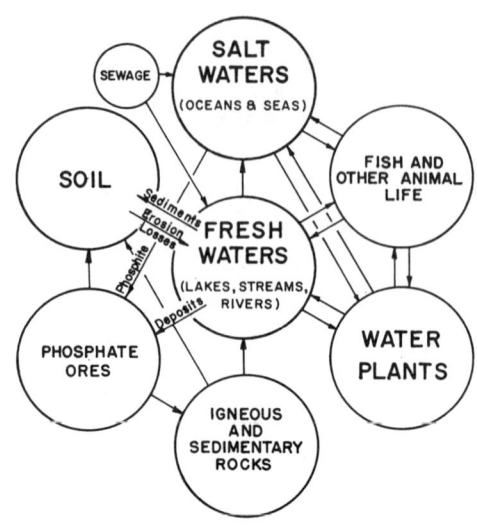

FIG. 5. Phosphorus cycle in fresh and salt waters.

The phosphorus in the final deposits in the ocean is derived from: dissolved phosphorus of eroded soil; rocks and minerals over which streams and rivers flow; marine life, e.g., aquatic plants, fish bones, crustacean shells; microorganisms; and other organic debris of the biosphere and zoosphere, including sewage. The biochemical cycle in the sea as developed by Redfield (1958), shows atom ratios of elements, phosphorus, nitrogen, carbon and oxygen, in the synthesis and decomposition of organic matter of the biosphere. The concentration of phosphorus in the sea and ocean cycle are postulated to depend almost strictly on these processes. The data also show that the sedimentary phosphate rock deposited on the ocean floors contains about 40,000 atoms of phosphorus for every atom in solution. The importance of plankton in this cycle is emphasized both with respect to their P–N–C ratios, which is about 1:16:106, and carbon dioxide production which acts in the solubilization processes making phosphorus available for assimilation by biological systems.

PHOSPHORUS CYCLE

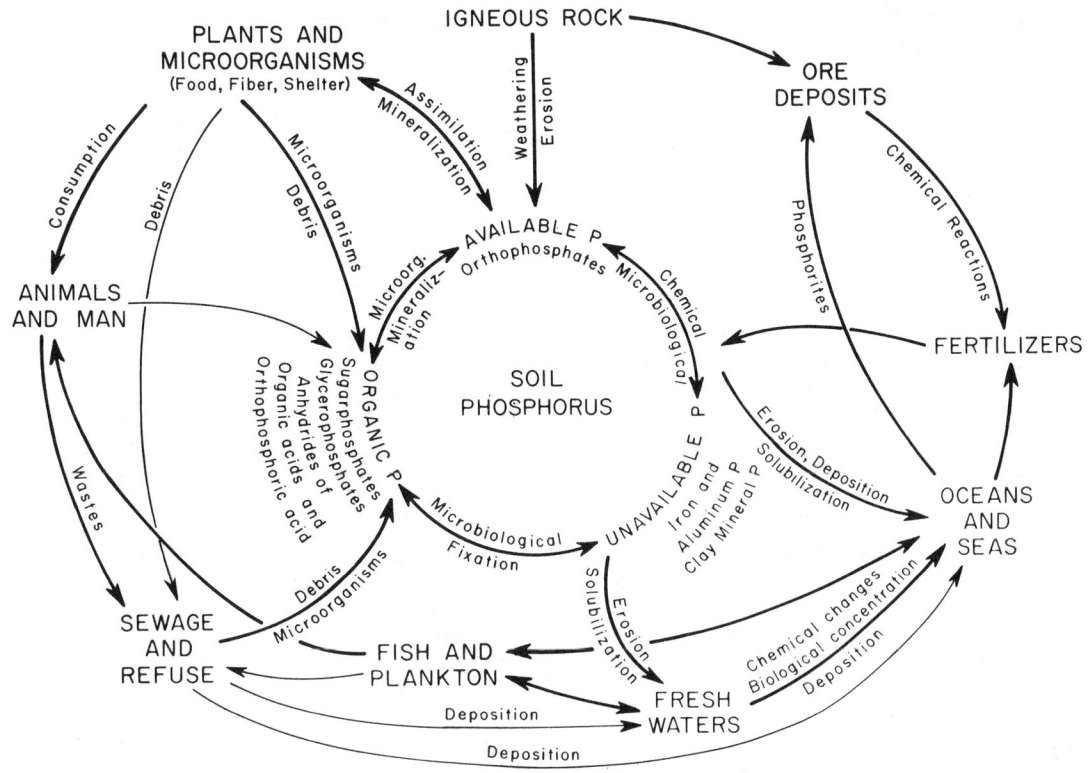

FIG. 6. The phosphorus cycle of the earth.

An idealized phosphorus cycle in seas and oceans has already been shown previously in Fig. 5 along with that of the fresh-water cycle. The hydrosphere cycle is so closely integrated with respect to phosphorus, at least, that it is not feasible to present the fresh- and salt-water cycles separately. Combining the individual cycles into a single cycle, the phosphorus cycle of the earth can be visualized as in Fig. 6.

WALLACE H. FULLER

References

Dietz, D. S., Emery, K. O., and Shepard, F. P., 1942, "Phosphorite deposits on the sea floor off Southern California," *Geol. Soc. Am. Bull.*, **53**, 815-847.

Fuller, W. H., 1965, "Basic concepts of nitrogen, phosphorus and potassium in calcareous soil," *Univ. Ariz. Agr. Expt. Sta. and Ext. Ser. Bull.*, **A-42**, 1-36.

Gulick, A., 1955, "Phosphorus a factor in the origin of life," *Am. Sci.*, **43**, 479-489.

*Hutchinson, H., 1957, "Phosphorus Cycle in Lakes," in, "A Treatise on Limnology," same author, Vol. 1, New York, John Wiley & Sons, pp. 727-752..

*Katchman, B. J., 1961, "Phosphate in Life Processes," in, "Phosphorus and Its Compounds," Vol. 2, (Van Wazer, J. R. editor), New York, Interscience Publ., pp. 1281-1343.

*Pierre, W. H., 1948, "The phosphorus cycle in soil fertility," *J. Am. Soc. Agron.*, **40**, 1-14.

*Rankama, K., and Sahama, T. G., 1950, "Geochemistry," Univ. Chicago Press, 912pp.

*Redfield, A. C., 1958, "Biological control of chemical factors in the environment," *Am. Sci.*, **46**, 205-221.

Truog, Emil, 1953, "Liming in Relation of Native and Applied Phosphorus," in (Norman, H. G., and Pierre, W. H., editors), "Soil and Fertilizer Phosphorus in Crop Nutrition," New York, Academic Press, pp. 291-292.

*Van Wazer, R., 1958 and 1961, "Phosphorus and Its Compounds," Vol. 1 (1958) Vol. 2 (1961), New York, Interscience Publ.

*Additional bibliographic references may be found in this work.

Cross-references: *Biogeochemistry; Calcium Cycle; Cycles—Geochemical; Iron; Lake Geochemistry; Phosphatization; Phosphorus: Element and Geochemistry; River Geochemistry; Seawater, Chemistry; Weathering, Chemical.* Vol. IVB: *Phosphates and Guano.* Vol. VI: *Limnology; Pedology; Soils.*

PHOTOLUMINESCENCE

PHOTOLUMINESCENCE—*See* Vol. IV B, LUMINESCENCE

PHOTOSYNTHESIS

Green plant photosynthesis is the process by which stable inorganic compounds such as CO_2 and water are converted into complex, unstable, energy-rich organic compounds in the plant. The complete process is an endergonic one, with light as the initial source of energy and the chlorophyll molecule of the green portions of the plant mediating the initial light-capturing process. The photosynthetic equation is commonly given as

$$6CO_2 + 12H_2O \xrightarrow[\text{chlorophyll}]{\text{light}} C_6H_{12}O_6 + 6O_2$$

but the process is not actually so simple or direct. Generally, the process can be thought of as consisting of two fairly distinct subprocesses, the so-called "light" and "dark" reactions, which operate at different rates.

Arnon (1961) has equated the "light" reaction with the production of chemical energy in the form of ATP, and "reducing power" in the form of $NADPH_2$, through processes called cyclic and noncyclic photophosphorylation. ATP and $NADPH_2$ together supply the energy requirements for manufacturing hexose sugars, and subsequently other carbohydrates, from CO_2 and water. In both cyclic and noncyclic photophosphorylation light energy is transformed, via chlorophyll and certain cytochromes, into chemical energy, the reaction being simply written as

$$nADP + nP \xrightarrow{\text{light}} nATP$$

The cyclic process can proceed anaerobically and does not evolve oxygen. The noncyclic process, in addition to performing the above reaction, also reduces NADP to $NADPH_2$ and evolves O_2. The composite reaction is

$$nNADP + nH_2O + nADP + nP \xrightarrow{\text{light}}$$
$$nNADPH_2 + nATP + nO_2$$

The "dark" reaction is actually a series of chemical reactions in which CO_2 is incorporated into the plant through a five-carbon compound, to form a six-carbon intermediate. This intermediate is split into two three-carbon compounds, which are subsequently reduced to hexose phosphates. These energy-rich compounds are then combined with other compounds in various ways to yield different carbohydrates. The "dark" reaction is slower than the "light" reaction, and thus is the rate-controlling reaction for the complete photosynthetic equation. The path of carbon in photosynthesis has been elucidated by Bassham and Calvin (1957), and can be simplified as shown in Fig. 1.

Various types of photosyntheses occur in almost every conceivable environment pervaded by light, and in many organisms besides green plants. Purple, red, and sulfur bacteria, red, coralline, and "snow" algae, and many other forms perform certain types of photosynthesis. Not always is CO_2 reduced or hexose sugar formed, nor is O_2 always evolved or chlorophyll necessarily the only pigment utilized. The only universal requirement of photosynthesis, as the name implies, is light energy. Different plants may best utilize different wavelengths of light, each wavelength having a characteristic energy manifestation, but all photosynthesizing plants of the earth are dependent upon some wavelength associated with the visible or near-visible spectrum (4000–7000 Å).

Green plant photosynthesis is by far the most prevalent type of photosynthesis on the earth. Through this mechanism, all green plants become part of a giant chemical industry that modifies the earth's atmosphere and hydrosphere. Yearly figures estimate that plants of the earth combine 150 billion tons of carbon with 25 billion tons of hydrogen to set free 400 billion tons of oxygen (Rabinowitch, 1955). If the total carbon reserve in the hydrosphere and atmosphere is 17×10^{12} tons, it would take the plants of the earth slightly less than 100 years to assimilate this quantity. If oxygen in the atmosphere amounts to 12×10^{14} tons, this supply would be renewed by photosynthesis every 3000 years. Perhaps up to 90% of this carbon fixation and oxygen evolution is accomplished under the surface of the ocean by phytoplankton algae, and only 10% on land.

The efficiency with which plants utilize solar energy falling on the earth's surface is quite small. Of the 73×10^8 kilocalories falling on an acre of land per year in temperate latitudes, only about 0.1 to 0.5% is fixed as organic material. In the sea and terrestrial croplands the average efficiency may be somewhat greater, on the order of 1 or 2%. Even though efficiency is low, the energy converted by photosynthesis annually is about 100 times greater than the heat of combustion of all the coal mined on the earth in one year, and it is about 10^4 times greater than all the energy derived by man from water power during one year (Rabinowitch, 1945). The theoretical

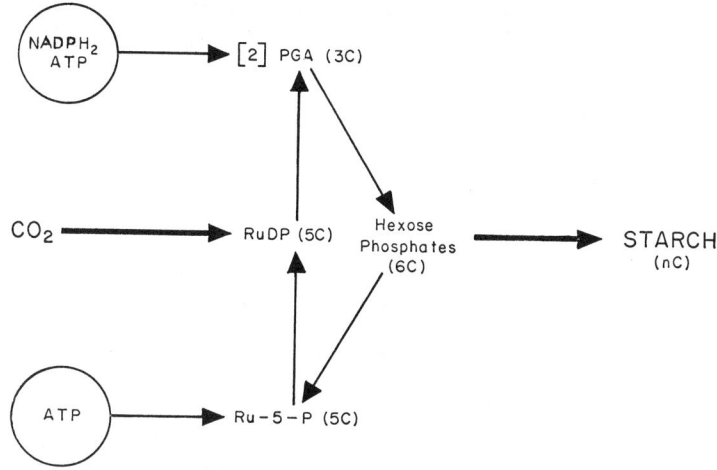

Fig. 1. Diagram of the reductive carbohydrate cycle ("dark reaction") in plant chloroplasts. Ribulose-5-phosphate (Ru-5-P) is phosphorylated (P added from ATP) to ribulose diphosphate (RuDP), which then accepts a molecule of CO_2 and is cleaved into two molecules of phosphoglyceric acid ([2]PGA). PGA is then reduced in several steps, upon input of ATP and $NADPH_2$, to hexose phosphates, which then can be converted into storage carbohydrate (starch) or back to Ru-5-P by several pathways. The symbols in parentheses (5C, 3C, 6C) refer to the number of carbon atoms per compound (after Arnon, 1961).

maximum photosynthetic efficiency of a plant cell, based on the inherent ability of the photosynthetic mechanism to capture and utilize light energy, is around 20%, and Tamiya (1957) and others have approached this value in mass cultures of unicellular algae. In the natural environment, high efficiencies of photosynthetic production of organic matter have been observed for sugar cane and for plant communities around coral reefs, in a turtle grass flat, and in certain polluted waters. Efficiencies less than maximum may be caused by a number of factors, including inability of plant chloroplasts to absorb all available light, CO_2 levels which are too low and light levels which are too high for optimum functioning of the photosynthetic mechanism, and inorganic nutrient deficiency (particularly nitrogen).

Factors which control a plant population can affect photosynthetic production of organic matter on an areal basis simply by limiting the number and types of plants growing in the area. Such factors as amount or seasonality of rainfall (in the case of terrestrial plants), salinity (in the case of marine plants), temperature extremes, and phosphorus availability fall into this category. Photosynthetic production in any area of the biosphere is thus controlled by factors affecting the photosynthetic mechanism of individual plants and the distribution and abundance of those plants. Because of this, large differences in photosynthetic productivity are found on both a temporal and spatial basis. Table 1 depicts daily photosynthetic production of organic carbon in several very different environments, and estimated average annual production for several different ecological regimes.

An estimate of annual production in marine environments by Steeman-Nielsen (1954) is $12-15 \times 10^9$ tons C/yr, or about an order of magnitude lower than the estimate in Table 1 (Riley, 1944). Ryther (1959) estimates about 30×10^9 tons C/yr produced in the world oceans, which is approximately twice the value of Steeman-Nielsen. Riley's estimates were earlier than the latter two, and necessarily based on less data for the world oceans. Steeman-Nielsen and Ryther also had the advantage of a fairly refined C^{14} technique for measuring net photosynthetic production. On these bases, perhaps the latter two estimates are more accurate, and oceanic production is one or two times the production on land rather than greater by a factor of six or seven. On an areal basis, Riley (1944) estimates the average annual production of organic matter in terrestrial environments is 160 tons C/km^2, and 340 tons/km^2 for open oceanic waters. Schroeder's (1919) annual average for land is 113 tons C/km^2. From this we gather that

PHOTOSYNTHESIS

TABLE 1. PHOTOSYNTHETIC PRODUCTION OF ORGANIC CARBON[a]

System	Gross (g C/m²/day)	Net (g C/m²/day)	Annual Net (metric tons C/yr)
Total earth	—	—	142.6×10^9
Total land(S)	—	—	16.6×10^9
Forest(S)	—	—	11.0×10^9
Pine forests (best growing years)	—	2.7*	—
Cultivated land(S)	—	—	4.3×10^9
Sugar cane	—	8.3*	—
Rice	—	4.1*	—
Wheat	—	2.1*	—
Grassland(S)	—	—	1.1×10^9
Tall prairie	—	1.4*	—
Short prairie	—	0.23*	—
Desert(S)	—	0.09*	0.2×10^9
Total ocean(R)	—	—	126.0×10^9
Coral reef	10.8+	4.3+	—
Polluted estuary	5.0+	3.6+	—
Grand Banks (April)	4.9+	2.9+	—
Continental shelf (May)	2.7+	1.7+	—
Continental shelf	0.33°	0.18°	—
Long Island Sound	0.95°	0.41°	—
Sargasso Sea (April)	1.8+	1.3+	—
Sargasso Sea	0.40°	0.18°	—
Mass outdoor *Chlorella* culture	—	12.6 (max)	—
Mass outdoor *Chlorella* culture	—	5.6 (mean)	—

[a] Gross production refers to net production plus respiration loss. Organic carbon was assumed to be 45% of dry weight to arrive at several values. (After Ryther, 1959; Schroeder, 1919; and Riley, 1944.)
* = maximum production, entire growing seasons.
+ = maximum production, single days.
° = annual mean production.
S = estimate of Schroeder (1919).
R = estimate of Riley (1944).

production per km² in the ocean is about two or three times that on land.

Of the total organic material synthesized by the plants of the earth, a very small amount is utilized by animals as food. A much greater fraction is used in the respiration and general metabolism of the plants themselves. The greatest part decomposes into CO_2, water, and inorganic material. Under certain geologic or climatic conditions decay is halted, and huge masses of plant material accumulate for millions of years under a protective layer of silt or rock, eventually to become peat, coal, or a petroleum reserve. Fig. 2 depicts the role played by photosynthesis in the biogeochemical carbon cycle (after Goldschmidt, 1958).

The effect of photosynthesis on CO_2 balance between the atmosphere and hydrosphere often is quite profound. Kalle (1943) reports, for example, that the partial pressure of CO_2 in the Gulf of Finland and Baltic Sea dropped from 3×10^{-4} to 2×10^{-4} atm during a spring diatom bloom.

On a much greater scale, comparisons of CO_2 partial pressures in Antarctic waters and tropical waters point up the influence of photosynthesis on CO_2 balance. In Antarctic waters, characteristically large populations of phytoplankton assimilate great quantities of CO_2 in the upper water layers. Because of this the partial pressure of CO_2 is almost always less in these waters than in the atmosphere above it, and CO_2 enters the water. At the Antarctic convergence, cells are moved deeper into the water column, out of the photic zone, to be picked up by the northward-moving Antarctic intermediate waters. No photosynthesis takes place here, but respiration continues. As these waters move northward they become warmer and more saline, and respiration proceeds at an increased rate. Calcium carbonate desposition also increases to take some of the CO_2, but by the time these deep waters approach the equator they are well charged with CO_2. These CO_2-rich waters come to the surface via divergences along the African and South American Coasts, and move westward along the surface. In these areas partial pressure of CO^2 is greater in the water than in the atmosphere above it, and some CO_2 is lost to the atmosphere, perhaps to be recovered again through photosynthetic fixation in Antarctic waters.

It is obvious that photosynthesis is the major

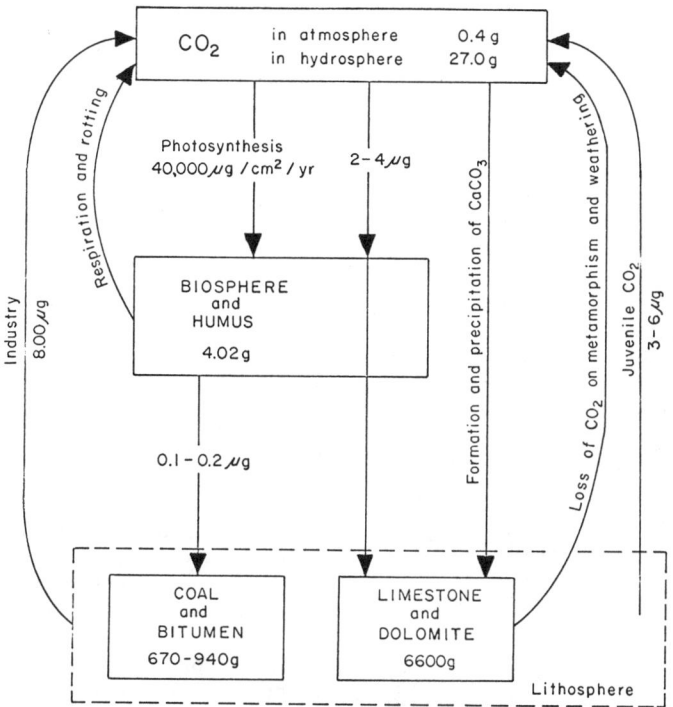

FIG. 2. The carbon cycle. Annual turnover of CO_2/cm^2 of the earth's surface.

mechanism for assimilating CO_2 from the atmosphere and hydrosphere and concentrating it as carbon in the biosphere. Kalle (1943) estimates nearly 100% of the CO_2 consumed in all environments is assimilated by plants carrying out photosynthesis. This estimate is likely too high in view of the vast areas of land and ocean where $CaCO_3$ precipitation is taking place, both organically and inorganically; nevertheless, the percentage of CO_2 consumption by green plant photosynthesis is great. Photosynthesis is the primary mechanism for recycling oxygen among the atmosphere, hydrosphere and biosphere, also. As Rabinowitch (1955) states: "In endlessly repeated cycles the atoms of carbon, oxygen, and hydrogen come from the atmosphere into the biosphere. After a tour of duty which may last seconds or millions of years in the unstable organic world, they return to the stable equilibrium of inorganic nature."

LAWRENCE F. SMALL

References

Arnon, D. I., 1961, "Cell-free Photosynthesis and the Energy Conversion Process," in (McElroy, W. D., and Glass, B., editors), "Life and Light," Baltimore, Johns Hopkins Press, pp. 480–569.

Bassham,, J. A., and Calvin, M., 1957, "The Path of Carbon in Photosynthesis," Englewood Cliffs, N.J., Prentice-Hall.

Clayton, R. K., 1965, "The biophysical problems of photosynthesis," *Science,* 149(3690), 1346–1354.

Goldschmidt, V. M., 1958, "Geochemistry," Oxford, Oxford Univ. Press.

Kalle, L., 1943, "Der Stoffhaushalt des Meeres," Leipzig, Akad. Verlag.

Rabinowitch, E. I., 1945, "Photosynthesis and Related Processes," Vol. I., New York, Interscience.

Rabinowitch, E. I., 1955, "Photosynthesis," in "The Physics and Chemistry of Life," New York, Simon & Schuster.

Riley, G. A., 1944, "The carbon metabolism and the photosynthetic efficiency of the earth as a whole," *Am. Sci.,* **32,** 129–134.

Ryther, J. A., 1959, "Potential productivity of the sea." *Science,* **130,** 602–608.

Schroeder, H., 1919, "Die jährliche Gesamtproduktion der grünen Pflanzendecke der Erde," *Naturwissenschaften,* **7,** 8–12, 23–29.

Steeman-Nielsen, E., 1954, "On organic production in the oceans," *J. Conseil, Perm. Intern. Exploration Mer,* **19,** 309–328.

Tamiya, H., 1957, "Mass culture of algae," *Ann. Rev. Plant Physiol.,* **8,** 309–334.

Cross-references: *Carbon Cycle; Chlorophyll; Organic Geochemistry; Oxygen Cycle.* Vol. II: *Solar Energy.*

PHREATIC WATER

According to Vollmer (1967) phreatic water is groundwater occurring in a zone of saturation having a water table, as distinguished from artesian, or confined water. Phreatic water also may be called free, or unconfined groundwater. This is the definition which will be used here. The word "phreatic" is derived from a Greek word meaning "well," and some writers have considered phreatic water to be any water recoverable from a well, and hence any *groundwater* (q.v.).

The confining effect observed in some groundwater systems is the result of layers of rock with low permeability to water which lie stratigraphically above more permeable beds that are fully saturated. Water encountered by a well penetrating this confining layer to the saturated material below will rise above the level where it is first encountered, to a height above the land surface in some systems. The level to which the artesian water will rise depends upon the amount of head in the hydraulic system.

Phreatic water may not have any chemical features distinguishing it from confined groundwater. However the relatively unhampered connection between phreatic water and the atmosphere can be expected to have several important effects that could be reflected in chemical composition. Phreatic water, especially that within and just below the range of seasonal rise and fall of the water table, can be expected generally to contain oxygen and to have an oxidizing effect on minerals or organic debris. Ferrous mineral species such as pyrite would not be stable in contact with such a solution, for example. Also the contents of calcium and bicarbonate in phreatic water cannot rise above levels that would be stable at the partial pressure of carbon dioxide in the gas phase in the air above the saturated zone.

It would appear, therefore, that in general phreatic water should not usually exhibit properties such as a low redox potential, or a tendency to be strongly supersaturated with respect to calcium carbonate so that this material is deposited where the water reaches the land surface. These properties are frequently found in confined groundwater. Analysis 10 under the topic *Aqueous Solution* (q.v.), and analyses 5 and 6 under the topic *Water—Nonmarine* (q.v.) are common types of phreatic water.

JOHN D. HEM

Reference

Vollmer, Ernst, 1967, "Encyclopaedia of Hydraulics, Soil and Foundation Engineering," New York, Elsevier Publ. Co., 219pp.

Cross-references: *Aqueous Solutions; Artesian Water; Groundwater; Water—Nonmarine; Water Table.*

PLANKTONIC PHOTOSYNTHESIS—See Vol. IV B

PLATINUM: ELEMENT AND GEOCHEMISTRY

Properties

The element platinum (symbol Pt) is one of the group of six platinum metals which are commonly associated together. Although "platina" had been known as an unwanted adjunct of gold from Spanish America in the 18th century, its composition was not worked out until the beginning of the 19th century by British and French scientists. Pure platinum was first produced by Wollaston in 1803.

The element has atomic number 78 and atomic weight 195.09. There are five stable isotopes (Wright and Fleischer, 1965) which have mass numbers and abundances as follows: 192 (0.78%), 194 (32.8%), 195 (33.7%), 196 (25.4%), and 198 (7.23%). The melting point of the pure metal is 1769.3°C and the density 21.45 g/cc at 20°C; the values of these and other physical properties are quite sensitive to impurities. The element crystallizes with a face-centered cubic lattice, with the lattice constant $a = 3.9231$ Å. The atomic radius for 12-coordination in the metal is 1.385 Å.

Occurrence

Platinum is a noble metal and shows considerable reluctance to enter into direct chemical combination with many other elements, and many of its compounds are decomposed at higher temperatures. Accordingly, the element is most commonly found uncombined, though a few mineral compounds are known and are described below. The usual form in which the metal occurs naturally is in the alloy known as *native platinum*. In this, platinum is the principal component generally with small amounts of the other platinum metals, appreciable amounts of iron, and often copper, nickel, or silver. Varying amounts of platinum are also present in other platinum-metal alloys such as iridosmine.

The commonest mineral species are *sperrylite*, $PtAs_2$, found in the Sudbury district in Ontario, and *cooperite*, PtS, and *braggite*, $(Pt, Pd, Ni)S$, both found especially in the Bushveld complex and in the Potgietersrust district of the Transvaal. Several other mineral forms have been newly identified and are described in a recent review (Wright and Fleischer, 1965). Platinum is found in concentrations of a few parts per million in some minerals, especially those like pyrrhotite, chalcopyrite, and pentlandite which form part of nickeliferous ore bodies. At such a level of attenuation it is really quite impossible to say in what chemical form the metal occurs in these minerals.

Abundance and Geochemical Features

The abundance of platinum in the earth's crust has been estimated at 0.005 ppm. In meteorites its abundance is much greater; for instance, over 10 ppm in siderites and about 1 ppm in chondrites. Estimates of cosmic abundance suggest 1.3 to 1.6 atoms per million atoms of silicon.

All evidence points to platinum being strongly siderophilic. It concentrates in a metallic phase during smelting of ores (Hampel, 1968), and in the iron-nickel phase of chondrites (meteorites). It is believed to have concentrated preferentially in the earth's iron-nickel core, and such a view is consistent with the generally low concentration level in crustal bodies. A marked preference is shown for ultrabasic rock bodies and associated minerals in preference to silicic rocks, and this has led to the theory that the metal was concentrated early in the sequence of magmatic differentiation (Rankama and Sahama, 1950). This is discussed more fully in the entry *Platinum Metals*.

Economic Geology

Platinum is economically the most important of the platinum metals both for its resistance to chemical attack and for its considerable catalytic powers. Practically all new platinum is produced from extensive deposits in the Sudbury district, Canada; the Bushveld Igneous Complex in South Africa; and the Noril'sk district, U.S.S.R. (Hampel, 1968).

W. A. E. McBryde

References

Hampel, C. A. (editor), 1961, "Rare Metals Handbook," Second edition, New York, Reinhold Publ. Corp., pp. 304–335.
Hampel, C. A. (editor), 1968, "Encyclopedia of the Chemical Elements," New York, Reinhold Publ. Corp., pp. 533–539.
Standen, A., (editor), Kirk-Othmer "Encyclopedia of Chemical Technology," Second ed., New York, John Wiley & Sons, Vol. 15 pp. 832–878.
Rankama, K., and Sahama, Th. G., 1950, "Geochemistry," Chicago, University of Chicago Press, 912pp.
Wright, T. L., and Fleischer, M., 1965, "Geochemistry of the Platinum Metals," *U.S. Geol. Surv. Bull.*, **1214-A**.

Cross-references: *Elements: Planetary Abundances and Distribution; Geochemical Classification of Elements; Geochemical Evolution of the Core, Mantle, and Crust; Platinum Metals.*

PLATINUM METALS

Characteristics

The platinum metals, along with iron, cobalt, and nickel occupy the eighth group of the periodic table in the following relationship to each other and to the coinage metals of group IB.

Iron	Cobalt	Nickel	Copper
Ruthenium	Rhodium	Palladium	Silver
Osmium	Iridium	Platinum	Gold

These are all transition elements in the sense of being differentiated chemically to a large extent by the number of d electrons in the penultimate shell of their atoms. The most evident kinship among the platinum metals is seen in the vertically arranged pairs, but significant differences exist between these and the ferrous metals lying above them in the table.

A distinctive chemical attribute of these metals is their nobility, manifest as a marked reluctance toward oxidation. Because of this the majority of natural occurrences of the platinum group are as native metal. All known mineral compounds are extremely rare.

Occurrences

The native metal is almost always found in the form of alloys, the composition of which may show considerable variation. A number of these alloys, of which the list in Table 1 is representative, bear special names according to their principal composition.

The limits of composition of these natural

TABLE 1. ALLOYS OF PLATINUM

Name	Crystal Form	Principal Composition
Platinum	Cubic	Pt
Platiniridium	Cubic	(Pt, Ir)
Palladiplatinum	Cubic	(Pd, Pt)
Palladium	Cubic	Pd
Allopalladium	Hexagonal	Pd
Iridosmine	Hexagonal	(Ir, Os)
Osmiridium	Cubic	(Os, Ir)

alloys are not clearly defined. Chemical analysis for these elements has traditionally presented difficulties, and there are also doubts concerning the homogeneity of many samples that have been analysed. In addition to other members of the group, native platinum-metal alloys may contain gold, silver, copper, iron, nickel, and other metals in varying proportions (Wright and Fleischer, 1965). Extensive or complete miscibility occurs among members of the platinum group of metals as well as with a number of elements nearby in the periodic table (Hampel, 1968). The metallic radii of the six elements in the platinum group lie within a range of 0.05 Å, and such a likeness in atomic size is recognized as predisposing to ample mutual solubility.

The existence of intermetallic compounds in which the platinum metals are constituents is not yet well established except in the case of the mineral potarite (PdHg or Pd_3Hg_2) found mainly in Guyana.

As early as 1922 Goldschmidt (1954) advanced the hypothesis that the scarcity of platinum metals in the earth's crust implied a strong siderophile character on the part of these elements. They were considered to have concentrated during the earth's formative period in the liquid iron-nickel core and accordingly to have become depleted from the lithosphere, much as they would become distributed between iron and slag in metallurgical operations. These suppositions appear to have gained strong support from subsequent analyses of meteoric materials, some of the results of which are included below.

The platinum metals, some more than others, also show chalcophile character. This can be inferred, first, from the presence of appreciable amounts of osmium and ruthenium and smaller amounts of the other metals in the sulfide (troilite) phase of chondrites; and, secondly, from the occurrence of certain of the platinum metals in sulfide minerals. Of the latter, particular mention may be made of *laurite,* (Ru, Os) S_2, *cooperite,* (Pt, Pd)S, and *braggite,* (Pt, Pd, Ni)S. The common occurrence of the platinum metals in association with sulfide minerals is presumably related to their chalcophile character.

Although mineral compounds of the platinum metals are quite rare, a number have been well established for some time. Quite a few new mineral compounds have been identified in recent years, particularly by Russian workers, as a result of the application of new physical methods of analysis such as x-ray spectroscopy and electron-probe analysis (Wright and Fleischer, 1965). In addition to the sulfide minerals mentioned previously, the following arsenides and antimonides are well characterized:

Sperrylite ($PtAs_2$)   Geversite ($PtSb_2$)

Arsenopalladinite (Pd_3As)   Stibiopalladinite (Pd_3Sb)

Some of the more recently identified minerals include appreciable amounts of tin, bismuth, and tellurium. The composition of these shows some variation from sample to sample, but its determination is subject to the same constraints that apply to analyses of the native metals.

A good deal of information is now available on the occurrence of platinum metals as trace constituents in minerals. The content of these elements rarely exceeds 10–20 ppm, and the commonest and generally greatest occurrence is in sulfides, selenides, tellurides, and arsenides. Some representative data in Table 2 indicate the average content and distribution among the three chief elements for representative sulfide minerals from Canada and Russia (Wright and Fleischer, 1965, and references therein).

The platinum metals have been found in comparable concentrations in oxide minerals such as pyrolusite, psilomelane, cassiterite, columbite, and chromite; and in silicates such as zircon, gadolinite, and thortveitite. Occurrence in a number of rock-forming silicates is at a much lower concentration.

It is difficult to place much geochemical significance upon these findings since it is difficult to interpret from the available information whether the platinum metals are present as distinct mineral species or in solid solution. It is evident that advantage should be taken of several new, much better analytical techniques to reexamine the data on the occurrence of platinum metals in rocks and minerals at the trace level, and, with regard to minerals, to ensure that homogeneous material is selected for analysis.

There are still some unresolved problems concerning the origin and distribution of platinum metals in rocks. In general, because of the siderophile character of these elements, their overall content in the lithosphere tends to be very low. But abundance determinations show clearly that the platinum metals show a marked preference for the ultrabasic compared to more silicic rocks. These are the rocks believed to have crystallized early in the sequence of magmatic differentiation. One view (Rankama and Sahama, 1950) is that in consequence of their nobility and high melting points the platinum metals underwent substantial deposition in the early stages of magmatic

TABLE 2. AVERAGE OF PLATINUM, PALLADIUM, AND RHODIUM IN SULFIDE MINERALS

Mineral	Platinum Metals (total ppm)	Percentage of Total Platinum Metals		
		Pt	Pd	Rh
Sudbury, Ontario				
Pyrite (FeS$_2$)	2.60	54.9	41.6	3.5
Pyrrhotite (FeS)	1.50	46.0	42.4	11.6
Pentlandite ((Fe,Ni)$_9$S$_8$)	6.70	31.0	60.8	8.2
Chalcopyrite (CuFeS$_2$)	7.74	19.0	80.2	0.8
Noril'sk, U.S.S.R.				
Pyrrhotite (veins)	19	10.6	52.5	36.9
Pyrrhotite (incrustations)	20.6	12.6	72.7	14.5
Chalcopyrite (veins)	4.5	17.0	83.0	—
Chalcopyrite (incrustations)	5.14	9.7	89.5	0.8

crystallization. Accordingly they are found, occasionally with content up to 1 ppm, in dunites, pyroxenites, and serpentines, and also concentrated in chromite. As the silica content increases in later crystallizing rocks the platinum metal content becomes very small. According to this view the sulfides undergoing earliest crystallization, pyrrhotite and nickeliferous pentlandite, contain the highest concentration of platinum metals, while diminishing amounts occur in the sulfides that crystallize later, pentlandite and then chalcopyrite. Finally, after the main stage of differentiation a different set of conditions provides for the possibility of further deposition of residual platinum metals but with quite different associated minerals. The residual magmatic mother liquors containing the concentrated volatile constituents are capable of retaining in the melt a small fraction of the platinum metals originally in the magma, together with various other rare elements. The final stages of crystallization produce niobate, tantalate, zirconate, and titanate minerals in which significant amounts of platinum metals have been found. Examples of such minerals include columbite, tantalite, samarksite and gadolinite.

It should be mentioned that the information outlined above has received a somewhat different interpretation by Goldschmidt (1954).

In summary, the following abundances for platinum metals in rocks have been given by Wright and Fleischer (1965), though these authors caution that the values could be in error by up to an order of magnitude.

Ultramafic rocks	0.05	ppm
Mafic igneous rocks	0.02	
Silicic and intermediate igneous rocks	0.005	

Data are too scanty to permit reliable estimates for the abundances in sedimentary and metamorphic rocks, but in general the platinum metals occur at a very low level except in some copper-bearing shales.

Abundances

Estimates of abundance in the earth's crust suffer from the limitations on the available analytical data for the platinum metal content of rocks, especially granitic rocks which comprise the largest part of the crust. The following are generally regarded as the best estimates at present

Palladium	0.01	ppm
Rhodium	0.001	
Ruthenium	0.001	
Platinum	0.005	
Iridium	0.001	
Osmium	0.001	

Considerably more data are available on the abundance of these elements in meteorites. Table 3 summarizes average values taken from recent work. The generally high concentrations in siderites as compared to chondrites bear out the siderophile character of the elements.

Evidence concerning the relative siderophile/chalcophile character has been based on the distribution of the elements between the nickel-iron and troilite phases in chondrites. Table 4

TABLE 3. AVERAGE PLATINUM METAL CONTENT OF METEORITES

Element	Siderites (ppm)	Chondrites (ppm)	Overall (ppm)
Ruthenium	7.3	1.0	1.6
Rhodium	2.3	0.2	0.6
Palladium	3.7	1.2	1.4
Osmium	4.3	0.7	1.7
Iridium	2.7	0.5	0.7
Platinum	10.1	1.5	2.3

TABLE 4. DISTRIBUTION OF PLATINUM METALS IN PHASES OF CHONDRITES[a]

Element	Nickel-Iron	Troilite (FeS)
Ruthenium	10	9
Rhodium	5	0.4
Palladium	9	2
Osmium	8	9
Iridium	4	0.4
Platinum	20	2

[a] Goldschmidt, 1954; see also Wright and Fleischer, 1965, Table 16.

shows that in platinum, iridium, and rhodium siderophile character predominates, but ruthenium and osmium are about equally siderophile and chalcophile.

Cosmic abundance of the platinum metals has been estimated on the basis of estimated meteoric and solar composition, with certain adjustments to these estimates to fit various theories of nucleogenesis. Three sets of such estimated values, in the form *atoms per million atoms of silicon,* are given in Table 5.

Economic Geology

The mineral deposits that lead to the production of economically significant quantities of the platinum metals are almost entirely primary deposits of two main types. A third type of primary deposit found in South Africa is included in this description. Formerly considerable amounts of platinum metals were found in placer deposits, but these have been largely worked out, and have declined to quite minor economic significance.

The first type of primary deposit consists of disseminations or local concentrations of the metals in olivine-rich rocks, especially in dunite and often associated with chromite. For instance, in the Lydenburg district at the eastern end of the Bushveld igneous complex just north of Pretoria in South Africa, the platinum metals are found in steeply descending pipes in hortonalite-dunite. In this deposit the metal is mainly native platinum, but some sperrylite and antimony-bearing minerals are also found. In Russia extensive deposits occur in dunite masses, the largest of which occurs at Nizhni Tagil on the eastern slopes of the Urals. In this case the dome-like masses of dunite surrounded by pyroxenite in turn surrounded by gabbro, contain the metals chiefly in the central core of dunite. Similar, though economically insignificant, deposits occur in western Ethiopia.

The second, and by far the most economically important, primary deposits occur in magmatic nickel-copper sulfide deposits generally associated with norite. The three main sources of present world production of platinum metals are of this general type.

In South Africa, for instance, the principal deposits occur in the Merensky Reef within the Bushveld Igneous Complex (Cousins, 1959). The platinum-bearing reef lies near the top of the so-called Critical Zone of the Complex. The Critical Zone is an unusual inclined rock formation, from 4000 to 5000 feet thick, containing a sequence of layered igneous rocks which show characteristics typical of a regular sedimentary series. Alternate layers of pyroxenites, anorthosites, and norites have interspersed between them seams of chrome ore. The Merensky Reef itself is made up of a coarse aggregate of pyroxene and feldspar, with sulfide mineralization. The platinum metals are found together with platinum predominating, either as native metal alloys which invariably contain iron, or as the minerals sperrylite and cooperite. The ore also contains a few percent of iron, copper, and nickel as sulfides. The main deposits being worked are near Rustenberg and to the north, in the western section of the Bushveld Complex. Similar deposits occur in the Potgietersrust and Lydenburg districts of this region, and again with comparable associations in the Great Dyke of Southern Rhodesia.

In Canada the production of platinum metals has been related to the output of nickel. The great nickel-producing area has been the Sudbury basin, in the marginal areas of which copper-nickel ores, pyrrhotite, pentlandite, chalcopyrite, and cubanite, are distributed in a rock mass made up of norite and a less basic micropegmatite (Hawley, 1962). Associated with the ore bodies is a good deal of quartz diorite. These ores are believed to be of magmatic origin, though the question of their origin remains not fully answered. In these ores the concentration of platinum metals is low

TABLE 5. COSMIC ABUNDANCE OF PLATINUM METALS

Element	Estimated Cosmic Abundance[a]		
	A	B	C
Ruthenium	1.5	0.9	0.9
Rhodium	0.2	0.15	0.2
Palladium	0.7	0.7	0.6
Osmium	1.0	0.6	0.8
Iridium	0.8	0.5	0.5
Platinum	1.6	1.3	1.6

[a] In atoms per million atoms of silicon.
(A) Suess and Urey, 1956, *Rev. Mod. Phys.* 28, 53.
(B) Cameron, 1959, *Astrophys. J.* 129, 676.
(C) Aller, 1961, "Abundances of the Elements," New York, Interscience Publ.

(0.6–0.9 ppm) and the exact form in which they occur is not fully known. Sperrylite, michenerite, and froodite are mineral forms that have been identified. Newer Canadian sources of nickel at Lynn Lake (Cousins, 1959) and Thompson Lake in Manitoba are not believed to be as well endowed with associated platinum metals.

Since 1965 Russian production of platinum metals has outstripped that of South Africa or Canada. There is much less published information about the sources of this production, but it is understood that much of it comes from the Noril'sk region in Siberia. Here the platinum metals occur at a relatively low concentration associated with copper-nickel sulfides in basic and ultrabasic rock formations similar to the deposits in South Africa. In addition, the sulfide deposits of the Kola peninsula (e.g., near Petsamo) containing pyrrhotite, pentlandite, and chalcopyrite, are believed to be the source of a substantial output of platinum metals.

The average concentration of platinum metals in these sulfide ore bodies is of the order of 4 to 10 ppm in the Merensky Reef, <1 ppm in the Sudbury basin, and probably intermediate between these levels in the Noril'sk deposits. The proportion of platinum to palladium is roughly 7:2, 1:1, and 1:2, respectively, and in all cases the other platinum group metals rarely exceed 10% of the total for the six elements.

Geologically, one other type of primary deposit has been recorded in parts of the Potgietersrust district of the northern arm of the Bushveld Igneous Complex. These are contact-metasomatic deposits where the floor of the norite is partly dolomite and banded ironstone. Localized metamorphosis has occurred, and invasion by emanations that introduced sulfides of iron, copper, and nickel, along with which platinum minerals such as sperrylite, cooperite, and braggite have been found.

Placer deposits formerly accounted for most of the world's supply of the platinum metals. The greatest occurrence of these was in the Ural mountains region in Russia, especially the districts around Issovsk and Nighni Tagil. These appear to have arisen through erosion of primary deposits in dunite previously described. The placer metal is found in river beds, past or present, and in terraces; it consists of native metal alloys, principally platinum with iron and copper, and small amounts of the other platinum metals.

The first platinum to be identified came from South America, and was brought to Europe in the 17th and 18th centuries along with gold and silver from the New World.

Today some placer platinum is still mined in Columbia; it too seems to have originated from ultrabasic rocks. Still smaller quantities of placer platinum come from Ethiopia, the Goodnews Bay district, Alaska, and Tasmania. The last mentioned deposits are of special interest since they contain much osmiridium, and appear to owe their origin to primary deposits in serpentine.

W. A. E. MCBRYDE

References

Cousins, C. A., 1959, "The Bushveld igneous complex," *Platinum Metals Rev.*, **3**, 94.

Goldschmidt V. M., 1954, "Geochemistry," London, Oxford Univ. Press.

Hampel, C. A. (editor), 1968, "Encyclopedia of the Chemical Elements," New York, Reinhold Publ. Corp.

Hawley, J. E., 1962, "The Sudbury ores: their mineralogy and origin," *Can. Mineralogist*, **7,** 1.

Lamey, C. A., 1966, "Metallic and Industrial Mineral Deposits," New York, McGraw-Hill Book Co., 567pp.

Rankama, K., and Sahama, Th. G., 1950, "Geochemistry," Chicago, Univ. of Chicago Press, 912pp.

Ware, G. C., 1965, "Platinum-Group Metals," U.S. Bur. Mines Min. Yearbook.

Wright, T. L., and Fleischer, M., 1965, "Geochemistry of Platinum Metals," *U.S. Geol. Surv. Bull.*, **1124-A**.

Cross-references: *Electron Probe Microanalysis; Elements: Planetary Abundances and Distribution; Geochemical Evolution of the Core, Mantle, and Crust; Platinum: Element and Geochemistry; Sulfides; X-Ray Spectroscopy*. Vol. IVB: *Mineral and Ore Deposits; Placer Deposits;* see also *Individual Elements and Minerals*.

PLUTONIUM: ELEMENT AND GEOCHEMISTRY*

Fifteen isotopes of plutonium are known at present; they range in mass number from 232 to 246, and all are radioactive. The longest-lived of these is Pu^{244}, with a half-life of 7.6×10^7 years. If we take 3.3×10^9 years as the age of the oldest terrestrial rocks, we find that only one atom in 10^{13} primordial Pu^{244} atoms might have survived to the present time. Furthermore, because Pu^{244} is produced through successive neutron capture by uranium, its concentration probably has never been high. Accordingly, it is not surprising that naturally occurring Pu^{244} has not been found in the few searches made; it seems reasonable that pri-

* Work performed under the auspices of the United States Atomic Energy Commission.

mordial plutonium no longer exists in presently detectable amounts in the earth's crust. The natural occurrence of Pu^{239} (half-life 24,360 years), however, as a secondary product in various uranium and thorium ores in concentrations approximating 10^{-14} to 10^{-12} has been reported (see references). The natural existence of minute quantities of this relatively unstable isotope depends on a complex dynamic equilibrium between its formation from U^{238} by neutron capture and β decay

$$(U^{238} + n \rightarrow U^{239} \xrightarrow{\beta-} Np^{239} \xrightarrow{\beta-} Pu^{239})$$

and its disappearance by decay

$$(Pu^{239} \rightarrow U^{235} + \alpha)$$

by neutron capture

$$(Pu^{239} + n \rightarrow Pu^{240})$$

or by fission. Therefore, the geochemistry of naturally occurring plutonium is that of uranium, but the amounts of plutonium are so minute that they cannot be considered significant.

F. W. SCHONFELD

References

Garner, C. S., Bonner, N. A., and Seaborg, G. T., 1948, "Search for elements 94 and 93 in nature. Presence of 94^{239} in carnotite," *J. Am. Chem. Soc.,* **70,** 3453–3454.

Levine, C. A., and Seaborg, G. T., 1951, "The occurrence of plutonium in nature," *J. Am. Chem. Soc.,* **73,** 3278–3283.

Peppard, D. F., Studier, M. H., Gergel, M. V., Mason, G. W., Sullivan, J. C., and Mech, J. F., 1951, "Isolation of microgram quantities of naturally-occurring plutonium and examination of its isotopic composition," *J. Am. Chem. Soc.,* **73,** 2529–2531.

Seaborg, G. T., and Perlman, M. L., 1948, "Search for elements 94 and 93 in nature. Presence of 94^{239} in pitchblende," *J. Am. Chem. Soc.,* **70,** 1571–1573.

Cross-references: *Radioactive Isotopes; Uranium.*

POLLUTION—See ENVIRONMENTAL POLLUTION

POLONIUM: ELEMENT AND GEOCHEMISTRY

Polonium (atomic number 84), symbol Po, is a radioactive metal lying just below tellurium in group VI of the Periodic Table. The element exists in two crystalline modifications: α-Po, cubic, and β-Po, rhombohedral. The transition point is about 36°C. The density of polonium is 9.4, whereas the melting point is close to 255°C. Its abundance in the earth's crust is on the order of 10^{-13} percent.

Polonium was discovered in 1898 by Madame Curie in the course of investigating the radioactivity of various uranium and thorium minerals.

Occurrence

Polonium occurs in nature in uranium and thorium minerals, such as pitchblende (uraninite), thorianite, carnotite, monazite, pilbarite, autunite, and thorite (orangite). Isotopes of polonium occur in the decay schemes of uranium-238, uranium-235, and thorium-232. These respective decay schemes are shown in Tables 1, 2, and 3. In these tables the isotopes of polonium are indicated by asterisks. Although there are seven naturally occurring isotopes of polonium, the only one occurring to any extent is Po^{210} (RaF) which decays by α-emission with a half-life of 138.4 days. The remaining isotopes have half-lives so short that their abundance in nature is negligible. One thousand kilograms of Joachimstal pitchblende containing 65% uranium contains about 0.05 mg of polonium.

Separation

Polonium can be extracted from aged radium salts, which contain about 0.2 mg of polonium per gram of radium at equilibrium. Extraction reagents include thenoyltrifluoroacetone (TTA) in benzene, acetylacetone, ether, and isobutylketone. Prior to the production of Po^{210} in nuclear reactors by the following scheme

$$_{83}Bi^{209} \; (n,\gamma) \; _{83}Bi^{210} \xrightarrow[5d]{\beta} \; _{84}Po^{210}$$

trace amounts of polonium were usually obtained from aged radon tubes, which contained an active deposit consisting of the radium-D-E-F series (Table 1).

M. H. LIETZKE

References

Bagnall, K. W., 1957, "Chemistry of the Rare Radioelements," London, Butterworths Scientific Publ.

Brasted, R. C., 1961, "Comprehensive Inorganic Chemistry," New York, D. Van Nostrand Co.

Friedlander, G. and Kennedy, J. W., 1962, "Nuclear and Radiochemistry," New York, John Wiley & Sons.

Cross-references: *Lead; Radium; Thorium; Uranium.*

POLONIUM: ELEMENT AND GEOCHEMISTRY

TABLE 1. THE U^{238} DISINTEGRATION SERIES

Elements	Symbol, Mass Number, Radiation	Half-Life
Uranium	U^{238}	4.51×10^9 yr
	↓ alpha	
Thorium	Th^{234}	24.1 days
	↓ beta	
Protoactinium	Pa^{234}	1.18 min
	↓ beta	
Uranium	U^{234}	2.48×10^5 yr
	↓ alpha	
Thorium	Th^{230}	7.6×10^4 yr
	↓ alpha	
Radium	Ra^{226}	1,622 yr
	↓ alpha	
Emanation (Radon)	Em^{222}	3.825 days
	↓ alpha	
*Polonium (RaA)	Po^{218}	3.05 min
	99+% alpha ↙ ↘ beta 0.1%	
Astatine	At^{218}	1.3 sec
Lead (RaB)	Pb^{214} ↙ alpha	26.8 min
	beta ↘	
Bismuth (RaC)	Bi^{214}	19.7 min
	99.96% beta ↙ ↘ alpha 0.04%	
*Polonium (RaC')	Po^{214}	1.6×10^{-4} sec
Thallium (RaC")	Tl^{210}	1.32 min
	alpha ↘ ↙ beta	
Lead (RaD)	Pb^{210}	22.2 yr
	↓ beta	
Bismuth (RaE)	Bi^{210}	5.02 days
	↓ beta	
*Polonium (RaF)	Po^{210}	138.4 days
	↓ alpha	
Lead (RaG)	Pb^{206} (Stable)	—

TABLE 2. THE U^{235} DISINTEGRATION SERIES

Elements	Symbol, Mass Number, Radiation	Half-Life
Uranium	U^{235}	7.13×10^8 yr
	↓ alpha	
Thorium	Th^{231}	25.6 hr
	↓ beta	
Protoactinium	Pa^{231}	3.43×10^4 yr
	↓ alpha	
Actinium	Ac^{227}	22.0 yr
	↓ beta	
Thorium	Th^{227}	18.6 days
	↓ alpha	
Radium	Ra^{223}	11.1 days
	↓ alpha	
Emanation (Actinon)	Em^{219}	3.92 sec
	↓ alpha	
*Polonium	Po^{215}	1.83×10^{-3} sec
	↓ alpha	
Lead	Pb^{211}	36.1 min
	↓ beta	
Bismuth	Bi^{211}	2.16 min
	99.68% alpha ↙ ↘ beta 0.32%	
*Polonium	Po^{211}	0.52 sec
Thallium	Tl^{207}	4.79 min
	beta ↘ ↙ alpha	
Lead	Pb^{207} (Stable)	—

POLONIUM: ELEMENT AND GEOCHEMISTRY

TABLE 3. THE Th^{232} DISINTEGRATION SERIES

Elements	Symbol, Mass Number, Radiation	Half-Life
Thorium	Th^{232}	1.39×10^{10} yr
	↓ alpha	
Radium	Ra^{228}	6.7 yr
	↓ beta	
Actinium	Ac^{228}	6.13 hr
	↓ beta	
Thorium	Th^{228}	1.90 yr
	↓ alpha	
Radium	Ra^{224}	3.64 days
	↓ alpha	
Emanation (Thoron)	Em^{220}	54.5 sec
	↓ alpha	
*Polonium	Po^{216}	0.158 sec
	↓ alpha	
Lead	Pb^{212}	10.6 hr
	↓ beta	
Bismuth	Bi^{212}	60.5 min
	66.3% beta ↙ ↘ alpha 33.7%	
*Polonium	Po^{212}	3.0×10^{-7} sec
Thallium	Tl^{208}	3.1 min
Lead	alpha ↘ ↙ beta	—
	Pb^{208} (Stable)	

POROSITY AND PERMEABILITY

Porosity

The percentage of pore space in the total volume of a rock, namely the total volume minus the volume of solid mineral matter, is termed porosity. Porosity can be defined as follows:

$$\phi = \frac{V_b - V_s}{V_b}$$

where ϕ = porosity, % or fraction; V_b = bulk volume; and V_s = sand-grain (solid) volume, in the same units as V_b.

Bulk volume is usually determined by the displacement of mercury. This may be accomplished gravimetrically with a pycnometer or volumetrically with some device (e.g., Beeson, 1950) designed for the purpose. Either method depends upon the fact that mercury does not enter the pore channels of most core samples at the moderate pressures imposed in the tests.

Sand-grain volume necessary for determining total porosity may be obtained by measuring the volume occupied by the particles of a crushed sample. Another method involves weighing the sample and measuring the density of particles from the crushed solid.

Sand-grain volume for determining effective porosity is most readily determined by a Boyle's-law porosimeter (Beeson, 1950) which is operated with a gas, such as helium, that is not appreciably adsorbed on the core solids. In this method, gas may be expanded or compressed, and the volume occupied by the solids is determined from the effect on some such reading as pressure, due to the displacement of gas by the solids. An excellent discussion on collection of unconsolidated materials for porosity determination and methods of measurement is presented by Krumbein and Pettijohn (1938).

Two types of porosity may be determined, the effective porosity and the total porosity. The pore volume involved in measuring the latter includes the disconnected voids. Fluids generally do not flow from or through these voids. Consequently, any hydrocarbons present in them do not contribute to the reserves recoverable by the usual methods of production.

Fluid content usually is reported as percent or fraction of the pore space which is occupied by the fluid, and it is designated by the term saturation. For instance, a core sample which had been found to contain oil equal to one-fifth the pore volume and water equal to one-fourth the pore volume would be reported to have an oil saturation S_o of 20%, a water saturation S_w of 25%, and a gas saturation S_g of 55%.

Thus to convert a fluid content to a saturation it is necessary to determine the pore volume. A knowledge of the proportion of the bulk volume that is porous or void is also useful in estimating the amount of oil or gas present in the reservoir.

It is possible to classify reservoir rocks on the basis of their effective porosity: > 25%

very high porosity; 15–25% high porosity; 8–15% medium porosity; 5–8% low porosity; and <5% very low porosity.

Permeability

To help answer the question about the rate of flow, methods have been developed for measuring a property called permeability which is designated by the symbol k. Credit for the early work in this field is usually given to Henry Darcy (1856). The American Petroleum Institute Code No. 27, entitled "Standard Procedure for Determining Permeability of Porous Media," discusses many special cases of measuring permeability.

The simplest expression for Darcy's law of flow through porous media neglects the effect of gravity and refers to a porous medium of constant cross section. All forms of Darcy's law are restricted to the nonturbulent or viscous region of flow. With these limitations in mind, Darcy's law may be expressed as:

$$q = \frac{kA\Delta p}{\mu L} \quad (1)$$

where q = rate of flow of one homogeneous, single-phase liquid (cm³/sec); k = permeability (darcys); A = cross-sectional area of porous medium (cm²); Δp = pressure drop along length of sample (atm); μ = viscosity of liquid (centipoise); and L = length of sample (cm).

Work by Darcy and others showed that the rate of flow would be directly proportional to the area and the pressure drop, and inversely proportional to the viscosity and the length. The proportionality constant, k, has been called the permeability and the unit has been given the name *darcy* when the other units are as indicated above. A more common unit is the millidarcy (md) which is equal to 0.001 darcy.

Reservoir rocks can be classified according to their permeability as follows (also see Eremenko, 1960, p. 372):

> 1.0 darcy   very high permeability
0.1 to 1.0 darcy   high permeability
0.01 to 0.1 darcy   medium permeability
0.001 to 0.01 darcy   low permeability
0.1 to 1.0 millidarcy   very low permeability
< 0.1 millidarcy   practically impermeable

Darcy's law also may be used in the above form for the flow of a gas through a porous medium, provided the rate of flow is considered to be the mean rate of flow. That is, a gas expands as it flows, so the volumetric rate of flow is greater downstream than it is upstream, even under steady-state conditions when the mass rate of flow is constant. This presents no difficulty if the volumetric rate of flow is con-

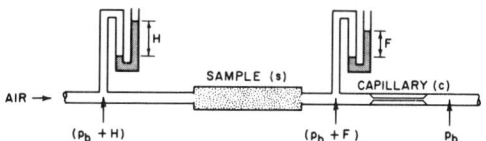

Fig. 1. Apparatus for measuring air permeability.

sidered to be measured at the arithmetic average of the upstream and downstream pressures.

Derivation of Equation for Calculating Air Permeability. An apparatus for measuring air permeability is shown diagrammatically in Fig. 1. As indicated, air flows from left to right through a sample and a capillary tube. The upstream and downstream pressures for the sample are noted as $p_b + H$ and $p_b + F$, where H and F are manometer deflections (pressures) expressed in atmospheres, in excess of the barometric pressure P_b. As the restriction to flow between the sample and the capillary tube is negligible compared to the restriction to flow in the capillary tube, the downstream pressure for the sample is substantially the same as the upstream pressure for the capillary. The downstream pressure for the capillary tube is almost equal to barometric pressure.

Using the above notation, one may apply Darcy's law to the system depicted in Fig. 1. With the subscript s referring to the sample, the subscript c referring to the capillary tube, and the bar indicating mean value:

$$\bar{q}_s = \frac{k_s A_s \Delta p_s}{\mu_s L_s} \qquad k_s = \frac{\bar{q}_s \mu_s L_s}{A_s \Delta p_s} \quad (2)$$

$$\bar{q}_c = \frac{k_c A_c \Delta p_c}{L_c \mu_c} \quad \text{or} \quad \bar{q}_c = k_c' \frac{\Delta p_c}{\mu_c} \quad (3)$$

where

$$k_c' = \frac{k_c A_c}{L_c} = \frac{\bar{q}_c}{\Delta p_c/\mu_c} \text{ in darcy-cm} \left(= \frac{\text{cm}^3/\text{sec}}{\text{atm/cp}}\right)$$

Substituting ($\Delta p_c = p_b + F - p_b = F$) into Eq. (3) gives

$$\bar{q}_c = \frac{k_c' F}{\mu_c} \quad (4)$$

Applying the gas law, $pV = zNRT$, to the low pressures involved here, the compressibility factor Z is substantially equal to unity. For isothermal flow, the absolute temperature T is constant. Under steady-state conditions, the number of moles N per unit of time is constant. Therefore, the product of the pressure p times the volume V per unit of time also is constant. This yields a relation similar to Boyle's law which is applicable to the isothermal, steady-state flow of compressible fluids:

$$\bar{p}_1 \bar{q}_1 = \bar{p}_2 \bar{q}_2 \quad (5)$$

Equation (5) may be used to compute the rate of flow in the sample from the rate of flow through the capillary tube as follows:

$$\overline{p}_s \overline{q}_s = \overline{p}_c \overline{q}_c \quad \text{or} \quad \overline{q}_s = \frac{\overline{p}_c}{\overline{p}_s} \overline{q}_c \quad (6)$$

Substituting $\Delta p_s = p_b + H - (p_b + F) = H - F$ into Eq. (2) gives

$$k_s = \frac{\overline{q}_s \mu_s L_s}{(H-F) A_s} \quad (7)$$

The values for $\overline{p}_c$ and $\overline{p}_s$ may be substituted into Eq. (5), along with the value of $\overline{q}_c$ from Eq. (3), to give

$$\overline{q}_s = \frac{2p_b + F}{2p_b + H + F} \cdot \frac{k_c' F}{\mu_c} \quad (8)$$

Substituting the value for $\overline{q}_s$ from Eq. (8) into Eq. (7) yields,

$$k_s = \frac{(2p_b + F) k_c' F}{(2p_b + H + F) \mu_c} \cdot \frac{\mu_s L_s}{(H-F) A_s} \quad (9)$$

For the usual case of substantially isothermal flow, the viscosity μ_s of the gas in the sample is practically equal to the viscosity μ_c of the gas in the capillary. Accordingly, one may simplify and rearrange Eq. (8) to give

$$k_s = k_c' \cdot \frac{(2p_b + F) F}{(2p_b + H + F)(H-F)} \cdot \frac{L_s}{A_s} \quad (10)$$

Equation (10) may be used for calculating air permeability. The capillary constant k_c' must be determined for the given capillary tube. The second factor following the equal sign depends upon p_b, H, and F, but it may be applied equally well to any capillary tube for which the range of F lies in the viscous region. Values of this factor may be compiled and tabulated for the chosen values of p_b, H, and F. A table which applies uniquely to the given capillary tube may then be prepared from the product of the capillary constant k_c' and the factor involving p_b, H, and F. Thus, determining the permeability of a sample to any gas may be accomplished by simply measuring the magnitude of F which results from impressing a suitable pressure H upon the upstream face of the sample, locating in the table the corresponding value of the product of the capillary constant and the factor involving p_b, H, and F, and then multiplying this value by the length-to-area ratio of the sample.

Estimating Gas Permeability at Infinite Mean Pressure. There is however, one difficulty in the determination of gas permeability which was brought to light by Klinkenberg (1941). He showed that the error due to slip may be appreciable in the range of permeabilities encountered in economically productive reservoirs. He also demonstrated that a plot of gas permeability versus reciprocal mean pressure yielded a straight line which crossed the permeability axis at a value equal to the liquid permeability. That is, the gas permeability is proportional to the reciprocal mean pressure, and as the mean pressure approaches infinity or the reciprocal mean pressure approaches zero, the gas permeability approaches the value measured with a liquid which does not react with the rock components.

The American Petroleum Institute established a project at the Pennsylvania State College to investigate questions which had been raised concerning the validity of Klinkenberg's conclusions (Grunberg and Nissan, 1943). The final report of the project in 1949 completely substantiated Klinkenberg's conclusions after a thorough study of 11 synthetic cores and 164 natural cores from various localities (American Petrol. Inst., 1949). The report also included a plot of the fractional error in the gas permeability at a mean pressure of one atmosphere versus the liquid permeability, on a log-log scale. Although there was considerable spread in the data, there was a definite trend toward decreasing error with increasing permeability. The straight line drawn through the points gave very close to the theoretical relation of the error, varying inversely with the square root of the permeability; as the error was 250% for a liquid permeability of 0.1 millidarcy and 4% for a liquid permeability of 1000 md.

For precise determinations of liquid permeability, rather than use the relation discussed above, it is preferable to measure the permeability with a liquid which does not react with the rock components or to measure the air permeability at varying mean pressures. A hydrocarbon would be a suitable liquid except that a core is seldom completely saturated with a hydrocarbon, and the inclusion of an extra step for this purpose usually would be too time-consuming. Water is too likely to react with clays and other solids contained in the core; this ordinarily eliminates water for measuring absolute permeability. For example, on adding 5% montmorillonite clay to clean quartz sand (0.20–0.25 mm grain diameter) having a permeability of 60 darcys, the permeability is lowered 30 times (Tsvetkova, 1949, in Strakhov, 1957, p. 300). Consequently, core laboratories usually have equipment for determining gas permeabilities at high pressures.

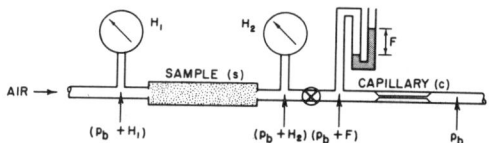

FIG. 2. Apparatus for measuring gas permeability at high pressures.

An apparatus for measuring gas permeability at high pressure is depicted in Fig. 2. It is apparent that the gas flow can be throttled between the downstream end of the sample and the upstream end of the capillary tube. Also, the pressure can be measured at each end of the sample, as well as at the upstream end of the capillary tube.

By the method used under the section on Derivation of Equation for Calculating Air Permeability, the following relation is readily derived for the case of throttling the flow of gas between the sample and the capillary tube:

$$k_s = k_c' \cdot \frac{(2p_b + F)F}{(2p_b + H_1 + H_2)(H_1 - H_2)} \cdot \frac{L_s}{A_s} \quad (11)$$

where H_1 and H_2 are the inlet and outlet pressures for the sample, respectively, and the other symbols have the same meaning as in Eq. (10).

Effective Permeability and Relative Permeability. The preceding discussion of permeability have involved the flow of a single, homogeneous fluid. When two or more immiscible fluids are present in a porous solid, the permeability to any of the fluids is termed the effective permeability of the solid to that fluid. The fluids of interest are oil, water, and gas. The symbols used for these fluids are k_{eo}, k_{ew}, and k_{eg}, respectively. The units are the same as for single-phase, homogeneous permeability.

It often is convenient to convert effective permeabilities into permeabilities relative to some standard base, in which case they are called relative permeabilities. The symbols for the relative permeabilities to oil, water, or gas are k_{ro}, k_{rw}, and k_{rg}, respectively. With these symbols, $k_{ro} = k_{eo}/k$, $k_{rw} = k_{ew}/k$, or $k_{rg} = k_{eg}/k$, where k is some standard permeability, often the single-phase, liquid permeability. Relative permeabilities generally are expressed in per cent or fraction.

It has been found that the sum of the relative permeabilities to the fluids contained in a porous solid seldom equal 100%. The case where a core sample contains only oil and water is shown in Fig. 3. The dashed diagonal lines indicate a practical lower limit of relative permeabilities for a bundle of capillary tubes of uniform size and length. For that ideal case, the sum of the relative permeabilities would equal 100%. For actual core samples, however, each relative permeability curve is considerably below the diagonal for most values of saturation.

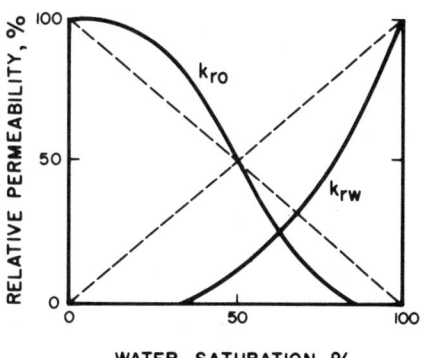

FIG. 3. Relative permeabilities to oil and water as a function of water saturation.

The relative permeability to the nonwetting phase can be above the diagonal, as shown at lower water saturations in Fig. 3. In fact, the actual curve can even rise above 100%. Yuster (1951) has developed for capillaries of constant cross section the relation:

$$k_{ro} = 2S_o(1 - S_o)\frac{\mu_o}{\mu_w} + S_o^2 \quad (12)$$

where k_{ro} is the fractional relative permeability to oil (the nonwetting phase); S_o is the fractional oil saturation; μ_o is the viscosity of the oil; and μ_w is the viscosity of the water (the wetting phase).

Figure 4, after Odeh and Yuster (1953), shows that the theoretical permeability curve, for oil (the nonwetting phase) in a bundle of capillaries of constant cross section and length, lies above the diagonal for oil-water viscosity

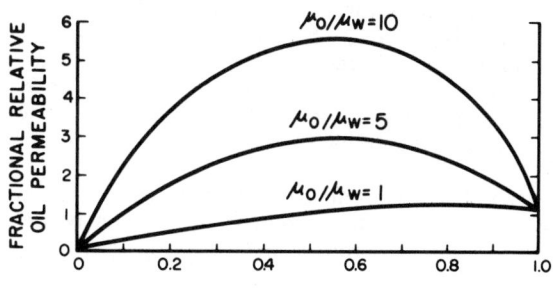

FIG. 4. Theoretical relationship between viscosity ratio and relative permeability in a circular capillary. (After Odeh and Yuster, 1953. *Courtesy SPE of AIME.*)

POROSITY AND PERMEABILITY

ratio of unity or above. The validity of this relationship, however, is questioned by some scientists (e.g., Baker, 1960).

The fact that the relative-oil permeability may be above the diagonal is due to the lower viscosity of the wetting phase which allows the water to move more rapidly than the oil, for the same pressure gradient, which in turn gives the oil an added impetus. The reason relative permeabilities are usually below the diagonal is related to surface phenomena.

Relative Permeability Determination. Reservoir performance can be usually predicted on the basis of core properties measured in the laboratory. Of these, the measurement of relative permeability is essential and routine. Many methods of obtaining relative permeability data on reservoir core samples have been proposed; however, many of these techniques are too tedious and time-consuming for practical application.

Hassler et al. (1936) measured the relative air permeability by blowing oil out of oil-saturated cores at various time intervals. Wyckoff and Botset (1936) measured relative gas and liquid permeabilities of carbon dioxide-brine systems using unconsolidated sand packs having different permeabilities. On flowing mixtures of oil and brine through unconsolidated sand packs, Leverett (1939) concluded that relative permeability is a function of pore size distribution, displacement pressure, pressure gradient, and water saturation. Leverett and Lewis (1941) extended their research to three-phase flow through unconsolidated sand packs. Morse et al. (1947) developed a technique for determination of the relative permeabilities of small core samples. Brownscombe et al. (1949) used the capillary pressure curves to obtain the relative permeability curves.

The research work of Buckley and Leverett (1942) was expanded by Welge (1952), who developed a method for determining the ratio of relative permeabilities. In his method, the core is first saturated with oil and then gas or water is injected. The produced oil and gas or water are collected and their volume is measured. On knowing the pore volume of the core, corresponding gas or water saturation can be computed as follows (Fig. 5):

(1) The oil produced, expressed as a fraction of cumulative production (F_o), can be determined from a plot of log q_t versus q_o:

$$F_o = \frac{1}{2.303 \, mq_t} \quad (13)$$

where m = slope, q_o = cumulative oil production, and q_t = cumulative total production.

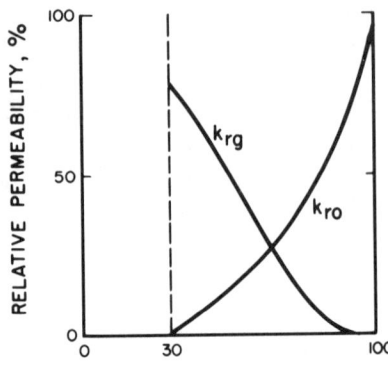

FIG. 5. Relative permeabilities to oil and gas as a function of total liquid saturation.

(2) The k_{rg}/k_{ro} or k_{rw}/k_{ro} ratios at the end of the sample can be obtained on using the following formulas:

$$k_{rg}/k_{ro} = \frac{1 - F_o}{F_o} \times \frac{\mu_g}{\mu_o} \quad (14)$$

$$k_{rw}/k_{ro} = \frac{1 - F_o}{F_o} \times \frac{\mu_w}{\mu_o} \quad (15)$$

where k_{rg} = relative permeability to gas, k_{ro} = relative permeability to oil, k_{rw} = relative permeability to water, μ_g = gas viscosity, and μ_o = oil viscosity.

(3) The corresponding S_g or S_w at the effluent end of the core can be determined from the following equation:

$$S = \frac{q_o}{PV} - F_o q_t \quad (16)$$

where PV = pore volume of core, $q_t = q_g + q_o$ or $q_t = q_w + q_o$, and S = saturation.

Johnson et al. (1959) developed a method for determining the individual relative permeability curves. This method is far less time-consuming than other methods.

Babalyan (1956) and Sinnokrot and Chilingar (1961) presented data on effect of polarity of oil and presence of carbonate particles on relative permeability of rocks. The relative permeability curves of oil + alkaline and oil + hard water were also presented; the former lie above the latter.

In the case of polar oils, the residual oil and water saturations also shift due to the chemistry of water. This is due to the fact that the following is true in the case of alkaline waters: (1) low interfacial tension between oil and water, (2) low values of contact angle, (3)

POROSITY AND PERMEABILITY

slow coalescence of oil droplets in water, and (4) greater degree of dispersion of oil in water. In the case of hard waters, on the other hand, the oil becomes a dispersed phase at higher water saturations of porous medium than in alkaline waters. Intensity of transformation of oil into a dispersed phase is greater in alkaline than in hard waters. In the case of nonpolar oil, the attachment of oil to solid surfaces is negligible in the presence of both alkaline and hard waters.

Relationship Between Porosity and Permeability

Permeability is governed by (1) manner of packing of grains during deposition, (2) compaction history of sediment, (3) grain size distribution, (4) shape of grains, and (5) grain size. With the exception of (5) porosity is also controlled by the same factors.

Porosity can be easily related to petrography and in many reservoir sands porosity is related to permeability (exponential function). According to Griffiths (1958, p. 15), when permeability is plotted on a logarithmic scale and porosity on an arithmetic scale, a linear relationship emerges. These relationships, however, are not exact as can be seen from the scatter around the trend lines (Griffiths, 1958, p. 16, Fig. 1).

Some data on the relationship between porosity and permeability had been assembled by Muskat (1949, p. 168–176). Khanin (1956; in Chilingar, 1957) conducted a comprehensive statistical study on relationship between the effective porosity and permeability, and found that the granulometric composition of sandstones has to be considered in order to establish correlation between these two variables. Khanin (1958) also started to study relationship between the porosity and permeability of unconsolidated and weakly cemented sands.

Chilingar (1963) showed that on considering grain size distribution, there is a correlation between the porosity and permeability of sandstones, *containing an irreducible minimum of interstitial formation water,* and unconsolidated sands. He plotted porosity versus permeability (Fig. 6) for (1) very coarse-grained,

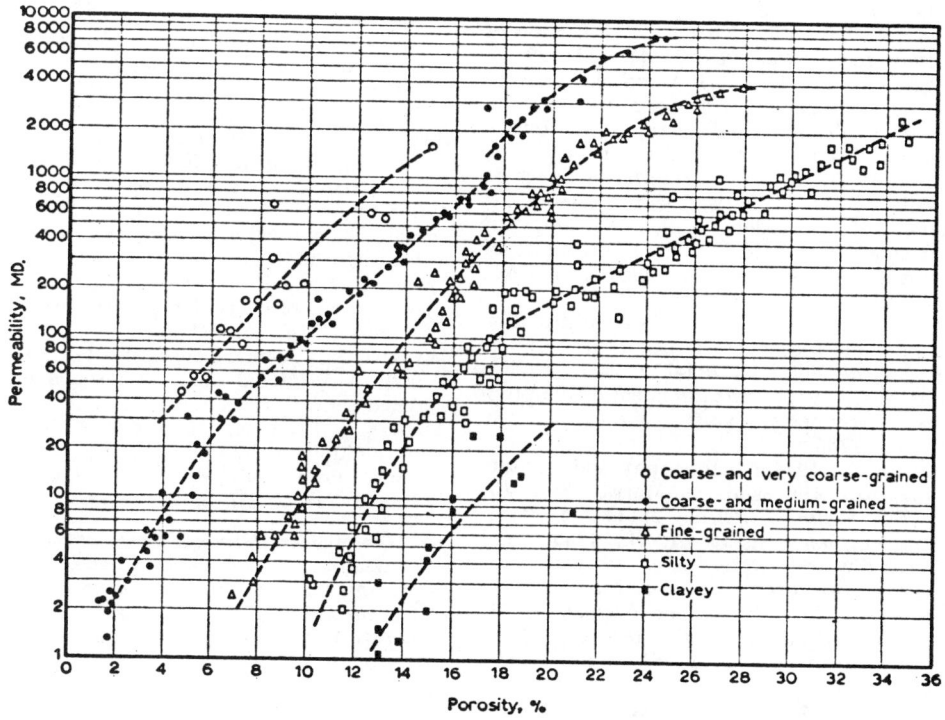

FIG. 6. Relationship between porosity and permeability of (1) very coarse-grained, (2) coarse- and medium-grained, (3) silty, and (4) clayey sandstones. (After Chilingar, 1964; in: Van Straaten, 1964. *Courtesy Elsevier Publ. Co.*) Very coarse-grained—more than 50% of 1–2 mm fraction; coarse-grained—more than 50% of 0.5–1 mm fraction: medium-grained—more than 50% of 0.25–0.5 mm fraction; fine-grained—more than 50% of 0.1–0.25 mm fraction; silty—more than 10% silt (<0.1 mm); and clayey—clay (<0.004 mm) content in excess of 7%.

(2) coarse- and medium-grained, (3) fine-grained, (4) silty, and (5) clayey sandstones.

GEORGE V. CHILINGARIAN
CARROL M. BEESON
IRAJ ERSHAGHI

References

American Petroleum Institute, 1949, Circ. D-319, July 25.
Babalyan, G. A., 1956, "Questions on Mechanism of Oil Recovery," Aznefteizdat, Baku, 254pp.
Baker, P. E., 1960, "Discussion of 'Effect of viscosity ratio on relative permeability,'" *Trans. AIME*, 219, 404–405.
Beeson, C. M., 1950, "The Kobe porosimeter and the Oilwell research porosimeter," *Trans. AIME*, 189, 313.
Beeson, C. M., 1953, "Core Analysis Principles and Experiments," Univ. South. Calif. Publ., 55pp.
Brownscombe, E. R., Terwilliger, P. L., and Coudle, B. H., 1949, "Relative permeability of cores desaturated by capillary pressure method," Am. Petrol. Inst. Spring Meet., Pacific Coast District, No. 801-25G, May 12–13.
Buckley, S. E., and Leverett, M. C., 1942, "Mechanism of fluid displacement in sands," *Trans. AIME*, 146, 107–116.
Chilingar, G. V., 1957, "Khanin's classification of reservoir rocks," *Compass*, 34(4), 335–339.
Chilingar, G. V., 1963, "Relationship between porosity, permeability, and grain size distribution of sands and sandstones," Proc. Intern. Sediment. Congress, Amsterdam-Antwerp.
Darcy, Henry, 1856, "Les Fontaines Publiques de la Ville de Dijon," Paris, Victor Dalmont.
Eremenko, N. A. (editor), 1960, "Geology of Petroleum" (handbook), Vol. 1, "Principles of Petroleum Geology," Moscow, Gos. Top. Tekh. Izd., 592pp.
Griffiths, J. C., 1952a, "Grain size distribution and reservoir rock characteristics," *Am. Assoc. Petrol. Geol. Bull.*, 36, 205–229.
Griffiths, J. C., 1952b, "Measurement of the properties of sediments" (abstr.), *Geol. Soc. Am. Bull.*, 63, 1256.
Griffiths, J. C., 1958, "Petrography and porosity of the Cow Run sand, St. Marys, West Virginia," *J. Sediment. Petrol.*, 28(1), 15–30.
Grunberg, L., and Nissan, A. H., 1943, "The permeability of porous solids to gases and liquids," *J. Inst. Petrol. Technol.*, 29, 236.
Hassler, G. L., Rice, R. R., and Leeman, E. H., 1936, "Investigations on the recovery of oil from sandstones by gas drive," *Trans. AIME*, 118, 116–137.
Heid, J. G., McMahon, J. J., Nielsen, R. F., and Yuster, S. T., 1950, "Study of the permeability of rocks to homogeneous fluids," *A.P.I. Drill. Prod. Prac.*, 230.
Johnson, E. F., Bossler, D. P., and Naumann, V. O., 1959, "Calculation of relative permeability from displacement experiments," *J. Petrol. Technol.*, Jan., p. 61.
Khanin, A. A., 1956, "About classification of petroleum and natural gas reservoir rocks," *Razvedka i Okhrana Nedr*, 1, 7–16.
Khanin, A. A., 1958, "Toward question of determining reservoir properties of non-cemented sandstones," *Tr. VNII Gas*, vypusk 4(12), 7pp.
Klinkenberg, L. J., 1941, "The permeability of porous media to liquids and gases," *A.P.I. Drill. Prod. Prac.*, 200.
Kotyakhov, F. I., 1949, "Interrelationship between major physical parameters of sandstones," *Neft. Khoz.*, 12, 29–32.
Krumbein, W. C., and Pettijohn, F. J., 1938, "Manual of Sedimentary Petrography," New York, Appleton-Century-Crofts, Inc., 549pp.
Leverett, M. C., 1939, "Flow of oil-water mixtures through unconsolidated sands," *Trans. AIME*, 132, 149–171.
Leverett, M. C., and Lewis, W. B., 1941, "Steady flow of gas-oil-water mixtures through unconsolidated sands," *Trans. AIME*, 142, 107–116.
Morse, R. A., Terwilliger, P. L., and Yuster, S. T., 1947, "Relative permeability measurements on small core samples," *Oil Gas J.*, 46, 109.
Muskat, M., 1949, "Physical Principles of Oil Production," New York, McGraw-Hill Book Co., 922pp.
Odeh, A. S., and Yuster, S. T., 1953, "The effect of viscosity on relative permeability," paper 264-G, Am. Inst. Min. Met. Engrs. Meet., Los Angeles, Feb. 14.
Sinnokrot, A., and Chilingar, G. V., 1961, 'Effect of polarity of oil and presence of carbonate particles on relative permeability of rocks," *Compass*, 38(2), 115–120.
Strakhov, N. M. (Editor), 1957, "Methods of Studying Sedimentary Rocks," Vol. 1, Moscow, Gos. Geol. Tekh. Izd., 611pp.
Van Straaten, L. M. J. U. (editor), 1964, "Deltaic and Shallow Marine Deposits," Amsterdam, Elsevier Publ. Co., 464pp.
Welge, H. J., 1952, "A simplified method for computing oil recovery by gas or water drive," *Trans. AIME*, 195, 91–98.
Wyckoff, R. D., and Botset, H. G., 1936, "The flow of gas-liquid mixtures through unconsolidated sands," *Physics*, 7, 325–345.
Yuster, S. T., 1951, "Theoretical considerations of multiphase flow in idealized capillary systems," *Proc. Third World Petrol. Congr.*, The Hague, Section III, pp. 437–455.

Cross-references: *Groundwater; Interstitial Water in Sediments; Sea Water, Chemistry; Thermogravimetric Analysis; Water—Nonmarine—Geochemistry.* Vol. I: *Sediment Coring Techniques.* Vol. VI: *Grain Size Studies; Sedimentary Petrography.*

POTASSIUM: ELEMENT AND GEOCHEMISTRY

Properties

Potassium (symbol K), of atomic number 19, is an alkali element, and one of the major rock-forming elements of the earth's crust. It was isolated by H. Davy in 1807. In pure form it is a silver-colored metal of atomic weight 39.100, density 0.86 (20°C), melting point 63.4°C, and boiling point 757°C. In compounds, potassium forms univalent cations of ionic radius 1.33Å. Bonds with oxygen and chlorine are essentially ionic in character.

Three natural isotopes of potassium are known. These, together with their relative abundances are: K^{39}, 93.1%; K^{40} 0.0119%; and K^{41} 6.9%.

K^{40} is naturally radioactive, and decomposes to form Ca^{40} and Ar^{40}. Its half-life is 1.27×10^9 years. The reaction has produced considerable heat in the earth, and provides a method of determining the absolute age of minerals and rocks.

TABLE 1. TERRESTRIAL POTASSIUM ABUNDANCE (from Heier and Adams, 1964)

	K %
Whole Earth	0.085
Crust	2.1
Oceanic crust	0.87
Continental crust	2.6
Mantle	0.11
Granite	3.40
Syenite	3.73
Nepheline syenite	4.53
Diorite	1.75
Gabbro	0.74
Dunite	0.033
Shale	2.66
Sandstone	1.07
Limestone	0.27

Minerals

The following list is a selection of some of the more common of the many existing potassium minerals:

potash feldspar	$KAlSi_3O_8$
leucite	$KAlSi_2O_4$
muscovite	$KAl_2AlSi_3O_{10}(OH)_2$
biotite	$K(Mg,Fe)_3AlSi_3O_{10}(OH)_2$
glauconite	$K_{1.5}(Fe,Mg,Al)_4(Si,Al)_8O_{20}(OH)_4$
sylvite	KCl
carnallite	$KMgCl_3 \cdot 6H_2O$
kainite	$KMg(SO_4)Cl \cdot 3H_2O$

Abundance

Potassium is a mineral-forming element in a large variety of crustal rocks. In granitic rocks for example, it is present in 2–5 wt % K_2O and is concentrated principally in muscovite, biotite, and potash feldspar. The average concentration in the earth's crust has been estimated at 2.1% K. Meteorites contain much lower concentrations; for example, the average K content of chondrites is 0.085% (see Table 1) (Edwards and Urey, 1955).

Geochemical Behavior

The process of magmatic differentiation causes an increase in potassium in the last stages of the process. Thus the differentiation of certain gabbroic rocks has evidently produced granophyre, which contains potassic feldspar.

Potassium is locally introduced to metamorphic rocks, a process that is referred to as potash metasomatism. The resulting rocks are veined gneisses, containing porphyroblasts of potash feldspar and veins of potash feldspar and quartz, or more homogeneous granitic rocks. This type of metasomatism is most common in rocks of the amphibolite facies.

Aureoles about pegmatite bodies are locally enriched in potassium, which is manifest in the crystallization of muscovite.

Studies have been made of the distribution of potassium in zoned pegmatite bodies. Solodov (1960) found an increase in the potassium content of microcline and albite and a decrease in the potassium content of muscovite from the periphery to the center of a zoned pegmatite body.

Although potassium appears to be relatively mobile at temperatures and pressures of rock formation, little is known regarding the kinetics of its transfer through solid rocks.

During weathering of crystalline rocks, potassium is dissolved with relative ease. However, in contrast to sodium, the potassium ions are more easily removed from solution, and adsorbed on clay minerals or contribute to the formation of glauconite and sericite. Under special conditions, potassium may be precipitated from sea water as a chloride or sulfate (carnalite, kainite, sylvite) together with other salts.

Potassium is found in soils, and in plants and animals, and is essential to life. Over 90% of the world's output of potassium salts goes into the manufacture of fertilizer.

RALPH KRETZ

References

Day, F. H., 1963, "The Chemical Elements in Nature," New York, Reinhold Publ. Corp., London, George G. Harrap & Co.

Edwards, G., and Urey, H. C., 1955, "Determination of alkali metals in meteorites by a distillation process," *Geochim. Cosmochim. Acta*, **7**, 154–168.

Heier, K. S., and Adams, J. A. S., 1964, "The Geochemistry of the Alkali Metal," in (Ahrens, L., et al., editors), "Physics and Chemistry of

POTASSIUM: ELEMENT AND GEOCHEMISTRY

the Earth," Vol. 5, New York, McGraw-Hill Book Co., pp. 255–381.
Rankama, K., and Sahama, T. G., 1950, "Geochemistry," Chicago, Univ. of Chicago Press, 912pp.
Solodov, N. A., 1960, "Distribution of alkali metals and beryllium in the minerals of a zoned pegmatite in the Mongolian Altai," *Geochemistry*, No. 8, 874–885, (translation of Russian.
Clark, S. P., Jr., Peterman, Z. E., and Heier, K. S., 1966, "Abundances of uranium, thorium, and potassium," in "Handbook of Physical Constants," *Geol. Soc. Am. Mem.*, **97**, Sec. 24, 521–541.
Verniani, F., 1966, "The total mass of the Earths atmosphere," *J. Geophys. Res.*, **71**(2), 385–391.
Wasserburg, G. J., et al., 1964, "Relative contributions of uranium, thorium, and potassium to heat production in the Earth,' *Science* **143** (3605), 465–467.

Cross-references: *Elements: Planetary Abundances and Distribution; Potassium/Rubidium Ratio in Geology; Radioactivity in Rocks; Weathering, Chemical. Vol. IVB: Potash; Origin and Economic Geology. Vol. V: Magmatic Differentiation; Metasomatism; Porphyroblast.*

POTASSIUM-ARGON AGE DETERMINATION

General Description

The K–Ar method of age determination has been successfully applied to samples ranging from a few hundred thousand years old to ones approximately 4500 million years old (meteorites, the oldest samples known). K^{40} which is present in all naturally occurring K ($K^{39}:K^{40}:K^{41} = 93.08:0.0119:6.91$) decays to Ca^{40} and Ar^{40} with a half-life of 1,310 million years. Decay by β^- emission to produce Ca^{40} is 8.1 times as probable as decay by K capture (the capture of an orbiting electron by the nucleus) to produce Ar^{40} (Fig. 1).

Three features of this reaction make possible the K–Ar method, one of the most widely applicable age dating methods used: *(1)* K is one of the most abundant and ubiquitous elements; *(2)* the half-life of K^{40} is on the order of magnitude of the age of the earth and its elements; and *(3)* Ar does not normally enter into chemical combination with other elements.

Many of the common rock-forming minerals contain K and have been used for dating by the K–Ar method. These include muscovite, biotite, lepidolite, orthoclase, sanidine, sylvite, glauconite, and clays. Some minerals which have been used successfully but which are more difficult because of their low potassium concentrations are amphiboles and pyroxenes. Determinations are sometimes made on whole rock samples which contain some of the minerals mentioned above. Obsidian has also been found to yield satisfactory age determinations by the K–Ar method.

The half-life of K^{40} is sufficiently long so that it has not yet decayed away since the time of its creation. It is sufficiently short so that the reaction proceeds fast enough to produce detectable amounts of Ar^{40} and Ca^{40} in geologically significant time periods.

Both $Ca^{40}:K^{40}$ and $Ar^{40}:K^{40}$ ratios increase with time after a potassium-bearing mineral forms (becomes a closed system). Even though Ca^{40} is produced at a greater rate than Ar^{40}, the $Ar^{40}:K^{40}$ ratio is the more suitable indication of elapsed time. Ar^{40} constitutes 99.6% of the total natural Ar and Ca^{40} constitutes 97.0% of total natural Ca. Therefore, natural Ar or Ca incorporated in a sample or acquired at some time during the life of the sample obscures the radiogenic Ar^{40} or Ca^{40}. Although a correction can be made for the natural isotopes in a sample, the results become less reliable as the correction becomes larger. Ar is an inert element and therefore generally is not taken up by a mineral at the time it forms. Hence, radiogenic Ar^{40} is much less obscured than Ca^{40}. There is evidence that some Ar may be incorporated in a mineral if it forms under unusually high pressure.

The chemical behavior of Ar has a disadvantage as well. Because it is not chemically bound within a mineral, it is rather easily lost. The extent of loss of radiogenic Ar depends on the type of mineral and on its history. Of the minerals most commonly used, muscovite generally retains radiogenic Ar the best, biotite less well, and feldspar rather poorly. Even mild heating of these minerals during their history promotes Ar loss. Ar loss is also enhanced by ion exchange and alteration within biotite.

Heating and alteration intense enough to be termed metamorphism apparently drives almost all of the Ar out of the potassium minerals.

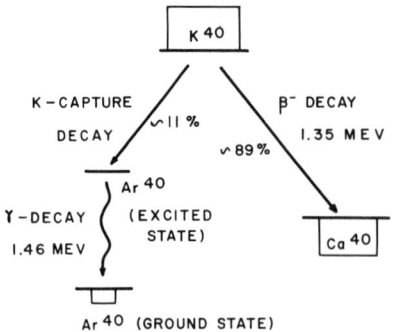

FIG. 1. The decay scheme for K^{40}.

Hence, a K–Ar determination on a metamorphosed rock generally gives results which indicate the time elapsed since the time of metamorphism.

History

The K–Ar method has its origins in 1935 when A. O. Nier published evidence for the existence of the radioactive isotope K^{40}. Weizsacker in 1937 suggested that Ar^{40} formed from K^{40} and that most of the Ar in the earth's atmosphere which has an isotopic composition of $Ar^{36}:Ar^{38}:Ar^{40} = 0.377:0.063:99.60$ formed from the decay of K^{40} in the rocks. He also predicted that old potassium minerals should have radiogenic argon trapped in them. Measurements by Aldrich and Nier in 1948 confirmed this prediction. Before the $Ar^{40}:K^{40}$ ratio in minerals could be used for determining ages, the decay rates of the two branches of the reaction needed to be determined with accuracy. In the early 1950s several workers made determinations employing natural samples, which had been dated by means of the U–Pb method applied to associated uranium minerals. By 1957 the decay rates had been determined with considerable accuracy by direct counting of the emissions from K^{40} (Weatherhill, 1957). During the past decade, thousands of samples have been dated by government, university, and private laboratories around the world.

Measurements

The ratio of K^{40} to total K is very constant throughout natural samples. Therefore, it is not necessary to measure K^{40} directly. It is sufficient to measure the total K content of a sample and multiply by the ratio of K^{40} to total K which equals 0.000119. The potassium concentration may be determined by a number of methods such as flame photometry, gravimetry, and x-ray spectrometry. Some investigators prefer to use mass spectrometry to measure the K^{40} directly.

Measurement of radiogenic Ar^{40}, the Ar^{40} produced by decay of K^{40} in the sample, is less direct and more difficult. For example, a typical muscovite sample a hundred million years old contains approximately 8% K but only 0.06 parts per million radiogenic argon by weight. In addition to measuring this very small quantity accurately, it is necessary to determine its isotopic composition. For this task a mass spectrometer is necessary.

The sample is melted in an evacuated chamber and the gases driven out of the sample are passed over a hot metal such as titanium to remove the chemically reactive elements and leave the inert argon as the major constituent of the gas. The sample is then introduced into the ionization chamber of the mass spectrometer tube where individual atoms are given a positive charge so that they can be propelled in a stream through a magnetic field which causes them to separate into beams according to mass. The intensities of these beams of ions are measured by the electric charges on the plates which they strike. When the signal is amplified and read out on a strip chart recorder, each isotope appears as a peak whose height indicates the relative abundance of that isotope. In order to derive absolute abundances from these relative quantities, it is necessary to compare these signals to a signal produced by a known quantity of an argon isotope. For this reason a known quantity of Ar^{38} (called a spike) is introduced into the gas that was driven out of the sample (Fig. 2). The signals from the Ar^{40} and Ar^{38} are compared and absolute quantities are determined.

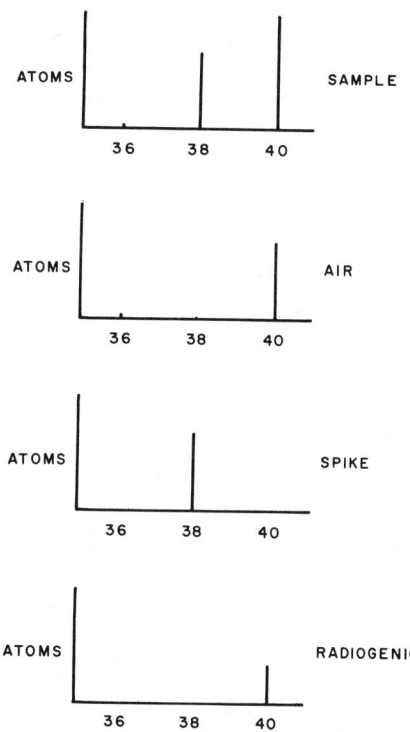

Fig. 2. The isotopic composition of Ar in a typical sample run is shown at the top. The isotopic compositions of the Ar contributed by the three sources, air, spike, and radiogenic, are shown. The objective is to determine the amount of radiogenic Ar^{40} in the sample. This is easily done when the compositions of the air and spike Ar are known and a known amount of spike is metered into the sample.

POTASSIUM-ARGON AGE DETERMINATION

The very isotope of Ar (Ar^{40}) which is to be measured is the most abundant isotope of Ar in the atmosphere. Consequently, contamination of the sample by air Ar presents a serious problem. It is possible, however, to correct for this contamination when the isotopic composition is known. Air contains 296 times as much Ar^{40} as Ar^{36} (Fig. 2). Therefore, air Ar may be distinguished from radiogenic Ar by means of the amount of Ar^{36} present, and a correction can be made.

Calculations

When the abundances of radiogenic Ar and the radioactive K in a sample are known, the age of the sample can be calculated from the following equation:

$$t = \frac{1}{\lambda} \ln \left(1 + \frac{\lambda_\beta + \lambda_k}{\lambda_k} \times \frac{N'}{N}\right)$$

where t = time elapsed since the mineral or rock became a closed system; λ = the rate of decay of K^{40} = 5.30×10^{-10} atoms per atom per year; λ_β = the rate of decay of K^{40} to Ca^{40} = 4.72×10^{-10} atoms per atom per year; λ_k = the rate of decay of K^{40} to Ar^{40} = 0.585×10^{-10} atoms per atom per year; N' = the number of atoms of Ar^{40} in the sample; and N = the number of atoms of K^{40} in the sample. The decay constants given here are the ones used by most workers in the United States and Canada.

Conclusions

The chemical behavior of K and of Ar and the rate of decay of K^{40} combine to provide one of the most widely applicable means for determining the age of a mineral or rock. However, it is rare that the method can be applied in a direct and simple manner. The results for a suite of potassium minerals should be compared and, where possible, the K-Ar results should be compared with measurements by other methods. The mineralogy, alteration, and evidence for metamorphism should be studied and conclusions about the age should be based on as much of this information as possible. Perhaps the K-Ar method along with other isotope methods should not be considered an age dating method but rather a measurement which can be interpreted in such a way as to yield information about the time which has elapsed since the formation or metamorphism of a mineral or rock.

WILLIAM A. BASSETT

References

Armstrong, R. L., 1966, "K-Ar Dating of Plutonic and Volcanic Rocks in Orogenic Belts," in (Schaeffer, O. A. and Zähringer, J., editors), "Potassium Argon Dating," Berlin, Springer-Verlag, p. 117–133.

Curtis, G. H., 1966, "The Problem of Contamination in Obtaining Accurate Dates of Young Geologic Rocks," in *ibid.*, p. 151–162.

Dalrymple, G. B., and Lanphere, M. A., 1969, "Potassium-Argon Dating," W. H. Freeman and Co.

Faul, H. (editor), 1954, 'Nuclear Geology," New York, John Wiley & Sons, 414pp.

Hamilton, E. I., and Farquhar, R. M. (editors), 1968, "Radiometric Dating for Geologists," New York, Interscience Publ., 506pp.

Hurley, P. M., 1959, "How Old is the Earth?," New York, Anchor Books, Doubleday and Co., 160pp.

Krueger, H. W., and Freedman, R. O., 1962, "K^{40}-Ar age determinations aid subsurface correlation," *World Oil*, April, 119–121.

Kulp, J. L. (editor), 1961, "Geochronology of rock systems," *Ann. N.Y. Acad. Sci.*, **91**, Part 2, 159–594.

Müller, O., 1966, 'Potassium Analysis," in (Schaeffer, O. A. and Zähringer, J., editors), "Potassium Argon Dating," Berlin, Springer-Verlag, p. 40–67.

Newell, N. D., 1962, "Geology's time clock; radioactive minerals and fossils both aid in decipherment of earth history," *Am. Mus. Nat. History*, **71**(5), 32–37.

Rancitelli, L., Fisher, D. E., Funkhouser, J., and Schaeffer, O. A., 1967, "Potassium: Argon dating of iron meteorites," *Science*, **155**(3765), 999–1000.

Rankama, K., 1954, "Isotope Geology," New York, McGraw-Hill Book Co., 535pp.

Schaeffer, O. A. and Zähringer, J. (editors), 1966, "Potassium-Argon Dating," Berlin, Springer-Verlag, 234pp.

Weatherill, G. W., 1957, "Radioactivity of potassium and geologic time," *Science*, **126**, 545–549.

Zykov, S. E., et al., 1964, "The age of the oldest formations of the Kola Peninsula," *Geokhymia*, **4**(transl. in *Geochem. Internat.*, **2**, 262).

Cross-references: *Flame Spectroscopy; Geochronometry; Mass Spectrometry; Uranium-Thorium-Lead Dating; X-Ray Diffraction Analysis.* Vol. IVB: *Radioactive Isotope Tracer Technology.*

POTASSIUM/RUBIDIUM RATIO IN GEOLOGY

Potassium and rubidium, due to their close chemical similarity, form a coherent pair in geochemical processes. Rb forms no minerals of its own, but substitutes in small amounts for K in the crystal structures of a variety of rock-forming minerals, notably the micas and feldspars. Geochemically, the most important difference between the two elements is the slightly larger size of the Rb^+ ion

TABLE 1. K/Rb Ratios in Meteorites, Tektites, and the Earth

	K (ppm)	Rb (ppm)	K/Rb	Source
Whole earth	140 ± 30	0.4 ± 0.1	350 ± 50	Hurley (1968)
Mantle and core	84 ± 30	0.17 ± 0.07	500	Hurley (1968)
Crust	13500 ± 2300	57 ± 10	237	Hurley (1968)
Crust	21000	91	231	Heier and Adams (1964)
Chondrites	850	2.8	303	Adams (1964)
Achondrites	330	0.25	1325	Gast (1965)
Tektites	19700	93.7	218	Heier and Adams (1964)
Tektites	25000	126	199	Pinson et al. (1965)

(radius 1.47 Å compared with 1.33 Å for K^+). This size difference leads to a measurable fractionation of K and Rb between coexisting minerals, with Rb concentrated relative to K in the phase with the larger available lattice sites. Partitioning between crystals and a melt will in most cases concentrate Rb in the latter. These relationships make the K/Rb ratio a useful indicator of such geologic processes as magmatic differentiation, metamorphic differentiation, and fractional melting. For a general discussion of the K–Rb association and other coherent element groups, see Ahrens (1965).

Meteorites, Tektites, and the Earth

Table 1 reviews analytical data for meteorites and tektites, together with computed estimates for the earth. Relative to chondritic meteorites, the earth shows a considerable depletion in K and Rb and a slightly higher K/Rb ratio. Taylor (1964) estimates that 62% and 72% of the earth's K and Rb, respectively, are concentrated in the crust; Hurley (1968) estimates 39% and 56% concentrations. Zones with anomalously high K/Rb ratio may exist in the uppermost mantle (Gast, 1968) and in the lower crust (Heier, 1966, Lambert, 1967). The low K/Rb ratios in tektites are compatible with an origin by impact fusion of crustal rocks. Significant variations in the K/Rb ratio both within and between tektite-strewn fields have been noted by Pinson et al. (1965).

Minerals

Data from numerous sources (Hart, 1966; Heier, 1966; Griffin, 1967; Lang, 1967; Steuber, 1967; White, 1966) on coexisting mineral pairs indicate that among common rock-forming silicates with large cation sites (coordination > 8) the K/Rb ratio decreases (i.e., the relative tolerance for Rb increases) roughly in the following sequence: hornblende > plagioclase > orthoclase > muscovite > biotite. In olivines, pyroxenes, and garnets, K contents are extremely low and K/Rb ratios are scattered but commonly lower than in coexisting hornblendes. A significant fraction of the K and Rb in minerals without large cation sites may be present in impurities, cracks, and lattice defects rather than in solid solution.

Igneous Rocks

Fractional crystallization of silicate melts commonly results in an absolute enrichment of both K and Rb in the liquid phase concurrent with a slight relative enrichment of Rb. As a result, differentiated igneous rock suites tend to show a gradual decrease in the K/Rb ratio with increasing K content from early to late fractions. Shaw (1968) computed an average trend from a variety of igneous suites, giving values of K/Rb of 433, 332, 254, and 195 at concentrations of 0.01, 0.1, 1.0, and 10% K, respectively.

The high and scattered K/Rb ratios in abyssal basalts, ranging from near 400 to over 3000 (Gast, 1965; Tatsumoto, 1965; Bence, 1966), if not a result of seawater contamination, may reflect an origin of these rocks by partial melting of upper mantle material previously depleted in Rb (Gast 1968). Basalts from oceanic islands and continental areas show values close to the average trend.

Granitoids show a wide range of K/Rb ratios with a mean probably close to 200. Beus and Oyzerman (1965) note that the strong positive correlation between K and Rb in most igneous rocks is weaker in granites and reversed in those granitic rocks which show evidence of metasomatic or postmagmatic alteration. Minerals from large pegmatites commonly show abnormally low K/Rb ratios (Heier and Adams, 1964) indicating strong enrichment of Rb in late stage and postmagmatic aqueous fluids.

Sedimentary Rocks

Fractionation of K and Rb between clays and water during weathering and sedimentation leads to low K/Rb ratios in argillaceous sediments (mean close to 150) and a very high ratio (3167) in sea water (Heier and

Adams, 1964). Data for sandstones are scattered and reflect both the provenance and mineralogy of the clastic material.

Metamorphic Rocks

K/Rb ratios in low- and medium-grade metamorphic rocks probably do not vary greatly from those of the parent igneous or sedimentary rocks. With intense metamorphism, local migration of alkalis combined with fractionation of K and Rb amongst biotite, feldspar, and hornblende may lead to mineralogical control of the K/Rb ratio in heterogeneous metamorphic rocks. Granitic veins in some migmatites show ratios considerably higher than the host rock, due to K/Rb fractionation between feldspar in the veins and biotite the host (White, 1966). Some granulite facies rocks show increased K/Rb on a regional scale, perhaps due to fractionation accompanying breakdown of the micas (Lambert, 1967, Heier, 1966).

PHILIP R. WHITNEY

References

Ahrens, L. H., 1965, "Distribution of the Elements in Our Planet," New York, McGraw-Hill Book Co.

Gast, P. W., 1968, "Trace element fractionation and the origin of tholeiitic and alkaline magma types," *Geochim. Cosmochim. Acta,* **32,** 1057–1086.

Heier, K. S., and J. A. S. Adams, 1964, "Geochemistry of the alkali metals," in (Ahrens, Press and Runcorn, editors), "Physics and Chemistry of the Earth," Vol. 5, New York, McGraw-Hill Book Co., pp. 253–381.

Hurley, P. M., 1968, "Correction to: Absolute abundance and distribution of Rb, K and Sr in the earth," *Geochim. Cosmochim. Acta,* **32,** 1025–1030.

Pinson, W. H. Jr., Philpotts, J. A., and Schnetzler, C. C., 1965, "K/Rb ratios in tektites," *J. Geophys. Res.,* **70,** 2889–2894.

Shaw, D. M., 1968, "A review of K-Rb fractionation trends by covariance analysis," *Geochim. Cosmochim. Acta,* **32,** 573–602.

*For other citations, see references in Shaw (1968).

Cross-references: *Geochemical Evolution of the Core, Mantle, and Crust; Isotope Fractionation; Potassium: Element and Geochemistry; Radioactivity in Rocks; Rubidium: Element and Geochemistry.* Vol. II: *Meteorites; Tektites.* Vol. IVB: *Rock-Forming Minerals.* Vol. V: *Magmatic Differentiation.*

PRASEODYMIUM: ELEMENT AND GEOCHEMISTRY

Von Welsbach, in 1885, first isolated praseodymium from other rare earths; he separated the element out of didymia, an intermediate rare earth concentrate of cerite, the first known light rare earth mineral, discovered by Cronstedt in 1751.

Properties

Elemental praseodymium (Pr) is a very reactive, silvery-white *lanthanide* (q.v.), or *rare earth* (q.v.) metal with atomic number 59 and outer electronic structure $4f^36s^2$. It occurs as one stable isotope Pr^{141}.

Praseodymium is determined analytically by neutron activation, x-ray spectrography, and by absorption spectrometry (it has very sharp spectral lines). In geologic environments, praseodymium forms highly ionic bonds. It forms weak complexes. The oxidation state is tripositive, with ionic radius 1.013 Å; tetravalent praseodymium is known in a few solid compounds; the system praseodymium-oxygen has a very complex phase diagram. Generally, chemical behavior is similar to the alkaline earths, and to other light rare earths.

Minerals

Praseodymium occurs in highest concentration in rare earth minerals that favor light rare earths, generally nitrates, sulfates, carbonates, niobates, and tungstates with high cation-coordination number, often containing structural halogens and alkali metals. Typical examples, with cation coordination number 10 and 11, are:

Monazite (light rare earths, Th) PO_4
Bastnaesite (light rare earths, Th) FCO_3

Geochemistry

Schmitt et al. (1964) have established a value 0.12 ppm (parts per million) praseodymium in chondrites and 1.4 ppm in eucrite achondrites. The standard deviation within a meteorite class is only about 20%. In meteoritic matter volatilization may have redistributed alkali metals but not rare earths: praseodymium has a boiling point of 3127°C.

The praseodymium concentration in average basalt is 8.5 ppm, but the range is over a factor of ten (Haskin et al. 1966). Alkaline oceanic basalts and most continental basalts have more praseodymium than oceanic tholeites and some continental diabases. Partial melting of mantle peridotite yields a basaltic liquid with enriched praseodymium and a mantle source area depleted in praseodymium.

The praseodymium concentration in granites with less than 60% silica averages 9.8 ppm, a value that is higher, but also more variable, than for basalts.

Igneous differentiation series tend toward higher praseodymium concentrations. The relative abundance of praseodymium and other rare earths changes. Haskin et al. (1966) report 14.2 ppm praseodymium in a Gough Island basalt and 54 ppm praseodymium in a trachyte differentiate. Towell et al. (1965) found 5.64 ± 0.62 ppm in the Rubidoux Mountain Leucogranite phase of the Southern California batholith, probably a differentiate (by crystal settling) of the San Marcos Gabbro, with 2.15 ± 0.24 ppm. About half of the praseodymium in the gabbro is in rock-forming minerals (plagioclase, pyroxene, hornblende) probably substituting for calcium. The remainder is in accessory minerals like apatite. The accessory minerals (probably apatite, zircon, sphene, and monazite) hold most of the praseodymium in the leucogranite.

Praseodymium forms complexes in alkali, carbonate, and halogen-rich hydrothermal fluid, but they are not as stable as heavy rare earth complexes. Pagmatoid and metasomatic deposits formed from the hydrothermal fluids often have low praseodymium to heavy rare earth ratios compared to granites.

Praseodymium has a 180-year residence time in seawater, about the same as other light rare earths. Probably it is complexed. Removal by absorption on clay particles, as well as depletion in surface water in areas of organic productivity are likely processes. Today, high secondary praseodymium concentrations in phosphatic fish remains occur only in deep water; Mesozoic fish remains also have high secondary praseodymium. Manganese nodules have 3600 ppm praseodymium whereas Pacific Ocean water has only 0.64×10^{-6} ppm (Goldberg et al., 1963).

In soils, the relative amounts of praseodymium in detrital, clay, and carbonate fractions is not well known. Sedimentary rocks have about 10 ppm praseodymium, with a variation of a factor of five. In a study of rare earths in Russian platform soils, Balashov et al. (1964) found higher concentrations in alkaline soils, the praseodymium precipitating as hydroxide; lower concentrations are in acid soils, probably due to removal as complexes.

Biologic concentration of rare earths has been found in hickory leaves; rare earths have been used to trace opium.

Economic Geology

The major source of praseodymium has been from monazite in beach sands. The accessory monazite originally present in granitic rocks resists weathering and is concentrated by sedimentary processes. Australia, Brazil, Ceylon, India, Malaysia, and South Africa mine the largest quantities. Usually, about 2% (by weight) of monazite is praseodymium. Recently, the bastnaesite deposit at Mountain Pass, California has been an important source of praseodymium.

High purity praseodymium has limited application in electronics and in optical filters. Most praseodymium is sold unpurified as misch metal (a mixed rare earth alloy), and as oxide polishing powders. Impure rare earths are also used as petroleum cracking catalysts and in arc carbons.

ROBERT KAY

References

Balashov, Y. A., et al., 1964, "The effects of climate and facies environment on the fractionation of the rare earths during sedimentation," *Geochem. Intern. No. 1,* **10,** 951–969 (English transl.).

Fisk, Z., and Matthias, B. T., 1969, "Rare-earth elements and high pressures," *Science,* **165** (3890), 279–280.

Goldberg, E. D., et al., 1963, "Rare-earth distributions in the marine environment," *J. Geophys. Res.,* **68,** 4209–4217.

Haskin, L. A., et al., 1966, "Meteoric, Solar, and Terrestrial Rare-Earth Distributions," 167-321 in (Ahrens, L. H., et al., editors), "Physics and Chemistry of the Earth," Vol. 7, Oxford, Pergamon Press.

Rankama, K., and Sahama, T. G., 1950, "Geochemistry," Chicago, University of Chicago Press, 912pp.

Schmitt, R. A., et al., 1964, 'Rare-earth, yttrium and scandium abundances in meteoric and terrestrial matter—II," *Geochim. Cosmochim. Acta,* **28,** 67–86.

Towell, D. C., et al., 1965, "Rare-earth distributions in some rocks and associated minerals of the batholith of Southern California," *J. Geophys. Res.,* **70,** 3485–3496.

Cross-references: *Complexes; Lanthanides; Neutron Activation Analysis; Rare Earths; X-Ray Spectroscopy.*

PRECAMBRIAN ATMOSPHERE—GEOCHEMICAL HISTORY

The evolution of the earth's atmosphere during Precambrian times (from about 4.55 AE to 0.6 AE ago; following a suggestion by H. C. Urey, I shall use as a unit of time the aeon, which is equivalent to 10^9 years and abbreviated AE), entailed highly significant changes in its composition and, presumably, structure. These changes, of a more revolutionary character by far than any which have occurred since, influenced profoundly the origin and development of life on this planet. They may also have had marked influences on the geochemical cycle, particularly the processes of weathering and sedimentation, but most records of such effects seem to have been erased subsequently. Fortunately, the slate has not been wiped quite clean, so that by using the extant geological record, reasonable inferences on the origin of the earth and on the origin of life, and data on the present composition of the atmosphere, it is possible to construct geochemical models of the Precambrian atmosphere. These models, while moderately detailed and relatively self-consistent, are as yet far from being accepted as unique representations of the actual events. Indeed, this article itself shall treat of three variant histories, which oddly enough are dictated largely by different sets of hypotheses for the origin of the moon. There are, however, certain common features of these models which seem to be so strongly demanded by much of what we know at present that these features are likely to persist throughout future modifications and improvements of the models. In this sense a consensus has been achieved.

Boundary Conditions

There are three major lines of argument which may be used to establish boundary conditions for the models we shall consider. Each of these depends upon sophisticated interpretations of complex observations, so that the arguments cannot be presented in any detail. However, the conclusions which I shall utilize seem to be as certain as might be expected, considering their derivation from observations of complex natural systems.

Constancy of Atmospheric Composition. Extensive geological, geochemical and paleontological evidence indicates that the abundances of the major components of the atmosphere have not changed markedly for approximately the latest aeon (or longer) of the earth's history. Some minor changes, such as the addition of Ar^{40} (a decay product of long-lived K^{40}) or fluctuations in the CO_2 content are either known or suspected to have occurred. The evidence in support of the assertion of the constancy of atmospheric composition has been discussed by Rubey (see reprint of Rubey's 1951 paper in Brancazio and Cameron, 1964). This classic paper is now somewhat outdated, largely due to advances in our understanding of the very early history of the earth, but nevertheless treats the problem under discussion with wit and rare good sense. Constancy of atmospheric composition implies feedback mechanisms, at least partly biological in origin, to stabilize the system. Another implication is that we must construct models which (1) yield approximately the contemporary composition at about 3 AE after t_0 (t_0 being the time of origin of the earth, approximately 4.55 AE ago), and (2) allow the feedback mechanisms to dominate since $t_0 + \sim 3\ AE$. These requirements comprise our terminal boundary condition.

Primitive Atmosphere. The initial boundary condition is imposed, though in a less unambiguous manner, by knowledge of the probable composition of the primitive atmosphere of the earth, that which it had during and shortly after its formation. If we could assert with plausibility that this atmosphere was closely similar to the gaseous portion of the primitive solar nebula from which the planets evolved (see Vol. II), its composition would be well defined. In fact, we would then have a very close analog in the present atmospheres of Saturn and especially Jupiter, which seem not to have changed significantly during the preceding 4.5 AE. However, as pointed out by Suess and by Brown (see references in Urey, 1959), the noble gas contents of the terrestrial atmosphere indicate that at most about 10^{-7} of any primitive nebular atmosphere has been retained by the earth. Arguments have been presented for the conclusion that the fraction retained may have been even smaller than 10^{-7} (see paper by Cameron in Brancazio and Cameron, 1964), so that, e.g., the overall hydrogen content of the atmosphere at the stage from which we can follow its evolution by means of detailed models remains uncertain by several orders of magnitude. Despite this ambiguity, we may use as our initial boundary condition an atmospheric composition which is much more highly reduced than the present atmosphere; indeed, it seems likely that H_2 or hydrogenated compounds such as CH_4 made up the bulk of the primitive atmosphere.

The Origin of Life. The presently available data on the composition of the primitive atmosphere have had a marked influence in determining the nature of some of the early, highly relevant experimentation on syntheses of crucial building blocks (e.g., amino acids) for living organisms. However, as Urey (1959) discusses in his review article, studies of the biochemical and geochemical evidence relating to the origin of life are now beginning to repay the debt. Thus, it is now possible to utilize what is known about the origin of life on this planet to place some restrictions on the initial composition of the earth's atmosphere and on the time scales for the earliest stages of its evolution. These restrictions parallel, to some extent, those imposed by knowledge of the composition of the primitive atmosphere but in addition suggest that a reducing atmosphere was present for at least some tens of millions of years early in the earth's history.

Input-Output Mechanisms

The evolutionary history we are concerned with was dictated by processes which fed material into the atmosphere or removed material from it. These input-output mechanisms may be classified as follows:

Input Mechanisms
(a) Capture from the primitive solar nebula.
(b) Outgassing of the earth's interior.
(c) Chemical reactions with surface materials (including biochemical interactions).
(d) Capture from the solar wind.
(e) Outgassing of interplanetary materials (e.g., comets, meteorites, cosmic dust) during impact or weathering.

Output Mechanisms
(a) Thermal escape from the exosphere.
(b) Chemical reactions with surface materials (including biochemical interactions).
(c) "Sweeping" by the solar wind or by other magnetohydrodynamic interactions.

In addition, Cameron (see preface to Brancazio and Cameron, 1964) has recently suggested that rotational instability, such as that which has been postulated to account for the origin of the moon, should be considered as an output mechanism.

The first of these mechanisms no longer operates. For the others, with the exception of output via rotational instability (which as yet remains speculative), we can infer much about their influences by observing their present action. Rubey (op. cit.) has discussed in some detail the evidence for outgassing of the earth's interior. His conclusion, which more recent work has tended to confirm, is that this is the predominant input mechanism and has probably been so during most of the earth's history. The predominant output mechanism, at least during the later Precambrian, has been thermal escape from the exosphere.

The Three-Stage Model

A self-consistent model based upon the boundary conditions and input-output mechanisms discussed above has been devised by Holland (see Holland's paper in Brancazio and Cameron, 1964). This model, which we shall take as the framework for further discussion, defines three stages in the evolution of the terrestrial atmosphere. Stage I is a tenuous atmosphere of CH_4, N_2, NH_3, and some H_2, the highly reduced state being imposed by the assumed $Fe^{\circ}-Fe^{2+}$ buffer system present in the upper mantle. It is further assumed that the atmosphere is in equilibrium with this buffer system. As Holland points out, if no significant quantities of metallic Fe had been present in the crust or upper mantle, Stage I would not have existed. (It should be noted that there is an implicit Stage 0, that in which nebular gases, with their very high abundances of H_2 and He, dominate. Holland assumes that this stage may be neglected, which is equivalent to assuming either that capture from the nebula was very inefficient or that the solar wind "sweeping" output mechanism (see above) or the speculative rotational instability output, or both, were very efficient before the beginning of stage I.).

Stage II is an atmosphere dominated by volcanic gases in equilibrium with an $Fe^{2+}-Fe^{3+}$ buffer system. The major component is N_2, with principal minor components of SO_2, H_2O, CO_2, and Ar. Hydrogen may be lost via thermal escape at a high enough rate to give rise to trace amounts of O_2, but free oxygen is essentially absent. Such volcanic gases are the predominant components of present-day volcanic exhalations, but their influence on the composition of the present atmosphere, except for localized effects, is minor. The present atmosphere is representative of Stage III, in which the oxidation state is determined principally by photosynthesis. Initially, Stage III must have been characterized by minor amounts of free oxygen in a predominantly N_2 atmosphere, but the O_2 partial pressure seems to have grown monotonically toward the present value (Rutten, 1966; Cloud, 1968).

Chronology

Holland suggests that Stage I occupied approximately the initial 0.5 AE of the earth's history, Stage II from 0.5 to, very roughly, 2 to 3 AE after t_0, yielding then to Stage III. This chronology is critically dependent upon the time scale for the growth of the core of the earth, which in turn has been associated, according to several hypotheses currently under discussion, with the origin of the Moon. Three possibilities may be distinguished, at least in broad outline:

The Urey Hypothesis. One widely accepted idea of the origin of the moon is that it is an exceedingly primitive body, one of Urey's primary objects which by accident survived violent collisional events and was captured by the earth. Under this hypothesis, the Moon's capture would have occurred very soon after

t_0. As has been discussed by Urey, Elasser, and Runcorn, the core of the earth could then have been grown slowly throughout the earth's history, so that metallic Fe would have been available in the upper mantle for perhaps 0.5 AE. This set of hypotheses thus is consistent with Holland's suggested chronology.

The MacDonald Hypothesis. On the basis of estimates of the moon-earth tidal interactions and their effects on the orbit of the moon, MacDonald has suggested that the moon originated via capture or via collisional breakup of preexisting satellites at about $t_0 + 2$ AE. This cataclysm, in one form or another, would probably have greatly accelerated the formation of the core. Under this hypothesis, a presumably highly evolved atmosphere, perhaps with characteristics intermediate between Stage 0 and Stage I, might have been completely lost. The extensive internal reorganization of the Earth would have induced efficient outgassing (see *Outgassing of the Planet Earth*), leading to a Stage II atmosphere shortly after the cataclysm. Thus, this hypothesis implies that we shall probably never be able to recognize evidence of a Stage I atmosphere, but rather must begin at about $t_0 + 2$ AE with Stage II. Evidence now available on the ages of rocks from lunar maria makes this hypothesis implausible.

The Cameron Hypothesis. Using arguments based in part on comparisons of the atmospheres of the earth and Venus, Cameron has suggested that the earth-moon fission via rotational instability occurred at about $t_0 + 0.05$ AE; i.e., very early in the history of the earth. This hypothesis has two important consequences for our discussion: First, it implies that any Stage 0 atmosphere would have been lost early in the game. Second, since no metallic Fe would be left in the upper mantle after the earth-moon fission, Stage I never existed and Stage II would have begun very early.

Conclusions

From this brief and sketchy overview of alternative histories of the Precambrian atmosphere, it is possible to isolate elements of a consensus. Clearly, at least Stages II and III of Holland's model must have been significant. There is some uncertainty about the time at which the change from one to the other occurred, but there seems to be agreement that it happened during the Precambrian. Whether or not Stage I existed, and the extent to which it was important geochemically, are questions which cannot be resolved without more detailed information on the origin of the moon and the growth of the core of the earth.

GORDON GOLES

References

Brancazio, P. J., and Cameron, A. G. W. (editors), 1964, "The Origin and Evolution of Atmospheres and Oceans," New York, John Wiley & Sons.

Cloud, P. E., 1968, "Atmospheric and hydrospheric evolution on the primitive Earth," *Science*, **160**(3829), 729–736.

Eugster, H. P., and Munoz, J., 1966, "Ammonium micas: possible sources of atmospheric ammonia and nitrogen," *Science*, **151**(3711), 683–686.

Rutten, M. G., 1966, "Geologic data on atmospheric history," *Palaeogeogr., Palaeoclimat., Palaeogeol.*, **2**(1),

Urey, H. C., 1959, "The atmospheres of the planets," in (S. Flügge, editor) "Astrophysics III; The Solar System," Vol. 52 in "Handbuch der Physik," Berlin, Springer-Verlag, pp. 363–418.

Cross-references: *Biogeochemistry; Buffer Systems; Gases—Volcanic; Geochemical Evolution of the Core, Mantle, and Crust; Organic Geochemistry; Outgassing of the Planet Earth; Oxygen—Its Evolution in the Earths Atmosphere; Potassium-Argon Age Determination. Vol. II: Moon—Theory of Origin; Planet Earth—Origin and Evolution; Planetary Atmospheres; Solar System—Origin; Solar Wind. Vol. VII: Life—Its Origin.*

PRECAMBRIAN HYDROCARBONS

Precambrian hydrocarbons have been studied as a means of determining the time of the origin of life on earth. The abundance of morphological fossils decreases markedly in sediments older than the Cambrian. The study of chemical fossils, or biological markers, was made to extend our knowledge of biological evolution into the Precambrian with the hope that at some point in time we would be able to see a gross change or absence of biological markers. The concentration of extractable *hydrocarbons* (q.v.) from Precambrian sediments is usually less than 100 parts per million and only recently has the equipment and techniques become available to isolate and identify hydrocarbons at the microgram level. The two prime tools are moderate- and high-resolution gas-liquid *chromatography* (q.v.) and *mass spectrometry* (q.v.).

The characteristics which biological markers should have are: (*1*) ubiquity in living systems, especially in primitive forms such as alga, phytoplankton, bacteria, etc.; (*2*) high expectation of longevity (on the order of billions of years) which means low chemical reactivity; (*3*) a high degree of structural specificity and complexity so that it may be recognized from the products of degradation or abiogenic synthesis; (4) presence in great excess above the proportion expected from the possible number of isomers at a given carbon number (biological systems have very high selectivity); (5) low probability of being formed readily by reasonable abiogenic processes.

Amino acids, nucleic acids, carbohydrates, and lipids are found in large amounts in all living organisms but most of these compounds are degraded by microorganisms or thermal processes soon after deposition. Much of the remainder polymerizes to become the insoluble organic polymer which comprises the bulk of all organic matter in sediments (and carbonaceous chondrites). Traces of amino acids, fatty acids, and hydrocarbons are found in most sediments examined although only certain hydrocarbons meet all of the requirements set forth as biological markers. Amino acids are readily formed by abiotic reactions (primitive atmosphere experiments) and fatty acids are common constituents of Fischer-Tropsch reactions (CO, H_2, and metal oxides as catalysts). Of the lipids, the saturated hydrocarbons, especially the isoprenoid types, fill all the requirements. Porphyrins, which are the chemical remnants of vital pigment molecules such as *chlorophyll* (q.v.) and hemoglobin, are also very stable, complex, and ubiquitous in living systems, but, they are not found in all sediments. Their presence, however, is strong evidence of biological origin and it is usually searched for in ancient deposits which have a low-temperature history.

Saturated hydrocarbons usually comprise the bulk of the organic extract of Precambrian sediments, are the easiest to isolate and identify, and were the first compounds studied. The saturated hydrocarbons may be divided into three groups: the straight-chained, or, normal; the branched-chain; and the cyclic. The normal hydrocarbons are presumed to arise from chemically changed fatty acids and are a distinctive and common group found in most sediments. They are usually present in a continuous homologous series with smoothly varying concentration between adjacent chains of increasing length. The range of carbon atoms in the molecules varies between about 6 and 35. Unfortunately, these compounds are not very complex in structure and are also formed in high abundance and in approximately the same distribution as the products of abiotic Fischer-Tropsch reactions, which have been postulated as a possible source of some of the organic matter originally formed in the chemical evolution period of the earth's history. If, however, there is marked alternation in concentration of adjacent straight-chained hydrocarbons, and contamination by much younger petroleum can be firmly ruled out, then this would be strong presumptive evidence of biological origin since this is a very common characteristic of straight-chained fatty acids and hydrocarbons in living organisms. Marked alternation of these hydrocarbons has not been found in Precambrian sediment extracts. Loss of alternation seems to be a result of normal chemical changes that occur in most deposits of marine origin.

Among the branched-chain hydrocarbons which have been found in ancient sediments are the singly branched iso- and anteiso-normal compounds. The methyl group is located at the second or third carbon of the chain. The compounds found contain about 18 carbon atoms. Some branched-chain fatty acids exist in living organisms, and singly branched hydrocarbons are commonly found in the waxy cuticle of terrestrial plants. Most of the arguments made for the normal hydrocarbons such as simplicity, alternation, and abiogenesis by Fischer-Tropsch reactions hold for these compounds, also.

The isoprenoid hydrocarbons (Fig. 1) with their single-carbon branching at a regular spacing of four carbons along the main chain are currently the prime biological markers. The C_{19} isoprenoid, pristane, is probably the single most ubiquitous saturated hydrocarbon in marine organisms ranging from zooplankton to whales. It is most likely passed along through the marine food chain with little digestive breakdown because of its low chemical reactivity. It is also often the single most abundant saturated hydrocarbon in the branched-cyclic fraction of Precambrian and younger sedimental extracts. It is not found in phytoplankton at the beginning of the food chain but is believed to be derived from the C_{20} isoprenoid alcohol, phytol, which is part of the chlorophyll molecules contained in phytoplankton. Pristane and phytane have not yet been identified in the various abiotic reactions related to the origin of biological precursors and still stand as valid biological markers. They have been found in all Precambrian sediments examined to date, including the oldest ones from South Africa which have been dated at about 3.3 billion years. Other isoprenoids are usually found along with pristane and phytane and are probably breakdown products. They include: C_{15}, farnesane; C_{16}; C_{18}; and occasionally, C_{21}.

The cyclic and branched cyclic compounds are present in much greater variety but in much lower individual abundance than the straight-chained and isoprenoid compounds in the organic extracts. The smaller and simpler molecules are most likely degradation products

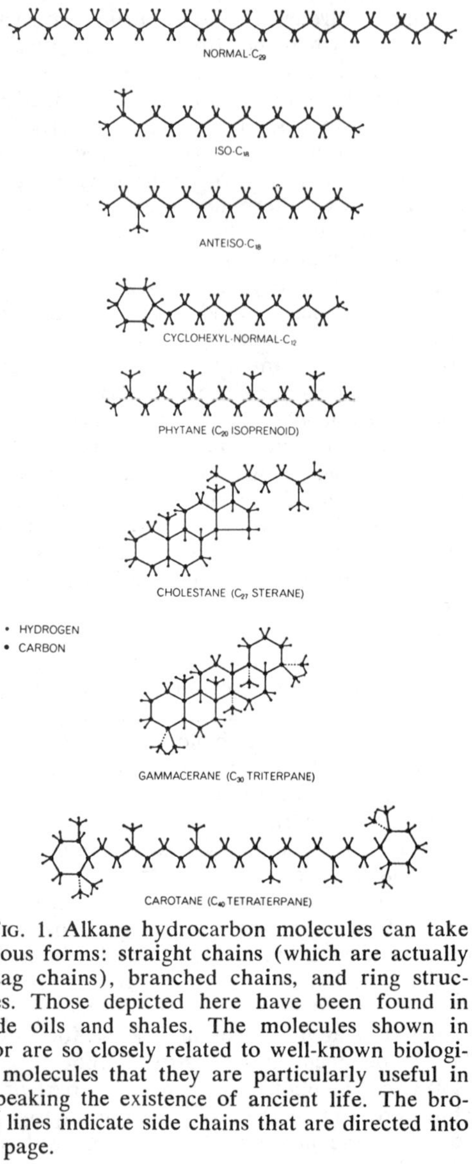

FIG. 1. Alkane hydrocarbon molecules can take various forms: straight chains (which are actually zigzag chains), branched chains, and ring structures. Those depicted here have been found in crude oils and shales. The molecules shown in color are so closely related to well-known biological molecules that they are particularly useful in bespeaking the existence of ancient life. The broken lines indicate side chains that are directed into the page.

believed to be present in the one-billion-year-old Nonesuch shale from Michigan, and in the 2.7-billion-year-old Soudan shale from Minnesota, although there is some question in the latter case as to when the steranes were incorporated in the sediment. The high-temperature history of the shale would seem to indicate that the steranes could not have survived and had entered the rock after cooling. Just when the sediment was at the elevated temperature is not known. Lithification is usually complete in the first 10 or 20 million years and we have an uncertainty in age of about 200 million years. Even if the hydrocarbons entered when the shale was 200 million years old we would still have a finding of importance. This would mean that the complex biochemical mechanisms required for the synthesis of steroids were already in existence in the first half of the 4.8 billion-year-old earth. Similarly, the finding of pristane and phytane in sediment with an age of about 3.3 billion years means that their biosynthetic pathways were developed during the first 1.5 billion years if these molecules are, indeed, biological remnants.

Two additional characteristics of biologically formed compounds which would be strong evidence of their biogenesis are stereospecificity in structure and C^{12}–C^{13} isotope fractionation. The former is caused by the high stereospecificity of the proteinaceous enzymes which control and catalyze most biochemical transformations and syntheses. The synthesized molecules possess varying degrees of right- or life-handedness which can often be determined if sufficient quantities are available. In the case of Precambrian hydrocarbons the amounts of pure compounds are about 1/1000th the amount needed. A similar situation holds with the carbon isotope measurements. Here, only the total extract and the insoluble polymer, *kerogen* (q.v. Vol. VI), can be measured and compared. Generally, there is good agreement between them although a few exceptions have been found in some of the Precambrian samples. Agreement indicates that they are related and probably were laid down at the same time. The values show a marked fractionation in favor of the lighter isotope and parallels those values obtained from contemporary plant hydrocarbons and fatty acids. Attempts are being made to develop new or finer techniques in order to measure both the carbon isotope ratio and stereospecificity on minute amounts of pure Precambrian compounds.

The geochemical and micropaleontological analyses of Precambrian sediments have been complementary and mutually supporting. Chemical fossils have been found in every

since they are not common in living organisms. This is probably true also for the monocyclic compounds with side chains of various lengths attached. Of interest in this group are the steranes—the hydrocarbon skeletons of the tetracyclic and pentacyclic steroids which are biochemically derived from isoprenoids and which are common to many plants and animals. Steranes have only infrequently been isolated from sediments of any age. Recently, cholestane, sitostane, ergostane, and gammacerane were found to be present in the 50 million-year-old Green River shale of Colorado and Wyoming. Another unusual find was the C_{40} saturated isoprenoid carotane. Steranes are sample in which morphological fossils have been found. The reverse is not always true since physical distortion of the rock may easily destroy the morphological but not the chemical remains. The morpohological analysis also runs into difficulty in the very old sediments because the apparent microfossils are much simpler than in the younger ones. The distinction between primitive life forms and organic coacervate droplets becomes less certain. A similar difficulty has not yet arisen in the chemical analyses. Extreme care has had to be used in both the chemical and paleontological studies because of the far reaching conclusions which are likely to be drawn from the results. A similar statement can be made for the forthcoming extraterrestrial samples for which these Precambrian studies have been the logical prerequisite.

THEODORE BELSKY

References

Clark, G. L. (editor), 1966, "Encyclopedia of Chemistry," New York, Reinhold Publ. Corp., p. 424.

Eglinton, G., Murphy, Sister Mary (editors), 1968, "Organic Geochemistry," New York, Springer-Verlag.

Eglington, G., and Calvin, M., 1967, "Chemical Fossils," *Sci. Am.*, **216**(1), 32–43.

Meinschein, W. G., 1965, 'Soudan formation: organic extracts of Early Precambrian rocks," *Science*, **150**(3696), 601–605.

Nagy, B., and Colombo, U. (editors), 1967, "Fundamental Aspects of Petroleum Geochemistry," New York, Elsevier.

Cross-references: *Chlorophyll; Chromatography; Hydrocarbons; Mass Spectrometry; Organic Geochemistry.* Vol. II: *Extraterrestrial Life:* Vol. IVB: *Green River Mineralogy.* Vol. VI: *Kerogen; Petroleum.* Vol. VII: *Life—Its Origin; Paleontology.*

PRIMARY PRODUCTION—See Vol. I

PROMETHIUM: ELEMENT AND GEOCHEMISTRY

Promethium (Pm), is a *lanthanide* (q.v.), or *rare earth* (q.v.) metal, first isolated in 1947 by Marinsky et al. by separation of radioactive Pm^{147} (2.6-yr half-life) on an ion-exchange column. Only milligram quantities have been recovered. Pm^{145}, the longest-lived isotope, has a half-life of 25 yr. The chemical properties of promethium are almost certainly intermediate between *samarium* (q.v.) and *neodymium* (q.v.), its neighboring rare earths.

ROBERT KAY

References

Haskin, L. A., et al., 1966, "Meteoric, solar and terrestrial rare-earth distribution," in (Ahrens, L. H., et al., editors), "Physics and Chemistry of the Earth," Vol. 7, Oxford, Pergamon Press, pp. 167–321.

Marinsky, J. A., Glendenin, L. E., and Coryell, C. D., 1947, "The chemical identification of radioisotopes of neodymium and of element 61," *J. Am. Chem. Soc.*, **69**, 278.

Rankama, K., and Sahama, Th. G., 1950, "Geochemistry," Chicago, Univ. of Chicago Press, 912pp.

Cross-references: *Lanthanides; Neodymium; Rare Earths; Samarium.*

PROTACTINIUM: ELEMENT AND GEOCHEMISTRY

Protactinium (symbol Pa), a naturally occurring radioactive element, atomic number 91, was discovered in 1917 independently by O. Hahn and L. Meitner, by K. Fajans, and by F. Soddy, J. Cranston, and A. Flack. In its pure form, first prepared by A. V. Grosse in 1927, protactinium is a gray metal with a density of 15.37. Because of its extreme rarity (average crustal abundance of $10^{-10}\%$), it is not found naturally as a pure compound or mineral.

Two isotopes of protactinium, one having two isomers, are found in nature, generally in radioactive equilibrium with U^{238} and U^{235} (Table 1).

Geochemical Behavior

Of the three protactinium species listed above, only Pa^{231} has the sufficiently long lifetime that makes possible its differentiation from other elements by geological processes. For most crustal materials Pa^{231} is in radioactive equilibrium with its parent, U^{235}. There is one major exception to this general observation, illustrated in Fig. 1. Uranium-235 and uranium-238, present in solution in seawater, undergo radioactive decay producing the nuclides shown. Pa^{231} and Th^{230}, with long lifetimes, do not remain in solution when formed but are adsorbed on particulate materials and carried to the ocean floor. Moore and Sackett (1964) report that each of these nuclides is present in amounts less than 1% of the amount that could be in equilibrium with the uranium present. This deficiency in seawater is matched by a corresponding excess in deep-sea sediments which is largest at the sediment-water interface and decreases exponentially with sediment depth. The distribution of Pa^{231} and/or Th^{230} in sediments has been used by many investigators to determine deposition rates over

PROTACTINIUM: ELEMENT AND GEOCHEMISTRY

TABLE 1. PROTACTINIUM ISOTOPE DATA

Isotope	Uranium Series	Half-life	Decay Mode	Particle Energy (MeV)
$_{91}Pa^{231}$	U^{235}	3.248×10^4 yr	Alpha	5.001(24%), 5.017(23%) 4.938(22%), and others
$_{91}Pa^{234m}$	U^{238}	1.18 min	Beta isomeric transition	0.58(99%), 2.31(1%)
$_{91}Pa^{234}$	U^{238}	6.66 hr	Beta	0.23–1.35

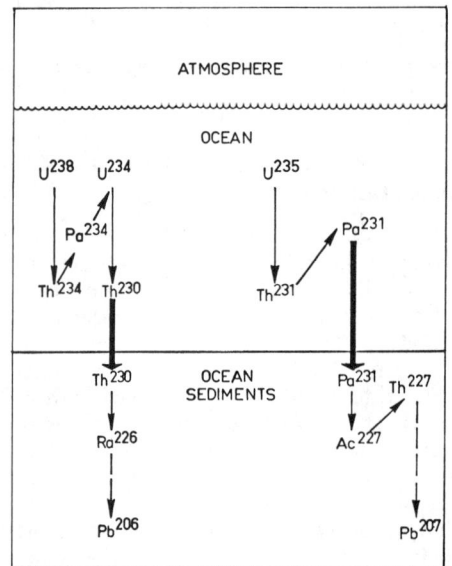

FIG. 1. Behavior of uranium series nuclides in the ocean (from Sackett, 1964).

the ocean floor. The half-lives of these nuclides limit these dating methods to the last 400,000 years.

WILLIAM M. SACKETT

References

Blackman, A., and Somayajulu, B. L. K., 1966, "Pacific Pleistocene cores: Faunal analyses and geochronology," *Science,* **154**(3751), 886–889.

Grosse, A. V., 1930, "The Rarest Metal Yet Obtained," *Sci. Am.,* **142**, 42–44.

Hahn, O., and Meitner, L., 1921, "Veber die Eigerschaften des Protoaktiniums," *Ber.,* **54**, 69–77.

Moore, W. S., and Sackett, W., 1964, "Uranium and thorium series inequilibrium in sea water," *J. Geophys. Res.,* **69**, 5401–5405.

Rosholt, J. N., Emiliani, C., Geiss, J., Koczy, F. F., and Wangersky, P. J., 1961, "Absolute dating of deep-sea cores by the Pa^{231}/Th^{230} Method," *J. Geol.,* **69**, 162–185.

Sackett, W. M., 1960, "Protactinium –231 content of ocean water and sediments," *Science,* **132**, 1761–1762.

Sackett, W. M., 1964, "Measured deposition rates of marine sediments and implications for accumulation rates of extraterrestrial dust," *Ann. N.Y. Acad. Sci.,* **119**, 340–350.

Sackett, W. M., 1965, "Deposition Rates by the Protactinium Method, "University of Rhode Island Occasional Pub. No. 3-1965.

Cross-references: *Geochemistry of Sediments; Protactinium-Thorium Dating; Seawater, Chemistry; Thorium: Element; Uranium: Element.* Vol. I: *Radionuclides in Oceans and Sediments.* Vol. VI: *Pelagic Sediments.*

PROTACTINIUM–THORIUM DATING*

Radioactive daughter products from uranium decay have been used in attempts to date archeological, geological, and oceanographic environments represented by samples of bone, wood, charcoal, continental and marine carbonates, and marine sediments. Two basically different techniques using isotopes of thorium and protactinium have been attempted for methods of Pleistocene dating. One depends on the measurement of the growth of Pa^{231} and Th^{230} following the deposition of uranium, and the other depends on the measurement of the radioactive decay of Pa^{231} and Th^{230} following their separation, in nature, from uranium parents.

Measurement of Pa^{231} and Th^{230} Growth in Fossils

This method covers a time range of about 200,000 years. To be datable by the method, a small amount of uranium of known isotopic abundance must have been incorporated in the material during or shortly after its formation. Pa^{231} is produced from the decay of U^{235}, and Th^{230} (commonly called ionium) is produced from the decay of U^{234} and U^{238}. The amounts of both daughter products can be defined by similar exponential relationships, of the form $U(1-e^{-\lambda t})$, where λ is the decay constant of the daughter product and t is the time elapsed since the incorporation of uranium. The ratio of Th^{230}/U^{234} in the material must be determined to obtain the age. The deter-

---

* Publication authorized by the Director, U.S. Geological Survey.

mination of the Pa^{231}/U^{235} ratio will provide an additional parameter for confirmation of the age. If the result is to represent the true age, three assumptions must be valid: (1) uranium was introduced into virtually nonuraniferous material over an interval of time that was short compared to the age of the daughter products; (2) no measurable amount of external contamination occurred; (3) measurable uranium or daughter products were not subsequently leached from the material. These conditions are rarely fulfilled in natural environments. The Pa^{231}/U^{235} ratio in coral limestones appears to meet the conditions most nearly (Ku, 1968), provided the initial aragonitic calcium carbonate has not recrystallized to calcite.

One of the chief difficulties with this method is the uncertainty of the isotopic ratio of U^{234} to U^{238} at the time of deposition of uranium. Prior to 1962, the radioactivity ratio of U^{234} to U^{238} was assumed to be equal to 1. Recently activity ratios of U^{234}/U^{238} greater than 2.5 have been measured in fossil carbonates of continental origin. Determination of the Pa^{231}/U^{235} ratio is required to estimate the original U^{234}/U^{238} ratio. This comparison should be valid because significant deviations from the normal U^{238}/U^{235} ratio have not been found in natural uranium-bearing materials.

Measurement of Radioactive Decay of Pa^{231} and Th^{230}

The second method useful for geochronological studies of the Pleistocene is based on the differential rates of decay of Pa^{231}, Th^{230}, and Th^{232}. One variation of this technique requires measurements of the uranium-unsupported Pa^{231}/Th^{230} ratios in deep-sea sediments. A second variation requires measurements of the Th^{230}/Th^{232} ratios in authigenic minerals (presumably derived from seawater) contained in deep-sea sediments.

Pa^{231}/Th^{230} **Method.** This method of dating, covering a time range of about 150,000 years, has been used for some selected deep-sea cores from Globigerina-ooze sediments. It has been demonstrated that uranium is soluble in seawater; the radioactive daughters Pa^{231} and Th^{230} are insoluble and are concentrated in the sediments in a distribution which is not in radioactive equilibrium with their respective parents, U^{235} and U^{234}. The chemical behavior of both Pa^{231} and Th^{230} in seawater and sediments is nearly identical, and presumably the two isotopes produced are adsorbed on particles settling to the bottom; these particles are incorporated in the bottom sediment in less than 1000 years. Over this time interval the Th^{230}/Pa^{231} radioactivity ratio in the settling particles will be constant at 10:1. This is equal to their production ratio in seawater with a U^{234}/U^{238} radioactivity ratio of 1.14. The production ratio will be the initial activity ratio at the time of accumulation of undisturbed sediments and it will be independent of both the uranium concentration in seawater and the sedimentation rate; however, it is dependent on the isotopic composition of uranium in seawater. The uranium-unsupported Pa^{231} and Th^{230} will decay in the sediments with half-lives of 32,500 and 75,200 years, respectively; after 57,000 years their activity ratio will be 20:1, thus increasing as a direct function of time only. The mathematical expressions used in this method are given by Rosholt et al. (1961).

Th^{230}/Th^{232} **Method (Ionium/Thorium Method).** This technique which is sometimes useful for obtaining rates of accumulation of deep-sea sediments, over a time range of about 300,000 years, is based upon the simultaneous removal from seawater of Th^{232} and Th^{230} to one or more mineral components of the sediments. Th^{232}, with an extremely long half-life, decayed very slightly during the Pleistocene Epoch, while Th^{230} decayed with a 75,200-year half-life. After deposition of the sediment, the Th^{230}/Th^{232} ratio is assumed to vary only because of the shorter half-life of Th^{230}. In this method, the main source of uranium-unsupported Th^{230} is that produced from the radioactive disintegration of U^{234} dissolved in seawater; and Th^{232} comes from continental or volcanic sources. The measured Th^{230}/Th^{232} ratio should decrease with depth in the deposit column. The age at any level should be reflected by comparison of the Th^{230}/Th^{232} ratio to the value of that ratio at the sediment surface. The values of the ratios in the deposit column should vary by the exponential relationship $e^{-\lambda t}$, where λ is the decay constant of Th^{230}. Descriptions of different types of depth profiles of the Th^{230}/Th^{232} ratio in sediments and mathematical interpretations are given by Goldberg and Koide (1962). This method depends on the presumption that (1) analyzed materials do not contain detrital substances of continental or volcanic origin with significant contributions of Th^{230} or Th^{232}, and (2) the Th^{230}/Th^{232} ratio in the water mass adjacent to the sediment in a given oceanic basin has remained constant over the dating interval sediment accumulation.

Although these protactinium-thorium methods are used in many laboratories in the United States and abroad, some controversial issues exist such as problems of selection of datable samples, sample preparation, radioactivity

PROTACTINIUM–THORIUM DATING

counting techniques, and interpretation of results. A summary of these controversial issues has been published (Rona, 1964).

JOHN N. ROSHOLT

References

Goldberg, E. D., and Koide, M., 1962, "Geochronological studies of deep sea sediments by the ionium/thorium method," *Geochim. Cosmochim. Acta,* **26,** 417–435.

Kaufman, A., and Broecker, W., 1965, "Comparison of Th^{230} and C^{14} ages for carbonate materials from Lakes Lahontan and Bonneville," *J. Geophys. Res.,* **70,** 4039–4054.

Ku, Teh-Lung, 1968, "Protactinium 231 method of dating coral from Barbados Island," *J. Geophys. Res.,* **73,** 2271–2276.

Rona, E., 1964, "Geochronology of marine and fluvial sediments," *Science,* **144,** 1595–1597.

Rosholt, J. N., Emiliani, C., Geiss, J., Koczy, F. F., and Wangersky, P. J., 1961, "Absolute dating of deep-sea cores by the Pa^{231}/Th^{230} method," *J. Geology,* **69,** 162–185.

Sackett, W. M., 1966, "Manganese nodules: thorium-230: protactinium-231 ratios," *Science,* **154**(3749), 646–647.

Cross-references: *Authigenesis of Minerals— Marine; Geochronometry; Ionium-Thorium Dating; Seawater, Chemistry; Uranium-Thorium-Lead Age Determination.*

PROTIUM—See **HYDROGEN**

PSYCHOGEOLOGY—See **WATER DIVINING**

QUATERNARY PERIOD—See Vol. III   QUICKSILVER—See MERCURY

RADIOACTIVE ISOTOPES

Henri Becquerel, in 1896, reported the first evidence of radioactivity. He had exposed photographic plates to uranium sulfate crystals and found that, regardless of their previous exposure to sunlight, the crystals darkened the plates. Other uranium salts, solutions, and uranium metal acted in the same manner, darkening the plates in proportion to the uranium content. Pierre and Marie Curie introduced the name "radioactivity" to apply to those phenomena associated with uranium and apparently unrelated to the chemical or physical state of the element. Rutherford and Soddy suggested that radioactivity was due to the spontaneous transformation of one element to another during which certain penetrating rays were emitted.

Radioactive Decay and Growth

Working with a ton of residue from a uranium mine, Mme. Curie isolated two substances which had considerably higher radioactivity than uranium. She named them polonium and radium (qq.v. *Polonium* and *Radium*). Purified radium was found to emit a gas that was itself radioactive. By isolating quantities of this so-called emanation and observing its activity with time, these early workers found that as the gas was allowed to stand, its activity decreased to a point where it was no longer detectable. They also observed that a radium solution would generate a new generation of the emanation. Such studies served as the foundation for an equation of radioactive decay and growth.

A given radioactive species decays according to an exponential law:

$$N = N_0 e^{-\lambda t} \qquad (1)$$

where N represents the number of atoms at time $= t$, N_o the number of atoms at time $(t) = 0$, and λ the decay constant. λ is expressed in units of inverse seconds and is a measure of the probability of a particular isotope decaying in a given time period. For each radioactive isotope, λ has a unique value which does not depend on the physical or chemical state of that isotope. By setting $N = \frac{1}{2} N_o$ in Eq. (1), an extremely useful term may be derived. This term is known as the half-life ($t_{1/2}$) and is a measure of the time required for the original number of atoms to decrease by one half.

$$t_{1/2} = \frac{\ln 2}{\lambda} = \frac{0.693}{\lambda} \qquad (2)$$

Eq. (1) may be expressed in an alternate form:

$$\frac{dN}{dt} = -\lambda N \qquad (3)$$

The term dN/dt, which is also known as the activity, represents the rate of change of the number of atoms (N) with time. This equation emphasizes the fact that the activity is proportional to the number of atoms present.

In a radioactive decay sequence the original or decaying nuclide is known as the parent, and each product or residual nuclide is called a daughter. If the daughter is radioactive, and if its half-life is shorter than its parent's, a condition known as transient equilibrium may be achieved. During this state, the activity of the parent is exactly equal to that of the daughter:

RADIOACTIVE ISOTOPES

$$\lambda_1 N_1 = \lambda_2 N_2 \qquad (4)$$

where 1 represents the parent isotope and 2 the daughter. This assumes that parent and daughter are in a closed system. In the case of radium decaying to radon (a rare gas), a state of equilibrium may not be reached due to escape of the gas.

Mass-Energy Relationships

Albert Einstein formulated a very important relationship in nuclear decay:

$$E = \Delta m\, c^2 \qquad (5)$$

where E is the energy released, c the speed of light, and Δm the change in mass.

Radiation Emission and Detection

There is no way to distinguish a radioisotope from any other isotope until it decays and emits an energetic particle. Therefore, a study of the particles is necessary to understand radioisotopes. Three types of particles, called alpha (α), beta (β), and gamma (γ) rays, are of prime importance. Alpha rays are helium nuclei ($_2He^4$); therefore, in α decay the atomic weight decreases by 4 units and the atomic number by 2; e.g.

$$_{92}U^{238} \rightarrow\, _{90}Th^{234} +\, _2He^4$$

Alpha decay only occurs for elements with high atomic weight. Beta particles are high-energy electrons emitted from the nucleus of an atom. Beta decay causes the atomic number to increase one unit. To produce a β-particle, a neutron must be converted to a proton and an electron: $n \rightarrow p^+ + e^-$; e.g.,

$$_{90}Th^{234} \rightarrow\, _{91}Pa^{234} +\, _{-1}\beta^0$$

Gamma rays are electromagnetic radiations similar to x rays; their emission causes no change in atomic number or atomic weight. Frequently, alpha or beta decay leaves the nucleus in an excited state; by gamma ray emission the energy is removed and the excited state decays to the ground state, e.g.:

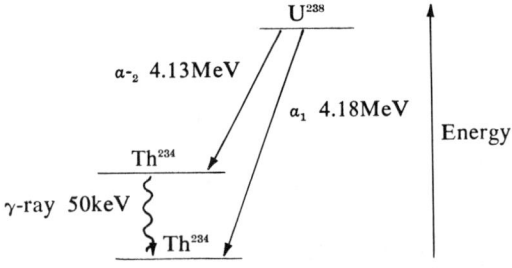

These particles are detectable when they interact with matter and dissipate their energy. There are three primary methods of quantitative radioactive measurement:

1. Using a gas, measure the conductivity resulting from ionizations due to the passage of high energy particles. This method uses ionization, Geiger, and proportional counting.
2. Measure light flashes from a fluorescent screen of zinc sulfide resulting from the particles striking the screen. This process is known as *scintillation counting*.
3. Measure the energy given off by defect crystals struck by particles. These energy pulses are directly proportional to the energy of the incident particles. This method allows particles to be measured as a function of their energy; thus radioisotopes of the same element can be distinguished since each isotope emits particles of characteristic energies. This method of discrimination is called *solid-state spectrometry*.

The easiest method used to distinguish between particles is absorption. Alpha particles may be stopped by only a few sheets of paper. Beta particles penetrate paper but may be absorbed by a thin sheet of metal such as a coin. Gamma rays, being much more penetrative than either alpha or beta particles, require thick lead sheets to stop them.

Natural Radioisotopes

All isotopes of atomic number greater than 83 (bismuth) are radioactive. These belong to one of three decay series (Fig. 1), each characterized by a long-lived parent: U^{238}, U^{235} or Th^{232}, and a number of short-lived daughters. These series all end with stable lead isotopes.

In addition to the uranium and thorium decay series, there are other radioactive isotopes that exist in the earth. These are summarized in Table 1 (see *Radioactivity of Rocks*).

Besides the radioisotopes which exist in the earth's rocks, certain other radioisotopes are naturally present on the earth as a result of the interaction of cosmic rays and atmospheric gases. Cosmic rays strike atoms in the upper atmosphere to produce a variety of particles including mesons, neutrons, electrons, and protons. These particles in turn interact with other gas atoms to produce radioisotopes. Among the most geologically important cosmic-ray-produced radionuclides is C^{14}, formed when a neutron strikes a N^{14} atom. (See also: *Radionuclides: Cosmic-Ray-Produced*.)

Artificial Radioisotopes

Nuclear explosions are fission and fusion reactions. Both reactions produce radioisotopes. When an atom bomb explodes (fission), the U^{235} nucleus splits into two isotopes. The

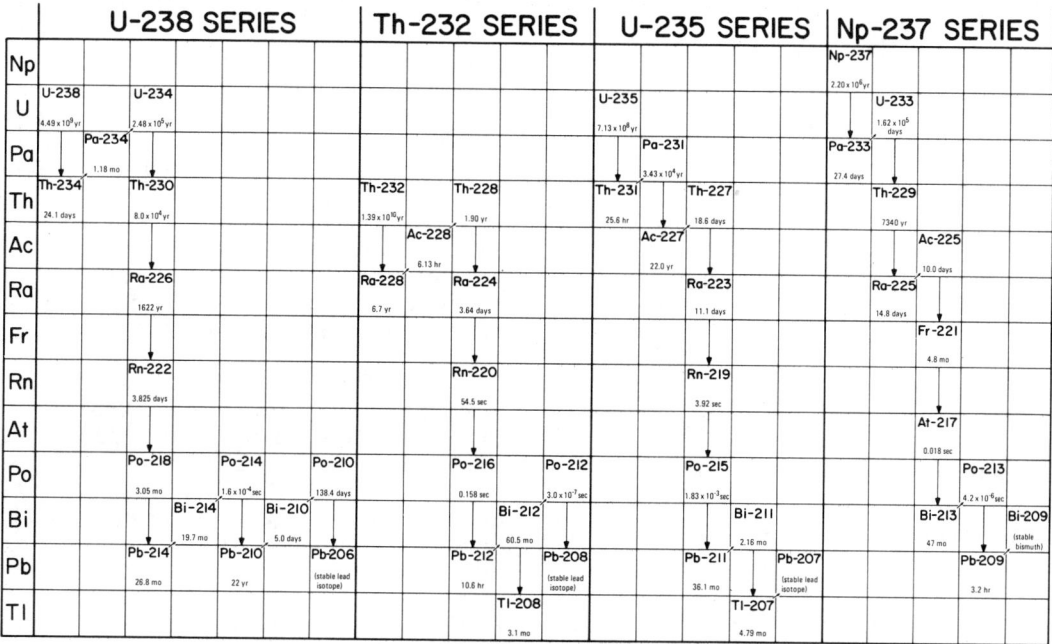

FIG. 1. Natural decay series of U-238, U-235, and Th-232, and artificial Np-237 series. Vertical arrows denote alpha decay; diagonal arrows represent beta decay.

most prominent are Sr^{90} and Cs^{137}. These isotopes are removed from the atmosphere by precipitation and fall to the earth; hence the name *fallout isotopes*. Hydrogen bombs (fusion) produce a high flux of neutrons in the atmosphere. These neutrons act the same as those produced by cosmic rays and thus produce many of the same radioisotopes. Since the advent of hydrogen bombs, atmospheric levels of tritium (H^3) have increased by orders of magnitude, while the level of C^{14} has about doubled (see Vol. 1, *Radionuclides in Oceans and Sediments*).

Health problems from radioactive fallout merit special concern. Although natural radioactivity rarely reaches levels considered harmful to the health, artificially produced radioisotopes change man's environment and may be quite detrimental. Strontium-90 produced in fission bombs is especially dangerous. The chemistry of strontium is much like calcium, a major constituent of bones and teeth. There is strong evidence that accumulations of Sr^{90} in bones lead to harmful effects such as cancer. It is interesting but unconsoling to contrast the radioactivity produced by atomic bombs with natural levels of radioactivity. Thirty years after the explosion of a one-megaton bomb, the amount of Sr^{90} remaining is radioactively equivalent to 100 kilograms of pure Ra^{226}. One-hundred kilograms of Ra^{226} could be extracted from about 10^{20} grams rock.

TABLE 1. NATURALLY OCCURRING RADIOISOTOPES OTHER THAN MEMBERS OF THE URANIUM AND THORIUM DECAY SERIES

Radioisotope	Type of Decay	Half-Life (yr)	Stable Product
K^{40}	β, EC[a]	1.2×10^9	Ca^{40}, Ar^{40}
Rb^{87}	β	6.2×10^{10}	Sr^{87}
In^{115}	β	6×10^{14}	Sn^{115}
La^{138}	EC, β	$\sim 2 \times 10^{11}$	Ba^{138}, Ce^{138}
Nd^{144}	α	$\sim 5 \times 10^{15}$	Ce^{140}
Sm^{147}	α	1.3×10^{11}	Nd^{143}
Lu^{176}	β	4.6×10^{10}	Hf^{176}
Re^{187}	β	$\sim 5 \times 10^{10}$	Os^{187}
Pt^{190}	α	$\sim 10^{12}$	Os^{186}

[a] EC stands for electron capture, a decay process that decreases the atomic number one unit and leaves the atomic weight unchanged.

TABLE 2. FALLOUT RADIOISOTOPES THAT MAY CONSTITUTE HEALTH HAZARDS

Part of Body Attacked	Isotope	Half-Life
Skeleton	Ca^{45}	164 days
	Sr^{89}	54 days
	Sr^{90}	20 yr[a]
	Ba^{140}	13 days
	La^{140}	2 days
	U^{235}	7×10^8 yr[a]
Thyroid	I^{131}	8 days
	I^{133}	1 day
Liver	Mn^{54}	1 yr
	Co^{60}	5.2 yr[a]
	Ce^{141}	33 days
	Ce^{144}	282 days
	Pr^{143}	2 wk
	Nd^{147}	11 days
Whole body	Cs^{137}	33 yr[a]
	C^{14}	5700 yr[a]

[a] Significant concentrations remaining from 1945–1956 bomb tests.

One of the most frightening aspects of fallout is that its effect cannot be evaluated for several generations. Man may be altering his evolution by the gradual accumulation of fallout isotopes in his body. Table 2 lists isotopes that may constitute health hazards and shows where in the body they accumulate.

Bombarding a nucleus with neutrons also produces artificial radioisotopes. This process, which is of growing importance as an analytical tool, is called *neutron activation* (q.v.). It is used to detect submicrogram quantities of radioactive or stable elements. The artificial radioisotopes produced in this process are quantitatively identified on the basis of their chemical and radioactive properties and thus give a measure of the amount of the element that was irradiated. Elements vary in their sensitivity to neutron activation analysis. Some commonly determined this way are shown in Table 3.

A third type of radioisotope production is with high-energy accelerators. Using Van de Graaff generators, cyclotrons, synchrotrons, betarons, and similar tools of high-energy physicists, particles may be accelerated to billions of electron volts. In this manner, charged particles such as alphas and protons as well as uncharged neutrons may be injected through the electron shell into the nucleus. Uranium has the highest atomic number found in nature (92) but by additions to its nucleus, elements with atomic numbers up to 101 (mendeluvium) have been built. Higher atomic numbers are possible; however, the extreme instability of the nucleus in the +100 atomic number range makes such work quite tedious.

Uses of Radioactive Isotopes

Radioactive isotopes are quite important to earth scientists because they make possible the measurement of time. The absolute dating of rocks and minerals (see *Fission Track Dating; Potassium-Argon Age Determination; Rubidium-Strontium Dating Method; Uranium-Helium Isotopic Age Method;* and *Uranium-Thorium-Lead Age Determination*) has placed much of geology on an absolute time scale (see *Geochronometry*). The dating of ocean sediments (see *Carbon-14 Dating; Ionium-Thorium Dating;* and *Protactinium-Thorium Dating Method*) has allowed marine geologists and geochemists to evaluate quantitatively the passage of minerals and ions through the ocean. Radioisotopes have proved useful in tracing ocean currents and studying oceanic mixing rates (see *Carbon-14 Dating; Radium in the Ocean;* also Vol. I, *Radionuclides: Their Application in Oceanography*).

Present Status

Certainly the discovery, production, and use of radioisotopes have produced a revolution in man's culture. Nuclear fuels are becoming competitive in cost with fossil fuels; treaties to halt the spread of nuclear weapons are being signed while more nations build atomic bombs. Man must answer the question: "Where will our radioisotopes take us from here?"

WILLARD S. MOORE

TABLE 3. SENSITIVITY OF VARIOUS ELEMENTS TO NEUTRON ACTIVATION ANALYSIS USING A NEUTRON FLUX OF 10^{12} NEUTRONS/CM^2/SEC

Lower Limit of Detection (g)	Elements
10^{-11}–10^{-12}	Eu, Dy
10^{-10}–10^{-11}	In, Ir, Mn, Sm
10^{-9}–10^{-10}	Cu, Na, W, Sb
10^{-8}–10^{-9}	Se, Ni, Zn, Co
10^{-7}–10^{-8}	Ag, Sn, Zr, Hg, Cr
10^{-6}–10^{-7}	Bi, Ca, Fe, S, Si

References

Choppin, Gregory, 1964, "Nuclei and Radioactivity," New York, W. A. Benjamin, Inc.

Friedlander, G., Kennedy, J., and Miller, J., 1964, "Nuclear and Radiochemistry," New York, John Wiley & Sons.

Fowler, J. M., 1960, "Fallout," New York, Basic Books, Inc.

Price, W. J., 1958, "Nuclear Radiation Detection," New York, McGraw-Hill Book Co.

Cross-references: *Carbon-14 Dating; Electron Capture; Fission Track Dating; Geochemistry: Ionization Potentials; Geochronometry; Ionium-*

Thorium Dating; Neutron Activation Analysis; Polonium; Potassium-Argon Age Determination; Protactinium-Thorium Dating Method; Radioactivity in Rocks; Radionuclides; Radium; Radium in the Oceans; Rubidium-Strontium Dating Method; Spectrophotometry; Strontium; Tritium; Uranium-Helium Isotopic Age Method; Uranium-Thorium-Lead Age Determination. Vol. I: *Radionuclides in Oceans and Sediments; Radionuclides—Their Application in Oceanography.* Vol. II: *Cosmic Rays.*

RADIOACTIVE ISOTOPE DATING—See GEOCHEMISTRY

RADIOACTIVE ISOTOPE TRACER TECHNOLOGY—See Vol. VI

RADIOACTIVE ORE DEPOSITS—See THORIUM; URANIUM

RADIOACTIVE WASTE IN THE OCEAN—See Vol. I

RADIOACTIVITY IN ROCKS

Types of Radioactive Isotopes

Isotopes are or are not radioactive by virtue of the intrinsic stability of their nuclei. For a given number of protons there are certain numbers of neutrons which render the combination stable. Other numbers of neutrons result in combinations which tend to break apart at specific rates; the least stable combinations break apart fastest or have the shortest half-life.

Thus, six protons together with five neutrons (C^{11}) are very unstable (half-life of 20 min), six protons and eight neutrons (C^{14}) are somewhat more stable (half-life of 5560 yr), whereas C^{12} and C^{13} are entirely stable (have no measurable tendency to break apart). Furthermore, some isotopes would be so unstable (e.g., C^9 which would have six protons and three neutrons) that they don't exist (or at least they break apart so quickly that we can't be sure that they were ever put together in the first place).

In considering the distribution of radioactive isotopes in the earth, then, all isotopes fall naturally into two groups on the basis of their half-lives alone: (*1*) *primordial isotopes*—those which have such long half-lives that if they were present when the earth formed, they would still be present. (*2*) *regenerated isotopes*—those which have such short half-lives that they could not have persisted from the earth's beginning till the present. If any of these isotopes are present today there must be some process which has created them anew.

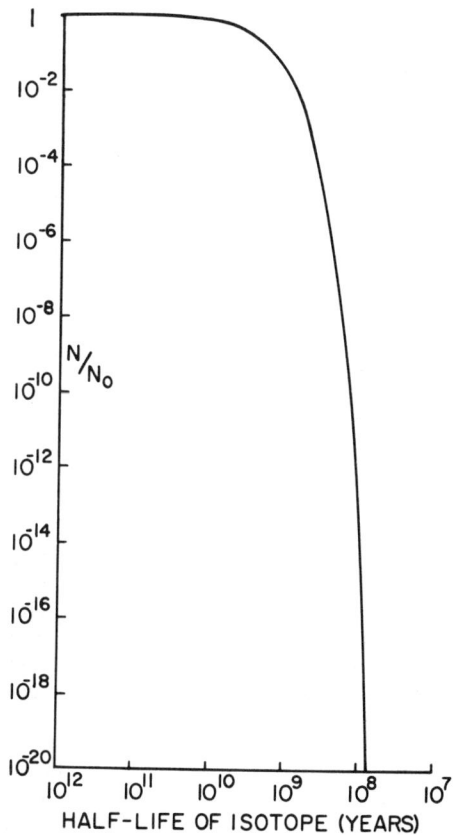

Fig. 1. Degree of reduction of abundance of radioactive isotopes during the 4.6 billion year history of the earth as a function of the half-life of the radioactive isotope. N/N_o is the ratio of the number of atoms presently existing to that which existed when the earth was formed.

Since the age of the earth is believed to be 4.6 billion years, an isotope with a 100-million-year half-life would be 10^{-14} as concentrated today as it was when the earth was formed (see Fig. 1). If we define a primordial isotope as one which is at least one-millionth as abundant today as then, Fig. 1 shows that it must have a half-life which is no less than about 300,000,000 years.

Distribution of Primordial Isotopes

Table 1 lists all isotopes which are definitely radioactive yet which have half-lives greater than 300 million years. Many of these are either so rare or so long-lived that very little is known about them. The five which are marked with asterisks are the only ones which are abundant enough and radioactive enough that they are both easily measured and have significant consequences with regard to the history of the earth.

RADIOACTIVITY IN ROCKS

TABLE 1. LONGEST-LIVED RADIOACTIVE ISOTOPES

Isotope	Half-Life (yr)	% of Total Element
Nd^{144}	5×10^{15}	23.87
In^{115}	6×10^{14}	95.8
Pt^{190}	10^{12}	0.012
La^{138}	2×10^{11}	0.089
Sm^{147}	1.3×10^{11}	15.07
*Rb^{87}	6.2×10^{10}	27.9
Re^{187}	5×10^{10}	62.93
Lu^{176}	4.6×10^{10}	2.60
*Th^{232}	1.4×10^{10}	100
*U^{238}	4.51×10^{9}	99.3
*K^{40}	1.2×10^{9}	0.0119
*U^{235}	7.13×10^{8}	0.715

TABLE 3. ESTIMATES OF CONCENTRATIONS OF PRIMORDIAL RADIOACTIVE ISOTOPES IN EARTH'S INTERIOR

Methods of Estimation	% K	ppm Rb	ppm U	ppm Th
Earth is chondritic	0.080	2.5	0.006	0.012
Mantle is dunitic	0.001	—	0.001	—
Maximum for non-melting	0.05	—	0.05	0.18
Concentrations in chondrites	0.086	2.8	0.014	0.04

Concentration in Rocks. Since samples of rock from the interior of the earth cannot be acquired, estimates of its radioactive content are quite uncertain. Samples from various components of the crust of the earth, on the other hand, have been analyzed in very large numbers. In the continents, the crust is believed to grade from granitic at the surface to basaltic at the base, while the oceanic crust is predominantly composed of basalt. A thin veneer of sediments covers both.

Table 2 lists the concentrations of the four main radioactive elements (U^{235} is always 0.71% as abundant as U^{238}) in the most important rock types. Sedimentary rocks, mainly shale, together comprise only about six percent of the mass of the crust. It should be remembered that the concentrations given are of the element and that to convert to concentrations of specific isotopes references should be made to Table 1.

Two generalities may be noticed in the igneous data in Table 2. First, all radioactive isotopes are enriched in granitic relative to basaltic rocks. Second, the proportions are maintained fairly constant so that K/Rb is about 300, K/U is about 10,000, and K/Th is about 3,000.

Estimates of the concentrations of the radioactive elements in the interior of the earth are made in several ways (see Table 3). First, we may accept the fairly plausible but recently challenged hypothesis that the average earth's composition is closely approximated by that of chondritic meteorites. If we subtract that portion of the radioactive elements known to be in the crust, we can compute the concentration in the interior.

Another estimate is based on the hypothesis that dunites represent mantle material because they originate at greater depths than any other rocks and because they have the seismic properties of mantle material.

Finally, an upper limit of the concentrations of radioactive elements can be estimated from the fact that the mantle is unmolten. Because of the low conductivity of rocks, the heat generated by radioactive elements would have melted the earth if their concentrations were greater than certain values. For this latter estimate, the assumption is made that K:U:Th in the interior of the earth is 10,000:1:3.5 (Rb is neglected because of the small amount of heat produced by its decay) as it is in the crust.

Concentrations in Minerals of Common Rocks. The magmas from which igneous rocks are derived usually contain about seven major elements (concentration greater than 1%) and many less abundant ones. The bulk of any single rock is generally made up of four or fewer mineral phases; none of the others comprises more than about 2% of the total rock mass. Hence, the dominant fraction of any major element is usually present in one or two major minerals. Potassium, being the only major radioactive element is therefore located mainly in one or two major minerals. For granitic rocks these are mica and K-feldspar, and for basaltic rocks, plagioclase.

Trace elements, on the other hand, may be either dispersed in small concentrations in the major minerals, or concentrated largely in the accessory minerals. Rubidium, being chemically

TABLE 2. CONCENTRATIONS OF PRIMORDIAL RADIOACTIVE ISOTOPES IN MAJOR CRUSTAL ROCK TYPES

Rock	% K	ppm Rb	ppm U	ppm Th
Igneous and metamorphic				
Granitic	3.2	140	3.0	11
Intermediate	2.7	110	1.8	6.3
Basaltic	.95	30	1.2	4.2
Sedimentary				
Sandstone	1.2	60	.45	1.8
Shale	2.9	140	3.7	12
Limestone	0.3	3	2.2	1.7
Average crust	1.9	74	1.7	6.0

TABLE 4. THORIUM AND URANIUM IN
ACCESSORY GRANITIC MINERALS

Accessory Mineral	ppm U	ppm Th	Th/U
Zircon	100–6000	50–4000	0.2–1
Apatite	5–150	20–250	1–25
Sphene	100–700	100–600	1–2
Allanite	30–700	500–20,000	5–50
Monazite	500–3000	25,000–200,000	25–50

similar to potassium is usually distributed among the same major minerals as is potassium.

The bulk of the uranium and thorium, however, is not present in the major minerals, though perhaps 20–50% may be. There are two other locations where these elements are often concentrated: in accessory minerals; and along mineral boundaries. Which is more important is not certain but they are at least roughly comparable.

Accessory minerals which incorporate significant amounts of uranium and thorium in granitic rocks (not much is known of their specific locations in other rocks) are listed in Table 4. It should be noted that though granites rarely have Th/U ratios very far from about 3.5, these accessory minerals have quite a range in this ratio.

Concentrations in Ore Deposits. Although the vast majority of the earth's content of radioactive elements is located in the rocks mentioned above, it is more profitable to recover what man needs from the much rarer rocks described below. The major ore of potassium is produced during the late stages of marine evaporation. Here are found several complex salts whose cations are K and Mg and whose anions are Cl and SO_4. These minerals have from 10 to 63% K, but the mined mass is usually 5 to 20% K.

The most important source of rubidium is the lithium-rich micas of granite pegmatites which have over 1% rubidium, this being recovered as a byproduct of lithium processing.

Because of the recent "uranium boom," a wide variety of types of sources of this metal was discovered. Concurrently, potential thorium deposits were revealed.

The most abundant uranium deposits are formed in the pegmatites of the late stages of magmatism. At this time, in the chemical evolution of a magma, many trace metals which had been discriminated against by the earlier forming crystals become greatly concentrated in the liquid. At the same time, the fluid becomes much less viscous due to its enrichment in volatile components such as H_2O, CO_2 sulfides, and halides. As a result, the potential for the formation of minerals whose compositions are very different from the parent magma is realized.

The increased abundances of such metals as Ni, Co, Ag, Cu, U, and Li together with the greater diffusion rates associated with low viscosity permits the formation of rather exotic minerals. The result is usually a series of veins where minerals such as quartz, calcite, pyrite, and fluorite are impregnated by these rare minerals which include uraninite and thorite where radioactive cations dominate, and a host of others where these elements may be present to the extent of several percent.

There are three types of sedimentary environments where U and Th may be greatly concentrated. In phosphorites and black shales, it appears that the very process of sediment deposition caused the concentration of uranium due to the chemical affinity of this element for the phosphate ion and organic compounds. In some stratified continental clastic sediments, notably in the Colorado Plateau, uranium is often concentrated by the postdepositional replacement of organic matter or the filling of pore spaces by U-rich cementing substances. Finally, there are placer deposits (in river beds and on beaches) where weathering processes concentrate the densest, most resistant materials from igneous rocks such as the thorium-rich mineral, monazite.

Distribution of Regenerated Isotopes

As mentioned above, of all the radioactive isotopes found in significant abundances, only K^{40}, Rb^{87}, Th^{232}, U^{235}, and U^{238} were here since the formation of the earth. If we find any others, there must be some processes which continually replenish them. Two such processes are the dominant ones: regeneration by decay of the primordial isotopes and regeneration by nuclear reactions involving cosmic irradiation.

Isotopes Regenerated by Primordial Parents. Of the five primordial isotopes, K^{40} and Rb^{87} decay immediately to stable isotopes (Ar^{40} and Sr^{87}, respectively) and hence do not regenerate any radioactive isotopes. The other three are actually each the first in a long chain of isotopes which are themselves radioactive and stability is only achieved when the lead isotopes are formed as shown in Table 5. Note that the well-known radium (Ra^{226}) is in the U^{238} chain and has so short a half-life that it must be constantly regenerated.

In order to understand the general distributions of all the daughter isotopes shown in Table 5, one must first understand two mathematical relationships governing the ratios of the concentrations of any radioactive mother-

RADIOACTIVITY IN ROCKS

TABLE 5. THE THREE NATURAL RADIOACTIVE DECAY CHAINS

U^{238} Series		
Isotope	Particle	Half-Life
U^{238}	α	4.51×10^9 yr
Th^{234}	β	24.1 day
Pa^{234}	β	1.18 min
U^{234}	α	2.48×10^5 yr
Th^{230}	α	7.52×10^4 yr
Ra^{226}	α	1622 yr
Rn^{222}	α	3.83 day
Po^{218}	α	3.05 min
Pb^{214}	β	26.8 min
Bi^{214}	β	19.7 min
Po^{214}	α	1.64×10^{-6} sec
Pb^{210}	β	19.4 yr
Bi^{210}	β	5.01 day
Po^{210}	α	138 day
Pb^{216}	stable	

U^{235} Series		
Isotope	Particle	Half-Life
U^{235}	α	7.13×10^8 yr
Th^{231}	β	25.6 hr
Pa^{231}	α	3.24×10^4 yr
Ac^{227}	β	21.6 yr
Th^{227}	α	18.2 day
Ra^{223}	α	11.7 day
Rn^{219}	α	3.92 sec
Po^{215}	α	1.83×10^{-3} sec
Pb^{211}	β	36.1 min
Bi^{211}	α	2.16 min
Tl^{207}	β	4.79 sec
Pb^{207}	stable	

Th^{232} Series		
Isotope	Particle	Half-Life
Th^{232}	α	1.39×10^{10} yr
Ra^{228}	β	6.7 yr
Ac^{228}	β	6.13 hr
Th^{228}	α	1.91 yr
Ra^{224}	α	3.64 day
Rn^{220}	α	51.5 sec
Po^{216}	α	0.158 sec
Pb^{212}	β	10.6 hr
Bi^{212}	β	60.5 min
Po^{212}	α	3.04×10^{-7} sec
Pb^{208}	stable	

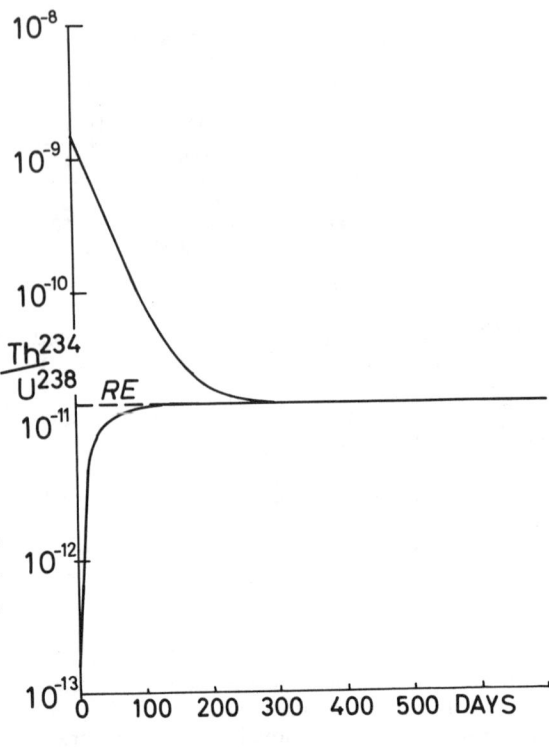

FIG. 2. The ratio of the number of Th^{234} atoms to the number of U^{238} atoms (U^{238} is the mother of Th^{234}) as a function of time for two cases. In the upper curve it is assumed that initially there was 100 times as much Th^{234} as there would be at equilibrium and in the lower curve that there was 1/100 as much. R_e, the equilibrium ratio equals the ratio of the half-life of Th^{234} to that of U^{238} and turns out to be 1.5×10^{-11}. Note that equilibrium is virtually achieved in 300 days in the upper case and 100 days in the lower case.

daughter pair where the mother is much longer lived than the daughter. If we take the example of U^{238} and its daughter Th^{234}, Fig. 2 illustrates these relationships. First, there is a specific ratio which the daughter will assume with respect to the mother if we wait long enough. This is called the equilibrium ratio or R_E in Fig. 2, and turns out to equal the ratio of the half-lives of daughter to mother (in the case of Th^{234}/U^{238}, $R_{EQU} = 24.1$ days/4.51×10^9 years or about 1.46×10^{-11}). Second, this ratio is approached, no matter what the starting ratio, at such a rate that the gap between the ratio at any time and the equilibrium ratio is reduced by one half during a half-life of the daughter.

The outcome of these rules is that in almost all rocks, the concentrations of all the daughter isotopes are proportional to the concentrations of U^{238}, U^{235}, or Th^{232}; if Rock A has ten times as much U^{238} as Rock B, it will have ten times as much Ra^{226}, etc., as Rock B, too. The only exceptions are rocks less than two million years old (which is eight times the longest daughter half-life appearing in Table 5) which were formed under the fairly rare conditions where the equilibrium ratios were disturbed. Two such cases will be described because they provide the bases for methods of radioactive age determination.

The first exception is the Th^{230}/U^{238} ratio in recent marine sediments. Because U^{238} is very soluble in seawater and Th^{230} is not, the Th^{230} produced from the U^{238} in solution tends to become adsorbed on the solid particles suspended in the water. As the latter settle out of the water and accumulate on the ocean floor, a high Th^{230}/U^{238} ratio is built up in the bottom sediments. Since the uranium within the particles themselves varies geographically as does the rate of particle accumulation the sediments do not all have the same ratio. The highest ratios, about 200 times the equilibrium ratio, are found on the floor of the South Pacific, whereas ratios as low as 10 are found in the equatorial Atlantic.

Another type of young rock in which ratios far from equilibrium are found is coral fossils. A living coral appears to include within its skeleton a host of trace metals in approximately the proportion in which they are dissolved in seawater. Since the Th^{230} produced in ocean water is removed by suspended particles, the Th^{230}/U^{238} ratio remaining is low and is incorporated into the skeleton as such. Thus, ratios as low as 0.003 of the equilibrium ratio have been found in the shells of recent coral.

Isotopes Regenerated by Cosmic Irradiation. There are a great many isotopes produced by the interaction of cosmic rays with the earth's atmosphere. Most of these are of such short half-life or are produced in such low quantities that they are of no interest to the present discussion.

One very interesting isotope produced in this manner is C^{14}. When a neutron strikes a N^{14} atom, the former is incorporated into its nucleus and a proton is emitted; the resulting nucleus has six protons and eight neutrons, hence C^{14}. Because of the high abundance of N^{14} in the atmosphere and the high probability of this interaction, the C^{14} in the atmosphere is easily measured. The newly formed C^{14} atom proceeds to enter CO_2 molecules and to mix with the nonradioactive CO_2 in the atmosphere.

This C^{14} may be incorporated in a carbonate rock in the following way. These rocks are generally formed either from the skeletons of plants or animals which live in water or from carbonate salts (usually $CaCO_3$) which are precipitated chemically from bodies of water (oceans and lakes). These processes usually involve the dissolved CO_3^{2-} ion, and it turns out that by the process of gas exchange, large amounts of C^{14} are injected into these carbonate ions from the atmosphere; their place is taken by equal amounts of nonradioactive carbon (C^{12}) which leave the water and go into the atmosphere.

The rate at which this gas exchange occurs is so fast that the carbonate ions in many water bodies have almost the same ratio of C^{14} to C^{12} as does the atmosphere, about 1 in 10^{12}. As a result, most presently forming carbonate rocks have a similar ratio and the C^{14} age determination method, as applied to rocks and shells, involves finding out how much of this original C^{14} has decayed by its own radioactivity. Because the half-life of C^{14} is only about 5,600 years, a rock as young as 100,000 years old would have so little C^{14} left that it would be virtually unmeasurable.

Another isotope constantly regenerated by cosmic irradiation is Be^{10} (half-life 2,700,000 years). This isotope, however, does not form a gaseous compound as does C^{14} and therefore eventually enters rocks in a manner very different from that of the latter. It is washed out of the atmosphere by rain and then is either adsorbed on soils or carried into the ocean by streams. In the latter case it would be adsorbed on suspended particles and accumulate in ocean bottom sediments. In any event it would only be found in rocks say twenty million years old or younger. Its presence has been detected in ocean bottom sediments, but too little study has been performed as yet to generalize on its distribution.

AARON KAUFMAN

References

Friedlander, G., Kennedy, J. W., Miller J. M., 1964, "Nuclear and Radiochemistry," New York, John Wiley & Sons, 585pp.

Gast, P. W., 1965, "Terrestrial ratio of K to Rb and the compositions of the earth's mantle," *Science* **147**, 858–860.

Hamilton, E. I., 1965, "Applied Geochronology," London, Academic Press, 267pp.

Heinrich, E. W., 1958, "Mineralogy and Geology of Radioactive Raw Material," New York, McGraw-Hill Book Co., 654pp.

Hurley, P. M., 1968, "Absolute abundances and distribution of Rb, K, and Sr in the earth," *Geochim. Cosmochim. Acta,* **32,** 273–283.

Wasserburg, G. T., Macdonald, G. J. F., Hoyle, F., and Fowler, W. A., 1964, "Relative contributions of U, Th, and K to heat production in the earth," *Science,* **143,** 465–467.

Cross-references: *Calcium Carbonate; Carbon Isotope Fractionation; Elements: Planetary Abundances and Distribution; Geochemical Evolution of the Core, Mantle, and Crust; Geochronometry; Isotopic Variation in Mineral Deposits; Potassium; Radioactive Isotopes; Rubidium; Thorium; Trace Elements; Trace Metals; Uranium.* Vol. II: *Atmospheric Chemistry; Cosmic Rays; Planet Earth—Origin and Evolution.* Vol. IVB: *Placer Deposits.* Vol. V: *Magma.* Vol. VI: *Pelagic Sediments.*

RADIOACTIVITY IN SEDIMENTS—See Vol. VI

RADIOCARBON DATING—See CARBON-14 DATING

RADIONUCLIDES: COSMIC-RAY-PRODUCED

All matter surrounding us contains traces of radioactive elements of widely varying half-lives and chemical nature. An appreciable fraction of these elements have half-lives much shorter than 10^7 years so that their present abundance on the earth must be ascribed to either a "continuous" or to a "recent" artificial production mechanism. Any continual production of radioactivity on a global scale must necessarily be implemented by natural agencies. Two known mechanisms responsible for this are: (1) natural radioactive disintegrations of long-lived radionuclides, e.g., U^{238} and Th^{232}, which have survived since the time of nucleosynthesis, leading to a continuous production of radioactive daughter nuclides, and (2) nuclear interactions in terrestrial materials due to energetic particles of "cosmic radiation" leading to the formation of a variety of end products, comprising all types of nuclear matter of mass lighter than the target nucleus. Besides these natural processes, radioactive species found on the earth today are those that have been produced due to nuclear weapon explosions in the atmosphere during the last decade or so. The global scale dispersion of artificial radioactivity due to this cause greatly exceeds that due to production in accelerators and nuclear reactors.

Any matter introduced in the atmosphere or in the upper layers of the earth, which are in continuous motion, migrates in complex trajectories from the point of its injection due to a variety of geophysical and geochemical processes. The most active zones from the point of view of rapid mixing are the atmosphere and the hydrosphere. The removal of cosmic-ray-produced or artificially injected activity introduced in the atmosphere occurs from the lower regions of the troposphere due to scavenging by droplets during condensations leading to wet precipitations. Any material present in the upper regions of the atmosphere has first to be brought to the lower layers by convective, diffusive, or advective mixing processes before its removal. Thus, all elements which can be removed along with water droplets are finally removed from the atmosphere and introduced in the upper layers of the lithosphere and the hydrosphere. The gaseous elements, e.g., CO_2, are removed at a slower rate from the atmosphere by molecular exchange processes at the atmosphere-hydrosphere interface.

Deeper down in the lithosphere, mixing occurs much more slowly, except when sporadic outbursts, e.g., volcanic eruptions, occur; convection currents in the mantle produce only a slow mixing between the crust and the mantle. Therefore, the radioactive elements found in the lithosphere, even in the uppermost layers, are mainly those which derive from decay of primordial radionuclides, e.g., U^{238}. The atmosphere and the hydrosphere, on the other hand, do receive appreciable quantities of decay products due to disintegration of primordial radionuclides present in the lithosphere. Exhalations of radioactive gaseous elements from the upper layers of the soil, e.g., Rn^{222}, and weathering of surface rocks introduce radioactive elements in the atmosphere and in the hydrosphere.

Thus, the entire atmosphere and the hydrosphere, and upper layers of the lithosphere, including the biosphere contains traces of radioactive elements having different origin and physical and chemical properties. The distribution of these elements is indicative of the characteristics of a wide variety of geophysical and geochemical processes. Nature, as a partial compensation for its intricacies, has provided us with several clues for unravelling its geological history and mysteries. With rapid developments in nuclear technology, the knowledge of natural processes and geological events is being acquired at a fast rate.

In this article we will confine our attention chiefly to radionuclides which are produced by cosmic rays. A total of twenty-one isotopes have already been detected; a few useful ones have yet to be detected. The life history of the radionuclides, since their formation in energetic nuclear reactions produced by cosmic ray particles, will be traced. Major emphasis will be laid on outlining the utility of these isotopes for studying geophysical and geochemical processes rather than presenting a detailed account of the various results obtained to date. Only a few exemplary illustrations of results showing the manner in which these isotopes serve either as "tracers" of mass motion of molecules in the atmosphere and the hydrosphere, or for "dating" certain geophysical events, will be given.

Isotopic changes produced directly by interactions of cosmic rays in the lithosphere, studies of the prehistoric cosmic-ray intensity based on terrestrial abundance of certain cosmic-ray-produced radionuclides, and rates of influx of extraterrestrial dust by observation on isotopes accreted on the earth with the dust grains, are some examples of topics which will not be covered in this article. Several less-pronounced cosmic-ray-produced isotopic changes, e.g., production of isotopes in the upper atmosphere by particles emitted from sun, and in the lithosphere by nuclear interactions of cosmic ray particles surviving to these depths, changes in the production rate due to solar effects, will also not be considered in this article; the geophysical implications of such processes have only recently begun to be studied.

For a detailed account of the various applications which the cosmic ray produced isotopes are finding for the study of geophysical and astrophysical problems, reference is made to recent review articles by Lal and Peters (1967) and Honda and Arnold (1967).

Cosmic-Ray-Produced Nuclides on the Earth

The energetic particles of the cosmic radiation incident on the earth interact with the atmospheric nuclei giving rise to a variety of products: nucleons, subnuclear particles (e.g., π and K-mesons), and complex nuclear aggregates. Nucleons and nuclear interacting particles emitted from these interactions lead to further interactions from which similar products are emitted. Thus, traversal of cosmic rays through the atmosphere leads to a nuclear cascade resulting in a significant modification in the composition and energy spectrum of the primary cosmic rays, and various complex nuclear fragments, lighter than the atmospheric nuclei, are formed. Much of the cosmic ray energy is dissipated in the atmosphere itself, which represents a vertical traverse length of more than ten mean free paths for nuclear-interacting particles. Except in the upper layers of the atmosphere where nuclear cascades lead to an increase in the particle flux due to multiplication, the total flux of primary and secondary cosmic ray particles, and thus also the rate of production of complex nuclear fragments, is a monotonically decreasing function of the atmospheric pressure.

Of the various nuclear fragments produced, those which are radioactive, can be detected more easily. The stable end products are difficult to detect for two reasons: (1) their rate of production by cosmic rays is small, and (2) they mix with terrestrial matter containing appreciable quantities of their kind and become indistinguishable. The stable isotope, He^3, is an exception; it is a noble gas element and accumulates in the atmosphere where the abundance of helium is comparatively small not to mask its detection.

With the exception of He^3, all cosmic-ray-produced isotopic changes detected on the earth refer to radioactive nuclei. Twenty radionuclides detected to date are listed in Table 1, in order of decreasing half-lives. Their half-lives, prominent decay modes, and the atmospheric target nuclei responsible for their formation have also been listed in Table 1.

It should be noted that except for Kr^{81}, no other products of interaction with permanent gases of abundance lower than argon, e.g., krypton and xenon, have been detected as yet.

TABLE 1. COSMIC-RAY-PRODUCED RADIONUCLIDES DETECTED ON THE EARTH

Nuclide	Half-Life	Main Radiation and Energy (keV)	Main Target Nuclide(s)
Be^{10}	2.7×10^6 yr	β^-–550	N, O
Al^{26}	7.4×10^5 yr	β^+–1170	Ar
Cl^{36}	3.1×10^5 yr	β^-–714	Ar
Kr^{81}	2.1×10^5 yr	K –X-ray	Kr
C^{14}	5730 yr	β^-–156	N, O
Si^{32}	500 yr	β^-–100	Ar
Ar^{39}	270 yr	β^-–565	Ar
H^3	12.3 yr	β^-–18	N, O
Na^{22}	2.6 yr	(β^+–540 (γ –1300	Ar
S^{35}	87 days	β^-–167	Ar
Be^7	53 days	γ –480 (11%)	N, O
P^{33}	25 days	β^-–250	Ar
P^{32}	14.3 days	β^-–1700	Ar
Mg^{28}	21 hr	β^-–420	Ar
Na^{24}	15 hr	β^-–1400	Ar
S^{38}	2.9 hr	β^-–1100	Ar
Si^{31}	2.6 hr	β^-–1470	Ar
Cl^{39}	55 min	β^-–1900	Ar
Cl^{38}	37 min	β^-–4800 (53%)	Ar
Cl^{34m}	32 min	β^-–4500 (47%)	Ar

RADIONUCLIDES: COSMIC-RAY-PRODUCED

Their detection involves more sensitive and refined techniques than are available at present. Even the detection of the nuclides listed in Table 1 requires unconventional techniques which are practiced only in a few laboratories. The very first significant advance in counting techniques was made by W. F. Libby and co-workers of the United States, who developed a method for suppressing the cosmic-ray background in connection with his successful efforts to detect the cosmic-ray-produced radiocarbon on the earth. Subsequent to the discovery of C^{14} in 1949, many improvements and new techniques were developed which led to a detection of a large number of cosmic-ray-produced nuclides during 1955–60 (Lal and Peters, 1967). Further advances in technology will make it possible to detect a few more geophysically useful cosmic ray produced radionuclides.

Global Dispersion and Applications of Cosmic-Ray-Produced Radioactivity

The most significant production of radionuclides by cosmic rays occurs in the atmosphere; isotopic changes directly induced in the crust are small and their studies have not led to any tangible results as yet.

The various nuclides listed in Table 1, except for He^3, are oxidized quickly after their

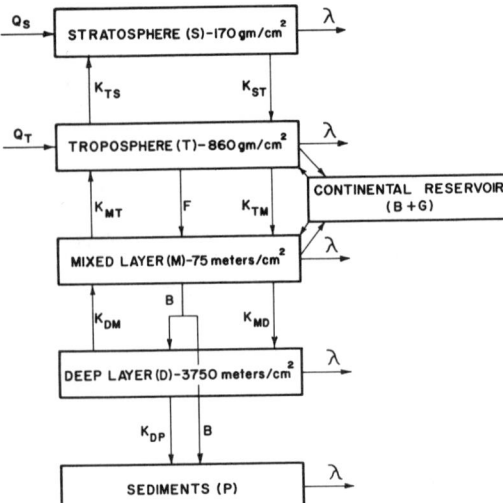

FIG. 1. A six-reservoir exchange model for the circulation of cosmic-ray-produced isotopes. Q denotes the production in the atmosphere. Other symbols associated with arrows represent decay or exchange constants.

TABLE 2. STEADY STATE FRACTIONAL INVENTORIES OF COSMIC-RAY PRODUCED RADIOISOTOPES IN THE EXCHANGE RESERVOIRS

Exchange Reservoir	Radioisotope					
	Be^{10}	Al^{26}	Cl^{36}	C^{14}	Si^{32}	Ar^{39}
Stratosphere (S)	3.7×10^{-7}	1.3×10^{-6}	10^{-6}	3×10^{-3}	1.9×10^{-3}	0.16
Troposphere (T)	2.3×10^{-3}	7.7×10^{-8}	6×10^{-8}	1.6×10^{-2}	1.1×10^{-4}	0.83
Continental reservoir (B + G)	0.29^a	0.29^a	0.29^a		0.29^a	0
Mixed oceanic layer (M)	5.7×10^{-6}	1.4×10^{-5}	1.4×10^{-2}	2.2×10^{-2}	3.5×10^{-3}	2×10^{-4}
Deep oceanic layer (D)	10^{-4}	7×10^{-5}	0.69	0.92	0.68	3×10^{-3}
Oceanic sediments (P)	0.71	0.71	0	4×10^{-3}	2.8×10^{-2}	0
λ (yr^{-1})	2.8×10^{-7}	9.4×10^{-7}	2.2×10^{-6}	1.24×10^{-4}	1.4×10^{-3}	2.3×10^{-3}

TABLE 2. (Continued)

Exchange Reservoir	Radioisotope						
	H^3	Na^{22}	S^{35}	Be^7	Ar^{37}	P^{33}	P^{32}
Stratosphere (S)	6.8×10^{-2}	0.25	0.57	0.60	0.63	0.64	0.60
Troposphere (T)	4×10^{-3}	1.7×10^{-2}	8×10^{-2}	0.11	0.37	0.16	0.24
Continental reservoir (B + G)	0.27	0.21	0.10	0.08	0	5.6×10^{-2}	4.7×10^{-2}
Mixed oceanic layer (M)	0.35	0.44	0.24	0.2	0	0.13	0.11
Deep oceanic layer (D)	0.3	8×10^{-2}	4×10^{-3}	2×10^{-3}	0	7×10^{-4}	10^{-4}
Oceanic sediments (P)	0	0	0	0	0	0	0
λ (yr^{-1})	5.6×10^{-2}	0.27	2.9	4.8	7.2	10	17.7

[a] Part of this inventory may in fact be carried as silt or dust into the oceans before decay.

formation and attached to particles of aerosols in sufficiently short periods. Radiocarbon behaves differently as it forms carbon dioxide, a "permanent" gas of the atmosphere. After attachment to aerosols, the various nuclides partake in the large-scale atmospheric circulation till they reach the lower layers of the troposphere, where they are efficiently removed by the scavenging action during wet condensations leading to precipitations. They thus get introduced into the geological domains, lithosphere and the hydrosphere. Radiocarbon is removed from the atmosphere primarily by molecular exchange processes across the atmosphere–hydrosphere interface: this exchange introduces radiocarbon in the oceans.

The quantitative features in the global dispersion of a cosmic-ray-produced element are determined by its half-life as well as by its physical and chemical properties. With the help of oversimplified exchange models, confined to a few geophysically significant reservoirs, or zones, which can be considered more or less typical with respect to the exchange of material within and between the adjacent reservoirs, Lal (1963) has calculated the expected fractional inventories of several radionuclides. Fig. 1 shows the six-reservoir exchange model, comprising the atmosphere, lithosphere, and the marine reservoir including the sediments, adopted for these calculations.

The reservoirs are assumed to be well mixed internally and the rates of flow of isotopes across the interfaces have been assumed to depend on first-order kinetic processes. The sizes of the reservoirs and the exchange coefficients $K_{ij}, K_{ji} \ldots$, effective for the transfer of material from the reservoir i to j, j to $i \ldots$ respectively (Fig. 1) have been based on limited data available from the physical and radioactive tracer studies. The mean residence time of material in a reservoir i against removal to all adjacent reservoirs is given by $1/\sum_j K_{ij}$. It is the time in which $[1 - (1/e)]$ of the material present is replaced by materials from the adjacent reservoirs.

The expected steady-state global fractional inventories of several radioactive isotopes of half-lives longer than a day listed in Table 1 are given in Table 2 for the six reservoirs. The isotope, Ar^{37}, which has not been detected as yet, has been included in this table for the sake of completeness. In view of the uncertainties of the various data used and the simplification made, the fractional inventories are only approximate. In fact, it can be considered to be one of the main tasks of the isotope work to determine the sizes of the reservoirs and the rates of exchange of material between them. At present, we make use of the rough values of these parameters to discuss the manner in which the various isotopes are distributed on the earth and to determine their useful applications as tracers in geophysical studies.

It becomes apparent from Table 2 that the steady state distribution of isotopes in the five reservoirs is different for the various isotopes. Extreme cases are Be^{10} (Al^{26}) and $A^{37,39}$ where most of their inventories are piled up in the sediments and in the atmosphere, respectively. When the exchange processes are rapid compared to the half-life of the isotopes, they behave as stable tracers and if the exchange of material can be considered to be essentially unidirectional, their inventories can be used to determine the average hold-up times in these reservoirs. If the flow of matter between adjacent reservoirs is cyclic in nature, the study of the rate of exchange of material has to be based on isotopes whose half-lives are comparable to the time scales involved. Isotopes which reach the biosphere, oceanic sediments or other materials, and are isolated from the exchange reservoirs, find applications for dating the intervals of time elapsed since they ceased to partake in the cycle.

Thus, the isotopes Be^{10}, Al^{26}, C^{14}, and Si^{32} find applications for determining the chronologies of oceanic sediments, whereas C^{14}, and Si^{32}, H^3 are suitable for dating archaeological organic remains and snow deposits in the polar ice caps, respectively. In glaciers, C^{14} determination on the small amounts of atmospheric CO_2 trapped during snow formation have also been used for dating. The isotopes C^{14}, Si^{32}, and Ar^{39} find application in the study of large-scale oceanic circulation (and the dating of polar ice and glaciers), whereas the isotopes H^3 and Be^7 are useful for studying the short-term vertical mixing in the upper layers of the oceans. H^3 finds application in the study of several problems in the field of hydrology, e.g., the time scales involved in the mixing and circulation of water between the atmosphere, continents, and oceans, and the characteristics and the sizes of the subterranean water reservoirs. Of all the isotopes, C^{14} finds maximum applications: in archaeology, glaciology, hydrology, and oceanography, because it partakes in the natural carbon cycle which plays an important role in many earth phenomena.

Other isotopes, Na^{22}, S^{35}, Be^7, P^{33} and Ar^{37}, P^{32} are useful for meteorological investigations. The shorter lived of these find applications in the studies of the tropospheric circulation and scavenging processes, whereas the longer lived ones are useful for investigating the advection and mixing processes operative in the stratosphere.

Seven short-lived radioisotopes of half-lives ranging from about half an hour to one day, listed in Table 1, decay primarily in the atmosphere before reaching the continental or the oceanic reservoir; their fractional inventories in the atmosphere (not listed in Table 2) are expected to be essentially proportional to their production rates in the stratosphere and troposphere. Their usefulness lies primarily in the study of wash-out processes occurring in the lower troposphere. They can also serve as indicators of temporal changes in the cosmic ray flux in the upper regions of the atmosphere.

Source and Sink-Functions of Cosmic-Ray-Produced Activity

We have so far discussed the migration of isotopes on the earth with the help of a few geophysical reservoirs without any reference to production rates of isotopes in the atmosphere. The fractional inventories given in Table 2 depend on a knowledge of the relative production rates of isotopes in the troposphere and the stratosphere; the production of all isotopes is confined chiefly within the atmosphere and direct production in the lithosphere or the hydrosphere is negligible except in the case of Cl^{36}, which is produced in appreciable quantities in the oceans due to neutron capture by the chlorine present. The absolute inventories in different reservoirs as well as the detailed distribution of radioisotopes in the atmosphere, however, depend sensitively on the source functions, i.e., the rate of production of isotopes in different regions of the atmosphere. The sink-functions, i.e., the space and time rate of removal if isotopes from the atmosphere become the source functions for the (B + G) and M reservoirs (Fig. 1). These functions, which have to be based on observational data, are a prerequisite in the application of cosmic-ray-produced radioactivity to geophysical problems.

In view of the importance of the knowledge of the source functions, concentrated efforts have been made to evaluate these accurately for the various isotopes using all available data based on observations of cosmic-ray fluxes in the atmosphere, and measurements of formation cross sections of isotopes in interactions produced by artificially accelerated particles, and of radioactivities produced by exposing targets to cosmic rays. The results of such calculations have been summarized by Lal and Peters (1967).

Some of the important properties of the atmospheric source functions of cosmic-ray-produced isotopes are briefly summarized below:

1. Isotope production is a sensitive function of latitude and altitude in the atmosphere; the

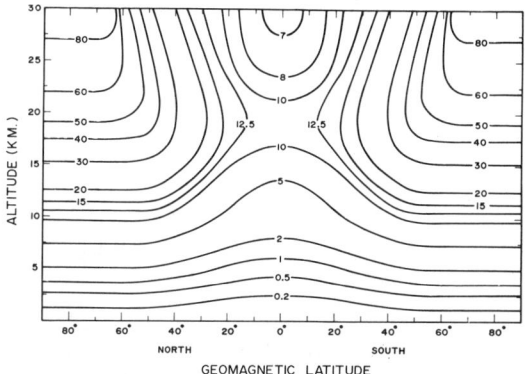

FIG. 2. A north-south cross section showing the position of lines of equal production rates of the radioisotope, Be^7. The numbers on the curves represent the production rate per minute in a cubic meter of air at S.T.P.

production being identical in the two hemispheres. For a given geographic latitude, the production rate changes only slightly with the longitude; the variation reflects the change in the geomagnetic coordinate system (the primary cosmic-ray flux, at the top of the atmosphere, is only a function of the geomagnetic latitude).

Fig. 2 shows the calculated production rate of Be^7 in the atmosphere. The strong altitude-latitude dependence can be easily seen from this figure. Similar altitude-latitude variation curves have been evaluated for all other isotopes.

2. A few altitude-latitude curves suffice to describe the variations in the production rates of all isotopes in the atmosphere. The *relative* production rates of H^3, Be^7, and Be^{10} are, for instance, identical throughout the atmosphere. Similarly Na^{22} and Al^{26} are expected to be produced in uniform ratios everywhere. The relative variations arise because the shape of the energy spectrum of cosmic ray particles changes in the atmosphere.

3. The integrated isotope production in the troposphere is practically independent of latitude. This arises because the increase in production rate with latitude due to increase in the primary cosmic ray flux (the earth's magnetic field acts as an analyzer and excludes lower-energy particles from entering the earth's atmosphere at lower latitudes) is effectively compensated by the decrease in the height of the tropopause.

The integrated production rate within the stratosphere is, however, strongly dependant on latitude; it is higher at the poles than at the equator by a factor of about 12.

Such a behavior is clearly seen from Fig. 3,

which shows the integrated production rate of Be^7 in the total atmosphere as well as the part which is produced in the troposphere alone.

4. The rate of isotope production is continuous and essentially constant. All time changes observed so far are correlated directly or indirectly with solar processes. The semi-diurnal, diurnal, and the 27-day cosmic ray intensity variations, changes due to the eleven-year solar cycle, and sporadic emission of particles during solar flares have been evaluated; except at high latitudes near the top of the atmosphere, the effects are small and can be neglected to a first approximation. The long-term changes in the cosmic-ray intensity, the primary galactic component itself, have been deduced to be small on the basis of study of isotopic changes in meteorites. These data suggest that during the past few million years, the cosmic-ray intensity averaged over time intervals, as defined by the half-lives of the isotopes studied, Mn^{53}, Be^{10}, Al^{26}, Cl^{36}, Si^{32}, Ti^{44}, has remained essentially the same as it is today (Honda and Arnold, 1967).

In most isotope applications, it thus suffices to use time-averaged production rates.

The calculated production rates of several isotopes of half-lives greater than a day, listed in Table 1, are summarized in Table 3 separately for the stratosphere and the troposphere.

The atmospheric sink functions of cosmic-ray-produced isotopes which, in fact, are source

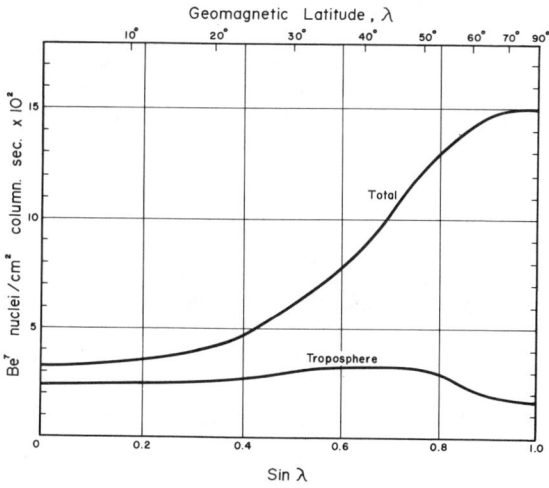

FIG. 3. The total rate of production of Be^7 in a 1-cm^2 cross section atmospheric column, per unit time, as a function of geomagnetic latitude, is given by the upper curve. The lower curve gives the corresponding production in the troposphere.

RADIONUCLIDES: COSMIC–RAY–PRODUCED

TABLE 3. COSMIC-RAY-PRODUCED RADIOISOTOPES[a]

Radio-isotope	Half-Life	Production Rate (atoms–cm^{-2}–sec^{-1})		Global Inventory
		Troposphere	Total Atmosphere	
Be^{10}	2.5×10^6 yr	1.5×10^{-2}	4.5×10^{-2}	430 tons
Al^{26}	7.4×10^5 yr	3.8×10^{-5}	1.4×10^{-4}	1.1 tons
Cl^{36}	3.1×10^5 yr	4×10^{-4}	1.1×10^{-3}	15 tons[b]
C^{14}	5730 yr	1.1	2.5	75 tons
Si^{32}	500 yr	5.4×10^{-5}	1.6×10^{-4}	1.4 kg
H^3	12.5 yr	8.4×10^{-2}	0.25	3.5 kg
Na^{22}	2.6 yr	2.6×10^{-5}	8.7×10^{-5}	1.8 g
S^{35}	87 days	4.9×10^{-4}	1.4×10^{-3}	4.5 g
Be^7	53 days	2.7×10^{-2}	8.1×10^{-2}	3.2 g
P^{33}	25 days	2.2×10^{-4}	6.8×10^{-4}	0.6 g
P^{32}	14.3 days	2.7×10^{-4}	8.1×10^{-4}	0.4 g

[a] Radioisotopes of half-life > 1 day, excluding Ar^{39} and Kr^{81}.
[b] Inventory includes a rough, estimate of Cl^{36} produced by the capture of neutrons at the earth's surface.
[b] Inventory includes a rough estimate of Cl^{36} produced by the capture of neutrons at the earth's surface.

functions for the lithosphere and the hydrosphere, depend on the rate and pattern of their removal from the atmosphere. These depend on the half-life and chemical nature of the isotope and the gross atmospheric circulation and tropospheric scavenging processes. The fractional inventories in Table 2 show the expected wide disparity in fall-out rates for various isotopes. The exact latitudinal fall-out of isotopes as a function of latitude have to be based on direct measurements. The limited available data show that the fall-out of all isotopes which can be removed effectively in wet precipitations is either a practically independent function of latitude (for isotopes of short half-life, e.g., P^{32} and Be^7) or a broad-peaked distribution, with maximum at 30–40° latitude where most of the stratosphere-troposphere exchange of air occurs. The latter distribution is applicable for long-lived isotopes, e.g., Na^{22}, Si^{32}, and Be^{10}, which have half-lives comparable to or longer than the stratosphere-troposphere exchange time. The experimental data on the fall-out of various isotopes have been discussed by Lal and Peters (1967).

Sufficient information is thus available on the source and sink functions of cosmic ray produced activity; this fact alone considerably enhances the utility of cosmic-ray-produced activity as tracers for geophysical studies over other natural or artificial tracers.

Observations on Cosmic-Ray-Produced Radioactivity

So far we have discussed the gross features in the circulations of isotopes on the earth with a view to illustrate their applications as tracers for geophysical studies. For this purpose, the atmosphere and the oceans were subdivided into zones which were considered to be internally well mixed (Fig. 1). The transport and mixing processes in these zones are known to be quite different and therefore such a characterization is not unmeaningful; nevertheless, such subdivisions may be grossly inapplicable for some isotopes. A treatment of this kind is particularly useful when only limited isotope data are available. The usefulness of such models was clearly demonstrated by Arnold, Craig, Revelle, and Suess (see Lal and Peters, 1967) in the case of oceanic circulation studies and air-sea exchange using C^{14} as a tracer.

In recent years, a large number of isotope measurements have become available for different regions of the atmosphere and of the oceans, and these make it possible to study the structure of these two important geophysical reservoirs, with respect to their stability and processes operative for mixing and transport of material. We will now briefly discuss those observations of isotope concentrations in the atmosphere, biosphere, hydrosphere and the marine sediments, which are relevant to geophysical studies. It is clearly beyond the scope of this article to discuss the data or the interpretations in any detail. For a detailed account of the investigations, reference is made to Lal and Peters (1967), and Israel and Krebs (1962).

It should be mentioned at this point that the natural distributions of cosmic-ray-produced isotopes has often been disturbed for short or long periods, depending on the half-life of the isotope in question, due to large-scale nuclear-weapons testing. For instance, the largest part of the present day inventories of C^{14} and H^3 in the atmosphere are due to man-made injections. The same was true for Na^{22} during 1963–1966. The discussions which follow are based on data which are believed to be free from errors due to any contaminations by artificially produced isotopes.

Atmosphere. With the development of high altitude submicron aerosol collection techniques, permitting quantitative sampling of radioactivity present in up to 30 tons of air, it became possible to measure systematically the distribution of the concentration of five of the short-lived cosmic-ray produced isotopes, Na^{22}, S^{35}, Be^7, P^{33}, and P^{32} up to altitudes of about 20 km (Feely et al., 1963; Bhandari et al., 1966). Above 20 km, where sampling is as yet possible only with balloons, the data are limited.

In the stratosphere, except in polar regions, and at altitudes close to the tropopause, the activities due to the short-lived isotopes S^{35}, Be^7, P^{33}, and P^{32} are found to be in near equilibrium with production. Of these isotopes, Be^7 has been most extensively studied in the stratosphere. The absolute concentrations of this isotope, as well as the ratios Be^7/P^{32}, P^{33}/P^{32}, clearly show that air motions, organized or turbulent, do not occur in most of the stratosphere on short time scales as compared to the half-lives of these isotopes, such that near equilibrium concentrations can be reached. The observed isoconcentration lines of Be^7 and the distribution expected for the case of a motionless atmosphere are shown in Fig. 4. The activity of the longer-lived isotope, Na^{22}, is however found to be considerably undersaturated in all regions of the stratosphere: order of magnitude departure from the secular equilibrium are found to occur. The measured disintegration ratios, Na^{22}/Be^7, in the tropospheric and stratospheric air samples are shown in Fig. 5. The activity-ratio expected in the stratosphere at secular equilibrium, i.e., in the case of a stagnant atmosphere, is about 125×10^{-5}. The observed values are usually less than 50×10^{-5}. These results have been discussed in terms of apparent irradiation ages and atmospheric circulation models (Bhandari et al., 1966). The calculated values of the coefficient of vertical eddy diffusivity in the stratosphere range from 2×10^3 to 3×10^4 cm^2-sec^{-1}.

Close to the tropopause, and in the polar regions, the mixing processes are time dependent and the air-motions are easily measurable during certain periods from the observations of Be^7 and P^{32}. Reference is made to Bhandari et al. (1966) for a detailed discussion of these observations.

In the troposphere, the distributions Na^{22}, S^{35}, Be^7, P^{33}, and P^{32} clearly suggest a lack of complete mixing, and variability in isotope concentrations as well as in ratios. The data indicate that the removal of these activities from the troposphere occurs in periods of the order of one month.

Fall-out in Wet Precipitations. The concentrations of the short-lived isotopes are very variable in rains. The fall-out of isotopes Be^7 and P^{32} is found to be practically independent of the latitude. This observation, which considered along with the observed absolute fall-out rates, indicates that most of the activity produced in the stratosphere decays there because of slow-exchange (cf. Fig. 3 which shows that the production in the troposphere is nearly independent of latitude). The observed ratios Be^7/P^{32} in rains indicate a mean wash-out period of the order of one month. The results based on other short-lived isotopes support such conclusions (Lal and Peters, 1967).

The fall-out of Si^{32}, whose half-life is long

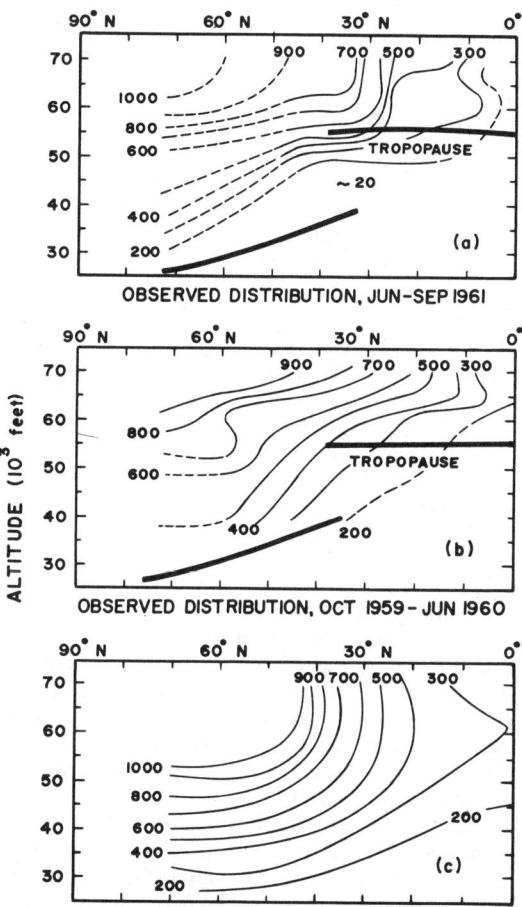

FIG. 4. Mean observed Be^7 isoconcentration lines in the atmosphere during two periods are given in (a) and (b). The expected distribution (Lal and Peters, 1967) for a motionless atmosphere is given by (c). The units are dpm/10^3 S.C.F. The thick lines show the position of mean tropopause. The figure is redrawn from Freely et al. (1963).

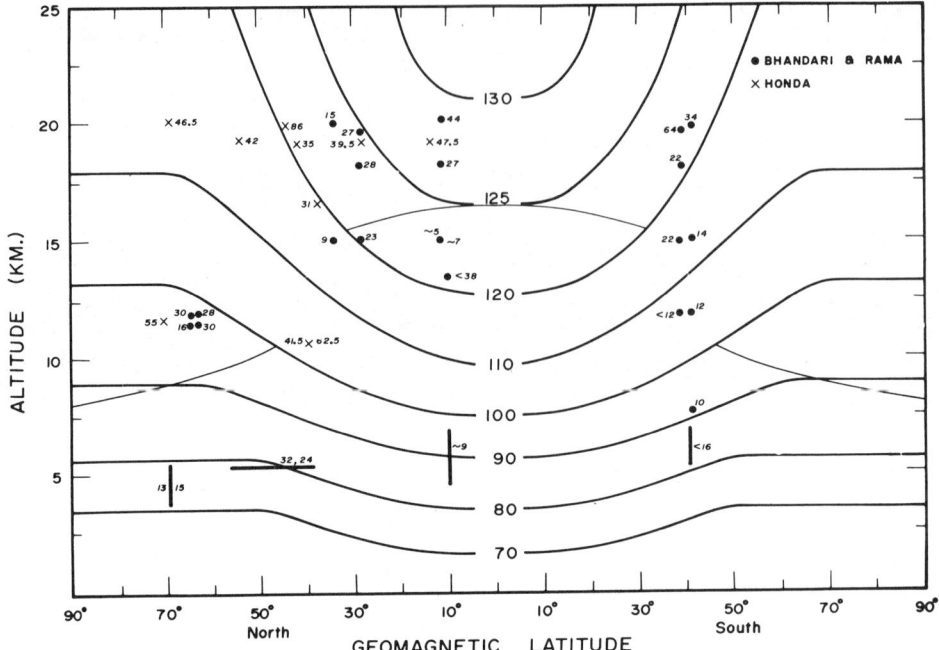

Fig. 5. The observed values of the ratios Na^{22}/Be^7 in the atmosphere. The numbers should be multiplied by 10^{-5} to obtain the ratio of disintegrations of these activities (after Bhandari et al., 1966).

compared to the time scales in atmospheric mixing processes, is found to be latitude dependent. The observations on the fall-out of Si^{32} are useful for understanding the nature of stratospheric-tropospheric exchange, and the global fall-out patterns. The concentrations of other longer-lived isotopes have not yet been measured in rain water.

Biosphere. So far only C^{14} has been studied in the continental biosphere. In the marine biosphere, C^{14} and Si^{32} have been studied. The specific activity of C^{14} in the land biosphere, prior to nuclear weapons tests, has been deduced to be independent of latitude. The values of the atmospheric ratio, C^{14}/C^{12}, have been experimentally measured to have remained the same within a few percent during the last 4000 years or so. These results form the basis of the well known archaeological dating method (Libby, 1965). The radiocarbon method is applicable for dating organic matter on the land or in the oceanic sediments, provided it is younger than about 50,000 years.

Oceans. The activities of Be^7, H^3, Si^{32}, and C^{14} have been measured in ocean water. The shorter-lived isotopes Be^7 and H^3 have been measured in order to study the rates of vertical mixing in the turbulent regions of the ocean above the thermocline. The distributions of the longer-lived isotopes have been investigated to understand the nature of large-scale mixing and transport processes within and between the oceans. Study of these isotopes essentially provide the only estimates of time scales available to date. The measurements are most extensive in the case of C^{14}, thanks to the work of Broecker and of Suess, and their collaborators.

The measurements of the concentration of radiocarbon in the atmosphere and the surface ocean waters have led to an evaluation of the time of exchange of carbon dioxide between the atmosphere and the ocean. The mean residence time of CO_2 in the atmosphere is determined to be about seven years. This corresponds to a mean global exchange rate of about 20 moles CO_2/m^2–yr.

The observed distribution of C^{14} in the principal oceans lead to a picture of large scale water circulation which is in conformity with the oceanographic evidence. The apparent ages and the average residence times of water in the oceans have been calculated. The residence times depend on the assumed model of mixing. The C^{14} data, however, restrict the choice of model (Broecker, 1963).

Fig. 6 shows a world ocean model which satisfies the main features of the observed C^{14} distribution, including the high C^{14} concentra-

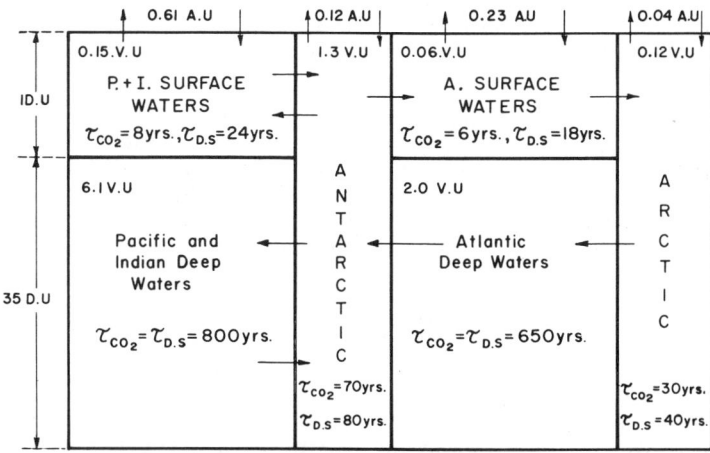

FIG. 6. A world-ocean model, separating Pacific plus Indian oceans from the Atlantic ocean by the vertical reservoir, Antarctic ocean. The directions of exchange of water, or of CO_2, are shown by arrows. A.U., V.U., and D.U. represent the fractional areas, volumes, and depths of the oceanic reservoirs. The model and the computed values of the residence times are after Broecker (1963).

tion in the Arctic and intermediate values in the Antarctic (Broecker, 1963). The computed residence times for CO_2 and dissolved solids (D.S.) are given in this figure. It should be pointed out here that whereas the apparent age of the Pacific deep waters is found to be older by 800 years compared to the Atlantic deep waters, the residence time in the Pacific is greater by only about 25% on this model. This is so because the C^{14} content of source waters are very different in the two cases.

The available Si^{32} data are largely confined to the surface waters. Reservoir model calculations, using the estimated production rate of Si^{32} and the observed specific activities of silicon in surface waters of the Pacific and the Atlantic oceans, yield values for the residence times which are consistent with results based on C^{14} data.

The distribution of both C^{14} and Si^{32} in the oceans is controlled in an important way by complex biological processes; dissolution of gravitationally settling calcareous and siliceous biogenic particles at depths, and at the ocean floor (Lal, 1962). The in situ biological term has to be implicitly considered in the analysis of C^{14} and Si^{32} data for the evaluation of diffusion/advection rates. Experimental data on the concentrations of both stable and radioactive isotopes are now becoming available.

Marine Sediments. In oceanic sediments, three cosmic-ray-produced isotopes have been detected so far: Si^{32}, C^{14}, and Be^{10}. These isotopes permit a study of chronology in the time interval of a few hunderd to a few million years in the past. The sediments contain a record of the earth's climatological and geological history and dating with the help of these isotopes is therefore proving very valuable.

In this section, we have attempted a very brief sketch of some of the observations made to date on the terrestrial distribution of cosmic-ray-produced isotopes. The material covered has been chosen to represent the types of geophysical and geochemical studies which are becoming possible using radioactive tracers which are continually produced by cosmic rays on the earth.

D. LAL

References

Bhandari, N., Lal, D. and Rama, 1966, "Stratospheric circulation studies based on natural and artificial radioactive tracer elements," *Tellus*, **18**, pp. 391–406.

Broecker, W. S., 1963, "Radioisotopes and Large Scale Oceanic Mixing," in (Hill, M. N., editor), "The Sea," New York, Interscience Publishers, pp. 88–108.

Feely, H. W., Davidson, B., Friend, J. P., Lagomarsino, R. J., and Leo, M. W. M., 1963, "Project Star Dust, DASA 1309," Isotope Inc., New Jersey.

Honda, M., Arnold, J. R., 1967, "Effects of Cosmic Rays on Meteorites," in "Handbuch der Physik," Berlin, Springer-Verlag, pp. 613–632.

Israel, H., and Krebs, A., 1962 editors), "Nuclear Radiation in Geophysics," Berlin, Springer-Verlag.

Lal, D., 1962, "Cosmic ray produced radionuclides in the sea," *J. Oceanogr. Soc. Japan,* 20th Anniv. Vol., pp. 600–614.

Lal, D., 1963, "On the Investigations of Geophysical Processes Using Cosmic Ray Produced Radioactivity," in (Geiss, J., and Goldberg, E. D., editors), "Earth Science and Meteoritics," Amsterdam, North Holland Publ. Co., pp. 115–142.

Lal, D., and Peters, B., 1967, "Cosmic Ray Produced Radioactivity on the Earth," in "Handbuch der Physik," Berlin, Springer-Verlag, pp. 551–612.

Libby, W. F., 1965, "Radiocarbon Dating," Second edition, Chicago, University of Chicago Press.

Cross-references: *Carbon Cycle; Carbon-14 Dating; Elements: Planetary Abundances and Distribution; Isotope Geology; Radioactive Isotope; Radioactivity in Rocks; Seawater, Chemistry; Silica Cycle; Trace Elements.* Vol. I: *Carbon Dioxide in Sea and Atmosphere; Radionuclides in Oceans and Sediments; Radionuclides: Their Applications in Oceanography.* Vol. II: *Aerosols; Atmosphere; Cosmic Rays; Tropopause; Sunspots and Solar Activity.* Vol. V: *Geomagnetism.*

RADIONUCLIDES IN OCEANS AND SEDIMENTS—See Vol. I

RADIUM: ELEMENT AND GEOCHEMISTRY

Physical Properties

Radium (L. *radius,* ray), symbol Ra, is a radioactive metal of Group IIA of the periodic table. Its atomic weight is 226 and its atomic number is 88. Its electronic configuration [Rn core] is $7s^2$. The melting point is 700°C and the boiling point is 1150°C. The specific gravity (at 20°C) is 6.0 and its valence is +2. Isotopes are given in Table 1 (see also Fig. 1). In general the name radium refers to Ra^{226}. To denote other isotopes, the common names or mass numbers are used.

The pure metal, prepared by electrolysis of a fused salt, is brilliant white but blackens in air. Both the metal and its salts exhibit a strong luminescence.

The primary method of measuring Ra^{226} is to seal the sample in a flask and allow its daughter, Rn^{222}, to grow to equilibrium value. The radon is then swept from the flask and its activity measured.

Minerals

The radium isotopes owe their existence on the earth to their continuing production from uranium and thorium isotopes. Thus uranium- and thorium-bearing minerals also contain the natural radium isotopes and it is not so much the chemistry of radium as it is the chemistry of uranium and thorium that determine in what minerals radium will be found.

Chemistry

Radium is a member of the alkaline earth family and, like barium, strontium, and calcium, it forms insoluble sulfate, carbonate, and chromate salts. The chloride, bromide, nitrate, and hydroxide salts of radium are soluble in water. In general, radium salts are less soluble than are corresponding barium salts, and the first method of separation was based on fractional crystallization processes. Recently, methods of separation based on ion-exchange and solvent extraction have made the isolation of radium relatively simple.

Radium was first isolated from a ton of processed uranium residues by Mme. Curie in what became a classical separation. She thus proved that radioactive substances other than uranium existed in uranium minerals. Soon after radium was isolated, it was found to emit a radioactive gas similar in properties to a certain thorium emanation called thoron. This gas in turn gave rise to a series of radioactive products called radium-A, -B, and -C. Through such studies the foundation was laid for later work linking many different isotopes in the uranium and thorium decay series (see Fig. 1).

Geochemistry

The geochemistry of radium depends on the isotope and the time scale of interest. If no separation takes place for 10,000 years, the geochemistry of Ra^{226} is entirely governed by its parent, Th^{230}; likewise separations must occur in less than 50 years for Ra^{228} not to be in radioactive equilibrium with its parent, Th^{232}. It is thus only in recent volcanic rocks, weathered and altered materials, natural waters, recent sediments and evaporites, and

TABLE 1. NATURALLY OCCURRING ISOTOPES

Mass	Common Name	Half-Life	Nature and Energy (MeV) of Decay
228	Mesothorium	6.7 yr	$\beta-$, 0.012
226	Radium	1622 yr	α, 4.79
224	Thorium-X	3.64 days	α, 5.68, 5.45
223	Actinium-X	11.2 days	α, several 5.4–5.7

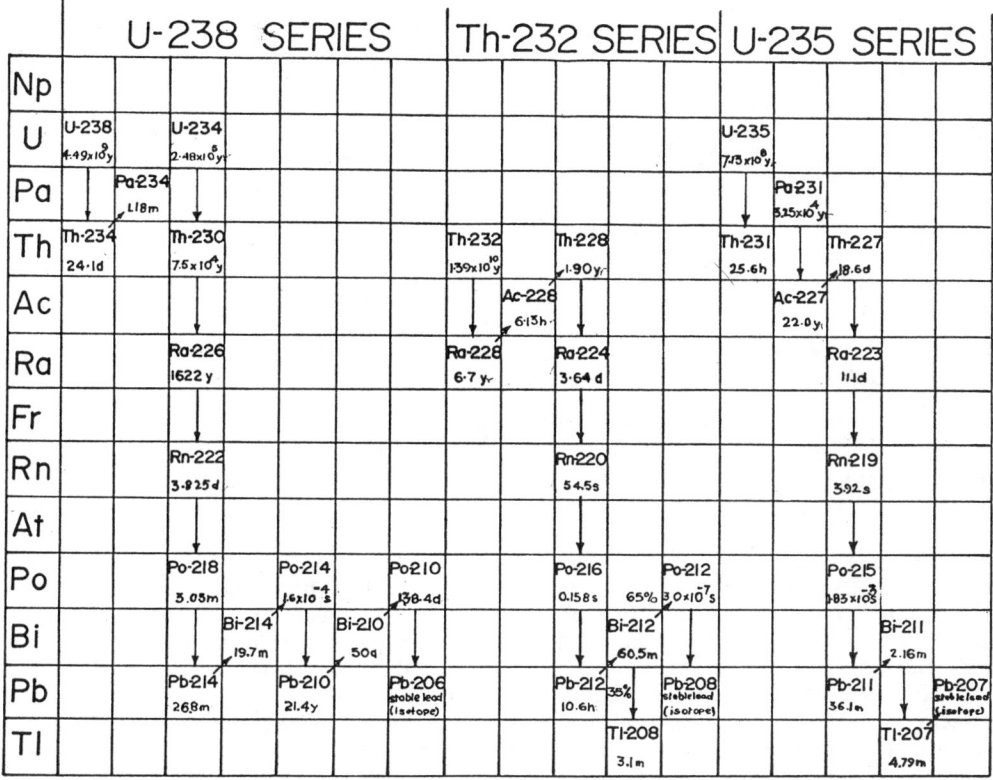

FIG. 1. Position of radium in the natural decay series of U-238, Th-232, and U-235. Vertical arrows indicate alpha decay; diagonal arrows represent beta decay.

the biosphere that a separation of radium isotope from its parent may be expected.

Perhaps the most interesting phase of the geochemistry of radium is in the ocean. Radium-226 exists in greater than tenfold excess over that which can be supported by Th^{230} in seawater. Koczy pointed this out and hypothesized that the excess radium was being supplied by diffusion from deep-sea sediments. He further documented this theory by showing a deficiency of Ra^{226} in deep-sea sediments. Moore has shown that Ra^{228} also exists in excess of its parent, Th^{232}, in the ocean. This was first suggested by high Th^{228}/Th^{232} activity ratios in seawater.

The potentiality of these two isotopes to oceanic circulation studies is immense. Of the radioactive tracers used to study large-scale oceanic mixing processes, only radium is injected directly into bottom waters from underlying sediments. Thus using these two isotopes, deep-mixing processes on the time scale of tens of years as well as thousands of years may be investigated.

Radium also plays a role in radiometric dating methods. It was first used as a measure of the Th^{230} concentration of deep-sea cores; however the discovery of its migration within the sediment has invalidated such measurements. In dating calcium carbonate materials, the existence of radioactive equilibrium between Th^{230} and Ra^{228} is used to help evaluate whether the material has been a closed system during the past 5000 years.

WILLARD S. MOORE

References

Blanchard, R. L., and Oakes, D., 1965, "Relationship between uranium and radium in coastal marine shells and their environment," *J. Geophys. Res.*, **70**, 2911-2921.

Kaufman, A., and Broecker, W., 1965, "Comparison of Th^{230} and C^{14} ages for carbonate materials from Lakes Lahontan and Bonneville," *J. Geophys. Res.*, **70**, 4039-4054.

Koczy, F. F., 1958, "Natural Radium as a tracer in the ocean," Second U.N. International Conference on Peaceful Uses of Atomic Energy, **18**, 351-357.

Moore, W. S., and Sackett, W. M., 1964, "Uranium and thorium series in equilibrium in sea water," *J. Geophys. Res.*, **69**, 5401-5405.

Cross-references: *Geochronometry; Radioactive Isotopes; Radioactivity in Rocks; Radium in the Oceans; Seawater, Chemistry; Seawater, Geochemical Balances; Thorium; Uranium; Uranium-Thorium-Lead Age Determination.*

RADIUM IN THE OCEANS

The element radium is the heaviest of the alkaline earth group of metals. Of the reported thirteen radium isotopes with mass numbers between 213 and 230, four occur naturally. These four isotopes are intermediate members of the three naturally occurring radioactive series:

$$U^{238} \xrightarrow{4.5 \times 10^9 \text{ yr}} Th^{234} \xrightarrow{24.1 \text{ days}} Pa^{234} \xrightarrow{6.7 \text{ hr}} U^{234} \xrightarrow{2.5 \times 10^5 \text{ yr}}$$

$$Th^{230} \xrightarrow{7.52 \times 10^4 \text{ yr}} Ra^{226}(Ra) \xrightarrow{1620 \text{ yr}} Rn^{222} \cdots Pb^{206}.$$

$$U^{235} \xrightarrow{7.1 \times 10^8 \text{ yr}} Th^{231} \xrightarrow{25.6 \text{ hr}} Pa^{231} \xrightarrow{3.43 \times 10^4 \text{ yr}} Ac^{227} \xrightarrow{22 \text{ yr}}$$

$$Th^{227} \xrightarrow{18.6 \text{ hr}} Ra^{223}(AcX) \xrightarrow{11.2 \text{ day}} Rn^{219} \cdots Pb^{207}.$$

$$Th^{232} \xrightarrow{1.4 \times 10^{10} \text{ yr}} Ra^{228}(MsTh) \xrightarrow{6.7 \text{ yr}} Ac^{228} \xrightarrow{6.13 \text{ hr}} Th^{228} \xrightarrow{1.9 \text{ yr}}$$

$$Ra^{224}(ThX) \xrightarrow{3.64 \text{ day}} Rn^{220} \cdots Pb^{208}.$$

Because of its relatively long half-life only the isotope Ra^{226} has been of interest in marine geochemistry. Not until recently has the distribution of Ra^{228} in the ocean begun to receive attention from geochemists.

As early as 1908, J. Joly reported the enrichment of Ra^{226} in the deep-sea sediments acquired during the *Challenger Expedition* (1872–1876). The impact of this finding was a series of investigations on the distribution of radium, as well as its radioactive parents, uranium and ionium (Th^{230}), in the marine environment beginning in the 1930s (Faul, 1954; Rankama, 1963). From this emerged the so-called "disequilibrium studies" of the members of the uranium and thorium radioactive series in the field of geochemistry. Such studies not only help to elucidate some geochemical processes taking place in the hydrosphere and lithosphere, but also introduce a time parameter for the study of late Pleistocene geological events (Faul, 1954; Rankama, 1963; Broecker, 1965a; and references therein).

Distribution of Ra^{226} in Seawater

As an intermediate member of the decay chain, the Ra^{226} present in the ocean has a concentration level primarily controlled by that of its immediate parental nuclide, Th^{230}.

The concentration of the dissolved uranium in seawater is rather uniform, having an average value of about 3.3×10^{-6} g/l. The amount of Th^{230} and Ra^{226} in radioactive equilibrium with this uranium should be 5.4×10^{-11} g/l and 1.1×10^{-12} g/l, respectively. Measurements showed that the Th^{230} concentration in the open ocean ranged from 1 to 4×10^{-14} g/l (Moore and Sackett, 1964). This indicates that the precipitation of Th^{230} as it is formed from uranium dissolved in the sea is essentially quantitative; less than 0.1% of the Th^{230} produced remaining in the solution. The rapid removal of Th^{230} (in a time on the order of 100 years) to the ocean bottom thus also leads to a low Ra^{226} content in the sea.

Table 1 summarizes the Ra^{226} distribution in the world oceans. The Ra^{226} content varies between 2 and 17×10^{-14} g/l. The general vertical distribution pattern shows an increase of Ra^{226} concentration toward the bottom. Surface values are commonly around 4 to 5×10^{-14} g/l, whereas values close to the bottom often range from 6 to 16×10^{-14} g/l. Fig. 1 gives some of the profiles. Ra^{226} content in the deep Pacific is higher than that in other parts of the ocean. There is an indication of a sharp change across the main thermocline. Below 3000 m the Ra^{226} content is essentially constant.

It is clear that although there is less Ra^{226} in the ocean than would be expected from the amount of uranium, there is more than can be supported by its immediate parent Th^{230}. The fact that the activity ratio of Ra^{226}/Th^{230} ranges from about twenty to several hundred, compared to unity under radioactive equilibrium conditions, means that the amount of Th^{230} found in seawater can account for only about 1–5% of the Ra^{226} in the sea. If the unsupported Ra^{226} in the ocean were not being continuously supplied by other sources, such as diffusion from deep-sea sediments, river influx, submarine volcanism, and particulate matter containing Th^{230}, it would disappear. Ra^{226} supplied by river influx and maintained in the surface water of the ocean is estimated to be on the order of 1×10^{-15} g/l, which also accounts for less than 4% of the oceanic content (Moore, 1967). As no evidence as to the con-

TABLE 1. WORLDWIDE DISTRIBUTION OF Ra^{226} IN SEAWATER ($\times 10^{-14}$ g/l)[a]

	0–500 m	500–2000 m	2000 m to max. Depth
North Pacific	2–10	4–13	6–16
South Pacific	2–10	5–15	8–17
North Atlantic	2–8	4–9	5–12
Indian Ocean	3–6	5–9	6–12
Mediterranean Sea	~6	5–12	~10
Caribbean Sea	2–6	3–7	4–7
Providence Channel	3–6	3–6	4–7
Tongue of the Ocean	3–6	3–6	
Florida Straits	3–5	~5	
Black Sea	7–12	~10	
Red Sea	4–6	~6	
Baltic Sea	3–15		

[a] From Szabo (1967).

tribution from the submarine volcanic emanation and the particulate matter has been gathered, Ra^{226} originating from these sources, if any, is considered to be of minor importance. Thus the major source of the Ra^{226} supply (>90%) must come from the sediments. Taking the residence time of Ra^{226} in the ocean to be its mean life (2300 years), we can estimate that, in order to maintain the observed amount, the upward flux of Ra^{226} from sediments is on the order of 10^{-10} g/m²-yr, or 10^{-21} g/cm²-sec. Observations on the diffusion of Ra^{226} in deep-sea cores will be treated below.

If Ra^{226} concentration in the sea has reached steady state, its distribution should be related to the rate of oceanic mixing. Theoretically, in an area where no vertical advection occurs and where the Ra^{226} distribution is controlled by diffusion only, the eddy diffusion coefficient of deep water can be derived from the steady state Ra^{226} distribution. Koczy (1958) has made calculations for five stations in the low-latitude region where the influence of convection and horizontal gradients are presumably small. He found that the vertical eddy diffusion coefficient increased approximately in an exponential way with depth; the maximum values near the bottom are about 3–30 cm²/sec. From the vertical eddy diffusivity, Koczy and Szabo (1962) have also attempted to determine the renewal time of bottom water masses.

However, calculations of this nature have recently been challenged by the observation that Ra^{226} in the ocean may be distributed through mechanisms other than diffusion, such as organic particle settling. It has been shown that in order to explain a fourfold increase of Ra^{226} downward in the Pacific and a twofold increase in the Atlantic (see Fig. 1) a mean eddy diffusion coefficient of much less than 0.1 cm²/sec across the main thermocline

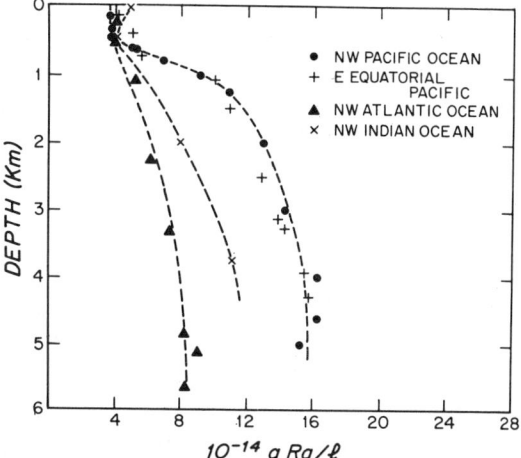

FIG. 1. Vertical profiles of Ra^{226} distribution in the oceans. Data of the Pacific and Atlantic are from Broecker et al., 1967; those of the Indian Ocean from Patterson, 1955 (See Koczy and Szabo, 1962).

is required. The mixing rates derived therefrom are an order of magnitude lower than those which fit the vertical distribution of natural radiocarbon. On the other hand, Broecker, et al. (1967) stressed that the distribution profiles in the Pacific and the Atlantic can be interpreted by the settling of radium-carrying particles at the same rate from surface to depth in the two oceans and the threefold longer residence time of water in the deep Pacific than in the deep Atlantic. Recent work by Szabo (1967) demonstrated that Ra^{226}, as well as barium, is preferentially taken up by plankton. Hence the application of Ra^{226} distribution for tracing movements appears to be complicated by biological removal of Ra^{226} from surface water and a net downward transport of radium-carrying particles.

Distribution of Ra²²⁶ in Deep-Sea Sediments

As a result of Th^{230} precipitation, deep-sea sediments are greatly enriched in Th^{230} with respect to its parent uranium. Recently deposited sediments contain Th^{230} in the range between 1×10^{-10} and 5×10^{-9} g/g, which is about 3 to 150 times greater than can be supported by the uranium in sediments (sediments contain about 2×10^{-6} g U/g; the equivalent amount of Th^{230} supported by this uranium is 3.3×10^{-11} g/g). The unsupported Th^{230} will decay with a half-life of 75,200 years. Thus in an ocean core, if both the accumulation for bulk sediment and Th^{230} remained constant with time, the Th^{230} concentration should assume an exponential decrease with depth (Fig. 2a). Ra^{226} is low in concentration in seawater, and is nonreactive with respect to oceanic precipitation processes. Its presence in sediments is therefore mainly due to decay of Th^{230}. An ideal case for Ra^{226}

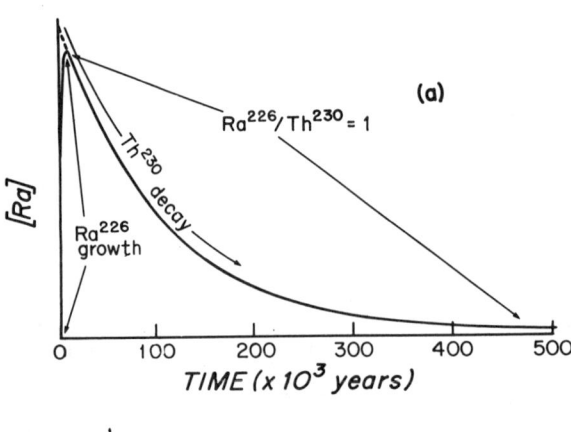

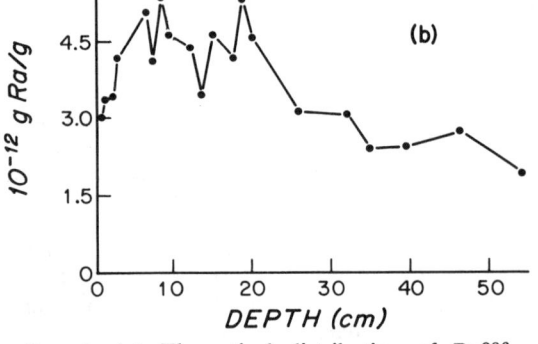

FIG. 2. (a) Theoretical distribution of Ra^{226} in the sediments when it is assumed that precipitation of both bulk sediment and Th^{230} are constant and that no post-depositional migration of Th^{230} and Ra^{226} occurs. (b) The Ra^{226} distribution in Core 86 (of the Swedish Expedition) taken from the equatorial Pacific measured by Kröll (See Koczy, 1958).

TABLE 2. Ra^{226} AND Th^{230} DISTRIBUTION IN CORE V10-95[a]

Depth (cm)	0 -6	30 -35	75 -80	92 -97	147 -150	210 -215	250 -255	330 -335	390 -395	470 -475	550 -555	720 -725	880 -885	1112 -1117
$\dfrac{Th^{230}}{U^{234}}$	37.8	16.9	7.72	4.38	2.02	1.10	1.01	1.07	0.89	1.00	0.93	1.03	1.02	0.98
$\dfrac{Ra^{226}}{Th^{230}}$	0.40	0.89	1.06	0.95	1.16	0.89	1.33	1.07	1.33	1.09	1.11	1.09	0.92	1.61

[a] From Ku (1965). Core location: 26°31′N, 51°47′W; Water depth: 5190 m.

distribution in a core is also shown in Fig. 2a: increase with depth near the top (growth with its 1620 years half-life), reaching a maximum at about 9000 years, then in radioactive secular equilibrium with and thus following the distribution of Th^{230}. Another assumption involved here is that sediments act as a chemically closed system for these two radionuclides after their deposition.

The ideal situation given in Fig. 2a provides a basis for the determination of sedimentation rates in a core, as has been attempted by a number of investigators (Pettersson, Piggot, Urry, Kröll, Sanderman, Utterback, Bernert, Volchok, and Kulp; see Faul, 1954, and Rankama, 1963). This so-called "radium-ionium" method utilized the assumption that Ra^{226} could be used as an index of Th^{230}. Hence by measuring the Ra^{226} along a core, the rate of deposition could be calculated from the decay rate of Th^{230}. The method was rather short-lived due to Kröll's extensive analyses. Fig. 2b, for example, shows the results on one of the many cores analyzed by Kröll. Whereas the general trend of Ra^{226} growth and decay (cf. Fig. 2a) is maintained, there are two or more maxima in the vertical distribution. To explain this, Kröll questioned the validity of the assumptions regarding constancy of both Th^{230} and sediment deposition and the "closed system" for Ra^{226} and Th^{230}. We now understand, through measurements of Ra^{226} and Th^{230} in seawater and in sediments, that in many cases Ra^{226} cannot serve as an index for Th^{230} because of the fact that Ra^{226} is subject to postdepositional migration. Table 2 lists data on a uniformly deposited red-clay core from the Atlantic. Whereas exponential decrease of Th^{230} (denoted by ratio Th^{230}/U^{234}) with depth is exhibited, Ra^{226} is not always in secular equilibrium with Th^{230} (indicated by ratio Ra^{226}/Th^{230}) in deeper sections of the core.

Koczy and Bourret (1958) first attempted to calculate the diffusion rate of Ra^{226} which is presumably present in ionic form in the sediments. They visualized that diffusion of Ra^{226} was primarily controlled by the concentration gradients on either side of the maximum near the core top (see Fig. 1). A difference between the two gradients caused the level of maximum Ra^{226} concentration to shift. The Ra^{226} deficit with respect to Th^{230} observed at top portions of cores (e.g., data in Table 2) is a combined effect of two factors: growth of Ra^{226} from Th^{230}, and diffusion of Ra^{226} from sediments into the sea. Knowing the sedimentation rate (e.g., from the Th^{230} decay) and the Ra^{226} distribution in that part of core above its highest concentration, we can estimate the Ra^{226} diffusion rate in the following way.

By assuming that rates for sediment accumulation, Th^{230} precipitation, and Ra^{226} diffusion are constant, the equation governing the steady-state Ra^{226} distribution in a core is

$$D\frac{\partial^2 c}{\partial x^2} - S\frac{\partial c}{\partial x} + \lambda_1 N_0 e^{-\lambda_1 x/S} - \lambda_2 c = 0 \quad (1)$$

where D = diffusion coefficient of Ra^{226}; S = sediment accumulation rate; x = depth in core; c = concentration of Ra^{226}; N_0 = concentration of Th^{230} in newly deposited sediments; and λ_1, λ_2 = decay constants (0.693/ half-life) of Th^{230} and Ra^{226}, respectively.

The first term of the equation, expressing the one-dimensional case of Fick's diffusion law, accounts for the changes in the concentration of Ra^{226} due to diffusion; the second term the changes resulting from the thickening of the sediment (mathematically equivalent to downward displacement of Ra^{226} at a constant rate, S); the third term the production rate of Ra^{226} from Th^{230}, and the fourth, the rate of Ra^{226} decay.

For the boundary conditions that Ra^{226} in freshly deposited sediments and at infinite depth are both zero, Eq. (2) can be written.

Fig. 3 shows a set of curves that are derived by taking $S = 1$ cm/10^3 yr, $N_0 = 1.7 \times 10^{-10}$ g Th^{230}/g sediment, and different rates for Ra^{226} diffusion. From experimentally determined S, N_0 and the Ra^{226} distribution data, workers have estimated values of D's to be on the order of 10^{-8} to 10^{-9} cm^2/sec. These values are much lower than the diffusion coefficient of electrolytes in water ($\sim 10^{-5}$ cm^2/sec), suggesting that Ra^{226} migration in sedimentary column may be strongly impeded by adsorption on surfaces of solids. The irregular pattern of Ra^{226}/Th^{230} distribution in the deeper portions of core as shown in Table 2 is not quite clearly understood; however, it presumably reflects the variations in the surface sorption properties of sediment mineral grains.

Ra^{226} in Marine Carbonates and Biosphere

Since radium is chemically similar to calcium, it is expected that the Ra/Ca ratio in contemporary marine carbonates should have the same value as that in surface seawater ($\sim 1.2 \times 10^{-13}$). This is probably true for aragonitic corals and inorganically precipitated

$$c(x) = \frac{\lambda_1 N_0}{D(\lambda_1/S)^2 + \lambda_1 - \lambda_2} \left\{ \exp\left[-\left(\sqrt{\frac{S^2}{4D^2} + \frac{\lambda_2}{D}} - \frac{S}{2D}\right)x\right] - \exp\left(\frac{-\lambda_1 x}{S}\right) \right\} \quad (2)$$

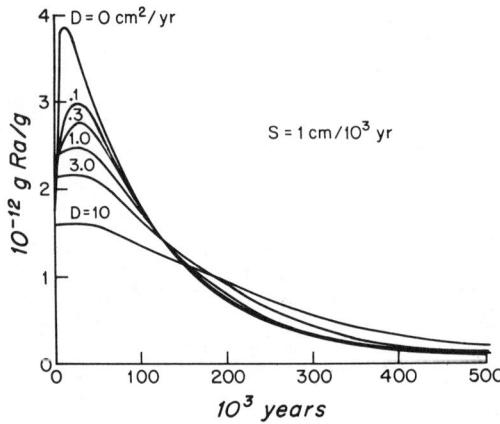

FIG. 3. Ra^{226} distribution in sediments deposited at a constant rate of 1 cm/10^3 yr. Assumptions: (1) a constant rate of Th^{230} precipitation corresponding to the total Th^{230} produced in a water column of average ocean depth (3800 m) containing 3×10^{-6} gU/l; and (2) a constant diffusion coefficient (D) for Ra^{226}. The different curves represent different values of D (after Koczy and Bourret, 1958).

oolites. For other marine organisms, a considerable range of Ra/Ca has been found. The discrimination factors of Ra^{226} with respect to Ca, i.e., $DF_{Ca}^{Ra} = (Ra/Ca)_{sample}/(Ra/Ca)_{seawater}$, for the various marine biological materials are summarized in Table 3. They appear to fall into three groups. For calcareous shells DF_{Ca}^{Ra} appears to be less than 10, with most of the values ranging from 0.1 to 4; for plankton, fish, and algae DF_{Ca}^{Ra} ranges from 20 to 60. A plankton sample of mainly siliceous diatoms indicates a high DF_{Ca}^{Ra} of 600. The high value suggests that plankton preferentially selects radium to calcium.

As there is no stable isotope of radium, the importance of the transport of radium by organisms cannot be readily estimated. Chow and Goldberg (1960) have suggested barium as a possible alternative. They have explained the observed enrichment of barium in the deep equatorial Pacific by release of barium by solution of organisms which incorporate the element in the surface water. Szabo (1967) analyzed Ra^{226} and other alkaline earth elements of six plankton samples from the Bahamas re-

TABLE 3. Ra^{226} CONTENT OF VARIOUS BIOLOGICAL MATERIALS[a]

Material	Investigator	Ra^{226} Content $\times 10^{-13}$ g/g ash	Average DF_{Ca}^{Ra}
	Plankton		
Calanus finmarchicus	Brunovski (1932)	2.0	55
Schizopod crustaceans	Evans et al. (1938)	4.6	
Diatoms	Föyn et al. (1939)	20.0	
Mixed plankton (mainly diatoms)	Koczy and Titze (1958)	2.8	600
Mixed plankton	Cerrai et al. (1964)	18.1	57
Mixed plankton	Szabo (1967)	2.2	19
	Fish		
Cottus gobia	Burkser et al. (1929)	2.7	
Various	Kunasheva (1944)	0.4–1.6	40
	Algae		
Brownalgae	Burkser et al. (1931)	5.6	30
Algae	Brunovski (1932)	2.0	
Algae	Wiesner (1938)	0.5–7.0	
Kelp	Evans et al. (1938)	2.0	
	Crustacea		
Balanus balanoides	Kunasheva (1944)	0.2	1
Various	Koczy and Titze (1958)	0.6–2.0	1.6–6.5
Barnacles	Broecker (1963)	1.0–3.3	2–6.6
	Mollusca		
Various	Koczy and Titze (1958)	0.2–1.4	0.4–2.4
Pelecypods	Broecker (1963)	0.1–2.5	0.2–5.0
Gastropods	Broecker (1963)	0.02–0.3	0.04–0.52
Pelecypods	Blanchard and Oakes (1965)	0.003–0.14	0.04–1.03

[a] Adapted from Szabo (1967).

gion. The concentration of alkaline earth elements was found to decrease in the order $Ca > Mg > Sr > Ba > Ra$. But the discrimination factors for these elements showed the following sequence of selectivity: $Ra > Ba > Sr > Ca > Mg$, or $19 > 18 > 1.2 > 1.0 > 0.10$. Radium and barium were strongly preferred relative to calcium. Also, since the sequence is the same as that found for the polystyrene type cation exchangers, such as Dowex 50, the suggestion has been made that the uptake of these elements is by an ion exchange process. Further investigation of the biological processes affecting the radium distribution in the ocean may be accomplished by simultaneous measurements of barium and radium.

Ra^{228} in the Ocean

Th^{228}, a member of the Th^{232} series, was found in excess of the amount presumed to be in radioactive equilibrium with the Th^{232} in both surface and deep ocean water. The ratio Th^{228}/Th^{232} observed generally varies between 10 and 25. High Th^{228}/Th^{232} was also detected in recent sea shells (Moore and Sackett, 1964). These somewhat unexpected results have been explained as being due to presence of an excess of Ra^{228}. From the consideration of material balance, it can be shown that rivers do not contain enough Ra^{228} to support all that is observed in the surface ocean. Whereas the excess of Ra^{228} in deep waters can be attributed to diffusion of radium from deep-sea sediments, in view of the short half-life of Ra^{228} ($t_{1/2} = 6.7$ years) compared to the mean residence time of water in the deep ocean (about 1500 years), the observed excess in surface ocean is difficult to explain. It has been tentatively suggested to result from leaching of Ra^{228} from nearshore and continental shelf sediments, and from the particulate phases containing Th^{232}.

Because of its relatively short half-life, Ra^{228} is potentially a useful tracer for studying the oceanic mixing on a time and spatial scale much smaller than that that can be dealt with by other radionuclides, such as C^{14}, Ra^{226}, Sr^{90}, and Cs^{137}. Furthermore, since radium is suspected to be subjected to biological transport in the ocean, the ratio of Ra^{228}/Ra^{226} would be unaffected by this transfer but will be a function of time the two species travelled together.

Rn^{222}/Ra^{226} Ratio in Seawater

Attention has recently been given to the 3.85-day half-life daughter of Ra^{226}, Rn^{222} (radon), in connection with problems in oceanography. The concentration of this inert gas in seawater well away from the sediment-sea and atmosphere-sea interfaces should be controlled by the amount of Ra^{226} dissolved in the water (i.e., the activity ratio $Rn^{222}/Ra^{226} = 1$). Thus a measurement of the Rn^{222} content of such a water sample is a measurement of its Ra^{226} content. As the partial pressure of Rn^{222} in the atmosphere over the oceans is negligible compared to that in the surface sea Rn^{222} continually escapes from the sea. From the vertical distribution of the Rn^{222} deficiency ($Rn^{222}/Ra^{226} < 1$) in the surface ocean it is possible to determine both the vertical mixing rate in these waters and the gas exchange rate at the air-sea interface. Pore water in deep sea sediments contains 10^4 to 10^5 times more Rn^{222} than the overlying seawater. Hence Rn^{222} diffuses from the sediments into the sea. The vertical distribution of the excess Rn^{222} ($Rn^{222}/Rn^{226} > 1$) in near bottom water allows estimates of the coefficient of vertical eddy diffusion in the deep sea to be made. Studies of this nature by shipboard measurements of the ratio Rn^{222} to Ra^{226} concentration have been carried out (Broecker, 1965b; Broecker et al., 1967).

TEH-LUNG KU

References

Broecker, W. S., 1965a, "Isotope Geochemistry and the Pleistocene Climatic Record," in (Wright, H., and Frey, D. G., editors), "Quaternary of the United States," Princeton Univ. Press, pp. 735–753.

Broecker, W. S., 1965b, "An Application of Natural Radon to Problems in Ocean Circulation," in (Ichiye, T., editor) "Symposium on Diffusion in Oceans and Fresh Waters," Palisades, N.Y., Lamont Geological Observatory, pp. 116–145.

Broecker, W. S., Li, Y. H., and Cromwell, J., 1967, "Ra^{226} and Rn^{222}: Concentration in Atlantic and Pacific Oceans," *Science,* **158**(3806), 1307–1310.

Chow, T. J., and Goldberg, E. D., 1960, "On the Marine Geochemistry of Barium," *Geochim. Cosmochim. Acta,* **20,** 192–198.

Faul, H. (editor), 1954, "Nuclear Geology," New York, John Wiley & Sons, 414pp.

Koczy, F. F., 1958, "Natural Radium as a Tracer in the Ocean," *Proc. Intern. Conf. Peaceful Uses At. Energy,* 2nd, Geneva, **18,** 351–357.

Koczy, F. F., and Bourret, R., 1958, "Radioactive Nuclides in Ocean Water and Sediments," Progress Report, The Marine Laboratory, Univ. of Miami.

Koczy, F. F., and Szabo, B. J., 1962, "Renewal time of bottom water in the Pacific and Indian Oceans," *J. Oceanogr., Soc. Japan,* 20th Anniv. Vol., pp. 590–599.

Ku, T. L., 1965, "An evaluation of the U^{234}/U^{238} method as a tool for dating pelagic sediments," *J. Geophys. Res.,* **70,** 3457–3474.

Moore, W. S., 1967, "Amazon and Mississippi River concentrations of uranium, thorium, and radium isotopes," *Earth and Planet. Sci. Letters*, **2**, 231–234.

Moore, W. S., and Sackett, W. M., 1964, "Uranium and thorium series in equilibrium in sea water," *J. Geophys. Res.*, **69**, 5401–5405.

Rankama, K., 1963, "Progress in Isotope Geology," New York, Interscience Publishers, 705pp.

Szabo, B. J., 1967, "Radium content in plankton and sea water in the Bahamas," *Geochim. Cosmochim. Acta*, **31**, 1321–1331.

Cross-references: *Geochronometry; Protactinium-Thorium Dating Method; Radioactive Isotopes; Radionuclides; Radium; Radon; River Geochemistry; Seawater, Chemistry; Thorium; Uranium. Vol. I: Radionuclides in the Oceans; Radionuclides: Their Application in Oceanography. Vol. VI: Pelagic Sediments.*

RADON: ELEMENT AND GEOCHEMISTRY

Physical Properties

Radon (from Radium), symbol Rn, is one of the rare or "noble" gases. Its atomic weight is about 222, and its atomic number is 86. Its electronic configuration [Xe core] is $4f^{14}5d^{10}6s^26p^6$. The melting point is $-71°C$ and the boiling point is $-62°C$. The density (gas) is 9.7 gm/liter, and specific gravity (of liquid at $-62°C$) is 4.4. In general, the name *radon* refers to Rn^{222}; other isotopes are further denoted by their common names or mass numbers (see Table 1; see also *Radium*).

The gas is readily absorbed on charcoal or silica gel and is fairly soluble in water. At its freezing point, radon exhibits a brilliant phosphorescence; however, the short half-life and intense radioactivity make it extremely difficult to study quantities of the gas or solid.

No minerals or compounds of radon are definitely known; however, the fact that xenon does form compounds would suggest that were enough radon gas available, compounds could be prepared and observed.

Geochemistry

In a closed system, on a time scale of greater than one month, the geochemistry of radon is entirely the geochemistry of its parent, radium. That is if there is no separation, a sample of Ra^{226} will produce an equilibrium amount of Rn^{222} in about five radon half-lives, or twenty days. Such radon is said to be radium-supported since it is in radioactive equilibrium with its parent.

Where there is a separation of radon from radium, radioactive equilibrium does not exist and several geochemical processes may be studied. Since the half-life of Rn^{222} is far greater than the half-lives of the other radon isotopes, it is the easiest to measure and hence the most useful for such studies. Being a gas, such a separation is often observed in nature. Cherdyntsev et al. have studied the emanation of radon from radium-bearing minerals and found that after heating and annealing these minerals, radon emanation may decrease by factors of ten or twenty. This is probably due to a partial restoration of the crystal lattice, and such studies may be useful in determining the thermal histories of mineral samples.

River and groundwaters are generally enriched in Rn^{222} over its parent Ra^{226}. This observation has been used to estimate an "age" for the water, i.e., the time elapsed since it was in contact with Ra^{226}-bearing rocks. Such ages are, at the best, only order of magnitude approximations, but they are useful in studying the movement of underground waters.

Similar "ages" may be applied to air masses. Most radon enters the atmosphere from the land surface, therefore the decrease in radon in a vertical profile should give the rate of air ascent. Both Rn^{222} and Rn^{220} have been applied to such studies. Since air over continents contains more radon than oceanic air, the source and movement of air masses near a land-sea boundary may be suggested from the radon content of the air.

Radon also has potential application to oceanic studies. Broecker (1964) has found large-scale discontinuities in radon concentration at air-sea and sea-sediment interfaces. In the surface ocean, measurable deficiencies of radon exist. The degree to which radon is depleted depends on the rate of gas exchange across the air-sea interface since the air has negligible radon compared to the sea. By measuring the degree to which radon is depleted in

TABLE 1. NATURALLY OCCURRING ISOTOPES

Mass	Common Name	Half-Life	Nature and Energy (MeV) of Decay
222	Radon	3.825 days	α, 5.48
220	Thoron	54.5 seconds	α, 6.28
219	Actinon	3.92 seconds	α, 6.82, 6.56, 6.43

the surface waters, the rate of gas exchanged and the depth of immediate mixing may be determined. Near the ocean bottom, radon is found in excess of that which is supported by Ra^{226} in the water. This is due to a leakage of radon from deep-sea deposits high in radium. Thus a sampling of very-near-bottom waters may help in answering questions on the vertical circulation of such waters.

To date, excesses or deficiencies of radon have been considered geochemical anomalies. Now that methods are available to sample and measure radon in the field, its potential as a geochemical tracer may be realized.

WILLARD S. MOORE

References

Broecker, W. S., 1964, "Symposium on Diffusion in Oceans and Fresh Water," Palisades, N.Y., Lamont Geological Observatory.

Cherdyntsev, V. V., 1961, "Abundance of the Chemical Elements," University of Chicago Press (transl).

Israel, H., Horbert, M., and Israel, G. W., 1966, "Results of continuous measurements of radon and its decay products in the lower atmosphere," *Tellus,* **18,** 638–641.

Jacobi, W., and Andre, K., 1963, "The vertical distribution of Rn^{222}, Rn^{220} and their decay products in the atmosphere," *J. Geophys. Res.,* **68,** p. 3799–3814.

Cross-references: *Radium; Rare Gases; Seawater, Chemistry.*

RAINWATER

Rainwater is meteoric water that is a dilute mixed electrolyte containing cations and anions derived from ocean surfaces, land surfaces, and man-made industries. In clouds, rainwater is in equilibrium with the CO_2 of the atmosphere and has a pH of 5.7. Analyses of rainwater show that the recorded pH range is from 3.0 to 9.8, the lowest in industrial areas, due to SO_2, H_2SO_4, and HCl. Both the pH and the composition of rainwater vary seasonally and geographically. A global average rainwater cannot be computed, but analyses from northern Europe (62 stations, 30 months) show average annual rainfall to be 560 mm; average pH 5.47; ions, in parts per million (ppm): Na, 2.05; K, 0.35; Ca, 1.42; Mg, 0.39; Cl, 3.47; SO_4, 2.19; NO_3, 0.27; NH_4, 0.41.

Rainwater adds important amounts of plant nutrient elements to the weathering surface of the earth and assists in maintaining microclimates. The elements added may accumulate as salt deposits in arid climates. The relation between the composition of rainwater and the exchangeable cations of clay minerals has not yet been assessed in detail.

Affect on Rock Weathering

There are now enough chemical analyses of rainwater from various parts of the world to permit an evaluation of its role in the chemical weathering of rocks and in soil development. An almost unexplored field in geochemistry is the interpretation of the effect of different types of rainwater on the composition of soil water, on the exchangeable cations of the soil clay minerals, and on the relation between exchangeable cations in soils, their addition by rain and from weathering rocks, and their redistribution by plants and animals. It is probable that the pH of rainwater, as well as the amount of rain and its contained cations and anions, has an important influence on the weathering of minerals and rocks in any locality.

Rainwater contains some constituents of local origin, and some that have been transported by winds from elsewhere. Even during rainless periods there is precipitation of mineral and organic dusts. Chemical constituents in both rain and dry precipitation are added continually to any area of the earth's crust to become part of the chemical weathering environment. Eriksson (1958, p. 177) calls this the *chemical climate.*

Chemical Composition

The major cations and anions of rainwater come from several sources. Salts are picked up by winds blowing across large stretches of open ocean. Rain deposited near a coast by onshore winds has a composition similar to that of diluted seawater. During winter rains (April to August 1952) at Perth, southwestern Australia, the following amounts of constituents were deposited by the rain, according to Turton (1953) (where *1* applies to rain water and *2* to seawater diluted to the same chloride content).

	Chemical Constituents (lb/acre)			
Chloride	Sodium	Magnesium	Calcium	Potassium
1 38	19.6	3.8	7.0	2.3
2 38	21	2.5	0.8	0.8

In rain deposited farther inland, the ratios $Cl^-:Na^+$, $K^+:Na^+$, and $SO_4^{2-}:Cl^-$, expressed as parts per million, differ from those in seawater (see Table 1). A decrease in chloride content of rainwater with distance from the coast has been observed in many countries.

Differences in composition are caused by several factors, principally by the amount of

TABLE 1. VARIATION IN COMPOSITION OF RAINWATER, IN EQUIVALENTS PER MILLION, WITH DISTANCE FROM COAST IN SOUTHEASTERN AUSTRALIA[a]

Distance from Coast (miles)	Average Annual Rainfall (mm)	Sodium Chloride (Average)	Sodium Calcium (Average)	Sodium Average	Sodium Range
1	800	0.9	6 8	13.8	10.1–21.1
20	828	1.0	7.8	5.5	3.9–6.4
65	617	1.0	3.9	2.5	1.8–3.2
120	447	1.5	1.4	.9	0.9–1.1
160	335	1.5	.9	1.1	0.7–1.4
200	264	1.5	.6	2.3	0.4–16.6

[a] After Hutton, 1958b.

rainfall, nearness to the ocean, and dust from the atmosphere (see Table 2).

The chloride content of rainwater shows a marked seasonal trend; and, as mentioned above, inland rain generally contains less chloride than coastal rain. In arid countries, however, the influence of cyclic salt is important. Hutton and Leslie (1958, p. 504) emphasize the fact that the surface of dry soil in southern Australia usually contains a quantity of sodium chloride amounting to about half the total soluble salts. They consider that it is not possible to use rainwater analyses to evaluate the amount of salt accumulating year by year from the amount received in rain. In saline inland areas—for example, Lake Eyre, central Australia (Bonython, 1958)—there is a recycling of salt from one land surface to another. The various salts in the basin are picked up as mineral particles by winds and deposited elsewhere.

Calcium, magnesium, and potassium ions in rainwater come both from oceanic salts and from land surfaces. Both calcium and magnesium increase with respect to sodium and chloride over land areas, and calcium increases more than magnesium. Calcareous dunes, as reported by Hingston (1958), are a greater source of calcium than seawater. Hutchinson (1957) points out that when sea spray evaporates, two kinds of solid particles form in the atmosphere. Calcium sulfate (gypsum) first crystallizes, and then later both sodium and magnesium chlorides crystallize. Thus sulfate and chloride become separated in the atmosphere. Fine-grained gypsum can also be readily picked up from playas. Salt rains containing sodium chloride crystals are not uncommon in the Salt Lake City–Ogden area, Utah (J. H. Feth, U.S. Geological Survey, written communication, 1960).

Sulfate ions in rainwater come from several sources. In industrial areas the principal source is the combustion of fuels containing sulfur. In other areas sulfate comes from shallow-water marine environments, particularly tidal estuaries and lagoons, from fresh-water lakes from salt flats, and from ocean waters. Some sediments contain pyrite that oxidizes slowly to SO_2. Sea salts supply some sulfate as gypsum. The amount of sulfate in rainwater varies greatly. In Uganda, Visser (1961) reports a range of sulfate from 0 to 68 ppm in samples of rainwater analyzed during one year. The source of this sulfate is Lake Victoria which is only a few miles from where the samples were collected. Rainwater on the Island of Hawaii analyzed by Eriksson (1957, p. 520) contained from 6.4 ppm sulfate near sea level to 0.8 ppm at 5550 feet above sea level. The interrelation of factors of height, prevailing winds, and volcanic emanations are probably responsible for these variations. Bermuda, the only other oceanic island for which figures are available, has rain with an average of 2.12 ppm sulfate (Junge and Werby, 1958, p. 422). The Australian rainwater samples are somewhat low in sulfate in comparison with amounts found elsewhere.

Nitrogenous compounds are always present in the atmosphere and are carried down in rainwater. Recent analyses record the nitrogen in three forms—NH_4, NO_3, and total nitrogen—depending on the method of analysis. Ammonia and nitrate are of terrestrial origin and are in particles of organic matter in dust and in soils. The amounts in rainwater are variable (see Table 3). The influence of vegetation on the nitrogen content of rainwater is shown by analyses of water from the open and from beneath trees. (It may be two to three times higher from beneath trees.)

Nitrification is a well known process in soil formation, and nitrogen is a necessary plant food. The nitrate ion is readily leached out of soils and weathering products. Ammonium can proxy for potassium in the mica structure, and becomes "fixed" in many soils; in a "fixed"

TABLE 2. CHEMICAL COMPOSITION OF RAINWATER FROM SEVERAL LOCALITIES IN THE CONTERMINOUS UNITED STATES[a]

Locality	Distance from Sea (miles)	Average Annual Rainfall (mm)	Constituents						
			Sodium (Na) (ppm)	Potassium (K) (ppm)	Calcium (Ca) (ppm)	Chloride (Cl) (ppm)	Sulfate (SO_4) (ppm)	Nitrate (NO_3) (ppm)	Ammonia (NH_4) (ppm)
Cape Hatteras, N.C.	0	1370	4.49	0.24	0.44	6.50	0.88	1.03	0.11
San Diego, Calif.	0	277	2.17	0.21	0.67	3.31	1.66	3.13	1.15
Brownsville, Tex.	1	635	22.30	1.00	6.50	21.96	5.34	1.76	0.28
Akron, Ohio	27[b]	889	0.10	0.10	0.69	0.17	1.62	4.68	0.38
Tallahassee, Fla.	37	1397	0.53	0.13	0.43	0.66	0.48	0.72	0.07
Greenville, N.C.	50	1194	0.18	0.07	0.31	0.13	0.57	2.97	0.14
Tacoma, Wash.	75	2032	14.50	0.59	0.73	22.58	1.69	0.99	0.05
Urbana, Ill.	85[b]	940	0.90	0.07	—	0.69	1.20	1.27	0.09
Washington, D.C.	85	1052	0.23	0.18	0.23	0.35	1.33	2.14	0.43
Fresno, Calif.	112	240	0.30	1.11	0.37	0.35	0.54	2.94	2.21
Indianapolis, Ind.	128[b]	995	0.26	0.12	0.69	0.18	4.00	2.06	0.27
Albany, N.Y.	150	914	0.21	0.09	0.43	0.23	0.10	4.05	0.21
Roanoke, Va.	200	1270	0.22	0.13	0.32	0.23	1.33	3.12	0.24
Ely, Nev.	410	381	0.69	0.14	3.79	0.30	1.05	0.81	0.35
Amarillo, Tex.	540	534	0.22	0.23	2.17	0.14	0.03	1.64	0.28
Glasgow, Mont.	625	380	0.40	0.26	1.72	0.17	1.30	1.82	0.75
Grand Junction, Colo.	650	226	0.69	0.17	3.41	0.28	2.37	2.63	0.33
Columbia, Mo.	650	1016	0.33	0.31	2.18	0.15	1.20	3.81	0.44

[a] After Junge and Werby, 1958. [b] Distance from fresh-water lake system.

chloride ions. There is a cyclic circulation of iodine between soils and air.

TABLE 3. AMOUNTS OF NITRATE AND AMMONIA IN RAINWATER IN PARTS OF THE U.S.

Locality	NO_3 (ppm)	NH_4 (ppm)
Cape Hatteras, N.C.	1.03	0.11
San Diego, Calif.	3.13	1.15
Brownsville, Texas	1.76	0.28
Akron, Ohio	4.68	0.38
Tallahassee, Fla.	0.72	0.07
Greenville, N.C.	2.97	0.14
Tacoma, Wash.	0.99	0.05
Urbana, Ill.	1.27	0.09
Washington, D.C.	2.14	0.43
Fresno, Calif.	2.94	2.21
Indianapolis, Ind.	2.06	0.27
Albany, N.Y.	4.05	0.21
Roanoke, Va.	3.12	0.24
Ely, Nev.	0.81	0.35
Amarillo, Texas	1.64	0.28
Glasgow, Mont.	1.82	0.75
Grand Junction, Colo.	2.63	0.33
Columbia, Mo.	3.81	0.44

pH of Rainwater

Water in clouds is assumed to be in equilibrium with the carbon dioxide of the atmosphere. The pH of rainwater in equilibrium with atmospheric carbon dioxide at 25°C is 5.7. Barrett and Brodin (1955) consider a pH of 5.7 as the neutral point for atmospheric water, not in a chemical sense, however, but as a reference point from which to discuss changes that may take place by the addition of cations and anions. Water of pH 5.7 is acid, and therefore rainwater in clouds is acid.

Rainwater having a pH below 5.7 contains gases or acid such as SO_2, H_2SO_4, or HCl. The bicarbonate ion content is extremely low.

Geochemical Implications of the Chemical Composition of Rainwater

In humid, well-drained regions, water selectively removes cations from weathering rocks and soils, but in areas of limited rainfall and stream activity the continued addition of water of definite chemical composition changes the composition of the soil water, which, under these circumstances, is only present during rainy periods. This effect is a result of the accumulation of sodium chloride deposited by rain.

The amounts of chloride in certain Western Australian soils from humid and arid areas illustrate this point (Teakle, 1929, 1937):

condition it is either not available to plants, or only very slowly available. The accumulation of nitrate deposits can occur only under very arid conditions.

Bromine, iodine, and boron also are constituents of seawater, and salts containing these elements can be expected to accompany other salts in rainwater derived from the oceans. Data on bromine and iodine are scanty. Hutchinson (1957) suggests that these constituents will show a seasonal variation similar to that of

RAINWATER

Locality	Distance from ocean (miles)	Annual rainfall (in.)	Sodium chloride deposited by rain (lb/acre/yr)	Chloride in soil solution (ppm dry soil)
Bakers Hill	50	45	25	40
Salmon Gums	100	12.5	27	570
Merredin	250	10.5	16	200

The cations of the soluble salts in saline soils occupy the exchange positions of the clay minerals and modify or prevent development of soil profiles, particularly with regard to the movement of clay. Sodium in exchange positions may cause a flocculated, jellylike mass of clay similar to a sodium bentonite when wet. Eluviation of sodium clay produces a columnar structure in a soil profile. This structure is due to the shrinking and swelling of the clay on change from a dry to a wet condition (Byers and others, 1938, pp. 976–977). Such soils are characteristically developed in saline areas and are known as *solonchaks*.

The effect of the cations in rainwater on the exchangeable cations of soil clay minerals and on chemical weathering is difficult to determine in the laboratory, but the overall effect is a regional pattern that has developed as a result of the chemical climate in which soil formation and weathering occurs.

The cations adsorbed by the exchange complex of soils, clay minerals, or minerals in weathering rocks are dependent on several factors—availability of different kinds of cations; their concentration and proportion in a soil solution or leaching water; the nature and number of exchange sites on the exchange com-

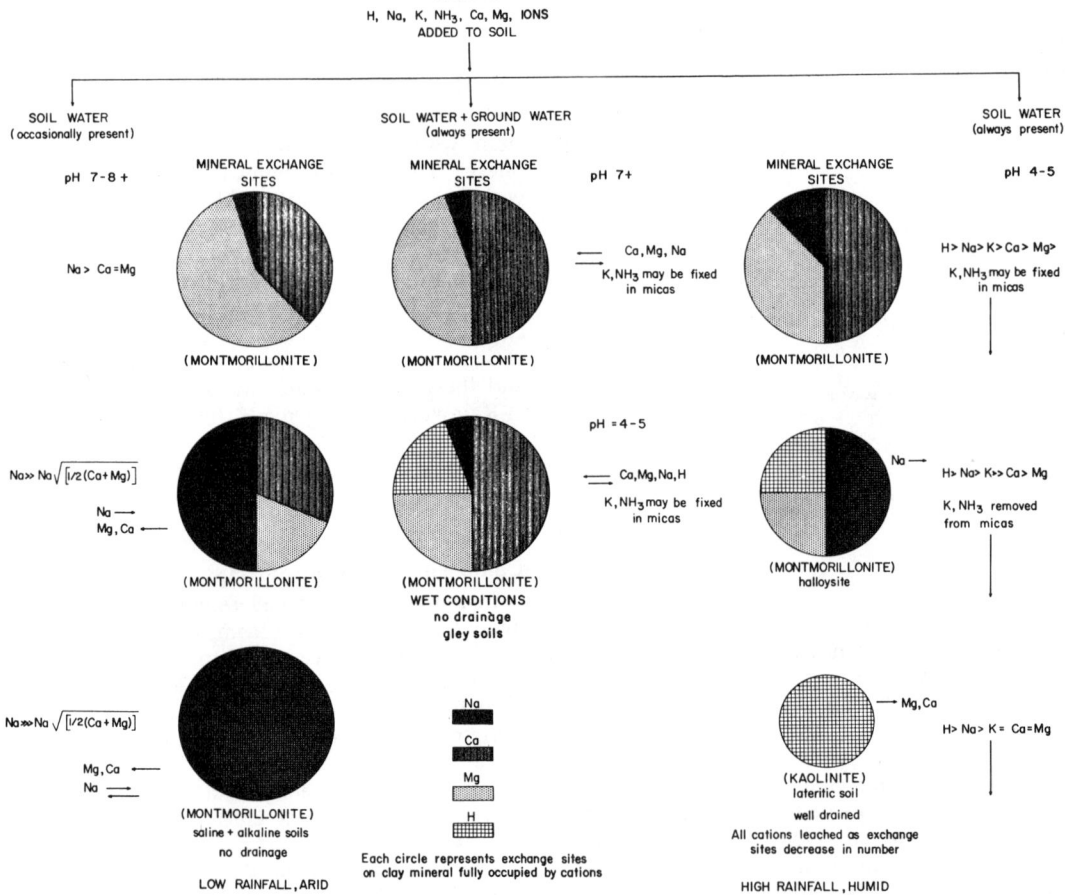

FIG. 1. Schematic representation of the relation of rainwater to soil formation and rock weathering (from Carroll, 1962).

plex; and the volume of water that is in contact with the exchange complex.

In Fig. 1 the accumulation of salts in the soil solution under arid conditions with no removal of cations added by rainwater is shown on the left. The exchange sites of montmorillonite originally contain calcium, magnesium, and a little sodium. As the soil water becomes saturated with sodium, the calcium and magnesium are gradually replaced by sodium until all the exchange sites are filled. The concentration of sodium in the soil solution, which is only intermittently present after showers, is much greater than that of calcium and magnesium. The clay mineral does not break down or alter.

In the center of Fig. 1, the relation for saturated soils (gleyed soils) at a pH of 7 and over and at a pH of 4–5 is shown. In both instances the cations in the exchange positions are in chemical equilibrium with those in the soil and groundwater present. This is a situation similar to that which one obtains in a laboratory experiment. Cations added by rainwater to the surrounding water of the clay mineral will establish equilibrium with those in the exchange positions, and little weathering or removal of cations can take place.

The leaching of cations from soil or weathering rock under medium to heavy rainfall in well-drained situations is shown on the right-hand side of Fig. 1. Here the cations added by rainwater have little influence. If the rainwater is alkaline, the leaching solution will not be as effective as if it were acid. The net effect is to remove cations from the exchange sites of the clay minerals, and to make original montmorillonite unstable, thereby changing it first to halloysite and later to kaolinite. The reduction in size of the circles indicates that the exchange capacity of the minerals formed is less than that of the original montmorillonite, that is, the number of exchange sites is reduced. These exchange sites are, in kaolinite, not present in sufficient numbers to retain all the cations being removed by leaching rainwater, and H^+ ions occupy them. This is the situation in lateritic weathering in which the soil minerals have an acid reaction.

No allowance has been made on Fig. 1 for the very considerable effects of the presence of organic matter either in the production of acids or in the complexing of Ca^{2+}, Mg^{2+}, and other cations by organic solutions.

The role of potassium, and to a lesser extent of ammonia, seems to be more complex than those of Na^+, Ca^{2+}, or Mg^{2+}. Potassium ions may react somewhat like Na^+, but K^+ ions may be largely withdrawn from circulation by fixation in micaceous clay minerals. The K^+ ion enters the interlayer positions of these minerals, thereby partially or wholly reconstituting them to true micas and preserving them in a relatively unweathered condition by maintaining their original uncharged state. The abundance of "illite" in soils and sedimentary rocks is due to its stability.

Dorothy Carroll (deceased)

(The above entry is based largely upon the article by the author, 1962, cited below.)

References

Alexander, F. E. S., 1959, "Observations on tropical weathering—a study of the movement of iron, aluminum, and silicon in weathering rocks in Singapore," *Geol. Soc. London Quart. J.,* **115,** 123–144.

Barrett, E., and Brodin, C., 1955, "The acidity of Scandinavian precipitation," *Tellus,* **7,** 251–257.

Bonython, C. W., 1058, "The salt of Lake Eyre—its occurrence in Madigan Gulf and its possible origin," *Royal Soc. S. Australia Trans.,* **79,** 66–92.

Bower, C. A., 1959, "Cation exchange equilibria in soils affected by sodium salts," *Soil Sci.,* **88,** 32–35.

Byers, H. G., Kellogg, C. E., Anderson, M. S., and Thorp, J., 1938, "Formation of soil," in "Soils and Men," *U.S. Dept. Agr. Yrbk.,* pp. 948–992.

Carroll, D., 1962, "Rainwater as a chemical agent of geologic processes—A review," *U.S. Geol. Surv. Water-Supply Paper,* **1535-G,** 18pp.

Eriksson, E., 1957, "The chemical composition of Hawaiian rainfall," *Tellus,* **9,** 509–520.

Eriksson, E., 1958, "The chemical climate and saline soils in the arid zone," *Arid Zone Res.,* **10,** UNESCO, 147–180.

Eriksson, E., 1959, "The yearly circulation of chloride and sulfur in nature; meteorological, geochemical and pedological implication," Pt. 1, *Tellus,* **11,** 375–403; Pt. 2, *ibid.,* **12,** 63–109.

Gorham, E., 1961, "Factors in influencing supply of major ions to inland waters, with special reference to the atmosphere," *Geol. Soc. Am. Bull.,* **72,** 795–840.

Hingston, F. J., 1958, "The major ions in Western Australian rainwaters," *Comm. Sci. Indus. Res. Org.* (Australia), *Div. Soils, Rept.* **1/58.**

Hutchinson, G. E., 1957, "A Treatise on Limnology," Vol. 1, New York, John Wiley & Sons, 1015pp.

Hutton, J. T., and Leslie, T. I., 1958, "Accession of non-nitrogenous ions dissolved in rainwater to soils in Victoria," *Australian J. Agr. Res.,* **9,** 492–507.

Ingham, G., 1950, "Effect of materials absorbed from the atmosphere in maintaining soil fertility," *Soil Sci.,* **70,** 202–212.

Junge, C. E., and Werby, R. T., 1958, 'The concentration of chloride, sodium, potassium, calcium and sulfate in rainwater over the United States," *J. Meteorol.,* **15,** 417–425.

Nash, V. E., and Marshall, C. E., 1956, "The surface reactions of silicate minerals, Pt. 2,

Reactions of feldspar surfaces with salt solutions," *Bull. Univ. Missouri Coll. Agri. Res.,* **614**, 36pp.

Teakle, L. J. H., 1929, "The water extracts of Western Australian soils No. 1, Studies on soils from Merredin, Ghooli, Salmon Gums, Wongan Hills, Chapman, Baker's Hill and Lake Brown," *J. Royal Soc. W. Australia,* **15**, 115–123.

Teakle, L. J. H., 1937, "The salt (sodium chloride) content of rainwater," *J. Dept. Agri. W. Australia,* **14**, 115–123.

Turton, A. G., 1953, "Atmospheric accessions," *Australian Conf. Soil Sci., Adelaide, Comm. Sci. Indus. Res. Org., Div. Soils,* **2**, 6.14.1–6.14.4.

Visser, S., 1961, "Chemical composition of rainwater in Kampala, Uganda, and its relation to meteorological and topographical conditions," *J. Geophys. Res.,* **66**, 3759–3766.

Cross-references: *Cyclic Salts; Groundwater; Ion Exchange; Meteoric Water; Pedology; pH-Eh; Seawater, Chemistry; Weathering—Chemical.* See also *Individual Elements.* Vol. IVB: *Clays and Clay Minerals.* Vol. VI: *Leaching; Soil Genesis.*

RARE EARTHS (LANTHANIDE SERIES)

The rare earth group of elements comprises the fifteen elements from lanthanum ($Z = 57$) to lutetium ($Z = 71$). These are also known as the *Inner Transition Elements,* the *Lanthanide Series,* or the *Lanthanons.* Yttrium ($Z = 39$) is usually included by geochemists in the group both because it is a member of Group III A of the periodic table, and because it is associated with the heavier rare earths (Gd–Lu) which are commonly known as the yttrium earths. Scandium ($Z = 21$), the other Group III A element, is not commonly included, since its geochemical behavior is in contrast to the rare earth group proper (Frondel 1968) mainly on account of its smaller ionic radius.

A list of the elements, with atomic number, atomic weight (1959), year of discovery, and origin of the element name is given in Table 1.

Nuclides

A list of naturally occurring rare earth nuclides is given in Table 2. The relative abundance of the isotopes of each element is given. $_{59}$Pr, $_{65}$Tb, $_{67}$Ho, and $_{69}$Tm are monoisotopic, $_{57}$La and $_{71}$Lu have one dominant isotope, and only $_{63}$Eu among the rare earths of odd atomic number has two abundant isotopes. $_{61}$Pm has no stable isotopes and the half-life of the longest-lived isotope is 2.26 years. Sm^{149}, Eu^{151}, Eu^{153}, Gd^{155}, Gd^{157}, and Yb^{168} have exceptionally large thermal neutron capture cross sections and are thus sensitive indicators of the presence of a strong neutron flux. The similarity in the ratios of these nuclides to other rare earth nuclides in terrestrial and meteoritic samples provides no evidence for exposure of terrestrial and meteoritic parent material to a different neutron flux.

Naturally Occurring Radioactive Nuclides. La^{138}, Ce^{142}, Sm^{147}, Sm^{148}, Sm^{149}, Gd^{152}, and Lu^{176} are unstable and decay with very long half-lives (Table 2) ranging from 3.6×10^{10} to 5×10^{15} years. For reasons such as long half-life, low abundance of parent or daughter product, swamping of daughter product with a common nuclide, etc., these decays have not yet been utilized for determinations of geological age.

Origin of Rare Earth Nuclides. Currently accepted theories, principally due to E. M. Bur-

TABLE 1. RARE EARTH ELEMENTS, WITH SCANDIUM AND YTTRIUM

Element	Symbol	Atomic Number	Atomic Weight (1959)	Year Discovered	Origin of Name
Scandium	Sc	21	44.96	1879	Scandinavia
Yttrium	Y	39	88.91	1843	Ytterby, Sweden
Lanthanum	La	57	138.92	1839	Latin—to lie hidden
Cerium	Ce	58	140.13	1803	Ceres (asteroid)
Praesodymium	Pr	59	140.92	1885	Greek—green twin
Neodymium	Nd	60	144.27	1885	Greek—new twin
Prometheum	Pm	61	(147)	1947	Prometheus
Samarium	Sm	62	150.35	1879	Samarskite
Europium	Eu	63	152.0	1896	Europe
Gadolinium	Gd	64	157.26	1880	Finnish chemist Gadolin
Terbium	Tb	65	158.93	1843	Ytterby, Sweden
Dysprosium	Dy	66	162.51	1886	Greek—difficult of access
Holmium	Ho	67	164.94	1879	Holmia = Stockholm
Erbium	Er	68	167.27	1843	Ytterby, Sweden
Thulium	Tm	69	168.94	1879	Thule
Ytterbium	Yb	70	173.04	1878	Ytterby, Sweden
Lutetium	Lu	71	174.99	1907	Lutetia = Paris

Table 2. Rare Earth Nuclides

Atomic Number	Nuclide	% Natural Abundance	Half-Life (in years)	Process of Origin[a]
21	Sc^{45}	100	—	s
39	Y^{89}	100	—	s
57	La^{138}	0.089	1.1×10^{11}	p
	La^{139}	99.911	—	s
58	Ce^{136}	0.193	—	p
	Ce^{138}	0.250	—	p
	Ce^{140}	88.48	—	s
	Ce^{142}	11.07	5×10^{15}	r
59	Pr^{141}	100	—	s
60	Nd^{142}	27.11	—	s
	Nd^{143}	12.17	—	s
	Nd^{144}	23.85	5×10^{15}	s
	Nd^{145}	8.30	—	s
	Nd^{146}	17.22	—	s
	Nd^{148}	5.73	—	r
	Nd^{150}	5.62	—	r
62	Sm^{144}	3.09	—	p
	Sm^{147}	14.97	1.06×10^{11}	rs
	Sm^{148}	11.24	1.2×10^{13}	s
	Sm^{149}	13.83	4×10^{14}	rs
	Sm^{150}	7.44	—	s
	Sm^{152}	26.72	—	rs
	Sm^{154}	22.71	—	r
63	Eu^{151}	47.82	—	rs
	Eu^{153}	52.18	—	rs
64	Gd^{152}	.20	1.1×10^{14}	ps
	Gd^{154}	2.15	—	s
	Gd^{155}	14.73	—	r
	Gd^{156}	20.47	—	r
	Gd^{157}	15.68	—	r
	Gd^{158}	24.87	—	r
	Gd^{160}	21.90	—	r
65	Tb^{159}	100	—	r
66	Dy^{156}	.052	—	p
	Dy^{158}	.090	—	p
	Dy^{160}	2.29	—	s
	Dy^{161}	18.88	—	r
	Dy^{162}	25.53	—	r
	Dy^{163}	24.97	—	r
	Dy^{164}	28.18	—	r
67	Ho^{165}	100	—	r
68	Er^{162}	.136	—	p
	Er^{164}	1.56	—	p
	Er^{166}	33.41	—	r
	Er^{167}	22.94	—	r
	Er^{168}	27.07	—	r
	Er^{170}	14.88	—	r
69	Tm^{169}	100	—	r
70	Yb^{168}	.135	—	p
	Yb^{170}	3.03	—	s
	Yb^{171}	14.31	—	sr
	Yb^{172}	21.82	—	sr
	Yb^{173}	16.13	—	sr
	Yb^{174}	31.84	—	sr
	Yb^{176}	12.73	—	r
71	Lu^{175}	97.41	—	sr
	Lu^{176}	2.59	3.6×10^{10}	p

[a] E. M. Burbidge et al., 1957, *Rev. Mod. Phys.* **29**, 547.

RARE EARTHS (LANTHANIDE SERIES)

TABLE 3. ELECTRONIC CONFIGURATION OF THE RARE EARTHS

	Z	1s	2s	2p	3s	3p	3d	4s	4p	4d	4f	5s	5p	5d	6s
Y	39	2	2	6	2	6	10	2	6	1		2			
La	57	2	2	6	2	6	10	2	6	10		2	6	1	2
Ce	58	2	2	6	2	6	10	2	6	10	1	2	6	1	2
Pr	59	2	2	6	2	6	10	2	6	10	3	2	6		2
Nd	60	2	2	6	2	6	10	2	6	10	4	2	6		2
Pm	61	2	2	6	2	6	10	2	6	10	5	2	6		2
Sm	62	2	2	6	2	6	10	2	6	10	6	2	6		2
Eu	63	2	2	6	2	6	10	2	6	10	7	2	6		2
Gd	64	2	2	6	2	6	10	2	6	10	7	2	6	1	2
Tb	65	2	2	6	2	6	10	2	6	10	9	2	6		2
Dy	66	2	2	6	2	6	10	2	6	10	10	2	6		2
Ho	67	2	2	6	2	6	10	2	6	10	11	2	6		2
Er	68	2	2	6	2	6	10	2	6	10	12	2	6		2
Tm	69	2	2	6	2	6	10	2	6	10	13	2	6		2
Yb	70	2	2	6	2	6	10	2	6	10	14	2	6		2
Lu	71	2	2	6	2	6	10	2	6	10	14	2	6	1	2

bidge et al. (1957) and Cameron (1959) relate the origin of the elements to continuous synthesis during stellar evolution. The rare earth elements are considered to be produced by three processes, s, r, and p. s-Process nuclides are produced by neutron capture on a slow time scale with 100 to 10^5 years between each neutron capture. This process occurs during the red giant stage of stellar evolution. r-Process nuclides are produced by neutron capture on a fast time scale (0.01–10 sec) with intervening β decays. This process is considered to occur in Type I supernovae. The p process, responsible for the origin of a number of proton-rich isotopes of low abundance, is thought to occur in Type II supernovae.

The process assigned for the origin of each nuclide by Burbidge et al., is given in Table 2. The most significant geochemical conclusion is that the rare earth nuclides are produced mainly by the s and r processes. These are independent processes and material from outside the solar system will probably possess different ratios of s- and r-produced nuclides. From Table 2 it can be seen that if minor contributions from the p-process nuclides are neglected, the isotopes of $_{21}Sc$, $_{39}Y$, $_{57}La$, $_{58}Ce$, $_{59}Pr$, and $_{60}Nd$ are dominantly produced by the s proc-

TABLE 4. TRIVALENT IONIC RADII[a] (r), IONIC POTENTIAL (i), IONIZATION POTENTIALS[b] (I^1, ETC.), AND ELECTRONEGATIVITIES[c] (e) OF THE RARE EARTHS, SCANDIUM, AND YTTRIUM

		r^{3+} (1)	r^{3+} (2)	i	I^1	I^2	I^3	e
39	Y	0.92	—	3.26	6.38	12.24	20.52	1.2
57	La	1.14	1.06	2.63	5.58	11.06	19.18	1.1
58	Ce	1.07	1.03	2.80	5.47	10.85	20.20	1.1
59	Pr	1.06	1.01	2.83	5.42	10.55	21.62	1.1
60	Nd	1.04	1.00	2.88	5.49	10.72	—	1.2
61	Pm	—	—	—	5.55	10.90	—	—
62	Sm	1.00	0.96	3.00	5.63	11.07	—	1.1
63	Eu	0.98	0.95	3.06	5.67	11.25	—	1.2
64	Gd	0.97	0.94	3.09	6.14	12.1	—	1.2
65	Tb	0.93	0.92	3.23	5.85	11.52	—	1.2
66	Dy	0.92	0.91	3.26	5.93	11.67	—	1.2
67	Ho	0.91	0.89	3.30	6.02	11.80	—	1.2
68	Er	0.89	0.88	3.37	6.10	11.93	—	1.2
69	Tm	0.87	0.87	3.45	6.18	12.05	23.71	1.1
70	Yb	0.86	0.86	3.49	6.25	12.17	25.2	1.2
71	Lu	0.85	0.85	3.53	5.43	13.9	—	—

[a] Ionic radii (1) from L. H. Ahrens (1952); (2) from Templeton and Dauben (1954).
[b] Ionization potentials from C. E. Moore, 1970, U.S. Natl. Bur. Stds. Circ. 34.
[c] Electronegativity values from W. Gordy and W. J. Thomas, 1956, J. Chem. Phys., 24, 439.

ess, those of $_{62}$Sm and $_{63}$Eu are of mixed origin, those of $_{64}$Gd, $_{65}$Tb, $_{66}$Dy, $_{67}$Ho, $_{68}$Er, and $_{69}$Tm are dominantly produced by the r process, and the isotopes of $_{70}$Yb and $_{71}$Lu are of mixed origin. Thus the element ratios of Sc, Y, La, Ce, Pr, and Nd to Gd, Tb, Dy, Ho, Er, and Tm are sensitive to processes of element formation.

Electronic Structure. The electronic configuration of the rare earths is given in Table 3, adapted from Topp (1965). The 4f shell is half filled at europium and filled at ytterbium. The stable trivalent ions result from losing the $6s^2$ electrons and usually one 4f electron.

Oxidation States. The outstanding feature of the geochemistry of the rare earths (and Y) is their oxidation states of +3. La^{3+}, Gd^{3+}, and Lu^{3+} are particularly stable. The only other oxidation states important in geological processes are Ce^{4+} and Eu^{2+}, although other valency states have been reported. Sm^{2+} and Yb^{2+} are known to exist in solution (see Topp, 1965, p. 69, Table 14).

Ionic Radii and Lanthanide Contraction. The trivalent ionic radii are given in Table 4. A regular decrease in radius with increase of atomic number is observed from La^{3+} (r = 1.14A) to Lu^{3+} (r = 0.85A). This is the well-known *lanthanide contraction*, which results from the filling of the 4f subshell. The addition of electrons to an inner shell and the increasing nuclear charge leads to the progressive reduction in ionic radius. The geochemical effect of the lanthanide contraction is to cause the heavier rare earths to be similar in ionic radius to much lighter elements (e.g., Y). This results in a geochemical coherence between $_{39}$Y and the rare earths from $_{64}$Gd to $_{71}$Lu, and so called "yttrium earths." The large rare earth cations $_{57}$La to $_{63}$Eu are commonly referred to as the "cerium earths."

The effect of the lanthanide contraction extends to elements such as $_{72}$Hf, $_{73}$Ta, etc., and is responsible for their similarity in ionic radius to $_{40}$Zr, $_{41}$Nb, etc., leading to the observed geochemical coherence of Zr^{4+}–Hf^{4+} and Nb^{5+}–Ta^{5+}.

The radii of other rare earth ions are Ce^{4+}, 0.94 Å; Eu^{2+}, 1.24 Å; Sm^{2+}, 1.30 Å; and Yb^{2+}, 1.06 Å.

Ionic Potential. The values for the function i, ionic charge (or valency)/ionic radius, are given in Table 4 for the trivalent ions using the Ahren's radii. These values indicate that the rare earths exist in magmas mainly as free ions. In the sedimentary environment, the rare earths enter the hydrolysate minerals (clays).

Ionization Potential. Values are given in Table 4. The data are very sparse, with incomplete values for the third I.P.

Electronegativity. Values are given in Table 4. The rare earth elements show few significant differences.

Abundances

A basic observation is that the rare earths of even atomic number are more abundant than their neighbors of odd atomic number. In Fig. 1, the chondritic abundances are plotted to show this effect.

Universe. Mainly through the production of s-process nuclides in red giants, r-process nuclides in supernovae, and the dispersion of this material in the interstellar medium, the abundance of the rare earth elements in particular galaxies increases with time so that younger stars contain a higher proportion than older stars (e.g., Population II). Compared with the solar system, high abundances of Y, La, Ce, Pr, and Nd are observed in s-type stars. Barium stars also show "anomalous" concentrations of these elements. These enhanced abundances are attributed to formation by the s process, during the red giant stage. The Peculiar A stars (apparently young Population I) contain increased concentrations of La, Ce, Pr, Nd, Sm, Eu, Gd, and Dy. These over abundances are associated with strong surface magnetic fields by Burbidge et al. (1957), Cameron (1959), and Aller (1961).

Sun. Values for solar abundances from Wallerstein (1966) and Grevesse and Blanquet (in press) are given in Table 5, expressed as log numbers of atoms relative to H = 10^{12} atoms (Log N_H = 12.00). There are considerable discrepancies between these values

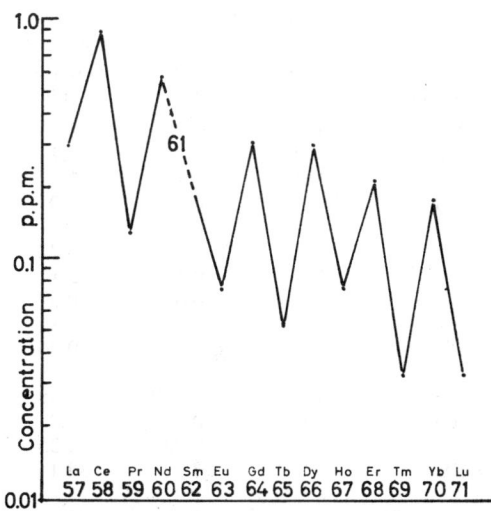

FIG. 1. Abundance, in parts per million, of the rare earths in chondritic meteorites, illustrating the alternation in the abundances of the even-Z and odd-Z elements

RARE EARTHS (LANTHANIDE SERIES)

TABLE 5. RARE EARTH SOLAR ABUNDANCES (ATOMIC) [ALL VALUES AS LOG N, RELATIVE TO H = 12.00 (10^{12} ATOMS H)]

		Sun[a]		Chondrites[b]
		(1)	(2)	
57	La	1.92	1.81	1.11
58	Ce	1.53	1.88	1.62
59	Pr	1.37	1.88	0.78
60	Nd	1.78	1.82	1.44
62	Sm	1.27	1.66	0.91
63	Eu	0.97	0.49	0.51
64	Gd	0.99	1.12	1.15
65	Tb	—	—	0.27
66	Dy	—	1.11	1.11
67	Ho	—	—	0.50
68	Er	—	0.76	0.87
69	Tm	—	0.43	0.09
70	Yb	—	0.81	0.87
71	Lu	—	0.84	0.09

[a] (1) Wallerstein, 1966.
(2) Grevesse and Blanquet (in press).
[b] Chondritic meteorites.

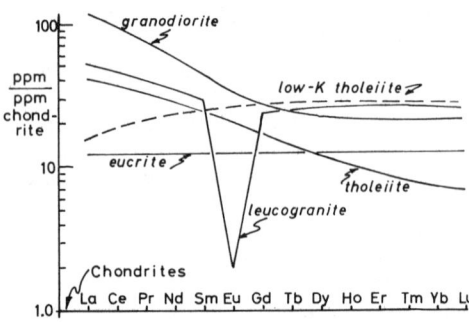

FIG. 2. Rare earth abundance patterns for igneous rocks and meteorites, plotted relative to chondrites. Note strong depletion in Eu in the leucogranite (data from Tables 4 and 5).

for the solar atmosphere and those for the meteorites. In particular, the lighter rare earths (notably La, Pr, Sm) appear to be enriched in the sun relative to chondritic meteories. Much of this difference may be due to analytical uncertainty in the solar values. However, if they are confirmed by future work, then it will be invalid to continue to use the relative meteoritic rare earth abundances as representative of the solar system nebula abundances.

TABLE 6. RARE EARTH METEORITE ABUNDANCES[a]

		Chondrites		Achondrites (Eucrites) (ppm)
		Atomic ($Si = 10^6$ atoms)	Wt (ppm)	
57	La	0.38	0.30	3.7
58	Ce	1.05	0.84	9.7
59	Pr	0.15	0.12	1.4
60	Nd	0.71	0.58	7.0
62	Sm	0.24	0.21	2.3
63	Eu	0.085	0.074	0.72
64	Gd	0.36	0.32	3.1
65	Tb	0.054	0.049	0.57
66	Dy	0.33	0.31	3.8
67	Ho	0.077	0.073	0.80
68	Er	0.22	0.21	2.3
69	Tm	0.033	0.033	0.38
70	Yb	0.17	0.17	2.0
71	Lu	0.031	0.031	0.35
Σ REE		—	3.32	38.1
39	Y	3.5	1.8	22
Σ REE + Y		—	5.12	60
La/Ce		0.36	0.36	0.38
Eu/Gd		0.24	0.24	0.23

[a] Data from Haskin et al., 1966, Table 8.

Meteorites. Abundances in *chondrites* are given in Table 6, expressed both in parts per million by weight (ppm) and as numbers of atoms relative to $Si = 10^6$ atoms. The rare earths show no real variation among the various classes of chondrites, outside of ± 15–20% variations usually ascribed to analytical error. The similarity in the relative abundances has led to the widespread use of the chondritic values as a standard against which to compare abundances in terrestrial materials.

Abundances in one class of the Ca-rich *achondrites (eucrites)* are also given. The absolute abundances are about twelve times those of the chondrites, but the relative abundance patterns are parallel (Fig. 2). Interest attaches to these eucritic meteorites since their bulk composition resembles that of the Surveyor analyses of the lunar maria surface.

Terrestrial Abundances

Abundances of the rare earth elements in common igneous rocks, sedimentary rocks and tektites are given in Tables 6 and 7 and are discussed in succeeding sections.

Presentation of Rare Earth Data

A major problem in comparing rare earth abundance data is to overcome the alternation in even Z–odd Z element abundances (Fig. 1) (the *Oddo-Harkins rule*, due to the increased stability of nuclei with even numbers of protons and neutrons). Plotting of raw abundance data produces a zig-zag effect which complicates comparisons. Many schemes have been tried (e.g., separate plotting of even-Z and odd-Z elements). The most useful method is to compare data relative to the chondritic or sedimentary rock patterns, dividing the abundances element by element, by the corresponding meteoritic or sedimentary rock abundances. This procedure, illustrated in Figs. 2, 3 and 6, was first suggested by Coryell, Chase, and

RARE EARTHS (LANTHANIDE SERIES)

TABLE 7. RARE EARTH IGNEOUS ROCK ABUNDANCES (ppm—wt)

		Perido-tite[a]	Low-K Tholeiite[a]	Hawaiian Tholeiite[a]	Calc-alkaline Andesite[b]	Grano-diorite[b]	Leuco-granite[b]	Nepheline Syenite[a]
57	La	0.005	4.7	10.5	11.9	36	17	360
58	Ce	0.006	16	35	24	47	31.5	670
59	Pr	0.001	2.4	3.9	3.2	8.5	5.2	74
60	Nd	0.042	14	18	13	26	17	270
62	Sm	0.033	4.9	4.2	2.9	6.8	11.2	48
63	Eu	0.02	1.9	1.3	1.0	1.2	0.17	10
64	Gd	0.07	7.0	4.7	3.3	7.4	6.8	42
65	Tb	0.08	1.37	0.66	0.68	1.3	1.6	9.3
66	Dy	—	5	3	—	—	—	40
67	Ho	0.05	1.7	0.64	0.71	1.6	2.4	8.8
68	Er	0.17	4.9	1.7	2.1	4.8	5.9	21
69	Tm	0.03	0.8	0.21	0.30	0.5	0.66	3.5
70	Yb	0.27	3.9	1.11	1.9	3.6	5.4	18
71	Lu	0.02	0.7	0.2	—	—	—	3.4
Σ REE		0.79	69	85	65	145	105	1578
39	Y	1.21	43	26	21	30	46	210
Σ REE + Y		2	112	111	86	175	151	1788
La/Ce		0.80	0.29	0.30	0.50	0.77	0.54	0.53
Eu/Gd		0.28	0.27	0.28	0.30	0.16	0.03	0.24

[a] Haskin et al. (1966).
[b] Taylor (unpublished data).

Winchester (1963) and has the following advantages: *(1)* It removes the even-Z odd-Z variations. *(2)* It enables direct comparison both of relative patterns and concentrations, compared to the standard.

When mineral analyses are available, ratios of the values for the elements to the total rock analysis can be computed (Figs. 4 and 5).

Normalization to one element (e.g., La, or Yb) was popular at one time but this procedure masks important information on the

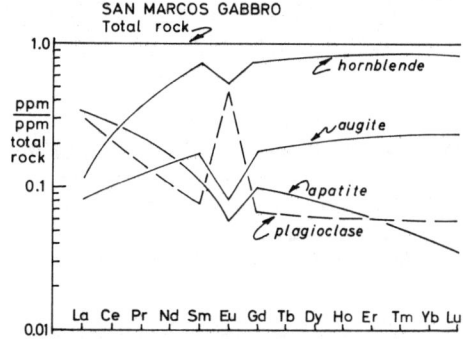

FIG. 4. Rare earth distribution in minerals of the San Marcos gabbro, relative to the total rock abundances (data from Towell et al., 1965).

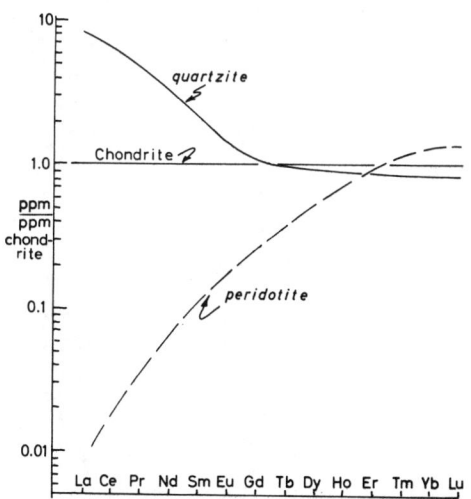

FIG. 3. Rare earth abuandance patterns, relative to chondrites in peridotite and quartzite (data from Haskin et al., 1966).

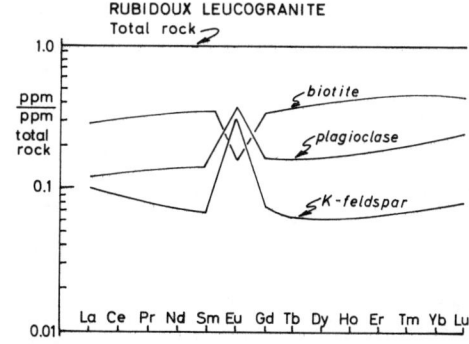

FIG. 5. Rare earth distribution in the minerals of the Rubidoux Mountain leucogranite (data from Towell et al., 1965).

RARE EARTHS (LANTHANIDE SERIES)

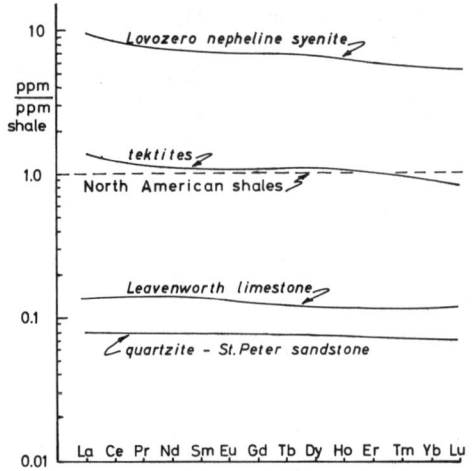

FIG. 6. Rare earth abundance patterns, relative to shales for nepheline syenite, tektites, limestones, and quartzites (data from Tables 5 and 6).

relative abundances, and places an undue weight on the reliability data for the chosen element.

Abundances may be plotted against ionic radii or atomic number (Z): the order is the same.

Relative Abundance Patterns

One of the most significant geochemical results of recent rare earth research has been the confirmation of fractionation of the individual elements during formation of igneous rocks. Separation of the elements during pegmatite formation has long been known, but it was previously thought that the chemical difficulty in separating the rare earths was such as to cause them to behave as a coherent group in natural processes less extreme than those existing during pegmatite formation. Earlier work by Noddack on meteorites, and Minami on shales, had revealed differing patterns but the belief that such fractionation was unlikely had led to a general rejection of Noddack's data. The work of Schmitt et al. (1963) on chondrites and Haskin and Gehl (1963) on sediments confirmed the earlier relative abundances.

Two basic abundance patterns have been established: (1) The chondritic pattern, similar for all classes of chondrites. Data are given in Table 6 and are plotted in Fig. 1; (2) The sedimentary rock pattern, characteristic of most types of sedimentary rocks (Table 8). Compared to the chondritic pattern, there is a relative depletion of the heavy rare earths, or enrichment of the lighter ones.

Igneous rocks do not possess unique patterns, but in general those for basic rocks lie between the chondritic and sedimentary patterns, and the granitic rocks show increased depletion in the heavier rare earths, compared to the sedimentary rock abundances, although there is great variation in detail. The pattern for tektites closely resembles that for sedimentary rocks.

During the formation of igneous rocks and pegmatites rare earth-rich and rare earth minerals form and exhibit a wide variety of abun-

TABLE 8. SEDIMENTARY ROCKS AND TEKTITES (ppm—wt)

		Shale[a]	Quartzite[a]	Limestone[a]	Sub-greywacke[b]	Tektite[b]
57	La	39	2.5	3.8	50	50
58	Ce	76	—	10	75	76
59	Pr	10.3	0.56	1.3	12	9
60	Nd	37	2.3	3.6	32	28
62	Sm	7.0	—	0.64	5.3	4.7
63	Eu	2.0	0.09	0.14	1.3	1.3
64	Gd	6.1	0.43	0.79	5.2	4.7
65	Tb	1.3	—	0.084	1.1	0.86
66	Dy	—	—	—	3.9	4.2
67	Ho	1.4	0.05	0.26	1.3	1.1
68	Er	4.0	0.20	0.38	2.8	2.2
69	Tm	0.58	—	0.063	0.5	0.35
70	Yb	3.4	0.28	0.40	2.6	1.9
71	Lu	0.6	0.03	0.07	—	0.36
Σ REE		189	6.2	21.5	193	189
39	Y	35	1.8	4.5	30	30
Σ REE + Y		224	8	26	223	219
La/Ce		0.51	—	0.38	0.66	0.65
Eu/Gd		0.33	0.21	0.17	0.25	0.28

[a] Haskin et al., (1966).
[b] Taylor (unpublished data).

dance patterns. These were originally studied by Goldschmidt and Thomassen. Later workers include Butler (1957, 1960), Murata et al, (1957), and Semenov and Barinskii (1958). A number of different types may be distinguished:

1. Apatite type: The lighter elements (La–Nd) predominate.
2. Yttrofluorite type: Intermediate and heavy elements predominate.
3. Monazite type: Mainly La–Sm.
4. Allanite type: Mainly La–Nd.
5. Thalenite type: Maximum at Dy.
6. Thortvetite type: High abundance of Yb and Lu.
7. Xenotime type: Predominance of Er and Yb.
8. Wiikite type: Yttrium earths predominate over cerium earths.
9. Davidite type: Depletion of intermediate members.

Behavior of Rare Earths in Igneous Processes

The ionic radii of the trivalent cations (1.14–0.85 Å) restrict substitution among the major cations to Ca^{2+} (0.99 Å). Less abundant cations with suitable properties are Zr^{4+} (0.79 Å) U^{4+} (0.97 Å), and Th^{4+} (1.02 Å).

Although the classical capture principle of Goldschmidt might be expected to apply to the substitution for Ca^{2+} resulting in preferential entry of the rare earths, the RE–O bonds are more covalent than the Ca–O bonds. This factor, coupled with difficulty in charge balance for trivalent ions in eight-fold coordination sites in minerals, tends to offset the advantage of the extra formal charge and leads to the concentration of the rare earths in later crystal fractions and residual melts.

Among the rare earths themselves, the effect of the lanthanide contraction is responsible for the preferential entry of the smaller cations (e.g., Gd^{3+}–Lu^{3+}) into Ca^{2+} positions. The larger cations (La^{3+}–Sm^{3+}) are accordingly concentrated in later fractions. Two effects are thus superimposed: a general enrichment of the entire group, and a preferential enrichment of the lighter elements. The behavior of yttrium is similar to that of the heavier "Yttrium earths." Scandium, on the other hand, is much smaller (r^{3+} = 0.75 Å; Frondel 1968) and is captured mainly in early Fe^{2+} positions, resulting in a decrease in concentration during fractional crystallization. The rare earth elements are widely fractionated among igneous rocks, so that a wide variety of patterns is observed. Some typical examples are shown in Figs. 2–6.

The high-temperature peridotites (e.g., Lizard, Venezuela) (Fig. 3), show extreme depletion of the large rare earths relative to the chondrites. Low-K tholeiites (Fig. 2) show a slight relative depletion. Hawaiian tholeiites show marked relative enrichment of the large rare earths relative to chondrites (Fig. 2) whereas granodiorites show a pattern subparallel to that of sedimentary rocks (Fig. 2). This is not unexpected if the average composition of the crust exposed to weathering is close to granodioritic composition. Some granites (e.g., G-1) show extreme enrichment of the large rare earths. Others, such as the leucogranite (Table 6, Fig. 2) show less enrichment but depletion in Eu.

Very high concentrations are observed in nepheline syenites (e.g., Lovozero Massif) and commonly, but not universally, in carbonatites. The rare earths are accompanied by high concentrations of the large, highly charged cations such as Th^{4+}, U^{4+}, Zr^{4+}, Hf^{4+}, Nb^{5+}, etc., and by a wide variety of accessory minerals containing these elements.

In the Lovozero alkaline complex, the rare earth elements are concentrated in eudialite, $(NaCeFe)_6$ $Zr(OHCl)$ $(SiO_3)_6$, loparite, $(NaCeCa)$ $(TiNb)_2$ O_6, titanite (sphene), and apatite. Eudialite concentrates the "yttrium earths" (the group Gd–Lu similar in size to Y^{3+}) and loparite, the "cerium earths" (La–Eu). The deeper, more slowly cooled parts of the intrusion showed sharper fractionation of the rare earths between the minerals, and the total concentration of the group increased with depth. The data support geological evidence of cooling from the top, and a single period of intrusion.

Rock-Forming Minerals. Typical distributions in a gabbro (San Marcos) and a leucogranite (Rubidoux Mt.) from the Southern California Batholith are shown in Figs. 4 and 5 (Towell et al., 1965). Similar studies have been made by Gavrilova and Turanskaya (1958).

Europium. The behavior of this element in igneous rocks may be attributed partly to its presence as Eu^{2+} (r, 1.25 Å). This ion is close in size to Sr^{2+} (1.18 Å) and these two elements exhibit parallel behavior.

Towell et al. (1965) showed that Eu was concentrated in plagioclase, in the San Marcos gabbro (Southern California batholith), relative to the other rare earths, and was present in plagioclase and K-feldspar in the Rubidoux Mountain granodiorite. Relative depletion of Eu was observed in hornblende, augite and apatite in the San Marcos Gabbro, and in biotite in the leucogranite. Eu was slightly concentrated in the K-feldspar relative to coexisting plagioclase (see Figs. 4, 5). The closest parallel to this behavior among other elements is strontium, which distributes itself about equally between coexisting K-feldspar and plagioclase, and does not readily enter the mica structure. The author showed that Eu, Ba, and Sr showed parallel behavior during fractionation and the extreme depletion of Ba and Sr in leucogranites was paralleled by the distribution of Eu. A useful measure of the variation of Eu relative to the other rare earths is provided by the Eu/Gd ratio.

Rare Earths in Sedimentary Rocks. The outstanding feature of the behavior of the rare earths in sedimentary processes is the uniformity of the abundance patterns observed for diverse rock types. The data given for sediments in Table 8 are derived for shales, sandstones, quartzites, and limestones (Haskin et al., 1966).

These are plotted in Fig. 6 relative to the shale pattern.

The uniformity of these patterns indicates that only minor fractionation takes place and that the dominant process is one of remixing of the diverse patterns produced by fractionation in igneous processes. The uniform sedimentary pattern indicates thorough remixing and can be interpreted as the relative abundance pattern typical of the continental crust.

Metamorphic Rocks. No modern data are available, but in the highest grades of metamorphism, some fractionation may be predicted to occur.

Earth's Crust. There is a marked concentration of the rare earths in the crust relative to chondritic abundances, and to presumed total earth compositions. The similarity in overall chemistry between granodiorites and the abundances obtained by widespread sampling of the crust suggest that the rare earth abundances of the accessible crust are similar to those observed in granodiorites and that the relative pattern is the same as that observed in sedimentary rocks.

Seawater. The abundance values are given in Table 9. Cerium is depleted relative to La, probably due to oxidation and removal by incorporation in Mn modules, where this element is concentrated. Wildeman and Haskin (1965) noted that the total content of rare earths in the ocean was equivalent to that in the upper 0.2 mm of oceanic sediment.

Tektites. Average values for australites are given in Table 8. These show similar relative and absolute rare earth abundances to those of terrestrial sedimentary rocks and this must be considered as supporting evidence for a terrestrial origin of tektites.

Rare Earth Minerals. The concentration of the rare earth elements during igneous rock formation (particularly granites and nepheline syenites) and in pegmatites leads to the crystallization of many rare earth minerals. Among the more important are the following:

TABLE 9. RARE EARTHS IN SEAWATER

		Seawater—Pacific Ocean ppm ($\times 10^6$)	Residence Time in Seawater (yr)[b]
57	La	2.9	210
58	Ce	1.3	48
59	Pr	0.64	180
60	Nd	2.3	180
62	Sm	0.44	180
63	Eu	0.11	160
64	Gd	0.61	280
65	Tb	—	—
66	Dy	0.73	440
67	Ho	0.22	—
68	Er	0.61	430
69	Tm	0.13	630
70	Yb	0.52	390
71	Lu	0.12	570
Σ REE		10.2	
39	Y	—	
Σ REE + Y		—	
La/Ce		2.2	
Eu/Gd		0.18	

[a] Goldberg et al., (1963).
[b] Wildeman and Haskin (1965).

Yttrofluorite	(CaF_2, YF_3)
Bastnaesite	$(CeF)Co_3$
Allanite	$(CaFeCeAl)$ silicate
Gadolinite	$Be_2Fe_2Si_2O_{10}$
Thalenite	$Y_2Si_2O_7$
Thortvetite	$(ScY)_2Si_2O_7$
Loparite	$(CeCaNa)$ titanite
Yttrotantalite	$(FeCaYErCe)$ tantalate
Samarskite	$(FeCa,UCeY)$ Niobate
Euxenite	$(YCeU)$ niobate-tantalate
Monazite	$(CeLa)PO_4$
Xenotime	YPO_4

Economic Sources. The principal source of the rare earths is the mineral monazite, occurring in beach sand deposits. Recovery of the rare earths is a byproduct of the extraction of thorium.

S. R. TAYLOR

References

Ahrens, L. H., 1952, *Geochim. Cosmochim. Acta*, **2**, 155.
Aller, L. H., 1961, "The Abundance of the Elements," New York, Interscience Publ.
Burbidge, E. M., Burbidge, G. R., Fowler, W. A., and Hoyle, F., 1957, "Synthesis of the elements in stars," *Rev. Mod. Phys.*, **29**, 547–650.

Butler, J. R., 1957, "Rare earths in Yttrotungstite," *Geochim. Cosmochim. Acta,* **12**, 190–194.

Butler, J. R., 1958, "Rare earths in some niobate-tantalates," *Mineral. Mag.,* **31**, 736–780.

Cameron, A. G. W., 1959, "The origin of the elements," *Phys. Chem. Earth,* **3**, Chap. 4, 199.

Coryell, C. D., Chase, J. W., and Winchester, J. W., 1963, "A procedure for geochemical interpretation of terrestrial rare-earth abundance patterns," *J. Geophys. Res.,* **68**, 559–566.

Fisk, Z., and Matthias, B. T., 1969, "Rare-earth elements and high pressures," *Science,* **165** (3890), 279–280.

Frondel, C., 1968, "Crystal chemistry of scandium as a trace element in minerals," *Z. Krist.,* **127**, 121.

Gavrilova, L. K., and Turanskaya, N. V., 1958, "Distribution of rare earths in rock-forming and accessory minerals of certain granites," *Geokhimiya,* **1958**(2), 163–170.

Goldberg, E. D., Koide M., Schmitt, R. A., and Smith, R. H., 1963, "Rare-earth distribution in the marine environment," *J. Geophys. Res.,* **68**, 4209–4217.

Grevesse, N., and Blanquet, G., "Abundances of the rare earths in the sun," *Solar Physics* (in press).

Haskin, L. A., Frey, F. A., Schmitt, R. A., and Smith, R. H., 1966, "Meteoritic, solar and terrestrial rare-earth distributions," *Phys. Chem. Earth,* **7**, 169–321.

Haskin, L., and Gehl, M. A., 1963, "The rare-earth distribution in tektites," *Science,* **139**, 1056–1058.

Haskin, L. A., Wildeman, T. R., Frey, F. A., Collins, K. A., Keedy, C. R., and Haskin, M. A., 1966, "Rare earths in sediments," *J. Geophys. Res.,* **71**, 6091–6105.

Murata, K. J., Rose, H. J., Jr., Carron, M. K., and Glass, J. J., 1957, "Systematic variation of rare earths in cerium-earth minerals," *Geochim. Cosmochim. Acta,* **11**, 141–164.

Schmitt, R. A., Smith, R. H., Lasch, J. E., Mosen, A. W., Olehy, D. A., and Vasilevskis, J., 1963, "Abundances of the fourteen rare-earth elements, scandium and yttrium in meteoritic and terrestrial matter," *Geochim. Cosmochim. Acta,* **27**(6), 577–622.

Semenov, E. I., and Barinskii, R. L., 1958, "The composition characteristics of the rare earths in minerals," *Geochem.,* **1958**, 398–419.

Templeton, D. H., and Dauben, C. H., 1954, "Lattice parameters of some rare-earth compounds and a set of crystal radii," *J. Am. Chem. Soc.,* **76**, 5239–5337.

Topp, N. E., 1965, "The Chemistry of the Rare Earth Elements," Amsterdam, Elsevier.

Towell, D. G., Winchester, J. W., and Spirn, R. V., 1965, *J. Geophys. Res.,* **70**, 3485.

Wallerstein, G., 1966, "A preliminary analysis of the abundances of the rare earths in the sun," *Icarus,* **5**, 75–78.

Wildeman, T. R., and Haskin, L. A., 1965, "Rare-earth elements in ocean sediments," *J. Geophys. Res.,* **70**, 2905–2910.

Wybourne, B. G., 1965, "Spectroscopic Properties of Rare Earths," New York, Interscience Publ., 236pp.

Cross-references: *Earth's Crust Geochemistry; Electron Capture; Electronegativity; Elements: Planetary Abundances and Distribution; Isotope Fractionation; Radioactive Isotopes; Rare Earths in Basalts; Seawater, Chemistry.* Vol. II: *Cosmochemistry; Cosmogony; Gravitational Collapse of a Star; Meteorites; Tektites.* Vol. IVB: *Rock-Forming Minerals.* Vol. V: *Carbonatites; Gabbro; Granite; Pegmatite; Peridotite; Syenite; Tholeiite.*

RARE EARTHS IN BASALTS

Elements of the lanthanide transition series are known to have very homologous chemistry because of the deeply buried 4f electronic transition shell, leaving the outer shells, the ones responsible for chemical behavior, unchanged for the complete series. Because of these electronic characteristics, separation and isolation of these elements have challenged chemists for a long time. They were among the last elements to have been discovered, although Mendeleev had predicted their place in the periodic table of elements. It is only through complex fractional crystallization schemes that chemists had succeeded in isolating these elements before ion-exchange techniques of separation were developed.

All the rare earths are present in terrestrial and cosmic materials but usually only in trace amounts. The even atomic numbers of these elements are more abundant by a factor of 6 to 10 relative to their odd atomic neighbors because of the greater number of even stable isotopes produced during nucleosynthesis (see Vol. II, *Nucleogenesis*). Thus, a zig-zag abundance pattern results as shown in Fig. 1 when reporting the concentration of these elements as a function of the atomic number for a basalt from Hawaii and the average of twelve chondritic meteorites (Haskin et al., 1966).

Until a decade ago, the relative rare earth abundances of cosmic matter were thought to be preserved in terrestrial rocks because of the homologous nature of the rare earths, although the absolute abundances would vary depending on the material. It is only since *neutron activation analyses* (q.v.) and, more recently, *mass spectrometry* (q.v.) techniques were developed for analyzing and separating the rare earths that much information has been gained in terrestrial and cosmic matter and that these earlier views had been revised.

The rare earths in most rocks of the continental crust are enriched relative to chondritic meteorites, with the light rare earths

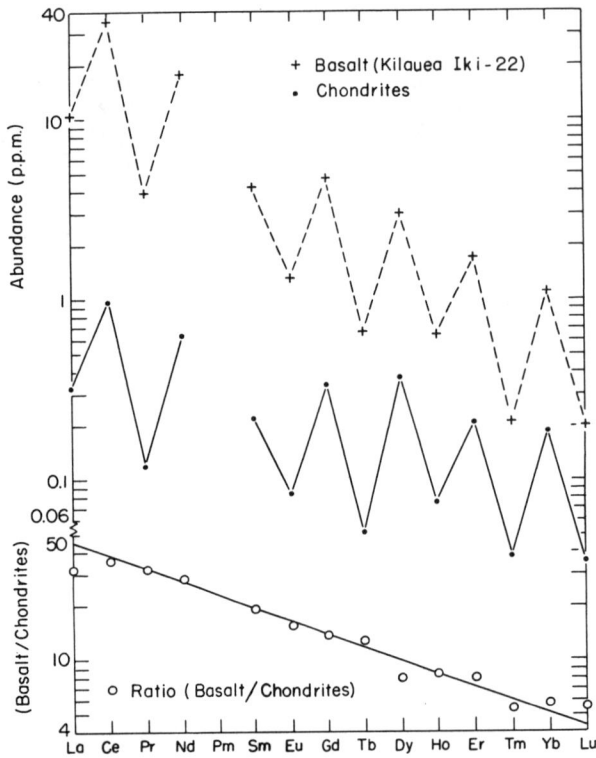

Fig. 1. Abundance of the rare-earth elements in the basalt Kilauea Iki-22 (broken line) and mean of twenty chondrites (solid line) plotted on a logarithmic scale as a function of the atomic number (no stable isotope of promethium (Pm) occurs in nature). The lower curve gives the ratios of the rare-earth abundances in basalt to the twenty chondrites, i.e., the enrichment factor relative to chondrites (Masuda-Coryell's plot).

preferentially so (Fig. 1). More importantly, the relative enrichment invariably varies smoothly as a function of the atomic number or the inverse of the ionic radius (which amounts to the same thing because the inverse of the ionic radius increases linearly as a function of the atomic number for the lanthanide series). Only cerium and europium may depart from this smooth fraction curve depending on the oxidation or reduction condition of the environment as they, respectively, can be oxidized or reduced relatively easily in nature.

Whatever the complexities of the actual history of the earth, the crust of the earth is enriched relative to chondritic meteorites (and presumably relative to the earth interior) in trace elements which do not fit easily into the structure of common rock-forming minerals of the mantle. This effect is more pronounced for the light rare earths or other electropositive trace elements of similar size such as uranium, thorium, potassium, rubidium, cesium, stron-

tium, and barium, and is probably determined by the large ionic size rather than the atomic weight. That size controls the enrichment factor is best demonstrated by the behavior of yttrium which is not truly a rare earth but which has an equivalent size and charge to erbium. Despite a twenty-nine mass unit difference between these two elements, their enrichment in the crust is almost invariably identical for both elements. Because the rare earths have a very smooth decreasing ionic radius with increasing atomic number the relative abundances of these elements and their absolute enrichment or depletion should be sensitive to the details of operative processes in the earth.

Rare earth studies of volcanic rocks, particularly basalts which are more likely to have derived directly from the mantle, are useful indicators of processes within the mantle of the earth as well as for shedding light on basalt genesis (Balashov, 1966; Schilling and Winchester, 1967).

Basalt Classification

Because of the widespread occurrence of basaltic rocks in both oceanic and continental regions, the term "basalt" has been widely used, often too loosely. A brief review of the classification and nomenclature of basaltic rocks seems in order. The Yoder and Tilley classification (see Vol. V, *Basalt*) will be adopted here and the reader is further referred to these authors for an excellent summary of previous and present views on basalt classification.

"Basalts" consist principally of plagioclase and clinopyroxene. In addition, they may carry olivine, nepheline, melilite, hypersthene, or quartz, as well as other minor phases. When lavas are chilled so quickly that a glass is formed, these minerals either may not be observable or may only crystallize as small metastable phases. A convenient way to describe the chemical composition of basalts is by giving the relative proportions of such minerals, whether present or not, as calculated from the bulk chemical analyses of the lava according to an arbitrary scheme developed by Cross, Iddings, Pirsson and Washington, the so-called *CIPW norm* (e.g., normative nepheline indicates calculated nepheline). The basalt tetrahedron of Yoder and Tilley (Fig. 2) is a convenient and simple illustration of this method of classification of basaltic rocks. These classes are:

1. *Theoleiite or tholeiitic suite* (oversaturated)—basalts characterized by the presence of normative hypersthene.

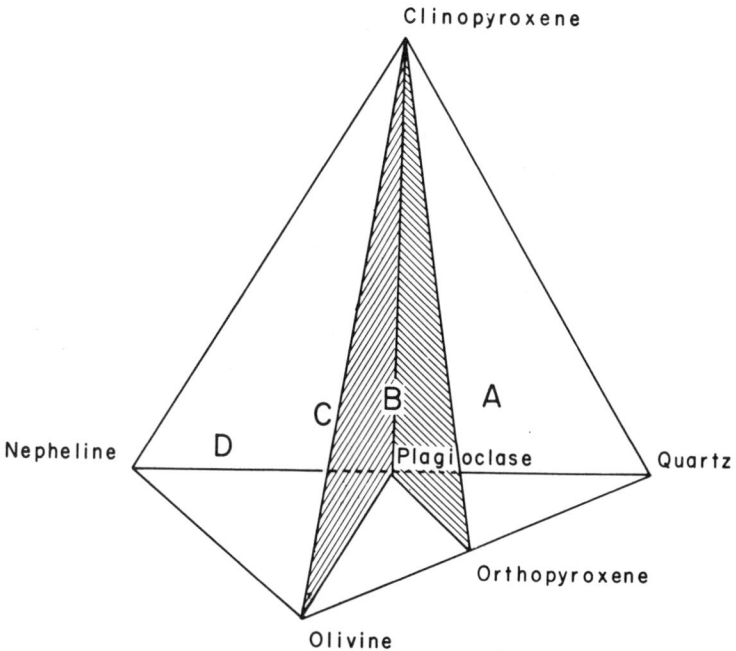

Fig. 2. Major mineralogy of basalts represented by the "Yoder and Tilley basalt tetrahedron." The plane of olivine-clinopyroxene-plagioclase is referred to as the "critical plane of silica undersaturation" and the plane clino-pyroxene-plagioclase-orthopyroxene as "plane of silica saturation." A, field of quartz tholeiites; B, field of olivine theoleiites; C, field of alkali olivine basalts; D, field of olivine basanites.

2. *Quartz tholeiite*—basalt with normative quartz and hypersthene.
3. *Olivine tholeiite*—basalt with normative olivine and hypersthene; this includes basalts falling within this group but having low (0–3%) normative hypersthene content.
4. *Alkali olivine basalt*—basalt characterized by the presence of normative olivine and nepheline but with normative nepheline less than 5%.
5. *Basanite*—basalt characterized by normative olivine and nepheline and with more than 5% nepheline.
6. *Olivine nephelinite*—basalt-like rocks without normative albite and typically with olivine, diopside, and nepheline as the major normative minerals.

Petrochemical Relationships

The first systematic study of rare earths in a series of petrochemically well documented and spatially and genetically closely related basalts was undertaken for the Hawaiian Islands by Schilling and Winchester (1969). A close quantitative correlation between absolute and relative rare earth contents and the petrochemistry of the basalts was demonstrated. This study will now be used here to illustrate the potentiality of rare earth geochemistry in basalt genesis.

On a Coryell-Masuda plot (Fig. 1), the rare earth pattern of each of the twenty-one lavas are found to be smoothly and progressively enriched relative to the chondritic meteorites. The lighter rare earths are preferred while the concentrations in the heavy region stay relatively more constant. Lanthanum shows the maximum range of variation and was used to illustrate quantitatively these variations with respect to the petrochemistry of these rocks. Fig. 3a shows the La abundance as a function of the commonly used differentiation index $SiO_2/3 + K_2O-MgO-FeO-CaO$. Three main groupings are apparent: *(1)* In the low La concentration region without apparent relationship to the index of differentiation, are the tholeiites. *(2)* At intermediate La content and systematically and logarithmically related to the index of differentiation, are the lavas of the alkali series. *(3)* At high La content and low values of index of differentiation, are the strongly silica undersaturated nepheline-melilite basalts.

These three groupings, corresponding to three major lava series, suggest an independ-

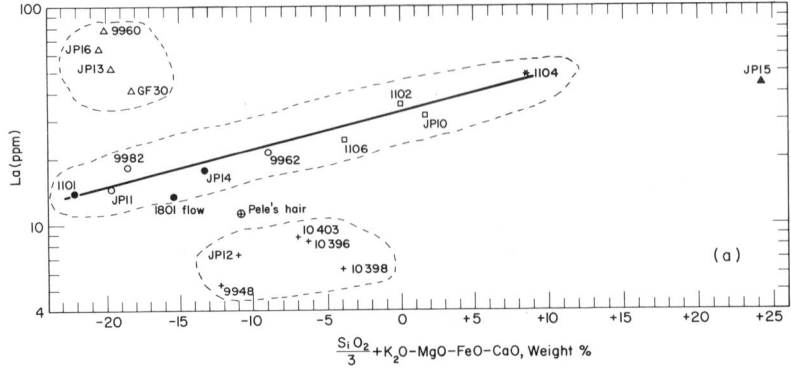

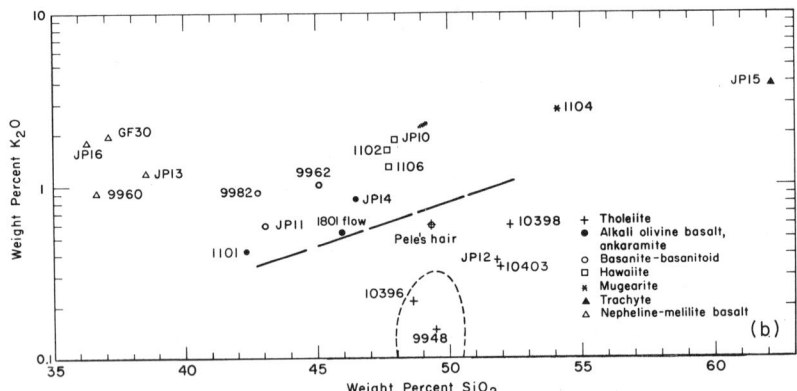

FIG. 3. Variation diagram of: (a) La content (in ppm) of twenty Hawaiian lavas and a diabase plotted as a function of index of fractionation of the $SiO_2/3 + K_2O-MgO-FeO-CaO$ (wt %). (b) K_2O content (wt %) of the same Hawaiian lavas as in (a) plotted as a function of SiO_2 content (wt %). Field above the dashed line represents the alkalic series and field below represents the tholeiitic series. The small field delineated by the dotted curve represents the submarine basalt (low K_2O) field.

ent origin for each group rather than being interrelated by a process of fractional crystallization. Within the alkali series, however, a genetic relationship between the lavas that compose this series was implied.

The similarity of Fig. 3a and 3b for the La and K_2O content of these lavas testifies to the similar geochemical behavior of these two trace elements in basalt genesis.

The same groupings corresponding to mineralogically well-identified basalt types are obtained from the rare earth patterns as a whole (Fig. 4). These groups are again: *(1)* The tholeiites with the lowest rare earth content, followed by the rocks of *(2)* the alkali series showing a *break* in the middle of the patterns and *(3)* the nepheline-melilite basalts with the light rare earths strongly enriched and fractionated relative to tholeiites.

In summary, Fig. 3 and 4 show a systematic relationship between the absolute or relative rare earth abundances and the petrochemistry of these rocks. They demonstrate the ability of the rare earth fractionation to reflect systematically the past chemical history of volcanic rocks.

Mineral Selectivity

Incorporation of rare earths into silicate minerals can happen very selectively because of lattice size restrictions of crystals.

This mineral selectivity is illustrated by rare earth data on common rock silicate minerals, or by minerals from ultramafic nodules found in basalts, or by rare earth partitioning factors between phenocrysts-volcanic matrix (-glass) applicable to basalts genesis (see references in Nagasawa et al., 1969). Some of these data are shown in Fig. 5. The partition coefficients are particularly consistent despite large varia-

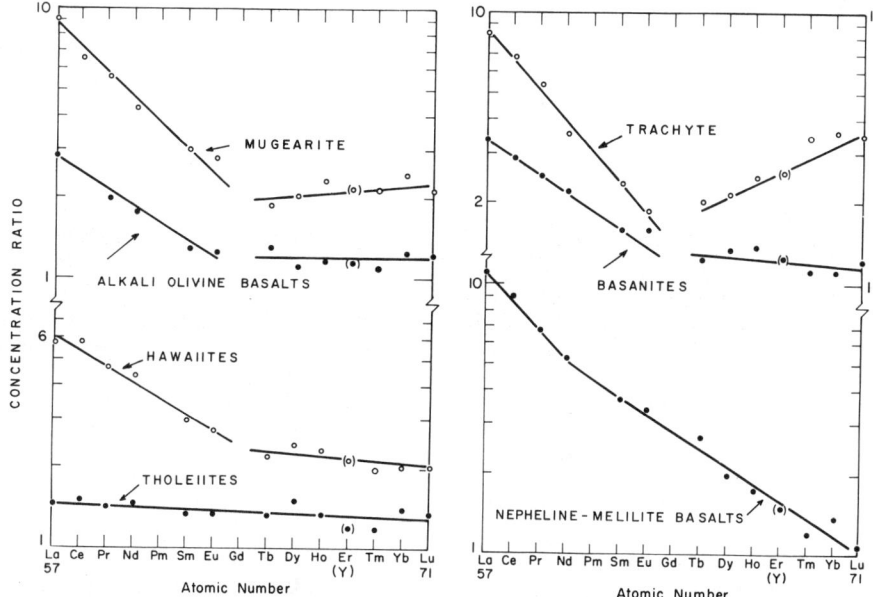

FIG. 4. Rare earth abundance averages of mineralogically recognized rock groups relative to an olivine tholeiite, Koolau basalt series, Oahu Island (Hawaii).

tions in temperature and pressure of formation and in bulk composition of the lavas. They tend to indicate an approach to equilibrium conditions. Extraction of such minerals from a primary melt may indeed affect the

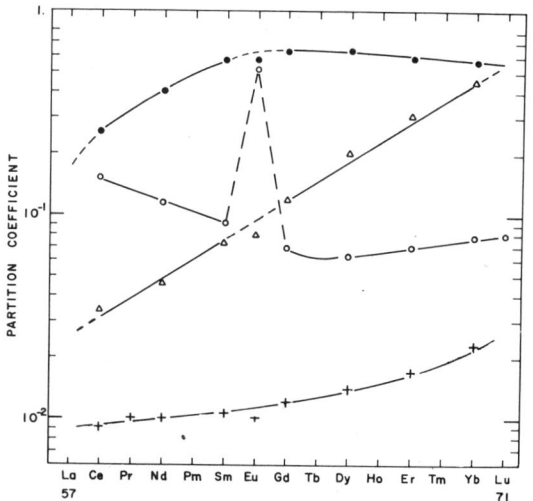

FIG. 5. Average rare earth partition coefficients of phenocrysts (ppm concentration in mineral phrenocryst divided by ppm concentration in matrix): closed circle, Ca-clinopyroxene; open circle, plagioclase feldspar; open triangle, orthopyroxene; cross, Mg-rich olivine. Note the europium enrichment in feldspar. (Data from Schnetzler and Philpotts, personal communication.)

rare earth abundance pattern of the residual melt and thus leave fingerprints of the process involved previous to the eruption of the lava.

Tests of models can be made for linking the three major basalts types of Hawaii, i.e., tholeiite, alkalic olivine basalt, or the nepheline-melilite basalt, by fractional crystallization of various phases satisfying: (a) the major element chemistry, (b) evidence from experimental petrology under controlled conditions, and (c) rare earth mineral data and crystal chemistry theories.

For instance, the distribution of rare earths with a break in the middle (Fig. 4) and a horizontal trend for heavy elements may be produced in the residual liquid by crystallization of calcium minerals such as clinopyroxenes. Substitution of trivalent rare earths for calcium with a preference for the heavier elements is plausible in the light of ionic radii. Goldschmidt gives the following radii: La^{3+} 1.22, Gd^{3+} 1.11, Lu^{3+} 0.99, Ca^{2+} 1.06, Mg^{2+} 0.78, and Fe^{2+} 0.83 Å. Rare earths with radii similar to and smaller than the radii of calcium may substitute for Ca^{2+} without fractionation, but larger ions may be discriminated against to a degree increasing with increasing radius. As crystallization proceeds and as clinopyroxenes, olivine, and limited amounts of plagioclase are removed from the liquid, the residual liquid should develop the pattern of basanites and alkali olivine basalts. Furthermore, the general anomalous enrich-

RARE EARTHS IN BASALTS

ment of europium in feldspars can put some stringent limits on how much of this mineral can be extracted from the primary melt without producing a noticeable depletion of this element in the residual melts (Fig. 5).

The calcium-crystallization model does not account for the nepheline-melilite basalt pattern which shows no break in the middle (Fig. 4). Removal of olivine and orthopyroxene from olivine tholeiitic magma to produce silica-undersaturated, nepheline-normative residual liquid, has been suggested. Fractional crystallization of such Mg^{2+} and Fe^{2+} minerals would not produce a break in the middle of the series (Fig. 4). Because Mg and Fe cation sites of orthopyroxene are smaller than the radius of the smallest rare earths, progressive discrimination against rare earths with increasing radii is expected over the entire range from lutecium to lanthanum (Fig. 5). The residual liquid should be enriched with rare earths to a degree which increases with increasing ionic radius as observed in nepheline-melilite basalt patterns. Such residual patterns can, however, also be produced by extraction from a tholeiitic melt at depth of garnet whose lattice strongly discriminates against the light rare earths. These two possibilities cannot yet be distinguished with our present limited knowledge. These few examples, however, demonstrate some of the potentiality of rare earths in basalt genesis.

Volcanic Provinces

Other volcanic regions show the same qualitative rare earth pattern variation for the principal basalt types as previously discussed for Hawaii (Fig. 6a). These variations are shown graphically in Fig. 6a-d. For the Japanese Islands (Fig. 6b), the progressive increase of rare earth content and degree of fractionation with alkali content of the basalts is also accompanied by a simple zonal arrangement of these lava series parallel to the Japanese Island Arc (Masuda, 1968). Going westward from the Pacific side of Japan, these are: the *tholeiitic zone*, the *high-aluminum basalt zone*, and the alkali-rich basalt zone near the Japanese Sea. In the Atlantic (Fig. 6d) again the average of subarial alkali-rich basalts of the

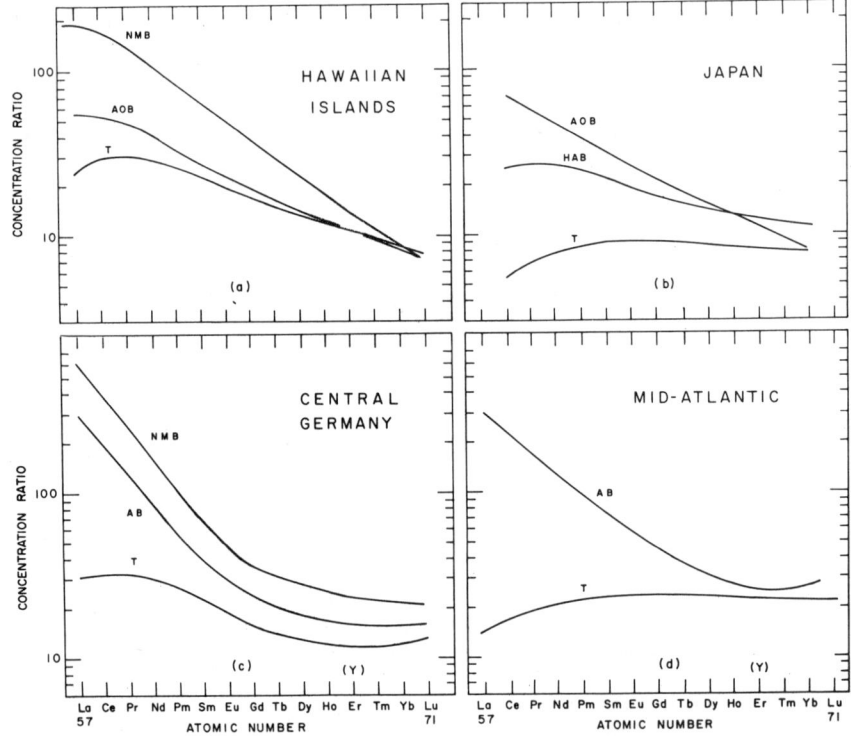

FIG. 6. Average rare earth fractionation patterns for the major basalt types, i.e., tholeiite (T), alkali olivine basalt (AOB), nepheline-melilite basalts (NMB) for: (a) Hawaiian Islands; (b) Japan (HAB stands for high-aluminum basalt); (c) Central Germany; and (d) Mid-Atlantic Ocean (T stands for submarine theoleiites from the mid-Atlantic Ridge and AB for lavas of the alkali series from Gough and Ascension Islands).

Ascension and Gough Islands on the mid-Atlantic Ridge show a preferential enrichment of the light rare earths with respect to the spatially related submarine tholeiites (low in K_2O) of this ridge (Haskin et al., 1966). In addition, one observes the same rare earth-tholeiitic, alkalic, and nepheline-melilite basalt relationships for continental basalts of Central Germany (Herrman, 1968) as that for oceanic regions, i.e., the light rare earths and degree of alkalinity of the basalts both increase simultaneously (Fig. 6c).

These results thus consistently show a regular increase of the light rare earths (larger ions) with increasing degree of alkalinity of the lavas, regardless of the degree of silica saturation or undersaturation.

Considering the simplified basalt tetrahedron of Yoder and Tilley (Fig. 2) the light rare earth content appears to increase regardless of whether the bulk chemistry of the melt evolved:

1. In the undersaturated block "nepheline-olivine-plagioclase-clinopyroxene toward nepheline enrichment, i.e., from basanite to nephelinite.
2. Near or within the critical plane of silica undersaturation for the line of descent of alkali olivine basalt-hawaiite-mugearite and to possibly trachyte (see also Fig. 3).
3. Or toward quartz enrichment in the block "olivine-plagioclase-clinopyroxene-quartz," as for the tholeiitic series.

However, for all these three cases the content of light rare earths increases with increasing alkalinity of the melts, although at different rates depending on the mineralogy involved in each block. The increase is most pronounced for the largest rare earth ion, i.e., lanthanum, and decreases regularly with decreasing trivalent ionic size towards the heavy rare earths (the smallest rare earth ions) which stay relatively more constant.

To summarize, Figs. 6a-d shows:

1. Each petrochemical group is characterized by a distinct rare earth abundance pattern.
2. For each volcanic region, the light rare earth abundances increase progressively from tholeiitic to more alkali-rich basalts. The rate of increase is variable from one series to another. It is most pronounced for the largest ions at the light rare earth end of the series and decreases regularly with decreasing trivalent ionic size. This produces a convergence of rare earth patterns towards the heavy rare earths, whose concentrations stay relatively more constant or in some rare cases, may even slightly decrease (e.g., for one or two nephelinites from Hawaii, but not all of them.
3. Tholeiites, the basalts with lowest alkali content, have the lowest rare earth concentrations and are the least fractionated. Their rare earth patterns are identical within any one volcanic region but differ from region to region as shown in Fig. 6a-d. The level of convergence of the patterns towards the heavy rare earths differs from region to region.

Theoretical Models

Theoretical models on the fractionation of rare earths in basalts by fractional crystallization have been explored by Masuda (for other magmatic processes see references in Schilling and Winchester, 1967). Similar patterns as those observed in basalts have been simulated. Promises for distinguishing between these processes have resulted but need to be experimentally verified. For instance, independent derivation of the three magma types shown in Fig. 6a can apparently also be obtained by different degrees of melting of the same mantle material, presumably at different depths. The tholeiite would be derived by a relatively larger degree of melting than the alkali olivine basalt and the nepheline-melilite basalt by a very small degree of melting.

Tectonic Relationships

A noticeable rare earth pattern variation is apparent for tholeiites of various volcanic provinces (Fig. 6a-d) despite that qualitatively the same systematic rare earth pattern variation from tholeiitic to alkalic and to nepheline-melilite basalts exists for these volcanic regions. For the sake of a better comparison these rare earth variations for only tholeiitic basalts are regrouped in Fig. 7.

Of particular significance is the light rare earth depletion of the submarine tholeiites erupted along the mid-oceanic ridges relative to usual crustal rocks which are light rare earth enriched. The rare earth fractionation pattern is remarkably uniform along each mid-ocean ridge (mid-Atlantic Ridge, Reykjanes Ridge, Juan de Fuca Ridge, East Pacific Rise, Chile Rise, Pacific-Antartic Ridge, mid-Indian Ridge, Carlsberg Ridge, as well as the Red Sea Trough; see references in Schilling, 1971). However, they clearly contrast with tholeiitic lavas from plateau basalts which resemble other continental rocks (Fig. 8). These rare earth differences contrast two very large and

RARE EARTHS IN BASALTS

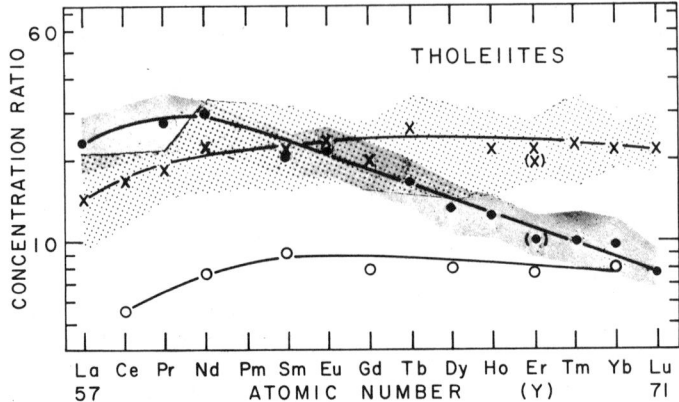

Fig. 7. Rare earth fractionation patterns in tholeiites from: mid-Atlantic Ridge (cross); Hawaiian Islands (closed circle); Izu Peninsula, Japan (open circle). The shaded areas represent the maximum deviations from the mean.

important tholeiitic volcanic provinces, i.e., oceanic versus continental areas. If one compares the Deccan tholeiitic plateau basalt to the Columbia tholeiitic plateau basalt, both geographically widely separated, no noticeable rare earth variations are observed; they form a relatively uniform group.

If one compares submarine tholeiites from a particular mid-ocean ridge (e.g., mid-Atlantic) with tholeiites from another widely separated ridge (e.g., East Pacific Rise) only small rare earth variations are noticed, thus forming again a remarkably uniform group, but different this time than from plateau basalts.

Work on the Columbia Plateau Basalt shows some systematic variation of the rare earth abundances between the lower and upper stratigraphic horizons of the Picture Gorge flows and Yakima basalts (Osawa and Goles 1970). The light rare earth abudances in these basalts are somewhat lower than those shown in Fig. 8, but remain fractionated and enriched relative to chondrites. The comments on oceanic versus the continental tholeiitic volcanism remain valid, however.

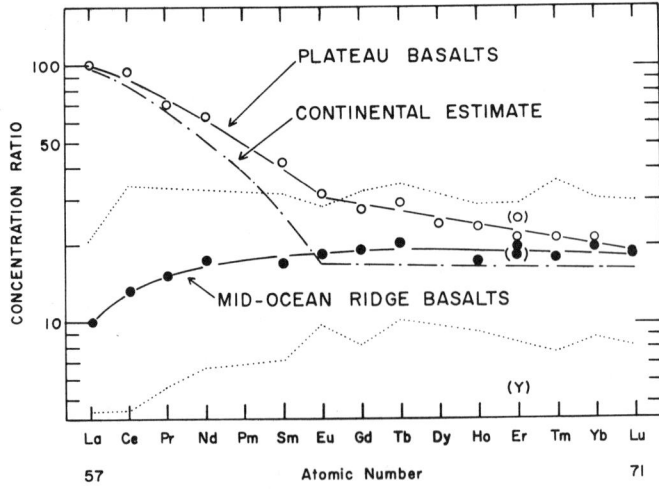

Fig. 8. Rare earth fraction pattern of tholeiitic basalts from: closed circle, mid-ocean ridges (East Pacific Rise, mid-atlantic, Carlsberg Ridge, and Red Sea Axial Trough); the dotted lines delineate the extreme deviations from the mean pattern; open circle, plateau basalts (Columbia, U.S.A. and Deccan, India); dot-dash line, estimated rare earth pattern for the upper part of the continental crust.

These two types of oceanic and continental volcanism are related to quite different types of structural and tectonic settings. Plateau basalts erupt along tensional fractures at the onset of continental break up and drift and may well represent an early magmatic phase of continental drift and spreading of the sea floor; whereas submarine ridge tholeiites are produced by eruption along the crest of mid-ocean ridges, partly healing the gap produced by the divergence of rigid plates of lithosphere, i.e., at the loci of the most recent oceanic crust formation produced by spreading of the sea floor.

It is thus tempting to relate such tectonic variations to the rare earth pattern variations of these two types of tholeiites. However, to fully understand these major rare earth differences, one needs to comprehend better factors influencing *"whole rock"* rare earth patterns in lavas. A rare earth synthesis of the Hawaiian work and thoretical considerations by Schilling and Winchester (1969) indicates that: *"whole rock"* rare earth patterns of basalts can significantly be influenced by *(1)* rare earth composition of source rocks in the mantle (i.e., *"source effect"*), *(2)* minerals involved during processes of partial melting and ascent of lavas to the surface (i.e., *"mineral effect"*), *(3)* mechanism of magma formation, fractionation and evolution (i.e., *"dynamical or kinetic effects"*), and possibly, but to a lesser extent, by *(4)* selective formation and transport of volatile rare earth complexes (i.e., *"complexing effect"*).

Unraveling of these factors and determining their relative importance is a difficult task. However, limitation of the comparison, e.g., to tholeiites only, may minimize variations due to "mineral effect" and presumably "complexing effect." Tholeiites are formed under a limited range of temperature and pressure conditions, and the small bulk chemical variations of tholeiites only are usually attributed to addition or removal mainly of olivine; which does not fractionate the rare earths significantly. Thus, these rare earth variations can either be related to variations of the source of these lavas presumably in the mantle or to tectonic variations or to the dynamics of vulcanic processes.

The great uniformity of mid-ocean ridge tholeiites requires a rather unique and simple magmatic process. It tends to speak against convection cells and partial melting underneath mid-ocean ridges but rather for the migration of homogeneous interstitial melts from the low velocity layer (Schilling, 1971).

The level of enrichment of the heavy rare earths is particularly useful in understanding regional and lateral inhomogeneities of the mantle source. We have already seen that within a single volcanic region, the heavy rare-earth abundances vary little despite large bulk chemical compositional variations from tholeiitic to alkali and nepheline-melilite basalts. Strong evidence from experimental petrology indicates that these three basalt types have been derived from different depths in the mantle underneath these volcanic regions and were produced by different degrees of melting of an ultramafic mantle. The tholeiites are derived from shallow depths by a large degree of melting, the alkali olivine basalts at deeper depth by a lesser degree of melting and, finally, the nepheline-melilite basalts from a very much smaller degree of melting at much greater depth (see Vol. V, *Basalts*). Thus, despite the large variations of melting conditions required to produce the major type of basalts, the heavy rare earths vary only little; yet from region to region more important variations are observed in the tholeiites only. These relatively large variations of the heavy rare earths can only be related to important lateral inhomogeneities of the mantle at shallow depth from where tholeiitic basalts were derived; and thus by interference they are related to a certain extent to the past chemical history of these shallow regions of the mantle.

The degree of enrichment or depletion of the light rare earths, which because of their large size are more sensitive to fractionation processes, may better reflect the past chemical history of these mantle sources. The light rare earth depletion of the submarine mid-oceanic ridge tholeiites is consistent with continental drift and spreading of sea floors. This depletion is also accompanied by a depletion of other large trace elements of similar behavior, to a degree depending on their ionic size for equal charge (e.g., Cs–Rb–K, Th–U, and Ba–Sr). Presumably, these large ions have been previously extracted from the mantle and concentrated in the growing continental crust before the opening and development of a new ocean. Such large ions do not fit into the dense minerals of the mantle and tend to concentrate in intercrystalline zones of the mantle. Upon melting of the mantle, these large electropositive ions were incorporated in first-melting fractions, transported upward and concentrated in the continental crust during probably several cycles of upper-mantle flows, leaving with time a mantle progressively more depleted in these ions (the larger the ion, the greater the depletion).

The age, spatial distribution, and light rare earth enrichment of plateau basalts are also consistent with such a model. Plateau basalts which were erupted at the onset of continental

RARE EARTHS IN BASALTS

drift in this case have probably been derived from an advecting mantle which at the time was not as much depleted in light rare earths as the mantle from which mid-oceanic ridge tholeiites were derived later.

Finally, of additional interest are tholeiites of *island arcs* which in part are produced by melting processes during the consumption of old oceanic crust thrust along the Gutenberg-Benioff seismic place in the mantle beneath islands arcs. Although the relative rare earth patterns of these tholeiites are related to those of mid-ocean ridges, their absolute values are much lower for the Japanese Arc (Masuda, 1968) and the South Sandwich Arc and more variable for the Mariana Arc (Schilling, unpublished). In producing a tholeiite from a tholeiitic oceanic crust, complete melting should result. The lower rare earth content of these tholeiites indicates a need for dilution with material impoverished in these elements. Incomplete melting of the crust with incorporation of material of more ultramafic derivation seems required to satisfy both the bulk and rare earth element chemistry. The problem of island arcs are discussed, however, in greater detail by Taylor (see *Rare-Earths*). The state of the art in these tectonically very active regions is still in an early stage.

Shield Volcanoes—Mid-Oceanic Ridge Relationships

Noticeable differences of the light rare earth content of *tholeiites* derived from subaerial shield volcanoes or seamounts relative to spatially related submarine mid-oceanic ridge *tholeiites* can be observed, while the heavy rare earth content stays nearly identical. These features are illustrated in Fig. 9a-c for the submarine basalts from the Red Sea Trough and Jebel Teir Island (a shield volcano at the southern end of the Red Sea Trough), and for Culpepper Island (Galapagos) and the East Pacific Rise, respectively. The heavy rare earths of the tholeittes building these volcanoes are almost identical to their spatially related mid-oceanic ridge tholeiites, but the light rare earths are enriched in the shield volcanoes. The building of these volcanoes results from an unusual accumulation of melt due to additional structural or tectonic controls; presumably superimposed on the rather steady-state vulcanic process associated with the crest of mid-ocean ridges and spreading of the sea floor. Melts in the magma reservoir at depth, intermittently feeding the overlying volcanic edifice, and slow cooling during the dormant periods produce some fractional crystallization which result in an increase of the light rare earths. For such a model to be in agreement

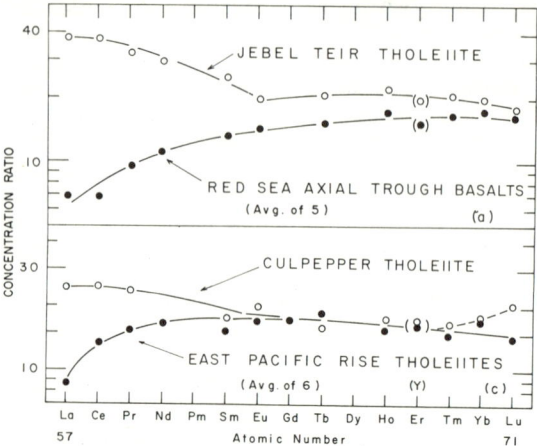

FIG. 9. Rare earth fractionation patterns of tholeiitic rocks contrasting shield volcanoes and related mid-ocean ridges. (a) Jebel Teir (Southern Red Sea) and Red Sea Axial Trough; (c) Culpepper Island (Galapagos) and East Pacific Rise.

with the bulk chemistry, i.e., melts of tholeiitic composition, a relationship between the volume of the volcanic edifice, the previous degree of melting in the mantle before accumulation of melts in the magma chamber, and light rare-earth enrichment is required and needs to be tested.

Conclusions

The same systematic rare earth variations have been observed between major basalt types for every volcanic region studied to date. These variations can easily be related to different degrees of melting of an ultramafic mantle at various depths below the volcanic province and to some limited extent of fractional crystallization during the ascent of lavas to the surface.

In addition, tholeiitic basalt types from various continental and oceanic regions characterized by different tectonic settings are clearly distinguished on the basis of rare earth abundance data, but not readily apparent from the major element chemistry of these basalts. Thus, on the basis of rare earth abundances and presumably other trace elements of similar behavior, "there are *tholeiites* and there are *tholeiites*." Such differences for tholeiitic basalts can be related to tectonic processes, associated modes of vulcanicity, and to the past chemical history of the source region in the mantle from which these lavas have derived.

JEAN-GUY SCHILLING

References

Balashov, Yu. A., 1966, "Differentiation of rare-earth elements during magmatic processes," in

(Vinogradov, A. P., editor), "Chemistry of the Earth's Crust," Vol. 1, pp. 272–287.
Graham, A. L., and Nicholls, G. D., 1969, "Mass spectrographic determinations of lanthanide element contents in basalts," *Geochim. Cosmochim. Acta,* **33,** 555–568.
Haskin, L. A., Frey, F. A., Schmitt, R. A., and Smith, R. H., 1966, "Meteoritic, solar, and terrestrial rare-earth distribution," *Phys. Chem. Earth,* **7,** 167–321.
Herrmann, A. G., 1968, "Die Verteilung der Lanthaniden in basaltischen Gesteinen," *Contr. Mineral. Petrol.,* **17,** 275–314.
Masuda, A., 1968, "Geochemistry of Lanthanides in Basalts of Central Japan," *Earth Planet. Sci. Letters,* **4,** 284–292.
Nagasawa, H., Wakita, H., Higuchi, H., and Onuma, N., 1969, "Rare earths in peridotite nodules: an explanation of the genetic relationship between basalt and peridotite nodules," *Earth Planet. Sci. Letters,* **5,** 377–381.
Osawa, M., and Goles, G. G., 1970, "Trace element abundances in Columbia River basalts," in (Gilmore, E. H., and Stradling, D., editors). *Proc. Second Columbia R. Basalt Symposium,* Eastern Washington College Press, pp. 55–71.
Schilling, J. G., 1971, "Sea-floor evolution: rare-earth evidence," *Phil. Trans. Roy. Soc. London, Ser. A,* **268,** 663–706.
Schilling, J-G., 1969, "Red Sea Floor origin: rare-earth evidence," *Science,* **165** 1357–1360.
Schilling, J-G., and Winchester, J. W., 1967, "Rare earth fraction and magmatic processes," in Runcorn, S. K., editor) "Mantles of the Earth and Terrestrial Planets," London, John Wiley & Sons, 267–283.
Schilling, J-G., and Winchester, J. W., 1969, "Rare earth contribution to the origin of Hawaiian lavas," *Contr. Mineral. Petrol.,* **23,** 27–37.
Schnetzler, C. C., and Philpotts, J. A., 1968, "Partition Coefficients of Rare Earth Elements and Barium Between Igneous Matrix Material and Rock Forming Mineral Phenocrysts," in (Ahrens, L. H., editor), "Origin and Distribution of the Elements," Oxford, Pergamon Press, pp. 929–939.
Tatsumoto, M., et al., 1965, "Potassium, rubidium, strontium, thorium, uranium, and the ratio of strontium-87 to strontium-86 in oceanic tholeiitic basalt," *Science,* **150,** 886–888.

Cross-references: *Atomic Number and Periodic Table; Elements: Planetary Abundances and Distribution; Erbium; Geochemical Evolution of the Core, Mantle, and Crust; Lanthanum; Lutetium; Mantle Geochemistry; Mass Spectrometry; Neutron Activation Analysis; Oxidation and Reduction; Partition Coefficients; Rare Earths; Trace Elements; Yttrium.* Vol. I: *Island Arcs; Mid-Oceanic Ridge.* Vol. II: *Cosmochemistry; Meteorites; Nucleogenesis.* Vol. IVB: *Clinopyroxenes; Mantle Mineralogy; Orthopyroxenes; Plagioclase Feldspurs.* Vol. V: *Basalt; Magma; Sea-Floor Spreading; Tholeiites.*

RARE (NOBLE, INERT) GASES

The rare gases are those elements that constitute Group O in the Periodic Table. They have the symbols He, Ne, Ar, Kr, Xe, and Rn. All are characterized by closed shells or subshells of electrons. There is no consistency about naming this group; it includes helium, for example, that is far from rare. "Noble" is simply a translation of German *edel* meaning "precious" or "rare." And a recent pioneering discovery by Neil Bartlett at the University of British Columbia showed that these elements are not, after all, inert. Their existence was first suspected by Cavendish in 1785 who found that there was a trace (nearly 1%) of some nonreactive gas in the earth's atmosphere, but the existence of a family of such elements was not appreciated until demonstrated by Rayleigh and Ramsay in 1894.

In the atmosphere the rare gases have the following concentrations in parts per million:

Argon
(Ar, atomic number 18) 9,340
Neon
(Ne, atomic number 10) 18.18
Helium
(He, atomic number 2) 5.24
Krypton
(Kr, atomic number 36) 1.14
Xenon
(Xe, atomic number 54) 0.086
Radon
(Rn, atomic number 86) ...... 6×10^{-14}

The rare gases are all products of radioactive decay. For example, argon (named from the Greek word for "lazy") owes its origin in the atmosphere to the slow decay of potassium-40 and has been gradually accumulating through time. Helium, on the other hand, is too light to be retained by a planet of the earth's mass and is therefore constantly being lost to outer space; its concentration in the atmosphere therefore represents an equilibrium between its emission and its diffusion rates. It is found in relatively high concentrations in some natural gas fields.

All the rare gases are colorless, odorless, and tasteless. They are all partly soluble in water, the more soluble ones having the higher molecular weights. They liquify at low temperatures, and solidify at extremely low temperatures (Table 1).

Commercially, mixtures of the rare gases are produced by fractionation of liquid air and used for filling electric light bulbs, in food preservation, and insulation.

RHODES W. FAIRBRIDGE

RARE (NOBLE, INERT) GASES

Table 1. Physical Properties of Rare Gas Elements[a]

Property		He³	He	Ne	Ar	Kr	Xe	Rn
Atomic weight	$C^{12} = 12.0000$	3.01603	4.0026	20.183	39.948	83.80	131.30	(222)
Gas density (0°C and 1 atm)	g/l		0.17847	0.89994	1.78403	3.7493	5.8971	9.73
Triple point	°K			24.55	83.78	115.95	161.3	202
Pressure at triple point	torr			324	516	548	612	~500
Heat of fusion at triple point	cal/mole		44.5[b]	80.1	281	390.7	548.5	
Solid density at triple point	g/cc			1.444	1.623	2.826	3.540	
Boiling point	°K		4.215	27.07	87.27	119.8	165.05	211
Heat of vaporization at boiling point	cal/mole		19.4	414	1557.5	2158	3020	
Liquid density at boiling point	g/cc		.1249	1.207	1.3798	2.413	3.06	
Critical temperature	°K		5.3	44.5	150.9	209.4	289.71	378
Critical pressure	atm		2.26	26.9	48.3	54.3	57.64	62
Critical density	g/cc		0.0693	0.484	0.536	0.908	1.100	
Solubility in water (0°C)	cc(STP)/1000g		9.78	14.0	52.4	99.1	203.2	510
Dielectric constant (25°C and 1 atm)			1.0000639	1.0001229	1.000768	1.001238	1.0005085	

[a] After Hyman, 1968. [b] Value given is at 25°K.

References

Asimov, I., 1966, "The Noble Gases," New York, Basic Books.
Bartlett, N., 1964, *Endeavour*, **23**, 3.
Hyman, H. H., 1968, "Noble Gases," in (Hampel, C. A., editor), "The Encyclopedia of the Chemical Elements," p. 461, New York, Reinhold Book Corp.
Mazor, E., Wasserburg, G. J., and Craig, H., 1964, "Rare gases in Pacific Ocean water," *Deep-Sea Res.*, **11**, 929-933.
Pepin, R. O., and Signen, P., 1965, "Primordial rare gases in meteorites," *Science*, **149**(3681), 253.
Tilles, D., 1965, "Atmospheric noble gases: Solar-wind bombardment of extraterrestrial dust as a possible source mechanism," *Science*, **148** (3673), 1085.

Cross-references: *Atomic Number and Periodic Table; Elements: Planetary Abundances and Distribution; Molecular Weights.* See also *Individual Elements.*

REACTION RIM—See Vol. V

REDUCTION—See **OXIDATION AND REDUCTION**

REGOLITH AND SAPROLITE—See Vol. III

RESIN AND AMBER—See Vol. VI

RHENIUM: ELEMENT AND GEOCHEMISTRY

Properties

Rhenium (Re), a rare metallic element, atomic number 75, was discovered by W. Noddack, I. Tacke, and O. Berg in 1925. In its elemental form it is a silvery-white metal with a close-packed hexagonal structure, a specific gravity of 21.02 (20°C), a melting point of 3180°C, and an estimated boiling point of 5627°C. The most important oxidation states are +7 (ionic radius 0.56Å) and +4 (0.72Å) but +6, +5, +3, +2, +1, −1 are also known.

The atomic weight is 186.2 and there are two naturally occurring isotopes—Re^{185} (37.07%) and Re^{187} (62.93%). Re^{187} is naturally radioactive ($t_{1/2} = 4.3 \pm 0.5 \times 10^{10}$yr) emitting a β^- particle (energy ≤ 0.008MeV) to form Os^{187}. This decay scheme forms the basis of the Re/Os geochronological dating technique.

Minerals

Rhenium is a "dispersed element" and occurs primarily in minerals of other elements. Molybdenite contains rhenium in amounts from ≤ 1 ppm up to 1.88%. An imperfectly known rhenium mineral, dzhezkazganite ($CuReS_4$?), has been reported. Other possible minerals are ReS_2(?) in the Mansfield copper shales and Re_2O_7(?) in sedimentary rocks associated with the uraninite deposits in Coconino County, Arizona.

Abundance

Rhenium is an extremely rare element, the average abundance in igneous rocks is estimated at about 50×10^{-4} ppm (Morris and Short, 1966). Sedimentary rocks usually contain $<5 \times 10^{-4}$ ppm but black pyritic shales may be relatively enriched in rhenium (about 825×10^{-4} ppm). Tektite abundances average about 0.56×10^{-4} ppm (Lovering and Morgan, 1964). Chondritic meteorite abundances average 580×10^{-4} ppm (or 0.053 Re atoms per 10^6 Si atoms), in good agreement with estimated "cosmic abundances" of 0.054 atoms per 10^6 Si (Morgan and Lovering, 1967). Iron meteorite abundances (Herr et al., 1961) are higher but very variable (0.002–4.8 ppm). Achondritic stony meteorite abundances are low ($\leq 7 \times 10^{-4}$ ppm).

Geochemical Behavior

In the reducing meteorite environment (and possibly also in the earth's deep interior), rhenium is strongly siderophilic since its metallic radius (in twelve-coordination) is 1.37 Å whereas that of iron is 1.27 Å. In the earth's crust, it is often chalcophilic and concentrates in molybdenite due to the similarity in the ionic radii between Re^{4+} (0.72 Å) and Mo^{4+} (0.70 Å). In more oxidizing (e.g., acid pegmatite) environments, the similarity of the ionic radii of Re^{4+} and Re^{6+} (i.e., 0.6–0.7 Å) enables rhenium to enter niobium (0.69 Å) and tantalum (0.60 Å) minerals such as wolframite. Rhenium has also been found in gadolinite, hellandite and zircon.

Little is known about the behavior of rhenium during igneous differentiation, metamorphism, weathering, or soil formation. Nothing is known about possible rhenium fractionation during sedimentary processes but some black pyritic shales apparently concentrate rhenium (Lovering and Morgan, 1964).

Economic Geology

Commercial sources of rhenium are (a) the concentration of volatile Re_2O_7 in the flue dust from molybdenite smelting, or (b) the concentration with the platinum metals in the "anode sludge" during electrolytic copper refining.

JOHN F. LOVERING

References

Herr, W., Hoffmeister, W., Hirt, B., Geiss, J., and Houtermans, F. G., 1961, "Versuch zur Datierung von Eisenmeteoriten nach der Rhenium-Osmium-Methode," *Z. Naturforschg.*, **16a**, 1053–1058.

Lovering, J. F., and Morgan, J. W., 1964, "Rhenium and osmium abundances in tektites," *Geochim. Cosmochim. Acta*, **28**, 761–768.

Morgan, J. W., and Lovering J. F., 1967, "Rhenium and osmium abundances in chondritic meteorites," *Geochim. Cosmochim. Acta*, **31**, 1893–1909.

Morris, D. F. C., and Short, E. L., 1966, "Minerals of rhenium," *Mineral. Mag.*, **35**, 871–873.

Stevens, R. F., Jr., 1965, "Rhenium," U.S. Bur. Mines Min. Yearbook.

Cross-references: *Elements: Planetary Abundances and Distribution.* Vol. II: *Meteorites.* Vol. IVB: *Meteoritic Minerals.*

RHODIUM: ELEMENT AND GEOCHEMISTRY

Properties

Rhodium, symbol Rh, is a member of the platinum family of metallic elements which occur in the eighth group of the periodic table. The element was first identified by W. H. Wollaston in 1803–1804. Its atomic number is 45 and the atomic weight 102.905. Alone among the platinum metals rhodium is anisotopic, all atoms being of mass number 103. The crystalline metal is structured on a face-centered cubic lattice with $a = 3.8031$ Å. The atomic radius for twelve-coordination in the metal is 1.34 Å. The melting point of the pure metal is 1960°C; this is appreciably higher than that of either platinum or palladium. Rhodium is also harder and less subject to chemical attack than the latter elements. The density of the pure compact metal is 12.41 g/cc at 20°C.

Occurrence

The platinum group elements are typically noble metals which resist direct chemical attack by many reagents. In consequence of this nobility and of the instability of their compounds at high temperature, most of the natural occurrences of these metals are as native metal alloys. In these they are found together and with certain base metals such as iron, copper, and nickel. There is no naturally occurring alloy in which rhodium is a major component, but it is found to varying small extents (usually less than 10%) in native platinum, palladium, iridosmine, and sometimes gold.

The only platinum-metal mineral in which rhodium occurs other than as a minor impurity (e.g., in sperrylite) is the recently characterized arsenosulfide *hollingworthite* (Wright and Fleischer, 1965).

Abundance and Geochemical Features

The abundance of all the platinum metals in the earth's crust is low, and since platinum and palladium together make up about 9/10 of the native alloys, rhodium is a relatively scarce element. Its crustal abundance is given as 0.001 ppm. The typical distribution is siderophilic, but some chalcophilic character is revealed by the less common occurrence in sulfides, tellurides, or arsenides. Thus platinum metals have been found to the extent of a few ppm in various minerals, especially sulfides and arsenides. Data are available from both Russian and Canadian sources showing that rhodium makes up a significant fraction of this platinum metal content, especially in the minerals pyrrhotite and pentlandite.

The rhodium content of igneous rocks is very low, of the order of parts per billion. In meteorites it has been found at concentrations of about 3 ppm in siderites and about 0.2 ppm in chondrites, and the rhodium in the latter is concentrated in the metal phase (see also *Platinum Metals*).

Economic Geology

The economic importance of rhodium is principally as an alloying element for platinum whose strength and hardness are thereby increased. Such platinum-rhodium alloys find numerous applications at high temperatures. Among the platinum metals rhodium ranks third in economic importance following platinum and palladium. The sources of platinum metals today are mainly primary deposits in northern Siberia, U.S.S.R., Northern Ontario, Canada, and the Transvaal, South Africa. (See *Platinum Metals*).

W. A. E. McBryde

References

Hampel, C. A. (editor), 1968, "Encyclopedia of the Chemical Elements," New York, Reinhold Publ. Corp., pp. 598–604.
Hampel, C. A. (editor), 1961, "Rare Metals Handbook," Second edition, New York, Reinhold Publ. Corp., pp. 304–335.
Rankama, K., and Sahama, Th. G., 1950, "Geochemistry," Chicago, University of Chicago Press.
Standen, A. (editor), Kirk-Othmer "Encyclopedia of Chemical Technology," Second ed., New York, John Wiley & Sons, Vol. 15, pp. 832–878.
Wright, T. L., and Fleischer, M., 1965, "Geochemistry of the Platinum Metals," *U.S. Geol. Surv. Bull.*, **1214-A**.

Cross-references: *Palladium; Platinum: Element and Geochemistry; Platinum Metals; Sulfides.*

RIVER GEOCHEMISTRY: ENVIRONMENTAL FACTORS

In discussing the geochemistry of rivers it is usual to subdivide the total material carried by a river into three categories: (*1*) the dissolved salts, (*2*) the suspended material, and (*3*) bed load. It must, however, be remembered that within the total material a complete gradation exists.

The dissolved salts of a river are those materials in true solution and, therefore, once mixed, remain uniformly distributed in the water mass, even when the water mass stops flowing.

The suspended materials of a river are solid particles ranging in size from colloids to coarse sand (1 mm) kept suspended within the water mass by the turbulent eddies of the flowing water. The coarseness of the material that can be carried increases with greater turbulence. Likewise, as the turbulence of a water mass decreases, the material settles out with the coarsest material settling first and the finest material remaining until turbulence has nearly ceased.

Bedload of a river is variously defined as (*1*) that material moved along the bottom by sliding and rolling, and/or (*2*) that material carried near, i.e., within 10 to 50 cm, the bottom. This material is always the most coarse material transported by the river. It is also the first material to be deposited when a decrease in the velocity of flow occurs. It should be noted, therefore, that the geochemical character of the total solid load of a single river can be drastically changed along the length of the river by changes of flow regime, such as a sluggish or accelerated flow span, a slow span through a lake or through a reservoir.

Factors

This discussion of the environmental factors involved in determining the geochemistry of rivers considers the factors related to natural, undisturbed river systems only. Since rivers are major suppliers of sediment to the ocean, it is this data of a river that is needed to help interpret the geological past. The effect of man on the natural system of a river can be tremendous, causing, for example, variation of erosion rates by many orders of magnitude.

The factors that control the overall geochemistry of a river are numerous. Of these, the major factors involved will be discussed here. It should be kept in mind constantly that the factors discussed are not always entirely independent of one another. For example, rainfall and vegetation may be related, as may be relief and vegetation.

The major environmental factors of a river basin can be grouped into four broad categories: climate, parent rock, relief, and vegetation, as briefly outlined in Table 1. A detailed summary of the various factors can be found in Leopold, Wolman, and Miller (1964).

Climate. The ultimate source of the water that forms rivers and lakes is precipitation over land areas. When the precipitation runs off the ground surface into rivers, it is called runoff. When it percolates into the ground to seep more slowly to the rivers, it is called groundwater. Precipitation is not pure H_2O. It contains gases absorbed from the atmosphere, sea salts derived from sea spray and carried over land (Gorham 1961), as well as solid particles carried as the nucleus of raindrops and snow crystals. This solid material is an insignificant source of solid materials compared to those materials derived from the land area drained by the river. Evidence of sea salt as a source of dissolved salts is observed in the decrease of the salinity of the precipitation inland from the sea. However, sea salts become a major source of dissolved salts in rivers in rare cases only, as in low salinity rivers (< 10 ppm) near the ocean.

The effect of precipitation on the geochemistry of the total material derived from a river basin is very complex and is not completely understood. In numerous studies it has been difficult to isolate the effect of precipitation from the other environmental factors, particularly that of relief. Discussion of some of the effects of precipitation on the geochemistry of rivers follows.

In order to isolate the relationship between annual precipitation, or annual runoff and annual dissolved load, further investigation is needed. However, the present state of knowledge permits several generalizations.

According to Leopold's (1964) observations concerning river loads of the United States, the dissolved load increases as annual runoff increases to approximately ten inches; whereas, in the case of annual runoff greater than ten inches, the dissolved load is approximately 125 tons per square mile.

The relationship between annual precipitation, or annual runoff, and annual sediment load given in Douglas (1967) can be seen in Fig. 1, along with the data of Langbein and Schumm (1958). The similar shapes of the curves should be noted as having their maximums in the low precipitation areas corresponding to areas of limited protective vegetation. The high erosion rates of Langbein and Schumm probably reflect the extensive effect of man on a number of the streams studied. As more data become available the relationship between runoff and suspended load will more likely be represented by a family of curves for varying relief of river basins, as is suggested from the data of Gibbs (1967).

The effect of different annual precipitations influences the ratio of dissolved salts to suspended solids in the following manner. With low precipitation, the suspended load, as a percentage of total load, can be high (up to 91%); in areas having high precipitation, the dissolved salts as a percentage of total load can be high

TABLE 1. MAJOR ENVIRONMENTAL FACTORS CONTROLLING RIVER GEOCHEMISTRY AND THEIR MEASURABLE PARAMETERS

Climate
Precipitation
   Total annual amount
   Form (snow or rain)
   Distribution through time
   Intensity of rain
   Pattern within basin
Temperature
   Mean annual
   Seasonal and daily variations
   Pattern within basin
Wind
   Intensity
   Direction and frequency
Humidity
   Amount and distribution

Parent Material
Type of rock exposed in basin
Distribution of rocks in basin
Coverage of rocks by soil and/or vegetation
Condition of rocks (massive, fractured, etc.)

Relief
Total relief of basin
Distribution of relief within basin
Slope

Vegetation
Total amount
Type of vegetation
Distribution within basin

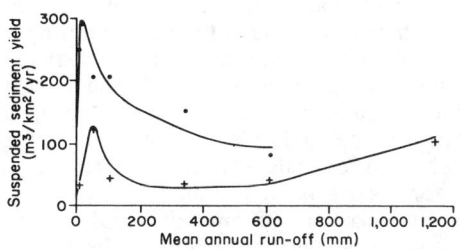

FIG. 1. Relationship between mean annual runoff (mm) and suspended sediment yield.

(up to 85%) (Leopold et al., 1964, p.77) and (Gibbs, 1967). In areas of high precipitation, particularly areas also having low relief, dense vegetation develops which protects the soil from erosion. Conversely, in regions of low precipitation, the sparsity of vegetation offers the soil little protection from erosion. It can be observed that these effects tend to cancel one another.

As runoff from seasonal precipitation increases, it produces a decrease in the salinity of a stream due to its diluting effect. Increased seasonal runoff also produces an increase in the concentration of the suspended material because of the generally higher velocities and related turbulence.

The distribution of precipitation can be significant. A long, light, steady drizzle will be far less effective in eroding solids than will a short, heavy downpour. Similarly, the distribution of precipitation throughout the year is significant. An area having its total annual precipitation concentrated in a rainy season, with a dry season in which vegetation growth is markedly restricted will have a much higher erosion rate than will another area having approximately the same annual precipitation evenly distributed throughout the year. This distribution of precipitation through time also influences the proportion of river water derived from runoff and from groundwater, these waters generally having different concentrations and compositions with their resulting effects on the geochemistry of the river.

Inasmuch as temperature controls whether H_2O is in the form of ice or water, it obviously influences the erosive processes. The type of weathering dominant in an environment is dictated, in part, by temperature, as in the case of tropical region chemical weathering and of Arctic region physical weathering.

Climate, to a great extent, dictates the overall dissolved salts composition of a river. For example, the water of humid regions are generally of the Ca–HCO_3^- type, whereas the waters of dry regions are generally of the Cl^- and SO_4^{2-} types.

Parent Material. The material which comprises the drainage basin of a river controls, to a certain extent, both the concentration of the dissolved and solid constituents carried in the river and their composition. The concentration of the sediment carried can vary, according to the ability of the material to be eroded. Erosion rates of solid materials are high for unconsolidated sediments and low for resistant rocks such as quartzites and granites. The concentrations of dissolved salts likewise vary with rock type, as shown in the work of Miller (1961) in studying the dissolved salts drained from areas characterized by three distinct rock types: quartzite, granite and sandstone-shale intermixed with a small amount of limestone. The average solute concentrations of the waters draining these respective areas are given, in proportion, as 1:2.5:10. The relationship of salinity and "basic" rocks in the drainage basins of rivers has been discussed by Gorham (1961) and by Gibbs (1967).

Few studies relating the overall composition of the sediment load to parent rock have provided sufficient control for isolating the effect of the parent material from climatic, topographic and vegetal factors. In a study of the tributaries of the Amazon River (Gibbs, 1967), the importance of parent rock was indicated by the observation that the majority of the sediments of the Amazon are simply physically broken-down preexisting minerals derived from the Andes Mountains. In the extensive chemical weathering of the Amazon rain forest basins, where the majority of the sediments are products of chemical processes, the parent material is again important. While kaolinite is the dominant clay mineral in these systems, the percentage of "calcic" rocks in each basin controls the percentage of montmorillonite eroded. It should be noted, as a generalization, that under differing climatic conditions the same parent material can weather to different end products.

The composition of the dissolved salts is related to the parent material. For example, in basins having a high percentage of limestone, the amount of Ca is high. In those having a high percentage of dolomite, the amounts of Mg and Ca are high. High percentages of evaporites provide high amounts of Cl^- and SO_4^{2-}. Numerous other more subtle relationships can be observed.

Relief. The topography of a river basin governs the potential energy of the water flowing in its streams. The steeper the slope, the faster the water moving down it flows and the greater is its erosive force. In addition, never is the slope of a river basin uniform. Rather, slope variation from place to place within a single river valley can be great. Actual average slope is impossible to measure practically, therefore, other related parameters are measured, such as (1) the ratio of maximum relief (vertical distance between greatest height and basin mouth) to maximum length, (2) the maximum valley-side slope, (3) an operationally defined mean slope, and (4) the hypsometric curve properties and integrals.

The relationship between the area-weighted mean relief above base level and the dissolved salts and suspended sediment concentrations of the tributaries of the Amazon River is shown

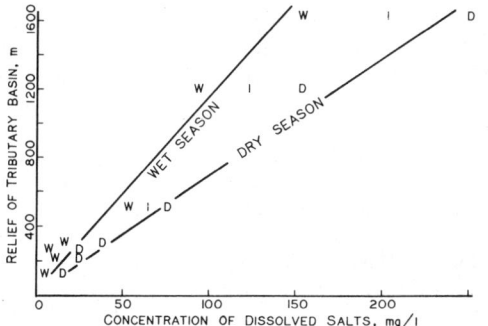

FIG. 2. Variation of dissolved salts of tributaries with relief: (W), wet-season value; (I) intermediate value; and (D), dry-season value.

in Figs. 2 and 3. It will be noted that the relief of a tributary basin dictates its possible range of concentration; seasonal precipitation governs the position of concentration within this range. The sediment load is generally an increasing percentage of the total load with increasing relief of a river basin.

Vegetation. The vegetal factor is so closely linked to climate (precipitation, temperature, etc.) that it is difficult to isolate its effect on the geochemistry of a river. In a study of soil erodibility in northern California wildlands, André and Anderson (1961) found that erodibility was greatest beneath brush, next highest beneath trees and least beneath grass. In addition to the mainly physical relationship of vegetation to erodibility of soils, the chemical effects may be equally important, since the type of vegetation influences the amount of organic material in the soil as well as pH and the amounts of organic acids.

Summary

Several major factors that control the geochemistry of rivers have been discussed in-

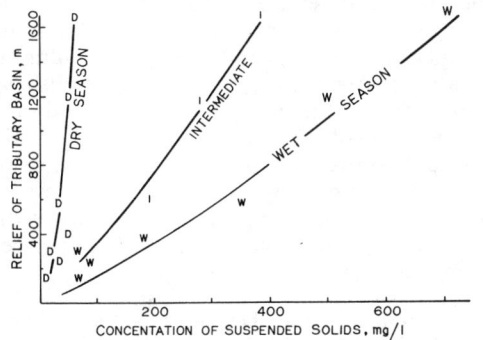

FIG. 3. Variation of suspended solids concentration of tributaries with relief: (W), wet-season value; (I), intermediate value; and (D), dry-season value.

RIVER GEOCHEMISTRY: REGIONAL

dividually. The problem of studying them arises in trying to isolate the effect of each in an environment in which they are all actively present and interrelated to varying degrees. The current state of knowledge regarding this problem does not yet allow each factor to be stated as a quantitative law or rule. Many generalizations concerning the factors can, however, be made based on investigations to date. Of the major factors, climate and relief appear to attain greatest significance in the control of the geochemistry of rivers.

RONALD J. GIBBS

References

André, J. E., and Anderson, H. W., 1961, "Variation of soil erodibility with geology, geographic zone, elevation, and vegetation type in northern California wildlands," *J. Geophys. Res.*, **66**, 3351–3358.

Douglas, I., 1967, "Man, vegetation and the sediment yields of rivers," *Nature*, **215**, 925–928.

Gibbs, R. J., 1967, "The geochemistry of the Amazon River System: Part I. The factors that control the salinity and the composition and concentration of the suspended solids," *Geol. Soc. Am. Bull.*, **78**, 1203–1232.

Gorham, E., 1961, "Factors influencing supply of major ions to inland waters with special reference to atmosphere," *Geol. Soc. Am. Bull.*, **72**, 795–840.

Langbein, W. B., and Schumm, S. A., 1958, "Yield of sediment in relation to mean annual precipitation," *Am. Geophys. Union Trans.*, **39**, 1076–1084.

Leopold, L. B., Wolman, M. G., and Miller, J. P., 1964, "Fluvial Processes in Geomorphology," San Francisco, W. H. Freeman & Co., 522pp.

Miller, J. P., 1961, "Solutes in small streams draining single rock types, Sangre de Cristo Range, New Mexico., *U.S. Geol. Surv. Water-supply Paper*, **1535-F**, 23pp.

Cross-references: *Colloids; Groundwater; River Geochemistry: Regional; Runoff; Weathering, Chemical.* Vol. III: *Anthropogenic Influences in Geomorphology; Rivers; Sediment Transport—Fluvial and Marine; Slopes; Weathering.* Vol. VI: *Weathering—Organic.*

RIVER GEOCHEMISTRY: REGIONAL

During the past decade, the chemistry of fresh waters has become an area of increasingly active research. This interest is derived, partially, from concern with water pollution. A thorough examination of the geochemistry of the world's rivers is made difficult since the waters of many of the large rivers of the world have never undergone geochemical analysis and the data for others are totally inadequate for any comprehensive evaluation.

Ideal data for geochemical research on a

river are its discharge-weighted salinity and its composition, based on samples taken at intervals dictated by the variability of the particular river. Monthly sampling is usually adequate for larger rivers. The sampling points selected should be free from contamination of seawater and of industrial, agricultural, and residential pollutants. Numerous published geochemical analyses of major rivers are based on a single sample which represents the river at a single seasonal state. Unfortunately, the resulting data present poor approximations of true values.

The factors that control the geochemistry of a river are discussed in detail in the article *River Geochemistry: Environmental Factors* appearing in this volume. These factors can be grouped into two categories: those that are variable with time and those that are relatively unchanged over a seasonal cycle. The factors that do not vary over a period of a year are relief, type of rock, shape of the river basin, vegetation, etc. These factors pose no sampling problem inasmuch as they are not the causes of seasonal variations in a river's geochemistry. It is those environmental factors which vary with time that pose difficulties in sampling rivers. These factors are, obviously, the climatic factors including temperature, precipitation, and evaporation. A river system responds to variations in these factors and thus its concentration and compositional parameters can vary through the year.

Geochemical Data Variability

The environmental factors that vary with time, and thus influence variations of the geochemical state of a river with time, must be reflected in the analytical results of sampling that is unbiased not only with respect to location but also seasonally. The major factors that influence the geochemistry of rivers during a seasonal cycle are the climatic factors, including *temperature, precipitation,* and *evaporation.*

The very form of the water of a river, whether flowing or frozen, is determined by differences in *temperature,* both seasonally and with respect to geographic locale. The form and mobility of the climatic factor precipitation is also affected by these differences in temperature.

Evaporation influences the amount of water supplied to a river. Variations in evaporation are due to the variable states of such meteorologic factors as wind, cloud cover, and humidity.

The amount and the seasonal and areal distribution of *precipitation* help make this climatic factor the dominant factor among those which change seasonally and thus influence the geochemistry of rivers with time.

The composition of the precipitation that falls on a river basin can vary, as Gorham (1958) has pointed out, with the presence or absence of dissolved salts from sea salt or from industrial pollution carried above the river basin by wind. When precipitation falls upon the basin, reactions with rocks and soils add soluble salts. Little soluble salts are added to the salts of precipitation by the soils of some tropical basins that have been thoroughly leached and have large volumes of water passing through them. For example, the Xingu, a tributary of the Amazon River, shows a remarkably low erosion rate by solution (Table 3). In river basins where rocks and soils have not been leached to such an extent, salts are readily available to be dissolved in the water.

The variability of a river's salinity and composition can also be related to the routes the major portion of the precipitation takes through the river basin on its way to the river at different times of the year. The precipitation that falls upon a river basin reaches the river by two means. The first of these is the direct runoff of the precipitation across the surface into the streams and the river. The second means is the indirect percolation of the precipitation as groundwater through the soils and rocks of the basin before it enters the streams and the river. Obviously, direct runoff allows time for minimum reaction with soils and rock, while groundwater allows time for maximum reaction. For most river systems, the groundwater flow is a more constant source of supply than is runoff, which can, of course, have wide seasonal variation. A striking example of the seasonal variation of the effect of precipitation on a river's salinity and composition are the temperature and cold climate river basins where winter surface runoff is negligible or absent and the streams are supplied solely from groundwater, while at the time of the spring melting of frozen precipitation, runoff becomes the dominant source of supply. The winter groundwater will generally have a higher salinity than the spring runoff waters, indicating the greater reaction time allowed by the slower percolation of the precipitation as groundwater. Recogntion of the several means by which precipitation reaches a river, in addition to the effect of seasonal variations, again reinforces the need for year-long sampling in order that the data obtained can be meaningful in a geochemical evaluation.

Geochemical Data that do not Vary Significantly with Time

As noted above, the environmental factors that affect the geochemistry of a river that do not vary significantly through the year are the geometric and compositional factors of the river basin. These geometric factors include relief, slope and basin shape while the compositional factors include the rocks and soils of the basin and the vegetation that covers the basin. Since these environmental factors do not vary significantly seasonally or throughout the year, they have little direct effect on the changes that occur in the geochemistry of a particular river basin during these spans of time. The range and effect of these factors on river geochemistry is discussed in detail in the entry *River Geochemistry: Environmental Factors*.

World River Geochemistry

The world average composition values given in Table 1 (from Livingstone, 1963) are discharge-weighted mean composition values obtained from data that range in quality from excellent to very poor. The world average salinity value given in Table 1 is a discharge-weighted mean salinity derived from the data of Livingstone (1963) incorporated with recent data on the Amazon River from Gibbs (1967). Many of the important rivers for which only poor quality data are available include, in South America: the Paraná, Orinoco, Magdalena, Tocantins, and São Francisco Rivers; in Africa: the Congo, Zambezi, Niger and Nile; and in Eurasia: the Yangtze, Bramaputra, Ganges, Lena, Irrawaddy, and Indus. On the whole, better data are available for the rivers of North America than for those of other continents. However, for some of the North American rivers limited analyses only are available with no discharge-weighted mean values based on regular sampling over a period of time.

The rivers of the world can be classified generally into three groups according to the chemical composition of their major cations and anions: (1) those rivers high in Ca–HCO$_3$, (2) those high in Ca–SO$_4$, and (3) those high in Na–Cl. A general increase in the total amount of dissolved salts can be observed, progressing from rivers high in Ca–HCO$_3$ to those high in Ca–SO$_4$ and on to those waters high in Na–Cl. This increase in salinity from the Ca and HCO$_3$, apexes to the Na–Cl apexes can be seen in Table 1 as progressing from the dilute Ca–HCO$_3$ river-type to hypersaline waters with a 29000-fold increase in Cl, an 18000-fold increase in Na, and a 5600-fold increase in total dissolved salts. It will be noted that the world river discharge-weighted mean chemical composition and total dissolved salts values lie between the dilute Ca–HCO$_3$ river type and the Ca–HCO$_3$ river type, indicating that by far the vast majority of river discharge is of the Ca–HCO$_3$ type.

TABLE 1. DISSOLVED SOLIDS: COMPOSITION OF VARIOUS RIVERS AND OTHER WATERS

	HCO$_3$	SO$_4$	Cl	Na	Mg	Ca	Fe	SiO$_2$	Total Dissolved Solids
World average	58.4	11.2	7.8	6.3	4.1	15.0	0.67	13.1	114
Dilute Ca-HCO$_3$-type river[a]	17	3.0	1.7	1.8	1.1	4.3	0.05	7.0	36
Ca-HCO$_3$-type river[b]	101	41	15	11	7.6	34	0.02	5.9	221
Ca-SO$_4$-type river[c]	183	289	113	124	30	94	0.01	14	853
Na-Cl-type river[d]	238	174	473	253	71	80	—	—	1310
Seawater[e]	139.7	2649	18980	10556	1272	400.1	0.002-02	0.02-4	34,481
Hypersaline water[f]	T	6680	55480	33170	2760	160	—	—	203,490

[a] Amazon River (Gibbs, 1967; and Gibbs, in preparation).
[b] Mississippi River near Baton Rouge, La. (Livingstone, 1963).
[c] Colorado River at Yuma, Arizona (Livingstone, 1963).
[d] Jordan River at Jericho (Livingstone, 1963).
[e] Average seawater (Sverdrup et al., 1942).
[f] Great Salt Lake (Clarke, 1924).

TABLE 2. DATA ON THE WORLD'S TWENTY-FIVE LARGEST RIVERS

	Discharge[a] (10^{10} m³/yr)	Area[b] (10^3/km²)	Solution (10^{12} g/yr)	Solids (10^{12} g/yr)	Annual Erosion[d] Solution (10^6 g/km²/yr)	Solids (10^6 g/km²/yr)	Total (10^6 g/km²/yr)	% Erosion by Solution
1. Amazon	552.02	6300	232[e]	499[e]	36.8[e]	79.2[e]	116.6[e]	21.7[e]
2. Congo	125.10	3885.0	98.5[e]	31[e]	25.4[e]	8[e]	33.4[e]	76[e]
3. Yangtze	68.80	1942.5		500[f]		257[f]		
4. Brahmaputra	62.55	934.9		726[f]		776.6[f]		
5. Ganges	58.97	1059.3	89.6[e]	1600[f]	84.6[e]	1510[f]	1594.6	5.3
6. Yenisei	54.86	2590.0	30.9	10.5	11.4	4.0	15.4	74
7. Mississippi	54.59	3221.9	118[e]	213[e]	36.6[e]	66.1[e]	102.7	36
8. Orinoco	53.61	880.6	29.0[c]	86.3[g]	32.9[c]	98[g]	130.87	25.2
9. Lena	48.88	2424.2	69.9		28.9[c]			
10. Parana	47.00	2305.1	63.4[c]	121[f]	27.5[c]	52.5[f]	90.0	31
11. St. Lawrence	44.68	1289.8	71.9[c]	3	5.6[c]	4	9.6	58.3
12. Irrawaddy	42.80	429.9		300[f]		698[f]		
13. Ob	39.40	2483.8	29.6	14.2	12.2	6	18.2	67.0
14. Mekong	34.85	802.9	60.6[e]	1000	75[c]	1200	1275	5.8
15. Tocantins	34.69	906.5						
16. Amur	34.67	1844.0	18.6	52	10.1	28	38.1	26.5
17. MacKenzie	25.02	1805.2	58.7	9.02	32.5	5	37.5	86.8
18. Magdalena	23.68	240.8						
19. Columbia	22.87	668.2	43.7[c]	36	65.4[c]	47	112.4	58.2
20. Zambezi	22.34	1295.0		100		75		
21. Danube	19.48	815.8	9.1[c]	5.5	11[c]	101	112	62.5
22. Niger	19.21	1113.7	7.8[c]	35	3.7[c]	32	35.7	10.5
23. Indus	17.51	927.2	45.9[c]	400	49.5[c]	420	469.5	10.5
24. Yukon	16.08	932.4	42.8[c]	88	46.2[c]	103	149.2	31
25. Pechora	12.86	326.3	5.6	6.5	17.0	20	37.0	45.9

[a],[b] From Am. Geol. Inst., 1964.
[c] Calculated from data in Livingstone, 1963.
[d] From Strakhov, 1967 except as noted.
[e] From Gibbs, 1967.
[f] From Holeman, 1968.

For purposes of comparison the composition of seawater is given in Table 1. The composition of seawater in the various oceans and seas of the world is constant relative to the widely varying chemical compositions of fresh water. The shift in the compositions of the waters from the Ca and HCO_3 apexes to the Na and Cl apexes is caused mainly by the precipitation of $CaCO_3$, with the resulting concentration of the solution by evaporation.

Rivers and streams have generally higher silica content values (world average 13.1 ppm) than the oceans (seawater average 0.02–4 ppm) (Table 1). The lower silica content value of the oceans is customarily attributed to the diatoms and other creatures that secrete silica shells thereby removing soluble silica from' the ocean water.

The low values of iron content in both fresh and ocean water (Table 1) is due to the fact that the major portion of the iron is in its ferric form, precipitating as ferric hydroxide. The low concentrations of other metals in waters are explained similarly.

The pH of rivers varies from a low pH value of 4 in some tropical streams having high amounts of organic acids, with an even lower pH value in some streams that drain mining areas in which pyrite or metallic sulfides oxidize to form sulfuric acid to a high pH value of well above 9. The average pH value, however, is between 6 and 8.

In Table 2 are tabulated data on the twenty-five largest rivers of the world ranked according to discharge. It will be noted that although the Amazon River is, by far, the river having the greatest discharge in total volume, several other rivers carry more total sediment per year. These are the Yangtze, the Brahmaputra, the Ganges, and the Mekong. The high quantity of sediment discharged from the Brahmaputra and Ganges is derived from the large percent of their basins draining the highly mountainous Himalayan area with its abundant glaciers. The Yangtze and Mekong drain areas of extensive cultivation and abundant, easily eroded loess deposits. The Amazon River System has 82% of its basin as low-relief tropical rain forest with only 18%, approximately, in the Andean mountainous environment. With regard to annual dissolved salts erosion rates, however, the Amazon River ranks first with the Mississippi River second.

The annual erosion rates per square kilometer provide an interesting study inasmuch as these values take into account the sizes of the river basins. The trend of river systems that have a large percentage of their basins in highly mountainous areas also having the higher rates of erosion is again correlated. River basins having high relief (i.e., highly mountainous basins) have high erosion rates for both solid materials and dissolved salts and generally low ($<50\%$) percent erosion by solution compared to low relief (nonmountainous) river basins, which have erosion values by solution generally above 50%.

In studying the data of large river systems, it must be remembered that each river system includes a wide range of environments which tend to average out extreme erosion rates. For example, in Table 3 are given data for two high-relief (mountainous) tributaries and two low-relief (nonmountainous) tributaries of the Amazon River.

It is observed that within the Amazon River System the dissolved salts erosion rate varies by more than 50-fold and the solid material erosion rate varies by about 340-fold. The solid material erosion rates for the large rivers of the world vary from one river basin to the next and, with time, vary far more than do the dissolved salts erosion rates.

The reader is referred to Livingstone (1963) and to Hutchinson (1957) for the more detailed information on the geochemistry of the major rivers of the world which cannot be examined thoroughly in a short article. The factors that control the geochemistry of rivers is discussed in greater detail in the article *River*

TABLE 3. EROSION RATES FOR TRIBUTARIES OF THE AMAZON RIVER

	Dissolved Salts Erosion Rate (10^6 g/km^2)	Solid Material Erosion Rate (10^6 g/km^2/yr)	Total	Percent of Total Erosion by Solution
		High-Relief Tributaries		
Maranon	92.8	251.5	344.3	27
Ucayali	152.0	307.1	459.1	33
		Low-Relief Tributaries		
Tapajos	3.8	1.2	5.0	76
Xingu	2.8	0.9	3.7	75
Amazon	36.8	79	115.8	32

Geochemistry: Environmental Factors in the present volume.

RONALD J. GIBBS

References

American Geological Institute, 1964, "Principal Rivers," *Am. Geol. Inst. Data Sheet No.* **32**, Washington, D.C.

Clarke, F. W., 1924, "The data of geochemistry," Fifth edition, *U.S. Geol. Surv. Bull.*, **770**, 841pp.

Gibbs, R. J., 1967, "The geochemistry of the Amazon River System, Part I, The factors that control the salinity and the composition and concentration of the suspended solids," *Geol. Soc. Am. Bull.*, **78**, 1203–1232.

Gibbs, R. J., "The geochemistry of the Amazon River System: Part II, The chemistry of and the factors that control the composition and concentration of the dissolved salts," (in preparation).

Gorham, E., 1958, "Factors influencing supply of major ions to inland waters, with special reference to the atmosphere," *Geol. Soc. Am. Bull.*, **72**, 795–840.

Holeman, J. N., 1968, "The sediment yield of major rivers of the world," *Water Resources Res.*, **4**, 737–747.

Hutchinson, G. E., 1957, "A treatise on Limnology," Vol. 1: "Geography, physics and chemistry." New York, John Wiley & Sons, 1015pp.

Livingstone, D. A., 1963, "Chemical composition of rivers and lakes," *U.S. Geol. Surv. Profess. Paper* **440-G**, 64pp.

Strakhov, N. M., 1967, "Principles of Lithogenesis," New York, Consultants Bureau, 245pp.

Sverdrup, H. U., Johnson, M. W., and Fleming, R. H., 1942, "The Oceans," Englewood Cliffs, N.J., Prentice-Hall, 1087pp.

van Andel, T. H., 1967, "The Orinoco Delta," *J. Sed. Petrol.* **37**, 297–310.

Cross-references: *Groundwater; Hydrologic Cycle; Hydrology; pH-Eh; Pollution; River Geochemistry: Environmental Factors; Runoff; Seawater, Chemistry; Weathering, Chemical. Vol. I: Salinity in the Ocean. Vol. III: Continental Erosion; Denudation.*

ROCK–FORMING MINERALS—See Vol. IV B

RUBIDIUM: ELEMENT AND GEOCHEMISTRY

Physical Properties. Rubidium, symbol Rb; atomic weight 85.48; atomic number 37; electron configuration (Kr core) $4s^2 4p^6 5s^1$; m.p. 38.5°C; b.p. 700°C; sp.gr. 1.475 (38.5°C); valence +1. Rubidium is a soft, white, silvery metal.

Natural Rb is composed of two isotopes, Rb^{85} (72.15%) and Rb^{87} (27.85%). Rb^{87} is radioactive and is transformed by beta decay to Sr^{87}. This scheme forms the basis of the important method of Rb–Sr geochronology (see *Rubidium-Strontium Dating Method*).

Minerals. Rubidium is camouflaged by potassium in potassium minerals and forms no minerals of its own. It is concentrated, often to the extent of several percent, in lepidolite, amazonite, and the cesium mineral pollucite. Muscovite and biotite are the most important rubidium carriers in common rocks. These micas often contain several thousand ppm rubidium. Potassium feldspars are poorer in rubidium by a factor of two or three.

Chemistry. The chemical properties of rubidium are very similar to those of potassium. Ionic radii and electronegatives are similar; both have a noble gas electron configuration, and both are limited to a single valence state, +1. Rubidium tends to be enriched in structures with higher coordination numbers because of its slightly greater ionic radius.

Elemental rubidium is a very active metal. Of the naturally occurring elements, only cesium has more pronounced metallic properties. Rubidium decomposes water with the evolution of hydrogen. In solutions, it forms nonhydrated univalent ions. Its common salts are soluble ionic compounds.

Geochemistry. The geochemistry of rubidium is dominated by its extensive diadochic relation with potassium. For this reason, rubidium is usually discussed in terms of its abundance with respect to potassium (i.e., the K/Rb ratio). Indeed, the geochemical coherence of these two elements is so definite that for some problems, data on K–Rb fractionation may be as instructive as data on the fractionation of stable isotope pairs.

Extensive work by Ahrens and others (see Heier and Adams, 1964 for a modern review and bibliography) has established that the K/Rb ratio in most rocks lies between 160 and 300. The mean value for crustal materials is 240. Different K/Rb ratios outside of these limits (and perhaps within them) probably represent real differences that are related to geological processes or conditions.

Chondritic meteorites have K/Rb ratios that are within the limits established for crustal rocks. These data support the view that chondrites are derived from a planet whose general composition and differentiation history parallels that of the earth. Some authorities have suggested that the earth's mantle has a chondritic composition. However, recent work shows that some Hawaiian basalts of deep origin are depleted in rubidium; their K/Rb ratios (which are in the vicinity of 500) are closer to those reported for achondritic meteorites. These same basalts commonly have lower Sr^{87}/Sr^{86} ratios than do crustal rocks. Because Sr^{87} is produced by the decay of Rb^{87}, the lower strontium isotope ratios are consistent with an assumption of rubidium impoverishment at depth. Consequently, it seems plausible that a disproportionate share of the earth's rubidium is concentrated in the crust.

The geochemistry of rubidium in igneous rocks is quite well understood. Igneous differentiation causes a slight relative enrichment in rubidium, therefore K/Rb ratios in basalts are usually somewhat higher than are K/Rb ratios in associated differentiates. Pegmatites are extremely enriched in rubidium. Pegmatitic micas and potassium feldspars contain among the lowest K/Rb ratios found in natural materials. Their K/Rb ratios may be as low as 10 and values near 50 are common.

Most low-grade metamorphic rocks have K/Rb ratios of approximately 240, indicating a retention of potassium and rubidium from normal premetamorphic crustal materials. scanty data suggest, however, that rocks from the amphibolite and particularly the granulite facies are impoverished in rubidium. Hence the K/Rb ratio might be, to some extent, an index of metamorphic grade. This interesting possibility must be substantiated by further research.

The behavior of rubidium in sedimentary processes is controlled largely by adsorption on clay minerals. Illites and montmorillonites are the most common clays with large cation exchange capacities and both adsorb rubidium more strongly than potassium. This behavior leads to a temporary decrease in K/Rb during the weathering process; presumably, potassium is more quickly leached away than rubidium. Intermediate weathering products may have K/Rb ratios of 30 to 80. Eventually, both potassium and rubidium are removed.

Rubidium and potassium are largely removed from seawater by adsorption on clays. Again, the exchange equilibria favor preferential removal of rubidium; the K/Rb ratio in seawater is 3130, whereas most sediments have values close to the crustal average of 240. Undoubtedly, unweathered clastic debris (e.g., micas, potassium feldspars) partially controls the K/Rb ratio, but studies of authigenic illite and glauconite indicate that these phases too have K/Rb ratios near 240. It appears that potassium and rubidium additions are balanced by sedimentary removals. The fact that old Procambrian sediments also have K/Rb ratios near 240 indicates that this balance has been in existence for a significant portion of geologic time.

R. C. REYNOLDS

RUBIDIUM: ECONOMIC GEOLOGY

References

*Heier, K. S., and Adams, J. A. S., 1964, "The Geochemistry of the Alkali Metals," in (Ahrens, L., et al., editors), "Physics and Chemistry of the Earth," Vol. 5, New York, Macmillan, p. 253.

Tatsumoto, M., et al., 1965, "Potassium, rubidium, strontium, thorium, uranium, and the ratio of strontium-87 to strontium-86 in oceanic tholeiitic basalt," *Science,* 150(3698), 886–888.

Vlasov, K. A. (editor), 1966, "Geochemistry and Mineralogy of Rare Elements," Jerusalem, Israel Program for Sci. Transl. (from the Russian), 2 vols.

*Additional bibliographic references may be found in this work.

Cross-references: *Earth's Crust Geochemistry; Mantle Geochemistry; Potassium: Element and Geochemistry; Potassium-Rubidium Ratio in Geology; Rubidium-Strontium Dating Method; Seawater, Chemistry; Weathering, Chemical.* Vol. V: *Metamorphic Grade.*

RUBIDIUM: ECONOMIC GEOLOGY

Although rubidium is widely distributed in the earth's crust, and actually more abundant than lead, copper, or zinc, high concentrations are rather uncommon. It is found, in trace concentrations, in a great many rocks (usually 0.02–0.09%), including granites, basalts, shales, and limestones. In the oceans its range is only about 1.2–3.5 g/ton (or hundred micrograms/liter); this low concentration may be because these ions are strongly absorbed by the clay minerals. In the "fresher" seas, such as the Black Sea the concentration is 4.5 g/ton and in the Caspian 570 g/ton. Some mineral springs measure up to 1.1 mg/l. Rb is preferentially concentrated in some of the potash evaporites. In the carnallite, $KMgCl_3 \cdot 6H_2O$, of Stassfurt, East Germany, there is 0.037–0.15% Rb.

Rubidium metal was first recognized spectroscopically from a sample of lepidolite from Saxony (later analyzed as containing 0.24% Rb_2O). There is no mineral with Rb as its principal component and most commercial sources of Rb today come from lepidolite, a type of lithium mica, $LiKAl_2(OH,F)_2(Si_2O_5)_2$, commonly found in "greisen" or stringers in pegmatites, i.e., late-stage crystallization products of granitic emplacement. Useful Rb-bearing lepidolites range from about 1.0 to 4.6% Rb_2O, but it should be remembered that the lepidolite is never found in rock-forming concentrations, but dispersed with other minerals (quartz, feldspars, biotite, etc.) within the pegmatite dike.

Because of its similarities (including ionic radius) to potassium the Rb tends to occupy spaces within crystal lattices normally occupied by K, and in some potash feldspars Rb can reach 25,000 ppm.

Other Rb source minerals include the cesium ore pollucite $(Cs,Na)AlSi_2O_6 \cdot nH_2O$ (with 0.06–1.15% Rb) from Maine, Manitoba, and Rhodesia; and green microcline, sometimes known as amazonstone or amazonite (0.87–1.13% Rb) from Colorado, East Germany, and Madagascar.

Uses. The lack of specific Rb minerals makes this alkali metal an expensive one to produce and uses up till recently have been limited mainly to experimental work. It has been considered for fueling an ion propulsion engine for space vehicles, but cesium is probably better.

Its major commercial use has been, thanks to its photoelectric properties, for vacuum tubes and photocells. The same is true for heat conversion to electricity using the thermionic process. Rubidium carbonates have been used for making some special glasses.

Rubidium metal can cause skin burns, and some of its common salts are highly toxic, e.g., the fluoride, cyanide, and hydroxide of rubidium.

The only producer of primary rubidium in the United States (as of 1967) was Penn Rare Metals Division of the Kawecki Chemical Co. (Revere, Pa.) Rubidium is normally handled along with cesium. High-grade Rb metal costs up to $4.00/g, the carbonate salt only about 30¢.

R.W.F.

References

Eilertsen, D. E., 1965, "Rubidium," in "Mineral Facts and Problem," Washington, *U.S. Bur. Mines, Bull.,* **630**.

Mosheim, C. E., 1968, "Rubidium," in (Hampel, C. A. editor), "The Encyclopedia of the Chemical Elements," New York, Reinhold Book Corp., 604–610.

Perel'man, F. H., 1965, "Rubidium and Cesium," New York, Macmillan & Co.

Stevens, R. E., and Schelle, W. T., 1942, "The rare alkalis in mica," *Am. Min.,* **27**, 525–537.

U.S. Bureau of Mines, 1968, Minerals Yearbook 1967, Washington.

Vlasov, K. A. (editor), 1966, "Geochemistry and Mineralogy of Rare Elements," Jerusalem, Israel Program for Sci. Transl. (from the Russian), 2 vols.

Cross-references: *Cesium; Earth's Crust Geochemistry; Potassium;* Vol. V: *Pegmatite.*

RUBIDIUM–STRONTIUM DATING METHOD

The rubidium-strontium method is used primarily to measure the ages of rocks and minerals. The underlying principle makes use of the fact that rubidium-87 undergoes radioactive decay by emission of a beta particle to produce strontium-87. Any rubidium-bearing material will thus experience a gradual decrease in its content of rubidium-87 and a parallel increase in its strontium-87 content with time. The age of the specimen can be calculated from the measured ratio of daughter to parent nuclide and a knowledge of the decay rate.

The Rb–Sr method has been extended beyond rock dating to include studies on the age and origin of meteorites and the earth; the development of the earth's crust and the composition of the mantle; and the gain or loss of alkalis and alkaline earths in minerals during certain geological processes.

Table 1 shows the isotopic compositions of Rb and Sr; the Sr abundances are only approximate because of variable quantities of radiogenic Sr in different geologic materials. The enrichment by radiogenic Sr^{87} is small in most samples, resulting in a practical lower limit on the ages that can be measured accurately. The chief analytical uncertainty is usually related to the ratio of "common" Sr^{87} (present when the rock formed) to radiogenic Sr^{87}. For samples with excessive common Sr or for very young samples, the radiogenic increment is relatively minute and, hence, undetectable. The youngest published age is seven million years on extremely Rb-rich pollucite and lepidolite from Elba (Ferrara et al., 1961). The K–Ar method has far greater potential for the measurement of young ages; values of only a few thousand years have been determined by this method.

The application of the Rb–Sr method to the dating of geological events depends on the suitability of the material to be dated. This relates to the geochemical behavior of Rb and Sr. Because the most favorable cases are those where the Rb/Sr ratio is large, potassium minerals are used primarily. There are no Rb minerals. Measurements have commonly been made on biotite, lepidolite, phlogopite, glauconite, mus-

TABLE 1. ISOTOPIC COMPOSITION OF RUBIDIUM AND STRONTIUM

Isotope	Abundance (atom fraction)	
	Rubidium	Strontium
84	—	0.0056 (stable)
85	0.722 (stable)	—
86	—	0.099 (stable)
87	0.278 (radio-active, $\beta-$)	0.070 (stable)
88	—	0.826 (stable)

covite, and the potassium feldspars. Other minerals which have been dated include: alkali amphiboles, sylvite, illite, and various phases of stony meteorites. In addition, samples of whole rocks have also been dated.

A rubidium-strontium age is calculated from the relationship:

$$t = \frac{1}{\lambda} \ln\left(1 + \frac{Sr^{87*}}{Rb^{87}}\right) \quad (1)$$

where Rb^{87} and Sr^{87*} are respectively the number of atoms of Rb^{87} and radiogenic Sr^{87} (measured at the present), λ is the radioactive decay constant ($\lambda = 0.693$/half-life), and t is time. The value of λ is difficult to obtain by direct laboratory measurements because of the large proportion of low energy beta particles in the Rb^{87} decay energy spectrum. One half-life adopted by many investigators is 4.7×10^{10} yr ($\lambda = 1.47 \times 10^{-11}$ yr^{-1}) determined by liquid scintillation counting techniques. Another much-used value, 5.0×10^{10} yr ($\lambda = 1.39 \times 10^{-11}$ yr^{-1}), was calculated from Rb–Sr analyses of minerals from pegmatites of known age. In spite of a number of recent determinations, the half-life discrepancy has not yet been resolved. Ages quoted in this article are recalculated with $\lambda = 1.47 \times 10^{-11}$ yr^{-1}. A possible future revision of the accepted half-life will not invalidate already determined daughter/parent ratios. The ages may be simply recalculated.

Method of Isotopic Analysis

Isotopic abundances are measured by isotope dilution using a mass spectrometer. Fig. 1 shows a typical Sr analysis as a specific example of the procedure. In the mass spectrum of common Sr (part A), the peak heights correspond to the relative abundances of Sr isotopes seen in Table 1. A portion of the Sr^{87} peak consists of radiogenic strontium. A known quantity of "spike" (artificially enriched in Sr^{84}, part B) is added and thoroughly mixed with the unknown quantity of Sr in the sample. The spectrum shown in part C is the total of contributions by common, radiogenic, and spike strontium. It is apparent that if the number of atoms of spike Sr^{84} is known, the quantities of all the other components can be obtained by comparison. For Rb analyses, a spike enriched in Rb^{87} is used.

Isotope ratios can be measured routinely with a precision of $\pm 0.05\%$ or better, and quantities of Rb and Sr to $\pm$ 1 to 2%. The measured Sr isotope ratios are corrected for mass-dependent fractionation in the mass spectrometer. The assumption is made that Sr^{86}/Sr^{88} is invariant in nature, and measured ratios are normalized to a standard reference $Sr^{86}/Sr^{88} = 0.1194$.

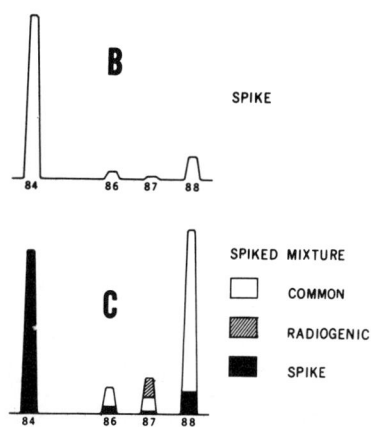

FIG. 1. Mass spectra of common and radiogenic Sr (A), spike Sr (B), and a mixture containing common, radiogenic, and spike Sr (C). (See text for discussion.)

Application of the Rb–Sr Method

Perhaps the most important problem in the determination of Rb–Sr ages is whether or not the system has remained closed to Sr and Rb exchange during its history. This is primarily a geological question. Factors affecting this assumption are: temperature, pressure, hydrothermal or groundwaters, and the activity of components in such waters. These in turn relate to depth of burial, metamorphism, metasomatism, groundwater alteration, and weathering. Some of these effects are readily observed in hand specimen or thin section. Others are subtle and have thus far escaped a means of detection which is independent of the measured isotopic abundances.

It is clear that the simplest age measurement would be for a mineral with a high Rb/Sr ratio which has experienced no significant external effects during its existence. The age obtained would then be the time when the mineral formed. An example might be that of a mica which formed from a cooling high level intrusive melt and which experienced no later events such as metamorphism. Normally, it is assumed that the date also gives the age of the intrusion. A second example would be the time of formation of a new mineral as a result of regional or contact metamorphism. The date obtained would be the date of the metamorphism or, more precisely, when the intensity (temperature?) of the metamorphism had decreased sufficiently to allow the minerals to become systems closed to Rb and Sr.

Numerous examples exist where this comparatively simple geological history has resulted in rocks whose ages have been determined with good accuracy, for example Kulp et al. (1960). However, frequently more than one thermal event has affected the rock or mineral under study (Giletti et al., 1961). The complex history is reflected in discordant (inconsistent) ages for different minerals in the rock. They may also be discordant with other isotopic age measurements or with a stratigraphical assignment. Some minerals may be completely exchanged to yield the age of the later event, whereas others may remain essentially closed systems and yield older ages, perhaps the age of formation of the rock. Still others may gain radiogenic nuclides to yield ages greater than the age of the rock. A simple example of age resetting occurs in one case of contact metamorphism (see Hart, 1964).

Rb–Sr Isochron

A major advance in the application of the Rb-Sr method came with the realization that two independent variables pertain to any system of related rocks or minerals (Compston and Jeffery, 1961; Nicolaysen, 1961 and reviewed by Faul, 1966). One variable is the age, or length of time that a mineral or rock has remained a closed chemical system with respect to Rb and Sr (no gain or loss of isotopes except by radioactive decay). The second variable is the composition of common Sr (expressed as Sr^{87}/Sr^{86}) that existed when $t = 0$. These quantities can be determined by plotting a so-called isochron diagram. The isotopic relationships of Rb and Sr can be expressed thus:

$$\frac{Sr^{87})}{Sr^{86})} = \frac{Sr^{87})_{initial} + Sr^{87})_{radiogenic}}{Sr^{86}}$$

$$= \frac{Sr^{87}}{Sr^{86}}\bigg)_{initial} + \frac{Rb^{87}(e^{\lambda t} - 1)}{Sr^{86}} \quad (2)$$

Eq. (2) is of the form $y = mx + b$, where x and y respectively are Rb^{87}/Sr^{86} and Sr^{87}/Sr^{86} (present day ratios); the y intercept (b) is the initial value of Sr^{87}/Sr^{86}; and the slope (m) equals $e^{\lambda t} - 1$, from which the age is readily calculated. The straight line (isochron) is fitted through a series of (Sr^{87}/Sr^{86}, Rb^{87}/Sr^{86}) val-

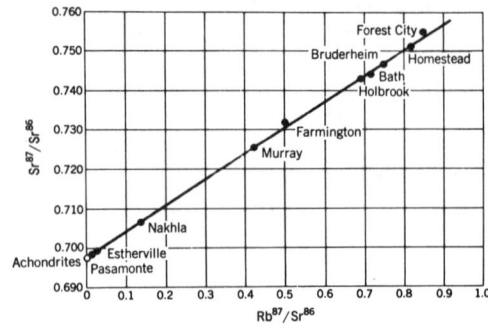

FIG. 2. Isochron diagram for stony meteorites. Achondrites fall into the open circle at the left, and chondrites are solid dots (from Faul, 1966; after Pinson et al., 1965).

ues from samples of differing Rb/Sr ratios. Care must be used to choose samples that appear on geologic grounds to be parts of a single "system" that experienced homogenization of Sr isotopes at a unique time.

Isochron Studies of Meteorites

The isochron concept has been applied successfully to studies of meteorites and a great variety of terrestrial rocks. Fig. 2 is an isochron diagram that presents some important information concerning the history of Rb and Sr in the solar system. The points refer to analyzed chondrites and achondrites (varieties of stony meteorite). The y-intercept indicates an initial $Sr^{87}/Sr^{86} = 0.698$, and the slope is equivalent to a 4.3 billion year age of homogenization of Sr isotopes. This age is close to that found by an independent but similar analysis of Pb isotopic composition in meteorites. Several lines of evidence suggest that the earth belongs to the same physico-chemical system as the meteorites. The meteoritic initial Sr^{87}/Sr^{86} is lower than in any known terrestrial rocks (all of which were formed at least 1000 million years later than the earth itself), although some terrestrial initial ratios closely approach the meteoritic value.

Isochron Studies of Terrestrial Rocks

Age. The mechanisms by which igneous, sedimentary, and metamorphic rocks are formed must be evaluated critically in the application of the isochron concept to these rocks. In principle, the most straightforward interpretation applies to igneous rocks. That is, the rock may have originated from a well mixed melt that crystallized over a geologically "short" interval. This simple picture is supported by the majority of isochron data from igneous rocks.

The isochron approach is less suitable to the

precise dating of sedimentary rocks due to the presence of detritus obtained from source regions with inhomogeneous Sr^{87}/Sr^{86}. The initially nonuniform initial ratios are manifested today as a scatter of points. Some striking exceptions have been found, in which the measured points fit very closely to a straight line. At present, the interpretation of the linear relationship is not entirely clear, and deserves further investigation.

One of the greatest achievements of the use of isochrons has been in the study of metamorphic rocks. Isochrons often can establish both the primary age and the age of latest metamorphism. A "whole rock" (usually a hand specimen but occasionally much larger) may remain chemically closed during metamorphism, whereas mineral grains adjacent to each other within the rock might experience complete homogenization of Sr isotopes. As a result of this episodic homogenization, a plot of the isotopic composition of individual minerals and of the whole rock becomes a horizontal isochron (Fig. 3), the "initial" Sr^{87}/Sr^{86} for the system of minerals simply being the whole rock value at that time. Fig. 4 shows that the slope of the isochron through whole rock values for a number of samples allows calculation of the primary age (t_p), and that the mineral subsystems form parallel isochrons corresponding to the age of metamorphism (t_m). Multiple histories have been documented for the Carn Chuinneag granite, Scottish Highlands (t_p = 530 million years; t_m = 390 million years; Long, 1964), granitic gneisses in central Colorado (t_p = 1560 million years; t_m = 1280 million years; Wetherill and Bickford, 1965), the Rotondo granite, Swiss Alps

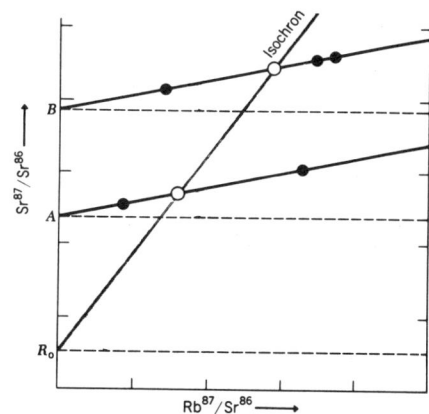

FIG. 4. Two-stage isochron diagram for whole rocks (open circles) from the same geologic system, and minerals (solid circles) comprising the whole rocks. The whole-rock isochron corresponds to the primary age, the mineral isochrons to the age of metamorphism. The initial $Sr^{87}Sr/^{86}$ for the whole rock is R_o, and Sr^{87}/Sr^{86} in individual whole-rock specimens at the later time of metamorphism is denoted by A and B (modified from Faul, 1966; after Lanphere et al., 1964).

(t_p = 260 million years; t_m = 13 million years; Jäger and Niggli, 1964), and elsewhere.

If thermal metamorphism causes only partial homogenization of the mineral isotopic ratios, the isochron diagram both for minerals and in some cases for whole rocks, will contain a considerable scatter of points. The effect of partial metamorphic overprinting has been studied in detail by Aldrich, Hart, and others. As a generalization, the stability of minerals against loss of Sr is (from lowest to highest): biotite, muscovite, feldspar. Some exceptions to this pattern have been observed.

For those rocks which have undergone three or more events involving isotope exchange subsequent to their formation the interpretation of the data becomes very difficult. In fact, unless specimens may be obtained from regions where fewer events occurred to the rock, correct interpretation of the timing of the events may be impossible.

It is best to obtain data for a terrane by use of several methods. In regions of complex history, especially, data from the potassium-argon and the uranium-thorium-lead methods are essential.

Initial Sr^{87}/Sr^{86}. Initial Sr^{87}/Sr^{86} ratios provide information about the prehistory of a rock. The interpretation of these ratios is based upon the belief that there is a decrease of Rb/Sr with increasing depth in the earth, especially between the crust and upper mantle. Rocks derived by partial or complete melting within these zones would inherit Sr^{87}/Sr^{86} ratios that depend upon the time of melting

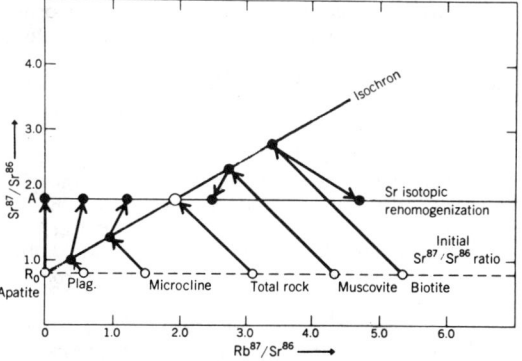

FIG. 3. Isochron diagram showing the effect of episodic homogenization of Sr. The total rock and component minerals had an initial Sr^{87}/Sr^{86} indicated by R_o. Later, a metamorphism caused the Sr composition in the minerals to rehomogenize to the total rock ratio (A) existing at that time (modified from Faul, 1966; after Lanphere et al., 1964).

and upon Rb/Sr in the source. Initial Sr^{87}/Sr^{86} from nearly all igneous rocks lies within the restricted range 0.699 to 0.710—a total variation of about 1.5%. With care, the ratios can be measured with a precision of ± 0.0003, enough to make useful distinctions between samples of similar composition.

Evidence suggests that oceanic basalts are derived from the upper mantle. Initial Sr^{87}/Sr^{86} for these rocks groups tightly at 0.702 to 0.705, indicating that the mantle must have rather uniform Rb/Sr of about 0.02. Detailed studies have demonstrated systematic variation of Sr^{87}/Sr^{86} within oceanic islands or between neighboring islands, suggesting local minor chemical inhomogeneities in the mantle, or perhaps some contamination with crustal Sr. Many granites and gneisses have present-day Sr^{87}/Sr^{86} of 0.8 or higher. Clearly, contamination of a mantle-derived magma by such a rock would alter Sr^{87}/Sr^{86} in the magma significantly. Oceanic basalts have the least chance of being contaminated. Continental basalts have a wider range of initial ratios, presumably due to sialic contamination.

Seawater Sr has a consistent ratio (0.709) derived from contributions by rocks of both continental and oceanic affinity. The seawater value appears to be buffered by exchange with Sr contained in marine sediments.

Initial ratios in alkalic igneous rocks (whether feldspathoid-bearing or not) are mostly in the range 0.703 to 0.708. The isotopic and elemental abundance data signify derivation from the mantle or deep crust; contamination by upper crustal materials is very minor.

A surprising result is the low initial value for most granites (0.704 to 0.711). If the average age of the upper crust is 2 billion years, if the crust began with $Sr^{87}/Sr^{86} = 0.704$ (typical of basalt), and if Rb/Sr = 0.25, then the average Sr^{87}/Sr^{86} today should be about 0.725. This estimate has received support by an actual measurement of 0.722 on composites of Paleozoic shales (Faure and Hurley, 1963). Granites of Phanerozoic age, if derived by melting of crustal materials, should thus inherit Sr^{87}/Sr^{86} of 0.72 or higher. The granites discovered to have initial ratios in this range are notably few. The ultimate origin of granite magma is one of the outstanding current problems being studied from the viewpoint of Sr isotope distributions.

LEON E. LONG
BRUNO J. GILETTI

References

Arriens, P. A., Brooks C., Bofinger, V. M., and Compston, W., 1966, "The discordance of mineral ages in granitic rocks resulting from the redistribution of rubidium and strontium," *J. Geophys. Res.*, **71**(20), 4981–4994.

Compston, W., and Jeffery, P. M., 1961, "Metamorphic chronology by the rubidium-strontium method," *Ann. N.Y. Acad. Sci.*, **91**, 185–191.

Faul, H., 1966, "Ages of rocks, planets, and stars," New York, McGraw-Hill Book Co., 109pp.

Faure, G., and Hurley, P. M., 1963, "The isotopic composition of strontium in oceanic and continental basalts: application to the origin of igneous rocks," *J. Petrol.*, **4**, 31–50.

Ferrara, G., Hirt, B., Marinelli, G., and Tongiorgi, E., 1961, "Primi risultati sulla determinazione con il metodo del rubidio-stronzio dell' eta di alcuni minerali dell' Isola d'Elba," *Boll. Soc. Geol. Ital.*, **80**, 1–6.

Gast, P. W., 1960, "Limitations on the composition of the upper mantle," *J. Geophys. Res.*, **65**, 1287–1297.

Giletti, B. J., Moorbath, S., and Lambert, R. St/J, 1961, "A geochronological study of the metamorphic complexes of the Scottish Highlands," *Quart. J. Geol. Soc. London*, **117**, 233–272.

Hart, S. R., 1964, "The petrology and isotopic-mineral age relations of a contact zone in the Front Range, Colorado," *J. Geol.*, **72**, 493–525.

Jäger, E., and Niggli, E., 1964, "Rubidium-Strontium-Isotopenalysen an Mineralien und Gesteinen des Rotondogranites und ihre geologische Interpretation," *Schweiz. Mineral. Petrog. Mitt.*, **44**, 61–81.

Kulp, J. L., Long, L. E., Giffin, C. E., Mills A. A., Lambert, R. St/J, Giletti, B. J., and Webster, R. K., 1960, "Potassium-argon and rubidium-strontium ages of some granites from Britain and Eire," *Nature*, **185**, 495–497.

Lanphere, M. A., Wasserburg, G., Albee, A. L., and Tilton, G. R., 1964, "Redistribution of strontium and rubidium isotopes during metamorphism, World Beater Complex, Panamint Range, California," in (Craig, Miller, and Wasserburg, editors). "Isotopic and Cosmic Chemistry," Amsterdam, North-Holland Publ. Co., pp. 269–320.

Long, L. E., 1964, "Rb-Sr chronology of the Carn Chuinneag intrusion, Ross-shire, Scotland," *J. Geophys. Res.*, **69**, 1589–1597.

Nicolaysen, L. O., 1961, "Graphic interpretation of discordant age measurements on metamorphic rocks," *Ann. N.Y. Acad. Sci.*, **91**, 198–206.

Pinson, W. H., Schnetzler, C. C., Beiser, E., Fairbairn, H. W., and Hurley, P. M., 1965, "Rb-Sr age of stony meteorites," *Geochim. Cosmochim Acta*, **29**, 455–466.

Wetherill, G. W., and Bickford, M. E., 1965, "Primary and metamorphic Rb-Sr chronology in central Colorado," *J. Geophys. Res.*, **70**, 4669–4686.

Cross-references: *Geochronometry; Potassium-Argon Age Determination; Radioactive Isotopes; Radioactivity in Rocks; Uranium-Thorium-Lead Dating.* Vol. II: *Meteorites.* Vol. V: *Magma; Metamorphism.*

RUNOFF

Runoff and stream flow are synonymous. Surface runoff, interflow, and groundwater flow in varying proportions make up the total runoff in stream channels.

The direct runoff is that runoff which enters the stream promptly after rainfall or melting of snow. It is equal to the surface runoff, the prompt subsurface runoff, plus the channel precipitation which falls directly on the water surfaces of lakes and streams. Surface runoff is commonly represented in the form of a hydrograph similar to Fig. 1. A hydrograph is a time record of runoff at a given cross section of the stream.

Interflow is that part of the precipitation which infiltrates the surface soil, moves laterally through the upper soil horizons as ephemeral, shallow, perched groundwater above the main groundwater level, and reaches the stream before it reaches the water table. This lateral movement results from the presence of relatively impervious horizons near the surface.

Precipitation which infiltrates the zone of saturation flows as groundwater and gradually contributes to runoff where the water table is above the surface of the stream.

Runoff from a drainage basin is influenced by climatic and physiographic factors. Climatic factors include the effects of various forms of precipitation, interception, evaporation, and transpiration, all of which exhibit seasonal variations in accordance with the climate.

An idea of the variation in precipitation in the United States may be gained by comparing the Red River of the North at Grand Forks, N. D., which has an annual precipitation of 18.5 in. and a runoff of 0.24 in., and the Chattahoochee River of Georgia, which has a precipitation of about 60 in. and a runoff of 11.6 in. The rivers of the western side of the Olympic Peninsula of Washington have runoffs of well over 100 in.

The physiographic factors influencing runoff are basin characteristics and channel characteristics. Basin characteristics include size, shape, and slope of drainage area, permeability, capacity, and orientation of the geologic formations, and land use. Streams fed primarily by direct surface runoff are always flashy. Streams fed largely by groundwater flow are well sustained throughout the year, although they may not escape floods produced by direct surface runoff during periods of intense precipitation (Meyer, 1942). Channel characteristics are related mostly to hydraulic properties of the channel which govern the movement of water and determine channel storage.

Most large drainage areas respond to precipitation events differently than most small drainage areas. Consequently, drainage basins may be classified as large and small, not on the basis of size alone, but on the basis of this response. Frequently two basins of nearly the same size may have entirely different runoff phenomena. One drainage basin may show prominent channel storage effects, like most large basins, whereas the other may be strongly influenced by the land use, like most small basins. A distinct characteristic of small basins is that the effect of overland flow rather than the effect of channel flow is a dominating factor affecting the shape of the hydrograph. Thus small basins will respond rapidly and strongly to high-intensity rainfalls of short duration. On large basins, channel storage both delays and diminishes the response. Therefore, small drainage basin may be defined as one that is so small that its response to high-intensity rainfalls of short durations and to land use is not suppressed by the channel characteristics (Chow, 1964).

The runoff from all drainage basins can be visualized as a cycle comprised of the following five phases described by Hoyt. (1) Just prior to the beginning of rainfall after an extended dry period, the groundwater table is low and with continued losses to stream flow its elevation continues to decrease gradually. Intermittent channels are dry (Fig. 2A). (2) As the rain starts, it is divided among channel precipitation, interception by vegetation, infiltration into the soil, and temporary retention in surface depressions (Fig. 2B). (3) As the rain continues, the capacities of interception and retention by vegetation and surface depressions are reached, and the excess rain be-

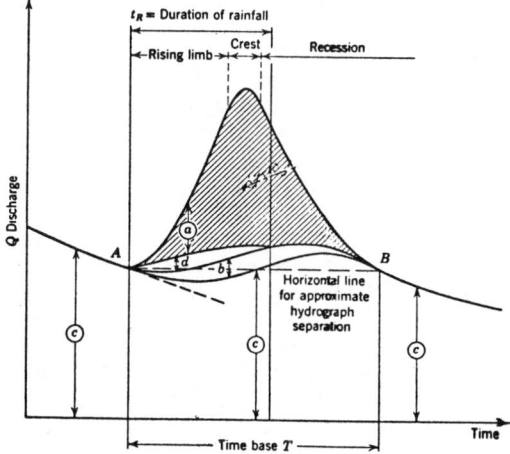

Fig. 1. Hydrograph parts and flow contributions (after Davis and DeWiest, 1966). Key: (a) surface runoff; (b) interflow; (c) groundwater flow; (d) channel precipitation.

RUNOFF

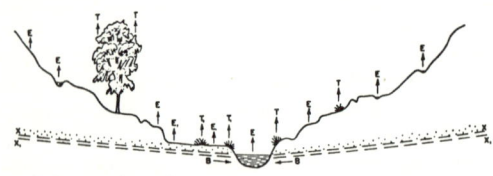

A. Rainless period.

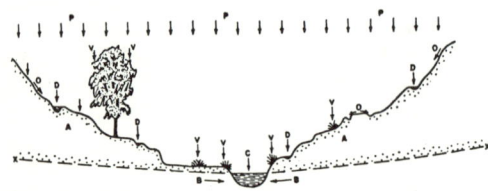

B. Initial rain period.

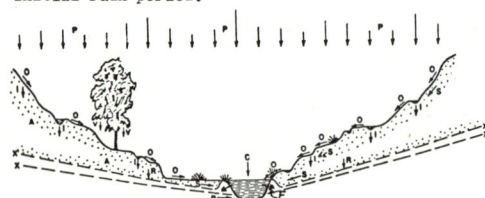

C. Continuation of rain.

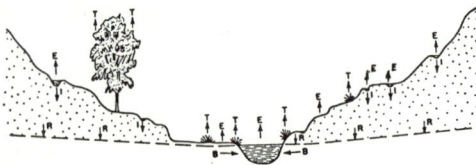

D. Following cessation of rain.

FIG. 2. The Runoff cycle (after Hoyt, 1942). Key: A, soil moisture; B, seepage to stream; C, channel Precipitation; D, depressed storage; E, Evaporation; F, bank storage; I, infiltration; O, overland flow; P, precipitation; R, recharge; S, interflow; T, transportation; V, interception; X, groundwater table; X^1, groundwater table later.

comes a source of runoff and detention storage on land surfaces and in channels (Fig. 2C). Overland flow occurs when the intensity of rain exceeds the infiltration rate. The infiltrated water will saturate the upper part of the zone of aeration which has been depleted in the first phases and will then move down to the water table. (4) If rain continues, the water table will rise and the groundwater contribution to stream flow will increase. If the stage of flow in the channels rises rapidly and becomes higher than the relatively slowly rising groundwater table the streams will change from effluent streams to influent streams, contributing to the groundwater and developing bank storage of water. (5) Following cessation of the rain, infiltration from channel storage and surface retention continues at a gradually diminishing rate until these sources are exhausted

and the conditions of the first phase are reached.

ROBERT K. FAHNESTOCK

References

Chow, V. T., 1964, "Handbook of Applied Hydrology," New York, McGraw-Hill Book Co., Section 14, 53pp.

Davis, S. N., and DeWiest, R. J. M., 1966, "Hydrogeology," New York, John Wiley & Sons, 463pp.

Hoyt, W. G., 1942, "The Runoff Cycle," in (Meinzer, D. E., ed.), "Hydrology," New York, McGraw-Hill Book Co., pp. 507–513.

Linsley, R. K., Jr., Kohler, M. A., and Paulhus, J. L. H., 1949, "Applied Hydrology," New York, McGraw-Hill Book Co., 689pp.

Meyer, A. F., 1942, "Runoff—Introduction," in (Meinzer, D. E., ed.), "Hydrology," New York, McGraw-Hill Book Co., pp. 478–485.

Cross-references: *Groundwater; Water Table.* Vol. III: *Drainage Basin; Slopes; Stream Channel Characteristics.*

RUTHENIUM: ELEMENT AND GEOCHEMISTRY

Properties

Ruthenium (Ru) is a member of the platinum group of metals, the six elements which are classified along with iron, cobalt, and nickel in group VIII of the periodic table. The element was identified and named by K. K. Klaus in 1844. Its atomic number is 44 and atomic weight 101.07. There are seven stable isotopes of ruthenium with mass numbers and relative abundances as follows (Wright and Fleischer, 1965): 96 (5.5%), 98 (1.9%), 99 (12.7%), 100 (12.6%), 101 (17.1%), 102 (31.5%), and 104 (18.6%). The metal crystallizes with a hexagonal close-packed lattice whose characteristic constants are $a = 2.7056$ Å and $c/a = 1.5820$. The average atomic radius for twelve-coordination in the metal is 1.335 Å. The compact metal has a density of 12.45 g/cc at 20°C.

The pure metal is remarkably hard and has a high melting point, 2310°C. In spite of some earlier thermal evidence to the contrary, ruthenium does not appear to exist in allotropic modification (Hampel, 1968). Chemically the element, like osmium, forms a volatile tetroxide, but this, in contrast to the osmium compound, does not form directly from the elements. The metal is not subject to attack by acids, but does react appreciably with alkalis under oxidizing conditions.

Occurrence

All six platinum metals are noble; i.e., they show considerable reluctance to unite with other elements. Accordingly their natural occurrence is chiefly uncombined in alloys with each other and with certain base metals such as iron or copper. Ruthenium is not a major component in these alloys, but it is usually found in iridosmine and osmiridium, and sometimes as a minor component in native palladium.

There is a mineral species *laurite,* RuS_2, often containing some osmium, found principally in Borneo and the Transvaal. More recently one or two other sulfide or arsenosulfide minerals have been described in which ruthenium is a significant, though not the exclusive, metallic constituent (Wright and Fleischer, 1965). The element is distributed at very low concentrations in some other minerals, particularly sulfides and arsenides.

Abundance and Geochemical Features

It is interesting that, although ruthenium is estimated to comprise 0.001 ppm in the earth's crust, and by some authorities to be the least abundant of the platinum metals there, it ranks next only to platinum in estimates of abundance in meteorites and of cosmic abundance. Like all of the platinum metals, the element is predominantly siderophilic, though less so than platinum (see also *Platinum Metals*).

Economic Geology

Economically valuable deposits of the platinum metals generally contain all of these elements together, with platinum and palladium by far the most abundant. The deposits are nearly always within or near ultrabasic rocks, and the metals are closely associated with sulfides (e.g., pyrrhotite or pentlandite) or with chromite. The commercial importance of ruthenium is mainly as a hardener for platinum or palladium (see *Platinum Metals*).

W. A. E. McBryde

References

Hampel, C. A. (editor), 1968, "Encyclopedia of the Chemical Elements," New York, Reinhold Publ. Corp., pp. 610–615.

Hampel, C. A. (editor), 1961, "Rare Metals Handbook," Second edition, New York, Reinhold Publ. Corp., pp. 304–335.

Rankama, K., and Sahama, Th. G., 1950, "Geochemistry," Chicago, University of Chicago Press, 912pp.

Standen, A. (editor), Kirk-Othmer, "Encyclopedia of Chemical Technology," Second ed., New York, John Wiley & Sons, Vol. 15, pp. 832–878.

Wright, T. L., and Fleischer, M., "Geochemistry of the Platinum Metals. Washington, U.S. Gov't. Print. Office, 1965.

Cross-reference: *Platinum Metals.*

S

SALINIZATION OF SOILS AND GROUNDWATER—*See* Vol. VI

SALT NUCLEI—*See* **CYCLIC SALTS:** *also* Vol. II, **SALT NUCLEI**

SAMARIUM: ELEMENT AND GEOCHEMISTRY

Properties

Elemental samarium (Sm) is a very reactive, silvery white *lanthanide* (q.v.) or *rare-earth* (q.v.) metal with atomic number 62 and outer electronic structure $4f^6\,6s^2$.

It occurs as four stable isotopes:

Sm^{144}	3.1%	natural abundance
Sm^{150}	7.4%	
Sm^{152}	26.8%	
Sm^{154}	22.7%	

Three isotopes are weakly radioactive, emitting alpha particles: Sm^{147} (15.0%, 1.14×10^{11} year half-life), Sm^{148} (11.2%, 1.2×10^{13} year half-life), and Sm^{149} (13.8%, 4×10^{14} year half-life).

Samarium is determined analytically by neutron activation (its thermal neutron cross section, 5600 barn, is high), x-ray spectrography, absorption spectrometry (it has very sharp spectral lines), and by isotope dilution mass spectrometry. In geologic environments, samarium forms highly ionic bonds and weak complexes. The oxidation state is tripositive, with ionic radius 0.064Å; dipositive samarium is oxidized by water and probably is unstable in most crystal lattices. Generally, chemical behavior is similar to the alkaline earths, and more similar to the heavy rare earths than to other light rare earths.

DE Boisbaudran (in 1879) first isolated samarium from other rare earths; he separated the element out of didymia, an intermediate rare earth concentrate of cerite, the first known light rare earth mineral, discovered by Cronstedt in 1751.

Minerals

Samarium occurs in highest concentration in rare earth minerals that favor light rare earths, generally nitrates, sulfates, carbonates, niobates, and tungstates with high cation-coordination number, often containing structural halogens and alkali metals. Typical examples, with cation-coordination numbers 10 and 11, are:

Monazite (light rare earths, Th) PO_4
Bastnaesite (light rare earths, Th) FCO_3

Geochemistry

Schmitt et al. (1964) have established a value 0.21 ppm (parts per million) samarium in chondrites and 2.3 ppm in eucrite achondrites. The standard deviation within a meteorite class is only about 20%. In meteoritic matter, volatilization may have redistributed alkali metals but not rare earths: samarium has a boiling point of 1900°C.

The samarium concentration in average basalt is 6.9 ppm, but the range is over a factor of ten (Haskin et al., 1966). Alkaline oceanic basalts and most continental basalts have more samarium than oceanic tholeites and some continental diabases. Partial melting of mantle peridotite yields a basaltic liquid with enriched samarium and a mantle source area depleted in samarium.

The samarium concentration in granites with 60 to 70% silica averages 8.5 ppm, a value that is higher, but also more variable, than for basalts.

Igneous differentiation series tend toward higher samarium concentrations. The relative abundance of samarium and other rare earths changes. Haskin et al. (1966) report 15 ppm samarium in a Gough Island basalt and 24 ppm in a trachyte differentiate. Towell et al. (1965) found 3.78 ± 0.04 ppm in the Rubidoux Mountain Leucogranite phase of the Southern California batholith, probably a differentiate (by crystal settling) of the San Marcos Gabbro, with 2.17 ± 0.02 ppm. About half of the samarium in the gabbro is in rock-forming minerals (plagioclase, pyroxene, hornblende) probably substituting for calcium. The remainder is in accessory minerals like apatite. The accessory minerals (probably apatite, zircon, sphene, and monazite) hold most of the samarium in the leucogranite.

Samarium forms complexes in alkali, carbonate, and halogen-rich hydrothermal fluids, but they are not as stable as heavy rare earth complexes. Pegmatoid and metasomatic deposits formed from the hydrothermal fluids often have low samarium- to-heavy-rare earth ratios compared to granites.

Samarium has a 180-year residence time in seawater, about the same as other light rare earths. Probably it is complexed. Removal by absorption on clay particles, as well as depletion in surface water in areas of organic productivity are likely processes. Today, high secondary samarium concentrations in phosphatic fish remains occur only in deep water; Mesozoic fish remains also have high secondary samarium. Manganese nodules have 6400 ppm samarium whereas Pacific Ocean water has only 0.44×10^{-6} ppm (Goldberg et al., 1963).

In soils, the relative amounts of samarium in detrital, clay, and carbonate fractions is not well known. Sedimentary rocks have about 7 ppm samarium, with a variation of a factor of five. In a study of rare earths in Russian platform soils, Balashov et al. (1964) found higher concentrations in alkaline soils, the samarium precipitating as hydroxide; lower concentrations are in acid soils, probably due to removal as complexes.

Biologic concentration of rare earths has been found in hickory leaves; rare earths have been used to trace opium.

Economic Geology

The major source of samarium has been from monazite in beach sands. The accessory monazite originally present in granitic rocks resists weathering and is concentrated by sedimentary processes. Australia, Brazil, Ceylon, India, Malaysia, and South Africa mine the largest quantities. Usually, about one percent (by weight) of monazite is samarium. Recently, the bastnaesite deposit at Mountain Pass, California has been an important source of samarium.

High purity samarium has limited application in electronics and in optical filters. Most samarium is sold unpurified as misch metal (a mixed rare earth alloy), for use as oxide polishing powders. Impure rare earths are also used as petroleum cracking catalysts and in arc carbons.

ROBERT KAY

References

Balashov, Y. A., et al., 1964, "The effects of climate and facies environment on the fractionation of the rare earths during sedimentation," *Geochem. Intern. No. 1,* **10,** 951–969 (English transl.).

Goldberg, E. D., et al., 1963, "Rare-earth distributions in the marine environment," *J. Geophys. Res.,* **68,** 4209–4217.

Haskin, L. A., et al., 1966, "Meteoric, Solar, and Terrestrial Rare-Earth Distributions," in (Ahrens L. H., et al., editors), "Physics and Chemistry of the Earth," Vol. 7, Oxford, Pergamon Press, pp. 167–321.

Rankama, K., and Sahama, T. G., 1950, "Geochemistry," Chicago, University of Chicago Press, 912pp.

Schmitt, R. A., et al., 1964, "Rare-earth, yttrium and scandium abundances in meteoric and terrestrial matter—II," *Geochim Cosmochim. Acta,* **28,** 67–86.

Towel, D. C., et al., 1965, "Rare-earth distributions in some rocks and associated minerals of the batholith of Southern California," *J. Geophys. Res.,* **70,** 3485–3496.

Cross-references: *Complexes; Lanthanides; Mass Spectrometry; Neutron Activation Analysis; Rare Earths; X-Ray Spectroscopy.*

SAPROPEL—See LIMNOLOGY; also Vol. VI

SCANDIUM: ELEMENT AND GEOCHEMISTRY

Physical and Chemical Properties

Scandium (Sc) was discovered in 1879 by Nilson and is an electropositive metal similar in physical properties to the rare earth elements. Only one stable isotope (Sc^{45}) is known in nature, and with electronic configuration $1s^2 2s^2 2p^6 3s^2 3p^6 3d^1 4s^2$ the element (atomic number 21) represents the first member of the transition series.

The chemistry of scandium is dictated by the fact that only one oxidation state exists, the Sc^{3+} ion being formed by the ready loss of the $3d$ and $4s$ electrons. None of the properties normally associated with transition elements thus appear and the element behaves in a way intermediate between aluminum (a typical representative element) and the rare earth elements (all characterized by the great stability of the tripositive ionic state). In the past, this latter similarity has been overstressed, but Goldschmidt correctly emphasized the fact that, since Sc^{3+} has an ionic radius (0.81 Å) considerably smaller than those of the rare earth elements, its crystal chemistry should be markedly different. In fact, isomorphous substitution for either Al^{3+} or the tripositive rare earths is the exception, substitution for Fe^{2+} in common rock-forming minerals being the rule. In the unusual assemblages of certain pegmatites, substitution for such ions as Zr^{4+}, Sn^{4+}, Nb^{5+}, Hf^{5+}, etc., occurs, further confirming Goldschmidt's hypothesis.

Geochemistry

Evidence gathered from the analysis of meteoritic material suggests that the bulk of the scandium in the earth is concentrated in the silicate phases of the mantle and crust. There is little difference between the average abundance computed for the earth's crust as a whole (22 ppm) and the silicate phase of meteorites (10 ppm). Thus, little differentiation apparently occurred during the formation of the mantle and crust. This would indicate that scandium is captured widely as a trace constituent of silicate minerals.

The process of magmatic differentiation apparently leads to a maximum concentration of about 30 ppm in rocks of basaltic composition. This is presumably due to the fact Sc^{3+} competes more effectively with Fe^{2+} for vacant lattice sites in the pyroxene phase than in the more compact olivine lattice. In more acid rocks the scandium content drops off to quite low levels. Only in the rich assemblages of certain rare pegmatites is there any evidence of further concentration in the magma. Such minerals as wiikite, wolframite, and cassiterite then may show considerable concentrations, and very occasionally a discrete scandium mineral, thortveitite, $(Sc, Y)_2 Si_2O_7$, is found.

In the processes of weathering and sedimentation scandium should behave as a typical hydrolysate element (ionic potential = ionic charge/ionic radius = 3.7). This is confirmed by the relatively high concentration found in shales (13 ppm) and deep-sea clays (19 ppm) as compared with the average composition of other sediment types (sandstone = 1 ppm; limestone = 1 ppm; deep-sea carbonates = 2 ppm). Concentrations of up to 100 ppm reported for some sedimentary iron ores cannot arise from the normal oxidative mechanisms since the solubility of scandium is unaffected by redox potential. Presumably coprecipitation of hydroxides is responsible. The fact that sedimentary manganese deposits show no parallel concentrations would tend to confirm this hypothesis.

The relative ionic potentials of Al^{3+} (6.5) and Sc^{3+} (3.7) are such that scandium should be preferentially leached at normal pH values during bauxite formation. This qualitatively explains the low Sc/Al ratios found in these deposits.

Although only one or two discrete scandium minerals are known (these occurring exceedingly infrequently), the element is not in itself as rare as was once believed. This is because the vast bulk of the world's scandium is dissipated throughout the common rock-forming minerals as a trace constituent.

Scandium metal is produced in small quantities, usually as a by-product of thorium or rare earth extraction.

C. D. CURTIS

References

*Goldschmidt, V. M., 1954, "Geochemistry," Oxford, Clarendon, 730pp.
*Taylor, S. R., 1964, "Abundance of chemical elements in the continental crust: a new table," Geochim. Cosmochim. Acta, **28,** 1273–1285.

*Additional bibliographic references may be found in this work.

Cross-references: *Atomic Number and Periodic Table; Earth's Crust Geochemistry; Mantle Geochemistry; Rare Earths; Weathering, Chemical.* Vol. IVB: *Rock-Forming Minerals.*

SEAWATER, CHEMISTRY

Seawater is a remarkable homogeneous substance consisting of about 96.5% water, 3.5% salt, small amounts of particulate matter, dissolved gases, and organic compounds. Some 98% of the hydrosphere is seawater. The dissolved salts, or salinity of surface water ranges between 33.99‰ at 50°S and 35.79‰ at 25°N. The average salinity for all ocean water is about 35‰. High salinities (up to 55‰) are observed in subtropical oceans and their landlocked extensions (e.g., Red Sea), whereas low salinities are found where seawater is diluted by continental rivers and melting ice. In Polar seas, the freezing of flow ice or freezing beneath ice shelves may lead to quite a strong brine in places. In some smaller landlocked embayments (e.g., Laguna Madre in the Gulf of Mexico), salinities as high as 130‰ have been observed.

It was not until the German chemist Professor William Dittmar aboard the British ship H.M.S. *Challenger* in 1872–1876 made a detailed analysis of the now famous "seventy samples" taken from most of the world's oceans that the true composition of the major ionic components of seawater was established. Not only did the results of these analyses fix the concentration of the major ionic components, but they demonstrated conclusively that within narrow limits, the ratios of these components are the same for all oceanic waters. Hence, seawater can be physically described by only three parameters: temperature, pressure, and a single number which will establish the concentrations of all the major components.

Considerable difficulty was encountered in obtaining a single number interpretable as the total salt composition, largely due to analytical difficulties, which led an International Commission composed of three Scandinavian oceanographers, Forch, Knudsen, and Sorensen, in 1902 to standardize on a technique yielding reproducible results which does not represent the true quantity of total solids but does represent a quantity slightly less, that is closely related, and by definition is called salinity of the water. *Salinity is defined as the total amount of solid material in grams contained in 1 kg of seawater when all the carbonate has been converted to oxide, the bromine and iodine replaced by chlorine, and all organic matter completely oxidized.* Using this definition, the concept of constancy of composition and the fact that accurate analysis of chloride can reflect the quantities of the other elements led the International Commission to establish the empirical relation:

Salinity = 0.03 + 1.805 × Chlorinity

to evaluate salinity. The chlorinity in this equation, also a defined term, is the number in grams per kilogram of a seawater sample which is identical with the number giving the mass in grams of "atomic weight silver" just necessary to precipitate the halogens in 0.3285233 kg of the seawater sample. Chlorinity is usually determined by titration with silver nitrate using a colorimetric or potentiometric end point.

To provide a primary standard for determination of salinity usable by all laboratories, the so-called Normal Water prepared by the Hydrographical Laboratories in Copenhagen, Denmark is used. Such water has been found to have a chlorinity of 19.381‰ according to the above definition. This water or a substandard referred to it is also used in standardizing physical measurements used in determining salinity.

In recent years most analyses for total salt content have been made by instrumental methods in which a property of ionic solutions such as their conductivity or refractive index is assessed. Data obtained by the conductivity methods give both greater sensitivity and accuracy (± 0.005‰ against ± 0.02‰ for titration) than the older volumetric procedures and allow for strong arguments for eliminating the salinity concept in favor of a physically defined parameter. However, equally strong arguments have prevailed to convert the instrumental values to salinity in order to make easy comparison between new and old oceanographic data possible.

The constancy of composition concept used in estimating the major salt components present in seawater does not apply to a large number of the elements, the dissolved gases, organic compounds, and particulate matter present in trace quantities not greatly affecting the physical properties of the water but yet having great biological and geochemical importance. Because of the variability of these components in ocean water and the difficult chemistries involved, most of the effort in marine chemistry has recently focused in these areas. Understanding is yet incomplete concerning many problems, yet discussion will be made in light of the knowledge now available.

Composition of Seawater

Major Components. Eight ions and largely undissociated boric acid contribute 99.95% of the total salt found in seawater. Values for these components are presented in Table 1.

Values for H_3BO_3 are probably the least reliable of those presented. Boric acid does not show a constant ratio to chlorinity with depth since variable amounts (0–10%) are found complexed to organic matter. Inorganic B/Cl ratios are, however, constant except at the oxygen minimum zone where the complexed boron reaches minimum values.

Changes in the ion ratios and thereby an error in salinity measurements may be brought about by land drainage, variation in nutrients and organic matter, changes in alkalinity and pH and sedimentation processes. Careful studies have indicated that a salinity effect of only 0.023‰ occurs in the extremes between water mass differences in the Pacific Ocean. Statistical evaluation of available data show the probability of a single salinity value falling within ± 0.024‰ is 0.99; ± 0.017 is 0.95; and ± 0.015 is 0.90. However, the variability in dissolved gas composition, the isotope variations and the variability in organic content makes a single element-to-salinity relation in the ± 0.005 accuracy range impossible.

Minor Components. Into this group fall essentially all the elements in the periodic table other than those listed above. These elements may be arbitrarily subdivided into those of intermediate concentration (1×10^{-4} to $1 \times 10^{-6} M$) and those present in trace or micro-

TABLE 1. MAJOR SALT CONSTITUENTS OF SEAWATER[a]

Component	Concentration (‰)	% of Total Salt
Cl^-	18.980	55.04
Na^+	10.543	30.61
SO_4^{2-}	2.465	7.68
Mg^{2+}	1.272	3.69
Ca^{2+}	0.400	1.16
K^+	0.380	1.10
HCO_3^-[b]	0.140	0.41
Br^-[b]	0.065	0.19
H_3BO_3	0.024	0.07
Total	34.455	99.95

[a] Values in grams per kilogram (‰) based on chlorinity of 19‰.
[b] Varies to give equivalent CO_3^{--} depending on pH. Value given is essentially true for pH 7.50 at 20°C.
[c] Corresponds to a salinity of 34.325‰.

TABLE 2. ELEMENTS OF INTERMEDIATE CONCENTRATION IN SEAWATER

Component	Concentration	Probable Species
Sr	8.0	Sr^{2-}; $SrSO_4$
O	4.6–7.5 (surface)	O_2 (gas)
Si	3.0	$Si(OH)_4$
F	1.3	F^- and metal complexes
N	0.5	N_2, NH_4^+, NO_2^-, NO_3^- and organic compounds
Ar	0.5	Ar(gas)
Li	0.17	Li^+
P	0.07	HPO_4^{2-}, H_2PO_4 and organic compounds
I	0.06	I^-, IO_3^-, organic compounds

quantities (less than $1 \times 10^{-6}M$). Although the composition of the major ions is considered conservative in seawater, the components included in these groups, particularly the latter, vary widely and in many cases, the quantity may be independent of the salt content. Individual analysis must be made for each species, and under widely varying oceanic environments, to gain an understanding of the distribution and the many factors controlling the observed distribution. Except in a very few cases, information in this area is incomplete and therefore subject to change as more, and perhaps better, data become available.

Intermediate Group. The intermediate group is composed of nine elements, including the three gases N_2, O_2, and argon and the three important nutrients, nitrogen, silicon and phosphorus compounds. The concentrations of this group of elements and their probable contributing species are given in Table 2.

Lithium has no known biological function, is not active geochemically (residence time in the ocean of 2.0×10^7 years) and tends to be conservative in distribution.

Nitrogen is present in multiple forms in the sea. Nitrogen gas, except under anaerobic conditions, is unreactive and shows only minor variations from saturation at *in situ* temperature and atmospheric pressure. The biologically important fixed nitrogen compounds vary widely in concentration with location, depth and season. Nitrate is the end product of the biological cycle and is the predominant form at depths below 150 meters in most seas. Organic nitrogen compounds have been found in all waters so far examined, but nitrite and ammonia are found in maximum concentrations at intermediate depths in middle latitudes during the spring and autumn. Nitrite concentration varies from none in many waters to as much as 50 µg/liter in regions of high biological production. Ammonia is distributed similarly and varies between 0 and 75 µg/liter. These are intermediates in the oxidative cycle of nitrogen, and their distribution and quantity mark the place and intensity of biological degradation. Nitrate concentration in the ocean varies between 1 µg/liter in some surface waters to as much as 600 µg/liter in deep Indian Ocean water.

Oxygen, while present in water and several anions, is most significant to the chemistry of seawater as the dissolved gas. Oxygen enrichment occurs only at the sea surface in exchange with the atmosphere or in the photic zone by photosynthesis. Depletion results primarily from biological activity, including the degradation of organic debris. In the photic zone, the oxygen content may increase above that found at the surface and has been found to reach values of 130% of saturation at in situ temperature and atmospheric pressure. At depths down to the compensation point (point at which respiration exceeds photosynthesis), oxygen production exceeds utilization but below this depth, a loss of oxygen is realized. The production of oxygen in the ocean ranges between 0.82×10^3 ml/m²/yr in Helsingör Sound to 1.86×10^6 ml/m²/yr in Long Island Sound with open ocean values being intermediate. Consumption of oxygen in the Atlantic has been found to be 0.21 ml/liter/yr near the surface and 1.3×10^{-3} ml/liter/yr at 2500 meters. The oxygen content of the bottom water of the eastern Atlantic does not fall below 60% saturation at in situ temperature and atmospheric pressure with a concentration of 4.9 ml/liter at 50°S increasing gradually to 6.4 ml/liter at 50°N. The western section is of the same general pattern with about 0.2 ml/liter less throughout. In the Pacific, subsurface waters are generally lower in oxygen concentration than the Atlantic. A typical vertical section showing the oxygen concentration along the west coast of North and Central America is shown in Fig. 1 (Sverdrup et al., 1942). Similar distribution is found along the coast of South America to at least 40°S. In these areas an intensely deoxygenated zone extends from about 100–1000 meters deep. Going eastward there is a distinct increase in the oxygen content of the water column, but only at great depths (4000 meters) does the water contain as much as 4 ml/liter.

The widespread occurrence of the oxygen minimum zone at intermediate depths is a very striking characteristic of oxygen distribution. Several theories and much experimental evidence have been presented to explain this un-

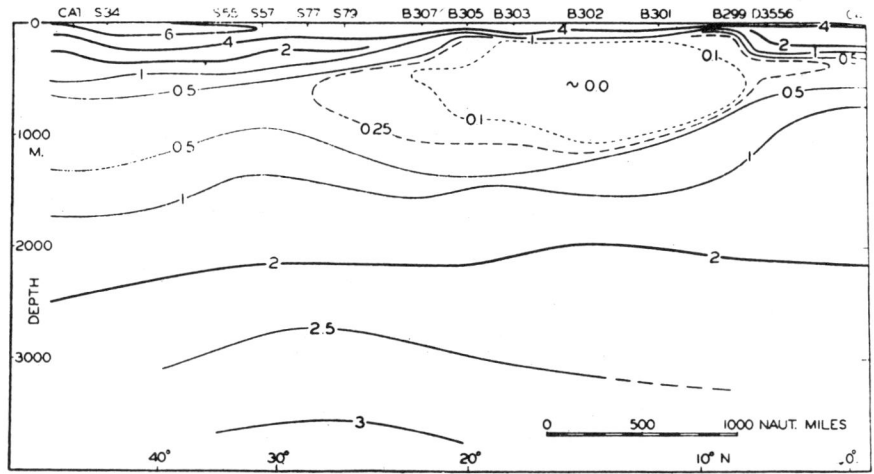

FIG. 1. Oxygen distribution (ml/liter) in a vertical section along the west coast of North and Central America at a distance of a few hundred miles from the coast (mainly based on observations of E. W. Scripps and Bushnell).

usual feature. There is some support for each of the three most often proposed models: the minima dynamically conditioned by circulatory processes, often referred to as the resting boundary theory; the minima biochemically conditioned and represents the depth at which the specific gravity relationships between seawater and sinking detritus are such that detritus accumulates, resulting in excessive consumption; and the minima circulation conditioned by sinking of a layer of water containing maximal amounts of organic material in high latitudes. The water density is such that it moves at intermediate depths, and oxidation of the organic matter results in a lower oxygen value than the waters above or below.

Silicon occurs in seawater as the monomeric silicic acid in concentrations between 0.02 and 3 ppm at the surface to 5–10 ppm in deep water. This is well below saturation equilibrium with any of the silica bearing minerals, including quartz (7–14 ppm). Since rivers are continually adding water with silica concentrations many times these values (30–35 ppm), a very effective, but yet unknown, removal process must be functional. The silica cycle in the sea is not well understood. Diatoms and other silica-secreting organisms form opaline tests that are very resistant to solution as attested by the abundance of frustules found in the sediment, yet convincing evidence has been presented showing a rapid regeneration of silica in surface waters during high organic production. It is apparent that a portion of the tests are decomposed at the surface by the action of bacteria and digestive enzymes of plankton-feeding animals. Fluctuation of silica concentration in surface waters usually exceeds that of nitrate and phosphate, probably because regeneration depends on the activity of grazers whose population is known to be patchy.

Phosphorus in seawater having a normal pH of between 7.8 and 8.2 is present largely as a mixture of the monobasic and dibasic ions $H_2PO_4^-$ and HPO_4^{2-}. In surface waters, a large portion of the total may be present as unregenerated organic phosphorus (0.0–15 μg/liter dissolved and 0–35 μg/liter particulate), and lipids containing organic phosphorus have been found in deep water of the Gulf of Mexico, Antarctic and Pacific Oceans. While other factors appear to exist, the phytoplankton productivity in the oceans is most often limited by compounds of phosphorus, nitrogen, and silicon. Large regions of the surface water of the oceans become almost completely exhausted of these nutrients. The phosphorus concentration increases with depth and generally shows a maximum in the oxygen minimum zone. The phosphorus content of the Indian and Pacific Ocean deep water (60–110 μg/liter) is about twice as high as that found in the Atlantic (35–45 μg/liter). The differences in the two oceans are related to the nature of the deep water circulation. Phosphorites ($Ca_{10}(PO_4)_6 F_{2-3}$) occur in extensive deposits in water 100–1000 feet deep along the California coast, which is thought to occur from supersaturation of calcium phosphate under low pH and oxygen tension conditions. More detailed discussion of the nutrient elements may be found in *Nutrients of the Sea*.

Argon, the most abundant of the inert gases, is found nearly saturated in seawater at in situ temperatures and atmospheric pressure. Small fluctuations that do occur are thought to be caused by barometric pressure differences at the surface and slight production from K^{40} decay and temperature changes in the water after sinking to the ocean depths. It is useful to use the ratio of this gas to other gases in valuating their stability in the ocean. Such studies with N_2 show a ratio of N_2/Ar of 37–39 to persist, indicating the conservative nature of N_2 gas.

Strontium may be considered a conservative element but has received much recent attention because of the Sr^{90} pollution problem resulting from nuclear fission reactions. Because of the high concentration factor of strontium in the bones of many animals over that in seawater (50), the long half-life (28 years) and the energy of the beta emission (1.46 MeV), this element is considered one of the most critical as an index of radioactive contamination. The ratio of Ca/Sr in the shells of organisms is

TABLE 3. COMPONENTS OF MICROCONCENTRATION IN SEAWATER[a]

Component	Concentration (μg/liter)	Probable Species
	(π/liter)	
He	0.005	He (gas)
Be	0.0005	
C	560 (g); 200–3000[a]	CO_2 (gas),[b] diss. organic compounds[d]
Ne	0.1	Ne (gas)
Al	1.0–10.0	
Sc	0.04	
Ti	0.02	
V	2.0	$VO_2(OH)_3^{2-}$
Cr	0.13–0.25	
Mn	0.1–8.0	Mn^{2+}; $MnSO_4$[c]
Fe	1.7–150	$Fe_2O_3 \cdot 3H_2O$[c]
Co	0.2–0.7	CO^{2+}; $CoSO_4$
Ni	2.0	Ni^{2+}; $NiSO_4$
Cu	0.5–3.5	Cu^{2+}; $CuSO_4$[c]
Zn	1.5–10.0	Zn^{2+}; $ZnSO_4$[c]
Ga	0.007–0.03	
Ge	0.07	$Ge(OH)_4$; $Ge(OH)_3O^-$
As	3.0	$HAsO^{2-}$; $H_2AsO_4^-$; AsO_4^{3-}; H_3AsO_3
Se	4.0–6.0	SeO_4^{2-}
Kr	0.3	Kr (gas)
Rb	120	Rb^+
Y	0.3	
Nb	0.01–0.02	
Mo	4.0–12.0	MoO_4^{2-}
Ag	0.145	$AgCl_2^-$; $AgCl_3^{2-}$
Cd	0.11	Cd^{2+}
Sn	0.3	
Sb	0.5	
Xe	0.1	Xe (gas)
Cs	0.5	Cs^+
Ba	10–63	Ba^{2+}; $BaSO_4$
La	0.3	
Ce	0.4	
W	0.12	WO_4^{2-}
Au	0.015–0.4	$AuCl_4^-$ [c]
Hg	0.15–0.27	$HgCl_3^-$; $HgCl_4^{2-}$
Pb	0.6–1.5	Pb^{2+}; $PbSO_4$
Bi	0.02	
Rn	0.6 (10^{-12})	Rn (gas)
Ra	1.0 (10^{-7})	Ra^{2+}; $RaSO_4$
Th	0.05	
Pa	2.0 (10^{-6})	
U	3.0	$UO_2(CO_3)^{4-}$

[a] From Goldberg, 1963; Hood, 1963.
[b] In presence of other CO_2 compounds.
[c] Evidence exists for organic complexed metal.
[d] Consisting of several hundred organic compounds.

proportional to the concentration in the water in which the organisms grew. The atomic size of the strontium ion (1.27Å) is such that strontium is usually found in the crystal lattice of deposited aragonite but not in calcite.

Iodine, based on thermodynamic considerations, should be present in seawater as the IO_3^- ion; however, I^- is the predominant species. This anomaly may be caused by plant and animal utilization of iodine in the reduced forms in compounds such as thyroxine and iodotyrosine and thus inhibit the thermodynamic equilibrium.

Microcomponents. Although the importance of the micro-components to the biochemistry of seawater has long been established, the concentration and chemical forms of these components are often in doubt. The concentration of many elements has been based on a single analysis, and where more analyses have been made, results may vary by several orders of magnitude. Most of the problem arises from the difficulty of analysis in the less than $10^{-6}M$ region, both because the analytical methods are usually at the lower limit of detectability and random contamination is difficult to control. Improvement in analytical techniques by innovation of x-ray fluorimetry, activation analysis, flame and atomic absorption spectrophotometry, improved polarography and new methods in colorimetric analysis and better sampling and laboratory conditions have much improved the situation in recent years.

The concentrations and probable contributing species of the micro-components that have been determined in seawater are presented in Table 3.

Aluminium data in the literature give values ranging between 1 μg to 1 mg/liter. Recent data for deep stations in the Pacific Ocean, Weddell Sea and in nearshore water along the California coast gave dissolved aluminum values of between 1 and 2 μg/liter. Particulate aluminum was variable, the concentration depending on location and depth, but values of over 3 μg/liter were not found except in nearshore water where higher values reflect land drainage.

Manganese occurs in seawater largely as the divalent ion, part of which appears to be complexed with organic matter. Particulate manganese is not found except in surface waters, and no evidence for the colloidal MnO_2 has been found in the water column, yet it occurs widely in ferromanganese minerals of the pelagic sediments. Whether chemical or biological oxidative processes are primarily responsible for this oxidation is still open to question.

Iron is present in seawater in the oxidized state in multiple physical forms. Very little is present in true solution because of the instability of the ferrous ion and the insolubility of the ferric hydroxides, yet stable colloids of hydrated ferric oxide occur at all depths of the ocean. The active surfaces of this colloid undoubtedly play an important role in geochemical and sedimentary processes of other microelements. The wide range of concentrations observed may partly result from the random distribution of the floccular iron colloids.

Copper and *zinc* have been investigated in seawater by many different analysts and analytical techniques. A wide disparity in values has resulted, but more recent work considering various physical and chemical states of the elements has provided more consistent data. Table 4 shows data for these elements for a typical station in the Gulf of Mexico. Similar distribution, but generally lower concentrations were found in the Antarctic and South Pacific Oceans. At intermediate depths, 20–80% of these elements is held in the nondialyzable fraction with lesser amounts in the shallower and deeper waters. The significance of this apparently organic complexed portion to the chemistry of the sea has not yet been determined.

Barium is present in surface waters in the

TABLE 4. DISTRIBUTION WITH DEPTH OF COPPER AND ZINC IN THE GULF OF MEXICO (95° 53'W; 24° 27'N)

Depth (m)	Extractable[a] (μg/liter)		Nondialyzable[b] (μg/liter)		Particulate[c] (μg/liter)		Total (μg/liter)	
	Cu	Zn	Cu	Zn	Cu	Zn	Cu	Zn
10	2.3	8.8	0.5	2.1	0.8	1.3	3.5	10.5
310	0.5	2.7	0.1	1.1	0.2	0.7	1.6	3.5
900	1.9	7.3	0.9	5.3	0.1	0.6	2.8	9.7
1200	0.5	4.7	0.2	4.1	0.6	2.3	1.4	5.0
2250	0.3	3.5	0.1	1.1	0.4	1.2	0.3	3.6
3400	0.3	3.5	0.1	1.1	0.1	0.5	0.6	1.6

[a] That fraction forming a solvent-soluble chelate with diethyldithiocarbamate, probably divalent.
[b] Fraction passing through 0.45μ millipore filter but not passing through cellulose acetate dialysis membrane (2 mμ average pore size).
[c] Material not passing through 0.45μ millipore filter.

range of 10 μg/liter and steadily increases to as much as 65 μg/liter in deep ocean water. Barite is found in ocean sediments and deep seawater may be in a state of supersaturation with respect to barium sulfate.

Radium shows a similar distribution to that of barium, but the concentration of Ra^{226}, which has a radiogenic origin, is about sixfold greater than can be supported by its parent Th^{230}. It is apparent that radium diffuses into the seawater subsequent to its origin from ionium in the sediments. This, coupled with biological incorporation at the surface and subsequent release at depth during decomposition processes, may explain the radium distribution found.

Other trace elements do not lend themselves to detailed considerations, especially vertical since in most cases relatively few samples, largely from the surface, have been analyzed.

Organic Constituents. Our knowledge of the kinds, distribution and significance of organic compounds in seawater is currently in a rapid stage of development. Organic matter in seawater is divided into particulate and dissolved on the basis of filtering through 0.45μ millipore filters with the soluble portion passing through the paper and the particulate remaining on the filter.

The quantity of dissolved organic carbon exceeds that of particulate organic carbon by a factor of 7–8 in waters rich in phytoplankton and in deep water by a factor of 1000. The amount of total organic carbon in the open ocean varies generally from 0.2–2.7 mg/liter. Higher values are found in landlocked areas; 3.3 mg/liter of carbon in the Black Sea, 4.6 in the Baltic; 6.0 in the Sea of Azov, and 8.0 in the landlocked coastal area of the Dutch Wadden Sea. Organic solute distribution is not homogeneous either horizontally or vertically, but there is a minimum of organic carbon in the oxygen minimum zone. The highest organic carbon concentrations are in waters of high phytoplankton productivity or recent productivity. Further details of the kind, amount and distribution of specific organic compounds, as presently understood are given in *Seawater, Geochemical Balance*.

Physical Chemistry of Seawater

Thermodynamics. Most of the dissolved constituents of seawater have had many thousands of years to reach equilibrium, yet thermodynamically unstable species apparently persist. Can this be attributed to lack of reaction sites, as is often suggested, or lack of knowledge concerning the complex interaction between these apparently unstable components and other components of seawater? While many efforts have been made to approach the problem theoretically, little can be gained by such an exercise until an exact knowledge of contributing forms is known. In the case of trace components in the concentration range less than 10^{-7} or $10^{-8}M$, quite enough organic material exists in seawater to form compounds requiring different thermodynamic consideration than for the inorganic ions alone. Our knowledge of the organic components is yet too limited to make this a worthwhile consideration. Except for the major ions, members of the carbon dioxide system, and perhaps some others, few experimental data exist that permit computation of the activity coefficients or reaction equilibrium constants for the components present. For the major ions, the problem is more easily approached.

By extension of the Lewis and Randall concept of activities of ions in dilute solution to seawater (ionic strength 0.7), remarkably good agreement between calculated and measured values has resulted. Using this and other methods for activity coefficient measurements, Garrels and Thompson (1963) have produced activity data for the free ions of the major components of seawater. These are: $A_{Na} = 0.356$; $A_{Mg}^2 = 0.0169$; $A_{Ca}^2 = 0.00264$; $A_K = 0.0063$; $A_{SO4}^2 = 1.79 (10^{-3})$; $A_{HCO3} = 9.74 (10^{-4})$; $A_{CO3}^2 = 4.7 (10^{-6})$. From this work the picture that emerges is that the major cations exist chiefly as uncomplexed species. Calcium and magnesium are only 10–15% complexed, and magnesium is the most active in complexing anions. The anions, except for chloride, are strongly paired with cations (most of carbonate, one-third of bicarbonate and nearly half of sulfate).

Solubility relations, closely allied with the above, are uncertain in seawater. It is, however, clear that the concentrations of the tabulated metal ions, including the alkaline earths, cannot be controlled by the solubility equilibria but by other more complicated processes.

pH. The pH is affected in seawater by temperature and salinity changes, photosynthesis and respiration, deposition of ions of the buffer system, mainly carbonate, and gaseous exchange with the atmosphere. The pH observed is usually between 7.8 and 8.3 in surface waters (being highest in the subtropics), the range depending on the combined effect of the above, but it is usually controlled to these limits by the natural buffer systems present. However, in estuaries, values as high as 9.0 have been observed at times of high photosynthesis. pH measurements may be used as a powerful tool in the interpretation of the above processes, but because of the buffer action of the carbonic ($2.4 \times 10^{-3}M$) and boric ($4.3 \times 10^{-4}M$)

TABLE 5. VERTICAL DISTRIBUTION OF CO_2 COMPONENTS AT A SOUTH PACIFIC STATION (21° 75'S; 72° 44'W)

Depth (m)	Cl (°/$_{00}$)	Temperature (°C)	pH	Alkalinity (eq./liter × 10⁻³)	pCO_2 ($A \times 10^{-4}$)	HCO_3^{2-} (moles/liter × 10⁻³)	CO_3^{2-} (moles/liter × 10⁻⁴)
0	19.33	17.66	8.21	2.36	2.79	1.74	2.64
31	19.30	17.47	8.21	2.36	2.82	1.74	2.61
88	19.12	12.56	7.76	2.34	9.05	2.11	0.99
176	19.22	11.63	7.60	2.35	13.63	2.19	0.69
356	19.19	9.27	7.57	2.35	14.00	2.20	0.62
632	19.08	5.92	7.71	2.34	9.54	2.16	0.75
1562	19.15	2.78	7.70	2.39	9.74	2.23	0.68
4535	19.20	1.73	7.36	2.37	21.91	2.30	0.30

acid systems, small differences in pH must be accurately detected if this parameter is to be used to its fullest potential. Devices are now available to detect voltage changes between a glass and calomel electrode to give a sensitivity to 0.0025 pH unit. In practice, however, better than ±0.02 pH unit is seldom achieved. The difficulty usually arises from poor sampling and electrode effects, largely caused by use of fresh-water buffer systems for calibrating the instruments for seawater measurements. Recent work has been done to help resolve both these problems.

Harvey (1957) has summarized the many measurements made of the pH values of seawater. The distribution of pH with depth at a South Pacific station is shown in Table 5 and Fig. 2. In middle and lower latitudes, as in this station, the pH of surface waters down to the thermocline is uniformly higher than in deeper water. It then undergoes a sharp decrease in the decomposition zone and rises slowly to values of 7.2–7.9 in the bottom water. The variation of pH in deep water suggests the use of this parameter in water mass studies.

Oxidation Potential. Oxidation potential in the ocean is controlled by the oxygen half-cell reaction, and deviation from 0.43 volt occurs only in anaerobic conditions in some static basins, severe oxygen-depleted intermediate depths, and at or below the sediment interface.

Carbon Dioxide System. Understanding the complex dynamic relationships of carbon dioxide in the air, sea and sediments is probably the most difficult problem in marine chemistry. Many of the difficulties arise from the nonequilibrium status of many of the reactions in the system and of incomplete information on many of the others.

Studies on the rate of exchange between the air and the sea have shown that equilibrium is seldom reached. In general, most of the ocean surface contains a lower concentration of gaseous carbon dioxide than the overlying atmosphere. Exceptions to this occur under conditions of local cooling and in regions of upwelling. The causes of this nonequilibrium are not clear, but it appears to be closely associated with the reaction between gaseous carbon dioxide and water: $H_2O + CO_2 \rightarrow H_2CO_3$. Only recently the forward reaction rates for this reaction were measured in seawater and found to be 3.3×10^{-2}/sec. The ionic components of the system may be calculated by use of equations summarized by Harvey

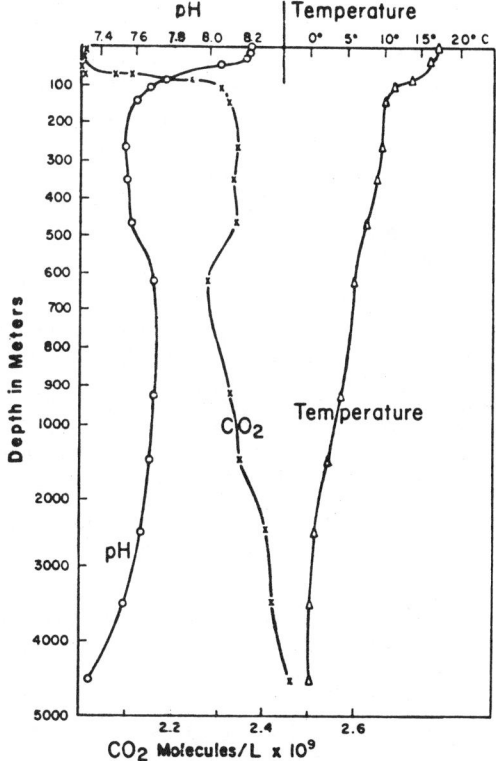

FIG. 2. The distribution of pH and temperature with depth. Measured in South Pacific.

(1957). Based on these calculations, data obtained from a representative station in the South Pacific (Table 5) show that the pCO_2 concentration increases rapidly below the thermocline, and the carbonate ion concentration goes through a corresponding decrease. The alkalinity remains essentially constant with depth, showing small deviations from chlorinity. Details of this relation are discussed more fully in *Alkalinity of Seawater* in Vol. I.

The solubility relations of $CaCO_3$ in seawater point to the surface waters of the ocean being almost universally supersaturated with respect to all the crystal forms of calcium carbonate. Yet only in limited local areas (e.g., Bahama Banks) is evidence for chemical precipitation existent. Deeper waters are in some cases undersaturated, and in these, tests of calcareous organisms are dissolved while settling toward the bottom leaving a sediment devoid of calcium carbonate. Except in limited areas, the deep waters also are nearly saturated. Supersaturation appears to exist because of the slow rate of formation of carbonate crystals in regions of relatively high dissolved organic content and where crystal sites are lacking.

The system is heavily influenced by biological factors which impose relatively short-term stresses (often diurnal) that are reflected in a nonequilibrium status for many of the components under wide environmental situations.

DONALD W. HOOD

References

Garrels, R. M., and Thompson, M. E., 1962, *Am. J. Sci.*, **260**, 57–66.
Goldberg, E. D., 1963, in (Hill, W. N., editor) "The Seas," Vol. 2, New York, Interscience Publishers, 554pp.
Harvey, H. W., 1957, "Chemistry and Fertility of Sea Waters," Cambridge, Cambridge University Press, 243pp.
Hood, D. W., 1963, "Chemical Oceanography," *Oceanogr. Marine Biol. Ann. Rev.*, **1**, 129–155.
Pytkowicz, R. M., 1965, "Calcium Carbonate Saturation in the Ocean," Oregon State Univ. Dept. of Oceanography, Ref. no. 65–16, 6pp.
Revelle, R., and Fairbridge R., 1957, in (Hedgpeth, J. W. editor) "Treatise on Marine Ecology and Paleocology," Vol. 1, *Geol. Soc. Am. Mem.*, **67**, 1296pp.
Sverdrup, H. V., Johnson, M. V., and Fleming, R. H., 1946, "The Oceans," New York, Prentice-Hall, Inc., 10–77.

SEAWATER, GEOCHEMICAL BALANCE

Geochemical balance of seawater is the material balance between the amount of substances present in seawater and the amounts introduced and removed from the sea. In order to draw a geochemical balance of seawater it is necessary to know (*1*) the amount of material present in seawater, the main components of which are water and substances dissolved and suspended in it, (*2*) the amount supplied to the ocean during its history, and (*3*) the amount removed by sedimentary processes.

The primary source of the major components of seawater are the igneous, metamorphic, and sedimentary rocks of the earth's crust. The volume of the ocean (1.37×10^9 km^3) is approximately 1/7 of the volume of the whole crust, and it is 1.4–2 times greater than the volume of all the sediments. Although crystalline rocks are considered as comprising more than 95% of the earth's crust, it has been estimated (Goldschmidt, 1954; Rubey, 1951) that their weathering alone could not have supplied all of H_2O, CO_2, chlorine, sulfur, nitrogen, boron, and some other elements present in the hydrosphere, atmosphere, and sedimentary rocks. These "excess volatiles," as they are called, were probably liberated in the process of degassing of the earth's interior. The important question of the time and rate of addition of H_2O and other volatiles to the earth's surface has been treated in a number of hypotheses and models of the evolution of the ocean and atmosphere (references in Mason, 1966).

Sources of Dissolved Components

Rivers of the world are the main avenues of supply of dissolved and suspended matter to the ocean. The rivers deliver to the ocean annually 3.33×10^{16} to 3.8×10^{16} kg of water, 3.85×10^{12} kg of dissolved matter, and 8.3×10^{12} to 32.5×10^{12} kg of suspended material (Mackenzie and Garrels, 1966; Nace, 1967). Additional input sources are groundwaters discharging through the continental shelf, submarine springs of volcanic and deeper crustal origin, and the atmosphere through which airborne detritus enters the ocean and the gases of which exchange with the surface seawater.

Eight chemical species are the major dissolved components of river and seawater: Na^+, Mg^{2+}, K^+, Ca^{2+}, Cl^-, SO_4^{2-}, HCO_3^-, and $Si(OH)_4^0$. Their mean proportions in the river and seawater, however, are very different: Fig. 1 shows the ratios of the molar concentration of each component to the Cl^--ion concentration. The similarity of the Na^+/Cl^- molar ratio in the river and seawater is pronounced whereas the other ratios for seawater are much lower. Approximately 50% of Na^+ and probably all of Cl^- in rivers are derived from the sea spray and recycled through the atmosphere, which accounts for the similarity of the Na^+/Cl^- ratios in the two habitats (Living-

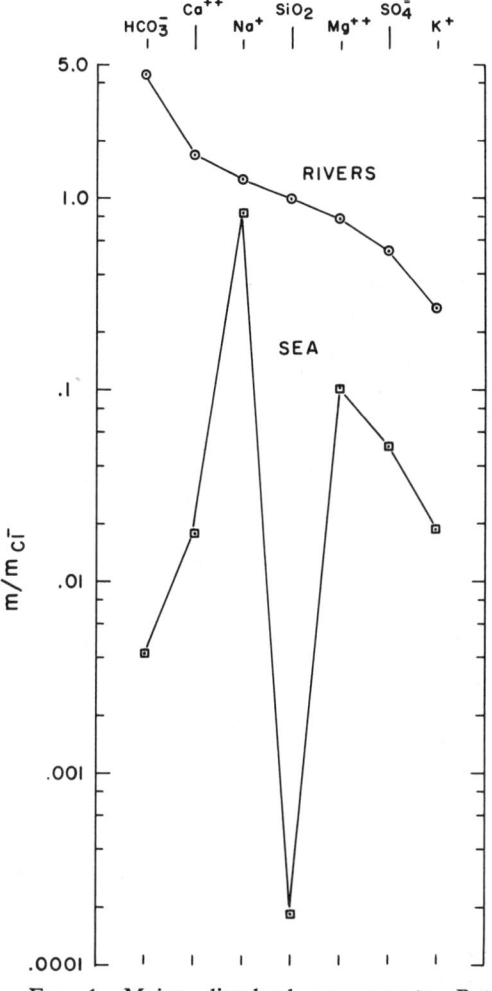

FIG. 1. Major dissolved components. Ratios of molar concentrations of seven dissolved species to the chloride ion molar concentration (m/m_{Cl^-}) in average river and seawater (from data in Mackenzie and Garrels, 1966).

The amount of water evaporating annually from the ocean surface is approximately ten times greater than the amount delivered to the ocean by rivers. Thus only near 10% of the oceanic water vapor is being recycled through the continents. The rate of turnover of the continental water is relatively rapid, as may be deduced from Nace's (1967) data on the water balance of the earth (Table 1).

The mean residence time of liquid water on the continents may be estimated as $8500/40 \simeq 200$ years. A more meaningful figure would probably be on the order of 50 years, assuming that only one-quarter of all the groundwater is open to the atmospheric input and outflow into the ocean.

Much slower is the recycling of matter through the sedimentation on the sea floor. Marine sediments near the continental margins may be raised above the sea level by tectonic processes and they may be metamorphosed and incorporated in the crystalline continental crust. By either mechanism, reintroduction of marine sediments into the weathering cycle may be a very slow process.

Atmospheric gases enter the ocean by direct exchange through the air-sea interface, with atmospheric precipitation and river water. The balance between the ocean and atmosphere for the gases N_2, O_2, He, Ar, Ne, Kr, and Xe is heavily weighed in favor of the atmosphere: the total amount of these gases in the ocean is a small fraction (0.3–3%) of their amount in the atmosphere. Of these only the isotopes He^4 and Ar^{40} are being produced in seawater and sediments by the radioactive decay of, respectively, the uranium series and K^{40}. The production rates are approximately thirty He^4 atoms and fifty A^{40} atoms per minute per liter of seawater (Revelle and Suess, 1962). For the CO_2, however, the balance is heavily on the ocean side: the amount of CO_2 in the ocean is approximately fifty-nine times its amount in the atmosphere. CO_2 reacts with water producing the bicarbonate and carbonate ions which control the buffer mechanism of seawater. Concentrations of carbon dioxide,

stone, 1963; Goldberg, 1965). For other components, the lower ratios in seawater reflect the different mechanisms and rates of removal from the ocean.

TABLE 1. WATER BALANCE OF THE EARTH

Amount of Continental Water Excluding Saline Lakes and Inland Seas (unit: 10^3 km³)		Runoff to Oceans from Continents (unit: 10^3 km³/year)	
Fresh water lakes	125	From rivers and icecaps	38
Rivers (mean instantaneous volume)	1.25	Groundwater outflow (assumed	
Soil moisture and vadose water	67	5% of above)	1.6
Groundwater (to depth 4000 m)	8350	Total	40
Total	8500		
Icecaps and glaciers	29200		
Total	37700		

nitrogen, and oxygen in the surface seawater are close to equilibrium with the atmosphere. Deviations in either direction from equilibrium are commonly due to organic activity.

Stationary Ocean

The idea that the volume and composition of the ocean have been stationary for a geologically long period of time is favored by many (references in Mackenzie and Garrels, 1966). The concept of a stationary or steady state implies that the water and dissolved components introduced into the ocean are being continually removed at the same rate as their rate of input. The removal of dissolved matter from seawater by sedimentation includes the following processes: (*1*) entrapment of seawater in the pore space of the sediment; (*2*) formation of mineral phases by chemical and biochemical reactions in seawater; (*3*) exchange on clays; (*4*) precipitation of evaporitic minerals.

On the assumption that the ocean has remained in a steady state for at least 10^8 years Mackenzie and Garrels (1966) have drawn a balance for the major species carried by the rivers into the ocean. In their calculation the excess of dissolved constituents in river water (Fig. 1) was disposed of by assuming "reversed weathering" reactions between degraded alumino-silicates carried by rivers and the silica, bicarbonate and cations present in seawater:

degraded Al-silicate + SiO_2 + HCO_3^- + cations = cation–Al–silicate + CO_2 + H_2O

The products of such reversed weathering reactions are aluminosilicate minerals stoichiometrically identical with montmorillonite, chlorite, and illite. The proportions of the cations and silica in the products, as deduced from the material balance, are similar to their proportions in deep-sea argillaceous sediments and "average" shales. Other important reactions removing the excess of dissolved components from sea water are: for Mg^{2+} and Ca^{2+}—formation of magnesian calcites, calcite, and aragonite; for SO_4^{2-}—formation of pyrite and calcium sulfate; for Na^+—80% removed in 1:1 ratio with Cl^- (interstitial water, precipitation of halite) and the remaining 20% exchanged on montmorillonite.

Residence Times

For a stationary ocean the residence time of a component is the number of years it takes to renew the total amount of it in the ocean, when the rates of supply and removal remain equal and constant. The residence time is defined as

$$\tau = \frac{A}{dA/dt}$$

where A is the amount of a component in the ocean and dA/dt is its rate of supply or removal (Barth, 1961; Goldberg, 1965). For fifty-one elements in seawater the residence times are in the range 80–2.6×10^8 years (Goldberg, 1965). Alkali and alkaline earth cations and chloride are characterized by long residence times of hundreds of years are of metals as Fe, Ti, Al, Th, and rare earths which form poorly soluble oxides, hydroxides, and silicates are less than 1000 years. Such short residence times of hundreds of years are of the same order as the time of passage of suspended material through the oceanic water column: passage time of less than 100 years for material coarser than 0.5μ, and less than 600 years for particles smaller than 0.5μ has been reported (Arrhenius, 1963).

The concentrations of such trace metals as Pb, Ni, Co, Cu, Zn, characterized by residence times longer than 1000 years, are unlikely to be controlled by equilibria with their solid carbonate and hydroxide phases: their equilibrium ionic activity products are not attained in seawater (Goldberg, 1965). Many trace metals are removed from seawater in detrital silicate minerals and clays (Arrhenius, 1963; Turekian, 1965).

Ronov (1968) has made new estimates of the total amount and composition of sediments deposited during the last "1.5×10^9 years." He has calculated the residence times of four cations—Na^+, Mg^{2+}, K^+, Ca^{2+}—in the ocean as $\tau = A/$(total amount in sediments/1.5×10^9). Ronov's values are higher by a factor of 3–23 than the values based on the direct river input (Table 2).

Diffusion from Sediment

The sediment is not only an outlet valve for the excess of dissolved material brought into the ocean. Diagenetic processes in the sediment and interstitial water, particularly those involving organic matter, alter the chemical nature of some of the constituents with the result that newly formed species may react to produce mineral phases and/or they may diffuse through the pore space of the sediment and into the overlying seawater. The necessary, although not the only, condition for diffusion out of the sediment is the existence of concentration gradients such that the concentration increases from the sediment-water interface for some distance down in the sediment.

The important biochemical reactions which produce species diffusing out of the sediment are the formation of ammonia from organic

TABLE 2. RESIDENCE TIMES OF MAJOR SPECIES IN SEAWATER
(MILLIONS OF YEARS)

	River Input (Goldberg, 1965)	Sedimentation (Goldberg, 1965)	River Input (Mackenzie and Garrels, 1966)	Total Sediments in 1.5 Billion Years (Ronov, 1968)
Na^+	210	260	70	1,600
Mg^{2+}	22	45	14	85
K^+	10	11	6.8	22
Ca^{2+}	1	8	1.2	4.5
Cl^-			103	
SO_4^{2-}			10	
HCO_3^-			0.1	
SiO_2	0.035	0.01	0.019	

nitrogen, formation of H_2S by sulfate reduction, and release of phosphate by decaying organic matter (Rittenberg et al., 1955; Richards, 1965). Ammonia and hydrogen sulfide are oxidized in seawater to nitrate and sulfate, and the phosphate released from the sediment is reutilized in the biological cycle. Reactions of this type are characteristic of anoxic basins with sediments rich in organic matter.

Of the inorganic processes, diffusion of Ra^{226} (half-life 1620 years) from the sediment results in its concentration in seawater (10^{-13} g/l) about ten times higher than the concentration at equilibrium with its parent Th^{230} (= ionium, 10^{-14} g/l; half-life 76,000 years) (Koczy, 1958). In the uppermost layers of oceanic sediments the concentration of Ra^{226} is lower than the concentration at equilibrium with Th^{230}. Th^{230} does not migrate through the sediment whereas Ra^{226} diffuses through the pore space at the rate characterized by the diffusion coefficient on the order of 10^{-9} cm^2/sec (Goldberg and Koide, 1963). This value is approximately three orders of magnitude lower than the values of molecular diffusion coefficients in water-filled porous media, which probably indicates that Ra^{226} interacts with the sediment in the process of diffusion.

Nonstationary Conditions

In the computation of the geochemical balance of a stationary ocean the nature of the distribution of the dissolved constituents concentrations is immaterial. The distribution of the concentrations through the oceanic water column is controlled by the rate of input, transport by diffusion and advection, and the rate of removal. A steady-state distribution of concentration as a function of depth retains no memory of the earlier transient states from which the steady-state had been reached. Major climatic changes, massive proliferation or extinction of groups of organisms, and major volcanic events may temporarily perturb the stationary conditions with respect to some of the constituents of oceanic water and sediments. If the new conditions imposed on the older environment persist, a new stationary state would be attained. Knowledge of the length of time required to develop a new steady-state in the water column and sediment is germane to the interpretation of the historical record of oceanic sediments.

A major feature of the present-day oceanic circulation which controls the distribution of dissolved species as a function of depth is the sinking of the cold surface water in the polar regions, the equatorward flow of deep cold water and its slow rising in the central oceanic region. The upper several hundred meters of the ocean are well mixed, but the exchange of water and dissolved substances with the underlying water masses is more restricted. The organic productivity cycle in the upper layers of the ocean is responsible for the seasonal fluctuations in the concentration of nitrate, phosphate, organic phosphates, and oxygen. The production and consumption of oxygen in seawater are the result, respectively, of production and oxidation of organic matter. Oxidation of organic matter releases phosphate and CO_2 into the water such that the concentration profiles of phosphate, oxygen, nitrate, and CO_2 as a function of depth in the ocean are to a large extent interrelated. Below the euphotic zone oxygen is only being consumed by the organic matter settling through the water. The deep bottom water masses forming in the polar regions are relatively rich in oxygen, and their advective upwelling replenishes the oxygen consumed. The balance of oxygen at intermediate depths is thus maintained by the supply from slowly rising deep water, eddy diffusional transport, and consumption, all of which processes result in a steady-state distribution of oxygen below the euphotic zone (Sverdrup, et al., 1942, p. 161; Munk, 1966). Munk (1966) has shown that in the layer between 1 km and 4 km depth the vertical distribution of temperature, salinity and such dissolved species as

SEAWATER, GEOCHEMICAL BALANCE

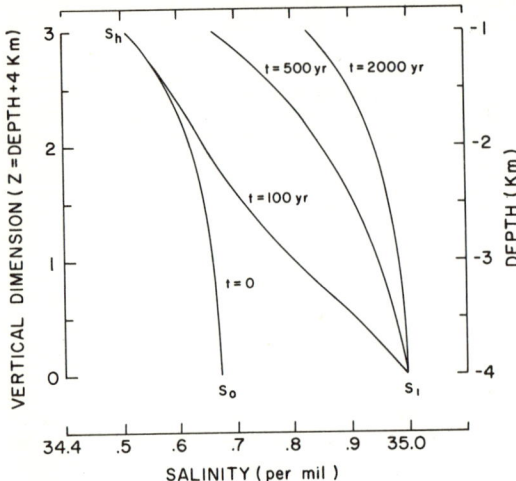

FIG. 2. Rate of change in salinity of the oceanic layer at a depth of 1–4 km due to "instantaneous" increase in salinity at 4-km depth from 34.67 permil to 35.00 permil. Profile $S_o - S_h$ at $t = 0$ from Munk (1966). Profiles at 100, 500, and 2000 years calculated for an eddy diffusion coefficient of 1.3 cm^2/sec, a rate of deep-water upwelling of 1.4×10^{-5} cm/sec, and a constant salinity ($S_1 = 35.00$ permil) at the lower boundary ($z = 0$).

O_2, C^{14}, and Ra^{226} may be regarded as stationary and maintained by eddy diffusion (constant eddy diffusion coefficient of approximately 1.3 cm^2/sec) and rising deep water (upwelling rate of 1.4×10^{-5} cm/sec). The stationary profiles for salinity and oxygen concentration as a function of depth, fitted by Munk (1966) to the literature data, are shown in Figs. 2 and 3, labelled $t = 0$. With reference to the present-day salinity profile $S_0 - S_h$ in Fig. 2, if the salinity at 4-km depth increased from $S_0 = 34.67$ permil to a constant value $S_1 = 35.00$ permil, a new salinity profile virtually parallel to the present-day profile would have been established in 2000 years (at the rate of upwelling and eddy diffusion unchanged). For oxygen (Fig. 3), the present-day concentration changes from 3.5 ml/l at 4-km depth to 0.6 ml/l at 1 km, the concentration-depth profile being stationary (curve $C_0 - C_h$). Increase in the oxygen concentration to 5.0 ml/l at 4-km depth would have resulted in a new steady-state attained in approximately 2000 years. Thus under the physical conditions in the Recent ocean the times of transition from one stationary distribution to a new one are short, and these are of the same order as the "mixing time" of the deep ocean estimated at 800–1000 years from C^{14} ages (Broecker, 1963). Time to steady state is of the order of 500 years when the concentration at 1-km depth remains constant (S_h and C_h, Figs. 2 and 3) but the concentration at 4-km depth increases as shown in Figs. 2 and 3.

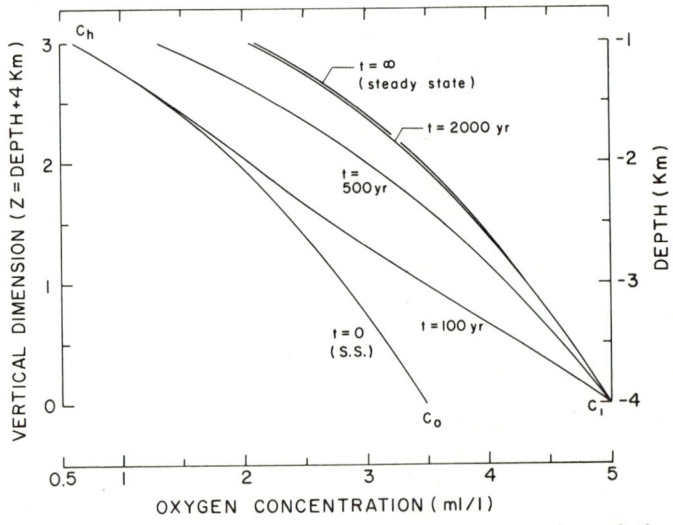

FIG. 3. Rate of change in the oxygen concentration of the oceanic layer at a depth of 1–4 km due to "instantaneous" increase in oxygen concentration at 4-km depth ($z = 0$) from 3.5 ml O_2/liter to 5.0 ml O_2/liter. Eddy diffusion coefficient = 1.3 cm^2/sec; upwelling rate = 1.4×10^{-5} cm/sec; oxygen consumption rate = 8.5×10^{-11} ml O_2/l/sec ($=0.0027$ ml O_2/l/yr). $C_o - C_h$ profile from Munk (1966). Oxygen concentration values after 2000 years are within 1% of the steady-state values ($t = \infty$).

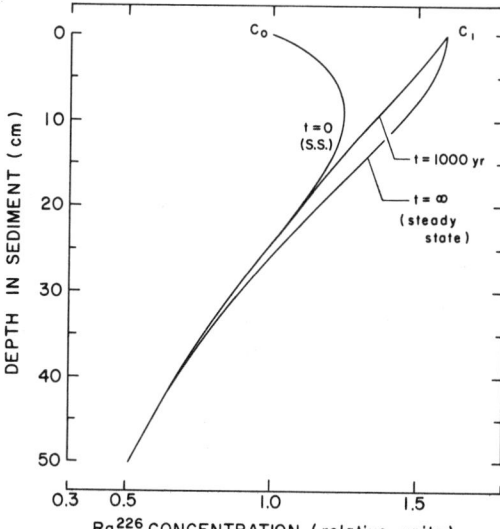

FIG. 4. Rate of change in Ra^{226} concentration in sediment due to an "instantaneous" increase in the Ra^{226} concentration at the sediment-water interface from $C_0 = 1$ to $C_1 = 1.6$ (arbitrary units). Th^{230} concentration in settling sediment = 100 (arbitrary units), sedimentation rate = 3×10^{-4} cm/yr; diffusion coefficient of Ra^{226} in sediment = 1×10^{-9} cm²/sec. Stationary concentration-depth profile ($t = \infty$) established in less than 5000 years.

The rates of change in a stationary distribution within the sediment may be illustrated for the case of Ra^{226}. A characteristic profile of Ra^{226} as a function of depth in the sediment is shown in Fig. 4, labelled $t = 0$. The Ra^{226} concentration scale was arbitrarily adjusted to the value of $C_0 = 1$ at the sediment-water interface. The present-day stationary profile is maintained by a balance between the decay of Th^{230} added to the sediment at constant concentration and constant sedimentation rate, constant concentration of Ra^{226} at the interface, and diffusion of Ra^{226} through the pore space of the sediment. The stationary distribution of Ra^{226} may be perturbed by any combination of such factors as a change in the sedimentation rate; change in the Th^{230} concentration in the settling sediment; change in the radium concentration at the interface. The consequences of an increase of 60% in the radium concentration at the interface (from $C_0 = 1$ to $C_1 = 1.6$, Fig. 4), keeping the Th^{230} and sedimentation rate unchanged, are shown in Fig. 4. The Ra^{226} profile changes drastically after as little as 1000 years. In less than 5000 years the profile would be virtually indistinguishable from the new steady-state profile ($t = \infty$). Under new conditions there would be no radium flux from the sediment insofar as the Ra^{226} concentration continually decreases from the interface down.

ABRAHAM LERMAN

References

Arrhenius, G., 1963, "Pelagic sediments," in (Hill, M. N., editor), "The Sea," Vol. 3, New York, Interscience Publ., pp. 655–727.
Barth, T. F. W., 1961, "Abundance of elements, areal averages and geochemical cycles," Geochim. Cosmochim. Acta, **23**, 1–8.
Broecker, W., 1963, "Radioisotopes and large-scale oceanic mixing," in (Hill, M. N., editor), "The Sea," Vol. 2, New York, Interscience Publ., pp. 88–108.
Goldberg, E. D., 1965, "Minor elements in sea water," in (Riley, J. P., and Skirrow, G., editors), "Chemical Oceanography," Vol 1, New York, Academic Press, pp. 163–196.
Goldberg, E. D., and Koide, M., 1963, "Rates of sediment accumulation in the Indian Ocean," in (Geiss, J., and Goldberg, E. D., editors), "Earth Science and Meteoritics," Amsterdam, North-Holland Publishing Co., pp. 91–102.
Goldschmidt, V. M., 1954, "Geochemistry," London, Oxford Univ. Press.
Koczy, F. F., 1958, "Natural radium as a tracer in the ocean," Proc. 2nd U.N. Intern. Conf. Peaceful Uses of Atomic Energy, **18**, 351–375, United Nations, Geneva.
Livingstone, D. A., 1963, "The sodium cycle and the age of the ocean," Geochim. Cosmochim. Acta, **27**, 1055–1069.
*Mackenzie, F. T., and Garrels, R. M., 1966, "Chemical mass balance between rivers and oceans," Am. J. Sci., **204**, 507–525.
*Mason, B., 1966, "Principles of Geochemistry," Third ed., New York, John Wiley & Sons.
Munk, W. H., 1966, "Abyssal recipes," Deep-Sea Res., **13**, 707–730.
Nace, R. L., 1967, "Water resources: a global problem with local roots," Environ. Sci. Technol., **1**, 550–560.
Revelle, R., and Suess, H. E., 1962, "Gases," in (Hill, M. N., editor), "The Sea," Vol. 1, New York, Interscience Publ., pp. 313–321.
Richards, F. A., 1965, "Anoxic basins and fjords," in (Riley, J. P., and Skirrow, G., editors), "Chemical Oceanography," Vol. 1, New York, Academic Press, pp. 611–645.
Rittenberg, S. C., Emery, K. O., and Orr, W. L., 1955, "Regeneration of nutrients in sediments of marine basins," Deep-Sea Res., **3**, 23–45.
Ronov, A. B., 1968, "Probable changes in the composition of sea water during the course of geological time," Sedimentol., **10**, 25–43.
Rubey, W. W., 1951, "Geologic history of sea water," Bull. Geol. Soc. Am., **62**, 1111–1147.
Sverdrup, H. U., Johnson, M. W., and Fleming, R. H., 1942, "The Oceans," Englewood Cliffs, N.J. Prentice-Hall.
Turekian, K. K., 1965, "Some aspects of the geochemistry of marine sediments," in (Riley, J. P., and Skirrow, G., editors), "Chemical Oceanog-

raphy," Vol. 2, New York, Academic Press, pp. 81–126.

*Additional bibliographic references may be found in this work.

Cross-references: *Buffer Systems; Calcium Cycle; Cycles—Geochemical; Groundwater; Outgassing of the Planet Earth; Oxygen Cycle; River Geochemistry; Seawater, Chemistry; Stoichiometry; Sulfate Reduction Microbial; Weathering, Chemical.* Vol. I: *Marine Sediments; Ocean-Atmosphere Interaction; Oceanic Circulation; Salinity in the Ocean; Upwelling.* Vol. V: *Volatiles.*

SEAWATER, HISTORY

The chemical composition of seawater was first studied on a world wide base by the *Challenger* expedition of 1877. Since that time, no systematic changes have been observed. Speculation about the history of seawater must, therefore, be based entirely on clues derived from the geologic record combined with calculations of the relevant chemical equilibria. Since Late Precambrian time (approximately 600 million years), the chemical composition of seawater is at least in part recorded in the extensive sequences of chemical and biochemical sediments of the stratigraphic column, and limits can be set for the possible variations of a number of important constituents. Sedimentary records of the seas of earlier times are more difficult to interpret.

Post-Precambrian History

Salt deposits' and limestones are the sediments which most directly reflect the composition of the seawater from which they were deposited. Salt deposits are known from every period of the geologic column ever since Cambrian times. No systematic differences in mineralogy and chemistry between deposits of different ages seem to exist, indicating that the relative abundance of cations and anions in seawater must have remained fairly constant. Furthermore, since Late Precambrian time, the ratio of marine limestones to all other marine sediments deposited during a particular time interval has remained nearly constant. Rubey (1951) has discussed CO_2 equilibria in seawater and the effects of changes in P_{CO_2}. Because of the precipitation of $CaCO_3$, P_{CO_2} of seawater is well buffered with respect to large additions of CO_2. This buffering capacity must have existed ever since limestones have been deposited. Loss of CO_2 from seawater would lead to high pH values which cannot be tolerated by most forms of life. For these reasons it is likely that P_{CO_2} of the atmosphere-ocean system has not changed very greatly since Late Precambrian time.

However, if the addition of CO_2 to the oceans through weathering processes is compared with the loss caused by the biochemical and chemical precipitation of carbonates, a deficit exists which has been estimated by Rubey (1951) to be as large as 10^{14} grams of CO_2 per year. This deficit can be balanced only by assuming a slow and continuous degassing of the earth's interior.

A number of other volatile constituents require the existence of the same source to balance similar deficits. All such volatiles are called "excess" volatiles. Some, like argon (see Damon and Kulp, 1958), probably were degassed rapidly at an early stage of the earth's history, while others, like H_2O, were retained in melts and solids and might have appeared at the earth's surface more gradually.

Excess Volatiles

The amounts of excess volatiles present in the hydrosphere and atmosphere have been calculated by a number of workers, using the geochemical balance approach suggested by V. M. Goldschmidt. Goldschmidt assumed that the early crust consisted of igneous rocks similar in composition to igneous rocks exposed today. Erosion of a certain amount of this crust yielded the sediments and the solutes in the oceans. Balances are usually established on the basis of sodium and potassium. Barth (1960) has criticized this approach, because it does not take into account the recycling of elements (erosion of older sediments) and the

TABLE 1. EXCESS VOLATILE CONTENTS (units of 10^{20} g)[a]

Volatile	Igneous Rocks, i	Sediments, s	Oceans, o	Atmosphere, a	Excess, $e=s+o+a-i$	Excess % $\frac{i}{s+o+a} \times 100$
H_2O	62	260	14,000	0.06	14,200	0.42
C (includes) CO_2	3.3	620	1.5	0.023	618	1.9
Cl	1.7	38	280	—	316	0.6
S	3.4	45	14	—	56	5.8
N	0.2	5.2	0.3	38.6	44	0.5
B	0.1	1.4	0.1	—	1.4	6.6

[a] From Wedepohl, 1963.

storage in interstitial brines. Nevertheless, useful comparisons can be obtained in this way. Wedepohl (1963) has assembled recent values for the six most important volatiles (Table 1). For each of these volatiles, weathering of igneous rocks accounts for less than 10%, and in some cases less than 1%, of the amount present in sediments, hydrosphere and atmosphere. Degassing of the earth's core and mantle is assumed to be the source for the remaining amounts. Channels for this degassing can be volcanic exhalations or hot springs. Comparisons of the composition of the "excess" volatiles with that of volcanic gases are not meaningful, because such gases are largely contaminated with surface materials and the "excess" volatiles themselves during their passage through the crust. Isotopic analyses have shown that primary H_2O formed by degassing of the mantle constitutes certainly less than, and probably much less than, 5% of the H_2O of volcanic gases and hot springs.

Early History of the Atmosphere and Hydrosphere

Much less specific information is available with respect to the early history of seawater (Early Precambrian). The early hydrosphere may have formed in part by condensation from the early atmosphere, and speculations about the early history of seawater, therefore, center around the history of the early atmosphere. Urey compared terrestrial with cosmic abundances and came to the conclusion that on the earth H, O, N, C and S are enriched greatly over Ne, Ar, Kr and Xe with respect to the universe. He argues that these volatiles could not have been retained by the earth's gravitational field if the earth formed at a high temperature and concludes that the earth must have formed by agglomeration of cosmic dust at low temperature. Radioactive heat subsequently raised the average temperature and made the separation of core, mantle and crust possible. The early atmosphere was probably very reducing and consisted either predominantly of CH_4, H_2 and NH_3 or H_2, H_2O and CO (Holland, 1962). This assumption is attractive because of the much greater cosmic abundance of hydrogen, because of the CH_4–NH_3 atmosphere of the outer planets (Kuiper, 1952, reference in Holland, 1962) and because a reducing environment is more favorable to the formation of the complex organic molecules necessary for the origin of life. The abundance of the individual gas species in the early atmosphere depends primarily upon the temperature at which this gas mixture was equilibrated with the earth's crust. The presence of metallic iron in the undifferentiated earth makes it likely that some of the carbon was present as graphite. If a gas phase of the system H–C–O is equilibrated with graphite, its composition can be calculated for a given temperature, provided that P_{O_2} is known. For an early crust, two mineral assemblages defining an upper and a lower limit with respect to P_{O_2} can be postulated: fayalite (Fe_2SiO_4) + magnetite (Fe_3O_4) + quartz (SiO_2) for high P_{O_2} and iron + magnetic (Fe_3O_4) below 560°C and iron + wüstite (FeO) above 560°C for the lower P_{O_2} limit. The gas pressure at which equilibration between solids and gas took place also influences the composition (though to a much lesser degree than temperature does). For reasons of reaction kinetics, a pressure of 100 bars has been assumed in the calculations. Such a pressure can exist a few hundred yards below the surface.

Figure 1 shows two sets of curves, one for low P_{O_2} and the other for high P_{O_2}. According to this model, an early cool atmosphere consisted essentially of CH_4 and NH_3. As heating of the earth proceeded, primarily through

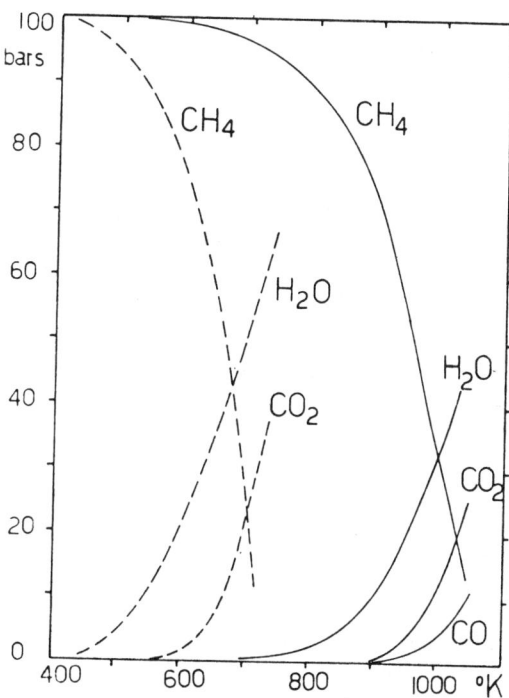

FIG. 1. Temperature (in degrees Kelvin)-pressure (in bars) diagram of a gas phase in the system C–H–O at a total gas pressure of 100 bars. Solid curves are for a gas phase equilibrated with the assemblage iron + magnetite + graphite (below 830°K) or iron + wustite + graphite (above 830°K). Dashed curves are for a gas phase equilibrated with the assemblage fayalite + magnetite + quartz + graphite.

radioactive decay of K^{40} to Ar^{40}, the composition of gases escaping from deeper levels of the earth, and with it the composition of the atmosphere, changed gradually to CO_2, CO and N_2. This model gives water vapor pressures at 25°C between 10^{-9} atm (low P_{O_2}) and 10^{-4} atm (high P_{O_2}). Therefore, liquid water could not condense during this period (P_{H_2O} at 25°C must be 3×10^{-2} atm).

A second step in the development of the atmosphere was initiated by the photodissociation of H_2O to H_2 and O_2 in the upper atmosphere. Loss of hydrogen by escape from the gravitational field was responsible for a slow, but steady, increase in the P_{O_2} at the earth's surface. Assuming a gas pressure of 1 atm, a temperature of 25°C and graphite still present, P_{O_2} would have to rise only a very small amount above that of a fayalite + magnetite + quartz assemblage (from 10^{-78} to 10^{-73} atm) for P_{H_2O} to reach a value of 3×10^{-2} atm and for liquid water to form.

Early condensates were probably accumulated in a number of isolated, closed basins. Because of the large amount of CO_2 present in the atmosphere at that time, these proto-oceans must have been quite acid. Exposed rocks would weather quickly and a large amount of solutes would be carried into the basins in a short time. Early anions were probably predominantly CO_3^{2-} and HCO_3^- rather than Cl^-. $CaCO_3$ can precipitate from such waters but, because of the high P_{CO_2}, only at high values of Ca^{2+}. Some of the oldest limestones have been formed earlier than 2560 million years ago (see Goldich, 1961) and might have precipitated in evaporating basins isolated from such oceans.

Intense evaporation must have been widespread, and salt deposits must have formed very frequently. Continuing evaporation would probably have yielded dolomite, anhydrite, Mg sulfates and finally nahcolite ($NaHCO_3$), rather than the sequences of younger salt deposits, in which halite (NaCl) is an important constituent. Because of the complex subsequent history, none of these early salt deposits can be expected to have been preserved.

During the period just discussed, P_{O_2} was still too low for hematite to be stable. Holland (1962), from the occurrence of sedimentary uraninite deposits 1800 million years old, has concluded that P_{O_2} of the atmosphere at that time was still very low.

The very gradual increase in P_{O_2} was in part due to the fact that a large amount of the oxygen produced by photodissociation of H_2O was consumed in the oxidation of CH_4, NH_3, graphite, and iron or ferrous ions. However, in the time span between 2500 and 1000 million years ago, P_O must have risen sufficiently for life to begin. As biological activity expanded, CO_2 was consumed and O_2 was produced at a much more rapid rate. (Today's production of oxygen by photosynthesis far outstrips that produced by photodissociation.)

Throughout this time H_2O continued to condense and the proto-oceans merged into oceans. As P_{CO_2} decreased, the pH of seawater rose to near its present level. Saturation with respect to $CaCO_3$ was reached at much lower concentrations of Ca^{2+} and large-scale precipitation of limestones was initiated, much of it probably aided by biological processes. From this time on (perhaps 1000 million years ago), a considerable portion of the Precambrian sediments consist of carbonate rocks. Partially as a consequence of this development, chlorine became the dominant anion and seawater had probably reached a composition very near that of today's seawater.

H. P. EUGSTER

References

Barth, T. F. W., 1960, "Abundance of elements, areal averages and geochemical cycles," *Geochim. Cosmochim. Acta*, 23, 1–8.

Damon, P. E., and Kulp, J. L., 1958, "Inert gases and the evolution of the atmosphere: Geochim. et Cosmochim.," *Acta*. v. 13, 280–292.

Derpgol'ts, V. F., "The principal planetary source of natural waters," *International Geol. Rev.*, Aug. 1964, v. 6, 1433–1444.

Goldich, S. S., et al., 1961, "The Precambrian geology and geochronology of Minnesota," *Minn. Geol. Surv. Bull.*, 41, 154.

Goldschmidt, V. M., 1938, "Geochemische Verteilungsgesetze der Elemente IX. Die Mengenverhältnisse der Elemente und der Atom-Arten," *Skrifter Norske Videnskaps-Akad.* Oslo, 1, Mat. Natur. Klasse 1937 No. 3, 1–148.

Goldschmidt, V. M., 1954, "Geochemistry," (ed. Alex. Muir), Oxford, Clarendon Press, 730pp.

*Holland, H. D., 1962, "Model for the evolution of the earth's atmosphere," Buddington Volume, *Geol. Soc. Am.*, 447-477.

*Rubey, W. W., 1951, "Geologic history of seawater," *Bull. Geol. Soc. Am.*, 62, 1111–1248.

Sillén, L. G., 1963, "How has seawater gotten its present composition?" *Svensk Kemisk Tidskrift*, 75(4), 161–177.

Urey, H. C., 1952, "The Planets, their origin and development," New Haven, Yale Univ. Press, 245pp.

Weber, J. N., "Chloride ion concentration in liquid inclusions of carbonate rocks as a possible environmental indicator," *Journal of Sed. Pet.*, Sept. 1964, v. 34, 677–680.

Wedepohl, K. H., 1963, "Einige Uberlegungen zur Geschichte des Meerswassers," *Fortschr. Geol. Rheinland Westfalen*, 10, 129–150.

SEDIMENTARY PARAGENESIS—See **PARAGENESIS**; also Vol. VI

SEDIMENTOLOGY—See Vol. VI

SELENATES AND TELLURATES; SELENITES AND TELLURITES

The minerals in these classes are relatively rare and found mostly as alteration products in the zone of oxidation. In some classifications these minerals, along with the arsenites and iodates, are included as a special division of the oxide class.

SELENATES AND TELLURATES

Montanite	$(BiO_2)(TeO_4) \cdot 2H_2O$
Ferrotellurite	$Fe(TeO_4)$?
Magnolite	$Hg_2(TeO_4)$?
Schmeiderite	$(Pb,Cu)_2SeO_4(OH)_2$

SELENITES AND TELLURITES

A(XO$_3$) xH$_2$O type

Chalcomenite	$Cu(SeO_3)\ 2H_2O$	
Teineite	$Cu(TeO_3)\ 2H_2O$	
Ahlfeldite	$Ni(SeO_3)\ 2H_2O$	⎫ A series of Co–Ni selenites
Cobaltomenite	$Co(SeO_3)\ 2H_2O$	⎭

A$_2$(XO$_3$)$_3$ · xH$_2$O type

Emmonsite	$Fe_2(TeO_3)_3 \cdot 2H_2O$
Mackayite	$Fe_2(TeO_3)_3 \cdot xH_2O$
Blakeite	(Ferric tellurite)

Miscellaneous

Guilleminite	$Ba(UO_2)_3(SeO_3)_2(OH)_4\ 3H_2O$
Molybdomenite	$PbSeO_3$
Spiroffite	$(Mn,Zn)_2\ Te_3O_8$

LLOYD W. STAPLES

References

Dana, J. W., and Dana, E. S., 1951, "The System of Mineralogy," Seventh ed., Vol. II, 635, (revised by Palache, C., Berman, H., and Frondel C.), New York, John Wiley & Sons.

Frondel, C., and Pough, F. H., 1944, "Two new tellurites of iron: mackayite and blakeite," *Am. Mineralogist*, **29**, 211.

Cross-references: *Iodates; Phosphates, Arsenates, and Vandates.*

SELENITES—See SELENATES AND TELLURATES; SELENITES AND TELLURITES

SELENIUM: ELEMENT AND GEOCHEMISTRY

Properties

The Swedish chemist Berzelius discovered selenium (Se) in 1817 while working on pyrite taken from copper mines at Fahlun. Its name is derived from the Greek word *selene*, for moon, because of its chemical similarity to tellurium (named after the Greek word for earth). The atomic number of selenium is 34, its atomic weight is 78.96. Elemental selenium exists in a number of polymorphs, two monoclinic, one hexagonal, and one amorphous. Metallic (hexagonal) selenium has a density of 4.81 g/cm³, melts at 220°C, and boils at 685°C.

In compounds with metals, selenium has a negative valence of two (radius 1.98 Å); in oxygen compounds, positive valences of four ($r = 0.69$ Å), and six ($r = 0.35$ Å). The atomic radius of selenium is 1.16 Å and the covalent radius is 1.14 Å (Pauling).

Seventeen isotopes of selenium are known. The six stable isotopes have the following relative abundances: Se^{74} 0.87%; Se^{76} 9.02%; Se^{77} 7.58%; Se^{78} 23.52%; Se^{80} 49.82%; and Se^{82} 9.19%.

Minerals

Approximately fifty selenium minerals are known; many of these have been discovered and described only recently. Nearly all selenium minerals occur in small quantities and in very small crystals. Selenides are normally structurally simple, dominately covalently bonded minerals containing high-atomic-number metals. Many of the selenides (e.g., berzelianite) appear to form only in hydrothermal deposits depleted or devoid of sulfur. In addition to selenides and selenites, selenium is found in a wide variety of sulfides and sulfosalts, particularly sulfides of nickel, cobalt, molybdenum, and copper. Among the most common (none are common) selenium minerals are:

Clausthalite	PbSe
Ferroselite	$FeSe_2$
Berzelianite	$Cu_{2-x}Se$
Umangite	$Cu_{4-x}Se$
Challomenite	$CuSeO_3 \cdot 2H_2O$

Abundance

Recent studies put the concentration of selenium in the crust between 1.4×10^{-5} and $5 \times 10^{-6}\%$ (0.14–0.05 ppm). There seems to be little difference in Se concentration in silicic and mafic rocks. Cosmic abundance of Se appears to be 67.6 atoms per 10^6 atoms of silicon.

Geochemical Behavior

The close crystallochemical behavior of selenium and sulfur ties much of the geochemistry of selenium to the more abundant sulfur. Selenium readily enters high-temperature sulfide mineral structures. Isomorphous selenides of many heavy metal sulfides are known. In magmatic, pegmatitic, and hydrothermal processes selenium follows closely the behavior of sulfur. However, in supergene processes, the high stability of S^{6+} and relatively low stability of Se^{6+} (relative to Se^{4+} or Se^0) results in a geochemical separation. Eh-pH studies have shown that elemental selenium is the most stable form for most of the range of natural, near-surface conditions.

Selenium is found in toxic concentration in plants in areas of North and South America, Africa, Australia, Spain, France, and Germany. The "loco weed" of western lore derived its name from the effects of selenium poisoning on animals. Seleniferous soils of the semiarid western United States often support heavy growths of selenium-indicator plants such as *Astragalus*. Some Late Paleozoic and Mesozoic shales have concentrations in the 50-ppm range. Plants growing on these formations may contain 5,000–10,000 ppm Se.

Economic Geology

Selenium production comes from the electrolytic copper refineries. There are no areas known where the concentration and quantity of selenium and selenium-bearing minerals are sufficient to allow profitable extraction. Selenium is present in the dispersed state in all hypogene sulfide deposits but rarely in concentrations greater than hundredths of a percent. Concentrations of as much as 0.1% selenium are not uncommon in the immediate vicinity of some oxidized sandstone-type uranium deposits.

Selenium deposits of diverse genetic types are found in the Baltic Shield; in zones of Paleozoic and Tertiary folding (Urals, Harz mountains); on the border of the Canadian Shield (Sudbury, Noranda, Flin-Flon); in zones of Mesozoic folding in South America (Pacajake, Bolivia); and extensively in the African shield.

GEORGE RAPP, JR.

References

Akaiwa, H., 1966, "Abundances of Selenium, Tellurium, and Indium in Meteorites," *J. Geophys. Res.*, **71**(7), 1919–1923.

Collins, A. G., Waters, C. J., and Pearson, C. J., 1964, "Methods of analyzing oilfield waters: selenium and tellurium," *U.S. Bur. Mines, Rept.*, **6474**, 19pp.

Luttrell, G. W., 1959, "Annotated bibliography on the geology of selenium," *U.S. Geol. Surv. Bull.*, **1019-M.**

Rosenfeld, I., and Beath, O. A., 1964, "Selenium; Geobotany, Biochemistry, Toxicity, and Nutrition," New York, Academic Press

Sindeeva, N. D., 1964, "Mineralogy and Types of Deposits of Selenium and Tellurium," New York, John Wiley & Sons (translated from the Russian), 363pp.

Cross-references: *Hypogene; pH-Eh; Selenates and Tellurates; Sulfides; Sulfosalts; Sulfur; Tellurium.*

SEMIARID REGION GROUNDWATER —See HYDROLOGY, SEMIARID REGIONS

SILICA—BIOGEOCHEMICAL CYCLE

Silica is a major constituent in many plants and in a few animal species, serving in most cases as a structural component in the formation of a hard skeleton. In the terrestrial environment certain plants, particularly grasses, can play an important role in the rate of rock weathering. For example, Lovering and Engel (1967) have calculated that the translocation of silica from rock into grasses removed 3.4×10^5 grams silica/acre/year from basalt.

It is in the marine environment that plants have the largest effect on the geochemical cycle of silica. Each year rivers and streams draining the continents contribute 4.3×10^{14} grams of dissolved silica to the oceans. At the present influx rate the total amount of silica in the ocean would double in only twenty thousand years if there were no processes actively removing silica from the oceanic reservoir. One of the major mechanisms for silica removal from the ocean is biological uptake and subsequent removal by sedimentation of biogenic siliceous oozes over large areas of the ocean floor.

Biological Utilization of Silica in the Sea

The principal groups of organisms which extract dissolved silica from seawater are the diatoms, radiolarians, silicoflagellates, and siliceous sponges. The diatoms and radiolarians contribute much larger quantities of skeletal material to marine sediments than silicoflagellates or sponges at the present time (Riedel, 1959).

Diatoms are unicellular algae which live primarily in the lighted surface waters of the oceans. They secrete hydrated amorphous silica in their cell wall. The diatoms presently predominate in the colder regions of the world ocean, particularly the Antarctic Ocean. They are common in the geological record back to the Triassic (Vol. I, *Phytoplankton*). Lewin (1962) has presented a detailed discussion of the biochemical aspects of silicification of algae.

The radiolarians are marine protozoans found widely distributed throughout the ocean. These organisms also secrete skeletal material composed of amorphous silica, with the exception of one group, the *Acantharia*, which have been reported to have skeletons composed of strontium sulfate and calcium and aluminum silicates.

The fossil record suggests that silicoflagellates and siliceous sponges were more abundant in past geological times and may therefore have played a more important role in the oceanic silica cycle.

Biogeochemical Cycle

The basic biogeochemical cycle of silica in the sea is outlined in Fig. 1. Biological utilization depletes the surface layers of the ocean in silica. The minimum silica concentration necessary for diatom growth is thought to be approximately 0.2 mg/l dissolved silica. Most analyses of the surface layers of the ocean range from 0.1 to 2.0 mg/l. Upon death the organic remains of organisms sink through the water column and undergo microbial decomposition and dissolution of the siliceous skeletal material. The regeneration of silica, phosphorus, and nitrogen from decomposing phytoplankton in seawater has been discussed in detail by Grill and Richards (1964) and Redfield et al. (1965). The dissolution of siliceous skeletal material takes place at a slower rate than the oxidation of organic matter and dissolved silica generally increases in concentration with increasing depth to values of 5 to 10 mg/l dissolved silica at the sea floor, while phosphorus and nitrogen generally show maximum values at intermediate depths (Vol. I, *Nutrients in the Sea*). The equilibrium solubility of biogenic silica in seawater is similar to inorganic amorphous silica gels (120 mg/l). The only natural environment in which equilibrium is observed between amorphous silica and seawater is sediment interstitial water. The

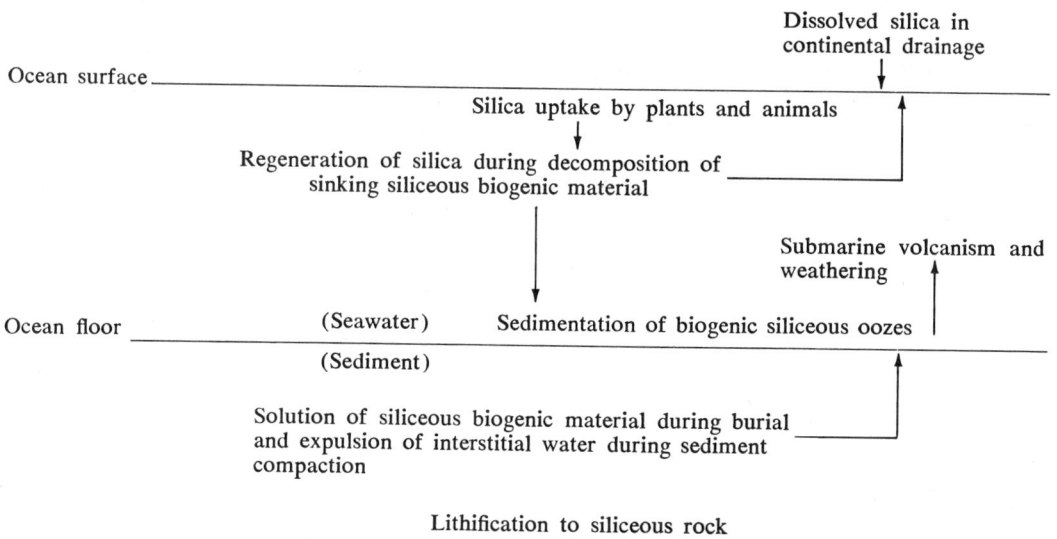

Fig. 1. Silica cycle in the sea.

SILICA—BIOGEOCHEMICAL CYCLE

TABLE 1. SILICA BUDGET FOR THE WORLD OCEAN

	Silicon Dioxide (g/yr)
Input	
Dissolved silica in continental drainage	4.3×10^{14}
Silica from submarine weathering	3.0×10^{12}
Removal	
Silica removed as biogenic sediment	5.0×10^{14}

downward flux of biogenic silica from the surface of the sea is thus balanced by an upward flux of dissolved silica derived from solution of sinking siliceous skeletal remains and an influx of dissolved silica in continental drainage water. Dissolution of siliceous skeletons can continue after burial below the sediment-water interface until equilibrium is attained. During compaction interstitial water is driven out of the sediment and further contributes to the upward flux of silica to the depleted surface layer.

Silica Budget in the Sea

Calculation of a silica budget for the oceans provides useful quantitative information about the influence of organisms on the concentration of oceanic silica (Table 1). The methods used for making such calculations have been discussed by Harriss (1966). It is apparent that the amount of silica removed from the oceans annually as biogenic oozes slightly exceeds the input of dissolved silica from continental drainage and submarine weathering. The influx of dissolved silica from submarine volcanism could possibly bring the budget close to a balance. It can be concluded from Table 1 that organisms, primarily diatoms and radiolaria, are responsible for controlling the concentration of dissolved silica in the modern ocean.

ROBERT C. HARRISS

References

Armstrong, F. A. J., 1965, "Silicon," in (Riley, J. and Skirrow, G., editors), "Chemical Oceanography," Vol. 1, New York, Academic Press, pp. 409–432.

Bogoyavlensky, A., 1967, "Distribution and imigration of dissolved silica in the oceans," *Intern. Geol. Rev.*, **9**, 133–154.

Grill, E. V., and Richards, F. A., 1964, "Nutrient regeneration from phytoplankton decomposing in seawater," *J. Marine Res.*, **22**, 51–69.

Harriss, R. C., 1966, "Biological buffering of oceanic silica," *Nature*, **212**, 275–276.

Lewin, J. C., 1962, "Silicification," in (Lewin, R. A., editor) "Physiology and Biochemistry of Algae," New York, Academic Press, pp. 445–455.

Lisitsyn, A. P., 1967, "Basic relationships in distribution of modern siliceous sediments and their connection with climatic zonation," *Intern. Geol. Rev.*, **9**, 1114–1128.

Lovering T., and Engel, C., 1967, "Translocation of silica and other elements from rock into *Equisetum* and three grasses," *U.S. Geol. Surv. Profess. Paper*, **594-B**, 17pp.

Mackenzie, F. T., Garrels, R. M., Bricker, O. P., and Bickley, F., 1967, "Silica in sea water: control by silica minerals," *Science*, **155**(3768), 1404–1405.

Redfield, A. C., Ketchum, B., and Richards, F. A., 1965, "The influence of organisms on the composition of seawater," in (Hill, M. N., editor), "The Sea," Vol. 2, New York, John Wiley & Sons, pp. 26–77.

Riedel, W. R., 1959, "Siliceous organic remains in pelagic sediments," in (Ireland, H. A., editor), "Silica in Sediments," *Soc. Econ. Paleontol. Mineral.* (Tulsa) *Spec. Publ. No.* **7**, pp. 80–91.

Sosman, R. B., 1965, "The Phases of Silica," New Brunswick, N.J., Rutgers Univ. Press, 388pp.

Cross-references: *Buffer Systems; Cycles—Geochemical; Geochemistry of Sediments; Interstitial Water in Sediments; River Geochemistry; Seawater, Chemistry; Silica Cycle. Vol. I: Nutrients in the Sea; Phytoplankton. Vol. VI: Pelagic Sediments.*

SILICA CYCLE

Silica is the second most abundant element in the earth's crust, and silicate minerals make up a very large part of the rocks at or near the earth's surface. Therefore, a knowledge of the solubility and mobility of silica is critical for an understanding of chemical weathering, soil formation, and the distribution of elements in natural waters. Also, recent work has shown that many silicate minerals, particularly clays, are important in controlling the important properties of pH and buffering capacity in the oceans and fresh-water bodies. These properties are respectively a measure of the amount of hydrogen ion in solution which determines the acid or alkaline character of water and of the ability of a water to take additional amounts of acid or alkaline materials without undergoing an appreciable change in pH.

General data on solubility, abundance, and occurrence of silica are given in Table 1. Silica is usually reported in chemical analyses and the literature as SiO_2 or, less commonly, as Si or other form, but this does not mean silica is actually present in that form in the natural solution. All silica values in this article with one exception in Table 1 are given as SiO_2.

TABLE 1. GENERAL DATA ON SILICA

	Solubility (mg SiO_2/liter)
Quartz	6 at 5°C
	8 at 25°C
Amorphous Silica	60–80 at 0°C
	100–140 at 25°C
	300–380 at 90°C

	Abundance (ppm Si)
Crust	28.2×10^4
Granite	32.3×10^4
Basalt	24.0×10^4
Shale	23.8×10^4
Seawater	3.0

Average Values in Natural Waters (ppm SiO_2)	
Oceans	4–6
Surface waters	0.5–2
Bottom waters	4
Streams	14
Lakes	1–48
Alkaline lakes	> 100
Ground water	17
In carbonate rocks	7
In rhyolitic ash	85
Thermal waters	> 100
Connate waters	20–60
Rainfall and snow	0.1–0.2

Solubility

Quartz is the most stable form of silica at ordinary pressures and temperatures; however, reactions between quartz and water proceed so slowly under normal conditions that the metastable phase of amorphous silica is also important. Without going into detail, it can be said that in the absence of biologic activity or organic reactions which remove silica or higher temperatures or higher pH which enhance solubility, most waters have a silica content between 10 ppm (quartz) and 120 ppm (amorphous silica). The solubilities, particularly of amorphous silica, are temperature sensitive (Table 1), but pH does not affect solubility until values are in excess of pH 9. Moreover, the solubility of silica is a function in part at least of the reactions of silicate minerals such as feldspar and clay. It is worth noting that the resulting solutions have an alkaline pH but not much higher than 9. Therefore, acid waters will tend to become alkaline in the presence of silicate minerals. Recent work on silica by Krauskopf, Iler, Alexander, and others is summarized well by Siever (1962).

Form of Silica in Solution

Work by Krauskopf (1967) and others (Siever, 1962) has shown that in the pH range of about 2 to 9, silica is in the form of dissolved but undissociated monosilicic acid, H_4SiO_4. At pH values in excess of 9, silicate ions become important. The controlling reactions are: $SiO_2 + H_2O = H_4SiO_4$ and $H_4SiO_4 = H^+ + H_3SiO_4^-$. Silica is probably not present in colloidal form except under local conditions that lead to high concentrations and saturated conditions.

Silica Cycle

The silica cycle as used herein is taken to mean the movement of silica, mainly in aqueous solutions between the land and the oceans in a zone at or near the earth's surface (Fig 1). Therefore, silica in deep-seated rocks or magmas is not considered part of this cycle unless the rocks become subject to erosion or the magmas make a direct contribution of silica, for example, during volcanism.

Land and Fresh Waters. A simple way to understand the process is with a schematic equation of the chemical weathering of silicate

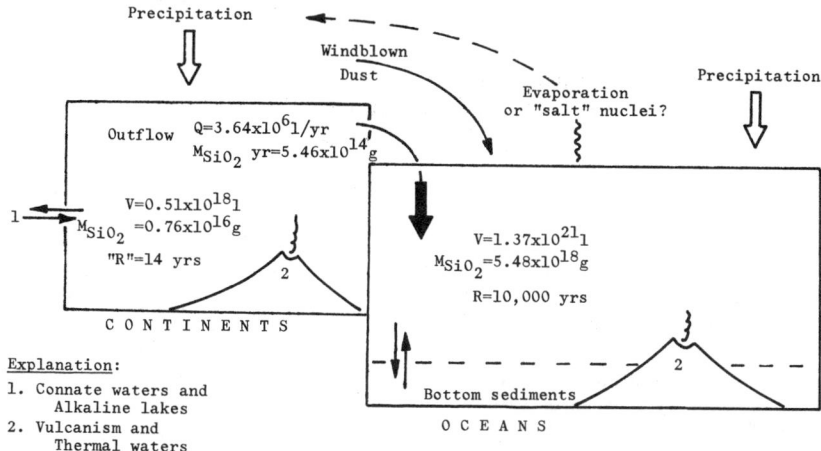

FIG. 1. Generalized scheme of the silica cycle.

rocks as illustrated by feldspar. The cycle begins with rain or snowfall on the earth's surface, which provides (a) water for hydrolysis, (b) some hydrogen ions, due to relatively low pH, from carbonic acid and other reactions, and (c) a low background of SiO_2 on the order of 0.1–0.2 ppm.

Feldspar + Water + Carbonic acid → Intermediate products → Clay + Dissolved ions + Dissolved silica

Some noteworthy features of this type of reaction are that although many such systems can be studied using the methods of classical thermodynamics, most of them are actually open and irreversible. Also, the intermediate products may not be greatly different than the final ones or those from similar reactions; so it can be quite difficult to determine if equilibrium is ever achieved. Furthermore, some silicate minerals such as quartz, pyroxene, and amphibole are so stable that hydrolysis or other chemical reactions proceed very slowly.

The weathering process provides silica which moves into groundwater, streams, and lakes. The only sources of silica in addition to rock weathering would appear to be small amounts in precipitation and perhaps windblown dust. Some silica may be removed or temporarily bypassed by uptake by vegetation. Reactions involving organic materials, aluminum, and iron may also slow down or stop the movement of silica in the soil. Chemical reactions continue during movement of the silica through the unsaturated zone, in groundwater, and with bottom and suspended sediments in streams and lakes. Finally, silica may be affected by organic reactions in swamps and bogs, and it will be taken up by phytoplankton, mainly diatoms, in streams and lakes. The latter effect in lakes can be very pronounced indeed, with silica levels being held at 1 or 2 ppm although silica inflow is much greater. Biologic uptake of silica introduces a seasonal variation into this part of the cycle.

Some typical values for silica in fresh waters are given in Table 1. Rock composition and mineral type can have a strong influence as shown by a SiO_2 content of 7 ppm in carbonate rocks as opposed to 85 ppm in unaltered rhyolitic ash. A range of 7–85 ppm may seem great, but silica tends to be one of the least variable constituents in groundwater and streams. Other than the influence of rock and mineral type, however, it is very difficult to determine the effects of environmental factors. For example, as shown by Davis (1964) there seems to be little or no compelling evidence for influence of climate or temperature. This is of particular interest because Davis doubts that there is a more rapid removal of silica in moist tropical climates as is generally postulated in most theories of laterization or formation of tropical soils. Furthermore, he suggests that silica may be removed as rapidly or possibly more so in moist temperate and moist subarctic climates.

The occurrence and quantities of silica in fresh waters has been discussed for groundwater and streams by Davis (1964) and for lakes by Hutchinson (1957). In addition, Bogomolov et al., (1966) have summarized data for silica in many waters of the Soviet Union and other countries.

Oceans. The silica in fresh waters moves eventually toward the oceans. Much of the silica is in dissolved form, but it is transported also in various kinds of suspended materials. For example, both clay particles and diatoms may be parts of the suspended load of a stream. Important reactions take place as the fresh waters encounter seawater, usually in a delta or estuary. A common assumption has been that electrolytes in seawater precipitate silica (see Bien et al., as discussed in Siever, 1962). This assumes that the silica is in colloidal form, which is now suspect. Furthermore, work by Krauskopf (1967) has shown that seawater does not affect solutions of silicic acid. On the other hand, studies by Garrels and others (Sillén, 1967) show that seawater will remove silica from clays.

It seems likely, with possible exceptions, that most of the dissolved silica in streams moves out into the oceans although much of the suspended material will be deposited in or near the estuary or delta. Other possible sources of ocean silica are from windblown dust off the lands and gases and hot waters resulting from volcanic activity. The first is likely to be widespread, but its importance is hard to assess. Silica from volcanic activity can be important locally, but its overall affect on the silica cycle is also difficult to determine. Over geologic time, however, this has probably been a major source of silica.

The distribution of silica in the oceans and the path of the cycle are hard to follow. In contradistinction to most fresh waters, silica tends to have a wider range of variation in ocean waters than any other element. Silica has a relatively short residence time in the oceans of about 10,000 years (calculated as ratio of weight of silica in the oceans to annual weight inflow from the continents) but the concentrations in seawater are much lower than in most fresh waters (Table 1). With the large input into the oceans of dissolved silica and suspended silicates over geologic time, the question arises as to why the silica content is so low. One obvious factor is removal in the surface layer by radiolaria and by phytoplankton such as diatoms and silicoflagellates. Siliceous sponges also precipitate silica. As in the case of fresh waters, biologic uptake introduces a seasonal variation in concentration in surface waters. Harriss (see *Silica—Biogeochemical Cycle*) presents evidence that uptake by organisms, mainly diatoms and radiolaria, followed by deposition of the biogenic material on the sea floor are the major controls on content of silica in seawaters. Other possible mechanisms include deposition in silicate fractions of clastic sediment and inorganic precipitation, but these would seem to be minor in comparison to the biogenic processes described by Harriss.

Silica in the oceans has been reviewed comprehensively by Armstrong (1965). Sillén (1967) has discussed the role of silicates in influencing pH and composition of seawater, and he has summarized the important work of others such as Garrels, Holland, and Mackenzie.

Other Waters. Brief mention should be made of several byways of the silica cycle that can be locally important. The most common types are *connate waters* (waters that have been out of contact with the atmosphere for a substantial amount of geologic time), *thermal waters* (not necessarily associated with volcanic activity), and *alkaline lakes* (usually in closed drainage basins). Some general values for SiO_2 are given in Table 1. Silica concentration in connate waters is not very different from that in groundwater but tends to be considerably higher in thermal waters and alkaline lakes.

Return to the Continents. Rough calculations indicate that silica in fresh waters takes on the average about fourteen years to reach the sea (calculated as ratio of weight of silica in fresh waters to annual weight outflow per year to the oceans), whereas the residence time in the oceans is about 10^4 years (Fig. 1). The oceans, therefore, introduce a rather large time lag in the silica cycle because, as opposed to a volatile constituent such as chloride (in cyclic salts), very little silica appears to be returned to the continents except by uplift of marine sediments with or without an intermediate step of metamorphism. To this extent then, the silica cycle is not really "cyclical." Nevertheless, it is convenient to consider silica in aqueous solutions and associated mineral phases as comprising a cycle while recognizing the presence of a large time lag or lack of completeness.

F. R. HALL

References

Armstrong, F. A. J., 1965, "Silicon" in (Riley, J. P., and Skirrow, G., editors), "Chemical Oceanography," Vol. I, New York, Academic Press.

Bissell, H. J., 1959, "Silica in sediments of the Upper Paleozoic of the Cordilleran area," in "Silica in Sediments," *Soc. Econ. Paleontol. Mineral. Spec. Publ.*, **7**, 150–185.

Bogomolov, G. V., Plotnikova, G. N., and Titova, E. A., 1966, "Silica in the subterranean waters of certain regions of the USSR and other countries," *Bull. Intern. Assoc. Sci. Hydrol.*, **11**(1), 24–33.

Davis, S. N., 1964, "Silica in streams and ground water," *Am. J. Sci.*, **262**, 870–891.

Hutchinson, G. E., 1957, "A Treatise on Limnology," I, "Geography, Physics, and Chemistry," New York, John Wiley & Sons.

Krauskopf, K. B., 1967, "Introduction to Geochemistry," New York, McGraw-Hill Book Co., 721pp.

MacKenzie, F. T., and Garrels, R. M., 1966, "Silicabicarbonate balance in the ocean and early diagenesis," *J. Sediment. Petrol.*, **38**(4), 1075–1084.

*Siever, R., 1962, "Silica solubility, 0°–200°C., and the diagenesis of siliceous sediments," *J. Geol.* **70**, 127–150.

*Sillén, L. G., 1967, "The ocean as a chemical system," *Science*, **156**, 1189–1197.

*Additional bibliographic references may be found in this work.

Cross-references: *Buffer Systems; Connate Water; Cycles—Geochemical; Electrolytes; Geochemistry of Sedimentary Silica; Lake Geochemistry; Magmatic Water; pH-Eh; Seawater, Chemistry; Silica—Biogeochemical Cycle; Silica Solubility; Weathering, Chemical.* Vol. I: *Nutrients in the Sea.* Vol. VI: *Weathering, Organic.*

SILICA GEOCHEMISTRY—See GEOCHEMISTRY OF SEDIMENTARY SILICA

SILICA SOLUBILITY

Experimental Data

There are more experimental data on the solubility of the various forms of silica in water than on any other single mineral, and they cover by far the widest range of both pressure and temperature. These data are summarized in Fig. 1 (see also Table 1). The stable form of SiO_2 in all parts of this diagram is quartz (except, of course, at temperatures above the "melting curve," where it is liquid silica). The separate fields of α and β quartz have not been indicated since no break in the solubility curves has been detected at the α–β transition, and, indeed, none should be ex-

SILICA SOLUBILITY

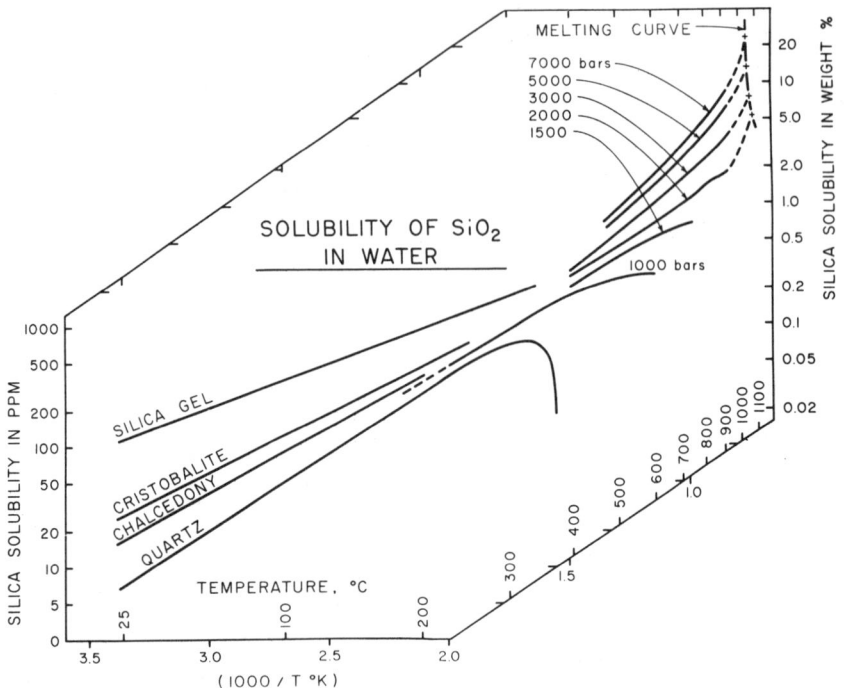

FIG. 1. Solubilities of various forms of silica in water as a function of temperature and pressure. Silica gel, chalcedony, and cristobalite are metastable with respect to quartz. Solubilities above 400°C are for quartz.

pected because of the small free energy difference between the two structural types.

The solubilities of quartz and the metastable forms of SiO_2 at the vapor pressure of the solutions are shown by the curves extending from 25 to 200–350°C. At temperatures of 350°C and below, the conversion of the metastable forms to quartz is sufficiently slow that their solubilities are important and may readily be measured. At 25°C, stable equilibrium is very difficult to achieve, some solutions having been agitated for more than a year without achieving chemical equilibrium (Morey, Fournier, and Rowe, 1962, 1964). Only one sample of chalcedony has been run (R. O. Fournier, unpublished) so the variation in solubilities is not known, but the range of values for the various forms of amorphous silica at any one temperature covers at least ± 50 ppm (see summary in Morey, Fournier, and Rowe, 1964). The curves for chalcedony and silica gel should thus be considered as zones of uncertain width rather than as lines.

The drop in solubilities of quartz between 350 and 371°C (the lower critical end point for SiO_2–H_2O; pressure 220 bars), is due to

TABLE 1.[a]

	Solubilities (ppm)			ΔH[b] (cal/mole)	Variation of Solubility with T[c]
	25°C	100°C	200°C		
Quartz	6	49	268	3350	$\log C = \dfrac{-1310}{T} + 0.42$
Chalcedony	~17	~83	~322	4560	$\log C = \dfrac{-1032}{T} - 0.09$
Cristobalite	27	125	465	4720	$\log C = \dfrac{-1000}{T} - 0.0$
Silica gel	~115	~360	~930	6000	$\log C = \dfrac{-731}{T} - 0.26$

[a] Unpublished equations and ΔH values courtesy of R. O. Fournier.
[b] Differential heat of solution in pure water, as measured by the slope of the curves in Fig. 1.
[c] C = silica concentration in moles/1000 g; T = temperature in Kelvins.

the very rapid decrease in the density of water in this region, which lowers the solubility more than the increasing temperature raises it (Kennedy, 1950).

At higher temperatures (400–900°C) the curves labeled 1500 bars–7000 bars indicate the solubility of quartz at these pressures and temperatures (Anderson and Burnham, 1965; Weill and Fyfe, 1964), and at still higher temperatures (1050–1150°C) the "melting curve," or upper three-phase curve, indicates the SiO_2 content of the aqueous solutions in equilibrium with both quartz and water-saturated liquid silica. This curve continues up to a value of 75 wt% at 9700 bars and 1080°C, the upper critical end point for SiO_2–H_2O, where the aqueous and liquid silica phases become identical (Kennedy, Wasserburg, Heard, and Newton, 1962).

Nature of Silica in Solution

On dissolving in water, it is known that SiO_2 complexes in some way with H_2O, but the stoichiometry of the aqueous species at any temperature and pressure has yet to be conclusively demonstrated. There are various reasons for suspecting that the complex is $SiO_2 \cdot 2H_2O$ or $Si(OH)_4$, the principal one being that Si shows a strong tendency to surround itself with four oxygens in silicate crystals, and it is reasonable to assume that it does the same in solution. Its monomeric nature is established by its rapid reaction with ammonium molybdate to form silicomolybdates.

Solubility in Acidic, Basic and Salt Solutions

The silica-water complex shows at least two ionization reactions in alkaline solutions. Assuming the complex to be $Si(OH)_4$, the reactions and their equilibrium constants at 25°C are

$H_4SiO_4 \rightleftharpoons H_3SiO_4^- + H^+ \quad K = 10^{-9.5}$

$H_3SiO_4^- \rightleftharpoons H_2SiO_4^= + H^+ \quad K = 10^{-12.7}$

These reactions result in greatly increased solubilities in solutions of pH greater than 9.5. The analogous reactions for the dissociation of OH^- ions in acid solutions proceed to such a slight extent as to be as yet unmeasurable, indicating the tenacity with which silicon bonds with its oxygen neighbors. Thus change in pH below 9.5 does not affect silica solubility, indicating that the silica-water complex is unchanged.

At room temperature and pressure, alkali chlorides have a negligible effect on silica solubility.

Silica in Natural Waters

The silica content of natural waters has an

TABLE 2.

Type of Water	Approximate SiO_2 range[a] (ppm)
Groundwater	5–60
Oil field brines	5–60
Rivers and lakes	5–25
Hot springs and geysers	100–600
Seawater	0.01–7

[a] Sources: Davis (1964), Livingstone (1963), White et al., (1963).

approximate range of from 0.1 ppm in melted snow to 4000 ppm in a California mineral spring. However, it is less variable than any other major dissolved constituent of natural waters.

Normal ranges of concentrations of SiO_2 in natural waters are indicated in Table 2. Except for seawater, these concentrations indicate that silica in natural waters is almost always in concentrations greater than would be in equilibrium with quartz. However, the silica has been shown to be usually not colloidal but in true solution, probably in the monomeric forms discussed above. Very slow reaction and recrystallization rates are thus indicated at normal temperatures, and this has been verified by laboratory experiments. In thermal waters, Ellis and Mahon (1964) have noted that at temperatures of 150 to 200°C and higher, silica concentrations correspond closely to those to be expected in solutions in equilibrium with quartz. At depth within the crust, hydrothermal solutions undoubtedly contain up to several weight percent dissolved silica.

The low silica content of seawater has always posed something of a problem, but is generally ascribed to the silica-consuming action of small organisms (diatoms, radiolaria, etc.), and to a lesser extent to the formation of authigenic silicates.

G. M. ANDERSON

References

Anderson, G. M., and Burnham, C. W., 1965, "The solubility of quartz in supercritical water," *Am. J. Sci.*, **263**, 494–511.

Davis, S. N., 1964, "Silica in streams and groundwater," *Am. J. Sci.*, **262**, 870–891.

Ellis, A. J., and Mahon, W. A. J., 1964, "Natural hydrothermal systems and experimental hot-water/rock interactions," *Geochim. Cosmochim. Acta*, **28**, 1323–1357.

Jones, B. F., Rettig, S. L., and Eugster, H. P., 1967, "Silica in Alkaline Brines," *Science*, **158**(3806), 1310–1314.

Kennedy, G. C., 1950, "A portion of the system silica-water," *Econ. Geol.*, **45**, 629–653.

Kennedy, G. C., Wasserburg, G. J., Heard, H. C.,

and Newton, R. C., 1962, "The upper three-phase region in the system SiO_2-H_2O," *Am. J. Sci.*, **260**, 501–521.

Livingstone, D. A., 1963, "Chemical composition of rivers and lakes," in (Fleischer, M., editor), "The Data of Geochemistry," *U.S. Geol. Surv. Profess. Paper* **440-G**.

Mackenzie, F. T., and Garrels, R. M., 1965, "Silicates: Reactivity with sea water," *Science*, **150**(3692), 57–58.

Morey, G. W., Fournier, R. O., and Rowe, J. J., 1962, "The solubility of quartz in water in the temperature interval from 25° to 300°C," *Geochim. Cosmochim. Acta*, **26**, 1029–1043.

Morey, G. W., Fournier, R. O., and Rowe, J. J., 1964, "The solubility of amorphous silica at 25°C," *J. Geophys. Res.*, **69**, 1995–2002.

Weill, D. F., and Fyfe, W. S., 1964, "The solubility of quartz in H_2O in the range 1000–4000 bars and 400°550°C," *Geochim. Cosmochim. Acta*, **28**, 1243–1256.

White, D. E., Hem., J. D., and Waring, G. A., 1963, "Chemical composition of subsurface waters," in (Fleischer, M., editor), "The Data of Geochemistry," *U.S. Geol. Surv. Profess. Paper* **440-F**.

Cross-references: *Complexes; Lake Geochemistry; pH-Eh; River Geochemistry; Seawater, Chemistry; Silica—Biogeochemical Cycle; Stoichiometry.* Vol. IVB: *Quartz.*

SILICATE MINERALS—See MINERAL CLASSES: SILICATES

SILICATES—HYDROTHERMAL SYSTEMS

History

Systematic, quantitative, studies of silicate systems with water as a component began with the studies of Morey and Fenner in 1917 (cited by Wyllie, 1963) at the Geophysical Laboratory, Carnegie Institution of Washington. Investigators at this laboratory have continued extensive and valuable studies under hydrothermal conditions since that time. In the years following World War II a number of additional laboratories, devoted in large part to these studies, have been established at universities and industrial organizations, both here and abroad, and in several federal agencies including the U.S. Geological Survey.

Objectives

The principal geologic objectives of hydrothermal studies in silicate systems include a better understanding of possible processes, and the interplay of such processes in the formation of crustal and upper mantle rock, and the characterization of specified rock occurrences in terms of the intensive parameters of state significant in their origin and development.

Thermodynamic Basis

The interpretation of the experimental results rests in very large part on the concept of heterogeneous equilibria in polycomponent systems as developed by J. Willard Gibbs (1876, 1878). The majority of the studies to date have been in terms of pressure, temperature, and composition as independent variables. The results are presented in terms of a pressure-temperature projection showing $P-T$ conditions under which specified invariant (point), univariant (curves), and bivariant (field) assemblages are thermodynamically stable. A $P-T$ projection may be used to illustrate aspects of the phase relations in systems of many components, if the reactions can be appropriately defined. However, this projection contains no information on the composition of the coexisting phases under invariant, univariant, or bivariant conditions. To illustrate the composition relationships under invariant and univariant conditions, an analogous projection, the pressure-composition or temperature-composition $(T-X)$, may be used. Similarly, if pressure and temperature are held constant (isobaric-isothermal) the bivariant phase relations may be illustrated. In a system of unit mass composed of n components, $n-1$ spatial dimensions are required to illustrate the general composition relations. Thus the technique using either $P-X$, $T-X$ projections, or isobaric-isothermal diagrams becomes unwieldy in systems with greater than four components, and various methods based on n–dimensional or projective geometry must be resorted to. An alternative method is to represent in analytical form the elements of the classical $P-T-X$ phase diagram. It is also possible to treat the heterogeneous equilibria in silicate-water systems in terms of volume-temperature-composition $(V-T-X)$, entropy-pressure-composition $(S-P-X)$, or entropy-volume-composition $(S-V-X)$. These forms of analysis have not been used extensively, probably due in large part to the ready availability of P and T "meters," and the lack of experimentally convenient S and V "meters." It is also possible to treat heterogeneous equilibria in terms of the chemical potentials (μ) (zeta potential of Gibbs) of the individual components as independent variables.

Experimental Techniques

Starting Materials. The more common starting materials used in hydrothermal studies include crystalline and amorphous (glass or gel) materials, depending on the type of information sought. It has been demonstrated that the nature of the starting materials frequently bias the results of the experimental studies. Methods of preparation of starting material have been presented by Schairer, Roy and others (Wyllie, 1963). The nature of the bias imposed upon the results by the choice of starting materials and problems of equilibria have been discussed by Fyfe and others (Wyllie, 1963).

Apparatus. The equipment available for hydrothermal studies at elevated pressure and temperature has been reviewed by Wyllie (1963) and will not be discussed here. It is sufficient to note that a variety of apparatus is available which permit investigations at pressures and temperatures corresponding to depths of 200 to 300 km in the earth, depending on the geothermal gradient believed to exist.

Experiments. Although many of the early hydrothermal studies were performed in large-volume, low-pressure devices developed by Morey (Wyllie, 1963), or in precious metal envelopes or capsules open to the pressure media, the more common technique in use at the present time involves sealed precious metal capsules containing the charge and subjected to an external pressure (solid, liquid, or gas) media. Precious metal (gold, platinum, silver, or silver palladium alloys) capsules are used principally to reduce interaction of the capsule material with the charge. The capsules are sealed in order to provide a closed (partially or completely) chemical environment. Typically a charge of silicate material and water, or an aqueous solution containing other constituents such as CO_2, F, Cl, etc., in known proportions are sealed in the capsule by electric arc welding. These capsules are then placed in a pressure vessel and subjected to external pressure and temperature (either by a furnace within or external to the pressure vessel.) Pressure in the capsule is generated by the change in volume of the water, or aqueous solution, and then transmitted to silicate phases. Due to the low strength, at high temperature, of the precious metal capsule, and the confining pressure exerted by the external pressure media, the internal volume of the capsule will change. If an appropriate amount of solution has been used, the sealed capsule will collapse until the internal pressure generated by the expansion of the solution is equal to the external confining pressure. A modification of the sealed-capsule technique has been developed by Eugster and Wones (Wyllie, 1963), whereby the capsule is permeable to hydrogen, but not to other components, permitting control of oxygen fugacity (f_{O_2}) of the vapor phase by means of an external buffer. A number of buffers are available, and are based on the fact that for a bivariant assemblage in an n-component system at constant pressure and temperature, the composition of each of the coexisting phases is fixed. Thus the f_{O_2} of the gas phase in the three phase assemblage magnetite-hematite-gas in the ternary system Fe-H-O may be used to "buffer" the f_{O_2} of the assemblage in the system of interest. Eugster and co-workers have developed techniques by which it is possible to control f_{HF}, f_{CO_2}, and f_{S_2}. Shaw, using a mixed gas technique, has been able to control f_{H_2O}.

Pressure, P_{H_2O}, P_{fluid}, P_{total}. The majority of the hydrothermal studies to date have taken place in what the authors refer to as "the presence of excess water." Depending on the particular study, this statement may either mean that the amount of H_2O present is more than sufficient to: (1) fill the pore space in the charge, or (2) saturate the silicate melt present. In both of these cases, the pressure variable is commonly referred to as P_{H_2O}. This is a consequence of the fact that the pressure exerted on the silicate phases is by a dominantly aqueous, supercritical phase. Rigorously, this notation is incorrect, inasmuch as the vapor phase (gas, fluid) coexisting with the silicate material (crystalline, or liquid, or both) contains notable amounts (Kennedy et al., cited by Wyllie, 1963) of dissolved silicate. In terms of the experimental studies, the subscripts on the pressure variables are not necessary if we can properly define the thermodynamic system.

Static and Dynamic Methods. An extension of the "quenching" technique, developed by the investigators at the Geophysical Laboratory for the study of dry silicate equilibria, has proven extremely useful in the hydrothermal studies of silicates. This method is based on the assumption that equilibria obtained at elevated pressures and temperatures can be "frozen in" by rapid cooling and decrease in pressure. Even if the equilibria attained at elevated pressures and temperatures is not completely "frozen in" on the quench, in many cases useful information can be obtained by careful examination of the textural features exhibited by the quenched material. Dynamic methods have not been used extensively; however, high pressure-high temperature differential thermal analysis (DTA) methods (Wyllie, 1963) show considerable promise in the study of reactions which are kinetically favorable. Another dynamic method used by Morey and co-workers (Wyllie, 1963) is based on analysis of the vapor (fluid, gas) phase as sampled at high pressure and temperature.

Examination of the Products. The petrographic microscope has proven to be an indispensable tool in the examination of the experimental products, both for the characterization of the phases and for the determination of the textural relations. X-ray powder diffraction techniques are used extensively in the identification and characterization of the crystalline materials. In certain cases it is necessary to use somewhat more sophisticated techniques such as infrared spectrometry and electron microprobe analysis. Burnham (Wyllie, 1963) has devised a technique which permits quantitative analysis of the vapor (gas, fluid) phase coexisting with silicates (crystalline or molten) at high pressure and temperature, after the material has been quenched.

Results of Hydrothermal Studies in Silicate Systems

Minerals. The stability relations in terms of pressure, temperature, and composition of a large number of hydrous silicates such as amphiboles and layer silicates have been determined (Deer, et al. 1962, 1963). The influence of f_O on the stability of many of the biotites and iron-bearing amphiboles has been studied in some detail. In a number of studies water is not an essential constituent of any of the phases encountered (other than the vapor phase) but simply provides a diffusion media, which consequently enhances reaction times. The studies on the alkali-feldspars by Tuttle and Bowen, Yoder, Stewart, and Smith, Mackenzie, and others (Wyllie, 1963) at temperatures below the solidus are typical examples.

Rocks. The studies of Tuttle and Bowen on the granitic rocks in terms of the simplified model system $NaAlSi_3O_8$–$KAlSi_3O_8$–SiO_2–H_2O, on the pyroxenites and peridotites in terms of the simplified model system MgO–SiO_2–H_2O, by Yoder on metamorphic assemblages typified by examples in the system MgO–Al_2O_3–SiO_2–H_2O, by Fyfe and Coombs on the zeolitic facies, and by Yoder and Tilley on the basalts, represent a sample of the major petrologic contribution (Wyllie, 1963). Winkler and co-workers, Yoder and Tilley, Wyllie and Tuttle, and others have been concerned with relationships using rocks (powdered) as starting material.

Metasomatism. Orville (Wyllie, 1963) has been very much concerned with the role of metasomatism in rock-forming processes. His work has been mainly concerned with K and Na metasomatism in terms of alkali feldspar ion exchange studies in aqueous alkali halide solutions (see Vol. V, *Metasomatism*).

Melting Relations. One of the most pronounced effects caused by the presence of water under pressure is the pronounced lowering of the isobaric solidus temperature shown by many silicate-water systems. Thermodynamically this is quite reasonable in view of the low molecular weight of water and the low melting point of ice in the one-component system H_2O. Although most silicate melts at pressures less than 10 kb show limited solubility of H_2O in the melt, some compositions, particularly those containing more Na_2O or K_2O than the 1:1 molecular ratio $Na_2O + K_2O:Al_2O_3$, show continuous solubility of H_2O in the melt.

Studies Other Than Phase Equilibria. Transport properties of silicate melts containing dissolved H_2O which have been studied include: viscosity, thermal and electrical conductivity, and diffusion rates. Rose, Smith, Schloemer, and others have been concerned with the devitrification of natural and synthetic glasses under hydrothermal conditions. Taylor and co-workers evaluated oxygen isotope (O^{18}, O^{16}) fractionation among coexisting silicates and the aqueous vapor (fluid, gas) phase (Eitel, 1966).

Future Research

Phase Equilibria. Although the results of a large number of systematic studies of phase equilibria in silicate-water systems are available, much of great interest remains to be done. Petrologically, we are less interested in the ultimate stability of a single phase than in the stability range of an *assemblage* of phases. Little is known concerning phase relations in systems where H_2O is a component, but the melt phase is undersaturated with respect to H_2O. Much more careful work needs to be done in terms of accurate determination of equilibria.

Kinetics. Very little has been done regarding rate studies in synthetic systems. Although we can duplicate the pressures and temperatures in the earth's crust and outer mantle, geologic times are not available for the studies.

Properties of Materials. *Chemical.* If we are to attain predictive abilities in silicate-water systems, thermodynamic properties, including mixing terms, are required for many phases, crystalline, liquid, and vapor.

Physical. In addition to properties such as thermal and electrical conductivity of the melt and vapor phase, seismic wave velocity and attenuation in polyphase silicate-water systems must be known for a better understanding of crustal rock-forming processes.

Review Articles. Many recent review articles are cited in the book by Eitel (1966) with extensive reviews of the original investigations. Applications of these studies to a number of petrologic problems can be found in Wyllie (1963), and in terms of specified minerals in the first four volumes of Deer, Howie, and Zussman (1962, 1963).

WILLIAM C. LUTH

References

Deer, W. A., Howie, R. A., and Zussman, J., 1962, 1963, "Rock-Forming Minerals," New York, J. Wiley & Sons, 5 Vols.
*Eitel, W. 1966, "Silicate Science," Vol. IV, "Hydrothermal Silicate Systems," New York, Academic Press.
Gibbs, J. Willard, 1876, 1878, "On the Equilibrium of Heterogeneous Substances," *Trans. Conn. Acad.*, **3**, 108–248, 343–524.
*Wyllie, P. J., 1963, "Applications of High Pressure Studies to the Earth Sciences," in (Bradley, R. S., editor), "High Pressure Physics and Chemistry," Vol. 2, Chap. 6, pp. 2–89.

*Additional bibliographic references mentioned in text may be found in this work.

Cross-references: *Differential Thermal Analysis; Infrared Analysis; Isotope Fractionation; Mineral Classes—Silicates; Phase Equilibria; Thermodynamics; X-Ray Diffraction Analysis. Vol. IVB: Electron Probe Microanalysis; Glass: Devitrification. Vol. V: Metasomatism.*

SILICEOUS DEPOSITS—See Vol. IV B; also Vol. VI

SILICON: ELEMENT AND GEOCHEMISTRY

Properties

Silicon, a major element, symbol Si, of atomic number 14, was first isolated by J. J. Berzelius in 1823. In its pure form, the element is dark, steel-gray when crystalline, and dark brown when amorphous. Its atomic weight is 27.7, density 2.4 (20°C), melting point 1410°C, and boiling point about 2680°C. In compounds, silicon is a tetravalent cation of ionic radius 0.42Å. Bonds with oxygen are partly covalent.

Three isotopes of silicon have been found. In olivine of dunite from Jackson County, North Carolina, the relative concentration of these isotopes is as follows: Si^{28} (92.18%); Si^{29} (4.71%); and Si^{30} (3.12%).

Minerals

A large number of silicate minerals are known, and in all of these, the fundamental structural unit is the SiO_4 tetrahedron in which a silicon ion is surrounded by four oxygen ions. Some of the common rock-forming silicate minerals are listed below.

quartz SiO_2
olivine $(Mg,Fe)_2SiO_4$
orthopyroxene $(Mg,Fe)SiO_3$
actinolite $Ca_2(Mg,Fe)_5Si_8O_{22}(OH)_2$
biotite $K(Mg,Fe)_3AlSi_3O_{10}(OH)_2$
potash feldspar $KAlSiO_3$
plagioclase $(NaSi,CaAl)AlSi_2O_8$
almandine garnet $Fe_3Al_2Si_3O_{12}$
sillimanite Al_2SiO_5
kaolinite $Al_4Si_4O_{10}(OH)_8$

Abundance

Silicon is second in abundance of all elements, and is estimated to have an average crustal abundance of 27.7% by weight.

Nearly all volcanic, plutonic, and metamorphic rocks, and some meteorites, are composed mainly of silicate minerals. The estimated proportions of the common rock-forming silicates in the earth's crust are: feldspars 60%, quartz 12%, amphiboles 17%, pyroxene 17%, and biotite 4%.

The frequency distribution of SiO_2 content of igneous rocks is evidently bimodal (Richardson and Sneesby, 1922; Ahrens, 1964) as shown in Fig. 1. The peak on the left corresponds to basaltic rocks and that on the right to granitic rocks.

Geochemical Behavior

Wager and Deer (1939) have shown that the differentiation of a gabbroic magma does not produce a notable alteration in the silica content. In magmas of more granitic composition, an enrichment of silica is expected in the last-formed products of the differentiation process.

During rock metamorphism, SiO_2 may become segregated and concentrated in the form of quartz veins and pods.

Silicon is introduced to some metamorphic rocks. Thus certain skarns have evidently resulted from the introduction of SiO_2 to dolomitic limestones, resulting in the crystallization of calc-silicate minerals.

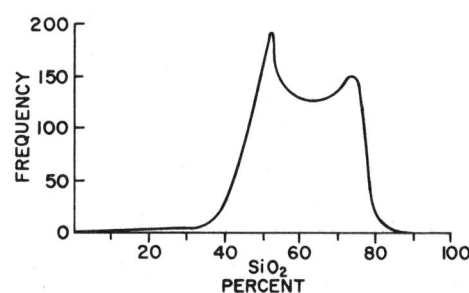

FIG. 1. Frequency distribution of silica in igneous rocks (after Richardson and Sneesby, 1922).

SILICON: ELEMENT AND GEOCHEMISTRY

The replacement of preexisting rocks by silica (silicification) is commonly associated with the formation of ore deposits.

During weathering of crystalline rocks, olivine, calcic plagioclase, and nepheline are decomposed relatively rapidly, whereas alkali feldspars, quartz, and micas are more resistant to dissolution. Under tropical conditions, silica may be preferentially leached from rocks, resulting in lateritic soils. The breakdown of many silicate minerals leads to the formation of clay minerals.

During sedimentation processes, silica may become concentrated in the form of sandstone. Some silica in solution reaches the sea where it is precipitated by diatoms, radiolarians, and sponges, leading to concentrations of silica in the form of chert and siliceous ooze.

RALPH KRETZ

References

Ahrens, L. H., 1964, "Element distribution in igneous rocks–VII. A reconnaissance survey of the distribution of SiO_2 in granitic and basaltic rocks," *Geochim. Cosmochim. Acta,* **28,** 271–290

Day, F. H., 1963, "The Chemical Elements in Nature," George G. Harrap & Co.

Rankama, K., and Sahama, Th. G., 1950, "Geochemistry," Chicago, University of Chicago Press, 912pp.

Richardson, W. A., and Sneesby, G., 1922, "The frequency-distribution of igneous rocks. I Frequency-distribution of the major oxides in analyses of igneous rocks." *Mineral. Mag.* **19,** 303–313.

Wager, L. R., and Deer, W. A., 1939, "The petrology of the Skaergaard Intrusion, Kangerdlugssuaef, East Greensland," *Med. Grønland.* **105,** No. 4.

Thatcher, J. W., 1965, "Silicon," U.S. Bur. Mines Minerals Yearbook.

Cross-references: *Mineral Classes — Silicates; Weathering, Chemical.* Vol. IVB: *Clays and Clay Minerals; Rock-Forming Minerals.* See also *Individual Minerals.* Vol. V: *Magmatic Differentiation; Skaegaard Intrusion; Skarn.* Vol. VI: *Pelagic Sediments.*

SILVER: ELEMENT AND GEOCHEMISTRY

Properties. Silver (symbol Ag), an element of atomic number 47 atomic weight 107.88, occurs in trace amounts in the earth. When pure, it is soft (Mohs hardness 3 and Brinell hardness 25 kg/cm^2), malleable, and ductile. Its density is 10.53, melting point 961°C, and boiling point 2180°C. In compounds it is usually univalent, but rarely is divalent. Its atomic radius in metallic bonding is 1.40 Å for eightfold coordination and 1.44 for twelvefold. The ionic radius when univalent is 1.31 Å for eightfold coordination and 1.34 for tenfold. Isotopes of mass numbers 107 and 109 are known and have relative abundances of 51.4% and 48.6%, respectively.

Minerals. Silver occurs as a major constituent of several minerals and as trace or minor amounts in others. Native silver is found at a number of localities and is never pure, usually containing several percent of gold, copper, and other metals in solid solution. Some of the commoner silver minerals are listed below:

argentite	Ag_2S (high-temperature structure)
acanthite	Ag_2S (low-temperature structure)
pyrargyrite	Ag_3SbS_3
proustite	Ag_3AsS_3
stephanite	Ag_5SbS_4
polybasite	$(Ag,Cu)_{16}Sb_2S_{11}$
miargyrite	$AgSbS_2$
dyscrasite	Ag_3Sb
argentian tetrahedrite	$(Cu,Fe,Ag)_{12}Sb_4S_{13}$
stromeyerite	$AgCuS$
matildite	$AgBiS_2$
andorite	$PbAgSb_3S_6$
sylvanite	$AuAgTe_2$
hessite	Ag_2Te
petzite	Ag_3AuTe_2
silver	Ag
cerargyrite	$AgCl$
argentojarosite	$AgFe_3(SO_4)_2(OH)_6$

Impurities of acanthite or matildite are common in galena from some districts and constitute an important source of silver.

Abundance. Few reliable figures are available for estimating the abundance of silver. Available data indicate concentrations of about 18 ppm in meteoritic troilite, 3.3 ppm in meteoritic iron-nickel, and 0.2 ppm in average crustal rocks. This places silver in about the 72nd place in the abundance list of elements in the earth's crust.

Geochemical Behavior. Silver is strongly chalcophile in character, as indicated by the abundance data already given. During differentiation of magmas, the silver is concentrated into late-stage minerals or deposited from associated hydrothermal solutions. Silver minerals commonly oxidize to silver sulfate during weathering and the silver may be dissolved and carried away by runoff and groundwater. Supergene enrichment of the upper portions of sulfide deposits is common. In the presence of chloride ion, deposits rich in cerargyrite are frequently formed. Native silver is rare in placers.

PAUL L. CLOKE

References

Boyle, R. W., 1968, "Geochemistry of silver and its deposits," *Geol. Surv. Canada Bull.,* **160**, 264pp.

Kerschagl, R., 1961, "Die Metallischen Rohstoffe, 13. Band, Silber," Stuttgart, Ferdinand Enke Verlag.

Rankama, K., and Sahama, Th. G., 1950, "Geochemistry." Chicago, Univ. of Chicago Press, 912pp.

Cross-references: *Elements: Planetary Abundances and Distribution; Gold; Economic Deposits; Mineralogy; Silver Ore Deposits; Supergene; Trace Elements.* Vol. IVB: *Mineral and Ore Deposits; Rare Metal Ore Deposits.* See also *Individual Minerals.*

SILVER: ECONOMIC DEPOSITS

Types of Deposits. Silver minerals occur in a large variety of associations making a detailed classification difficult. Simple schemes with few categories necessarily group together deposits which differ from each other in important particulars. The subdivisions based primarily on structure, given by Bateman (1950), namely, cavity fillings, replacement deposits, contact metasomatic deposits, and supergene sulfide enrichments, are reasonably satisfactory, although they do not distinguish important chemical differences or include types such as the Kupferschiefer or deposits in sandstones. Most districts differ appreciably in regard to structure, mineralogy, and/or country rock from others, giving most areas certain distinguishing characteristics.

The majority of the major silver deposits are or were located near the west coasts of North and South America. These are in, or

FIG. 1. Important former or present sources of silver in North America.

SILVER: ECONOMIC DEPOSITS

FIG. 2. Important former or present sources of silver in South America.

associated with, Tertiary volcanics and intrusives. Gold is usually present in significant amounts. Districts include (a) epithermal, supergene-enriched lode and vein deposits of acanthite (see *Silver* for chemical formulas of minerals), with lesser amounts of polybasite and other silver sulfosalts at Tonopah and the Comstock Lode, Nevada, the latter having intensive propylitic and sericitic alteration of the volcanic country rock; (b) mesothermal limestone replacement deposits at Tintic, Utah, where ores include siliceous silver—bearing pyrite—enargite, lead-silver, and pyritic silver bodies; (c) mesothermal lead-zinc-silver (contained in tetrahedrite) fissure fillings in quartzite at Coeur d'Alene, Idaho; (d) epithermal fissure fillings in volcanics, as at Pachuca and Real del Norte, Mexico, where the principal silver mineral is acanthite and wall rocks are propylitized, and at Red Mountain, Colorado, where most of the silver is in stromeyerite and wall rocks are silicified; (e) tin-silver deposits of Bolivia, which in the northern part of the country are meso- or hypothermal, but at Potosi are shallow with complex mineralogy, some zoning, and the high- to low-temperature assemblages typical of xenothermal ores; and (f) hypothermal quartzite replacement deposit of fine-grained sphalerite and argentiferous galena at the Sullivan Mine, British Columbia.

The cobalt-bismuth-silver-uranium fissure veins of Cobalt, Ontario, are notable for the large masses of native silver and the unusual assemblage of arsenides. Similar deposits occur at Great Bear Lake, Canada, and in Germany. The Kupferschiefer (see *Copper Deposits*) is an important source of by-product silver, as are many other types of copper ores; among these are the low-grade "porphyry copper" bodies, e.g., Bingham, Utah, and hypothermal vein ores, such as at Butte, Montana. Broken Hill, Australia, is a major silver producer; lead, zinc, and lesser amounts of other metals are also produced there. The massive ore occurs in distinct beds near the axes of folds in strongly metamorphosed and contorted metasedimentary rocks. Opinion is sharply divided between a hypothermal (i.e., hydrothermal) and a metamorphosed syngenetic origin.

Much silver has also been produced from cerargyrite and minor amounts of other silver minerals deposited in sandstone, often in association with uranium and vanadium minerals, as in the Silver Reef District, Utah. Silver rarely occurs in placers.

Geography. About three-fourths of the world production of silver comes from the Western Hemisphere, mostly from Mexico, the United States, and Canada. Localities of present or former importance are shown on Figs. 1 and 2 for North and South America, respectively. Localities in other parts of the world are shown in Figs. 3–6, although little is known about new deposits or production figures in Russia or Communist China.

PAUL L. CLOKE

References

Bateman, A. M., 1950, "Economic Mineral Deposits," Second ed., New York, John Wiley & Sons.

Kerschagl, R., 1961, "Die Metallischen Rohstoffe, 13. Band, Silber," Stuttgart, Ferdinand Enke Verlag.

Lindgren, W., 1933, "Mineral Deposits," Fourth ed., New York, McGraw-Hill Book Co.

Park, C. F., Jr., and MacDiarmid, R. A., 1964, "Ore Deposits," San Francisco, W. H. Freeman & Co.

Proctor, P. D., 1953, "Geology of the Silver Reef (Harrisburg) Mining District," *Utah Geol. Mineral. Surv., Bull.* 44.

Cross-references: *Copper; Gold Ore Deposits; Hydrothermal Solutions—Sulfide Transport; Silver: Element; Sulfosalts; Supergene.* Vol. IVB: *Economic Geology; Mineral and Ore Deposits; Mining and Mineral Centers of the World.*

FIG. 3. Important former or present sources of silver in Europe.

FIG. 4. Important former or present sources of silver in Australia.

FIG. 5. Important former or present sources of silver in Africa.

SILVER: ECONOMIC DEPOSITS

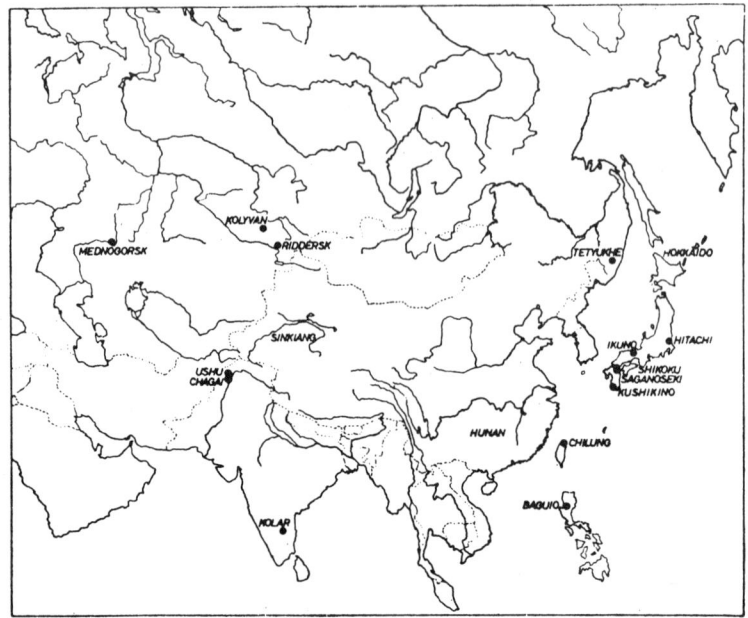

FIG. 6. Important former or present sources of silver in Asia.

SMOG—*See* Vol. II, **FOG, SMOG, MIST**
SNOW—*See* Vol. V

SODIUM: ELEMENT AND GEOCHEMISTRY

Chemical Properties

Sodium (Engl., soda; Latin, *Natrium*) has symbol Na, atomic number 11, atomic weight 22.9898 (C^{12} = 12), density 0.9712 g/cm^3 (20°C), melting point 97.82°C, boiling point 892°C, valence electron $3s^1$, chemical valence 1, and atomic diameter 3.83Å. Its crystal structure is body-centered cubic. The element was isolated by Davy in 1807; it is a Group I alkali metal.

Isotopic abundances are as follows: Na^{23} = 100%; Na^{22} is unstable with a half-life of 2.60 yr; and Na^{24} is unstable with a half-life of 15.0 hr.

The extreme chemical reactivity of sodium can be related to the ease with which its valence electron is lost. For this reason sodium occurs only in combined form in nature. Metallic sodium is an excellent conductor of electricity. It is excited in the flame and emits a bright yellow line at 5900 Å. The metal is soft and readily fusible. In air, sodium rapidly oxidizes to the oxide:

$$2Na + \tfrac{1}{2}O_2 = Na_2O$$

In excess oxygen, the oxide yields the peroxide:

$$Na_2O + \tfrac{1}{2}O_2 = Na_2O_2$$

Sodium decomposes water violently:

$$2Na + 2H_2O = 2NaOH + H_2$$

Similarly, it dissolves in liquid ammonia and releases hydrogen:

$$2Na + 2NH_3 = 2NaNH_2 + H_2$$

Sodium reacts with the halogens to form sodium halides such as sodium chloride or common salt. It also forms a hydride, NaH, carbonates Na_2CO_3 and $NaHCO_3$, sulfates Na_2SO_4 and $NaHSO_4$, and the nitrate $NaNO_3$ or "Chile saltpeter," all of which are very water-soluble.

Mineralogy

Sodium is one of the most abundant of the elements in the earth's crust and consequently occurs in a large variety of minerals. The three major modes of occurrence of sodium minerals in the lithosphere are: (*1*) Soluble salts found chiefly in evaporite deposits which are of economic importance, e.g., halite, natron, trona. (*2*) Complex rock-forming silicates of major geological interest, e.g., albite, nepheline, jadeite. (*3*) Rare, unusual, simple halides or aluminohalides which are scientific curiosities, e.g. villaumite, cryolite.

Some important sodium minerals are listed below.

halite	NaCl
villaumite	NaF
cryolite	Na_3AlF_6
natron	$Na_2CO_3 \cdot 10H_2O$
trona	$Na_2CO \cdot NaHCO_3 \cdot 2H_2O$
nahcolite	$NaHCO_3$
gay-lussite	$CACO_3 \cdot Na_2CO_3 \cdot 5H_2O$
anorthoclase	$(Na,K)AlSi_3O_8$
plagioclase group	$NaAlSi_3O_8$ to $CaAl_2Si_2O_8$
aegirine	$NaFeSi_2O_6$
jadeite	$NaAlSi_2O_6$
pectolite	$NaCa_2(Si_3O_8)OH$
glaucophane	$Na_2Mg_3Al_2(Si_8O_{22})(OH,F)_2$
riebeckite	$Na_2Fe_3^{II}Fe_2^{III}(Si_8O_{22})(OH)_2$
nepheline	$NaAlSiO_4$
sodalite	$3Na_2Al_2Si_2O_8 \cdot 2NaCl$
lazurite	$3Na_2Al_2Si_2O_8 \cdot 2Na_2S$
natrolite	$Na_2Al_2Si_3O_{10} \cdot 2H_2O$
analcite	$NaAlSi_2O_6 \cdot H_2O$
Chile saltpeter	$NaNO_3$
borax	$Na_2B_4O_7 \cdot 10H_2O$
glauberite	$Na_2SO_4 \cdot CaSO_2$
mirabilite	$Na_2SO_4 \cdot 10H_2O$
thenardite	Na_2SO_4

Geochemistry

Sodium is the sixth most abundant element in the earth's crust after oxygen, silicon, aluminum, iron, and calcium. The average amount of sodium in magmatic rocks is 2.83% by weight. In the hydrosphere, it is the most abundant cation with a concentration of 10,556 g/ton.

The geochemical behavior of sodium in *igneous rocks* is controlled to a large extent by its ionic radius (0.98 Å) which is closer in size to that of calcium (0.99 Å) than either to potassium (1.33 Å) or lithium (0.68 Å). Therefore sodium is frequently associated with calcium in silicate minerals rather than with the other alkali elements. Examples of complete solid solution or isomorphous substitution of Na for Ca are the plagioclase feldspars and the diopside $CaMgSi_2O_6$–aegerine $NaFeSi_2O_6$ series of the pyroxenes. On the other hand, Na and potassium (K) exhibit only limited miscibility except at elevated temperatures. Thus anorthoclase $(Na,K)AlSi_3O_8$ unmixes below 660°C to an intergrowth of albite and orthoclase called *perthite*.

The amount of alkalis in igneous rocks increases toward the later stages of fractional crystallization, resulting in sodium enrichment especially in the feldspathoid-bearing rocks (e.g., nepheline syenite) and to a lesser extent in granites. This is illustrated by the familiar zoning of plagioclase phenocrysts and alkali pyroxenes in igneous rocks. Residual hydrothermal solutions associated with magmatic activity have had a fairly high concentration of sodium, judging by the effects of hydrothermal alteration, which range from deposition of albite and zeolites in cavities to myrmekitization (replacement of K-feldspar by an aggregate of Na-plagioclase and quartz). Some perthitic intergrowths may be the result of sodium alteration of orthoclase, rather than exsolution.

Regional metamorphism frequently involves a form of sodium metasomatism on a tremendous scale. During the transformation of argillaceous sediments such as shales into their metamorphic equivalent (phyllite, mica schists, gneisses), aluminum undergoes a change of coordination from six, as in micas, to four as in feldspars. Mineral assemblages change toward those requiring increased amounts of sodium, such as a quartz-sericite-biotite phyllite going to albite-bearing mica schist through oligoclase-biotite gneiss and finally to Na-plagioclase and K-feldspar gneiss. The difficulty in distinguishing the ultimate products of regional metamorphism from presumed magmatic granites which often are gneissoid and contain garnet-bearing xenoliths, has led many geologists to believe that many so-called granites are really produced by alkali metasomatism of metamorphosed sediments. This process has been called *"granitization."*

Formation of glaucophane schists is another process requiring sodium addition on a vast scale. Glaucophane schists are restricted to orogenic belts of Mesozoic age or younger (e.g., southern Alps, Japan, New Zealand, California), and are characterized by the presence of glaucophane, jadeite, lawsonite, pumpellyite, epidote, chlorite, and almandine. P–T equilibrium data suggest that high pressures (5–10 kbars) and low temperatures (below 300°C) were necessary to produce the observed assemblages. Yet geological evidence does not indicate deep burial. The requirement of high pressure could be lifted if the glaucophane schists formed by Na and Al metasomatism.

Sodium readily forms soluble salts during weathering and has accumulated in the ocean to a greater extent than potassium, rubidium, and cesium which have been adsorbed by clay minerals in sediments. The amount of sodium in the ocean has been used as a measure of geological time and of the total amount of sedimentation, but such calculations have failed to consider the recycling of sodium leached from marine sediments (see *Sodium Cycle*) (see also Table 1).

Although the amounts of Na and K are nearly equal in igneous rocks, the K/Na ratio of river water is 0.40 and that of ocean water

SODIUM: ELEMENT AND GEOCHEMISTRY

TABLE 1. CHEMICAL COMPOSITION OF WATER FROM (1) OCEAN, (2) GREAT SALT LAKE, (3) KARA BOGAZ GULF, (4) DEAD SEA (in mg/l)

	1	2	3	4
Na^+	10,561	67,498	52,250	34,940
K^+	380	3,378	4,820	7,560
Rb^+	.2	—	—	60
Ca^{2+}	400	326	tr.	15,800
Mg^{2+}	1,272	5,616	33,830	41,960
Cl^-	18,980	112,896	132,730	208,020
Br^-	64.6	—	380	5,920
SO_4^{2-}	2,648.6	13,593	67,830	540
HCO_3^-	139.7	183	670	240
Total	34,446.1	203,490	292,510	315,040

only 0.0362. In contrast to sodium, potassium is preferentially adsorbed by the clay minerals of marine sediments. Silicate minerals formed during early diagenesis (glauconite, illite, phillipsite), or altering from preexisting clays (montmorillonite to mixed-layer illite), incorporate K in preference to Na despite the greater abundance of sodium in seawater. The sorption of cations from seawater by river clays depends on the energy of solution and bonding energy between clay and cation. The bonding energy in turn depends upon the nature of the cation and the presence and concentration of other ions. Generally, divalent cations are more tightly held than monovalent ones. Experimentally, the bonding energy of K is higher when undersaturated (Keller, 1963). Thus, Ca-rich clays entering the ocean first adsorb Na, Mg, and K in that order of abundance (determined by the relative concentration of the cation and the change in free energy for the reaction), but later Mg and K become tightly incorporated into the clay structure instead of Na. The average Na_2O content of marine shales is 1.3% whereas K_2O content is 3.25%.

The evaporation of seawater in shallow marginal basins of restricted circulation in hot, arid climates leads to the concentration and precipitation of various soluble salts of the alkali and alkali earth elements. Halite (NaCl) is the most common sodium salt; glauberite, $Na_2SO_4 \cdot CaSO_4$, trona, $Na_2CO_3 \cdot NaHCO_3 \cdot 2H_2O$, and natron, $Na_2CO_3 \cdot 10H_2O$, are less common, but of economic importance. The sequence of precipitation of salts from seawater during evaporation is determined by the initial concentration of the brine, the temperature, and the relative solubilities of the salts. At 25°C, the usual sequence is calcite (dolomite), gypsum, halite, epsomite, kieserite, kainite and carnallite, bischofite. The frequency of occurrence and total thickness of any evaporite salt is inversely proportional to its solubility (the more insoluble, the more abundant). Nonmarine evaporites form as marginal salt pan deposits, salina deposits, or lagoonal deposits; they also form in lakes without external drainage whose salt content is influenced by river drainage and high rates of evaporation.

Economic Geology

The economic geology of sodium depends upon the occurrence of evaporite deposits. Recent evaporites form in hot, arid climates in a belt 23° N and S of the equator. The distribution of evaporites has varied considerably in the past, due to climatic changes and other causes. While modern evaporites are predominantly nonmarine, those of the past were mostly marine. Important examples of bedded marine salt deposits are: Zechstein deposits in Germany, Salina Formation in Michigan, and West Permian Basin in Texas–New Mexico.

Salt undergoes plastic deformation easily with rising temperature and increased stress exerted by the overburden, depending on the depth of burial. Such deformation may ultimately lead to the development of salt domes ("diapirs") which pierce through overlying sediments. Salt domes are abundant in the Gulf Coast region of the U.S., in northern Germany, Spain, Rumania, South Russia, Iran, and North Africa. They are of considerable economic importance because in addition to halite, they are associated with large deposits of petroleum and sulfur. In the U.S.A., salt domes are the major commercial source of NaCl. Salt is removed from the earth by underground mining and by pumping water down to force up the brines. Salt is also obtained by the direct solar evaporation of seawater. Salt or alkali lakes and playas of the southwestern U.S.A. are sources of many useful salts such as mirabilite, thenardite, natron, trona, borax, ulexite, and glauberite.

Halite is the chief raw material from which numerous industrially important chemicals are manufactured. Sodium hydroxide, sodium carbonate, sodium sulfate, and sodium metal are all derived from halite. Sodium hydroxide in turn is used in the manufacture of other chemicals, cellulose and rayon, soap, petroleum refining, and by the pulp and paper industry. Sodium carbonate finds its major use in the glass industry and also in the manufacture of soap, detergents, and cleansers. The pulp and paper industry is the major consumer of sodium sulfate. Sodium metal is important in the synthesis of tetraethyl lead, an antiknock agent for automotive gasolines. Salt is used as a food preservative and flavoring agent, but this accounts for only a small percentage of NaCl consumption, most of which enters the industrial processes outlined above.

Major salt producers are: the U.S.A., U.S.S.R, England, Germany, Canada, and France.

VIVIEN GORNITZ

References

Goldschmidt, V. M., 1958, "Geochemistry," Oxford, Clarendon Press, 730 pp.
Govett, C. J. S., 1958, "Sodium Sulfate Deposits in Alberta," *Res. Council Alberta, Geol. Div., Prelim. Rpt.* **58-5**, 34pp.
Borchert, H., and Muir, R. O., 1964, "Salt Deposits," London, New York, Van Nostrand, 338pp.
Keller, W. D., 1963, "Diagenesis in Clay Minerals—A Review," Proc. 11th Natl. Conf. Clays and Clay Minerals, New York, Macmillan.
Kerns, W. H., 1965, "Sodium and Sodium Compounds," U.S. Bur. Mines Minerals Yearbook.
Rankama, K., and Sahama, T. G., 1950, "Geochemistry," Chicago, Univ. Chicago Press, 912pp.

Cross-references: *Bonding; Earth's Crust Geochemistry; Evaporite Processes; Exsolution; Hydrothermal Alteration; Lagoon Geochemistry; Mineral Classes—Silicates; Natural Brines; River Geochemistry; Seawater, Chemistry; Sodium Cycle; Weathering, Chemical.* Vol. III: *Great Salt Lake; Kara-bogaz Gulf.* Vol. IVB: *Plagioclase Feldspars; Rock-Forming Minerals; Salt Economy;* See also *Individual Minerals.* Vol. V: *Granitization; Metamorphism.* Vol. VI: *Diagenesis; Evaporites.*

SODIUM CYCLE

Before the advent of radiometric dating the sodium cycle was used to estimate the length of geologic time. Clarke (1924) compared the rate of sodium transfer by rivers with the amount dissolved in the sea and concluded that the process had been going on for about 100 million years. This is a much shorter period of time than radiometric estimates suggest. Perhaps the ocean is younger than the oldest rocks, which have an age greater than 4000 million years, but it is not likely to be younger than the oldest known fossils. Although there is no direct evidence that the present configuration of the ocean basins antedates the Cretaceous, it is clear from the fossil record that most major groups of marine organisms were evolved before the beginning of the Cambrian, and such Precambrian fossils as those of the Gunflint Formation strongly suggest the presence of an extensive permanent aqueous phase in the biosphere 1600 million years ago.

Conway (1943) sought to reconcile the geochemical and radiometric estimates of geologic time by considering the geologic peculiarities of the presence and return of sodium to the land through the atmosphere. Generous allowances for the erosional effects of agricultural man and for river sampling at times of low-water stage enabled him to raise the geochemical age to about 1000 million years.

More complete data based on systematic sampling at all water stages (Livingstone, 1963a) have shown that the discharge effect is a small one. If discharge as well as concentration is considered, it appears that the Amazon, which Conway used as an example of a river little affected by agriculture, actually delivers more sodium than the average.

Sodium Budget

About 14.1×10^{15} metric tons of sodium are dissolved in the sea. The rate of delivery of rivers is well known, except in the tropics, and further data are not likely to demand a major change in the present estimate of 20.5×10^7 tons, of which three percent is of human origin, carried to the sea annually.

Recycling through the atmosphere can be estimated by attributing a marine origin to all the sodium that falls in the rain. This suggests that about one-third of the river sodium has been cycled through the atmosphere, and a generous estimate for salt transferred in crystalline form might increase the cyclic portion of river sodium to almost one-half. This would reduce the net rate of river transfer to about 10.7×10^7 tons per year. At such a rate 64.2×10^{15} tons would have been transferred in the 600 million years since the beginning of the Cambrian, or more than four times as much as is actually present in the sea.

Part of the missing sodium may be sought in the sediments as rock salt, as ions sorbed to the sedimentary particles, and as salt solution in the pore space. Assuming a sorptive capacity equal to that of finely divided marine clay, half saturated with sodium, and a pore space of twenty percent filled with triple strength seawater, one might account for 10.5×10^{15} tons of sodium in sediments now under the sea and an additional 2.6×10^{15} tons in the sediments on land. The rock salt deposits account for only 0.4×10^{15} metric tons. This leaves the budget for post-Algonkian time seriously out of balance:

	metric tons
Carried by rivers to the sea	119.4×10^{15}
Returned by atmosphere to land	55.2×10^{15}
	64.2×10^{15}
Dissolved in seawater	14.1×10^{15}
In suboceanic sediments	10.5×10^{15}
In continental sediments	3.0×10^{15}
	27.6×10^{15}

In terms of time, all of the oceanic sodium is accountable for in 280 million years, using very generous estimates for atmospheric return and sedimentary sodium. More reasonable estimates would reduce the geochemical age of the ocean to 200 million years, or only twice what Clarke made it.

A large part of the discrepancy, and perhaps all of it, springs from the failure to take into account the recycling of continental sediment. The amount of marine sediment that is found on the continents today may be a very small part of the total that has been uplifted and subjected to erosion during post-Algonkian time.

There is no completely satisfactory way of estimating the rate of sedimentary recycling. Livingstone (1963b), whose treatment is otherwise followed here, employed an indirect argument based on Ronov's measurements of sedimentary masses, which are probably not suitable for the purpose. A more direct approach is to compare the present rate of delivery of sediment by rivers with estimates of the quantity of marine sediment in different parts of the lithosphere. Adjusting Kuenen's (1950) estimate of 12 km³ per year to 10 km³ for pore space, and multiplying by the length of post-Algonkian time, indicates that some 6000×16^6 km³ of sediment has been deposited in the ocean. Subtracting from this the sediment that lies beneath the surface of the ocean we are left with 5280×10^6 km³, or about 30 times the volume of post-Algonkian sediments still on the land. This is more than enough to balance the sodium budget.

A more conservative procedure is to treat the shallow-water shelf sediments as part of the continental ones, and only the pelagic sediments as marine. On such a basis one would expect a tenfold recycling of both continental and shelf sediments, which would also be more than enough to balance the budget for post-Algonkian time. On either basis about 1000 million years would be required for the total sodium transfer at the present rate.

Amount of Sedimentary Rocks

Igneous rocks contain about two percent sodium and sedimentary rocks that are available for analysis only about half as much. Assuming that the oceanic sodium can account for the difference, Clarke computed the mass of sedimentary rocks. With current data such a computation yields 700×10^{15} metric tons, substantially less than Poldervaart's direct estimate of 1702×10^{15} tons and very much less than the mass of the sediments if the second seismic layer under the deep sea is sedimentary. It is much closer to the direct measurement of continental sediments, which is 480×10^{15} tons. The computation makes no allowance for the considerable mass of metamorphosed Precambrian sediment.

The sodium content of metamorphic rocks is about 1.72%, or closer to the accepted value for igneous rocks than that for sediments. Probably most analyses of sedimentary rocks are unrealistically low because the continental sediments have been leached by groundwater. The mean sodium content of all sedimentary rocks may be much higher, perhaps even higher than that of metamorphic rocks, which may have been partially leached before metamorphosis.

If good figures were available for the sodium content of each of the major classes of rocks, and also for the mass of the sediments, it would be possible to calculate the mass of the metasediments in a manner analogous to that which Clarke used to compute the mass of sediments. Rubey (cited in Livingstone, 1963b) has made such a computation, using reasonable assumptions where accurate data were not available, and concluded that the mass of metasediments in a manner analogous to that unaltered sedimentary rocks. The actual result depends very much on the assumptions of sedimentary sodium content. Good data for the chemical composition of unleached suboceanic sediment are needed as badly for this purpose as they are for drawing up a reliable budget for sodium transport since the beginning of the Cambrian.

The available data give little reason to doubt that the sodium cycle is now in a steady state. If Barth (1961) is right in believing that the so-called igneous rocks are really metasediments, the steady state has probably lasted throughout the accessible part of geologic time.

D. A. LIVINGSTONE

References

Barth, T. F. W., 1961, "Ideas on the interrelation between igneous and sedimentary rocks," *Comp. Rend. Soc. Geol. Finlande*, **33**, 321–326.

*Clarke, F. W., 1924, "The Data Geochemistry," Fifth edition *U.S. Geol. Surv. Bull.*, **70**, 841pp.

*Conway, E. J., 1943, "The chemical evolution of the ocean," *Proc. Roy. Irish Acad. Sect. B*, **48**, 161–212.

*Kuenen, P. H., 1950, "Marine Geology," New York, John Wiley & Sons, 568pp.

Livingstone, D. A., 1963a, "The chemical composition of lakes and rivers," *U.S. Geol. Surv. Profess. Paper* **440-G**, 64pp.

*Livingstone, D. A., 1963b, "The sodium cycle and the age of the ocean," *Geochim. Cosmochim. Acta*, **27**, 1055–1069.

*Additional bibliographic references may be found in this work.

Cross-references: *Cycles—Geochemical; Lake Geochemistry; River Geochemistry; Seawater, Chemistry; Seawater, History; Sodium: Element and Geochemistry.* Vol. VI: *Sedimentation.*

SOIL CHEMISTRY—See Vol. VI

SOIL EROSION

"Erosion" is the general term applied to the loosening and transportation of rock and soil materials by weathering, moving water, glaciers, gravity (mass movement), and wind (Howell et al., 1957, p. 99). The term "soil erosion" is usually applied to erosion of soil that has been accelerated by the activities of man. Moving water removes soil by rain-splash and sheet-wash, and by rill and gully formation. Erosion of soil by wind is known as *deflation* (Howell et al., 1957, p. 75).

Balance of Nature

Under most natural conditions, where soils are protected by vegetation, the rate of rock weathering at the base of the soil may exceed or may be about equal to the rate of erosion at the surface and the total thickness of soil will either increase gradually or will remain approximately constant. But as soon as the natural vegetative cover has been destroyed or substantially modified by deforestation, plowing, over-grazing, or planted tilled crops, the precarious balance of nature is upset. Beating rains loosen soil particles and destroy the structure of exposed soils; much of the rainwater runs off instead of being absorbed; and flood waters carry away the upper and most fertile layers of soil. Turbulent winds pick up silt and clay particles and carry them away, and shift the sand fraction into small drifts and large dunes.

History of Soil Erosion

Accelerated soil erosion has plagued agricultural and pastoral man ever since his increasing numbers forced him from the level plains and gentle slopes to the hilly lands where rain-splash and runoff scour away the exposed soil at an alarming rate.

Soil erosion has been a problem for several thousand years in Mediterranean lands, in the loess region of northern and northwestern China, and in the hills of India. Farmers in ancient times built thousands of miles of terraces in their attempts to stem the tide of soil erosion and to make "shelf gardens" out of the small patches of land between gullies. Many sedentary people lost their battles with erosion when roving nomads or the armies of hostile civilizations laid their lands waste. Thus, a substantial part of accelerated soil erosion is due to political instability, both in ancient and in recent times; economic instability can have similar bad effects.

Recently, soil erosion has damaged severely vast areas of farm and grazing land in all the continents, especially throughout the uplands of China (Thorp, 1936), in much of the United States (Stallings, 1957), in Australia (Holmes, 1938; Soil Conservation Authority, 1957) and New Zealand, in many African countries, and in much of South America (Baldwin et al., 1954).

Conditions Conducive to Erosion

Soil erosion by water is especially severe in sparsely vegetated arid regions, and in those parts of semiarid, subhumid, and humid regions where much of the rainfall is torrential or where heavy rains come immediately after long dry periods. The problem is less troublesome where the rains are gentle and well distributed throughout the year, as in the British Isles.

Wind erosion is a special hazard in semiarid regions where winds are strong and rainfall is unreliable; and it takes a heavy toll both in cultivated areas and on land that has been overgrazed. Even in humid regions wind erosion is a problem where very sandy soils and muck beds have been plowed. Some of the sand plains of Michigan have been converted to dune fields; and muck farmers must plant wind-breaks to protect their fluffy soils from destruction by the wind.

Some of the soil material eroded from the uplands is carried away by rivers or by the wind and is lost in the sea. Some of it collects on the flood plains of streams or settles on the land from the air. Fine mineral sediments and organic matter, so deposited, may enrich the soils of both valleys and uplands; but where

SOIL EROSION

the streams dump sand and gravel on good alluvial soils, the latter are ruined or greatly reduced in value.

Erosion Controls

Modern programs of erosion control have been organized on a large scale in the United States, Australia, New Zealand, and in several countries in Africa, south of the Sahara. In fact nearly all countries where agriculture is important have recognized the problem and are working to control erosion.

The cultivation of irrigated rice in oriental countries reduces erosion to an absolute minimum on nearly all of the land actually used for rice (Thorp, 1936). In China and India and in many smaller countries, however, much of the land between the irrigated paddy fields has lost its upper layers of soil to sheet erosion and is scored by deep gullies. This kind of problem has been alleviated in Japan by intensive reforestation and by other erosion-control measures on land used for unirrigated crops.

Erosion in many European countries has been kept at a low level for several generations through enlightened land-use programs and intensive cultivation and fertilization; but much of the upland soil bordering the Mediterranean Sea has been lost to erosion. Fertile floodplains around the Mediterranean have been badly damaged by deposition of coarse debris washed from cultivated uplands.

Modern soil conservationists recommend using the land in ways that will produce the greatest income consistent with the least loss of soil. They endeavor to protect sloping lands by planting soil-conserving crops, by contour cultivation, by alternation of strips of close-growing crops with clean-cultivated ones, and by using graded terraces and grass-covered waterways to control the rate of flow of runoff water. Shallow gullies may be filled by plowing, and deep gullies may be controlled by check dams or vegetative cover. But structures like terraces and dams may do more harm than good, if they are not well maintained. The large volume of water collected behind terraces and check dams sometimes breaks through the barriers, sweeps away large volumes of soil, and scours out new gullies. Ditches that were dug to drain wet lowland soils for farming have provided lower base levels for water flow, and as a consequence gullies have eaten their way back from the ditches high into the adjacent uplands. Many examples of this exist in eastern Nebraska and western Iowa. The balance of nature is indeed a delicate one!

The bad effects of wind erosion (deflation) are ameliorated on cultivated land by planting strips of close-growing crops in alternation with strips of sod or fallow land arranged at right angles to the prevailing wind. Stubble mulch in semiarid wheat land is highly effective (Stallings, 1957, pp. 293–312). List-furrowing across the wind is effective in some areas, and "shelterbelts" of trees and shrubs have been helpful to a limited extent. Wind erosion of muck and peat beds can be greatly reduced by planting wind-breaks at right angles to the prevailing wind.

Measures for erosion control most likely to succeed are based primarily on good agronomic practice. Maintenance of an optimum level of organic matter and good soil structure through the use of organic manures, soil amendments, fertilizers, and adapted crops will go a long way toward reducing soil erosion to a minimum. Close-growing crops with leafy tops will protect the soil from the impact of rain drops.

The conservationist's goal is to manage soils in a way that will maintain something comparable to the balance of nature between the gradual loss of soil from the surface and the gradual replacement of soil material through rock weathering or surface deposition. If there is a net loss of soil each year the land will, in time, inevitably become a barren waste. Society as a whole has a vital stake in the conservation of the soil and can make its greatest contribution by maintaining stable government and programs of technical assistance for those who derive their living from the land.

JAMES THORP

References

Baldwin, M., et al., 1954, "Soil Erosion Survey of Latin America," The Conservation Foundation and The Food and Agriculture Organization of U.N., *Soil and Water Conservation*, July, Sept., and Nov., 31pp.

Holmes, M., 1938, "The Meaning of Soil Erosion," *Geography*, No. 1 (Univ. of Sydney, N.S.W., Australia) 38pp.

Hovell, J. V., et al., 1957, "Glossary of Geology and Related Sciences," *Am. Geol. Inst. Natl. Acad. Sci., Natl. Res. Council*, 325pp.

Soil Conservation Authority, Victoria, 1957, 8th Ann. Rept., Gov't Printer, Melbourne, Australia, 42pp.

Stallings, J. H., 1957, "Soil Conservation," Englewood Cliffs, N.J., Prentice-Hall, 575pp. (This is an excellent general reference on soil erosion and its control.)

Thorp, J., 1936, "Geography of the Soils of China," *Nat. Geol. Surv. China*, 552pp.

Cross-references: *Conservation; Hydrology, Semiarid Regions; Pedology; Runoff; Weathering, Chemical.* Vol. III: *Anthropogenic Influence in Geomorphology; Deflation; Denudation; Ero-*

sion; Gully Erosion; Mass Wasting; Sheet Erosion; Weathering; Wind Action. Vol VI: Agricultural Geology; Agrogeology; Agronomy.

SOIL SALINITY AND ALKALINITY

Generally speaking, the arid and semiarid regions of every continent contain important acreages of soils that are termed "salt affected." This classification refers to soils which have significant amounts of free salts or which show the chemical and physical effects of salts present at some past period of their development. It has been estimated that about 30% of the arable land of California is salt affected and similar percentages have been reported for the western and southwestern states of the United States in general.

The free salts found in soils are the result of the weathering of parent rocks. However, the salts present in most saline soils have been transported from elsewhere by surface and subsurface drainage. That is, salt-affected soils are usually found in the less elevated parts of the landscape where salts accumulate as a result of the evaporation of surface drainage waters or of the capillary rise and evaporation of subsurface drainage waters (Kelley, 1951).

The salts of saline soils are most often those of sodium. However, excessive amounts of the soluble salts of potassium, calcium, or magnesium occur in some areas. When these elements are present in the form of soluble chlorides or sulfates, the soils may be neutral or even acid in reaction. When the elements occur as the bicarbonates and carbonates or when appreciable amounts of precipitated calcium and magnesium carbonates are present, the soils are alkaline in reaction; that is, the pH of the soil pastes may vary from 8.3 to 10 or over.

Because of osmotic and specific ion effects, the presence of excessive amounts of salts in soil is injurious to crops. As the result of many years of research, workers at the U.S. Salinity Laboratory in Riverside, California, have developed a system of classification of saline soils according to their suitability for crops (U.S. Dept. Agriculture, 1954). The system is based on the conductivity of the filtrate obtained from a so-called "saturated paste." The preparation of the saturated paste has been prescribed in detail (U.S. Dept. Agriculture, 1954). The moisture content of the paste is approximately twice that of the well-drained soil a few days after irrigation. Where the conductivity of the saturation extract is measured in millimhos per cm, Table 1 is valid: In general, purely saline soils can be reclaimed

TABLE 1. EFFECT OF SALINITY AND ALKALINITY ON CROPS

Conductivity of Saturation Extract (millimhos per cm at 25°C)	Effect on Crops
0–2	Salinity effect mostly negligible
2–4	Yields of very sensitive crops may be restricted
4–8	Yields of many crops restricted
8–16	Only tolerant crops yield satisfactorily
16	Only a few very tolerant crops yield satisfactorily

for agriculture by the establishment of good drainage followed by heavy irrigation to carry the salts below the root zone of plants (Thomas, 1936).

Because soils have ionic-exchange characteristics, the presence of soluble salts frequently gives rise to still another deleterious soil condition; that is, a high level of alkali cations (most commonly sodium) in the adsorbed form. Normally, agricultural soils contain predominantly calcium and magnesium in the adsorbed state.

Soils having undesirable percentages of sodium in their exchange complexes are designated as "sodic." By and large, sodic soils are unsuitable for agriculture and have very poor physical properties. As a rough rule, reductions in crop yields can be expected when the equivalent percentage of sodium in the exchange complex (i.e., the exchangeable sodium percentage = ESP) exceeds about 15%. In a recent study of a nonsaline sodic soil of the Fresno series, the correlation between ESP and the yield of bush beans was established in the Department of Soils and Plant Nutrition at Berkeley (Fig. 1).

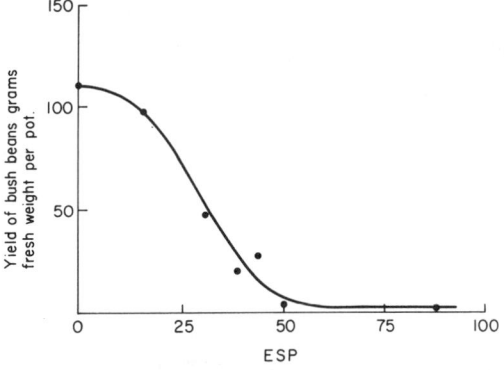

FIG. 1. Correlation between exchangeable sodium percentage (ESP) and the yield of bush beans.

SOIL SALINITY AND ALKALINITY

The reclamation of a sodic soil also requires the establishment of good drainage. In addition, an amendment must be added that will supply calcium in a soluble form (for example, gypsum). The adsorbed sodium is thereby released to the soil solution in exchange for Ca. After the addition of the amendment, the soil must be heavily and frequently irrigated to obtain the necessary leaching (Overstreet et al., 1951).

ROY OVERSTREET (deceased)

References

Kelley, W. P., 1951, "Alkali Soils; Their Formation, Properties, and Reclamation," New York, Reinhold Publ. Corp.

Overstreet, Roy, Martin, J. C., and King, H. M., 1951, "Gypsum, sulfur, and sulfuric acid for reclaiming an alkali soil of the Fresno series," *Hilgardia*, **21**(5), 113–27.

Thomas, E. E., 1936, "Reclamation of white alkali soils in the Imperial Valley," *Calif. Agr. Expt. Sta. Bull.* **601**, 1–15.

U.S. Department of Agriculture, 1954, "Saline and Alkali Soils," Agriculture Handbook No. 60.

Cross-references: *Evaporite Processes; Ion Exchange; Pedology; pH-Eh; Sodium; Sulfates; Weathering, Chemical.* Vol. VI: *Salinization; Salt Deposits.*

SOIL SCIENCE—See PEDOLOGY; also Vol. VI

SOLAR ACTIVITY—See Vol. II, SUNSPOTS AND SOLAR ACTIVITY

SOLAR CYCLE, RADIATION; SOLAR-CLIMATIC RELATIONSHIPS; SUNSPOT CYCLES—See Vol. II

SOLAR ENERGY—See Vol. II

SOLID SOLUTION

The concept of solid solution derives from phase-rule studies in which the existence of homogeneous solid phases over a range of compositions plays almost the same phenomenological role as does the existence of liquid solutions. Just as liquids may be either miscible in all proportions or may have restricted ranges of solubility of one in another, so crystalline solids may form either a continuous solid solution series or may form solid solutions of each in the other to a limited extent. The components in a solid solution series are frequently referred to as *end-members*, to indicate that these are the extremes in the system. Although there are differences between the way in which solid solutions and liquid solutions are usually discussed, this arises not so much from differences in the phenomena as from the fact that we understand the solid state much better than the liquid state. It is accordingly possible to discuss solid solutions in detail in terms of the phenomena involved at the atomic level.

Ideal and Nearly Ideal Solid Solutions

Rigorous definitions of the criteria for the existence of an ideal solid solution are given in standard works on thermodynamics (e.g., Guggenheim, 1967). They require that the formation of the solution from the pure components shall involve no volume change or enthalpy change, and that the configurational entropy change shall be a maximum—that is that there shall be no tendency in the solution for relative ordering of the structural entities of the two components. Such ideal solid solutions can only be formed between components which have very closely similar formulas, and crystal structures containing atoms or ions of very closely similar size and chemical properties; that is, very closely isomorphous substances (see *Isomorphism*). Ideal solid solution necessarily implies the existence of a complete range of solid miscibility in all proportions.

Lack of a volume change on mixing implies that the volume of the unit cell of the ideal solid solution should be a linear function of the mole fractions of the components. Because the other criteria can be satisfied only if the atoms or ions in the structure do not differ very much in size, it follows that the range in cell volume will be small and consequently that the unit cell dimensions will also be linear functions of composition to within a high degree of approximation.

Whether any solid solution series satisfies the ideality conditions perfectly is perhaps a matter for speculation. The greater the precision with which such systems are investigated the more likely is it that departures from ideality will be detected. However, it may safely be said that examples of solid solution series with minimal departures from ideality are forsterite-fayalite and diopside-hedenbergite.

In such mineral systems, in which there is a substantial degree of ionic character in the bonding, the phenomena can be better understood if the solid solution is regarded as a case of isomorphous substitution of certain ions by others within a framework provided by the remainder of the structure. Thus the diopside-hedenbergite solutions are most usefully to be regarded as varying degrees of substitution of Fe for Mg in $CaMgSi_2O_6$ or of Mg for Fe in $CaFeSi_2O_6$, rather than of $CaFeSi_2O_6$ in $CaMgSi_2O_2$. Thus a 50:50 solution is best represented by $CaMg_{0.5}Fe_{0.5}Si_2O_6$.

Nonideal Solid Solutions with Complete Miscibility

For the existence of a complete range of solid miscibility in all proportions between two substances it is usually necessary that they should be isomorphous. In systems involving minerals whose structures can be approached from an ionic viewpoint the lack of ideality can often be traced to one or another of the following factors:

1. Difference in size of the ions involved leads to local strains in the structure around the ion from the minor component (solute). This leads to an enthalpy of mixing.
2. Difference in size or in electronic characteristics (e.g. covalent stereochemistry, crystal field effects (see *Crystal Field Theory*) or magnetic moment) between these ions may lead to preferential occupation of a certain fraction of the crystallographic sites by one ion. This results in a configurational entropy of solution that is less than the maximum. The system enstatite-orthoferrosilite is a case in point, where the departure from ideality arises from crystal field effects.
3. If the substitution involves two pairs of ions of different, but balancing, charges (e.g., $Na^+, Si^{4+} \rightleftharpoons Ca^{2+}, Al^{3+}$ as in the plagioclase feldspars) then the tendency to maintain local electrostatic neutrality will promote a tendency to short-range ordering of these pairs of ions, and again the configurational entropy of solution will be less than the maximum.

Complete miscibility between nonisomorphous species is rare, but may occur when two substances have very closely related structures differing in crystallographic symmetry as a result of a small structural distortion. In such circumstances the higher symmetry may survive over an appreciable range of composition as a result of the averaging of opposite distortions in different cells. An example is provided by the complete miscibility between high albite (triclinic) and high sanidine (monoclinic).

Partial Solid Miscibility Between Substances with Similar Structures

Any departure from the conditions for ideality in a solid solution must lead to a nonzero value for ΔH, the enthalpy of mixing. If this is negative then an intermediate compound may be formed between the substances, and it is then more convenient to discuss the phenomena in terms of separate systems each involving this compound and the original components. If ΔH is positive, then the solid solution will be stable or unstable with respect to its components according as

$$\Delta G = \Delta H - T\Delta S \lessgtr 0$$

Since ΔS increases with the mole fraction of the minor component there will always be some temperature below which a solid solution will be unstable if it contains more than some limited proportion of solute. Thus all components forming nonideal solid solutions are only partially miscible below some particular temperature and the extent of miscibility will decrease below this temperature. If a solution is cooled to a temperature at which it is unstable, the components segregate by diffusion to give domains of the two conjugate solutions stable at that temperature (see *Exsolution*). However, if this temperature is low the rate of diffusion may be negligible and the solution remains metastable. Thus many components forming nearly ideal systems are, for all practical purposes, completely miscible at all temperatures.

Conversely, although there is in principle a temperature above which any components should be completely miscible, this temperature may be so high that they decompose, or otherwise transform into other phases, before it is reached. In this case they are only partially miscible at all attainable temperatures.

Solid Solubility Between Phases of Different Structural Types

Solid solution series covering a wide range of compositions are nearly always between components of closely similar structure, and may be understood in terms of isomorphous substitution of one ion by another, or of a pair of ions by another pair. In such cases the solid solution has a crystal structure that is intermediate between the structures of the components. There are however many pairs of components which are of totally dissimilar structure and yet form restricted (often very restricted) ranges of solid solution. In such cases the crystal structure is that of the major component and has no relationship to that of the solute.

The possibility of variable chemical composition may arise by the omission of certain atoms or ions to leave vacancies, by the insertion of interstitial atoms or ions into the structure, by the replacement of one atom or ion by another in a site in which it is not strictly appropriate, or by some combination of these processes. In many cases the detailed mechanisms are incompletely known, but to explain a solid solution of silica in a silicate one may hypothesize that a metal cation and an adjacent oxygen ion are occasionally omitted from the structure. The conjugate solution of a silicate in silica might occur to a slight degree by replacement of an occasional silicon ion by the metal cation, and the preservation of neutrality by omission of an adjacent oxygen. Such processes lead to a substantial increase of enthalpy, but they also lead to an increase in entropy; they may therefore have a limited stability at a sufficiently high temperature.

An unusually extensive example of this type of solid solution that has been investigated in some detail is the solubility of sulfur in troilite (FeS). This is effected by omission of a proportion of the Fe^{2+} ions balanced by conversion of others to Fe^{3+}. At certain proportions it is possible for the ferric ions and vacancies to be ordered in ways which give rise to energy minima. In such circumstances it is possible for such a solid solution series to be punctuated by the formation of a series of compounds of the end-members.

E. J. W. WHITTAKER

References

Guggenheim B. A., 1967, "Thermodynamics," Fifth edition, Amsterdam, North-Holland Publ. Co.

Vanders, I. and Kerr, P. F., 1967, "Mineral Recognition," New York, John Wiley & Sons, pp. 53–56.

Cross-references: *Crystal-Field Theory; Enthalpy or Heat Content; Entropy; Exsolution; Phase Equilibria; Thermodynamics.* Vol. IVB: *Crystallography; Defects in Crystals; Isomorphism.* See also *Individual Minerals.*

SOLUBILITY

The concentration of an element or ion that can occur in aqueous solution has an upper limit, or maximum solubility. This limiting solubility is a function of temperature, pressure, and the concentrations of other solutes present. Supersaturation is sometimes attained in natural solutions, usually owing to environmental changes taking place during the movement of the water. A supersaturated solution is unstable and tends to return to equilibrium by precipitation of a solid phase, but the process of restoration of equilibrium may be slow.

The solubility of a solid mineral is a slightly different concept. In general this concept is applied in the literature to indicate not only the amount of the mineral that can be dissolved before saturation is reached, but also the rate at which chemical attack by water can take place. The chemical alteration of the mineral may yield solutes and a new solid phase.

The concentrations of major dissolved ions of natural water are controlled by different factors. The reversible chemical reactions whereby minerals are brought into solution range from simple disruption of the crystal lattice as in the solution of sodium chloride, through reversible alteration processes of a more complex nature such as the hydrolysis in the solution of carbon dioxide and carbonates, and the oxidation or reduction processes involved in the solution and deposition of iron minerals. The solution of minerals may also take place during an irreversible decomposition in which the original crystal lattice cannot be restored without drastic alteration of the conditions of the reaction. Crystalline silicates, for example, generally decompose to yield ions that go into solution, hydrolysis products such as clay minerals or metal hydroxides of low solubility, and silica.

Ions which participate in oxidation or reduction reactions, or form sparingly soluble compounds in response to pH or availability of other ions, are the most likely to be present in natural waters in amounts that might be predicted from chemical equilibrium considerations.

When the ions in solution are by-products of the breakdown of silicate minerals in irreversible reactions, the concentrations may be related to reaction rates and the length of time the solutions and source minerals were in contact. In all instances the amount of water available and the rate at which it circulates are important factors. In well-watered areas major dissolved constituents may be continually below saturation levels.

Saturation with respect to highly soluble substances such as sodium chloride is attained most commonly in environments where water is depleted by evaporation or other processes, and solutes are abundantly available.

The calculation of equilibrium solubility for a particular ion in mixed solutions of the type dealt with in natural water chemistry is based on the *law of mass action*. The mass law states that at equilibrium in the reaction

$$aA + bB = cC + dD$$

the equation

$$\frac{[C]^c [D]^d}{[A]^a [B]^b} = K$$

holds. The quantities in square brackets represent effective concentrations of reactants and products, in moles per liter, and K is the

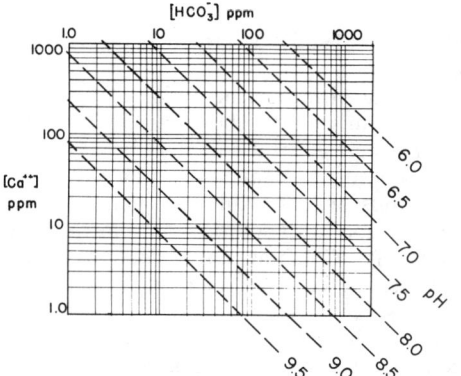

FIG. 1. Equilibrium pH in relation to calcium and bicarbonate activities in solutions in contact with calcite. Total pressure, 1 atm; temperature, 25°C.

equilibrium constant. The effective concentration of an ion is smaller than the measured, or stoichiometric, concentration because the charged ions cannot move about freely in the presence of one another. Methods for converting stoichiometric to effective concentrations, or activities, are given by Garrels (1960). For solutions of relatively constant composition the equilibrium constant may be adjusted so that stoichiometric concentrations can be used instead of activities.

The mass law applied to a solution of calcite in water containing H^+ gives the expression

$$\frac{[Ca^{2+}][HCO_3^-]}{[H^+][CaCO_3]} = K$$

By convention, the activity of $CaCO_3$ a solid, is taken as 1. The equilibrium constant for this reaction is 0.97×10^2 (Hem, 1961). Fig. 1 shows the relationship in graphical form. The pH at which a solution would be in equilibrium with calcite is indicated by the dashed lines. In this system a gas phase is assumed to be absent, the condition one would anticipate in groundwater at a considerable depth below the water table. Systems exposed to air should also adjust to equilibria in which gaseous carbon dioxide participates. However, some of the stepwise chemical reactions that are involved in attaining full equilibrium are not very rapid, and surface waters not infrequently become supersaturated with respect to calcite.

In the chemical analysis of natural water, the concentrations determined for the elements or ions in solution generally represent the total amount present. Aside from the effects of ion activities, which must be taken into account, the possible effects of different ionic species which may go to make up the total amount of the determined concentration must also be considered in calculating the concentrations to be expected at equilibrium. Some of the dissolved ions in water tend to become associated in complexes. For example, ferric iron forms hydroxide complexes of the form $FeOH^{2+}$ and $Fe(OH)_2^+$, depending on pH, all of which may affect the total iron content at equilibrium in some solutions. Manganese may form complexes with bicarbonate and sulfate ions (Hem, 1963) and in more concentrated solutions, such as seawater, a large number of complexes may occur, involving major as well as minor constituents (Garrels and Thompson, 1962). Because the chemistry of some of these complex ions is not known the calculation of maximum or equilibrium solubility of many of the ions in natural water, particularly minor constituents, is not always practicable.

The silicate minerals which are most readily attacked by water in weathering reactions are those of the inosilicate type in which much of the structure is made up of metal-oxygen bonding. The more stable forms are those in which larger proportions of the crystal structure are maintained by silicon-oxygen bonding. In hydrothermal environments, alterations such as olivene → serpentine may take place with liberation of cations and silica. Surficial weathering commonly produces reactions of the type

feldspar + H^+ + H_2O = clay
+ silicic acid + metal ions

In some environments, the clay may be altered to metal hydroxides, releasing additional silicic acid. Concentrations of silica may sometimes be controlled by precipitation of amorphous

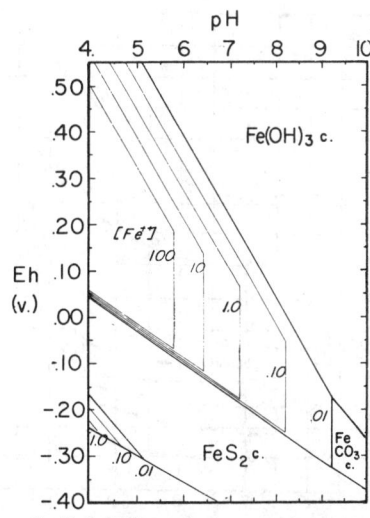

FIG. 2. Solubility of ferrous iron in a system containing dissolved activities of 100 ppm HCO_3 and 10 ppm SO_4, and stability fields of pyrite, siderite, and ferric hydroxide (Hem, 1960).

silica, and concentrations of cations by precipitation of carbonates or oxides, but generally the amounts present depend on reaction rates, total surface area where water-solid contact occurs, and amount of water and its rate of movement. The solution of silica has been extensively studied and many results have been summarized by Siever (1962).

Generalizations regarding the behavior of major dissolved constituents of natural water form only a very primitive basis on which to build an understanding of the complexities of solution geochemistry. Many methods of classification of water analyses based on the relative concentrations of dissolved ions are described in the literature, and are particularly abundant in the extensive Soviet literature in the field of groundwater geochemistry. The classification schemes may use ion ratios, or graphical techniques. A number of these have been discussed by Hem (1970). Although these procedures provide some help in the systematic understanding of solution geochemistry, the factors which control the occurrence of any single element generally can only be explored by study of equilibrium solubility and reaction rates and mechanisms, taking into account the various solute and solid species that are present. Fundamental chemical thermodynamic data which may be used for calculating solubilities are widely scattered in published literature. A compilation of equilibrium constants by Sillén and Martell (1964) however contains many thousands of these values. Garrels and Christ (1965) described many calculation methods.

The concentrations of major components of natural waters tend to be controlled in different ways by their general chemical behavior. Thus, some abundant elements which form many compounds of low solubility, such as iron, are generally found in solution only in minor concentrations. Calcium concentrations may be controlled by equilibrium precipitaton of moderately soluble substances such as calcite or gypsum. Chloride, and to a somewhat lesser extent sodium, form few compounds which are not readily soluble, and in the waters which have attained a high dissolved solids concentration, a large, and commonly predominating, residue of sodium and chloride remains in solution.

JOHN D. HEM

References

Clark, S. P., Jr., 1966, "Solubility," in "Handbook of Physical Constants," *Geol. Soc. Am. Mem.*, **97**, 415–436.
Garrels, R. M., 1960, "Mineral Equilibria at Low Temperature and Pressure," New York, Harper and Bros. 254pp.
Garrels, R. M., and Thompson, M. E., 1962, "A chemical model for sea water at 25° C. and one atmosphere total pressure," *Am. J. Sci.*, **260**, 57–60.
Garrels, R. M., and Christ, C. L., 1965, "Solutions, Minerals and Equilibria," New York, Harper and Row, 450pp.
Hem, J. D., 1970, "Study and interpretation of chemical characteristics of natural water," *U.S. Geol. Surv. Water Supply Paper*, **1473**, 363pp.
Hem, J. D., 1960, "Chemical equilibrium diagrams for ground-water systems," *Bull. Intern. Assoc. Sci. Hydrol.*, **19**, 45–53.
Hem, J. ,D., 1961, "Calculation and use of ion activity," *U.S. Geol. Surv. Water Supply Paper*, **1535 C**, cl-c17.
Hem, J. D., 1963, "Manganese complexes with bicarbonate and sulfatae in natural water," *J. Chem. Eng. Data*, **8**, 99–101.
Okamoto, G., Okura, T., and Goto, K., 1957, "Properties of silica in water," *Geochim. Cosmochim. Acta*, **12**, 123–132.
Siever, R., 1962, "Silica solubility 0°–200°C. and the diagenisis of siliceous sediments," *J. Geol.*, **70**, p. 127.
Sillén, L. G., and Martell, A. E., 1964, "Stability constants of metal-ion complexes," *Chem. Soc. (London) Spec. Publ.*, **17**, 754pp.

Cross-references: *Aqueous Solutions; Complexes; Crystal Chemistry; Mass Action and Equilibrium Constant; Oxidation-Reduction; pH–Eh; Phase Equilibria; Seawater, Chemistry; Silica Solubility; Solubility Product Constant; Stoichiometry; Weathering, Chemical.*

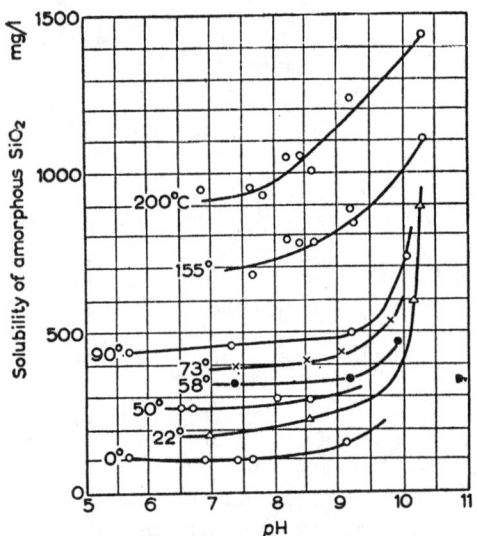

FIG. 3. Solubility of amorphous silica in relation to temperature and pH (Okamoto, Okura, and Goto, 1957).

SOLUBILITY PRODUCT CONSTANT

The solubility product constant, or K_{sp}, is one type of equilibrium constant which is applied to the study of sparingly soluble materials and their saturated solutions. When slightly soluble, strong electrolytes, as in seawater, are in equilibrium with a saturated solution, e.g., $CaCO_3$, the soluble phase consists of ions. The product of the ion concentration in the saturated solution has a fixed value at a given temperature; e.g.,

$$BaSO_4 \rightleftharpoons Ba^{2+} + SO_4^{2-} \text{ (satd. sol.)}$$
barite

$$K_{sp} = [Ba^{2+}][-SO_4^{2-}] = 1 \times 10^{-10}$$

where the ionic concentrations are measured in moles per liter. From this equation, it is apparent that the concentration of barium ion is inversely proportional to the concentration of SO_4^{2-}. If both $BaSO_4$ and $FeSO_4$ are present in a solution, the solubility of $BaSO_4$ varies inversely with the total concentration of the common ion. The repression of the solubility of a salt by adding a common ion is called the *common-ion effect*, and is employed extensively in qualitative analytical chemistry for precipitation purposes. In nature the solubility of $CaCO_3$ can be lowered, e.g., over the Bahama Banks, by evaporation, leading to high salinity.

The K_{sp} varies with changing temperature. The solubility of most salts increases with a rise in temperature, but some like $CaCO_3$ (calcite and aragonite) become less soluble with increasing temperature. Solubility and solubility product constants depend also on grain size of the precipitate. Kolthoff has calculated that $BaSO_4$ crystals with a diameter of 4×10^{-6} cm are 1000 times as soluble as coarsely crystalline barite.

Slightly soluble salts show an increase in solubility when in solutions containing other salts (*salt effect*). Interionic forces are believed to be responsible for the increased solubility. At the same molarity, salts of high ionic charges are more effective than salts with low charges. Because the K_{sp} of a sparingly soluble salt varies slightly when other salts are present in solution, the ionic concentration must be replaced by a more exact quantity—the *ion activity*. The activity of an ion is defined as the product of the molar concentration C and an activity coefficient γ:

$$a = C\gamma$$

For example, for fluorite, CaF_2

$$CaF_2 \rightleftharpoons Ca^{2+} + 2F^-$$

$$K_{sp} \text{ (correct)} = a_{Ca^{2+}} a_{F^-}^2 = [Ca^{2+}][F^-]^2 \cdot \gamma_{Ca^{2+}} \gamma_{F^-}^2$$

The activity coefficient is an index of the deviation from ideal behavior. As the solution becomes more and more dilute, the activity approaches the solution concentration as a limit. (They become identical at infinite dilution.)

$$\lim_{C \to 0} \gamma = 1$$

In dilute solutions, the activity coefficient of a given strong electrolyte is the same for all solutions of the same ionic strength. The ionic strength μ, introduced by Lewis and Randall, measures the effect of ionic interactions on a slightly soluble salt. It is defined as

$$\mu = 1/2 \sum_i C_i Z_i^2$$

where μ = ionic strength; C_i = concentration (moles/liter) of the ith ion; Z_i = charge on the ith ion; and the summation includes all the ions in solution. The ionic strength is a property of the solution and not of any particular ion.

Example problem:

Find μ for a solution of $0.01M$ NaCl and $0.01M$ Na_2SO_4.

Concn. Na^+ = 0.03 (0.01 from NaCl and 0.02 from Na_2SO_4)

Concn. Cl^- = 0.01

Concn. SO_4^{2-} = 0.01

$\mu = 1/2 (0.03 \times 1^2 + 0.01 \times 1^2 + 0.01 \times 2^2) = 0.04$

The ionic strength of seawater is about 0.7.

The concept of K_{sp} is useful in geological systems involving the formation of mineral deposits. An example is the evaporation of saline solutions. The K_{sp} determines the sequence of crystallization of salts from evaporating seawater in a shallow basin. The order of precipitation roughly follows the solubility, with $CaCO_3$ (the least soluble) coming out first, then gypsum, then halite, with a residue of the very soluble Mg,K,Na salts. In nature, the situation is complicated by the many constituents dissolved, the temperature fluctuations, and other environmental changes during deposition. Additional applications of the K_{sp} principle to the transportation of ore solutions and origin of hydrothermal deposits may be found in Barnes (1967). A compilation of K_{sp} values is given in Sillén (1964).

VIVIEN GORNITZ

References

Barnes, H. L., ed., 1967, "Geochemistry of Hydrothermal Ore Deposits," New York, Holt, Rhinehart and Winston, 670pp.

SOLUBILITY PRODUCT CONSTANT

Garrels, R. M., and Christ, C. L., 1965, "Solutions, Minerals, and Equilibria," New York, Harper and Row, 450pp.
Hem, J. D., 1959, "Study and interpretation of chemical characteristics of natural water," U.S. Geol. Surv. Water Supply Paper **1473**, 269pp.
Lewin, S., 1960, "The Solubility Product Principle: An Introduction to Its Uses and Limitations," New York, Interscience Publ., 116pp.
Sillén, L. G., 1964, "Stability constants of metal-ion complexes; section I: Inorganic ligands," Chemical Society, London, Spec. Publ. **17**.

Cross-references: *Calcium Carbonate; Electrolytes; Evaporite Processes; Hydrothermal Solutions—Sulfide Transport; Seawater, Chemistry; Sea-Water, Geochemical Balance; Solubility.* Vol. IVB: *Gypsum; Halite.*

SOLUSPHERE—See Vol. VI

SOLUTIONS—See HYDROTHERMAL SOLUTIONS; also Vol. V, WATER—SOLUBILITY IN ROCKS

SONOCHEMICAL PROCESSES IN THE OCEAN

When water undergoes mechanical agitation of sufficiently high intensity (>0.1 Watt/cm^2) and frequency (>1 kc/sec), it undergoes *cavitation*. Cavitation is the spontaneous formation of microscopic voids in the agitated liquid. As soon as they are formed, these cavities become filled with water vapor as well as with molecules of any volatile solute present. During the cycle of agitation these gas-filled cavities or bubblets undergo extreme changes in size and as a consequence the gas inside them undergoes compression to *high pressures* (on the order of hundreds of atmospheres). These fast compressions are adiabatic in nature and result, in the presence of *noncondensable gases*, in the rapid development of extremely *high temperatures* (6000–9000°C). The high temperatures induce extensive chemical reactions between the gaseous constituents of the system. A large fraction of the products of these reactions diffuse into the bulk of the liquid phase before undergoing further reactions. In other words, the cavitation process which is induced by mechanical agitation of water results in high-temperature gas phase reactions, the products of which are frozen at the ambient temperatures of the liquid. These chemical changes are called *sonochemical* processes, because they are induced by high-frequency agitation in the sonic or ultrasonic range.

The products of the sonochemical reactions depend on the nature of the noncondensable gas. Thus oxygen-saturated water produces mainly hydrogen peroxide; nitrogen-saturated water mainly ammonia and hydrogen; whereas air-saturated water produces primarily nitrous and nitric acids. The latter finding implies that if water is agitated to a sufficient extent under natural air-saturated conditions sonochemical reactions would take place resulting in nitrogen fixation.

Little is known about the occurrence of cavitation under natural conditions. Cavitation undoubtedly occurs when waves break at sea cliffs or when seawater surges through narrows and especially over the sharp edges of submerged rocks. Laboratory experiments have demonstrated the occurrence of cavitation when water impinges on water at velocities in the range of rolling waves. Whenever cavitation takes place in nature it will be accompanied by sonochemical reactions and first of all in nitrogen fixation. As there is at present no quantitative information on the extent of occurrence of cavitation in the ocean, it is impossible to assess the contribution of sonochemical processes to the nitrogen cycle.

MICHAEL ANBAR

References

Anbar, M., 1965, "Chemical reactions induced by sound," *New Scientist*, **30**, 365.
Anbar, M., 1968, "Cavitation during impact of liquid water on water: Geochemical implications," *Science*, **161**, 1343.
Anbar, M., and Pecht, I., 1964, 1967, "The sonochemical formation of hydrogen peroxide in water and the sonochemical decomposition of organic solutes in dilute aqueous solutions," *J. Phys. Chem.*, **68**, 352, 1460, 1462; **71**, 1246.
Elpiner, I. E., 1964, "Ultrasound, Physical, Chemical, and Biological Effects," New York, Consultant Bureau.

Cross-reference: *Nitrogen Cycle in the Oceans.*

SOROSILICATES—See CRYSTAL CHEMISTRY; MINERAL CLASSES: SILICATES

SPA—See MEDICINAL SPRINGS

SPECIFIC GRAVITY

Specific gravity (G) or *relative density* of a mineral is a number that expresses the ratio between the weight of the mineral and the weight of an equal volume of water at 4°C. For example, for quartz G has a value of 2.65. This means that a given specimen of quartz

SPECIFIC GRAVITY

weighs 2.65 times as much as an equivalent volume of water.

Although the terms density and specific gravity are often used interchangeably, a distinction exists between them. *Density* is defined as the ratio of the mass of a substance to the volume of the substance, or "mass per unit volume." Density values must always include specification of units, e.g., grams per cubic centimeter, or pounds per cubic foot. When density is expressed in cgs units it is numerically equal to specific gravity.

Specific gravity determinations of a mineral are an important tool in specimen identification, except where composition (and hence G) is highly variable. Determination of G does not injure the specimen, and thus is particularly helpful when dealing with fine crystals or gemstones. Table 1 groups a number of minerals according to values of G. Mursky and Thompson (1958) published a list of specific gravity data for more than 1400 minerals.

Several factors determine specific gravity in a mineral: crystal structure and chemical composition are the most important. If a substance of variable composition crystallizes in a specific structure, the variation in density will depend essentially on the mass of the individual atoms. Thus *fayalite* is heavier than *forsterite* because iron is heavier than magnesium, and anorthite is heavier than albite (with labradorite in between) because $CaAl_2Si_2O_8$ (anorthite) contains atoms of more total mass than $NaAlSi_3O_8$ (albite).

TABLE 1. MINERALS ARRANGED ACCORDING TO INCREASING SPECIFIC GRAVITY[a]

G	Name	G	Name	G	Name
0.917	Ice	2.2–2.65	Serpentine	2.76–3.1	Muscovite
1.6	Carnallite	2.42	Colemanite	2.8–2.95	Prehnite
1.7	Borax	2.42	Petalite	2.8–3.0	Datolite
1.75	Epsomite	2.4–2.45	Lazurite	2.8–3.0	Lepidolite
1.95	Kernite	2.45–2.50	Leucite	2.89–2.98	Anhydrite
1.96	Ulexite	2.2–2.8	Garnierite	2.9–3.0	Boracite
1.99	Sylvite	2.54–2.57	Microcline	2.95	Aragonite
		2.57	Orthoclase	2.95	Erythrite
2.0–2.19				2.8–3.2	Biotite
		2.6–2.79		2.95–3.0	Cryolite
2.0–2.55	Bauxite			2.97–3.00	Phenacite
2.0–2.4	Chrysocolla	2.55–2.65	Nepheline		
2.0	Sepiolite	2.6–2.63	Kaolinite	3.0–3.19	
2.05–2.09	Sulfur	2.62	Albite		
2.05–2.15	Chabazite	2.60–2.66	Cordierite	2.97–3.02	Danburite
1.9–2.2	Opal	2.58–2.68	Vivianite	2.85–3.2	Anthophyllite
2.09–2.14	Niter	2.65	Oligoclase	3.0–3.1	Amblygonite
2.1–2.2	Stilbite	2.65	Quartz	3.0–3.1	Lazulite
2.16	Halite	2.69	Andesine	3.0–3.2	Magnesite
2.12–2.30	Chalcanthite	2.6–2.8	Alunite	3.0–3.1	Margarite
2.18–2.20	Heulandite	2.6–2.8	Turquois	3.0–3.25	Tourmaline
		2.71	Labradorite	3.0–3.3	Tremolite
2.2–2.39		2.65–2.74	Scapolite	3.09	Lawsonite
		2.72	Calcite	3.1–3.2	Autunite
2.12–2.30	Chalcanthite	2.6–2.9	Chlorite	3.1–3.2	Chondrodite
2.0–2.4	Chrysocolla	2.62–2.76	Plagioclase	3.15–3.20	Apatite
2.2–2.65	Serpentine	2.6–2.9	Collophane	3.15–3.20	Spodumene
2.25	Natrolite	2.74	Bytownite	3.16–3.20	Andalusite
2.26	Tridymite	2.7–2.8	Pectolite	3.18	Fluorite
2.27	Analcime	2.7–2.8	Talc		
2.29	Soda niter	2.70–2.85	Glauberite	3.2–3.39	
2.30	Cristobalite	2.76	Anorthite		
2.30	Sodalite	2.75–2.8	Beryl	3.1–3.3	Scorodite
2.3	Graphite	2.78	Polyhalite	3.2	Hornblende
2.32	Gypsum			3.23	Sillimanite
2.33	Wavellite	2.8–2.99		3.2–3.3	Diopside
2.3–2.4	Apophyllite			3.2–3.4	Augite
2.39	Brucite	2.6–2.9	Collophane	3.25–3.37	Clinozoisite
		2.8–2.9	Pyrophyllite	3.26–3.36	Dumortierite
2.4–2.59		2.8–2.9	Wollastonite	3.27–3.35	Axinite
		2.85	Dolomite	3.27–4.37	Olivine
2.0–2.55	Bauxite	2.86	Phlogopite	3.2–3.5	Enstatite

SPECIFIC GRAVITY

TABLE 1. (Continued)

G	Name	G	Name	G	Name
3.4–3.59		3.7–4.7	Psilomelane	5.68	Zincite
		4.18–4.25	Rutile	5.7	Arsenic
3.27–4.27	Olivine	4.3	Manganite	5.5–6.0	Jamesonite
3.3–3.5	Jadeite	4.3	Witherite	5.3–7.3	Columbite
3.35–3.45	Diaspore	4.37	Goethite		
3.35–3.45	Epidote	4.35–4.40	Smithsonite	5.8–5.99	
3.35–3.45	Idocrase				
3.4–3.5	Hemimorphite	4.4–4.59		5.8–5.9	Bournonite
3.45	Arfvedsonite			5.85	Pyrargyrite
3.40–3.55	Aegirite	4.4	Stannite		
3.4–3.55	Sphene	4.43–4.45	Enargite	6.0–6.49	
3.48	Realgar	4.5	Barite		
3.42–3.56	Triphylite	4.55	Gahnite	5.9–6.1	Crocoite
3.49	Orpiment	4.52–4.62	Stibnite	5.9–6.1	Scheelite
3.4–3.6	Topaz			6.0	Cuprite
3.5	Diamond			6.07	Arsenopyrite
3.45–3.60	Rhodochrosite	4.6–4.79		6.0–6.2	Polybasite
3.5–4.3	Garnet			6.2–6.3	Stephanite
		3.7–4.7	Psilomelane	6.2–6.4	Anglesite
3.6–3.79		4.6	Chromite	5.3–7.3	Columbite
		4.58–4.65	Pyrrhotite	6.33	Cobaltite
3.27–4.37	Olivine	4.7	Ilmenite		
3.5–4.2	Allanite	4.75	Pyrolusite	6.5–6.99	
3.5–4.3	Garnet	4.6–4.76	Covellite		
3.6–4.0	Spinel	4.62–4.73	Molybdenite	6.5	Skutterudite
3.56–3.66	Kyanite	4.68	Zircon	6.55	Cerussite
3.58–3.70	Rhodonite			6.78	Bismuthinite
3.65–3.75	Staurolite	4.8–4.99		6.5–7.1	Pyromorphite
3.7	Strontianite			6.8	Wulfenite
3.65–3.8	Chrysoberyl	4.6–5.0	Pentlandite	6.7–7.1	Vanadinite
3.75–3.77	Atacamite	4.6–5.1	Tetrahedrite-Tennantite	6.8–7.1	Cassiterite
3.77	Azurite	4.89	Marcasite		
		4.9	Greenockite	7.0–7.49	
3.8–3.99					
		5.0–5.19		7.0–7.2	Mimetite
3.7–4.7	Psilomelane			7.0–7.5	Wolframite
3.6–4.0	Spinel	5.02	Pyrite	7.3	Argentite
3.6–4.0	Limonite	4.8–5.3	Hematite		
3.83–3.88	Siderite	5.06–5.08	Bornite	7.5–7.99	
3.5–4.2	Allanite	5.15	Franklinite	7.4–7.6	Galena
3.5–4.3	Garnet	5.0–5.3	Monazite	7.3–7.9	Iron
3.9	Antlerite	5.18	Magnetite	7.78	Niccolite
3.9–4.03	Malachite				
3.95–3.97	Celestite	5.2–5.39		>8.0	
4.0–4.19		5.4–5.59		8.0–8.2	Sylvanite
				8.10	Cinnabar
3.9–4.1	Sphalerite	5.5	Millerite	8.9	Copper
4.02	Corundum	5.5±	Cerargyrite	9.0–9.7	Uraninite
3.9–4.2	Willemite	5.55	Proustite	9.35	Calaverite
				9.8	Bismuth
4.2–4.39		5.6–5.79		10.5	Silver
				15.0–19.3	Gold
4.1–4.3	Chalcopyrite	5.5–5.8	Chalcocite	14–19	Platinum

^a Hurlbut, 1966, p. 585.

Similarly in a group of isomorphous compounds the value of G is directly related to the mass of the atoms present, as seen in Table 2.

Crystal structure affects specific gravity largely through packing of the constituent ions, with more closely packed lattices producing denser substances than loosely packed lattices (see Table 3). If the dimensions of the unit cell have been measured, and the number and kinds of atoms in the unit cell are known, it is possible to calculate the value of G with great accuracy.

TABLE 2. SPECIFIC GRAVITY CHANGES WITH CHANGES IN CATIONS IN ORTHORHOMBIC CARBONATES[a]

Mineral	Composition	Atomic Weight of Cation	G
Aragonite	$CaCO_3$	40.08	2.95
Strontianite	$SrCO_3$	87.63	3.78
Witherite	$BaCO_3$	137.36	4.31
Cerussite	$PbCO_3$	207.21	6.55

[a] Modified from Hurlbut, 1966, p. 155.

TABLE 3. RELATIONSHIP BETWEEN PACKING INDEX (DERIVED FROM CRYSTAL STRUCTURE) AND DENSITY IN POLYMORPHIC MINERALS[a]

Composition	Polymorph	Packing Index	Density ($=G$)
TiO_2	Rutile	6.6	4.25
	Brookite	6.4	4.14
	Anatase	6.3	3.90
Al_2SiO_5	Kyanite	7.0	3.63
	Sillimanite	6.2	3.24
	Andalusite	6.0	3.15

[a] Modified from Berry and Mason, 1959.

Determination of specific density of a specimen requires considerable care in selection of method, preparation of specimen of homogeneous material, avoidance of air bubbles, and so forth. Berry and Mason (1959) and Hurlbut (1966) provide a brief synthesis of the available methods, whereas Muller (1967) provides a more technical account, with direct references to the literature.

One type of method of determination of G is to measure the apparent loss of weight when a weighed fragment of a mineral is immersed in a suitable liquid, usually water or organic solutions. The Jolly balance, the Berman balance, and the ordinary beam balance can be used in this method, which is termed *hydrostatic weighing*. Accuracy of better than a hundredth of one percent is obtainable with the more refined techniques of this method.

In the simplest case, the specific gravity of the solid is determined by Archimedes' Principle, $G = W_A L / W_A - W_L$, where L is density of the liquid, W_A is the weight of the solid in air, and W_L is the weight of the solid when immersed in the liquid.

A second technique involves use of a small stoppered bottle called a *pycnometer*. Four weighings are required: one weights the empty bottle (W_1); then the bottle with mineral (W_2); the bottle with mineral and liquid of density L (W_3); and finally the bottle with liquid alone (W_4). Then

$$G = L \frac{W_2 - W_1}{(W_4 - W_1) - (W_3 - W_2)}$$

A third method involves suspension of minerals in liquids of known density. If a mineral neither sinks nor floats in a liquid, it is the specific gravity (density) of that liquid. Liquid density is determined by a Westphal balance. Combinations of bromoform (G–2.89), methylene iodide ($G=3.33$), clerici solution ($G=4.2$), water ($G=1$), and acetone ($G=0.79$) are used to provide a wide range of values of G. Accuracies to a fraction of a percent are obtainable with the second and third methods listed above.

As seen in Table 1, the most common rock-forming minerals (quartz, feldspars, calcite) have G values in the range 2.65–2.75. Metallic minerals are appreciably heavier, with pyrite ($G=5.0$) representing an average for this group.

LEE WILSON

References

Berry, L. G., and Mason, B., 1959, *Mineralogy*, San Francisco, W. H. Freeman & Co.

Hurlbut, C. S., 1966, *Dana's Manual of Mineralogy*, 17th ed., New York, John Wiley & Sons.

Muller, L. D., 1967, "Density determinations," in (Zussman, J., editor), *Physical Methods in Determinative Mineralogy*, London, Academic Press, pp. 459–466.

Musrsky, G. A., and Thompson, R. M., 1958, "A specific gravity index for minerals," *Can. Mineralogist*, **6**, 273.

Cross-references: *Crystal Chemistry*. Vol. IVB: *Crystallography—Morphological; Jolly Balance;* see also *Individual Minerals*.

SPECTROPHOTOMETRY

Spectrophotometry, or absorption spectroscopy, is one of the most important methods of chemical analysis. Spectrophotometry is based on the fact that radiant energy is absorbed differentially by substances; a given substance will absorb particular wavelengths in a given way and to a particular extent. The range of absorption provides qualitative information as to the nature of the substance; the extent of absorption is a quantitative measure of the amount of the substance present.

A wide range of the electromagnetic spectrum may be involved—from short x rays to long microwaves. Fig. 1 shows the spectral distribution of radiant energy. Thus far, the principal analytical applications of spectrophotometry are in the wavelength range of 200 to 25000 mμ.

Long-path ultraviolet absorption spectrophotometry involves the spectral region from 220 to 400 mμ (2200–4000 Å), which is associated with energy-level transitions of outer electrons, and has been used for measuring the presence and concentration of gases and vapors in trace amounts.

A new spectrometric method receiving much attention recently is *atomic absorption*. The entry in this volume on *Flame Spectroscopy* contains a discussion of atomic absorption. Although flame photometry is an emission spectrochemical method, the instrumentation used can be adapted to atomic absorption. Slavin (1968) has published a monograph on Atomic Absorption Spectroscopy which includes a chapter on Geochemical Applications. Atomic absorption covers the spectral range from the far ultraviolet through the visible (200–750 mμ).

A specialized version of visible-light spectrophotometry is *Colorimetry* (q.v.), which is applicable to solutions of colored atoms, ions, and molecules.

Molecules or ions in which the constituent atoms are covalently bound (e.g., CO_3^{2-}) absorb energies in the long-wavelength regions between 1000 and 25000 mμ. The vibration-rotation spectra thus produced are characteristic of the functional groups present, and may be determined through *Infrared Analysis* (q.v.)

Theory

There are two fundamental laws for the absorption of radiant energy by homogeneous, transparent media.

1. The *Bouguer-Lambert Law* expresses the relationship between absorptive capacity and thickness of an absorbing medium:

$$T = 10^{-ab}$$

where T is the transmittance, 10 is the base of common logarithms, a is the absorbance index, and b is the thickness of the absorbing medium. There are no known exceptions to this law among homogeneous systems.

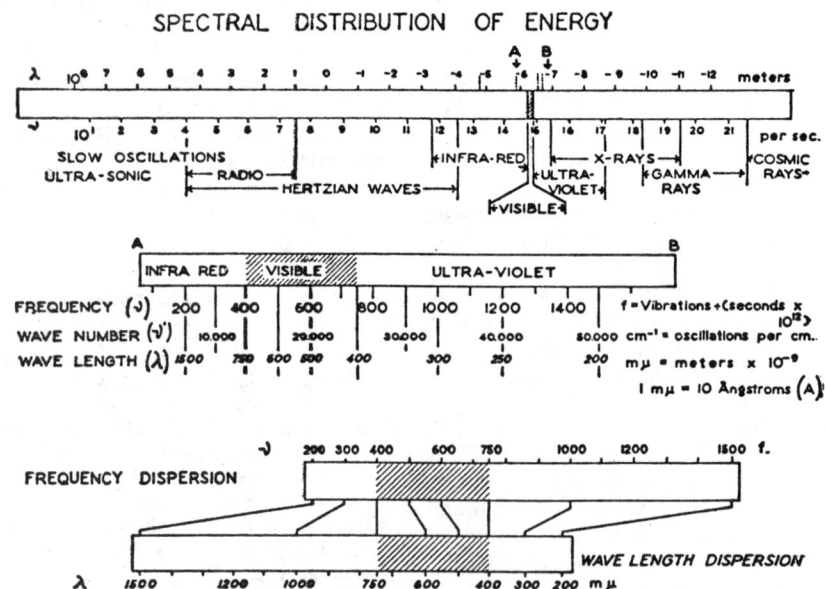

FIG. 1. Spectral distribution of radiant energy. In the lower portion of this figure the region from A to B is shown on a linear scale in frequency (above) and on a linear scale in wavelength (below) (from Brode, 1962, p. 3).

SPECTROPHOTOMETRY

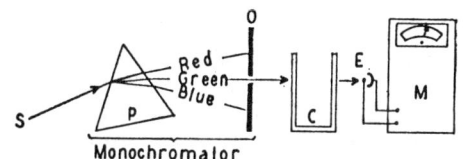

FIG. 2. Essential components of a photoelectric spectrophotometer (from Clark, 1960, p. 3).

2. *Beer's Law* expresses the relationship between absorptive capacity and concentration of the solute in a solution. The absorptive capacity is directly proportional to the number of absorbing entities (ions or molecules). Beer's Law is usually stated:

$$I = I_0 \, 10^{-abc}$$

where I_0 is the intensity of the incident light, I is its intensity after passage through b cm of the given material, and c is the concentration of the solution. Most solutions obey Beer's Law, if dilute and the law applies if the radiant energy is approximately monochromatic.

These two laws may be combined in the expression

$$T = 10^{-abc}$$

or

$$c = A/ab$$

where $A = \log_{10}(1/T)$ = absorbancy of solute.

Instrumentation

Instruments designed for measuring absorption of radiant energy are variously referred to as spectrometers, photometers, absorptimeters and spectrophotometers.

The essential components of a spectrometer are (1) a source of radiant energy, (2) a detector for unabsorbed radiant energy, (3) a monochromator for the isolation of the desired spectral band of radiant energy, (4) an absorption cell, if the sample is gas or liquid, and (5) a photometer. Fig. 2 is a schematic diagram of the relationship of these components. As this figure indicates, spectrophotometers are relatively simple. However, in practice, and considering only the major variations that can be used in the spectral, geometrical and photometric parts of the instruments, over 72 arrangements are possible. Fig. 3 illustrates some possibilities. Commercial instruments are available for ultraviolet, visible, and infrared regions.

Applications

Spectrophotometry has a wide range of qualitative and quantitative applications (see *Colorimetry, Infrared Analysis*, and the discussion of atomic absorption in *Flame Spectroscopy*).

ISABELLA DREW

References

Brode, Wallace R., 1962, "Chemical Spectroscopy," Second ed., New York, John Wiley & Sons.

Clark, G. L., (editor), 1960, "The Encyclopedia of Spectroscopy," New York, Reinhold Publ. Co.

Hershenson, H. M., 1961, "Ultraviolet and Visible Absorption Spectra," New York, Academic Press.

Slavin, W., 1968, "Atomic Absorption Spectroscopy," New York, Interscience Publ. (see Chapter VII, "Geochemical Applications").

Cross-references: *Colorimetry; Emission Spectrography; Flame Spectroscopy; Infrared Analysis.*

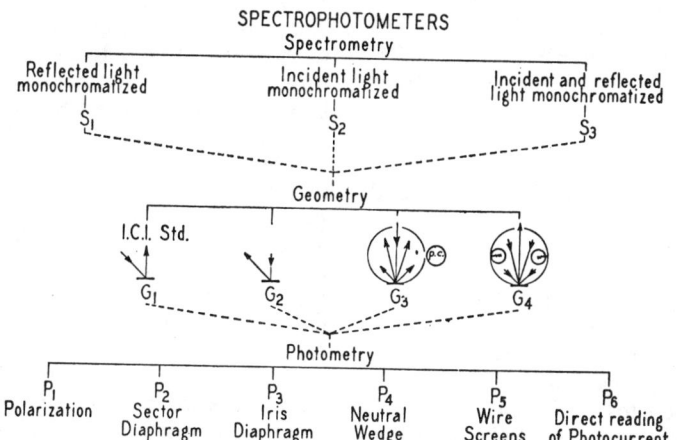

FIG. 3. Diagram showing various spectrophotometric combinations (from Clark, 1960, p. 4).

SPRINGS

A spring is a place where, without the agency of man, water flows from a rock or soil upon the land or into a body of surface water (Meinzer, 1923, p. 48). Throughout the world, springs are important sources of domestic and stock water, and some are large enough to supply cities, industries, or irrigation projects or to form attractive recreational areas. Their flow, along with groundwater seepage too diffuse to be called a spring, is the principal source of dry-weather flow in streams.

Bryan (1919, pp. 559–561) divided springs into (1) those whose water is of meteoric origin and flows under the influence of gravity, and (2) those whose water is of deep origin (including connate, juvenile, and deeply circulating meteoric water) and, in part, is brought to the land surface by forces in addition to gravity. The gravity springs he divided into four groups; as redefined by Meinzer (1942, p. 419), these are:

1. Depression springs, which are due to the water table intersecting the land surface in permeable rocks.

2. Contact springs, which are due to an outcrop of permeable water-bearing rock overlying relatively impermeable rock.

3. Artesian springs, which are due to water rising from a permeable water-bearing bed, confined between relatively impermeable beds, either at an outcrop or through a fissure penetrating the upper confining bed.

4. Springs that issue from tubular openings or fractures in otherwise impermeable rocks.

The springs of deep origin Bryan divided into two groups:

1. Volcanic springs, which are due to or associated with volcanism or volcanic rocks.

2. Fissure springs, which are due to fractures extending deep into the earth's crust.

As a result of the size of springs, Meinzer (1923, p. 53) proposed a classification based on discharge, expressed in units used in the United States:

Magnitude	Discharge
First	100 cfs (cubic feet per second) or more
Second	10 to 100 cfs
Third	1 to 10 cfs
Fourth	100 gal/min to 1 cfs (448.8 gal/min)
Fifth	10 to 100 gal/min
Sixth	1 to 10 gal/min
Seventh	1 pint to 1 gal/min
Eighth	Less than 1 pint/min (less than 180 gallons, or about 5 barrels, per day)

In the United States, according to a study by Meinzer (1927), about sixty-five springs of the first magnitude are known. Of these, thirty-eight rise in volcanic rocks or associated gravel, twenty-four rise in limestone, and three rise in sandstone. Among the largest springs, according to records of the United States Geological Survey, are Malad Springs, Idaho (average discharge 1250 cfs in 1956); Thousand Springs, Idaho (average discharge 1250 cfs in 1956); Big Springs, Missouri (average discharge 434 cfs for forty-one years of record); Comal Springs, Texas (average discharge 289 cfs for thirty years of record); Silver Springs, Florida (average discharge 816 cfs for thirty years of record); and Giant Springs, Montana (discharge about 600 cfs). There are innumerable springs of lesser magnitude, including perhaps hundreds of second magnitude and thousands of third magnitude.

The flow of springs fluctuates. Some springs are perennial; others are intermittent; a geyser, for example, is a special kind of intermittent spring whose discharge is caused by the expansive force of highly heated steam. Some springs fluctuate seasonally or over prolonged periods of time in accordance with seasonal and long-term variations in groundwater recharge. Some are periodic, i.e., they fluctuate at intervals related not to variations in recharge but, rather, to such causes as diurnal changes in transpiration and evaporation, variations in atmospheric pressure, tidal fluctuations, and perhaps even special local conditions that produce an intermittent siphon action. Some fluctuate, or are permanently changed, as a consequence of developments by man. For example, Kissengen Spring, formerly one of the large springs of the Florida Peninsula, ceased to flow in 1950 owing to the heavy draft from wells in the surrounding region. In contrast, a large increase—from about 3800 cfs in 1902 to about 5900 cfs in 1956—in the discharge of springs to the Snake River, Idaho, between Milner and Bliss, was due to the increase in recharge from irrigation on the Snake River Plain.

Springs are classified also as thermal and nonthermal (Meinzer, 1923, pp. 54–55). Thermal springs are those whose water has a temperature noticeably above the mean annual air temperature for their localities. They may be divided into hot springs, whose water has a higher temperature than that of the human body, and warm springs, whose water has a lower temperature than that of the human body.

Thermal springs commonly are associated with areas of present or geologically recent volcanic activity, such as Yellowstone National Park in Wyoming; the lava areas of Idaho, eastern Oregon, and northern California; the lava of the Auvergne region in France and of areas of volcanic rocks in Central Europe, Italy and the circum-Pacific belt, notably Japan, New Zealand and Chile. They are associated also with areas in which the rocks have been intensely folded and faulted, such as the Alps and Pyrenees and the mountains of the western United States (Waring, 1965). More than 1000 thermal springs are known in the United States, and several thousand in the world.

The principal mineral substances in water from springs are generally the same as those in other natural groundwater. Springs whose water contains unusually large quantities of mineral salts in solution are known as mineral springs. Many of these are thermal, and some have gained widespread notice because of the therapeutic value attributed to their water. Some present a striking appearance because of their relation to colorful terraces formed by minerals precipitated from the spring water, for example, at Mammoth Hot Springs in Yellowstone National Park.

O. M. HACKETT

References

Bryan, Kirk, 1919, "Classification of springs," *J. Geol.*, **27**, 522–561.

Meinzer, O. E., 1923, "Outline of ground-water hydrology, with definitions," *U.S. Geol Surv. Water Supply Paper* **494**, 48–56 (71pp).

Meinzer, O. E., 1927, "Large springs in the United States," *U.S. Geol. Surv. Water Supply Paper* **557**, 94pp.

Meinzer O. E. (editor), 1942, "Hydrology," in "Physics of the Earth Series," Vol. 9, pp 416–432, New York, McGraw-Hill Book Co., 712pp.

Stearns, Norah D., Stearns, H. T., and Waring, G. A., 1937, "Thermal springs in the United States," *U.S. Geol. Surv. Water Supply Paper* **679B**, 59–206.

Tolman, C. F., 1937, "Ground Water," pp. 435–466, New York, McGraw-Hill Book Co., 593pp.

Waring, G. A., 1965, "Thermal springs of the United States and other countries of the world," revised by R. R. Blankenship and Ray Bentall, *U.S. Geol. Surv. Profess. Paper* **492**, 383pp.

White, D. E., and Brannock, W. W., 1950, "The sources of heat and water supply of thermal springs, with particular reference to Steamboat Springs, Nevada," *Am. Geophys. Union Trans.*, **31**, 566–574.

Cross-references: *Artesian Water; Connate Water; Groundwater; Juvenile Water; Medicinal Springs; Meteoric Water; Water Table.* Vol. V: *Geyser.*

STATISTICS—CONTINENTS AND OCEANS—See Vol. III, CONTINENTS AND OCEANS—STATISTICS OF AREA, VOLUME, RELIEF

STEREOCHEMISTRY AND ENANTIOMORPHISM

Stereochemistry is the branch of chemistry concerned with the spatial arrangement of atoms within molecules, complex ions and infinite three-dimensional networks, and the chemical and physical properties resulting from such arrangement.

Classical stereochemistry was formerly restricted to the study of organic molecules and complex ions, but with the development of solid-state chemistry, the scope of stereochemistry has been extended to include the configuration of atoms in inorganic crystals.

Stereoisomerism in organic compounds refers to isomers that differ only in the orientation of the constituent atoms in space. (Isomers are organic compounds having the same empirical formula, but differing in molecular structure.) Optical isomers are stereoisomers that can exist in two forms related to each other as mirror images (enantiomorphs). They have the ability to rotate the plane of polarized light in opposite directions and are therefore described as *optically active* (q.v.). The phenomenon of optical activity was discovered by J. B. Biot (1815) and studied by Pasteur (1848) in tartrates, and was applied to compounds of carbon by van't Hoff and Le Bel (1874–1875). It can be used as a test for natural hydrocarbons.

Enantiomorphism (the ability of a substance to exist in left-handed and right-handed forms) is also exhibited by inorganic crystals. This property is related to molecular and crystal symmetry. If the molecule or crystal lacks a center or plane of symmetry then it will be enantiomorphous. Of the thirty-two crystal classes, eleven are enantiomorphic:

1, 2, 3, 4, 6, 23, 222, 32, 42, 62, 432.

A solid which crystallizes in one of these classes will be enantiomorphous, and if suitable faces develop in the crystal it can be identified as such. The crystal is also optically active, but this property disappears in the molten state. Common minerals which display this

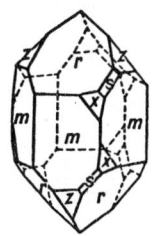 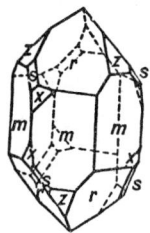

Fig. 1. Stereoisomers of quartz: right-handed quartz (left); left-handed quartz (right).

STEREOCHEMISTRY AND ENANTIOMORPHISM

property include quartz and cinnabar (HgS).

In quartz, the two forms s and z, if present on the crystal, reveal the right- and left-handed nature of this substance (Fig. 1).

VIVIEN GORNITZ

References

Eliel, E. L., 1962, "Stereochemistry of Carbon Compounds," New York, McGraw-Hill Book Co., 486pp.

Wells, A. F., 1962, "Structural Inorganic Chemistry," Oxford, Clarendon Press, 1055pp.

Cross-references: *Crystal Chemistry; Hydrocarbons; Optical Activity.* Vol. IVB: *Cinnabar; Quartz.*

STOICHIOMETRY

Stoichiometry is defined as the mathematics of chemical reactions and processes. The calculations are based on two laws: (a) the law of conservation of mass and energy, which states that mass and energy are neither created nor destroyed in a chemical reaction; and, (b) the law of definite composition, which states that a chemical compound is always composed of the same elements combined in definite and constant proportions by weight. The relations expressed in a chemical equation are used to compute weights and volumes of substances involved in the reaction. It follows that the quantities of reacting materials are proportional to their molecular weights. This forms the basis for solving many problems in geochemistry.

Weight-to-weight problems are those in which the weight of one material is given and the weight of another must be found. These are derived from mole ratios in the balanced equation. A mole is defined as the quantity of material having a mass in grams equivalent to the atomic or molecular weight.

Problem 1. A sample of impure magnetite, Fe_3O_4, weighing 1.542 g is dissolved, the iron is oxidized to Fe^{3+} and precipitated as $Fe(OH)_3$. The precipitate is heated ("ignited") to form hematite, Fe_2O_3, weighing 1.485 g. Calculate the percentage of Fe_3O_4 in the ore.

In order to solve problems of this type, one finds a gravimetric factor relating the weight of the substance sought to the weight of the given substance.

Wt substance sought =

$$\left(\frac{X \text{ mol wt substance sought}}{Y \text{ mol wt substance given}}\right) \text{wt substance given}$$

where X and Y are in the proportions determined by the chemical reaction. In this case the gravimetric factor is $2Fe_3O_4/3Fe_2O_3$ because two moles of Fe_3O_4 will yield three moles of Fe_2O_3 when oxidized. (The number of Fe atoms must be equal.)

$$\text{wt } Fe_3O_4 = (1.485 \text{ g}) \frac{(2 \text{ } Fe_3O_4)}{(3 \text{ } Fe_2O_3)}$$

$$= (1.485) \frac{(2)(231.55)}{(3)(159.7)}$$

$$= 1.437 \text{ g}$$

$$\% \text{ } Fe_3O_4 = \frac{(1.437)}{(1.542)} (100) = 93.1\%$$

Problem 2. An impure sample of pyrite, FeS_2, weighing 0.850g, is fused and the sulfur oxidized to SO_4^{2-}. The iron sulfate is then precipitated by a Ba salt as $BaSO_4$ weighing 1.43 g. What is the percentage of pyrite in the rock?

$$\text{wt } FeS_2 = (\text{wt } BaSO_4) \frac{(FeS_2)}{(2 \text{ } BaSO_4)}$$

$$= (1.430 \text{ g}) \frac{(119.97)}{(2)(233.42)}$$

$$= 0.3675 \text{ g } FeS_2$$

$$\% \text{ } FeS_2 = \frac{(0.3675)(100)}{(0.850)}$$

$$= 43.23\%$$

Since many geochemical reactions take place in solution, it is often necessary to find the concentration of a given product. Ordinary concentration units are expressed in weight % and volume %, but the chemist prefers a unit based upon a chemical quantity. An important chemical concentration unit is *molarity*, defined as the number of moles of substance dissolved in one liter of solution.

Problem 3. Find the molarity (M) of 10 g of chalcanthite, $CuSO_4 \cdot 5H_2O$, dissolved in 50 ml. of water.

First find the molecular weight of chalcanthite:

$$\begin{aligned} Cu &= 63.5 \\ S &= 32 \\ 4 \times O &= 64 \\ \underline{5 \times H_2O} &= \underline{90} \\ \text{mol wt} &= 249.5 \end{aligned}$$

The number of moles = wt given/mol wt

$$= 10\text{g}/249.5$$
$$= 0.0401 \text{ moles}$$

0.0401 moles dissolved in 50 ml or 0.05 liter is equivalent to 0.805 moles/liter; or molarity = 0.805.

Normality is another frequently encountered unit of concentration. It is defined as the number of equivalents per liter of solution. In this method, equal volumes of equal concentration are chemically equivalent.

$$\text{equivalent weight} = \frac{\text{mol wt}}{\text{combining capacity}}$$

For acids, the combining capacity is the number of H^+ atoms replaced in neutralization; e.g.,

$HCl + NaOH \rightarrow NaCl + H_2O$
combining capacity = 1

$H_2SO_4 + 2\,NaOH \rightarrow NaSO_4 + 2H_2O$
combining capacity = 2

$H_3PO_4 + 3\,NaOH \rightarrow Na_3PO_4 + 3H_2O$
combining capacity = 3

For bases, it is the number of OH^- groups replaced in a reaction.

$2HCl + Ca(OH)_2 \rightarrow CaCl_2 + 2H_2O$
combining capacity = 2

For salts, it is the product of the oxidation numbers; e.g., for $Al_2^{III}(SO4)_3^{(-2)}$ $n=6$ or (3)(2).

In oxidation-reduction reactions, it is the change in oxidation number/atom.

$n = 5 \quad Mn^{(VII)}O_4^- \rightarrow Mn^{(II)2+}$
$n = 2 \quad Cu^\circ \rightarrow Cu^{2+}$

(The symbol $^\circ$ indicates zero valence, i.e., an element in the "native" or uncombined state.)

Problem 4. Compute the normality of sulfuric acid with density 1.622 g/ml and 71% H_2SO_4 by weight.

0.71×1.622 g/ml = 1.152 g H_2SO_4/ml
= 1152 g H_2SO_4/1000 ml or 1 liter

N = eq./liter = (wt. in g/eq wt)/liter

eq. wt = 1/2 mol wt for H_2SO_4
= 1/2 (98) = 49 g

$\dfrac{1152}{49}$ = 23.5 eq/liter

Normality = 23.5 N

Oxidation-reaction equations are important in geochemistry. Some sample problems are given below.

Problem 5. Write the equation for the oxidation of magnetite to hematite in terms of water, hydrogen ions, and electrons.

$$\begin{array}{l}
2\,Fe_3O_4\,(c) + 1/2\,O_2\,(g) \rightleftharpoons 3\,Fe_2O_3\,(c) \\
\underline{H_2O\,(e) \rightleftharpoons 2\,H^+\,(aq.) + 1/2\,O_2\,(g) + 2e^-} \\
2\,Fe_3O_4\,(c) + H_2O\,(e) \rightleftharpoons 3\,Fe_2O_3\,(c) + 2H^+\,(aq.) + 2e^-
\end{array}$$

Problem 6. What is the equivalent weight of siderite ($FeCO_3$) in the oxidation-reduction reaction of siderite to hematite? The reaction is

$2FeCO_3(c) + H_2O(e) \rightleftharpoons Fe_2O_3(c) + 2CO_2(g) + 2H^+\,(aq.) + 2e^-$

The change in oxidation state of iron is from Fe^{II} to Fe^{III}, which corresponds to a change of one per Fe atom, but two for the reaction

$$\text{eq. wt} = \frac{\text{mol wt FeCO}_3}{2} = \frac{115.86}{2}\text{ g}$$
$$= 57.93\text{ g}$$

Additional Stoichiometric Problems

Problem 7. Calculation of composition.

In the Bayer process for recovering aluminum from siliceous ores, some aluminum is always lost because of the formation of an unworkable "mud" having the formula

$3\,Na_2O \cdot 3Al_2O_3 \cdot 5SiO_2 \cdot 5H_2O$

Since Al and Na ions are present in excess in the solution from which this precipitate is formed, the precipitation of Si in the mud is complete. A certain ore contained 13% by weight kaolin ($Al_2O_3 \cdot 2SiO_2 \cdot 2H_2O$) and 87% gibbsite ($Al_2O_3 \cdot 3H_2O$). (a) Find the % of total Al in this ore recoverable in the Bayer process. (b) How much Al can be recovered from 100 g of ore?

The g–mol wt of kaolin, $Al_2O_3 \cdot 2SiO_2 \cdot 2H_2O$, is 258. The g–mol wt of gibbsite, $Al_2O_3 \cdot 3H_2O$, is 156. In 100 g of ore there are 13 g kaolin and 87 g gibbsite.

wt Al in 13 g kaolin =
$$(13\text{ g kaolin})\frac{(2)\,(\text{g--mol wt Al})}{(\text{g--mol wt kaolin})}$$
$$= \frac{13 \times 54.0}{258}$$
$$= 2.7\text{ g Al}$$

wt Al in 87 g gibbsite =
$$(87\text{ g gibbsite})\frac{(2)\,(\text{at. wt Al})}{(\text{g--mol wt gibbsite})}$$
$$= 87 \times \frac{54.0}{156}$$
$$= 30.1\text{ g Al}$$

STOICHIOMETRY

In 100 g ore there are $2.7 + 30.1 = 32.8$ g Al. In kaolin there is an equal number of Al and Si atoms, and 13 g kaolin contains 2.7 g Al. The mud takes 6 Al atoms for 5 Si atoms. The precipitation of all the Si from 13 g kaolin involves the loss of $6/5 \times 2.7$ g $= 3.2$ g Al. Thus:

(a) % Al recoverable $= \dfrac{\text{recoverable Al}}{\text{total Al}}$

$= \dfrac{32.8 - 3.2 \text{ g}}{32.8 \text{ g}}$

$= 0.90$

$= 90\%$

(b) wt Al recoverable from 100 g ore

$= 32.8 - 3.2$ g

$= 30$ g

Problem 8. Calculation of Formulas. Determine the empirical formula of a uranium oxide ore which weighed 2.949 g and contained 2.500 g uranium.

In 2.949 g of the oxide there are 2.500 g uranium and $(2.949 - 2.500) = 0.449$ g oxygen.

g-atoms U $= \dfrac{2.500 \text{ g}}{238.1 \text{ (at. wt)}}$

$= 0.0105$ g—atoms U

g-atoms O $= \dfrac{0.449 \text{ g}}{16.00 \text{ (at. wt)}}$

$= 0.02806$ g—atoms O

The ratio of U to O atoms in this compound is 0.0105:0.02806. To express this ratio in terms of small whole numbers, divide each term in the ratio by the smaller number 0.01050.

no. of atoms U $= \dfrac{0.01050}{0.01050} = 1$

no. of atoms O $= \dfrac{0.02806}{0.01050} = 2.672$

Find the smallest multiplying integer that will give whole numbers. This is 3.

U:O = 1 atom : 2.67 atoms
    = 3 atoms : 8.02 atoms

The empirical formula is U_3O_8.

Limitations of Stoichiometry as Applied to Natural Systems

The chemist deals with reactions that obey stoichiometric principles. The geochemical reactions which occur in the earth and which have produced rocks and minerals are extremely complex and cannot always be treated in this manner. For example, a reaction like $CaCO_3 + SiO_2 \leftrightarrows CaSiO_3 + CO_2$ if carried out in a laboratory is subject to quantitative relations. If one starts with a specified amount of $CaCO_3$ one can calculate the amount of $CaSiO_3$ present after the completion of the reaction. This is not always so in natural systems. The supply of SiO_2 may be continuous over long periods of geologic time, and the CO_2 may escape altogether. The replacement of one mineral by another is carried out by solutions filling very small capillary openings. Minerals are dissolved to make way for the new material. The volume of rock remains constant throughout the process. Replacement occurs volume for volume and not mole for mole so that ordinary stoichiometric equations are not really applicable.

Furthermore, there are many inorganic compounds, both naturally occurring and artificial, the properties of which do not correspond to a compound with a simple and characteristic formula. Such substances are known as *nonstoichiometric* or *berthollide compounds*.

Nonstoichiometric compounds are solid phases which deviate from the law of constant composition by having variable chemical composition. These compounds arise from crystal defects (q.v.), solid solutions, and structures with interstices and intergrowths, and are quite common in nature. In some crystals, there is an excess or deficiency of one type of atom over another, or interstitial ions occupy spaces in the crystal lattice. An example of this is the mineral pyrrhotite which has an Fe deficiency and an apparent composition of Fe_6S_7 but is really $Fe_{1-x}S$.

Solid solutions of two phases that have the same crystal structures and are miscible in all proportions also cause nonstoichiometric compounds to form (e.g., olivines, plagioclase series). Substances with layer lattices can accommodate foreign atoms between the layers (clays, graphite). Other materials having open-framework structures can trap impurities—gases, ions—resulting in a berthollide compound (e.g., zeolites). Two phases of different composition but similar in structure can coexist in an intimate intergrowth (e.g., perthite).

VIVIEN GORNITZ

References

Pierce, G., and Smith, R. N., 1968, "General Chemistry Workbook: How to Solve Chemistry Problems," Second ed., San Francisco, Freeman and Co., 243pp.

Wells, A. F., 1962, "Structural Inorganic Chemistry," Third ed., Oxford, Clarendon Press, 1055pp.

STRONTIUM: ELEMENT AND GEOCHEMISTRY

Cross-references: *Molecular Weights; Oxidation-Reduction; Solid Solution.* Vol. IVB: *Defects in Crystals.*

STRONTIUM: ELEMENT AND GEOCHEMISTRY

Physical Properties

Strontium (Sr) has atomic number 38 and atomic weight 87.63. Its electronic configuration is $1s^2 2s^2 2p^6 3s^2 3p^6 3d^{10} 4s^2 4p^6 5s^2$ and its ionic radius is 1.12A. It has a melting point of 704°C, a boiling point of 1638°C, and a density of 2.6g/cm^3.

Geochemistry

Strontium is the least abundant of the alkaline earth metals, occurring to the extent of 0.042% in the earth's crust. The main ores are celestite, $SrSO_4$, and strontianite, $SrCO_3$. The colors of these ores vary from red to green to colorless, depending on the impurities, which are usually calcium, aluminum, and iron. Strontium exists in the geologic realm in several isotopic forms. Sr^{88} is the most abundant, making up 82.5% of the total. Sr^{87} is 7.02%, Sr^{86} is 9.86% in abundance. The Sr^{87}/Sr^{86} ratio is used in age determinations, since Rb^{87} decays to Sr^{87}. The recent presence of bomb-produced Sr^{90}, with a radioactive half-life of 28 years and a β^- emission energy of 1.46 MeV has added significantly to the understanding of oceanic circulation and of the patterns of distribution of strontium in natural systems.

The distribution of strontium in minerals, rocks, sediments, and water is affected to a certain extent by the presence of calcium, which has an ionic radius of 0.99 Å, very close to that of strontium, 1.12 Å. Strontium is usually found in greatest amounts in calcium-rich minerals, and to a lesser extent in potassium-rich minerals. This phenomenon is explained by Goldschmidt's "admittance" theory, which says that Sr^{2+} is "admitted" into the smaller calcium lattice but is "entrapped" in the potassium lattice which has a larger radius (1.33 Å).

In basaltic rocks, very little coherence of strontium with calcium is seen. This is due to the fact that calcic plagioclase is a major constituent of basaltic rocks, so there are a large number of sites into which strontium can enter with reasonably low potential energy. Hence, since strontium never approaches saturation, there is no obvious coherence with calcium. Differing values for strontium concentrations can be assigned to regional variations in the strontium content of the source magma, and to fractional crystallization and differentiation during formation. The average concentration of strontium in basaltic rocks, as found by Turekian (1956), is about 465 ppm. Other workers report different values, again suggesting the lack of coherence in basaltic rocks.

In contrast to basaltic rocks, which show a large range of strontium values for a given calcium concentration, granitic rocks show a definite coherence between the two elements, with increasing calcium content implying increasing strontium concentration. Here the limited number of easily available sites for strontium forces coherence, which far outweighs the effects of regional variations. Coherence is implemented by the high degree of mobility of calcium and strontium which would be implied by the intense metamorphism and metasomatism required to produce granitic rocks. Turekian reports a value of 282 ppm for the average strontium content of all granitic rocks but the interpretation of this value depends on the model of the continental crust chosen and on the relative sampling of high- and low-calcium rocks.

Just as with the basaltic rocks, there are regional variations in the strontium content of granitic rocks. Samples from Africa show 160 ppm Sr, whereas ones from the Canadian shield show 305 ppm, and values as high as 340 ppm are found in samples from the Mesozoic belts of western North America.

The behavior of strontium in sedimentary rocks is almost unpredictable because of the many influences on Sr content in low temperature deposition. The ratio of Sr to Ca in carbonate shells and carbonate-bearing sediments is a function of: (*1*) The Sr/Ca ratio in the liquid phase from which the solid phase is derived; (*2*) The vital effect of the organism; (*3*) The particular polymorph, calcite or aragonite, into which the strontium is incorporated; (*4*) The temperature; (*5*) The salinity of the liquid phase. Diagenesis and groundwater can further alter the Sr/Ca relation in sediments. The average Sr/1000 Ca atomic ratio of 155 limestones studied by Turekian was 0.71, which corresponds to a Sr content of 610 ppm for a "pure $CaCO_3$" limestone. Shales seem to have an ability to concentrate strontium due to the ion-exchange properties of the clay minerals, and shales have an average strontium concentration of 298 ppm, but the interpretation of this value is difficult since there is evidence that the strontium content of various limestones may have been changing with time. Shales and limestones, then, do not seem to show strontium coherence with calcium, just as in basaltic rocks, but recent deep-sea sediments do seem to show coherence, suggesting that weathering of limestones may have a significant effect on the calcium-strontium of the original sediment.

Strontium and calcium show little or no coherence in metamorphic schists. This suggests that the degree of metamorphism is not high enough to allow mass migration and equilibration of the Sr–Ca system. Strontium contents for various minerals have been determined, and the highest values occur in the plagioclases and feldspars, with values as high as 1600 ppm for the feldspars.

Strontium exists in seawater to the extent of 8 mg/liter. In deep-sea sediments, the Sr content of the carbonate-free fraction is generally less than 150 ppm. In samples of sediment cores with less than 2% Ca, the average Sr content is about 130 ppm. The range of Sr concentrations of low-Ca cores includes values equal to or lower than the shale content of 300 ppm Sr. The major contributors of strontium to the deep-sediments are the carbonate-forming coccoliths and foraminifera. Most of these organisms have Sr contents around 1200 ppm. Coccoliths and foraminifera seem to show large variations around the average value, indicating the possibility of significant regional variations in the strontium content of the marine environment.

The introduction of Sr^{90} into the atmosphere and ocean by nuclear testing has increased man's knowledge of oceanographic circulation and of the distribution of strontium in the atmosphere-ocean ecosystem. Bones of animals and man have the ability to concentrate strontium as much as fifty times over seawater, so the physiological implications of Sr^{90} contamination cannot be over-emphasized.

JOHN WEHMILLER

References

Collin, R. L., 1966, "Precipitate formation in the strontium-phosphate system," *Science,* **151** (3716), 1386–1388.

Mackenzie, F. T., 1964, "Strontium content and variable strontium-chlorinity relationship of Sargasso Sea water," *Science,* **146**, 517–518.

Powell, J. L., and DeLong, S. E., 1966, "Isotopic composition of strontium in volcanic rocks from Oahu," *Science,* **153**(3741), 1239–1242.

Rankama, K., and Sahama, T. G., 1950, "Geochemistry," Chicago, Univ. of Chicago Press, 912pp.

Turekian, K., 1956, "The geochemistry of strontium," *Geochim. Cosmochim. Acta,* **10**, 145–196.

Turekian, K., 1964, "The marine geochemistry of strontium," *Geochim. Cosmochim. Acta,* **28**, 1479–1496.

Cross-references: *Calcium; Rubidium; Rubidium-Strontium Dating Method; Strontium Cycle;*
Weathering, Chemical. Vol. VI: *Diagenesis; Pelagic Sediments.*

STRONTIUM CYCLE

The cycle of strontium to and through the sea has been the subject of much conjecture. Turekian (1956) originally proposed that the emergence of many aragonitic corals and molluscs since the early Mesozoic has been caused by either intense, low-strontium calcite deposition by brachiopods or else by a new high Sr/Ca source of weathering, namely granites. Odum (1957) suggests that starting with a basaltic crust, and the weathering of $CaCO_3$ with a low Sr/Ca ratio, the Sr/Ca ratio of the sea is variable and increasing until the first sedimentary uplift after which weathering and recycling brings the Sr/Ca ratio of seawater to a constant value.

The constancy of the Sr^{87}/Sr^{86} ratio in limestones of Paleozoic age and younger is a good reason for believing that the Sr/Ca ratio of the oceans has been constant during the Phanerozoic. If the mean Sr/Ca ratio of streams draining limestones of various ages was constant at any given time, and if removal rates equalled supply rates to the ocean, then the oceanic Sr/Ca ratio would remain constant. Implicit in this model is a significant amount of strontium loss from limestones by diagenesis and weathering. If the Sr/Ca ratio of seawater remains constant, then there must be a representative recycling of all sediments deposited. If any independent sink for calcium or strontium develops which does not recycle, then the oceanic Sr/Ca ratio would respond to the new conditions.

Because of preferential loss of strontium by limestones during diagenesis, no recycling of sediments would mean that the net Sr/Ca ratio supplied to the oceans would decrease, if Turekian's (1964) conclusion that the $CaCO_3$ deposition in the oceans equals that supplied by streams is assumed to be correct. Thus the Sr/Ca ratio in seawater would decrease exponentially with time, if Sr/Ca ratios in supply and removal are identical.

The value for the oceanic Sr/Ca ratio through time cannot be determined very well because fossil shells which are well-preserved through the Cenozoic show strontium contents quite similar to contemporary shells, indicating a long time constant for the decrease in Sr/Ca, if there is one at all. The proven ability of well-preserved shells to add various trace elements, including strontium, further adds to the difficulty in understanding the strontium cycle. Mechanisms are known by which the ocean can alter its strontium concentration, but there are no unequivocal ways of determining the extent of any changes from the fossil record.

JOHN WEHMILLER

References

Odum, H. T., 1951, "The stability of the world strontium cycle," *Science*, **114**, 407–411.
Odum, H. T., 1957, "Biogeochemical deposition of strontium," *Publ. Inst. Marine Sci.*, **4**, 39–106.
Turekian, K., 1956, "The geochemistry of strontium," *Geochim. Cosmochim. Acta*, **10**, 145–196.
Turekian, K., 1964, "The marine geochemistry of strontium," *Geochim. Cosmochim. Acta*, **28**, 1479–1496.

Cross-references: *Cycle, Geochemical; Seawater, Chemistry; Seawater, Geochemical Balance; Strontium: Element and Geochemistry; Weathering, Chemical.* Vol. VI: *Diagenesis.* Vol. VII: *Paleontology.*

SUBMARINE SPRINGS—*See* Vol. I

SUBSTITUTION, ATOMIC—*See* TRACE ELEMENTS IN SILICATE MINERALS

SUBSURFACE WATER—*See* GROUNDWATER; HYDROLOGY

SULFATES

The fundamental unit is the $(SO_4)^{2-}$ anionic group producing anisodesmic compounds, sulfur being in symmetrical tetrahedral coordination with four oxygen atoms. Many of the groups have had their structures worked out in considerable detail. The sulfates are one of the commonest and most important classes of minerals.

This class is usually divided into two large divisions, the anhydrous and the hydrous sulfates. Gypsum and anhydrite are the most abundant sulfates, and gypsum is probably the most important sulfate economically. Barite is probably the most important of the anhydrous sulfates. Most sulfates are white or colorless, nonmetallic, and relatively soft. (See table below.)

ANHYDROUS ACID SULFATES

Mercallite	$KHSO_4$	
Misenite	$K_8H_6(SO_4)_7$	
Letovicite	$(NH_4)_3H(SO_4)_2$	

A_2XO_4 Type

Mascagnite	$(NH_4)_2SO_4$	*Mascagnite group:* orthorhombic; complete substitution of K, Rb, Tl and Cs in artificial compounds, and probably in minerals
Arcanite	K_2SO_4	
Taylorite	$(K,NH_4)_2SO_4$ (?)	
Aphthitalite	$(K,Na)_3Na(SO_4)_2$	
Palmierite	$(K,Na)_2Pb(SO_4)_2$	
Thenardite	Na_2SO_4	

AXO_4 Type

Barite	$BaSO_4$	*Barite group:* orthorhombic; complete series between barite and celestite, but less substitution of Ba or Sr for Pb in anglesite; little substitution of Ca for Ba or Sr
Celestite	$SrSO_4$	
Anglesite	$PbSO_4$	
Anhydrite	$CaSO_4$	
Chalcocyanite	$CuSO_4$	

$A_mB_n(XO_4)_p$ Type

Vanthoffite	$Na_6Mg(SO_4)_4$	
Glauberite	$Na_2Ca(SO_4)_2$	
Langbeinite	$K_2Mg_2(SO_4)_3$	*Langbeinite group:* isometric, tetartohedral
Manganolangbeinite	$K_2Mn_2(SO_4)_3$	

HYDRATED ACID AND NORMAL SALTS

Hydrous Acid

Rhomboclase	$FeH(SO_4)_2 \cdot 4H_2O$
Minasragrite	$(VO)_2H_2(SO_4)_3 \cdot 15H_2O$

SULFATES

$A_2(XO_4) \cdot xH_2O$ Type

Lecontite	$Na(NH_4,K)(SO_4) \cdot 2H_2O$
Mirabilite	$Na_2SO_4 \cdot 10H_2O$

$A_2B(XO_4)_2 \cdot xH_2O$ Type

Syngenite	$K_2Ca(SO_4)_2 \cdot H_2O$	
Koktaite	$(NH_4)_2Ca(SO_4)_2 \cdot H_2O$	
Kroehnkite	$Na_2Cu(SO_4)_2 \cdot 2H_2O$	
Loeweite	$Na_2Mg(SO_4)_2 \cdot 2\frac{1}{2}H_2O$	
Bloedite	$Na_2Mg(SO_4)_2 \cdot 4H_2O$	*Bloedite group:* monoclinic; not isostructural with each other, but leonite is isostructural with many other salts
Leonite	$K_2Mg(SO_4)_2 \cdot 4H_2O$ (?)	
Wattevilleite	$Na_2Ca(SO_4)_2 \cdot 4H_2O$ (?)	
Picromerite	$K_2Mg(SO_4)_2 \cdot 6H_2O$	*Picromerite group:* monoclinic, but pseudo cubic; an artificial series called "Tuttons Salts" contains many salts with formula $A_2B(XO_4)_2 \cdot 6H_2O$
Cyanochroite	$K_2Cu(SO_4)_2 \cdot 6H_2O$	
Boussingaultite	$(NH_4)_2Mg(SO_4)_2 \cdot 6H_2O$	

$A_mB_n(XO_4)_p \cdot xH_2O$, with $(m + n):p < 3:2$ and $> 1:1$ Type

Ferrinatrite	$Na_3Fe(SO_4)_3 \cdot 3H_2O$
Polyhalite	$K_2Ca_2Mg(SO_4)_4 \cdot 2H_2O$
Leightonite	$K_2Ca_2Cu(SO_4)_4 \cdot 2H_2O$

$AB(XO_4)_2 \cdot xH_2O$ Type

Krausite	$KFe(SO_4)_2 \cdot H_2O$	
Voltaite	$(K,Fe^{II})_3Fe^{III}(SO_4)_3 \cdot 4H_2O$ (?)	
Tamarugite	$NaAl(SO_4)_2 \cdot 6H_2O$	*Tamarugite group:* monoclinic
Amarillite	$NaFe(SO_4)_2 \cdot 6H_2O$	
Mendozite	$NaAl(SO_4)_2 \cdot 11H_2O$	*Mendozite group:* monoclinic; fibrous alums
Kalinite	$KAL(SO_4)_2 \cdot 11H_2O$	
Potash Alum	$KAL(SO_4)_2 \cdot 12H_2O$	*Alum group:* isometric alums; many artificial salts but only these are known in nature
Soda Alum	$NaAl(SO_4)_2 \cdot 12H_2O$	
Ammonia Alum	$(NH_4)Al(SO_4)_2 \cdot 12H_2O$	

$A(XO_4) \cdot xH_2O$ Type

Bassanite	$2CaSO_4 \cdot H_2O$	
Kieserite	$MgSO_4 \cdot H_2O$	*Kieserite group:* monoclinic and isostructural
Szomolnokite	$FeSO_4 \cdot H_2O$	
Szmikite	$MnSO_4 \cdot H_2O$	
Gypsum	$CaSO_4 \cdot 2H_2O$	
Ilesite	$MnSO_4 \cdot 4H_2O$ (?)	
Chalcanthite	$CuSO_4 \cdot 5H_2O$	*Chalcanthite group:* triclinic and isostructural with each other
Siderotil	$FeSO_4 \cdot 5H_2O$	
Pentahydrite	$MgSO_4 \cdot 5H_2O$	
Hexahydrite	$MgSO_4 \cdot 6H_2O$	*Hexahydrite group:* monoclinic; also isostructural with sulfates and selenates of Mg, Co, Ni, and Zn
Bianchite	$ZnSO_4 \cdot 6H_2O$	
Retgersite	$NiSO_4 \cdot 6H_2O$	
Melanterite	$FeSO_4 \cdot 7H_2O$	*Melanterite group:* monoclinic; isostructural and forming partial or complete series; all water-soluble
Pisanite	$(Fe,Cu)SO_4 \cdot 7H_2O$	
Kirovite	$(Fe,Mg)SO_4 \cdot 7H_2O$	
Boothite	$CuSO_4 \cdot 7H_2O$	
Bieberite	$CoSO_4 \cdot 7H_2O$	

SULFATES

Mallardite	$MnSO_4 \cdot 7H_2O$	*Melanterite group* (continued)
Zinc-melanterite	$(Zn,Cu)SO_4 \cdot 7H_2O$	
Epsomite	$MgSO_4 \cdot 7H_2O$	*Epsomite group:* orthorhombic, disphenoidal; partial or complete substitution of divalent cations
Goslarite	$ZnSO_4 \cdot 7H_2O$	
Morenosite	$NiSO_4 \cdot 7H_2O$	
Tauriscite	$FeSO_4 \cdot 7H_2O$	

$A_2B(XO_4)_4 \cdot xH_2O$ Type

Ransomite	$Cu(Fe,Al)_2(SO_4)_4 \cdot 7H_2O$	
Roemerite	$Fe^{II}Fe_2^{III}(SO_4)_4 \cdot 14H_2O$	
Pickeringite	$MgAl_2(SO_4)_4 \cdot 22H_2O$	
Halotrichite	$Fe^{II}Al_2(SO_4)_4 \cdot 22H_2O$	*Halotrichite group:* monoclinic; complete substitution in pickeringite and halotrichite and partial substitution in others
Apjohnite	$Mn^{II}Al_2(SO_4)_4 \cdot 22H_2O$	
Dietrichite	$ZnAl_2(SO_4)_4 \cdot 22H_2O$	
Bilinite	$Fe^{II}Fe_2^{III}(SO_4)_4 \cdot 22H_2O$	
Redingtonite	$(Fe^{II},Mn,Ni)(Cr,Al)_2(SO_4)_4 \cdot 22H_2O$ (?)	

$A_2(XO_4)_3 \cdot xH_2O$ Type

Lausenite	$Fe_2(SO_4)_3 \cdot 6H_2O$
Kornelite	$Fe_2(SO_4)_3 \cdot 7H_2O$
Coquimbite	$Fe_2(SO_4)_3 \cdot 9H_2O$
Paracoquimbite	$Fe_2(SO_4)_3 \cdot 9H_2O$
Quenstedtite	$Fe_2(SO_4)_3 \cdot 10H_2O$
Alunogen	$Al_2(SO_4)_3 \cdot 18H_2O$

ANHYDROUS SULFATES WITH HYDROXYL OR HALOGEN

$A_m(XO_4)_pZ_q$ with $m:p > 2:1$ Type

Brochantite	$Cu_4(SO_4)(OH)_6$
Antlerite	$Cu_3(SO_4)(OH)_4$
Caracolite	Na,Pb chloride-sulfate
Chlorothionite	$K_2Cu(SO_4)Cl_2$
Schairerite	$Na_3(SO_4)(F,Cl)$
Sulfohalite	$Na_6ClF(SO_4)_2$

$A_2(XO_4)Z_q$ Type

Lanarkite	$Pb_2(SO_4)O$	
Dolerophanite	$Cu_2(SO_4)O$	
Linarite	$PbCu(SO_4)(OH)_2$	
Alunite	$KAl_3(SO_4)_2(OH)_6$	
Natroalunite	$NaAl_3(SO_4)_2(OH)_6$	
Jarosite	$KFe_3(SO_4)_2(OH)_6$	*Alunite group:* hexagonal (rhombohedral) with general formula $AB(SO_4)_2(OH)_6$; mostly only partial substitution
Ammoniojarosite	$(NH_4)Fe_3(SO_4)_2(OH)_6$	
Natrojarosite	$NaFe_3(SO_4)_2(OH)_6$	
Argentojarosite	$AgFe_3(SO_4)_2(OH)_6$	
Carphosiderite	$(H_2O)Fe_3(SO_4)_2[(OH)_5 \cdot H_2O]$	
Beaverite	$Pb(Cu,Fe,Al)_3(SO_4)_2(OH)_6$	
Plumbojarosite	$PbFe_6(SO_4)_4(OH)_{12}$	

HYDROUS SULFATES WITH HYDROXYL OR HALOGEN

$A_mB_n(XO_4)_pZ_q \cdot xH_2O$ with $(m+n):p > 4:1$ Type

Connellite	$Cu_{19}(SO_4)(OH)_{32}Cl_4 \cdot 3H_2O$	*Connellite group:* hexagonal; substitution of (SO_4) and (NO_3)
Buttgenbachite	$Cu_{19}(NO_3)_2(OH)_{32}Cl_4 \cdot 3H_2O$	
Glaucocerinite	$Zn_{13}Al_8Cu_7(SO_4)_2(OH)_{60} \cdot 4H_2O$	
Mooreite	$(Mg,Mn,Zn)_8(SO_4)(OH)_{14} \cdot 4H_2O$	
Torreyite	$(Mg,Mn,Zn)_7(SO_4)(OH)_{12} \cdot 4H_2O$	

SULFATES

Spangolite	$Cu_6Al(SO_4)(OH)_{12}Cl \cdot 3H_2O$
Cyanotrichite	$Cu_4Al_2(SO_4)(OH)_{12} \cdot 2H_2O$
Zincaluminite	$Zn_3Al_3(SO_4)(OH)_{13} \cdot 2\frac{1}{2}H_2O$
Woodwardite	$Cu_4Al_2(SO_4)(OH)_{12} \cdot 2\text{–}4H_2O$ (?)
Chalcoalumite	$CuAl_4(SO_4)(OH)_{12} \cdot 3H_2O$
Uranopilite	$(UO_2)_6(SO_4)(OH)_{10} \cdot 12H_2O$
Meta-uranopilite	$(UO_2)_6(SO_4)(OH)_{10} \cdot 5H_2O$

$$A_4(XO_4)Z_q \cdot xH_2O \text{ Type}$$

Langite	$Cu_4(SO_4)(OH)_6 \cdot H_2O$ (?)
Felsobanyaite	$Al_4(SO_4)(OH)_{10} \cdot 5H_2O$ (?)
Basaluminite	$Al_4(SO_4)(OH)_{10} \cdot 5H_2O$
Hydrobasaluminite	$Al_4(SO_4)(OH)_{10} \cdot 36H_2O$ (?)
Glockerite	$Fe_4(SO_4)(OH)_{10} \cdot nH_2O$ (?)

$$A_mB_n(XO_4)_pZ_q \cdot xH_2O \text{ with } (m+n):p \text{ from 5:2 to 3:1 Type}$$

Kamarezite	$Cu_3(SO_4)(OH)_4 \cdot 6H_2O$ (?)
Ettringite	$Ca_6Al_2(SO_4)_3(OH)_{12} \cdot 26H_2O$
Devillite	$Cu_4Ca(SO_4)_2(OH)_6 \cdot 3H_2O$
Arnimite	$Cu_5(SO_4)_2(OH)_6 \cdot 3H_2O$ (?)
Serpierite	$(Zn,Cu,Ca)_5(SO_4)_2(OH)_6 \cdot 3H_2O$ (?)

$$(AB)_2(XO_4)Z_q \cdot xH_2O \text{ Type}$$

Kainite	$KMg(SO_4)Cl \cdot 3H_2O$
Ungemachite	$Na_9K_3Fe(SO_4)_6(OH)_3 \cdot 9H_2O$
Zippeite	$(UO_2)_2(SO_4)(OH)_2 \cdot 4H_2O$
Aluminite	$Al_2(SO_4)(OH)_4 \cdot 7H_2O$

$$A_3(XO_4)_2Z_q \cdot xH_2O \text{ Type}$$

Natrochalcite	$NaCu_2(SO_4)_2(OH) \cdot H_2O$
Metasideronatrite	$Na_4Fe_2(SO_4)_4(OH)_2 \cdot 3H_2O$
Sideronatrite	$Na_2Fe(SO_4)_2(OH) \cdot 3H_2O$
Johannite	$Cu(UO_2)_2(SO_4)_2(OH)_2 \cdot 6H_2O$
Vernadskite	$Cu_4(SO_4)_3(OH)_2 \cdot 4H_2O$

$$A(XO_4)Z_q \cdot xH_2O \text{ Type}$$

Metahohmannite	$Fe_2(SO_4)_2(OH)_2 \cdot 3H_2O$
Butlerite	$Fe(SO_4)(OH) \cdot 2H_2O$
Parabutlerite	$Fe(SO_4)(OH) \cdot 2H_2O$
Amarantite	$Fe(SO_4)(OH) \cdot 3H_2O$
Hohmannite	$Fe_2(SO_4)_2(OH)_2 \cdot 7H_2O$
Fibroferrite	$Fe(SO_4)(OH) \cdot 5H_2O$ (?)
Botryogen	$MgFe(SO_4)_2(OH) \cdot 7H_2O$
Guildite	$Cu_3Fe_4(SO_4)_7(OH)_4 \cdot 15H_2O$
Metavoltine	$(K,Na,Fe)_5Fe_3^{III}(SO_4)_6(OH)_2 \cdot 9H_2O$ (?)
Slavikite	$Na_2Fe_{10}(SO_4)_{13}(OH)_6 \cdot 63H_2O$ (?)
Copiapite	$Fe^{II}Fe_4^{III}(SO_4)_6(OH)_2 \cdot 20H_2O$
Magnesiocopiapite	$MgFe_4^{III}(SO_4)_6(OH)_2 \cdot 20H_2O$
Cuprocopiapite	$CuFe_4^{III}(SO_4)_6(OH)_2 \cdot 20H_2O$

Copiapite group: triclinic; complete substitution of Mg, Fe, and probably Cu and Zn

COMPOUND SULFATES

Miscellaneous

Hanksite	$Na_{22}K(SO_4)_9(CO_3)_2Cl$
Caledonite	$Cu_2Pb_5(SO_4)_3(CO_3)(OH)_6$
Wherryite	$Pb_4Cu(CO_3)(SO_4)_2(OH,Cl)_2O$ (?)
Burkeite	$Na_6(SO_4)_2(CO_3)$

LLOYD W. STAPLES

References

Dana, J. W., and Dana, E. S., 1951, "The System of Mineralogy," Seventh ed., Vol. II, 390, (revised by Palache, C., Berman, H., and Frondel, C.), New York, John Wiley & Sons.
Dean, G. B., and Brobst, D. A., 1955, "Annotated Bibliography and Index Map of Barite Deposits in United States," *U.S. Geol. Surv. Bull.*, **1019C**.
Douglas, C. V., and Goodman, N. R., 1957, "The deposition of gypsum and anhydrite," *Econ. Geol.*, **52**, 831.
Holser, W. T., and Kaplan, I. R., 1966, "Isotope geochemistry of sedimentary sulphates," *Chem. Geol.*, 1(2).
MacDonald, G. J. F., 1953, "Anhydrite-gypsum equilibrium relations," *Am. J. Sci.*, **251**, 884.

Cross-references: *Mineral Classes-Nonsilicates.* Vol. IVB: *Anhydrite; Barite; Gypsum.*

SULFATE REDUCTION—MICROBIAL

Many microbes obtain the sulfur needed for biosyntheses from the inorganic sulfate ion, and in doing so they reduce the sulfur atom to a low valency state. Green plants obtain their sulfur in a similar fashion. The major products of this kind of sulfate reduction are proteins incorporating sulfur-containing amino acids such as cysteine and methionine. The process is known as *assimilatory sulfate reduction,* because sulfur becomes assimilated into the cell structure. A specialized group of bacteria exists, known as "the sulfate-reducing bacteria," which conduct a more rapid and large-scale reduction of sulfate and which form equivalent quantities of sulfide as the end product. For these organisms sulfate is the terminal oxidizing agent for their respiration: like oxygen, which is the terminal oxidizing agent for air-breathing organisms, sulfate accepts the electrons generated during the metabolism of carbon compounds that leads to growth. Little, if any, of the sulfur enters into the structure of the organisms; this process is known as *dissimilatory* or *respiratory sulfate reduction.*

Assimilatory Sulfate Reduction

A wide variety of molds, yeasts, algae, and both aerobic and anaerobic bacteria can satisfy their growth requirements for sulfur from the sulfate ion. The biochemical pathway by which sulfate is reduced has largely been elucidated by the study of mutant strains of microbe that have been induced, by physical or chemical maltreatment, to require various intermediates (they have been called "parathiotrophic" mutants). Most of this work has involved mutants of two molds, *Neurospora* or *Aspergillus*, or of the *Escherichia coli;* more recently some enzymological work on the primary activation of sulfate by yeasts and bacteria has become available. There is no reason to believe that all organisms use similar pathways, but evidence is accumulating that most pathways have many intermediates in common. Research on the status of possible intermediates in this pathway is made difficult by the fact that many possible compounds are unstable in physiological conditions and change readily into one another, hence doubt exists about several steps in schemes that have been proposed. Among the well-established inorganic intermediates are the sulfite and sulfide ions. Thiosulfate, though frequently utilized by parathiotrophic mutants, is less well-established because it reacts readily with many components of biological systems. Ions such as tetrathionate and sulfoxylate are of more dubious status. The sulfur atom appears to enter organic combination permanently at the sulfide level of oxidation via the enzyme serine sulfhydrase, which catalyses the formation of cysteine from serine and sulfide; proposals, based on studies with molds, that it enters at the sulfite level (e.g., as cysteic, sulfinylpyruvic, or cysteine-S-sulfonic acids) seem less soundly based but cannot be excluded. Species may differ in their behavior. Subsequent to cysteine, a condensation with homoserine to yield cystathionine and thence homocysteine followed by methionine is belived to occur, thus supplying the major sulfur-containing amino acids in bacterial protein.

Initial activation of the sulfate ion takes place by formation of "active sulfate," which is a nucleotide (phosphoadenosine-phosphosulfate or PAPS). In bacteria, cysteine inhibits the synthesis of this material, thus preventing the organism from conducting unnecessary sulfate reduction if cysteine is already available.

Dissimilatory Sulfate Reduction

The sulfate-reducing bacteria proper are all obligate anaerobes (unable to grow in the presence of oxygen) yet their physiology is essentially oxidative and they show several physiological analogies to aerobic microbes.

Classification

Two major genera exist, distinguished fundamentally by whether or not they form spores. An early belief that members of the two groups are interconvertible by training is mistaken. The main taxonomic characteristics of the two groups follow.

Desulfotomaculum are spore-forming sulfate-reducing bacteria. Marine strains of this group have not been reported.

Desulfotomaculum nigrificans are multiflagellate, rod-shaped bacteria. They are thermophilic (grow best between 45 and 70°C) but can be "trained" to grow at lower temperatures. These bacteria are found in hot springs, hot water installations, and soils.

Desulfotomaculum orientis are multiflagellate, curved bacteria, and are nonthermophilic. So far they have been found only in soils from Singapore.

Desulfotomaculum ruminis are multiflagellate, rod-shaped bacteria that are nonthermophilic but otherwise are like *Dm. nigrificans*. They are found in the rumina of ruminant mammals.

Desulfovibrio are nonsporulating sulfate-reducing bacteria found in soils, marine environments, and brackish waters. Nearly all have a porphyroprotein pigment (desulfoviridin) which causes them to fluoresce red in ultraviolet light if they are treated with alkali.

Desulfovibrio desulfuricans and *vulgaris* are singly flagellated curved rods that are nonthermophilic. The two differ in certain biochemical details. They are found in a wide variety of environments, particularly in sulfur-bearing or polluted waters and muds. Marine varieties exist; a species with an obligate requirement for salt water has been recognized (*Desulfovibrio salexigens*).

Desulfovibrio gigas and *africanus* are sigmoid organisms resembling spirilla, with a single polar bundle of flagella.

The generic names of these bacteria have changed in the past fifty years: *Spirillum, Microspira, Vibrio,* and *Sporovibro* have been used in place of *Desulfovibrio; Dm nigrificans* was earlier *Clostridium nigrificans,* and *Dm. orientis* was first classified with the genus *Desulfovibrio.* Certain organisms which are mentioned in the literature (e.g., *Desulfovibrio rubentschikii, Desulforistella*) are not generally accepted.

Physiology

Sulfate-reducing bacteria metabolize relatively few carbon compounds for growth and sulfate reduction. These become oxidized to fatty acids; typical metabolic reactions are the formation of acetate from lactate (Eq. (1)) or of *n*-butyrate from *n*-butanol (Eq. (2)).

$$2CH_3CH(OH)COO^- + SO_4^{2-} \rightarrow 2CH_3COO^- + 2CO_2 + 2H_2O + S^{2-} \quad (1)$$

$$2C_3H_7CH_2OH + SO_4^{2-} \rightarrow 2C_3H_7COOH + 2H_2O + S^{2-} \quad (2)$$

Certain types (e.g., *Dm. nigrificans, D. desulfuricans*) can grow at the expense of a fermentative reaction which does not involve sulfate, analogous to the facultatively anaerobic bacteria. The reaction in Eq. 3 requires pyruvate to be the substrate and yields gaseous hydrogen.

$$CH_3-\overset{\overset{O}{\|}}{C}-COO^- + H_2O \rightarrow CH_3COO^- + CO_2 + H_2 \quad (3)$$

With certain strains of *Desulfovibrio*, choline or fumarate can be substrates for sulfate-free growth. The belief that *Desulfovibrio* could grow autotrophically, by reducing CO_2, as do green plants and certain chemotrophic bacteria, is probably mistaken. The belief was based on the observation that cultures provided with hydrogen, CO_2, sulfate, but no organic matter grew better than comparable cultures with nitrogen in place of hydrogen. Experiments with radioactive CO_2, however, showed that little of the cell carbon came from this source; the explanation seems to be that hydrogen acts as an energy source and promotes the more efficient utilization of organic impurities which are usually present in laboratory reagents and water. Isobutanol and oxamate can, for certain strains, perform a comparable function: they can be utilized for sulfate reduction but not for growth, hence they provide energy for the assimilation of other substrates but do not themselves contribute to the cell material.

Sulfate-reducing bacteria contain enzyme proteins of the iron-porphyrin class (cytochromes) which, before they were observed in the bacteria, were considered to be unique to aerobic organisms. The two genera have different types of cytochrome: the first group, the spore formers, have insoluble protoporphyrin derivatives, the second group has soluble haematoporphyrin compounds called cytochromes c_3. One of the latter has been purified, crystallized, and its structure determined. These cytochromes probably play a part in the generation of metabolic energy by the cells, and one important energy-rich compound in this process is the nucleotide adenosine phosphosulfate (APS). The organisms expend energy to convert sulfate to APS before reducing it; subsequently they make a net 'profit' in energy from the reduction of APS coupled to oxidation of carbon compounds. Interesting from the point of view of comparative biochemistry is the fact that a similar reaction in reverse provides energy from the oxidation of sulfur compounds by the aerobic sulfur bacteria *Thiobacillus,* a second group of microbes whose metabolism is based on large-scale turn-

SULFIDES (WITH SELENIDES, TELLURIDES, ARSENIDES, ANTIMONIDES)

over of the element sulfur. (Assimilatory sulfate reduction also passes through APS, but this is converted to PAPS, which has an extra phosphate group, before reduction.) The sulfite ion, or something closely related to it biochemically, is an intermediate after APS in the dissimilatory reduction of sulfate; other stages are not known. The pathways of carbon metabolism are probably fairly conventional except for the arrest at the fatty acid level. Flavoproteins, in addition to cytochromes, are involved in electron transport; so are nonhaem iron proteins of the ferredoxin and rubredoxin classes. A few strains of *D. desulfuricans* fix atmospheric nitrogen; most sulfate-reducing bacteria metabolize gaseous hydrogen readily, though not all use sulfate as the hydrogen acceptor.

Ecology and Economic Activities

Despite their anaerobic habit and peculiar metabolism, sulfate-reducing bacteria may be found in a remarkably wide variety of habitats, including most soils and waters as well as the intestinal tracts of insects and ruminant mammals. They are rarely encountered in airborne dust. Sewage is a plentiful source of these bacteria; sulfur springs, polluted waters, muds, and sands (notably estuarine sand) are usually rich in these bacteria, whose presence is indicated by a black coloration and a smell of sulfide. Even ordinary well-aerated soils usually contain some sulfate-reducing bacteria including, for no clear reason, the thermophile *Dm. nigrificans*. Deep-sea sediments, brackish waters, brines, salt pans, deep telluric waters associated with hot springs or oil deposits, supercooled natural brines in the Antarctic; these all contain representatives of the sulfate-reducing bacteria, whose collective tolerance of extremes of salinity and temperature is thus remarkable.

Massive bacterial formation of sulfide, in nature, leads to evolution of H_2S, to alkalinity in the soil or water, to precipitation of heavy metal ions as sulfides, and sometimes, in conjunction with other sulfur bacteria, to formation of acid zones where free H_2SO_4 is present. H_2S is toxic to most other forms of life; it also removes oxygen so that aerobic organisms that are immune to its toxicity may die of anoxia. Thus these bacteria create a anaerobic environment that favors the survival of themselves, of some other sulfur bacteria and a few resistant microbes. Such an ecosystem has been called a *sulfuretum*. Economic consequences of the establishment of whole or partial sulfureta can be considerable: the Texas, Louisiana, and Gulf sulfur deposits are of biogenic origin; anaerobic corrosion of iron pipes results, by an oblique mechanism, from bacterial sulfate reduction. Water pollution, some types of water calamity, plugging of injection systems in oil wells; spoilage of stored petroleum and emulsions, of canned and sealed foods; formation of natural soda deposits and deposition of metal sulfide ores usually involve, directly or indirectly, the sulfate-reducing bacteria. A review of the topic listed ten main economic fields in which these bacteria were involved. Biogenic formation of reduced sulfur may be detected readily since these organisms fractionate the natural sulfur isotopes by reducing the light isotope somewhat the more readily; in nonindustrial areas most of the atmospheric sulfur has been shown to be of biogenic origin.

JOHN POSTGATE

References

Jensen, M. L. (editor), 1963, "Biogeochemistry and Sulfur Isotopes," Symposium, Yale Univ., 193pp.
Postgate, J., 1960, *Progr. Ind. Microbiol.*, **2**, 49.
Postgate, J., 1965, *Bacteriol. Rev.*, **29**, 425.
Trudinger, P. A., 1969, *Advan, Microbial Physiol.*, **3**, 111.

Cross-references: *Biogeochemistry; Microorganisms; Organic Geochemistry; Sulfides; Sulfide Mineral Oxidation—Microbial; Sulfides in Sediments; Sulfur Isotope Fractionation; Sulfur Oxidation—Bacterial.*

SULFIDES (WITH SELENIDES, TELLURIDES, ARSENIDES, ANTIMONIDES)

These are arranged in decreasing A:X ratio where A is one of the metals and X is sulfur, selenium, tellurium, arsenic, or, rarely, antimony or bismuth.

Many of the most important ore minerals are sulfides and they are often, but not always, opaque or metallic. They frequently are the result of hydrothermal deposition.

Considerable work has been done on the structures of the sulfides, and satisfactory determinations have been made in most cases. The bonding may be simple ionic bonding, covalent, or even partial metallic bonding.

Secondary enrichment frequently is responsible for the formation of sulfide ore bodies, especially in the case of copper.

The selenide and telluride minerals are very rare as compared with the sulfides, and the selenides almost always occur in small amounts. (See table below.)

SULFIDES (WITH SELENIDES, TELLURIDES, ARSENIDES, ANTIMONIDES)

A_mX_n Type (with m:n > 3:1)

Tellurobismuthite	Bi_2Te_3	*Tetradymite group:* very similar minerals, usually distinguished from each other by x rays; all with perfect basal cleavage; hexagonal; gray metallic luster
Tetradymite	Bi_2Te_2S	
Grüenlingite	Bi_4TeS_3	
Nagyagite	$Pb_5Au(Te,Sb)_4S_{5-8}$	
Algodonite	Cu_6As	*Copper arsenide group*
Domeykite	Cu_3As	

A_3X Type

Dyscrasite	Ag_3Sb	*Antimonides*
Stibiopalladinite	Pd_3Sb	

A_2X Type

Argentite	Ag_2S	
Aguilarite	$Ag_4(Se,S)_2$	
Naumannite	Ag_2Se	*Argentite group:* sulfides, selenides, tellurides of silver and copper, usually isometric when formed but often inverting to orthorhombic crystals
Digenite	$Cu_{2-x}S$	
Berzelianite	Cu_2Se	
Eucairite	$CuAgSe$	
Hessite	Ag_2Te	
Petzite	Ag_3AuTe_2	
Chalcocite	Cu_2S	*Chalcocite group:* nonisometric species, formed at lower temperature
Stromeyerite	$AgCuS$	
Acanthite	Ag_2S	

A_3X_2 Type

Mauchierite	$Ni_{12}As_8$
Umangite	Cu_3Se_2
Bornite	Cu_5FeS_4
Oregonite	Ni_2FeAs_2

A_4X_3 Type

Dimorphite	As_4S_3
Rickardite	Cu_4Te_3

AX Type

Galena	PbS	*Galena group:* face-centered isometric structure; lead minerals are metallic, others are nonmetallic
Clausthalite	PbSe	
Altaite	PbTe	
Alabandite	MnS	
Oldhamite	CaS	
Sphalerite	ZnS	*Sphalerite group:* tetrahedral (isometric) structure is common, indicating tetrahedral coordination similar to diamond
Metacinnabar	$(Hg,Fe,Zn)S$	
Tiemannite	HgSe	
Coloradoite	HgTe	
Chalcopyrite	$Cu_2Fe_2S_4$	*Chalcopyrite group:* tetragonal species closely related to the sphalerite group in structure
Stannite	Cu_2FeSnS_4	
Wurtzite	ZnS	*Wurtzite group:* hexagonal minerals, with wurtzite a dimorph of sphalerite
Greenockite	CdS	
Voltzite	$Zn(As,S)$	
Pyrrhotite	$Fe_{1-x}S$	*Niccolite group:* hexagonal crystals, with a similar structure (nickel-arsenide structure)
Niccolite	NiAs	
Breithauptite	NiSb	

SULFIDES (WITH SELENIDES, TELLURIDES, ARSENIDES, ANTIMONIDES)

Millerite	NiS	
Pentlandite	$(Fe,Ni)_9S_8$	
Cubanite	$CuFe_2S_3$	
Sternbergite	$AgFe_2S_3$	
Covellite	CuS	*Covellite group:* the remainder of the minerals do not fit well into any group, but are placed together because they are AX-type compounds
Klockmannite	CuSe	
Cinnabar	HgS	
Realgar	AsS	
Cooperite	PtS	
Braggite	$(Pt,Pd,Ni)S$	
Empressite	AgTe	
Muthmannite	$(Ag,Au)Te$	

A_3X_4 Type

Linnaeite	Co_3S_4	
Siegenite	$(Co,Ni)_3S_4$	*Linnaeite group:* a group of relatively rare minerals which are isometric with the spinel structure; a series
Carrollite	Co_2CuS_4	
Violarite	Ni_2FeS_4	
Polydymite	Ni_3S_4	
Daubreelite	Cr_2FeS_4	
Smythite	Fe_3S_4	

A_2X_3 Type

Orpiment	As_2S_3	
Stibnite	Sb_2S_3	*Stibnite group:* these are salts of arsenic, antimony, and bismuth characterized by good cleavage and flexibility
Bismuthinite	Bi_2S_3	
Guanajuatite	Bi_2Se_3	
Kermesite	Sb_2S_2O	
Montbrayite	Au_2Te_3	

AX_2 Type

Pyrite	FeS_2	
Bravoite	$(Ni,Fe)S_2$	
Laurite	RuS_2	*Pyrite group:* hard, metallic minerals with pyrite (diploidal) structure
Sperrylite	$PtAs_2$	
Hauerite	MnS_2	
Penroseite	$(Ni,Cu,Pb)Se_2$	
Froodite	$PdBi_2$	
Michenite	$PdBi_2$	
Aurostibite	$AuSb_2$	
Geversite	$PtSb_2$	
Cobaltite	CoAsS	*Cobaltite group:* similar to pyrite in structure, with sulfur pairs replaced by As–S or Sb–S
Gersdorffite	NiAsS	
Ullmannite	NiSbS	
Loellingite	$FeAs_2$	*Loellingite group:* orthorhombic arsenides of iron, cobalt, and nickel
Safflorite	$(Co,Fe)As_2$	
Rammelsbergite	$NiAs_2$	
Pararammelsbergite	$NiAs_2$	
Marcasite	FeS_2	
Arsenopyrite	FeAsS	*Arsenopyrite group:* now believed to be monoclinic or triclinic species
Glaucodot	$(Co,Fe)AsS$	
Gudmundite	FeSbS	

SULFIDES (WITH SELENIDES, TELLURIDES, ARSENIDES, ANTIMONIDES)

Lautite	CuAsS	
Molybdenite	MoS_2	*Molybdenite group:* notable because of their perfect basal cleavage; hexagonal
Tungstenite	WS_2	
Krennerite	$AuTe_2$	*Krennerite group:* tellurides of gold, silver, and nickel
Calaverite	$AuTe_2$	
Sylvanite	$(Ag,Au)Te_2$	
Melonite	$NiTe_2$	

AX_3 Type

Skutterudite	$(Co,Ni)As_3$	*Skutterudite group:* this is a series of isometric arsenides of cobalt and nickel where isomorphous substitution of the metals forms a continuous series
Smaltite	$(Co,Ni)As_{3-x}$	
Nickel-Skutterudite	$(Ni,Co)As_3$	
Chloanthite	$(Ni,Co)As_{3-x}$	

LLOYD W. STAPLES

References

Berry, L. G., 1965, "Recent advances in sulfide mineralogy," *Am. Mineralogist*, **50**, 301.

Dana, J. W., and Dana, E. S., 1944, "The System of Mineralogy," Seventh ed., Vol. I, 155, (revised by Palache, C., Berman, H., and Frondel, C), New York, John Wiley & Sons.

Earley, J. W., 1950, "Description and synthesis of the selenide minerals," *Am. Mineralogist*, **35**, 337.

Hellner, E., 1958, "A structural scheme for sulfide minerals," *J. Geol.*, **66**, 503.

Kostov, I., 1964, "Paragenetic Analysis and Classification of the Sulphide Minerals," Chap. 17, in (Battey and Tomkeieff, editors) "Aspects of Theoretical Mineralogy in the U.S.S.R.," New York, MacMillan.

Kullerud, G., 1959, "Sulfide Systems as Geological Thermometers," in (Abelson, P. H., editor), "Researches in Geochemistry," New York, John Wiley & Sons, 301.

Kullerud, G., 1966, "Phase Relations in Sulfide-type Systems." in (Clark, S. P., editor), "Handbook of Physical Constants," *Geol. Soc. Am. Mem.*, **97**, 323–343.

Thompson, R. M., 1949, "The telluride minerals and their occurrence in Canada," *Am. Mineralogist*, **34**, 342.

Cross-references: *Bonding; Hydrothermal Solutions—Sulfide Transport; Mineralogy. Vol. IVB: Mineral and Ore Deposits.* See also *Individual Minerals*.

SULFIDE MINERAL OXIDATION—MICROBIAL

Metal sulfides in the presence of oxygen and water will oxidize in time to the corresponding metal sulfates and sulfuric acid. However, oxidation rates can be accelerated greatly by the activities of acidophilic iron-oxidizing bacteria.

Acidophilic Iron-Oxidizing Bacteria

Three species have been described: *Thiobacillus ferrooxidans, Ferrobacillus ferrooxidans,* and *Ferrobacillus sulfooxidans* (reviewed by Silverman and Ehrlich, 1964). Physiologically, these organisms are chemoautotrophs; that is, they require no preformed organic compounds for growth but synthesize their cell substance from carbon dioxide, water, and minerals using the energy derived from the oxidation of inorganic compounds. All three species are obligate aerobes (require molecular oxygen) and grow well at about 25°C in acidic inorganic media (pH 2.5–4.5) with ferrous iron as the sole energy source, but are differentiated according to their ability to use inorganic sulfur compounds as alternate energy sources. They are resistant to, or can adapt to, normally toxic concentrations of metal ions. For example, up to 15,000 ppm Cu, 40,000 ppm Fe, and 40,000 ppm Zn are tolerated.

The acidophilic iron-oxidizing bacteria are indigenous to aqueous habitats in which sulfide minerals and oxygen are present under acid conditions. Thus they have been found in the acid waters of metal sulfide deposits and sulfide-bearing coal deposits in Canada, America, Mexico, Scotland, Portugal, Spain, Germany, Denmark, Sweden, Russia, and Japan. They are often accompanied by the non-iron-oxidizing sulfur-oxidizing chemoautotrophs, *Thiobacillus thiooxidans* and *T. concretivorus,* and other acid-tolerant microorganisms.

Mechanisms of Bacterial Sulfide-Mineral Oxidation

The acidophilic iron-oxidizing bacteria markedly accelerate the rate of oxidation of many naturally occurring sulfide minerals (Table 1) as well as synthetic sulfides of copper, antimony, iron and titanium (reviewed by Kuznetsov et al., 1963; Silverman and Ehrlich,

TABLE 1. NATURALLY OCCURRING SULFIDE MINERALS OXIDIZED BY ACIDOPHILIC IRON-OXIDIZING BACTERIA

Simple Sulfides		Mixed Sulfides	
chalcocite	Cu_2S	arsenopyrite	$FeS_2 \cdot FeAs_2$
covellite	CuS	bornite	Cu_5FeS_4
galena	PbS	chalcopyrite	$CuFeS_2$
marcasite	FeS_2	enargite	$3Cu_2S \cdot As_2S_5$
millerite	NiS	tetrahedrite	$Cu_8Sb_2S_7$
molybdenite	MoS_2		
orpiment	As_2S_3		
pyrite	FeS_2		
sphalerite	ZnS		

1964). One means by which they accomplish this is indirect and occurs during the oxidation of metal sulfides (MS) by acid ferric sulfate solutions (Eqs. (1), (2), (3), and (4)).

$$MS + Fe_2(SO_4)_3 \rightarrow MSO_4 + 2FeSO_4 + S \quad (1)$$

$$S + 1\tfrac{1}{2}O_2 + H_2O \rightarrow H_2SO_4 \quad (2)$$

$$MS + Fe_2(SO_4)_3 + 1\tfrac{1}{2}O_2 + H_2O \rightarrow MSO_4 + 2FeSO_4 + H_2SO_4 \quad (3)$$

$$2FeSO_4 + H_2SO_4 + \tfrac{1}{2}O_2 \rightarrow Fe_2(SO_4)_3 + H_2O \quad (4)$$

Here the initial oxidative attack on a metal sulfide by ferric sulfate results in the production of metal sulfate (MSO_4) and elemental sulfur, and the reduction of ferric sulfate to the ferrous salt (Eq. (1)). *Thiobacillus thiooxidans, T. concretivorus,* and sulfur-oxidizing species of the iron-oxidizing bacteria oxidize the elemental sulfur to sulfuric acid (Eq. (2)) which aids in maintaining acid conditions. The net result (Eq. (3)) is the oxidation of metal sulfide to the corresponding metal sulfate and sulfuric acid. In the absence of bacteria, reaction (1) would soon cease upon exhaustion of available ferric sulfate. But the acidophilic iron-oxidizing bacteria rapidly oxidize ferrous sulfate to ferric sulfate (Eq. (4)), thereby regenerating the primary oxidant and permitting metal sulfide oxidation to continue.

When sulfide minerals contain iron, e.g., bornite, chalcopyrite, arsenopyrite, or are accompanied by pyrite or marcasite, an increase in available iron sulfate occurs. For example, the oxidation of pyrite by ferric sulfate results in a net increase in ferrous sulfate (Eq. (5)).

$$FeS_2 + Fe_2(SO_4)_3 \rightarrow 3FeSO_4 + 2S \quad (5)$$

Subsequent bacterial oxidation of ferrous sulfate provides more ferric sulfate than was originally present. Increased quantities of ferric sulfate in turn oxidize additional iron sulfide to produce even greater quantities of ferrous sulfate. Thus an ever-increasing cycle of oxidations and reductions ensues, resulting in the oxidation of large quantities of metal sulfides.

In addition to their role in regenerating ferric sulfate from spent ferrous sulfate solutions, the acidophilic iron-oxidizing bacteria are involved in direct oxidative attack on metal sulfide minerals independent of the action of ferric sulfate. This has been demonstrated with the iron-free minerals covellite, chalcocite, tetrahedrite, molybdenite, orpiment, and synthetic Cu_2S, ZnS, TiS_2, and Sb_2S_3. The exact mechanism remains unknown.

Geological Importance

The acidophilic iron-oxidizing bacteria play an important role in solubilizing the metals of sulfide ores, making them available for deposition elsewhere as secondary ore deposits (reviewed by Kuznetsov et al., 1963). Conditions favoring bacterial activity in ore deposits are moderate temperatures, the presence of iron sulfides as well as fissures, cracks and dislocations in the ore, good irrigation, and the absence of alkaline materials in associated rocks. Conversely, the absence of iron sulfides or cracks and fissures in massive sulfide deposits, poor irrigation, immobilization of water in permafrost regions, and the presence of alkaline rocks are unfavorable for bacterial activity.

Economic Importance

The oxidative capabilities of the acidophilic iron-oxidizing bacteria can be exploited for the benefit of man. The bacteria are employed in mining operations to regenerate spent ferric sulfate leach solutions and have opened the way for profitable leaching of waste rocks and sulfide ores of marginal grade (Malouf and Prater, 1961). These microorganisms may also find application in removing pyritic sulfur from coal (Silverman et al., 1963).

The activities of the acidophilic iron-oxidizing bacteria can also be detrimental. During the mining of coal, the associated pyritic material becomes exposed to the action of oxygen and water. Under these conditions pyrite oxidizes to sulfuric acid and ferric sulfate. These corrosive products eventually enter streams and rivers and constitute a serious pollution problem. The bacteria cause a manyfold increase in the rate of acid production and contribute significantly to the magnitude of the problem (Leathen et al., 1953).

MELVIN P. SILVERMAN

References

*Kuznetsov, S. I., Ivanov, M. V., and Lyalikova, N. N., 1963, "Introduction to Geological Microbiology," New York, McGraw-Hill Book Co., 252pp.

Leathen, W. W., Braley, S. A., and McIntyre, L. D., 1953, "The role of bacteria in the formation of acid from certain sulfuritic constituents associated with bituminous coal. II. Ferrous iron-oxidizing bacteria," *Appl. Microbiol.*, **1**, 65–68.

Malouf, E. E., and Prater, J. D., 1961, "Role of bacteria in the alteration of sulfide minerals," *J. Metals*, **13**, 353–356.

*Silverman, M. P., and Ehrlich, H. L., 1964, "Microbial Formation and Degradation of Minerals," in "Advances in Applied Microbiology," Vol. 6, 153–206.

Silverman, M. P., Rogoff, M. H., and Wender, I., 1963, "Removal of pyritic sulphur from coal by bacterial action," *Fuel (London)*, **42**, 113–124.

*Additional bibliographic references may be found in this work.

Cross-references: *Microorganisms; Organic Geochemistry; Oxidation and Reduction; Pollution; Sulfate Reduction—Microbial; Sulfides in Sediments; Sulfur Oxidation—Bacterial. Vol. IVB: Individual Minerals. Vol. VI: Leaching.*

SULFIDES IN SEDIMENTS

Sulfides are widespread in sedimentary rocks and these undoubtedly contain the major part of the sulfide of the earth's crust. The iron sulfide pyrite (cubic FeS_2) is by far the most common and may occur in all concentrations from sparse disseminations to layers of almost pure pyrite. Pyrrhotite, sphalerite, chalcopyrite, and galena are not so abundant or widespread but are sometimes locally prominent and may attain quite high concentrations over restricted areas. All of these sulfides, with various minor associates, are found in sedimentary rocks ranging from Recent unmetamorphosed materials right through to very old and often highly metamorphosed sedimentary sequences.

The most frequent host rocks are black shales containing organic carbon or its derivative graphite, though pyrite is often present in silty layers and carbonates are frequent associates. Copper sulfides are often associated with dolomitic carbonate, and the pyrite of iron formations with sideritic material.

Sulfide in sedimentary and particularly metasedimentary rocks is not necessarily of sedimentary origin. Disseminated iron sulfide in older sediments, particularly when of disseminated form, is generally assumed to be of sedimentary origin by analogy with its form and mode of occurrence in modern ones, and there seems little question that the high concentrations of pyrite in "bedded iron formations" are sedimentary also. However, where it attains high concentrations in old rocks—particularly where these have been involved in orogenic processes and where sulfides of other metals occur—later deposition by replacement is frequently suspected.

The principal requirements for the formation and accumulation of sulfides during sedimentation are (a) a source of sulfur for the production of the sulphide ion (H_2S), (b) readily mobilizeable metals, (c) conditions under which the relevant metallic sulfides are stable, and (d) bottom conditions sufficiently quiet and stable to allow the precipitating sulfide to accumulate.

The chief source of sulfur appears to be sulfate derived from the body of the oceans or, locally, more or less directly from volcanic sources. Sulfites, polythionates, and other sulfur compounds may also be important in the latter. Reduction to H_2S seems to be almost exclusively bacterial, the anaerobic bacteria *Desulphovibrio* being the chief agent concerned. The chemical and physical conditions of activity of this organism have been studied in some detail by Baas Becking (1961) and his associates, and many of its characteristics are now known in quite a quantitative way. It cannot occur at a pH under 4.2, but it has been found at pH's well over 10 (seawater about 8.0). The highest Eh at which it may develop is +110 millivolts, and it may generate potentials as low as −500 millivolts. It is apparently highly thermotolerant but its development is inhibited by brines rich in magnesium or selenates. They are dependent on a sulfate source as shown by their sparseness or absence from bogs of low sulfate concentration and subsequent development in these when sulfate is added. Although generally regarded as heterotrophic, some strains are able to persist and develop in an inorganic medium if hydrogen is provided. Baas Becking and Moore suggest that such an autotrophic existence may be more common than has been assumed, though observations of the natural environment indicate that the great bulk of the genus is heterotrophic and depends upon decaying organic matter as an energy source.

Most experimental work has been carried out using an organic medium such as lactate or acetate to simulate the natural organic matter of the sea or lake floor. Utilization of this may be represented by the generalized equation:

$$2CH_2O + SO_4^{2-} \rightarrow 2H_2O + 2CO_2 + S^{2-}$$

two atoms of carbon being oxidized for each atom of sulfur reduced (Kaplan et al., 1963). The bacteria reproduce prolifically under suitable conditions and 10^3 to 10^5 sulfur reducers per gram of sediment have been found under natural reducing conditions. The limits of the "natural" environment as defined by variation in hydrogen ion concentration and oxidation-reduction potential (i.e., total spread of many measurements in sea, lake, and river waters and sediments, bogs, soils, geothermal waters, and other milieu) and the characteristics of these in which sulfate reducers have been found (Baas Becking et al., 1960) are shown in Fig. 1.

As the sulfate reducers require an organic substrate their maximum development is expectably confined to areas of sedimentation in which large quantities of organic matter accumulate and decay. Two principal environments are involved: (a) bottoms of substantially enclosed and hence stagnant bodies of water, such as the Black Sea and many fiords, where euxinic conditions develop and all organic matter reaching the bottom is shielded from oxygen and decays, and (b) fairly shallow, open, areas where, due to a high rate of photosynthesis, a fairly short water column, and rapid burial, large quantities of organic matter reach the sea floor and are buried before significant oxidation takes place.

In the first instance, sulfate reducers are abundant on the floor itself, H_2S is contributed to the bottom waters quite freely and may remain unoxidized to considerable heights (as in the Black Sea where free H_2S has been found to persist from the sea floor at 6000 feet to within about 600 feet of the surface) and metallic sulfide precipitates directly onto the sea floor (Goldschmidt, 1954). In the second, the production of sulfide is substantially an authigenic process.

The reduction of sulfate to other sulfur compounds—and particularly iron sulfide—by authigenic processes in near-shore marine muds has recently been investigated by Kaplan and others off the coast of southern California (Kaplan et al., 1963). The areas concerned were the Santa Barbara, Santa Monica, and Santa Catalina Basins and the San Diego Trough in all of which sulfide has long been known to be forming. Sulfate contents of the overlying water are about 2.64‰ (which compares with the mean value of about 2.65‰ for the open ocean) and there appear to be no notable differences above and below the sill in each case. From various studies on nutrient regeneration, oxygen content, sediment stratification and sulfate sulfur isotope fractionation it is concluded that the time of water renewal in the deeper parts was on the order of two to twenty years either by regular short period exchange or perhaps by continuous renewal and although oxygen contents as low as 0.04 ml/l are found, complete stagnation with the evolution of H_2S in the water has not been observed. The sediments have been cored to about four meters in each case and Ep, pH, sulfate, organic sulfur, sulfide, and other constituents determined at half-meter intervals to indicate any systematic changes with depth below the mud-water interface. Some sediments are highly reduced, containing H_2S from the surface down, whereas others are mildy reducing to slightly oxidized. Sediments of Santa Barbara Basin are oxidized as far down as two meters, below which they are reducing. Reduction of sulfate therefore occurs in all cases, position of commencement varying from the mud-water interface itself down to some two meters below this. The concentration of sulfate in the interstitial water of the sediments decreases markedly with depth in most cases indicating that it is being reduced. Where reduced conditions begin at the mud-water interface the decrease in sulfate starts here and is essentially complete at 1.0 to 1.5 meters. However, sulfate has in no case been found to

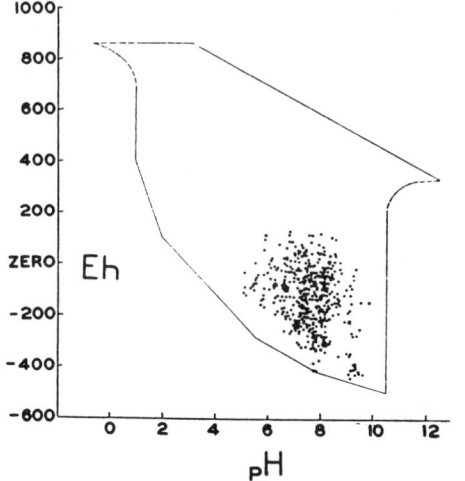

FIG. 1. Limits of the "natural environment" and of bacterial sulfate reduction (from Backing et al., 1960).

completely disappear, due possibly to the cessation of sulfate reduction following exhaustion of utilizable organic matter; (the investigators concerned note, on the other hand, that this could be an analytical effect resulting from the oxidation of a small amount of pyrite sulfur to sulfate during manipulation). The product of sulfate reduction is free sulfide which transforms in the sediment to elemental sulfur, organic sulfur, and iron sulfide, the latter being the dominant form of sulfur and comprising mainly pyrite, but also metastable iron sulfides.

This sulfide produced in euxinic bottom waters and/or during the diagenesis of putrid organic sediments is then available for the precipitation of any available metallic ions capable of forming sulfides that are stable under the prevailing conditions.

There is now a substantial body of information on the stabilities of sulfides and their associates in low-temperature aqueous environments, due mainly to Garrels and his co-workers, and the stage has been reached where it is possible to understand at least a number of sedimentary (chemical) mineral assemblages on the basis of Eh, pH, and temperature, and of CO_2 and sulfide concentrations. On the assumption of equilibrium conditions, it is possible to define the Eh-pH conditions under which various associations of

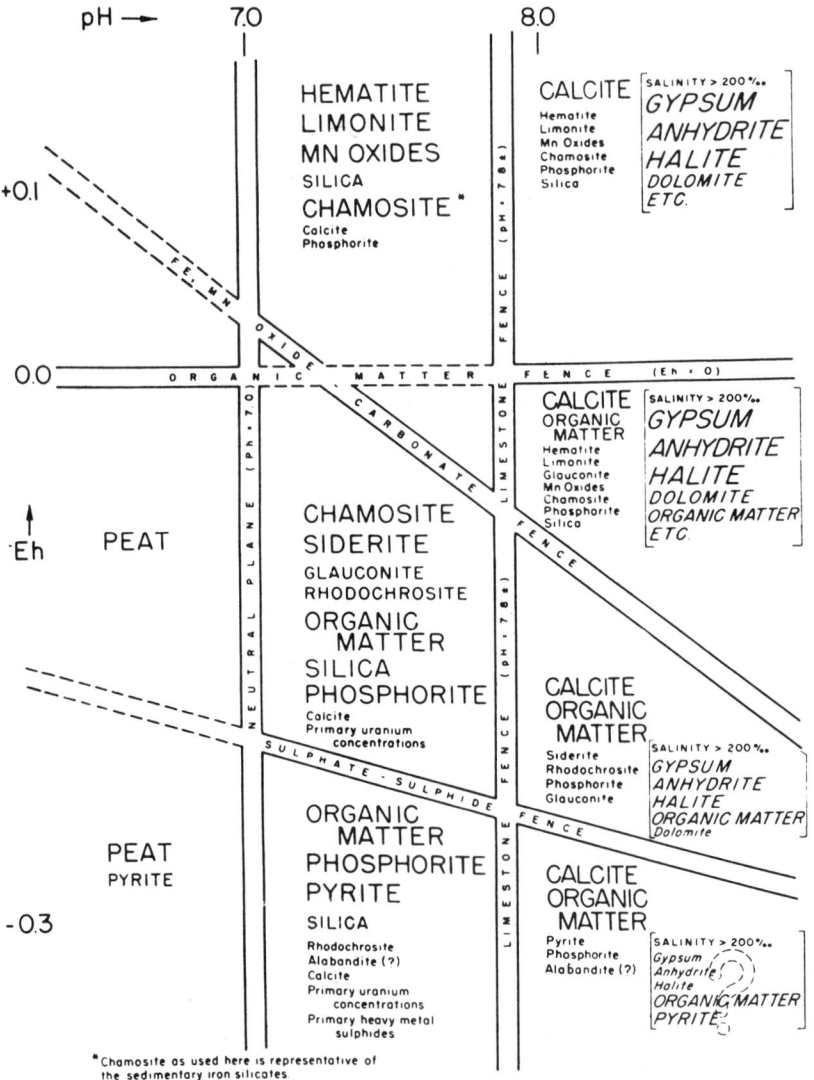

FIG. 2. Sedimentary chemical associations and their relations to environmental limitations imposed by selected Eh and pH values (from Garrels, 1960).

chemical sediments may be expected to be stable (Garrels, 1960), and to show that a number of well-known environmental groupings such as those of sedimentary iron formations and phosphorite beds can be accounted for in good approximation by the respective Eh-pH stability fields of the substance concerned (Fig. 2). A large number of diagrams are now available showing the stability relations of numerous substances frequently formed as chemical sediments, under conditions approximating various natural sedimentary environments. In Fig. 3 are given, in simplified form, the Eh-pH limits of Baas Becking's "total natural milieu" already referred to, the limits of bacterial sulfate reduction, the approximate normal limits of such reduction, and generalized sulfate-sulfide stability boundaries for the more common sulfide-forming metals. It will be seen that all of the latter form sulfides that may be expected to be stable under natural conditions of sulfate reduction.

Sulfides of all of the principal "base" metals have now been produced by bacterial sulfate reduction in the laboratory, the principal modern worker in this field again being Baas Becking. Ferrous and ferric sulfides have been produced repeatedly in this way, though the precise constitution and mechanisms of formation of the different minerals have not been determined finally. It has been suggested (Baas Becking, 1956) that the first product in the interaction of iron oxide, hydroxide, or carbonate with H_2S in a marine environment is hydrotroilite, essentially $Fe(OH)(SH)$. Further reaction with H_2S gives $Fe(SH)_2$—disulphydryl iron—which in turn may react to give FeS_2 as follows:

$$Fe(SH)_2 + S \rightarrow FeS_2 + H_2S$$

$$Fe(SH)_2 + \tfrac{1}{2}O_2 \rightarrow FeS_2 + H_2O$$

$$Fe(SH)_2 + 2Fe(OH)_3 \rightarrow FeS_2 + 2Fe(OH)_2 + 2H_2O$$

Other compounds identified are $Fe_3S_2O_2$ (sulphomagnetite), possibly Fe_2S_3 (though this has been reported to break down to FeS_2 and FeS at temperatures over 20°C) and Fe_3S_4 (greigite). Undoubted pyrrhotite $(Fe_{1-x}S)$ does not seem to have been produced experimentally. The precise nature of the FeS_2 as formed is also somewhat conjectural. At low temperatures, marcasite is the product of acid conditions, pyrite of alkaline ones, so that marcasite is normally restricted to fresh-water sediments. Most FeS_2 collected from modern sediments is pyrite though the possibility of the occurrence of the slightly more iron-rich $(Fe_5S_7?)$ form, melnikovite, is often referred to. Low-temperature experiments (Lepp, 1957) have yielded sooty black iron sulfide with a composition corresponding to $13FeS:87FeS_2$, giving a pyrite x-ray diffraction pattern but showing microscopic pyrrhotite after heating. However "melnikovite" yielding quite a distinctive x-ray pattern has been produced by adding NH_4HS to dilute $FeSO$ (Lepp, 1957); this is black, magnetic, soluble in dilute HCl, and appears to yield pyrite and $Fe_2O_3 \cdot H_2O$ when slightly oxidized. Thus there is apparently quite a complex of iron sulfides, usually formed as short-lived intermediate products in the development of pyrite, and whose precise nature is dictated by delicate variations in the chemistry of the precipitating medium.

Investigations of the bacterial precipitation of lead, zinc, copper, and other metallic sulfides are fewer and more recent (Baas Becking and Moore, 1961), and have been provoked by the modern resurgence of the sedimentary theory of stratiform ore formation. A possible major problem here is the toxicity of some of the non-ferrous metals—particularly copper—to the bacteria producing the H_2S. This caused earlier investigations to be designed around metallic source compounds of very low solubility—in the case of copper, such as malachite $(CuCO_3 \cdot Cu(OH)_2)$, chrysocolla $(CuO \cdot SiO_2 \cdot H_2O)$, and tenorite (CuO). Using malachite and *Desulfovibrio* in both lactate and acetate cultures; metallic copper has been pro-

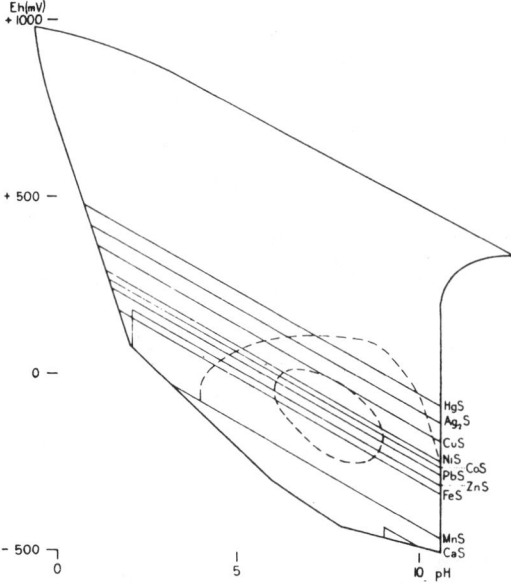

Fig. 3. Eh–pH limits of the "natural aqueous milieu" (outer boundary), bacterial sulfate reduction (dashed line), bacterial sulfate reduction as normally encountered (area enclosed by dashed line), and generalized sulfate-sulfide equilibrium lines (from Becking and Moore, 1961).

duced within three days of incubation at 30°C, and a black precipitate, identified by x-ray powder pattern as covellite (CuS), after five to seven days; galena from both the carbonate and basic carbonate after four days at 30°C in media containing hydrogen, lactate or acetate; and sphalerite, argentite, and digenite have similarly all been produced by bacterial reduction, and all confirmed by x-ray diffraction analysis. There is therefore little question that these sulfides may be precipitated by bacteria under natural conditions provided the source material is not so soluble as to yield toxic quantities of the metallic ions, or, rather, that solution does not proceed faster than the microbes are able to dispose of the ions by reducing them to sulfide.

Recently, this problem of copper toxicity has been investigated quantitatively by following the effect of increasing concentrations of copper sulfate on sulfate reduction by *Desulfovibrio* (Temple and LeRoux, 1964). As this genus can use molecular hydrogen to reduce sulfate to sulfide according to the reaction

$$SO_4{}^{2-} + 4H_2 \rightarrow S^{2-} + 4H_2O$$

the course of sulfate reduction using such a hydrogen source can be followed manometrically without disturbing the system.

Similar cultures, to all of which was added a trace element solution containing, among a group of compounds, 2.0 mg% of $CuSO_4 \cdot 5H_2O$, were placed in five flasks. To these were added different quantities of $CuSO_4 \cdot 5H_2O$ over a three-week period, and the activity of the bacteria followed by resulting variation in the rate of hydrogen uptake as indicated in Fig. 4. It was found that complete inhibition of sulfate reduction by *Desulfovibrio* required the addition of enough copper sulfate to give between 0.249 and 0.289% solution. Temporary inhibition without destroying the bacteria resulted from copper concentrations of 0.129–0.249% and lesser concentrations had no effect other than that of supplying additional sulfate and of precipitating copper sulfide. It therefore appears that a sulfate-reducing organism protects itself from metal ion toxicity by precipitating—and hence rendering innocuous—such metals as insoluble sulfide. Presumably this principle would apply to any biological sulfate-reducing agent and to any metal that formed an insoluble sulfide; whether or not an influx of salt is toxic depends on the balance between rate of supply and rate of production of sulfide from sulfate. Thus sulfate-reducing bacterial cultures are unaffected by copper unless the amount added exceeds that necessary to precipitate all of the sulfide present.

Sulfides in Modern Sediments

The occurrence of sulfide in present-day sediments has been recognized for a very long time, particularly in euxinic environments. However, apart from very restricted areas of volcanic contribution, such as that of the island of Vulcano in Italy where sulfides of iron, lead, zinc, and copper, derived from the heavy metal content of the geothermal matters, are being precipitated in shallow waters round the shore, and in the geothermal basins of the Red Sea, only iron sulfides have been observed. Of these, pyrite is by far the dominant species in marine sediments, whereas marcasite may be abundant in acid fresh-water lakes. So far there have been no high concentrations found, the maximum being on the order of 1% of the sediment concerned. The highest pyrite sulfur contents found in the Southern Californian corings were about 0.9%, giving a pyrite content of about 1.5% of the dry weight of the containing sediment (Kaplan et al., 1963). Probably the most spectacular marine sulfide occurrence so far described is that of the Kuriles (Ostroumov and Shilov, 1956).

Sulfides in Older Sediments

Occurrences in older rocks cover a very much wider range of type. Most are in marine sediments, though there are important occurrences in lake, stream, and alluvial fan accumulations. Those in marine sediments show enormous variation in concentration and constitution; the

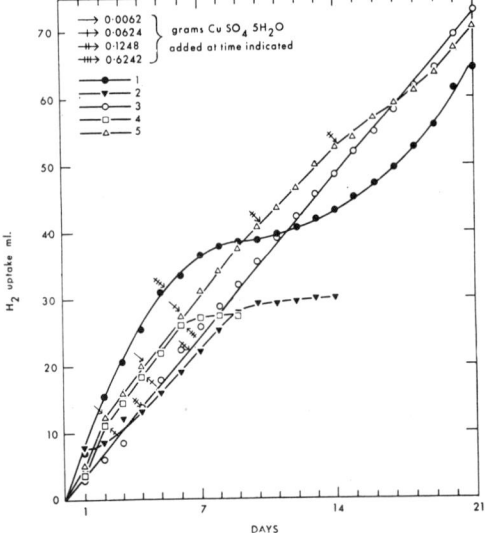

FIG. 4. Relation between H_2 uptake and $CuSO_4$ addition in the experiments of Temple and LeRoux (from Temple and LeRoux, 1964).

former ranges from widespread sparse disseminations through deposits such as those of the Roan, Sullivan, and Lake Superior pyritic iron formations to the massive deposits at Broken Hill (New South Wales) and Balmat (New York). Many of the disseminated occurrences are regarded with little doubt as true sedimentary products, but as sulfide content increases opinion diverges, and, apart from indubitable iron formations, high concentrations have generally been regarded as epigenetic, derived from igneous intrusions and deposited by selective replacement. At what particular sulfide content an occurrence is more, or less, likely to be syn- or epigenetic has not been stated, and there is now increasing emphasis on the possibility of the development of high concentrations during sedimentation. In order of approximately increasing concentration of sulfide, well-known examples of non-ferrous metal bearing stratiform sulfide occurrences in older sedimentary rocks are (a) the Permian Marl Slate of N.E. England, (b) the related Permian Mansfeld Kupferschiefer of Germany, (c) the Roan copper deposits of Zambia and Katanga, (d) the Precambrian Mount Isa (Queensland) and Sullivan (British Columbia) lead-zinc deposits, and (e) the Precambrian Broken Hill (New South Wales) and Balmat (New York) lead-zinc and zinc deposits.

The major mineralogy of most of the deposits is fairly simple. The sulfide facies of bedded iron formations appear to be simple low Eh-low pH analogs of the oxide and carbonate portions of a given iron formation sequence and their sulfide fraction is almost entirely pyrite. Any other sulfides occur as no more than traces. Mount Isa, Sullivan, and innumerable other lead-zinc and copper ores of their class are also highly pyritic and may contain substantial quantities of pyrrhotite. In fact the "ore bodies" are often (Mount Isa) no more than comparatively small parts of very large volumes of pyritic sediments in which the valuable metals happen to reach local concentrations that are sufficiently high for profitable working. Others such as those of the Roan (i.e., away from the pyritic zones) and cupriferous red-beds are composed of chalcopyrite, bornite, covellite, and chalcocite with much less iron sulfide, and ores such as Broken Hill and Balmat contain little discrete (i.e., as distinct from that contained in sphalerite and other base metal minerals) iron sulfide.

Calculation of the atomic proportions of copper, lead, zinc in a number of sulfide occurrences (of quite widely different sulfide content) yields the field shown in Fig. 5 (Stanton, 1958). Clearly there may be wide deviation from average crustal proportions, though the field is not unlike that found for a range of igneous rocks, indicating the possibility of some kind of connection.

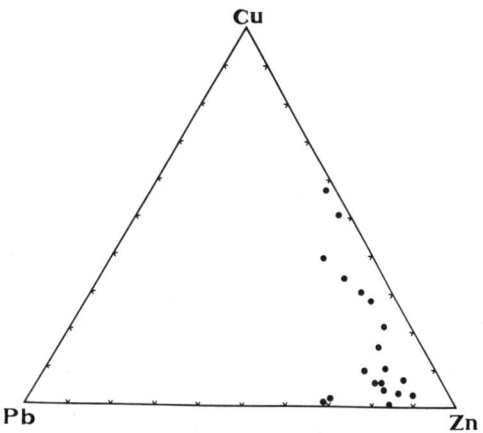

FIG. 5. Atomic proportions of copper, zinc, and lead in some stratiform sulfide concentrations in old sediments (from Stanton, 1958).

Isotopic Constitution of the Lead and Sulfur

Mass spectrometric measurements of both lead and sulfur have now been carried out on large numbers of samples of sulfides found in sedimentary and metasedimentary rocks. As discussed elsewhere, various attempts have been made to use the isotopic data as an indication of origin and mode of deposition of those occurrences whose histories are not clear and some valuable information has been obtained.

As discussed in the section on lead isotopes, any naturally occurring sulfide lead consists of primaeval (non-radiogenic) Pb^{204}, together with various amounts of Pb^{206}, Pb^{207}, and Pb^{208}, derived by radioactive disintegration from U^{238}, U^{235}, and Th^{232}, respectively. Using certain assumptions and various values for constants, various "models" for terrestrial lead production have been proposed, and these may be used to determine the "model" age of a given lead whose isotopic constitution is known. A large number of galenas from sedimentary rocks have now been measured but the isotopic constitution of their leads falls into no simple pattern. Many are demonstrably younger than the enclosing sediments and some even have an apparent negative age (i.e., are younger, according to their Pb isotopes, than today). A few are older than the enclosing rocks. Some can be accounted for mathematically on the basis of a "two-stage" history, i.e., generation of lead in a source essentially uniform in U–Th–Pb ratios (such as, perhaps, the mantle or deeper regions of the earth), extraction from this, and deposition in the crust without mixing with any other lead already in the latter. Others require a "three-stage" history for mathematical solution, e.g., generation in, and initial extraction from, such as the above source, mixing with other, substantially radiogenic lead already in the crust, and then deposition. Some of these leads may, in fact, have had an even more complex history than this. It has been recognized fairly recently (Stanton and Russell, 1959) that all of a number of stratiform galena concentrations in sediments yield leads that approximate closely to a simple two-stage "growth curve," and are isotopically indistinguishable within any one deposit (this latter characteristic is demonstrated quite spectacularly by the Broken Hill deposit, the isotopic condition of whose lead is uniform over a strike length of almost four miles). This is in sharp contrast to many vein leads which may show great variability within a deposit and which generally require a three- (or more) stage history to explain the ratios of the isotopes. It is also different from the leads of other sulfide occurrences in sediments (e.g., of limestone lead-zinc deposits) and those of recent sediments and of the present oceans, all of which appear to have undergone crustal mixing and to have histories more complex than the simple "two-stage" one. This apparent marked simplicity of the leads of the Broken Hill type of deposit has been interpreted as indicating their possible direct derivation from the mantle via volcanism, involving minor to almost infinitesimal contamination of crustal materials, with emission and rapid precipitation by bacterial sulfate reduction on the sea floor. Certainly many of the deposits concerned occur in highly carbonaceous layers within volcanic sequences.

As is pointed out in the section on sulfur isotopes, sulfides in sediments show a greater range in S^{32}/S^{34} ratios than any other class of naturally occurring sulfides so far subjected to mass spectrometric analysis. The present known spread of this ratio for sulfides in sediments is from $-45‰$ to $+40‰$ or some 8.5%. This has presumably been induced by fractionation during bacterial reduction, such fractionation tending to concentrate the lighter isotope in the sulfide as opposed to the sulfate, and varying substantially with variation in the conditions under which the bacteria operated. The resulting sulfide sulfur is thus usually "lighter" than the average for the crust, but very variable.

This is shown particularly by disseminated sulfides in modern and older shaly sediments and by many of those in limestones. However, it is interesting to note that the class of sulfide concentration found to yield "two-stage" leads, has so far always been found to show very uniform S^{32}/S^{34} ratios, invariably *heavier* than crustal sulfur. The relevant values appear to cluster about that found for the total sulfur (ie., sulfate, sulfite, polythionate, native sulfur, and sulfide) of several volcanic emanations, which suggests, again, that they may have been derived from submarine volcanism and deposited rapidly and under fairly constant conditions, without significant seawater contamination, on the nearby sea floor.

R. L. STANTON

References

Baas Becking, L. G. M., 1956, "Biological processes in the estuarine environment. N The state of the iron in the estuarine mud iron sulphides," *Koninkl. Ned. Akad. Wetenschap.* **59**, 181–189.

Baas Becking, L. G. M., Kaplan, I. R., and Moore, D., 1960, "Limits of the natural environment in terms of pH and oxidation-reduction potentials," *J. Geol.*, **68**, 243–284.

Baas Becking, L. G. M., and Moore, D., 1961, "Biogenic sulfides," *Econ. Geol.*, **56**, 259–272.

Garrels, R. M., 1960, "Mineral Equilibria at Low Temperature and Pressure," New York, Harper & Brothers, 254pp.

Germanov, A. I., 1963, "Role of organic substances in the formation of hydrothermal sulfide deposits," *Intern. Geol. Rev.* **5**, 379–394.

Goldschmidt, V. M., 1954, "Geochemistry," Oxford, Clarendon Press, 730pp.

Kaplan, I. R., Emery, K. O., and Rittenberg, S. C., 1963, "The distribution and isotopic abundance of sulphur in recent marine sediments off southern California," *Geochim. Cosmochim. Acta*, **27**, 297–331.

Lepp, H., 1957, "The synthesis and probable geologic significance of melnikovite," *Econ. Geol.*, **52**, 528–535.

Ostroumov, E. A., and Shilov, V. M., 1956, "Distribution of ferrous sulphide and hydrogen sulphide in deposits of the northwestern part of the Pacific Ocean," *Geokhimya*, **7**, 25–38.

Stanton, R. L., 1958, "Abundances of copper, zinc and lead in some sulfide deposits," *J. Geol.*, **66**, 484–502.

Stanton, R. L., and Russell, R. D., 1959, "Anomalous leads and the emplacement of lead sulfide ores," *Econ. Geol.*, **54**, 588–607.

Temple, K. L., and LeRoux, N. W., 1964, "Syngenesis of sulfide ores: sulfate reducing bacteria and copper toxicity," *Econ. Geol.*, **59**, 271–278.

Cross-references: *Authigenesis of Minerals; Earth's Crust Geochemistry; Interstitial Waters in Sediments; Mass Spectrometry; Oxidation and Reduction; pH-Eh; Seawater, Chemistry; Sulfates; Sulfate Reduction—Microbial; Sulfides; Sulfide Mineral Oxidation—Microbial; Sulfur; Sulfur Isotope Fractionation; Sulfur Oxidation—Bac-*

terial; Syngenesis; X-Ray Diffraction Analysis. Vol. I: Black Sea. Vol. IVB: Mineral and Ore Deposits; see also Individual Minerals. Vol. VI: Diagenesis; Euxinic Basins; Sedimentation.

SULFIDE SOLUTIONS—See HYDROTHERMAL SOLUTIONS—SULFIDE TRANSPORT

SULFOSALTS

Sulfosalts are minerals of the $A_m B_n X_p$ type, where A is Ag, Cu, Pb, or Sn; B is Sb, As, Bi, or Sn; and X is S.

In the sulfosalts, a semimetal (Sb, As, Bi) acts like a metal in the structure. Although the structures of the sulfides have now been worked out in most cases, the sulfosalt structures are less well understood. In general, the sulfur in sulfides occurs either singly or in pairs, whereas in the sulfosalts the sulfur atoms are arranged with antimony, arsenic, or bismuth in sheets or chains. In the sulfosalts, lead and silver are the commonest metals; copper and iron are next in importance.

The ruby silvers, pyrargyrite and proustite, and enargite and tetrahedrite are the best known of the sulfosalts. In some classifications the sulfosalt minerals are included with the sulfides, as complex sulfides. (See table below.)

$A_m B_n X_p$ Type $(m + n : p > 4 : 3)$

Polybasite	$(Ag,Cu)_{16}Sb_2S_{11}$	Polybasite group: monoclinic, black metallic silver salts
Pearceite	$(Ag,Cu)_{16}As_2S_{11}$	
Argyrodite	Ag_8GeS_6	Argyrodite group: this is an isomorphous series with substitution of Ge by Sn; isometric
Canfieldite	Ag_8SnS_6	
Stephanite	Ag_5SbS_4	

A_3BX_3 Type

Pyrargyrite	Ag_3SbS_3	Ruby silver group: hexagonal, red silver minerals, usually in low temperature zone of silver deposits
Proustite	Ag_3AsS_3	
Pyrostilpnite	Ag_3SbS_3	Dimorphous with ruby silvers: monoclinic or triclinic
Xanthoconite	Ag_3AsS_3	
Tetrahedrite	$(Cu,Fe)_{12}Sb_4S_{13}$	Tetrahedrite group: isometric; forms an isomorphous series
Tennantite	$(Cu,Fe)_{12}As_4S_{13}$	

A_3BX_4 Type

Sulvanite	Cu_3VS_4	Sulvanite group: isometric minerals
Germanite	Cu_3GeS_4 ?	
Colusite	$Cu_3(Sn,Te,Fe,V,As)S_4$	
Famatinite	Cu_3SbS_4	Enargite group: probably orthorhombic; frequently intergrown
Enargite	Cu_3AsS_4	
Beegerite	$Pb_6Bi_2S_9$	
Samsonite	$Ag_4MnSb_2S_6$	
Gratonite	$Pb_9As_4S_{15}$	
Lengenbachite	$Pb_6(Ag,Cu)_2As_4S_{13}$	
Jordanite	$Pb_{14}As_7S_{24}$	
Geocronite	$Pb_5(Sb,As)_2S_8$	
Guitermanite	$Pb_{10}As_6S_{19}$	
Meneghinite	$Pb_{13}Sb_7S_{23}$	
Lillianite	$Pb_3Bi_2S_6$	

A_2BX_3 Type

Bournonite	$PbCuSbS_3$	Bournonite group: orthorhombic gray metallic minerals
Seligmannite	$PbCuAsS_3$	
Aikinite	$PbCuBiS_3$	
Diaphorite	$Pb_2Ag_3Sb_3S_8$	
Freieslebenite	$Pb_3Ag_5Sb_5S_{12}$	
Klaprothite	$Cu_6Bi_4S_9$?	

SULFOSALTS

ABX_2 Type $(A:B \sim 1:1)$

Boulangerite	$Pb_{20}Sb_{16}S_{44}$ to $Pb_{18}Sb_{18}S_{44}$
Miargyrite	$AgSbS_2$
Aramayoite	$Ag(Sb,Bi)S_2$
Matildite	$AgBiS_2$
Smithite	$AgAsS_2$
Chalcostibite	$CuSbS_2$ } *Chalcostibite group:* orthorhombic,
Emplectite	$CuBiS_2$ } gray, metallic
Lorandite	$TlAsS_2$
Teallite	$PbSnS_2$
Benjaminite	$Pb_2(Cu,Ag)_2Bi_4S_9$

$A_2B_2X_5$ Type $(A:B \sim 1:1)$

Dufrenoysite	$Pb_2As_2S_5$
Cosalite	$Pb_2Bi_2S_5$
Franckeite	$Pb_5Sn_3Sb_2S_{14}$
Jamesonite	$Pb_4FeSb_6S_{14}$
Rathite	$Pb_{13}As_{18}S_{40}$

$A_2B_3X_6$ Type $(A + B:X \sim 5:6)$

Andorite	$PbAgSb_3S_6$ } *Andorite group:* lead gray, metallic
Lindstromite	$PbCuBi_3S_6$ }
Baumhauerite	$Pb_4As_6S_{13}$
Fuloppite	$Pb_3Sb_8S_{15}$ }
Plagionite	$Pb_5Sb_8S_{17}$ } *Plagionite group:* monoclinic; gray to
Heteromorphite	$Pb_7Sb_8S_{19}$ } black, metallic
Semseyite	$Pb_9Sb_8S_{21}$ }

AB_2X_4 Type $(A:B \sim 1:2)$

Hutchinsonite	$(Pb,Tl)_2(Cu,Ag)As_5S_{10}$
Galenobismutite	$PbBi_2S_4$
Zinkenite	$Pb_6Sb_{14}S_{27}$
Sartorite	$PbAs_2S_4$
Berthierite	$FeSb_2S_4$
Cylindrite	$Pb_3Sn_4Sb_2S_{14}$
Gladite	$PbCuBi_5S_9$
Vrbaite	$TlAs_2SbS_5$

AB_4X_7 Type

Livingstonite	$HgSb_4S_7$

LLOYD W. STAPLES

References

Berry, L. G., 1965, "Recent advances in sulfide mineralogy," *Am. Mineralogist,* **50,** 301.

Born, L., and Heller, E., 1960, "A structural proposal for boulangerite," *Am. Mineralogist,* **45,** 1266.

Dana, J. W., and Dana, E. S., 1944, "The System of Mineralogy," Seventh ed., Vol. I, 348, (revised by Palache, C., Berman, H., and Frondel, C.), New York, John Wiley & Sons.

Hellner, E. 1958, "A structural scheme for sulfide minerals," *J. Geol.,* **66,** 503.

Kostov, I., 1964, "On isomorphism amongst minerals of the group of sulphosalts," Chap. 18, in (Battey and Tomkeieff, editors), "Aspects of Theoretical Mineralogy in the USSR" by MacMillan, N.Y., 1964.

Cross-references: *Mineral Classes–Non-Silicates; Sulfides; Sulfur: Element and Geochemistry.*

SULFUR: ELEMENT AND GEOCHEMISTRY

Sulfur, known to ancient civilizations and sometimes referred to as "brimstone" or the stone that burns, is widely dispersed as an important minor constituent of the biosphere, hydrosphere, and nearly all major rock types

SULFUR: ELEMENT AND GEOCHEMISTRY

of the lithosphere. Early usage for bleaching, fumigation, and medicinal purposes in Egyptian and Greek cultures by 2000 BC arose from the remarkable chemical properties of this element and its not infrequent occurrence in the native state. Today, because of the enormous quantities used in the agricultural, chemical, and manufacturing industries, the per capita consumption of sulfur is a reliable index of the industrial activity of a nation.

Properties

Sulfur, having atomic number 16 and atomic weight 32.064, occupies group VI A in the Periodic Table of the elements along with oxygen, selenium, tellurium, and polonium. It possesses four stable isotopes, S^{32}, S^{33}, S^{34}, and S^{36}, for which the terrestrial abundances, although variable, are approximately 95.1, 0.74, 4.2, and 0.02%, respectively. In naturally occurring sulfur-bearing compounds, however, the S^{34}/S^{32} ratio varies by more than 11%.

Under standard conditions, elemental sulfur exists as yellow orthorhombic crystals. The crystals have a resinous luster, white streak, hardness of 1.5 to 2.5, and density of 2.07. When heated, orthorhombic sulfur inverts to a monoclinic form at 96°C. The melting and boiling points of sulfur are 113 and 445°C, respectively. Liquid sulfur melts become increasingly more viscous as the temperature is raised above the melting point. Investigations of molecular structure reveal that sulfur in crystalline form and in low-temperature melts has a ring-type structure with the general formula S_8. At higher temperatures, the ring structures break to form chains which may then polymerize and thereby increase the viscosity of the liquid.

In addition to the elemental state, sulfur displays -2, $+4$, and $+6$ oxidation states in combination with other elements. The most important of these with respect to the common sulfur-bearing minerals are the S^{-2} sulfide anion (ionic radius 1.84Å) and the S^{+6} cation (ionic radius 0.30Å) of sulfate. Atomic bonding of sulfur to other elements in compounds may be ionic, covalent, intermediate types thereof, or partly metallic.

Minerals

The sulfur-bearing minerals, although mineralogically diverse and of great economic importance, are quantitatively insignificant in the earth's crust relative to the silicates, carbonates, and oxides. They may be conveniently subdivided into two classes; the sulfides, including the sulfosalt subgroup, and the sulfates. A number of the more common representatives of these classes, including their composition, crystal system, and sulfur content, are listed in Table 1.

By convention, the sulfides may be represented by the general formula A_mX_p, where X is one or more of the larger sulfur atoms, or less commonly arsenic, antimony, bismuth, selenium, or tellurium, and A is one or more of the smaller transition metal atoms (Mason and Berry, 1968). Similarly, the sulfosalts may be depicted by the formula $A_mB_nX_p$, or double sulfide $A_mX_q \cdot B_nX_{p-q}$, where X is sulfur, B is arsenic, antimony, bismuth, or tin, and A is copper, lead, or silver. Both sulfides and sulfosalts are listed in order of decreasing metal/sulfur ratios in Table 1. Bond types among these minerals are normally complex mixtures of covalent-ionic character and some with high luster; for example, chalcopyrite and galena, exhibit metallic bonding as well. The most important factors that control the deposition of sulfides, apart from gross composition of the systems, are fugacity of sulfur and temperature (Barton and Skinner, 1967). Both selenium and tellurium may substitute for sulfur in the sulfides and sulfosalts. The sulfides are comparatively more abundant than the sulfosalts in most natural occurrences.

The sulfates are normally subdivided into anhydrous, hydrated, and hydroxyl-bearing varieties (Mason and Berry, 1968). Representative members are listed in Table 1. In sulfates, the sulfur atom is covalently bonded to four oxygen atoms to form the sulfate (SO_4^{-2}) radical. The sulfate radical, in turn, is joined by ionic bonds to an alkali, alkaline earth, or transition metal. Water molecules, or water of crystallization, are normally coordinated to both metallic cations and sulfate radical in the hydrous sulfates.

Sulfur is also present in organic matter as proteins of plants and animal tissue. Upon hydrolysis, the proteins break down to form amino acids of which cysteine, $HSCH_2CH(NH_2)COOH$, cystine, $HOOCH(NH_2)CH_2SCCH_2CH(NH_2)COOH$, and methionine $CH_3SCH_2CH_2CH(NH_2)COOH$ are sulfur-bearing. Apparently sulfur is one of the essential elements to the vital processes of organisms.

Abundance

With the exception of restricted concentrations such as in the cap rock of salt domes, in sulfide mineral deposits, and in sedimentary evaporites, sulfur is normally a minor constituent of the earth's crust. Average concentrations of sulfur in selected rock types and waters of the crust are listed in Table 2 (see Krauskopf, 1967, and listed references). Comparison of these data to reliable estimates of cosmic, solar, and meteoritic abundances of sulfur re-

SULFUR: ELEMENT AND GEOCHEMISTRY

TABLE 1. COMMON SULFUR-BEARING MINERALS SHOWING COMPOSITION, CRYSTAL SYSTEM, AND PERCENTAGE SULFUR

Class-Mineral	Composition	System	S (%)
Sulfides			
Argentite	Ag_2S	Isometric	12.9
Chalcocite	Cu_2S	Orthorhombic	20.2
Bornite	Cu_5FeS_4	Isometric	25.5
Galena	PbS	Isometric	13.4
Sphalerite	ZnS	Isometric	33.0
Chalcopyrite	$CuFeS_2$	Tetragonal	35.0
Pyrrhotite	$Fe_{(1-x)}S$	Hexagonal	~36.5
Pentlandite	$(Fe,Ni)_9S_8$	Isometric	~36.0
Covellite	CuS	Hexagonal	33.6
Cinnabar	HgS	Trigonal	13.8
Realgar	AsS	Monoclinic	29.9
Orpiment	As_2S_3	Monoclinic	39.0
Stibnite	Sb_2S_3	Orthorhombic	28.6
Bismuthinite	Bi_2S_3	Orthorhombic	18.8
Pyrite	FeS_2	Isometric	53.4
Marcasite	FeS_2	Orthorhombic	53.4
Arsenopyrite	$FeAsS$	Orthorhombic	19.7
Molybdenite	MoS_2	Hexagonal	40.1
Sulfosalts			
Pyrargyrite	Ag_3SbS_3	Trigonal	17.8
Proustite	Ag_3AsS_3	Trigonal	19.4
Tetrahedrite	$(Cu,Fe,Zn,Ag)_{12}Sb_4S_{13}$	Isometric	~25.0
Tennantite	$(CuFe,Zn,Ag)_{12}As_4S_{13}$	Isometric	~28.2
Enargite	Cu_3AsS_4	Orthorhombic	32.6
Bournonite	$PbCuSbS_3$	Orthorhombic	19.7
Boulangerite	$Pb_5Sb_4S_{11}$	Monoclinic	18.9
Native sulfur	S	Orthorhombic	100.0
Sulfates			
Barite	$BaSO_4$	Orthorhombic	13.7
Anglesite	$PbSO_4$	Orthorhombic	10.6
Anhydrite	$CaSO_4$	Orthorhombic	23.5
Gypsum	$CaSO_4 \cdot 2H_2O$	Monoclinic	18.6
Polyhalite	$K_2Ca_2Mg(SO_4)_4 \cdot 2H_2O$	Triclinic	21.2
Chalcanthite	$CuSO_4 \cdot 5H_2O$	Triclinic	12.8
Melanterite	$FeSO_4 \cdot 7H_2O$	Monoclinic	11.5
Epsomite	$MgSO_4 \cdot 7H_2O$	Orthorhombic	13.0
Brochantite	$Cu_4(SO_4)(OH)_6$	Monoclinic	7.1
Alunite	$KAl_3(SO_4)_2(OH)_6$	Trigonal	15.4
Jarosite	$KFe_3(SO_4)_2(OH)_6$	Trigonal	12.8

TABLE 2. ABUNDANCE OF SULFUR IN VARIOUS NATURAL ENVIRONMENTS[a]

Locale	S (ppm)
Crustal average	260
Ultramafic	300
Basalt	250
Granite	270
Shale	2400
Sandstone	240
Carbonate	1200
Deep-sea sediment	1300
Soils	850
Terrestrial plants	500
Seawater	885
Freshwater	5.5

[a] After Krauskopf, 1967, Appendix III; and other sources.

veals a probable deficiency of this element in the crust. For example, the sulfur content of chondritic meteorites is 1.93%, chiefly as troilite (FeS), and its relative cosmic and solar abundances are approximately 37.5 and 63% that of silicon, whereas the concentration of sulfur rarely exceeds 0.03% in common rock types of the crust (Krauskopf, 1967, Table 21-1 and references). The apparent deficit presumably resides in the earth's core where sulfur, by virtue of its geochemical behavior, and larger quantities of iron and nickel were concentrated at the time of primary differentiation.

Geochemical Behavior

Although quantitatively of minor abundance, sulfur is widely distributed as a principal con-

SULFUR: ELEMENT AND GEOCHEMISTRY

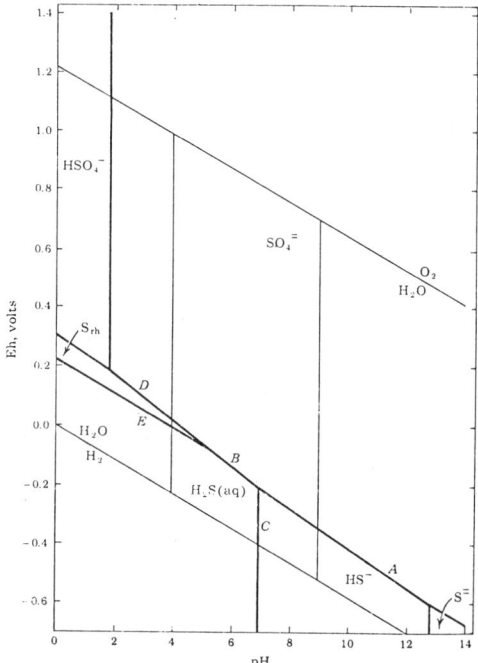

FIG. 1. Eh–pH diagram for stable sulfur species at 25°C and 7 atm total pressure; assumed total concentration of dissolved sulfur species 0.001M (Krauskopf, 1967, Fig. 10-2).

stituent of minerals that have formed throughout a broad spectrum of geologic environments and which in turn are representative of diverse physiochemical conditions. The chemical versatility of sulfur may be attributed to its ability to exhibit -2, 0, $+4$, and $+6$ valence states and to combine with both metals and nonmetals via either covalent or ionic bonds. Moreover, many of these reactions will take place readily under standard conditions. This behavior is evident from the stability fields of common sulfur species illustrated in Fig. 1. Under most natural conditions only the -2 and $+6$ valence states are important. These form the sulfide anion and sulfate radical that in combination with metals are the basis of the two great classes of sulfur-bearing minerals. The sulfide minerals are diagnostic of reducing (commonly deep-seated) conditions whereas the sulfates are representative of oxidizing (near-surface) conditions. Sulfide-sulfur is the essential component of the chalcophile groups in Goldschmidt's *geochemical classification of the elements* (q.v.). However, in oxidizing environments the chalcophile behavior of sulfur becomes less important owing to the large solubility of the sulfate radical.

In the majority of high-temperature plutonic, volcanic, and hydrothermal environments sulfur is present in the reduced state as a sulfide melt, as aqueous or gaseous hydrogen sulfide, as bisulfide or sulfide ions, or as polysulfides, the relative stabilities of which are determined by Eh–pH, temperature, pressure, and compositional factors. Experimental studies and detailed examinations of meteorites, smelter products, and plutonic and volcanic rocks indicate that an immiscible sulfide melt will separate from a magma provided that the concentration of sulfur reaches a certain threshold value. The sulfide melt may then serve as an effective scavenger for the transition elements, in conformance with the chalcophile behavior of sulfur, ultimately to form economic concentrations of base, ferroalloy, and precious metals. In hydrothermal systems the sulfide precipitant is probably hydrogen sulfide. Less commonly, sulfates such as anhydrite, gypsum, or barite are deposited under near-surface magmatic or hydrothermal conditions in response to decreasing pH and/or increasing Eh.

Because sulfide stability necessitates reducing conditions, sulfates are normally the stable sulfur compounds in near-surface and sedimentary environments. Less commonly, native sulfur is deposited with incomplete oxidation of hydrogen sulfide from biogenic or volcanic sources. The sulfates are formed in surficial environments either through oxidation of pre-existing sulfide minerals or by direct precipitation in restricted marine imbayments to produce extensive evaporite deposits that contain carbonates and chlorides as well. However, the oceans, by virtue of their volume, constitute a vast residuum of sulfate owing to the relatively high solubility of sulfate minerals and particularly that of the sulfate radical. Because of its high mobility the sulfate ion is used as a pathfinder in geochemical exploration.

The presence of native sulfur in the cap rock of salt domes and of sulfides in sediments and sedimentary rock was for many years a geologic enigma. Combined investigations of carbon and sulfur isotope distributions in these minerals and associated compounds, together with laboratory study of bacterial cultures have conclusively demonstrated the role of anaerobic bacteria in localizing these deposits via the low-temperature reduction of sulfate to hydrogen sulfide (Thode, 1963; Feely and Kulp, 1957; and later investigators). The biogenic hydrogen sulfide was either oxidized to form native sulfur or served as the sulfide precipitant. Measured isotopic fractionation effects were in close agreement with those predicted by theory and later verified by experiment.

The isotopic composition of most magmatic

and hydrothermal sulfide-sulfur exhibits a relatively narrow dispersion about the zero permil value for meteoritic (troilite) sulfur. In contrast, dissolved sulfate in ocean water is enriched in S^{34} by approximately +20 permil, relative to the meteoritic standard, and contemporary evaporite sulfates are isotopically similar. Moreover, sedimentary sulfides are depleted in S^{34} by as much as −50 permil as a consequence of fractionation effects attendant with the bacterial reduction of sulfate to hydrogen sulfide. Holser and Kaplan (1966) have shown that the isotopic composition of seawater sulfate, as deduced from evaporite sulfates, has ranged from approximately +30 to +10 permil since Precambrian time. These variations are attributed to differing rates of influx and sedimentation between the isotopically heavy sulfates and light sulfides.

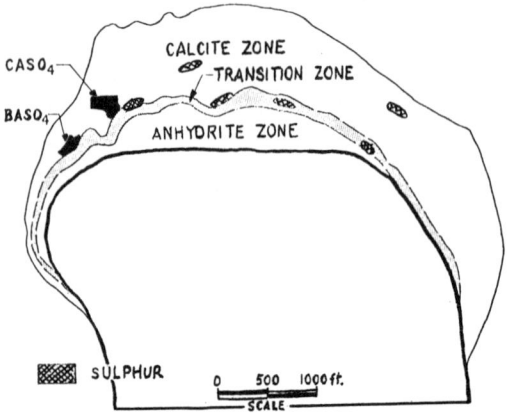

Fig. 2. Cross section of Jefferson Island dome, Louisiana, showing three mineralogical zones typical of cap rock on many domes (Halbouty, 1967, Fig. 4–2).

Economic Geology

Sulphur is an essential mineral to the industrial economy of a nation. According to Ambrose (1965), total production of sulfur in the United States for 1963 amounted to 6,643,802 long tons, valued at $125,000,000, of which a small amount was available for export. Other important producers include Mexico, France, Russia, and Canada. Approximately 85% of the domestic sulfur production is converted to sulfuric acid for use in agricultural, chemical, and related industries. Sulfur resources of the United States are more than ample for domestic needs as the estimated reserves of native sulfur alone exceed 200 million tons. Moreover, to this may be added other sources such as from sulfide ores, petroleum, oil shale, and coal which collectively render the potential reserves enormous.

In terms of commercial recovery, the principal geologic sources include native sulfur deposits associated with salt domes, sedimentary rocks, and volcanic exhalations and by-product recovery from the roasting or smelting of sulfide ores or from "sour" natural and refinery gases. Deposits of native sulfur in salt domes and in sedimentary rock are quantitatively most important.

Salt domes are widely distributed along the Gulf Coast of the United States, particularly in Louisiana and Texas, and Mexico. Elsewhere, they are found in West Germany, Rumania, Russia, and Iran. The domes consist of a vertical cylinder of salt that has intruded the enclosing sediments from an evaporite source bed at depth. With time, and in response to meteoric waters, a cap rock may develop on top of the salt plug. If fully developed, the cap rock will consist of three mineralogically distinct units: from bottom to top—anhydrite, gypsum, and calcite. The salient features of a salt dome are illustrated in Fig. 2. Native sulfur, if present, is normally found in the porous gypsum or calcite units of the cap rock. The lowermost anhydrite unit represents a residuum of sulfate that originally was carried as inclusions in the salt plug. Reaction with ground waters converted the anhydrite to gypsum. Carbon and sulfur isotope studies have demonstrated that the native sulfur formed by partial oxidation of hydrogen sulfide that in turn was derived from the bacterial reduction of sulfate (see Thode, 1963, and previous investigators). Additionally calcite-carbon of the upper cap rock layer is depleted in C^{13} relative to sedimentary carbonate. As the calcite is isotopically similar to the carbon in nearby petroleum, it may be inferred that the anaerobic sulfate-reducing bacteria utilized the hydrocarbon as an energy source. Carbon dioxide or methane evolved by this process, and depleted in C^{13}, then combined with dissolved calcium of the cap rock. A similar origin has been postulated for the native sulfur deposits in Tertiary sedimentary rocks that are common to Sicily, Poland, Russia, and Gulf Coast region of the United States.

Prior to 1913 Sicily, by virtue of a worldwide monopoly, was the leading producer of sulfur. Development of the Frasch process, by H. Frasch in 1895, led to the efficient and profitable extraction of native sulfur from buried salt domes. This ingenious method, as illustrated in Fig. 3, utilizes the low melting point of sulfur (113°C) and a system of three coaxial pipes. Compressed air and superheated water (about 160°C) are forced into the cap rock or sediment reservoir. The heated water

SULFUR: ELEMENT AND GEOCHEMISTRY

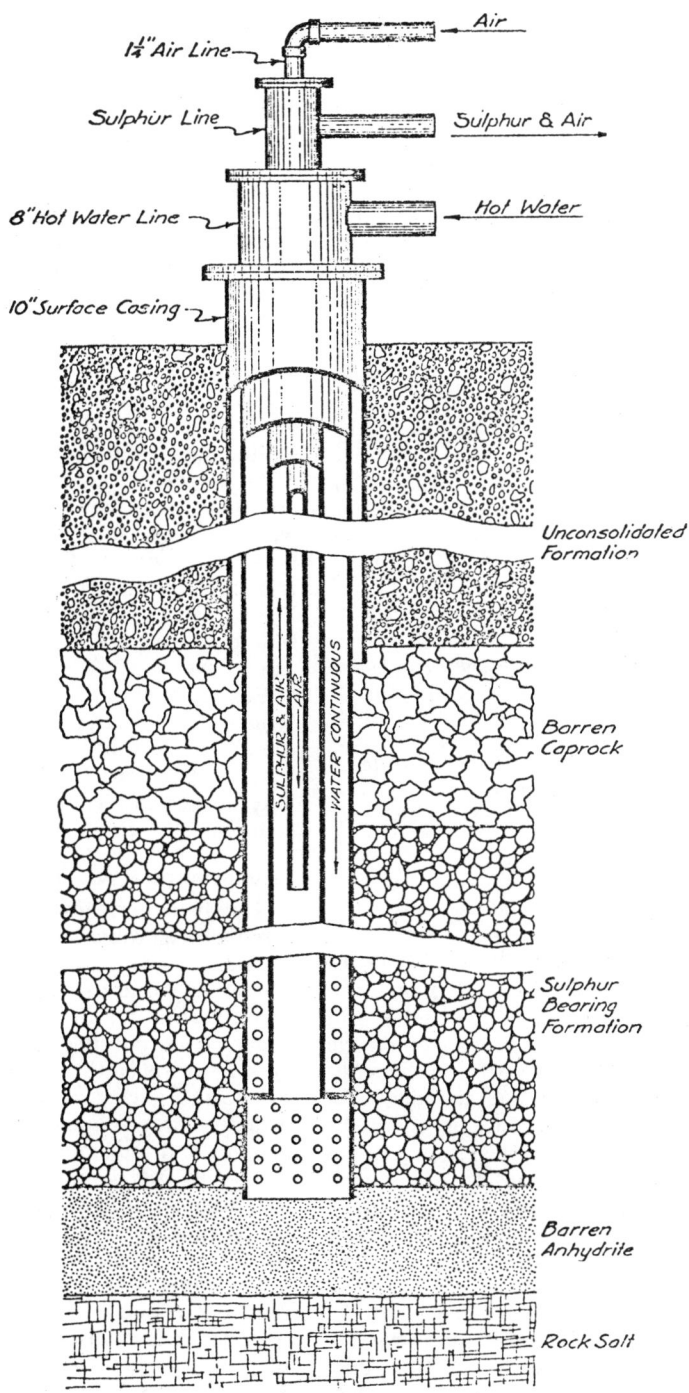

FIG. 3. Frasch method of sulfur extraction (Lundy, 1949, Fig. 3).

melts the native sulfur and it flows as a liquid to storage vats at the surface. Several hundred tons of sulfur and having a purity of 99.5% may be produced daily from a single well.

CYRUS W. FIELD

References

Ambrose, P. M., 1965, "Sulfur and Pyrites," in "Mineral Facts and Problems," *U.S. Bur. Mines Bull.*, **630**, 901–917.

Barton, P. B., Jr., and Skinner, B. J., 1967, "Sulfide Mineral Stabilities," in (Barnes, H. L., editor) "Geochemistry of Hydrothermal Ore Deposits," New York, Holt, Rinehart, and Winston, pp. 236-333.

Feely, H. W., and Kulp, J. L., 1957, "The origin of the Gulf Coast salt dome sulphur deposits," *Bull. Am. Assoc. Petrol. Geol.*, **41**, 1802-1853.

Halbouty, M. T., 1967, "Salt Domes," Houston, Gulf Publishing Co., 425pp.

Holser, W. T., and Kaplan, I. R., 1966, "Isotope geochemistry of sedimentary sulfates," *Chem. Geol.*, **1**, 93-135.

Krauskopf, K. B., 1967, "Introduction to Geochemistry," New York, McGraw-Hill Book Co., 721pp.

Lundy, W. T., 1949, "Sulphur and Pyrites," in (Dolbear, S. H., and Bowles, O., editors), "Industrial Minerals and Rocks," New York, Am. Inst. Mining Metallurgical Engineers, pp. 989-1017.

Mason, B., and Berry, L. G., 1968, "Elements of Mineralogy," San Francisco, W. H. Freeman and Co., 550pp.

Thode, H. G., 1963, "Sulphur Isotope Geochemistry," in (Shaw, D. M., editor) "Studies in Analytical Chemistry," Royal Soc. Canada Spec. Publ. No. 6, pp. 25-41.

Cross-references: *Elements: Planetary Abundances and Distribution; Geochemical Classification of the Elements; Hydrothermal Solutions—Sulfide Transport; Isotopic Variations in Mineral Deposits; Seawater, Chemistry; Sulfides; Sulfides in Sediments; Sulfosalts; Sulfur Isotope Fractionation; Sulfur Oxidation—Bacterial.* Vol. V: *Salt Domes.*

SULFUR CYCLE

Chemistry and Cosmic Distribution of Sulfur

Sulfur is the tenth most abundant element in the universe (Mason, 1958, p. 22) and eighth most abundant in the solar atmosphere. In the earth's crust, represented largely by igneous rocks, it is the fourteenth most abundant element, but in sediments its place is eighth or ninth (see Table 1). Meteorites show a relatively high enrichment of sulfur, where it is concentrated largely as troilite (FeS), both in iron and stony meteorites, and where it is the fifth most abundant element.

On earth, sulfur is strongly chalcophilic, forming sulfide minerals with silver and the base metals. An analogy has been drawn between meteoritic and terrestrial sulfur abundances, and it has been hypothesized that zones of the lower mantle may contain high concentrations of metal sulfides, particularly troilite. Sulfur reacts with oxygen in fourfold coordination to make sulfate, a soluble double charged anion (SO_4^{2-}).

Sulfur Deposits

An estimate of the sulfur deposits on the earth's crust taken from Holser and Kaplan (1966) is given in Table 1. Since ultrabasic (mafic) rocks comprise the bulk of the crust, it is not surprising that this constitutes one of the largest reservoirs for sulfur. Sedimentary rocks and the sea (9100×10^{12} tons), however, constitute the largest reservoir. In mafic, metamorphic or volcanic rocks, sulfur exists almost exclusively as sulfide. Hydrothermal processes often concentrate the sulfides and deposit them as copper, lead, zinc, silver, mercury, or iron minerals (see *Sulfides in Sediments*).

The estimates indicate that gypsum ($CaSO_4 \cdot 2H_2O$), anhydrite ($CaSO_4$), and other evaporite sulfates are the most abundant form of sedimentary sulfur. Next in abundance are metal sulfides, mainly pyrite (FeS_2), in shales, limestone, and sandstone. The ocean contains half as much sulfur (1280×10^{12} tons, entirely in the form of dissolved sulfate ion) as is present in reduced form in all the sediments. Fresh water and the atmosphere have much smaller reservoirs that are continually being turned over.

Elemental sulfur is generally present in most bituminous shales, probably associated with the organic matter, but quantitatively unimportant. Sedimentary traps, such as salt domes, may concentrate elemental sulfur and make it economically profitable to mine.

Geochemical Cycle of Sulfur

The high enrichment of sulfur in the oxidized phase, which was most probably absent in the early stage of the earth's history, suggests that sulfur is being released from depth and concentrated at the surface. Two mechanisms are responsible for this transfer; (a) degassing of the earth, along rift zones, supported by violent volcanism and accelerated by times of active orogeny, and (b) leaching of surface igneous and metamorphic rocks. In both cases the solubilized component is transferred into the ocean, either by streams or rain, during which time they are oxidized to sulfate. In the ocean the soluble ion undergoes two possible fates, either it is reduced by bacteria at the mud-water interface and converted to pyrite, or it enters a semiisolated basin in an arid zone of the earth and is precipitated as calcium sulfate (gypsum) after evaporation has concentrated it to four times its normal seawater concentration (2.64 g/l). Holser and Kaplan (1966) calculated the residence time of sulfate in the ocean to be 21×10^6 years.

In periods of marine regression and tectonic uplift, sedimentary pyrite and gypsum are weathered out and transferred back to the ocean in the oxidized form. Burial of the gyp-

TABLE 1. GEOCHEMICAL INVENTORY OF SULFUR IN THE EARTH'S CRUST

	Mass of Rock (10^{15}T)	Mean S Content in %, and Standard Deviation of the Mean	Mass of Sulfur (10^{12}T)	Mean δS^{34} (‰)
Deep oceanic rocks				
Sediments	300. ± 75	0.025 ± 0.003	75 ± 20	+20. ± 1
Mafic rocks (to Moho)	4400. ± 1100	0.053 ± 0.013	2300 ± 800	+ 1. ± 2
Sedimentary rocks				
Sandstone	280. ± 70	0.090 ± 0.01	250 ± 60	−12. ± 5
Shale	750. ± 190	0.27 ± 0.04	2000 ± 580	−12. ± 5
Limestone	290. ± 80	0.13 ± 0.01	380 ± 110	−12. ± 5
Evaporites	30. ± 12	17. ± 4	5100 ± 1600	+17. ± 2
Volcanics	120. ± 30	0.04 ± 0.01	50 ± 18	+ 5. ± 5
Connate water	140. ± 30	0.019 ± 0.02	27 ± 5	+17. ± 2
Total evaporites and connate water	170. ± 30		5100 ± 1600	+17. ± 2
Total other sediments	1440. ± 220	0.19 ± 0.05	2700 ± 600	−11.8 ± 5.5
Total all sediments	1600. ± 220	0.49 ± 0.13	7800 ± 1700	+7.1 ± 4.5
Fresh water	0.3 ± 0.2	0.0011 ± 0.0001	0.003 ± 0.002	+10. ± 5
Ice	35.	0.00003 ± 0.00001	0.006 ± 0.002	+ 5. ± 5
Atmosphere	5.3		(3.6×10^6)	+ 5. ± 5
Sea	1420 ± 40	0.090 ± 0.003	1280 ± 55	+20. ± 0.5
Total in sedimentary rocks and sea			9100 ± 1700	+8.5 ± 4.0
Continental igneous and metamorphic rocks (to Moho)				
Granitic	10500. ± 2600	0.021 ± 0.003	2200 ± 600	+10. ± 5
Mafic	8700. ± 220	0.053 ± 0.013	4600 ± 1600	+ 1. ± 2
Total	21000. ± 3000	0.032 ± 0.005	6800 ± 1700	+ 4. ± 3
Total in Crust			18200 ± 2500	+ 6. ± 2.3

SULFUR CYCLE

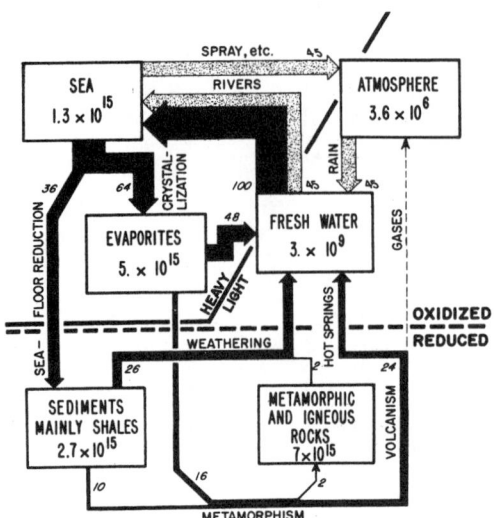

FIG. 1. Geochemical cycle of sulfur. Masses are in millions of metric tons. Most material above the dashed line is oxidized to sulfate, most below this line is reduced to sulfide. Long-term (dark) and short-term fluxes of sulfur between reservoirs are indicated on the basis of the long-term component of fresh-water sulfate flowing to the sea (from Holser and Kaplan, 1966).

sum to a depth of 800 m or greater will cause it to be dehydrated and converted to anhydrite.

The cycling process is shown in Fig. 1. The relative magnitude of the flows are calculated from the size of the sulfur reservoir and from the proportion of rock type exposed, assuming that the sulfur content of the ocean is neither increasing or diminishing. Weathering and volcanism contribute 5.4 to 6.1×10^7 tons per year of sulfur dissolved in fresh water and transferred ultimately to the ocean. Hot springs account for about 1.2×10^7 tons per year of the above estimates. The atmospheric cycle has a very high frequency and at the present time contributes over 4×10^7 tons per year of sulfur. A quantitative evaluation of the sulfur cycle has become increasingly difficult, since the start of the Industrial Revolution. A calculation on the basis of 1958 statistics shows that liberation of industrial wastes and sewage into streams, and sulfur dioxide through fuel combustion into the atmosphere (Eriksson, 1959/60) equaled the annual contribution of sulfur by streams to the ocean.

Biological Transformations of Sulfur

The microorganisms involved in the sulfur cycle offer an outstanding example of the interaction between biology and geology. Bacteria are the most important members of this group, representing several different families. In this group of organisms are two families of photosynthetic bacteria, *Chlorobacteriaceae* and *Thiorhodaceae*, green and purple sulfur bacteria. Here also are the chemoautotrophic bacteria *Thiobacillus*, capable of using CO_2 as the only carbon source, and deriving energy from the oxidation of reduced sulfur compounds. The group known as *Beggiatoaceae*, comprise several genera of filamentous bacteria capable of oxidizing sulfide and storing the elemental sulfur as globules within their cells. These organisms are closely related to the blue green algae, and probably represent an evolutionary link between true bacteria and true algae (using chlorophyll A for photosynthesis). The last components of the sulfur bacteria and perhaps the most important in nature are the sulfate reducers, with *Desulfovibrio desulfuricans* as the best known species.

In the cycling process shown in Fig. 2a, the starting point is usually sulfate, the most abundant sulfur ion dissolved in water. Sulfide may be important in geothermal environments as the primary source of sulfur. Bacterial reduction proceeds directly to hydrogen sulfide. The liberated gas (or sulfhydryl ion, SH^-) is then oxidized in three ways: (a) inorganically by atmospheric oxygen, probably in the presence of trace metals as catalysts, (b) by sulfur bacteria of the *Thiobacillus* genus requiring high

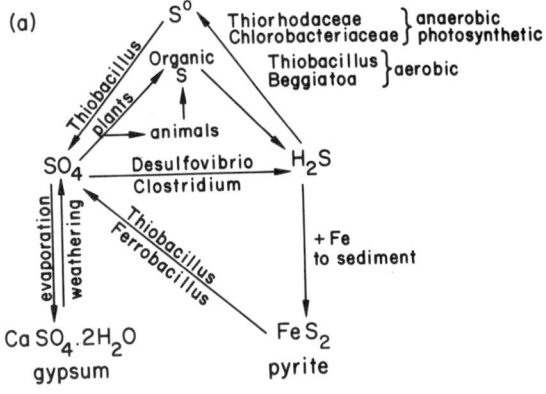

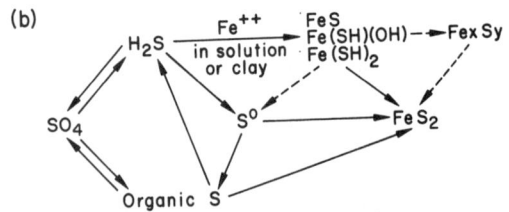

FIG. 2. Pathways of sulfur transformation: (a) the biological sulfur cycle; (b) reactions leading to pyrite formation in marine sediments.

SULFUR CYCLE

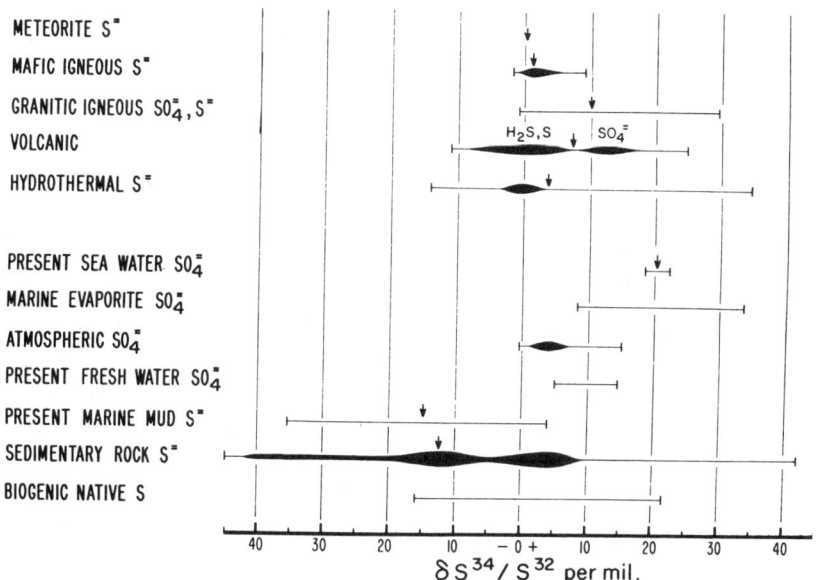

Fig. 3. S^{34}/S^{32} variation in different sulfur-containing components. Arrows denote mean values with meteoritic sulfur taken as zero. The number of measurements for any single reading is indicated by the thickness of the line.

concentrations of oxygen relative to sulfide, or by filamentous sulfur bacteria that can metabolize at relatively low oxygen tensions, and (c) by photosynthetic sulfur bacteria which oxidize sulfide or elemental sulfur in the *absence* of oxygen only, by the following reactions:

$$CO_2 + 2H_2S \xrightarrow{light} (CH_2O)_n + 2S + H_2O$$
$$\text{cell material}$$

$$2CO_2 + H_2S + 2H_2O \xrightarrow{light} 2(CH_2O)_n + H_2SO_4$$
$$\text{cell material}$$

The biological cycle has two leaks of geological importance. The first is the removal of soluble sulfate from solution by evaporation to form calcium sulfate or gypsum (and later anhydrite). The second is the formation of iron disulfide or pyrite. Fig. 2b shows the pathway through which pyrite probably forms in marine sediments. The first mineral to form is troilite or hydrotroilite (FeS or FeS·OH). This is then oxidized in situ by reaction with colloidal sulfur and perhaps organic sulfur to form metastable iron sulfides and eventually pyrite.

Isotopic Abundance of Sulfur

As a result of its existence in several valence states and because sulfur is continually being cycled, it has a total spread of S^{34}/S^{32} ratio of about ±4.5% from a mean value. This value is close to the S^{34}/S^{32} ratio of meteoritic sulfur. Sulfate reducing bacteria are able to fractionate the isotopes during metabolism, with an enrichment of up to about 45‰ in S^{32} in the released sulfide relative to the starting S^{34}/S^{32} ratio of the sulfate (Kaplan and Rittenberg, 1964). Hence sedimentary sulfides are generally enriched in the light isotope (Fig. 3) whereas the residual sulfate remaining becomes enriched in the S^{34}. Fig. 3 shows that assuming the earth had an original isotopic composition close to that of meteoritic sulfur, or mafic igneous sulfur, the greatest differentiation is between the reduced sedimentary pyrite and oxidized gypsum in evaporites. Volcanic sulfur seems to equal the mean terrestrial values.

Using the isotopic effect as a mechanism, several studies (Thode et al., 1954; Ault and Kulp, 1959) have been made to understand the origin of mineral deposits. Recent studies on the variation of S^{34}/S^{32} with age in marine evaporitic sulfur deposits (Holser and Kaplan, 1966) suggest that the sulfur content of the ocean may have periodically changed since the early Paleozoic.

I. R. KAPLAN

References

Ault, W. U., and Kulp, J. L., 1959, "Isotopic geochemistry of sulphur," *Geochim. Cosmochim. Acta*, **16**, 201–235.

*Eriksson, E., 1959/60, "The yearly circulation of chloride and sulfur in nature; meteorological,

geochemical and pedological implications," *Tellus,* **12,** 63–109.
*Holser, W. T., and Kaplan, I. R., 1966, "Isotope geochemistry of sedimentary sulfates," *Chem. Geol.,* **1,** 93–135.
Kaplan, I. R., and Rittenberg, S. C., 1964, "Microbial fractionation of sulfur isotopes," *J. Gen. Microbiol.,* **34,** 195–212.
La Riviere, J. W. M., 1966, "The microbial sulfur cycle and some of its implications for the geochemistry of sulfur isotopes," *Geol. Rundschau,* **55**(3), 568–582.
*Mason, B., 1958, "Principles of Geochemistry," New York, John Wiley & Sons.
Nielsen, H., 1966, "Schwefelisotope im marinen Kreislauf und das $\delta^{34}S$ der früheren Meere," *Geol. Rundschau,* **55**(1), 160–172.
Thode, H. G., Wanless, R. K., and Wallouch, R., 1954, "The origin of native sulphur deposits from isotope fractionation studies," *Geochim. Cosmochim. Acta,* **5,** 286–298.

*Additional bibliographic references may be found in this work.

Cross-references: *Cycles—Geochemical; Earth's Crust Geochemistry; Evaporite Processes; Gases—Volcanic; Hydrothermal Solutions—Sulfide Transport; Outgassing of the Planet Earth; Seawater, Chemistry; Seawater, History; Sulfate Reduction—Microbial; Sulfide Mineral Oxidation—Microbial; Sulfides in Sediments; Sulfur: Element and Geochemistry; Sulfur Isotope Fractionation; Sulfur Oxidation—Bacterial; Weathering, Chemical.* Vol. IVB: *Exhalative Sedimentary Ore Deposits.* Vol. VI: *Evaporites; Hot Springs; Salt Domes.*

SULFUR ISOTOPE FRACTIONATION—IN BIOLOGICAL PROCESSES

Equilibrium Isotope Effects

The isotopes of the light elements differ in their chemical properties and are fractionated in chemical reactions, both in the laboratory and in nature. The theory of equilibrium isotope effects is now well understood and the simple process enrichment factor for an isotope exchange process may be calculated using the well-known methods of statistical mechanics. These methods require a knowledge of the molecular vibrational frequencies for the molecules involved which are obtained from spectroscopic data (Urey and Greiff, 1935; Bigeleisen and Mayer, 1947; Urey, 1947).

Effect of Enzymes

The equilibrium isotope effect or the isotope fractionation which occurs in an equilibrium process will, of course, not depend on the method or path by which equilibrium is established. The presence of enzymes in biological systems may speed up isotope exchange for a reaction, but will not affect the extent of isotope fractionation at equilibrium. For example, in the isotopic exchange reaction:

$$C^{13}O_2 \text{ (gas)} + HC^{12}O_3^- \text{ (soln)} \leftrightarrows$$
$$C^{12}O_2 \text{ (gas)} + C^{13}O_3^{2-} \text{ (soln)}$$

the equilibrium constant k turns out from both theory and experiment to be ~ 1.007, favoring C^{13} in the bicarbonate (Thode et al., 1964). This means that in a single equilibrium stage there will be 7‰ more in C^{13} in HCO_3^- (soln) than in the CO_2 (gas). The equilibrium isotope effect is defined as $(k-1)\,1000 = 7\,‰$.

This isotope exchange takes place through the hydration-dehydration of CO_2 gas which is a relatively slow process (Mills and Urey, 1940). However, the enzyme carbonic anhydrase which is found in the blood of mammals speeds up this CO_2 hydration-dehydration process (speeds up the evolution of CO_2 in the lungs and prevents suffocation) and therefore also speeds up the carbon isotopic exchange between CO_2 and HCO_3^-. Enzyme systems may therefore facilitate isotopic exchange processes but will not alter the exchange constant or equilibrium isotope fractionation.

Kinetic Isotope Effects

In the case of kinetic isotope effects where we are concerned with the relative rates of two competing isotopic reactions, the situation is somewhat different. Here the isotope effect or extent of isotope fractionation depends on the mechanism of the reaction. Thus, the kinetic isotope effect may be quite different for a reaction in a biological system catalyzed by enzymes than in a straight chemical system. In general, a primary kinetic isotope effect results when a chemical bond involving the isotopes in question is broken in the rate-determining step.

There is, of course, a theoretical basis for this since the substitution of a heavy isotope in a molecule decreases the classical vibrational frequencies involved and thereby stabilizes the molecule, making the heavy isotope species less reactive than the light species. If the bonds involving these frequencies are broken in the rate-determining step, then the light species will react faster and the first product to form will be richer in the light isotope as compared to the reactant. The theory of kinetic isotope effects has been reviewed by Bigeleisen and Wolfsberg, 1958.

For two competing isotope reactions:

$$A_1 \xrightarrow{k_1} B_1$$
$$A_2 \xrightarrow{k_2} B_2$$

where the subscripts 1 and 2 refer to the light and heavy isotope species, respectively, and k_1 and k_2 are the specific rate constants for the two reactions, the kinetic isotope effect in ‰ is defined in terms of specific rate constants and is given by the relation

$$\alpha = \left(\frac{k_1}{k_2} - 1\right) 1000$$

Thus, in biological or microbial processes, the kinetic isotope effects will depend on the mechanism of the enzyme-catalyzed process and the extent to which the bond-breaking step is rate controlling. In general, one might expect the bond breaking step to be speeded up in the presence of the enzyme and where this is one of a sequence of reactions, the others may then become rate controlling. In such cases the kinetic isotope effect may vary depending on conditions and may actually drop to zero in going from a straight chemical reaction to an enzyme catalyzed process. In a complex sequence of reactions there may be a combination of an equilibrium and kinetic isotope effect. For example, if in a stepwise process there is an equilibrium step followed by a rate-controlling unidirectional step, then the overall isotope fractionation or apparent kinetic isotope effect will be a combination of the equilibrium and the kinetic effects.

Sulfur Isotope Fractionation

Early studies of the distribution of the sulfur isotopes in nature showed wide variations in the S^{34} content of sulfur found in various forms in rocks and minerals (Thode, Macnamara and Collins, 1949; Trofimov, 1949; Ault and Kulp, 1959; Vinogradov, 1958; Thode, Monster and Dunford, 1961). In general, it was found that in sedimentary rocks sulfates are enriched and sulfides depleted in S^{34} with respect to ogneous rock and meteoritic sulfur (see Fig. 1). The extent of enrichment or depletion is expressed in terms of δS^{34}, where

$$\delta S^{34} = \frac{(S^{34}/S^{32})_{sample} - (S^{34}/S^{32})_{standard}}{(S^{34}/S^{32})_{standard}} \times 1000 \text{ ‰}$$

In isotope fractionation studies standards appropriate to the experiment under consideration are used. The question arose as to the nature of the processes which brought about this fractionation of the sulfur isotopes.

The exchange constant K for the isotope exchange reaction

$$S^{32}O_4^{2-} \text{ (soln)} + H_2S^{34} \text{ (gas)} \rightleftharpoons S^{34}O_4^{2-} \text{ (soln)} + H_2S^{32} \text{ (gas)}$$

was calculated to be 1.071, indicating a 71‰ enrichment of S^{34} in SO_2^{2-} under equilibrium conditions (Tudge and Thode, 1950). The establishment of isotopic equilibrium between sulfate and sulfide would therefore lead to S^{34} enrichment in sulfate. However, SO_4^{2-} is not easily reduced and in the absence of enzyme systems, isotopic exchange between SO_4^{2-} and S^{2-} does not take place.

Biological Processes

It was suggested quite early that the oxidation and reduction of sulfate in the biological sulfur cycle might well be responsible for the S^{34} enrichment found in sedimentary sulfates in nature and that biological processes might indeed provide a mechanism for isotopic exchange. A number of sulfur isotope fractionation studies have been carried out involving the various stages of the biological sulfur cycle (see Fig. 2) (Thode, Kleerekoper and McElcheran, 1951; Jones and Starkey, 1957; Harrison and Thode, 1957, 1958; Kaplan and Rittenberg, 1964). It was found that considerable sulfur isotope fractionation did occur in the bacterial reduction of sulfate and that this step alone could account for much of the sulfur isotope fractionation in nature.

The results of the first experiments showed that the sulfide released during sulfate reduction by growing cells of *Desulfovibrio desulfuricans* was depleted in S^{34} by a factor of 10–12‰ with respect to initial sulfate. Subsequent experiments were carried out using both growing culture media of *Desulfovibrio desulfuricans* and resting cell suspensions harvested from these media. The results obtained over a

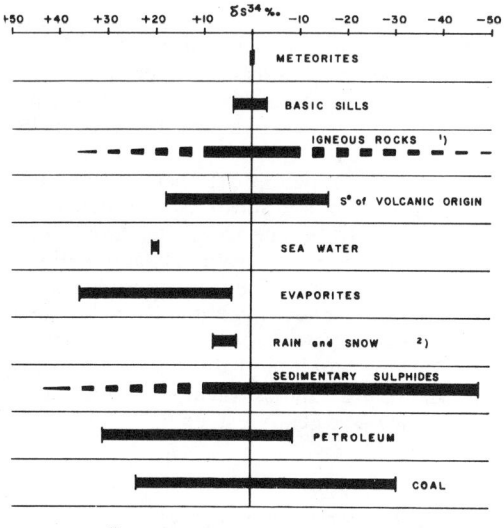

FIG. 1. The distribution of S^{34} in nature.

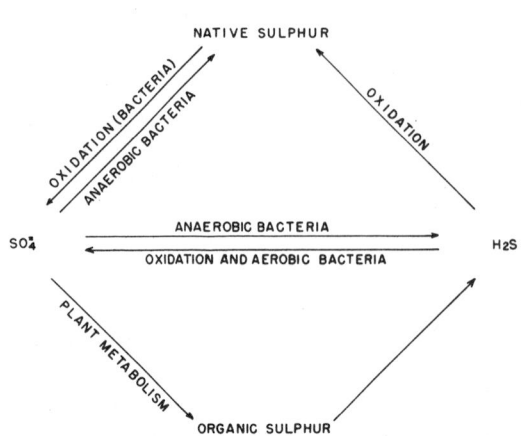

Fig. 2. Sulfur cycle in the sea.

wide range of temperatures and a wide range of SO_4^{2-} and lactate nutrient concentrations, showed that the magnitude of the sulfur isotope effect in the reduction of SO_4^{2-} by these microorganisms is dependent on the overall rate of sulfate reduction. This was found to be the case regardless of whether this rate is controlled by temperature, concentration of metabolites or conditions of growth. The sulfur isotope fractionation was found to vary from ~0‰, at extremely rapid metabolic rates or at extremely low SO_4^{2-} concentrations, to ~25‰ at moderate metabolic rates. However, for a wide range of intermediate rates, the H_2S produced was found to be depleted in S^{34} by from 10 to 20‰.

This fractionation phenomenon has been critically studied by Harrison and Thode, 1958, and Kaplan and Rittenberg, 1964. Experiments were carried out to determine whether this fractionation was due to equilibrium between pairs of reduction products in the stepwise reduction of SO_4^{2-} or due to kinetic isotope effects, or a combination of both.

Chemical Reduction of Sulfate

In the straight chemical reduction of SO_4^{2-} by a mixture of HI, H_3PO_2 and HCl, there is no evidence of isotope exchange and the isotope fractionation which occurs is obviously due to a kinetic isotope effect (Harrison and Thode, 1957). For the competing isotopic reactions

$$S^{32}O_4^{2-} \xrightarrow{HCl,\ HI,\ H_3PO_2}_{k_1} H_2S^{32}$$

$$S^{34}O_4^{2-} \xrightarrow{}_{k_2} H_2S^{34}$$

The first H_2S produced (1–3% reaction) is found to be depleted in S^{34} with respect to initial SO_4^{2-} by 23‰, indicating that the $S^{32}O_4^{2-}$ reacts 23‰ faster than the $S^{34}O_4^{2-}$ species. There is considerable evidence to suggest that the chemical reduction of SO_4^{2-} to H_2S is stepwise and that the rate controlling step in the process is the initial S–O bond breaking in the step SO_4^{2-} to SO_3^{2-}. It is therefore assumed that the isotopic fractionation occurs in this step and that the SO_3^{2-} which is more rapidly reduced ends as H_2S without any further isotope fractionation. In this regard, estimates from theory of the kinetic isotope effect expected in the SO_4^{2-} to SO_3^{2-} step are in reasonable agreement with the experimental value and for the competing isotopic reactions above

$$\alpha = \frac{k_1}{k_2} = \sim 1.023$$

Bacterial Reduction of Sulfate

In the case of the bacterial or enzyme-catalyzed reduction of sulfate where the extent of isotope fractionation depends on the rate of sulfate reduction or metabolic rate, the situation is obviously more complex. The reduction rates are, in general, slower and the mechanism of the microbial process obviously involves the attachment of SO_4^{2-} to an enzyme and its subsequent reduction in steps. However, under the conditions of experiments ranging from very rapid to slow metabolic rates (extreme slow rates excluded) there is little evidence of sulfur isotope exchange between reactant and reduction products in the stepwise reduction and the sulfur isotope fractionation is assumed to be due to kinetic processes only.

The variation of isotope fractionation with rate can be explained by postulating a reaction sequence involving two consecutive steps, either of which could be rate controlling.

The magnitude and direction of the isotope effect found under various conditions indicates two consecutive steps as follows:

$$SO_4^{2-}\ (\text{soln}) + \text{enzyme} \underset{k_2}{\overset{k_1}{\rightleftarrows}} \begin{array}{l} SO_4^{2-} \cdot \text{enzyme complex} \\ \downarrow \quad \text{II (S—O bond breaking step)} \\ SO_3^{2-} \cdot \text{enzyme complex} \\ \downarrow \quad (\text{rapid reduction relative to I and II}) \\ H_2S \end{array}$$

I

Either steps I or II may be rate controlling. If the former is rate controlling then a small sulfur isotope effect will result with the $S^{34}O_4^{2-}$ species reacting faster. Under these conditions step I is an equilibrium process involving the attachment of SO_4^{2-} to an enzyme. The small exchange constant expected and found favors $S^{34}O^{2-}$ attached to enzyme. If, however, the latter step is rate controlling, there will be a large isotope effect with the lighter species $S^{32}O_4^{2-}$ reacting faster. Intermediate values of the isotope effect will be obtained when both steps simultaneously influence the rate of sulfate reduction.

The results obtained by Jones and Starkey, 1957, Harrison and Thode, 1958, and Kaplan and Rittenberg, 1964, at different metabolic rates under various conditions of temperature and nutrient concentrations, would all seem to confirm the above postulate of two possible rate-determining steps in the bacterial reduction of sulfate. At slow metabolic rates the S–O bond breaking step becomes rate controlling and the isotope effect approaches 25‰, the value obtained in the straight chemical reduction. At very rapid rates, step I, the diffusion and attachment of SO_4^{2-} to an enzyme, would become rate determining and the isotope effect resulting would be quite small. Over a wide range of metabolic rates, the two steps simultaneously influence the rate and intermediate isotope effects are obtained in the neighborhood of 15‰.

It is interesting to point out here that no sulfur isotope fractionation occurs in the plant reduction of sulfate in contrast to the large isotope effect in the bacterial reduction (Ishii, 1953; Kaplan and Rittenberg, 1964). In view of the high specificity of enzymes and the universality of biochemical processes, one might expect enzyme systems for the reduction of sulfate in plants to be similar to those for microbial processes. The lack of any sulfur isotope effect in the plant metabolism of sulfate may be simply a matter of relative reaction rates. For example, if in the plant metabolism of sulfate, step I in the mechanism proposed above were always rate controlling, then no isotope fractionation would ever occur.

When bacterial reduction of SO_4^{2-} is carried out at extremely slow rates, such as in the absence of added nutrient lactate (Ford, 1957) or in the presence of ethanol as an electron donor (Kaplan and Rittenberg, 1964), much higher isotope fractionation is obtained between the SO_4^{2-} nutrient and the H_2S produced. Values up to 46‰ have been reported. Under these extreme conditions not all the reduced sulfur analyzed is in the form of H_2S and there is evidence of a buildup of intermediates in the stepwise reduction of SO_4^{2-}. The overall sulfur isotope fractionation obtained by comparing the isotope ratios of the first H_2S to form with that of the original nutrient is therefore not due to a simple isotope effect in a single step, but is the result of a combination of isotope effects, both equilibrium and kinetic effects. The isotope effects involved under these extreme conditions are being studied further. Isotope studies of ocean bottom sediments indicate that these extreme conditions, extremely slow rates of sulfate reduction, can prevail in nature (Vinogradov et al., 1962; Kaplan et al., 1963).

Bacterial Fractionation in Nature

The processes of sulfate reduction in nature are widespread, sulfide in sedimentary rocks and many elementary sulfur deposits are formed by bacterial reduction of sulfates. Isotope distribution studies have shown that these sulfide and elemental sulfur deposits are depleted in S^{34}, compared to sulfate in qualitative agreement with laboratory reduction experiments.

However, the dependence of isotope fractionation on metabolic rate and therefore environmental factors suggest a range of simple isotope fractionation factors is to be expected in natural processes. However, barring extreme conditions, there is a wide range of conditions and metabolic rates which give intermediate isotope effects in the neighborhood of 12 to 18‰.

In nature, the simple isotope fractionation factor or isotope effect involved in the bacterial reduction of SO_4^{2-} may be multiplied many times by a batch-type process. Just as in the case of a batch distillation process, the residual liquid becomes richer and richer in the heavy component as the distillation proceeds, so in the bacterial reduction of a limited reservoir of $SO_4^{2-}{}_{sol}$, the SO_4^{2-} residual in solution becomes richer and richer in S^{34} as a larger and larger fraction of the SO_4^{2-} is reduced to H_2S: this type of process undoubtedly accounts for the variable and often very high S^{34} enrichments in gypsum and anhydrite evaporites which are deposited from inland seas or basins (Thode and Monster, 1963).

However, in the case of sulfate reduction in contact with a very large reservoir such as the open ocean, the isotope ratio of the ocean sulfate will not change perceptibly as reduction takes place and the S^{2-} produced will be depleted in S^{34} with respect to ocean sulfates. The S^{2-} in ocean bottom sediments will reflect isotope fractionation both in sulfate reduction at ocean interface (open system), and in sulfate reduction after burial (closed system or batch process). The extent of S^{34} depletion in sediment S^{2-} will therefore depend on the proportion of each.

Ocean water sulfate, which is today enriched in S^{34} by $\sim 20\%_0$ with respect to meteoritic and igneous rock sulfur is remarkably uniform over its breadth and depth in sulfur isotope content (Ault and Kulp, 1959; Thode, Monster, and Dunford, 1961). It is significant that reduced sulfur partially derived from ocean water sulfate, such as SO_2 in rainwater formed from H_2S by photochemical processes in the upper atmosphere and sulfides in recent sediments formed by bacterial reduction of seawater sulfates in shallow water, are depleted in S^{34} by about $15\%_0$ with respect to present-day ocean sulfate. Also, studies of the sulfur isotope distribution in petroleum and associated evaporites suggest that the S^{34} content of the oceans has been changing in a somewhat cyclic fashion throughout geological time and that the S^{34} contents of the petroleum deposits, in general, are displaced $\sim 15\%_0$ with respect to the associated anhydrite and gypsum deposits. This indicates, as one might expect, that the sulfur in petroleum has been derived from seawater sulfate and is first reduced by bacterial action before being introduced into the petroleum or the proto-petroleum (Thode and Monster, 1963).

It is clear from studies of the distribution of the sulfur isotopes in the salt domes of Louisianna and Texas, including the cores of the sulfur wells, that the elemental sulfur was indeed formed by the bacterial reduction of sulfate in the gypsum and anhydrite cap rock (Thode, Wanless, and Wallouch, 1954). The elemental sulfur is depleted in S^{34} with respect to associated sulfate, indicating that the sulfate has been reduced by bacterial action in a batch-type process. The larger is the fraction of SO_4^{2-} reduced in the sulfur well core samples, the greater will be the overall fractionation. Carbon isotope ratios of the secondary carbonate in which the elemental sulfur is found also confirms this view. The C^{13} content of this carbonate was up to $50\%_0$ light as compared to standard limestone carbonate, indicating that it was of organic origin and that further fractionation of carbon isotopes occurred in the conversion of the original organic matter to CO_2 and H_2O by the bacteria.

Sulfite Reduction

The sulfur isotope fractionation which occurs in the reduction of sulfite (SO_3^{2-}) by cell suspensions of *Desulfovibrio desulfuricans* has also been measured (Harrison and Thode, 1958; Kaplan and Rittenberg, 1964) at 30°C with sulfite and lactate concentrations similar to that used in sulfate reduction. The sulfur isotope fractionation factor was found to be 1.011 on the average, which is identical with the fractionation found for the reduction of sulfate under similar conditions (same rate of H_2S production). Using rapidly growing cells, it was found that increased rates of reduction of sulfite were accompanied by a marked decrease in isotope fractionation. Thus the magnitude of the sulfur isotope effect and its dependence on metabolic rates is identical in the bacterial reduction of both sulfite and sulfate.

This suggests a mechanism for the reduction of SO_3^{2-} similar to that proposed above for SO_4^{2-} reduction, namely an equilibrium step with little or no isotope effect, followed by a unidirectional S–O bond breaking step with a large isotope effect. The overall isotope fractionation would therefore depend on the relative rates of these two steps which, in turn, would depend on nutrient concentrations and metabolic rates.

Sulfide Oxidation

Laboratory experiments have indicated little sulfur isotope fractionation in either the chemical or biological oxidation of sulfide to elemental sulfur. *Thiobacillus concretivorus* and *Chromatium sp* are capable of oxidizing sulfide to sulfur, which is stored internally until sulfide supply is exhausted. In the second step, the internally stored sulfur is oxidized to sulfate. The isotope effect in both these steps has been studied using growing cultures and in some cases, resting cell suspensions. Little or no sulfur isotope fractionation was found in the oxidation of sulfide to sulfur or in step I. In most cases, some slight depletion of S^{34} in the first S^0 to be formed was reported, although the maximum isotope effect measured was usually less than $2\%_0$ (Ford, 1957; Kaplan and Rafter, 1958; Kaplan and Rittenberg, 1964).

In the oxidation of internally stored sulfur to sulfate by growing thiobacillus and Chromatium, similar results have been reported. The fractionation factor expressed in terms of the ratio of rate constants for the reactions

$$\text{Chromatium}$$
$$S^{32}{}^\circ + 4[0] \xrightarrow{k_1} S^{32}O_4{}^{2-}$$

$$S^{34}{}^\circ + 4[0] \xrightarrow{k_2} S^{34}O_4{}^{2-}$$

or k_1/k_2 was obtained from the isotope ratio of the initial sulfur and product sulfate and the extent of reaction in the usual manner. The fractionation factors reported indicated isotope effects of less than ±2‰ (McElcheran, 1951; Jones and Starkey; 1957, Ford, 1957).

In recent experiments of Kaplan and Rittenberg (1964), growing cultures of thiobacillus and Chromatium are used to oxidize sulfide under conditions which yield both S^0 and SO_4^{2-}, as well as intermediate polythionates ($S_xO_6^{2-}$). In thiobacillus experiments, the SO_4^{2-} product was found to be depleted in S^{34} by $\sim$ (10–15‰) and the $S_xO_6^{2-}$ fraction enriched in this isotope by about the same amount. Since the ratio of SO_4^{2-}-sulfur to $S_xO_6^{2-}$-sulfur was nearly unity, the results suggest some sulfur isotope disproportionation between the oxidation products under the conditions of their experiments. Under conditions where oxidation intermediates build up in the nutrient solutions, such isotope disproportionation can be expected. For example, in the decomposition of thiosulfate $S_2O_3^{2-}$ to form S^0 and SO_3^{2-}, large kinetic isotope effects result.

However, with the exception of the possibility of some isotope disproportionation between oxidation products under certain circumstances, there would appear to be little isotope fractionation in microbial processes involving sulfide or sulfur oxidation. Certainly, in nature the primary process responsible for sulfur isotope fractionation must be the bacterial reduction of sulfate where very large isotope effects are involved.

H. G. THODE

References

Ault, W. V., and Kulp, J. L., 1959, "Isotopic geochemistry of sulphur," *Geochim. Cosmochim. Acta*, **16**, 201–235.

Bigeleisen, J., and Wolfsberg, M., 1958, "Aspects of Isotope Effects in Chemical Kinetics," Vol. 1 in (Prigogine, I., editor), "Advances in Chemical Physics," New York, Interscience Publishers, pp. 15–76.

Bigeleisen, J., and Mayer, M. G., 1947, "Calculation of equilibrium constants for isotopic exchange reactions," *J. Chem. Phys.*, **15**, 216–267.

Ford, R. W., 1957, "Sulphur Isotope Effects in Chemical and Biological Processes," Ph. D. Thesis, McMaster University, Hamilton, Canada.

Harrison, A. G., and Thode, H. G., 1958, "Mechanism of the bacterial reduction of sulphate from isotope fractionation studies," *Trans. Faraday Soc.*, **54**, 84–92.

Harrison, A. G., and Thode, H. G., 1957, "The Kinetic Isotope Effect in the Chemical Reduction of Sulphate," *Trans. Faraday Soc.*, **53**, 1–4.

Ishii, M., 1953, "The Fractionation of Sulfur Isotopes in the Plant Metabolism of Sulphates," Master's Thesis, McMaster University, Hamilton, Canada.

Jones, G. E., and Starkey, R. L., 1957, "Fractionation of Stable Isotopes of Sulfur by Microorganisms and their Role in Deposition of Native Sulfur," *Appl. Microbiol.*, **5**, 111–115.

Kaplan, I. R., Emery, K. O., and Rittenberg, S. C., 1963, "The distribution and isotopic abundance of sulphur in Recent marine sediments off Southern California," *Geochim. Cosmochim. Acta*, **27**, 297–331.

Kaplan, I. R., and Rittenberg, S. C., 1964, "Microbiological Fractionation of Sulfur Isotopes," *J. Gen. Microbiol.*, **34**, 195–212.

Kaplan, I. R., and Rafter, T. A., 1958, "Fractionation of Stable Isotopes of Sulfur by Thiobacilli," *Science*, **127**, 517–518.

McElcheran, D. E., 1951, "Fractionation of Sulphur Isotopes by Bacterial Processes," Master's Thesis, McMaster University, Hamilton, Canada.

Mills, G. A., and Urey, H. C., 1940, "The Kinetics of Isotopic Exchange Between Carbon Dioxide, Bicarbonate Ion, Carbonate Ion and Water," *J. Am. Chem. Soc.*, **62**, 1019–1026.

Thode, H. G., Shima, M., Rees, C. E., and Krishnamurty, K. V., 1964, unpublished work, McMaster University, Hamilton, Canada.

Thode, H. G., and Monster, J., 1963, "The Sulphur Isotope Abundances in Evaporites and in the Ancient Oceans," *Vernadsky Memorial Volume*, II, Moscow.

Thode, H. G., Monster, J., and Dunford, H. B., 1961, "Sulphur Isotope Geochemistry," *Geochim. Cosmochim. Acta*, **25**, 159–174.

Thode, H. G., Kleerekoper, H., and McElcheran, D., 1951, "Isotope Fractionation in the Bacterial Reduction of Sulphate," *Research (London)*, **4**, 581–582.

Thode, H. G., Macnamara, J., and Collins, C. B., 1949, "Natural Variations in the Isotopic Content of Sulphur and their Significance," *Can. J. Res.*, **B27**, 361–375.

Thode, H. G., Wanless, R. K., and Wallouch, R., 1954, "The origin of native sulphur deposits from isotope fractionation studies," *Geochim. Cosmochim. Acta*, **5**(6), 286–298.

Trofimov, A., 1949, "Isotopic composition of sulphur in meteorites and in terrestrial objects," *Dokl. Akad. Nauk S.S.S.R.*, **66**, 181–184.

Tudge, A. P., and Thode, H. G., 1950, "Thermodynamic Properties of Isotopic Compounds of Sulphur," *Can. J. Res.*, **B28**, 567–578.

Urey, H. C., 1947, "Thermodynamic properties of isotopic substances," *J. Chem. Soc.*, **1947**, 562–581.

Urey, H. C., and Greiff, L. J., 1935, "Isotopic exchange equilibria," *J. Am. Chem. Soc.*, **57**, 321–327.

Vinogradov, A. P., 1958, "Isotopic composition of sulphur in meteorites and in the earth," in (Extermann, R. C., editor), "Radioisotopes in Scientific Research," Vol. II, New York, Pergamon Press, pp. 581–591.

Vinogradov, A. P., Grinenko, V. A., and Ustinov, V. I., 1962, "Isotopic composition of sulphur compounds in the Black Sea," *Geokhimiya*, **10**, 871–873.

Cross-references: *Anhydrite; Carbon; Carbon Isotope Fractionation; Cycles—Geochemical; Sea-*

water, Chemistry; Seawater, History; Sulfate Reduction—Microbial; Sulfide Mineral Oxidation—Microbial; Sulfides in Sediments; Sulfur: Element and Geochemistry; Sulfur Cycle; Sulfur Oxidation—Bacterial. Vol. IVB: *Isotopic Variations in Mineral Deposits.* Vol. V: *Salt Domes.* Vol. VI: *Petroleum—Origin.*

SULFUR OXIDATION—BACTERIAL

Geologically the most important group of sulfur-oxidizing bacteria are the Thiobacilli. Different species in this group vary as to the preferred conditions for sulfur oxidation. The most frequently encountered members of this group are *Thiobacillus thiooxidans, T. ferrooxidans, T. thioparus,* and *T. denitrificans.* These organisms all grow autotrophically on elemental sulfur, i.e., they oxidize sulfur to sulfuric acid for energy and reducing power, which they use for CO_2 assimilation in protoplasmic synthesis. *T. thiooxidans* and *T. ferrooxidans* grow, and oxidize sulfur best, at pH 2.0 to 3.5. They may lower the pH of their growth medium to 1.0 or less. *T. thiooxidans* has an optimal rH_2 of 24 to 26 (Eh = +362 to +620 MV). Such an rH_2 range is probably also optimal for *T. ferrooxidans* when it oxidizes sulfur. *T. thioparus* and *T. denitrificans* grow and oxidize sulfur best around pH 7, but the growth of *T. thioparus* is slow. They lower the pH of the medium during growth to about 5. *T. thioparus* has an optimum rH_2 of around 20 to 26 when oxidizing sulfur in contrast to an optimum rH_2 of 10 to 16 when oxidizing hydrogen sulfide or thiosulfate. All foregoing Thiobacilli, except *T. denitrificans,* require O_2 for sulfur oxidation. *T. denitrificans* can substitute nitrate for O_2 as electron acceptor. Some other Thiobacilli can also oxidize sulfur, but their importance in nature remains to be established. All Thiobacilli can oxidize thiosulfate, and some can oxidize hydrogen sulfide and tetrathionate. The oxidation of hydrogen sulfide to sulfur by *T. thioparus* is of major importance in most epigenetic and syngenetic sulfur depositions in nature. *T. novellus* is unable to oxidize hydrogen sulfide, sulfur, or tetrathionate. It, unlike other Thiobacilli, oxidizes thiosulfate only very slowly. For a more complete survey of sulfur-oxidizing Thiobacilli see the discusson by Sokolova and Karavaiko (1968).

The mechanism of sulfur oxidation by Thiobacilli is not completely understood. However, recent work has shown that *T. thiooxidans* contains an enzyme system that oxidizes sulfur to sulfite, and another which oxidizes sulfite to sulfate (Suzuki and Silver, 1966). Although it was once postulated that, before being oxidized, sulfur first had to be reduced to sulfide by the cell, more recent evidence suggests that the sulfur combines with glutathione to form a polysulfide which then is oxidized to sulfite and glutathione by the sulfur oxidase enzyme system (*ibid.*). Similar enzyme systems are used in thiosulfate oxidation (*ibid.*).

In nature, Thiobacilli are found in soil, sulfur deposits, and fresh and saline waters in contact with oxidizable inorganic sulfur compounds (Silverman and Ehrlich, 1964; Sokolova and Karavaiko, 1968). Prevailing Eh and pH conditions as well as the form of the oxidizable sulfur compounds determine which of the Thiobacilli shall be active.

Sulfur oxidation is not restricted to Thiobacilli. Green and purple sulfur bacteria can oxidize sulfur as can some heterotrophic microorganisms. However, the extent of their activity with respect to sulfur oxidation is generally thought to be negligible in nature.

The activity of sulfur-oxidizing bacteria is detrimental in the exploitation of sulfur deposits because of their destructive activity, which is initiated as soon as a given deposit comes in contact with air in a moist environment (Sokolova and Karavaiko, 1968). The amount of sulfuric acid produced by *T. thiooxidans* may be great enough to result in major alteration of the sulfur-associated host rock. Calcite may be decomposed and aluminum in aluminosilicates may be mobilized (*ibid.*).

H. L. EHRLICH

References

*Silverman, M. P., and Ehrlich, H. L., 1964, "Microbial formation and degradation of minerals," *Advan. Appl. Microbiol.,* **6,** 153–206.
*Sokolova, G. A., and Karavaiko, G. I., 1968, "Physiology and Geochemical Activity of Thiobacilli," U.S. Department of Commerce, Clearing House for Federal Scientific and Technical Information, Springfield, Va., English transl.
Suzuki I., and Silver, M., 1966, "The initial product and properties of the sulfur-oxidizing enzyme of Thiobacilli," *Biochim. Biophys. Acta,* **122,** 22–33.

*Additional bibliographic references may be found in this work.

Cross-references: *pH-Eh; Sulfide Mineral Oxidation—Microbial; Sulfides in Sediments; Sulfur: Element and Geochemistry.*

SUPERGENE

Supergene is a term used in geology to denote at or near-surface locus of mineral deposition.

1. Epigenetic or syngenetic mineral formation caused by descending meteoric water, connate water, weathering, precipitation from solution, biological action, etc. Examples include: (a) *epigenetic-supergene:* Supergene enrichment forming chalcocite in preexisting copper veins near the water table; (b) *syngenetic-supergene:* Supergene formation of manganese nodules on the sea floor.
2. Opposite to *hypogene* (q.v.).
3. Broadly equivalent to *epigene* (q.v.).

RAYMUNDO J. CHICO

References

Schieferdecker, A. A. G. (editor), 1959 "Geological Nomenclature," Royal Geol. and Mining Soc. of the Netherlands, 523pp.

Stamp, L. D. (editor), 1961, "A Glossary of Geographical Terms," London, Longmans, Green & Co., 539pp.

Cross-references: *Connate Water; Epigenesis; Hypogene; Meteoric Water;* Vol. I: *Manganese Nodules.* Vol. IVB: *Economic Geology; Mineral Deposits—Classification.*

SUSPENDED WATER

Water that occurs between the land surface and the zone of saturation was termed "vadose water" by Meinzer (1923). Most writers have considered the terms "vadose water" and "suspended water" to be synonymous. Meinzer's definition, however, is concerned primarily with the physical location of the water. Suspended water is commonly transported to the water table by the processes of recharge, and Meinzer considered the movement downward through the vadose zone as a transient condition of saturation. Details of the processes by which suspended or vadose water moves down to the water table are not completely known. The descending solutions carry with them solutes derived by reactions with minerals in the soil zone, and probably are altered in composition by various factors during transit. In the more arid climatic zones a deposit of calcium carbonate and other minerals may be formed a short distance below the land surface as a result of the precipitation of solutes from descending water. These deposits can cause a nearly impermeable layer to be formed, composed mostly of mineral grains cemented by calcium carbonate, and thus prevent direct recharge to groundwater over large areas. The deposits are commonly called "caliche" in the western United States (see Vol. VI, *Caliche*).

Where permeability is not significantly impaired, the vadose or suspended water must represent the vehicle whereby solutes derived from weathering reactions, or added through the activities of man, are carried to the groundwater reservoir. The residual solutes from irrigation water that are left in the soil, for example, must be kept from rising in concentration to levels where they would interfere with plant growth. The usual method of control is to add sufficient surplus water to the soil so that some percolates below the root zone, carrying the surplus solutes with it. Excess quantities of readily soluble fertilizer, such as liquid ammonia, or ammonium nitrate may also be carried through the zone of suspended water.

Very little information has been obtained on the actual composition of suspended water. In one of the few detailed studies that have been published, Stewart and others (1967) found rather large amounts of dissolved nitrogen species in water in the vadose zone beneath irrigated fields and feed lots in the South Platte River valley of Colorado. The composition of solutions draining from saturated soil obviously would depend on many factors. Some indication of composition of such solutions is given in Table 1. Analysis 1 is for a water draining from a saturated soil in Kentucky, presumably to become part of the runoff in a nearby stream. This solution is probably higher in dissolved salts than might be expected where rocks are less soluble and vegetation less abundant. Analysis 2 represents the extract from a saturated sample of irrigated soil in the Pecos River valley near Roswell, New Mexico. The abundant solutes in this extract are partly from the rather highly mineralized groundwater (dissolved solids concentration over 3000 ppm) that was used to irrigate the soil, and partly from the solution of gypsum which was present in the soil.

Analyses 3 and 4 represent two wells of different depth in the alluvium of the Rio Grande valley near Albuquerque, N.M. The area had a high water table, only two feet below the land surface in the shallow well represented by analysis 3. The analysis indicates the higher solute concentration sometimes observed in the upper part of very shallow groundwater bodies, owing to extraction of water by evaporation and use of water by vegetation, leaving residual solutes behind. The concentration of solutes in the main groundwater body at greater depth is indicated by analysis 4. The latter represents a nearby well thirty-five feet deep which is supplied from water occurring farther below the surface.

JOHN D. HEM

SUSPENDED WATER

Table 1. Composition of Soil Water and Groundwater (ppm)

Source	1	2	3	4
Silica (SiO_2)	10	—	21	24
Iron (Fe)	.05	—	—	—
Manganese (Mn)	.01	—	—	—
Calcium (Ca)	54	610	130	97
Magnesium (Mg)	25	161	27	12
Sodium (Na)	8.7	311	218	29
Potassium (K)	1.5	0	—	—
Bicarbonate (HCO_3)	240	—	394	201
Carbonate (CO_3)	0	—	—	—
Sulfate (SO_4)	41	2,000	445	144
Chloride (Cl)	9.5	404	86	28
Fluoride (F)	.1	—	.4	0
Nitrate (NO_3)	14	—	0	0
Dissolved solids	288	3,560	1,160	456
pH	7.3	—	—	—
Specific conductance (micromhos/cm at 25°C)	486	4,210	1,620	671

1. Drainage from saturated soil near Goose Creek, Jefferson County, Kentucky (Hendrickson and Krieger, 1964).
2. Solution extracted from saturated soil in filter press at 140 psi. Soil from 0-58 in. in productive cotton field near Roswell, N.M. Irrigated with groundwater having specific conductance of 3270 micromhos/cm at 250°C; May 24, 1940 (Natural Resources Planning Board, 1942).
3. Water from observation well in Sec. 9, T. 11 N., R. 3 E., Bernalillo County, N.M., north of Albuquerque. Water about 2 ft below land surface; Oct. 28, 1936 (Scofield, 1938).
4. Water from domestic well 35 ft deep, Sec. 9, T. 11 N., R. 3 E., Bernalillo County, N.M.; Nov. 3, 1936 (Scofield, 1938).

References

Hendrickson, G. E., and Krieger, R. A., 1964, "Geochemistry of natural waters of the Blue Grass Region, Kentucky," *U.S. Geol. Surv. Water Supply Paper* **1700**, 135pp.

Meinzer, O. E., 1923, "Outline of ground-water hydrology with definition of terms," *U.S. Geol. Surv. Water Supply Paper* **394**.

National Resources Planning Board, 1942, "The Pecos River joint investigation-reports of the participating agencies," Washington, D.C., *Natl. Res. Plan. Bd.*, p. 270.

Scofield, C. S., 1938, "Quality of water of the Rio Grande basin above Fort Quitman, Texas," *U.S. Geol. Surv. Water Supply Paper* **839**, 244–245, 254–255.

Stewart, B. A., Viets, F. G., Jr., Hutchinson, G. L., and Kemper, W. D., 1967, "Nitrate and other water pollutants under fields and feed lots," *Environ. Sci. Technol.*, **1**, 736–739.

Cross-references: *Groundwater; Porosity and Permeability; Runoff; Vadose Water; Water Table; Weathering, Chemical.*

SYNGENESIS

Noun. 1. In rock and ore genesis processes, minerals or mineral deposits are syngenetic if their formation (deposition, crystallization, precipitation, etc.) took place before or during the formation, crystallization, precipitation, etc., of the enclosing rock; the term syngenesis refers only to a time relation and does not say anything about the source of materials (Fersman, 1922).

2. Syngenetic deposits are "those formed by processes similar to those which have formed the enclosing rocks and, in general, simultaneously with them" (Lindgren, 1930).

3. Sometimes a synonym of congruency (Amstutz and Chico, 1959); e.g., ore bodies' congruency with submarine slumping features (Chico, 1963).

4. Opposite to epigenesis [see *Ore Genesis*, or *Katagenesis* (of Fersman)].

5. In sedimentology (lithification and authigenesis), the almost equivalent term is *syndiagenesis*.

Adjective. *Syngenetic:* Of, relating to, or produced by syngenesis. Examples: the *syngenetic* nature of chert development; the possible *syngenetic* nature of chromite in the Bushveld Complex, etc.

Compound adjective. (a) *Syngenetic-supergene* (e.g., formation of iron pan in rice paddies; i.e., all mineral matter forms at the same time due to local interactions, weathering, or descending fluid action. Compare with caliche and tin alluvial deposits).

(b) *Syngenetic-hypogene* (e.g., formation of tantalite clusters within pegmatite; i.e., all mineral matter forms at the same time but from an ascending source).

Raymundo J. Chico

References

Amstutz, G. C., 1961, "Syngenesis and epigenesis in petrography and the study of mineral deposits," Pt. 1, *Intern. Geol. Rev.*, **3**, 119–140; Pt. 2, *ibid.*, **3**, 202–226.

Chico, R. J., 1963, "Submarine slumping and ore genesis discussion," *Econ. Geol.*, **58**(7), 1193 (abstract).

Fersman, A. E., 1922, "The Geochemistry of Russia, I," Leningrad, Goskhimizdat.

Lindgren, W., 1930, "Pseudo-eutetic textures," *Econ. Geol.*, **25**, 1–13.

Lovering, T. S., 1963, "Epigenetic, diplogenetic, syngenetic and lithogene deposits," *Econ. Geol.*, **58**(3), 315–331.

Cross-references: Vol. IVB: *Economic Geology, Ore Genesis; Mineral and Ore Deposits.*

T

TANTALUM: ELEMENT AND GEOCHEMISTRY

Tantalum (Ta) was discovered in 1802 by the Swedish chemist Ekeberg. It is listed in group VB of the periodic table.

Properties

Some of the important properties of tantalum are listed below:

Atomic number: 73
Atomic weight: 180.95
Atomic radius: 1.41Å coordination of 8
              1.47Å coordination of 12
Ionic radius: 0.68Å coordination of 6
Electron configuration: $1s^2, 2s^2, 2p^6, 3s^2, 3p^6, 3d^{10}, 4s^2, 4p^6, 4d^{10}, 4f^{14}, 5s^2, 5p^6, 5d^3, 6s^2$
Valence: 5
Isotopic abundance: $Ta^{181} = 100\%$ (alternatively, the only stable isotope is Ta^{181})
Structure: body-centered cubic $a_0 = 3.306$Å
Density: 16.6 g/cc
Melting point: ~3000°C
Boiling point: ~5400°C
Electrical conductivity: 13.5×10^{-6} ohm-cm at 20°C
Thermal expansion: 70.9×10^{-7} per°C (25–1000°C)

Minerals

Tantalum usually occurs in complex oxides combined with Nb, Ti, Y and the lanthanides. Most of the industrial needs for the metal are supplied by the minerals tantalite and columbite. High-grade tantalite contains up to approximately 84% Ta_2O_5 and columbite about 77% Nb_2O_5; between these maxima there is a whole solid solution series of minerals varying between tantalum and niobium.

Minerals containing tantalum are listed below:

Tantalite	$(Fe,Mn)(Ta,Nb)_2O_6$
Columbite	$(Fe,Mn)(Nb,Ta)_2O_6$
Mossite	$(Fe,Mn)(Nb,Ta)_2O_6$
Tapiolite	$(Fe,Mn)(Ta,Nb)_2O_6$
Microlite	$(Ca,Na)_2Ta_2O_6(O,OH,F)$
Euxenite	(Y,Er,Ce,U,Pb,Ca) $[(Ti,Nb,Ta)_2(O,OH)_6]$
Blomstrandite	(Y,Ce,Ca,Na,Th,U) $[(Nb,Ta,Ti)_2O_6]$
Fergusonite	$Y(Nb,Ta)O_4$
Stibiotantalite	$SbTaO_4$
Bismutotantalite	$BiTaO_4$
Samarskite	$(Y,Er)_4[(Nb,Ta)_2O_7]_3$
Yttrotantalite	$Y_4(Ta_2O_7)_3$
Simpsonite	$Al_2Ta_2O_8$

Abundance

The average content of tantalum in igneous rocks, according to Rankama, is 2.1 g/ton. The average tantalum contents in the various groups of igneous rocks quoted by Rankama and Sahama are listed in Table 1.

Rankama also reported that silicate meteorites contain 0.38 g/ton of tantalum and meteoritic irons contain 0.06 g/ton.

Geochemistry

In the earth, tantalum is strongly enriched in the uppermost part of the lithosphere and is closely associated with niobium, due to the similarities in chemistry and ionic radii. Tantalum-containing minerals are concentrated in the late stages of magmatic differentiation, and the most typical geochemical feature of tantalum is its concentration in pegmatites.

In the processes of weathering and sediment deposition, a small fraction of tantalum remains in resistant minerals like cassiterite, columbite, tantalite, and rutile. They are deposited in residual sediments such as sands and sandstones. However, most of the tantalum is precipitated with hydrolyzate sediments, such as clays and bauxites. Only bauxite showed an appreciable amount of tantalum among the sedimentary rocks examined by Rankama.

Economic Geology

Much tantalite is produced as a by-product from cassiterite placer mining. Tantalite de-

TABLE 1. TANTALUM CONTENT IN IGNEOUS ROCKS

Rock	Ta (g/ton)
Monomineralic rocks	0.7
Ultrabasic rocks	1.0
Eclogites	0.7
Gabbros	1.1
Diorites	0.7
Granites	4.2
Syenites	2.0
Nepheline syenites	0.8
Basic alkalic rocks	1.2

posits in pegmatites, such as those in Brazil and the Congo, are the most important sources of tantalum. Tantalum, unlike niobium, does not occur in significant amounts in carbonatites and alkaline igneous rocks. The major producers of tantalum-containing minerals (1965) are Brazil, Mozambique, Democratic Republic of the Congo (Kinshasa-Leopoldville), Rhodesia, Portugal, Nigeria, and Australia. (See also *Niobium—Economic Geology*.)

The major consumption of tantalum falls in the following areas: (*1*) ferro-tantalum-niobium steel; (*2*) pure metal in chemical equipment; (*3*) pure metal in electronic applications; (*4*) tantalum carbide; (*5*) special nonferrous alloys; and (*6*) pure metal for surgical, dental, and miscellaneous purposes.

R. C. DOMAN

References

Barton, W. H., 1962, "Columbium and Tantalum, A Materials Survey," *U.S. Bur. Mines Inf. Circ.*, **8120**, 110pp.
Goldschmidt, V. M., 1958, "Geochemistry," Oxford Clarendon Press, 730pp.
Hampel, C. A., 1968, "Encyclopedia of the Chemical Elements," New York, Reinhold Publ. Corp.
Lyman, T. (editor), 1961, "Metals Handbook, Properties and Selection of Metals," Vol. I, Eighth ed., The American Society for Metals, 1300pp.
Miller, G. L., 1959, "Tantalum and Niobium," New York, Academic Press, 767pp.
Quarrell, A. G. (editor), 1961, "Niobium, Tantalum, Molybdenum and Tungsten," New York, Elsevier Publ. Co., 413pp.
Rankama, K., and Sahama, Th. G., 1950, "Geochemistry," Chicago, University of Chicago Press, 912pp.
Stevens, R. F., Jr., 1965, "Columbium and Tantalum," *U.S. Bur. Mines Min. Yearbook.*
Van der Veen, A. H., 1963, "A Study of Pyrochlore," *K. Nederl. Geol.-Mijn., Geol. Ser.* **dl 22**, 188pp. (English).

Cross-references: *Lanthanides; Niobium (Columbium); Solid Solution. Vol. V: Magmatic Differentiation.*

TAR SANDS—*See Vol. IVB*

TECHNETIUM: ELEMENT AND GEOCHEMISTRY

Primordial technetium (atomic number 43) does not occur in nature since all the isotopes of the element undergo radioactive decay with half-lives that are short relative to the age of the earth. Minute quantities of one isotope, that of mass number 99, however, are found in uranium ores where it is produced by spontaneous fission. Traces of Tc^{99} and other long-lived technetium isotopes also may be formed by the interaction of cosmic rays with ores such as molybdenite, etc. The recent discovery in the light from S-type and N-type stars of lines belonging to the optical spectrum of Tc–I has been of great significance, as it may be inferred that nucleogenesis is occurring continuously in some parts of the universe.

Three very long-lived technetium isotopes have been identified (Table 1), and all knowledge about the chemical behavior of ponderable quantities of the element has been derived from researches on one of these.

The weakly β^--active isotope Tc^{99} is formed in relatively large yield in the fission of uranium in nuclear reactors. Therefore, as a consequence of the large-scale use of nuclear energy in the future, multikilogram quantities of the element should become available.

Elemental fission-product technetium is a light silver gray metal with an atomic weight of 98.913 (measured mass spectrometrically). The weakly paramagnetic metal crystallizes in an hexagonal close-packed arrangement with a calculated density of 11.50 g/cm^3 at 25°C. The melting point of technetium is 2250 ± 50°C, and its boiling point has been estimated as ~ 4700°C. Metallic technetium, either alone or in alloys with molybdenum or niobium, exhibits electrical superconductivity at low temperatures.

The chemistry of technetium is much more like that of its cogener, rhenium, than like manganese. In its compounds, Tc is mainly heptavalent, although tetravalent and pentavalent states of combination exist as do less stable compounds of the divalent and trivalent oxidation states. The heptavalent state in small concentrations in aqueous solutions as pertechnetate ion, TcO_4^-, dramatically inhibits the corrosion of soft iron. A nuclear isomer may be produced in technetium–99 which decays with a six-hour half-life and low-energy gamma rays. This isotope Tc^{99m} is widely used medically for the diagnosis of thyroid disorders in human patients.

GEORGE E. BOYD

References

Boyd, G. E., 1959, "Technetium and Promethium," *J. Chem. Ed.*, **36**, 3–14.
Colton, R., 1965, "The Chemistry of Rhenium and Technetium," New York and London, John Wiley & Sons, 185pp.
Kotegov, K. V., Pavlov, O. N., and Shvedov, V. P., 1965, "Technetium," Moscow, Atomizdat (for English translation, see "Advances in Inorganic Chemistry and Radiochemistry," Vol. 11, New York, Academic Press, Inc., 1968).

TECHNETIUM: ELEMENT AND GEOCHEMISTRY

TABLE 1. LONG-LIVED TECHNETIUM RADIOISOTOPES

Mass Number	Half-Life (years)	Decay Mode
Tc^{97}	2.6×10^6	Orbital electron capture
Tc^{98}	1.5×10^6	Beta-ray emission
Tc^{99}	2.2×10^5	Beta-ray emission

Peacock, R. D., 1966, "The Chemistry of Technetium and Rhenium," Amsterdam, Elsevier Publ. Co.

Tribalat, S., 1957, "Rhénium et Technétium," Paris, Gauthier-Villars.

Cross-references: *Niobium (Columbium); Radioactive Isotopes.* Vol. II: *Nucleogenesis.*

TEKTITES—*See* Vol. II

TELLURATES, TELLURITES—*See* **SELENATES AND TELLURATES: SELENITES AND TELLURITES**

TELLURIUM: ELEMENT AND GEOCHEMISTRY

Properties

Tellurium (Te), a minor element, atomic number 52, was discovered in 1782 by Franz Müller von Reichenstein and first isolated by Klaproth in 1798. In stable form it is a bright silvery hexagonal crystalline metalloid having trigonal symmetry. Its atomic weight is 127.60, density 6.24 (25°C), melting point 450°C, and boiling point 990°C. As a member of Group VIA of the Periodic Table, tellurium bears a definite resemblance to selenium and sulfur in a number of its properties, but is more metallic in nature than either of these elements.

Eight stable isotopes of tellurium are known having mass numbers of 120, 122, 123, 124, 125, 126, 128, and 130 and abundances of 0.08, 2.46, 0.87, 4.61, 6.99, 18.71, 31.79, and 34.49%, respectively.

Hexagonal crystalline tellurium is characterized by a marked anisotropy in many of its physical properties including the following: electrical conductivity, linear compressibility, linear coefficient of thermal expansion, optical absorption and index of refraction with respect to polarized light, and galvanomagnetic properties.

Geochemistry

Tellurium is closely associated in nature with sulfur and selenium and is commonly found in pyritic ores. In contrast to selenium which occurs isomorphously with sulfur in the sulfide lattice, tellurium forms distinct mineral species which are present as micro segregations in the sulfide mineral. Although the concentrations of selenium and tellurium are variable and generally quite low in sulfide minerals, such minerals, notably copper sulfides, constitute a major source of both elements which are recovered as by-products in the refining process.

The most extensive data on the occurrence and abundance of tellurium in the earth's crust are given by Sindeeva (1964). On the basis of recent mineralogical evidence Vinogradov (1956) has given the relative terrestrial abundance of tellurium and selenium to be $10^{-7}\%$ and $6 \times 10^{-5}\%$, respectively.

Thirty-nine tellurium minerals have been identified the most common being hessite, Ag_2Te; nagyagite, $Au(Pb,Sb,Fe)_3(TeS)_{11}$; sylvanite, $AuAgTe_4$, and tetradymite, Bi_2Te_2S. Native tellurium has been encountered in small quantities.

The deposition of tellurium (and selenium) in concentrations of industrial significance in primary mineral-forming processes is associated with magmatic and postmagmatic sulfide formations chiefly at the hydrothermal stage. According to Markham (1960), there are three basic environments in which distinctive telluride minerals occur:

Telluride mineralization *(1)* with tertiary volcanic rocks; *(2)* within Precambrian rocks; and *(3)* in contact metamorphic deposits.

Further studies are required particularly on the crystallo-chemistry of tellurium to define more accurately the occurrence and distribution of tellurium.

W. CHARLES COOPER

References

Akaiwa, H., 1966, "Abundances of Selenium, Tellurium, and Indium in Meteorites," *J. Geophys. Res.*, **71**(7), 1919–1923.

Collins, A. G., Waters, C. J., and Pearson, C. J., 1964, "Methods of analyzing oilfield waters: selenium and tellurium," *U. S. Bur. Mines, Rept. Invest.*, 6474, 19pp.

Markham, N. L., 1960, *Econ. Geol.*, **55**, parts 1 and 2, 1148, 1460.

Sindeeva, N. D., 1964, "Mineralogy and Types of Deposits of Selenium and Tellurium," New York, Interscience Publ., 363pp.

Vinogradov, A. P., 1956, "Laws of Distribution of Chemical Elements in the Earth's Crust," *Geokhimiya (Geochemistry)*, No. 1.

TERBIUM: ELEMENT AND GEOCHEMISTRY

Cross-references: *Anisotropism; Selenium; Sulfides; Sulfur.* Vol. V: *Magmatism.*

Properties

Elemental terbium (Tb) is a very reactive, silvery white *lanthamide* (q.v.), or *rare earth* (q.v.), metal with atomic number 65 and outer electronic structure $4f^96s^2$. It occurs as one stable isotope, Tb^{159}.

Terbium is determined analytically by neutron activation, x-ray spectrography, and by absorption spectrometry (it has very sharp spectral lines). In geologic environments, terbium forms highly ionic bonds. It forms weak complexes. The oxidation state is tripositive, with ionic radius 0.923 Å; tetrapositive and tripositive terbium occur in the oxide Tb_4O_7, but no evidence for tetrapositive terbium has been found in geologic environments. Generally chemical behavior is similar to the alkaline earths, and almost identical to other heavy rare earths.

Delafontaine and de Marignac (in 1878) were the first to isolate terbium from other rare earths; they separated it from "old erbia" which they renamed "new terbia." Terbium is named after Ytterby, Sweden, where Arrhenius discovered gadolinite, the first heavy rare earth mineral.

Minerals

Terbium occurs in highest concentration in rare-earth minerals that favor heavy rare earths, generally nitrates, sulfates, carbonates, niobates, and tungstates with cation-coordination number of six to eight, often containing structural halogens and alkali metals. Typical examples are:

Xenotime (Y, heavy rare earths) PO_4

Fergusonite (Y, heavy rare earths) (Nb, Ta) O_4

Geochemical Behavior

Schmitt et al. (1964) have established a value of 0.049 ppm (parts per million) terbium in chondrites and 0.57 ppm in eucrite achondrites. The standard deviation within a meteorite class in only about 20%. In meteoritic matter, volatization may have redistributed alkali metals but not rare earths: terbium has a boiling point of 2800°C.

The terbium concentration in average basalt is 1.08 ppm, and the range is less than a factor of ten (Haskin et al., 1966). Oceanic and continental basalts have similar ranges. Partial melting of mantle peridotite yields a basaltic liquid with enriched terbium and a mantle source area depleted in terbium.

Igneous differentiation series may tend toward higher or lower terbium concentrations, although the total variation is less than for light rare earths. The relative abundance of terbium and other heavy rare earths remains almost unchanged. Haskin et al (1966) report 2.0 ppm terbium in an Ascension Island basalt and 1.13 ppm terbium in a trachyte differentiate. Towell et al. (1965) found 0.633 ± 0.057 ppm in the Rubidoux Mountain Leucogranite phase of the Southern California batholith, probably a differentiate (by crystal settling) of the San Marcos Gabbro, with 0.478 ± 0.043 ppm.

Average granite with 60 to 70% silica has 1.02 ppm, but the range of values for granites is even greater than for basalts.

In most basaltic, syenitic and granitic rocks, terbium is mainly in mafic rock forming minerals, probably substituting for divalent calcium (ionic radius 0.99 Å).

Terbium and other heavy rare earths form complexes and are transported in alkali, halogen, and carbon dioxide-rich hydrothermal fluids; metasomatic and pegmatoid deposits formed from such fluids may have high concentrations of terbium and higher ratios of terbium to light rare earths than granites.

Terbium probably has about a 400-year residence time in seawater (which is alkaline), about the same as other heavy rare earths. Removal by adsorption on clay particles, as well as depletion in surface water in areas of high organic productivity, are likely processes. High secondary terbium concentrations in phosphatic fish remains today occur only in deep water; paleozoic fish remains also have high rare earths. Manganese nodules have 690 ppm, whereas Pacific Ocean water has only about 0.2×10^{-6} ppm (Goldberg et al., 1963).

In soils, the relative amounts of terbium in detrital, clay, and carbonate fractions is not well known. Sedimentary rocks have about 1.30 ppm, with a variation of a factor of five. In a study of rare earths in Russian platform soils, Balashov et al. (1964) found higher terbium in alkaline soils, the terbium in groundwater precipitating as hydroxide; lower concentrations are in acid soils, probably due to removal of complexes.

Biologic concentration of rare earths has been found in hickory leaves; rare earths have been used to trace opium.

Economic Geology

The major source of terbium has been from a light rare earth mineral, monazite in beach sands; the accessory monazite originally present in granitic rocks resists weathering and is concentrated by sedimentary processes. Australia, Brazil, Ceylon, India, Malaysia and South Africa mine the largest quantities. Less than one per cent of the monazite is terbium. Recently, the bastnaesite deposit at Mountain Pass, California has been an important source of terbium, again in low concentration.

High purity terbium has a very limited market. Most terbium is sold unpurified in rare earth concentrates used for polishing, petroleum cracking catalysis, lighter flints, arc carbons, and in alloys.

ROBERT KAY

References

Balashov, Y. A., et al., 1964, "The effects of climate and facies environment on the fractionation of the rare earths during sedimentation," *Geochem. Intern. No. 1*, **10**, 951–969 (English transl.).

Goldberg, E. D., et al., 1963, "Rare-earth distributions in the marine environment," *J. Geophys. Res.*, **68**, 4209–4217.

Haskin, L. A., et al., 1966, "Meteoric, Solar, and Terrestrial Rare-Earth Distributions," in (Ahrens, L. H., et al., editors), "Physics and Chemistry of the Earth," Vol. 7, Oxford, Pergamon Press, pp. 167–321.

Rankama, K., and Sahama, T. G., 1950, "Geochemistry," Chicago, University of Chicago Press, 912pp.

Schmitt, R. A., et al., 1964, "Rare-earth, yttrium and scandium abundances in meteoric and terrestrial matter—II," *Geochim. Cosmochim. Acta*, **28**, 67–86.

Towell, D. C., et al., 1965, "Rare-earth distributions in some rocks and associated minerals of the batholith of Southern California," *J. Geophys. Res.*, **70**, 3485–3496.

Cross-references: *Complexes; Hydrothermal Solutions; Rare Earths.*

TERRACES, FLUVIAL—ENVIRONMENTAL CONTROLS—See Vol. III

THALLIUM: ELEMENT AND GEOCHEMISTRY

Thallium (Tl) was discovered in 1861 by Sir William Crookes when examining a spectrum of a sample of the residue from a sulfuric acid plant. The name is from Latin *thallus* a green twig, referring to the green line in the spectrum observed. In the freshly prepared elemental condition thallium has a bright silvery appearance; however, on exposure to the atmosphere it becomes coated with a black oxide. Its atomic number is 81; atomic weight 204.39; density 11.85 (20°C); melting point 303°C; boiling point 1457°C. In the Periodic Table it comes in Group IIIA, between indium and lead. The element and many of its compounds are toxic and care should be taken in handling them.

Geochemistry

Thallium exhibits the properties of both a lithophylic and a chalcophylic element. As a lithophylic element it is found concentrated late in the magmatic crystallization of potassium minerals such as feldspars and micas. As a chalcophylic element it is found in nature as independent minerals and as a substitute element in minerals such as galena.

Estimates regarding its abundance in the earth's crust vary by a factor of ten. Goldschmidt reported a value of 0.3 g/T (ppm), Mason 0.6 to 1.0 g/T, and Ahrens reported 3 g/T. On the basis of a ranking of the relative abundance of the element, Mason lists thallium as fifty-eighth.

Some limited data show the presence of thallium in meteorites in the range of 0.15 to 0.3 ppm. Thallium has not been found to be present in the solar atmosphere. Its presence has been observed in seawaters as well as in mineral waters.

The following minerals of thallium have been found in nature:

Mineral	Formula	Thallium Content (%)
Crookesite	$(Cu,Tl,Ag)_2Se$	16–19
Hutchinsonite	$PbS(Tl,Ag)_2S \cdot 2As_2S_3$	18–30
Verbaite	$Tl_3S \cdot 3(As,Sb)_2S_3$	29–30
Lorandrite	$Tl_2S \cdot As_2S_3$	59–60
Avicennite	$7Tl_2O_3 \cdot Fe_2O_3$	—

The ionic and atomic radii of thallium and its relationship to some associated elements of geochemical interest have been summarized by Goldschmidt and are listed as follows:

Ionic radii

Tl^+	1.49Å	Tl^{3+}	1.05Å
Rb^+	1.49	Y^{3+}	1.05
K^+	1.33	In_3^+	0.92

Metallic radii		*Covalent radii*	
Tl	1.71Å	Tl	1.47Å
Pb	1.75	Pb	1.46
In	1.57	In	1.44
Zn	1.37	Zn	1.31

From a comparison of the univalent ionic radius of thallium with those of univalent potassium and rubidium, its presence as a substituted element for K+ in potash and plagioclase feldspars can be understood as well as its accompaniment with rubidium with which it is concentrated during the late stages of crystallization in igneous rocks.

Economics

The most important economic source of the element is thalliferous galena in which thallium is present as a replacement element for lead. It is in the treatment of these types of ores that the thallium is sufficiently concentrated to be recovered as a by-product metal.

Thallium, being volatile at lead and zinc smelting temperatures, can be collected from the flue dusts, usually as an oxide or sulfate. In these dusts it often is in association with cadmium, indium, selenium and tellurium.

The uses of thallium are somewhat restricted because of its toxicity to man. On the other hand, it is an effective rat killer and ant killer, but lacking an effective antidote, this application is not recommended. Within strict limits it can be used in pharmaceuticals. Certain thallium salts have unique photosensitivity, particularly in the long-wavelength, infrared range. Alloys of thallium are employed in the electrical industry for switches and fuses. A binary alloy with mercury freezes at only $-59°C$ (approaching minimal polar or stratosphere temperatures). A ternary alloy of thallium with mercury and indium with a freezing point of $-82°F$ ($63.33°C$) has been patented. Some organic compounds of thallium are liquid at room temperature. The compounds have a high density and are used for sink-float separations. Thallium nitrate may be added to glass to produce lenses with increased index of refraction. It is also used as an additive to counter electrode alloys used in the manufacture of selenium rectifiers.

S. C. CARAPELLA, JR.

References

Ahrens, L. H., 1947, "The abundance of thallium in the earth's crust," *Science*, **106**, 268.
Goldschmidt, V. M., 1954, "Geochemistry," London, Oxford University Press.
Howe, H. E., 1968, "Thallium," in (Hampel, C., editor), "Encyclopedia of the Chemical Elements," New York, Reinhold Book Corp., 706–711.
Klienert, R. Z., 1963, "Thallium, a rare metal—an impurity metal," *Z. Erzbergbau Metall huethenw.*, **16**, 67–76.
Mason, B. H., 1966, "Principles of Geochemistry," Third ed., New York, John Wiley & Sons, 329pp.
Vlasov, K. A. (editor), 1966, "Geochemistry and Mineralogy of Rare Elements," Jerusalem, Israel Program Sci. Transl., 2 vols. (transl. by Lerman, Z.).

Cross-references: *Geochemical Classification of Elements; Hydrothermal Solutions—Sulfide Transport; Lead.* Vol. IVB: *Plagioclase Feldspars; Rare Metal Ore Deposits.*

THERAPEUTIC ROLE OF GROUNDWATER—See MEDICINAL SPRINGS

THERMAL ANALYSIS (THERMOANALYSIS)

Thermal methods of analysis have been defined as including only those techniques in which some physical parameter of a system is measured as a function of temperature. To yield useful information the property measured must change significantly as a function of temperature and must therefore be a dynamic function of temperature (Wendlandt, 1964).

Thermal analysis or thermoanalysis is, therefore, the general term used to describe collectively the group of independent techniques by which some physical parameter is measured. Usually this data is recorded and plotted, either automatically or manually, as a continuous variation curve. The particular parametral variations measured are attendant upon chemical and physical changes, which occur in single or multicomponent samples as their temperature is varied, at a constant predetermined rate or in accordance with a more complex temperature variation program.

The individual methods comprising thermal analysis are summarized in Table 1.

Excellent overall coverages of thermal analysis techniques are given in the books by Wendlandt (1964) and Garn (1965); those of Smothers and Chiang (1966) and Mackenzie (1957) are devoted almost exclusively to *differential thermal analysis* (q.v.). In particular, differential thermal analysis (DTA) has been placed on an excellent basis by the recent publication of a punch card index containing all available DTA data for both minerals and chemical compounds (Mackenzie, 1962).

Of the seventeen thermal analysis methods listed in Table 1 some have direct and marked application to mineralogical, geochemical, and petrological studies, while others are of little or no value in these "geological" investigations. The geologically valuable methods have been defined and discussed in detail elsewhere under the following eight headings, which here are accompanied only by brief definitions.

(1) Differential Thermal Analysis (DTA). The measurement of energy changes (heat effects) produced in a substance by chemical or physical changes, as its temperature is raised at a uniform rate. These are measured as a

THERMAL ANALYSIS (THERMOANALYSIS)

Table 1. Thermal Analysis Methods[a]

Method	Instrument Used	Parameter Measured as a Function of Temperature
Thermal analysis	Calorimeter; thermal analysis unit	Temperature vs. time or heat content
Differential thermal analysis (DTA)	Differential thermal analysis unit	Temperature difference between sample and inert reference material
Derivative differential thermal analysis (DDTA)	Differential thermal analysis unit ± derivative computor	First derivative of temperature difference obtained from DTA
Thermogravimetric analysis (TGA) or thermogravimetry (TG))	Thermobalance	Change in mass
Differential thermogravimetric analysis (DTGA)	Thermobalance ± derivative computor or differential thermobalance	First derivative of change in mass obtained from TGA
Gas evolution analysis (GEA) or effluent gas analysis (EGA)	Gas thermal conductivity or density cell (± mass spectrograph or gas chromatograph, etc.)	Gas thermal conductivity or density indicating presence of evolved gas (± actual gas composition)
X-ray diffraction with sample temperatures above or below ambient	X-ray diffractometer; powder and single crystal cameras	Change in d-spacings
Thermoluminescence	Thermoluminescence unit employing a photomultiplier	Light emission
Oxyluminescence	Same as for thermoluminescence but with sample in oxygen	Light emission
Dynamic reflectance spectroscopy (DRS)	Spectroreflectometer	Sample reflectance
Electrical conductivity	Resistance bridge	Change in electrical resistance
Dilatometry	Dilatometer	Change in volume
Specific heat measurement	Differential calorimeter	Specific heat
Pyrolysis	Mass or infrared spectrometer or gas chromatograph	Pyrolysis fractions
Differential scanning calorimetric analysis	Differential calorimeter	Heat change supplied to sample
Thermometric titrimetry	Thermometric titrimeter	Temperature change vs. time or volume of titrant
Derivative thermometric titrimetry	Thermometric titrimeter ± derivative computor	First (± second) derivative of temperature change as obtained from thermometric titrimetry

[a] Modified after Wendlandt, 1964.

differential temperature (ΔT) and continuously recorded graphically as a function of temperature or time to produce a DTA curve or thermogram.

(2) Derivative Differential Thermal Analysis (DDTA). The continuous plot representing the first calculus derivative of the differential temperature (ΔT) data obtained by DTA.

(3) Thermogravimetric Analysis (TGA). The continuous measurement of weight variation as a function of temperature or time as a sample is heated at a uniform rate.

(4) Differential Thermogravimetric Analysis (DTGA). The continuous plot representing the first calculus derivative of the weight variation data obtained by TGA.

(5) Gas Evolution Analysis (GEA) or Effluent Gas Analysis (EGA). The measurement of density, volume or thermal conductivity of the gas liberated from a sample as its temperature is raised at a uniform rate.

(6) Variable Temperature X-Ray Diffraction. The x-ray diffraction study of materials whose temperatures may be raised, lowered, maintained or made to follow some more complex temperature variation program.

(7) Thermoluminescence. The emission of visible light by certain substances, when heated at a slow constant rate to temperatures below that of incandescence, the luminescent light energy being measured and recorded as a function of temperature to produce a "glow curve."

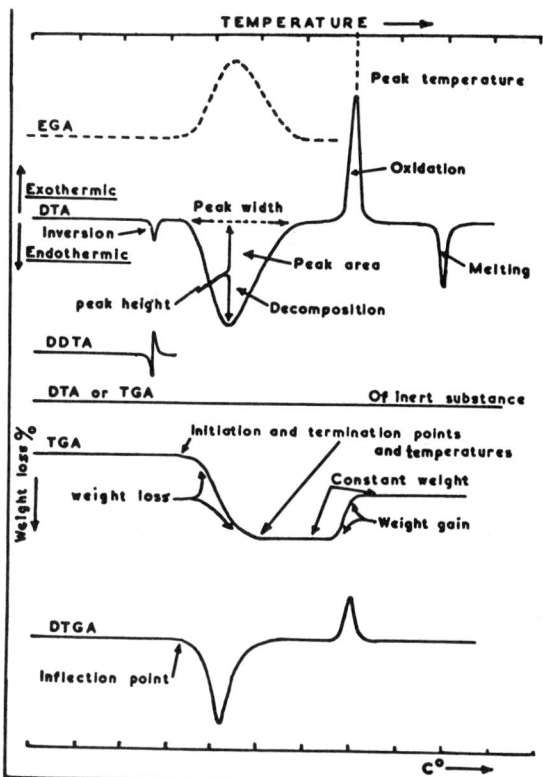

FIG. 1. A diagrammatic comparison of thermal analysis curves produced by methods 1 to 5 described in the text.

(8) *Oxyluminescence.* Similar to thermoluminescence except that light emission will take place only if oxygen is in contact with the surface of the material under test. This necessitates heating in an atmosphere of oxygen or air.

To assist in the clarification and evaluation of the data displayed by the thermal analysis curves produced by methods 1 to 5 above, and to highlight their interdependent and complementary nature, a comparison of curves produced by these methods has been made in Fig. 1.

Repeated reference to this figure in connection with the detailed discussion of these five methods of thermal analysis (which occur elsewhere) will prove of considerable value and assistance.

S. ST. J. WARNE

References

Garn, P. D., 1965, "Thermoanalytical Methods of Investigation," New York, Academic Press, 606pp.
Mackenzie, R. C. (editor), 1957, "The Differential Thermal Investigation of Clays," *Min. Soc. London* (Clay Minerals Group), 456pp.
Mackenzie, R. C. (compiler), 1962, "Scifax Differential Thermal Analysis Data Index," London, Cleaver-Hume Press Ltd. (plus first supplement, 1964).
Smothers, W. J., and Chiang, Y., 1966, "Differential Thermal Analysis," Second edition, Chemical Publ. Co., 633pp.
Wendlandt, W. Wm., 1964, "Thermal Methods of Analysis," New York, Interscience Publ., 424pp.

Cross-references: *Derivative Differential Thermal Analysis; Differential Thermal Analysis; Differential Thermogravimetric Analysis; Effluent Gas Analysis; Gas Evolution Analysis; Oxyluminescence; Thermogravimetric Analysis; Thermoluminescence; X-Ray Diffraction—Variable Temperature.*

THERMAL ENERGY—See Vol. IV B, GEOTHERMAL ENERGY

THERMAL POLLUTION

Thermal pollution is defined as heat added to natural bodies of water by the activities of man, which cause undesirable effects to the biota. It is sometimes called thermal enhancement if the ecological change proves desirable for the waters under consideration. Increased heat changes the physical properties of the water and thus the biota. It may reduce the oxygen content below levels required for some desirable organisms or for some major uses. A very slight increase in temperature may trigger excessive growth of undesirable organisms such as algae, may render infections lethal to some forms of aquatic life, may have profound effects on eggs, larvae, or fry, and may depress reproduction rates of fish or shellfish. Heat may adversely effect sedimentation and the ability of the water body to assimilate organic wastes, or it may inhibit the natural turnover of a lake.

The degree of thermal pollution (or enhancement) depends on: the type of receiving water, i.e., marine estuary, fresh-water lake, or river; the shape and volume of the receiving water; velocity; current; the presence or absence of stratification, particularly in lakes or estuaries; regional weather conditions; the chief uses of water in the area and their relative importance; and the location of the facility in the drainage basin.

Fisheries and preservation of the base of the food pyramid are affected most by pollution. Oyster cultures and some types of fin fish catches may be improved by the additions of heat.

Thermal pollution became significant early

THERMAL POLLUTION

in the twentieth century with the development of the steam-river generator and the chemical industry. Thermal pollution of the Columbia River by the nuclear industrial activity at Hanford, Washington triggered lethal infections with respect to a species of commercial salmon, thus providing an early documented ecological problem, which subsequent investigations indicate was far more complex than originally believed.

Approximately 80% of man-made heat is contributed by electrical generating stations. Fossil fuel plants dissipate about 10% of their heat to the atmosphere. Nuclear plants (at present operating at lower pressure and efficiency) reject about 50% more heat per kilowatt than the modern fossil fuel high-pressure plants. Condensor water discharge temperatures range from 10 to 20° (average 13–15°) above ambient temperature of the receiving water. This thermal pollution can be reduced, as at Oyster Creek, New Jersey, by mixing condensor water with volumes of bypass water. Cooling ponds or one of several kinds of cooling towers have been used where land values or climate permit.

The modern nuclear generating stations, which are two to five times the size of fossil fuel plants, require such large amounts of cooling water that thermal pollution will present increasingly difficult problems in water resource management. The planned Forked River, New Jersey plants will discharge through a pipeline to the ocean. The Newbold Island, New Jersey plant on the Delaware estuary will require cooling towers because of insufficient river flow. These latter solutions to thermal pollution may create environmental problems which have not been anticipated.

Radiation to the air and mixing with receiving waters rapidly dissipates thermal pollution effects. Model studies show that mixing zones need to be from one to five miles long with temperatures usually less than 5° in a small area around the discharge points before the receiving water returns to normal.

The Federal Water Pollution Control Act, as amended by the Water Quality Act, gives states the authority to establish water quality standards, which must be approved by the Secretary of the Interior. Thermal pollution is to be judged by several criteria, using the most sensitive criteria as the governing factors. Under most circumstances, temperatures that are only three to five degrees above ambient will be accepted. and all standards presented for approval require maximum temperatures of less than 100°F. Most states require maximum temperatures between 70 and 90°F.

The literature is fragmented, but becoming increasingly voluminous. Most information is now found in AEC documents or in publications related to the power industry.

KEMBLE WIDMER

References

Anon., 1970, *Proceedings of the Conference on International and Interstate Regulation of Water Pollution,* Columbia University, School of Law, March 12–13, 1970.

Bloom, S. C., 1970, "Environment Reporter," Monograph #4, Bureau of National Affairs, Inc., Vol. 1, No. 1.

Parker, F. L., and Krenkel, P. A., 1969 "Physical and Engineering Aspects of Thermal Pollution," Cleveland, Ohio, CRC Monoscience Series, 100pp.

Cross-references: *Ecology; Environmental Science; Environmental Pollution.*

THERMAL SPRINGS—See SPRINGS

THERMOCHEMISTRY

Thermochemistry is a branch of physical chemistry concerned with the heat effects which accompany chemical reactions, phase transformations (fusion of a crystalline solid, change from one crystalline modification to another, vaporization of a liquid), and solutions. When these processes evolve heat they are said to be *exothermic*. Concersely, those absorbing heat are *endothermic*. The heat effects are measured in calories, rather than joules, the fundamental unit of energy. A calorie is defined as being equal to 4.1840 joules.

Thermochemistry is based on the first law of thermodynamics, i.e., the law of conservation of energy. $\Delta E = q - w$ where ΔE is the change in internal energy of a system, q is the heat absorbed by the system, and w is the work done by the system. By convention, the heat given off by a system is negative, whereas the heat absorbed is positive. Similarly, the work done to a system is negative and that done by the system is positive. Both q and w depend upon the path of the reaction or the steps in which it was carried out, but ΔE is independent of path. ΔE is the same for the same initial and final states regardless of how one gets from one to the other.

Generally, the only kind of work performed by chemical reactions is pressure-volume work (except for the electrical work done by galvanic cells). If the volume remains constant, no work is done. Under conditions of constant volume, $\Delta E = q$.

The heat of a reaction equals the change in internal energy, volume remaining fixed. Usually reactions occur at constant pressure open to the atmosphere. Work is done against a confining pressure, P external.

$$w = \int_{V_1}^{V_2} P_{\text{ext.}}\, dV$$

At constant pressure

$$w = P_{\text{ext.}}(V_2 - V_1) = P_{\text{ext.}}\, \Delta V$$

$$\Delta E = q - P\Delta V$$

or rearranging

$$q = \Delta E + P\Delta V$$

A new function convenient for dealing with constant pressure processes can be introduced. This function, the *enthalpy, H,* is defined as

$$H = E + PV$$

Differentiating

$$dH = dE + PdV + VdP$$

At constant pressure $dP = 0$, so

$$dH = dE + PdV$$

or

$$\Delta H = \Delta E + P\Delta V = q$$

Therefore, the heat of a reaction at constant pressure equals the enthalpy.

The heats of reaction can be measured experimentally by calorimetric methods. Only a few types of chemical reactions can be determined in this way. A reaction, in order to be suitable for precise study must be rapid, complete, and without side reactions. In the field of organic chemistry, combustion reactions are particularly appropriate. A compound of C, H and O, when burned, usually forms only H_2O and CO_2, e.g.,

$$CH_4\, (g) + 2O_2\, (g) \rightarrow 2H_2O\, (l) + CO_2\, (g)$$
$$\Delta H°_{25°C} = -212{,}800 \text{ cal}$$

The heat of combustion $\Delta H°_{25°C}$ refers to the heat liberated by one mole of the substance burned at 25°C. Heats of inorganic reactions (heats of neutralization, solution, and complex formation) can be measured in a calorimeter open to the atmosphere.

The heat of a reaction is recorded by writing the balanced chemical reaction for the process, and the ΔE or ΔH value. The reaction for the production of water from the elements is: $H_2(g) + 1/2 O_2(g) \rightarrow H_2O(l)$ at 25°C and 1 atm ($\Delta H°_{25°C} = -68{,}317$ cal). The letters in parentheses refer to the physical states of matter. Thus O_2 and H_2 are gases, and H_2O is a liquid at room temperature and one atmosphere. The standard states are the common forms of the elements at one atmosphere and a given temperature (usually 25°C), and unit activity. When dealing with solids, one must specify the crystalline modification. For example, carbon exists as both graphite and diamond. The standard state of C is taken as graphite because it is the stable form at room temperature. The enthalpies of elements in their standard states at 25°C are zero. Thus:

$$C \text{ (graphite)} + O_2\, (g) \rightarrow CO_2\, (g)$$
$$\Delta H° = -94{,}052 \text{ cal}$$

$$C \text{ (diamond)} + O_2\, (g) \rightarrow CO_2\, (g)$$
$$\Delta H° = -94{,}505 \text{ cal}$$

Note the difference in the heats of combustion of the two polymorphs. The enthalpy content of diamond is 453 cal above that of graphite.

Hess's Law

Frequently one must know the heat of a reaction which cannot be determined experimentally. The desired heat effect may be found by combining chemical equations and corresponding $\Delta H°$ values. Hess's law permits $\Delta H°$ to be calculated regardless of the intermediate steps, because it states that the change in enthalpy is independent of path.

Example 1. Find $\Delta H°$ for the reaction

$$2Fe_2O_3\, (s) + 3C \text{ (graphite)} \rightarrow$$
$$4\,Fe\, (s) + 3CO_2\, (g)$$

at standard conditions. In tables such as those in the "Handbook of Chemistry and Physics," the heats of formation of CO_2 and Fe_2O_3 are given.

1. $C \text{ (graphite)} + O_2\, (g) \rightarrow CO_2\, (g)$
$$\Delta H°(CO_2) = -94{,}052 \text{ cal}$$

2. $2Fe\, (s) + 3/2\, O_2\, (g) \rightarrow Fe_2O_3\, (s)$
$$\Delta H°\, (Fe_2O_3) = -196{,}500 \text{ cal}$$

Multiply Eq. (1) by 3 to get 3 moles of CO_2. Then multiply Eq. (2) by 2 to get 2 moles of Fe_2O_3, reverse the equation and change the sign of $\Delta H(Fe_2O_3)$. Add the two equations algebraically. The final equation is:

$$2Fe_2O_3 + 3C \rightarrow 4Fe + 3CO_2$$

$\Delta H°$ for the reaction is $= 3\Delta H(CO_2) - 2\Delta H\,(Fe_2O_3)$

$$\Delta H°_{\text{reaction}} = 3(-94{,}052) - 2(-196{,}500)$$
$$= -282{,}156 + 393{,}000$$
$$= 110{,}844 \text{ cal}$$

Since $\Delta H°$ is positive, heat is absorbed and the reaction is endothermic. This reaction occurs in the blast furnace where hematite, Fe_2O_3, is reduced by carbon to pig iron, Fe.

THERMOCHEMISTRY

Example 2. Calculate the heat of reaction for the oxidation of pyrite, FeS_2. The balanced equation for the reaction is:

$$4FeS_2 \text{ (s)} + 11O_2 \text{ (g)} \rightarrow 2Fe_2O_3 \text{ (s)} + 8SO_2 \text{ (g)}$$

$$\begin{aligned}
\Delta H°_{reaction} &= 2\Delta H(Fe_2O_3) + 8\Delta H(SO_2) \\
&\quad - 4\Delta H(FeS_2) \\
&= 2(-196{,}500) + 8(-70{,}960) \\
&\quad - 4(-42{,}520) \\
&= -393{,}000 - 567{,}680 \\
&\quad + 170{,}080 \\
&= -790{,}600 \text{ cal}
\end{aligned}$$

Example 3. Cement is made from ground limestone and clay. The raw materials are heated to form many complex calcium silicates and aluminates. When water is added to cement, a series of reactions take place. Gypsum is added to control the speed of hydration. During the period of setting, reactions such as this hydrolysis slowly proceed:

$$Ca_2SiO_4 \text{ (s)} + H_2O \text{ (l)} \rightarrow CaSiO_3 \text{ (s)} + Ca(OH)_2 \text{ (aq)}$$

$$\begin{aligned}
\Delta H°_{reaction} &= \Delta H(CaSiO_3) + \Delta H(Ca(OH)_2) \\
&\quad - \Delta H(Ca_2SiO_4) - \Delta H(H_2O) \\
&= -377{,}400 - 239{,}680 \\
&\quad - (-538{,}000) - (-68{,}320) \\
&= -617{,}080 + 606{,}320 \\
&= -10{,}760 \text{ cal}
\end{aligned}$$

This is an exothermic reaction.

Spontaneous Combustion

Certain substances undergo slow oxidation at room temperature with the liberation of heat. If the material is confined in an enclosed space, the heat released by oxidation cannot escape and gradually causes the temperature to rise. When the temperature reaches the kindling point, combustion occurs. It has been suggested that spontaneous combustion is responsible for some explosions in coal mines. When any combustible substance is finely subdivided into a powder, the rate of oxidation increases tremendously. Thus a lump of coal may take a few hours to burn, whereas the same amount of coal ground to a fine powder will burn in sereval minutes. The slow oxidation of pyrite, FeS_2 (see example 2), associated with coal dust in the mine may generate enough heat to raise the temperature to the kindling point and set off a violent explosion.

The Temperature Dependence of ΔH

Up to this point, only ΔH for processes at 25°C has been considered. To compute ΔH for different temperatures, *Kirchhoff's law* may be utilized:

$$\Delta H \text{ at } T_2 - \Delta H \text{ at } T_1 = \int_{T_1}^{T_2} C_p \, dT$$

where C_p stands for the difference in heat capacities of the products and reactants of a reaction at constant pressure. For small ranges of temperature (20 to 30 degrees), C_p is assumed to be relatively constant. Then

$$\Delta H(T_2) - \Delta H(T_1) = \Delta C_p (T_2 - T_1)$$

If temperature ranges are larger, one may calculate C_p for each substance in the equation by a series of the form $C_p = a + bT + cT^2 + \ldots$, where a, b, and c are constants. For the whole reaction $\Delta C_p = \Delta a' + \Delta b'T + \Delta c'T^2 + \ldots$, where $\Delta a'$ is the sum of a terms for the products minus the sum of a terms for the reactants, and so on.

The calculation of enthalpies as shown for heats of reaction can be applied as well to changes of physical state (phase transformations) and to solutions.

VIVIEN GORNITZ

References

Garrels, R. M., and Christ, C. L., 1965, "Solutions, Minerals, and Equilibria," New York, Harper and Row, 450pp.

Hall, H. T., 1966, "Application of thermochemical data to problems of ore deposition," *Econ. Geol.*, **61,** 622–623.

Holland, H. D., 1965, "Some applications of thermochemical data to problems of ore deposits. II. Mineral assemblages and the composition of ore-forming fluids," *Econ. Geol.*, **60,** 1101–1166.

Krauskopf, K. B., 1967, "Introduction to Geochemistry," New York, McGraw-Hill Book Co., 721pp.

Moore, W. J., 1954, "Physical Chemistry," New York, Prentice-Hall, 592pp.

Robie, R. A., 1965, "Heat and free energy of formation of herzenbergite, troilite, magnesite, and rhodochrosite, calculated from equilibrium data," *U. S. Geol. Surv. Profess. Paper* **525-D,** 65–72.

Cross-references: *Enthalpy or Heat Content; Thermodynamics.*

THERMOCLINE

The thermocline (or metalimnion) is an intermediate water layer developed in a thermally stratified lake or reservoir, where the temperature decreases rapidly and density increases rapidly with depth. It lies between the *epilimnion* (q.v.) and the *hypolimnion* (q.v.). The thermocline, in natural lakes, is often observed at depths of 8–13 m below the surface, and has a thermal gradient in excess of 1°C/m. In impounded water bodies the position of the thermocline is dependent on the depth of the water outlet and upon the withdrawal rates.

In lakes which do not freeze over, both those in mild climates and those which are of very great size, this stratification is poorly or not at all developed, since circulation is good. In a climate with freezing winters, the thermocline begins to develop after the spring overturn and becomes increasingly well-defined and sharp through the summer. In late autumn, as the surface water becomes chilled and dense, there is another overturn and the thermocline is once again destroyed, until the next spring.

B.C.S.

References

Chow, Ven Te, 1964, "Handbook of Applied Hydrology," New York, McGraw-Hill Book Co., 1445pp.
Hutchinson, G. E., 1957, "A Treatise on Limnology," New York, John Wiley & Sons, Vol. 1, 1015pp.
Welch, P. S., 1952, "Limnology," Second ed., New York, McGraw-Hill Book Co.

Cross-references: *Epilimnion; Hypolimnion; Limnology.*

THERMODYNAMICS

Thermodynamics is a branch of physical science which is concerned with the relationships between all forms of energy such as heat, chemical energy, mechanical or electrical work, and the macroscopic properties of material systems. Because thermodynamics is concerned with the general laws governing the transformations of macroscopic systems, without consideration of the microscopic mechanisms (at the atomic or molecular level), the results are widely applicable and independent of the microscopic explanations, and thermodynamics is relevant to physics, chemistry, or biology as well as to geology, the applied sciences and engineering.

Classical thermodynamics is an exact science and the phenomenological relations are derived from the fundamental laws and from one another by straightforward mathematical arguments. The thermodynamic functions, such as enthalpy, entropy, or free energy, which are introduced to describe the properties of the systems, are defined unambiguously, but they cannot be predicted by theory and they must be determined experimentally. The definitions and relations of thermodynamics can be applied with confidence and exactness; the measured values of the thermodynamic functions, however, are subject to experimental errors, and the applications to practical problems are often limited by the lack of reliable data.

The interpretation on a molecular level of the experimental properties of macroscopic systems is the object of *statistical thermodynamics*. By statistical quantum mechanics, the properties of individual molecules and the nature of interactions among them can be related to the fundamental thermodynamic equations.

The conditions for equilibrium and the criteria for a spontaneous or natural process result from classical thermodynamics. The theories of *thermodynamics of irreversible processes* extend the range of validity to phenomena proceeding under steady-state conditions slightly removed from equilibrium.

The Laws of Thermodynamics

The fundamental laws of thermodynamics are postulates supported by experimental evidence. They do not require to be proven, but their acceptance is based on the fact that the conclusions derived from them are consistent with the results of all known experiments.

The *first law of thermodynamics* is the law of *conservation of energy* within an isolated system. For a system interacting with the surroundings, *the change in internal energy equals the heat received minus the work performed by the system.* A consequence of the first law is that the energy is defined as a thermodynamic function which depends only on the state of the system as described by its temperature, pressure, chemical composition, etc. The *enthalpy* (or heat content) and the heat capacity (or specific heat) are defined on the basis of the first law (see *Enthalpy or Heat Content*).

The *second law of thermodynamics* expresses the concept that natural processes occur in a given direction and cannot be reversed without producing some changes in the surroundings. For instance, heat always flows spontaneously from a warmer body to a colder one. By introducing a new thermodynamic function, called entropy which is defined in such a way that its change equals the heat exchanged divided by the absolute temperature, the second law can be stated as, *the entropy of an isolated system undergoing a natural or spontaneous transformation cannot decrease.* The microscopic interpretation of entropy relates it to a measure of the disorder of a system at the atomic level. From a practical viewpoint, the combined statements of the first and second law lead to the conclusion that, although the energy of the universe remains constant, all actual transformations from one form into another are accompanied by an

THERMODYNAMICS

irreversible degradation of the energy which makes it less available for further processes (see article on *Entropy*).

The *third law of thermodynamics,* also called the Nernst heat theorem, states that at *the absolute zero of temperature the entropy of a perfect crystalline substance becomes zero.* The postulate of the existence of temperature, which is fundamental for the significance of the other laws, is sometimes referred to as the *zeroth law of thermodynamics.*

Chemical Thermodynamics

The object of chemical thermodynamics is the application of the fundamental principles to systems of variable composition. It is concerned with the criteria for spontaneous processes and for chemical equilibria, the energy changes accompanying chemical reactions or state transformations, and the relations between the properties of materials and the thermodynamic variables such as temperature, pressure, and chemical composition.

Additional thermodynamic functions are introduced for the purpose of expressing the consequences of the first and second law in a more convenient form. The *Helmholtz free energy* and the *Gibbs free energy* (or free enthalpy) are defined respectively as the internal energy and the enthalpy minus the product of the entropy by the absolute temperature. They combine in a single relation the tendency towards lower energy and higher entropy, and they exhibit a minimum for equilibrium (or reversible) processes and they decrease for spontaneous transformations (see *Free Energy and Free Enthalpy*).

For dealing with phases of variable composition other functions must also be defined such as the *partial molal free energy* (or free enthalpy), also called *chemical potential,* which can be described as the "contribution of one constituent to the total free energy or free enthalpy of the system," and the *activity,* which is an "effective concentration." The introduction of these functions simplifies the statement of the conditions of chemical equilibrium, which can be expressed by the *law of mass action* relating the activities of the various constituents of a chemical system in equilibrium to a thermodynamic function called the equilibrium constant (see *Mass Action and Equilibrium Constant*).

Several relations, which are direct consequences of the fundamental laws are of particular importance for the geologist. The *Gibbs phase rule* defines the number of variables such as temperature, pressure, or concentrations which can be arbitrarily fixed when a given system is in equilibrium. It is useful in establishing phase diagrams (see *Phase Equilibria—Principles*). *Clapeyron's equation* is also of particular interest since it predicts the pressure temperature dependence of a phase transformation (see *Clapeyron's Equation*).

Application of Thermodynamics to Geology

Thermodynamics provides the mathematical relations for the quantitative study of geological problems involving phase transformations or chemical reactions such as those encountered in petrology or geochemistry. The physicochemical model which can be conceived by the geochemist to describe a geological problem always involves an idealization of the actual system. Thermodynamics provides the equations to verify that the model is consistent with the fundamental laws of nature. The same equations can then be used to extrapolate the experimental data and to predict with confidence the behavior of the system outside of the range of physical observations.

Several limitations must be kept in mind in order to make an optimum use of thermodynamics. Most of the relations to be applied by the geochemist are based on the assumption that the system under consideration is in a state of chemical equilibrium, and this condition is obviously not always realized for natural systems. Furthermore, systems of geological interest are usually very complex and simplifying assumptions have to be made in order to analyze them. Last but not least, experimental data are always needed, and sometimes they are not readily available or they are poorly reliable; in those cases approximations have to be made to obtain values of the functions which are involved in thermodynamic calculations.

Examples of the use of thermodynamics for the study of problems of geological interest are presented in the articles previously referred to, as well as in others, and they are discussed at greater length in some of the references listed below.

PAUL DUBY

References

Barth, T. F. W., 1962, "Theoretical Petrology," Second ed., New York, John Wiley & Sons, Inc., 416pp.

Garrels, R. M., and Christ, C. L., 1965, "Solutions, Minerals and Equilibria," New York, Harper and Row, 450pp.

Helgeson, H. C., 1969, "Thermodynamics of Hydrothermal Systems at Elevated Temperatures and Pressures," *Am. J. Sci.,* **267,** 729–804.

Kern, R., and Weisbrod, A., 1967, "Thermodynamics for Geologists," San Francisco, Freeman, Cooper and Co., 304pp; translated by Duncan McKie from "Thermodynamique de

Base pour Minéralogistes, Pétrographes et Géologues," Paris Masson et Cie, 1964.
Klotz, I. M., 1964, "Basic Chemical Thermodynamics," New York, W. A. Benjamin, Inc., 468pp.
Krauskopf, K. B., 1967, "Introduction to Geochemistry," New York, McGraw-Hill Book Co., 721pp.
Lewis, G. N., and Randall, M., 1961, "Thermodynamics," Second ed., revised by Pitzer, K. S., and Brewer, L., New York, McGraw-Hill Book Co., 723pp.
Mohan, B. H., 1963, "Elementary Chemical Thermodynamics," New York, W. A. Benjamin, Inc., 155pp.
Moore, W. J., 1962, "Physical Chemistry," Englewood Cliffs, N.J., Prentice-Hall, Inc., Third ed., 844pp.
Waser, J., 1966, "Basic Chemical Thermodynamics," New York, W. A. Benjamin, Inc., 278pp.

Cross-references: *Clapeyron's Equation; Enthalpy or Heat Content; Entropy; Free Energy and Free Enthalpy; Geochemistry; Geologic Thermometry; Hydrothermal Solutions—Sulfide Transport; Mass Action and Equilibrium Constant; pH-Eh Diagrams; Phase Equilibria; Thermochemistry.*

THERMOGRAVIMETRIC ANALYSIS

Thermogravimetric analysis (TGA) or thermogravimetry (TG) is the technique whereby the *weight* and temperature of a sample under test are continuously and preferably automatically recorded as it is heated up at a constant and linear rate by a *thermobalance*.

The heating, usually within the range ambient to 1100 or 1500°C, is achieved by means of a furnace which surrounds, but does not touch, the sample. Thus the sample remains freely suspended from the balancing mechanism which is actuated as the sample mass alters in response to chemical reactions produced as its temperature is progressively increased.

Thus variations in weight (w) are continuously recorded as a function of temperature (T) or time (t), i.e.,

$$w = f(T \text{ or } t)$$

These data are recorded as a TGA curve, thermogram (sometimes called a thermogravigram), or weight loss curve, preferably on either an X–time or X–Y recorder. (The term "weight loss curve" should be used only where *no weight gains* are recorded, even though the overall effect is one of weight loss.) For a detailed comparison and evaluation of these time- and temperature-based recording techniques see Garn (1965, p. 127).

The major components of a generalized thermobalance, as used by Wendlandt (1964), are shown schematically in Fig. 1.

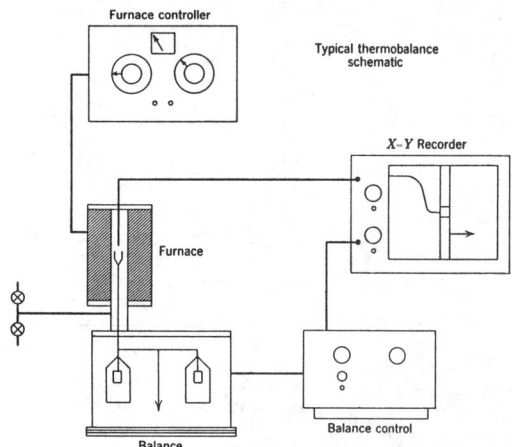

FIG. 1. The schematic layout of a typical automatic recording thermobalance (Wendlandt 1964).

Temperature ranges over which no weight changes occur are recorded as straight lines (conventionally oriented "horizontally" across the thermogram), whereas weight losses or gains are represented by curved departures toward the top and bottom of the thermogram, respectively (Fig. 2).

Typically, therefore, TGA thermograms exhibit, due to the alternation of these constant-weight "flats" or "plateaus" and weight variation "curves," a stepped configuration, the definition of which degenerates where separate weight-varying reactions closely follow each other. In this case the "flat" is replaced by a point of inflection (see Fig. 3 and Warne and Bayliss, 1962).

Quantitative weight changes, between two constant-weight "flats" or any other points of interest on the TGA curve, are obtained by reference to the weight-variation axis (ordinate) (see Fig. 3).

The points where TGA curves first depart from and return to a "horizontal" trace are called *initiation* and *termination temperatures*, respectively. They indicate the temperatures at which the weight varying reactions start and stop. Furthermore, the sloping portions of the curve are also indicative of reaction rate, i.e., the steeper the curve the faster the rate of weight variation and the reaction or process causing it.

Considerable importance has been placed on the accurate determination of various reference points, particularly the *initiation* temperature, although its actual location for any given substance depends upon a number of factors such as furnace and sample holder type and geometry, heating rate, furnace atmosphere, and recorder sensitivity. Even though these factors

THERMOGRAVIMETRIC ANALYSIS

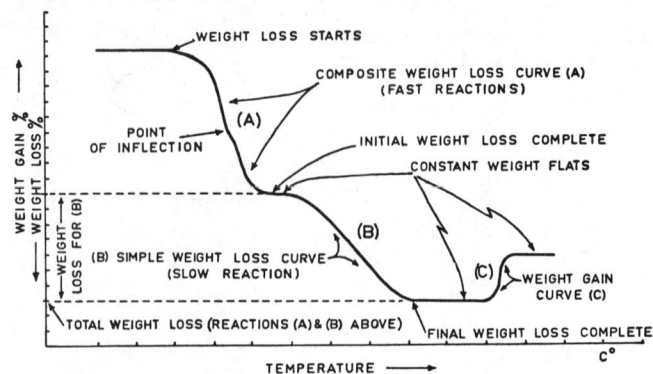

Fig. 2. Diagrammatic TGA curve to illustrate various features and configurations.

vary from unit to unit, the detailed characterization and comparison of the thermograms produced from a particular thermobalance under reproducible conditions still remains of paramount importance. To this end, the accurate determination of the weight-variation data, particularly the location of specific reference points, such as the initiation temperature, is often greatly enhanced by plotting the first derivative of the TGA curve. (For further details see *Derivative Thermogravimetric Analysis*).

History

The development of thermobalances and TGA has been reviewed in detail by Duval (1951, 1953, and 1963 (Second ed.)); Lukaszewski and Redfern 1961 (which contains Vigneron, 1949; and Jacque, Guiochon, and Gendrel, 1961, see in Wendlandt, 1964).

Excellent coverages of TGA and its instrumentation are given in the books by Duval (1963), Wendlandt (1964), and Garn (1965), the very comprehensive reviews by Gordon and Campbell (1960), Coats and Redfern (1963), and the four-part paper by Lukaszewski and Redfern (1961).

The first automatically recording balances were produced by Kuhlmann in 1910 and Abderhalden in 1913, whereas the technique of TGA and the term *thermobalance* appear to originate from Honda (1915). However, for approximately the next twenty-five years this technique attracted little attention and was kept viable mainly by the work of Guichard (1925) and Noshida (1927).

The details of the automatically recording thermobalance were published by Chevenard, Wacké, and de la Tullaye in 1944. The establishment of TGA on a sound, quantitative basis by Peltier and Duval (1947), the application to the thermogravimetric studies of some 1200 compounds by Duval (1953), (Second ed. 1963), and their critical examination by Newkirk (1960), were all major steps in the very rapid expansion in the use, applications, and acceptance of this technique which have occurred since 1947.

Instrumentation and Theory

The various commercial and noncommercial automatically operating and recording thermobalances have collectively been reviewed, in considerable detail by Duval (1951), Gordon and Campbell (1960), Jacque, Guiochon, and Gendrel (1961), Lewin (1962) and in books by Duval (1963), Wendlandt (1964), and Garn (1965).

Balance Types. In operation, thermobalances fall basically into two groups, the *null-point*

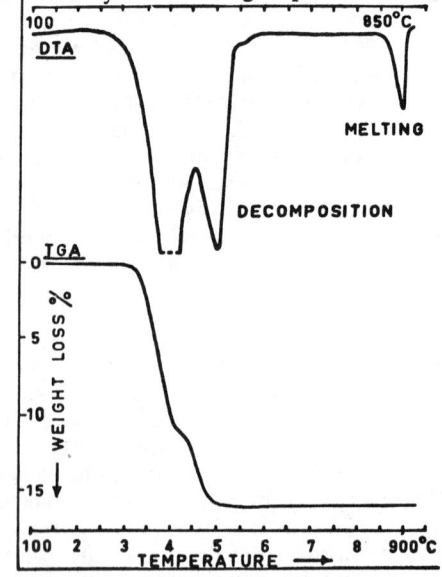

Fig. 3. Comparison of DTA and TGA curves of the same specimen of cerussite with their respective multipeak and multicurve configurations which indicate the presence of more than one reaction.

THERMOGRAVIMETRIC ANALYSIS

and the *deflection types* of instruments, while further subdivisions are based on the actual mechanisms used to obtain the weight variations, (Gordon and Campbell, 1960). (See Figs. 1, 4, 5, 6, and 7 for diagrammatic representations.)

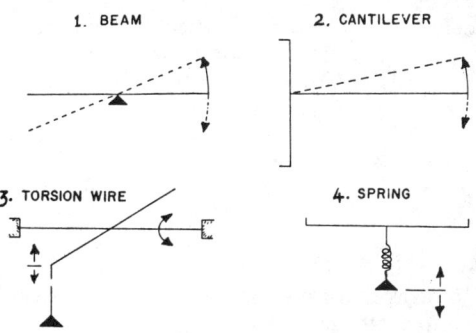

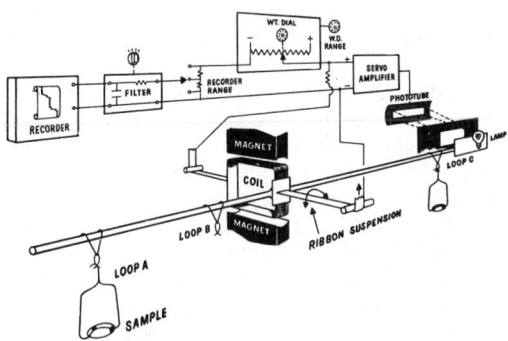

Fig. 4. The null-type balance. The restoring force, which is controlled by the null detector, N.D., is proportional to change in sample weight (Gordon and Campbell, 1960).

Fig. 6. Types of Deflection Balances. Beam (1), Cantilever (2), Torsion wire (3), Spring (4), where the actual deflection is proportional to sample weight changes (Gordon and Campbell 1960).

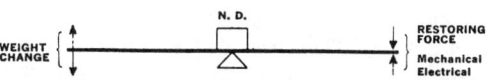

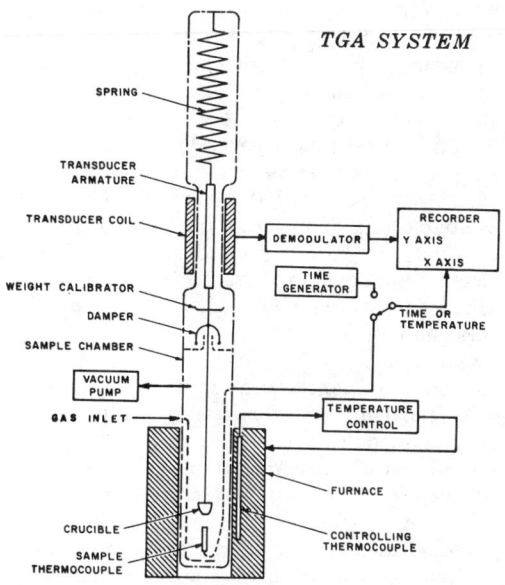

Fig. 5a. Detailed schematic diagram of the null-point Cahn Electrobalance. For operational details and thermobalance conversion see Wendlandt (1964, p. 58). *Courtesy Cahn Instrument Co.*

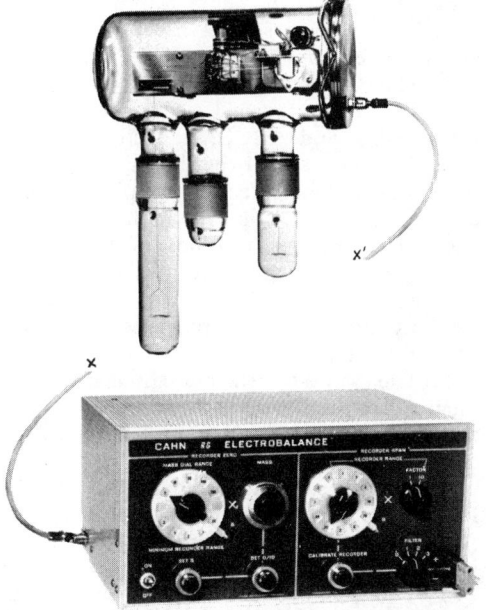

Fig. 7. Detailed schematic diagram of the helical spring, "Thermo-Grav," thermobalance. For operational details see Wendlandt (1964, p. 60). *Courtesy American Instrument Co.*

For *null-point* balances the restoring force may be exerted mechanically or electrically, whereas the null detector may be operated by means of mechanical, electrical, electromechanical, electrooptical, or radiation detection devices (see Fig. 4). In all cases, however, the weight of the sample is directly proportional to the restoring force, the varying magnitude of which is recorded to give the required TGA curve (Fig. 5).

Fig. 5b. Photo of Cahn Electrobalance (connection is broken at x and x'). *Courtesy Cahn Instrument Co.*

The *deflection type* balances involve the conversion, to a graphical representation (TGA curve), of the progressive balance "beam"

1177

deflections which are caused by, and are proportional to, changes in sample weight.

The *four* main deflection balance types, schematically shown in Fig 6, are:

1. *The beam balance*, where the deflections of a balance beam about a fulcrum are proportional to changes in sample weight (see also Fig. 1).

2. *The cantilevered balance*, one end of which is fixed, while the other is free to undergo deflection caused by weight variations in the sample attached to it.

3. *The torsion wire*, which is taut and rigidly fixed at both ends to serve as a fulcrum for an attached beam, the deflections of which are proportional to the torsional characteristics of the wire and changes in sample weight. For the use of this principle, *but* converted to a null-point type, see Fig. 5.

4. *The helical spring* or other appropriately mounted strain gages that stretch or contract in proportion to weight changes, the measurement of which is often achieved with suitable transducers (Fig. 7).

Thermobalance Operation Mechanisms. In respect to thermobalances in general, Gordon and Campbell (1960) have tabulated *three* valuable sets of information.

First, the various methods employed to detect any deviation from the null-point or changes in weight with *deflection type* balances, namely:

Optical: (1) light source-mirror-photographic paper; *(2)* Light source-shutter-photocell.

Electronic: (3) Capacitance bridge; *(4)* Mutual inductance: coil-plate, coil-coil; *(5)* Differential transformer or variable permeance transducer; *(6)* Radiation detector (Geiger tube); *(7)* Strain gauge.

Mechanical: (8) Pen electromechanically linked to balance beam or coulometer.

Secondly, the various methods used to apply the restoring force to *null-type* balances.

Mechanical: (1) Addition or removal of discrete weights, or beam rider positioning; *(2)* Incremental or continuous application of torsional or helical spring force; *(3)* Incremental or continuous chainomatic operation; *(4)* Incremental addition or withdrawal of liquid (buoyancy); *(5)* Incremental increase or decrease of pressure (hydraulic).

Electromagnetic interaction: (6) Coil-armature; *(7)* Coil-magnet; *(8)* Coil-coil.

Electrochemical: (9) Coulometric dissolution or deposition of metal at electrode suspended from balance beam or coulometer.

Thirdly, weight change recording techniques used in *automatic* balances.

Mechanical: (1) Pen linked to potentiometer slider; *(2)* Pen linked to chain-restoring drum; *(3)* Pen or electric arcing-point on end of beam; *(4)* Pen(s) linked to servo-driven photoelectric beam-deflection follower.

Photographic: (5) Light source-mirror-photographic paper using either a drum, time base, or flat bed, temperature-base mirror galvanometer.

Electronic: (6) Current generated in a transducing circuit, e.g., photocell, differential transformer, variable permeance transducer, strain gauge, bridge, radiation detector, capacitor, or inductor; *(7)* Current passing through the coil of an electromagnet.

Thermobalance Requirements. The details and relative merits of the individual thermobalances have been covered by the above reviewers; however, many very desirable features are now available. For example, automatic weight loading together with simultaneously recorded changes of both sample temperature and weight, with preselected continuously variable linear, nonlinear, or more complex heating rates and programs, including periods at constant temperature with furnace atmosphere conditions ranging from vacuum to static or dynamic oxidizing, reducing, or inert gases, at ambient or elevated pressures.

Furthermore, Lukaszewski and Redfern (1961) have listed the following criteria which are invaluable when considered in relation to the investigation requirements of any thermobalance under consideration.

1. The thermobalance should be capable of recording the variations in sample weight as a function of time and temperature.

2. The furnace should be capable of operation between ambient and 1500°C.

3. The sample weight variation and temperature should be recorded with accuracies of at least 0.01% and ±1°C, respectively.

4. The accuracy of the thermobalance determinations is not affected by physical effects caused by its normal functioning, e.g., radiation and convection currents and furnace winding magnetic effects. Not only should the latter be negligible, but no reaction between them and any conducting or magnetic materials under study should take place.

5. The sample crucible should always be identically positioned within the furnace, so that the sample and recorded temperature correspond.

6. The instrument should be designed to allow the heating of samples in various gaseous furnace atmospheres vacuum is also advantageous.

7. The instrument should be versatile in respect to heating rates, automatic control of temperature programs and the ease of their modification and change.

8. The balance should be adequately protected from the furnace and the weighing accuracy assured by reducing the wear on the knife edges and other moving parts to a minimum.

9. The thermobalance calibration can be easily checked periodically to ensure operational accuracy.

10. The chart speed of the continuous recorder can be varied with provision for the accurate recording of suitable time intervals.

Not all these features are included or required in every commercial thermobalance.

Factors Affecting Thermogravimetric Analyses

The actual configuration and data portrayed by the TGA curve may be influenced by two groups of factors, most of which are dealt with in detail by Wendlandt (1964) and Garn (1965), namely:

(1) Instrumental factors: (a) Furnace geometry and constitution; (b) Furnace heating rate and method; (c) Furnace atmosphere type and conditions; (d) Furnace atmosphere control method; (e) Sample holder, geometry, size, and composition; (f) Sample holder, thermal characteristics; (g) Sample holder, open "loosely" covered or sealed; (h) Random fluctuations in balance and recording mechanisms; (i) Recording mechanism sensitivity; (j) Electrostatic effects on balance mechanism; (k) Condensation on sample support; (l) Chart and recording speed; (m) Chart paper rulings; (n) Sample "buoyancy" due to gas flow and convection currents; (o) Temperature and weight measurement calibration.

(2) Sample Characteristics: (a) Form and nature; (b) Size and volume; (c) Particle size and size distribution; (d) Packing density; (e) Thermal conductivity; (f) Heat capacity; (g) Expansion or contraction effects on thermal characteristics; (h) Thermal characteristics of contaminants or dilutants; (i) Heat of reactions; (j) Speed of reactions; (k) Solubility of evolved gas in sample; (l) The influence of furnace and evolved gases on sample reaction; (m) Reaction between sample and sample holder.

Effects of Major Controllable Variables

A great number of these factors are inherent in the actual type of instrument, sample, and sample preparation technique used. Others, classifiable as *controllable variables,* may be adjusted by the operator to give the best results for individual samples and research aims, while the remaining factors are kept as reproducibly constant as possible. Of these, heating rate, recording or chart speed, and the

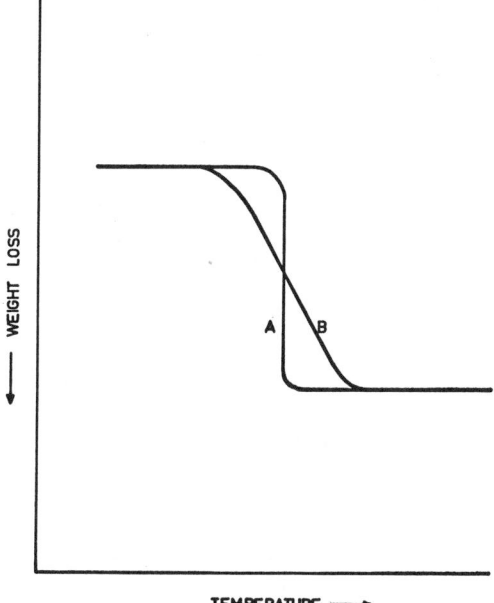

Fig. 8. An illustration of the simple steepening effects produced on the weight loss portion of the TGA curves obtained from identical samples by rapid (A) and slow (B) heating rates (after Lukaszewski and Redfern, 1961).

nature and type of the furnace atmosphere are most important.

Heating Rate or Program. Detailed studies have been made by Duval (1963), Newkirk (1960), Rynasiewicz and Flagg (1954), Lukaszewski and Redfern (1961), and Fruchart and Michel (1958).

The most fundamental effect is that the recorded reaction temperatures increase with increasing heating rates, but the reaction temperature range decreases. However, for fast irreversible reactions the associated weight variation curve is little affected by different heating rates.

Increased heating rates, where single reactions are involved, also result in a simple steepening of the slope of the corresponding weight variation portion of the TGA curve (Fig. 8). This is particularly important because the degree of slope is relatable to reaction rate. However, where several closely adjacent reactions occur, the weight variation "curves" progressively overlap as one reaction does not have time to reach completion before the next starts. For this reason the vital quantitative measurement, the constant weight "flat", first loses its definition (but may still be detected by an inflection in the curve) and finally may completely disappear (Fig. 9).

Among the wide variety of heating programs

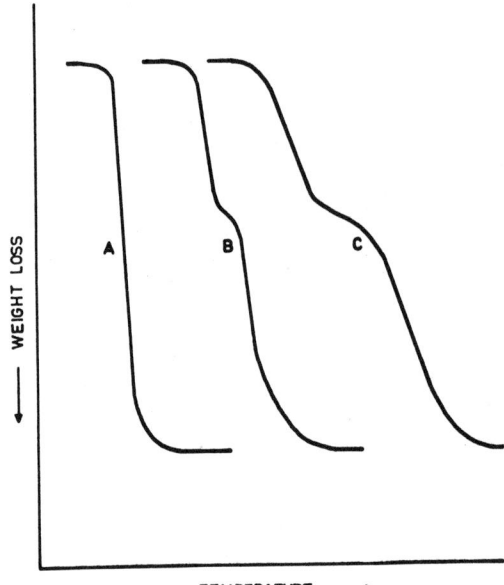

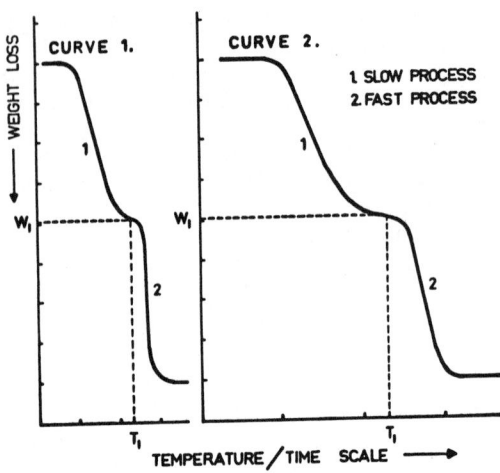

Fig. 10. The influence of chart speed on the TGA curve shapes for slow processes (1) and fast processes (2). Curves 1 and 2 recorded with slow and fast chart speeds, respectively (after Luzaszewski and Redfern, 1961).

Fig. 9. The effect of heating and recording rates on the separation and definition of closely adjacent but separate weight-loss reactions. Curves obtained from identical samples, with fast heating rate (A), slow heating rate (B) (both with identical slow recording rates), and a slow heating rate with fast recording rate (C) (after Lukaszewski and Redfern, 1961).

used, one of the most useful is the interruption of a uniform linear *"dynamic"* heating program, by maintaining a steady temperature for an appropriate period of time before resuming the original heating rate or terminating the run. This technique, termed *quasi-static* is particularly useful, where slow reactions are involved, and for the resolution of complex TGA curves. According to Lukaszewski and Redfern (1961) such curves are in general steeper than their *"dynamic"* counterparts.

Where the prime object is not how the sample attains a predetermined temperature, but its maintenance at this temperature for a set period of time during which the weight changes of interest occur and are recorded, the terms *static* or *isothermal* thermogravimetry are used. By contrast, all other heating programs *may* generally be termed *nonisothermal*.

Recording or Chart Speed. The actual speed at which the weight variation data is recorded may greatly affect the configuration of the resultant TGA curve, particularly the resolution of closely adjacent successive decomposition "steps." Lukaszewski and Redfern (1961) have clearly demonstrated that TGA curves: (a) for both slow and fast reactions, are flattened by increasing chart speed (see Fig. 9); (b) for slow reactions, suffer greater flattening with increased chart speed, than those from fast reactions (cf curves 1 and 2 Fig. 10); (c) where slow and fast reactions follow each other, decreased chart speeds cause a deterioration in the resolution of the individual weight variations and vice versa (cf curves 1 and 2, Fig. 10).

These authors have also described the very useful concept of the *resolution factor* (R) and defined it as follows:

"For a recording chart speed of 1 cm/min and a weight loss at time t represented on the chart by dw cm/min. The shape of the slope at time t depends on both chart speed and the weight scale in g/cm; thus,

$$R = \tan \alpha = \frac{dw}{dl}$$

which is independent of t. The limitations on R are the chart speed and the weight scale, a value for α around 45° or $dw/dl = 1$ being desirable for the mid point of a reaction." These various functions are shown by Fig. 11. Furthermore, these authors recommend chart speeds of 6–12 in. with heating rates of from 1 to 6°C/min for most investigations.

One remaining fact must never be overlooked; that even with perfectly clean and empty sample holders, apparent weight changes are recorded during most heating programs. This results from the interaction of several factors; furnace atmosphere type, convection and buoyancy effects, sample and furnace geometry, radiation and the internal furnace tem-

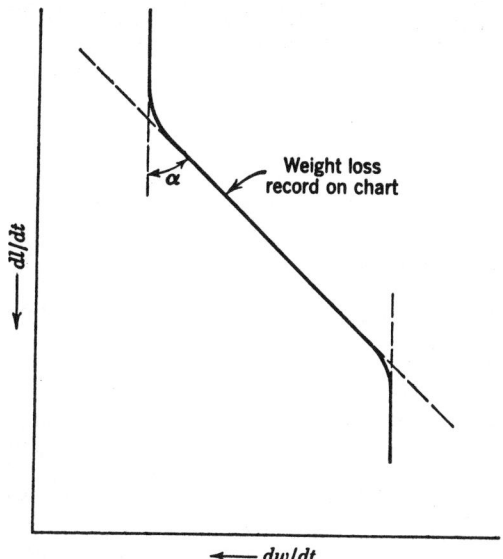

FIG. 11. The resolution factor, R; where ideally $a = 45°$ (Lukaszewski and Redfern, 1961; after Wendlandt, 1964).

perature gradients which may affect both sample sample holder and its support.

Furnace Atmosphere Conditions. The terms commonly used are *static, dynamic, vacuum,* and *pressure,* the latter two of which may also be either *static* or *dynamic*.

Static. Where the furnace atmosphere (assume air) is in good connection with the outside atmosphere into which any gaseous decomposition products may freely escape and be replaced by air, to maintain a furnace atmosphere of fairly constant composition.

Dynamic. When a particular gas is made to flow through the furnace; thus for nitrogen or oxygen, inert and oxidizing conditions are maintained, respectively. This is achieved by the "flushing" action of the gas which also has the advantage of effectively removing sample decomposition gases, which, because they cause additional unwanted effects, are deleterious or corrosive, or are wanted for collection and identification.

In the latter case, the plot of the amount of the evolved gas against temperature and the actual identity of this gas may be determined automatically and to considerable advantage by the technique of EGA (see *Effluent Gas Analysis*). Under these condtions the flushing gas is termed the *carrier gas* which is usually inert and must be of different composition than the evolved gases.

Vacuum. Once established, it is usual to maintain the vacuum during a TGA by keeping the pump in operation to give *dynamic vacuum* conditions. Otherwise the vacuum would be reduced somewhat by the evolved decomposition gas. In this case the term vacuum would apply.

Pressure. The amount of gas liberated by the sample is relatively insignificant compared with the volume of the furnace and, usually, the enclosing pressure vessel. Furthermore, the magnitude of this effect decreases with increasing pressure. Thus, once achieved, the pressure is little altered by the evolved gas so that *static pressure* conditions predominate. However, for the reasons mentioned above, unwanted evolved gases may still be removed by the flushing technique. Thus once the desired pressure has been attained the amount of gas entering and leaving the pressure vessel is kept constant to give *dynamic pressure* conditions.

By controlling and altering furnace atmosphere conditions, individual sample reactions, for example, oxidation, may be accelerated (in oxygen gas), inhibited (in air), and stopped or removed (in enert gases, i.e., argon). Furthermore, certain reactions may be delayed. For example, the temperature at which the reaction $CaCO_3 \rightarrow CaO + CO_2 \uparrow$ occurs is considerably raised if the partial pressure of CO_2 in the furnace atmosphere is increased, i.e., by replacing air by CO_2, or even more so, by CO_2 under pressure (Wolf, Easton and Warne, 1967, Fig. 18).

By judicious manipulation of furnace atmosphere, not only may certain types of reactions be identified but others may be accelerated, retarded, or completely removed to give greatly improved and controlled TGA curve modification, definition, configuration, and identifcation, particularly of intermediate reactions.

To aid the TGA of samples in atmospheres of their own gaseous decomposition products, an ingeneous piston type sample holder has been described by Garn and Kessler (1960) (see Fig. 12). For the improved curve definition obtainable with it, see Fig. 13.

Presentation and Recording of Thermogravimetric Data

The continuous weight variation data is best portrayed as a TGA curve or thermogram. The redrafting of these for publication should be as faithful a reproduction of the original recording as possible so that information, no matter how apparently trivial, is available to other workers. Furthermore, for the full value of published thermograms to be maintained, Newkirk and Simons (1963) have strongly suggested that eight pieces of supporting information should accompany each thermogram: *(1)* Identification of the substance examined by

THERMOGRAVIMETRIC ANALYSIS

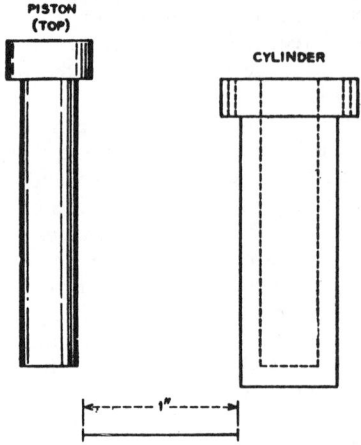

FIG. 12. The sample holder described by Garn and Kessler (1960) for thermogravimetry in self-generating atmospheres. For details see also Wendlandt (1964, p. 15).

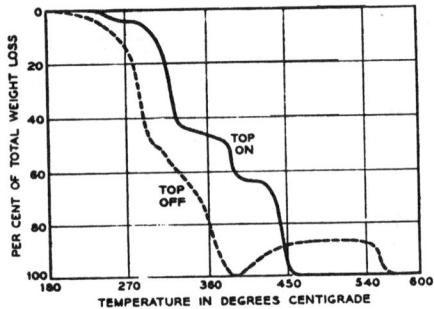

FIG. 13. The improved definition of TGA curves of lead carbonate determined with (top on) and without (top off) the presence of a self-generating atmosphere (Garn and Kessler 1960).

a definitive name, or an empirical formula, or equivalent compositional data; and the source of the substance; (2) The sample weight, and a weight scale for the ordinate; (3) Furnace heating rate; (4) Atmosphere; (5) Size, shape, and material of the container; (6) Methods used to identify intermediates and final products; (7) Identification of the thermobalance, including the location of the thermocouple used for temperature measurements; (8) Identification of the abscissa scale in terms of time, furnace temperature, or sample temperature.

Thermogram Interpretation

Most aspects of this complex subject have been evaluated and discussed as a guide to the interpretation of thermogravimetric measurements in relation to new studies of calcium oxalate monohydrate by Simons and Newkirk (1964).

Applications of Thermogravimetric Analysis

General. Thermogravimetric analysis yields accurate information on the size and rate of weight changes and the temperature range over which these occur. Indirectly however, this information may be applied to studies of rates of reaction and energies of activation for the reaction, sublimation, or vaporization of chemical compounds (including minerals) singly or in mixtures.

Weight gains, for example, are caused by the adsorption and absorption of gases by solid samples, direct reaction as in oxidation, corrosion, recarbonation, and hydration, or the reaction of gases to produce solids.

Conversely *weight losses* are produced by the desorption of gases, dehydration, vaporization, sublimation, gaseous desolvation, and gas-liberating reactions of both organic and inorganic substances.

Within these general aspects of research lie most of the applications of TGA.

Specific. A wide coverage of specific uses of TGA has been made by Gordon and Campbell (1960), Lukaszewski and Redfern (1961), Coats and Redfern (1963), Wendlandt (1964), and Garn (1965). Examples of these, which are important in mineralogy and geochemistry follow. This list is not fully exhaustive nor are the individual topics applicable to all investigations.

1. Automatic thermogravimetric analysis.
2. The elucidation of the thermal decomposition and recombination characteristics of minerals and resultant chemical compounds together with investigations into the kinetics and mechanisms of the reactions involved.
3. The identification of minerals from these characteristic TGA curve configurations.
4. The thermal decomposition and recombination behavior of minerals in mixtures with other minerals or chemical compounds.
5. The thermal stability of minerals and mineral mixtures.
6. The pyrolysis of coals, petroleum, and wood.
7. The analysis of soils.
8. The intensity of thermal reactivity of minerals and mineral mixtures.
9. The modes of mineral decomposition.
10. The size and rate of weight varying mineralogical reactions.
11. Roasting and calcining investigations.
12. The rate and conditions of formation of certain mineral products, slags, etc.
13. Thermochemical reactions of ceramics and cements.

14. Dehydration hygroscopicity studies.
15. Carbon dioxide and carbonate contents of rocks.
16. Solid-state reactions.
17. Determinations of volatile, moisture, or ash contents and combustion studies.
18. The absorption, adsorption, and desorption properties of minerals.
19. Automatic heating to constant weight determinations.
20. Rates of crystallization and diffusion.
21. Changes in specific gravity or gas density.
22. Effect of radiation on various substances.
23. Weight of effluent from chromatographic or other separatory systems.
24. Corrosion studies.
25. Determination of ignition temperatures.
26. The temperature and conditions of oxygen liberation from oxidizing agents such as nitrates, chlorates, bromates, stannates, and chromates.
27. The effects of oxidizing agents and catalysts.
28. Phase relationships.
29. Rates and conditions of evaporation, distillation, and sublimation.
30. The study of homologous series of inorganic complexes.
31. The detection of short-lived unstable intermediate compounds.

In many cases TGA gives the maximum information when used in conjunction with additional techniques such as x-ray diffraction, differential thermal analysis, and infrared spectroscopy.

S. St. J. Warne

References

Coats, A. W., and Redfern, J. P., 1963, "Thermogravimetric analysis," *The Analyst*, **88**, p. 906–924.
*Duval, C., 1953, "Inorganic Thermogravimetric Analysis," First ed., Amsterdam, Elsevier, 531pp. (second ed., 1963).
*Garn, P. D., 1965, "Thermoanalytical Methods of Investigation," New York, Academic Press, 606pp.
Gordon, S., and Campbell, C., 1960, "Automatic recording balances," *Anal. Chem.*, **32**, 271R–289R.
*Lukaszewski, G. M., and Redfern, J. P., 1961, "Thermogravimetric analysis," Parts I to IV, *Anal. Chem.*, **10**, 469–473, 552–555, 630–632, 721–724.
Newkirk, A. E., and Simons, E. L., 1963, "Suggestions for reporting dynamic thermogravimetric Data," *Talanta*, **10**, 1199.
Simons, E. L., and Newkirk, A. E., 1964, "New studies on calcium oxalate, monohydrate—a guide to the interpretation of thermogravimetric measurements," *Talanta*, **11**, 549–571.
Warne, S. St. J., and Bayliss, P., 1962, "The Differential Thermal Analysis of Cerussite," *Am. Mineralogist*, **47**, 1011–1023.
*Wendlandt, W. Wm., 1964, "Thermal Methods of Analysis," New York, Interscience Publishers, 424pp.
*Wolf, K. H., Easton, A. J., and Warne, S. St. J., 1967, "Techniques of Examining and Analyzing Carbonate Skeletons, Minerals and Rocks," Chap. 8, in "Carbonate Rocks," part B, Amsterdam, Elsevier, 413pp.

*Additional bibliographic references mentioned in text may be found in this work.

Cross-references: *Derivative Differential Thermal Analysis; Differential Thermal Analysis; Differential Thermogravimetric Analysis; Effluent Gas Analysis; Infrared Analysis; Mineralogy; x-Ray Diffraction Analysis. Vol. VI: Coal; Pedology; Petroleum.*

THERMOLUMINESCENCE—See Vol. IV B

THORIUM: ELEMENT AND GEOCHEMISTRY

Physical Properties. Thorium (Thor, God of War), symbol Th, has atomic weight 232.05 and Atomic Number 90. Its electronic configuration is (Rn Core), $6d^2$, $7s^2$ or (Rn Core) $5f^1$ $6d^1$ $7s^2$. Th has a melting point of 1650–1800°C, a boiling point of 3000–5200°C, and a density of 11.1–11.7 g/cm³. Its valence is 4+ and ionic radius (Th^{4+}) = 0.99Å. It exists in face-centered cubic crystal forms, transforming to body-centered near the melting point. Iso-

TABLE 1. Isotopes of Thorium

Mass	Common Name	Half-Life	Nature and Energy (MeV) of decay
234	UX$_1$	24.1 days	$\beta-$ (0.193)
232	Thorium	1.39×10^{10} yr	α (4.00, 3.95)
231	UY	25.6 hr	$\beta-$ (0.302, 0.216, 0.094)
230	Ionium	$\begin{cases} 8 \times 10^4 \text{ yr} \\ 7.6 \times 10^4 \text{ yr} \end{cases}$	α (4.685, 4.619)
228	Radio thorium	1.91 yr	α (5.421, 5.338)
227	Radio actinium	18.17 days	α (Several 5.65–6.03)

THORIUM: ELEMENT AND GEOCHEMISTRY

topes (Th^{232} = 100% by weight) are listed in Table 1 (see also *Radium*, Fig. 1).

Pure thorium is a silvery white metal that is spontaneously flammable in air; however, a few percent of ThO$_2$ stabilizes the metal. Thus the physical properties of the metal are a strong function of the oxide content.

Thorium is mainly used as a 'fertile' nuclear fuel for the production of fissionable U^{233} by neutron capture by the reaction Th^{232} (n, γ)→ Th^{233} $\xrightarrow{\beta^-}$ Pa^{233} $\xrightarrow{\beta^-}$ U^{233}.

Minerals

Thorium occurs in commercial concentrations mainly in monazite where it substitutes for the trivalent rare earths. The thorium content of this mineral is about 8–10%. In some of the rare earth silicates, such as orthite and gadolinite, the thorium concentration is of the order of one percent; whereas in the niobates and titanoniobates of rare earths, thorium is concentrated to a very large extent, in some cases up to 25%. The following list gives some of the important ores of thorium:

Ore	Composition	Th Content
Monazite	(Ce, Y, La, Th) (PO$_4$)	~10%
Cheralite	(Th, Ca, Ce) (PO$_4$, SiO$_2$)	~30%
Thorite	ThSiO$_4$	~80%
Thorianite	ThO$_2$	~90%
Pilbarite	ThO$_2$·UO$_3$·PbO·2SiO$_2$·4H$_2$O	~30%

Chemistry

Before the discovery of transuranium elements, thorium was placed in the secondary subgroup IV elements with titanium, zirconium, and hafnium. Although it does resemble these elements in chemical properties, all elements of $Z \geq 89$ are now placed in the actinide series. These elements are filling 5f shells and form a series similar to the lanthanide series where elements are filling 4f shells. Actually it is not known at which of the actinide elements the development of the 5f shell begins, but on the basis of similarities in chemical, physical, and spectroscopic properties, all elements from actinium ($Z = 89$) to lawrencium ($Z = 103$) are placed in the actinide series.

The geochemistry of thorium is simplified by the existence of but one valence state, 4+. Although other states have been prepared in the laboratory, this is the only one known in nature.

Common insoluble compounds used in thorium separation are the hydroxide, fluoride, iodate, chromate, phosphate, and oxalate. Soluble compounds include the chloride, nitrate, and sulfate.

Because thorium in solution is such a small, highly charged cation, it undergoes extensive interaction with water and many anions. Thus the solution chemistry of thorium is primarily a study of its complex ions. The following agents form strong complexes with thorium: Cl$^-$, NO$_3^-$, ClO$_3^-$, BrO$_3^-$, Cl$_2$CHCOOH, IO$_3^-$, HSO$_4^-$, H$_3$PO$_4$, and HF as well as ions of many organic acids: oxalate, citrate, pthalate, maleate, and EDTA. All of these are positively charged or cation complexes. In nitric acid solutions of >3M, thorium forms an anion complex. At pH values greater than 3, thorium undergoes hydrolysis in aqueous solution.

In separation involving gram quantities of thorium, precipitation methods based on the insoluble compounds listed above may be employed. When working with trace amounts, separations based on the complex ions must be used, such as ion-exchange resins and solvent extractions. Often a carrier, that is an element that approximates the chemistry of thorium, must be employed.

Geochemistry

The geochemistry of the quadrivalent thorium ion is characterized by its ionic radius 0.99Å and its high charge. It is concentrated in the late stage of magmatic crystallization where it substitutes for rare earths in monazite, zirconium in zircon, and calcium in apatite and sphene. Uranium follows a similar path in crystallizing magmas; however, notable separations occur and high uranium low thorium pitchblendes result. This may be due to the oxidation of U^{+4} to U^{+6} in an intermediate step, resulting in its separation from thorium, and then a subsequent reduction of U^{6+} to U^{4+}.

Abundance

The thorium concentrations in ordinary chondrites is of the order of 0.04 ppm, Wood (1963). The average Th/U ratio for chondrites is 3.6.

Since no samples are available, the thorium concentration in the upper mantle and oceanic crust may only be calculated. A Th/U ratio of 4 is generally assumed. This number is based primarily on the value of 4.1 calculated by Cannon et al. (1961) who made the assumption that lead ore bodies have been derived from the upper mantle. The other major heat producing isotope is K^{40}, and once its distribution is fixed, models for the crustal and upper mantle structure may be constructed.

One such model is that of Clark and Ringwood (1964), Table 2.

The concentration of Th in different types of rocks varies widely, and the following table lists the average value of Th concentration:

Rock	Concentration (ppm)	Reference
Granite	18.5	Heier et al. (1963)
Basalt	2.7	Heier et al. (1963)
Eclogite	0.18–0.45	Tilton and Reed (1963); Heier (1963)
Shales	12	Adams (1959)

Pb^{208}/Th^{232} Method of Age Determination

Thorium has been used to date the time of crystallization of certain minerals. In a closed system, the amount of Th^{232} that has decayed may be determined by the amount of radiogenic Pb^{208} in the mineral. Using the equation

$$Pb^{208}(t) = Th^{232}(t)(e^{\lambda 232 t} - 1)$$

the time of mineral formation may be obtained. It is important to discriminate between that Pb^{208} originally incorporated into the mineral (hence nonradiogenic) and that Pb^{208} produced by Th^{232} decay since the formation of the mineral. This common lead correction may be obtained by measuring the Pb^{204} in the mineral. It is obvious that high-thorium, low-lead minerals are the most useful for age determination.

Similar methods may be used to obtain a Pb^{260}/U^{238} and Pb^{207}/U^{235} age on the mineral. The fact that U^{235} and U^{238} occur in a constant ratio in nature allows the calculation of yet another age based on the $Pb^{207/206}$ ratio. This lead-lead age is the least affected by weathering processes and hence the standard against which other ages are compared. In general, the Pb–U ages are closer to the Pb–Pb age than is the Pb–Th age; however both Pb–U and Pb–Th ages are usually less than Pb–Pb ages. This is probably due to a partial lead loss during the history of the mineral. The fact that Pb–U ages are better than Pb–Th ages may reflect a difference in lattice sites for U and Th. If Th is nearer the grain boundary, the chance is greater that the Pb^{208} produced would be leached out easily, which would result in a lower Pb–Th age.

Th in the Weathering Cycle

Monazite, the primary thorium ore and a very resistate mineral, is concentrated by weathering, wave action, and near-shore currents in commercial quantities in sands along rivers and beaches near areas of the primary host rocks. In these deposits, monazite typically occurs as small, round, translucent grains associated with minerals like ilmenite, garnet, magnetite, and zircon.

No important source of thorium minerals other than of detrital material has yet been found in sedimentary rocks.

Because of the high ionic potential of Th^{4+} ion, the thorium which is brought into solution is quickly adsorbed or precipitated as hydrolysates. Thus the primary transport of thorium from the continents to the ocean follows the detrital phase. The available data for the thorium content of river water is restricted to several major rivers. The Th^{232} concentration is of the order of 10^{-5} micrograms/liter. The Th^{228}/Th^{232} activity ratio is slightly above unity, and the Th^{230}/Th^{232} activity ratio is close to natural abundance ratio (0.8) for both the water and the sediment. The concentration of Th^{232} in the suspended load of rivers is about 10 ppm, a value close to shales and pelagic sediments, (Moore, 1967). The amount of thorium isotopes present in ground waters has been analyzed by V. S. Dement'yev et al. (1965) who find that a large part of the thorium is transported as colloidal and organic anion complexes. They pointed out a dependance of Th isotopic ratios on the particle size of the suspended load.

Th in the Marine Environment

The concentration of thorium in ocean water is extremely low and only recently with the help of highly specific analytical and counting techniques has it been possible to obtain reliable limits of its concentration. The determination of Th^{232} in ocean water was tried many times previously, but only an upper limit of the concentration could be obtained. The first attempts for determining the thorium concentration of ocean water were made by Foyn et al. (1939), and Th^{230} values much below that which should be in equilibrium with dissolved uranium were found. Petterson (1937) had earlier hypothesized, from the low Ra^{226} content of sea water, that precipitation of Th^{230} took place soon after its formation. Koczy et al. (1957) used photographic emulsion techniques to determine the different thorium isotopes separated from sea water. Sackett et al (1958) used *alpha proportional*

TABLE 2. DISTRIBUTION OF RADIOACTIVE ELEMENTS CALCULATED FROM HEAT-FLOW DATA ASSUMING AN ECLOGITE MODEL[a]

	Surface Heat Flow (cal/cm²·sec)	Depth Interval (km)	U (ppm)	Th (ppm)	K (%)	Heat Production (×10⁻¹³ cal/cm³ sec)
Oceanic	1.2	5–12	0.42	1.68	0.69	0.714
		12–400	0.13–0.03	0.52–0.12	0.22–0.05	0.270–0.068
Shield	1.0	0–16	1.00	4.00	1.63	1.67
		16–37	0.37	1.48	0.61	0.63
		37–200	0.01	0.04	0.02	Zero
		200–400	0.05	0.20	0.10	Effectively zero
Continental	1.2	0–16	1.32	5.28	2.15	2.2
		16–37	0.42	1.68	0.69	0.714
Continental	1.5	0–16	1.87	7.48	3.05	3.125
		16–37	0.57	2.28	0.93	0.95

[a] Clark and Ringwood, 1964.

counting techniques to obtain the isotopic thorium concentrations.

Recent works of Moore and Sackett (1964) and Somayajulu and Goldberg (1966) has given a good idea of the concentrations of different thorium isotopes in ocean water. They have studied the three isotopes of thorium, namely Th^{232}, Th^{230}, and Th^{228}. Moore and Sackett did the work by collecting 130–190 liters of water at different depths to determine the isotopic ratios after separating the particulate matter by centrifuging. Somayajulu and Goldberg used an in situ method of extraction of the thorium isotopes on cleaned sponges loaded with $Fe(OH)_3$ from a few thousand liters of water. Some of these values of thorium isotopes are given in Table 3.

From the concentrations of Th^{230} in seawater, its residence time has been calculated. Moore and Sackett have given a value of less than fifty years, while Somayajulu and Goldberg has quoted 70 ± 40 years, which is further supported by a value of 60 years, obtained by their calculations from sedimentation rates. Because of this short residence time of Th^{230}, it is doubtful whether there is a complete mixing of thorium isotopes in ocean water.

The excess of Th^{228} activity over the parent Th^{232} activity in ocean water has recently become a problem of great interest. This was first observed by Moore and Sackett (1964), in the North Atlantic and Caribbean Sea, where Th^{228}/Th^{232} is fifteen times greater than the expected equilibrium ratio.

Somayajulu and Goldberg also obtained a 10–25 fold excess of Th^{228}/Th^{232} in the Pacific Ocean. The excess Th^{228} over Th^{232} must reflect a similar excess of Ra^{228} in the ocean.

Excluding Th^{232}, all the other isotopes of thorium are produced in situ in the ocean, and river input is negligible for them. The presence of four isotopes, with half-lives varying between a month and 80,000 years may help in the study of various oceanic processes.

Another important application of the different thorium isotopes in the marine environment is the dating of oceanic sediments. A detailed account of the various methods based on the Th isotopes is given in Vol. I (see *Radionuclides in Ocean and Sediments*) and only a brief outline is mentioned here. First attempts to use the decay of the unsupported Th^{230} activity for dating deep-sea sediments was done by Piggot and Urry (1942). They measured the Ra^{226} concentration of sediment cores as an index of Th^{230} concentration and suggested that by measuring Ra^{226} and U^{238} it would be possible to get the "excess" Th^{230}, i.e., that precipitated from sea water, and this "excess" could be used for obtaining the sedimentation

TABLE 3. CONCENTRATIONS OF DIFFERENT ISOTOPES OF THORIUM IN OCEAN WATER

Origin	Depth (m)	Th^{232} (g/l)	Th^{230}/Th^{232a}	Th^{228}/Th^{232a}	Reference
North Atlantic	0	$0.64 \pm 0.2 \times 10^{-3}$	2.7 ± 1.1	16 ± 5	Moore and Sackett (1964)
Caribbean	0	$0.64 \pm 0.2 \times 10^{-3}$	2.7 ± 1.1	14 ± 5	Moore and Sackett (1964)
Caribbean	800	$0.36 \pm 0.17 \times 10^{-3}$	3.5 ± 2.1	12 ± 6	Moore and Sackett (1964)
North Pacific	0	$0.33 \pm 0.04 \times 10^{-3}$	6.8 ± 0.9	25 ± 3	Somayajulu and Goldberg (1966)
North Pacific	2500	$0.65 \pm 0.09 \times 10^{-3}$	1.01 ± 1.4	10 ± 1.5	Somayajulu and Goldberg (1966)

^a Activity ratio.

rate. But the measurements of Kroll (1954) showed that the Ra^{226} does not follow the ideal decay of Th^{230}, and hence the validity of the assumptions of Piggot and Urry were doubted. Kroll suggested that Th^{230} should be measured directly, and that Ra^{226} is not an index of Th^{230}.

Picciotto and Wilgain (1954) suggested the possibility of using the Th^{230}/Th^{232} ratio for dating oceanic sediments. Much work has been done for dating sediments using Th^{230}/Th^{232} ratios by Goldberg and his colleagues (1962, 1964). In all these cases, the most important assumption is that Th^{230} and Th^{232} have the same chemical speciation in water and that there is no fractionation of thorium isotopes between sea water and precipitated phases. They also have assumed that, by leaching of sediments, only the "authigenic" part is dissolved and detrital fraction is left behind. The validity of these assumptions has been widely criticized, mainly because Th^{230} is produced in situ from dissolved uranium and Th^{232} transported from continents (Koczy, 1965; Sackett, 1964; Broecker, 1965).

Another isotope which is used to normalize for Th^{230} is Pa^{231}, suggested independently by Sackett (1960) and Rosholt et al (1961). Even though they are isotopes of two different elements, the geochemical behavior of Pa^{231} is similar to Th^{230} in that it is also precipitated to sediments in a short time compared to its half-life. A number of sedimentation rates has been calculated using this ratio by Sackett, Rosholt, and Ku (1966). However, a large number of measurements have been reported in which the Th^{230}/Pa^{231} ratios at the tops of cores are much higher than the values calculated from the U^{234}/U^{235} ratio of seawater. Ku (1966) has considered the possibility that significant Th^{230} is transported with the particulate matter. He also pointed out the possibility of volcanism as another alternative for the production of excess of Th^{230}/Pa^{231} ratios in deep-sea sediments. Recently Sackett (1966) has observed Th^{230}/Pa^{231} ratios in manganese nodules less than the calculated values, suggesting that these two isotopes are not removed by the same process.

Because of the deficiency of Th^{230} relative to U^{234} in natural waters, carbonate materials growing in these waters incorporate uranium almost free of thorium. If the system that incorporates uranium remains closed, the growth to equilibrium of Th^{230} may be used to determine the age of the carbonate material in the range 10,000 to 300,000 yrs B.P. The first measurements were carried out by Barnes et al (1956) on coral materials and the assumption appeared justified. However, further work revealed that few carbonates incorporate significant uranium during the life of the organism. With the exception of coral, most uranium in fossil carbonates appears to be secondary. Whether shells are an open system to uranium, or whether there is a period after the death of the organism when uranium is locked into the shell structure has not been resolved. However, high U^{234}/U^{238} ratios for shells million of years old would indicate that, at least in some instances, shells are quite open to uranium.

Coral, on the other hand, grows with about 3 ppm uranium and negligible thorium, and the Th^{230}/U^{234} method applied to corals from raised shorelines has given us an account of eustatic sea level changes over the past 200,000 years. Veeh (1960) has assigned an age of 120,000 years BP to a eustatic high sea stand, possibly an interglacial stage of the Pleistocene.

WILLARD S. MOORE
S. KRISHNA SWAMI

References

Adams, J. A. S., Osmond, J. K., and Rodgers, J. J. W., 1959, "The geochemistry of uranium and thorium;" *Physics and Chemistry of Earth* Vol. III, Pergamon Press.

Barnes, J. W., Lang, E. J., and Potratz, H. A., 1956, "Ratio of ionium to uranium in coral limestone," *Science*, **124**, 175–176.

Broecker, W. S., 1965, "Isotope geochemistry and the Pleistocene climatic record," *Quarternary of the U.S. Review*, Volume VII, INQUA Congress.

Cannon, R. S., Pierce, A. P., Antweiler, J. C., and Buck, K. C., 1961, "The data of lead isotope geology related problems of ore genesis;" *Econ. Geol.*, **56**, 1–38.

Clark, S. P., and Ringwood, A. E., 1964, "Density distribution and constitution of the mantle;" *Rev. Geophys.*, **2**, 35–88.

Dementyev, V. S., and Syromyatnikov, N. G., 1965, "Mode of occurrence of thorium isotopes in ground waters," *Geokhimiya*, **2**, 211–218.

Foyn, E., Karlik, B., Petterson, H., and Rona, E., 1939, "The Radioactivity of sea water," *Nature* **143**, 275–276.

Goldberg, E. D., and Koide, M., 1962, "Geochronological studies of deep sea sediments by ionium-thorium method," *Geochim Cosmochim Acta*, **26**, 417–443.

Goldberg, E. D., and Griffin, J., 1964, "Sedimentation rates and minerology in South Atlantic," *J. Geophys. Res.*, **69**, 4293–4309.

Heier, K. S., 1963,, "Uranium, thorium, potassium in eclogite rocks," *Geochim. Cosmochim. Acta*, **27**, 849–860.

Heier, K. S., and Rogers, J. J. M., 1963, "Radiometric determination of thorium, uranium and potassium in basalts and in two magmatic differentiation series," *Geochim. Cosmochim. Acta*, **27**, 137–154.

Koczy, F. F., Picciotto, E., Poulaert, G., and Wilgain, S., 1957, "Measure des isotopes du thorium dansl'eaudemer," *Geochim. Cosmochim. Acta*, **11**, 103–129.

Koczy, F. F., 1965, "Remarks on the age determination in deep-sea sediments," in *Progress in Oceanography* Vol. III, Pergamon Press.

Kroll, V. St., 1954, "On the age determination in deep-sea sediments by radium measurements," *Deep Sea Res.*, **1**, 211.

Ku T. L., 1966, "Uranium series disequilibrium in deep-sea sediments," Thesis, Columbia University.

Moore, W. S., 1967, "Amazon and Mississippi River Concentrations of Uranium, Thorium and Radium Isotopes," *Earth Planetary Sci. Letters* **2**, 231–234.

Moore, W. S., and Sackett, W. M., 1964, "Uranium and Thorium Series inequilibrium in sea water; *J. Geophys, Res.*, **69**, 5401–5405.

Picciotto, E., and Wilgain, S., 1954, "Thorium determination in deep-sea sediments," *Nature*, **173**, 632–633.

Piggot, C. S., and Urry, W. D., 1942, "Time relations in Ocean Sediments," *Geol. Soc. Amer. Bull.*, **53**, 1187–1210.

Rosholt, J. N., Emiliani, C., Geiss, J., Koczy, F. F., and Wangersky, P. J., 1961, "Absolute dating of deep sea cores by the Pa^{231}/Th^{230} method;" *J. Geol.*, **69**, 162–185.

Sackett, W. M., 1964, "Measured deposition rates of marine sediments and implication for accumulation of extraterrestrial dust, *Ann. N.Y. Acad.*, **119**, 339–346.

Sackett, W. M., 1966, "Manganese Nodules: Thorium 230: Pa-231 ratios," *Science*, **154**, 646–647.

Sackett, W. M., Portratz, H. A., and Goldberg, E. D., 1958, "Thorium content of ocean water," *Science*, **128**, 204–205.

Somayajulu, B. L. K., and Goldberg, E. D., 1966, "Thorium and uranium isotopes in sea water and sediments," *Earth Planetary Sci. Letters*, **1**, 102–106.

Tilton, G. R., and Reed, G. W., 1963, "Radioactive heat production in eclogites and some ultramafic rocks," in *Earth Science and Meteoritics*, Amsterdam, North Holland Publishing Co.

Urry, W. D., 1949, "Radioactivity of ocean sediments," *J. Marine Res.*, **7**, 618–634.

Veeh, H. H., 1966, "Th^{230}/U^{238} and U^{234}/U^{238} ages of Pleistocene high sea level stands," *J. Geophys. Res.*, **71**, 3379–3386.

Wood, J. A., 1963, "Physics and Chemistry of Meteorites," *The Moon Meteorites, and Comets*, University of Chicago Press.

Clark, S. P., Jr., Peterman, Z. E., and Heier, K. S., 1966, "Abundances of Uranium, Thorium, and Potassium," in (Clark, S. P., editor), "Handbook of Physical Constants," *Geol. Soc. Am. Mem.*, **97**, 521–541.

Wasserburg, G. J., MacDonald, G. J. F., Hoyle, F., and Fowler, W. A., 1964, "Relative contributions of uranium, thorium, and potassium to heat production in the earth," *Science*, **143** 465–467.

Cross-references: *Actinide Series; Complexes; Manganese Nodules: Geochemistry; Radium; Rare Earths; River Geochemistry; Seawater, Chemistry; Seawater, Geochemical Balance; Thorium Ore Deposits; Uranium-Thorium-Lead Age Determination. Vol. I: Radionuclides in Oceans and Sediments. Vol. IVB: Monazite.*

THORIUM: ECONOMIC DEPOSITS

Mineralogy. Thorium occurs in more than 100 minerals, commonly in association with zirconium, hafnium, uranium, and the rare earth metals. World production of thorium has come almost exclusively from monazite, although deposits of thorite ($ThSiO_4$, containing 25.2 to 62.7% of ThO_2) in Colorado and Idaho are considered to be potential commercial sources. Monazite is a complex phosphate of the rare earth metals in which thorium substitutes in solid solution for cerium. The usual content is 8–10% ThO_2, but some specimens contain as much as 30% ThO_2.

Types of Deposits. Monazite occurs widely as an accessory mineral in granites, gneisses, aplites, and pegmatites, usually in concentrations far too limited to be of commercial interest. Monazite has been produced from a few vein deposits, notably in the Republic of South Africa. The major source, however, has been beach and stream placers in various parts of the world, where heavy minerals have been concentrated in poorly consolidated sediments by stream and wave action. Many of these deposits have been worked for ilmenite, cassiterite, gold, or zircon with monazite as a by-product.

Description of Mining Districts. *India.* Thorium was discovered in monazite sand of Travancore in 1909. Between 1911 and 1945, India supplied 40% of the world's production, mainly from the Kerola area. The beaches and dunes of the Malabar and Coromandel coasts are the main commercial source of monazite, which is associated with ilmenite, rutile and zircon. Other beaches are also known to contain monazite and the reserves are among the largest in the world. Recognizing thorium as a potential source of fissionable material, India banned exports in 1946.

Brazil. First explored in 1895, Brazilian placer deposits produced half of the world's supply of monazite during the first half of the 20th century. The mineral is concentrated in stream bars and beach sands mainly between Cape Frio and Recife. The best deposits are about 10 m wide, 1 km long, and 1/2 m thick.

TABLE 1. LOCATION OF THORIUM RESERVES

Country	Recoverable Short Tons of ThO_2
India and Ceylon[a]	250,000
Canada[b]	200,000
United States[c]	100,000
Brazil[a]	10,000
Australia and S.E. Asia[a]	10,000
West Africa[a]	15,000
East Africa[a]	20,000
South Africa[a]	10,000
Egypt[a]	5,000

[a] U.S. Atomic Energy Comm., 1960.
[b] Can. Dept. Mines and Tech. Surv., 1963.
[c] U.S. Atomic Energy Comm., 1964.

Since January 1, 1951, Brazil has restricted exports.

United States. Monazite was mined on a small scale from colluvial and alluvial placers in North and South Carolina from 1887 and 1917. After foreign imports were curtailed in 1950, dredging and recovery of monazite was started in Bear Valley, Idaho.

Republic of South Africa. The van Rhynsdorp deposit is a vein up to 14 ft. wide which contains zones of massive monazite, up to 2 ft. wide, along the foot wall. Since 1950 this has been an important source.

Other Countries. Monazite placers are widely distributed at places throughout the world, including Nigeria, Malay, Taiwan, Australia, Ceylon, Indonesia, and Korea. Uranium deposits of the Blind River district, Canada, contain probably the largest supply of thorium in North America but the grade is low (0.05% ThO_2).

Reserves. Free world reserves of thorium recoverable at $10.00 (U.S.) per pound ThO_2, or less, are estimated in Table 1.

ROBERT J. WRIGHT

References

Bateman, A. M., 1950, "Economic Mineral Deposits," Second ed., New York, John Wiley & Sons, 916pp.
Baroch, C. T., 1965, "Thorium," U.S. Bur. Mines Min. Yearbook.
Lamey, C. A., 1966, "Metallic and Industrial Mineral Deposits," New York, McGraw-Hill Book Co., 567pp.
Soister, P. E., and Conklin, D. R., 1959, "Bibliography of U.S.G.S. Reports on Uranium and Thorium, 1942 through May, 1958," *U.S. Geol. Surv. Bull.*, **1107-A**, 167pp.

Cross-references: *Economic Geology; Hafnium; Mining and Mineral Centers of the World; Monazite; Placer Deposits; Rare Earths; Thorium: Element and Geochemistry; Uranium; Zirconium.*

THULIUM: ELEMENT AND GEOCHEMISTRY

THORIUM DATING METHOD—See PROTACTINIUM–THORIUM DATING METHOD

THULIUM: ELEMENT AND GEOCHEMISTRY

Properties

Elemental thulium is a very reactive silvery white *lanthanide* (q.v.), or *rare earth* (q.v.) metal with atomic number 69 and outer electronic structure $4f^{13} 6s^2$. It occurs as one stable isotope, Tm^{169}.

Thulium is determined analytically by neutron activation, x-ray spectrography, and by absorption spectrometry (it has very sharp spectral lines). In geologic environments, thulium forms highly ionic bonds and weak complexes. The oxidation state is tripositive, with ionic radius 0.869Å. Generally, chemical behavior is similar to the alkaline earths, and almost identical to other heavy rare earths.

Cleve, in 1879, was the first to isolate thulium from other rare earths. Thulium was named after Thule, the ancient name of Scandinavia.

Minerals

Thulium occurs in highest concentration in rare earth minerals that favor heavy rare earths, generally nitrates, sulfates, carbonates, niobates, and tungstates with cation coordination number of six to eight, often containing structural halogens and alkali metals. Typical examples are:

Xenotime (Y, heavy rare earths)PO_4

Fergusonite (Y, heavy rare earths) $(Nb, Ta)O_4$

Geochemical Behavior

Schmitt et al. (1964) have established a value of 0.033 ppm (parts per million) thulium in chondrites and 0.38 ppm in eucrite achondraites. The standard deviation within a meteoric class is only about 20%. In meteoritic matter, volatization may have redistributed alkali metals but not rare earths: thulium has a boiling point of 1727°C.

The thulium concentration in average basalt is 0.44 ppm, and the range is less than a factor of ten (Haskin et al. 1966). Oceanic and continental basalts have similar ranges. Partial melting of mantle peridotite yields a basaltic liquid with enriched thulium and a mantle source area depleted in thulium.

Igneous differentiation series may tend toward higher or lower thulium concentration, although the total variation is less than for light rare earths. The relative abundance of thulium and other heavy rare earths remains unchanged. Haskin et al (1966) report 0.73 ppm thulium in an Ascersion Island basalt and 0.52 ppm thulium in a trachyte differentiate. Towell et al. (1965) found 0.263 ± 0.018 ppm in the Rubidoux Mountain Leucogranite phase of the Southern California batholith, probably a differentiate (by crystal settling) of the San Marcos Gabbro, with 0.248 ± 0.017 ppm.

Average granite with less than 60 per cent silica has 0.52 ppm, but the range of value for granites is even greater than for basalts.

In most basaltic, syenitic and granitic rocks, thulium is mainly in mafic rock-forming minerals, probably substituting for divalent calcium (ionic radius 0.99Å.)

Thulium and other heavy rare earths form complexes and are transported in alkali, halogen, and carbon dioxide-rich hydrothermal fluids; metasomatic and pegmatoid deposits formed from such fluids may have high concentrations of thulium and higher ratios of thulium to light rare earths than granites.

Thulium has a 630-year residence time in seawater (which is alkaline), about the same as other heavy rare earths. Removal by adsorption on clay particles, as well as depletion in surface water in areas of high organic productivity, are likely processes. High secondary thulium concentrations in phosphatic fish remains today occur only in deep water; paleozoic fish remains also have high rare earths. Manganese nodules have 170 ppm, whereas Pacific Ocean water has only 0.13×10^{-6} ppm (Goldberg et al., 1963).

In soils, the relative amounts of thulium in detrital, clay, and carbonate fractions is not well known. Sedimentary rocks have about 0.58 ppm, with a variation of a factor of five. In a study of rare earths in Russian platform soils, Balashov et al. (1964) found higher rare earths in alkaline soils, the thulium in groundwater probably precipitating as hydroxide; lower concentrations are in acid soils, probably due to removal of complexes.

Biologic concentration of rare earths has been found in hickory leaves; rare earths have been used to trace opium.

Economic Geology

The major source of thulium has been from a light rare earth mineral, monazite, in beach sands; the accessory monazite originally present in granitic rocks resists weathering and is concentrated by sedimentary processes. Australia, Brazil, Ceylon, India, Malaysia and South Africa mine the largest quantities. Less than one per cent of the monazite is thulium. Recently, the bastnaesite deposit at Mountain Pass, California has been an important source of thulium, again in low concentration.

High purity thulium has a very limited market. Most thulium is sold unpurified in rare earth concentrates used for polishing, petroleum cracking catalysis, lighter flints, are carbons, and in alloys.

ROBERT KAY

References

Balashov, Y. A., et al., 1964, "The effects of climate and facies environment on the fractionation of the rare earths during sedimentation," *Geochem. Intern. No. 1*, **10**, 951–969 (English transl.).

Goldberg, E. D., et al., 1963, "Rare-earth distributions in the marine environment," *J. Geophys. Res.*, **68**, 4209–4217.

Haskin, L. A., et al., 1966, "Meteoric, Solar, and Terrestrial Rare-Earth Distributions," in (Ahrens, L. H. et al., editors), "Physics and Chemistry of the Earth," Vol. 7, Oxford, Pergamon Press, pp. 167–321.

Rankama, K., and Sahama, T. G., 1950, "Geochemistry," Chicago, University of Chicago Press, 912pp.

Schmitt, R. A., et al., 1964, "Rare-earth, yttrium and scandium abundances in meteoric and terrestrial matter—II," *Geochim. Cosmochim. Acta*, **28**, 67–86.

Towell, D. G., et al., 1965, "Rare-earth distributions in some rocks and associated minerals of the batholith of Southern California," *J. Geophys. Res.*, **70**, 3485–3496.

Cross-references: *Hydrothermal Solutions; Neutron Activation Analysis; Rare Earths; Seawater, Chemistry.*

TIN: ELEMENT AND GEOCHEMISTRY

Tin is a valuable metal element, symbol Sn, atomic number 50, and atomic weight 118.69. It has been found as an alloy in bronze articles dating back to 5500 years. It played an important part in the history of mining and trade in northwestern Europe; the Phoenicians carried it from mines in Spain and in Cornwall, England, to the metal-working centers of the eastern Mediterranean. Its principal virtues are its anticorrosive, nonrusting characteristics, its malleability, and its low melting point. The use of tin alloyed with lead to make an easily melting *solder* was mentioned in AD 79 by Pliny. The Romans tin-plated and soldered copper utensils. Tin-plated steel was developed in Britain and in Saxony in the mid-17th century.

There are two allotropic forms of the metal, the familiar silvery *"white tin"* (beta form) which is sometimes found in nature in the native elemental state and crystallizes in the body-centered tetragonal system, and the rare *"gray tin"* (alpha form), which forms only at very low temperatures and has a diamond cubic structure. The melting point is 231.9°C, and boiling point 2270°C. Density of beta (white) tin is 7.29 and of alpha (gray) tin 5.77.

Chemically tin is nontoxic and nonreactive. Actually on exposure to the air it develops a very thin (invisible) layer of oxide which forms a stable coating. It is thus stable in the presence of distilled water, nitrogen, hydrogen, CO_2, and NH_3. On the other hand it reacts with chlorine, bromine, and iodine, and with strong acids and strong bases.

Tin has ten stable isotopes in nature, the largest number of any element. As with tellurium there has probably been both slow and rapid neutron capture. These isotopes are:

$Sn^{112} = 0.95\%$	$Sn^{118} = 24.01\%$
$Sn^{114} = 0.65\%$	$Sn^{119} = 8.58\%$
$Sn^{115} = 0.34\%$	$Sn^{120} = 32.97\%$
$Sn^{116} = 14.24\%$	$Sn^{122} = 4.71\%$
$Sn^{117} = 7.57\%$	$Sn^{124} = 5.98\%$

Geochemistry

Tin is relatively rare in the earth's crust (0.004% or 3 g/ton), following such relatively "exotic" elements as hafnium and dysprosium. Its cosmic abundance is 0.013 (atoms/10,000 atoms Si, according to Suess and Urey, 1956). In meteorites it is commonly concentrated in the iron-rich phases, but also enriched in sulfide and silicate phases. In seawater its concentration is 0.003 g/ton, following iron. It is only a microconstituent in natural organisms. It appears to have little biological significance although it is sometimes used in therapeutics.

In nature tin is markedly siderophile, but also shows thiophile and lithophile characteristics. The most common mineral by far is the oxide, cassiterite.

Its concentration in the earth's crust is usually in association with granites, but is rather enigmatic, for it is quite abundant in a few places, but extremely rare elsewhere. It is estimated that since about 500 BC Cornwall has furnished over 3 million tons of tin, but some other granites supply none at all. For example there is a belt of granites, often called the "tin granites" that runs through S.E. Asia, from Yunnan (China), through Malaya, Indonesia, eastern Australia to Tasmania (Devonian to Triassic in age). In Europe tin granites are found in Saxony, Czechoslovakia (Bohemia), England, France, and Spain; in South America, in Bolivia; and in Africa, Nigeria, and the Congo. Yet in the vast granite areas of western North America, the Canadian Shield, Brazil, South Africa, Western Australia, the Urals, and Siberia it is virtually absent.

TIN: ELEMENT AND GEOCHEMISTRY

TABLE 1. TIN MINERALS[a]

Name	Formula	Occurrence
Stannite	Cu_2SnFeS_4	Hydrothermal veins, Cornwall, etc.
Teallite	$PbSnS_2$	
Montesite	$PbSn_4S_5$	Silver-tin veins, Bolivia
Cylindrite	$Pb_3Sn_4Sb_2S_{14}$?	
Franckeïte	$Pb_5Sn_3Sb_2S_{14}$?	
Cassiterite	SnO_2	Hydrothermal and pneumatolytic deposits, widespread
Nordenskiöldine	$CaSnB_2O_6$	South Norway
Stokesite	$CaSnSi_3O_9 \cdot 2H_2O$	S.W. Africa
Arandisite	$Sn_5Si_3O_{16} \cdot 4H_2O$	

[a] From Day, 1963.

According to one theory tin is concentrated in the mantle, transported up through the crust by the volatile fluorides or chlorides as SnF_4 or $SnCl_4$ and near the surface hydrolysed to the stable form SnO_2. In Bolivia there are sulfide ores associated with silver deposits. While the average tin in basalt in only 4 g/ton, average granite carries 80 g/ton and in greisens there is up to 8000 g/ton.

Tin does not readily associate with other elements, so the list of possible minerals is short (Table 1), and after cassiterite, most of them are rare. The ionic radius of Sn^{4+} 0.71Å would not be favorable for the replacement of silicon in the common silicate minerals. It can, however, replace iron, scandium, and titanium.

Cassiterite is an extremely durable mineral and it is commonly found in weathering crusts, saprolites and placer accumulations, both alluvial and littoral. For example, much of the production in Malaya and Indonesia comes from stream gravels that accumulated during the low sea-level phases of the Quaternary glaciations which were semiarid in the tropics, and so the streams tended to slit up instead of carrying their bed loads out to the ocean. Since sea level was lower then some of these alluvial tracts may be followed out onto what is now the continental shelf.

Economics

World production of tin runs about 150,000–200,000 tons of metal annually. The commodity is thus rather scarce, the demand is fairly high and the price is generally attractive, so that placers containing only 1 part in 10,000 of cassiterite, or veins with 1% ore are still profitable to work. In Malaya both alluvial placers and primary emplacements (veins or stockworks) are exploited, some 500 mines being operable. In Nigeria it is associated with columbite in a deeply weathered granitic saprolite. In Cornwall it is found in dikes and veins, penetrating the slates surrounding the granites; near the surface there are silver-lead ores, then copper, and finally, deepest, there is tin associated with wolfram and fluoride minerals. The Bolivia development is in veins associated with sulfides ores such as marcasite, bismuthite, pyrrhotite etc.

The principal uses of tin are as the metal, for coatings, for alloys, and in chemical compounds. The largest world consumer is the United States with an annual demand of about 60,000 tons (from the ore), in addition to about 25,000 tons obtained from scrap. Prices fluctuate but an approximation of $2/lb, or $4000/ton gives the scale.

R.W.F.

References

Day, F. H., 1963, "The Chemical Elements in Nature," London, George G. Harrap & Co. (also New York, Reinhold Book Corp.), 372pp.

Engel, G. T., 1965, "Tin," U.S. Bur. Mines Min. Yearbook.

Faulkner, C. J., 1954, "The Properties of Tin," Columbus, Ohio, Tin Res. Inst. Publ. No. 218.

Hedges, E. S., 1960, "Tin and Its Alloys," London, E. Arnold, Ltd.

International Tin Council, 1967, "A Technical Conference on Tin," London, 2 vols, 641pp.

MacIntosh, R. M., 1968, "Tin," in (Hampel, C. A., editor), "Encyclopedia of the Chemical Elements," New York, Reinhold Book Corp., pp. 722–732.

Mantell, C. L., 1949, "Tin: Its Mining, Production and Technology," Second edition, New York, Reinhold Publ. Corp.

Rankama, K., and Sahama, T. G., 1950, "Geochemistry," Second edition, Chicago, Univ. of Chicago Press, 912pp.

Cross-references: *Earth's Crust Geochemistry; Isotope Geology—Stable; Lead;* Vol. III: *Regolith and Saprolite.* Vol. IVB: *Metallogenetic or Minerogenetic Provinces.*

TITANIUM: ELEMENT AND GEOCHEMISTRY

History

By analyzing ilmenite in 1789, W. Gregor discovered titanium oxide. Several years later, M. Klaproth studied rutile ores, and gave the element the name titanium. J. Berzelius isolated the element in 1825.

Properties and Characteristics

Titanium (symbol Ti) is in group IV B of Periodic Table. It has the following properties: atomic number 22; atomic weight 47.90; atomic radii 1.41Å (eight-fold coordination) and 1.45Å (twelve-fold coordination); ionic radii, Ti^{3+} 0.76Å (six-fold coordination) and Ti^{4+} 0.68Å (six-fold coordination); electron configuration $1s^2, 2s^2, 2p^6, 3s^2, 3p^6, 3d^2, 4s^2$; electron shells 2, 8, 10, 2; valence 2, 3, 4 (but mainly tetravalent); crystal structure hexagonal close packed, where $c = 4.729$Å, $a = 2.953$Å, and body-centered cubic above 880°C, where $a = 3.32$Å; isotopes 48 (73%), 46 (8%), 47 (8%), 50 (5%), 49 (5%); specific gravity 4.507 (20°C); hardness 60 (Rockwell A); melting point 1668°C ± 10°C; boiling point 3260°C (1 atm); coefficient of thermal expansion 8.2×10^{-6} per°C (0–300°C); and (α form) 10.5×10^{-6} per°C (25–500°C); thermal conductivity 0.037 $(cal/_{sec})(sq\ cm)(°C/_{cm})$ at room temperature; electrical resistivity $= \sim 42 \times 10^{-6}$ ohm–cm at 20°C; heat of atomization of solid titanium $= \sim 112.7$ kcal/g-atom at 298.15°K.

Abundance, Mineralogical, and Geochemical Characteristics

Titanium is the ninth most abundant element in the earth's crust. The average content of titanium in igneous rocks is 0.44% (4400 g/ton Ti), which makes it the most common trace element in igneous rocks. It is much more abundant than zirconium and hafnium. More TiO_2 is found in mafic igneous rocks (1.8%) than in silicic igneous rocks (0.5%).

Titanium concentrates (1800 g/ton Ti) in the silicate phase of meteorites, whereas no titanium has been found in the sulfide phase of meteorites. A small amount of titanium (100 g/ton Ti, primarily as TiC and TiN) is present in the metal phase of meteorites. The above information shows the strong lithophile characteristic of titanium.

In the earth's crust titanium combines with oxygen to form stable oxides. Sometimes these oxides are combined with silica to form silicate minerals such as sphene ($CaTiSiO_5$). However, the most common minerals of titanium do not contain silica (ilmenite, $FeTiO_3$; rutile TiO_2; titanomagnetite, $Fe(Fe,Ti)_2O_4$; perovskite, $CaTiO_3$; anatase, TiO_2; and brookite, TiO_2).

Titanium is often found in solid solution in common silicate minerals such as amphiboles, pyroxenes, and biotites. Titanium can be found as either Ti^{3+} or Ti^{4+} (much more common). In general, Ti^{3+} is only present when all of the iron has been reduced to Fe^{2+} or metallic iron. Laboratory work done at the Corning Glass Works, Corning, New York, has shown that when Ti is present in the form of Ti^{3+} large quantities (>10%) of it can go into solid solution in periclase at high temperatures by forming a cation defect structure, whereas, only <2% of Ti^{4+} can go into solid solution in periclase. The Ti^{3+}-containing phase is violet to purple in color. The Ti^{4+} phases are cream color. In the femic minerals, depending on the oxidation potential, Ti^{3+} or Ti^{4+} probably substitutes for ions such as Ca^{2+}, Mg^{2+}, Fe^{2+}, Fe^{3+}, Cr^{3+}, Al^{3+}, Mn, etc. When this occurs, often Group IA ions (R^{1+}) substitute for Group IIA ions (R^{2+}) which balances the charge. Probably, in some cases, the charge is balanced by forming a cation defect.

In addition to the minerals mentioned above, there are many other titanium-containing minerals of much less importance, such as cerium titanosilicates and the rare earth niobate-titanates. Also, there are several calcium zirconium titanates and numerous combinations of Group IA and/or IIA titanosilicates. Also, manganese-containing titanium minerals exist. No binary phase of only TiO_2 and SiO_2 exists. This is probably due to the small differences in their ionic field strength.

Because many titanium-containing minerals, such as ilmenite and rutile, are resistant to weathering, they are sometimes found associated with other heavy, resistant minerals that remain largely in the resistates. The titanium contained in the structure of femic minerals is dissolved in water during weathering, but then is promptly hydrolyzed and carried into the hydrolzates. Bauxites and laterites can contain as much as 4% titanium which is present in the form of sphene, ilmenite, brookite, and anatase.

Economic Geology

The most important ore minerals of titanium are ilmenite ($FeTiO_3$), rutile (TiO_2), and, to a lesser degree, sphene, also called titanite ($CaTiSiO_5$). Because these minerals are resistant to weathering and have a high specific gravity, they accumulate in placer deposits. Rutile ores can be purchased that contain 92–98% TiO_2. Commonly, monazite and zircon are associated with rutile in these placers.

Ilmenite contains 52.7% TiO_2 and sphene, 41%.

Besides placer deposits in which titanium accessory minerals have been concentrated, there are many magmatic deposits of economic value. The most important ore deposits of magmatic orgin resulted from differentiation of basic magmas. Titanomagnetite, ilmenite, and rutile are associated commonly with magnetite and ferromagnesian minerals, sometimes with hematite and calcic plagioclase, sometimes with chromite, and sometimes in dikes with apatite. Important deposits are found in North America, India, Norway, and Australia.

Because titanium dioxide has a high refractive index, and a high dielectric constant, it is used in pigments (paints, glazes, enamels, textiles, inks, plastics, paper, etc.) and in ceramics for special electrical purposes. Titania is also being used successfully as a catalyst to increase the rate and amount of devitrification of glass for the formation of glass-ceramics. Additions of titania to basic refractories increase their spall resistance. The halides of titanium are useful in many applications such as sky writing, dyeing leather, and bleaching and conditioning textiles. Titanium and its alloys are lighter weight and more refractory than steel alloys. Therefore, they are being used in aerospace applications. Titanium-containing steel alloys have high strength and abrasion resistance, and titanium-aluminum alloys have greater ability to withstand attack by organic acids and salt solutions than pure aluminum. Titanium carbides are useful as cutting tools and as refractory materials.

A. M. ALPER

References

Alper, A. M., 1970, "Oxides of rare earths, titanium, zirconium, hafnium, niobium and tantalum," in "High Temperature Oxides," New York, Academic Press, Part II.

Barksdale, J., 1949, "titanium," New York, Ronald Press Company, 591pp.

Bateman, A. M., 1949, "Economic Mineral Deposits," John Wiley & Sons, 898pp.

Goldschmidt, V. M., 1954, "Geochemistry," London, Oxford University Press, 730pp.

Lindgren, W., 1933, "Mineral Deposits," Fourth ed., New York, McGraw-Hill Book Co., 930pp.

Lyman, T., 1961, "Metals Handbook—Properties and Selection of Metals," Eighth ed., Vol. I, Am. Soc. Metals, Metals Park, Ohio, 1236pp.

Poldervaart, A., 1955, "Crust of the Earth," *Geol. Soc. Spec. Paper*, **62**, 762pp.

Rankama, K., and Sahama, Th. G., 1950, "Geochemistry," University of Chicago Press, 912pp.

Rose, A. & Rose, E., 1956, "The Condensed Chemical Dictionary," Fifth ed., New York, Reinhold Publ. Corp., 1200pp.

Shaffer, Peter T. B., 1964, "High Temperature Materials," No. I, Materials Index, New York, Plenum Press, 740pp.

Stecher, P. G., 1960, "The Merck Index of Chemical and Drugs, Rakeway, N. J., Merck and Co., 1639pp.

Czamanske, G. K., and Porter, S. C., 1965, "Titanium dioxide in pyroclastic layers from volcanoes in the Cascade Range," *Science*, **150** (3699), 1022–1025.

Howie, R. A., and Woolley, A. R., 1968, "The role of titanium and the effect of TiO_2 on the cell-size, refractive index, and specific gravity in the andradite-melanite-schorlomite series," *Mineral. Mag.*, **36**(282), 775–790.

Stamper, J. W., 1965, "Titanium," U. S. Bur. Mines Min. Yearbook.

Cross-references: *Earth's Crust Geochemistry; Solid Solution; Trace Elements.* Vol. IVB: *Ceramic Technology; Glass: Devitrification of Volcanic Glass; Placer Deposits; Refractive Index; Refractory Minerals*; Vol. V: *Magmatic Differentiation.*

TRACE ELEMENTS: GEOCHEMISTRY

Trace elements are those comprising less than two percent by weight of a rock or specimen of mineral matter. For the upper lithosphere, this generally means the elements other than the eight major constituents O, Si, Al, Fe, Mg, Ca, Na, and K. Concentrations of the more common minor constituents S, Ti, Mn, P, Cl, F, C, and H are high enough that in many rocks they assume major status. Also, in meteorites K is to be regarded as a trace element whereas Ni is usually a major component. Similarly there are numerous terrestrial rock types such as pegmatites, which have high concentrations of very rare elements. In terms of the effect of trace elements on the stability relations of mineral assemblages, the classification by weight percentage is somewhat misleading insomuch as the chemistry is governed by molar concentrations. Consequently such light elements as H and C may behave as major components although they occur in trace weight percentages.

Interest in the minor elements may be divided into the following categories:

1. Utilization in the compositional characterization of rocks, waters, and gases in the interpretation of geologic environments and the history of seawater and the atmosphere. This also includes such uses as geochemical indicator aids to prospecting and the nutritional evaluation of soils.

2. Sources of information on the intrinsic properties of elements such as ionic radii and other characteristics related to position in the periodic table.

3. Determination and utilization of thermodynamic parameters such as the distribution constants which yield information on the temperature, pressure, and chemical environment of crystallization as well as criteria for the attainment of equilibrium in natural systems. In order to quantify these data, use must be made of solution models which have their source in statistical mechanics.

Although these categories are closely related, it is important not to confuse them.

Category *1* is of great interest to geologists because the minor element content of rocks and waters is usually closely related to their mode of origin. For example, the Precambrian iron formations, which are inferred to be chemical sediments, have characteristically low values of Ti, V, Ni, Co, Cr, and other metals, whereas laterites tend to be enriched in many of these elements. Again the degree of concentration of ancient brines of marine origin can be deduced from the Br/Cl ratio of salt beds precipitated from them. Also, in meteoritics, certain classes of chondrites may be distinguished on the basis of the elements Bi, Pb, Tl, and Hg (Mason, 1962). However, some geologists have attempted to use the absolute amount of certain elements in rocks as indicators of the grade of metamorphism in a sense which confuses categories *1* and *3*.

Category *2* has in the past received much attention chiefly through the impetus of Goldschmidt (1954), Ahrens (1952), and Ramberg (1953). This theory, which treats ions and atoms as entities with properties relatively independent of the phase in which they occur, has served as a valuable first approximation in explaining the gross distribution of elements and their occurrence in different crystal structures. By these means we can explain such features as the relative concentration of Ba in feldspars or qualitative aspects of the distribution of Fe, Mg, and Mn among certain silicates.

In Category *3* we find the verification of the principles of Gibbs in the most complex systems of nature. Thermodynamics, with its adjunct solution theory, leads to quantitative expressions for the distribution of elements among coexisting minerals. The great order of these natural distributions was not generally anticipated before it was demonstrated in metamorphic rocks by Kretz (1959). Fig. 1 shows one of the simpler distribution relations found by him. The slope of the line, which is 1.2 in this case, is simply related to the thermodynamic distribution constant.

The thermodynamic approach to the distribution of the elements is the most general one, and there is no fundamental difference between the treatment of minor and major constituents.

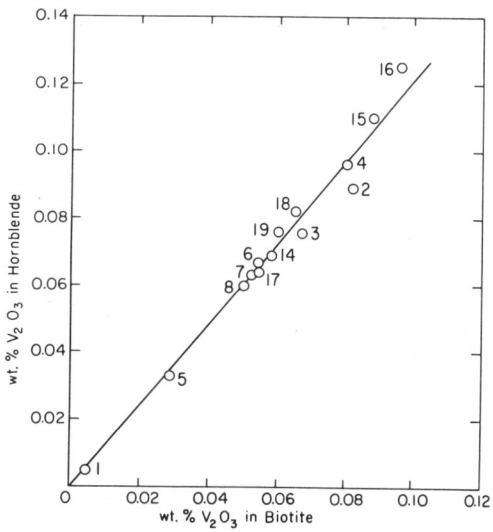

FIG. 1. The distribution of vanadium between coexisting biotites and hornblendes from the metamorphic gneisses of Quebec (Kretz, 1959). Each point represents a mineral pair. This illustrates one of the simpler types of distributions commonly observed. The orderly adherence of the points to the line indicates the attainment of a high degree of local equilibrium. In principle the position and form of the curve, which are functions of the thermodynamic distribution constant, shift with changes in pressure and temperature. *Courtesy University of Chicago Press.*

The distribution relations, which form an important class within the general equilibria, may be approached formally through two types of reactions: (a) transfer reactions and (b) exchange reactions.

As an illustration of equilibria governed by transfer reactions we have the distribution of Co, Ni, and Fe among the α (Kamacite) and γ (taenite) metallic phases of meteorites. The reactions are

$$\overset{\alpha}{\text{Co}} \rightleftharpoons \overset{\gamma}{\text{Co}} \qquad (a)$$

$$\overset{\alpha}{\text{Ni}} \rightleftharpoons \overset{\gamma}{\text{Ni}} \qquad (b)$$

$$\overset{\alpha}{\text{Fe}} \rightleftharpoons \overset{\gamma}{\text{Fe}} \qquad (c)$$

Each of these reactions has its corresponding equation of equilibrium, which may be written down directly from the *principle of mass action*:

$$K_a = \frac{a_{\text{Co}}^\gamma}{a_{\text{Co}}^\alpha} = \frac{X_{\text{Co}}^\gamma f_{\text{Co}}^\gamma}{X_{\text{Co}}^\alpha f_{\text{Co}}^\alpha} \qquad (1)$$

$$K_b = \frac{a_{\text{Ni}}^\gamma}{a_{\text{Ni}}^\alpha} = \frac{X_{\text{Ni}}^\gamma f_{\text{Ni}}^\gamma}{X_{\text{Ni}}^\alpha f_{\text{Ni}}^\alpha} \qquad (2)$$

$$K_c = \frac{a_{Fe}{}^\gamma}{a_{Fe}{}^\alpha} = \frac{X_{Fe}{}^\gamma f_{Fe}{}^\gamma}{X_{Fe}{}^\alpha f_{Fe}{}^\alpha} \quad (3)$$

In these equations, the K's are equilibrium (distribution) constants, which are functions solely of the temperature and pressure. The a's are relative activities or thermodynamic concentrations, and the X's are the corresponding atomic fractions or molar concentrations. For example, $X^\gamma{}_{Co} = N_{Co}/(N_{Co} + N_{Ni} + N_{Fe})$ in taenite, where the N's represent the numbers of moles of each component in this phase. The f's are the activity coefficients which measure the deviations from ideal solutions for which they reduce to unity. Generally the f's are functions of the temperature, pressure, and the total composition expressed as the X's. Deducing values of the f's from empirical data or giving them explicit functional form is generally difficult. However, this may at times be accomplished by applying solution model theory to natural or experimental distributions.

An additional example of a transfer reaction is that which governs the distribution of Ba between coexisting feldspars:

Plagioclase   Microcline
$BaAl_2Si_2O_8 \rightleftharpoons BaAl_2Si_2O_8$ (d)

The choice of this complex component is based on its homology with anorthite and albite for which similar transfer reactions have proved useful in treating the phase relations.

In general simple transfer reactions cannot be applied to the distribution relations among complex minerals since they require considerable similarity in the coexisting phases. Of wider application are the exchange reactions which govern the distribution of many elements between various silicates, A specific example is the distribution of Mn and Mg among coexisting pyroxenes and olivines:

Pyroxene   Olivine
$MgSiO_3 + \tfrac{1}{2} Mn_2SiO_4 \rightleftharpoons$

Pyroxene   Olivine
$MnSiO_3 + \tfrac{1}{2} Mg_2SiO_4$ (e)

The units of exchange are Mn^{2+} and Mg^{2+} ions which are chosen because they are also the units of mixing in solid solutions of this type. A formalism based on such units as $Mn_2{}^{2+}$, $Mg_2{}^{2+}$, etc., which from time to time has appeared in the literature, introduces needless complications into the calculations and has no rationale whatever.

Also we note that the standard states are chosen as the pure substances Mg_2SiO_4, Mn_2SiO_4, $MgSiO_3$, and $MnSiO_3$. The first three of these are well defined compounds whereas the last apparently is not. However this does not effect our treatment since it does not depend on the existence of stable end members.

In terms of the phase rule, the system represented by Eq. (e) consists of two phases and three components, one choice of these components being Mg_2SiO_4, Mn_2SiO_4, and SiO_2. It thus retains one degree of freedom under isothermal-isobaric conditions. The equation of equilibrium corresponding to Eq. (e) may be written as

$$K_e = \frac{X_{Mn}{}^{Px} X_{Mg}{}^{Ol} f_{Mn}{}^{Px} f_{Mg}{}^{Ol}}{X_{Mg}{}^{Px} X_{Mn}{}^{Ol} f_{Mg}{}^{Px} f_{Mn}{}^{Ol}} \quad (4)$$

which is also general enough for a system of any number of components. For the ternary system, the solutions are quasi-binary and the X's represent the atomic fractions; thus, $X_{Mn}{}^{Px} = N_{Mn}/(N_{Mn} + N_{Mg})$. The f's are again the activity coefficients as previously defined. Where more components (for example Fe_2SiO_4) are present, an equation similar to Eq. (4) is written for each pair of exchangeable ions.

When Mn is a trace element we have $X_{Mg}{}^{Ol} \rightarrow X_{Mg}{}^{Px} \rightarrow 1$, $f_{Mg}{}^{Ol} \rightarrow f_{Mg}{}^{Px} \rightarrow 1$, and $f_{Mn}{}^{Px}$ and $f_{Mn}{}^{Ol}$ become constants. Then eq. (4) reduces to

$$K_{e'} = \frac{X_{Mn}{}^{Px}}{X^{Ol}{}_{Mn}}$$

where

$$K_{e'} = K_e \left(\frac{f_{Mn}{}^{Ol}}{f_{Mn}{}^{Px}} \right) \quad (5)$$

which is seen to be similar in form to Eqs. (1) to (3). This type of linear relation between the concentrations appears to hold for a number of systems and is illustrated in Fig. 1. However, simple relations of this type can hold only for systems which are quasi-binary or in which there is no interference from additional components. For these more complex situations, Eq. (5) may not be satisfied and the f's do not necessarily reduce to constants but may vary widely as other components vary in concentration. Under these circumstances the distribution diagrams may show a great apparent scatter. This case is illustrated by Fig. 2 which shows the effect of Ca variation in garnet on the distribution of vanadium.

Functions of the type in Eqs. (1) to (5) and analogous relations for the major components provide a variety of potential temperature and pressure indicators. Usually the effect of temperature and pressure may be evaluated qualitatively or semiquantitatively by the use of the general thermodynamic relations

$$\left(\frac{\partial \ln K}{\partial T} \right)_p = \frac{\Delta H}{RT^2} \quad \text{and} \quad \left(\frac{\partial \ln K}{\partial P} \right)_T = \frac{-\Delta V}{RT} \quad (6)$$

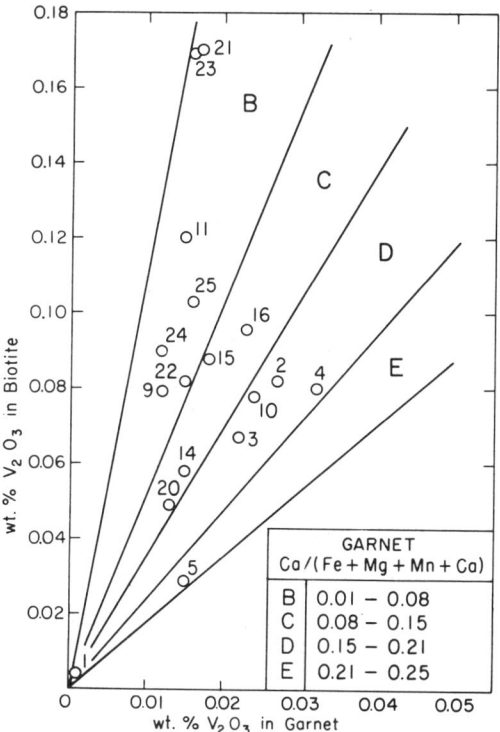

FIG. 2. The distribution of vanadium between coexisting biotites and garnets from the metamorphic gneisses of Quebec (Kretz, 1959). Many of the mineral pairs represented by the points came from the same specimens as in Fig. 1, as indicated by the numbers. Note that in this case the distribution is strongly affected by the Ca variation in garnet, which apparently entirely obscures the effects of pressure and temperature. *Courtesy University of Chicago Press.*

in which ΔH and ΔV are the differences in enthalpy and volume between the pure reactants and products for any transfer or exchange reaction, and R and T are the gas constant and absolute temperature, respectively. Unfortunately in many cases ΔH can only be estimated and ΔV is not known as a function of the temperature and pressure. With the determination of the distribution constant K as a function of T and P by any means it then becomes possible in principle to apply the distribution relations of a number of elements to a metamorphic rock in order to fix the physicochemical environment of crystallization by the method of cross checks. However, in relations of the type shown in Fig. 2 the pressure and temperature effects are apt to be completely obscured by overriding compositional variations.

When the coexisting minerals and the exchangeable particles are similar in nature with respect to size, oxidation state, etc., then the distribution constants will have values near unity. This is the case for our example of the distribution of Mn^{2+} between olivine and pyroxene. When, on the other hand, the coexisting phases differ greatly in character, the distribution may be very unequal corresponding to very small or very large distribution constants. Such is the case for the distribution of Ni and Co between meteoritic iron and the coexisting olivines. Although iron is usually in high concentration in both phases, Ni and Co resist oxidation to a far greater degree and so are concentrated in the metal phases. The magnitude of the effect, which was determined by E. J. Olsen and the writer, restricts the concentration of Ni and Co in olivine to less than 30 ppm. By contrast, the olivine of terrestrial rocks, which have formed under more oxidizing conditions frequently contain much higher concentrations of these elements. As an example, the olivines of the Stillwater basic complex of Montana contain up to 2500 ppm NiO and 200 ppm CoO as determined by us.

Another condition which results in very unequal distribution of the elements is a large deviation from ideal solutions in a sense which makes the activity coefficients large, thus resulting in an immiscibility gap. Usually this behavior is confined to major constituents since fairly high concentrations are required for saturation. But for some common trace elements saturation occurs although the total concentration is only a few hundreds of ppm or less. This apparently is the case for zirconium silicate, which is almost always present as zircon ($ZrSiO_4$), and thus indicates a supersaturation with respect to other silicates. Another example of a slightly different character is to be found in the behavior of TiO_2, which appears to reach a saturation concentration in olivine at about 100 ppm as determined from the Stillwater rocks, although it is as high as 3000 ppm in the coexisting Ca-pyroxenes.

According to the phase rule, saturation imposes an additional condition on the system by giving rise to a new phase. Such systems present us with potential pressure and temperature indicators of a somewhat different type than those discussed earlier. For example, the activity of $ZrSiO_4$ in a silicate coexisting with pure zircon is given by

$$K = a_{ZrSiO_4}$$

where K is the equilibrium constant for the transfer reaction. The use to which equilibria of this kind may be put is, however, limited by the great possibility of contamination by the exsolved phase.

Thermodynamic relations similar to those just discussed apply quite generally to the distributions of the elements, and equations of the same form govern relations between silicate minerals, magmas, aqueous solutions, and gases. Some of the most interesting geochemical problems concern the fractionation of the elements between the lithosphere and hydrosphere. The character if this fractionation is established by a multiplicity of factors in addition to thermodynamic considerations, but an important role may be attributed to the chemical interaction of the waters with lithosphere minerals. The chemical content of the waters is to a large extent controlled by the compositional variables that determine the types of minerals equilibrated with them. Two of the most frequently referred to variables are the oxidation potential or Eh and the negative logarithm of the hydrogen ion activity or pH, since all minerals are restricted by these parameters. The type of major phase which is stabilized then exerts its influence on the minor element distributions. One of the best illustrations of this is the elemental composition of sea water. The pH of seawater is near 8.2 and deep-sea waters tend to be oxidizing as is shown by the widespread occurrence of iron and manganese oxides in pelagic sediments. These conditions do not favor the solubility of heavy metals (Garrels, 1960), and as a consequence Mn, Ni, Co, V, Sc, and other elements occur in seawater in concentrations less than $10^{-3}\%$ of the amounts to be expected by weathering processes throughout geologic time (Rankama and Sahama, 1950). Of course, many elements released by weathering never actually reach the sea but are retained in residues such as laterites.

In contrast, the minor element Br has been highly concentrated in seawater relative to the lithosphere. Its behavior, which might have been anticipated from its position in the periodic table, is thus similar to, Cl whose abundance in the oceans is attributable to volcanic and other emanations from the earth's interior. We are thus led to consider the problem of differentiation in the earth's crust and the concentration of trace elements involved in the genesis of certain rare rocks as well as ore deposits. The problem is essentially one of the distribution of the elements between solid, liquid, and vapor phases which participate in magmatic and metamorphic processes. The question of the concentration of various elements in a vapor in equilibrium with a solidifying magma has been treated in some detail by Krauskopf (1957)) who calculated the partial pressures of such metals as Hg, Sb, Pb, Sn, Zn, Co, etc., above the corresponding chlorides, oxides, and sulfides. As might be expected from kinetic considerations, volatile elements such as Hg and Sb, which are concentrated in the gas phase, are readily transported great distances from the igneous source, whereas less volatile elements like Co and Ni are less mobile. Similar factors are involved in the genesis of the rare metal pegmatites which are rich in Li, Be, and B: The sources of these elements are not known, but it is possible that they are segregated from the surrounding rocks by some chemical mechanism. A clue may be the coarse crystallinity of the deposits, which indicates a high mobility for the elements which comprise them. It seems plausible that many of the elements are brought into high concentrations in a vapor phase by the formation of stable molecular complexes of fluorine, chlorine, and other elements. This problem will be amendable to attack when thermochemical data for the rare molecular complexes become available.

In summary we may say that the laws of energetics which govern the distribution of major and trace elements are identical except as they are modified by the lower concentrations of the latter. The most general formulation of these laws are the equations of thermodynamics which always take a similar form whether the coexisting phases be crystalline, liquid, or gaseous. Recent studies have shown that a high degree of order prevails in the distribution of the elements between coexisting minerals of metamorphic rocks, indicating the attainment of a high degree of local equilibrium. Indeed, the observed distributions are good criteria for equilibrium as well as potential pressure and temperature indicators. In certain rocks, where equilibrium was not attained, the major and minor elemental distributions serve to distinguish disruptive influences such as crystal settling or other mechanical movements. The element fractionation between such large units as the hydrosphere and lithosphere and between different lithospheric layers are only partly attributable to chemical factors; yet the major trends in the distributions may be explained by thermodynamic and kinetic theory. Also the distribution of a given element may be highly dependent on compositional variations of a phase. This is particularly true in systems involving liquids and gases but is also exhibited by certain crystalline phases.

ROBERT F. MUELLER

References

Ahrens, L. H., 1952, "The use of ionization potentials," parts 1 and 2, *Geochim. Cosmochim. Acta*, **2**, 168; **3**, 1.

Garrels R. M. 1960, "Mineral Equilibria at Low

Temperature and Pressure," New York, Harper & Bros.
Goldschmidt, V. M., 1954, "Geochemistry," Oxford, Clarendon Press.
Kamen, M. D., 1949, "Tracers," *Sci. Am.,* February.
Krauskopf, K. B., 1957, "Heavy metal content of magmatic vapor at 600°C.," *Econ. Geol.* **52,** 786.
Kretz, R., 1959, "Chemical study of garnet, biotite, and hornblende from gneisses of southwestern Quebec, with emphasis on distribution of elements in coexisting minerals," *J. Geol.* **67,** 371.
Mason, B., 1962, "Meteorites," New York, John Wiley & Sons.
Ramberg, H., 1953, "Relationships between heats of reaction among solids and properties of the constituent ions, and some geochemical implications," *J. Geol.* **61,** 318.
Rankama, K., and Sahama, T. G., 1950, "Geochemistry," Chicago, University of Chicago Press, 912pp.
Shaw, D. M., 1964, "Interprétation Geochimique des Eléments en Traces dans les Roches Cristallines," Paris, Masson et Cie, 237pp.

Cross-references: *Earth's Crust Geochemistry; Elements: Planetary Abundances and Distribution; Enthalpy; Free Energy; pH-Eh; Seawater, Chemistry; Seawater, History; Thermochemistry; Seawater, History; Thermochemistry; Thermodynamics; Trace Elements in Silicate Minerals—Substitution; Water.* Vol. IVB: *Geochemical Prospecting.* Vol. V: *Volatiles.* Vol. VI: *Laterites.*

TRACE ELEMENTS IN PLANTS

A definite relationship exists between soil, plants, animals, and man. Plants are good indicators of the health of the soil in which they are growing. If the soil is in a state of imbalance, either naturally or due to abuse in cultivation, plants will indicate it by their appearance, by a change in their biochemical makeup, or both. Animals are dependent upon plants or upon the flesh of herbivorous animals for their survival. Therefore, man's health depends directly or indirectly upon plants.

Plants have been used for healing purposes since antiquity. Modern biological and chemical assays have made us aware of the highly complex makeup of plants. The value of a plant as a curative agent is commonly associated with its chief constituent(s) such as glycosides or alkaloids; however, if the whole picture is considered, trace elements gain in significance (Koffler, 1964). Our studies showed that some plants usually classified as weeds have a high content of some trace elements. Examples are: *Chelidonium majus* (Greater Celandine), *Cirsium arvense* (Canadian Thistle), *Melilotus officinalis* and *alba* (Yellow and White Sweet Clover), and *Phytolacca americana* (Poke Root). Some of these weeds have been, and still are, well-known as medicinal plants. Some plants have roots which reach deeply into the earth and accumulate trace elements not found in the shallower layers of the soil. When these plants die, the upper layers of the soil become enriched with these elements. The delicate interrelationship of soil, plant, animal, and human health was discussed in an Interdisciplinary Symposium in the Earth and Medical Sciences during the AAAS convention at Montreal, December 1964 (Warren, 1965; for further information see Warren, 1960).

Our knowledge of the importance of trace elements in plant physiology has increased considerably in the last years. To mention but a few examples, manganese plays an important role as an activator of enzymes catalyzing various stages of plant respiration; it also plays a role in the nitrogen metabolism of plants. Zinc, too, is a catalyst in some enzyme systems. Boron plays a role in the carbohydrate metabolism and is also involved in the synthesis of ribonucleic acid in plants. B, Mn, Zn, Cu, Mo, and Co were found to increase the rate of photosynthesis (Stiles, 1961).

Ecological findings of recent years have pointed to a close interrelationship between plants in a given area and between plants and the microflora connected with them. Trace elements enter into all these interrelationships. Minute changes may have far-reaching effects. Applications of trace elements to soils or as foliar sprays may greatly affect the morphology and physiology of plants.

ANNA H. KOFFLER

References

Koffler, A. H., 1964, "Trace element investigations of medicinal plants." *J. Am. Inst. Homeopathy,* **56,** 279–282; **57,** 48–54, 193–196, 262–264.
Stiles, W., 1961, "Trace Elements in Plants," Third ed. Cambridge, The University Press.
Warren, H. V., 1960, "Health and geology," *West. Miner, Oil Rev.* (August 1960).
Warren, H. V., 1965, "Reports of sections and societies: Symposium on medical geology and geography," *Science* **147,** 918–919.

Cross-references: *Medical Geography; Medical Geology; Medical Geology—Trace Metals in Mammals; Pedology; Photosynthesis.* Vol. VI: *Soil Mineralogy.*

TRACE ELEMENTS IN SILICATE MINERALS—SUBSTITUTION

Several processes result in the occurrence of trace elements in minerals. These include ion exchange, surface adsorption, occlusion during mineral growth, and solid solution. Solid solution at relatively high temperatures may also be followed by exsolution as the temperature, and consequently the heat of mixing, is lowered. In general, the subject is very complex and the actual processes resulting in the distribution of a particular trace element in silicate minerals have frequently not been adequately ascertained. It is difficult to obtain pure minerals free of admixed phases; consequently most studies of trace elements, however accurate the analysis, have involved inadequately purified minerals. The increasing availability of electron probes now makes it possible to determine the distribution of many trace elements in the individual minerals of igneous rocks without the necessity of mineral separation. Surface adsorption and occlusion can be distinguished from solid solution with the aid of the electron probe.

This article will be restricted to the phenomena of solid solution in igneous minerals. The references below contain information on the other phenomena and also adequate treatment of the thermodynamic approach to solid solution involving the Berthelot-Nernst distribution law (partition coefficients).

There are certain properties of trace element cations which allow for more or less accurate prediction of their distribution in igneous silicate minerals. These properties are cationic charge (Z_c), radius (R_c), and electronegativity (X_c). They are important because the growth of stable ionic silicate minerals requires efficient cation-anion packing, electrical neutrality, and adequate bond strength (see Vol. V, *Silicate Melting*).

By attention to the results of phase equilibria experiments, thermodynamic relationships, and trace element analyses for natural and synthetic silicate minerals, it is possible to set up rules, such as Goldschmidt's rules, for the substitution relationships. Goldschmidt's rules were based on charge and radius relationships alone and were surprisingly useful considering that pure ionic bonding was assumed. Ionic bonding is dominant in silicate minerals, at least for the network modifiers, (see Vol. V, *Silicate Melting*), but there is always a covalent contribution. Because of this dominantly ionic structure, silicate minerals are destabilized when the ionic contribution to the mixed-bond energy decreases.

In order to include the effect of mixed bonding, the following relationship for the difference in bond energy ($\Delta\delta$) of a trace element relative to a major element can be used as a first approximation (see Vol. V, *Silicate Melting*, eqs. 5 and 6):

$$\Delta\delta(\%) = 100 \frac{(\delta_t - \delta_m)}{\delta_m}$$

$$= 100 \left[\frac{F_t Z_t (R_m + R_a)}{F_m Z_m (R_t + R_a)} - 1 \right] \quad (1)*$$

Where the subscripts t and m refer to the trace element and major element respectively, and F is the fractional ionic character of the bond $[1 - e^{-0.25} (X_a - X_c)^2]$. It is also convenient to define certain useful terms:

1. *Capture:* A trace cation is said to be captured by a major cation position if it is preferentially incorporated in the mineral.
2. *Camouflage:* A trace cation is said to be camouflaged by a major element if it is neither preferentially incorporated or merely admitted, i.e., if its relative concentration in the mineral is nearly equal to its relative abundance in the melt.
3. *Admission:* If a trace element is neither captured or camouflaged but significantly substitutes for the major element, it is said to be admitted.
4. *Rejection:* If there is not significant substitution of the trace element for the major element, it is said to be rejected.

With these terms in mind, the following set of substitution rules appear to apply successfully to trace element substitution:[‡]

1. Electrical neutrality must be maintained. For example, when a divalent cation substitutes for a monovalent cation, a concomitant substitution must take place. Frequently this consists of the substitution of Al^{3+} for Si^{4+}. Substitution becomes less likely as the charge difference increases.
2. The trace element cation must be close to or within the limits of radius ratio R_c/R_a for the coordination of the major cation. For fourfold (tetrahedral) coordination the limits are 0.22–0.41, for sixfold (octahedral) coordination the limits are 0.41–0.73, and for coordination greater than six the radius ratio must be greater than 0.73.
3. If the electronegativity of the cation (X_t) is greater than 1.9, the cation will be rejected from all silicate minerals. For example, the noble metals (Ru, Rh, Pd, Os, Ir, Pt, Au)

---

*Actually, when the substitution involves a charge difference, prediction would be better with the following relationship for which there is only a qualitative theoretical justification:

$$\Delta\delta'(\%) = 100 \left[\frac{Z_t X_m^2 R_m}{Z_m X_t^2 R_t} - 1 \right] \quad \ldots\ldots\ldots\ldots (1)^1$$

‡When $\Delta\delta'(\%)$ is used, the rules below are the same except for Rule 5. $\Delta\delta'$ must be less than -75% for rejection. Also, the limits for camouflage are: $\Delta\delta' = \pm 15\%$.

remain as metals or accumulate in the sulfide phase with As, Se, Sb, and Te.

4. If the radius difference between the trace and major cation is greater than 40%, the trace cation will be rejected. This difference is given by the relationship:

$$\Delta R(\%) = 100 \frac{[R_t - R_m]}{R_m} \quad (2)$$

5. If $\Delta\delta$ is less than -35%, the trace cation will be rejected.

6. If the cation is not rejected by the above considerations and the radius difference is less than 20%, it may be captured, camouflaged, or admitted. It will be captured if the ionic bond energy $\Delta\delta$, is greater than $+10\%$ and admitted if $\Delta\delta$ is less than -10%. If $\Delta\delta$ falls between the limits of $\pm 10\%$, it will be camouflaged.

7. If the cation is not rejected by the above considerations and the radius difference is between 20% and 40%, the cation may be admitted, but not captured or camouflaged.

There are some interesting exceptions to these rules. For example, Zr^{4+} ($R_c=0.83$Å) is well within the coordination limits for sixfold coordination, but its coordination in zircon is eightfold. Consequently U^{4+}, Th^{4+}, and the rare earths are admitted for zirconium in zircon (Rule 6 applies). With the exception of zirconium, the above rules appear to hold quite well for cations with a closed shell electronic configuration. For transition elements, another factor must be introduced. This is the crystal-field stabilization energy which results from the energy splitting upon removal of electron degeneracy in the electric field of a tetrahedral or octahedral site. This stabilization energy amounts to 29.3 kcal/mole for Ni^{2+} and, despite the higher electronegativity of Ni^{2+} ($X_t=1.7$, $R_t=0.72$Å) relative to Mg^{2+} ($X_m=1.2$, $R_m=0.65$Å), it results in nickel being captured by Mg^{2+} or Fe^{2+} in olivine. Compressibility relationships at very high pressures and substitutional disorder at high temperatures also have not been discussed. The fact is that the cations are not simple, charged billiard balls and, ultimately, a precise solution of the Schrödinger wave equation would be required for precise prediction. Unfortunately the computations are too complex and laborious for all but simple chemical relations. Consequently, silicate geochemistry remains essentially an experimental-observational science.

PAUL E. DAMON

References

Ahrens, L. H., 1964, "The Significance of the Chemical Bond for Controlling the Geochemical Distribution of the Elements," Part I, in (Ahrens, L. H., et al., editors), "Physics of the Earth," Vol. 5, New York, Pergamon Press, pp. 1–54.

Curtis, C. D., 1964, "Applications of the crystal-field theory to the inclusion of trace transition elements in minerals during magmatic differentiation," *Geochim. Cosmochim. Acta,* **28,** 389–403.

Kirkinskii, V. A., 1963, "On polarity in isomorphism," *Geochemistry,* No. 2, pp. 133–144 (Translated from the Russian and produced by Scripta Technica, Inc.).

McIntire, W. L., 1963, "Trace element partition coefficients—a review of theory and applications to geology," *Geochim. Cosmochim. Acta,* **27,** 1209–1264.

Rankama, K., and Sahama, Th. G., 1955, "Geochemistry," Chicago, Univ. of Chicago Press, 912pp.

Cross-references: *Crystal Chemistry; Crystal-Field Theory; Electronegativity; Exsolution; Geochemistry of Sedimentary Silica; Ion Exchange; Phase Equilibria; Rare Earths; Silica Solubility; Solid Solution; Trace Elements: Geochemistry.* Vol. IVB: *Electron Probe Microanalysis.* Vol. V: *Silicate Melting.* Vol. VI: *Radioactive Isotope Tracer Technology.*

TRACE METALS IN SEDIMENTS, OILS, AND ALLIED SUBSTANCES

The accumulation of trace elements is subject to other causes in addition to those active on major elements, such as sorption, capture of ions in crystal lattices, and ineffectiveness of precipitation when the solubility products and/or the saturation points are not attained. The distribution and proportion of trace elements may be as characteristic for the geological environment as the distribution and proportion of major elements. Only ubiquitous elements (e.g., V), however, show regular associations with petrology or facies (although multivalent elements may accumulate in more than one environment), whereas detrital (Ti) and rare elements (U) tend to occur in patches. Even in seawater the distribution of many trace elements, depending on terrestrial, volcanic, and planktonic contributions, is variable; contents in plankton and derived muds seem more regular.

This article is concerned only with metals, yet nonmetallic elements will be mentioned where they are characteristic or essential for the environment of trace metals. Early work on this subject was done by V. M. Goldschmidt and his followers. The number of publications is considerable: Zulfugarly lists 730. Important workers include Fester, Hodgson, Katchenkov, Leutwein, Schroll, Vinogradov, and Zulfugarly.

TRACE METALS IN SEDIMENTS, OILS, AND ALLIED SUBSTANCES

Sediments

Abiogenic Sediments. Abiogenic sediments have been formed by forces not essentially connected with life.

Clastic sediments. Clastic sediments are formed by the transportation and deposition of fragments of rocks obtained by weathering. Resistance to physical and chemical destruction causes the accumulation of residual minerals which may contain typical elements (extremes: laterites and bauxites with Ti, Th, V, Nb, Ta, Cr, Co, and Ni; eluvial placers with Au, Sn, Pt-metals, and diamonds). Further accumulation is due to grinding of softer minerals during transport, and deposition of heavy minerals in traps, together with coarser grains of lighter minerals (sands with magnetite, ilmenite, pyrite; extreme: alluvial placers with Au, Fe, Ti, Sn, Th, rare earth metals, U, Pt-metals, and diamonds; Th:U>7). Such accumulations may be found in very thin layers rich in heavy minerals; with regard to the whole rock, the elements will be present in the quantity of traces and only the association of placer elements will point to the nature of the accumulation process. (For a list of trace metals in sediments, see Table 1.)

Accumulations of Ti and Th seem to mark sediments covering surfaces, or derived from surfaces, which have undergone longer periods of sub-aerial weathering. Accumulations of Au and Sn in sedimentary iron ores point to clastic rather than chemical accumulation of the iron. With regard to the contents of seawater, pelitic sediments are enriched in K and Sr (factor 10^2), Rb (10^3), Ba (10^4), and Cs (10^5). Sediments with U>4 ppm may tentatively be regarded as continental; U decreases with distance from continents toward the deep sea. Distinctive accumulations of Ga have been reported from fresh-water sediments by some authors, whereas others indicate that B, Cr, Cu, Ga, Ni, and V are significantly more abundant in marine than in fresh-water argillaceous sediments.

Chemical Sediments. In evaporites, the concentration of trace metals seems to correspond to the concentration of the salts. Gold (with more than 10% Pt-metals), Cu, Zn, Pb, and Mn are concentrated by a factor of 100 in German salts. Within marine limestones, Mn seems to prefer well-aerated waters, whereas V prefers less well-aerated waters (bituminous limestones). Sr increases from back-reef to fore-reef but is influenced by dolomitization. Sr accumulates in aragonite and anhydrite, Mg and Mn in calcite and gypsum. Phosphatic deposits are frequently enriched in U, sometimes together with V. An association of Ge, P, and V is also found in some sedimentary oxidic iron ores. The relationship of Fe:Mn in subaquatic deposits depends, at a given pH, on the content of O_2. At pH 7, Fe is soluble up to 0.5 mg O_2/dm^3, Mn up to 1 mg/dm^3. MnO_2 may accumulate Li, K, Ba, Mo, and Co by adsorption or as manganites.

Biogenic Sediments. Biogenic sediments are formed by parts or derivates of parts of the bodies, or by the action of organisms, or by the interaction of organic and inorganic matter. Some elements are selectively accumulated by organisms. For example, marine algae are enriched in Au (up to 10^3 times), Pb and Mo (up to 10^4), and Cu and Ni (up to 10^5) with regard to seawater. Some organisms even form skeletons composed of compounds of rare elements, e.g., xenophyophora ($BaSO_4$).

Anorganic biogenic sediments are formed by skeletons, or parts of skeletons, composed of carbonates, phosphates, or silica. Aragonitic and calcitic shells are characterized by enrichments in Sr and Mg, respectively. Phosphatic and calcitic skeletons seem to be relatively more frequent in Paleozoic sediments, thus simulating an "age effect" in limestones but not in shales. Siliceous rocks show the same enrichments as shales of the same facies, but to a lesser degree. Diatoms accumulate Mn, radiolaria Sr.

Organo-mineralic sediments are formed in regions that have little or no oxygen. In sub-aerial deposits and in well-aerated waters only small quantities of the most resistant organic substances are preserved. In terrestrial deposits, waxes and resins are preserved; for subaquatic deposits, remains of skleroproteins and chitin are more characteristic. Since these latter two have C/N ratios of 3 and 6 respectively, recent subaquatic deposits of this type have C/N ratios of about 10 to 12, fossil ones of about 14 to 15. The deposits are anorganic (abiogenic or biogenic). In regions that have little or no oxygen, several types of deposits may be distinguished:

Gyttja is formed in regions with bottom waters poor in O_2, where the production of organic substances is greater than the rate of oxidation during sedimentation. More labile substances, such as soft proteins, are destroyed; skleroproteins and chitin, as well as carbohydrates and fats are preserved, mostly after transformations. Since carbohydrates and fats make up most of the organic substances, and contain no N, fossil gyttjas have C/N ratios ranging from 50 to 350, whereas recent ones (with some proteins still preserved) have ratios down to 8. In gyttja, there are all transitions from mineral deposits of well aerated waters, to sapropels of hydrogen sulfide waters. Re-

TABLE 1. TRACE METALS IN SEDIMENTS (IN PPM)

Formation	Locality	Rock	Facies	NiO	Co_3O_4	Cr_2O_3	V_2O_3	MoO_3
—	—	Magmatic rocks[a]	Eruptive	130	50	300	200	20
Neogene	Rumania	6 sands		10–50	10–50	<10	<100	n.f.
Neogene	Germany	45 sandstones		10–50	10–100	10–50	<100–100	n.f.
Neogene	Rumania	6 clays	Mineralic	50–100	10–50	10–50	100–500	n.f.
Recent	Oahu, Hawaii	Reef limestones		50–500	10–50	50–100	<100	n.f.
Neogene	Rumania	4 limestones		<5–10	<10–50	n.f.	<100	n.f.
Oligocene	Rumania	Rock salt		n.f.[c]	n.f.	<10	<100	n.f.
U. Permian	Fallstein, Germany	Anhydrite + dolomite		n.f.	n.f.	n.f.	n.f	n.f.
Pliocene	Rumania	7 carbonaceous rocks		100–500	<50–100	<50–500	<100–500	<<100–100
Miocene	Beuern, Germany	6 carbonaceous rocks	Coals	100–500	50–100	100–500	100–500	n.f.
L. Eocene	Geiseltal, Germany	6 lignites		<5–50	<50–100	<<50–50	<100–100	n.f.
Carboniferous	—	Ashes of coals[a]		(20–500)	400	500	(<<100–100)	300
Recent	Germany	7 muds (tidal flats)		10–50	10–50	10–100	<100–100	n.f.
U. Eocene	Rumania	4 clays (flysch)		100–500	50–100	50–100	100–500	n.f.
Carboniferous	Germany	17 greywacks	Gyttjas	10–50	50–100	10–50	100	n.f.
Wealden	Obernkirchen, Germany	2 bituminous clays		50–100	10–100	100–500	100–500	n.f.
M. Rhetic	Edesse, Germany	4 bituminous clays		10–100	50–100	50–100	100–500	n.f.
L. Devonian	Bundenbach, Germany	18 shales		50–500	50–100	100–500	100–500	n.f.
M. Eocene	Messel, Germany	3 oilshales		100–800	50–100	100–1000	100–500	n.f.
Tertiary	Marahu, Brasil	Marahuite		6–10	n.f.	70–150	7–15	8–15
Ordovician	Kohtla, Estonia	Kukkersite	Sapropelites	50	n.f.	n.f.	<100	<100
Ordovician	Kohtla, Estonia	Tar residue		50	n.f.	100–500	<100	n.f.
Sub-Recent	Black Sea	Black mud[b]		n.d.[d]	n.d.	n.d.	600–900	n.d.
Cornu shales	Rumania	2 bituminous clays		100–500	50–100	50–100	500–5000	<100–500
Oligocene	Rumania	6 bituminous clays		<5–100	n.f.	10–50	<100–500	<<100–100
Cretaceous	Venezuela	La Luna limestone		10–50	<10–50	50	1000–5000	100–500
Barremian	Wenden, Germany	Foliaceous shales		100	10–50	50–100	500	<100–100
Lias E	Germany	2 Posid. shales		10–500	<50–50	<50–100	500–1000	100
U. Triassic	Schröfeln, Germany	2 bituminous marls		500	50	100–500	1000–5000	100–500
Triassic	Meride, Switz.	Bituminous marl		100–500	50–100	n.f.	1000–5000	50–100
Hauptdol.	Volkenroda, Germany	Black clay		60–100	16–80	150–700	150–700	15–75
Zechstein	Mansfeld, Germany	Kupferschiefer[a]		500	100	50	1000–5000	100–500
Silurian	Hohenleuben, Germany	Alune shales[a]		25	n.d.	45	1.500	330
Silurian	Weckersdorf, Germany	Alune shales[a]		60	n.d.	90	800	200
Ordovician	N. Europe	Dictyonema shale		10–500	<<50[e]	<50–100	1000–5000	<<100–500
Jatulian	Suojärvi, Karel. ASSR	Shungite		100–500	<50–50	100–500	1000	n.f.
Jatulian	Shunga, Gouv. Olonetsk	Shungite		600–1300	trace	trace	750–1500	75–15

[a] Average. [b] Contains 100 ppm Cu. [c] Not found. [d] Not determined. [e] Much less than 50.

cent gyttja consists of an oxidized, light-colored upper layer ranging in thickness from millimeters to decimeters, and a reducing, dark-colored layer with H_2S. At higher redox potentials Mn is accumulated in the mineral deposits; Cr seems characteristic for most gyttjas and Io (ionium) has also been reported. Fe ranges from oxides and oxide-hydrates at higher Eh potentials to carbonates and silicates to sulfides on and below the zero level; P is accumulated at the lower Eh values and U straddles the limit to anaerobic conditions. There are movements of elements as the conditions change while sediment is piling up, and the reducing environment moves upward.

Peats (and *coals*) (see Table 2) are formed in an amphibious environment somewhat similar to that of gyttja. The groundwater level is near the surface; more highly situated pores are partly filled by capillary humic waters. These, as well as bodies of open water are poor in, or devoid of, oxygen. Organic matter passes through a stage of aerobic decomposition when labile substances are destroyed; the accumulation of lignine and/or cellulose leads to a high C/N ratio of <15 in recent and about 50–100 in fossil deposits. Remains of higher plants (fusite, vitrite) are mixed with those of lower plants and with detritus (durite). Subaquatic deposits are dy (colloidal humic substances) or dygyttja (cannel coal with spores or pollens; bogheads with algae). Dygyttja seems enriched in Sn; humic coals in Be, Ge, and As. A loose inverse correlation of Ge and

TABLE 2. ELEMENTS ACCUMULATED IN ASHES OF COALS (PPM)

No.	Element	Earth's Crust (average)	Pelitic Sediments (average)	Ashes of Coals (max. content)
3	Li	65	46	960
4	Be	6	<4	2,800
5	B	10	310	8,600
21	Sc	5	6.5	400
22	Ti	4400	4300	20,000
23	V[a]	150	120	11,000[d]
24	Cr	200	550	1,200
25	Mn	1000	620	22,000[d]
27	Co	40	8	2,000
28	Ni[b]	100	24	16,000[d]
29	Cu	70	192	4,000[d]
30	Zn	40	200–1000	10,000[d]
31	Ga	19	50	6,000
32	Ge	7	7	90,000
33	As	5	~5	8,000
37	Rb	280	300	33
39	Y	28	28	800
40	Zr	220	120	5,000
41	Nb	20	—	2
42	Mo	2.3	—	2,000[d]
47	Ag[c]	0.02	0.05	5–10
48	Cd	0.18	0.3	80
49	In	0.1	0.5	2
50	Sn	40	40	6,000
51	Sb	1	3	3,000
53	I	3	0.3	950
55	Cs	3.2	12	4
57	La	18	18	31
73	Ta	2.1	—	0.1
78	Pt	0.005	—	0.7
79	Au[c]	0.001	—	0.2–0.5
80	Hg	0.5	0.3	50
81	Tl	1.3	2	25
82	Pb	16	20	1,000[d]
83	Bi	0.2	1	200
92	U	4	1.2	600[d]

[a] Several percents were reported from ashes of Uralian and Bashkirian coals.
[b] Eight percent Ni was indicated in vitrinite.
[c] Ashes from Cambria, Wy. contained $2 Au and $0.28 Ag per ton.
[d] Higher values are probably due to secondary enrichment, which is not excluded even for the numerically given values.

As to Ga and V was reported from Germany; a direct correlation of Ge with Mo, Co, and Ni from the Donetz Basin. The greatest amount (average 80%) of Ge is found in the organic substance of coals (Ge-humacite). Be, Cr, and Zr are bound to the organic substances of coals. As, B, Cu, Mo, Ni, Ti, S, and Y are more abundant in the organic substances; Co, Ga, La, P, Pb, Sb, V, and Zn are more abundant in the minerals. Al, Ba, Ca, Fe, K, Mg, Mn, Na, and Si occur in greatest part in the mineral substances.

Sapropels are formed in water free of oxygen. The limit between O_2 and H_2O (the zero level of the redox potential), which, with gytta, lies within the sediment, is here above the sediment in free water. Organic matter entering the H_2S zone is transformed anaerobically, but its substance is largely preserved, including labile proteins. The ratio C/N may be as low as 3 in recent sediments, and is usually 20–30 in fossil sediments. Bacterial activity is high, depending on the C/N ratio; in soils microorganisms thrive only at ratios lower than 20 and are increasingly inactive above 30. In the free water, bacteria are most frequent at the limit O_2/H_2S; where this limit cuts into the coastal sediments there is a maximum of sulfides. P escapes from the bacterial mud and is found in the water above it. Elements which would not precipitate because of dilution are adsorbed by clay or organic particles or dragged down with precipitating iron (or other metal) sulfides (e.g., Mo). Uranium accumulated near the zero potential; Pb and Zn at lower, and Cu, Ag, V, Mo, and Ni at higher negative redox potentials.

The Ni:Co ratio in plants is mostly 5–10; in gyttjas, ~ 1; in coals, mostly >1 (in humic coals higher than in sapropelic); in cannel coals and bogheads (dygyttja deposits), 1–10; in true sapropels, >1. The V:Mo ratio in gyttja is <5, in sapropel, >5. The V:Cr ratio is about 1 in gyttja, >1 in sapropels. The V:Ni ratio is about 1 in gyttjas, 2–10 in sapropels. The Cr:Ni ratio is >1 in gyttja, <1 in mineralic and sapropelic deposits. The Mn:Ni ratio is high (up to 10) in well-aerated environments and low in low O_2 environments, (even <1).

Secondary Enrichments. These are caused by the physical or chemical nature of sediments and may complicate the picture. Sands may be percolated by waters, and the resulting solutions precipitated on included or adjacent organic matter or clays. Such action, effective even on a small scale, may lead to considerable accumulations of Cu, Ag, V, and U. Although fresh oils will not stick to sand grains, oxidized oils contain compounds that will be strongly adsorbed, forming coatings enriched in trace metals (German salts have acquired Pb from oilfield brines). Metasomatic replacements in limestones are well known in commercial deposits of Fe, Zn, Pb and Ag. Organic substances, present as remains in clastic sediments, carbonates, and especially in phosphates, or as integral components in coals, gyttjas, and sapropels, act as adsorbents and precipitators. In clastic sediments, such inclusions may be accompanied by V and U; coals may be secondarily enriched in Cu, Ag, Au, U, Pb, and Zn, and minute cracks in gyttjas and sapropels contain galena and sphalerite. Peat and lignite accumulate copper from copper-bearing solutions (up to 4% of their own weight). Most Mo is found in pyrite; its accumulation depends largely on highly negative redox potential. During low degrees of metamorphism, Ag, Au, Zn, Pb, and U may accumulate in coals; during higher degrees (as in anthracites), Ge is partly lost.

Oil and Bitumen

Oil and bitumen contain optically active substances, porphyrins, and probably steroids, all of which are only formed in organisms; they are thus biogenic products. Oil contains substances such as adamantanes, alkynaphthobenzenes, polycyclic aromatic hydrocarbons, etc., which are not found in living organisms. Organic substances have, therefore, been transformed. The transformation of porphyrins, and the change of small oxygen-bearing molecules into large aromatic hydrocarbons, has been followed step by step. Oil is thus formed after the death and burial of organisms, i.e., within sediments, under reducing conditions. (For a list of trace metals in oil and bitumen, see Tables 3 and 4).

Oil, like any liquid or gas, has the ability to migrate. The more common hydrocarbons may be formed in many ways, but porphyrins are found primarily in algal gyttja and sapropels. A certain family of trace metals connects oils, asphalts, and ozokerites with sapropels. The content of trace metals in ashes of oils is frequently in inverse relation to the amount of ash; it is thus probable that the trace metals are present in organic compounds or in solution rather than as solid impurities. Porphyrins, which quickly lose their central ions (Mg^{2+} or $(FeCl)^{2+}$) after sedimentation, acquire new ones consisting of, or containing, the metals Fe, Ni, V, and Ga; other organometallic compounds (or clay-organometallic compounds) seem to be found among the asphaltenes. Since these are large molecules they accumulate in heavy oils and asphalts (light oils are poor in metals), and in the residues of distillation;

TABLE 3. TRACE METALS IN TARS FROM OIL SHALES, IN OILS, ASPHALTS AND OZOKERITES (PPM)

Formation	Locality		Substance	Cu	Sn	V	Cr	Mo	Co	Ni
M. Eocene	Messel, Germany	Oilshale	tar	n.d.[a]	n.d.	200	80	n.d.	n.d.	200
M. Eocene	Messel, Germany	Oilshale	paraffin	n.d.	n.d.	140	60	n.d.	n.d.	750
Ordovician	Kohtla, Estonian ASSR	Oilshale	tar	n.d.	40–80	≪100	70–350	n.f.	n.f.	40
Ordovician	Kohtla, Estonian ASSR	Oilshale	blow pitch	n.d.	n.f.[b]	<70–70	40–70	n.f.	≪40	10–40
Pliocene	Nagy Lengyel, Hungary	Naph. oils		n.d.	n.d.	300.000	n.d.	n.d.	n.d.	157,000
U. Miocene	Gösting I, Austria	Naph. oils	0.2% ash	n.d.	n.f.	70	>30–30	n.f.	40–70	1,600
U. Miocene	Gösting II, Austria	Naph. oils	0.04% ash	n.d.	80–400	350–700	30–70	n.f.	400	8,000–80,000
Miocene	Maidan, Iran	Oils		n.d.	n.d.	50,000	n.d.	n.f.	n.d.	27,000
Oligo-Miocene	La Rosa, Venezuela	Oils		n.d.	<40	>3,400	<30	n.f.	40	1,600–3,200
Oligo-Miocene	Lagunillas, Venezuela	Oils		n.d.	n.f.	>3,400	<30	n.f.	40–70	3,200–8,000
Permian	NE Caucasus, φ 13	Oils		4,100	n.d.	300,000	100	n.d.	n.f.	72,000
Carboniferous	Wolga- φ 17	Oils		4,800	n.d.	170,000	200	n.d.	n.d.	77,000
Carboniferous	Ural φ 27	Oils		4,300	n.d.	176,000	200	n.d.	n.d.	59,000
Devonian	Oklahoma	Oils		n.d.	n.d.	124,000	n.d.	n.d.	n.d.	46,000
Recent	Dead Sea		Manjak	n.d.	n.f.	>3,500	n.f.	700–7000	n.f.	4,000–8,000
U. Pliocene	Matiza, Rumania		Asphaltite, 2.9% ash	n.d.	n.f.	350–700	70–350	n.f.	<40–40	600–1,600
Pliocene	Selenica, Albania		Asphalt	n.d.	n.f.	>3,500	35–70	<70–70	7–40	400–800
Neogene	Boeton,[d] Celebes		Limestone-asphalt	n.d.	n.f.	100–500	trace	n.f.	5–10	50–100
Hauterive	Bentheim, Germany	Albertites	0.9% ash	n.d.	n.f.	70–350	35–70	n.f.	40	~8000
Cretaceous	Yauli, Peru	Albertites	11.7% ash	n.d.	n.f.	3,500	n.f.	350	n.f.	600–1600
Cretaceous	Huari, Peru	Albertites	0.9% ash	n.d.	n.f.	>3,500	35	n.f.	n.f.	>8000
Cretaceous	S. Martin, Argentina[c]	Albertites		n.d.	n.f.	27,000	n.d.	n.f.	n.f.	5,000
Cretaceous	Auca Mahuida, Argentina	Albertites	Rafaelite	n.d.	n.f.	195,000	n.d.	n.f.	n.d.	73,600
M. Triassic	Mt. Coldai	Ozokerite	Anthraxolite	n.d.	n.f.	89,000	<35–35	350–700	n.f.	8–40
Jatulian	Shunga	Ozokerite	Shungite	240	n.f.	3,000	n.d.	1,200	n.d.	35,000
Miocene	Cheleken, Kaspi	Ozokerite		n.d.	n.f.	<70–70	30	n.f.	40–70	600
Miocene	Boryslaw, Ukraine	Ozokerite		n.d.	n.f.	70–340	<30–30	n.f.	40–70	400–600
Miocene	Moinesti, Rumania	Ozokerite		n.d.	4–8	<70–70	7–30	n.f.	7–40	800
Eocene	Wigan, England	Ozokerite	Elaterite	n.d.	n.f.	<70	<30–30	70	40–70	80–400
Eocene	Montrelais, France	Ozokerite	Elaterite	n.d.	n.f.	<70	<30–30	<70–70	40–70	400–600

[a] n.d. = not determined. [b] n.f. = not found. [c] φ (13 = average of 13 analyses). [d] Boeton asphalt has 60–75% ash, shungite 35%. Ash of shungite has up to 9000 ppm V.

such residues may contain up to 97% V_2O_5 or 90% NiO. (Under the name of *asphalt* we combine all substances, from the soft manjak wholly soluble in organic solvents to the nearly or totally insoluble albertites, impsonites, and anthraxolites. Insolubility may be caused by exposure).

Oils are usually poor in metals of the 6-valent group: Cr, Mo, W, and U. This is expected with regard to Cr, which is not accumulated in sapropels, but with regard to their electronegativity we would expect Mo to follow V geochemically. We would also expect Cr to be accumulated in tars obtained from gyttja oil shales; but some such tars are poor in Cr. However, Mo accumulates in some asphalts, as does U (Wichita-Amarillo). U also accumulates in the Swedish Kolm and in the "Pechhieken" of the Kupferschiefer. It seems that 6-valent metals belong to the least mobile compounds, since they are found accumulated in veins of (predominantly hard) asphalts connected not with oils, but with bituminous rocks (Mt. Coldai, Tirol; Argentina), supposedly expelled by volcanic heat. Mo, together with V and Ni, is also accumulated in the soft manjak of the Dead Sea. It seems that the patronite (VS_4) of Minasragras is similarly a product of contact metamorphosis of an asphaltic vein, containing also quisqueite (C_4S), possibly a kind of thiophen graphite.

Many elements have been reported from crude oils (Table 4); of these, Co, Cr, Cu, Mo, Ni, Pb, V, and Zn are most consistently present. The ratio V:Cr is equal to or greater than 10 in most oils, and from 10 to 3000 in asphalts; V:Ni is from 0.01 to 10 in oils, about 0.1 in Ozokerites, and from 0.01 to several thousand in asphalts. The ratio V:Ni may be very constant within certain horizons and/or regions. Mo is generally far below 100 ppm in oils, but may reach from zero to several permils in asphalts. V:Mo varies from near infinity to 1 in asphalts. In the Ural-Volga region, ratios in oils follow the sequence V>Ni>Cu>Cr; these are the usual ratios found elsewhere, but in some oils and asphalts Ni may equal or surpass V. In ozokerites, Ni>V>Cr. Most crude oils contain less U than does seawater, but several times more is found in oilfield brines. Paraffinic oils contain below average amounts of U; enrichments have been reported from naphthenic oils.

In smaller concentrations, V, Mo, and Ni have an inverse linear relationship to ash, whereas in very pure asphalts the increase in trace metals is greater than the decrease in ash. The separation of the trace metals is probably not only due to the size and charge of the ions but also to the pH and Eh of the solution from which the elements are captured or exchanged (e.g., V against Fe in reducing environments). The process may be comparable to the action of some selective organic reagents.

The fact that ashes of oils and asphalts may be composed of up to 97% V_2O_5, up to 50–95% oxides of Cu + V + Ni, and contain >10% U points to secondary enrichments. That is, V might be taken up by porphyrins without central ions, or, under reducing conditions, in exchange for Fe, etc.; all might precipitate in reducing environments. Oils produced by secondary recovery methods (extracted by solvents from rocks, etc.), are enriched in U, which probably comes from solidly adsorbed coatings on sand grains, clay particles, etc. U is most easily taken up from $CaCl_2$ waters, least easily from $NaHCO_3$ waters. Reduction by organic matter, formation of insoluble organouranyl compounds, and adsorption and bacterial action seem to be responsible for the enrichment in U.

The elements most consistently present in oil and bitumen are Cu, Pb, Cr, Mo, V, Co, and Ni; in coal, Be, B, Ge, V, and As. Even with differential movement of elements during metamorphosis, it is often possible to attribute graphites either to the geochemical group of coals or of bitumen.

Oilfield Brines

Typical oilfield waters from deeper horizons are chloridic brines having little or no sulfates, with irregular enrichments in NH_3, Na, K, Ba, Ra, B, Cl, Br, and I (see Table 5, p. 1210).

Young (Pliocene and Pleistocene) sediments are sometimes encountered in mines below groundwater level in dry areas, with clays adhering to the tongue (the pores of sands being "empty"). Oilfield brines are found in freshwater sediments, even if these sediments have never been covered by later seas.

Dry organic substances contain more than 30% oxygen, which disappears during oil genesis eventually as H_2O and CO_2. It seems that oil field brines are by-products of oil genesis and products of primarily marine organic muds.

Original brines may be diluted by surface waters, which are nonchloridic in humid climates; we therefore regard the relation of an element to Cl as characteristic. Original brines may be diluted with chloridic surface waters of arid regions and especially with waters from saline formations; we think, therefore, of the lowest ratio (Cl: trace elements) as the most characteristic. Thus it may be seen which elements have relationships similar to those found in biochemistry; e.g., enrichments in K, B,

TABLE 4. MAXIMAL ACCUMULATION OF TRACE METALS IN OILS AND ASPHALTS (PPM)

		Oil[a]		Asphalts		Contents in	
No.	Element	Total	Ash	Total	Ash	Earth's Crust	Pelitic Sediments
3	Li	—	750	—	—	65	46
4	Be	—	10	—	(X0)	6	<4
11	Na	+	228,000	—	(X00,000)	28,300	9,700
12	Mg	700	70,000	5,300	35,000	20,900	14,800
13	Al	+	210,000	5,400	(X00,000)	81,300	81,900
14	Si	100	280,000	24,000	155,000	280,000	240,000
19	K	+	15,000	—	(X0,000)	25,900	27,000
20	Ca	180	186,000	12,000	(X00,000)	36,300	22,300
21	Sc	—	(X00)	—	(X0)	5	6.5
22	Ti	+	8,000	—	(X0,000)	4,400	4,300
23	V	260	460,000	23,400	530,000	150	120
24	Cr	3	450	6	330	200	540
25	Mn	330	33,000	0.4	(X0,000)	1,000	620
26	Fe	850	530,000	420	250,000	50,000	47,300
27	Co	1	10,000	0.14	70	40	8
28	Ni	240	158,000	3,000	200,000	100	24
29	Cu	19	130,000	1,660	108,000	70	192
30	Zn	0.6	18,000	100	1,000	40	600
31	Ga	4	3,800	—	(X00)	19	50
32	Ge	0.7	(X00)	—	(X00)	7	7
38	Sr	+	10,000	—	(X,000)	150	170
39	Y	—	(X00)	—	(X,000)	28	28
40	Zr	—	100	—	(X,000)	220	120
42	Mo	3	1,300	40	4,000	2.3	—
46	Pd	+	+	—	—	0.01	—
47	Ag	+	(X0)	—	(X00)	0.02	0.05
48	Cd	—	(X0)	—	—	0.18	0.03
50	Sn	0.1	500	—	(X00)	40	40
51	Sb	1	(X)	11	(X,000)	1	3
55	Cs	—	+	—	—	3.2	12
56	Ba	+	10,000	—	(X,000)	430	460
57	La	—	(X0,000)	—	(X,000)	18	—
58	Ce	—	(X0,000)	—	(X,000)	42	—
60	Nd	—	—	—	—	24	—
70	Yb	—	(X,000)	—	—	3	—

TRACE METALS IN SEDIMENTS, OILS, AND ALLIED SUBSTANCES

Br, and I. in oilfield brines match those in marine algae.

Many trace elements are dissolved from the country rock (see above, enrichments in pelitic rocks); with regard to CuS and $RaSO_4$, solubility products are not attained within the contents observed. Due to formation of complex ions the solubility products may be transgressed: at 75°C and 20 atm, 10 mg ZnS/dm^3 are soluble in H_2S-saturated water; this is a million times higher than the solubility product.

With regard to the high contents of very common elements, such as Fe and Al, it may be remembered that two wells at Teis, Rumania produced oil with 40% clay in inseparable emulsion; emulsion of oils and brines are well known.

KARL KREJCI-GRAF

References

Erickson, R. L., Myers, A. T., Horr, C. A., 1954, "Association of uranium and other metals with crude oil, asphalt, and petroligerous rock," *Bull. Am. Assoc. Petrol. Geologists,* **38**(10), 2200–2218.

Goldschmidt, V. M., 1954, "Geochemistry," Oxford, Clarendon Press, 730pp.

Goldschmidt, V. M., Krejci-Graf, K., Witte, H., 1948, "Spurenmetalle in Sedimenten," *Nachr. Akad. Göttinger math. phys,* Kl, 1948, pp. 35–52.

Hesemann, J. (editor), 1963, "Unterscheidungsmöglichkeiten mariner und nichtmariner Sedimente," *Fortschr. Geol. Rheinld. Westf.,* **10**, XII + 482pp., 10 plates, 93 figs., 93 tables.

Krejci-Graf, K., 1964, "Geochemical diagnosis of facies," *Proc. Yorkshire Geol. Soc.,* **34**, 469–521.

Rankama, K., and Sahama, Th. G., 1950, "Geochemistry," Chicago, Univ. of Chicago Press, 912pp.

Zulfugarly, D. I., 1960, "Distribution of microelements in kaustobioliths, organisms, sediments and oilfield waters," *Akad. Nauk SSSR Inst. Geol. Razrab. Gor. Iskop. Baku.,* 230pp. (in Russian).

Cross-references: Carbonates; Hydrocarbons; Natural Brines; Organic Geochemistry; pH-Eh; Phosphates; Precambrian Hydrocarbons; Solubility Product Constant; Trace Elements; Weathering, Chemical. Vol. VI: *Asphalt, Asphaltites, Asphaltoids; Coal; Crude Oil; Dy; Petroleum; Sapropel.*

TABLE 4. *(Continued)*

No.	Element	Oil[a] Total	Oil[a] Ash	Asphalts Total	Asphalts Ash	Earth's Crust	Contents in Pelitic Sediments
74	W	—	(X)	—	—	1	2
76	Os	—	(X)	—	—	—	—
77	Ir	—	(0,X)	—	—	0.001	—
78	Pt	—	(X)	—	—	0.005	—
79	Au	0.005	+	0.001	0.3	0.001	0.4
80	Hg	27	(X,000)	—	(X0)	0.1	2
81	Tl	—	+	—	2.8	1.3	2
82	Pb	—	6,000	—	(X,000)	16	20
83	Bi	—	(X0)	—	(X0)	0.2	1
90	Th	0.004	(X)	—	—	12	10
92	U	2.1	4,600	>60,000	>100,000	4	12

[a] Amounts for ashes and oils or bitumen are mostly from different samples. Values in brackets are less reliable. Averages are meaningful only for oils of specified character and density within specified horizons and regions. Also, they should be stated in relation to a standard element (as Cl in brines) if we could find such. The range goes down from the max. to values below detection.

TABLE 5. ENRICHMENT OF ELEMENTS IN OILFIELD WATERS, SEAWATER, ASHES OF PLANKTON AND BENTHONIC ALGAE

Element	Sea (ppm)	Oilfield Waters (max., mg/l)	Plankton (max., ppm)	Ratios (weight)	Sea	Oilfield Waters Min. Ratio	Oilfield Waters Max. Enrichment	Benthonic Algae Min. Ratio	Benthonic Algae Max. Enrichment
Li	0.180	140	—	Cl:Li	106,000	90	1,200	—	—
Na	10,470	110,000	150,000	Cl:Na	1.85	1.2[a]	1.5[a]	~1	~2
K	380	16,050	20,000	Cl:K	50	1.8	28	1–2	25–50
Rb	0.2	21	—	Cl:Rb	97,000	8,000	12	(750)[e]	(130)[e]
Cu	0.005	0.34	2,700	Cl:Cu	3,900,000	450,000	9	(6000)[e]	(650)[e]
Mg	1,280	6,400	(20,000)[f]	Cl:Mg	15	12	1.2	~1	~15
Ca	410	23,000	40,000	Cl:Ca	48	8	6	~1	~50
Sr	8.1	1,400	—	Cl:Sr	2,400	120	20	(100)[e]	(24)[e]
Ba	0.05[g]	28	—	Cl:Ba	390,000	30,000	13	(1000)[e]	(400)[e]
Ra	10⁻¹⁰	10⁻⁶	—	Cl:Ra	—	—	—	—	—
B	4.6	200	760	Cl:B	4,100	46	87	~400	~10
Ga	0.00003	0.0023	—	Cl:Ga	640,000,000	820,000	800	(200,000)[e]	(~30)[e]
Sn	0.003	0.18	90	Cl:Sn	6,400,000	40,000	160	(9,000)[e]	(600)[e]
Pb	0.003–0.005	0.40	1,300	Cl:Pb	5,500,000	180,000	30	15[c,d]	250,000 or 25,000[c,d]
NH₃	(1[d])	—	—	Cl:NH₃	3,900,000 to 390,000	45[a]	85,000[a] or 8,500[a]	—	—
V	0.005–0.05	300	85	Cl:V	65,000,000	—	—	(6,000)[e]	(~10,000)[e]
U	0.0003	>0.02	—	Cl:U	10,000,000	—	—	—	—
Cl	19,345	170,000	—						
Br	65	6,500	—	Cl:Br	300	80	3.6	~100, ~67[c]	~3, ~4[c]
J	0.05	100	620	Cl:J	390,000	71, 37[b]	5,400, 10,000[c]	~71, ~31[c]	~3,800, ~12,000[c]
Ni	0.0004	40	480	Cl:Ni	50,000,000	—	—	(~30,000)[e]	(~1,700)[e]

[a] Deep-seated waters only. [b] Waters from Flysch and Molasse. [c] Tang meadows of California. [d] Cl:N. [e] Related to average Cl-contents of the species. [f] Referring to dry weight. [g] According to V. Engelhardt; Bowen gives 0.006 ppm.

TRACER TECHNOLOGY—See Vol. VI, RADIOACTIVE ISOTOPE TRACER TECHNOLOGY

TRANSURANIUM SERIES—See ACTINIDE SERIES

TRITIUM: ELEMENT AND GEOCHEMISTRY

Properties

Tritium (H^3 or T) is the isotope of hydrogen with three neutrons in the nucleus. It is radioactive and decays by beta radiation (18.6 keV) with a half-life of 12.26 r. Tritium was first identified as a product of deuterium bombardment by Rutherford in 1934. Its presence in nature was discovered by Faltings and Harteck in 1950 and by von Grosse, Libby, and colleagues in 1951.

Natural tritium is produced in the upper atmosphere by cosmic ray spallation and by the interaction of fast neutrons with nitrogen, $N^{14}(n, H^3) C^{12}$. Measurements of samples not affected by fallout tritium indicate a natural production rate on the order of 0.5 atoms per square centimeter of the earth's surface per second.

The mass ratio of 3:1 between tritium and protium introduces significant differences in the physical properties. Several of the properties are compared below:

	H_2	T_2
Triple point (°K)	13.96	20.62
Boiling point (°K)	20.39	25.04
Heat of vaporization (cal/mole)	216	333

Abundance

Since tritium oxidizes rapidly to HTO it exists essentially as water and its distribution among the oceanic, terrestrial and atmospheric reservoirs is controlled by the operation of the hydrological cycle. Tritium abundance is usually expressed in tritium units (T.U. = T/10^{18}H) and was on the order of 1 T.U. or less for surface ocean water before thermonuclear testing began. Other natural abundance values were approximately 10 T.U. for northern hemisphere continental midlatitude precipitation, 1 T.U. for equatorial and oceanic precipitation, and essentially zero for deep ocean water and ground water.

In 1952 thermonuclear testing released the first artificial tritium to the atmosphere. By 1954 this had raised precipitation levels in the Northern Hemisphere to several hundred T.U. Later test series up to the moratorium in 1963 added a total of approximately 100 kilograms of tritium.

Natural tritium values primarily for North America, were reviewed by Kaufman and Libby (1954). A world review of tritium fallout is available in the report of USAEC contract AT (30–1) 3162.

Geophysical Behavior

Tritium circulation in nature is governed primarily by two effects, the stratospheric input and the circulation in the hydrological cycle. This is true for both artificial and natural tritium since the former is carried through the stratosphere by the jet stream and other mechanisms before it enters the weather zone. Hence tritium, like other stratospheric fallout isotopes, is injected into the troposphere by the annual "spring leak" and reaches its highest concentration in precipitation in late spring. The stratospheric residence time, according to Leventhal and Libby (1968) is 1.5 yr.

Tritium fallout is maximum in the northern latitudes (above about 40°); the concentrations are highest in the continental air masses, particularly the polar air masses, and are lowest in oceanic air masses. Hence, tritium isoconcentration lines are found to circle the earth approximately along latitude parallels with strong distortions along coastlines and southward penetrations over continents.

Tritium measurements confirmed the long suspected isolation of the southern hemisphere from the air masses of the northern hemisphere. Thermonuclear testing was confined to the northern hemisphere until 1967 and tritium in the southern hemisphere remained relatively very low.

Tritium has been a particularly important tool in oceanography for the investigation of mixing across the thermocline and identification of downwelling and upwelling. Results to date suggest the mixing rates show great variations as tritium has been found as deep as 500 meters in the Atlantic but is usually confined to the top 100 meters. Suess (1969) has reported the discovery of tritium rich ocean layers that indicate fast moving currents coming under the surface from northern latitudes.

Tritium has been particularly fruitful in hydrologic investigations. These were pioneered by Begemann and Libby (1957) in a study of the northern Mississippi Valley. The most rigorous application of tritium is the detection of recent recharge, Thatcher et al. (1961). Hydrologic investigations based on tritium include base flow in the Hudson by Giletti et al. (1958), groundwater stratification, Carlston et al. (1960), and runoff relations in the Ottawa River by Brown (1961). Recent applications of tritium to hydrological investigations have been reviewed by Payne (1967).

LELAND L. THATCHER

References

Begemann, F., and Libby, W. F., 1957, "Continental water balance, ground water inventory and storage times, surface ocean mixing rates and world wide circulation patterns from cosmic ray and bomb tritium," *Geochim. Cosmochim. Acta*, **12**, 277.
Brown, R. M., 1961, "Hydrology of tritium in the Ottawa Valley," *Geochim. Cosmochim. Acta*, **21**, 214.
Carlston, C. W., Thatcher, L., and Rhodehamel, E. C., 1960, "Tritium as a Hydrologic Tool," The Wharton Tract Study, Publ. No. 52, IASH Committee on Subterranean Waters, p. 503.
Giletti, B., Bazan, F., and Kulp, L. J., 1958, "Tritium geochemistry," *Trans. Am. Geophys. Union*, **39**, 807.
Kaufman, S., and Libby, W. F., 1954, "The natural distribution of tritium," *Phys. Rev.* **93**, 1337.
Leventhal, J., and Libby, W. F., 1968,, "Tritium geophysics from 1961-62 tests," *J. Geophys. Res.*, **73**(8).
Payne, B. R., 1967, "Isotope techniques in the hydrologic cycle," *Geophys. Monograph Series*, **11**, 62.
Suess, H. E., 1969, "Tritium geophysics as an international research project," *Science*, **163** (3874), 1405.
Thatcher, L., Rubin, M., and Brown, G., 1961, "Dating desert ground water," *Science*, **134** (3472), 105.

Cross-references: *Deuterium; Hydrogen; Hydrologic Cycle. Vol. I: Upwelling. Vol. II: Jet Streams; Stratosphere; Thoposphere.*

TUNGSTATES—See MOLYBDATES AND TUNGSTATES

TUNGSTEN: ELEMENT AND GEOCHEMISTRY

Tungsten is a valuable metal, especially remarkable for its high melting point. The name tungsten, first used by A. F. Cronstedt in 1755, comes from the Swedish *tung* (heavy) and *sten* (stone), the miners' term for scheelite $CaWO_4$, first analyzed in 1781 by the Swedish chemist C. W. Scheele. The symbol W comes from the mineral wolfram or wolframite (Fe, Mn)WO_4, first described by L. Ecker in 1574 and so-called on account of its "wolfish" behavior in replacing tin in some ores. The actual metal is not found in nature, but was isolated by two Spaniards, J. J. and F. de Elhuyar in 1783, and was experimentally used by them to harden steel; it was only in the present century, however, that it was so employed by industry.

Physical Properties and Characteristics

Tungsten, symbol W, has the following properties: atomic number 74; atomic weight 183.92; atomic radii—1.26Å in fourfold coordination, 1.31Å in eightfold coordination, and 1.41Å in twelvefold coordination; ionic radii—0.70Å for valence 4 in sixfold coordination, 0.62Å for valence 6 in sixfold coordination, and 0.59Å for valence 6 in fourfold coordination; electron shells 2, 8, 18, 32, 12, 2; valences 2,4,5,6; electron configuration $1s^2$, $2s^2$, $2p^6$, $3s^2$, $3p^6$, $3d^{10}$, $4s^2$, $4p^6$, $4d^{10}$, $4f^{14}$, $5s^2$, $5p^6$, $5d^4$, $6s^2$; type of lattice: body-centered cubic; $\alpha = 3.159$Å; isotopes and abundance—$180 = 0.122\%$; $182 = 25.77\%$; $183 = 14.24\%$; $184 = 30.68\%$; and $186 = 29.17\%$; specific gravity $= 19.35$; atomic volume $= 9.50$ (atomic weight/specific gravity); hardness $= 6.5$ to 7.5 (Mohs scale); melting point $= 3410°C \pm 20°C$; boiling point $= 5900°C$ (approx.); coefficient of thermal expansion $= 4.6 \times 10^{-6}$ cm/cm/°C (0 to 500°C); thermal conductivity $= 0.35$ cal/sec/cm^2/°C–cm; electrical resistivity $= 5.5$ microhm–cm at $20°C$; 108.5 at $3227°C$.

Minerals

Wolframite group minerals include the following:

Wolframite—$FeWO_4$ with 20% to 80% $MnWO_4$, i.e., (Fe,Mn)WO_4, with two end members, as follows:
Ferberite—$FeWO_4$ with less than 20% $MnWO_4$
Hubnerite—$MnWO_4$ with less than 20% $FeWO_4$

Other minerals of this group, but economically unimportant are:

Reinite—$FeWO_4$, pseudomorph after Scheelite
Ferritungstite—$FeO \cdot WO_3 \cdot 6H_2O$
Tungstite—$WO_3 \cdot H_2O$ or H_2WO_4 (Meymacite)
Thorotungstite—$[2\ WO_3 \cdot H_2O + (ThO_2, ClO_2, ZrO_2)H_2O]$
Hydrotungstite—$H_2WO_2 \cdot H_2O$
Limonite—$[(Fe_2O_3)_n WO_3(H_2O)_n]$
Russellite—$BiO_3 \cdot WO_3$
Tungomelane—1 to 7% WO_3 with oxides of Mn, Ba, K, and water

Scheelite group involves essentially scheelite —$CaWO_4$ (80.6% WO_3, 63.9% W). Other minerals of this group, but economically unimportant are:

Powellite—$CaMoO_4$
Stolzite and Raspite—$PbWO_4$
Wulfenite—$PbMoO_4$
Chillagite—$[Pb,(W,Mo)O_4]$
Cuproscheelite and cuprotungstite—
  $[(Ca,Cu)WO_4]$
Tungstenite—WS_2
Hollandite—barium manganate with up to .6% WO_3

Geochemistry

Time-stable isotopes of tungsten are known in nature: all but W^{180} are common. The element is very stable in the presence of mineral acids due to the formation of a passive state. An important feature of tungsten chemistry is the stability of the W^{VI} state, marked especially by the acid H_2WO_4 (a crystalline solid) and a vast number of complicated salts with very variable acid/base proportions. Even more complex are the "heteropoly acids" in which tungstic acid is linked with some other highly oxidized acid residue (e.g., as in the phosphotungstates). The free acid, when liberated from salts, has a pronounced tendency to adopt colloidal properties.

The geochemistry of this element is not particularly well documented. Tungsten is concentrated in the silicate phase of meteorites. Its mean concentration in the earth's crust is of the order 1 ppm.

Ore-level concentrations are found in a number of different geological situations, but all may be linked with granitic intrusions. The element (presumably as $WO_4{}^{2-}$) appears to concentrate within residual granitic magmas. This is reasonable; the stereochemistry of $WO_4{}^{2-}$ is such that direct incorporation in rock-forming silicate structures is unlikely.

Geochemically, tungsten is strongly lithophile and weakly siderophile, with no chalcophile tendency. The volatile constituents of the parent granitic magma, water and compounds of boron, fluorine, and chlorine, act as powerful fluxes (mineralizers) and carriers for the deposition of tungsten. In the absence of these mineralizers, sulfides may act similarly. As the granitic magma cools, it can become an aqueous igneous residual magma from which pegmatites form; then finally veins can develop from the aqueous tungsten-forming solution. Tungsten is formed under conditions of medium-to-high temperatures by gaseous emanations or hydrothermal solutions.

Unlike molybdenum, tungsten is rarely found as the sulfide, almost always occurring as metal tungstates. Wolframite (Fe, Mn) WO_4 and scheelite $CaWO_4$ are the most common ore minerals.

Surface weathering of W-ores does not result in secondary enrichment zones. It is probable that acid leaching liberates tungstic acid and that this is fairly readily removed as a colloid. It has been suggested that further reaction with Ca^{2+}-rich waters leads to secondary precipitation of scheelite.

Economic Aspects

Tungsten is of major economic importance (Table 1). The tungsten-bearing areas of the world, in order of their approximate economic importance are: Sino-Malaysian province 50%; North American—Cordilleran province 12%; Andean province 12%; Australian province 4%; Japan and Korea province 5%; Iberian peninsula and Cornwall province 9%; other regions 7%. Tables 2 and 3 summarize geologic data. E. W. Pehrson estimates that world reserves of tungsten ore will last 125 years at present rate of production.

Uses of Tungsten

The approximate distribution of tungsten in various industrial applications in the United States is:

Tools, steels, dies, hot work, punches, etc.	45%
Military requirements	25%
Carbide inserts for machine tools, wire dies, drills, etc.	19%
Powder for wire, rod, and sheet	9%
Miscellaneous (inks, fluorescent powder, chemicals, etc.)	2%

This article is based partly on information from The American Chemical Society Monograph No. 94 (Monograph Series No. 130) by

TABLE 1. WORLD PRODUCTION OF TUNGSTEN, 1905 TO 1952 ACCUMULATIVE

Locality	Metric Tons, 60% WO_3	% of World Total
North America	138,089	13.48
South America	135,341	13.21
Europe	157,003	15.31
Asia	533,365	52.05
Africa	16,656	1.62
Oceania	44,296	4.33

TUNGSTEN: ELEMENT AND GEOCHEMISTRY

TABLE 2. CONTENT OF TUNGSTEN IN IGNEOUS ROCKS

Type of Rock	Tungsten (G/ton)
Gabbros and norites (Hevesy and Hobbie, 1933)	24
Basic rocks, central Roslagen, Sweden (Lundegardh, 1946)	10
Silicic and intermediate igneous rocks (Sandell, 1946)	1.5
Granite, Schwarzwald, Germany (Hevesy and Hobbie, 1933)	83
Acidic Rocks, central Roslagen, Sweden (Lundegardh, 1946)	7

TABLE 3. GENETIC CLASSIFICATION OF MINERAL DEPOSITS

Temperature of Formation	Zones of Formation	Forms		Examples
Atmospheric	Surface	Eluvial		Placer deposit of scheelite at Atolia, California
50–200°C	Epithermal	Quartz vein		Hubnerite veins in Tonopah, Nevada, and Cripple Creek, Colorado
200–300°C	Mesothermal	Quartz vein deposits	Replacement Deposits	Boulder County, Colorado; Kiangsi deposits, China
300–500°C	Hypothermal (a) Lower limit (subzone)	Quartz vein deposits		Kiangsi deposits, China; Bolivia deposits; Portugal deposits; Australia deposits
	(b) Upper limit	Quartz vein deposits		Malay deposits; Burma deposits; Kiangsi deposits
500–800°C	Pyro-metasomatic	Contact metamorphic deposits		Scheelite deposits in California, Utah
575–1000°C	Orthotectic (a) Lower limit (b) Upper limit	Pegmatoid deposits; Segregation deposits (dissemination deposits)		Pegmatites in Burma, China, U.S.; Granite at Mawchi, Burma

K. C. Li and Chung Yu Wang, to whom credit is due and gratefully acknowledged.

R. K. SMITH
C. D. CURTIS

References

Bateman, A. M., 1950, "Economic Mineral Deposits," Second ed., New York, John Wiley & Sons, 898pp.

Goldschmidt, V. M., 1954, "Geochemistry," Oxford, Clarendon Press, 730pp.

Li, K. C., and Wang, Chung Yu, 1955, "Tungsten," Third edition, New York, Reinhold Publ., 506pp.

Pehrson, E. W., 1951, "Estimates of Selected World Mineral Suppliers by Cost Range," Proc. U.N. Scient. Conf. Cons. & Util. Res., 1949, New York, pp. 2–4.

Rankama, K., and Sahama, T. G., 1950, "Geochemistry," Chicago, Univ. of Chicago Press, 912pp.

Sidgwick, N. V., 1950, "The Chemical Elements and Their Compounds," Oxford, Clarendon Press, 1703pp.

Stevens, R. F., Jr., 1965, "Tungsten," U.S. Bur. Mines Min. Yearbook.

Cross-references: *Hydrothermal Solutions; Weathering, Chemical.* Vol. IVB: *Ore-Forming Fluids; Ore Genesis; Scheelite.*

U

UNITS, NUMBERS, CONSTANTS, SYMBOLS—*See* Vol. II

UPWELLING—*See* Vol. I

URANIUM: ELEMENT AND GEOCHEMISTRY

General Properties. Uranium (L. *uranus*, heaven) was first recognized as an element in 1789 by H. M. Klaproth and first chemically isolated by E. M. Peligot in 1841. A member of the actinide series and designated U, uranium has atomic number 92 and atomic mass 238.07. Elemental uranium is not stable in the geologic realm but in this form is a white, ductile, and malleable metal with a density of 19.04 g/cc and Brinell hardness of 187 (< glass) at room temperature. Metallic uranium exhibits phase transitions at 660 and 770°C, melts at 1330°C, and boils at 3818°C. Uranium is a poor conductor of electricity, is pyrophoric as a powder, and is unstable due to spontaneous chemical and nuclear reactions.

The metallic properties of uranium have little economic use, however its nuclear properties are exploited in radiation therapy, commercial and military power systems, and military ordnance. Uranium salts are used as colorants in ceramics (see *Uranium Ore Deposits*).

Chemical Properties. The electronic configuration of neutral gaseous uranium is [Rn] $5f^3$, $6d$, $7s^2$, and the common ionic species are U(III), U(IV), U(V), and U(VI). The $5f$ electrons are those principally involved in chemical bonding and in the formation of aqueous ions. Fig. 1 contains oxidation potentials of uranium species, the U(O)–U(III) couple of +1.8V being close to aluminum and beryllium in the electrochemical series.

Uranium (III) slowly evolves hydrogen from aqueous solutions and may be precipitated from acid solution by fluoride and oxalate ions. Uranium (IV) is precipitated by iodates, substituted arsenates, and cupferron anions. Uranium (V) is precipitated as a potassium salt from strong carbonate solutions. Uranium (VI) is the only stable state in aerated solutions. Peroxide precipitates uranium (VI) and also oxidizes and precipitates uranium (IV) from moderately acid solutions. Hydroxides of all oxidized states of uranium are insoluble.

For a given anion the order of stability of uranium complexes is: $U(IV) > U(VI)O_2^{2+} > U(III) > U(V)O_2^{1+}$. The order of stability of anion complexes is (-1) fluoride > nitrate > chloride > bromide > iodine > perchlorate and (-2) carbonate > oxalate > sulfate. Uranium forms numerous organic complexes. (See Table

$$U(O) \xrightarrow{+1.80V} U(III) \xrightarrow{+0.63V} U(IV) \xrightarrow{-0.58V} UO_2(V)^+ \xrightarrow{-0.06V} UO_2(VI)^{2+}$$

$$-0.32V$$
irreversible

Fig. 1. Oxidation potentials in $1M$ perchloric acid solution at 25°C.

TABLE 1. URANIUM CATION PROPERTIES IN AQUEOUS SOLUTION

Species	Ionic Radius	Color	Preparation	Stability
U(III)	1.03 Å	Red-brown	Na or Zn/Hg on U(VI)O$_2^{2+}$	Slowly oxidized by H$_2$O, rapidly by air to U(IV)
U(IV)	0.93 Å	Green	Air or O$_2$ on U(III)	Stable, slowly oxidized by air to U(VI)O$_2^{2+}$
U(V)O$_2^+$ Uranous complex	0.80 Å as U(V)	?	Transient species	Greatest at pH 2–4, disproportionates to U(IV) and U(VI)O$_2^{2+}$
U(VI)O$_2$ Uranyl complex	3.4 Å as the uranyl complex	Yellow	Oxidize U(IV) with HNO$_3$	Very stable, difficult to reduce

TABLE 2. NUCLEAR PROPERTIES OF NATURALLY OCCURRING URANIUM ISOTOPES[a]

Isotope	% Abundance	Z	A	Thermal Neutron Cross Section (barns)	Half-Life (yr)	Decay Modes Energies (MeV)
U-234	0.0056	92	234.0409	105. (absorption) ≤ 0.65 (nf) 90. (activation)	2.48×10^5 (alpha) 1.6×10^{16} (SF)	Alpha: 4.77, 4.72 $e-$ (γ) 0.053 SF
U-235 Parent of $4N+3$ Series	0.72	92	235.0439	678. (absorption) 580. (nf) 101. (activation)	7.13×10^8 (alpha) 1.9×10^{17} (SF)	Alpha: 4.39 γ: 0.18, 0.14, 0.10 SF
U-238 Parent of $4N+2$ Series	99.27	92	238.0508	2.71 (absorption) $<.5 \times 10^{-4}$ (nf)	4.51×10^9 (alpha) 8.04×10^{15} (SF)	Alpha: 4.19 γ: 0.045 SF

[a] SF, spontaneous fission; nf, thermal neutron fission; γ, gamma emission.

1 for cation properties in aqueous solution.)

Nuclear Properties (see Tables 2–4). Elemental uranium consists of three naturally occurring isotopes in the proportion 99.27% U–238, 0.72% U–235, and 0.0056% U–234. Eleven other transient species have been identified but their short half-lives preclude natural occurrence. The decay schemes of U–235 and U–238 include decay modes of alpha, beta, gamma, thermal energy emission, and spontaneous fission; all but the latter terminate in stable lead products.

Analytical Techniques. The most common analytical techniques of uranium are listed in Table 5. In special situations the detection limits can be lowered and the precision improved by significant amounts.

Mineralogy (see Table 6). Classification of uranium-bearing minerals is based on a chemical scheme and includes eight major groups which are differentiated by the major anion (see Frondel, 1958). Approximately 103 uranium minerals have been confirmed to which over 200 names have been applied. In addition there are approximately forty minerals with minor amounts of uranium and at least ten minerals containing localized concentrations of uranium as impurities or intergrowths. Part of the multiplicity in mineral names arises from the secondary alteration nature of many uranium minerals. Uraninite, ideally UO_2, is the most common mineral and is found in pegmatites and granites. Pitchblende occurs in hydrothermal sulfide veins. The balance of uranium minerals are not major rock-forming minerals and usually occur as hydrothermal or weathering alteration products of accessory minerals.

Distribution and Abundance. The distribution and abundance of uranium have been studied extensively and current understanding includes some uncertainty inherent in sampling, analysis, and interpretation by different sources. Approximate ranges of uranium concentrations in various materials are presented in Table 7 (on p. 1222).

Speculations of the cosmic abundance of uranium are inferred from meteorite values, theoretical considerations of nuclear stability, and direct measurements of terrestrial rocks. One popular nucleosynthesis model postulates rapid neutron absorption relative to intervening beta emission in stars. Under these conditions, a value not less than 0.14 ppm is deduced for the primordial uranium concentration of the inner solar system. If no subsequent uranium removal has occurred, the present day concentration is then 0.054 ppm.

The uranium concentrations of meteorites occupy a wide range of values. Although overlap exists, iron meteorites have considerably less uranium than stony meteorites. As a group, meteorites contain less uranium than the average of the earth's crust, 3.4 ppm. Tektites usually contain more uranium than do meteorites. The similarity of concentrations to crustal materials and dissimilarity to meteorite concentrations is often cited as evidence for the terrestrial origin of tektites.

Uranium exists in the (IV) state under most magmatic conditions and appears in igneous rock suites as an oxyphile element. Uranium and thorium are geochemically coherent due to similarities in oxidation state and ionic radius. The same two properties prohibit geochemical coherence of uranium with the more common elements that appear in ultrabasic and basic rocks. In intermediate and granitic magmas, substitution does occur with zirconium in zircon and calcium in apatite and sphene. Coordination requirements of the calcium lattice site in the plagioclase series prevent extensive isomorphic substitution of uranium. A direct relation exists between the silica and uranium content of igneous rocks.

The occurrence of primary uranium minerals in igneous rocks is partially controlled by the oxidation state of uranium in the melt. Uranium (IV) compounds are insoluble in water and crystallize as primary minerals from wet granitic magmas. Under oxidizing conditions, as in near surface environments, uranium exists in the (VI) state. The higher valence produces soluble complexes which are highly mobile and unlikely to sustain crystal growth. Leaching experiments indicate that significant amounts of uranium are superficially incorporated within some igneous rocks. Autoradiograph and induced fission-track studies reveal localized distribution patterns within cracks and along grain boundaries.

Uranium minerals are chemically weathered in humid climates to soluble U(VI) complexes and removed by river water. Under arid conditions, uranium becomes complexed as oxides, hydroxides, sulfates, carbonates, phosphates, vanadates, and arsenates. Uranium remains soluble in the sea as carbonate and other complexes. Methods of removal include adsorption on clays, substitution for calcium in authigenic apatite, complexing by humic and sapropelic matter, and incorporation in skeletal material of organisms.

The uranium concentration of sediments is controlled by provenance of the source materials and the geochemical properties of the depositional environment. Terrigenous sands contain an average of 3 ppm of uranium. Exceptions exist where primary or accessory minerals are included in the sediment supply.

The oxidation of organic matter in clays creates a reducing environment which precipitates uranium (IV). As a consequence, black, carbonaceous-rich shales usually contain more uranium than red, gray, or green shales. The incorporation of uranium in carbonate sediments is selectively controlled in early stages by biological factors. Skeletal material of corals contain approximately four times as much uranium as mollusc shells from identical marine environments. Lithification of marine carbonate sediments by fresh water removes uranium from the carbonate system in conjunction with the replacement of metastable carbonate minerals by calcite. Diagenesis exerts numerous other effects on the uranium in sediments which can disperse or concentrate the element.

As in the case of most trace elements, systematic effects of metamorphism on uranium concentration and distribution are relatively unknown. Mobilization has been demonstrated in some rock assemblages and not in others. Generally the lower grade metamorphic processes do not significantly alter uranium concentrations. Higher-grade metamorphism causes alterations as there is evidence that some rocks of the granulite facies have undergone uranium loss. The effects of high grade metamorphism are difficult to categorize because of uncertainty in initial uranium concentration, initial rock texture, and active metamorphic conditions.

Oceanic waters contain a relatively uniform amount of uranium and are usually saturated in uranium with respect to local geochemical environments. Uranium concentrations of oceans commonly range from 0.001 to 0.003 ppm. Variations exist in response to local effects of undersaturated tributaries. Variations of 20% may also exist within a depth profile due to zonation in Eh and pH conditions by pelagic organisms. Uranium concentrations in continental waters show wide ranges due to the nature of the rocks forming the drainage basin. Concentrations are usually in the tenths of part per billion, but may be a few tenths of part per million in waters of mineralized aquifers.

Radiometric Age Determination. The nuclear decay schemes of uranium provide means of absolute age determinations. There are three basic types of dating. The most widely used type involves evaluation of the uranium isotope parent and corresponding lead isotope product with the appropriate decay constant. An age may be calculated for a single decay scheme; however routine isotopic analysis of uranium and lead provides data for both uranium decay schemes. Age interpretations are commonly made by means of a *concordia plot*.

TABLE 3. THE URANIUM $4N+2$ SERIES

			U^{234}, U_{II} (uranium II) 2.48×10^5 yr							
		Pa^{234}, UX_2 1.18 min Pa^{234}, UZ 6.7 hr	β (99.85%) I.T.(0.15%) β α							
	U^{238}, U_I (uranium I) 4.51×10^9 yr	α β								
		Th^{234}, UX_1 (uranium X_1) 24.1 days		Th^{230}, Io (ionium) 7.52×10^4 yr	α					
					Ra^{226}, Ra (radium) 1622 yr	α				
							Rn^{222}, Rn (radon) 3.825 days	α		
									At^{218} 1.3 sec β (0.02%) α	
								Po^{218}, RaA (radium A) 3.05 min		Po^{214}, RaC' (radium C') 1.6×10^{-4} sec	Po^{210}, RaF (polonium) 138.4 days

92	U
91	Pa
90	Th
89	Ac
88	Ra
87	Fr
86	Rn
85	At
84	Po

URANIUM: ELEMENT AND GEOCHEMISTRY

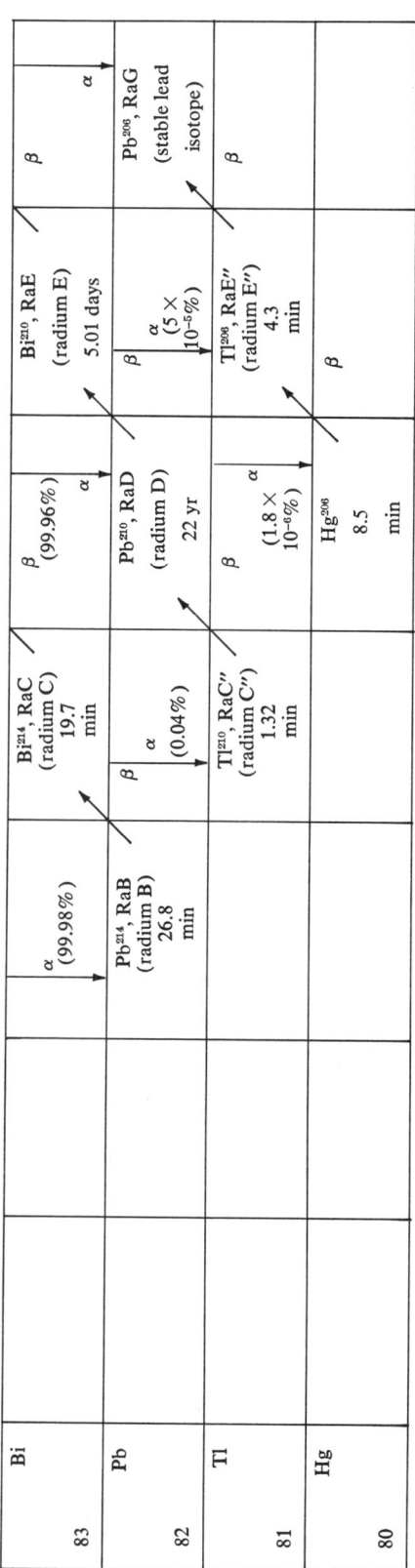

The plot is defined by lead-uranium isotope ratios for each scheme. The concordia curve itself is the locus of points having equal ages according to each scheme. Departure of measured values from the concordia indicates the influence of some geologic event on the isotopic or chemical system. The time of mineral crystallization and intervening event are then interpreted by the relation of the measured values to the concordia. The technique has been effectively applied to uraninite, pitchblende, zircon, xenotime, and monazite, among others (see *Uranium- Thorium-Lead Age Determination*).

Another basic type of dating is an evaluation of the uranium decay series disequilibrium. At the time uranium is incorporated in a host, its decay products begin to accumulate. Secular equilibrium is attained when a given intermediate member decays at the rate in which it is produced. If the intermediate members are in disequilibrium, the age of the host can be computed. Accurate age results depend on correct assumptions of parent-daughter abundances and deposition rates at the time of isolation. The assumptions are simplified by evaluating two independent nuclides that are deposited in fixed proportions. Age determinations of sediments have been made using ratios of thorium-230 (ionium)-, thorium-232-, and thorium-230-protactinium-231.

The third type of dating deals with time dependent radiation damage within mineral lattices. Uranium-238 spontaneously undergoes fission producing two heavy fission products. The host mineral absorbs the energetic projectiles which result in linear trails of damage. Treatment of a given mineral under proper etching conditions enlarges these trails to provide a record of fossil fission events. Thermal neutron fission of uranium-235 produces induced fission products which are similarly recorded and revealed. The fossil-track density records time dependent events and induced track density is a measure of the present uranium concentration. Calculation of the track retention age of a mineral includes the fossil and induced track densities and the spontaneous fission decay constant of uranium-238. Although fission tracks may be annealed, reliable age determinations have been made of micas, sphene, zircon, apatite, hornblende, and natural and synthetic glasses (see *Fission Track Dating*).

Pleochroic halos (q.v.) form from the radioactivity effects of minute inclusions of uranium in certain minerals, notably micas. The neighboring crystal lattice absorbs alpha radiation to form spheres of darkened color. The radius of the sphere is related to unique alpha decay

URANIUM: ELEMENT AND GEOCHEMISTRY

TABLE 4. THE URANIUM $4N+3$ SERIES

U 92	U^{235}, AcU (actinouranium) 7.13×10^8 yr					
Pa 91	↓ α	Pa^{231}, Pa (protactinium) 3.48×10^4 yr ↗				
Th 90	Th^{231}, UY (uranium Y) 25.6 hr	β → ↓ α	Th^{227}, RdAc (radioactinium) 18.17 days ↗			
Ac 89		Ac^{227}, Ac (actinium) 22.0 yr	β (98.8%) → ↓ α			
Ra 88		α (1.2%) ↓	Ra^{223}, AcX (actinium X) 11.7 days ↗			
Fr 87		Fr^{223}, AcK (actinium K) 22 min	β → ↓ α			
Rn 86		α $(4 \times 10^{-3}\%)$ ↓	Rn^{219}, An (actinon) 3.92 sec ↗			
At 85		At^{219} 0.9 min	β (3%) → ↓ α	At^{215} 10^{-4} sec ↗		
Po 84		α (97%) ↓	Po^{215}, AcA (actinium A) 1.83×10^{-3} sec ↗	β $(5 \times 10^{-4}\%)$ → ↓ α	Po^{211}, AcC′ (actinium C′) 0.52 sec ↗	
Bi 83			Bi^{215} 8 min	β → ↓ α	Bi^{211}, AcC (actinium C) 2.15 min ↗	β (0.32%) → ↓ α
Pb 82			Pb^{211}, AcB (actinium B) 36.1 min ↗	β → α (99.68%) ↓		Pb^{207}, AcD (stable lead isotope) ↗
Tl 81					Tl^{207}, AcC″ (actinium C″) 4.79 min ↗	β

schemes and the darkness is related to the exposure time. Theoretically it is possible to obtain an absolute age by determining the number of alpha events by comparison with standard materials. Unfortunately the darkening process is reversible and the method is not considered very versatile.

DAVID S. HAGLUND

References

Adams, J. A. S., Osmond, J. K., and Rodgers, J. J. W., 1959, "The geochemistry of thorium and uranium," in (Ahrens, L. H., Press, F. Rankama, K. and Runcorn, S. K., (editors), "Physics and Chemistry of the Earth," B, Vol. 3, New York, Pergamon Press, pp. 298–348.

Aswathanarayana, U., 1961, "Present status of knowledge on the geochemistry of uranium,"

TABLE 5. ANALYTICAL TECHNIQUES FOR URANIUM

Name	Principle[a]	Approximate Detection Limit, (μg/g)	Probable Error (%)
Fission track	NA, ND	10^{-4}	5.–10.
Isotope dilution	MS	10^{-4}	0.5–5.
Neutron activation	NA, ND	10^{-4}	2.–5.
Delayed neutron counting	NA, ND	10^{-3}	3.
Emission spectroscopy	ET	5×10^{-2}	1.–10.
Fluorescence (ultraviolet and x-ray)	ET	10^{-1}	5.–50.
Chromatography	WC	10^{-1}	10.
Gamma-ray spectrometry	ND	10^{-1}	10.
Autoradiography	ND	1.	1.–10.
Titrimetry	WC	1.	0.5–5.
Colorimetry	WC	10.	1.–3.
Atomic absorption	ET	5×10^{1}	2.–5.
Alpha counting (electronic)	ND	5×10^{1}	1.–10.
Polarography	EC	10^{2}	2.–5.
Potentiometry	EC	2×10^{2}	1.–5.
Gravimetry	WC	5×10^{4}	0.1–2.

[a] ND, nuclear decay; ET, electron transitions; WC, wet chemical; EC, electrochemical; NA, neutron activation; MS, mass spectrometry.

TABLE 6. URANIUM MINERALS

Name	Chemical Group	Composition	Uranium Content (%)
Uraninite	Oxide	$[U(IV)_{1-x}U(VI)_x]O_{2+x}$	46.5–88.2
Pitchblende (var.) a fine grained colloform habit			
Gummite (var.) a silicate, phosphate, and oxide alteration product of uraninite			
Rutherfordine	Carbonate	$(UO_2)(CO_3)$	72.1
Johannite	Sulfate	$(Cu(UO_2)_2(SO_4)_2(OH)_2 \cdot 6H_2O$	50.8
Umohoite	Molybdate	$(UO_2)(MoO_4) \cdot 4H_2O$	57.
Autunite	Phosphate	$Ca(UO_2)_2(PO_4)_2 \cdot nH_2O$	60.
Carnotite	Vanadate	$K_2(UO_2)_2(UO_4)_2 \cdot nH_2O$	52.–55.
Coffinite	Silicate	$U(SiO_4)_{1-x}(OH)_{4x}$	60.
Uranophane	Silicate	$Ca(UO_2)_2(SiO_3)_2(OH)_2 \cdot 5H_2O$	55.6
Betafite	Multiple oxide	$(U,Ca)(Nb,Ta,Ti)_3O_9 \cdot nH_2O$	12.–23.

in (Krishnan, M. S., editor), "A Collection of Geological Papers," Mahadevan Volume, New Delhi, Today and Tomorrow's Book Agency, pp. 148–158.

Blanchard, R. L., 1965, "U234/U238 ratios in coastal marine waters and calcium carbonates," *J. Geophys. Res.*, **70**, 4055–4061.

Chamberlain, J. A., 1964, "A field method for determining uranium in natural waters," *Geol. Surv. Can., Dept. Mines and Tech. Surv., Field and Lab. Methods* No. 7.

Clark, S. P., Jr., Peterman, Z. E., and Heier, K. S., 1966, "Abundances of uranium, thorium, and potassium," in (Clark, S. P., Jr., editor), "Handbook of Physical Constants," Rev. ed. *Geol. Soc. Am. Mem.*, **97**.

Frondel, Clifford, 1958, "Systematic mineralogy of uranium and thorium," *U.S. Geol. Surv. Bull.*, **1064**.

Hamilton, E. I., 1966, "Distribution of uranium in some natural minerals," *Science*, **151**(3710), 570–572.

Hoyle, F., and Fowler, W. A., 1964, "On the abundances of uranium and thorium in solar system material," in (Craig, H., Miller, S. L., and Wasserburg, G. J., editors), "Isotopic and Cosmic Chemistry," Amsterdam, North-Holland Co., pp. 516–529.

Katz, J. J., and Rabinowitch, E., 1951, "The Chemistry of Uranium. Part I, The Element, Its Binary and Related Compounds," New York, McGraw-Hill Book Co.

Ku, T.-L., 1965, "An evaluation of the U-234/-U-238 method as a tool for dating pelagic

URANIUM: ELEMENT AND GEOCHEMISTRY

TABLE 7. ABUNDANCES OF URANIUM IN NATURAL MATERIALS

Material	Reported Concentration Ranges (ppm)[a]	
Cosmos	0.01–0.1	
(r model)	(0.054)	
Meteorites		
Stony achondrites	$0.15–14.5 \times 10^{-2}$	
Stony chondrites	$0.62–2.8 \times 10^{-2}$	
Irons	$<0.003–1.7 \times 10^{-2}$	
Tektites	1.2–3.0	
Igneous rocks		
Ultramafic (dunites and peridotites)	0.001–0.8	
Mafic (gabbro and diabase)	0.3–3.4	Median 0.5
Intermediate (diorite and quartz diorite)	0.1–11.	Median 1.7
Sialic (granite, syenite, and monzonite)	0.15–21.	Median 3.9
Pegmatites	1.–4.	(see primary minerals of uranium)
Sedimentary rocks		
Black shales	3.–250.	Median 8.
Red, gray, and green shales	1.2–12.	Median 3.2
Orthoquartzite	0.2–0.6	Median 0.45
Limestone and dolomite	0.1–9.	Median 2.2
Bentonite	1.–21.	Median 5.
Bauxite	3.–27.	Median 8.
Halite	0.01–0.02	Median 0.013
Anhydrite	0.25–0.43	Median 0.37
Metamorphic rocks		
Marble	0.11–0.24	
Slate	1.2–6.1	
Phyllite	1.0–2.7	
Schist	1.8–2.9	
Gneiss	4.5–15.	
Amphibolite	2.6–4.1	
Granulite	3.2–7.	

[a] From analyses compiled by Adams et al., 1959; Clark et al., 1966; and Hoyle and Fowler, 1964. The reader is directed to these compilations for specific references.

sediments," *J. Geophys. Res.*, **70**(14), 3457–3474.

Wasserburg, G. J., et al., 1964, "Relative contributions of uranium, thorium, and potassium to heat production in the earth," *Science*, **143** (3605), 465–467.

Cross-references: *Actinide Series; Chromatography; Colorimetry; Emission Spectrography; Earth's Crust Geochemistry; Fission Track Dating; Isotope Geology—Stable; Mass Spectrometry; Neutron Activation Analysis; Protactinium-Thorium Dating Method; Radioactive Isotopes; Radioactivity in Rocks; Seawater, Chemistry; Trace Elements; Uranium-Helium Isotopic Age Method; Uranium Ore Deposits; Uranium-Thorium-Lead Age Determination; Weathering, Chemical; X-ray Diffraction Analysis. Vol. II: Meteorites; Nucleogenesis; Tektites.*

URANIUM: ECONOMIC DEPOSITS

With the discovery (in 1898) of radium, a market was established for uranium ore from which radium, a radioactive decay product, could be extracted. Early needs were supplied by pitchblende-bearing veins of Jachymov (Joachimsthal), Czechoslovakia. Interest then shifted to carnotite ore of southwestern Colorado and southeastern Utah, which was the world's major source of uranium from 1911 to 1923. In 1923, the large, high-grade pitchblende deposit at Shinkolobwe, Belgian Congo, was brought into production and it dominated world markets until 1936. During 1930, a new find of pitchblende was made at a remote locality near Great Bear Lake, Canada; it was worked from 1933 to 1940.

With the United States' decision, in 1942, to develop an atomic bomb, an unprecedented demand was created for uranium ores, because the U^{235} isotope is naturally fissionable, and the more abundant U^{238} isotope can be transformed into fissionable plutonium by treatment in a reactor. The Shinkolobwe, Great Bear Lake and Colorado-Utah mines were reactivated, and a massive search for uranium was carried on throughout the world, largely under the auspices of various governments. An important uranium mining industry developed in a number of countries.

Uranium has been identified in some 115 different mineral species and is an essential component in approximately 80. Among these are nine which form most of the ore in mines throughout the world: uraninite, UO_2; pitchblende (variety of UO_2); coffinite, $U(Si,H_4)O_4$; brannerite, $(U,Ca,Fe,Y,Th)_3Ti_5O_{16}$; uranothorite, $(Th,U)SiO_4$; uranophane, $CaU_2O_3Si_2O_8 \cdot 7H_2O$; davidite, mixture; carnotite, $K_2(UO_2)_2(VO_4)_2 \cdot 3H_2O$; and tyuyamuite, $Ca(UO_2)_2(VO_4)_2 \cdot nH_2O(?)$.

Description of Deposits. *Veins.* Before World War II uranium production was mainly from veins containing pitchblende and uraninite. The deposits occupy fractures, more or less complex, and the ore shows evidence of open-space filling. The favored host rocks are granitic intrusives and metamorphosed sedimentary rocks.

Important deposits include:

Jachymov (Joachimsthal), Czechoslovakia—for centuries a source of lead, silver, nickel, cobalt, and uranium.

Shinkolobwe, Belgian Congo—uraninite and various cobalt, nickel, gold and silver minerals distributed along fracture and bedding planes in the Precambrian Mines series of carbonate rocks.

Great Bear Lake, Canada—nearly east-west steeply dipping veins containing pitchblende and other minerals in the Precambrian Echo Bay complex.

Lake Athabaska, Canada—veins and shear zones containing pitchblende and calcite.

Radium Hill, Australia—davidite, along with biotite, rutile, and ilmenite in fracture planes in Precambrian gneiss.

Douro, Portugal—simple veins in granite with quartz, jasper, pitchblende, and minor sulfides.

Pitchblende veins have been mined in France, the United States, and in various other places.

Deposits in Precambrian Conglomerate. In two major uranium districts of the world—Blind River, Canada, and the Rand, South Africa—the deposits are bedded features in Precambrian conglomerate. At Blind River, Canada, the ore consists of rounded quartz pebbles in an arkosic matrix containing abundant pyrite, brannerite (the main ore mineral), uraninite, and monazite. Ore averages 0.1% U_3O_8 and 0.05% ThO_2. The conglomerate "reefs" of the Rand, Republic of South Africa, are the world's main source of gold, with which uranium is associated. Pebbles of quartz are cemented by a matrix of fine-grained quartz, sericite, and chlorite which contains native gold, uraninite, and thucolite (a radioactive hydrocarbon mixture). Uranium is recovered as a by-product of gold, but some mines are operated mainly for uranium.

Deposits in Younger Sedimentary Rocks. Sandstone ore: Most of the uranium found in the United States has been found in fluvial sandstones, a type of deposit uncommon elsewhere. Main producers are the Morrison formation (Jurassic), Chinle formation (Triassic), Shinarump conglomerate (Triassic), and Wind River formation (Eocene). Production is centered in the Colorado Plateau area of Colorado, New Mexico, Utah, and Arizona, and similar deposits are being mined in Wyoming and South Dakota.

Mineralogically, the ores are of two types: *(1) unoxidized*—containing uraninite, montroseite, coffinite, and *(2) oxidized*—containing carnotite, tyuyamunite, hewettite. Ore minerals fill the pores of the sandstone and also replace fossil wood. Carnotite ore contains several times more vanadium than uranium and has been the main source of vanadium in the United States.

The deposits are mainly tabular, forming layers from a few feet to as much as sixty feet in thickness. The layers are generally parallel to the bedding, which is horizontal or slightly inclined, but in detail the mineralization cuts across bedding planes. A common feature is "rolls," sharply curving forms that cut across bedding, separating ore from waste. Many deposits are elongate parallel to the course of the stream which deposited the sandstone. Thousands of ore bodies are known but most are small. Individually they range in size from a few tons to more than 100,000 tons.

Limestone ore: Production of uranium from limestone has been relatively minor. At Tyuya-Muyun, Fergana district, Uzbeck, U.S.S.R., tyuyamunite and other secondary minerals form encrustations up to 1.5 m thick in caves and solution channels in Paleozoic metamorphic limestone. At Grants, New Mexico, the Todilto limestone (Jurassic) contains deposits of uraninite, carnotite, and tyuyamunite associated with fluorite. In the Bighorn-Pryor Mountains, Wyoming-Montana, tyuyamunite has been mined from solution cavities at the top of the Madison limestone (Mississippian).

Black carbonaceous shales: The alum shale of Sweden (Cambrian-Ordovician) contains about 100 grams of uranium per metric ton. The Chattanooga black shale (Devonian) in Tennessee contains 0.005–0.008% U_3O_8 in beds 15 ft thick. Reserves are enormous (estimated to contain 5 million metric tons of uranium) but sub-commercial.

Lignites: Uranium is intimately associated with organic material in flat-lying lignite beds

URANIUM: ECONOMIC DEPOSITS

TABLE 1. URANIUM RESERVES IN VARIOUS COUNTRIES

Country	Reserves (short tons U_3O_8)
Canada	270,000
U.S.A.[a]	160,000
Republic of South Africa	150,000
France	40,000
Australia	12,000
Other	25,000

[a] U.S.A. figures are for January 1, 1964; all others are for January 1, 1963.

of North Dakota and South Dakota. Although widespread, the units are thin, averaging 0.2–0.4% U_3O_8 over thicknesses of 1.5 ft.

Reserves. Free world uranium reserves, at $8.00 (U.S.) per pound, were estimated by U.S. Atomic Energy Commission (Table 1).

Communist controlled uranium reserves were estimated in 1960 to contain 100,000–370,000 metric tons of uranium metal.

ROBERT J. WRIGHT

References

Bateman, A. M., 1950, "Economic Mineral Deposits," Second ed., New York, John Wiley & Sons, 916pp.

Bernard, J. H., Rösler, H. J., and Baumann, L., 1968, "Hydrothermal Ore Deposits of the Bohemian Massif," XXIII Internat'l. Geol. Congress, Prague (Guide to Excursion 22AC).

Kerr, P. F., et al., 1957, "Marysvale, Utah, Uranium Area," *Geol. oc. Am. Spec. Paper* **64**, 212pp.

Lamey, C. A., 1966, "Metallic and Industrial Mineral Deposits," New York, McGraw-Hill Book Co., 567pp.

Soister, P. E., and Conklin, D. R., 1959, "Bibliography of U.S.G.S. Reports on Uranium and Thorium, 1942 through May, 1958," *U.S. Geol. Surv. Bull.*, 1107-A, 167pp.

Cross-references: *Uranium: Element and Geochemistry.* Vol. IVB: *Economic Geology; Mineral and Ore Deposits; Mining and Mineral Centers of the World; see also Individual Minerals.*

URANIUM-HELIUM ISOTOPIC AGE METHOD

Helium–4 is produced in rocks by the decay of uranium, thorium, and alpha-emitting daughters in their decay chains:

$$Th^{232} \rightarrow Pb^{208} + 6He^4$$
$$U^{235} \rightarrow Pb^{207} + 7He^4$$
$$U^{238} \rightarrow Pb^{206} + 8He^4$$

If the decay rate of the intermediate daughters in the U and Th series has reached equilibrium, the U,Th–He age may be stated as a function of the ratio (radiogenic He)/(U + Th). In 1905, Rutherford proposed the use of helium accumulation in rocks as a means of age determination, and Strutt reported the pioneer measurements in 1908. Since then, the He method has undergone many vicissitudes of abandonment and revival. Its history of development has been marred by the use of improper standards and inadequate laboratory techniques; today the method is little used.

Analytical Techniques

Helium concentrations are most satisfactorily determined by isotope dilution. A tracer consisting of virtually pure He^3 is added to He^4 released from the sample, and the He^4/He^3 ratio determined by a mass spectrometer. Since the amount of tracer He^3 is known, the amount of He^4 from the sample may be calculated. Contamination by atmospheric He is not a serious problem because of the very low concentration of He in air (5.4 ppm).

Uranium and thorium are commonly determined by isotope dilution, fluorimetry, or by alpha counting.

Discussion

The most serious failure of the He age method was soon found to be incomplete retention of He by nearly all rocks and minerals. Typical retentivities are 50% or less, and metamict zircon may have a retentivity of even less than 1%. Theoretically, at room temperature a helium atom should not diffuse appreciably through a perfect crystal over intervals of millions of years. However, alpha particles of the U and Th series characteristically bombard the crystal with energies of 4 to 7 × 10^6 electron volts. The high energy particles destroy the crystal structure along paths 10 to 40 microns long; He can diffuse rapidly through these defects. The retentivity problem is compounded by the tendency of radioactive materials to concentrate along grain boundaries where parent and daughter isotopes have rather high and variable mobility. Of the common minerals, magnetite may on occasion retain He quantitatively, and therefore it is of some potential use for age determination.

Another prerequisite for valid He ages is that the mineral must form with an absence of initial, or "excess" He^4. Damon and Kulp found that beryl, cordierite, and other ring silicates whose structures have continuous channels could contain up to 10^3 times as much helium and argon as can be accounted for by the decay of uranium and potassium. The channels are sites where gas atoms may be trapped during mineral formation. Other minerals with more closely packed structures accept excess He to a far smaller extent, but may still contain some He in fluid inclusions.

LEON E. LONG

References

Damon, P. E., and Green, W. D., 1963, "Investigations of the helium age dating method by stable isotope-dilution technique," *Radioactive Dating* (Intern. Atomic Energy Agency, Vienna), pp. 55–71.

Fanale, F. P., and Schaeffer, O. A., 1965, "Helium-uranium ratios for Pleistocene and Tertiary fossil aragonites," *Science,* **149,** 312–317.

Hurley, P. M., 1954, "The helium age method and the distribution and migration of helium in rocks," in (Faul, H., editor), "Nuclear Geology," New York, John Wiley & Sons, pp. 301–329.

Cross-references: *Geochronometry; Helium: Element and Geochemistry; Radioactivity in Rocks; Uranium: Element and Geochemistry.*

URANIUM–THORIUM–LEAD AGE DETERMINATION

Naturally occurring uranium consists of two isotopes, U^{238} and U^{235}, with relative abundances at the present time of 99.27% and 0.72%, respectively. Both of these isotopes are unstable and give rise to radioactive decay series which have as their end products Pb^{206} and Pb^{207}. In the course of decay from U^{238} to Pb^{206} there are fourteen intervening unstable nuclides; from U^{235} to Pb^{207} there are twelve. Of the two parent nuclides, U^{235} decays more rapidly, any given amount halving itself in 713 million years, whereas U^{238} has a longer "half-life" of 4510 million years. Since U^{235} decays more rapidly than U^{238} the ratio $U^{235}:U^{238}$ has decreased throughout geological time down to the present day relative abundance of 1:137.7. Naturally occurring thorium consists of a single isotope Th^{232} with a half-life of 14,200 million years. There are ten intervening nuclides between Th^{232} and its stable end product Pb^{208}.

The decay of an atom of U^{238} to its final end product involves the production of eight alpha particles (helium nuclei), similarly the complete decay of an atom of U^{235} produces seven helium nuclei and Th^{232} produces six. Thus both uranium and thorium differ from the other two elements of interest to geochronology, namely potassium and rubidium, in that their radioactive decay passes through a large number of intervening stages. Decay products begin to accumulate from the moment of formation of a radioactive mineral, but since there is a series of unstable isotopes with varying half-lives, there is a time interval before one disintegration of a parent atom is accompanied by one disintegration of each successive decay product. When this occurs the whole series is said to be in equilibrium. The time required for equilibrium to be reached by uranium minerals is about a million years, but by thorium minerals only about thirty years. It is possible to use simple radiometric equipment to determine whether a uranium mineral is in equilibrium since the radioactivity of natural uranium (or thorium) is contributed by both the parent element and intermediate decay products, each with β and γ radiation in different proportions. For a decay series as a whole, the proportion of $\beta:\gamma$ radiation is fixed as long as it is in equilibrium, and since radiometric counting equipment can be used to count γ and $\beta+\gamma$ radiation separately the state of equilibrium can be determined. Uranium that has only recently migrated to its present position, or uranium that has recently lost some of its decay products can be readily distinguished in this way. This is a convenient means of eliminating at an early stage of an investigation material which is unsuitable for age determination.

For any decay series in equilibrium the relationship between the number of parent atoms (P) at the present time and the number of daughter atoms (D) produced in time t is given by

$$t = \frac{1}{\lambda} \log_e \left(1 + \frac{D}{P}\right)$$

where λ is the decay constant of the parent isotope, which is related to the half-life $T_{1/2}$ as follows

$$T_{1/2} = \frac{\log_e 2}{\lambda} = \frac{0.6931}{\lambda}$$

This relationship covers the decay of U^{238} to Pb^{206}, U^{235} to Pb^{207}, and Th^{232} to Pb^{208}, and enables an age to be derived from any of these pairs, which is reliable provided there has been no loss or gain of parent or daughter isotopes. Because of the different rates at which U^{235} and U^{238} decay, the ratio of the end products Pb^{207} and Pb^{206} is itself a measure of age. Fig. 1 shows the values of the ratios N_{206}/N_{238}, N_{207}/N_{235}, N_{207}/N_{206}, and N_{208}/N_{232} over the period 0–600 million years where N is the number of atoms of the various isotopes. The diagram illustrates the uncertainty in the determination of the decay constants, also the sensitivity of the ages to changes of one unit

URANIUM-THORIUM-LEAD AGE DETERMINATION

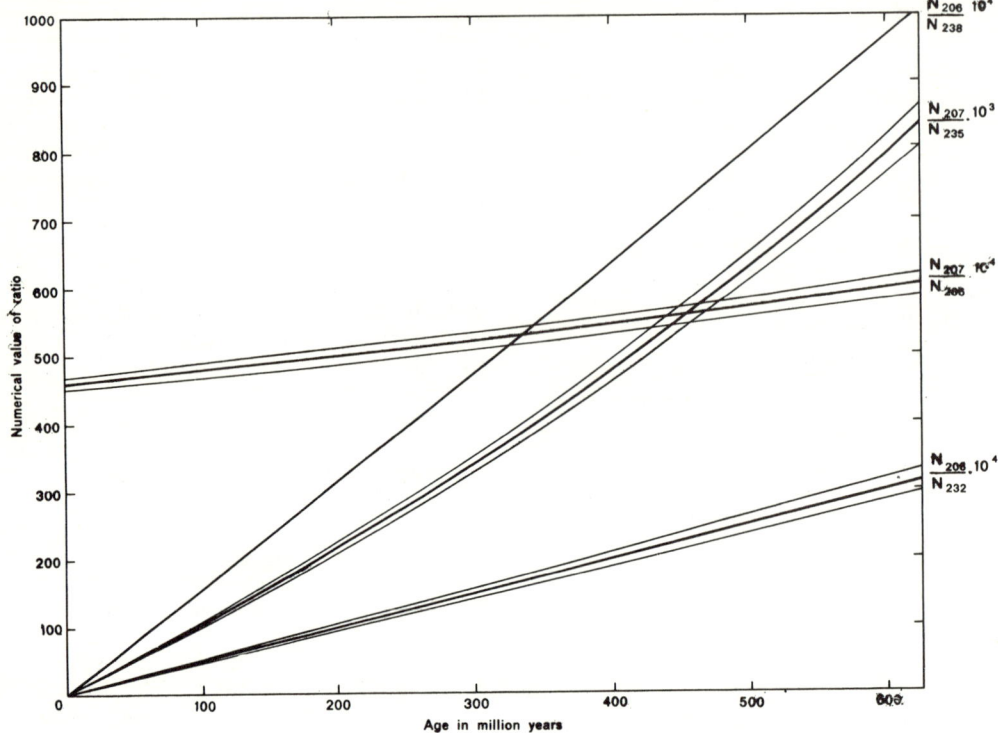

Fig. 1. Isotopic ratios plotted against age. Outer lines indicate limits due to uncertainty with which decay constants are known.

in the third significant figure of the ratios. It can be seen that the N_{206}/N_{238} ratio has negligible uncertainty and a satisfactory slope, whereas N_{207}/N_{206} is unsatisfactory in both respects. In addition to these two criteria, others also enter into assessing the relative reliability of the different ratios, in particular the absolute amount of each lead isotope present. The most abundant isotope can be determined with the greatest relative accuracy and is least sensitive to the common-lead correction (discussed below). For a uranium-bearing mineral which crystallized in the period 0–600 million years, Pb^{206} constitutes about 95% of the uranium-lead formed, Pb^{207} only about 5%. Thus Pb^{206} can be determined with much greater precision than can Pb^{207}, particularly in young materials where the total amount of lead present is small. In such material this fact enhances the reliability of the N_{206}/N_{238} ratio and diminishes that of N_{207}/N_{235} and especially N^{207}/N^{206}. Although there is no clear-cut dividing line below which it can be said that N_{207}/N_{206} ages are completely unreliable, they do become progressively less reliable from Precambrian time onward. Conversely N_{207}/N_{206} ratios become increasingly reliable with increasing age. U^{235} has a much shorter half life than U^{238}; consequently, minerals that formed early in the earth's history accumulated Pb^{207} at an appreciably higher rate than in more recent geological times. This accounts for the increasing reliability of N_{207}/N_{206} with age, and also results in N_{207}/N_{235} ratios from *old* materials being as reliable as N_{206}/N_{238}.

Ages based on the ratio N_{206}/N_{210} where N_{210} is the number of Pb^{210} atoms merit consideration. Pb^{210} is an unstable intermediate nuclide in the decay of U^{238} to Pb^{206}. Since for a decay series in equilibrium the quantities of intermediate decay members are proportional to the amount of the parent remaining, the ratio of any intermediate member to the stable end product is related to the age of the system. Only a comparatively small number of N_{206}/N_{210} ages have been reported, although the measurement of N_{210} can be carried out radiometrically. Both N_{210}/N_{206} and N_{207}/N_{206} ages (sometimes known as lead-lead ages) can be obtained without analyzing for total uranium, thorium or lead since it is necessary to determine only the isotopic composition of a separated portion of lead.

Average crustal lead consists of four isotopes Pb^{204}, Pb^{206}, Pb^{207}, and Pb^{208}. Pb^{204} is not derived from any radioactive parent so far as is known. It is generally accepted that the composition of average lead at the time of the Earth's formation was the same as that now found in those meteorites which have a negligible uranium and thorium content, such as the Canyon Diablo meteorite, which has a composition of $Pb^{204} = 2.0$, $Pb^{206} = 18.8$, $Pb^{207} = 20.6$, and $Pb^{208} = 58.6$ at. %. Throughout the history of the earth the composition of the original average lead has changed progressively with the addition of Pb^{206}, Pb^{207}, and Pb^{208} from uranium and thorium also present in the earth. There is evidence for supposing that the isotopic composition of crustal lead is established at a depth below that at which any exposed igneous rocks were emplaced. It is only possible to envisage the continuous homogenization of existing lead with new radiogenic lead at very deep levels in the crust or in the upper part of the mantle. Because this common lead is continuously changing in isotopic composition with time it is theoretically possible to determine its age. To do this it is necessary to choose four parameters in addition to those required for dating uranium and thorium minerals. These cover the composition of original lead and the age of its formation (considered by some authors to be the age of the Earth, by others the age of the crust). From this basic data the Holmes-Houterman method of calculation is independent of variations in the U:Th ratio of parent material whereas other methods are not. This leads to different assessments, according to the method used, of what constitutes an "anomalous" result, but these are beyond the scope of this review. Russell (1963) provides a more extended introduction to the problems concerned. It will be apparent, however, that ages obtained from common lead differ in certain important respects from the ages derived from the analysis of uranium or thorium minerals. Firstly, there is a wider choice of parameters that can be used in calculating an age and for purposes of comparison it is important to know which have been used. Secondly, the method is very dependent upon a high level of analytical precision. Thirdly, the age does not date the crystallization of the mineral but (provided it has not suffered contamination from lead of a different composition) the time at which the lead separated from its source. It is possible to find examples of galena where the difference between model age and time of crystallization established from geological evidence is several hundred million years.

Most uranium minerals contain some thorium, excepting those formed at low temperatures, and most thorium minerals contain some uranium so that to obtain an age determination it is normal practice to analyze for both uranium and thorium as well as for lead. Ideally, measurement of these three elements alone is sufficient to calculate the age of a mineral. When this is done by chemical methods this may be referred to as a "chemical age." Wickman (1944) published graphs which enable a chemical age to be obtained directly from a chemical analysis. The lead-alpha method developed by Larsen and co-workers about 1952, and applied to zircon, monazite, thorite, and xenotime relies on the radiometric determination of the combined U+Th content and lead determination by optical emission spectrography. For this method it is necessary to assume the value of the U:Th ratio in any particular concentrate and choose the appropriate constant for calculation of the age.

Both the chemical age and the lead-alpha age may be grossly misleading since there is no means of knowing the extent to which lead was present in the mineral at the time of its formation, nor is there anything to indicate whether or not a mineral has been a chemically closed system since its formation. Many uranium and thorium minerals contain a significant percentage of common lead. In the case of low-temperature pitchblende it may be 50% of the total lead present. This lead may be visible as galena inclusions or it may be invisible at the highest magnifications. Isotopic analyses of the lead present in a mineral determines the percentage abundance of Pb^{204}, Pb^{206}, Pb^{207}, and Pb^{208}. If the mineral is completely free of common lead, as may happen with high-temperature uraninite, then Pb^{204} will not be found. If it is present then not all of the Pb^{206}, Pb^{207}, and Pb^{208} has been produced in the lifetime of the mineral and a correction based on Pb^{204} must be made. Quite often low temperature uranium minerals with high common lead contents have a negligible Th content, in which case all the Pb^{208} is attributable to common lead. Since the ratio of Pb^{204}:Pb^{208} in common lead varies between about 1:33 and 1:39 according to its age, a common lead correction based on Pb^{208} is preferable. The choice of common lead composition used for the correction depends on circumstances. If galena is present in the same specimen or the same vein or horizon as provided the radioactive mineral, then this should be analyzed and used for the correction. Failing this, judgment must be exercised in selecting a composition corresponding to that of common lead in the region at the time the radioactive mineral crystallized. However, even in the most unfavorable circumstance of a very young mineral with large common lead contamination, the precise value used for common lead is not critical, since changes in common-lead composition with time are small. Of course, if there was anything abnormal about the composition of the lead incorporated in a mineral at its formation, then the problem of correction becomes much more difficult. Further details of the method of calculation of ages from basic analytical data, and tables for the rapid conversion of ratios into ages are published by the United States Geological Survey (Stieff, Stern, Oshiro, and Senftle, 1959).

After an appropriate correction for common lead, and provided the mineral has not been altered, the isotopic ratios should all be in agreement. If there are discrepancies between the isotopic ages, the term "apparent age" is sometimes used where there is uncertainty as to the significance of the figures obtained to emphasize that a calculated age is not necessarily the true age.

The simplest examples of discordance between calculated ages are brought about by loss or gain of uranium or lead through recent supergene action. Appreciable supergene alteration of a high grade uranium mineral normally results in distinct mineralogical changes both in the mineral itself and its surroundings so that discordance of the ratios would merely confirm what was obvious. In such circumstances only the N_{207}/N_{206} age is likely to be of any value, and if the mineral is post-Cambrian this is not likely to be reliable either. A mineral may not necessarily show much outward sign of alteration; if it has been in a permeable environment for an appreciable proportion of its history it may have suffered loss of the gas radon. Radon is an intermediate decay product in the decay of U^{238}, and its loss would result in a deficiency of Pb^{206}. In such a case only the N_{207}/N_{235} age could be significant. The problems presented by discordant ages, especially when they are believed to have occurred at some intermediate period in a mineral's history are too complex for further consideration in this article and reference should be made to sources given in the accompanying bibliography.

A. G. Darnley

References

Aldrich, L. T., 1956, "Measurement of the radioactive ages of rocks," *Science*, **123**, 871–875.

Aldrich, L. T., and Wetherill, G. W., 1958, "Geochronology by radioactive decay," *Ann. Rev. Nuclear Sci.* **8**, 257–298.

Cobb, J. C., and Kulp, J. L., 1961, "Isotopic geochemistry of uranium and lead in the Swedish Kolm and its associated shale," *Geochim. Cosmochim. Acta*, **34**, 226–249.

Darnley, A. G., 1964, "Uranium-thorium-lead age determinations with respect to the Phanerozoic time-scale," *Quart. J. Geol. Soc. London*, **120s**, 73–86.

Giletti, B. J., and Kulp, J. L., 1955, "Radon leakage from radioactive minerals," *Am. Mineralogist*, **40**, 481–496.

Kulp, J. L., and Eckelmann, W. R., 1957, "Discordant U-Pb ages and mineral type," *Am. Mineralogist*, **42**, 154–164.

Larsen, E. S., Gottfried, D., Jaffe, H. W., and Waring, C. L., 1958, "Lead-alpha ages of the Mesozoic batholiths of Western North America," *Bull. U.S. Geol. Surv.*, **1070-A**.

Miller, D. S., and Kulp, J. L., 1963, "Isotopic evidence on the origin of the Colorado Plateau ores," *Bull. Geol. Soc. Am.*, **74**, 609–630.

Osmond, J. K., Carpenter, J. R., and Windom, H. L., 1965, "Th^{230}/U^{234} age of the Pleistocene corals and oölites of Florida," *J. Geophys. Res.*, **70(8)**, 1843–1847.

Russell, R. D., 1963, "Some recent researches on lead isotope abundances." in (Geiss J., and Goldberg, G. D., editors) "Earth Science and Meteoritics, Amsterdam, pp. 44–73.

Russell, R. D., and Farquhar, R. M., 1960, "Lead Isotopes in Geology," New York and London.

Silver, L. T., and Deutsch, S., 1963, "Uranium-lead variations in zircons: a case study," *J. Geol.*, **71**, 721–758.

Steiger, R. H., and Wasserburg, G. J., 1966, "Systematics in the Pb^{208}-Th^{232}, Pb^{207}-U^{235}, and Pb^{206}-U^{238} systems," *J. Geophys. Res.*, **71** (24), 6065–6090.

Stieff, S. R., Stern, T. W., Seiki Oshiro, and Senftle, F. E., 1959, "Tables for the calculation of lead isotope ages," *U.S. Geol. Surv. Profess. Paper*, **334-A**.

Tilton, G. R., and Hart, S. R., 1963, "Geochronology," *Science*, **140**, 357–366.

Wetherill, G. W., 1956, "Discordant uranium-lead ages, I" *Trans. Am. Geophys. Union*, **37**, 320–326.

Wickman, F. E., 1944, "A graph for the calculation of the age of minerals according to the lead method," *Sveriges Geol. Undersokn. Arsbok*, **37**, 1–6.

Zykov, S. I., et al., 1964, "The age of the oldest formations of the Kola Peninsula," *Geokhymia* (4), transl. in *Geochem. Internat.* (2), p. 262.

Cross-references: *Geochronometry; Geologic Time Scale; Radioactivity in Rocks; Supergene.*

URBAN AIR POLLUTION

Man's most vital resource is the air he breathes. An average man can live on 4½ lb of water a day and 4 lb of food. But to stay alive, he requires approximately 30 lb of air. He can select his food and water; its contamination is often revealed by taste, odor, or appearance. But he has no choice about the air he breathes. He must take it as it comes to him, even though it may be laden with pollutants.

Air pollution is ubiquitous. It is primarily a problem of municipalities where people and sources of air pollution tend to concentrate. These sources are many and varied. They derive from our expanding needs for transporting ourselves and our goods, from industrial activities which add copious amounts of gaseous and particulate contaminants to the atmosphere. Expansion of our science and technology brings about new kinds of pollutants which are added to the old. These pollutants derive from combustion of fuels for heat and power, from the processing of materials, and from the disposal of wastes.

All of these trends are increasing rapidly. At the same time, population and urbanization are increasing. In 1920, for example, less than half the country's 100 million people lived in cities. Forty years later, the urban population accounted for 70% of the nation's 179 million. By the end of the century 95% of an estimated 280 million in the United States will live in urban areas.

While these and other similar trends of contemporary life continue to rise, one critical factor, the available supply of air, remains constant. The more we concentrate our population into small portions of the total land mass, the less of the total air mass we have available for our use.

Nature and Source of Pollution

The earth's atmosphere is divided into three regions characterized by progressively lower air pressures and widely fluctuating temperatures. The troposphere, which rises to a height of 5–12 miles, contains 95% of the earth's total air mass and provides the locale for nearly all meteorological phenomena. The stratosphere stretches beyond, to about 60 miles and finally, the ionosphere, containing mainly charged particles, reaches out to some 650 miles.

Clean, dry air contains 78.09% nitrogen by volume, and 20.94% oxygen. The remaining 0.97% of the gaseous constituents of dry air include small amounts carbon dioxide, helium, argon, krypton, and xenon, as well as very small amounts of other inorganic and organic gases whose concentration may differ with time and place. Water vapor and carbon dioxide, which are essentially harmless though present in relatively large amounts, are not generally regarded as part of the pollution problem.

The air also contains aerosols, dispersed solid or liquid particles. The size of the individual species range from several angstroms (10^{-8} cm) to hundreds of microns. Suspensions of particles that settle out by virtue of their large size (in excess of $\sim 50\mu$) are customarily classified as particulates, whereas stable suspensions containing smaller particles are referred to as aerosols. Dust particles occur in a size range (0.1–1000μ) that spans the boundary between aerosols and particulates.

Formation of air pollutants can be caused by natural processes or by human activities. Typical of the former are wind erosion, volcanic eruptions, evaporation of sea spray, plant and animal discharges, and atmospheric reactions engendered by solar radiation and lightning. Some thirty million tons of particulates due to natural processes settle out annually over the United States alone.

Human activities contributing to air pollution include mechanized attrition, combustion, and vaporization. Mechanical attrition comprises a large variety of demolishing, crushing, grinding, cutting, mixing, and sweeping operations, a common aspect of metropolitan and industrial activities. The effective extraction of energy from nature by man has significantly contributed to atmospheric pollution. Combustion is utilized primarily to provide energy for motive power, for industrial processes, for electrical power plants, for space heating, and also as a means of refuse disposal.

Emission of the principal pollutants in the atmosphere in the United States totals about 142 million tons per year (Table 1). Analysis of Table 1 shows:

1. Internal-combustion-powered transportation to be the major source of carbon monoxide, the oxides of nitrogen, and the hydrocarbons.

2. Central-station generation of electricity to

TABLE 1. NATIONAL AIR POLLUTANT EMISSIONS (MILLIONS OF TONS PER YEAR, 1965)

	Carbon Monoxide	Sulfur Oxides	Hydrocarbons	Nitrogen Oxides	Particles	Totals	% of Totals
Automobiles	66	1	12	6	1	86	60
Industry	2	9	4	2	6	23	17
Electric power plants	1	12	1	3	3	20	14
Space heating	2	3	1	1	1	8	6
Refuse disposal	1	1	1	1	1	5	3
Totals	72	26	19	13	12	142	72

URBAN AIR POLLUTION

Table 2. Major Pollutants in New York City Air

Type of Pollutant	Tons per Year
Carbon monoxide	1,536,000
Sulfur dioxide	750,000
Hydrocarbons	567,000
Nitrogen oxides	298,000
Particulates	230,000

be principally responsible for the oxides of sulfur.

3. Industry to be the principal source of particulate matter.

Though the problem of atmospheric pollution is ubiquitous, it is primarily a problem of municipalities, where pollutants are concentrated and often become entrained in relatively small geographical areas that are also densely populated. For example, the major pollutants found in New York City's air are given in Table 2. Analysis of Table 2 is identical to the analysis of Table 1.

The average New Yorker is forced to contend with approximately 870 pounds of pollutants each year. But even this large amount of noxious and obnoxious airborne materials is not enough to cause a major problem. There are other factors which greatly effect a city's air pollution problem such as location, topography, and climate.

By climate is meant the net result of several interacting variables, including temperature, the amount of water vapor in the air, the speed of the wind, the amount of solar radiation, and the amount of precipitation.

Weather and Air Quality

Air quality is strongly dependent upon weather. When waste products have been discharged into the atmosphere, their subsequent chronological history is a meteorological problem. Even though pollutants are put into the atmosphere at a constant rate, the condition of the weather will determine whether these contaminants accumulate to the point where discomfort and damage result or are thinned out enough so that no problem results. Weather conditions also determine whether such a problem will be localized in one small section of the community or will be widespread.

The weather elements of primary concern in air pollution are (1) air temperature, (2) instability and turbulence, (3) inversions, and (4) wind direction and speed.

Air Temperature. Within the troposphere, the stability of the air can be characterized in terms of its temperature lapse rate, usually shortened to lapse rate. The lapse rate is the rate of decrease of temperature with height. The equilibrium condition for nonsaturated air is $-1°C$ per 100 meters or $-5.4°F$ per 1000 feet. This is the rate at which nonsaturated air would cool if displaced upward, or heat if displaced downward, and is called the adiabatic lapse rate. Thus, if a theoretical parcel of air is moved from a low altitude to a high one and there is no exchange of heat with its environment, it becomes colder at a given rate as the pressure on it decreases and allows it to expand. The reverse is also true: if the air parcel is moved to a lower altitude, the pressure increases, the air is compressed, and its temperature rises.

Instability and Turbulence. If the gradient of temperature in the atmosphere exceeds the adiabatic lapse rate, it is obvious that a parcel of air, displaced upward by an infinitesimal amount from a level at which it had the same temperature and pressure as the surrounding atmosphere, will be at a higher temperature than the environment at the new level and will therefore be of lower density than the surrounding air. The force of buoyancy which must result from this condition means that the volume is likely to continue ascending, so that such an atmosphere must be classed as *statically unstable*. On the other hand, in an atmosphere whose gradient of temperature falls below the adiabatic lapse rate a mass of air forced upward will be denser than its environment and will tend to sink back to its old level, a necessary condition for *static stability*. The same result holds if the displacements are downward, and therefore generally, provided that the changes are always adiabatic.

The vertical temperature gradient controls the up and down motions of the air and thus the rate at which diffusion in the vertical direction occurs. For example, in a sunlit level grassland, the warm air near the ground tends to rise, and the cold air aloft to sink, resulting in rapid convective overturning and thorough mixing of the air near the ground with clean air from upper levels. In this way a mixing process, called *thermal turbulence* is effected and helps to dilute pollutants.

Inversions. Temperature inversion is a meteorological phenomenon which, when occurring over large cities, can have very serious consequences. Essentially, temperature inversion is an atmospheric condition in which the air temperature increases with height above the earth's surface.

Temperature inversions may be brought about in several ways, and occur more often in the fall and during the early morning hours. The most frequent cause of temperature inversions is radiative cooling of the air layer near the surface of the earth. At night, the earth radiates its warmth into space and the ground cools. Air passing over the ground is cooled and sinks. By morning the lowest layers of air are considerably cooler than the air above, thus producing a ground inversion.

A temperature inversion can also be produced by radiative cooling of a cloud bank or a dust layer. In this case, the cool air sinks and is warmed at the adiabatic lapse rate of 5.4°F per 1000 feet. This sinking air will produce a layer of air warmer than the layer of air that is at the earth's surface, thus forming a temperature inversion.

Temperature inversions may also be produced by warm or cold fronts. These frontal inversions are caused by a warm air mass riding over a cold air mass. The reason the warm air rises is the difference in density between warm and cold air.

The final and most efficient inversion producer is a high stationary atmosphere pressure system or anticyclone. Wind currents within the anticyclone spiral slowly downward, in and around the center of the air mass. The subsiding currents, compressed by the weight of the air above, are both warmed and dried, and result in a cloudless sky. The sinking air is compressed further causing the formation of warm air layers over the cooler air mass. The resulting subsidence inversions will then act independently or together with typically formed ground inversions to inhibit the dispersal of air pollutants.

All inversions have one particular feature in common—they are stable. The warm air acts like the lid on a pot, inhibiting the natural upward movement of air; this restricts vertical air currents and air pollutants become trapped in this layer. Unless the inversion is broken the air pollutants will not be dispersed and an air pollution problem or episode will exist.

Persistent thermal inversions have been experienced in New York City. Under such conditions, lethal layers of sulfur dioxide, carbon monoxide, nitrogen oxides, hydrocarbons, particulates, and many other pollutants have been statically entrapped for days at a time. Either alone or when combined, these pollutants have proved to be detrimental to health, economic welfare, and aesthetics.

Wind Speed and Direction. Inversions imply a lack of air movement, not just the convection-caused vertical movement of air called thermal turbulence. Inversions also imply a lack of wind.

Wind is the horizontal movement of the air from one place to another. It is the means by which contaminants are transported through the atmosphere. The wind speed tells how fast the pollutants are being carried. In addition, the wind speed determines the volume of air into which the pollutant is injected per unit time, and the degree of turbulent diffusion. The stronger the wind, the more energy goes into turbulent fluctuations which spread the pollution laterally and vertically. Wind blowing over level ground ordinarily moves smoothly. Wind blowing over rough and hilly terrain, or around and over tall buildings, stirs up eddies and cross-currents. These swirls are referred to as *mechanical turbulence*. The pollution tends to be less concentrated if the wind is strong, more concentrated if it is light or calm.

The total accumulated downwind dosage is a function of the variability of wind direction. Naturally, the degree of air pollution you are exposed to depends on whether the wind is blowing pollutants from its source toward you or away from you.

Heat Island Effect

Cities differ from the countryside not only in their temperature but in all other aspects of climate. The city itself is the cause of these differences. Because of its preponderous high perpendicular buildings and its ravine-like streets and the properties of the materials used in these structures and the surrounding streets, the city absorbs more energy during the day and retains it longer at night. The many vertical surfaces tend to reflect solar radiation to the ground instead of the sky. Since air is heated almost entirely by contact with warmer surfaces rather than by direct radiation, a city provides a highly efficient system for using sunlight to heat large volumes of air. In addition, the city's many structures have a braking effect on the wind, thereby reducing surface wind speed and the amount of heat carried away. The city also produces an appreciable amount of energy generated internally through space heating and other activities. This also contributes to heating of the air.

The resulting phenomenon is the *heat island effect:* warm air tends to concentrate near the structural center of the city. This warm air rises, carrying with it its burden of pollution; then it expands, flows outward over the edges of the city, and, cooling, sinks. Cooler air from the edge of the city flows into the center to replace the rising air and is followed by the now cooled dirty city air. A self-contained circulatory system has been set up.

Over a long period of time the continuous introduction and movement of particles creates a dome-shaped layer of haze over the city. In the absence of a prevailing wind, or a heavy rain to clear away the dust dome, this system can become highly stabilized and can trap pollutant emissions in a closed system. A characteristic haze or dust dome is created over the city that tends to be self-perpetuating, since it serves to reflect and back-scatter appreciable quantities of solar energy before it reaches the surface.

In winter, since less and less sunshine penetrates the dome to warm the city naturally, more and more fuel is burned to make up the difference. The combustion contributes further to the processes that build up smog. It is in this gradual but inexorable way that the smog problem has attained serious dimensions in many large cities.

LAWRENCE SLOTE

References

Anon., 1966, "Waste management and Control," *Natl. Acad. Sci.—Natl. Res. Council, Washington, D.C., Publ.* **1400**.

Anon., June 20, 1966, "Freedom to Breathe," Rept., Mayor's Task Force on Air Pollution in the City of New York.

Anon., 1966, "The Sources of Air Pollution and their Control," Public Health Publication No. **1548**, Gov. Ptg. Off., Washington, D.C.

Anon., 1969, "Cleaning Our Environment—The Chemical Basis For Action," *Am. Chem. Soc.*, Washington, D. C.

Bouman, D. J., and Schmidt, F. H., 1961, "On the growth of ground concentration of atmospheric pollution in cities during stable atmospheric conditions," *Beitr. Phys. Atm.*, **33**.

Chow, T. J., and Earl, J. L., 1970, "Lead aerosols in the atmosphere: increasing concentrations," *Science*, **169**(3945), 577–580.

Corman, R., 1969, "Air Pollution Primer," Natl. Tuberculosis Respiratory Disease Assoc., New York.

DeMarrais, G. A., 1961, "Vertical temperature difference observed over an urban area," *Bull. Am. Meteorol. Soc.*, **42**.

Duckworth, F. S., and Sandberg, J. S., 1954, "The effect of cities on horizontal and vertical temperature gradients," *Bull. Am. Meteorol. Soc.*, **35**.

Ferrand, E., June 1969, "Urban air," *Sci. Technol.*

Friedlander, G. D., October 1965, "Airborne Asphyxia—an international problem," *IEEE Spectrum*.

Landsberg, H. E., 1962, "City Air—Better or Worse," Symposium "Air Over Cities." U.S. Public Health Report A 62–5, Taft Sanitary Engineering Center.

Landsberg, H. E., September 1963, "Air Pollution and Urban Climate," Proceedings Third Internat. Biometeorol. Congr., Pau, S. France.

Landsberg, H. E., 1966, "Dust," "Encyclopedia Britannica," Vol. 7.

Lowry, W. P., August 1967, "The climate of cities," *Sci. Am.*

McCormick, R. A., and Kurfis, K. R., 1965, "Vertical diffusion of aerosols over a city," *Quart. J. Roy. Meteorol. Soc.*, **92**.

Mitchell, J. M., Jr., August 1953, "On the causes of instrumentally observed secular temperature trends," *Jour. Meteorol.*

Pack, D. H., 1966, "Air pollution, where and when?" *Ann. N. Y. Acad. Sci.*, **136**.

Panofsky, H. A., 1969, "Air pollution meteorology," *Am. Scientist*, **57**.

Sundborg, A., August 1950, "Local climatological studies of the temperature conditions in an urban area," *Tellus*.

Sutton, O. G., 1953, "Micrometeorology," New York, McGraw-Hill Book Co.

Tilson, S., June 1965, "Air pollution," *Intern. Sci. and Technol.*

Cross-references: *Air Pollution and Urban Climate; Argon; Environmental Pollution; Helium; Krypton; Nitrogen; Oxygen; Sulfur; Xenon.* Vol. II: *Aerosols; Atmospheric Nuclei and Dusts; Inversion; Ionosphere; Troposphere; Stratosphere.*

V

VADOSE WATER

For the purpose of this discussion the terms vadose water and suspended water are considered to synonymous. Vadose water is all the water held stationary or which is moving through the soil, subsoil, and underlying rock down to the upper boundary of the zone of saturation. It does not include either water at the surface of the land or the groundwater body below the water table. Chemical composition of vadose water is rather poorly known. Some indication may be gained from analyses and discussion under the topic *Suspended Water* in this volume.

JOHN D. HEM

Cross-references: *Groundwater; Hydrology; Suspended Water; Water Table.*

VANADATES—See PHOSPHATES, ARSENATES, AND VANADATES

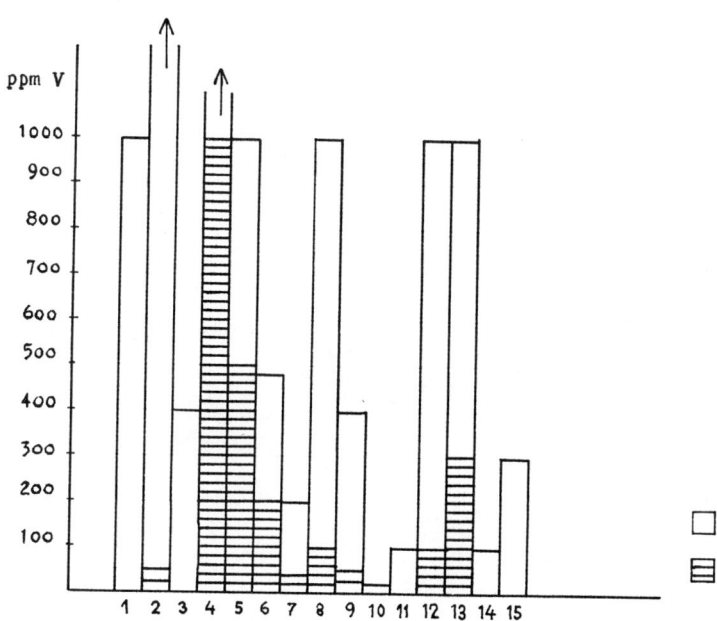

FIG. 1. Distribution of vanadium in some minerals (shaded blocks, minimum average; unshaded blocks, maximum average):

- (1) chromite
- (2) magnetite up to 1000 ppm V and more
- (3) spinel
- (4) ilmenite up to 200 ppm V
- (5) aegirine
- (6) augite
- (7) hypersthene
- (8) hornblende
- (9) biotite
- (10) plagioclase
- (11) sphalerite
- (12) chalcopyrite
- (13) pyrite and marcasite
- (14) pyrrhotite
- (15) arsenopyrite

TABLE 1. PRINCIPAL ORE MINERALS OF VANADIUM

Mineral	Composition	Crystal Structure	Physical and Optical Properties	Association	Principal Occurrences
Patronite	VS_4	C_{2h}–$2/m$ $a_0=6.78$ Å; $\beta=100.8°$ $b_0=10.42$ Å $c_0=12.11$ Å	$H=2$ $d=2.81$ Black greenish; in polish section microcrystals	Bituminous sedimentary deposit	Mina Ragra, Peru Karatau, Russia (?)
Sulvanite	Cu_3VS_4	T_d–$I\bar{4}3m$ (?) $a_0=10.77$ Å	$H=3.5$ $d=4.0$ Opaque; metallic white	Hydrothermal	Burra, Australia Kazakhstan, Russia Katanga, Africa Mercur Mine, Utah
Colusite	$Cu_3(As,Fe,Sn,V,Te)S_4$	T_d–$I\bar{4}3m$ $a_0=10.61$ Å	$H=3-4$ $d=4.5$ Opaque; bronze-yellow-black	Hydrothermal	Butte, Montana Red Mountain, Colorado Hokkaido, Japan
Coulsonite	$(Fe,V)_3O_4$	O_h–$Fd3m$ $a_0=8.49$ Å (synth. crystals)	$H=5.5-6.5$ $d=?$ (natural material) Opaque; brown-red-black	Magnetite differentiations	Shingbhum and Mayurbhang, India Taberg, Sweden Otanmäki, Finland
Vanadinocker	V_2O_5	D_{2h}–$Pmmm$ $a_0=11.52$ Å $b_0=4.37$ Å $c_0=3.56$ Å	$H=?$ $d=4.87$ Earthy yellow powder	Oxidation zone	Cliff Mine, Lake Superior
Montroseite	$(V,Fe)OOH$	D_{2h}–$Pbnm$ $a_0=4.54$ Å $b_0=9.97$ Å $c_0=3.03$ Å	$H=?$ $d=?$ Black masses	Oxidation zone	Creek Mine, Colorado
Vanadinite	$Pb_5Cl(VO_4)_3$	C_{6h}–$P6_3/m$ $a_0=10.49$ Å $c_0=7.44$ Å	$H=3$ $d=6.8-7.0$ Resinous luster; yellow-brown	Oxidation zone	Abenab, SWA Tsumeb, SWA Sierra de Cordoba, Argentina
Descloizite	$Pb(Zn,Cu)(VO_4)(OH)$	D_2–$P222$ $a_0=6.06$ Å $b_0=9.41$ Å $c_0=7.58$ Å	$H=3.5$ $d=5.5-6.2$ Radical aggregates; brown-red-black	Oxidation zone	Abenab, SWA Broken Hill, Rhodesia
Mottramite	$Pb(Cu,Zn)(VO_4)(OH)$	Related to descloizite after powder diagrams	Related to descloizite	Oxidation zone	Tsumeb, SWA
Carnotite	$K_2(UO_2)_2(VO_4)_2 \cdot 3H_2O$	C_{2h}–$P2_1/a$ $a_0=10.47$ Å $b_0=8.41$ Å $c_0=6.91$ Å	$H=4$ $d=4.5$ Mainly yellow corns	In pores of sandstones	Colorado Plateau, U.S.A. Utah, U.S.A. Katanga, Africa Turkestan

VANADIUM: ELEMENT AND GEOCHEMISTRY

Vanadium (V) is a metallic element (at.wt. = 50.95, $Z = 23$, $d = 5.98$, mp = 1715°C, bp ~3000°C) which was discovered in 1830 by Seifström in ores from Taberg (Sweden).

Vanadium occurs with valences $2+$, $3+$, $4+$, and $5+$. The hydroxide of the $2+$ and $3+$ are basic, those with higher positive valences are amphoteric. The vanadium compounds are the most stable in correspondence with their position in group VB of the periodic table. The colors of the compounds depend on the oxidation state.

Most of the V-containing minerals are of secondary origin. The important ones are listed in Table 1.

Characteristics of V^{2+} to V^{5+} Oxidation States

Vanadium, probably in V^{2+} state of oxidation, has been shown to occur in the FeS phase in Fe-meteorites (Goldschmidt, 1958; Sarma and Mayeda, 1961). V^{3+} (ionic radius 0.65, 4-coordination; 0.74 Å 6-coordination (Strunz, 1957)) substitutes for Fe^{3+} (ionic radius 0.67; 0.64 Å) in minerals of magmatic origin (see Fig. 1). Experimental mineral syntheses have shown that V^{3+} can substitute for 70% of the Al^{3+} in chrysoberyll (Sarazin and Forestier, 1959); in $FeFe_2O_4$, V^{3+} can substitute completely for Fe^{3+}, but shows limited substitution for Al^{3+} and Cr^{3+} (Trojer, 1963).

V^{4+} (ionic radius 0.61; 0.63 Å (Strunz, 1957)) substitutes for Ti^{4+} (ionic radius 0.64; 0.68 Å) in rutile and ilmenite (Fig. 1, p. 1233) formed from gabbroic magma (Goldschmidt, 1958). In reducing conditions in sedimentary deposits, such as Mina Ragra, V^{4+} occurs in patronite with porphyrin containing Ni,Fe,Mo, and Sr (Baumann, 1964; Allmann et al., 1964).

Oxidation to V^{5+} takes place during the oxidation cycle of weathering; $[VO_4]^{3-}$ is soluble over a wide pH range (in contrast, V^{3+} forms insoluble hydroxides when pH is low). When cations of Pb, Zn, and Cu are present, local precipitation of vanadates (vanadinite, descloizite, mottramite) occurs. Isomorphic relations are known between V^{5+}, P^{5+}, As^{5+}, and Mo^{5+}. The presence of $(UO_2)^{2+}$ cation leads to local precipitation of carnotite and tyuyamunite. The $[VO_4]^{3-}$ ion has adsorptive properties and thus occurs with Fe–Al–Mn hydroxides in laterites, bauxites (Goldschmidt, 1958), and Fe–Mn nodules (Wakell and Riley, 1961). Such adsorption has been demonstrated experimentally (Krauskopf, 1958). Adhesion of $[VO_4]^{3-}$ to clay minerals containing Fe^{3+} lead to considerable

TABLE 1. (Continued)

Mineral	Composition	Crystal Structure	Physical and Optical Properties	Association	Principal Occurrences
Tyuyamunite	$Ca(UO_2)_2(VO_4)_2 \cdot nH_2O$	rhomb. (?) $a_0 = 10.63$ Å $b_0 = 8.36$ Å $c_0 = 20.40$ Å	$H=2$ $d=3.7$–4.35 Microcrystals or plates; yellow to greenish	In pores and fractures in limestones	Tyuya Muyan, Russia Ferghana, Russia Turkestan Colorado Plateau, U.S.A.
Roscoelite	$KV_2(AlSi_3O_{10})(OH,F)_2$	Mica group lattice constant between muscovite and biotite	$H=2.5$ $d=2.90$–2.94 Micaceous; greenish-gray-black	In pores of sandstones In gold and gold telluride veins	San Miguel, Coloradeau Cripple Creek, Colorado Kalgoorlie, Australia

TABLE 2. THE V–FE RATIO OF THE
MAIN GROUPS OF ROCKS

Rocks	V (ppm)	V:Fe_2O_3
Ultrabasic rocks	40	1:3300
Basaltic rocks	250	1: 500
Ca-rich granites	88	1: 470
Ca-poor granites	44	1: 450
Syenites	30	1:1750
Shales	130	1: 500
Sandstones	20	1: 700
Limestones	20	1: 270

concentration of V in many sediments (Leutwein, 1951).

The geochemical distribution of V is indicated in Table 2, which also shows the V–Fe ratio (Hartmann, 1963; Turekian and Wedepohl, 1961).

Vanadium is concentrated in many bituminous sediments, sapropelites, oils, and coals, In organic compounds (e.g., porphyrins) vanadyl replaces Mg, (e.g., in chlorophyll) Blumer, 1950; Thomas and Blumer, 1964. Analyses of algae gave $2 \times 10^{-4}\%$ V (Black and Mitchell, 1952), of oceanic microorganisms $8.5 \times 10^{-3}\%$ V (Noddack, 1943). Analyses of crude oils (Park, 1961) are listed in Table 3.

Vanadium is concentrated in basic rocks such as gabbros, norites and anorthites (see Table 2). In ore deposits, its concentration varies; in India titanomagnetite ores, which contain 1.4–1.9% coulsonite on the average, the maximum amount of V_2O_3 is 8.8%; Canadian ores average 0.3–0.4; Swedish 0.5; Finish 0.23; U.S. and Russian ores contain less than 0.5% V_2O_3. Average values of 1.5–2.0% V_2O_5 have been found to occur in the ores of the Bushveld complex.

In hydrothermal ores V is not usually enriched due to the low solubility of V^{3+}.

The vanadium concentration in sediments is economically important; Jurassic iron ores from Cleveland Hills, England contain 0.08–0.10% V_2O_5, minette ores and oolitic iron ores from Salzgitter, Germany contain 0.02–0.2%. Vanadium is produced from phosphates as a by-product in the western U.S.

In the past, the bituminous sediments from Mina Ragra supplied important quantities of V from patronite. Today most of the V left over is concentrated in beds, which at the present time are below ore grade despite an average V content of 1% V_2O_3. Ashes of petroleum and asphalts from various places may contain as much as 40% V (Goldschmidt, 1958).

The origin of V in the oxidation zones of Pb–Zn–Cu ores in limestones and dolomites USA, Mexico, Africa, Australia (Skul, 1934; Verwoerd, 1957) is uncertain. Mining of vanadinite, descloizite, and mottramite is taking place in the Otavi district (SWA), Broken Hill (Rhodesia), and Mammoth area (Arizona).

Vanadium is also obtained from supergene ores (roscoelite, vanadates, and other V-oxides) which also contain Cu and U, for instance on the Colorado Plateau (USA) and Tyuya Muyan (USSR). Such deposits contain 1.5–2% V_2O_5 on the average.

INGRID BURKART-BAUMANN

References

Altmann, R., Baumann, I., Kutoglu, A., Rösch, H., and Hellner, E., 1964, "Die Kristallstruktur des Patronits $V(S_2)_2$," *Naturwissenschaften,* **51**, 11, 263–264.

Baumann, I., 1964, "Patronit, VS_4 und die Mineralparagenese von Minasragra, Peru," *N. Jb. Miner. Abh.,* **101**, 1, 97–108.

Black, W. A. P., and Mitchell, H. L., 1952, "Trace elements in the common brown algae and in seawater," *J. Marine Biol. Assoc. U.K.,* **30**, 575.

Blumer, M., 1950, "Porphyrin-Metallkomplexe in schweizerischen Bitumina," *Helv. Chim. Acta,* **33**, 1627–1637.

Goldschmidt, V. M., 1958, "Vanadium, Geochemistry," Oxford Press, pp. 485–499.

Hartmann, H., 1963, "Einige geochemische Untersuchungen an Sandsteinen aus Perm und Trias," *Geochim. Cosmochim. Acta,* **27**, 459–499.

Krauskopf, K. B., 1956, "Factors controlling the concentration of thirteen rare metals in seawater," *Geochim. Cosmochim. Acta,* **9**, 1–32 B.

Leutwein, F., 1951, "Geochemical studies of the alum and silica shales of Thuringia," *Arch. Lagerstättenforsch.* **82**, 1–45, L 51.

Noddack, I. and W., 1943, "Die Häufigkeiten von Schwermetallen in Meerestieren," *Zbl. Miner., 1* 173.

Park, R., 1961, "Stable carbon isotope studies of crude oils and their porphyrin aggregates," *Geochim. Cosmochim. Acta,* **22**, 99–105.

Sarazin, J. A., and Forestier, H., 1959, "Etude de la substitution d'ions V^{3+} aux ions Al^{3+} dans le chrysoberyl," *Compt. Rendus,* **248**(13), 2208–2210

TABLE 3. ANALYSES OF CRUDE OILS

Localities	Porphyrin (ppm)	Vanadium (ppm)	Nickel (ppm)
Boscan, Venez.	1700	900	70
Lagunillas, Venez.	170	320	40
Santa Maria, Valley/Calif.	300	280	130
Tatums, Okla.	160	150	70

Sarma, D. V., and Mayeda, T., 1961, "Meteorite analysis: the search for diamonds," *Geochim. Cosmochim. Acta,* **22,** 166–175.
Skerl, A. C., 1934, "Vanadium at the Rhodesia Broken Hill," *Mineral. Mag.,* **50,** 280–283.
Strunz, H., 1957, "Mineralogische Tabellen," *Akad. Verlagsgesellschaft.* 30–32.
Thomas, D. W., and Blumer, M., 1964, "Porphyrin Pigments of Triassic Sediment," *Geochim. Cosmochim. Acta,* **28,** 1147–1154.
Trojer, T., 1963, "Die oxydischen Kristallphasen der anorg. Industrie-produkte," Stuttgart, Schweizerbartsche Verlagsbuchhandlung, 167pp.
Turekian, K. K., and Wedepohl, K. H., 1961, "Distribution of the elements in some major units of the earth crust," *Bull. Geol. Soc. Am.,* **72,** 145–192.
Verwoerd, W. J., 1957, "The Mineralogy and Genesis of the Lead-Zinc-Vanadium Deposit of Abenab West in the Otavi Mountains, South-West Africa," *Ann. Univ. Stellenbosch,* **33A,** 235–328.
Wakeel, S. K., and Riley, J. P., 1961, "Chemical and mineralogical studies of deep-sea sediments," *Geochim. Cosmochim. Acta,* **25,** 110–146.

Cross-references: *Weathering, Chemical. Vol. IVB: Economic Geology Mineral and Ore Deposits.*

VANADIUM OXYSALTS

This class contains the oxysalts of vanadium which do not have discrete $(VO_4)^{-3}$ anisodesmic groups and which, therefore, have not been included with the phosphates and arsenates. It does not include the simple oxides or hydroxides of vanadium. For many of the vanadium oxysalts included in this section the structures are not well known, although considerable progress has been made in recent years in determining some of the structures. The minerals are arranged, in general, according to decreasing ratio of the cations to vanadium. (See table below.)

Carnotite	$K_2(UO_2)_2(VO_4)_2 \cdot 3H_2O$
Tyuyamunite	$Ca(UO_2)_2(VO_4)_2 \cdot 5-8H_2O$
Metatyuyamunite	$Ca(UO_2)_2(VO_4)_2 \cdot 3-5H_2O$
Sengierite	$Cu_2(UO_2)_2(V_2O_8)(OH)_2 \cdot 6H_2O$
Ferghanite	$U_3(VO_4)_2 \cdot 6H_2O$
Kolovratite	Ni vanadate ?
Fervanite	$Fe_4V_4O_{16} \cdot 5H_2O$
Steigerite	$Al_2(VO_4)_2 \cdot 6\frac{1}{2}H_2O$
Pucherite	$Bi(VO_4)$
Coulsonite	FeV_2O_4
Brackebuschite	$Pb_4MnFe(VO_4)_4 \cdot 2H_2O$
Chervetite	$Pb_2V_2O_7$
Pintadoite	$Ca_2V_2O_7 \cdot 9H_2O$
Delrioite	$CaSrV_2O_7 \cdot 3H_2O$
Rossite	$CaV_2O_6 \cdot 4H_2O$
Metarossite	$CaV_2O_6 \cdot 2H_2O$
Pascoite	$Ca_2V_6O_{17} \cdot 11H_2O$
Uvanite	$U_2V_6O_{21} \cdot 15H_2O$
Sincosite	$Ca(VO)_2(PO_4)_2 \cdot 5H_2O$
Hummerite	$K_2Mg_2(V_{10}O_{28}) \cdot 16H_2O$
Grantsite	$(Na,Ca)_2(V_6O_{16}) \cdot 4H_2O$
Barnesite	$Na_2(V_6O_{16}) \cdot 3H_2O$
Sherwoodite	$Ca_3(V_8O_{22}) \cdot 15H_2O$
Rauvite	$CaU_2V_{12}O_{36} \cdot 20H_2O$
Melanovanadite	$Ca_2V_4V_6O_{25}$
Hewettite	$CaV_6O_{16} \cdot 9H_2O$
Metahewettite	$CaV_6O_{16} \cdot 9H_2O$
Fernandinite	$CaO \cdot V_2O_4 \cdot 5V_2O_5 \cdot 14H_2O$

Lloyd W. Staples

References

Dana, J. W., and Dana, E. S., 1951, "The System of Mineralogy," Seventh ed., Vol. II, 1402, (revised by Palache, C., Berman, H., and Frondel, C.), New York, John Wiley & Sons.
Fischer, R. P., 1942, "Vanadium deposits of Colorado and Utah," *U.S. Geol. Soc. Bull.,* **936P.**
Weeks, A. D., and Thompson, M. E., 1954, "Identification and occurrence of uranium and vanadium minerals from the Colorado Plateau," *U.S. Geol. Soc. Bull.,* **1009-B.**

Cross-references: *Mineral Classes-Non-Silicates; Vanadium.*

VAN DER WAALS FORCE

The Van der Waals force is the residual attractive force that exists between all molecules after bonding forces due to electrostatics have been subtracted. In 1873 van der Waals postulated the existence of such a force between gas molecules so that he might justify his equation of state for real gases. The equation of state for ideal gases is

$$PV = nRT$$

where R is the gas constant and n the number of moles. Clausius had modified this equation to account for the actual volume occupied by the gas molecules, hence to

$$P(V - nb) = nRT$$

where b is a constant for a particular gas. However, gases did not behave exactly as this equation predicted. Although some, like helium, were quite close, others, like water vapor, were considerably off. Van der Waals reasoned that there remained an intermolecular attractive force much greater than that due to gravity alone and that this force was quite variable for different gases.

To understand the existence of this attrac-

VAN DER WAALS FORCE

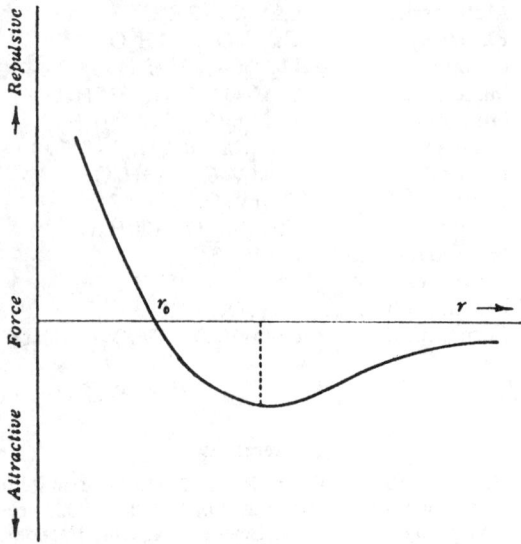

FIG. 1. The force between two molecules as a function of the separatoin of their centers (from Fleagle and Businger, 1963).

tive force, we must realize that, although a molecule may be electrically neutral, it is composed of electrically charged particles, and it is through the interaction of these charged units of molecules that the van der Waals force arises. The force decreases roughly as the seventh power of the distance of separation and is thus only measurable between a molecule and its nearest neighbor. Molecules of a gas in a container are on the average equally attracted in all directions; however, those in the outermost layer experience a net inward attraction. This leads to a lowering of the number of collisions with the walls of the container. Thus, the gas will exhibit less pressure than the ideal gas equation predicts. Since only nearest neighbor interactions are important, only the molecules in the outer two layers need be considered. The density of each of these layers should be equal, so the correction term will be proportional to the square of the density. The pressure reduction term is thus $a(n/V)^2$ where a is a constant. The van der Waals equation is

$$\left(P + \frac{an^2}{V^2}\right)(V - nb) = nRT$$

The attractive term is positive because a pressure reduction term has been subtracted. The constants a and b vary for different gases and are almost independent of temperature and pressure as the equation assumes.

In solids the van der Waals force is usually obscured by the much stronger polar, covalent, and metallic bonds. These bonds act between atoms and are much stronger than the van der Waals bond that exists between molecules. However, in crystals of Group VIII gases, and in certain organic crystals, a bond is observed that cannot be explained by any interatomic bond type. Studies of these crystals have given us a partial understanding of the van der Waals bond. I+ is a very weak bond and is similar to the metallic bond in that it has no directional properties; that is, the molecule may be linked to all surrounding molecules in a spatially unoriented manner. The structure of the solid is thus determined by geometry. There is an optimum distance of approach of molecules where the attraction of the van der Waals force is balanced by the repulsion of interpenetration of electronic shells. One-half this distance is called the van der Waals radius for the molecule. This is the radius the molecule will exhibit in a crystal lattice of neutral particles.

Van der Waals crystals have characteristic properties. Since there are no strong electronic links, the crystals are insulators. They are soft and plastic and exhibit high compressibility and low melting points.

WILLARD S. MOORE

References

Evans, R. C., 1964, "An Introduction to Crystal Chemistry," Second ed., Cambridge University Press, 410pp.
Fleagle, R. G., and Businger, J. A., 1963, "An Introduction to Atmospheric Physics," New York, Academic Press, 346pp.
Pauling, L., 1953, "General Chemistry," Second ed., San Francisco, W. H. Freeman & Co.
Smith, F. G., 1963, "Physical Geochemistry," Reading, Mass., Addison-Wesley Publ. Co., 624pp.
Widom, B., 1967, "Intermolecular forces and the nature of the liquid state," *Science*, **157**(3787), 375–382.

Cross-references: *Bonding; Crystal Chemistry.*

VEGETATION INDICATORS

Ecosystems

The use of vegetation as an indicator is based on the theory that the vegetation and the site on which it grows form a unit called an *ecosystem*. A description of an ecosystem includes all plant and animal species (even microbial ones) as well as climatic conditions and soil characteristics. Every landscape consists of a large number of individual ecosystems, each one with its own distinctive features.

Ecosystems are exceedingly complex and

have so far defied all attempts at precise description. And yet, an intelligent land-use requires an intimate knowledge of all site qualities if maximum productivity is to be maintained on a sustained yield basis without reducing the capability of the land. Although an ecosystem may not be expressed by a mathematical formula, it is nevertheless possible to obtain a deep and quite adequate insight into its nature because of the vegetation on it (for references, see Küchler, 1967).

Phytocenoses

The distribution of plant species in the landscape is governed (1) by the ranges of tolerance of each species with regard to every individual feature of its environment, and (2) by the competing species. The limited ranges of tolerance and the competition result in the fact that on any given site only a given combination of species will grow naturally. Such a particular combination of species is called a natural plant community or *phytocenose*. It is adjusted to the combination of site qualities and is, in fact, their integrated visible expression. Where the site qualities (or any one of them) change, the combination of plant species changes as well. As a result of this close interrelationship between natural phytocenoses and their sites, the plant communities "indicate" the qualities of their sites.

Natural communities are becoming less common because of man's increasing utilization of the land. A wheat field is also a phytocenose though obviously not natural. Such man-made phytocenoses are called "cultural." Man is not free to plant any crop anywhere; he plants those crops of which he can reasonably expect the greatest returns. Such returns are best assured by selecting crops that fit most precisely the conditions prevailing in his fields. Thus a hay meadow may promise good returns on wet ground where wheat may fail easily, whereas on a well-drained site a hay meadow may be less economical than wheat. The weeds among the crop plants also form communities of considerable indicator value, revealing the site qualities usually more precisely than cultivated crop plants do.

Indication

Plants are rooted in the soil and exposed to the daily weather conditions of the seasons; they can therefore indicate the nature of their environment much more comprehensively than any instrument. The presence or absence of given species, their appearance and vitality, the patterns of their distribution, their associations with other species, and many other features tell the story of the site on which they grow.

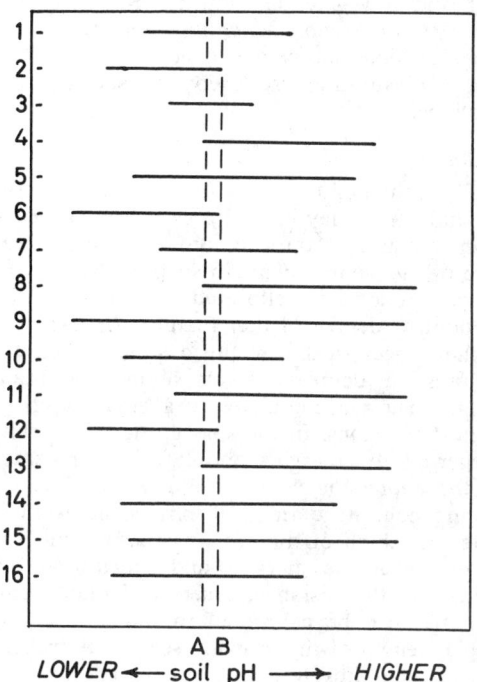

FIG. 1. Indicator value of a phytocenose (for explanation, see text).

Individual species can be used as indicators of certain environmental conditions but they are much less reliable than phytocenoses because their range of tolerance is usually wider than that of a community as a unit, and because a given species may behave (i.e., "indicate") differently in different regions and with different competitors. On the other hand, the indicator value of a phytocenose may be remarkably accurate. In Fig. 1, the horizontal lines may represent the ranges of tolerance of sixteen species concerning the degree of acidity or pH of the soil of a given site. Some species have a much narrower tolerance than others, but all have a range so wide as to preclude the accurate determination of the pH by a single species. Yet the phytocenose as a unit reveals that the pH value must lie between points A and B because a higher pH would result in the disappearance of species 2, 6, and 12, whereas a lower pH would eliminate species 4, 8, and 13. It is therefore the community rather than an individual species that indicates the pH most accurately.

Some environmental features may be so forceful as to overshadow most or all the others of a given site. Yet, the landscape consists of many unlike sites, and the dominant site quality may therefore vary within short distances. A landscape analysis may utilize this fact and a vegetation map of the area (see Vol. VI, *Vegetation Mapping*) will reveal the type, location, and extent of plant communities that indicate an assortment of specific site qualities.

Calibration

From the point of view of a particular form of land-use, it may be desirable to have a maximum amount of information on a single feature of the environment. In such a case, phytocenoses may be calibrated to indicate the amount or degree of that feature. An example is the concentrically arranged series of phytocenoses in depressions on alpine mountain ranges. Snow accumulates to a great depth in such depressions. In the spring, the snow melts earlier at the margins and last in the center of the depression where it is deepest. The resulting concentric rings of phytocenoses indicate the length of the growing season which is longest near the margins and shortest at the center. In this instance, a series of plant communities can be calibrated to indicate a particular length of the growing season. A trained person can thus interpret the vegetation and tell the duration of the snow-free period at any given spot simply by analyzing the phytocenoses growing there. Similar calibrations can be made for a host of other environmental features.

The various qualities of environmental features indicated by plant communities may be arranged in three major groups of climate, soil, and water. Each of these is enormously complex but a few selected examples will suffice to illustrate the indicator relations between plant communities and their environment.

Climate

The intimate relation between vegetation and climate is perhaps best illustrated in landscapes with high mountains. The temperature usually decreases with altitude, the wind increases as often does the relative humidity. The upper atmosphere is so much thinner that the sun's rays are more powerful with a higher portion of shorter waves. The vegetation reacts to all these climatic features and reveals them at every altitude through its structure and floristic composition (i.e., the particular combination of species). One of the most obvious vegetational changes resulting from climatic change with altitude occurs at "timberline," i.e., the elevation up to which forests can grow, and above which the vegetation is limited to phytocenoses of herbaceous plants, dwarf shrubs, and the like.

In addition, vegetation indicates the effects of different exposures to the sun and to rain-bearing winds. Thus in the latitudes of the United States, vegetation boundaries such as timberline are higher on south-facing slopes than on north-facing slopes because they receive more heat from the sun on the south side of the mountains.

On the other hand, rain-bearing winds result in vegetation contrasts between the windward and leeward sides of mountain ranges. Thus the north east slopes of the Blue Mountains in Jamaica face the tradewinds and are covered with a lush evergreen forest, indicating a heavy rainfall in all seasons. In the lee of these same mountains, the vegetation consists of thorny deciduous scrub with many cacti, indicating hot and dry conditions with very light rainfall.

A whole field of investigation called *phenology* has developed, exploring the behavior of vegetation and of individual species as they indicate the progressive climatic changes through the seasons. More recently, lichen communities have been discovered to be remarkably faithful indicators of air pollution. From unpolluted areas with a rich lichen flora, the number of species decreases and the communities are progressively impoverished as air pollution grows. In severely polluted areas as in many large cities, the lichen communities have completely disappeared even in large parks.

Soil

The very close relationship between phytocenoses and soil types (cf. Fig. 1) becomes obvious when one considers that, on the one hand, plants take their nutrients from the soil and, on the other hand, are in part responsible for the quality of the soil by contributing its humus content. As a result, the correlation between vegetation and soil is often so close that soils can be mapped by observing the extent and distribution of plant communities. Where the soils are thin, the quality of the rocks underneath is also indicated by the vegetation for it makes a great difference to phytocenoses whether the substrate consists of coarse sand, finely textured compact clay, limestone, or granite, etc. The floristic composition of the phytocenoses indicates such differences unfailingly. There have even been efforts to prospect for minerals, e.g., uranium, by using vegetation as an indicator but the usefulness of this method is limited (see Vol. IV B, *Geobotanical and Biogeochemical Methods of Prospecting for Minerals*).

Water

Water and soil often go hand in hand because it is the water-holding capacity of the soil that may determinte the amount of water available to plants. A well structured soil with much humus can hold much water and yet permit a normal air circulation in the soil. Where the water table is close to the surface of the soil, the vegetation indicates the water-logged condition very clearly by the species combination because for most plants a water logged soil is beyond their range of tolerance. In areas where the water table has a tendency to be high, phytocenoses can be calibrated to indicate the depth of the groundwater with remarkable accuracy; they will even indicate the seasonal fluctuations of the water table. The chemical composition of the water is also reflected in the vegetation since the soil water carries a great variety of compounds in solution, and plants absorb their nutrients in dissolved form.

These examples can only hint at the complexity of an ecosystem, but they also illustrate how plant communities vary their species combinations and thereby reflect environmental conditions as these change from place to place. Phytocenoses indicate the character of their sites most accurately as can be observed readily. The problem for man is to interpret his observations correctly, and that is sometimes very difficult. The indicator value of phytocenoses is often of decisive importance in such fields as forestry, agriculture, range management, etc., and continues to be explored in ever greater detail.

A. W. KÜCHLER

References

Küchler, A. W., 1967, "Vegetation Mapping," New York, Ronald Press, 472pp.

Cross-references: *Pedology; pH-Eh; Water Table.* Vol. II: *Climatic Classification; Vegetation Classification and Description.* Vol. IVB: *Geobotanical and Biogeochemical Methods of prospecting for Minerals.* Vol. VI: *Vegetation Mapping.*

VERSENE (EDTA) SOLUTION STUDIES

The application of the chelating agent, EDTA, Versene, or ethylenediaminetetraacetic acid, to the solution of carbonate rocks replacing the time-honored acid solution techniques was first reported in the early 1960s, almost simultaneously by Hill and Runnels (1960) and Glover (1961). (Both groups of researchers working independently.) The basic research was originally aimed at improving the recovery of noncarbonate fractions of carbonate rocks for detailed study.

It has long been realized that the terms acid-insoluble residue, and noncarbonate rocks are not synonymous. When carbonate rocks are dissolved in organic or mineral acids, the solution of other accessory minerals also takes place, and clay mineral structures are broken up and either ion exchanged or dissolved.

Remedial tactics to improve the recovery of these minerals from the breakdown of carbon-rocks has trended toward weaker acids and greater dilutions. These techniques have not solved the problem.

The use of chelating agents to dissolve calcium and magnesium carbonates is of value in a number of ways:

1. The solution of the alkaline earth carbonates is carried out at a neutral or basic pH.
2. There is no effervescence or foaming as with acid solution techniques. In fact, there are no fumes or gas liberation from the reaction vessel. Therefore, neither fume hoods nor special ventilation are needed.
3. Large numbers of the solution reactions can be carried out simultaneously if desired.

The use of chelating agents in solution studies offers a new approach to the study of carbonate diagenesis.

Glover (1961) reported the results of his work with commercially available EDTA compounds to obtain residues from carbonate rocks. He specifically reported the effects of concentration, pH, temperature, and particle size on the solution rate. The data presented by Hill and Goebel (1963) showed the effects of various substituted EDTA solutions on the solubility of limestone blocks.

These initial studies pointed out not only the lack of adequate information on the solubility of various minerals in EDTA solutions but the lack of acid solution data on the same minerals. Hill and Evans (1965) initiated studies into the solubility of a number of minerals normally found as secondary or accessory minerals in limestones and dolomites. More detailed evaluations of the solubility of these and other minerals in EDTA solutions and other chelating agents is necessary to fully develop the effectiveness of carbonate rock solution studies. Studies into the beneficiation of some industrial minerals and some possible bulk applications of EDTA solutions to effect an improvement of porosity and permeability in sedimentary rocks are underway.

WALTER E. HILL, JR.

References

Ellingboe, J., and Wilson, J., 1964, "A quantitative separation of non-carbonate minerals," *J. Sediment Petrol.* **34**(2), 412–418.

Glover, E. D., 1961, "Method of solution of calcareous materials using the complexing agent, EDTA," *J. Sediment. Petrol.*, **31**(4), 622–626.

VERSENE (EDTA) SOLUTION STUDIES

Hill, W. E., Jr., and Evans, D. R., 1965, "Solubility of twenty minerals in selected versene (EDTA) solutions," *Bull. Kansas Geol. Surv.*, **175**, pt. 3, 22pp.

Hill, W. E., Jr., and Goebel, E. D., 1963, "Rates of solution of limestone using the chelating properties of Versene (EDTA) compounds," *Kansas Geol. Survey Bull.*, **165**, pt. 7, p. 1–15.

Hill, W. E., Jr., and Runnels, R. T., 1960, "Versene, a new tool for the study of carbonate rocks," *Am. Assoc. Petrol. Geol Bull.*, **44**(5), 632–633.

Palache, C., Berman, H., and Frondel, C., 1944, "The system of mineralogy of J. D. Dana and E. S. Dana," Vol. 2, Seventh ed., New York, John Wiley & Sons, 1124pp.

Thompson, T. L., 1965, "Conodonts from the Meramecian Stage (Upper Mississippian) of Kansas," unpublished Ph.D. Dissertation, Univ. of Iowa, 205pp.

Welch, R. G., Hill, F. E., Jr., and Ireland, H. A., 1964, "A contin[u]ous extraction technique for insoluble residues and phosphatic microfossils," *Trans. Kan. Acad. Sci.*, **67**(3), 553–555.

Cross-references: *Calcium Carbonate Geochemistry;-Chelation.* Vol. VI: *Diagenesis; Limestone.*

VOLATILES—See Vol. V

VOLCANIC ENERGY—See Vol. IV B, GEOTHERMAL ENERGY

VOLCANIC GASES—See GASES—VOLCANIC

VOLCANIC TERRAIN GROUNDWATER —See HYDROLOGY, VOLCANIC TERRAIN

W

WATER—NONMARINE

Water at or near the earth's surface is partly held in chemical combination in the structure of rock minerals and is partly present as free H_2O, in vapor, liquid, or solid form. The combined water is not considered here. The major reservoirs of free water are the oceans. Large quantities also are stored as ice in the polar regions.

The water which is considered here is the portion which is circulated from the oceans through the atmosphere and precipitated on the land surfaces to produce the fresh water supplies that are vital to living organisms and the processes that shape the land surface. As water circulates it becomes charged with impurities carried in solution and suspension. This discussion is concerned with amounts and kinds of dissolved impurities in the circulating fresh water that is encountered in various parts of the hydrologic cycle apart from the ocean. Information is available for composition of liquid water in the atmosphere, soil moisture, surface water of rivers and lakes, and underground water from the zone of saturation.

Some processes by which the composition of natural water in these forms is attained are described elsewhere in this volume (see *Aqueous Solutions; Oxidation and Reduction; Solubility*).

Liquid water in the atmosphere is present as suspended droplets in clouds and fog and in the larger drops of rain, or the ice crystals of snow and other solid species of precipitation. Air-borne dust and debris, salt particles from the ocean, and soluble vaporized material all constitute sources of solutes in precipitation. The composition of atmospheric water is discussed by Junge (1963). Near the ocean the principal components of dissolved matter in rain or snow are usually sodium and chloride but further inland the principal anion is usually sulfate. Analysis 1 in Table 1 represents the average composition of rain for five months at Menlo Park, Calif. (Whitehead and Feth 1964). Analysis 2 is the average of twenty samples of snow collected in Utah in 1959 (Feth, Rogers, and Roberson 1964). A considerable part of the load of ions carried in runoff from some stream basins can be accounted for in the rainfall.

TABLE 1. COMPOSITION OF RAINFALL, RIVER, AND UNDERGROUND WATERS (PPM)

Source	1	2	3	4	5	6
Silica (SiO_2)	0.29	—	13.1	10	44	12
Aluminum (Al)	—	—	—	—	.05	.2
Iron (Fe)	—	—	.67	—	.03	.06
Manganese (Mn)	—	—	—	—	.00	.00
Calcium (Ca)	.77	2.23	15	790	54	50
Magnesium (Mg)	.43	.33	4.1	424	20	6.0
Sodium (Na)	2.24	.60	6.3	3,160	51	2.4
Potassium (K)	.35	.47	2.3	—	7.2	3.0
Bicarbonate (HCO_3)	4.95	6.29	58.4	116	242	184
Carbonate (CO_3)	—	—	—	—	—	—
Sulfate (SO_4)	1.76	2.25	11.2	3,310	61	2.1
Chloride (Cl)	3.76	.97	7.8	5,690	46	1.8
Fluoride (F)	—	—	—	—	.4	0
Nitrate (NO_3)	.15	—	1	—	6.0	6.5
Dissolved solids	12.4	10.6	90	13,900	532	268
pH	—	—	—	—	7.7	7.4

1. Mean composition of rain collected Dec. 1957 to April, 1958 at Menlo Park, California.
2. Mean composition of 20 samples of snow collected at various points in Utah, January and February, 1959.
3. Mean composition of river water of the world (after Livingstone 1963); dissolved solids recomputed.
4. Discharge-weighted average of daily samples Pecos River near Girvin, Pecos County Texas for period Oct. 1, 1962 to Sept. 30, 1963.
5. Well 380' deep in Snake River Basalt at Eden, Jerome County, Idaho.
6. Spring 1 mile east of Melbourne, Izard County, Arkansas. Water from St. Peter Sandstone.

Although all geochemists are aware of the importance of reactions taking place in the soil in supplying solutes to water, especially bicarbonate and associated hydrogen ions, most published work in the field contains little actual information on the composition of soil moisture, from which many solutes that later appear in river water or ground water are obtained. An exception is the work by Schoeller (1962) who cited a considerable number of analyses from pertinent published sources and studies of the relation of composition of soil moisture to that of ground water in specific areas.

River and lake waters obviously have a wide range in composition. Geochemists have long been interested in determining the quantities of solute ions carried to the ocean by rivers each year. As an outgrowth of such studies, average analyses have been published by several writers, to show the composition of runoff reaching the ocean. These averages are weighted by discharge and are strongly affected by the largest and most dilute rivers. The composition of most of these large rivers, for example the Amazon, is not well known. The most recent such average analysis is no. 3 in Table 1 and was taken from Livingstone (1963). Where soluble rocks are abundant in the drainage basin the runoff may be very high in dissolved solids as shown by analysis 4, representing the average composition of the Pecos River near Girvin, Texas, for the year ended Sept. 30, 1963 (U.S. Geological Survey, 1966). Alekin & Brazhnikova (1964) have summarized the composition of river waters in the U.S.S.R.

Obviously the composition of ground water has a very wide range depending on the nature of the rock with which the water has been in contact and the extent to which many other factors may have influenced chemical reactions that brought mineral matter into solution. White, Hem and Waring (1963) have cited many analyses for different geologic terranes. Analyses 5 and 6, taken from that paper, represent water from an igneous terrane and one from a sandstone. Both are dilute waters of types widely used for water supplies. More concentrated water is commonly associated with salt deposits and with sediments that are deeply buried and through which circulation of water occurs very slowly. Some of the solutes in such waters are connate, representing marine salt trapped in the sediments when they were deposited. The brines associated with petroleum often are near saturation with respect to sodium chloride. The concentration of salt in deep waters of this kind may also result from selective permeability of rock strata toward ions and water molecules and related processes which are not yet fully understood by geochemists.

JOHN D. HEM

References

Alekin, O. A., and Brazhnikova, L. V., 1964, "Stok rastvorennikh veshchestv s territorii S.S.S.R.," Moskva, Izdatel'stvo "Nauka," 143pp.

Feth, J. R., Rogers, S. M., and Roberson, C. E., 1964, "Chemical composition of snow in the northern Sierra Nevada and other areas," *U.S. Geol. Surv. Water Supply Paper* **1535J**, J 18.

Junge, C. E., 1963, "Air Chemistry and Radioactivity," New York, Academic Press, 382pp.

Krejci-Graf, K., Appelt, W., and Kreher, A., 1966, "Zur Geochemie des Wiener Beckens," *Geol. Mitt.*, **7**, 49–108.

Livingstone, D. A., 1963, "Chemical composition of rivers and lakes," *U.S. Geol. Surv. Profess. Paper,* **440G**, G 41.

Schoeller, H., 1962,, "Les Eaux Souterraines," Paris, Masson et Cie, pp. 334–350.

U.S. Geological Survey, 1966, "Quality of surface waters of the United States 1963," *U.S. Geol. Surv. Water Supply Paper,* **1950**, 588pp.

White, D. E., Hem, J. D., and Waring, G. A., 1963, "Chemical composition of subsurface waters," *U.S. Geol. Surv. Profess. Paper,* **440F**.

White, D. E., 1965, "Saline waters of sedimentary rocks," in "Fluids in subsurface environments—A symposium," *Am. Assoc. Petrol. Geol. Mem.* **4**, 342–366.

Whitehead, H. C., and Feth, J. H., 1964, "Chemical composition of rain, dry fallout and bulk precipitation at Menlo Park, California, 1957–1959," *J. Geophys. Res.,* **69**, 3326.

Cross-references: *Aqueous Solutions; Connate Water; Groundwater; Hydrologic Cycle; Lake Geochemistry; Natural Brines; Oxidation and Reduction; River Geochemistry; Runoff; Solubility.* Vol. II: *Atmospheric Nuclei and Dust.*

WATER—SUBSTANCE AND SOLVENT

Structure

The two protons of each water molecule (molecular weight, 18.015) are deeply embedded in the oxygen atom as shown in Fig. 1. Ice, up to about 2 kbs, has an open structure of tetrahedrally coordinated H_2O molecules, with a density of 0.9167 g/cc at 0°C and 1 atm. On melting (heat of fusion, 1.436 kcal/mole at 0°C), expanded, ice-like clusters of H_2O molecules form surrounded by unordered monomeric molecules. Dodecahedral cavities in the clusters are sufficiently large to accept an interstitial H_2O molecule which accounts for water having a greater density (0.999841 g/cc) than ice at 0°C and 1 atm. The increase with temperature of the fraction of cavities filled by interstitial molecules counteracts thermal expansion of the clusters and leads to an in-

WATER—SUBSTANCE AND SOLVENT

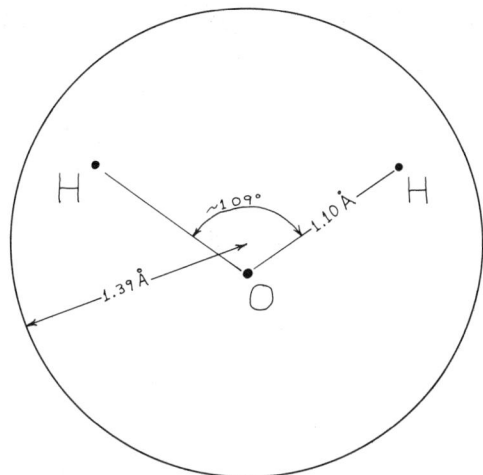

FIG. 1. Geometry of a water molecule in the liquid at 25°C (schematic) showing the atomic nuclei and the "hard-sphere" radius which is nearly identical to that of oxygen (data from Narten, Danford, and Levy, 1967).

crease in density to a maximum at 3.98°C and 1 atm. Above 3.98°C about half the cavities remain filled and normal thermal expansion causes a decrease in water density with increasing temperature.

The abnormally wide range of thermal stability for liquid water at 1 atm, and its high surface tension (71.97 dynes/cm at 25°C) reflect the strength of hydrogen-bonding between water molecules in the clusters. These bonds are continuously being broken and reestablished so that the average lifetime of a cluster is about 10^{-10} seconds. The concentration and distribution of cluster sizes are uncertain (see Kavanau, 1964) but both are affected by solutes and temperature. Pressure may have only a minor effect on water structure because of water's low compressibility (4.57×10^{-11} cm²/dyne at 25°C). Some properties of water along the liquid-vapor curve are given in Table 1. P–V–T properties at other pressures and temperatures are tabulated in Clark (1966, p. 371). For properties at 25 and 100°C see Table 2.

Solutes effect the stability of liquid water. Solids are generally more soluble in water than in ice or water vapor and, therefore, tend to increase water stability, shifting the ice-liquid curve to lower temperatures and pressures and the liquid-vapor curve to higher temperatures and pressures. Because gases are more soluble in water vapor than in liquid water, and in liquid water than in ice, gases in solution tend to shift both the liquid-vapor and ice-liquid curves to lower temperatures and pressures. Consistent with this behavior, salts generally increase the critical temperature (T_c) and decrease the ice point of water, whereas gases decrease T_c and lower the temperature of the ice point. For examples, an NaCl solution saturated at 25°C (6.18 molal, 26.4 wt%) has a T_c of 700°C (1237 bars), whereas T_c for a 27 mole % CO_2 solution is 275°C (885 bars).

Water as a Solvent

The unequalled ability of water as a natural solvent for ionic solutes reflects its dipolar nature and strong molecular polarizability, and in acid or alkaline solutions, the capacity of H^+ or OH^- ions to form complexes with solute species. The dielectric constant of water, ϵ (Table 1), is a measure of the combined effects of its dipolar and polarizing character, and is higher than for any other inorganic liquid. The dependence of ionic solubilities on ϵ in dilute solutions is given by the Debye-Hückel Limiting Law (Robinson and Stokes, 1959, p. 230), which for an ionic species, i, at constant temperature and ionic strength may be written

TABLE 1. PROPERTIES OF LIQUID WATER IN EQUILIBRIUM WITH VAPOR

	Temp (°C)	Pressure (bars)	Specific Volume (cc/g)	Dielectric Constant	-Log K_w[a]
Ice point	0	0.006107	1.0002	87.74	14.944
Triple point[b]	0.01	0.006112	1.0002	87.74	14.944
Maximum density point	3.98	0.008129	1.0000	86.20	14.793
	25	0.031663	1.0030	78.30	13.997
Boiling point	100	1.01325	1.0435	55.32	12.254
	200	15.551	1.1565	34.51	11.254
	300	85.92	1.4036	19.55	11.034
	350	165.37	1.741	12.55	11.422
Critical point	374.15	221.2	3.17	9.01	11.997

[a] $K_w = (a_{H^+})(a_{OH^-})$.
[b] A standard point on the temperature scale.

TABLE 2. PROPERTIES OF WATER

Property	At 25°C	At 100°C
Dipole moment (Debye units)	1.87	—
Enthalpy (kcal/mole)	−68.315	−67.747
Entropy (cal/mole-°K)	16.71	20.76
Gibbs free energy (kcal/mole)	−56.688	−53.820
Heat capacity (cal/mole-°K)	17.995	18.15
Heat of vaporization (kcal/mole)	—	9.717
Index of refraction, Na light		
absolute	1.33287	1.31819
in air	1.33251	1.31783
Resistivity, intrinsic (ohms)	18.24×10^6	(4.97×10^6 at 55°C)
Specific volume (cc/g)		
1 bar	1.0030	1.0435
1,000 bars	1.0000	0.9999
10,000 bars	0.807	0.8389
Surface tension, in air (dynes/cm)	71.97	58.9
Thermal conductivity (cal/cm-sec-°C)	0.00145	0.00160
Viscosity (poises)	0.008904	0.002790

[a] At 25 and 100°C and 1 bar (0.986923 atm), except where otherwise noted.

$$\log (1/\gamma_i) = k\epsilon^{3/2}$$

where k is a constant, and γ_i is the individual ion activity coefficient, a factor which is an inverse function of the solubility of salts forming solute ions. The force of attraction between two ionic species in solution is proportional to the reciprocal of the dielectric constant. Thus forces between solute ions in water are relatively small because of the ease of polarization of water molecules about an ion, an effect causing ion hydration and complexing. The dielectric constant of water decreases with rising temperature (Table 1), but increases with increasing pressure or density below the critical point.

A second important mechanism of solution in water is complex formation and ion association between H^+ and OH^- and solute species. This mechanism is, of course, more important in acidic or alkaline solutions. Most minerals, including those of the feldspar and carbonate groups, are salts of weak acids and strong bases. Their attack by water produces cations, OH^- ions, and other species. Consequently, they are most soluble under acid conditions. In contrast, the solubility of silica, which hydrates to form the weak acid H_4SiO_4, increases under alkaline conditions by the formation of H^+ ions plus anionic species such as $H_3SiO_4^-$ and $H_2SiO_4^{2-}$. A number of multivalent metal ions, such as Fe^{+3} and Al^{3+} which form relatively insoluble oxyhydroxides in near-neutral solutions, are amphoteric, exhibiting high cationic solubilities in acid waters and high anionic solubilities in alkaline waters.

The acidity or alkalinity of an aqueous solution at any temperature or pressure is measured by its pH, defined as the log of the reciprocal of the hydrogen ion activity. The pH scale is temperature dependent as is evident from the increase in the activity product of water with temperature (Table 1); the scale is also slightly pressure dependent. The neutral point, where the activities of H^+ and OH^- are equal, decreases from pH = 7.00 at 25°C to pH = 5.52 at 300°C for solutions along the liquid vapor curve. Barnes and Ellis (in Barnes, 1967) give pH data for other conditions to 700°C and over 2000 bars.

At a given temperature in dilute solutions, the solubility of minerals which dissolve to form ions increases with ionic strength, i.e., on the addition of other ions. The solubility of gases and solids which form molecular species in solution is controlled by the hydration process and generally decreases slightly with increasing ionic strength.

The effects on solubility of pressure and temperature variations are complex. Increasing pressure favors solubility to a minor extent at low temperatures for salts where ΔV_{soln} and solution compressibility are small, and to a major extent for gases at higher temperatures where both V_{soln} and solution compressibility are large. In the subcritical region, minerals *which react with a gas on dissolving,* such as calcite, with CO_2, usually decrease in solubility with increasing temperature; exceptions are the solubilities of cinnabar and acanthite in solutions with H_2S, where solubilities of both minerals pass through a maximum at about 100°C. Minerals such as halite or quartz, *which do not involve gas species in their solution reactions,* generally increase in solubility with rising temperature in subcritical solutions; however some such minerals, as anhydrite and fluorite, show decreasing solubility. In the supercritical region, solubility is often a direct function of solution density. Summaries of high temperature solubilities are given in Barnes (1967, chapt. 8, 9, and 11).

Geologic Distribution

Water content of the crust of the earth is $2.2\text{–}2.6 \times 10^{18}$ metric tons distributed among the atmosphere, 1.3×10^{13} tons, the hydrosphere, 1.4×10^{18} tons, and the lithosphere (Poldervaart, 1955). Water of the hydrosphere and atmosphere is believed to have originated by the continuous dehydration through geologic time of the hydrous minerals of crustal and subcrustal igneous rocks (Rubey, 1955). Igneous rocks contain an average of 1.15% water by weight, primarily as structural water of hydration in such mineral groups as the amphiboles and micas. The extent of hydration of crustal rocks depends on the thermodynamic water pressure or fugacity of water, a property which is a function of temperature, pressure, and water purity. Fugacity data for water are presented by Anderson (in Barnes, 1967).

Composition of Natural Waters

Average rain or snow contains about 10 ppm dissolved solids, chiefly at Na^+, Ca^{2+}, Cl^-, and SO_4^{2-}, and has a pH of about 5.5 due to carbonic acid formed by solution of atmospheric CO_2. Rain falling near ocean coastlines or saline deposits occasionally has more than 100 ppm dissolved solids due to solution of windblown salts (Carroll, 1962).

Ca^{2+} and HCO_3^- are the predominant ions in most surface and groundwaters. The pH of most natural waters is between 6 and 8, buffered by the dissociation of carbonic acid and hydrolysis reactions with carbonate and silicate minerals. The lowest pH's, near 1, occur in surface waters which contain H_2SO_4 or rarely HCl. High pH's of 9–10, or exceptionally 12, may be found in alkali lakes and springs, in the presence of carbonate and evaporate minerals.

The dissolved solids content of surface waters is highly variable, ranging from less than 50 ppm in small mountain streams on siliceous rocks, to as much as 50,000 ppm in streams flowing over saline deposits, and 400,000 ppm in some closed soda alkali lakes. Average river water contains 120 ppm solutes, including 58.4 HCO_3^-, 11.2 SO_4^{2-}, 7.8 Cl^-, 1 NO_3^-, 15 Ca^{2+}, 4.1 Mg^{2+}, 6.3 Na^+, 2.3 K^+, 0.67 total iron, and 13.1 SiO_2 (Livingstone, 1963). Ocean water contains 34,500 ppm solutes as follows: 18.980 Cl^-, 10,560 Na^+, 2,650 SO_4^{2-}, 1,270 Mg^{2+}, 400 Ca^{2+}, 380 K^+, 140 HCO_3^-, 65 Br^-, and others less than 10 ppm.

The dissolved solids content of ground water generally exceeds that of surface water (White, Hem, and Waring, 1963), and ranges from less than 100 ppm in shallow groundwaters from siliceous rocks in areas of high rainfall, to 300,000–600,000 ppm in some saline groundwaters. Compositions of thermal springs, which vary even more widely, have been compiled by Waring (1965).

Hardness of water reflects chiefly its Ca^{2+} and Mg^{2+} content. Generally waters with dissolved solids less than about 120 ppm are considered soft; from 120 to 350 ppm moderately hard to hard; and greater than 350 ppm very hard.

Potable waters in the United States are usually expected to contain less than 500 ppm dissolved solids, although in some western states where supplies are high in solutes, they contain 500–1000 ppm. Half the United States population served by public supplies receives water containing less than 150 ppm, whereas 90% receives water with less than 550 ppm (Durfor and Becker, 1964).

The U.S. Public Health Service states that a water is toxic and a health hazard, and should not be used for drinking if it contains substances in excess of the following mg/1 concentrations: As 0.05, Ba 1.0, Cd 0.01, Cr^{6+} 0.05, CN 0.2, F 1.5 (approx.), Pb 0.05, Se 0.01, and Ag 0.05. The Health Service further recommends that when more suitable supplies are available, water should not be used if concentrations in mg/1 exceed the following: alkyl benzenesulfonate (from detergents) 0.5, Cl^- 250, Cu 1.0, Fe 0.3, Mn 0.05, NO_3^- 45, phenols 0.001 SO_4^{2-} 250, total dissolved solids 500, and Zn 5.0.

DONALD LANGMUIR
H. L. BARNES

References

Barnes, H. L., (editor), 1967, "Geochemistry of Hydrothermal Ore Deposits," New York, Holt., Rinehart, and Winston, Inc., 670pp.

Carroll, D., 1962, "Rainwater as a chemical agent of geologic processes-a review," *U.S. Geol. Surv. Water-Supply Paper*, **1535G**, 18pp.

Clark, S. P., Jr. (editor), 1966, "Handbook of Physical Constants," *Geol. Soc. Am. Mem.*, **97**, 587pp.

Durfor, C. N., and Becker, E., 1964, "Chemical quality of public water supplies of the United States and Puerto Rico, 1962," *U.S. Geol. Surv. Hydrologic Investig. Atlas*, **HA-200**.

WATER—SUBSTANCE AND SOLVENT

Kavanau, J. L., 1964, "Water and Solute-Water Interactions," San Francisco, Holden-Day, Inc., 101pp.

Livingstone, D. A., 1963, "Chemical composition of rivers and lakes," in (Fleischer, M., editor), "Data of Geochemistry," *U.S. Geol. Surv. Profess. Paper*, **440-G**, 64pp.

Narten, A. H., et al., 1967, "X-ray diffraction study of liquid water in the temperature range 4–200°C," *Discussions Faraday Soc.*, **43**, 97–107.

Poldervaart, A., 1955, "Chemistry of the earth's crust," in (Poldervaart, A., editor), "Crust of the Earth," *Geol. Soc. Am. Spec. Paper*, **62**, 119–144.

Robinson, R. A., and Stokes, R. H., 1959, "Electrolyte Solutions," Second ed., New York, Academic Press, 559pp.

Rubey, W. W., 1955, "Development of the Hydrosphere and Atmosphere," in (Poldervaart, A., editor), "Crust of the Earth," *Geol Soc. Am. Spec. Paper*, **62**, 631–650.

Waring, G. A., 1965, "Thermal Springs of the United States and other countries of the world—a summary," *U.S. Geol. Surv. Prof, Paper* **492**, 383pp. (revised by R. A. Blankenship and R. Bentall).

White, D. E., 1965, "Saline waters of sedimentary rocks," in "Fluids in subsurface environments—A symposium," *Am. Assoc. Petrol. Geol. Mem.*, **4**, 342–366.

White, D. E., et al., 1963, "Chemical composition of subsurface waters," in (Fleischer, M., editor), "Data of Geochemistry," *U.S. Geol. Surv. Prof. Paper* **440-F**, 67pp.

Cross-references: *Acids and Bases; Complexes; Enthalpy; Entropy; Free Energy; Lake Geochemistry; pH-Eh; River Geochemistry; Seawater, Chemistry; Solubility. Vol. VI: Hydrosphere; Ice.*

WATER, SUBSURFACE—See GROUNWATER; HYDROLOGY; CONNATE WATER; JUVENILE WATER; MAGMATIC WATER; METEORIC WATER; SUSPENDED WATER; VADOSE WATER

WATER BALANCE

As defined by Thornthwaite and Mather (1957), "... the term water balance refers to the balance between the income of water from precipitation and the outflow of water by evapotranspiration."

By comparing monthly or seasonal values of precipitation with evapotranspiration, other associated moisture parameters such as water surplus, water deficit, soil moisture storage and water runoff may be measured.

One can distinguish between the water balance at a locality and the water balance of the world. The latter will be discussed in the last section of this article.

Seasonal Change of Evapotranspiration and Precipitation

The Thornthwaite system of determining water balance represents a valuable contribution to the field of climatology. His method begins with a monthly accounting of the precipitation and potential evapotranspiration (PE). When precipitation exceeds PE, there is a net gain in soil moisture for that month. If the soil is at field capacity (i.e., its saturation limit), then the difference between excess precipitation and PE is water runoff. As long as the soil remains at field capacity, evapotranspiration will continue at the potential rate.

The Thornthwaite system of hydroclimatic bookkeeping is shown in Figs. 1 and 2 for two examples: Seabrook, N.J. and Bismarck, N.D. The water holding capacity in the root zone of the soil is 300 mm for Seabrook and 200 mm for Bismarck. Note four different moisture conditions on these diagrams.

(a) *Water surplus*—when the soil is at field capacity and precipitation exceeds PE. Note

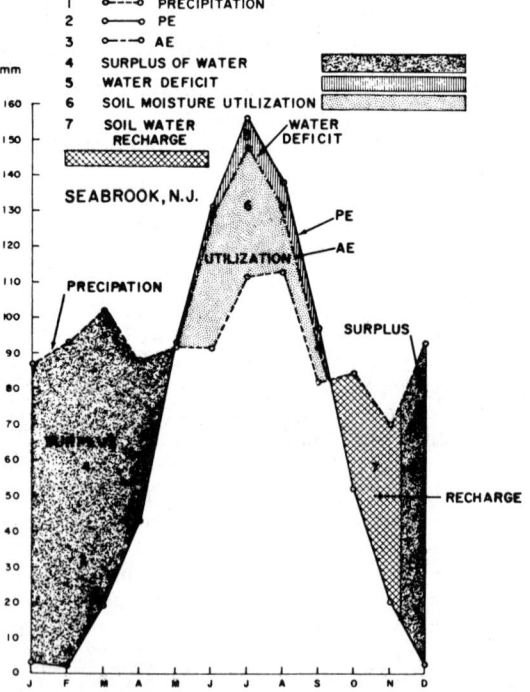

Fig. 1. Thornthwaite system of hydroclimatic bookkeeping for Seabrook, N.J. Surplus of water at Seabrook (No. 4 above) equals 374mm; water deficit (No. 5 above) equals 22mm.

WATER BALANCE

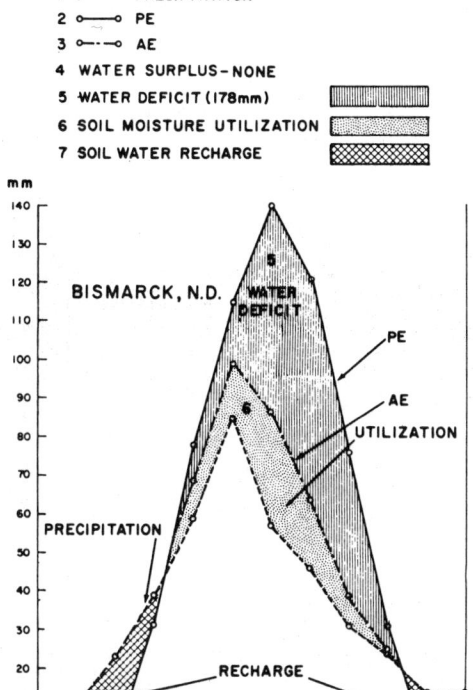

FIG. 2. Thornthwaite system of hydroclimatic bookkeeping for Bismarck, N.D.

that Seabrook has 374 mm of surplus water compared to none for Bismarck.

(b) *Soil moisture utilization*—when precipitation is less than PE, and plants draw on soil moisture.

(c) *Water deficiency*—when precipitation is again less than PE, and there is no longer any available soil moisture. Note the large water deficit of 178 mm for Bismarck.

(d) *Soil moisture recharge*—when precipitation exceeds PE, and there is a deficiency in soil moisture.

Information Necessary to Determine the Water Balance

The water balance at a locality may be determined when the following information is available:

(1) Mean monthly or daily air temperature.
(2) Mean monthly or daily precipitation.
(3) Conversion and computational tables to reduce the complication in the relationship between evaporation, temperature, latitude, and length of day.
(4) The depth of the root zone of the soil (and thus the water holding capacity, which may vary with soil type and vegetation).

Information on the last item is the most difficult to establish, as the water holding capacity of the soil depends on: (a) soil type and structure; (b) type of vegetation growing on the surface. For example, a sandy soil will hold only 1–2 cm of moisture per 30 cm depth of soil while a silt or clay may hold 10 cm of water in the same depth. Also, the amount of water in the root zone of a soil at field capacity can vary from a few milimeters on shallow sand to over 400 mm on a deep, well-aerated silt loam (Thornthwaite and Mather, 1955).

Thornthwaite's Bookkeeping Procedures

Sample monthly water balance computation for Seabrook, N.J. and Bismarck, N.D. are indicated in Table 1. The bookkeeping procedures will be briefly discussed line by line. Note that PE is potential evapotranspiration.

Line 1. $T°C$. Record the mean monthly air temperature on line 1.

Line 2. Heat Index I. Obtain the heat index i from tables and record on line 2. Summation of the twelve monthly values yields the index I. Note that i is zero when the mean monthly temperature is below 0°C, as the logarithm of a negative number is indeterminate.

Line 3. Unadjusted PE. Obtain the unadjusted daily PE from tables and record on line 3. Note that PE is zero below 0°C.

Line 4. PE. Multiply unadjusted PE by the appropriate month and day length correction factor (obtained from tables) for the station's latitude. Record this new adjusted monthly value of PE on line 4.

Line 5. Precipitation. Record the mean monthly precipitation on line 5.

Line 6. Precipitation (PE). Determine the difference between precipitation and PE and record on line 6. A negative value indicates a period of moisture deficiency, while a positive value indicates water available for soil moisture recharge and runoff. Note that Seabrook has an annual moisture excess of 352 mm whereas Bismarck has an annual moisture deficiency of 178 mm.

Line 7. Accumulated Potential Water Loss. The accumulated sum of the negative precipitation–PE values are recorded on line 7. Since the annual value of precipitation–PE is positive for Seabrook, the value of accumulated potential water loss with which one starts accumulating the negative values of line 6 is zero. In the case of Bismarck, the annual value of precipitation–PE is negative. Therefore, it is necessary to refer to the tables and by a series of successive approximations determine the accumulated potential water loss.

Line 8. ST: Storage. Values of soil moisture storage are recorded on line 8 after they are

TABLE 1. SAMPLE MONTHLY BALANCE COMPUTATIONS[a]

Seabrook, New Jersey
(All values except T and I in mm. Water holding capacity in root zone of soil is 300 mm)

	J	F	M	A	M	J	J	A	S	O	N	D	Y
1. T°C[a]	0.9	1.2	5.9	11.3	17.5	22.3	24.7	23.7	20.2	14.0	7.6	2.3	
2. I	0.07	0.12	1.29	3.44	6.66	9.62	11.23	10.55	8.28	4.75	1.89	0.31	58.21
3. Unadjusted PE	0.1	0.1	0.6	1.3	2.5	3.5	4.1	3.9	3.1	1.8	0.8	0.1	
4. PE	3	2	19	43	93	131	156	138	97	52	20	2	756
5. P	87	93	102	88	92	91	112	113	82	85	70	93	1108
6. P-PE	84	91	83	45	−1	−40	−44	−25	−15	33	50	91	352
7. Accumulated Potential WL					−1	−41	−85	−110	−125				
8. ST	300	300	300	300	299	261	225	207	197	230	280	300	
9. ΔST	0	0	0	0	−1	−38	−36	−18	−10	+33	+50	+20	
10. AE	3	2	19	43	93	129	148	131	92	52	20	2	734
11. D	0	0	0	0	0	2	8	7	5	0	0	0	22
12. S	84	91	83	45	0	0	0	0	0	0	0	71	374
13. RO	59	76	79	62	31	15	8	4	2	1	1	36	374
14. SMRO	0	0	0	0	0	0	0	0	0	0	0	0	
15. Tot. RO	59	76	79	62	31	15	8	4	2	1	1	36	374
16. DT	360	375	379	362	330	277	233	211	199	231	280	335	

(Snow 0)

Bismarck, North Dakota
(All values except T and I in mm. Water holding capacity in root zone of soil is 200 mm)

	J	F	M	A	M	J	J	A	S	O	N	D	Y
1. T°C[a]	−13.4	−12.1	−4.3	5.6	12.5	17.6	21.0	19.6	14.5	7.2	−1.9	−9.6	
2. I	0	0	0	1.19	4.00	6.72	8.78	7.91	5.01	1.74	0	0	35.35
3. Unadjusted PE	0	0	0	0.9	2.0	2.9	3.5	3.3	2.4	1.1	0	0	
4. PE	0	0	0	31	78	115	140	121	76	31	0	0	592
5. P	11	11	23	39	59	85	57	46	31	24	14	14	414
6. P-PE	11	11	23	8	−19	−30	−83	−75	−45	−7	14	14	−178
7. Accumulated Potential WL				(−116)	−135	−165	−248	−323	−368	−375			
8. ST	69	80	103	111	101	87	57	39	31	30	44	58	
9. ΔST	11	11	23	8	−10	−14	−30	−18	−8	−1	14	14	
10. AE	0	0	0	31	69	99	87	64	39	25	0	0	414
11. D	0	0	0	0	9	16	53	57	37	6	0	0	178
12. S	0	0	0	0	0	0	0	0	0	0	0	0	
13. RO	0	0	0	0	0	0	0	0	0	0	0	0	
14. SMRO	0	0	0	0	0	0	0	0	0	0	0	0	
15. Tot. RO	0	0	0	0	0	0	0	0	0	0	0	0	
16. DT	69	80	103	111	101	87	57	39	31	30	44	58	

(Snow 73 mm)

[a] Abbreviations: T, mean air temperature; I, heat index; Unadj. PE unadjusted potential evapotranspiration; PE, potential evapotranspiration; P, precipitation; P-PE, precipitation minus the potential evapotranspiration; Acc. Pot. WL, accumulated potential water loss (accumulated sum of the negative P-PE values); ST, storage ΔST, change in soil moisture; AE, actual evapotranspiration; D, moisture deficit; S, moisture surplus; RO, water runoff; SMRO, snow melt runoff; Tot. RO, total runoff; DT, total moisture detention.

determined by reference to tables and lines 1, 6 and 7.

Line 9. ΔST: Change in Soil Moisture. Determine the difference in soil moisture storage from one month to the next and record on line 9.

Line 10. AE: Actual Evapotranspiration. When the precipitation > PE, the soil remains at field capacity and AE approximates PE. When precipitation < PE, the soil begins to dry out and AE < PE. In those months AE = precipitation + | ΔST |. AE is recorded on line 10.

Line 11. D: Moisture Deficit. D = PE − AE and is recorded on line 11.

Line 12. S: Moisture Surplus. Any excess precipitation after the soil reaches field capacity is recorded on line 12 as moisture surplus which can become runoff.

Line 13. RO: Water Runoff. For large catchment areas, only about 50% of the surplus water available for runoff in any month actually does run off. The rest is detained the following month when another 50% of the surplus water will run off.

Line 14. SMRO: Snow Melt Runoff. Since ST < water holding capacity in the case of Bismarck, that is no SMRO as it is assumed it will go into soil storage.

Line 15. Tot. RO: Total Runoff. Tot. RO = Σ RO + Σ SMRO.

Line 16. DT: Total Moisture Detention. The moisture detention represents the total of the water stored in the soil, the snow remaining on the soil surface and the surplus water which has been detained for a month and is in the process of running off.

Note A: A daily water balance can also be determined for a station by computing procedures similar to the one already discussed for the monthly water balance.

Note B: Seabrook, N.J. and Bismarck, N.D. represent stations with one "wet" and one "dry" season. A monthly water balance can also be computed for stations with multiple wet and dry seasons by more detailed procedures described in Thornthwaite and Mather (1957).

The Water Balance of the Earth

Although precipitation equals evaporation on a worldwide basis, the amounts vary considerably from region to region (Gentilli, 1958).

Table 2 indicates that the area poleward of 40°N has precipitation exceeding evaporation by 25,900 km³/yr. Most of this precipitation is caused by the polar front between 40 and 60°N. In contrast, the area between 10 and 40°N has evaporation exceeding precipitation by 42,300 km³. The belt between the equator and 10°N gains 19,300 km³ of water, the greatest amount of any 10° latitudinal belt on earth.

Thus, the middle and middle-low latitudes of the northern hemisphere (10–40°N) lose a tremendous amount of water to the higher latitudes and to the intertropical front. The greater land masses in the northern hemisphere cause the intertropical front to be generally located north of the equator. Consequently, the northern hemisphere obtains approximately 300 km³ of water from the southern hemisphere.

TABLE 2. WATER BALANCE OF THE EARTH[a]
Precipitation minus Evaporation

Latitude	Oceans Precip.-Evap. (cm)	Oceans Precip.-Evap. (km³ × 10⁻³)	Continents Precip.-Evap. (cm)	Continents Precip.-Evap. (km³ × 10⁻³)	World Precip.-Evap. (cm)	World Precip.-Evap. (km³ × 10⁻¹)
90–80°N	(+10)	(+0.3)	(+29)	(+0.1)	(+12)	(+0.4)
80–70°N	(+20)	(+1.7)	(+17)	(+0.6)	(+20)	(+2.3)
70–60°N	+36	+2.0	(+23)	(+3.1)	(+27)	(+5.0)
60–50°N	+56	+6.0	+14	+2.1	+31	+8.1
50–40°N	+47	+7.1	+18	+2.9	+32	+10.1
40–30°N	−45	−9.3	+14	+2.2	−20	−7.1
30–20°N	−93	−23.4	+29	+4.3	−48	−19.0
20–10°N	−58	−18.1	+16	+1.8	−38	−16.2
10–0°	40	+13.5	+57	+5.8	+44	+19.3
0–10°S	−19	−6.2	+59	+6.1	0	−0.2
10–20°S	−54	−17.9	+20	+1.8	−37	−16.0
20–30°S	−61	−18.7	+23	+2.2	−42	−16.5
30–40°S	−1	−0.2	+6	+0.2	0	+0.1
40–50°S	+34	+10.3	+37	+0.4	+34	+10.7
50–60°S	+47	+11.9	+82	+0.2	+47	+12.0
60–70°S	(+20)	(+3.5)	(+20)	(+0.1)	(+19)	(+3.6)
70–80°S	(+10)	(+0.3)	(+25)	(+2.2)	(+19)	(+2.5)
80–90°S	(0)	(0.0)	(+25)	(+1.0)	(+25)	(+1.0)
World	−10.0	−37.1	+24.9	+37.1	0	0.0

[a] After Wüst (1922)—less reliable estimates in brackets.

WATER BALANCE

As shown in Table 3, the area between the equator and 30°S loses 32,700 km³ of water per year, 30,000 of which go to the higher latitudes with the westerlies, the other 3000 crossing the equator northward.

The prevailing westerlies of both hemispheres transport the net vapor equivalent of nearly 56,000 km³ of water toward the polar fronts. A tremendous amount of heat is needed to evaporate this water into water vapor. When the water vapor condenses and returns to the earth in liquid form as precipitation, a tremendous amount of heat is liberated. Thus, water in the air plays a large role in the heat transfer of the earth (see Vol. II, *Energy Budget of Earth's Surface*).

It is more difficult to determine the water balance of the tropical and equatorial regions from Table 3. Vertical air movements predominate over horizontal movements in low latitudes. Therefore, enormous quantities of water vapor which are absorbed by the tropical easterlies over the oceans return to earth within a narrow latitudinal belt along the intertropical front, and also where a coastline intercepts the easterlies.

Oceanic evaporation is much more important than evaporation from continents in terms of the contribution to the total moisture balance of the atmosphere. Precipitation exceeds evaporation over the continents by 37,100 km³ of water a year. This amount is then returned to the oceans as runoff. Note also the considerable transfer of heat from the oceans to the continents by the transfer of vapor which condenses later over the land.

ROBERT M. HORDON

References

Gentilli, J., 1958, "A Geography of Climate," Second ed. revised, Perth, Australia, University of Western Australia Press.

Prescott, J. A., 1958, "Climatic indices in relation to the water balance," *Climatology and Microclimatology* (Canberra Symposium), Paris, UNESCO, 48–51.

Thornthwaite, C. W., "An approach toward a rational classification of climate," *Geograph. Rev.*, **38**, 55–94.

Thornthwaite, C. W., and Mather, J. R., 1955, "The water balance," *Johns Hopkins Univ., Laboratory in Climatology, Publ. in Climat.*, **8**, No. 1.

Thornthwaite, C. W., and Mather, J. R., 1957, "Instructions and tables for computing potential evapotranspiration and the water balance," *Johns Hopkins Univ., Laboratory in Climatology, Publ. in Climat.*, **10**, No. 3

Wust, G., 1922, *Z. Ges. Erdk., Berlin*, 35.

Cross-references: *Hydrologic Cycle; Hydrology; Hydrology, Semiarid Regions; International Hydrological Decade; Runoff.* Vol. II: *Energy Budget of the Earth's Surface; Evapotranspiration; Hydroclimate; Precipitation.*

WATER CYCLE—See HYDROLOGIC CYCLE

WATER DIVINING (DOWSING)

Introduction and Nomenclature

The word "divining" is derived from the verb "to divine," and was used originally to describe the capacities of some human subjects who claimed to be able to predict certain processes or phenomena in the nonperceptible world. This capacity was considered by the believers to be a heavenly (i.e., divine) gift. However, in recent years divining phenomena have been defined as a group of complex physical and physiological phenomena associated with living organisms and unconsciously perceptible by many subjects (Tromp, 1949). The transformation of the unconscious stimuli into conscious motor contractions of the forearm muscles (in the case of dowsing) requires a physiological amplifier in certain parts of the human brain. This may explain the observation that these unconsciously recorded stimuli are experienced consciously only in some subjects and not in others.

Divining phenomena comprise all those parapsychological phenomena known as *water-divining, dowsing,* or *muscle tone reflex* (Tromp, 1956). According to the claims of the water diviner or dowser, he is able to locate underground water, ore deposits, etc., without the use of geophysical instruments. He experiences certain bodily changes in muscular tone (usually in his arms) if he walks over such a location. He uses a simple object, a forked twig, wire, etc., to indicate clearly this change in muscular tone that he experiences (see Fig. 1a–f).

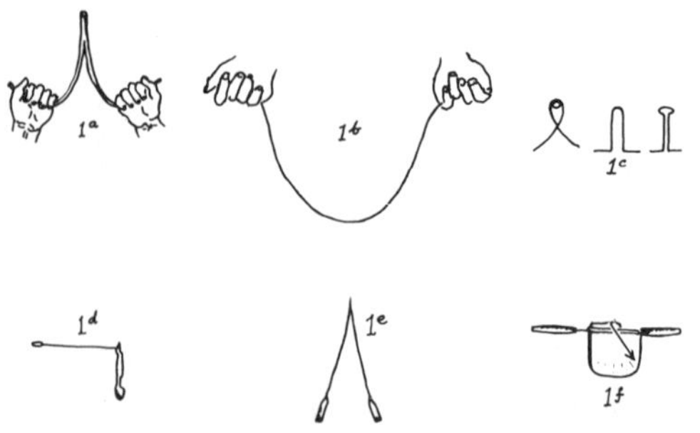

FIG. 1. Different types of dowsing rods.

On the basis of the objects used in water divining the subject is classified into two major fields:

1. *Rhabdomancy,* using different types of rods. The name is derived from the Greek words *rhabdos* = rod and *manteia* = divination.

2. *Radiesthesia* or *Pallomancy,* using different types of pendulums. The first name indicates the sensitivity to radiations: pallomancy derives from the Greek words *pallo* = to shake and *manteia* = divination.

Historical Material

The origin of the dowsing rod is not known although there is some evidence that it has been in use for over 7000 years. Several passages in the Bible suggest the use of divining rods in olden times, e.g., the "smiting of the rock" by Moses. Herodotus mentioned the use of the rod by Scythians, Persians, and Medes. It was used for searching for water, lost objects, or precious metals, for forecasting events, and for other occult practices. The old Chinese also used the divining rod, for example, to search for a safe locality for their houses. The first published description was found in the book *"De Re Metallica"* (1556) by Georgius Agricola. Other historical facts can be found in *"Psychical Physics"* (Tromp, 1949) which contains an extensive list of references.

The use of the pendulum and its practical application started probably only in the beginning of the 19th century. This first "magic" pendulum consisted of a finger ring or piece of metal and was used mainly for locating wells. A special group of these pallomantic phenomena is the basis for the study of so-called "human radiations" and was used chiefly by Catholic priests in France as a means of medical diagnosis. It was only in 1930 that the word "radiesthesia" was created by the Abbot Bouly. In 1939 the "Medical Society for the Study of Radiesthesia" was founded in England, followed by the creation of similar societies in France and Belgium. At present two branches of radiesthesia are known:

1. *Teleradiesthesia* or *Intuitive* or *Psychical Radiesthesia.* The persons using this method claim that with a pendulum above a map or photograph they can indicate the location of water, ore deposits, and certain diseases, hundreds of miles away. However, in no critical tests has it been possible to confirm these claims.

2. *Physical Radiesthesia.* In this case the dowser operates in the field in direct relation to a geophysical discontinuity in the soil. The rotation or swinging of the pendulum is explained by physical and physiological laws.

Types of Divining Rods

The dowser, walking over a piece of land with a forked twig in his hands (see Fig. 1a), experiences at certain places a contraction in his forearm muscles which forces the forked twig to turn either upward or downward. Alternatively, a pendulum attached to a wire between two fingers deviates from its original direction of swing or may even start to rotate. This physiological phenomenon was given the name *muscle tone reflex,* MTR (Tromp, 1956).

Since the time of the first divining rods described by Agricola (1556) and Valentine (1651) a great number of different rods have been used (see Fig. 1) which consist either of nonconducting materials (whale baleen, oxhorn, ivory, wood, etc.) or of conducting material (copper wire, steel springs, etc.). The oldest and most commonly used material is wood and as a rule twigs are used either from the peach, willow, hazel, or witchhazel. The shape of the divining rod is usually *forked* or *loopshaped,* in the latter case usually a wire or spring being used. The loop-shaped rod does not turn upward and downward but from the left to right or vice versa.

A pendulum may be attached to a 10–20 cm long chain, the chain being held usually between the thumb and forefinger 30–40 cm in front of the body. The dowser walks with the chain and pendulum, either kept nonswinging or swung in a plane perpendicular to the body at the beginning of the experiment. In both cases, at certain places, the dowser experiences a deviation of the pendulum from the perpendicular plane, or the pendulum may rotate.

Most trained dowsers agree that neither the shape of the dowsing rod nor the material is of fundamental importance. The same is true for the material and weight of the pendulum and the material and length of the wire or chain holding the pendulum. However, each of these factors affects the speed and intensity of a dowsing reaction as anyone can try out for himself by turning the rod or swinging the pendulum purposely. Material of low elasticity will cause the rod to be too stiff. The pendulum will react considerably faster than any dowsing rod and a short chain with heavy pendulum will deviate more easily than one with a long chain or wire. In other words, in any controlled scientific dowsing experiment for psychological reasons the dowser should be free as much as possible to follow his own method. A comparison of the results of different people using different methods, or even of the same person changing his method from time to time, is extremely difficult and a perfect coincidence of the dowsing spots or the reacting zones is excluded. A perfect coincidence between muscle tone reflex (MTR) zones determined by different dowsers would be comparable with the illusion of a physicist who wants to measure the same electric current with a number of ampere meters of different sensitivity and expects the same scale reading on all his meters at the same moment.

Experiments on the Physiological Mechanisms of Dowsing

Research work in the field of dowsing (Tromp, 1956) has shown that in order to reproduce the observations of dowsers a number of precautionary measures have to be taken, such as similarity in speed of movement, direction of movement, etc., of the dowsers. As in practically all tests carried out previously no such precautionary measures have been taken, it is not surprising that in only a few instances were statistically significant reproducible results obtained.

A second major cause of misunderstanding is that a dowser is always tested on his claims and not on his actual physiological performance. In other words a dowser may claim, on the basis of the turning of his dowsing rod, that at this particular site water, gas, or ore deposits are present. If this claim is tested the result is usually negative. However, if the dowser is tested on his physiological performance (MTS) only, one should test whether below the locality where the dowser experiences a contraction of the muscles of his forearm there is indeed some kind of geological discontinuity (e.g., change in subsurface rock composition, change in groundwater, etc., referred to here as "MTR Zones"). As practically every shallow subsurface geological discontinuity can be established by an ordinary resistivity soil survey, the best method for testing dowsing phenomena is to bring the dowser to a new, preferably flat, area unknown both to the dowser and observers. The surface soil should not show any obvious changes in lithology. The vegetation should be uniform. In other words, there should be no clues concerning subsurface conditions. The dowser is made to walk slowly along a marked distance and each time the rod turns or the pendulum deviates the exact point is recorded. Along exactly the same line an electrogeophysical profile is measured at several depths (1 or 2m, etc.). The various dowsing zones, previously established by the dowser, should coincide with the geological discontinuities. If such a coincidence is found repeatedly and the number of statistically significant guesses exceeds the number of chance guesses, there can be little doubt that the dowsing phenomenon is some sort of a true physiological phenomenon and not due only to autosuggestion (as often assumed by many scientists), despite the fact that the physiological mechanisms involved are still relatively unknown. One is inclined to assume that the subsurface discontinuity, through certain physical properties of the rock or soil, either directly or indirectly, affects the autonomic nervous system of the human body which, in turn, stimulates certain parts of the brain, including the cortex, which is responsible for the change in muscle tone of the forearm that has been holding a dowsing rod in muscular balance.

Tests to Demonstrate the MTR Reaction

A number of tests have been devised to demonstrate the MTR reaction. They can be classified into three main groups: magnetic, electrocardiograph, and soil resistivity experiments.

Magnetic Experiments. Magnetic experiments include the following: (a) using artificial magnetic fields (created by a ringshaped tangent galvanometer); (b) using local disturbances in the earth's magnetic field (indicated by an ordinary compass or magnetometer). The experiments were carried out both in buildings and fields.

Electrocardiograph Experiments. These experiments include: (a) measuring electrical variations in skin potentials of dowsers walking over MTR zones; or alternatively with the forearms attached to a string galvanometer; (b) indicating the existence of MTR zones over electromagnetic low-frequency fields in modern buildings (as a result of buried conductors); in this case, forearms of dowsers were connected with a cathode-ray-type electrocardiograph.

(a) Experiments Using Artificial Magnetic Fields. An artificial constant magnetic field was developed by a tangent galvanometer consisting of a ring (one meter in diameter) and a single electric coil; an electric current of 10 amp developed a field strength of 0.125 Gauss in the center and 0.001 Gauss at a distance of 230 cm from the ring. This field could be sensed by a blindfolded dowser holding a loop-shaped rod, whenever sudden variations in the field strength were set up. These variations were obtained either by movement of the dowser from a point outside the field into the field or by a sudden change of the angle of the ring of the tangent galvanometer (the dowser standing still in front of the ring), by switching the current on and off or by changing the direction of the current. In all those experiments neither the dowser nor the person who registered the dowser's reactions knew whether the current was on or off. The contacts

WATER DIVINING (DOWSING)

were completely noiseless and the handling of the switch by a third person could not be seen although all three persons were in the same room. As an extra precaution thick cotton-wool was placed in the ears of the dowser.

It was found that several dowsers (not all of them) were able to register these changes in the magnetic field. Apparently they were not able to register the field strength as such, only the changes therein. A difference between a field created by a 5- or 10-amp current could not be distinguished; only a sudden change from one field to another. No reaction was found for a field strength of less than 1 amp. The experiments showed that magnetic gradients of less than 0.001 Gauss/cm could be registered by sensitive dowsers.

(b) Experiments using Local Disturbances in the Earth's Magnetic Field. Two types of experiments were carried out, in buildings and in the field.

Experiments in buildings: In most buildings, particularly modern ones, often narrow MTR zones are located by dowsers which seem to be related to water pipes, central heating pipes, reinforced concrete beams, etc., below the floor. It was found that considerable deviations of the magnetic needle of an ordinary geological compass could be observed above those places indicated by dowsers (see Fig. 2). Deviations of up to 30° are not uncommon. As these dowsing zones were established by dowsers before any magnetic measurement was done, any possibility of autosuggestion can be excluded. The deviation of the needle is maximal at floor level; in the case of a 30° deviation caused by a steel beam the deviation was only 15° at a height of 30cm. At the level of the arms of a person holding a dowsing rod, the needle no longer deviates. A great number of such experiments have been carried out in which the dowser, walking along a straight line, first indicated the dowsing zones. A magnetic survey was made afterward along the same line with an ordinary compass and excellent results were obtained.

Experiments in the field: During geophysical work in mountainous areas it was noticed that a relationship seems to exist between dowsing zones and zones of strong local disturbances of the earth's magnetic field. The first systematic studies were carried out by Wüst (1936) and Petschke in Germany, measured with magnetometers.

In Holland in flat open country with horizontal sedimentary formations the variations in the magnetic field are negligible (only 1 or 2 gammas). The dowsing observations show that many MTR zones occur, which suggests that the effect may be due to some cause other than the geomagnetic field and that the correlation with magnetic phenomena observed in mountainous areas is only a secondary feature.

(c) Electrocardiographic experiments with a string galvanometer. Experiments were carried out with a physiological millivolt meter, to register the changes in electric skin potentials of a moving dowser. The electrocardiograph,

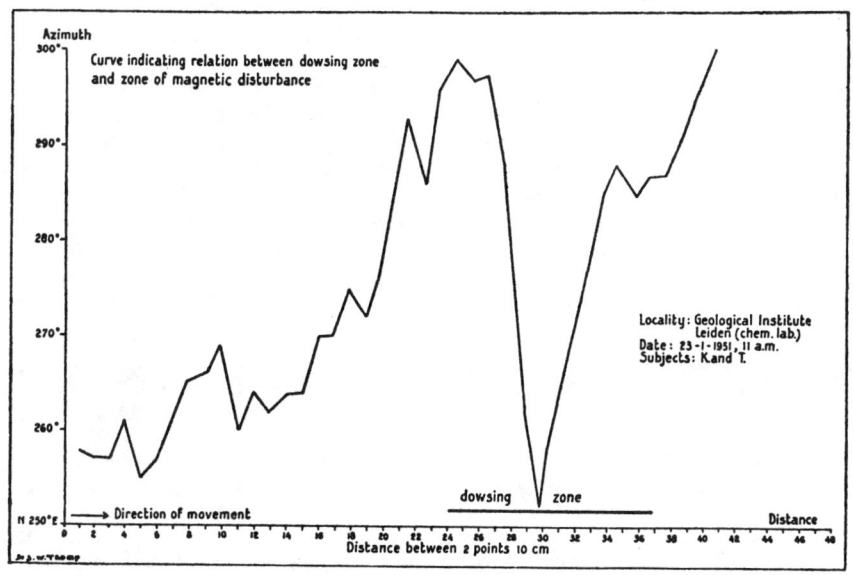

Fig. 2. Changes in azimuth of the compass needle, along a straight line of about five meters, in relation to a predetermined dowsing zone, which coincides with the drastic change in azimuth created by a pipe below the floor of the room.

1255

WATER DIVINING (DOWSING)

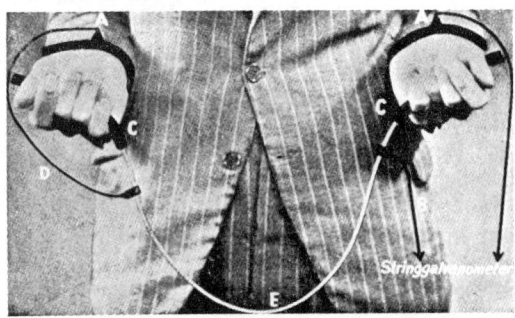

FIG. 3. Arrangement of a loop-shaped dowsing rod connected to a string galvanometer (after Einthoven): A, metal grips placed on pulse; B, electrodes leading to string galvanometer; C, insulated grips; D, copper wire connecting metal rod with metal grip on pulse; E, steel divining rod.

invented by the Nobel prizewinner, Prof. Einthoven in Leiden, proved to be extremely useful inasmuch as it is not easily disturbed by external electromagnetic fields. It was found that a loop-shaped elastic metal rod does not turn in the hands of a dowser if the rod endings are placed in two insulated grips in which they can turn freely (Fig. 3). A current is made to flow through the body and the rod which makes it possible to register fluctuations of skin potentials if the dowser moves through a MTR zone without turning the rod. Experiments with a moving dowser outside a dowsing zone showed that the new circuit does not change the electrocardiograms and that neither standing nor walking changes the level of the peaks or the frequency in the electrocardiograms. The skin potentials and electrocardiograms were next studied with the dowser moving through a strong MTR zone. The electrocardiograms show considerable departures of the general curve (not in the heart frequency) as soon as the dowser enters the MTR zone (Fig. 4) and becomes normal again after the dowser leaves the zone (Tromp, 1949).

Similar electrocardiographic phenomena in such cases were observed with people who are not sensitive to dowsing, but here the departure of the curve was less pronounced. In all, over 500 experiments were carried out. A careful study of the various electrocardiograms showed that the departures of the cardiographic curve were not due to a psychogalvanic reflex.

In another experiment it was found that if a dowser moves his hands above a reclining person interesting diagrams are obtained that were different from those in the MTR zones. The Q-peak level (without changes in frequency) jumped upward or downward depending on whether the dowser's forearms were above the upper or lower part of the body. The central part of the body was neutral and the Q-level was the same at a distance from the person. In other words, the human body showed a polarity in the cardiograms which was also experienced by the dowser.

Similar electrocardiographic experiments in the field were carried out later in the United States. On September 8, 1948, James L. Jenks, of the Sandborn Electrocardiograph Co. (Cambridge, Mass.) tested a well-known dowser, Henry Gross, with a cathode-ray-type electrocardiograph. If the rod was turned purposely by Mr. Gross no large departures of the electrocardiogram were observed. However, as soon as Mr. Gross crossed a MTR zone very large fluctuations were noticed indicating large electric potential fluctuations on the skin. On April 25, 1952, four electronic engineers in New York, H. Gallay, H. Cohen, A. Goldschmidt, and J. Levin, using a specially constructed very sensitive microvoltmeter, observed the same phenomena when they tested Mr. Gross in a similar way.

Experiments with an Elmquist electrocardiograph. It is well-known to most cardiologists that at certain spots in modern buildings no electrocardiogram can be taken because of electrical disturbances, known as the "humming effect." However, the actual causes of this phenomenon were not studied. Studies by Tromp, in the Netherlands, indicated that the "humming effect" was caused by the presence

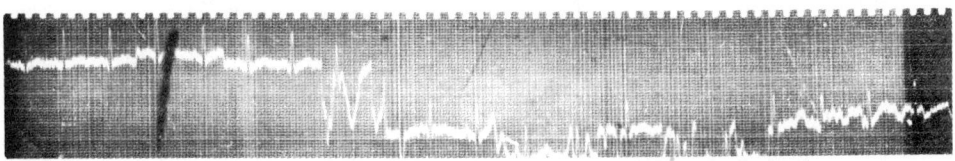

FIG. 4. Dowser walks slowly (electrode circuit as in Fig. 3) from a position outside the dowsing zone (left part of the cardiogram) into the zone of disturbance, perpendicular to the longest axis of the zone (right part of the cardiogram). Immediately upon entering the dowsing zone, the whole curve shows a considerable excursion downwards. It becomes normal again after the dowser leaves the zone. The subsidence of the curve took place immediately and not after one or more seconds, as we may observe in the case of psychogalvanic reflexes.

WATER DIVINING (DOWSING)

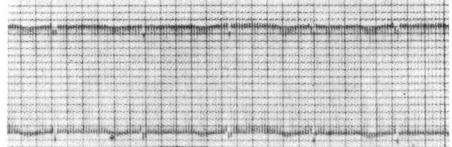

Fig. 5a. Subject stands in the center of the dowsing zone. Lack of the PQRST curve is due to the humming effect.

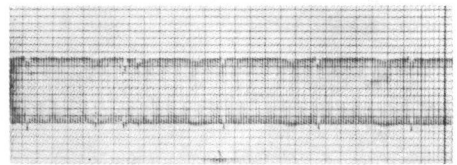

Fig. 5b. Subject stands near the edge of the dowsing zone.

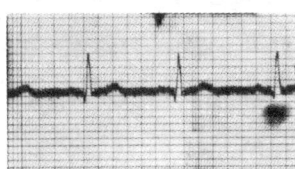

Fig. 5c. Normal cardiogram (electrodes attached to both pulses) of a dowser, recorded with an Elmquist cathode ray type of electrocardiograph, standing outside a dowsing zone caused by a metal pipe below the floor in a house.

of reinforced concrete beams or pipes beneath the floors which seemed to concentrate the electromagnetic fields (created by the lighting circuits) into narrow zones (Fig. 5). These zones were also characterized by strong local disturbances of the earth's magnetic field which showed up in sudden changes in azimuth, up to 50°, of the magnetic needle of an ordinary geologic compass. It is interesting that these zones of electromagnetic and magnetic disturbances could be located by dowsers before the instrumental measurements were taken, which indicated that autosuggestion was excluded in this experiment.

Soil Resistivity Experiments. In geophysics various electric methods are used to determine the subsurface soil resistivity. A modified version of the well-known Shepard Resistivity meter was designed, using only two electrodes, through which a 600-Hz alternating current is passed. With a number of resistances in the instrument it is possible to neutralize this current and to establish by the zero-method the soil resistivity (or its inverse value, the soil conductivity). It was found that a geological discontinuity, for example an undulating peat layer under a surface clay or a shallow buried valley, will show up in the soil resistivity.

In order to study the possible relationship between soil resistivity and MTR zones the following procedure was followed: first the dowser walked along a 20-m tape over a piece of uniform land. In view of its greater sensitivity the pendulum method was used in most of the experiments. The dowser walked slowly along the tape and each deviation was recorded, including the direction and rate of deviation, and weather conditions.

Around 1950 over 200 field sections were carefully studied and both MTR zones and changes in soil resistivity, along the same profile, were recorded. In all these experiments first the MTR zones were established by a dowser and then the soil resistivity was measured in order to exclude any possible suggestive factor. In most sections an excellent, or at least statistically significant, correlation was found between MTR zones and zones of low soil resistivity (Tromp, 1956). Deviations proved to be due to experimental errors such as walking too quickly with the rod, or the lapse of too great a time interval between the dowsing experiment and the subsequent geophysical survey. For example, a rainfall changes both the dowsing and soil resistivity patterns. The MTR zones were found to correspond mainly to buried stream channels (with sands), undulating peat deposits, under surface clay, subsurface faults near the surface, etc.

The experiments described strongly suggest that the MTR phenomenon is a true physiological phenomenon not caused by autosuggestion. The mechanisms involved are not yet known although some evidence has been found that infrared thermal radiation causing minor disturbances of the hypothalamic thermoregulation mechanism in the dowser may be involved. Further reseearch is required to give a definite answer to this physiological aspect of dowsing.

SOLCO W. TROMP

Editorial Note: The question of water divining (dowsing) has been one of the most controversial problems in water search. Undoubtedly many of the self-styled water diviners are "confidence men" and charlatans, or at least totally naive persons. They often claim to be able to locate deep-seated artesian water by the use of a divining rod (or through map dowsing), or to locate missing documents in a building. Critical tests in the past have shown beyond doubt that these claims cannot be confirmed. On the other hand, many dowsers with long local experience may perform a useful service by finding shallow water; in such cases it is not easy to judge to what extent their success comes from a certain unknown physiological sensitivity or simply from good powers of observation. In view of the studies of Dr. Tromp (a professional geologist, well-known for his studies

WATER DIVINING (DOWSING)

in structural geology), who spent more than twenty-five years testing dowsers with orthodox geophysical methods, the editor feels that this review paper should be included in this volume.

R.W.F.

References

Barton, D. C., 1926, "The wigglestick," *Bull. Am. Assoc. Petrol. Geol.*, **10**, 312–313.

Jenny, E., 1947, "Experimental-biologischen Untersuchungen zum Erdstrahlenproblem," *Gesund. Wohl.*, **1**, 1–40.

Jenny, E., Oehler, A., and Stauffer, H., 1935/-1936, "Experimentelle Untersuchungen über biologische Wirkungen der sogenannten Erdstrahlen," *Schweiz Med. Wochenschr.*, **39**, 947; **24**, 572.

Maby, J. C., and Franklin, T. B., 1939, "The Physics of the Divining Rod." London, Bell Co.

Prokop, O., 1955, "Wünschelrute Erdstrahlen and Wissenschaft," Stuttgart, Ferd. Enke Verlag, 193pp.

Trinder, W. H., 1939, "Dowsing," London, Camelot Press 137pp.

Tromp, S. W., 1949, "Psychical Physics," Amsterdam, Elsevier Publ. Co., 534pp.

Tromp, S. W., 1954, "Experimente auf Zonen Konzentrierter niederfrequenter Wechselfelder in Häusern und ihre möglichen biologischen Wirkungen," *Hippokrates*, **7**, 199–206.

Tromp, S. W., 1956, "Experiments on the Possible Relationship between Soil Resistivity and Dowsing Zones," Oegstgeest, Leiden (Foundation for Study of Psycho-Physics), 35pp.

Wilhelmi, G., 1948, "Weitere experimentell-biologische Untersuchungen zum Erdstrahlenproblem," *Gesund. und Wohl.*, **2**.

Wüst, J., 1936, "Weitere Versuche zur Klärung der physikalischen Seite des Wüschelruten problem," *Zeit. f. Wünschel.*, **17**, 2.

Cross-references: *Geophysical Methods for Hydrologic Search; Groundwater; Hydrology.*

WATER GAS—See HYDROGEN

WATER MOLECULE—STRUCTURE

Composition of Water

Water is a compound of the elements hydrogen and oxygen. This may be shown by decomposing a sample of water by electrolysis and identifying the constituent gases. When this is done it is always found that 11% of the weight of the water is hydrogen and 89% is oxygen. The same percentages are obtained if part of the water is frozen, or if some of the water is evaporated by boiling, and only the liquid water remaining is analyzed. In other words, water, ice, and water vapor all have the same composition, and a molecule of water must consist of a combination of hydrogen and oxygen atoms.

In the synthesis of water vapor from hydrogen and oxygen, it is found that if the volumes of the gases are measured at the same temperature and pressure, two liters of hydrogen combine with one liter of oxygen to form two liters of water vapor. From Avogadro's hypothesis this implies that two molecules of hydrogen and one molecule of oxygen combine to form two molecules of water vapor, that is

$$2H_2 + O_2 \rightarrow 2H_2O$$

If the atomic weight of oxygen is taken as 16, that of hydrogen is 1.0080, and the molecular weight of water is 18.0160. The mass of the water molecule is obtained by dividing its molecular weight by Avogadro's number (6.002×10^{23}), which gives 29.92×10^{-24} gram.

Geometry of the Water Molecule

The geometrical configuration of the water molecule in the vapor phase is shown in Fig. 1. The distance from the oxygen atom to a hydrogen atom is 0.958 Å, and the bond angle is 104° 31'. This arrangement of the atoms gives the water molecule a permanent dipole moment of 1.831×10^{-18} esu. It is clear also that the configuration of the three centers of masses is such that the water molecule possesses three principal moments of inertia. The water molecule is therefore an "asymmetric top" which may absorb or emit radiation when it rotates in an applied electric field. In addition to three rotational degrees of freedom, the atoms of the water molecule may vibrate in three normal modes about their equilibrium positions (Fig. 2). The rotational and vibrational

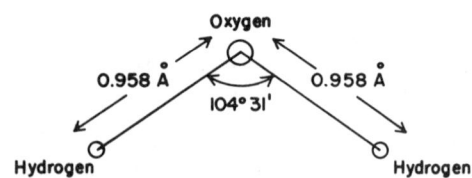

FIG. 1. Geometrical configuration of the water molecule in the vapor phase.

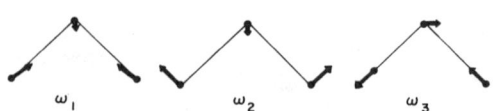

FIG. 2. Normal modes of vibration of the water molecule (from Wilson et al., 1955). To represent actual relative motions in space, the arrows representing the displacements of the oxygen atom should be only one-fourth as long as shown here.

modes of oscillation of the water molecule produce the complex infrared spectra of water vapor, liquid water, and ice. In ice the bond angle is about 109° 28′ and the bond length is 1.01 Å; these modifications, together with the polarization of the molecule by the electric fields of adjacent molecules, produce an average molecular dipole moment of 2.60×10^{-18} esu.

Bonding of the Water Molecule

The hydrogen atoms are bound to the oxygen atom in a water molecule by so-called *valence* forces which are fairly strong. Thus the total energy necessary to dissociate a water molecule into its constituent atoms is 218.9 kcal/mole. The nature of the bonding can be understood in simple terms as follows. The hydrogen atom consists of a central positive nucleus and a single orbital electron which, in its lowest energy state, is in the first shell ($1s$) characterized by a principal quantum of 1 and an *s*-type atomic orbital. Surrounding the nucleus of the oxygen atom, on the other hand, are eight electrons; two of them fill the first shell $(1s)^2$, two are in the s orbital of the second shell $(2s)^2$, and the remaining four are in the p orbital of the second shell $(2p)^4$. The maximum number of electrons that the first two shells can accommodate are two in the first shell, and, in the second shell, two s electrons and six p electrons. Now, if each hydrogen atom shared an electron-pair with the p orbital of the oxygen atom, this would result in the first shell of each hydrogen atom being filled with two electrons and the second shell of the oxygen atom being occupied by the maximum number of eight electrons. The atoms may then be considered to be held together by virtue of the sharing of the electron-pairs.

Although the simple model described above is conceptually helpful in understanding the nature of the bonding in the water molecule, closer analysis reveals that it cannot be entirely correct. For example, if the water molecule were held together by pure *p*-bonding, the bond angle would be about 95° rather than 104° 31′. In the modern valence theory this problem is overcome by considering that hybridized orbitals exist in which both the s and p orbitals of the oxygen atom enter into the bonding between the three atoms. The water molecule is then viewed as consisting of three nuclei surrounded by ten electrons, two of these electrons circle the $1s$ shell around the oxygen nucleus, and the remaining eight electrons are in pairs occupying four directed orbitals which partake of both s and p character. Two of these four orbitals are directed towards the two hydrogen nuclei and are known as

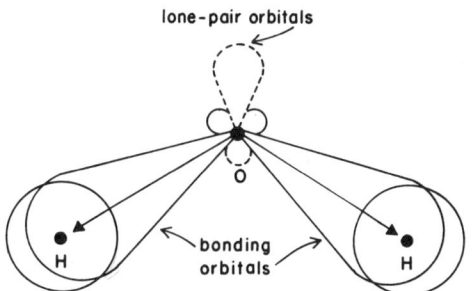

FIG. 3. Hybridization in H_2O (from Coulson, 1952).

the *bonding orbitals*, whereas the remaining two are on the other side of the oxygen atom from the hydrogen nuclei and are referred to as the *lone-pair orbitals* (Fig. 3). This arrangement of the electric charges along four directed orbitals plays an important role in the bonding together of water molecules in ice and liquid water.

PETER V. HOBBS

References

Coulson, C. A., 1952, "Valence," Oxford Univ. Press.
Debye, P., 1929, "Polar Molecules," New York, Chemical Catalog Co., Reinhold Publ. Corp.
Pauling, L., 1960, "The Nature of the Chemical Bond," Ithaca, N.Y., Cornell Univ. Press.
Wilson, E. B., Decius, J. C., and Cross, P. C., 1955, "Molecular Vibrations," New York, McGraw-Hill Book Co.

Cross-references: *Bonding; Hydrothermal Solutions; Molecular Weights; Water—Substance and Solvent.* Vol. II: *Avogadro's Law.* Vol. VI: *Ice: Hydrogen Bonding; Ice: Structure and Properties.*

WATER POLLUTION—See ENVIRONMENTAL POLLUTION

WATER POWER—See Vol. VI

WATER OF RETENTION—See CONNATE WATER

WATER SATURATION

This term refers to the degree to which a rock or soil is saturated with respect to water, i.e., if all of its interstices are filled with water, it is saturated. An unconfined column of saturated material which is both porous and permeable will drain in response to gravity; however, traces of water will remain as adherent droplets or as wetted surfaces. If the system is open and the atmosphere unsaturated, even this residue will disappear by evaporation. If the system is porous but relatively impermeable, as in the case of clay sediments, the water will remain relatively fixed in the system and only externally applied pressures, gradual drying from the outside (with related cracking and deformation, creating permeability), or thermal and chemical changes will allow removal of this *interstitial water* (q.v.). In the case of porous and permeable saturated materials, in which drainage is precluded, water removal may nontheless occur by surface evaporation and upward movement to the surface through capillary action.

Occasionally, as a relatively loose dry sand (both porous and permeable) becomes wet the phenomenon of *bulking* takes place, in which the total volume of the partially wetted material will become as much as 20–25% greater than in either the dry or saturated state. As the moisture content approaches saturation, this gradually disappeears, vanishing by 100% water content. The cause of this phenomenon is that moisture "hulls" or films form around each grain, adsorbed to the grain surface by surficial tension, causing the grains to occupy a greater volume than when in the dry statae. When totally saturated, this surface tension is removed and the extra volume occupied by the grain-plus-water unit is eliminated.

B. C. S.

References

Davis, S., and DeWiest, R., 1966, "Hydrogeology," New York, John Wiley & Sons, 463pp.

Feret, R., 1892, "Sur la compacité des mortiers hydrauliques," *Ann. Ponts et Chaussées,* Paris, IV, 5–164.

Jumikis, A., 1962, "Soil Mechanics," Princeton, N.J., Van Nostrand, 791pp.

Cross-references: *Groundwater; Interstitial Water in Sediments; Phreatic Water; Suspended Water.*

WATER SUPPLY: ECONOMICS

General History

The economics of water supply always has been a determining factor in water-resource planning. Most early efforts were on an ad hoc basis, and general theory is of recent origin. Earlier planning often was along the line of "What is the problem, how can we solve it, what will it cost, can we afford it?" During the 19th century, the Federal government began to enter into water-resources planning in the United States in both flood control and navigation (Ackerman and Löf, 1959, pp. 508–509, 530–531). With the Reclamation Act of 1902, the Federal government assumed a major role in irrigation problems of the West.

In 1927, the Corps of Engineers began its extensive series of "308 reports" (Ackerman and Löf, pp. 472–473) which were the first large-scale Federal entry into integrated river basin studies, covering power, irrigation, flood control, fishing, and navigation, and their integrated economic feasibility and impact. There were no generally accepted standards for economic analysis, however, so that the basin studies still were done on an ad hoc basis. Although not designed for use as a basis for construction projects, in some cases they were so used when the Flood Control Act of 1936 became law.

The large public works programs of the 1930s brought a realization that water resource planning involved the allocation of scarce resources (available water and money) among several competing claims. Therefore, some standardization of judging feasibility was seen as desirable. The requirement of the Flood Control Act of 1936 that benefits must exceed costs for projects to be authorized, led to the widespread use of the benefit-cost ratio as a criterion for project feasibility.

Benefit-Cost Ratio

Although the various Federal agencies are not entirely consistent in detailed application of benefit-cost analyses [Eckstein, 1958; U.S. (IACWR), 1958], they agree on certain general principles for judging feasibility and relative ranking of water resources projects. For an individual project, "the optimum scale of development is that at which the net benefits are at a maximum ... The ratio of benefits to costs ... is the recommended basis for comparison of projects" [U.S. (IACWR), 1958]. The reasoning involved is shown in Fig. 1. For a given project, point 1, at which the benefit-cost ratio is a maximum, represents that point at which the ratio of the incremental benefit to the incremental cost is equal to the benefit-cost ratio at that point. Point 2 is that point at which the incremental benefit is equal to the incremental cost. Therefore, at point 2 all increments for which benefits exceed cost are included. For a given excess of benefits over cost, that project with minimum cost will have the maximum benefit-cost ratio and, thus, by this criterion will be considered the most efficient.

Operations Research Methods

Recent developments in the optimal design of water-resource systems have used the tools of operations research and of simulation (Chow, 1964; Maass, 1962). The hydrology and eco-

WATER SUPPLY: ECONOMICS

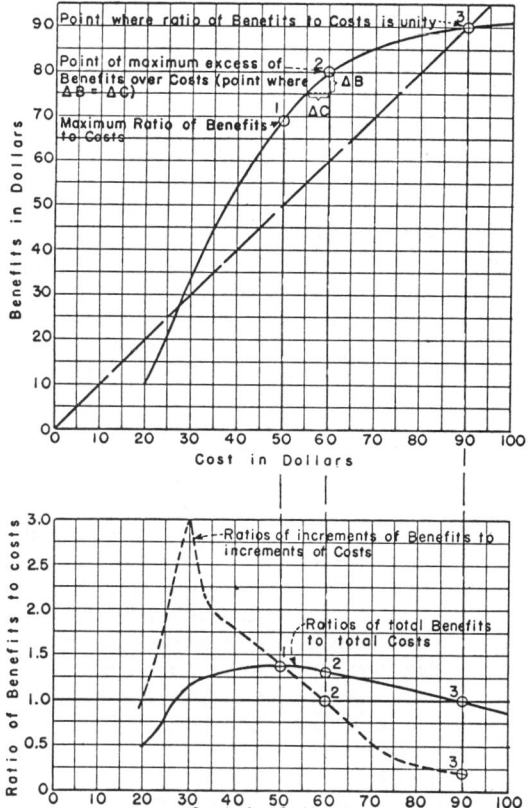

Fig. 1. Relationships between benefits and costs for varying scales of development (U.S. Govt., IACWR, 1958).

nomics of the river basin system are described by mathematical programming, a standard tool of operations research. An objective function is stated which is to be maximized (say a description of net benefits). Certain constraints then are stated, which may include inequalities (cost cannot exceed so many dollars, the flood pool must not fall below so many acre-feet of volume, etc.). If all constraints and the objective function are linear, the problem may be solved directly by linear programming. If only the objective function is nonlinear, certain types of nonlinearities may be solved by nonlinear programming. If conditions change with time, dynamic programming can be used.

Simulation techniques are used for more complicated techniques (which includes most realistic river basin studies). Both the hydrology and the economics of the basin are modeled mathematically. In addition, various synthetic histories of events can be included in the simulation. Through the use of high-speed computers many alternative designs can be studied and compared in the search for an optimal system.

For any model, the "most efficient design" which results will be related to the statement of the objective function. Much of the discussion concerning the benefit-cost criteria results from a lack of a clearly defined objective function or disagreement as to what it should include. The statement of the objective function is a policy matter (some constraints also may have policy implications). Operations research, by stating the problem mathematically, focuses attention properly on policy assumptions. It also measures the cost to any one objective which occurs as the result of achieving another objective. Thus, the relative price of each constraint is presented to the policy maker to aid him to evaluate the implications of his policy decisions. "An objective contains, explicitly or implicitly, the policy-maker's attitudes toward fundamental matters such as the relative values to be placed on outputs received in the present and in the future, which is expressed finally in the interest or discount rate; the necessity for a constraint on budget or expenditure and how it should vary, if at all, over successive time periods; and the degree of certainty desired in meeting planned outputs" (Maas, 1962, p. 4).

Determination of Costs and Benefits

Any study of feasibility is complicated by the determination of costs and benefits, both direct and indirect, tangible and intangible [Dixon, 1964; Eckstein, 1958; U.S. (IACWR), 1958]. Certain direct benefits, such as those resulting from hydroelectric power revenue, and direct costs, such as construction costs and direct operation and maintenance expenses, are relatively easily estimated. Others, such as reduction of flood damages, depend upon hydrologic data, and are thus more liable to random error. "The benefits derived from a flood-control reservoir are measured by the reduction in flood damage by the project. The flood damage can be evaluated by a flood survey. Flood occurrence being stochastic in nature, a probability analysis of the flood flow is necessary" (Chow, 1964). The results of such a probability analysis are shown in Fig. 2. "Curve AB shows the flood-damage probability function under natural conditions. Curve CD shows the residual flood-damage probability function when floods of a given probability, say 30%, and smaller floods are completely controlled. Thus the area ABCD represents the average annual gross benefits of the project. From a flood-frequency curve, the peak flood flow for the given probability can be obtained. Hence the annual benefits can be plotted against peak flood flows for given probabilities" (Chow, 1964). The combination of the flood-benefit function with the corresponding cost function is shown in

1261

WATER SUPPLY: ECONOMICS

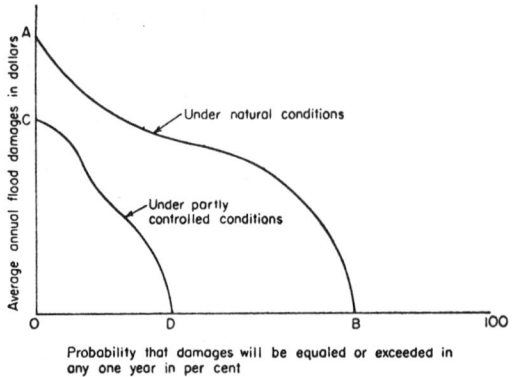

FIG. 2. Flood-damage frequency curves (Chow, 1964).

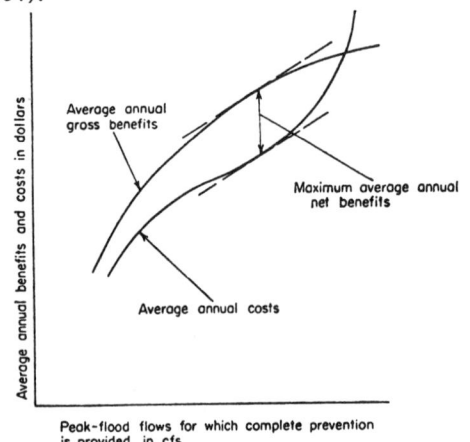

FIG. 3. Determination of optimal scale of development for a flood-control reservoir (Chow, 1964).

Fig. 3. The "maximum annual net benefits" shown correspond to point 2 on Fig. 1. That is, total net benefits are maximized at that point.

Indirect and intangible benefits often can be given monetary value only on an arbitrary basis, and thus are difficult to fit into the framework of benefit-cost analysis.

Some direct benefits which seemingly are easily amenable to measurement are affected by government policy in other areas. Thus, monetary worth of crops grown on irrigated land at least partly is determined by farm price-support subsidies. A change in policy on price-support levels therefore may affect the measure of feasibility of a project unless the true nature of the flexibility of policy in this area is considered in measuring benefits. Similarly, a change in flood-plain zoning policy may affect estimates of future flood damages.

Importance of Investments Assumptions

Perhaps as important as any facet of the benefit-cost analysis, is the dependence of relative efficiency upon the investment assumptions. "The monetary values of benefits and costs that accrue at varying times are comparable only if all are adjusted to a uniform time basis. The use of interest rates provides a means for converting estimates to a common time point or period" [U.S. (IACWR), p. 23]. The U.S. Government Inter-Agency Committee on Water Resources recommends the use of the average yield on long-term Federal bonds as an appropriate interest rate with which to convert to present worth all benefits and costs for federally financed projects. Krutilla and Eckstein (1958) stress the dependence of relative efficiency upon investment assumptions, and give several illustrative case studies for which "most efficient" depends upon imputed interest rates and amortization schedules.

The competitive model of the market would maximize the rate of return on investment. This would be a major consideration for private in-

TABLE 1. BENEFIT-COST RATIOS AND RATES OF RETURN FOR SELECTED CAPITAL INTENSITIES AND PERIODS OF ANALYSIS[a]

Benefit-Cost Ratio	Rates of Return			
	Period of 50 yr		Period of 100 yr	
	$O/K = 0.01$	$O/K = 0.10$	$O/K = 0.01$	$O/K = 0.10$
0.8	0.015	−0.02	0.020	−0.01
1.0	0.030	0.030	0.030	0.030
1.2	0.044	0.064	0.039	0.058
1.4	0.054	0.092	0.047	0.088
1.6	0.065	0.121	0.057	0.111
1.8	0.076	0.149	0.065	0.137
2.0	0.086	0.177	0.073	0.163
2.5	0.111	0.247	0.094	0.229
3.0	0.136	0.317	0.115	0.295
4.0	0.185	0.456	0.157	0.427
5.0	0.234	0.594	0.198	0.558

[a] From Eckstein, 1958.

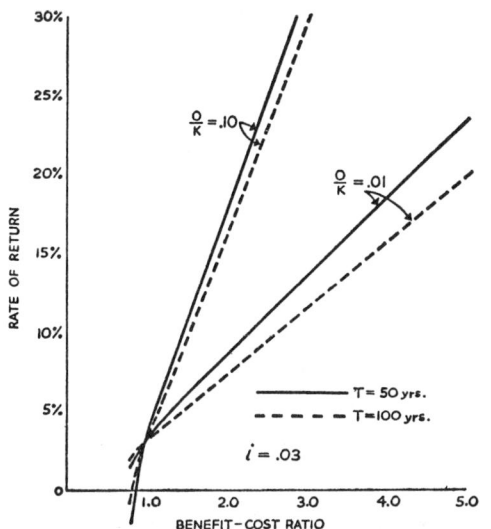

Fig. 4. Benefit-cost ratio and rate of return (Eckstein, 1958). (For explanation, see text.)

vestment in water resources development. Benefit-cost ratio, imputed interest rate, and rate of return are integrally related (Eckstein, 1958). Table 1 and Fig. 4 show a typical relation for an imputed interest rate of 3%. The ratio O/K is the ratio of annual operation and maintenance costs to fixed capital investment, and thus is a determining factor in computing rate of return. If imputed interest is equal to rate of return, a cost-benefit ratio of 1.0 results. Any lower imputed interest rate gives a benefit-cost ratio greater than 1.0. Eckstein's conclusion on the basis of his study of this interrelationship is that "the government (should) use a relatively low interest rate for the design and evaluation of projects, but let projects be considered justified only if the benefit-cost ratio is well in excess of 1.0" (Eckstein, 1958, p. 101).

Summary of Problems

McKean (1958, pp. 97–99) summarizes many of the problems inherent in the economics of water system design as follows:

General Criterion Difficulties. 1. Guard against particularly treacherous tests. One ubiquitous and untrustworthy candidate is the maximization of the ratio of gain (i.e., effectiveness) to cost.

2. Use the generally suitable form of criterion, the maximization of gain *minus* cost, if both can be expressed in the same unit. . . . Also, it should go without saying that the adoption of these general forms does not *insure* that the test will be an appropriate one.

3. Examine the criterion for consistency with higher-level criteria.

4. Guard against tests which involve erroneous concepts of cost or gain, such as treatment of valuable inputs as though they were free goods, or the inclusion of historical costs.

The Appropriate Alternatives. 1. Decide on the scope of the systems to be compared in conjunction with the selection of the criterion.

2. Give careful attention to the devising of the alternative courses of action to be compared. Sound models and criteria will not result in picking out good policies if only poor ones are considered.

3. Watch out for possible effects of adopting one policy upon the costs or gains from other policies, especially if they are to be ranked.

Intangibles. 1. Explore the possibilities of measuring effects which seem at first glance to be intangibles—that is, of measuring those effects in the same units of gain or cost used for the principal estimates.

2. In some special cases, show the value of the intangibles that would be implied by preference for one course of action over another.

3. Try to devise indicators of the magnitude and nature of major intangibles, using units of measure other than those adopted for the principal estimates.

Uncertainty. If practicable, show ranges of outcomes to which roughly the same degree of confidence can be attached.

Time Streams. 1. If gains and costs can both be measured in monetary units, discount the streams to their present values and choose the set of investments that yields maximum present worth for the investment budget.

2. If gains and costs cannot be expressed in the same unit, the best that can be done usually is to specify the time path of the task as a "requirement" and to discount the cost stream.

Higher Criteria

Most of the emphasis in the economics of water resource system design is aimed toward quantification of decision criteria. There is a possible inherent problem involved in the philosophy of this approach, in and of itself. "There may be higher criteria than efficiency criteria . . . River basin programs may be undertaken to increase the national product (consistent with efficiency considerations), but also may be undertaken for strategic, social, and perhaps other objectives, which may not be compatible with maximum efficiency in terms of the relatively narrow definition of efficiency employed . . . Where projects are undertaken for the latter type of goals, there will be a smaller net economic gain than otherwise would be possible—national income will be smaller than if conditions of maximum efficiency were to prevail. This is not meant to

WATER SUPPLY: ECONOMICS

imply that such objectives are unworthy, and that ... efficiency considerations provide the preferable course of action. Social, strategic, and other objectives may be preferred, and may be undertaken with sanction of collective choice expressed through the political process in a representative government" (Krutilla and Eckstein, 1958, p. 12).

DAVID R. DAWDY

References

*Ackerman, E., and Löf, G., 1959, "Technology in American Water Development: Baltimore," Johns Hopkins University Press, 710pp.
*Chow, V. T., 1964, "System Design by Operations Research," in "Handbook of Applied Hydrology," New York, McGraw-Hill Book Co., Section 26–II, pp. 30–47.
*Dixon, J. W., 1964, "Water Resources Planning and Development," in "Handbook of Applied Hydrology," New York, McGraw-Hill Book Co., Section 26–I, pp. 1–29.
*Eckstein, O., 1958, "Water-Resources Development, the Economics of Project Evaluations," Cambridge, Harvard University Press.
Krutilla, J., and Eckstein, O., 1958, "Multiple Purpose River Development: Studies in Applied Economic Analysis," Baltimore, Johns Hopkins University Press.
Maass, A., et al., 1963, "Design of Water-Resource Systems," Cambridge, Harvard University Press.
*McKean, R., 1958, "Efficiency in Government Through Systems Analysis, with Emphasis on Water Resources Development," New York, John Wiley & Sons.
U.S. Govt. (Inter-Agency Comm. on Water Resources), 1958, "Proposed Practices for Economic Analysis of River Basin Projects: A Report," Subcomm. on Evaluation Standards.

*Additional bibliographical references may be found in this work.

Cross-references: *Hydrology; Natural Resources; Pollution.* Vol. VI: *Water Power.*

WATER TABLE

The *water table* is the upper water surface in unconfined material, a surface which separates the groundwater (phreatic) zone and the zone of vadose or *suspended water* (q.v.). The base of the latter is often marked by a *capillary fringe* in which upwards migration of moisture occurs during dry seasons. The water table tends to follow the contours of the overlying ground surface, although as a subdued reflection. In valleys it is nearer the ground surface than it is in the highlands; the surfaces of most lakes, permanent streams, or swamps mark the position of the water table at that particular place. At the water table the hydrostatic pressure is equal to the atmospheric pressure.

The amplitude of the relief of the water table depends largely upon the texture of the material comprising the zone of saturation (Ward, 1967). The more truly the groundwater surface reflects the topography, the slower the seepage of water through the material. With low precipitation the water table level tends to drop and even out. A *perched water table* is one which lies above the regional water table due to the presence of impervious layers or lenses of limited extent.

R. W. F.

References

Davis, S. N., and DeWiest, R. J. M., 1966, "Hydrogeology," New York, John Wiley & Sons, 463pp.
Linsley, R. K., Jr., Kohler, M. A., and Paulhus, J. L. H., 1958, "Hydrology for Engineers," New York, McGraw-Hill Book Co., 340pp.
Ward, R. C., 1967, "Principles of Hydrology," London, McGraw-Hill Publ. Co. Ltd., 403pp.

Cross-references: *Groundwater; Phreatic Water; Suspended Water.*

WEATHERING—See Vol. III

WEATHERING—CHEMICAL

The chemical reactions coupled with the physical reactions *between* the elemental components of the "weather" *and* the elemental components of earth's crust yield (*1*) the soils on land and the sediments in the oceans and lakes; (*2*) the precipitated deposits of oxides, hydroxides, carbonates, phosphates, and sulfates on and within the soils and sediments; and (*3*) the soluble salts of the waters of the oceans, lakes, rivers, and underground reservoirs.

The elemental components of the weather are those of the atmosphere (oxygen, nitrogen, and carbon dioxide), the hydrosphere (water, vapor, and ice), and the biosphere (plants, animals, and microbes) coupled with the fluctuations of temperature, pressure, and sunlight on earth.

The elemental components of the earth's crust are the solid rocks and volcanic ash and the minerals within them; the soils and sediments and the minerals within them; and the organic residues of plants and animals.

Chemical weathering has played an important role in the evolution and maintenance of life on earth through the conversion of the elements essential to life from an unavailable to an available form and in creating a vast reservoir of these elements. They are stored: (*1*) in the soils and sediments in an exchangeable

TABLE 1. CHEMICAL COMPOSITION OF THE SAND AND SILT OF VARIOUS IGNEOUS ROCKS BEFORE AND AFTER WEATHERING AND OF THE CLAY THAT FORMED IN LATERITIC RED EARTH SOILS OF THE REPUBLIC OF SOUTH AFRICA[a]

	Granite			Dolerite		
	Before	After	Clay (1)[b]	Before	After	Clay (2)[b]
SiO_2	72.20	82.20	48.28	46.60	67.00	38.66
Fe_2O_3	0.20	0.00	7.74	19.77	22.70	18.97
Al_2O_3	17.80	9.48	40.12	17.20	5.20	38.30
TiO_2	0.29	0.34	0.70	1.96	6.05	0.77
Mn_2O_4	Tr	Tr	Tr	0.21	0.22	0.13
CaO	0.02	0.00	0.71	4.59	0.90	0.23
MgO	0.30	0.00	0.75	6.88	0.00	2.73
Na_2O	1.87	1.80	0.23	0.88	0.22	0.13
K_2O	5.83	5.43	1.57	0.76	0.15	0.37
P_2O_5	0.06	0.06	0.15	0.09	0.13	0.23
H_2O	3.46	1.94	15.90	8.90	15.25	21.87

[a] Data from Vander Merwe, 1941.
[b] Amount of clay that formed from 100 g of rock to bring about the change in chemical composition of the rocks: (1) = 44.2 g, (2) = 42.5 g.

form and in solids (having a solubility sufficient for maintaining plant needs); (2) in the waters of the earth in soluble form; and (3) in salt deposits of the earth (as potential fertilizers).

The activities of "Life" in turn have accelerated and "directed" chemical weathering through the carbon, nitrogen, phosphorus and sulfur cycles.

Major changes of rocks induced by chemical weathering are best described in terms either of changes in mineralogical composition or in chemical composition.

Changes in mineralogical composition are expressed by gradual disappearance of the nonresistant minerals and the accumulation of the resistant minerals, by the formation of clay minerals and uncombined oxides, and by the alteration of the particle-size distribution of the nonresistant minerals in the direction of an increase in fineness.

Changes in chemical composition reflect the changes in mineralogical composition brought about by clay formation and by clay migration. The loss of bases and the gain in silica characterizes the formation of clay, whereas the loss or gain of sesquioxides characterizes the loss or gain of clay particles by migration; that is, by the movement into or out of a given matrix of soil material. Tables 1 and 2 illustrate the chemical changes occurring to various rocks upon weathering and the chemical composition of clay that formed from these rocks.

History of Chemical Weathering

Chemical weathering must have begun as soon as water vapor could be absorbed by the surfaces of the rocks and minerals to cause hy-

TABLE 2. CHEMICAL COMPOSITION OF THE SAND AND SILT OF VARIOUS IGNEOUS ROCKS, BEFORE AND AFTER WEATHERING, AND OF THE CLAY THAT FORMED IN SUBTROPICAL BLACK CLAY SOILS OF THE REPUBLIC OF SOUTH AFRICA[a]

	Basalt			Norite			Peridotite		
	Before	After	Clay (1)[b]	Before	After	Clay (2)[b]	Before	After	Clay (3)[b]
SiO_2	54.20	91.3	64.08	54.70	85.8	56.19	55.45	80.40	63.33
Fe_2O_3	8.58	3.15	6.39	15.28	3.77	10.19	10.76	3.38	17.33
Al_2O_3	18.22	0.49	22.12	11.23	4.40	27.15	1.40	3.16	9.23
TiO_2	1.06	0.36	1.22	0.40	1.04	0.45	Tr	0.49	0.55
Mn_3O_4	0.16	0.36	0.14	0.30	0.52	0.19	0.18	0.35	0.18
CaO	8.94	1.74	2.08	4.26	1.07	2.43	1.80	0.62	0.63
MgO	7.01	0.29	3.07	13.35	4.74	2.99	32.03	9.85	9.17
Na_2O	1.35	0.00	0.46	0.46	0.35	0.06	0.25	0.20	0.12
K_2O	0.52	0.27	0.89	0.16	0.60	0.13	0.07	0.31	0.50
P_2O_5	0.12	0.00	0.09	0.02	0.02	0.05	0.03	0.02	0.05
H_2O	4.20	1.50	11.93	5.70	0.00	16.09	0.12	0.00	15.11

[a] Data from Vander Merwe, 1941.
[b] Amount of clay formed from 100 g of rock to bring about the change in chemical composition of the rocks: (1) 59.0 g, (2) 49.4 g, (3) 57.6 g.

dration of surface ions. It accelerated greatly upon the appearance of liquid water which is the keystone to all chemical weathering. In the younger days of the earth, the highest rate of chemical activity must have occurred wherever disaggregated rocks and minerals were present in the form of volcanic ash. These areas must have been the home of the first land-supported life, not only because of stored available essential elements but also because of their capacity to store water.

Due to the presence of an essentially CO_2 atmosphere on the young earth and the emanation by volcanic activity of large quantities of Cl_2, SO_2, and H_2S gases, the water on earth was acid (see Vol. 2, *Earth—Geology of the Planet*). In acids the rate of dissolution of the silicate minerals rich in bases and aluminum was high. Through the release of these bases into the waters, the waters of the earth were neutralized and the adsorbed H^+ became incorporated in the products as crystal structured water. In other words, through chemical weathering the water on earth was neutralized and thereby influenced the evolution of plants and animals. In turn, through the *photosynthesis* (q.v.) of the plants, the carbon dioxide of the atmosphere was used up and oxygen liberated. Thus, in time chemical weathering altered from that in a reducing environment to an oxidizing environment and with it the rate and the nature of the chemical reactions. A recent finding, by Barshad and Kishk, demonstrated that the ease of replaceability of potassium from biotite is much greater when its iron is in the Fe^{2+} than the Fe^{3+} state, suggesting that the rate of weathering of all the minerals containing Fe^{2+} iron must have been greater in the reducing acid environment of the young earth than in the present oxidizing environment.

From a study (Barshad, unpublished) of the weathering of recent lava flows and volcanic ash, it was discovered that the rate of weathering of rocks is directly proportional to their water holding capacity. Early stages of weathering of rocks consists primarily of corrosion and is strongly aided by lichens which have the facility to establish themselves even on "solid" surfaces. As soon as fine grained material accumulates in crevices and cracks within rocks, higher plants establish themselves and help secure this material from being eroded away. With time these "pockets" of soil grow in size and eventually envelop a whole area. Because of the absence of an abundant amount of clay minerals in the young stages of a soil, plant residues play a very important role in building up a reservoir of essential elements for maintaining plant life. As the soil becomes enriched with clay minerals, their capacity to store the *essential plant nutrients* and *water* increases and thereby also increases their capacities to enhance and maintain all the biological cycles of the biosphere.

This enhancement of storage, in turn, enables higher plant forms to sustain and to prolong their growth and thus anchor a soil to a given place for relatively long periods of time and thereby to promote its development in depth and to reach a state of equilibrium with its environment as circumscribed by the Factors of Soil Formation.

Chemical Reactions in Weathering

The most important reactants in chemical weathering are the silicate and the aluminosilicate minerals, water, and numerous acids dissolved in the water (CO_2, HNO_3, H_3PO_4, H_2SO_4, and various organic acids) and organic chelating compounds in the organic matter of soils. The products resulting from chemical weathering are the clay minerals, hydroxides and carbonates and salts of low solubility, oxides and bases, and salts of high solubility. The clay minerals, insoluble oxides, etc., accumulate within the weathered crust whereas the higher soluble bases and salts are removed from the site of the reactions by water moving over and through the weathering crust.

The major chemical reactions occurring in weathering are: (*1*) hydrolysis, (*2*) solution, (*3*) cation interchange, (*4*) cation exchange, and (*5*) oxidation and reduction.

Hydrolysis, Solution, and Cation Interchange

Because the aluminosilicate and silicate minerals are primarily salts of weak acids and bases and weak acids and strong bases (i.e., silicic acid, $Al(OH)_3$, $Fe(OH)_2$, and bases of the alkali and alkaline earth metals) the hydrolysis reaction, whereby the ionic species of water (H^+ and OH^-) become incorporated into the structural framework of the minerals, is one of the major reactions responsible for solubilization and/or decomposition. However, to decompose a mineral completely the hydrolysis must be coupled with the interchange reaction whereby the adsorbed H^+ on the surface of a crystal diffuses to the interior; in turn an interior cation diffuses outward to occupy the position of the cation which was replaced by the H^+. Hydrolysis and the interchange reaction repeatedly can occur until the mineral is completely decomposed.

For a mineral like a feldspar, for example, orthoclase, having the structural formula $K(AlSi_3O_8)$, two types of hydrolysis reactions occur in the course of complete decomposition. In one type the feldspar behaves like a salt of a weak acid and a strong base during which the products of hydrolysis are an undissociated insoluble fraction, $H(AlSi_3O_8)$, and a double base, KOH, i.e.,

$$K(AlSi_3O_8) + H^+ + OH^- \longrightarrow$$
$$(H_2O)$$

$$H(AlSi_3O_8) + K^+ + OH^-$$
$$\text{(insoluble)} + \text{(soluble)}$$

After the interchange reaction has taken place; that is, $H(AlSi_3O_8) \rightarrow Al_{1/3}(HAl_{2/3}Si_3O_8)$, the insoluble fraction will behave as a salt of a weak acid and a weak base; that is, silicic acid and aluminum hydroxide and the second type of hydrolysis takes place during which both the H^+ and the OH^- ions of water are being adsorbed by the mineral fraction and resulting in silicic acid and aluminum hydroxide. Thus

$$Al_{1/3}(HAl_{2/3}Si_3O_8) + 3H^+ + 3OH^- (H_2O)$$
$$\longrightarrow Al(OH)_3 + (H_4Si_3O_8)$$

Some of these products which result from the hydrolysis and the interchange reaction react with each other to form the clay minerals, some are left as oxides, hydroxides and carbonates, and some are removed from the site of reaction by water moving over and through the weathering material. The relative abundance and nature of each of these secondary products is determined by the conditions which exist at the site of the reaction. These conditions are determined to a large extent by the Factors of Soil Formation: Parent Material, Topography, Climate, Organisms, and the Duration of the Reaction.

A cation-exchange reaction is a reaction whereby the surface-held cations on primary minerals are replaced by other cations in solution. This action is similar to the hydrolysis reaction with H^+ ion as the replacing cation but it plays an entirely different role when cations in solution are other than $H^+(H_3O^+)$.

The presence of H^+ in solution is partly due to solution of CO_2 of the atmosphere in water that enters or is present in the weathered crust and partly due to the biological activities involved in the carbon, nitrogen, and sulfur cycles. These cycles furnish to the solution not only carbonic acid but also nitric, sulfuric and phosphoric acids and various organic acids. These cycles are also largely responsible for the composition of the cation in solution other than H^+ through their absorption by plants and later by their release into solution by decomposition of the organic matter.

The effects on the mineral of the replacement of surface cation by exchange with H^+ are identical to those from hydrolysis discussed previously.

The effect, however, of the replacement of surface cation by cations other than H^+ ions is determined by the nature of these cations and the nature of the minerals themselves. Thus the mica minerals weather very little if the solution cations are rich in K^+ since the surface adsorbed cations are also K^+. But if the solution cations are rich in Na^+, Ca^{2+}, and/or Mg^{2+} then the mica alters to vermiculite. The formation of vermiculite hastens the physical breakdown of the micas into clay size particles and also preserves the content of these cations in a form available for plant uptake since these ions are readily exchangeable.

Because of the variation in the relative replacing power of the solution cations for adsorbed cations, the chemical composition of the soil solution in contact with a weathering mineral affects the direction and extent of weathering. It is for this reason that a variation occurs in the weathering of a given mineral depending on its position in the weathering crust (or soil) and on the minerals with which it is associated. Thus mica minerals present in subsoil horizons persist longer than those present in surface horizons since the K^+ released by hydrolysis or exchange at the surface are leached downward to enrich the solution with K^+. Consequently, the replacement of K^+ from the mica in subsoil material is reduced or eliminated and thereby prevents weathering. Similarly, it was found that mica minerals which are present in granites in association with orthoclase alter less than those present in granodiorites and diorites in association with plagioclase feldspars. In other words, the release of K^+ by hydrolysis or exchange with H^+ from orthoclase feldspar helps to preserve the mica whereas the release of Na^+ and Ca^{2+} from plagioclase feldspars enhances the weathering of mica.

The variation of the cation-exchange properties of the various cations in solution and their variation in composition and in concentration in soil solution appears to affect markedly the nature of the clay minerals which form by synthesis from the insoluble weathering products; that is, from the alumina and the silica gels. Thus when K^+ predominates, illite minerals tend to form and when Ca^{2+} and Mg^{2+} predominate, montmorillonite minerals tend to form. But when Al^{3+} predominates over all the other cations then the kaolinite minerals tend to form. When, however, both Ca^{2+} and Mg^{2+} and Al^{3+} and Fe^{3+} are equally abundant, then vermiculites tend to form.

The oxidation reaction plays an important role in the minerals containing Fe^{2+} iron—the mafic minerals—the olivines, pyroxenes, amphiboles, and biotites. As a response to oxidation, which causes an increase in positive electrical charge in the crystal structure, the olivines, pyroxenes and amphiboles decompose completely whereas the biotites tend partly to break down in size to form the mica and vermiculite clays and partly to decompose to accommodate the increase in positive charge.

Because of the higher degree of basicity of the mafic minerals, they affect strongly the nature of clay mineral formation in association with the products of decomposition of the feldspars by affecting the degree of basicity and salinity of the environment in which clay mineral formation takes place.

Rate of Chemical Weathering

The rate of weathering, as measured by various indexes, is an extremely variable quantity and is determined by all the factors of soil formation; namely, parent material, climate, topography, organisms, and time. Taking the loss of Na_2O from rocks as an index by which to measure the rate of weathering; that is, the *thickness of rock* which loses a given amount of Na_2O per year, it may be possible to calculate a "mean rate" of weathering of the earth surface as a whole from beginning of time until now (by assuming a uniform parent rock and uniform weathering conditions). This rate may be calculated from the Na^+ content of the oceans; since it is the only major element in the ocean which accumulated there primarily from weathering and is not lost subsequently.

By expressing the Na^+ content of the oceans [which is equal to 1.0556 g/100 ml (see Vol. 2, *Earth—Geology of the Planet*] as Na_2O [it is equal to 1.425 g/100 ml or 1.425% (by wt)] and from the total mass of the ocean waters (which is $14{,}130 \times 10^{20}$ g) the total Na_2O content in the ocean is equal to $(14{,}130)(10^{20})(1.425)/10^2 = 20{,}103 \times 10^{18}$ g. To calculate the mass of rocks which could have supplied such an amount of Na_2O, we may further assume that this amount of Na_2O represents a loss of either 1%, 2%, or 3% (by weight) of the Na_2O content of the rocks. Therefore the total mass of the rocks which weathered would equal to either $20{,}013 \times 10^{20}$ g, $10{,}007 \times 10^{20}$ g, or $6{,}667 \times 10^{20}$ g, respectively. Assuming that the density of such rocks is approximately 2.73 g/ml then the volumes of these masses is equal to either $7{,}380 \times 10^{20}$ ml; $3{,}670 \times 10^{20}$ ml; or $2{,}442 \times 10^{20}$ ml, respectively. In terms of km³, they are equal either to $7{,}380 \times 10^5$ km³, $3{,}670 \times 10^5$ km³, or $2{,}422 \times 10^5$ km³ since 10^{15} ml = 1.0 km³.

To estimate what thickness of rock these masses represent, it is necessary to divide them by an area which may be considered to have been subjected to weathering. Here again it is necessary to make further assumptions. For the sake of simplicity and contrast, it may be assumed that either the whole of the surface of the earth has been subjected to weathering; namely, 514×10^6 km² or only an area equal to the present land surface area; namely 153×10^6 km² (see Vol. 2 *Earth—Geology of the Planet*). By dividing the volume of the rock masses which have weathered, as given above, by the land surface alone we obtain the following possible thicknesses of weathered rock: 14.36 km, 7.16 km, and 4.76 km; or 48.20 km, 24.95 km, and 15.59 km.

To calculate the rate of weathering in terms of rock thickness per year it is necessary to divide the entire weathered rock thickness given above by the length of the period which we may assume to have induced weathering. If we assume a period of about 0.5×10^9 yr (which is approximately the post-Cambrian period) the rates would be 28×10^{-6} m, 14×10^{-6} m, and 9.54×10^{-6} m; or 96.4×10^{-6} m, 50×10^{-6} m, or 31.2×10^{-6} m/yr. If we assume a weathering period of 2×10^9 yr (beginning of life) the rate would be 7×10^{-6} m, 3.5×10^{-6} m, and 2.38×10^{-6} m; or 24.1×10^{-6} m, 12.5×10^{-6} m, and 7.8×10^{-6} m/yr.

It is seen, therefore, that depending on the assumptions made, the "mean" rate of weathering as measured by loss of 1% to 3% of Na_2O could have varied between about 2×10^{-6} m and 100×10^{-6} m of rock thickness per year or between 2 and 100 m per million years.

An examination, however, of the thicknesses of weathered crust (soils) on primary rocks which have lost 1% to 3% Na_2O reveals that even under the most favorable weathering conditions such thickness never exceeds 2 to 3 m. The absence, therefore, of great thicknesses of weathered crust on primary rocks showing large losses of Na_2O content indicates that the *rate of erosion*, that is, the removal by water, by wind, and/or by gravity of the weathering crust in which such losses did occur, must be at a rate which is only slightly less than the rate of weathering.

Because of the near equality of weathering and erosion, it is necessary to conclude that to account for the loss of 1–3% of Na_2O by the rocks, several cycles of weathering and erosion must have occurred—a phenomenon which is well recognized.

Weathering and Erosion and Deposition

Erosion and deposition affect chemical weathering as follows:

1. On the eroding surface, erosion continually brings "fresh" minerals into the cycles of weathering.

2. Erosion brings about a segregation of the reactants from the products, thus enabling the formation of sandstone, shale, and deposits of various textures ranging from pure sands to pure clays. This segregation represents the greatest contribution to chemical differential in the weathering cycle.

3. The weathering of nonresistant minerals to completion is possible primarily through erosion and redeposition in a new environment where they are subjected to a new cycle of weathering.

ISAAC BARSHAD

References

Barshad, I., 1964, "Chemistry of soil development" in (Bear, F. E., editor), "Chemistry of the Soil," Reinhold Publishing Corp., New York.

Barshad I., and Kishk, F. M., 1968, "Oxidation of ferrous to ferric iron in vermiculite and biotite alters fixation and replaceability of potassium," *Science*, **162**, 1401–1402.

Jackson, M. L., and G. D. Sherman, 1953, "Chemical Weathering of Minerals in Soils," *Advan. Agron.*, **5**, 219.

Jackson M. L., 1964, "Chemical composition of soils" in "Chemistry of the Soil," Reinhold Publishing Corp., New York.

Jenny, H., 1941, "Factors of Soil Formation," McGraw-Hill Book Co., New York.

Keller, W. D., 1966, "Geochemical weathering of rocks: source of raw materials for good living," *J. Geol. Ed.*, **14**(1), 17–22.

Loughhann, F. C., 1969, "Chemical Weathering of the Silicate Minerals," Amsterdam, Elsevier, 154pp.

Polynov, B. B., 1937, "The Cycle of Weathering," translated by A. Muir, Nordemann Publishing Company, New York.

Rankama, K., and T. G., Sahama, 1950, "Geochemistry," the University of Chicago Press, Chicago, Ill.

Van der Merwe, C. R., 1941, "Soil Groups and Sub-Groups of South Africa," Govt. Printer, Union of South Africa.

Cross-references: *Acids and Bases; Cycles—Geochemical; Gases—Volcanic; Ion Exchange; Outgassing of the Planet Earth; Oxygen—Its Evolution in the Earth's Atmosphere; Pedology; Photosynthesis; River Geochemistry; Seawater, Chemistry; Seawater, History; Water.* Vol. II: *Earth—Geology of the Planet; Hydrosphere.* Vol. III: *Denudation; Erosion.* Vol. VI: *Soil Genesis; Weathering, Organic.*

WEATHERING, ORGANIC—See Vol. VI

WEATHERING TROPICAL—See Vol. III, **TROPICAL WEATHERING**

WELLS, WATER—See Vol. VI

WINDS—PRINCIPLES, LOCAL; WIND ACTION—See resp. Vol. II and Vol. III

X

XENON: ELEMENT AND GEOCHEMISTRY

Properties

Xenon, a rare or "noble" gas of atomic number 54, was discovered as a residue in liquid air by Ramsay and Travers in 1898. Contrary to earlier beliefs, xenon has been observed to form stable compounds with fluorine and other elements. The chemistry of the heavy noble gases is in fact becoming recognized as quite complex (Selig et al., 1964). Xenon is colorless and odorless with a melting point of −112°C and a boiling point of −107.1°C. Its atomic weight is 131.3, the density of the liquid is 3.52 g/cc (−109°C), and the density of the solid is 2.7 g/cc (−140°C). The ionization potential of xenon is 12.08 eV.

Abundance

There are three to four atoms of xenon in the solar system for every 10^6 atoms of silicon. The inner planets, depleted in volatile elements, contain even less xenon so that only about four atoms of xenon exist terrestrially for every 10^{13} atoms of silicon. Its occurrence on earth is thought to be limited to the atmosphere and crustal rocks, with the bulk of the xenon in the atmosphere.

Meteorites are, in general, richer in xenon than is the earth. A particular class of them, the carbonaceous chondrites, contain up to 740 times as much xenon per gram as the earth, and define a type of xenon that is isotopically somewhat different from terrestrial xenon. Neglecting for the moment Xe^{129}, the light isotopes in these meteorites are enriched relative to terrestrial xenon (see Table 1). The recurring isotope pattern and relative abundance have led many scientists to conclude that carbonaceous chondrite xenon represents a true primordial xenon, trapped by the parent bodies of meteorites as they cooled in the primitive solar system.

Iodine-Xenon Dating

The *iodine-xenon formation* interval for a sample refers to the time interval between the isolation of the material from galactic element formation and the cooling of the object to a point where it can quantitatively retain xenon (Reynolds, 1963). This contrasts with the potassium-argon age, where the interval *begins* with rare gas retention.

A record of I^{129}, now extinct, is preserved in the isotopic ratios of xenon in some meteorites. This radionuclide decays with a 17-million-year half-life to Xe^{129}, often creating a measurable excess of this isotope. The ratio of I^{129} to I^{127} when xenon retention began, together with an estimate of this ratio when the solar nebula became isolated from nucleosynthesis, permits

TABLE 1. ABUNDANCE AND ISOTOPE RATIOS OF TERRESTRIAL[a] AND CARBONACEOUS CHONDRITE[b] XENON

	Terrestrial	Carbonaceous Chondrite
Abundance	0.087 ppm by vol. (atmosphere)[c] 3.4×10^{-10} cc/g (at STP) (old granite, Halfway House)[d]	4.2×10^{-8} cc/g (at STP) (Murray)
Isotopic Ratios[e]	Xe^{124} 0.003575	0.00486
	Xe^{126} 0.003331	0.00439
	Xe^{128} 0.07137	0.0830
	Xe^{129} 0.9832	—
	Xe^{130} 0.1515	0.1619
	Xe^{131} 0.7876	0.820
	Xe^{132} 1.0000	1.0000
	Xe^{134} 0.3882	0.082
	Xe^{136} 0.3298	0.322

[a] Nier, 1950.
[b] Krummenacher et al., 1962.
[c] Townsend, 1960.
[d] Butler et al., 1963.
[e] The isotopic ratios are given relative to Xe^{132}, defined as unity.

a calculation of the formation interval for a meteorite. The latter ratio must be inferred from nuclear theory and current theories of element formation. The former can be measured in the laboratory from present-day xenon and iodine in the sample. A bulk measurement of excess Xe^{129} and I^{127} in the meteorite does not allow an accurate determination of this ratio because some xenon has undoubtedly been lost over the intervening 4.6×10^9 years due to effects of metamorphism and diffusion. Only a measurement of excess Xe^{129} and I^{127} at mineral sites where xenon has been quantitatively retained yields the correct value for the ratio of I^{129} to I^{127} at the onset of xenon retention. This measurement is best accomplished by a neutron irradiation of the sample, which implants Xe^{128} at the iodine sites (by neutron absorption in iodine) and an analysis of the xenon evolved as the meteorite is heated stepwise to progressively higher temperatures. Only at higher temperatures is the xenon from very retentive sites released, and those that have retained all of their xenon release excess Xe^{129} in a single fixed proportion to the implanted Xe^{128}. The ratio of these two isotopes permits a determination of the ratio of I^{129} to I^{127} present in these minerals when they began retaining xenon.

Because the I^{129} to I^{127} ratio at the time of isolation from galactic nucleosynthesis depends upon details of element formation which are not known with certainty, the formation interval, in the absolute sense, can be computed only for idealized models. But the *relative* formation intervals for two meteorites, the difference in the times at which they began to retain xenon, can be measured quite accurately and independently of any particular model for nucleosynthesis. All stone meteorites examined to date which contain an appreciable excess of Xe^{129}, began retaining xenon within one or two million years of each other (Hohenberg et al., 1967. Using any reasonable model for nucleosynthesis, the formation interval for these meteorites is surprisingly short, ranging from 60 to 180 million years (Turner, 1965; Hohenberg, 1969).

CHARLES M. HOHENBERG

References

Butler, W. A., Jeffery, P. M., Reynolds, J. H., and Wasserburg, G. J., 1963, "Isotopic variations in terrestrial xenon," *J. Geophys. Res.*, **68**, 3283–3291.

Hohenberg, C. M., Podosek, F. A., and Reynolds, J. H., 1967, "Xenon-iodine dating: Sharp isochronism in chondrites," *Science*, **156**, 233–235.

Hohenberg, C. M., 1969, "Radioisotopes and the history of nucleosynthesis in the Galaxy," *Science*, **166**, 212–215.

Krummenacher, D., Merrihue, C. M., Pepin, R. O., and Reynolds, J. H., 1962, "Meteoritic krypton and barium versus the general isotopic anomalies in meteoritic xenon," *Geochim. Cosmochim. Acta*, **26**, 231–249.

Nier, A. O., 1950, "A redetermination of the relative abundances of the isotopes of neon, krypton, rubidium, xenon, and mercury," *Phys. Rev.*, **79**, 450–454.

Reynolds, J. H., 1963, "Xenology," *J. Geophys. Res.*, **68**, 2939–2956.

Selig, H., Malm, J. G., Claasen, H. H., 1964, "The chemistry of the noble gases," *Sci. Am.*, **210**, 66–77.

Townsend, J. W., 1960, "Composition of the Upper Atmosphere," in "The Physics and Medicine of the Atmosphere and Space," New York, John Wiley & Sons, 112–133.

Turner, G., 1965, "Extinct iodine 129 and trace elements in chondrites," *J. Geophys. Res.*, **70**, 5433–5445.

Takagi, J., Sakamoto, K., and Tanaka, S., 1967, "Terrestrial xenon anomaly and explosion of our galaxy," *J. Geophys. Res.*, **72**(8), 2267–2270.

Verniani, F., 1966, "The total mass of the Earth's atmosphere," *J. Geophys. Res.*, **71**(2), 385–391.

Wasserburg, G. J., and Mazor, E., 1965, "Spontaneous Fission Xenon in Natural Gases," in "Fluids in Subsurface Environments," *Am. Assoc. Petrol. Geol. Mem.*, **4**.

Cross-references: *Iodine; Potassium-Argon Age Determination; Rare Gases.* Vol. II: *Carbonaceous Meteorites.*

X-RAY DIFFRACTION ANALYSIS

X-ray diffraction analysis dates back to 1912, when the first attempt was made to use a crystal as a diffraction grating for x rays. The regular configuration of atoms in space, which was hypothesized by many early crystallographers because of the symmetry of crystals, was confirmed by the first experiments. Since then, the techniques and methods of x-ray diffraction analysis have been applied to many varied investigations ranging from routine identification to structure analysis of complicated compounds.

Wavelengths of x-rays employed for diffraction analysis range from 0.7 Å to 2.3 Å. Because interatomic distances are of the same magnitude, crystals act as three-dimensional diffraction gratings when placed in the path of x rays. Therefore, as in optical diffraction, where, if the wavelength of the light is known, the spacing of the diffraction grating can be found, so in the case of x-ray diffraction, the interatomic spacings can be determined if the wavelength of the x radiation is known. Because a set of interatomic distances is deter-

X-RAY DIFFRACTION ANALYSIS

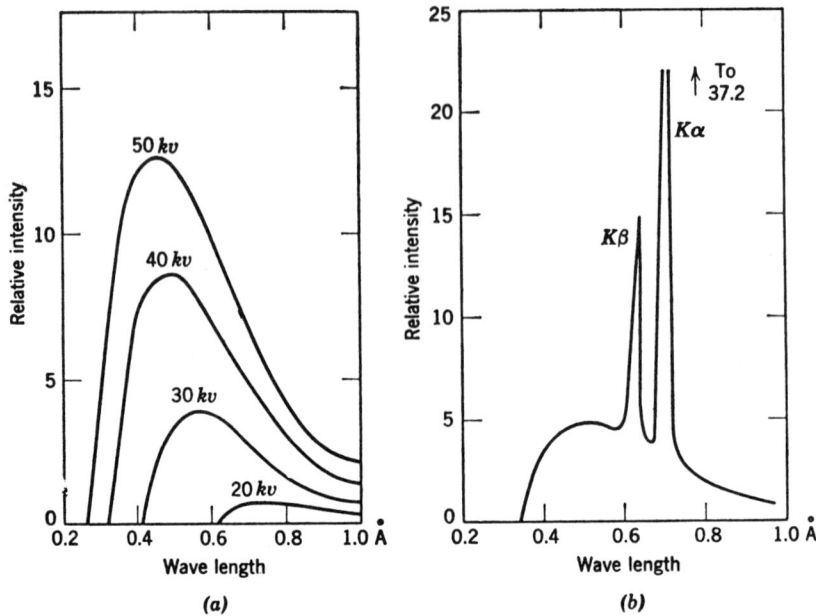

Fig. 1. X-ray spectra: (a) Continuous spectra. (b) A characteristic spectrum superimposed on a continuous spectrum. (After Uhrey, *Phys. Rev.*, **11**, 401.)

mined by the kinds of atoms present and the directions and lengths of their bonds, it characterizes a compound and may be used for identification.

Although x-ray diffraction data may be obtained easily, the derivation of the crystal structure, i.e., the determination of atomic positions, is a tedious and difficult task, requiring time-consuming computations and frequently taxing the ingenuity of the investigator. As will be pointed out below, x-ray diffraction data are not sufficiently complete to reduce the task of solving a structure problem to processing and evaluating a large amount of data. However, computations, which previous to the advent of high-speed computers required an inordinate amount of time, can now be carried out quickly and relatively easily.

Physics of X rays

When electrons that are accelerated in a high-potential field strike any substance, x rays may be produced. In practice, x rays are generated in an x-ray tube containing a cathode, the source of electrons, and an anode, a metallic target toward which the electrons are accelerated. The energies or wavelengths of the x rays emanating from the tube are distributed over two superimposed spectra: (a) a continuous spectrum and (b) a characteristic or line spectrum (Fig. 1). The continuous spectrum is due to x radiation emitted by electrons that interact with the coulomb fields of force of nuclei in the target material. The minimum wavelength of this spectrum is a function of the accelerating potential only and is independent of the target material. Therefore, an increase in the voltage across the tube elements causes the minimum wavelength to shift toward a smaller value. When operated at a sufficiently high voltage, an x-ray tube emits, in addition to the continuous spectrum, a line spectrum which is characteristic of the element of the target. Some high-energy electrons, upon striking the anode or target, interact with electrons of the target substance and may cause them to occupy higher energy levels or be ejected from the atom. The removal of an electron creates a vacancy which is quickly filled by another electron from a higher energy level. The transition from a higher to a lower energy level is accompanied by the emission of a photon whose wavelength is a function of the energy difference of the respective levels and, therefore, is characteristic of the element used as the anode (Fig. 2). Elements such as Mo, Cu, Ni, Co, Fe, W, and Cr are used as anodes in x-ray tubes because the wavelengths of the radiation emitted when a transition to the lowest energy level occurs are within the desired range, i.e., 0.7–2.3 Å. Instead of reporting the radiation employed in a given analysis by its wavelength, it is customary to designate the source of the radiation. For example, Cu K_α radiation is due

X-RAY DIFFRACTION ANALYSIS

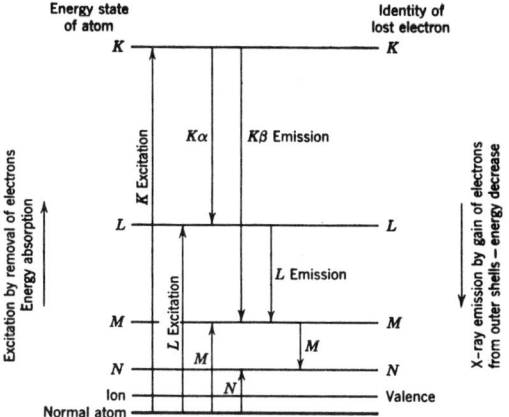

FIG. 2. Schematic energy-level diagram for an atom (from Alexander, 1954).

to electron transitions from the L level to the K level of copper, the anode material (Fig. 2).

The physics of emission of x-ray photons is identical to that of photons in the region of visible light, the former accompany inner orbital transitions which require large amounts of energy while the latter involve loosely held, outer orbital electrons which may be activated by relatively low thermal energy.

Almost all techniques of analysis require x radiation of essentially one wavelength only, i.e., monochromatic radiation. Most radiation of undesirable wavelengths can be removed by placing appropriate filters in the path of the beam. Filters, usually metal foils, are selected because of their absorption characteristics; that is, they must be almost transparent to radiation of the desired wavelength and almost opaque to the unwanted portion of the spectrum.

Diffraction

When a substance is exposed to x rays its electrons are accelerated and reradiate energy at the frequency of the incident radiation. Because atoms in a crystalline solid are arranged in an orderly array or pattern, and because the wavelengths of x rays employed in analysis are of the same magnitude as distances between atoms, diffraction effects may be observed. An approximate optical analog would be a multiple-slit screen exposed to monochromatic light. Each slit corresponds to all the electrons in the unit cell, and the identity distance between the slits to the distance between equivalent lattice points. However, a multiple slit screen or diffraction grating is essentially a two-dimensional entity; a crystal lattice, on the other hand, represents a three-dimensional array of diffraction centers.

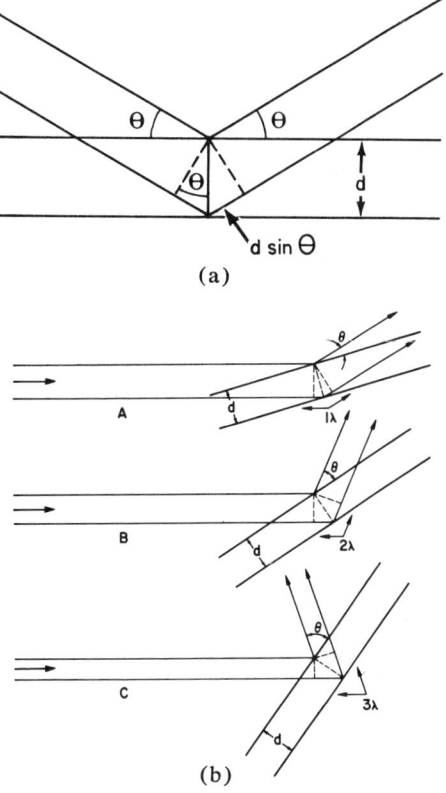

FIG. 3. Construction upon which the Bragg equation is based: (a) from Kittel, 1966; (b) from Buerger, 1942.

In order to relate the positions of diffraction maxima to the interatomic distances of a crystal, W. L. Bragg, in 1913, developed a model which assumes that x rays are reflected from parallel, semitransparent planes of atoms. Figs. 3a and 3b illustrate the model upon which Eq. (1) is based.

$$n\lambda = 2d \sin \theta \qquad (1)$$

where n = an integer; λ = the wavelength; d = the distance between equivalent planes, the so-called d-spacing; and θ = the angle between these planes and the x-ray beam. Eq. (1), known as Bragg's law or Bragg's equation, and Fig. 3 show that when x radiation is scattered by points lying in a plane the scattered ray elements interfere with each other destructively in all directions except in those where they are in phase, i.e., where θ is such that the path difference is a whole number of wavelengths. In these directions, the beam elements reinforce each other, resulting in a relatively strong signal.

The number of wavelengths of the path dif-

1273

X-RAY DIFFRACTION ANALYSIS

ference, n in Eq. (1), is the *order* of the diffraction; for example, the second-order diffraction maximum for a given d-spacing coincides with the first-order maximum from a d-spacing one half as wide. This makes possible an alternate expression of Bragg's law, namely

$$\lambda = 2d_{hkl} \sin \theta_{hkl} \qquad (2)$$

where d_{hkl} and θ_{hkl} are the d-spacing and angle of a particular set of planes (hkl). Although lattice planes, unlike mirror surfaces, do not reflect the incident rays at any angle, there is a superficial similarity between the construction upon which Bragg's equation is based and those illustrating the laws of reflection. Because of this similarity, it is common practice among crystallographers to speak of *reflection* when referring to diffraction phenomena, a practice which will be followed here. For example, the 110 reflection refers to the diffraction maximum associated with (110) planes of the crystal.

Reciprocal Lattice

The understanding of diffraction phenomena is greatly aided by the concept of the reciprocal lattice or of reciprocal space. The construction shown in Fig. 4 illustrating conditions for reflection for a set of planes (hkl) serves to illustrate the fundamentals of reciprocal space. The construction consists of a sphere of radius $1/\lambda$, drawn around the crystal at its center. It can be seen from Eq. (2) and Fig. 4 that

$$\frac{1}{d_{hkl}} = \frac{2}{\lambda} \sin \theta_{hkl} = S_{hkl} \qquad (3)$$

where S_{hkl} is the vector originating at O, the origin, and terminating on the sphere. Vectors S_{hkl} are perpendicular to the crystallographic planes (hkl) and of magnitude $1/d_{hkl}$. Because

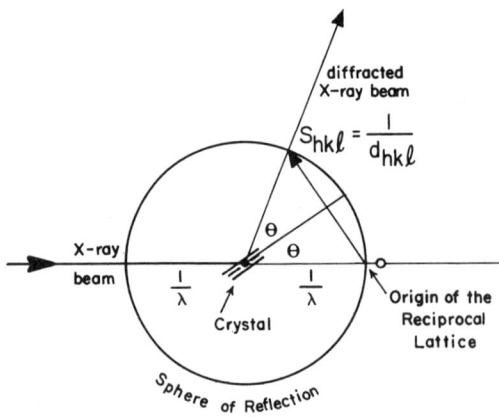

Fig. 4. The sphere of reflection and a reciprocal lattice vector, S_{hkl}.

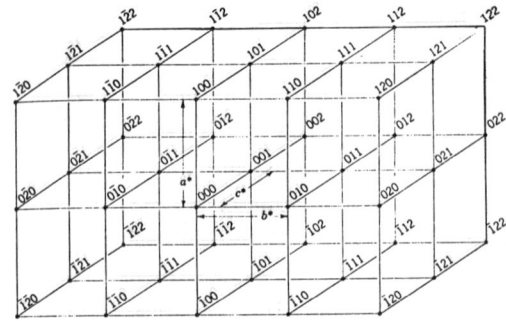

Fig. 5. Reciprocal lattice for an orthorhombic crystal.

the vectors are perpendicular to the planes (hkl) and their lengths are $1/d_{hkl}$, they form a lattice that is reciprocal to the crystal lattice. If a^*, b^*, and c^* are the unit lengths of the reciprocal lattice, their relation to the real or crystal lattice may be expressed as follows:

$$a^* = \frac{1}{d_{100}} = \frac{bc}{V} \sin \alpha \qquad (4a)$$

and is perpendicular to the bc plane

$$b^* = \frac{1}{d_{010}} = \frac{ac}{V} \sin \beta \qquad (4b)$$

and is perpendicular to the ac plane

$$c^* = \frac{1}{d_{001}} = \frac{ab}{V} \sin \gamma \qquad (4c)$$

and is perpendicular to the ab plane, where V is the volume of the unit cell and a, b, c, α, β, γ, are the parameters of the unit cell in real space.

Fig. 5 illustrates a portion of the reciprocal lattice of an orthorhombic crystal. It should be noted that all points of this lattice can be derived from a^*, b^*, and c^*. The multiples are the coordinates of the points as well as their indices. For example, $1a^*$, $2b^*$, $0c^*$, or 120 represents the location of the reciprocal space vector S_{120} whose length is $1/d_{120}$ and whose direction is perpendicular to the (120) plane of the crystal lattice. The reciprocal lattice, whose origin is fixed by the location of the crystal and by the wavelength employed, occupies a position in space determined by that of the crystal lattice. As the crystal is made to rotate or oscillate in the x-ray beam, the reciprocal lattice may be pictured to perform the same motion about its origin. Conditions for reflection are met when a reciprocal lattice vector, S_{hkl}, touches the sphere of radius $1/\lambda$, or the sphere of reflection. Therefore, in all methods of analysis employing monochromatic radiation, the crystal is made to move so that many reciprocal lattice points may come in contact with the sphere of reflection. From Fig. 4 it is also clear that only points lying within the distance of $2/\lambda$ from the origin can be made to diffract radiation of wavelength λ.

Powder Diffraction Analysis

The most widely used x-ray method is based on the analysis of diffraction patterns obtained from powdered samples. The pattern may be recorded on film or, with the aid of proper instrumentation, on a strip chart. The former is carried out with simple equipment and yields accurate results but the film must be developed and dried before the pattern can be measured and analyzed. The latter is more convenient and rapid but is carried out with complicated electronic equipment.

In the common film method, the sample is mounted on a rotating sample holder which occupies the center of a cylindrical camera. The camera is equipped with a collimator for the entering primary beam and a beam stop for the nondiffracted portion of the beam. A length of 35-mm wide film is held against the inner surface of the camera. Common camera diameters are 57.3 mm and 114.6 mm. Conditions for reflection are met by all planes (hkl) making angles θ_{hkl} with the primary beeam. Because the sample is rotated during exposure and because of the random orientation of the crystals, a large number of planes can be brought into reflecting positions. The diffracted beams form generators of coaxial cones whose apex angles, 4θ, are located at the point where the x-ray beam and the planes intersect. The diffraction maxima, therefore, are on the intersections formed by these cones and the cylinder formed by the film (Fig. 6). Alternatively, one may picture a reciprocal lattice associated with each of the randomly oriented crystals. When and where a reciprocal lattice vector S_{hkl} touches the sphere of radius $1/\lambda$ diffraction occurs (Fig. 4). From the distances between diffraction maxima, which appear in the form of arcs (Fig. 6), the interplanar spacings, or d-spacings, are computed. An unknown substance is identified by finding a matching set of d-spacings among those of previously identified substances. Diffraction data of thousands of compounds are compiled by the American Society for Testing Materials and are published on convenient cards. The cards are indexed according to the d-spacings of the three most intense lines.

Powder diffraction analysis is well suited for mineralogical work where single crystals are not always available and where rapid identification of several samples is often called for. Beyond mere identification, powder data may be analyzed for other information. For example, techniques have been developed for the identification of mineral species of some isomorphous series based on the correlation of chemical composition and d-spacings of specific planes. Another technique yielding chemical information which is often useful in cases

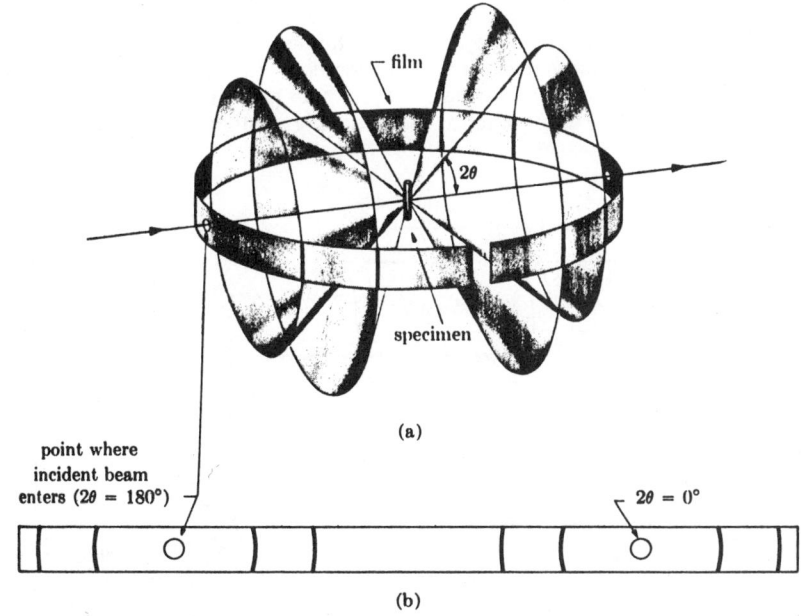

FIG. 6. Orientation of specimen and film for the powder method. (Cullity, 1956).

where analysis is difficult, is based on the comparison of the actual and theoretical densities. The computation of the theoretical density is carried out by taking into account the volume of the unit cell and the assumed chemical composition. If the measured and theoretical densities are in agreement, the assumed composition is taken to be correct. Because of the inherent simplicity, rapidity of application, and the wealth of information it yields, the analysis of powdered samples is the basis of most x-ray diffraction analytical work.

Single Crystal Diffraction Analysis

The locations of diffraction maxima are functions of the *d*-spacings or, in a more general sense, of the shape of the unit cell. The intensities of the maxima, on the other hand, are determined mainly by the amount and distribution of diffracting matter, i.e., electrons, in the unit cell. An approximate optical analog for unit cells as diffracting centers for x-rays would be the grooves of diffraction gratings. Assuming perfect spacing of the grooves, the intensity of diffraction maxima is to a large extent a function of the quality of the grooves. It may be apparent that in order to derive the crystal structure from its x-ray diffraction pattern, one must obtain data on the intensity of the diffraction maxima. Patterns obtained from single crystals furnish the required information.

In a common experimental technique, the crystal, supported on a fiber, is placed in the center of a long cylindrical camera and the film is fastened against the inner surface of the camera (Fig. 7). The fiber supporting the crystal is mounted on a goniometer head assembly, a clamp which permits two translatory and two rotary motions. By manipulating these, it is possible to orient the crystal so that one of its major crystallographic axes coincides with the axis of the cylindrical camera. While being exposed to x rays, the crystal is made to rotate or oscillate. Unlike a powder pattern, the diffraction pattern obtained from a single crystal consists of spots. The correspondence between diffraction spots and reflecting planes is best understood in terms of reciprocal space. We may picture the reciprocal lattice located with its origin on the sphere of reflection (Fig. 4). Each spot of the diffraction pattern corresponds to a point of the reciprocal lattice. For example, if the axis of rotation of an orthorhombic crystal crystal is the c-axis (Fig. 5), all points of the reciprocal lattice with indices $hk0$, which form a layer in reciprocal space, will register on the film in a line, the so-called 0-layer line (Fig. 8). The layers above and below it are due to $hk1$, $hk2$ and $hk\bar{1}$, $hk\bar{2}$ points of the lattice. For a given wavelength and camera size, the distance between successive layer lines in the example above is directly related to the identity distance in the c direction, or d_{001}. The identity distances along the other coordinates may be found by rotating the crystal about the a and b axes.

In order to facilitate the indexing of the pattern, moving film techniques are employed. In the camera shown in Fig. 7 a screen containing a 2 mm. slit is placed between the crystal and the film, permitting only points lying in one layer of the reciprocal lattice, such as points with indices $hk0$, to be recorded on the film. While the screen is kept stationary, the film holder is made to move parallel to the camera's cylindrical axis as the crystal is rotated, the two movements being synchronized by a gear system. The effect of this displace-

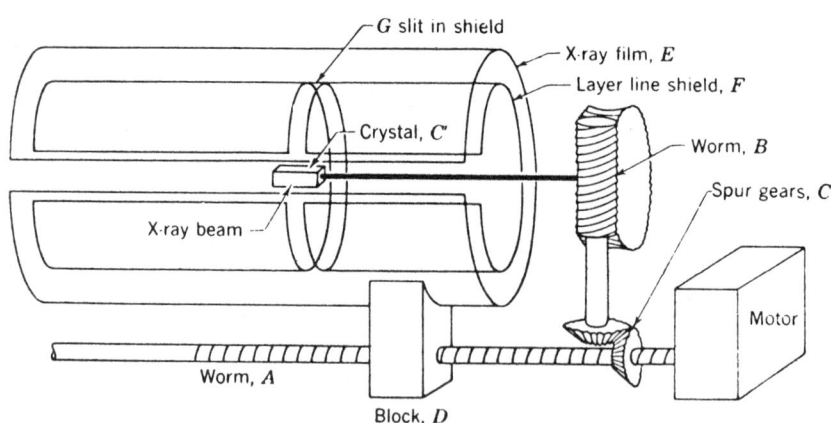

Fig. 7. Schematic drawing of a single-crystal camera (Weissenberg camera) (from McLachlan, 1957).

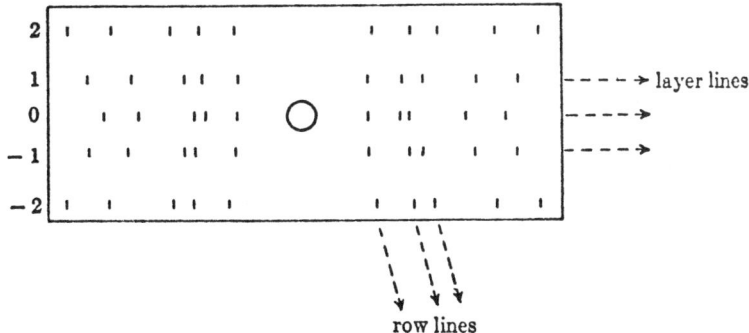

FIG. 8. Diffraction pattern of a rotating crystal (Gould, 1955).

ment is that the points of a single reciprocal layer which, had the film been stationary, would appear as closely spaced spots on a single line (Fig. 8) are spread out over the entire film. The moving film has the further advantage of separating spots which would be superimposed on a stationary film, i.e., spots representing points in reciprocal space which are equidistant from the origin. With the aid of templates, it is relatively simple to index the spots. From absent reflections, it is possible to deduce among others the presence of such symmetry elements as glide planes and screw axes. This information, as well as optical and other data, are then utilized to allocate the symmetry of the structure under investigation to one of the 230 space groups. The identification of the crystal's symmetry is an essential preliminary step in structure analysis.

Structure Factor Equation

The procedure of structure analysis is to either derive the atomic positions from the intensities of diffraction maxima or to postulate an arrangement of atoms in space and, using this model, to calculate the intensities of the maxima. If the calculated and observed intensities are equal within reasonable limits, the model is considered correct. Relative intensities of the spots recorded on film may be estimated by comparing them with some standard, or they may be measured with a densitometer. A technique yielding accurate intensity data utilizes a Geiger counter or scintillation counter mounted on a movable arm which permits it to receive the diffracted beam. Corrections must be made for several geometric and physical factors before the observed intensities can be utilized in analysis.

The corrected relative intensity of a diffraction spot is mainly a function of (a) the kinds of atoms, (b) the number of atoms per unit cell, and (c) their arrangement within the unit cell. Because the size of an atom is of the same magnitude as the wavelength of the radiation, its electrons scatter rays in phase only in the forward direction (Fig. 9). In any other direction there will be a phase difference among the scattered wavelets and consequently a reduction in intensity of the diffracted beam. Therefore, the efficiency with which an atom scatters radiation, expressed by the atomic scattering factor f, is a function of the number of electrons, i.e., the atomic number, the angle θ_{hkl}, and the wavelength. Atomic scattering factors for all atoms have been computed and are available in texts and standard reference works on structure analysis.

The implied assumption of the Bragg equation (Eq. 1) is that all diffracting matter, i.e., all electrons, lie in the reflecting planes. This simplified model which serves to explain the angular relationship between d_{hkl} and θ_{hkl} is shown in Fig. 10A. Fig. 10B illustrates a more realistic model containing atoms 1, 2, and 3. The interplanar distance d, is of course the same for any two adjacent planes of similar atoms and, therefore, each set of similar planes meets the conditions of the Bragg equation for reflection. For the first order of diffraction, the waves scattered from adjacent similar

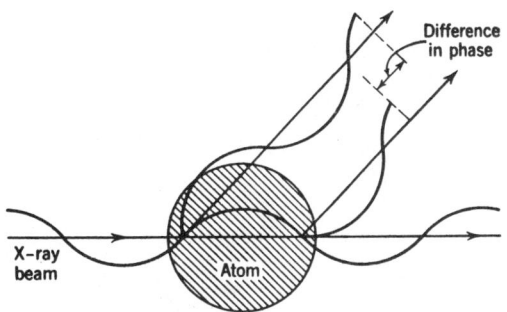

FIG. 9. Phase difference due to scattering from different parts of the atom (Klug and Alexander, 1954).

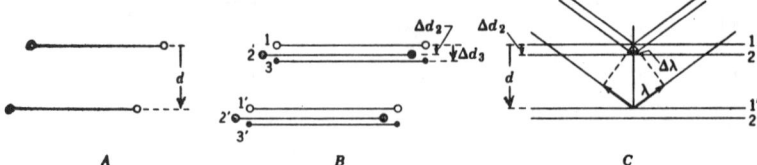

FIG. 10. X radiation scattered by a stack of lattice planes (Buerger, 1960).

planes, separated by distance d, scatter x rays so that there is a path difference of one wavelength, λ, and a phase difference of 2π. Planes of dissimilar atoms, for example planes of atoms 1 and 2 in Fig. 10B and 10C being separated by Δd, scatter x rays with a path difference of $\Delta\lambda$ and a phase difference of ϕ, so that

$$\frac{\phi}{2\pi} = \frac{\Delta\lambda}{\lambda} = \frac{\Delta d}{d}$$

or

$$\phi = 2\pi \frac{\Delta d}{d} \quad (5)$$

and for the second order of diffraction

$$\phi_2 = 4\pi \frac{\Delta d}{d}$$

or, generally

$$\phi_n = n[2\pi] \frac{\Delta d}{d} \quad (6)$$

where n is the order of diffraction. As will be shown below, the net effect of the mutual interference of the wavelets is a wave whose amplitude and phase equal the vector sum of the individual wavelets. Fig. 11 shows three wavelets with amplitudes f and phases ϕ and the resultant wave with amplitude F, which was obtained by compounding the wavelets. It is more convenient to represent a wave as a vector in the complex plane where the length of the vector equals the wave's amplitude, f, and the angle it makes with the x axis equals the phase angle, ϕ. The vectors in Fig. 12 represent the wavelets of Fig. 11. Here, the amplitude and the phase of the resultant wave were found by adding vectors f. Vectors in the complex plane can be resolved into their real and imaginary components, the real components having the form $f \cos \phi$ and the imaginary components having the form $f \sin \phi$. Vectors of the type shown in Fig. 11 may be represented by complex numbers where

$$e^{i\phi} = \cos \phi + i \sin \phi \quad (7)$$

Using this notation and rearranging the terms shown in Fig. 12 results in the expression

$$F = f_1 [\cos \phi_1 + i \sin \phi_1] + f_2 [\cos \phi_2 + i \sin \phi_2] + f_3 [\cos \phi_3 + i \sin \phi_3] \quad (8)$$

or, equivalently

$$F = f_1 e^{i\phi_1} + f_2 e^{i\phi_2} + f_3 e^{i\phi_3}$$
$$= \sum_j f_j e^{i\phi_j} \quad (9)$$

The term ϕ_j, representing the phase angle of the wavelet scattered by atom j, may be expressed in terms of the d-spacing and the order of diffraction [Eq. (6)] so that

$$F = \sum_j f_j e^{n2\pi i \left[\frac{d}{\Delta d}\right]_j} \quad (10)$$

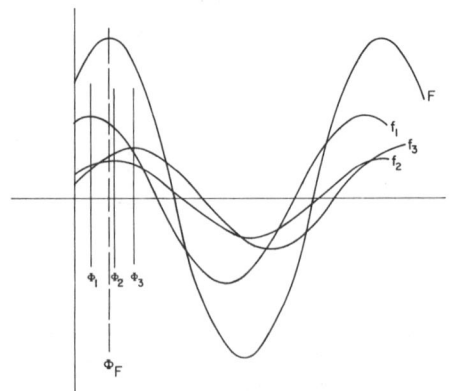

FIG. 11. A wave whose amplitude, F, and phase ϕ, are due to the interference of three waves with amplitudes f_1, f_2, and f_3, and phases ϕ_1, ϕ_2, and ϕ_3.

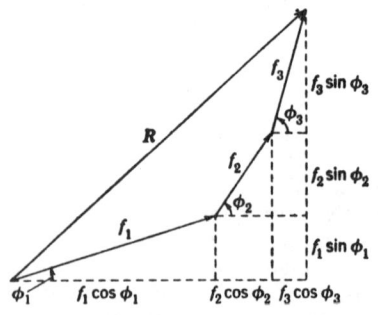

FIG. 12. Vector sum representing waves of different phases and amplitudes (Buerger, 1960).

Written in terms of its components,

$$\left[\frac{\Delta d}{d}\right]_j = \frac{x_j}{a} + \frac{y_j}{b} + \frac{z_j}{c} \quad (11)$$

where a, b, and c are the lengths of the unit cell edges and x_j, y_j, and z_j the coordinates of atom j. The term n in Eq. (10), denoting the order of diffraction, is equivalent to the Miller index of the reflecting plane (Eq. 2). Writing Eq. (10) in terms of the atomic coordinates and the Miller indices results in the useful expression

$$F_{hkl} = \sum_{j=1}^{N} f_j e^{2\pi i \left[\frac{hx_j}{a} + \frac{ky_j}{b} + \frac{lz_j}{c}\right]} \quad (12)$$

where N is the number of atoms per unit cell. Eq. (12) is known as the *structure factor equation* and F_{hkl} as the *structure factor*.

It is possible to calculate the structure factors of a trial structure, that is, the F's of a postulated arrangement of atoms. If the observed F's obtained from the intensity data, agree within the limits of unavoidable errors with the calculated F's then the postulated model is considered correct.

A formal derivation of the distribution of electrons in the unit cell from the structure factors is beyond the scope of this article. But, the relationship may be grasped intuitively if one recalls that F_{hkl} is the amplitude of a wave. We may imagine a wave associated with every diffraction spot of the pattern. The distribution of electrons in the cell may be closely approximated by the sum of these waves. The physical meaning of this is illustrated in Fig. 13, where two waves of amplitudes F_{020} and F_{300} are added. The sum is a function which is represented in the form of a warped surface and a contour map, showing positive and negative maxima. The latter, presenting the projection of the structure on a plane, is a form commonly employed by crystallographers. The expression for the summation of structure factors is

$$\zeta(xyz) = \frac{V}{1} \sum_{h} \sum_{k} \sum_{l}^{+\infty}_{-\infty} F_{hkl}\, e^{-2\pi i \left[\frac{hx}{a} + \frac{ky}{b} + \frac{lz}{c}\right]} \quad (13)$$

where $\zeta(xyz)$ is the electron density for an element of volume, and V is the volume of the unit cell. The summation may be carried out if the magnitudes and phase angles of the structure factors, i.e., the F_{hkl}'s of Eqs: (12) and (13) are known, which in turn requires a knowledge of the atomic positions (x_j, y_j, z_j). The summation above is a Fourier series and the F_{hkl}'s are its constants. The structure factor

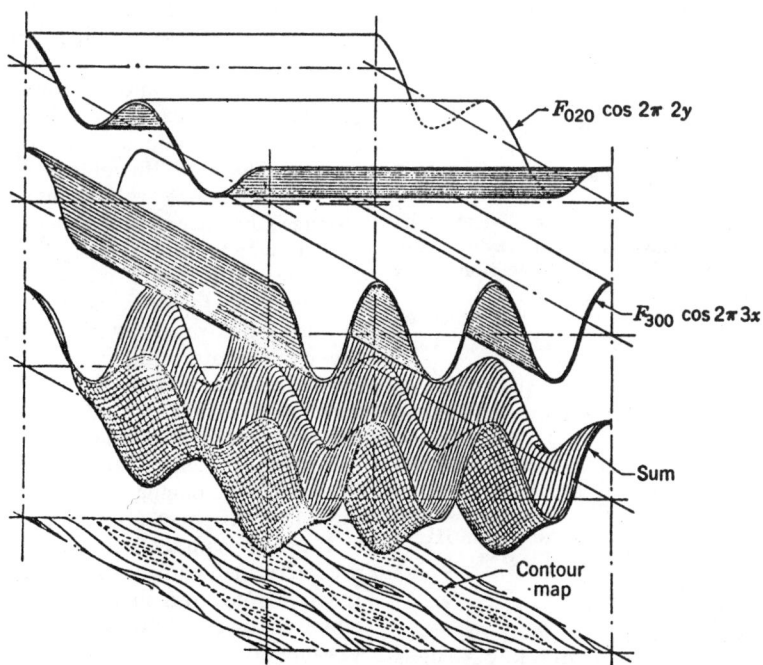

FIG. 13. The addition of two waves and the contour map (McLachlan, 1957).

Eq. (12), being an expression for these constants, is a Fourier transform of the series.

Although the absolute magnitudes of the structure factors may be derived from the experimentally obtained intensity data, the derivation of the structure cannot be carried out in a straightforward manner because information about the phase angles (ϕ) is not furnished by x-ray data. This problem, the so-called phase problem, is a major obstacle to structure-analytical work. A few ingenious, indirect techniques and methods are available but none reduce the task to mere computational labor. Structure-analytical work still requires the ingenuity and diligence of the researcher who must develop a complete model from incomplete data.

EDWARD STURM

References

Aoyagi, K., 1967, "Mineralogical study of sedimentary rocks in the oil fields of Japan by the X-ray diffraction method," *Clay Sci.*, 3(1–2), 37–53.
*Azaroff, L. V., and M. J. Buerger, 1958, "The Powder Method," New York, McGraw-Hill Book Co.
*Barrett, C. S., 1952, "Structure of Metals, Crystallographic Methods, Principles and Data," New York, McGraw-Hill Book Co.
Barrett, C. S., and Massalski, T. B., 1966, "Structure of Metals," Third ed., New York, McGraw-Hill Book Co., 654pp.
*Buerger, M. J., 1942, "X-Ray Crystallography," New York, John Wiley & Sons.
*Buerger, M. J., 1941, reprinted in 1953, "Numerical Structure Factor Tables," Geological Society of America Special Paper 33.
*Buerger, M. J., 1960, "Crystal Structure Analysis," New York, John Wiley & Sons.
*Bunn, C. W., 1961, "Chemical Crystallography," Oxford, Clarendon Press.
Cullity, B. D., 1956, "Elements of X-Ray Diffraction," Reading, Mass. Addison-Wesley Publ.
Gould, E. S., 1955, "Inorganic Reactions and Structure," New York, Holt & Co.
James, R. W., 1962, "The Optical Principles of the Diffraction of X-Rays" Ithaca, N.Y., Cornell Univ. Press, 664pp.
Klug, H. P., and Alexander, L. E., 1954, "X-Ray Diffraction Procedures," New York, John Wiley & Sons.
Kittel, C., 1966, "Solid State Physics," Second edition, New York, John Wiley & Sons.
*Lipson H., and Cochran, W., 1953, "The Determination of Crystal Structures," Vol. III in (Bragg, L., editor) "The Crystalline State," G. Bell and Sons, London.
*McLachlan, Dan, 1957, "X-Ray Crystal Structure," New York, McGraw-Hill Book Co.
Mirkin, L. I., 1964, "Handbook of X-Ray Analysis of Polycrystalline Materials," New York, Consultants Bureau, 731pp. (transl. from Russian).

*Additional bibliographic references may be found in this work.

Cross-references: Crystal Chemistry; Differential Thermal Analysis; Mineralogy; X-Ray Diffraction—Variable Atmosphere; X-Ray Diffraction—Variable Temperature; X-Ray Spectroscopy. Vol. IVB: Crystallography—Morphological; Dispersion—Optical; Electron Probe Microanalysis; Optical Mineralogy.

X-RAY DIFFRACTION—VARIABLE ATMOSPHERE

The basic techniques of "X-Ray Diffraction Analysis" and "X-ray Diffraction—Variable Temperature" are discussed under these headings elsewhere.

The combined application of these two techniques is often greatly expanded when the gaseous atmosphere conditions surrounding and often reacting with the diffracting sample are capable of controlled modification.

By altering the internal gaseous environmental conditions of the diffraction cameras with suitable single or multicomponent gases, oxidizing, reducing or inert conditions may be produced. These conditions may be maintained, promoted, or altered at will before or during individual variable temperature x-ray diffraction determinations.

Cameras (for a good example see Fig. 1). For detailed references to camera types, see under "X-Ray Diffraction—Variable Temperature," but in general they operate on: (1) a dynamic gas flow, purging effect, where the main concern is to maintain the desired atmosphere conditions and remove efficiently any gaseous decomposition products, or (2) under gas pressures, static or dynamic, above or below normal (see Buwalda, 1966, particularly for high-pressure references). To date the emphasis has been on the low-pressure vacuum aspect, which has the added advantage of increasing diffraction line definition and decreasing film fogging and exposure times, due to reduced x-ray scattering effects caused by the camera atmosphere.

In the second case the predetermined pressure controls the partial pressure of the gas present, which if critical, will increase or decrease the reaction temperature, speed and intensity, i.e., magnesite or calcite in a carbon dioxide atmosphere (Bayliss and Warne, 1962).

General Application. Thus by the use of predetermined camera "atmosphere" conditions, reaction rates and their products (see Fig. 2), which only occur under specific gaseous environments, can be studied with confidence. Conversely it may prevent some unwanted reactions, which may be confirmed by the absence

X-RAY DIFFRACTION—VARIABLE TEMPERATURE

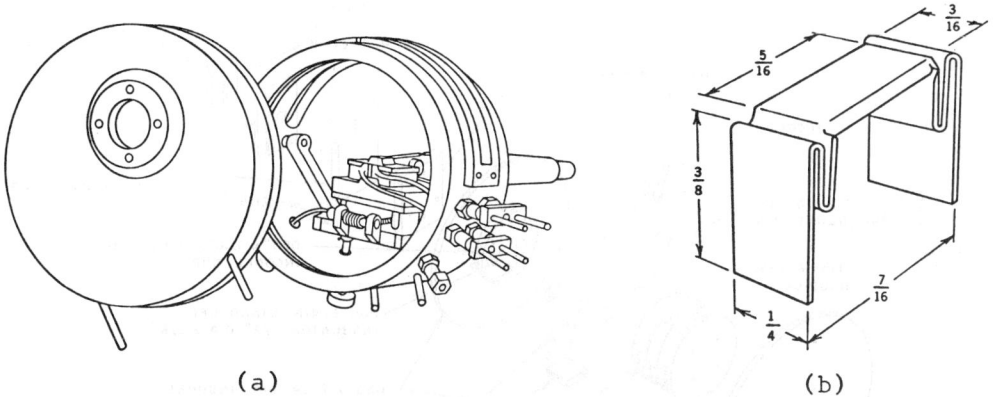

FIG. 1. (a) Schematic view of the high-temperature variable atmosphere x-ray camera of Intrator and Hurwitt (1961). (b) Detail of the platinum-40% rhodium heater strip (from Wendlandt 1964).

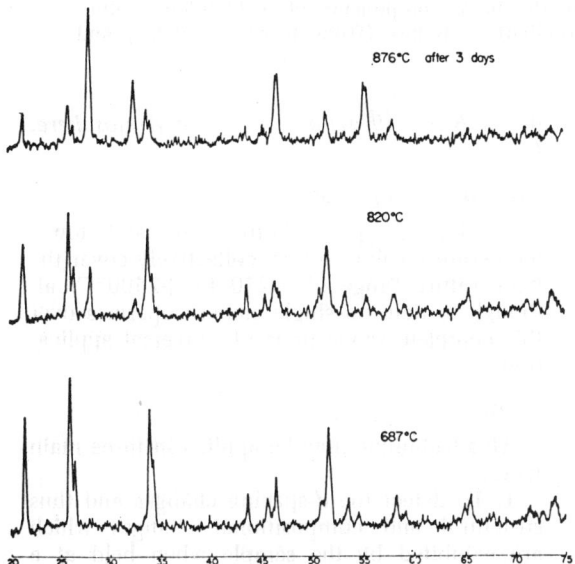

FIG. 2. Pressure-temperature relations of U_3O_{8-y}. The three patterns obtained at an O_2 pressure of 10^{-3} mm Hg show the appearance of the face-centered cubic UO_{2+x} phase at 830°C. Its appearance is gradual as indicated by the upper pattern taken three days later. The evidence indicates that the 10^{-3} O_2 isobar crosses the two-phase region U_3O_{8-y}–UO_{2+x} below 820°C (from Smith 1963).

of their reaction products, but will not stop solid-state reactions. It is also widely used in conjunction with controlled atmosphere *differential thermal* and *thermogravimetric* analysis, to elucidate thermogram modifying reactions from their products.

S. St. J. Warne

References

Bayliss, P., and Warne, S. St. J., 1962, "The effects of controllable variables on differential thermal analysis," *Am. Mineralogist*, **47**, 775–778.
Buwalda, J. (editor), 1966, "Review of Literature," Third ed., Vol. XRD, Netherlands, Philips, 78pp.
Smith, D. K., 1963, "Techniques of high-temperature X-ray diffraction using metal ribbon furnaces," *Norelco Reporter*, pp. 19–29.
Wendlandt, W. Wm., 1964, "Thermal Methods of Analysis," New York, Interscience Publ., 424pp.

Cross-references: *Differential Thermal Analysis; Thermal Analysis; Thermogravimetric Analysis; X-Ray Diffraction Analysis; X-Ray Diffraction—Variable Temperature.*

X-RAY DIFFRACTION—VARIABLE TEMPERATURE

The basic technique is covered elsewhere under the heading of *X-Ray Diffraction Analysis*. However, the usual diffraction pattern determination conditions, i.e., in air at normal temperatures and pressures, are not suitable for certain types of investigations. For example, thermally unstable materials or products which outside specific temperature ranges suffer further decomposition, recombination, phase or crystallographic change, or inversion-reversion reactions. Furthermore, often these are only produced under certain "atmospheric" conditions, i.e., oxidizing, reducing, and inert (see under *X-Ray Diffraction—Variable Atmosphere*).

The investigation of such materials requires the direct determination of their x-ray diffraction pattern while the sample is held at, or

X-RAY DIFFRACTION—VARIABLE TEMPERATURE

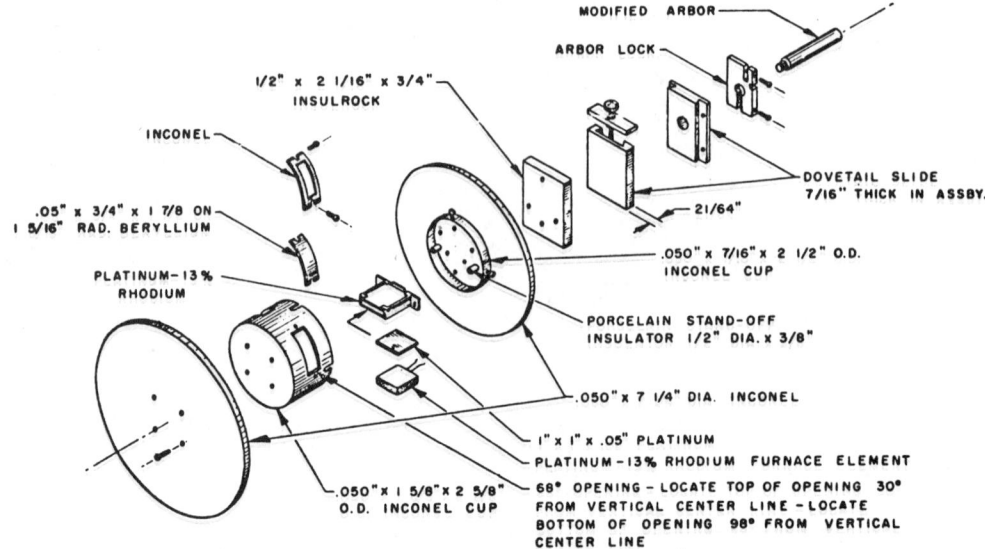

FIG. 1. "Exploded" representation, showing the basic components of a high-temperature x-ray diffraction camera—in this case for oscillation studies (from Rowland, Weiss, and Lewis, 1959).

passed through, a predetermined temperature or temperature range, often under controlled gaseous atmosphere conditions.

The employment of these additional variable parameters gives rise to the two independent, but complementary, techniques of *variable temperature* and *variable atmosphere x-ray diffraction*, which usually utilize a single dual purpose unit. The former is also a recognized thermal analysis method (see *Thermal Analysis*).

The range of application and usefulness of x-ray diffraction may therefore be greatly increased if the sample temperature can be varied as required.

Cameras

The technique has expanded rapidly with the development and description of many camera types and modifications, e.g. Goldschmidt (1955); Fridrichsons (1956); Austin, Richard, and Schwartz (1956); Jetter, McHargue, Williams, and Yokel, (1957); Goon, Mason, and Gibb (1957); Katz and Kay (1957); Markowitz, Kishel, and Cree (1958); Bond (1958); Horne, Croft, and Smith (1959); Intrater and Hurwitt (1961); Intrater (1961); Shimura (1961); Intrater and Appel (1961) (see in Wendlandt, 1964); Weiss and Rowland (1956); Rowland, Weiss, and Lewis (1959) (see in Garn, 1965); Bassett and Lapham (1957), and Skinner, Stewart, and Morgenstern (1962). These basically utilize powder (e.g. Debye-Scherrer and diffractometer) and single-crystal techniques. (For examples of basic camera components and layout, see Fig. 1 and under *X-Ray Diffraction—Variable Atmosphere*, Fig. 1a.)

Operating Temperatures

The known types of both "high-" and "low-" temperature cameras now collectively cover the temperature range of -270 to $+2500°C$, although to date no single camera operates over this complete range or is of universal application.

Methods

This technique may be applied in three main ways:

1. To detect the *d*-spacing changes and thus structural and compositional changes which are exhibited by the sample when held at a specific static temperature (see Fig. 2 under *X-Ray Diffraction—Variable Atmosphere*) or a series of constant temperatures (see Fig. 2).

2. To continuously study and record the

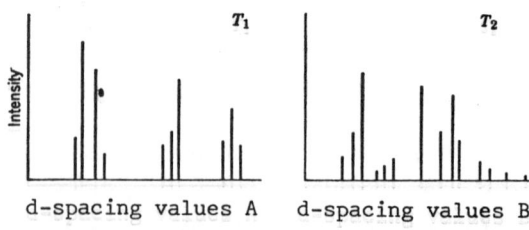

FIG. 2. Static-temperature method. Two different sets of *d*-spacing values A and B, produced by changing the static examination temperatures from T_1 to T_2 (from Wendlandt, 1964).

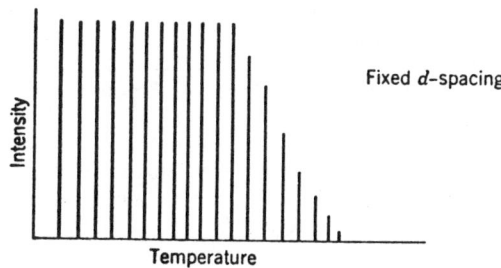

Fig. 3. Dynamic-temperature method. The intensity of a specific single d-space value remains static or changes over particular temperature ranges through which the sample is heated (from Wendlandt, 1964).

effect of progressively increasing temperature on a single critical or diagnostic diffraction peak (d-spacing value) (see Fig. 3).

3. For single crystals, to detect and measure, usually on a series of separately exposed photographs, the progressive movement, appeearance or disappearance of particular reflections in response to temperature variations.

In the first method the sample, having been raised to the required temperature, is maintained there while its diffraction pattern is determined. By repeating this procedure at set intervals the temperature range within which the particular reaction occurs can be established and the new product or products identified. Subsequently by repeating this process, within the now established temperature range, using smaller and smaller temperature increments, the "reaction" temperature can eventually be established within acceptably close and accurate limits (see Fig. 4).

In the second method the diffractometer goniometer is: (a) located in a suitable stationary position so that the presence and intensity of a diagnostic peak of the initial or derived product, and thus the amount of this material present, may be continuously evaluated with respect to progressively increasing sample temperature (see Fig. 3); and (b) made to oscillate across a selected x-ray diffraction maximum (Weiss and Rowland, 1956, see Fig. 5. and for their camera, Fig. 1; for a recent appraisal, see Garn, 1965).

In the third case a number of methods are available for the initial determinations, e.g., the Laue, Buerger Precession, Rotation and Oscillation, Equi-inclination Weissenberg and Single Crystal Diffractometry (see Nuffield, 1966). One major advantage is that weak reflections which may distinguish structural types are often not apparent on powder photographs (see Smith and Gay, 1958).

The Laue photograph, obtained with the crystal stationary is difficult to relate to the dimensions of the unit cell or the atomic distribution within the cell although crystal symmetry can often be obtained. If however, once correctly oriented, the crystal is rotated or oscillated, the resultant diffraction pattern may be directly related to cell dimensions and the indexes of reflecting planes obtained.

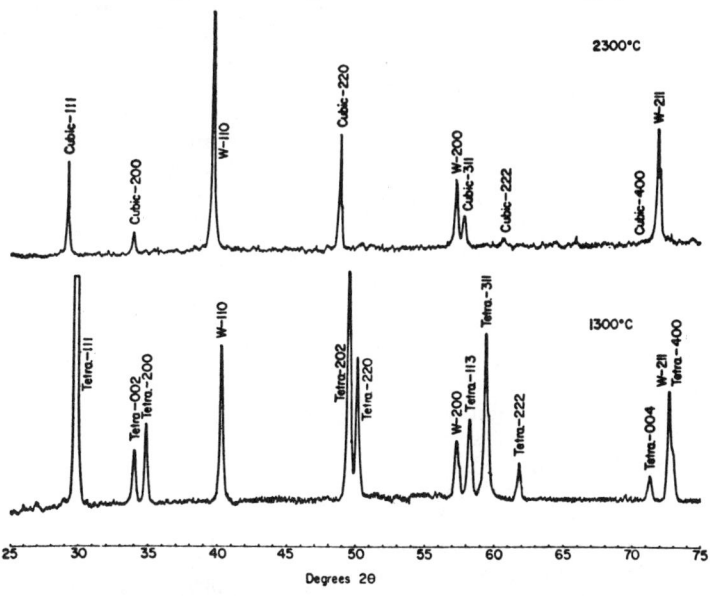

Fig. 4. Diffractometer patterns of two polymorphic forms of ZrO_2 obtained at 1300 and 2300°C. The face-centered pattern is observed in the higher-temperature pattern. Reflections resulting from the tungsten heater are also present (from Smith, 1963).

X-RAY DIFFRACTION—VARIABLE TEMPERATURE

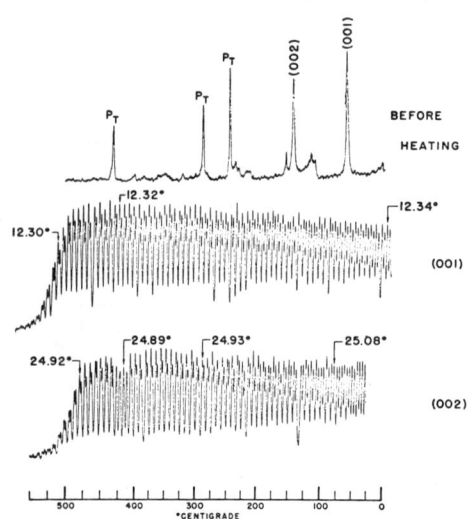

FIG. 5. Heating-oscillation patterns of the first and second order of the basal spacing of kaolinite (from Weiss and Rowland, 1956).

Commonly used is the Weissenberg method which employs an oscillating crystal and a film movement synchronous with and in a direction parallel to the crystals axis of rotation.

The comprehensive annotated bibliography "High Temperature X-ray Diffraction Techniques" by Goldschmidt (1964), covers this whole topic in detail, and is complemented by "Review of Literature" (Buwalda, 1966).

Detection Objectives

In most cases the principal objectives are the detection of the appearance, disappearance, and intensity variations of specific diffraction peaks or reflections which indicate the presence, absence, or content of a particular sample constituent; sudden peak position shifts due to state change; or the more gradual peak migrations used by Skinner (1962) to calculate thermal lattice expansions and molar volumes.

Heating Methods

The form in which the sample is heated varies from a small rod, virtually the whole surface of which is exposed to a much larger flat layer, where all but the top surface is obscured (i.e., less than 50% surface exposed). The sample volume and geometry thus have considerable bearing on the actual sample temperature response to the heating mechanism and the thermal gradients and variations produced within the sample. This causes some difficulty in measuring accurately the temperature of the actual portion of the sample producing the x-ray diffraction effects.

There are two alternatives:

1. The sample is heated by the direct conduction of heat from the source, usually some form of strip heater (Fig. 6). Upon this the sample rests with only one side (lower) in thermal contact with the heater. As the heat losses by radiation from the upper unheated face of the sample are not fully compensated by heat transfer from below and within the sample, a temperature gradient results which becomes progressively more pronounced with increasing temperature. In another case the sample was mounted on a wire which facilitated centering and heating when an electric current was passed along it.

2. The sample is heated by radiation from a heat source with which it is not in contact and which heats the upper or diffracting surface or even completely surrounds the sample rod (Bond 1958).

In this way, as the x-ray beam penetration of the sample surface is minimal, the actual diffraction producing material receives the maximum unmodified heat direct from the source. Furthermore, its temperature is essentially unaffected by variable heat gradients particularly if the sample is kept thin and supported on insulating material.

For investigations below ambient temperature the first method is modified by employing a circulating coolant such as precooled nitrogen gas.

Additional details are given in Wendlandt (1964) who has described a further complication. If during heating the sample suffers volume changes, then diffractometer determined diffraction patterns may be misleading due to sample surface changes causing increased x-ray scattering. This may be overcome by re-smoothing the surface before each new diffraction determination.

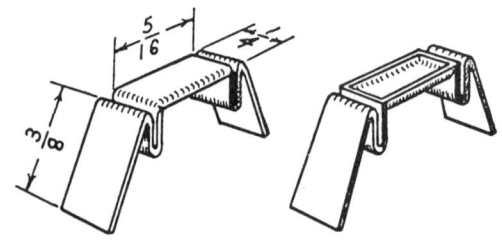

FIG. 6. Strip heater styles: (1) for use with thick pastes; (2) for samples suspended in volatile liquids or solid samples which would slide off the heater at high diffraction angles (from Smith 1963).

General Applications

Applications include the following:
1. Structural changes and crystallographic inversions and reversions.
2. Phase changes.
3. Lattice defects.
4. Lattice expansions and molar volumes.
5. The detection of materials, polymorphs, and breakdown and reaction products, usually not produced *or* unstable at normal temperatures.
6. The detection of transition reaction products.
7. As a major supporting technique to other thermal analysis techniques, in particular, differential thermal and thermogravimetric analysis.
8. In a vast number of economic and industrial applications where the complete and sequential temperature controlled reaction mechanisms and products produced by a given process or set of conditions, are required.

In connection with *differential thermal analysis* (DTA) and *thermogravimetric analysis* (TGA; for details see under these headings elsewhere) variable temperature x-ray diffraction is an invaluable aid in establishing the actual reactions responsible for unknown thermogram modifications by the identification of the constituents present before and after reaction (or change) has taken place. In such cases the sample may be heated up to just above the reaction temperature as shown by DTA or TGA and the new constituents identified from the resultant diffraction pattern *or* the DTA/TGA uniform heating rate or variable heating program is reproduced in the x-ray camera so that the advent, disappearance, or stable temperature range of a particular compound may be established.

S. St. J. Warne

References

Bassett, W. A., and Lapham, D. M., 1957, "A Thermal Increment Diffractometer," *Am. Mineralogist,* **42,** 548–555.
Buwalda, J., (editor), 1966, "Review of Literature," Third ed., Vol. XRD, Netherlands, Philips, 78pp.
*Garn, P. D., 1965, "Thermoanalytical Methods of Investigation," New York, Academic Press, 606pp.
Goldschmidt, H. J., (editor), 1964, "High Temperature X-ray Diffraction Techniques," International Union of Crystallography.
*Nuffield, E. W., 1966, "X-ray Diffraction Methods," New York, John Wiley & Sons, 409pp.
Rowland, R. A., Weiss, E. V., and Lewis, D. R., 1959, "Apparatus for the oscillating-heating method of X-ray powder diffraction," *J. Am. Ceram. Soc.,* **42,** 133–138.
Skinner, B. J., Stewart, D. B., and Morgenstern, J. C., 1962, "A New Heating Stage for the X-ray Diffractometer," *Am. Mineralogist,* **47,** 962–967.
Skinner, B. J., 1962, "Thermal Expansion of Ten Minerals," *U.S. Geol. Surv. Profess. Paper,* 450-D.
Smith, D. K., 1963, "Techniques of high-temperature X-ray diffraction using metal ribbon furnaces," *Norelco Reporter,* 19–29.
Smith, J. V., and Gay, P., 1958, "The Powder Patterns and Lattice Parameters of Plagioclase Feldspars," II, *Mineral. Mag.* **31,** 744–762.
Weiss, E. J., and Rowland, R. A., 1956, "Oscillating reacting X-ray diffractometer studies of clay mineral dehydroxylation," *Am. Mineralogist,* **41,** 117–126.
*Wendlandt, W. Wm., 1964, "Thermal Methods of Analysis," New York, Interscience Publishers, a div. of John Wiley & Sons, 424pp.

Cross-references: *Differential Thermal Analysis; Thermal Analysis; Thermogravimetric Analysis; X-Ray Diffraction Analysis; X-Ray Diffraction —Variable Atmosphere.*

X-RAY EMISSION ANALYSIS—See X-RAY SPECTROSCOPY

X-RAY SPECTROSCOPY

Scientists and engineers need a rapid, highly specific method of elemental analysis. X-ray fluorescence achieves this in many cases. It was introduced by Coster and von Hevesy in 1923 and at once proved its value by showing the presence of hafnium in Norwegian zircons. Quantitative analysis of major and minor elements, and some trace elements by x-ray spectroscopy finds increasing use in geologic research and ore exploration. Semiquantitative techniques are applied in such areas as oceanography where many samples have to be evaluated rapidly.

X-rays impinging on a sample target dislodge electrons from the inner shells of the target matter. The vacancy thus produced is rapidly filled by another electron from an outer shell. Quantum jumps caused by this electron displacement give rise to secondary x rays of wavelengths characteristic of the target matter. Because only a few energy steps are possible, the x-ray fluorescence spectra have fewer lines and are easier to interpret than emission spectra. The intensity of secondary radiation of characteristic wavelengths is a measure of the amount of an atomic species present in the sample target. The method is nondestructive and rapid. It consists of exposing a sample

X-RAY SPECTROSCOPY

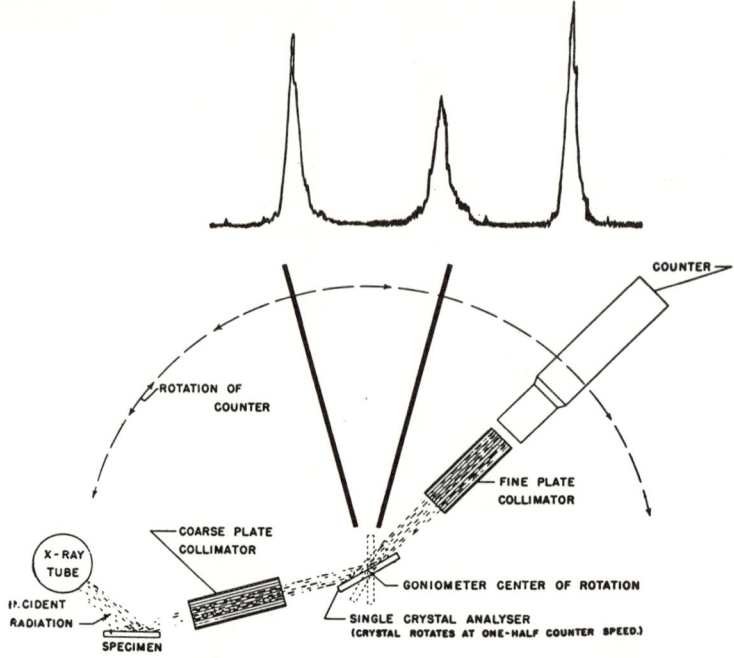

Fig. 1. Diagrammatic representation of x-ray apparatus and (above) the resultant trace. *Courtesy Philips Electronic Instruments.*

material to x rays and measuring the secondary x rays emitted by the sample. Since the wavelengths of these secondary x rays are characteristic for the elements which cause them, it is only necessary to measure the radiation intensity of specific wavelengths excluding as much as possible all other radiation. This can be achieved by collimating the radiation given off by the sample material and directing it to an analyzing crystal. A crystal is chosen for this purpose which has an interatomic spacing of the proper magnitude to satisfy the Bragg equation for the wavelength of the characteristic radiation sought.

As an example: the wavelength of the K_a radiation of cobalt is 1.7853 Å. A lithium fluoride analyzing crystal has a d-spacing of 4.028 Å between the crystallographic (200) planes. The crystal will therefore reflect characteristic Co-radiation that strikes it at an incident angle of 12°48′. A radiation detector placed at the corresponding exit angle will register radiation with a wavelength of 1.78 Å, and to a lesser extent the higher-order waves of 0.89 Å, 0.59 Å, etc. Radiation of other wavelengths is almost completely excluded from the detector. Characteristic radiation emitted by other elements is measured similarly, by placing the analyzing crystal and the detector in position to satisfy Bragg's law. X radiation registered by the detector can be amplified and counted by means of an electronic rate meter and scaler (Fig. 1).

A radioisotope source of suitable wavelength can be substituted for the x-ray generator, and portable x-ray fluorescence analyzers are available. Electrons produced by an electron gun may also be used to excite secondary x rays, as in electron probes.

Absorption Edge and Background

The nature and generation of x rays as well as Bragg's law are explained in the entry on x ray diffraction analysis of this encyclopedia.

When absorption of energy is plotted against wavelength each element shows sharp breaks at certain wavelengths called the *absorption edge*. For most efficient production of secondary radiation a tube emitting x-rays with a peak wavelength slightly shorter than the absorption edge of the element sought should be chosen.

In addition to the characteristic radiation the detector will register scattered, or background radiation. Coherent scattering results from elastic collision of x-ray quanta. Compton or inelastic scattering is due to low-atomic number atoms such as oxygen, carbon, hydrogen. It appears as a broad hump in the small-angle portion of the spectrum. About half the

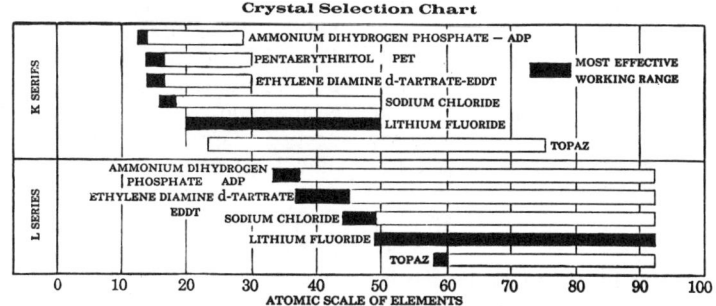

FIG. 2. Commonly used crystals, and their effective range.

background is caused by scattering of the secondary radiation by the analyzing crystal. This portion of the background can be eliminated by pulse height discrimination.

The Analyzing Crystal

Wavelength range, intensity, and resolution are the main criteria for choosing an analyzing crystal. Figure 2 shows the most commonly used crystals and their effective range.

In addition, graphite crystals are coming into use as highly efficient analyzers in the short-wave spectrum, and multiple layer aggregates, such as barium stearate for the very long wavelength range ($d = 100.4$ Å).

Instead of a flat analyzing crystal, a curved or "focusing" crystal may be used. Several different geometrical arrangements are possible with the crystal either in diffraction or in transmission position. Secondary radiation emitted from a very small specimen may be measured, or conversely, a low concentration of an element may be detected by focusing the rays from a large sample area. Adler and Axelrod (1957) used and described various techniques involving curved crystal optics. It is also possible to use a relatively thin edge (0.125 mm) of a crystal (LiF) for simultaneous recording of the complete spectrum. A homogeneous sample specimen is required and the spectrum may be recorded on film without rotating the crystal (Birks and Brooks, 1955).

Detectors and Pulse Height Discrimination

Any detector of radiation may be used to register secondary radiation given off by the sample. Because proportional counters and scintillation counters are capable of distinguishing energy (wavelength) in addition to intensity, these two are most common at present. These counters register the number of x-ray quanta per unit time, and their energy, in the form of electric pulses of different intensity. By means of an appropriate electronic circuit it is thus possible to eliminate radiation of wavelength other than that of interest. Pulse height discriminators suppress high-order radiation contributed by interfering elements, and increase the signal-to-background in trace element analysis.

Nondispersive Spectrometers

In recent years solid-state detectors, usually consisting of a lithium-drifted silicon or germanium crystal were developed. In combination with multichannel analyzers they can measure x-ray intensities directly while discriminating between wavelengths. In such an instrument analyzing crystals are not needed, and a complete wavelength spectrum can be obtained in seconds (Aitkins, 1968; Slivinsky and Ebert, 1969). Nondispersive spectrometers with a radioactive isotope to excite characteristic radiation can be made very compact. A portable fluorescence spectrometer, set on the characteristics wavelength of lead, can be hand-carried to instantaneously detect poisonous lead paint in slum areas.

Qualitative Analysis

X-ray fluorescence analysis is nondestructive, it requires relatively little sample, and results are inherently reproducible. The spectra contain fewer lines than light optical emission spectra, which makes identification easier. Qualitative analysis is usually made by rapid scanning and recording of 2θ against intensity. A rate meter with a logarithmic scale is useful because it helps to keep peaks on scale even at high-sensitivity settings. For a complete scan two different analyzing crystals and two detectors are needed. For the light elements from Al to Ca, an EDTA (ethylenediamine tartrate) or a PET (pentaerythritol) is preferable with a flow-proportional counter as detector. In this range air absorption is strong, and either a helium-pass or a vacuum spectrometer are needed. For the elements heavier than potassium a lithium fluoride crystal and a scintillation counter as the detector are most efficient.

A thin-window tungsten tube will cover the range of elements from Al upward.

Comparison of spectra obtained by scanning from 90° to 3°2θ with and without pulse height discrimination, and with the use of thin metal filters permits identification of elements $Z = 83$ to $Z = 13$. Templates are available for rapid comparison.

Semiquantitative Analysis

Relative intensities of identified peaks can be related to abundance of a given element. Absorption and enhancement depend on elemental composition and also contribute to the measured intensities. Comparing an unknown with a known standard of different composition thus may become rather inaccurate. These matrix effects can be partly overcome by using scattered radiation (Andermann and Kemp, 1958).

Quantitative Analysis

The intensity of radiation of a specific reflection is a function of the abundance of the excited element and of several other factors. It is very difficult, and usually impractical to compute concentration directly from intensity. Comparison of measured intensities with those of standards of known composition will render quantitative results, if the other parameters are taken into account. Mass absorption, self absorption, or enhancement effects produced by other elements in the sample affect intensities to varying degree. Several methods are available to compensate for these effects: (*1*) use of standards with an elemental composition similar to the unknown; (*2*) dilution technique; (*3*) addition of a heavy absorber; (*4*) addition of an internal standard; (*5*) compensation of absorption-enhancement effects by means of a computer (Kodama et al., 1967).

Sample Preparation

Particle size, flatness, density, and thickness of the specimen affect intensity measurements. Claisse (1956) thoroughly evaluated these variables and indicated ways to deal with the problems. One method frequently used is to fuse the weighted sample with a flux, such as lithium tetraborate and then either cast the molten mass onto a flat, heated plate, or let it cool and grind it to prepare a pressed pellet (Rose et al., 1963, and Van den Heuvel, 1965).

Geological Applications

Addition of lanthanum oxide as a heavy absorber, lithium borate fusion, grinding, and pressing a pellet will produce straight-line calibration plots over a wide range of elemental composition. The major elements in silicates such as Si, Al, Ca, K, Fe, Ti can be routinely determined by this method with good accuracy. For Na and Mg a separate sample pellet is pressed without fusion or absorber, and a thin-window flow counter and a Cr–X-ray tube are needed. The light elements, Na, F, O, N, C, and B have been determined with a Henke tube (1963). This tube has an aluminum target and is operated at high current (150 milliamp) and low potential (10 kV).

Major elements in sulfide ores can be analyzed by fusing with an excess of K-pyrosulfate which acts as a diluent (Cullen, 1960).

X-ray fluorescence is useful in trace analysis of elements difficult to measure by other methods, such as zirconium-hafnium (J. Parks et al., 1966). Rubidium, strontium, and barium were analyzed (Price and Angell, 1968) by thin film techniques. Tantalum-columbium in ores, uranium and thorium in monazite, and many rare earths as well as nonmetals such as sulfur, selenium, bromine have been successfully determined in a variety of minerals. The limits of detection are generally in the parts-per-million range. Pulse-height discrimination is essential to eliminate the effects of interfering elements and higher-order reflections.

X-Ray Macroprobe

A macroprobe using a curved analyzing crystal can be mounted on a standard x-ray spectrograph. It is capable of analyzing elemental concentration in a sample specimen with an area of as little as 100-microns diameter. Changes in elemental composition within the specimen may be measured (Hermes and Ragland, 1967).

Electron Microprobes

In the electron microprobe, unlike x-ray fluorescence, the elements in the sample are brought into an excited state by means of high-energy electrons. The sample target, which must be made electrically conductive, is exposed to a stream of electrons of sufficient energy to dislodge valence electrons from the inner shells of the sample material. This process gives rise to characteristic x radiation. The effect is similar to that produced by bombardment with x rays. Here, too, the x rays produced are characteristic of the type and abundance of the elements present in the sample, and they may be treated in a similar fashion as secondary x rays in fluorescence. The stream of electrons may be produced by an electron gun, or by a radio isotope emitting beta rays. In either case it is necessary to evacuate the sample chamber to avoid absorption by air. Beta probes are capable of rapid analysis of light elements. Fluorine, oxygen, carbon, and boron

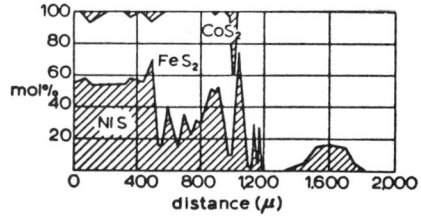

FIG. 3. Electron-probe measurements on natural zoned bravoite (after Springer and Long, 1963).

are within the range of this method, but too light for x-ray fluorescence.

Electrons can be focused almost as precisely as visible light using electronic lenses. It is thus possible to concentrate an electron beam and to direct it to a very small sample area. The highly concentrated electron beam may be focused on an area as small as one micron (0.0001 cm) in diameter and produce an almost instantaneous reading of the amount of a desired element present. The electron beam can also be made to scan the surface of a sample specimen, such as the scanning which produces a picture in a television tube. It is thus possible to obtain a line profile of the changes in concentration of one element in a sample, and all this with high resolution. A series of parallel scans will render a two-dimensional picture of elemental distribution.

By varying the angular positions of the analyzing crystal and the detector, the distribution of other elements may be determined in the same fashion. Figure 3 shows line profiles of the iron, cobalt and nickel content in a specimen of zoned bravoite. Elements of low atomic number can be analyzed because electron-excited x-ray fluorescence is a more efficient process than x-ray excited fluorescence. Electron-excited fluorescence also lends itself to mathematical treatment more readily than x-ray excited fluorescence and, therefore, element concentration can be often directly computed from x-ray intensity of the characteristic radiation. Calibration curves and standards are needed mostly as a check.

The advantages of electron-excited analysis are balanced by the higher cost of equipment and maintenance which may be as much as ten times that of equivalent x-ray excited fluorescence, and by the need to operate at high vacuum. In spite of these drawbacks electron-probe analysis is an important tool where high resolution elemental analysis is required, such as in the study of fine detail of ore mineral intergrowth and alteration boundaries, crystal inclusion, zoning, etc. A more detailed treatment of the electron microprobe, its theory, instrumentation, and applications is found in Adler (1966).

WILLIAM LODDING

References

Adler, I., 1966, "X-ray emission spectrography in geology," New York, Elsevier Publ. Co., 258pp.

Adler, I., and Axelrod, J. M., 1957, "Reflecting curved-crystal X-ray spectrograph," *Econ. Geol.*, 52(6), 694–701.

Aitkins, D. W., 1968, "High resolution with non-dispersive X-ray spectrometers," *Office Aerospace Research Rev.*, 7, 9, 1.

Andermann, G., and Kemp, J. W., 1958, "Scattered X-rays as internal standards in X-ray emission spectroscopy," *Anal. Chem.* 30, 1306–1309.

Birks, L. S., and Brooks, E. J., 1955, "Applications of curved crystal x-ray spectrometers," *Anal. Chem.* 27, 437–440.

Claisse, F., 1956, "Accurate X-ray fluorescence analysis without internal standards," *Quebec Dept. Mines, Prelim. Report*, 32, 16.

Coster, D., and Von Hevesy, G., 1923, "The new element hafnium," *Nature*, 111, 182.

Cullen, T. J., 1960, "K-pyrosulfate fusion technique," *Anal. Chem.* 32, 516–517.

Henke, B. L., 1963, "Production, detection and application of ultra-soft X-rays," in "X-ray Optics and X-ray Microanalysis," New York, Acad. Press, 157–172.

Hermes, O. D., and Ragland, P. C., 1967, "Quantitative chemical analysis of minerals in thin section with the X-ray macro-probe," *Am. Mineralogist*, 52(3–4), 493–508.

Kodama, H., Brydon, J. E., Stone, B. C., 1967, "X-ray spectrochemical analysis of silicates using synthetic standards with a correction for interelemental effects by a computer method," *Geochim. Cosmochim. Acta*, 31, 649–659.

Price, N. B., and Angell, G. R., 1968, "Determination of minor elements in rocks by thin film X-ray fluorescence techniques," *Anal. Chem.* 40, 660–663.

Rose, H. J., Adler, I., and Flanagan, F. J., 1963, "Use of X-ray fluorescence Analysis of the light elements in rocks and minerals," *Appl. Spectr.* 17(4), 81–85.

Slivinsky, V. W., and Ebert, P. J., 1969, "Efficiency calibration of a Ge-Li detector from 8–98 keV," *Nuclear Instrum. Methods*, 71, 346.

Springer, G., and Long, J. V. P., 1963, "Electron-probe analysis of minerals in the system FeS2-CoS2-NiS2," in "X-ray Optics and X-ray Microanalysis," New York, Acad. Press, 611–617.

Van den Heuvel, R. C., 1965, "X-ray emission spectrography," *Methods Soil Anal., Am. Soc. Agron.*, 786–820.

Cross-references: *Atomic Number and Periodic Table; Crystal Chemistry; Elements; Radioactive Isotopes; Spectrophotometry; Trace Elements; Versene (EDTA) Solution Studies; X-Ray Diffraction Analysis. See also Individual Elements.* Vol. IVB: *Crystallography; Electron Probe Microanalysis; Luminescence.*

Y

YTTERBIUM: ELEMENT AND GEOCHEMISTRY

De Marignac (in 1878) first isolated ytterbium from other rare earths; he separated the element out of ytterbia, a heavy rare earth oxide mixture. Ytterbium was named after Ytterby, Sweden, where Arrhenius discovered gadolinite, the first heavy rare earth mineral.

Properties

Elemental ytterbium (Yb) is a very reactive, silvery white *lanthanide* (q.v.), or *rare earth* (q.v.) metal with atomic number 70 and outer electronic structure $4f^{14}\,6s^2$. It occurs as seven stable isotopes:

Yb^{168}	0.14% natural abundance
Yb^{170}	3.03
Yb^{171}	14.3
Yb^{172}	21.8
Yb^{173}	16.2
Yb^{174}	31.8
Yb^{176}	12.7

Ytterbium is determined analytically by neutron activation, x-ray spectrography, and isotope dilution mass spectrometry. Unlike other rare earths, tripositive ytterbium has a broad infrared absorption band. In geologic environments, ytterbium forms highly ionic bonds and weak complexes. The oxidation state is tripositive, with ionic radius 0.858Å; dipositive ytterbium is oxidized by water, and probably is unstable in most crystal lattices. Generally, chemical behavior is similar to the alkaline earths, and almost identical to other heavy rare earths.

Minerals

Ytterbium occurs in highest concentration in rare earth minerals that concentrate heavy rare earths. Generally nitrates, sulfates, carbonates, niobates and tungstates with cation-coordination number of six to eight favor the heavy rare earths, and often contain structural halogens and alkali metals. Typical examples are:

Xenotime (Y, heavy rare earths) PO_4
Fergusonite (Y, heavy rare earths) $(Nb, Ta)O_4$

Geochemical Behavior and Abundance

Schmitt et al. (1964) have established a value of 0.17 ppm (parts per million) ytterbium in chondrites and 2.0 ppm in eucrite achondrites. The standard deviation within a meteorite class is only about 20%. In meteoritic matter, volatization may have redistributed alkali metals but not most rare earths: ytterbium, however has a boiling point of 1427°C, the lowest of any rare earth.

The ytterbium concentration in average basalt is 2.7 ppm, and the range is less than a factor of ten (Haskin et al., 1966). Oceanic and continental basalts have similar ranges. Partial melting of mantle peridotile yields a basaltic liquid with enriched ytterbium and a mantle source area depleted in ytterbium.

Igneous differentiation series may tend toward higher or lower ytterbium concentrations, although the total variation is less than for light rare earths. The relative abundance of ytterbium and other heavy rare earths remains almost unchanged. Haskin et al. (1966) report 3.0 ppm ytterbium in a Gough Island basalt and 8.5 ppm ytterbium in a trachyte differentiate Towell et al. (1965) found 1.30 ± 0.10 ppm in the Rubidoux Mountain Leucogranite phase of the Southern California batholith, probably a differentiate (by crystal settling) of the San Marcos Gabbro, with 1.70 ± 0.14 ppm.

Average granite with 60 to 70% silica has 4.0 ppm but the range of values for granites is even greater than for basalts.

In most basaltic, syenitic, and granitic rocks, ytterbium is mainly in mafic rock-forming minerals, probably substituting for divalent calcium (ionic radius 0.99Å).

Ytterbium and other heavy rare earths form complexes and are transported in alkali, halogen, and carbon dioxide-rich hydrothermal fluids; metasomatic and pegmatoid deposits formed from such fluids may have high concentrations of ytterbium and higher ratios of ytterbium to light rare earths than granites.

Ytterbium has a 390-year residence time in seawater (which is alkaline), about the same as other heavy rare earths. Removal by adsorption on clay particles, as well as depletion in surface water in areas of high organic productivity, are likely processes. High secondary ytterbium concentrations in phosphatic fish remains today occur only in deep watery Paleozoic fish remains also have high rare earth concentrations. Manganese nodules have 460 ppm, whereas Pacific Ocean water has only 0.52 × 10^{-6} ppm (Goldberg et al., 1963).

In soils, the relative amounts of ytterbium in detrital, clay, and carbonate fractions is not well known. Sedimentary rocks have about 3.4 ppm, with a variation of a factor of five. In a study of rare earths in Russian platform soils, Balashov et al. (1964) found higher ytterbium in alkaline soils, the ytterbium groundwater precipitating as hydroxide; lower concentrations are in acid soils, probably due to removal of complexes.

Biologic concentration of rare earths has been found in hickory leaves; rare earths have also been used to trace opium.

Economic Geology

The major source of ytterbium has been from a light rare earth mineral, monazite in beach sands; the accessory monazite originally present in granitic rocks resists weathering and is concentrated by sedimentary processes. Australia, Brazil, Ceylon, India, Malaysia, and South Africa mine the largest quantities. Less than one percent of the monazite is ytterbium. Recently, the bastnaesite deposit at Mountain Pass, California has been an important source of ytterbium, again in low concentration.

High-purity ytterbium has a very limited market. Most ytterbium is sold unpurified in rare earth concentrates used for polishing, petroleum cracking catalysis, lighter flints, arc carbons, and in alloys.

ROBERT KAY

References

Balashov, Y. A., et al., 1964, "The effects of climate and facies environment on the fractionation of the rare earths during sedimentation," *Geochem. Intern. No. 1*, **10**, 951–969 (English transl.).

Goldberg, E. D., et al., 1963, "Rare-earth distributions in the marine environment," *J. Geophys. Res.*, **68**, 4209–4217.

Haskin, L. A., et al., 1966, "Meteoric, Solar, and Terrestrial Rare-Earth Distributions," in (Ahrens, L. H., et al., editors), "Physics and Chemistry of the Earth," Vol. 7, Oxford, Pergamon Press, pp. 167–321.

Rankama, K., and Sahama, T. G., 1950, "Geochemistry," Chicago, University of Chicago Press, 912pp.

Schmitt, R. A., et al., 1964, "Rare-earth, yttrium and scandium abundances in meteoric and terrestrial matter—II," *Geochim. Cosmochim. Acta*, **28**, 67–86.

Towell, D. C., et al., 1965, "Rare-earth distributions in some rocks and associated minerals of the batholith of Southern California," *J. Geophys. Res.*, **70**, 3485–3496.

Cross-references: *Mass Spectrometry; Neutron Activation Analysis; Rare Earths; X-Ray Spectroscopy.*

YTTRIUM: ELEMENT AND GEOCHEMISTRY

Mosander (in 1843) was the first to isolate yttrium from the other rare earth elements; he separated it from yttria, a heavy rare earth concentrate. Yttrium was named after Ytterby, Sweden, where Arrhenius found gadolinite, the first heavy rare earth, or *lanthanide* mineral.

Properties

Elemental yttrium (Y) is a very reactive silvery white *rare earth* (q.v.) metal with atomic number 39 and outer electronic structure $4d^1 5s^2$. Yttrium occurs as one stable isotope, Y^{89}.

Yttrium is determined analytically by neutron activation and by x-ray spectrography. In geologic environments, yttrium forms highly ionic bonds; it forms weak complexes. The oxidation state is tripositive, with ionic radius 0.88Å. Generally, chemical behavior is similar to the alkaline earths and to other heavy rare earths or *lanthanides* (q.v.); commonly the chemical behavior of *erbium* (q.v.) is assumed to be almost identical.

Minerals

Yttrium occurs in highest concentration in rare earth minerals that concentrate heavy rare earths. Generally nitrates, sulfates, carbonates, niobates, and tungstates, with cation-coordination number of six to eight favor the heavy rare earths, and often contain structural halogens and alkali metals. Typical examples are:

Xenotime (Y, heavy rare earths) PO_4
Fergusonite (Y, heavy rare earths) $(Nb, Ta)O_4$

Geochemical Behavior and Abundance

Schmitt et al. (1964) have established a value of 1.8 ppm (parts per million) yttrium in chondrites and 2.2 ppm in eucrite achondrites. The standard deviation within a meteorite class is only about 20%. In meteoritic matter, volatization may have redistributed alkali metals but not rare earths; yttrium has a boiling point of 2927°C.

The yttrium concentration in average basalt is 25 ppm, and the range is less than a factor of ten (Haskin et al., 1966). Oceanic and continental basalts have similar ranges. Partial melting of mantle peridotite yields a basaltic liquid with enriched yttrium and a mantle source area developed in yttrium.

Igneous differentiation series tend toward higher or lower yttrium concentrations, although the total variation is less than for light rare earths. The relative abundance of yttrium and other heavy rare-earths remains almost unchanged. Haskin et al. (1966) report 56 ppm yttrium in an Ascension Island basalt and 34 ppm yttrium in a trachyte differentiate; differentiation of a Gough Island basalt with 30 ppm yttrium yields a trachyte with 101 ppm. Towell et al. (1965) found 17.3 ppm in the Rubidoux Mountain Leucogranite phase of the Southern California batholith, probably a differentiate (by crystal settling) of the San Marcos Gabbro, with 15.5 ppm.

Average granite with 60 to 70% silica has 44 ppm but the range of values for granites is even greater than for basalts.

In most basaltic, syenitic and granitic rocks, yttrium is mainly in mafic rock-forming minerals, probably substituting for divalent calcium (ionic radius 0.99Å).

Yttrium and other heavy rare earths form complexes and are transported in alkali, halogen, and carbon dioxide-rich hydrothermal fluids; metasomatic and pegmatoid deposits formed from such fluids may have high concentrations of yttrium and higher ratios of yttrium to light rare earths than granites.

Probably, yttrium has a 500-year residence time in sea water (which is alkaline), about the same as heavy rare earths. Removal by adsorption on clay particles, as well as depletion in surface water in areas of high organic productivity, are likely processes. High secondary yttrium concentrations in phosphatic fish remains today occur only in deep water. Paleozoic fish remains also have high rare earths. Manganese nodules have about 14,000 ppm, whereas Pacific Ocean water has only about 6×10^{-6} ppm; these values are interpolated from data of Goldberg et al. (1965), assuming a constant ratio of yttrium to heavy rare earths.

In soils, the relative amounts of yttrium in detrital, clay, and carbonate fractions is not well known. Sedimentary rocks have about 35 ppm, with a variation of a factor of five. In a study of rare earths in Russian platform soils, Balashov et al. (1964) found higher yttrium in alkaline soils, the yttrium in groundwater precipitating as hydroxide; lower concentrations are in acid soils, probably due to removal of complexes.

Biologic concentration of rare earths has been found in hickory leaves; rare earths have also been used to trace opium.

Economic Geology

The major source of yttrium has been from monazite (a light rare earth mineral) in beach sands. The accessory monazite originally present in granitic rocks resists weathering and is concentrated by sedimentary processes. Australia, Brazil, Ceylon, India, Malaysia, and South Africa mine the largest quantities. Less than one percent (by weight) of monazite is yttrium. Recently, the bastnaesite deposit at Mountain Pass, California has been an important source of yttrium, again in low concentration.

High-purity yttrium has a limited market; recently yttrium has been used in garnet lasers and television phosphors. Most yttrium is sold unpurified, in rare earth concentrates used for polishing, petroleum cracking catalysis, lighter flints, arc carbons, and in alloys.

ROBERT KAY

References

Balashov, Y. A., et al., 1964, "The effects of climate and facies environment on the fractionation of the rare earths during sedimentation," *Geochem. Intern. No. 1,* **10,** 951–969. (English transl.).

Goldberg, E. D., et al., 1963, "Rare-earth distributions in the marine environment," *J. Geophys. Res.,* **68,** 4209–4217.

Haskin, M. A., and Haskin, L. A., 1966, "Rare earths in European shales: a redetermination," *Science,* **154**(3748), 507–509.

Haskin, L. A., et al., 1966, "Meteoric, Solar, and Terrestrial Rare-Earth Distributions," in (Ahrens, L. H., et al., editors), "Physics and Chemistry of the Earth," Vol. 7, Oxford, Pergamon Press, pp. 167–321.

Rankama, K., and Sahama, T. G., 1950, "Geochemistry," Chicago, University of Chicago Press, 912pp.

Schmitt, R. A., et al., 1964, "Rare-earth, yttrium and scandium abundances in meteoric and terrestrial matter—II," *Geochim. Cosmochim. Acta,* **28,** 67–86.

Towell, D. C., et al., 1965, "Rare-earth distributions in some rocks and associated minerals of the batholith of Southern California," *J. Geophys. Res.,* **70,** 3485–3496.

Cross-references: *Erbium; Rare Earths.*

ZINC: ELEMENT AND GEOCHEMISTRY

Physical and Chemical Properties

Zinc (atomic number 30) is a lustrous white metal with the unusually low melting point (for heavy metals) of 419°C. In this respect it is similar to the other members of group IIB, cadmium and mercury. Five stable isotopes are known to exist in nature:

$Zn^{64} = 48.89\%$ $Zn^{68} = 18.56\%$
$Zn^{66} = 27.81\%$ $Zn^{70} = 0.62\%$
$Zn^{67} = 4.11\%$

With electronic structure $1s^2 2s^2 2p^6 3s^2 3p^6 3d^{10} 4s^2$, zinc follows immediately after the first row of the transition elements. The element's chemical behavior is dictated by the existence of a single oxidation state. In this state (Zn^{2+}) the $3d$ level is fully occupied, and chemical reaction is uncomplicated by the several physical and chemical factors associated with an incomplete shell. The halides and oxyacid salts of Zinc^{II} are readily soluble in water, as would be expected from a charge/radius value (ionic potential) of 2.7. The formation of complexes is commonly met with in aqueous solution, but these are never particularly stable or inert.

The crystal chemistry of zinc is complicated by the tendency of the element to form bonds which are considerably covalent in character and which show a distinct preference for tetrahedral coordination. When octahedral coordination does occur (not uncommonly), simple packing principles generally suffice to explain behavior.

Geochemistry

Rather surprisingly, recent analyses of meteoritic material demonstrate that zinc has no marked affinity for the sulfide phase, the element being comparatively uniformly distributed throughout silicate (50 ppm), iron (20 ppm), and sulfide (50 ppm). On the other hand, in the earth's crust, ore-level concentrations are almost exclusively restricted to sulfide bodies. The mean continental crustal abundance has been estimated at 70 ppm zinc.

In the sequence of igneous rocks, the bulk of the total zinc occurs as a trace constituent of common minerals. The Zn^{2+} ion is normally found wtih ferrous iron, and apparently competes with it during fractionation for vacant octahedral lattice sites. Certain magmatic iron ores (ilmenite, magnetite) have been reported to contain concentrations of the order of 10^3 ppm. Both ions (Fe^{2+}, Zn^{2+}) have the same ionic radius (0.74Å). Differences in the behavior of these ions during fractionation probably arise from zinc's preference for tetrahedral coordination.

The sulfide magmas generally believed to separate at early stages in magmatic differentiation are always very poor in zinc. Only in sulfides associated with very late hydrothermal bodies are rich zinc deposits found, usually together with copper and lead (sphalerite, ZnS along with galena PbS and various complex sulfides). This provides further evidence that the affinity of zinc for the sulfide phase is not general and manifests itself only in certain circumstances.

Much of the present knowledge concerning zinc in the processes of weathering and sedimentation has come from studies of the oxidative weathering of sulfide ore bodies. There, the formation of the very soluble sulfate allows for rapid removal by transport. The dipositive ion may then be removed by reaction with carbonates, or by precipitation in reduzate sediments as sulfide. Much remains in solution and presumably reaches the sea, where concentrations as high as 0.01 ppm have been reported. Further evidence of this mass transport comes from the composition of deep-sea clays, which show zinc concentrations well over twice those of the estimated mean crustal abundance or mean shale composition.

Resistate sediments sometimes show high concentrations, but these are almost always due to the presence of zinc-rich magnetite or ilmenite grains of magmatic origin.

Zinc is an important essential element in both the plant and animal kingdoms. The high mobility of zinc during weathering processes causes the not-infrequent appearance of deficiency diseases.

Metallic zinc is produced by very many of the world's nations, it being comparatively easily won from rich sulfide ores and it is of considerable technical importance. The present world production is approximately 3.3 million tons each year.

C. D. CURTIS

ZINC: ELEMENT AND GEOCHEMISTRY

References

*Goldschmidt, V. M., 1954, "Geochemistry," Oxford, Clarendon Press, 730pp.
*Turekian, K. K., and Wedepohl, K. H., 1961, "Distribution of the elements in some major units of the earth's crust," *Bull. Geol. Soc. Am.*, **72**, 175–192.
Schroeder, H. J., 1965, "Zinc," U.S. Bur. Mines Min. Yearbook.

*Additional bibliographic references may be found in this work.

Cross-references: *Copper: Economic Geology; Hydrothermal Solutions—Sulfide Transport; Sulfides.*

ZINC—ECONOMIC DEPOSITS—See LEAD AND ZINC: ECONOMIC DEPOSITS

ZIRCONIUM: ELEMENT AND GEOCHEMISTRY

History

Zirconium stones have been known from ancient times, but it was in 1789 that M. H. Klaproth reported that zircon contained the oxide (ZrO_2). Elemental zirconium was first prepared in 1824 by J. J. Berzelius.

Properties

Zirconium, symbol Zr, has an atomic number of 40, an atomic weight of 91.22, and electron configuration $1s^2$, $2s^2$, $2p^6$, $3s^2$, $3p^6$, $4s^2$, $3d^{10}$, $4p^6$, $5s^2$, $4d^2$. Its electron shells contain 2, 8, 18, 10, and 2 electrons and it has valences of 2, 3, and 4. Zirconium exists as close-packed hexagonal crystals at 25°C ($a = 3.2312$Å, $c = 5.1477$Å) and body-centered cubic crystals at 862°C ($a = 3.6090$Å). The ionic radius for Zr^{4+} is 0.79 for six-fold coordination and 0.82 for eight-fold coordination. The atomic radius is 1.55 for eight-fold coordination and 1.60 for twelve-fold coordination.

Isotopes exist in the following abundances: 90 = 51.46%; 91 = 11.23%; 92 = 17.11%; 94 = 17.40%; 96 = 2.80%.

Other physical properties include: specific gravity 6.4; hardness (Rockwell B) 25–30; melting point 1850°C; boiling point 3500°C; coefficient of expansion: 298–1143°K, along a axis = 5.5×10^{-6} per °K; along c axis = 10.8×10^{-6} per °K; β, 1143–1600°K = 9.7×10^{-6} per °K; thermal conductivity, at 125°C in (cal/sec) (cm²) (°C/cm) = 0.035 ± 5%; electrical resistivity at room temperature = 40 microhm-cm.

Minerals

Zirconium is found principally in the mineral zircon ($ZrSiO_4$—tetrahedral), which ideally contains 67.2% ZrO_2 and 32.8% SiO_2. It is also found in minerals such as baddeleyite (ZrO_2) and the complex silicates, such as eudialite (Na, Ca, Fe)$_6$Zr [OH, Cl | Si_3O_9)$_2$], låvenite (Na, Ca, Mn)$_3$rZ[F | (SiO_4)$_2$] and catapleite (Na, Zr [Si_3O_9]·H_2O. At least thirty-five minerals contain zirconium. All are oxyphile and lithophyle, but zirconium does not occur in the main rock-forming minerals because of its high ionic radius (0.79Å).

Occurrence

The cosmic abundance of zirconium is 0.55 atoms per 10,000 atoms of silicon (Suess and Urey) but only 0.23 in a more recent estimate (Cameron). The average amount of zirconium reported to be in the earth's crust is 220 g/ton, which is three to four times higher than zinc and copper, for example. Hafnium is always present in nature with zirconium. The average concentration in granite is said to be 460 g/ton, and may be concentrated in laterite or bauxite. Zircon and baddeleyite are the important ore minerals of zirconium; zircon is derived from the chemical weathering of granitic rocks and, thanks to its hardness and density, forms lange placers, either in stream alluvium or on ocean beaches in Australia, India, Brazil, and elsewhere. Baddeleyite is found in alluvial pebbles in Brazil. Zircon is generally mined in these so-called *"black sands,"* along with ilmenite ($FeTiO_3$) or rutile (TiO_2). Those deposits associated with hafnium, as in the mineral cyrtolite, are particularly interesting.

Zircon is present as an accessory mineral in all types of igneous rocks, in particular the more silicic types as granite, granodiorite, syenite, and monzonite. Also, it is present in many metamorphic and sedimentary rocks.

Gem zircons come from placer deposits in Matura, Ceylon, Ural Mountains, Australia, and Madagascar. In the United States, they are present in Maine, New York, and North Carolina.

Geochemistry

Since zirconium is closely related to hafnium it is difficult chemically to determine the percent of each in a given mineral; therefore, a total of the two is generally given. Zircon is one of the first phases to crystallize from an igneous magma. Zirconium and hafnium are concentrated in "residual" magmas in rock series which are believed to result from fractional crystallization. This is particularly true in the residual magmas which have a concentration of the alkali minerals (alkali feldspars). Zircon, the greatest source of zirconium, shows great stability toward ordinary hydrothermal solutions and the crystal remains unmodified during rock metamorphism. Zirconium is also found in the most insoluble residual sediments, but most zirconium is probably found in hydrolysate sediments.

Because of its high degree of stability during sedimentation and metamorphism, zircon has been found to be useful in distinguishing igneous granitic rocks from granitic rocks formed from metamorphic processes. Since zircon has a very high melting point, and its shape is dependent on the physicochemical environment of the magma, it is helpful in determining the sequence of magmatic activity and in correlating shallow intrusives with their related associated extrusives.

Uses

Zirconium is used primarily for atomic energy purposes in nuclear reactor chambers and space applications such as parts in re-entry vehicles; when it is free of hafnium, zirconium is corrosion resistant and has low absorption for neutrons. It is also used in high-vacuum work, high-intensity electric arc lights, and as an alloying agent.

Zirconia is one of the highest-melting oxides and is used as a refractory itself, in the manufacture of crucibles and of ceramic raw materials. It has such diverse uses as pigment for paints, and as an opacifier in white glass for indirect electric lighting.

Zircon is important primarily as a source of zirconium and zirconia, but is also used as a refractory, an abrasive, and in enamels.

Zirconium metal has a very low order of toxicity, but certain dangers exist from explosions owing to the metal having a high affinity for oxygen.

Zirconium is not known to play any significant role in biological metabolism.

ROBERT N. McNALLY

References

Alper, A. M., and Poldervaart, A., 1957," Zircons from the Animas Stock and associated rocks," New Mexico," *Econ. Geol.*, **52,** 952–971.
Bateman, A. M., 1949, "Economic Mineral Deposits," New York, John Wiley & Sons, 898pp.
Blumenthal, W. B., 1958, "The Chemical Behavior of Zirconium," New York, D. Van Nostrand Co., 398pp.
Goldschmidt, V. M., 1954, "Geochemistry," London, Oxford University Press, 730pp.
Hampel C. A. (editor), 1968, "The Encyclopedia of the Chemical Elements," New York, Reinhold Publ. Corp., 849pp.
Hurlbut, C. S., Jr., 1959, "Dana's Manual of Mineralogy," New York, John Wiley & Sons, 609pp.
Lyman, T., 1961, "Metals Handbook—Properties and Selection of Metals," Vol. 1, Eighth ed., Metals Park, Ohio, American Society for Metals, 1236pp.
Mason, B., 1952, "Principles of Geochemistry," New York, John Wiley & Sons, 276pp.
Rankama, K., and Sahama, Th. G., 1950, "Geochemistry," The University of Chicago Press, 912pp.

Cross-references: *Hafnium.* Vol. IVB: *Alkali Feldspars; Baddeleyite; Placer Deposits; Refractory Minerals; Zircon.* Vol. II: *Nucleogenesis.*

INDEX*

Aa basalt, 555
Abiogenic sediments, 1202
Abioliths, 825
Abiotic environment, 338
Abiotic reactions, 981
Abney, 579
ABRASION pH, **1**, 650
Abrasives, 24
Absolute age, 446
Absolute Age Determination. See GEOCHRONOMETRY. Also individual dating entries
Absolute entropy, 307
Absolute temperature, 305
Absolute time scale, 454
Absorption, 583
Absorption edge, 1286
Absorption lines, 277
Absorption spectography, 388
Absorption spectrometry, 343, 384, 494, 1114
Abundances of elements, 402
Abyssal hills, 597. See also Vol. I
Abyssinian plateau, 474
Acantharia, 1081
Acanthite, 1130, 1092
Acanthocephalans, 153
Accelerators, 996
Accessory minerals, 101, 993
Acetic acids, 3
Acetylene, 130
Achondrite model, 683
Achondrites, 18, 35, 146, 241, 343, 1024
Acids, 428, 718, 924
ACIDS AND BASES, **2**, 259, 924
Acid-base catalaysis, 144
Acid bogs, 931
Acid gases, 573
Acidic granitoids, 245
Acid leaching, 23
Acid mine drainage, 312
Acid mine waste seepage, 324
"Acid ocean," 122
Acid(ic) oxides, 48
Acid salts of silicic acids, 499, 718
Acid soils, 350, 385, 495
Acid solutions, 429
Acid-type hot springs, 869
Acidity, 4
Acidophilic iron-oxidizing bacteria, 1132
Acmite, 582
Acquitards, 552
Acratopegs, 703
Actinians, 35
ACTINIDE SERIES, **5**, 8, 44, 47
ACTINIUM: ELEMENT AND GEOCHEMISTRY, 6, **9**, 16, 44, 278
Actinolite, 1, 1091
Actinon, 1014
Actinouranium, 9
Activation analysis, 445
Activation energy, 234
Activity, 924
Activity coefficients, 573, 685, 1109
Adamite, 937
Adamantane, 496
Adenosine phosphosulfate (APS), 1128
Adenosine triphosphate (ATP), 126
Adirondack, 606
Adsorption, 33, 63, 74, 79, 160, 259, 729, 770
Adelite, 936
Advertisement wastes, 321
Aegirine, 1097
Aeolian (eolian) erosion, 545. See also Vol. III
Aeon, 977
Aeration zone, 470. See also Vol. VI
Aerial photo analysis, 460. See also Vol. VI
Aerobic, 1127
Aerobic organisms, 70
Aerodynamic effect, 16
Aerolites, 679. See also Vol. II
Aerosols, 259, 273, 409, 1229, 1003
Aesthetics, 1231
Affinity, 684
Afghanistan, 551. See also Vol. VIII
Africa, 474. See also Vol. VIII
African shield, 475
Ag, 573

Age determination, 455, 994, 1185
Age of the earth, 991
Agents, 695
"Ages," 1014
Agglutination, 284
Agricola, G., 732, 740
Agricultural wastes, 318
Agriculture, 593
Agriculture and forestry resources, 774
Agriculture and soil service, 775
Aguilarite, 1130
Ahlfeldite, 179
A horizon, 75, 914
Ahrens and Morris, 415
Ain Figeh, 548
Air, 600. See also Vol. II
Air masses, 1014. See also Vol. II
Air pollutants, 11, 312
Air pollution, 312, 314, 340, 732. See also POLLUTION
AIR POLLUTION AND GLOBAL CLIMATE, **11**
AIR POLLUTION AND URBAN CLIMATE, **14**
Air quality, 593
Air-borne dust, 1243. See also Vol. II
Aircraft pollution, 311, 324
Air-sea interface, 1014, 1071
Air-sea sediment carbonate equilibria, 98
Akrochordite, 938
"Alabamine," 492, 1130
Alanine, 72, 814
Albedo, 13, 341, 543, 893, 895. See also Vol. II, ALBEDO AND REFLECTIVITY
Albertites, 827, 1206
Albite, 1, 17, 22, 60, 203, 361, 900
Albite porphyroblasts, 19
Albitization, 19
Alcohols, 71, 77, 815
Aldehydes, 71
Alfalfa, 549
Alfalfa grass (Medicago sativa), elemental abundance, 76
Algae, 102, 111, 113, 114, 118, 151, 158, 353, 358, 437, 801, 802, 808, 818, 952, 1012, 1081
Algae (marine), 137, 180, 325, 333
Algal activity, 77
Algal blooms, 312
Algeria, 542. See also Vol. VIII
Algodonite, 42, 1130
Aliphatic hydrocarbons, 495
Alkali, alkalies, 350, 430, 573, 593, 606
ALKALIS, ALKALI METALS, AND ALKALINE EARTH METALS, 2, **16**, 62, 67, 74, 83, 88, 119, 147, 178, 241, 245, 343, 349, 384, 445, 494, 411
Alkali basalts, 18–173
Alkalies, 430
Alkali earth elements, 445
Alkali earth metals, 395
Alkali feldspar, 361, 365, 593
Alkali halides, 237
Alkali Lake (Oregon), 765
Alkali lakes, 1247
Alkali silicate, 32
Alkalic rocks, 156
Alkaline brines, 436
Alkaline earth, 349, 593, 1006
Alkaline earth chlorides, 573
Alkaline igneous rocks, 795
Alkaline lakes, 1085
Alkalis, 593
Alkaline soils, 350, 385, 485, 495
Alkaline solutions, 429
Alkalinity, 4, 121, 424, 587, 1070, 1246
Alkanes, 498
Alkemade line, 918
Alkemade, Van Rijn van, 918
Alkylation, catalytic reaction, 144
Alkyl groups, 498
Allactite, 936
Allanite, 149, 993, 1027
Allemontite, 763
Allergic reactions, 312
Allopalladium, 897, 957
Allophane, 227, 505, 744
Allothigenic minerals, 57
Alloys, 21, 146, 241, 599

Alluaudite, 934
Alluvial plains, 474
Almaden, 707
Almandine, 904
Alpha decay, 8, 492
Alpha emitting nuclides, 9
Alpha particles, 1225
Alpha particle decay, 366
Alpha radiation, 148
Alpha rays, 988
Alpha-recoil dating method, 149
Alpha spectrometry, 595
Alpine Mt. belt, 169
Alstonite, 65, 142
Altaite, 705, 1130
Atlantic, 587, See also Vol. I
Alteration products, 37
Alteration zones, 5
Alums, 1, 21
Alum shales, 815, 1223
Alumina, 5, 21, 23, 248, 464
Alumina hydrates, 24, 26
Aluminite, 1126
Aluminosilicates, 22, 25, 122, 436, 591, 1072
Aluminosis, 735
ALUMINUM: ELEMENT AND GEOCHEMISTRY, 17, 20, **21**, 38, 44, 68, 76, 106, 169, 178, 187, 210, 273, 278, 300, 365, 368, 386, 396, 422, 430, 433, 437, 440, 465, 475, 600, 735, 1067, 1119
Aluminum hydroxide, 5, 260, 423
Aluminum industry, 378
ALUMINUM ORE DEPOSITS, **23**, 310
Aluminum oxides, 24, 231, 432, 437, 562
Aluminum phosphates, 23
Aluminum silicates, 23
Aluminum sulphates, 23
Aluminum-thermal vacuum retorts method, 104
ALUMINUM TOXICITY, **27**
Aluminous mud, 606
Alumohydrocalcite, 143
Alune shales, 1203
Alunite, 23, 900, 1125
Alvite, 488
Alunogen, 1125
Amarantite, 1126, 1124
Amazon, 1044, 1046, 1047, 1099
Amazon stone, 1052
Amazonite, 17, 1052
Amber, 155, 811, 812, 826, 1040
Amblygonite, 17, 663, 936
Amebae, 154
Amenities, 771
Americium, 6, 44, 278
Amigdules, 111
Amines, 71, 74, 77, 83
Amino, 981
Amino Acids, 29, 70, 79, 126, 151, 154, 180, 231, 428, 416, 798, 801, 820, 850, 978, 1127
Amino acid substrates, 499
Amino compounds, 72
Amino sugars, 72, 154
Aminobutyric acid, 72
Ammonia, 320, 427, 587, 808, 863, 1017A, 1072
Ammonia alum, 1124
Ammonia (liquid), 3
AMMONIA, IN MINERALS AND EARLY ATMOSPHERE, **29**, 126, 272. See also HYDROGEN; NITROGEN
Ammonia oxidation, 144
"Ammoniacal Nitrogen," 797
Ammonification, 801
Ammonioborite, 87
Ammoniojarosite, 1125
Ammonium, 1016
Ammonium micas, 30
Ammonium molybdate, 436
Ammonium muscovite, 29
Ammonium phlogopite, 29
Ammonium silicates, 30
Ammonium sulfate, 273
Amorphous, 744
Amorphous silica, 435, 1108
Amosite, 734
Ampere, 377
Amphiboles, 17, 22, 57, 88, 101, 155, 202, 206, 378, 432, 425, 429, 579, 599, 605, 1084

*Words listed in small capital letters represent titles of articles; boldface numbers refer to pages on which articles begin.

INDEX

Amphibolite, 904
Amphibolite facies, 20, 156, 248, 711
Amphineura, 114. See also Vol. VII
Amphoterism, 21
Amphoteric elements, 85
Amur, 1048
Anadiagenesis, 179
Anaerobic bacteria, 126, 848, 1127
Anaerobic degradation, 335
Anaerobic sewers, 334
Analcime, 58, 61
Analcite, 1097
Analeite, 147
Analysis, 369
   minerals, 444, 579
   rocks, 444, 579
Analyte, 369
Anapaite, 935
Anatase, 58, 844, 1193
Anatexis, 709. See also Vol. V
Anauxite, 227
Ancylite, 143
Andalusite, 22, 562, 904
Andersonite, 143
Andes, 12
Andesine, 566
Andesite, 19, 156, 248, 273, 555, 849. See also Vol. V
Andesitic volcanics, 465
Andorite, 1092
Andrewsite, 936
Anemias, 699
Angara-Lena basin, 767, See also Vol. VIII
Anglesite, 50, 1123
Anhydrite, 1, 58, 77, 101, 103, 112, 354, 383, 427, 734, 902, 1078, 1123, 1148, 1247
Aniline, 67
Animals, 541
Animal behavior; patterns analysis, 258
Animal husbandry, 313
Anions, 37, 198, 719
Anion affinity, 412
Anion exchange, 591. See also ION EXCHANGE
Anionic radii, 414
Anisodesmic compounds, 33, 159, 934
Anisodesmic crystal, 201
Ankerite, 58, 64, 142, 604
Annabergite, 790, 935
Annelids, 102, 114, 118, 153. See also Vol. VII
Anode sludge, 1041
Anorthite, 17, 22, 100, 386, 429
Anorthoclase, 899, 1097
Anorthosite, 18, 872
Anoxygenic atmosphere, 850
Antarctica, 351. See also Vol. VIII
Antarctic convergence, 954. See also Vol. I
Antarctic Ocean sediments, 77. See also Vol. I
Anthoinite, 752
Anthraxolite, 1206
Anthropogenic influences, 13, 186, 326, 335, 340. See also Vol. III
Anthrax, 692
Anthraxalite, 579, 827
Anthophyllite, 734
Antibodies, 695
Antigorite, 790
Antilles, 474. See also Vol. VIII
ANTIMONATES, ANTIMONITES, AND ARSENITES, 33, 717
ANTIMONY: ELEMENT AND GEOCHEMISTRY, 33, 44, 82, 112, 151, 191, 278, 300, 763
Antimony-lead alloys, 35
Antiseptic, 89
Antlerite, 195, 1125
Apatite, 18, 83, 99, 101, 103, 146, 149, 367, 378, 425, 432, 449, 462, 567, 698, 899, 900, 940, 993, 1027, 1184, 1217
Apatitized wood, 941
Aphthitalite, 1123
Apjohnite, 1125
Aplites, 22
Apollo mission, 149
Apophyllite, 203
Apparent age, 1228
Apple cuticle wax, 499
Applied geochemistry, 585
AQUEOUS SOLUTIONS, 36, 52, 185, 259, 304, 309, 436, 538, 592, 600, 686, 900, 557, 561, 572, 586, 716
Aqueous mineral synthesis, 743
Aquiclude, 39, 473, 539, 552
AQUIFER, 39, 42, 240, 457, 472, 478, 485, 552, 593
Aquitard, 39
Ar⁴⁰, 449
Arabia, 471, 542, 548. See also Vol. VIII
Arable land, 1103

Aragonite, 20, 51, 58, 63, 77, 100, 104, 108, 114, 142, 153, 204, 359, 424, 579, 408, 587
Aragonite sediments, 433
Aragonite-strontianite system, 107
Aragonitic corals, 1122
Aramayoite, 82
Arandisite, 1192
Arcanite, 1123
Archaeocyathids, 6, 118. See also Vol. VII
Archaeology, 133, 451
Archeozoic, 850. See also Vol. VII
Archimedes Principle. See also SPECIFIC GRAVITY, 1113
Arco Reactor (Idaho), 557
Ardealite, 939
Arenaceous rocks, 90, 245, 579. See also Vol. VI
Arfvedson, A., 661
Argentian tetrahedrite, 1092
Argentite, 900, 1092, 1130
Argentojarosite, 1092, 1125
Argillaceous sediments, 90, 119, 579. See also Vol. VI
Argillic alteration, 567
Argillite, 273
Arginine, 72
ARGON: ELEMENT AND GEOCHEMISTRY, 39, 44, 201, 261, 272, 277, 278, 396, 400, 404, 411, 418, 427, 492, 837, 849, 1039, 1065
Argyrodite, 462
Arid climates, 351, 541, 1103
Arid coasts, 536
Arid lands, 470, 876
Arid soils, 178, 215
Arid zones, Mediterranean, 544
Aridity, 1246
Aristotle, 219
Arizonite, 846
Arkose, 270, 432, 447, 449. See also Vol. VI
Arkositic sandstones, 548
Armangite, 34
Arnimite, 1126
Aromatic hydrocarbons, 496
Arquerite, 704
Arrhenius theory, 2
Arrojadite, 934
Arsenates, 33, 41, 717. See also PHOSPHATES, ARSENATES AND VANADATES
ARSENIC: ELEMENT AND GEOCHEMISTRY, 34, 41, 44, 79, 82, 112, 151, 191, 198, 278, 300, 319, 320, 425, 433, 605, 735, 763
Arsenic sulfide, 41, 900
Arsenic trisulfide, 41
Arsenides, 41
Arseniopléite, 937
Arseniosiderite, 938
Arsenites, 33, 589
Arsenobismite, 937
Arsenoclasite, 936
Arsenolite, 41, 42, 834
Arsenopalladinite (Pd₃As), 958
Arsenopyrite, 44, 42, 468, 900, 1131
Arteriosclerosis, 691
Artesian aquifers, 471
Artesian basins, 536
ARTESIAN FLOW, 239
Artesian springs, 1116
ARTESIAN WATER, 42, 555, 956
Artesian wells, 551
Arthrites, 703
Arthropods, 102, 153. See also Vol. VII
Artois, 551
Artificial radioisotopes, 988
Artificial "reefs," 774
Artificially produced isotopes, 1002
Artinite, 143
As, 571
Asama, Japan, 12
Asbestos, 103, 314, 731. See also actinolite, chrysotile, crocidolite (blue asbestos), tremolite, other fibrous amphiboles, serpentine
Asbestosis, 730
Asbestos spray insulation, 739
Ascariassis, 693
Ascension Island, 384, 494. See also Vol. VIII
Ash, 271, 330, 465, 555, 823, 1203, 1236
Ash-fall deposits, 414. See also Vol. VI
Askja (Vatna Jökull), Iceland, 12
Aspartic acid, 72
Aspergillus niger, 387
Asphalt, asphaltites, waxes, 150, 812, 1206. See also Vol. IVB
Asphaltene, 498
Asphaltite, 824, 827. See also Vol. IVB
Assemblage, 726
Assimilation, 74
Association, 726
Astatine, 10, 44, 278, 492, 963
Asteroid, 791. See also Vol. II

"Asthma," 732
Aston, F. W., 454
Astragalus, 76, 1080
Asymmetry potential, 173
Atacama Desert, Peru, 216
Atacamite, 155, 191, 195, 490
Atelestite, 936
Atherosclerosis, 700
Atlantic coast, 537. See also Vol. III
Atlantic cores, 597
Atlantic deep water, 121. See also Vol. I
Atlantic Ocean clays, 275
Atlantic Ocean sediments, 52, 111, 121, 275. See also Vol. I
Atlantis Deep, 571
Atlas, 775. See also Vol. VIII
Atmophile elements, 47, 212, 396
Atmosphere, 74, 123, 156, 272, 396, 408, 428, 451, 492, 495, 498, 601, 849, 850, 861, 996, 1014, 1077, 1243. See also Vol. II, ATMOSPHERIC AIR
Atmosphere, anoxygenic, 850
Atmosphere-biosphere-surface ocean reservoir, 451
Atmosphere, evolution and history, 29, 126, 271, 387, 397, 836, 849. See also Vol. II
Atmosphere, oxygenic, 271
Atmosphere, prebiological, 29, 126
Atmosphere, primeval, 122, 210, 271
Atmospheric carbon dioxide, 102
Atmospheric chemistry, 69, 96. See also Vol. II
Atmospheric circulation, 11. See also Vol. II
Atmospheric contaminants, 326
Atmospheric gases, 389, 409. See also Vol. II
Atmospheric nitrogen, 799
Atmospheric nitrogen fixation, 144
Atmospheric nuclei and dust. See Vol. II
Atmospheric particulates, 15
Atmospheric pollution, 324. See also POLLUTION
Atom, 45, 84, 197. See also ENCYCLOPEDIA OF THE CHEMICAL ELEMENTS
Atomic absorption, 1114
Atomic absorption spectroscopy, 369, 742
Atom bomb, 988, 1222
Atomic charge, 410, 411
Atomic dispersion theory, 590
Atomic donors, 84
Atomic equilibrium locations, 234
Atomic generating plants, 314
Atomic mass, 750
ATOMIC NUMBER AND PERIODIC TABLE, 43, 263
Atomic reactors, 350
Atomic receptors, 84
Atomic volume, 47
Atomic wastes, 547
Atomic weight, 43
ATP (adenosine triphosphate), 126
Attapulgite, 178
Au, 407, 573
Aufwuchs, 660
Augelite, 937
Augite, 1, 101, 430, 899
Aureoles, 68, 706, 709, 971
Auric chloride complex, 468
Aurichalcite, 143
Aurostibite, 1131
Austinite, 936
Australia, 474, 548. See also Vol. VIII
Australian artesian basins, 537
Australian shield, 475. See also Vol. VIII
Australite, 148. See also Vol. II
Australopithicus, 731. See also Vol. VII
Authigenic clays, 213
Authigenic feldspars, 422, 427
Authigenic minerals, 674, 880, 985
AUTHIGENESIS OF MINERALS (MARINE), 48
Authigenic sediments, 433. See also Vol. VI
Authigenic silicates, 1087
Authigenesis, 91, 1160
AUTHIGENESIS OF MINERALS (NON-MARINE), 57
Autohydrothermal deposits, 190
Autometamorphic processes, 88
Autotrophs, 256
Autumn turnover, lakes, 342
Autunite, 939, 962
Average seawater, 1047. See also Vol. I
Avogadrite, 147, 491
Avogadro's number, 94, 235, 307, 750
Avogadro's principle, 1258
Awaruite, 791
Aivu (Awoe), 12
Axinite, 88
Ayres, 392
Azonal soil, 912. See also Vol. VI
Azotobacter, 75, 802, 809
Azurite, 58, 143, 191

B, 403, 571
Back scattering, 12

INDEX

Bacteria, Bacteriology, 37, 70, 74, 77, 111, 126, 136, 159, 180, 312, 333, 356, 499, 540, 670, 672, 729, 764, 801, 802, 1065, 1150, 1158. See also Vol. I, MARINE MICROBIOLOGY
Bacterial oxidation, 271, 587
BACTERIAL SULFATE REDUCTION. See SULFATE REDUCTION MICROBIAL
BACTERIAL SULFUR OXIDATION. See SULFUR OXIDATION BACTERIAL
Baddeleyite, 488, 844, 1294
Bagdad Crater, Calif., 103
Bahamas, 51, 102, 108, 111, 112. See also Vol. VIII
Bahama eolianites, 114
Baja California, 360. See also Vol. VIII
Bakerite, 87
Bakery pollutants, commercial, 321
Baking powder, 89
"Balance of nature," 772, 1102
Balmat (New York), 1139
Balneo-geohydrology, 702
Baltic Sea, 587, 819, 1068. See also Vols. I and VIII
Baltic shield, 475. See also Vol. VIII
Banalsite, 62
Bananas, 28
Bandai San, Japan, 12
Banded iron ores, 271
Bandylite, 87
Bankets, 469
Barada River, 548
Bararite, 491
Barbertonite, 845
Barima River, British Guiana, 536
Barite, 433, 900, 1123
Barite ore deposits, 64
Barite (barytes), 1, 17, 50, 58, 62, 78, 99
BARIUM: ELEMENT AND GEOCHEMISTRY, 16, 44, 50, **62,** 74, 106, 112, 210, 247, 270, 274, 278, 300, 320, 404, 406, 1012, 1067
BARIUM: ECONOMIC GEOLOGY, **63**
Barium carbonate, 64
Barium chloride, 62
Barium feldspar, 62
Barium hydroxide, 63
Barium, metallic, 62
Barium nitrate, 66
Barium oxide, 66
Barium peroxide, 66
Barium silicates, 62
Barium sulfate, 63, 271
Barium titanate, 66
Barnacles, 118
Barnesite, 1237
Barred basin, 353
Barrier islands, 344. See also Vol. III
Bartlett, Neil, 1039
Bartram, William, 187
Barstow Formation (Miocene), 882
Barysphere, 791
Baryta water, 63
Barytes, 605
Baryтоanglesite, 62
Barytocalcite, 142
Basal spacing, 30
Basalts, 18, 19, 83, 93, 99, 146, 156, 248, 273, 343, 350, 384, 401, 430, 449, 488, 494, 555, 582, 669, 1029, 1050, 1056, 1060, 1080, 1185. See also Vol. V
Basalt aquifers, 474
Basaltic magma, 391, 680, 1062
Basaltic melts, 375
Basaltic reservoirs, 547
Basaltic rocks, 992, 1062, 1091
Basaluminite, 1126
Basanite, 1031
BASE—CHEMICAL, **67**
Base Exchange, 450. See also ION EXCHANGE
Bases, 924
Base-line location studies, 219
Base metal ores, 469
Base metal refining, 462
Basic salts of silicic acid, 718
Basic oxides, 48
Basic rocks, 245. See also Vol. V
Basic salts of silicic acids, 499, 718
Basins, 351, 547. See also Vol. V
Bass (fish), 660
Bassanite, 1124
Bassetite, 939
Bastnaesite, 143, 146, 241, 343, 349, 350, 385, 495, 976, 1028, 1060, 1166
"Batch" operation analysis, 164
Batholiths, 434, 467
Bauxite, 5, 21, 23, 25, 68, 75, 423, 432, 734, 901, 902, 1062, 1235
Bayer Process, 25, 1119

Bay of Fundy, 348. See also Vol. I
Bayldonite, 938
Bay-mouth bar, 348. See also Vol. II'
Bayerite, 25
Bayleyite, 143
Beaches, 1189. See also Vol. III
Beach placers, 467
Beach rock, 116. See also Vol. III
Beach sands, 146, 241, 343, 432, 489, 495, 1061. See also Vol. VI
Beartooth Mountains, 711
Beaverite, 1125
Beckerite, 826
Becker, G. F., 454
Becking, Baas, 933
Becquère, Henri, 987
Becquerelite, 844
Bed load, 1042. See also Vol. III
Bedded ore deposits, 64
Bedding, 886. See also Vol. VI
Bedrock, 458
Beer's Law, 1115
Beetle wing, fossil, 155
Beggiatoaceae, 1150
Beidellite. See montmorillonite
Belemnites, 134, 893
Bellingerite, 590
Bendigo, Australia, 468
Benefit-cost ratio, 1260
Benitoite, 62, 202
Bens, 392
Benstonite, 142
Benthic zones, 660. See also Vol. I
Benthonic fauna, 48, 74, 136. See also Vol. I
Benthos, 660
Bentonite, 437. See also montmorillonite
Bentonite membrane, 174
Benzene, 79, 130, 496
Benzoic acids, 3
Beraunite, 939
Bergman, 369
Bering Sea, 436. See also Vol. I
Berkelium, 6, 44, 278
Berlinite, 935
Berman Balance, 1113
Bermanite, 939
Bermuda, 587. See also Vol. VIII
Bermuda eolianites, 114
Bernal, 415, 850
Bernalite, 817, 828
Berthelot-Nernst distribution law, 905, 909
Berthollide Compounds, 1120
Beryl, 17, 40, 68, 147, 201, 493, 899
Beryl-molybdenite veins, 148
Berylliosis, 735
BERYLLIUM: ELEMENT AND GEOCHEMISTRY, 16, 44, **68**, 79, 147, 150, 247, 266, 274, 278, 300, 422, 440, 489, 735
"Beryllometers," 785
Beryllonite, 68, 934
Berzelianite, 1079, 1080, 1130
Berzeliite, 934
Berzelius-Dana plan, 717
Berzelius, Jons Jakob, 152, 740
Bessemer ore, 605
Beta decay, 8, 40, 129, 267
Beta emission, 100
Beta particles, 988
Betafite, 846
Beudantite, 939
Beyerite, 143
Bezruhkov, 436
Bezymyannaya, Kamchatka (U.S.S.R.), 12
B horizon, 914. See also Vol. VI
Bianchite, 1124
Bicarbonate, 37, 38, 121, 126, 129, 134, 374, 388, 475, 548, 561, 573, 606, 1244
Bicarbonate alkalinity, 587
Bicarbonate solutions, 423
Bieberite, 1124
Big Springs, Missouri, 1116
Bilinite, 1125
Bilgewater pollution, 311
Biliprotein (chlorophyll b), 157
Bindheimite, 34
Binding power (of cations), 412
Bingham, Utah, 469, 1094
Biocatalysis, 145
BIOCHEMICALS, **68**, 325, 319
Biochemical cycle, 69
Biochemical inorganic processes, 595
Biochemical reactions, 1072
Biochemical silica, 441
Biochemistry, 409
Biocides, 318, 325
Bioclastic deposits, 114, 812, 1202
Biogenic (-ous) oozes, 52, 433, 1082
Biogenic carbonates, 113

Biogenic cycles, 587
Biogenic precipitation, 437
Biogenic sediments, 114, 812, 1202. See also Vol. VI
Biogenous minerals, 48
Biogeochemical cycles, 338
Biochemical prospecting. See Vol. IVB, GEOBOTANICAL AND BIOGEOCHEMICAL METHODS OF PROSPECTING FOR MINERALS
BIOGEOCHEMISTRY, 23, 52, **74**, 102, 253
Biogeochemistry, Pelagic, See Vol. I
Bioliths, 79, 825
Biologic activity, 36, 210, 430
Biological cycle, 946, 1073
Biological oxidation, 334
Biological oxygen demand (BOD), 312, 334
Biological pollutants, 312
Biomass, 76, 862
Biometeorology. See Vol. II
Biomobilisates, 827
Biophile elements, 88, 396
Biophillic systems, 944
Biophyll elements, 813
Biosphere, 74, 123, 209, 253, 338, 495, 498, 600, 800, 861, 1004. See also Vol. VII
Biosphere (marine), 74, 102
Biosphere (terrestrial), 76
Biosynthesis, 157, 499
Biotic environment, 338
Biotin, 821
Biotite, 1, 5, 19, 22, 29, 40, 63, 99, 147, 426, 447, 448, 449, 450, 734, 899, 901, 902, 904, 971
Birnessite, 49, 674
Bird droppings (See also guano), 941; also Vol. VI
Bird habitat, 333
Bischofite, 155, 490
Bismite, 844
Bismoclite, 490
BISMUTH: ELEMENT AND GEOCHEMISTRY, 34, 44, 79, **82**, 151, 191, 198, 267, 278, 763, 900
Bismuth (RaC), 963
Bismuth ochre, 83
Bismuthates, 83
Bismuthinite, 82, 1131
Bismutotantalite, 82, 846, 1162
Bisulfates (in volcanic gases), 388
Bisulfides (in sulfide transport), 574
Biosulfide solutions, 561
Bitter taste, 2
Bitterns, 66, 93, 353
Bitumen, Bituminous Sediments, 151, 817, 823, 824, 1205. (See also Vol. IVB
Bituminous shales, 71, 426, 1236
Bixbyite, 207, 844
Black Ash, 64
Black Hills (S. Dakota), 441, 448, 663
Black-eye peas, 28
Black marble, 120
Blake Plateau, 49, 52. See also Vol. I
Black Sea, 51, 271, 587, 754, 809, 819, 1051, 1068, 1135. See also Vols. I and VI
Black shales, 35, 150, 194, 195, 268, 356, 579, 993, 1223. See also Vol. VI
Blakeite, 1079
Blanc fixe, 64
Bloedite, 1124
Blomstrandite, 1162
Blood groups, 691
Blooms, 808
Blowpipe analysis, 152, 369, 443, 444, 742. See also Vol. IVB
Blue mud, 587
Blue-green algae, 76
BOD (biological oxygen demand), 312, 334
Boehmite, 21, 423, 844
Boehmitic bauxite, 23, 26
Body fluids, 905
Bog iron ore, 78, 729, 601
Boisse, 679
Boléite, 490
Bolivia, 35, 1192. See also Vol. VIII
Boltwood, B. B., 454
Boltzmann constant, 235, 301
Boltzmann's Relation, 306
Bomb-produced radioactivity, 409
Bond, covalent, 84, 422
Bond, homopolar, 84
Bond, hydrogen, 84
Bond, ionic, 84, 422
Bond, metallic, 84
Bond, metal-oxygen, 413
Bond, polar, 84
Bond, van der Waals, 84
Bonding, aluminum-oxygen, 579
BONDING, **84,** 145, 154, 184, 193, 197, 261, 422, 1237

1299

INDEX

in X-ray diffraction analysis, 1272
Bonding orbitals, 1259
Bonding, mixed, 200
Bone, 378, 452, 689, 815, 822, 989, 1122
Bone material, 75
Bonsen, 368
Boothite, 1124
Boracite, 87
BORATES, **87**, 231, 374, 388, 583, 717
Borax, 87, 902, 1097
Borax beads, 276
Borax Lake, Calif., 89
Boreal region, 432. *See also* Vols. II and VII
Boreholes, 544, 547, 573
Borehole waters, 573
Boric acid, 88, 121, 1063
Borickite, 938
Borislavite, 827
Bornite, 58, 191, 195, 1130
BORON: ELEMENT AND GEOCHEMISTRY, 38, 44, 75, 79, **88**, 149, 151, 245, 266, 278, 300, 420, 430, 433, 440, 475, 886, 1017
Boron fixation, 432
BORON GEOCHEMISTRY IN MARINE ENVIRONMENTS, **90**
Borosilicates, 88
Bosmina (microfossils), 660
Boss, 555. *See also* Vol. V
Bossons Glacier, Chamonix, France, 465
Botallackite, 490
Botany, 187
Botryogen, 1126
Bouguer-Lambert Law, 1114
Boulangerite, 34
Boundary layer. *See* Vol. II
Bournonite, 35, 192
Boussingaultite, 1124
Bover Glacier, Lom-Skjolden, Norway, 465
Boyle, Robert, 373
Bowen, N. L., 898
Bowen's reaction series, 417
Boxwork. *See* Vol. VI
B. P. (Before Present), 131. *See also* Vol. III
Brachiaria purpurascens, 28
Brachiopods, 79, 102, 113, 118. *See also* Vol. VII
Brackebuschite, 1237
Bradleyite, 143
Bragg, W. L., 741
Bragg equation, 1277
Braggite, 956, 958, 1131
Brahmaputra, 1048. *See also* Vol. VIII
"Brain Drain," 775
Branched hydrocarbons, 495
Brandtite, 935
Brannerite, 844, 846, 1223
Bravais, A., 741
Bravaisite. *See* Illite
Bravoite, 1131
Brazil, 474
Brazilian shield, 475
Brazilianite, 937
Breccia, 436
Breislak, 373
Breithauptite, 790, 1130
Brewer's processing pollutants, 321
Brewster, 373
Bridges (engineering problems), 478
Brimstone, 1142
Brines, 89, 93, 103, 113, 155, 174, 185, 319, 321, 353, 561, 571, 587, 562, 666, 764, 869, 884, 1244
Brine refrigeration, 104
Bristol Lake, Calif., 103
Britannia metal, 35
British Guiana (Guyana), 28. *See also* Vol. VIII
Brochantite, 195, 1125
Broken Hill (New South Wales), Australia, 1094
Bromellite, 68
Bromide, 38, 185, 388
BROMINE: ELEMENT AND GEOCHEMISTRY, 44, 75, **92**, 185, 216, 278, 360, 489, 492, 886, 1017
Bromlite, 62, 843
Bromyrite, 93, 490
Bronchitis, 734
Bronsted-Lowry acid, 2, 144
Bronsted-Lowry Theory, 2, 67
Bronze Age, 189. *See also* Vol. VII
Bronzite, 149
Brookite, 58, 794, 844, 1193
Brown algae, 114
Brown clays, 64
Brown Coal, 826. *See also* Vol. VI
Brown hematite, 604
Brown marble, 120
Brownian motion, 94, 373. *See also* Vol. II
Brownlee Dam, 556
Brucillosis, 692
Brucite, 114, 667, 844

Brugnatellite, 845
Brushite, 698, 935, 940
Bryozoa, 102, 113, 118, 153. *See also* Vol. VII
BUBBLE MOTION IN FLUID INCLUSIONS, **93**
Bubonic plague, 694
Buddingtonite, 29, 31
Buehler, 392
Buetschliite, 142
Buffer mechanism, 1071
BUFFER SYSTEMS, **95**, 818, 979, 1068
Buffering, 122, 842, 1082
Buildings, 478
Building materials, 116, 118. *See also* Vol. VI
Bulk volume, 964
Bull Domingo, Colorado, 468
Bunsenite, 843
Buried valleys, 460
Burkeite, 1126
Burrowing organisms, 428, 596 914
"Bush sickness," 75
Bushveld Complex, S. Africa, 169, 709, 957, 960.
See also Vol. VIII
Butane, 165, 496, 812
Butlerite, 1126
Butte, Montana, 1094
Buttgenbachite, 1125
Butylenes, 165
Bystromite, 34

C, 404
C-12, C-13, C-14; *See* CARBON, 12, 13, 14
Ca, 404, 547, 600
Cobalt, 310
Cacoxenite, 939
Cactus (isoprenoid hydrocarbons), 499
Cactus leaves (alkanes), 499
CADMIUM: ELEMENT AND GEOCHEMISTRY, 44, 75, 79, **99**, 106, 150, 278, 300, 320, 735
Cadmoselite, 99
Cadwaladerite, 490
Cahnite, 87
Calamine, 99
Calaveras formation, Calif., 65
Calaverite, 466, 468, 1132
Calc-alkaline volcanics, 273
Calcareous algae, 111, 122
Calcareous argillite, 465
Calcareous dunes (*See also* eolianite), 1016
Calcareous foraminifera, 353
Calcareous ooze (*See also* biogenic ooze). 424
Calcareous shells, 1012
Calcification, 100
Calcining, 1182
Calcioferrite, 939
Calciovolborthite, 936
Calcite, 1, 17, 20, 51, 58, 63, 77, 88, 97, 100, 102, 104, 108, 142, 149, 153, 185, 200, 201, 231, 424, 429, 433, 475, 557, 579, 587, 900, 901, 902, 904, 941, 993
Calcite, high-low magnesium, 101, 118
Calcite-dolomite, 112
Calcite-siderite, 112
Calcite-smithsonite system, 107
Calcite- strontianite, 112
CALCIUM: ELEMENT AND GEOCHEMISTRY, 16, 20, 37, 38, 44, 75, **100**, 126, 146, 178, 185, 200, 210, 216, 245, 261, 278, 296, 300, 320, 350, 354, 369, 374, 384, 406, 408, 432, 433, 464, 465, 475, 494, 561, 580, 587, 589, 600, 954, 1017
ECONOMIC GEOLOGY, **103**
Calcium acid sulfide, 100
Calcium carbide, 100, 118
CALCIUM CARBONATE: GEOCHEMISTRY, 20, 52, 59, 77, 96, 101, **104**, 153, 349, 354, 408, 424, 432, 437, 540, 595
Calcium carbonate cave deposits, 452
CALCIUM CARBONATE: ECONOMIC GEOLOGY, **118**
Calcium carbonate hexahydrate, 104
Calcium carbonate, melting relations, 111
  terrace deposits, 452
Calcium chloride, 103
Calcium cyanamide, 100
CALCIUM CYCLE, **120**, 212
Calcium hydroxide, 100
Calcium magnesium chloride (tachydrite), 103
Calcium metal, 104
Calcium nitride, 100
Calcium oxide, 100, 111, 843
Calcium phosphate, 425
Calcium sulfate, 273, 352, 354
Calculi, 697
Caledonian Mt. belt, 169. *See also* Vol. VIII
Caledonite, 1126
Caliche, 100, 548, 802, 1159. *See also* Vol. VI
Californium, 6, 44, 278
Calomel, 490, 704
Calc, chalk, 100

Cambrian, 454, 857, 1099
Cambrian period fossils, 155. *See also* Vol. VII
Campbell, 392
Campeche Bank, 115
Camphor, 828
Canada, 957
Canadian Shield, 475
Cancer, 696
Candy manufacturing pollutants, 321
Canfieldite, 462
Cannel coal, 1204
Canyon Diable meteorite, 1227
CAPILLARITY, 470, 543. *See also* Vol. II
Capillary concentrates, 829
Capillary fringe, 470, 1264
CAPILLARY WATER. *See* Vol. VI
Caracolite, 1125
Carbohydrates, 70, 136, 231, 312, 815, 821, 981
C-12/C-13 ratio, 133
C-13, 863
C-14, 123, 447, 894
CARBON-14 DATING, 108, **129**, 451, 544, 597
CARBON CYCLE, 81, 123, **124**, 210, 255
CARBON: ELEMENT AND GEOCHEMISTRY, 44, 70, 75, 92, 95, **123**, 198, 267, 275, 278, 296, 320, 338, 391, 422, 446, 495, 600, 819
Carbon dioxide, 11, 14, 69, 77, 95, 124, 126, 129, 272, 326, 347, 374, 427, 429, 437, 451, 475, 547, 587, 601, 606, 837, 869, 1069, 1229. *See also* Vol. II
Carbon dioxide balance, 339
CARBON DIOXIDE CYCLE IN THE SEA AND ATMOSPHERE, 121, 131, **340**. *See also* Vols. I and II
Carbon dioxide, photosynthetic cycle, 466
Carbon fixation, 952
CARBON ISOTOPE FRACTIONATION, **133**, 452. *See also* SULFUR ISOTOPE FRACTIONATION
Carbon monoxide, 272, 311, 312, 326, 945
CARBON/NITROGEN RATIO, **136**, 797, 819, 1202
Carbon, organic, inorganic, 92, 95, 134, 349
Carbon pollutants, 313
Carbon reservoir, 127
Carbon, vegetable, 124
Carbonaceous chondrites (meteorites) 40, 134, 226, 275, 493, 680. *See also* Vol. II
Carbonaceous complexes, 813
Carbonaceous fuel, 848
Carbonaceous limestone, 119
CARBONACEOUS METEORITES, 151. *See also* Vols. II and IVB
Carbonaceous shale, 35, 430
CARBONATES, 38, 51, 59, 80, 89, 101, 106, 114, 121, 128, 135, **142**, 200, 231, 241, 246, 271, 274, 343, 349, 353, 374, 384, 408, 419, 424, 430, 432, 433, 435, 467, 468, 494, 548, 559, 579, 583, 588, 812, 954. *See also* Vol. VI
Carbonate apatite, 52, 425, 937
Carbonate aquifer systems, 540
Carbonate equilibria, 96
Carbonate-fluoride-barite-sulfide associations, 99
Carbonate geochemistry, 425, 861
CARBONATE HYDROLOGY. *See* HYDROLOGY, COASTAL TERRAIN
Carbonate magmas, 111
Carbonate rhombs, 438
Carbonate rocks, 118, 194, 245, 539. *See also* Vol. VI
Carbonate thermometry, 745
Carbonate-water system, 892
Carbonation, 903
Carbonatites, 18, 111, 274, 590, 794. *See also* Vol. V
Carbonatitic explosions, 274
Carbonic acids, 59, 101, 121, 126
Carbonic anhydrase, 151
Carboniferous, 454, 886, 887. *See also* Vol. VII
Carboniferous period fossils, 155. *See also* Vol. VII
Carbonium ion theory, 144
Carbophyll elements, 813
Carbosphere, 823
Carboxyl groups, 70, 151
Carboxylic acids, 182
Carcinogenic responses, 738
Cardiovascular disease, 696
Caribbean, 894. *See also* Vols. I and VIII
Carminite, 938
Carnallite, 93, 103, 155, 354, 491, 667, 971, 1051
Carnivores, 256, 661
Carnotite, 754, 962, 1235, 1237
Carnotization, 342
Carotenoids, 71, 816, 830
Carphosiderite, 1125
Carrier ions (Berthelot-Nerst distribution law), 909
Carrollite, 1131

INDEX

Cartiladge, 412
Caryinite, 934
Caspian Sea, 352, 587, 589, 1051
Cassava, 28
Cassiterite, 148, 578, 794, 844, 899, 900, 1062, 1192
Cassiterite-lead-zinc ores, 578
Cassiterite-polymetallic ores, 578
Cassiterite-sulfide skarns, 578
Castella, Calabria, (Italy), 894
Castile formation, 358
Castor bean, 28
Castrations, 1062
"Cat clays," 28
Catacombs, 70
CATALYSIS, 67, **143**, 670
catalyst, 600, 981, 1150, 1199
Catapleiite, 488, 1294
Cations, 37, 198, 719
Cation exchange, 591, 1266. *See also* ION EXCHANGE
Cation substitution. *See* TRACE ELEMENTS IN SILICATE MINERALS, SUBSTITUTION
Cationic forces, 412
Cationic radii, 414
Caucasus Mts., U.S.S.R., 418. *See also* Vol. VIII
Cave deposits, 108, 112. *See also* Vol. III
Cavendish, 1039
Caves, prehistoric, 70
Cavitation, 1110
Cavities (hydrothermal deposits in), 900
Cavity-filling ore deposits, 64, 342
Cawk (barite), 63
Cd, 578
Ceboruco, Mexico, 12
Cedarite, 826
Celadonite, 30
Celestite, 17, 50, 58, 1123
Cellulose, 73, 79, 153, 815, 824
Cellulose acetate, 222
Celsian, 62
Cement, 103, 118, 1182. *See also* Vol. VI, BUILDING MATERIALS
Cenozoic, 877, 885, 893. *See also* Vol. VII
Central America, 474. *See also* Vol. VIII
Centrarchids (fish), 660
Centrifugal force, 688
Cephalopods, 114. *See also* Vol. VII
Cerargyrite, 93, 155, 490, 1092
Ceramics, 378, 923, 1182
Ceramic glazes, 66, 118
Ceramic industries, 364
Cereals, 190
Cerite, 145, 1060
CERIUM: ELEMENT AND GEOCHEMISTRY, 7, 44, **145**, 278, 399, 1030
Ceruléite, 938
Ceruleolactite, 939
Cerussite, 58, 107, 142, 204
Cervantite, 844
CESIUM: ELEMENT GEOCHEMISTRY, 16, 20, 44, **147**, 274, 278, 300, 462, 489, 1052
Chabazite, 58, 592, 749
Chain coordination, 88
Chains-of-wells, 547
Chain structure, linear, 198, 202
Chalcanthite, 191, 1118, 1124
Chalcedony, 35, 58, 422, 435, 438, 731, 900, 1086. *See also* quartz, silica
Chalcoalumite, 1126
Chalcocite, 191, 195, 234, 750, 900, 1130
Chalcocyanite, 1123
Chalcomenite, 1079
Chalcophanite, 846
Chalcophile elements, 35, 47, 83, 99, 212, 395
Chalcophile groups, 1145
Chalcophyllite, 939
Chalcopyrite, 58, 64, 190, 191, 195, 355, 569, 578, 900, 959, 1130
Chalcosiderite, 938
Chalcosphere, 396
Chalk, 114, 551
Challenger Expedition, 674, 1008, 1062, 1076
Challomenite, 1080
Chalybite, 604
Chamosite, 58, 61, 599, 601, 605, 606
Chaoborus (larvae), 660
Chao Phraya River, Thailand, 474. *See also* Vol. VIII
Charcoal, 447, 452
Charge balance reaction, 127
CHARGED-PARTICLE TRACKS, **148**. *See also* FISSION TRACK DATING
Charred bone (dating), 447
Chattahoochee River, 1057
Chelatable metals, 268
CHELATION, 28, **149**, 180, 190, 416, 671, 833, 1241
Chemical age, 1227

Chemical analysis, 1114
Chemical base. *See* BASE, CHEMICAL
Chemical bonding, 406
Chemical climate, 1015
Chemical deposition (precipitation), 118
CHEMICAL MINERALOGY, **152**
Chemical-physical weathering, 606
Chemical pollutants, 312
Chemical potential, 382, 684
Chemical precipitation, 272, 351, 548
Chemical reactions, 891, 1118, 1170
Chemical sediments, 901. *See also* PARAGENESIS; *also* Vol. VI
Chemical weathering. *See* WEATHERING, CHEMICAL
Chemiluminescence. *See Vol.* VI, LUMINESCENCE
Chemisorption, 160
Chemistry, 409
Chemistry, methods, etc. *See Encyclopedia of Chemistry* (2nd ed., Reinhold Publ. Corp.)
Chemoautotrophs, 1132
Chemosynthesis, 850
Chenevixite, 937
Cheralite, 1184
Chert, 59, 64, 422, 436, 438, 600, 734. *See also* Vol. VI
Cherty limestone, 119
Chervetite, 1237
Childrenite, 938
Chilean Desert, 802
Chile saltpeter, 1097
Chillagite, 754, 1213
China, 35, 378. *See also* Vol. VIII
China clay. *See* kaolinite
Chinchona bark, 772
Chindwin River, 474
Chinle Formation (Triassic), 1223
Chironomidae, 660
Chironomid larvae, 660
Chironomus plumosus, 660
CHITIN AND CHITINOUS CUTICLES, 75, **153**, 1202
Chitinases, 154
Chitosan, 154
Chloanthite, 1132
Chloraluminite, 490
Chlorapatite, 155, 937
Chlorides, 37, 38, 83, 93, 185, 271, 320, 475, 571, 574, 589, 1017, 1017A, 1192. *See also* Vol. I
Chloride complexes, 574
Chloride type waters, 868
Chlorinated hydrocarbon insecticides, 340
Chlorination, 328
CHLORINE: ELEMENT AND GEOCHEMISTRY, 36, 44, 75, **155**, 200, 216, 225, 245, 270, 274, 278, 374, 377, 408, 489, 492
Chlorinity, 121, 1063
Chlorins, 832
Chlorite, 22, 55, 58, 91, 168, 178, 203, 230, 260, 438, 568, 587, 605, 790, 901, 902
Chloritoid, 23, 904
Chlormanganokalite, 491
Chlorabacteriaceae, 1150
Chlorobium chlorophyll, 156
Chlorocalcite, 490
Chloromagnesite, 490
CHLOROPHYLL, 71, 150, **156**, 600, 831, 1236
Chlorophyll-degradation products, 71
Chlorothionite, 1125
Chloroxiphite, 490
Cholera, 692
Cholesterol, 700
Chondrites, 19, 93, 99, 146, 241, 255, 343, 349, 384, 462, 494, 599, 664, 679, 810, 865, 959, 971, 1060, 1184, 1270. *See also* Vol. II
Chondritic earth, 838
Chondritic meteorites, 18, 35, 398, 401, 402, 582, 992, 1024, 1029, 1041, 1050
Chondrules, 276
CHROMATES, **159**, 583, 717
Chromatite, 159
Chromatium, 1156
Chromite, 898, 899
Chromatogram, 163
Chromatographic procedures (hydrocarbon analysis), 496
Chromatographic "plate" theory, 770
CHROMATOGRAPHY, 71, **159**, 409, 496, 769
Chromian spinel group, 167
Chromite, 167, 845, 1059
CHROMIUM: ELEMENT AND GEOCHEMISTRY, 44, 106, 151, 159, **167**, 274, 278, 296, 300, 319, 320, 423, 599, 701, 736
Chrysoberyl, 22, 68, 845, 1235
Chrysocolla, 195, 744, 1137
Chrysotile, 734, 791. *See also* serpentine, asbestos
Churchite, 936

Chydoridae, 660
Chydorus spaericus, 660
Cinnabar, 35, 462, 573, 704, 732, 810, 900, 1131
Cirrepods, 114. *See also* Vol. VII
Circumpolar Waters, Antarctic, 121. *See also* Vol. I
Citrus, 28
Cladocera, 660
CLAPEYRON'S EQUATION, 4, **170**, 382, 1174. *See also* Vol. IVB
Clarke, F. W., 454
"Clarke Value," 99
Clarkeite, 844
Clathrate compounds, 201
Claudetite, 41, 42, 844
Clausius, 305
Clausius-Clapeyron equation, 171
Clausthalite, 705, 1080, 1130
Clays and clay minerals, 5, 21, 22, 25, 29, 35, 54, 59, 64, 68, 70, 88, 91, 144, 147, 148, 153, 155, 181, 203, 213, 226, 238, 245, 246, 260, 275, 321, 349, 419, 430, 433, 458, 459, 462, 472, 475, 480, 495, 565, 587, 591, 593, 600, 601, 901, 902, 1018, 1051, 1162. *See also* Vols. IVB and VI
Clay-bound ammonium, 797
Clay lenses, 552
Clay materials, 438
CLAY MEMBRANE PHENOMENA, **172**, 768
CLAY MINERALS—BASE EXCHANGE, **176**
Clay mineraloid, 745
Clay particles, 20
Clay petrology, 420
Clay sediments, deep-sea, 35, 52, 180, 190, 246. *See also* Vols. I and VI
Cleanser, 89
Cleavelandite, 19
Cleve, 494
Cliachite, 24
Climate, 428, 471, 544, 773, 1043, 1240
Climate, global, 11, 338
Climatic fluctuation, 14, 133, 876. *See also* Vol. II, CLIMATIC VARIATIONS
Climatic variations. *See* Vol. II
Climatic zonation, 435
Climatology, 187
Climax, Colorado, 756
Climax vegetation, 255
Clinoptilolite, 53
Clinopyroxene, 17, 168, 273, 1030
Clinopyroxene insolicates, 582
Clinton Ores, New York, 606
Closed (endorheic) drainage, 547; *see also* Vol. III
Cloth dating, 447
Clouds (rainwater content), 1015
Cloud condensation nuclei, 15. *See also* Vol. II
Cloud seeding, 549
Cloudburst, 545
Cloudiness, 13. *See also* Vol. II, CLOUDINESS AND RAINFALL
Clover leaves, 499
C/N ratio. *See* CARBON-NITROGEN RATIO
Co, 412
Coal, 68, 78, 88, 128, 135, 151, 332, 335, 409, 427, 462, 734, 812, 824, 1203. *See also* Vols. IVB and VI
Coal burning, 732
Coal-dust aerosols, 739
Coal engine pollution, 323
Coal flue dust, 387
Coal formation, 427. *See also* Vol. VI
Coal gasification and liquefaction. *See* Vol. IVB
Coal measures, 886
Coal tars, 427
Coastal dunes, 536
Coastal plains, 474
Coastal terrain groundwater. *See* HYDROLOGY, COASTAL TERRAIN
Coastal waters, 595
COBALT: ELEMENT AND GEOCHEMISTRY, 41, 44, 49, 75, 79, 82, 106, 151, **179**, 190, 193, 274, 278, 296, 300, 320, 426, 433, 599, 699, 736
Cobalt, Ontario, 1094
Cobaltite, 42, 1131
Cobaltocalcite, 142
Cobaltomenite, 1079
Coccidiomycosis, 312
Coccinite, 490, 704
Coccolith, Coccolithoforidate, 49, 115. *See also* Vol. VII
Cochina, 114
Cocoa, 28
Coconuts, 28
Coefficient of thermal expansion, 374
Coelenterata, 113, 153. *See also* Vol. VII
Coelenterate perisare, 155

1301

INDEX

Coesite, 274, 579, 580
Coeur d'Alene, Idaho, 35, 1094
Coeur d'Alene lead ores, 449
Coffee, 28
Coffinite, 1223
Coherent exsolution, 364
Coinage metals, 411
Cold gas trap, 496
Colemanite, 87, 101
Coliform bacteria, 320
Colima, Mexico, 12
Collagen, 154, 730
Collenia, 122. See also Vol. VII
Colligative properties, 751
Collinsite, 935
Colloform, 744
Collogen fibers, 114
COLLOIDS, 20, **180**, 259, 319, 574, 1042, 1067
Colloidal clays, 430
Colloidal particles, suspensions, systems, 259, 902, 911, 948
Colloidal silica, 436
"Collomorphic" aggregates (colloidal genesis), 744
Collophane, 52, 58, 101, 425, 902, 941, 944
Color (of minerals), 208. See also Vol. IVB
Colorado, 378, 547
Colorado Plateau, 152, 993, 1223
Colorado River, 1047
Coloradoite, 704, 1130
COLORIMETRY, **182**, 374, 409, 436, 444, 1114
Columbia, 1048
Columbia Basalt, 1036
Columbia River Plateau, Washington, 474, 555
Columbite, 489, 794, 846, 1162
Columbium, 909. See also NIOBIUM
Columbus Limestone, Ohio, 120
Columbian Marble, Vermont, 120
Column chromatography, 160, 496
Comal Springs, Texas, 1116
Combining capacity of acids, 1119
Combustion, 15, 848
Combustion of fuels, 1016, 1229
Commission for Spectroscopy (Berlin), 583
Common ion effect, 1109
Compaction (carbonate rocks), 539
Compaction (muds), 435
Compatibility triangles, 918
Compensation depth, 126. See also Vol. I, CALCIUM CARBONATE COMPENSATION
Compensation point, 77
COMPLEXES, 3, 71, 74, 83, 108, 126, 136, 150, 157, 180, **184**, 190, 259, 378, 384, 558, 575, 816, 957
Complex formation, 573
Complex oxyanions, 847
Complexes, "Alpine," 169
Complexes, bicarbonate, 576
Complexes, bisulfide; hydrogen sulfide, 575
Complexes, stratiform, 169
Complexation (in chromatographic separation), 160
Complexing agents, 572
Complexing of ions, 1068, 1246
Components, 917
Composition triangles, 918
Compounds, 3, 83
Composition of the earth, 264
Compounds, metal-inorganic, 149
Compounds, metal-organic, 149
Compressibility of aquifers, 552
Compressional waves, 459
Computer application in earth sciences. See Vol. VI
Comstock Lode, Nevada, 469
Concentrations of substances in solution, 924, 1119
Conchineal layer, 114
Concordia curve, plot, 448, 1217
Concrete treatment, 104
Concretion (in mineral development), 709
Concretions, oceanic, 49
Condensation, 547
Condensation nuclei, 545
Conducting gelatine, 482
Conducting papers, 482
Conduction, Conductivity, 543, 545, 988, 992
Cone of depression, 239
Confining layer (phreatic water), 956
Congo, 1048. See also Vol. VIII, ZAIRE
Congruent melting point, 918
Conichalcite, 1106
Coniferous forest, carbon cycle, 128
Coniology. See Vol. II, KONIOLOGY
Conjugate solution, 1106
CONNATE WATER, **184**, 359, 470, 485, 547, 769, 1085
Connellite, 1125
CONSERVATION, **185**, 772

Conservationists, 1102
Conservative ions, 669
Consolidation, effect on permeability of carbonate rocks, 539
Construction, 478, 593
Construction equipment, noise, 315
Construction industry, 314
CONSTRUCTION MATERIALS. See Vol. VI
Contact springs, 1116
Containers, urban waste, 311, 317
Contamination, 949, 1002, 1056
Continents, 210, 992. See also Vol. V
Continents and oceans—statistics of area, volume, relief. See Vol. III
Continental crust, 17, 244, 402. See also Vol. V
Continental deposits, 358. See also Vol. VI
Continental ice, 894. See also Vol. II
Continental shelf, 435. See also Vol. I
Continental shelf sediments, 54, 102. See also Vol. VI
Continental slope, 588. See also Vol. I
Continental structure, 682
Continuous reactions (in paragenesis), 898
Control aquifers, 549
Control runoff, 549
Convection, 129, 314, 401, 545, 572. See also Vols. II and V
Cooper Marl, South Carolina, 539
Cooperite, 956, 958, 1131
Coordinate bond, 198
Coordination number, 86, 88, 199
Coorong lagoon, South Australia, 441, 764
Copalite, 826
Copepods, 74, 137. See also Vol. VII
Copiapite, 1126
COPPER: ELEMENT AND GEOCHEMISTRY, 34, 41, 44, 49, 75, 106, 130, 150, **189**, 197, 201, 231, 278, 296, 300, 320, 360, 396, 411, 426, 433, 466, 467, 573, 574, 587, 598, 605, 697, 699, 763, 1067
Copper deficiency, 696
COPPER DEPOSITS, 191, **193**, 310
Copper-lead-zinc ore, 462, 468
COPPER: MINERALOGY OF COMPOUNDS, **190**
Copper oxysalts, 190
Copper reserves, 195
Copper toxicity, 1138
Coprogenic ooze, 660
Corals, 102, 113, 118, 126. See also Vol. VII
Coral fossils, 994
Coral reefs, Recent (Holocene), 114
Coralline algae, 118
Cordierite, 904
Cordilleran Mt. belt, 169. See also Vol. VIII
Cordylite, 143
Core, 394, 397, 429, 600, 678, 791. See also Vol. V
Core drilling, 550
CORE GEOCHEMISTRY, 32, **195**, 263
Core-mantle interface, 195. See also Vol. V
Cores, Pacific Ocean, 597. See also Vol. I
Corkite, 939
Cornetite, 936
Cornwall, 1192
Cornwallite, 938
Corona, 706. See also Vol. V
Coronadite, 846
Coroongite, 829
Corrosion, 334, 598, 1183, 1266
Corrosion-free alloys, 897
Corundum, 21, 22, 27, 844
Cosmic abundances, 1041
Cosmic abundance curve, 265, 266
Cosmic dust, 397, 545, 812. See also Vol. II
Cosmic irradiation, 995
Cosmic radiation, 129, 149, 326, 996
Cosmic rays, 398, 451, 493, 988
Cosmic-ray-produced radionuclides, 40, 68, 409. See also RUDIONUCLIDES, COSMIC-RAY-PRODUCED
Cosmic-ray spallation, 1211
Cosmochemistry, 402
"Cosmogenic" argon, 40
Cosmogenous particles, 49
Coster, 488
Cotopaxi, Equador, 12. See also Vol. VIII
Cottonwood, 549
Cotunnite, 892
Coulsonite, 1236, 1237
Coulomb interaction, 761
Counterions, 259
Covalent bond, 84, 197, 259, 262, 495
Covalent crystals, 414
Covellite, 191, 195, 1131, 1138
Cr, 404
Cracking, catalytic reaction, 144, 146, 241, 343
Crandallite, 937, 941

Crassulacea, 499
Crater Lake, Oregon, 555
Crednerite, 845
Creedite, 491
Cretaceous chalk of England, 474
Cretaceous greensands, 605
Cretaceous period, 102, 454. See also Vol. VII
Crimea, U.S.S.R., 545. See also Vol. VIII
Crinoids, 70, 118
Cripple Creek, Colorado, 468, 469
Cristobalite, 59, 203, 438, 579, 734, 915, 1086
Crocidolite, 734
Crocoite, 159, 168
Cronstedt, Axel Frederich von, 152, 777, 792, 976
Crude oil, composition and migration, 421, 427, 498, 770, 799, 831, 1236. See also Vol. IVB
Crushed stone, 118
Crust, 394, 397, 1267, 600, 678, 764, 861, 992, 1056. See also EARTH'S CRUST, GEOCHEMISTRY; GEOCHEMICAL EVOLUTION OF THE CORE, MANTLE, AND CRUST; also Vol. V
Crustacea, 74, 660. See also Vol. VII
Crustacean shells, 950
Crustal deformation, 186
Crustal fusion, 40
Crustal history, 448
Cryolithionite, 491
Cryolite, 21, 24, 378, 491, 1097
Cryoscopic constant, 751
Cryptohalite, 491
Cryptomelane, 846
Cryptoperthite, 238
CRYSTAL CHEMISTRY, **197**, 374, 740, 906
CRYSTAL FIELD THEORY, 193, **204**, 205, 416
Crystal, growth, 363, 573
 morphology, 153, 201, 583
 segregations, 898
 solubility, 923
 solution, 919
 symmetry, 1117
Crystalline rocks, schists, 427, 465
Crystallization, 342, 400, 898, 923
Crystallization, fractional, 18
Crystallographic, changes, 217
 inversions, 394
Crystallography, 740. See also Vol. IVB
Cs, 406
Cu, 573, 578, 589
Cuba, 474. See also Vol. VIII
Cubanite, 191, 1131
Cultural environment, 690
Cumengite, 490
Cuprite, 191, 195, 843
Cuprocopiapite, 1126
Cuproscheelite, 1213
Cuprotungstite, 752, 1213
Curie, Marie, 987, 1006
Curie point, 231, 746
Curite, 844
Curium, 6, 44, 278. See also ACTINIDE SERIES
Current base. See Vol. VI
Curtisite, 226, 811, 828
Cuticular laminae, fossil, 155
Cyanide, 320
Cyanochroite, 1124
Cyanotrichite, 1126
Cycles, biological, 80, 124
CYCLES, GEOCHEMICAL, 81, 102, 124, 193, **208**
Cycle, iron, 600
Cycle of weathering, 1269
Cyclic hydrocarbons, 498
Cyclic salts, 216, 351, 359, 669, 1016, 1043. See also Vol. VI, SALTS—CYCLIC
Cycloalkanes, 498
Cyclobutane, 496
Cyclohexane, 496
Cyclopentane, 496
Cyclopentaphenanthrenes, 498
Cyclosilicates, 201, 720. See also CRYSTAL CHEMISTRY; MINERAL CLASSES: SILICATES
Cyclothems, 886
Cylindrite, 578, 1192
Cyrtolite, 488
Cysteic acid, 72
Cystine, 72
Cytherean atmosphere, 33
Czechoslovakia, 35. See also Vol. VIII

Dacite, 555. See also Vol. V
Dahllite, 52, 58, 698
Dakota Sandstone, 474
Dams, 478. See also Vol. VI
Dana, J. D., 454
Dana's System of Mineralogy, 717
Danburite, 88
Danube River, 1048
Danube River Basin, 418
Darcy, Henry, 476, 551, 965

INDEX

Darcy velocity, 551
Darcy's law, 372, 471, 480, 551
Darcy's law of flow, 965
Darwin, Charles, 337, 454
Data processing, 774
Dating, 594, 996, 1052
Dative bond, 198
Datolite, 88
Daubrée, G. A., 399, 476, 679
Daubréeite, 490
Daubréelite, 168, 1131
Daughter, minerals, 374
　phases (exsolutions), 361
Davidite, 1027, 1223
Davis, 436
Davisonite, 21, 23, 143, 938
Davy, Sir Humphrey, 88, 100, 155, 373, 377
Day, Arthur, 153
DC arc techniques, 445
DDT, 311, 318, 327, 340
DDTA (derivative differential thermal analysis), 217
Dead Sea, 754, 764, 765, 1098. *See also* Vols. III and VIII
Deafness, 315
Death (organic), 451, 946
Death Valley, California, 89
Debienne, A., 9
De Boisbaudran, 385
Debye-Hückel equation, 110, 592
Debye-Waller factor, 759
n-Decane, 496
Decapods, 114. *See also* Vol. VIII
Decarboxylation, 79, 816, 824
Decay, radioactive, 446, 780, 984, 987, 1053
Decay series (U^{238}, Th^{232}, U^{235}), 1007
Deccan Plateau, India, 474, 1036. *See also* Vol. VIII
De Chancourtois, B., 43
Deciduous forest, carbon cycle, 128
Decomposing agents, (biological), 104, 127, 661
Decomposition (organic), 69, 217, 226, 349, 1182
Decrepitation, 373
Deep-sea, clays, 180, 1062. *See also* Vols. I and VI
　cores, 596, 894
　sediments, 48, 55, 63, 68, 102, 110, 434, 588, 590, 594, 1007, 1010, *See also* Vols. I and VI
Deep waters, (isotopic carbon composition), 134
Deep-well, disposal, 540
　turbine pump, 551
Defect crystals, 204
Deflation, 1101. *See also* Vol. III
Defluidation cycle, 284
Defoliants, 325
Deforestation, 1101
Degassing, 428, 499, 837, 1077
Degradation, anaerobic, 335
　of mineral accumulations by microbes, 729
Degrees of freedom, 917
De Hevesy, 488
Dehrnite, 937
Dehydration reactions, 104, 217, 226, 231, 274, 562, 824, 1183
Dehydrogenation, 71, 945
Delafossite, 845
De L'Isle, R., 741
Delrioite, 1237
Deltaite, 937
Delvauxite, 938
Demand, 772
Demarcay, 349
Dendritic crystal growth, 373
Dengue (fever), 693
Denhardite, 826
Density, 1111
Denitrification, 806
Denudation, 418. *See also* Vol. III
Deoxidation, 104
Dysentery, 692
"Depoisoning," 151
Depolymerization, 815
Deposition, 57, 77, 342, 420, 428, 454, 901
　estuarine, 348. *See also* Vol. VI
Deposits, 670, 1080
Deposits, abysso-pelagic, 432. *See also* Vol. VI
Depression springs, 1116
Derbylite, 34
DERIVATIVE DIFFERENTIAL THERMAL ANALYSIS, 217, 1168
DESALINATION PROCESSES—U.S. DESALTING PROGRAM, 219, 332
Desalting, 549
Deserts, 75, 128, 541. *See also* Vol. III
Desert-grassland transition, 542
Desert vegetation, 545
Deschutes Basin, 556
Desulfotomaculum, 729, 1127

Desulfovibrio, 729, 1128, 1134, 1137, 1150, 1156
Descloizite, 936, 1235, 1236
Detection, radiation, 988
Detergents, 317, 475, 1247
Detrital sediments, 901. *See also* Vol. VI
DEUTERIUM, 225, 266, 572, 868
Deuterium/hydrogen ratio, 225
Deuteron bombardment, 7
Developed resource, 773
Development of life, 454
Devillite, 1126
Devil's River, Texas, 542
Devitrification, 59, 439, 829
Devonian, 454. *See also* Vol. VII
Dew point, 544. *See also* Vol. VII
Dewindtite, 937
Diabase, 146, 156, 465. *See also* Vol. V
Diaboleite, 490
Diadochite, 490
Diadochy. *See* TRACE ELEMENTS IN SILICATE MINERALS
Diagenesis, 22, 49, 51, 69, 89, 114, 118, 151, 178, 208, 349, 418, 435, 441, 485, 586, 588, 669, 670, 715, 744, 872, 901, 902, 1072, 1121, 1217. *See also* PARAGENESIS; also Vol. VI
Diaminopimelic acid, 72
Diamond, 85, 123, 134, 197, 201, 275, 382, 763, 899. *See also* Vol. IVB
Diapir (salt domes), 1098
Diaspore, 21, 27, 86, 423, 845
Diatoms, 53, 70, 92, 113, 137, 216, 422, 433, 437, 1081. *See also* Vol. VII
Diatom bloom, 954
Diatomaceous earth (opal), 734
Diatomite, 80, 422, 440
Diatremes, 681. *See also* Vol. V
Diastrophism, 576. *See also* Vol. V
Dickinsonite, 935
Dickite, 565. *See also* KAOLINITE, Vol. II
Diesel engine pollution, 323
Dietrichite, 1125
Dietzeite, 590
Dietzel, 412
Differential entropy change, 305
DIFFERENTIAL THERMAL ANALYSIS, 217, 226, 258, 391, 743, 1167
Differential thermobalance, 232
DIFFERENTIAL THERMOGRAVIMETRIC ANALYSIS, 231, 1168
Differentiation, 18, 21, 68, 190, 401, 709
Diffraction, 301
Diffraction grating, 1271
Diffusion-geological role, 160, 234, 274, 573, 574, 709, 1008, 1072, 1183
Digenite, 191, 1130
Digital models (aquifer performance), 477
Dihydroretene, 499
Dijon, France, 551
Dikes, 478, 555
Dimension stone (architectural), 118
Dimethyl glyoxime, 150
Dimethylphenanthrene, 499
2, 2-Dimethylpropane, 496
Dimorphite, 42, 1130
Dinoflagellates, 76
Dinwoody Glacier (Wind River Mts.), Wyoming, 465
Diopside, 1, 100, 120, 149, 202, 399, 582
Dioptase, 191
Diorite, 19, 101, 156, 194, 248. *See also* Vol. V
Diorthosilicates, 202
Diphenyl-thiocarbazone, 150, 444
Diphtheria, 692
Direct current arc excitation, 301
Direct interchange diffusion, 236
Discontinuous reaction (paragenesis), 898
Discharge, 554, 1116
Discharge zone, 539
Disease, 690
Disease-carrying pollutants, 312
Disease patterns, 738
Disequilibrium studies, 1008
Disintegration ratios (Na_{22}/Be_7), 1003
Dislocations, lattice, 105, 237
Dispersed element, 1040
Dispersion patterns, 706, 998
Dispersion of pollutants, 333
Dissimilation, 74
Dissociation constants, 573, 925
Dissolved components, sources, 1070
Dissolved salts, 669, 1042, 1046
Dissolved solids, 37, 38
Distillation methods (hydrocarbon analysis), 496
Distillation process (LTV), 220
Distribution constant (trace elements), 907
Divergence theorem, 230
Doerner-Hoskins heterogeneous relationship, 907

Djalmaite, 846
Djerissa, Tunisia, 606. *See also* Vol. VIII
Dokuchaiev, 912
Dolerite, 83, 556. *See also* Vol. V
Dolerophanite, 1125
Dolomite, 1, 17, 20, 52, 58, 64, 100, 103, 107, 112, 118, 120, 142, 156, 356, 378, 423, 424, 441, 480, 557, 587, 667, 734, 900, 941, 1078, 1241
Dolomitization, 114. *See also* Vol. VI
Domains (crystals), 906
Domestic waters, 317, 322
Domesticated flocks, 541
Domeykite, 42
Donnay, 438
Donor-acceptor bond, 198
Doppler effect, 759
Dopplerite, 829
Double bond, 71
Double-layer theory, 177
Double tetrahedra structures, 201
Douglasite, 491
Dover chalk (Cretaceous), 114. *See also* Vol. VI
Dowsing. *See* WATER DIVINING
Dra Valley (Morocco), 549
Drains and Drainage, 430, 478. *See also* Vol. VI
Drainage basins, 478, 480, 484, 876, 1057
DRAWDOWN, CONE OF DEPRESSION, 239, 552
Dredging, 313
Drill-core sediments, 427
Driling fluids, technology. *See* Vol. VI
Drilling, hydraulic, 551
Drinking water standards, 320, 702
Dropping method, 304
Drought, *See also* Vol. II
Drusy quartz, 65
Dry ice, 549
DTA (differential thermal analysis), 217, 226, 258
DTG (derivative thermogravimetry), 231
DTGA (differential thermogravimetric analysis), 231
Duboscq colorimeter, 182
Dufrenite, 937
Duftite, 936
Dumontite, 938
Dumortierite, 23, 88
Dundasite, 143
Dundie Limestone, Mich., 120
Dunes, 360. *See also* Vol. III
Dunite, 19, 62, 101, 156, 248, 273, 401, 582, 680, 992. *See also* Vol. V
Dupuit, Jules, 476
Durangite, 937
Duricrust, 215. *See also* Vol. III
Dussertite, 937
Dusts, 697, 1015. *See also* Vol. II
Dust control, 104
Dust counter. *See* KONIOLOGY; ATMOSPHERIC NUCLEI AND DUST, Vol. II
Dust cover, 546
"Dust disease," 731
Dust pollution, 314, 341
Dust storms, 312, 545
Dutch Antilles dolomite, 112. *See also* Vols. VI and VIII
Dygyttja, 1204. *See also* GYTTJA, Vol. VI
Dynamic metamorphism, 903
Dynamo Theory, 196
Dyscrasite, 1092, 1130
DYSPROSIUM: ELEMENT AND GEOCHEMISTRY, 44, 240, 278
Dzhezkazganite, 1041

Earth, 394, 402. *See also* Vol. II
　atmospheric interface, 337, 341
　biological history, 68, 186. *See also* Vol. VII
　composition, 264, 273
　See EARTH'S CRUST GEOCHEMISTRY; also GEOCHEMICAL EVOLUTION OF THE CORE, MANTLE, AND CRUST
　shells, 429
Earth's, basaltic shell, 246
　core, 462. *See also* Vol. V
　crust, 405, 428, 432, 470, 489, 492, 571, 599, 606, 718, 1052, 1070, 1082
EARTH'S CRUST GEOCHEMISTRY, 29, 32, 36, 80, 95, 100, 103, 180, 189, 190, 243, 270
　crust, origin, 270
　granitic shell, 245; *See also* Vol. V
　interior, 273
　magnetism, 196
　mantle, 462. *See also* Vol. V
　resources, 310
　sedimentary shell, 244
Earth sciences, 186. *See also* Vol. VI
Earthworms, 154
East African rift valleys, 795. *See also* Vol. V

1303

INDEX

East Pacific Rise, 434, 1035
 sediments, 50, 271
Ebeco Volcano, Kurile Islands, 765
Ebulliscopic constant, 751
Ecdemite, 34
Echinoderms, 35, 102, 113, 114, 118. *See also* Vol. VII
Echinoidea, 113
Eclogite, 18, 83, 264, 273, 710, 904, 1185. *See also* Vol. V
"Ecologic balance," 772
Ecological change, 1169
 niche, 257
Ecologist, 690
ECOLOGY, 74, 185, **254**, 324, 544, 856
 tarine. *See* Vols. I and VII
 terminology. *See* Vol. VII
Economic, expansion, 187
 geology, mineralogy. *See* Vol. IVB
Ecosystems, 186, 326, 1238
Eddies in fluid motion. *See* FLUID MECHANICS, Vols. II and IV
Edible oils hydrogenation, 144
Eddy, conductivity. *See* Vol. VI
 diffusion, 1074
 diffusion coefficient, 1009
EDTA, 416, 444. *See also* VERSENE (EDTA) SOLUTION STUDIES
Edwards limestone, 539
Eelworms, 154
Effective permeability, 967
EFFLUENT GAS ANALYSIS, **258**, 391, 1168
EGA, 258
Eglestonite, 490, 704
Egueiite, 938
Egypt, 548, 551. *See also* Vol. VIII
Egyptian mummies, 731
Eh, 84, 374, 416, 435, 576, 672, 1136
Eh–pH, geochemical application. *See* pH–Eh DIAGRAMS
Elba, Italy, 1052
n-Eicosane, 496
Einstein, A., 94, 988
Einsteinium, 6, 44, 278
EK value, 412
Ekman dredge, 660
El Laco, Northern Chile, 606
Elasticity of rocks, 459
Elastomers, synthetic, 144
Elaterite, 828, 1206
Elbe River, Germany, 473
Elburz Mountains, Iran, 551
Electric, analog model, 472
 double layer, 259
 field, 457
 logging, 460
 generating station, 1170
 groundwater models, 482
 power plants, 1229
 resistivity methods, 457
Electrocardiagraph, 1254
Electrodialysis, 221, 374
Electroluminescence. *See* Vol. IVB LUMINESCENCE
Electrolysis, 21, 24, 104
Electrolytes, 437, 482, 574, 606, 687, 924, 1015, 1084, 1109
ELECTROLYTES: FLOCCULATION OF COLLOIDS, 103, **259**
Electrolytic conductance, 2
Electromagnetic, pulses, 460
 spectrum, 579
Electromotive force, 380
Electron, 3, 43, 85, 266, 411, 839, 932
ELECTRON CAPTURE, **260**, 840, 989
"Electron cloud," 85, 262
Electron diffraction, 741
Electron, holes, 145
 microscope, 438
 microscopy, 115, 149. *See also* Vol. IVB
 probe, 445
 probe microanalysis, 409, 419, 742
Electronegativity, 3, 48, 86, 200, **261**, 415, 847
Electronic energy, 3, 891
Electroosmosis, 172. *See also* Vol. VI
Electroplating, 99
Electropositive oxides, 394
Electrostatic, attraction, 84, 591
 bond, 259
 field, 86, 416
 neutrality, 17
 potential, 261
 shift, 761
Electrovalent bond, 178
ELEMENTS, 210, 253, **263**, 398, 406, 443, 579, 680
Elements, geochemical classification. *See* GEOCHEMICAL CLASSIFICATION OF ELEMENTS
Elements, origin, 266

ELEMENTS: PLANETARY ABUNDANCES AND DISTRIBUTION, 255, **268**. *See also* Vol. IVB
Element synthesis, 266
Ellestadite, 937
Elpasolite, 491
Elpidite, 488
Embolite, 93
Emery, 23
Eminence Dolomite, Missouri, 65
EMISSION SPECTROGRAPHY, **300**, 388, 409, 742. *See also* Vol. IV
Emmons Glacier, Mt. Rainier, Washington, 465
Emmonsite, 1079
Emotionalists, 772
Empirical mineral synthesis, 153
Empressite, 1131
"Empty Quarter," Arabia, 542. *See also* Vol. VIII
Enamels, 66
Enantiomers, 810
Enantiomorphism. *See* Vol. IVB, *also* STEREOCHEMISTRY AND ENANTIOMORPHISM
Enargite, 42, 192, 195, 462, 900
End-members (in solid solutions), 1104
Endellite. *See* halloysite
Endogenic Cycle, 602
Endoskeleton, 893
Endothermic reactions, 217, 226, 743, 1170. *See also* THERMOCHEMISTRY
Enelectrite, 829
Energetic particles, 996
Energy, 234, 1170
 balance, 546
 budget of earth's surface, 337. *See also* Vols. I and II
 flow, 254, 339
 level, 45, 1272
 of expansion, 593
 resources, 187, 774
 wastes, 317
Engineering, 478
England, 551. *See also* Vol. VIII
Englishite, 938
Enhydros, 94. *See also* Vol. IVB
Enstatite, 1, 149, 277, 667
ENTHALPY OR HEAT CONTENT, 184, **302**, 303, 304, 684, 1104, 1171, 1173. *See also* Vol. IVB
Entropy, 1104, 1173
ENTROPY, 170, 184, **305**, 1104, 1173. *See also* Vol. IVB
Environment, 481, 690, 902
Environments, ancient, 80
Environmental factors, 696, 1046
Environmental geochemistry symposium, 296
Environmental indicators, 57
ENVIRONMENTAL POLLUTION, 76, 296, **309**. *See also* Vol. IVB
Environmental pollution—global effects. *See* AIR POLLUTION AND GLOBAL CLIMATE; AIR POLLUTION AND URBAN CLIMATE
Environmental resources, 311
ENVIRONMENTAL SCIENCE, 187, **337**. *See also* Vol. III, ANTHROPOGENIC INFLUENCES IN GEOMORPHOLOGY
Enzymes, 74, 154, 334, 699, 857, 1065, 1128, 1152, 1155, 1158, 1199
Enzyme activator, calcium as, 103
Eocene, 454, 588, 881. *See also* Vol. VII
Eocene fossils, 155
Eolian erosion. *See* "Aeolian . . .", 545
Eolianites, 114. *See also* Vol. III
Eosphorite, 938
Epicuticle, 154
Epidemiology, 738
Epidiagenesis, 178, 342
Epidote, 367, 567, 568, 901, 904
Epidote amphibolite, 904
Epidote group, 23, 101, 149
Epidotehornfels, 904
EPIGENESIS, **342**, 422, 441, 1159, 1160. *See also* Vol. IVB
EPILIMNION, **342**, 577, 1172. *See also* Vol. VI
Epsomite, 667, 1125
Equation of continuity, 478
Equations, Navier-Stokes, 551
Equation of state, 478
Equatorial sediments, 77
Equilibrium, 2, 37, 133, 170, 213, 266, 274, 379, 684, 725, 909, 917, 1106, 1152, 1173
Equilibrium assemblage, 726
Equilibrium constant, 383, 561, 685, 927, 1107, 1109. *See also* MASS ACTION AND EQUILIBRIUM CONSTANT
Equilibrium rate, 994
Equivalent weight, 1119
ERBIUM: ELEMENT AND GEOCHEMISTRY, 44, 278, 300, **342**. *See also* Vol. IV
Erinite, 936
Eriochalcite, 490

Erionite, 749
Erodibility of soils, 1045
Erosion, 20, 40, 186, 313, 325, 338, 454, 669, 876, 901, 1268. *See also* Vol. III
Erosion controls, 1102
Erosion, estuarine, 348
Erosion rates, 1043, 1046
Erythrite, 935
Erythrosiderite, 491
Eschynite, 846
Eskolite, 168
Essential to life, 971
Esterification, 157
Esters, 815
Estheriid shells, 882. *See also* Vol. VII
Estuarine environments, 842
ESTUARINE HYDROLOGY, 265, **344**. *See also* Vol. VI
Estuarine sediments, 348, 587. *See also* Vol. VI
Etching, 366
Etch pits, 148, 373. *See also* 148; *also* Vol. IVB
Etch tubes, 373
Ethane, 496, 812
Ethylene, 130
Ethylene bromide, 93
Ethylenediamine, 67
Ethylphenanthrene, 499
Ettringite, 1126
Eucairite, 1130
Euchroite, 938
Eucrite achondrites, 146, 241, 350, 384, 494
Eucrites, 1024
Eucritic meteorites, 275
Eucryptite, 661
Eudialite, endialyte, 1027, 488, 1294
Euglena gracilis, 157
Euhedra, 60
Euphotic zone, 74, 1073
Euphrates River, Iraq, 474, 547
EUROPIUM: ELEMENT AND GEOCHEMISTRY, 44, 278, 300, **349**, 439, 1030. *See also* Vol. IVB
Eustatic sea level changes, 537. *See also* Vols. I, III, and VII
Eutectics, 561, 918, 923
Eutrophication, 660, 312, 345
Euxenite, 846, 1028, 1162
Euxenic conditions, environments, 1135, 1138. *See also* BLACK SEA, Vols. I and VI
Evaluation of constants, 927
Evansite, 938
Evaporite deposits, rocks, 353, 358, 699, 764, 768, 1096
Evaporite minerals, 351
Evaporation, 121, 344, 351, 437, 471, 481, 536, 541, 544, 545, 554, 876, 902, 1078, 1046, 1103, 1159
Evaporating pits, 103
Evaporites, 17, 78, 85, 88, 101, 112, 213, 245, 419, 426, 432, 433, 462, 588, 1202
Evaporites—physiochemical condition of origin. *See* Vol. VI
EVAPORITE PROCESSES, 103, 113, **351**. *See also* Vol. IVB
Evaporite sulfates (Maine), 80
Evapotranspiration, 473, 543, 878, 1248
Evenkite, 811
Everglades, Florida 538
Evolution (of life), 29, 122, 186, 258, 428, 850, 854. *See also* Vol. VII
Evolved gas detection, 391
Excess volatiles, 837, 1070, 1076
Exhalations, 996, 1077
Exchange of cations, 908
Exchange reactions, 420
Exchangeable cations, 949, 1018
Exchangeable reservoirs, 407
Excitation process (spectrochemical analysis), 301
Excited nucleii (fission), 8
Exhaust gasses (air pollution), 311, 315
Exobiology, 82
Exocuticle, 154
Exogenic Cycle, 600
Exothermic reaction, 217, 226, 743, 1170. *See also* THERMOCHEMISTRY
Exothermic oxidation processes, 547
Experimental mineralogy, 740
Experimental Mohole, 588
Exploitation, 772
Exploitation, fresh water, 537
Explosives (pollution), 314
"Exposure Ages," 493
EXSOLUTION, 117, **361**, 561, 727. *See also* Vol. IVB
Extraterrestrial, abundances, elements, 275
 chemistry, 585. *See also* Vol. II
 dust, 996
 environments, 123
 life, 810. *See also* Vol. II

INDEX

F, 571
Face mask, 739
Facultative anaerobes, 806
Fairchildite, 142
Fairfieldite, 935
Falcon Island, Tonga, 12. *See also* Vol. VIII
Fallout, 130, 1004, 1211
Fallout isotopes, 989
Fallow (land), 1102
False equilibrium, 725
FAO, 774
Faraday constant, 4, 380
Farnesane, 498
Farnesyl pyrophosphate, 500
Fars Formation (Iraq, Iran, Syria), 548. *See also* Vol. VIII
Fats, 231
Fatty acids, 79, 312, 428, 981
Fatty alcohols, 547
Faujasite, 749
Faunal distribution, 541. *See also* Vol. VII
Fayalite, 31, 201
Fe, 403, 404, 412, 578. *See also* iron
Feasibility (resource development), 1260
Feature (natural resource), 771
Fecal contamination, 312
Fecal pellets, 55
Feigl, 444
Feldspar, 17, 22, 29, 49, 53, 57, 153, 203, 386, 421, 429, 462, 573, 591, 735, 899, 971, 1051, 1084, 1091, 1107, 1166. *See also* Vol. IVB
Feldspathization, 714
Feldspathoids, 23, 203
Feldspathoid-bearing rocks, 1097
Felsic minerals, 898
Felsobanyaite, 1126
Felton, 392
Ferberite, 752, 1212
Ferghanite, 1237
Fergusonite, 241, 343, 384, 494, 664, 846, 1162, 1165, 1190, 1290, 1291
Fermentation, 77
Fermium, 6, 44, 278
Fermorite, 937
Fernandina I., Galapagos, 12. *See also* Vol. VIII
Fernandinite, 1237
Ferric hydroxide, 181, 260, 437
Ferric iron, 61, 356, 430
Ferric oxide, 600
Ferricrete, 215
Ferrides, 599
Ferrimolybdite, 752
Ferrinatrite, 1124
Ferri-sicklerite, 934
Ferrite. *See* Vol. VI
Ferritungstite, 1212
Ferrobacillus, 729, 1132
Ferromagnesian minerals, 19, 669
Ferromagnetism, 191
Ferromanganese oxides, 430, 673
Ferromanganese precipitates, 430
Ferrous bicarbonate, 601
Ferrous carbonate, 601
Ferrous iron, 106
Ferrous minerals, 600
Ferroselite, 1080
Ferrotellurite, 1079
Ferruccite, 491
Fersman, 412
FERTILITY OF THE OCEAN. *See* Vol. I
Fertility, soil (clay minerals), 593
Fertilization, artificial, 311
Fertilizer, 971
Fertilizer, agricultural, 103, 312, 318
Feruginous limestone, 119
Fervanite, 1237
Festings, 579
Fibroferrite, 1126
Fichtelite, 811, 829
Fick's diffusion law, 234, 1011
Fiedlerite, 490
Field capacity, 1248
Filariasis, 693
Filling temperature, 375
Fillowite, 935
Filter-feeders, 819
Filter pressing, 606
Filter sands, 551
Fingernail clams, 660
Finnemanite, 341
Fjords, 808. *See also* Vol. III
First law of thermodynamics, 303, 379
Fischer-Tropsch, 981
Fish, 660, 774. *See also* Vol. VII
Fish boxes, 950
Fish remains, 778
Fisheries, 1169

Fisheries and Wildlife Service, 775
Fission, 8, 783, 962, 988
Fission products, 1219
Fission products, air-borne, 130
Fission tracks, 148
FISSION TRACK DATING, 149, **366**. *See also* Vol. IVB
Fissure Springs, 1116
Fissure vein, 83
Flagellates (toxin production), 822
Flagstaff limestone (Paleocene and Eocene), 881
Flame emission spectroscopy, 369
Flame photometer, 368
Flame photometry, 409. *See also* FLAME SPECTROSCOPY
Flame spectrometry, 445
FLAME SPECTROSCOPY, **367**. *See also* Vol. IV
Flanders, 551. *See also* Vol. VIII
Flash distillation plant, 221
Flavinoid pigments, 816
Flavonoids, 833
Fleischerite, 462
Flinkite, 936
Flint, 438, 735. *See also* Vol. VI
Flint clays, 60
Floating debris, 314
Flocculation, 259, 820
Flood control, 1260
Flood-damage, 1261
Floodplain sediments, 480
Floods, 547, 1057
Florencite, 937
Florida, 378, 587. *See also* Vol. VIII
Florida sediments, 51, 112, 114
Floridian Aquifer, 540
Florissant Lake Beds (Oligocene), 882. *See also* Vol. VIII
Flow breccias, 474
Flowering plants, 877
Flowers, 499
FLOW NETWORK, **372**
Flows, 555
Flowstone, 110. *See also* Vol. III
Flow system, 484
Flue dusts, 1167
Fluellite, 491
Fluid, conductivity, 552
 discharge, 480
 inclusions, 40, 93, **373**, 558, 561, 571, 768. *See also* Vol. IVB
 mechanics, 476. *See also* Vol. II
 motion, 478
Fluidity, 898
Fluoborite, 87
Fluocerite, 490
Fluorapatite, 378, 937, 944
Fluorens, 498
Fluorescence. *See* Vol. IVB, LUMINESCENCE
Fluorescent screen, 988
Fluoride(s), 37, 38, 99, 320, 374, 377, 475, 1192, 1215
 toxic, 317, 322
Fluorimetry, 445
FLUORINE: ELEMENT AND GEOCHEMISTRY, 44, 147, 245, 261, 271, 278, 300, 462, 489, 492, 697, **377**. *See also* Vol. IV
Fluorite, 64, 85, 100, 201, 303, 308, 375, 377, 468, 490, 735, 900, 993, 1109, 1247
Fluorspar, 103
Flushing process, 184
Fluvioglacial deposits, 91
Fog, smog, mist, 259, 315. *See also* Vol. II, FOG, SMOG, MIST; BIOMETEOROLOGY
Fogarras, 547
Fongo, 702
Food chains, 325, 338. *See also* Vols. I and VII
Food preservatives (borax), 89
Food pyramid, 1169
Foraminifera, 49, 102, 118, 126, 451, 595, 605. *See also* Vol. VII
Foraminiferal ooze, 52. *See also* biogenic ooze; deep-sea ooze; pelagic sediment
Forbesite, 935
Forchheimer, P., 551
Forced-circulation vapor compression process, 221
"Foreign ion effect," 110
Forest, clearing, 313
 fires, man-made, 311, 326
Forested regions, 876
Formaldehyde, 272
Formation fluids, 769
Formation of ores, 377
Formanite, 846
Formula calculation. *See* MOLECULAR WEIGHTS
Formyl group, 156
Forsterite, 683
Fossil, corals shells, 493
 debris, 59
 fish, 881

fuels, 11, 14, 68, 71, 78, 127, 131, 187, 311, 326, 340, 774. *See also* Vol. IVB
 groundwater, 547
 microbes. *See* Vol. VII MICROBES
 seawater, 184
 sedimentary basins, 587
 shells, 893
 teeth, 378
 water, 544
 wood, 80
Fossil-fuel power plants, 315
Fossilization, 857
Fossils, 73, 79, 136, 155. *See also* Vol. VII
Fountain of Vaucluse, France, 540
"Fountain of water," 42
Fourmarierite, 844
Foxite, 828
Fractional crystallization, 18
Fractional inventories, 999
Fractional melting, 680
Fractionation, 18, 124, 134, 401, 595, 892, 1026, 1050
Fractionation series, 828
Framework silicates, 580
Framework structures (tektosilicates), 203
France, 474, 545. *See also* Vol. VIII
Franciscan formation, 436, 441
Francium, 10, 16, 44, 278
Francolite, 52, 698
Frankeite, 1192
Frankel defect, 236
Frankenheim, M. S., 741
Franklinite, 845
Frasch process, 1146
FREE ENERGY AND FREE ENTHALPY, 105, 109, 153, 170, 305, 308, **379**, 592, 927, 1173
Freezing-point, 751
Freezing-thawing, 914
Freirinite, 938
Fresh water, 219, 425, 459, 498
 carbon dioxide, 102
 level, 554
 cycle, 949
 sediments, 71, 78, 420
 snails, 452
Freidel-Grafts reactions, 144
Frondelite, 937
Froodite, 897, 1131
Froth flotation, 64
Fruits, 499
Fuchsite, 168
Fuels, 772. *See also* Vol. IVB
Fuel combustion, 1150
Fuel-electric power, 476
Fugacity, 558, 1247
Fuller's earth, 734. *See also* Vol. VIII
Fulvic acids, 28, 70. *See also* Vols. VI and Vol, I
Fumaroles, 88, 112, 388
Funafuti Atoll, Gilbert and Ellice Is., 116. *See also* Vols. III and VIII
Fungi, 77, 153, 312, 672
Fusates, 827
Fused silicon, 581
Fusion, 170, 226, 273, 303, 308, 988, 1170

Ga, 578
Gabbro, 19, 21, 93, 101, 146, 156, 169, 248, 350, 582, 1060. *See also* Vol. V
Gadolinite, 1028, 1041, 1165
GADOLINIUM: ELEMENT AND GEOCHEMISTRY, 44, 278, 300, 350, **384**, 489
Gahn, 369
Gahnite, 845
Galactosamine, 72
Galapagos Islands, 1038. *See also* Vol. VIII
Galaxite, 845
Galaxy, 397
Galena, 35, 58, 64, 83, 99, 296, 448, 489, 574, 900, 1092, 1130, 1166
Galleries, 547
GALLIUM: ELEMENT AND GEOCHEMISTRY, 44, 79, 151, 273, 277, 278, 300, **385**, 423, 433, 578
Galvanometer, 368
Gamma hematite, 604
Gamma-methane bridge carbon, 156
Gamma radiation, 130
Gamma-ray spectrometry, 780, 782
Gamma rays, 988
Gamow, 266
Ganges River, 1048
Gangue (minerals), 35, 64, 467, 468, 605. *See also* Vol. IVB
Garbage pollution, 311, 314
Garnet, 462, 680, 750, 1091
 group, 17, 23, 101, 168, 201
 peridotite, 146
Garnierite, 790, 791
Gas bubbles, 94, 374, 571

1305

INDEX

Gas, chromatography, 160, 230, 388, 493
　density detector, 392, 751
　detection curve, 391
　drilling, 551
　drilling wastes, 321
　exchange, 995, 1014
　field brines, 185
　industry. See Vol. IVB, NATURAL GASES
　liquid chromatography, 165, 497
　phase reactions, 143
　reservoirs, 478. See also Vol. VI
　saturation, 964
　solid chromatography, 165
GAS-EVOLUTION ANALYSIS, 259, **391**, 1168
Gaseous wastes, 315
Gases, 495, 547, 1039
GASES, VOLCANIC, 88, 272, **387**
Gasoline engine fuel pollution, 311, 312, 323
Gasoline synthesis, 144
Gastropods, 114, 353. See also Vol. VII
Gauging stations, 544
Gause equations, 257
Gause's axiom, 257
Gaylussite, 142, 884, 1097
GEA. See GAS EVOLUTION ANALYSIS
Gearksutite, 491
Geiger, 988
Geiger counter, 130, 780, 1277
Geikie, Archibald, 454
Geikielite, 844
Gel, gel minerals, 436, 744. See also COLLOIDS, MINERALOIDS
Gemology. See Vol. IVB
Gems, 579
Genesis (ore), 605
Genotypes, 691. See also Vol. VII
Geobarometers, 377
Geobotanical and biogeochemical method of prospecting for minerals. See Vol. IVB
Geochemical abundance categories, 269
GEOCHEMICAL CLASSIFICATION OF ELEMENTS, **394**
Geochemical crises, 945
Geochemical cycles, 429, 977
Geochemical documentation, 585
GEOCHEMICAL EVOLUTION OF THE CORE, MANTLE, AND CRUST, 264, 270, **397**
Geochemical nomenclature, 585
Geochemical prospecting, 99, 268, 284, see also Vol. IVB
Geochemical separation, 41
GEOCHEMISTRY, 296, **402**
Geochemistry of the earth, 74. See also EARTH'S CRUST GEOCHEMISTRY; GEOCHEMICAL EVOLUTION OF THE CORE, MANTLE, AND CRUST
GEOCHEMISTRY: IONIZATION POTENTIALS, **410**
Geochemistry of Pelagic Sediments, Antarctic, 417. See also Vol. I, ANTARCTIC PELAGIC SEDIMENTS
Geochemistry of Precambrian Atmosphere and Ocean, 417. See also PRECAMBRIAN ATMOSPHERE, GEOCHEMISTRY
Geochemistry of sediments, 585
GEOCHEMISTRY OF SEDIMENTS (ANCIENT), **417**
GEOCHEMISTRY OF SEDIMENTS (MODERN), **428**
GEOCHEMISTRY OF SEDIMENTARY SILICA, **434**
GEOCHEMISTRY: TESTING FOR ELEMENTS, **443**
GEOCHEMISTRY OF TRACE ELEMENTS. See TRACE ELEMENTS, GEOCHEMISTRY.
Geochronology, 454, 455. See also Vol. VIII
GEOCHRONOMETRY, **446**
Geodynamic processes, 186
Geogens, 692
Geographic situation resources, 771
Geohydrology. See Vol. VI
Geologic environment and health, 696
Geologic maps, 476, 481. See also Vol. VI
Geologic surveys, 187, 775. See also Vol. VIII
Geologic thermometry, 374, 815. See also Vol. IVB
Geology, 402, 471, 480
GEOLOGIC TIME SCALE, **453**. See also Vol. VII
Geomicirite (Geomycrite), 826
Geomorphology. See also Vol. III
Geophysical exploration, 481. See also Vol. V
GEOPHYSICAL METHODS FOR HYDROLOGIC SEARCH, **456**
Geophysics. See Vol. V
Georgiadesite, 936
Geospheres, 268, 278, 281
Geosynclines, 246, 434, 903. See also Vols. V and VII
Geosynclinal sediments, 115, 193, 245. See also Vol. V
Geotectonic cycles, 433
Geothermal gradient, 547
Geothermal energy, 571, 868. See also Vol. IVB
Geothermal waters, 868
Gerling, E. K., 455

German Lahn-Dill ores, 606
Germanates, 462
Germanite, 386, 462
GERMANIUM: ELEMENT AND GEOCHEMISTRY, 44, 79, 106, 151, 273, 277, 278, 300, 399, 415, 422, **461**
Germany, Mansfeld (copper shale), 754. See also Kupferschiefer
Germicides, 156
Gersdorffite, 42, 790, 1131
Geversite, 958, 1131
Geyserite, 77
Geyserite deposits, 436
Geysers, 572, 869, 1087
Ghanats, 474
Ghyben-Herzberg Theory, 554, 557. See also GROUNDWATER MOTION: HYDROLOGY OF SUBSURFACE WATERS; HYDROLOGY, COASTAL TERRAIN
Giant Springs, Montana, 1116
Gibbs, J. Willard, 153, 917
Gibbs' free energy, 105, 170, 305, 308, 363, 379, 684, 1174. See also Vol. V.
Gibbs function, 170, 379
Gibbs-Helmholtz equation, 305
Gibbs phase rule, 426
Gibbsite, 21, 25, 423, 430, 435, 562
Gibbsitic bauxite, 23
Gilbert, G. K., 454, 880
Gillespite, 62
Gilsonite, 827
Ginorite, 87
Glacial deposits, 460. See also Vol. VI
Glacial drift, 449. See also Vol. III
Glacial geology: periglacial and global effects. See Vols. II and III
Glacial-interglacial changes, periods, 544, 894
GLACIAL MILK, (meltwater), **463**
Glacial till, 449, 459, 475, 915
Glaciations, 544, 893, 895
Glaciation, cyclic, 339
Glacier Bay National Monument, 188
Glacier National Park, 188
Gladstone and Dale law, 445
Glaserite, 826
Glass, 66, 118, 149, 156, 744. See also Vol. V.
Glass, devitrification of. See Vol. IVB
Glass-membrane electrodes, 181
Glauberite, 101, 1097, 1123
Glaucocerinite, 1125
Glaucodot, 1131
Glauconite, 1, 29, 53, 58, 61, 89, 421, 427, 447, 449, 450, 599, 605, 971, 1051. See also Vol. VI
Glaucophane, 22, 904, 1097
Glaucophane schists, 112, 904, 1097. See also Vol. V.
GLC (Gas-liquid chromatography), 165
Gley, 28. See also Vol. VI
Glimmerton. See Illite
Global climate, 11
Global hydrology, 585
Global temperature, 340
Globar, 227
Globigerina ooze, 985. See also Foraminiferal Ooze; Deep-Sea Ooze
Globerinoides rubra, 894
Glockerite, 1126
Glucosamine, 72
Glutamic acid, 72
Glycine, 72, 157
Glycogen, 154
Glycoprotein, 154
Gmelinite, 749
Gneiss, 23, 156, 248, 448, 449, 465, 713. See also Vol. V
Goethite, 49, 386, 430, 599, 600, 604, 673, 674, 791, 845, 929
Gold, 35, 44, 64, 83, 157, 197, 275, 278, 300, 320, 394, 411, 432, 574, 763, 839, 897, 900
GOLD: ECONOMIC DEPOSITS, **467**
GOLD: ELEMENT AND GEOCHEMISTRY, **466**
Gold-quartz veins, 41
Gold tellurides, 900
Goldschmidt, V. M., 93, 905
Goldschmidt's "admittance" theory, 1121
Goldschmidt's geochemical classification of the elements, 1145
Goldschmidt's hypothesis, 1061
Goldschmidt's rules, 176, 1200
Goniolithon, 114
Gonorrhea, 692
Gorceixite, 937
Gordon, 392
Gordonite, 939
Gorgonia, 113
Gorshkova, 587

Goslarite, 578, 1125
Goslute Lake, 881
Gossan, 467
Gough Island, 384. See also Vol. VIII
Goyazite, 937
Graftonite, 934
Grain, dating of, 447
Grand Canyon, (of Colorado River), 188
Granite, 17, 32, 62, 93, 101, 156, 194, 248, 343, 350, 384, 401, 432, 448, 459, 465, 471, 494, 572, 582, 734, 971, 1026, 1056, 1060, 1091, 1185, 1192. See also Vol. V
Granitic crust, shell, 243, 250, 448, 449. See also Vol. V
Granitic melts, 375
Granitization, 20, 1097
Granitoids, acidic, 245
Granodiorites, 18, 35, 93, 101, 156, 248, 401, 465. See also Vol. V
Granophyric rocks, 169. See also Vol. V
Grantsite, 1237
Granulite, 904. See also Vol. V
Granulite facies suite, 18, 710. See also Vol. V
Graphite, 32, 123, 134, 201, 204, 382, 735, 763, 812
Grass Valley, Nevada City, California, 469
Grassland, carbon cycle, 128
Grasses, 877
Gratonite, 42
Gravel, 459, 471, 480. See also Vol. VI
Gravimetric, 742
Gravitational convection, 587. See also Vol. V
Gravitational field, 492. See also Vol. V
Gravity, 457, 460, 472
　flow, 36
　separation, 606
Graywacke (greywacke), 245, 270. See also Vol. VI
Graywacke-argillite sequence, 273
Great American Desert, 542
Great Basin (U.S.), 884
Great Basin Desert, 133
Great Bear Lake, Canada, 1094, 1222
Great Dyke, Rhodesia, 169, 960
Great Salt Lake, 764, 765, 880, 884, 1047, 1098. See also Vol. III
Great soil group, 913
Green algae, 114, 158. See also Vols. I and VIII
Green marble, 120
Green pellets (glauconite), 53
Green plants, 813, 952
Green River shale formation (Wyoming), 23, 815, 881, 982
Greenalite, 58, 599, 605
Greenhouse effect, 12, 15, 326, 858. See also Vol. II
Greenland, 273, 378, 757. See also Vol. VIII
Greenochite, 99
Greenockite, 1130
Greenschist, 89, 904. See also Vol. V
Greenstone, 465. See also Vol. V
Greigite, 1137
Greisen, 17, 147, 1051, 1192. See also Vol. V
Grinding-induced strain, 110
Grindstones, 733
Grinnell Glacier, Glacier National Park, Montana, 465
Griphite, 937
Grossularite, 101, 904
GROUNDWATER, 25, 28, 36, 42, 78, 148, 174, 194, 313, 317, 358, 372, 446, 447, 450, 452, 456, **470**, 495, 536, 539, 544, 546, 547, 555, 576, 842, 879, 925, 956, 1043, 1046, 1057, 1070, 1087, 1159, 1211, 1233, 1247, 1264
　action, discharge, flow, 352, 480, 481, 587
　in coastal, karst, semiarid, volcanic terrain. See resp. HYDROLOGY, COASTAL TERRAIN; HYDROLOGY, KARST TERRAIN; HYDROLOGY, SEMIARID REGIONS; HYDROLOGY, VOLCANIC TERRAIN
　hydraulics, 476, 551
　mathematical models, 482
　membrane models, 482
GROUNDWATER MOTION IN DRAINAGE BASINS, **478**
Groundwater seepage, 1116
Groundwater, therapeutic role. See MEDICINAL SPRINGS
Groutite, 845
Grunerite, 605
Grünlingite, 1130
GSC (gas-solid chromatography), 165
Guanajuatite, 82, 1131
Guano, 100, 829, 941. See also Vol. IVB
Guatemala basin, 50. See also Vol. I
Gudmundite, 1131
Guianas, 538. See also Vol. VIII
Guildite, 1126

1306

INDEX

Guilleminite, 1079
Gulf coast sediments, 91
Gulf of Karaboghaz (Turkmen SSR), 352, 765, 1098. *See also* Vol. III
Gulf of Mexico, 358, 558. *See also* Vol. I limestones, 119
Gullies, 1102. *See also* Vol. III
Gummite, 844
Gunung Agung, Bali, 12
Guyana, 28, 958. *See also* Vol. VIII
Guyot, 52. *See also* Vol. I
Gypsum, 1, 58, 77, 100, 103, 108, 112, 231, 353, 383, 427, 475, 734, 750, 881, 900, 902, 1016, 1124, 1148
Gyttja, 660, 1202, 1203. *See also* Vol. VI

HAFNIUM: ELEMENT AND GEOCHEMISTRY, 7, 44, 274, 278, 300, **488**
Hagen, 551
Haidingerite, 935
Haüy, 741
Haiti, 474. *See also* Vol. VIII
Haldane, 850
Half-life, 8, 987, 994, 1002, 1014, 1020, 1053
HALIDES, 198, 231, 237, 415, 432, **489**, 583, 717, 1096
Halite, 17, 58, 85, 112, 155, 200, 352, 353, 475, 489, 490, 548, 900, 902, 1078, 1097, 1246
Halophyte vegetation, 549
Halley, Edmund, 476
Halloysite, 25, 91, 179, 227
Halmeic minerals, 49. *See also* Vol. I
Halmyrolysis, 49, 54. *See also* Vol. VI
Halogen gases, 391
HALOGENS, 82, 87, 92, 241, 270, 343, 350, 377, 384, 406, 489, **492**, 494, 1063
Halotrichite, 1125
Hamersley range (Western Australia), 606
Hannayite, 935
Hanford plant, Washington, 557
Hanskite, 1126
Hard acids, 3
Hardness of water, 1247
Harzburgite, 681
Harmotome, 53
Hartite, 811, 829
Hatchettite, 827
Hauerite, 1131
Hausmannite, 49, 207, 845
Havana, Cuba, 548
Hawaii (Lanai), 556
Hawaii Volcanoes National Park, 188
Hawaiian Islands, 26, 273, 388, 434, 474, 537. *See also* Vol. VIII
Hawleyite, 99
Hayden Expedition, 188
He, 397. *See also* Helium
Health, 969
Health hazards (radioactive), 990
Heat, 541, 544, 1173
Heat balance, 545
Heat balance sheet, 543
Heat budget, planetary, 000
Heat capacity, 303, 307
Heat conductivity, 197
Heat content. *See* ENTHALPY
Heat energy, 545, 709
Heat-exchange medium, 386
Heat flow, 680
Heat of hydration, 206
Heat of reaction, 305
"Heat Island effect," 15, 1231
Heat transfer of the earth, 1252
Heavy metals, 41
Heavy metal poisoning, 150, 319
Heavy metal sulfides, 586
Heavy minerals, 432
Heavy spar (barite), 63
Heavy water. *See* DEUTERIUM
Hectorite, 227. *See* montmorillonite
Hedenbergite, 582
Hedyphane, 937
Height-discharge relationship, 544
Hekla, Iceland, 12, 869. *See also* Vol. VIII
Heliophyllite, 34
HELIUM: ELEMENT AND GEOCHEMISTRY, 44, 123, 165, 201, 263, 272, 279, 396, 405, 411, 427, **492**, 454, 837, 838, 988, 997, 1039
Helium bombardment, 7
Helium oil chemistry. *See* Vol. IVB
Hellandite, 1041
Helmholtz free energy, 379, 1174
Hemafibrite, 938
Hematite, 26, 31, 58, 120, 558, 599, 600, 601, 604, 791, 844, 900, 901, 902, 927, 936, 1078
Hematophanite, 845
Hemihedrite, 159

Hemimorphite, 202
Hemin, 150, 157
Hemocyanin, 74
Hemoglobin, 150, 190
Hemovanadin, 74
Henderson, Colorado, 757
Heptane, 496
Herbivores, 159, 256
Hercynite, 845
Herderite, 936
Hermesite, 705
Herschel, W., 579
Hess's Law, 000
Hessite, 1092, 1130, 1164
Hetaerolite, 207, 845
Heterocyclic phenols, 75
Heterodesmic crystal, 201
Heterogeneous systems, catalysis, 144
Heteronuclear bond, 262
Heteropolar bond, 198
Heteropoly acids, 1213
Heterosite, 934
Heterotrophic bacteria, organisms, 661, 806
Heulandite, 58
Heulandite-Clinoptilolite, 61
Hewettite, 1237
Hexacorals, 113
Hexadecanol, 549
Hexahydrite, 354, 1124
Hexane, 496, 498
HF (hydrofluoric acid), 415. *See also* Fluorine
Hickory leaves, 146, 343, 350
Hieratite, 491
High-energy radiation, 863
High-pressure mineral synthesis, 743
Highgate resin, 826
High-temperature mineral synthesis, 743
High-temperature solutions, 898
Highway construction (pollution), 313
Highway de-icing, 104
Hilgardite, 87
Hill, 392
Hillslopes, 879
Hinsdalite, 939
Histidine, 72
Hoar-frost, 545. *See also* Vol. II
Hoegbomite, 845
Hoff equation (Van't Hoff), 685
Hohmannite, 1126
Holdenite, 936
Hollandite, 846, 1213
Hollingworthite, 1042
Holmia, 240, 494
HOLMIUM: ELEMENT AND GEOCHEMISTRY, 44, 279, 300, **494**
Holocene ("Recent"), 586. *See also* Vol. III
Holothurian, 70. *See also* Vol. VII
Holston Orthomarble, Tenn., 120
Homodesmic crystal, 201
Homogeneous distribution law, 907
Homogeneous systems, catalysis, 144, 728
Homogenization of isotopes, 1054
Homogenization temperature, 375
Homonuclear covalent bond energy, 262
Homopolar bond, 84, 197
Homosphere, 948
Honolulu, Hawaii, 556
Hoofworm, 700
Hopeite, 935
Hornblende, 1, 22, 40, 101, 120, 146, 153, 367, 447, 449, 451, 899, 901, 904
Hornblende granulite facies, 274. *See also* Vol. V
Hornblende-hornfels, 904
Hornstone, 438. *See also* Vol. V
Host materials, 940
Hot springs, 436, 571, 572, 666, 900, 1077, 1087. *See also* SPRINGS
Hot spring activity, 387
Hot spring deposits, 29, 97, 101, 112
Howlite, 87
Huebnerite, 752, 1212
Huhnerkobelite, 934
Hulsite, 87
Human activity, 13, 430
Human disease, 730
"Human resources," 771
Humboldt, 337
Humboldt current, 216
Humboldtine, 811, 824
Humic acid, 70, 75, 151, 539
Humic organic acids, 606
Humic waters, 1204
Humidity, 541, 544
Huminite, 825
Hummerite, 1237
Humose soils, 601. *See also* Vol. VI
Humus, terrestrial, 69, 70, 75, 127. *See also* Vol. VI

Hund's Rule, 204
Hund's stabilization energy, 205
Huntite, 142
Hureaulite, 935
Hurlbutite, 934
Hybrid analog-digital models, 477
Hydrates, 231
Hydration, 160, 562, 1246
Hydraulic conductance, 480
Hydraulic gradient, 471, 480, 481
Hydraulic head, 372, 485
Hydraulic system, 956
Hydrobasaluminite, 1126
Hydroboracite, 87
HYDROCARBONS, 70, 78, 92, 126, 151, 165, 311, 312, 335, 427, 478, 485, **495**, 811, 823, 830, 980, 981, 1205, 1231
Hydrocarbonates, 83
Hydrocerussite, 143
Hydrochemical zoning, 594
Hydrochemistry, 547
Hydrochloric acid, 3, 764
Hydroclimate. *See* Vol. II
Hydrocracking, catalytic reaction, 144
Hydrodynamics. *See* Vol. II
Hydroelectric energy. *See* WATER POWER, Vol. VI
Hydroelectric potential, 771
Hydrofluoric acid, 378. *See also* Fluorine
HYDROGEN: ELEMENT AND GEOCHEMISTRY, 2, 31, 44, 75, 96, 122, 123, 126, 165, 199, 225, 263, 266, 272, 279, 397, 405, 427, 461, 492, 495, **503**, 561, 572, 573, 593, 686, 1129, 1215
Hydrogen bond, 84, 154, 200
Hydrogen chloride, 296
Hydrogen electrode, 925
Hydrogen fusion, 492
Hydrogen ion concentration (pH), 77
Hydrogen metasomatism, 562. *See also* Vol. V
Hydrogen sulfide, 77, 185, 315, 321, 335, 374, 427, 475, 1073. *See also* Vol. IVB
Hydrogenous minerals, 49
Hydrogenous sediments, 433
HYDROGEOLOGY, 478, **508**
Hydrograph, 1057
Hydrologic associations. *See* Vol. VI
HYDROLOGIC CYCLE, 338, 470, **515**. 542, 585, 716, 1211, 1243
HYDROLOGIC MAPS, **519**
HYDROLOGY, 430, 476, 480, 481, 530, 585, 1000, 1260
HYDROLOGY, COASTAL TERRAIN, **535**
Hydrology, dissolved substances. *See* Vol. VI
HYDROLOGY, LIMESTONE TERRAIN, **538**
HYDROLOGY, SEMIARID REGIONS, **541**
HYDROLOGY, SUBSURFACE WATERS, **550**
HYDROLOGY, VOLCANIC TERRAIN, **555**
Hydrolysate sediments, 387
Hydrolysis, 2, 9, 37, 108, 160, 562, 729, 1084, 1106, 1266
Hydrolyzates, 23, 72, 83, 432, 433, 829, 1162
Hydromagnesite, 58, 143
Hydromica, 22. *See also* ILLITE
Hydronium. *See* Hydrogen
Hydrophilite, 490
Hydrosols, 606
Hydrosphere, 29, 36, 74, 123, 156, 210, 245, 270, 378, 428, 495, 498, 716, 861, 949, 996, 1062, 1077, 1247. *See also* Vol. II
Hydrostatic head, 42, 353
Hydrostatic pressure, 57, 126, 1264
Hydrostatic theory, 554
Hydrostatic weighing, 1113
Hydrosulfide complex, 468
Hydrothermal activity, submarine, 49
Hydrothermal alteration, 5, 1216
HYDROTHERMAL ALTERATION—NONSILICATE, **557**
HYDROTHERMAL ALTERATION—SILICATE ROCKS—GENERAL PRINCIPLES, **561**
Hydrothermal conditions, 1088
Hydrothermal deposits, 99, 112, 467, 670, 900, 1109, 1293
Hydrothermal phase, process, 10, 37, 578, 1148
Hydrothermal replacement, mineralization, 576, 902
HYDROTHERMAL SOLUTIONS, FLUIDS, 146, 156, 238, 343, 350, 384, 468, 494, 558, **571**. 768, 1060, 1097, 1213
HYDROTHERMAL SOLUTIONS—SULFIDE TRANSPORT, **572**
Hydrothermal sulfide deposits, 35, 191
Hydrothermal systems—silicates. *See* SILICATES: HYDROTHERMAL SYSTEMS
Hydrothermal vapors, 19
Hydrotroilite, 51, 425, 1151
Hydrotungstite, 1212
Hydrous aluminosilicates, 22
Hydrous minerals, 155

1307

INDEX

Hydrous oxides, 35
Hydroxides, 350, 385, 422, 430, 495, 559, 583, 717, 843. *See also* OXIDES AND HYDROXIDES
Hydroxides, aluminum, 430
Hydroxylapatite, 378, 937
Hydroxy bonds, 415
Hydroxyl groups, 2, 87, 144, 151
Hydroxyl hydrogen, 226
Hydroxyproline, 72
Hydrozincite, 143
Hygroscopic salt, 103
Hypersthene, 149, 274, 899, 904
Hypocupremia, 699
HYPOGENE, **576.** 578
HYPOLIMNION, **577,** 944, 1172
Hypsometric curve, 1044

IAGC, 584
Ianthinite, 844
Iberian peninsula, 474
Ice, 86, 429, 459, 843, 1000, 1243, 1258. *See also* Vol. VI
Ice Ages, 409, 858. *See also* Vols. II and III
Ice-age theory, meteorology. *See* Vol. II
Ice caps, 716
Ice crystals, 1243
Ice nuclei, 15
Ice sheets, 340
Iceland, 388. *See also* Vol. VIII
Iceland spar, 103
Idaho, Big Springs, 556
  Twin Falls, 555
Idarado, Colorado, 468
Idrialite, 828
Igneous, 957
  intrusions, 903. *See also* Vol. V
  paragenesis, 898. *See also* PARAGENESIS; also Vol. V
  rocks, 19, 80, 88, 93, 101, 111, 147, 156, 193, 273, 278, 378, 430, 462, 573, 590, 599
  metamorphic rocks, 573
IHD (International Hydrological Decade), 585
Ilesite, 1124
Illinois, 378
Illites, 22, 29, 58, 89, 91, 176, 230, 260, 421, 435, 566, 582, 591, 886, 1051, 1019. *See also* Vol. VI
Ilmenite, 362, 432, 489, 599, 604, 794, 844, 899, 1193, 1235
Ilsemannite, 735, 844
Immiscibility, 919
Imperial Valley, California, 545
Impermeable rocks, 1116
Impervious horizons, 1057
Impurities, 907
In situ mineral growth, 433. *See also* AUTHIGENESIS
In situ weathering, 430
Incinerator pollution, 313
Inclusions, 373, 747
Inclusion complexes, 748
Incongruent melting point, 918
Indan, 496
Indanes, 498
Inderborite, 87
Inderite, 87
"Index" trace elements, 111
Indian Ocean, 587. *See also* Vol. I
  Sediments, 50, 121
INDIUM: ELEMENT AND GEOCHEMISTRY, 44, 279, 300, **578**
Indo-Gangetic plain, 474. *See also* Vol. VIII
Induced polarization, 459
Indus River, Pakistan, 547, 1048. *See also* Vol. VIII
Industrial activity, 771
Industrial center, 11
Industrial complexes, 315
Industrial expansion, 187
Industrial poison, 701
Industrial pollutants, 409
  concentrations, 321
Industrial pollution, 312
Industrial Revolution, 733
Industrial wastes, 318, 319, 322
Industry interest in hydrology, 548
Inert, 1039
Inert gases, 406, 579. *See also* RARE GASES
Inert gas configuration, 93
Infiltration, 480, 547, 876
Infiltration galleries, 550
Infiltration rates, 1058
Influenza, 692
Information Gap, 775
Infrared absorption, 498, 741
INFRARED ANALYSIS, 29, **579,** 1114
Infrared measurements, 460

Infrared spectrophotometry, 579
Infrared spectroscopy, 230, 261, 301, 392
Ingramite, 827
Inhalation, 730
"Initial" argon, 40
Inner transition elements, 6, 47, 1020
Inorganic acids, 312
Inorganic compounds, 319
Inorganic pollutants, 312
Inorganic precipitation, 000
Inorganic salts, 579
Inosilicates, 101, 202, 462, 720, 1107. *See also* CRYSTAL CHEMISTRY; MINERAL CLASSES: SILICATES
Insects, 73, 154, 333, 660. *See also* Vol. VII
Insecticides, 93, 340
Insolation, 338. *See also* Vol. II
Insoluble minerals, 901
Instability, 1230
Insulators, electrical, 85
Insulators, thermal, 85
Insuzi Series, Swaziland, 27. *See also* Vol. VIII
Interatomic distances, 1271
Interflow, 1057
Interfluve areas, 879
Intergranular fluids, 116
Internal energy, 593, 1173
Internal gas proportional counter, 451
INTERNATIONAL ASSOCIATION OF GEOCHEMISTRY AND COSMOCHEMISTRY (IAGC), **584**
International Atomic Energy Agency (IAEA), 222, 585
International Council of Scientific Unions (ICSU), 586
"International Geological Correlation Programme" (IGCP), 773
INTERNATIONAL HYDROLOGICAL DECADE (IHD), 226, 542, **585**
Interstadials, 895
Interstellar matter, 267. *See also* Vol. II
Interstitial cement, 108, 422, 587. *See also* Vol. VI
Interstitial clays, 555
Interstitial diffusion, 236
Interstitial pore fluids, 458
Interstitial sediments, 587
Interstitial solutions, 49
Interstitial water, 49, 427, 435, 458, 1072, 1081, 1260
  formation, 969
INTERSTITIAL WATER IN SEDIMENTS, **586**
Intertidal environments, zones, 102, 538. *See also* Vols. I and VI
Intrazonal soil, 912
Intrusives, 555
Invariant point, 919
Inventory, 695, 774
"Inverse" spinels, 207
Inversion, 314, 1230, *see* Vol. II
  mineral, 59
Invertebrates, 137. *See also* Vol. VII
Invisible gold, 468
Inyoite, 87
IODINE: ELEMENT AND GEOCHEMISTRY, 44, 75, 272, 279, 489, 492, **590,** 1017, 1067
Iodine-xenon formation, 1270
IODATES, **589,** 717
Iodide, 185
Iodobromite, 93
Iodyrite, 490
Ions, 1105, 1106. *See also* TRACE ELEMENTS IN SILICATE MINERALS
Ion activity, 1109
Ion engine research, 148
ION EXCHANGE, 79, 103, 151, 160, 589, **591,** 597, 749
Ion-exchange, capacity, 591
  membranes, 593
  processes, 908
Ion ratios, 1063
Ionic bond, 84, 160, 198
Ionic compounds, 719
Ionic leakage, 174
Ionic migration, 85
Ionic radius, 20, 30, 198, 1023
Ionic salts, 924
Ionic solubilities, 1245
Ionium, 6
IONIUM—THORIUM DATING, **594.** 985
Ionization, 3, 84, 988
Ionization Potential, 20, 47, 410. *See also* GEOCHEMISTRY: IONIZATION POTENTIALS
Ionizing radiation, 326
Iran, 542, 548, 551. *See also* Vol. VIII
Iranite, 159
Iraq, 548. *See also* Vol. VIII
IRIDIUM: ELEMENT AND GEOCHEMISTRY, 44, 275, 279, 300, **598,** 959
Iridosmine, 599, 763, 836, 897, 957, 1059

IRON: ELEMENT AND GEOCHEMISTRY, 20, 32, 38, 40, 44, 49, 61, 75, 107, 150, 157, 169, 178, 179, 190, 196, 210, 248, 264, 266, 272, 279, 300, 320, 360, 395, 399, 422, 426, 430, 433, 475, 569, 598, **599,** 699, 736, 763, 842, 1067, 1139
IRON: ECONOMIC DEPOSITS, **603**
Iron, carbonates, 268
  conglomerates, 606
  Cycle, 82, 151, 215. *See also* IRON: ELEMENT AND GEOCHEMISTRY
  Hydroxide, 5, 25, 579
  metallic, 47
  meteorites, 9, 263, 394, 462, 493
  nitride, 32
  ores, 83, 310, 432, 433, 603
  ore formations, 423, 600
  oxides, 231, 268, 360, 432, 600
  precipitation, 49
  sands, 606
  sedimentation, 601
  silicates, 601
  sulfides, 425, 558, 573, 578
  sulfide deposits, 99, 268
  bearing oxides, 898
  magnesium silicates, 429
  manganese ores, 586
  nickel core, 599
Iron King, Arizona, 468
IRON OXIDATION AND REDUCTION—MICROBIAL, **610**
Ironstones, 600, 601
Irradiation, 326
Irradiation ages, 1003
Irrawaddy River, Burma, 1048. *See also* Vol. VIII
Irreversible processes, 1173
Irrigation, 475, 476, 478, 548, 1103, 1116
Island-arcs, 433, 1034. *See also* Vol. V and VIII
Isobar, 918
Isobutane, 165, 496
Isochemical metamorphism, 709
Isochron, 450, 1054
Isoclasite, 938
Isodesmic crystal, 201
Isoleucine, 72
Isomorphic compounds, 41
Isomorphous crystal, 204
Isomorphous substances, miscibility, 1105
Isomorphous substitution, 1097
Isopentane, 165
Isopleth, 918
Isoprenoids, 80
Isoprenoid hydrocarbons, 498
Isopropylphphenanthrene, 499
"Iso" terms. *See* Vol. II
Isotherm. 918
Isothermal flow, 551
Isotopes, 403, 812, 974, 987, 991, 996, 1006, 1014, 1225
  tagged, 107
Isotope dilution method, 447
ISOTOPE FRACTIONATION, 79, 212, **612,** 1153
Isotope fractionation in the ocean. *See* Vol. I
Isotope geochemistry, 585
ISOTOPE GEOLOGY: STABLE, **618**
Isotope glaciology. *See* Vol. VI
Isotope pairs, 1050
Isotope tracer technology. *See* Vol. VI, RADIOACTIVE ISOTOPE TRACER TECHNOLOGY
Isotopic abundance, 688
Isotopic age, 1228
Isotopic analysis, 1053
Isotopic equilibrium, 892
Isotopic neutron sources, 784
Isotopic paleotemperatures. *See* PALEOTEMPERATURES—ISOTOPIC DETERMINATION
Isotopic variations in Mineral deposits. *See* Vol. IVB
Isotropism, 274
Italy, 388, 474, 551. *See also* Vol. VIII
Itoite, 462
Ixolite (Ixolyte), 829

Jacksonburg limestone, Pa., 120
Jachymov, Czechoslovakia, 1222
Jacobsite, 845
Jadeite, 22, 904, 1097
Jagger, 388
Jahn-Teller effect, 189, 193, 206
Jamaica, 474. *See also* Vol. VIII
Jamesonite, 34
Japan, 388, 587, 474. *See also* Vol. VIII
Japanese, 1034
Jarlite, 91
Jarosite, 604, 1125. *See also* Vol. VI
Jasper, 438; *See also* Vol. IV
Jeremejevite, 87
Jet aircraft noises, 311
Jet Stream, 1211. *See also* Vol. II

INDEX

Ježekite, 936
Jigging (barite recovery), 64
Jimboite, 87
Joachimstal pitchblende, 962
Johannite, 1126
JOIDES (Joint Oceanographic Institutions Deep Earth Sampling Program), 588
Joly, John, 454
Jolly Balance, 1113
Juice and fruit drink bottling pollutants, 321
Julienite, 829
Junk automobile pollution, 322
Jupiter, 33. See also Vol. II
Jurassic, 454, 880. See also Vol. VII
Jurassic oolitic iron ores, 605
Jute, 28
JUVENILE WATER, 450, 470, **623**, 666, 702. See also Vol. VI

K, 547. See also Potassium
"K" capture, 493, 500. See also ELECTRON CAPTURE, Vol. IV
Kainite, 667, 971, 1126
Kalahari Desert, S. Africa, Botswana, 548. See also Vols. III and VIII
Kaliborite, 87
Kalicinite, 142
Kalinite, 1124
Kamacite, 763, 791, 1195
Kamarezite, 1126
Kamchatka, U.S.S.R., 388. See also Vol. VIII
Kamcite, 99
Kämmererite, 168
Kanalstrahlen, 688
Kanats, 550
Kandite (the kaolin group of minerals). See Kaolinite
Kanthal, 227
Kaolin, 178, 735. See also Kaolinite
Kaolinite, 1, 5, 22, 25, 55, 58, 176, 178, 181, 218, 227, 230, 429, 435, 562, 582, 591, 1091
K-Ar, 367, 447, 448, 449, 455, 473, 972
Kara-Bogaz Gulf, U.S.S.R., 352, 764, 765, 1098. See also Vol. III
Karelia, U.S.S.R., 449, 661
Karst, 26, 480, 538. See also Vol. III
Karst basins, 547
Karst-limestone, 26
Karst terrain, groundwater, hydrology. See HYDROLOGY, LIMESTONE TERRAIN
Karst waters, 26
Katagenesis, 1160
Katangan Copper belt (Zaïre, Zambia), 180. See also Vol. VIII
Katmai National Monument, Alaska, 12, 188. See also Vol. VIII
Katoptrite, 34
Keith, 440
Kelvin (Lord), 454
Kempite, 490
Kentucky, 378. See also Vol. VIII
Kenyaite, 436
Kermesite, 34, 1131
Kernite, 87
Kerogen, 68, 136, 798, 817, 825, 982. See Vol. IVB; also Vol. VI
Ketones, 77
Keweenawan lavas, Michigan, 190, 193. See also Vols. VII and VIII
Kherazes, 547
Khorat plateau, Thailand, 474. See also Vol. VIII
Kidney ore, 604
Kidney stone, 697
Kieserite, 354, 667, 1124
Kilauea (Hawaii), 273, 388, 555. See also Vol. VIII
Kimberlite, 681. See also Vol. V
Kinetic Isotope Effects, 1152
Kinetic processes, 133, 145, 219
Kings Canyon National Park, 188. See also Vol. VIII
Kings Mountain, 663
Kink bands, 274. See also Vol. V
Kirchoff, 368
Kirkendall effect, 236
Kirovite, 1124
Kiruna, Sweden, 606. See also Vol. VIII
Kiruna ores, 602
Kirunavaara, 432
Kiscellite, 825
Kizer, 411. See also Ahrens
Klaus, K. K., 1058
Kleinite, 491, 704
Klockmannite, 1131
Kluchev, Kamchatka, U.S.S.R., 12. See also Vol. VIII
Koechlinite, 752
Koenenite, 491

Koettigite, 935
Koktaite, 1124
Kola Peninsula, U.S.S.R., 273. See also Vol. VIII
Kolbeckite, 939
Kolm (Sweden), 1207
Kolovratite, 1237
Kongsbergite, 704
Koniology (coniology). See Vol. II
Kornelite, 1125
Kotoite, 87
Krakatoa, Indonesia, 12, 802. See also Vol. VIII
Krakatoa Winds, See Vol. II
Kramer Lake, California, 89
Kratochwilite, 811
Krausite, 1124
Kremersite, 491
Krennerite, 468, 1132
Kribergite, 939
KRYPTON: ELEMENT AND GEOCHEMISTRY, 44, 272, 279, 411, **624**, 849, 1039. See also Vol. IV INERT GASES
Kukkersite, 1203
Kullenberg, 587
"Kupfernickel," 792
Kupferschiefer copper shales, Mansfeld, Germany, 150, 190, 194, 754, 1139, 1203, 1207
Kurile Islands, 26. See also Vol. VIII
Kurnakovite, 87
Kutnahorite, 142
Kuwait, 549. See also Vol. VIII
Kwashiorkor, 699
Kyanite, 22, 31, 735, 904. See also Vol. V

Labradorite, 1, 23
Lacroixite, 936
Lagoon, 351, 359. See also Vols. III and VI
LAGOON GEOCHEMISTRY, 344, 548, **629**
Laguna Madre, Texas, 764, 1062
Lakes, 36, 378, 437, 470, 482, 539, 577, 601, 880, 944, 949, 1042, 1087, 1172, 1243. See also Vols. III and VI
Lake Athabasca, Canada, 1223
Lake Bonneville, Utah, 880, 884
Lake Bonney, Antarctica, 965
Lake Erie, 949
Lake Erie pollution, 312, 325
LAKE GEOCHEMISTRY, 101, 133, 148, 312, 342, **634**. See also Vol. VI
Lake Lahonton, Nevada, 880
Lake Magadi, Kenya, 436, 765. See also Vol. VIII
Lake San Augustin, New Mexico, 884
Lake stratigraphy, 660
Lake Superior, 424, 603, 1139
Lake Uinta, 881
Lamarck, 337
Lambert-Beer law, 182
Lamellibranch (pelecypod), 74. See also Vol. VII
Lamprophyllite, 273
Laminar flow, 471
Lanarkite, 1125
Landergren, 603
Landesite, 935
Landscape, 771, 1238. See also Vol. III
Land subsidence, 344, 476. See also Vol. VI
Land-use, 1239
Langbeinite, 667, 1123
Langite, 1124
Lansfordite, 142
LANTHANIDE, 6, 44, 47, 145, 151, 241, 342, 349, 494, **640**, 664, 777, 976, 1020, 1030, 1060, 1291. See also Vol. IVB
Lanthanide contraction, 7
Lanthanide rare earths, 406
Lanthanite, 143
Lanthanons, 1020
LANTHANUM: ELEMENT AND GEOCHEMISTRY, 44, 53, 279, 300, **641**. See also Vol. IVB
Lanthinite, 844
Laplace's equation, 372, 480
de Lapparent, A., 454
Larderellite, 87
Lark, Utah, 468
Las Plumas, California, 440
Latent heat of evaporation, 545
Latent heat of transformation, 303
Latent heat of vaporization, 221, 543
Laterite, 5, 21, 168, 215, 430, 600, 604, 1235. See also Vol. VI
Lateritic alteration, weathering, 178, 600
Lateritic bauxite, 171
Lateritic soil, 556, 1092
Lateritic surfaces, 792. See also Vol. III
Laterization, 25, 75, 296
Latite, 555
Lattice distortions, 105
Lattice energy, 206
Lattice sites, 18
Lattice substitutions, 591

Laubmannite, 936
Laumontite, 58, 422, 749
Laundry pollutants, 321
Laurionite, 490
Laurite, 836, 958, 1059, 1131
Lausenite, 1125
Lautarite, 590
Lautite, 42, 1132
Lava tubes, 474
Lavenite, 1294
Law of conservation of energy, 1170, 1173
Law of mass action, 684, 1106
"Law of Octaves," 45
Lawrencite, 490
Lawrencium, 6, 44
Layered silicates, 53
Lazulite, 937
Lazurite, 1097
Leachates, 768
Leaching, 26, 28, 60, 178, 184, 216, 548, 568, 670, 671, 764, 932, 1019, 1104. See also Vol. VI
LEAD-ELEMENT AND GEOCHEMISTRY, 20, 35, 44, 49, 75, 79, 82, 106, 150, 159, 191, 270, 273, 279, 296, 300, 312, 319, 320, 396, 400, 404, 425, 430, 433, 454, 455, 575, 587, 605, **642**, 697, 699, 736, 763, 963. See also COPPER, LEAD, ZINC, ORE DEPOSITS, Vol. I
LEAD AND ZINC: ECONOMIC GEOLOGY, 99, 194, 310, **645**. See also Vol. VIII
LEAD: INTERPRETATION OF STABLEISOTOPE ABUNDANCES, **646**. See also Vol. IVB
Lead isotopes, 1139
Lead ore deposits, 296. See also Copper, lead, and zinc ore deposits
Leadhillite, 143
Lead poisoning, 701
Lead, South Dakota, 469
Lead-uranium isotope ratios, 1219
Lead-zinc ores, 468
Leaves, 452
Lebanon, 542. See also Vol. VIII
LeChatelier, 686
Lecontite, 1124
Legrandite, 939
Legumes, 75
Leguminous plants, 803
Lehiite, 938
Leightonite, 1124
Leishmaniasis, 695
Lemma minor, 387
Lena River, U.S.S.R., 1048
Leonite, 1124
Lepidocrocite, 604, 844
Lepidolite, 17, 147, 447, 450, 462, 661, 663, 899, 1051, 1052
Leprosy, 692
Les Diablerets, Switzerland, 465
Lesse river, 540
Lethal cosmic radiation, 862
Letovicite, 1123
Leucine, 72
Leucite, 1, 22, 899, 971
Leucochalcite, 938
Leucogranite, 146, 241, 350, 494, 1028. See also Vol. V
Leucophosphite, 938
Leucosphenite, 62
Leucoxene, 58, 567
Levyite, 749
Lewis acid, 3
Lewistonite, 937
Lherzolite, 681
Li, 403, 571. See also LITHIUM
Libethenite, 937
Libya, 549. See also Vol. VIII
Lichen, 152
Libigite, 143
LIESEGANG RINGS, 600, **648**. See also Vol. VI
Life sciences, 186
Ligand, 149, 204, 574
Ligand-field theory, 416
Lightest structural metal, 668
Lignin, 79, 151, 815
Lignin-humus organic complexes, 136
Lignite, 70, 335, 423
Lime, 23, 111, 127, 319, 604. See also Vol. VI
Lime breccia, 119
Lime, slaked, 100, 103
Limestone, 29, 77, 100, 103, 118, 147, 156, 375, 378, 423, 430, 432, 459, 462, 465, 480, 488, 548, 559, 576, 606, 670, 734, 1122, 1241. See also Vol. VI
Limestone aquifers, 474
Limestone, high-calcium, 118
Limestone, magnesian, 118
Limestone schists, 465

1309

INDEX

Limestone terrains, 538
LIMNOLOGY, **650**, 660, 881. *See also* Vol. VI
Limonite, 58, 120, 604, 605, 901, 902, 1212
Linarite, 1125
Lindackerite, 939
Lindgrenite, 752
Lineages, 906
Linnaeite, 1131
Lipids, 70, 134, 136, 495, 815, 821
Liquid-gas colloidal system, 259
Liquid-gas inclusions, 40
Liquid-liquid colloidal systems, 259
Liquid-phase reactions, 143
Liquid scintillation counting, 451
Liquid segregations, 898
Liquid wastes, 315, 333
Liquidus, 918, 923
Liroconite, 938
Liskeardite, 938
Lithomarge, 843
Lithification, 91, 115, 151, 208, 342, 1160
Lithiophilite, 661, 934
Lithiophosphatite, 935
LITHIUM: ELEMENT AND GEOCHEMISTRY, 16, 68, 147, 149, 185, 266, 274, 279, 300, 420, 440, 462, **661**, 1051, 1064
LITHIUM: ECONOMIC DEPOSITS, 44, **662**. *See also* LITHIUM, Vol. IVB
Lithium-rich micas, 993
Lithology, 886. *See also* Vol. VI
Lithophile elements, 10, 47, 88, 155, 212, 247, 396
Lithopone chemicals, 64
Lithos, 661
Lithosphere, 21, 74, 123, 243, 378, 396, 405, 466, 495, 498, 600, 664, 996. *See also* Vol. V
Littoral zone, 660. *See also* Vols. I and VI
Living organisms, 946
Livingstonite, 704
Lizardite. *See* serpentine
"Loco weed," 1080
Lode gold, 468
Lodestone, 604
Loellingite (Löllingite), 42, 1131
Loess, 915
Loeweite, 1124
Logarithimic distribution law, 909
Lonsdaleite, 763
Longshore currents, 348. *See also* Vols. I and VI
Loparite, 1028
Lopezite, 159, 168
Lorandite, 42
Lorettoite, 490
Lorraine, France, 604. *See also* Vol. VIII
Loseyite, 143
Lost rivers, 544
Lovozero alkaline complex, 1027
Lower Carboniferous fossils, 115
Lower Cretaceous, 881
Ludlamite, 938
Ludwigite, 87
Luminescence, 664. *See also* Vol. IVB
Lunar abundances, elements, 275. *See also* Vol. II
Lunar glass, 149
Lunar mare analysis, 254. *See also* Vol. II
Lunar minerals, 149. *See also* Vol. II
Lunar soil contamination, 324
Lungs, 730
Lusczynski, 554
Lussatine, lussatite, 438
Lutecine, lutecite, 438
LUTETIUM: ELEMENT AND GEOCHEMISTRY, 44, 279, 300, **664**. *See also* Vol. IVB
Lydite, 438
Lyell, Charles, 454
Lysine, 72

Mackayite, 1079
Macroflora, 25
Mafic minerals, Mafics, 350, 465, 898
Mafic rocks, 25, 35, 156. *See also* Vol. V
Magadiite, 436
Mackenzie River, Canada, 1948
Magdalena River, Colombia, 1048
Maghemite, 50, 604, 845
Magma, 17, 63, 83, 111, 147, 168, 208, 272, 375, 391, 555, 593, 605, 666, 709, 898, 910. *See also* PARAGENESIS, Vol. V
Magmatic crystallization, 10, 19, 88, 101, 190. *See also* Vol. V
Magmatic differentiation, 21, 68, 83, 273, 387, 603, 670, 1062. *See also* Vol. V
Magmatic fluids, 57, 427, 571
Magmatic gases, 387, 493, 572
Magmatic rocks, 17, 83, 244, 248. *See also* Vol. V
Magmatic sulfides, 41
MAGMATIC WATER, 226, 436, 470, 547, **666**

Magmatism, 576, 993. *See also* Vol. V
Magnesia, 430
Magnesian calcite, 441
Magnesian limestone, 118. *See also* Vol. VI
Magnesiocopiapite, 1126
Magnesite, 1, 17, 58, 107, 142, 441, 559, 667, 941
Magnesiochromite, 845
Magnesioferrite, 845
MAGNESIUM: ELEMENT AND GEOCHEMISTRY, 16, 21, 37, 38, 44, 75, 88, 106, 113, 126, 151, 157, 178, 185, 196, 248, 264, 267, 279, 296, 300, 320, 360, 374, 396, 424, 433, 464, 475, 569, 580, 587, 599, **666**
Magnesium carbonate, 600
Magnesium chloride, 354
Magnesium chlorophoenicite, 936
MAGNESIUM CYCLE, 213, **669**
Magnesium feldspar, 200
Magnesium oxide, 196
Magnesium silicates, 55
Magnesium salts, 353
Magnesium sulfate, 354
Magnetic properties, minerals, 208
Magnetic field, 457, 1255. *See also* Vol. V
Magnetic reversals, 597. *See also* Vol. V
Magnetite, 31, 424, 432, 493, 558, 599, 601, 604, 606, 845, 898, 899, 900, 927, 1118
Magnetite-ulvospinel solid solution, 362
Magnetoplumbite, 845
Magnolite, 704, 1079
Maize, 28
Makjan, Molucca Is., 12
Malachite, 58, 143, 191, 195, 207, 1137
Malacon, 488
Malad Springs, Idaho, 1116
Maladjustment, 691
Malaria, 693
 prevention, 327
Malaya, 1192. *See also* Vol. VIII
Malayan ores, 606
Maldonite, 763
Malladrite, 491
Mallardite, 1125
Mammoth Hot Springs, 1117
Man (*Homo sapiens*), 76, 103, 137
Man-made industries, 1015
Mangan-alluaudite, 934
Manganberzeliite, 934
MANGANESE: ELEMENT AND GEOCHEMISTRY, 38, 44, 49, 75, 106, 150, 279, 284, 300, 320, 360, 414, 433, 475, 587, 599, 604, 605, **670**, 699, 736, 842, 1067
MANGANESE CYCLE, **671**
MANGANESE NODULES, GEOCHEMISTRY, 49, 63, 78, 146, 241, 275, 343, 350, 384, 462, 495, 665, 672, **673**, 792, 1166, 1190
Manganese ores, 433
Manganese oxides, 49, 207, 231, 422
Manganese precipitation, 49
Manganese residence time in seawater, 50
Manganite, 207, 674, 845
Manganocalcite, 52
Manganolangbeinite, 1123
Manganosite, 843
Mangroves, 536. *See also* Vols. VI and VII
Manjak, 1207
Man's activities, 541
Mansfeld Kupferschiefer (Copper Shales), 35
Mansfieldite, 935
Mantle, 394, 397, 429, 448, 493, 573, 600, 666, 678, 992, 1030, 1037, 1052, 1056. *See also* Vol. V
MANTLE GEOCHEMISTRY, 17, 32, 195, 264, 270, 273, **677**. *See also* Vol. V
Manto deposits, 190
Manuscript dating, 452
Map making, 690. *See also* Vol. VI
Marahuite, 1203
Maranon, 1049
Marble, 18, 101, 103, 112, 120, 248, 489, 734
Marcasite, 51, 58, 61, 425, 547, 900
Margarite, 30
Mariana Arc, 1038. *See also* Vol. VIII
Marine chemistry, 36. *See also* Vol I
Marine deposits, 358, 480. *See also* Vol. VI
Marine deserts, 818
Marine ecology, 683. *See also* Vols. I and VII
Marine environment, 79. *See also* Vol. I
Marine food chain, 325
Marine fossils, 79
Marine geochemistry. *See* SEA-WATER CHEMISTRY; *also* Vol. I
Marine microbiology, 78. *See also* Vol. I
Marine organisms, 425
Marine protozoans, 50
Marine scavengers, 462
Marine sediments, 71, 78, 90, 102, 110, 135, 138,
 184, 407, 420, 425, 587, 995, 1005. *See also* Vol. VI
Marine sedimentation rates, 10
Mariotte, Edmé, 476
Marmora, Ontario, Canada, 606
Mars, 848
Marsh, 79, 325, 471, 538, 548, 944. *See also* Vol. VI
Marsh, George Perkins, 187
Marshite, 490
Martite, 604
Mascagnite, 1123
MASS ACTION AND EQUILIBRIUM CONSTANT, 3, 96, 177, 181, **684**
Mass budget, 337
Mass law, 1107
Mass spectrographic analysis, 455
Mass spectrometric measurments, 455
MASS SPECTROMETRY, 145, 343, 349, 374, 384, 392, 409, 497, **688**, 1029
Mass spectroscopy, 230, 742
Massicot, 843
Matildite, 82, 1092
Matlockite, 490
Matrix errors, 369
Maucherite, 42, 1130
Maui, Hawaii, 556. *See also* Vol. VIII
Mauna loa, Hawaii, 555. *See also* Vol. VIII
Mayflower, Utah, 468
Mean sea level, 554. *See also* Vol. I
Measles, 693
Meat processing pollutants, 321
Mechanical, weathering/erosion, 428. *See also* Vol. VI
MEDICAL GEOGRAPHY, **690**
MEDICAL GEOLOGY, **696**
MEDICAL GEOLOGY—TRACE METALS IN MAMMALS, **699**
Medical research, radioisotopes, 326
Medical therapy, radioisotopes, 326
Medicinal plants, 1199
MEDICINAL SPRINGS, **702**
Mediterranean, 474
 climatic regions, 548
 semiarid climate, 543
Meerschaum. *See* sepiolite
Megaw, 415
Mekong River, Laos, Cambodia, Vietnam, 474, 1048. *See also* Vol. VIII
Melanostibian, 34
Melanovanadite, 1237
Melanterite, 1, 1124
Mellite, 829
Melnikovite, 1137
Melonite, 1132
Melting, 217, 602, 1085
Melting point, 47, 85, 918
Membranes, 589, 704. *See also* CLAY MEMBRANE PHENOMENA
Mendelevium, 6, 44. 279, 990
Mendeleyev (Mendeleeff; Mendeleev), Dimitri, 45, 385, 1029
Mendipite, 490
Mendozite, 1124
Mercallite, 1123
Mercaptans, 104
MERCURY: ELEMENT AND GEOCHEMISTRY, 35, 44, 106, 150, 270, 277, 279, 300, 319, 433, 573, **704**, 763, 828
Meso-compounds, 810
Mesocuticle, 154
Mesodesmic crystal, 201
Meson, 130
Mesophere, 409. *See also* Vol. II
Mesozoic, 878, 893. *See also* Vol. VII
Mesozoic bauxites, 26
Mesquite, 549
Meta-autunite, 939
Metabitumite, 828
Metabolic end products, 729
Metabolic rates, 1156
Metabolism, 74, 77, 660, 812, 834
Metabolites (organic), 818
Metacinnabar, metacinnabarite, 573, 704, 1130
Metahewettite, 1237
Metahohmannite, 1126
Metals, metallurgy, mettallogenesis, 66, 70, 84, 406, 558, 709, 763. *See also* Vol. IVB
Metal biogeochemistry, 415
Metal complexes, 70, 831
Metal containers, pollution, 311
Metal hydroxides, 37
Metal meteorites, 398. *See also* Vol. II
Metals, noble, 1058. *See also* gold, platinum, silver
Metal ore deposits, 214
Metalimestone, 18
Metalimnion, 709, 1172. *See also* THERMOCLINE

INDEX

Metallic bonds, 84, 197, 413
Metallic compounds, 15
Metallic iron phase, 189
Metallic oxides, 59, 847
Metallic sulfide, 1135
Metalloid, 33
Metallurgical flux, 119
Metallurgy, metallurgic industries, 364, 558
Metamict minerals, 446
Metamict substance, 744
METAMORPHIC ENVIRONMENTS—CHEMICAL MOBILITY, **709**. See also Vol. V
Metamorphic facies, grade, 428, 709, 746, 903. See also Vol. V
Metamorphic parageneses, 898
Metamorphic rocks, 19, 111, 156, 244, 248, 274, 418, 459, 561, 590. See also Vol. V
Metamorphism, 19, 40, 57, 89, 101, 115, 117, 127, 134, 208, 113, 387, 451, 467, 561, 572, 576, 602, 1054. See PARAGENESIS; also Vol. V
Metarossite, 1237
Metasedimentary rocks, 245
Metasideronatrite, 1126
Metasomatism, 5, 19, 57, 99, 342, 350, 602, 904, 1097. See also Vol. V
Metasomatic deposits, 1093
Metastable equilibrium, 725
Metastable mineral phases, 427
Metastibnite, 34
Metastrengite, 936
Metatorbernite, 939
Metatyuyamunite, 1237
Meta-uranopilite, 1126
Metavariscite, 936
Metavauxite, 939
Metavoltine, 1126
Metazeunerite, 939
Metazoans, 76
Meteoric dust, 792
METEORIC WATER, 78, 184, 470, 547, 571, 572, 606, **716**, 1015
Meteorites, 8, 18, 21, 40, 74, 82, 100, 149, 179, 189, 190, 226, 263, 386, 398, 448, 455, 466, 493, 494, 578, 599, 664, 679, 779, 865, 869, 897, 923, 957, 971, 975, 1024, 1052, 1054, 1195, 1216. See also METEORITES Vol. II
Meteorite abundances, elements, 275
Meteorites—organic constituents. See CARBONACEOUS METEORITES, Vol. II
Meteoritic minerals, 149, 670. See also Vol. IVB
Meteorological cycles. See Vol. II
Meteorology, 402, 409, 546. See also Vol. II
Meteorology service, 775
Methane, 32, 77, 79, 126, 130, 226, 272, 335, 427, 451, 496, 863. See also Vol. VIB
Method of mixtures, 304
Methionine, 72
Methyl group, 156, 495, 499
Methylbutane, 496
Methylcyclohexane, 496
Methylamine, 67
Methylpentane, 496
Methylphenanthrene, 499
Mexico, 35, 378. See also Vol. VIII
Meyerhofferite, 87
Mg, 547, 600. See also magnesium
Miargyrite, 1092
Mica group. See mineral groups; SILICATES
Mica, 17, 22, 88, 148, 149, 155, 168, 176, 203, 367, 378, 429, 433, 449, 599, 887, 900
Mica schist, 22. See also Vol. V
Micelles, 436
Michenerite, 897, 1131
Microbenthos, 660
Microbes, 428, 672, 914, 1127
Microbial iron oxidation and reduction. See IRON OXIDATION AND REDUCTION—MICROBIAL
Microbial mineral accumulation, 729
Microbial mineral genesis and transformation. See MINERAL GENESIS AND TRANSFORMATION—MICROBIAL
Microbial mineral sulfide oxidation. See SULFIDE MINERAL OXIDATION—MICROBIAL
Microbial oxidation and reduction of iron. See IRON OXIDATION AND REDUCTION—MICROBIAL
Microbial oxidative activity, 334
Microbial sulfate reduction. See SULFATE REDUCTION—MICROBIAL
Microbial sulfate reduction, 425
Microbiology. See MARINE MICROBIOLOGY, Vol. I
Microchemical analysis, 444
Microclimatic environment, 878. See also Vol. II
Microcline, 1, 60, 117, 447, 448, 450, 579, 901, 902, 904, 1052
Micrococcus, 806

Microflora, 25, 75
Microlite, 794, 846, 1162
Microorganisms, microbes, 77, 180, 334, 356, 729. See also Vol. VII
Microperthite, 234
Microprobe, 908, 565, 1288
Microscope freezing stage, 374
Microscopy. See *Encyclopedia of Microscopy* (G. L. Clark, editor, Reinhold Publishing Co.)
Microtektites, 276
Mid-Atlantic Ridge, 434, 1035. See also Vols. I, III, and V
Mid-Atlantic Ridge Valleys, 597
Middle East, 548. See also Vol. VIII
Midge larvae, 660
Midocean ridge, 260, 1035. See also Vols. I, III, and V
Mid-ocean rift valleys, 247
Miersite, 490
Migmatization, 709
Migration, 709, 771. See also CRUDE OIL COMPOSITION AND MIGRATION, Vol. IVB
Milk processing pollutants, 321
Miller, W. H., 741
Millerite, 790, 1131
Millisite, 938
Mimetite, 937
Minasragrite, 1123
Mine gases, 493
Mine water, 77
Miner's consumption, 734
Mineral, 583, 702, 740. See also Vol. IVB
Mineral chemistry, 153
Mineral classes, 599
MINERAL CLASSES: NONSILICATES, **717**
MINERAL CLASSES: SILICATES, 53, 83, 88, 191, 204, 206, 231, 234, 237, 260, 264, **718**
Mineral collecting, 743
Mineral cycling, 339
Mineral deposits, resources, 409, 457, 574, 773. See also Vol. IVB
Mineral fuels. See Vol. IVB
MINERAL GENESIS—EQUILIBRIUM, 342, **725**. See also ISOTOPIC VARIATION IN MINERAL GENESIS, Vol. IVB
MINERAL GENESIS AND TRANSPORTATION—MICROBIAL, **729**
Mineral groups (organic). See ORGANIC MINERALOIDS
MINERAL PARTICLES AND HUMAN DISEASE, **730**
Mineral prospecting. See Vol. IVB
Mineral salts, 485
Mineral solubilities, 573
Mineral springs. See MEDICINAL SPRINGS
Mineral sulfide oxidation—microbial. See SULFIDE MINERAL OXIDATION—MICROBIAL
MINERAL THERMOMETRY, **745**
Mineral waxes. See ASPHALT, ASPHALTITES, ASPHALTOIDS AND MINERAL WAXES, Vol. IVB
Mineralization, 334
Mineralization, groundwater, 548
MINERALOGY, 405, 444, 579, **740**. See also Vol. IVB
MINERALOIDS, **744**, 811, 823
Mines and mining, 486, 489, 555, 606. See also Vol. VI
Mines, dewatering, 478
Minguzzite, 811
Mining industry pollution, 312, 316, 419, 320
Minium, 843
Minnesotaite, 599, 605
Minyulite, 939
Miocene, 555, 882, 454. See also Vol. VII
Miogeosynclinal facies, 115. See also Vol. VII
Mirabilite, 1097, 1124
Miscibility, 1104
Misenite, 1123
Missile firing pollution, 324
Mississippi Delta, 437
Mississippi River, 1047, 1048
Mississippi Valley, 474
Mitridatite, 938
Mitscherlichite, 491
Mixed layer clays, 421
Mixing, 1003, 1007
Mixing time, 1074
Mixite, 938
Mn, 406, 412. See also Manganese
Mobile phase, chromatography, 160
Mobilisates, 825
Mobility, 906
Models. See Vol. II
Mohole type sediment cores, 427
Mohorovičic discontinuity, 244, 264, 679. See also Vol. V
Molal free energy, 1174
Molalities, 687
Molarity, 1118

Molecular attraction, 474
Molecular bond, 84
Molecular energy, 891
Molecular formation. See MOLECULAR WEIGHT
Molecular-kinetic theory, 94
MOLECULAR SIEVES, 496, **748**
MOLECULAR WEIGHTS, **750**
Molecules, 1237
Mollusks, mollusca, 102, 113, 118, 126, 135, 153. See also Vol. VII
MOLYBDATES AND TUNGSTATES, 101, 241, 343, 583, 717, **752**
Molybdenite, 148, 204, 753, 899, 1132, 1041
MOLYBDENUM: ELEMENT AND GEOCHEMISTRY, 20, 44, 75, 79, 106, 150, 193, 274, 279, 399, 423, 697, 699, **753**
Molybdite, 753
Molybdomenite, 1079
Molysite, 490
Monazite, 146, 241, 343, 349, 350, 385, 489, 495, 665, 876, 934, 944, 962, 993, 1027, 1028, 1189, 1219, 1292, 1165, 1184, See also Vol. IVB
Monetite, 934, 940
Monomeric silica, 435
Monsoon semiarid climate, 542. See also Vol. II
Mont Pelée, Martinique, 12. See also Vol. VIII
Montanite, 1079
Montbrayite, 1131
Monte Amiata, 707
Montebrasite, 936
Monteponite, 843
Monteporrite, 99
Monterey series, 441
Montesite, 1192
Montgomeryite, 939
Montmorillonite, 5, 22, 55, 58, 91, 181, 227, 230, 421, 429, 435, 562, 587, 591, 604, 1019, 1051
Montroydite, 704, 843
Montroseite, 845
Monzonite, 18, 1060
Moon—theories of origin, 149, 270. See also Vol. II
Mooreite, 1125
Morbidity (disease), 738
Mordenite, 749
Morenosite, 1125
Morinite, 936
Morrison formation (Jurassic), 1223
Morocco, 549. See also Vol. VIII
Mortality (death), 738
Moschellandsbergite, 704
Mosesite, 491, 704
MOSSBAUER EFFECT, **759**
Mossite, 846, 1162
Mother Lode, California, 468
Motion sickness, 690
Mottramite, 936, 1235, 1236
Mount Enid, 606
Mount Isa, Queensland, Australia, 1139
Mt. Hood, Oregon, 555
Mt. Lassen, California, 555
Mount Magnitnaya, 606
Mt. Rainier National Park, 188
Mt. Ranier, Washington, 555
Mt. Spurr, Alaska, 12
Mountain belts, 169
Mountain Pass, California, 385
Mountain sickness, 690
Mowry formation, 441
Muck, 28
Mud baths, 703
Muds, 433, 702
Mud-feeding organisms, 54
Mullite, 22
Multiple sclerosis, 697
Multiplet theory, 145
Mummy cloth dating, 452
Muscle tone reflex, 1252
Muscovite, 22, 29, 40, 55, 58, 63, 203, 435, 447, 448, 449, 450, 562, 735, 899, 901, 902, 904, 971
Muskox Intrusion, Canada, 169
Mutagenic factors, 691
Mutations, 326, 691
Muthmannite, 1131
Mycotic infections, 312
Myristic, 499
Myrmekitization, 1097. See also Vol. V

NMR, 498
Na, 547. See also sodium
NaCl (sodium chloride), 383, 489, 492, 572, 588
Nacreous layer, 114
Nacrite. See Kaolinite
Nadorite, 34
Naëgite, 488

1311

INDEX

Nagelschmidt, 438
Nagyagite, 1130, 1164
Nahcolite, 111, 142, 884, 1078, 1097
Namib desert. See NAMIBIA, Vol. VIII
Nannoplankton, 74
Nantokite, 490
Naphthalene, 496
Naphthalenic ring systems, 498
Nari (calcrete duricrust), 548. See also Vols. III and VI
Native elements, 717
Native platinum, 956
National Audubon Society, 188
National forests, 188
National parks, 188, 311
National registry of natural landmarks, 188
NATIVE ELEMENTS, 763. See also under individual names
Natroalunite, 1125
Natrochalcite, 1126
Natrojarosite, 1125
Natrolite, 58, 749, 1097
Natromontebrasite, 936
Natron, 58, 142, 1097
Natrophilite, 934
NATURAL BRINES, 764
Natural chromatography. See Vol. IVB
NATURAL CHROMATOGRAPHY IN SEDIMENTARY ROCKS, 769
"Natural" community, environment, 257, 1135
Natural gas, 80, 124, 226, 335, 493, 827. See also Vols. IVB and VI
Natural glasses, 591
Natural parks, 772
NATURAL RESOURCES, 186, 771
Natural resources service, 775
Natural waters, 575, 1247
Nature Conservancy, 188
Naumannite, 1130
Navier-Stokes equations, 479
Navigation, 1260
NBS oxalic acid standard, 452
Near-shore sediments, 111
Necking down, 376
Negev, Israel, 545
Nekton, 74, 660. See also Vol. I
Nematodes, 153, 660
NEODYMIUM: ELEMENT AND GEOCHEMISTRY, 44, 279, 300, 777
Neogaea, 246
Neolithic man, 730
NEON: ELEMENT AND GEOCHEMISTRY, 21, 44, 269, 272, 396, 411, 840, 1039, 778
Nepalite, 828
Nepheline, 1, 23, 735, 899, 1097
Nepheline syenite, 18, 156, 248, 273. See also Vol. V
Nephelinite, 1031
Nephelometry, 183
Nephthelosis, 730
Nepouite, 790, 791
Neptunium, 6, 44, 279
Nernst equation, 173, 307
Nernst heat theorem, 1174
NESOSILICATES, 100, 201, 462, 720. See also CRYSTAL CHEMISTRY; MINERAL CLASSES: SILICATES
Nesquehonite, 58, 142
Nessler tubes, 182
Netherlands Antilles, 116. See also Vol. VIII
Neurological disorders, 738
Neurotoxin, 822
Neuston, 660
Neutron, 43, 266, 989, 991, 1211
NEUTRON ACTIVATION ANALYSIS, 145, 275, 343, 349, 374, 384, 409, 443, 494, 590, 990, 779, 1029
Neutron capture, 7, 267, 405, 962, 1020
Neutron diffraction, 191
Neutron fluxes, 784
Neutron well-logging, 785
Nevyanskite, 836
New Caledonia, 792. See also Vol. VIII
New Idria (California), 707
New Zealand, 571, 573. See also Vol. VIII
Newlands, John A. R., 43
New Market Limestone, Va., 120
New Mexico, 1092. See also Vol. VIII
New York bight area dumping, 313, 316
Newark Canyon Formation (Lower Cretaceous), 881
Newberyite, 935
Ni, 404, 412. See also nickel
Niacin, 821
Niafu, Tonga Islands, 12. See also Vol. VIII
Niccolite, 42, 790, 1130
Niche, 690
Nichrome, 227

NICKEL: ELEMENT AND GEOCHEMISTRY, 34, 41, 44, 49, 75, 79, 106, 150, 179, 190, 193, 274, 279, 296, 300, 395, 399, 411, 426, 430, 433, 598, 599, 605, 700, 736, 790, 831
Nickel-cadmium batteries, 99
Nickel-iron. See Iron
Nickel-iron alloy, 680, 763, 791
Nickel-iron phase, 599. See also Iron
Nickel-skutterudite, 1132
Nier, A. O., 455
NIFE (core), 791. See also Vol. V
Nigards Glacier, Gaupne, Norway, 465
Niger Delta, 887
Niger River, Nigeria, 1048
Nigeria, 489, 1192
Niggli numbers, 712
Nile River, Egypt, Sudan, 473, 547
Niobates, 241, 343, 349, 384, 494
NIOBIUM (COLUMBIUM): ELEMENT GEOCHEMISTRY, 35, 44, 245, 274, 279, 300, 423, 793
Nisqually Glacier, Mt. Rainier, Wash., 465
Niter, 29
Nitrate, 37, 38, 155, 216, 231, 241, 320, 343, 349, 384, 388, 475, 494, 579. 583, 941, 1017, 1064, 1073
Nitrate fertilizers, 312
Nitric acids, 3
Nitric oxide, 326
Nitrides, 29, 100
Nitrification, 801
Nitrite, 1064
Nitrobacter, 805
Nitrobarite, 62
NITROGEN: ELEMENT AND GEOCHEMISTRY, 29, 32, 44, 69, 70, 75, 123, 126, 129, 150, 165, 272, 279, 396, 408, 493, 795, 819, 863, 1016, 1064
NITROGEN CYCLE, 210, 253, 338, 801
NITROGEN CYCLE IN THE OCEAN, 807
Nitrogen dioxide, 15, 272
Nitrogen donor complexes, 190
Nitrogen fixation, 29, 801
Nitrogen isotopes, 796
Nitrogen metabolism, 1199
Nitrogen, organic, 349
Nitrogen oxides, 272, 312
Nitrogenous compounds, 1016
Nitrosomonas, 805
Nobelium, 6, 44, 279
Noble gases, 492, 1014. See also RARE GASES, also under specific names (Neon, Argon, etc.)
Noble metals, 839, 956, 1039, 1041
Nocerite, 490
Nodules, oceanic, 49, 63
Nodular phosphorite, 950
Noise pollution, 311, 314
Nonane, 496
Nonbiogenic nitrogen, 29
Non-Brownian mobility, 94
Noncondensable gases, 1110
Nondestructive analysis of rocks, 789
Nonequilibrium formula, 476
Nonequilibrium theories, 266
Nonesuch shale, 982
Non-metals, 763
Non-metallic minerals. See MINERAL CLASSES: SILICATES
Nonrenewable resources, 187
Nonstoichiometric compounds, 362
Nonsilicates, 717
Nontronite, 55, 227, 604. See also montmorillonite
Nordenskiöldine, 87, 1192
Nordstrandite, 25
Noril'sk district, U.S.S.R., 957
Normal water, 1063
Normality, 1119
Norsethite, 142
North Africa, 378, 474, 548. See also Vol. VIII
North America, 548. See also Vol. VIII
Nose-cone reentry, 579
North German plain sediments, 91. See also Vol. III
Northupite, 143
Nova, 397. See also Vol. II
Nostocaceae, 802
Novaculite, 65, 438. See also Vol. IV
Nubian system, 474. See also Vol. VIII
Nuclear bomb contamination, 296
Nuclear decay, 988, 1217
Nuclear explosions, 314, 326, 988
Nuclear fuels, 990. See also Vol. IVB
Nuclear generating stations, 1170
Nuclear interactions, 996
Nuclear magnetic resonance analysis, 498
Nuclear magnetic resonance logging, 461
Nuclear processes, 267
Nuclear reactions, 6, 84, 123, 267, 488

Nuclear reactors, 99, 996
Nuclear reactor fuel, 9
Nuclear weapons, 6
Nuclear weapons testing, 409, 1002
Nucleation, 363
Nucleic acids, 981
Nucleosynthesis, 263, 405, 996, 1216, 1270
Nucleotides, 126
Nuclides, 7, 398, 784, 1020
Null-point balances, 1177
Nutrient concentrations, 1156
Nutrient elements, 209
Nutrients in the sea, 349, 1065. See also Vol. I
Nutrient salts, 586
Nutritional supplements, 699

O, 404. See also oxygen
Oahu, Hawaii, 556
Oasis, 548
Ob River, U.S.S.R., 1048
Obesity, 691
Obsidian, 582, 972. See also Vol. V
Oceans, 432, 471, 594. See also Vol. I
Ocean-atmosphere interaction, 271, 333. See also Vol. I
Ocean biomass, 74. See also Vol. I
Ocean bottom, spreading (sea-floor spreading), 367. See also Vol. V
Ocean drilling, 888
Ocean evolution, 387
Ocean pollution, 312
Ocean productivity, 74. See also Vol. I
Ocean salt, spray, 454, 669. See also cyclic salts
Ocean waters, 389, 447, 452, 1004. See also Vol. I
Oceanic basalts, 18. See also Vol. V
Oceanic circulation, 1000, 1073. See also Vol. I
Oceanic crust, 244, 682, 1037. See also Vol. V
Oceanic islands, 1056. See also Vol. III
Oceanic resources, 774. See also Vol. I
OCEANOGRAPHY, 74, 402. See also Vol. I
Ochsenius, bar theory, 359. See also evaporites
Octacorals, 113
Octane, 496
Oddo-Harkins rule, 266, 1024
Oeschger counter, 130
Offshore acid waste dumps, 316
Ohio River, 473
Oil, 812, 1205. See also Vol. VI
Oil brines, 20, 185
Oil composition and migration. See CRUDE OIL COMPOSITION AND MIGRATION, Vol. IVB
Oil deposits, 457
Oil drilling, 551
Oil drilling wastes, 321
Oil fields, 549
Oil field brines, 1087
Oil field waters, 88, 498
Oil gas deposits, 586
Oil (palm), 28
Oil pollution, 33, 321
Oil reservoirs, 116, 478
Oil saturation, 964
Oil shale, 23, 71, 831, 1206. See also Vols. IVB and VI
Oldhamite, 101, 1130
Olduvai Gorge, Tanzania, 23, 367. See also Vols. VII and VIII
Olefins, 144, 496
Olefinite (bernalite), 828
Oligist (iron), 604
Oligocene, 454, 882. See also Vol. VII
Olivenite, 937
Olivine, 1, 17, 101, 149, 180, 190, 196, 201, 206, 273, 579, 418, 429, 462, 599, 667, 679, 680, 735, 750, 898, 899, 901, 1091, 1196
Olivine basalts, 384. See also Vol. V
Olivine nodules, 401
Olivine-spinel transition, 462
Omphacite, 582
Oolites, 111, 115, 604, 1012. See also Vol. VI
Oolith, 51. See also Vol. VI
Opal, 53, 58, 77, 204, 433, 435, 438, 900, 902. See also Vols. IVB and VI
Opaline silica, 430, 468
Oparin, 850
Open-pit mining, 64
Operations research, 1260
Operculum, 114. See also Vol. VII
Ophiolite, 465. See also Vol. V
Opium, 350, 385
Opium tracing, 146, 241, 343
OPTICAL ACTIVITY, 810. See also STEREOCHEMISTRY AND ENANTIOMORPHISM
Optical density, 302
Optical isomers, 810
Optical microscopes, 741
Optical pyrometers, 921

INDEX

Optical rotation, 810
Orangite, 962
Orbitals, 45
Orbital hybridization, 415
Orcelite, 790
Order-disorder. See Vol. IVB
Ordoñezite, 34
Ordovician, 454. See also Vol. VIII
Ores, ore genesis, 296, 342, 1160. See also, Vol. IVB
Ore deposits, 5, 23, 214, 268, 273, 296, 377, 387, 561, 572, 773, 909, 993
Ore-forming fluids, solutions, 377, 389, 561, 576, 1109
Ore-forming metals, 296
Oregonite, 1130
Organic acids, 59
Organic carbon, 438
Organic chemistry, 812
ORGANIC COMPOUNDS, 69, 75, 77, 319, 408, 717, **811**, 953
Organic cosmochemistry. See Vol. II
Organic debris, detritus, 349, 427
Organic fixation, 669
ORGANIC GEOCHEMISTRY, 69, 79, 150, 254, 274, **812**
ORGANIC GEOCHEMISTRY OF SEAWATER, **818**. See also SEAWATER, CHEMISTRY OF, Vol. I
Organic ligands, 74, 416, 468
ORGANIC MINERALOIDS, 817, **823**
ORGANIC PIGMENTS, **830**
Organic pollutants, 312
Organic reef, 353
Organic sediments. See Vol. VI
Organic soils, 28
Organisms, estuarine, 349
Organisms, marine, 1012
Organisms, photoautotrophic, 325
Organized elements, 82
Organo-clays, 421
Organofluoride compounds, 313
Organometallic complexes, 151
Organometallic systems, catalysis, 144
Organo-mineralic sediments, 1202
Orgueil (meteorite), 865
Origin of elements, 266
Origin of life, 73, 978. See also Vol. VII
Origin of petroleum, 427
Orinoco River, Venezuela, 1048
Ornithine, 72
Orogenic belts, 1097. See also Vol. V
Orogenesis, 208
Orpiment, 41, 42, 573, 900, 1131
Orsat apparatus, 388
Orthobismuthates, 83
Orthoclase (feldspars), 22, 58, 62, 429, 579, 904, 972
Orthomarble, 120
Orthophosphate, 80
Orthopyroxene, 17, 1091
Orthosilicates. See CRYSTAL CHEMISTRY
Osbornite, 29
Osmiridium, 599
Osiridium, 1059
Osmiridium, 836, 957
OSMIUM: ELEMENT AND GEOCHEMISTRY, 44, 279, 300, 599, **836**, 959, 1059
Osmosis, chemical, 172, 185
Osmosis, reverse (RO), 222
Osmotic excretion, 122
Osmotic pressure, 94, 172, 321
Ostracods, 114, 118, 660. See also Vol. III
Otavite, 99, 142
Othmer, K., 599
Ouenza, Algeria, 606. See also Vol. VIII
Outer space, 493. See also Vol. II
OUTGASSING OF THE PLANET EARTH, 29, 40, 126, 245, 324, 768, **836**, 979
Overite, 939
Overgrazing, 338
Owens Lake, Calif., 884
Oxalates, 811, 1215
Oxammite, 811
Oxanite, 829
Oxidants, 924
Oxidates, 433
Oxidate ores, 180
OXIDATION AND REDUCTION, 8, 32, 37, 71, 77, 104, 121, 217, 226, 334, 573, 600, 606, **839**, 932, 1132, 1266
OXIDATION AND REDUCTION OF IRON—MICROBIAL. See IRON OXIDATION AND REDUCTION—MICROBIAL
Oxidation cycle, 848
Oxidation potential, state, 41, 77, 601, 1023, 1069
Oxidation-reduction potential, 932
Oxidative degradation, 333
OXIDES AND HYDROXIDES, 21, 41, 49, 59, 100, 107, 146, 191, 231, 237, 260, 343, 419, 422, 430, 558, 583, 717, **843**
Oxidizing agents, 377, 861
Oxychloride, 83
OXYGEN: ELEMENT AND GEOCHEMISTRY, 17, 44, 69, 75, 123, 150, 196, 199, 262, 264, 267, 272, 279, 312, 325, 347, 390, 395, 396, 422, 429, 440, 466, 492, 558, 572, 587, 599, 600, 686, **847**, 952, 1064
OXYGEN CYCLE, 81, 210, 338, 600, **861**
OXYGEN: EVOLUTION IN THE EARTH'S ATMOSPHERE, 126, 271, **849**
OXYGEN ISOTOPE GEOCHEMISTRY, 108, **864**
Oxygen-isotope thermometer, 745
OXYLUMINESCENCE, **874**, 1169
Oxysalts, 142, 191
Oysters, 821
Ozocerite, 817, 827
Ozokerite, 1206
Ozone, 15, 126, 272, 311, 324, 326, 324, 847, 863. See also Vol. II
Ozonosphere, 847. See also Vol. II

Pachnolite, 491
Pachuca, 1094
Pacific Coast, 537
Pacific Ocean, 587
Pacific Ocean clays, 275
Pacific Ocean sediments, 49, 111, 121, 275
Packaging material wastes, 322
Paddy fields, 1102
Paddy muds, 587
Padina, 114
Pahoehoe, 555
Paigeite, 87
PAL (Present Atmosphere Level), 853
Paleobiochemistry, 73
Paleocene, 881
Paleochannel discharge, 879
Paleoclimates, 423
Paleoecology, 113, 885
Paleoenvironments, 180, 425
Paleogeography, 419, 449
PALEOHYDROLOGY, 539, **876**
PALEOLIMNOLOGY, **880**
Paleontology, 428, 597. See also Vol. VII
Paleopathology. See Vol. VII
PALEOSALINITY, 84, **885**. See also SEA WATER, HISTORY
Paleotectonics, 419
PALEOTEMPERATURES, ISOTOPIC DETERMINATIONS, 79, 422, 871, **891**, 893, 910
Paleozoic bauxites, 27
Palingenesis, 246
Palisades, 556
Palissy, Bernard, 476
Palladiplatinum, 957
Palladium, 411, 599, 763, 836, 957, 959, 1059
PALLADIUM: ELEMENT AND GEOCHEMISTRY, 44, 273, 279, 300, **897**
Palladium group, 394
Pallasites, 865
Pallomancy, 1253
Palmierite, 1123
Palynology, 133
Pampas, 542
Panama Basin, 50
Pantocycle, 219
Paper chromatography, 160
Paper coal, 826
Parabutlerite, 1126
Paracelsus, 732
Paracoquimbite, 1125
Paraffinic hydrocarbons, 79
Paraffins, 816, 827, 1206
PARAGENESIS, 148, **898**
Paragonite, 30
Parahilgardite, 87
Parahopeite, 935
Paralaurionite, 490
Parameleaconite, 843
Paramontroseite, 845
Pararammelsbergite, 42, 1131
Parasites, 690
Paratacamite, 490
Paratenorit, 843
Paravauxite, 939
Parent material, 915, 1043, 1267
Parisite, 143
Parsonsite, 938
Partial molal free energy, 382
Partial molal free enthalpy, 382
Particle tracks. See CHARGED PARTICLE TRACKS
Particulate contaminants, 1229
Particulate matter, 316, 499
Particulates, atmospheric, 11, 15, 312
Particulates in water, 313
Partition, chromatographic, 160

PARTITION COEFFICIENTS, 886, **905**
Partridgeite, 207
Pascoite, 1237
Past environments, 876
Pasteur level, 863
Pasteur point, 272
Pasturing animals, 548
Paternoite, 87
Pathfinder element, 99
Pathogenic agents, 690, 692, 730
Patronite, 1207, 1236
Pauling scale electronegativities, 261
Pauling's rules, 106, 199
Pb, 493, 578
Pb 207-Pb 206, 448
Peak configuration studies, 219
Pearceite, 42
Pearly aragonite, 114
Peat bogs, 28, 78, 133, 226, 335, 447, 452, 702, 1204
Pechora, 1048
Pectolite, 1097
PEDOLOGY (SOIL SCIENCE), 74, **911**
Pedosphere, 947, 949
Peedée formation, South Carolina, 134
Pegmatic/muscovite, 449
Pegmatite bodies, 971
Pegmatites, 156, 448, 661, 662, 715, 794, 825, 975, 1026, 1051, 1062, 1162, 1213. See also Vol. V
Pegmatitic solution, 238
Pegmatitization, 17, 22, 68, 83, 99
Pegmatoid deposits, 350
Pelagic carbonates, 275
Pelagic clays, 275
Pelagic sediments, Geochemistry, 49, 50, 246, 271, 275, 387, 433. See also ANTARCTIC PELAGIC SEDIMENTS, PELAGIC BIOCHEMISTRY, Vol. I; PELAGIC SEDIMENTS, Vol. VI
Pelecypods, 114
Peloids, 702
Pencil ore, 604
Pendletonite, 811, 828
Pendulums, 1253
Penfieldite, 490
Peninsular shield, 475
Penroseite, 705, 1131
Pentahydrite, 1124
Pentahydrocalcite, 142
Pentane, 496
Pentlandite, 35, 190, 599, 790, 959, 1042, 1059, 1131
Peptides, 126, 151, 427
Perched flow, 539
Perched water table, 555, 1264
Perchloric acids, 3
Percolation, 1046
Percylite, 490
"Perfect gas" state, 304
Periclase, 843
Peridinians, 137
Peridotite, 146, 156, 248, 264, 343, 350, 384, 494, 606, 680, 1027, 1090
Periodic law, 45
Periodic table. See ATOMIC NUMBER AND PERIODIC TABLE
Periodicity, 45
Perkins-Elmer infrared spectrophotometer, 582
Permeability, 39, 458, 472, 480, 481, 539, 552, 555, 557, 965, 1159. See also POROSITY AND PERMEABILITY
Permeable barrier, 353
Permeable rocks, 1116
Permian, 351, 454, 858, 880
Permian algal reefs, 114
Permian salt deposits, 156
Perovskite, 794, 845, 1193
Perrault, Pierre, 476
Persian Gulf, 102, 358, 545
Persilicic glass, 59
Perthite, 361, 1097
Peru (Humboldt) current, 216
Perylene, 496
Pesticides, 311, 313, 317, 318, 325
Petalite, 17, 462, 661, 663
Petrochemical industry, 144
Petrogenesis, 417
Petrographic microscope, 445, 451, 565
Petroleum, 68, 71, 77, 104, 124, 128, 135, 150, 158, 185, 335, 375, 409, 421, 493, 771, 772, 798, 810, 824, 830, 1156. See also Vol. IVB
Petroleum exploration, 268, 444
Petroleum geochemistry, 274
Petroleum industry, 144
Petroleum industry wastes, 321
Petrologic terminology, 5
Petrology, 152, 405. See also Vol. V

1313

INDEX

Petzite, 468, 1092, 1130
Pewter, 35
pH, 374, 416, 429, 435, 464, 539, 571, 573, 587, 606, 672, 908, 1015, 1068, 1082, 1106, 1136, 1198, 1239, 1246
pH and Eh, 422, 432
pH-Eh DIAGRAMS, 153, 181, **924**
pH-Eh field in natural environments, 933
pH-Eh RELATIONS, 1, 37, 38, 70, 77, 96, 101, 121, 151, 153, 178, 190, **932**
Phanerites, 464
Phanerozoic time scale, 453
Pharmaceutical pollutants, 321
Pharmacolite, 935
Pharmacosiderite, 939
Phase assemblages, 923
Phase contrast microscopy, 366
Phase diagram, 922
PHASE EQUILIBRIA, 559, 561, **917**, 1090
Phase rule, 561, 917, 1174. *See also* FREE ENERGY; *also* FREE ENERGY AND THE GIBBS PHASE RULE, Vol. V
Phase transformations, 273, 1170
Phenolic compounds, 320
Phenology, 1240
Phenols, 70, 77
Phenylalanine, 72
Pheophorbides, 158
Pheophytins, 158
Philippine Islands, 474
Phillipsite, 53, 58, 61, 433, 462, 674, 749
Phlogopite, 1, 17, 30, 447, 562, 735
Phoenicians, 1191
Phoenicochroite, 159
Phorbides, 159
Phosgenite, 143
Phosphate cycle. *See* PHOSPHORUS CYCLE
Phosphate and guano; *See* Vol. IVB
Phosphate fertilizers, 312, 318
Phosphate replacement, 52
Phosphate rock, 100, 950
Phosphates, 378, 425, 475, 579, 583, 717, 1073, 1189
PHOSPHATES, ARSENATES, AND VANADATES, 33, 52, 59, 78, 100, 181, 231, **934**
Phosphatic fish, 495, 1190
Phosphatic fish remains, 343, 350
PHOSPHATIZATION, **940**
Phosphine, 944
Phosphoferrite, 935
Phosphoguanidine, 945
Phosphophyllite, 935
Phosphorescence. *See* Vol, IVB, LUMINESCENCE
Phosphoric acid, 941
Phosphorites, 49, 52, 101, 425, 433, 586, 993, 1065
Phosphorröslerite, 935
PHOSPHORUS: ELEMENT AND GEOCHEMISTRY, 44, 69, 75, 247, 271, 279, 300, 349, 408, 465, 600, 605, 736, 819, 940, **942**, 1065
PHOSPHORUS CYCLE, 210, **946**
Phosphorus deficiency, 28
Phosphuranylite, 937
Photic zone, 954
Photocell manufacture, 148
Photochemical processes, 1156
Photodissociation, 32, 126, 850, 863, 1078
Photoelectric sensing, 301
Photoelectrometers, 444
Photoexcitation, 850
Photoluminescence. *See* LUMINESCENCE, Vol. IVB
Photometers, 1115
Photoreduction, 857
Photosensitization, 830
PHOTOSYNTHESIS, 51, 68, 70, 77, 121, 124, 129, 134, 157, 272, 334, 340, 347, 812, 857, 862, 863, **952**, 1064, 1266
Photosynthetic pigments, 79
PHREATIC WATER, **956**
Phreatic zone, 1264
Phreatophytes, 549
Phreatophytic vegetation, 485
Phthanite, 438
Phycobilins, 832
Phyllite, 449
Phylloerythorin, 159
Phylloretite, 811
Phyllosilicates, 720. *See also* CRYSTAL CHEMISTRY; MINERAL CLASSES: SILICATES; *See also* Vol. IVB
Physical chemical weathering, 427
Physical pollutants, 313
Phytane, 498
Phytoceroses, 1239
Phytochrome, 834
Phytohormones, 822
Phytological desert, 877

Phytoplankton, 69, 136, 157, 340, 437, 808, 952, 1065
Phytyl alcohol, 157
Phytyl esters, 158
Pickelmeer (Netherlands Antilles) sediments, 116
Pickeringite, 1125
Picrolite. *See* Serpentine
Picromerite, 1124
Picropharmacolite, 935
Piedmont deposits, 544
Piedmont Plateau, 475
Pietricikite, 827
Piezometric contour map, 43
Piezometric head, 372
Piezometric surface, 485, 537, 539. *See also* GROUNDWATER
Pig iron, 605
Pigeonite, 899
Pigments, 151, 816, 830, 981
Pikes Peak, Colo., 448
Pilbarite, 962, 1184
Pillow lava, 555
Pimelite, 790
Pinakiolite, 87
Pinewood tar, 499
Pinnoite, 87
Pintadoite, 1237
Pirssonite, 142
Pisanite, 1124
Pisidium, 660
Piston corer, 587
Pitchblende, 9, 447, 779, 962, 1184, 1219, 1223, 1227
Pitticite, 939
Placer, placer deposits, 168, 466, 467, 489, 961, 993, 1093, 1189, 1192
Plagioclase (feldspar), 17, 58, 101, 103, 146, 149, 203, 566, 579, 679, 899, 901, 902, 904, 1030
Plagioclase group, 1097
Planar structures, 274
Plane-polarized light, 810
Planetary defluidization, 271, 276
Planets, 398, 405
Planktogenic ooze, 660
Plankton, 48, 76, 135, 822, 1012. *See also* Vol. I
Plankton bloom, 818
Planktonic photosynthesis. *See* Vol. I and Vol IVB
Plant community, 1239
Plant cover, 548
Plant ecology, 541, 544
Plant indicators, 2
Plant nutrients, 464
Plant nutrition, 593
Plant wax, 499
Plantains, 28
Plants, 498, 1199, 1239
Plaster of Paris, 103
Plastic containers, pollution, 311, 317
Platform facies, 425
Platform sediments, 245, 246
Platiniridium, 763, 957
Platinoids, 399
Platinum, 411, 736, 763, 836, 839, 897, 899, 957, 959
PLATINUM: ELEMENT AND GEOCHEMISTRY, 44, 197, 227, 279, 296, 300, **956**
Platinum group, 394, 466, 598, 1041
PLATINUM METALS, **957**, 1059. *See also* individual entries.
Plattnerite, 844
Playas, 89, 103, 178, 216, 359, 485, 1016. *See also* Vol. III
Pleistocene, 454, 597, 894
Pleistocene bauxites, 26
Pleistocene dating, 984
Pleistocene deposits, 102
Pleistocene lakes, 880, 884
Pleochroic halos, 1219
Pleochroism, 208
Pliocene, 454, 555, 894
Plugs, 555
Plumboferrite, 845
Plumbogummite, 937
Plumbojarosite, 1125
Plutonic iron deposits, 603
Plutonic rocks, 361
Plutonium, 783, 1222
PLUTONIUM: ELEMENT AND GEOCHEMISTRY, 6, 44, 148, 279, **961**
Pluvial epoch, 226, 547
Pluvials, 544
Pm, 403
Pneumatolytic deposits, 10, 41, 83
Pneumatolytic emanations, 147
Pneumoconiosis, 697, 730, 731, 734
Pucherite, 1237
Pocos de Caldas plateau, 489

Podzolic soil, 913
Podsolization, 25. *See also* Vol. VI
Podzols, 28. *See also* Vol. VI.
Poiseuille, 551
"Poising," 842
Poisons, 317, 823
Polar bond, 84, 198
Polar ice, 225
Polarizability, 3
Polarization, 810
Polarizing power, 412
Polarography, 444
Polders, 538, 774
Poleyave, N. I., 455
Policing, 772
Poliomyelitis, 692
Pollen, 858
Pollen, air enrichment, 312, 314
Pollen analysis, 133
Pollucite, 17, 147, 462, 1052
Pollutants, 311, 312, 313, 540
Polluted waters, 1129
Pollution, 309, 772, 1134. *See also* Environmental Pollution
Pollution, atmospheric, 11
Pollution, environmental. *See* ENVIRONMENTAL POLLUTION
Pollution, industrial, 97
Pollution, urban, 15, 99
Pollution effect, 16
Pollution losses, 948
Pollution prevention, 327
Polonium, 963, 987
POLONIUM: ELEMENT AND GEOCHEMISTRY, 44, 279, **962**
Polybasite, 1092, 1094
Polycrase, 846
Polydymite, 1131
Polygorskite, 53, 230
Polyhalite, 101, 667, 1124
Polyhydroxy polycarboxylates, 152
Polymerization, 70, 439
Polymers, 144, 153, 231
Polymignite, 488, 846
Polymorphism, 153, 170, 204, 308, 918
Polymorphs, 373, 683, 1171
Polyphenols, 154
Polyprotic acid, 3
Polysaccharide, 73, 427
Polysulfide solutions, 561
Polysulfides, 185, 575
Polythionates, 575
Pools, 703
Population growth, 187, 254, 310, 690
Population management, 257
Porcupine, Ontario, 468
Pore canals, fossil, 155
Pore fluids, 59, 586
Pore water, 433
Porifera, 113. *See also* Vol. VII
Porosity, 458, 471, 539, 551
POROSITY AND PERMEABILITY, **965**
Porous media, 965
Porphobilinogen, 157
Porphyrin pigments, 71, 150
Porphyrins, 814, 831, 1236
Porphyroblasts, 89
Porphyry copper deposits, 190, 194, 238, 468, 566, 758
Portland cement, 103, 118
Portlandite, 844
Positron emission, 260
Postdiagenesis, 114
Postmagmatic solutions, 572
Potable waters, 1247
Potarite, 897, 958
Potash, 23, 605, 971
Potash alum, 1124
Potash salts, 356
Potassic alteration, 567
POTASSIUM: ELEMENT AND GEOCHEMISTRY, 16, 37, 38, 40, 44, 63, 75, 147, 178, 185, 216, 245, 247, 279, 300, 365, 369, 374, 418, 433, 446, 461, 465, 475, 587, 588, 593, 837, **970**, 992, 1017, 1019, 1051, 1225
POTASSIUM/ARGON AGE DETERMINATION, 40, 149, 261, 418, **972**
Potassium/argon/rubidium/strontium/uranium/thorium/lead methods, 409
Potassium chloride, 259, 296, 320, 561
Potassium feldspar, 19, 450
POTASSIUM/RUBIDIUM RATIO IN GEOLOGY, 20, 247, **974**
Potassium salts, 353
Potassium/sodium ratios, 273
Potato chip manufacturing pollutants, 321
Potential resource, 772
Potentiometer, 217, 227

Potentiometric surface, 174, 473
Potosi dolomites, Missouri, 65
"Potters asthma," 734
Powder diffraction analysis, 1275
Powell, Major John Wesley, 187
Powellite, 752, 753, 1213
Power generating plants, industrial, 311, 314, 323
Power generation, 549
Power plant effluents, 314, 316
PRASEODYMIUM: ELEMENT AND GEOCHEMISTRY, 44, 279, 300, **976**
Prebiological atmosphere, 29, 126
"Prebiological" equilibrium, 812
Prebiological history, 800
PRECAMBRIAN ATMOSPHERE, GEOCHEMICAL HISTORY, **977**
Precambrian bauxites, 27
PRECAMBRIAN HYDROCARBONS, **980**
Precambrian iron formations, 601, 1195
Precambrian oceans, 270
Precambrian oxidative environment, 333
Precambrian-Paleozoic transition, 272
Precambrian time scale, 455
Precipitation, 351, 361, 409, 471, 481, 539, 541, 544, 1003, 1043, 1046, 1047, 1057, 1071, 1248
Precipitation, chemical, 52, 118
Precipitation chromatography, 160
Precipitation, mineral, 4
Predators, 127
Prehistoric caves, 70
Prehnite, 88
Pre-pleistocene deposits, 104
Pressure, 551, 562
Pressure changes, 37
Pressure head, 42
Pressure sickness, 690
Prevegetation time, 878
Prey-predator equations, 257
Priceite, 87
Primary condensates, 825
Primary consumers, 661
Primary igneous rock, 669
Primary inclusions, 373
Primary production. See Vol. I
Primary volatiles, 702
Primitive atmosphere, 862, 978
Primitive earth, 408
Primitive man, 186, 340
"Primordial" argon, 40
Primordial atmosphere, 849
Primordial helium, 493
Primordial hydrocarbons, 499
Primordial isotopes, 991
Primordial volatiles, 837
Primordial xenon, 1170
Priorite, 846
Pristane, 498
Probertite, 87
Procuticle, 154
Productivity, 946
Profundal zone, 660
Proline, 72
PROMETHIUM: ELEMENT AND GEOCHEMISTRY, 44, 279, **983**
Propane, 165, 496, 812
Proportional counting, 988
Propylene, 165
Propylitic alterations, 567
Prosopite, 491
Prospecting, 706. See also Vol. IVB
PROTACTINIUM: ELEMENT AND GEOCHEMISTRY, 6, 44, 279, 963, **983**
Protactinium/ionium dating method, 597
PROTACTINIUM/THORIUM DATING METHOD, 597, 894, **984**
Proteases, 154
Proteins, 29, 73, 136, 154, 157, 231, 312, 416, 1127, 1202
Proterozoid, 850
Protista, 113. See also Vol. VII
Protium. See hydrogen
Protoatmosphere, 799
Protochordates, 114. See also Vol. VII
Protoearth, 32, 399
Proton transfer, 3
Protons, 43, 266
Proto-petroleum, 1156
Protoporphyrin, 157
Protoplanets, 398
Protosun, 398
Protozoa, 77, 136, 159, 213, 660
Proustite, 42, 1092
Psammon, 660
Psartite, 829
Pseudoboléite, 490
Pseudobrookite, 846
Pseudocotunnite, 491

Pseudomalachite, 936
Pseudomorphic replacement, 940
Psilomelane, 49, 58, 63, 674, 845
Psychology. See water divining
Psychogeology. See water divining
PT, 407
Pteridines, 834
Ptygmatic folding, 112
Puerto Rico, 474
Pumice, 555, 735
Puntiagudo, Chile, 12
Pure solutions, 907
Purine bases, 71
Purpurite, 934
Pyconometer, 964, 1113
Pyrargyrite, 35, 1092
Pyrethrum wax, 499
Pyridine, 67
Pyrimidine bases, 71
Pyrite, 35, 51, 58, 61, 64, 77, 190, 197, 208, 231, 358, 425, 433, 438, 448, 468, 547, 558, 599, 604, 725, 900, 931, 959, 993, 1016, 1079, 1118, 1131, 1133, 1134, 1148, 1172
Pyritic shales, 1041
Pyritic gold-quartz, 468
Pyritic sandstones, 271
Pyrobelonite, 936
Pyrobitumens, 824
Pyrochlore, 794, 846
Pyrochroite, 844
Pyroclastics, 555
Pyrolite, 680
Pyrolusite, 58, 844, 902
Pyrolysis, 1182
Pyromorphite, 937, 944
Pyrope, 904
Pyrophanite, 844
Pyrophyllite, 1, 86, 562
Pyropissite, 826, 829
Pyroretine, 827
Pyrosilicates, 202
Pyroxenes, 17, 22, 49, 57, 101, 146, 180, 190, 202, 206, 386, 426, 429, 449, 462, 579, 582, 599, 679, 680, 683, 901, 1084, 1196
Pyroxenehornfels, 904
Pyroxenites, 156, 1090
Pyrrhotite, 35, 190, 198, 208, 362, 558, 599, 725, 900, 959, 1042, 1059, 1130, 1134, 1137
Pyrrhotite-pyrite thermometer, 747
Pyrrole ring, 150, 156

Qatarra Depression, 548
Quantum mechanics, 1173
Quartz, 1, 19, 31, 35, 58, 64, 130, 196, 197, 203, 234, 274, 373, 418, 429, 432, 435, 436, 438, 462, 468, 562, 576, 579, 580, 591, 735, 810, 899, 900, 901, 902, 904, 993, 1084, 1086, 1091, 1246
Quartz-beryl veins, 148
Quartz-monozonite, 194, 238, 566
Quartz-muscovite, 562
Quartz porphyry, 488, 606
Quartz-wolframite veins, 148
Quartzine, 438
Quartzite, 471, 489, 734
Quartzose clay, 25
Quasicratonic crust, 248
Quaternary bauxites, 26
Quaternary period, 339, 544, 547, 555, 894, 987. See also Vol. III
Quebec-Labrador iron belt, 603
Queensland, 360
Queen's metal, 35
Quenselite, 845
Quenstedtite, 1125
Quick clays, 260. See also Vol. VI
Quicklime, 100
Quicksilver, 987. See also mercury
Quinones, 834
Quisqueite, 1207

Ra-493
Racine Dolomite, Illinois, 120
Radiant energy, 36, 1114
Radiation, 457, 543, 847
Radiation damage study, 231
Radiation emission, 988
Radiation logging, 461
Radiesthesia, 1253
Radioactinium, 10
Radioactive "age," 595
Radioactive contamination, 1066
Radioactive decay, 18, 32, 129, 454, 594, 983, 1054, 1071, 1222, 1225
Radioactive gas, 1006
Radioactive heat, 1077
RADIOACTIVE ISOTOPES, **987**, 991

Radioactive isotope dating, 991. See also GEOCHEMISTRY
Radioactive isotope tracer technology. See also Vol. VI
Radioactive nuclides, 1020
Radioactive ore deposits, 991. See also thorium; uranium
Radioactive pollutants, 314
Radioactive tracers, 1007
Radioactive waste, 326
Radioactive waste in the ocean, 991. See also Vol. I
Radioactivity, 398, 780, 987
RADIOACTIVITY IN ROCKS, **991**
RADIOACTIVITY IN SEDIMENTS. See Vol. VI
Radioactivity in surface waters, 326
Radiocarbon, 998
Radiocarbon dating, 451, 996. See also CARBON-14 DATING
"Radiogenic" argon, 40
"Radiogenic" helium, 493
"Radiogenic" nuclides, 1054
Radiogenic strontium, 1053
Radioisotopes, 1002
Radiolaria, 53, 113, 216, 422, 432, 437, 1081. See also Vol. VII
Radiometric age determination, 1217
Radiometric time scale, 454
Radionuclides, 988
RADIONUCLIDES: COSMIC-RAY-PRODUCED, 326, **996**. See also RADIONUCLIDES IN THE OCEANS, Vol. I
Radionuclides in oceans and sediments, See Vol. I
RADIUM: ELEMENT AND GEOCHEMISTRY, 16, 44, 62, 106, 148, 279, 319, 320, 326, 963, 987, 993, **1006**, 1068, 1222
Radium Hill, Australia
RADIUM IN THE OCEANS, **1008**
Radium-ionium method, 1011
Radium sulfate, 320
RADON: ELEMENT AND GEOCHEMISTRY, 44, 315, 279, 411, 963, 1039, **1014**, 1228
Rafaelite, 1206
Rain, 1243
Raindrops, 409
Rainfall, 15, 430, 437, 546, 1057. See also PRECIPITATION (RAINFALL), Vol. II
Rain-shadow deserts, 545
Rain-splash, 1101
RAINWATER, 271, 572, 716, **1015**
Ralstonite, 491
Raman spectroscopy, 498
Rammelsbergite, 42, 790, 1131
Ramsdellite, 49, 674
Rancieite, 49
Ranco Puyehe, Chile, 12
Rand, South Africa, 1223
Ransomite, 1125
RARE EARTHS (LANTHANIDE SERIES), 6, 45, 53, 64, 106, 111, 145, 240, 245, 270, 342, 349, 401, 425, 488, 494, 664, 777, 976, **1020**, 1060
RARE EARTHS IN BASALTS, 146, **1029**
RARE (NOBLE, INERT) GASES, 21, 39, 45, 261, 396, 492, 849, **1039**
Rare gas retention, 1270
Raspite, 752, 1213
Rate of erosion, 1268
Rate phenomena, 274
Rationalists, 772
Rauvite, 1237
Rayleigh, Lord, 454
Rb, 406, 493
Rb-Sr, 455
Rb-Sr dating, 449
Rb^{87}-Sr^{87}, 447, 448
Re^{187}-Os^{187} dating, 446
Reaction rates, 728
Reaction rim. See Vol. V
Reaction series, 898
  Bowen, 430
Reactivity, 1096
Real del Norte, Mexico, 1094
Realgar, 42, 900, 1131
Recent (Holocene), 351
Recent bauxites, 26
Recent coral reefs, 114
Recent sediments, 101, 104
Recharge waters, 548, 1159
Reciprocal lattice, 1274
Recombination, chemical, 226
Recording thermometers, 375
Recreational areas, 1116
Recrystallization, 52, 59, 116, 226, 539, 557
Recycling, 661, 669, 955, 1016, 1099
Recycling, environmental, 332
Red algae, 113, 114
Red-bed copper deposits, 190, 194

INDEX

Red blood worms, 660
Red clay, 64, 587
Red deep-sea clays, 35, 52, 246
Red marble, 120
Red Mountain, Colorado, 1094
Red ochre, 604
Red Sea, 49, 112, 360, 544, 571, 587, 869, 1035, 1062, 1138
Red Sea (Atlantis II Deep), 766
Red Sea brines, 271
Red shales, 268
Red sunsets, 545
Red tide, 822
Reddingite, 935
Redingtonite, 1125
Redox potential, 37, 423, 432, 840, 932, 1205
Reduction, 356, 672, 839, 932, 1040, 1145, 1266.
    See also OXIDATION AND REDUCTION
Reduceates, 433
Reducing atmosphere, 32, 849
Reductants, 924
Reduzates, 83, 180
Reefs, 114, 480
Reef limestones, 1203
Reflection, 583
Reflectivity, 543
Reforming, catalytic reaction, 144
Refraction index, 85
Refractories, 24
Refractory coatings, 579
Refractory material, 119
Refractory minerals, 898
Refuse disposal emissions, 316
Regenerated isotopes, 991
Regional facies, 904
Regional metamorphism, 709, 903
Regional tectonics, 420
Regolith, 911, 1040
Regolith and saprolite. *See* Vol. III
Reinite, 1212
Reich, 578
Relative density (specific gravity), 1110
Relative permeability, 968
Relict seas, 351
Renardite, 938
"Renewable" resources, 187, 547
Renewable and unrenewable resources, 771
Renierite, 462
Replacement deposits, 1093
Replacement reactions, 558
Reproduction, 946
Republic of South Africa, 35
Reservoirs, 478, 557
Reservoir rocks, 70, 116, 965
Resetting of isotopic "clock," 1054
Residence times, 11, 75, 211, 1004, 1009, 1071
Residual bond, 200
Residual deposits, 64, 670, 901
Residual fluids, 88
Residual sediments, 432
Resin, 498, 811, 826, 828, 1040
Resin and amber, 176. *See also* Vol. VI
Resistant minerals, 902, 1265
Resistates, 432, 1293
Resistate sediments, 180, 430
Resonance spectroscopy, 261
Respiration, 37, 68, 77, 108, 848, 854, 862, 739
Restricted basin ("barred" basin), 353
Retene, 499
Retgersite, 1124
Retzian, 936
Reverse osmosis (RO), 222
Reversed weathering, 1072
Reversible transformation, 379
Rhabdomancy, 1253
Rhabdophane, 936
RHENIUM: ELEMENT AND GEOCHEMISTRY, 44, 279, 300, **1040**
Rheumatism, 703
Rhine River, 473
Rhodesia, 960
RHODIUM: ELEMENT AND GEOCHEMISTRY, 44, 277, 279, 300, 959, **1041**
Rhodizite, 87, 147
Rhodochrosite, 58, 142, 900
Rhodovioloscin, 71
Rhomboclase, 1123
Rhone Glacier, Chamonix, France, 465
Rhone River, 473
Rhyolite, 83, 248, 273, 391, 555, 582
Rias, 344. *See also* Vol. III
Rice, 28
Richellite, 938
Rickardite, 1130
Ridgidity of rocks, 459
Riebeckite, 1097
Ring diffusion, 236
Ring structure, 201

Ringwood, 415
Rinneite, 491
Ritter Island, 12
Rivers, 378, 436, 470, 595, 667, 949, 1087, 1243
River-basin phenomena, 585
River-flow, 345
River geochemistry, 101, 148, 155, 271
RIVER GEOCHEMISTRY: ENVIRONMENTAL FACTORS, 312, **1042**
RIVER GEOCHEMISTRY: REGIONAL, 36, **1045**
River waters, 606, 670, 1071, 1244
Rn, 493
Roads, 478
Robe River, 606
Rock alterations, 575
Rock crystallization, 387
Rock dust disease, 697
Rock flour, 178, 463
Rock-forming minerals, 23, 36, 99, 146, 418, 458, 485, 489. *See also* Vol. IVB
Rock forming silicates, 21
Rock magnetism, 460
Rock salt, 155
Rockbridgeite, 937
Roemerite, 1125
Roentgenite, 143
Roesslerite, 935
Rogers, 392
Rogers City Limestone, Michigan, 120
Rome formation, Georgia, 65
Roméite, 34
Rooseveltite, 935
Roots, 499, 914
Root zone, 1249
Rope, 452
Rosasite, 143
Roscherite, 939
Roscoelite, 1236
Roselite, 935
Rosenbuschite, 488
Rose petal wax, 499
Rossite, 1237
Rotary drilling, 551
Rotary polarization, 810
Rotational energy, 891
Rotifers, 660
Row crops, 313
Roweite, 87
Roy and Roy, 440
Rubble, 459
RUBIDIUM: ELEMENT AND GEOCHEMISTRY, 16, 44, 76, 147, 247, 277, 279, 300, 420, 446, 992, **1050**, 1225
RUBIDIUM: ECONOMIC GEOLOGY, **1051**
Rubidium muscovite, 30
RUBIDIUM-STRONTIUM DATING METHOD, 19, 149, **1052**
Rubidoux Mountain (granodiorite), (California), 146, 241, 494, 665, 777, 1027
Ruby, 23
Rumen, 159
Ruminants, 75
RUNOFF, 121, 313, 351, 481, 546, 555, 595, 716, 876, 1043, **1057**, 1101, 1159, 1211, 1248
Russell, I. C., 880
Russellite, 844, 1212
Russian platform, 418
Rust, 930
RUTHENIUM: ELEMENT AND GEOCHEMISTRY, 44, 277, 279, 300, 959, **1058**
Rutherford, Lord, 454, 795
Rutherfordine, 143
Rutile, 58, 489, 683, 794, 844, 899, 1162, 1193, 1235

S, 407
Sabkha, 359, 536
Saddle reefs, 468
"Safe levels" (radioactive elements in water), 703
Safflorite, 42, 1131
Sahara Desert, 542, 547
Sahlinite, 936
Sakurashima, Japan, 12
Sal ammoniac, 155, 490
Saléeite, 939
Salem limestone, Indiana, 120
Salesite, 590
Salina, 359
Salina formation, 1098
Salinas, 351, 548
Saline areas, 1018
Saline compounds, 426
Saline groundwaters, 593
Saline lakes, 587
Saline marshes, 548
Saline soils, 1103
Salines-geochemistry, 274
Saline Water Act, U.S., 219

Saline water conversion solar still (diagram), 220
Salinity, 90, 110, 122, 344, 424, 458, 1046, 1062
gradients, 588
Salinization (of soils and groundwater). *See* Vol. VI
Salmonid fish, 660
Salmonsite, 935
Saltation, 548
Salton sea, California, 511, 573, 766, 869
Salt deposits, 2, 103, 156, 427, 430, 1078
Salt Desert iron, 542
Salt domes, 78, 214, 1098, 1143, 1146
Salt lakes, 548
Salt nuclei, 351, 359. *See* CYCLIC SALTS; *Also* SALT NUCLEI, Vol. II
Salt pans, 351, 359
Salt particles, 1243
Salt plugs, 588
Salts, 353, 374, 482, 492, 547, 1015
Salt sieving, 172
Salt-water encroachment, 458, 476, 554
Salt-water wedge, 346
SAMARIUM: ELEMENT AND GEOCHEMISTRY, 44, 279, 300, 350, 489, **1060**
Samarskite, 846, 1028, 1162
Sampleite, 938
Sanbornite, 62
Sand, 432, 458, 459, 471, 480, 601
blasting, 314
dune belt, 536
and gravel aquifers, 473
Sandstone, 29, 80, 90, 100, 128, 147, 156, 246, 274, 387, 432, 440, 447, 449, 459, 488, 552, 606, 734
aquifers, 474
Sanidine, 361, 447, 904
Sanidinite, 904
San Marcos gabbro, 146, 494, 665, 1027
Sanmartinite, 752
Santa Barbara basin, 1135
Santa Barbara oil seepage, 321
Santa Maria, Guatemala, 12
Saponite, 790. *See also* montmorillonite
Sapphirine, 845
Sapropel, 335, 1205. *See also* Vol. VI
Saprolite, 430, 475, 1040
Sapropelites, 754, 1203, 1236
Sarcopside, 937
Sargasso Sea, 818, 822
Sarkinite, 937
Sarmientite, 939
Sartorite, 42
Sassolite, 88
Satellite construction, 579
Satellites, 544
Saturometry, 587
Saturated zones, 470
Sauconite. *See* montmorillonite
Saudi Arabia, 547
Sauerstoff, 2
Savanna, 542
SCANDIUM: ELEMENT AND GEOCHEMISTRY, 44, 279, 300, 1020, **1061**
Scacchite, 490
Scaphopods, 114. *See also* Vol. VII
Scapolite, 23, 155
Scarlet fever, 692
Scatter (heat), 12, 15
Scavenger, refining, 104
Scavengers, marine, 70
Schacher, 392
Schafarzikite, 34
Schairerite, 1125
Scharizite, 829
Scheele, K. S., 753
Scheele, C., 377, 795
Scheelite, 101, 752, 754, 900, 1212
Schérerite, 828
Schists, 23, 156, 168, 248, 449, 464, 713
Schistosomiasis, 693, 700
Schizophrenia, 691
Schmeider, 486, 487
Schmeiderite, 1079
Schneeberg, 792
Schreibersite, 679
Schlumberger configuration, 458
Schoepite, 844
Schottky defect, 236
Schroeckingerite, 143
Schuettite, 704
Schuchert, Charles, 454
Schultenite, 934
Schulze-Hardy rule, 260
Schwartzembergite, 590
Schwazite, 705
Scientific Committee on Water Research, 586
Scintillation, 780
Scintillation counting, 988

INDEX

Sclerites, 154
Scleretinite, 826
Sclerotization, 154
Scorodite, 935
Scorpion, fossil, 155
Scorzalite, 937
Scrap materials, 311
Se, 407
Sea, 420, 818, 822
Sea fans, 113
Sea floor spreading, 271
Sea level, 539
Sea level rise, 344
Sea salt, 433, 547, 1046
Sea spray, 271, 1043
Seamanite, 87
Searles Lake, Calif., 89, 133, 664, 884
Seasonal cycles, 1046
Seasonal precipitation, 879
SEAWATER, CHEMISTRY, 20, 35, 38, 41, 49, 50, 63, 69, 72, 75, 78, 88, 93, 95, 101, 108, 113, 121, 124, 146, 150, 155, 174, 184, 190, 209, 219, 241, 260, 271, 343, 350, 353, 354, 378, 384, 407, 435, 459, 462, 466, 495, 498, 587, 601, 606, 667, 765, 1008, 1028, 1056, 1061, **1062**, 1109, 1110, 1122
SEAWATER, GEOCHEMICAL BALANCE, **1070**
SEAWATER, HISTORY, **1076**
Seawater sulfates, 79, 425
Sea wax, 826
Sebca, 536
Sebkha, 359, 548, 764
Second law of thermodynamics, 305, 379
Secondary enrichments, 1129, 1205, 1213
Secondary inclusions, 373
Secondary minerals, 101
Sedimentation rates, 1011
Sediments, 20, 90, 95, 313, 409, 417, 420, 430, 435, 498, 595
Sediments, arenaceous, 90
Sediments, argillaceous, 90
Sediments, cosmogenous, 433
Sediments (tropical region), 432
Sediment barriers, 313
Sediment squeezers, 587
Sediment-water interface, 356, 436, 1075
Sediment-water-plant interaction, 586
Sediment yield, 876
Sedimentary cycle, 415
Sedimentary deposits, 599
Sedimentary metamorphic rocks, 586
Sedimentary ore deposits, 268
Sedimentary paragenesis, 898. *See also* PARAGENESIS; *also* Vol. VI
Sedimentary rocks, 69, 80, 146, 156, 243, 244, 274, 278, 343, 350, 378, 428, 462, 495, 498, 572, 590, 600, 1026, 1100
Sedimentary rock resistivity, 458
Sedimentary series, average composition, 246
Sedimentary silica, 434
Sedimentation, 20, 186
  rates, 10, 16, 596, 1075
Seeds, 499
Segregation, mineral, 119
Seiches, 348. *See also* Vol. I
Seismic refraction methods, 457
Seismic shear waves, 459
Seismology, 459
Selectivity orders, 593
Selenites, 589, 717. *See also* SELENATES AND TELLURATES; SELENITES AND TELLURITES
SELENIUM: ELEMENT AND GEOCHEMISTRY, 41, 44, 79, 83, 198, 204, 280, 300, 320, 736, 763, **1079**
Seligmannite, 42
Sellaite, 490
Semenenko, N. P., 455
Semiarid zones, Mediterranean, 544
Semiarid climate, 877
Semiarid soils, 429
Semiarid region groundwater. *See* HYDROLOGY, SEMIARID REGIONS
Semiarid regions, 541, 1101, 1103
Semi-artesian water, 42
Semimetals, 198, 763
Semipermeable membranes, 594
Senarmontite, 34, 844
Sengierite, 1237
Sepiolite, 53, 178, 230
Sequestration, 150
Seral succession (ecological, environmental), 252
Sericite, 5, 238, 468, 565. *See also* Illite
Serine, 72
Serpentine, 55, 76, 120, 230, 605, 667, 904. *See also* Chrysotile
Serpentinized ultrabasics, 88
Serpierite, 1126
Serpulids, 114

Sesame (aluminum tolerance), 28
Sesquioxides, 913
Seston, 71, 158
Setae, fossil, 155
"Seventh Approximation," 913
Sewage, 949, 1129, 1150
  disposal, 948
  pollution, 76, 312
  treatment, 317, 319, 326
Shady Formation, Georgia, 65
Shales, 20, 23, 29, 35, 80, 89, 100, 128, 147, 150, 156, 168, 180, 245, 274, 427, 432, 435, 459, 475, 552, 593, 799
Shale membranes, 174
Shaler, Nathaniel Southgate, 187
Shan Plateau, Burma, 474
Sharpite, 143
Shatter cones, 274
Shatter zones, 65
Shchukarev, 587
Shear waves, 195
Shearing stress, 57, 903
Sheet silicates, 22, 101, 176
Sheet structures, 203
Shell hash, 120
Shells, 447
Sherwoodite, 1237
Shields, continental, 245, 248, 250
Shield volcanoes, 1038
Shinarump conglomerate (Triassic), 1223
Skinkolobue, Belgian Congo, 1222
Shock mechanism, 274
Shortite, 111, 142
Shrinkage stopping, 64
Shtyubelya, Kamchatka, 12
Shungite, 828, 1203
Sial, 243
Sialic crust, 21
Sickle cell anemia, 691
Sicklerite, 934
Siderazot, 29
Siderite, 1, 35, 52, 58, 61, 64, 112, 142, 599, 601, 604, 606, 679, 753, 900, 1119
Siderolites, 679
Sideronatrite, 1126
Siderophile elements, 47, 212, 395, 399, 411
Siderosis, 736
Siderosphere, 395
Siderotil, 1124
Siegburgite, 826
Siegenite, 1131
Sierra Club, 188
Silcrete, 216
Silexite, 438
Silica, 37, 38, 53, 245, 248, 260, 262, 320, 343, 384, 422, 429, 430, 462, 464, 475, 555, 587, 580, 1107
Silica alumina, 605
SILICA—BIOGEOCHEMICAL CYCLE, **1080**
Silica budget, 1082
SILICA CYCLE, 216, 1065, 1081, **1082**
Silica gel, 5, 496, 1086
Silica geochemistry. *See* GEOCHEMISTRY OF SEDIMENTARY SILICA
Silica glass, 591
Silica minerals, 231
Silica, organic, 92
SILICA SOLUBILITY, 436, **1085**
Silicates, 399, 419, 433, 462, 467, 557, 572, 573, 578, 583, 599, 718
SILICATES, HYDROTHERMAL SYSTEMS, **1088**
Silicate analysis, 409
Silicates of iron, 604
Silicate melts, 375
Silicate minerals, 600. *See also* MINERAL CLASSES: SILICATES
Silicate networks, 591
Silicate rocks, 18, 405, 561
Silicate—sulfide deposits, 99
Silicatosis, 730
Siliceous deposits, 1091. *See also* Vols. IVB, VI
Siliceous oozes, 1081
Siliceous sponges, 437
Silicic acid, 5, 703, 718, 1065, 1084
Silicic rocks, 35
Silicification, 570, 1092
Silicified wood, 440
Silico flagellates, 1081
Silicomolybates, 436
SILICON: ELEMENT AND GEOCHEMISTRY, 20, 36, 44, 75, 178, 196, 247, 264, 280, 300, 365, 396, 399, 422, 432, 440, 465, 600, 718, 1065
Silicon—oxygen bonds, 579
Silicosis, 697, 730, 734
Sillénite, 844
Sillimanite (mullite), 22, 735, 904, 1091
Sillimanite zone, 89
Sills, 555

Silting, 311, 313, 475, 480
Siltstone, 447, 449
Silurian, 272, 352, 454, 858
Silurian fossils, 155
Silurian salt deposits, 156
SILVER: ELEMENT AND GEOCHEMISTRY, 34, 44, 64, 75, 79, 82, 106, 191, 280, 300, 301, 320, 394, 396, 399, 411, 466, 574, 737, 763, 900, **1092**
SILVER, ECONOMIC DEPOSITS, **1093**
Silver iodide, 549
Silver salts, 319
Silver Springs, Florida, 1116
Silver sulfosalts, 900
Sima, 243, 666
Simonellite, 811
Simpsonite, 846, 1162
Sincosite, 1237
Sink-functions, 1000
Sinkholes, 539
Sinter deposits, 436
Sinter feed (pyrites), 604
Sintering, 923
Si, 02, 579
Siserskite, 763, 836
Sittampundi complex, 711
Skaergaard intrusion, 90, 662
Skarniron ores, 68
Skarns, 559, 709, 1091
Skeletal crystal growth, 373
Sjögrenite, 845
Skutterudite, 42, 790, 1132
Slag, 605, 1182
Slate, 489, 713, 734
Slavikite, 1126
Slichter, 551
Slaughter house pollutants, 321
Sludge, 330
Slurres, 319
Smallpox, 693
Smaltite, 1132
Smectite, 53, 176. *See also* montmorillonite
Smith, 572
Smithsonite, 107, 142
Smithstone, 578
Smog. *See* FOG, SMOG, MIST, Vol. II
Smoke, 259
SMOW (Standard Mean Ocean Water), 225, 864
Smythite, 131
Sn, 403
Snails, 154
Snake River Plain of Idaho, 474, 555
Snow, 225, 545, 1000. *See also* Vol. VI
Snowflakes, 409
Soap holes, 485
Soapstone. *See* Talc
Soda, 23, 83, 605
Soda alum, 1124
Soda ash, 156
Sodalite, 155, 1097
Soda-niter, 29, 902
SODIUM: ELEMENT AND GEOCHEMISTRY, 16, 37, 38, 44, 75, 126, 185, 200, 216, 234, 245, 271, 280, 300, 320, 365, 369, 374, 396, 408, 433, 434, 437, 465, 475, 561, 569, 589, 593, 1017, **1096**
  bromide, 354
  budget, 1099
  carbonate, 111, 119, 121
  chloride, 185, 197, 236, 259, 272, 296, 374, 485
  fluoride, 377
SODIUM CYCLE, **1099**
Sodium/potassium ratio, 247, 275
Sodium sulfate, 320, 485
Soerensenite, 274
Soft acids, 3
"Soft" corals, 113
Soft drink bottling pollutants, 321
Soil, 350, 384, 427, 430, 466, 471, 495, 498, 539, 541, 586, 911, 1017, 1044, 1057, 1199, 1233, 1240, 1266
  analysis, 1182
  Chemistry, 27, 71, 78, 101, 108, 146, 178, 180, 190, 215, 231, 343, 547. *See also* Vol. VI
  classifications, 912
  fertility, 672
  formation, 75, 548, 1267
  genesis, 913. *See also* SOIL, Vol. VI
  management, 912
  map, 911
  microflora, 75
  mixture, 1243
  moisture, 471, 482, 544, 546, 1248
  moisture storage, 477
  organisms, 914
  profile, 911
  resistivity, 1257

1317

INDEX

resources, 774
science, 587, 911. See also PEDOLOGY; also Vol. VI
scientists, 546
subtropical, 5
types, 482, 913
water, 1160
SOIL EROSION, 338, 548, **1101**
SOIL SALINITY AND ALKALINITY, **1103**
Sol, 259
Solar abundance, elements, 277
Solar activity. See SUNSPOTS AND SOLAR ACTIVITY, Vol. II
Solar atmosphere, 263
Solar cycle, radiation; solar-climatic relationships; sunspot cycles, 12, 15, 272, 341, 344, 1001. See also Vol. II
Solar energy. See Vol. II
Solar energy spectrum, 579
Solar flare, 812, 1001
Solar insolation, 541
Solar nebula, 979
Solar photosphere, 851
Solar system, 149, 263, 265, 397, 402, 493, 1170. See also Vol. II
Solar wind, 132, 149, 493, 837, 979, See also Vol. II
Solder, 1191
Solids, 36, 495
Solid diffusion, 239, 424, 709
Solid exchange, 592
Solid-gas colloidal system, 259
SOLID SOLUTION, 35, 106, 204, 361, 430, 582, 593, 746, 919, 958, **1104**, 1120
Solid-state diffusion, 365
Solid-state reactions, 143, 361
Solid-state spectrometry, 988
Solid-state track detectors, 148
Solid-solid phase changes, 226
Solidus, 918
Solid waste, 314, 316
Solid waste emissions, 316
Sollas, W. J., 454
Solonchak soils, 548, 1018
SOLUBILITY, 59, 108, 574, 1068, **1106**
SOLUBILITY PRODUCT CONSTANT, 2, 49, 126, 687, **1109**
Solubility of silica, 1083
SOLUSPHERE. See Vol. VI
Solutes, 36, 39, 1245
Solution chemistry, 36
Solutions, 906, 924, 1109, 1266. See also HYDROTHERMAL SOLUTIONS; also WATER SOLUBILITY IN ROCKS, Vol. V
Soluviation, 28
Solvation hulls, 177
Solvay process, 103
Solvent, 1245
Solvus, 362, 364
Sonic Booms, 311
SONOCHEMICAL PROCESSES IN THE OCEAN, **1110**
Soot, 341
Sorby, H. C., 741
Sorosilicates, 101, 201, 720. See also CRYSTAL CHEMISTRY; MINERAL CLASSES: SILICATES
Sorption capacity, 91
Sorting coefficient, 90
Soudan shale, 982
Soufrière, St. Vincent, 12
Sound energy, 315, 317
Sour taste, 2
Source aspects, 419
Souss valley, 549
South Africa, Republic of, 35, 548, 957
South Australia, 441
Souzalite, 938
Sövite, 432
SPA, see Medicinal Springs, 702
Spa therapy, 703
Space exploration, 82
Space heating emissions, 316
Spain, 474
Spallation reactions, 493
Spallation recoils, 149
Spangolite, 1126
Spate-breaker, 549
Spathose iron, 604
Specific discharge, 551
SPECIFIC GRAVITY, 650 **1110**
Specific heat, 307
Specific yield, 39, 472
Spectral analysis, 263, 579
Spectrochemical analysis, 76, 153, 302
Spectrochemical series, 206
Spectrofluorometry, 498
Spectrophotometers, 444
SPECTROPHOTOMETRY, 182, **1114**
Spectroscopic analysis, 367, 444, 497, 688

Specularite, 604
Spencerite, 938
Sperrylite, 42, 956, 958, 1041, 1131
Sphalerite, 58, 64, 99, 365, 462, 557, 578, 705, 900, 1130, 1134
Sphalerite-Pyrrhotite thermometer, 746
Sphene, 17, 149, 367, 567, 899, 993, 1184, 1193, 1217
Spherulitic obsidian, 556
Spike, 973, 1053
Spinel, 17, 21, 167, 207, 604, 683, 845
Spinodes, 364
Spindle top oil field, 551
Spiral arrangement of atoms, 810
Spiroffite, 1079
Spodumene, 17, 22, 462, 661, 663
Spores, 858
Sponges, 102, 113, 118, 216, 422, 437, 1081. See also Vol. VII
Spongy chert, 438
Spontaneous combustion, 1172
Spontaneous fission, 8, 149, 446
Spontaneous fussion, 366
"Spontaneous" nucleation, 51
Spontaneous transformation, 379
Spray, 359
SPRINGS, 20, 26, 37, 77, 88, 101, 126, 226, 326, 470, 482, 485, 548, 549, 555, 702, 944, **1116**
Spring turnover, lakes, 342, 577
SST's, 311, 324
Stabilities of minerals, 923
Stability-crystals solutions, 406
Stability fields, 558
Stable isotope studies, 419
Stable mineral phases, 427
Stalactites and stalagmites, 110. See also Vol. III
Standard enthalpy of formation, 304
Standard-state solubilities, 108
Stanley Shale, Arkansas, 65
Stannite, 191, 578, 1130, 1192
Stantienite, 826
Starches, 231
Starik, I. Ye, 455
Starikova, 587
Stars, 267, 492. See also Vol. II
Stassfurt, Germany, 104, 1051
Statistical thermodynamics, 1173
Statistics—Continents and Oceans. See CONTINENTS AND OCEANS—STATISTICS OF AREA VOLUME, RELIEF, Vol. III
St. Augustine, Alaska, 12
Staurolite, 22, 904
Steady-state conditions, 1173
Steamboat springs, 869
Steatite. See Talc
Steam vents, 900
Steno, N., 740
Stepanovite, 811, 1092
Steel, 605
Steel industry, 378
Steigerite, 1237
Stellar abundances, elements, 277
Stellar evolution, 1022
Stellar interiors, 267. See also Vol. II
Stenothermic organisms, 314
Steppe, 542
STEREOCHEMISTRY AND ENANTIOMORPHISM, 500, **1117**
Stereoisomerism, 1117
Sterline, 35
Sternbergite, 1131
Stercorite, 935
Steroids, 71, 79, 815
Sterols, 154, 500
Sterrettite, 939
Stewartite, 935
Stibiconite, 844
Stibine, 34
Stibiocolumbite, 846
Stibiopalladinite, 897, 958, 1130
Stibiotantalite, 846, 1162
Stibnite, 34, 82, 573, 900, 1131
Stichtite, 168
Stillwater Complex, Montana, 169, 711
Stilpnomelane, 599, 605
Stilpnosiderite, 604
Stishovite, 196, 201, 579, 580
St. Lawrence, 1048
Stocks, 467
STOICHIOMETRY, 436, **1118**. See also Vol. IV
Stokesite, 1192
Stolzite, 752, 1213
Stone, 392
Stone Age, 466
"Stone Age Cultures," 731
Stone, building, construction, statuary, 118
Stony meteorites, 9, 40, 134, 226, 263, 395, 493

Storage coefficient, 39
Storage runoff, 549
Stottite, 462
Stratification, 552
"Stratisphere," (earth's crust), 243
Stratosphere, 311, 326, 408, 1003, 1229. See also Vol. II
Streams, 437, 471, 482, 949
Streamflow, 1057
Strengite, 935
String bean wax, 499
Stromatolites, 126. See also Vol. VII
Stromatolitic algae, 863
Stromatoporids, 118
Stromeyerite, 1092, 1130
Strontian aragonite, 112
Strong acid, 932
Strontianite, 17, 58, 64, 107, 112, 142
STRONTIUM: ELEMENT AND GEOCHEMISTRY, 16, 20, 38, 44, 62, 106, 113, 210, 247, 273, 280, 296, 300, 314, 320, 400, 406, 424, 425, 433, 475, 989, 1051, 1066, **1121**
STRONTIUM CYCLE, **1122**
Structure of the earth, 677
Struvite, 698, 935
Subcrustal fusion, 40
Subhumid belts, 541
Subgeospheres, 268
Sublimation, 170, 303, 308
Sublittoral zones, 660
Submarine aquifers, 588
Submarine exhalations, 606
Submarine slumping, 1160
Submarine springs, 436, 1070. See also Vol. I
Subnitrates, 82
Subnuclear particles, 997
Subsoil, 1233, 1267
Substitution, atomic, 909, 1104, 1200. See also TRACE ELEMENTS IN SILICATE MINERALS, SUBSTITUTION
Subsurface geology, 481
Subsurface water, 471, 474, 550. See also GROUNDWATER; HYDROLOGY
Subtropical soils, 5
Succinyl coenzyme A, 157
Sudbury, Ontario, 193, 792, 957
Suess, 666
Sugar cane, 28, 556
Sugar cane cuticle wax, 499
Sugars, 71, 428, 815
Sulfate, 354, 360, 374, 561, 573, 606, 1017A, 1073
SULFATES, 21, 37, 38, 50, 59, 63, 80, 159, 231, 241, 271, 273, 320, 343, 349, 384, 475, 494, 547, 573, 579, 583, 718, 869, **1123**
SULFATE REDUCTION—MICROBIAL, 51, 77, 80, 185, 214, 1073, **1127**
Sulfide transport, 572
SULFIDES (WITH SELENIDES, TELLURIDES, ARSENIDES, ANTIMONIDESQ, 35, 41, 51, 59, 77, 83, 99, 185, 190, 198, 214, 231, 237, 335, 356, 386, 399, 419, 423, 425, 433, 462, 467, 468, 547, 558, 572, 573, 575, 587, 704, **1129**, 1134
Sulfide bonds, 414
Sulfide complexes, 575
Sulfide-copper ores, 180, 189
Sulfide enrichments, 1093
Sulfide minerals, 574
SULFIDE MINERAL OXIDATION—MICROBIAL, **1132**
Sulfide ores, 578
Sulfide rocks, 599
Sulfide segregates, 791
SULFIDES IN SEDIMENTS, **1134**
Sulfide ore deposits, 396
Sulfide ores, 365
Sulfide-oxide layer, 396
Sulfide-oxide sphere, 395
Sulfide solutions. See HYDROTHERMAL SOLUTIONS—SULFIDE TRANSPORT
Sulfidic lead, 42
Sulfidization, 576
Sulfites, 583
Sulfoborite, 87
Sulfohalite, 1125
SULFOSALTS, 41, 191, 578, **1141**
SULFUR: ELEMENT AND GEOCHEMISTRY, 41, 44, 58, 69, 75, 77, 82, 99, 112, 150, 159, 180, 185, 190, 245, 277, 280, 320, 332, 390, 391, 395, 424, 425, 466, 558, 561, 569, 572, 604, 753, 763, 813, 900, 1016, **1142**
Sulfur bacteria, 77
SULFUR CYCLE, 81, 214, 870, **1148**, 1153
Sulfur dioxide, 272, 312, 316, 332, 334, 1231
Sulfuretum, 1129
Sulfuric acid, 571
SULFUR-ISOTOPE FRACTIONATION IN BIOLOGICAL PROCESSES, **1152**

INDEX

Sulfur oxidation—bacterial, **1158**
Sulfur trioxide, 334
Sulfuric acid, 334, 931, 1158
Sullivan (British Columbia), 1139
Sulphomagnetite, 1137
Sulphur springs, .1129
Sun, 397, 402, 492, 1023
Sun geochemistry, 100, 123
Sunburn, 858
Sunfishes, 660
Sunspots, 132, 326, See also Vol. II
Supergene, 99, 191, 195, 342, **1158**
Supergene enrichment, 1092
Supergene processes, 578
Supernova, 397, 1022
Supernova explosion, 7
Super-oxides, 41
Supersaturation, 1106
Supratidal environments, 102
Surface mapping, 456
Surface runoff, 546
Surface tension, 36
Surficial equilibrium, 727
Surtsey, Iceland, 12
Susannite, 143
Suspended solids, 321
Suspended water, **1159**, 1233, 1264
Sussexite, 87
Svabite, 937
Svanbergite, 939
Swamp, 536
Swamps, 325, 601
Swartzite, 143
Sweden, 602
Swedenborgite, 34
Sweet potatoes, 28
Swiss chard, 499
Syenitic granitic rocks, 494
Syenites, 18, 62, 83, 93, 156, 248, 343, 606
Sylvanite, 466, 468, 1092, 1132, 1164
Sylvite, 17, 421, 447, 490, 902, 971
Symbiotic Nitrogen Fixation, 803
Synadelphite, 936
Synchisite, 143
Syndiagenesis, 49, 178, 1160
Syngenesis, 178, 342, 418, 441, **1160**
Syngenetic-supergene, 1159
Syngenite, 1124
Synthesizing, 923
Synthetic oil, 810
Syphilis, 693
Syria, 548
System, 818, 917, 1173
Szaibelyite, 87
Szmikite, 1124
Szomolnokite, 1124

Taal, Luzon, 12
Table salts-purification, 320
Tabling, 64
Tachydrite, 103
Tachyhydrite, 491
Taconic Mts. belt, 169
Taconite, 435, 603
Taenite, 763, 791, 1195
Tagged isotopes, 107
Tagilite, 938
Tailings, pollution, 320
Talc, 1, 86, 120, 203, 568, 605, 735, 904
Talcosis, 340
Tamarugite, 1124
Tantalite, 794, 846, 1162
Tantalum: element and geochemistry, 35, 44, 147, 274, 280, 300, 489, 763, 909, **1162**
Tanytarsus, 660
Tapajos, 1049
Tapiolite, 846
Tar, 70, 1203
Tar sands. See Vol. IVB
Tar spills, 321
Taranakite, 939
Tarapacaite, 159, 168
Tarawera, N. Z., 12
Tarbuttite, 937
Tauriscite, 1125
Tavistockite, 937
Taylor Glacier, West end of Lake; Bonney, Antarctica, 465
Taylorite, 1123
Tc, 403, 407
Teallite, 1192
Technetium: element and geochemistry, 44, 280, **1163**
Technological Gap, 775
Tectonic silt, 353
Tectosilicates, 62, 101, 176, 203, 462, 720. See also Crystal chemistry; Mineral classes: silicates

Teepleite, 87
Teeth, 697, 698, 989
Teineite, 1079
Tektites, 148, 226, 275, 367, 582, 865, 975, 1026, 1041. See also Vol. II
Tektites, Australasian, 277
Tektites, Ivory Coast, 277
Tektosilicates, 580
Tellurates, tellurites. See Selenates and tellurates: selenites and tellurites
Tellurates, 717
Telluric screw, 43
Tellurides, 466
Tellurite, 589, 717, 844
Tellurium: element and geochemistry, 41, 44, 82, 280, 300, 466, 468, 736, **1164**
Tellurobismuthite, 1130
Temperature changes (fluctuation), 11, 37, 57, 541, 894
Temperature coefficient, 910
Temperature differential, 217
Temperature, mean annual, 11
Temperature measurement, 921
Temperature, urban, 15
Tennant, S., 598
Tennantite, 42, 195, 705
Tenorite, 191, 195, 207, 843, 1137
Teorell-Meyer-Siever Theory, 174
Terbium: element and geochemistry, 44, 280, 300, **1165**
Terlinguaite, 490, 704
Ternary system, 918
Terpenes, 79, 498
Terpinoids, 815
Terra rossa bauxites, 25. See also Vol. VI
Terra rossa soils, 548
Terraces, fluvial-environmental controls. See Vol. III
Terrameter, 459
Terrigenous minerals, 48
Terrigenous sediments, 433
Tertiary, 893
Tertiary bauxites, 26
Tertiary pisolitic limonites, 606
Tertiary sediments, 102
Teschemacherite, 142
Test drilling, 456
Teton Glacier Wind River Mts., Wyoming, 465
Tetracorals, 118
Tetradecane, 496
Tetradymite, 82, 1130, 1164
Tetrahedral coordination, 85, 88
Tetrahedrite, 35, 42, 191, 195, 705, 900
Tetralins, 498
Tetrapyrole pigment, 156
Texas, 551
Textile fiber production, 144
Textural equilibrium, 724
Th, 493
Th^{232} to Pb^{208}, 447
Th^{232} disintegration series, 964
Thailand, 474
Thalenite, 1027
Thallium (RaC"), 963
Thallium: element and geochemistry, 44, 280, 300, 414, **1166**
Thalweg, 483
Theis, 551
Theoleiites, 1027, 1030, See also Tholeiites
Thenardite, 1097, 1123
Therapeutic, 703
Therapeutic role of groundwater. See Medicinal springs
Thermal analysis (thermoanalysis), 153, **1167**
Thermal balance, 543
Thermal conductivity detector, 392
Thermal degradation, 428
Thermal effect, 16
Thermal energy, 317, 593. See also Geothermal energy, Vol, IVB
Thermal facies, 904
Thermal gradient detector, 94
Thermal indicator minerals, 422
Thermal ionizing radiation, 317
Thermal logging, 461
Thermal metamorphism, 903
Thermal neutron cross section, 384
Thermal pollution, 15, 311, 314, **1169**
Thermal radiation, 12
Thermal springs, 359, 436, 561, 702, 1085, 1247. See also Springs
Thermal stability, 30, 1182
Thermistors, 227
Thermobalance, 1175, 1176
Thermobaric gradient, 105
Thermochemistry, **1170**
Thermocline, 577, 1008, **1172**
Thermodynamics, 170, 296, 303, 305, 307, 380,

382, 684, 923, 1068, 1088, 1104, 1170, **1173**
Thermodynamic concentrations, 592
Thermodynamic equilibrium constant, 108, 306
Thermodynamic potentials, 379
Thermogravimetric analysis (tga), 227, 230, 1168, **1175**
Thermoluminescence, 874, 1168. See also Vol. IVB
Thermonatrite, 142
Thermonuclear testing, 1211
Thermo-osmosis, 173
Th-He dating, 446
Thiem, Adolph, 476
Thiamine, 821
Thin-layer chromatography (TLC), 160
Thiobacillus, 77, 729, 806, 1128, 1132, 1150, 1158
Thioelaterite, 829
Thiophenes, 104
Thiorhodaceae, 1150
Thiosulfate, 575
Third law of thermodynamics, 307
Tholeiites, 18, 146, 1037, See also Theoleiites
Tholeiitic basalt, 247, 273, 432
Thomsenolite, 491
Thomsonite, 61
Thoreaulite, 846
Thorianite, 844, 962, 1184
Thorite, 962, 993, 1184
Thorium: element and geochemistry, 6, 44, 145, 149, 245, 247, 267, 280, 300, 488, 594, 783, 963, 993, 1006, **1183**
Thorium dating method. See Protractinium– thorium dating method
Thorium: economic deposits, **1189**
Thornthwaite system, 1248
Thoron, 1006, 1014
Thorotungstite, 1212
Thortveitite, 1027, 1062
Thousand Springs, Idaho, 555, 1116
Threonine, 72
Throat breccias, 555
Thucholite, 779, 825
Thulium: element and geochemistry, 44, 280, 300, **1190**
Thuringite, 605
Ti, 404, 412, 415, 578
Tidal bore, 348
Tidal currents, 346
Tidal current energy, 90
Tidal exchange, 344, 346
Tidal flats, 348
Tiemannite, 704, 1130
Tiff, 63
Tigris River, 474, 547
Tikhomirava, 587
Tilasite, 937
Timberline, 1240
Tin: element and geochemistry, 20, 44, 76, 82, 106, 191, 273, 280, 300, 310, 489, 578, 605, 737, 763, **1191**
Tin granites, 1191
Tin greisens, 578
Tin veins, 41
Tincalconite, 87
Tintic, Utah, 1094
Tinticite, 939
Tire weighting, 104
Titanite, 1193
Titanium: element and geochemistry, 25, 44, 76, 151, 247, 280, 300, 433, 440, 599, 600, 604, 603, **1193**
Titanium dioxide, 67
Titanomagnetite, 1193
Titration, 67, 121, 388
TLC (Thin-Layer Chromatography), 165
Tobacco leaves, 499
Tocantins, 1048
Todilto limestone (Jurassic), 880
Todorokite, 49, 674, 846
Toluene, 496, 498
Tongue of the Ocean (Bahamas), 108
Tonopah, 1094
Topaz, 22, 462, 900
Topographic maps, 480
Topography, 430, 471, 480, 481, 775, 915
Torbernite, 939
Torreyite, 1125
Torry Canyon oil spill, 321
Tourmaline, 58, 88, 430, 432, 462, 468, 886, 899, 900
Toxic concentration, 1080
Toxicity, 1167
Toxins, 76, 818, 822
Trace constituents, 909, 1293
Trace elements: geochemistry, 62, 74, 111, 148, 150, 190, 209, 273, 277, 420, 430, 579, 587, 696, 886, 905, 992, **1194**
Trace elements in plants, 876, **1199**

1319

INDEX

Trace elements in silicate minerals-substitution, 88, 91, **1199**
Trace element studies, 443
Trace metals, 572, 1072
Trace metals in sediments, oils, and allied substances, **1201**
Tracers, 996, 1015
Tracer technology. *See* Radioactive isotope tracer technology, Vol. VI
Trachyte, 350, 384, 494, 555
Transcaucasus, 418
Transient equilibrium, 987
Transition elements, 45, 74, 179, 189, 204
Transition metals, 75
Translational energy, 891
Transmissivity, 552
Transpiration, 28, 108, 471, 481, 545
Transplutonium elements, 7
Transportation minerals, 4
Transportation wastes, 323
Transported sediment, 901
Transuranium series. *See* Actinide series
Trapping temperature, 375
Travertine, 101, 110, 112
Trechmannite, 42
Tree-ring analysis, 131, 135, 451, 905. *See also* Vol. II and VII
Tremolite, 1, 100, 202
Trevorite, 845
Triangular coordination, 87
Triassic, 454, 556
Triassic salt deposits, 156
Trichalcite, 935
Tridymite, 203, 439, 579, 735
Trient, Switzerland, 465
Trigonite, 34
Trihalides, 83
Trihydrocalcite, 142
Trilobites, 155
Trinidad Lake asphalt, 827
Trioxides, 82
Triphylite, 934
Triplite, 937
Triploidite, 937
Tripositive state, 7
Trippkeite, 34
Tripuhyite, 34
Triterpanes, 498
Tritium: element and geochemistry, 315, 409, 572, 989, **1211**
Troegerite, 939
Troilite, 99, 179, 263, 275, 396, 679, 1092, 1146, 1148
Troilite phase, 41, 448
Trona, 884, 1097
Trondhjemite, 714
Tropical forest, carbon cycle, 128
Tropical rain forest, 536
Tropical semiarid climate, 542
Tropical weathering, 23. *See also* Vol. III
Troposphere, 129, 408, 1229. *See also* Vol. II
Trucial Coast, 359
Trudellite, 491
Trypanosomiais, 693
Tryptophane, 72
Tschiatura Kutais (U.S.S.R.), 424
Tsumebite, 938
Tsumeh, Southwest Africa, 386
Tuberulosis, 692
Tubicolous annelids, 118
Tubificid oligochaete worms, 660
Tucholites, 814
Tufas, 77, 101, 447
Tuffs, 449, 593
Tularemia, 694
Tundra, 542
Trunda, carbon cycle, 128
Tunell, 572
Tungomelane, 1212
Tungstates, 349, 384, 494, 583, 717, 752. *See also* Molybdates and tungstates
Tungsten, 565, 737
Tungsten: element and geochemistry, 44, 280, 300, 565, 737, **1212**
Tungstenite, 1132, 1213
Tungstite, 844, 1212
Tunicates, 74. *See also* Vol. VII
Turanite, 936
Turbidimetry, 183
Turbidites, 108. *See also* Vol. VI
Turbulent molecular diffusion, 545
Turgarinov, A. I., 455
Turquois, 938, 944
Tuttle, 440
Two-feldspar method, 746
Tychite, 143
Tycho, 275
Typhoid, 692

Typhus, 694
Tyrolite, 938
Tyrosine, 72
Tyuyamunite, 1223, 1235, 1237

U^{235} disintegration series, 963
U^{238} disintegration series, 963
U-He dating, 446
Ucayali, 1049
Uixan, Morocco, 606
Ukranian shield, 418
Ulexite, 87
Ullmannite, 790, 1131
Ultrabasic rocks, 18, 245, 590, 599
Ultrabasite, 462
Ultracentrifugation, 752
Ultrafiltration, 172, 185
Ultramafic rocks, 156, 168, 274, 296
Ultrametamorphism, 246
Ultraviolet absorption, 1114
Ultraviolet radiation, 326, 334
Ultraviolet wavelength, 301
Umangite, 1080, 1130
Underground storage groundwater, 549
Underground waters, 36, 1014
Undersea springs, 460
Ungemachite, 1126
United Nations, 585
United Nations Dept. of Economic and Social Affairs, (Resources and Transport Div.), 222
UNESCO, 585, 773
United Nations Food and Agriculture Organization (FAO), 585
United Nations Resources and Transport Division, 774
U.S.S.R., 378, 957
  ore deposits, 27
United States, 378, 474, 542
U.S. Bureau of Mines Bulletin, 489
U.S. Geological Survey (USGS), 187
U.S. Geological Survey Water-Supply Paper, 486, 487
U.S. Public Health Service, 326
Units, numbers, constants symbols. *See* Vol. II
Universe, 397, 402, 1023
Unmixing, 361
Unrecrystallized marine shells, 452
Unsaturated hydrocarbons, 495
Unstable equilibrium, 725
U^{235}-Pa^{231} Dating, 446
U^{235}-Pb^{206}, 447
U^{235}-Pb^{207}, 448
U^{238}-Pb^{206}, 447
U^{283}-Pb^{206}, 448
U-Pb age calculations, 454
U-Pb Dating, 447
Upper atmosphere, 11
Upper crust, 462
Upwelling, 111, 121, 210, 216, 941, 1211. *See also* Vol. I
Ural Mts., 606
Uraninite, 447, 448, 844, 993, 1078, 1216, 1223
Uranite, 1041
Uranium, 366, 404, 425, 430, 433, 446, 454, 489, 493, 594, 696, 737, 783, 963, 987, 1184
Uranium decay, 984
Uranium: element and geochemistry, 6, 16, 20, 44, 78, 147, 148, 245, 247, 267, 274, 280, 300, **1215**
  Economic deposits, 76, 271, **1222**
Uranium-helium isotopic age method, **1224**
Uranium mining, 320
Uranium oxide, 78
Uranium/thorium/lead age determination, 149, **1225**
Uranium-vanadium ore deposition, 842
Uranocircite, 939
Uranopilite, 1126
Uranospinite, 939
Uranothorite, 1223
Urban air pollution, 15, **1228**
Urban climate, 14
Urban construction, 313
Urban development, 548, 1229
U^{238}-Th^{230} dating, 446
U-Th-Pb age determinations, 455
U^{238}-U^{234} dating, 446
Urey, H. C., 403, 977
Urey effect, 851
Urey hypothesis, 979
UV spectrum, 851
Uvanite, 1126
Uvarovite, 101, 168

Vacancy diffusion, 236
Vacuum tube degassing, 148
Vadose water, 470, 702, 1159, **1233**

Valence, 3, 84, 410, 928, 1259
Valence bond theory, 415
Valentine limestone, Pennsylvania, 120
Valentinite, 844
Valine, 72
Valle Grande Caldera, New Mexico, 555
Vanadates, 83, 583, 717, 1236. *See also* Phosphates, arsenates, and vanadates
Vanadinite, 937, 1235, 1236
Vanadium: element and geochemistry, 20, 44, 74, 75, 106, 150, 280, 296, 300, 420, 430, 433, 599, 605, 699, 737, **1235**
Vanadium oxysalts, **1237**
Vanadyl, 831
Vandenbrandeite, 844
Van de Graaff generators, 786
Van der Waals force, 84, 160, 181, 196, 200, 259, **1237**
Vanduse Spring, France, 540
Van't Hoff, Jacobus H., 153
Van't Hoff equation, 4, 685
Vanthoffite, 1123
Vapor bubbles, 374
Vapor condensation, 311
Vapor domes, 221
Vapor pressure, 751
Vaporization, 170, 221, 226, 303, 308, 1170
Vapor-phase gas chromatography, 392
Variscan Mt. belt, 169
Variscite, 935
Varulite, 934
Varves, 881
Vascular plants, 252
Vaseline, 828
Vashegyite, 939
Vassallo, 392
Vaterite, 100, 104
Vauquelinite, 159
Vauxite, 939
Veatchite, 87
Vectors, 695
Vegetation, 37, 76, 135, 252, 338, 471, 482, 541, 876, 1043, 1057, 1238
Vegetative effects, 915
Vegetation indicators, **1238**
Vegetative cover, 1101
Vein-filling ore deposits, 64
Venus, 33, 272, 837, 848, *See also* Vol. II
Vermiculites, 176, 230
Vermicular crystals, 60
Vernadskite, 1126
Vernadsky, 587
Versene (EDTA) solution studies, **1241**
Vertebrates, 137
Vesuvius, 12, 493
Veszelyite, 938
Vibrational energy, 891
Vibration frequencies, 45
Villiaumite, 377, 490, 1097
Vinogradov, A. P., 455, 587
Vinyl groups, 156
Vicking Sandstone, Canada, 174
Violarite, 1131
Viruses, water-borne, 312
Viscosity, 95. *See also* Vol. II
Vitamins, 71, 818, 821
Vitamin B_{12}, 151
Vitamin E, 701
Vitrain, 462
Vitravius, Marcus, 476
Vivianite, 935, 944
Voglite, 143
Volatiles, 276, 396, 493, 709. *See also* Vol. V
Volatiles, excess, 245
Volatile acid gases, 122
Volatilization, 93, 270
Volborthite, 936
Volcanic activity, 408
Volcanic activity, submarine, 49
Volcanic ash, 436, 449
Volcanic carbon dioxide, 102, 126
Volcanic chronology, 12
Volcanic domes, 26
Volcanic dust, 11
Volcanic emanations, 156
Volcanic energy. *See* Vol. IVB Geothermal Energy
Volcanic eruptions, 12, 1229
Volcanic exhalations, 26, 88, 122, 194, 676, 979
Volcanic gases, 493, 686. *See* Gases—volcanic
Volcanic glass, 433
Volcanic glass, 53, 433
Volcanic provinces, 1034
Volcanic rocks, 245, 361
Volcanic sediments, 246
Volcanic springs, 1116
Volcanic sublimations, 41
Volcanic terrain, 555

INDEX

Volcanic terrain, groundwater. *See* HYDROLOGY, VOLCANIC TERRAIN
Volcanic tuffs, 547
Volcanism, 127, 134, 428, 576, 1008, 1150
Volcanoes, 837
Volcanogenous minerals, 49
Volkonskoite. *See* montmorillonite
Volta potential, 173
Voltaite, 1124
Volterra prey-predator equations, 257
Voltzite, 1130
Volume of the ocean, 1070
Volumetric analysis, 742
von Laue, M., 741
Vrbaite, 42
Vugs, 111, 238

Wad, 58
Wadden Sea, 819, 1068
Wadis, 544, 549
Wagnerite, 937
Wairakei, New Zealand, 573
Walcott, C. D., 454
Walpurgite, 936
Walsh-Healy Public Contracts Act, 315
Wardite, 938
Warm/hot springs, Hamme, Yarmouh, 547
Warwickite, 87
Wash-out period, 1003
Waste disposal, 310, 313, 328
Waste disposal, incineration at sea, 331
Waste disposal, piping to open ocean, 331
Waste disposal, underground, 319, 332
Waste gases, 549
Waste treatment, 319, 326
Waste waters, 315, 316
Water, 427, 429, 457, 543, 572, 600, 702, 769, 837, 876, 1107, 1116
Water-air surface film, 660
Water-air transport, 432
Water analysis, 369
WATER BALANCE, 232, 542, 544, 1071, **1248**
Water-bearing rocks, 476
Water collecting galleries, 537
Water column, 49
Water, combined, 248
Water of crystallization, 850
Water cycle, 427, 585. *See also* HYDROLOGIC CYCLE
Water deficiency, 545, 1249
Water desalination in developing countries, 224
Water desalination (symposium), 222
WATER DIVINING (DOWSING), **1252**
Water gas. *See* hydrogen
Water drops, 549
Water ionization, 4
Water losses, 876
Water management, 473
WATER—MOLECULE AND STRUCTURE, 36, 85, 181, **1258**
WATER—NONMARINE, **1243**
Water, oxygen-depleted, 315
Water pollution, 1045. *See also* ENVIRONMENTAL POLLUTION
Water power. *See* Vol. VI
Water, pure, 96
Water Quality Act, 1170
Water recovery, 332
Water resources, 585, 773
Water resource planning, 1260
Water of retention. *See* CONNATE WATER
Water saturation, 964, **1259**. *See also* POROSITY
Water science technology, 585
Water-sediment interface, 50, 70, 74
Water storage, 547
WATER-SUBSTANCE AND SOLVENT, 36, 39, 338, **1244**
Water (subsurface). *See* GROUNDWATER, HYDROLOGY; CONNATE WATER; JUVENILE WATER; MAGMATIC WATER; METEORIC WATER; SUSPENDED WATER; VADOSE WATER
Water supply, 470
WATER SUPPLY: ECONOMICS, **1260**
Water surplus, 545
WATER TABLE, 191, 239, 458, 470, 481, 484, 537, 538, 539, 554, 556, 1057, 1233, **1264**
Water table (aquifer), 473
Water temperature, 547
Water vapors, 15, 315, 545, 686, 898, 1071, 1252, 1258
Water vapor, oceanic, 225
Watercraft pollution, 323
Wattevillite, 1124
Waves, 348. *See also* Vols. I and III
Waves break, 1110
Wavellite, 22, 939, 941, 944
Wave propagation velocities, 457

Waves, 498, 824. *See also* Vol. IVB
Weather, 1230
Weather, modification, 549
Weathered surface material, 459
Weathering, 20, 36, 115, 186, 418, 434, 593, 594, 600, 670, 671, 791, 901, 911, 971, 1015, 1070, 1082, 1092, 1107, 1150, 1162, 1185, 1193, 1213, 1216, 1293. *See also* Vol. III
WEATHERING-CHEMICAL, 2, 22, 25, 28, 36, 101, 115, 121, 129, 152, 156, 191, 213, 272, **1264**
Weathering rates, 877
Weathering, organic. *See* Vol. VI
Weathering processes, 948
Weathering residues of igneous rocks, 418
Weathering, Tropical, 23. *See also* TROPICAL WEATHERING, Vol. III
Weaver, 420
Weberite, 491
Weddellite, 698, 811, 829
Wegscheiderite, 142
Weight, 1182
Weinschenkite, 936
Weisner Formation, Georgia, 65
Welded tuffs, 555
Wells, 470, 478, 537, 544, 547
Well construction, 550
Well-drilling muds, 66
Well logging, 785
Wellington Formation (Permian), 880
Wells, Water. *See* Vol. VI
Wells, water. *See* Vol. VI
Wenner configuration, 458
Werner, A. G., 740
Westphal balance, 1113
"Wet-chemicals" qualitative analysis, 443
Wet granite, 562, 563
Wetting-drying, 914
Wherryite, 1126
Whewellite, 698, 811, 829
White copper, 792
White Sands National Monument
Whitlockite, 149, 934, 940
Whole metamorphic rock (geochronometry), 447
Whole rocks, 1053, 1055
Whole volcanic rock, 447
Widmanstätten figures, 791
Wiechert-Gutenberg discontinuity, 679
Wiikite, 1027
Wilberforce, Ontario (Grenville Province), 448
Wilkeite, 937, 1062
Wind, 429, 545
Wind erosion, 541, 1101, 1229
Windmills, 771
Winds-principles, local; wind action, 344. *See* resp. Vols. II and III
Wind River Formation (Eocene), 1223
Wind strength, 544
Wind wells, 545
Wine, 446
Witherite, 17, 58, 62, 107, 142
Witwatersrand districts, 469
Wöhler, Friedrich, 152
Wöhlerite, 488
Wolfeite, 937
Wolfram. *See* Tungsten
Wolframite, 148, 752, 794, 899, 900, 1062, 1212
Wolaston, W. H., 741, 897, 956, 1041
Wollastonite, 1, 101
Wood, 447, 452
Wood analysis, 131
Woodruffite, 846
Woodwardite, 1126
World Health Organization (WHO), 326, 585
World mapping, 773
World Meteorological Organization, 585
Worms, 70, 914
Woodhouseite, 939
Work, 1170
Work content, 379
Worms, parasitic, 312
Wright Glacier, Wright tongue of Wilson Piedmont Glacier, Antarctica, 465
Wulfenite, 752, 753, 1213
Wurtzite, 1130
Wurzilite, 827
Wustite, 604, 842

Xanthoconite, 42
Xanthocroite, 99
Xanthophylls, 830
Xanthoxenite, 939
XENON: ELEMENT AND GEOCHEMISTRY, 8, 44, 272, 280, 411, 849, 1039, **1270**
Xenophyophora, 50, 1202

Xenotime, 241, 343, 384, 481, 664, 934, 1027, 1165, 1190, 1219, 1290, 1291
Xingu, 1049
X-ray, 718, 759, 988, 1114
  analysis, 584
  crystallography, 498
  diffraction, 438, 445, 489, 740, 752, 921, 1168
  diffractometer, 583
  emission analysis. *See* X-RAY SPECTROSCOPY
  fluorescence, 409, 742, 912
  fluorescence spectrometer, 445
  medical, 326
  spectrography, 145, 343, 384, 494
X-RAY DIFFRACTION ANALYSIS, 89, 230, 565, **1271**
X-RAY DIFFRACTION—VARIABLE ATMOSPHERE, **1280**
X-RAY DIFFRACTION—VARIABLE TEMPERATURE, **1281**
X-RAY SPECTROSCOPY, **1285**
X-Y recorder, 227

Yams, 28, 693
Yangtze, 1048
Yampi Sound, Western Australia, 606
Yasuda, 392
Yellow fever, 694
Yellow ochre, 604
Yellowstone National Park, 188, 556, 571, 869, 1117
Yenisei, 1048
Yosemite, 188
YTTERBIUM: ELEMENT AND GEOCHEMISTRY, 44, 53, 280, 300, **1290**
YTTRIUM: ELEMENT AND GEOCHEMISTRY, 44, 53, 83, 280, 300, 1020, **1291**
Yttrium-europium oxide red television phosphors, 350
Yttrium lanthanides, 83
Yttrofluorite, 1027
Yttrotantalite, 846, 1028, 1162
Yucatan Peninsula, 540
Yugoslavia, 35, 474
Yukon, 1048
Yunnan, 791

Zaitseva, 387
Zambezi River, 1048
Zambian (Rhodesian) Copper Belt, 194
Zaratite, 143
Zaura, Libya, 536
Zechstein, 1098
Zechstein lagoons, 358
Zeeman splitting, 760
Zenotime, 944
Zeolites, 53, 58, 61, 88, 176, 203. *See also* Vol. IVB
Zeolites, 412, 591, 593, 748, 900
Zeolitic facies, 57
Zero-order reactions, 219
Zermatt, Switzerland, 465
Zeroth law of thermodynamics, 1174
Zeta potential, 181
Zeunerite, 939
Ziegler-Natta catalysts, 144
ZINC: ELEMENT AND GEOCHEMISTRY, 20, 35, 44, 75, 82, 99, 106, 150, 190, 191, 274, 280, 300, 320, 360, 386, 392, 462, 587, 605, 697, 699, 737, 763, 1067, **1293**
Zinc, economic deposits. *See* LEAD AND ZINC: ECONOMIC DEPOSITS
Zinc carbonate, 99
Zincaluminite, 1126
Zincblende, 191, 386
Zinc-melanterite, 1125
Zincite, 843
Zippeite, 1126
Zircon, 149, 201, 367, 419, 432, 447, 448, 451, 455, 462, 488, 493, 993, 1041, 1184, **1217**, 1219, 1294
ZIRCONIUM: ELEMENT AND GEOCHEMISTRY, 44, 145, 151, 245, 247, 274, 280, 300, 415, 423, 488, **1294**
Zirkelite, 491, 846
Zn, 573, 578
Zonal fusion, 402
Zonal soil, 912
Zone depression, 554
Zones-high pressure, 545
Zone of saturation, 1233, 1243
Zoned equilibrium, 727
Zoned feldspars, 593
Zoned pegmatite, 971
Zoology, 187
Zooplankton, 70
Zoosphere, 947
Zunyite, 155

Ref
QE
515
F24
1972 b